T0181515

Concise Encyclopedia
of Coding Theory

Concise Encyclopedia of Coding Theory

Edited by

W. Cary Huffman
Loyola University Chicago, USA

Jon-Lark Kim
Sogang University, Republic of Korea

Patrick Solé
University Aix-Marseille, Marseilles, France.

CRC Press
Taylor & Francis Group
Boca Raton London New York

CRC Press is an imprint of the
Taylor & Francis Group, an **informa** business

First edition published 2021
by CRC Press
6000 Broken Sound Parkway NW, Suite 300, Boca Raton, FL 33487-2742

and by CRC Press
2 Park Square, Milton Park, Abingdon, Oxon, OX14 4RN

ISBN: 9781138551992 (hbk)
ISBN: 9781315147901 (ebk)

To a new generation, my grandchildren

Ezekiel James Huffman — 20 November 2019

Eliana Rei Beaton — 24 June 2020

— W. C. H.

To my daughter *Sylvie Jinna Kim*

To the memory of my advisor

Vera Pless — 1931–2020

— J.-L. K.

To the memory of my advisor

Gérard D. Cohen — 1951–2018

— P. S.

Contents

II   Families of Codes   283

14 Coding Theory and Galois Geometries   285

Leo Storme

15 Algebraic Geometry Codes and Some Applications   307

Alain Couvreur and Hugues Randriambololona

17 Constacyclic Codes over Finite Commutative Chain Rings 385

Hai Q. Dinh and Sergio R. López-Permouth

18 Weight Distribution of Trace Codes over Finite Rings 429

Minjia Shi

19 Two-Weight Codes 449

Andries E. Brouwer

20 Linear Codes from Functions

Sihem Mesnager

III   Applications   553

Bibliography

Index

Preface

The 1948 publication of the paper *A mathematical theory of communication* by Claude E. Shannon is considered to be the genesis of the now vast area of coding theory. In the over 70 years since this monumental work first appeared, coding theory has grown into a discipline intersecting mathematics, computer science, and engineering with applications to almost every area of communication and data storage and even beyond. Given a communication channel on which data is transmitted or a storage device on which data is kept, that data may be corrupted by errors or erasures. In what form do you put that data so that the original information can be recovered and how do you make that recovery? Shannon's paper showed that coding theory provides an answer to that question. The *Concise Encyclopedia of Coding Theory*, somewhat in the spirit of the 1998 *Handbook of Coding Theory* [1521], examines many of the major areas and themes of coding theory taking the reader from the basic introductory level to the frontiers of research.

The authors chosen to contribute to this encyclopedia were selected because of their expertise and understanding of the specific topic of their chapter. Authors have introduced the topic of their chapter in relationship to how it fits into the historical development of coding theory and why their topic is of theoretical and/or applied interest. Each chapter progresses from basic to advanced ideas with few proofs but with many references to which the reader may go for more in-depth study. An attempt has been made within each chapter to point the reader to other chapters in the encyclopedia that deal with similar or related material. An extensive index is provided to help guide the reader interested in pursuing a particular concept.

The *Concise Encyclopedia of Coding Theory* is divided into of three parts: Part I explores the fundamentals of coding theory, Part II examines specific families of codes, and Part III focuses on the practical application of codes.

The first thirteen chapters make up Part I of this encyclopedia. This part explores the fundamental concepts involved in the development of error-correcting codes. Included is an introduction to several historically significant types of codes along with some of their natural generalizations. The mathematical theory behind these codes and techniques for studying them are also introduced. Readers of this encyclopedia who are new to coding theory are encouraged to first read Chapter 1 and then consider other chapters that interest them. More advanced readers may wish to skim Chapter 1 but then move to other chapters.

Chapter **1**, written by the editors of this encyclopedia, is an introduction to the *basic concepts of coding theory* and sets the stage with notation and terminology used throughout the book. The chapter starts with a simple communication channel moving then to the definition of linear and nonlinear codes over finite fields. Basic concepts needed to explore codes are introduced along with families of classical codes: Hamming, Reed–Muller, cyclic, Golay, BCH, and Reed–Solomon codes. The chapter concludes with an brief introduction to encoding, decoding, and Shannon's Theorem, the latter becoming the justification and motivation for the entire discipline.

Chapter **2**, written by Cunsheng Ding, describes two fundamental constructions of *cyclic codes* and the BCH and Hartmann–Tzeng Bounds on cyclic codes. The main task of this chapter is to introduce several important families of cyclic codes, including irreducible cyclic codes, reversible cyclic codes, BCH codes, duadic codes, punctured generalized Reed–Muller codes, and a new generalization of the punctured binary Reed–Muller codes.

Shannon's proof of the existence of good codes is non-constructive and therefore of little use for applications, where one needs one or all codes with specific (small) parameters. Techniques for *constructing and classifying codes* are considered in Chapter **3**, written by Patric R. J. Östergård, with an emphasis on computational methods. Some classes of codes are discussed in more detail: perfect codes, MDS codes, and general binary codes.

Self-dual codes form one of the important classes of linear codes because of their rich algebraic structure and their close connections with other combinatorial configurations like block designs, lattices, graphs, etc. Topics covered in Chapter **4**, by Stefka Bouyuklieva, include construction methods, results on enumeration and classification, and bounds for the minimum distance of self-dual codes over fields of size 2, 3, and 4.

Combinatorial designs often arise in codes that are optimal with respect to certain bounds and are used in some decoding algorithms. Chapter **5**, written by Vladimir D. Tonchev, summarizes links between combinatorial designs and perfect codes, optimal codes meeting the restricted Johnson Bound, and linear codes admitting majority logic decoding.

Chapters 1–5 explore codes over fields; Chapter **6**, by Steven T. Dougherty, introduces *codes over rings*. The chapter begins with a discussion of quaternary codes over the integers modulo 4 and their Gray map, which popularized the study of codes over more general rings. It then discusses codes over rings in a very broad sense describing families of rings, the Chinese Remainder Theorem applied to codes, generating matrices, and bounds. It also gives a description of the MacWilliams Identities for codes over rings.

Quasi-cyclic codes form an important class of algebraic codes that includes cyclic codes as a special subclass. Chapter **7**, coauthored by Cem Güneri, San Ling, and Buket Özkaya, focuses on the algebraic structure of *quasi-cyclic codes*. Based on these structural properties, some asymptotic results, minimum distance bounds, and further applications, such as the trace representation and characterization of certain subfamilies of quasi-cyclic codes, are elaborated upon.

Cyclic and quasi-cyclic codes are studied as ideals in ordinary polynomial rings. Chapter **8**, by Heide Gluesing-Luerssen, is a survey devoted to *skew-polynomial rings and skew-cyclic block codes*. After discussing some relevant algebraic properties of skew polynomials, the basic notions of skew-cyclic codes, such as generator polynomial, parity check polynomial, and duality are investigated. The basic tool is a skew-circulant matrix. The chapter concludes with results on constructions of skew-BCH codes.

The coauthors Jürgen Bierbrauer, Stefano Marcugini, and Fernanda Pambianco of Chapter **9** develop the theory of *cyclic additive codes*, both in the permutation sense and in the monomial sense when the code length is coprime to the characteristic of the underlying field. This generalizes the classical theory of cyclic and constacyclic codes, respectively, from the category of linear codes to the category of additive codes. The cyclic quantum codes correspond to a special case when the codes are self-orthogonal with respect to the symplectic bilinear form.

Up to this point the codes considered are block codes where codewords all have fixed length. That is no longer the case in Chapter **10**, coauthored by Julia Lieb, Raquel Pinto, and Joachim Rosenthal. This chapter provides a survey of *convolutional codes* stressing the connections to module theory and systems theory. Constructions of codes with maximal possible distance and distance profile are provided.

Chapter **11**, written by Elisa Gorla, provides a mathematical introduction to *rank-metric codes*, beginning with the definition of the rank metric and the corresponding codes, whose elements can be either vectors or matrices. This is followed by the definition of code equivalence and the notion of support for a codeword and for a code. This chapter treats some of the basic concepts in the mathematical theory of rank-metric codes: duality, weight enumerators and the MacWilliams Identities, higher rank weights, MRD codes, optimal anticodes, and q-polymatroids associated to a rank-metric code.

The final two chapters of Part I deal with the important technique of linear programming and a related generalization to produce bounds. As described in Chapter 12, coauthored by Peter Boyvalenkov and Danyo Danev, general *linear programming* methods imply universal bounds for codes and designs. The explanation is organized in the Levenshtein framework extended with recent developments on universal bounds for the energy of codes, including the concept of universal optimality. The exposition is done separately for codes in Hamming spaces and for spherical codes.

Linear programming bounds, initially developed by Delsarte, belong to the most powerful and flexible methods to obtain bounds for extremal problems in coding theory. In recent years, after the pioneering work of Schrijver, *semidefinite programming* bounds have been developed with two aims: to strengthen linear programming bounds and to find bounds for more general spaces. Chapter 13, by Frank Vallentin, introduces semidefinite programming bounds with an emphasis on error-correcting codes and its relation to semidefinite programming hierarchies for difficult combinatorial optimization problems.

The next eight chapters make up Part II of the *Concise Encyclopedia of Coding Theory*, where the focus is on specific families of codes. The codes presented fall into two categories, with some overlap. Some of them are generalizations of classical codes from Part I. The rest have a direct connection to algebraic, geometric, or graph theoretic structures. They all are interesting theoretically and often possess properties useful for application.

There are many problems in coding theory which are equivalent to geometrical problems in Galois geometries. Certain formulations of some of the classical codes have direct connections to geometry. Chapter 14, written by Leo Storme, describes a number of the many links between *coding theory and Galois geometries*, and shows how these two research areas influence and stimulate each other.

Chapter 15, written by Alain Couvreur and Hugues Randriambololona, surveys the development of the theory of *algebraic geometry codes* since their discovery in the late 1970s. The authors summarize the major results on various problems such as asymptotic parameters, improved estimates on the minimum distance, and decoding algorithms. In addition, the chapter describes various modern applications of these codes such as public-key cryptography, algebraic complexity theory, multiparty computation, and distributed storage.

Very often the parameters of good/optimal linear codes can be realized by *group codes*, that is, ideals in a group algebra $\mathbb{F}G$ where G is a finite group and $\mathbb{F}$ is a finite field. Such codes, the topic of Chapter 16 written by Wolfgang Willems, carry more algebraic structure than only linear codes, which leads to an easier analysis of the codes. In particular, the full machinery of representation theory of finite groups can be applied to prove interesting coding theoretical properties.

Chapter 17, coauthored by Hai Q. Dinh and Sergio R. López-Permouth, discusses foundational and theoretical aspects of the role of finite rings as alphabets in coding theory, with a concentration on the class of *constacyclic codes* over finite commutative chain rings. The chapter surveys both the simple-root and repeated-root cases. Several directions in which the notion of constacyclicity has been extended are also presented.

The next three chapters focus on codes with few weights; such codes have applications delineated throughout this encyclopedia. As described in Chapter 18, written by Minjia Shi, one important construction technique for few-weight codes is to use *trace codes*. For example the simplex code, a one-weight code, can be constructed as a trace code by using finite field extensions. In recent years, this technique has been refined by using ring extensions of a finite field coupled with a linear Gray map. Moreover, these image codes can be applied to secret sharing schemes.

Codes with few weights often have an interesting geometric structure. Chapter **19**, written by Andries E. Brouwer, focuses specifically on *codes with exactly two nonzero weights*. The chapter discusses the relationship between two-weight linear codes, strongly regular graphs, and 2-character subsets of a projective space.

Functions in general and more specifically cryptographic functions, that is highly nonlinear functions (PN, APN, bent, AB, plateaued), have important applications in coding theory since they are used to construct optimal linear codes and linear codes useful for applications such as secret sharing, two-party computation, and storage. The ultimate goal of Chapter **20**, by Sihem Mesnager, is to provide an overview and insights into *linear codes* with good parameters that are *constructed from functions* and polynomials over finite fields using multiple approaches.

Chapter **21** by Christine A. Kelley, the concluding chapter of Part II, gives an overview of *graph-based codes* and iterative message-passing decoding algorithms for these codes. Some important classes of low-density parity check codes, such as finite geometry codes, expander codes, protograph codes, and spatially coupled codes are discussed. Moreover, analysis techniques of the decoding algorithm for both the finite length case and the asymptotic length case are summarized. While the area of codes over graphs is vast, a few other families such as repeat accumulate and turbo-like codes are briefly mentioned.

The final chapter of Part II provides a natural bridge to the applications in Part III as codes from graphs were designed to facilitate communication. The thirteen chapters of Part III examine several applications that fall into two categories, again with some overlap. Some of the applications present codes developed for specific uses; other applications use codes to produce other structures that themselves become the main application.

The first chapter in Part III is again a bridge between the previous and successive chapters as it has a distinct theoretical slant but its content is useful as well in applications. Chapter **22**, written by Marcelo Firer, gives an account of many *metrics* used in the context of coding theory, mainly for decoding purposes. The chapter tries to stress the role of some metric related invariants and aspects that are eclipsed at the usual setting of the Hamming metric.

Chapter **23**, written by Alfred Wassermann, examines *algorithms for computer construction* of "good" linear codes and methods to determine the minimum distance and weight enumerator of a linear code. For code construction the focus is on the geometric view: a linear code can be seen as a suitable set of points in a projective geometry. The search then reduces to an integer linear programming problem. The chapter explores how the search space can be much reduced by prescribing a group of symmetries and how to construct special code types such as self-orthogonal codes or LCD codes.

In Chapter **24** by Swastik Kopparty we will see some algorithmic ideas based on *polynomial interpolation* for decoding algebraic codes, applied to generalized Reed–Solomon and interleaved generalized Reed–Solomon codes. These ideas will power decoding algorithms that can decode algebraic codes beyond half the minimum distance.

The theory of pseudo-noise sequences has close connections with coding theory, cryptography, combinatorics, and discrete mathematics. Chapter **25**, coauthored by Tor Helleseth and Chunlei Li, gives a brief introduction of two kinds of *pseudo-noise sequences*, namely sequences with low correlation and shift register sequences with maximum periods, which are of particular interest in modern communication systems.

Lattice coding is presented in Chapter **26**, written by Frédérique Oggier, in the context of Gaussian and fading channels, where channel models are presented. Lattice constructions from both quadratic fields and linear codes are described. Some variations of lattice coding are also discussed.

A bridge between classical coding theory and quantum error control was firmly put in place via the stabilizer formalism, allowing the capabilities of a quantum stabilizer code to be inferred from the properties of the corresponding classical codes. Well-researched tools and the wealth of results in classical coding theory often translate nicely to the design of good *quantum codes*, the subject of Chapter **27** by Martianus Frederic Ezerman. Research problems triggered by error-control issues in the quantum setup revive and expand studies on specific aspects of classical codes, which were previously overlooked or deemed not so interesting.

Chapter **28** on *space-time coding*, written by Frédérique Oggier, defines what space-time coding actually is and what are variations of space-time coding problems. The chapter also provides channel models and design criteria, together with several examples.

Chapter **29**, by Frank R. Kschischang, describes *error-correcting network codes* for packet networks employing random linear network coding. In such networks, packets sent into the network by the transmitter are regarded as a basis for a vector space over a finite field, and the network provides the receiver with random linear combinations of the transmitted vectors, possibly also combined with noise vectors. Unlike classical coding theory—where codes are collections of well-separated *vectors*, each of them a point of some ambient vector space—here codes are collections of well-separated *vector spaces*, each of them a subspace of some ambient vector space. The chapter provides appropriate coding metrics for such subspace codes, and describes various bounds and constructions, focusing particularly on the case of constant-dimension codes whose codewords all have the same dimension.

Erasure codes have attained a position of importance for many streaming and file download applications on the internet. Chapter **30**, written by Ian F. Blake, outlines the development of these codes from simple erasure correcting codes to the important Raptor codes. Various decoding algorithms for these codes are developed and illustrated.

Chapter **31**, with coauthors Vinayak Ramkumar, Myna Vajha, S. B. Balaji, M. Nikhil Krishnan, Birenjith Sasidharan, and P. Vijay Kumar, deals with the topic of designing reliable and efficient *codes for the storage and retrieval* of large quantities of data over storage devices that are prone to failure. Historically, the traditional objective has been one of ensuring reliability against data loss while minimizing storage overhead. More recently, a third concern has surfaced, namely, the need to efficiently recover from the failure of a single storage unit corresponding to recovery from the erasure of a single code symbol. The authors explain how coding theory has evolved to tackle this fresh challenge.

Polar codes are error-correcting codes that achieve the symmetric capacity of discrete input memoryless channels with a polynomial encoding and decoding complexity. Chapter **32**, coauthored by Noam Presman and Simon Litsyn, provides a general presentation of *polar codes* and their associated algorithms. At the same time, most of the examples in the chapter use the basic Arıkan's $(u + v, v)$ original construction due to its simplicity and wide applicability.

While one thinks of coding theory as the major tool to reveal correct information after errors in that information have been introduced, the final two chapters of this encyclopedia address the opposite problem: using coding theory as a tool to hide information. Chapter **33**, by Cunsheng Ding, first gives a brief introduction to *secret sharing schemes*, and then introduces two constructions of secret sharing schemes with linear codes. It also documents a construction of multisecret sharing schemes with linear codes. Basic results about these secret sharing schemes are presented in this chapter.

Chapter **34**, written by Philippe Gaborit and Jean-Christophe Deneuville, gives a general overview of basic tools used for *code-based cryptography*. The security of the main difficult problem for code-based cryptography, the Syndrome Decoding problem, is considered, together with its quasi-cyclic variations. The current state-of-the-art for the cryptographic primitives of encryption, signature, and authentication is the main focus of the chapter.

The editors of the *Concise Encyclopedia of Coding Theory* thank the 48 other authors for sharing their expertise to make this project come to pass. Their cooperation and patience were invaluable to us. We also thank Gayle Imamura-Huffman for lending her transparent watercolor *Coded Information* and opaque watercolor *Linear Subspaces* for use on the front and back covers of this encyclopedia. Additionally we thank the editorial staff at CRC Press/Taylor and Francis Group: Sarfraz Khan, who helped us begin this project; Callum Fraser, who became the Mathematical Editor at CRC Press as the project progressed; and Robin Lloyd-Starkes, who is Project Editor. Also with CRC Press, we thank Mansi Kabra for handling permissions and copyrights and Kevin Craig who assisted with the cover design. We thank Meeta Singh, Senior Project Manager at KnowledgeWorks Global Ltd., and her team for production of this encyclopedia. And of course, we sincerely thank our families for their support and encouragement throughout this journey.

– W. Cary Huffman
– Jon-Lark Kim
– Patrick Solé

Contributors

S. B. Balaji
Qualcomm
Bangalore, India
balaji.profess@gmail.com

Jürgen Bierbrauer
Professor Emeritus
Department of Mathematical Sciences
Michigan Technological University
Houghton, Michigan, USA
jbierbra@mtu.edu

Ian F. Blake
Department of Electrical and Computer
    Engineering
University of British Columbia
Vancouver, British Columbia, Canada
ifblake@ece.ubc.ca

Stefka Bouyuklieva
Faculty of Mathematics and Informatics
St. Cyril and St. Methodius University
Veliko Tarnovo, Bulgaria
stefka@ts.uni-vt.bg

Peter Boyvalenkov
Institute of Mathematics and Informatics
Bulgarian Academy of Sciences
Sofia, Bulgaria
and Technical Faculty
South-Western University
Blagoevgrad, Bulgaria
peter@math.bas.bg

Andries E. Brouwer
Department of Mathematics
Eindhoven University of Technology
Eindhoven, Netherlands
Andries.Brouwer@cwi.nl

Alain Couvreur
Inria *and* LIX

École Polytechnique
Palaiseau, France
alain.couvreur@inria.fr

Danyo Danev
Department of Electrical Engineering and
    Department of Mathematics
Linköping University
Linköping, Sweden
danyo.danev@liu.se

Jean-Christophe Deneuville
Ecole Nationale de l'Aviation Civile
University of Toulouse, France
jean-christophe.deneuville@enac.fr

Cunsheng Ding
Department of Computer Science and
    Engineering
The Hong Kong University of Science and
    Technology
Hong Kong, China
cding@ust.hk

Hai Q. Dinh
Department of Mathematical Sciences
Kent State University
Warren, Ohio, USA
hdinh@kent.edu

Steven T. Dougherty
Department of Mathematics
University of Scranton
Scranton, Pennsylvania, USA
prof.steven.dougherty@gmail.com

Martianus Frederic Ezerman
School of Physical and Mathematical
    Sciences
Nanyang Technological University
Singapore
fredezerman@ntu.edu.sg

Marcelo Firer
Department of Mathematics
State University of Campinas (Unicamp)
Campinas, Brasil
mfirer@ime.unicamp.br

Philippe Gaborit
XLIM-MATHIS
University of Limoges, France
gaborit@unilim.fr

Heide Gluesing-Luerssen
Department of Mathematics
University of Kentucky
Lexington, Kentucky, USA
heide.gl@uky.edu

Elisa Gorla
Institut de Mathématiques
Université de Neuchâtel
Neuchâtel, Switzerland
elisa.gorla@unine.ch

Cem Güneri
Faculty of Engineering and Natural Sciences
Sabancı University
Istanbul, Turkey
guneri@sabanciuniv.edu

Tor Helleseth
Department of Informatics
University of Bergen
Bergen, Norway
Tor.Helleseth@uib.no

W. Cary Huffman
Department of Mathematics and Statistics
Loyola University
Chicago, Illinois, USA
whuffma@luc.edu

Christine A. Kelley
Department of Mathematics
University of Nebraska-Lincoln
Lincoln, Nebraska, USA
ckelley2@unl.edu

Jon-Lark Kim
Department of Mathematics
Sogang University
Seoul, South Korea
jlkim@sogang.ac.kr

Swastik Kopparty
Department of Mathematics
Rutgers University
Piscataway, New Jersey, USA
swastik.kopparty@gmail.com

M. Nikhil Krishnan
Department of Electrical and Computer
    Engineering
University of Toronto
Toronto, Ontario, Canada
nikhilkrishnan.m@gmail.com

Frank R. Kschischang
Department of Electrical and Computer
    Engineering
University of Toronto
Toronto, Ontario, Canada
frank@ece.utoronto.ca

P. Vijay Kumar
Department of Electrical Communication
    Engineering
Indian Institute of Science
Bangalore, India
and
Ming Hsieh Department of Electrical and
Computer Engineering
University of Southern California
Los Angeles, California, USA
pvk@iisc.ac.in, vijayk@usc.edu

Chunlei Li
Department of Informatics
University of Bergen
Bergen, Norway
chunlei.li@uib.no

Julia Lieb
Institut für Mathematik
Universität Zürich
Zürich, Switzerland
julia.lieb@math.uzh.ch

San Ling
School of Physical and Mathematical
    Sciences
Nanyang Technological University
Singapore
lingsan@ntu.edu.sg

Simon Litsyn
Department of Electrical
  Engineering-Systems
Tel Aviv University
Ramat Aviv, Israel
litsyn@tauex.tau.ac.il

Sergio R. López-Permouth
Department of Mathematics
Ohio University
Athens, Ohio, USA
lopez@ohio.edu

Stefano Marcugini
Dipartimento di Matematica e Informatica
Università degli Studi di Perugia
Perugia, Italy
stefano.marcugini@unipg.it

Sihem Mesnager
Department of Mathematics
University of Paris VIII, 93526 Saint-Denis,
  France
and
Laboratoire de Géométrie, Analyse et
Applications, UMR 7539, CNRS
University Sorbonne Paris Cité, 93430
Villetaneuse, France
and
Télécom Paris
91120 Palaiseau, France
smesnager@univ-paris8.fr

Frédérique Oggier
Division of Mathematical Sciences
Nanyang Technological University
Singapore
frederique@ntu.edu.sg

Patric R. J. Östergård
Department of Communications and
  Networking
Aalto University
Espoo, Finland

Buket Özkaya
School of Physical and Mathematical
  Sciences
Nanyang Technological University
Singapore
buketozkaya@ntu.edu.sg

Fernanda Pambianco
Dipartimento di Matematica e Informatica
Università degli Studi di Perugia
Perugia, Italy
fernanda.pambianco@unipg.it

Raquel Pinto
Departamento de Matemática
Universidade de Aveiro
Aveiro, Portugal
raquel@ua.pt

Noam Presman
Department of Electrical
  Engineering-Systems
Tel Aviv University
Ramat Aviv, Israel
presmann@gmail.com

Vinayak Ramkumar
Department of Electrical Communication
  Engineering
Indian Institute of Science
Bangalore, India
vinram93@gmail.com

Hugues Randriambololona
ANSSI – Laboratoire de Cryptographie
Paris, France
and
Télécom Paris
Palaiseau, France
hugues.randriam@ssi.gouv.fr

Joachim Rosenthal
Institut für Mathematik
Universität Zürich
Zürich, Switzerland
rosenthal@math.uzh.ch

Birenjith Sasidharan
Department of Electronics and
  Communication Engineering
Govt. Engineering College, Barton Hill
Trivandrum, India
birenjith@gmail.com

Minjia Shi
School of Mathematical Sciences
Anhui University
Hefei, 230601, China
smjwcl.good@163.com

Patrick Solé
Lab I2M
CNRS, Aix-Marseille Université, Centrale
   Marseille
13 009 Marseilles, France
patrick.sole@telecom-paris.fr

Leo Storme
Department of Mathematics: Analysis,
   Logic and Discrete Mathematics
Ghent University
9000 Gent, Belgium
Leo.Storme@ugent.be

Vladimir D. Tonchev
Department of Mathematical Sciences
Michigan Technological University
Houghton, Michigan, USA
tonchev@mtu.edu

Myna Vajha
Department of Electrical Communication
   Engineering
Indian Institute of Science

Bangalore, India
mynaramana@gmail.com

Frank Vallentin
Mathematisches Institut
Universität zu Köln
Köln, Germany
frank.vallentin@uni-koeln.de

Alfred Wassermann
Mathematisches Institut
Universität Bayreuth
95440 Bayreuth, Germany
alfred.wassermann@uni-bayreuth.de

Wolfgang Willems
Fakultät für Mathematik
Otto-von-Guericke Universität
Magdeburg, Germany
and
Departamento de Matemáticas y
Estadística
Universidad del Norte
Barranquilla, Colombia
willems@ovgu.de

Part I

Coding Fundamentals

Chapter 1

Basics of Coding Theory

W. Cary Huffman

Loyola University, Chicago

Jon-Lark Kim

Sogang University

Patrick Solé

CNRS, Aix-Marseille Université

1.1　Introduction

Coding theory had it genesis in the late 1940s with the publication of works by Claude Shannon, Marcel Golay, and Richard Hamming. In 1948 Shannon published a landmark

FIGURE 1.1: A simple communication channel

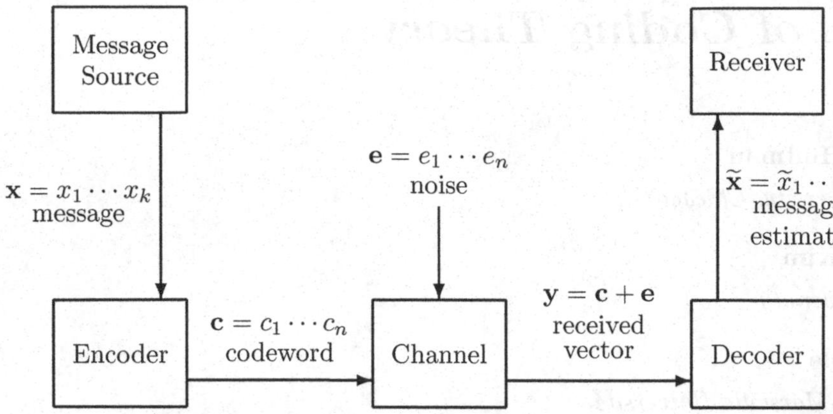

paper *A mathematical theory of communication* [1661] which marked the beginning of both information theory and coding theory. Given a communication channel, over which information is transmitted and possibly corrupted, Shannon identified a number called the 'channel capacity' and proved that arbitrarily reliable communication is possible at any rate below the channel capacity. For example, when transmitting images of planets from deep space, it is impractical to retransmit the images that have been altered by noise during transmission. Shannon's Theorem guarantees that the data can be encoded before transmission so that the altered data can be decoded to the original, up to a specified degree of accuracy. Other examples of communication channels include wireless communication devices and storage systems such as DVDs or Blue-ray discs. In 1947 Hamming developed a code, now bearing his name, in an attempt to correct errors that arose in the Bell Telephone Laboratories' mechanical relay computer; his work was circulated through a series of memoranda at Bell Labs and eventually published in [895]. Both Shannon [1661] and Golay [820] published Hamming's code, with Golay generalizing it. Additionally, Golay presented two of the four codes that now bear his name. A monograph by T. M. Thompson [1801] traces the early development of coding theory.

A simple **communication channel** is illustrated in Figure 1.1. At the source a **message**, denoted $\mathbf{x} = x_1 \cdots x_k$ in the figure, is to be sent. If no modification is made to $\mathbf{x}$ and it is transmitted directly over the channel, any noise would distort $\mathbf{x}$ so that it could not be recovered. The basic idea of coding theory is to embellish the message by adding some redundancy so that hopefully the original message can be recovered after reception even if noise corrupts the embellished message during transmission. The redundancy is added by the encoder and the embellished message, called a **codeword** $\mathbf{c} = c_1 \cdots c_n$ in the figure, is sent over the channel where noise in the form of an **error vector** $\mathbf{e} = e_1 \cdots e_n$ distorts the codeword producing a received vector $\mathbf{y}$.[1] The received vector is then sent to be decoded where the errors are removed. The redundancy is then stripped off, and an estimate $\widetilde{\mathbf{x}}$ of the original message is produced. Hopefully $\widetilde{\mathbf{x}} = \mathbf{x}$. (There is a one-to-one correspondence

---

[1]Generally message and codeword symbols will come from a finite field $\mathbb{F}$ or a finite ring R. Messages will be 'vectors' in $\mathbb{F}^k$ (or R^k), and codewords will be 'vectors' in $\mathbb{F}^n$ (or R^n). If $\mathbf{c}$ entered the channel and $\mathbf{y}$ exited the channel, the difference $\mathbf{y} - \mathbf{c}$ is what we have termed the error vector $\mathbf{e}$ in Figure 1.1. While this is the normal scenario, other ambient spaces from which codes arise occur in this encyclopedia.

between codewords and messages. In many cases, the real interest is not in the message $\mathbf{x}$ but the codeword $\mathbf{c}$. With this point of view, the job of the decoder is to obtain an estimate $\widetilde{\mathbf{y}}$ from $\mathbf{y}$ and hope that $\widetilde{\mathbf{y}} = \mathbf{c}$.) For example in deep space communication, the message source is the satellite, the channel is outer space, the decoder is hardware at a ground station on Earth, and the receiver is the people or computer processing the information; of course, messages travel from Earth to the satellite as well. For a DVD or Blue-ray disc, the message source is the voice, music, video, or data to be placed on the disc, the channel is the disc itself, the decoder is the DVD or Blue-ray player, and the receiver is the listener or viewer.

Shannon's Theorem guarantees that the hope of successful recovery will be fulfilled a certain percentage of the time. With the right encoding based on the characteristics of the channel, this percentage can be made as high as desired, although not 100%. The proof of Shannon's Theorem is probabilistic and nonconstructive. No specific codes were produced in the proof that give the desired accuracy for a given channel. Shannon's Theorem only guarantees their existence. In essence, the goal of coding theory is to produce codes that fulfill the conditions of Shannon's Theorem and make reliable communication possible.

There are numerous texts, ranging from introductory to research-level books, on coding theory including (but certainly not limited to) [170, 209, 896, 1008, 1323, 1505, 1506, 1520, 1521, 1602, 1836]. There are two books, [169] edited by E. R. Berlekamp and [212] edited by I. F. Blake, in which early papers in the development of coding theory have been reprinted.

1.2   Finite Fields

Finite fields play an essential role in coding theory. The theory and construction of finite fields can be found, for example, in [1254] and [1408, Chapter 2]. Finite fields, as related specifically to codes, are described in [1008, 1323, 1602]. In this section we give a brief introduction.

Definition 1.2.1 A **field** $\mathbb{F}$ is a nonempty set with two binary operations, denoted $+$ and $\cdot$, satisfying the following properties.

(a) For all $\alpha, \beta, \gamma \in \mathbb{F}$, $\alpha + \beta \in \mathbb{F}$, $\alpha \cdot \beta \in \mathbb{F}$, $\alpha + \beta = \beta + \alpha$, $\alpha \cdot \beta = \beta \cdot \alpha$, $\alpha + (\beta + \gamma) = (\alpha + \beta) + \gamma$, $\alpha \cdot (\beta \cdot \gamma) = (\alpha \cdot \beta) \cdot \gamma$, and $\alpha \cdot (\beta + \gamma) = \alpha \cdot \beta + \alpha \cdot \gamma$.

(b) $\mathbb{F}$ possesses an **additive identity** or **zero**, denoted 0, and a **multiplicative identity** or **unity**, denoted 1, such that $\alpha + 0 = \alpha$ and $\alpha \cdot 1 = \alpha$ for all $\alpha \in \mathbb{F}_q$.

(c) For all $\alpha \in \mathbb{F}$ and all $\beta \in \mathbb{F}$ with $\beta \neq 0$, there exists $\alpha' \in \mathbb{F}$, called the **additive inverse of** α, and $\beta^* \in \mathbb{F}$, called the **multiplicative inverse of** β, such that $\alpha + \alpha' = 0$ and $\beta \cdot \beta^* = 1$.

The additive inverse of α will be denoted $-\alpha$, and the multiplicative inverse of β will be denoted β^{-1}. Usually the multiplication operation will be suppressed; that is, $\alpha \cdot \beta$ will be denoted $\alpha\beta$. If n is a positive integer and $\alpha \in \mathbb{F}$, $n\alpha = \alpha + \alpha + \cdots + \alpha$ (n times), $\alpha^n = \alpha\alpha \cdots \alpha$ (n times), and $\alpha^{-n} = \alpha^{-1}\alpha^{-1} \cdots \alpha^{-1}$ (n times when $\alpha \neq 0$). Also $\alpha^0 = 1$ if $\alpha \neq 0$. The usual rules of exponentiation hold. If $\mathbb{F}$ is a finite set with q elements, $\mathbb{F}$ is called a **finite field of order** q and denoted $\mathbb{F}_q$.

Example 1.2.2 Fields include the rational numbers $\mathbb{Q}$, the real numbers $\mathbb{R}$, and the complex numbers $\mathbb{C}$. Finite fields include $\mathbb{Z}_p$, the set of integers modulo p, where p is a prime.

The following theorem gives some of the basic properties of finite fields.

Theorem 1.2.3 *Let $\mathbb{F}_q$ be a finite field with q elements. The following hold.*

(a) $\mathbb{F}_q$ *is unique up to isomorphism.*

(b) $q = p^m$ *for some prime p and some positive integer m.*

(c) $\mathbb{F}_q$ *contains the subfield $\mathbb{F}_p = \mathbb{Z}_p$.*

(d) $\mathbb{F}_q$ *is a vector space over $\mathbb{F}_p$ of dimension m.*

(e) $p\alpha = 0$ *for all $\alpha \in \mathbb{F}_q$.*

(f) *If $\alpha, \beta \in \mathbb{F}_q$, $(\alpha + \beta)^p = \alpha^p + \beta^p$.*

(g) *There exists an element $\gamma \in \mathbb{F}_q$ with the following properties.*

    (i) $\mathbb{F}_q = \{0, 1 = \gamma^0, \gamma, \ldots, \gamma^{q-2}\}$ *and $\gamma^{q-1} = 1$,*

    (ii) $\{1 = \gamma^0, \gamma, \ldots, \gamma^{m-1}\}$ *is a basis of the vector space $\mathbb{F}_q$ over $\mathbb{F}_p$, and*

    (iii) *there exist $a_0, a_1, \ldots, a_{m-1} \in \mathbb{F}_p$ such that*

$$\gamma^m = a_0 + a_1\gamma + \cdots + a_{m-1}\gamma^{m-1}. \tag{1.1}$$

(h) *For all $\alpha \in \mathbb{F}_q$, $\alpha^q = \alpha$.*

Definition 1.2.4 In Theorem 1.2.3, p is called the **characteristic** of $\mathbb{F}_q$. The element γ is called a **primitive element** of $\mathbb{F}_q$.

Remark 1.2.5 Using Theorem 1.2.3(f), the map $\sigma_p : \mathbb{F}_q \to \mathbb{F}_q$ given by $\sigma_p(\alpha) = \alpha^p$ is an automorphism of $\mathbb{F}_q$, called the **Frobenius automorphism of $\mathbb{F}_q$**. Once one primitive element γ of $\mathbb{F}_q$ is found, the remaining primitive elements of $\mathbb{F}_q$ are precisely γ^d where $\gcd(d, q-1) = 1$.

The key to constructing a finite field is to find a primitive element γ in $\mathbb{F}_q$ and the equation (1.1). We do not describe this process here, but refer the reader to the texts mentioned at the beginning of the section. Assuming γ is found and the equation (1.1) is known, we can construct addition and multiplication tables for $\mathbb{F}_q$. This is done by writing every element of $\mathbb{F}_q$ in two forms. The first form takes advantage of Theorem 1.2.3(g)(ii). Every element $\alpha \in \mathbb{F}_q$ is written uniquely in the form

$$\alpha = a_0\gamma^0 + a_1\gamma + a_2\gamma^2 + \cdots + a_{m-1}\gamma^{m-1} \text{ with } a_i \in \mathbb{F}_p = \mathbb{Z}_p \text{ for } 0 \le i \le m-1,$$

which we abbreviate $\alpha = a_0 a_1 a_2 \cdots a_{m-1}$, a vector in $\mathbb{Z}_p^m$. Addition in $\mathbb{F}_q$ is accomplished by ordinary vector addition in $\mathbb{Z}_p^m$. To each $\alpha \in \mathbb{F}_q$, with $\alpha \ne 0$, we associate a second form: $\alpha = \gamma^i$ for some i with $0 \le i \le q-2$. Multiplication is accomplished by $\gamma^i\gamma^j = \gamma^{i+j}$ where we use $\gamma^{q-1} = 1$ when appropriate. We illustrate this by constructing the field $\mathbb{F}_9$.

Example 1.2.6 The field $\mathbb{F}_9$ has characteristic 3 and is a 2-dimensional vector space over $\mathbb{Z}_3$. One primitive element γ of $\mathbb{F}_9$ satisfies $\gamma^2 = 1 + \gamma$. Table 1.1 gives the two forms of all elements. The zero element is $0\gamma^0 + 0\gamma = 00$; the unity element is $1 = 1\gamma^0 + 0\gamma = 10$. Now $\gamma = 0\gamma^0 + 1\gamma = 01$, $\gamma^2 = 1 + \gamma = 1\gamma^0 + 1\gamma = 11$, $\gamma^3 = \gamma\gamma^2 = \gamma(1+\gamma) = \gamma + \gamma^2 = \gamma + (1+\gamma) = 1\gamma^0 + 2\gamma = 12$, and $\gamma^4 = \gamma\gamma^3 = \gamma(1+2\gamma) = \gamma + 2\gamma^2 = \gamma + 2(1+\gamma) = 2\gamma^0 + 0\gamma = 20$. Note $\gamma^4 = -1$. $\gamma^5, \gamma^6, \gamma^7$ are computed similarly. As an example, we compute $(\gamma^5 - 1 + \gamma^6)/(\gamma^5 + \gamma^3 + 1)$ as follows. First $\gamma^5 - 1 + \gamma^6 = 02 - 10 + 22 = 11 = \gamma^2$, and $\gamma^5 + \gamma^3 + 1 = 02 + 12 + 10 = 21 = \gamma^7$. So $(\gamma^5 - 1 + \gamma^6)/(\gamma^5 + \gamma^3 + 1) = \gamma^2/\gamma^7 = \gamma^{-5} = \gamma^3$ since $\gamma^8 = 1$.

Tables 1.2, 1.3, and 1.4 give addition and multiplication tables for $\mathbb{F}_4$, $\mathbb{F}_8$, and $\mathbb{F}_{16}$, respectively. These fields have characteristic 2. Notice that $\mathbb{F}_{16}$ contains the subfield $\mathbb{F}_4$ where $\omega = \rho^5$.

TABLE 1.1: $\mathbb{F}_9$ with primitive element γ where $\gamma^2 = 1 + \gamma$ and $\gamma^8 = 1$

vector	power of γ	vector	power of γ	vector	power of γ
00	—	11	γ^2	02	γ^5
10	$\gamma^0 = 1$	12	γ^3	22	γ^6
01	γ	20	$\gamma^4 = -1$	21	γ^7

TABLE 1.2: $\mathbb{F}_4$ with primitive element ω where $\omega^2 = 1 + \omega$ and $\omega^3 = 1$

vector	power of ω	vector	power of ω	vector	power of ω	vector	power of ω
00	—	10	$\omega^0 = 1$	01	ω	11	ω^2

1.3 Codes

In this section we introduce the concept of codes over finite fields. We begin with some notation.

The **set of n-tuples with entries in** $\mathbb{F}_q$ forms an n-dimensional vector space, denoted $\mathbb{F}_q^n = \{x_1 x_2 \cdots x_n \mid x_i \in \mathbb{F}_q, \ 1 \leq i \leq n\}$, under componentwise addition of n-tuples and componentwise multiplication of n-tuples by scalars in $\mathbb{F}_q$. The vectors in $\mathbb{F}_q^n$ will often be denoted using bold Roman characters $\mathbf{x} = x_1 x_2 \cdots x_n$. The vector $\mathbf{0} = 00 \cdots 0$ is the **zero vector** in $\mathbb{F}_q^n$.

For positive integers m and n, $\mathbb{F}_q^{m \times n}$ denotes **the set of all $m \times n$ matrices with entries** in $\mathbb{F}_q$. The matrix in $\mathbb{F}_q^{m \times n}$ with all entries 0 is the **zero matrix** denoted $\mathbf{0}_{m \times n}$. The **identity matrix** of $\mathbb{F}_q^{n \times n}$ will be denoted I_n. If $A \in \mathbb{F}_q^{m \times n}$, $A^\mathsf{T} \in \mathbb{F}_q^{n \times m}$ will denote the **transpose** of A. If $\mathbf{x} \in \mathbb{F}_q^m$, $\mathbf{x}^\mathsf{T}$ will denote $\mathbf{x}$ as a column vector of length m, that is, an $m \times 1$ matrix. The column vector $\mathbf{0}^\mathsf{T}$ and the $m \times 1$ matrix $\mathbf{0}_{m \times 1}$ are the same.

If S is any finite set, its **order** or **size** is denoted $|S|$.

Definition 1.3.1 A subset $\mathcal{C} \subseteq \mathbb{F}_q^n$ is called a **code of length n over** $\mathbb{F}_q$; $\mathbb{F}_q$ is called the **alphabet** of $\mathcal{C}$, and $\mathbb{F}_q^n$ is the **ambient space** of $\mathcal{C}$. Codes over $\mathbb{F}_q$ are also called q-**ary codes**. If the alphabet is $\mathbb{F}_2$, $\mathcal{C}$ is **binary**. If the alphabet is $\mathbb{F}_3$, $\mathcal{C}$ is **ternary**. The vectors in $\mathcal{C}$ are the **codewords** of $\mathcal{C}$. If $\mathcal{C}$ has M codewords (that is, $|\mathcal{C}| = M$) $\mathcal{C}$ is denoted an $(n, M)_q$ code, or, more simply, an (n, M) code when the alphabet $\mathbb{F}_q$ is understood. If $\mathcal{C}$ is a linear subspace of $\mathbb{F}_q^n$, that is $\mathcal{C}$ is closed under vector addition and scalar multiplication, $\mathcal{C}$ is called a **linear code of length n over** $\mathbb{F}_q$. If the dimension of the linear code $\mathcal{C}$ is k, $\mathcal{C}$ is denoted an $[n, k]_q$ code, or, more simply, an $[n, k]$ code. An $(n, M)_q$ code that is also linear is an $[n, k]_q$ code where $M = q^k$. An $(n, M)_q$ code may be referred to as an **unrestricted code**; a specific unrestricted code may be either linear or nonlinear. When referring to a code, expressions such as (n, M), $(n, M)_q$, $[n, k]$, or $[n, k]_q$ are called the **parameters** of the code.

Example 1.3.2 Let $\mathcal{C} = \{1100, 1010, 1001, 0110, 0101, 0011\} \subseteq \mathbb{F}_2^4$. Then $\mathcal{C}$ is a $(4, 6)_2$ binary nonlinear code. Let $\mathcal{C}_1 = \mathcal{C} \cup \{0000, 1111\}$. Then $\mathcal{C}_1$ is a $(4, 8)_2$ binary linear code. As $\mathcal{C}_1$ is a subspace of $\mathbb{F}_2^4$ of dimension 3, $\mathcal{C}_1$ is also a $[4, 3]_2$ code.

Remark 1.3.3 Basic development of linear codes is found in papers by D. Slepian [1722, 1723, 1724]. In some chapters of this book, codes will be considered where the alphabet is

TABLE 1.3: $\mathbb{F}_8$ with primitive element δ where $\delta^3 = 1 + \delta$ and $\delta^7 = 1$

vector	power of δ	vector	power of δ	vector	power of δ	vector	power of δ
000	—	010	δ	110	δ^3	111	δ^5
100	$\delta^0 = 1$	001	δ^2	011	δ^4	101	δ^6

TABLE 1.4: $\mathbb{F}_{16}$ with primitive element ρ where $\rho^4 = 1 + \rho$ and $\rho^{15} = 1$

vector	power of ρ	vector	power of ρ	vector	power of ρ	vector	power of ρ
0000	—	0001	ρ^3	1101	ρ^7	0111	ρ^{11}
1000	$\rho^0 = 1$	1100	ρ^4	1010	ρ^8	1111	ρ^{12}
0100	ρ	0110	ρ^5	0101	ρ^9	1011	ρ^{13}
0010	ρ^2	0011	ρ^6	1110	ρ^{10}	1001	ρ^{14}

not necessarily a field but rather a ring R. In these situations, the vector space $\mathbb{F}_q^n$ will be replaced by an R-module such as $R^n = \{x_1 x_2 \cdots x_n \mid x_i \in R,\ 1 \le i \le n\}$, and a code will be considered *linear* if it is an R-submodule of that R-module. See for example Chapters 6, 17, and 18.

1.4 Generator and Parity Check Matrices

When choosing between linear and nonlinear codes, the added algebraic structure of linear codes often makes them easier to describe and use. Generally, a linear code is defined by giving either a generator or a parity check matrix.

Definition 1.4.1 Let $\mathcal{C}$ be an $[n, k]_q$ linear code. A **generator matrix** G for $\mathcal{C}$ is any $G \in \mathbb{F}_q^{k \times n}$ whose row span is $\mathcal{C}$. Because any k-dimensional subspace of $\mathbb{F}_q^n$ is the kernel of some linear transformation from $\mathbb{F}_q^n$ onto $\mathbb{F}_q^{n-k}$, there exists $H \in \mathbb{F}_q^{(n-k) \times n}$, with independent rows, such that $\mathcal{C} = \{\mathbf{c} \in \mathbb{F}_q^n \mid H\mathbf{c}^{\mathsf{T}} = \mathbf{0}^{\mathsf{T}}\}$. Such a matrix, of which there are generally many, is called a **parity check matrix of $\mathcal{C}$**.

Example 1.4.2 Continuing with Example 1.3.2, there are several generator matrices for $\mathcal{C}_1$ including

$$G_1 = \begin{bmatrix} 1 & 0 & 0 & 1 \\ 0 & 1 & 0 & 1 \\ 0 & 0 & 1 & 1 \end{bmatrix}, \ G_1' = \begin{bmatrix} 1 & 1 & 1 & 1 \\ 1 & 1 & 0 & 0 \\ 0 & 1 & 1 & 0 \end{bmatrix}, \ \text{and } G_1'' = \begin{bmatrix} 1 & 1 & 0 & 0 \\ 0 & 1 & 1 & 0 \\ 0 & 0 & 1 & 1 \end{bmatrix}.$$

In this case there is only one parity check matrix $H_1 = \begin{bmatrix} 1 & 1 & 1 & 1 \end{bmatrix}$.

Remark 1.4.3 Any matrix obtained by elementary row operations from a generator matrix for a code remains a generator matrix of that code.

Remark 1.4.4 By Definition 1.4.1, the rows of G form a basis of $\mathcal{C}$, and the rows of H are independent. At times, the requirement may be relaxed so that the rows of G are only required to span $\mathcal{C}$. Similarly, the requirement that the rows of H be independent may be dropped as long as $\mathcal{C} = \{\mathbf{c} \in \mathbb{F}_q^n \mid H\mathbf{c}^{\mathsf{T}} = \mathbf{0}^{\mathsf{T}}\}$ remains true.

Theorem 1.4.5 ([1323, Chapter 1.1]) *Let* $G \in \mathbb{F}_q^{k \times n}$ *and* $H \in \mathbb{F}_q^{(n-k) \times n}$ *each have independent rows. Let* $\mathcal{C}$ *be an* $[n, k]_q$ *code. The following hold.*

(a) *If* G, *respectively* H, *is a generator, respectively parity check, matrix for* $\mathcal{C}$, *then* $HG^T = \mathbf{0}_{(n-k) \times k}$.

(b) *If* $HG^T = \mathbf{0}_{(n-k) \times k}$, *then* G *is a generator matrix for* $\mathcal{C}$ *if and only if* H *is a parity check matrix for* $\mathcal{C}$.

Definition 1.4.6 Let $\mathcal{C}$ be an $[n, k]_q$ linear code with generator matrix $G \in \mathbb{F}_q^{k \times n}$. For any set of k independent columns of G, the corresponding set of coordinates forms an **information set** for $\mathcal{C}$; the remaining $n - k$ coordinates form a **redundancy set** for $\mathcal{C}$. If G has the form $G = [I_k \mid A]$, G is in **standard form** in which case $\{1, 2, \ldots, k\}$ is an information set with $\{k+1, k+2, \ldots, n\}$ the corresponding redundancy set.

Theorem 1.4.7 ([1602, Chapter 2.3]) *If* $G = [I_k \mid A]$ *is a generator matrix of an* $[n, k]_q$ *code* $\mathcal{C}$, *then* $H = [-A^T \mid I_{n-k}]$ *is a parity check matrix for* $\mathcal{C}$.

Example 1.4.8 Continuing with Examples 1.3.2 and 1.4.2, the matrix G_1 is in standard form. Applying Theorem 1.4.7 to G_1, we get the parity check matrix H_1 of Example 1.4.2. The matrices G_1' and G_1'' both row reduce to G_1; so all three are generator matrices of the same code, consistent with Remark 1.4.3. Any subset of $\{1, 2, 3, 4\}$ of size 3 is an information set for $\mathcal{C}_1$. The fact that $HG_1^T = HG_1'^T = HG_1''^T = \mathbf{0}_{1 \times 3}$ is consistent with Theorem 1.4.5. Finally, let $\mathcal{C}_2 = \{0000, 1100, 0011, 1111\}$ be the $[4, 2]_2$ linear subcode of $\mathcal{C}_1$. $\mathcal{C}_2$ does not have a generator matrix in standard form; the only information sets for $\mathcal{C}_2$ are $\{1, 3\}$, $\{1, 4\}$, $\{2, 3\}$, and $\{2, 4\}$.

Example 1.4.9 Generator and parity check matrices for the $[7, 4]_2$ binary linear Hamming code $\mathcal{H}_{3,2}$ are

$$
G_{3,2} = \left[\begin{array}{cccc|ccc}
1 & 0 & 0 & 0 & 0 & 1 & 1 \\
0 & 1 & 0 & 0 & 1 & 0 & 1 \\
0 & 0 & 1 & 0 & 1 & 1 & 0 \\
0 & 0 & 0 & 1 & 1 & 1 & 1
\end{array}\right] \quad \text{and} \quad H_{3,2} = \left[\begin{array}{cccc|ccc}
0 & 1 & 1 & 1 & 1 & 0 & 0 \\
1 & 0 & 1 & 1 & 0 & 1 & 0 \\
1 & 1 & 0 & 1 & 0 & 0 & 1
\end{array}\right],
$$

respectively. $G_{3,2}$ is in standard form. Two information sets for $\mathcal{H}_{3,2}$ are $\{1, 2, 3, 4\}$ and $\{1, 2, 3, 5\}$ with corresponding redundancy sets $\{5, 6, 7\}$ and $\{4, 6, 7\}$. The set $\{2, 3, 4, 5\}$ is not an information set. More general Hamming codes $\mathcal{H}_{m,q}$ are defined in Section 1.10.

1.5 Orthogonality

There is a natural inner product on $\mathbb{F}_q^n$ that often proves useful in the study of codes.[2]

Definition 1.5.1 The **ordinary inner product**, also called the **Euclidean inner product**, on $\mathbb{F}_q^n$ is defined by $\mathbf{x} \cdot \mathbf{y} = \sum_{i=1}^n x_i y_i$ where $\mathbf{x} = x_1 x_2 \cdots x_n$ and $\mathbf{y} = y_1 y_2 \cdots y_n$. Two vectors $\mathbf{x}, \mathbf{y} \in \mathbb{F}_q^n$ are **orthogonal** if $\mathbf{x} \cdot \mathbf{y} = 0$. If $\mathcal{C}$ is an $[n, k]_q$ code,

$$
\mathcal{C}^\perp = \{\mathbf{x} \in \mathbb{F}_q^n \mid \mathbf{x} \cdot \mathbf{c} = 0 \text{ for all } \mathbf{c} \in \mathcal{C}\}
$$

---

[2]There are other inner products used in coding theory. See for example Chapters 4, 5, 7, 11, and 13.

is the **orthogonal code** or **dual code** of C. C is **self-orthogonal** if $C \subseteq C^\perp$ and **self-dual** if $C = C^\perp$.

Theorem 1.5.2 ([1323, Chapter 1.8]) *Let C be an $[n, k]_q$ code with generator and parity check matrices G and H, respectively. Then $C^\perp$ is an $[n, n-k]_q$ code with generator and parity check matrices H and G, respectively. Additionally $(C^\perp)^\perp = C$. Furthermore C is self-dual if and only if C is self-orthogonal and $k = \dfrac{n}{2}$.*

Example 1.5.3 C_2 from Example 1.4.8 is a $[4, 2]_2$ self-dual code with generator and parity check matrices both equal to

$$\begin{bmatrix} 1 & 1 & 0 & 0 \\ 0 & 0 & 1 & 1 \end{bmatrix}.$$

The dual of the Hamming $[7, 4]_2$ code in Example 1.4.9 is a $[7, 3]_2$ code $\mathcal{H}_{3,2}^\perp$. $H_{3,2}$ is a generator matrix of $\mathcal{H}_{3,2}^\perp$. As every row of $H_{3,2}$ is orthogonal to itself and every other row of $H_{3,2}$, $\mathcal{H}_{3,2}^\perp$ is self-orthogonal. As $\mathcal{H}_{3,2}^\perp$ has dimension 3 and $(\mathcal{H}_{3,2}^\perp)^\perp = \mathcal{H}_{3,2}$ has dimension 4, $\mathcal{H}_{3,2}^\perp$ is not self-dual.

1.6 Distance and Weight

The error-correcting capability of a code is keyed directly to the concepts of Hamming distance and Hamming weight.[3]

Definition 1.6.1 The **(Hamming) distance** between two vectors $\mathbf{x}, \mathbf{y} \in \mathbb{F}_q^n$, denoted $d_H(\mathbf{x}, \mathbf{y})$, is the number of coordinates in which $\mathbf{x}$ and $\mathbf{y}$ differ. The **(Hamming) weight** of $\mathbf{x} \in \mathbb{F}_q^n$, denoted $wt_H(\mathbf{x})$, is the number of coordinates in which $\mathbf{x}$ is nonzero.

Theorem 1.6.2 ([1008, Chapter 1.4]) *The following hold.*

(a) (nonnegativity) $d_H(\mathbf{x}, \mathbf{y}) \geq 0$ *for all* $\mathbf{x}, \mathbf{y} \in \mathbb{F}_q^n$.

(b) $d_H(\mathbf{x}, \mathbf{y}) = 0$ *if and only if* $\mathbf{x} = \mathbf{y}$.

(c) (symmetry) $d_H(\mathbf{x}, \mathbf{y}) = d_H(\mathbf{y}, \mathbf{x})$ *for all* $\mathbf{x}, \mathbf{y} \in \mathbb{F}_q^n$.

(d) (triangle inequality) $d_H(\mathbf{x}, \mathbf{z}) \leq d_H(\mathbf{x}, \mathbf{y}) + d_H(\mathbf{y}, \mathbf{z})$ *for all* $\mathbf{x}, \mathbf{y}, \mathbf{z} \in \mathbb{F}_q^n$.

(e) $d_H(\mathbf{x}, \mathbf{y}) = wt_H(\mathbf{x} - \mathbf{y})$ *for all* $\mathbf{x}, \mathbf{y} \in \mathbb{F}_q^n$.

(f) *If* $\mathbf{x}, \mathbf{y} \in \mathbb{F}_2^n$, *then*

$$wt_H(\mathbf{x} + \mathbf{y}) = wt_H(\mathbf{x}) + wt_H(\mathbf{y}) - 2wt_H(\mathbf{x} \star \mathbf{y})$$

where $\mathbf{x} \star \mathbf{y}$ *is the vector in* $\mathbb{F}_2^n$ *which has 1s precisely in those coordinates where both* $\mathbf{x}$ *and* $\mathbf{y}$ *have 1s.*

(g) *If* $\mathbf{x}, \mathbf{y} \in \mathbb{F}_2^n$, *then* $wt_H(\mathbf{x} \star \mathbf{y}) \equiv \mathbf{x} \cdot \mathbf{y} \pmod{2}$. *In particular,* $wt_H(\mathbf{x}) \equiv \mathbf{x} \cdot \mathbf{x} \pmod{2}$.

---

[3]There are other notions of distance and weight used in coding theory. See for example Chapters 6, 7, 10, 11, 17, 18, 22, and 29.

(h) *If* $\mathbf{x} \in \mathbb{F}_3^n$, *then* $\mathrm{wt}_\mathrm{H}(\mathbf{x}) \equiv \mathbf{x} \cdot \mathbf{x} \pmod 3$.

Remark 1.6.3 A distance function on a vector space that satisfies parts (a) through (d) of Theorem 1.6.2 is called a **metric**; thus d_H is termed the **Hamming metric**. Other metrics useful in coding theory are examined in Chapter 22.

Definition 1.6.4 Let $\mathcal{C}$ be an $(n, M)_q$ code with $M > 1$. The **minimum (Hamming) distance** of $\mathcal{C}$ is the smallest distance between distinct codewords. If the minimum distance d of $\mathcal{C}$ is known, $\mathcal{C}$ is denoted an $(n, M, d)_q$ code (or an $[n, k, d]_q$ code if $\mathcal{C}$ is linear of dimension k). The **(Hamming) distance distribution** or **inner distribution** of $\mathcal{C}$ is the list $B_0(\mathcal{C}), B_1(\mathcal{C}), \ldots, B_n(\mathcal{C})$ where, for $0 \le i \le n$,

$$B_i(\mathcal{C}) = \frac{1}{M} \sum_{\mathbf{c} \in \mathcal{C}} |\{\mathbf{v} \in \mathcal{C} \mid d_\mathrm{H}(\mathbf{v}, \mathbf{c}) = i\}|.$$

The **minimum (Hamming) weight** of a nonzero code $\mathcal{C}$ is the smallest weight of nonzero codewords. The **(Hamming) weight distribution** of $\mathcal{C}$ is the list $A_0(\mathcal{C}), A_1(\mathcal{C}), \ldots, A_n(\mathcal{C})$ where, for $0 \le i \le n$, $A_i(\mathcal{C})$ is the number of codewords of weight i. If $\mathcal{C}$ is understood, the distance and weight distributions of $\mathcal{C}$ are denoted $B_0, B_1, \ldots, B_n$ and $A_0, A_1, \ldots, A_n$, respectively.

Example 1.6.5 Let $\mathcal{C}$ be the $(4, 6)_2$ code in Example 1.3.2. Its distance distribution is $B_0(\mathcal{C}) = B_4(\mathcal{C}) = 1$, $B_2(\mathcal{C}) = 4$, $B_1(\mathcal{C}) = B_3(\mathcal{C}) = 0$, and its minimum distance is 2. In particular $\mathcal{C}$ is a $(4, 6, 2)_2$ code. The weight distribution of $\mathcal{C}$ is $A_2(\mathcal{C}) = 6$ with $A_i(\mathcal{C}) = 0$ otherwise; its minimum weight is also 2. Let $\mathcal{C}' = 1000 + \mathcal{C} = \{0100, 0010, 0001, 1110, 1101, 1011\}$. The distance distribution of $\mathcal{C}'$ agrees with the distance distribution of $\mathcal{C}$ making $\mathcal{C}'$ a $(4, 6, 2)_2$ code. However, the weight distribution of $\mathcal{C}'$ is $A_1(\mathcal{C}') = A_3(\mathcal{C}') = 3$ with $A_i(\mathcal{C}') = 0$ otherwise; the minimum weight of $\mathcal{C}'$ is 1.

Theorem 1.6.6 ([1008, Chapter 1.4]) *Let $\mathcal{C}$ be an $[n, k, d]_q$ linear code with $k > 0$. The following hold.*

(a) *The minimum distance and minimum weight of $\mathcal{C}$ are the same.*

(b) $A_i(\mathcal{C}) = B_i(\mathcal{C})$ *for* $0 \le i \le n$.

(c) $\displaystyle\sum_{i=0}^{n} A_i(\mathcal{C}) = q^k$.

(d) $A_0(\mathcal{C}) = 1$ *and* $A_i(\mathcal{C}) = 0$ *for* $1 \le i < d$.

(e) *If* $q = 2$ *and* $\mathbf{1} = 11 \cdots 1 \in \mathcal{C}$, *then* $A_i(\mathcal{C}) = A_{n-i}(\mathcal{C})$ *for* $0 \le i \le n$.

(f) *If* $q = 2$ *and* $\mathcal{C}$ *is self-orthogonal, every codeword of $\mathcal{C}$ has even weight and* $\mathbf{1} \in \mathcal{C}^\perp$.

(g) *If* $q = 3$ *and* $\mathcal{C}$ *is self-orthogonal, every codeword of $\mathcal{C}$ has weight a multiple of 3.*

Remark 1.6.7 Analogous to Theorem 1.6.6(c) and (d), if $\mathcal{C}$ is an $(n, M, d)_q$ code, then

$$\sum_{i=0}^{n} B_i(\mathcal{C}) = M \text{ with } B_0(\mathcal{C}) = 1 \text{ and } B_i(\mathcal{C}) = 0 \text{ for } 1 \le i < d.$$

Binary vectors possess an important relationship between weights and inner products. If $\mathbf{x}, \mathbf{y} \in \mathbb{F}_2^n$ and each have even weight, Theorem 1.6.2(f) implies $\mathbf{x} + \mathbf{y}$ also has even weight. If $\mathbf{x}, \mathbf{y} \in \mathbb{F}_2^n$ are orthogonal and each have weights a multiple of 4, Theorem 1.6.2(f) and (g) show that $\mathbf{x} + \mathbf{y}$ has weight a multiple of 4. This leads to the following definition for binary codes.

Definition 1.6.8 Let $\mathcal{C}$ be a binary linear code. $\mathcal{C}$ is called **even** if all of its codewords have even weight. $\mathcal{C}$ is called **doubly-even** if all of its codewords have weights a multiple of 4. An even binary code that is not doubly-even is **singly-even**.

Remark 1.6.9 By Theorem 1.6.6(e), self-orthogonal binary linear codes are even. The converse is not true; code $\mathcal{C}_1$ from Example 1.3.2 is even but not self-orthogonal. Doubly-even binary linear codes must be self-orthogonal by Theorem 1.6.2(f) and (g). There are self-orthogonal binary codes that are singly-even; code $\mathcal{C}_2$ from Examples 1.4.8 and 1.5.3 is singly-even and self-dual.

Example 1.6.10 Let $\mathcal{H}_{3,2}$ be the $[7,4]_2$ binary Hamming code of Example 1.4.9. With $A_i = A_i(\mathcal{H}_{3,2})$, $A_0 = A_7 = 1$, $A_3 = A_4 = 7$, and $A_1 = A_2 = A_5 = A_6 = 0$, illustrating Theorem 1.6.6(d) and (e), and showing $\mathcal{H}_{3,2}$ is a $[7,4,3]_2$ code. The $[7,3]_2$ dual code $\mathcal{H}_{3,2}^{\perp}$ is self-orthogonal by Example 1.5.3 and hence even. Also by self-orthogonality, $\mathcal{H}_{3,2}^{\perp} \subseteq (\mathcal{H}_{3,2}^{\perp})^{\perp} = \mathcal{H}_{3,2}$; the weight distribution of $\mathcal{H}_{3,2}$ shows that the 8 codewords of weights 0 and 4 must be precisely the codewords of $\mathcal{H}_{3,2}^{\perp}$. In particular, $\mathcal{H}_{3,2}^{\perp}$ is a doubly-even $[7,3,4]_2$ code. $\mathcal{H}_{3,2}^{\perp}$ is called a **simplex** code, described further in Section 1.10.

The minimum weight of a linear code is determined by a parity check matrix for the code; see [1008, Corollary 1.4.14 and Theorem 1.4.15].

Theorem 1.6.11 *A linear code has minimum weight d if and only if its parity check matrix has a set of d linearly dependent columns but no set of $d-1$ linearly dependent columns. Also, if $\mathcal{C}$ is an $[n, k, d]_q$ code, then every $n - d + 1$ coordinate positions contain an information set; furthermore, d is the largest number with this property.*

1.7 Puncturing, Extending, and Shortening Codes

There are several methods to obtain a longer or shorter code from a given code; while this can be done for both linear and nonlinear codes, we focus on linear ones. Two codes can be combined into a single code, for example as described in Section 1.11.

Definition 1.7.1 Let $\mathcal{C}$ be an $[n, k, d]_q$ linear code with generator matrix G and parity check matrix H.

(a) For some i with $1 \le i \le n$, let $\mathcal{C}^*$ be the codewords of $\mathcal{C}$ with the ith component deleted. The resulting code, called a **punctured code**, is an $[n - 1, k^*, d^*]$ code. If $d > 1$, $k^* = k$, and $d^* = d$ unless $\mathcal{C}$ has a minimum weight codeword that is nonzero on coordinate i, in which case $d^* = d - 1$. If $d = 1$, $k^* = k$ and $d^* = 1$ unless $\mathcal{C}$ has a weight 1 codeword that is nonzero on coordinate i, in which case $k^* = k - 1$ and $d^* \ge 1$ as long as $\mathcal{C}^*$ is nonzero. A generator matrix for $\mathcal{C}^*$ is obtained from G by deleting column i; G^* will have dependent rows if $d^* = 1$ and $k^* = k - 1$. Puncturing is often done on multiple coordinates in an analogous manner, one coordinate at a time.

(b) Define $\widehat{\mathcal{C}} = \{c_1 c_2 \cdots c_{n+1} \in \mathbb{F}_q^{n+1} \mid c_1 c_2 \cdots c_n \in \mathcal{C} \text{ where } \sum_{i=1}^{n+1} c_i = 0\}$, called the **extended code**. This is an $[n + 1, k, \widehat{d}]_q$ code where $\widehat{d} = d$ or $d + 1$. A generator matrix $\widehat{G}$ for $\widehat{C}$ is obtained by adding a column on the right of G so that every row sum in this $k \times (n+1)$ matrix is 0. A parity check matrix $\widehat{H}$ for $\widehat{C}$ is

$$\widehat{H} = \left[\begin{array}{ccc|c} 1 & \cdots & 1 & 1 \\ \hline & & & 0 \\ & H & & \vdots \\ & & & 0 \end{array}\right].$$

(c) Let S be any set of s coordinates. Let $C(S)$ be all codewords in C that are zero on S. Puncturing $C(S)$ on S results in the $[n-s, k_S, d_S]_q$ **shortened code** C_S where $d_S \geq d$. If $C^\perp$ has minimum weight $d^\perp$ and $s < d^\perp$, then $k_S = k - s$.

Example 1.7.2 Let $\mathcal{H}_{3,2}$ be the $[7,4,3]_2$ binary Hamming code of Examples 1.4.9 and 1.6.10. Extending this code, we obtain $\widehat{\mathcal{H}}_{3,2}$ with generator and parity check matrices

$$\widehat{G}_{3,2} = \left[\begin{array}{cccc|cccc} 1 & 0 & 0 & 0 & 0 & 1 & 1 & 1 \\ 0 & 1 & 0 & 0 & 1 & 0 & 1 & 1 \\ 0 & 0 & 1 & 0 & 1 & 1 & 0 & 1 \\ 0 & 0 & 0 & 1 & 1 & 1 & 1 & 0 \end{array}\right] \text{ and } \widehat{H}_{3,2} = \left[\begin{array}{cccc|cccc} 1 & 1 & 1 & 1 & 1 & 1 & 1 & 1 \\ 0 & 1 & 1 & 1 & 1 & 0 & 0 & 0 \\ 1 & 0 & 1 & 1 & 0 & 1 & 0 & 0 \\ 1 & 1 & 0 & 1 & 0 & 0 & 1 & 0 \end{array}\right],$$

respectively. Given the weight distribution of $\mathcal{H}_{3,2}$ found in Example 1.6.10, the weight distribution of $\widehat{\mathcal{H}}_{3,2}$ must be $A_0(\widehat{\mathcal{H}}_{3,2}) = A_8(\widehat{\mathcal{H}}_{3,2}) = 1$, $A_4(\widehat{\mathcal{H}}_{3,2}) = 14$, and $A_i(\widehat{\mathcal{H}}_{3,2}) = 0$ otherwise, implying $\widehat{\mathcal{H}}_{3,2}$ is doubly-even and self-dual; see Remark 1.6.9. Certainly if $\widehat{\mathcal{H}}_{3,2}$ is punctured on its right-most coordinate, the resulting code is $\mathcal{H}_{3,2}$.

There is a relationship between punctured and shortened codes via dual codes.

Remark 1.7.3 If C is a linear code over $\mathbb{F}_q$ and S a set of coordinates, then $(C^\perp)_S = (C^S)^\perp$ and $(C^\perp)^S = (C_S)^\perp$ where C^S and $(C^\perp)^S$ are C and $C^\perp$ punctured on S; see [1008, Theorem 1.5.7].

1.8 Equivalence and Automorphisms

Two vector spaces over $\mathbb{F}_q$ are considered the *same* (that is, isomorphic) if there is a nonsingular linear transformation from one to the other. For linear codes to be considered the *same*, we want these linear transformations to also preserve weights of codewords. In Theorem 1.8.6, we will see that these weight preserving linear transformations are directly related to monomial matrices. This leads to two different concepts of code equivalence for linear codes.

Definition 1.8.1 If $P \in \mathbb{F}_q^{n \times n}$ has exactly one 1 in each row and column and 0 elsewhere, P is a **permutation matrix**. If $M \in \mathbb{F}_q^{n \times n}$ has exactly one nonzero entry in each row and column, M is a **monomial matrix**. If C is a code over $\mathbb{F}_q$ of length n and $A \in \mathbb{F}_q^{n \times n}$, then $CA = \{cA \mid c \in C\}$. Let C_1 and C_2 be linear codes over $\mathbb{F}_q$ of length n. C_1 is **permutation equivalent** to C_2 provided $C_2 = C_1 P$ for some permutation matrix $P \in \mathbb{F}_q^{n \times n}$. C_1 is **monomially equivalent** to C_2 provided $C_2 = C_1 M$ for some monomial matrix $M \in \mathbb{F}_q^{n \times n}$.

Remark 1.8.2 Applying a permutation matrix to a code simply permutes the coordinates; applying a monomial matrix permutes and re-scales coordinates. Applying either a permutation or monomial matrix to a vector does not change its weight. Also applying either a permutation or monomial matrix to two vectors does not change the distance between these two vectors. There is a third more general concept of equivalence, involving semi-linear transformations, where two linear codes C_1 and C_2 over $\mathbb{F}_q$ are **equivalent** provided one can be obtained from the other by permuting and re-scaling coordinates and then applying an automorphism of the field $\mathbb{F}_q$. Note that applying such maps to a vector or to a pair of vectors preserves the weight of the vector and the distance between the two vectors, respectively; see [1008, Section 1.7] for further discussion of this type of equivalence. There are other concepts of equivalence that arise when the code may not be linear but has some specific algebraic structure (e.g., additive codes over $\mathbb{F}_q$ that are closed under vector addition but not necessarily closed under scalar multiplication). The common theme when defining equivalence of such codes is to use a set of maps which preserve distance between the two vectors, which preserve the algebraic structure under consideration, and which form a group under composition of these maps. We will follow this theme when we define equivalence of unrestricted codes at the end of this section.

Remark 1.8.3 Let C_1 and C_2 be linear codes over $\mathbb{F}_q$ of length n. Define $C_1 \sim_P C_2$ to mean C_1 is permutation equivalent to C_2; similarly define $C_1 \sim_M C_2$ to mean C_1 is monomially equivalent to C_2. Then both $\sim_P$ and $\sim_M$ are equivalence relations on the set of all linear codes over $\mathbb{F}_q$ of length n; that is, both are reflexive, symmetric, and transitive. If $q = 2$, the concepts of permutation and monomial equivalence are the same; if $q > 2$, they may not be. Furthermore, two permutation or monomially equivalent codes have the same size, weight and distance distributions, and minimum weight and distance. If two linear codes are permutation equivalent and one code is self-orthogonal, so is the other; this may not be true of two monomially equivalent codes.

Row reducing a generator matrix of a linear code to reduced echelon form and then permuting columns yields the following result.

Theorem 1.8.4 *Let C be a linear $[n, k, d]_q$ code with $k \geq 1$. There is a code permutation equivalent to C with a generator matrix in standard form.*

Example 1.8.5 Let C be an $[8, 4, 4]_2$ binary linear code. By Theorem 1.8.4, C is permutation equivalent to a code with generator matrix $G = [I_4 \mid A]$. A straightforward argument using minimum weight 4 shows that columns of A can be permuted so that the resulting generator matrix is $\widehat{G}_3$ from Example 1.7.2. This verifies that C is permutation equivalent to $\widehat{\mathcal{H}}_{3,2}$.

The following is a generalization of a result of MacWilliams [1318]; see also [229, 1876]. This result motivated Definition 1.8.1.

Theorem 1.8.6 (MacWilliams Extension) *There is a weight preserving linear transformation between equal length linear codes C_1 and C_2 over $\mathbb{F}_q$ if and only if C_1 and C_2 are monomially equivalent. Furthermore, the linear transformation agrees with the associated monomial transformation on every codeword in C_1.*

Definition 1.8.7 Let C be a linear code over $\mathbb{F}_q$ of length n. If $CP = C$ for some permutation matrix $P \in \mathbb{F}_q^{n \times n}$, then P is a **permutation automorphism** of C; the set of all permutation automorphisms of C is a group under matrix multiplication, denoted $\mathrm{PAut}(C)$. Similarly, if $CM = C$ for some monomial matrix $M \in \mathbb{F}_q^{n \times n}$, then M is a **monomial automorphism** of C; the set of all monomial automorphisms of C is a matrix group, denoted $\mathrm{MAut}(C)$. Clearly $\mathrm{PAut}(C) \subseteq \mathrm{MAut}(C)$.

We now consider when two unrestricted codes are equivalent. It should be noted that, in this definition, a linear code may end up being equivalent to a nonlinear code. See Chapter 3 for more on this general equivalence.

Definition 1.8.8 Let $\mathcal{C}_1$ and $\mathcal{C}_2$ be unrestricted codes of length n over $\mathbb{F}_q$ of the same size. Then $\mathcal{C}_1$ is **equivalent** to $\mathcal{C}_2$ provided the codewords of $\mathcal{C}_2$ are the images under a map of the codewords of $\mathcal{C}_1$ where the map is a permutation of coordinates together with n permutations of the alphabet $\mathbb{F}_q$, independently within each coordinate.[4]

1.9 Bounds on Codes

In this section we present seven bounds relating the length, dimension or number of codewords, and minimum distance of an unrestricted code. The first five are considered upper bounds on the code size given length, minimum distance, and field size. By this, we mean that there does not exist a code of size bigger than the upper bound with the specified length, minimum distance, and field size. The last two are lower bounds on the size of a linear code. This means that a linear code can be constructed with the given length and minimum distance over the specified field having size equalling or exceeding the lower bound. We also give asymptotic versions of these bounds. Some of these bounds will be described using $A_q(n,d)$ and $B_q(n,d)$, which we now define.

Definition 1.9.1 For positive integers n and d, $A_q(n,d)$ is the **largest number of codewords in an** $(n,M,d)_q$ **code**, linear or nonlinear. $B_q(n,d)$ is the **largest number of codewords in a** $[n,k,d]_q$ **linear code**. An $(n,M,d)_q$ code is **optimal** provided $M = A_q(n,d)$; an $[n,k,d]_q$ linear code is **optimal** if $q^k = B_q(n,d)$. The concept of 'optimal' can also be used in other contexts. Given n and d, $k_q(n,d)$ denotes the largest dimension of a linear code over $\mathbb{F}_q$ of length n and minimum weight d; an $[n,k_q(n,d),d]_q$ code could be called 'optimal in dimension'. Notice that $k_q(n,d) = \log_q B_q(n,d)$. Similarly, $d_q(n,k)$ denotes the largest minimum distance of a linear code over $\mathbb{F}_q$ of length n and dimension k; an $[n,k,d_q(n,k)]_q$ code may be called 'optimal in distance'. Analogously, $n_q(k,d)$ denotes the smallest length of a linear code over $\mathbb{F}_q$ of dimension k and minimum weight d; an $[n_q(k,d),k,d]_q$ code might be called 'optimal in length'.[5]

Clearly $B_q(n,d) \le A_q(n,d)$. On-line tables relating parameters of various types of codes are maintained by M. Grassl [845].

The following basic properties of $A_q(n,d)$ and $B_q(n,d)$ are easily derived; see [1008, Chapter 2.1].

Theorem 1.9.2 *The following hold for* $1 \le d \le n$.

(a) $B_q(n,d) \le A_q(n,d)$.

(b) $B_q(n,n) = A_q(n,n) = q$ *and* $B_q(n,1) = A_q(n,1) = q^n$.

---

[4]In a more general setting, unrestricted codes do not have to have $\mathbb{F}_q$ as an alphabet. If $\mathfrak{A}$ is the alphabet, the permutations within each coordinate are permutations of $\mathfrak{A}$.

[5]Further restrictions might be placed on a family of codes when discussing optimality. For example, given n, a self-dual $[n, \frac{n}{2}, d]_q$ code over $\mathbb{F}_q$ with largest minimum weight d is sometimes called an 'optimal q-ary self-dual code of length n'. Optimal codes are explored in chapters such as 2–5, 11, 12, 16, 18, 20, and 23.

(c) $B_q(n, d) \leq qB_q(n - 1, d)$ and $A_q(n, d) \leq qA_q(n - 1, d)$ when $1 \leq d < n$.

(d) $B_q(n, d) \leq B_q(n - 1, d - 1)$ and $A_q(n, d) \leq A_q(n - 1, d - 1)$.

(e) If d is even, $B_2(n, d) = B_2(n - 1, d - 1)$ and $A_2(n, d) = A_2(n - 1, d - 1)$.

(f) If d is even and $M = A_2(n, d)$, then there is an $(n, M, d)_2$ code such that all codewords have even weight and the distance between all pairs of codewords is also even.

1.9.1   The Sphere Packing Bound

The Sphere Packing Bound, also called the Hamming Bound, is based on packing $\mathbb{F}_q^n$ with non-overlapping spheres.

Definition 1.9.3 The **sphere of radius** r **centered at** $\mathbf{u} \in \mathbb{F}_q^n$ is the set $S_{q,n,r}(\mathbf{u}) = \{\mathbf{v} \in \mathbb{F}_q^n \mid d_H(\mathbf{u}, \mathbf{v}) \leq r\}$ of all vectors in $\mathbb{F}_q^n$ whose distance from $\mathbf{u}$ is at most r.

We need the size of a sphere, which requires use of binomial coefficients.

Definition 1.9.4 For a, b integers with $0 \leq b \leq a$, $\binom{a}{b}$ is the number of b-element subsets in an a-element set. $\binom{a}{b} = \dfrac{a!}{b!(a - b)!}$ and is called a **binomial coefficient**.

The next result is the basis of the Sphere Packing Bound; part (a) is a direct count and part (b) follows from the triangle inequality of Theorem 1.6.2.

Theorem 1.9.5 *The following hold.*

(a) For $\mathbf{u} \in \mathbb{F}_q^n$, $|S_{q,n,r}(\mathbf{u})| = \sum_{i=0}^{r} \binom{n}{i}(q - 1)^i$.

(b) If $\mathcal{C}$ is an $(n, M, d)_q$ code and $t = \lfloor \frac{d-1}{2} \rfloor$, then spheres of radius t centered at distinct codewords are disjoint.

Theorem 1.9.6 (Sphere Packing (or Hamming) Bound) *Let* $d \geq 1$. *If* $t = \lfloor \frac{d-1}{2} \rfloor$, *then*

$$B_q(n, d) \leq A_q(n, d) \leq \frac{q^n}{\sum_{i=0}^{t} \binom{n}{i}(q - 1)^i}.$$

Proof: Let $\mathcal{C}$ be an $(n, M, d)_q$ code. By Theorem 1.9.5, the spheres of radius t centered at distinct codewords are disjoint, and each such sphere has $\alpha = \sum_{i=0}^{t} \binom{n}{i}(q - 1)^i$ total vectors. Thus $M\alpha$ cannot exceed the number q^n of vectors in $\mathbb{F}_q^n$. The result is now clear. $\qquad\square$

Remark 1.9.7 The Sphere Packing Bound is an upper bound on the size of a code given its length and minimum distance. Additionally the Sphere Packing Bound produces an upper bound on the minimum distance d of an $(n, M)_q$ code in the following sense. Given n, M, and q, compute the smallest positive integer s with $M > \dfrac{q^n}{\sum_{i=0}^{s} \binom{n}{i}(q - 1)^i}$; for an $(n, M, d)_q$ code to exist, $d < 2s - 1$.

Definition 1.9.8 If $\mathcal{C}$ is an $(n, M, d)_q$ code with $M = \dfrac{q^n}{\sum_{i=0}^{t} \binom{n}{i}(q-1)^i}$ (that is, equality holds in the Sphere Packing Bound), $\mathcal{C}$ is called a **perfect code**. Perfect codes are precisely those $(n, M, d)_q$ codes where the disjoint spheres of radius $t = \left\lfloor \frac{d-1}{2} \right\rfloor$ centered at all codewords fill the entire space $\mathbb{F}_q^n$. Perfect codes are discussed in Sections 3.3.1 and 5.3.

Example 1.9.9 The code $\mathcal{H}_{3,2}$ of Examples 1.4.9 and 1.6.10 is a $[7, 4, 3]_2$ code. So in this case $t = \left\lfloor \frac{3-1}{2} \right\rfloor = 1$ and $\frac{q^n}{\sum_{i=0}^{t} \binom{n}{i}(q-1)^i} = \frac{2^7}{1+7} = 2^4$ yielding equality in the Sphere Packing Bound. So $\mathcal{H}_{3,2}$ is perfect.

1.9.2 The Singleton Bound

The Singleton Bound was formulated in [1717]. As with the Sphere Packing Bound, the Singleton Bound is an upper bound on the size of a code.

Theorem 1.9.10 (Singleton Bound) *For $d \leq n$, $A_q(n, d) \leq q^{n-d+1}$. Furthermore, if an $[n, k, d]_q$ linear code exists, then $k \leq n - d + 1$; i.e., $k_q(n, d) \leq n - d + 1$.*

Remark 1.9.11 In addition to providing an upper bound on code size, the Singleton Bound yields the upper bound $d \leq n - \log_q(M) + 1$ on the minimum distance of an $(n, M, d)_q$ code.

Definition 1.9.12 A code for which equality holds in the Singleton Bound is called **maximum distance separable (MDS)**. No code of length n and minimum distance d has more codewords than an MDS code with parameters n and d; equivalently, no code of length n with M codewords has a larger minimum distance than an MDS code with parameters n and M. MDS codes are discussed in Chapters 3, 6, 8, 14, and 33.

The following theorem is proved using Theorem 1.6.11.

Theorem 1.9.13 *$\mathcal{C}$ is an $[n, k, n-k+1]_q$ MDS code if and only if $\mathcal{C}^\perp$ is an $[n, n-k, k+1]_q$ MDS code.*

Example 1.9.14 Let $\mathcal{H}_{2,3}$ be the $[4, 2]_3$ ternary linear code with generator matrix

$$G_{2,3} = \begin{bmatrix} 1 & 0 & 1 & 1 \\ 0 & 1 & 1 & -1 \end{bmatrix}.$$

Examining inner products of the rows of $G_{2,3}$, we see that $\mathcal{H}_{2,3}$ is self-orthogonal of dimension half its length; so it is self-dual. Using Theorem 1.6.2(h), $A_0(\mathcal{H}_{2,3}) = 1$, $A_3(\mathcal{H}_{2,3}) = 8$, and $A_i(\mathcal{H}_{2,3}) = 0$ otherwise. In particular $\mathcal{H}_{2,3}$ is a $[4, 2, 3]_3$ code and hence is MDS.

1.9.3 The Plotkin Bound

The Binary Plotkin Bound [1527] is an upper bound on the size of an unrestricted binary code of length n and minimum distance d provided d is close enough to n.

Theorem 1.9.15 (Binary Plotkin Bound) *Let $2d > n$. Then*

$$A_2(n, d) \leq 2 \left\lfloor \frac{d}{2d - n} \right\rfloor.$$

This result is generalized in [230] to unrestricted codes over $\mathbb{F}_q$.

Theorem 1.9.16 (Generalized Plotkin Bound) *If an* $(n, M, d)_q$ *code exists, then*

$$M(M-1)d \leq 2n \sum_{i=0}^{q-2} \sum_{j=i+1}^{q-1} M_i M_j$$

where $M_i = \left\lfloor \dfrac{M+i}{q} \right\rfloor.$

Example 1.9.17 The Sphere Packing Bound yields $A_2(17,9) \leq \frac{131\,072}{3\,214}$ and $A_2(18,10) \leq \frac{262\,144}{4\,048}$; so $A_2(17,9) \leq 40$ and $A_2(18,10) \leq 64$. The Singleton Bound produces $A_2(17,9) \leq 512$ and $A_2(18,10) \leq 512$. The Binary Plotkin Bound gives $A_2(17,9) \leq 18$ and $A_2(18,10) \leq 10$. Using Theorem 1.9.2(e), the Plotkin Bound is best with $A_2(18,10) = A_2(17,9) \leq 10$. According to [845], there is a $(18,10,10)_2$ code implying $A_2(18,10) = A_2(17,9) = 10$.

1.9.4 The Griesmer Bound

The Griesmer Bound [855] is a lower bound on the length of a linear code given its dimension and minimum weight.

Theorem 1.9.18 (Griesmer Bound) *Let* $\mathcal{C}$ *be an* $[n, k, d]_q$ *linear code with* $k \geq 1$. *Then*

$$n \geq \sum_{i=0}^{k-1} \left\lceil \frac{d}{q^i} \right\rceil.$$

Remark 1.9.19 One can interpret the Griesmer Bound as an upper bound on the code size given its length and minimum weight. Specifically, $B_q(n, d) \leq q^k$ where k is the largest positive integer such that $n \geq \sum_{i=0}^{k-1} \lceil \frac{d}{q^i} \rceil$. This bound can also be interpreted as a lower bound on the length of a linear code of given dimension and minimum weight; that is, $n_q(k, d) \geq \sum_{i=0}^{k-1} \lceil \frac{d}{q^i} \rceil$. Finally, the Griesmer Bound can be understood as an upper bound on the minimum weight given the code length and dimension; given n and k, $d_q(n, k)$ is at most the largest d for which the bound holds.

Example 1.9.20 Suppose we wish to find the smallest code length n such that an $[n, 4, 3]_2$ code can exist. By the Griesmer Bound $n \geq \lceil \frac{3}{1} \rceil + \lceil \frac{3}{2} \rceil + \lceil \frac{3}{4} \rceil + \lceil \frac{3}{8} \rceil = 3 + 2 + 1 + 1 = 7$. Note that equality in this bound is attained by the $[7, 4, 3]_2$ code $\mathcal{H}_{3,2}$ of Examples 1.4.9 and 1.6.10.

1.9.5 The Linear Programming Bound

The Linear Programming Bound is a result of the work of P. Delsarte in [517, 519, 521]. This is generally the most powerful of the bounds but does require setting up and solving a linear program involving Krawtchouk polynomials.

Definition 1.9.21 For $0 \leq k \leq n$, define the **Krawtchouk polynomial** $K_k^{(n,q)}(x)$ **of degree** k to be

$$K_k^{(n,q)}(x) = \sum_{j=0}^{k} (-1)^j (q-1)^{k-j} \binom{x}{j} \binom{n-x}{k-j}.$$

An extensive presentation of properties of the Krawtchouk polynomials can be found in [1229, 1365] and in Section 12.1. A simple proof of the following result, known as the Delsarte–MacWilliams Inequalities, is found in [1885].

Theorem 1.9.22 (Delsarte–MacWilliams Inequalities) *Let $\mathcal{C}$ be an $(n, M, d)_q$ code with distance distribution $B_i(\mathcal{C})$ for $0 \leq i \leq n$. Then for $0 \leq k \leq n$*

$$\sum_{i=0}^{n} B_i(\mathcal{C}) K_k^{(n,q)}(i) \geq 0.$$

Let $\mathcal{C}$ be an $(n, M, d)_q$ code with distance distribution $B_i(\mathcal{C})$ for $0 \leq i \leq n$. By Remark 1.6.7, $M = \sum_{i=0}^{n} B_i(\mathcal{C})$, $B_0(\mathcal{C}) = 1$, and $B_i(\mathcal{C}) = 0$ for $1 \leq i \leq d-1$. Although $B_i(\mathcal{C})$ may not be an integer, $B_i(\mathcal{C}) \geq 0$. By the Delsarte–MacWilliams Inequalities, we also have $\sum_{i=0}^{n} B_i(\mathcal{C}) K_k^{(n,q)}(i) \geq 0$ for $0 \leq i \leq n$. As $K_0^{(n,q)}(i) = 1$, the 0th Delsarte–MacWilliams Inequality is merely $\sum_{i=0}^{n} B_i(\mathcal{C}) \geq 0$, which is clearly already true. If $q = 2$, there are additional inequalities that hold. When $q = 2$, it is straightforward to show that $B_n(\mathcal{C}) \leq 1$. Furthermore when $q = 2$ and d is even, we may also assume that $B_i(\mathcal{C}) = 0$ when i is odd by Theorem 1.9.2(f). Properties of binomial coefficients show that $K_k^{(n,2)}(i) = (-1)^i K_{n-k}^{(n,2)}(i)$; thus the kth Delsarte–MacWilliams Inequality is the same as the $(n-k)^{\text{th}}$ Delsarte–MacWilliams Inequality because $B_i(\mathcal{C}) = 0$ when i is odd. This discussion leads to the linear program that is set up to establish an upper bound on $A_q(n, d)$.

Theorem 1.9.23 (Linear Programming Bound) *The following hold.*

(a) *When $q \geq 2$, $A_q(n, d) \leq \max \{\sum_{i=0}^{n} B_i\}$ where the maximum is taken over all B_i subject to the following conditions:*

   (i) *$B_0 = 1$ and $B_i = 0$ for $1 \leq i \leq d-1$,*

   (ii) *$B_i \geq 0$ for $d \leq i \leq n$, and*

   (iii) *$\sum_{i=0}^{n} B_i K_k^{(n,q)}(i) \geq 0$ for $1 \leq k \leq n$.*

(b) *When d is even and $q = 2$, $A_2(n, d) \leq \max \{\sum_{i=0}^{n} B_i\}$ where the maximum is taken over all B_i subject to the following conditions:*

   (i) *$B_0 = 1$ and $B_i = 0$ for $1 \leq i \leq d-1$ and all odd i,*

   (ii) *$B_i \geq 0$ for $d \leq i \leq n$ and $B_n \leq 1$, and*

   (iii) *$\sum_{i=0}^{n} B_i K_k^{(n,2)}(i) \geq 0$ for $1 \leq k \leq \lfloor \frac{n}{2} \rfloor$.*

Sometimes additional constraints can be added to the linear program and reduce the size of $\max \{\sum_{i=0}^{n} B_i\}$. Linear Programming Bounds will be considered in more detail in Chapters 12 and 13.

1.9.6 The Gilbert Bound

The Gilbert Bound [806] is a lower bound on $B_q(n, d)$ and hence a lower bound on $A_q(n, d)$.

Theorem 1.9.24 (Gilbert Bound)

$$B_q(n, d) \geq \frac{q^n}{\sum_{i=0}^{d-1} \binom{n}{i}(q-1)^i}.$$

1.9.7   The Varshamov Bound

The Varshamov Bound [1844] is similar to the Gilbert Bound; asymptotically they are the same as stated in Section 1.9.8.

Theorem 1.9.25 (Varshamov Bound)

$$B_q(n,d) \geq q^{n-\lceil \log_q(1+\sum_{i=0}^{d-2} \binom{n-1}{i}(q-1)^i) \rceil}.$$

1.9.8   Asymptotic Bounds

We now describe what happens to the bounds, excluding the Griesmer Bound, as the code length approaches infinity; these bounds are termed **asymptotic bounds**. We first need some terminology.

Definition 1.9.26 The **information rate**, or simply **rate**, of an $(n, M, d)_q$ code is defined to be $\dfrac{\log_q M}{n}$. If the code is actually an $[n, k, d]_q$ linear code, its rate is $\dfrac{k}{n}$, measuring the number of information coordinates relative to the total number of coordinates. In either the linear or nonlinear case, the higher the rate, the higher the proportion of coordinates in a codeword that actually contain information rather than redundancy. The ratio $\dfrac{d}{n}$ is called the **relative distance** of the code; as we will see later, the relative distance is a measure of the error-correcting capability of the code relative to its length.

Each asymptotic bound will be either an upper or lower bound on the largest possible rate for a family of (possibly nonlinear) codes over $\mathbb{F}_q$ of lengths going to infinity with relative distances approaching δ. The function, called the **asymptotic normalized rate function**, that determines this rate is

$$\alpha_q(\delta) = \limsup_{n \to \infty} \frac{\log_q A_q(n, \delta n)}{n}.$$

As the exact value of $\alpha_q(\delta)$ is unknown, we desire upper and lower bounds on this function. An upper bound would indicate that all families with relative distances approaching δ have rates, in the limit, at most this upper bound. A lower bound indicates that there exists a family of codes of lengths approaching infinity and relative distances approaching δ whose rates are at least this bound. Three of the bounds in the next theorem involve the entropy function.

Definition 1.9.27 The **entropy function** is defined for $0 \leq x \leq r = 1 - q^{-1}$ by

$$H_q(x) = \begin{cases} 0 & \text{if } x = 0, \\ x \log_q(q-1) - x \log_q x - (1-x) \log_q(1-x) & \text{if } 0 < x \leq r. \end{cases}$$

Discussion and proofs of the asymptotic bounds can be found in [1008, 1323, 1505, 1836]. The **MRRW Bound**, named after the authors of [1365] who developed the bound, is the Asymptotic Linear Programming Bound. The MRRW Bound has been improved by M. Aaltonnen [2] in the case $q > 2$.

Theorem 1.9.28 (Asymptotic Bounds) *Let $q \geq 2$ and $r = 1 - q^{-1}$. The following hold.*

(a) (Asymptotic Sphere Packing) $\alpha_q(\delta) \leq 1 - H_q(\delta/2)$ *if $0 < \delta \leq r$.*

(b) (Asymptotic Singleton) $\alpha_q(\delta) \leq 1 - \delta$ *if* $0 \leq \delta \leq 1$.

(c) (Asymptotic Plotkin) $\alpha_q(\delta) = 0$ *if* $r \leq \delta \leq 1$ *and* $\alpha_q(\delta) \leq 1 - \frac{\delta}{r}$ *if* $0 \leq \delta \leq r$.

(d) (MRRW)

    (i) (First MRRW)
$$\alpha_q(\delta) \leq H_q\left(\frac{1}{q}\left(q - 1 - (q-2)\delta - 2\sqrt{(q-1)\delta(1-\delta)}\right)\right) \text{ if } 0 < \delta < r.$$

    (ii) (Second MRRW) *Let* $g(x) = H_2((1 - \sqrt{1-x})/2)$.
$$\alpha_2(\delta) \leq \min_{0 \leq u \leq 1-2\delta}\{1 + g(u^2) - g(u^2 + 2\delta u + 2\delta)\} \text{ if } 0 < \delta < 1/2.$$

(e) (Asymptotic Gilbert–Varshamov) $1 - H_q(\delta) \leq \alpha_q(\delta)$ *if* $0 < \delta \leq r$.

1.10   Hamming Codes

A binary code permutation equivalent to the code of Example 1.4.9 was discovered in 1947 by R. W. Hamming while working at Bell Telephone Laboratories. Because of patent considerations, his work was not published until 1950; see [895]. This Hamming code actually appeared earlier in C. E. Shannon's seminal paper [1661]. It was also generalized to codes over fields of prime order by M. J. E. Golay [820].

Given a positive integer m, if one takes an $m \times n$ binary matrix whose columns are nonzero and distinct, the binary code with this parity check matrix must have minimum weight at least 3 by Theorem 1.6.11. Binary Hamming codes $\mathcal{H}_{m,2}$ arise by choosing an $m \times n$ parity check matrix with the maximum number of columns possible that are distinct and nonzero.

Definition 1.10.1 Let $m \geq 2$ be an integer and $n = 2^m - 1$. Let $H_{m,2}$ be an $m \times n$ matrix whose columns are all $2^m - 1$ distinct nonzero binary m-tuples. A code with this parity check matrix is called a **binary Hamming code**. Changing the column order of $H_{m,2}$ produces a set of pairwise permutation equivalent codes. Any code in this list is denoted $\mathcal{H}_{m,2}$ and is a $[2^m - 1, 2^m - 1 - m, 3]_2$ code.

The code $\mathcal{H}_{3,2}$ of Example 1.4.9 is indeed a binary Hamming code. These codes are generalized to Hamming codes $\mathcal{H}_{m,q}$ over $\mathbb{F}_q$, all with minimum weight 3 again from Theorem 1.6.11.

Definition 1.10.2 Let $m \geq 2$ be an integer and $n = (q^m - 1)/(q-1)$. There are a total of n 1-dimensional subspaces of $\mathbb{F}_q^m$. Let $H_{m,q}$ be an $m \times n$ matrix whose columns are all nonzero m-tuples with one column from each of the distinct 1-dimensional subspaces of $\mathbb{F}_q^m$. A code with this parity check matrix is called a **Hamming code over** $\mathbb{F}_q$. Re-scaling columns and/or changing column order of $H_{m,q}$ produces a set of pairwise monomially equivalent codes. Any code in this list is denoted $\mathcal{H}_{m,q}$ and is a $\left[(q^m-1)/(q-1), (q^m-1)/(q-1)-m, 3\right]_q$ code. The code $\mathcal{H}_{m,q}^\perp$ is called a **simplex code**.

Example 1.10.3 The parity check matrix of the code in Example 1.9.14 is
$$\begin{bmatrix} -1 & -1 & 1 & 0 \\ -1 & 1 & 0 & 1 \end{bmatrix}.$$

This code satisfies the definition of a Hamming $[4, 2, 3]_3$ code, and so $\mathcal{H}_{2,3}$ is the appropriate labeling of this code.

The parameters of the Hamming codes in fact determine the code. That $\mathcal{H}_{m,q}$ is perfect follows by direct computation from Definition 1.9.8.

Theorem 1.10.4 *The following hold.*

(a) *If $\mathcal{C}$ is a $[2^m - 1, 2^m - 1 - m, 3]_2$ binary linear code, then $\mathcal{C}$ is permutation equivalent to $\mathcal{H}_{m,2}$.*

(b) *If $\mathcal{C}$ is a $[(q^m - 1)/(q-1), (q^m - 1)/(q-1) - m, 3]_q$ linear code, then $\mathcal{C}$ is monomially equivalent to $\mathcal{H}_{m,q}$.*

(c) *$\mathcal{H}_{m,q}$ is perfect.*

The weight distribution of $\mathcal{H}_{3,2}^\perp$ was given in Example 1.6.10. The following generalizes this; for a proof see [1008, Theorem 2.7.5].

Theorem 1.10.5 *The nonzero codewords of the $[(q^m - 1)/(q-1), m]_q$ simplex code over $\mathbb{F}_q$ all have weight q^{m-1}.*

1.11 Reed–Muller Codes

In 1954 the binary Reed–Muller codes were first constructed and examined by D. E. Muller [1409], and a majority logic decoding algorithm for them was described by I. S. Reed [1581]. The non-binary Reed–Muller codes, called generalized Reed–Muller codes, were developed in [1089, 1887]; see also Example 16.4.11 and Section 2.8. We define binary Reed–Muller codes recursively based on the $(\mathbf{u} \mid \mathbf{u} + \mathbf{v})$ construction; see [1323]. Other constructions of Reed–Muller codes can be found in Chapters 2, 16, and 20.

Definition 1.11.1 For $i \in \{1, 2\}$, let $\mathcal{C}_i$ be linear codes both of length n over $\mathbb{F}_q$. The $(\mathbf{u} \mid \mathbf{u} + \mathbf{v})$ **construction** produces the linear code $\mathcal{C}$ of length $2n$ given by $\mathcal{C} = \{(\mathbf{u}, \mathbf{u} + \mathbf{v}) \mid \mathbf{u} \in \mathcal{C}_1, \mathbf{v} \in \mathcal{C}_2\}$.

Remark 1.11.2 Let $\mathcal{C}_i$, for $i \in \{1, 2\}$, be $[n, k_i, d_i]_q$ codes with generator and parity check matrices G_i and H_i, respectively. $\mathcal{C}$ obtained by the $(\mathbf{u} \mid \mathbf{u} + \mathbf{v})$ construction is a $[2n, k_1 + k_2, \min\{2d_1, d_2\}]_q$ code with generator and parity check matrices

$$G = \left[\begin{array}{c|c} G_1 & G_1 \\ \hline \mathbf{0}_{k_2 \times n} & G_2 \end{array} \right] \quad \text{and} \quad H = \left[\begin{array}{c|c} H_1 & \mathbf{0}_{(n-k_1) \times n} \\ \hline -H_2 & H_2 \end{array} \right]. \tag{1.2}$$

We now define the binary Reed–Muller codes.

Definition 1.11.3 Let r and m be integers with $0 \le r \le m$ and $1 \le m$. The rth **order binary Reed–Muller (RM) code of length** 2^m, denoted $\mathcal{RM}(r, m)$, is defined recursively. The code $\mathcal{RM}(0, m) = \{\mathbf{0}, \mathbf{1}\}$, the $[2^m, 1, 2^m]_2$ **binary repetition code**, and $\mathcal{RM}(m, m) = \mathbb{F}_q^{2^m}$, a $[2^m, 2^m, 1]_2$ code. For $1 \le r < m$, define

$$\mathcal{RM}(r, m) = \{(\mathbf{u}, \mathbf{u} + \mathbf{v}) \mid \mathbf{u} \in \mathcal{RM}(r, m-1), \mathbf{v} \in \mathcal{RM}(r-1, m-1)\}.$$

Remark 1.11.4 Let $G(r, m)$ be a generator matrix of $\mathcal{RM}(r, m)$. By Definition 1.11.3, $G(0, m) = \begin{bmatrix} 1 \ 1 \ \cdots \ 1 \end{bmatrix}$ and $G(m, m) = I_{2^m}$. By Definition 1.11.3 and (1.2), for $1 \leq r < m$,

$$G(r, m) = \left[\begin{array}{c|c} G(r, m-1) & G(r, m-1) \\ \hline O & G(r-1, m-1) \end{array} \right]$$

where $O = \mathbf{0}_{k \times 2^{m-1}}$ with k the dimension of $\mathcal{RM}(r-1, m-1)$.

Example 1.11.5 We give generator matrices for $\mathcal{RM}(r, m)$ with $1 \leq r < m \leq 3$:

$$G(1, 2) = \left[\begin{array}{cc|cc} 1 & 0 & 1 & 0 \\ 0 & 1 & 0 & 1 \\ 0 & 0 & 1 & 1 \end{array} \right], \quad G(1, 3) = \left[\begin{array}{cccc|cccc} 1 & 0 & 1 & 0 & 1 & 0 & 1 & 0 \\ 0 & 1 & 0 & 1 & 0 & 1 & 0 & 1 \\ 0 & 0 & 1 & 1 & 0 & 0 & 1 & 1 \\ 0 & 0 & 0 & 0 & 1 & 1 & 1 & 1 \end{array} \right],$$

$$G(2, 3) = \left[\begin{array}{cccc|cccc} 1 & 0 & 0 & 0 & 1 & 0 & 0 & 0 \\ 0 & 1 & 0 & 0 & 0 & 1 & 0 & 0 \\ 0 & 0 & 1 & 0 & 0 & 0 & 1 & 0 \\ 0 & 0 & 0 & 1 & 0 & 0 & 0 & 1 \\ \hline 0 & 0 & 0 & 0 & 1 & 0 & 1 & 0 \\ 0 & 0 & 0 & 0 & 0 & 1 & 0 & 1 \\ 0 & 0 & 0 & 0 & 0 & 0 & 1 & 1 \end{array} \right].$$

From these generator matrices, we see that $\mathcal{RM}(1, 2)$ and $\mathcal{RM}(2, 3)$ consist of all even weight binary vectors of lengths 4 and 8, respectively. Also $\mathcal{RM}(1, 3)$ is an $[8, 4, 4]_2$ code, which by Example 1.8.5 must be $\widehat{\mathcal{H}}_{3,2}$.

Using the definition of Reed–Muller codes and properties from the $(\mathbf{u} \mid \mathbf{u} + \mathbf{v})$ construction, along with induction, the following hold; see [1008, Theorem 1.10.1].

Theorem 1.11.6 *Let r and m be integers with $0 \leq r \leq m$ and $1 \leq m$. The following hold.*

(a) $\mathcal{RM}(i, m) \subseteq \mathcal{RM}(j, m)$ *if $0 \leq i \leq j \leq m$.*

(b) *The dimension of $\mathcal{RM}(r, m)$ equals $\binom{m}{0} + \binom{m}{1} + \cdots + \binom{m}{r}$.*

(c) *The minimum weight of $\mathcal{RM}(r, m)$ equals 2^{m-r}.*

(d) $\mathcal{RM}(m, m)^{\perp} = \{\mathbf{0}\}$, *and if $0 \leq r < m$, then $\mathcal{RM}(r, m)^{\perp} = \mathcal{RM}(m - r - 1, m)$.*

Remark 1.11.7 Theorem 1.11.6(a) is sometimes called the **nesting property** of Reed–Muller codes. As observed in Example 1.11.5, $\mathcal{RM}(1, 3) = \widehat{\mathcal{H}}_{3,2}$. Using Theorem 1.11.6(d), it can be shown that $\mathcal{RM}(m-2, m) = \widehat{\mathcal{H}}_{m,2}$; see [1008, Exercise 61]. By Theorem 1.11.6(d), $\mathcal{RM}(m-1, m) = \mathcal{RM}(0, m)^{\perp}$. Since $\mathcal{RM}(0, m) = \{\mathbf{0}, \mathbf{1}\}$, $\mathcal{RM}(m-1, m)$ must be all even weight vectors in $\mathbb{F}_2^{2^m}$, a fact observed for $m = 2$ and $m = 3$ in Example 1.11.5.

1.12 Cyclic Codes

The study of cyclic codes seems to have begun with a series of four Air Force Cambridge Research Laboratory (AFCRL) technical notes [1536, 1537, 1538, 1539] by E. Prange from

1957 to 1959. The 1961 book by W. W. Peterson [1505] compiled extensive results about cyclic codes and laid the framework for much of the present-day theory. In 1972 this book was expanded and published jointly by Peterson and E. J. Weldon [1506].

Up to this point, the coordinates of $\mathbb{F}_q^n$ have been denoted $\{1, 2, \ldots, n\}$. For cyclic codes, the coordinates of $\mathbb{F}_q^n$ will be denoted $\{0, 1, \ldots, n-1\}$.

Definition 1.12.1 Let $\mathcal{C}$ be a code of length n over $\mathbb{F}_q$. $\mathcal{C}$ is **cyclic** provided that for all $\mathbf{c} = c_0 c_1 \cdots c_{n-1} \in \mathcal{C}$, the cyclic shift $\mathbf{c}' = c_{n-1} c_0 \cdots c_{n-2} \in \mathcal{C}$.

Remark 1.12.2 The *cyclic shift* described in Definition 1.12.1 is cyclic shift to the *right* by one position with wrap-around. The code $\mathcal{C}$ is cyclic if and only if $P \in \mathrm{PAut}(\mathcal{C})$ where the permutation matrix $P = [p_{i,j}]$ is defined by $p_{i,i+1} = 1$ for $0 \le i \le n-2$, $p_{n-1,0} = 1$, and $p_{i,j} = 0$ otherwise. Cyclic codes are closed under cyclic shifts with wrap-around of *any* amount and in *either* the *left* or *right* directions.

Example 1.12.3 Let $\mathcal{C}$ be the $[7, 4]_2$ code with generator matrix

$$G = \begin{bmatrix} 1 & 1 & 0 & 1 & 0 & 0 & 0 \\ 0 & 1 & 1 & 0 & 1 & 0 & 0 \\ 0 & 0 & 1 & 1 & 0 & 1 & 0 \\ 0 & 0 & 0 & 1 & 1 & 0 & 1 \end{bmatrix}.$$

Labeling the rows of G as $\mathbf{r}_1, \mathbf{r}_2, \mathbf{r}_3, \mathbf{r}_4$ top to bottom, we see that $\mathbf{r}_5 = 1000110 = \mathbf{r}_1 + \mathbf{r}_2 + \mathbf{r}_3$, $\mathbf{r}_6 = 0100011 = \mathbf{r}_2 + \mathbf{r}_3 + \mathbf{r}_4$, and $\mathbf{r}_7 = 1010001 = \mathbf{r}_1 + \mathbf{r}_2 + \mathbf{r}_4$. Since $\mathcal{C}$ is spanned by $\{\mathbf{r}_1, \mathbf{r}_2, \ldots, \mathbf{r}_7\}$ and this list is closed under cyclic shifts, $\mathcal{C}$ must be a cyclic code. By row reducing G, we obtain another generator matrix

$$G' = \begin{bmatrix} 1 & 0 & 0 & 0 & 1 & 1 & 0 \\ 0 & 1 & 0 & 0 & 0 & 1 & 1 \\ 0 & 0 & 1 & 0 & 1 & 1 & 1 \\ 0 & 0 & 0 & 1 & 1 & 0 & 1 \end{bmatrix}.$$

Label the columns of G' left to right as $0, 1, \ldots, 6$. If P is the 7×7 permutation matrix induced by the permutation that sends column 0 to column 2, column 2 to column 3, column 3 to column 1, column 1 to column 0, and fixes columns 4, 5, and 6, then $G'P$ has the same rows as $G_{3,2}$, the generator matrix of $\mathcal{H}_{3,2}$ in Example 1.4.9. Therefore, by ordering the coordinates of $\mathcal{H}_{3,2}$ appropriately, we see that $\mathcal{H}_{3,2}$ is a cyclic code.

While cyclic codes can be nonlinear, throughout this section we will examine only those that are linear. To study linear cyclic codes it is useful to consider elements of $\mathbb{F}_q^n$ as polynomials inside a certain quotient ring of polynomials. In that framework, linear cyclic codes are precisely the ideals of that quotient ring. We now establish the framework.

Definition 1.12.4 Let R be a commutative ring with identity. A subset I of R is an **ideal of** R if for all $a, b \in I$ and $r \in R$, then $a - b \in I$ and $ra \in I$. The ideal I is a **principal ideal** if there exists $a \in I$ such that $I = \{ra \mid r \in R\}$; a is a **generator** of I and I is denoted $\langle a \rangle$.[6] The ring R is an **integral domain** if whenever $a, b \in R$ and $ab = 0$, either $a = 0$ or $b = 0$. R is a **principal ideal domain (PID)** if it is an integral domain and all its ideals are principal. The **quotient ring of** R **by the ideal** I, denoted R/I, is the set of cosets $\{a + I \mid a \in R\}$ with addition and multiplication of cosets given by $(r + I) + (s + I) = (r + s) + I$ and $(r + I)(s + I) = rs + I$; two cosets $a + I$ and $b + I$ are equal if and only if $a - b \in I$.

---

[6]In some chapters of this books, the principal ideal generated by a will be denoted (a).

Definition 1.12.5 Let x be an indeterminate over $\mathbb{F}_q$. The set

$$\mathbb{F}_q[x] = \{a_0 + a_1 x + \cdots + a_m x^m \mid a_i \in \mathbb{F}_q \text{ for } 0 \le i \le m \text{ and some } m\}$$

is the **ring of polynomials with coefficients in** $\mathbb{F}_q$. Let $a(x) = a_0 + a_1 x + \cdots + a_m x^m \in \mathbb{F}_q[x]$ be a nonzero polynomial where $a_m \ne 0$. The **degree of** $a(x)$, denoted $\deg(a(x))$, is m. If $a_m = 1$, $a(x)$ is **monic**. If $\deg(a(x)) \ge 1$ and there do not exist $b(x), c(x) \in \mathbb{F}_q[x]$ with $\deg(b(x)) < \deg(a(x))$ and $\deg(c(x)) < \deg(a(x))$ such that $a(x) = b(x)c(x)$, then $a(x)$ is **irreducible over** $\mathbb{F}_q$.

The following result is standard; see for example [751].

Theorem 1.12.6 *The following hold.*

(a) $\mathbb{F}_q[x]$ *is a commutative ring with identity 1.*

(b) (Division Algorithm) *For $a(x), b(x) \in \mathbb{F}_q[x]$ with $b(x)$ nonzero, there exists unique $s(x), t(x) \in \mathbb{F}_q[x]$ such that $a(x) = b(x)s(x) + t(x)$ where either $t(x) = 0$ or $\deg(t(x)) < \deg(b(x))$.*

(c) $\mathbb{F}_q[x]$ *is a PID.*

(d) (Unique Factorization) *Let $p(x) \in \mathbb{F}_q[x]$ with $\deg(p(x)) \ge 1$. There exists a unique set $\{f_1(x), f_2(x), \ldots, f_t(x)\} \subseteq \mathbb{F}_q[x]$, a unique list $n_1, n_2, \ldots, n_t$ of positive integers, and a unique $\alpha \in \mathbb{F}_q$ where each $f_i(x)$ is monic and irreducible over $\mathbb{F}_q$ such that*

$$p(x) = \alpha f_1(x)^{n_1} f_2(x)^{n_2} \cdots f_t(x)^{n_t}.$$

(e) (Unique Coset Representatives) *Let $p(x) \in \mathbb{F}_q[x]$ be nonzero. The distinct cosets of the quotient ring $\mathbb{F}_q[x]/\langle p(x)\rangle$ are uniquely representable as $a(x) + \langle p(x)\rangle$ where $a(x) = 0$ or $\deg(a(x)) < \deg(p(x))$; $\mathbb{F}_q[x]/\langle p(x)\rangle$ has order $q^{\deg(p(x))}$. The quotient ring $\mathbb{F}_q[x]/\langle p(x)\rangle$ is also a vector space over $\mathbb{F}_q$ of dimension $\deg(p(x))$.*

(f) *If $p(x)$ is irreducible over $\mathbb{F}_q$, then $\mathbb{F}_q[x]/\langle p(x)\rangle$ is a field.*

The map $\mathbf{a} = a_0 a_1 \cdots a_{n-1} \mapsto a(x) + \langle x^n - 1\rangle$ where $a(x) = a_0 + a_1 x + \cdots + a_{n-1}x^{n-1}$ is a vector space isomorphism from $\mathbb{F}_q^n$ onto $\mathbb{F}_q[x]/\langle x^n - 1\rangle$. We denote this map by $\mathbf{a} \mapsto a(x)$, dropping the '$+ \langle x^n - 1\rangle$'. Thus a linear code $\mathcal{C}$ of length n can be viewed equivalently as a subspace of $\mathbb{F}_q^n$ or as an $\mathbb{F}_q$-subspace of $\mathbb{F}_q[x]/\langle x^n - 1\rangle$. Notice that if $\mathbf{a} \mapsto a(x)$, then $\mathbf{a}' = a_{n-1}a_0 \cdots a_{n-2} \mapsto xa(x)$ as $x^n + \langle x^n - 1\rangle = 1 + \langle x^n - 1\rangle$. So $\mathcal{C}$ is a cyclic code in $\mathbb{F}_q^n$ if and only if $\mathcal{C} \mapsto I$ where I is an ideal of $\mathbb{F}_q[x]/\langle x^n - 1\rangle$. Therefore we study cyclic codes as ideals of $\mathbb{F}_q[x]/\langle x^n - 1\rangle$.

To find the ideals of $\mathbb{F}_q[x]/\langle x^n - 1\rangle$ requires factorization of $x^n - 1$. From the theory of finite fields (see [170, 1254, 1362]), there is an extension field of $\mathbb{F}_q$ that contains all the roots of $x^n - 1$. The smallest such field, called a **splitting field of** $x^n - 1$ **over** $\mathbb{F}_q$, is $\mathbb{F}_{q^t}$ where t is the smallest integer such that $n \mid (q^t - 1)$. When $\gcd(n, q) = 1$, there exists $\alpha \in \mathbb{F}_{q^t}$, called a **primitive n^{th} root of unity**, such that the n distinct roots of $x^n - 1$ (called the **roots of unity**) are $\alpha^0 = 1, \alpha, \alpha^2, \ldots, \alpha^{n-1}$; alternately if γ is a primitive element of $\mathbb{F}_{q^t}$, one choice for α is $\gamma^{(q^t - 1)/n}$. When $\gcd(n, q) \ne 1$, $x^n - 1$ has repeated roots. **For the remainder of this section, we assume $\gcd(n, q) = 1$.**[7]

---

[7]The theory of cyclic codes when $\gcd(n, q) \ne 1$ has some overlap with the theory when $\gcd(n, q) = 1$, but there are significant differences. When $\gcd(n, q) \ne 1$, cyclic codes are called **repeated-root cyclic codes**. Repeated-root cyclic codes were first examined in their most generality in [369, 1835].

Definition 1.12.7 Let s be an integer with $0 \le s < n$. The **q-cyclotomic coset of s modulo n** is the set

$$C_s = \{s, sq, \ldots, sq^{r-1}\} \bmod n$$

where r is the smallest positive integer such that $sq^r \equiv s \pmod{n}$. The distinct q-cyclotomic cosets modulo n partition the set of integers $\{0, 1, 2, \ldots, n-1\}$.

Remark 1.12.8 A splitting field of $x^n - 1$ over $\mathbb{F}_q$ is $\mathbb{F}_{q^t}$ where t is the size of the q-cyclotomic coset of 1 modulo n.

Theorem 1.12.9 ([1323, Chapter 7.5]) *Let α be a primitive n^{th} root of unity in the splitting field $\mathbb{F}_{q^t}$ of $x^n - 1$ over $\mathbb{F}_q$. For $0 \le s < n$ define $M_{\alpha^s}(x) = \prod_{i \in C_s}(x - \alpha^i)$. Then $M_{\alpha^s}(x) \in \mathbb{F}_q[x]$ and is irreducible over $\mathbb{F}_q$. Furthermore, the unique factorization of $x^n - 1$ into monic irreducible polynomials over $\mathbb{F}_q$ is given by $x^n - 1 = \prod_s M_{\alpha^s}(x)$ where s runs through a set of representatives of all distinct q-cyclotomic cosets modulo n.*

Example 1.12.10 The 2-cyclotomic cosets modulo 7 are $C_0 = \{0\}$, $C_1 = \{1, 2, 4\}$, and $C_3 = \{3, 6, 5\}$. By Remark 1.12.8, $\mathbb{F}_{2^3} = \mathbb{F}_8$ is the splitting field of $x^7 - 1$ over $\mathbb{F}_2$. In the notation of Table 1.3, $\alpha = \delta$ is a primitive 7th root of unity. In the notation of Theorem 1.12.9, $M_{\alpha^0}(x) = -1 + x = 1 + x$, $M_\alpha = (x - \alpha)(x - \alpha^2)(x - \alpha^4) = 1 + x + x^3$, $M_{\alpha^3} = (x - \alpha^3)(x - \alpha^6)(x - \alpha^5) = 1 + x^2 + x^3$, and $x^7 - 1 = M_{\alpha^0}(x)M_\alpha(x)M_{\alpha^3}(x)$.

Using Theorem 1.12.9, we have the following basic theorem [1008, Theorem 4.2.1] describing the structure of cyclic codes over $\mathbb{F}_q$. We remark that all of this theorem except part (g) is valid when $\gcd(n, q) \ne 1$. We note that if $a(x), b(x) \in \mathbb{F}_q[x]$, then $a(x)$ **divides** $b(x)$, denoted $a(x) \mid b(x)$, means that there exists $c(x) \in \mathbb{F}_q[x]$ such that $b(x) = a(x)c(x)$.

Theorem 1.12.11 *Let C be a nonzero linear cyclic code over $\mathbb{F}_q$ of length n viewed as an ideal of $\mathbb{F}_q[x]/\langle x^n - 1 \rangle$. There exists a polynomial $g(x) \in C$ with the following properties.*

(a) *$g(x)$ is the unique monic polynomial of minimum degree in C.*

(b) *$C = \langle g(x) \rangle$ in $\mathbb{F}_q[x]/\langle x^n - 1 \rangle$.*

(c) *$g(x) \mid (x^n - 1)$.*

With $k = n - \deg(g(x))$, let $g(x) = \sum_{i=0}^{n-k} g_i x^i$ where $g_{n-k} = 1$. Then

(d) *the dimension of C is k and $\{g(x), xg(x), \ldots, x^{k-1}g(x)\}$ is a basis for C,*

(e) *every element of C is uniquely expressible as a product $g(x)f(x)$ where $f(x) = 0$ or $\deg(f(x)) < k$,*

(f) *a generator matrix G of C is*

$$G = \begin{bmatrix} g_0 & g_1 & g_2 & \cdots & g_{n-k} & \cdots & \cdots & 0 \\ 0 & g_0 & g_1 & \cdots & g_{n-k-1} & g_{n-k} & \cdots & 0 \\ \vdots & & & & & & & \vdots \\ 0 & 0 & 0 & g_0 & \cdots & & \cdots & g_{n-k} \end{bmatrix}$$

$$\leftrightarrow \begin{bmatrix} g(x) & & & \\ & xg(x) & & \\ & & \ddots & \\ & & & x^{k-1}g(x) \end{bmatrix},$$

and

(g) *if α is a primitive n^{th} root of unity in the splitting field $\mathbb{F}_{q^t}$ of $x^n - 1$ over $\mathbb{F}_q$, then*

$$g(x) = \prod_s M_{\alpha^s}(x)$$

where the product is over a subset of representatives of distinct q-cyclotomic cosets modulo n.

Definition 1.12.12 The polynomial $g(x)$ in Theorem 1.12.11 is the **generator polynomial** of $\mathcal{C}$. By convention, the cyclic code $\mathcal{C} = \{\mathbf{0}\}$ has generator polynomial $g(x) = x^n - 1$.

The following are immediate consequences of Theorem 1.12.11.

Corollary 1.12.13 *There are 2^m linear cyclic codes of length n (including the zero code) over $\mathbb{F}_q$ where m is the number of q-cyclotomic cosets modulo n.*

Corollary 1.12.14 *If $g_1(x)$ and $g_2(x)$ are generator polynomials of $\mathcal{C}_1$ and $\mathcal{C}_2$, respectively, and if $g_1(x) \mid g_2(x)$, then $\mathcal{C}_2 \subseteq \mathcal{C}_1$.*

By Theorem 1.12.11, when $\gcd(n, q) = 1$, a linear cyclic code of length n over $\mathbb{F}_q$ is uniquely determined by its generator polynomial. This in turn is determined by its roots in the splitting field of $x^n - 1$ over $\mathbb{F}_q$. This leads to the following definition.

Definition 1.12.15 Let $\mathcal{C}$ be a linear cyclic code of length n over $\mathbb{F}_q$ with generator polynomial $g(x) \mid (x^n - 1)$. Let α be a fixed primitive n^{th} root of unity in a splitting field of $x^n - 1$ over $\mathbb{F}_q$. By Theorems 1.12.9 and 1.12.11, $g(x) = \prod_s \prod_{i \in C_s} (x - \alpha^i)$ where s runs through some subset of representatives of the q-cyclotomic cosets C_s modulo n. Let $T = \bigcup_s C_s$ be the union of these q-cyclotomic cosets. The roots of unity $\{\alpha^i \mid i \in T\}$ are called the **zeros** of $\mathcal{C}$; $\{\alpha^i \mid 0 \le i < n, \ i \notin T\}$ are the **nonzeros** of $\mathcal{C}$. The set T is called the **defining set** of $\mathcal{C}$ relative to α.

Remark 1.12.16 In Definition 1.12.15, if you change the primitive n^{th} root of unity, you change the defining set T; so T is computed relative to a fixed primitive root of unity.

Remark 1.12.17 Corollary 1.12.14 can be translated into the language of defining sets: If T_1 and T_2 are defining sets of $\mathcal{C}_1$ and $\mathcal{C}_2$, respectively, relative to the same primitive root of unity, and if $T_1 \subseteq T_2$, then $\mathcal{C}_2 \subseteq \mathcal{C}_1$.

Example 1.12.18 Continuing with Example 1.12.10, Table 1.5 describes the $2^3 = 8$ binary cyclic codes of length 7. The code with $g(x) = 1 + x + x^3$ is $\mathcal{H}_{3,2}$ as discussed in Example 1.12.3. The code with $g(x) = 1 + x^2 + x^3$ is permutation equivalent to $\mathcal{H}_{3,2}$. The code of dimension $k = 1$ is the binary repetition code $\{\mathbf{0}, \mathbf{1}\}$.

The dual code of a cyclic code is also cyclic. We can determine its generator polynomial and defining set; see [1323, Chapter 7.4].

Theorem 1.12.19 *Let $\mathcal{C}$ be an $[n, k]_q$ cyclic code with generator polynomial $g(x)$. Define $h(x) = \dfrac{x^n - 1}{g(x)}$. Then $\mathcal{C}^\perp$ is cyclic with generator polynomial $g^\perp(x) = \dfrac{x^k h(x^{-1})}{h(0)}$. Let α be a primitive n^{th} root of unity in a splitting field of $x^n - 1$ over $\mathbb{F}_q$. If T is the defining set of $\mathcal{C}$ relative to α, the defining set of $\mathcal{C}^\perp$ is $T^\perp = \{0, 1, \ldots, n-1\} \setminus (-1)T \bmod n$.*

TABLE 1.5: The $[7, k, d]_2$ cyclic codes with generator polynomial $g(x)$ and defining set T relative to α

k	d	$g(x)$	T
0	—	$1 + x^7 = 0$	$\{0, 1, 2, 3, 4, 5, 6\}$
1	7	$1 + x + x^2 + x^3 + x^4 + x^5 + x^6$	$\{1, 2, 3, 4, 5, 6\}$
3	4	$1 + x^2 + x^3 + x^4$	$\{0, 1, 2, 4\}$
3	4	$1 + x + x^2 + x^4$	$\{0, 3, 5, 6\}$
4	3	$1 + x + x^3$	$\{1, 2, 4\}$
4	3	$1 + x^2 + x^3$	$\{3, 5, 6\}$
6	2	$1 + x$	$\{0\}$
7	1	1	$\emptyset$

Example 1.12.20 Continuing with Example 1.12.18, the dual of any code in Table 1.5 must be a code in the table. By comparing dimensions, the codes of dimension 0 and 7 in the table are duals of each other as are the codes of dimension 1 and 6; this is confirmed by examining the defining sets and using Theorem 1.12.19. By this theorem, $\{0, 1, 2, 3, 4, 5, 6\} \setminus (-1)\{1, 2, 4\} \bmod 7 = \{0, 1, 2, 3, 4, 5, 6\} \setminus \{6, 5, 3\} = \{0, 1, 2, 4\}$ showing that the codes with defining sets $\{0, 1, 2, 4\}$ and $\{1, 2, 4\}$ are duals of each other. By Remark 1.12.17, as $\{1, 2, 4\} \subseteq \{0, 1, 2, 4\}$, $\langle 1 + x + x^3 \rangle^\perp = \langle 1 + x^2 + x^3 + x^4 \rangle \subseteq \langle 1 + x + x^3 \rangle$, a fact we already observed in Example 1.6.10. Similarly, the codes with defining sets $\{0, 3, 5, 6\}$ and $\{3, 5, 6\}$ are duals of each other.

The following is a somewhat surprising fact about cyclic self-orthogonal binary codes; see [1008, Theorem 4.4.18].

Theorem 1.12.21 *A cyclic self-orthogonal binary code is doubly-even.*

Example 1.12.22 As detailed in Examples 1.6.10 and 1.12.20, $\mathcal{H}_{3,2}^\perp$ is self-orthogonal and doubly-even, illustrating Theorem 1.12.21.

Definition 1.12.23 Quasi-cyclic codes are a natural generalization of cyclic codes. Let $\mathcal{C}$ be a code of length n and ℓ a positive integer dividing n. $\mathcal{C}$ is ℓ-**quasi-cyclic** provided whenever $c_0 c_1 \cdots c_{n-1} \in \mathcal{C}$ then $c_{n-\ell} c_{n-\ell+1} \cdots c_{n-1} c_0 \cdots c_{n-\ell-2} c_{n-\ell-1} \in \mathcal{C}$. Cyclic codes are 1-quasi-cyclic codes. Quasi-cyclic codes will be studied in Chapter 7.

See Chapter 2 and Sections 8.6 and 20.5 for more on cyclic codes over fields.

1.13   Golay Codes

In the same remarkable one-half page 1949 paper [820] in which Golay generalized the Hamming codes, he also introduced what later became known as the $[23, 12, 7]_2$ binary Golay code and the $[11, 6, 5]_3$ ternary Golay code. There are many ways to present these two codes; one way is to describe them as cyclic codes.

Example 1.13.1 There are three 2-cyclotomic cosets modulo 23 with sizes 1, 11, and 11: $C_0 = \{0\}$, $C_1 = \{1, 2, 4, 8, 16, 9, 18, 13, 3, 6, 12\}$, $C_5 = \{5, 10, 20, 17, 11, 22, 21, 19, 15, 7, 14\}$. Theorem 1.12.9 implies that, over $\mathbb{F}_2$, $x^{23} - 1 = x^{23} + 1$ factors into 3 monic irreducible polynomials of degrees 1, 11, and 11. These irreducible factors are $b_0(x) = 1 + x$, $b_1(x) = 1 + x + x^5 + x^6 + x^7 + x^9 + x^{11}$, and $b_2(x) = 1 + x^2 + x^4 + x^5 + x^6 + x^{10} + x^{11}$. There are 8 binary linear cyclic codes of length 23 by Corollary 1.12.13. By Theorem 1.12.11(d), the codes $\mathcal{C}_1 = \langle b_1(x) \rangle$ and $\mathcal{C}_2 = \langle b_2(x) \rangle$ are $[23, 12]_2$ codes. The map that fixes coordinate 0 and switches coordinates i and $23 - i$ for $1 \le i \le 11$ leads to a permutation matrix P where $\mathcal{C}_1 P = \mathcal{C}_2$. Any code permutation equivalent to $\mathcal{C}_1$ is termed the $[23, 12]_2$ **binary Golay code of length 23** and is denoted $\mathcal{G}_{23}$. The splitting field of $x^{23} - 1$ over $\mathbb{F}_2$ is $\mathbb{F}_{2^{11}}$. In $\mathbb{F}_{2^{11}}$ there is a primitive 23rd root of unity α where the defining sets of $\mathcal{C}_1$ and $\mathcal{C}_2$ are C_1 and C_5, respectively, relative to α. Another primitive 23rd root of unity is $\beta = \alpha^5$; relative to β, the defining sets of $\mathcal{C}_1$ and $\mathcal{C}_2$ are C_5 and C_1, respectively. By Theorem 1.12.19 the defining set of $\mathcal{C}_1^\perp$ relative to α is $C_0 \cup C_1$ implying by Remark 1.12.17 that $\mathcal{C}_1^\perp \subseteq \mathcal{C}_1$ showing $\mathcal{C}_1^\perp$ is self-orthogonal. Using Theorem 1.12.21, $\mathcal{C}_1^\perp$ is the doubly-even $[23, 11]_2$ code consisting of all codewords in $\mathcal{C}_1$ of even weight.

Example 1.13.2 The 3-cyclotomic cosets modulo 11 are $C_0 = \{0\}$, $C_1 = \{1, 3, 9, 5, 4\}$, and $C_2 = \{2, 6, 7, 10, 8\}$ of sizes 1, 5, and 5, respectively. Theorem 1.12.9 implies that, over $\mathbb{F}_3$, $x^{11} - 1$ factors into 3 monic irreducible polynomials of degrees 1, 5, and 5. These irreducible factors are $t_0(x) = -1 + x$, $t_1(x) = -1 + x^2 - x^3 + x^4 + x^5$, and $t_2(x) = -1 - x + x^2 - x^3 + x^5$. There are 8 ternary linear cyclic codes of length 11 by Corollary 1.12.13. By Theorem 1.12.11(d), the codes $\mathcal{C}_1 = \langle t_1(x) \rangle$ and $\mathcal{C}_2 = \langle t_2(x) \rangle$ are $[11, 6]_3$ codes. The map that fixes coordinate 0 and switches coordinates i and $11 - i$ for $1 \le i \le 5$ leads to a permutation matrix P where $\mathcal{C}_1 P = \mathcal{C}_2$. Any code monomially equivalent to $\mathcal{C}_1$ is termed the $[11, 6]_3$ **ternary Golay code of length 11** and is denoted $\mathcal{G}_{11}$. The splitting field of $x^{11} - 1$ over $\mathbb{F}_3$ is $\mathbb{F}_{3^5}$. In $\mathbb{F}_{3^5}$ there is a primitive 11th root of unity α where the defining sets of $\mathcal{C}_1$ and $\mathcal{C}_2$ are C_1 and C_2, respectively, relative to α. Another primitive 11th root of unity is $\beta = \alpha^2$; relative to β, the defining sets of $\mathcal{C}_1$ and $\mathcal{C}_2$ are C_2 and C_1, respectively.

Definition 1.13.3 $\mathcal{G}_{23}$ can be extended as in Section 1.7 to a $[24, 12]_2$ code $\widehat{\mathcal{G}}_{23}$, denoted $\mathcal{G}_{24}$, and called the **binary Golay code of length 24**. Similarly $\mathcal{G}_{11}$ can be extended to a $[12, 6]_3$ code $\widehat{\mathcal{G}}_{11}$, denoted $\mathcal{G}_{12}$, and called the **ternary Golay code of length 12**.

Remark 1.13.4 The automorphism groups of the four Golay codes involve the **Mathieu groups** M_p for $p \in \{11, 12, 23, 24\}$ discovered by Émile Mathieu [1352, 1353]. These four permutation groups on p points are 4-fold transitive, when $p \in \{11, 23\}$, and 5-fold transitive, when $p \in \{12, 24\}$, simple groups. Properties of these groups and their relationship to Golay codes can be found in [442].

The following two theorems give basic properties of the four Golay codes. Parts (a), (b), and (c) of each theorem can be found in most standard coding theory texts. The uniqueness of these codes in part (d) of each theorem is a culmination of work in [525, 1514, 1518, 1732] with a self-contained proof in [1008, Chapter 10]. Part (e) of each theorem follows by direct computation from Definition 1.9.8. The automorphism groups in part (f) of each theorem can be found in [434], which is also [442, Chapter 10].

Theorem 1.13.5 *The following properties hold for the binary Golay codes.*

(a) $\mathcal{G}_{23}$ *has minimum distance 7 and weight distribution* $A_0 = A_{23} = 1$, $A_7 = A_{16} = 253$, $A_8 = A_{15} = 506$, $A_{11} = A_{12} = 1288$, *and* $A_i = 0$ *otherwise.*

(b) $\mathcal{G}_{24}$ *has minimum distance 8 and weight distribution* $A_0 = A_{24} = 1$, $A_8 = A_{16} = 759$, $A_{12} = 2576$, *and* $A_i = 0$ *otherwise.*

(c) $\mathcal{G}_{24}$ *is doubly-even and self-dual.*

(d) *Both a $(23, M)_2$ and a $(24, M)_2$, possibly nonlinear, binary code each containing $\mathbf{0}$ with $M \geq 2^{12}$ codewords and minimum distance 7 and 8, respectively, are unique up to permutation equivalence. They are the $[23, 12, 7]_2$ and $[24, 12, 8]_2$ binary Golay codes.*

(e) *$\mathcal{G}_{23}$ is perfect.*

(f) *$\mathrm{PAut}(\mathcal{G}_{23}) = \mathrm{M}_{23}$ and $\mathrm{PAut}(\mathcal{G}_{24}) = \mathrm{M}_{24}$.*

Theorem 1.13.6 *The following properties hold for the ternary Golay codes.*

(a) *$\mathcal{G}_{11}$ has minimum distance 5 and weight distribution $A_0 = 1$, $A_5 = A_6 = 132$, $A_8 = 330$, $A_9 = 110$, $A_{11} = 24$, and $A_i = 0$ otherwise.*

(b) *$\mathcal{G}_{12}$ has minimum distance 6 and weight distribution $A_0 = 1$, $A_6 = 264$, $A_9 = 440$, $A_{12} = 24$, and $A_i = 0$ otherwise.*

(c) *$\mathcal{G}_{12}$ is self-dual.*

(d) *Both a $(11, M)_3$ and a $(12, M)_3$, possibly nonlinear, ternary code each containing $\mathbf{0}$ with $M \geq 3^6$ codewords and minimum distance 5 and 6, respectively, are unique up to monomial equivalence. They are the $[11, 6, 5]_3$ and $[12, 6, 6]_3$ ternary Golay codes.*

(e) *$\mathcal{G}_{11}$ is perfect.*

(f) *$\mathrm{MAut}(\mathcal{G}_{11}) = \widetilde{M}_{11}$ and $\mathrm{MAut}(\mathcal{G}_{12}) = \widetilde{M}_{12}$ where $\widetilde{M}_{11}$ and $\widetilde{M}_{12}$ are isomorphic to the double covers, or the non-splitting central extensions by a center of order 2, of M_{11} and M_{12}.*

Remark 1.13.7 If there is equality for given parameters of a code in a bound from Section 1.9, we say the code **meets** the bound. Perfect codes are those meeting the Sphere Packing Bound; MDS codes are those meeting the Singleton Bound. Using Theorem 1.10.5, direct computation shows that the $[(q^m - 1)/(q - 1), m, q^{m-1}]_q$ simplex code meets the Griesmer Bound, as do $\mathcal{G}_{11}$ and $\mathcal{G}_{12}$. Neither $\mathcal{G}_{23}$ nor $\mathcal{G}_{24}$ meet the Griesmer Bound.

1.14   BCH and Reed–Solomon Codes

There is a lower bound, presented in Theorem 1.14.3, on the minimum distance of a cyclic code based on the defining set of the code. BCH codes take advantage of this bound. The binary BCH codes were discovered by A. Hocquenghem [968] and independently by R. C. Bose and D. K. Ray-Chaudhuri [248, 249], and were generalized to all finite fields by D. C. Gorenstein and N. Zierler [840]. Some properties of BCH codes are given in Section 2.6.

Definition 1.14.1 Let $\mathcal{N} = \{0, 1, \ldots, n-1\}$ and $T \subseteq \mathcal{N}$. We say T contains a set of $s \leq n$ **consecutive** elements provided there exists $b \in \mathcal{N}$ such that

$$\{b, b+1, \ldots, b+s-1\} \bmod n \subseteq T.$$

Remark 1.14.2 When considering the notion of *consecutive*, wrap-around is allowed. For example if $n = 10$, $\{8, 9, 0, 1\}$ is a consecutive set in $T = \{0, 1, 2, 5, 8, 9\}$.

TABLE 1.6: The $[7, k, d]_2$ BCH codes with defining set T relative to α, b, and Bose distance δ

k	d	T	b	δ
0	—	$\{0,1,2,3,4,5,6\} = C_1 \cup C_2 \cup \cdots \cup C_6 \cup C_0$	1	—
1	7	$\{1,2,3,4,5,6\} = C_1 \cup C_2 \cup \cdots \cup C_6$	1	7
3	4	$\{0,1,2,4\} = C_0 \cup C_1 \cup C_2$	0	4
3	4	$\{0,3,5,6\} = C_5 \cup C_6 \cup C_0$	5	4
4	3	$\{1,2,4\} = C_1 \cup C_2$	1	3
4	3	$\{3,5,6\} = C_5 \cup C_6$	5	3
6	2	$\{0\} = C_0$	0	2

Rather surprisingly, the existence of consecutive elements in the defining set of a cyclic code determines a lower bound, called the *BCH Bound*, on the minimum distance of the code. A proof of the following can be found in [1323, Chapter 7.6].

Theorem 1.14.3 (BCH Bound) *Let $\mathcal{C}$ be a linear cyclic code of length n over $\mathbb{F}_q$ and minimum distance d with defining set T relative to some primitive n^{th} root of unity. Assume T contains $\delta - 1$ consecutive elements for some integer $\delta \geq 2$. Then $d \geq \delta$.*

Definition 1.14.4 Let δ be an integer with $2 \leq \delta \leq n$. A **BCH code over $\mathbb{F}_q$ of length n and designed distance δ** is a linear cyclic code with defining set

$$T = C_b \cup C_{b+1} \cup \cdots \cup C_{b+\delta-2}$$

relative to some primitive n^{th} root of unity where C_i is the q-cyclotomic coset modulo n containing i. As T contains the consecutive set $\{b, b+1, \ldots, b+\delta-2\}$, this code has minimum distance at least δ by the BCH Bound. If $b = 1$, the code is **narrow-sense**; if $n = q^t - 1$ for some t, the code is **primitive**.

Definition 1.14.5 Sometimes a BCH code can have more than one designed distance; the largest designed distance is called the **Bose distance**.

Example 1.14.6 Consider the eight $[7, k, d]_2$ binary cyclic codes from Example 1.12.18 and presented in Table 1.5. All except the code with defining set $T = \emptyset$ are BCH codes as seen in Table 1.6. As $7 = 2^3 - 1$, all these BCH codes are primitive. Technically, the zero code is primitive with designed distance 8; of course distance in the zero code is meaningless. Of the six remaining codes, three are narrow-sense. Notice that the code with defining set $\{1, 2, 4\}$ is narrow-sense with two designed distances 2 and 3 as $\{1, 2, 4\} = C_1 = C_1 \cup C_2$; the Bose distance is 3. The code with defining set $\{1, 2, 3, 4, 5, 6\}$ is narrow-sense with designed distances 4 through 7 as $\{1, 2, 3, 4, 5, 6\} = C_1 \cup C_2 \cup C_3 = C_1 \cup C_2 \cup C_3 \cup C_4 = C_1 \cup C_2 \cup C_3 \cup C_4 \cup C_5 = C_1 \cup C_2 \cup C_3 \cup C_4 \cup C_5 \cup C_6$; the Bose distance is 7. The Bose designed distance and the true minimum distance are the same for the seven nonzero BCH codes.

Example 1.14.7 In the notation of Example 1.13.1, $\mathcal{G}_{23}$ has defining set $T = C_1$ which contains 4 consecutive elements $\{1, 2, 3, 4\}$. By the BCH Bound, $\mathcal{G}_{23}$ has minimum weight at least 5; its true minimum weight is 7 from Theorem 1.13.5(a). As the defining set of $\mathcal{G}_{23}$ is $T = C_1 = C_1 \cup C_2 \cup C_3 \cup C_4$, $\mathcal{G}_{23}$ is a narrow-sense[8] BCH code of Bose designed distance

---

[8]$\mathcal{G}_{23}$ is permutation equivalent to the BCH code with designed distance 5 and defining set $C_5 = C_{19} = C_{19} \cup C_{20} \cup C_{21} \cup C_{22}$; in this formulation $\mathcal{G}_{23}$ is not narrow-sense.

$\delta = 5$ with $b = 1$. Similarly, $\mathcal{G}_{11}$ of Example 1.13.2 is a BCH code viewed in several ways. $\mathcal{G}_{11}$ is a narrow-sense BCH code with $b = 1$, $\delta = 2$ and defining set $C_1 = \{1, 3, 4, 5, 9\}$. It is also a BCH code with $b = 3$, $\delta = 2, 3,$ or 4 as $\{1, 3, 4, 5, 9\} = C_3 = C_3 \cup C_4 = C_3 \cup C_4 \cup C_5$. The Bose distance of $\mathcal{G}_{11}$ is 4 while its true minimum distance is 5 from Theorem 1.13.6(a).

At about the same time as BCH codes appeared in the literature, I. S. Reed and G. Solomon [1582] published their work on the codes that now bear their names. These codes, which are now commonly presented as a special case of BCH codes, were actually first constructed by K. A. Bush [319] in 1952 in the context of orthogonal arrays. Because of their burst error-correction capabilities, Reed–Solomon codes are used to improve the reliability of compact discs, digital audio tapes, and other data storage systems.

Definition 1.14.8 A **Reed–Solomon (RS) code**[9] **of length** n **over** $\mathbb{F}_q$ is a primitive BCH code of length $n = q - 1$.

When $n = q - 1$, the q-cyclotomic coset modulo n containing s is $C_s = \{s\}$. So if $\mathcal{C}$ is an $[n, k, d]_q$ Reed–Solomon code, its defining set T has size $n - k$ and must be $\{b, b+1, \ldots, b + (n - k - 1)\} \bmod n$ for some b. By the BCH Bound, $d \geq n - k + 1$. By the Singleton Bound, $d \leq n - k + 1$. Therefore $d = n - k + 1$ and $\mathcal{C}$ is MDS; in particular the designed distance $\delta = n - k + 1$ equals the true minimum distance. In general the dual code of an MDS code is also MDS by Theorem 1.9.13. The dual code of a BCH code may not be BCH; however the dual of a Reed–Solomon code is Reed–Solomon as follows. By Theorem 1.12.19, $\mathcal{C}^\perp$ has defining set $T^\perp = \mathcal{N} \setminus (-1)T \bmod n$ where $\mathcal{N} = \{0, 1, \ldots, n - 1\}$. Since $(-1)T \bmod n$ consists of $n - k$ consecutive elements of $\mathcal{N}$, $T^\perp$ is the remaining k elements $\mathcal{N}$, which clearly must be consecutive (recalling that *wrap-around* is allowed in consecutive sets modulo n). This discussion yields the following result.

Theorem 1.14.9 *Let $\mathcal{C}$ be a Reed–Solomon code over $\mathbb{F}_q$ of length $n = q - 1$ and designed distance δ. The following hold.*

(a) *$\mathcal{C}$ has defining set $T = \{b, b+1, \ldots, b + \delta - 2\}$ for some integer b.*

(b) *$\mathcal{C}$ has minimum distance $d = \delta$ and dimension $k = n - d + 1$.*

(c) *$\mathcal{C}$ is MDS.*

(d) *$\mathcal{C}^\perp$ is a Reed–Solomon code of designed distance $k + 1$.*

Example 1.14.10 Using Table 1.1, γ is both a primitive element of $\mathbb{F}_9$ and a primitive 8th root of unity. Let $\mathcal{C}$ be the narrow-sense Reed–Solomon code over $\mathbb{F}_9$ of length 8 and designed distance $\delta = 4$. Then $\mathcal{C}$ has defining set $\{1, 2, 3\}$ relative to γ and generator polynomial $g(x) = (x - \gamma)(x - \gamma^2)(x - \gamma^3) = \gamma^2 + \gamma x + \gamma^3 x^2 + x^3$. $\mathcal{C}$ is an $[8, 5, 4]_9$ code. By Theorem 1.12.19, $\mathcal{C}^\perp$ has defining set $T^\perp = \{0, 1, \ldots, 7\} \setminus (-1)\{1, 2, 3\} \bmod 8 = \{0, 1, \ldots, 7\} \setminus \{7, 6, 5\} = \{0, 1, 2, 3, 4\}$ and hence generator polynomial $g^\perp(x) = (x - 1)(x - \gamma)(x - \gamma^2)(x - \gamma^3)(x - \gamma^4) = (x^2 - 1)g(x) = \gamma^6 + \gamma^5 x + \gamma^5 x^2 + \gamma^7 x^3 + \gamma^3 x^4 + x^5$. So $\mathcal{C}^\perp$ is an $[8, 3, 6]_9$ non-narrow-sense Reed–Solomon code with $b = 0$ and designed distance 6, consistent with Theorem 1.14.9(d). As $T \subseteq T^\perp$, $\mathcal{C}^\perp \subseteq \mathcal{C}$ by Remark 1.12.17.

The original formulation of Reed and Solomon for the narrow-sense Reed–Solomon codes is different from that of Definition 1.14.8. This alternative formulation of narrow-sense Reed–Solomon codes is of particular importance because it is the basis for the definitions

---

[9]While this is a common definition of Reed–Solomon codes, there are other codes of lengths different from $q - 1$ that are also called Reed–Solomon codes. See Remark 15.3.21.

of generalized Reed–Solomon codes, Goppa codes, and algebraic geometry codes; see Chapters 15 and 24.

For this formulation, let $\mathcal{P}_{k,q} = \{p(x) \in \mathbb{F}_q[x] \mid p(x) = 0 \text{ or } \deg(p(x)) < k\}$ when $k \geq 0$. See [1008, Theorem 5.2.3] for a proof of the following.

Theorem 1.14.11 *Let $n = q - 1$ and let α be a primitive n^{th} root of unity in $\mathbb{F}_q$. For $0 < k \leq n = q - 1$, let $\mathcal{RS}_k(\boldsymbol{\alpha}) = \{(p(\alpha^0), p(\alpha), \ldots, p(\alpha^{q-2})) \in \mathbb{F}_q^n \mid p(x) \in \mathcal{P}_{k,q}\}$. Then $\mathcal{RS}_k(\boldsymbol{\alpha})$ is the narrow-sense $[q - 1, k, q - k]_q$ Reed–Solomon code.*

In general, extending an MDS code may not produce an MDS code; however extending a narrow-sense Reed–Solomon code does produce an MDS code. With the notation of Theorem 1.14.11, $\mathbb{F}_q = \{0, 1 = \alpha^0, \alpha, \alpha^2, \ldots, \alpha^{q-2}\}$ and, when $q \geq 3$, $\sum_{i=0}^{q-2} \alpha^i = 0$. Using this, it is straightforward to show that if $q \geq 3$, $k < q - 1$, and $p(x) \in \mathcal{P}_{k,q}$, then $\sum_{\beta \in \mathbb{F}_q} p(\beta) = 0$. This leads to the following result.

Theorem 1.14.12 *With the notation of Theorem 1.14.11 and $0 < k < n = q - 1$, $\widehat{\mathcal{RS}}_k(\boldsymbol{\alpha}) = \{(p(\alpha^0), p(\alpha), \ldots, p(\alpha^{q-2}), p(0)) \in \mathbb{F}_q^n \mid p(x) \in \mathcal{P}_{k,q}\}$ is a $[q, k, q - k + 1]_q$ MDS code.*

Remark 1.14.13 The code $\mathcal{RS}_{q-1}(\boldsymbol{\alpha})$ omitted from consideration in Theorem 1.14.12 equals $\mathbb{F}_q^{q-1}$. Its extension is not as given in Theorem 1.14.12; however $\widehat{\mathcal{RS}}_{q-1}(\boldsymbol{\alpha})$ is still a $[q, q - 1, 2]_q$ MDS code.

1.15 Weight Distributions

The weight distribution of a linear code determines the weight distribution of its dual code via a series of equations, called the MacWilliams Identities or the MacWilliams Equations. They were first developed by F. J. MacWilliams in [1319]. There are in fact several equivalent formulations of these equations. Among these are the Pless Power Moments discovered by V. S. Pless [1513]. The most compact form of these identities is expressed in a single polynomial equation relating the weight distribution of a code and its dual.

Definition 1.15.1 Let $\mathcal{C}$ be a linear code of length n over $\mathbb{F}_q$ with weight distribution $A_i(\mathcal{C})$ for $0 \leq i \leq n$. Let x and y be independent indeterminates over $\mathbb{F}_q$. The **(Hamming) weight enumerator of $\mathcal{C}$** is defined to be

$$\mathrm{Hwe}_{\mathcal{C}}(x, y) = \sum_{i=0}^{n} A_i(\mathcal{C}) x^i y^{n-i}.$$

The formulation of the Pless Power Moments involves Stirling numbers.

Definition 1.15.2 The **Stirling numbers $S(r, \nu)$ of the second kind** are defined for nonnegative integers r, ν by the equation

$$S(r, \nu) = \frac{1}{\nu!} \sum_{i=0}^{\nu} (-1)^{\nu-i} \binom{\nu}{i} i^r;$$

$\nu! S(r, \nu)$ is the number of ways to distribute r distinct objects into ν distinct boxes with no box left empty.

The next theorem gives six equivalent formulations of the **MacWilliams Identities** or **MacWilliams Equations**. The fourth in the list involves the Krawtchouk polynomials; see Definition 1.9.21. The last two are the **Pless Power Moments**. One proof of the equivalence of these identities is found in [1008, Chapter 7.2].

Theorem 1.15.3 (MacWilliams Identities and Pless Power Moments) *Let C be a linear $[n, k]_q$ code and $C^\perp$ its $[n, n - k]_q$ dual code. Let $A_i = A_i(C)$ and $A_i^\perp = A_i(C^\perp)$, for $0 \le i \le n$, be the weight distributions of C and $C^\perp$, respectively. The following are equivalent.*

(a) $\displaystyle\sum_{j=0}^{n-\nu} \binom{n-j}{\nu} A_j = q^{k-\nu} \sum_{j=0}^{\nu} \binom{n-j}{n-\nu} A_j^\perp$ *for $0 \le \nu \le n$.*

(b) $\displaystyle\sum_{j=\nu}^{n} \binom{j}{\nu} A_j = q^{k-\nu} \sum_{j=0}^{\nu} (-1)^j \binom{n-j}{n-\nu} (q-1)^{\nu-j} A_j^\perp$ *for $0 \le \nu \le n$.*

(c) $\mathrm{Hwe}_{C^\perp}(x, y) = \dfrac{1}{|C|} \mathrm{Hwe}_C(y - x, y + (q-1)x).$

(d) $A_j^\perp = \dfrac{1}{|C|} \displaystyle\sum_{i=0}^{n} A_i K_j^{(n,q)}(i)$ *for $0 \le j \le n$.*

(e) $\displaystyle\sum_{j=0}^{n} j^r A_j = \sum_{j=0}^{\min\{n,r\}} (-1)^j A_j^\perp \left(\sum_{\nu=j}^{r} \nu! S(r, \nu) q^{k-\nu} (q-1)^{\nu-j} \binom{n-j}{n-\nu} \right)$ *for $0 \le r$.*

(f) $\displaystyle\sum_{j=0}^{n} (n-j)^r A_j = \sum_{j=0}^{\min\{n,r\}} A_j^\perp \left(\sum_{\nu=j}^{r} \nu! S(r, \nu) q^{k-\nu} \binom{n-j}{n-\nu} \right)$ *for $0 \le r$.*

Remark 1.15.4 In the case of a linear code, the Delsarte–MacWilliams Inequalities of Theorem 1.9.22 follow from Theorem 1.15.3(d).

Example 1.15.5 In Theorem 1.13.6(b), we gave the weight distribution of $\mathcal{G}_{12}$. Using the Pless Power Moments of Theorem 1.15.3(e), we can verify this weight distribution using only the fact that $\mathcal{G}_{12}$ is a $[12, 6, 6]_3$ self-dual code. As $\mathcal{G}_{12}$ is self-dual, $A_i^\perp = A_i$ and $A_i = 0$ when $3 \nmid i$ by Theorem 1.6.2(h). As $A_0 = 1$ and the minimum weight of $\mathcal{G}_{12}$ is 6, the only unknown A_i are A_6, A_9, and A_{12}. We can find these from the first three power moments in Theorem 1.15.3(e). For a general $[n, k]_q$ code, as $A_0^\perp = 1$, the first three power moments are

$$\sum_{j=0}^{n} A_j = q^k,$$

$$\sum_{j=0}^{n} j A_j = q^{k-1}(qn - n - A_1^\perp), \text{ and}$$

$$\sum_{j=0}^{n} j^2 A_j = q^{k-2}\big((q-1)n(qn - n + 1) - (2qn - q - 2n + 2)A_1^\perp + 2A_2^\perp\big).$$

Applied specifically to $\mathcal{G}_{12}$, these become

$$1 + A_6 + A_9 + A_{12} = 729$$
$$6A_6 + 9A_9 + 12A_{12} = 5\,832$$
$$36A_6 + 81A_9 + 144A_{12} = 48\,600.$$

The unique solution to this system is $A_6 = 264$, $A_9 = 440$, and $A_{12} = 24$. Thus the weight enumerator of $\mathcal{G}_{12}$ is

$$\text{Hwe}_{\mathcal{G}_{12}}(x, y) = y^{12} + 264x^6 y^6 + 440 x^9 y^3 + 24 x^{12}.$$

Example 1.15.6 Let $\mathcal{C}$ be the $\left[(q^m - 1)/(q - 1), m, q^{m-1}\right]_q$ simplex code of Theorem 1.10.5; by this theorem, with $n = (q^m - 1)/(q - 1)$,

$$\text{Hwe}_{\mathcal{C}}(x, y) = y^n + (q^m - 1)x^{q^{m-1}} y^{n - q^{m-1}}.$$

By Definition 1.10.1, $\mathcal{C}^\perp = \mathcal{H}_{m,q}$. Using Theorem 1.15.3(d),

$$A_j(\mathcal{H}_{m,q}) = \frac{1}{q^m}\left(K_j^{(n,q)}(0) + (q^m - 1)K_j^{(n,q)}(q^{m-1})\right) \text{ for } 0 \le j \le n$$

noting that

$$K_j^{(n,q)}(0) = (q - 1)^j \binom{n}{j} \text{ and}$$

$$K_j^{(n,q)}(q^{m-1}) = \sum_{i=1}^{j}(-1)^i(q-1)^{j-i}\binom{q^{m-1}}{i}\binom{n - q^{m-1}}{j - i}.$$

The MacWilliams Identities can be used to find the weight distribution of an MDS code as found, for example, in [1323, Theorem 6 of Chapter 11]. A resulting corollary gives bounding relations on the length, dimension, and field size.

Theorem 1.15.7 Let $\mathcal{C}$ be an $[n, k, d]_q$ MDS code over $\mathbb{F}_q$. The weight distribution of $\mathcal{C}$ is given by $A_0(\mathcal{C}) = 1$, $A_i(\mathcal{C}) = 0$ for $1 \le i < d = n - k + 1$ and

$$A_i(\mathcal{C}) = \binom{n}{i}\sum_{j=0}^{i-d}(-1)^j\binom{i}{j}\left(q^{i+1-d-j} - 1\right)$$

for $d \le i \le n$.

Corollary 1.15.8 Let $\mathcal{C}$ be an $[n, k, d]_q$ MDS code over $\mathbb{F}_q$.

(a) If $2 \le k$, then $d = n - k + 1 \le q$.

(b) If $k \le n - 2$, then $k + 1 \le q$.

This corollary becomes a foundation for the MDS Conjecture 3.3.21 in Chapter 3.

1.16 Encoding

Figure 1.1 shows a simple communication channel that includes a component called an *encoder*, in which a message is encoded to produce a codeword. In this section we examine two encoding processes.

As in Figure 1.1, a message is any of the q^k possible k-tuples $\mathbf{x} \in \mathbb{F}_q^k$. The encoder will convert $\mathbf{x}$ to an n-tuple $\mathbf{c}$ from a code $\mathcal{C}$ over $\mathbb{F}_q$ with q^k codewords; that codeword will then be transmitted over the communication channel.

Suppose that C is an $[n, k, d]_q$ linear code with generator matrix G and parity check matrix H. We first describe an encoder that uses the generator matrix G. The most common way to encode the message $\mathbf{x}$ is as $\mathbf{x} \mapsto \mathbf{c} = \mathbf{x}G$. If G is replaced by another generator matrix, the encoding of $\mathbf{x}$ will, of course, be different. A nice relationship exists between message and codeword if G is in standard form $[I_k \mid A]$. In that case the first k coordinates of the codeword $\mathbf{c}$ are the information symbols $\mathbf{x}$ in order; the remaining $n - k$ symbols are the parity check symbols, that is, the redundancy added to $\mathbf{x}$ in order to help recover $\mathbf{x}$ if errors occur during transmission. A similar relationship between message and codeword can exist even if G is not in standard form. Specifically, suppose there exist column indices $i_1, i_2, \ldots, i_k$ such that the $k \times k$ matrix consisting of these k columns of G is the $k \times k$ identity matrix. In that case the message is found in the k coordinates $i_1, i_2, \ldots, i_k$ of the codeword scrambled but otherwise unchanged; that is, the message symbol x_j is in component i_j of the codeword. If this occurs where the message is embedded in the codeword, we say that the encoder is a **systematic encoder of** C. We can always force an encoder to be systematic. For example, if G is row reduced to a matrix G_1 in reduced row echelon form, G_1 remains a generator matrix of C by Remark 1.4.3; the encoding $\mathbf{x} \mapsto \mathbf{c} = \mathbf{x}G_1$ is systematic as G_1 has k columns which together form I_k. Another way to force an encoder to be systematic is as follows. By Theorem 1.8.4, C is permutation equivalent to an $[n, k, d]_q$ code C' with generator matrix G' in standard form. If the code C' is used in place of C, the encoder $\mathbf{x} \mapsto \mathbf{x}G'$ is a systematic encoder of C'.

Example 1.16.1 Let C be the $[7, 4, 3]_2$ binary Hamming code $\mathcal{H}_{3,2}$ with generator matrix $G_{3,2}$ from Example 1.4.9. Encoding $\mathbf{x} = x_1 x_2 x_3 x_4 \in \mathbb{F}_2^4$ as $\mathbf{x}G_{3,2}$ produces the codeword $\mathbf{c} = x_1 x_2 x_3 x_4 (x_2 + x_3 + x_4)(x_1 + x_3 + x_4)(x_1 + x_2 + x_4)$.

Example 1.16.2 Let C be an $[n, k, d]_q$ cyclic code with generator polynomial $g(x)$ and generator matrix G obtained from cyclic shifts of $g(x)$ as in Theorem 1.12.11(f). Suppose the message $\mathbf{m} = m_0 m_1 \cdots m_{k-1}$ is to be encoded as $\mathbf{c} = \mathbf{m}G$. Using the polynomial $m(x) = m_0 + m_1 x + \cdots + m_{k-1} x^{k-1}$ to represent the message $\mathbf{m}$ and $c(x) = c_0 + c_1 x + \cdots + c_{n-1} x^{n-1}$ to represent the codeword $\mathbf{c}$, it is a simple calculation to show $c(x) = m(x)g(x)$. Generally, this encoding is not systematic. Recall from Examples 1.12.3 and 1.12.18 that the Hamming $[7, 4, 3]_2$ code $\mathcal{H}_{3,2}$ has a cyclic form with generator polynomial $g(x) = 1 + x + x^3$. The nonsystematic encoder $m(x) \mapsto c(x) = m(x)g(x)$ yields $c(x) = m_0 + (m_0 + m_1)x + (m_1 + m_2)x^2 + (m_0 + m_2 + m_3)x^3 + (m_1 + m_3)x^4 + m_2 x^5 + m_3 x^6$.

The second method to encode uses the parity check matrix H. This is easiest to do when G is in standard form $[I_k \mid A]$; in this case $H = [-A^\mathsf{T} \mid I_{n-k}]$ by Theorem 1.4.7. Suppose that $\mathbf{x} = x_1 x_2 \cdots x_k$ is to be encoded as the codeword $\mathbf{c} = c_1 c_2 \cdots c_n = \mathbf{x}G$. As G is in standard form, $c_1 c_2 \cdots c_k = x_1 x_2 \cdots x_k$. So we need to determine the $n - k$ redundancy symbols $c_{k+1} c_{k+2} \cdots c_n$. Because $\mathbf{0}^\mathsf{T} = H\mathbf{c}^\mathsf{T} = [-A^\mathsf{T} \mid I_{n-k}]\mathbf{c}^\mathsf{T}$, we have $A^\mathsf{T}\mathbf{x}^\mathsf{T} = c_{k+1} c_{k+2} \cdots c_n^\mathsf{T}$, or equivalently $c_{k+1} c_{k+2} \cdots c_n = \mathbf{x}A$. This process can be generalized when $\mathbf{x} \mapsto \mathbf{x}G$ is a systematic encoder.

Example 1.16.3 Continuing with Example 1.16.1, we can encode $\mathbf{x} = x_1 x_2 x_3 x_4$ using $H_{3,2}$ from Example 1.4.9. Here $c_5 c_6 c_7 = \mathbf{x}A$ where

$$A = \begin{bmatrix} 0 & 1 & 1 \\ 1 & 0 & 1 \\ 1 & 1 & 0 \\ 1 & 1 & 1 \end{bmatrix}.$$

Thus $c_5 = x_2 + x_3 + x_4$, $c_6 = x_1 + x_3 + x_4$, and $c_7 = x_1 + x_2 + x_4$, consistent with Example 1.16.1.

Example 1.16.4 Let C be an $[n, k, d]_q$ cyclic code with generator polynomial $g(x)$. In Example 1.16.2 a nonsystematic encoder was described that encodes a cyclic code using $g(x)$. There is a systematic encoder of C using the generator polynomial $g^{\perp}(x)$ of $C^{\perp}$. By Theorem 1.12.19, $g^{\perp}(x) = x^k h(x^{-1})/h(0) = h'_0 + h'_1 x + \cdots + h'_{k-1} x^{k-1} + h'_k x^k$ where $h(x) = (x^n - 1)/g(x)$ and $h'_k = 1$. Let H, which is a parity check matrix for C, be determined from the shifts of $g^{\perp}(x)$ as follows:

$$H = \begin{bmatrix} h'_0 & h'_1 & h'_2 & \cdots & h'_k & \cdots & \cdots & 0 \\ 0 & h'_0 & h'_1 & \cdots & h'_{k-1} & h'_k & \cdots & 0 \\ \vdots & & & & & & & \vdots \\ 0 & 0 & 0 & h'_0 & \cdots & \cdots & \cdots & h'_k \end{bmatrix}$$

$$\leftrightarrow \begin{bmatrix} g^{\perp}(x) & & & \\ & xg^{\perp}(x) & & \\ & & \ddots & \\ & & & x^{n-k-1} g^{\perp}(x) \end{bmatrix}.$$

Examining the generator matrix G for C in Theorem 1.12.11(f), $\{0, 1, \ldots, k-1\}$ is an information set for C. Let $\mathbf{c} = c_0 c_1 \cdots c_{n-1} \in C$; so $c_0 c_1 \cdots c_{k-1}$ can be considered the associated message. The redundancy components $c_k c_{k+1} \cdots c_{n-1}$ are determined from $H\mathbf{c}^{\mathsf{T}} = \mathbf{0}^{\mathsf{T}}$ and can be computed in the order $i = k, k+1, \ldots, n-1$ where

$$c_i = -\sum_{j=0}^{k-1} h'_j c_{i-k+j}. \tag{1.3}$$

Example 1.16.5 We apply the systematic encoding of Example 1.16.4 to the cyclic version of the Hamming $[7, 4, 3]_2$ code $\mathcal{H}_{3,2}$ with generator polynomial $g(x) = 1 + x + x^3$; see Example 1.12.18. By Example 1.12.20, $g^{\perp}(x) = 1 + x^2 + x^3 + x^4$ and (1.3) yields $c_4 = c_0 + c_2 + c_3$, $c_5 = c_1 + c_3 + c_4$, and $c_6 = c_2 + c_4 + c_5$. In terms of the information bits $c_0 c_1 c_2 c_3$, we have $c_4 = c_0 + c_2 + c_3$, $c_5 = c_0 + c_1 + c_2$, and $c_6 = c_1 + c_2 + c_3$.

As discussed in Section 1.1, sometimes the receiver is interested only in the sent codewords rather than the sent messages. However, if there is interest in the actual message, a question arises as to how to recover the message from a codeword. If the encoder $\mathbf{x} \mapsto \mathbf{x}G$ is systematic, it is straightforward to recover the message. What can be done otherwise? Because G has independent rows, there is an $n \times k$ matrix K such that $GK = I_k$; K is called a **right inverse for** G. A right inverse is not necessarily unique. As $\mathbf{c} = \mathbf{x}G$, the message $\mathbf{x} = \mathbf{x}GK = \mathbf{c}K$.

Example 1.16.6 In Example 1.16.2, we encoded the message $m_0 m_1 m_2 m_3$ using the $[7, 4, 3]_2$ cyclic version of $\mathcal{H}_{3,2}$ with generator polynomial $g(x) = 1 + x + x^3$. The resulting codeword was $\mathbf{c} = (m_0, m_0 + m_1, m_1 + m_2, m_0 + m_2 + m_3, m_1 + m_3, m_2, m_5)$. The generator matrix G obtained from $g(x)$ as in Theorem 1.12.11(f) has right inverse

$$K = \begin{bmatrix} 1 & 0 & 0 & 0 \\ 0 & 1 & 0 & 0 \\ 0 & 1 & 0 & 0 \\ 0 & 1 & 0 & 0 \\ 0 & 1 & 0 & 0 \\ 0 & 0 & 1 & 0 \\ 0 & 0 & 0 & 1 \end{bmatrix}.$$

Computing $\mathbf{c}K$ gives $m_0 m_1 m_2 m_3$ as expected.

FIGURE 1.2: Binary symmetric channel with crossover probability ϱ

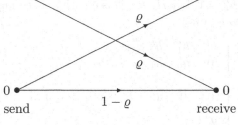

1.17 Decoding

Decoding is the process of determining which codeword **c** was sent when a vector **y** is received. Decoding is generally more complex than encoding. Decoding algorithms usually vary with the type of code being used. In this section we discuss only the basics of hard-decision decoding. Decoding is discussed more in depth in Chapters 15, 21, 24, 28–30, and 32.

Definition 1.17.1 A **hard-decision decoder** is a decoder that inputs 'hard' data from the channel (e.g., elements from $\mathbb{F}_q$) and outputs hard data to the receiver. A **soft-decision decoder** is one which inputs 'soft' data from the channel (e.g., estimates of the symbols with attached probabilities) and generally outputs hard data.

Initially we focus our discussion to the decoding of binary codes. To set the stage for decoding, we begin with one possible mathematical model of a channel that transmits binary data. Before stating the model, we establish some notation. If E is an event, $\mathrm{Prob}(E)$ is the probability that E occurs. If E_1 and E_2 are events, $\mathrm{Prob}(E_1 \mid E_2)$ is the conditional probability that E_1 occurs given that E_2 occurs. The model for transmitting binary data we explore is called the **binary symmetric channel (BSC) with crossover probability** ϱ as illustrated in Figure 1.2. In a BSC, we have the following conditional probabilities: For $y, c \in \mathbb{F}_2$,

$$\mathrm{Prob}(y \text{ is received} \mid c \text{ is sent}) = \begin{cases} 1 - \varrho & \text{if } y = c, \\ \varrho & \text{if } y \neq c. \end{cases} \tag{1.4}$$

In a BSC we also assume that the probability of error in one bit is independent of previous bits. We will assume that it is more likely that a bit is received correctly than in error; so $\varrho < \frac{1}{2}$.[10]

Let $\mathcal{C}$ be a binary code of length n. Assume that $\mathbf{c} \in \mathcal{C}$ is sent and $\mathbf{y} \in \mathbb{F}_2^n$ is received and decoded as $\widetilde{\mathbf{c}} \in \mathcal{C}$. Of course the hope is that $\widetilde{\mathbf{c}} = \mathbf{c}$; otherwise the decoder has made a **decoding error**. So $\mathrm{Prob}(\mathbf{c} \mid \mathbf{y})$ is the probability that the codeword **c** is sent given that **y** is received, and $\mathrm{Prob}(\mathbf{y} \mid \mathbf{c})$ is the probability that **y** is received given that the codeword **c**

---

[10]While ϱ is usually very small, if $\varrho > \frac{1}{2}$, the probability that a bit is received in error is higher than the probability that it is received correctly. One strategy is to then immediately interchange 0 and 1 at the receiving end. This converts the BSC with crossover probability ϱ to a BSC with crossover probability $1 - \varrho < \frac{1}{2}$. This of course does not help if $\varrho = \frac{1}{2}$; in this case communication is not possible.

is sent. These probabilities are related by Bayes' Rule

$$\text{Prob}(\mathbf{c}\,|\,\mathbf{y}) = \frac{\text{Prob}(\mathbf{y}\,|\,\mathbf{c})\text{Prob}(\mathbf{c})}{\text{Prob}(\mathbf{y})}$$

where $\text{Prob}(\mathbf{c})$ is the probability that $\mathbf{c}$ is sent and $\text{Prob}(\mathbf{y})$ is the probability that $\mathbf{y}$ is received. There are two natural means by which a decoder can make a choice based on these two probabilities. First the decoder could decode $\mathbf{y}$ as $\widetilde{\mathbf{c}} \in \mathcal{C}$ where $\text{Prob}(\widetilde{\mathbf{c}}\,|\,\mathbf{y})$ is maximum; such a decoder is called a **maximum *a posteriori* probability (MAP) decoder**. Symbolically a MAP decoder makes the decision

$$\widetilde{\mathbf{c}} = \arg\max_{\mathbf{c}\in\mathcal{C}} \text{Prob}(\mathbf{c}\,|\,\mathbf{y}).$$

Here $\arg\max_{\mathbf{c}\in\mathcal{C}} \text{Prob}(\mathbf{c}\,|\,\mathbf{y})$ is the argument $\mathbf{c}$ of the probability function $\text{Prob}(\mathbf{c}\,|\,\mathbf{y})$ that maximizes this probability. Alternately the decoder could decode $\mathbf{y}$ as $\widetilde{\mathbf{c}} \in \mathcal{C}$ where $\text{Prob}(\mathbf{y}\,|\,\widetilde{\mathbf{c}})$ is maximum; such a decoder is called a **maximum likelihood (ML) decoder**. Symbolically an ML decoder makes the decision

$$\widetilde{\mathbf{c}} = \arg\max_{\mathbf{c}\in\mathcal{C}} \text{Prob}(\mathbf{y}\,|\,\mathbf{c}).$$

We further analyze ML decoding over a BSC. If $\mathbf{y} = y_1 y_2 \cdots y_n$ and $\mathbf{c} = c_1 c_2 \cdots c_n$,

$$\text{Prob}(\mathbf{y}\,|\,\mathbf{c}) = \prod_{i=1}^{n} \text{Prob}(y_i\,|\,c_i)$$

since bit errors are independent. By (1.4) $\text{Prob}(y_i\,|\,c_i) = \varrho$ if $y_i \neq c_i$ and $\text{Prob}(y_i\,|\,c_i) = 1 - \varrho$ if $y_i = c_i$. Therefore

$$\text{Prob}(\mathbf{y}\,|\,\mathbf{c}) = \varrho^{d_{\mathrm{H}}(\mathbf{y},\mathbf{c})}(1-\varrho)^{n-d_{\mathrm{H}}(\mathbf{y},\mathbf{c})} = (1-\varrho)^n \left(\frac{\varrho}{1-\varrho}\right)^{d_{\mathrm{H}}(\mathbf{y},\mathbf{c})}. \tag{1.5}$$

Since $0 < \varrho < \frac{1}{2}$, $0 < \frac{\varrho}{1-\varrho} < 1$. Thus maximizing $\text{Prob}(\mathbf{y}\,|\,\mathbf{c})$ is equivalent to minimizing $d_{\mathrm{H}}(\mathbf{y},\mathbf{c})$; so a ML decoder finds the codeword $\mathbf{c}$ closest to the received vector $\mathbf{y}$ in Hamming distance.

Definition 1.17.2 If a decoder decodes a received vector $\mathbf{y}$ as the codeword $\mathbf{c}$ with $d_{\mathrm{H}}(\mathbf{y},\mathbf{c})$ minimized, the decoder is called a **nearest neighbor decoder**.

From this discussion, on a BSC, maximum likelihood and nearest neighbor decoding are the same. We can certainly perform nearest neighbor decoding on any code over any field.

Before presenting an example of a nearest neighbor decoder, we need to establish the relationship between the minimum distance of a code and the error-correcting capability of the code under nearest neighbor decoding. Notice this theorem is valid for any code, linear or not, over any finite field.

Theorem 1.17.3 *Let $\mathcal{C}$ be an $(n, M, d)_q$ code and $t = \lfloor \frac{d-1}{2} \rfloor$. If a codeword $\mathbf{c} \in \mathcal{C}$ is sent and $\mathbf{y}$ is received where t or fewer errors have occurred, then $\mathbf{c}$ is the unique codeword closest to $\mathbf{y}$. In particular nearest neighbor decoding uniquely and correctly decodes any received vector in which at most t errors have occurred in transmission.*

Proof: By definition $\mathbf{y} \in S_{q,n,t}(\mathbf{c})$, the sphere of radius t in $\mathbb{F}_q^n$ centered at $\mathbf{c}$. By Theorem 1.9.5(b), spheres of radius t centered at codewords are pairwise disjoint; hence if $\mathbf{y} \in S_{q,n,t}(\mathbf{c}_1)$ with $\mathbf{c}_1 \in \mathcal{C}$, then $\mathbf{c} = \mathbf{c}_1$. $\qquad \square$

Definition 1.17.4 A code C is a t-**error-correcting code** provided that whenever any $\mathbf{c} \in C$ is transmitted and $\mathbf{y} \in \mathbb{F}_q^n$ is received, where $\mathbf{y}$ differs from $\mathbf{c}$ in at most t coordinates, then every other codeword in C differs from $\mathbf{y}$ in more than t coordinates.

Remark 1.17.5 By Theorem 1.17.3, an $(n, M, d)_q$ code C is t-error-correcting for any $t \le \lfloor \frac{d-1}{2} \rfloor$. Furthermore, when $M > 1$ and $t = \lfloor \frac{d-1}{2} \rfloor$, there exist two distinct codewords such that the spheres of radius $t + 1$ about them are not disjoint; if this were not the case, the minimum distance of C is in fact larger than d. Thus when $M > 1$ and $t = \lfloor \frac{d-1}{2} \rfloor$, C is not $(t + 1)$-error-correcting.[11]

The nearest neighbor decoding problem for an $(n, M, d)_q$ code becomes one of finding an efficient algorithm that will correct up to $t = \lfloor \frac{d-1}{2} \rfloor$ errors. An obvious decoding algorithm is to examine all codewords until one is found with distance t or less from the received vector. This is a realistic decoding algorithm only for M small. Another obvious algorithm is to make a table consisting of a nearest codeword for each of the q^n vectors in $\mathbb{F}_q^n$ and then look up a received vector in the table to decode it. This is impractical if q^n is very large.

For an $[n, k, d]_q$ linear code, we can devise an algorithm using a table with q^{n-k} rather than q^n entries where one can find the nearest codeword by looking up one of these q^{n-k} entries. This general nearest neighbor decoding algorithm for linear codes is called *syndrome decoding*, which is the subject of the remainder of the section.

Definition 1.17.6 Let C be an $[n, k, d]_q$ linear code. For $\mathbf{y} \in \mathbb{F}_q^n$, the **coset of** C **with coset representative** $\mathbf{y}$ is $\mathbf{y} + C = \{\mathbf{y} + \mathbf{c} \mid \mathbf{c} \in C\}$. The **weight** of the coset $\mathbf{y} + C$ is the smallest weight of a vector in the coset, and any vector of this smallest weight in the coset is called a **coset leader**.

The next result follows from the theory of finite groups as a linear code is a group under addition.

Theorem 1.17.7 *Let C be an $[n, k, d]_q$ linear code. The following hold for $\mathbf{y}, \mathbf{y}', \mathbf{e} \in \mathbb{F}_q^n$.*

(a) $\mathbf{y} + C = \mathbf{y}' + C$ *if and only if* $\mathbf{y} - \mathbf{y}' \in C$.

(b) *Cosets of C all have size q^k.*

(c) *Cosets of C are either equal or disjoint. There are q^{n-k} distinct cosets of C and they partition $\mathbb{F}_q^n$.*

(d) *If $\mathbf{e}$ is a coset representative of $\mathbf{y} + C$, then $\mathbf{e} + C = \mathbf{y} + C$. In particular, if $\mathbf{e}$ is a coset leader of $\mathbf{y} + C$, then $\mathbf{e} + C = \mathbf{y} + C$.*

(e) *Any coset of weight at most $t = \lfloor \frac{d-1}{2} \rfloor$ has a unique coset leader.*

Let C be an $[n, k, d]_q$ code; fix a parity check matrix H of C. For $\mathbf{y} \in \mathbb{F}_q^n$, $\mathrm{syn}(\mathbf{y}) = H\mathbf{y}^\mathsf{T}$ is called the **syndrome of** $\mathbf{y}$. Syndromes are column vectors in $\mathbb{F}_q^{n-k}$. The code C consists of all vectors whose syndrome equals $\mathbf{0}^\mathsf{T}$. As H has rank $n - k$, every column vector in $\mathbb{F}_q^{n-k}$ is a syndrome. From Theorem 1.17.7, if $\mathbf{y}, \mathbf{y}' \in \mathbb{F}_q^n$ are in the same coset of C, then $\mathbf{y} - \mathbf{y}' = \mathbf{c} \in C$. Therefore $\mathrm{syn}(\mathbf{y}) = H\mathbf{y}^\mathsf{T} = H(\mathbf{y}' + \mathbf{c})^\mathsf{T} = H\mathbf{y}'^\mathsf{T} + H\mathbf{c}^\mathsf{T} = H\mathbf{y}'^\mathsf{T} + \mathbf{0}^\mathsf{T} = \mathrm{syn}(\mathbf{y}')$. Conversely, if $\mathrm{syn}(\mathbf{y}) = \mathrm{syn}(\mathbf{y}')$, then $H(\mathbf{y} - \mathbf{y}')^\mathsf{T} = \mathbf{0}^\mathsf{T}$ and so $\mathbf{y} - \mathbf{y}' \in C$. Thus we have the following theorem.

---

[11]In the trivial case where $M = 1$, C is n-error-correcting as every received vector decodes to the only codeword in C. However, since the information rate (Definition 1.9.26) of such a code is 0, it is never used in practice.

Theorem 1.17.8 *Two vectors belong to the same coset if and only if they have the same syndrome.*

Hence there is a one-to-one correspondence between cosets of C and syndromes. For $\mathbf{s} \in \mathbb{F}_q^{n-k}$, denote by $C_\mathbf{s}$ the coset of C consisting of all vectors in $\mathbb{F}_q^n$ with syndrome $\mathbf{s}^\mathsf{T}$. Also let $\mathbf{e_s}$ be a coset leader of $C_\mathbf{s}$. Thus $C_\mathbf{s} = \mathbf{e_s} + C$.

Suppose a codeword sent over a communication channel is received as a vector $\mathbf{y}$. In nearest neighbor decoding we seek a vector $\mathbf{e}$ of smallest weight such that $\mathbf{y} - \mathbf{e} \in C$. So nearest neighbor decoding is equivalent to finding a coset leader $\mathbf{e}$ of the coset $\mathbf{y} + C$ and decoding the received vector $\mathbf{y}$ as $\mathbf{y} - \mathbf{e}$. The *Syndrome Decoding Algorithm* is the following implementation of nearest neighbor decoding.

Algorithm 1.17.9 (Syndrome Decoding)

Use the above notation.

Step 1: For each syndrome $\mathbf{s} \in \mathbb{F}_q^{n-k}$, choose a coset leader $\mathbf{e_s}$ of the coset $C_\mathbf{s}$. Create a table pairing the syndrome with the coset leader.

Step 2: After receiving a vector $\mathbf{y}$, compute $\mathbf{s} = \text{syn}(\mathbf{y})$.

Step 3: Decode $\mathbf{y}$ as the codeword $\mathbf{y} - \mathbf{e_s}$.

Step 1 of this algorithm can be somewhat involved, but it is a one-time preprocessing task that is carried out before received vectors are analyzed. We briefly describe this table creation. Begin with all vectors in $\mathbb{F}_q^n$ of weight $t = \lfloor \frac{d-1}{2} \rfloor$ or less and place them in the table paired with their syndromes; by Theorem 1.17.7(e), no syndrome is repeated. If all syndromes have not been accounted for, place all vectors in $\mathbb{F}_q^n$ of weight $t + 1$, one at a time, paired with their syndromes into the table as long as the syndrome is not already in the table. If all syndromes have still not been accounted for, repeat this procedure with vectors in $\mathbb{F}_q^n$ of weight $t + 2$, then weight $t + 3$, and continue inductively. End the process when all syndromes are in the table.

Example 1.17.10 Let C be the $[6, 3, 3]_2$ binary code with parity check matrix

$$H = \begin{bmatrix} 0 & 1 & 1 & 1 & 0 & 0 \\ 1 & 0 & 1 & 0 & 1 & 0 \\ 1 & 1 & 0 & 0 & 0 & 1 \end{bmatrix}.$$

The table of Step 1 in the Syndrome Decoding Algorithm is the following.

leader	syndrome	leader	syndrome	leader	syndrome	leader	syndrome
000000	000^T	010000	101^T	000100	100^T	000001	001^T
100000	011^T	001000	110^T	000010	010^T	100100	111^T

Notice that the coset with syndrome 111^T has weight 2 and does not have a unique coset leader. This coset has two other coset leaders: 010010 and 001001. All other cosets have unique coset leaders by Theorem 1.17.7(e). We analyze three received vectors.

- Suppose $\mathbf{y} = 110110$ is received. Then $\text{syn}(\mathbf{y}) = H\mathbf{y}^\mathsf{T} = 000^\mathsf{T}$ and $\mathbf{y}$ is decoded as $\mathbf{y}$. $\mathbf{y}$ was the sent codeword provided 2 or more errors were not made.

- Now suppose that $\mathbf{y} = 101000$ is received. Then $\text{syn}(\mathbf{y}) = 101^{\mathsf{T}}$ and $\mathbf{y}$ is decoded as $\mathbf{y} - 010000 = 111000$. This was the sent codeword provided only 1 error was made.

- Finally suppose that $\mathbf{y} = 111111$ is received. Then $\text{syn}(\mathbf{y}) = 111^{\mathsf{T}}$ and $\mathbf{y}$ is decoded as $\mathbf{y} - 100100 = 011011$ and at least 2 errors were made in transmission. If exactly 2 errors were made, and we had chosen one of the other two possible coset leaders for the table, $\mathbf{y}$ would have been decoded as $\mathbf{y} - 010010 = 101101$ or $\mathbf{y} - 001001 = 110110$.

For this code, any received vector where 0 or 1 errors were made would be decoded correctly. If 2 errors were made, the decoder would decode the received vector to one of three possible equally likely codewords; there is no way to determine which was actually sent. If more than 2 errors were made, the decoder would always decode the received vector incorrectly.

Example 1.17.11 Nearest neighbor decoding of the binary Hamming code $\mathcal{H}_{m,2}$ is particularly easy. The parity check matrix for this code consists of the $2^m - 1$ nonzero binary m-tuples of column length m; these can be viewed as the binary expansions of the integers $1, 2, \ldots, 2^m - 1$. Choose the parity check matrix H for $\mathcal{H}_{m,2}$ where column i is the associated binary m-tuple expansion of i. Step 1 of the Syndrome Decoding Algorithm 1.17.9 can be skipped and the algorithm becomes the following: If $\mathbf{y}$ is received, compute $\mathbf{s} = \text{syn}(\mathbf{y})$. If $\mathbf{s} = \mathbf{0}^{\mathsf{T}}$, decode $\mathbf{y}$ as the codeword $\mathbf{y}$. Otherwise $\mathbf{s}$ represents the binary expansion of some integer i; the nearest codeword $\mathbf{c}$ to $\mathbf{y}$ is obtained from $\mathbf{y}$ by adding 1 to the ith bit.

As an illustration, the parity check matrix to use for $\mathcal{H}_{4,2}$ is

$$
H = \begin{bmatrix}
0 & 0 & 0 & 0 & 0 & 0 & 0 & 1 & 1 & 1 & 1 & 1 & 1 & 1 & 1 \\
0 & 0 & 0 & 1 & 1 & 1 & 1 & 0 & 0 & 0 & 0 & 1 & 1 & 1 & 1 \\
0 & 1 & 1 & 0 & 0 & 1 & 1 & 0 & 0 & 1 & 1 & 0 & 0 & 1 & 1 \\
1 & 0 & 1 & 0 & 1 & 0 & 1 & 0 & 1 & 0 & 1 & 0 & 1 & 0 & 1
\end{bmatrix}.
$$

Suppose $\mathbf{y} = 100110001111000$ is received. Then $\text{syn}(\mathbf{y}) = 0100^{\mathsf{T}}$, which is column 4 of H. Hence 1 is added to coordinate 4 of $\mathbf{y}$ to yield the codeword $\mathbf{c} = 100010001111000$.

1.18   Shannon's Theorem

Shannon's Channel Coding Theorem [1661] guarantees that good codes exist making reliable communication possible. We will discuss this theorem in the context of binary linear codes for which maximum likelihood decoding over a BSC is used. Note however that the theorem can be stated in a more general setting.

Assume that the communication channel is a BSC with crossover probability ϱ and that syndrome decoding is used as the implementation of ML decoding to decode an $[n, k, d]_2$ code $\mathcal{C}$. The **word error rate** P_{err} for this channel and decoding scheme is the probability that the decoder makes an error, averaged over all codewords of $\mathcal{C}$; for simplicity assume that each codeword of $\mathcal{C}$ is equally likely to be sent. A decoder error occurs when $\widetilde{\mathbf{c}} = \arg\max_{\mathbf{c} \in \mathcal{C}} \text{Prob}(\mathbf{y} \mid \mathbf{c})$ is not the originally transmitted codeword $\mathbf{c}$ when $\mathbf{y}$ is received. The syndrome decoder makes a correct decision if $\mathbf{y} - \mathbf{c}$ is a coset leader. The probability that the decoder makes a correct decision is

$$
\varrho^{\text{wt}_{\text{H}}(\mathbf{y}-\mathbf{c})}(1-\varrho)^{n-\text{wt}_{\text{H}}(\mathbf{y}-\mathbf{c})}
$$

by (1.5). Therefore the probability that the syndrome decoder makes a correct decision averaged over all equally likely transmitted codewords is $\sum_{i=0}^{n} \alpha_i \varrho^i (1 - \varrho)^{n-i}$ where α_i is the number of coset leaders of weight i. Thus

$$P_{\mathrm{err}} = 1 - \sum_{i=0}^{n} \alpha_i \varrho^i (1 - \varrho)^{n-i}. \qquad (1.6)$$

Example 1.18.1 Suppose binary messages of length k are sent unencoded over a BSC with crossover probability ϱ. This in effect is the same as transmitting codewords from the $[k, k, 1]_2$ code $\mathcal{C} = \mathbb{F}_2^k$. This code has a unique coset, the code itself, and its leader is the zero codeword of weight 0. Hence $\alpha_0 = 1$ and $\alpha_i = 0$ for $i > 0$. Therefore (1.6) shows that the probability of decoder error is

$$P_{\mathrm{err}} = 1 - \varrho^0 (1 - \varrho)^k = 1 - (1 - \varrho)^k.$$

This is precisely what we expect as the probability of no decoding error is the probability $(1 - \varrho)^k$ that the k bits are received without error. For instance if $\varrho = 0.01$ and $k = 4$, P_{err} without coding the length 4 messages is 0.03940399.

Example 1.18.2 We compare sending $2^4 = 16$ binary messages unencoded to encoding using the $[7, 4, 3]_2$ binary Hamming code $\mathcal{H}_{3,2}$. By Theorem 1.17.7(c), there are $2^{7-4} = 8$ cosets of $\mathcal{H}_{3,2}$ in $\mathbb{F}_2^7$. Since $\mathbb{F}_2^7$ has 1 vector of weight 0 and 7 vectors of weight 1, these must be the coset leaders for all 8 cosets of $\mathcal{H}_{3,2}$ in $\mathbb{F}_2^7$ by Theorem 1.17.7(e). Thus $\alpha_0 = 1$, $\alpha_1 = 7$, and $\alpha_i = 0$ for $i > 1$. Hence the probability of decoder error is

$$P_{\mathrm{err}} = 1 - (1 - \varrho)^7 - 7\varrho(1 - \varrho)^6$$

by (1.6). For example if $\varrho = 0.01$, $P_{\mathrm{err}} = 0.00203104\cdots$, significantly lower than the word error rate for unencoded transmissions of binary messages of length 4 found in Example 1.18.1. For comparison, when transmitting 10 000 unencoded binary messages each of length 4, one can expect about 394 to be received in error. On the other hand, when transmitting 10 000 binary messages each of length 4 encoded to length 7 codewords from $\mathcal{H}_{3,2}$, one can expect about 20 to be decoded in error.

In order to state Shannon's Theorem, we need to define the channel capacity.

Definition 1.18.3 For a BSC with crossover probability ϱ, the **capacity of the channel** is

$$C(\varrho) = 1 + \varrho \log_2 \varrho + (1 - \varrho) \log_2(1 - \varrho).$$

The capacity $C(\varrho) = 1 - H_2(\varrho)$ where $H_2(\varrho)$ is the binary entropy function defined in more generality in Section 1.9.8.[12] See Figure 1.3.

The next theorem is Shannon's Theorem for binary symmetric channels. Shannon's original theorem was stated for nonlinear codes but was later shown to be valid for linear codes as well. The theorem also holds for other channels provided channel capacity is appropriately defined. For discussion and proofs of various versions of Shannon's Theorem, see for example [467, 1314]. For binary symmetric channels, Shannon's Theorem is as follows.

Theorem 1.18.4 (Shannon) *Let $\delta > 0$ and $R < C(\varrho)$. Then for large enough n, there exists an $[n, k]_2$ binary linear code $\mathcal{C}$ with $\frac{k}{n} \geq R$ such that $P_{\mathrm{err}} < \delta$ when $\mathcal{C}$ is used for communication over a BSC with crossover probability ϱ. Furthermore no such code exists if $R > C(\varrho)$.*

---

[12] When $q = 2$, the domain of the entropy function $H_2(x)$ can be extended from $0 \leq x < \frac{1}{2}$ to $0 \leq x < 1$.

FIGURE 1.3: Channel capacity for a BCS with crossover probability ρ

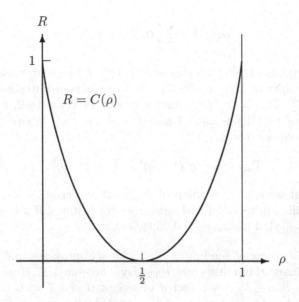

Remark 1.18.5 When the crossover probability is $\rho = \frac{1}{2}$, $C\left(\frac{1}{2}\right) = 0$. In this case Shannon's Theorem indicates that communication is not possible. This is not surprising; when $\rho = \frac{1}{2}$, whether a binary symbol is received correctly or incorrectly is essentially determined by a coin flip. See Footnote 10.

Remark 1.18.6 Recall that $\frac{k}{n}$ is the information rate of the code as in Definition 1.9.26. The proof of Shannon's Theorem is nonconstructive, but the theorem does guarantee that good codes exist with information rates just under channel capacity and decoding error rates arbitrarily small; unfortunately these codes may have to be extremely long. The objective becomes to find codes with a large number of codewords (to send many messages), large minimum distance (to correct many errors), and short length (to minimize transmission time or storage space). These goals conflict as seen in Section 1.9.

Chapter 2

Cyclic Codes over Finite Fields

Cunsheng Ding

The Hong Kong University of Science and Technology

2.1   Notation and Introduction

A brief introduction to cyclic codes over finite fields was given in Section 1.12. The objective of this chapter is to introduce several important families of cyclic codes over finite fields. We will follow the notation of Chapter 1 as closely as possible.

By an $[n, \kappa, d]_q$ code, we mean a linear code over $\mathbb{F}_q$ with length n, dimension κ and minimum distance d. Notice that the minimum distance of a linear code is equal to the minimum nonzero weight of the code. By the parameters of a linear code, we mean its length, dimension and minimum distance. An $[n, \kappa, d]_q$ code is said to be **distance-optimal** (respectively **dimension-optimal**) if there is no $[n, \kappa, d+1]_q$ (respectively $[n, \kappa+1, d]_q$) code. By the best known parameters of $[n, \kappa]$ linear codes over $\mathbb{F}_q$ we mean an $[n, \kappa, d]_q$ code with the largest known d reported in the tables of linear codes maintained at [845].

In this chapter, we deal with cyclic codes of length n over $\mathbb{F}_q$ and always assume that $\gcd(n, q) = 1$. Under this assumption, $x^n - 1$ has no repeated factors over $\mathbb{F}_q$. Denote by C_i the q-cyclotomic coset modulo n that contains i for $0 \leq i \leq n - 1$. Put $m = \text{ord}_n(q)$, and let γ be a generator of $\mathbb{F}_{q^m}^* := \mathbb{F}_{q^m} \setminus \{0\}$. Define $\alpha = \gamma^{(q^m - 1)/n}$. Then α is a primitive n^{th} root of unity. The canonical factorization of $x^n - 1$ over $\mathbb{F}_q$ is given by

$$x^n - 1 = M_{\alpha^{i_0}}(x) M_{\alpha^{i_1}}(x) \cdots M_{\alpha^{i_t}}(x), \tag{2.1}$$

where $i_0, i_1, \ldots, i_t$ are representatives of the q-cyclotomic cosets modulo n, and

$$M_{\alpha^{i_j}}(x) = \prod_{h \in C_{i_j}} (x - \alpha^h),$$

which is the minimal polynomial of α^{i_j} over $\mathbb{F}_q$ and is irreducible over $\mathbb{F}_q$.

Throughout this chapter, we define $\mathcal{R}_{(n,q)} = \mathbb{F}_q[x]/\langle x^n - 1 \rangle$ and use $\mathrm{Tr}_{q^m/q}$ to denote the trace function from $\mathbb{F}_{q^m}$ to $\mathbb{F}_q$ defined by $\mathrm{Tr}_{q^m/q}(x) = \sum_{j=0}^{m-1} x^{q^j}$. The ring of integers modulo n is denoted by $\mathbb{Z}_n = \{0, 1, \ldots, n - 1\}$.

Cyclic codes form an important subclass of linear codes over finite fields. Their algebraic structure is richer. Because of their cyclic structure, they are closely related to number theory. In addition, they have efficient encoding and decoding algorithms and are the most studied linear codes. In fact, most of the important families of linear codes are either cyclic codes or extended cyclic codes.

2.2 Subfield Subcodes

Let $\mathcal{C}$ be an $[n, \kappa]_{q^t}$ code. The **subfield subcode** $\mathcal{C}|_{\mathbb{F}_q}$ of $\mathcal{C}$ with respect to $\mathbb{F}_q$ is the set of codewords in $\mathcal{C}$ each of whose components is in $\mathbb{F}_q$. Since $\mathcal{C}$ is linear over $\mathbb{F}_{q^t}$, $\mathcal{C}|_{\mathbb{F}_q}$ is a linear code over $\mathbb{F}_q$.

The dimension, denoted κ_q, of the subfield subcode $\mathcal{C}|_{\mathbb{F}_q}$ may not have an elementary relation with that of the code $\mathcal{C}$. However, we have the following lower and upper bounds on κ_q.

Theorem 2.2.1 *Let $\mathcal{C}$ be an $[n, \kappa]_{q^t}$ code. Then $\mathcal{C}|_{\mathbb{F}_q}$ is an $[n, \kappa_q]$ code over $\mathbb{F}_q$, where $\kappa \geq \kappa_q \geq n - t(n - \kappa)$. If $\mathcal{C}$ has a basis of codewords in $\mathbb{F}_q^n$, then this is also a basis of $\mathcal{C}|_{\mathbb{F}_q}$ and $\mathcal{C}|_{\mathbb{F}_q}$ has dimension κ.*

Example 2.2.2 The Hamming code $\mathcal{H}_{3,2^2}$ over $\mathbb{F}_{2^2}$ has parameters $[21, 18, 3]_4$. The subfield subcode $\mathcal{H}_{3,2^2}|_{\mathbb{F}_2}$ is a $[21, 16, 3]_2$ code with parity check matrix

$$\begin{bmatrix}
1 & 0 & 0 & 1 & 1 & 0 & 0 & 1 & 1 & 0 & 0 & 1 & 1 & 1 & 1 & 1 & 0 & 0 & 1 & 1 & 0 & 1 \\
0 & 1 & 0 & 0 & 1 & 0 & 1 & 1 & 0 & 0 & 1 & 1 & 0 & 1 & 0 & 0 & 1 & 1 & 0 & 0 & 1 \\
0 & 0 & 1 & 1 & 0 & 0 & 1 & 1 & 0 & 0 & 1 & 1 & 0 & 0 & 1 & 1 & 0 & 0 & 1 & 1 & 0 \\
0 & 0 & 0 & 0 & 0 & 1 & 1 & 1 & 1 & 0 & 0 & 0 & 0 & 0 & 0 & 0 & 1 & 1 & 1 & 1 \\
0 & 0 & 0 & 0 & 0 & 0 & 0 & 0 & 0 & 1 & 1 & 1 & 1 & 1 & 1 & 1 & 1 & 0 & 0 & 0 & 0
\end{bmatrix}.$$

In this case, $n = 21$, $\kappa = 18$, and $n - t(n - \kappa) = 15$. Hence $\kappa_q = 16$, which is very close to $n - t(n - \kappa) = 15$.

The following is called **Delsarte's Theorem**, which exhibits a dual relation between subfield subcodes and trace codes. This theorem is very useful in the design and analysis of linear codes.

Theorem 2.2.3 (Delsarte) *Let $\mathcal{C}$ be a linear code of length n over $\mathbb{F}_{q^t}$. Then*

$$(\mathcal{C}|_{\mathbb{F}_q})^{\perp} = \mathrm{Tr}_{q^t/q}(\mathcal{C}^{\perp}),$$

where $\mathrm{Tr}_{q^t/q}(\mathcal{C}^{\perp}) = \left\{ \left(\mathrm{Tr}_{q^t/q}(v_1), \ldots, \mathrm{Tr}_{q^t/q}(v_n) \right) \mid (v_1, \ldots, v_n) \in \mathcal{C}^{\perp} \right\}$.

Theorems 2.2.1 and 2.2.3 work for all linear codes, including cyclic codes. Their proofs could be found in [1008, Section 3.8]. We shall need them later.

2.3 Fundamental Constructions of Cyclic Codes

In Section 1.12, it was shown that every cyclic code of length n over $\mathbb{F}_q$ can be generated by a generator polynomial $g(x) \in \mathbb{F}_q[x]$. The objective of this section is to describe several other fundamental constructions of cyclic codes over finite fields. By a fundamental construction, we mean a construction method that can produce every cyclic code over any finite field.

An element e in a commutative ring $\mathcal{R}$ is called an **idempotent** if $e^2 = e$. The ring $\mathcal{R}_{(n,q)}$ has in general quite a number of idempotents. Besides its generator polynomial, many other polynomials can generate a cyclic code $\mathcal{C}$. Let $\mathcal{C}$ be a cyclic code over $\mathbb{F}_q$ with generator polynomial $g(x)$. It is easily seen that a polynomial $f(x) \in \mathbb{F}_q[x]$ generates $\mathcal{C}$ if and only if $\gcd(f(x), x^n - 1) = g(x)$.

If an idempotent $e(x) \in \mathcal{R}_{(n,q)}$ generates a cyclic code $\mathcal{C}$, it is then unique in this ring and called the **generating idempotent**. Given the generator polynomial of a cyclic code, one can compute its generating idempotent with the following theorem [1008, Theorem 4.3.3].

Theorem 2.3.1 *Let $\mathcal{C}$ be a cyclic code of length n over $\mathbb{F}_q$ with generator polynomial $g(x)$. Let $h(x) = (x^n - 1)/g(x)$. Then $\gcd(g(x), h(x)) = 1$, as it was assumed that $\gcd(n, q) = 1$. Employing the Extended Euclidean Algorithm, one computes two polynomials $a(x) \in \mathbb{F}_q[x]$ and $b(x) \in \mathbb{F}_q[x]$ such that $1 = a(x)g(x) + b(x)h(x)$. Then $e(x) = a(x)g(x) \bmod (x^n - 1)$ is the generating idempotent of $\mathcal{C}$.*

The polynomial $h(x)$ in Theorem 2.3.1 is called the **parity check polynomial** of $\mathcal{C}$. Given the generating idempotent of a cyclic code, one obtains the generator polynomial of this code as follows [1008, Theorem 4.3.3].

Theorem 2.3.2 *Let $\mathcal{C}$ be a cyclic code over $\mathbb{F}_q$ with generating idempotent $e(x)$. Then the generator polynomial of $\mathcal{C}$ is given by $g(x) = \gcd(e(x), x^n - 1)$, which is computed in $\mathbb{F}_q[x]$.*

Example 2.3.3 The cyclic code $\mathcal{C}$ of length 11 over $\mathbb{F}_3$ with generator polynomial $g(x) = x^5 + x^4 + 2x^3 + x^2 + 2$ has parameters $[11, 6, 5]$ and parity check polynomial $h(x) = x^6 + 2x^5 + 2x^4 + 2x^3 + x^2 + 1$.

Let $a(x) = 2x^5 + x^4 + x^2$ and $b(x) = x^4 + x^3 + 1$. It is then easily verified that $1 = a(x)g(x) + b(x)h(x)$. Hence

$$e(x) = a(x)g(x) \bmod (x^{11} - 1) = 2x^{10} + 2x^8 + 2x^7 + 2x^6 + 2x^2,$$

which is the generating idempotent of $\mathcal{C}$. On the other hand, we have $g(x) = \gcd(e(x), x^{11} - 1)$.

A generator matrix of a cyclic code can be derived from its generating idempotent as follows [1008, Theorem 4.3.6].

Theorem 2.3.4 *If $\mathcal{C}$ is an $[n, \kappa]$ cyclic code with generating idempotent $e(x) = \sum_{i=0}^{n-1} e_i x^i$, then the $\kappa \times n$ matrix*

$$\begin{bmatrix} e_0 & e_1 & e_2 & \cdots & e_{n-2} & e_{n-1} \\ e_{n-1} & e_0 & e_1 & \cdots & e_{n-3} & e_{n-2} \\ \vdots & \vdots & \vdots & \vdots & \vdots & \vdots \\ e_{n-\kappa+1} & e_{n-\kappa+2} & e_{n-\kappa+3} & \cdots & e_{n-\kappa-1} & e_{n-\kappa} \end{bmatrix}$$

is a generator matrix of $\mathcal{C}$.

Let $f(x) = f_0 + f_1 x + \cdots + f_\ell x^\ell$ be a polynomial over a field with $f_\ell \neq 0$. Then the **reciprocal** of f is defined by

$$f^\perp(x) = f_\ell^{-1} x^\ell f(x^{-1}).$$

Another fundamental construction of cyclic codes over finite fields is the following trace construction [1899], which is a direct consequence of Theorem 2.2.3.

Theorem 2.3.5 *Let $\mathcal{C}$ be a cyclic code of length n over $\mathbb{F}_q$ with parity check polynomial $h(x)$. Let J be a subset of $\mathbb{Z}_n$ such that*

$$h^\perp(x) = \prod_{j \in J} M_{\alpha^j}(x),$$

where $h^\perp(x)$ is the reciprocal of $h(x)$. Then $\mathcal{C}$ consists of all the following codewords:

$$c_a(x) = \sum_{i=0}^{n-1} \mathrm{Tr}_{q^m/q}(f_a(\alpha^i)) x^i$$

where

$$f_a(x) = \sum_{j \in J} a_j x^j \ \text{ for } a_j \in \mathbb{F}_{q^m}.$$

The trace and generator polynomial approaches are the most popular and effective ways for constructing and analysing cyclic codes. In particular, the trace approach allows the use of exponential sums for the determination of the weight distribution of cyclic codes. A lot of progress in this direction has been made in the past decade [560]. A less investigated fundamental approach to cyclic codes is the q-polynomial method developed in [562]. Another fundamental construction uses sequences as described in Section 20.5.

The following theorem says that every projective linear code is a punctured code of a special cyclic code [1914].

Theorem 2.3.6 *Every linear code $\mathcal{C}$ of length n over $\mathbb{F}_q$ with minimum distance of $\mathcal{C}^\perp$ at least 3 is a punctured code of the cyclic code*

$$\left\{ \left(\mathrm{Tr}_{q^m/q}(a\gamma^0), \mathrm{Tr}_{q^m/q}(a\gamma^1), \ldots, \mathrm{Tr}_{q^m/q}(a\gamma^{q^m-2}) \right) \mid a \in \mathbb{F}_{q^m} \right\}.$$

for some integer m, where γ is a generator of $\mathbb{F}_{q^m}^*$.

2.4   The Minimum Distances of Cyclic Codes

The length of a cyclic code is clear from its definition. However, determining the dimensions and minimum distances of cyclic codes is nontrivial. If a cyclic code $\mathcal{C}$ of length n is defined by its generator polynomial $g(x)$, then the dimension of $\mathcal{C}$ equals $n - \deg(g)$. But it may be hard to find the degree of $g(x)$ when $g(x)$ is given as the least common multiple of a number of polynomials. If a cyclic code is defined in the trace form, it may also be difficult to determine the dimension. Determining the exact minimum distance of a cyclic code is more difficult. In the case that the minimum distance of a cyclic code cannot be settled, the best one could expect is to develop a good lower bound on the minimum distance. Unlike many other subclasses of linear codes, cyclic codes have some lower bounds on their minimum distances. Some of the bounds are easy to use, while others are hard to employ. Below we introduce a few effective lower bounds on the minimum distances of cyclic codes.

Let $\mathcal{C}$ be a cyclic code of length n over $\mathbb{F}_q$ with generator polynomial

$$g(x) = \prod_{i \in T}(x - \alpha^i)$$

where T is the union of some q-cyclotomic cosets modulo n, and is called the **defining set of $\mathcal{C}$ relative to** α. The following is a simple but very useful lower bound ([248] and [968]).

Theorem 2.4.1 (BCH Bound) *Let $\mathcal{C}$ be a cyclic code of length n over $\mathbb{F}_q$ with defining set T and minimum distance d. Assume T contains $\delta - 1$ consecutive integers for some integer δ. Then $d \geq \delta$.*

The BCH Bound depends on the choice of the primitive n^{th} root of unity α. Different choices of the primitive root may yield different lower bounds. When applying the BCH Bound, it is crucial to choose the right primitive root. However, it is open how to choose such a primitive root. In many cases the BCH Bound may be far away from the actual minimum distance. In such cases, the lower bound given in the following theorem may be much better. It was discovered by Hartmann and Tzeng [914]. To introduce this bound, we define

$$A + B = \{a + b \mid a \in A,\ b \in B\},$$

where A and B are two subsets of the ring $\mathbb{Z}_n$, n is a positive integer, and $+$ denotes the integer addition modulo n.

Theorem 2.4.2 (Hartmann–Tzeng Bound) *Let $\mathcal{C}$ be a cyclic code of length n over $\mathbb{F}_q$ with defining set T and minimum distance d. Let A be a set of $\delta - 1$ consecutive elements of T and $B(b, s) = \{jb \bmod n \mid 0 \leq j \leq s\}$, where $\gcd(b, n) < \delta$. If $A + B(b, s) \subseteq T$ for some b and s, then $d \geq \delta + s$.*

When $s = 0$, the Hartmann–Tzeng Bound becomes the BCH Bound. Other lower bounds can be found in [1838]. As cyclic codes are a special case of quasi-cyclic codes, bounds on such codes found in Section 7.4 can be applied to cyclic codes.

2.5 Irreducible Cyclic Codes

Let $\mathcal{C}(q, n, i)$ denote the cyclic code of length n over $\mathbb{F}_q$ with parity check polynomial $M_{\alpha^i}(x)$, which is the minimal polynomial of α^i over $\mathbb{F}_q$, and where α is a primitive n^{th} root of unity over an extension field of $\mathbb{F}_q$. These $\mathcal{C}(q, n, i)$ are called **irreducible cyclic codes.** Since the ideals $\langle (x^n - 1)/M_{\alpha^i}(x) \rangle$ of $\mathcal{R}_{(n,q)}$ are minimal, these $\mathcal{C}(q, n, i)$ are also called **minimal cyclic codes.**

By Theorem 2.3.5, $\mathcal{C}(q, n, i)$ has the following trace representation:

$$\mathcal{C}(q, n, i) = \left\{ \left(\text{Tr}_{q^{m_i}/q}(a\beta^0), \text{Tr}_{q^{m_i}/q}(a\beta), \dots, \text{Tr}_{q^{m_i}/q}(a\beta^{n-1}) \right) \mid a \in \mathbb{F}_{q^{m_i}} \right\},$$

where $\beta = \alpha^{-i} \in \mathbb{F}_{q^{m_i}}$ and $m_i = |C_i|$.

Example 2.5.1 Let $n = (q^m - 1)/(q - 1)$ and $\alpha = \gamma^{q-1}$, where γ is a generator of $\mathbb{F}_{q^m}^*$. If $\gcd(q - 1, m) = 1$, then $\mathcal{C}(q, n, 1)$ has parameters $[n, m, q^{m-1}]$ and is equivalent to the simplex code whose dual is the Hamming code. Hence, when $\gcd(q-1, m) = 1$, the Hamming code is equivalent to a cyclic code.

Example 2.5.2 The celebrated Golay codes introduced in Section 1.13 are also irreducible cyclic codes and the binary [24, 12, 8] extended Golay code was used on the Voyager 1 and Voyager 2 missions to Jupiter, Saturn, and their moons.

By definition, the dimension of $C(q, n, i)$ equals $\deg(M_{\alpha^i}(x))$, which is a divisor of $m := \mathrm{ord}_n(q)$. The determination of the weight enumerators of irreducible cyclic codes is equivalent to the evaluation of Gaussian periods, which is extremely difficult in general. However, in a small number of cases, the weight enumerator of some irreducible cyclic codes is known. One-weight, two-weight and three-weight irreducible cyclic codes exist. It is in general hard to determine the minimum distance of an irreducible cyclic code. A lower bound on the minimum distances of irreducible cyclic codes has been developed. The reader is referred to [568] for detailed information.

Irreducible cyclic codes are very important for many reasons. First of all, every cyclic code is the direct sum of a number of irreducible cyclic codes. Secondly, the automorphism group of some irreducible codes (Golay codes) has high transitivity. Thirdly, some irreducible codes can be employed to construct maximal arcs, elliptic quadrics (ovoids), inversive planes, and t-designs. Hence, irreducible cyclic codes are closely related to group theory, finite geometry and combinatorics. In addition, irreducible cyclic codes also have a number of applications in engineering.

2.6 BCH Codes and Their Properties

BCH codes are a subclass of cyclic codes with special properties and are important in both theory and practice. Experimental data shows that binary and ternary BCH codes of certain lengths are the best cyclic codes in almost all cases; see [549, Appendix A]. BCH codes were briefly introduced in Section 1.14. This section treats BCH codes further and summarizes their basic properties.

Let δ be an integer with $2 \leq \delta \leq n$ and let b be an integer. A **BCH code** over $\mathbb{F}_q$ of length n and **designed distance** δ, denoted by $C_{(q,n,\delta,b)}$, is a cyclic code with defining set

$$T(b, \delta) = C_b \cup C_{b+1} \cup \cdots \cup C_{b+\delta-2} \tag{2.2}$$

relative to the primitive n^{th} root of unity α, where C_i is the q-cyclotomic coset modulo n containing i.

When $b = 1$, the code $C_{(q,n,\delta,b)}$ with defining set in (2.2) is called a **narrow-sense** BCH code. If $n = q^m - 1$, then $C_{(q,n,\delta,b)}$ is referred to as a **primitive** BCH code. The Reed–Solomon code introduced in Section 1.14 is a primitive BCH code.

Sometimes $T(b_1, \delta_1) = T(b_2, \delta_2)$ for two distinct pairs (b_1, δ_1) and (b_2, δ_2). The **maximum designed distance** of a BCH code is defined to be the largest δ such that the set $T(b, \delta)$ in (2.2) defines the code for some $b \geq 0$. The maximum designed distance of a BCH code is also called the **Bose distance**.

Given the canonical factorization of $x^n - 1$ over $\mathbb{F}_q$ in (2.1), we know that the total number of nonzero cyclic codes of length n over $\mathbb{F}_q$ is $2^{t+1} - 1$. Then the following two natural questions arise:

1. How many of the $2^{t+1} - 1$ cyclic codes are BCH codes?

2. Which of the $2^{t+1} - 1$ cyclic codes are BCH codes?

The first question is open. Regarding the second question, we have the next result whose proof is straightforward.

Theorem 2.6.1 *A cyclic code of length n over $\mathbb{F}_q$ with defining set $T \subseteq \mathbb{Z}_n$ is a BCH code if and only if there exists an integer δ with $2 \le \delta \le n$, an integer b with $-(n-1) \le b \le n-1$, and an element $a \in \mathbb{Z}_n$ such that $\gcd(n, a) = 1$ and*

$$aT \bmod n = \bigcup_{i=0}^{\delta-2} C_{i+b}.$$

In general it is not easy to use Theorem 2.6.1 to check if a cyclic code is a BCH code. In addition, it looks hard to answer the first question above with Theorem 2.6.1.

Definition 2.6.2 A family of codes is **asymptotically good**, provided that there exists an infinite subset of $[n_i, k_i, d_i]$ codes from this family with $\lim_{i \to \infty} n_i = \infty$ such that both $\liminf_{i \to \infty} k_i/n_i > 0$ and $\liminf_{i \to \infty} d_i/n_i > 0$.

The family of primitive BCH codes over $\mathbb{F}_q$ is asymptotically bad in the following sense [1262].

Theorem 2.6.3 *If C_i are $[n_i, k_i, d_i]_q$ codes for $i = 1, 2, \ldots$ with $\lim_{i \to \infty} n_i = \infty$, then either $\liminf_{i \to \infty} k_i/n_i = 0$ or $\liminf_{i \to \infty} d_i/n_i = 0$.*

Despite this asymptotic property, binary primitive BCH codes of length up to 127 are among the best linear codes known [549]. It is questionable if the definition of asymptotic badness above makes sense for applications.

2.6.1 The Minimum Distances of BCH Codes

It follows from Theorem 2.4.1 that a cyclic code with designed distance δ has minimum weight at least δ. It is possible that the actual minimum distance is equal to the designed distance. Sometimes the actual minimum distance is much larger than the designed distance.

A codeword $(c_0, \ldots, c_{n-1})$ of a linear code C is **even-like** if $\sum_{j=0}^{n-1} c_j = 0$, and **odd-like** otherwise. The weight of an even-like (respectively odd-like) codeword is called an **even-like weight** (respectively **odd-like weight**). Let C be a primitive narrow-sense BCH code of length $n = q^m - 1$ over $\mathbb{F}_q$ with designed distance δ. The defining set is then $T(1, \delta) = C_1 \cup C_2 \cup \cdots \cup C_{\delta-1}$. The following theorem provides useful information on the minimum weight of narrow-sense primitive BCH codes.

Theorem 2.6.4 *Let C be the narrow-sense primitive BCH code of length $n = q^m - 1$ over $\mathbb{F}_q$ with designed distance δ. Then the minimum weight of C is its minimum odd-like weight.*

The coordinates of the narrow-sense primitive BCH code C of length $n = q^m - 1$ over $\mathbb{F}_q$ with designed distance δ can be indexed by the elements of $\mathbb{F}_{q^m}^*$, and the extended coordinate in the extended code $\widehat{C}$ can be indexed by the zero element of $\mathbb{F}_{q^m}$. The general affine group $\mathrm{GA}_1(\mathbb{F}_{q^m})$ then acts on $\mathbb{F}_{q^m}$ and also on $\widehat{C}$ doubly transitively, where

$$\mathrm{GA}_1(\mathbb{F}_{q^m}) = \{ax + b \mid a \in \mathbb{F}_{q^m}^*, \ b \in \mathbb{F}_{q^m}\}.$$

Since $\mathrm{GA}_1(\mathbb{F}_{q^m})$ is transitive on $\mathbb{F}_{q^m}$, it is a subgroup of the permutation automorphism group of $\widehat{C}$. Theorem 2.6.4 then follows.

In the following cases, the minimum distance of the BCH code $C_{(q,n,\delta,b)}$ is known. We first have the following [1323, p. 260].

Theorem 2.6.5 *For any h with $1 \le h \le m - 1$, the narrow-sense primitive BCH code $\mathcal{C}_{(q,q^m-1,q^h-1,1)}$ has minimum distance $d = q^h - 1$.*

It is easy to prove the following result [1247], which is a generalization of the classical result for the narrow-sense primitive case.

Theorem 2.6.6 *The code $\mathcal{C}_{(q,n,\delta,b)}$ has minimum distance $d = \delta$ if δ divides $\gcd(n, b - 1)$.*

The following is proved in [1245], which is a generalization of a classical result of Kasami and Lin for the case $q = 2$.

Theorem 2.6.7 *Let $m \ge 3$ for $q = 2$, $m \ge 2$ for $q = 3$, and $m \ge 1$ for $q \ge 4$. For the narrow-sense primitive BCH code $\mathcal{C}_{(q,q^m-1,\delta,1)}$ with $\delta = q^m - q^{m-1} - q^i - 1$, where $(m-2)/2 \le i \le m - \lfloor m/3 \rfloor - 1$, the minimum distance $d = \delta$.*

Although it is notoriously difficult to find out the minimum distance of a BCH code in general, in a small number of cases other than the cases dealt with in the three theorems above, the minimum distance of the BCH code $\mathcal{C}_{(q,n,\delta,b)}$ is known. In several cases, the weight distribution of some BCH codes are known. Detailed information can be found in [551, 555, 1245, 1246, 1247, 1277].

2.6.2 The Dimensions of BCH Codes

The dimension of the BCH code $\mathcal{C}_{(q,n,\delta,b)}$ with defining set $T(b, \delta)$ in (2.2) is $n - |T(b, \delta)|$. Since $|T(b, \delta)|$ may have a very complicated relation with n, q, b and δ, the dimension of the BCH code cannot be given exactly in terms of these parameters. The best one can do in general is to develop tight lower bounds on the dimension of BCH codes. The next theorem introduces such bounds [1008, Theorem 5.1.7].

Theorem 2.6.8 *Let $\mathcal{C}$ be an $[n, \kappa]$ BCH code over $\mathbb{F}_q$ of designed distance δ. Then the following statements hold.*

(a) $\kappa \ge n - \mathrm{ord}_n(q)(\delta - 1)$.

(b) *If $q = 2$ and $\mathcal{C}$ is a narrow-sense BCH code, then δ can be assumed odd; furthermore if $\delta = 2w + 1$, then $\kappa \ge n - \mathrm{ord}_n(q)w$.*

The bounds in Theorem 2.6.8 may not be improved for the general case, as demonstrated by the following example. However, in some special cases, they could be improved.

Example 2.6.9 Note that $m = \mathrm{ord}_{15}(2) = 4$, and the 2-cyclotomic cosets modulo 15 are

$$C_0 = \{0\}, \ C_1 = \{1, 2, 4, 8\}, \ C_3 = \{3, 6, 9, 12\},$$
$$C_5 = \{5, 10\}, \ C_7 = \{7, 11, 13, 14\}.$$

Let γ be a generator of $\mathbb{F}_{2^4}^*$ with $\gamma^4 + \gamma + 1 = 0$ and let $\alpha = \gamma^{(2^4-1)/15} = \gamma$ be the primitive 15th root of unity.

When $(b, \delta) = (0, 3)$, the defining set $T(b, \delta) = \{0, 1, 2, 4, 8\}$, and the binary cyclic code has parameters $[15, 10, 4]$ and generator polynomial $x^5 + x^4 + x^2 + 1$. In this case, the actual minimum weight is more than the designed distance, and the dimension is larger than the bound in Theorem 2.6.8(a).

When $(b, \delta) = (1, 3)$, the defining set $T(b, \delta) = \{1, 2, 4, 8\}$, and the binary cyclic code has parameters $[15, 11, 3]$ and generator polynomial $x^4 + x + 1$. It is a narrow-sense BCH code. In this case, the actual minimum weight is equal to the designed distance, and the dimension reaches the bound in Theorem 2.6.8(b).

When $(b, \delta) = (2, 3)$, the defining set $T(b, \delta) = \{1, 2, 3, 4, 6, 8, 9, 12\}$, and the binary cyclic code has parameters $[15, 7, 5]$ and generator polynomial $x^8 + x^7 + x^6 + x^4 + 1$. In this case, the actual minimum weight is more than the designed distance, and the dimension achieves the bound in Theorem 2.6.8(a).

When $(b, \delta) = (1, 5)$, the defining set $T(b, \delta) = \{1, 2, 3, 4, 6, 8, 9, 12\}$, and the binary cyclic code has parameters $[15, 7, 5]$ and generator polynomial $x^8 + x^7 + x^6 + x^4 + 1$. In this case, the actual minimum weight is equal to the designed distance, and the dimension is larger than the bound in Theorem 2.6.8(a). Note that the three pairs $(b_1, \delta_1) = (2, 3), (b_2, \delta_2) = (2, 4)$ and $(b_3, \delta_3) = (1, 5)$ define the same binary cyclic code with generator polynomial $x^8 + x^7 + x^6 + x^4 + 1$. Hence the maximum designed distance of this $[15, 7, 5]$ cyclic code is 5.

When $(b, \delta) = (3, 4)$, the defining set $T(b, \delta) = \{1, 2, 3, 4, 5, 6, 8, 9, 10, 12\}$, and the binary cyclic code has parameters $[15, 5, 7]$ and generator polynomial $x^{10} + x^8 + x^5 + x^4 + x^2 + x + 1$. In this case, the actual minimum weight is more than the designed distance, and dimension is larger than the bound in Theorem 2.6.8(a).

The following is a general result on the dimension of BCH codes [47].

Theorem 2.6.10 *Let* $\gcd(n, q) = 1$ *and* $q^{\lfloor m/2 \rfloor} < n \leq q^m - 1$, *where* $m = \mathrm{ord}_n(q)$. *Let* $2 \leq \delta \leq \min\{\lfloor nq^{\lceil m/2 \rceil}/(q^m - 1)\rfloor, n\}$. *Then*

$$\dim(\mathcal{C}_{(q,n,\delta,1)}) = n - m\lceil (\delta - 1)(1 - 1/q) \rceil.$$

Theorem 2.6.10 is useful when $n = q^m - 1$ or $n = (q^m - 1)/(q - 1)$, but may not be useful in some cases as the range for δ may be extremely small. It is in general very difficult to determine the dimensions of BCH codes. In a very small number of cases, the dimension of the BCH code $\mathcal{C}_{(q,n,\delta,1)}$ is known. For further information, the reader is referred to [551, 555, 1245, 1246, 1247, 1277].

2.6.3 Other Aspects of BCH Codes

The automorphism groups of BCH codes in most cases are open, but are known in some cases [161]. The weight distributions of the cosets of some BCH codes were considered in [386, 387, 388]. This problem is as hard as the determination of the weight distributions of BCH codes. The dual of a BCH code may not be a BCH code. An interesting problem is to characterise those BCH codes whose duals are also BCH codes.

Almost all references on BCH codes are about the primitive case. Only a few references on BCH codes with lengths $n = (q^m - 1)/(q - 1)$ or $n = q^\ell + 1$ exist in the literature [1246, 1247, 1277]. Most BCH codes have never been investigated. This is due to the fact that the q-cyclotomic cosets modulo n are very irregular and behave very badly in most cases. For example, in most cases it is extremely difficult to determine the largest coset leader, not to mention the dimension and minimum distance of a BCH code. This partially explains the difficulty in researching into BCH codes. A characteristic of BCH codes is that it is hard in general to determine both the dimension and minimum distance of a BCH code.

2.7 Duadic Codes

Duadic codes are a family of cyclic codes and are generalizations of the quadratic residue codes. Binary duadic codes were defined in [1220] and were generalized to arbitrary finite fields in [1517, 1519]. Some duadic codes have very good parameters, while some have very bad parameters. The objective of this section is to give a brief introduction of duadic codes.

As before, let n be a positive integer and q a prime power with $\gcd(n,q) = 1$. Let S_1 and S_2 be two subsets of $\mathbb{Z}_n$ such that

- $S_1 \cap S_2 = \emptyset$ and $S_1 \cup S_2 = \mathbb{Z}_n \setminus \{0\}$, and

- both S_1 and S_2 are a union of some q-cyclotomic cosets modulo n.

If there is a unit $\mu \in \mathbb{Z}_n$ such that $S_1\mu = S_2$ and $S_2\mu = S_1$, then (S_1, S_2, μ) is called a **splitting** of $\mathbb{Z}_n$.

Recall that $m := \mathrm{ord}_n(q)$ and α is a primitive n^{th} root of unity in $\mathbb{F}_{q^m}$. Let (S_1, S_2, μ) be a splitting of $\mathbb{Z}_n$. Define

$$g_i(x) = \prod_{i \in S_i}(x - \alpha^i) \ \ \text{and} \ \ \widetilde{g}_i(x) = (x-1)g_i(x)$$

for $i \in \{1, 2\}$. Since both S_1 and S_2 are unions of q-cyclotomic cosets modulo n, both $g_1(x)$ and $g_2(x)$ are polynomials over $\mathbb{F}_q$. The pair of cyclic codes $\mathcal{C}_1$ and $\mathcal{C}_2$ of length n over $\mathbb{F}_q$ with generator polynomials $g_1(x)$ and $g_2(x)$ are called **odd-like duadic codes**, and the pair of cyclic codes $\widetilde{\mathcal{C}}_1$ and $\widetilde{\mathcal{C}}_2$ of length n over $\mathbb{F}_q$ with generator polynomials $\widetilde{g}_1(x)$ and $\widetilde{g}_2(x)$ are called **even-like duadic codes**.

By definition, $\mathcal{C}_1$ and $\mathcal{C}_2$ have parameters $[n, (n+1)/2]$ and $\widetilde{\mathcal{C}}_1$ and $\widetilde{\mathcal{C}}_2$ have parameters $[n, (n-1)/2]$. For odd-like duadic codes, we have the following result [1008, Theorem 6.5.2].

Theorem 2.7.1 (Square Root Bound) *Let $\mathcal{C}_1$ and $\mathcal{C}_2$ be a pair of odd-like duadic codes of length n over $\mathbb{F}_q$. Let d_o be their (common) minimum odd-like weight. Then the following hold.*

(a) *$d_o^2 \geq n$.*

(b) *If the splitting defining the duadic codes is given by $\mu = -1$, then $d_o^2 - d_o + 1 \geq n$.*

(c) *Suppose $d_o^2 - d_o + 1 = n$, where $d_o > 2$, and assume that the splitting defining the duadic codes is given by $\mu = -1$. Then d_o is the minimum weight of both $\mathcal{C}_1$ and $\mathcal{C}_2$.*

Example 2.7.2 Let $(n, q) = (49, 2)$. Define

$$S_1 = \{1,2,4,8,9,11,15,16,18,22,23,25,29,30,32,36,37,39,43,44,46\} \cup \{7,14,28\}$$

and

$$S_2 = \{1,2,\dots,48\} \setminus S_1.$$

It is easily seen that $(S_1, S_2, -1)$ is a splitting of $\mathbb{Z}_{49}$. The pair of odd-like duadic codes $\mathcal{C}_1$ and $\mathcal{C}_2$ defined by this splitting have parameters $[49, 25, 4]$ and generator polynomials

$$x^{24} + x^{22} + x^{21} + x^{10} + x^8 + x^7 + x^3 + x + 1$$

and

$$x^{24} + x^{23} + x^{21} + x^{17} + x^{16} + x^{14} + x^3 + x^2 + 1$$

respectively. The minimum weight of the two codes is even (i.e., 4), while the minimum odd-like weight in the two codes is 9. Note the lower bound on d_o given in Theorem 2.7.1 is 7.

Duadic codes of prime lengths are of special interest as they include the quadratic residue codes, which are defined as follows.

Definition 2.7.3 Let n be an odd prime and q be a quadratic residue modulo n. Denote by S_1 and S_2 the set of quadratic residues and the set of quadratic non-residues, respectively. Let μ be an element of S_2. Then (S_1, S_2, μ) is a splitting of $\mathbb{Z}_n$. The corresponding four duadic codes $\mathcal{C}_1, \mathcal{C}_2, \widetilde{\mathcal{C}}_1, \widetilde{\mathcal{C}}_2$ are called **quadratic residue codes.**

It is known that the automorphism groups of the extended odd-like quadratic residue codes are transitive. Hence, their minimum weight codewords must be odd-like. We then have the following.

Theorem 2.7.4 (Square Root Bound) *Let n be an odd prime and q be a quadratic residue modulo n. Let $\mathcal{C}_1$ and $\mathcal{C}_2$ be a pair of odd-like quadratic residue codes of length n over $\mathbb{F}_q$. Let d be their (common) minimum weight. Then the following hold.*

(a) $d^2 \geq n$.

(b) *If -1 is a quadratic non-residue, then $d^2 - d + 1 \geq n$.*

(c) *Suppose $d^2 - d + 1 = n$, where $d > 2$, and assume that -1 is a quadratic non-residue. Then d is the minimum weight of both $\mathcal{C}_1$ and $\mathcal{C}_2$.*

The Golay codes $\mathcal{C}_1$ and $\mathcal{C}_2$ introduced in Section 1.13 of Chapter 1 are the odd-like quadratic residue binary codes of length 23 and have parameters $[23, 12, 7]_2$. The corresponding even-like quadratic residue codes have parameters $[23, 11, 8]_2$. The ternary Golay codes described in Section 1.13 are also quadratic residue codes.

It is very hard to determine the minimum distance of quadratic residue codes. However, the Square Root Bound on their minimum distances is good enough. Quadratic residue codes are interesting partly because their extended codes hold 2-designs and 3-designs. Quadratic residue codes of length the product of two primes were introduced in [546].

Under certain conditions the extended odd-like duadic codes are self-dual [1220]. Duadic codes have a number of interesting properties. For further information on the existence, constructions, and properties of duadic codes, the reader is referred to [1008, Chapter 6], [546], [559], and [565].

2.8 Punctured Generalized Reed–Muller Codes

Binary Reed–Muller codes were introduced in Section 1.11. It is known that these codes are equivalent to the extended codes of some cyclic codes. In other words, after puncturing the binary Reed–Muller codes at a proper coordinate, the obtained codes are permutation equivalent to some cyclic codes. The purpose of this section is to introduce a family of cyclic codes of length $n = q^m - 1$ over $\mathbb{F}_q$ whose extended codes are the generalized Reed–Muller code over $\mathbb{F}_q$.

Let q be a prime power as before. For any integer $j = \sum_{i=0}^{m-1} j_i q^i$, where $0 \leq j_i \leq q-1$ for all $0 \leq i \leq m-1$ and m is a positive integer, we define

$$\omega_q(j) = \sum_{i=0}^{m-1} j_i,$$

where the sum is taken over the ring of integers, and is called the q-**weight** of j.

Let ℓ be a positive integer with $1 \leq \ell < (q-1)m$. The ℓth order **punctured generalized Reed–Muller code** $\mathcal{RM}_q(\ell, m)^*$ over $\mathbb{F}_q$ is the cyclic code of length $n = q^m - 1$ with generator polynomial

$$g(x) = \sum_{\substack{1 \leq j \leq n-1 \\ \omega_q(j) < (q-1)m-\ell}} (x - \alpha^j),$$

where α is a generator of $\mathbb{F}_{q^m}^*$. Since $\omega_q(j)$ is a constant function on each q-cyclotomic coset modulo $n = q^m - 1$, $g(x)$ is a polynomial over $\mathbb{F}_q$.

The parameters of the punctured generalized Reed–Muller code $\mathcal{RM}_q(\ell, m)^*$ are known and summarized in the next theorem [71, Section 5.5].

Theorem 2.8.1 *For any ℓ with $0 \leq \ell < (q-1)m$, $\mathcal{RM}_q(\ell, m)^*$ is a cyclic code over $\mathbb{F}_q$ with length $n = q^m - 1$, dimension*

$$\kappa = \sum_{i=0}^{\ell} \sum_{j=0}^{m} (-1)^j \binom{m}{j} \binom{i - jq + m - 1}{i - jq}$$

and minimum weight $d = (q - \ell_0)q^{m-\ell_1-1} - 1$, where $\ell = \ell_1(q-1) + \ell_0$ and $0 \leq \ell_0 < q-1$.

Example 2.8.2 Let $(q, m, \ell) = (3, 3, 3)$, and let α be a generator of $\mathbb{F}_{3^3}^*$ with $\alpha^3 + 2\alpha + 1 = 0$. Then $\mathcal{RM}_3(3, 3)^*$ is a ternary code with parameters $[26, 17, 5]$ and generator polynomial

$$g(x) = x^9 + 2x^8 + x^7 + x^6 + x^5 + 2x^4 + 2x^3 + 2x^2 + x + 1.$$

The dual of the punctured generalized Reed–Muller code is described in the following theorem [71, Corollary 5.5.2].

Theorem 2.8.3 *For $0 \leq \ell < m(q-1)$, the code $(\mathcal{RM}_q(\ell, m)^*)^\perp$ is the cyclic code with generator polynomial*

$$h(x) = \sum_{\substack{0 \leq j \leq n-1 \\ \omega_q(j) \leq \ell}} (x - \alpha^j),$$

where α is a generator of $\mathbb{F}_{q^m}^*$. In addition,

$$(\mathcal{RM}_q(\ell, m)^*)^\perp = (\mathbb{F}_q\mathbf{1})^\perp \cap \mathcal{RM}_q(m(q-1) - 1 - \ell, m)^*,$$

where $\mathbf{1}$ is the all-one vector in $\mathbb{F}_q^n$ and $\mathbb{F}_q\mathbf{1}$ denotes the code over $\mathbb{F}_q$ with length n generated by $\mathbf{1}$.

Example 2.8.4 Let $(q, m, \ell) = (3, 3, 3)$, and let α be a generator of $\mathbb{F}_{3^3}^*$ with $\alpha^3 + 2\alpha + 1 = 0$. Then $(\mathcal{RM}_3(3, 3)^*)^\perp$ is a ternary code with parameters $[26, 9, 9]$ and generator polynomial

$$\begin{aligned} g^\perp(x) = {}& x^{17} + 2x^{16} + 2x^{15} + x^{14} + x^{13} + x^{11} + 2x^{10} + 2x^9 + \\ & x^8 + 2x^7 + 2x^5 + x^4 + 2x^3 + 2x + 2. \end{aligned}$$

The codes $\mathcal{RM}_q(\ell, m)^*$ have a geometric interpretation. Their extended codes hold 2-designs for $q > 2$, and 3-designs for $q = 2$. The reader is referred to [71, Corollary 5.2] for further information.

2.9 Another Generalization of the Punctured Binary Reed–Muller Codes

The punctured generalized Reed–Muller codes are a generalization of the classical punctured binary Reed–Muller codes, and were introduced in the previous section. A new generalization of the classical punctured binary Reed–Muller codes was given recently in [561]. The task of this section is to introduce the newly generalized cyclic codes.

Let $n = q^m - 1$. For any integer a with $0 \leq a \leq n - 1$, we have the following q-adic expansion

$$a = \sum_{j=0}^{m-1} a_j q^j,$$

where $0 \leq a_j \leq q - 1$. The Hamming weight of a, denoted by $\mathrm{wt_H}(a)$, is the number of nonzero coordinates in the vector $(a_0, a_1, \ldots, a_{m-1})$.

Let α be a generator of $\mathbb{F}_{q^m}^*$. For any $1 \leq h \leq m$, we define a polynomial

$$g_{(q,m,h)}(x) = \prod_{\substack{1 \leq a \leq n-1 \\ 1 \leq \mathrm{wt_H}(a) \leq h}} (x - \alpha^a).$$

Since $\mathrm{wt_H}(a)$ is a constant function on each q-cyclotomic coset modulo n, $g_{(q,m,h)}(x)$ is a polynomial over $\mathbb{F}_q$. By definition, $g_{(q,m,h)}(x)$ is a divisor of $x^n - 1$.

Let $\mho(q, m, h)$ denote the cyclic code over $\mathbb{F}_q$ with length n and generator polynomial $g_{(m,q,h)}(x)$. By definition, $g_{(q,m,m)}(x) = (x^n - 1)/(x - 1)$. Therefore, the code $\mho(q, m, m)$ is trivial, as it has parameters $[n, 1, n]$ and is spanned by the all-1 vector. Below we consider the code $\mho(q, m, h)$ for $1 \leq h \leq m - 1$ only.

Theorem 2.9.1 *Let $m \geq 2$ and $1 \leq h \leq m - 1$. Then $\mho(q, m, h)$ has parameters $[q^m - 1, \kappa, d]$, where*

$$\kappa = q^m - \sum_{i=0}^{h} \binom{m}{i}(q-1)^i$$

and

$$\frac{q^{h+1} - 1}{q - 1} \leq d \leq 2q^h - 1. \tag{2.3}$$

When $q = 2$, the code $\mho(q, m, h)$ clearly becomes the classical punctured binary Reed–Muller code $\mathcal{RM}(m-1-h, m)^*$. Hence, $\mho(q, m, h)$ is indeed a generalization of the original punctured binary Reed–Muller code. In addition, when $q = 2$, the lower bound and the upper bound in (2.3) become identical. It is conjectured that the lower bound on d is the actual minimum distance.

Example 2.9.2 The following is a list of examples of the code $\mho(q, m, h)$.

1. When $(q, m, h) = (3, 3, 1)$, $\mho(q, m, h)$ has parameters $[26, 20, 4]$, and is distance-optimal.

2. When $(q, m, h) = (3, 4, 1)$, $\mho(q, m, h)$ has parameters $[80, 72, 4]$, and is distance-optimal.

3. When $(q, m, h) = (3, 4, 2)$, $\mho(q, m, h)$ has parameters $[80, 48, 13]$, and its minimum distance is one less than that of the best code with parameters $[80, 48, 14]$.

4. When $(q, m, h) = (3, 4, 3)$, the code $\mho(q, m, h)$ has parameters $[80, 16, 40]$, which are the best parameters known.

5. When $(q, m, h) = (4, 3, 1)$, the code $\mho(q, m, h)$ has parameters $[63, 54, 5]$, which are the best parameters known.

An interesting fact about the family of newly generalized codes $\mho(q, m, h)$ is the following.

Corollary 2.9.3 *Let $m \geq 2$. Then the ternary code $\mho(3, m, 1)$ has parameters $[3^m - 1, 3^m - 1 - 2m, 4]$ and is distance-optimal.*

We have also the next two special cases in which the parameters of the code $\mho(q, m, h)$ are known.

Theorem 2.9.4 *Let m be even. Then the cyclic code $\mho(q, m, 1)$ has parameters $[q^m - 1, q^m - 1 - m(q - 1), q + 1]$.*

Theorem 2.9.5 *Let $m \geq 2$. Then the cyclic code $\mho(q, m, m - 1)$ has parameters $[q^m - 1, (q - 1)^m, (q^m - 1)/(q - 1)]$.*

The following theorem gives information on the parameters of the dual code $\mho(q, m, h)^{\perp}$.

Theorem 2.9.6 *Let $m \geq 2$ and $1 \leq h \leq m - 1$. The dual code $\mho(q, m, h)^{\perp}$ has parameters $[q^m - 1, \kappa^{\perp}, d^{\perp}]$, where*

$$\kappa^{\perp} = \sum_{i=1}^{h} \binom{m}{i} (q - 1)^i.$$

The minimum distance $d^{\perp}$ of $\mho(q, m, h)^{\perp}$ is bounded below by

$$d^{\perp} \geq q^{m-h} + q - 2.$$

When $q = 2$, the lower bound on the minimum distance $d^{\perp}$ of $\mho(q, m, h)^{\perp}$ given in Theorem 2.9.6 is achieved. Experimental data shows that the lower bound on $d^{\perp}$ in Theorem 2.9.6 is not tight for $q > 2$. It is open how to improve it or determine the exact minimum distance.

The code $\mho(q, m, h)$ is clearly different from the punctured generalized Reed–Muller code. It is open if $\mho(q, m, h)$ has a geometric interpretation. For a proof of the results introduced above and further properties of the cyclic code $\mho(q, m, h)$, the reader is referred to [561].

2.10   Reversible Cyclic Codes

Definition 2.10.1 A linear code $\mathcal{C}$ is **reversible**[1] if $(c_0, c_1, \ldots, c_{n-1}) \in \mathcal{C}$ implies that $(c_{n-1}, c_{n-2}, \ldots, c_0) \in \mathcal{C}$.

---

[1]A linear code of length n over $\mathbb{F}_q$ is called an **LCD code (linear code with complementary dual)** if $\mathcal{C} \cap \mathcal{C}^{\perp} = \{0\}$, which is equivalent to $\mathcal{C} \oplus \mathcal{C}^{\perp} = \mathbb{F}_q^n$. Reversible cyclic codes are in fact LCD codes.

Reversible cyclic codes were considered in [1346, 1347]. A cryptographic application of reversible cyclic codes was proposed in [353]. A well rounded treatment of reversible cyclic codes was given in [1236]. The objective of this section is to deliver a basic introduction to reversible cyclic codes.

Definition 2.10.2 A polynomial $f(x)$ over $\mathbb{F}_q$ is called **self-reciprocal** if it equals its reciprocal $f^{\perp}(x)$.

The conclusions of the following theorem are known in the literature [1323, page 206] and are easy to prove.

Theorem 2.10.3 *Let $\mathcal{C}$ be a cyclic code of length n over $\mathbb{F}_q$ with generator polynomial $g(x)$. Then the following statements are equivalent.*

(a) *$\mathcal{C}$ is reversible.*

(b) *$g(x)$ is self-reciprocal.*

(c) *β^{-1} is a root of $g(x)$ for every root β of $g(x)$ over the splitting field of $g(x)$.*

Furthermore, if -1 is a power of q mod n, then every cyclic code over $\mathbb{F}_q$ of length n is reversible.

Now we give an exact count of reversible cyclic codes of length $n = q^m - 1$ for odd primes m. Recall the q-cyclotomic cosets C_a modulo n given in Definition 1.12.7. It is straightforward that $-a = n - a \in C_a$ if and only if $a(1 + q^j) \equiv 0 \pmod{n}$ for some integer j. The following two lemmas are straightforward and hold whenever $\gcd(n, q) = 1$.

Lemma 2.10.4 *The irreducible polynomial $M_{\alpha^a}(x)$ is self-reciprocal if and only if $n - a \in C_a$.*

Lemma 2.10.5 *The least common multiple $\mathrm{lcm}(M_{\alpha^a}(x), M_{\alpha^{n-a}}(x))$ is self-reciprocal for every $a \in \mathbb{Z}_n$.*

Definition 2.10.6 The least nonnegative integer in a q-cyclotomic coset modulo n is called the **coset leader** of this coset.

By Lemma 2.10.4, we have that

$$\mathrm{lcm}(M_{\alpha^a}(x), M_{\alpha^{n-a}}(x)) = \begin{cases} M_{\alpha^a}(x) & \text{if } n - a \in C_a, \\ M_{\alpha^a}(x)M_{\alpha^{n-a}}(x) & \text{otherwise.} \end{cases}$$

Let $\Gamma_{(n,q)}$ denote the set of coset leaders of all q-cyclotomic cosets modulo n. Define

$$\Pi_{(n,q)} = \Gamma_{(n,q)} \setminus \{\max\{a, \mathrm{leader}(n - a)\} \mid a \in \Gamma_{(n,q)}, n - a \notin C_a\},$$

where $\mathrm{leader}(i)$ denotes the coset leader of C_i. Then $\{C_a \cup C_{n-a} \mid a \in \Pi_{(q,n)}\}$ is a partition of $\mathbb{Z}_n$.

The following conclusion then follows directly from Lemmas 2.10.4, 2.10.5, and Theorem 2.10.3.

Theorem 2.10.7 *The total number of reversible cyclic codes over $\mathbb{F}_q$ of length n is equal to $2^{|\Pi_{(q,n)}|}$, including the zero code and the code $\mathbb{F}_q^n$. Every reversible cyclic code over $\mathbb{F}_q$ of length n is generated by a polynomial*

$$g(x) = \prod_{a \in S} \mathrm{lcm}\left(M_{\alpha^a}(x), M_{\alpha^{n-a}}(x)\right),$$

where S is a (possibly empty) subset of $\Pi_{(q,n)}$.

Example 2.10.8 Let $(n, q) = (15, 2)$. The 2-cyclotomic cosets modulo 15 are

$$C_0 = \{0\}, \ C_1 = \{1, 2, 4, 8\}, \ C_3 = \{3, 6, 9, 12\}, \ C_5 = \{5, 10\}, \ \text{and} \ C_7 = \{7, 11, 13, 14\}.$$

We also have

$$x^{15} - 1 = M_{\alpha^0}(x) M_{\alpha^1}(x) M_{\alpha^3}(x) M_{\alpha^5}(x) M_{\alpha^7}(x),$$

where

$$M_{\alpha^0}(x) = x + 1,$$
$$M_{\alpha^1}(x) = x^4 + x + 1,$$
$$M_{\alpha^3}(x) = x^4 + x^3 + x^2 + x + 1,$$
$$M_{\alpha^5}(x) = x^2 + x + 1,$$
$$M_{\alpha^7}(x) = x^4 + x^3 + 1.$$

Note that $M_{\alpha^i}(x)$ are self-reciprocal for $i \in \{0, 3, 5\}$ while $M_{\alpha^1}(x)$ and $M_{\alpha^7}(x)$ are reciprocals of each other. In this case,

$$\Gamma_{(n,q)} = \{0, 1, 3, 5, 7\}.$$

But

$$\Pi_{(n,q)} = \{0, 1, 3, 5\}.$$

Hence, there are 16 reversible binary cyclic codes of length 15, including the zero code and the code $\mathbb{F}_2^{15}$.

Corollary 2.10.9 *Let q be an even prime power and $n = q^m - 1$. If m is odd, then the only self-reciprocal irreducible divisor of $x^n - 1$ over $\mathbb{F}_q$ is $x - 1$. If m is an odd prime, then the total number of reversible cyclic codes of length n over $\mathbb{F}_q$ is equal to $2^{\frac{q^m + (m-1)q}{2m}}$, including the zero code and the code $\mathbb{F}_q^n$.*

Corollary 2.10.10 *Let q be an odd prime power and $n = q^m - 1$. If m is odd, then the only self-reciprocal irreducible divisors of $x^n - 1$ over $\mathbb{F}_q$ are $x - 1$ and $x + 1$. If m is an odd prime, then the total number of reversible cyclic codes of length n over $\mathbb{F}_q$ is equal to $2^{\frac{q^m + (m-1)q + m}{2m}}$, including the zero code and the code $\mathbb{F}_q^n$.*

The following three theorems follow directly from Theorem 2.10.3 and the definition of BCH codes, and can be viewed as corollaries of Theorem 2.10.3.

Theorem 2.10.11 *The BCH code $\mathcal{C}_{(q,n,\delta,b)}$ is reversible if $b = -t$ and the designed distance is $\delta = 2t + 2$ for any nonnegative integer t.*

Theorem 2.10.12 *The BCH code $\mathcal{C}_{(q,n,\delta,b)}$ is reversible if n is odd, $b = (n - t)/2$ and the designed distance is $\delta = t + 2$ for any odd integer t with $1 \leq t \leq n - 2$.*

Theorem 2.10.13 *The BCH code $\mathcal{C}_{(q,n,\delta,b)}$ is reversible if n is even, $b = (n - 2t)/2$ and the designed distance is $\delta = 2t + 2$ for any integer t with $0 \leq t \leq n/2$.*

The dimensions of some of the reversible BCH codes described in Theorems 2.10.11, 2.10.12, and 2.10.13 were determined in [1236]. Two families of reversible BCH codes were studied in [1247]. Non-primitive reversible cyclic codes were also treated in [1236]. There are many reversible cyclic codes and it is easy to construct them. However, determining their parameters is a difficult problem in general. There are clearly reversible cyclic codes with both good and bad parameters.

Chapter 3

Construction and Classification of Codes

Patric R. J. Östergård

Aalto University

3.1 Introduction

Shannon's seminal work [1661] showed in a nonconstructive way that good codes exist, as discussed in Section 1.1, and marked the birth of coding and information theory. One of the main goals of coding theory is to construct codes that are as good as possible for given parameters and types of codes. There is a mathematical as well as an engineering dimension of the construction problem. Whereas mathematicians may wish to study arbitrary parameters and the behavior of codes when the length tends to infinity, the parameters of a code used in some engineering applications are necessarily fixed and bounded. Depending on one's philosophical viewpoint, one may say that codes are discovered rather than constructed. Hence, perhaps the most fundamental question in coding theory can also be phrased in terms of **existence**: Does a code with given parameters exist or not?

If the answer to an existence question is affirmative, one may further wish to carry out a **classification** of those codes, that is, to determine all possible such codes up to symmetry (equivalence, isomorphism). Also classification has both a theoretical and a practical side. Classified codes can be studied to gain more insight into codes, to state or refute conjectures, and so on. But they also form an exhaustive set of candidate codes, whose performance can be tested in actual applications.

Remark 3.1.1 A nonexistence result is intrinsically a classification result, where the outcome is the empty set.

A lot of (manual and/or computational) resources may be required for the study of existence or classification of codes with certain parameters. However, whereas verifying correctness of a (positive) outcome is fast and straightforward for a constructed code, it can be as time-consuming as the original work for a set of classified codes.

Validation of classification results is considered in [1090, Chap. 10]. Especially double-counting techniques have turned out to be useful. If we know the total number of codes—for example, through a *mass formula*—then we can check whether the orbit-stabilizer theorem applied to a set of classified codes leads to the same number; see for example Theorem 4.5.2. This approach can even be used to show that a set of codes found in a non-exhaustive search is complete. Lam and Thiel [1192] showed how double-counting can be integrated into classification algorithms when no mass formula is available (which is the typical situation).

The main classical reference in coding theory, *The Theory of Error-Correcting Codes* by MacWilliams and Sloane [1323], considers the existence problem extensively and is still, many decades after the publication of the first edition, a good source of information. The theme of classifying codes, on the other hand, has flourished much later, in part because of its high dependence on computational resources; the monograph [1090] provides an in-depth treatment of this theme.

In this chapter, a general overview of existence and classification problems will be given, and some highlights from the past will be singled out. The chapter touches upon several of the problems and problem areas presented in the list of major open problems in algebraic coding theory in [630].

3.2   Equivalence and Isomorphism

The concepts of equivalence and isomorphism of codes are briefly discussed in Section 1.8. Generally, the term *symmetry* covers both of those concepts, especially when considering maps from a code onto itself, that is, automorphisms. Namely, such maps lead to groups under composition, and groups are essentially about symmetries. The group formed by all automorphisms of a code is, whenever the type of automorphisms is understood, simply called the **automorphism group** of the code. A subgroup of the automorphism group is called a **group of automorphisms**.

Symmetries play a central role when constructing as well as classifying codes: several types of constructions are essentially about prescribing symmetries, and one core part of classification is about dealing with maps and symmetries.

On a high level of abstraction, the same questions are asked for linear and unrestricted codes and analogous techniques are used. On a detailed level, however, there are significant differences between those two types of codes.

Consider codes of length n over $\mathbb{F}_q$. We have seen in Definition 1.8.8 that equivalence of unrestricted codes is about permuting coordinates and the elements of the alphabet, individually within each coordinate. All such maps form a group that is isomorphic to the wreath product $S_q \wr S_n$. For linear codes on the other hand, the concepts of permutation equivalence, monomial equivalence, and equivalence lead to maps that form groups isomorphic to S_n, $\mathbb{F}_q^* \wr S_n$, and the semidirect product $(\mathbb{F}_q^* \wr S_n) \rtimes_\theta \mathrm{Aut}(\mathbb{F}_q)$, respectively, where $\mathbb{F}_q^*$ is the multiplicative group of $\mathbb{F}_q$ and $\theta : \mathrm{Aut}(\mathbb{F}_q) \to \mathrm{Aut}(\mathbb{F}_q^* \wr S_n)$ is a group homomorphism.

Remark 3.2.1 For *binary* linear codes, all three types of equivalence coincide.

3.2.1 Prescribing Symmetries

A code of size M is a subset of M vectors from the n-dimensional vector space over $\mathbb{F}_q$ which fulfills some requirements depending on the type of code. The number of ways to choose M arbitrary vectors from such a space is $\binom{q^n}{M}$, which becomes astronomically large already for rather small parameters. (This is obviously the total number of $(n, M)_q$ codes.) Although no general conclusion regarding the hardness of solving construction and classification problems can be drawn from this number, the number does give a clue that the limit of what is feasible might be reached quite early. Indeed, this is what happens, but perhaps not as early as one would think.

Example 3.2.2 In some special cases—in particular, for perfect codes—quite large unrestricted codes have been classified, such as the $(23, 4096, 7)_2$ code (the binary Golay code is unique [1732]; see also [525]) and the $(15, 2048, 3)_2$ codes (with the parameters of a Hamming code; there are 5983 such codes [1472]).

But what can be done if we go beyond parameters for which the size of an optimal code can be determined and the optimal codes can be classified? Analytical upper bounds and constructive lower bounds on the size of codes can still be used. One way to speed up computer-aided constructive techniques—some of which are discussed in Chapter 23—is to restrict the search by imposing a structure on the codes. This is a two-edged sword: the search space is reduced, but good codes might not have that particular structure. Hence some experience is of great help in tuning the search. A very common approach is that of prescribing symmetries (automorphisms).

Remark 3.2.3 In the discussion of groups in the context of automorphism groups of codes, we are not only interested in the abstract group but in the group and its action. This is implicitly understood in the sequel when talking about one particular group or all groups of certain orders. For example, "prescribing a group" means "prescribing a group and its action" and "considering all groups" means "considering all groups and all possible actions of those groups".

By prescribing a group G, the n-dimensional vector space is partitioned into orbits of vectors. The construction problem then becomes a problem of finding a set of those orbits rather than finding a set of individual vectors. It must further be checked that the orbits themselves are feasible; an orbit whose codewords do not fulfill the minimum distance criterion can be discarded immediately.

Remark 3.2.4 An $[n, k]_q$ linear code can be viewed as an unrestricted code which contains the all-zero codeword and has a particular group of automorphisms G of order q^k, which only permutes elements of the alphabet, individually within each coordinate.

Example 3.2.5 Consider the $[4, 2]_2$ binary linear code generated by

$$G = \begin{bmatrix} 1 & 0 & 1 & 0 \\ 0 & 1 & 1 & 1 \end{bmatrix}.$$

If this code contains some word $\mathbf{c}$, then it also contains, for example, $\mathbf{c} + 1010$. Adding 0 to a coordinate value means applying a value permutation that is the identity permutation, and adding 1 to a coordinate value means applying the value permutation $0 \leftrightarrow 1$. Permuting elements of the alphabet, individually within each coordinate, is indeed covered by the definition of equivalence of unrestricted codes.

Unrestricted codes that consist of cosets of a linear code can in a similar manner be considered in the framework of prescribed groups of automorphisms. Let H be an $(n-k) \times n$ parity check matrix for an $[n, k, d']_q$ linear code[1], let $\mathbf{x}, \mathbf{y} \in \mathbb{F}_q^{n-k}$, and let $\mathcal{S} \subseteq \mathbb{F}_q^{n-k}$. Then, if $\mathrm{wt_H}(\mathbf{t})$ is the Hamming weight of $\mathbf{t}$, we define

$$d^H(\mathbf{x}, \mathbf{y}) = \min \{ \mathrm{wt_H}(\mathbf{t}) \mid H\mathbf{t}^\mathsf{T} = (\mathbf{x} - \mathbf{y})^\mathsf{T}, \ \mathbf{t} \in \mathbb{F}_q^n \},$$

$$d^H(\mathcal{S}) = \min_{\mathbf{a}, \mathbf{b} \in \mathcal{S}, \mathbf{a} \neq \mathbf{b}} d^H(\mathbf{a}, \mathbf{b}).$$

The following theorem [1470, Theorem 1] shows that the situation here is a generalization of the standard approach of finding arbitrary codes with minimum distance $2r+1$ by packing Hamming balls of radius r. Here we are packing more general geometrical objects given by the columns of H, and we get the standard approach when $k = 0$ and thereby $H = I_n$.

Theorem 3.2.6 *Let H be an $(n - k) \times n$ parity check matrix for an $[n, k, d']_q$ linear code, and let $\mathcal{S} \subseteq \mathbb{F}_q^{n-k}$. Then the code $\mathcal{C} = \{ \mathbf{c} \in \mathbb{F}_q^n \mid H\mathbf{c}^\mathsf{T} \in \mathcal{S} \}$ has minimum distance $\min \{ d^H(\mathcal{S}), d' \}$.*

When searching for unrestricted codes, one may consider any subgroup of $\mathrm{S}_q \wr \mathrm{S}_n$. Due to the large number of (conjugacy classes of) subgroups, it is typically necessary to restrict the set of subgroups considered. The subgroups should not be too small, whence the search would not be limited enough, and not too big either. Experiments and experience, especially regarding automorphism groups of known good codes, are helpful in the process of sifting candidate groups. For unrestricted codes, Theorem 3.2.6 has turned out to be very useful. Groups that act transitively on the n coordinates or even on the qn coordinate–value pairs have also been considered with success [1186].

Arguably the principal automorphism for good linear as well as unrestricted codes is a cyclic permutation of the coordinates. Codes with such an automorphism are called cyclic. Cyclic *linear* codes—which have a nice algebraic structure as they can be viewed as ideals of certain quotient rings—are discussed in Section 1.12 and Chapter 2.

We have seen a generalization of cyclic codes to l-quasi-cyclic codes in Definition 1.12.23, and these are covered in Chapter 7; cyclic codes are 1-quasi-cyclic. Other possible automorphisms that are generalizations of the cyclic ones and that are common amongst good codes are given by the following definition. The definition applies to both linear and unrestricted codes. For linear codes, these types of codes can be considered algebraically. See also Chapter 17.

Definition 3.2.7 Fix $\alpha \in \mathbb{F}_q^*$. A code $\mathcal{C}$ is called α-**constacyclic** (or α-**twisted**) if $c_0 c_1 \cdots c_{n-1} \in \mathcal{C}$ implies that $(\alpha c_{n-1}) c_0 c_1 \cdots c_{n-2} \in \mathcal{C}$. A 1-constacyclic code is cyclic. A code $\mathcal{C}$ is (α, l)-**quasi-twisted** if l divides n and $c_0 c_1 \cdots c_{n-1} \in \mathcal{C}$ implies that $(\alpha c_{n-l})(\alpha c_{n-l+1}) \cdots (\alpha c_{n-1}) c_0 c_1 \cdots c_{n-l-1} \in \mathcal{C}$. An $(\alpha, 1)$-quasi-twisted code is α-constacyclic (or α-twisted).

Important types of codes with a large automorphism group include quadratic residue codes.

Definition 3.2.8 Let p be an odd prime, let Q be the set of quadratic residues modulo p, let $q \in Q$ be a prime, and let ζ be a primitive p^{th} root of unity in some finite extension field of $\mathbb{F}_q$. A **quadratic residue code of length p over $\mathbb{F}_q$** is a cyclic code with generator polynomial

$$f(x) = \prod_{j \in Q} (x - \zeta^j).$$

---

[1]This means that H has full rank, which is a reasonable assumption. We get essentially the same main theorem when H does not have full rank, but then some details regarding distances have to be tuned.

Remark 3.2.9 The quadratic residue codes in Definition 3.2.8 can be generalized to codes whose alphabet size is a prime power and to codes whose length is a prime power.

The dimension of a quadratic residue code of length p is $(p+1)/2$. The minimum distance is known to be at least $\sqrt{p}$, which can be further strengthened for various parameters; see Theorem 2.7.4. The true minimum distance typically has to be determined on a case-by-case basis.

Example 3.2.10 The $[7,4,3]_2$ binary Hamming code, the $[23,12,7]_2$ binary Golay code, and the $[11,6,5]_3$ ternary Golay code are quadratic residue codes.

Gleason and Prange showed that the automorphism group of an extended quadratic residue code is rather large. The result was originally published only in a laboratory research report, but it has later been discussed in most coding theory textbooks, including [1323, Chap. 16]. See also [210, 1002]. When extending quadratic residue codes, the standard Definition 1.7.1 can be used for $q = 2, 3$, but a different definition has to be used for general values of q; see, for example, [1323, Chap. 16] for details.

Theorem 3.2.11 (Gleason–Prange) *Every extended quadratic residue code of length p has a group of automorphisms that is isomorphic to the projective special linear group* $\mathrm{PSL}_2(p)$.

Remark 3.2.12 The extensions of the three codes in Example 3.2.10 are exceptional examples of extended quadratic residue codes whose automorphism group has $\mathrm{PSL}_2(p)$ as a proper subgroup.

There is a two-way interaction between codes and groups: known good codes (or families of good codes) can be studied to find their automorphism groups, and groups can be prescribed to find good codes with such automorphisms.

A search for codes with prescribed automorphisms that has a negative outcome leads to results of scientific value only if the search is exhaustive and the parameters are of general interest. For the most important open existence problems, smaller and smaller orders of automorphisms and groups of automorphisms are commonly considered in a sequence of publications. The hope in each and every such study is clearly to find codes with the prescribed groups of automorphisms.

Remark 3.2.13 If the answer to a general existence problem is negative, then even a proof that a code cannot have nontrivial automorphisms does not essentially bring us closer to a nonexistence proof. However, this is not work in vain, but forms a strong basis for making a nonexistence conjecture.

Example 3.2.14 Neil Sloane [1726] asked in 1973 whether there is a self-dual doubly-even $[72, 36, 16]_2$ code. This would be the third code in a sequence of self-dual doubly-even $[24m, 12m, 4m + 4]_2$ codes, which begins with the extended binary Golay code ($m = 1$) and the extended binary quadratic residue code of length 48 ($m = 2$). The latter code is a unique such code [991]; the extended binary Golay code is unique as a general linear code [1514] and even as an unrestricted code [1732].

Jessie MacWilliams, in her review of [1726] for *Mathematical Reviews*, mentions that the author had offered \$10 for a solution and added another \$10 to the prize money. It was thus clear from the very beginning that this is an important and interesting problem. Almost half a century later, the problem is still open, no less fascinating, and arguably the most important open specific case in the theory of linear codes.

Conway and Pless [437] showed that the possible prime orders for automorphisms of a self-dual doubly-even $[72, 36, 16]_2$ code are 2, 3, 5, 7, 11, 17, and 23. Starting from the biggest order, the largest four orders have been eliminated in [1516], [1525], [1010], and [723], respectively. After further elimination of various groups of orders $2^i 3^j 5^k$, $i, j, k \geq 0$ in [238, 239, 241, 274, 280, 1422, 1922, 1930, 1931], the groups of order at most 5, except for the cyclic group of order 4, remain. Moreover, automorphisms of order 2 and 3 cannot have fixed points [273, 274], and automorphisms of order 5 must have exactly two fixed points [437]. See also [240] and Section 4.3.

3.2.2   Determining Symmetries

The obvious recurrent specific questions when studying equivalence (or isomorphism) of (linear and unrestricted) codes and the symmetries of such codes are the following:

1. Given two codes, $\mathcal{C}_1$ and $\mathcal{C}_2$, are these equivalent (isomorphic) or not?

2. Given a code $\mathcal{C}$, what is the automorphism group of $\mathcal{C}$?

The two questions are closely related, since if we are able to find all possible maps between two codes, $\mathcal{C}_1$ and $\mathcal{C}_2$, we can answer both of them (the latter by letting $\mathcal{C}_1 = \mathcal{C}_2 = \mathcal{C}$).

Invariants can be very useful in studying the first question.

Definition 3.2.15 An **invariant** is a property of a code that depends only on the abstract structure, that is, two equivalent (isomorphic) codes necessarily have the same value of an invariant.

Remark 3.2.16 Two inequivalent (non-isomorphic) codes may or may not have the same value of an invariant.

Example 3.2.17 The distance distribution is an invariant of codes. This invariant can be used to show that the two unrestricted $(4, 3, 2)_3$ codes $\mathcal{C}_1 = \{0000, 0120, 2121\}$ and $\mathcal{C}_2 = \{1000, 1111, 2112\}$ are inequivalent. Actually, to distinguish these codes an even less sensitive invariant suffices: the number of pairs of codewords with mutual Hamming distance 3 is 0 for $\mathcal{C}_1$ and 1 for $\mathcal{C}_2$.

Since invariants are only occasionally able to provide the right answer to the first question above, alternative techniques are needed for providing the answer in all possible situations. One such technique relies on producing canonical representatives of codes.

Definition 3.2.18 Let S be a set of possible codes, and let $r : S \to S$ be a map with the properties that (i) $r(\mathcal{C}_1) = r(\mathcal{C}_2)$ if and only if $\mathcal{C}_1$ and $\mathcal{C}_2$ are equivalent (isomorphic) and (ii) $r(\mathcal{C}) = r(r(\mathcal{C}))$. The **canonical representative** (or **canonical form**) of a code $\mathcal{C} \in S$ with respect to this map is $r(\mathcal{C})$.

To test whether two codes are equivalent (isomorphic), it suffices to test their canonical representatives for equality.

Remark 3.2.19 The number of equivalence (isomorphism) classes that can be handled when comparing canonical representatives is limited by the amount of computer memory available. However, in the context of classifying codes, there are actually methods that do not require any comparison between codes; see [1090].

For particular codes with special properties, algebraic and combinatorial techniques can be used to determine the automorphism group of a code.

Example 3.2.20 The automorphism group of the $[2^m - 1, 2^m - r - 1, 3]_2$ binary Hamming code is isomorphic to the general linear group $\mathrm{GL}_m(2)$.

Remark 3.2.21 Determining the automorphism group essentially consists of two parts: finding the set of all automorphisms and proving that the set is complete. The latter task is trivial in cases where the automorphism group is a maximal subgroup of the group of all possible symmetries.

In the general case, algorithmic tools are required to answer the questions above. In the early 1980s, Leon [1217] published an algorithm for computing automorphism groups of linear codes over arbitrary fields, considering monomial equivalence. More recently, Feulner [721] developed an algorithm for computing automorphism groups *and* canonical representatives of linear codes, considering equivalence.

Algorithms similar to those developed by Leon and Feulner could also be developed for unrestricted codes. However, the task of developing such tailored algorithms is rather tedious. An alternative, convenient approach for unrestricted codes and their various subclasses is to map the codes to colored graphs, which can be then be considered in the framework of graph isomorphism. In particular, the **nauty** graph automorphism software [1370] can then be used.

Maps from codes to colored graphs are discussed in [1090, Chap. 3]. For unrestricted codes over $\mathbb{F}_q$ with length n and size M, one may consider the following graph with $M + qn$ vertices. Take one vertex for each codeword and a complete graph with q vertices for each coordinate. Insert edges so that the neighbors of a codeword vertex show the values in the respective coordinates. One may color the graph so that no automorphism can map a vertex of one of the two types to the other (although the structure of the graph is such that for nontrivial codes no such maps are possible even if the graph is uncolored).

Example 3.2.22 For the two codes in Example 3.2.17, one may construct the graphs in Figure 3.1. The leftmost vertex in each triangle corresponds to value 0, and the other two vertices correspond to 1 and 2 in a clockwise manner. Now we can also use graph invariants, such as the distribution of the degrees of the vertices, to see that the graphs are non-isomorphic, which in turn implies that the codes are inequivalent.

Remark 3.2.23 The number of codewords of linear codes grows exponentially as a function of the dimension of the codes. Hence we do not wish to map individual codewords to vertices of some graph, so an approach similar to that for unrestricted codes is not practical. However, various other approaches have been proposed which partly rely on **nauty**; see [483, 1467, 1608] and [1090, Chap. 7]. Further algorithms for classifying linear codes are presented in [190].

3.3 Some Central Classes of Codes

By Definition 1.9.1, the maximum size of error-correcting codes with length n and minimum distance d are given by the functions $A_q(n, d)$ and $B_q(n, d)$ for unrestricted and linear codes, respectively. Most general bounds on these functions, such as those in Section 1.9,

FIGURE 3.1: Graphs of two ternary codes

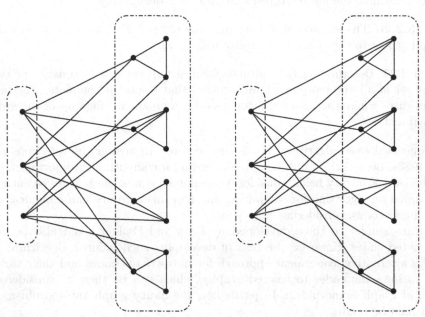

consider upper bounds and are about nonexistence of codes. Lower bounds, on the other hand, are typically obtained by constructing explicit codes. Especially for small parameters, many best known codes have been obtained on a case-by-case basis. One possible approach for finding such codes is that of prescribing symmetries—as discussed in Section 3.2.1—and carrying out a computer search; see Chapter 23.

In some rare situations, there exist codes that attain some general upper bounds. For such parameters, the problem of finding the size of an optimal code is then settled. When this occurs and the upper bound is the Sphere Packing Bound, we get perfect codes (Definition 1.9.8), and when the upper bound is the Singleton Bound, we get maximum distance separable (MDS) codes (Definition 1.9.12). In this section we will take a glance at these two types of codes as well as general binary linear and unrestricted codes.

3.3.1 Perfect Codes

Perfect codes were considered in the very first scientific papers in coding theory. We have already seen two types of perfect codes in Sections 1.10 and 1.13. Hamming codes [895] have parameters

$$[n = (q^m - 1)/(q - 1), n - m, 3]_q \tag{3.1}$$

and exist for $m \geq 2$ and prime powers q. Golay codes [820] have parameters

$$[23, 12, 7]_2 \quad \text{and} \quad [11, 6, 5]_3. \tag{3.2}$$

There are also some families of *trivial* perfect codes: codes containing one word, codes containing all codewords in the space, and $(n, 2, n)_2$ codes for odd n. If the order of the alphabet q is a prime power, these are in fact the only sets of parameters for which (linear and unrestricted) perfect codes exist [1805, 1949].

Theorem 3.3.1 *The nontrivial perfect linear codes over* $\mathbb{F}_q$, *where q is a prime power, are precisely the Hamming codes with parameters* (3.1) *and the Golay codes with parameters* (3.2). *A nontrivial perfect unrestricted code (over* $\mathbb{F}_q$, *q a prime power) that is not equivalent to a linear code has the same length, size, and minimum distance as a Hamming code* (3.1).

Although the remarkable Theorem 3.3.1 gives us a rather solid understanding of perfect codes, there are still many open problems in this area, including the following (a code with different alphabet sizes for different coordinates is called **mixed**):

Research Problem 3.3.2 Solve the existence problem for perfect codes when the size of the alphabet is not a prime power.

Research Problem 3.3.3 Solve the existence problem for perfect mixed codes.

Research Problem 3.3.4 Classify perfect codes, especially for the parameters covered by Theorem 3.3.1.

Since Theorem 3.3.1 covers alphabet sizes that are prime powers, that is, exactly the sizes for which finite fields and linear codes exist, Research Problems 3.3.2 to 3.3.4 are essentially about unrestricted codes (although many codes studied for Research Problem 3.3.3 have clear algebraic structures and close connections to linear codes).

Constructing Perfect q-Ary Codes

The following conjecture is related to Research Problem 3.3.2.

Conjecture 3.3.5 The only nontrivial $(n, M)_q$ perfect codes are those in Theorem 3.3.1.

A survey of known results for Research Problem 3.3.2 and Conjecture 3.3.5 can be found in [930]. It seems to make sense to try to generalize techniques used in the various proofs of Theorem 3.3.1 to the case when the alphabet size is not a prime power. Indeed, a lot can be proved in this manner, but a full proof is still missing. Results obtained so far [136, 185, 975, 1185, 1215, 1588, 1806] show that minimum distance of such a nontrivial perfect code is necessarily 3 or 5, that is, it would be 1-error-correcting or 2-error-correcting. For 2-error-correcting codes, nonexistence proofs have further been obtained for various alphabet sizes [136, 1588, 1832]. But not a single nontrivial perfect code is known when the alphabet is not a prime power, and this supports Conjecture 3.3.5.

Remark 3.3.6 Van Lint [1836, Chap. 7] writes that "The problem of the possible existence of unknown 2-error-correcting codes over alphabets Q with $|Q|$ not a prime power seems to be very hard but probably not impossible. The case [of 1-error-correcting codes] looks hopeless."

When the alphabet size is $q = p^i$, p a prime, then by the Sphere Packing Bound a necessary condition for the existence of a perfect 1-error-correcting code is

$$1 + n(q - 1) \mid q^n = p^{ni};$$

that is, $1 + n(q - 1)$ should be a power of p. However, one may prove an even stronger necessary condition, namely that $1 + n(q-1)$ should be a power of q; see [1836, Chap. 7]. This condition shows that for alphabet sizes that are prime powers, the only possible parameters for unrestricted perfect 1-error-correcting codes equal those in (3.1), that is,

$$\left((q^m - 1)/(q - 1), q^{(q^m - 1)/(q-1) - m}, 3\right)_q \quad \text{for } m \geq 2. \tag{3.3}$$

The parameters (3.3) are feasible also when the alphabet size is not a prime power, but for those alphabets there are also other parameters to consider.

Example 3.3.7 The Sphere Packing Bound cannot be used to prove nonexistence of perfect $(19, 2^{14}3^{18}, 3)_6$ codes. To prove nonexistence in this case, we may use the necessary condition [931, Theorem 1] that

$$1 + n(q-1) \mid q^{1+(n-1)/q}.$$

The tricky instances for general alphabet sizes seem to be those of type (3.3), and this is perhaps the reason behind the pessimism in the quote of Remark 3.3.6. In fact, a nonexistence proof [826, Theorem 6] for $(7, 6^5, 3)_6$ codes is the only result so far for (3.3). The next open case for $m = 2$, that is, $n = q + 1$, is here singled out as a particularly interesting question.

Research Problem 3.3.8 Solve the existence problem for $(11, 10^9, 3)_{10}$ codes.

Remark 3.3.9 Codes with $m = 2$ ($n = q+1$) in (3.3) are of special interest because if there exists such a code, then there exists an infinite sequence of perfect codes with parameters (3.3) over the same alphabet [1509].

Constructing Perfect Mixed Codes

For Research Problem 3.3.3, there is obviously a greater variety of possible parameters than for Research Problem 3.3.2. For mixed codes, the main focus has been on cases when the minimum distance is 3 and all coordinates have alphabets that are powers of some prime p. One useful property of such codes is that the they can be used to construct codes over other alphabets. The following theorem is central in such constructions [1068].

Theorem 3.3.10 *Let q be a prime power, and assume that there is a perfect code with minimum distance 3 and length n over $\mathbb{F}_{q^m}\mathbb{F}^{(2)}\cdots\mathbb{F}^{(n)}$. Then there is a perfect code with minimum distance 3 and length $n - 1 + (q^m - 1)/(q-1)$ over $\mathbb{F}_q^{(q^m-1)/(q-1)}\mathbb{F}^{(2)}\cdots\mathbb{F}^{(n)}$.*

Proof: Since q is a prime power, there is an $[n = (q^m - 1)/(q-1), n - m, 3]_q$ Hamming code $\mathcal{C}_0$ (3.1). The q^m cosets of that code partition all words in the space,

$$\mathbb{F}_q^n = \mathcal{C}_0 \cup \mathcal{C}_1 \cup \cdots \cup \mathcal{C}_{q^m-1}.$$

For each codeword $x_1 x_2 \cdots x_n \in \mathbb{F}_{q^m}\mathbb{F}^{(2)}\cdots\mathbb{F}^{(n)}$ in the original code, include the words $c x_2 x_3 \cdots x_n$, $\mathbf{c} \in \mathcal{C}_{x_1}$ in the new code. To show that the obtained code is perfect, it suffices to observe that the minimum distance of the new code is 3 (there are two cases to consider: codewords that come from different codewords, respectively the same codeword of the original code) and that its size attains the Sphere Packing Bound. $\square$

Example 3.3.11 Starting from a perfect Hamming code over $\mathbb{F}_4^5$ and applying Theorem 3.3.10 repeatedly, we get perfect codes with minimum distance 3 over $\mathbb{F}_2^3\mathbb{F}_4^4$, $\mathbb{F}_2^6\mathbb{F}_4^3$, $\mathbb{F}_2^9\mathbb{F}_4^2$, $\mathbb{F}_2^{12}\mathbb{F}_4$, and $\mathbb{F}_2^{15}$. The last code again has the parameters of a Hamming code.

In Example 3.3.11, we started from a well-known perfect code, but in this way only specific families of perfect mixed codes can be obtained. What about the general case? In the general case, there are both existence and nonexistence results. For nonexistence results, see, for example, [1588, 1842] and their references.

When the alphabet sizes are powers of a prime power q, mixed perfect codes with minimum distance 3 are known to exist for a wide range of parameters. Actually, for all those parameters, such perfect codes can be constructed via a vector space partition.

Definition 3.3.12 Let q be a prime power. A collection $\{\mathcal{C}_1, \mathcal{C}_2, \ldots, \mathcal{C}_m\}$ of subspaces of $\mathbb{F}_q^n$ is called a **partition** of $\mathbb{F}_q^n$ if (i) $\mathbb{F}_q^n = \cup_{i=1}^m \mathcal{C}_i$ and (ii) $\mathcal{C}_i \cap \mathcal{C}_j = \{\mathbf{0}\}$ when $i \neq j$.

Let $\{C_1, C_2, \ldots, C_m\}$ be a partition of $\mathbb{F}_q^n$. We now define a map φ from the direct product $C_1 \times C_2 \times \cdots \times C_m$ to $\mathbb{F}_q^n$ by

$$\varphi : (c_1, c_2, \ldots, c_m) \mapsto c_1 + c_2 + \cdots + c_m.$$

The following theorem is proved in [949].

Theorem 3.3.13 *Let $\{C_1, C_2, \ldots, C_m\}$ be a partition of $\mathbb{F}_q^n$, where C_i is an $[n, k_i]_q$ code. Then the kernel of φ is a perfect code with minimum distance 3 over $\mathbb{F}_{q^{k_1}} \mathbb{F}_{q^{k_2}} \cdots \mathbb{F}_{q^{k_m}}$.*

Example 3.3.14 Consider partitions of $\mathbb{F}_2^3$. By taking the seven $[3, 1]_2$ codes in the partition, Theorem 3.3.13 gives the binary Hamming code of length 7. With $C_1 = \{000, 100, 010, 110\}$, $C_2 = \{000, 001\}$, $C_3 = \{000, 101\}$, $C_4 = \{000, 011\}$, and $C_5 = \{000, 111\}$, we get a perfect code over $\mathbb{F}_4 \mathbb{F}_2^4$.

Remark 3.3.15 The term *partition* in Definition 3.3.12 is justifiable when the all-zero word is ignored. Such partitions into subspaces (linear codes) should not be confused with partitions into arbitrary subsets (unrestricted codes), such as those considered in the proof of Theorem 3.3.10.

Remark 3.3.16 Partitions of vector spaces into t-dimensional subspaces are known as t-*spreads* in finite geometry. Partitions of vector spaces are further related to q-analogs of combinatorial designs.

Partitions of vector spaces have been studied extensively—see, for example, [667, 669, 1214, 1637] for selected results on this problem obtained during the last decade. Any existence results in this context give perfect codes, but our current knowledge does not allow us to use nonexistence results to deduce nonexistence of mixed codes with the related parameters. Mixed codes obtained by Theorem 3.3.13 have a "linear" structure, and one central question is about the existence of such codes. The next problem is [929, Problem 4].

Research Problem 3.3.17 Are there parameters for which perfect mixed codes with minimum distance 3 exist but which cannot be obtained via partitions of vector spaces (Theorem 3.3.13)?

Classifying Perfect Codes

Classifying perfect codes goes even beyond constructing them; so Research Problem 3.3.4 is in general a very challenging problem.

The Hamming codes are the unique *linear* codes with those parameters, as we have seen in Theorem 1.10.4. However, for unrestricted perfect codes with the same parameters, there are more codes for all but the smallest parameters. Perfect $(q + 1, q^{q-1}, 3)_q$ codes are MDS codes; we will get back to those in Section 3.3.2 and show some classification results. The only other classification results obtained so far are Zaremba's [1939] uniqueness proof of the $(7, 16, 3)_2$ (Hamming) code in 1952 and the classification results for the $(15, 2048, 3)_2$ codes and the Golay codes, mentioned in Examples 3.2.2 and 3.2.14. As for the $(15, 2048, 3)_2$ codes, the 5983 codes with these parameters were obtained and validated in an extensive computational study [1472] and later reproduced with a different computational approach in [1166]; an analytical proof seems far out of reach.

Little has been done on classifying mixed perfect codes, but see [1473].

3.3.2 MDS Codes

Maximum distance separable (MDS) codes are not only of theoretical interest, but rather important families of codes are of this type, such as Reed–Solomon codes (Section 1.14). An entire chapter is devoted to MDS codes in the book by MacWilliams and Sloane [1323, Chap. 11].

MDS codes are closely connected to many other structures in combinatorics and geometry. For example, an $[n, k, n-k+1]_q$ MDS code with dimension $k \geq 3$ corresponds to an n-arc in the projective geometry $\mathrm{PG}(k-1, q)$; see Chapter 14. Finite geometry is indeed a commonly used framework for studying MDS codes. In combinatorics, MDS codes correspond to certain orthogonal arrays.

Definition 3.3.18 An **orthogonal array** of *size* N, with m *constraints*, s *levels*, and *strength* t, denoted $\mathrm{OA}(N, m, s, t)$, is an $m \times N$ matrix with entries from $\mathbb{F}_s$, having the property that in every $t \times N$ submatrix, every $t \times 1$ column vector appears $\lambda = N/s^t$ (called the *index*) times.

Theorem 3.3.19 *An* $n \times q^k$ *matrix with columns formed by the codewords of a linear* $[n, k, n-k+1]_q$ *MDS code or an unrestricted* $(n, q^k, n-k+1)_q$ *MDS code is an* $\mathrm{OA}(q^k, n, q, k)$, *which has index* $\lambda = 1$.

Remark 3.3.20 As the codewords of an MDS code with dimension k form an orthogonal array with strength k and index 1, such codes are systematic and any k coordinates can be used for the message symbols.

In a paper [319] published by Bush in 1952, the framework of orthogonal arrays is used to construct objects that we now know as Reed–Solomon codes. In that study it is also shown that for linear codes over $\mathbb{F}_q$ with $k > q$, $n \leq k+1$ is a necessary condition for an $[n, k, n-k+1]_q$ MDS code to exist, and that there are $[k+1, k, 2]_q$ MDS codes. Such codes, and generally codes with parameters $[n, 1, n]_q$, $[n, n-1, 2]_q$, and $[n, n, 1]_q$, are called *trivial* MDS codes.

For $k \leq q$, on the other hand, the following MDS Conjecture related to a question by Segre [1638] in 1955 is still open.

Conjecture 3.3.21 (MDS) If $k \leq q$, then a linear $[n, k, n-k+1]_q$ MDS code exists exactly when $n \leq q+1$ unless $q = 2^h$ and $k = 3$ or $k = q-1$, in which case it exists exactly when $n \leq q+2$.

Remark 3.3.22 MDS codes are typically discussed in the linear case, but the parameters of the codes in Conjecture 3.3.21 conjecturally also cover the parameters for which unrestricted MDS codes exist.

The MDS Conjecture has throughout the years been proved for various sets of parameters k and q; see [962] for a survey. A major breakthrough was made in 2012 by Ball [104], who proved that the MDS Conjecture holds when q is a prime; see Theorem 14.2.17. The conjecture is still open for non-prime q; see [104, 106] and their references for some related results. Much less is known in the more difficult case of unrestricted MDS codes, where the MDS Conjecture has been proved for $q \leq 8$ [1142, 1143].

MDS codes, linear and unrestricted, have been classified for various parameters. For example, a general result is that for prime q, $[q+1, k, q-k+2]_q$ MDS codes—which have the maximum value of n given k and q [104]—are Reed–Solomon codes. However, when q is not a prime, examples of $[q+1, k, q-k+2]_q$ MDS codes that are not equivalent to Reed–Solomon codes are known [811, 957].

Unrestricted MDS codes have been classified for small values of n and q. For the smallest nontrivial case, $(4, 9, 3)_3$ MDS codes, Taussky and Todd [1794] claimed uniqueness of the Hamming code in 1948 but omitted the details—see [1074] for a complete proof. That code is the first nontrivial member of a family of perfect 1-error-correcting codes with varying q and smallest possible length, that is, with parameters $(q + 1, q^{q-1}, 3)_q$. We have earlier seen these parameters in Research Problem 3.3.8.

Actually, for $q \leq 5$ there is a unique unrestricted MDS code for each set of admissible parameters. The number of equivalence classes of unrestricted $(n, 7^k, d)_7$ MDS codes and $(n, 8^k, d)_8$ MDS codes are given in Tables 3.1 and 3.2, respectively, for $n \geq k + 2$. See [35, 664, 1142, 1143] and their references for details.

TABLE 3.1: Equivalence classes of $(n, 7^k, n - k + 1)_7$ MDS codes

$n\backslash k$	2	3	4	5	6
4	4				
5	1	1			
6	1	3	1		
7	1	1	1	1	
8	1	1	1	1	1

TABLE 3.2: Equivalence classes of $(n, 8^k, n - k + 1)_8$ MDS codes

$n\backslash k$	2	3	4	5	6	7
4	2165					
5	39	12484				
6	1	39	14			
7	1	2	2	8		
8	1	2	1	2	4	
9	1	2	1	1	2	4
10		1				1

Remark 3.3.23 An $(n, q^2, n - 1)_q$ MDS code corresponds to a set of $n - 2$ **mutually orthogonal Latin squares (MOLS)** of side n. Two Latin squares $L = (a_{i,j})$ and $L' = (b_{i,j})$ with elements from $S = \{0, 1, \ldots, n - 1\}$ are *orthogonal* if every element in $S \times S$ occurs exactly once among the pairs $(a_{i,j}, b_{i,j})$, $1 \leq i, j \leq n$. In particular, $(3, q^2, 2)$ MDS codes correspond to Latin squares. Moreover, $(n, q^{n-1}, 2)$ MDS codes correspond to Latin hypercubes.

Remark 3.3.24 The existence of a $(q + 1, q^2, q)_q$ MDS code is equivalent to the existence of a projective plane of order q and an affine plane of order q.

3.3.3   Binary Error-Correcting Codes

All codes considered in this section are assumed to be binary, unless otherwise mentioned. We first look at unrestricted codes.

Binary Unrestricted Codes

The function $A_2(n, d)$ is perhaps the most studied function in combinatorial coding theory. Plotkin introduced this function in his M.Sc. thesis [1526] from 1952 and in the related journal paper [1527], which, however, did not appear until 1960. Binary codes had earlier been studied by Hamming, but the function defined in [895] specifically concerns *systematic* codes (so it can take only values of the form 2^i).

As for determining values of $A_2(n, d)$ and classifying optimal codes, there has been considerable progress since the first edition of the classical reference *The Theory of Error-Correcting Codes* [1323] appeared. Whereas in the early 1990s not even the exact value of $A_2(10, 3)$ was known, the values of $A_2(n, d)$ are now known for all parameters when $n \leq 15$ and all those optimal codes have been classified. These results are summarized in Tables 3.3 and 3.4; Table 3.4 updates [1090, Table 7.2] with results from [1166, 1472].

TABLE 3.3: The values of $A_2(n, d)$ for $n \leq 15$ and odd $d \leq 11$

$n \backslash d$	3	5	7	9	11
1	1	1	1	1	1
2	1	1	1	1	1
3	2	1	1	1	1
4	2	1	1	1	1
5	4	2	1	1	1
6	8	2	1	1	1
7	16	2	2	1	1
8	20	4	2	1	1
9	40	6	2	2	1
10	72	12	2	2	1
11	144	24	4	2	2
12	256	32	4	2	2
13	512	64	8	2	2
14	1024	128	16	4	2
15	2048	256	32	4	2

Remark 3.3.25 Since $A_2(n - 1, d - 1) = A_2(n, d)$ when d is even (Theorem 1.9.2(e)), $A_2(n, d)$ is known for all even d when $n \leq 16$.

One may focus on either of the parities of d—even or odd—not only when determining the values of $A_2(n, d)$ but also when classifying codes. Namely, codes with odd minimum distance can be *extended* to codes with even minimum distance, and codes with even minimum distance can be *punctured* to codes with odd minimum distance. These operations can be carried out in several ways (but may still lead to equivalent codes).

When puncturing an arbitrary binary code, the minimum distance decreases by at most one—if the length of the original code is n, this can be done in n ways. One may find all possible ways of extending a code with odd minimum distance by finding all proper 2-colorings of a certain graph; see [1090, Sect. 7.1.4] for details. In particular, one possible extension is obtained by adding a parity check symbol. Even parity is typically considered, but odd parity gives equivalent codes. Codes extended with a parity check symbol further have even distance between all pairs of codewords, cf. Theorem 1.9.2(f).

For parameters outside the range of Table 3.3, the exact value of $A_2(n, d)$ is known only for sporadic parameters, including those of Hamming codes, the binary Golay code, and Preparata codes [1541]. For more on Preparata codes, see Chapters 5 and 6.

TABLE 3.4: Equivalence classes of $(n, A_2(n, d), d)_2$ codes for $n \leq 15$ and odd $d \leq 11$

$n \backslash d$	3	5	7	9	11
1	1	1	1	1	1
2	1	1	1	1	1
3	1	1	1	1	1
4	2	1	1	1	1
5	1	1	1	1	1
6	1	2	1	1	1
7	1	3	1	1	1
8	5	1	2	1	1
9	1	1	3	1	1
10	562	1	4	2	1
11	7398	1	1	3	1
12	237610	2	9	4	2
13	117823	1	6	5	3
14	38408	1	10	1	4
15	5983	1	5	9	5

It is known that the binary Golay code leads to optimal codes even after shortening it six times [1469].

Proposition 3.3.26 $A_2(n, 7) = 2^{n-11}$ *for* $17 \leq n \leq 23$.

Since we further know [1468] that $A_2(16, 7) = 36$, we actually know $A_2(n, 7)$ for all $n \leq 23$ (and consequently $A_2(n, 8)$ for all $n \leq 24$).

In the general case, we have lower and upper bounds on $A_2(n, d)$. For codes with parameters just outside the range of Table 3.3, examples of techniques for finding these bounds include, for lower bounds, conducting a computer search for codes with prescribed automorphisms [1186] and, for upper bounds, employing semidefinite programming [804, 1634]. See Chapter 13 for details of the latter method.

For fixed (small) d, the relative distance $d/n \to 0$ as $n \to \infty$. In this case, the asymptotic behavior is conveniently discussed in terms of the density of a code.

Definition 3.3.27 The **density** of an $(n, M, 2t + 1)_q$ code is

$$\frac{M \sum_{i=0}^{t} \binom{n}{i} (q - 1)^i}{q^n}.$$

Remark 3.3.28 A perfect code has density 1.

The asymptotic behavior of 1-error-correcting codes has been solved for prime-power alphabets [1068]. For clarity we focus on the binary case.

Theorem 3.3.29 *The density of optimal* $(n, M, 3)_2$ *codes tends to* 1 *as* $n \to \infty$.

Outline of proof: The proof contains five ingredients.

1. Hamming codes over $\mathbb{F}_q$, where q is a prime power. The parameters of these are given by (3.1).

2. The following construction of a code C' over $\mathbb{F}_{2^i}$ from a code C over $\mathbb{F}_q$, where $q > 2^i$. Ignoring the algebraic structure of the alphabet, label the elements of $\mathbb{F}_{2^i}$ so that they form a subset of the elements of $\mathbb{F}_q$, and take exactly those codewords of C that have all coordinate values in $\mathbb{F}_{2^i}$.

3. Transform a code over $\mathbb{F}_{2^i}$ to a code over $\mathbb{F}_2$ using the construction in the proof of Theorem 3.3.10.

4. Lengthen an $(n, M, 3)_2$ code to an $(2n + 1, 2^n M, 3)_2$ code.

5. Shorten codes to obtain codes with all possible parameters.

The Hamming codes have density 1, the constructions (3) and (4) do not affect the density of a code, and constructions (2) and (5) decrease the density (in the way they are used in practice). By putting the pieces together carefully, the desired result is reached. □

For 2-error-correcting codes, that is, codes with minimum distance 5, the density of Preparata codes tends to 1 as $n \to \infty$. However, this only shows that the limit superior is 1.

For a relative distance approaching δ as $n \to \infty$, the asymptotic behavior of $A_2(n, \delta n)$ is considered in Section 1.9.8 in terms of upper bounds on the asymptotic normalized rate function. It was not until 1972 that Justesen [1063] published the first construction of codes—now called *Justesen codes*—with a lower bound on the asymptotic normalized rate function that is greater than 0. All these bounds apply to linear as well as unrestricted codes.

Binary Linear Codes

Classification results have been published for many types of linear codes, including general linear codes [190], rate-1/2 linear codes, self-dual linear codes [1005], and so on. If we have codes with parameters $[n-1, k, d]_2$, $[n+1, k+1, d]_2$, or $[n+1, k, d+1]_2$, then $[n, k, d]_2$ codes can be constructed. Hence some studies focus explicitly on parameters $[n, k, d]_2$ for which codes cannot be constructed from other codes in this way (and even use the term 'optimal' for those codes) [1031]. However, in this situation we cannot always get a full classification of one family of codes from a classification of one of the other mentioned families; so it still makes sense to handle sets of parameters individually.

By Definition 1.9.1, the maximum dimension of q-ary linear error-correcting codes with length n and minimum distance d is given by the function $k_q(n, d)$. Values of $k_2(n, d)$ for small values of n and d and the number of optimal codes are given in Tables 3.5 and 3.6, respectively. The entries of the two tables are compiled from [190, Table 9.6]. Note that no further restrictions are assumed in the tables here. For example, we allow coordinates with value 0 in all codewords. Due to the connection between the odd-distance and even-distance cases, which is analogous to that in the unrestricted situation, we restrict the tables to odd d. Only entries for codes with dimension greater than 0 are included.

Classification results for linear codes are known beyond the range of Tables 3.5 and 3.6. The threshold between what has been done and the smallest open cases is not as abrupt for linear codes as for unrestricted codes. Hence there are possibilities for fruitful research for the classification problem as well as the existence problem.

Research Problem 3.3.30 Collect and extend classification results for binary linear codes.

Research Problem 3.3.31 Determine the value of $k_2(n, d)$ for the smallest open case.

At the time of writing, $n = 32$ is the shortest length with open cases [845], which are $k_2(32, 7)$ and $k_2(32, 9)$.

TABLE 3.5: The values of $k_2(n,d)$ for $n \leq 17$ and odd $d \leq 17$

$n\backslash d$	3	5	7	9	11	13	15	17
3	1							
4	1							
5	2	1						
6	3	1						
7	4	1	1					
8	4	2	1					
9	5	2	1	1				
10	6	3	1	1				
11	7	4	2	1	1			
12	8	4	2	1	1			
13	9	5	3	1	1	1		
14	10	6	4	2	1	1		
15	11	7	5	2	1	1	1	
16	11	8	5	2	1	1	1	
17	12	9	6	3	2	1	1	1

TABLE 3.6: Equivalence classes of $[n, k_2(n,d), \geq d]_2$ codes for $n \leq 17$ and odd $d \leq 17$

$n\backslash d$	3	5	7	9	11	13	15	17
3	1							
4	2							
5	1	1						
6	1	2						
7	1	3	1					
8	6	1	2					
9	5	4	3	1				
10	4	2	4	2				
11	3	1	1	3	1			
12	2	14	4	4	2			
13	1	15	1	5	3	1		
14	1	11	1	1	4	2		
15	1	6	1	4	5	3	1	
16	145	1	7	9	6	4	2	
17	129	1	3	2	1	5	3	1

Chapter 4

Self-Dual Codes

Stefka Bouyuklieva

Veliko Tarnovo University

4.1   Introduction

Definition 4.1.1 If C is an $[n, k]_q$ linear code over the field $\mathbb{F}_q$, its **dual** or **orthogonal code** $C^\perp$ is the set of vectors which are orthogonal to all codewords of C:

$$C^\perp = \{\mathbf{u} \in \mathbb{F}_q^n \mid (\mathbf{u}, \mathbf{v}) = 0 \text{ for all } \mathbf{v} \in C\},$$

where $(\mathbf{u}, \mathbf{v})$ is an inner product in the vector space $\mathbb{F}_q^n$.

Definition 4.1.2 A linear code C is called **self-orthogonal** if $C \subseteq C^\perp$, and **self-dual** if $C = C^\perp$.

Remark 4.1.3 The most important inner products used in the theory of self-dual codes are:

- **Euclidean inner product** or **ordinary inner product**

$$(\mathbf{u}, \mathbf{v}) = \mathbf{u} \cdot \mathbf{v} = \sum_{i=1}^{n} u_i v_i \ \text{ for } \mathbf{u} = u_1 u_2 \cdots u_n, \ \mathbf{v} = v_1 v_2 \cdots v_n \in \mathbb{F}_q^n,$$

    where $q = p^m$ for an arbitrary prime p and an arbitrary integer $m \geq 1$, and

- **Hermitian inner product**

$$(\mathbf{u}, \mathbf{v}) = \sum_{i=1}^{n} u_i \overline{v}_i \ \text{ for } \mathbf{u} = u_1 u_2 \cdots u_n, \ \mathbf{v} = v_1 v_2 \cdots v_n \in \mathbb{F}_q^n,$$

    where q is an even power of an arbitrary prime, and $\overline{x} = x^{\sqrt{q}}$ for $x \in \mathbb{F}_q$.

Remark 4.1.4 Any self-dual code of length n over a finite field $\mathbb{F}_q$ has dimension $k = \frac{n}{2}$. Therefore self-dual codes over a field exist only for even lengths.

Definition 4.1.5 A code $\mathcal{C}$ is divisible by Δ provided all codewords have weights divisible by an integer Δ, called a **divisor** of $\mathcal{C}$; the code is called **divisible** if it has a divisor $\Delta > 1$.

Definition 4.1.6 If all codewords in a linear code have even weights, the code is **even**. If all codewords in a binary linear code have weights divisible by 4, the code is **doubly-even**. Binary self-orthogonal codes, which are not doubly-even, are called **singly-even**.

Remark 4.1.7 Any binary self-orthogonal code is even but there are even binary codes which are not self-orthogonal. It is easy to see that any binary doubly-even code is self-orthogonal. If $\mathcal{C}$ is a binary singly-even self-orthogonal code, then $\mathcal{C}$ is even but it contains at least one codeword of weight $\equiv 2 \pmod 4$.

Remark 4.1.8 Definition 4.1.6 repeats Definition 1.6.8 but we give it here for completeness. Remark 1.6.9 and Example 1.6.10 also give some information about binary self-orthogonal codes.

Theorem 4.1.9 (Gleason–Pierce–Ward [1008]) *Let $\mathcal{C}$ be an $[n, n/2]$ divisible code over the field $\mathbb{F}_q$ with divisor $\Delta \geq 2$. Then one (or more) of the following holds.*

(a) $q = 2$, $\Delta = 2$, or

(b) $q = 2$, $\Delta = 4$, and $\mathcal{C}$ is self-dual, or

(c) $q = 3$, $\Delta = 3$, and $\mathcal{C}$ is self-dual, or

(d) $q = 4$, $\Delta = 2$, and $\mathcal{C}$ is Hermitian self-dual, or

(e) $\Delta = 2$, and $\mathcal{C}$ is equivalent to the code over $\mathbb{F}_q$ with generator matrix $\begin{bmatrix} I_{n/2} & I_{n/2} \end{bmatrix}$.

Remark 4.1.10 Note that in case (b), the code is doubly-even self-dual, and these codes are also called **Type II** codes; the singly-even self-dual codes are called **Type I** codes.

Remark 4.1.11 Theorem 4.1.9 generalizes the Gleason–Pierce–Turyn Theorem presented and proved in [1323].

Theorem 4.1.12 ([1008, Theorem 9.1.3]) *There exists a Euclidean self-dual code over $\mathbb{F}_q$ of even length n if and only if $(-1)^{n/2}$ is a square in $\mathbb{F}_q$. Furthermore, if n is even and $(-1)^{n/2}$ is not a square in $\mathbb{F}_q$, then the dimension of a maximal self-orthogonal code of length n is $(n/2) - 1$. If n is odd, then the dimension of a maximal self-orthogonal code of length n is $(n-1)/2$.*

Theorem 4.1.13 ([1008, Corollary 9.2.2]) *A self-dual doubly-even binary code of length n exists if and only if $8 \mid n$; a self-dual ternary code of length n exists if and only if $4 \mid n$; and a Hermitian self-dual code over $\mathbb{F}_4$ of length n exists if and only if n is even.*

The structure of the self-dual codes plays a major role in their research. The Balance Principle gives very important information on this structure. Theorem 4.1.14 is the base for most of the construction methods presented in Section 4.4.

Theorem 4.1.14 ([1008, Theorem 9.4.1]) *Let $\mathcal{C}$ be a self-dual code of length $n = n_1 + n_2$. Let $\mathcal{B}$, respectively $\mathcal{D}$, be the largest subcode of $\mathcal{C}$ whose support is contained entirely in the left n_1, respectively right n_2 coordinates. Suppose $\mathcal{B}$ and $\mathcal{D}$ have dimensions k_1 and k_2, respectively. Then the following hold:*

(a) (Balance Principle)

$$k_1 - \frac{n_1}{2} = k_2 - \frac{n_2}{2}.$$

(b) $\mathcal{C}$ *has a generator matrix of the form*

$$G = \begin{bmatrix} B & O \\ O & D \\ E & F \end{bmatrix}$$

where B is a $k_1 \times n_1$ matrix with $\begin{bmatrix} B & O \end{bmatrix}$ a generator matrix of $\mathcal{B}$, D is a $k_2 \times n_2$ matrix with $\begin{bmatrix} O & D \end{bmatrix}$ a generator matrix of $\mathcal{D}$, O is the appropriate size zero matrix, and $\begin{bmatrix} E & F \end{bmatrix}$ is a $k_3 = (n/2 - k_1 - k_2) \times n$ matrix.

(c) *Let $\mathcal{B}^*$ be the code of length n_1 generated by B, $\mathcal{B}_E$ the code of length n_1 generated by the rows of B and E, $\mathcal{D}^*$ the code of length n_2 generated by D, and $\mathcal{D}_F$ the code of length n_2 generated by the rows of D and F. Then*

$$\mathcal{B}_E^\perp = \mathcal{B}^* \quad \text{and} \quad \mathcal{D}_F^\perp = \mathcal{D}^*.$$

4.2   Weight Enumerators

The Hamming weight enumerator is defined in Definition 1.15.1 in Chapter 1. Recall that

$$\mathrm{Hwe}_\mathcal{C}(x, y) = \sum_{i=0}^{n} A_i(\mathcal{C}) x^i y^{n-i}.$$

Definition 4.2.1 A linear code $\mathcal{C}$ is called **formally self-dual** if $\mathcal{C}$ and its dual code $\mathcal{C}^\perp$ have the same weight enumerator, $\mathrm{Hwe}_\mathcal{C}(x, y) = \mathrm{Hwe}_{\mathcal{C}^\perp}(x, y)$. A linear code is **isodual** if it is equivalent to its dual code.

Remark 4.2.2 Any isodual code is also formally self-dual, but there are formally self-dual codes that are neither isodual nor self-dual. The smallest length for which a formally self-dual code is not isodual is 14, and there are 28 such codes amongst 6 weight enumerators [867]. Any self-dual code is also isodual and formally self-dual.

Example 4.2.3 The $[6, 3, 3]$ binary code $\mathcal{C}$ with a generator matrix

$$\begin{bmatrix} 100 & 111 \\ 010 & 110 \\ 001 & 101 \end{bmatrix}$$

is isodual. Its weight enumerator is $\mathrm{Hwe}_\mathcal{C}(x, y) = y^6 + 4x^3y^3 + 3x^4y^2$, and its automorphism group has order 24. Obviously, this code is not self-dual as it contains codewords with odd weight.

Theorem 4.2.4 (Gleason [1008]) *Let C be an $[n, n/2]$ code over the field $\mathbb{F}_q$, and let*

$$g_1(x, y) = y^2 + x^2,$$
$$g_2(x, y) = y^8 + 14x^4y^4 + x^8,$$
$$g_3(x, y) = y^{24} + 759x^8y^{16} + 2576x^{12}y^{12} + 759x^{16}y^8 + x^{24},$$
$$g_4(x, y) = y^4 + 8x^3y,$$
$$g_5(x, y) = y^{12} + 264x^6y^6 + 440x^9y^3 + 24x^{12},$$
$$g_6(x, y) = y^2 + 3x^2, \quad and$$
$$g_7(x, y) = y^6 + 45x^4y^2 + 18x^6.$$

The following hold.

(a) *If $q = 2$ and C is a formally self-dual even code, then*

$$\mathrm{Hwe}_C(x, y) = \sum_{i=0}^{\lfloor \frac{n}{8} \rfloor} a_i g_1(x, y)^{\frac{n}{2} - 4i} g_2(x, y)^i.$$

(b) *If $q = 2$ and C is a doubly-even self-dual code, then*

$$\mathrm{Hwe}_C(x, y) = \sum_{i=0}^{\lfloor \frac{n}{24} \rfloor} a_i g_2(x, y)^{\frac{n}{8} - 3i} g_3(x, y)^i.$$

(c) *If $q = 3$ and C is a self-dual code, then* $\mathrm{Hwe}_C(x, y) = \displaystyle\sum_{i=0}^{\lfloor \frac{n}{12} \rfloor} a_i g_4(x, y)^{\frac{n}{4} - 3i} g_5(x, y)^i.$

(d) *If $q = 4$ and C is a Hermitian self-dual code, then*

$$\mathrm{Hwe}_C(x, y) = \sum_{i=0}^{\lfloor \frac{n}{6} \rfloor} a_i g_6(x, y)^{\frac{n}{2} - 3i} g_7(x, y)^i.$$

In all cases, all coefficients a_i are rational and $\sum_i a_i = 1$.

Remark 4.2.5 In Theorem 4.2.4(a), if we let $g_2'(x, y) = x^2y^2(x^2 - y^2)^2$, then we can use $g_2'(x, y)$ in place of $g_2(x, y)$; so

$$\mathrm{Hwe}_C(x, y) = \sum_{i=0}^{\lfloor \frac{n}{8} \rfloor} a_i g_1(x, y)^{\frac{n}{2} - 4i} g_2(x, y)^i$$

$$= \sum_{i=0}^{\lfloor \frac{n}{8} \rfloor} c_i (x^2 + y^2)^{\frac{n}{2} - 4i} (x^2y^2(x^2 - y^2)^2)^i.$$

The usage of $g_2'(x, y)$ in place of $g_2(x, y)$ is much more useful in many applications due to the fact that the coefficients c_i are integers.

Remark 4.2.6 The polynomials $g_1, g_2, \ldots, g_7$ presented in Gleason's Theorem are all weight enumerators of certain self-dual codes and are called **Gleason polynomials**. The polynomial $g_1(x, y)$ is the weight enumerator of the binary repetition code of length 2, and $g_2(x, y)$ is the weight enumerator of the $[8, 4, 4]$ extended binary Hamming code $\widehat{\mathcal{H}}_{3,2}$. The weight enumerators of the extended binary and ternary Golay codes are $g_3(x, y)$ and $g_5(x, y)$. The ternary $[4, 2, 3]$ Hamming code $\mathcal{H}_{2,3}$ of Examples 1.9.14 and 1.10.3 (often called the **tetracode**) has weight enumerator $g_4(x, y)$, the repetition code of length 2 over $\mathbb{F}_4$ has weight enumerator $g_6(x, y)$, and the $[6, 3, 4]$ **hexacode** over $\mathbb{F}_4 = \{0, 1, \omega, \omega^2\}$ with generator matrix

$$
\begin{bmatrix}
1 & 0 & 0 & 1 & \omega & \omega \\
0 & 1 & 0 & \omega & 1 & \omega \\
0 & 0 & 1 & \omega & \omega & 1
\end{bmatrix}
$$

has weight enumerator $g_7(x, y)$.

Remark 4.2.7 The Gleason polynomials g_3, g_5, and g_7 can be replaced by $g_3'(x, y) = x^4 y^4 (x^4 - y^4)^4$, $g_5'(x, y) = x^3 (x^3 - y^3)^3$, and $g_7'(x, y) = x^2 (x^2 - y^2)^2$ in the corresponding formulas in Gleason's Theorem.

Lemma 4.2.8 ([1555, Lemma 1]) *Let $\mathcal{C}$ be a binary singly-even self-orthogonal code, and let $\mathcal{C}_0$ be its subset of doubly-even codewords. Then $\mathcal{C}_0$ is a linear subcode of index 2 in $\mathcal{C}$.*

Definition 4.2.9 The **shadow** $\mathcal{S}$ of a self-orthogonal binary code $\mathcal{C}$ is

$$
\mathcal{S} = \begin{cases} \mathcal{C}_0^\perp \setminus \mathcal{C}^\perp & \text{if } \mathcal{C} \text{ is singly-even,} \\ \mathcal{C}^\perp & \text{if } \mathcal{C} \text{ is doubly-even.} \end{cases}
$$

Remark 4.2.10 The term 'shadow' was introduced by Conway and Sloane [440] with the purpose to prove a new upper bound for the minimum distance of the binary singly-even self-dual codes as well as some restrictions on their weight enumerators.

Theorem 4.2.11 ([440]) *Let $\mathcal{S}$ be the shadow of the singly-even self-dual code $\mathcal{C}$ of length n and minimum distance d. The dual code $\mathcal{C}_0^\perp$ of the doubly-even subcode $\mathcal{C}_0$ of $\mathcal{C}$, is a union of four cosets of $\mathcal{C}_0$, i.e., $\mathcal{C}_0^\perp = \mathcal{C}_0 \cup \mathcal{C}_1 \cup \mathcal{C}_2 \cup \mathcal{C}_3$, where $\mathcal{C} = \mathcal{C}_0 \cup \mathcal{C}_2$. The following hold.*

(a) $\mathcal{S} = \mathcal{C}_0^\perp \setminus \mathcal{C} = \mathcal{C}_1 \cup \mathcal{C}_3$.

(b) *The sum of any two vectors in $\mathcal{S}$ is a codeword in $\mathcal{C}$. More precisely, if $\mathbf{u}, \mathbf{v} \in \mathcal{C}_1$, then $\mathbf{u} + \mathbf{v} \in \mathcal{C}_0$; if $\mathbf{u} \in \mathcal{C}_1$, $\mathbf{v} \in \mathcal{C}_3$, then $\mathbf{u} + \mathbf{v} \in \mathcal{C}_2$; if $\mathbf{u}, \mathbf{v} \in \mathcal{C}_3$, then $\mathbf{u} + \mathbf{v} \in \mathcal{C}_0$.*

(c) *Let $\mathrm{Hwe}_\mathcal{S}(x, y) = \sum_{r=0}^{n} A_r(\mathcal{S}) x^r y^{n-r}$ be the weight enumerator of the shadow $\mathcal{S}$. Then*

$$
\mathrm{Hwe}_\mathcal{S}(x, y) = \mathrm{Hwe}_\mathcal{C}\left(i \frac{y - x}{\sqrt{2}}, \frac{y + x}{\sqrt{2}}\right),
$$

where $\mathrm{Hwe}_\mathcal{C}(x, y)$ is the weight enumerator of $\mathcal{C}$ and $i = \sqrt{-1}$. Furthermore,

(i) $A_r(\mathcal{S}) = A_{n-r}(\mathcal{S})$ *for $r = 0, 1, \ldots, n$;*

(ii) $A_r(\mathcal{S}) = 0$ *for any $r \not\equiv n/2 \pmod{4}$;*

(iii) $A_0(\mathcal{S}) = 0$;

(iv) $A_r(\mathcal{S}) \leq 1$ *for $r < d/2$; and*

(v) $A_{d/2}(\mathcal{S}) \leq 2n/d$.

Theorem 4.2.12 ([440]) *If the weight enumerator of a self-dual code $\mathcal{C}$ of length n is given by*

$$
\mathrm{Hwe}_\mathcal{C}(x, y) = \sum_{j=0}^{\lfloor \frac{n}{8} \rfloor} c_j (x^2 + y^2)^{\frac{n}{2} - 4j} (x^2 y^2 (x^2 - y^2)^2)^j,
$$

then the weight enumerator of its shadow is

$$\text{Hwe}_{\mathcal{S}}(x,y) = \sum_{j=0}^{\lfloor \frac{n}{8} \rfloor} (-1)^j 2^{\frac{n}{2}-6j} c_j (xy)^{\frac{n}{2}-4j} (x^4 - y^4)^{2j}.$$

Example 4.2.13 By [440, Section V], a consequence of Theorem 4.2.12 is that the largest minimum weight of a self-dual code of length 42 is 8. Additionally, a self-dual $[42, 21, 8]$ code $\mathcal{C}$ and its shadow $\mathcal{S}$ have the following weight enumerators, respectively,

$$\begin{cases} \text{Hwe}_{\mathcal{C}}(x,y) = y^{42} + (84 + 8\beta)x^8 y^{34} + (1449 - 24\beta)x^{10}y^{32} + \cdots, \\ \text{Hwe}_{\mathcal{S}}(x,y) = \beta x^5 y^{37} + (896 - 8\beta)x^9 y^{33} + \cdots, \end{cases} \tag{4.1}$$

$$\begin{cases} \text{Hwe}_{\mathcal{C}}(x,y) = y^{42} + 164 x^8 y^{34} + 697 x^{10}y^{32} + 15088 x^{12}y^{30} + \cdots, \\ \text{Hwe}_{\mathcal{S}}(x,y) = xy^{41} + 861 x^9 y^{33} + \cdots, \end{cases} \tag{4.2}$$

where β is a nonnegative integer [440]. A self-dual $[42, 21, 8]$ code with weight enumerator $\text{Hwe}_{\mathcal{C}}(x,y)$ given by (4.1) exists if and only if $\beta \in \{0, 1, \ldots, 22, 24, 26, 28, 32, 42\}$; see [279]. A self-dual $[42, 21, 8]$ code with weight enumerator $\text{Hwe}_{\mathcal{C}}(x,y)$ given by (4.2) is also known; see [1005].

Remark 4.2.14 Similar definitions for a shadow can be given in some other cases. For example, a shadow is defined if $\mathcal{C}$ is an additive trace-Hermitian self-orthogonal code over $\mathbb{F}_4$ and $\mathcal{C}_0$ is its subcode with even Hamming weights, or if $\mathcal{C}$ is a self-orthogonal code over $\mathbb{Z}_m$ (m even) and $\mathcal{C}_0$ is the subcode with Euclidean norms divisible by $2m$. Definitions and theorems for these codes and their shadows are given in [1555].

Remark 4.2.15 Extremality for some of the parameters always gives interesting problems. In [86] the minimum weights of a code and its shadow are considered simultaneously. In [283], the authors study binary self-dual codes having a shadow with the smallest possible minimum weight.

Theorem 4.2.16 ([86]) *Let $\mathcal{C}$ be a self-dual binary code, assumed not to be doubly-even, of minimum weight d, and let $\mathcal{S}$ be its shadow, of minimum weight s. Then $2d + s \le 4 + n/2$, unless $n \equiv 22 \pmod{24}$ and $d = 4\lfloor n/24 \rfloor + 6$, in which case $2d + s \le 8 + n/2$.*

Definition 4.2.17 ([86]) Let $\mathcal{C}$ be a binary self-dual $[n, n/2, d]$ code whose shadow $\mathcal{S}$ has minimum distance s. If $2d + s = 4 + n/2$ or $2d + s = 8 + n/2$, $\mathcal{C}$ is said to be *s-extremal*.

Definition 4.2.18 ([283]) Let $\mathcal{C}$ be a binary self-dual code of length $n = 24m + 8l + 2r$, $l = 0, 1, 2$, $r = 0, 1, 2, 3$. Suppose the shadow of $\mathcal{C}$ has minimum weight s. Then $\mathcal{C}$ is a code with **minimal shadow** if: (i) $s = r$ when $r > 0$, and (ii) $s = 4$ when $r = 0$.

Remark 4.2.19 Some s-extremal codes and codes with minimal shadow support 1- and 2-designs; see Chapter 5 for appropriate terminology. More properties of the s-extremal codes are given in [86]. Connections between self-dual codes with minimal shadow, combinatorial designs and secret sharing schemes are presented in [282].

4.3 Bounds for the Minimum Distance

Theorem 4.3.1 ([1008, 1555]) *Let $\mathcal{C}$ be an $[n, n/2, d]$ code over $\mathbb{F}_q$, $q = 2, 3, 4$.*

TABLE 4.1: Extremal even formally self-dual binary codes

n	d	♯ sd	♯ even fsd (not sd)	n	d	♯ sd	♯ even fsd (not sd)	n	d	♯ sd	♯ even fsd (not sd)
2	2	1	-	12	4	1	2	22	6	1	41519
4	2	1	-	14	4	1	9	24	8	1	-
6	2	1	1	16	6	-	-	26	8	-	-
8	4	1	-	18	6	-	1	28	8	-	1
10	4	-	1	20	6	-	7	30	8	-	42

(a) *If $q = 2$ and C is formally self-dual and even, then $d \leq 2\lfloor n/8 \rfloor + 2$.*

(b) *If $q = 2$ and C is self-dual doubly-even, then $d \leq 4\lfloor n/24 \rfloor + 4$.*

(c) *If $q = 3$ and C is self-dual, then $d \leq 3\lfloor n/12 \rfloor + 3$.*

(d) *If $q = 4$ and C is Hermitian self-dual, then $d \leq 2\lfloor n/6 \rfloor + 2$.*

In all cases, if equality holds in the bounds, the weight enumerator of C is unique.

Remark 4.3.2 Recall from Remark 4.1.10 that Type I codes are binary singly-even self-dual codes, and Type II codes are binary doubly-even self-dual codes. We say that a ternary self-dual code is **Type III**, and a code over $\mathbb{F}_4$ that is self-dual under the Hermitian inner product is **Type IV**.

Theorem 4.3.3 (Zhang [1555, 1943]) *Let $\mathrm{Hwe}_C(x, y)$ be the weight enumerator of a formally self-dual code C of Type I, or a self-dual code C of Type II, III or IV meeting the corresponding bound from Theorem 4.3.1. Let $c = 2, 4, 3, 2$, respectively, and $\mu = \lfloor n/8 \rfloor$, $\lfloor n/24 \rfloor$, $\lfloor n/12 \rfloor$, $\lfloor n/6 \rfloor$. Then the coefficient $A_{c(\mu+2)}$ in $\mathrm{Hwe}_C(x, y)$ is negative if and only if*

Type I: $n = 8i$ $(i \geq 4)$, $n = 8i + 2$ $(i \geq 5)$, $n = 8i + 4$ $(i \geq 6)$, $n = 8i + 6$ $(i \geq 7)$.

Type II: $n = 24i$ $(i \geq 154)$, $n = 24i + 8$ $(i \geq 159)$, $n = 24i + 16$ $(i \geq 164)$. *In particular, C cannot exist for $n > 3928$.*

Type III: $n = 12i$ $(i \geq 70)$, $n = 12i + 4$ $(i \geq 75)$, $n = 12i + 8$ $(i \geq 78)$. *So code C cannot exist for $n > 932$.*

Type IV: $n = 6i$ $(i \geq 17)$, $n = 6i + 2$ $(i \geq 20)$, $n = 6i + 4$ $(i \geq 22)$. *In particular, C cannot exist for $n > 130$.*

Theorem 4.3.4 ([1008]) *There exist self-dual binary codes of length n meeting the bound of Theorem 4.3.1(a) if and only if $n = 2$, 4, 6, 8, 12, 14, 22, and 24. There exist even formally self-dual binary codes of length n meeting the bound of Theorem 4.3.1(a) if and only if n is even with $n \leq 30$, $n \neq 16$, and $n \neq 26$. For all these codes, the weight enumerator is uniquely determined by the length.*

Remark 4.3.5 The formally self-dual even codes meeting the bound of Theorem 4.3.1(a) are called **extremal**. These extremal codes are classified. The number of all inequivalent extremal formally self-dual even codes is given in Table 4.1 ([275, Table 1]). Table 1 in [912] lists the number of all inequivalent extremal even codes, but for length 30 the number is given by ≥ 6; the classification of the $[30, 15, 8]$ formally self-dual even codes is completed in [275]. In [867], Gulliver and Östergård classified all binary optimal linear $[n, n/2]$ codes up to length 28. Moreover, they proved that at least one linear $[n, n/2, d_{max(n)}]$ code is formally

TABLE 4.2: Type I and II codes of length $2 \leq n \leq 72$

n	d_I	num$_I$	d_{II}	num$_{II}$	n	d_I	num$_I$	d_{II}	num$_{II}$
2	2^O	1			38	8^E	2744		
4	2^O	1			40	8^E	10 200 655	8^E	16 470
6	2^O	1			42	8^E	$\geq 16\,607$		
8	2^O	1	4^E	1	44	8^E	$\geq 395\,555$		
10	2^O	2			46	10^E	1		
12	4^E	1			48	10°	≥ 74	12^E	1
14	4^E	1			50	10°	$\geq 2\,910$		
16	4^E	1	4^E	2	52	10°	≥ 499		
18	4^E	2			54	10°	≥ 54		
20	4^E	7			56	$10(12)$	?	12^E	$\geq 1\,151$
22	6^E	1			58	10°	≥ 101		
24	6°	1	8^E	1	60	12^E	≥ 18		
26	6°	1			62	12^E	≥ 20		
28	6°	3			64	12^E	≥ 22	12^E	$\geq 3\,270$
30	6°	13			66	12^E	≥ 3		
32	8^E	3	8^E	5	68	12^E	≥ 65		
34	6°	938			70	$12(14)$	?		
36	8^E	41			72	$12(14)$	?	$12(16)$	?

self-dual for lengths $n \leq 64$. The smallest length for which a formally self-dual code is not isodual is 14, and there are 28 such codes amongst 6 weight enumerators. Simonis [1715] showed that the extended quadratic residue code of length 18 is the unique $[18, 9, 6]$ code, up to equivalence. Since its dual distance is 6, this code is isodual. Jaffe [1031] proved that there is a unique linear $[28, 14, 8]$ code. In [728] the authors showed that there are exactly 7 inequivalent $[20, 10, 6]$ even extremal formally self-dual binary codes and that there are over 1000 inequivalent $[22, 11, 6]$ even extremal formally self-dual binary codes. The classification for length 22 was completed in [912]. The formally self-dual codes of smaller lengths were constructed and classified in [82, 188, 727, 1105].

Conway and Sloane in [440] obtained a new upper bound for the minimum distance of the singly-even self-dual codes, which is updated by Rains in [1553].

Theorem 4.3.6 (Rains [1553]) *If C is a binary self-dual $[n, n/2, d]$ code then*

$$d \leq 4\lfloor \tfrac{n}{24} \rfloor + 4 \quad \text{if } n \not\equiv 22 \pmod{24},$$
$$d \leq 4\lfloor \tfrac{n}{24} \rfloor + 6 \quad \text{if } n \equiv 22 \pmod{24}.$$

If n is multiple of 24 and $d = 4\lfloor n/24 \rfloor + 4$, the code C is doubly-even.

Definition 4.3.7 A self-dual code is called **extremal** if it meets the bound in Theorem 4.3.6. The code is called **optimal** if there exists no self-dual code with larger minimum distance for the same length.

Remark 4.3.8 All extremal self-dual codes are optimal but the opposite is not true. For example, there are no extremal self-dual codes of length $n = 26$. The situation with the optimal self-dual binary codes of length n for $n = 2, 4, \ldots, 72$ is presented in Table 4.2 which updates [1008, Table 9.1], [1423, Table 11.2], and [1555, Table X]. In the table, d_I is the largest minimum weight for which a Type I code is known to exist while d_{II} is the largest minimum weight for which a Type II code is known to exist. The superscript E indicates that the code is extremal; the superscript O indicates the code is not extremal but optimal. The number of inequivalent Type I and II codes of the given minimum weight is listed under num_I and num_{II}, respectively. The optimal codes of length 34 have been classified in [205] and [21] independently, the extremal codes of length 36 have been classified in [21]. The papers [22] and [277] have presented the classification of the $[38, 19, 8]$ codes using different methods. The singly-even and doubly-even extremal self-dual codes of length 40 have been classified in [277] and [189], respectively. For lengths 42 and 44, the extremal self-dual codes having automorphisms of odd prime order have been classified (see [284, 285]). For the larger lengths we give the same bound as in [1423]; nevertheless many new codes have been constructed so far.

Conjecture 4.3.9 Extremal binary self-dual codes of lengths $n \equiv 2, 4, 6$, and 10 (mod 24) do not exist.

Remark 4.3.10 ([1008]) The $[8, 4, 4]$ extended Hamming code $\widehat{\mathcal{H}}_{3,2}$ of Example 1.7.2 is the unique Type II code of length 8 indicated in Table 4.2. Two of the five $[32, 16, 8]$ Type II codes are extended quadratic residue and Reed–Muller codes. The $[24, 12, 8]$ and $[48, 24, 12]$ binary extended quadratic residue codes are extremal Type II codes, but the binary extended quadratic residue code of length 72 has minimum weight 12 and hence is not extremal. The first two lengths for which the existence of an extremal Type II code is undecided are 72 and 96. For lengths beyond 96, extremal Type II codes are known to exist only for lengths 104 and 136.

Remark 4.3.11 The extremal self-dual codes of length a multiple of 24 are of particular interest. According to Theorem 4.3.6 these codes are doubly-even. Moreover, for any nonzero weight w in such a code, the codewords of weight w hold a 5-design; see Chapter 5. The extended Golay code $\mathcal{G}_{24}$ is the only $[24, 12, 8]$ code and the extended quadratic residue code q_{48} is the only $[48, 24, 12]$ self-dual doubly-even code. The automorphism group of $\mathcal{G}_{24}$ is the 5-transitive Mathieu group M_{24} of order $2^{10} \cdot 3^3 \cdot 5 \cdot 7 \cdot 11 \cdot 23$; see Theorem 1.13.5. The automorphism group of q_{48} is isomorphic to the projective special linear group $\text{PSL}_2(47)$ and has order $2^4 \cdot 3 \cdot 23 \cdot 47$. Both M_{24} and $\text{PSL}_2(47)$ are non-abelian simple groups, and so in particular are not solvable. These two codes are the only known extremal self-dual codes of length a multiple of 24. The existence of an extremal code of length 72 is a long-standing open problem; see Example 3.2.14 for a discussion. A series of papers investigates the structure of its automorphism group excluding most of the subgroups of the symmetric group S_{72} [238, 239, 241, 273, 274, 280, 437, 723, 1010, 1422, 1516, 1525, 1930, 1922, 1931]. The following theorem summarizes the works of Pless, Conway, Thompson, Huffman, Yorgov, Bouyuklieva, O'Brien, Willems, Yankov, Feulner, Nebe, Borello, and Dalla Volta.

Theorem 4.3.12 *Let $\mathcal{C}$ be a self-dual $[72, 36, 16]$ code. Then $\text{PAut}(\mathcal{C})$ is trivial or it is isomorphic to C_2, C_3, $\text{C}_2 \times \text{C}_2$, or C_5 where C_i is the cyclic group of order i.*

Remark 4.3.13 The possible automorphism groups of a putative extremal self-dual code of length 72 are abelian and very small. So this code is almost a rigid object (i.e., without symmetries) and it might be very difficult to find, if it exists.

Remark 4.3.14 Extremal ternary self-dual (Type III) codes do not exist for lengths $n = 72, 96, 120$, and all $n \geq 144$. Extremal Type III codes of length n are known to exist for all n divisible by 4 with $4 \leq n \leq 64$; all other lengths through 140 except 72, 96, and 120 remain undecided. The known results about the extremal ternary codes are summarized in [1005, Table 6].

Remark 4.3.15 By [1943], no extremal Type IV codes exist for lengths $n = 102$, 108, 114, 120, 122, 126, 128, and $n \geq 132$. The existence of extremal codes of even lengths $32 \leq n \leq 130$ except the lengths given above is unknown. We refer to [1005] for more details about the extremal and optimal Type IV codes.

For the definition of t-designs used in the next theorem, see Chapter 5.

Theorem 4.3.16 ([1008]) *The following results on t-designs hold in extremal codes of Types II, III, and IV.*

(a) *Let $\mathcal{C}$ be a $[24m+8\mu, 12m+4\mu, 4m+4]$ extremal Type II code for $\mu = 0$, 1, or 2. Then codewords of any fixed weight except 0 hold t-designs for the following parameters:*

    (i) $t = 5$ *if $\mu = 0$ and $m \geq 1$,*

    (ii) $t = 3$ *if $\mu = 1$ and $m \geq 0$, and*

    (iii) $t = 1$ *if $\mu = 2$ and $m \geq 0$.*

(b) *Let $\mathcal{C}$ be a $[12m + 4\mu, 6m + 2\mu, 3m + 3]$ extremal Type III code for $\mu = 0$, 1, or 2. Then codewords of any fixed weight i with $3m + 3 \leq i \leq 6m + 3$ hold t-designs for the following parameters:*

    (i) $t = 5$ *if $\mu = 0$ and $m \geq 1$,*

    (ii) $t = 3$ *if $\mu = 1$ and $m \geq 0$, and*

    (iii) $t = 1$ *if $\mu = 2$ and $m \geq 0$.*

(c) *Let $\mathcal{C}$ be a $[6m + 2\mu, 3m + \mu, 2m + 2]$ extremal Type IV code for $\mu = 0$, 1, or 2. Then codewords of any fixed weight i with $2m + 2 \leq i \leq 3m + 2$ hold t-designs for the following parameters:*

    (i) $t = 5$ *if $\mu = 0$ and $m \geq 2$,*

    (ii) $t = 3$ *if $\mu = 1$ and $m \geq 1$, and*

    (iii) $t = 1$ *if $\mu = 2$ and $m \geq 0$.*

4.4 Construction Methods

4.4.1 Gluing Theory

Remark 4.4.1 Gluing is a method to construct self-dual codes from shorter self-orthogonal codes systematically. The construction begins with a direct sum $\mathcal{C}_1 \oplus \mathcal{C}_2 \oplus \cdots \oplus \mathcal{C}_t$ of self-orthogonal codes which forms a subcode of the self-dual code $\mathcal{C}$. If $\mathcal{C}$ is equal to such a direct sum, it is called **decomposable**. Otherwise, additional vectors, called **glue vectors**, have to be added. A generator matrix G for a code formed by gluing components $\mathcal{C}_1, \ldots, \mathcal{C}_t$ together has the form

$$G = \left[\begin{array}{cccc} G_1 & 0 & \cdots & 0 \\ 0 & G_2 & \cdots & 0 \\ & & \ddots & \\ 0 & 0 & \cdots & G_t \\ \hline & & X & \end{array} \right],$$

where G_i is a generator matrix of $\mathcal{C}_i$, $1 \leq i \leq t$, and X denotes the rest of the generator matrix G. The glue vectors are rows of the matrix X. Gluing theory is described in detail in many books; see, for example, [1008, 1555].

Example 4.4.2 Construct the self-dual $[14, 7, 4]$ code using gluing. We choose two gluing components $\mathcal{C}_1 \cong \mathcal{C}_2 \cong e_7$ where e_7 is the $[7, 3, 4]$ code $\mathcal{H}_{3,2}^\perp$, the dual of the Hamming code $\mathcal{H}_{3,2}$; see Example 1.5.3. We need one glue vector. Since $(0101010) + e_7$ is a coset in $e_7^\perp$ with minimum weight 3, we can take for a glue vector $\mathbf{v} = (0101010|0101010)$. So we have a generator matrix of the code in the form

$$G = \left[\begin{array}{c|c}
0001111 & 0000000 \\
0110011 & 0000000 \\
1010101 & 0000000 \\
\hline
0000000 & 0001111 \\
0000000 & 0110011 \\
0000000 & 1010101 \\
\hline
0101010 & 0101010
\end{array} \right].$$

4.4.2 Circulant Constructions

Remark 4.4.3 An $m \times m$ matrix A is **circulant** provided

$$A = \left[\begin{array}{cccc}
a_1 & a_2 & \cdots & a_m \\
a_m & a_1 & \cdots & a_{m-1} \\
& & \vdots & \\
a_2 & a_3 & \cdots & a_1
\end{array} \right].$$

An $(m + 1) \times (m + 1)$ matrix B is **bordered circulant** if

$$B = \left[\begin{array}{cc}
\alpha & \beta\beta\cdots\beta \\
\gamma & \\
\vdots & A \\
\gamma &
\end{array} \right],$$

where A is a circulant matrix.

Definition 4.4.4 We say that a code has a **double circulant generator matrix** or a **bordered double circulant generator matrix** provided it has a generator matrix of the form

$$[I_m \quad A] \quad \text{or} \quad [I_{m+1} \quad B],$$

where A is an $m \times m$ circulant matrix and B is an $(m + 1) \times (m + 1)$ bordered circulant matrix, respectively. A code has a **double circulant construction** or a **bordered double circulant construction** provided it has a double circulant or bordered double circulant generator matrix, respectively.

The proof of the following result is straightforward from the definitions.

Proposition 4.4.5 *A code with a double circulant construction is isodual. A code with a bordered double circulant construction is isodual provided the bordered matrix used in the construction satisfies either $\beta = \gamma = 0$ or both β and γ are nonzero.*

Remark 4.4.6 All extremal binary self-dual codes of lengths up to 90 with circulant and bordered circulant constructions are classified in a series of papers by Gulliver and Harada [864, 865, 905].

Remark 4.4.7 For the structure of the quasi-cyclic self-dual codes we refer to the tetralogy of papers of Ling, Niederreiter, and Solé [1268, 1269, 1271, 1272] and Chapter 7.

Remark 4.4.8 Double negacirculant (DN) codes are the analogs in odd characteristic of double circulant codes. Four-negacirculant codes are introduced in [906]. Self-dual negacirculant and other generalizations of quasi-cyclic codes are studied by many authors; see, for example, [29, 866, 906].

4.4.3   Subtraction Procedure

Proposition 4.4.9 *Let C be a binary self-dual $[n, k, d]$ code with $n = 2k > 2$ and $C_0 = \{x_1 x_2 \cdots x_n \in C \mid x_{n-1} = x_n\}$. If the last two coordinates of C are not equal and C_1 is the punctured code of C_0 on the coordinate set $\{n-1, n\}$, then C_1 is a self-dual $[n-2, k-1, d_1]$ code with $d_1 \geq d - 2$.*

Remark 4.4.10 Any two coordinates of a self-dual code of length $n > 2$ and minimum distance $d > 2$ are not equal. But even if a self-dual code of length $n > 2$ has minimum distance $d = 2$, it has two coordinates which are not equal.

Remark 4.4.11 Any self-dual $[n, n/2, 2]$ code C_n for $n > 2$ is decomposable as $C_n = i_2 \oplus C_{n-2}$ where C_{n-2} is a self-dual code of length $n - 2$ and i_2 is the binary repetition code of length 2. Furthermore, up to equivalence, the number of the self-dual $[n, n/2, 2]$ codes is equal to the number of all self-dual $[n - 2, n/2 - 1]$ codes for $n > 2$)

4.4.4   Recursive Constructions

There are several methods to construct self-dual codes of length $n + 2$ from self-dual codes of length n. In [22], the authors describe three such methods calling them the **recursive construction**, the **Harada–Munemasa Construction**, and the **Building-up Construction**. The first one (recursive construction) gives all inequivalent self-dual $[n+2, n/2+1, d+2]$ codes starting from the self-dual $[n, n/2, d]$ codes [910]. The other two constructions give all self-dual codes of length $n + 2$ from the self-dual codes of length n, which we now describe.

Proposition 4.4.12 (Harada–Munemasa Construction [910]) *Let G_1 be a generator matrix of a binary self-dual $[n, n/2, d]$ code C_1. Then the matrix*

$$G_2 = \begin{bmatrix} a_1 & a_1 & \\ \vdots & \vdots & G_1 \\ a_{n/2} & a_{n/2} & \end{bmatrix}$$

where $a_i \in \mathbb{F}_2$, $1 \leq i \leq n/2$, generates a binary self-orthogonal $[n+2, n/2, \geq d]$ code C_2. Moreover, if $z_2 \in C_2^\perp \setminus C_2$, then $C = C_2 \cup (z_2 + C_2)$ is a self-dual code of length $n + 2$.

Remark 4.4.13 The Harada–Munemasa Construction is a binary version of Huffman's construction presented in [1003]. Similar constructions can be used for codes over different fields.

Proposition 4.4.14 (Huffman Construction [1003]) *Let G_1 be a generator matrix of a Hermitian self-dual $[n, n/2, d]_4$ code C_1. Then the matrix*

$$G_2 = \begin{bmatrix} a_1 & a_1 & \\ \vdots & \vdots & G_1 \\ a_{n/2} & a_{n/2} & \end{bmatrix}$$

where $a_i \in \mathbb{F}_4$, $1 \le i \le n/2$, generates a Hermitian self-orthogonal $[n + 2, n/2, \ge d]_4$ code C_2. Moreover, if $\mathbf{z_2} \in C_2^\perp \setminus C_2$, then $C = C_2 \cup (\mathbf{z_2} + C_2)$ is a Hermitian self-dual code of length $n + 2$.

Theorem 4.4.15 (Harada Construction [904]) *Let $\mathbf{x} = x_1 x_2 \cdots x_n$ be a vector in $\mathbb{F}_2^n$ such that $\mathrm{wt_H}(\mathbf{x}) \equiv n + 1 \pmod d$, and $G_0 = \begin{bmatrix} I_n & A \end{bmatrix}$ be a generator matrix of a binary self-dual code C_0 of length $2n$. Then the matrix*

$$G = \begin{bmatrix} 1 & 0 & x_1 \cdots x_n & 1 \cdots 1 \\ y_1 & y_1 & & \\ \vdots & \vdots & I_n & A \\ y_n & y_n & & \end{bmatrix},$$

where $y_i = x_i + 1 \pmod 2$, $i = 1, \ldots, n$, generates a self-dual code C of length $2n + 2$.

Theorem 4.4.16 (Building-up Construction [1119]) *Let $G_0 = [\mathbf{r_i}]$ be a generator matrix (may not be in standard form) of a binary self-dual code C_0 of length n, where $\mathbf{r_i}$ is a row of G_0 for $1 \le i \le n/2$. Let $\mathbf{x} \in \mathbb{F}_2^n$ be a binary vector with an odd weight. Define $y_i := \mathbf{x} \cdot \mathbf{r_i}$ for $1 \le i \le n/2$, where $\cdot$ denotes the standard inner product. Then the matrix*

$$G = \begin{bmatrix} 1 & 0 & \mathbf{x} \\ y_1 & y_1 & \\ \vdots & \vdots & G_0 \\ y_n & y_n & \end{bmatrix}$$

generates a self-dual binary code C of length $n + 2$.

Theorem 4.4.17 ([1119]) *Any self-dual code C over $\mathbb{F}_2$ of length n with minimum weight $d > 2$ is obtained from some self-dual code C_0 of length $n - 2$ (up to equivalence) by the construction in Theorem 4.4.16.*

Remark 4.4.18 Choosing G_0 as $G_0 = \begin{bmatrix} I_n & A \end{bmatrix}$ and $\mathbf{x} = x_1 x_2 \cdots x_{n/2} 1 1 \cdots 1$ in Theorem 4.4.16, we obtain the Harada Construction as a corollary.

Remark 4.4.19 Let C be a binary self-dual code and C_d be its subcode generated by the set of codewords with minimum weight d. If G_d is a generator matrix of C_d with rank k, and G_E is an $(n/2 - k) \times n$ matrix such that $G = \begin{bmatrix} G_d \\ G_E \end{bmatrix}$ is a generator matrix of C, then the matrix

$$\begin{bmatrix} \begin{array}{cc|c} 1 & 0 & \mathbf{x} \\ \hline 1 & 1 & \\ \vdots & \vdots & G_d \\ 1 & 1 & \\ \hline a_1 & a_1 & \\ \vdots & \vdots & G_E \\ a_{n/2-k} & a_{n/2-k} & \end{array} \end{bmatrix},$$

where $a_i \in \mathbb{F}_2$, $1 \leq i \leq n/2 - k$, and the even weight vector $10\mathbf{x} \in \mathbb{F}_2^{n+2}$ is orthogonal to all other rows of the matrix, generates a self-dual code. To obtain only $[n + 2, n/2 + 1, d + 2]$ codes, the authors of [22] take only those vectors $\mathbf{x} \in \mathcal{C}^\perp \setminus \mathcal{C}$ for which the constructed code has minimum distance $d + 2$.

4.4.5 Constructions of Codes with Prescribed Automorphisms

Huffman and Yorgov (see [999, 1928, 1929]) developed a method for constructing binary self-dual codes via an automorphism of odd prime order. Their method was extended by other authors for automorphisms of odd composite order and for automorphisms of order 2 [272, 281, 616].

Huffman has also studied the properties of the linear codes over $\mathbb{F}_q$, having an automorphism of prime order p coprime with q [1000]. Further, he has continued with Hermitian and additive self-dual codes over $\mathbb{F}_4$ [1001, 1006], and with self-dual codes over rings [1004, 1007].

Let $\mathcal{C}$ be a binary self-dual code of length n with an automorphism σ of prime order $p \geq 3$ with exactly c independent p-cycles and $f = n - cp$ fixed points in its decomposition. We may assume that

$$\sigma = (1, 2, \cdots, p)(p+1, p+2, \cdots, 2p) \cdots ((c-1)p+1, (c-1)p+2, \cdots, cp), \qquad (4.3)$$

and say that σ is of **type** $p\text{-}(c, f)$. We present the main theorems about the structure of such a code. This structure has been used by many authors in order to construct optimal self-dual codes with different parameters.

Theorem 4.4.20 ([999]) *Let $\mathcal{C}$ be a binary $[n, n/2]$ code with automorphism σ from (4.3). Let $\Omega_1 = \{1, 2, \ldots, p\}, \ldots, \Omega_c = \{(c-1)p+1, (c-1)p+2, \ldots, cp\}$ denote the cycles of σ, and let $\Omega_{c+1} = \{cp+1\}, \ldots, \Omega_{c+f} = \{cp+f = n\}$ be the fixed points of σ. Define*

$$F_\sigma(\mathcal{C}) = \{\mathbf{v} \in \mathcal{C} \mid \sigma(\mathbf{v}) = \mathbf{v}\},$$
$$E_\sigma(\mathcal{C}) = \{\mathbf{v} \in \mathcal{C} \mid \mathrm{wt_H}(\mathbf{v}_{|\Omega_i}) \equiv 0 \pmod 2, \ i = 1, 2, \ldots, c + f\},$$

where $\mathbf{v}_{|\Omega_i}$ is the restriction of $\mathbf{v}$ on Ω_i. Then $\mathcal{C} = F_\sigma(\mathcal{C}) \oplus E_\sigma(\mathcal{C})$, $\dim(F_\sigma(\mathcal{C})) = \frac{c+f}{2}$, and $\dim(E_\sigma(\mathcal{C})) = \frac{c(p-1)}{2}$.

Theorem 4.4.21 ([1928]) *Let $\mathcal{C}$ be a binary $[n, n/2]$ code with automorphism σ from (4.3). Let $\pi : F_\sigma(\mathcal{C}) \to \mathbb{F}_2^{c+f}$ be the projection map, where, for $\mathbf{v} \in F_\sigma(\mathcal{C})$, $(\pi(\mathbf{v}))_i = v_j$ for some $j \in \Omega_i$, $i = 1, 2, \ldots, c + f$. Let $\mathcal{E}$ (respectively $\mathcal{P}$) be the set of all even-weight vectors in $\mathbb{F}_2^p$ (respectively even-weight polynomials in $\mathbb{F}_2[x]/\langle x^p - 1 \rangle$). Define $\varphi' : \mathcal{E} \to \mathcal{P}$ by $\varphi'(v_0 v_1 \cdots v_{p-1}) = v_0 + v_1 x + \cdots + v_{p-1} x^{p-1}$. Let $E_\sigma(\mathcal{C})^*$ be $E_\sigma(\mathcal{C})$ punctured on all the fixed points of σ. Define $\varphi : E_\sigma(\mathcal{C})^* \to \mathcal{P}^c$ by $\varphi(\mathbf{v}) = (\varphi'(\mathbf{v}_{|\Omega_1}), \varphi'(\mathbf{v}_{|\Omega_2}), \ldots, \varphi'(\mathbf{v}_{|\Omega_c}))$ for $\mathbf{v} \in E_\sigma(\mathcal{C})^* \subseteq \mathcal{E}^c$. Then $\mathcal{C}$ is self-dual if and only if the following two conditions hold:*

(a) *$\mathcal{C}_\pi = \pi(F_\sigma(\mathcal{C}))$ is a binary self-dual code of length $c + f$, and*

(b) *for every two vectors $\mathbf{u}, \mathbf{v} \in \mathcal{C}_\varphi = \varphi(E_\sigma(\mathcal{C})^*)$, we have $\sum_{i=1}^{c} u_i(x) v_i(x^{-1}) = 0$ where $u_i(x) = \varphi'(\mathbf{u}_{|\Omega_i})$ and $v_i(x) = \varphi'(\mathbf{v}_{|\Omega_i})$ for $i = 1, 2, \ldots, c$.*

Theorem 4.4.22 ([999]) *Let 2 be a primitive root modulo p. Then the binary code $\mathcal{C}$ with an automorphism σ is self-dual if and only if the following two conditions hold:*

(a) *$\mathcal{C}_\pi$ is a self-dual binary code of length $c + f$, and*

(b) $C_\varphi = \varphi(E_\sigma(C)^*)$ *is a self-dual code of length c over the field* $\mathcal{P}$ *under the inner product*

$$(\mathbf{u}, \mathbf{v}) = \sum_{i=1}^{c} u_i v_i^{2^{(p-1)/2}}.$$

Example 4.4.23 Consider the extended Hamming code $\widehat{\mathcal{H}}_{3,2}$ of Example 1.7.2. It has parameters $[8, 4, 4]$ and automorphism group of order $7 \cdot 3 \cdot 2^6$. Let σ be an automorphism of $\widehat{\mathcal{H}}_{3,2}$ of type 3-(2,2). Then C_π is the binary self-dual $[4, 2, 2]$ code with generator matrix $\begin{bmatrix} 1 & 0 & 1 & 0 \\ 0 & 1 & 0 & 1 \end{bmatrix}$, and C_φ is the $[2, 1, 2]_4$ code with generator matrix $\begin{bmatrix} e & e \end{bmatrix}$, where e is the identity element of the field $\mathcal{P} \cong \mathbb{F}_4$. Here $\mathcal{P} = \{0, x + x^2, 1 + x^2, 1 + x\}$ is the field $\mathbb{F}_4 = \{0, 1, \omega, \omega^2\}$ with $1 \leftrightarrow e = x + x^2$, $\omega \leftrightarrow 1 + x^2$, and $\omega^2 \leftrightarrow 1 + x$. Hence we can write a generator matrix for $\widehat{\mathcal{H}}_{3,2}$ in the form

$$\left[\begin{array}{cc|cc} 111 & 000 & 1 & 0 \\ 000 & 111 & 0 & 1 \\ \hline 011 & 011 & 0 & 0 \\ 101 & 101 & 0 & 0 \end{array}\right].$$

The first two rows of this matrix come from the generator matrix for C_π with the left two columns replaced by 000 and 111 corresponding to the two 3-cycles of σ. The bottom two rows come from the row of the generator matrix for C_φ followed by that row multiplied by $\omega \leftrightarrow 1 + x^2$.

4.5 Enumeration and Classification

Remark 4.5.1 The main tool to classify self-dual codes is based on the so-called *mass formula* which gives the possibility of checking whether the classification is correct. The number of the self-dual binary codes of even length n is $N(n) = \prod_{i=1}^{n/2-1}(2^i + 1)$. If $\mathcal{C}$ has length n, then the number of codes equivalent to $\mathcal{C}$ is $n!/|\mathrm{PAut}(\mathcal{C})|$. To classify binary self-dual codes of length n, it is necessary to find inequivalent self-dual codes $\mathcal{C}_1, \ldots, \mathcal{C}_r$ so that the following mass formula holds:

$$N(n) = \sum_{i=1}^{r} \frac{n!}{|\mathrm{PAut}(\mathcal{C}_i)|}.$$

There are such formulas for all families of self-dual and also of self-orthogonal codes. Detailed information is presented in [1008, 1555]. See also Proposition 7.5.1.

Theorem 4.5.2 *We have the following mass formulas.*

(a) *For self-dual binary codes of even length* n,

$$\sum_j \frac{n!}{|\mathrm{PAut}(\mathcal{C}_j)|} = \prod_{i=1}^{n/2-1} (2^i + 1).$$

(b) *For doubly-even self-dual binary codes of length* $n \equiv 0 \pmod 8$,

$$\sum_j \frac{n!}{|\mathrm{PAut}(\mathcal{C}_j)|} = \prod_{i=1}^{n/2-2} (2^i + 1).$$

(c) *For self-dual ternary codes of length* $n \equiv 0 \pmod 4$,

$$\sum_j \frac{2^n n!}{|\mathrm{MAut}(\mathcal{C}_j)|} = 2 \prod_{i=1}^{n/2-1} (3^i + 1).$$

(d) *For Hermitian self-dual codes over* $\mathbb{F}_4$ *of even length* n,

$$\sum_j \frac{2 \cdot 3^n n!}{|\Gamma\mathrm{Aut}(\mathcal{C}_j)|} = \prod_{i=1}^{n/2-1} (2^{2i+1} + 1).$$

In each case, the summation is over all j, *where* $\{\mathcal{C}_j\}$ *is a complete set of representatives of inequivalent codes of the given type. The automorphism group* $\Gamma\mathrm{Aut}(\mathcal{C}_j)$ *is the set of all semi-linear monomial transformations from* $\mathbb{F}_4^n$ *to* $\mathbb{F}_4^n$ *that fix* $\mathcal{C}_j$; *see* [1008, Section 1.7].

Example 4.5.3 Consider the binary self-dual codes of length 8. The codes i_2^4, which is the direct sum of 4 $[2,1,2]_2$ repetition codes, and the extended Hamming code $\widehat{\mathcal{H}}_{3,2}$ are of this type. Since $\mathrm{PAut}(i_2^4) \cong \mathbb{Z}_2^4 \wr S_4$ and $|\mathrm{PAut}(\widehat{\mathcal{H}}_{3,2})| = 7 \cdot 3 \cdot 64$, we have

$$\frac{8!}{2^4 \cdot 4!} + \frac{8!}{7 \cdot 3 \cdot 64} = 105 + 30 = 135 = (2+1)(4+1)(8+1).$$

This proves, by Theorem 4.5.2(a), that these two codes are all of the inequivalent binary self-dual codes of length 8.

Lemma 4.5.4 (Thompson [1800]) *Let* n *be an even positive integer. Let* $\mathrm{Hwe}_{\mathcal{C}}(x) = \mathrm{Hwe}_{\mathcal{C}}(x, 1)$ *where* $\mathrm{Hwe}_{\mathcal{C}}(x, y)$ *is the weight enumerator of a code* $\mathcal{C}$. *Then*

$$\sum_{\mathcal{C}} \mathrm{Hwe}_{\mathcal{C}}(x) = (1 + x^n) \prod_{i=1}^{n/2-1} (2^i + 1) + \sum_{j=1}^{n/2-1} \binom{n}{2j} \prod_{i=1}^{n/2-2} (2^i + 1) x^{2j},$$

where $\mathcal{C}$ *runs through the set of all self-dual codes of length* n.

Theorem 4.5.5 ([910]) *Let* n *and* d *be even positive integers and let* U *be a family of inequivalent binary self-dual codes of length* n *and minimum weight at most* d. *Then* U *is a complete set of representatives for equivalence classes of self-dual codes of length* n *and minimum weight at most* d *if and only if*

$$\sum_{\mathcal{C} \in U} \frac{n!}{|\mathrm{PAut}(\mathcal{C})|} |\{\mathbf{v} \in \mathcal{C} \mid \mathrm{wt}_{\mathrm{H}}(\mathbf{v}) = d\}| = \binom{n}{d} \prod_{i=1}^{n/2-2} (2^i + 1).$$

Remark 4.5.6 The last formula gives us the possibility of checking the result even if only the self-dual codes of length n and minimum distance $\leq d < 4\lfloor n/24 \rfloor + 4$ are constructed.

The classification of binary self-dual codes began in the seventies in the work of Vera Pless [1515], where she classified the codes of length $n \leq 20$. The self-dual codes of lengths 22 and 24 were classified by Pless and Sloane [1523], and for lengths $n = 26, 28, 30$, by Conway and Pless [436]. The classification of the doubly-even self-dual codes of length 32 was presented in the paper [436]. All these results are summarized in the survey of Conway, Pless and Sloane [439]. To construct the codes, the authors used gluing theory.

TABLE 4.3: Binary self-dual codes of length $n \leq 40$

n	$\sharp_I$	$\sharp_{II}$	$d_{max,I}$	$\sharp_{max,I}$	$d_{max,II}$	$\sharp_{max,II}$
2	1		2	1		
4	1		2	1		
6	1		2	1		
8	1	1	2	1	4	1
10	2		2	2		
12	3		4	1		
14	4		4	1		
16	5	2	4	1	4	2
18	9		4	2		
20	16		4	7		
22	25		6	1		
24	46	9	6	1	8	1
26	103		6	1		
28	261		6	3		
30	731		6	13		
32	3 210	85	8	3	8	5
34	24 147		6	938		
36	519 492		8	41		
38	38 682 183		8	2 744		
40	8 250 058 081	94 343	8	10 200 655	8	16 470

The classification of binary self-dual codes of length 34 was given in [205]. Harada and Munemasa in [910] completed the classification of the self-dual codes of length 36 and created a database of self-dual codes [908]. The doubly-even self-dual codes of length 40 were classified by Betsumiya, Harada and Munemasa [189]. Moreover, using these codes, they classified the optimal self-dual $[38, 19, 8]$ codes. All binary self-dual codes of lengths 38 and 40 were classified in [276] and [269], respectively.

Table 4.3 summarizes the classification of binary self-dual codes of length $n \leq 40$. The number of all inequivalent singly-even (Type I) and doubly-even (Type II) codes is denoted by $\sharp_I$ and $\sharp_{II}$, respectively. In the table $d_{max,I}$ ($d_{max,II}$) is the largest minimum distance for which singly-even (doubly-even) self-dual codes of the corresponding length exist, and $\sharp_{max,I}$ ($\sharp_{max,II}$) is their number.

Remark 4.5.7 Actually, the construction itself is easy; the equivalence test is the difficult part of the classification. Because of that, for larger lengths the recursive constructions that are used are preferably heuristic, for building examples for codes with some properties. There are also many partial classifications, namely classifications of self-dual codes with special properties, for example self-dual codes invariant under a given permutation, or self-dual codes connected with combinatorial designs with given parameters. These classifications are not recursive but they use codes with smaller lengths; that is why the full classification is very important in these cases, too.

Remark 4.5.8 In [276], the authors present an algorithm for generating binary self-dual codes which gives as output exactly one representative of every equivalence class. To develop this algorithm, the authors use an approach introduced by Brendan McKay known as **isomorph-free exhaustive generation** [1369]. The constructive part of the algorithm is not different from the other recursive constructions for self-dual codes, but to take only one representative of any equivalence class, we use a completely different technique. This approach changes extremely the speed of generation of the inequivalent codes of a given length. Its special feature is that there is practically no equivalence test for the objects.

Conjecture 4.5.9 Denote by SD_k the number of all inequivalent binary self-dual codes of dimension k. The sequence $\{a_k = SD_k(2k)!/\prod_{i=1}^{k-1}(2^i + 1)\}$ is decreasing for $k \geq 10$.

Remark 4.5.10 For classification results for families of self-dual codes over larger fields we refer to [1005, 1008, 1555].

Chapter 5

Codes and Designs

Vladimir D. Tonchev

Michigan Technological University

5.1   Introduction

Combinatorial designs often arise in codes that are optimal with respect to certain bounds. This chapter summarizes some important links between codes and combinatorial designs. Other references that discuss various relations between codes and designs are [71, 1008, 1814, 1816].

In Section 5.2, designs associated with supports of codewords of given nonzero weight and how the regularity of such designs is related to transitivity properties of the automorphism group of the code are discussed.

In Section 5.3, a necessary and sufficient condition for a code to be perfect, that is, to attain the Sphere Packing Bound, is given in terms of a design supported by the codewords of minimum weight. In particular, the codewords of minimum weight in any binary perfect single-error-correcting code (linear or nonlinear) that contains the zero vector, support a Steiner triple system, while the minimum weight codewords in the extended code support a Steiner quadruple system.

The celebrated Assmus–Mattson Theorem that provides sufficient conditions for the codewords of given weight in a linear code to support a t-design is discussed in Section 5.4, together with examples of designs obtained from extremal self-dual codes.

Section 5.5 contains Delsarte's generalization of the Assmus–Mattson Theorem for non-linear codes. Delsarte's Theorem yields designs from nonlinear binary codes meeting the Johnson Bound, and 3-designs from the Preparata, Kerdock and Goethals codes in partic-ular. Two infinite classes of 4-designs obtained from Preparata codes are also presented.

The main topic of Section 5.6 is combinatorial designs supported by linear codes that can be used for majority logic decoding of their dual codes. Hamada's Conjecture about the minimum p-rank of designs derived from a finite projective or affine geometry, as well as the proved cases of the conjecture and the known counter-examples, are discussed.

The last Section 5.7 gives references to other links between codes and designs that are not covered in this chapter.

Definition 5.1.1 A **combinatorial design** D (or a **design** for short) is a pair of a finite set $X = \{x_i\}_{i=1}^v$ of **points** and a collection $\mathcal{B} = \{B_j\}_{j=1}^b$ of subsets $B_j \subseteq X$ called **blocks** [187].

Definition 5.1.2 A design $D = (X, \mathcal{B})$ is a **t-(v, k, λ) design**, where $v \geq k \geq t$ and λ are nonnegative integers, if every t-subset of X is contained in exactly λ blocks.

Any t-(v, k, λ) design is also an s-(v, k, λ_s) design for every $0 \leq s \leq t$ where λ_s is given by

$$\lambda_s = \frac{\binom{v-s}{t-s}}{\binom{k-s}{t-s}} \lambda. \tag{5.1}$$

In particular, the number of blocks b is equal to

$$b = \lambda_0 = \frac{\binom{v}{t}}{\binom{k}{t}} \lambda. \tag{5.2}$$

The **incidence matrix** of a design is a $(0, 1)$-matrix $A = (A_{i,j})$, with rows labeled by the blocks and columns labeled by the points, such that $A_{i,j} = 1$ if the ith block contains the jth point, and $A_{i,j} = 0$ otherwise. A design is **simple** if there are no repeated blocks, or equivalently, if all rows of its incidence matrix are distinct.

Two designs are **isomorphic** if there is a bijection between their point sets that maps the blocks of the first design to blocks of the second design. An **automorphism** of a design is a permutation of the points that preserves the collection of blocks. The set of all automorphisms of a design D is a permutation group called the **automorphism group** of D.

The **complementary** design $\overline{D}$ of a design D is obtained by replacing every block of D by its complement.

A **Steiner design** (or **Steiner system**) is a t-design with $\lambda = 1$. An alternative notation used for a Steiner t-$(v, k, 1)$ design is $\mathrm{S}(t, k, v)$. A **Steiner triple system** $\mathrm{STS}(v)$ is a Steiner 2-design with block size 3. A **Steiner quadruple system** $\mathrm{SQS}(v)$ is a Steiner 3-design with block size 4.

5.2   Designs Supported by Codes

The **support** of a nonzero vector $\mathbf{x} = x_1 \cdots x_n \in \mathbb{F}_q^n$ is the set of indices of its nonzero coordinates: $\mathrm{supp}(\mathbf{x}) = \{i \mid x_i \neq 0\}$.

Definition 5.2.1 A design D is **supported** by a block code $\mathcal{C}$ of length n if the points of D are labeled by the n coordinates of $\mathcal{C}$, and every block of D is the support of some nonzero codeword of $\mathcal{C}$.

Remark 5.2.2 If $\mathcal{C}$ is a linear code over a finite field of order $q > 2$, and $\mathbf{c}$ is a codeword of weight $w > 0$, all $q - 1$ nonzero scalar multiples of $\mathbf{c}$ have the same support. To avoid repeated blocks, we associate only one block with all scalar multiples of $\mathbf{c}$. Suppose that D is a t-(n, w, λ) design supported by a linear q-ary code $\mathcal{C}$. It follows that the number of blocks b of D is smaller than or equal to $A_w/(q-1)$, where A_w is the number of codewords of weight w. If the support of every codeword of weight w is a block of D, then we have

$$b = A_w/(q-1), \tag{5.3}$$

and the parameter λ can be computed using (5.2) and (5.3):

$$\lambda = \frac{A_w}{q-1} \cdot \frac{\binom{w}{t}}{\binom{n}{t}}. \tag{5.4}$$

Theorem 5.2.3 *If a code is invariant under a monomial group that acts t-transitively or t-homogeneously on the set of coordinates, the supports of the codewords of any nonzero weight form a t-design.*

Corollary 5.2.4 *If C is a cyclic code of length n, the supports of all codewords of any nonzero weight w form a 1-design.*

Example 5.2.5 The binary Hamming code of length 7 is a linear cyclic code with generator polynomial $g(x) = 1 + x + x^3$ and weight distribution

$$A_0 = A_7 = 1, \ A_3 = A_4 = 7.$$

By Corollary 5.2.4 and equations (5.1) and (5.4), the sets of codewords of weight 3 and 4 support 1-designs with parameters 1-$(7,3,3)$ and 1-$(7,4,4)$, respectively. The full automorphism group of the code is of order 168 and acts 2-transitively on the set of coordinates. By Theorem 5.2.3, the 1-designs supported by the codewords of weight 3 and 4 are actually 2-designs, with parameters 2-$(7,3,1)$ and 2-$(7,4,2)$, respectively. The incidence matrix of the 2-$(7,3,1)$ design supported by the codewords of weight 3 is

$$\begin{bmatrix} 1 & 1 & 0 & 1 & 0 & 0 & 0 \\ 0 & 1 & 1 & 0 & 1 & 0 & 0 \\ 0 & 0 & 1 & 1 & 0 & 1 & 0 \\ 0 & 0 & 0 & 1 & 1 & 0 & 1 \\ 1 & 0 & 0 & 0 & 1 & 1 & 0 \\ 0 & 1 & 0 & 0 & 0 & 1 & 1 \\ 1 & 0 & 1 & 0 & 0 & 0 & 1 \end{bmatrix}.$$

Example 5.2.6 The $[24, 12, 8]$ extended binary Golay code $\mathcal{G}_{24}$ is invariant under the 5-transitive Mathieu group M_{24}, while the $[12, 6, 6]$ extended ternary Golay code $\mathcal{G}_{12}$ is invariant under $\widetilde{M}_{12}$, a central extension of the Mathieu group M_{12} that acts 5-transitively on the coordinates; see Section 1.13. Both codes support 5-designs by Theorem 5.2.3.

Example 5.2.7 The binary Reed–Muller code $\mathcal{RM}(r, m)$ of length 2^m ($m \geq 2$) and order r ($1 \leq r \leq m-1$) is invariant under the 3-transitive group of all affine transformations of the 2^m-dimensional binary vector space. Consequently, the codewords of any nonzero weight $w < 2^m$ support a 3-design.

5.3 Perfect Codes and Designs

The size of a code $\mathcal{C}$ of length n and minimum Hamming distance $d = 2e + 1$ over an alphabet of size q is bounded from above by the Sphere Packing Bound:

$$|\mathcal{C}| \leq \frac{q^n}{\sum_{i=0}^{e} \binom{n}{i}(q-1)^i}. \tag{5.5}$$

A code meeting the equality in (5.5) is called **perfect**.

Trivial examples of perfect codes are: the zero vector of length n $(e = n)$; the n-dimensional vector space $\mathbb{F}_q^n$ over a field $\mathbb{F}_q$ of order q $(e = 0)$; the zero vector together with the all-one vector of odd length n over $\mathbb{F}_2$ $(e = (n-1)/2)$.

If q is a prime power, any nontrivial perfect code with $d = 3$ over $\mathbb{F}_q$ is either a linear Hamming code $\mathcal{H}_{m,q}$ with parameters $\left[\frac{q^m-1}{q-1}, \frac{q^m-1}{q-1} - m, 3\right]$, or a nonlinear code having the same length, size, and minimum distance as $\mathcal{H}_{m,q}$. Up to equivalence, the only nontrivial perfect codes over a finite field having minimum distance $d > 3$ are the $[23, 12, 7]$ binary Golay code $\mathcal{G}_{23}$ and the $[11, 6, 5]$ ternary Golay code $\mathcal{G}_{11}$; see [820, 1805, 1833, 1949].

For constructions of binary nonlinear perfect single-error-correcting codes see Chapter 3, as well as [139, 694, 929, 946, 1507, 1508, 1847], and for non-binary codes see [1265, 1630]. Every linear q-ary code with the parameters of $\mathcal{H}_{m,q}$ is monomially equivalent to $\mathcal{H}_{m,q}$. Up to equivalence, the binary linear Hamming code $\mathcal{H}_{3,2}$ of length 7 is the only nontrivial binary perfect single-error-correcting code of this length that contains the zero vector. All nonlinear binary perfect single-error-correcting codes of length 15 were classified in [1472].

The following theorem, due to Assmus and Mattson [73], characterizes perfect codes in terms of designs supported by the codewords of minimum weight.

Theorem 5.3.1 *The following hold.*

(a) *A linear $[n, k, d = 2e + 1]$ code over $\mathbb{F}_q$ is perfect if and only if the design supported by the codewords of minimum weight is an $(e+1)$-$(n, 2e+1, (q-1)^e)$ design.*

(b) *A binary (linear or nonlinear) code of length n and minimum distance $d = 2e+1$ that contains the zero vector is perfect if and only if the design supported by the codewords of minimum weight is a Steiner $(e+1)$-$(n, 2e+1, 1)$ design. In addition, the codewords of minimum weight in the extended code support a Steiner $(e+2)$-$(n+1, 2e+2, 1)$ design.*

Theorem 5.3.1 implies that the codewords of minimum weight in the Hamming code $\mathcal{H}_{m,q}$ support a 2-$((q^m - 1)/(q - 1), 3, q - 1)$ design. If $q = 2$, the binary Hamming code of length $2^m - 1$, as well as any other binary perfect single-error-correcting code that contains the zero vector, supports a Steiner triple system $STS(2^m - 1)$, while the extended code supports a Steiner quadruple system $SQS(2^m)$.

If $q = 2$ and $e = 3$, Theorem 5.3.1 implies that the minimum weight codewords of the perfect binary $[23, 12, 7]$ Golay code $\mathcal{G}_{23}$ support a Steiner 4-$(23, 7, 1)$ design, while the extended $[24, 12, 8]$ Golay code $\mathcal{G}_{24}$ supports a Steiner 5-$(24, 8, 1)$ design.

If $q = 3$ and $e = 2$, Theorem 5.3.1 implies that the 132 codewords of minimum weight in the ternary perfect $[11, 6, 5]$ Golay code $\mathcal{G}_{11}$ support a 3-$(11, 5, 4)$ design D. Since $\mathcal{G}_{11}$ is invariant under the 4-transitive group $\widetilde{M}_{11}$, a central extension of the Mathieu group M_{11}, the design D is actually a 4-design by Theorem 5.2.3, with $\lambda_4 = 1$ (calculated by using $A_5 = 132$ and (5.4)); that is, D is a Steiner 4-design.

As we mentioned in Example 5.2.6, the codewords of weight 6 in the extended ternary $[12, 6, 6]$ Golay code $\mathcal{G}_{12}$ support a 5-design D' by Theorem 5.2.3. Since the number of codewords of weight 6 is 264, it follows from (5.4) that $\lambda_5 = 1$; that is, D' is a Steiner 5-$(12, 6, 1)$ design.

5.4   The Assmus–Mattson Theorem

The following theorem, due to Assmus and Mattson [72], gives sufficient conditions for a linear code to support t-designs.

Theorem 5.4.1 (Assmus–Mattson) *Let C be an $[n, k, d]$ linear code over $\mathbb{F}_q$, and let $C^\perp$ be the dual $[n, n - k, d^\perp]$ code. Denote by n_0 the largest integer smaller than or equal to n such that $n_0 - \frac{n_0 + q - 2}{q - 1} < d$, and define $n_0^\perp$ similarly for the dual code $C^\perp$. Suppose that for some integer t with $0 < t < d$, there are at most $d - t$ nonzero weights w in $C^\perp$ such that $w \leq n - t$. Then*

(a) *the codewords in C of any weight u with $d \leq u \leq n_0$ support a t-design;*

(b) *the codewords in $C^\perp$ of any weight w with $d^\perp \leq w \leq \min\{n - t, n_0^\perp\}$ support a t-design.*

Remark 5.4.2 Note that $n_0 = n$ if $q = 2$, $n_0 = 2d$ if $q = 3$, and $n_0 = \lceil 3d/2 \rceil$ if $q = 4$.

The largest value of t for any known t-design supported by a code via the Assmus–Mattson Theorem is $t = 5$. All such 5-designs come from self-dual codes.

An $[n, k]$ code C is **self-orthogonal** if $C \subseteq C^\perp$ and **self-dual** if $C = C^\perp$. The length of a self-dual code is even and $n = 2k$. A code is **even** if all weights are even, and **doubly-even** if it is binary and all weights are divisible by 4. A binary code is **singly-even** if it is even but not doubly-even.

The next theorem gives an upper bound on the minimum distance of a self-dual code over $\mathbb{F}_q$ for $q = 2$, 3, and 4. We note that the ordinary inner product is used to define the dual code of a binary or ternary code, while when $q = 4$, we use the **Hermitian inner product** $(\mathbf{x}, \mathbf{y})$, defined by the equation

$$(\mathbf{x}, \mathbf{y}) = x_1 y_1^2 + \cdots + x_n y_n^2 \text{ for } \mathbf{x} = x_1 \cdots x_n, \ \mathbf{y} = y_1 \cdots y_n.$$

A code over $\mathbb{F}_4$ is self-orthogonal with respect to the Hermitian inner product if and only if it is even.

Theorem 5.4.3 *The minimum distance d of a self-dual $[n, \frac{n}{2}]$ code C satisfies*

$$d \leq \begin{cases} 2 \left\lfloor \frac{n}{8} \right\rfloor + 2 & \text{if } q = 2 \text{ and } C \text{ is singly-even,} \\ 4 \left\lfloor \frac{n}{24} \right\rfloor + 4 & \text{if } q = 2 \text{ and } C \text{ is doubly-even,} \\ 3 \left\lfloor \frac{n}{12} \right\rfloor + 3 & \text{if } q = 3, \\ 2 \left\lfloor \frac{n}{6} \right\rfloor + 2 & \text{if } q = 4 \text{ and } C \text{ is even.} \end{cases}$$

A self-dual code whose distance d achieves equality in Theorem 5.4.3 is called **extremal**.[1] Extremal self-dual codes can exist only for finitely many values of n, but the spectrum is unknown; see Theorem 4.3.3. The smallest $n \equiv 0 \pmod{24}$ for which the existence of a binary doubly-even self-dual code is unknown is $n = 72$. A data base of self-dual codes is available online at [908]; see also Chapter 4.

Self-dual extremal codes yield t-designs with $t \leq 5$ via the Assmus–Mattson Theorem 5.4.1; see also Theorem 4.3.16.

---

[1]With improved bounds, another meaning of the term 'extremal' has developed over time for binary self-dual codes; see Section 4.3 and, in particular, Definition 4.3.7 for this alternate meaning of 'extremal'.

Theorem 5.4.4 *The following hold.*

(a) *An extremal binary singly-even $[n, \frac{n}{2}]$ code supports 3-, 2-, or 1-designs provided that $n \equiv 0, 2,$ or $4 \pmod 8$, respectively.*

(b) *An extremal binary doubly-even $[n, \frac{n}{2}]$ code supports 5-, 3-, or 1-designs provided that $n \equiv 0, 8,$ or $16 \pmod{24}$, respectively.*

(c) *An extremal ternary code with $n \equiv 0 \pmod{12}$ supports 5-designs.*

(d) *An extremal even code over $\mathbb{F}_4$ with $n \equiv 0 \pmod 6$ supports 5-designs.*

Remark 5.4.5 The weights of codewords that support designs depend on the value of n_0; see Remark 5.4.2.

Example 5.4.6 The nonzero weights in the $[24, 12, 8]$ extended binary self-dual Golay code $\mathcal{G}_{24}$ are 8, 12, 16, and 24. The nonzero weights in the $[12, 6, 6]$ extended ternary self-dual Golay code $\mathcal{G}_{12}$ are 6, 9, and 12. Both codes support 5-designs by the Assmus–Mattson Theorem 5.4.1.

A table with parameters of t-designs obtained from self-dual codes is given in [1816, Table 1.61, page 683].

Remark 5.4.7 There are extremal binary self-dual codes that do not satisfy the conditions of the Assmus–Mattson Theorem 5.4.1 or Theorem 5.4.4, but some of their punctured codes support 2-designs. An example of this phenomenon is a $[50, 25, 10]$ extremal binary code $\mathcal{C}_{50}$ discovered in [1009] with Hamming weight enumerator

$$\mathrm{Hwe}_{\nabla\prime} = 1 + 196x^{10}y^{40} + 11368x^{12}y^{38} + 31752x^{14}y^{36} + \cdots. \qquad (5.6)$$

All minimum weight codewords of $\mathcal{C}_{50}$ share one nonzero position, and the codewords of minimum weight of the punctured code $\mathcal{C}_{50}^*$ with respect to this position support a 2-$(49, 9, 6)$ design D, having the additional property that every two distinct blocks intersect each other in either one or three points.

A 2-design that has only two distinct block intersection numbers is called **quasi-symmetric**. Quasi-symmetric designs are related to other interesting combinatorial structures, such as strongly regular graphs and association schemes; see [308, Section VII.11] and [1814, Section 6].

More $[50, 25, 10]$ binary self-dual codes with weight enumerator (5.6) whose punctured codes support quasi-symmetric 2-$(49, 9, 6)$ designs were later found in [278, 909].

5.5 Designs from Codes Meeting the Johnson Bound

In this section we consider some nonlinear binary codes that meet the Johnson Bound, and support t-designs with $t = 2$ or 3.

Recall that we denote by $(n, M, d)_2$ the parameters a binary code $\mathcal{C}$ of length n, size $M = |\mathcal{C}|$, and minimum distance d; see Section 1.3. Also recall from Section 1.6 that the **distance distribution** of an $(n, M, d)_2$ code $\mathcal{C}$ is the sequence $\{B_i\}_{i=0}^n$, where B_i is the number of ordered pairs of codewords at distance i apart, divided by M. Note that the distance and weight distribution of a linear code coincide.

The **MacWilliams transform** $\{B_i'\}_{i=0}^n$ of the distance distribution $\{B_i\}_{i=0}^n$ of an $(n, M, d)_2$ code $\mathcal{C}$ is defined by the identity

$$2^n \sum_{j=0}^n B_j x^j = M \sum_{i=0}^n B_i'(1+x)^{n-i}(1-x)^i.$$

Let $\sigma_1 < \sigma_2 < \cdots < \sigma_{s'}$ be the nonzero subscripts i for which $B_i' \neq 0$. Then $d' = \sigma_1$ is called the **dual distance** of $\mathcal{C}$, and s' is the **external distance** of $\mathcal{C}$.

Remark 5.5.1 If $\{B_i\}_{i=0}^n$ is the distance distribution of a binary *linear* code $\mathcal{C}$, then $\{B_i'\}_{i=0}^n$ is the weight distribution of the dual code $\mathcal{C}^\perp$.

The following theorem, due to Delsarte, generalizes the Assmus–Mattson Theorem 5.4.1 to nonlinear codes containing the zero vector $\mathbf{0}$.

Theorem 5.5.2 (Delsarte [521]) *Let s be the number of distinct nonzero distances in an $(n, M, d)_2$ code $\mathcal{C}$ with distance distribution $\{B_i\}_{i=0}^n$ such that $\mathbf{0} \in \mathcal{C}$. Let $\bar{s} = s$ if $B_n = 0$ and $\bar{s} = s - 1$ if $B_n = 1$.*

(a) *If $\bar{s} < d'$, then the codewords of any weight $w \geq d' - \bar{s}$ in $\mathcal{C}$ support a $(d' - \bar{s})$-design.*

(b) *If $d - s' \leq s' < d$, the codewords of any fixed weight support a $(d - s')$-design.*

The next theorem is a special case of the Johnson Bound.

Theorem 5.5.3 *For any $(n, M, d = 2e + 1)_2$ code*

$$M \leq \frac{2^n}{\sum_{i=0}^e \binom{n}{i} + \frac{1}{\lfloor n/(e+1) \rfloor} \binom{n}{e} \left(\frac{n-e}{e+1} - \lfloor \frac{n-e}{e+1} \rfloor \right)}.$$

A code for which equality holds in Theorem 5.5.3 is called **nearly perfect**. It follows from Delsarte's Theorem 5.5.2 that nearly perfect codes support designs.

Theorem 5.5.4 *The following hold.*

(a) *The minimum weight codewords in a nearly perfect $(n, M, d = 2e + 1)_2$ code $\mathcal{C}$ that contains the zero vector support a e-$(n, 2e + 1, \lfloor (n - e)/(e + 1) \rfloor)$ design.*

(b) *The minimum weight codewords in the extended code $\widehat{\mathcal{C}}$ support a $(e + 1)$-$(n + 1, 2e + 2, \lfloor (n - e)/(e + 1) \rfloor)$ design.*

Example 5.5.5 ([1446]) The Nordstrom–Robinson code $\mathcal{NR}$ is a nonlinear $(16, 256, 6)_2$ code obtained as the union of eight carefully chosen cosets of the first order Reed–Muller code $\mathcal{RM}(1, 4)$ of length 16. For a detailed description of how this code can be obtained from the extended binary Golay code $\mathcal{G}_{24}$, see [1008, Section 2.3.4]. The weight and distance distribution of $\mathcal{NR}$ is

$$B_0 = B_{16} = 1, \ B_6 = B_{10} = 112, \ B_8 = 30,$$

and it coincides with its MacWilliams transform: $B_i' = B_i$ for $0 \leq i \leq 16$. Hence, $d = d' = 6$ and $s = s' = 4$; by Theorem 5.5.2(a) (or Theorem 5.5.4(b)), the nonzero codewords support 3-designs with parameters 3-(16,6,4), 3-(16,8,3), and 3-(16,10,24).

The punctured code $\mathcal{NR}^*$ of length 15 and minimum distance 5 meets the Johnson Bound 5.5.3, and supports 2-designs by Theorem 5.5.4(a).

TABLE 5.1: Some 3-designs from nonlinear binary codes

(n, M, d)	Weight w	Comment
$(4^m, 2^{4^m - 4m}, 6)$	even $w \geq 6$	Preparata code [1541]
$(4^m, 2^{4m}, 2^{2m-1} - 2^{m-1})$	$2^{2m-1} \pm 2^{m-1}$	Kerdock code [1107]
$(4^m, 2^{4^m - 6m+1}, 8)$	even $w \geq 8$	Goethals code C [816]
$(4^m, 2^{6m-1}, 2^{2m-1} - 2^m)$	$2^{2m-1} \pm 2^m$ $2^{2m-1} \pm 2^{m-1}$	Goethals code D [816]

Remark 5.5.6 The Nordstrom–Robinson code is the first member of two infinite classes of nonlinear binary codes that support 3-designs: the **Preparata codes** [1541], and the **Kerdock codes** [1107] (see also Section 20.4.4). The Preparata code $\mathcal{P}_m$ is of length $n = 2^{2m}$ ($m \geq 2$), contains 2^{n-4m} codewords, and has minimum distance $d = 6$. The Kerdock code $\mathcal{K}_m$ is of length $n = 2^{2m}$ ($m \geq 2$), contains 2^{4m} codewords and has minimum distance $2^{2m-1} - 2^{m-1}$. The codes $\mathcal{P}_m$ and $\mathcal{K}_m$ contain the zero vector, and their distance distributions are related by the MacWilliams transform as if the codes were duals of each other. Two decades after the discovery of these codes, as well as some nonlinear binary codes found by Goethals [816], it was proved by Hammons, Kumar, Calderbank, Sloane, and Solé [898] that appropriate versions of $\mathcal{P}_m$ and $\mathcal{K}_m$, as well as the Goethals codes, can be obtained from pairs of dual codes that are linear over the ring $\mathbb{Z}_4$; see Chapter 6.

Parameters of 3-designs supported by Preparata, Kerdock, and Goethals codes are listed in Table 5.1.

The 3-designs supported by the minimum weight codewords in the Preparata code $\mathcal{P}_m$ have been used for the construction of infinite classes of 4-designs and Steiner 3-designs.

Theorem 5.5.7 ([81]) *Let P_m be the 3-$(4^m, 6, \frac{1}{3}(4^m - 4))$ design supported by the codewords of minimum weight in the Preparata code $\mathcal{P}_m$.*

(a) *The 4-subsets of coordinate indices that are not covered by any block of P_m are the blocks of a Steiner system S_m with parameters 3-$(4^m, 4, 1)$.*

(b) *The 3-$(4^m, 4, 2)$ design S_m^* consisting of two identical copies of S_m is extendable to a 4-$(4^m + 1, 5, 2)$ design D_m by adding one new point to all blocks of S_m^*, and adding as further new blocks all 5-subsets that are contained in blocks of P_m.*

Remark 5.5.8 The class of 4-$(4^m + 1, 5, 2)$ designs D_m ($m \geq 2$) of Theorem 5.5.7 is the only known infinite class of 4-(v, k, λ) designs derived from codes. Although these designs contain repeated blocks, they provide an infinite class of 4-designs with the smallest known λ ($= 2$). It is known that a 4-$(17, 5, 1)$ design does not exist [1471].

Research Problem 5.5.9 The existence of a 4-$(4^m + 1, 5, 1)$ design is an open problem for every $m \geq 3$.

The Preparata codes yield yet another infinite class of 4-designs, namely Steiner 4-designs with blocks of two different sizes.

Definition 5.5.10 Given nonnegative integers v, t, λ, and k_i, $1 \leq i \leq s$, where $v \geq k_i \geq t$, a **t-wise balanced design** with parameters t-(v, K, λ) is a set V of v points together with a collection of subsets of V, called blocks, where each block size belongs to the set $K = \{k_1, k_2, \ldots, k_s\}$, and every t-subset of V is contained in exactly λ blocks.

Theorem 5.5.11 ([1813]) *The minimum weight vectors in the Preparata code* $\mathcal{P}_m$ *together with the blocks of the Steiner system* S_m *from Theorem 5.5.7 extended by one new point, yield a Steiner* 4-$(4^m, \{5, 6\}, 1)$ *design.*

5.6 Designs and Majority Logic Decoding

A linear code $\mathcal{C}$ whose dual code $\mathcal{C}^\perp$ supports a t-design admits one of the simplest decoding algorithms, known as **majority logic decoding**. In a nutshell, for each symbol y_i of the received codeword $\mathbf{y} = y_1 y_2 \cdots y_n$, a set of values $y_i^{(j)}$, $j = 1, 2, \ldots, r$, of linear functions defined by the blocks of the design are computed, and the true value of y_i is decided to be the one that appears most frequently among $y_i^{(1)}, y_i^{(2)}, \ldots, y_i^{(r)}$.

Theorem 5.6.1 (Rudolph [1609]) *If the dual code* $\mathcal{C}^\perp$ *of a linear code* $\mathcal{C}$ *supports a* 2-(n, w, λ) *design, then* $\mathcal{C}$ *can correct by majority logic decoding up to* e *errors where*

$$e \leq \left\lfloor \frac{r + \lambda - 1}{2\lambda} \right\rfloor$$

with $r = \lambda(n - 1)/(w - 1)$.

Example 5.6.2 The maximum value of the minimum distance of a binary linear code of length 28 and dimension 9 is 10 [845]. A $[28, 9, 10]$ binary linear code $\mathcal{C}$ can be obtained as the null space of the 63×28 incidence matrix A of a 2-$(28, 4, 1)$ design D, known as the **Ree unital** [306], or the design associated with a maximal $(28, 4)$-arc in the projective plane of order 8 [1817]. The rank over $\mathbb{F}_2$ (or the 2-rank) of A is 19; so any set of 19 linearly independent rows of A is a parity check matrix of $\mathcal{C}$. Since $\mathcal{C}^\perp$ supports D, the code $\mathcal{C}$ can correct by majority logic decoding up to $e \leq \lfloor (9 + 1 - 1)/2 \rfloor = 4$ errors, which is the maximum number of errors that a code with minimum distance 10 can correct.

Remark 5.6.3 There are more than 4000 known non-isomorphic 2-$(28, 4, 1)$ designs [428, 1168], with 2-ranks ranging from 19 to 26 [1033]. However, up to isomorphism, the Ree unital is the unique 2-$(28, 4, 1)$ design of minimum 2-rank 19 [1368].

Theorem 5.6.1 can be improved if the code supports a t-design with $t > 2$.

Theorem 5.6.4 (Rahman–Blake [1552]) *Assume that the dual code of a linear code* $\mathcal{C}$ *supports a* t-(n, w, λ) *design* D *with* $t \geq 2$. *Let* $A_l = \sum_{j=0}^{l-1} (-1)^j \binom{l-1}{j} \lambda_{j+2}$, *where* $\lambda_i = \lambda \binom{n-i}{t-i} / \binom{w-i}{t-i}$ *is the number of blocks of* D *containing a set of* i *fixed points for* $0 \leq i \leq t$. *Define further*

$$A_l' = \begin{cases} A_l & \text{if } l \leq t - 1, \\ A_{t-1} & \text{if } l > t - 1. \end{cases}$$

Then $\mathcal{C}$ *can correct up to* l *errors by majority logic decoding, where* l *is the largest integer that satisfies the inequality*

$$\sum_{i=1}^{l} A_i' \leq \lfloor (\lambda_1 + A_l' - 1)/2 \rfloor.$$

A straightforward way of finding a linear code $\mathcal{C}$ over a finite field of characteristic p whose dual code $\mathcal{C}^\perp$ supports designs is by using the same approach as in Example 5.6.2, that is, by defining $\mathcal{C}$ as the null space of the incidence matrix A of a t-(n, w, λ) design D. The number of errors correctable by majority decoding is then determined by the parameters of D, while the dimension of $\mathcal{C}$ is equal to $n - \mathrm{rank}_p A$, where $\mathrm{rank}_p A$ (or p-**rank** of A) is the rank of A over $\mathbb{F}_p$. We will refer to $\mathrm{rank}_p A$ as the p-**rank** of D.

To maximize the dimension of $\mathcal{C}$, one needs to choose a t-(n, w, λ) design whose incidence matrix has the minimum possible p-rank among all designs with parameters t, n, w, λ.

Theorem 5.6.5 ([889]) *Let A be the incidence matrix of a 2-(n, w, λ) design. Let $r = \lambda(n-1)/(w-1)$ and p be a prime.*

(a) *If p does not divide $r(r - \lambda)$, then $\mathrm{rank}_p A = n$.*

(b) *If p divides r but does not divide $r - \lambda$, then $\mathrm{rank}_p A \geq n - 1$.*

(c) *If $\mathrm{rank}_p A < n - 1$, then p divides $r - \lambda$.*

Example 5.6.6 Up to isomorphism, there are exactly four 2-$(8, 4, 3)$ designs [428, page 27]. Since $r = 3(8 - 1)/(4 - 1) = 7$ and $r - \lambda = 7 - 3 = 4$, the only primes p for which the p-rank of a 2-$(8, 4, 3)$ design can be smaller than 8 are $p = 7$ and $p = 2$. The 2-ranks of the four 2-$(8, 4, 3)$ designs are 4, 5, 6, and 7. The binary code spanned by the incidence matrix of the design of minimum 2-rank 4 is a self-dual $[8, 4, 4]$ code equivalent to the first order Reed–Muller code $\mathcal{RM}(1, 3)$ of length 8.

Most of the known majority-logic decodable codes are based on designs arising from finite geometry, a notable class of such codes being the Reed–Muller codes. We refer to any design having as points and blocks the points and subspaces of a given dimension of a finite affine or projective geometry (see Chapter 14) as a **geometric design**. We denote by $PG_d(n, q)$ (respectively $AG_d(n, q)$) the geometric design having as blocks the d-dimensional subspaces of the n-dimensional projective space $\mathrm{PG}(n, q)$ (respectively the n-dimensional affine space $\mathrm{AG}(n, q)$) over $\mathbb{F}_q$. $PG_d(n, q)$ is a 2-(v, k, λ) design with parameters

$$v = \frac{q^{n+1} - 1}{q - 1}, \ k = \frac{q^{d+1} - 1}{q - 1}, \ \lambda = \begin{bmatrix} n - 1 \\ d - 1 \end{bmatrix}_q,$$

where

$$\begin{bmatrix} n \\ i \end{bmatrix}_q = \frac{(q^n - 1) \cdots (q^{n-i+1} - 1)}{(q^i - 1) \cdots (q - 1)}.$$

$AG_d(n, q)$ is a 2-(v, k, λ) design with parameters

$$v = q^n, \ k = q^d, \ \lambda = \begin{bmatrix} n - 1 \\ d - 1 \end{bmatrix}_q.$$

In the special case $q = 2$, $AG_d(n, 2)$ is also a 3-design (see, e.g., [1812, Theorem 1.4.13]) with parameters

$$v = 2^n, \ k = 2^d, \ \lambda_3 = \begin{bmatrix} n - 2 \\ d - 2 \end{bmatrix}_2.$$

In particular, the planes in $\mathrm{AG}(n, 2)$ are the blocks of a Steiner quadruple system $\mathrm{SQS}(2^n)$.

Remark 5.6.7 The codewords of minimum weight in the binary Reed–Muller code $\mathcal{RM}(r, n)$ of length 2^n and order r with $1 \leq r \leq n - 1$ support a 3-design isomorphic to $AG_{n-r}(n, 2)$ [71, Section 5.3].

Theorem 5.6.8 *Let q be a prime power.*

(a) **([1056])** *The number of non-isomorphic Steiner designs having the parameters 2-$((q^{n+1} - 1)/(q - 1), q + 1, 1)$ of $PG_1(n, q)$ grows exponentially with linear growth of n.*

(b) **([1059])** *The number of non-isomorphic designs having the parameters of $PG_d(n, q)$, $2 \leq d \leq n - 1$, grows exponentially with linear growth of n.*

(c) **([411])** *The number of non-isomorphic designs having the parameters of $AG_d(n, q)$, $2 \leq d \leq n - 1$, grows exponentially with linear growth of n.*

Remark 5.6.9 The 2-$(8, 4, 3)$ design of minimum 2-rank 4 from Example 5.6.6 is in fact a 3-$(8, 4, 1)$ design and is isomorphic to the geometric design $AG_2(3, 2)$ of the points and planes in $AG(3, 2)$.

If $q = p^s$ where p is a prime number, the q-rank of a $(0, 1)$-matrix is equal to its p-rank. The p-ranks of the incidence matrices of geometric designs were computed in the 1960s and 1970s, starting with $PG_1(2, p^s)$ (Graham and MacWilliams [844], MacWilliams and Mann [1322]), $PG_{n-1}(n, q)$ and $AG_{n-1}(n, q)$ (Smith [1731], Goethals and Delsarte [818]), and a formula for the p-rank of $PG_d(n, q)$ and $AG_d(n, q)$, for any d in the range $1 \leq d \leq n - 1$ was found by Hamada [889].

In [889], Hamada made the following conjecture.

Conjecture 5.6.10 (Hamada) The p-rank of any design D having the parameters of a geometric design G in $PG(n, q)$ or $AG(n, q)$ $(q = p^m)$ is greater than or equal to the p-rank of G, with equality if and only if D is isomorphic to G.

Remark 5.6.11 Hamada's Conjecture has been proved in the following cases:

- $PG_{n-1}(n, 2)$ and $AG_{n-1}(n, 2)$ (Hamada and Ohmori [892]);

- $PG_1(n, 2)$ and $AG_1(n, 3)$ (Doyen, Hubaut and Vandensavel [640]);

- $AG_2(n, 2)$ (Teirlinck [1795]).

In all proven cases of the conjecture, the geometric designs not only have minimum p-rank, but are also the unique designs, up to isomorphism, of minimum p-rank for the given parameters.

Although there are no known designs that have the same parameters, but smaller p-rank than a geometric design, there are non-geometric designs that have the same p-rank as a geometric design with the same parameters; hence they provide counter-examples to the "only if" part of Hamada's Conjecture 5.6.10. Throughout the remainder of this section, we list all known designs that are counter-examples to Hamada's Conjecture and examine related questions.

Example 5.6.12 ([1811]) The codewords of minimum weight 8 in a $[32, 16, 8]$ binary doubly-even self-dual code support a 3-design by Theorem 5.4.4 with parameters 3-$(32, 8, 7)$. There are exactly five inequivalent such codes [436], and each code is generated by its set of codewords of weight 8. Thus, the five 3-$(32, 8, 7)$ designs supported by the five $[32, 16, 8]$ extremal self-dual codes have 2-rank 16 and are pairwise non-isomorphic. One of these designs, namely the design supported by the second order Reed–Muller code $\mathcal{RM}(2, 5)$, is isomorphic to the geometric design $AG_3(5, 2)$.

Any $[31, 16, 7]$ code punctured from a $[32, 16, 8]$ doubly-even self-dual code supports a 2-$(31, 7, 7)$ design and is spanned by its set of 155 codewords of minimum weight. Since all $[32, 16, 8]$ doubly-even self-dual codes admit transitive automorphism groups, their punctured codes support a total of five pairwise non-isomorphic 2-$(31, 7, 7)$ designs, all having 2-rank 16. One of these designs, namely the one supported by the punctured Reed–Muller code $\mathcal{RM}(2, 5)^*$, is isomorphic to the geometric design $PG_2(4, 2)$.

Example 5.6.13 ([907]) A symmetric $(4, 4)$-**net** is a 1-$(64, 16, 16)$ design $N = (V, \mathcal{B})$ with the following properties:

- The point set V is partitioned into 16 disjoint subsets of size 4 (called **point groups**) so that every two points that belong to different groups appear together in four blocks, while two points belonging to the same group do not appear together in any block.

- The block set $\mathcal{B}$ is partitioned into 16 disjoint subsets of size 4 (called **block groups**) so that every two blocks that belong to different groups share exactly four points, while any two blocks from the same group are disjoint.

All non-isomorphic symmetric $(4, 4)$-nets admitting an automorphism group of order 4 that acts regularly on every point group and every block group were enumerated in [907], and the binary linear codes of length 64 spanned by their incidence matrices were analyzed. Three of these codes have parameters $[64, 16, 16]$ and support three non-isomorphic 2-$(64, 16, 5)$ designs, each having 2-rank 16. One of these three designs is isomorphic to the geometric design $AG_2(3, 4)$, while the other two designs are counter-examples to the "only if" part of Hamada's Conjecture 5.6.10.

Alternative constructions of the two non-geometric 2-$(64, 16, 5)$ designs with 2-rank 16 were given later in [1355, 1817].

Definition 5.6.14 A **generalized incidence matrix** of a design D over $\mathbb{F}_q$ is any matrix obtained from the $(0, 1)$-incidence matrix A of D by replacing nonzero entries of A with nonzero elements from $\mathbb{F}_q$.

A revised version of Hamada's Conjecture that uses generalized incidence matrices instead of $(0,1)$-incidence matrices, was proved by Tonchev [1815] for the complementary designs of the classical geometric designs with blocks being hyperplanes in a projective or affine space.

Theorem 5.6.15 ([1815]) *The q-rank d of a generalized incidence matrix over $\mathbb{F}_q$ of a* 2-$(\frac{q^{n+1}-1}{q-1}, q^n, q^{n-1})$ *design $(n \geq 2)$ is greater than or equal to $n+1$. The equality $d = n+1$ holds if and only if the design is isomorphic to the complementary design of the geometric design $PG_{n-1}(n, q)$.*

Theorem 5.6.16 ([1815]) *The q-rank d of a generalized incidence matrix over $\mathbb{F}_q$ of a* 2-$(q^n, q^n - q^{n-1}, q^n - q^{n-1} - 1)$ *design $(n \geq 2)$ is greater than or equal to $n+1$. The equality $d = n+1$ holds if and only if the design is isomorphic to the complementary design of the geometric design $AG_{n-1}(n, q)$.*

An infinite class of non-geometric designs that share their parameters and p-rank with geometric designs was discovered in [1058] by generalizing some properties of one of the 2-$(31, 7, 7)$ designs with 2-rank 16 from Example 5.6.12. These designs were constructed by using polarities in $PG(2d - 1, q)$ and are called **polarity designs**.

Theorem 5.6.17 ([1058]) *For every prime $p \geq 2$ and every integer $d \geq 2$, there exists a polarity 2-design P having the same parameters and the same p-rank as $PG_d(2d, p)$, but P is not isomorphic to $PG_d(2d, p)$.*

Theorem 5.6.18 ([412]) *The linear code over $\mathbb{F}_p$ that is the null space of the incidence matrix of the polarity design P from Theorem 5.6.17 corrects by majority logic decoding the same number of errors as the code that is the null space of the incidence matrix of $PG_d(2d, p)$.*

The following theorem describes an infinite class of non-geometric designs having the same parameters and 2-rank as certain designs in binary affine geometry.

Theorem 5.6.19 *The following hold.*

  (a) ([410]) *For every integer $r \geq 2$, there exists a 3-design P_r having the same parameters and 2-rank as $AG_{r+1}(2r+1, 2)$, but is not isomorphic to $AG_{r+1}(2r+1, 2)$.*

  (b) ([412]) *The binary code spanned by the incidence matrix of P_r is a self-dual code having the same parameters and minimum distance as the Reed–Muller code $\mathcal{RM}(r, 2r+1)$ of length $n = 2^{2r+1}$ and order r, and corrects by majority logic decoding the same number of errors as the Reed–Muller code.*

Remark 5.6.20 The designs from Theorem 5.6.19 generalize properties of one of the five 3-(16, 8, 7) designs from Example 5.6.12, and were constructed by modifying the method from [1058].

The first few codes of the infinite class of codes from Theorem 5.6.19(b) have the same weight distribution as the Reed–Muller code $\mathcal{RM}(r, 2r+1)$ of length 2^{2r+1} and order r [911]. This motivates the following.

Conjecture 5.6.21 For all r, the code from Theorem 5.6.19(b) has the same weight distribution as the Reed–Muller code $\mathcal{RM}(r, 2r+1)$ of length 2^{2r+1} and order r.

Conjecture 5.6.21 is true for $r \leq 3$ [911]. The smallest open case is $r = 4$.

5.7 Concluding Remarks

In this chapter, we reviewed several important links between combinatorial designs and codes.

Section 5.1 contains the definitions of various types of combinatorial designs considered in this chapter. A basic sufficient condition for a code to support a t-design, based on transitivity properties of the code automorphism group is given in Section 5.2. Another fundamental sufficient condition for a linear code to support t-designs is the Assmus–Mattson Theorem 5.4.1 (Section 5.4), and its generalization for nonlinear codes due to Delsarte (Theorem 5.5.2), with an application to nonlinear codes meeting the Johnson Bound (Section 5.5). The Assmus–Mattson characterization of perfect codes (Theorem 5.3.1) in terms of designs supported by the codewords of minimum weight, is presented in Section 5.3. Section 5.6 reviews the application of t-designs supported by linear codes for majority-logic decoding of their dual codes, The classical examples of majority logic decodable codes based on designs defined by the subspaces in a finite affine or projective geometry, and Hamada's Conjecture 5.6.10 about the minimum p-rank of these designs are reviewed in Section 5.6.

There are other interesting links between codes and designs beside the ones considered in this chapter. Designs based on Hadamard matrices are used for the construction of binary codes meeting the Plotkin Bound (see, for example [1814, Section 2]). The equivalence between 2-designs and certain binary constant weight codes that meet the restricted Johnson Bound, as well as the equivalence of Steiner systems and certain constant weight binary codes that meet the unrestricted Johnson Bound, are reviewed in [1814, Section 3]. A thorough discussion of quasi-symmetric designs and their links to self-orthogonal and self-dual codes is given in [1814, Section 9]).

Chapter 6

Codes over Rings

Steven T. Dougherty

University of Scranton

6.1   Introduction

In this chapter, we expand the collection of acceptable alphabets for codes and study codes over finite commutative rings. Although there were some early papers about codes over some finite commutative rings, a major interest in codes over rings did not really develop until the landmark paper about quaternary codes, *The $\mathbb{Z}_4$-linearity of Kerdock, Preparata, Goethals, and related codes* [898]. In that paper, the authors showed that some well known nonlinear binary codes could actually be understood as the images of linear codes over the ring $\mathbb{Z}_4$ under a nonlinear Gray map. This explained why certain binary codes acted like linear codes in so many ways yet were not linear in the classical sense for binary codes. Following this important paper, a flurry of activity occurred studying codes over the ring $\mathbb{Z}_4$. For example, see Wan's text on codes over $\mathbb{Z}_4$ [1861]. The next step was to consider codes over the other rings of order 4 together with their respective Gray maps; see [621, 622]. Other papers followed that found interesting applications for codes over rings. For example, in [236] the notion of Type II codes over $\mathbb{Z}_4$ is introduced and a very simple construction of the Leech lattice is given, and in [115] self-dual codes over $\mathbb{Z}_{2k}$ were used to construct real unimodular lattices which in turn were used to construct modular forms. In [332], codes over the p-adics were considered and a unified approach to cyclic codes was given. The acceptable alphabet for codes continued to grow to chain rings [979, 980], principal ideal rings [1449], Galois rings [1862, 1863], and finally to finite Frobenius rings [1905, 1906].

In extending coding theory to codes over rings, several important things must be considered. To begin, linear codes over fields are vector spaces, and the most important weight is the Hamming weight. When switching to codes over rings, we must make new definitions for *linear* codes and determine which *weight* is the most important one for the particular application. For example, when considering quaternary codes which give rise to interesting binary codes via the Gray map, the Lee weight is the most important weight (see [898]), and the homogeneous weight generalizes this weight to chain rings. However, when considering self-dual codes over $\mathbb{Z}_{2k}$ used to produce real unimodular lattices, the Euclidean weight is the most important weight; see [115]. Additionally, when determining a generator matrix for a code over a field and determining the cardinality of a code from that matrix, the standard definition of linear independence is used along with the classical results of linear algebra. For codes over rings, we need new results to determine when a set of generators is minimal and to determine the cardinality of the code generated by them. Moreover, we must determine what is the largest class of rings that we can use as alphabets for codes. While any set can be used as an alphabet for coding theory, in general, we want alphabets for which the fundamental theorems of coding theory apply, like the MacWilliams theorems (Theorem 1.8.6 and Theorem 1.15.3). Therefore, we generally restrict ourselves to rings for which we have analogs of these theorems. In [1906], the class of rings for which both MacWilliams theorems hold were shown to be the class of finite Frobenius rings.

In [1906], codes over both commutative and noncommutative rings were considered. For noncommutative rings the situation is much different than for commutative rings. For example, there is not a unique orthogonal. Rather, there is a left orthogonal and a right orthogonal. Moreover, a linear code can either be a left submodule of the ambient space or a right submodule of the ambient space. In this chapter, we shall restrict ourselves to commutative rings since codes over commutative rings have been widely studied while the study of codes over noncommutative rings is still in its infancy.

Throughout this chapter, we shall take the definition of ring that says a ring has a multiplicative unity.

6.2 Quaternary Codes

We begin by studying codes over the ring $\mathbb{Z}_4$, which is the ring that has received the most attention so far in the study of codes over rings. As $\mathbb{Z}_4$ is not a field we cannot say that a linear code is a vector subspace of the ambient space. Rather, we say that a code $\mathcal{C}$ is a **linear quaternary code** if $\mathcal{C}$ is a submodule of $\mathbb{Z}_4^n$. Specifically, this means that if $\mathbf{v}, \mathbf{w} \in \mathcal{C}$ and $\alpha, \beta \in \mathbb{Z}_4$, then $\alpha\mathbf{v} + \beta\mathbf{w} \in \mathcal{C}$. As an example of how different this can be from the classic case, consider the code $\mathcal{C}$ of length 1 defined by $\mathcal{C} = \{0, 2\}$. This code is a nontrivial linear code of length 1, which cannot occur for codes over fields.

It is straightforward to see that any linear code over $\mathbb{Z}_4$ is permutation equivalent to a code with generator matrix of the form

$$\begin{bmatrix} I_{k_0} & A & B \\ \mathbf{0}_{k_1 \times k_0} & 2I_{k_1} & 2C \end{bmatrix}$$

where A and C are matrices over $\mathbb{Z}_2$ and B is a matrix over $\mathbb{Z}_4$. It follows immediately that if $\mathcal{C}$ is generated by this matrix, then $|\mathcal{C}| = 4^{k_0}2^{k_1}$. In this case, we say that the linear code has **type** $4^{k_0}2^{k_1}$. We note that for codes over fields of cardinality q, all linear codes have cardinality of the form q^k for some integer k, but not all quaternary linear codes have cardinality of the form 4^k. For example, the code $\mathcal{C} = \{0, 2\}$ of type 4^02^1 has cardinality 2, which is not an integer power of 4.

The usual inner product is attached to the ambient space and is defined as $\mathbf{v} \cdot \mathbf{w} = \sum v_i w_i$. The **orthogonal** is defined as $\mathcal{C}^\perp = \{\mathbf{v} \mid \mathbf{v} \cdot \mathbf{w} = 0 \text{ for all } \mathbf{w} \in \mathcal{C}\}$. It follows immediately that if $\mathcal{C}$ is a linear code of length n of type $4^{k_0} 2^{k_1}$, then $\mathcal{C}^\perp$ is a linear code of type $4^{n-k_0-k_1} 2^{k_1}$. Additionally, we have that $|\mathcal{C}||\mathcal{C}^\perp| = 4^n$.

6.3 The Gray Map

The first important application for linear quaternary codes was in producing interesting binary codes using the following Gray map, defined as $\phi : \mathbb{Z}_4 \to \mathbb{F}_2^2$ where

$$\phi(0) = 00, \quad \phi(1) = 01, \quad \phi(2) = 11, \quad \phi(3) = 10.$$

Note that this is not the usual representation of quaternary elements in binary notation. This map is nonlinear since $\phi(1) + \phi(3) = 11$ whereas $\phi(1+3) = \phi(0) = 00$.

Extend this Gray map coordinate-wise to $\mathbb{Z}_4^n$ by $\phi(v_1 v_2 \cdots v_n) = \phi(v_1)\phi(v_2) \cdots \phi(v_n)$. Unlike for binary codes, the most important weight for quaternary codes is not the Hamming weight but rather the Lee weight, where the **Lee weight** is defined as

$$\text{wt}_L(\mathbf{v}) = \text{wt}_H(\phi(\mathbf{v}))$$

where wt_H denotes the Hamming weight, that is, the number of nonzero elements in the vector.

It is immediate from the definition that if $\mathcal{C}$ is a linear quaternary code of length n, type $4^{k_0} 2^{k_1}$, and minimum Lee weight d, then $\phi(\mathcal{C})$ is a binary possibly nonlinear code of length $2n$, with cardinality $2^{2k_0+k_1}$, and minimum Hamming weight d. Specifically, it is a $(2n, 2^{2k_0+k_1}, d)_2$ code. While the image is not necessarily linear, it often occurs that the image is linear. For example, consider the code $\{0, 2\}$ over $\mathbb{Z}_4$. The image of this code under ϕ is $\{00, 11\}$ which is a linear binary code.

In [519], Delsarte defined additive codes as subgroups of the underlying abelian group in a translation association scheme. In the binary Hamming scheme, that is, when the underlying abelian group is of order 2^n, the only structures for the abelian group are those of the form $\mathbb{Z}_2^\alpha \times \mathbb{Z}_4^\beta$ with $\alpha + 2\beta = n$. This gives that subgroups of $\mathbb{Z}_2^\alpha \times \mathbb{Z}_4^\beta$ are the only additive codes in a binary Hamming scheme. These codes are called *translation-invariant propelinear codes*. With this in mind, it should have been understood 20 years earlier that codes over $\mathbb{Z}_4$ should be studied for precisely the reason that they were studied later – namely, that interesting binary codes could be produced from them. Specifically, what was shown in [898] were the following:

- The binary Reed–Muller codes $\mathcal{RM}(r, m)$, with $m \geq 1$, of length $n = 2^m$ and orders $0 \leq r \leq m$ are the images of linear quaternary codes under the Gray map.

- The nonlinear binary Kerdock code $\mathcal{K}_m$ of length 2^{2m} containing 2^{4m} codewords with minimum distance $2^{2m-1} - 2^{m-1}$ is the image of a linear quaternary cyclic code under the Gray map. The Gray image of the dual of this code has the same weight enumerator and parameters as the nonlinear binary Preparata code $\mathcal{P}_m$.

- The nonlinear Nordstrom–Robinson code $\mathcal{NR}$ of length 16 is the image of the linear quaternary octacode of length 8 under the Gray map.

For an early description of the Reed–Muller codes see [1581]. Descriptions of the nonlinear binary Kerdock code, the nonlinear Preparata code and the nonlinear Nordstrom–Robinson code can be found in [1107], [1541], and [1446], respectively. For a description of the octacode, see [898]. See Chapters 3, 5, 12, and 20 for more on the Kerdock and Preparata codes.

6.3.1 Kernels of Quaternary Codes

In this subsection, $\mathcal{C}$ will always denote a linear quaternary code of length n. There are two important quaternary codes, $\mathcal{K}(\mathcal{C})$ and $\mathcal{R}(\mathcal{C})$, related to $\mathcal{C}$. Before defining these codes, we need to define two concepts. First $\mathrm{Ker}(\phi(\mathcal{C})) = \{\mathbf{v} \in \phi(\mathcal{C}) \mid \mathbf{v} + \phi(\mathcal{C}) = \phi(\mathcal{C})\}$. Second, if $\mathcal{S}$ is a set of binary, respectively quaternary, vectors of fixed length, then $\langle \mathcal{S} \rangle$ denotes the binary, respectively quaternary, linear span of the vectors in $\mathcal{S}$, that is, the intersection of all binary linear codes, respectively quaternary linear codes, that contain $\mathcal{S}$.

With this notation, the **kernel** $\mathcal{K}(\mathcal{C})$ of $\mathcal{C}$ is

$$\mathcal{K}(\mathcal{C}) = \{\mathbf{v} \in \mathcal{C} \mid \phi(\mathbf{v}) \in \mathrm{Ker}(\phi(\mathcal{C}))\}.$$

The code $\mathcal{R}(\mathcal{C})$ is defined by

$$\mathcal{R}(\mathcal{C}) = \{\mathbf{v} \in \mathbb{Z}_4^n \mid \phi(\mathbf{v}) \in \langle \phi(\mathcal{C}) \rangle\}.$$

For vectors $\mathbf{v}, \mathbf{w} \in \mathbb{Z}_4^n$, define the Hadamard product $\mathbf{v} \star \mathbf{w} = (v_1 w_1, v_2 w_2, \ldots, v_n w_n)$, that is, the component-wise product of the two vectors. The following lemma can be proved by simply evaluating the possible cases for the elements of $\mathbb{Z}_4$. It appeared first in [898].

Lemma 6.3.1 *Let $\mathbf{v}, \mathbf{w}$ be vectors in $\mathbb{Z}_4^n$. Then*

$$\phi(\mathbf{v} + \mathbf{w}) = \phi(\mathbf{v}) + \phi(\mathbf{w}) + \phi(2\mathbf{v} \star \mathbf{w}). \tag{6.1}$$

The next result can be proved from this lemma and can be found in [718].

Theorem 6.3.2 *Let $\mathcal{C}$ be a linear quaternary code.*

(a) *The code $\phi(\mathcal{C})$ is linear if and only if $2\mathbf{v} \star \mathbf{w} \in \mathcal{C}$ for all $\mathbf{v}, \mathbf{w} \in \mathcal{C}$.*

(b) *The kernel of $\mathcal{C}$ is*

$$\mathcal{K}(\mathcal{C}) = \{\mathbf{v} \in \mathcal{C} \mid 2\mathbf{v} \star \mathbf{w} \in \mathcal{C} \text{ for all } \mathbf{w} \in \mathcal{C}\}.$$

(c) *The code $\mathcal{R}(\mathcal{C})$ is*

$$\mathcal{R}(\mathcal{C}) = \langle \mathcal{C}, 2\mathbf{v} \star \mathbf{w} \mid \mathbf{v}, \mathbf{w} \in \mathcal{C} \rangle.$$

Theorem 6.3.2(a) is important both for its use in proving various results for quaternary codes but also as an easy computational technique to determine if the image of the quaternary code under the Gray map is going to be linear. Namely, if you can construct a generator matrix for the code, where if $\mathbf{v}$ and $\mathbf{w}$ are rows of the matrix such that $2\mathbf{v} \star \mathbf{w} \in \mathcal{C}$, then you can be sure that the Gray image of the code is linear. As an easy example of this, consider any code with a generator matrix of the form $\begin{bmatrix} 2I_k & 2A \end{bmatrix}$. The rows of this generator matrix satisfy this since $2\mathbf{v} \star \mathbf{w} = \mathbf{0}$ for any two rows $\mathbf{v}, \mathbf{w}$ in the matrix. Therefore, the image of the code generated by this matrix is a linear binary code.

The following results appear in [718, 719]. They follow easily from (6.1).

Theorem 6.3.3 *Let $\mathcal{C}$ be a linear quaternary code and let $\mathbf{v} \in \mathcal{C}$. Then $\mathbf{v} \in \mathcal{K}(\mathcal{C})$ if and only if $2\mathbf{v} \star \mathbf{w} \in \mathcal{C}$ for all $\mathbf{w} \in \mathcal{C}$.*

The next theorem shows the relationship between $\mathcal{C}$, $\mathcal{K}(\mathcal{C})$, and $\mathcal{R}(\mathcal{C})$.

Theorem 6.3.4 *Let $\mathcal{C}$ be a linear quaternary code. Then $\mathcal{K}(\mathcal{C})$ and $\mathcal{R}(\mathcal{C})$ are linear codes and $\mathcal{K}(\mathcal{C}) \subseteq \mathcal{C} \subseteq \mathcal{R}(\mathcal{C})$.*

We note that if $\mathcal{K}(\mathcal{C}) = \mathcal{C}$, then $\mathcal{R}(\mathcal{C}) = \mathcal{C}$ as well. Any code generated by a matrix of the form $\begin{bmatrix} 2I_k & 2A \end{bmatrix}$ satisfies this relation.

6.4 Rings

Following an intense study of quaternary codes, the question arose as to what is the largest class of rings that can be used as alphabets for algebraic coding theory. Specifically, the question is: what is the largest class of rings such that both MacWilliams theorems (Theorem 1.8.6 and Theorem 1.15.3) have analogs for codes over this class of rings. Without these theorems many of the techniques used in coding theory, which make it a powerful tool in mathematics, would not work. Therefore, we really want to restrict ourselves to rings where we have these theorems. This was answered in the landmark paper [1906] by J. A. Wood. The class of rings which were identified as the answer to this question is the class of Frobenius rings. Namely, Wood showed that if a ring is Frobenius then analogs of both MacWilliams theorems hold for codes over that ring. Since that paper, a great deal of work has been done studying codes over finite Frobenius rings and, in general, it is a tacit assumption in papers about codes over rings that the alphabets will be Frobenius rings.

6.4.1 Codes over Frobenius Rings

We begin with a description of Frobenius rings. In order to do that we require an understanding of characters.

Let G be a finite abelian group. A **character** is a homomorphism $\chi : G \to \mathbb{C}^*$ where $\mathbb{C}^* = \mathbb{C} \setminus \{0\}$, the nonzero complex numbers. We note that sometimes characters are defined as maps into $\mathbb{Q}/\mathbb{Z}$ but we will not use this definition here. As G is abelian, the set of characters $\widehat{G}$ is a group under function composition. The group $\widehat{G}$ is isomorphic to G but not in a canonical manner. Given a finite ring R, R is a finite abelian group under addition, whether R is commutative or noncommutative, and hence has a character group $\widehat{R}$.

The ring R is both a left and right module over itself. We denote by ${}_R R$ the left module of R over itself and by R_R the right module of R over itself. Moreover, $\widehat{R}$ is a module over R as well.

The definition of Frobenius rings as given by Nakayama is quite complex and is given for all rings, finite or infinite, commutative or noncommutative. We shall define them by the characterization of a Frobenius ring in the following theorem which can be found in [1906].

Theorem 6.4.1 *Let R be a finite ring. The following are equivalent.*

(a) *The ring R is Frobenius.*

(b) *As a left module, $\widehat{R} \cong {}_R R$.*

(c) *As a right module $\widehat{R} \cong R_R$.*

Examples of Frobenius rings include $\mathbb{Z}_n$, chain rings, and Galois rings. Not all finite commutative rings are Frobenius. For example, the ring $\mathbb{F}_2[u, v]/\langle u^2, v^2, uv \rangle$ is not Frobenius. We shall show why this is true in Example 6.5.6.

We shall give a further characterization of Frobenius rings in terms of generating characters. We restrict ourselves to commutative rings. In this case, we have that $\widehat{R} \cong {}_R R$ and $\widehat{R} \cong R_R$ are the same and we do not have to distinguish between left and right modules. Let R be a Frobenius ring and let $\phi : R \to \widehat{R}$ be the module isomorphism. Set $\chi = \phi(1)$ so that $\phi(r) = \chi_r$ for $r \in R$, where $\chi_r(a) = \chi(ra)$. This character χ is called a **generating character** for $\widehat{R}$. The following theorem gives a characterization for Frobenius rings in terms of the generating character. See [1906] for a proof.

Theorem 6.4.2 *The finite commutative ring R is Frobenius if and only if $\widehat{R}$ has a generating character.*

Example 6.4.3 Consider the Frobenius ring $\mathbb{Z}_n$. Let $\xi = e^{\frac{2\pi i}{n}}$, a complex primitive n^{th} root of unity. The generating character for $\widehat{\mathbb{Z}_n}$ is given by $\chi(a) = \xi^a$.

The following is shown in [1906] and is a straightforward way of determining if a character is in fact a generating character.

Theorem 6.4.4 *The character χ for the finite commutative ring R is a generating character if and only if the kernel of χ contains no nontrivial ideals.*

As an example, consider the non-Frobenius ring $R = \mathbb{F}_2[u,v]/\langle u^2, v^2, uv \rangle$ mentioned previously. One might naively think that $\chi : R \to \mathbb{C}^*$, defined by $\chi(a+bu+cv) = (-1)^{a+b+c}$, might be a generating character. However, the kernel of this character contains the ideal $\{0, u+v\}$ and so is not a generating character. In fact, as previously asserted, there is no character of R with kernel that does not contain a nontrivial ideal.

This theorem also shows that the generating character is not unique. Consider a finite field; since every nonzero element is a unit, there are no nontrivial ideals. Therefore any character of the field can serve as the generating character.

When a ring can be decomposed, we can use this decomposition to construct a generating character for the ring from the generating characters of the component rings.

Theorem 6.4.5 ([619]) *Let R be a finite commutative Frobenius ring with $R \cong R_1 \times R_2 \times \cdots \times R_s$, where each R_i is a Frobenius ring. Let χ_{R_i} be the generating character for R_i. Then the character χ for R defined by*

$$\chi(a) = \prod_{i=1}^{s} \chi_{R_i}(a_i),$$

where a is the element of R corresponding to $(a_1, a_2, \ldots, a_s)$, is a generating character for R.

For a finite Frobenius ring R, we can make a character table by labeling the rows and columns of the matrix with the elements of R and in the ath row and b^{th} column putting in the value of $\chi(ab)$ where χ is the generating character. Given that χ is a generating character this value corresponds to $\chi_a(b)$ where χ_a is the character of $\widehat{R}$ that corresponds to the element $a \in R$. We can give the character table for $\mathbb{Z}_4$, where i is the complex fourth root of unity, that is $i = \sqrt{-1}$:

	0	1	2	3
0	1	1	1	1
1	1	i	-1	$-i$
2	1	-1	1	-1
3	1	$-i$	-1	i

Compare this to the character table for $\mathbb{F}_4 = \{0, 1, \omega, 1+\omega\}$ with $\omega^2 = 1 + \omega$, which is:

	0	1	ω	$1+\omega$
0	1	1	1	1
1	1	-1	-1	1
ω	1	-1	1	-1
$1+\omega$	1	1	-1	-1

We note that the character table for a ring is not unique as we can take different generating characters for the ring. This non-uniqueness will not pose any problems in their coding theory applications.

For this general class of rings we make the following definitions. Let R be a finite commutative Frobenius ring. Then a **linear code** over R of length n is an R-submodule of R^n. The ambient space R^n is attached with the usual inner product, namely $\mathbf{v} \cdot \mathbf{w} = \sum v_i w_i$. The orthogonal defined with respect to this inner product is $\mathcal{C}^\perp = \{\mathbf{w} \mid \mathbf{w} \cdot \mathbf{v} = 0 \text{ for all } \mathbf{v} \in \mathcal{C}\}$. The code $\mathcal{C}^\perp$ is linear even if $\mathcal{C}$ is not. If $\mathcal{C} \subseteq \mathcal{C}^\perp$, the code is said to be **self-orthogonal** and if $\mathcal{C} = \mathcal{C}^\perp$, then the code is said to be **self-dual**. For a description of self-dual codes over Frobenius rings see [628], and for an encyclopedic description of self-dual codes see [1555].

Two codes $\mathcal{C}$ and $\mathcal{D}$ are said to be **equivalent** if $\mathcal{D}$ can be formed from $\mathcal{C}$ by a combination of permutations of the coordinates and multiplication of coordinates by units of R. Multiplication by non-units does not give an equivalent code; for example consider the code of length 2 with generator matrix $\begin{bmatrix} 1 & 1 \end{bmatrix}$ over $\mathbb{Z}_9$. This code has 9 codewords and is $\{aa \mid a \in \mathbb{Z}_9\}$. Multiplying the first and second coordinate by 3 gives the code $\{00, 33, 66\}$ which is substantially different as the cardinality is different.

6.4.2 Families of Rings

Recall that $\mathfrak{a}$ is an **ideal** in a commutative ring R if $\mathfrak{a}$ is an additive subgroup of R and for all $b \in R$, we have $b\mathfrak{a} \subseteq \mathfrak{a}$. An ideal $\mathfrak{m}$ is **maximal** if there is no ideal $\mathfrak{b}$ with $\mathfrak{m} \subset \mathfrak{b} \subset R$, where the containments are proper. If R is a finite commutative ring and $a \in R$, then the **principal ideal generated by** a is

$$\langle a \rangle = \{ba \mid b \in R\}.$$

In terms of coding theory, any ideal is a linear code of length 1 over the ring.

Recall the following definitions from the theory of rings. These definitions apply to all rings, but we are only concerned with commutative rings. Moreover, they apply to non-Frobenius and Frobenius rings as well; for example a local ring may or may not be Frobenius. However, in terms of coding theory we are only concerned with Frobenius rings.

- A **local ring** is a ring with a unique maximal ideal.

- A **chain ring** is a local ring where the ideals are linearly ordered.

- A **principal ideal ring** is a ring where every ideal $\mathfrak{a}$ is generated by a single element; that is, $\mathfrak{a} = \langle a \rangle$ for some $a \in R$.

A chain ring R is necessarily a local principal ideal ring and its maximal ideal is $\mathfrak{m} = \langle a \rangle$ for some $a \in R$. It follows that the ideals of R are

$$\{0\} \subset \langle a^{e-1} \rangle \subset \langle a^{e-2} \rangle \subset \cdots \subset \langle a^2 \rangle \subset \langle a \rangle \subset R.$$

Here, the integer e is known as the **index of nilpotency**. If p is a prime, then $\mathbb{Z}_{p^e}$ is a chain ring with maximal ideal $\langle p \rangle$ and index of nilpotency e. In this ring, the ideals are of the form $\langle p^i \rangle$ for $i = 0, 1, 2, \ldots, e$, where $i = 0$ and $i = e$ give trivial ideals, that is, the entire ring and the ideal consisting of only zero. Here the ideal $\langle p^i \rangle$ has orthogonal $\langle p^{e-i} \rangle$.

Example 6.4.6 The ring $\mathbb{Z}_n$ is a principal ideal ring for all $n > 1$. Consider the ring $\mathbb{Z}_{12}$. This ring has ideals $\langle 0 \rangle, \langle 2 \rangle, \langle 3 \rangle, \langle 4 \rangle, \langle 6 \rangle, \langle 1 \rangle = \mathbb{Z}_{12}$ with $\langle 0 \rangle^\perp = \mathbb{Z}_{12}$, $\langle 2 \rangle^\perp = \langle 6 \rangle$, and $\langle 3 \rangle^\perp = \langle 4 \rangle$. The ring $\mathbb{Z}_{12}$ is neither a local ring nor a chain ring.

Example 6.4.7 Consider the ring $R = \mathbb{F}_2[u,v]/\langle u^2, v^2, uv - vu \rangle$. The ring R has 16 elements with maximal ideal $\mathfrak{m} = \{0, u, v, uv, u+v, u+uv, v+uv, u+v+uv\}$. The ring is a local ring that is not a chain ring nor a principal ideal ring, since the maximal ideal is $\langle u, v \rangle$ which is not principal.

There are four rings of order 4; they are all commutative and Frobenius. Namely, they are $\mathbb{Z}_4$ which is a chain ring, $\mathbb{F}_4$ which is a finite field, $\mathbb{F}_2[u]/\langle u^2 \rangle$ which is a chain ring, and $\mathbb{F}_2[v]/\langle v^2 + v \rangle \cong \mathbb{F}_2 \times \mathbb{F}_2$ which is not a local ring but is a principal ideal ring. Each of these rings is equipped with a Gray map to $\mathbb{F}_2^2$ and they are given below:

$\mathbb{Z}_4$	$\mathbb{F}_4$	$\mathbb{F}_2[u]/\langle u^2 \rangle$	$\mathbb{F}_2[v]/\langle v^2 + v \rangle$	$\mathbb{F}_2^2$
0	0	0	0	00
1	1	1	v	01
2	$1 + \omega$	u	1	11
3	ω	$1 + u$	$1 + v$	10

The Lee weight is defined for each of these rings according to the Hamming weight of its image under the respective Gray map. We note that only the Gray map for $\mathbb{Z}_4$ is nonlinear.

Each of these rings generalizes to a family of Frobenius rings with a corresponding family of Gray maps. For $\mathbb{F}_4$ the generalization is the natural generalization to finite fields $\mathbb{F}_{2^r}$ where the Gray map is simply writing each element of the finite fields in its canonical representation as an element in the vector space $\mathbb{F}_2^r$.

The generalization of $\mathbb{Z}_4$ is the family of rings $\mathbb{Z}_{2^k}$. These rings are all chain rings. We shall describe the Gray map as given in [620]. Let $\mathbf{1}_i$ denote the all-one vector of length i and let $\mathbf{0}_i$ denote the all-zero vector of length i. Then we define the Gray map $\phi_k : \mathbb{Z}_{2^k} \to \mathbb{Z}_2^{2^{k-1}}$ by

$$\phi_k(i) = \begin{cases} \mathbf{0}_{2^{k-1}-i} \mathbf{1}_i & \text{if } 0 \leq i \leq 2^{k-1}, \\ \mathbf{1}_{2^{k-1}} + \phi_k(i - 2^{k-1}) & \text{if } i > 2^{k-1}. \end{cases}$$

We note that this map is a nonlinear map.

Any linear code over $\mathbb{Z}_{2^k}$ is equivalent to a code with a generator matrix of the following form, where the elements in $A_{i,j}$ are from $\mathbb{Z}_{2^k}$:

$$G = \begin{bmatrix} I_{\delta_0} & A_{0,1} & A_{0,2} & A_{0,3} & \cdots & & \cdots & A_{0,k} \\ 0 & 2I_{\delta_1} & 2A_{1,2} & 2A_{1,3} & \cdots & & \cdots & 2A_{1,k} \\ 0 & 0 & 4I_{\delta_2} & 4A_{2,3} & \cdots & & \cdots & 4A_{2,k} \\ \vdots & \vdots & & 0 & \ddots & \ddots & & \vdots \\ \vdots & \vdots & \vdots & & \ddots & \ddots & \ddots & \vdots \\ 0 & 0 & 0 & \cdots & & 0 & 2^{k-1}I_{\delta_{k-1}} & 2^{k-1}A_{k-1,k} \end{bmatrix}. \quad (6.2)$$

The matrix G is said to be in **standard form** and such a code $\mathcal{C}$ is said to have **type** $\{\delta_0, \delta_1, \ldots, \delta_{k-1}\}$. It is immediate that a code $\mathcal{C}$ with this generator matrix has $\prod_{i=0}^{k-1}(2^{k-i})^{\delta_i}$ vectors. It follows that the type of $\mathcal{C}^\perp$ is $\{n - \sum_{i=0}^{k-1} \delta_i, \delta_{k-1}, \delta_{k-2}, \ldots, \delta_1\}$, where n is the length of the code. The following can be found in [620].

Theorem 6.4.8 *Let $\mathcal{C}$ be a linear code over $\mathbb{Z}_{2^k}$ of type $\{\delta_0, \delta_1, \ldots, \delta_{k-1}\}$. If $m = \dim(\mathrm{Ker}(\phi_k(\mathcal{C})))$, then*

$$m \in \left\{ \sum_{i=0}^{k-1} \delta_i, \sum_{i=0}^{k-1} \delta_i + 1, \ldots, \sum_{i=0}^{k-1} \delta_i + \delta_{k-2} - 2, \sum_{i=0}^{k-1} \delta_i + \delta_{k-2} \right\}.$$

Moreover, there exists such a code C for any m in the interval. If C has length n, let
$s = n - \sum_{i=0}^{k-1} \delta_i$. *If r is the dimension of* $\langle \phi_k(C) \rangle$, *then*

$$r \in \left\{ \sum_{i=0}^{k-1} 2^{k-(i+1)} \delta_i, \sum_{i=0}^{k-1} 2^{k-(i+1)} \delta_i + 1, \ldots, 2^{k-1} \delta_0 + (2^{k-1} - 1)((\sum_{i=1}^{k-1} \delta_i) + s) \right\}.$$

The ring $\mathbb{F}_2[u]/\langle u^2 \rangle$ can be extended to an infinite family of rings R_k as in [626, 638, 639]. Let $R_k = \mathbb{F}_2[u_1, u_2, \ldots, u_k]/\langle u_i^2 = 0, u_i u_j = u_j u_i \rangle$; for $k \geq 1$, the ring R_k is a finite commutative ring.

We can describe the representation of the elements in the ring R_k. Take a subset $A \subseteq \{1, 2, \ldots, k\}$ and let

$$u_A := \prod_{i \in A} u_i$$

with the convention that $u_\emptyset = 1$. Any element of R_k can be represented as

$$\sum_{A \subseteq \{1, \ldots, k\}} c_A u_A \text{ with } c_A \in \mathbb{F}_2.$$

We note that the element is a unit if and only if $c_\emptyset = 1$. The ring R_k is a Frobenius local ring with maximal ideal $\langle u_1, u_2, \ldots, u_k \rangle$ and $|R_k| = 2^{2^k}$. This ring is neither a principal ideal ring nor a chain ring when $k \geq 2$.

The Gray map ψ_k on R_k can be defined inductively, building on the Gray map given previously, recognizing R_1 as $\mathbb{F}_2[u_1]/\langle u_1^2 \rangle$ with Gray map ψ_1. Recall that $\psi_1(\alpha + \beta u_1) = (\beta, \alpha + \beta)$ for $\alpha, \beta \in \mathbb{F}_2$. Then define

$$\psi_k(\alpha + \beta u_k) = (\psi_{k-1}(\beta), \psi_{k-1}(\alpha) + \psi_{k-1}(\beta))$$

for $\alpha, \beta \in R_{k-1}$. This map is a linear map and in fact preserves orthogonality.

A binary self-dual code is called **Type II** if all of its codewords have Hamming weights congruent to 0 (mod 4). If a binary self-dual code is not Type II, it is **Type I**. Generalizing these concepts, a self-dual code over R_k is **Type II** if all of its codewords have Lee weights congruent to 0 (mod 4); otherwise it is **Type I**. Here, the Lee weight of a vector in R_k^n is the Hamming weight of its image under the Gray map ψ_k. The following can be found in [639].

Theorem 6.4.9 *If C is a self-dual code over R_k, then $\psi_k(C)$ is a binary self-dual code of length 2^k. If C is a Type II code, then $\psi_k(C)$ is Type II, and if C is Type I, then $\psi_k(C)$ is Type I.*

The following can be found in [626]. We note that quasi-cyclic codes are introduced in Definition 1.12.23 and studied in Chapter 7.

Theorem 6.4.10 *Let C be a cyclic code of length n over the ring R_k. Then $\psi_k(C)$ is a 2^k-quasi-cyclic binary linear code of length $2^k n$.*

The ring $\mathbb{F}_2[v]/\langle v^2 + v \rangle$ can be extended to an infinite family of rings A_k as in [373]. Let $A_k = \mathbb{F}_2[v_1, v_2, \ldots, v_k]/\langle v_i^2 = v_i, v_i v_j = v_j v_i \rangle$. For all k, the ring A_k is a finite commutative Frobenius ring.

We can describe the representation of the elements in the ring A_k. Take a subset $B \subseteq \{1, 2, \ldots, k\}$; then let $v_B = \prod_{i \in B} v_i$, with the convention that $v_\emptyset = 1$. Any element of A_k can be represented as

$$\sum_{B \subseteq \{1, 2, \ldots, k\}} \alpha_B v_B \text{ with } \alpha_B \in \mathbb{F}_2.$$

The ring A_k has characteristic 2 and cardinality 2^{2^k} and is not a local ring. For example, if $k = 1$ then $\langle v_1 \rangle \neq \langle 1 + v_1 \rangle$ and both ideals are maximal.

The Gray map Θ_k on A_k can be defined inductively in a manner similar to the case for R_k, building on the Gray map given previously, recognizing A_1 as $\mathbb{F}_2[v_1]/\langle v_1^2 + v_1 \rangle$ with Gray map Θ_1. Recall that $\Theta_1(\alpha + \beta v_1) = (\alpha, \alpha + \beta)$ for $\alpha, \beta \in \mathbb{F}_2$. Then define

$$\Theta_k(\alpha + \beta v_k) = \big(\Theta_{k-1}(\alpha), \Theta_{k-1}(\alpha) + \Theta_{k-1}(\beta)\big)$$

for $\alpha, \beta \in A_{k-1}$. This map is a linear map.

Theorem 6.4.11 ([373]) *If C is a self-dual code over A_k, then $\Theta_k(C)$ is a binary self-dual code.*

6.4.3   The Chinese Remainder Theorem

Perhaps the most powerful tool in studying codes over commutative rings is the Chinese Remainder Theorem. This theorem allows us to restrict ourselves to a much smaller class of rings where the results can easily be extended to whole classes of Frobenius rings.

Let R be a finite commutative ring. Since the ring is finite, there is a finite set of maximal ideals of R. Recall that the product of ideals $\mathfrak{a}$ and $\mathfrak{b}$ is the ideal $\mathfrak{ab}$ generated by all products of the form ab where $a \in \mathfrak{a}$ and $b \in \mathfrak{b}$. The ideal $\mathfrak{a}^k$ is the k-fold product of $\mathfrak{a}$ with itself. For any ideal $\mathfrak{m}$, the chain of ideals $\mathfrak{m} \supset \mathfrak{m}^2 \supset \mathfrak{m}^3 \supset \cdots$ necessarily stabilizes. The smallest $t \geq 1$ such that $\mathfrak{m}^t = \mathfrak{m}^{t+i}$ for $i \geq 0$ is called the **index of stability** of $\mathfrak{m}$. When $\mathfrak{m}$ is nilpotent, that is $\mathfrak{m}^e = \{0\}$ for some e, then the smallest $t \geq 1$ such that $\mathfrak{m}^t = \{0\}$ is called the **index of nilpotency** of $\mathfrak{m}$. When this occurs, the index of stability of $\mathfrak{m}$ coincides with the index of nilpotency.

We can now state the Chinese Remainder Theorem. A proof of the theorem can be found in most advanced algebra books. In a coding theory setting, it can be found in [619].

Theorem 6.4.12 (Chinese Remainder Theorem) *Let R be a finite commutative ring with maximal ideals $\mathfrak{m}_1, \ldots, \mathfrak{m}_s$. Let the index of stability of $\mathfrak{m}_i$ be e_i, for $1 \leq i \leq s$. Then the map $\Psi : R \to \prod_{i=1}^s R/\mathfrak{m}_i^{e_i}$, defined by $\Psi(x) = (x + \mathfrak{m}_1^{e_1}, \ldots, x + \mathfrak{m}_k^{e_k})$, is a ring isomorphism.*

The natural consequence of this theorem is the following result which decomposes rings into rings which are more easily studied.

Theorem 6.4.13 *Let R be a commutative ring.*

(a) *If R is a principal ideal ring, then R is isomorphic via the Chinese Remainder Theorem to a product of chain rings.*

(b) *If R is a finite Frobenius ring, then R is isomorphic via the Chinese Remainder Theorem to a product of local Frobenius rings.*

Example 6.4.14 Let n be a positive integer with $n > 1$. The Fundamental Theorem of Arithmetic gives that $n = p_1^{e_1} p_2^{e_2} \cdots p_s^{e_s}$ where p_i is a prime, $p_i \neq p_j$ when $i \neq j$, and $e_i \geq 1$. Then the Chinese Remainder Theorem shows that

$$\mathbb{Z}_n \cong \mathbb{Z}_{p_1}^{e_1} \times \mathbb{Z}_{p_2}^{e_2} \times \cdots \times \mathbb{Z}_{p_s}^{e_s}.$$

Let R be a finite ring and let $R \cong R_1 \times R_2 \times \cdots \times R_s$, via the Chinese Remainder Theorem, with isomorphism Ψ. Let CRT denote the inverse isomorphism of Ψ. Therefore, $R = \text{CRT}(R_1, R_2, \ldots, R_s)$.

Example 6.4.15 Let $R = \mathbb{Z}_{60} \cong \mathbb{Z}_{2^2} \times \mathbb{Z}_3 \times \mathbb{Z}_5$. Then $\mathrm{CRT}(2, 0, 2) = 42$ as $\Psi(42) = (42 + \langle 2^2 \rangle, 42 + \langle 3 \rangle, 42 + \langle 5 \rangle) = (2 + \langle 2^2 \rangle, 0 + \langle 3 \rangle, 2 + \langle 5 \rangle)$. Similarly, $\mathrm{CRT}(1, 2, 2) = 17$ and $\mathrm{CRT}(3, 2, 3) = 23$. We see that each triple $(a, b, c) \in \mathbb{Z}_{2^2} \times \mathbb{Z}_3 \times \mathbb{Z}_5$ corresponds to a unique element in $\mathbb{Z}_{60}$.

With the notation above, let n be a positive integer. Extend the definition $\mathrm{CRT} : R_1 \times R_2 \times \cdots \times R_s \to R$ to $\mathrm{CRT} : R_1^n \times R_2^n \times \cdots \times R_s^n \to R^n$ as follows. Let $\mathbf{v}_i = v_{i,1} v_{i,2} \cdots v_{i,n} \in R_i^n$ for $1 \leq i \leq s$; thus $(\mathbf{v}_1, \mathbf{v}_2, \ldots, \mathbf{v}_s) \in R_1^n \times R_2^n \times \cdots \times R_s^n$. Set $\mathbf{w}_j = (v_{1,j}, v_{2,j}, \ldots, v_{s,j}) \in R_1 \times R_2 \times \cdots \times R_s$ for $1 \leq j \leq n$; hence $\mathrm{CRT}(\mathbf{w}_j) \in R$. Define

$$\mathrm{CRT}(\mathbf{v}_1, \mathbf{v}_2, \ldots, \mathbf{v}_s) = (\mathrm{CRT}(\mathbf{w}_1), \mathrm{CRT}(\mathbf{w}_2), \ldots, \mathrm{CRT}(\mathbf{w}_n)) \in R^n.$$

Finally, let $\mathcal{C}_i$ be a code of length n over R_i. Define the code $\mathrm{CRT}(\mathcal{C}_1, \mathcal{C}_2, \ldots, \mathcal{C}_s)$ over R of length n by

$$\mathrm{CRT}(\mathcal{C}_1, \mathcal{C}_2, \ldots, \mathcal{C}_s) = \{\mathrm{CRT}(\mathbf{v_1}, \mathbf{v_2}, \ldots, \mathbf{v_s}) \mid \mathbf{v}_i \in \mathcal{C}_i \text{ for all } i, 1 \leq i \leq s\}.$$

The **rank** of a code $\mathcal{C}$ over a ring R is the minimum number of generators of $\mathcal{C}$. A code $\mathcal{C}$ is **free** over a ring R if it is isomorphic to R^k for some k. The following theorems appear in [627].

Theorem 6.4.16 *Let R_i be finite commutative rings and let*

$$R = \mathrm{CRT}(R_1, R_2, \ldots, R_s).$$

Let $\mathcal{C}_i$ be a code over R_i with $\mathcal{C} = \mathrm{CRT}(\mathcal{C}_1, \mathcal{C}_2, \ldots, \mathcal{C}_s)$. The following hold.

(a) $|\mathcal{C}| = \prod_{i=1}^{s} |\mathcal{C}_i|$.

(b) $\mathrm{rank}(\mathcal{C}) = \max\{\mathrm{rank}(\mathcal{C}_i) \mid i = 1, 2, \ldots, s\}$.

(c) $\mathcal{C}$ *is free if and only if $\mathcal{C}_i$ is free for all i each of the same rank.*

Theorem 6.4.17 *Let R_i be finite commutative rings and let*

$$R = \mathrm{CRT}(R_1, R_2, \ldots, R_s).$$

Let $\mathcal{C}_i$ be a code over R_i with $\mathcal{C}_i^{\perp}$ its orthogonal in R_i^n. Let $\mathcal{C} = \mathrm{CRT}(\mathcal{C}_1, \mathcal{C}_2, \ldots, \mathcal{C}_s)$.

(a) *We have that $\mathcal{C}^{\perp} = \mathrm{CRT}(\mathcal{C}_1^{\perp}, \mathcal{C}_2^{\perp}, \ldots, \mathcal{C}_s^{\perp})$.*

(b) *If each $\mathcal{C}$ is self-dual, that is $\mathcal{C}_i = \mathcal{C}_i^{\perp}$ for all i, then $\mathcal{C} = \mathcal{C}^{\perp}$.*

Given these results, it is clear that in most cases, one need only study codes over local rings to get general results for codes over all finite commutative rings.

6.5 The MacWilliams Identities

One of the most significant results in coding theory is the MacWilliams Identities. These identities state that the weight enumerator of a linear code completely determines the weight enumerator of its orthogonal. F. J. MacWilliams proved the relation for binary codes in [1318]. These were easily extended to codes over arbitrary finite fields; see [1323] for a complete description. An analog for the MacWilliams Identities exists for codes over rings if the ring is Frobenius. Without this fundamental result many of the techniques that give coding theory power are no longer applicable. Therefore, we generally only study codes over Frobenius rings.

We begin with the definition of various weight enumerators to give analogs to Theorem 1.15.3(c).

Definition 6.5.1 Let C be a code over an alphabet $A = \{a_0, a_1, \ldots, a_{r-1}\}$ with the convention that if A is a finite abelian group, $a_0 = 0$. The **complete weight enumerator** for the code C is defined as

$$\text{cwe}_C(x_{a_0}, x_{a_1}, \ldots, x_{a_{r-1}}) = \sum_{\mathbf{c} \in C} \prod_{i=0}^{r-1} x_{a_i}^{n_i(\mathbf{c})},$$

where there are $n_i(\mathbf{c})$ occurrences of a_i in the vector $\mathbf{c}$.

Let G be a finite abelian group, and define the equivalence relation $\equiv$ on G by $g \equiv g'$ if and only if $g = \pm g'$. Let $g_0 = 0, g_1, \ldots, g_{s-1}$ be a list of distinct representatives of each class. The **symmetrized weight enumerator** of a code C over a group G is given by

$$\text{swe}_C(x_{g_0}, x_{g_1}, \ldots, x_{g_{s-1}}) = \sum_{\mathbf{c} \in C} \text{swt}(\mathbf{c}),$$

where $\text{swt}(\mathbf{c}) = \prod_{i=0}^{s-1} x_{g_i}^{\beta_i}$ with β_i the number of times $\pm g_i$ appears in the codeword $\mathbf{c}$.

The **Hamming weight enumerator** of a code C of length n, extending Definition 1.15.1, is

$$\text{Hwe}_C(x, y) = \sum_{\mathbf{c} \in C} x^{\text{wt}_H(\mathbf{c})} y^{n - \text{wt}_H(\mathbf{c})} = \text{cwe}_C(y, x, x, \ldots, x).$$

The following analogs of the MacWilliams Identities can be found in [1906].

Theorem 6.5.2 *Let C be a linear code over a finite commutative Frobenius ring $R = \{a_0 = 0, a_1, a_2, \ldots, a_{r-1}\}$. Let χ be a fixed generating character associated with R. Define the matrix T indexed by elements of R as $T_{a_i, a_j} = \chi(a_i a_j)$. Then*

$$\text{cwe}_{C^{\perp}}(x_{a_0}, x_{a_1}, \ldots, x_{a_{r-1}}) = \frac{1}{|C|} \text{cwe}_C(T \cdot (x_{a_0}, x_{a_1}, \ldots, x_{a_{r-1}})^T).$$

Let $S = \{g_0 = 0, g_1, \ldots, g_{s-1}\}$ be a set of distinct representatives of each equivalence class $[g]$ of the equivalence relation $\equiv$ on R given by $g \equiv g'$ if and only if $g = \pm g'$. Define the matrix S indexed by equivalence classes as $S_{[g_i][g_j]} = \sum_{g \in [g_j]} \chi(g_i g)$. Then

$$\text{swe}_{C^{\perp}}(x_{g_0}, x_{g_1}, \ldots, x_{g_{s-1}}) = \frac{1}{|C|} \text{swe}_C(S \cdot (x_{g_0}, x_{g_1}, \ldots, x_{g_{s-1}})^T).$$

Note that we are not saying that there is a unique way to express the MacWilliams Identities since it depends on the generating character which is not unique for a given ring. However, different matrices will still give the same weight enumerator for the orthogonal.

Example 6.5.3 We shall give the specific MacWilliams Identities for codes over $\mathbb{Z}_4$ as an example. Let C be a linear code over $\mathbb{Z}_4$. Then

$$\text{cwe}_{C^{\perp}}(x_0, x_1, x_2, x_3) = \frac{1}{|C|} \text{cwe}_C(x_0 + x_1 + x_2 + x_3, x_0 + ix_1 - x_2 - ix_3,$$
$$x_0 - x_1 + x_2 - x_3, x_0 - ix_1 - x_2 + ix_3),$$

and

$$\text{swe}_{\mathcal{C}^\perp}(x_0, x_1, x_2) = \frac{1}{|\mathcal{C}|}\text{swe}_{\mathcal{C}}(x_0 + 2x_1 + x_2, x_0 - x_2, x_0 - 2x_1 + x_2).$$

The following was first proved by MacWilliams in [1318, 1319] for codes over finite fields. Here we can extend the result to codes over finite commutative Frobenius rings.

Theorem 6.5.4 *Let R be a finite commutative Frobenius ring with $|R| = r$. Let $\mathcal{C}$ be a linear code over R. Then*

$$\text{Hwe}_{\mathcal{C}^\perp}(x, y) = \frac{1}{|\mathcal{C}|}\text{Hwe}_{\mathcal{C}}(y - x, y + (r - 1)x). \tag{6.3}$$

One of the most significant corollaries of the MacWilliams Identities is the following.

Corollary 6.5.5 *Let $\mathcal{C}$ be a linear code of length n over a finite commutative Frobenius ring R. Then $|\mathcal{C}||\mathcal{C}^\perp| = |R|^n$.*

Proof: In (6.3), let $x = y = 1$. Then $|\mathcal{C}^\perp| = \text{Hwe}_{\mathcal{C}^\perp}(1, 1) = \frac{1}{|\mathcal{C}|}\text{Hwe}_{\mathcal{C}}(0, r) = \frac{1}{|\mathcal{C}|}r^n$, which gives the result. $\square$

It follows immediately that if $\mathcal{C}$ is a linear code over a Frobenius ring, then $(\mathcal{C}^\perp)^\perp = \mathcal{C}$. In general, one of the simplest ways of showing a ring is not Frobenius is to find an ideal $\mathfrak{a}$ in the ring with $|\mathfrak{a}||\mathfrak{a}^\perp| \neq |R|$.

Example 6.5.6 Consider the ring $R = \mathbb{F}_2[u, v]/\langle u^2, v^2, uv\rangle$, which we claimed was not Frobenius after Theorem 6.4.1. This ring has 8 elements $\{0, 1, u, v, u+v, 1+u, 1+v, 1+u+v\}$. Consider the ideal $\langle u, v\rangle = \mathfrak{a}$. Then $\mathfrak{a} = \{0, u, v, u + v\}$. Its orthogonal is $\mathfrak{a}^\perp = \mathfrak{a}$. But $4(4) = 16 \neq 8$; so the ring R is not Frobenius.

6.6 Generating Matrices

For codes over fields, it is easy to show that any code is permutation equivalent to a code with generator matrix of the form $\begin{bmatrix} I_k & A \end{bmatrix}$ where k is the dimension of the code. For codes over rings this is not necessarily true. For example, the code $\{0, 2, 4, 6\}$ of length 1 over $\mathbb{Z}_8$ has generator matrix $\begin{bmatrix} 2 \end{bmatrix}$ which is not of this form. Moreover, for codes over rings it is not as easy to tell when vectors are a minimal generating set.

Example 6.6.1 Consider the code $\mathcal{C}$ over $\mathbb{Z}_{10}$ generated by $\begin{bmatrix} 2 & 0 \\ 0 & 5 \end{bmatrix}$. This code is $\mathcal{C} = \{00, 20, 40, 60, 80, 05, 25, 45, 65, 85\}$ and has 10 codewords. However, the vectors 20 and 05 do not constitute a minimal generating set. Consider the vector 25. We have that $\alpha 25 \neq \beta 25$ if $\alpha \neq \beta$ for all $\alpha, \beta \in \mathbb{Z}_{10}$. Therefore the code $\mathcal{C}$ is generated by a single vector. Simply because the generator matrix was in diagonal form and each diagonal element was an integer prime did not guarantee that the set of vectors was a minimal generating set.

We shall give definitions for *modular independent* and *independent* which will be used to describe a basis for a code over a ring, namely a minimal generating set for the code. These definitions first appeared in [1490] for codes over $\mathbb{Z}_n$ and were expanded to codes over Frobenius rings in [632]. We begin with the definition for modular independence.

Definition 6.6.2 Let R be a finite local commutative ring with unique maximal ideal $\mathfrak{m}$, and let $\mathbf{v}_1, \mathbf{v}_2, \ldots, \mathbf{v}_k$ be vectors in R^n. Then $\mathbf{v}_1, \mathbf{v}_2, \ldots, \mathbf{v}_k$ are **modular independent** if and only if $\sum_{j=1}^{k} \alpha_j \mathbf{v}_j = \mathbf{0}$ implies that $\alpha_j \in \mathfrak{m}$ for all j.

The importance of this definition is in the following result.

Theorem 6.6.3 ([632]) *Let $\mathbf{v}_1, \mathbf{v}_2, \ldots, \mathbf{v}_k \in R^n$ where R is a finite local commutative ring. Then $\mathbf{v}_1, \mathbf{v}_2, \ldots, \mathbf{v}_k$ are modular dependent if and only if some $\mathbf{v}_j$ can be written as a linear combination of the other vectors.*

It follows that for a finite chain ring (which is local by definition), this definition is enough to find a minimal generating set. Moreover, we can put the generator matrix in a form quite similar to the standard form for codes over fields. Namely, we have the following theorem which is a generalization of (6.2). See [980, 1447, 1448] for a complete description.

Theorem 6.6.4 *Let R be a finite commutative chain ring with maximal ideal $\langle \gamma \rangle$ and index of nilpotency e. Let $\mathcal{C}$ be a linear code over R. Then $\mathcal{C}$ is permutation equivalent to a linear code over R with generator matrix*

$$\begin{bmatrix} I_{k_0} & A_{0,1} & A_{0,2} & A_{0,3} & \cdots & & \cdots & A_{0,e} \\ 0 & \gamma I_{k_1} & \gamma A_{1,2} & \gamma A_{1,3} & \cdots & & \cdots & \gamma A_{1,e} \\ 0 & 0 & \gamma^2 I_{k_2} & \gamma^2 A_{2,3} & \cdots & & \cdots & \gamma^2 A_{2,e} \\ \vdots & \vdots & 0 & & \ddots & \ddots & & \vdots \\ \vdots & \vdots & \vdots & & \ddots & \ddots & \ddots & \vdots \\ 0 & 0 & 0 & \cdots & 0 & \gamma^{e-1} I_{k_{e-1}} & & \gamma^{e-1} A_{e-1,e} \end{bmatrix}$$

where the $A_{i,j}$ are arbitrary matrices with elements from the ring R, and I_{k_j} is the $k_j \times k_j$ identity matrix.

A code with generator matrix of this form is said to have **type** $\{k_0, k_1, \ldots, k_{e-1}\}$. A simple counting argument gives the following corollary.

Corollary 6.6.5 *Let R be a finite chain ring with maximal ideal $\langle \gamma \rangle$. Let $\mathcal{C}$ be a code over R of type $\{k_0, k_1, \ldots, k_{e-1}\}$. Then*

$$|C| = |R/\langle \gamma \rangle|^{\sum_{i=0}^{e-1} (e-i)k_i}.$$

For codes over chain rings, the situation is quite similar to codes over finite fields. Namely, we have a standard form for the generator matrix and from that matrix we can directly compute the cardinality of the code. For codes over other rings the situation is not as simple.

We now use the previous definition of modular independence for local rings to give a definition for the most general class of rings for which we define codes.

Definition 6.6.6 Let R be a finite commutative Frobenius ring with

$$R = \mathrm{CRT}(R_1, R_2, \ldots, R_s).$$

Since $R \cong R_1 \times R_2 \times \cdots \times R_s$, let $\Psi_i : R \to R_i$ be the projection of R onto R_i. The vectors $\mathbf{v}_1, \mathbf{v}_2, \ldots, \mathbf{v}_k \in R^n$ are **modular independent** if and only if $\Psi_i(\mathbf{v}_1), \Psi_i(\mathbf{v}_2), \ldots, \Psi_i(\mathbf{v}_k)$ are modular independent for some i with $1 \leq i \leq s$.

It is very important to note here that we are not saying that the projection to the local components of the ring must be modular independent for *all* i but rather for *some* i. We only need one projection to be modular independent for the definition to be fulfilled.

Example 6.6.7 Let us return to the codewords in Example 6.6.1. The vectors 20 and 05 project to 00 and 01 over $\mathbb{Z}_2$ and to 20 and 00 over $\mathbb{Z}_5$. Hence these vectors are not modular independent over $\mathbb{Z}_{10}$.

We now present a concept needed to formulate the definition for a basis of a code over a ring.

Definition 6.6.8 Let R be a finite commutative Frobenius ring. Let $\mathbf{v}_1, \mathbf{v}_2, \ldots, \mathbf{v}_k$ be nonzero vectors in R^n. Then $\mathbf{v}_1, \mathbf{v}_2, \ldots, \mathbf{v}_k$ are **independent** if $\sum_{j=1}^{k} \alpha_j \mathbf{v}_j = \mathbf{0}$ implies that $\alpha_j \mathbf{v}_j = \mathbf{0}$ for all j.

Notice that if R is a field and $\mathbf{v}_j$ is nonzero, then $\alpha_j \mathbf{v}_j = \mathbf{0}$ implies $\alpha_j = 0$. Therefore Definition 6.6.8 generalizes the concept of independence of vectors over fields. However, this differs from the standard definition of linear independence for vectors over fields, since we are not simply saying that each α_j must be 0, but rather that the vector $\alpha_j \mathbf{v}_j$ must be the zero vector. As an example, consider the vectors $\{(2,0), (0,2)\} \in \mathbb{Z}_4^2$. These vectors are not *linearly independent* since $2(2,0) + 2(0,2) = (0,0)$. However, they are *independent* since if $\alpha(2,0) + \beta(0,2) = (2\alpha, 2\beta) = (0,0)$, then α and β must either be 0 or 2. In either case, both $\alpha(2,0)$ and $\beta(0,2)$ are the zero vector.

Note as well that modular independence does not imply independence nor does independence imply modular independence. We first give an example where modular independence does not imply independence.

Example 6.6.9 Via the Chinese Remainder Theorem, $\mathbb{Z}_{45}$ is isomorphic to $\mathbb{Z}_9 \times \mathbb{Z}_5$. Consider the vectors $(44, 7)$ and $(5, 7)$ in $\mathbb{Z}_{45}^2$. Projecting to $\mathbb{Z}_9$, these vectors are $(8, 7)$ and $(5, 7)$ which are modular independent as follows. If $\alpha_1(8, 7) + \alpha_2(5, 7) = (0, 0)$ in $\mathbb{Z}_9^2$, then $7\alpha_1 + 7\alpha_2 = 0$ and $8\alpha_1 + 5\alpha_2 = 0$ implying $\alpha_1 = -\alpha_2$ and $3\alpha_1 = 0$. Thus $\alpha_1, \alpha_2 \in \langle 3 \rangle$. As $\langle 3 \rangle$ is the maximal ideal of $\mathbb{Z}_9$, the vectors are modular independent over one projection and hence over $\mathbb{Z}_{45}$. However, $15(44, 7) + 3(5, 10) = (0, 0)$, with neither $15(44, 7)$ nor $3(5, 10)$ the zero vector, and so these vectors are not independent.

Next we give an example where independence does not imply modular independence.

Example 6.6.10 Consider the vectors $(9, 0)$ and $(0, 5)$ in $\mathbb{Z}_{45}^2$. Suppose $\alpha_1(9, 0) + \alpha_2(0, 5) = (0, 0)$ with $\alpha_1, \alpha_2 \in \mathbb{Z}_{45}$. Then $9\alpha_1 = 0$ and $5\alpha_2 = 0$ implying α_1 is a multiple of 5 and α_2 is a multiple of 9. In either case, both $\alpha_1(9, 0)$ and $\alpha_2(0, 5)$ are the zero vector. Hence these vectors are independent. However, projecting to $\mathbb{Z}_9$, these vectors are $(0, 0)$ and $(0, 5)$ and projecting to $\mathbb{Z}_5$, these vectors are $(4, 0)$ and $(0, 0)$. Neither set is modular independent. Therefore the vectors are not modular independent over $\mathbb{Z}_{45}$.

This leads us to the following definition.

Definition 6.6.11 Let $\mathcal{C}$ be a code over a finite commutative Frobenius ring R. The codewords $\mathbf{v}_1, \mathbf{v}_2, \ldots, \mathbf{v}_k$ are called a **basis** of $\mathcal{C}$ if they are independent, modular independent, and generate $\mathcal{C}$.

Theorem 6.6.12 ([632]) *Let R be a finite commutative Frobenius ring.*

(a) *If $\mathbf{v}_1, \mathbf{v}_2, \ldots, \mathbf{v}_k$ are modular independent over R and $\sum_{j=1}^{k} \alpha_j \mathbf{v}_j = \mathbf{0}$, then α_j is not a unit in R for all j.*

(b) *If $\mathbf{v}_1, \mathbf{v}_2, \ldots, \mathbf{v}_k$ are independent over R and $\alpha \mathbf{w} \notin \langle \mathbf{v}_1, \mathbf{v}_2, \ldots, \mathbf{v}_k \rangle$, for any $\alpha \neq 0$, then $\mathbf{v}_1, \mathbf{v}_2, \ldots, \mathbf{v}_k, \mathbf{w}$ are independent.*

(c) *If R is a principal ideal ring and C is an arbitrary code over R, then any basis for C contains exactly r codewords, where r is the rank of C.*

Note that over a local ring, the definition of independence implies modular independence. Namely, if $\mathbf{v}_1, \mathbf{v}_2, \ldots, \mathbf{v}_k$ are independent over a local ring R with maximal ideal $\mathfrak{m}$, then $\sum_{j=1}^{k} \alpha_j \mathbf{v}_j = \mathbf{0}$ implies each $\alpha_j \mathbf{v}_j = \mathbf{0}$. Note that the non-units of R are precisely the elements in $\mathfrak{m}$. If α_j is a unit of R, then $\mathbf{v}_j = \mathbf{0}$, a contradiction. So α_j is a not a unit of R implying that $\alpha_j \in \mathfrak{m}$ which shows $\mathbf{v}_1, \mathbf{v}_2, \ldots, \mathbf{v}_k$ are modular independent.

6.7 The Singleton Bound and MDR Codes

The combinatorial version of the Singleton Bound Theorem 1.9.10 applies to all alphabets, even if they do not have an algebraic structure; namely it states that any code C of length n over an alphabet of size q satisfies

$$d_H(C) \leq n - \log_q(|C|) + 1$$

where $d_H(C)$ is the minimum Hamming distance of C. Codes meeting this bound are called **maximum distance separable (MDS)** codes. See Singleton's early paper [1717] for details. The classification of such codes is at present an open question and has proven to be quite intractable; see [630] for a complete description. For linear codes over fields this bound becomes $d_H(C) \leq n - k + 1$ where k is the dimension of the code as a vector space. Since a code over a ring is not a vector space, this algebraic version does not apply. An algebraic version of this has been proven for codes over principal ideal rings. There are additional results for more general classes of rings but they are not as easy to state. Recall that the **rank of a code over a principal ideal ring** is the minimum number of generators of that code.

Theorem 6.7.1 ([1682]) *Let C be a linear code over a finite commutative Frobenius principal ideal ring. Then*
$$d_H(C) \leq n - \mathrm{rank}(C) + 1.$$

A code meeting this bound is said to be a **maximum distance with respect to rank (MDR)** code.[1] An MDR code is not necessarily an MDS code as in the following example.

Example 6.7.2 Let R be a finite principal ideal ring. Let $\mathfrak{a}$ be any nontrivial ideal in R. We know that the ideal has a single generator since all the ideals are principally generated. Therefore the rank of $\mathfrak{a}$ is 1. Its minimum Hamming weight and length are also 1. Then since $1 = 1 - 1 + 1$ the code is an MDR code. This code is, however, not an MDS code since it does not have cardinality $|R|$.

Recall that a code C is a **free** code if it is isomorphic to R^k for some natural number k. The following theorem determines when an MDR code is MDS.

Theorem 6.7.3 ([627]) *Let R be a finite commutative Frobenius ring and let C be a code over R. The code C is an MDS code if and only if C is an MDR code and C is free.*

---

[1]The concept of MDR codes is not to be confused with MRD codes arising in the study of rank-metric codes where MRD codes also satisfy an appropriate Singleton-type bound; see Chapter 11.

One can also use the Chinese Remainder Theorem to construct MDR codes as explained in the following theorem.

Theorem 6.7.4 ([627]) *Let $R = \mathrm{CRT}(R_1, R_2, \ldots, R_s)$ and let $\mathcal{C}_i$ be a code over R_i for $1 \leq i \leq s$. If $\mathcal{C}_i$ is an MDR code for each i, then $\mathcal{C} = \mathrm{CRT}(\mathcal{C}_1, \mathcal{C}_2, \ldots, \mathcal{C}_s)$ is an MDR code. If $\mathcal{C}_i$ is an MDS code of the same rank for each i, then $\mathcal{C} = \mathrm{CRT}(\mathcal{C}_1, \mathcal{C}_2, \ldots, \mathcal{C}_s)$ is an MDS code.*

Various other results can be found for different weights in terms of what constitutes a maximal code; see [634] for examples. These results indicate that we are trying to find codes which are maximal for certain weights given their parameters. Specifically, we are trying to find codes over a ring R which have the largest minimum weight (for a given weight) given its length and type. Essentially, this is a broader question than the main problem of coding theory which seeks to find the largest minimum Hamming distance for a given length and dimension over a field. In terms of rings, this is generalized for various weights and for various forms of generators.

6.8 Conclusion

Codes over rings have become an integral part of algebraic coding theory. Its study started with the realization that certain important binary codes were the images of linear quaternary codes under a Gray map and expanded greatly when it was shown that both MacWilliams theorems generalized to codes over Frobenius rings.

In this chapter, we have laid out foundational results for the study of codes over rings that parallel classical coding theory. These results make it possible to expand the uses of classical coding theory in applications as well as in number theory, the geometry of numbers, ring theory and discrete mathematics. The interested reader can find a complete description of codes over rings in [619, 1668, 1735].

Chapter 7

Quasi-Cyclic Codes

Cem Güneri

Sabancı University

San Ling

Nanyang Technological University

Buket Özkaya

Nanyang Technological University

7.1 Introduction

Cyclic codes are among the most useful and well-studied code families for various reasons, such as effective encoding and decoding. These desirable properties are due to the algebraic structure of cyclic codes, which is rather simple. Namely, a cyclic code can be viewed as an ideal in a certain quotient ring obtained from a polynomial ring with coefficients from a finite field.

It is then natural for coding theorists to search for generalizations of cyclic codes, which also have nice properties. This chapter is devoted to one of the first such generalizations, namely the family of quasi-cyclic (QC) codes. Algebraically, QC codes are modules rather than ideals. As desired, QC codes turned out to be a very useful generalization and their investigation continues intensively today.

There is a vast literature on QC codes and it is impossible to touch upon all aspects in a chapter. This chapter is constrained to some of the most fundamental aspects to which the authors also made contributions over the years. We start with the algebraic structure of QC codes, present how the vectorial and algebraic descriptions are related to each other, and also discuss the code generators. Then we focus on the decomposition of QC codes, via the Chinese Remainder Theorem (CRT decomposition) and concatenation. In fact, these decompositions turn out to be equivalent in some sense, which is also mentioned in the chapter. Decomposition of a QC code yields a trace representation. Moreover, the dual QC code can be described in the decomposed form, which has important consequences of its own. For instance, self-dual and linear complementary dual QC codes can be characterized in terms of their CRT decompositions. Moreover, the asymptotic results presented in the chapter heavily rely on the decomposed structure of QC codes. Three different general minimum distance bounds on QC codes are presented here, with fairly complete proofs or detailed elaborations. A relation between QC codes and convolutional codes is also shown in the end.

7.2 Algebraic Structure

Let $\mathbb{F}_q$ denote the finite field with q elements, where q is a prime power, and let m and ℓ be two positive integers. A linear code $\mathcal{C}$ of length $m\ell$ over $\mathbb{F}_q$ is called a **quasi-cyclic (QC) code of index** ℓ if it is invariant under shift of codewords by ℓ positions and ℓ is the minimal number with this property. Note that if $\ell = 1$, then $\mathcal{C}$ is a cyclic code. If we view codewords of $\mathcal{C}$ as $m \times \ell$ arrays as follows

$$
\mathbf{c} = \begin{bmatrix} c_{0,0} & \cdots & c_{0,\ell-1} \\ \vdots & & \vdots \\ c_{m-1,0} & \cdots & c_{m-1,\ell-1} \end{bmatrix}, \tag{7.1}
$$

then being invariant under a shift by ℓ units amounts to being closed under the row shift where each row is moved downward one row and the bottom row moved to the top.

Consider the principal ideal $I = \langle x^m - 1 \rangle$ of $\mathbb{F}_q[x]$ and define the quotient ring $R := \mathbb{F}_q[x]/I$. If T represents the shift-by-1 operator on $\mathbb{F}_q^{m\ell}$, then its action on $\mathbf{v} \in \mathbb{F}_q^{m\ell}$ will be denoted by $T \cdot \mathbf{v}$. Hence, $\mathbb{F}_q^{m\ell}$ has an $\mathbb{F}_q[x]$-module structure given by the multiplication

$$
\begin{aligned}
\mathbb{F}_q[x] \times \mathbb{F}_q^{m\ell} &\longrightarrow \mathbb{F}_q^{m\ell} \\
(a(x), \mathbf{v}) &\longmapsto a(T^\ell) \cdot \mathbf{v}.
\end{aligned}
$$

Note that, for $a(x) = x^m - 1$, we have $a(T^\ell) \cdot \mathbf{v} = (T^{m\ell}) \cdot \mathbf{v} - \mathbf{v} = 0$. Hence, the ideal I fixes $\mathbb{F}_q^{m\ell}$ and we can view $\mathbb{F}_q^{m\ell}$ as an R-module. Therefore, a QC code $\mathcal{C} \subseteq \mathbb{F}_q^{m\ell}$ of index ℓ is an R-submodule of $\mathbb{F}_q^{m\ell}$.

To an element $\mathbf{c} \in \mathbb{F}_q^{m \times \ell} \cong \mathbb{F}_q^{m\ell}$ as in (7.1), we associate an element $\mathbf{c}(x)$ of R^ℓ as

$$
\mathbf{c}(x) := (c_0(x), c_1(x), \ldots, c_{\ell-1}(x)) \in R^\ell,
$$

where, for each $0 \leq j \leq \ell - 1$,

$$c_j(x) := c_{0,j} + c_{1,j}x + c_{2,j}x^2 + \cdots + c_{m-1,j}x^{m-1} \in R.$$

Thus, the following map is an R-module isomorphism:

$$\phi : \quad \mathbb{F}_q^{m\ell} \quad \longrightarrow \quad R^\ell$$

$$\mathbf{c} = \begin{bmatrix} c_{0,0} & \cdots & c_{0,\ell-1} \\ \vdots & & \vdots \\ c_{m-1,0} & \cdots & c_{m-1,\ell-1} \end{bmatrix} \quad \longmapsto \quad \mathbf{c}(x). \tag{7.2}$$

Note that, for $\ell = 1$, this amounts to the classical polynomial representation of cyclic codes. Observe that the T^ℓ-shift on $\mathbb{F}_q^{m\ell}$ corresponds to the componentwise multiplication by x in R^ℓ. Therefore, a q-ary QC code $\mathcal{C}$ of length $m\ell$ and index ℓ can be considered as an R-submodule of R^ℓ.

Lally and Fitzpatrick proved in [1191] that every quasi-cyclic code, viewed as an R-submodule in R^ℓ, has a generating set in the form of a reduced Gröbner basis. In order to explain their findings, we need to fix some further notation first.

Consider the following ring homomorphism:

$$\Psi : \mathbb{F}_q[x]^\ell \quad \longrightarrow \quad R^\ell$$

$$(f_0(x), f_1(x), \ldots, f_{\ell-1}(x)) \quad \longmapsto \quad (f_0(x) + I, f_1(x) + I, \ldots, f_{\ell-1}(x) + I).$$

Given a QC code $\mathcal{C} \subseteq R^\ell$, it follows that the preimage $\Psi^{-1}(\mathcal{C}) = \widetilde{\mathcal{C}}$ of $\mathcal{C}$ in $\mathbb{F}_q[x]^\ell$ is an $\mathbb{F}_q[x]$-submodule containing $\widetilde{K} = \{(x^m - 1)\mathbf{e}_j \mid 0 \leq j \leq \ell - 1\}$, where $\mathbf{e}_j$ denotes the standard basis vector of length ℓ with 1 at the coordinate j and 0 elsewhere. Throughout, the tilde ~ represents structures over $\mathbb{F}_q[x]$.

Since $\widetilde{\mathcal{C}}$ is a submodule of the finitely generated free module $\mathbb{F}_q[x]^\ell$ over the principal ideal domain $\mathbb{F}_q[x]$ and it contains $\widetilde{K}$, it has a generating set of the form

$$\{\mathbf{u}_1, \ldots, \mathbf{u}_p, (x^m - 1)\mathbf{e}_0, \ldots, (x^m - 1)\mathbf{e}_{\ell-1}\},$$

where p is a nonnegative integer and $\mathbf{u}_b = (u_{b,0}(x), \ldots, u_{b,\ell-1}(x)) \in \mathbb{F}_q[x]^\ell$ for each $b \in \{1, \ldots, p\}$. Hence, the rows of the matrix

$$M = \begin{bmatrix} u_{1,0}(x) & \cdots & u_{1,\ell-1}(x) \\ \vdots & & \vdots \\ u_{p,0}(x) & \cdots & u_{p,\ell-1}(x) \\ x^m - 1 & \cdots & 0 \\ \vdots & \ddots & \vdots \\ 0 & \cdots & x^m - 1 \end{bmatrix}$$

generate $\widetilde{\mathcal{C}}$. By using elementary row operations, we may triangularize M so that another generating set can be obtained from the rows of an upper-triangular $\ell \times \ell$ matrix with entries in $\mathbb{F}_q[x]$ as follows:

$$\widetilde{G}(x) = \begin{bmatrix} g_{0,0}(x) & g_{0,1}(x) & \cdots & g_{0,\ell-1}(x) \\ 0 & g_{1,1}(x) & \cdots & g_{1,\ell-1}(x) \\ \vdots & \vdots & \ddots & \vdots \\ 0 & 0 & \cdots & g_{\ell-1,\ell-1}(x) \end{bmatrix}, \tag{7.3}$$

where $\widetilde{G}(x)$ satisfies the following conditions (see [1191, Theorem 2.1]):

(1) $g_{i,j}(x) = 0$ for all $0 \leq j < i \leq \ell - 1$.

(2) $\deg(g_{i,j}(x)) < \deg(g_{j,j}(x))$ for all $i < j$.

(3) $g_{i,i}(x) \mid (x^m - 1)$ for all $0 \leq i \leq \ell - 1$.

(4) If $g_{i,i}(x) = x^m - 1$, then $g_{i,j}(x) = 0$ for all $i \neq j$.

Note that the rows of $\widetilde{G}(x)$ are nonzero, and each nonzero codeword of $\widetilde{\mathcal{C}}$ can be expressed in the form $(0, \ldots, 0, c_j(x), \ldots, c_{\ell-1}(x))$ where $j \geq 0$, $c_j(x) \neq 0$, and $g_{j,j}(x) \mid c_j(x)$. This implies that the rows of $\widetilde{G}(x)$ form a Gröbner basis of $\widetilde{\mathcal{C}}$ with respect to the position-over-term (POT) order in $\mathbb{F}_q[x]$, where the standard basis vectors $\{\mathbf{e}_0, \ldots, \mathbf{e}_{\ell-1}\}$ and the monomials x^i are ordered naturally in each component. Moreover, the second condition above implies that the rows of $\widetilde{G}(x)$ form a reduced Gröbner basis of $\widetilde{\mathcal{C}}$, which is uniquely defined up to multiplication by constants with monic diagonal elements.

Let $G(x)$ now be the matrix with the rows of $\widetilde{G}(x)$ under the image of the homomorphism Ψ. Clearly, the rows of $G(x) := \widetilde{G}(x) \mod I$ form an R-generating set for $\mathcal{C}$. When $\mathcal{C}$ is the zero code of length $m\ell$, we have $p = 0$ which implies $G(x) = \mathbf{0}_{\ell \times \ell}$. Otherwise we say that $\mathcal{C}$ is an r-**generator QC code** (generated as an R-submodule of R^ℓ) if $G(x)$ has r (nonzero) rows. The $\mathbb{F}_q$-dimension of $\mathcal{C}$ is given by (see [1191, Corollary 2.4] for the proof)

$$m\ell - \sum_{i=0}^{\ell-1} \deg(g_{i,i}(x)) = \sum_{i=0}^{\ell-1} (m - \deg(g_{i,i}(x))).$$

7.3 Decomposition of Quasi-Cyclic Codes

From this section on, we assume that $\gcd(m, q) = 1$.

7.3.1 The Chinese Remainder Theorem and Concatenated Decompositions of QC Codes

We now describe the decomposition of a QC code over $\mathbb{F}_q$ into shorter codes over extension fields of $\mathbb{F}_q$. We follow the brief presentation in [869] and refer the reader to [1269] for details. Let the polynomial $x^m - 1$ factor into irreducible polynomials in $\mathbb{F}_q[x]$ as

$$x^m - 1 = f_1(x)f_2(x)\cdots f_s(x). \tag{7.4}$$

Since m is relatively prime to q, there are no repeating factors in (7.4). By the Chinese Remainder Theorem (CRT), given for finite commutative rings in Theorem 6.4.12, we have the following ring isomorphism:

$$R \cong \bigoplus_{i=1}^{s} \mathbb{F}_q[x]/\langle f_i(x)\rangle. \tag{7.5}$$

Since each $f_i(x)$ divides $x^m - 1$, its roots are powers of some fixed primitive m^{th} root of unity ξ in an extension field of $\mathbb{F}_q$. For each $i = 1, 2, \ldots, s$, let u_i be the smallest nonnegative integer such that $f_i(\xi^{u_i}) = 0$. Since the $f_i(x)$'s are irreducible, the direct summands in (7.5) are field extensions of $\mathbb{F}_q$. If $\mathbb{E}_i := \mathbb{F}_q[x]/\langle f_i(x)\rangle$ for $1 \leq i \leq s$, then we have

$$\begin{aligned} R &\cong \mathbb{E}_1 \oplus \cdots \oplus \mathbb{E}_s \\ a(x) &\mapsto \left(a(\xi^{u_1}), \ldots, a(\xi^{u_s})\right) \end{aligned} \tag{7.6}$$

This implies that
$$R^\ell \cong \mathbb{E}_1^\ell \oplus \cdots \oplus \mathbb{E}_s^\ell.$$

Hence, a QC code $\mathcal{C} \subseteq R^\ell$ can be viewed as an $(\mathbb{E}_1 \oplus \cdots \oplus \mathbb{E}_s)$-submodule of $\mathbb{E}_1^\ell \oplus \cdots \oplus \mathbb{E}_s^\ell$ and **decomposes** as
$$\mathcal{C} \cong \mathcal{C}_1 \oplus \cdots \oplus \mathcal{C}_s, \tag{7.7}$$

where $\mathcal{C}_i$ is a linear code of length ℓ over $\mathbb{E}_i$, for each i. These length ℓ linear codes over various extension fields of $\mathbb{F}_q$ are called the **constituents** of $\mathcal{C}$. Let $\mathcal{C} \subseteq R^\ell$ be r-generated as an R-module by

$$\left\{ \left(a_{1,0}(x), \ldots, a_{1,\ell-1}(x)\right), \ldots, \left(a_{r,0}(x), \ldots, a_{r,\ell-1}(x)\right) \right\} \subseteq R^\ell.$$

Then for $1 \le i \le s$, we have

$$\mathcal{C}_i = \mathrm{span}_{\mathbb{E}_i} \left\{ \left(a_{b,0}(\xi^{u_i}), \ldots, a_{b,\ell-1}(\xi^{u_i})\right) \mid 1 \le b \le r \right\}. \tag{7.8}$$

Note that each extension field $\mathbb{E}_i$ above is isomorphic to a minimal cyclic code of length m over $\mathbb{F}_q$, namely, the cyclic code whose check polynomial is $f_i(x)$. If we denote by θ_i the generating primitive idempotent for the minimal cyclic code in consideration (see Section 2.3), then the isomorphism is given by the maps

$$
\begin{array}{ccc}
\varphi_i : \langle \theta_i \rangle & \longrightarrow & \mathbb{E}_i \\
a(x) & \longmapsto & a(\xi^{u_i})
\end{array}
\quad \text{and} \quad
\begin{array}{ccc}
\psi_i : \mathbb{E}_i & \longrightarrow & \langle \theta_i \rangle \\
\delta & \longmapsto & \sum_{k=0}^{m-1} a_k x^k
\end{array}, \tag{7.9}
$$

where
$$a_k = \frac{1}{m} \mathrm{Tr}_{\mathbb{E}_i/\mathbb{F}_q}(\delta \xi^{-k u_i}).$$

Here, $\mathrm{Tr}_{\mathbb{E}_i/\mathbb{F}_q}$ denotes the trace map from $\mathbb{E}_i$ onto $\mathbb{F}_q$; if $[\mathbb{E}_i : \mathbb{F}_q] = e_i$, then $\mathrm{Tr}_{\mathbb{E}_i/\mathbb{F}_q}(x) = \mathrm{Tr}_{q^{e_i}/q}(x) = x + x^q + \cdots + x^{q^{e_i-1}}$. If $\mathcal{C}_i$ is a length ℓ linear code over $\mathbb{E}_i$, we denote its concatenation with $\langle \theta_i \rangle$ by $\langle \theta_i \rangle \square \mathcal{C}_i$ and the concatenation is carried out by the map ψ_i, extended to $\mathbb{E}_i^\ell$. In other words, ψ_i is applied to each symbol of the codeword in $\mathcal{C}_i$ to produce an element of $\langle \theta_i \rangle^\ell$.

Jensen gave the following **concatenated description** for QC codes.

Theorem 7.3.1 ([1042])

(a) *Let $\mathcal{C}$ be an R-submodule of R^ℓ (i.e., a QC code). Then for some subset $\mathcal{I}$ of $\{1, \ldots, s\}$, there exist linear codes $\mathcal{C}_i$ of length ℓ over $\mathbb{E}_i$, which can be explicitly described, such that*
$$\mathcal{C} = \bigoplus_{i \in \mathcal{I}} \langle \theta_i \rangle \square \mathcal{C}_i.$$

(b) *Conversely, let $\mathcal{C}_i$ be a linear code in $\mathbb{E}_i^\ell$ for each $i \in \mathcal{I} \subseteq \{1, \ldots, s\}$. Then*
$$\mathcal{C} = \bigoplus_{i \in \mathcal{I}} \langle \theta_i \rangle \square \mathcal{C}_i$$

is a q-ary QC code of length $m\ell$ and index ℓ.

It was proven in [869, Theorem 4.1] that, for a given QC code $\mathcal{C}$, the constituents $\mathcal{C}_i$'s in (7.7) and the outer codes $\mathcal{C}_i$'s in the concatenated structure of Theorem 7.3.1 are the same.

7.3.2 Applications

In this section, we present some constructions and characterizations of QC codes using their CRT decomposition.

7.3.2.1 Trace Representation

By (7.9) and Theorem 7.3.1, an arbitrary codeword $\mathbf{c} \in \mathcal{C}$ can be written as an $m \times \ell$ array in the form (see [1269, Theorem 5.1])

$$
\mathbf{c} = \frac{1}{m}
\begin{bmatrix}
\left(\sum\limits_{i=1}^{s} \mathrm{Tr}_{\mathbb{E}_i/\mathbb{F}_q} \left(\lambda_{i,t} \xi^{-0u_i} \right) \right)_{0 \le t \le \ell-1} \\
\vdots \\
\left(\sum\limits_{i=1}^{s} \mathrm{Tr}_{\mathbb{E}_i/\mathbb{F}_q} \left(\lambda_{i,t} \xi^{-(m-1)u_i} \right) \right)_{0 \le t \le \ell-1}
\end{bmatrix}
\tag{7.10}
$$

where $\lambda_i = (\lambda_{i,0}, \dots, \lambda_{i,\ell-1}) \in \mathcal{C}_i$ for all i. Since $m\mathcal{C} = \mathcal{C}$, every codeword in $\mathcal{C}$ can still be written in the form of (7.10) with the constant $\frac{1}{m}$ removed. Note that the row shift invariance of codewords amounts to being closed under multiplication by ξ^{-1} in this representation.

Let $\mathbb{F} := \mathbb{F}_q(\xi^{u_1}, \dots, \xi^{u_s})$ be the splitting field of $x^m - 1$ (i.e., the smallest field containing all the $\mathbb{E}_i$'s), and let $k_1, \dots, k_s \in \mathbb{F}$ with $\mathrm{Tr}_{\mathbb{F}/\mathbb{E}_i}(k_i) = 1$ for each i. We can now unify the traces and rewrite $\mathbf{c} \in \mathcal{C}$ as follows (see [869]):

$$
\mathbf{c} = \left[\left(\mathrm{Tr}_{\mathbb{F}/\mathbb{F}_q} \left(\sum_{i=1}^{s} k_i \lambda_{i,t} \xi^{-ju_i} \right) \right)_{\substack{0 \le t \le \ell-1 \\ 0 \le j \le m-1}} \right].
\tag{7.11}
$$

Note that the case $\ell = 1$ gives us the trace representation of a cyclic code of length m, in the sense of the following formulation:

Theorem 7.3.2 ([1899, Proposition 2.1]) *Let* $\gcd(m, q) = 1$ *and let* ξ *be a primitive* m^{th} *root of unity in some extension field* $\mathbb{F}$ *of* $\mathbb{F}_q$. *Assume that* $u_1, \dots, u_s$ *are nonnegative integers and let* $\mathcal{C}$ *be a q-ary cyclic code of length m such that the **basic set of zeros** of its dual is* $BZ(\mathcal{C}^\perp) = \{\xi^{u_1}, \xi^{u_2}, \dots, \xi^{u_s}\}$ *(i.e., the generator polynomial of $\mathcal{C}^\perp$ is* $\prod_i m_i(x)$, *where $m_i(x) \in \mathbb{F}_q[x]$ is the minimal polynomial of ξ^{u_i} over $\mathbb{F}_q$). Then*

$$
\mathcal{C} = \left\{ \left(\mathrm{Tr}_{\mathbb{F}/\mathbb{F}_q}(c_1 \xi^{ju_1} + \dots + c_s \xi^{ju_s}) \right)_{0 \le j \le m-1} \mid c_1, \dots, c_s \in \mathbb{F} \right\}.
$$

Hence, the columns of any codeword in the QC code $\mathcal{C}$ viewed as in (7.11) lie in the cyclic code $D \subseteq \mathbb{F}_q^m$ with $BZ(D^\perp) = \{\xi^{-u_1}, \dots, \xi^{-u_s}\}$; see [868, Proposition 4.2].

Example 7.3.3 Let $m = 3$ and $q \equiv 2 \pmod{3}$ such that $x^3 - 1$ factors into $x - 1$ and $x^2 + x + 1$ over $\mathbb{F}_q$. By (7.5) and (7.6), we obtain

$$
R = \mathbb{F}_q[x]/\langle x^3 - 1 \rangle \cong \mathbb{F}_q[x]/\langle x - 1 \rangle \oplus \mathbb{F}_q[x]/\langle x^2 + x + 1 \rangle \cong \mathbb{F}_q \oplus \mathbb{F}_{q^2}.
$$

Therefore, any QC code $\mathcal{C}$ of length 3ℓ and index ℓ has two linear constituents $\mathcal{C}_1 \subseteq \mathbb{F}_q^\ell$ and $\mathcal{C}_2 \subseteq \mathbb{F}_{q^2}^\ell$ such that $\mathcal{C} \cong \mathcal{C}_1 \oplus \mathcal{C}_2$. By (7.11), the codewords of $\mathcal{C}$ are of the form (see [1269, Theorem 6.7])

$$
\begin{bmatrix}
\mathbf{z} + 2\mathbf{a} - \mathbf{b} \\
\mathbf{z} - \mathbf{a} + 2\mathbf{b} \\
\mathbf{z} - \mathbf{a} - \mathbf{b}
\end{bmatrix},
$$

where $\mathbf{z} \in \mathcal{C}_1$, $\mathbf{a} + \beta\mathbf{b} \in \mathcal{C}_2$ with $\mathbf{a}, \mathbf{b} \in \mathbb{F}_q^\ell$ and $\beta \in \mathbb{F}_{q^2}$ such that $\beta^2 + \beta + 1 = 0$.

In particular, let $q = 2^t$, where t is odd (otherwise, $2^t \equiv 1 \pmod{3}$ if t is even and $x^3 - 1$ splits into linear factors). By using the fact that $2\mathbf{a} = 2\mathbf{b} = \mathbf{0}_\ell$ in even characteristic, we can simplify the above expression further and write $\mathcal{C}$ as

$$\mathcal{C} = \left\{ \begin{bmatrix} \mathbf{z} + \mathbf{b} \\ \mathbf{z} + \mathbf{a} \\ \mathbf{z} + \mathbf{a} + \mathbf{b} \end{bmatrix} \;\middle|\; \mathbf{z} \in \mathcal{C}_1, \; \mathbf{a} + \beta\mathbf{b} \in \mathcal{C}_2 \right\}. \tag{7.12}$$

Example 7.3.4 Let $m = 5$ and let q be such that $x^4 + x^3 + x^2 + x + 1$ is irreducible over $\mathbb{F}_q$. Let $\alpha \in \mathbb{F}_{q^4}$ such that $\alpha^4 + \alpha^3 + \alpha^2 + \alpha + 1 = 0$, and let $\text{Tr} = \text{Tr}_{\mathbb{F}_{q^4}/\mathbb{F}_q}$. Let $\mathcal{C}_1 \subseteq \mathbb{F}_q^\ell$ and $\mathcal{C}_2 \subseteq \mathbb{F}_{q^4}^\ell$ be two linear codes. Then the code (see [1269, Theorem 6.14])

$$\mathcal{C} = \left\{ \begin{bmatrix} \mathbf{z} + \text{Tr}(\mathbf{y}) \\ \mathbf{z} + \text{Tr}(\alpha^{-1}\mathbf{y}) \\ \mathbf{z} + \text{Tr}(\alpha^{-2}\mathbf{y}) \\ \mathbf{z} + \text{Tr}(\alpha^{-3}\mathbf{y}) \\ \mathbf{z} + \text{Tr}(\alpha^{-4}\mathbf{y}) \end{bmatrix} \;\middle|\; \mathbf{z} \in \mathcal{C}_1, \; \mathbf{y} \in \mathcal{C}_2 \right\}$$

is a QC code of length 5ℓ and index ℓ over $\mathbb{F}_q$.

In particular, let $q = 2^t$ and set $\mathbf{y} = \mathbf{c} + \alpha\mathbf{d} + \alpha^2\mathbf{e} + \alpha^3\mathbf{f}$ for some $\mathbf{c}, \mathbf{d}, \mathbf{e}, \mathbf{f} \in \mathbb{F}_q^\ell$. We can rewrite the codewords in $\mathcal{C}$ as

$$\begin{bmatrix} \mathbf{z} + \mathbf{d} + \mathbf{e} + \mathbf{f} \\ \mathbf{z} + \mathbf{c} + \mathbf{e} + \mathbf{f} \\ \mathbf{z} + \mathbf{c} + \mathbf{d} + \mathbf{f} \\ \mathbf{z} + \mathbf{c} + \mathbf{d} + \mathbf{e} \\ \mathbf{z} + \mathbf{c} + \mathbf{d} + \mathbf{e} + \mathbf{f} \end{bmatrix}$$

where $\mathbf{z} \in \mathcal{C}_1$ and $\mathbf{c} + \alpha\mathbf{d} + \alpha^2\mathbf{e} + \alpha^3\mathbf{f} \in \mathcal{C}_2$.

7.3.2.2 Self-Dual and Complementary Dual QC Codes

Observe that the monic polynomial $x^m - 1$ is self-reciprocal. We rewrite the factorization into irreducible polynomials given in (7.4) as follows, which is needed for the dual code analysis (see also [1269]):

$$x^m - 1 = g_1(x) \cdots g_n(x) h_1(x) h_1^*(x) \cdots h_p(x) h_p^*(x) \tag{7.13}$$

where $g_i(x)$ is self-reciprocal for all $1 \leq i \leq n$, and $h_t^*(x) = x^{\deg(h_t)} h_t(x^{-1})$ denotes the reciprocal polynomial of $h_t(x)$ for all $1 \leq t \leq p$. Note that $s = n + 2p$ since (7.4) and (7.13) must be identical.

Let $\mathbb{G}_i := \mathbb{F}_q[x]/\langle g_i(x)\rangle$, $\mathbb{H}_t' := \mathbb{F}_q[x]/\langle h_t(x)\rangle$, and $\mathbb{H}_t'' := \mathbb{F}_q[x]/\langle h_t^*(x)\rangle$ for each i and t. By the CRT, the decomposition of R in (7.5) now becomes

$$R \cong \left(\bigoplus_{i=1}^{n} \mathbb{G}_i \right) \oplus \left(\bigoplus_{t=1}^{p} \left(\mathbb{H}_t' \oplus \mathbb{H}_t'' \right) \right),$$

which implies

$$R^\ell \cong \left(\bigoplus_{i=1}^{n} \mathbb{G}_i^\ell \right) \oplus \left(\bigoplus_{t=1}^{p} (\mathbb{H}_t')^\ell \oplus (\mathbb{H}_t'')^\ell \right).$$

Hence, a QC code $C \subseteq R^\ell$ viewed as an R-submodule of R^ℓ now decomposes as (see (7.7))

$$C \cong \left(\bigoplus_{i=1}^{n} C_i \right) \oplus \left(\bigoplus_{t=1}^{p} \left(C_t' \oplus C_t'' \right) \right), \qquad (7.14)$$

where the C_i's are the $\mathbb{G}_i$-linear constituents of C of length ℓ for all $i = 1, \ldots, n$, and the C_t''s and C_t'''s are the $\mathbb{H}_t'$-linear and $\mathbb{H}_t''$-linear constituents of C of length ℓ, respectively, for all $t = 1, \ldots, p$. By fixing roots corresponding to the irreducible factors $g_i(x), h_t(x), h_t^*(x)$ and via the CRT, one can write explicitly the constituents, as in (7.8), in this setting as well. Observe that $\mathbb{H}_t' = \mathbb{H}_t''$ since $\mathbb{F}_q(\xi^a) = \mathbb{F}_q(\xi^{-a})$ for any $a \in \{0, 1, \ldots, m - 1\}$.

Since $g_i(x)$ is self-reciprocal, the cardinality of $\mathbb{G}_i$, say q_i, is an even power of q for all $1 \leq i \leq n$ with two exceptions. One of these exceptions, for all m and q, is the field coming from the irreducible factor $x - 1$ of $x^m - 1$. When q is odd and m is even, $x + 1$ is another self-reciprocal irreducible factor of $x^m - 1$. In these cases, $q_i = q$. Except for these two cases, we equip each $\mathbb{G}_i^\ell$ with the inner product

$$\langle \mathbf{c}, \mathbf{d} \rangle := \sum_{j=0}^{\ell-1} c_j d_j^{\sqrt{q_i}}, \qquad (7.15)$$

where $\mathbf{c} = (c_0, \ldots, c_{\ell-1}), \mathbf{d} = (d_0, \ldots, d_{\ell-1}) \in \mathbb{G}_i^\ell$. This is the Hermitian inner product. For the two exceptions, in which case the corresponding field $\mathbb{G}_i$ is $\mathbb{F}_q$, we equip $\mathbb{G}_i^\ell$ with the usual Euclidean inner product. With the appropriate inner product, Hermitian or Euclidean, on $\mathbb{G}_i$, $\perp_{\mathbb{G}}$ denotes the dual on $\mathbb{G}_i^\ell$. For each $1 \leq t \leq p$, $\mathbb{H}_t'^\ell = \mathbb{H}_t''^\ell$ is also equipped with the Euclidean inner product; $\perp_e$ denotes the dual on $\mathbb{H}_t'^\ell = \mathbb{H}_t''^\ell$.

The dual of a QC code is also QC and the following result is immediate.

Proposition 7.3.5 ([1269, Theorem 4.2]) *Let C be a QC code with CRT decomposition as in (7.14). Then its dual code $C^\perp$ (under the Euclidean inner product) is of the form*

$$C^\perp = \left(\bigoplus_{i=1}^{n} C_i^{\perp_{\mathbb{G}}} \right) \oplus \left(\bigoplus_{t=1}^{p} \left(C_t''^{\perp_e} \oplus C_t'^{\perp_e} \right) \right). \qquad (7.16)$$

Recall that a linear code C is said to be **self-dual** if $C = C^\perp$, and C is called a **linear code with complementary dual (LCD)** if $C \cap C^\perp = \{\mathbf{0}\}$. We now characterize self-dual and LCD QC codes (QCCD codes) via their constituents; see [871, 1270].

Theorem 7.3.6 *Let C be a q-ary QC code of length $m\ell$ and index ℓ with a CRT decomposition as in (7.14).*

(a) *C is (Euclidean) self-dual if and only if $C_i = C_i^{\perp_{\mathbb{G}}}$ for all $1 \leq i \leq n$, and $C_t'' = C_t'^{\perp_e}$ over $\mathbb{H}_t' = \mathbb{H}_t''$ for all $1 \leq t \leq p$.*

(b) *C is (Euclidean) LCD if and only if $C_i \cap C_i^{\perp_{\mathbb{G}}} = \{\mathbf{0}\}$ for all $1 \leq i \leq n$, and $C_t' \cap C_t''^{\perp_e} = \{\mathbf{0}\}$, $C_t'' \cap C_t'^{\perp_e} = \{\mathbf{0}\}$ over $\mathbb{H}_t' = \mathbb{H}_t''$ for all $1 \leq t \leq p$.*

Proof: The proof is immediate from the CRT decompositions of C in (7.14) and of its dual $C^\perp$ in (7.16). $\qquad \square$

The following special cases are easy to derive from Theorem 7.3.6.

Corollary 7.3.7

(a) *If the CRT decomposition of C is as in (7.14) with $C_i = C_i^{\perp_{\mathbb{G}}}$ for all $1 \leq i \leq n$, and $C_t' = \{\mathbf{0}\}$, $C_t'' = (\mathbb{H}_t'')^\ell$ (or $C_t'' = \{\mathbf{0}\}$, $C_t' = (\mathbb{H}_t')^\ell$) for all $1 \leq t \leq p$, then C is self-dual.*

(b) *If the CRT decomposition of C is as in (7.14) with $C_i \cap C_i^{\perp_{\mathbb{G}}} = \{\mathbf{0}\}$ for all $1 \leq i \leq n$, and Euclidean LCD codes $C_t' = C_t''$ over $\mathbb{H}_t' = \mathbb{H}_t''$ for all $1 \leq t \leq p$, then C is LCD.*

7.4 Minimum Distance Bounds

Given a q-ary code $\mathcal{C}$ that is not the trivial zero code, the minimum (Hamming) distance of $\mathcal{C}$ is defined as $d_H(\mathcal{C}) := \min\{\text{wt}_H(\mathbf{c}) \mid \mathbf{0} \neq \mathbf{c} \in \mathcal{C}\}$, where $\text{wt}_H(\mathbf{c})$ denotes the number of nonzero entries in $\mathbf{c} \in \mathcal{C}$. Using the tools described in Sections 7.2 and 7.3, we address some lower bounds on the minimum distance of a given QC code in this section.

7.4.1 The Jensen Bound

The concatenated structure in Theorem 7.3.1 yields a minimum distance bound for QC codes ([1042, Theorem 4]). This bound is a consequence of the work of Blokh and Zyablov in [222], which holds for general concatenated codes.

Theorem 7.4.1 (Jensen Bound) *Let $\mathcal{C}$ be a q-ary QC code of length $m\ell$ and index ℓ with the concatenated structure*

$$\mathcal{C} = \bigoplus_{t=1}^{g} \langle \theta_t \rangle \square \mathcal{C}_t,$$

where, without loss of generality, we have chosen the indices t such that $0 < d_H(\mathcal{C}_1) \leq d_H(\mathcal{C}_2) \leq \cdots \leq d_H(\mathcal{C}_g)$. Then we have

$$d_H(\mathcal{C}) \geq \min_{1 \leq e \leq g} \left\{ d_H(\mathcal{C}_e) d_H\left(\langle \theta_1 \rangle \oplus \cdots \oplus \langle \theta_e \rangle \right) \right\}.$$

Proof: Let $\mathbf{c}$ be a nonzero codeword in $\mathcal{C}$. By the concatenated structure of Section 7.3.1 and (7.9),

$$\mathbf{c} = \sum_{i=1}^{g} \psi_i(\mathbf{c}_i),$$

with $\mathbf{c}_i \in \mathcal{C}_i$ for all $1 \leq i \leq g$. Assume that e is the maximal index such that $\mathbf{c}_e$ is a nonzero vector and $\mathbf{c}_{e+1}, \ldots, \mathbf{c}_g$ are all zero vectors. Hence,

$$\mathbf{c} = \psi_1(\mathbf{c}_1) + \cdots + \psi_e(\mathbf{c}_e),$$

and each $\psi_i(\mathbf{c}_i) \in \langle \theta_i \rangle^\ell$. Since the sum of $\langle \theta_i \rangle$'s is direct, the codeword $\mathbf{c}$ belongs to $(\langle \theta_1 \rangle \oplus \cdots \oplus \langle \theta_e \rangle)^\ell$. Note that $\mathbf{c}_e$ has at least $d_H(\mathcal{C}_e)$ nonzero coordinates, all of which are mapped to a nonzero vector in $\langle \theta_e \rangle$ by ψ_e. This implies that $\mathbf{c}$ has at least $d_H(\mathcal{C}_e)$ nonzero coordinates, since each coordinate of $\mathbf{c}$ belongs to the direct sum $\langle \theta_1 \rangle \oplus \cdots \oplus \langle \theta_e \rangle$. Hence the result follows. $\qquad\square$

7.4.2 The Lally Bound

Let $\mathcal{C}$ be a QC code of length $m\ell$ and index ℓ over $\mathbb{F}_q$. Let $\{1, \alpha, \ldots, \alpha^{\ell-1}\}$ be some fixed choice of basis of $\mathbb{F}_{q^\ell}$ as a vector space over $\mathbb{F}_q$. We view the codewords of $\mathcal{C}$ as $m \times \ell$ arrays as in (7.1) and consider the following map:

$$\Phi: \quad \mathbb{F}_q^{m\ell} \quad \longrightarrow \quad \mathbb{F}_{q^\ell}^m$$

$$\mathbf{c} = \begin{bmatrix} c_{0,0} & \cdots & c_{0,\ell-1} \\ \vdots & & \vdots \\ c_{m-1,0} & \cdots & c_{m-1,\ell-1} \end{bmatrix} \longmapsto \begin{bmatrix} c_0 \\ \vdots \\ c_{m-1} \end{bmatrix}$$

where $c_i = c_{i,0} + c_{i,1}\alpha + \cdots + c_{i,\ell-1}\alpha^{\ell-1} \in \mathbb{F}_{q^\ell}$ for all $0 \leq i \leq m-1$.

Clearly $\Phi(\mathbf{c})$ lies in some cyclic code for any $\mathbf{c} \in \mathcal{C}$. We now define the smallest such cyclic code as $\widehat{\mathcal{C}}$. First, we equivalently extend the map Φ above to the polynomial description of codewords as in (7.2):

$$\Phi : \mathbb{F}_q[x]^\ell \longrightarrow \mathbb{F}_{q^\ell}[x]$$

$$\mathbf{c}(x) = (c_0(x), \ldots, c_{\ell-1}(x)) \longmapsto c(x) = \sum_{j=0}^{\ell-1} c_j(x)\alpha^j. \tag{7.17}$$

If $\mathcal{C}$ has generating set $\{\mathbf{f}_1, \ldots, \mathbf{f}_r\}$ where $\mathbf{f}_k = \left(f_0^{(k)}(x), \ldots, f_{\ell-1}^{(k)}(x)\right) \in \mathbb{F}_q[x]^\ell$ for each $k \in \{1, \ldots, r\}$, then $\widehat{\mathcal{C}} = \langle \gcd(f_1(x), \ldots, f_r(x), x^m - 1) \rangle$ with $f_k = \Phi(\mathbf{f}_k) \in \mathbb{F}_{q^\ell}[x]$ for all $k \in \{1, \ldots, r\}$ [1189].

Next, we consider the q-ary linear code of length ℓ that is generated by the rows of the codewords in $\mathcal{C}$, which are represented as $m \times \ell$ arrays as in (7.1). Namely, let $\mathcal{B}$ be the linear block code of length ℓ over $\mathbb{F}_q$ generated by $\{\mathbf{f}_{k,i} \mid k \in \{1, \ldots, r\}, i \in \{0, \ldots, m-1\}\} \subseteq \mathbb{F}_q^\ell$, where each $\mathbf{f}_{k,i} := \left(f_{i,0}^{(k)}, \ldots, f_{i,\ell-1}^{(k)}\right) \in \mathbb{F}_q^\ell$ is the vector of the ith coefficients of the polynomials $f_j^{(k)}(x) = f_{0,j}^{(k)} + f_{1,j}^{(k)}x + \cdots + f_{m-1,j}^{(k)}x^{m-1}$, for all $k \in \{1, \ldots, r\}$ and $j \in \{0, \ldots, \ell-1\}$.

Since the image of any codeword $\mathbf{c}(x) \in \mathcal{C}$ under the map Φ is an element of the cyclic code $\widehat{\mathcal{C}}$ over $\mathbb{F}_{q^\ell}$, there are at least $d_H(\widehat{\mathcal{C}})$ nonzero rows in each nonzero codeword of $\mathcal{C}$. For any $i \in \{0, \ldots, m-1\}$, the ith row $\mathbf{c}_i = (c_{i,0}, \ldots, c_{i,\ell-1})$ of a codeword $\mathbf{c} \in \mathcal{C}$ can be viewed as a codeword in $\mathcal{B}$; therefore, a nonzero $\mathbf{c}_i$ has weight at least $d_H(\mathcal{B})$. Hence we have shown the following.

Theorem 7.4.2 (Lally Bound [1189, Theorem 5]) *Let $\mathcal{C}$ be an r-generator QC code of length $m\ell$ and index ℓ over $\mathbb{F}_q$ with generating set $\{\mathbf{f}_1, \ldots, \mathbf{f}_r\} \subseteq \mathbb{F}_q[x]^\ell$. Let the cyclic code $\widehat{\mathcal{C}} \subseteq \mathbb{F}_{q^\ell}^m$ and the linear code $\mathcal{B} \subseteq \mathbb{F}_q^\ell$ be defined as above. Then*

$$d_H(\mathcal{C}) \geq d_H(\widehat{\mathcal{C}})d_H(\mathcal{B}).$$

7.4.3 Spectral Bounds

In [1645], Semenov and Trifonov developed a spectral theory for QC codes by using the upper triangular polynomial matrices (7.3) given by Lally and Fitzpatrick [1191]; this gives rise to a BCH-like minimum distance bound. Their bound was improved by Zeh and Ling in [1940] by using the Hartmann–Tzeng (HT) Bound for cyclic codes.[1] Before proceeding to the spectral theory of QC codes and their minimum distance bounds, we explore relevant minimum distance bounds for cyclic codes that generalize the BCH and HT Bounds, namely the Roos Bound and the Shift Bound.

7.4.3.1 Cyclic Codes and Distance Bounds From Their Zeros

Recall from Section 1.12 and Chapter 2 that a cyclic code $\mathcal{C} \subseteq \mathbb{F}_q^m$ can be viewed as an ideal of R. Since R is a principal ideal ring, there exists a unique monic polynomial $g(x) \in R$ such that $\mathcal{C} = \langle g(x) \rangle$; i.e., each codeword $c(x) \in \mathcal{C}$ is of the form $c(x) = a(x)g(x)$ for some $a(x) \in R$. The polynomial $g(x)$, which is a divisor of $x^m - 1$, is called the **generator polynomial** of $\mathcal{C}$. For any positive integer p, let $\mathbf{0}_p$ denote throughout the all-zero vector of length p. A cyclic code $\mathcal{C} = \langle g(x) \rangle$ is $\{\mathbf{0}_m\}$ if and only if $g(x) = x^m - 1$.

---

[1]The BCH Bound is found in [248, 968] and in Theorem 1.14.3; the HT Bound, which generalizes the BCH Bound, is found in [914, Theorem 2] and Theorem 2.4.2.

The roots of $x^m - 1$ are of the form $1, \xi, \ldots, \xi^{m-1}$ where ξ is a fixed primitive m^{th} root of unity. Henceforth, let $\Omega := \{\xi^k \mid 0 \leq k \leq m - 1\}$ be the set of all m^{th} roots of unity. Recall that the splitting field of $x^m - 1$, denoted by $\mathbb{F}$, is the smallest extension field of $\mathbb{F}_q$ that contains Ω. Given the cyclic code $\mathcal{C} = \langle g(x) \rangle$, the set $L := \{\xi^k \mid g(\xi^k) = 0\} \subseteq \Omega$ of roots of its generator polynomial is called the **zeros** of $\mathcal{C}$. Note that $\xi^k \in L$ implies $\xi^{qk} \in L$ for each k. A nonempty subset $E \subseteq \Omega$ is said to be **consecutive** if there exist integers e, n, and δ with $e \geq 0$, $\delta \geq 2$, $n > 0$, and $\gcd(m,n) = 1$ such that

$$E := \{\xi^{e+zn} \mid 0 \leq z \leq \delta - 2\}. \tag{7.18}$$

We now describe the Roos Bound for cyclic codes. For $\emptyset \neq P \subseteq \Omega$, let $\mathcal{C}_P$ denote any cyclic code of length m over $\mathbb{F}_q$ whose set of zeros contains P. Let d_P denote the least possible amount that the minimum distance of any $\mathcal{C}_P$ can have. In particular, $d_P = d_H(\mathcal{C}_P)$ when P is the set of zeros of $\mathcal{C}_P$.

Theorem 7.4.3 (Roos Bound [1594, Theorem 2]) *Let N and M be two nonempty subsets of Ω. If there exists a consecutive set M' containing M such that $|M'| \leq |M| + d_N - 2$, then we have $d_{MN} \geq |M| + d_N - 1$ where $MN := \{\varepsilon\vartheta \mid \varepsilon \in M, \vartheta \in N\}$.*

If N is consecutive like in (7.18), then we get the following.

Corollary 7.4.4 ([1594, Corollary 1]) *Let N, M, and M' be as in Theorem 7.4.3, with N consecutive. Then $|M'| < |M| + |N|$ implies $d_{MN} \geq |M| + |N|$.*

Remark 7.4.5 In particular, the case $M = \{1\}$ in Corollary 7.4.4 yields the BCH Bound for the associated cyclic code. Taking $M' = M$ in the corollary yields the Hartmann–Tzeng Bound.

Another improvement to the Hartmann–Tzeng Bound for cyclic codes, known as the Shift Bound, was given by van Lint and Wilson in [1838]. To describe the Shift Bound, we need the notion of an *independent set*, which can be constructed over any field in a recursive way.

Let S be a subset of some field $\mathbb{K}$ of any characteristic. One inductively defines a family of finite subsets of $\mathbb{K}$, called **independent with respect to** S, as follows:

Condition 1: $\emptyset$ is independent with respect to S.

Condition 2: If $A \subseteq S$ is independent with respect to S, then $A \cup \{b\}$ is independent with respect to S for all $b \in \mathbb{K} \setminus S$.

Condition 3: If A is independent with respect to S and $c \in \mathbb{K}^* = \mathbb{K} \setminus \{0\}$, then cA is independent with respect to S.

We define the **weight** of a polynomial $f(x) \in \mathbb{K}[x]$, denoted by $\text{wt}_{\text{H}}(f)$, as the number of nonzero coefficients in $f(x)$.

Theorem 7.4.6 (Shift Bound [1838, Theorem 11]) *Let $0 \neq f(x) \in \mathbb{K}[x]$ and let $S := \{\theta \in \mathbb{K} \mid f(\theta) = 0\}$. Then $\text{wt}_{\text{H}}(f) \geq |A|$ for every subset A of $\mathbb{K}$ that is independent with respect to S.*

The minimum distance bound for a given cyclic code follows by considering the weights of its codewords $c(x) \in \mathcal{C}$ and the independent sets with respect to subsets of its zeros L. Observe that, in this case, the universe of the independent sets is Ω, not $\mathbb{F}$, because all of the possible roots of the codewords are contained in Ω. Moreover, we choose b from $\Omega \setminus P$ in Condition 2 above, where $P \subseteq L$, and c in Condition 3 is of the form $\xi^k \in \mathbb{F}^*$, for some $0 \leq k \leq m - 1$.

Remark 7.4.7 In particular, $A = \{\xi^{e+zn} \mid 0 \leq z \leq \delta - 1\}$ is independent with respect to the consecutive set E in (7.18), which gives the BCH Bound for $\mathcal{C}_E$. Let $D := \{\xi^{e+zn_1+yn_2} \mid 0 \leq z \leq \delta - 2, \ 0 \leq y \leq s\}$, for integers $b \geq 0$, $\delta \geq 2$ and positive integers s, n_1, and n_2 such that $\gcd(m, n_1) = 1$ and $\gcd(m, n_2) < \delta$. Then, for any fixed $\zeta \in \{0, \ldots, \delta - 2\}$, $A_\zeta := \{\xi^{e+zn_1} \mid 0 \leq z \leq \delta - 2\} \cup \{\xi^{e+\zeta n_1+yn_2} \mid 0 \leq y \leq s+1\}$ is independent with respect to D, and we get the Hartmann–Tzeng Bound for $\mathcal{C}_D$.

7.4.3.2 Spectral Theory of QC Codes

Semenov and Trifonov [1645] used the polynomial matrix $\widetilde{G}(x)$ in (7.3) to develop a spectral theory for QC codes, a topic we now explore.

Given a QC code $\mathcal{C} \subseteq R^\ell$, let the associated $\ell \times \ell$ upper triangular matrix $\widetilde{G}(x)$ be as in (7.3) with entries in $\mathbb{F}_q[x]$. The determinant of $\widetilde{G}(x)$ is

$$\det(\widetilde{G}(x)) := \prod_{j=0}^{\ell-1} g_{j,j}(x),$$

and we define an **eigenvalue** of $\mathcal{C}$ to be a root β of $\det(\widetilde{G}(x))$. Note that, since $g_{j,j}(x) \mid (x^m - 1)$ for each $0 \leq j \leq \ell - 1$, all eigenvalues of $\mathcal{C}$ are elements of Ω; i.e., $\beta = \xi^k$ for some $k \in \{0, \ldots, m - 1\}$. The **algebraic multiplicity** of β is the largest integer a such that $(x - \beta)^a \mid \det(\widetilde{G}(x))$. The **geometric multiplicity** of β is the dimension of the null space of $\widetilde{G}(\beta)$. This null space, denoted by $\mathcal{V}_\beta$, is called the **eigenspace** of β. In other words,

$$\mathcal{V}_\beta := \{\mathbf{v} \in \mathbb{F}^\ell \mid \widetilde{G}(\beta)\mathbf{v}^\mathsf{T} = \mathbf{0}_\ell^\mathsf{T}\},$$

where $\mathbb{F}$ is the splitting field of $x^m - 1$, as before, and T denotes the transpose. It was shown in [1645] that, for a given QC code and the associated $\widetilde{G}(x) \in \mathbb{F}_q[x]^{\ell \times \ell}$, the algebraic multiplicity a of an eigenvalue β is equal to its geometric multiplicity $\dim_{\mathbb{F}}(\mathcal{V}_\beta)$.

Lemma 7.4.8 ([1645, Lemma 1]) *The algebraic multiplicity of any eigenvalue of a QC code $\mathcal{C}$ is equal to its geometric multiplicity.*

From this point on, we let $\overline{\Omega} \subseteq \Omega$ denote the nonempty set of all eigenvalues of $\mathcal{C}$ where $|\overline{\Omega}| = t > 0$. Note that $\overline{\Omega} = \emptyset$ if and only if the diagonal elements $g_{j,j}(x)$ in $\widetilde{G}(x)$ are constant and $\mathcal{C}$ is the trivial full space code. Choose an arbitrary eigenvalue $\beta_i \in \overline{\Omega}$ with multiplicity n_i for some $i \in \{1, \ldots, t\}$. Let $\{\mathbf{v}_{i,0}, \ldots, \mathbf{v}_{i,n_i-1}\}$ be a basis for the corresponding eigenspace $\mathcal{V}_i$. Consider the matrix

$$V_i := \begin{bmatrix} \mathbf{v}_{i,0} \\ \vdots \\ \mathbf{v}_{i,n_i-1} \end{bmatrix} = \begin{bmatrix} v_{i,0,0} & \cdots & v_{i,0,\ell-1} \\ \vdots & \vdots & \vdots \\ v_{i,n_i-1,0} & \cdots & v_{i,n_i-1,\ell-1} \end{bmatrix} \tag{7.19}$$

having the basis elements as its rows. We let

$$H_i := (1, \beta_i, \ldots, \beta_i^{m-1}) \otimes V_i \text{ and}$$

$$H := \begin{bmatrix} H_1 \\ \vdots \\ H_t \end{bmatrix} = \begin{bmatrix} V_1 & \beta_1 V_1 & \cdots & \beta_1^{m-1} V_1 \\ \vdots & \vdots & \vdots & \vdots \\ V_t & \beta_t V_t & \cdots & \beta_t^{m-1} V_t \end{bmatrix}. \tag{7.20}$$

Observe that H has $n := \sum_{i=1}^t n_i$ rows. We also have that $n = \sum_{j=0}^{\ell-1} \deg(g_{j,j}(x))$ by Lemma 7.4.8. The rows of H are linearly independent, a fact shown in [1645, Lemma 2]. This proves the following lemma.

Lemma 7.4.9 *The matrix H in (7.20) has rank $m\ell - \dim_{\mathbb{F}_q}(\mathcal{C})$.*

It is immediate to confirm that $H\mathbf{c}^\mathsf{T} = \mathbf{0}_n^\mathsf{T}$ for any codeword $\mathbf{c} \in \mathcal{C}$. Together with Lemma 7.4.9, we obtain the following.

Proposition 7.4.10 ([1645, Theorem 1]) *The $n \times m\ell$ matrix H in (7.20) is a parity check matrix for $\mathcal{C}$.*

Remark 7.4.11 Note that if $\overline{\Omega} = \emptyset$, then the construction of H in (7.20) is impossible. Hence, we have assumed $\overline{\Omega} \neq \emptyset$ and we can always say $H = \mathbf{0}_{m\ell}$ if $\mathcal{C} = \mathbb{F}_q^{m\ell}$. The other extreme case is when $\overline{\Omega} = \Omega$. By using Lemma 7.4.9 above, one can easily deduce that a given QC code $\mathcal{C} = \{\mathbf{0}_{m\ell}\}$ if and only if $\overline{\Omega} = \Omega$, each $V_i = \mathbb{F}^\ell$ (equivalently, each $V_i = I_\ell$, where I_ℓ denotes the $\ell \times \ell$ identity matrix), and $n = m\ell$ so that we obtain $H = I_{m\ell}$. On the other hand, $\overline{\Omega} = \Omega$ whenever $(x^m - 1) \mid \det(\widetilde{G}(x))$, but $\mathcal{C}$ is nontrivial unless each eigenvalue in Ω has multiplicity ℓ.

Definition 7.4.12 Let $\mathcal{V} \subseteq \mathbb{F}^\ell$ be an eigenspace. We define the **eigencode** corresponding to $\mathcal{V}$ by

$$\mathbb{C}(\mathcal{V}) = \mathbb{C} := \left\{ \mathbf{u} \in \mathbb{F}_q^\ell \,\middle|\, \sum_{j=0}^{\ell-1} v_j u_j = 0 \text{ for all } \mathbf{v} \in \mathcal{V} \right\}.$$

When we have $\mathbb{C} = \{\mathbf{0}_\ell\}$, it is assumed that $d_H(\mathbb{C}) = \infty$.

The BCH-like minimum distance bound of Semenov and Trifonov for a given QC code in [1645, Theorem 2] was expressed in terms of the size of a consecutive subset of eigenvalues in $\overline{\Omega}$ and the minimum distance of the common eigencode related to this consecutive subset. Zeh and Ling generalized their approach and derived an HT-like bound in [1940, Theorem 1] without using the parity check matrix in their proof. The eigencode, however, is still needed. In the next section, we will prove the analogs of these bounds for QC codes in terms of the Roos and Shift Bounds.

7.4.3.3 Spectral Bounds for QC Codes

First, we establish a general spectral bound on the minimum distance of a given QC code. Let $\mathcal{C} \subseteq \mathbb{F}_q^{m\ell}$ be a QC code of index ℓ with nonempty eigenvalue set $\overline{\Omega} \subsetneq \Omega$. Let $P \subseteq \overline{\Omega}$ be a nonempty subset of eigenvalues such that $P = \{\xi^{u_1}, \xi^{u_2}, \ldots, \xi^{u_r}\}$ where $0 < r \leq |\overline{\Omega}|$. We define

$$\widetilde{H}_P := \begin{bmatrix} 1 & \xi^{u_1} & \xi^{2u_1} & \cdots & \xi^{(m-1)u_1} \\ \vdots & \vdots & \vdots & \vdots & \vdots \\ 1 & \xi^{u_r} & \xi^{2u_r} & \cdots & \xi^{(m-1)u_r} \end{bmatrix}.$$

Recall that d_P denotes a positive integer such that any cyclic code $\mathcal{C}_P \subseteq \mathbb{F}_q^m$, whose zeros contain P, has a minimum distance at least d_P. We have $\widetilde{H}_P \mathbf{c}_P^\mathsf{T} = \mathbf{0}_r^\mathsf{T}$ for any $\mathbf{c}_P \in \mathcal{C}_P$. In particular, if P is equal to the set of zeros of $\mathcal{C}_P$, then $\widetilde{H}_P$ is a parity check matrix for $\mathcal{C}_P$.

Let $\mathcal{V}_P$ denote the common eigenspace of the eigenvalues in P, and let V_P be the matrix, say of size $t \times \ell$, whose rows form a basis for $\mathcal{V}_P$ (compare (7.19)). If we set $\widehat{H}_P = \widetilde{H}_P \otimes V_P$, then $\widehat{H}_P \mathbf{c}^\mathsf{T} = \mathbf{0}_{rt}^\mathsf{T}$, for all $\mathbf{c} \in \mathcal{C}$. In other words, $\widehat{H}_P$ is a submatrix of some H of the form in (7.20) if $\mathcal{V}_P \neq \{\mathbf{0}_\ell\}$. If $\mathcal{V}_P = \{\mathbf{0}_\ell\}$, then $\widehat{H}_P$ does not exist. We first handle this case separately so that the bound is valid even if we have $\mathcal{V}_P = \{\mathbf{0}_\ell\}$, before the cases where we can use $\widehat{H}_P$ in the proof.

Henceforth, we consider the quantity $\min\{d_P, d_H(\mathbb{C}_P)\}$, where $\mathbb{C}_P$ is the eigencode corresponding to $\mathcal{V}_P$. We have assumed $P \neq \emptyset$ so that $\widetilde{H}_P$ is defined, and we also have $P \neq \Omega$

as $P \subseteq \overline{\Omega} \subsetneq \Omega$ to make d_P well-defined. Since $|P| \geq 1$, the BCH Bound implies $d_P \geq 2$. Hence, $\min\{d_P, d_H(\mathbb{C}_P)\} = 1$ if and only if $d_H(\mathbb{C}_P) = 1$. For any nonzero QC code $\mathcal{C}$, we have $d_H(\mathcal{C}) \geq 1$. Therefore, whenever $d_H(\mathbb{C}_P) = 1$, we have $d_H(\mathcal{C}) \geq \min\{d_P, d_H(\mathbb{C}_P)\}$. In particular, when $\mathcal{V}_P = \{\mathbf{0}_\ell\}$, which implies $\mathbb{C}_P = \mathbb{F}_q^\ell$ and $d_H(\mathbb{C}_P) = 1$, we have $d_H(\mathcal{C}) \geq 1 = \min\{d_P, d(\mathbb{C}_P)\}$.

Now let $\emptyset \neq P \subseteq \overline{\Omega} \subsetneq \Omega$ and let $d_H(\mathbb{C}_P) \geq 2$. Assume that there exists a codeword $\mathbf{c} \in \mathcal{C}$ of weight ω such that $0 < \omega < \min\{d_P, d_H(\mathbb{C}_P)\}$. For each $0 \leq k \leq m-1$, let $\mathbf{c}_k = (c_{k,0}, \ldots, c_{k,\ell-1})$ be the kth row of the codeword $\mathbf{c}$ given as in (7.1), and we set $\mathbf{s}_k := V_P \mathbf{c}_k^\mathsf{T}$. Since $d_H(\mathbb{C}_P) > \omega$, we have $\mathbf{c}_k \notin \mathbb{C}_P$, and therefore $\mathbf{s}_k = V_P \mathbf{c}_k^\mathsf{T} \neq \mathbf{0}_t^\mathsf{T}$ for all $\mathbf{c}_k \neq \mathbf{0}_\ell$ with $k \in \{0, \ldots, m-1\}$. Hence, $0 < |\{\mathbf{s}_k \mid \mathbf{s}_k \neq \mathbf{0}_t^\mathsf{T}\}| \leq \omega < \min\{d_P, d_H(\mathbb{C}_P)\}$. Let $S := [\mathbf{s}_0\ \mathbf{s}_1 \cdots \mathbf{s}_{m-1}]$. Then $\widetilde{H}_P S^\mathsf{T} = 0$, which implies that the rows of the matrix S lie in the right kernel of $\widetilde{H}_P$. But this is a contradiction since any row of S has weight at most $\omega < d_P$, showing the following.

Theorem 7.4.13 ([700, Theorem 11]) *Let $\mathcal{C} \subseteq R^\ell$ be a QC code of index ℓ with nonempty eigenvalue set $\overline{\Omega} \subsetneq \Omega$. Let $P \subseteq \overline{\Omega}$ be a nonempty subset of eigenvalues, and let $\mathcal{C}_P \subseteq \mathbb{F}_q^m$ be any cyclic code with zeros $L \supseteq P$ and minimum distance at least d_P. We define $\mathcal{V}_P := \bigcap_{\beta \in P} \mathcal{V}_\beta$ as the common eigenspace of the eigenvalues in P, and let $\mathbb{C}_P$ denote the eigencode corresponding to $\mathcal{V}_P$. Then*

$$d_H(\mathcal{C}) \geq \min\{d_P, d_H(\mathbb{C}_P)\}.$$

Theorem 7.4.13 allows us to use any minimum distance bound derived for cyclic codes based on their zeros. The following special cases are immediate after the preparation that we have done in Section 7.4.3.1; compare this to Theorems 7.4.3 and 7.4.6.

Corollary 7.4.14 ([700, Corollary 12]) *Let $\mathcal{C} \subseteq R^\ell$ be a QC code of index ℓ with $\overline{\Omega} \subsetneq \Omega$ as its nonempty set of eigenvalues.*

(a) *Let N and M be two nonempty subsets of Ω such that $MN \subseteq \overline{\Omega}$, where $MN := \{\varepsilon\vartheta \mid \varepsilon \in M, \vartheta \in N\}$. If there exists a consecutive set M' containing M with $|M'| \leq |M| + d_N - 2$, then $d_H(\mathcal{C}) \geq \min\{|M| + d_N - 1, d_H(\mathbb{C}_{MN})\}$.*

(b) *For every $A \subseteq \Omega$ that is independent with respect to $\overline{\Omega}$, we have $d_H(\mathcal{C}) \geq \min\{|A|, d_H(\mathbb{C}_{T_A})\}$ where $T_A := A \cap \overline{\Omega}$.*

Proof: For part (a), let $N = \{\xi^{u_1}, \ldots, \xi^{u_r}\}$ and $M = \{\xi^{v_1}, \ldots, \xi^{v_s}\}$ be such that there exists a consecutive set $M' = \{\xi^z \mid v_1 \leq z \leq v_s\} \subseteq \Omega$ containing M with $|M'| \leq |M| + d_N - 2$. We define the matrices

$$\widetilde{H}_N := \begin{bmatrix} 1 & \xi^{u_1} & \xi^{2u_1} & \cdots & \xi^{(m-1)u_1} \\ \vdots & \vdots & \vdots & \vdots & \vdots \\ 1 & \xi^{u_r} & \xi^{2u_r} & \cdots & \xi^{(m-1)u_r} \end{bmatrix} \quad \text{and} \quad \widetilde{H}_M := \begin{bmatrix} 1 & \xi^{v_1} & \xi^{2v_1} & \cdots & \xi^{(m-1)v_1} \\ \vdots & \vdots & \vdots & \vdots & \vdots \\ 1 & \xi^{v_s} & \xi^{2v_s} & \cdots & \xi^{(m-1)v_s} \end{bmatrix}.$$

Consider the joint subset $MN = \{\xi^{u_i + v_j} \mid 1 \leq i \leq r, 1 \leq j \leq s\} \subseteq \overline{\Omega}$. Let B_k be the kth column of $\widetilde{H}_N$ for $k \in \{0, \ldots, m-1\}$. We create the joint matrix

$$\widetilde{H}_{MN} = \begin{bmatrix} B_0 & \xi^{v_1} B_1 & \xi^{2v_1} B_2 & \cdots & \xi^{(m-1)v_1} B_{m-1} \\ \vdots & \vdots & \vdots & \vdots & \vdots \\ B_0 & \xi^{v_s} B_1 & \xi^{2v_s} B_2 & \cdots & \xi^{(m-1)v_s} B_{m-1} \end{bmatrix}.$$

Now let $\mathcal{V}_{MN} := \bigcap_{\beta \in MN} \mathcal{V}_\beta$ denote the common eigenspace of the eigenvalues in MN, and let V_{MN} be the matrix whose rows form a basis for $\mathcal{V}_{MN}$, built as in (7.19). Let $\mathbb{C}_{MN}$ be the eigencode corresponding to $\mathcal{V}_{MN}$. Setting $\widehat{H}_{MN} := \widetilde{H}_{MN} \otimes V_{MN}$ implies $\widehat{H}_{MN} \mathbf{c}^{\mathsf{T}} = \mathbf{0}^{\mathsf{T}}$ for all $\mathbf{c} \in \mathcal{C}$. The rest of the proof is identical with the proof of Theorem 7.4.13, where P is replaced by MN, and the result follows by the Roos Bound in Theorem 7.4.3.

For part (b) if $A \subseteq \Omega$ is an independent set with respect to $\overline{\Omega}$, let $T_A = \{\xi^{w_1}, \xi^{w_2}, \ldots, \xi^{w_y}\} = A \cap \overline{\Omega}$. Since $\overline{\Omega}$ is a proper subset of Ω, a nonempty T_A can be obtained by the recursive construction of A. We define

$$\widetilde{H}_{T_A} = \begin{bmatrix} 1 & \xi^{w_1} & \xi^{2w_1} & \cdots & \xi^{(m-1)w_1} \\ \vdots & \vdots & \vdots & \vdots & \vdots \\ 1 & \xi^{w_y} & \xi^{2w_y} & \cdots & \xi^{(m-1)w_y} \end{bmatrix}.$$

Let V_{T_A} be the matrix corresponding to a basis of $\mathcal{V}_{T_A}$, which is the intersection of the eigenspaces belonging to the eigenvalues in T_A. Let $\mathbb{C}_{T_A}$ be the eigencode corresponding to the eigenspace $\mathcal{V}_{T_A}$. We again set $\widehat{H}_{T_A} := \widetilde{H}_{T_A} \otimes V_{T_A}$ and the result follows in a similar way by using the Shift Bound in Theorem 7.4.6. $\qquad\square$

Remark 7.4.15 We can obtain the BCH-like bound in [1645, Theorem 2] and the HT-like bound in [1940, Theorem 1] by using Remarks 7.4.5 and 7.4.7.

7.5 Asymptotics

Let $(\mathcal{C}_i)_{i \geq 1}$ be a sequence of linear codes over $\mathbb{F}_q$, and let N_i, d_i and k_i denote respectively the length, minimum distance, and dimension of $\mathcal{C}_i$, for all i. Assume that $\lim_{i \to \infty} N_i = \infty$. Let

$$\delta := \liminf_{i \to \infty} \frac{d_i}{N_i} \quad \text{and} \quad \mathcal{R} := \liminf_{i \to \infty} \frac{k_i}{N_i}$$

denote the relative distance and the relative rate of the sequence $(\mathcal{C}_i)_{i \geq 1}$. Both $\mathcal{R}$ and δ are finite as they are limits of bounded quantities. If $\mathcal{R}\delta \neq 0$, then $(\mathcal{C}_i)_{i \geq 1}$ is called an **asymptotically good sequence** of codes.

We will require the celebrated entropy function

$$H_q(y) = y \log_q(q-1) - y \log_q(y) - (1-y) \log_q(1-y),$$

defined for $0 < y < \frac{q-1}{q}$ and of constant use in estimating binomial coefficients of large arguments [1323, pages 309–310]. The asymptotic Gilbert–Varshamov Bound (see [1323, Chapter 17, Theorem 30] or Theorem 1.9.28) states that, for every q and $0 < \delta < 1 - \frac{1}{q}$, there exists an infinite family of q-ary codes with limit rate

$$\mathcal{R} \geq 1 - H_q(\delta).$$

The existence of explicit families of QC codes satisfying a modified Gilbert Varshamov Bound were shown in [393], and then this result was improved in [1088]. Below, we focus on self-dual and LCD families of QC codes.

7.5.1 Good Self-Dual QC Codes Exist

In this section, we construct families of binary self-dual QC codes. We assume that all binary codes are equipped with the Euclidean inner product and all the $\mathbb{F}_4$-codes are equipped with the Hermitian inner product (7.15). Self-duality in the following discussion is with respect to these respective inner products. A binary self-dual code is said to be of **Type II** if and only if all its weights are multiples of 4 and of **Type I** otherwise. We first recall some background material on mass formulas for self-dual codes over $\mathbb{F}_2$ and $\mathbb{F}_4$. See also Theorem 4.5.2.

Proposition 7.5.1 *Let ℓ be an even positive integer.*

(a) *The number of self-dual binary codes of length ℓ is given by*

$$N(2,\ell) = \prod_{i=1}^{\frac{\ell}{2}-1}(2^i+1).$$

(b) *Let $\mathbf{v} \in \mathbb{F}_2^\ell$ with even Hamming weight, other than 0 and ℓ. The number of self-dual binary codes of length ℓ containing $\mathbf{v}$ is given by*

$$M(2,\ell) = \prod_{i=1}^{\frac{\ell}{2}-2}(2^i+1).$$

(c) *The number of self-dual $\mathbb{F}_4$-codes of length ℓ is given by*

$$N(4,\ell) = \prod_{i=0}^{\frac{\ell}{2}-1}(2^{2i+1}+1).$$

(d) *The number of self-dual $\mathbb{F}_4$-codes of length ℓ containing a given nonzero codeword of length ℓ and even Hamming weight is given by*

$$M(4,\ell) = \prod_{i=0}^{\frac{\ell}{2}-2}(2^{2i+1}+1).$$

Proof: Parts (a) and (c) are well-known facts, found for example in [1008, 1555]. Part (b) is an immediate consequence of [1324, Theorem 2.1] with $s = 2$, noting that every self-dual binary code must contain the all-one vector $\mathbf{1}$. Part (d) follows from [438, Theorem 1] with $n_1 = \ell$ and $k_1 = 1$. □

Proposition 7.5.2 *Let ℓ be a positive integer divisible by 8.*

(a) *The number of Type II binary self-dual codes of length ℓ is given by*

$$T(2,\ell) = 2\prod_{i=1}^{\frac{\ell}{2}-2}(2^i+1).$$

(b) *Let $\mathbf{v} \in \mathbb{F}_2^\ell$ with Hamming weight divisible by 4, other than 0 and ℓ. The number of Type II binary self-dual codes of length ℓ containing $\mathbf{v}$ is given by*

$$S(2,\ell) = 2\prod_{i=1}^{\frac{\ell}{2}-3}(2^i+1).$$

Proof: Part (a) is found in [1008, 1555], and part (b) is exactly [1324, Corollary 2.4]. □

Let C_1 denote a binary code of length ℓ and let C_2 be a code over $\mathbb{F}_4$ of length ℓ. We construct a binary QC code C of length 3ℓ and index ℓ whose codewords are of the form given in (7.12). It is easy to check that C is self-dual if and only if both C_1 and C_2 are self-dual, and C is of Type II if and only if C_1 is of Type II and C_2 is self-dual.

We assume henceforth that C is a binary self-dual QC code constructed in the above way. Any codeword $\mathbf{c} \in C$ must necessarily have even Hamming weight. Suppose that $\mathbf{c}$ corresponds to the pair $(\mathbf{c}_1, \mathbf{c}_2)$, where $\mathbf{c}_1 \in C_1$ and $\mathbf{c}_2 \in C_2$. Since C_1 and C_2 are self-dual, it follows that $\mathbf{c}_1$ and $\mathbf{c}_2$ must both have even Hamming weights. When $\mathbf{c} \neq \mathbf{0}_{3\ell}$, there are three possibilities for $(\mathbf{c}_1, \mathbf{c}_2)$:

Case 1: $\mathbf{c}_1 \neq \mathbf{0}_\ell$, $\mathbf{c}_2 \neq \mathbf{0}_\ell$;

Case 2: $\mathbf{c}_1 = \mathbf{0}_\ell$, $\mathbf{c}_2 \neq \mathbf{0}_\ell$; and

Case 3: $\mathbf{c}_1 \neq \mathbf{0}_\ell$, $\mathbf{c}_2 = \mathbf{0}_\ell$.

We count the number of codewords $\mathbf{c}$ in each of these cases for a given even weight d.

For Case 2, if the Hamming weight of $\mathbf{c}$ is d, then $C \cong C_1 \oplus C_2$ implies that $\mathbf{c}_2$ has Hamming weight $d/2$. Since $\mathbf{c}_2$ has even Hamming weight, it follows that d is divisible by 4 in order for this case to occur. It is easy to see that the number $A_2(\ell, d)$ of such words $\mathbf{c}$ is bounded above by $\binom{\ell}{d/2}3^{d/2}$ where $4 \,|\, d$. For d not divisible by 4, set $A_2(\ell, d) = 0$.

The argument to obtain the number of words of Case 3 is similar. It is easy to show that the number $A_3(\ell, d)$ of such words is bounded above by $\binom{\ell}{d/3}$ where $6 \,|\, d$. When d is not divisible by 6, set $A_3(\ell, d) = 0$.

The total number of vectors in $\mathbb{F}_2^{3\ell}$ of weight d is $\binom{3\ell}{d}$; so for Case 1, we have

$$A_1(\ell, d) \leq \binom{3\ell}{d} - A_2(\ell, d) - A_3(\ell, d).$$

In particular, $A_1(\ell, d)$ is bounded above by $\binom{3\ell}{d}$.

Combining the above observations and Proposition 7.5.1, the number of self-dual binary QC codes of length 3ℓ and index ℓ with minimum weight less than d is bounded above by

$$\sum_{\substack{e < d \\ e:\,\text{even}}} \left(A_1(\ell, e)M(2, \ell)M(4, \ell) + A_2(\ell, e)N(2, \ell)M(4, \ell) + A_3(\ell, e)M(2, \ell)N(4, \ell) \right).$$

Theorem 7.5.3 ([1270, Theorem 3.1]) *Let ℓ be an even integer and let d be the largest even integer such that*

$$\sum_{\substack{e < d \\ e \equiv 0 \bmod 2}} \binom{3\ell}{e} + (2^{\frac{\ell}{2}-1} + 1)\left(\sum_{\substack{e < d \\ e \equiv 0 \bmod 4}} \binom{\ell}{e/2} 3^{e/2} \right) + (2^{\ell-1} + 1)\left(\sum_{\substack{e < d \\ e \equiv 0 \bmod 6}} \binom{\ell}{e/3} \right)$$

$$< (2^{\frac{\ell}{2}-1} + 1)(2^{\ell-1} + 1).$$

Then there exists a self-dual binary QC code of length 3ℓ and index ℓ with minimum distance at least d.

Proof: Multiplying both sides of the inequality by $M(2, \ell)M(4, \ell)$, and applying the above upper bounds for $A_i(\ell, d)$ $(i = 1, 2, 3)$, we see that the inequality in Theorem 7.5.3 implies that the number of self-dual binary QC codes of length 3ℓ and index ℓ with minimum distance $< d$ is strictly less than the total number of self-dual binary QC codes of length 3ℓ and index ℓ.   □

If we are interested only in Type II QC codes, using Proposition 7.5.2, we easily see that the number of Type II binary QC codes of length 3ℓ and index ℓ with minimum weight is $< d$ is bounded above by

$$\sum_{\substack{e < d \\ e \equiv 0 \bmod 4}} \left(A_1(\ell, e)S(2, \ell)M(4, \ell) + A_2(\ell, e)T(2, \ell)M(4, \ell) + A_3(\ell, e)S(2, \ell)N(4, \ell)\right).$$

Using an argument similar to that for Theorem 7.5.3, we obtain the following result.

Theorem 7.5.4 ([1270, Theorem 3.2]) *Let ℓ be divisible by 8 and let d be the largest multiple of 4 such that*

$$\sum_{\substack{e < d \\ e \equiv 0 \bmod 4}} \binom{3\ell}{e} + (2^{\frac{\ell}{2}-2} + 1)\left(\sum_{\substack{e < d \\ e \equiv 0 \bmod 4}} \binom{\ell}{e/2} 3^{e/2}\right) + (2^{\ell-1} + 1)\left(\sum_{\substack{e < d \\ e \equiv 0 \bmod 12}} \binom{\ell}{e/3}\right)$$

$$< (2^{\frac{\ell}{2}-2} + 1)(2^{\ell-1} + 1).$$

Then there exists a Type II binary QC code of length 3ℓ and index ℓ with minimum distance at least d.

We are now in a position to state and prove the asymptotic version of Theorems 7.5.3 and 7.5.4.

Theorem 7.5.5 ([1270, Theorem 4.1]) *There exists an infinite family of binary self-dual QC codes $\mathcal{C}_i$ of length $3\ell_i$ and of distance d_i with $\lim_{i \to \infty} \ell_i = \infty$ such that $\delta = \liminf_{i \to \infty} \dfrac{d_i}{3\ell_i}$ exists and is bounded below by*

$$\delta \geq H_2^{-1}(1/2) = 0.110 \cdots.$$

Proof: The right-hand-side of the inequality of Theorem 7.5.3 is plainly of the order of $2^{3\ell/2}$ for large ℓ. We compare this in turn to each of the three summands on the left-hand-side (at the price of a more stringent inequality, congruence conditions on the summation range are neglected). By [1323, Chapter 10, Corollary 9], for large ℓ (with $\mu = \delta$ and $n = \ell$), the first and third summands are of order $2^{3\ell H_2(\delta)}$ and $2^{\ell + \ell H_2(\delta)}$, respectively. They both are of the order of the right-hand-side for $H_2(\delta) = 1/2$. By [1323, Chapter 10, Lemma 7], for large ℓ (with $\lambda = \delta$ and $n = \ell$), the second summand is of order $2^{\ell f(3\delta/2)}$ for $f(t) := 0.5 + t \log_2(3) + H_2(t)$, which is of the order of the right-hand-side for

$$\delta = 0.1762 \cdots,$$

a value $> H_2^{-1}(1/2)$.   □

Similarly, for Type II codes, we have the following.

Theorem 7.5.6 *There exists an infinite family of Type II binary QC codes $\mathcal{C}_i$ of length $3\ell_i$ and of distance d_i with $\lim_{i \to \infty} \ell_i = \infty$ such that $\delta = \liminf_{i \to \infty} \dfrac{d_i}{3\ell_i}$ exists and is bounded below by*

$$\delta \geq H_2^{-1}(1/2) = 0.110 \cdots.$$

Proof: Since we neglected the congruence conditions in the preceding analysis, the calculations are exactly the same but using Theorem 7.5.4.   □

7.5.2 Complementary Dual QC Codes Are Good

It is known that both Euclidean and Hermitian (using the inner product (7.15)) LCD codes are asymptotically good; see [1347, Propositions 2 and 3] and [871, Theorem 3.6], respectively. Recall that the second part of Corollary 7.3.7 suggests an easy construction of QCCD codes from LCD codes. In [871], this idea was used to show the existence of good long QCCD codes, as shown in the next result, together with the help of Theorem 7.4.1.

Theorem 7.5.7 ([871, Theorems 3.3 and 3.7]) *Let q be a power of a prime and let $m \geq 2$ be relatively prime to q. Then there exists an asymptotically good sequence of q-ary QCCD codes.*

Proof: Let ξ be a primitive m^{th} root of unity over $\mathbb{F}_q$. There are two possibilities.

  (1) 1 and -1 are not in the same q-cyclotomic coset modulo m.

  (2) 1 and -1 are in the same q-cyclotomic coset modulo m.

In case (1), ξ and ξ^{-1} have distinct minimal polynomials $h'(x)$ and $h''(x)$, respectively, over $\mathbb{F}_q$. Let $\mathbb{H}' = \mathbb{F}_q[x]/\langle h'(x) \rangle$ and $\mathbb{H}'' = \mathbb{F}_q[x]/\langle h''(x) \rangle$. Recall that these fields are equal: $\mathbb{H}' = \mathbb{F}_q(\xi) = \mathbb{F}_q(\xi^{-1}) = \mathbb{H}''$. We denote both by $\mathbb{H}$. Let θ' and θ'' denote the primitive idempotents corresponding to the q-ary length m minimal cyclic codes with check polynomials $h'(x)$ and $h''(x)$, respectively. Let $(\mathcal{C}_i)_{i \geq 1}$ be an asymptotically good sequence of (Euclidean) LCD codes over $\mathbb{H}$ (such a sequence exists by [1347]), where each $\mathcal{C}_i$ has parameters $[\ell_i, k_i, d_i]$. For all $i \geq 1$, define the q-ary QC code $\mathcal{D}_i$ as

$$\mathcal{D}_i := \mathcal{C}_i \oplus \mathcal{C}_i = \big(\langle \theta' \rangle \square \mathcal{C}_i \big) \oplus \big(\langle \theta'' \rangle \square \mathcal{C}_i \big) \subset \mathbb{H}^{\ell_i} \oplus \mathbb{H}^{\ell_i}.$$

By Corollary 7.3.7, each $\mathcal{D}_i$ with $i \geq 1$ is a QCCD code of index ℓ_i. If $e := [\mathbb{H} : \mathbb{F}_q]$, then the length and the $\mathbb{F}_q$-dimension of $\mathcal{D}_i$ are $m\ell_i$ and $2ek_i$, respectively. By the Jensen Bound of Theorem 7.4.1, the minimum distance of $\mathcal{D}_i$ satisfies

$$d_H(\mathcal{D}_i) \geq \min \big\{ d_H(\langle \theta' \rangle) d_i, d_H(\langle \theta' \rangle \oplus \langle \theta'' \rangle) d_i \big\} \geq d_H(\langle \theta' \rangle \oplus \langle \theta'' \rangle) d_i.$$

For the sequence of QCCD codes $(\mathcal{D}_i)_{i \geq 1}$, the relative rate is

$$\mathcal{R} = \liminf_{i \to \infty} \frac{2ek_i}{m\ell_i} = \frac{2e}{m} \liminf_{i \to \infty} \frac{k_i}{\ell_i},$$

and this quantity is positive since $(\mathcal{C}_i)_{i \geq 1}$ is asymptotically good. For the relative distance, we have

$$\delta = \liminf_{i \to \infty} \frac{d_H(\mathcal{D}_i)}{m\ell_i} \geq d_H(\langle \theta' \rangle \oplus \langle \theta'' \rangle) \liminf_{i \to \infty} \frac{d_i}{\ell_i}.$$

Note again that δ is positive since $(\mathcal{C}_i)_{i \geq 1}$ is asymptotically good.

In case (2), we clearly have that n and $-n$ are always in the same q-cyclotomic coset modulo m. Therefore every irreducible factor of $x^m - 1$ over $\mathbb{F}_q$ is self-reciprocal. As $m \geq 2$, we can choose such a factor $g(x) \neq x - 1$. Let $\mathbb{G} = \mathbb{F}_q[x]/\langle g(x) \rangle$. We denote by θ the primitive idempotent corresponding to the q-ary minimal cyclic code of length m with check polynomial $g(x)$. Let $(\mathcal{C}_i)_{i \geq 1}$ be an asymptotically good sequence of Hermitian LCD codes over $\mathbb{G}$. Such a sequence exists by [871, Theorem 3.6]. Assume that each $\mathcal{C}_i$ has parameters $[\ell_i, k_i, d_i]$. For each $i \geq 1$, define the q-ary QC code $\mathcal{D}_i$ as the QC code with one constituent:

$$\mathcal{D}_i := \langle \theta \rangle \square \mathcal{C}_i.$$

If $e := [\mathbb{G} : \mathbb{F}_q]$, then the length and the dimension of $\mathcal{D}_i$ are $m\ell_i$ and ek_i, respectively. By the Jensen Bound of Theorem 7.4.1, we have

$$d_H(\mathcal{D}_i) \geq d_H(\langle\theta\rangle)d_H(\mathcal{C}_i).$$

As in part (a) we conclude that $(\mathcal{D}_i)_{i\geq 1}$ is asymptotically good since $(\mathcal{C}_i)_{i\geq 1}$ is asymptotically good. $\qquad\square$

7.6   Connection to Convolutional Codes

An (ℓ, k) convolutional code $\mathcal{C}$ over $\mathbb{F}_q$ is defined as a rank k $\mathbb{F}_q[x]$-submodule of $\mathbb{F}_q[x]^\ell$, which is necessarily a free module since $\mathbb{F}_q[x]$ is a principal ideal domain. The weight of a polynomial $c(x) \in \mathbb{F}_q[x]$ is defined as the number of nonzero terms in $c(x)$ and the weight of a codeword $\mathbf{c}(x) = (c_0(x), \ldots, c_{\ell-1}(x)) \in \mathcal{C}$ is the sum of the weights of its coordinates. The free distance $d_{free}(\mathcal{C})$ of the convolutional code $\mathcal{C}$ is the minimum weight among its nonzero codewords. See Chapter 10 and also [1363] for additional information.

Remark 7.6.1 An encoder of an (ℓ, k) convolutional code $\mathcal{C}$ is a $k \times \ell$ matrix G of rank k with entries from $\mathbb{F}_q[x]$. In other words,

$$\mathcal{C} = \left\{(u_0(x), \ldots, u_{k-1}(x))\,G \mid (u_0(x), \ldots, u_{k-1}(x)) \in \mathbb{F}_q[x]^k\right\}.$$

Remark 7.6.2 Note that a convolutional code is in general defined as an $\mathbb{F}_q(x)$-submodule (subspace) of $\mathbb{F}_q(x)^\ell$. However, this leads to codewords with rational coordinates and infinite weight. From the practical point of view, there is no reason to use this as the definition; see [809, 1190]. Note that even if $\mathcal{C}$ is defined as a subspace of $\mathbb{F}_q(x)^\ell$, it has an encoder which can be obtained by clearing off the denominators of all the entries in any encoder. Moreover, it is usually assumed that G is noncatastrophic in the sense that a finite weight codeword $\mathbf{c}(x) \in \mathcal{C}$ can only be produced from a finite weight information word $\mathbf{u}(x)$. In other words, an encoder G is said to be noncatastrophic if, for any $\mathbf{u}(x) \in \mathbb{F}_q(x)^k$, $\mathbf{u}(x)G$ has finite weight implies that $\mathbf{u}(x)$ also has finite weight. Hence, with a noncatastrophic encoder G, all finite weight codewords are covered by the $\mathbb{F}_q[x]$-module structure. Noncatastrophic encoders exist for any convolutional code; see [1363].

Let $R = \mathbb{F}_q[x]/\langle x^m - 1\rangle$ as before and consider the projection map

$$\begin{aligned} \Psi : \mathbb{F}_q[x] &\longrightarrow R \\ f(x) &\longmapsto f'(x) := f(x) \bmod \langle x^m - 1\rangle. \end{aligned}$$

It is clear that for a given (ℓ, k) convolutional code $\mathcal{C}$ and any $m > 1$, there is a natural QC code $\mathcal{C}'$, of length $m\ell$ and index ℓ, related to it as shown below. Note that we denote the map from $\mathcal{C}$ to $\mathcal{C}'$ also by Ψ.

$$\begin{aligned} \Psi : \mathcal{C} &\longrightarrow \mathcal{C}' \\ \mathbf{c}(x) = (c_0(x), \ldots, c_{\ell-1}(x)) &\longmapsto \mathbf{c}'(x) = (c_0'(x), \ldots, c_{\ell-1}'(x)). \end{aligned}$$

Lally [1190] showed that the minimum distance of the QC code $\mathcal{C}'$ above is a lower bound on the free distance of the convolutional code $\mathcal{C}$.

Theorem 7.6.3 ([1190, Theorem 2]) *Let $\mathcal{C}$ be an (ℓ, k) convolutional code over $\mathbb{F}_q$ and let $\mathcal{C}'$ be the related QC code in R^ℓ. Then $d_{free}(\mathcal{C}) \geq d_H(\mathcal{C}')$.*

Proof: Let $\mathbf{c}(x)$ be a codeword in $\mathcal{C}$ and set $\mathbf{c}'(x) = \Psi(\mathbf{c}(x)) \in \mathcal{C}'$. We consider the two possibilities. First, if $\mathbf{c}'(x) \neq \mathbf{0}$, then $\text{wt}_H(\mathbf{c}(x)) \geq \text{wt}_H(\mathbf{c}'(x))$. For the second possibility, suppose $\mathbf{c}'(x) = \mathbf{0}$. Let $\gamma \geq 1$ be the maximal positive integer such that $(x^m - 1)^\gamma$ divides each coordinate of $\mathbf{c}(x)$. Write $\mathbf{c}(x) = (x^m - 1)^\gamma (v_0(x), \ldots, v_{\ell-1}(x))$ and set $\mathbf{v}(x) = (v_0(x), \ldots, v_{\ell-1}(x))$. Then $\mathbf{v}(x)$ is a codeword of $\mathcal{C}$, and for $\mathbf{v}'(x) \in \mathcal{C}' \setminus \{\mathbf{0}\}$, we have $\text{wt}_H(\mathbf{c}(x)) \geq \text{wt}_H(\mathbf{v}'(x))$ by the first possibility. Combining the two possibilities, for any $\mathbf{c}(x) \in \mathcal{C}$, there exists a $\mathbf{v}'(x) \in \mathcal{C}'$ such that $\text{wt}_H(\mathbf{c}(x)) \geq \text{wt}_H(\mathbf{c}'(x))$. This proves that $d_{free}(\mathcal{C}) \geq d_H(\mathcal{C}')$. $\qquad\square$

Remark 7.6.4 Note that Lally uses an alternative module description of convolutional and QC codes in [1190]. Namely, a basis $\{1, \alpha, \ldots, \alpha^{\ell-1}\}$ of $\mathbb{F}_{q^\ell}$ over $\mathbb{F}_q$ is fixed and the $\mathbb{F}_q[x]$-modules $\mathbb{F}_q[x]^\ell$ and $\mathbb{F}_{q^\ell}[x]$ are identified via the map Φ in (7.17). With this identification, a length ℓ convolutional code is viewed as an $\mathbb{F}_q[x]$-module in $\mathbb{F}_{q^\ell}[x]$ and a length $m\ell$, index ℓ QC code is viewed as an $\mathbb{F}_q[x]$-module in $\mathbb{F}_{q^\ell}[x]/\langle x^m - 1 \rangle$. However, all of Lally's findings can be translated to the module descriptions that we have been using for convolutional and QC codes, and this is how they are presented in Theorem 7.6.3.

Acknowledgement

The third author thanks Frederic Ezerman, Markus Grassl, and Patrick Solé for their valuable comments and discussions. The authors are also grateful to Cary Huffman for his great help in improving this chapter. The second and third authors are supported by NTU Research Grant M4080456.

Chapter 8

Introduction to Skew-Polynomial Rings and Skew-Cyclic Codes

Heide Gluesing-Luerssen

University of Kentucky

8.1 Introduction

In classical block coding theory, cyclic codes are the most studied class of linear codes with additional algebraic structure. This additional structure turns out to be highly beneficial from a coding-theoretical point of view. Not only does it allow the design of codes with large minimum distance, it also gives rise to very efficient algebraic decoding algorithms. For further details we refer to Chapter 1 of this encyclopedia, to [1008, Ch. 4 and 5] in the textbook by Huffman/Pless, and to the vast literature on this topic.

Initiated by Boucher/Ulmer in [256, 258, 259], the notion of cyclicity has been generalized in various ways to skew-cyclicity during the last decade. In more precise terms the quotient space $\mathbb{F}[x]/(x^n - 1)$, which is the ambient space for classical cyclic codes, is replaced by $\mathbb{F}[x; \sigma]/{}^\bullet(x^n - 1)$, where $\mathbb{F}[x; \sigma]$ is the skew-polynomial ring induced by an automorphism σ of $\mathbb{F}$ (see Definition 8.2.1), and ${}^\bullet(x^n - 1)$ is the left ideal generated by $x^n - 1$. Further generalizations are obtained by replacing the modulus $x^n - 1$ by $x^n - a$, leading to skew-constacyclic codes, or even more general polynomials f of degree n. In any case, the quotient is isomorphic as a left $\mathbb{F}$-vector space to $\mathbb{F}^n$, and thus we may consider linear codes in $\mathbb{F}^n$ as subspaces of the quotient.

This allows us to define skew-cyclic codes. A linear code in $\mathbb{F}^n$ is (σ, f)-skew-cyclic if it is a left submodule of $\mathbb{F}[x; \sigma]/{}^\bullet(f)$. As in the classical case, every such submodule is generated by a right divisor of the modulus f. If $f = x^n - a$ or even $f = x^n - 1$, the resulting codes are called (σ, a)-skew-constacyclic or σ-skew-cyclic, respectively. The first most striking difference when compared to the classical case is that the modulus $x^n - 1$ has in general far more right divisors in $\mathbb{F}[x; \sigma]$ than in $\mathbb{F}[x]$. As a consequence, a polynomial may have more roots than its degree suggests. All of this implies that the family of skew-cyclic codes of given length is far larger than that of cyclic codes.

While these basic definitions are straightforward, a detailed study of the algebraic and coding-theoretic properties of skew-cyclic codes requires an understanding of the skew-polynomial ring $F[x; \sigma]$. In Sections 8.2, 8.4, and 8.5 we will present the theory of skew polynomials as it is needed for our study of skew-cyclic codes. This entails division properties in the ring $F[x; \sigma]$, evaluations of skew polynomials and their (right) roots, and algebraic sets along with a skew version of Vandermonde matrices. Division and factorization properties were studied in detail by Ore, who introduced skew-polynomial rings in the 1930's in his seminal paper [1466]. Evaluations of skew polynomials were first considered by Lam [1193] in the 1980's and then further investigated by Lam and Leroy; see [1193, 1194, 1195, 1196, 1223]. For Sections 8.2, 8.4, and 8.5, we will closely follow these sources. We will add plenty of examples illustrating the differences to commutative polynomial rings. In Section 8.3 we will briefly present the close relation between skew polynomials over a finite field and linearized polynomials, which play a crucial role in the area of rank-metric codes.

The material in Sections 8.2, 8.4, and 8.5 provides us with the right toolbox to study skew-cyclic codes and their generalizations. In Sections 8.7 and 8.8 we will derive the algebraic theory of (σ, f)-skew-cyclic codes and specialize to skew-constacyclic codes whenever necessary. We will do so by introducing skew circulant matrices because their row spaces are the codes in question. As a guideline to skew circulants we give a brief approach to classical cyclic codes via classical circulants in Section 8.6. Among other things we will see in Section 8.8 that the dual of a $(\sigma, x^n - a)$-skew-constacyclic code is $(\sigma, x^n - a^{-1})$-skew-constacyclic, and a generator polynomial arises as a certain kind of reciprocal of the generator polynomial of the primal code. This result appeared first in [258] and was later derived in more conceptual terms in [736].

In Section 8.9 we will report on constructions of skew-cyclic codes with designed minimum distance. The results are taken from the very recent papers [260, 828, 1791]. They amount to essentially two kinds of skew-BCH codes. For the first kind the generator polynomial has right roots that appear as consecutive powers of a suitable element in a field extension (similar to classical BCH codes), whereas for the second kind it has right roots that are consecutive Frobenius powers of a certain element. Both cases can be generalized to Hartmann–Tzeng form as for classical cyclic codes. The theory of skew Vandermonde matrices will be a natural tool in the discussion.

We wish to stress that in this survey we restrict ourselves to skew-cyclic codes derived from skew polynomials of automorphism type over fields. More general situations have been studied in the literature; see Remark 8.2.2. In addition, some results have been obtained for quasi-skew-cyclic codes. We will not survey this material either. Finally, we also do not discuss decoding algorithms for the codes of Section 8.9 in this survey.

8.2   Basic Properties of Skew-Polynomial Rings

In this section we introduce skew-polynomial rings with coefficients in a field. These rings were considered and studied first by Ore in [1466]. We will give a brief account of the ring-theoretic results from [1466] insofar as they are important for our later discussions of skew-cyclic codes.

Definition 8.2.1 Let F be any field and $\sigma \in \mathrm{Aut}(F)$. The **skew-polynomial ring** $F[x; \sigma]$ is defined as the set $\left\{ \sum_{i=0}^{N} f_i x^i \mid N \in \mathbb{N}_0, \ f_i \in F \right\}$ endowed with usual addition, i.e., coefficient-wise, and multiplication given by the rule

$$xa = \sigma(a)x \quad \text{for all } a \in F \tag{8.1}$$

along with distributivity and associativity. Then $(F[x; \sigma], +, \cdot)$ is a ring with identity $x^0 = 1$. Its elements are called **skew polynomials** or simply polynomials.

If $\sigma = \text{id}$, then $F[x; \sigma] = F[x]$, the classical commutative polynomial ring over F. We refer to this special case as the *commutative case* and *commutative polynomials*. In the general case, the additive groups of $F[x; \sigma]$ and $F[x]$ are identical, whereas multiplication in $F[x; \sigma]$ is given by

$$\Big(\sum_{i=0}^{N} f_i x^i \Big) \Big(\sum_{j=0}^{M} g_j x^j \Big) = \sum_{i,j} f_i \sigma^i(g_j) x^{i+j}.$$

Note that the set of skew polynomials may also be written as $\{ \sum_{i=0}^{N} x^i f_i \mid N \in \mathbb{N}_0, \, f_i \in F \}$, i.e., with coefficients on the right of x. The only rule we have to obey is to apply σ when moving coefficients from the right to the left of x, and thus σ^{-1} for the other direction. We will always write polynomials as $\sum_{i=0}^{N} f_i x^i$, and coefficients are meant to be left coefficients. As a consequence, the **leading coefficient** of a polynomial is meant to be its *left* leading coefficient.

Note that $F[x; \sigma]$ is a left and right vector space over F, but these two vector space structures are not identical.

Remark 8.2.2 Skew polynomial rings are commonly introduced and studied in much more generality. One may replace the coefficient field F by a division algebra or even a noncommutative ring; one may consider (ring) endomorphisms σ instead of automorphisms; and one may introduce a σ-derivation, say δ, which then turns (8.1) into $xa = \sigma(a)x + \delta(a)$. All of this is standard in the literature of skew-polynomial rings. For simplicity of this presentation we restrict ourselves to skew-polynomial rings as in Definition 8.2.1. However, we wish to point to the article [260] for examples showing that a σ-derivation may indeed lead to skew-cyclic codes with better minimum distance than what can be achieved with the aid of an automorphism alone. In addition, we will not discuss skew-polynomial rings with coefficients from a finite ring.

Remark 8.2.3 Consider the skew-polynomial ring $F[x; \sigma]$, and let $K \subseteq F$ be the fixed field of σ. If σ has finite order, say m, the center of $F[x; \sigma]$ is given by the commutative polynomial ring $K[x^m]$. This is easily seen by using the fact that any f in the center satisfies $xf = fx$ and $af = fa$ for all $a \in F$. If σ has infinite order, the center is K.

Example 8.2.4 Consider the field $\mathbb{C}$ of complex numbers, and let σ be complex conjugation. Then the center of $\mathbb{C}[x; \sigma]$ is the commutative polynomial ring $\mathbb{R}[x^2]$. Furthermore, $\mathbb{R}[x]$ is a subring of both $\mathbb{C}[x; \sigma]$ and $\mathbb{C}[x]$, which shows that a skew-polynomial ring may be a subring of skew-polynomial rings with different automorphisms.

Later we will restrict ourselves to skew-polynomial rings over finite fields. The following situation essentially covers all such cases because each automorphism is a power of the Frobenius automorphism over the prime field. Throughout, the automorphism $\mathbb{F}_{q^m} \to \mathbb{F}_{q^m}$ given by $c \mapsto c^q$ is simply called the q-**Frobenius**.

Example 8.2.5 Consider the skew-polynomial ring $\mathbb{F}[x; \sigma]$ where $\mathbb{F} = \mathbb{F}_{q^m}$ and $\sigma \in \text{Aut}(\mathbb{F})$ is the q-Frobenius. Then σ has order m and fixed field $\mathbb{F}_q$, and the center of $\mathbb{F}[x; \sigma]$ is $\mathbb{F}_q[x^m]$.

Let us return to general skew-polynomial rings $F[x; \sigma]$ and fix the following standard notions.

Definition 8.2.6 The **degree** of skew polynomials is defined in the usual way as the largest exponent of x appearing in the polynomial, and $\deg(0) := -\infty$. This does not depend on the side where we place the coefficients because σ is an automorphism, and we obtain

$$\deg(f + g) \leq \max\{\deg(f), \deg(g)\} \quad \text{and} \quad \deg(fg) = \deg(f) + \deg(g).$$

As a consequence, the group of units of $F[x; \sigma]$ is given by $F^* = F \setminus \{0\}$. A nonzero polynomial is **monic** if its leading coefficient is 1. Again, this does not depend on the sidedness of the coefficients because $\sigma(1) = 1$. We say that g is a **right divisor** of f and write $g\,|_r\,f$ if $f = hg$ for some $h \in F[x; \sigma]$. A polynomial $f \in F[x; \sigma] \setminus F$ is **irreducible** if all its right (hence left) divisors are units or polynomials of the same degree as f. Clearly, polynomials of degree 1 are irreducible.

Example 8.2.7 Let $F = \mathbb{F}_4 = \{0, 1, \omega, \omega^2\}$, where $\omega^2 = \omega + 1$, and let σ be the 2-Frobenius. Then $\sigma^{-1} = \sigma$. In $\mathbb{F}_4[x; \sigma]$ we have:

(1) $x^2 + 1 = (x+1)(x+1) = (x+\omega^2)(x+\omega) = (x+\omega)(x+\omega^2)$. Thus, a skew polynomial may have more linear factors than its degree suggests.

(2) $(x^2 + \omega x + \omega)(x + \omega) = x^3 + \omega^2 x + \omega^2$, and thus $x + \omega$ is a right divisor of $x^3 + \omega^2 x + \omega^2$. It is not a left divisor. This is easily seen by computing

$$(x + \omega)(f_2 x^2 + f_1 x + f_0) = f_2^2 x^3 + (\omega f_2 + f_1^2) x^2 + (\omega f_1 + f_0^2) x + \omega f_0$$

and comparing coefficients with those of $x^3 + \omega^2 x + \omega^2$.

(3) The polynomial $x^{14} + 1 \in \mathbb{F}_4[x; \sigma]$ has 599 nontrivial monic right divisors. On the other hand, in the commutative polynomial ring $\mathbb{F}_4[x]$ the same polynomial has only 25 nontrivial monic right divisors.

Just as for commutative polynomials, one can carry out division with remainder in $F[x; \sigma]$ if one takes sidedness of the coefficients into account. This is spelled out in (a) below. The proof is entirely analogous to the commutative case. Indeed, if $\deg(f) = m \geq \deg(g) = \ell$ and the leading coefficients of f and g are f_m and g_ℓ, respectively, then the polynomial $f - f_m \sigma^{m-\ell}(g_\ell^{-1}) x^{m-\ell} g$ has degree less than m. This allows one to proceed until a remainder of degree less than ℓ is obtained. The rest of the theorem formulates the familiar consequences of division with remainder.

Theorem 8.2.8 ([1466, pp. 483–486]) $F[x; \sigma]$ *is a left Euclidean domain and a right Euclidean domain. More precisely, we have the following.*

(a) **Right division with remainder:** *For all $f, g \in F[x; \sigma]$ with $g \neq 0$ there exist unique polynomials $s, r \in F[x; \sigma]$ such that $f = sg + r$ and $\deg(r) < \deg(g)$. If $r = 0$, then g is a right divisor of f.*

(b) *For any two polynomials $f_1, f_2 \in F[x; \sigma]$, not both zero, there exists a unique monic polynomial $d \in F[x; \sigma]$ such that $d\,|_r\,f_1$, $d\,|_r\,f_2$ and such that whenever $h \in F[x; \sigma]$ satisfies $h\,|_r\,f_1$ and $h\,|_r\,f_2$ then $h\,|_r\,d$. The polynomial d is called the **greatest common right divisor** of f_1 and f_2, denoted by $\operatorname{gcrd}(f_1, f_2)$. It satisfies a **right Bezout identity**, that is,*

$$d = u f_1 + v f_2 \quad \text{for some } u, v \in F[x; \sigma].$$

*We may choose u, v such that $\deg(u) < \deg(f_2)$ and, consequently, $\deg(v) < \deg(f_1)$. This is a consequence of the Euclidean algorithm; see also [802, Sec. 2]. If $d = 1$, we call f_1, f_2 **relatively right-prime**.*

(c) *For any two nonzero polynomials f_1, $f_2 \in F[x; \sigma]$, there exists a unique monic polynomial $\ell \in F[x; \sigma]$ such that $f_i \mid_r \ell$, $i = 1, 2$, and such that whenever $h \in F[x; \sigma]$ satisfies $f_i \mid_r h$, $i = 1, 2$, then $\ell \mid_r h$. The polynomial ℓ is called the **least common left multiple** of f_1 and f_2, denoted by $\mathrm{lclm}(f_1, f_2)$. Moreover, $\ell = u f_1 = v f_2$ for some $u, v \in F[x; \sigma]$ with $\deg(u) \leq \deg(f_2)$ and $\deg(v) \leq \deg(f_1)$.*

(d) *For all nonzero f_1, $f_2 \in F[x; \sigma]$*

$$\deg(\mathrm{gcrd}(f_1, f_2)) + \deg(\mathrm{lclm}(f_1, f_2)) = \deg(f_1) + \deg(f_2).$$

Analogous statements hold true for the left-hand side.

We refer to [356, 802] for various algorithms for fast computations in $\mathbb{F}[x; \sigma]$ for a finite field $\mathbb{F}$, in particular for factoring skew polynomials into irreducibles.

Exactly as in the commutative case, the above leads to the following consequence.

Theorem 8.2.9 *Let $I \subseteq F[x; \sigma]$ be a left ideal (that is, $(I, +)$ is a subgroup of $(F[x; \sigma], +)$ and I is closed with respect to left multiplication by elements from $F[x; \sigma]$). Then I is principal; i.e., there exist $f \in I$ such that $I = F[x; \sigma]f = \{gf \mid g \in F[x; \sigma]\}$. For brevity we will use the notation $^{\bullet}(f)$ for $F[x; \sigma]f$ and call it the **left ideal generated by** f. An analogous statement is true for right ideals. Thus, $F[x; \sigma]$ is a **left principal ideal ring** and a **right principal ideal ring**.*

Recall the center of $F[x; \sigma]$ from Remark 8.2.3. It is clear that for any polynomial f in the center, the ideal $^{\bullet}(f)$ is **two-sided**, i.e., a left ideal and a right ideal. Polynomials generating two-sided ideals are closely related to central elements.

Remark 8.2.10 Let σ have order m. An element $f \in F[x; \sigma]$ is called **two-sided** if the ideal $^{\bullet}(f)$ is two-sided, i.e., $^{\bullet}(f) = (f)^{\bullet}$. It is not hard to see [1029, Thm. 1.1.22] that the two-sided elements of $F[x; \sigma]$ are exactly the polynomials of the form $\{cx^t g \mid c \in F, t \in \mathbb{N}_0, g \in Z\}$, where $Z = K[x^m]$ is the center of $F[x; \sigma]$. As a special case, for any $a \in F^*$, the polynomial $x^n - a$ is two-sided if and only if it is central if and only if $m \mid n$ and $\sigma(a) = a$.

With respect to many properties addressed thus far, the skew-polynomial ring $F[x; \sigma]$ behaves similarly to the commutative polynomial ring $F[x]$. However, a main difference is that in $F[x; \sigma]$, where $\sigma \neq \mathrm{id}$, polynomials do not factor uniquely (up to order) into irreducible polynomials. We have seen this already in Example 8.2.7(1). Of course, irreducible factorizations of non-units still exist, which is a consequence of the boundedness of the degree by zero. In order to formulate a uniqueness result, we need the notion of similarity defined in [1466]. The equivalence of (a) and (b) below is straightforward. The equivalence to (c) can be found in [1029, Prop. 1.2.8].

Definition 8.2.11 Let $f, g \in F[x; \sigma]$. The following are equivalent.

(a) There exist $h, k \in F[x; \sigma]$ such that $\mathrm{gcrd}(f, h) = 1$, $\mathrm{gcld}(g, k) = 1$, and $gh = kf$.

(b) There exist $h \in F[x; \sigma]$ such that $\mathrm{gcrd}(f, h) = 1$ and $\mathrm{lclm}(f, h) = gh$.

(c) The left $F[x; \sigma]$-modules $F[x; \sigma]/^{\bullet}(f)$ and $F[x; \sigma]/^{\bullet}(g)$ are isomorphic.

If (a), hence (b) and (c), holds true, the polynomials f, g are called (left) **similar**.

In the commutative ring $F[x]$, two polynomials are thus similar if and only if they differ by a constant factor. In general, similar polynomials have the same degree (see Proposition 8.2.8(d)). Part (c) shows that similarity is indeed an equivalence relation on $F[x; \sigma]$ and does not depend on the sidedness; i.e., f, g are right similar if and only if they are left similar (see also [1466, Thm. 13, p. 489] without resorting to (c)). Part (c) above is the similarity notion for left ideals as introduced and discussed by Cohn [427, Sec. 3.2]. It leads to a simple criterion for similarity, which we will present next.

For a monic polynomial $f = \sum_{i=0}^{n-1} f_i x^i + x^n \in F[x; \sigma]$ define the ordinary **companion matrix**

$$
C_f = \begin{bmatrix}
 & 1 & & & \\
 & & \ddots & & \\
 & & & 1 & \\
 & & & & 1 \\
-f_0 & -f_1 & \cdots & -f_{n-2} & -f_{n-1}
\end{bmatrix} \in F^{n \times n}. \tag{8.2}
$$

Consider the map L_x given by left multiplication by x in the left $F[x; \sigma]$-module $\mathcal{M}_f :=$ $F[x; \sigma]/^\bullet(f)$. This map is σ-semilinear, i.e., $L_x(at) = \sigma(a) L_x(t)$ for all $a \in F$ and $t \in \mathcal{M}_f$. Furthermore, the rows of C_f are the coefficient vectors of $L_x(x^i)$ for $i = 0, \ldots, n-1$. All of this shows that $L_x(\sum_{i=0}^{n-1} a_i x^i) = \sum_{i=0}^{n-1} b_i x^i$, where $(b_0, \ldots, b_{n-1}) = (\sigma(a_0), \ldots, \sigma(a_{n-1})) C_f$. In this sense, C_f is the matrix representation of the semi-linear map L_x with respect to the basis $\{1, x, \ldots, x^{n-1}\}$.

If g is another monic polynomial of degree n, then both $\mathcal{M}_f$ and $\mathcal{M}_g$ are n-dimensional over F and thus isomorphic F-vector spaces. They are isomorphic as left $F[x; \sigma]$-modules if we can find a left $F[x; \sigma]$-linear isomorphism. Along with Definition 8.2.11(c) this easily leads to the following criterion for similarity (see also [1196, Thm. 4.9]).

Proposition 8.2.12 *Let $f, g \in F[x; \sigma]$ be monic polynomials of degree n. Then f, g are similar if and only if there exists a matrix $B \in \mathrm{GL}_n(F)$ such that $C_g = \sigma(B) C_f B^{-1}$.*

In [356, Prop. 2.1.17] a different criterion is presented for finite fields F in terms of the reduced norm.

Now we are ready to formulate the uniqueness result for irreducible factorizations.

Theorem 8.2.13 ([1466, p. 494]) *Let $f_1, \ldots, f_r, g_1, \ldots, g_s$ be irreducible polynomials in $F[x; \sigma]$ such that $f_1 \cdots f_r = g_1 \cdots g_s$. Then $r = s$ and there exists a permutation π of $\{1, \ldots, r\}$ such that $g_{\pi(i)}$ is similar to f_i for all $i = 1, \ldots, r$. In particular, $\deg(f_i) = \deg(g_{\pi(i)})$ for all i.*

The reader should be aware that the converse of the above statement is not true: it is not hard to find (monic) polynomials $f_i, g_i, i = 1, 2$, such that f_i and g_i are similar for $i = 1, 2$, but $f_1 f_2$ and $g_1 g_2$ are not similar (and thus certainly not equal).

We close this section with the following example illustrating yet some further challenging features in factorizations of skew polynomials.

Example 8.2.14 Consider $\mathbb{F}_4[x; \sigma]$ as in Example 8.2.7. In (1) we had seen that $x^2 + 1 = (x + \omega)(x + \omega^2)$. Right-multiplying the first factor by ω^2 and left-multiplying the second one by ω does not change the product and thus

$$
x^2 + 1 = (x + \omega)(x + \omega^2) = (\omega x + 1)(\omega x + 1).
$$

Hence we have two factorizations of $x^2 + 1$ into linear polynomials. While in the first factorization the two factors are relatively right-prime, the two factors of the second factorization are identical. In the first factorization the linear factors are monic, whereas in the second one they are normalized such that their constant coefficients are 1. The reader is invited to check that Theorem 8.2.13 is indeed true: choose $\pi = \mathrm{id}$ and $h_1 = \omega$, $h_2 = \omega x$. Examples like this will make it impossible to define multiplicities of roots for skew polynomials in a meaningful way. We will comment on this at the end of Section 8.5.

We refer to [1466] for further results on decompositions of skew polynomials, most notably completely reducible polynomials, i.e., polynomials that arise as the least common left multiple of irreducible polynomials — and thus generalize square-free commutative polynomials.

8.3 Skew Polynomials and Linearized Polynomials

In this short section we discuss the relation between skew-polynomial rings and the ring of linearized polynomials over finite fields. The latter play an important role in the study of rank-metric codes.

Consider the ring $\mathbb{F}[x; \sigma]$, where $\mathbb{F} = \mathbb{F}_{q^m}$ and σ is the q-Frobenius; see Example 8.2.5. In the commutative polynomial ring $\mathbb{F}[y]$ define the subset

$$\mathcal{L} := \mathcal{L}_{q^m, q} := \Big\{ \sum_{i=0}^{N} f_i y^{q^i} \;\Big|\; N \in \mathbb{N}_0, f_i \in \mathbb{F} \Big\}.$$

Polynomials of this type are called q-**linearized** because for any $f \in \mathcal{L}$ the associated map $\mathbb{F} \longrightarrow \mathbb{F}$, $a \longmapsto f(a)$ is $\mathbb{F}_q$-linear. Linearized polynomials have been well studied in the literature and a nice overview of the basic properties can be found in [1254, Ch. 3.4]. In particular, $(\mathcal{L}, +, \circ)$ is a (noncommutative) ring, where $+$ is the usual addition and $\circ$ the composition of polynomials. The rings $\mathbb{F}[x; \sigma]$ and $\mathcal{L}$ are isomorphic in the obvious way. Indeed, the map

$$\Lambda : \mathbb{F}[x; \sigma] \longrightarrow \mathcal{L} \quad \text{where} \quad \sum_{i=0}^{N} g_i x^i \longmapsto \sum_{i=0}^{N} g_i y^{q^i} \tag{8.3}$$

is a ring isomorphism between $\mathbb{F}[x; \sigma]$ and $(\mathcal{L}, +, \circ)$. The only interesting part of the proof is the multiplicativity of Λ. But that follows from $\Lambda(ax^i bx^j) = \Lambda(a\sigma^i(b)x^{i+j}) = \Lambda(ab^{q^i} x^{i+j}) = ab^{q^i} y^{q^{i+j}} = ay^{q^i} \circ by^{q^j}$ for all $a, b \in \mathbb{F}$ and $i, j \in \mathbb{N}_0$. As a consequence, $\mathcal{L}$ inherits all the properties of $\mathbb{F}[x; \sigma]$ presented in the previous section. We can go even further. The polynomial $y^{q^m} - y \in \mathcal{L}$ induces the zero map on $\mathbb{F} = \mathbb{F}_{q^m}$. Its pre-image under Λ is $x^m - 1$, which is in the center of $\mathbb{F}[x; \sigma]$; see Example 8.2.5. As a consequence, the left ideal generated by $y^{q^m} - y$ is two-sided and gives rise to the quotient ring $\mathcal{L}/(y^{q^m} - y)$. Since the latter ring has cardinality q^{m^2}, this tells us that the quotient is isomorphic to the space of all $\mathbb{F}_q$-linear maps on $\mathbb{F}_{q^m}$. Thus,

$$\mathbb{F}[x; \sigma]/(x^m - 1) \cong \mathcal{L}/(y^{q^m} - y) \cong \mathbb{F}_q^{m \times m}.$$

Clearly, the second map is given by $g + (y^{q^m} - y) \longmapsto [g]_B^B$, where $[g]_B^B$ denotes the matrix representation of the map g with respect to a chosen basis B of $\mathbb{F}_{q^m}$ over $\mathbb{F}_q$. Fix the basis $B = (b_0, \ldots, b_{m-1})$ and define its **Moore matrix** $S = \big(b_j^{q^i}\big)_{i,j=0}^{m-1}$ (called the **Wronskian** in [1194, (4.11)]). The linear independence of $b_0, \ldots, b_{m-1}$ implies that S is in $\mathrm{GL}_m(\mathbb{F}_{q^m})$ (see

[1254, Cor. 2.38]). Furthermore, in [1909, Lem. 4.1] it is shown that for any $g = \sum_{i=0}^{m-1} g_i y^{q^i}$ we have $S\,[g]_B^B\,S^{-1} = D_g$, where

$$
D_g = \begin{bmatrix}
g_0 & g_1 & \cdots & g_{m-2} & g_{m-1} \\
g_{m-1}^q & g_0^q & \cdots & g_{m-3}^q & g_{m-2}^q \\
\vdots & \vdots & & \vdots & \vdots \\
g_1^{q^{m-1}} & g_2^{q^{m-1}} & \cdots & g_{m-1}^{q^{m-1}} & g_0^{q^{m-1}}
\end{bmatrix}
\tag{8.4}
$$

is the **Dickson matrix** of g. This matrix is also known as q-**circulant** and generalizes the notion of a classical circulant matrix; see (8.8). The isomorphisms above allow us to define the Dickson matrix of $g \in \mathbb{F}[x;\sigma]$ as $D_g := D_{\Lambda(g)}$ so that we obtain the ring embedding

$$
\mathbb{F}[x;\sigma]/(x^m - 1) \longrightarrow \mathbb{F}^{m \times m} \quad \text{where} \quad g + (x^m - 1) \longmapsto D_g.
$$

Note that the ith row of D_g is given by the coefficient vector of $x^i g \in \mathbb{F}[x;\sigma]$ reduced modulo $x^m - 1$ via right division. This will be made precise in the realm of skew circulants in Section 8.7 (see Definition 8.7.3 and the paragraph thereafter).

Linearized polynomials and their kernels play a crucial role in the study of the rank distance. We refer to the vast literature on rank-metric codes, initiated by [760]; see Chapter 11. Most closely related to this survey are the articles [260, 389, 1339, 1340] on the rank distance of skew-cyclic codes.

8.4 Evaluation of Skew Polynomials and Roots

In this section we provide an overview of evaluating skew polynomials at field elements and roots of skew polynomials. Ore in his seminal work [1466] did not define these concepts. They were in fact introduced much later in 1986 by Lam, and the material of this and the next section is taken from the work of Lam and Leroy [1193, 1194, 1195, 1196]. Roots of skew polynomials as defined below in Definition 8.4.1 also appeared previously in the monograph [427, Sec. 8.5] by Cohn in 1985. (However, what is now commonly called a 'right root' is considered a 'left root' by Cohn.)

Throughout, we fix the skew-polynomial ring $F[x;\sigma]$. Consider a polynomial $f = \sum_{i=0}^{N} f_i x^i \in F[x;\sigma]$. Clearly, the usual notion of evaluating f at a point $a \in F$, that is, $f(a) = \sum_{i=0}^{N} f_i a^i$ is not well-defined if $\sigma \neq \text{id}$ because x does not commute with the field elements. For instance, for the polynomial $f = bx = x\sigma^{-1}(b)$, where b is not fixed by σ, substituting a nonzero element a for x would lead to the contradiction $ba = a\sigma^{-1}(b)$.

Circumventing this issue by requiring that coefficients be on the left when substituting a for x does not solve the problem because it does not lead to a nice remainder theory. Take for instance $f = x^3 + \omega \in \mathbb{F}_4[x;\sigma]$, where $\mathbb{F}_4[x;\sigma]$ is as in Example 8.2.7. Substituting ω for x yields ω^2, and thus ω is not a root in this naive sense. Yet, one easily verifies that $x - \omega$ is a right divisor (and a left divisor) of f.

A meaningful notion of evaluation of skew polynomials at field elements and roots of skew polynomials is obtained by making use of division with remainder by the associated linear polynomials. In the following we define right evaluation and right roots, and in this survey 'root' will always mean 'right root'. The left-sided versions are analogous.

Definition 8.4.1 Let $f \in F[x;\sigma]$ and $a \in F$. We define $f(a) = r$, where $r \in F$ is the remainder upon right division of f by $x - a$; that is, $f = g \cdot (x - a) + r$ for some $g \in F[x;\sigma]$. If $f(a) = 0$, we call a a **(right) root** of f. Thus a is a root of f if and only if $(x - a)\,|_r\,f$.

Example 8.2.7(1) shows that a polynomial of degree N may have more than N roots. If the field F is infinite, it may even have infinitely many roots. For instance, in $\mathbb{C}[x;\sigma]$, where σ is complex conjugation, the polynomial $f = x^2 - 1$ splits as $(x + \bar{a})(x - a)$ for any a on the unit circle, and thus has exactly the complex numbers on the unit circle as roots. The next example shows yet another surprising phenomenon, namely even if f splits into linear factors, it may only have one root.

Example 8.4.2 The polynomial $(x - \alpha^2)(x - \alpha) \in \mathbb{F}_8[x;\sigma]$, where $\alpha^3 + \alpha + 1 = 0$ and σ is the 2-Frobenius, has the sole root α in $\mathbb{F}_8$. Extending σ to the 2-Frobenius on $\mathbb{F}_{8^2}$, results in two additional roots of $f \in \mathbb{F}_{8^2}[x;\sigma]$ in $\mathbb{F}_{8^2} \setminus \mathbb{F}_8$.

The evaluation $f(a)$ can be computed explicitly without resorting to division with remainder.

Definition 8.4.3 For any $i \in \mathbb{N}_0$ define $N_i : F \longrightarrow F$ as $N_0(a) = 1$ and $N_i(a) = \prod_{j=0}^{i-1} \sigma^j(a)$ for $i > 0$. We call N_i the ith **norm on** F.

Thus, $N_1(a) = a$ and $N_{i+1}(a) = N_i(a)\sigma^i(a)$ for all $a \in F$. For $\mathbb{F} = \mathbb{F}_{q^m}$ and σ the q-Frobenius, N_m is simply the field norm of $\mathbb{F}$ over $\mathbb{F}_q$. Note that in the commutative case, i.e., $\sigma = \text{id}$, we simply have $N_i(a) = a^i$, and thus the following result generalizes evaluation of commutative polynomials.

Proposition 8.4.4 ([1194, Lem. 2.4] or [1193, Eq. (11) and Thm. 3]) *Let* $f = \sum_{i=0}^N f_i x^i \in F[x;\sigma]$ *and* $a \in F$. *Then*

$$f(a) = \sum_{i=0}^N f_i N_i(a).$$

Now that we have a notion of roots for skew polynomials $f \in F[x;\sigma]$ we may wonder about the relation to the roots of the associated linearized polynomial $\Lambda(f)$ introduced in Section 8.3. The latter are simply commutative polynomials and thus the ordinary notion of roots applies.

Remark 8.4.5 Consider $F[x;\sigma] = \mathbb{F}[x;\sigma]$ as in Example 8.2.5, and let $g \in \mathbb{F}[x;\sigma]$ and $\Lambda(g) \in \mathbb{F}[y]$ be as in (8.3). Note that $\Lambda(g)(0)$ is always 0.

(a) For any q we have the following relation between the roots of g and $\Lambda(g)$. For any $b \in \mathbb{F}^*$

$$g(b^{q-1}) = 0 \Longleftrightarrow \Lambda(g)(b) = 0;$$

see also [828, Lem. A.3]. In order to see this, note that $\Lambda(g)(b) = 0$ implies $\Lambda(g)(\alpha b) = 0$ for all $\alpha \in \mathbb{F}_q$, which means that the linearized polynomial $y^q - b^{q-1}y$ is a divisor of $\Lambda(g)$ in $(\mathbb{F}[y], +, \cdot)$. But then it is also a divisor of $\Lambda(g)$ in the ring $(\mathcal{L}, +, \circ)$ (see [1254, Thm. 3.62]), i.e., $\Lambda(g) = G \circ (y^q - b^{q-1}y)$ for some $G \in \mathcal{L}$. Applying the ring homomorphism Λ^{-1} shows that $x - b^{q-1}$ is a right divisor of g.

(b) For $q = 2$, part (a) shows that the set of nonzero roots of $\Lambda(g)$ coincides with the set of nonzero roots of g.

(c) If $q \neq 2$, the roots of g do not agree with the roots of $\Lambda(g)$. To see this, take for instance $g = x - a$ for some nonzero $a \in \mathbb{F}_{q^m}$. Then a is a left and right root of g. The associated linearized polynomial is $\Lambda(g) = y^q - ay = y(y^{q-1} - a)$, and it may or may not have nonzero roots. For instance, if $\mathbb{F}_{q^m} = \mathbb{F}_{3^2}$, the equation $y^2 = a$ has two distinct roots for four values of a (the nonzero squares) and no roots for the other four values of a. The discrepancy between roots of skew polynomials and roots of linearized polynomials is, of course, also related to the fact that multiplication in $\mathcal{L}$ is composition whereas a root c corresponds to a factor $y - c$ in the ordinary sense (which is not even a linearized polynomial).

There is an obvious relation between the right roots of a skew polynomial over a finite field and the roots of a different associated commutative polynomial. Consider again the skew-polynomial ring $\mathbb{F}[x;\sigma]$ as in Example 8.2.5. Then the ith norm is given by

$$N_i(a) = a^{q^0 + q^1 + \cdots + q^{i-1}} = a^{[\![i]\!]} \quad \text{where} \quad [\![i]\!] := \frac{q^i - 1}{q - 1} \text{ for } i \geq 0. \tag{8.5}$$

Proposition 8.4.4 allows us to translate the evaluation of skew polynomials into evaluation of commutative polynomials (see also [389, p. 278]).

Remark 8.4.6 Define the map

$$\mathbb{F}[x;\sigma] \longrightarrow \mathbb{F}[y] \quad \text{where} \quad \sum_{i=0}^{n} f_i x^i \longmapsto P_f := \sum_{i=0}^{n} f_i y^{[\![i]\!]} \in \mathbb{F}[y].$$

Then $f(a) = P_f(a)$.

Properties of the polynomial P_f can be found in [1223, Sec. 2]. Unfortunately, the map $f \longmapsto P_f$ does not behave well under multiplication and therefore the above result is of limited use.

Let us return to the general case with an arbitrary field F. Having defined evaluation of polynomials, one may study properties of the map

$$\text{ev}_a : F[x;\sigma] \longrightarrow F \quad \text{where} \quad f \longmapsto f(a).$$

Clearly, this map is additive and left F-linear, but unlike in the commutative case, it is not multiplicative. As an extreme case of this non-multiplicativity note that $f = x - a$ satisfies $f(a) = 0$, whereas $(fb)(a) \neq 0$ for any $b \in \mathbb{F}$ not fixed by σ. Yet, evaluation is close to being multiplicative. Before we can make this precise we need the following definition.

Definition 8.4.7 Let $a \in F$. For $c \in F^*$ we define $a^c := \sigma(c)ac^{-1}$. We say that $a, b \in F$ are σ-**conjugate** if $b = a^c$ for some $c \in F^*$. The σ-**conjugacy class** of a is

$$\Delta(a) = \{a^c \mid c \in F^*\}.$$

Since the automorphism is fixed throughout this text, we will drop the prefix σ. The reader should be aware of the ambiguity of the notation. For instance, for $c = -1 \in F$, the notation a^c does not represent the inverse of a; in fact, in this case $a^c = a$. More generally, $a^{-c} = a^c$ for any $c \in F^*$.

It is easy to see that conjugacy defines an equivalence relation. Moreover, $\Delta(0) = \{0\}$, and if $\sigma = \text{id}$, then $\Delta(a) = \{a\}$ for all $a \in F$. Furthermore, if $a \neq 0$, then $a^c = a$ if and only if c is in the fixed field of σ. The relevance of conjugacy for us stems from the equivalence

$$b = a^c \Longleftrightarrow (x - b)c = \sigma(c)(x - a)$$

and the identities [1195, (3.3)]

$$\text{lclm}(x - a, x - b) = (x - b^{b-a})(x - a) = (x - a^{a-b})(x - b) \quad \text{for any } a \neq b, \tag{8.6}$$

which tell us that, up to conjugacy, linear factors can be reordered. Furthermore, as a special case of Proposition 8.2.12 we have

$$\left.\begin{array}{r} x - a \text{ and } x - b \text{ are similar} \\ \text{in the sense of Definition 8.2.11} \end{array}\right\} \Longleftrightarrow a \text{ and } b \text{ are conjugate.}$$

Example 8.4.8

(a) For any finite field $F = \mathbb{F}_{q^m}$ with q-Frobenius σ, the identity $a^c = c^{q-1}a$ implies that the nonzero conjugacy classes are given by the cosets of $\Delta(1) = \{c^{q-1} \mid c \in F^*\}$ in F^*. Thus, for $q = 2$ the conjugacy classes are $\{0\}$ and $\Delta(1) = F^*$, whereas for $q = 3$ there are two nonzero conjugacy classes, one of which consists of the squares of F^* and the other of the non-squares.

(b) For $\mathbb{C}$ with complex conjugation, the nonzero conjugacy classes are exactly the circles about the origin.

Remark 8.4.9 Obviously, the conjugacy classes are the orbits of the group action $F^* \times F \longrightarrow F$ where $(c, a) \longmapsto a^c$, and the stabilizer of any nonzero a is the multiplicative group of the fixed field of σ. Therefore, in the case where $F = \mathbb{F}_{q^m}$ and σ is the q-Frobenius, the nonzero conjugacy classes have size $(q^m - 1)/(q - 1)$. This also shows that there are q conjugacy classes (including $\{0\}$).

Now we can formulate the product theorem. It appeared first in [1193, Thm. 2] and was later extended to more general skew-polynomial rings in [1194, Thm. 2.7]. The proof follows by direct computations using Proposition 8.4.4 and properties of the norms N_i.

Theorem 8.4.10 *Let $f, g \in F[x; \sigma]$ and $a \in F$. Then*

$$(fg)(a) = \begin{cases} 0 & \text{if } g(a) = 0, \\ f(a^{g(a)})g(a) & \text{if } g(a) \neq 0. \end{cases}$$

In particular, if a is a root of fg, but not of g, then the conjugate $a^{g(a)}$ is a root of f.

We have seen already that the number of roots of a skew polynomial may vastly exceed its degree. Taking conjugacy into account, however, provides us with the following generalization of the commutative case. It follows quickly from the previous result by inducting on the degree.

Theorem 8.4.11 ([1193, Thm. 4]) *Let $f \in F[x; \sigma]$ have degree N. Then the roots of f lie in at most N distinct conjugacy classes. Furthermore, if $f = (x - a_1) \cdots (x - a_N)$ for some $a_i \in F$ and $f(a) = 0$, then a is conjugate to some a_i.*

Note that for $F = \mathbb{F}_{q^m}$ with $q = 2$, the theorem does not provide any insight because in this case there is only one conjugacy class. This also shows that the converse of Theorem 8.4.11 is not true: not every conjugate of some a_i is a root of f (take $N = 1$ for instance).

Even more, the above theorem does not state that every conjugacy class $\Delta(a_i)$ contains a root of f. This is indeed in general not the case, and an example can be found in the skew-polynomial ring $\mathbb{Q}(t)[x; \sigma]$, where σ is the $\mathbb{Q}$-algebra automorphism given by $t \mapsto t + 1$. However, for finite fields the last statement is in fact true.

Theorem 8.4.12 *Consider the ring $\mathbb{F}[x; \sigma]$ as in Example 8.2.5, and let $f = (x - a_1) \cdots (x - a_N)$ for some $a_i \in \mathbb{F}^*$. Then each conjugacy class $\Delta(a_i)$ contains a root of f.*

Proof: It suffices to show that $\Delta(a_1)$ contains a root of f, which means that f is of the form $f = (x - b_1) \cdots (x - b_{N-1})(x - a_1^c)$ for some $b_1, \ldots, b_{N-1}, c \in \mathbb{F}^*$. Since $(a^{c_1})^{c_2} = a^{c_1 c_2}$, the case $N = 2$ is sufficient. Thus, let $f = (x - b)(x - a)$ for some $a, b \in \mathbb{F}^*$. If $a \in \Delta(b)$ there is nothing to prove. Thus let $a \notin \Delta(b)$. Let now $c \in \mathbb{F}^*$. Theorem 8.4.10 implies $f(b^c) = ((b^c)^{b^c - a} - b)(b^c - a)$, and thus $f(b^c) = 0$ if and only if $(b^c)^{b^c - a} = b$. It is easy to see that the latter is equivalent to $\sigma(\sigma(c)b - ac) = \sigma(c)b - ac$, which in turn is equivalent to $\sigma(c)b - ac \in \mathbb{F}_q$. Thus we have to establish the existence of some $c \in \mathbb{F}^*$ such that $\sigma(c)b - ac \in \mathbb{F}_q$. Consider the $\mathbb{F}_q$-linear map $\Psi_{a,b} : \mathbb{F} \longrightarrow \mathbb{F}$ where $c \longmapsto \sigma(c)b - ac$. If we can show that $\Psi_{a,b}$ is injective, then it is bijective and we are done. Suppose there exists $d \in \ker \Psi_{a,b} \setminus \{0\}$. Then $\sigma(d)b - ad = d^q b - ad = 0$; hence $d^{q-1} = a/b$. But then $1 = d^{q^m - 1} = (d^{q-1})^{[\![m]\!]} = (a/b)^{[\![m]\!]}$. This shows that $\Psi_{a,b}$ is injective if $(a/b)^{[\![m]\!]} \neq 1$. On the other hand, if $(a/b)^{[\![m]\!]} = 1$, then the order of a/b in $\mathbb{F}^*$, say t, is a divisor of $[\![m]\!]$. Furthermore, $a/b = \omega^{k(q^m - 1)/t}$ for some k and a primitive element ω. Writing $ts = [\![m]\!]$, we conclude $a/b = (\omega^{ks})^{q-1}$. But this means that a and b are conjugate (see Example 8.4.8(a)), a contradiction. $\qquad \square$

Similar reasoning shows that the previous result is also true in the skew-polynomial ring $\mathbb{C}[x; \sigma]$ with complex conjugation σ.

8.5   Algebraic Sets and Wedderburn Polynomials

We now further the theory of right roots of skew polynomials by introducing minimal polynomials and algebraic sets and presenting some of their properties. The material is again from [1193, 1195]. Throughout, we fix a skew-polynomial ring $F[x; \sigma]$.

Definition 8.5.1 For a polynomial $f \in F[x; \sigma]$ denote by $V(f)$ its set of (right) roots in F; thus $V(f) = \{a \in F \mid f(a) = 0\}$. We call $V(f)$ the **vanishing set** of f. A subset $A \subseteq F$ is called σ-**algebraic** if there exists some nonzero $f \in F[x; \sigma]$ such that $A \subseteq V(f)$; that is, f vanishes on A. In this case, the monic polynomial of smallest degree, say f, such that $A \subseteq V(f)$, is uniquely determined by A and called the σ-**minimal polynomial** of A, denoted by m_A. The degree of m_A is called the σ-**rank** of A, denoted by $\mathrm{rk}(A)$.

Again, we will drop the prefix σ as there will be no ambiguity. That the minimal polynomial is well-defined is a consequence of Theorem 8.4.10 along with the fact that $F[x; \sigma]$ is a left principal ideal ring.

The reader may wonder why an algebraic set has to be merely contained in a vanishing set, but not necessarily be a vanishing set itself. The latter would be too restrictive for a meaningful theory because in general the minimal polynomial m_A of a set A has additional roots outside A (see Example 8.5.7(b)). This phenomenon gives rise to the **closure** of A, defined as the vanishing set $V(m_A)$. However, we do not need this notion and therefore will not discuss it in further detail.

We begin with discussing the vanishing sets $V(f)$ for given polynomials. First of all,

$$V(f) \subseteq V(g) \implies V(fh) \subseteq V(gh)$$

for all $f, g, h \in F[x; \sigma]$. On first sight the implication may feel counterintuitive because $V(f)$ denotes the set of right roots. Yet the reader can readily verify that it is just a simple consequence of Theorem 8.4.10. On the other hand, the analogous statement with left factors h is not true in general; that is, there exists $f, g, h \in F[x; \sigma]$ such that

$$V(f) \subseteq V(g) \quad \text{and} \quad V(hf) \nsubseteq V(hg).$$

Example 8.5.2 Consider $\mathbb{F}_4[x;\sigma]$ from Example 8.2.7. Let $f = h = x + 1$ and $g = x^2 + \omega^2 x + \omega$. Then one easily checks that $V(f) = \{1\} = V(g)$. In Example 8.2.7(1) we have seen that $V(hf) = \{1, \omega, \omega^2\}$. However, $hg = x^3 + \omega^2 x^2 + \omega$ has sole root 1.

We now turn to algebraic sets. Obviously, in the commutative case, i.e., $\sigma = \text{id}$, algebraic sets are exactly the finite sets. This is not the case for skew polynomials as we have seen right after Definition 8.4.1 for $\mathbb{C}[x;\sigma]$ with complex conjugation σ. From (8.6) we deduce that every set $A = \{a, b\}$ of cardinality 2 has rank 2, whereas Example 8.2.7(1) provides us with a set of cardinality 3 and rank 2. In general, a finite set $A = \{a_1, \ldots, a_n\}$ has minimal polynomial $m_A = \text{lclm}(x - a_1, \ldots, x - a_n)$. Using induction on the cardinality and the degree formula in Proposition 8.2.8(d), one obtains immediately:

Proposition 8.5.3 (see also [1193, Prop. 6]) *Let $A = \{a_1, \ldots, a_n\} \subseteq F$. Then A is algebraic and $\text{rk}(A) =: r \leq |A|$. Furthermore, there exist distinct $b_1, \ldots, b_r \in A$ such that $m_A = \text{lclm}(x - b_1, \ldots, x - b_r)$.*

Example 8.5.4 ([1223, Rem. 2.4]) Let $A = \mathbb{F}_{p^r}$, where p is prime and $r \in \mathbb{N}$. Then $m_A = \text{lclm}(x - a \mid a \in \mathbb{F}_{p^r}) = x^{r(p-1)+1} - x$. Thus, $\text{rk}(\mathbb{F}_{p^r}) = r(p-1) + 1$.

Here is a particularly interesting example. The proof is analogous to the commutative case.

Example 8.5.5 ([260, Prop. 4]) Let $\mathbb{F}_{q^s}$ be an extension field of $\mathbb{F} = \mathbb{F}_{q^m}$ and consider $\mathbb{F}_{q^s}[x;\sigma]$ with q-Frobenius σ. Fix an element $a \in \mathbb{F}_{q^s}$ and set $A = \{\tau(a) \mid \tau \in \text{Aut}(\mathbb{F}_{q^s} \mid \mathbb{F}_{q^m})\}$. Then the minimal polynomial m_A is in $\mathbb{F}[x;\sigma]$ and is the nonzero monic polynomial of smallest degree in $\mathbb{F}[x;\sigma]$ with (right) root a. It is called the σ-**minimal polynomial** of a over $\mathbb{F}$.

The minimal polynomial of an algebraic set always factors linearly. The following result comes closest to the commutative case.

Proposition 8.5.6 ([1193, Lem. 5]) *Let $A \subseteq F$ be an algebraic set of rank r. Then its minimal polynomial is of the form $m_A = (x - a_1) \cdots (x - a_r)$, where each a_i is conjugate to some $a \in A$.*

This result is indeed the best possible in the sense that the roots of the linear factors need not be in A.

Example 8.5.7

(a) Consider $\mathbb{F}_{3^3}[x;\sigma]$ with 3-Frobenius σ and primitive element β satisfying $\beta^3 + 2\beta + 1 = 0$. Let $A = \{\beta^{14}, \beta^{25}\}$. Then $m_A = x^2 + \beta x + \beta = (x - \beta^{13})(x - \beta^{14}) = (x - \beta^2)(x - \beta^{25})$, and thus m_A is not the product of two linear terms with roots in A. With the aid of Example 8.4.8 one concludes that β^{13} is conjugate to β^{25} and β^2 is conjugate to β^{14}. Finally, $A = V(m_A)$; that is, m_A has no further roots in $\mathbb{F}_{3^3}$.

(b) The linear factors $x - a_i$ in Proposition 8.5.6 need not be distinct. Consider for instance $\mathbb{F}_{2^4}[x;\sigma]$ with 2-Frobenius σ and primitive element γ satisfying $\gamma^4 + \gamma + 1 = 0$. The polynomial

$$f = (x - \gamma^2)(x - \gamma^{12})(x - \gamma^2) = (x - \gamma^3)(x - \gamma^{14})(x - \gamma^{14})$$
$$= x^3 + \gamma^7 x^2 + \gamma^3 x + \gamma$$

is the minimal polynomial of $A = \{1, \gamma^2, \gamma^3, \gamma^6, \gamma^8, \gamma^{13}, \gamma^{14}\}$ and in fact $V(f) = A$. In order to illustrate Proposition 8.5.3 we mention that $f = \mathrm{lclm}(x - 1, x - \gamma^2, x - \gamma^3)$, which shows that the set $B = \{1, \gamma^2, \gamma^3\}$ is algebraic, but not a vanishing set: every polynomial vanishing on B has additional roots in $\mathbb{F}_{2^4}$. On the other hand, $f \neq \mathrm{lclm}(x - 1, x - \gamma^2, x - \gamma^8)$. The latter polynomial is given by $g = x^2 + \gamma^5 x + \gamma^{10}$.

The rank of a finite algebraic set can be determined via the 'skew version' of the classical Vandermonde matrix. The **skew Vandermonde matrix** has been introduced by Lam in [1193, p. 194]. For $a_1, \ldots, a_r \in F$ it is the matrix in $F^{n \times r}$ defined as

$$V_n^\sigma(a_1, \ldots, a_r) := V_n(a_1, \ldots, a_r) = \begin{bmatrix} 1 & \cdots & 1 \\ N_1(a_1) & \cdots & N_1(a_r) \\ \vdots & & \vdots \\ N_{n-1}(a_1) & \cdots & N_{n-1}(a_r) \end{bmatrix}. \tag{8.7}$$

The skew Vandermonde matrix depends on σ (because the norms do). Proposition 8.4.4 shows that for $g = \sum_{i=0}^{n-1} g_i x^i \in F[x; \sigma]$ we have

$$(g(a_1), \ldots, g(a_r)) = (g_0, \ldots, g_{n-1}) V_n^\sigma(a_1, \ldots, a_r).$$

Using (8.5), we conclude that for the skew-polynomial ring $\mathbb{F}[x; \sigma]$, where σ is the q-Frobenius, the skew Vandermonde matrix evaluates the powers $x^{[0]}, x^{[1]}, \ldots, x^{[n-1]}$ at $a_1, \ldots, a_r$. This matrix must not be confused with the Moore matrix of $a_1, \ldots, a_r$, which evaluates the powers $x^{q^0}, x^{q^1}, \ldots, x^{q^{n-1}}$ at $a_1, \ldots, a_r$. The relation between the Moore matrix and an associated Vandermonde matrix is spelled out in Example 8.5.10 below. The following results are not hard to show (in [1193] a bit more work is required because the coefficients are from a division ring).

Theorem 8.5.8 ([1193, Thm. 8]) *Let $A = \{a_1, \ldots, a_n\} \subseteq F$. Then*

$$\mathrm{rk}(A) = \mathrm{rank}(V_n(a_1, \ldots, a_n)).$$

*As a consequence, if $\mathrm{rk}(A) = |A|$ (such a set is called **P-independent**), then $\mathrm{rk}(B) = |B|$ for every subset $B \subseteq A$.*

Example 8.5.9

(a) Consider $\mathbb{C}[x; \sigma]$ with complex conjugation σ. If $A = \{a_1, \ldots, a_n\} \subseteq \mathbb{C}$, where $a_1, \ldots, a_n$ are not all equal and $|a_1| = \cdots = |a_n| =: c$, then $N_2(a_i) = a_i \sigma(a_i) = c^2$ for all i. This implies that $V_n(a_1, \ldots, a_n)$ has rank 2, and this is consistent with the fact that $m_A = x^2 - c^2 = (x + a_i)(x - \overline{a_i})$.

(b) Consider Example 8.5.7(b). Then

$$V_3(1, \gamma^2, \gamma^8) = \begin{bmatrix} 1 & 1 & 1 \\ 1 & \gamma^2 & \gamma^8 \\ 1 & \gamma^6 & \gamma^9 \end{bmatrix}$$

has rank 2, consistent with the fact that $\mathrm{lclm}(x - 1, x - \gamma^2, x - \gamma^8)$ has degree 2.

Example 8.5.10 Consider $\mathbb{F}[x; \sigma]$ as in Example 8.2.5 where $\mathbb{F} = \mathbb{F}_{q^m}$. Suppose that $a_0, \ldots, a_{m-1}$ is a basis of $\mathbb{F}_{q^m}$ over $\mathbb{F}_q$. Then one easily verifies ([260, Eq. (4)] or [1282, Thm. 5]) that

$$V_m(a_0^{q-1}, \ldots, a_{m-1}^{q-1}) \begin{bmatrix} a_0 & & & \\ & a_1 & & \\ & & \ddots & \\ & & & a_{m-1} \end{bmatrix} = \begin{bmatrix} a_0 & \cdots & a_{m-1} \\ a_0^q & \cdots & a_{m-1}^q \\ \vdots & & \vdots \\ a_0^{q^{m-1}} & \cdots & a_{m-1}^{q^{m-1}} \end{bmatrix}.$$

Note that the matrix on the right hand side is the Moore matrix of $a_0, \ldots, a_{m-1}$ (see Section 8.3). Since the latter is invertible thanks to the linear independence of $a_0, \ldots, a_{m-1}$ ([1254, Cor. 2.38]), the same is true for the Vandermonde matrix on the left hand side. Let us now apply this to a normal basis $\{\gamma, \gamma^q, \ldots, \gamma^{q^{m-1}}\}$ of $\mathbb{F}_{q^m}$. Then the Vandermonde matrix above is $V_m(b, b^q, \ldots, b^{q^{m-1}})$, where $b = \gamma^{q-1}$. It is easy to see that $N_i(b^{q^j}) = (\gamma^{-1})^{q^j}\gamma^{q^{i+j}}$, which in turn yields that b^{q^j} is a right root of $x^m - 1 \in \mathbb{F}[x; \sigma]$ for all $j = 0, \ldots, m-1$. As a consequence, $x^m - 1 = \mathrm{lclm}(x - b, x - b^q, \ldots, x - b^{q^{m-1}})$ thanks to Theorem 8.5.8. Note that the obvious right root 1 of $x^m - 1$ does not appear in the list $b, b^q, \ldots, b^{q^{m-1}}$. Theorem 8.5.8 also shows for any subset $\{j_1, j_2, \ldots, j_r\} \subseteq \{0, 1, \ldots, m-1\}$ the polynomial $\mathrm{lclm}(x - b^{q^{j_1}}, x - b^{q^{j_2}}, \ldots, x - b^{q^{j_r}})$ has degree r (see also [827, Lem. 3.1]). Finally, consider a polynomial $x^m - a \in \mathbb{F}[x; \sigma]$ for some $a \in \mathbb{F}^*$ and suppose $c \in \mathbb{F}$ is a root of $x^m - a$. Using the multiplicity of the maps N_i, one easily deduces that $x^m - a = \mathrm{lclm}(x - cb, x - cb^q, \ldots, x - cb^{q^{m-1}})$.

We now turn to polynomials that occur as minimal polynomials of an algebraic set A.

Definition 8.5.11 A monic polynomial $f \in F[x; \sigma]$ is a **Wedderburn polynomial over** F or simply **W-polynomial** if $f = m_A$ for some $A \subseteq F$.

The polynomial $x^2 + 1$ is a W-polynomial over $\mathbb{F}_4$ as it is $m_{\{1, \omega, \omega^2\}}$ (see Example 8.2.7(1)), but it is not a W-polynomial over $\mathbb{F}_2$. Thus the field F matters in the definition of a W-polynomial. We will always assume that the field is the coefficient field of the skew polynomial ring under consideration. In general, $x^m - 1 \in \mathbb{F}_{q^m}[x; \sigma]$, where σ is the q-Frobenius, is a W-polynomial by Example 8.5.10.

The polynomial $x^2 - 1 \in \mathbb{C}[x; \sigma]$, where σ is complex conjugation, is the minimal polynomial of the unit circle (or of the set $\{1, -1\}$) and thus a W-polynomial. In $\mathbb{F}_4[x; \sigma]$ the polynomial $f = (x + 1)(x + \omega)$ is not a W-polynomial because $V(f) = \{\omega\}$.

If the algebraic set A is finite, say $A = \{a_1, a_2, \ldots, a_N\}$, then Theorem 8.4.10 implies that $f = m_A = \mathrm{lclm}(x - a_1, x - a_2, \ldots, x - a_N)$. This shows that in the commutative case, W-polynomials are simply the separable polynomials that factor linearly. One may also note that W-polynomials are a special case of completely reducible polynomials in the sense of [1466, p. 495] by Ore. The latter are defined as the least common left multiple of irreducible polynomials.

Since $m_{V(f)} \mid_r f$ for any $f \in F[x; \sigma]$, we observe that f is a W-polynomial if and only if $m_{V(f)} = f$. If $\deg(f) = N$, then this reads as "f is Wedderburn if and only if $\mathrm{rk}(V(f)) = N$", which may be understood as f having sufficiently many roots (see [1195, Prop. 3.4]).

Let us list some properties of W-polynomials.

Proposition 8.5.12 ([1195, Prop. 4.3]) *Let* $A, B \subseteq F$ *be algebraic sets.*

(a) $m_{A \cup B} = \mathrm{lclm}(m_A, m_B)$; *thus* $\mathrm{rk}(A \cup B) \le \mathrm{rk}(A) + \mathrm{rk}(B)$.

(b) $\mathrm{rk}(A \cup B) = \mathrm{rk}(A) + \mathrm{rk}(B) \iff \gcd(m_A, m_B) = 1 \iff V(m_A) \cap V(m_B) = \emptyset$.

Part (a) as well as the first equivalence in (b) are clear; see also Theorem 8.2.8(d). The second part in (b) requires more work. We now present some strong properties of W-polynomials. More machinery is needed to derive most of them, especially the Φ-transform and λ-transform introduced by Lam/Leroy in [1195], and we refer to the excellent presentation in [1195, Sec. 4 and 5] for further details.

Theorem 8.5.13 ([1195, Thms. 5.1, 5.3, 5.9, 5.10])

(a) *Let $f \in F[x; \sigma]$ be a monic polynomial of degree N. The following are equivalent.*

    (i) *f is a W-polynomial.*

    (ii) *$f = \mathrm{lclm}(x - a_1, \ldots, x - a_N)$ for some distinct elements $a_i \in F$.*

    (iii) *f splits completely and every monic factor of f is a W-polynomial.*

(b) *Let $f, g \in F[x; \sigma]$ be similar monic polynomials and f a W-polynomial. Then g is a W-polynomial.*

(c) *Let g, h be W-polynomials. The following are equivalent.*

    (i) *gh is a W-polynomial.*

    (ii) *$1 \in {}^\bullet(g) + (h)^\bullet$.*

    (iii) *$\{k \in F[x; \sigma] \mid gk \in {}^\bullet(g)\} \subseteq {}^\bullet(g) + (h)^\bullet$.*

The set on the left-hand side of (c)(iii) above is called the *idealizer* of ${}^\bullet(g)$. It is the largest subring of $F[x; \sigma]$ in which ${}^\bullet(g)$ is a two-sided ideal.

We close this section with a brief discussion of 'multiplicities of roots' and 'splitting fields' for skew polynomials.

It is tempting to define the multiplicity of a root a of the skew polynomial $f \in F[x; \sigma]$ as the largest exponent r for which $(x - a)^r$ is a right divisor of f. However, this defines the multiplicity based on the *monic* linear factor with root a. Unfortunately, rescaling a linear factor with root a from the left (or from the right) may change the exponent for the right divisor of f. Indeed, in Example 8.2.14 we have seen that $x + \omega^2$ appears with exponent 1 as a right divisor of $x^2 + 1$, whereas $\omega(x + \omega^2) = \omega x + 1$ appears with exponent 2. For this reason it is not meaningful to define the multiplicity of roots in this way. The reader may also note that over a finite field $\mathbb{F}$ the commutative polynomial P_f from Remark 8.4.6 has only simple roots (if $f_0 \neq 0$) and thus cannot serve for the definition of multiplicity either. To our knowledge, no notion of multiplicity of roots for skew polynomials has been discussed in the literature.

It should not come as a surprise that also the notion of a splitting field is questionable. First of all, when considering extension fields of F we also need to extend the automorphism σ. Of course, this is not unique in general. But even if we extend the q-Frobenius of $\mathbb{F}_{q^m}$ to the q-Frobenius on an extension field $\mathbb{F}_{q^M}$, we still may ask whether we want a splitting field to be a 'smallest' extension in which the given polynomial splits or whether we want it to be a 'smallest' one in which the polynomial has 'all its roots'? Example 8.4.2 has already shown that these two objectives are not identical. The latter objective is easy to achieve for the finite field case $\mathbb{F}_{q^m}[x; \sigma]$: from Remark 8.4.6 it is clear that the splitting field of the commutative polynomial $P_f \in \mathbb{F}_{q^m}[y]$ is the smallest field that contains all the roots of f. On the other hand, the polynomial $x^2 - i \in \mathbb{C}[x; \sigma]$ with complex conjugation σ has no roots in $\mathbb{C}$ (since $N_2(c) = |c|^2$ is real for any $c \in \mathbb{C}$), and thus its splitting field, if it exists, must be a (transcendental) field extension of $\mathbb{C}$. Again to our knowledge, no theory of splitting fields has been developed for skew polynomials.

8.6 A Circulant Approach Toward Cyclic Block Codes

We briefly summarize the algebraic theory of classical cyclic block codes with the aid of circulant matrices. This is standard material of any introductory course on block codes (see also Section 1.12 and Chapter 2 of this encyclopedia), and is presented here for the mere purpose to serve as a guideline for the skew case in the next sections.

Let us fix a finite field $\mathbb{F} = \mathbb{F}_q$ and a length n. Usually, one requires that q and n be relatively prime, but that is not needed for the algebraic theory. It only plays a role when it comes to distance considerations. We consider the quotient ring $\mathcal{R} = \mathbb{F}[x]/(x^n - 1)$, which as an $\mathbb{F}$-vector space is isomorphic to $\mathbb{F}^n$ via the map

$$\mathfrak{p} : \mathbb{F}^n \longrightarrow \mathcal{R} \quad \text{with} \quad (g_0, g_1, \ldots, g_{n-1}) \longmapsto \overline{\sum_{i=0}^{n-1} g_i x^i},$$

where $^-$ denotes cosets. Set $\mathfrak{v} := \mathfrak{p}^{-1}$ (and think of these maps as *vectorization* and *polynomialization*).

A **cyclic code** in $\mathbb{F}^n$ is, by definition, a subspace of the form $\mathcal{C} = \mathfrak{v}(I)$, where I is an ideal in $\mathcal{R}$. This means that $\mathcal{C}$ is invariant under the cyclic shift $(a_0, \ldots, a_{n-1}) \longmapsto (a_{n-1}, a_0, \ldots, a_{n-2})$. Usually, the code is identified with its corresponding ideal so that its codewords are polynomials of degree less than n. It follows from basic algebra that $\mathcal{R}$ is a principal ideal ring and every ideal is of the form $(\overline{g})$, where g is a monic divisor of $x^n - 1$. Such a generator is unique and called the **generator polynomial** of the cyclic code $(\overline{g})$. We conclude that the number of cyclic codes of length n over $\mathbb{F}$ equals the number of divisors of $x^n - 1$. If $gh = x^n - 1$, then h is the **parity check polynomial** of $\mathcal{C}$. For all of this see also Theorem 1.12.11 of Section 1.12 in Chapter 1 of this encyclopedia.

Before summarizing some well-known facts about cyclic codes we introduce circulant matrices. For any $g = (g_0, g_1, \ldots, g_{n-1}) \in \mathbb{F}^n$ define the **circulant**

$$\Gamma(g) := \begin{bmatrix} g_0 & g_1 & \cdots & g_{n-2} & g_{n-1} \\ g_{n-1} & g_0 & \cdots & g_{n-3} & g_{n-2} \\ \vdots & \vdots & & \vdots & \vdots \\ g_2 & g_3 & \cdots & g_0 & g_1 \\ g_1 & g_2 & \cdots & g_{n-1} & g_0 \end{bmatrix}. \tag{8.8}$$

Indexing by $i = 0, 1, \ldots, n-1$, we see that the ith row of $\Gamma(g)$ is given by $\mathfrak{v}(x^i \sum_{j=0}^{n-1} g_j x^j)$. The set $\text{Circ} := \{\Gamma(g) \mid g \in \mathbb{F}^n\}$ is an n-dimensional subspace of $\mathbb{F}^{n \times n}$. We also define

$$\rho : \qquad \mathbb{F}[x] \longrightarrow \mathbb{F}[x]$$
$$g = \sum_{i=0}^{r} g_i x^i \longmapsto x^r g(x^{-1}) = \sum_{i=0}^{r} g_i x^{r-i} \quad \text{(where } g_r \neq 0\text{)}. \tag{8.9}$$

The image $\rho(g)$ is called the **reciprocal** of g.

The following properties of circulant matrices are either trivial or well-known; see [1160, Thm. 4], [1323, p. 501] or [492] for a general reference on circulant matrix theory.

Remark 8.6.1

(a) The map $\text{Circ} \longrightarrow \mathcal{R}$, $\Gamma(g_0, \ldots, g_{n-1}) \longmapsto \overline{\sum_{i=0}^{n-1} g_i x^i}$, is an $\mathbb{F}$-algebra isomorphism. We may and will use the notation $\Gamma(\sum_{i=0}^{n-1} g_i x^i)$ for $\Gamma(g_0, \ldots, g_{n-1})$. Then the above tells us, among other things, that $\Gamma(\overline{gh}) = \Gamma(\overline{g})\Gamma(\overline{h}) = \Gamma(\overline{h})\Gamma(\overline{g})$.

(b) $\text{rank}(\Gamma(\overline{g})) = \deg \frac{x^n - 1}{\gcd(g, x^n - 1)} =: k$ (where the quotient is evaluated in $\mathbb{F}[x]$) and every set of k consecutive rows (respectively columns) of $\Gamma(\overline{g})$ is linearly independent.

(c) The map $\phi : \mathcal{R} \longrightarrow \mathcal{R}$ where $\overline{g} \longmapsto \overline{g(x^{n-1})}$ is a well-defined involutive $\mathbb{F}$-algebra automorphism corresponding to transposition in Circ, i.e., $\Gamma(\overline{g})^{\mathsf{T}} = \Gamma(\phi(\overline{g}))$.

(d) Let $g = \sum_{i=0}^{r} g_i x^i$ be of degree r. Then $\overline{x^r} \phi(\overline{g}) = \overline{\rho(g)}$ or, equivalently, $\phi(\overline{g}) = \overline{x^{n-r} \rho(g)}$. Hence $\Gamma(\overline{\rho(g)}) = \Gamma(\overline{x^r})\Gamma(\phi(\overline{g})) = \Gamma(\phi(\overline{g}))\Gamma(\overline{x^r})$ and since $\overline{x^r}$ is a unit, the circulants $\Gamma(\phi(\overline{g}))$ and $\Gamma(\overline{\rho(g)})$ have the same row space and the same column space. In fact, the left (respectively right) factor $\Gamma(\overline{x^r})$ simply permutes the rows (respectively columns) of $\Gamma(\phi(\overline{g}))$.

(e) If g is a divisor of $x^n - 1$, then so is $\rho(g)$. But the representative of $\phi(\overline{g})$ of degree less than n is in general not a divisor of $x^n - 1$. Thus, while involution $g(x) \longmapsto g(x^{n-1})$ is the appropriate map for transposition of circulants, it does not behave well when it comes to divisors of $x^n - 1$.

Now we review the basic algebraic properties of cyclic codes in the terminology of circulant matrices; for further details see for instance [1008, Sec. 4.1, 4.2] or Section 1.12 of this encyclopedia and specifically Theorem 1.12.11.

Remark 8.6.2 Let $x^n - 1 = hg$, where $g = \sum_{i=0}^{r} g_i x^i$, $h = \sum_{i=0}^{k} h_i x^i$ are monic of degree r and k, respectively. Let $\mathcal{C}$ be the cyclic code $\mathcal{C} = \mathfrak{v}((\overline{g})) \subseteq \mathbb{F}^n$.

(a) The ideal $(\overline{g})$ has dimension $k := n - r$ as an $\mathbb{F}$-vector space and $\overline{g}, \overline{xg}, \ldots, \overline{x^{k-1}g}$ is a basis. Thanks to the isomorphism $\mathfrak{v}$, this implies that $\mathcal{C}$ is the row space of the circulant $\Gamma(g)$ and actually of its first k rows (see also Remark 8.6.1(b)). Since $\deg(g) = r$, these first rows have the form

$$
G = \begin{bmatrix} \mathfrak{v}(\overline{g}) \\ \mathfrak{v}(\overline{xg}) \\ \vdots \\ \mathfrak{v}(\overline{x^{k-1}g}) \end{bmatrix} = \begin{bmatrix} g_0 & g_1 & \cdots & g_r & & & \\ & g_0 & g_1 & \cdots & g_r & & \\ & & \ddots & \ddots & & \ddots & \\ & & & g_0 & g_1 & \cdots & g_r \end{bmatrix},
$$

which is the well-known **generator matrix** of the cyclic code generated by g.

(b) Remark 8.6.1(a) yields $\Gamma(\overline{g})\Gamma(\overline{h}) = 0$ and thus $\Gamma(\overline{g})\Gamma(\phi(\overline{h}))^\mathsf{T} = 0$ for ϕ as in Remark 8.6.1(c). As a consequence, the code $\mathcal{C} = \mathfrak{v}((\overline{g}))$ satisfies

$$
\mathcal{C} = \{v \in \mathbb{F}^n \mid \Gamma(\phi(\overline{h}))v^\mathsf{T} = 0\}.
$$

By Remark 8.6.1(b), the last $n - k$ rows of $\Gamma(\phi(\overline{h}))$ form a basis of the row space of this matrix. Writing $h = \sum_{i=0}^{k} h_i x^i$, all of this implies $\mathcal{C} = \{v \in \mathbb{F}^n \mid Hv^\mathsf{T} = 0\}$ where

$$
H = \begin{bmatrix} h_k & h_{k-1} & \cdots & h_0 & & & \\ & h_k & h_{k-1} & \cdots & h_0 & & \\ & & \ddots & \ddots & & \ddots & \\ & & & h_k & h_{k-1} & \cdots & h_0 \end{bmatrix} \in \mathbb{F}^{(n-k)\times n},
$$

which is known as the **parity check matrix** of $\mathcal{C}$. This is also the submatrix consisting of the first $n - k$ rows of $\Gamma(\overline{\rho(h)})$, where $\rho(h)$ is the reciprocal of h.

(c) $\rho(h)\rho(g) = 1 - x^n$, and the dual code $\mathcal{C}^\perp$ (see Section 1.5 of Chapter 1) is cyclic with generator and parity check polynomial $\rho(h)/h_0$ and $\rho(g)/g_0$, respectively.

8.7 Algebraic Theory of Skew-Cyclic Codes with General Modulus

In this section we introduce the notion of skew-cyclic codes in most generality and present the basic algebraic properties. Later we will restrict ourselves to more special cases. The material of this section is drawn from [258, 262, 736]. We will put an emphasis on the generalization of the circulant.

From now on we consider the skew-polynomial ring $\mathbb{F}[x; \sigma]$ where $\mathbb{F} = \mathbb{F}_{q^m}$ and σ is the q-Frobenius. In order to generalize the last section we need to first generalize the quotient ring $\mathcal{R} = \mathbb{F}[x]/(x^n - 1)$. To do so, note that for any $f \in \mathbb{F}[x; \sigma]$ we obtain a left $\mathbb{F}[x; \sigma]$-module $\mathbb{F}[x; \sigma]/{}^{\bullet}(f)$ (we may, of course, also consider right ideals and right modules). It is known from basic module theory that this module is a ring if and only if f is a two-sided polynomial; see Remark 8.2.10. In this section, no ring structure is needed and thus we fix the following setting until further notice.

Let $f \in \mathbb{F}[x; \sigma]$ be a monic polynomial of degree n, which we call the **modulus**, and consider the left $\mathbb{F}[x; \sigma]$-module

$$\mathcal{R}_f = \mathbb{F}[x; \sigma]/{}^{\bullet}(f).$$

Note that the left module structure means that $z\overline{g} = \overline{zg}$ for any $z, g \in \mathbb{F}[x; \sigma]$, where $\overline{g}$ denotes the coset $g + {}^{\bullet}(f)$ in $\mathcal{R}_f$. As in the previous section we consider the map

$$\mathfrak{p}_f : \mathbb{F}^n \longrightarrow \mathcal{R}_f \quad \text{where} \quad (c_0, \ldots, c_{n-1}) \longmapsto \overline{\sum_{i=0}^{n-1} c_i x^i}.$$

It is crucial that the coefficients c_i appear on the left of x, because this turns $\mathfrak{p}_f$ into an isomorphism of (left) $\mathbb{F}$-vector spaces. This map will relate codes in $\mathbb{F}^n$ to submodules in $\mathcal{R}_f$. Again we set $\mathfrak{v}_f = \mathfrak{p}_f^{-1}$. The map $\mathfrak{v}_f$ coincides with the map ϕ given in [262, Prop. 3], where it is defined with the aid of a semi-linear map based on the companion matrix of f; see (8.2).

The following facts about submodules of $\mathcal{R}_f$ are straightforward generalizations of the commutative case, Theorem 1.12.11(a)–(c) in Chapter 1, and are proven in exactly the same way (with the aid of right division with remainder in $\mathbb{F}[x; \sigma]$). We use the notation ${}^{\bullet}(\overline{g})$ for the left submodule $\{z\overline{g} \mid z \in \mathbb{F}[x; \sigma]\}$ of $\mathcal{R}_f$ generated by $\overline{g}$.

Proposition 8.7.1 *Let M be a left submodule of $\mathcal{R}_f$. Then $M = {}^{\bullet}(\overline{g})$, where $g \in \mathbb{F}[x; \sigma]$ is the unique monic polynomial of smallest degree such that $\overline{g} \in M$. Alternatively, g is the unique monic right divisor of f such that ${}^{\bullet}(\overline{g}) = M$. Finally $g \mid_r h$ for any $h \in \mathbb{F}[x; \sigma]$ such that $\overline{h} \in M$.*

The following definition of skew-cyclic codes was first cast in [256] for the case where f is a central polynomial of the form $f = x^n - 1$, i.e., $\sigma^n = \text{id}$. In the form below, the definition appeared in [258]. A different, yet equivalent, definition was cast in [262, Def. 3]. Note the generality of the setting. It includes for instance the case $f = x^n$, for which even in the commutative case the terminology 'cyclic' may be questionable because reduction modulo $f = x^n$ simply means truncating the given polynomial at power x^n. Yet, the basic part of the algebraic theory, presented in this section, applies indeed to this generality, and only in the next section will we restrict ourselves to skew-constacyclic codes.

Definition 8.7.2 *A subspace $\mathcal{C} \subseteq \mathbb{F}^n$ is called (σ, f)-**skew-cyclic** if $\mathfrak{p}_f(\mathcal{C})$ is a submodule of $\mathcal{R}_f$. For $a \in \mathbb{F}^*$ the code $\mathcal{C} \subseteq \mathbb{F}^n$ is called (σ, a)-**skew-constacyclic** if it is $(\sigma, x^n - a)$-skew-cyclic. The code is called σ-**cyclic** if it is $(\sigma, 1)$-skew-constacyclic. We will also call the image $\mathfrak{p}_f(\mathcal{C})$ a cyclic code (of the same type).*

Thus, up to the isomorphism $\mathfrak{p}_f$, the skew-cyclic codes are the submodules of $\mathcal{R}_f$. In the literature, the above defined codes are often called *ideal σ-codes* if f generates a two-sided ideal and *module-σ-codes* otherwise. The q-cyclic codes introduced in [761] are the $(\sigma, 1)$-skew-constacyclic codes for the case where $m = n$.

Skew-constacyclic codes can easily be described in $\mathbb{F}^n$. Just as in the commutative case one observes that a subspace $\mathcal{C} \subseteq \mathbb{F}^n$ is (σ, a)-skew-constacyclic if and only if

$$(c_0, \dots, c_{n-1}) \in \mathcal{C} \implies (a\sigma(c_{n-1}), \sigma(c_0), \dots, \sigma(c_{n-2})) \in \mathcal{C}. \tag{8.10}$$

Indeed, the right-hand side is simply $\mathfrak{v}_{x^n-a}(\overline{x \sum_{i=0}^{n-1} c_i x^i})$. In other words, a (σ, a)-skew-constacyclic code is invariant under the σ-semilinear map induced by the companion matrix C_{x^n-a} (see (8.2) and the paragraph thereafter). This characterization also generalizes to (σ, f)-skew-cyclic codes; see [262, pp. 466].

Thanks to Proposition 8.7.1 every (σ, f) skew-cyclic code is generated by a single element in $\mathcal{R}_f$ (the analog of principal ideals), i.e., has a generator polynomial. As a consequence, the number of (σ, f)-skew-cyclic codes equals the number of monic right divisors of f. This leads in general to a significantly larger number of skew-cyclic codes than classical cyclic codes. For instance, in the constacyclic case where $f = x^{15} - \omega$ and $\omega \in \mathbb{F}_4$ satisfies $\omega^2 + \omega + 1 = 0$, the polynomial f has 8 monic divisors (including the trivial ones) in the commutative ring $\mathbb{F}_4[x]$, whereas it has 32 monic right divisors in the skew-polynomial ring $\mathbb{F}_4[x; \sigma]$.

Proposition 8.7.1 allows us to present generator matrices just as for the commutative case. Again we will do so with the aid of circulants. The definition of a skew circulant matrix is straightforward.

Definition 8.7.3 For $\overline{g} \in \mathcal{R}_f$ define the (σ, f)-**circulant**

$$\Gamma_f^\sigma(\overline{g}) := \begin{bmatrix} \mathfrak{v}_f(\overline{g}) \\ \mathfrak{v}_f(x\overline{g}) \\ \vdots \\ \mathfrak{v}_f(x^{n-2}\overline{g}) \\ \mathfrak{v}_f(x^{n-1}\overline{g}) \end{bmatrix} \in \mathbb{F}^{n \times n}.$$

In the case where $f = x^n - a$, $a \in \mathbb{F}^*$, we write Γ_a^σ instead of $\Gamma_{x^n-a}^\sigma$. We call any matrix of the form $\Gamma_f^\sigma(\overline{g})$ a **skew circulant**.

One may regard $\Gamma_f^\sigma(\overline{g})$ as the matrix representation of the left $\mathbb{F}$-linear map on $\mathcal{R}_f$ given by right multiplication by $\overline{g}$ with respect to the basis $\{\overline{x^0}, \dots, \overline{x^{n-1}}\}$. If $f = x^n - a$, the skew circulant of $\overline{g}$ can be given explicitly. For $g = \sum_{i=0}^{n-1} g_i x^i$ we have $\Gamma_a^\sigma(\overline{g}) =$

$$\begin{bmatrix} g_0 & g_1 & g_2 & \cdots & g_{n-2} & g_{n-1} \\ a\sigma(g_{n-1}) & \sigma(g_0) & \sigma(g_1) & \cdots & \sigma(g_{n-3}) & \sigma(g_{n-2}) \\ a\sigma^2(g_{n-2}) & \sigma(a)\sigma^2(g_{n-1}) & \sigma^2(g_0) & \cdots & \sigma^2(g_{n-4}) & \sigma^2(g_{n-3}) \\ \vdots & \vdots & & \ddots & \vdots & \vdots \\ a\sigma^{n-2}(g_2) & \sigma(a)\sigma^{n-2}(g_3) & \sigma^2(a)\sigma^{n-2}(g_4) & \cdots & \sigma^{n-2}(g_0) & \sigma^{n-2}(g_1) \\ a\sigma^{n-1}(g_1) & \sigma(a)\sigma^{n-1}(g_2) & \sigma^2(a)\sigma^{n-1}(g_3) & \cdots & \sigma^{n-2}(a)\sigma^{n-1}(g_{n-1}) & \sigma^{n-1}(g_0) \end{bmatrix}.$$

If $f = x^n - 1$, then $\Gamma_1^\sigma(\overline{g}) = D_g$, the Dickson matrix in (8.4), and this specializes to the classical circulant $\Gamma(g)$ in (8.8) if $\sigma = $ id. Furthermore, for any monic f the skew circulant

$\Gamma_f^\sigma(\overline{x})$ equals C_f, the companion matrix of f in (8.2). Note also that modulo $x^n - a$

$$\Gamma_a^\sigma(\overline{x}) = \begin{bmatrix} & 1 & & & \\ & & 1 & & \\ & & & \ddots & \\ & & & & 1 \\ a & & & & \end{bmatrix} \quad \text{and} \quad \Gamma_a^\sigma(\overline{x^2}) = \begin{bmatrix} & & 1 & & \\ & & & \ddots & \\ & & & & 1 \\ a & & & & \\ & \sigma(a) & & & \end{bmatrix}. \tag{8.11}$$

Example 8.7.4 Let $f = x^7 + \alpha \in \mathbb{F}_8[x; \sigma]$, where $\alpha^3 + \alpha + 1 = 0$ and σ is the 2-Frobenius. Let $g = x^4 + \alpha x^3 + \alpha^5 x^2 + \alpha$. Then g is a right divisor of f and

$$\Gamma := \Gamma_f^\sigma(\overline{g}) = \begin{bmatrix} \alpha & 0 & \alpha^5 & \alpha & 1 & 0 & 0 \\ 0 & \alpha^2 & 0 & \alpha^3 & \alpha^2 & 1 & 0 \\ 0 & 0 & \alpha^4 & 0 & \alpha^6 & \alpha^4 & 1 \\ \alpha & 0 & 0 & \alpha & 0 & \alpha^5 & \alpha \\ \alpha^3 & \alpha^2 & 0 & 0 & \alpha^2 & 0 & \alpha^3 \\ 1 & \alpha^6 & \alpha^4 & 0 & 0 & \alpha^4 & 0 \\ 0 & 1 & \alpha^5 & \alpha & 0 & 0 & \alpha \end{bmatrix}.$$

The first row is simply the vector of left coefficients of g. The second and third row of Γ are the cyclic shift of the previous row followed by the map σ applied entrywise. Thus, the first 3 rows do not depend on f. Only in the last 4 rows, where $\deg(x^i g)$ is at least 7, reduction modulo $^\bullet(f)$ kicks in.

Now one obtains the straightforward analog of Remark 8.6.2(a): every (σ, f)-skew-cyclic code has a generator matrix that reflects the skew-cyclic structure. Consider a general modulus f of degree n. For any matrix G we use the notation rowsp(G) for the row span of G.

Proposition 8.7.5 (see also [259], [736, Cor. 2.4]) *Let* $\mathcal{M} = {}^\bullet(\overline{g}) \subseteq \mathcal{R}_f$, *where* $g = \sum_{i=0}^r g_i x^i \in \mathbb{F}[x; \sigma]$ *has degree* r. *Then:*

(a) *For any* $u \in \mathbb{F}^n$ *we have* $\mathfrak{p}_f(u \Gamma_f^\sigma(\overline{g})) = \mathfrak{p}_f(u)\overline{g}$.

(b) $\mathfrak{v}_f(\mathcal{M}) = \text{rowsp}(\Gamma_f^\sigma(\overline{g}))$.

(c) *Suppose that* g *is a right divisor of* f *of degree* r. *Then* $\mathcal{M}$ *is a left* $\mathbb{F}$-*vector space of dimension* $k := n - r$ *with basis* $\{\overline{g}, \overline{xg}, \dots, \overline{x^{k-1}g}\}$. *As a consequence,* $\text{rank}(\Gamma_f^\sigma(\overline{g})) = k$ *and*

$$\mathfrak{v}_f(\mathcal{M}) = \text{rowsp}(G),$$

where $G \in \mathbb{F}^{k \times n}$ *consists of the first* k *rows of the skew circulant* $\Gamma_f^\sigma(\overline{g})$:

$$G = \begin{bmatrix} \mathfrak{v}_f(\overline{g}) \\ \mathfrak{v}_f(\overline{xg}) \\ \vdots \\ \mathfrak{v}_f(\overline{x^{k-1}g}) \end{bmatrix}$$

$$\tag{8.12}$$

$$= \begin{bmatrix} g_0 & g_1 & \cdots & & g_r & & \\ & \sigma(g_0) & \sigma(g_1) & \cdots & & \sigma(g_r) & \\ & & \ddots & \ddots & & & \ddots \\ & & & \sigma^{k-1}(g_0) & \sigma^{k-1}(g_1) & \cdots & \sigma^{k-1}(g_r) \end{bmatrix}.$$

If, in addition, g *is monic, we call it the* **generator polynomial** *of the* (σ, f)-*skew-cyclic code* $\mathcal{M}$.

(d) *Let $z \in \mathbb{F}[x; \sigma]$ and $g = \gcrd(z, f)$. Then $^\bullet(\bar{z}) = {}^\bullet(\bar{g})$ and thus $\mathrm{rowsp}(\Gamma_f^\sigma(\bar{z})) = \mathrm{rowsp}(\Gamma_f^\sigma(\bar{g}))$.*

In order to provide a feeling for the line of reasoning we provide a short proof.

Proof: (a): For any $u_i \in \mathbb{F}$ we have that $(u_0, \ldots, u_{n-1})\Gamma_f^\sigma(\bar{g}) = \sum_{i=0}^{n-1} u_i \mathfrak{v}_f(\overline{x^i g}) = \mathfrak{v}_f((\sum_{i=0}^{n-1} u_i x^i)\bar{g})$; hence $(\sum_{i=0}^{n-1} u_i x^i)\bar{g} = \mathfrak{p}_f((u_0, \ldots, u_{n-1})\Gamma_f^\sigma(\bar{g}))$, which proves the statement.

(b): '$\supseteq$' follows from (a). For '$\subseteq$' consider $\overline{zg} \in {}^\bullet(g)$, where $z \in \mathbb{F}[x; \sigma]$. Thanks to Theorem 8.2.8(c) there exist $u, v \in \mathbb{F}[x; \sigma]$ such that $ug = vf = \mathrm{lclm}(g, f)$ and $\deg(u) \leq n$. Right division with remainder of z by u provides us with polynomials $t, r \in \mathbb{F}[x; \sigma]$ such that $z = tu + r$ and $\deg(r) < \deg(u) \leq n$. Now we have $\overline{zg} = \overline{tvf + rg} = \overline{rg}$, and writing $r = \sum_{i=0}^{n-1} r_i x^i$, we conclude $\mathfrak{v}_f(\overline{zg}) = \mathfrak{v}_f(\overline{rg}) = (r_0, \ldots, r_{n-1})\Gamma_f^\sigma(\bar{g})$.

(c): Let $hg = f$. It suffices to show that every $\overline{zg} \in {}^\bullet(\bar{g})$ is of the form $\overline{rg}$, where $\deg r < k$. But this follows from the previous part because $hg = f = \mathrm{lclm}(g, f)$.

(d): $^\bullet(\bar{z}) \subseteq {}^\bullet(\bar{g})$ holds since $g \mid_r z$. For the other containment use a Bezout identity $g = uf + vz$ with $u, v \in \mathbb{F}[x; \sigma]$ (see Theorem 8.2.8(b)) and take cosets. $\qquad\square$

As we already noticed in Example 8.7.4, the matrix G in (8.12) above does not depend on the modulus f. The dependence materializes only through the fact that the code $\mathcal{M}$ is (σ, f)-skew-cyclic. As a consequence, a given subspace of $\mathbb{F}^n$ may be (σ, f)-skew-cyclic for various moduli f. This has been studied in further detail in [259, Sec. 2]. Therein, the authors discuss existence and degree of the smallest degree monic *two-sided* polynomial f such that $\overline{f} \in {}^\bullet(\bar{g})$; such f is called the *bound* of g (see also [1028, Ch. 3]). Its degree is the shortest length in which the given g generates an ideal-σ-code.

In this context we wish to remark that a code $\mathcal{C} \neq \mathbb{F}^n$ can only be (σ, a)-skew-constacyclic with respect to at most one polynomial $x^n - a$. Indeed, if a polynomial g is a right divisor of $x^n - a$ and $x^n - b$ in $\mathbb{F}[x; \sigma]$, then $a = b$ or $g \in \mathbb{F}^*$. However, it is possible that a code $\mathcal{C} \neq \mathbb{F}^n$ is (σ, a)-skew-constacyclic and (σ', b)-skew-constacyclic for some $\sigma \neq \sigma'$ and $a \neq b$. For instance, over the field $\mathbb{F}_4$ the polynomial $g = x + \omega$ is a right divisor of $x^2 - 1$ in $\mathbb{F}_4[x; \sigma]$ and a divisor of $x^2 - \omega^2$ in $\mathbb{F}_4[x]$. Hence $\mathrm{rowsp}\begin{bmatrix}\omega & 1\end{bmatrix} \subseteq \mathbb{F}_4^2$ is $(\sigma, 1)$-skew-constacyclic and (id, ω^2)-skew-constacyclic.

Let us now study the map induced by the skew circulant.

Remark 8.7.6 Consider the map $\Gamma_f^\sigma : \mathcal{R}_f \longrightarrow \mathbb{F}^{n \times n}$ where $\bar{g} \longmapsto \Gamma_f^\sigma(\bar{g})$.

(a) Γ_f^σ is injective and additive.

(b) $\Gamma_f^\sigma(c\bar{g}) = \Gamma_{f'}^\sigma(\bar{c})\Gamma_f^\sigma(\bar{g})$ for all $c \in \mathbb{F}$, $g \in \mathcal{R}_f$ and all monic $f' \in \mathbb{F}[x; \sigma]$ of degree n. This follows directly from the definition along with the fact that

$$\Gamma_{f'}^\sigma(\bar{c}) = \begin{bmatrix} c & & & \\ & \sigma(c) & & \\ & & \ddots & \\ & & & \sigma^{n-1}(c) \end{bmatrix}$$

for any monic f' of degree n. As a consequence, Γ_f^σ is not $\mathbb{F}$-linear (unless $\sigma = \mathrm{id}_\mathbb{F}$), but it is $\mathbb{F}_q$-linear (recall that $\mathbb{F}_q$ is the fixed field of σ).

(c) Γ_f^σ is not multiplicative, that is, $\Gamma_f^\sigma(\overline{gg'}) \neq \Gamma_f^\sigma(\bar{g})\Gamma_f^\sigma(\bar{g'})$ in general. This simply reflects the fact that $\mathcal{R}_f$ is not a ring.

By (c), the identity $hg = f$ does not imply $\Gamma_f^\sigma(\bar{h})\Gamma_f^\sigma(\bar{g}) = 0$. The situation becomes much nicer when f is two-sided. The following is obtained by applying Proposition 8.7.5(a) twice to $\mathfrak{p}_f(u\Gamma_f^\sigma(\bar{g})\Gamma_f^\sigma(\bar{g'}))$ for $u \in \mathbb{F}^n$.

Theorem 8.7.7 (see also [736, Thm. 3.6]) *Let $f \in \mathbb{F}[x; \sigma]$ be two-sided; thus $\mathcal{R}_f$ is a ring. Then*

$$\Gamma_f^\sigma(\overline{gg'}) = \Gamma_f^\sigma(\overline{g})\Gamma_f^\sigma(\overline{g'}) \text{ for all } g, g' \in \mathbb{F}[x; \sigma].$$

Hence Γ_f^σ is an $\mathbb{F}_q$-algebra isomorphism between $\mathcal{R}_f$ and the subring $\Gamma_f^\sigma(\mathcal{R}_f)$ of $\mathbb{F}^{n \times n}$ consisting of the (σ, f)-circulants.

This result does not generalize if f is not two-sided. For instance, (8.11) shows that $\Gamma_a^\sigma(\overline{x^2}) \neq \left(\Gamma_a^\sigma(\overline{x})\right)^2$ if $\sigma(a) \neq a$.

The following consequence for (σ, f)-skew-constacyclic codes is immediate. In [262, Cor. 1] the matrix $\Gamma_f^\sigma(\overline{h'})$ appearing below is called the *control matrix* of the code $\mathcal{C}$. This is not to be confused with the parity check matrix to which we will turn later.

Corollary 8.7.8 *Let $f \in \mathbb{F}[x; \sigma]$ be two-sided and $f = hg = gh'$ for some $g, h, h' \in \mathbb{F}[x; \sigma]$. Then $\Gamma_f^\sigma(\overline{g})\Gamma_f^\sigma(\overline{h'}) = 0$ and the code $\mathcal{C} = \mathfrak{v}_f({}^\bullet(\overline{g})) = \text{rowsp}(\Gamma_f^\sigma(\overline{g}))$ is the left kernel of the skew circulant $\Gamma_f^\sigma(\overline{h'})$.*

It is not hard to see that actually the two-sidedness of f along with $f = hg$ implies the existence of h' such that $f = gh'$.

Now that we have a natural notion of generator matrix for a skew-cyclic code it remains to discuss whether such a code also has a parity check matrix that reflects the skew-cyclic structure. We have shown in Remark 8.6.2(b) that in the commutative case the parity check matrix hinges on two facts: (i) the product of circulants is again a circulant, (ii) the transpose of a circulant is a circulant. Theorem 8.7.7 shows that property (i) carries through to the noncommutative case if the modulus is two-sided (and thus also to the commutative case for arbitrary moduli f instead of $x^n - 1$). But if f is not two-sided, then even in the skew-constacyclic case (i.e., moduli of the form $f = x^n - a$), the product of two (σ, f)-circulants is not a (σ, f)-circulant in general. However, we will encounter a proxy of such multiplicativity in the next section that fully serves our purposes.

Transposition of (σ, f)-circulants is an even bigger obstacle. For general modulus f the transpose of a (σ, f)-circulant need not be a (σ', f')-circulant for any automorphism σ' and any modulus f' of the same degree as f. This is actually not very surprising because even in the commutative case the transpose of a circulant in the sense of Definition 8.7.3 need not be a circulant. A trivial example is $f = x^n$ and $g = x^r$, but examples also exist for polynomials f with nonzero constant term. The following (noncommutative) example illustrates this.

Example 8.7.9 Consider $f = x^3 + x^2 + \omega^2$, $g = x^2 + \omega x + \omega \in \mathbb{F}_4[x; \sigma]$, where $\omega^2 + \omega + 1 = 0$ and σ is the 2-Frobenius. Then g is a right divisor of f and

$$G := \Gamma_f^\sigma(\overline{g}) = \begin{bmatrix} \omega & \omega & 1 \\ \omega^2 & \omega^2 & \omega \\ \omega & \omega & 1 \end{bmatrix}.$$

The matrix G has rank 1 and thus generates a 1-dimensional (σ, f)-skew-cyclic code $\mathcal{C} = \mathfrak{v}_f({}^\bullet(\overline{g}))$. Suppose the transpose G^T is a (σ', f')-circulant for some automorphism σ' and $f' \in \mathbb{F}[x; \sigma']$ of degree 3, say $G^\mathsf{T} = \Gamma_{f'}^{\sigma'}(\overline{g'})$. Then clearly g' is given by the first column of G; hence $g' = \omega + \omega^2 x + \omega x^2$. Using for instance SageMath one checks that $G^\mathsf{T} \neq \Gamma_{f'}^{\sigma'}(\overline{g'})$ for any automorphism σ' of $\mathbb{F}_4$ and any $f' \in \mathbb{F}_4[x; \sigma']$ of degree 3 (even non-monic). Furthermore, there exists no skew circulant $H = \Gamma_{f'}^{\sigma'}(\overline{h})$ such that $\text{rank}(H) = 2$ and $GH^\mathsf{T} = 0$. This means there is no analog of Remark 8.6.2(b),(c): $\mathcal{C}$ does not have a skew circulant parity check matrix and $\mathcal{C}^\perp$ is not (σ', f')-skew-cyclic for any (σ', f').

In the next section we restrict ourselves to skew-constacyclic codes and will see that in that case these obstacles can be overcome.

We close this section by presenting a different type of parity check matrix, namely a generalization of the Vandermonde type parity check matrix for classical cyclic codes. Recall W-polynomials from Section 8.5. The following result is immediate with the definition of the skew Vandermonde matrix in (8.7).

Theorem 8.7.10 ([262, Prop. 4]) *Let $f \in \mathbb{F}[x; \sigma]$ be any monic modulus of degree n and $g \in \mathbb{F}[x; \sigma]$ be a monic right divisor of f of degree r. Suppose g is a W-polynomial. Thus we may write $g = \mathrm{lclm}(x - a_1, \dots, x - a_r)$ for distinct $a_1, \dots, a_r \in \mathbb{F}$; see Theorem 8.5.13(a)(ii). Let $M = V_n(a_1, \dots, a_r) \in \mathbb{F}^{n \times r}$ be the skew Vandermonde matrix. Then the cyclic code $\mathcal{C} = \mathfrak{v}({}^\bullet(\overline{g}))$ is given by*

$$\mathcal{C} = \{(c_0, \dots, c_{n-1}) \mid (c_0, \dots, c_{n-1})M = 0\}.$$

8.8 Skew-Constacyclic Codes and Their Duals

We now restrict ourselves to skew-constacyclic codes, that is to modulus $x^n - a$. In this case we are able to obtain a parity check matrix, and thus a generator matrix of the dual, that reflects the skew-constacyclic structure. The material is mainly drawn from [258, 260, 736].

Throughout, we fix a modulus $f = x^n - a$ for some $a \in \mathbb{F}^*$. In order to formulate the main results we need, as in the commutative case, the reciprocal of a polynomial. In the noncommutative case this can be done in different ways depending on the position of the coefficients. The following left version of (8.9) will suffice for this survey. Let

$$\rho_l : \quad \mathbb{F}[x; \sigma] \longrightarrow \mathbb{F}[x; \sigma]$$

$$\sum_{i=0}^{r} g_i x^i \longmapsto \sum_{i=0}^{r} x^{r-i} g_i = \sum_{i=0}^{r} \sigma^i(g_{r-i}) x^i \quad (\text{where } g_r \neq 0).$$

Then $\rho_l(g)$ is called the **left reciprocal** of g. Furthermore, we extend the automorphism σ to the ring $\mathbb{F}[x; \sigma]$ via $\sigma(\sum_{i=0}^{r} g_i x^i) = \sum_{i=0}^{r} \sigma(g_i) x^i$. Then σ is a ring automorphism of $\mathbb{F}[x; \sigma]$ satisfying $xg = \sigma(g)x$ for all $g \in \mathbb{F}[x; \sigma]$.

The following partial product formula for skew circulants will be sufficient to discuss the duals of skew-constacyclic codes. Recall that for $f = x^n - a$ we denote the skew circulant Γ_f^σ by Γ_a^σ. We will have to deal with different moduli, $x^n - a$ and $x^n - c$, and of course the notation $\Gamma_c^\sigma(\overline{g})$ means that the coset of g is taken modulo ${}^\bullet(x^n - c)$.

Theorem 8.8.1 ([736, Thm. 5.3]) *Let $x^n - a = hg$. Set $c = \sigma^n(g_0) a g_0^{-1}$, where g_0 is the constant coefficient of g. Then $x^n - c = \sigma^n(g)h$ and*

$$\Gamma_a^\sigma(\overline{g'g}) = \Gamma_c^\sigma(\overline{g'})\Gamma_a^\sigma(\overline{g}) \quad \text{for any } g' \in \mathbb{F}[x; \sigma].$$

Note that c defined in the theorem is the conjugate a^{g_0} with respect to the automorphism σ^n in the sense of Definition 8.4.7. If $c = a$ (i.e., $\sigma^n(g_0) = g_0$) we have the much nicer formula $\Gamma_a^\sigma(\overline{g'g}) = \Gamma_a^\sigma(\overline{g'})\Gamma_a^\sigma(\overline{g})$, which may be regarded as a generalization of the two-sided case in Theorem 8.7.7. However, the above result holds true only for right divisors g

of $x^n - a$. Check, for instance, with the aid of (8.11) that $\Gamma_a^\sigma(\overline{x(x+1)}) \neq \Gamma_b^\sigma(\overline{x})\Gamma_a^\sigma(\overline{x+1})$ for any $b \neq 0$ unless $a = \sigma(a) = b$.

The above product formula plays a central role in the following quite technical result. It tells us that the transpose of a $(\sigma, x^n - a)$-circulant is a $(\sigma, x^n - a')$-circulant for a suitable constant a'.

Theorem 8.8.2 ([736, Thm. 5.6]) *Suppose $x^n - a = hg$ for some $g, h \in \mathbb{F}[x; \sigma]$ of degree r and $k = n - r$, respectively. Again set $c = \sigma^n(g_0)ag_0^{-1}$, where g_0 is the constant coefficient of g. Then*

$$\Gamma_a^\sigma(\overline{g})^\mathsf{T} = \Gamma_{c^{-1}}^\sigma(\overline{g^\#}) = \Gamma_{\sigma^k(c^{-1})}^\sigma(\overline{g^\circ})\Gamma_{c^{-1}}^\sigma(\overline{x^k}),$$

where $g^\# = a\sigma^k(\rho_l(g))x^k$ and $g^\circ = a\sigma^k(\rho_l(g))$. Furthermore, g° is a right divisor of the modulus $x^n - \sigma^k(c^{-1})$.

The result generalizes Remark 8.6.1(c) and (d): if $f = x^n - 1 = hg$ and $\sigma = \mathrm{id}$, then $c = 1$, $g^\circ = \rho(g)$ and thus $g^\# = \rho(g)x^k$. Furthermore, in general and analogously to Remark 8.6.1(e), g° is a right divisor of the modulus $x^n - \sigma^k(c^{-1})$, whereas the representative of $\overline{g^\#}$ of degree less than n is in general not a divisor of $x^n - c^{-1}$. This is the reason why we provide two formulas pertaining to the transpose of a skew circulant $\Gamma_a^\sigma(\overline{g})$. The first one above is interesting in itself as it tells us that the transpose is again a skew circulant. The second formula says that the skew-constacyclic code $^\bullet(\overline{g})$, i.e., the row space of $\Gamma_a^\sigma(\overline{g})$, equals the row space of a transposed skew circulant where the representing polynomial is a right divisor of the modulus. In all these cases it is crucial that g is a right divisor of the modulus $x^n - a$ for otherwise the transpose of $\Gamma_a^\sigma(\overline{g})$ is not a skew circulant in general [736, Ex. 5.7].

Now we are ready to derive a parity check matrix reflecting the skew-constacyclic structure of the code. The second part of the following theorem appeared first, proven differently, in [258, Thm. 8]. The mere (σ, a^{-1})-skew-constacyclicity of $\mathcal{C}^\perp$ can also be shown directly with the aid of (8.10); see [1827, Thm. 2.4].

Theorem 8.8.3 ([736, Cor. 4.4, Thm. 5.8 and Thm. 6.1]) *Let $x^n - a = hg$, where $\deg(g) = r$ and $\deg(h) = k = n - r$. Set $h^\circ := \rho_l(\sigma^{-n}(h))$. Then $h^\circ |_r (x^n - a^{-1})$. Consider the (σ, a)-skew-constacyclic code $\mathcal{C} = \mathfrak{v}_{x^n - a}(^\bullet(\overline{g}))$. Then $\Gamma_a^\sigma(\overline{g})\Gamma_{a^{-1}}^\sigma(\overline{h^\circ})^\mathsf{T} = 0$ and $\mathrm{rank}(\Gamma_{a^{-1}}^\sigma(\overline{h^\circ})) = n - k$. Hence*

$$\mathcal{C} = \mathrm{rowsp}(\Gamma_a^\sigma(\overline{g})) = \{c \in \mathbb{F}^n \mid \Gamma_{a^{-1}}^\sigma(\overline{h^\circ})c^\mathsf{T} = 0\},$$

and the $(n - k) \times n$-submatrix consisting of the first $n - k$ rows of $\Gamma_{a^{-1}}^\sigma(\overline{h^\circ})$ is a parity check matrix of $\mathcal{C}$. As a consequence, the dual code $\mathcal{C}^\perp$ is (σ, a^{-1})-skew-constacyclic with (non-monic) generator polynomial h° and generator matrix and parity check matrix given by the first $n - k$ rows of $\Gamma_{a^{-1}}^\sigma(\overline{h^\circ})$ and the first k rows of $\Gamma_a^\sigma(\overline{g})$, respectively.

Clearly, this parity check matrix of $\mathcal{C}$ has a form analogous to (8.12) and thus reflects the skew-constacyclic structure of $\mathcal{C}$. As in Proposition 8.7.5, its row space equals the row space of the entire skew circulant $\Gamma_{a^{-1}}^\sigma(\overline{h^\circ})$. In the commutative cyclic case, where $x^n - 1 = hg = gh$, we have $a^{-1} = a = 1$ and $h^\circ = \rho(h)$ and thus recover Remark 8.6.2(b).

Having an understanding of the dual of skew-constacyclic codes, we can address self-duality. The following corollary is immediate.

Corollary 8.8.4 ([258, Prop. 13], [261, Cor. 6], [736, Cor. 6.2]) *If there exists a (σ, a)-skew-constacyclic self-dual code of length n, then n is even and $a = \pm 1$. More specifically, let n be even and consider the modulus $x^n - \epsilon$, where $\epsilon \in \{1, -1\}$. Then there exists a self-dual skew-constacyclic code of length n if and only if there exists a polynomial $h \in \mathbb{F}[x; \sigma]$ such that $x^n - \epsilon = hh^\circ$. In this case the self-dual code is given by $\mathcal{C} = \mathfrak{v}_{x^n - \epsilon}(^\bullet(\overline{h^\circ}))$.*

In [255, 261] the identity $x^n - \epsilon = hh^\circ$ is exploited to enumerate or construct self-dual skew-constacyclic codes, some with very good minimum distance, e.g., [261, Ex. 30].

Let us briefly turn to the notion of check polynomials for skew-constacyclic codes. Recall that in the classical case, where $x^n - 1 = hg$ in $\mathbb{F}[x]$ we call h a check polynomial for the simple reason that $\overline{z} \in (\overline{g}) \Leftrightarrow \overline{zh} = 0$ for any $z \in \mathbb{F}[x]$. In other words $(\overline{h})$ is the annihilator ideal of $(\overline{g})$. The following generalization to $\mathbb{F}[x; \sigma]$ is based on the fact [736, Thm. 4.2] that $x^n - a = hg$ implies $x^n - \tilde{c} = g\sigma^{-n}(h)$ for $\tilde{c}$ defined below.

Theorem 8.8.5 ([736, Prop. 6.5]) *Let $x^n - a = hg$ and set $c = \sigma^n(g_0)ag_0^{-1}$, where g_0 is the constant coefficient of g. Define $\tilde{c} = \sigma^{-n}(c)$. Then the map*

$$\Psi : \mathbb{F}[x; \sigma]/^\bullet(x^n - a) \longrightarrow \mathbb{F}[x; \sigma]/^\bullet(x^n - \tilde{c}) \quad \text{where} \quad \overline{z} \longmapsto \overline{z\sigma^{-n}(h)}$$

*is a well-defined left $\mathbb{F}[x; \sigma]$-linear map with kernel $^\bullet(\overline{g})$. Therefore we call $\sigma^{-n}(h)$ the **check polynomial** of the code $\mathcal{C} = \mathfrak{v}_{x^n - a}(^\bullet(\overline{g}))$.*

One has to be aware that the check equation $\overline{z\sigma^{-n}(h)} = 0$ has to be carried out modulo $x^n - \tilde{c}$. The above generalizes [789, Thm. 2.1(iii)], where the modulus is central of the form $x^n - 1$. In that case we have $\sigma^n = \text{id}$ and $c = a$, and the above also reflects Corollary 8.7.8. Theorem 8.8.5 also extends [259, Lem. 8], where general two-sided moduli are considered.

We close this section by mentioning idempotent generators of skew-constacyclic codes. In [789] the authors consider central moduli of the form $x^n - 1$. Such polynomials have a factorization into pairwise coprime *two-sided maximal* polynomials [1029, Thm. 1.2.17'], which in turn gives rise to a decomposition of $\mathbb{F}[x; \sigma]/(x^n - 1)$ into a direct product of rings generated by central idempotents [789, Thm. 2.11]. As a consequence, just as in the classical cyclic case, a code $(\overline{g})$, where g is a central divisor of $x^n - 1$, has a unique central generating idempotent [735, Thm. 6.2.15]. In the thesis [735] a partial generalization to the non-central case is presented along with the obstacles that occur in this scenario; see [735, Ch. 6].

8.9   The Minimum Distance of Skew-Cyclic Codes

In this section we report on constructions of skew-cyclic codes with designed minimum distance. We only consider the Hamming distance; there also exist a few results on the rank distance in the literature. The results are from the papers [260, 828, 1791].

Throughout, $\mathbb{F} = \mathbb{F}_{q^m}$ and σ is the q-Frobenius. Furthermore, we consider a code

$$\mathcal{C} = \mathfrak{v}_f(^\bullet(\overline{g})) \text{ for some monic } f, g \in \mathbb{F}[x; \sigma] \text{ such that } \deg(f) = n \text{ and } g \mid_r f.$$

In the results below we present conditions on f, g that guarantee a desired minimum distance. In all interesting cases the generator g of the code in question will be the least common left multiple of linear factors over some extension field, and thus g is a W-polynomial over that extension field; see Theorem 8.5.13(a). In Theorem 8.7.10 we presented a parity check matrix of a skew-cyclic code generated by a W-polynomial in the form of a skew Vandermonde matrix. This matrix is the basis of the distance results in this section.

The first two results lead to what we will call skew-BCH codes of the first kind. They are based on generator polynomials with roots that are consecutive ordinary powers of some element. Thereafter, we present skew-BCH codes of the second kind, which are based on generator polynomials with roots that are consecutive q-powers of some element. We conclude with two examples illustrating the constructions.

We start with skew-BCH codes of the first kind.

Theorem 8.9.1 ([260, Thm. 4]) *Fix $b, \delta \in \mathbb{N}$. Suppose there exists some $\alpha \in \bar{\mathbb{F}}$ (the algebraic closure of $\mathbb{F}$) such that*

$$\alpha^{[0]}, \alpha^{[1]}, \dots, \alpha^{[n-1]} \text{ are distinct and } g(\alpha^{b+i}) = 0 \text{ for } i = 0, \dots, \delta - 2.$$

*Then the code $\mathcal{C} = \mathfrak{v}_f(^\bullet(\bar{g}))$ has minimum distance at least δ. If g is the smallest degree monic polynomial with roots $\alpha^b, \dots, \alpha^{b+\delta-2}$, then $\mathcal{C}$ is called an $(n, q^m, \alpha, b, \delta)$-**skew-BCH code of the first kind**.*

Corollary 8.9.2 ([260, Thm. 5]) *Consider the situation of Theorem 8.9.1 and where $\alpha \in \mathbb{F}$. Then the polynomial $g' := \mathrm{lclm}(x - \alpha^b, \dots, x - \alpha^{b+\delta-2})$ is in $\mathbb{F}[x; \sigma]$ and of degree $\delta - 1$. Thus for any left multiple f' of g' of degree n, the skew-cyclic code $\mathfrak{v}_{f'}(^\bullet(\bar{g'}))$ has dimension $n - \delta + 1$ and is MDS. It is called an $(n, q^m, \alpha, b, \delta)$-**skew-RS code of the first kind**.*

Theorem 8.9.1 follows immediately from the fact that the code is contained in the left kernel of the skew Vandermonde matrix (see (8.7))

$$V_n(\alpha^b, \dots, \alpha^{b+\delta-2}) = \begin{bmatrix} 1 & \cdots & 1 \\ (\alpha^{[1]})^b & \cdots & (\alpha^{[1]})^{b+\delta-2} \\ \vdots & & \vdots \\ (\alpha^{[n-1]})^b & \cdots & (\alpha^{[n-1]})^{b+\delta-2} \end{bmatrix}.$$

The columns of this matrix consist of consecutive $[\![i]\!]$-powers of $\alpha^b, \dots, \alpha^{b+\delta-2}$ (which simply accounts for skew-polynomial evaluation), while the rows consist of ordinary consecutive powers of $\alpha^{[0]}, \dots, \alpha^{[n-1]}$. The latter together with the fact that $\alpha^{[0]}, \alpha^{[1]}, \dots, \alpha^{[n-1]}$ are distinct, guarantees that any $(\delta - 1) \times (\delta - 1)$-minor of $V_n(\alpha^b, \dots, \alpha^{b+\delta-2})$ is nonzero, which establishes the stated designed distance.

There exist some precursors of Theorem 8.9.1. In [256, Prop. 2] the case where $f = x^m - 1$ (thus $n = m$), $q = 2$, $b = 0$, and α is a primitive element of $\mathbb{F}$ was considered. It was the first appearance of skew-BCH codes. Subsequently, in [389, Prop. 2] the modulus f was relaxed to a two-sided polynomial of degree n, and α to a primitive element of a field extension $\mathbb{F}_{q^s}$, where $n \leq (q-1)s$. In the same paper, examples of such codes were constructed by translating the above situation into the realm of linearized polynomials.

Theorem 8.9.1 has been generalized to the following form.

Theorem 8.9.3 ([1791, Thm. 4.10]) *Let f have a nonzero constant coefficient. Suppose there exist δ, $t_1, t_2 \in \mathbb{N}$ and $b, \nu \in \mathbb{N}_0$ and some $\alpha \in \bar{\mathbb{F}}$ such that*

(a) $g(\alpha^{b+t_1 i + t_2 j}) = 0$ *for $i = 0, \dots, \delta - 2$, $j = 0, \dots, \nu$,*

(b) $(\alpha^{t_\ell})^{[i]} \neq 1$ *for $i = 1, \dots, n - 1$, $\ell = 1, 2$ (if $\nu = 0$, the condition on α^{t_2} is omitted).*

*Then the code $\mathfrak{v}_f(^\bullet(\bar{g})) \subseteq \mathbb{F}^n$ has minimum distance at least $\delta + \nu$. It may also be called an $(n, q^m, \alpha, b, t_1, t_2, \delta)$-**skew-BCH code of the first kind**.*

Note that for $\nu = 0$ and $t_1 = 1$ this result reduces to Theorem 8.9.1 because Condition (b) is equivalent to $\alpha^{[0]}, \alpha^{[1]}, \dots, \alpha^{[n-1]}$ being distinct.

Since the constructed code has dimension $n - \deg(g)$, it remains to investigate how to find the smallest degree monic polynomial g satisfying (a) above (and has degree at most n). Recall from Proposition 8.7.5(c) that the modulus f does not play a role in the generator matrix of the resulting code $\mathfrak{v}_f(^\bullet(\bar{g})) \subseteq \mathbb{F}^n$. Thus, once g is found, any monic left multiple f of degree n suffices. The polynomial g is obtained as follows, which is a direct consequence of Example 8.5.5.

Remark 8.9.4 Consider the situation of Theorem 8.9.3 and suppose α is in the field extension $\mathbb{F}_{q^{ms}}$ of $\mathbb{F} = \mathbb{F}_{q^m}$. Set $T = \{b + t_1 i + t_2 j \mid i = 0, \ldots, \delta - 2, j = 0, \ldots, \nu\}$ and $A = \{\tau(\alpha^t) \mid \tau \in \mathrm{Aut}(\mathbb{F}_{q^{ms}} \mid \mathbb{F}), t \in T\}$. Then $g := m_A$ is in $\mathbb{F}[x; \sigma]$ and is the smallest degree monic polynomial satisfying (a) of Theorem 8.9.3.

Example 8.9.5 Consider the field extension $\mathbb{F}_{2^{12}} \mid \mathbb{F}_{2^6}$. Let α be the primitive element of $\mathbb{F}_{2^{12}}$ with minimal polynomial $x^{12} + x^7 + x^6 + x^5 + x^3 + x + 1$ and let $\gamma = \alpha^{65}$, which is thus a primitive element of $\mathbb{F}_{2^6}$. As always, σ is the 2-Frobenius. Let $b = 0, t_1 = 23, t_2 = 1, \delta = 4, \nu = 0$. Since $\nu = 0$, Condition (b) of Theorem 8.9.3 amounts to $(\alpha^{23})^{[i]} \neq 1$ for $i = 1, \ldots, n - 1$. Since $i = 12$ is the smallest positive integer such that $(\alpha^{23})^{[i]} = 1$, we can construct skew-BCH codes up to length 12. Condition (a) and Remark 8.9.4 shows that the desired g in $\mathbb{F}_{2^6}[x; \sigma]$ is given by $g = m_A$, where

$$A = \{\alpha^0, \alpha^{23}, \alpha^{46}, (\alpha^0)^{2^6}, (\alpha^{23})^{2^6}, (\alpha^{46})^{2^6}\} \subseteq \mathbb{F}_{2^{12}}.$$

Hence $g = \mathrm{lclm}(x - a \mid a \in A)$, and this results in $g = x^3 + \gamma^{47} x^2 + \gamma^{19} x + \gamma^{40}$. By construction, for any monic left multiple f of g of degree $3 \leq n \leq 12$, the skew-BCH code $\mathcal{C} = \mathfrak{v}_f(\bullet(g)) \subseteq \mathbb{F}_{2^6}^n$ has minimum distance at least 4 and dimension $n - 3$; thus it is MDS. It is interesting to observe that the polynomial $f = x^{12} - 1$ is a left multiple of g, and therefore for length $n = 12$ the code is σ-cyclic. The code is not σ-constacyclic for any length between $3 \leq n \leq 11$. Finally, note that by definition, g is a W-polynomial in $\mathbb{F}_{2^{12}}[x; \sigma]$; it is, however, not a W-polynomial in $\mathbb{F}_{2^6}[x; \sigma]$ (it is not the minimal polynomial of its vanishing set in $\mathbb{F}_{2^6}$).

We now turn to skew-BCH codes of the second kind. The codes presented next are σ-cyclic, i.e., skew-cyclic with respect to the central modulus $x^n - 1$, and *a priori* defined over a field extension $\mathbb{F}_{q^n}$ of $\mathbb{F} = \mathbb{F}_{q^m}$. The result generalizes the Hartmann–Tzeng Bound for classical cyclic codes (see Theorem 2.4.2).

Theorem 8.9.6 ([828, Thm. 3.3] and [1339, Cor. 5]) *Let σ be the q-Frobenius on $\mathbb{F}_{q^n}$. Let $f = x^n - 1$ and $g \in \mathbb{F}_{q^n}[x; \sigma]$ be a right divisor of f. Let $\alpha \in \mathbb{F}_{q^n}$ be such that $\alpha, \alpha^q, \ldots, \alpha^{q^{n-1}}$ is a normal basis of $\mathbb{F}_{q^n}$ over $\mathbb{F}_q$ and set $\beta = \alpha^{-1}\sigma(\alpha) = \alpha^{q-1}$. Suppose there exist $\delta, t_1, t_2 \in \mathbb{N}$ and $b, \nu \in \mathbb{N}_0$ such that $\gcd(n, t_1) = 1$ and $\gcd(n, t_2) < \delta$ and*

$$g(\beta^{q^{b+it_1+jt_2}}) = 0 \text{ for } i = 0, \ldots, \delta - 2, \ j = 0, \ldots, \nu.$$

Then the code $\mathfrak{v}_f(\bullet(\overline{g})) \subseteq \mathbb{F}_{q^n}^n$ has minimum distance at least $\delta + \nu$.

The version given in [1339, Cor. 5] is slightly different from the above. The difference is explained in [828, Rem. A.6].

We have seen already in Example 8.5.10 that $x^n - 1 = \mathrm{lclm}(x - \beta^{q^t} \mid t = 0, 1, \ldots, n - 1)$. Therefore the root condition on g does not clash with the condition that g be a right divisor of f.

Let us now assume that $n = ms$ so that $\mathbb{F}_{q^n}$ is a field extension of $\mathbb{F}_{q^m}$. In [828] it is shown how to obtain from the code of the previous theorem a code over the subfield $\mathbb{F} = \mathbb{F}_{q^m}$ with the same designed minimum distance $\delta + s$. Thus, as for skew-BCH codes of the first kind we want to find the smallest degree monic polynomial g in $\mathbb{F}_{q^m}[x; \sigma]$ with the desired roots. This can again be achieved with the aid of Remark 8.9.4, where we simply have to replace the set T by $\widetilde{T} = \{q^{b+t_1 i + t_2 j} \mid i = 0, \ldots, \delta - 2, j = 0, \ldots, \nu\}$. Noting that $\mathrm{Aut}(\mathbb{F}_{q^{ms}} \mid \mathbb{F}_{q^m}) = \{\tau_0, \ldots, \tau_{s-1}\}$, where $\tau_\ell(a) = a^{q^{\ell m}}$, we conclude that the set $A = \{\tau(\beta^t) \mid t \in \widetilde{T}, \tau \in \mathrm{Aut}(\mathbb{F}_{q^{ms}} \mid \mathbb{F}_{q^m})\}$ is given by

$$A = \{\beta^{q^{b+t_1 i + t_2 j + \ell m}} \mid i = 0, \ldots, \delta - 2, j = 0, \ldots, \nu, \ell = 0, \ldots, s - 1\}.$$

All of this can simply be described in terms of the q-exponents within the cyclic group C_{ms} of order ms. Consider C_s, the cyclic group of order s, as a subgroup of C_{ms}. Furthermore, let $X_0 = C_s, X_1, \ldots, X_{m-1}$ be the cosets of C_s in C_{ms}. Then Remark 8.9.4 and Theorem 8.9.6 lead to the following.

Theorem 8.9.7 ([828, Thm. 4.5]) *Let* $n = ms$ *and consider the situation of Theorem 8.9.6. Consider the set* $S = \{b + it_1 + jt_2 \mid i = 0, \ldots, \delta - 2, \ j = 0, \ldots, \nu\}$ *as a subset of* C_{ms} *(this is well-defined since* $\sigma^{ms} = $ *id). Define* $\overline{S}$ *as the smallest union of cosets* X_i *containing* S. *Then the polynomial* $g' = \mathrm{lclm}(x - \beta^{q^t} \mid t \in \overline{S})$ *is in* $\mathbb{F}[x; \sigma]$. *Thus it defines a* (σ, f)-*skew cyclic code* $\mathcal{C} = \mathfrak{v}_f(\bullet(\overline{g'}))$ *of length* $n = ms$ *over* $\mathbb{F}$. *The code* $\mathcal{C}$ *has minimum distance at least* $\delta + \nu$ *and is called an* $(n, q^m, \alpha, b, t_1, t_2, \delta)$-**skew-BCH code of the second kind.**

The last part follows from the fact that g' is a left multiple of g from Theorem 8.9.6 and thus generates a code contained in the code from that theorem.

There is a connection between the above and q-cyclotomic spaces defined in [1339, Sec. 3.2]. In particular, the q-polynomial of an element β defined in [1339, Lem. 3] is the linearized version of the σ-minimal polynomial of β as discussed earlier in Example 8.5.5. Further details on the connection are given in [828, Prop. A.7].

Example 8.9.8 As in Example 8.9.5, consider $\mathbb{F}_{2^{12}} \mid \mathbb{F}_{2^6}$ with the same primitive element α and the same data $\gamma = \alpha^{65}, b = 0, t_1 = 23, t_2 = 1, \delta = 4, \nu = 0$. The element α^5 generates a normal basis of $\mathbb{F}_{2^{12}}$ over $\mathbb{F}_2$. Thus, $\beta = \alpha^{-5}\sigma(\alpha^5) = \alpha^5$. We have to consider the set $S = \{b + it_1 \mid i = 0, 1, 2\} = \{0, 11, 10\}$ and find the smallest union of cosets of C_2 in C_{12} containing S. This is given by $\overline{S} = \{0, 6, 11, 5, 10, 4\}$. Then $\mathrm{lclm}(x - (\alpha^5)^{q^t} \mid t \in \overline{S}) = x^6 + \gamma^{61}x^5 + \gamma^{41}x^4 + \gamma^4 x^3 + \gamma^{20}x^2 + \gamma^{46}x + \gamma^7 \in \mathbb{F}_{2^6}[x; \sigma]$ generates a skew-cyclic code over $\mathbb{F}_{2^6}$ of length 12 and designed minimum distance 4. It has dimension 6 and its actual minimum distance is 6. Thus the code is not MDS.

To our knowledge, no general comparison of the two kinds of skew-BCH codes has been conducted so far.

Finally, we briefly address evaluation codes in the skew polynomial setting. Recall that the evaluation below, $p(\alpha_i)$, is carried out according to Definition 8.4.1.

Theorem 8.9.9 ([260, Prop. 2]) *Let* $k \in \{1, 2, \ldots, n - 1\}$ *and* $\alpha_1, \ldots, \alpha_n \in \mathbb{F}$ *be such that the skew-Vandermonde matrix* $V_n(\alpha_1, \ldots, \alpha_n) \in \mathbb{F}^{n \times n}$ *has rank* n. *Then the code*

$$\mathcal{E}_{\sigma, \alpha_1, \ldots, \alpha_n} := \{(p(\alpha_1), \ldots, p(\alpha_n)) \mid p \in \mathbb{F}[x; \sigma], \ \deg p \leq k - 1\} \subseteq \mathbb{F}^n$$

has dimension k *and minimum distance* $n - k + 1$, *and thus is MDS.*

By Theorem 8.5.8 the rank of the skew Vandermonde matrix equals the degree of the minimal polynomial of the set $\{\alpha_1, \ldots, \alpha_n\}$, which is given by $\mathrm{lclm}(x - \alpha_i \mid i = 1, 2, \ldots, n)$. Hence the rank condition above is equivalent to $\deg(\mathrm{lclm}(x - \alpha_i \mid i = 1, 2, \ldots, n)) = n$. In the classical case where $\sigma = $ id, this is equivalent to $\alpha_1, \ldots, \alpha_n$ being distinct, and the code $\mathcal{E}_{\mathrm{id}, \alpha_1, \ldots, \alpha_n}$ is a generalized $[n, k]$ Reed–Solomon code; see [1008, Sec. 5.3].

The proof of Theorem 8.9.9 follows easily as in the classical case of generalized Reed–Solomon codes with the aid of Theorem 8.5.8. It is well-known that in many cases classical generalized Reed–Solomon codes are cyclic, e.g., $\mathcal{E}_{\mathrm{id}, 1, \alpha, \ldots, \alpha^{n-1}}$ is cyclic if α is a primitive element of $\mathbb{F}$ and $n \leq |\mathbb{F}|$. However, no such statement holds true for skew-polynomial evaluation codes. Indeed, it is not hard to find examples of codes $\mathcal{E}_{\sigma, 1, \alpha, \ldots, \alpha^{n-1}}$, where α is a primitive element of $\mathbb{F}$, that are not (σ, f)-skew cyclic for any monic modulus f of degree n. The same is true for the evaluation codes $\mathcal{E}_{\sigma, 1, \alpha^{[1]}, \ldots, \alpha^{[n-1]}}$.

We close this chapter by mentioning that many of the papers cited above also present decoding algorithms for the codes constructed therein. We refer to the above literature on this important topic.

Acknowledgement

The author was partially supported by grant #422479 from the Simons Foundation.

Chapter 9

Additive Cyclic Codes

Jürgen Bierbrauer

Michigan Technological University

Stefano Marcugini

Università degli Studi di Perugia

Fernanda Pambianco

Università degli Studi di Perugia

9.1   Introduction

We develop the theory of additive cyclic codes in the case when the length of the code is coprime to the characteristic of the field. The definition of an additive code and some basic relations between a code defined over an extension field and its subfield and trace codes are recalled in Section 9.2. Section 9.3 discusses the notions of cyclicity of a code. In Section 9.4 we develop the theory of additive codes which are cyclic in the permutation sense. This contains the classical theory of cyclic linear codes. We discuss when cyclic codes are equivalent and consider the special case of cyclic quantum codes. The general theory of additive codes which are cyclic in the monomial sense is in Section 9.5. It is proved in Corollary 9.5.10 that each additive code over $\mathbb{F}_4$ which is cyclic in the monomial sense either is cyclic in the (more restricted) permutation sense or is linear over $\mathbb{F}_4$.

Special cases of this theory were published in earlier work [196, 197, 198, 199, 201, 663].

9.2 Basic Notions

We refer to the following family of codes collectively as **additive codes**.

Definition 9.2.1 A *q*-**linear** q^m-**ary** $[n, k]_{q^m}$ **code** is a km-dimensional $\mathbb{F}_q$-subspace $\mathcal{C} \subseteq E^n$, where $E = \mathbb{F}_q^m$. In particular $\mathcal{C}$ has q^{km} codewords.

Clearly this generalizes the notion of linear codes. Linear codes correspond to the special case of Definition 9.2.1 when $m = 1$. The alphabet is seen not as the field $\mathbb{F}_{q^m}$ but as an m-dimensional vector space over the subfield $\mathbb{F}_q$. Observe that the dimension k in Definition 9.2.1 is not necessarily an integer. It may have m in the denominator. As an example, we will encounter additive $[7, 3.5]_4$ codes over $\mathbb{F}_4$. Such a code has $4^{3.5} = 2^7$ codewords. We will describe the general theory of additive codes which are cyclic in the monomial sense under the general assumption that the characteristic p of the underlying field $\mathbb{F}_q$ is coprime to the length n of the code. The main effect of the assumption is that the action of a (cyclic) group of order n on a vector space over a field of characteristic p is completely reducible, by Maschke's Theorem; see Definition 9.3.5 and Theorem 9.3.6. We will make use of basic relations between linear codes over a finite field $\mathbb{F}$ and its trace codes and subfield codes over a subfield $\mathbb{K} \subset \mathbb{F}$. These basic facts and fundamental notions like the Galois group, cyclotomic cosets and Galois closedness are introduced in [199, Chapter 12]. Recall in particular Delsarte's Theorem [199, Theorem 12.14] relating codes over $\mathbb{F}$, their duals, the trace codes and the subfield codes as well as [199, Theorem 12.17] which states among other things that $\dim_{\mathbb{F}}(U) = \dim_{\mathbb{K}}(\mathrm{Tr}_{\mathbb{F}/\mathbb{K}}(U))$ provided the $\mathbb{F}$-linear code U is Galois closed with respect to the subfield $\mathbb{K}$.

9.3 Code Equivalence and Cyclicity

Definition 9.3.1 Codes $\mathcal{C}$ and $\mathcal{D}$ are **permutation equivalent** if there is a permutation π on n objects such that

$$(x_0, x_1, \ldots, x_{n-1}) \in \mathcal{C} \iff (x_{\pi(0)}, x_{\pi(1)}, \ldots, x_{\pi(n-1)}) \in \mathcal{D}$$

for all $\mathbf{x} = (x_0, x_1, \ldots, x_{n-1}) \in \mathcal{C}$.

The notion of permutation equivalence can be used for all codes of fixed block length n. It uses the symmetric group S_n as the group of motions. Two codes are equivalent if they are in the same orbit under S_n. The stabilizer under S_n is the permutation automorphism group of the code; see Section 1.8. An additive code is called *cyclic in the permutation sense* if it admits an n-cycle in its permutation automorphism group.

In Chapters 2 and 17, respectively, the notions of cyclic linear codes and the more general notion of constacyclic linear codes are studied. In the present chapter we consider generalizations of those concepts from the category of linear codes to the category of additive codes. Here the concept of an additive code which is cyclic in the permutation sense is a direct generalization of the concept of a cyclic linear code. The corresponding generalization of the concept of a constacyclic linear code is the notion of an additive code which is *cyclic in the monomial sense*. It is based on the following rather general notion of equivalence.

Definition 9.3.2 q-linear q^m-ary codes $\mathcal{C}$ and $\mathcal{D}$ are **monomially equivalent** if there exist a permutation π on n objects and elements $A_i \in \mathrm{GL}_m(q)$ such that

$$(x_0, x_1, \ldots, x_{n-1}) \in \mathcal{C} \iff (A_0 x_{\pi(0)}, A_1 x_{\pi(1)}, \ldots, A_{n-1} x_{\pi(n-1)}) \in \mathcal{D}$$

for all $\mathbf{x} = (x_0, x_1, \ldots, x_{n-1}) \in \mathcal{C}$. Here we view $\mathrm{GL}_m(q)$ as the group of nonsingular $\mathbb{F}_q$-linear transformations on $E = \mathbb{F}_q^m$ with $A_i x_{\pi(i)}$ the image $A_i(x_{\pi(i)})$ of $x_{\pi(i)}$ under A_i.[1]

The group of motions is the wreath product $\mathrm{GL}_m(q) \wr \mathrm{S}_n$ of $\mathrm{GL}_m(q)$ and S_n of order $|\mathrm{GL}_m(q)|^n \times n!$; two additive codes are equivalent in the monomial sense if they are in the same orbit under the action of this larger group. The elements of the wreath product are described by $(n+1)$-tuples $(A_0, \ldots, A_{n-1}, \pi)$ where the coefficients $A_i \in \mathrm{GL}_m(q)$ and the permutation $\pi \in \mathrm{S}_n$. Write the elements of the monomial group as $g(A_0, \ldots, A_{n-1}, \pi)$; observe the rule: first the entries are permuted according to permutation π and then the matrices A_i are applied in the coordinates.

Definition 9.3.3 A code $\mathcal{C}$ of length n is **cyclic in the permutation sense** if there is an n-cycle $\pi \in \mathrm{S}_n$ such that $\pi(\mathcal{C}) = \mathcal{C}$. A q-linear q^m-ary code $\mathcal{C}$ is **cyclic in the monomial sense** if it is invariant under an element $g(A_0, \ldots, A_{n-1}, \pi)$ of the monomial group where $\pi \in \mathrm{S}_n$ is an n-cycle.

As the n-cycles are conjugate in S_n, it is clear that each code which is cyclic in the permutation sense is permutation equivalent to a code which is invariant under the permutation $\pi = (0, 1, \ldots, n-1)$.[2] In what follows, n-tuples $(x_0, x_1, \ldots, x_{n-1})$ or $(x_1, x_2, \ldots, x_n)$ will often be denoted (x_i) where the subscripts are determined by the context.

Proposition 9.3.4 *Let $\mathcal{C}$ be a q-linear q^m-ary code of length n and cyclic in the monomial sense. Let I be the identity transformation of $\mathrm{GL}_m(q)$. Then the following hold.*

(a) *$\mathcal{C}$ is monomially equivalent to a code invariant under $g(I, \ldots, I, A, \pi)$ where $\pi = (0, 1, \ldots, n-1)$.*

(b) *If $\mathcal{C}$ is invariant under $g(A_0, A_1, \ldots, A_{n-1}, (0, 1, \ldots, n-1))$ and $A_0 A_1 \cdots A_{n-1} = I$, then $\mathcal{C}$ is monomially equivalent to a code that is cyclic in the permutation sense.*

(c) *If $\mathcal{C}$ is invariant under $g(I, \ldots I, A, (0, 1, \ldots, n-1))$ and the order of A is coprime to n, then $\mathcal{C}$ is monomially equivalent to a code that is cyclic in the permutation sense.*

Proof: As stated after Definition 9.3.3, we can assume $\mathcal{C}$ is invariant under $g = g(A_0, A_1, \ldots, A_{n-1}, \pi)$ where $\pi = (0, 1, \ldots, n-1)$. Let $B_i \in \mathrm{GL}_m(q)$ be defined inductively as follows:

$$B_0 = I \quad \text{and} \quad B_i = B_{i-1} A_{i-1} \text{ for } 1 \le i \le n-1.$$

Let

$$\mathcal{C}' = \{(B_0 x_0, B_1 x_1, \ldots, B_{n-1} x_{n-1}) \mid (x_0, \ldots, x_{n-1}) \in \mathcal{C}\}.$$

Suppose $(y_i) \in \mathcal{C}'$ where $y_i = B_i x_i$ with $(x_i) \in \mathcal{C}$. As $\mathcal{C}$ is invariant under g, $(A_i x_{i+1}) \in \mathcal{C}$,

---

[1] Elements of $\mathbb{F}_{q^m}$ are viewed as coordinate vectors in $E = \mathbb{F}_q^m$ by fixing a basis of $\mathbb{F}_{q^m}$ over $\mathbb{F}_q$.

[2] In our definition, a code is cyclic in the permutation sense if it is invariant under a cyclic shift to the left by one position with wrap-around. In Section 1.12, a code is cyclic if it is invariant under a cyclic shift to the right by one position with wrap-around; the definitions are equivalent by using π^{n-1} for the permutation. As discussed in Remark 1.12.2, the direction of shift does not matter.

where x_n means x_0. This implies that $(B_i A_i x_{i+1}) = (B_i A_i B_{i+1}^{-1} y_{i+1}) \in \mathcal{C}'$, where B_n means B_0 and y_n means y_0. By definition $B_i A_i B_{i+1}^{-1} = I$ for $i \neq n-1$ and

$$B_{n-1} A_{n-1} B_0^{-1} = B_{n-1} A_{n-1} = B_{n-2} A_{n-2} A_{n-1} = \cdots$$
$$= B_0 A_0 A_1 \cdots A_{n-1} = A_0 A_1 \cdots A_{n-1}.$$

It follows that $\mathcal{C}'$ is invariant under $g(I, \ldots, I, A, (0, 1, \ldots, n-1))$ where $A = A_0 A_1 \cdots A_{n-1}$, proving (a). Part (b) follows immediately from part (a). For part (c), suppose that $\mathcal{C}$ be invariant under $g = g(I, \ldots, I, A, (0, 1, \ldots, n-1))$ and $\mathrm{ord}(A) = u$ is coprime to n. Then $\mathcal{C}$ is invariant under g^u; g^u has u of its components equal to A and the others equal to I. Part (c) now follows from part (b).   □

The cyclic group $G = \langle g(A_0, A_1, \ldots, A_{n-1}, (0, 1, \ldots, n-1)) \rangle$ has order a multiple of n and acts on an nm-dimensional vector space V over $\mathbb{F}_q$. The cyclic codes are the G-submodules of V. The theory simplifies considerably if the action of G on V is completely reducible.

Definition 9.3.5 Let the group G act on a vector space V. An **irreducible submodule** $A \subseteq V$ is a nonzero G-submodule which does not contain a proper G-submodule (different from $\{0\}$ and from A itself). The action of G is **completely reducible** if for every G-submodule $A \subset V$ there is a direct complement B of A which is a G-submodule of V.

If the action is completely reducible, it suffices in a way to know the irreducibles, as every G-submodule is a direct sum of irreducibles. There is still work to do as the representation of a module as a direct sum of irreducibles may not be unique. However the situation is a lot simpler than in the cases where complete reducibility is not satisfied. This is where Maschke's Theorem comes in.

Theorem 9.3.6 (Maschke) *In the situation of Definition 9.3.5, let the order of G be coprime to the characteristic p of the underlying field. Then the action of G is completely reducible.*

See also [1203, page 666]. This fundamental theorem is the reason why in the theory of cyclic codes it is often assumed that $\gcd(n, p) = 1$. We will apply it in the study of additive codes which are cyclic in the permutation sense.

9.4   Additive Codes Which Are Cyclic in the Permutation Sense

Recall the situation: W is the cyclic group generated by the permutation $(0, 1, \ldots, n-1)$. It acts on the vector space $V = \mathbb{F}_q^n$ as a cyclic permutation of the coordinates. The W-submodules are precisely the cyclic linear codes over $\mathbb{F}_q$. We assume $\gcd(n, p) = 1$ where $\mathbb{F}_q$ has characteristic p. Because of Maschke's Theorem the action of W on $V = \mathbb{F}_q^n$ is completely reducible.

Throughout the remainder of this section we adopt the following notation. Let $r = \mathrm{ord}_n(q)$, the order of q when calculating mod n. Then $n \mid (q^r - 1)$ and r is the smallest natural number with this property. Let $F = \mathbb{F}_{q^r}$. There is a cyclic group $\langle \alpha \rangle$ of order n in $F^* = F \setminus \{0\}$. It will be profitable to identify W with this subgroup of F^*, and so let $W = \langle \alpha \rangle$.

As we are interested in cyclic *additive* codes, our alphabet is $E = \mathbb{F}_q^m$, and we study the action of W on $V_m = E^n = \mathbb{F}_q^{mn}$. Keep in mind however that we consider the elements of V_m (the codewords) not as nm-tuples over $\mathbb{F}_q$ but rather as n-tuples over E. Refer to the n entries of a codeword as the **outer coordinates** (in bijection with the elements α^i, $i = 0, 1, \ldots, n - 1$ of W) and to the m coordinates of the alphabet E as the **inner coordinates**. Let $V_{m,F} = V_m \otimes F = (F^m)^n$ be obtained by constant extension with the action of W on outer coordinates. We will relate the codes $\mathcal{C} \subseteq V_m = E^n = \mathbb{F}_q^{nm}$ to their constant extensions $\mathcal{C} \otimes F \subseteq V_{m,F}$ and use basic facts concerning relations between codes defined over a larger field F and a subfield $\mathbb{F}_q$. In particular we associate to each subcode $\mathcal{U} \subseteq V_{m,F}$ its trace code $\mathrm{Tr}_{F/\mathbb{F}_q}(\mathcal{U}) \subseteq V_m$ obtained by applying the trace to each inner coordinate. Recall that the trace function from an extension field $F = \mathbb{F}_{q^r}$ to its subfield $\mathbb{F}_q$ is defined by $\mathrm{Tr}_{F/\mathbb{F}_q}(x) = \mathrm{Tr}_{q^r/q}(x) = x + x^q + \cdots + x^{q^{r-1}}$.

Project to one of the m inner coordinates. We then obtain spaces $\mathbb{F}_q^n$ and F^n. The elements of F^n can be uniquely described by univariate polynomials $p(X) \in F[X]$ of degree $< n$, where the n-tuple in F^n defined by $p(X)$ is the **evaluation** $\mathrm{Ev}(p(X)) = (p(\alpha^i))$, $i = 0, 1, \ldots, n - 1$. Doing this for each inner coordinate we see that the elements of $V_{m,F}$ can be uniquely described by tuples $(p_1(X), p_2(X), \ldots, p_m(X))$ of polynomials $p_j(X) \in F[X]$ of degree $< n$, where the corresponding vector in $(F^m)^n$ is the evaluation $\mathrm{Ev}(p_1(X), p_2(X), \ldots, p_m(X)) = (p_1(\alpha^i), p_2(\alpha^i), \ldots, p_m(\alpha^i))_i$ (an n-tuple of m-tuples where each of the nm entries is in F).

Consider the Galois group $G = \mathrm{Gal}(F \mid \mathbb{F}_q)$ (cyclic of order r) and its orbits on $\mathbb{Z}/n\mathbb{Z}$, the q-cyclotomic cosets. We interpret the elements of $\mathbb{Z}/n\mathbb{Z}$ as the exponents of α in the description of the elements of the cyclic group W. For each polynomial $p(X) \in F[X]$ of degree $< n$, we consider the exponents of the monomials occurring in $p(X)$ and how they distribute on the q-cyclotomic cosets; see Definition 1.12.7.

Definition 9.4.1 Let $\mathcal{P}$ be the space of polynomials in $F[X]$ of degree $< n$ along with the zero polynomial. Then $\mathcal{P}$ is an n-dimensional F-vector space. Let $A \subseteq \mathbb{Z}/n\mathbb{Z}$ be a set of exponents. The F-vector space $\mathcal{P}(A)$ consists of the zero polynomial and the polynomials in $F[X]$ of degree $< n$ all of whose monomials have degrees in A. The **Galois closure** $\tilde{A}$ of A is the union of all q-cyclotomic cosets that intersect A nontrivially.

Observe that $\mathcal{P}(A)$ is an F-vector space of dimension $|A|$, and it is isomorphic to $\mathrm{Ev}(\mathcal{P}(A)) \subset V_{1,F}$. The terminology of Definition 9.4.1 is justified by the obvious fact that the Galois closure of the code $\mathrm{Ev}(\mathcal{P}(A)) \subset V_{1,F} = F^n$ is $\mathrm{Ev}(\mathcal{P}(\tilde{A}))$.

9.4.1 The Linear Case $m = 1$

We have $\mathcal{P} = \bigoplus_Z \mathcal{P}(Z)$ where Z varies over the q-cyclotomic cosets and correspondingly $V_{1,F} = F^n = \bigoplus_Z \mathrm{Ev}(\mathcal{P}(Z))$. As the $\mathrm{Ev}(\mathcal{P}(Z))$ are Galois closed we also have $V_1 = \mathbb{F}_q^n = \bigoplus_Z \mathrm{Tr}_{F/\mathbb{F}_q}(\mathrm{Ev}(\mathcal{P}(Z)))$ and $\dim_{\mathbb{F}_q}(\mathrm{Tr}_{F/\mathbb{F}_q}(\mathrm{Ev}(\mathcal{P}(Z)))) = \dim_F(\mathrm{Ev}(\mathcal{P}(Z)) = |Z|$; see [199, Theorem 12.17]. Let $V_1(Z) = \mathrm{Tr}_{F/\mathbb{F}_q}(\mathrm{Ev}(\mathcal{P}(Z)))$. It is our first task to identify the irreducible W-submodules, that is the **irreducible cyclic codes**, in V_1.

Theorem 9.4.2 *In case $m = 1$ with $\gcd(n, p) = 1$, the irreducible cyclic codes in the permutation sense are precisely the $V_1(Z) = \mathrm{Tr}_{F/\mathbb{F}_q}(\mathrm{Ev}(\mathcal{P}(Z)))$ of $\mathbb{F}_q$-dimension $|Z|$ where Z is a q-cyclotomic coset. Each cyclic code can be written as a direct sum of irreducibles in precisely one way. The total number of cyclic codes is 2^t, where t is the number of different q-cyclotomic cosets.*

Theorem 9.4.2 is of course well known in the classical theory of linear cyclic codes. In order to be self-contained we will prove it in the remainder of this subsection.

It follows from basic properties of the trace that in the description of a code $\mathcal{C} \subseteq V_1$ by codewords $\mathrm{Tr}_{F/\mathbb{F}_q}(\mathrm{Ev}(p(X)))$, it suffices to use polynomials of the form $p(X) = a_1 X^{z_1} + a_2 X^{z_2} + \cdots$ where $z_1, z_2, \ldots$ are representatives of different q-cyclotomic cosets. Such a polynomial is said to be in **standard form**, for the moment.

Lemma 9.4.3 *Let Z be a q-cyclotomic coset, $z \in Z$, $|Z| = s$, and $L = \mathbb{F}_{q^s}$. The $\mathbb{F}_q$-vector space $\langle W^z \rangle$ generated by the u^z where $u \in W$ is L. The codeword $\mathrm{Tr}_{F/\mathbb{F}_q}(\mathrm{Ev}(aX^z))$ where $a \in F$ is identically zero if and only if $a \in L^\perp$ where duality is with respect to the trace form. Here for a field extension $\mathbb{F}_q \subset F = \mathbb{F}_{q^r}$ the trace form is defined by $\langle x, y \rangle = \mathrm{Tr}_{F/\mathbb{F}_q}(xy)$ where $x, y \in F$.*

Proof: As $\mathrm{Ev}(\mathcal{P}(Z))$ is F-linear and Galois closed of dimension s, it follows that $\mathrm{Tr}_{F/\mathbb{F}_q}(\mathrm{Ev}(\mathcal{P}(Z)))$ is $\mathbb{F}_q$-linear of dimension s. Its generic codeword is $\mathrm{Tr}_{F/\mathbb{F}_q}(\mathrm{Ev}(aX^z))$. This codeword is identically zero if and only if $a \in \langle W^z \rangle^\perp$. Comparing dimensions shows $\dim(\langle W^z \rangle) = s$. Let $u \in W$. Then $u^{zq^s} = u^z$ by definition of a cyclotomic coset showing $W^z \subset L$. It follows that $\langle W^z \rangle = L$. □

Lemma 9.4.3 shows that we can assume $a_i \notin L_i^\perp$ for each i. Also observe that $\mathrm{Tr}_{F/\mathbb{F}_q}(\mathrm{Ev}(p(X))) = \mathrm{Tr}_{F/\mathbb{F}_q}(a_1 X^{z_1}) + \mathrm{Tr}_{F/\mathbb{F}_q}(a_2 X^{z_2}) + \cdots$ and the summands $\mathrm{Tr}_{F/\mathbb{F}_q}(a_i X^{z_i})$ are in different parts of the direct sum decomposition $V_1 = \bigoplus_Z V_1(Z)$.

Lemma 9.4.4 *Let $\mathcal{C} \subseteq V_1$ be a cyclic code, and let*

$$\mathcal{B} = \{p(X) \in \mathcal{P} \mid \mathrm{Tr}_{F/\mathbb{F}_q}(\mathrm{Ev}(p(X))) \in \mathcal{C}\}.$$

Let $p(X) = a_1 X^{z_1} + a_2 X^{z_2} + \cdots \in \mathcal{B}$ where $z_1, z_2, \ldots$ are representatives of different cyclotomic cosets. Then $p^{(l)}(X) = a_1 \alpha^{l z_1} X^{z_1} + a_2 \alpha^{l z_2} X^{z_2} + \cdots \in \mathcal{B}$ for all l. The smallest cyclic code containing $\mathrm{Tr}_{F/\mathbb{F}_q}(\mathrm{Ev}(p(X)))$ is the code spanned by the $\mathrm{Tr}_{F/\mathbb{F}_q}(\mathrm{Ev}(p^{(l)}(X)))$ for all l.

Proof: The entry of $\mathrm{Tr}_{F/\mathbb{F}_q}(\mathrm{Ev}(p(X)))$ in coordinate α^i is $\mathrm{Tr}_{F/\mathbb{F}_q}(a_1 \alpha^{i z_1} + a_2 \alpha^{i z_2} + \cdots)$. After a cyclic shift we obtain a codeword whose entry in coordinate α^i is $\mathrm{Tr}_{F/\mathbb{F}_q}(a_1 \alpha^{(i+1) z_1} + a_2 \alpha^{(i+1) z_2} + \cdots) = \mathrm{Tr}_{F/\mathbb{F}_q}(a_1 \alpha^{z_1} \alpha^{i z_1} + a_2 \alpha^{z_2} \alpha^{i z_2} + \cdots)$ which is the trace of the evaluation of $a_1 \alpha^{z_1} X^{z_1} + a_2 \alpha^{z_2} X^{z_2} + \cdots$. The first claim follows by repeated application, and the second is obvious. □

Let us complete the proof of Theorem 9.4.2. Let $\mathcal{C}$ be an irreducible cyclic code and $\mathrm{Tr}_{F/\mathbb{F}_q}(\mathrm{Ev}(p(X))) \in \mathcal{C}$ where $p(X)$ is in standard form. Assume $p(X)$ is not a monomial. Let $\mathcal{B}$ be defined as in Lemma 9.4.4 and $p(X) = a_1 X^{z_1} + a_2 X^{z_2} + \cdots \in \mathcal{B}$. Lemma 9.4.4 implies that $a_1 \alpha^{l z_1} X^{z_1} + a_2 \alpha^{l z_2} X^{z_2} + \cdots \in \mathcal{B}$ for all l. Assume at first $|Z_1| \neq |Z_2|$. Choose l such that $\alpha^{l z_1} = 1$ and $\alpha^{l z_2} \neq 1$. By subtraction we have $a_2(\alpha^{l z_2} - 1) X^{z_2} + \cdots \in \mathcal{B}$. The cyclic code generated by the trace-evaluation of this polynomial has trivial projection to $V_1(Z_1)$. As this is not true of $\mathcal{C}$ it follows from irreducibility that this codeword must be the zero codeword; hence $a_2(\alpha^{l z_2} - 1) \in L_2^\perp$. As $0 \neq \alpha^{l z_2} - 1 \in L_2$ (see Lemma 9.4.3), this implies $a_2 \in L_2^\perp$, a contradiction. Assume now $|Z_1| = |Z_2| = s$; let $L = \mathbb{F}_{q^s}$. Use Lemma 9.4.4. Let $c_l \in \mathbb{F}_q$ such that $\sum_l c_l \alpha^{l z_1} = 0$. The same argument as above shows that $\sum_l c_l \alpha^{l z_2} = 0$. We have $\beta = \alpha^{z_1} \in L$ and L is the smallest subfield of F containing β. Also $\alpha^{z_2} = \beta^j$ where j is coprime to the order of β. Let $\sum_{l=0}^s c_l \beta^l$ be the minimal polynomial of β. We just saw that $\sum_{l=0}^s c_l \beta^{jl} = 0$. This shows that the mapping $x \mapsto x^j$ is a field automorphism of L. This implies that z_1 and z_2 are in the same cyclotomic coset, a contradiction.

We have seen that a polynomial in standard form whose trace-evaluation generates an irreducible cyclic code is necessarily a monomial aX^z where $a \notin L^\perp$ using the by now standard terminology ($z \in Z$, $|Z| = s$, $L = \mathbb{F}_{q^s}$, and $\perp$ with respect to the trace form).

Lemma 9.4.4 shows that $a\alpha^{lz}X^z \in \mathcal{B}$ for all l. As the α^{lz} generate L (see Lemma 9.4.3), it follows that $auX^z \in \mathcal{B}$ for all $u \in L$. By Lemma 9.4.4 again we have in fact $\mathcal{C} = \mathrm{Tr}_{F/\mathbb{F}_q}(\mathrm{Ev}(aLX^z))$. Consider the map $u \mapsto \mathrm{Tr}_{F/\mathbb{F}_q}(\mathrm{Ev}(auX^z))$ from L onto $\mathcal{C}$. The kernel of this map is 0, and so $\mathcal{C}$ has dimension s. As it is contained in $V_1(Z)$ of dimension s, we have equality.

This completes the proof of Theorem 9.4.2. As a result we obtain an algorithmic description of all codewords of all cyclic linear codes in the permutation sense of length n, where $\gcd(n, p) = 1$. The codes are in bijection with sets of cyclotomic cosets. Let $Z_1 \cup \cdots \cup Z_\ell$ be such a set. Let $z_k \in Z_k$, $s_k = |Z_k|$, and $L_k = \mathbb{F}_{q^{s_k}}$. The dimension of the code is $\sum_{k=1}^{\ell} s_k$. Its generic codeword $w(x_1, \ldots, x_\ell)$ where $x_k \in L_k$ has entry

$$\sum_{k=1}^{\ell} \mathrm{Tr}_{L_k/\mathbb{F}_q}(x_k\alpha^{iz_k})$$

in coordinate i where $i = 0, 1, \ldots, n - 1$. The cyclic shift of the codeword $w(x_1, \ldots, x_\ell)$ is $w(x_1\alpha^{z_1}, \ldots, x_\ell\alpha^{z_\ell})$.

9.4.2 The General Case $m \geq 1$

Now let $m \geq 1$ be arbitrary. This is the case of additive not necessarily linear cyclic codes in the permutation sense. Recall that we still assume $\gcd(n, p) = 1$. We want to determine the irreducible cyclic codes in this case. The fact that we have already dealt with the case $m = 1$ will be helpful. Let $\mathcal{C}$ be an irreducible cyclic code and π_j, $j = 1, \ldots, m$, the projections to the inner coordinates. If $\pi_j(\mathcal{C})$ is not identically zero, then $\pi_j(\mathcal{C})$ is a linear irreducible cyclic code. Therefore by Theorem 9.4.2 it is described by a q-cyclotomic coset. It can be assumed that this happens for all j; otherwise we are dealing with a smaller value of m. Let $Z_1, \ldots, Z_m$ be the corresponding q-cyclotomic cosets. First we show that they are identical.

Lemma 9.4.5 *Let $\mathcal{C}$ be an irreducible additive cyclic code. The q-cyclotomic cosets determined by the irreducible linear cyclic codes $\pi_j(\mathcal{C})$ are all identical.*

Proof: We can assume $m = 2$ and $\mathrm{Tr}_{F/\mathbb{F}_q}(\mathrm{Ev}(p_1(X), p_2(X))) \in \mathcal{C}$ where $p_1(X) = a_1X^{z_1}$, $p_2(X) = a_2X^{z_2}$ and $a_i \notin L_i^\perp$ with the notation used in the previous subsection. The proof is similar to the main portion of the proof of Theorem 9.4.2. Lemma 9.4.4 shows that $(a_1\alpha^{lz_1}X^{z_1}, a_2\alpha^{lz_2}X^{z_2}) \in \mathcal{B}$ for all l, where $\mathcal{B}$ is the $\mathbb{F}_q$-linear space of tuples of polynomials whose trace-evaluation is in $\mathcal{C}$. Assume z_1, z_2 are in different q-cyclotomic cosets Z_1, Z_2, of lengths s_1, s_2. If $s_1 \neq s_2$, the same argument applies as in the proof of Theorem 9.4.2. If $s_1 = s_2$, the argument used in the proof of Theorem 9.4.2 shows $Z_1 = Z_2$, another contradiction. $\qquad\square$

Theorem 9.4.6 *Let Z be a q-cyclotomic coset. Each nonzero codeword in $V_m(Z) = \mathrm{Tr}_{F/\mathbb{F}_q}(\mathrm{Ev}(\mathcal{P}(Z), \ldots, \mathcal{P}(Z)))$ generates an irreducible cyclic code of dimension $s = |Z|$.*

Proof: Such a codeword can be written as $\mathrm{Tr}_{F/\mathbb{F}_q}(\mathrm{Ev}(a_1X^z, a_2X^z, \ldots, a_mX^z))$. The entry in outer coordinate i and inner coordinate j is $\mathrm{Tr}_{F/\mathbb{F}_q}(a_j\alpha^{iz})$. Lemma 9.4.4 shows that the cyclic code $\mathcal{C}$ generated by this codeword is the span of the codewords with entry $\mathrm{Tr}_{F/\mathbb{F}_q}(a_j\alpha^{lz}\alpha^{iz})$ in the same position. As the α^{lz} generate $L = \mathbb{F}_{q^s}$ (see Lemma 9.4.3) it follows that $\mathcal{C}$ consists of the codewords with entry $\mathrm{Tr}_{F/\mathbb{F}_q}(a_ju\alpha^{iz})$ in the (i, j)-coordinate, where $u \in L$. It is now clear that this code is the cyclic code generated by any of its nonzero codewords; in other words $\mathcal{C}$ is irreducible. By the transitivity of the trace we have $\mathrm{Tr}_{F/\mathbb{F}_q}(a_j u\alpha^{iz}) = \mathrm{Tr}_{L/\mathbb{F}_q}(b_j u\alpha^{iz})$ where $b_j = \mathrm{Tr}_{F/L}(a_j)$. The fact that the code is nonzero means that not all the b_j vanish. The map $u \mapsto \mathrm{Tr}_{F/\mathbb{F}_q}(\mathrm{Ev}(a_1 u X^z, \ldots, a_m u X^z))$ is injective, and so $\dim(\mathcal{C}) = s$. $\qquad\square$

Corollary 9.4.7 *The total number of irreducible permutation cyclic q-linear q^m-ary codes of length n coprime to the characteristic is $\sum_Z (q^{ms} - 1)/(q^s - 1)$ where Z varies over the q-cyclotomic cosets and $s = |Z|$.*

Proof: In fact, $V_m(Z)$ has $\mathbb{F}_q$-dimension ms, and each of the irreducible subcodes has $q^s - 1$ nonzero codewords. $\qquad\square$

Observe that this is true also in the linear case $m = 1$: the number of irreducible cyclic codes is the number of q-cyclotomic cosets in this case; see Theorem 9.4.2. More importantly we obtain a parametric description of the irreducible codes. Such a code is described by the following data:

1. A q-cyclotomic coset Z. Let $s = |Z|$ and $L = \mathbb{F}_{q^s}$; choose $z \in Z$.

2. A point $P = (b_1 : \cdots : b_m) \in \mathrm{PG}(m - 1, L)$, the $(m - 1)$-dimensional projective geometry over L.

The codewords are then parameterized by $x \in L$. The entry in outer coordinate i and inner coordinate j is $\mathrm{Tr}_{L/\mathbb{F}_q}(b_j x\alpha^{iz})$. Denote this code by $\mathcal{C}(z, P)$. This leads to a parametric description of all cyclic additive codes, not just the irreducible ones.

Definition 9.4.8 Let Z be a q-cyclotomic coset of length s, $z \in Z$, $L = \mathbb{F}_{q^s}$, and $U \subset V(m, L)$ a k-dimensional vector subspace (equivalently: a $\mathrm{PG}(k-1, L) \subset \mathrm{PG}(m-1, L)$). Let $P_1, \ldots, P_k$ be the projective points determined by a basis of U. Define $\mathcal{C}(z, U) = \mathcal{C}(z, P_1) \oplus \cdots \oplus \mathcal{C}(z, P_k)$.

We refer to the $\mathcal{C}(z, U)$ as **constituent codes**. Here $\mathcal{C}(z, U)$ has dimension ks if $U \subseteq L^m$ has dimension k. Use an **encoding matrix** $B = [b_{lj}]$, a $k \times m$ matrix with entries from L. The rows $\mathbf{z}_l$ of B form a basis of U. The codewords $w(\mathbf{x})$ are parameterized by $\mathbf{x} = (x_l) \in L^k$. The entry of codeword $w(\mathbf{x})$ in outer coordinate i and inner coordinate $j = 1, \ldots, m$ is

$$\mathrm{Tr}_{L/\mathbb{F}_q}((\mathbf{x} \cdot \mathbf{s}_j)\alpha^{iz}) \tag{9.1}$$

where $\mathbf{s}_j$ is column j of B and $\mathbf{x} \cdot \mathbf{s}_j$ is the ordinary dot product. The image of $w(\mathbf{x})$ under one cyclic shift is $w(\alpha^z \mathbf{x})$. What happens if we use a different basis $(\mathbf{z}_l')$ for the same subspace U? Then $\mathbf{z}_l' = \sum_{t=1}^k a_{lt}\mathbf{z}_t$ and $A = [a_{lt}]$ is invertible. The image of $w(\mathbf{x})$ under this operation (replacing $\mathbf{z}_l$ by $\mathbf{z}_l'$) has (i, j)-entry $\mathrm{Tr}_{L/\mathbb{F}_q}(\sum_{l=1}^k x_l \sum_{t=1}^k a_{lt}b_{tj}\alpha^{iz})$. This describes $w(\mathbf{x}')$ where $x_l' = \sum_{t=1}^k x_t a_{tl}$; in other words $\mathbf{x}' = \mathbf{x}A \in L^k$. We have seen that $\mathcal{C}(z, U)$ is indeed independent of the choice of an encoding matrix. Basic properties of the trace show that the dependence on the choice of representative $z \in Z$ is given by $\mathcal{C}(qz, U^q) = \mathcal{C}(z, U)$.

Here is a concrete expression for the weights of constituent codes: Codeword $w(\mathbf{x})$ has entry $w(\mathbf{x})_i = 0$ in outer coordinate i if and only if $\mathbf{x} \cdot \mathbf{s}_j \in (\alpha^{iz})^\perp$ for all j, with respect to the $\mathrm{Tr}_{L/\mathbb{F}_q}$-form.

Finally each cyclic code can be written in a unique way as the direct sum of its constituent codes: $\mathcal{C} = \bigoplus_Z \mathcal{C}(z, U_Z)$ where Z varies over the cyclotomic cosets and z is a fixed representative of Z.

Definition 9.4.9 Let $N(m, q)$ be the total number of subspaces of the vector space $\mathbb{F}_q^m$.

In particular $N(1, q) = 2$, $N(2, q) = q + 3$, and $N(3, q) = 2(q^2 + q + 2)$.

Corollary 9.4.10 *The total number of permutation cyclic q-linear q^m-ary codes of length n coprime to the characteristic is $\prod_Z N(m, q^{|Z|})$.*

Example 9.4.11 Let $q = 2$, $m = 2$, and $n = 7$. The cyclotomic cosets have lengths $1, 3, 3$ (representatives $0, 1, -1$). There are $3 + 9 + 9 = 21$ irreducible cyclic codes and $5 \times 11 \times 11 = 605$ cyclic codes total. A $[7, 3.5, 4]_4$ code is obtained as $\mathcal{C}(1, (1,0)\mathbb{F}_8) \oplus \mathcal{C}(-1, (0,1)\mathbb{F}_8) \oplus \mathcal{C}(0, (1,1)\mathbb{F}_2)$, where $\mathcal{C}(0, (1,1)\mathbb{F}_2)$ simply is the repetition code $\langle (11)^7 \rangle$. In the language of Definition 9.2.1 we have $km = 2k = 7$ and the $\mathbb{F}_4$-dimension is therefore $k = 3.5$. It may be checked that the minimum distance is indeed 4. This leads to the following generator matrix for our code:

$$
\left[
\begin{array}{c|c|c|c|c|c|c}
10 & 10 & 10 & 00 & 10 & 00 & 00 \\
10 & 10 & 00 & 10 & 00 & 00 & 10 \\
10 & 00 & 10 & 00 & 00 & 10 & 10 \\
\hline
01 & 00 & 00 & 01 & 00 & 01 & 01 \\
01 & 01 & 00 & 00 & 01 & 00 & 01 \\
01 & 01 & 01 & 00 & 00 & 01 & 00 \\
\hline
11 & 11 & 11 & 11 & 11 & 11 & 11
\end{array}
\right].
$$

Here the first section of 3 lines corresponds to $\mathcal{C}(1, (1,0)\mathbb{F}_8)$, the second section to $\mathcal{C}(-1, (0,1)\mathbb{F}_8)$; in both cases $k = 1$, $L = \mathbb{F}_8 = \mathbb{F}_2(\epsilon)$ defined by $\epsilon^3 + \epsilon^2 + 1 = 0$, and $\alpha = \epsilon$.

Example 9.4.12 In the case $q = 2$, $m = 2$, and $n = 13$ there are only two 2-cyclotomic cosets, of lengths 1 and 12. An optimal cyclic $[13, 6.5, 6]_4$ code is obtained. This was first described by Danielsen and Parker in [484]. A generator matrix can be written in the form $[\,I \mid A\,]$ where I is the 13×13 identity matrix and

$$
A = \left[
\begin{array}{c|c|c|c|c|c|c}
0 & 11 & 10 & 11 & 00 & 10 & 11 \\
1 & 10 & 01 & 00 & 00 & 01 & 11 \\
1 & 01 & 00 & 01 & 11 & 11 & 10 \\
1 & 00 & 01 & 10 & 00 & 11 & 01 \\
0 & 10 & 11 & 11 & 01 & 01 & 00 \\
1 & 11 & 01 & 01 & 10 & 01 & 00 \\
0 & 00 & 10 & 11 & 11 & 01 & 01 \\
0 & 01 & 11 & 01 & 01 & 10 & 01 \\
1 & 10 & 01 & 10 & 11 & 10 & 10 \\
1 & 10 & 00 & 11 & 01 & 00 & 01 \\
0 & 10 & 00 & 10 & 10 & 01 & 01 \\
1 & 11 & 11 & 00 & 11 & 00 & 11 \\
1 & 10 & 11 & 00 & 10 & 11 & 10
\end{array}
\right].
$$

Example 9.4.13 Let $q = 2$, $m = 2$, and $n = 15$. Representatives of the 2-cyclotomic cosets are 0 (length 1), 5 (length 2) and $1, 3, 14$ (length 4 each). There are $3 + 5 + 3 \times 17 = 59$ irreducible cyclic codes and a total of $5 \times 7 \times 19^3$ cyclic codes. A $[15, 4.5, 9]_4$ code is obtained as $\mathcal{C}(1, (\epsilon, 1)\mathbb{F}_{16}) \oplus \mathcal{C}(3, (1, \epsilon^2)\mathbb{F}_{16}) \oplus \mathcal{C}(0, (1,1)\mathbb{F}_2)$; see [197]. Here $F = \mathbb{F}_{16} = \mathbb{F}_2(\epsilon)$ and

$\epsilon^4 = \epsilon + 1$. The generator matrix is

$$
\left[
\begin{array}{c|c|c|c|c|c|c|c|c|c|c|c|c|c|c}
00 & 00 & 10 & 01 & 00 & 10 & 11 & 01 & 10 & 01 & 10 & 11 & 11 & 11 & 01 \\
00 & 10 & 01 & 00 & 10 & 11 & 01 & 10 & 01 & 10 & 11 & 11 & 11 & 01 & 00 \\
10 & 01 & 00 & 10 & 11 & 01 & 10 & 01 & 10 & 11 & 11 & 11 & 01 & 00 & 00 \\
01 & 00 & 10 & 11 & 01 & 10 & 01 & 10 & 11 & 11 & 11 & 01 & 00 & 00 & 10 \\
\hline
00 & 10 & 10 & 11 & 11 & 00 & 10 & 10 & 11 & 11 & 00 & 10 & 10 & 11 & 11 \\
01 & 01 & 11 & 01 & 10 & 01 & 01 & 11 & 01 & 10 & 01 & 01 & 11 & 01 & 10 \\
00 & 01 & 00 & 11 & 10 & 00 & 01 & 00 & 11 & 10 & 00 & 01 & 00 & 11 & 10 \\
10 & 10 & 11 & 11 & 00 & 10 & 10 & 11 & 11 & 00 & 10 & 10 & 11 & 11 & 00 \\
\hline
11 & 11 & 11 & 11 & 11 & 11 & 11 & 11 & 11 & 11 & 11 & 11 & 11 & 11 & 11 \\
\end{array}
\right].
$$

Example 9.4.14 In the case $q = 2$, $m = 2$, and $n = 21$ we have the following representatives for the 2-cyclotomic cosets: 0 (length 1), 7 (length 2), 3, 9 (length 3 each), and 1, 5 (length 6 each). Particularly interesting is a $[21, 10.5, 8]_4$ code. A generator matrix can be written $\left[\begin{array}{c|c} I & B \\ 0 & D \end{array}\right]$ where I is the 18×18 identity matrix and

$$
B = \left[
\begin{array}{c|c|c|c|c|c|c|c|c|c|c|c}
01 & 01 & 00 & 00 & 10 & 11 & 00 & 10 & 01 & 01 & 00 & 00 \\
01 & 01 & 00 & 11 & 01 & 01 & 10 & 10 & 10 & 10 & 10 & 00 \\
00 & 01 & 01 & 00 & 00 & 10 & 11 & 00 & 10 & 01 & 01 & 00 \\
00 & 01 & 01 & 00 & 11 & 01 & 01 & 10 & 10 & 10 & 10 & 10 \\
00 & 00 & 01 & 01 & 00 & 00 & 10 & 11 & 00 & 10 & 01 & 01 \\
01 & 01 & 01 & 01 & 10 & 00 & 01 & 11 & 11 & 11 & 10 & 10 \\
01 & 01 & 00 & 10 & 00 & 01 & 10 & 00 & 01 & 10 & 00 & 01 \\
01 & 00 & 01 & 01 & 11 & 01 & 00 & 11 & 10 & 10 & 11 & 10 \\
01 & 00 & 01 & 11 & 11 & 01 & 11 & 00 & 10 & 11 & 00 & 00 \\
01 & 00 & 00 & 01 & 11 & 00 & 01 & 10 & 10 & 11 & 10 & 11 \\
00 & 01 & 00 & 01 & 11 & 11 & 01 & 11 & 00 & 10 & 11 & 00 \\
00 & 01 & 00 & 11 & 10 & 01 & 10 & 01 & 01 & 01 & 01 & 10 \\
00 & 00 & 01 & 00 & 01 & 11 & 11 & 01 & 11 & 00 & 10 & 11 \\
01 & 01 & 01 & 00 & 01 & 01 & 01 & 00 & 00 & 00 & 01 & 01 \\
00 & 00 & 00 & 10 & 11 & 11 & 01 & 11 & 10 & 00 & 10 & 10 \\
01 & 00 & 01 & 10 & 01 & 00 & 11 & 11 & 10 & 10 & 10 & 01 \\
01 & 01 & 00 & 00 & 00 & 00 & 11 & 11 & 10 & 11 & 00 & 10 \\
01 & 00 & 00 & 10 & 11 & 00 & 10 & 01 & 01 & 00 & 00 & 10 \\
\end{array}
\right],
$$

$$
D = \left[
\begin{array}{c|c|c|c|c|c|c|c|c|c|c|c}
11 & 00 & 01 & 00 & 10 & 11 & 00 & 01 & 10 & 11 & 11 & 00 \\
00 & 11 & 00 & 01 & 00 & 10 & 11 & 00 & 01 & 10 & 11 & 11 \\
00 & 00 & 11 & 11 & 10 & 10 & 00 & 11 & 11 & 10 & 00 & 11 \\
\end{array}
\right].
$$

For small and moderate length, the parameters of the best *cyclic* additive codes over $\mathbb{F}_4$ have the tendency to coincide with the best parameters of additive codes over $\mathbb{F}_4$ in general; see [199, Section 18.3], and [200, 202].

9.4.3   Equivalence

When are two additive cyclic codes $\bigoplus_Z \mathcal{C}(z, U_Z)$ and $\bigoplus_Z \mathcal{C}(z, U_Z')$ equivalent? Two such situations are easy to see. Here is the first.

Proposition 9.4.15 *Let $C \in \mathrm{GL}_m(q)$. Then the additive cyclic code $\mathcal{C} = \bigoplus_Z \mathcal{C}(z, U_Z)$ is monomially equivalent to $\bigoplus_Z \mathcal{C}(z, U_Z C)$.*

Proof: Use the GL-part of monomial equivalence. An equivalent code is obtained if we apply the matrix $C = [c_{uj}]$ to each entry of $\mathcal{C}$. This means that for each constituent Z the entry $w(\mathbf{x})_i(j) = \mathrm{Tr}_{L/\mathbb{F}_q}((\mathbf{x} \cdot \mathbf{s}_j)\alpha^{iz})$ is replaced by $\sum_{u=1}^{m} c_{uj}w(\mathbf{x})_i(u)$. As $c_{uj} \in \mathbb{F}_q$ this amounts to replacing the encoding matrix B_Z by $B_Z C$, and therefore the subspace U_Z by $U_Z C$. $\square$

As a special case of Proposition 9.4.15 consider the case of irreducible codes for $m = 2$. We have a fixed q-cyclotomic coset Z of length $s = |Z|$ and $U = (b_1 : b_2)$ is a point of the projective line $\mathrm{PG}(1, L)$. Multiplication by C from the right amounts to applying a Möbius transformation, an element of $\mathrm{PGL}_2(q)$. The number of inequivalent irreducible codes belonging to the q-cyclotomic coset Z with respect to the equivalence described by Proposition 9.4.15 equals the number of orbits of $\mathrm{PGL}_2(q)$ on the projective line $\mathrm{PG}(1, q^s)$.

Here is the second such situation.

Proposition 9.4.16 *If* $\gcd(t, n) = 1$, *then the additive permutation cyclic code* $\mathcal{C} = \bigoplus_Z \mathcal{C}(z, U_Z)$ *is permutation equivalent to* $\bigoplus_Z \mathcal{C}(tz, U_Z)$.

Proof: The generic codeword $w(\mathbf{x})$ of the second code above has entry $\mathrm{Tr}_{L/\mathbb{F}_q}((\mathbf{x} \cdot \mathbf{s}_j)\alpha^{itz})$ in outer coordinate i, inner coordinate j (see Equation 9.1). This is the entry of the first code in outer coordinate $i/t \bmod n$ and inner coordinate j. $\square$

Proposition 9.4.16 describes an action of the group of units in $\mathbb{Z}/n\mathbb{Z}$. Here is an example.

Example 9.4.17 There are precisely three inequivalent length 7 irreducible additive codes over $\mathbb{F}_4$ which are cyclic in the permutation sense. This can be seen as follows. This is the case $q = 2$, $m = 2$, and $n = 7$. The total number of irreducible cyclic codes is $3 + 9 + 9 = 21$ (once $\mathrm{PG}(1, 2)$ and twice $\mathrm{PG}(1, 8)$). The number of inequivalent such codes is at most $1 + 2 = 3$. In fact, the three codes $\mathcal{C}(0, U)$ where $U \in \mathrm{PG}(1, 2)$ are all equivalent by Proposition 9.4.15. By Proposition 9.4.16 it suffices to consider $Z(1)$, the 2-cyclotomic coset containing 1, for the remaining irreducible codes. The number of possibly inequivalent such irreducible codes is the number of orbits of $\mathrm{PGL}_2(2)$ on $\mathrm{PG}(1, 8)$. There are two such orbits. The corresponding codes are $\mathcal{C}(1, (0, 1))$ and $\mathcal{C}(1, (1, \epsilon))$ where $\epsilon \notin \mathbb{F}_2$. It is in fact obvious that those irreducible codes are pairwise inequivalent. Clearly the first one is inequivalent to the others as it has binary dimension 1 (the repetition code $\langle (11)^7 \rangle$). The second code has $w(x)_i = (0, \mathrm{Tr}_{\mathbb{F}_8/\mathbb{F}_2}(x\epsilon^i))$, of constant weight 4. The third of those irreducible codes has $w(x)_i = (\mathrm{Tr}_{\mathbb{F}_8/\mathbb{F}_2}(x\epsilon^i), \mathrm{Tr}_{\mathbb{F}_8/\mathbb{F}_2}(x\epsilon^{i+1}))$, clearly not of constant weight.

9.4.4 Duality and Quantum Codes

Fix a non-degenerate bilinear form $\langle \cdot, \cdot \rangle$ on the vector space $E = \mathbb{F}_q^m$; extend it to $V_m = E^n$, to F^m and to $V_{m,F} = (F^m)^n$ in the natural way. Recall that a bilinear form $\langle \cdot, \cdot \rangle$ is non-degenerate if $\langle \mathbf{x}, \mathbf{y} \rangle = 0$ for all $\mathbf{y}$ implies $\mathbf{x} = \mathbf{0}$ (equivalently: $\langle \mathbf{x}, \mathbf{y} \rangle = 0$ for all $\mathbf{x}$ implies $\mathbf{y} = \mathbf{0}$); in other words, only the $\mathbf{0}$ vector is orthogonal to the whole space. In coordinates $\langle \cdot, \cdot \rangle$ is described by coefficients $a_{kl} \in \mathbb{F}_q$ where $k, l = 1, \ldots, m$ such that $\langle (x_k), (y_l) \rangle = \sum_{k,l} a_{kl}x_k y_l$. Duality will be with respect to this bilinear form. Basic information is derived from the following fact.

Lemma 9.4.18 *For* $q^r > 2$, *let* $W = \langle \alpha \rangle \subset F = \mathbb{F}_{q^r}$ *be the cyclic subgroup of order n, and let l be an integer not a multiple of n. Then* $\sum_{i=0}^{n-1} \alpha^{il} = 0$.

Proof: Let $S = \sum_{i=0}^{n-1} \alpha^{il}$. Then $\alpha^l S = S$ and $\alpha^l \neq 1$. It follows that $(1 - \alpha^l)S = 0$ which implies $S = 0$. $\square$

Proposition 9.4.19 *With the previous notation* $V_{m,F}(Z)^{\perp} = \bigoplus_{Z' \neq -Z} V_{m,F}(Z')$ *and* $V_m(Z)^{\perp} = \bigoplus_{Z' \neq -Z} V_m(Z')$.

Proof: As the dimensions are right, for the first equality, it suffices to show that, with respect to $\langle \cdot, \cdot \rangle$, $V_{m,F}(Z)$ is orthogonal to $V_{m,F}(Z')$ when $Z' \neq -Z$; an analogous statement holds for the second equality. Concretely this means $\sum_{k,l=1}^{m} a_{kl} \sum_{i=0}^{n-1} p_k(\alpha^i) q_l(\alpha^i) = 0$ where $p_k(X) \in \mathcal{P}(Z)$, $q_l(X) \in \mathcal{P}(Z')$. By bilinearity it suffices to prove this for fixed k, l and $p_k(X) = X^z$, $q_l(X) = X^{z'}$ where $z \in Z$, $z' \in Z'$. Let $j = z + z'$. Then j is not a multiple of n. We need to show $\sum_{i=0}^{n-1} \alpha^{iz+iz'} = 0$. This follows from Lemma 9.4.18, proving the first equality. The proof of the second equality is analogous, using the definition of the trace. $\square$

Observe that the result of Proposition 9.4.19 is independent of the bilinear form $\langle \cdot, \cdot \rangle$. The information contained in the bilinear form will determine the duality relations between subspaces of $V_{m,F}(Z)$ and $V_{m,F}(-Z)$ as well as between subspaces of $V_m(Z)$ and $V_m(-Z)$. In order to decide self-orthogonality, the following formula is helpful; see [199, Lemma 18.54].

Lemma 9.4.20 *Let Z be a nonzero q-cyclotomic coset, $|Z| = s$, $z \in Z$, and $L = \mathbb{F}_{q^s}$. Then*

$$\sum_{u \in W} \mathrm{Tr}_{F/\mathbb{F}_q}(au^z) \mathrm{Tr}_{F/\mathbb{F}_q}(bu^{-z}) = n \times \mathrm{Tr}_{F/\mathbb{F}_q}(a\mathrm{Tr}_{F/L}(b)).$$

Proof: Let S be the sum on the left. Writing out the definition of the trace, $S = \sum_{i,j=0}^{r-1} a^{q^i} b^{q^j} \sum_{u \in W} u^{(q^i - q^j)z}$. The inner sum vanishes if $(q^i - q^j)z$ is not divisible by n. In the contrary case the inner sum is n. The latter case occurs if and only if $i = j + \rho s$ where $\rho = 0, \ldots, s-1$. It follows that $S = n \sum_{i=0}^{r} a^{q^i} \sum_{\rho=0}^{s-1} b^{q^i q^{\rho s}} = n\mathrm{Tr}_{F/\mathbb{F}_q}(a\mathrm{Tr}_{F/L}(b))$. $\square$

An interesting special case concerns the cyclic quantum stabilizer codes over an arbitrary ground field $\mathbb{F}_q$. In this case $m = 2$, and the bilinear form is the symplectic form $\langle (x_1, x_2), (y_1, y_2) \rangle = x_1 y_2 - x_2 y_1$; in particular $\langle \mathbf{x}, \mathbf{x} \rangle = 0$ always. By definition such a code is a (cyclic) q-ary quantum stabilizer code if it is self-orthogonal with respect to the symplectic form. The basic result to decide self-orthogonality is the following.

Proposition 9.4.21 *When $m = 2$, $\gcd(n, p) = 1$, and $\langle \cdot, \cdot \rangle$ is the symplectic form, irreducible codes $\mathcal{C}(z, P)$ and $\mathcal{C}(-z, P')$ are orthogonal if and only if $P = P'$.*

Proof: Observe that Z and $-Z$ have the same length s, $L = \mathbb{F}_{q^s}$, and $P, P' \in \mathrm{PG}(1, L)$. It suffices to show that $\mathcal{C}(z, P)$ and $\mathcal{C}(-z, P)$ are orthogonal. Let $P = (v_1 : v_2)$. Writing out the symplectic product of typical vectors we need to show

$$\sum_{u \in W} \mathrm{Tr}_{F/\mathbb{F}_q}(av_1 u^z) \mathrm{Tr}_{F/\mathbb{F}_q}(bv_2 u^{-z}) = \sum_{u \in W} \mathrm{Tr}_{F/\mathbb{F}_q}(av_2 u^z) \mathrm{Tr}_{F/\mathbb{F}_q}(bv_1 u^{-z})$$

for $a, b \in F$. Lemma 9.4.20 shows that the sum on the left equals

$$n \times \mathrm{Tr}_{F/\mathbb{F}_q}(av_1 \mathrm{Tr}_{F/L}(bv_2)) = n \times \mathrm{Tr}_{F/\mathbb{F}_q}(av_1 v_2 \mathrm{Tr}_{F/L}(b))$$
$$= n \times \mathrm{Tr}_{L/\mathbb{F}_q}(v_1 v_2 \mathrm{Tr}_{F/L}(a) \mathrm{Tr}_{F/L}(b))$$

which by symmetry coincides with the sum on the right. $\square$

This implies in particular that the $[21, 10.5, 8]_4$ code of Example 9.4.14 is a $[\![21, 0, 8]\!]$ quantum code; see [331].

Theorem 9.4.22 *The number of additive cyclic q^2-ary quantum stabilizer codes is*

$$\prod_{Z = -Z,\, s=1} (q+2) \prod_{Z = -Z,\, s>1} (q^{s/2} + 2) \prod_{Z \neq -Z} (3q^s + 6).$$

The number of such codes that are self-dual is

$$\prod_{Z=-Z,\,s=1} (q+1) \prod_{Z=-Z,\,s>1} (q^{s/2}+1) \prod_{Z\neq-Z} (q^s+3).$$

Here $s = |Z|$ and the last product is over all pairs $\{Z, -Z\}$ of q-cyclotomic cosets such that $Z \neq -Z$.

Proof: Let $\mathcal{C} = \sum_Z S_Z$ where $S_Z = \mathcal{C}(z, U_Z)$. This is self-orthogonal if and only if S_Z and S_{-Z} are orthogonal for each Z. Consider first the generic case $Z \neq -Z$. If S_Z or S_{-Z} is $\{\mathbf{0}\}$, then there is no restriction on the other. If $S_Z = \mathcal{C}(z, L^2)$, then $S_{-Z} = \{\mathbf{0}\}$.

Consider the case $Z = -Z$ and $s > 1$. Then $s = 2i$. Either $S_Z = \{\mathbf{0}\}$ or $S_Z = \mathcal{C}(z, P)$ is a self-orthogonal irreducible code where $P \in \mathrm{PG}(1, L)$. The dual of S_Z in its constituent is $\mathcal{C}(-z, P)$ by Proposition 9.4.21. As $-z = z^{q^i}$, we have $\mathcal{C}(-z, P) = \mathcal{C}(z, P^{q^i})$. This equals S_Z if and only if $P \in \mathrm{PG}(1, L')$ where $L' = \mathbb{F}_{q^i}$. There are $q^i + 1$ choices for P. The case $Z = -Z$ and $s = 1$ contributes $q + 2$ self-orthogonal and $q + 1$ self-dual codes. $\qquad\square$

Example 9.4.23 In the case $q = 2$ and $n = 7$, we obtain $4 \times 30 = 120$ quantum codes total and $3 \times 11 = 33$ self-dual ones.

Example 9.4.24 For $q = 2$ and $n = 15$, the number of quantum codes is $4 \times 4 \times 6 \times 54$, and there are $3 \times 3 \times 5 \times 19$ self-dual quantum codes.

For more on quantum codes, see Chapter 27.

9.5 Additive Codes Which Are Cyclic in the Monomial Sense

Let $\mathcal{C}$ be a q^m-ary q-linear code of length n, which is cyclic in the monomial sense. Let $\gcd(n, p) = 1$, $m \geq 1$, and $A \in \mathrm{GL}_m(q)$ of order u where $\gcd(u, p) = 1$. Consider q-cyclotomic cosets $Z \subset \mathbb{Z}/un\mathbb{Z}$. For fixed Z of length s, let $z \in Z$ and $L = \mathbb{F}_{q^s}$. Also, let $\kappa = \mathrm{ord}_u(q)$ and $K = \mathbb{F}_{q^\kappa}$. It follows from Proposition 9.3.4 that by monomial equivalence it suffices to consider cyclic codes in the monomial sense fixed under the action of $g = g(I, \ldots I, A, (0, 1, \ldots, n-1))$ of order un. Such codes are also known as **constacyclic** codes. Here the matrix A plays the role of a constant factor. Let $G = \langle g \rangle$. Also, let β be a generator of the group of order un in an extension field of both F and K such that $\alpha = \beta^u \in F$. Then $\gamma = \beta^n$ has order u, and $\gamma \in K = \mathbb{F}_{q^\kappa}$.

Definition 9.5.1 Let $A \in \mathrm{GL}_m(q)$ with $A^u = I$. Define the **inflation map** $I_A : \mathbb{F}_q^{mn} \to \mathbb{F}_q^{umn}$ by

$$I_A(\underbrace{x_0, \ldots, x_{n-1}}_{\mathbf{x}}) = (\mathbf{x} | A\mathbf{x} | \cdots | A^{u-1}\mathbf{x}).$$

The inverse map, called **contraction**, from $I_A(\mathbb{F}_q^{mn})$ to $\mathbb{F}_q^{mn}$ is given by

$$\mathrm{contr}(\mathbf{x} | A\mathbf{x} | \cdots | A^{u-1}\mathbf{x}) = \mathbf{x}.$$

Lemma 9.5.2 *In the situation of Definition 9.5.1, the q-linear q^m-ary code $\mathcal{C}$ of length n is invariant under the (monomial) action of the cyclic group G of order un generated by $g(I, \ldots, I, A, (0, 1, \ldots, n-1))$ if and only if $I_A(\mathcal{C})$ is cyclic in the permutation sense under the permutation $(0, 1, \ldots, un-1)$. The contraction $\mathrm{contr}(\mathcal{C})$ is irreducible under the action of G if and only if the cyclic length un code $\mathcal{C}$ is irreducible.*

Proof: Observe that $\mathcal{C}$ and $I_A(\mathcal{C})$ have the same dimension and $\mathcal{C}$ is recovered from $I_A(\mathcal{C})$ as the projection to the first n coordinates. The claims are now obvious. $\qquad\square$

9.5.1 The Linear Case $m = 1$

In this special case we have $g = g(1, \ldots, 1, \gamma, (0, 1, \ldots, n-1))$ where the matrix A now is a constant $\gamma \in \mathbb{F}_q^*$ of order u where $u \mid (q-1)$. Write the elements of the ambient space $\mathbb{F}_q^{un}$ as $\mathbf{x} = (x_i)$ where $i \in \mathbb{Z}/un\mathbb{Z}$. Then $\mathbf{x} \in \mathbb{F}_q^{un}$ is in $I_\gamma(\mathbb{F}_q^n)$ if and only if $x_{i+n} = \gamma x_i$, with subscripts read modulo un, always holds.

Theorem 9.5.3 *Let* $\gcd(n, p) = 1$, *and let* $g = g(1, \ldots, 1, \gamma, (0, 1, \ldots, n-1))$ *where* $\gamma \in \mathbb{F}_q^*$ *with* $\mathrm{ord}(\gamma) = u$. *The codes which are invariant under the group generated by* g *are the contractions of the cyclic codes of length* un *in the permutation sense defined by cyclotomic cosets consisting of numbers which are* 1 *mod* u. *If there are* t *such cyclotomic cosets in* $\mathbb{Z}/un\mathbb{Z}$, *then the number of codes invariant under* g *is* 2^t.

Proof: Observe that $\gcd(un, p) = 1$. By Lemma 9.5.2 the codes in question are the contractions of the length un cyclic codes all of whose codewords (x_i) satisfy $x_{i+n} = \gamma x_i$ for all i. Let Z be a q-cyclotomic coset, $z \in Z$, $|Z| = s$, and $L = \mathbb{F}_{q^s}$. A typical codeword of the irreducible cyclic code defined by Z has entry $x_i = \mathrm{Tr}_{L/\mathbb{F}_q}(a\beta^{iz})$, where $a \in L$. As $\beta^n = \gamma$, it follows that $x_{i+n} = \mathrm{Tr}_{L/\mathbb{F}_q}(a\beta^{(i+n)z}) = \gamma^z x_i$. We must have $z \equiv 1 \pmod{u}$. $\square$

Here is an algorithmic view concerning linear length n q-ary codes which are invariant under $g = g(1, \ldots, 1, \gamma, (0, 1, \ldots, n-1))$ where $\mathrm{ord}(\gamma) = u$. Let $r' = \mathrm{ord}_{un}(q)$ and $F' = \mathbb{F}_{q^{r'}}$. Let $\beta \in F'$ have order un such that $\gamma = \beta^n$ and $\alpha = \beta^u$. Consider the q-cyclotomic cosets $Z_1, \ldots, Z_t$ in $\mathbb{Z}/un\mathbb{Z}$ whose elements are 1 mod u. Observe that $|Z_1 \cup \cdots \cup Z_t| = n$. Let $s_k = |Z_k|$, $z_k \in Z_k$, and $L_k = \mathbb{F}_{q^{s_k}} \subseteq F'$. Then $\mathcal{C} \subseteq \oplus \mathcal{C}_k$ where the $\mathcal{C}_k$ are the irreducible γ-constacyclic codes of length n with $\dim(\mathcal{C}_k) = s_k$. The codewords of $\mathcal{C}_k$ are $w(x)$ where $x \in L_k$ and the entry of $w(x)$ in coordinate i is $\mathrm{Tr}_{L_k/\mathbb{F}_q}(x\beta^{iz_k})$. The image of $w(x)$ under the monomial operation $g(1, \ldots, 1, \gamma, (0, 1, \ldots, n-1))$ is $w(\beta^{z_k}x)$. After n applications this leads to $w(\gamma^{z_k}x) = w(\gamma x) = \gamma w(x)$, as expected.

Example 9.5.4 Consider $q = 4$, $u = 3$, and $n = 15$. Then $F = \mathbb{F}_{16}$ whereas $F' = \mathbb{F}_{4^6}$. The 4-cyclotomic cosets in $\mathbb{Z}/45\mathbb{Z}$ consisting of numbers divisible by 3 are in bijection with the nine 4-cyclotomic cosets mod 15. The contractions of the cyclic length 45 codes defined by Z-cosets all of whose elements are divisible by 3 reproduce precisely the length 15 cyclic codes in the permutation sense. There are only three 4-cyclotomic cosets in $\mathbb{Z}/45\mathbb{Z}$ all of whose elements are 1 mod 3. They are

$$\{1, 4, 16, 19, 31, 34\}, \quad \{7, 13, 22, 28, 37, 43\}, \quad \{10, 25, 40\}.$$

It follows that there are $2^3 = 8$ constacyclic linear codes over $\mathbb{F}_4$ with constant $\gamma = \beta^{15}$ of order 3.

Example 9.5.5 Consider $q = 8$, $u = 7$, and $n = 21$. Then $F' = \mathbb{F}_{2^{42}} = \mathbb{F}_{8^{14}}$. The constant is $\gamma = \beta^{21}$ of order 7. We have $un = 7 \times 21 = 147$, and we have to consider the action of $q = 8$ (the Galois group $\mathrm{Gal}(F'|\mathbb{F}_8)$ has order 14) on the 21 elements which are 1 mod 7. The corresponding cyclotomic cosets are

$$Z(1) = \{1, 8, 64, 71, -20, -13, 43, 50, -41, -34, 22, 29, 85, -55\}$$

of length 14 and

$$Z(15) = \{15, -27, 78, 36, -6, -48, 57\}$$

of length 7. It follows that there are precisely two irreducible γ-constacyclic $\mathbb{F}_8$-linear codes of length 21, of dimensions 14 and 7.

Here are some more interesting and well-known examples.

Example 9.5.6 Let $q = 4$, $u = 3$, $Q = 2^f$ where f is odd, and $n = (Q+1)/3$. Then $F = \mathbb{F}_{Q^2}$. The irreducible constacyclic code over $\mathbb{F}_4$ defined by the 4-cyclotomic coset containing 1 has dimension f and dual distance 5. Its dual is therefore a linear $[(2^f+1)/3, (2^f+1)/3-f, 5]_4$ code. This is the first family from [801]; see also [199, Section 13.4].

Example 9.5.7 The second family of constacyclic codes over $\mathbb{F}_4$ from [801] occurs in the case $u = 3$ and $n = (4^f - 1)/3$ for arbitrary f where the 4-cyclotomic cosets are those generated by 1 and -2. This yields a $2f$-dimensional code over $\mathbb{F}_4$ of dual distance 5 and therefore $[(4^f - 1)/3, (4^f - 1)/3 - 2f, 5]_4$ codes for arbitrary f.

9.5.2 The General Case $m \geq 1$

Theorem 9.5.8 *With notation as introduced above, choose a representative z for each cyclotomic coset Z and U_Z the eigenspace of the eigenvalue γ^z of A in its action on L^m. Then each code stabilized by G is contained in $\bigoplus_Z \mathrm{contr}(\mathcal{C}(z, U_Z))$.*

Proof: The additive cyclic length un codes are direct sums over cyclotomic cosets Z of codes parameterized by subspaces $U_Z \subset L^m$, using the customary terminology. Such a code is in $I_A(\mathbb{F}_q^{mn})$ if and only if each codeword $\mathbf{x} = (x_i)$ (where $i = 0, \ldots, un - 1$ and $x_i \in \mathbb{F}_q^m$) satisfies $x_{i+n} = Ax_i$. Fix Z and ask for which subspace U this is satisfied. Let $A = [a_{jj'}] \in \mathrm{GL}_m(q)$ and $U = P = (b_1 : \cdots : b_m) \in \mathrm{PG}(m - 1, L)$ a point. The stability condition is

$$\mathrm{Tr}_{L/\mathbb{F}_q}(xb_j\beta^{(i+n)z}) = \sum_{j'=1}^m a_{jj'}\mathrm{Tr}_{L/\mathbb{F}_q}(xb_{j'}\beta^{iz})$$

for all i, all j, and $x \in L$. As $\beta^n = \gamma$, an equivalent condition is

$$\mathrm{Tr}_{L/\mathbb{F}_q}\left(x\beta^{iz}\left(\gamma^z b_j - \sum_{j'=1}^m a_{jj'}b_{j'}\right)\right) = 0$$

which is equivalent to $\gamma^z b_j = \sum_{j'=1}^m a_{jj'}b_{j'}$; in other words $\mathbf{b}$ (written as a column vector) is an eigenvector for the eigenvalue γ^z of A. □

Here is an algorithmic view again. The constacyclic length n additive codes are direct sums of their constituents, where the constituents correspond to q-cyclotomic cosets in $\mathbb{Z}/un\mathbb{Z}$. Let Z be such a cyclotomic coset, $z \in Z$, $|Z| = s$, $L = \mathbb{F}_{q^s}$, and β, α, γ as usual with $\gamma = \beta^n$. The irreducible subcodes are the contractions of $\mathcal{C}(z, P)$ where $P = (b_1, \ldots, b_m) \in L^m$ has to satisfy $AP = \gamma^z P$ (and we write P as a column vector). This irreducible length n constacyclic code $\mathrm{contr}(\mathcal{C}(z, P))$ has dimension s. Its codewords are $w(x)$ for $x \in L$. The entry of $w(x)$ in outer coordinate $i = 0, 1, \ldots, n-1$ and inner coordinate j is $\mathrm{Tr}_{L/\mathbb{F}_q}(xb_j\beta^{iz})$. The effect of the generator $g = g(I, \ldots, I, A, (0, \ldots, n-1))$ of the cyclic group G is $g(w(x)) = w(\beta^z x)$.

Proposition 9.5.9 *Let $\mathcal{C}$ be a q-linear q^m-ary length n code, which is invariant under $g(I, \ldots, I, A, (0, 1, \ldots, n - 1))$, where $\gcd(n, p) = 1$ and A of order $u = q^m - 1$ represents multiplication by a primitive element γ in $\mathbb{F}_{q^m}$. Then $\mathcal{C}$ is $\mathbb{F}_{q^m}$-linear.*

Proof: It suffices to prove this for irreducible constacyclic codes $\mathrm{contr}(\mathcal{C}(z, P))$. The generic codeword $w(x)$ has been described above. Applying matrix A in each outer coordinate $i = 0, \ldots, n - 1$ yields the codeword $w(\gamma^z x)$ which is still in the code. □

Proposition 9.5.9 applies in particular in the cases $q = 2$, $m = 2$, $u = 3$ and $q = 2$, $m = 3$, $u = 7$.

Corollary 9.5.10 *Each additive code over* $\mathbb{F}_4$, *which is cyclic in the monomial sense, either is linear over* $\mathbb{F}_4$ *or is equivalent to a cyclic code in the permutation sense.*

Proof: We use Proposition 9.3.4 to obtain a code which is invariant under $g = (I, \ldots, I, A, (0, \ldots, n-1))$ where $A \in \mathrm{GL}_2(2)$ and $u = \mathrm{ord}(A)$. If $u = 1$ or $u = 2$ or when $u = 3$ and n is not divisible by 3, then by Proposition 9.3.4 the additive code is cyclic in the permutation sense. In the case when $u = 3$, Proposition 9.5.9 shows that the code is a constacyclic code over $\mathbb{F}_4$. $\qquad\square$

This shows that in the $\mathbb{F}_4$ case the full theory of additive constacyclic codes does not produce anything useful as in each case we are reduced to a more elementary theory, either linear constacyclic over $\mathbb{F}_4$ or additive and cyclic in the permutation sense.

Chapter 10

Convolutional Codes

Julia Lieb

Universität Zürich

Raquel Pinto

Universidade de Aveiro

Joachim Rosenthal

Universität Zürich

10.1   Introduction

Convolutional codes were introduced by Peter Elias [674] in 1955. They can be seen as a generalization of block codes. In order to motivate this generalization, consider a $k \times n$ generator matrix G whose row space generates an $[n, k]$ block code $\mathcal{C}$. Denote by $\mathbb{F}_q$ the finite field with q elements. In case a sequence of message words $m_i \in \mathbb{F}_q^k, i = 1, \ldots, N$ has to be encoded one would transmit the sequence of codewords $c_i = m_i G \in \mathbb{F}_q^n, i = 1, \ldots, N$.

Using polynomial notation and defining

$$m(z) := \sum_{i=1}^{N} m_i z^i \in \mathbb{F}_q[z]^k \quad \text{and} \quad c(z) := \sum_{i=1}^{N} c_i z^i \in \mathbb{F}_q[z]^n,$$

the whole encoding process using the block code $\mathcal{C}$ would be compactly described as

$$m(z) \longmapsto c(z) = m(z)G.$$

Instead of using the constant matrix G as an encoding map, Elias suggested using more general polynomial matrices of the form $G(z)$ whose entries consist of elements of the polynomial ring $\mathbb{F}_q[z]$.

There are natural connections to automata theory and systems theory, and this was first recognized by Massey and Sain in 1967 [1351]. These connections have always been fruitful in the development of the theory on convolutional codes; the reader might also consult the survey [1596].

In the 1970s Forney [740, 741, 742, 743] developed a mathematical theory which allowed the processing of an infinite set of message blocks having the form $m(z) := \sum_{i=1}^{\infty} m_i z^i \in \mathbb{F}_q[[z]]^k$. Note that the quotient field of the ring of formal power series $\mathbb{F}_q[[z]]$ is the field of formal Laurent series $\mathbb{F}_q((z))$, and in the theory of Forney, convolutional codes were defined as k-dimensional linear subspaces of the n-dimensional vector space $\mathbb{F}_q((z))^n$ that also possess a $k \times n$ polynomial generator matrix $G(z) \in \mathbb{F}_q[z]^{k \times n}$. The theory of convolutional codes as first developed by Forney can also be found in the monograph by Piret [1511] and in the textbook by Johanesson and Zigangirov [1051]; McEliece also provides a survey [1363].

In this chapter our starting point is message words of finite length, i.e., polynomial vectors of the form $m(z) := \sum_{i=0}^{N} m_i z^i \in \mathbb{F}_q[z]^k$ which get processed by a polynomial matrix $G(z) \in \mathbb{F}_q[z]^{k \times n}$. The resulting code becomes then in a natural way a rank k module over the polynomial ring $\mathbb{F}_q[z]$. The connection to discrete time linear systems by duality is then also natural as was first shown by Rosenthal, Schumacher, and York [1597].

10.2   Foundational Aspects of Convolutional Codes

10.2.1   Definition of Convolutional Codes via Generator and Parity Check Matrices

Let $R = \mathbb{F}_q[z]$ be the ring of polynomials with coefficients in the field $\mathbb{F}_q$, and denote by $\mathbb{F}_q(z)$ the field of rational functions with coefficients in $\mathbb{F}_q$. R is a principal ideal domain (PID). Modules over a PID admit a basis and two different bases have the same number of elements, called the **rank** of the module.

Throughout this chapter, three notations will be used for vectors of polynomials in R^n. The usual n-tuple notation for $c(z) \in R^n$ will be used: $c(z) = (c_1(z), c_2(z), \ldots, c_n(z))$ where $c_i(z) \in R$ for $1 \leq i \leq n$. Related, $c(z)$ will be written as the $1 \times n$ matrix $c(z) = [c_1(z) \ c_2(z) \ \cdots \ c_n(z)]$. The **degree** of $c(z)$ is defined as $\deg(c(z)) = \max_{1 \leq i \leq n} \deg(c_i(z))$. The third more compact notation will be $c(z) = \sum_{i=0}^{\deg(c(z))} c_i z^i$ where $c_i \in \mathbb{F}_q^n$.

A **convolutional code** $\mathcal{C}$ of **rate** k/n is an R-submodule of R^n of rank k. A $k \times n$ matrix $G(z)$ with entries in R whose rows constitute a basis of $\mathcal{C}$ is called a **generator matrix** for $\mathcal{C}$.

Recall that a $k \times k$ matrix $U(z)$ with entries in R is a **unimodular matrix** if there is a $k \times k$ matrix $V(z)$ with entries in R such that

$$U(z)V(z) = V(z)U(z) = I_k.$$

By Cramer's rule and elementary properties of determinants, it follows that $U(z)$ is unimodular if and only if $\det(U(z)) \in \mathbb{F}_q^* := \mathbb{F}_q \setminus \{0\}$.

Assume that $G(z)$ and $\widetilde{G}(z)$ are both generator matrices of the same code $\mathcal{C} = \mathrm{rowspace}_R(G(z)) = \mathrm{rowspace}_R(\widetilde{G}(z))$. Then we can immediately show that there is a unimodular matrix $U(z)$ such that

$$\widetilde{G}(z) = U(z)G(z).$$

Note that this induces an equivalence relation on the set of $k \times k$ generator matrices: $G(z)$ and $\widetilde{G}(z)$ are **equivalent** if and only if $\widetilde{G}(z) = U(z)G(z)$ for some unimodular matrix $U(z)$. A canonical form for such an equivalence relation is the column Hermite form.

Definition 10.2.1 ([788, 1073]) Let $G(z) \in R^{k \times n}$ with $k \leq n$. Then there exists a unimodular matrix $U(z) \in R^{k \times k}$ such that

$$H(z) = U(z)G(z) = \begin{bmatrix} h_{11}(z) & h_{12}(z) & \cdots & h_{1k}(z) & h_{1,k+1}(z) & \cdots & h_{1n}(z) \\ & h_{22}(z) & \cdots & h_{2k}(z) & h_{2,k+1}(z) & \cdots & h_{2n}(z) \\ & & \ddots & \vdots & \vdots & & \vdots \\ & & & h_{kk}(z) & h_{k,k+1}(z) & \cdots & h_{kn}(z) \end{bmatrix}$$

where $h_{ii}(z)$, $i = 1, 2, \ldots, k$, are monic polynomials such that $\deg h_{ii} > \deg h_{ji}$ for $j < i$. $H(z)$ is the (unique) **column Hermite form** of $G(z)$.

Other equivalence relations are induced by right multiplication with a unimodular matrix or by right and left multiplication with unimodular matrices. Canonical forms of such equivalence relations are the row Hermite form and the Smith form, respectively.

Definition 10.2.2 ([788, 1073]) Let $G(z) \in R^{k \times n}$ with $k \leq n$. Then there exists a unimodular matrix $U(z) \in R^{n \times n}$ such that

$$H(z) = G(z)U(z) = \begin{bmatrix} h_{11}(z) & & & 0 & & 0 \\ h_{21}(z) & h_{22}(z) & & & \vdots & & \vdots \\ \vdots & \vdots & \ddots & & \vdots & & \vdots \\ h_{k1}(z) & h_{k2}(z) & \cdots & h_{kk}(z) & 0 & \cdots & 0 \end{bmatrix}$$

where $h_{ii}(z)$, $i = 1, 2, \ldots, k$, are monic polynomials such that $\deg h_{ii} > \deg h_{ij}$ for $j < i$. $H(z)$ is the (unique) **row Hermite form** of $G(z)$.

Definition 10.2.3 ([788, 1073]) Let $G(z) \in R^{k \times n}$ with $k \leq n$. Then there exist unimodular matrices $U(z) \in R^{k \times k}$ and $V(z) \in R^{n \times n}$ such that

$$S(z) = U(z)G(z)V(z) = \begin{bmatrix} \gamma_1(z) & & & 0 & \cdots & 0 \\ & \gamma_2(z) & & \vdots & & \vdots \\ & & \ddots & \vdots & & \vdots \\ & & & \gamma_k(z) & 0 & & 0 \end{bmatrix}$$

where $\gamma_i(z)$, $i = 1, 2, \ldots, k$, are monic polynomials such that $\gamma_{i+1}(z) \mid \gamma_i(z)$ for $i = 1, 2, \ldots, k-1$. These polynomials are uniquely determined by $G(z)$ and are called **invariant polynomials** of $G(z)$. $S(z)$ is the **Smith form** of $G(z)$.

Since two equivalent generator matrices differ by left multiplication with a unimodular matrix, they have equal $k \times k$ (full size) minors, up to multiplication by a constant. The maximal degree of the full size minors of a generator matrix (called its **internal degree**) of a convolutional code C is called the **degree** (or **complexity**) of C, and it is usually denoted by δ. A convolutional code of rate k/n and degree δ is also called an (n, k, δ) convolutional code. *Throughout this chapter* $L := \lfloor \frac{\delta}{k} \rfloor + \lfloor \frac{\delta}{n-k} \rfloor$.

For $i = 1, \ldots, k$, the largest degree of any entry in row i of a matrix $G(z) \in R^{k \times n}$ is called the ith **row degree** ν_i. It is obvious that if $G(z)$ is a generator matrix and $\nu_1, \nu_2, \ldots, \nu_k$ are the row degrees of $G(z)$, then $\delta \leq \nu_1 + \nu_2 + \cdots + \nu_k$. The sum of the row degrees of $G(z)$ is called its **external degree**. If the internal degree and the external degree coincide, $G(z)$ is said to be **row reduced**, and it is called a **minimal** generator matrix. Thus, the degree of the code can be equivalently defined as the external degree of a minimal generator matrix of C.

Lemma 10.2.4 ([741, 1073]) *Let $G(z) \in R^{k \times n}$ with row degrees $\nu_1, \nu_2, \ldots, \nu_k$, and let $[G]_{hr}$ be the highest row degree coefficient matrix defined as the matrix with the ith row consisting of the coefficients of z^{ν_i} in the ith row of $G(z)$. Then $\delta = \nu_1 + \nu_2 + \cdots + \nu_k$, i.e., $G(z)$ is row reduced, if and only if $[G]_{hr}$ is full row rank.*

Let $G(z)$ be a row reduced generator matrix with row degrees $\nu_1, \nu_2, \ldots, \nu_k$ and $c(z) = u(z)G(z)$ where $u(z) = [u_1(z) \ u_2(z) \ \cdots \ u_k(z)] \in R^k$. Obviously, $\deg c(z) \leq \max\limits_{i:u_i(z) \neq 0} \{\nu_i + \deg u_i(z)\}$. Let $\Lambda = \max\limits_{i:u_i(z) \neq 0} \{\nu_i + \deg u_i(z)\}$ and write $u_i(z) = \alpha_i z^{\Lambda - \nu_i} + r_i(z)$ with $\deg r_i(z) < \Lambda - \nu_i$ for $i = 1, 2, \ldots, k$. Then

$$c(z) = \left(\left[\ \alpha_1 z^{\Lambda - \nu_1} \ \ \alpha_2 z^{\Lambda - \nu_2} \ \ \cdots \ \ \alpha_k z^{\Lambda - \nu_k} \ \right] + \left[\ r_1(z) \ \ r_2(z) \ \ \cdots \ \ r_k(z) \ \right] \right) \times$$

$$\left(\begin{bmatrix} z^{\nu_1} & & & \\ & z^{\nu_2} & & \\ & & \ddots & \\ & & & z^{\nu_k} \end{bmatrix} [G]_{hr} + G_{rem}(z) \right),$$

where $G_{rem}(z) \in R^{k \times n}$ has i^{th} row degree smaller than ν_i for $i = 1, 2, \ldots, k$. The coefficient of $c(z)$ of degree Λ is given by $c_\Lambda = [\alpha_1 \ \alpha_2 \ \cdots \ \alpha_k][G]_{hr}$ which is different from zero since $\alpha_i \neq 0$ for some $i \in \{1, 2, \ldots, k\}$ and $[G]_{hr}$ is full row rank, i.e.,

$$\deg c(z) = \max\limits_{i:u_i(z) \neq 0} \{\nu_i + \deg u_i(z)\}. \tag{10.1}$$

Equality (10.1) is called the **predictable degree property**, and it is an equivalent characterization of the row reduced matrices [741, 1073].

Given a generator matrix $G(z)$, there always exists a row reduced generator matrix equivalent to $G(z)$ [1073]. That is, all convolutional codes admit minimal generator matrices. If $G_1(z)$ and $G_2(z)$ are two equivalent generator matrices, each row of $G_1(z)$ belongs to the image of $G_2(z)$ and vice-versa. Then, if $G_1(z)$ and $G_2(z)$ are row reduced matrices, the predictable degree property implies that $G_1(z)$ and $G_2(z)$ have the same row degrees, up to row permutation.

Another important property of polynomial matrices is left (or right) primeness.

Definition 10.2.5 A polynomial matrix $G(z) \in R^{k \times n}$, with $k \leq n$, is **left prime** if in all factorizations $G(z) = \Delta(z)\overline{G}(z)$ with $\Delta(z) \in R^{k \times k}$ and $\overline{G}(z) \in R^{k \times n}$, the left factor $\Delta(z)$ is unimodular.

Left prime matrices admit several very useful characterizations. Some of these characterizations are presented in the next theorem.

Theorem 10.2.6 ([1073]) *Let $G(z) \in R^{k \times n}$ with $k \le n$. The following are equivalent.*

(a) *$G(z)$ is left prime.*

(b) *The Smith form of $G(z)$ is $[I_k \quad \mathbf{0}_{k \times (n-k)}]$.*

(c) *The row Hermite form of $G(z)$ is $[I_k \quad \mathbf{0}_{k \times (n-k)}]$.*

(d) *$G(z)$ admits a right $n \times k$ polynomial inverse.*

(e) *$G(z)$ can be completed to a unimodular matrix, i.e., there exists $L(z) \in R^{(n-k) \times n}$ such that $\begin{bmatrix} G(z) \\ L(z) \end{bmatrix}$ is unimodular.*

(f) *The ideal generated by all the kth order minors of $G(z)$ is R.*

(g) *For all $u(z) \in \mathbb{F}_q((z))^k$, $u(z)G(z) \in R^n$ implies that $u(z) \in R^k$.*

(h) *$\operatorname{rank} G(\lambda) = k$ for all $\lambda \in \overline{\mathbb{F}}_q$, where $\overline{\mathbb{F}}_q$ denotes the algebraic closure of $\mathbb{F}_q$.*

Since generator matrices of a convolutional code $\mathcal{C}$ differ by left multiplication with a unimodular matrix, it follows that if a convolutional code admits a left prime generator matrix then all its generator matrices are also left prime. We call such codes **noncatastrophic** convolutional codes.

Example 10.2.7 Let us consider the binary field, i.e., $q = 2$. The convolutional code $\mathcal{C}$ of rate 2/3 with generator matrix

$$G(z) = \begin{bmatrix} 1 & 1 & z \\ z^2 & 1 & z+1 \end{bmatrix}$$

is noncatastrophic, since $G(z)$ is left prime by Theorem 10.2.6(d) as $G(z)$ admits the right polynomial inverse $\begin{bmatrix} 0 & 0 \\ z+1 & z \\ 1 & 1 \end{bmatrix}$. The highest coefficient matrix of $G(z)$, $[G]_{hr}$, is full row rank and, consequently, $G(z)$ is row reduced by Lemma 10.2.4. The degree of $\mathcal{C}$ is then equal to the sum of the row degrees of $G(z)$, which is 3. Therefore $\mathcal{C}$ is a $(3,2,3)$ binary convolutional code.

On the other hand, the convolutional code $\widetilde{\mathcal{C}}$ with generator matrix

$$\widetilde{G}(z) = \begin{bmatrix} 1+z & 1 \\ 0 & 1 \end{bmatrix} G(z) = \begin{bmatrix} 1+z+z^2 & z & 1+z^2 \\ z^2 & 1 & z+1 \end{bmatrix}$$

is a catastrophic convolutional code contained in $\mathcal{C}$ as the first equality of the preceding equation implies $\operatorname{rowspace}_R \widetilde{G}(z) \subsetneq \operatorname{rowspace}_R G(z)$. The matrix $[\widetilde{G}]_{hr} = \begin{bmatrix} 1 & 0 & 1 \\ 1 & 0 & 0 \end{bmatrix}$ has full row rank 2, making $\widetilde{G}(z)$ row reduced, and implying that the degree of $\widetilde{\mathcal{C}}$ is $2+2 = 4$. Hence $\widetilde{\mathcal{C}}$ is $(3,2,4)$ convolutional subcode of $\mathcal{C}$.

Let $\mathcal{C}$ be a noncatastrophic convolutional code with generator matrix $G(z) \in R^{k \times n}$. By Theorem 10.2.6(e), there exists a polynomial matrix $N(z) \in R^{(n-k) \times n}$ such that $\begin{bmatrix} G(z) \\ N(z) \end{bmatrix}$ is unimodular. Let $L(z) \in R^{k \times n}$ and $H(z) \in R^{(n-k) \times n}$ such that

$$\begin{bmatrix} G(z) \\ N(z) \end{bmatrix} \begin{bmatrix} L(z)^{\mathsf{T}} & H(z)^{\mathsf{T}} \end{bmatrix} = I_n. \tag{10.2}$$

One immediately sees that

$$c(z) \in \mathcal{C} \text{ if and only if } H(z)c(z)^{\mathsf{T}} = \mathbf{0}^{\mathsf{T}}.$$

$H(z)$ is called a **parity check matrix** of $\mathcal{C}$, analogous to the block code case.

It was shown that if a convolutional code is noncatastrophic, then it admits a parity check matrix. But the converse is also true.

Theorem 10.2.8 ([1932]) *Let $\mathcal{C}$ be a convolutional code of rate k/n. Then there exists a full row rank polynomial matrix $H(z) \in R^{(n-k) \times n}$ such that*

$$\mathcal{C} = \ker H(z) = \{c(z) \in R^n \mid H(z)c(z)^{\mathsf{T}} = \mathbf{0}^{\mathsf{T}}\},$$

i.e., a parity check matrix of $\mathcal{C}$, if and only if $\mathcal{C}$ is noncatastrophic.

Proof: Let us assume that $\mathcal{C}$ admits a parity check matrix $H(z) \in R^{(n-k) \times n}$ and let us write $H(z) = X(z)\widetilde{H}(z)$, where $X(z) \in R^{(n-k) \times (n-k)}$ is full rank and $\widetilde{H}(z) \in R^{(n-k) \times n}$ is left prime. Then there exists a matrix $L(z) \in R^{k \times n}$ such that $\left[\, L(z)^{\mathsf{T}} \;\; \widetilde{H}(z)^{\mathsf{T}} \,\right]$ is unimodular and therefore

$$\begin{bmatrix} G(z) \\ N(z) \end{bmatrix} \begin{bmatrix} L(z)^{\mathsf{T}} & \widetilde{H}(z)^{\mathsf{T}} \end{bmatrix} = I_n,$$

for some left prime matrices $N(z) \in R^{(n-k) \times n}$ and $G(z) \in R^{k \times n}$. Then $\widetilde{H}(z)G(z)^{\mathsf{T}} = \mathbf{0}_{(n-k) \times k}$ and consequently also $H(z)G(z)^{\mathsf{T}} = \mathbf{0}_{(n-k) \times k}$. It is clear that $\mathcal{C} = \mathrm{rowspace}_R G(z)$ and therefore $\mathcal{C}$ is noncatastrophic. $\qquad\square$

Remark 10.2.9 If $\widetilde{\mathcal{C}}$ is catastrophic (i.e., its generator matrices are not left prime), we can still obtain a right prime matrix $H(z) \in R^{(n-k) \times n}$ such that $\widetilde{\mathcal{C}} \subsetneq \ker_R H(z)$. If $\widetilde{G}(z)$ is a generator matrix of $\widetilde{\mathcal{C}}$, we can write $\widetilde{G}(z) = [\Delta(z) \;\; \mathbf{0}_{k \times (n-k)}]U(z)$ with $\Delta(z) \in R^{k \times k}$ where $[\Delta(z) \;\; \mathbf{0}_{k \times (n-k)}]$ is the row Hermite form of $\widetilde{G}(z)$ and $U(z)$ is a unimodular matrix. Then $\widetilde{G}(z) = \Delta(z)U_1(z)$ where $U_1(z)$ is the submatrix of $U(z)$ consisting of its first k rows. This means, by Theorem 10.2.6(e), that $U_1(z)$ is left prime. The matrix $H(z)$ is a parity check matrix of the convolutional code $\mathcal{C} = \mathrm{rowspace}_R U_1(z)$ and $\widetilde{\mathcal{C}} \subsetneq \mathcal{C}$.

Example 10.2.10 Consider the convolutional codes $\mathcal{C}$ and $\widetilde{\mathcal{C}}$ from Example 10.2.7. The matrix $H(z) = \begin{bmatrix} 1 & 1+z+z^3 & 1+z^2 \end{bmatrix}$ is a parity check matrix of $\mathcal{C}$. Since $\widetilde{\mathcal{C}}$ is catastrophic, it does not admit a parity check matrix. However, since $\widetilde{\mathcal{C}} \subsetneq \mathcal{C}$, it follows that $\widetilde{\mathcal{C}} \subsetneq \ker H(z)$.

Given a noncatastrophic code $\mathcal{C}$, we define the **dual** of $\mathcal{C}$ as

$$\mathcal{C}^\perp = \{y(z) \in R^n \mid y(z)c(z)^{\mathsf{T}} = 0 \text{ for all } c(z) \in \mathcal{C}\}.$$

The dual of a noncatastrophic convolutional code is also noncatastrophic. The left prime parity check matrices of $\mathcal{C}$ are the generator matrices of $\mathcal{C}^\perp$ and vice-versa. The degree of a noncatastrophic code and its dual are the same. This result is a consequence of the following lemma and Theorem 10.2.6(f).

Lemma 10.2.11 ([741]) *Let $H(z) \in R^{(n-k) \times n}$ and $G(z) \in R^{k \times n}$ be a left prime parity check matrix and a generator matrix of a noncatastrophic convolutional code, respectively. Given a full size minor of $G(z)$ consisting of the columns $i_1, i_2, \ldots, i_k$, let us define the complementary full size minor of $H(z)$ as the minor consisting of the complementary columns, i.e., by the columns $\{1, 2, \ldots, n\} \setminus \{i_1, i_2, \ldots, i_k\}$. Then the full size minors of $G(z)$ are equal to the complementary full size minors of $H(z)$, up to multiplication by a nonzero constant.*

Proof: For simplicity, let us consider $i_1 = 1, i_2 = 2, \ldots, i_k = k$, i.e., the full size minor of $G(z)$, $M_1(G)$, consisting of the first k columns:

$$M_1(G) = \det \begin{bmatrix} G(z) \\ \mathbf{0}_{(n-k)\times k} \quad I_{n-k} \end{bmatrix}.$$

Then considering $L(z)$ as in (10.2), we have that

$$\begin{bmatrix} G(z) \\ \mathbf{0}_{(n-k)\times k} \quad I_{n-k} \end{bmatrix} \begin{bmatrix} L(z)^\mathsf{T} & H(z)^\mathsf{T} \end{bmatrix} = \begin{bmatrix} I_k & \mathbf{0}_{k\times(n-k)} \\ Q(z) & \widetilde{H}(z) \end{bmatrix} \tag{10.3}$$

where $Q(z) \in R^{(n-k)\times k}$ and $\widetilde{H}(z)$ is the submatrix of $H(z)$ consisting of its last $n-k$ columns, i.e., $\overline{M}_1(H) = \det \widetilde{H}(z)$ is the complementary minor to $M_1(G)$, and from (10.3) we conclude that $M_1(G) = \alpha \overline{M}_1(H)$, where $\alpha = \det \begin{bmatrix} L(z)^\mathsf{T} & H(z)^\mathsf{T} \end{bmatrix}$ belongs to $\mathbb{F}_q^*$. Applying the same reasoning we conclude that all the full size minors of $G(z)$ are equal to α times the complementary full size minors of $H(z)$. $\square$

Therefore if $G(z)$ is a generator matrix and $H(z)$ is a parity check matrix of a noncatastrophic convolutional code $\mathcal{C}$, they have the same maximal degree of the full size minors, which means that $\mathcal{C}$ and $\mathcal{C}^\perp$ have the same degree.

10.2.2   Distances of Convolutional Codes

The minimum distance of a code is an important measure of robustness of the code since it provides a means to assess its capability to protect data from errors. Several types of distance can be defined for convolutional codes. We will consider the free distance and the column distances. To define these notions, one first has to define the distance between polynomial vectors.

Definition 10.2.12 The **(Hamming) weight** $\mathrm{wt_H}(c)$ of $c \in \mathbb{F}_q^n$ is defined as the number of nonzero components of c, and the **weight** of a polynomial vector $c(z) = \sum_{t=0}^{\deg(c(z))} c_t z^t \in R^n$ is defined as $\mathrm{wt_H}(c(z)) = \sum_{t=0}^{\deg(c(z))} \mathrm{wt_H}(c_t)$. The **(Hamming) distance** between $c_1, c_2 \in \mathbb{F}_q^n$ is defined as $\mathrm{d_H}(c_1, c_2) = \mathrm{wt_H}(c_1 - c_2)$; correspondingly the **distance** between $c_1(z), c_2(z) \in R^n$ is defined as $\mathrm{d_H}(c_1(z), c_2(z)) = \mathrm{wt_H}(c_1(z) - c_2(z))$,

Definition 10.2.13 The **free distance** of a convolutional code $\mathcal{C}$ is given by

$$d_{free}(\mathcal{C}) := \min_{c_1(z), c_2(z) \in \mathcal{C}} \{\mathrm{d_H}(c_1(z), c_2(z)) \mid c_1(z) \neq c_2(z)\}.$$

During transmission of information over a q-ary symmetric channel, errors may occur; i.e., information symbols can be exchanged by other symbols in $\mathbb{F}_q$ in a symmetric way. After channel transmission, a convolutional code $\mathcal{C}$ can detect up to s errors in any received word $w(z)$ if $d_{free}(\mathcal{C}) \geq s+1$ and can correct up to t errors in $w(z)$ if $d_{free}(\mathcal{C}) \geq 2t+1$, which gives the following theorem.

Theorem 10.2.14 *Let $\mathcal{C}$ be a convolutional code with free distance d. Then $\mathcal{C}$ can always detect $d-1$ errors and correct $\lfloor \frac{d-1}{2} \rfloor$ errors.*

As convolutional codes are linear, the difference between two codewords is also a codeword which gives the following equivalent definition of free distance.

Lemma 10.2.15 *The free distance of a convolutional code* $\mathcal{C}$ *is given by*

$$d_{free}(\mathcal{C}) := \min_{c(z) \in \mathcal{C}} \left\{ \sum_{t=0}^{\deg(c(z))} \mathrm{wt}_H(c_t) \ \Big| \ c(z) \neq \mathbf{0} \right\}.$$

Example 10.2.16 Consider the convolutional code $\mathcal{C}$ defined in Example 10.2.7. Since $[1 \ 1 \ z]$ is a codeword of $\mathcal{C}$ with weight 3, it follows that $d_{free}(\mathcal{C}) \leq 3$. Let $w(z) \in \mathcal{C}$ with $w(z) \neq \mathbf{0}$. So $w(z) = \sum_{i=\ell_0}^{\ell_1} w_i z^i$ for some $\ell_0, \ell_1 \in \mathbb{N}_0$ such that $w_{\ell_0} \neq 0$ and $w_{\ell_1} \neq 0$. Then $w_{\ell_0} \in \mathrm{rowspace}_R G(0)$, and consequently $w(z)$ has weight 2 or 3. Moreover, the predictable degree property (10.1) of $G(z)$ implies that $\ell_1 > \ell_0$, and therefore $w(z)$ must have weight at least 3. Hence, we conclude that $\mathcal{C}$ has free distance 3.

Besides the free distance, convolutional codes also possess another notion of distance, the so-called *column distances*. These distances have an important role in transmission of information over an erasure channel, which is a suitable model for many communication channels, in particular packet switched networks such as the internet. In this kind of channel, each symbol either arrives correctly or does not arrive at all, and it is called an erasure. In Section 10.5.1 the decoding over these type of channels will be analyzed in detail.

Column distances have important characterizations in terms of the generator matrices of the code, but also in terms of its parity check matrices if the code is noncatastrophic. For this reason, we will consider throughout this section noncatastrophic convolutional codes.

Definition 10.2.17 For $c(z) \in R^n$ with $\deg(c(z)) = \gamma$, write $c(z) = c_0 + \cdots + c_\gamma z^\gamma$ with $c_t \in \mathbb{F}_q^n$ for $t = 0, \ldots, \gamma$ and set $c_t = \mathbf{0} \in \mathbb{F}_q^n$ for $t \geq \gamma + 1$. We define the jth **truncation** of $c(z)$ as $c_{[0,j]}(z) = c_0 + c_1 z + \cdots + c_j z^j$.

Definition 10.2.18 ([809]) For $j \in \mathbb{N}_0$, the jth **column distance** of a convolutional code $\mathcal{C}$ is defined as

$$d_j^c(\mathcal{C}) := \min_{c(z) \in \mathcal{C}} \left\{ \mathrm{wt}_H(c_{[0,j]}(z)) \mid c_0 \neq \mathbf{0} \right\}.$$

Let $G(z) = \sum_{i=0}^{\mu} G_i z^i \in R^{k \times n}$ and $H(z) = \sum_{i=0}^{\nu} H_i z^i \in R^{(n-k) \times n}$ be a generator matrix and a parity check matrix, respectively, of the convolutional code $\mathcal{C}$. Note that $G_i \in \mathbb{F}_q^{k \times n}$ and $H_i \in \mathbb{F}_q^{(n-k) \times n}$. For $j \in \mathbb{N}_0$, define the **truncated sliding generator matrices** $G_j^c \in \mathbb{F}_q^{(j+1)k \times (j+1)n}$ and the **truncated sliding parity check matrices** $H_j^c \in \mathbb{F}_q^{(j+1)(n-k) \times (j+1)n}$ as

$$G_j^c := \begin{bmatrix} G_0 & G_1 & \cdots & G_j \\ & G_0 & \cdots & G_{j-1} \\ & & \ddots & \vdots \\ & & & G_0 \end{bmatrix} \quad \text{and} \quad H_j^c := \begin{bmatrix} H_0 & & & \\ H_1 & H_0 & & \\ \vdots & \vdots & \ddots & \\ H_j & H_{j-1} & \cdots & H_0 \end{bmatrix}.$$

Then if $c(z) = \sum_{i \in \mathbb{N}_0} c_i z^i$ is a codeword of $\mathcal{C}$, it follows that

$$[c_0 \ c_1 \cdots c_j] = [u_0 \ u_1 \cdots u_j] G_j^c \quad \text{for some } u_0, u_1, \ldots, u_j \in \mathbb{F}_q^k \tag{10.4}$$

and $H_j^c [c_0 \ c_1 \ \cdots \ c_j]^\mathsf{T} = \mathbf{0}^\mathsf{T}$. Note that since $G(z)$ is left prime, G_0 has full row rank, and therefore $c_0 \neq \mathbf{0}$ in (10.4) if and only if $u_0 \neq \mathbf{0}$. Consequently,

$$d_j^c(\mathcal{C}) = \min_{u_0 \neq \mathbf{0}} \left\{ \mathrm{wt}_H([u_0 \ \cdots \ u_j] G_j^c) \right\} = \min_{c_0 \neq \mathbf{0}} \left\{ \mathrm{wt}_H(c_{[0,j]}(z)) \mid H_j^c [c_0 \ \cdots \ c_j]^\mathsf{T} = \mathbf{0}^\mathsf{T} \right\},$$

and one has the following theorem.

Theorem 10.2.19 ([809]) *For $d \in \mathbb{N}$ the following statements are equivalent.*

(a) $d_j^c(\mathcal{C}) = d$.

(b) *None of the first n columns of H_j^c is contained in the span of any other $d-2$ columns and one of the first n columns is contained in the span of some other $d-1$ columns of that matrix.*

Example 10.2.20 The convolutional code $\mathcal{C}$ defined in Example 10.2.7 has column distances $d_0^c(\mathcal{C}) = \min_{u_0 \neq \mathbf{0}} \{\mathrm{wt}_H(u_0 G_0^c)\} = 2$, and for $j \geq 1$,

$$d_j^c(\mathcal{C}) = \min_{u_0 \neq \mathbf{0}} \{\mathrm{wt}_H([u_0 \ \cdots \ u_j] G_j^c)\}$$

$$= \mathrm{wt}_H([u_0 \ \cdots \ u_j] G_j^c) = 3 \ \text{ with } u_0 = [1 \ 0], \ u_1 = u_2 = \cdots = u_j = \mathbf{0}.$$

As for block codes, there exist upper bounds for the distances of convolutional codes.

Theorem 10.2.21 *Let $\mathcal{C}$ be an (n, k, δ) convolutional code. Then*

(a) ([1598]) $d_{free}(\mathcal{C}) \leq (n-k)\left(\lfloor \frac{\delta}{k} \rfloor + 1\right) + \delta + 1$, *and*

(b) ([809]) $d_j^c(\mathcal{C}) \leq (n-k)(j+1) + 1$ *for all $j \in \mathbb{N}_0$.*

The bound in (a) is called the **Generalized Singleton Bound** since for $\delta = 0$ one gets the Singleton Bound for block codes.

An (n, k, δ) convolutional code $\mathcal{C}$ such that

$$d_{free}(\mathcal{C}) = (n-k)\left(\left\lfloor \frac{\delta}{k} \right\rfloor + 1\right) + \delta + 1$$

is called a **maximum distance separable (MDS)** code [1598]. In [1598] it was proved that an (n, k, δ) convolutional code always exists over a sufficiently large field. In Section 10.3.1 constructions of such codes are presented.

The Generalized Singleton Bound has implications on the values that the column distances can achieve. Note that $0 \leq d_0^c(\mathcal{C}) \leq d_1^c(\mathcal{C}) \leq \cdots \leq d_{free}(\mathcal{C})$ and $d_{free}(\mathcal{C}) = \lim_{j \to \infty} d_j^c(\mathcal{C})$, which implies $d_j^c(\mathcal{C}) \leq (n-k)\left(\lfloor \frac{\delta}{k} \rfloor + 1\right) + \delta + 1$ for all $j \in \mathbb{N}_0$. Hence $j = L := \lfloor \frac{\delta}{k} \rfloor + \lfloor \frac{\delta}{n-k} \rfloor$ is the largest possible value of j for which $d_j^c(\mathcal{C})$ can attain the upper bound in Theorem 10.2.21(b). Moreover, the following lemma shows that maximal j^{th} column distance implies maximal column distance of all previous ones.

Lemma 10.2.22 ([809]) *Let $\mathcal{C}$ be an (n, k, δ) convolutional code. If $d_j^c(\mathcal{C}) = (n-k)(j+1) + 1$ for some $j \in \{1, \ldots, L\}$, then $d_i^c(\mathcal{C}) = (n-k)(i+1) + 1$ for all $i \leq j$.*

Definition 10.2.23 ([809]) An (n, k, δ) convolutional code $\mathcal{C}$ is said to be **maximum distance profile (MDP)** if

$$d_j^c(\mathcal{C}) = (n-k)(j+1) + 1 \ \text{ for } j = 0, \ldots, L = \left\lfloor \frac{\delta}{k} \right\rfloor + \left\lfloor \frac{\delta}{n-k} \right\rfloor.$$

Lemma 10.2.22 shows that it is sufficient to have equality for $j = L$ in Theorem 10.2.21(b) to get an MDP convolutional code.

A convolutional code $\mathcal{C}$ where $d_j^c(\mathcal{C})$ meets the Generalized Singleton Bound for the smallest possible value of j is called **strongly maximum distance separable (sMDS)**. Note that either $j = L$ or $j = L+1$. More precisely, an (n, k, δ) convolutional code is sMDS if

$$d_M^c(\mathcal{C}) = (n-k)\left(\left\lfloor \frac{\delta}{k} \right\rfloor + 1\right) + \delta + 1 \ \text{ where } M := \left\lfloor \frac{\delta}{k} \right\rfloor + \left\lceil \frac{\delta}{n-k} \right\rceil.$$

The next remark points out the relationship between MDP, MDS and strongly MDS convolutional codes.

Remark 10.2.24 ([1015])

(a) Each sMDS code is an MDS code.

(b) If $n - k$ divides δ, an (n, k, δ) convolutional code $\mathcal{C}$ is MDP if and only if it is sMDS.

In the following, we will provide criteria to check whether a convolutional code is MDP.

Theorem 10.2.25 ([809]) *Let $\mathcal{C}$ be a convolutional code with generator matrix $G(z) = \sum_{i=0}^{\mu} G_i z^i \in R^{k \times n}$ and with left prime parity check matrix $H(z) = \sum_{i=0}^{\nu} H_i z^i \in R^{(n-k) \times n}$. The following statements are equivalent.*

(a) $\mathcal{C}$ *is MDP.*

(b) $\mathcal{G}_L := \begin{bmatrix} G_0 & \cdots & G_L \\ & \ddots & \vdots \\ 0 & & G_0 \end{bmatrix}$, *with $G_i = \mathbf{0}_{k \times n}$ for $i > \mu$, has the property that every full size minor formed by rows with indices $1 \le j_1 < \cdots < j_{(L+1)k} \le (L+1)n$ which fulfills $j_{sk} \le sn$ for $s = 1, \ldots, L$ is nonzero.*

(c) $\mathcal{H}_L := \begin{bmatrix} H_0 & & 0 \\ \vdots & \ddots & \\ H_L & \cdots & H_0 \end{bmatrix}$, *with $H_i = \mathbf{0}_{(n-k) \times n}$ for $i > \nu$, has the property that every full size minor formed by columns with indices $1 \le j_1 < \cdots < j_{(L+1)(n-k)} \le (L+1)n$ which fulfills $j_{s(n-k)} \le sn$ for $s = 1, \ldots, L$ is nonzero.*

The property of being an MDP convolutional code is invariant under duality as shown in the following theorem.

Theorem 10.2.26 ([809]) *An (n, k, δ) convolutional code is MDP if and only if its dual code, which is an $(n, n - k, \delta)$ convolutional code, is MDP.*

MDP convolutional codes are very efficient for decoding over the erasure channel. Next we introduce two special classes of MDP convolutional codes, the reverse MDP convolutional codes and the complete MDP convolutional codes, which are especially suited to deal with particular patterns of erasures. Section 10.5.1 is devoted to decoding over this channel, and the decoding capabilities of these codes will be analyzed in more detail.

Definition 10.2.27 ([1014]) Let $\mathcal{C}$ be an (n, k, δ) convolutional code with left prime row reduced generator matrix $G(z) = [g_{ij}(z)]$. Set $\overline{g}_{ij}(z) := z^{\nu_i} g_{ij}(z^{-1})$ where ν_i is the ith row degree of $G(z)$. Then the code $\overline{\mathcal{C}}$ with generator matrix $\overline{G}(z) = [\overline{g}_{ij}(z)]$ is also an (n, k, δ) convolutional code, called the **reverse code** to $\mathcal{C}$. It follows that $c_0 + \cdots + c_d z^d \in \overline{\mathcal{C}}$ if and only if $c_d + \cdots + c_0 z^d \in \mathcal{C}$.

Definition 10.2.28 ([1810]) Let $\mathcal{C}$ be an MDP convolutional code. If $\overline{\mathcal{C}}$ is also MDP, $\mathcal{C}$ is called a **reverse MDP** convolutional code.

Remark 10.2.29 ([1810]) Let $\mathcal{C}$ be an (n, k, δ) MDP convolutional code such that $(n - k) \mid \delta$ and $H(z) = H_0 + \cdots + H_\nu z^\nu$, with $H_\nu \ne 0$, be a left prime and row reduced parity check matrix of $\mathcal{C}$. Then the reverse code $\overline{\mathcal{C}}$ has parity check matrix $\overline{H}(z) = H_\nu + \cdots + H_0 z^\nu$. Moreover, $\mathcal{C}$ is reverse MDP if and only if every full size minor of the matrix

$$\mathfrak{H}_L := \begin{bmatrix} H_\nu & \cdots & H_{\nu-L} \\ & \ddots & \vdots \\ 0 & & H_\nu \end{bmatrix}$$

formed from the columns with indices $j_1, \ldots, j_{(L+1)(n-k)}$ with $j_{s(n-k)+1} > sn$ for $s = 1, \ldots, L$ is nonzero.

Theorem 10.2.30 ([1810]) *An (n, k, δ) reverse MDP convolutional code exists over a sufficiently large base field.*

Definition 10.2.31 ([1810]) Let $H(z) = H_0 + H_1 z + \cdots + H_\nu z^\nu \in R^{(n-k) \times n}$, with $H_\nu \neq 0_{(n-k) \times n}$, be a left prime and row reduced parity check matrix of the (n, k, δ) convolutional code $\mathcal{C}$. Set $L = \lfloor \frac{\delta}{n-k} \rfloor + \lfloor \frac{\delta}{k} \rfloor$. Then

$$
\mathfrak{H} := \begin{bmatrix} H_\nu & \cdots & H_0 & & 0 \\ & \ddots & & \ddots & \\ 0 & & H_\nu & \cdots & H_0 \end{bmatrix} \in \mathbb{F}_q^{(L+1)(n-k) \times (\nu+L+1)n}
$$

is called a **partial parity check matrix** of the code. Moreover, $\mathcal{C}$ is called a **complete MDP** convolutional code if every full size minor of $\mathfrak{H}$ that is formed by columns $j_1, \ldots, j_{(L+1)(n-k)}$ with $j_{(n-k)s+1} > sn$ and $j_{(n-k)s} \leq sn + \nu n$ for $s = 1, \ldots, L$ is nonzero.

Remark 10.2.32

(a) [1810] Every complete MDP convolutional code is a reverse MDP convolutional code.

(b) [1255] A complete (n, k, δ) MDP convolutional code exists over a sufficiently large base field if and only if $(n - k) \mid \delta$.

10.3 Constructions of Codes with Optimal Distance

10.3.1 Constructions of MDS Convolutional Codes

In this section, we will present the most important known constructions for MDS convolutional codes. They differ in the constraints on the parameters and the necessary field size. The first two constructions that will be considered are for convolutional codes with rate $1/n$. The following theorem gives the first construction of MDS convolutional codes.

Theorem 10.3.1 ([1064]) *For $n \geq 2$ and $|\mathbb{F}_q| \geq n + 1$, set $s_j := \lceil (j-1)(|\mathbb{F}_q| - 1)/n \rceil$ for $j = 2, \ldots, n$ and*

$$
\delta := \begin{cases} \lfloor \frac{2}{9} |\mathbb{F}_q| \rfloor & \text{if } n = 2, \\ \lfloor \frac{1}{3} |\mathbb{F}_q| \rfloor & \text{if } 3 \leq n \leq 5, \\ \lfloor \frac{1}{2} |\mathbb{F}_q| \rfloor & \text{if } n \geq 6. \end{cases}
$$

Moreover, let α be a primitive element of $\mathbb{F}_q$, and set $g_1(x) := \prod_{k=1}^{\delta} (x - \alpha^k)$, $g_j(x) := g_1(x \alpha^{-s_j})$ for $j = 2, \ldots, n$. Then $G(z) = [g_1(z) \cdots g_n(z)]$ is the generator matrix of an $(n, 1, \delta)$ MDS convolutional code with free distance equal to $n(\delta + 1)$.

The second construction works for the same field size as the first one but contains a restriction on the degree of the code, which is different from the restriction on the degree in the first construction.

Theorem 10.3.2 ([807]) *Let $|\mathbb{F}_q| \geq n + 1$, $0 \leq \delta \leq n - 1$ and α be an element of $\mathbb{F}_q$ with order at least n. Then $G(z) = \sum_{i=0}^{\delta} z^i [1 \ \alpha \ \alpha^2 \ \cdots \ \alpha^{(n-1)i}]$ generates an $(n, 1, \delta)$ MDS convolutional code.*

Finally, we present a construction that works for arbitrary parameters but has a stronger restriction on the field size.

Theorem 10.3.3 ([1728]) *Let a, r be integers such that $a \geq \lfloor \frac{\delta}{k} \rfloor + 1 + \frac{\delta}{n-k}$ and $an = p^r - 1$. Let α be a primitive element of $\mathbb{F}_{p^r}$. Set $N = an$, $K = N - (n - k)\left(\lfloor \frac{\delta}{k} \rfloor + 1\right) - \delta$, $g(z) = (z - \alpha^0)(z - \alpha^1) \cdots (z - \alpha^{N-K-1})$ and write $g(z) = g_0(z^n) + g_1(z^n)z + \cdots + g_{n-1}(z^n)z^{n-1}$. Then*

$$G(z) = \begin{bmatrix} g_0(z) & g_1(z) & g_2(x) & \cdots & \cdots & \cdots & g_{n-1}(z) \\ zg_{n-1}(z) & g_0(z) & g_1(z) & \cdots & \cdots & \cdots & g_{n-2}(z) \\ zg_{n-2}(z) & zg_{n-1}(z) & g_0(z) & \cdots & \cdots & \cdots & g_{n-3}(z) \\ \vdots & \vdots & & \ddots & \ddots & & \vdots \\ zg_{n-k+1}(z) & zg_{n-k+2}(z) & \cdots & zg_{n-1} & g_0(z) & \cdots & g_{n-k}(z) \end{bmatrix}$$

is the generator matrix of an (n, k, δ) MDS convolutional code.

10.3.2 Constructions of MDP Convolutional Codes

MDP convolutional codes can be constructed by selecting appropriate columns and rows of so-called *superregular matrices*, which we define in the following.

Definition 10.3.4 *Let $l \in \mathbb{N}$.*

(a) Let $A = [a_{ij}]$ be a matrix in $\mathbb{F}_q^{l \times l}$. Define $\bar{a} = [\bar{a}_{ij}]$ where $\bar{a}_{ij} = 0$ if $a_{ij} = 0$ and $\bar{a}_{ij} = x_{ij}$ if $a_{ij} \neq 0$ and consider the determinant of $\bar{A}$ as an element of the ring of polynomials in the variables x_{ij}, $i, j \in \{1, \ldots, l\}$ and with coefficients in $\mathbb{F}_q$. One calls the determinant of A **trivially zero** if the determinant of $\bar{A}$ is equal to the zero polynomial. A is called **superregular** if all its not trivially zero minors are nonzero.

(b) A Toeplitz matrix of the form $\begin{bmatrix} a_1 & & 0 \\ \vdots & \ddots & \\ a_l & \cdots & a_1 \end{bmatrix}$ with $a_i \in \mathbb{F}_q$ for $i = 1, \ldots, l$ that is superregular is called a **lower triangular superregular** matrix.

In the following, we present two constructions for (n, k, δ) MDP convolutional codes using superregular matrices of different shapes. For the first construction, which is presented in the following theorem, it is required that $(n - k) \,|\, \delta$ and $k > \delta$, i.e., $L = \frac{\delta}{n-k}$.

Theorem 10.3.5 ([1810]) *Let $(n - k) \,|\, \delta$, $k > \delta$ and T be an $r \times r$ lower triangular superregular matrix with $r = (L + 1)(2n - k - 1)$. For $j = 0, \ldots, L$, let I_j and J_j be the following sets:*

$$I_j = \{(j + 1)n + j(n - k - 1), \ldots, (j + 1)(2n - k - 1)\},$$
$$J_j = \{jn + j(n - k - 1) + 1, \ldots, (j + 1)n + j(n - k - 1)\},$$

and let $I = \bigcup_{j=0}^{L} I_j$ and $J = \bigcup_{j=0}^{L} J_j$. Form $\mathcal{H}_L = \begin{bmatrix} H_0 & & 0 \\ \vdots & \ddots & \\ H_L & \cdots & H_0 \end{bmatrix}$ taking the rows and columns of T with indices in I and J, respectively. Then $H(z) = \sum_{i=0}^{L} H_i z^i$ is the parity check matrix of an MDP convolutional code.

The construction of the preceding theorem could be explained using the following steps:

Step 1: Construct the lower triangular superregular matrix T.

Step 2: Partition T into $L + 1$ blocks with $2n - k - 1$ rows each and delete the first $n - 1$ rows in each block. Define $\widehat{T}$ as the matrix consisting of the remaining rows.

Step 3: Partition $\widehat{T}$ into $L + 1$ blocks with $2n - k - 1$ columns each and delete the first $n - k - 1$ columns in each block. Define $\mathcal{H}_L$ as the matrix consisting of the remaining columns.

The following theorem provides a general construction for such a superregular matrix if the characteristic of the underlying field is sufficiently large.

Theorem 10.3.6 ([809]) *For every $b \in \mathbb{N}$ the not trivially zero minors of the Toeplitz*

$$
matrix \begin{bmatrix} 1 & 0 & \cdots & \cdots & 0 \\ \binom{b}{1} & \ddots & \ddots & & \vdots \\ \vdots & \ddots & \ddots & \ddots & \vdots \\ \binom{b}{b-1} & & \ddots & \ddots & 0 \\ 1 & \binom{b}{b-1} & \cdots & \binom{b}{1} & 1 \end{bmatrix} \in \mathbb{Z}^{b \times b} \ \text{are all positive. Hence for each } b \in \mathbb{N} \text{ there}
$$

exists a smallest prime number p_b such that this matrix is superregular over the prime field $\mathbb{F}_{p_b}$.

The second construction for MDP convolutional code also requires large field sizes but has the advantage that it works for arbitrary characteristic of the underlying field as well as for arbitrary code parameters.

Theorem 10.3.7 ([39]) *Let n, k, δ be given integers and $m := \max\{n - k, k\}$. Let α be a primitive element of a finite field $\mathbb{F}_{p^N}$, where p is prime and N is an integer, and define*

$$
T_i := \begin{bmatrix} \alpha^{2^{im}} & \alpha^{2^{im+1}} & \cdots & \alpha^{2^{(i+1)m-1}} \\ \alpha^{2^{im+1}} & \alpha^{2^{im+2}} & & \alpha^{2^{(i+1)m}} \\ \vdots & & \ddots & \vdots \\ \alpha^{2^{(i+1)m-1}} & \alpha^{2^{(i+1)m}} & \cdots & \alpha^{2^{(i+2)m-2}} \end{bmatrix} \ \text{for } i = 1, \ldots, L = \left\lfloor \frac{\delta}{k} \right\rfloor + \left\lfloor \frac{\delta}{n-k} \right\rfloor. \ \text{Define}
$$

$$
\mathcal{T}(T_0, \ldots, T_L) := \begin{bmatrix} T_0 & & 0 \\ \vdots & \ddots & \\ T_L & \cdots & T_0 \end{bmatrix} \in \mathbb{F}_{p^N}^{(L+1)m \times (L+1)m}.
$$

If $N \geq 2^{m(L+2)-1}$, then the matrix $\mathcal{T}(T_0, T_1, \ldots, T_L)$ is superregular over $\mathbb{F}_{p^N}$.

The following theorem provides a construction for (n, k, δ) MDP convolutional codes with $(n - k) \,|\, \delta$ using the superregular matrices from the preceding theorem.

Theorem 10.3.8 ([39]) *Let n, k, δ be integers such that $(n - k) \,|\, \delta$ with $\nu = \frac{\delta}{n-k}$, and let $T_l = [t^l_{ij}]$ for $i, j = 1, 2, \ldots, m$ and $l = 0, 1, 2, \ldots, L$ be the entries of the matrix T_l as in Theorem 10.3.7. Define $\overline{H}_l = [t^l_{ij}]$ for $i = 1, 2, \ldots, n - k$, $j = 1, 2, \ldots, k$ and $l = 0, 1, 2, \ldots, L$. If $q = p^N$ for $N \in \mathbb{N}$ and $|\mathbb{F}_q| \geq p^{2m(L+1)+n-2}$, let $\mathcal{C}$ be the convolutional code $\mathcal{C} = \ker_R[A(z) \ B(z)]$ where $A(z) = \sum_{i=0}^{\nu} A_i z^i \in R^{(n-k) \times (n-k)}$ and $B(z) = \sum_{i=0}^{\nu} B_i z^i \in R^{(n-k) \times k}$ are determined as follows:*

- $A_0 = I_{n-k}$ and $A_i \in \mathbb{F}_q^{(n-k)\times(n-k)}$, for $i = 1,\ldots,\nu$, are obtained by solving the equations

$$[A_\nu \;\cdots\; A_1] \begin{bmatrix} \overline{H}_{l-\nu} & \cdots & \overline{H}_1 \\ \vdots & & \vdots \\ \overline{H}_{L-1} & \cdots & \overline{H}_\nu \end{bmatrix} = -[\overline{H}_L \;\cdots\; \overline{H}_{\nu+1}].$$

- $B_i = A_0\overline{H}_i + A_1\overline{H}_{i-1} + \cdots + A_i\overline{H}_0$ for $i = 0,\ldots,\nu$.

Then $\mathcal{C}$ is an (n,k,δ) MDP convolutional code.

In [1414] this construction was generalized to arbitrary code parameters, where $(n-k)\,|\,\delta$ is not necessary. In this way, the authors of [1414] obtained constructions of (n,k,δ) convolutional codes that are both MDP and sMDS convolutional codes. Other constructions for convolutional codes can be found in [809]. There are two general constructions of complete MDP convolutional codes, similar to the constructions for MDP convolutional codes given in Theorem 10.3.5 and Theorem 10.3.8, presented in the following two theorems.

Theorem 10.3.9 ([1255]) *Let* $n,k,\delta \in \mathbb{N}$ *such that* $k < n$ *and* $(n-k)\,|\,\delta$, *and let* $\nu = \frac{\delta}{n-k}$. *Then* $H(z) = \sum_{i=0}^{\nu} H_i z^i$ *with*

$$H_0 = \begin{bmatrix} \binom{\nu n+k}{k} & \cdots & 1 & & 0 \\ \vdots & & & \ddots & \\ \binom{\nu n+k}{n-1} & & \cdots & & 1 \end{bmatrix},$$

$$H_i = \begin{bmatrix} \binom{\nu n+k}{in+k} & \cdots & \binom{\nu n+k}{(i-1)n+k+1} \\ \vdots & & \vdots \\ \binom{\nu n+k}{(i+1)n-1} & \cdots & \binom{\nu n+k}{in} \end{bmatrix} \quad \text{for } i = 1,\ldots,\nu-1, \quad \text{and}$$

$$H_\nu = \begin{bmatrix} 1 & & \cdots & & \binom{\nu n+k}{n-1} \\ & \ddots & & & \vdots \\ 0 & & 1 & \cdots & \binom{\nu n+k}{k} \end{bmatrix}$$

is the parity check matrix of an (n,k,δ) *complete MDP convolutional code if the characteristic of the base field is greater than* $\binom{\nu n+k}{\lfloor(\nu n+k)/2\rfloor}^{(n-k)(L+1)} \cdot ((n-k)(L+1))^{(n-k)(L+1)/2}$ *where* $L = \lfloor\frac{\delta}{k}\rfloor + \lfloor\frac{\delta}{n-k}\rfloor$.

Theorem 10.3.10 ([1255]) *Let* $n,k,\delta \in \mathbb{N}$ *with* $k < n$ *and* $(n-k)\,|\,\delta$. *With* $L = \lfloor\frac{\delta}{k}\rfloor + \lfloor\frac{\delta}{n-k}\rfloor$, *let* α *be a primitive element of a finite field* $\mathbb{F}_{p^N}$ *where* p *is a prime and* $N > (L + 1)\cdot 2^{(\nu+2)n-k-1}$. *Then* $H(z) = \sum_{i=0}^{\nu} H_i z^i$ *with* $H_i = \begin{bmatrix} \alpha^{2^{in}} & \cdots & \alpha^{2^{(i+1)n-1}} \\ \vdots & & \vdots \\ \alpha^{2^{(i+1)n-k-1}} & \cdots & \alpha^{2^{(i+2)n-k-2}} \end{bmatrix}$ *for*

$i = 0,\ldots,\nu = \frac{\delta}{n-k}$ *is the parity check matrix of an* (n,k,δ) *complete MDP convolutional code.*

10.4 Connections to Systems Theory

The aim of this section is to explain the correspondence between convolutional codes and discrete-time linear systems [1600] of the form

$$\begin{cases} x_{t+1} = x_t A + u_t B \\ y_t = x_t C + u_t D \\ c_t = [y_t \ u_t] \end{cases} \tag{10.5}$$

where $(A, B, C, D) \in \mathbb{F}_q^{s \times s} \times \mathbb{F}_q^{k \times s} \times \mathbb{F}_q^{s \times (n-k)} \times \mathbb{F}_q^{k \times (n-k)}$, $t \in \mathbb{N}_0$ and $s, k, n \in \mathbb{N}$ when $n > k$. The system (10.5) will be represented by $\Sigma = (A, B, C, D)$, and the integer s is called its **dimension**. We call $x_t \in \mathbb{F}_q^s$ the **state vector**, $u_t \in \mathbb{F}_q^k$ the **information vector**, $y_t \in \mathbb{F}_q^{n-k}$ the **parity vector** and $c_t \in \mathbb{F}_q^n$ the **code vector**. We consider that the initial state is the zero vector, i.e., $x_0 = \mathbf{0}$.

The input, state, and output sequences (trajectories), $\{u_t\}_{t \in \mathbb{N}_0}$, $\{x_t\}_{t \in \mathbb{N}_0}$, and $\{y_t\}_{t \in \mathbb{N}_0}$, respectively, can be represented as formal power series:

$$u(z) = \sum_{t \in \mathbb{N}_0} u_t z^t \in \mathbb{F}_q[[z]]^k,$$

$$x(z) = \sum_{t \in \mathbb{N}_0} x_t z^t \in \mathbb{F}_q[[z]]^s, \text{ and}$$

$$y(z) = \sum_{t \in \mathbb{N}_0} y_t z^t \in \mathbb{F}_q[[z]]^{n-k}.$$

A trajectory $\{x_t, u_t, y_t\}_{t \in \mathbb{N}_0}$ satisfies the first two equations of (10.5) if and only if

$$\begin{bmatrix} x(z) & u(z) & y(z) \end{bmatrix} E(z) = \mathbf{0}_{1 \times (n-k+s)},$$

where

$$E(z) := \begin{bmatrix} I_s - Az & -C \\ -Bz & -D \\ \mathbf{0}_{(n-k) \times s} & I_{n-k} \end{bmatrix}.$$

In order to obtain the codewords of a convolutional code by means of the system (10.5) we must only consider the polynomial input-output trajectories $c(z) = \begin{bmatrix} u(z) & y(z) \end{bmatrix}$ of the system. Moreover, we discard the input-output trajectories $c(z)$ with corresponding state trajectory $x(z)$ having infinite weight, since this would make the system remain indefinitely excited. Thus, we restrict to polynomial input-output trajectories with corresponding state trajectory also polynomial. We will call these input-output trajectories **finite-weight input-output trajectories**. The set of finite-weight input-output trajectories of the system (10.5) forms a submodule of R^n of rank k, and thus it is a convolutional code of rate $\frac{k}{n}$, denoted by $\mathcal{C}(A, B, C, D)$. The system $\Sigma = (A, B, C, D)$ is said to be an **input-state-output (ISO) representation** of the code.

Since $\mathcal{C}(A, B, C, D)$ is a submodule of rank k, there exists a matrix $G(z) \in R^{k \times n}$ such that $\mathcal{C}(A, B, C, D) = \text{rowspace}_R G(z)$. In fact, if $X(z) \in R^{k \times s}$ and $G(z) \in R^{k \times n}$ are such that $[X(z) \ G(z)]E(z) = \mathbf{0}_{k \times (n-k+s)}$ and $[X(z) \ G(z)]$ is left prime, then $G(z)$ is a generator matrix for $\mathcal{C}$.

Conversely, for each convolutional code $\mathcal{C}$ of rate $\frac{k}{n}$, there exists $(A, B, C, D) \in \mathbb{F}_q^{s \times s} \times \mathbb{F}_q^{k \times s} \times \mathbb{F}_q^{s \times (n-k)} \times \mathbb{F}_q^{k \times (n-k)}$ such that $\mathcal{C} = \mathcal{C}(A, B, C, D)$, if one allows permutation of the coordinates of the codewords.

If S is an invertible $s \times s$ constant matrix, the change of basis on the state vector, $x'(t) = x(t)S$, produces the algebraically equivalent system $\widetilde{\Sigma} = (S^{-1}AS, BS, S^{-1}C, D)$ with the same input-output trajectories as the system Σ. $\widetilde{\Sigma} = (S^{-1}AS, BS, S^{-1}C, D)$ is another ISO representation of $\mathcal{C}(A, B, C, D)$ of the same dimension.

However there are other ISO representations of the code with different dimension. We are interested in characterizing the ISO representations of the code with minimal dimension. These ISO representations will be called **minimal ISO representations**. The following properties of the system (10.5) have an important role not only in the characterization of the minimal ISO representations of the code, but also reflect on the properties of the code.

Given $A \in \mathbb{F}_q^{s \times s}$, $B \in \mathbb{F}_q^{k \times s}$, and $C \in \mathbb{F}_q^{s \times (n-k)}$, define the matrices

$$\Phi(A, B) = \begin{bmatrix} B \\ BA \\ \vdots \\ BA^{s-1} \end{bmatrix} \quad \text{and} \quad \Omega(A, C) = \begin{bmatrix} C & AC & \cdots & A^{s-1}C \end{bmatrix}.$$

Definition 10.4.1 ([1073]) Let $A \in \mathbb{F}_q^{s \times s}$, $B \in \mathbb{F}_q^{k \times s}$, and $C \in \mathbb{F}_q^{s \times (n-k)}$.

(a) The pair (A, B) is called **reachable** if rank $\Phi(A, B) = s$.

(b) The pair (A, C) is called **observable** if rank $\Omega(A, C) = s$.

The following lemma gives equivalent conditions for reachability and observability, and it is called the **Popov, Belevitch and Hautus (PBH) criterion**.

Lemma 10.4.2 ([1073]) *Let* $A \in \mathbb{F}_q^{s \times s}$, $B \in \mathbb{F}_q^{k \times s}$, *and* $C \in \mathbb{F}_q^{s \times (n-k)}$.

(a) (A, B) *is reachable if and only if* $\begin{bmatrix} z^{-1}I_s - A \\ B \end{bmatrix}$ *is right prime in the indeterminate* z^{-1}.

(b) (A, C) *is observable if and only if* $\begin{bmatrix} z^{-1}I_s - A & C \end{bmatrix}$ *is left prime in the indeterminate* z^{-1}.

If $\Sigma = (A, B, C, D)$ is not reachable, then rank $\Phi(A, B) = \delta < s$. Let $T = \begin{bmatrix} T_1 \\ T_2 \end{bmatrix} \in \mathbb{F}_q^{s \times s}$, be an invertible matrix such that the rows of T_1 form a basis of rowspace$_{\mathbb{F}_q} \Phi(A, B)$. Then $B = \begin{bmatrix} B_1 & \mathbf{0}_{k \times (s-\delta)} \end{bmatrix} T$, for some $B_1 \in \mathbb{F}_q^{k \times \delta}$. Moreover, since rowspace$_{\mathbb{F}_q} \Phi(A, B)$ is A-invariant (by the Cayley-Hamilton Theorem), it follows that the rows of T_1A belong to rowspace$_{\mathbb{F}_q} \Phi(A, B)$ and therefore $TA = \begin{bmatrix} A_1 & \mathbf{0}_{\delta \times (s-\delta)} \\ A_2 & A_3 \end{bmatrix} T$, for some $A_1 \in \mathbb{F}_q^{\delta \times \delta}$, $A_2 \in \mathbb{F}_q^{(s-\delta) \times \delta}$, and $A_3 \in \mathbb{F}_q^{(s-\delta) \times (s-\delta)}$. Then

$$S^{-1}AS = \begin{bmatrix} A_1 & \mathbf{0}_{\delta \times (s-\delta)} \\ A_2 & A_3 \end{bmatrix}, \quad BS = \begin{bmatrix} B_1 & \mathbf{0}_{k \times (s-\delta)} \end{bmatrix}, \quad S^{-1}C = \begin{bmatrix} C_1 \\ C_2 \end{bmatrix} \tag{10.6}$$

where $C_1 \in \mathbb{F}_q^{\delta \times (n-k)}$ and $S = T^{-1}$. Moreover, $\Phi(A, B)S = \begin{bmatrix} \Phi(A_1, B_1) & \mathbf{0}_{(k\delta) \times (s-\delta)} \\ B_1 A_1^\delta & \mathbf{0}_{k \times (s-\delta)} \\ \vdots & \vdots \\ B_1 A_1^{s-1} & \mathbf{0}_{k \times (s-\delta)} \end{bmatrix}$ has

rank δ. It follows from the Cayley-Hamilton Theorem that $\Phi(A_1, B_1)$ also has rank δ, i.e., the system $\Sigma_1 = (A_1, B_1, C_1, D)$ is reachable. The partitioning (10.6) is called the **Kalman (controllable) canonical form** [1073].

The system $\widetilde{\Sigma} = (S^{-1}AS, BS, S^{-1}C, D)$ has updating equations

$$
\begin{cases}
x_{t+1}^{(1)} = x_t^{(1)} A_1 + x_t^{(2)} A_2 + u_t B_1 \\
x_{t+1}^{(2)} = x_t^{(2)} A_3 \\
\phantom{x_{t+1}^{(2)}} y_t = x_t^{(1)} C_1 + x_t^{(2)} C_2 + u_t D
\end{cases}
$$

where $x_t = [x_t^{(1)} \ x_t^{(2)}]$, with $x_t^{(1)} \in \mathbb{F}_q^\delta$ and $x_t^{(2)} \in \mathbb{F}_q^{s-\delta}$. Since $x_0 = \mathbf{0}$, then $x_t^{(2)} = \mathbf{0}$ for all $t \in \mathbb{N}_0$. Therefore the system

$$
\begin{cases}
x_{t+1}^{(1)} = x_t^{(1)} A_1 + u_t B_1 \\
\phantom{x_{t+1}^{(1)}} y_t = x_t^{(1)} C_1 + u_t D
\end{cases}
$$

has the same finite-weight input-output trajectories. So $\mathcal{C}(A, B, C, D) = \mathcal{C}(A_1, B_1, C_1, D)$; i.e, $\Sigma_1 = (A_1, B_1, C_1, D)$ is another ISO representation of the code with smaller dimension. This means that a minimal ISO representation of a convolutional code must be necessarily reachable. But the converse is also true as stated in the next theorem.

Theorem 10.4.3 ([1600]) *Let $\Sigma = (A, B, C, D)$ be an ISO representation of a convolutional code $\mathcal{C}$ of degree δ. Then Σ is a minimal ISO representation of $\mathcal{C}$ if and only if it is reachable.*

Moreover, if Σ is a minimal ISO representation of $\mathcal{C}$, then it has dimension δ and all the minimal ISO representations of $\mathcal{C}$ are of the form $\widetilde{\Sigma} = (\widetilde{S}^{-1}A\widetilde{S}, B\widetilde{S}, \widetilde{S}^{-1}C, D)$, where $\widetilde{S} \in \mathbb{F}_q^{\delta \times \delta}$ is invertible.

The reachability together with the observability of the system influence the properties of the corresponding code as the next theorem shows.

Theorem 10.4.4 ([1600]) *Let $\Sigma = (A, B, C, D)$ be a reachable system. The convolutional code $\mathcal{C}(A, B, C, D)$ is noncatastrophic if and only if $\Sigma = (A, B, C, D)$ is observable.*

If $\Sigma = (A, B, C, D)$ is an ISO representation of a code $\mathcal{C}$ which is reachable and observable, then the polynomial input-output trajectories of the system coincide with the finite-weight input-output trajectories, which means that $\mathcal{C}$ is the set of the polynomial input-output trajectories of $\Sigma = (A, B, C, D)$.

The correspondence between linear systems and convolutional codes allows one to obtain further constructions of convolutional codes with good distance properties. The following theorem presents a construction for MDS convolutional codes with rate $1/n$ and arbitrary degree.

Theorem 10.4.5 ([1729]) *Let $|\mathbb{F}_q| \geq n\delta + 1$ and α be a primitive element of $\mathbb{F}_q$.*

Set $A = \begin{bmatrix} \alpha & & & 0 \\ & \alpha^2 & & \\ & & \ddots & \\ 0 & & & \alpha^\delta \end{bmatrix} \in \mathbb{F}_q^{\delta \times \delta}$, $B = [1 \cdots 1] \in \mathbb{F}_q^\delta$, $D = [1 \cdots 1] \in \mathbb{F}_q^{n-1}$ *and*

$C \in \mathbb{F}_q^{\delta \times (n-1)}$ *where the columns $c_1, c_2, \ldots, c_{n-1}$ of the matrix C are chosen such that* $\det(sI_\delta - (A - c_iB)) = \prod_{k=1}^{\delta}(s - \alpha^{r_i+k})$ *and r_i, $i = 1, \ldots, \delta$, are chosen such that* $\{\alpha^{r_i+1}, \alpha^{r_i+2}, \ldots, \alpha^{r_i+\delta}\} \cap \{\alpha^{r_j+1}, \alpha^{r_j+2}, \ldots, \alpha^{r_j+\delta}\} = \emptyset$ *for $i \neq j$. Then $\mathcal{C}(A, B, C, D)$ is an MDS convolutional code.*

ISO representations could also be helpful for the construction of MDP convolutional codes, using the following criterion for being MDP.

Theorem 10.4.6 ([1015, Corollary 1.1]) *The matrices (A, B, C, D) generate an MDP convolutional code if and only if the matrix* $\mathcal{F}_L :=$ $\begin{bmatrix} D & BC & \cdots & BA^{L-1}C \\ 0 & \ddots & \ddots & \vdots \\ \vdots & \ddots & \ddots & BC \\ 0 & \cdots & 0 & D \end{bmatrix}$ *with $L =$ $\lfloor \frac{\delta}{k} \rfloor + \lfloor \frac{\delta}{n-k} \rfloor$ has the property that every minor which is not trivially zero is nonzero.*

10.5 Decoding of Convolutional Codes

In this section, we will present decoding techniques for convolutional codes. The first part of this section describes the decoding of convolutional codes over the erasure channel (see Section 10.2.2), where simple linear algebra techniques are applied. The second part presents the most famous decoding algorithm for convolutional codes over the q-ary symmetric channel, the Viterbi Algorithm. Other decoding principles for this kind of channel such as sequential decoding, list decoding, iterative decoding and majority-logic decoding are explained in, e.g., [1051, 1260]; see also Chapters 21, 24, and 30.

10.5.1 Decoding over the Erasure Channel

To make it easier to follow, the decoding over an erasure channel will be explained first for $\delta = 0$ and afterwards for the general case.

10.5.1.1 The Case $\delta = 0$

Convolutional codes of degree zero are block codes. The only column distance that is defined for a block code $\mathcal{C}$ of rate k/n is the 0th column distance, which coincides with its minimum distance $d(\mathcal{C}) = \min_{\mathbf{c} \in \mathcal{C}} \{ \mathrm{wt}_H(\mathbf{c}) \mid \mathbf{c} \neq \mathbf{0} \}$. The minimum distance of a block code can be characterized by its parity check matrices.

Lemma 10.5.1 *Let $\mathcal{C}$ be a block code of rate k/n and let $H_0 \in \mathbb{F}_q^{(n-k) \times n}$ be a parity check matrix of $\mathcal{C}$. Then $\mathcal{C}$ has minimum distance d if and only if all the sets of $d-1$ columns of H_0 are linearly independent and there exist d linearly dependent columns of H_0.*

The next theorem establishes the number of erasures that can be corrected by a block code.

Theorem 10.5.2 *Let $\mathcal{C}$ be a block code with minimum distance d. Then $\mathcal{C}$ can correct at most $d-1$ erasures.*

Let $\mathcal{C}$ be a block code of rate k/n and minimum distance d with parity check matrix $H_0 \in \mathbb{F}_q^{(n-k) \times n}$. Let $\mathbf{c} \in \mathbb{F}_q^n$ be a received codeword of $\mathcal{C}$ after transmission over an erasure channel with at most $d-1$ erasures. Then $H_0 \mathbf{c}^\mathsf{T} = \mathbf{0}^\mathsf{T}$. Denote by $\mathbf{c}^{(e)}$ the vector consisting of the components of $\mathbf{c}$ that are erased during transmission and by $\mathbf{c}^{(r)}$ the vector consisting of the components of $\mathbf{c}$ that are received (correctly). Moreover, denote by $H_0^{(e)}$ the matrix consisting of the columns of H_0 whose indices correspond to the indices of the erased components of $\mathbf{c}$ and by $H_0^{(r)}$ the matrix consisting of the other columns of H_0. Then the equation $H_0 \mathbf{c}^\mathsf{T} = \mathbf{0}^\mathsf{T}$ is equivalent to the system of linear equations $H_0^{(e)} \mathbf{c}^{(e)\mathsf{T}} = -H_0^{(r)} \mathbf{c}^{(r)\mathsf{T}}$ with the erased components as unknowns. Since $H_0^{(e)}$ has full column rank, the system $H_0^{(e)} \mathbf{c}^{(e)\mathsf{T}} = -H_0^{(r)} \mathbf{c}^{(r)\mathsf{T}}$ has a unique solution and the erasures are recovered.

Clearly, the decoding is optimal if as many of these linear equations as possible are linearly independent for as many as possible erasure patterns. This is the case if all full size minors of H_0 are nonzero, which is true if and only if $\mathcal{C}$ is MDS, and the maximal number of erasures that a block code of rate k/n can correct is $n - k$.

10.5.1.2 The General Case

In this subsection, the considerations of the preceding subsection are generalized to noncatastrophic convolutional codes of rate k/n and arbitrary degree δ; see [1810]. Moreover, it will be seen that convolutional codes have a better performance than block codes in decoding over the erasure channel. This is due to the capability of considering particular parts (windows) of the sequence, of different sizes, through the decoding process and sliding along this sequence to choose the window to start decoding. Such capability is called the **sliding window property**, and it allows the decoder to adapt the process to the erasure pattern.

Let $\mathcal{C}$ be an (n, k, δ) noncatastrophic convolutional code and assume that for a codeword $c(z) = \sum_{i \in \mathbb{N}_0} c_i z^i$ of $\mathcal{C}$, the coefficients $c_0, \ldots, c_{t-1}$ are known for some $t \in \mathbb{N}_0$ and that there exists at least one erasure in c_t. Let $H(z) = \sum_{i=0}^{\nu} H_i z^i$ with $H_i \in \mathbb{F}_q^{(n-k) \times n}$ be a left prime parity check matrix of $\mathcal{C}$. Then, for each $j \in \mathbb{N}_0$ and

$$
\mathfrak{H}_j := \begin{bmatrix} H_\nu & \cdots & H_0 & & 0 \\ & \ddots & & \ddots & \\ 0 & & H_\nu & \cdots & H_0 \end{bmatrix} \in \mathbb{F}_q^{(j+1)(n-k) \times (\nu+j+1)n},
$$

one has $\mathfrak{H}_j [c_{t-\nu} \cdots c_{t+j}]^\mathsf{T} = \mathbf{0}^\mathsf{T}$, where $c_i = \mathbf{0}$ for $i \notin \{0, \ldots, \deg(c)\}$. Denote by $\mathfrak{H}_j^{(1)}$ the matrix consisting of the first νn columns of $\mathfrak{H}_j$. Then $\mathfrak{H}_j = [\mathfrak{H}_j^{(1)} \ H_j^c]$, and consequently, to recover the erasures in $c_t, \ldots, c_{t+j}$, one has to consider the window $[c_{t-\nu} \cdots c_{t+j}]$ and solve the linear system

$$
H_j^{(e)} [c_t^{(e)} \cdots c_{t+j}^{(e)}]^\mathsf{T} = -H_j^{(r)} [c_t^{(r)} \cdots c_{t+j}^{(r)}]^\mathsf{T} - \mathfrak{H}_j^{(1)} [c_{t-\nu} \cdots c_{t-1}]^\mathsf{T},
$$

where $c_i^{(e)}$ and $c_i^{(r)}$ denote the erased and correctly received components of c_i, respectively, and $H_j^{(e)}$ and $H_j^{(r)}$ denote the corresponding columns of H_j^c. The erasures are recovered if and only if the system has a unique solution, i.e., if and only if $H_j^{(e)}$ has full column rank.

Hence, bearing in mind Theorem 10.2.19, one obtains the following theorem, which relates the capability to correct erasures by forward decoding (i.e., decoding from left to right) with the column distances of the convolutional code.

Theorem 10.5.3 ([1810]) *Let $\mathcal{C}$ be an (n, k, δ) convolutional code. If in any sliding window of length $(j + 1)n$ at most $d_j^c(\mathcal{C}) - 1$ erasures occur, then full error correction from left to right is possible.*

Clearly, the best situation is if one has an MDP convolutional code, i.e., $d_j^c(\mathcal{C}) - 1 = (j + 1)(n - k)$ for $j = 0, \ldots, L$. Define the **recovery rate** as the ratio of the number of erasures and the total number of symbols in a window. In [1810], it has been shown that the recovery rate $\frac{n-k}{n}$ of an MDP convolutional code is the maximal recovery rate one could get over an erasure channel.

Reverse MDP convolutional codes have the MDP property forward and backward and hence also the erasure correcting capability described in Theorem 10.5.3 from left to right and from right to left, and therefore they can recover even more situations than MDP convolutional codes [1810]. There are bursts of erasures that cannot be forward decoded but could be skipped and afterwards be decoded from right to left (backward); see [1810] for examples.

If patterns of erasures occur that do not fulfill the conditions of Theorem 10.5.3, neither forward nor backward, one has a block that could not be recovered and gets lost in the recovering process. In order to continue recovering, one needs to find a block of νn correct symbols, a so-called guard space, preceding a block that fulfills the conditions of Theorem 10.5.3. Complete MDP convolutional codes have the additional advantage that to compute a guard space, it is not necessary to have a large sequence of correct symbols [1810]. Instead it suffices to have a window with a certain percentage of correct symbols as the following theorem states.

Theorem 10.5.4 ([1810]) *Let C be an (n, k, δ) complete MDP convolutional code and $L = \lfloor \frac{\delta}{k} \rfloor + \lfloor \frac{\delta}{n-k} \rfloor$. If in a window of size $(L + \nu + 1)n$ there are not more than $(L + 1)(n - k)$ erasures, and if they are distributed in such a way that between positions 1 and sn and between positions $(L + \nu + 1)n$ and $(L + \nu + 1)n - sn + 1$, for $s = 1, \ldots, L + 1$, there are not more than $s(n - k)$ erasures, then full correction of all symbols in this interval will be possible. In particular a new guard space can be computed.*

The way of decoding over the erasure channel described here could be used for any convolutional code. However, to correct as many erasures as possible it is optimal to use MDP or even complete MDP convolutional codes. For algorithms to do that we refer to [1810].

Remark 10.5.5 If $\widetilde{C}$ is a catastrophic convolutional code, it is possible to find a noncatastrophic convolutional code C with $\widetilde{C} \subsetneq C$ (see Remark 10.2.9). Let $H(z)$ be a left prime parity check matrix for C. Then $H(z)c(z)^\mathsf{T} = \mathbf{0}^\mathsf{T}$ for all $c(z) \in \widetilde{C}$. Hence, the decoding procedure described in this subsection could be applied to $\widetilde{C}$ by using $H(z)$.

MDP convolutional codes are able to recover patterns of erasures that MDS block codes, with the same recovery rate, cannot recover, as illustrated in the next example.

Example 10.5.6 ([1810]) Consider the received sequence $w = (w_0, w_1, \ldots, w_{100})$ with $w_i \in \mathbb{F}_q^2$ and with 120 erasures in w_i, $i \in \{0, 1, \ldots, 29\} \cup \{70, 71, \ldots, 99\}$. Let us assume that w is a received word of an MDS block code of rate $101/202$. Such a code can correct up to 101 symbols in a sequence of 202 symbols, i.e., it has a recovery rate of 50%. Thus, since w has 120 erasures it cannot be recovered.

Let us assume now that $w(z) = \sum_{i=0}^{100} w_i z^i$ is a received word of a $(2, 1, 50)$ MDP convolutional code C. C has the same recovery rate, but it is able to correct the erasures in $w(z)$. Note that $d_j^c(C) - 1 = j + 1$, for $j = 0, 1, \ldots, L$ with $L = 100$. To recover the whole sequence one can consider a window with the first 120 symbols (i.e., the sequence consisting of $w_0, w_1, \ldots, w_{59}$). Since $d_{59}^c(C) - 1 = 60$, the first 60 symbols can be recovered by applying the decoding algorithm described in this section. Afterwards, take another window with the symbols $w_{41}, w_{42}, \ldots, w_{100}$. Similarly, this window has 120 symbols and 60 erasures, which means that these erasures can also be recovered, and the whole corrected sequence is obtained.

10.5.2   The Viterbi Decoding Algorithm

The most commonly used error-correction decoding algorithm for convolutional codes is the Viterbi Decoding Algorithm, which was proposed by Viterbi in 1967 [1854]. It is a minimum-distance decoder, i.e., it computes all codewords of a certain length and compares the received word to each of them. Then, the codeword closest to the received word is selected (maximum-likelihood decoding). It could be understood as applying techniques from dynamic programming to the linear systems representation of the convolutional code, as explained next.

A **state-transition diagram** for the convolutional code can be defined from a minimal ISO representation of the code, $\Sigma = (A, B, C, D)$ of dimension δ, as a labeled directed graph with the set of nodes $X = \mathbb{F}_q^\delta$ such that (x_1, x_2), with $x_1, x_2 \in X$, is an arc (transition) of the graph if there exists $u \in \mathbb{F}_q^k$ such that $x_2 = x_1 A + uB$. In this case we assign the label (u, y) to the arc (x_1, x_2), where $y = x_1 C + uD$. The codewords of the code are obtained considering all closed walks on the state-transition diagram that start at the node $x = \mathbf{0}$ and end at the same node, and a codeword corresponding to such a walk is the sequence of the labels of the arcs that constitute this walk.

When we introduce the dimension of time, this state-transition diagram can be represented as a trellis diagram by considering a different copy of the set of nodes (states) of the state-transition diagram at each time instant (also called depth). For every $t \in \mathbb{N}_0$, we consider an arc e from a state x_1 in time instant t to another state x_2 at time instant $t + 1$, if there exists the arc (x_1, x_2) in the state-transition diagram. The label of the arc e is taken to be the same as the label of (x_1, x_2) in the state-transition diagram. The codewords of the code correspond to the paths on the trellis diagram that start and end at the zero state [533].

If $c^r(z) \in \mathbb{F}_q[z]^n$ is a received word, the Viterbi algorithm searches the paths of the trellis diagram starting at the state $x = \mathbf{0}$ and ending at the same state, $x = \mathbf{0}$, such that the corresponding codeword has minimum distance to $c^r(z)$. The decoding progress is simplified by breaking it down into a sequence of steps using the recursive relation given by the equations defining the linear system. In each step, the distance between the received word and the estimated word is minimized.

In the following, this process is explained in detail.

Algorithm 10.5.7 (Viterbi Decoding)

Assume that a convolutional code $\mathcal{C}$ and a received word $c^r(z) = \sum_{i \in \mathbb{N}_0} c_i^r z^i$, which should be decoded, are given. Take a minimal ISO representation $\Sigma = (A, B, C, D)$ of dimension δ of $\mathcal{C}$ and set $x_0 = \mathbf{0}$.

Step 1: Set $t = 0$, $d_{min} = \infty$ and $sp_{min} = \emptyset$, and assign to the initial node the label $(d = 0, sp = \emptyset)$. Go to Step 2.

Step 2: For each node x_2 at time instant $t + 1$ do the following: For each of the predecessors x_1 at time instant t with label (d, sp) and $d < d_{min}$, compute the sum $d + d((u, y), c_t)$, where (u, y) is the label of the arc (x_1, x_2) in the state-transition diagram, determine the minimum of these sums $\overline{d}$, and assign to x_2 the label $(\overline{d}, \overline{sp})$, where $\overline{sp}$ is the shortest path from $x = \mathbf{0}$, at time instant zero, to x_2. If there are several sums with the minimal value, then there are several paths from $x = \mathbf{0}$, at time instant zero, to x_2 with distance $\overline{d}$; in this case select randomly one of these paths. If $x_2 = \mathbf{0}$ and $\overline{d} < d_{min}$, then set $d_{min} = \overline{d}$ and $sp_{min} = \overline{sp}$. Go to Step 3.

Step 3: If at time instant $t + 1$, all the nodes x with label $(\bar{d}, \overline{sp})$ are such that $\bar{d} \geq d_{min}$, then STOP and the result of the decoding is the codeword corresponding to the path sp_{min}. Otherwise set $t = t + 1$ and go to Step 2.

The complexity of this algorithm grows with the number of states, at each time instant, in the trellis diagram. The set of states is $X = \mathbb{F}_q^\delta$, and therefore it has q^δ elements. This algorithm is practical only for codes with small degree and defined in fields of very small size.

Instead of using the linear systems representation, the Viterbi Decoding Algorithm could also be operated using the trellis of the convolutional code; see, e.g., [533].

10.6 Two-Dimensional Convolutional Codes

In this section we consider convolutional codes of higher dimension, namely the two-dimensional (2D) convolutional codes. Let $R = \mathbb{F}_q[z_1, z_2]$ be the ring of polynomials in the indeterminates z_1 and z_2, $\mathbb{F}_q(z_1, z_2)$ the field of rational functions in z_1 and z_2 with coefficients in $\mathbb{F}_q$, and $\mathbb{F}_q[[z_1, z_2]]$ the ring of formal power series in z_1 and z_2 with coefficients in $\mathbb{F}_q$.

10.6.1 Definition of 2D Convolutional Codes via Generator and Parity Check Matrices

A **two-dimensional (2D) convolutional code** $\mathcal{C}$ of **rate** k/n is a *free* R-submodule of R^n of rank k. A **generator matrix** of $\mathcal{C}$ is a full row rank matrix $G(z_1, z_2)$ whose rows constitute a basis of $\mathcal{C}$, and it induces an injective map between R^k and R^n. This is the main reason of the restriction of the definition of 2D convolutional codes to free submodules of R^n.

Analogously as defined in the 1D case, a matrix $U(z_1, z_2) \in R^{k \times k}$ is unimodular if there exists a $k \times k$ matrix $V(z_1, z_2)$ over R such that

$$U(z_1, z_2)V(z_1, z_2) = V(z_1, z_2)U(z_1, z_2) = I_k,$$

or equivalently, if $\det(U(z_1, z_2)) \in \mathbb{F}_q^*$. Using the same reasoning as in Section 10.2.1 for one-dimensional (1D) convolutional codes, the matrices $G(z_1, z_2)$ and $\widetilde{G}(z_1, z_2)$ in $R^{k \times n}$ are said to be equivalent if they are generator matrices of the same code, which happens if and only if

$$\widetilde{G}(z_1, z_2) = U(z_1, z_2)G(z_1, z_2)$$

for some unimodular matrix $U(z_1, z_2) \in R^{k \times k}$.

Complexity and degree are equivalent and important notions of 1D convolutional codes. They are one of the parameters of the Generalized Singleton Bound on the minimum distance of these codes, and they also provide a lower bound on the dimension of their ISO representations. To define similar notions for 2D convolutional codes, we consider the usual notion of (total) degree of a polynomial in two indeterminates $p(z_1, z_2) = \sum_{(i,j) \in \mathbb{N}_0} p_{ij} z_1^i z_2^j$

with $p_{ij} \in \mathbb{F}_q$ as $\deg(p(z_1, z_2)) = \max\{i + j \mid p_{ij} \neq 0\}$. We also define the **internal degree** of a polynomial matrix $G(z_1, z_2)$, denoted by $\delta_i(G)$, as the maximal degree of the full size minors of $G(z_1, z_2)$ and its **external degree**, denoted by $\delta_e(G)$, as $\displaystyle\sum_{i=1}^{k} \nu_i$, where ν_i is the maximum degree of the entries of the ith row of $G(z_1, z_2)$. Obviously $\delta_i(G) \leq \delta_e(G)$.

Since two generator matrices of a 2D convolutional code $\mathcal{C}$ differ by a unimodular matrix, their full size minors are equal, up to multiplication by a nonzero constant. The **complexity** of $\mathcal{C}$ is defined as the internal degree of any generator matrix of $\mathcal{C}$, and it is represented by δ_c. The **degree** of $\mathcal{C}$ is the minimum external degree of all generator matrices of $\mathcal{C}$, and it is represented by δ_d. Clearly the internal degree of any generator matrix is less than or equal to the corresponding external degree, and therefore $\delta_c \leq \delta_d$ [415, 1412].

The row reduced generator matrices of 1D convolutional codes are the ones for which the corresponding notions of internal degree and external degree coincide, and this is why the complexity and degree of a 1D convolutional code are the same. However, 2D convolutional codes do not always admit such generator matrices, and there are 2D convolutional codes such that $\delta_c < \delta_d$, as is illustrated in the following simple example.

Example 10.6.1 For any finite field, the 2D convolutional code with generator matrix

$$G(z_1, z_2) = \begin{bmatrix} 1 & z_1 & 0 \\ 1 & z_2 & 1 \end{bmatrix}$$

has complexity 1 and degree 2.

Another important property of a 1D convolutional code is the existence (or not) of prime generator matrices. When we consider polynomial matrices in two indeterminates, there are two different notions of primeness: factor-primeness and zero-primeness [1177, 1234, 1402, 1592, 1933].

Definition 10.6.2 A matrix $G(z_1, z_2) \in R^{k \times n}$, with $n \geq k$, is

(a) **left factor-prime** (ℓFP) if for every factorization

$$G(z_1, z_2) = T(z_1, z_2)\overline{G}(z_1, z_2),$$

with $\overline{G}(z_1, z_2) \in R^{k \times n}$ and $T(z_1, z_2) \in R^{k \times k}$, $T(z_1, z_2)$ is unimodular;

(b) **left zero-prime** (ℓZP) if the ideal generated by the $k \times k$ minors of $G(z_1, z_2)$ is $\mathbb{F}_q[z_1, z_2]$.

A matrix $G(z) \in R^{k \times n}$, with $k \geq n$, is **right factor-prime** (rFP) / **right zero-prime** (rZP) if its transpose is ℓFP / ℓZP, respectively. The notions (a) and (b) of the above definition are equivalent for polynomial matrices in one indeterminate (see Theorem 10.2.6). However, for polynomial matrices in two indeterminates, zero-primeness implies factor-primeness, but the contrary does not happen, as is illustrated in the following example.

Example 10.6.3 The matrix $\begin{bmatrix} z_1 & z_2 \end{bmatrix}$ is left factor-prime but it is not left zero-prime.

The following lemmas give characterizations of left factor-primeness and left zero-primeness. Compare them to Theorem 10.2.6.

Lemma 10.6.4 *Let* $G(z_1, z_2) \in R^{k \times n}$, *with* $n \geq k$. *Then the following are equivalent.*

(a) $G(z_1, z_2)$ *is* ℓFP.

(b) *There exist polynomial matrices $X_i(z_1, z_2)$ such that*

$$X_i(z_1, z_2)G(z_1, z_2) = d_i(z_i)I_k,$$

with $d_i(z_i) \in \mathbb{F}_q[z_i] \setminus \{0\}$, for $i = 1, 2$.

(c) *For all $u(z_1, z_2) \in \mathbb{F}_q(z_1, z_2)^k$, $u(z_1, z_2)G(z_1, z_2) \in R^n$ implies that $u(z_1, z_2) \in R^k$.*

(d) *the $k \times k$ minors of $G(z_1, z_2)$ have no common factor.*

Lemma 10.6.5 *Let $G(z_1, z_2) \in R^{k \times n}$, with $n \geq k$. Then the following are equivalent.*

(a) *$G(z_1, z_2)$ is ℓZP.*

(b) *$G(z_1, z_2)$ admits a polynomial right inverse.*

(c) *$G(\lambda_1, \lambda_2)$ is full column rank, for all $\lambda_1, \lambda_2 \in \overline{\mathbb{F}}_q$, where $\overline{\mathbb{F}}_q$ denotes the algebraic closure of $\mathbb{F}_q$.*

From the above lemmas, it immediately follows that if a 2D convolutional code admits a left factor-prime (left zero-prime) generator matrix then all its generator matrices are also left factor-prime (left zero-prime). 2D convolutional codes with left factor-prime generator matrices are called **noncatastrophic** and if they admit left zero-prime generator matrices, they are called **basic**.

A **parity check matrix** of a 2D convolutional code $\mathcal{C}$ is a full row rank matrix $H(z_1, z_2) \in R^{(n-k) \times n}$ such that

$$c(z_1, z_2) \in \mathcal{C} \text{ if and only if } H(z_1, z_2)c(z_1, z_2)^{\mathsf{T}} = \mathbf{0}^{\mathsf{T}}.$$

As for 1D convolutional codes, the existence of parity check matrices for a 2D convolutional code is connected with primeness properties of its generator matrices as stated in the following theorem.

Theorem 10.6.6 ([1826]) *A 2D convolutional code $\mathcal{C}$ admits a parity check matrix $H(z_1, z_2)$ if and only if it is noncatastrophic.*

The free distance of a 2D convolutional code was first defined in [1883], and it is a generalization of the free distance defined in the 1D case. The weight of a polynomial vector $c(z_1, z_2) = \sum_{(i,j) \in \mathbb{N}_0^2} c_{ij} z_1^i z_2^j$ is given by $\mathrm{wt_H}(c(z_1, z_2)) = \sum_{(i,j) \in \mathbb{N}_0^2} \mathrm{wt_H}(c_{ij})$ and the **free distance** of $\mathcal{C}$ is defined as

$$d_{free}(\mathcal{C}) = \min\{\mathrm{wt_H}(c_1(z_1, z_2) - c_2(z_1, z_2)) \mid c_1(z_1, z_2), c_2(z_1, z_2) \in \mathcal{C}, c_1(z_1, z_2) \neq c_2(z_1, z_2)\}$$
$$= \min\{\mathrm{wt_H}(c(z_1, z_2)) \mid c(z_1, z_2) \in \mathcal{C} \setminus \{\mathbf{0}\}\}.$$

The degree of a 2D convolutional code $\mathcal{C}$ is an important parameter for establishing an upper bound on the distance of $\mathcal{C}$.

Theorem 10.6.7 ([415]) *Let $\mathcal{C}$ be a 2D convolutional code of rate k/n and degree δ. Then*

$$d_{free}(\mathcal{C}) \leq n\frac{\left(\lfloor \frac{\delta}{k} \rfloor + 1\right)\left(\lfloor \frac{\delta}{k} \rfloor + 2\right)}{2} - k\left(\left\lfloor \frac{\delta}{k} \right\rfloor + 1\right) + \delta + 1.$$

This upper bound is the extension to 2D convolutional codes of the Generalized Singleton Bound for 1D convolutional codes (Theorem 10.2.21(a)), and it is called the **2D Generalized Singleton Bound**. Moreover, a 2D convolutional code of rate k/n and degree δ is said to be a **maximum distance separable** (MDS) 2D convolutional code if its free distance equals the 2D Generalized Singleton Bound. Constructions of MDS 2D convolutional codes can be found in [414, 415].

10.6.2 ISO Representations

Two-dimensional convolutional codes can also be represented by a linear system. Unlike the 1D case, there exist several state-space models of a 2D linear system, namely the Roesser model, the Attasi model, and the Fornasini–Marchesini model. The ISO representations of 2D convolutional codes investigated in the literature consider the Fornasini–Marchesini model [739]. In this model a first quarter plane 2D linear system is given by the updating equations

$$\begin{cases} x_{i+1,j+1} = x_{i,j+1}A_1 + x_{i+1,j}A_2 + u_{i,j+1}B_1 + u_{i+1,j}B_2 \\ y_{i,j} = x_{i,j}C + u_{i,j}D \\ c_{i,j} = \begin{bmatrix} y_{i,j} & u_{i,j} \end{bmatrix} \end{cases} \tag{10.7}$$

with $(A_1, A_2, B_1, B_2, C, D) \in \mathbb{F}_q^{s \times s} \times \mathbb{F}_q^{s \times s} \times \mathbb{F}_q^{k \times s} \times \mathbb{F}_q^{k \times s} \times \mathbb{F}_q^{s \times (n-k)} \times \mathbb{F}_q^{k \times (n-k)}$, $i, j \in \mathbb{N}_0$ and $s, k, n \in \mathbb{N}$ with $n > k$. The system (10.7) will be represented by $\Sigma = (A_1, A_2, B_1, B_2, C, D)$, and the integer s is called its **dimension**. We call $x_{i,j} \in \mathbb{F}_q^s$ the **local state vector**, $u_{i,j} \in \mathbb{F}_q^k$ the **information vector**, $y_{i,j} \in \mathbb{F}_q^{n-k}$ the **parity vector**, and $c_{i,j} \in \mathbb{F}_q^n$ the **code vector**. Moreover, the input and the local state have past finite support, i.e., $u_{i,j} = \mathbf{0}$ and $x_{i,j} = \mathbf{0}$, for $i < 0$ or $j < 0$, and we consider zero initial conditions, i.e., $x_{0,0} = \mathbf{0}$.

The input, the local state, and the output 2D sequences (trajectories) of the system, $\{u_{i,j}\}_{(i,j) \in \mathbb{N}_0^2}$, $\{x_{i,j}\}_{(i,j) \in \mathbb{N}_0^2}$, $\{y_{i,j}\}_{(i,j) \in \mathbb{N}_0^2}$, respectively, can be represented as formal power series in the indeterminates z_1, z_2:

$$u(z_1, z_2) = \sum_{(i,j) \in \mathbb{N}_0^2} u_{i,j} z_1^i z_2^j \in \mathbb{F}_q[[z_1, z_2]]^k,$$

$$x(z_1, z_2) = \sum_{(i,j) \in \mathbb{N}_0^2} x_{i,j} z_1^i z_2^j \in \mathbb{F}_q[[z_1, z_2]]^s, \text{ and}$$

$$y(z_1, z_2) = \sum_{(i,j) \in \mathbb{N}_0^2} y_{i,j} z_1^i z_2^j \in \mathbb{F}_q[[z_1, z_2]]^{n-k}.$$

The $\mathbb{F}_q[[z_1, z_2]]$-kernel of the matrix

$$E(z_1, z_2) = \begin{bmatrix} I_s - A_1 z_1 - A_2 z_2 & -C \\ -B_1 z_1 - B_2 z_2 & -D \\ \mathbf{0}_{(n-k) \times s} & I_{n-k} \end{bmatrix}$$

consists of the $(x(z_1, z_2), u(z_1, z_2), y(z_1, z_2))$ trajectories of the system, and we say that an input-output trajectory has corresponding state trajectory $x(z_1, z_2)$ if

$$\begin{bmatrix} x(z_1, z_2) & u(z_1, z_2) & y(z_1, z_2) \end{bmatrix} E(z_1, z_2) = \mathbf{0}_{1 \times (n-k+s)}.$$

For the same reasons stated for 1D convolutional codes in Section 10.4, we only consider the finite-weight input-output trajectories of the system (10.7) to obtain a 2D convolutional code, i.e, the polynomial trajectories $(u(z_1, z_2), y(z_1, z_2))$ with corresponding state trajectory $x(z_1, z_2)$ also polynomial.

Theorem 10.6.8 ([1412]) *The set of finite-weight input-output trajectories of the system* (10.7) *is a 2D convolutional code of rate* k/n.

The 2D convolutional code whose codewords are the finite-weight input-output trajectories of the system (10.7) is denoted by $\mathcal{C}(A_1, A_2, B_1, B_2, C, D)$. The system $\Sigma =$

$(A_1, A_2, B_1, B_2, C, D)$ is called an **input-state-output (ISO) representation** of the code $\mathcal{C}(A_1, A_2, B_1, B_2, C, D)$. If $L(z_1, z_2) \in R^{k \times s}$ and $G(z_1, z_2) \in R^{k \times n}$ are such that

$$\begin{bmatrix} L(z_1, z_2) & G(z_1, z_2) \end{bmatrix} E(z_1, z_2) = \mathbf{0}_{k \times (n-k+s)}$$

where $\begin{bmatrix} L(z_1, z_2) & G(z_1, z_2) \end{bmatrix}$ is left factor-prime, then $G(z_1, z_2)$ is a generator matrix of $\mathcal{C}(A_1, A_2, B_1, B_2, C, D)$.

Reachability and observability are properties of a linear system that also reflect on the corresponding convolutional code. However, 2D linear systems as in (10.7) admit different types of reachability and observability notions.

Definition 10.6.9 ([739]) Let $\Sigma = (A_1, A_2, B_1, B_2, C, D)$ be a 2D linear system with dimension s, and define the $s \times s$ matrices $^r\Delta^t$ by the following:

$$A_1 \, ^r\Delta^t A_2 = A_1(A_1 \, ^{r-1}\Delta^t A_2) + A_2(A_1 \, ^r\Delta^{t-1} A_2), \text{ for } r, t \geq 1,$$

$$A_1 \, ^r\Delta^0 A_2 = A_1^r \text{ and } A_1 \, ^0\Delta^t A_2 = A_2^t, \text{ for } r, t \geq 0, \text{ and}$$

$$A_1 \, ^r\Delta^t A_2 = 0, \text{ when either } r \text{ or } t \text{ is negative.}$$

(a) Σ is **locally reachable** if the reachability matrix

$$\mathcal{R} = \begin{bmatrix} R_1 \\ R_2 \\ R_3 \\ \vdots \end{bmatrix}$$

is full row rank, where R_k is the block matrix consisting of all

$$B_1(A_1 \, ^{i-1}\Delta^j A_2) + B_2(A_1 \, ^i\Delta^{j-1} A_2)$$

with $i + j = k$, for $i, j \geq 0$, i.e.,

$$R_k = \begin{bmatrix} B_1 A_1^{k-1} \\ B_1(A_1 \, ^{k-2}\Delta^1 A_2) + B_2(A_1 \, ^{k-1}\Delta^0 A_2) \\ B_1(A_1 \, ^{k-3}\Delta^2 A_2) + B_2(A_1 \, ^{k-2}\Delta^1 A_2) \\ \vdots \\ B_1(A_1 \, ^0\Delta^{k-1} A_2) + B_2(A_1 \, ^1\Delta^{k-2} A_2) \\ B_2 A_2^{k-1} \end{bmatrix}.$$

(b) Σ is **modally reachable** if the matrix

$$\begin{bmatrix} I_s - A_1 z_1 - A_2 z_2 \\ B_1 z_1 + B_2 z_2 \end{bmatrix}$$

is right factor-prime.

(c) Σ is **modally observable** if the matrix

$$\begin{bmatrix} I_s - A_1 z_1 - A_2 z_2 & C \end{bmatrix}$$

is left factor-prime.

There also exists a notion of local observability that will not be considered here. There are 2D systems that are locally reachable (observable) but not modally reachable (observable) and vice-versa [739]. The next lemma shows the influence of these properties on the corresponding convolutional code.

Lemma 10.6.10 ([416, 1412]) *Let* $\Sigma = (A_1, A_2, B_1, B_2, C, D)$ *be a modally reachable* 2D *linear system. Then* Σ *is modally observable if and only if* $\mathcal{C}(A_1, A_2, B_1, B_2, C, D)$ *is non-catastrophic.*

If $\Sigma = (A_1, A_2, B_1, B_2, C, D)$ is an ISO representation of a 2D convolutional code of dimension s, and S is an invertible $s \times s$ constant matrix, the algebraically equivalent system $\widetilde{\Sigma} = (S^{-1}A_1 S, S^{-1}A_2 S, B_1 S, B_2 S, S^{-1}C, D)$ is also an ISO representation of the code, i.e., $\mathcal{C}(A_1, A_2, B_1, B_2, C, D) = \mathcal{C}(S^{-1}A_1 S, S^{-1}A_2 S, B_1 S, B_2 S, S^{-1}C, D)$; see [1412]. Among the algebraically equivalent ISO representation of a code there exists the Kalman canonical form considered in the next definition.

Definition 10.6.11 ([739]) A 2D linear system $\Sigma = (A_1, A_2, B_1, B_2, C, D)$ with dimension s, k inputs and $n - k$ outputs is in **Kalman canonical form** if

$$
A_1 = \begin{bmatrix} A_{11}^{(1)} & \mathbf{0}_{\delta \times (s-\delta)} \\ A_{21}^{(1)} & A_{22}^{(1)} \end{bmatrix}, \quad A_2 = \begin{bmatrix} A_{11}^{(2)} & \mathbf{0}_{\delta \times (s-\delta)} \\ A_{21}^{(2)} & A_{22}^{(2)} \end{bmatrix},
$$

$$
B_1 = \begin{bmatrix} B_1^{(1)} & \mathbf{0}_{k \times (s-\delta)} \end{bmatrix}, \quad B_2 = \begin{bmatrix} B_1^{(2)} & \mathbf{0}_{k \times (s-\delta)} \end{bmatrix}, \text{ and } C = \begin{bmatrix} C_1 \\ C_2 \end{bmatrix},
$$

where $A_{11}^{(1)}, A_{11}^{(2)} \in \mathbb{F}_q^{\delta \times \delta}$, $B_1^{(1)}, B_1^{(2)} \in \mathbb{F}_q^{k \times \delta}$, $C_1 \in \mathbb{F}_q^{\delta \times (n-k)}$, with $s \geq \delta$ and the remaining matrices of suitable dimensions, and $\Sigma_1 = (A_{11}^{(1)}, A_{11}^{(2)}, B_1^{(1)}, B_1^{(2)}, C_1, D)$ a locally reachable system. Then Σ_1 is called the **largest locally reachable subsystem** of Σ.

Theorem 10.6.12 ([1412]) *Let* $\Sigma = (A_1, A_2, B_1, B_2, C, D)$ *be an ISO representation of a* 2D *convolutional code* $\mathcal{C}$. *Let* S *be an invertible constant matrix such that*

$$
\widetilde{\Sigma} = (S^{-1}A_1 S, S^{-1}A_2 S, B_1 S, B_2 S, S^{-1}C, D)
$$

is in Kalman reachability canonical form, and let

$$
\Sigma_1 = (A_{11}^{(1)}, A_{11}^{(2)}, B_1^{(1)}, B_1^{(2)}, C_1, D)
$$

be the largest locally reachable subsystem of $\widetilde{\Sigma}$. *Then* $\mathcal{C} = \mathcal{C}(A_{11}^{(1)}, A_{11}^{(2)}, B_1^{(1)}, B_1^{(2)}, C_1, D)$.

Unlike 1D convolutional codes, there are no characterizations of minimal ISO representations of a 2D convolutional code. However, Theorem 10.6.12 allows one to obtain a necessary condition for minimality; i.e., a minimal ISO representation of a 2D convolutional code must be locally reachable.

10.7 Connections to Symbolic Dynamics

In Section 10.4 we explained in detail a close connection between convolutional codes and linear systems. Concepts closely connected to convolutional codes appear also in automata theory and in the theory of symbolic dynamics. The survey article of Marcus [1329] provides details. In this section we describe the connection to symbolic dynamics. The reader will find more details on this topic in [1126, 1264, 1596]. In the sequel we closely follow [1596].

In symbolic dynamics one often works with a finite alphabet $\mathcal{A} := \mathbb{F}_q^n$ and then considers sequence spaces such as $\mathcal{A}^{\mathbb{Z}}$, $\mathcal{A}^{\mathbb{N}}$, or $\mathcal{A}^{\mathbb{N}_0}$. In order to be consistent with the rest of the chapter we will work in the sequel with $\mathcal{A}^{\mathbb{N}_0}$, i.e., with the *nonnegative time axis* $\mathbb{N}_0$.

Let k be a natural number. A **block** over the alphabet $\mathcal{A}$ is defined as a finite string $\beta = x_1 x_2 \cdots x_k$ consisting of the k elements $x_i \in \mathcal{A}$, $i = 1, \dots, k$. If $w(z) = \sum_i w_i z^i \in \mathcal{A}[[z]]$ is a sequence, one says that the block β occurs in w if there is some integer j such that $\beta = w_j w_{j+1} \cdots w_{k+j-1}$. If $X \subset \mathcal{A}[[z]]$ is any subset, we denote by $\mathscr{B}(X)$ the set of blocks which occur in some element of X.

As we will explain in this section one can view observable convolutional codes as the dual of linear, compact, irreducible and shift-invariant subsets of $\mathbb{F}_q^n[[z]]$. In order to establish this result we will have to explain the basic definitions from symbolic dynamics.

For this consider a set $\mathscr{F}$ of blocks. It is possible that this set is infinite.

Definition 10.7.1 The subset $X \subset \mathcal{A}[[z]]$ consisting of all sequences $w(z)$ which do not contain any of the (forbidden) blocks of $\mathscr{F}$ is called a **shift space**.

The left-shift operator is the $\mathbb{F}_q$-linear map

$$\sigma: \ \mathbb{F}_q[[z]] \longrightarrow \mathbb{F}_q[[z]] \ \text{ where } \ w(z) \longmapsto z^{-1}(w(z) - w(0)).$$

Let I_n be the $n \times n$ identity matrix acting on $\mathcal{A} = \mathbb{F}_q^n$. The shift map σ extends to the **shift map**

$$\sigma I_n: \ \mathcal{A}[[z]] \longrightarrow \mathcal{A}[[z]].$$

One says that $X \subset \mathcal{A}[[z]]$ is a **shift-invariant set** if $(\sigma I_n)(X) \subset X$.

Shift spaces can be characterized in a topological manner.

Definition 10.7.2 Let $v(z) = \sum_i v_i z^i$ and $w(z) = \sum_i w_i z^i$ be elements of $\mathcal{A}[[z]]$. Let $\mathrm{d_H}(x, y)$ be the Hamming distance between elements in $\mathcal{A}$. One defines the distance between the sequences $v(z)$ and $w(z)$ as

$$\mathrm{d}(v(z), w(z)) := \sum_{i \in \mathbb{N}_0} 2^{-i} \mathrm{d_H}(v_i, w_i).$$

Note that in this metric two elements $v(z), w(z)$ are 'close' if they coincide in a large number of elements in the beginning of the sequence. The function $\mathrm{d}(\cdot, \cdot)$ defines a metric on the sequence space $\mathcal{A}^{\mathbb{N}_0}$. The metric introduced in Definition 10.7.2 is equivalent to the metric described in [1264, Example 6.1.10]. The following result can be found in [1264, Theorem 6.1.21].

Theorem 10.7.3 *A subset of $\mathcal{A}[[z]]$ is a shift space if and only if it is shift-invariant and compact.*

Next we need the notion of irreducibility.

Definition 10.7.4 A shift space $X \subset \mathcal{A}[[z]]$ is called **irreducible** if for every ordered pair of blocks β, γ of $\mathscr{B}(X)$, there is a block μ such that the concatenated block $\beta\mu\gamma$ is in $\mathscr{B}(X)$.

Of particular interest are shift spaces which have a *kernel representation*. For this let $P(z)$ be an $r \times n$ matrix having entries in the polynomial ring $\mathbb{F}_q[z]$. Define the set

$$\mathcal{B} = \left\{ w(z) \in \mathcal{A}[[z]] \mid P(\sigma)w(z)^{\mathsf{T}} = \mathbf{0}^{\mathsf{T}} \right\}. \tag{10.8}$$

Subsets $\mathcal{B} \subseteq \mathcal{A}[[z]]$ having the particular form (10.8) can be characterized in a purely topological manner.

Theorem 10.7.5 *A subset $\mathcal{B} \subseteq \mathcal{A}[[z]]$ has a kernel representation of the form (10.8) if and only if $\mathcal{B}$ is a linear, irreducible, compact and shift invariant subset of $\mathcal{A}[[z]]$.*

Subsets of the form (10.8) also appear prominently in the behavioral theory of linear systems championed by Jan Willems. The following theorem was proven by Willems [1892, Theorem 5]. Note that a metric space is called complete if every Cauchy sequence converges in this space.

Theorem 10.7.6 *A subset $\mathcal{B} \subset \mathbb{F}_q^n[[z]]$ is linear, time-invariant, and complete if and only if $\mathcal{B}$ has a representation of the form (10.8).*

There is a small difference in the above notions as a subset $\mathcal{B} \subset \mathcal{A}[[z]]$ which is linear, irreducible, compact, and shift invariant is automatically also a 'controllable behavior' in the sense of Willems.

We are now in a position to connect to convolutional codes (polynomial modules) using Pontryagin duality. For this consider the bilinear form:

$$(\cdot,\cdot): \quad \mathbb{F}_q^n[[z]] \times \mathbb{F}_q^n[z] \quad \longrightarrow \quad \mathbb{F}_q$$
$$(w,v) \quad \longmapsto \quad \sum_{i=0}^{\infty} \langle w_i, v_i \rangle,$$

where $\langle \cdot, \cdot \rangle$ represents the standard dot product on $\mathcal{A} = \mathbb{F}_q^n$. As the sum has only finite many nonzero terms the bilinear form $(\cdot, \cdot)$ is well defined and nondegenerate. Using this bilinear form, for a subset $\mathcal{C}$ of $\mathbb{F}_q^n[z]$ one defines the **annihilator**

$$\mathcal{C}^\perp = \{w \in \mathbb{F}_q^n[[z]] \mid (w,v) = 0 \text{ for all } v \in \mathcal{C}\};$$

analogously the **annihilator** of a subset $\mathcal{B}$ of $\mathbb{F}_q^n[[z]]$ is

$$\mathcal{B}^\perp = \{v \in \mathbb{F}_q^n[z] \mid (w,v) = 0 \text{ for all } w \in \mathcal{B}\}.$$

The relation between these two annihilator operations is given by the following theorem which was derived and proven in [1597].

Theorem 10.7.7 *If $\mathcal{C} \subseteq \mathbb{F}_q^n[z]$ is a convolutional code with generator matrix $G(z)$, then $\mathcal{C}^\perp$ is a linear, left-shift-invariant, and complete behavior with kernel representation $P(z) = G(z)^T$. Conversely, if $\mathcal{B} \subseteq \mathbb{F}_q^n[[z]]$ is a linear, left-shift-invariant, and complete behavior with kernel representation $P(z)$, then $\mathcal{B}^\perp$ is a convolutional code with generator matrix $G(z) = P(z)^T$. Moreover $\mathcal{C} \subseteq \mathbb{F}_q^n[z]$ is noncatastrophic if and only if $\mathcal{C}^\perp$ is a controllable behavior.*

Acknowledgement

This work is supported by The Center for Research and Development in Mathematics and Applications (CIDMA) through the Portuguese Foundation for Science and Technology (FCT - Fundação para a Ciência e a Tecnologia), references UIDB/04106/2020 and UIDP/04106/2020, by the Swiss National Science Foundation grant n. 188430, and by the German Research Foundation grant LI 3101/1-1.

Chapter 11

Rank-Metric Codes

Elisa Gorla

Université de Neuchâtel

11.1 Definitions, Isometries, and Equivalence of Codes

Let q be a prime power and let $\mathbb{F}_q$ denote the finite field with q elements. Let m, n be positive integers and denote by $\mathbb{F}_q^{n \times m}$ the $\mathbb{F}_q$-vector space of matrices of size $n \times m$ with entries in $\mathbb{F}_q$.

In this chapter, we discuss the mathematical foundations of rank-metric codes. Rank-metric codes are also considered in Chapters 14 and 29. We restrict our attention to linear codes. All dimensions are over $\mathbb{F}_q$, unless otherwise stated.

Definition 11.1.1 For a matrix $A \in \mathbb{F}_q^{n \times m}$, we let rank$(A)$ denote the rank of A. The function

$$\mathrm{d_R} : \mathbb{F}_q^{n \times m} \times \mathbb{F}_q^{n \times m} \longrightarrow \mathbb{F}_q^{n \times m}$$
$$(A, B) \longmapsto \mathrm{rank}(A - B)$$

is a distance on $\mathbb{F}_q^{n \times m}$, which we call **rank distance** or simply **distance**. The rank is the corresponding **weight function**.

A (**matrix**) **rank-metric code** is an $\mathbb{F}_q$-linear subspace $\mathcal{C} \subseteq \mathbb{F}_q^{n \times m}$.

A class of rank-metric codes that has received a lot of attention is that of vector rank-metric codes, introduced independently by Gabidulin [760] and Roth [1601].

Definition 11.1.2 The **rank weight** rank(v) of a vector $v \in \mathbb{F}_{q^m}^n$ is the dimension of the $\mathbb{F}_q$-linear space generated by its entries. The function

$$\mathrm{d_R} : \mathbb{F}_{q^m}^n \times \mathbb{F}_{q^m}^n \longrightarrow \mathbb{F}_{q^m}^n$$
$$(u, v) \longmapsto \mathrm{rank}(u - v)$$

is a distance on $\mathbb{F}_{q^m}^n$, which we call **rank distance** or simply **distance**.[1]

A **vector rank-metric code** is an $\mathbb{F}_{q^m}$-linear subspace $C \subseteq \mathbb{F}_{q^m}^n$.

---

[1] In both Definitions 11.1.1 and 11.1.2, the distance $\mathrm{d_R}$ satisfies the triangle inequality and so is indeed a metric as in Definition 1.6.3, justifying the term 'rank-metric code'.

Every vector rank-metric code can be regarded as a rank-metric code, up to the choice of a basis of $\mathbb{F}_{q^m}$ over $\mathbb{F}_q$.

Definition 11.1.3 Let $\Gamma = \{\gamma_1, \gamma_2, \ldots, \gamma_m\}$ be a basis of $\mathbb{F}_{q^m}$ over $\mathbb{F}_q$ and let $v \in \mathbb{F}_{q^m}^n$. Define $\Gamma(v) \in \mathbb{F}_q^{n \times m}$ via the identity

$$v_i = \sum_{j=1}^{m} \Gamma_{ij}(v)\gamma_j \ \text{ for } i = 1, \ldots, n.$$

Let $C \subseteq \mathbb{F}_{q^m}^n$ be a vector rank-metric code. The set

$$\Gamma(C) = \{\Gamma(v) \mid v \in C\}$$

is the **rank-metric code associated to C with respect to Γ**.

Example 11.1.4 Let C be the vector rank-metric code $C = \langle (1, \alpha) \rangle \subseteq \mathbb{F}_8^2$. Let $\mathbb{F}_8 = \mathbb{F}_2[\alpha]/(\alpha^3 + \alpha + 1)$ and let $\gamma_1 = 1$, $\gamma_2 = \alpha$, $\gamma_3 = \alpha^2$. Then $\Gamma = \{\gamma_1, \gamma_2, \gamma_3\}$ is a basis of $\mathbb{F}_8$ over $\mathbb{F}_2$ and

$$\Gamma(1, \alpha) = \begin{bmatrix} 1 & 0 & 0 \\ 0 & 1 & 0 \end{bmatrix}, \ \Gamma(\alpha, \alpha^2) = \begin{bmatrix} 0 & 1 & 0 \\ 0 & 0 & 1 \end{bmatrix}, \ \Gamma(\alpha^2, \alpha + 1) = \begin{bmatrix} 0 & 0 & 1 \\ 1 & 1 & 0 \end{bmatrix}.$$

Hence

$$\Gamma(C) = \left\langle \begin{bmatrix} 1 & 0 & 0 \\ 0 & 1 & 0 \end{bmatrix}, \begin{bmatrix} 0 & 1 & 0 \\ 0 & 0 & 1 \end{bmatrix}, \begin{bmatrix} 0 & 0 & 1 \\ 1 & 1 & 0 \end{bmatrix} \right\rangle \subseteq \mathbb{F}_2^{2 \times 3}.$$

The image $\Gamma(C)$ of a vector rank-metric code C via Γ as defined above is a rank-metric code, whose parameters are determined by those of C. The proof of the next proposition is easy and may be found, e.g., in [842, Section 1].

Proposition 11.1.5 *The map $v \mapsto \Gamma(v)$ is an $\mathbb{F}_q$-linear isometry, i.e., it is a homomorphism of $\mathbb{F}_q$-vector spaces which preserves the rank. In particular, if $C \subseteq \mathbb{F}_{q^m}^n$ is a vector rank-metric code of dimension k over $\mathbb{F}_{q^m}$, then $\Gamma(C)$ is an $\mathbb{F}_q$-linear rank-metric code of dimension mk over $\mathbb{F}_q$.*

The following is the natural notion of equivalence for rank-metric codes.

Definition 11.1.6 An $\mathbb{F}_q$-**linear isometry** φ of $\mathbb{F}_q^{n \times m}$ is an $\mathbb{F}_q$-linear homomorphism $\varphi : \mathbb{F}_q^{n \times m} \to \mathbb{F}_q^{n \times m}$ such that $\text{rank}(\varphi(M)) = \text{rank}(M)$ for every $M \in \mathbb{F}_q^{n \times m}$.

Two rank-metric codes $\mathcal{C}, \mathcal{D} \subseteq \mathbb{F}_q^{n \times m}$ are **equivalent** if there is an $\mathbb{F}_q$-linear isometry $\varphi : \mathbb{F}_q^{n \times m} \to \mathbb{F}_q^{n \times m}$ such that $\varphi(\mathcal{C}) = \mathcal{D}$. If $\mathcal{C}$ and $\mathcal{D}$ are equivalent rank-metric codes, we write $\mathcal{C} \sim \mathcal{D}$.

Some authors define a notion of equivalence for vector rank-metric codes as follows.

Definition 11.1.7 An $\mathbb{F}_{q^m}$-**linear isometry** ψ of $\mathbb{F}_{q^m}^n$ is an $\mathbb{F}_{q^m}$-linear homomorphism $\psi : \mathbb{F}_{q^m}^n \to \mathbb{F}_{q^m}^n$ such that $\text{rank}(\psi(v)) = \text{rank}(v)$ for every $v \in \mathbb{F}_{q^m}^n$.

Two vector rank-metric codes $C, D \subseteq \mathbb{F}_{q^m}^n$ are **equivalent** if there is an $\mathbb{F}_{q^m}$-linear isometry $\psi : \mathbb{F}_{q^m}^n \longrightarrow \mathbb{F}_{q^m}^n$ such that $\psi(C) = D$. If C and D are equivalent vector rank-metric codes, we write $C \sim D$.

Notice however that Definition 11.1.3 allows us to apply the notion of equivalence from Definition 11.1.6 to vector rank-metric codes. It is therefore natural to ask whether the rank-metric codes associated to equivalent vector rank-metric codes are also equivalent. It is easy to show that the answer is affirmative.

Proposition 11.1.8 ([841], Proposition 1.15) *Let $C, D \subseteq \mathbb{F}_{q^m}^n$ be vector rank-metric codes. Let Γ and Γ' be bases of $\mathbb{F}_{q^m}$ over $\mathbb{F}_q$. If $C \sim D$, then $\Gamma(C) \sim \Gamma'(D)$.*

Linear isometries of $\mathbb{F}_q^{n \times m}$ and of $\mathbb{F}_{q^m}^n$ can be easily characterized. The following result was shown by Hua for fields of odd characteristic and by Wan for fields of characteristic 2.

Theorem 11.1.9 ([993, 1860]) *Let $\varphi : \mathbb{F}_q^{n \times m} \to \mathbb{F}_q^{n \times m}$ be an $\mathbb{F}_q$-linear isometry with respect to the rank metric.*

(a) *If $m \neq n$, then there exist matrices $A \in \mathrm{GL}_n(\mathbb{F}_q)$ and $B \in \mathrm{GL}_m(\mathbb{F}_q)$ such that $\varphi(M) = AMB$ for all $M \in \mathbb{F}_q^{n \times m}$.*

(b) *If $m = n$, then there exist matrices $A, B \in \mathrm{GL}_n(\mathbb{F}_q)$ such that either $\varphi(M) = AMB$ for all $M \in \mathbb{F}_q^{n \times n}$, or $\varphi(M) = AM^TB$ for all $M \in \mathbb{F}_q^{n \times n}$.*

The corresponding characterization of isometries of $\mathbb{F}_{q^m}^n$ was given by Berger.

Theorem 11.1.10 ([162]) *Let $\psi : \mathbb{F}_{q^m}^n \to \mathbb{F}_{q^m}^n$ be an $\mathbb{F}_{q^m}$-linear isometry with respect to the rank metric. Then there exist $\alpha \in \mathbb{F}_{q^m}^*$ and $B \in \mathrm{GL}_n(\mathbb{F}_q)$ such that $\psi(v) = \alpha vB$ for all $v \in \mathbb{F}_{q^m}^n$.*

Notation 11.1.11 For a vector rank-metric code $C \subseteq \mathbb{F}_{q^m}^n$ and $B \in \mathrm{GL}_n(\mathbb{F}_q)$, let

$$CB = \{vB \mid v \in C\} \subseteq \mathbb{F}_{q^m}^n.$$

For a rank-metric code $\mathcal{C} \subseteq \mathbb{F}_q^{n \times m}$, let

$$\mathcal{C}^\mathsf{T} = \{M^\mathsf{T} \mid M \in \mathcal{C}\} \subseteq \mathbb{F}_q^{m \times n}.$$

The MacWilliams Extension Theorem 1.8.6 is a classical result in the theory of linear block codes in $\mathbb{F}_q^n$ with the Hamming distance. It essentially says that any linear isometry of block codes can be extended to a linear isometry of the ambient space $\mathbb{F}_q^n$. It is natural to ask whether an analog of the MacWilliams Extension Theorem holds for rank-metric codes. In other words, given rank-metric codes $\mathcal{C}, \mathcal{D} \subseteq \mathbb{F}_q^{n \times m}$ and an $\mathbb{F}_q$-linear isometry $f : \mathcal{C} \to \mathcal{D}$, one may ask whether there exists an $\mathbb{F}_q$-linear isometry $\varphi : \mathbb{F}_q^{n \times m} \to \mathbb{F}_q^{n \times m}$ such that $\varphi |_\mathcal{C} = f$. The answer is no, as the next example shows. More counterexamples can be found in [131] and in the preprint [501, Section 7].

Example 11.1.12 ([131, Example 2.9(a)]) Denote by $\mathbf{0}_{2 \times 1}$ the zero matrix of size 2×1 and let

$$\mathcal{C} = \left\{ \begin{bmatrix} A & \mathbf{0}_{2 \times 1} \end{bmatrix} \mid A \in \mathbb{F}_2^{2 \times 2} \right\} \subseteq \mathbb{F}_2^{2 \times 3}.$$

Let $\varphi : \mathcal{C} \to \mathbb{F}_2^{2 \times 3}$ be defined by $\varphi(\begin{bmatrix} A & \mathbf{0}_{2 \times 1} \end{bmatrix}) = \begin{bmatrix} A^\mathsf{T} & \mathbf{0}_{2 \times 1} \end{bmatrix}$. Then φ is an $\mathbb{F}_2$-linear isometry defined on $\mathcal{C}$ which is not the restriction to $\mathcal{C}$ of an $\mathbb{F}_2$-linear isometry of $\mathbb{F}_2^{2 \times 3}$. In fact, there is no choice for $\varphi \left(\begin{bmatrix} 0 & 0 & 1 \\ 0 & 0 & 0 \end{bmatrix} \right)$ that preserves the property that φ is an $\mathbb{F}_2$-linear isometry.

11.2   The Notion of Support in the Rank Metric

In analogy with the notion of support of a codeword for linear block codes, one may define the support of a codeword in a vector rank-metric code. For a matrix $M \in \mathbb{F}_q^{n \times m}$, we denote by $\mathrm{colsp}(M) \subseteq \mathbb{F}_q^n$ the $\mathbb{F}_q$-vector space generated by the columns of M and by $\mathrm{rowsp}(M) \subseteq \mathbb{F}_q^m$ the $\mathbb{F}_q$-vector space generated by the rows of M.

Definition 11.2.1 ([1061, Definition 2.1]) Let $C \subseteq \mathbb{F}_{q^m}^n$ be a vector rank-metric code, and let $\Gamma = \{\gamma_1, \gamma_2, \ldots, \gamma_m\}$ be a basis of $\mathbb{F}_{q^m}$ over $\mathbb{F}_q$. The **support** of $v \in C$ is the $\mathbb{F}_q$-linear space

$$\mathrm{supp}(v) = \mathrm{colsp}(\Gamma(v)) \subseteq \mathbb{F}_q^n.$$

The **support** of a subcode $D \subseteq C$ is

$$\mathrm{supp}(D) = \sum_{v \in D} \mathrm{supp}(v) \subseteq \mathbb{F}_q^n.$$

Notice that $\mathrm{supp}(v)$ does not depend on the choice of the basis Γ, since if Γ' is another basis of $\mathbb{F}_{q^m}$ over $\mathbb{F}_q$, then there exists a $B \in \mathrm{GL}_m(\mathbb{F}_q)$ such that $\Gamma(v) = \Gamma'(v)B$. This also implies that $\mathrm{supp}(D)$ does not depend on the choice of Γ. See also [841, Proposition 1.13].

In the context of rank-metric codes, we define the support as follows.

Definition 11.2.2 Let $\mathcal{C} \subseteq \mathbb{F}_q^{n \times m}$ be a rank-metric code.
If $n \le m$ define the **support** of $M \in \mathcal{C}$ as the $\mathbb{F}_q$-linear space

$$\mathrm{supp}(M) = \mathrm{colsp}(M) \subseteq \mathbb{F}_q^n.$$

If $n > m$ define the **support** of $M \in \mathcal{C}$ as the $\mathbb{F}_q$-linear space

$$\mathrm{supp}(M) = \mathrm{rowsp}(M) \subseteq \mathbb{F}_q^m.$$

The **support** of a subcode $\mathcal{D} \subseteq \mathcal{C}$ is

$$\mathrm{supp}(\mathcal{D}) = \sum_{M \in \mathcal{D}} \mathrm{supp}(M) \subseteq \mathbb{F}_q^{\min\{m,n\}}.$$

Notice that, if $n \le m$, Definition 11.2.2 agrees with Definition 11.2.1, when restricted to rank-metric codes associated to vector rank-metric codes. Precisely, if $C \subseteq \mathbb{F}_{q^m}^n$ is a vector rank-metric code and $\Gamma = \{\gamma_1, \ldots, \gamma_m\}$ is a basis of $\mathbb{F}_{q^m}$ over $\mathbb{F}_q$, then

$$\mathrm{supp}(\Gamma(v)) = \mathrm{supp}(v)$$

for all $v \in C$, under the assumption that $n \le m$.

Remark 11.2.3 If $n > m$, the support of $v \in \mathbb{F}_{q^m}^n$ according to Definition 11.2.1 is $\mathrm{colsp}(\Gamma(v))$, while the support of $\Gamma(v) \in \mathbb{F}_q^{n \times m}$ according to Definition 11.2.2 is $\mathrm{rowsp}(\Gamma(v))$. In other words, Definition 11.2.2 for the elements of the rank-metric code associated to a vector rank-metric code does not coincide with Definition 11.2.1 for the elements of the vector rank-metric code. This will not create confusion, since our notation allows us to distinguish the two situations: $\mathrm{supp}(v) = \mathrm{colsp}(\Gamma(v))$ while $\mathrm{supp}(\Gamma(v)) = \mathrm{rowsp}(\Gamma(v))$ if $n > m$.

We wish to stress that, in the context of matrices, taking the support of the transpose yields a different, but well-behaved notion of support of a matrix. Below we make a few remarks on different possible notions of support and on why we choose to adopt Definition 11.2.2. It is clear that, depending on the application or on the information that one wishes to encode, one may also choose to work with different notions of support.

Remark 11.2.4 If $n = m$, then one may define a notion of support by considering row spaces instead of column spaces. This yields a different, but substantially equivalent notion of support. A different, but possibly interesting, notion of support for a square matrix would be defining the support of $M \in \mathbb{F}_q^{n \times n}$ to be the pair of vector spaces $(\mathrm{rowsp}(M), \mathrm{colsp}(M))$. This is connected to the definition of generalized weights (see Section 11.5) and to the approach taken in [841] for studying generalized weights of square matrices via q-polymatroids (see Section 11.6).

Remark 11.2.5 For any value of m, n, one may define a different notion of support as follows: For a rank-metric code $\mathcal{C} \subseteq \mathbb{F}_q^{n \times m}$ and for $M \in \mathcal{C}$, let

$$\mathrm{supp}(M) = \mathrm{rowsp}(M) \subseteq \mathbb{F}_q^m \quad \text{if } n \leq m \tag{11.1}$$

and let

$$\mathrm{supp}(M) = \mathrm{colsp}(M) \subseteq \mathbb{F}_q^n \quad \text{if } n > m. \tag{11.2}$$

Then the support of a subcode $\mathcal{D} \subseteq \mathcal{C}$ is

$$\mathrm{supp}(\mathcal{D}) = \sum_{M \in \mathcal{D}} \mathrm{supp}(M) \subseteq \mathbb{F}_q^{\max\{m,n\}}. \tag{11.3}$$

This definition yields a different notion of support from that of Definition 11.2.2; in particular it takes values in $\mathbb{F}_q^{\max\{n,m\}}$. One can check that both notions of support are regular in the sense of [1575]. However, the definition of support in (11.1), (11.2), and (11.3) for $n \neq m$ yields an empty extremality theory in the sense of [1575, Section 7] and a series of redundant MacWilliams Identities. Moreover, in [841] we showed that, for $n < m$, the q-polymatroid determined by the supports as in Definition 11.2.2 allows one to easily recover the generalized weights of the code, while the q-polymatroid determined by the supports as in (11.1), (11.2), and (11.3) does not. With these in mind, we choose to adopt Definition 11.2.2.

Remark 11.2.6 Some authors choose to work with a definition of support which is the same for every value of m, n; e.g., in [1341] the authors define

$$\mathrm{supp}(M) = \mathrm{colsp}(M) \quad \text{for any } m, n \text{ and any } M \in \mathbb{F}_q^{n \times m}. \tag{11.4}$$

This notion of support agrees with Definition 11.2.2 for $n \leq m$ and with the definition discussed in the previous remark for $n > m$. For all the reasons discussed in the previous remark, for $n > m$ we prefer Definition 11.2.2 to this definition. Notice however that the definition of support in (11.4) is compatible with the notion of generalized matrix weights as defined by Martínez-Peñas and Matsumoto (see Definition 11.5.12). In this chapter, however, we define generalized weights as in Definition 11.5.7, which is compatible with the notion of support as by Definition 11.2.2.

Given a definition of support, it is natural to consider the subcodes of a code which are supported on a fixed vector space.

Definition 11.2.7 Let $\mathcal{C} \subseteq \mathbb{F}_q^{n \times m}$ be a rank-metric code. Let $V \subseteq \mathbb{F}_q^{\min\{m,n\}}$ be a vector subspace. The **subcode of $\mathcal{C}$ supported on V** is

$$\mathcal{C}(V) = \{M \in \mathcal{C} \mid \mathrm{supp}(M) \subseteq V\}.$$

11.3 MRD Codes and Optimal Anticodes

The basic invariants of a rank-metric code $\mathcal{C} \subseteq \mathbb{F}_q^{n \times m}$ are n, m, the dimension of $\mathcal{C}$ as a vector space over $\mathbb{F}_q$, and its minimum distance.

Definition 11.3.1 The **minimum distance** of a rank-metric code $0 \neq \mathcal{C} \subseteq \mathbb{F}_q^{n \times m}$ is the integer
$$d_{\min}(\mathcal{C}) = \min \{\operatorname{rank}(M) \mid M \in \mathcal{C}, \ M \neq 0\}.$$
We define the minimum distance of the trivial code $\mathcal{C} = 0$ as
$$d_{\min}(0) = \min \{m, n\} + 1.$$

Sometimes, one is also interested in the maximum rank of an element of $\mathcal{C}$.

Definition 11.3.2 Let $\mathcal{C} \subseteq \mathbb{F}_q^{n \times m}$ be a rank-metric code. The **maximum rank** of $\mathcal{C}$ is
$$\max \operatorname{rank}(\mathcal{C}) = \max \{\operatorname{rank}(M) \mid M \in \mathcal{C}\}.$$

Analogous definitions can be given for vector rank-metric codes, using the corresponding rank distance.

Definition 11.3.3 The **minimum distance** of a vector rank-metric code $0 \neq C \subseteq \mathbb{F}_{q^m}^n$ is
$$d_{\min}(C) = \min \{\operatorname{rank}(v) \mid v \in C, \ v \neq 0\}.$$
The minimum distance of the trivial code $C = 0$ is
$$d_{\min}(0) = n + 1.$$

It follows from Proposition 11.1.5 that
$$d_{\min}(C) = d_{\min}(\Gamma(C))$$
for any vector rank-metric code $0 \neq C \subseteq \mathbb{F}_{q^m}^n$ and any basis Γ of $\mathbb{F}_{q^m}$ over $\mathbb{F}_q$.

Definition 11.3.4 Let $C \subseteq \mathbb{F}_{q^m}^n$ be a vector rank-metric code. The **maximum rank** of C is
$$\max \operatorname{rank}(C) = \max \{\operatorname{rank}(v) \mid v \in C\}.$$

Both the minimum distance and the maximum rank of a code are related to the other invariants of the code. The first inequality in the next theorem goes under the name of Singleton Bound and was proved by Delsarte in [523, Theorem 5.4]. The second goes under the name of Anticode Bound. The Anticode Bound was proved by Meshulam in [1375, Theorem 1] in the square case, but one can check that the proof also works in the rectangular case. The proof by Meshulam relies heavily on a result by König [1144] (see also [888, Theorem 5.1.4]). Finally, a coding-theoretic proof of the Anticode Bound was given by Ravagnani in [1574, Proposition 47].

Theorem 11.3.5 *Let $\mathcal{C} \subseteq \mathbb{F}_q^{n \times m}$ be a rank-metric code. Then*
$$\dim(\mathcal{C}) \leq \max \{n, m\}(\min \{m, n\} - d_{\min}(\mathcal{C}) + 1)$$
and
$$\dim(\mathcal{C}) \leq \max \{n, m\} \cdot \max \operatorname{rank}(\mathcal{C}).$$

Remark 11.3.6 For a vector rank-metric code $C \subseteq \mathbb{F}_{q^m}^n$, the Singleton Bound can be stated as

$$\dim_{\mathbb{F}_{q^m}}(C) \leq n - d_{\min}(C) + 1.$$

This bound appeared in [760, Corollary of Lemma 1], under the assumption that $n \leq m$. However, it is easy to check that the bound holds for any n, m.

Remark 11.3.7 For a vector rank-metric code $C \subseteq \mathbb{F}_{q^m}^n$ with $\dim_{\mathbb{F}_{q^m}}(C) \leq m$, the Anticode Bound can be stated as

$$\dim_{\mathbb{F}_{q^m}}(C) \leq \max \operatorname{rank}(C). \tag{11.5}$$

The bound was proved in [1573, Proposition 11], under the assumption that $n \leq m$. Using the same type of arguments however, one can easily prove the bound in the more general form stated here. Notice moreover that, if $\dim_{\mathbb{F}_{q^m}}(C) > m$, then

$$\dim_{\mathbb{F}_{q^m}}(C) > m \geq \max \operatorname{rank}(C).$$

In particular, the inequality (11.5) cannot hold if $\dim_{\mathbb{F}_{q^m}}(C) > m$.

The codes whose invariants meet the bounds of Theorem 11.3.5 go under the names of MRD codes and optimal anticodes, respectively. They have both been extensively studied.

Definition 11.3.8 A vector rank-metric code $C \subseteq \mathbb{F}_{q^m}^n$ is a **maximum rank distance (MRD) code** if

$$\dim_{\mathbb{F}_{q^m}}(C) = n - d_{\min}(C) + 1.$$

It is an **optimal vector anticode** if

$$\dim_{\mathbb{F}_{q^m}}(C) = \max \operatorname{rank}(C).$$

A rank-metric code $\mathcal{C} \subseteq \mathbb{F}_q^{n \times m}$ is a **maximum rank distance (MRD) code** if

$$\dim(\mathcal{C}) = \max\{n, m\}(\min\{m, n\} - d_{\min}(\mathcal{C}) + 1).$$

It is an **optimal anticode** if

$$\dim(\mathcal{C}) = \max\{n, m\} \cdot \max \operatorname{rank}(\mathcal{C}).$$

Remark 11.3.9 Let $\mathcal{C} \subseteq \mathbb{F}_q^{n \times m}$ be a rank-metric code and let $\mathcal{C}^\mathsf{T} \subseteq \mathbb{F}_q^{m \times n}$ be the code obtained from $\mathcal{C}$ by transposition. The following are immediate consequences of the definitions:

- $\mathcal{C}$ is MRD if and only if $\mathcal{C}^\mathsf{T}$ is MRD,

- $\mathcal{C}$ is an optimal anticode if and only if $\mathcal{C}^\mathsf{T}$ is an optimal anticode.

Several examples and constructions of MRD codes are given in Chapter 14.

Remark 11.3.10 Notice that 0 and $\mathbb{F}_{q^m}^n$ are the only MRD vector rank-metric codes, if $n > m$. In fact, if a vector rank-metric code C exists with

$$\dim_{\mathbb{F}_{q^m}}(C) = n - d_{\min}(C) + 1 < n,$$

then $d_{\min}(C) > 1$. Assume that $C \neq 0$ and let Γ be a basis of $\mathbb{F}_{q^m}$ over $\mathbb{F}_q$. By Proposition 11.1.5 $\dim(\Gamma(C)) = m \dim_{\mathbb{F}_{q^m}}(C)$ and $d_{\min}(C) = d_{\min}(\Gamma(C))$. Using the Singleton Bound, $\Gamma(C)$ is a rank-metric code of dimension

$$\dim(\Gamma(C)) = m(n - d_{\min}(\Gamma(C)) + 1) \leq n(m - d_{\min}(\Gamma(C)) + 1),$$

contradicting the assumption that $n > m$.

It is easy to produce examples of optimal anticodes and optimal vector anticodes.

Example 11.3.11 (Standard optimal anticodes) If $n \leq m$, let $C \subseteq \mathbb{F}_q^{n \times m}$ consist of the matrices whose last $n - k$ rows are zero. Then $\dim(C) = mk$ and $\max \operatorname{rank}(C) = k$; hence C is an optimal anticode.

If $n \geq m$, let $C \subseteq \mathbb{F}_q^{n \times m}$ consist of the matrices whose last $m - k$ columns are zero. Then $\dim(C) = nk$ and $\max \operatorname{rank}(C) = k$; hence C is an optimal anticode.

Example 11.3.12 (Standard optimal vector anticodes) Let $0 \leq k \leq \min\{m, n\}$ and let $C = \langle e_1, \ldots, e_k \rangle \subseteq \mathbb{F}_{q^m}^n$, where e_i denotes the ith vector of the standard basis of $\mathbb{F}_{q^m}^n$. Then $\dim_{\mathbb{F}_{q^m}}(C) = k$ and $\max \operatorname{rank}(C) = k$; hence C is an optimal vector anticode.

Remark 11.3.13 Notice that every optimal vector anticode $C \subseteq \mathbb{F}_{q^m}^n$ has

$$\dim_{\mathbb{F}_{q^m}}(C) = \max \operatorname{rank}(C) \leq \min\{m, n\}.$$

In particular, $\mathbb{F}_{q^m}^n$ is not an optimal vector anticode if $n > m$.

It is natural to ask whether the rank-metric code associated to an MRD vector rank-metric code or to an optimal vector anticode is an MRD rank-metric code or an optimal anticode, respectively. It is easy to show that, up to the trivial exceptions, this happens only if $n \leq m$.

Proposition 11.3.14 *Let $C \subseteq \mathbb{F}_{q^m}^n$ be a vector rank-metric code with $d_{\min}(C) = d$. Let Γ be a basis of $\mathbb{F}_{q^m}$ over $\mathbb{F}_q$ and let $\Gamma(C) \subseteq \mathbb{F}_q^{n \times m}$ be the rank-metric code associated to C with respect to Γ. If $n \leq m$, then:*

(a) C is MRD if and only if $\Gamma(C)$ is MRD.

(b) C is an optimal vector anticode if and only if $\Gamma(C)$ is an optimal anticode.

If $n > m$, then:

(c) The codes C and $\Gamma(C)$ are both MRD if and only if $C = 0$ or $C = \mathbb{F}_{q^m}^n$.

(d) C is an optimal vector anticode and $\Gamma(C)$ is an optimal anticode if and only if $C = 0$.

Proof: Notice that if $C = 0$ or $C = \mathbb{F}_{q^m}^n$, then both C and $\Gamma(C)$ are MRD for any n, m. Moreover, $C = 0$ is an optimal vector anticode and $\Gamma(C) = 0$ is an optimal anticode for any n, m. In the sequel, we suppose that $C \neq 0$. Recall that for any m, n one has $\dim(\Gamma(C)) = m \dim_{\mathbb{F}_{q^m}}(C)$, $d_{\min}(\Gamma(C)) = d$, and $\max \operatorname{rank}(\Gamma(C)) = \max \operatorname{rank}(C)$ by Proposition 11.1.5.

Suppose first that $n \leq m$. For part (a) the code C is MRD if and only if $\dim_{\mathbb{F}_{q^m}}(C) = n - d + 1$. The code $\Gamma(C)$ is MRD if and only if $\dim(\Gamma(C)) = m(n - d + 1)$. Then C is MRD if and only if $\Gamma(C)$ is MRD, giving (a). For (b) the code C is an optimal vector anticode if and only if $\dim_{\mathbb{F}_{q^m}}(C) = \max \operatorname{rank}(C)$. The code $\Gamma(C)$ is an optimal anticode if and only if $\dim(\Gamma(C)) = m \max \operatorname{rank}(C)$. Then C is an optimal vector anticode if and only if $\Gamma(C)$ is an optimal anticode, which is (b).

Suppose now that $n > m$. Then for (c) $C = 0$ and $C = \mathbb{F}_{q^m}^n$ are the only MRD vector rank-metric codes by Remark 11.3.10. The associated code $\Gamma(C)$ is MRD in both cases, giving (c). Finally for (d) the code C is an optimal vector anticode if and only if $\dim_{\mathbb{F}_{q^m}}(C) = \max \operatorname{rank}(C)$. The code $\Gamma(C)$ is an optimal anticode if and only if $\dim(\Gamma(C)) = n \max \operatorname{rank}(C)$. Then C is an optimal vector anticode and $\Gamma(C)$ is an optimal anticode if and only if $n \max \operatorname{rank}(C) = m \max \operatorname{rank}(C)$. Since $n > m$, this is equivalent to $\max \operatorname{rank}(C) = 0$, yielding (d). $\qquad \square$

Optimal anticodes were characterized by de Seguins Pazzis, who proved that, up to code equivalence, they are exactly the standard optimal anticodes of Example 11.3.11.

Theorem 11.3.15 ([511, **Theorem 4 and Theorem 6**]) *Recalling Definition* 11.2.7, *the optimal anticodes of* $\mathbb{F}_q^{n \times m}$ *with respect to the rank metric are exactly the following codes:*

- $\mathbb{F}_q^{n \times m}(V)$ *for some* $V \subseteq \mathbb{F}_q^{\min\{m,n\}}$, *if* $m \neq n$,

- $\mathbb{F}_q^{n \times n}(V)$ *and* $\mathbb{F}_q^{n \times n}(V)^T$ *for some* $V \subseteq \mathbb{F}_q^n$, *if* $m = n$.

In particular, every optimal anticode is equivalent to a standard optimal anticode.

Proof: The only part of the statement which is not contained in the proof of [511, Theorem 4 and Theorem 6] is the claim that every optimal anticode is equivalent to a standard optimal anticode. Notice that the standard optimal anticodes are $\mathbb{F}_q^{n \times m}(E_\ell)$ and $\mathbb{F}_q^{n \times n}(E_\ell)^T$, where $E_\ell = \langle e_1, \ldots, e_\ell \rangle$ and $\ell = 0, \ldots, \min\{m, n\}$.

If $n \leq m$, let $A \in \mathrm{GL}_n(\mathbb{F}_q)$ be a matrix whose first $k = \dim(V)$ columns are a basis of V and let $B \in \mathrm{GL}_m(\mathbb{F}_q)$ be any matrix. If $n > m$, let $A \in \mathrm{GL}_n(\mathbb{F}_q)$ be any matrix and let $B \in \mathrm{GL}_m(\mathbb{F}_q)$ be a matrix whose first $k = \dim(V)$ rows are a basis of V. Then

$$A\mathbb{F}_q^{n \times m}(E_k)B = \mathbb{F}_q^{n \times m}(V). \tag{11.6}$$

In fact, if $n \leq m$ and $\mathrm{colsp}(M) \subseteq E_k$, then

$$\mathrm{colsp}(AMB) = \mathrm{colsp}(AM) \subseteq V.$$

Similarly, if $n > m$ and $\mathrm{rowsp}(M) \subseteq E_k$, then

$$\mathrm{rowsp}(AMB) = \mathrm{rowsp}(MB) \subseteq V.$$

Therefore, $AMB \in \mathbb{F}_q^{n \times m}(V)$ for every $M \in \mathbb{F}_q^{n \times m}(E_k)$. Since the two vector spaces in (11.6) have the same dimension and one is a subset of the other, they must be equal. Hence $\mathbb{F}_q^{n \times m}(V)$ is equivalent to the standard optimal anticode $\mathbb{F}_q^{n \times m}(E_k)$. Moreover, if $n = m$, by taking the transpose of (11.6) one has

$$B^T \mathbb{F}_q^{n \times n}(E_k)^T A^T = \mathbb{F}_q^{n \times n}(V)^T.$$

Therefore $\mathbb{F}_q^{n \times n}(V)^T$ is equivalent to the standard optimal anticode $\mathbb{F}_q^{n \times n}(E_k)^T$. □

Optimal vector anticodes were characterized by Ravagnani in [1573, Theorem 18], under the assumption that $n \leq m$. One can also show that, up to code equivalence, optimal vector anticodes are exactly the standard optimal vector anticodes of Example 11.3.12. Notice that a vector rank-metric code $C \subseteq \mathbb{F}_{q^m}^n$ with $\dim_{\mathbb{F}_{q^m}}(C) > m$ cannot be an optimal vector anticode by Remark 11.3.13. Hence we may assume without loss of generality that $\dim_{\mathbb{F}_{q^m}}(C) \leq m$.

Theorem 11.3.16 *Let* $C \subseteq \mathbb{F}_{q^m}^n$ *be a vector rank-metric code with* $k = \dim_{\mathbb{F}_{q^m}}(C) \leq m$. *The following are equivalent.*

(a) *C is an optimal vector anticode,*

(b) *C has a basis consisting of vectors with entries in* $\mathbb{F}_q$,

(c) *$C \sim \langle e_1, e_2, \ldots, e_k \rangle$, where e_i denotes the ith vector of the standard basis of* $\mathbb{F}_{q^m}^n$.

Proof: Let $\phi : \mathbb{F}_{q^m}^n \to \mathbb{F}_{q^m}^n$ be the Frobenius endomorphism, defined by $\phi(v_1, v_2, \ldots, v_n) = (v_1^q, v_2^q, \ldots, v_n^q)$. Recall that a subspace $V \subseteq \mathbb{F}_{q^m}^n$ is fixed by ϕ if and only if it has an $\mathbb{F}_{q^m}$-basis that consists of vectors with entries in $\mathbb{F}_q$. Combining this fact with the argument in [1573, Theorem 18], one has that C is an optimal vector anticode if and only if C has a basis consisting of vectors with entries in $\mathbb{F}_q$. Although [1573, Theorem 18] is proved under the assumption that $n \leq m$, one can check that the proof works for arbitrary n, m, under the assumption that $\dim_{\mathbb{F}_{q^m}}(C) \leq m$. This proves that (a) and (b) are equivalent.

By Theorem 11.1.10, $C \sim \langle e_1, e_2, \ldots, e_k \rangle$ if and only if there exist $\alpha \in \mathbb{F}_{q^m}^*$ and $B \in GL_m(\mathbb{F}_q)$ such that

$$C = \alpha \langle e_1, e_2, \ldots, e_k \rangle B = \langle e_1, e_2, \ldots, e_k \rangle B.$$

Equivalence of (b) and (c) follows readily. □

11.4   Duality and the MacWilliams Identities

The usual scalar product for matrices induces a notion of dual for rank-metric codes.

Definition 11.4.1 The **dual** of $\mathcal{C} \subseteq \mathbb{F}_q^{n \times m}$ is

$$\mathcal{C}^\perp = \{M \in \mathbb{F}_q^{n \times m} \mid \mathrm{TR}(MN^{\mathsf{T}}) = 0 \text{ for all } N \in \mathcal{C}\},$$

where $\mathrm{TR}(\cdot)$ denotes the trace of a matrix.

The usual scalar product of $\mathbb{F}_{q^m}^n$ induces a notion of dual for vector rank-metric codes.

Definition 11.4.2 The **dual** of $C \subseteq \mathbb{F}_{q^m}^n$ is the vector rank-metric code

$$C^\perp := \{v \in \mathbb{F}_{q^m}^n \mid \langle v, w \rangle = 0 \text{ for all } w \in C\},$$

where $\langle \cdot, \cdot \rangle$ is the standard inner product of $\mathbb{F}_{q^m}^n$.

The two notions of dual code are compatible with the definition of associated rank-metric code with respect to a basis of $\mathbb{F}_{q^m}$ over $\mathbb{F}_q$, for a suitable choice of bases.

Definition 11.4.3 Two bases $\Gamma = \{\gamma_1, \ldots, \gamma_m\}$ and $\Gamma' = \{\gamma'_1, \ldots, \gamma'_m\}$ of $\mathbb{F}_{q^m}$ over $\mathbb{F}_q$ are **orthogonal** if

$$\mathrm{Tr}_{q^m/q}(\gamma_i \gamma'_j) = \begin{cases} 1 & \text{if } i = j \\ 0 & \text{if } i \neq j \end{cases}$$

where $\mathrm{Tr}_{q^m/q} : \mathbb{F}_{q^m} \to \mathbb{F}_q$ is the trace function given by $\mathrm{Tr}_{q^m/q}(x) = \sum_{i=0}^{m-1} x^{q^i}$.

Proposition 11.4.4 ([1574, Theorem 21]) *Let $C \subseteq \mathbb{F}_{q^m}^n$ be a vector rank-metric code and let Γ, Γ' be orthogonal bases of $\mathbb{F}_{q^m}$ over $\mathbb{F}_q$. Then*

$$\Gamma(C)^\perp = \Gamma'(C^\perp).$$

We continue with Example 11.1.4 to illustrate Proposition 11.4.4.

Example 11.4.5 Let C be the vector rank-metric code $C = \langle (1, \alpha) \rangle \subseteq \mathbb{F}_8^2$, where $\mathbb{F}_8 = \mathbb{F}_2[\alpha]/(\alpha^3 + \alpha + 1)$. Its dual code is $C^\perp = \langle (1, \alpha^2 + 1) \rangle \subseteq \mathbb{F}_8^2$. Let $\Gamma = \{\gamma_1 = 1, \gamma_2 = \alpha, \gamma_3 = \alpha^2\}$ be an $\mathbb{F}_2$-basis of $\mathbb{F}_8$. The rank-metric code associated to C with respect to Γ is

$$\Gamma(C) = \left\langle \begin{bmatrix} 1 & 0 & 0 \\ 0 & 1 & 0 \end{bmatrix}, \begin{bmatrix} 0 & 1 & 0 \\ 0 & 0 & 1 \end{bmatrix}, \begin{bmatrix} 0 & 0 & 1 \\ 1 & 1 & 0 \end{bmatrix} \right\rangle \subseteq \mathbb{F}_2^{2 \times 3}.$$

Its dual code is

$$\Gamma(C)^\perp = \left\langle \begin{bmatrix} 0 & 0 & 1 \\ 1 & 0 & 0 \end{bmatrix}, \begin{bmatrix} 1 & 0 & 1 \\ 0 & 1 & 0 \end{bmatrix}, \begin{bmatrix} 0 & 1 & 0 \\ 0 & 0 & 1 \end{bmatrix} \right\rangle \subseteq \mathbb{F}_2^{2 \times 3}.$$

The orthogonal basis of Γ is $\Gamma' = \{\gamma_1' = 1, \gamma_2' = \alpha^2, \gamma_3' = \alpha\}$. The rank-metric code associated to $C^\perp$ with respect to Γ' is

$$\Gamma'(C^\perp) = \left\langle \begin{bmatrix} 1 & 0 & 0 \\ 1 & 1 & 0 \end{bmatrix}, \begin{bmatrix} 0 & 0 & 1 \\ 1 & 0 & 0 \end{bmatrix}, \begin{bmatrix} 0 & 1 & 0 \\ 0 & 0 & 1 \end{bmatrix} \right\rangle.$$

It is easy to check that $\Gamma(C)^\perp = \Gamma'(C^\perp)$.

There are a number of interesting relations between the invariants of a code and those of its dual. The simplest one is probably the equality

$$\dim(\mathcal{C}) + \dim(\mathcal{C}^\perp) = mn,$$

which holds for any rank-metric code $\mathcal{C} \subseteq \mathbb{F}_q^{n \times m}$.

The minimum distances of $\mathcal{C}$ and $\mathcal{C}^\perp$ do not satisfy such a simple relation. Nevertheless, one can relate them through the next inequality, which follows easily from the Singleton Bound.

Proposition 11.4.6 ([1574, Proposition 43]) *Let $\mathcal{C} \subseteq \mathbb{F}_q^{n \times m}$ be a rank-metric code. Then*

$$d_{\min}(\mathcal{C}) + d_{\min}(\mathcal{C}^\perp) \leq \min\{m, n\} + 2$$

and equality holds if and only if $\mathcal{C}$ is MRD.

Using the Anticode Bound, one can produce an inequality which involves $\max\mathrm{rank}(\mathcal{C})$ and $\max\mathrm{rank}(\mathcal{C}^\perp)$.

Proposition 11.4.7 ([1574, Proposition 55]) *Let $\mathcal{C} \subseteq \mathbb{F}_q^{n \times m}$ be a rank-metric code. Then*

$$\max\mathrm{rank}(\mathcal{C}) + \max\mathrm{rank}(\mathcal{C}^\perp) \geq \min\{m, n\}$$

and equality holds if and only if $\mathcal{C}$ is an optimal anticode.

Finally, by combining the Singleton Bound and the Anticode Bound, one obtains an inequality which involves $d_{\min}(C)$ and $\max\mathrm{rank}(\mathcal{C}^\perp)$.

Proposition 11.4.8 ([1574, Proposition 49]) *Let $\mathcal{C} \subseteq \mathbb{F}_q^{n \times m}$ be a rank-metric code. Then*

$$d_{\min}(\mathcal{C}) \leq \max\mathrm{rank}(\mathcal{C}^\perp) + 1.$$

Notice that equality holds in Proposition 11.4.8 if and only if $\mathcal{C}$ is MRD and an optimal anticode. We will see in Corollary 11.5.23 that this is the case if and only if $\mathcal{C} = 0$ or $\mathcal{C} = \mathbb{F}_q^{n \times m}$.

Although the minimum distances of $\mathcal{C}$ and $\mathcal{C}^\perp$ do not determine each other, the weight distribution of $\mathcal{C}$ determines the weight distribution of $\mathcal{C}^\perp$, and vice versa. We now define the weight distribution, which is an important invariant of a rank-metric code.

Definition 11.4.9 Let $\mathcal{C} \subseteq \mathbb{F}_q^{n \times m}$ be a rank-metric code. The **weight distribution** of $\mathcal{C}$ is the collection of natural numbers

$$A_i(\mathcal{C}) = \left| \{ M \in \mathcal{C} \mid \mathrm{rank}(M) = i \} \right| \quad \text{for } i = 0, 1, \ldots, \min\{m, n\}.$$

Clearly $A_0(\mathcal{C}) = 1$, $d_{\min}(\mathcal{C}) = \min\{i \mid A_i(\mathcal{C}) \neq 0, i \neq 0\}$, and $\max \mathrm{rank}(\mathcal{C}) = \max\{i \mid A_i(\mathcal{C}) \neq 0\}$.

Definition 11.4.10 The **q-ary Gaussian coefficient** of $a, b \in \mathbb{Z}$ is

$$\begin{bmatrix} a \\ b \end{bmatrix}_q = \begin{cases} 0 & \text{if } a < 0,\ b < 0,\ \text{or } b > a, \\ 1 & \text{if } b = 0 \text{ and } a \geq 0, \\ \frac{(q^a - 1)(q^{a-1} - 1) \cdots (q^{a-b+1} - 1)}{(q^b - 1)(q^{b-1} - 1) \cdots (q - 1)} & \text{otherwise.} \end{cases}$$

The relations between the weight distribution of $\mathcal{C}$ and $\mathcal{C}^\perp$ go under the name of MacWilliams Identities and were first proved in [523, Theorem 3.3]. A different proof, inspired by [1575, Theorem 27], was given in [842, Theorem 2]. Another proof was given in [1683, Proposition 15]. The following formulation is analogous to Theorem 1.15.3(d).

Theorem 11.4.11 (MacWilliams Identities) *Let $\mathcal{C} \subseteq \mathbb{F}_q^{n \times m}$ be a rank-metric code. Let*

$$P_j^{(m,n,q)}(i) = \sum_{\ell = 0}^{\min\{m,n\}} (-1)^{i-\ell} q^{\max\{m,n\} \cdot \ell + \binom{i-\ell}{2}} \begin{bmatrix} \min\{m,n\} - \ell \\ \min\{m,n\} - i \end{bmatrix}_q \begin{bmatrix} \min\{m,n\} - j \\ \ell \end{bmatrix}_q.$$

Then for $i = 0, 1, \ldots, \min\{m, n\}$

$$A_i(\mathcal{C}^\perp) = \frac{1}{|\mathcal{C}|} \sum_{j=0}^{\min\{m,n\}} A_j(\mathcal{C}) P_j^{(m,n,q)}(i).$$

The following is an equivalent formulation of the MacWilliams Identities; for comparison, see Theorem 1.15.3(a). Identities of this form for vector rank-metric codes were proved in [782, Proposition 3]. The same identities were proved in [1574, Theorem 21] for rank-metric codes.

Theorem 11.4.12 *Let $\mathcal{C} \subseteq \mathbb{F}_q^{n \times m}$ be a rank-metric code. For $\ell = 0, 1, \ldots, \min\{m, n\}$,*

$$\sum_{i=0}^{\min\{m,n\}-\ell} A_i(\mathcal{C}) \begin{bmatrix} \min\{m,n\} - i \\ \ell \end{bmatrix}_q = \frac{|\mathcal{C}|}{q^{\max\{m,n\} \cdot \ell}} \sum_{j=0}^{\ell} A_i(\mathcal{C}^\perp) \begin{bmatrix} \min\{m,n\} - j \\ \ell - j \end{bmatrix}_q.$$

From the MacWilliams Identities, one can derive a number of nontrivial consequences. Here we give two relevant ones, starting with the celebrated result of Delsarte, which states that the dual of an MRD code is MRD. Compare this result to Theorem 1.9.13.

Theorem 11.4.13 ([523, Theorem 5.5]) *Let $\mathcal{C} \subseteq \mathbb{F}_q^{n \times m}$ be a rank-metric code. Then $\mathcal{C}$ is MRD if and only if $\mathcal{C}^\perp$ is MRD.*

The weight distribution of an MRD code was first computed by Delsarte in [523, Theorem 5.6] and can be derived from the MacWilliams Identities via a standard computation. An analogous result can be obtained for dually quasi-MRD codes.

Definition 11.4.14 Let $\mathcal{C} \subseteq \mathrm{Mat}_{n \times m}(\mathbb{F}_q)$ be a rank-metric code. $\mathcal{C}$ is **dually quasi-MRD** if

$$d_{\min}(\mathcal{C}) + d_{\min}(\mathcal{C}^\perp) = \min\{m, n\} + 1.$$

Discussing the family of dually quasi-MRD codes is beyond the scope of this chapter. The definition however is motivated by Proposition 11.4.6, which shows that dually quasi-MRD codes are exactly the non-MRD codes which maximize the quantity $d_{\min}(\mathcal{C}) + d_{\min}(\mathcal{C}^\perp)$. We refer the interested reader to [500] for a discussion of the properties of codes which are close to being MRD in this sense. The weight distribution of a dually quasi-MRD code was computed in [500, Corollary 28]. Below we give a statement that covers both MRD and dually quasi-MRD codes. Once again, compare this to Theorem 1.15.7.

Theorem 11.4.15 *Let $\mathcal{C} \subseteq \mathbb{F}_q^{n \times m}$ be an MRD or dually quasi-MRD rank-metric code. Let $d = d_{\min}(\mathcal{C})$. Then $A_0(\mathcal{C}) = 1$, $A_i(\mathcal{C}) = 0$ for $i = 1, \ldots, d-1$, and*

$$A_i(\mathcal{C}) = \begin{bmatrix} \min\{m,n\} \\ i \end{bmatrix}_q \sum_{j=0}^{i-d} (-1)^j q^{\binom{j}{2}} \begin{bmatrix} i \\ j \end{bmatrix}_q \left(q^{\dim(\mathcal{C}) - \max\{m,n\}(\min\{m,n\}+j-i)} - 1 \right)$$

for $i = d, \ldots, \min\{m, n\}$.

The analog of Theorem 11.4.13 for optimal anticodes was proved by Ravagnani.

Theorem 11.4.16 ([1574, Theorem 54]) *Let $\mathcal{C} \subseteq \mathbb{F}_q^{n \times m}$ be a rank-metric code. Then $\mathcal{C}$ is an optimal anticode if and only if $\mathcal{C}^\perp$ is an optimal anticode.*

Ravagnani also proved the next interesting result, relating MRD codes and optimal anticodes.

Proposition 11.4.17 ([1574, Proposition 53]) *Let $\mathcal{C} \subseteq \mathbb{F}_q^{n \times m}$ be a rank-metric code of dimension $\dim(\mathcal{C}) = k \cdot \max\{m, n\}$. Then $\mathcal{C}$ is an optimal anticode if and only if*

$$\mathcal{C} + \mathcal{D} = \mathbb{F}_q^{n \times m}$$

for every $\mathcal{D} \subseteq \mathbb{F}_q^{n \times m}$ MRD code of minimum distance $d_{\min}(\mathcal{D}) = k + 1$.

11.5 Generalized Weights

Generalized Hamming weights were introduced by Helleseth, Kløve, and Mykkeltveit in [939] for linear block codes. In [1882], Wei studied them in the context of wire-tap channels. Different definitions of generalized weights were given for vector rank-metric codes and rank-metric codes. In this section, we give the different definitions and compare them with each other.

In the context of vector rank-metric codes, generalized weights were first defined by Oggier and Sboui.

Definition 11.5.1 ([1459, Definition 1]) *Let $n \leq m$ and let $C \subseteq \mathbb{F}_{q^m}^n$ be a vector rank-metric code. For $i = 1, 2, \ldots, \dim_{\mathbb{F}_{q^m}}(C)$ the **generalized weights** of C are*

$$w_i(C) = \min_D \left\{ \max_v \{\dim \operatorname{supp}(v) \mid v \in D, v \neq 0\} \mid D \subseteq C, \dim_{\mathbb{F}_{q^m}}(D) = i \right\}.$$

A definition of relative generalized weights for vector rank-metric codes was given by Kurihara, Matsumoto, and Uyematsu. Let $\phi : \mathbb{F}_{q^m}^n \to \mathbb{F}_{q^m}^n$ be the **Frobenius endomorphism** defined by $\phi(v_1, \ldots, v_n) = (v_1^q, \ldots, v_n^q)$.

Definition 11.5.2 ([1178, Definition 2]) Let $C \subseteq \mathbb{F}_{q^m}^n$ be a vector rank-metric code, and let $D \subsetneq C$ be a proper subcode. Let $\phi : \mathbb{F}_{q^m}^n \to \mathbb{F}_{q^m}^n$ be the Frobenius endomorphism. For $i = 1, 2, \dots, \dim_{\mathbb{F}_{q^m}} (C) - \dim_{\mathbb{F}_{q^m}} (D)$ the **relative generalized weights** of C and D are

$$w_i(C, D) = \min \{\dim \operatorname{supp}(V) \mid V \subseteq \mathbb{F}_{q^m}^n, \ \phi(V) = V,$$
$$\dim_{\mathbb{F}_{q^m}} (C \cap V) - \dim_{\mathbb{F}_{q^m}} (D \cap V) \geq i\}.$$

In particular, for $i = 1, 2, \dots, \dim_{\mathbb{F}_{q^m}} (C)$ the relative generalized weights of C and 0 are

$$w_i(C, 0) = \min \{\dim \operatorname{supp}(V) \mid V \subseteq \mathbb{F}_{q^m}^n, \ \phi(V) = V, \ \dim_{\mathbb{F}_{q^m}} (C \cap V) \geq i\}.$$

In [645], Ducoat proposed and studied the following modification of Definition 11.5.1. For any $D \subseteq \mathbb{F}_{q^m}^n$, let $D^* = D + \phi(D) + \cdots + \phi^{m-1}(D)$, where ϕ denotes the Frobenius endomorphism. D^* is the smallest $\mathbb{F}_{q^m}$-linear space containing D which is fixed by ϕ.

Definition 11.5.3 Let $C \subseteq \mathbb{F}_{q^m}^n$ be a vector rank-metric code. For $i = 1, 2, \dots, \dim_{\mathbb{F}_{q^m}} (C)$ the **generalized weights** of C are

$$w_i(C) = \min_D \left\{ \max_v \{\dim \operatorname{supp}(v) \mid v \in D^*, v \neq 0\} \mid D \subseteq C, \dim_{\mathbb{F}_{q^m}} (D) = i \right\}.$$

Notice that, although the definition by Ducoat does not assume $n \leq m$, most of the results that he establishes in [645] do.

It was shown by Ducoat in [645, Proposition II.1] for $n \leq m$, and by Jurrius and Pellikaan in [1061, Theorem 5.4] for any n, m, that the relative generalized weights of C and 0 agree with the generalized weights of C, i.e.,

$$w_i(C, 0) = w_i(C) \text{ for } i = 1, \dots, \dim_{\mathbb{F}_{q^m}} (C).$$

Moreover, it follows from [1061, Theorem 5.2 and Theorem 5.8] that Definition 11.5.1 and Definition 11.5.3 are equivalent for $n \leq m$.

The next definition is the natural analog of the definition of generalized rank weights for linear block codes, endowed with the Hamming distance. It was given by Jurrius and Pellikaan, who in [1061, Theorem 5.2] proved that it is equivalent to Definition 11.5.1 if $n \leq m$. In [1061, Theorem 5.8] they proved that it is equivalent to Definition 11.5.3.

Definition 11.5.4 ([1061, Definition 2.5]) Let $C \subseteq \mathbb{F}_{q^m}^n$ be a vector rank-metric code. For $i = 1, 2, \dots, \dim_{\mathbb{F}_{q^m}} (C)$ the **generalized weights** of C are

$$w_i(C) = \min \{\dim \operatorname{supp}(D) \mid D \subseteq C, \dim_{\mathbb{F}_{q^m}} (D) = i\}.$$

The next result provides another equivalent definition of generalized weights for vector rank-metric codes. It was proved by Ravagnani under the assumption $n \leq m$, but it can easily be extended to arbitrary n, m as follows.

Theorem 11.5.5 ([1573, Corollary 19]) *Let $C \subseteq \mathbb{F}_{q^m}^n$ be a vector rank-metric code of dimension $\dim_{\mathbb{F}_{q^m}} (C) \leq m$. Then for $i = 1, 2, \dots, \dim_{\mathbb{F}_{q^m}} (C)$,*

$$w_i(C) = \min \{\dim_{\mathbb{F}_{q^m}} (A) \mid A \subseteq \mathbb{F}_{q^m}^n \text{ optimal vector anticode with } \dim_{\mathbb{F}_{q^m}} (C \cap A) \geq i\}.$$

Remark 11.5.6 Let $C \subseteq \mathbb{F}_{q^m}^n$ be a vector rank-metric code of dimension $\dim_{\mathbb{F}_{q^m}} (C) > m$ and let $m < i \leq \dim_{\mathbb{F}_{q^m}} (C)$. Then the quantity

$$\min \{\dim_{\mathbb{F}_{q^m}} (A) \mid A \subseteq \mathbb{F}_{q^m}^n \text{ optimal vector anticode with } \dim_{\mathbb{F}_{q^m}} (C \cap A) \geq i\}$$

cannot be equal to $w_i(C)$ since, by Remark 11.3.13, there exists no optimal vector anticode A with $\dim_{\mathbb{F}_{q^m}} (A) \geq \dim_{\mathbb{F}_{q^m}} (C \cap A) \geq i > m$.

The first definition of generalized weights for the larger class of rank-metric codes was given by Ravagnani.

Definition 11.5.7 ([1573, Definition 23]) Let $\mathcal{C} \subseteq \mathbb{F}_q^{n \times m}$ be a rank-metric code. For $i = 1, 2, \ldots, \dim(\mathcal{C})$ the **generalized weights** of $\mathcal{C}$ are

$$d_i(\mathcal{C}) = \frac{1}{\max\{m, n\}} \min\{\dim(\mathcal{A}) \mid \mathcal{A} \subseteq \mathbb{F}_q^{n \times m} \text{ optimal anticode with } \dim(\mathcal{C} \cap \mathcal{A}) \geq i\},$$

Remark 11.5.8 The characterization of optimal anticodes from Theorem 11.3.15, together with the observation that

$$\dim(\mathbb{F}_q^{n \times m}(V)) = \max\{n, m\} \cdot \dim(V),$$

implies that for $i = 1, 2, \ldots, \dim(\mathcal{C})$ one has

$$d_i(\mathcal{C}) = \min\{\dim(V) \mid V \subseteq \mathbb{F}_q^{\min\{n, m\}}, \ \dim(\mathcal{C}(V)) \geq i\} \quad \text{if } m \neq n$$

and

$$d_i(\mathcal{C}) = \min\{\dim(V) \mid V \subseteq \mathbb{F}_q^n, \ \max\{\dim(\mathcal{C}(V)), \dim(\mathcal{C}^{\mathsf{T}}(V))\} \geq i\} \quad \text{if } m = n.$$

Notice that, for $m = n$, this definition of generalized weights is coherent with the definition of support given in Remark 11.2.4.

The next statement appears as a theorem in [1573, Theorem 28]. However, the proof is incomplete. Therefore, we state it here as a conjecture.

Conjecture 11.5.9 ([1573, Theorem 28]) *Let $n \leq m$ and let $C \subseteq \mathbb{F}_{q^m}^n$ be a vector rank-metric code. Let $\Gamma = \{\gamma_1, \ldots, \gamma_m\}$ be a basis of $\mathbb{F}_{q^m}$ over $\mathbb{F}_q$. Then*

$$w_i(C) = d_{mi-e}(\Gamma(C))$$

for $i = 1, 2, \ldots, \dim_{\mathbb{F}_{q^m}}(C)$ and $e = 0, 1, \ldots, m - 1$.

Remark 11.5.10 In particular, under the assumptions of Conjecture 11.5.9, this conjecture implies that

$$d_{m(i-1)+1}(\Gamma(C)) = \cdots = d_{mi-1}(\Gamma(C)) = d_{mi}(\Gamma(C))$$

for $i = 1, 2, \ldots, \dim_{\mathbb{F}_{q^m}}(C)$.

One can easily find an example that shows that the equality in Conjecture 11.5.9 does not hold if $n > m$.

Example 11.5.11 Let $C = \langle(1, 0, 0)\rangle \subseteq \mathbb{F}_4^3$, where $\mathbb{F}_4 = \mathbb{F}_2[\alpha]/(\alpha^2 + \alpha + 1)$. Then

$$w_1(C) = d_{\min}(C) = 1.$$

Let $\Gamma = \{1, \alpha\}$ be an $\mathbb{F}_2$-basis of $\mathbb{F}_4$. Then

$$\Gamma(C) = \left\langle \begin{bmatrix} 1 & 0 \\ 0 & 0 \\ 0 & 0 \end{bmatrix}, \begin{bmatrix} 0 & 1 \\ 0 & 0 \\ 0 & 0 \end{bmatrix} \right\rangle$$

has $d_1(\Gamma(C)) = d_{\min}(\Gamma(C)) = 1$. Since $\Gamma(C)$ is not an optimal anticode, any nonzero optimal anticode $\mathcal{A} \supsetneq \Gamma(C)$. Hence $\mathcal{A}$ must have $\max \operatorname{rank}(\mathcal{A}) = 2$ and $\dim(\mathcal{A}) = 6$. Therefore, $d_2(\Gamma(C)) = 2$.

A definition of relative generalized weights for rank-metric codes was proposed by Martínez-Peñas and Matsumoto. This yields in particular a definition of generalized weights, which is different from Definition 11.5.7, as we discuss below. In order to avoid confusion, we call the weights defined by Martínez-Peñas and Matsumoto generalized matrix weights.

Definition 11.5.12 ([1341, Definition 10]) Let $\mathcal{C} \subseteq \mathbb{F}_q^{n \times m}$ be a rank-metric code, and let $\mathcal{D} \subsetneq \mathcal{C}$ be a proper subcode. Denote by

$$\mathbb{F}_q^{n \times m}{}_V^{\text{colsp}} = \{M \in \mathbb{F}_q^{n \times m} \mid \text{colsp}(M) \subseteq V\}.$$

For $i = 1, 2, \ldots, \dim(\mathcal{C}) - \dim(\mathcal{D})$ the **relative generalized matrix weight** of $\mathcal{C}$ and $\mathcal{D}$ is

$$\delta_i(\mathcal{C}, \mathcal{D}) = \min\left\{\dim(V) \mid V \subseteq \mathbb{F}_q^n,\ \dim\left(\mathcal{C} \cap \mathbb{F}_q^{n \times m}{}_V^{\text{colsp}}\right) - \dim\left(\mathcal{D} \cap \mathbb{F}_q^{n \times m}{}_V^{\text{colsp}}\right) \geq i\right\}.$$

The ith **generalized matrix weight** of $\mathcal{C}$ is the ith relative generalized matrix weight of $\mathcal{C}$ and 0; i.e., for $i = 1, 2, \ldots, \dim(\mathcal{C})$

$$\delta_i(\mathcal{C}) = \min\left\{\dim(V) \mid V \subseteq \mathbb{F}_q^n,\ \dim(\mathcal{C} \cap \mathbb{F}_q^{n \times m}{}_V^{\text{colsp}}) \geq i\right\}.$$

Generalized matrix weights measure the information leakage to a wire-tapper in a linearly coded network and, more generally, in a matrix-multiplicative channel. The model discussed in [1341] is not invariant with respect to transposition, since the wiretapper's observation is AM, where M is the codeword and A is the wiretap transfer matrix. Accordingly, in Definition 11.5.12 the authors consider the column space of the matrix independently of whether the matrix has more rows or columns. In particular, one should not expect Definition 11.5.12 to be equivalence-invariant; i.e., equivalent codes may not have the same generalized matrix weights. In Example 11.5.15 we show that this can in fact happen. Therefore, Definition 11.5.12 is not equivalence-invariant. The next proposition shows that Definition 11.5.7 is equivalence-invariant.

Proposition 11.5.13 ([841, Proposition 2.4]) *Let* $\mathcal{C}, \mathcal{D} \subseteq \mathbb{F}_q^{n \times m}$ *be rank-metric codes. If* $\mathcal{C} \sim \mathcal{D}$, *then*

$$d_i(\mathcal{C}) = d_i(\mathcal{D}) \quad \text{for } i = 1, 2, \ldots, \dim(\mathcal{C}).$$

The next result compares Definition 11.5.7 and Definition 11.5.12. It follows easily from Theorem 11.3.15 and appears in the literature as [1341, Theorem 9]. Notice that the assumption that $n \leq m$ is missing throughout [1341, Section VIII.C]. As a consequence, the statement of [1341, Theorem 9] claims that Definition 11.5.7 and Definition 11.5.12 agree for $m \neq n$; however the result is proved only for $n < m$.

Theorem 11.5.14 *For* $\mathcal{C} \subseteq \mathbb{F}_q^{n \times m}$ *a rank-metric code, we have*

(a) *if* $m > n$, *then* $d_i(\mathcal{C}) = \delta_i(\mathcal{C})$ *for* $i = 1, 2, \ldots, \dim(\mathcal{C})$, *and*

(b) *if* $m = n$, *then* $d_i(\mathcal{C}) \leq \delta_i(\mathcal{C})$ *for* $i = 1, 2, \ldots, \dim(\mathcal{C})$.

Proof: The thesis follows from Remark 11.5.8, after observing that for $n \leq m$ one has

$$\mathcal{C}(V) = \mathcal{C} \cap \mathbb{F}_q^{n \times m}{}_V^{\text{colsp}}.$$

$\square$

One can easily find examples that show that Definition 11.5.7 and Definition 11.5.12 do not agree in the case $m = n$.

Example 11.5.15 ([841, Example 2.10]) Let $\mathcal{C} \subseteq \mathbb{F}_2^{2 \times 2}$ be the code

$$\mathcal{C} := \left\{ \begin{bmatrix} a & a \\ b & b \end{bmatrix} \ \middle| \ a, b \in \mathbb{F}_2 \right\}.$$

Then $\mathcal{C}$ is an optimal anticode of dimension 2. Therefore $d_2(\mathcal{C}) = 1$. On the other hand, $\mathrm{supp}(\mathcal{C}) = \mathbb{F}_2^2$; hence $\delta_2(\mathcal{C}) = 2 \neq d_2(\mathcal{C})$.

Moreover, observe that $\mathcal{C} \sim \mathcal{C}^\mathsf{T}$. In particular, $d_2(\mathcal{C}) = d_2(\mathcal{C}^\mathsf{T}) = 1$. However, $\delta_2(\mathcal{C}) = 2$, while $\delta_2(\mathcal{C}^\mathsf{T}) = 1$.

In fact, one can also find examples that show that Definition 11.5.7 and Definition 11.5.12 do not agree in the case $m < n$. This implies in particular that the part of the statement of [1341, Theorem 9] concerning the case $m < n$ is incorrect.

Example 11.5.16 Let $\mathcal{C} \subseteq \mathbb{F}_2^{3 \times 2}$ be the code

$$\mathcal{C} := \left\{ \begin{bmatrix} a & a \\ b & b \\ c & c \end{bmatrix} \ \middle| \ a, b, c \in \mathbb{F}_2 \right\}.$$

Then $\mathcal{C}$ is an optimal anticode of dimension 3. Therefore $d_3(\mathcal{C}) = 1$. On the other hand,

$$\sum_{M \in \mathcal{C}} \mathrm{colsp}(M) = \mathbb{F}_2^3;$$

hence $\delta_3(\mathcal{C}) = 3$.

The code $\mathcal{C} \subseteq \mathbb{F}_2^{3 \times 2}$ of Example 11.5.16 has $d_3(\mathcal{C}) < \delta_3(\mathcal{C})$. One may wonder whether $d_i(\mathcal{C}) \leq \delta_i(\mathcal{C})$ for all i, for a rank-metric code $\mathcal{C} \subseteq \mathbb{F}_q^{n \times m}$ with $n > m$. The answer turns out to be negative, as the next example shows.

Example 11.5.17 Let $\mathcal{C} \subseteq \mathbb{F}_2^{3 \times 2}$ be the code

$$\mathcal{C} := \left\{ \begin{bmatrix} a & b \\ 0 & 0 \\ 0 & 0 \end{bmatrix} \ \middle| \ a, b \in \mathbb{F}_2 \right\}.$$

In Example 11.5.11 we showed that $d_2(\mathcal{C}) = 2$. On the other hand,

$$\sum_{M \in \mathcal{C}} \mathrm{colsp}(M) = \langle (1, 0, 0) \rangle;$$

hence $\delta_2(\mathcal{C}) = 1$.

In fact, as an easy consequence of Theorem 11.3.15 and of Remark 11.3.9, one obtains the following result.

Theorem 11.5.18 *For $\mathcal{C} \subseteq \mathbb{F}_q^{n \times m}$ a rank-metric code, we have*

(a) *if $m < n$, then $d_i(\mathcal{C}) = \delta_i(\mathcal{C}^\mathsf{T})$ for $i = 1, 2, \ldots, \dim(\mathcal{C})$, and*

(b) *if $m = n$, then $d_i(\mathcal{C}) = \min\{\delta_i(\mathcal{C}), \delta_i(\mathcal{C}^\mathsf{T})\}$ for $i = 1, 2, \ldots, \dim(\mathcal{C})$.*

Proof: Since $n \geq m$, then

$$\mathcal{C}^\mathsf{T}(V) = \mathcal{C}^\mathsf{T} \cap \mathbb{F}_q^{m \times n} \mathrm{colsp}_V. \tag{11.7}$$

If $n > m$, then

$$d_i(\mathcal{C}) = d_i(\mathcal{C}^\mathsf{T}) = \min\left\{\dim(V) \mid V \subseteq \mathbb{F}_q^m,\ \dim(\mathcal{C}^\mathsf{T}(V)) \geq i\right\} = \delta_i(\mathcal{C}^\mathsf{T}),$$

where the first equality follows from Proposition 11.5.13, the second from Remark 11.5.8, and the third from (11.7). If $n = m$, then

$$d_i(\mathcal{C}) = \min\left\{\dim(V) \mid V \subseteq \mathbb{F}_q^n,\ \max\left\{\dim(\mathcal{C}(V)), \dim(\mathcal{C}^\mathsf{T}(V))\right\} \geq i\right\} = \min\left\{\delta_i(\mathcal{C}), \delta_i(\mathcal{C}^\mathsf{T})\right\},$$

where the first equality follows from Remark 11.5.8 and the second from (11.7). $\square$

As in the case of generalized weights, one can relate the generalized weights of a vector rank-metric code and the generalized matrix weights of its associated rank-metric code. In fact more is true, since the relative versions of the weights can also be related, and the assumption $n \leq m$ is not needed.

We conclude this section with a few results on generalized weights. The next theorem establishes some properties of the sequence of generalized weights of a rank-metric code.

Theorem 11.5.19 ([1573, Theorem 30]) *Let $\mathcal{C} \subseteq \mathbb{F}_q^{n \times m}$ be a rank-metric code of dimension $\dim(\mathcal{C}) = \ell$. The following hold.*

(a) $d_1(\mathcal{C}) = d_{\min}(\mathcal{C})$,

(b) $d_\ell(\mathcal{C}) \leq \min\{m, n\}$,

(c) $d_i(\mathcal{C}) \leq d_{i+1}(\mathcal{C})$ *for* $i = 1, 2, \ldots, \ell - 1$, *and*

(d) $d_i(\mathcal{C}) \leq d_{i+\max\{m,n\}}(\mathcal{C})$ *for* $i = 1, 2, \ldots, \ell - \max\{m, n\}$.

Theorem 11.5.19 allows one to compute the generalized weights of MRD codes and optimal anticodes. In Section 11.4 we stated analogous results for the weight distribution of MRD codes and optimal anticodes.

Corollary 11.5.20 ([1573, Corollary 31]) *Let $\mathcal{C} \subseteq \mathbb{F}_q^{n \times m}$ be a rank-metric code of dimension $\dim(\mathcal{C}) = \ell = k \cdot \max\{m, n\}$. The following are equivalent.*

(a) $\mathcal{C}$ *is MRD.*

(b) $d_i(\mathcal{C}) = \min\{m, n\} - k + \lceil i / \max\{m, n\} \rceil$ *for* $i = 1, 2, \ldots, \ell$.

Corollary 11.5.21 ([1573, Corollary 32]) *Let $\mathcal{C} \subseteq \mathbb{F}_q^{n \times m}$ be a rank-metric code of dimension $\dim(\mathcal{C}) = \ell = k \cdot \max\{m, n\}$. The following are equivalent.*

(a) $\mathcal{C}$ *is an optimal anticode.*

(b) $d_\ell(\mathcal{C}) = k$.

(c) $d_i(\mathcal{C}) = \lceil i / \max\{m, n\} \rceil$ *for* $i = 1, 2, \ldots, \ell$.

An immediate consequence of Corollary 11.5.20 and Corollary 11.5.21 is the following.

Corollary 11.5.22 *Let $\mathcal{C} \subseteq \mathbb{F}_q^{n \times m}$ be a rank-metric code. Then $\mathcal{C}$ is both MRD and an optimal anticode if and only if $\mathcal{C} = 0$ or $\mathcal{C} = \mathbb{F}_q^{n \times m}$.*

A similar result can be obtained for dually quasi-MRD codes. It follows from [500, Corollary 18] that the dimension of a dually quasi-MRD code $\mathcal{C} \subseteq \mathbb{F}_q^{n \times m}$ is not divisible by $\max\{m, n\}$. Therefore, in the next result we make this assumption without loss of generality.

Corollary 11.5.23 ([500, Theorem 22]) *Let* $\mathcal{C} \subseteq \mathbb{F}_q^{n \times m}$ *be a rank-metric code of dimension* $\dim(\mathcal{C}) = \ell = k \cdot \max\{m, n\} + r$, *with* $k \geq 0$ *and* $0 < r < \max\{m, n\}$. *The following are equivalent.*

(a) $\mathcal{C}$ *is dually quasi-MRD.*

(b) $d_1(\mathcal{C}) = \min\{m, n\} - k$ *and* $d_{r+1}(\mathcal{C}) = \min\{m, n\} + 1 - k$.

Moreover, if $\mathcal{C}$ *is dually quasi-MRD, then its generalized weights are:*

$$d_1(\mathcal{C}) = \cdots = d_r(\mathcal{C}) = \min\{m, n\} - k,$$

$$d_{r+1+i \cdot \max\{m,n\}}(\mathcal{C}) = \cdots = d_{r+(i+1)\max\{m,n\}}(\mathcal{C})$$
$$= \min\{m, n\} + 1 + i - k \quad \text{for} \quad i = 0, 1, \ldots, k-2,$$

$$d_{r+1+(k-1)\max\{m,n\}}(\mathcal{C}) = \cdots = d_\ell(\mathcal{C}) = \min\{m, n\}.$$

We already observed that, while there is no easy relation between the minimum distance of $\mathcal{C}$ and $\mathcal{C}^\perp$, the weight distribution of $\mathcal{C}$ determines the weight distribution of $\mathcal{C}^\perp$. The next result shows that the generalized weights of $\mathcal{C}$ determine the generalized weights of $\mathcal{C}^\perp$.

Theorem 11.5.24 ([1573, Corollary 38]) *Let* $\mathcal{C} \subseteq \mathbb{F}_q^{n \times m}$ *be a rank-metric code of dimension* $\dim(\mathcal{C}) = \ell$. *Define the sets*

$$W_i(\mathcal{C}) = \{d_{i+j \cdot \max\{m,n\}}(\mathcal{C}) \mid j \geq 0, \ 1 \leq i + j \cdot \max\{m, n\} \leq \ell\}$$

and

$$\overline{W}_i(\mathcal{C}) = \{\min\{m, n\} + 1 - d_{i+j \cdot \max\{m,n\}}(\mathcal{C}) \mid j \geq 0, \ 1 \leq i + j \cdot \max\{m, n\} \leq \ell\}.$$

Then for $i = 1, 2, \ldots, \max\{m, n\}$

$$W_i(\mathcal{C}^\perp) = \{1, 2, \ldots, \min\{m, n\}\} \setminus \overline{W}_{i+\ell}(\mathcal{C}).$$

11.6 q-Polymatroids and Code Invariants

q-polymatroids are the q-analog of polymatroids. In this section we associate to every rank-metric code a q-polymatroid for $m \neq n$ and a pair of q-polymatroids for $m = n$. We then discuss how several invariants and structural properties of codes, such as generalized weights, the property of being MRD or an optimal anticode, and duality, are captured by the associated q-polymatroids. The material of this section is contained in [841, 1683], but the presentation we give differs at times from the original papers.

We start by giving the definition of a q-matroid, the q-analog of a matroid.

Definition 11.6.1 ([1062, Definition 2.1]) A **q-matroid** is a pair $P = (\mathbb{F}_q^\ell, \rho)$ where ρ is a function from the set of all subspaces of $\mathbb{F}_q^\ell$ to $\mathbb{Z}$ such that, for all $U, V \subseteq \mathbb{F}_q^\ell$:

(P1) $0 \leq \rho(V) \leq \dim(V)$,

(P2) if $U \subseteq V$, then $\rho(U) \leq \rho(V)$,

(P3) $\rho(U + V) + \rho(U \cap V) \leq \rho(U) + \rho(V)$.

q-polymatroids were defined independently by Shiromoto in [1683] and by Gorla, Jurrius, Lopez, and Ravagnani in [841]. The two definitions are essentially equivalent. Here we follow the approach of [841].

Definition 11.6.2 ([841, Definition 4.1]) A q-**polymatroid** is a pair $P = (\mathbb{F}_q^\ell, \rho)$ where ρ is a function from the set of all subspaces of $\mathbb{F}_q^\ell$ to $\mathbb{R}$ such that, for all $U, V \subseteq \mathbb{F}_q^\ell$:

(P1) $0 \leq \rho(V) \leq \dim(V)$,

(P2) if $U \subseteq V$, then $\rho(U) \leq \rho(V)$,

(P3) $\rho(U + V) + \rho(U \cap V) \leq \rho(U) + \rho(V)$.

Definition 11.6.2 is a direct q-analog of the definition of an ordinary polymatroid, with the extra property that $\rho(V) \leq \dim(V)$ for all $V \subseteq \mathbb{F}_q^\ell$. As in the ordinary case, a q-matroid is a q-polymatroid.

Remark 11.6.3 Definition 11.6.2 is slightly different from the definition of (q, r)-polymatroid given by Shiromoto in [1683, Definition 2]. However, a (q, r)-polymatroid (E, ρ) as defined by Shiromoto corresponds to the q-polymatroid $(E, \rho/r)$ according to Definition 11.6.2. Moreover, a q-polymatroid whose rank function takes values in $\mathbb{Q}$ corresponds to a (q, r)-polymatroid as defined by Shiromoto up to multiplying the rank function by an r which clears denominators.

We now give two simple examples of q-matroids.

Example 11.6.4 The pair $(\mathbb{F}_q^\ell, \dim(\cdot))$ is a q-matroid, where $\dim(\cdot)$ denotes the function that associates to a vector space its dimension.

Example 11.6.5 For a fixed $U \subseteq \mathbb{F}_q^\ell$, let

$$\rho_U(V) = \dim(V) - \dim(V \cap U^\perp)$$

for $V \subseteq \mathbb{F}_q^\ell$. The pair $(\mathbb{F}_q^\ell, \rho_U)$ is a q-matroid.

One has the following natural notion of equivalence for q-polymatroids.

Definition 11.6.6 ([841, Definition 4.4]) Let $(\mathbb{F}_q^\ell, \rho_1)$ and $(\mathbb{F}_q^\ell, \rho_2)$ be q-polymatroids. We say that $(\mathbb{F}_q^\ell, \rho_1)$ and $(\mathbb{F}_q^\ell, \rho_2)$ are **equivalent** if there exists an $\mathbb{F}_q$-linear isomorphism $\varphi : \mathbb{F}_q^\ell \to \mathbb{F}_q^\ell$ such that $\rho_1(V) = \rho_2(\varphi(V))$ for all $V \subseteq \mathbb{F}_q^\ell$. In this case we write $(\mathbb{F}_q^\ell, \rho_1) \sim (\mathbb{F}_q^\ell, \rho_2)$.

The following is the natural notion of duality for q-polymatroids.

Definition 11.6.7 ([841, Definition 4.5]) Let $P = (\mathbb{F}_q^\ell, \rho)$ be a q-polymatroid. For all subspaces $V \subseteq \mathbb{F}_q^\ell$ define

$$\rho^*(V) = \dim(V) - \rho(\mathbb{F}_q^\ell) + \rho(V^\perp),$$

where $V^\perp$ is the dual of V with respect to the standard inner product on $\mathbb{F}_q^\ell$. We call $P^* = (\mathbb{F}_q^\ell, \rho^*)$ the **dual** of the q-polymatroid P.

It is easy to show that P^* is indeed a q-polymatroid. The dual of a q-polymatroid satisfies the usual properties for a dual. Moreover, duality is compatible with equivalence.

Theorem 11.6.8 ([841, Theorem 4.6 and Proposition 4.7]) *Let P, Q be q-polymatroids. Then the following hold.*

(a) P^* *is a q-polymatroid.*

(b) $P^{**} = P$.

(c) *If $P \sim Q$, then $P^* \sim Q^*$.*

One can associate q-polymatroids to rank-metric codes as follows. In [841, Theorem 5.4] it is shown that these are indeed q-polymatroids according to Definition 11.6.2.

Definition 11.6.9 ([841, Notation 5.3]) Let $\mathcal{C} \subseteq \mathbb{F}_q^{n \times m}$ be a rank-metric code, and let $V \subseteq \mathbb{F}_q^{\min\{m,n\}}$. Define

$$\rho_{\mathcal{C}}(V) = \frac{1}{\max\{m,n\}} \left(\dim(\mathcal{C}) - \dim(\mathcal{C}(V^{\perp})) \right) \in \mathbb{Q}.$$

If $m \neq n$, we associate to $\mathcal{C}$ the q-polymatroid $P(\mathcal{C}) = \left(\mathbb{F}_q^{\min\{m,n\}}, \rho_{\mathcal{C}} \right)$. If $m = n$, we associate to $\mathcal{C}$ the pair of q-polymatroids $P(\mathcal{C}) = (\mathbb{F}_q^n, \rho_{\mathcal{C}})$, $P(\mathcal{C}^\mathsf{T}) = (\mathbb{F}_q^n, \rho_{\mathcal{C}^\mathsf{T}})$.

Notice that this is slightly different from what is done in [841], where a pair of q-polymatroids is associated to each rank-metric code. In this chapter, we choose to present the material of [841] differently, in order to stress the following facts (stated following the notation [841, Notation 5.3]):

(1) For $n < m$ the q-polymatroid that contains all the relevant information on $\mathcal{C}$ is $(\mathbb{F}_q^n, \rho_\mathrm{c}(\mathcal{C}, \cdot))$.

(2) For $n > m$ the q-polymatroid that contains all the relevant information on $\mathcal{C}$ is $(\mathbb{F}_q^m, \rho_\mathrm{r}(\mathcal{C}, \cdot))$.

(3) For $n = m$ one needs to consider both $(\mathbb{F}_q^n, \rho_\mathrm{c}(\mathcal{C}, \cdot))$ and $(\mathbb{F}_q^n, \rho_\mathrm{r}(\mathcal{C}, \cdot))$, at least if one wishes to have the property that equivalent codes have equivalent associated q-polymatroids.

Remark 11.6.10 In [1683, Proposition 3], Shiromoto associates a (q, m)-polymatroid to any rank-metric code with $n \leq m$. If $n < m$, his definition is equivalent to Definition 11.6.9, given what we observed in Remark 11.6.3. For $n = m$, Shiromoto's definition is not equivalent to Definition 11.6.9; in particular it is not equivalence-invariant (while Definition 11.6.9 is). Notice moreover that the original definition by Shiromoto does not contain the assumption that $n \leq m$. However, this hypothesis is used implicitly throughout his paper. Whenever stating the results from [1683], we always add the assumption $n \leq m$.

The code of [841, Example 2.10] shows that the definition of an associated (q, n)-polymatroid given by Shiromoto for a rank-metric code $\mathcal{C} \subseteq \mathrm{Mat}_{n \times n}(\mathbb{F}_q)$ is not equivalence-invariant.

Example 11.6.11 Let $\mathcal{C} \subseteq \mathbb{F}_2^{2 \times 2}$ be the code

$$\mathcal{C} := \left\{ \begin{bmatrix} a & a \\ b & b \end{bmatrix} \,\middle|\, a, b \in \mathbb{F}_2 \right\}.$$

Let $(\mathbb{F}_2^2, \rho_1)$ and $(\mathbb{F}_2^2, \rho_2)$ be the $(q, 2)$-polymatroids associated to $\mathcal{C}$ and $\mathcal{C}^\mathsf{T}$ respectively, according to [1683, Proposition 3]. By definition, for any $V \subseteq \mathbb{F}_2^2$

$$\rho_1(V) = 2 - \dim(\mathcal{C}(V^\perp)) = \dim(V)$$

and

$$\rho_2(V) = 2 - \dim(\mathcal{C}^\mathsf{T}(V^\perp)) = \begin{cases} \dim(V) & \text{if } V = 0, \langle(1,1)\rangle, \mathbb{F}_2^2, \\ 2 & \text{if } V = \langle(1,0)\rangle, \langle(0,1)\rangle. \end{cases}$$

The natural notion of equivalence for (q, r)-polymatroids is the following: $(\mathbb{F}_q^\ell, \rho_1)$ and $(\mathbb{F}_q^\ell, \rho_2)$ are equivalent if there exists an $\mathbb{F}_q$-linear isomorphism $\varphi : \mathbb{F}_q^\ell \to \mathbb{F}_q^\ell$ such that $\rho_1(V) = \rho_2(\varphi(V))$ for all $V \subseteq \mathbb{F}_q^\ell$. Clearly $(\mathbb{F}_2^2, \rho_1)$ and $(\mathbb{F}_2^2, \rho_2)$ are not equivalent with respect to such a notion of equivalence.

The interest in associating q-polymatroids to rank-metric codes comes from the fact that many invariants of rank-metric codes can be computed from the associated q-polymatroids. In fact, one could think of (equivalence classes of) q-polymatroids as invariants of the rank-metric codes to which they are associated, since equivalent codes are associated to equivalent q-polymatroids.

Proposition 11.6.12 ([841, Proposition 6.8]) *Let $\mathcal{C}, \mathcal{D} \subseteq \mathbb{F}_q^{n \times m}$ be rank-metric codes. Assume that $\mathcal{C} \sim \mathcal{D}$. If $m \neq n$, then $P(\mathcal{C}) \sim P(\mathcal{D})$. If $n = m$, then one of the following holds:*

- $P(\mathcal{C}) \sim P(\mathcal{D})$ *and* $P(\mathcal{C}^\mathsf{T}) \sim P(\mathcal{D}^\mathsf{T})$,

- $P(\mathcal{C}) \sim P(\mathcal{D}^\mathsf{T})$ *and* $P(\mathcal{C}^\mathsf{T}) \sim P(\mathcal{D})$.

One can also show that the q-polymatroid(s) associated to the rank-metric code $\Gamma(C)$ associated to a vector rank-metric code $C \subseteq \mathbb{F}_{q^m}^n$ do not depend on the choice of the basis Γ.

Proposition 11.6.13 ([1062, Corollary 4.7], [841, Proposition 6.10]) *Let $C \subseteq \mathbb{F}_{q^m}^n$ be a vector rank-metric code, and let Γ, Γ' be $\mathbb{F}_q$-bases of $\mathbb{F}_{q^m}$. Then*

$$P(\Gamma(C)) \sim P(\Gamma'(C)).$$

In the rest of this section, we discuss how to recover various invariants of rank-metric codes from the associated q-polymatroids. We start with the simplest invariants, namely the dimension and the minimum distance.

Proposition 11.6.14 ([841, Proposition 6.1 and Corollary 6.3]) *Let $\mathcal{C} \subseteq \mathbb{F}_q^{n \times m}$ be a rank-metric code. Then*

$$\dim(\mathcal{C}) = \max\{m, n\} \cdot \rho_{\mathcal{C}}\left(\mathbb{F}_q^{\min\{m,n\}}\right)$$

and

$$d_{\min}(\mathcal{C}) = \min\{m, n\} + 1 - \delta$$

where

$$\delta = \min\left\{ k \;\middle|\; \rho_{\mathcal{C}}(V) = \frac{\dim(\mathcal{C})}{\max\{m, n\}} \text{ for all } V \subseteq \mathbb{F}_q^{\min\{m,n\}} \text{ with } \dim(V) = k \right\}.$$

The next result shows how one can compute the generalized weights of a rank-metric code from its associated q-polymatroid(s).

Theorem 11.6.15 ([841, Theorem 7.1]) *Let $\mathcal{C} \subseteq \mathbb{F}_q^{n \times m}$ be a rank-metric code. For $i = 1, 2, \ldots, \dim(\mathcal{C})$ let*

$$d_i(P(\mathcal{C})) = \min \{m, n\} - \max \{\dim(V) \mid V \subseteq \mathbb{F}_q^{\min\{m,n\}},$$
$$\dim(\mathcal{C}) - \max \{m, n\} \cdot \rho_{\mathcal{C}}(V) \geq i\}.$$

If $n \neq m$, then

$$d_i(\mathcal{C}) = d_i(P(\mathcal{C})) \quad for \quad i = 1, 2, \ldots, \dim(\mathcal{C}).$$

If $n = m$, then

$$d_i(\mathcal{C}) = \min \{d_i(P(\mathcal{C})), \ d_i(P(\mathcal{C}^T))\} \quad for \quad i = 1, 2, \ldots, \dim(\mathcal{C}).$$

The associated q-polymatroid(s) also determine the weight distribution of a rank-metric code. The result is stated in terms of the rank weight enumerator of the code.

Definition 11.6.16 Let $\mathcal{C} \subseteq \mathbb{F}_q^{n \times m}$ be a rank-metric code. The **rank weight enumerator** of $\mathcal{C}$ is the polynomial

$$\mathrm{rwe}_{\mathcal{C}}(x, y) = \sum_{i=0}^{\min\{m,n\}} A_i(\mathcal{C}) x^i y^{\min\{m,n\}-i}.$$

The next theorem was proved by Shiromoto.

Theorem 11.6.17 ([1683, Theorem 14]) *Let $\mathcal{C} \subseteq \mathbb{F}_q^{n \times m}$ be a rank-metric code of dimension $\dim(\mathcal{C}) = \ell$. Assume that $n \leq m$. Then*

$$\mathrm{rwe}_{\mathcal{C}}(x, y) = x^{n - \ell/m} \left(\sum_{V \subseteq \mathbb{F}_q^n} (q^m x)^{\rho_{\mathcal{C}}(\mathbb{F}_q^n) - \rho_{\mathcal{C}}(V)} x^{-(\dim(V) - \rho_{\mathcal{C}}(V))} \right)^{\dim(V)-1} \prod_{i=0} (y - q^i x).$$

Finally, we state two results that show that the property of being MRD or an optimal anticode can be characterized in terms of the associated q-polymatroid(s).

Theorem 11.6.18 ([841, Theorem 6.4 and Corollary 6.6]) *Let $\mathcal{C} \subseteq \mathbb{F}_q^{n \times m}$ be a rank-metric code with minimum distance $d_{\min}(\mathcal{C}) = d$. The following are equivalent.*

(a) *$\mathcal{C}$ is MRD.*

(b) *$\rho_{\mathcal{C}}(V) = \dim(V)$ for all $V \subseteq \mathbb{F}_q^{\min\{m,n\}}$ with $\dim(V) \leq \min \{m, n\} - d + 1$.*

(c) *$\rho_{\mathcal{C}}(V) = \dim(V)$ for some $V \subseteq \mathbb{F}_q^{\min\{m,n\}}$ with $\dim(V) = \min \{m, n\} - d + 1$.*

In particular, if $\mathcal{C}$ is MRD with $d_{\min}(\mathcal{C}) = d$, then $P(\mathcal{C}) = \left(\mathbb{F}_q^{\min\{m,n\}}, \rho_{\mathcal{C}} \right)$ where

$$\rho_{\mathcal{C}}(V) = \begin{cases} \min \{m, n\} - d + 1 & \text{if } \dim(V) \geq \min \{m, n\} - d + 1, \\ \dim(V) & \text{if } \dim(V) \leq \min \{m, n\} - d + 1. \end{cases}$$

The corresponding result for optimal anticodes is the following. Notice that the q-polymatroids associated to MRD codes or optimal anticodes are in fact q-matroids.

Theorem 11.6.19 ([841, Theorem 7.2 and Corollary 6.6]) *Let $\mathcal{C} \subseteq \mathbb{F}_q^{n \times m}$ be a rank-metric code with $r = \max \mathrm{rank}(\mathcal{C})$. The following are equivalent.*

(a) *$\mathcal{C}$ is an optimal anticode.*

(b) $\{\rho_{\mathcal{C}}(V) \mid V \subseteq \mathbb{F}_q^{\min\{m,n\}}\} = \{0,1,\ldots,r\}$, or $\{\rho_{\mathcal{C}^T}(V) \mid V \subseteq \mathbb{F}_q^n\} = \{0,1,\ldots,r\}$ and $m = n$.

(c) $\rho_{\mathcal{C}}\big(\mathbb{F}_q^{\min\{m,n\}}\big) = r$, or $\rho_{\mathcal{C}^T}(\mathbb{F}_q^n) = r$ and $m = n$.

In particular, if $\mathcal{C}$ is an optimal anticode with $r = \max \operatorname{rank}(\mathcal{C})$, let

$$\rho(V) = \dim(V + \langle e_1, e_2, \ldots, e_{\min\{m,n\}-r}\rangle) - (\min\{m,n\} - r),$$

where e_i denotes the ith vector of the standard basis of $\mathbb{F}_q^{\min\{m,n\}}$. If $m \neq n$, then $P(\mathcal{C}) \sim (\mathbb{F}_q^{\min\{m,n\}}, \rho)$. If $m = n$, then either $P(\mathcal{C}) \sim (\mathbb{F}_q^n, \rho)$ or $P(\mathcal{C}^T) \sim (\mathbb{F}_q^n, \rho)$.

We conclude with a result on associated q-polymatroids and duality. The theorem as we state it was proved by Gorla, Jurrius, López, and Ravagnani in [841, Theorem 8.1 and Corollary 8.2]. Shiromoto also proved in [1683, Proposition 11] that $P(\mathcal{C})^* = P(\mathcal{C}^\perp)$ for a rank-metric code $\mathcal{C} \subseteq \mathbb{F}_q^{n \times m}$ with $n \leq m$.

Theorem 11.6.20 *Let $\mathcal{C} \subseteq \mathbb{F}_q^{n \times m}$ be a rank-metric code and let $C \subseteq \mathbb{F}_{q^m}^n$ be a vector rank-metric code. Let Γ be a basis of $\mathbb{F}_{q^m}$ over $\mathbb{F}_q$. Then*

$$P(\mathcal{C})^* = P(\mathcal{C}^\perp) \quad and \quad P(\Gamma(C))^* \sim P(\Gamma(C^\perp)).$$

Acknowledgement

Part of this chapter was written while the author was participating in the Nonlinear Algebra program at ICERM in Fall 2018. The author wishes to thank ICERM and Brown University for an excellent working environment.

Chapter 12

Linear Programming Bounds

Peter Boyvalenkov

Institute of Mathematics and Informatics, Bulgarian Academy of Sciences

Danyo Danev

Linköping University

12.1   Preliminaries – Krawtchouk Polynomials, Codes, and Designs

Definition 12.1.1 Let $n \geq 1$ and $q \geq 2$ be integers. The **Krawtchouk polynomials** are defined as

$$K_i^{(n,q)}(z) := \sum_{j=0}^{i} (-1)^j (q-1)^{i-j} \binom{n-z}{i-j}\binom{z}{j} \quad \text{for } i = 0, 1, 2, \ldots$$

where $\binom{z}{j} := z(z-1)\cdots(z-j+1)/j!$ for $z \in \mathbb{R}$.

Theorem 12.1.2 ([1229]) *The polynomials $K_i^{(n,q)}(z)$ satisfy the three-term recurrence relation*

$$(i+1)K_{i+1}^{(n,q)}(z) = \big(i + (q-1)(n-i) - qz\big)K_i^{(n,q)}(z) - (q-1)(n-i+1)K_{i-1}^{(n,q)}(z)$$

with initial conditions $K_0^{(n,q)}(z) = 1$ and $K_1^{(n,q)}(z) = n(q-1) - qz$.

Theorem 12.1.3 ([1229]) *The discrete measure*

$$d\mu_n(t) := q^{-n} \sum_{i=0}^{n} \binom{n}{i}(q-1)^i \delta(t-i)dt, \tag{12.1}$$

where $\delta(i)$ is the Dirac-delta measure at $i \in \{0, 1, \ldots, n\}$, and the form

$$\langle f, g \rangle = \int f(t)g(t)\,d\mu_n(t) \tag{12.2}$$

define an inner product over the class $\mathcal{P}_n$ of real polynomials of degree less than or equal to n.

Theorem 12.1.4 (Orthogonality Relations [1229]) *Under the inner product* (12.2) *the Krawtchouk polynomials satisfy*

$$\sum_{u=0}^{n} K_i^{(n,q)}(u) K_j^{(n,q)}(u)(q-1)^u \binom{n}{u} = \delta_{i,j} q^n (q-1)^i \binom{n}{i}$$

for any $i,j \in \{0,1,\ldots,n\}$, *where* $\delta_{i,j} = 1$ *if* $i = j$ *and* $\delta_{i,j} = 0$ *if* $i \neq j$. *Moreover,*

$$\sum_{u=0}^{n} K_i^{(n,q)}(u) K_u^{(n,q)}(j) = \delta_{i,j} q^n.$$

Theorem 12.1.5 (Expansion [1229]) *If* $f(z) = \sum_{i=0}^{n} f_i K_i^{(n,q)}(z)$, *then*

$$f_i = \left(q^n (q-1)^i \binom{n}{i} \right)^{-1} \sum_{u=0}^{n} f(u) K_i^{(n,q)}(u)(q-1)^u \binom{n}{u}$$

$$= q^{-n} \sum_{u=0}^{n} f(u) K_u^{(n,q)}(i).$$

Definition 12.1.6 Define $K_{n+1}^{(n,q)}(z) := \frac{q^{n+1}}{(n+1)!} \prod_{u=0}^{n}(u - z)$. Note that $K_{n+1}^{(n,q)}(z)$ is orthogonal to any polynomial $K_i^{(n,q)}(z)$, $i = 0, 1, \ldots, n$, with respect to the measure (12.1).

Theorem 12.1.7 (Krein Condition [1229]) *For any* $i,j \in \{0,1,\ldots,n\}$

$$K_i^{(n,q)}(z) K_j^{(n,q)}(z) = \sum_{u=0}^{n} p_{i,j}^u K_u^{(n,q)}(z) \pmod{K_{n+1}^{(n,q)}(z)}$$

with $p_{i,j}^u = 0$ *if* $i + j > u$, $p_{i,j}^u > 0$ *if* $i + j = u \leq n$, *and* $p_{i,j}^u \geq 0$ *otherwise.*

Definition 12.1.8 Denote

$$T_i^{n,q}(z,w) := \sum_{j=0}^{i} K_j^{(n,q)}(z) K_j^{(n,q)}(w) \left((q-1)^j \binom{n}{j} \right)^{-1}.$$

Theorem 12.1.9 (Christoffel–Darboux Formula [1229]) *For any real* z *and* w *and for any* $i = 0, 1, \ldots, n$

$$(w - z) T_i^{n,q}(z,w) = \frac{i+1}{q(q-1)^i \binom{n}{i}} \left(K_{i+1}^{(n,q)}(z) K_i^{(n,q)}(w) - K_{i+1}^{(n,q)}(w) K_i^{(n,q)}(z) \right).$$

Let $\mathcal{C} \subseteq F_q^n$ be a code, where $F_q = \{0, 1, \ldots, q-1\}$ is the alphabet of q symbols (so q is not necessarily a power of a prime). For $\mathbf{x}, \mathbf{y} \in F_q^n$, recall that $\mathrm{d_H}(\mathbf{x}, \mathbf{y})$ is the number of coordinates where $\mathbf{x}$ and $\mathbf{y}$ disagree.

Definition 12.1.10 The vector $B(\mathcal{C}) = (B_0, B_1, \ldots, B_n)$, where

$$B_i = \frac{1}{|\mathcal{C}|} \left| \{ (\mathbf{x}, \mathbf{y}) \in \mathcal{C}^2 \mid \mathrm{d_H}(\mathbf{x}, \mathbf{y}) = i \} \right| \quad \text{for } i = 0, 1, \ldots, n,$$

is called the **distance distribution** of $\mathcal{C}$. Clearly, $B_0 = 1$ and $B_i = 0$ for $i = 1, 2, \ldots, d-1$, where d is the minimum distance of $\mathcal{C}$.

Definition 12.1.11 The vector $B'(\mathcal{C}) = (B'_0, B'_1, \ldots, B'_n)$, where

$$B'_i = \frac{1}{|\mathcal{C}|} \sum_{j=0}^{n} B_j K_i^{(n,q)}(j) \ \text{ for } i = 0, 1, \ldots, n,$$

is called the **dual distance distribution** of $\mathcal{C}$ or the **MacWilliams transform** of $B(\mathcal{C})$. Obviously $B'_0 = 1$.

Theorem 12.1.12 ([519, 521]) *The dual distance distribution of $\mathcal{C}$ satisfies*

$$B'_i \geq 0 \ \text{ for } i = 1, 2, \ldots, n.$$

Theorem 12.1.13 ([1323]) *If q is a power of a prime and $\mathcal{C}$ is a linear code in $\mathbb{F}_q^n$, then $B'(\mathcal{C})$ is the distance distribution of the dual code $\mathcal{C}^\perp$.*

Definition 12.1.14 The smallest positive integer i such that $B'_i \neq 0$ is called the **dual distance** of $\mathcal{C}$ and is denoted by $d' = d'(\mathcal{C})$. Denote by $s = s(\mathcal{C})$ (respectively $s' = s'(\mathcal{C})$) the number of nonzero B_i's (respectively B'_i's), $i \in \{1, 2, \ldots, n\}$, i.e.,

$$s = \left| \{i \mid B_i \neq 0, i > 0\} \right| \ \text{ and } \ s' = \left| \{i \mid B'_i \neq 0, i > 0\} \right|.$$

The number s' is called the **external distance** of $\mathcal{C}$. Define $\delta = 0$ if $B_n = 0$ and $\delta = 1$ otherwise (respectively $\delta' = 0$ if $B'_n = 0$ and $\delta' = 1$ otherwise).

Definition 12.1.15 Let $\mathcal{C} \subseteq F_q^n$ be a code and M be a codeword matrix consisting of all vectors of $\mathcal{C}$ as rows. Then $\mathcal{C}$ is called a τ-**design** if any set of τ columns of M contains any τ-tuple of F_q^τ the same number of times, namely $\lambda := |\mathcal{C}|/q^\tau$. The largest positive integer τ such that $\mathcal{C}$ is a τ-design is called the **strength** of $\mathcal{C}$ and is denoted by $\tau(\mathcal{C})$. The number λ is called the **index** of $\mathcal{C}$.

Remark 12.1.16 A τ-design in F_q^n is also called an **orthogonal array of strength** τ [521] or a τ-**wise independent set** [41]; see also [928].

Theorem 12.1.17 ([519, 521]) *If $\mathcal{C} \subseteq F_q^n$ has dual distance $d' = d'(\mathcal{C})$, then $\mathcal{C}$ is a τ-design where $\tau = d' - 1$.*

Definition 12.1.18 For a real polynomial $f(z) = \sum_{i=0}^{n} f_i K_i^{(n,q)}(z)$, the polynomial

$$\widehat{f}(z) = q^{-n/2} \sum_{j=0}^{n} f(j) K_j^{(n,q)}(z)$$

is called the **dual** to $f(z)$. Note that the dual to $\widehat{f}(z)$ is $f(z)$ and also that $\widehat{f}(i) = q^{n/2} f_i$ for any $i \in \{0, 1, \ldots, n\}$.

12.2   General Linear Programming Theorems

Theorem 12.2.1 ([519, 521]) *For any code $\mathcal{C} \subseteq F_q^n$ that has distance distribution $(B_0, B_1, \ldots, B_n)$ and dual distance distribution $(B'_0, B'_1, \ldots, B'_n)$, and any real polynomial $f(z) = \sum_{i=0}^{n} f_i K_i^{(n,q)}(z)$, it is valid that*

$$f(0) + \sum_{i=1}^{n} B_i f(i) = |\mathcal{C}| \left(f_0 + \sum_{i=1}^{n} f_i B'_i \right).$$

In Definition 1.9.1 in Chapter 1, $A_q(n,d)$ is defined for codes over $\mathbb{F}_q$. We now extend that definition to codes over F_q as only the alphabet size is important and not its structure.

Definition 12.2.2 For fixed q, n, and $d \in \{1,2,\ldots,n\}$ denote

$$A_q(n,d) := \max\{|\mathcal{C}| \mid \mathcal{C} \subseteq F_q^n,\ d(\mathcal{C}) = d\}.$$

Definition 12.2.3 For fixed q, n, and $\tau \in \{1,2,\ldots,n\}$ denote

$$B_q(n,\tau) := \min\{|\mathcal{C}| \mid \mathcal{C} \subseteq F_q^n,\ \tau(\mathcal{C}) = \tau\}.$$

Theorem 12.2.4 (Linear Programming Bound for Codes [519, 521]) *Let the real polynomial* $f(z) = \sum_{i=0}^{n} f_i K_i^{(n,q)}(z)$ *satisfy the conditions*

(A1) $f_0 > 0$, $f_i \geq 0$ *for* $i = 1,2,\ldots,n$;

(A2) $f(0) > 0$, $f(i) \leq 0$ *for* $i = d, d+1, \ldots, n$.

Then $A_q(n,d) \leq f(0)/f_0$. *Equality holds for codes* $\mathcal{C} \subseteq F_q^n$ *with* $d(\mathcal{C}) = d$, *distance distribution* $(B_0, B_1, \ldots, B_n)$, *dual distance distribution* $(B_0', B_1', \ldots, B_n')$ *and polynomials* $f(z)$ *such that* $B_i f(i) = 0$ *and* $B_i' f_i = 0$ *for every* $i = 1,2,\ldots,n$.

Remark 12.2.5 The polynomial f from Theorem 12.2.4 is implicit in Theorem 1.9.23(a) from Chapter 1 as $\sum_{i=0}^{n} B_i K_i^{(n,q)}(i)$ and it is normalized for f_0 (or B_0) to be equal to 1; note also the normalization of the Krawtchouk polynomials. Thus the bound $f(1)/f_0$ appears as $\sum_{i=0}^{n} B_i$ in Chapter 1.

Theorem 12.2.6 (Linear Programming Bound for Designs [519, 521]) *Let the real polynomial* $f(z) = \sum_{i=0}^{n} f_i K_i^{(n,q)}(z)$ *satisfy the conditions*

(B1) $f_0 > 0$, $f_i \leq 0$ *for* $i = \tau+1, \tau+2, \ldots, n$;

(B2) $f(0) > 0$, $f(i) \geq 0$ *for* $i = 1,2,\ldots,n$.

Then $B_q(n,\tau) \geq f(0)/f_0$. *Equality holds for designs* $\mathcal{C} \subseteq F_q^n$ *with* $\tau(\mathcal{C}) = \tau$, *distance distribution* $(B_0, B_1, \ldots, B_n)$, *dual distance distribution* $(B_0', B_1', \ldots, B_n')$ *and polynomial* $f(z)$ *such that* $B_i f(i) = 0$ *and* $B_i' f_i = 0$ *for every* $i = 1,2,\ldots,n$.

Remark 12.2.7 The Rao Bound in Theorem 12.3.3 and Levenshtein Bound in Theorem 12.3.10 below are obtained by suitable polynomials in Theorems 12.2.4 and 12.2.6, respectively. Examples of codes attaining these bounds are listed in Table 12.1 presented later.

Theorem 12.2.8 (Duality [1231]) *A real polynomial* $f(z)$ *satisfies the conditions* (A1) *and* (A2) *if and only if its dual polynomial* $\widehat{f}$ *satisfies the conditions* (B1) *and* (B2). *Moreover,*

$$\frac{f(0)}{f_0} \cdot \frac{\widehat{f}(0)}{\widehat{f}_0} = q^n.$$

Remark 12.2.9 Rephrased, the duality means that for any polynomial f which is good for linear programming for codes, its dual $\widehat{f}$ is good for linear programming for designs (and conversely). Thus we obtain, in a sense, bounds for free. In particular, the duality justifies the pairs of bounds in Theorems 12.3.1, 12.3.3 and 12.3.10 below.

See also [309, 528, 813, 814, 1231].

Definition 12.2.10 For a code $\mathcal{C} \subseteq F_q^n$ and a function $h : \{1, 2, \dots, n\} \to \mathbb{R}$ the **potential energy** of $\mathcal{C}$ with respect to h is defined to be

$$E_h(\mathcal{C}) := \sum_{\mathbf{x}, \mathbf{y} \in \mathcal{C}, \mathbf{x} \neq \mathbf{y}} h(d_H(\mathbf{x}, \mathbf{y})).$$

Definition 12.2.11 For fixed q, n, h, and $M \in \{2, 3, \dots, q^n\}$ denote

$$E_h(n, M; q) := \min \{E_h(\mathcal{C}) \mid \mathcal{C} \subseteq F_q^n, \ |\mathcal{C}| = M\}.$$

Theorem 12.2.12 (Linear Programming Bound for Energy of Codes [426]) *Let n and q be fixed, $h : (0, n] \to (0, +\infty)$ be a function, and $M \in \{2, 3, \dots, q^n\}$. Let the real polynomial $f(z) = \sum_{i=0}^n f_i K_i^{(n,q)}(z)$ satisfy the conditions*

(D1) $f_0 > 0$, $f_i \geq 0$ *for* $i = 1, 2, \dots, n$;

(D2) $f(0) > 0$, $f(i) \leq h(i)$ *for* $i = 1, 2, \dots, n$.

Then $E_h(n, M; q) \geq M(f_0 M - f(0))$. Equality holds for codes $\mathcal{C} \subseteq F_q^n$ with distance distribution $(B_0, B_1, \dots, B_n)$, dual distance distribution $(B_0', B_1', \dots, B_n')$ and polynomials $f(z)$, such that $B_i [f(i) - h(i)] = 0$ and $B_i' f_i = 0$ for every $i = 1, 2, \dots, n$.

Remark 12.2.13 Theorems 12.2.4, 12.2.6 and 12.2.12 can be applied with the usual simplex method for quite large parameters. For instance, the website [1610] (Delsarte, a.k.a. Linear Programming (LP), upper bounds) offers a tool for computation of bounds via integer LP with Theorem 12.2.4. Several websites maintain tables of best known bounds (lower and upper) for codes of relatively small lengths; see, e.g., [305].

12.3 Universal Bounds

The Singleton Bound presented in Theorem 1.9.10 is an upper bound on the code cardinality, given q, n, and d. It is the upper bound in (12.3) below. Its proof by linear programming and the duality imply the lower bound in (12.3).

Theorem 12.3.1 (Singleton Bound [1717]) *For any code $\mathcal{C} \subseteq F_q^n$ with minimum distance d and dual distance d'*

$$q^{d'-1} \leq |\mathcal{C}| \leq q^{n-d+1}. \tag{12.3}$$

The bounds (12.3) can be attained only simultaneously and this happens if and only if $d + d' = n + 2$ and all possible distances are realized (so the attaining code is an MDS code).

Definition 12.3.2 For fixed q, n, and d denote by

$$V_k(n, q) := \sum_{i=0}^k \binom{n}{i} (q-1)^i \quad \text{for} \ 0 \leq k \leq n,$$

(the volume of a sphere of radius k in F_q^n) and

$$H^{n,q}(d) := q^\varepsilon V_k(n - \varepsilon, q),$$

where $d = 2k + 1 + \varepsilon$ with $\varepsilon \in \{0, 1\}$.

The Sphere Packing or Hamming Bound presented in Theorem 1.9.6 is another upper bound on the code size, given q, n, and d. Rao gave a lower bound on the code size, given q, n, and d'. These bounds are combined in the following theorem as they are connected by the duality.

Theorem 12.3.3 (Rao Bound [1566] and Hamming (Sphere Packing) Bound [895]) *For any code $\mathcal{C} \subseteq \mathbb{F}_q^n$ with minimum distance d and dual distance d'*

$$H^{n,q}(d') \leq |\mathcal{C}| \leq \frac{q^n}{H^{n,q}(d)}. \tag{12.4}$$

Definition 12.3.4 Codes attaining the upper bound in (12.4) are called **perfect**. Designs attaining the lower bound in (12.4) are called **tight**.

Recall the definitions of s, s', δ, and δ' in Definition 12.1.14.

Theorem 12.3.5 *Let $\mathcal{C} \subseteq \mathbb{F}_q^n$ with $|\mathcal{C}| > 1$. The following hold.*

(a) *$\mathcal{C}$ is a tight design if and only if $d' = 2s - \delta + 1$.*

(b) *$\mathcal{C}$ is a perfect code if and only if $d = 2s' - \delta' + 1$.*

See also [112, 652, 1231, 1566].

Definition 12.3.6 For fixed n, q, and i denote by $\xi_i^{n,q}$ the smallest root of the Krawtchouk polynomial $K_i^{(n,q)}(z)$. In the case of $i = 0$ set $\xi_0^{n,q} = n + 1$.

Lemma 12.3.7 ([1229]) *The following are valid.*

(a) *For $i = 1, 2, \ldots, n$, $\xi_i^{n-2,q} < \xi_i^{n-1,q} < \xi_{i-1}^{n-2,q}$.*

(b) *For $z \in [1, n]$, there exists a unique $k = k(z) \in \{1, 2, \ldots, n\}$ and a unique $\varepsilon = \varepsilon(z) \in \{0, 1\}$ such that*
$$\xi_k^{n-1-\varepsilon,q} + 1 < z \leq \xi_{k-1+\varepsilon}^{n-2+\varepsilon,q} + 1.$$

Example 12.3.8 One has $\xi_0^{n,q} = n + 1$, $\xi_1^{n,q} = \frac{(q-1)n}{q}$, and

$$\xi_2^{n,q} = \frac{2(q-1)n - q + 2 - \sqrt{4(q-1)n + (q-2)^2}}{2q}.$$

Definition 12.3.9 If $z \in [1, n]$, let $k = k(z)$ and $\varepsilon = \varepsilon(z) \in \{0, 1\}$ be as in Lemma 12.3.7(b). If $\xi_k^{n-1-\varepsilon,q} + 1 < z \leq \xi_{k-1+\varepsilon}^{n-2+\varepsilon,q} + 1$, define

$$D_k^{n,q}(z) := V_{k-1}(n,q) - (q-1)^k \binom{n}{q} \frac{K_{k-1}^{(n-1,q)}(z-1)}{K_k^{(n,q)}(z)} \quad \text{and}$$

$$L^{n,q}(z) := L_{2k-1+\varepsilon}^{n,q}(z) := q^\varepsilon D_k^{n-\varepsilon,q}(z).$$

Theorem 12.3.10 (Levenshtein Bound [1227, 1229, 1230]) *For a code $\mathcal{C} \subseteq \mathbb{F}_q^n$ with minimum distance d and dual distance d'*

$$\frac{q^n}{L^{n,q}(d')} \leq |\mathcal{C}| \leq L^{n,q}(d). \tag{12.5}$$

The lower (respectively upper) bound is attained if and only if $d \geq \max\{2s' - \delta', 2\}$ (respectively $d' \geq \max\{2s - \delta, 2\}$).

Example 12.3.11 In the first three relevant intervals, the Levenshtein (upper) Bound is given by

$$A_q(n,d) \leq \frac{qd}{qd - (q-1)n} = L_1^{n,q}(d)$$

(which is the Plotkin Bound discussed in Section 1.9.3) if $n - \frac{n-1}{q} = \xi_1^{n-1,q} + 1 \leq d \leq \xi_0^{n-2,q} + 1 = n$,

$$A_q(n,d) \leq \frac{q^2 d}{qd - (q-1)(n-1)} = L_2^{n,q}(d)$$

if $\xi_1^{n-2,q} + 1 \leq d \leq \xi_1^{n-1,q} + 1$, and

$$A_q(n,d) \leq \frac{qd(n(q-1)+1)(n(q-1)-qd+2-q)}{qd(2n(q-1)-q+2-qd) - (n-1)(q-1)^2} = L_3^{n,q}(d)$$

if $\xi_2^{n-1,q} + 1 \leq d \leq \xi_1^{n-2,q} + 1$.

Remark 12.3.12 It is also worth noting two important values of the Levenshtein Bound:

$$L^{n,q}(\xi_{k-1+\varepsilon}^{n-2+\varepsilon,q} + 1) = H^{n,q}(2k - 1 + \varepsilon) \quad \text{for } \varepsilon \in \{0,1\}.$$

So at the ends of the intervals $\left[\xi_k^{n-1-\varepsilon,q} + 1, \xi_{k-1+\varepsilon}^{n-2+\varepsilon,q} + 1\right]$ the Levenshtein Bound coincides with the corresponding Rao Bound.

Recall the kernels $T_i^{n,q}$ from Definition 12.1.8 and the parameters $k = k(d)$ and $\varepsilon = \varepsilon(d)$ from Lemma 12.3.7(b). The next theorem gives a Gauss–Jacobi quadrature formula (12.6), introduced by Levenshtein in [1228, 1229], which is instrumental in proofs of Theorem 12.3.16 and Theorem 12.3.21(a). Theorem 12.3.13 also introduces parameters needed for the universal bound (12.8) below.

Theorem 12.3.13 ([1229]) *For any $d \in \{1, 2, \ldots, n\}$ the polynomial*

$$g_d(z) = z(d-z)(n-z)T_{k-1}^{n-1-\varepsilon,q}(z-1, d-1),$$

where $k = k(d)$ and $\varepsilon = \varepsilon(d) \in \{0,1\}$, has $k + 1 + \varepsilon$ simple zeros

$$\alpha_0 = 0 < \alpha_1 = d < \cdots < \alpha_{k+\varepsilon} \leq n$$

with $\alpha_{k+\varepsilon} = n$ if and only if $\varepsilon = 1$ or $\varepsilon = 0$ and $d = \xi_{k-1}^{n-2,q} + 1$.

Moreover, for any real polynomial $f(z)$ of degree at most $2k - 1 + \varepsilon$ the following equality holds:

$$f_0 = \frac{f(0)}{L^{n,q}(d)} + \sum_{i=1}^{k+\varepsilon} \rho_i^{(d)} f(\alpha_i), \tag{12.6}$$

where all coefficients (weights) $\rho_i^{(d)}$, $i = 1, 2, \ldots, k$, are positive, and, in the case $\varepsilon = 1$, $\rho_{k+1}^{(d)} \geq 0$ with equality if and only if $d = \xi_k^{n-1,q} + 1$. We have

$$\rho_i^{(d)} = \frac{q^{-1-\varepsilon}(q-1)n(n-1)^\varepsilon}{\alpha_i(n-\alpha_i)T_{k-1}^{n-1-\varepsilon,q}(\alpha_i - 1, \alpha_i - 1)} \quad \text{for } i = 1, 2, \ldots, k$$

and

$$\rho_{k+1}^{(d)} = \frac{q^{-\varepsilon}K_k^{(n-\varepsilon,q)}(d)}{K_k^{(n-\varepsilon,q)}(d)K_{k-1}^{(n-1-\varepsilon,q)}(-1) - K_{k-1}^{(n-1-\varepsilon,q)}(d-1)K_k^{(n-\varepsilon,q)}(0)}.$$

Remark 12.3.14 Levenshtein used the polynomial

$$f(t) = (t - \alpha_{k+\varepsilon})^\varepsilon (t - \alpha_1) \prod_{i=1+\varepsilon}^{k-1+\varepsilon} (t - \alpha_i)^2 \qquad (12.7)$$

to obtain the bounds (12.5). It was shown in [289] that its zeros $\alpha_1, \ldots, \alpha_{k+\varepsilon}$ strongly suggest the optimal choice of nodes for the simplex method of Theorem 12.2.4 (equivalently, Theorem 1.9.23(a) from Chapter 1). Computational experiments show that simple replacement of any double zero $\alpha_i \in (j, j+1)$ of Levenshtein's polynomial (12.7) by two simple zeros j and $j+1$ gives in most cases (conjecture: for every sufficiently large rate d/n) the best result that can ever be obtained from Theorem 12.2.4.

Remark 12.3.15 The assertions of Theorem 12.3.13 remain true when $d \in [1, n]$ is a continuous variable. In particular, the quadrature formula (12.6) holds true for any real polynomial $f(z)$ of degree at most $2k - 1 + \varepsilon$ with well defined roots of $g_d(z)$ and corresponding positive $\rho_i^{(d)}$ for $i = 1, 2, \ldots, k+1$.

Theorem 12.3.16 ([1228, 1703]) *The upper bound in (12.5) cannot be improved by a real polynomial f of degree at most $2k - 1 + \varepsilon$ satisfying (A1) and the condition $f(z) \leq 0$ for $z \in [d, n]$.*

Definition 12.3.17 For $j \in \{1, 2, \ldots, n\}$ and $d \in (0, n]$ denote by

$$P_j^{n,q}(d) := \frac{K_j^{(n,q)}(0)}{L^{n,q}(d)} + \sum_{i=1}^{k+\varepsilon} \rho_i^{(d)} K_j^{(n,q)}(\alpha_i),$$

where $k = k(d)$, $\varepsilon = \varepsilon(d)$, and $\rho_i^{(d)}$ are as in Theorem 12.3.13. Note that $P_j^{n,q} = 0$ for $j \leq 2k - 1 + \varepsilon$. Note also that $K_j^{(n,q)}(0) = (q-1)^j \binom{n}{j}$.

Theorem 12.3.18 ([287, 1230]) *The upper bound in (12.5) can be improved by a real polynomial f of degree at least $2k + \varepsilon$ satisfying (A1) and the condition $f(z) \leq 0$ for $z \in [d, n]$ if and only if $P_j^{n,q}(d) < 0$ for some $j \geq 2k + \varepsilon$.*

Definition 12.3.19 A function $h : (0, n] \to (0, +\infty)$ is called **(strictly) completely monotone** if $(-1)^i h^{(i)}(z) \geq 0$ (> 0) for every nonnegative integer i and every $z \in (0, n]$. The derivatives $h^{(i)}(z)$ can be discrete (then $z \in \{1, 2, \ldots, n\}$) or continuous.

Theorem 12.3.20 (Universal Lower Bound on Energy [291]) *Let n, q and $M \in \{2, 3, \ldots, q^n\}$ be fixed and $h : (0, n] \to (0, +\infty)$ be a completely monotone function. If $M \in (H^{n,q}(2k - 1 + \varepsilon), H^{n,q}(2k + \varepsilon)]$ and $d = d(M) \in [1, n]$ is the smallest real root of the equation $M = L^{n,q}(d)$, then*

$$E_h(n, M; q) \geq M^2 \sum_{i=1}^{k+\varepsilon} \rho_i^{(d)} h(\alpha_i), \qquad (12.8)$$

where the parameters α_i and $\rho_i^{(d)}$ are determined as in Theorem 12.3.13. Equality is attained if and only if there exists a code with M codewords which attains the upper bound in (12.5).

Theorem 12.3.21 ([291]) *Let h be a strictly completely monotone function. The bound (12.8):*

(a) *cannot be improved by a real polynomial f of degree at most $2k - 1 + \varepsilon$ satisfying* (A1) *and $f(z) \le h(z)$ for $z \in [d, n]$;*

(b) *can be improved by a real polynomial f of degree at least $2k + \varepsilon$ satisfying* (A1) *and the condition $f(z) \le h(z)$ for $z \in [d, n]$ if and only if $P_j^{n,r}(d) < 0$ for some $j \ge 2k + \varepsilon$.*

Definition 12.3.22 A code $\mathcal{C} \subseteq F_q^n$ is called **universally optimal** if it (weakly) minimizes potential energy among all configurations of $|\mathcal{C}|$ codewords in F_q^n for each completely monotone function h.

The conditions for attaining the bounds (12.5) and (12.8) coincide. Thus, any code which attains the upper bound in (12.5) for its cardinality (and therefore the lower bound for (12.8) for its energy) is universally optimal; see Table 12.1. We summarize this in the next theorem.

Theorem 12.3.23 ([291, 426]) *All codes which attain the upper bound in (12.5) are universally optimal.*

Theorem 12.3.24 ([426]) *Let $\mathcal{C} \subseteq F_q^n$ be a code and $h : (0, n] \to \mathbb{R}$ be any function such that the bound in Theorem 12.2.12 is attained. Let $\mathbf{c} \in \mathcal{C}$. Then*

$$E_h(\mathcal{C} \setminus \{\mathbf{c}\}) = E_h(n, |\mathcal{C}| - 1; q).$$

In particular, if $\mathcal{C}$ is proved to be universally optimal by Theorem 12.2.12, then $\mathcal{C} \setminus \{\mathbf{c}\}$ is universally optimal for all $\mathbf{c} \in \mathcal{C}$ as well.

Example 12.3.25 The Kerdock codes $\mathcal{K}_m \subset \mathbb{F}_2^{2^{2m}}$ [1107] are nonlinear codes existing for lengths $n = 2^{2m}$. Their cardinality is $|\mathcal{K}_m| = n^2 = 2^{4m}$ and their distance (weight) distribution is as follows:

$$B_{2^{2m-1}-2^{m-1}} = 2^{2m}(2^{2m-1} - 1), \ B_{2^{2m-1}} = 2^{2m+1} - 2, \ B_{2^{2m-1}+2^{m-1}} = 2^{2m}(2^{2m-1} - 1),$$

$$B_0 = B_n = 1, \text{ and } B_i = 0 \text{ for } i \notin \{0, 2^{2m-1} - 2^{m-1}, 2^{2m-1}, 2^{2m-1} + 2^{m-1}\}.$$

It is easy to check that the Kerdock codes are asymptotically optimal with respect to the upper bound in (12.5) and the bound (12.8) as they are very close to the bounds already for small m.

Definition 12.3.26 For any code $\mathcal{C} \subseteq F_q^n$, $\rho(\mathcal{C}) := \max \{d_H(\mathbf{x}, \mathcal{C}) \mid \mathbf{x} \in F_q^n\}$ is called the **covering radius** of C. Here $d_H(\mathbf{x}, \mathcal{C}) := \min \{d_H(\mathbf{x}, \mathbf{c}) \mid \mathbf{c} \in \mathcal{C}\}$.

Theorem 12.3.27 (Delsarte Bound [521]) *For any code $\mathcal{C} \subseteq F_q^n$ with external distance s'*

$$\rho(\mathcal{C}) \le s'.$$

Theorem 12.3.28 (Tietäväinen Bound [1807, 1808]) *For any code $\mathcal{C} \subseteq F_q^n$ with dual distance $d' = 2k - 1 + \varepsilon$, where k is positive integer and $\varepsilon \in \{0, 1\}$,*

$$\rho(\mathcal{C}) \le \xi_k^{n-1+\varepsilon, q}.$$

See also [709, 1733].

Asymptotic versions of some of the bounds in this section can be found in Section 1.9.8 of Chapter 1; see [1, 2, 123, 1230, 1365].

TABLE 12.1: Parameters of known codes attaining (simultaneously) the upper bound in (12.5) (see also [286]) and the lower bound (12.8). This table appears for (12.5) in [1228]. Here the column for the external distance s' is added. All codes in the table are universally optimal.

n	q	$s'(\mathcal{C})$	$d'(\mathcal{C})$	Distances	$|\mathcal{C}|$	Comment
n	q	n	2	$\{n\}$	n	repetition code, $s'(\mathcal{C}) = \lfloor n/2 \rfloor$ for $q = 2$
n	q	$n-1$	2	$\{d\}$	$\frac{qd}{qd-n(q-1)}$	$n > d > \frac{(q-1)n+1}{q}$ coexistence with resolvable block designs $2\text{-}\big(|\mathcal{C}|, \frac{|\mathcal{C}|}{q}, n-d\big)$, [1643]
n	q	$n-2$	3	$\{\frac{(q-1)n+1}{q}\}$	$(q-1)n+1$	coexistence with affine resolvable block designs $2\text{-}\big(|\mathcal{C}|, \frac{|\mathcal{C}|}{q}, \frac{n-1}{q}\big)$, [1644]
$p^l q$	$q = p^m$	$n-2$	3	$\{n-p^l, n\}$	nq	$l, m = 1, 2, \ldots$, [1644]
$qh+h-q$	q	$n-2$	3	$\{n-h, n\}$	q^3	$2 \mid q, h \mid q, 2 < h < q$, [531]
q^2+1	q	$n-3$	4	$\{q^2-q, q^2\}$	q^4	ovoid in $\mathrm{PG}(3,q)$, [247, 1551]
56	3	53	4	$\{36, 45\}$	3^6	Projective cap, Hill [952]
78	4	75	4	$\{56, 64\}$	4^6	Projective cap, Hill [952]
$4l$	2	$2l-2$	4	$\{2l, 4l\}$	$8l$	Hadamard codes
$q+2$	q	$q-1$	4	$\{q, q+2\}$	q^3	$2 \mid q$, $s'(\mathcal{C}) = 2$ for $q = 4$; hyperoval in $\mathrm{PG}(2,q)$, [247]
11	3	5	5	$\{6, 9\}$	243	projection of Golay code
12	3	3	6	$\{6, 9, 12\}$	729	Golay code, [820]
22	2	10	6	$\{8, 12, 16\}$	1024	projection of Golay code
23	2	7	7	$\{8, 12, 16\}$	2048	projection of Golay code
24	2	4	8	$\{8, 12, 16, 24\}$	4096	Golay code, [820]
n	2	1	n	all even	2^{n-1}	even weight code

12.4 Linear Programming on $\mathbb{S}^{n-1}$

Definition 12.4.1 Let $\mathbb{S}^{n-1} = \{\mathbf{x} = (x_1, x_2, \ldots, x_n) \mid x_1^2 + x_2^2 + \cdots + x_n^2 = 1\}$ be the unit sphere in $\mathbb{R}^n$. The **Euclidean distance** between $\mathbf{x} = (x_1, x_2, \ldots, x_n)$ and $\mathbf{y} = (y_1, y_2, \ldots, y_n)$ is

$$d_E(\mathbf{x}, \mathbf{y}) := \sqrt{(x_1 - y_1)^2 + (x_2 - y_2)^2 + \cdots + (x_n - y_n)^2}.$$

The **inner product** is defined as

$$\langle \mathbf{x}, \mathbf{y} \rangle := x_1 y_1 + x_2 y_2 + \cdots + x_n y_n.$$

Note that on $\mathbb{S}^{n-1}$ the distance and the inner product are connected by

$$\langle \mathbf{x}, \mathbf{y} \rangle = 1 - \frac{d_E^2(\mathbf{x}, \mathbf{y})}{2}.$$

Definition 12.4.2 An (n, M, s)-**spherical code** is a nonempty finite set $\mathcal{C} \subset \mathbb{S}^{n-1}$ with cardinality $|\mathcal{C}| = M$ and maximal inner product

$$s = \max \{\langle \mathbf{x}, \mathbf{y} \rangle \mid \mathbf{x}, \mathbf{y} \in \mathcal{C}, \mathbf{x} \neq \mathbf{y}\}.$$

The minimum distance $d = d_E(\mathcal{C}) := \min \{d_E(\mathbf{x}, \mathbf{y}) \mid \mathbf{x}, \mathbf{y} \in \mathcal{C}, \mathbf{x} \neq \mathbf{y}\}$ and the maximal inner product are connected by

$$s = 1 - \frac{d^2}{2}.$$

Definition 12.4.3 For fixed n and s denote by

$$A(n, s) := \max \{|\mathcal{C}| \mid \mathcal{C} \text{ is an } (n, M, s)\text{-spherical code}\}.$$

Definition 12.4.4 For $a, b \in \{0, 1\}$ let $\{P_i^{a,b}(t)\}_{i=0}^{\infty}$ be the Jacobi polynomials $\{P_i^{\alpha, \beta}(t)\}_{i=0}^{\infty}$ ([8, Chapter 22], [1766]) with

$$(\alpha, \beta) = \left(a + \frac{n-3}{2}, b + \frac{n-3}{2}\right)$$

normalized by $P_i^{\alpha, \beta}(1) = 1$. When $(a, b) = (0, 0)$ we get the **Gegenbauer polynomials** and use the (n) indexing instead of $0, 0$. Denote by $t_i^{a,b}$ the greatest zero of the polynomial $P_i^{a,b}(t)$ and define $t_0^{1,1} = -1$. Let

$$r_i := \frac{2i + n - 2}{i + n - 2} \binom{i + n - 2}{i}.$$

Theorem 12.4.5 ([8, 1766]) *The Gegenbauer polynomials $\{P_i^{(n)}(t)\}_{i=0}^{\infty}$ satisfy the recurrence relations*

$$(i + n - 2)P_{i+1}^{(n)}(t) = (2i + n - 2)tP_i^{(n)}(t) - iP_{i-1}^{(n)}(t) \quad \text{for } i = 1, 2, \ldots$$

where $P_0^{(n)}(t) = 1$ and $P_1^{(n)}(t) = t$.

Theorem 12.4.6 ([8, 1766]) *The Gegenbauer polynomials are orthogonal on $[-1, 1]$ with respect to the measure*

$$d\mu(t) := \gamma_n (1 - t^2)^{\frac{n-3}{2}} dt \quad \text{for } t \in [-1, 1]$$

where $\gamma_n := \Gamma(\frac{n}{2})/\sqrt{\pi}\,\Gamma(\frac{n-1}{2})$ is a normalizing constant.

Theorem 12.4.7 (Linear Programming Bound for Spherical Codes [527, 1067]) *Let $n \geq 2$ and $f(t)$ be a real polynomial such that*

(A1) $f(t) \leq 0$ *for $-1 \leq t \leq s$, and*

(A2) *the coefficients in the Gegenbauer expansion $f(t) = \sum_{i=0}^{\deg(f)} f_i P_i^{(n)}(t)$ satisfy $f_0 > 0$, $f_i \geq 0$ for $i = 1, 2, \ldots, \deg(f)$.*

Then $A(n, s) \leq f(1)/f_0$.

Theorem 12.4.8 (Levenshtein Bound [1226, 1228]) *For the quantity $A(n, s)$ we have*

$$A(n, s) \leq L_\tau(n, s) := \left(1 - \frac{P_{k-1+\varepsilon}^{1,0}(s)}{P_k^{0,\varepsilon}(s)} \right) \sum_{i=0}^{k-1+\varepsilon} r_i \quad \text{for all } s \in I_\tau \tag{12.9}$$

where I_τ is the interval

$$I_\tau := \left[t_{k-1+\varepsilon}^{1,1-\varepsilon}, t_k^{1,\varepsilon} \right] \quad \text{for } \tau = 2k - 1 + \varepsilon \text{ and } \varepsilon \in \{0, 1\}.$$

Example 12.4.9 The first three bounds in (12.9) are $A(n, s) \leq (s-1)/s$ for $s \in [-1, -1/n]$,

$$A(n, s) \leq \frac{2n(1 - s)}{1 - ns}$$

for $s \in [-1/n, 0]$ and

$$A(n, s) \leq \frac{n(1 - s)(2 + (n + 1)s)}{1 - ns^2}$$

for $s \in \left[0, \frac{\sqrt{n+3}-1}{n+2} \right]$.

See also [684, 1227, 1230].

Definition 12.4.10 A spherical code $\mathcal{C} \subset \mathbb{S}^{n-1}$ is a **spherical τ-design** if and only if

$$\int_{\mathbb{S}^{n-1}} p(\mathbf{x}) \, d\sigma_n(\mathbf{x}) = \frac{1}{|\mathcal{C}|} \sum_{\mathbf{x} \in \mathcal{C}} p(\mathbf{x})$$

(σ_n is the normalized $(n-1)$-dimensional Hausdorff measure) holds for all polynomials $p(\mathbf{x}) = p(x_1, x_2, \ldots, x_n)$ of degree at most τ.

Theorem 12.4.11 (Linear Programming Bound for Spherical Designs [527]) *Let $n \geq 2$, $\tau \geq 1$ and $f(t)$ be a real polynomial such that*

(B1) $f(t) \geq 0$ *for $-1 \leq t \leq 1$, and*

(B2) *the coefficients in the Gegenbauer expansion $f(t) = \sum_{i=0}^{\deg(f)} f_i P_i^{(n)}(t)$ satisfy $f_0 > 0$, $f_i \leq 0$ for $i = \tau + 1, \ldots, \deg(f)$.*

Then any spherical τ-design $\mathcal{C} \subset \mathbb{S}^{n-1}$ has cardinality $|\mathcal{C}| \geq f(1)/f_0$.

Theorem 12.4.12 (Delsarte–Goethals–Seidel Bound [527]) *Any τ-design $\mathcal{C} \subseteq \mathbb{S}^{n-1}$ has cardinality*

$$|\mathcal{C}| \geq D(n,\tau) := \binom{n+k-2+\varepsilon}{n-1} + \binom{n+k-2}{n-1} \tag{12.10}$$

where $\tau = 2k - 1 + \varepsilon$ and $\varepsilon \in \{0,1\}$.

Definition 12.4.13 A spherical τ-design on $\mathbb{S}^{n-1}$ is called **tight** if it attains the bound (12.10).

Theorem 12.4.14 ([113, 114]) *Let $n \geq 3$. Tight spherical τ-designs on $\mathbb{S}^{n-1}$ exist for $\tau = 1$, 2 and 3 for every n, and possibly for $\tau = 4$, 5, 7, and 11. Tight spherical 4-designs on $\mathbb{S}^{n-1}$ exist for $n = 6$, 22, and possibly for $n = m^2 - 3$, where $m \geq 7$ is an odd integer. Tight spherical 5-designs on $\mathbb{S}^{n-1}$ exist for $n = 3$, 7, 23, and possibly for $n = m^2 - 2$, where $m \geq 7$ is an odd integer. Tight spherical 7-designs on $\mathbb{S}^{n-1}$ exist for $n = 8$, 23, and possibly for $n = 3m^2 - 4$, where $m \geq 4$ is an integer. Tight spherical 11-designs on $\mathbb{S}^{n-1}$ exist only for $n = 24$.*

Remark 12.4.15 ([527]) Tight spherical 4- and 5-designs coexist and are known for $m = 3$ and 5 only. Tight spherical 7-designs are known for $m = 2$ and 3 only.

See also [111, 112, 232] for general theory and [263, 709, 1734] for other bounds for designs.

Theorem 12.4.16 ([1226, 1227, 1228, 1230]) *The bounds (12.9) and (12.10) are related by the equalities*

$$L_{\tau-1-\varepsilon}(n, t^{1,1-\varepsilon}_{k-1-\varepsilon}) = L_{\tau-\varepsilon}(n, t^{1,1-\varepsilon}_{k-1-\varepsilon}) = D(n, \tau - \varepsilon) \quad \text{for } \varepsilon \in \{0,1\}$$

at the ends of the intervals I_τ ($\tau = 2k - 1 + \varepsilon$, $\varepsilon \in \{0,1\}$). In particular, if $\mathcal{C} \subseteq \mathbb{S}^{n-1}$ is a tight spherical τ-design, then it attains (12.9) in the left end of the interval I_τ.

Definition 12.4.17 Given an (extended real-valued) function $h : [-1,1] \to [0,+\infty]$, the h-**energy** (or **potential energy**) of $\mathcal{C}$ is given by

$$E_h(\mathcal{C}) := \sum_{\mathbf{x},\mathbf{y} \in \mathcal{C}, \mathbf{x} \neq \mathbf{y}} h(\langle \mathbf{x}, \mathbf{y} \rangle).$$

Definition 12.4.18 For fixed n, h, and $M \geq 2$ denote by

$$\mathcal{E}_h(n,M) := \inf \{E_h(\mathcal{C}) \mid \mathcal{C} \subset \mathbb{S}^{n-1}, \ |\mathcal{C}| = M\}.$$

Theorem 12.4.19 (Linear Programming Bound for Energy [1936]) *Let $n \geq 2$, $M \geq 2$, $h : [-1,1] \to [0,+\infty]$, and $f(t)$ be a real polynomial such that*

(C1) $f(t) \leq h(t)$ *for $t \in [-1,1]$, and*

(C2) *the coefficients in the Gegenbauer expansion $f(t) = \sum_{i=0}^{\deg(f)} f_i P_i^{(n)}(t)$ satisfy $f_0 > 0$, $f_i \geq 0$ for $i = 1, 2, \ldots, \deg(f)$.*

Then $\mathcal{E}_h(n,M) \geq M(f_0 M - f(1))$.

Definition 12.4.20 A real valued extended function $h : [-1,1] \to (0,+\infty]$ is called **absolutely monotone** if $h^{(k)}(t) \geq 0$, for every $t \in [-1,1)$ and every integer $k \geq 0$, and $h(1) = \lim_{t \to 1^-} h(t)$.

The following absolutely monotone potential functions are commonly used:

Newton potential
$$h(t) = (2 - 2t)^{-(n-2)/2} = d_E(\mathbf{x}, \mathbf{y})^{-(n-2)},$$

Riesz s-potential
$$h(t) = (2 - 2t)^{-s/2} = d_E(\mathbf{x}, \mathbf{y})^{-s},$$

and Gaussian potential
$$h(t) = \exp(2t - 2) = \exp(-d_E(\mathbf{x}, \mathbf{y})^2).$$

Definition 12.4.21 For $(a, b) = (0, 0), (1, 0)$, and $(1, 1)$ set

$$T_j^{a,b}(u, v) := \sum_{i=0}^{j} r_i^{a,b} P_i^{a,b}(u) P_i^{a,b}(v),$$

where $r_i^{a,b} = 1/c^{a,b} \int_{-1}^{1} \left(P_i^{a,b}(t) \right)^2 (1 - t)^a (1 + t)^b \, d\mu(t)$, $c^{1,0} = c^{0,0} = \gamma_n$, and $c^{1,1} = \gamma_{n+2}$ (see Theorem 12.4.6 for relevant notation). Note that $r_i^{0,0} = r_i$.

Let $\alpha_1 < \alpha_2 < \cdots < \alpha_{k+\varepsilon}$ be the roots of the polynomial used for obtaining the Levenshtein Bound $L_\tau(n, s)$ with $s = \alpha_{k+\varepsilon}$, $\tau = 2k - 1 + \varepsilon$, $\varepsilon \in \{0, 1\}$, and let

$$\rho_1 := \frac{T_k^{0,0}(s, 1)}{T_k^{0,0}(-1, -1) T_k^{0,0}(s, 1) - T_k^{0,0}(-1, 1) T_k^{0,0}(s, -1)} \quad \text{for } \varepsilon = 1 \text{ and}$$

$$\rho_{i+\varepsilon} := \frac{1}{c^{1,\varepsilon}(1 + \alpha_{i+\varepsilon})^\varepsilon (1 - \alpha_{i+\varepsilon}) T_{k-1}^{1,\varepsilon}(\alpha_{i+\varepsilon}, \alpha_{i+\varepsilon})} \quad \text{for } i = 1, 2, \ldots, k.$$

Theorem 12.4.22 ([290]) *Let $n \geq 2$, $\tau = 2k - 1 + \varepsilon \geq 1$, $\varepsilon \in \{0, 1\}$, and h be absolutely monotone in $[-1, 1]$. For $M \in (D(n, \tau), D(n, \tau + 1)]$ let s be the largest root of the equation $M = L_\tau(n, s)$ and $\alpha_1 < \alpha_2 < \cdots < \alpha_{k+\varepsilon} = s$, $\rho_1, \rho_2, \ldots, \rho_{k+\varepsilon}$ be as in Definition 12.4.21. Then*

$$\mathcal{E}_h(n, M) \geq M^2 \sum_{i=1}^{k+\varepsilon} \rho_i h(\alpha_i). \tag{12.11}$$

If an (n, M, s) code attains (12.11), then it is a spherical τ-design and its inner products form the set $\{\alpha_1, \alpha_2, \ldots, \alpha_{k+\varepsilon}\}$.

See also [245, Chapter 5]. Note that $\alpha_1, \alpha_2, \ldots, \alpha_{k+\varepsilon} = s$ are in fact the roots of the equation $M = L_\tau(n, s)$.

Remark 12.4.23 The conditions for attaining the bounds (12.9) and (12.11) coincide. Thus a spherical (n, M, s) code attains (12.9) if and only if it attains (12.11). In particular, every tight spherical design attains (12.11).

Theorem 12.4.24 ([1703] for (12.9), [290] for (12.11)) *The bounds (12.9) and (12.11) cannot be improved by using in Theorem 12.4.7 and 12.4.19, respectively, polynomials of the same or lower degree.*

Theorem 12.4.25 ([288] for (12.9), [290] for (12.11)) *For $j > \tau = 2k - 1 + \varepsilon$ and $\varepsilon \in \{0, 1\}$, let*

$$Q_j^{(n)} := \frac{1}{L_\tau(n, \alpha_{k+\varepsilon})} + \sum_{i=1}^{k+\varepsilon} \rho_i P_j^{(n)}(\alpha_i).$$

The bounds (12.9) and (12.11) (where $M = L_\tau(n, \alpha_{k+\varepsilon})$ for (12.11)) can be (simultaneously) improved if and only if $Q_j^{(n)} < 0$ for some $j > \tau$.

Definition 12.4.26 A spherical code $\mathcal{C} \subset \mathbb{S}^{n-1}$ is **universally optimal** if it (weakly) minimizes h-energy among all configurations of $|\mathcal{C}|$ points on $\mathbb{S}^{n-1}$ for each absolutely monotone function h.

Theorem 12.4.27 ([424]) *Every spherical code which is a spherical $(2k-1)$-design and which admits exactly k inner products between distinct points is universally optimal. The 600-cell (the unique $(4, 120, (1+\sqrt{5})/4)$-spherical code) is universally optimal.*

See also [109, 245, 425].

Remark 12.4.28 Any code which attains (12.9) (and (12.11)) is universally optimal. It is unknown if there exists a spherical code, apart from the 600-cell, which is universally optimal but does not attain (12.9) (and (12.11)). Tables with all known codes which attain (12.9) (and (12.11)) can be found in [424, 1228, 1230].

Example 12.4.29 Consider the case $(n, M) = (4, 24)$. The well known code D_4 (D_4 root system; equivalently, the set of vertices of the regular 24-cell) is optimal in the sense that it realizes the fourth kissing number [1411]. However, this code is not universally optimal [421], despite having energy which is very close to the bound (12.11). For example, with the Newtonian $h(t) = \frac{1}{2(1-t)}$ it has energy 334, while (12.11) gives 333 (which can be improved to ≈ 333.157).

Conjecture 12.4.30 Every universally optimal spherical code attains the Linear Programming Bound of Theorem 12.4.19.

Example 12.4.31 (Example 12.3.25 continued.) A standard construction (see [442, Chapter 5]) maps binary codes from $\mathbb{F}_2^n$ to the sphere $\mathbb{S}^{n-1}$ – the coordinates 0 and 1 are replaced by $\pm 1/\sqrt{n}$, respectively. Denote this map by $\mathbf{x} \to \overline{\mathbf{x}}$. The inner product $\langle \overline{\mathbf{x}}, \overline{\mathbf{y}} \rangle$ on $\mathbb{S}^{n-1}$ and the Hamming distance $d_H(\mathbf{x}, \mathbf{y})$ in $\mathbb{F}_2^n$ are connected by $\langle \overline{\mathbf{x}}, \overline{\mathbf{y}} \rangle = 1 - \frac{2d_H(\mathbf{x}, \mathbf{y})}{n}$. Thus the weights of the Kerdock code $\mathcal{K}_m$ correspond to the inner products $1, \frac{1}{\sqrt{n}}, 0, -\frac{1}{\sqrt{n}}, -1$, respectively.

The image $\overline{\mathcal{K}}_m \subset \mathbb{S}^{2^{2m}-1}$ of $\mathcal{K}_m$ is asymptotically optimal with respect to both bounds (12.9) and (12.11). For example, it has energy

$$E_h(\overline{\mathcal{K}}_m) = n^2[(2^{2m+1} - 2)h(0) + 2^{2m}(2^{2m-1} - 1)(h(2^{-m}) + h(-2^{-m})) + h(-1)].$$

When n tends to infinity, we obtain

$$E_h(\overline{\mathcal{K}}_m) \sim n^2\left((2^{4m} - 2)h(0) + h(-1)\right) \sim h(0)n^4,$$

which coincides with the asymptotic of (12.11) (obtained by a polynomial of degree 5).

12.5 Linear Programming in Other Coding Theory Problems

Remark 12.5.1 (Linear Programming in Johnson Spaces) All concepts and results from Sections 12.1-12.3 hold true for the Johnson spaces with changes corresponding to the role of the Hahn polynomials instead of the Krawtchouk polynomials. See [519, 528, 1228, 1230, 1231, 1365].

Theorem 12.5.2 (Linear Programming Bound for Binomial Moments [68]) *Let* $\mathcal{C} \subset \mathbb{F}_2^n$ *be a code with distance distribution* $(B_0, B_1, \ldots, B_n)$, $w \in \{1, 2, \ldots, n\}$, *and*

$$\mathcal{B}_w := \sum_{i=1}^{w} \binom{n-i}{n-w} B_i.$$

Let the real polynomial $f(z) = \sum_{i=0}^{n} f_i K_i^{(n,2)}(z)$ *satisfy the conditions*

(i) $f_i \geq 0$ *for* $i = 1, 2, \ldots, n$, *and*

(ii) $f(j) \leq \binom{n-j}{n-w}$ *for* $j = 1, 2, \ldots, n$.

Then $\mathcal{B}_w \geq f_0 |\mathcal{C}| - f(0)$.

The final bound in this chapter is an upper bound on the size of a quantum code. Quantum codes are the subject of Chapter 27.

Theorem 12.5.3 (Linear Programming Bound for Quantum Codes [70]; see also [1188]) *Let* $f(z) = \sum_{i=0}^{n} f_i K_i^{(n,4)}(z)$ *be a polynomial satisfying the conditions*

(i) $f_i \geq 0$ *for every* $i = 0, 1, \ldots, n$,

(ii) $f(z) > 0$ *for every* $z = 0, 1, \ldots, d-1$, *and*

(iii) $f(z) \leq 0$ *for every* $z = d, d+1, \ldots, n$.

Then every $((n, K))$ *quantum code of minimum distance* d *satisfies*

$$K \leq \frac{1}{2^n} \max_{j \in \{0,1,\ldots,d-1\}} \frac{f(j)}{f_j}.$$

See also [83, 125, 126, 422, 629].

Chapter 13 covers semidefinite linear programming, a related technique to linear programming, that can be used to develop bounds on codes.

Acknowledgement

A significant part of the work of the first author on this chapter was done during his stay (August-December 2018) as a visiting professor in the Department of Mathematical Sciences at Purdue University Fort Wayne. His research was supported, in part, by a Bulgarian NSF contract DN02/2-2016.

Chapter 13

Semidefinite Programming Bounds for Error-Correcting Codes

Frank Vallentin

Universität zu Köln

13.1   Introduction

Linear programming bounds belong to the most powerful and flexible methods to obtain bounds for extremal problems in coding theory as described in Chapter 12. Initially, Delsarte [519] developed linear programming bounds in the algebraic framework of association schemes. A central example in Delsarte's theory is finding upper bounds for the parameter $A_2(n, d)$, the largest number of codewords in a binary code of length n with minimum Hamming distance d.

The application of linear programming bounds led to the best known asymptotic bounds, such as the MRRW Bound [1365]. It was realized that linear programming bounds are also applicable to finite and infinite two-point homogeneous spaces [442, Chapter 9]. These are metric spaces in which the symmetry group acts transitively on pairs of points having the same distance. So one can treat metric spaces like the q-ary Hamming space $\mathbb{F}_q^n$, the sphere, real/complex/quaternionic projective space, or Euclidean space [423].

In recent years, semidefinite programming bounds have been developed with two aims: to strengthen linear programming bounds and to find bounds for more general spaces. Semidefinite programs are convex optimization problems which can be solved efficiently and which are a vast generalization of linear programs. The optimization variable of a semidefinite program is a positive semidefinite matrix whereas it is a nonnegative vector for a linear program.

Schrijver [1634] was the first who applied semidefinite programming bounds to improve the known upper bounds for $A_2(n, d)$ for many parameters n and d. The underlying idea is that linear programming bounds only exploit constraints involving pairs of codewords, whereas semidefinite programming bounds can exploit constraints between triples, quadruples, ... of codewords.

This chapter introduces semidefinite programming bounds with an emphasis on error-correcting codes. The structure of the chapter is as follows.

In Section 13.2 the basic theory of linear and semidefinite programming is reviewed in the framework of conic programming.

Semidefinite programming bounds can be viewed as semidefinite programming hierarchies for difficult combinatorial optimization problems. One can express the computation of $A_2(n, d)$ as finding the independence number of an appropriate graph $G(n, d)$ and apply the Lasserre hierarchy to find upper bounds for $A_2(n, d)$. This approach is explained in Section 13.3.

The graph $G(n, d)$ has exponentially many vertices and the Lasserre hierarchy for $G(n, d)$ employs matrices whose rows and columns are indexed by all t-element subsets of $G(n, d)$; so a computation of the semidefinite programs is not directly possible. However, the graph has many symmetries and these symmetries can be exploited to substantially reduce the size of the semidefinite programs. The technique of symmetry reduction is the subject of Section 13.4. There, this technique is applied to the graph $G(n, d)$ and the result of Schrijver is explained.

After Schrijver's breakthrough result, semidefinite programming bounds were developed for different settings. These developments are reviewed in Section 13.5.

13.2 Conic Programming

Semidefinite programming is a vast generalization of linear programming. Geometrically, both linear and semidefinite programming are concerned with minimizing or maximizing a linear functional over the intersection of a fixed convex cone with an affine subspace. In the case of linear programming the fixed convex cone is the nonnegative orthant and the resulting intersection is a polyhedron. In the case of semidefinite programming the fixed convex cone is the cone of positive semidefinite matrices and the resulting intersection is a spectrahedron. Linear and semidefinite programming belong to the field of conic programming.

Textbooks and research monographs dealing with semidefinite programming include: Wolkowicz, Saigal, and Vangenberghe (ed.) [1904], Ben-Tal and Nemirovski [160], de Klerk [498], Tunçel [1823], Anjos and Lasserre (ed.) [50], Gärtner and Matoušek [795], Blekherman, Parrilo, and Thomas (ed.) [221], and Laurent and Vallentin [1207].

13.2.1 Conic Programming and its Duality Theory

Conic programs are convex optimization problems. In general, conic programming deals with minimizing or maximizing a linear functional over the intersection of a fixed convex cone with an affine subspace. See Nemirovski [1429] and the references therein for a detailed overview of conic programming.

Let E be an n-dimensional real or complex vector space equipped with a real-valued inner product $\langle \cdot, \cdot \rangle_E : E \times E \to \mathbb{R}$.

Definition 13.2.1 A set $K \subseteq E$ is called a **(convex) cone** if for all $x, y \in K$ and all *nonnegative* numbers $\alpha, \beta \in \mathbb{R}_+$ one has $\alpha x + \beta y \in K$. A convex cone K is called **pointed** if $K \cap (-K) = \{0\}$. A convex cone is called **proper** if it is pointed, closed, and full-dimensional. The **dual cone** of a convex cone K is given by

$$K^* = \{ y \in E \mid \langle x, y \rangle_E \geq 0 \text{ for all } x \in K \}.$$

The simplest convex cones are **finitely generated cones**; the vectors $x_1, \ldots, x_N \in E$ determine the finitely generated cone K by

$$K = \text{cone}\{x_1, \ldots, x_N\} = \left\{ \sum_{i=1}^{N} \alpha_i x_i \;\middle|\; \alpha_1, \ldots, \alpha_N \geq 0 \right\}.$$

A pointed convex cone $K \subseteq E$ determines a partial order on E by

$$x \succeq_K y \text{ if and only if } x - y \in K.$$

To define a conic program, we fix the space E, a proper convex cone $K \subseteq E$, and an m-dimensional vector space F with a real-valued inner product $\langle \cdot, \cdot \rangle_F : F \times F \to \mathbb{R}$.

Definition 13.2.2 A linear map $A : E \to F$ and vectors $c \in E$, $b \in F$ determine a **primal conic program**
$$p^* = \sup \{ \langle c, x \rangle_E \mid x \in K, \, Ax = b \}.$$

The corresponding **dual conic program** is

$$d^* = \inf \{ \langle b, y \rangle_F \mid y \in F, \, \overline{A}^\mathsf{T} y - c \in K^* \},$$

where $\overline{A}^\mathsf{T} : F \to E$ is the usual adjoint of A.

The vector $x \in E$ is the **optimization variable** of the primal conic program; the vector $y \in F$ is the optimization variable of the dual. A vector x is called a **feasible solution** for the primal if $x \in K$ and $Ax = b$. It is called a **strictly feasible solution** if additionally x lies in the interior of K. It is called an **optimal solution** if x is feasible and $p^* = \langle x, c \rangle_E$. Similarly, a vector y is called feasible for the dual if $\overline{A}^\mathsf{T} y - c \in K^*$, and it is called strictly feasible if $\overline{A}^\mathsf{T} y - c$ lies in the interior of K^*. It is called optimal if y is feasible and $d^* = \langle b, y \rangle_F$.

The Bipolar Theorem (see for example Barvinok [134] or Simon [1714]) states that $(K^*)^* = K$ when K is a proper convex cone. From this it follows easily that taking the dual of the dual conic program gives a conic program which is equivalent to the primal.

Duality theory of conic programs looks at the (close) relationship between the primal and dual conic programs. In particular duality can be used to systematically find upper bounds for the primal program and lower bounds for the dual program.

Theorem 13.2.3 (Duality Theorem of Conic Programs)

(a) **weak duality**: $p^* \leq d^*$.

(b) **optimality condition/complementary slackness**: *Suppose that $p^* = d^*$. Let x be a feasible solution for the primal and let y be a feasible solution of the dual. Then x is optimal for the primal and y is optimal for the dual if and only if $\langle x, \overline{A}^\mathsf{T} y - c \rangle_E = 0$ holds.*

(c) **strong duality**: *Suppose that primal and dual conic programs both have a strictly feasible solution. Then $p^* = d^*$ and both primal and dual possess an optimal solution.*

13.2.2   Linear Programming

To specialize conic programs to linear programs we choose E to be $\mathbb{R}^n$ with standard inner product $\langle x, y \rangle_E = x^\mathsf{T} y$. For the convex cone K we choose the nonnegative orthant.

Definition 13.2.4 The **nonnegative orthant** is the following proper convex cone:

$$\mathbb{R}^n_+ = \{x \in \mathbb{R}^n \mid x_1 \geq 0, \ldots, x_n \geq 0\}.$$

The nonnegative orthant is self-dual, $(\mathbb{R}^n_+)^* = \mathbb{R}^n_+$. So, for a matrix $A \in \mathbb{R}^{m \times n}$, a vector $b \in \mathbb{R}^m$, and a vector $c \in \mathbb{R}^n$, we get the **primal linear program**

$$p^* = \sup\{c^\mathsf{T} x \mid x \geq 0, \ Ax = b\},$$

and its **dual linear program**

$$d^* = \inf\{b^\mathsf{T} y \mid y \in \mathbb{R}^m, \ A^\mathsf{T} y - c \geq 0\}.$$

Here we simply write $x \geq 0$ for the partial order $x \succeq_{\mathbb{R}^n_+} 0$.

Linear programming is a well established method, which is extremely useful in theory and practice; see for example Schrijver [1633], Grötschel, Lovász, and Schrijver [857] and Wright [1908]. The main algorithms to solve linear programs are the Simplex Algorithm, the Ellipsoid Algorithm, and the Interior-point Algorithm. Each one of these three algorithms has specific advantages: In practice, the Simplex Algorithm and the Interior-point Algorithm can solve very large instances. The Simplex Algorithm allows the computation of additional information which is useful for the broader class of mixed integer linear optimization problems, where some of the optimization variables are constrained to be integers. The Ellipsoid Algorithm and the Interior-point Algorithm are polynomial time algorithms. The Ellipsoid Algorithm is a versatile mathematical tool to prove the existence of polynomial time algorithms, especially in combinatorial optimization.

13.2.3   Semidefinite Programming

To specialize conic programs to semidefinite programs we choose E to be the $n(n+1)/2$-dimensional space $\mathcal{S}^n$ of real symmetric $n \times n$ matrices. This space is equipped with the **trace (Frobenius) inner product** $\langle \cdot, \cdot \rangle_E = \langle \cdot, \cdot \rangle_T$ defined by

$$\langle X, Y \rangle_T = \mathrm{TR}(Y^\mathsf{T} X) = \sum_{i=1}^n \sum_{j=1}^n X_{ij} Y_{ij}$$

where TR denotes the trace of a matrix, that is, the sum of the diagonal entries. For the convex cone K we choose the cone of positive semidefinite matrices.

Definition 13.2.5 The **cone of positive semidefinite matrices** is the proper convex cone

$$\mathcal{S}^n_+ = \{X \in \mathcal{S}^n \mid X \text{ is positive semidefinite}\}.$$

Let us recall that a matrix X is positive semidefinite if and only if for all $x \in \mathbb{R}^n$ we have $x^\mathsf{T} X x \geq 0$. Alternatively, looking at a spectral decomposition of X given by

$$X = \sum_{i=1}^n \lambda_i u_i u_i^\mathsf{T},$$

where λ_i, $i = 1, \ldots, n$, are the (real) eigenvalues of X and $\{u_i \mid i = 1, \ldots, n\}$ is an orthonormal basis consisting of corresponding eigenvectors, X is positive semidefinite if and only if all its eigenvalues are nonnegative: $\lambda = (\lambda_1, \ldots, \lambda_n) \in \mathbb{R}^n_+$. We write $X \succeq 0$ for $X \succeq_{\mathcal{S}^n_+} 0$.

The cone of positive semidefinite matrices is self-dual, $(\mathcal{S}^n_+)^* = \mathcal{S}^n_+$. So, for symmetric matrices $C, A_1, \ldots, A_m \in \mathcal{S}^n$ and a vector $b \in \mathbb{R}^m$, we get the **primal semidefinite program**

$$p^* = \sup\{\langle C, X \rangle_T \mid X \succeq 0, \langle A_1, X \rangle_T = b_1, \ldots, \langle A_m, X \rangle_T = b_m\}. \tag{13.1}$$

Its **dual semidefinite program** is

$$d^* = \inf\left\{ b^\mathsf{T} y \mid y \in \mathbb{R}^m, \sum_{j=1}^m y_j A_j - C \succeq 0 \right\}.$$

Restricting semidefinite programs to diagonal matrices, one recovers linear programming as a special case of semidefinite programming.

Definition 13.2.6 The set of feasible solutions of a primal semidefinite program

$$\mathcal{F} = \{X \in \mathcal{S}^n \mid X \succeq 0, \langle A_j, X \rangle_T = b_j \text{ for } j = 1, \ldots, m\}$$

is called a **spectrahedron**.

Spectrahedra are generalizations of polyhedra. They are central objects in convex algebraic geometry; see [221].

Under mild technical assumptions one can solve semidefinite programming problems in polynomial time. The following theorem was proved in Grötschel, Lovász, and Schrijver [857] using the Ellipsoid Algorithm and by de Klerk and Vallentin [499] using the Interior-point Algorithm.

Theorem 13.2.7 *Consider the primal semidefinite program* (13.1) *with rational input C, $A_1, \ldots, A_m$, and $b_1, \ldots, b_m$. Suppose we know a rational point $X_0 \in \mathcal{F}$ and positive rational numbers r, R so that*

$$B(X_0, r) \subseteq \mathcal{F} \subseteq B(X_0, R),$$

where $B(X_0, r)$ is the ball of radius r, centered at X_0, in the affine subspace

$$\{X \in \mathcal{S}^n \mid \langle A_j, X \rangle_T = b_j \text{ for } j = 1, \ldots, m\}.$$

For every positive rational number $\epsilon > 0$ one can find in polynomial time a rational matrix $X^* \in \mathcal{F}$ such that

$$\langle C, X^* \rangle_T - p^* \leq \epsilon,$$

where the polynomial is in n, m, $\log_2 \frac{R}{r}$, $\log_2(1/\epsilon)$, and the bit size of the data X_0, C, $A_1, \ldots, A_m$, and $b_1, \ldots, b_m$.

Sometimes—especially when dealing with invariant semidefinite programs or in the area of quantum information theory—it is convenient to work with complex Hermitian matrices instead of real symmetric matrices. A complex matrix $X \in \mathbb{C}^{n \times n}$ is called *Hermitian* if $X = X^*$, where $X^* = \overline{X}^\mathsf{T}$ denotes the conjugate transpose of X, i.e., $X_{ij} = \overline{X}_{ji}$. A Hermitian matrix is called *positive semidefinite* if for all vectors $x \in \mathbb{C}^n$ we have $x^* X x \geq 0$. The space of Hermitian matrices is equipped with the real-valued inner product $\langle X, Y \rangle_T = \mathrm{TR}(Y^* X)$. Now a **primal complex semidefinite program** is

$$p^* = \sup\{\langle C, X \rangle_T \mid X \succeq 0, \langle A_1, X \rangle_T = b_1, \ldots, \langle A_m, X \rangle_T = b_m\}, \qquad (13.2)$$

where $C, A_1, \ldots, A_m \in \mathbb{C}^{n \times n}$ are given Hermitian matrices, $b \in \mathbb{R}^m$ is a given vector, and $X \in \mathbb{C}^{n \times n}$ is the positive semidefinite Hermitian optimization variable (denoted by $X \succeq 0$).

One can easily reduce complex semidefinite programming to real semidefinite programming by the following construction: A complex matrix $X \in \mathbb{C}^{n \times n}$ defines a real matrix

$$X' = \begin{bmatrix} \mathrm{Re}(X) & -\mathrm{Im}(X) \\ \mathrm{Im}(X) & \mathrm{Re}(X) \end{bmatrix} \in \mathbb{R}^{2n \times 2n},$$

where $\mathrm{Re}(X) \in \mathbb{R}^{n \times n}$ and $\mathrm{Im}(X) \in \mathbb{R}^{n \times n}$ are the real, respectively the imaginary parts of X. Then X is Hermitian and positive semidefinite if and only if X' is symmetric and positive semidefinite.

13.3 Independent Sets in Graphs

13.3.1 Independence Number and Codes

In the following we are dealing with finite simple graphs. These are finite undirected graphs without loops and multiple edges. This means that the vertex set is a finite set and the edge set consists of (unordered) pairs of vertices.

Definition 13.3.1 Let $G = (V, E)$ be a simple finite graph with vertex set V and edge set E. A subset of the vertices $I \subseteq V$ is called an **independent set** if every pair of vertices $x, y \in I$ is not adjacent, i.e., $\{x, y\} \notin E$. The **independence number** $\alpha(G)$ is the largest cardinality of an independent set in G.

In the optimization literature, independent sets are sometimes also called **stable sets**, and the independence number is referred to as the **stability number**.

Frequently the largest number of codewords in a code with given parameters can be equivalently expressed as the independence number of a specific graph.

Example 13.3.2 Recall that $A_2(n, d)$ is the largest number M of codewords in a binary code of length n with minimum Hamming distance d. Consider the graph $G(n, d)$ with vertex set $V = \mathbb{F}_2^n$ and edge set $E = \{\{\mathbf{x}, \mathbf{y}\} \mid \mathrm{d_H}(\mathbf{x}, \mathbf{y}) < d\}$. Then independent sets in $G(n, d)$ are exactly binary codes $\mathcal{C}$ of length n with minimum Hamming distance d. Furthermore, $A_2(n, d) = \alpha(G(n, d))$.

The graph $G(n, d)$ can also been seen as a Cayley graph over the additive group $\mathbb{F}_2^n$. The vertices are the group elements and two vertices $\mathbf{x}$ and $\mathbf{y}$ are adjacent if and only if their difference $\mathbf{x} - \mathbf{y}$ has Hamming weight strictly less than d.

Computing the independence number of a given graph G is generally a very difficult problem. Computationally, determining even approximate solutions of $\alpha(G)$ is an NP-hard problem; see Håstad [923].

13.3.2 Semidefinite Programming Bounds for the Independence Number

One possibility to systematically find stronger and stronger upper bounds for $\alpha(G)$, which is often quite good for graphs arising in coding theory, is the *Lasserre hierarchy* of semidefinite programming bounds.

The Lasserre hierarchy was introduced by Lasserre in [1204]. He considered the general setting of 0/1 polynomial optimization problems, and he proved that the hierarchy converges in finitely many steps using Putinar's Positivstellensatz [1548]. Shortly after, Laurent [1205] gave a combinatorial proof, which we reproduce here.

The definition of the Lasserre hierarchy requires some notation. Let V be a finite set. By $\mathcal{P}_t(V)$ we denote the set of all subsets of V of cardinality at most t.

Definition 13.3.3 Let t be an integer with $0 \leq t \leq n$. A symmetric matrix $M \in \mathcal{S}^{\mathcal{P}_t(V) \times \mathcal{P}_t(V)}$ is called a **(combinatorial) moment matrix of order t** if

$$M_{I,J} = M_{I',J'} \quad \text{whenever } I \cup J = I' \cup J'.$$

A vector $y = (y_I) \in \mathbb{R}^{\mathcal{P}_{2t}(V)}$ defines a combinatorial moment matrix of order t by

$$M_t(y) \in \mathcal{S}^{\mathcal{P}_t(V) \times \mathcal{P}_t(V)} \quad \text{with } (M_t(y))_{I,J} = y_{I \cup J}.$$

The matrix $M_t(y)$ is called the **(combinatorial) moment matrix of order t of y**.

Example 13.3.4 For $V = \{1,2\}$, the moment matrices of order one and order two of y have the following form:

$$
M_1(y) =
\begin{array}{c}
\quad \\
\emptyset \\
1 \\
2
\end{array}
\begin{array}{c}
\begin{array}{ccc}
\emptyset & 1 & 2
\end{array} \\
\begin{bmatrix}
y_\emptyset & y_1 & y_2 \\
y_1 & y_1 & y_{12} \\
y_2 & y_{12} & y_2
\end{bmatrix}
\end{array}
\qquad
M_2(y) =
\begin{array}{c}
\quad \\
\emptyset \\
1 \\
2 \\
12
\end{array}
\begin{array}{c}
\begin{array}{cccc}
\emptyset & 1 & 2 & 12
\end{array} \\
\begin{bmatrix}
y_\emptyset & y_1 & y_2 & y_{12} \\
y_1 & y_1 & y_{12} & y_{12} \\
y_2 & y_{12} & y_2 & y_{12} \\
y_{12} & y_{12} & y_{12} & y_{12}
\end{bmatrix}
\end{array}.
$$

Here and in the following, we simplify notation and use y_i instead of $y_{\{i\}}$ and y_{12} instead of $y_{\{1,2\}}$. Note that $M_1(y)$ occurs as a principal submatrix of $M_2(y)$.

Definition 13.3.5 Let $G = (V, E)$ be a graph with n vertices. Let t be an integer with $1 \leq t \leq n$. The **Lasserre Bound of G of order t** is the value of the semidefinite program

$$\text{las}_t(G) = \max \left\{ \sum_{i \in V} y_i \;\middle|\; y \in \mathbb{R}_+^{\mathcal{P}_{2t}(V)}, \; y_\emptyset = 1, \; y_{ij} = 0 \text{ if } \{i,j\} \in E, \; M_t(y) \in \mathcal{S}_+^{\mathcal{P}_t(V) \times \mathcal{P}_t(V)} \right\}.$$

Theorem 13.3.6 *The Lasserre Bound of G of order t forms a hierarchy of stronger and stronger upper bounds for the independence number of G. In particular,*

$$\alpha(G) \leq \text{las}_n(G) \leq \cdots \leq \text{las}_2(G) \leq \text{las}_1(G)$$

holds.

Proof: To show that $\alpha(G) \leq \text{las}_t(G)$ for every $1 \leq t \leq n$ we construct a feasible solution $y \in \mathbb{R}_+^{\mathcal{P}_{2t}(V)}$ from any independent set I of G. This feasible solution will satisfy $|I| = \sum_{i \in V} y_i$ and the desired inequality follows. For this, we simply set y to be equal to the characteristic vector $\chi^I \in \mathbb{R}_+^{\mathcal{P}_{2t}(V)}$ defined by

$$\chi_J^I = \begin{cases} 1 & \text{if } J \subseteq I, \\ 0 & \text{otherwise.} \end{cases}$$

Clearly, y satisfies the conditions $y_\emptyset = 1$ and $y_{ij} = 0$ if i and j are adjacent. The moment matrix $M_t(y)$ is positive semidefinite because it is a rank-one matrix of the form (note the slight abuse of notation here, χ^I is now interpreted as a vector in $\mathbb{R}^{\mathcal{P}_t(V)}$)

$$M_t(y) = \chi^I (\chi^I)^\mathsf{T} \text{ where } M_t(y)_{J,J'} = y_{J \cup J'} = \chi_J^I \chi_{J'}^I \text{ and } \chi^I \in \mathbb{R}_+^{\mathcal{P}_t(V)}.$$

Since $M_t(y)$ occurs as a principal submatrix of $M_{t+1}(y)$, the inequality $\mathrm{las}_{t+1}(G) \leq \mathrm{las}_t(G)$ follows. $\qquad\square$

One can show, using the **Schur complement** for block matrices,

$$\text{for } A \text{ positive definite, then } \begin{bmatrix} A & B \\ B^\mathsf{T} & C \end{bmatrix} \succeq 0 \iff C - B^\mathsf{T} A^{-1} B \succeq 0,$$

that the first step of the Lasserre Bound coincides with the **Lovász ϑ-number** of G, a famous graph parameter which Lovász [1288] introduced to determine the Shannon capacity $\Theta(C_5)$ of the cycle graph C_5. Determining the Shannon capacity of a given graph is a very difficult problem and has applications to the zero-error capacity of a noisy channel; see Shannon [1663]. For instance, the value of $\Theta(C_7)$ is currently not known.

Theorem 13.3.7 *Let $G = (V, E)$ be a graph. We have $\mathrm{las}_1(G) = \vartheta'(G)$ where $\vartheta'(G)$ is defined as the solution of the following semidefinite program*

$$\vartheta'(G) = \max \left\{ \sum_{i,j \in V} X_{i,j} \;\middle|\; X \in \mathcal{S}_+^V, \; X_{i,j} \geq 0 \text{ for all } i,j \in V, \right.$$
$$\left. \mathrm{TR}(X) = 1, \; X_{i,j} = 0 \text{ if } \{i,j\} \in E \right\}.$$

Technically, the parameter $\vartheta'(G)$ is a slight variation of the original Lovász ϑ-number as introduced in [1288]. The difference is that in the definition of $\vartheta(G)$ one omits the nonnegativity condition $X_{i,j} \geq 0$ for all $i, j \in V$.

Schrijver [1632] and independently McEliece, Rodemich, and Rumsey [1364] realized that $\vartheta'(G)$ is nothing other than the Delsarte Linear Programming Bound in the special case of the graph $G = G(n, d)$, which was defined in Example 13.3.2. We will provide a proof of this fact in Section 13.4.3.

An important feature of the Lasserre Bound is that it does not lose information. If the step of the hierarchy is high enough, we can exactly determine the independence number of G.

Theorem 13.3.8 *For every graph G the Lasserre Bound of G of order $t = \alpha(G)$ is exact; that means $\mathrm{las}_t(G) = \alpha(G)$ for every $t \geq \alpha(G)$.*

Proof: (sketch) First we show that the hierarchy becomes stationary after $\alpha(G)$ steps. Let $J \subseteq V$ be a set of vertices which contains an edge $\{i, j\} \in E$ with $i, j \in J$. Let $y \in \mathbb{R}^{\mathcal{P}_{2t}(V)}$ be a feasible solution of $\mathrm{las}_t(G)$ with $2t \geq |J|$. Then $y_J = 0$, which can be seen as follows: Write $J = J_1 \cup J_2$ with $|J_1|, |J_2| \leq t$ and $\{i, j\} \subseteq J_1$. First, consider the following 2×2 principal submatrix of the positive semidefinite matrix $M_t(y)$

$$\begin{array}{c} \\ ij \\ J_1 \end{array} \begin{array}{cc} ij & J_1 \\ \begin{bmatrix} y_{ij} & y_{J_1} \\ y_{J_1} & y_{J_1} \end{bmatrix} \end{array} \succeq 0 \implies y_{J_1} = 0,$$

where we applied the constraint $y_{ij} = 0$. Then, consider the following 2×2 principal submatrix of $M_t(y)$

$$
\begin{array}{cc}
J_1 & J_2
\end{array}
$$
$$
\begin{array}{c}
J_1 \\
J_2
\end{array}
\begin{bmatrix}
y_{J_1} & y_J \\
y_J & y_{J_2}
\end{bmatrix}
\succeq 0 \implies y_J = 0.
$$

Hence,

$$
\text{las}_t(G) = \text{las}_{t+1}(G) = \cdots = \text{las}_n(G) \quad \text{for } t \geq \alpha(G).
$$

The next step is showing that vectors $y \in \mathbb{R}^{\mathcal{P}_n(V)}$, indexed by the full power set $\mathcal{P}_n(V)$, which determine a positive semidefinite moment matrix $M_n(y)$, form a finitely generated cone:

$$
M_n(y) \succeq 0 \iff y \in \text{cone}\{\chi^I \mid I \subseteq V\},
$$

where χ^I are the characteristic vectors. Sufficiency follows easily from $\chi^I_{J \cup J'} = \chi^I_J \chi^I_{J'}$. For necessity, we first observe that the characteristic vectors form a basis of $\mathbb{R}^{\mathcal{P}_n(V)}$. Let $(\psi^J)_{J \in \mathcal{P}_n(V)}$ be its dual basis; it satisfies $(\chi^I)^\mathsf{T} \psi^J = \delta_{I,J}$. Let y be so that $M_n(y)$ is positive semidefinite and write y in terms of the basis

$$
y = \sum_{I \in \mathcal{P}_n(V)} \alpha_I \chi^I \quad \text{with } \alpha_I \in \mathbb{R}.
$$

Since $M_n(y)$ is positive semidefinite, we have

$$
0 \leq (\psi^J)^\mathsf{T} M_n(y) \psi^J = \alpha_J.
$$

Now we finish the proof. Let $y \in \mathbb{R}^{\mathcal{P}_n(V)}$ be a feasible solution of $\text{las}_n(G)$. Then from the previous arguments we see

$$
y = \sum_{I \text{ independent}} \alpha_I \chi^I \quad \text{with } \alpha_I \geq 0.
$$

Furthermore, the semidefinite program is normalized by

$$
1 = y_\emptyset = \sum_{I \text{ independent}} \alpha_I,
$$

and the objective value of y equals

$$
\sum_{i \in V} y_i = \sum_{i \in V} \sum_{I \text{ independent}} \alpha_I \chi^I(i) = \sum_{I \text{ independent}} \alpha_I \sum_{i \in V} \chi^I(i) \leq 1 \cdot \alpha(G).
$$

$\square$

13.4 Symmetry Reduction and Matrix ∗-Algebras

One can obtain semidefinite programming bounds for $A_2(n, d)$ by using the Lasserre Bound of order t for the graph $G(n, d)$, defined in Example 13.3.2. Since the graph $G(n, d)$ has exponentially many vertices, even computing the first step $t = 1$ amounts to solving a semidefinite program of exponential size. On the other hand, the graph $G(n, d)$ is highly symmetric and these symmetries can be used to simplify the semidefinite programs considerably.

13.4.1   Symmetry Reduction of Semidefinite Programs

Symmetry reduction of semidefinite programs is most easily explained using complex semidefinite programs of the form (13.2). Let Γ be a finite group and let $\pi : \Gamma \to U(\mathbb{C}^n)$ be a **unitary representation** of Γ, that is, a group homomorphism from Γ to the group of unitary matrices $U(\mathbb{C}^n)$. Then Γ acts on the space of complex matrices by

$$(g, X) \mapsto gX = \pi(g)X\pi(g)^*.$$

A complex matrix X is called Γ-**invariant** if $X = gX$ holds for all $g \in \Gamma$. By

$$(\mathbb{C}^{n \times n})^\Gamma = \{X \in \mathbb{C}^{n \times n} \mid X = gX \text{ for all } g \in \Gamma\}$$

we denote the set of all Γ-invariant matrices.

Definition 13.4.1 Let Γ be a finite group. A complex semidefinite program is called Γ-**invariant** if for every feasible solution X and every $g \in \Gamma$, the matrix gX also is feasible and $\langle C, X \rangle_T = \langle C, gX \rangle_T$ holds. (Recall $\langle X, Y \rangle_T = \mathrm{TR}(Y^*X)$.)

Suppose that the complex semidefinite program (13.2) is Γ-invariant. Then we may restrict the optimization variable X to be Γ-invariant without changing the supremum. In fact, if X is feasible for (13.2), so is its Γ-**average**

$$\overline{X} = \frac{1}{|\Gamma|} \sum_{g \in \Gamma} gX.$$

Hence, (13.2) simplifies to

$$p^* = \sup\left\{\langle C, X \rangle_T \mid X \succeq 0, \ X \in (\mathbb{C}^{n \times n})^\Gamma, \ \langle A_1, X \rangle_T = b_1, \ldots, \langle A_m, X \rangle_T = b_m\right\}. \quad (13.3)$$

If we intersect the Γ-invariant complex matrices $(\mathbb{C}^{n \times n})^\Gamma$ with the Hermitian matrices we get a vector space having a basis $B_1, \ldots, B_N$. If we express X in terms of this basis, (13.3) becomes

$$p^* = \sup\{\langle C, X \rangle_T \mid x_1, \ldots, x_N \in \mathbb{C}, \ X = x_1 B_1 + \cdots + x_N B_N \succeq 0,$$
$$\langle A_1, X \rangle_T = b_1, \ldots, \langle A_m, X \rangle_T = b_m\}. \quad (13.4)$$

So the number of optimization variables is N. It turns out that we can simplify (13.4) even more by performing a simultaneous block diagonalization of the basis $B_1, \ldots, B_N$. This is a consequence of the main structure theorem of matrix $*$-algebras.

13.4.2   Matrix $*$-Algebras

Definition 13.4.2 A linear subspace $\mathcal{A} \subseteq \mathbb{C}^{n \times n}$ is called a **matrix algebra** if it is closed under matrix multiplication. It is called a **matrix $*$-algebra** if it is also closed under taking the conjugate transpose: if $A \in \mathcal{A}$, then $A^* \in \mathcal{A}$.

The space of Γ-invariant matrices $(\mathbb{C}^{n \times n})^\Gamma$ is a matrix $*$-algebra. Indeed, for Γ-invariant matrices X, Y and $g \in \Gamma$, we have

$$g(XY) = \pi(g)XY\pi(g)^* = (\pi(g)X\pi(g)^*)(\pi(g)Y\pi(g)^*) = (gX)(gY) = XY,$$

and

$$g(X^*) = \pi(g)X^*\pi(g)^* = (\pi(g)X\pi(g)^*)^* = (gX)^* = X^*.$$

The main structure theorem of matrix $*$-algebras—it is due to Wedderburn and it is well-known in the theory of C^*-algebras, where it can be also stated for the compact operators on a Hilbert space—is the following:

Theorem 13.4.3 *Let $\mathcal{A} \subseteq \mathbb{C}^{n \times n}$ be a matrix $*$-algebra. Then there are natural numbers d, $m_1, \ldots, m_d$ such that there is a $*$-isomorphism between $\mathcal{A}$ and a direct sum of full matrix $*$-algebras*

$$\varphi \colon \mathcal{A} \to \bigoplus_{k=1}^{d} \mathbb{C}^{m_k \times m_k}.$$

*Here a $*$-**isomorphism** is a bijective linear map between two matrix $*$-algebras which respects multiplication and taking the conjugate transpose.*

An elementary proof of Theorem 13.4.3, which also shows how to find a $*$-isomorphism φ algorithmically, is presented in [87]. An alternative proof is given in [1829, Section 3] in the framework of representation theory of finite groups; see also [84, 1828].

Now we want to apply Theorem 13.4.3 to block diagonalize the Γ-invariant semidefinite program (13.4). Let $\mathcal{A} = (\mathbb{C}^{n \times n})^{\Gamma}$ be the matrix $*$-algebra of Γ-invariant matrices. Let φ be a $*$-isomorphism as in Theorem 13.4.3; then φ preserves positive semidefiniteness. Hence, (13.4) is equivalent to

$$p^* = \sup \{ \langle C, X \rangle_T \mid x_1, \ldots, x_N \in \mathbb{C}, \ x_1 \varphi(B_1) + \cdots + x_N \varphi(B_N) \succeq 0,$$
$$X = x_1 B_1 + \cdots + x_N B_N, \ \langle A_1, X \rangle_T = b_1, \ldots, \langle A_m, X \rangle_T = b_m \}.$$

Thus, instead of dealing with one (potentially big) matrix of size $n \times n$ one only has to work with d (hopefully small) block diagonal matrices of size $m_1, \ldots, m_d$. This reduces the dimension from n^2 to $m_1^2 + \cdots + m_d^2$. Many practical semidefinite programming solvers can take advantage of this block structure and numerical calculations can become much faster. However, finding an explicit $*$-isomorphism is usually a nontrivial task, especially if one is interested in parameterized families of matrix $*$-algebras.

13.4.3 Example: The Delsarte Linear Programming Bound

Let us apply the symmetry reduction technique to demonstrate that the exponential size semidefinite program $\vartheta'(G(n, d))$ collapses to the linear size Delsarte Linear Programming Bound.

Since the graph $G(n, d)$ is a Cayley graph over the additive group $\mathbb{F}_2^n$, the semidefinite program $\vartheta'(G(n, d))$ is $\mathbb{F}_2^n$-invariant where the group is acting as permutations of the rows and columns of the matrix $X \in \mathbb{C}^{\mathbb{F}_2^n \times \mathbb{F}_2^n}$. The graph $G(n, d)$ has even more symmetries. Its automorphism group $\mathrm{Aut}(G(n, d))$ consists of all permutations of the n coordinates $\mathbf{x} = x_1 x_2 \cdots x_n \in \mathbb{F}_2^n$ followed by independently switching the elements of $\mathbb{F}_2$ from 0 to 1, or vice versa. So the semidefinite program $\vartheta'(G(n, d))$ is $\mathrm{Aut}(G(n, d))$-invariant. The $*$-algebra $\mathcal{B}_n$ of $\mathrm{Aut}(G(n, d))$-invariant matrices is called the **Bose–Mesner algebra (of the binary Hamming scheme)**. A basis $B_0, \ldots, B_n$ is given by zero-one matrices

$$(B_r)_{\mathbf{x}, \mathbf{y}} = \begin{cases} 1 & \text{if } \mathrm{d_H}(\mathbf{x}, \mathbf{y}) = r, \\ 0 & \text{otherwise,} \end{cases}$$

with $r = 0, \ldots, n$. So, $\vartheta'(G(n, d))$ in the form of (13.4) is the following semidefinite program in $n + 1$ variables:

$$\max \left\{ 2^n \sum_{r=0}^{n} \binom{n}{r} x_r \ \middle| \ x_0 = \frac{1}{2^n}, \ x_1 = \cdots = x_{d-1} = 0, \ x_d, \ldots, x_n \geq 0, \ \sum_{r=0}^{n} x_r B_r \succeq 0 \right\}.$$

Finding a simultaneous block diagonalization of the B_r's is easy since they pairwise commute and have a common system of eigenvectors. An orthogonal basis of eigenvectors is given by $\chi_{\mathbf{a}} \in \mathbb{C}^{\mathbb{F}_2^n}$ defined componentwise by

$$(\chi_{\mathbf{a}})_{\mathbf{x}} = \prod_{j=1}^{n} (-1)^{a_j x_j}.$$

Indeed,

$$(B_r \chi_{\mathbf{a}})_{\mathbf{x}} = \sum_{\mathbf{y} \in \mathbb{F}_2^n} (B_r)_{\mathbf{x},\mathbf{y}} (\chi_{\mathbf{a}})_{\mathbf{y}} = \sum_{\mathbf{y} \in \mathbb{F}_2^n} (B_r)_{\mathbf{x},\mathbf{y}} (\chi_{\mathbf{a}})_{\mathbf{y}-\mathbf{x}} (\chi_{\mathbf{a}})_{\mathbf{x}}$$

$$= \left(\sum_{\mathbf{y} \in \mathbb{F}_2^n, \, d_H(\mathbf{x},\mathbf{y})=r} (\chi_{\mathbf{a}})_{\mathbf{y}-\mathbf{x}} \right) (\chi_{\mathbf{a}})_{\mathbf{x}} = \left(\sum_{\mathbf{y} \in \mathbb{F}_2^n, \, d_H(\mathbf{0},\mathbf{y})=r} (\chi_{\mathbf{a}})_{\mathbf{y}} \right) (\chi_{\mathbf{a}})_{\mathbf{x}}.$$

The eigenvalues are given by the **Krawtchouk polynomials**

$$K_r^{(n,2)}(x) = \sum_{j=0}^{r} (-1)^j \binom{x}{j} \binom{n-x}{r-j}$$

through

$$\sum_{\mathbf{y} \in \mathbb{F}_2^n, \, d_H(\mathbf{0},\mathbf{y})=r} (\chi_{\mathbf{a}})_{\mathbf{y}} = K_r^{(n,2)}(d_H(\mathbf{0},\mathbf{a})).$$

Altogether, we have the $*$-algebra isomorphism

$$\varphi : \mathcal{B}_n \to \bigoplus_{r=0}^{n} \mathbb{C},$$

(so $m_0 = \cdots = m_n = 1$) defined by

$$\varphi(B_r) = (K_r^{(n,2)}(0), K_r^{(n,2)}(1), \ldots, K_r^{(n,2)}(n)).$$

So the semidefinite program $\vartheta'(G(n,d))$ degenerates to the following linear program

$$\max \left\{ 2^n \sum_{r=0}^{n} \binom{n}{r} x_r \;\middle|\; x_0 = \frac{1}{2^n}, \, x_1 = \cdots = x_{d-1} = 0, \, x_d, \ldots, x_n \geq 0, \right.$$

$$\left. \sum_{r=0}^{n} x_r K_r^{(n,2)}(j) \geq 0 \text{ for } j = 0, \ldots, n \right\}.$$

This is the Delsarte Linear Programming Bound; see also Theorem 1.9.23.

13.4.4   Example: The Schrijver Semidefinite Programming Bound

To set up a stronger semidefinite programming bound one can apply the Lasserre Bound directly, but also many variations are possible. These variations are crucial to be able to exploit the symmetries of the problem at hand. For instance, one can consider only "interesting" principal submatrices of the moment matrices to simplify the computation.

A rough classification for these variations can be given in terms of *k-point bounds*. This refers to all variations which make use of variables y_I with $|I| \leq k$. A k-point bound is capable of using obstructions coming from the local interaction of configurations having at most k points. For instance the Lovász ϑ-number is a 2-point bound and the tth step in

Lasserre's hierarchy is a $2t$-point bound. The relation between k-point bounds and Lasserre's hierarchy was first made explicit by Laurent [1206] in the case of bounds for binary codes; see also Gijswijt [803], who discusses the symmetry reduction needed to compute k-point bounds for block codes, and de Laat and Vallentin [506], who consider k-point bounds for compact topological packing graphs.

Schrijver's bound for binary codes [1634] is a 3-point bound. Essentially, it looks at principal submatrices $M_{\mathbf{a}} \in \mathbb{R}^{\mathbb{F}_2^n \times \mathbb{F}_2^n}$ of the matrix $M_2(y)$ defined by

$$(M_{\mathbf{a}}(y))_{\mathbf{b},\mathbf{c}} = y_{\{\mathbf{a},\mathbf{b},\mathbf{c}\}} \quad \text{with } \mathbf{a}, \mathbf{b}, \mathbf{c} \in \mathbb{F}_2^n.$$

The group which leaves the corresponding semidefinite program invariant is the stabilizer of a codeword in $\mathrm{Aut}(G(n,d))$, which is the symmetric group permuting the n coordinates of $\mathbb{F}_2^n$.

The algebra $\mathcal{A}_n \subseteq \mathbb{R}^{\mathbb{F}_2^n \times \mathbb{F}_2^n}$ invariant under this group action is called the **Terwilliger algebra of the binary Hamming scheme**. Schrijver determined a block diagonalization of the Terwilliger algebra which we recall here.

For nonnegative integers i, j, t, with $t \leq i, j$ and $i + j \leq n + t$, the matrices

$$(B_{i,j}^t)_{\mathbf{x},\mathbf{y}} = \begin{cases} 1 & \text{if } \mathrm{wt}_H(\mathbf{x}) = i, \ \mathrm{wt}_H(\mathbf{y}) = j, \ \mathrm{d}_H(\mathbf{x},\mathbf{y}) = i + j - 2t, \\ 0 & \text{otherwise} \end{cases}$$

form a basis of $\mathcal{A}_n$. Hence, $\dim \mathcal{A}_n = \binom{n+3}{3}$. The desired $*$-isomorphism

$$\varphi : \mathcal{A}_n \to \bigoplus_{k=0}^{\lfloor n/2 \rfloor} \mathbb{C}^{(n-2k+1) \times (n-2k+1)}$$

is defined as follows: Set

$$\beta_{i,j,k}^t = \sum_{u=0}^{n} (-1)^{u-t} \binom{u}{t} \binom{n-2k}{u-k} \binom{n-k-u}{i-u} \binom{n-k-u}{j-u}$$

so that

$$\varphi(B_{i,j}^t) = \left(\dots, \left(\binom{n-2k}{i-k}^{-1/2} \binom{n-2k}{j-k}^{-1/2} \beta_{i,j,k}^t \right)_{i,j=k}^{n-k}, \dots \right)_{k=0,\dots,\lfloor n/2 \rfloor}.$$

Schrijver determined the $*$-isomorphism from first principles using linear algebra. Later, Vallentin [1829] used representation theory of finite groups to derive an alternative proof. Here the connection to the orthogonal Hahn and Krawtchouk polynomials becomes visible. Another constructive proof of the explicit block diagonalization of $\mathcal{A}_n$ was given by Srinivasan [1748]; see also Martin and Tanaka [1334].

13.5 Extensions and Ramifications

Explicit computations of k-point semidefinite programming bounds have been done in a variety of situations, in the finite and infinite setting. Table 13.1 gives a guide to the literature.

TABLE 13.1: Computation of k-point bounds.

Problem	2-point bound	3-point bound	4-point bound
Binary codes	Delsarte [519]	Schrijver [1634]	Gijswijt, Mittelmann, Schrijver [804]
q-ary codes	Delsarte [519]	Gijswijt, Schrijver, Tanaka [805]	Litjens, Polak, Schrijver [1275]
Constant weight codes	Delsarte [519]	Schrijver [1634], Regts [1584]	Polak [1530]
Lee codes	Astola [75]	Polak [1529]	
Bounded weight codes	Bachoc, Chandar, Cohen, Solé Tchamkerten [85]		
Grassmannian codes	Bachoc [83]		
Projective codes	Bachoc, Passuello, Vallentin [88]		
Spherical codes	Delsarte, Goethals, Seidel [527]	Bachoc, Vallentin [89]	
Codes in $\mathbb{RP}^{n-1}$	Kabatianskii, Levenshtein [1067]	Cohn, Woo [425]	
Sphere packings	Cohn, Elkies [423]		
Binary sphere and spherical cap packings	de Laat, Oliveira, Vallentin [504]		
Translative body packings	Dostert, Guzmán, Oliveira, Vallentin [618]		
Congruent copies of a convex body	Oliveira, Vallentin [508]		

Semidefinite programming bounds have also been developed for generalized distances and list-decoding radii of binary codes by Bachoc and Zémor [90], for permutation codes by Bogaerts and Dukes [228], for mixed binary/ternary codes by Litjens [1274], for subsets of coherent configurations by Hobart [965] and Hobart and Williford [966], for ordered codes by Trinker [1820], for energy minimization on $\mathbb{S}^2$ by de Laat [503], and for spherical two-distance sets and for equiangular lines by Barg and Yu [129] and by Machado, de Laat, Oliveira, and Vallentin [505]. They have been used by Brouwer and Polak [311] to prove the uniqueness of several constant weight codes.

In extremal combinatorics, (weighted) vector space versions of the Erdős–Ko–Rado Theorem for cross intersecting families have been proved using semidefinite programming bounds by Suda and Tanaka [1760] and by Suda, Tanaka and Tokushige [1761]; see also the survey by Frankl and Tokushige [752].

Another coding theory application of the symmetry reduction technique includes new approaches to the Assmus–Mattson Theorem by Tanaka [1777] and by Morales and Tanaka [1399]

Acknowledgement

The author was partially supported by the SFB/TRR 191 "Symplectic Structures in Geometry, Algebra and Dynamics", funded by the DFG. He also gratefully acknowledges support by DFG grant VA 359/1-1. This project has received funding from the European Unions Horizon 2020 research and innovation programme under the Marie Skłodowska-Curie agreement number 764759.

Part II

Families of Codes

Chapter 14

Coding Theory and Galois Geometries

Leo Storme

Ghent University

14.1   Galois Geometries

14.1.1   Basic Properties of Galois Geometries

This section states the basic properties of Galois geometries; it is similar to [1408, Section 14.4]. We refer the reader to the standard references [958, 959, 963] for an in-depth study of Galois geometries.

Definition 14.1.1 Let $V = V(n+1, q)$ be the vector space of dimension $n+1$ over the finite field $\mathbb{F}_q$ of order q. Define the equivalence relation $\sim$ on the set of nonzero vectors of V: for $\mathbf{v}, \mathbf{w} \in V \setminus \{0\}$, $\mathbf{v} \sim \mathbf{w}$ if and only if $\mathbf{w} = \rho \mathbf{v}$ for some $\rho \in \mathbb{F}_q \setminus \{0\}$. The equivalence classes of this equivalence relation $\sim$ on $V \setminus \{0\}$ consist of the vector lines $\langle v \rangle = \{\rho \mathbf{v} \mid \rho \in \mathbb{F}_q\}$ of V, without the zero vector. The set of equivalence classes in V is called the **n-dimensional projective space $\mathrm{PG}(n, q)$ over the finite field $\mathbb{F}_q$**.

Definition 14.1.2

(a) The elements of $\mathrm{PG}(n, q)$ are called the **projective points**, or simply **points**, of $\mathrm{PG}(n, q)$.

(b) The equivalence class of the nonzero vector $\mathbf{v}$ is the point $P(\mathbf{v})$, or is simply denoted by P when the explicit reference to $\mathbf{v}$ is not required.

(c) The nonzero vector $\mathbf{v}$ is called a **coordinate vector** for $P(\mathbf{v})$. Note that also every nonzero multiple $\rho\mathbf{v}$, $\rho \in \mathbb{F}_q \setminus \{0\}$, is a coordinate vector for $P(\mathbf{v})$.

Definition 14.1.3 Consider $V(n+1, q)$ and its corresponding projective space $\mathrm{PG}(n, q)$.

(a) For any $m = -1, 0, 1, \ldots, n$, an m-**dimensional subspace**, also called m-**space**, of $\mathrm{PG}(n, q)$ is a set of points for which the union of all the corresponding coordinate vectors, together with the zero vector, form an $(m+1)$-dimensional vector subspace of $V(n+1, q)$. We will denote an m-space by Π_m.

(b) In particular, the -1-dimensional subspace is the empty set of $\mathrm{PG}(n, q)$, and a 0-dimensional subspace is a point. A 1-dimensional subspace is called a (**projective**) **line**, a 2-dimensional subspace is called a (**projective**) **plane**, and a 3-dimensional subspace is called a (**projective**) **solid**. An $(n-1)$-dimensional subspace of $\mathrm{PG}(n, q)$ is called a **hyperplane**. An $(n-r)$-dimensional subspace of $\mathrm{PG}(n, q)$ is also called a **subspace of codimension** r.

A lot of numerical data is known regarding finite projective spaces.

Definition 14.1.4

(a) Let $\theta_n = (q^{n+1} - 1)/(q - 1)$.

(b) Let

$$\begin{bmatrix} t \\ s \end{bmatrix}_q = \frac{(q^t - 1)(q^{t-1} - 1) \cdots (q^{t-s+1} - 1)}{(q^s - 1)(q^{s-1} - 1) \cdots (q - 1)}.$$

$\begin{bmatrix} t \\ s \end{bmatrix}_q$ is called a q-**ary Gaussian coefficient**.

Theorem 14.1.5

(a) *The projective space* $\mathrm{PG}(n, q)$ *contains* θ_n *points.*

(b) *The projective space* $\mathrm{PG}(n, q)$ *contains* $\begin{bmatrix} n+1 \\ t+1 \end{bmatrix}_q$ *different t-spaces.*

Remark 14.1.6 In $\mathrm{PG}(n, q)$, every hyperplane is a set of points $P(X)$ whose coordinate vectors $X = (X_0, \ldots, X_n)$ satisfy a linear equation

$$u_0 X_0 + u_1 X_1 + \cdots + u_n X_n = 0,$$

with $u_0 \cdots u_n \in \mathbb{F}_q^{n+1} \setminus \{0 \cdots 0\}$ fixed.

Definition 14.1.7

(a) If a point P lies in a subspace Π of $\mathrm{PG}(n, q)$, then we say that P is **incident** with Π, and vice versa, that Π is **incident** with P.

(b) If Π_r and Π_s are subspaces of $\mathrm{PG}(n, q)$, then the **intersection** $\Pi_r \cap \Pi_s$ of Π_r and Π_s is the set of points belonging to both the spaces Π_r and Π_s.

(c) If Π_r and Π_s are subspaces of $\mathrm{PG}(n, q)$, then the **span** $\langle \Pi_r, \Pi_s \rangle$ of Π_r and Π_s is the smallest subspace of $\mathrm{PG}(n, q)$ containing both Π_r and Π_s.

Substructures and subsets of a finite projective space sometimes are similar to each other. This is expressed via collineations of a finite projective space.

Definition 14.1.8 A **collineation** α of $PG(n, q)$, $n \geq 2$, is a bijection which preserves incidence: if $\Pi_r \subseteq \Pi_s$, then $\alpha(\Pi_r) \subseteq \alpha(\Pi_s)$.

Theorem 14.1.9 (Fundamental Theorem of Galois Geometry) *If α is a collineation of $PG(n, q)$, $q = p^h$, p prime, $h \geq 1$, then α is a semilinear bijective transformation of $V(n + 1, q)$; i.e., there exists a nonsingular $(n + 1) \times (n + 1)$ matrix A over $\mathbb{F}_q$ and an automorphism $\sigma : \mathbb{F}_q \to \mathbb{F}_q$, where $\sigma : x \mapsto \sigma(x) = x^{p^i}$ for some $0 \leq i \leq h - 1$, such that*

$$\alpha : \begin{bmatrix} x_0 \\ x_1 \\ \vdots \\ x_n \end{bmatrix} \mapsto A \begin{bmatrix} \sigma(x_0) \\ \sigma(x_1) \\ \vdots \\ \sigma(x_n) \end{bmatrix}.$$

Definition 14.1.10 Let $Z_{n+1}(q) = \{\alpha \mid x \mapsto \rho I_{n+1} x, \; \rho \in \mathbb{F}_q \setminus \{0\}, \; x \in V(n+1, q)\}$, where I_{n+1} is the $(n + 1) \times (n + 1)$ identity matrix.

(a) The **projective group** of $PG(n, q)$, $n \geq 1$, is the group

$$PGL_{n+1}(q) = GL_{n+1}(q)/Z_{n+1}(q).$$

(b) The **collineation group** of $PG(n, q)$, $n \geq 1$, is the group

$$P\Gamma L_{n+1}(q) = PGL_{n+1}(q) \rtimes \text{Aut}(\mathbb{F}_q).$$

Definition 14.1.11 Two sets S and S' of spaces contained in $PG(n, q)$ are called **projectively equivalent** to each other if and only if there is a collineation $\alpha \in P\Gamma L_{n+1}(q)$ which maps S onto S', i.e., $\alpha(S) = S'$.

14.1.2 Spreads and Partial Spreads

We now discuss a substructure of finite projective spaces which has been investigated for a long time. Its relevance increased when links to subspace codes were found (Theorem 14.5.6). Here, the problem of finding the maximal size for partial $(k - 1)$-spreads in $PG(n - 1, q)$ proved to be of great importance.

Definition 14.1.12

(a) A **partial $(k - 1)$-spread** in $PG(n - 1, q)$ is a set of pairwise disjoint $(k - 1)$-spaces.

(b) A **$(k - 1)$-spread** in $PG(n - 1, q)$ is a set of pairwise disjoint $(k - 1)$-spaces which partitions the point set of $PG(n - 1, q)$.

The question regarding the existence of $(k - 1)$-spreads in $PG(n - 1, q)$ has been completely solved.

Theorem 14.1.13 ([959, Theorem 4.1]) *A $(k - 1)$-spread in $PG(n - 1, q)$ exists if and only if k divides n.*

Regarding the other cases, where k does not divide n, so that $(k - 1)$-spreads in $PG(n - 1, q)$ do not exist, the following results on the largest partial $(k - 1)$-spreads in $PG(n - 1, q)$ have been found.

Theorem 14.1.14

(a) ([195]) *Let $n \equiv r \pmod{k}$. Then, for all q, $\mathrm{PG}(n-1,q)$ contains a partial $(k-1)$-spread of size at least $\frac{q^n - q^k(q^r - 1) - 1}{q^k - 1}$.*

(b) ([193, 194]) *Let $n \equiv 1 \pmod{k}$. Then, for all q, the largest partial $(k-1)$-spreads of $\mathrm{PG}(n-1,q)$ have size equal to $\frac{q^n - q^k(q-1) - 1}{q^k - 1}$.*

(c) ([1415]) *Let $n \equiv r \pmod{k}$. For all q, if $k > \frac{q^r - 1}{q - 1}$, then the largest partial $(k-1)$-spreads of $\mathrm{PG}(n-1,q)$ have size equal to $\frac{q^n - q^k(q^r - 1) - 1}{q^k - 1}$.*

(d) ([668]) *Let $n \equiv c \pmod{3}$ with $0 \leq c \leq 2$. Then the largest size of the partial 2-spreads in $\mathrm{PG}(n-1,2)$ is equal to $\frac{2^n - 2^c}{7} - c$.*

(e) ([1179]) *Let $n \equiv 2 \pmod{k}$. Then, for $k \geq 4$, $n \geq 2k + 2$, the largest partial $(k-1)$-spreads of $\mathrm{PG}(n-1,2)$ have size equal to $\frac{2^n - 3 \cdot 2^k - 1}{2^k - 1}$.*

14.2 Two Links Between Coding Theory and Galois Geometries

Many problems in coding theory can be retranslated into equivalent geometric problems on specific substructures in finite projective spaces. We now describe two of the most important ways to establish links between coding theoretic problems and geometric problems on substructures in finite projective spaces.

14.2.1 Via the Generator Matrix

Let $\mathcal{C}$ be an $[n, k]_q$ code. Let $G = \begin{bmatrix} G_1 & \cdots & G_n \end{bmatrix}$ be a $k \times n$ generator matrix of $\mathcal{C}$, described via the n columns $G_1, \ldots, G_n$ of this matrix. Assume that the matrix G does not have zero columns. Then the columns $G_1, \ldots, G_n$ define points of the projective space $\mathrm{PG}(k-1, q)$. If s columns out of $G_1, \ldots, G_n$ define the same projective point P, then we now consider this point P to have **weight** $w(P) = s$. Then a multiset $(K, w) = \{G_1, \ldots, G_n\}$ of size n of $\mathrm{PG}(k-1, q)$ is obtained, generating the space $\mathrm{PG}(k-1, q)$.

Similarly, from a multiset $(K, w) = \{G_1, \ldots, G_n\}$ of size n of $\mathrm{PG}(k-1, q)$ generating the space $\mathrm{PG}(k-1, q)$, it is possible to construct a generator matrix $G = \begin{bmatrix} G_1 & \cdots & G_n \end{bmatrix}$ of a linear $[n, k]_q$ code.

This first link leads to the following theorem.

Theorem 14.2.1 ([1198]) *Let $(K, w) = \{P_1, \ldots, P_n\}$ be a multiset of size n of $\mathrm{PG}(k-1, q)$ generating the space $\mathrm{PG}(k-1, q)$. Let $\mathcal{C}$ be the linear $[n, k]_q$ code with generator matrix $G = \begin{bmatrix} G_1 & \cdots & G_n \end{bmatrix}$ and minimum Hamming weight $d_H(\mathcal{C})$. Then $d_H(\mathcal{C}) = d$ if and only if every hyperplane of $\mathrm{PG}(k-1, q)$ contains at most $n - d$ points of the multiset (K, w), and at least one hyperplane of $\mathrm{PG}(k-1, q)$ contains exactly $n - d$ points of the multiset (K, w).*

14.2.2 Via the Parity Check Matrix

Let $\mathcal{C}$ be an $[n, k]_q$ code. Let $H = \begin{bmatrix} H_1 & \cdots & H_n \end{bmatrix}$ be an $(n-k) \times n$ parity check matrix of $\mathcal{C}$, described via the n columns $H_1, \ldots, H_n$ of this matrix. Assume that the matrix H does not have zero columns. Then the columns $H_1, \ldots, H_n$ define points of the projective space $\mathrm{PG}(n-k-1, q)$. If s columns out of $H_1, \ldots, H_n$ define the same projective point P, then we now consider this point P to have **weight** $w(P) = s$. Then a multiset $(K, w) = \{H_1, \ldots, H_n\}$ of size n of $\mathrm{PG}(n - k - 1, q)$ is obtained, generating the space $\mathrm{PG}(n - k - 1, q)$.

Similarly, from a multiset $(K, w) = \{H_1, \ldots, H_n\}$ of size n of $\mathrm{PG}(n - k - 1, q)$ generating the space $\mathrm{PG}(n - k - 1, q)$, it is possible to construct a parity check matrix $H = [\, H_1 \; \cdots \; H_n \,]$ of a linear $[n, k]_q$ code.

This second link leads to the following theorem, also mentioned in Theorem 1.6.11.

Theorem 14.2.2 *Let $(K, w) = \{H_1, \ldots, H_n\}$ be a multiset of size n of $\mathrm{PG}(n - k - 1, q)$, generating the space $\mathrm{PG}(n-k-1, q)$. Let $\mathcal{C}$ be the linear $[n, k]_q$ code with parity check matrix $H = [\, H_1 \; \cdots \; H_n \,]$. Then $d_H(\mathcal{C}) = d$ if and only if every $d - 1$ points of the multiset $(K, w) = \{H_1, \ldots, H_n\}$ are linearly independent, and there exists at least one subset of d points of the multiset $(K, w) = \{H_1, \ldots, H_n\}$ which are linearly dependent.*

14.2.3 Linear MDS Codes and Arcs in Galois Geometries

In Section 1.9.2, the Singleton Bound was presented. For a linear $[n, k, d]_q$ code, the Singleton Bound states that $k \leq n - d + 1$. A linear $[n, k, n - k + 1]_q$ code is called a Maximum Distance Separable (MDS) code.

Linear MDS codes proved to be equivalent to a specific substructure in finite projective spaces, called *arcs*.

Definition 14.2.3 An *n-arc* in $\mathrm{PG}(k - 1, q)$ is a set $K = \{P_1, \ldots, P_n\}$ of n points of $\mathrm{PG}(k - 1, q)$, every k of which are linearly independent.

Using Theorem 14.2.2, the definition of linear MDS codes and the definition of arcs in finite projective spaces led to the following equivalence between these codes in coding theory and these substructures in finite projective spaces.

Theorem 14.2.4 *A linear $[n, n - k]_q$ code $\mathcal{C}$ is an MDS code if and only if the n columns of a parity check matrix of $\mathcal{C}$ form an n-arc in $\mathrm{PG}(k - 1, q)$.*

This chapter focuses on the many links between coding theory and Galois geometries. We arrive to one of the most particular examples of these links. We now give the definition of a *normal rational curve* in Galois geometries. This is the classical example of an arc. This will correspond to the classical example of a linear MDS code, the *generalized doubly-extended Reed–Solomon* codes. Hence, the study of these objects in Galois geometries and in coding theory are in fact the study of the same objects.

Definition 14.2.5 A **normal rational curve** L in $\mathrm{PG}(k - 1, q)$, $2 \leq k \leq q - 1$, is a $(q + 1)$-arc projectively equivalent to the set of points

$$\{(1, t, \ldots, t^{k-1}) \mid t \in \mathbb{F}_q\} \cup \{(0, \ldots, 0, 1)\}.$$

A normal rational curve L of $\mathrm{PG}(2, q)$ is also called a **conic** of $\mathrm{PG}(2, q)$.

Definition 14.2.6 A **generalized doubly-extended Reed–Solomon (GDRS) code** is a $[q + 1, k, q + 2 - k]_q$ MDS code with a $(q + 1 - k) \times (q + 1)$ parity check matrix whose columns consist of the points of a normal rational curve of $\mathrm{PG}(q - k, q)$.

It is known that a linear $[n, k, d]_q$ code $\mathcal{C}$ is MDS if and only if its dual $[n, n - k, d']_q$ code $\mathcal{C}^\perp$ also is MDS (Theorem 1.9.13). Because of the link between linear MDS codes and arcs in finite projective spaces, the concept of dual arcs also was introduced in finite geometry.

Definition 14.2.7 An n-arc K in $\mathrm{PG}(k-1,q)$ and an n-arc K' in $\mathrm{PG}(n-k-1,q)$ are called **dual arcs** with respect to each other if and only if they define two linear MDS codes, via parity check matrices, which are dual codes to each other.

Note that because a generator matrix and a parity check matrix of a linear code are a parity check matrix and a generator matrix, respectively, of the dual linear code, two arcs are dual to each other if and only if they define two linear MDS codes, via generator matrices, which are dual codes to each other.

It is known that the dual code of a $[q+1,k,q+2-k]_q$ GDRS code is a $[q+1,q+1-k,k+1]_q$ GDRS code. This follows for instance from the standard form of a generator matrix of a GDRS code, described in [1603]. This translates to the following equivalent geometric result.

Theorem 14.2.8 *The dual $(q+1)$-arc to a normal rational curve in $\mathrm{PG}(k-1,q)$ is a normal rational curve of $\mathrm{PG}(q-k,q)$.*

Note that Section 1.14 also discusses Reed–Solomon codes. For instance, in Theorem 1.14.12, the Reed–Solomon code of length q defined by a generator matrix whose columns consist of the q points $(1,t,\ldots,t^{k-1})$, $t \in \mathbb{F}_q$, was discussed.

Many researchers in finite geometry have obtained results on arcs in finite projective spaces. These results on arcs translate immediately into corresponding results on linear MDS codes. Depending on the theorem, we present these results in the language of arcs or in the language of MDS codes, but we state the first theorems in both languages. The first result of the next theorem was proven by B. Segre. This characterization result was found as a theorem on arcs, and motivated many researchers to also start investigating problems in finite geometry. The links with the coding theoretic results on MDS codes then also inspired these researchers to work on coding theoretic problems, and, vice-versa, inspired researchers in coding theory to also investigate problems in finite geometry.

Theorem 14.2.9

(a) ([1639]) *Every $(q+1)$-arc in $\mathrm{PG}(2,q)$, q odd, and every $(q+1)$-arc in $\mathrm{PG}(q-3,q)$, q odd, is equal to a normal rational curve.*

    For q odd, every $[q+1,3,q-1]_q$ MDS code and every $[q+1,q-2,4]_q$ MDS code is GDRS.

(b) ([1638]) *Every $(q+1)$-arc in $\mathrm{PG}(3,q)$, $q > 3$ odd, and every $(q+1)$-arc in $\mathrm{PG}(q-4,q)$, $q > 3$ odd, is equal to a normal rational curve.*

    For $q > 3$ odd, every $[q+1,4,q-2]_q$ MDS code and every $[q+1,q-3,5]_q$ MDS code is GDRS.

(c) ([364]) *Every $(q+1)$-arc in $\mathrm{PG}(4,q)$, $q = 2^h$, $h \geq 3$, and every $(q+1)$-arc in $\mathrm{PG}(q-5,q)$, $q = 2^h$, $h \geq 3$, is equal to a normal rational curve.*

    For $q = 2^h$, $h \geq 3$, every $[q+1,5,q-3]_q$ MDS code and every $[q+1,q-4,6]_q$ MDS code is GDRS.

The only known example of a $(q+1)$-arc in $\mathrm{PG}(n,q)$, q odd, not equal to a normal rational curve is a 10-arc in $\mathrm{PG}(4,9)$, discovered by Glynn.

Theorem 14.2.10 (Glynn [811]) *Every 10-arc in $\mathrm{PG}(4,9)$ is projectively equivalent to a normal rational curve or to the 10-arc*

$$L = \{(1,t,t^2 + \eta t^6, t^3, t^4) \mid t \in \mathbb{F}_9\} \cup \{(0,0,0,0,1)\},$$

where $\eta \in \mathbb{F}_9$ with $\eta^4 = -1$.

We now present an important result on MDS codes over finite fields $\mathbb{F}_q$ of even order, and immediately discuss the geometrically equivalent result.

Theorem 14.2.11

(a) *For q even, there exist $[q+2,3,q]_q$ MDS codes.*

(b) *In $\mathrm{PG}(2,q)$, q even, there exist $(q+2)$-arcs.*

Definition 14.2.12 A **hyperoval** is a $(q+2)$-arc in $\mathrm{PG}(2,q)$, q even.

Definition 14.2.13 A **permutation polynomial** F over the finite field $\mathbb{F}_q$ is a polynomial for which the mapping $x \mapsto F(x)$, $x \in \mathbb{F}_q$, is a bijection over the finite field $\mathbb{F}_q$.

Theorem 14.2.14 (Segre [1640]) *Any $(q+2)$-arc of $\mathrm{PG}(2,q)$, $q = 2^h$ and $h > 1$, is projectively equivalent to a $(q+2)$-arc $\{(1,t,F(t)) \mid t \in \mathbb{F}_q\} \cup \{(0,1,0),(0,0,1)\}$, where F is a permutation polynomial over $\mathbb{F}_q$ of degree at most $q-2$ satisfying $F(0) = 0$, $F(1) = 1$, and such that for each s in $\mathbb{F}_q$*

$$F_s(X) = \frac{F(X+s) + F(s)}{X}$$

is a permutation polynomial over $\mathbb{F}_q$ satisfying $F_s(0) = 0$.

All polynomials $F(X)$ of this type are called **o-polynomials**. Examples of o-polynomials can be found in Section 20.4.15.

Remark 14.2.15 Let q be even. The classical example of a hyperoval in $\mathrm{PG}(2,q)$ is the *regular hyperoval*, which is defined in the following way. Every conic C in $\mathrm{PG}(2,q)$, q even, has a **nucleus** N. The nucleus N of a conic C is the unique point belonging to all the lines sharing one point with this conic C in $\mathrm{PG}(2,q)$. The union of a conic C and its nucleus N forms a hyperoval, called a **regular hyperoval**. In Theorem 14.2.14, $F(t) = t^2$ is an example of an o-polynomial defining a regular hyperoval.

Other infinite classes of hyperovals are known: the **translation** hyperovals, the **Segre** hyperoval, the two **Glynn** hyperovals, the **Payne** hyperoval, the **Cherowitzo** hyperoval, the **Subiaco** hyperovals, and also the **Adelaide** hyperoval. We refer to [961] for the corresponding o-polynomials. There is also one sporadic hyperoval in $\mathrm{PG}(2,32)$ known [1463].

Remark 14.2.16 The MDS Conjecture on linear MDS codes states that the maximal length n for a linear $[n,k,n-k+1]_q$ MDS code, with $2 \leq k \leq q-1$, is at most $q+1$, up to two exceptional cases. This MDS Conjecture is summarized in Table 14.1; see also Conjecture 3.3.21.

TABLE 14.1: The MDS Conjecture for linear $[n,k,n-k+1]_q$ MDS codes

Conditions	$n \leq$
$k \geq q$	$k+1$
$k \in \{3, q-1\}$ and q even	$q+2$
in all other cases	$q+1$

The next theorem, by S. Ball, proves the MDS Conjecture on linear MDS codes over prime fields, and is one of the nicest contributions of Galois geometries to coding theory. Because of its great relevance, we state the result both in the language of MDS codes and in the language of arcs.

Theorem 14.2.17 (Ball [104]) *Let q be an odd prime.*

(a) *Every $[q+1, k, q-k+2]_q$ MDS code is GDRS.*

(b) *Every $(q+1)$-arc in $\mathrm{PG}(k-1, q)$, $2 \le k \le q-1$, is a normal rational curve.*

We end this section with the complete classification of the $(q+1)$-arcs in $\mathrm{PG}(3, q)$, $q = 2^h$, $h \ge 3$.

Theorem 14.2.18 (Casse–Glynn [363]) *For $q = 2^h$, $h \ge 3$, every $[q+1, 4, q-2]_q$ MDS code arises from a $(q+1)$-arc of $\mathrm{PG}(3, q)$, projectively equivalent to a $(q+1)$-arc*

$$L_e = \{(1, t, t^e, t^{e+1}) \mid t \in \mathbb{F}_q\} \cup \{(0, 0, 0, 1)\},$$

where $e = 2^v$ and $\gcd(v, h) = 1$.

14.2.4   Griesmer Bound and Minihypers

The Griesmer Bound (Theorem 1.9.18) states that for every linear $[n, k, d]_q$ code,

$$n \ge \sum_{i=0}^{k-1} \left\lceil \frac{d}{q^i} \right\rceil = g_q(k, d).$$

Definition 14.2.19 Linear $[g_q(k, d), k, d]_q$ codes, i.e., meeting the Griesmer Bound, are called **Griesmer codes**.

We now present *minihypers*, a geometric structure linked to Griesmer codes.

Definition 14.2.20 An $(f, m; N, q)$-**minihyper** (F, w) is a set of points F of the projective space $\mathrm{PG}(N, q)$, with a weight function $w : \mathrm{PG}(N, q) \to \mathbb{N}$ where $P \mapsto w(P)$ satisfies the conditions:

1. $w(P) > 0$ if and only if $P \in F$,

2. $\sum_{P \in \mathrm{PG}(N,q)} w(P) = f$, and

3. $\min_{H \in \mathcal{H}} \{ \sum_{P \in H} w(P) \} = m$, with $\mathcal{H}$ the set of all the hyperplanes of $\mathrm{PG}(N, q)$.

Remark 14.2.21 An $(f, m; N, q)$-minihyper (F, w) is also known in the literature on Galois geometries under the name of **weighted m-fold blocking set of size f with respect to the hyperplanes of $\mathrm{PG}(N, q)$**.

The link between minihypers in $\mathrm{PG}(k-1, q)$ and linear $[g_q(k, d), k, d]_q$ Griesmer codes begins with the so-called *q-ary expansion* of d. Choose the unique positive integer s such that $(s-1)q^{k-1} < d \le sq^{k-1}$. Then there is a unique q-ary expansion $d = sq^{k-1} - \sum_{i=1}^{h} q^{\lambda_i}$ where

- $0 \le \lambda_1 \le \cdots \le \lambda_h < k-1$, and

- at most $q-1$ of the integers λ_i are equal to any given value.

Using this expression for d, the Griesmer Bound for a linear $[n, k, d]_q$ code is

$$n \geq s\theta_{k-1} - \sum_{i=1}^{h} \theta_{\lambda_i},$$

where $\theta_n = \frac{q^{n+1}-1}{q-1}$.

When $d = sq^{k-1} - \sum_{i=1}^{h} q^{\lambda_i}$, there is a one-to-one correspondence between the set of all inequivalent $[g_q(k, d), k, d]_q$ Griesmer codes and the set of all projectively distinct $\left(\sum_{i=1}^{h} \theta_{\lambda_i}, \sum_{i=1}^{h} \theta_{\lambda_i-1}; k-1, q\right)$-minihypers (F, w), where the maximum weight for a point P in $PG(k-1, q)$ is equal to s.

Belov, Logačev, and Sandimirov [157] gave a general construction method for Griesmer codes. We describe this construction method via the minihypers.

Construction 14.2.22 (Belov–Logačev–Sandimirov) Consider in $PG(k-1, q)$ a multiset of ϵ_0 points, ϵ_1 lines, ϵ_2 planes, ϵ_3 solids, $\ldots$, ϵ_{k-2} $(k-2)$-spaces, with $0 \leq \epsilon_i \leq q-1$, $i = 0, \ldots, k-2$. Then such a multiset defines a $\left(\sum_{i=0}^{k-2} \epsilon_i\theta_i, \sum_{i=0}^{k-2} \epsilon_i\theta_{i-1}; k-1, q\right)$-minihyper (F, w), where the weight $w(R)$ of a point R of $PG(k-1, q)$ equals the number of objects in the description of the multiset above in which it is contained.

One of the main contributions of finite geometries to the problem of linear codes meeting the Griesmer Bound is the characterization of Griesmer codes via the characterization of the corresponding minihypers. We mention two particular such characterization results.

Theorem 14.2.23 ([495]) *A $\left(\sum_{i=0}^{k-2} \epsilon_i\theta_i, \sum_{i=0}^{k-2} \epsilon_i\theta_{i-1}; k-1, q\right)$-minihyper (F, w), where $\sum_{i=0}^{k-2} \epsilon_i < \sqrt{q}+1$, is a multiset of ϵ_0 points, ϵ_1 lines, $\ldots$, ϵ_{k-3} $(k-3)$-spaces, ϵ_{k-2} hyperplanes, and so is of Belov–Logačev–Sandimirov type.*

Theorem 14.2.24 ([1199]) *An $\left(x\theta_{k-2}, x\theta_{k-3}; k-1, q\right)$-minihyper (F, w), $q = p^h$, p prime, $h \geq 1$, with $x \leq q - q/p$, is a multiset of x hyperplanes, and so is of Belov–Logačev–Sandimirov type.*

14.3 Projective Reed–Muller Codes

In Section 1.11, the binary Reed–Muller codes are investigated; Reed–Muller codes generalized to other fields are discussed in Chapters 2 and 16. In this section, we discuss related q-ary projective Reed–Muller codes.

Definition 14.3.1 Let $\mathcal{F}_d$ be the set of the homogeneous polynomials of degree d over the finite field $\mathbb{F}_q$ in the variables $X_0, \ldots, X_n$. An **algebraic hypersurface Φ of degree d in $PG(n, q)$** is the set of points of $PG(n, q)$ whose coordinates satisfy a homogeneous polynomial f in $\mathcal{F}_d$, i.e., $\Phi = \{P \in PG(n, q) \mid f(P) = 0\}$ for given $f \in \mathcal{F}_d$.

Definition 14.3.2 Let $PG(n, q) = \{P_0, \ldots, P_{\theta_n-1}\}$, where we normalize the coordinates of the points P_i by making the leftmost nonzero coordinate equal to one. Then the dth **order q-ary projective Reed–Muller code** $\mathcal{PRM}_q(d, n)$ is

$$\mathcal{PRM}_q(d, n) = \left\{\left(f(P_0), \ldots, f(P_{\theta_n-1})\right) \mid f \in \mathcal{F}_d \cup \{0\}\right\}.$$

We now present a geometric link between codewords of $\mathcal{PRM}_q(d,n)$ and algebraic hypersurfaces of degree d in $\mathrm{PG}(n,q)$. The weight of a codeword is the number of nonzero positions in this codeword. A point P_i of $\mathrm{PG}(n,q)$ belongs to an algebraic hypersurface Φ if and only if its coordinates satisfy the equation $f(X_0,\ldots,X_n) = 0$ of Φ. Hence, $P_i \in \Phi$ if and only if $f(P_i) = 0$. Since a nonzero codeword of minimum weight of $\mathcal{PRM}_q(d,n)$ is in fact a nonzero codeword of $\mathcal{PRM}_q(d,n)$ with the largest number of zeroes, the preceding equivalence leads to the following theorem.

Theorem 14.3.3 *The nonzero codewords of minimum weight of $\mathcal{PRM}_q(d,n)$ correspond to the algebraic hypersurfaces of degree d having the largest number of points in $\mathrm{PG}(n,q)$.*

In the context of projective Reed–Muller codes, finite geometries contributed to the characterization of algebraic hypersurfaces of the largest size. As an example, we present the following results.

Theorem 14.3.4

(a) **(Serre [1654])** *The minimum weight codewords of the code $\mathcal{PRM}_q(d,n)$, for $d \leq q - 1$, are defined by the algebraic hypersurfaces of degree d which are the union of d hyperplanes, passing through a common $(n-2)$-space of $\mathrm{PG}(n,q)$. So $d_H(\mathcal{PRM}_q(d,n)) = q^n - (d-1)q^{n-1}$ for $d \leq q-1$.*

(b) **(Sørensen [1741])** *Let $d - 1 = r(q-1) + s$, with $0 \leq s < q-1$. For $d \leq n(q-1)$, $d_H(\mathcal{PRM}_q(d,n)) = (q-s)q^{n-r-1}$.*

Theorem 14.3.5 (Sboui [1626])

(a) *The codewords of the second smallest nonzero weight of the code $\mathcal{PRM}_q(d,n)$, with $5 \leq d \leq \frac{q}{3} + 2$, are defined by the algebraic hypersurfaces $\mathcal{A}_2^d$ of degree d which are the union of d hyperplanes, $d-1$ of which meet in a common $(n-2)$-space Π_{n-2} and with the dth hyperplane not passing through this common $(n-2)$-space Π_{n-2}. This second smallest nonzero weight is equal to $q^n - (d-1)q^{n-1} + (d-2)q^{n-2}$.*

(b) *The codewords of the third smallest nonzero weight of the code $\mathcal{PRM}_q(d,n)$, with $5 \leq d \leq \frac{q}{3} + 2$, are defined by the algebraic hypersurfaces $\mathcal{A}_3^d$ of degree d which are the union of d hyperplanes, $d-2$ of which meet in a common $(n-2)$-space K_1, and where the last two hyperplanes H_{d-1} and H_d meet in an $(n-2)$-space K_2, different from K_1, such that K_2 is contained in exactly one of the $d-2$ hyperplanes passing through K_1. This third smallest nonzero weight is equal to $q^n - (d-1)q^{n-1} + 2(d-3)q^{n-2}$.*

Theorem 14.3.6 (Bartoli–Sboui–Storme [133]) *Let c be a nonzero codeword of the code $\mathcal{PRM}_q(d,n)$, where $d < \sqrt[3]{q}$, satisfying*

$$\mathrm{wt_H}(c) < \begin{cases} q^n - \left(\frac{r+d-4}{2}\right)q^{n-1} - ((d-r+1)^2 + d - 1 + 2\sqrt{q})q^{n-2} \\ \quad - (2d - r + 2\sqrt{q})q^{n-3} - \left(d-1+2\sqrt{q}\right)\left(\frac{q^{n-3}-1}{q-1}\right), \\ \quad \textit{when } d - r + 1 \textit{ is odd}, \\[2mm] q^n - \left(\frac{d+r-3}{2}\right)q^{n-1} - \left(\frac{(d-r+1)^2}{2} + d\right)q^{n-2} \\ \quad - \left(\frac{3d-r+1}{2}\right)q^{n-3} - d\left(\frac{q^{n-3}-1}{q-1}\right) + \frac{r+d}{2}, \\ \quad \textit{when } d - r + 1 \textit{ is even}. \end{cases}$$

Then c corresponds to an algebraic hypersurface of degree d in $\mathrm{PG}(n,q)$ containing at least r hyperplanes defined over $\mathbb{F}_q$.

14.4 Linear Codes Defined by Incidence Matrices Arising from Galois Geometries

Finite projective spaces led to particular linear codes. This was motivated by the fact that finite projective spaces are well-studied structures. A lot of properties of finite projective spaces are known. This makes it realistic that linear codes defined by finite projective spaces can be investigated in detail and can have interesting properties.

Definition 14.4.1 Let $s < t$. Consider an ordering $\Pi_i^{(s)}$, $i = 1, \ldots, \begin{bmatrix} n+1 \\ s+1 \end{bmatrix}_q$, of the s-spaces of $PG(n, q)$ and an ordering $\Pi_j^{(t)}$, $j = 1, \ldots, \begin{bmatrix} n+1 \\ t+1 \end{bmatrix}_q$, of the t-spaces of $PG(n, q)$. The **incidence matrix** $G_{s,t}$ of the s-spaces and t-spaces of $PG(n, q)$ is the matrix whose rows correspond to the t-spaces and whose columns correspond to the s-spaces of $PG(n, q)$, and where

$$(G_{s,t})_{i,j} = \begin{cases} 1 & \text{if } \Pi_j^{(s)} \subseteq \Pi_i^{(t)}, \\ 0 & \text{if } \Pi_j^{(s)} \not\subseteq \Pi_i^{(t)}. \end{cases}$$

Definition 14.4.2 Let $q = p^h$, p prime, $h \geq 1$, and let $s < t$. The code $\mathcal{C}_{s,t}(n, q)$ is the linear code over the prime field $\mathbb{F}_p$ generated by the incidence matrix $G_{s,t}$ of $PG(n, q)$.

Definition 14.4.3 Let $s < t$. Let $\Delta_{s,t}$ denote the **incidence system** whose points and blocks are the s-spaces and t-spaces in $PG(n, q)$, respectively, and the incidence is inclusion. This means that a block consists of all the s-spaces of $PG(n, q)$ contained in a fixed t-space of $PG(n, q)$.

The fundamental parameters of the linear codes $\mathcal{C}_{s,t}(n, q)$ have been determined.

Theorem 14.4.4

(a) *The length of $\mathcal{C}_{s,t}(n, q)$ is $\begin{bmatrix} n+1 \\ s+1 \end{bmatrix}_q$.*

(b) *([1024]) The dimension of the p-ary code $\mathcal{C}_{0,t}(n, q)$, $q = p^h$, p prime, $h \geq 1$, is*

$$1 + \sum_{i=1}^{n-t} \sum_{\substack{1 \leq r_1, \ldots, r_{h-1} \leq n-t \\ r_0 = r_h = i}} \prod_{j=0}^{h-1} \sum_{s=0}^{r_{j+1}-1} (-1)^s \binom{n+1}{s} \binom{n+pr_{j+1} - r_j - ps}{n}.$$

(c) *([818]) The dimension of the p-ary code $\mathcal{C}_{0,n-1}(n, q)$, $q = p^h$, p prime, $h \geq 1$, is*

$$\binom{n+p-1}{n}^h + 1.$$

Intensive research has been done in Galois geometries to characterize small weight codewords in the codes arising from the incidence matrices of finite projective spaces.

Theorem 14.4.5 (Bagchi–Inamdar [92]) *The minimum weight $d_H(\mathcal{C}_{s,t}(n, q))$ of the p-ary code $\mathcal{C}_{s,t}(n, q)$ is $\begin{bmatrix} t+1 \\ s+1 \end{bmatrix}_q$, and the minimum weight codewords are the scalar multiples of the incidence vectors of the blocks of $\Delta_{s,t}$.*

Theorem 14.4.6

(a) ([71]) *The minimum weight of the code* $C_{0,t}(n,q)$ *is* θ_t *and the minimum weight code-words are the scalar multiples of the incidence vectors of the* t*-spaces of* $\mathrm{PG}(n,q)$.

(b) ([1533]) *The second smallest nonzero weight of the code* $C_{0,n-1}(n,q)$ *is* $2q^{n-1}$ *and every codeword of* $C_{0,n-1}(n,q)$ *of weight* $2q^{n-1}$ *is a scalar multiple of the difference of two hyperplanes of* $\mathrm{PG}(n,q)$.

Remark 14.4.7 The code $C_{0,1}(2,p)$, p prime, contains a particular codeword c of weight $3p - 3$, which is not a linear combination of three lines; see [91]. Let (X, Y, Z) be the coordinates for the points of $\mathrm{PG}(2,p)$. For a point Q of $\mathrm{PG}(2,p)$, we denote the coordinate position of this point Q in a codeword c by c_Q.

Consider the three lines $\ell_1 : X = 0$, $\ell_2 : Y = 0$, and $\ell_3 : X = Y$ through the point $P(0,0,1)$. Consider also the line $m : Z = 0$ not through the point P. The support of the particular codeword c will consist of the points of $(\bigcup_{i=1}^{3} \ell_i) \setminus (m \cup \{P\})$. For this particular codeword c, the coordinate values are

$$c_Q = \begin{cases} t & \text{if } Q = (0,1,t) \text{ or } Q = (1,0,t), \ t \in \mathbb{F}_q, \\ -t & \text{if } Q = (1,1,t), \ t \in \mathbb{F}_q, \\ 0 & \text{for all the other points of } \mathrm{PG}(2,p). \end{cases}$$

Then c is a codeword of the code $C_{0,1}(2,p)$, p prime, of weight $3p - 3$. Up to projective equivalence, such a particular codeword is called a **special codeword**. Let c be such a special codeword, with its support contained in the lines ℓ_1, ℓ_2, ℓ_3. Then we call a codeword of the form $c + \sum_{i=1}^{3} \alpha_i \ell_i$, with $\alpha_i \in \mathbb{F}_p$, an **extended special codeword**.

Theorem 14.4.8 (Szőnyi–Weiner [1767]) *Let p be prime, $p \geq 19$. Let c be a codeword of the p-ary code $C_{0,1}(2,p)$ of weight* $\mathrm{wt_H}(c) \leq \max\{3q + 1, 4q - 22\}$. *Then this codeword c is a linear combination of at most three lines or an extended special codeword.*

Theorem 14.4.9 (Szőnyi–Weiner [1767])

(a) *Let c be a codeword of the p-ary code $C_{0,1}(2,q)$ with $q = p^h$, p prime, $h \geq 1$, and $q > 17$. If* $\mathrm{wt_H}(c) < \sqrt{\frac{q}{2}}(q+1)$, *then* $\mathrm{supp}(c)$ *can be covered by* $\lceil \frac{\mathrm{wt_H}(c)}{q+1} \rceil$ *lines.*

(b) *Let c be a codeword of the p-ary code $C_{0,1}(2,q)$ with $q = p^h$, p prime, $h > 2$, and $q > 27$. If* $\mathrm{wt_H}(c) < (\sqrt{q}+1)(q+1-\sqrt{q})$, *then c is a linear combination of exactly* $\lceil \frac{\mathrm{wt_H}(c)}{q+1} \rceil$ *lines.*

(c) *Let c be a codeword of the p-ary code $C_{0,1}(2,q)$ with $q = p^2$, p prime, and $q > 27$. If* $\mathrm{wt_H}(c) < \frac{(p-1)(p-4)(p^2+1)}{2p-1}$, *then c is a linear combination of exactly* $\lceil \frac{\mathrm{wt_H}(c)}{q+1} \rceil$ *lines.*

Remark 14.4.10 Table 14.2 presents known values and bounds on the minimum distance of linear codes arising from incidence matrices of projective spaces $\mathrm{PG}(n,q)$, $q = p^h$, p prime, $h \geq 1$. We refer the reader to [1208] for the exact references to the results in this table.

TABLE 14.2: Known values and bounds on the minimum distance of linear codes from projective spaces $\mathrm{PG}(n, q)$, $q = p^h$, p prime, $h \geq 1$

Code	Minimum distance d
$\mathcal{C}_{s,t}(n, q)$	$\begin{bmatrix} t+1 \\ s+1 \end{bmatrix}_q$
$\mathcal{C}_{0,t}(n, q)$	θ_t
$\mathcal{C}_{s,t}(n, q)^{\perp}$	$2\left(\frac{q^{n-s}-1}{q^{t-s}-1}\left(1 - \frac{1}{p}\right) + \frac{1}{p} \right) \leq d \leq 2q^{n-t}$
$\mathcal{C}_{s,s+1}(n, p)^{\perp}$	$2p^{n-s-1}$
$\mathcal{C}_{0,1}(2, q)^{\perp}$	$q + p \leq d$
$\mathcal{C}_{0,1}(2, q)^{\perp}$, $p > 2$	$4q/3 + 2 \leq d$
$\mathcal{C}_{0,t}(n, q)^{\perp}$, $p > 2$	$(4\theta_{n-t} + 2)/3 \leq d$
$\mathcal{C}_{0,t}(n, q)^{\perp}$	$(q + p)q^{n-t-1} \leq d \leq 2q^{n-t}$, $d = d_H(\mathcal{C}_{0,1}(n - t + 1, q)^{\perp})$
$\mathcal{C}_{0,t}(n, q)^{\perp}$, $p = 2$	$(q + 2)q^{n-t-1}$
$\mathcal{C}_{0,t}(n, p)^{\perp}$	$2p^{n-t}$
$\mathcal{C}_{0,t}(n, q)^{\perp}$	$d \leq 2q^{n-t} - q^{n-t-1}\frac{q-p}{p-1}$
$\mathcal{C}_{0,n-1}(n, q) \cap \mathcal{C}_{0,n-1}(n, q)^{\perp}$	$2q^{n-1}$

14.5 Subspace Codes and Galois Geometries

In [1156], subspace codes are defined; they are the subject of Section 29.5.1. Their relevance follows from the fact that they enable efficient transmission of information through wireless networks. These subspace codes are in fact sets of subspaces in a finite projective space. This implies that a lot of geometric methods can be used to investigate these subspace codes. This intensified the close relation between Galois geometries and coding theory.

14.5.1 Definitions

Definition 14.5.1 A **subspace code** $\mathcal{C}$ is a set of subspaces of a projective space $\mathrm{PG}(n - 1, q)$.

Definition 14.5.2

(a) The **subspace distance** $d_S(U, V)$ between two subspaces U and V of $\mathrm{PG}(n - 1, q)$ is equal to
$$d_S(U, V) = \dim U + \dim V - 2\dim(U \cap V).$$

(b) The **minimum subspace distance** $d_S(\mathcal{C})$ of a subspace code $\mathcal{C}$ is
$$d_S(\mathcal{C}) = \min\{d_S(U, V) \mid U, V \in \mathcal{C}, U \neq V\}.$$

Definition 14.5.3

(a) A **constant-dimension code**, also called a **Grassmannian code**, is a set of subspaces of the same dimension, i.e., $(k - 1)$-spaces of $\mathrm{PG}(n - 1, q)$.

(b) The minimum subspace distance $d_S(\mathcal{C})$ of a constant-dimension code is always even. We denote this by $d_S(\mathcal{C}) = 2\delta$. An $(n, 2\delta, k)_q$ constant-dimension code is a constant-dimension code of $(k-1)$-spaces in $\mathrm{PG}(n-1, q)$ with minimum subspace distance 2δ.

(c) The parameter $\mathcal{A}_q(n, 2\delta, k)$ denotes the largest number of codewords in an $(n, 2\delta, k)_q$ constant-dimension code.[1]

Theorem 14.5.4 (Johnson Bound [695])

(a) $\displaystyle \mathcal{A}_q(n, 2\delta, k) \leq \frac{\begin{bmatrix} n \\ k - \delta + 1 \end{bmatrix}_q}{\begin{bmatrix} k \\ k - \delta + 1 \end{bmatrix}_q}.$

(b) $\displaystyle \mathcal{A}_q(n, 2\delta, k) \leq \left\lfloor \frac{q^n - 1}{q^k - 1} \mathcal{A}_q(n - 1, 2\delta, k - 1) \right\rfloor.$

(c) $\displaystyle \mathcal{A}_q(n, 2\delta, k) \leq \left\lfloor \frac{q^n - 1}{q^k - 1} \left\lfloor \frac{q^{n-1} - 1}{q^{k-1} - 1} \cdots \left\lfloor \frac{q^{n-k+\delta} - 1}{q^\delta - 1} \right\rfloor \cdots \right\rfloor \right\rfloor.$

We now present a lower bound on $\mathcal{A}_q(n, 2\delta, k)$. This lower bound follows from the theory of rank-metric codes and the lifting procedure, described in Section 14.5.3, where we also state this lower bound in Theorem 14.5.22. We refer to Section 14.5.3 and to Theorem 14.5.22 for more detailed information on this lower bound.

Theorem 14.5.5 $\mathcal{A}_q(n, 2\delta, k) \geq q^{(n-k)(k-\delta+1)}$.

An $(n, 2k, k)_q$ subspace code is a set of $(k-1)$-spaces in $\mathrm{PG}(n-1, q)$ which are pairwise disjoint. Hence, the codewords of an $(n, 2k, k)_q$ code form a partial $(k-1)$-spread in $\mathrm{PG}(n-1, q)$. This leads to the following result linking the Johnson Bound for $(n, 2k, k)_q$ codes to the problem of finding the largest partial $(k-1)$-spreads in $\mathrm{PG}(n-1, q)$.

Theorem 14.5.6 *The value* $\mathcal{A}_q(n, 2k, k)$ *equals the size of the largest partial* $(k-1)$*-spread in* $\mathrm{PG}(n-1, q)$.

The preceding theorem has motivated a lot of recent research on partial $(k-1)$-spreads [668, 1179, 1415], mentioned in Section 14.1.2. It is another example of how the two research areas coding theory and Galois geometries interact with and influence each other.

---

[1]Chapter 29 examines subspace codes from the non-projective point of view. In Section 29.5.2, the subspace distance between non-projective spaces U and V is also defined as $d_S(U, V) = \dim U + \dim V - 2\dim(U \cap V)$. As the projective dimension of a subspace is 1 less than the non-projective dimension, the value of $d_S(U, V)$ is the same in either point of view. In Section 29.5.2, another distance, the injection distance $d_I(\cdot, \cdot)$ is defined. If U and V are of equal dimension, in either perspective, $d_S(U, V) = 2d_I(U, V)$ implying $d_S(\mathcal{C}) = 2d_I(\mathcal{C})$ for constant-dimension codes $\mathcal{C}$. In Section 29.6, $A_q(n, d, k)$ denotes the largest constant-dimension code in $\mathbb{F}_q^n$ where the codewords have dimension k and the code has minimum *injection* distance d. Therefore $\mathcal{A}_q(n, 2\delta, k) = A_q(n, \delta, k)$ and the Johnson Bound of Theorem 14.5.4(b) and Theorem 29.6.5 are identical. The reader might compare the lower bound of Theorem 14.5.5 to the upper bound of (29.11) in Chapter 29. (Note that the notation $(n, 2\delta, k)_q$ used in this section is the same as $(n, d, k)_q$ with $d = \delta$ in Section 29.6.)

14.5.2 Designs over $\mathbb{F}_q$

The study of the largest possible sizes for subspace codes led to a new link with the theory of designs in combinatorics.

Definition 14.5.7 A t-$(n, k, \lambda)_q$ **design** is a collection $\mathcal{B}$ of $(k-1)$-spaces of $PG(n-1, q)$, called **blocks**, such that each $(t-1)$-space of $PG(n-1, q)$ is contained in exactly λ blocks. If $\mathcal{B}$ consists of all the $(k-1)$-spaces of $PG(n-1, q)$, then $\mathcal{B}$ is called **trivial**.

Definition 14.5.8 A q-**Steiner system** is a t-$(n, k, 1)_q$ design.

The interest in coding theory for q-Steiner systems arises from the following link between q-Steiner systems and the Johnson Bound.

An $(n, 2(k-t+1), k)_q$ subspace code $\mathcal{C}$ is a set of $(k-1)$-subspaces of $PG(n-1, q)$, such that each $(t-1)$-space of $PG(n-1, q)$ is contained in at most one codeword of $\mathcal{C}$. This geometric property leads to the upper bound $\begin{bmatrix} n \\ t \end{bmatrix}_q \Big/ \begin{bmatrix} k \\ t \end{bmatrix}_q$ on the size of an $(n, 2(k-t+1), k)_q$ subspace code $\mathcal{C}$, stated as the Johnson Bound in Theorem 14.5.4(a). The counting arguments leading to this upper bound also show that equality in this upper bound occurs if and only if each $(t-1)$-space of $PG(n-1, q)$ is contained in exactly one codeword of $\mathcal{C}$. Hence, when this upper bound is attained, then the corresponding $(n, 2(k-t+1), k)_q$ subspace code $\mathcal{C}$ is in fact a t-$(n, k, 1)_q$ q-Steiner system. This leads to the following theorem.

Theorem 14.5.9 *A* t-$(n, k, 1)_q$ q-*Steiner system exists if and only if there exists an* $(n, 2(k-t+1), k)_q$ *subspace code of size* $\begin{bmatrix} n \\ t \end{bmatrix}_q \Big/ \begin{bmatrix} k \\ t \end{bmatrix}_q$ *, i.e., if and only if* $\mathcal{A}_q(n, 2(k-t+1), k) = \begin{bmatrix} n \\ t \end{bmatrix}_q \Big/ \begin{bmatrix} k \\ t \end{bmatrix}_q$.

As a particular application of this theorem, for $t = 1$, the preceding theorem implies that every projective point of $PG(n-1, q)$ belongs to exactly one codeword of this $(n, 2k, k)_q$ subspace code $\mathcal{C}$. Hence, the $(k-1)$-spaces, which form the codewords of this subspace code $\mathcal{C}$, form a $(k-1)$-spread of $PG(n-1, q)$, and this leads to the following theorem, also related to Theorem 14.5.6.

Theorem 14.5.10 *A* 1-$(n, k, 1)_q$ q-*Steiner system exists if and only if a* $(k-1)$-*spread in* $PG(n-1, q)$ *exists, i.e., if and only if* k *divides* n.

A main research question was if a nontrivial q-Steiner system exists. A major breakthrough in the theory of q-Steiner systems, and in the theory of subspace codes, was obtained when indeed the first example of a nontrivial q-Steiner system was found.

Theorem 14.5.11 ([295]) *There exists a* 2-$(13, 3, 1)_2$ *Steiner system; i.e.,* $\mathcal{A}_2(13, 4, 3) = \begin{bmatrix} 13 \\ 2 \end{bmatrix}_2 \Big/ \begin{bmatrix} 3 \\ 2 \end{bmatrix}_2 = 1\,597\,245$.

Designs arising from projective space are also examined in Section 5.6.

14.5.3 Rank-Metric Codes

We now present results on rank-metric codes. Rank-metric codes are the subject of Chapter 11; the codes described here are called (matrix) rank-metric codes in Chapter 11. Via the lifting procedure discussed in Definition 14.5.21, rank-metric codes lead to important classes of subspace codes. The second motivation for discussing rank-metric codes is the many connections to finite geometry. These will be discussed in the sections that follow.

Definition 14.5.12 Let $\mathbb{F}_q^{m \times n}$ be the set of $m \times n$ matrices over $\mathbb{F}_q$. For $A, B \in \mathbb{F}_q^{m \times n}$, the **rank distance** between the matrices A and B is $d_R(A, B) = \text{rank}(A - B)$.

Definition 14.5.13 An $[m \times n, k, d]_q$ **rank-metric code** $\mathcal{C}$ is a linear subspace of $\mathbb{F}_q^{m \times n}$ with $\mathbb{F}_q$-dimension k such that $d_R(A, B) \geq d$ for every two distinct codewords A and B of $\mathcal{C}$ and $d_R(A, B) = d$ for some two codewords A and B of $\mathcal{C}$.

Similar to $[n, k, d]_q$ linear codes, there exists a Singleton Bound for rank-metric codes; see also Theorem 11.3.5.

Theorem 14.5.14 (Singleton Bound) *For an* $[m \times n, k, d]_q$ *rank-metric code* $\mathcal{C}$,

$$k \leq n(m - d + 1) \text{ if } m \leq n \quad \text{and} \quad k \leq m(n - d + 1) \text{ if } n \leq m.$$

Definition 14.5.15 An $[m \times n, k, d]_q$ **rank-metric code** $\mathcal{C}$ is called a **Maximum Rank Distance (MRD)** code if $k = n(m - d + 1)$ when $m \leq n$ and $k = m(n - d + 1)$ when $n \leq m$.

Definition 14.5.16 Consider the set $G_{n,n,k}$ of linearized polynomials

$$G_{n,n,k} = \{a_0 x + a_1 x^q + a_2 x^{q^2} + \cdots + a_{k-1} x^{q^{k-1}} \mid a_0, a_1, \ldots, a_{k-1} \in \mathbb{F}_{q^n}\}.$$

Let B be a basis of $\mathbb{F}_{q^n}$ with respect to $\mathbb{F}_q$. We can write an element $a \in \mathbb{F}_{q^n}$ as a column vector $v(a)$ in $\mathbb{F}_q^n$ with respect to this basis B. Let $\{\beta_1, \ldots, \beta_n\}$ be a set of n linearly independent elements of $\mathbb{F}_{q^n}$ with respect to $\mathbb{F}_q$; then the **Gabidulin code** $\mathcal{G}_{n,n,k}$ is the set of $n \times n$ matrices

$$\mathcal{G}_{n,n,k} = \{[\, v(f(\beta_1)) \quad \cdots \quad v(f(\beta_n)) \,]^\mathsf{T} \mid f \in G_{n,n,k}\}.$$

Theorem 14.5.17 (Gabidulin [760]) *The Gabidulin codes* $\mathcal{G}_{n,n,k}$ *are* $[n \times n, nk, n - k + 1]_q$ *MRD codes.*

Definition 14.5.18 A **punctured** Gabidulin code $\mathcal{G}_{m,n,k}$, $m < n$, is obtained by restricting the linearized mappings of $G_{n,n,k}$ to an m-dimensional $\mathbb{F}_q$-subspace of $\mathbb{F}_{q^n}$.

Theorem 14.5.19 (Gabidulin [475]) *The punctured Gabidulin codes* $\mathcal{G}_{m,n,k}$, $m < n$, *are* $[n \times m, nk, m - k + 1]_q$ *MRD codes.*

We now present other examples of MRD codes via their corresponding sets of linearized polynomials.

Theorem 14.5.20 *The following sets of linearized polynomials define* $[n \times n, nk, n - k + 1]_q$ *MRD codes.*

(a) ([1171]) $\overline{G}_{k,s} = \{a_0 x + a_1 x^{q^s} + \cdots + a_{k-1} x^{q^{s(k-1)}} \mid a_0, \ldots, a_{k-1} \in \mathbb{F}_{q^n}\}$, with n, k, s positive integers and $\gcd(n, s) = 1$, defines a **generalized Gabidulin code**.

(b) ([1667]) $H_k(\nu, h) = \{a_0 x + a_1 x^q + \cdots + a_{k-1} x^{q^{k-1}} + \nu a_0^{q^h} x^{q^k} \mid a_0, \ldots, a_{k-1} \in \mathbb{F}_{q^n}\}$, with n, k, h positive integers where $k < n$ and $\nu \in \mathbb{F}_{q^n}$ such that $\nu^{\frac{q^n - 1}{q - 1}} \neq (-1)^{nk}$, defines a **twisted Gabidulin code**.

(c) ([1303, 1667]) $\overline{H}_{k,s}(\nu, h) = \{a_0 x + a_1 x^{q^s} + \cdots + a_{k-1} x^{q^{s(k-1)}} + \nu a_0^{q^h} x^{q^{sk}} \mid a_0, \ldots, a_{k-1} \in \mathbb{F}_{q^n}\}$, with n, k, s, h positive integers where $\gcd(n, s) = 1$, $k < n$, and $\nu \in \mathbb{F}_{q^n}$ such that $\nu^{\frac{q^{ns} - 1}{q^s - 1}} \neq (-1)^{nk}$, defines a **generalized twisted Gabidulin code**.

We now present the lifting procedure which transforms rank-metric codes into subspace codes. One of the most important consequences of this lifting procedure is the lower bound on the Johnson Bound, stated in Theorem 14.5.22, also mentioned in Theorem 14.5.5.

Definition 14.5.21 (Lifting Procedure [1711]) Let $m \leq n$. An $m \times n$ matrix A over $\mathbb{F}_q$ is **lifted** to the m-dimensional subspace of $\mathbb{F}_q^{m+n}$ whose generator matrix is $[\, I_m \;\; A \,]$, with I_m the $m \times m$ identity matrix. An $[m \times n, k, \delta]_q$ rank-metric code $\mathcal{C}$ is **lifted** to a subspace code $\mathcal{C}'$ of $(m-1)$-spaces in $\mathrm{PG}(m+n-1, q)$ by lifting all the codewords of $\mathcal{C}$, i.e., $\mathcal{C}' = \{\mathrm{rowspan}([\, I_m \;\; A \,]) \mid A \in \mathcal{C}\}$. This code $\mathcal{C}'$ is a subspace code with q^k codewords and minimum subspace distance 2δ. If a subspace code $\mathcal{C}'$ is constructed by lifting an MRD code $\mathcal{C}$, then $\mathcal{C}'$ is called a **Lifted MRD (LMRD)** code. (See Section 29.7.1 for more details.)

The known rank-metric codes, for instance mentioned in the preceding theorems, can now be used in this lifting procedure to construct subspace codes with specified minimum distance 2δ. This leads to the proof of Theorem 14.5.5, also stated in the next theorem.

Theorem 14.5.22 $A_q(n, 2\delta, k) \geq q^{(n-k)(k-\delta+1)}$.

14.5.4 Maximum Scattered Subspaces and MRD Codes

We now present a first link between finite geometry and MRD codes.

Definition 14.5.23 Let $\Lambda = \mathrm{PG}(r-1, q^n)$, $q = p^h$, p prime, $h \geq 1$, be defined by the r-dimensional vector space $V = V(r, q^n)$ over the finite field $\mathbb{F}_{q^n}$, and let L be a set of points of Λ. The set L is called an $\mathbb{F}_q$-**linear set** of Λ of rank k if it is defined by the nonzero vectors of an $\mathbb{F}_q$-vector subspace U of V of dimension k; i.e., $L = L_U = \{\langle \mathbf{u} \rangle_{\mathbb{F}_{q^n}} \mid \mathbf{u} \in U \setminus \{0\}\}$.

Remark 14.5.24 Note that the r-dimensional vector space V over the finite field $\mathbb{F}_{q^n}$ is isomorphic to the finite field $\mathbb{F}_{q^{rn}}$, considered as an r-dimensional vector space over the finite field $\mathbb{F}_{q^n}$. This is used in the next definition.

Definition 14.5.25 An $\mathbb{F}_q$-subspace U of $\mathbb{F}_{q^{rn}}$ is called **scattered** with respect to the finite field $\mathbb{F}_{q^n}$ if it defines a scattered $\mathbb{F}_q$-linear set in $\mathrm{PG}(r-1, q^n)$ defined by considering the finite field $\mathbb{F}_{q^{rn}}$ as an r-dimensional vector space over the finite field $\mathbb{F}_{q^n}$ where $\dim_{\mathbb{F}_q}(U \cap \langle \mathbf{x} \rangle_{\mathbb{F}_{q^n}}) \leq 1$ for each $\mathbf{x} \in \mathbb{F}_{q^{rn}} \setminus \{0\}$.

Definition 14.5.26 A scattered $\mathbb{F}_q$-linear set of Λ of the highest possible rank is called a **maximum scattered $\mathbb{F}_q$-linear set of Λ**.

Theorem 14.5.27 ([223]) *The rank of a scattered $\mathbb{F}_q$-linear set of $\mathrm{PG}(r-1, q^n)$, rn even, is at most $rn/2$.*

Theorem 14.5.28 ([474]) *Let U be an $rn/2$-dimensional $\mathbb{F}_q$-subspace of the r-dimensional vector space $V = V(r, q^n)$ over the finite field $\mathbb{F}_{q^n}$, where rn is even, and let $i = \max\{\dim_{\mathbb{F}_q}(U \cap \langle v \rangle_{\mathbb{F}_{q^n}}) \mid v \in V \setminus \{0\}\}$. Suppose $i < n$. For any $\mathbb{F}_q$-linear function $G : V \to W$, with $W = V(rn/2, q)$ such that $\ker G = U$, the pair (U, G) defines an $[(rn/2) \times n, rn, n-i]_q$ rank-metric code $\mathcal{C}_{U,G}$. Also, $\mathcal{C}_{U,G}$ is an MRD-code if and only if U is a maximum scattered $\mathbb{F}_q$-subspace with respect to $\mathbb{F}_{q^n}$.*

14.5.5   Semifields and MRD Codes

We now present links of MRD codes to algebraic structures, called quasifields and semi-fields [501].

Definition 14.5.29 A **finite (right) quasifield** $\mathbb{Q}$ is a finite algebra with at least two elements, and two binary operations $+$ and $\circ$, satisfying the following axioms:

1. $(\mathbb{Q}, +)$ is an abelian group with neutral element 0 which satisfies $0 \circ a = 0 = a \circ 0$ for all $a \in \mathbb{Q}$.

2. There exists an identity $e \in \mathbb{Q}$ such that $e \circ a = a \circ e = a$ for all $a \in \mathbb{Q}$.

3. For all $a, b \in \mathbb{Q}$ with $a \neq 0$ there exists exactly one $x \in \mathbb{Q}$ such that $a \circ x = b$.

4. For all $a, b, c \in \mathbb{Q}$ with $a \neq b$ there exists exactly one $x \in \mathbb{Q}$ such that $x \circ a = x \circ b + c$.

5. $(a + b) \circ c = a \circ c + b \circ c$ for all $a, b, c \in \mathbb{Q}$ (right distributivity).

The set $\ker(\mathbb{Q}) = \{c \in \mathbb{Q} \mid c \circ (a+b) = c \circ a + c \circ b, \text{ and } c \circ (a \circ b) = (c \circ a) \circ b \text{ for all } a, b \in \mathbb{Q}\}$ is called the **kernel** of the quasifield $\mathbb{Q}$.

Definition 14.5.30 A **finite semifield** $\mathbb{S}$ is a finite quasifield which also satisfies the left distributivity law $a \circ (b + c) = a \circ b + a \circ c$ for all $a, b, c \in \mathbb{S}$.

We identify the elements of a finite semifield, which is n-dimensional over some finite field $\mathbb{F}_q$, with the elements of the field extension $\mathbb{F}_{q^n}$.

Example 14.5.31 The **Dickson commutative semifields** $\mathbb{S}$ are defined in the following way. Consider the finite field $\mathbb{F}_{p^m}$ with p odd, $m > 1$, σ a nontrivial automorphism of $\mathbb{F}_{p^m}$ over $\mathbb{F}_p$, and f a non-square in $\mathbb{F}_{p^m}$. Then the Dickson semifield has elements $a + \lambda b$, $a, b \in \mathbb{F}_{p^m}$, where λ is a symbol not in $\mathbb{F}_{p^m}$. The addition in the Dickson semifield is componentwise and the multiplication is defined by

$$(a + \lambda b) \circ (c + \lambda d) = (ac + \sigma(b)\sigma(d)f) + \lambda(ad + bc).$$

Remark 14.5.32 For an element a of the semifield $\mathbb{S} = (\mathbb{F}_{q^n}, \circ)$, we can define the endo-morphism of left multiplication $M_a \in \mathrm{End}_{\mathbb{F}_q}(\mathbb{F}_{q^n})$ by $M_a(b) = a \circ b$. We then can define the **spread set** of the semifield $\mathbb{S}$ to be the set $\mathcal{C}(\mathbb{S}) = \{M_a \mid a \in \mathbb{S}\} \subset \mathrm{End}_{\mathbb{F}_q}(\mathbb{F}_{q^n})$. Since $\mathbb{S}$ is a finite-dimensional division algebra, every M_a, $a \in \mathbb{S} \setminus \{0\}$, is invertible. Moreover, $\mathcal{C}(\mathbb{S})$ is a $\mathbb{F}_q$-subspace of $\mathrm{End}_{\mathbb{F}_q}(\mathbb{F}_{q^n})$ since $M_{\lambda a + b} = \lambda M_a + M_b$, for every $\lambda \in \mathbb{F}_q$ and every $a, b \in \mathbb{F}_{q^n}$.

Theorem 14.5.33 ([501]) *MRD codes in $\mathbb{F}_q^{n \times n}$, which contain the zero and the identity matrix, with minimum distance n correspond to finite quasifields $\mathbb{Q}$ with $\mathbb{F}_q \leq \ker(\mathbb{Q})$ and $\dim_{\mathbb{F}_q}(\mathbb{Q}) = n$.*

Similarly, the following result is valid.

Theorem 14.5.34 ([501]) *Additively closed MRD codes in $\mathbb{F}_q^{n \times n}$, which contain the iden-tity matrix, with minimum distance n correspond to finite semifields $\mathbb{S}$ with $\mathbb{F}_q \leq \ker(\mathbb{S})$ and $\dim_{\mathbb{F}_q}(\mathbb{S}) = n$.*

14.5.6 Nonlinear MRD Codes

This section presents interesting links between MRD codes and classical substructures in finite geometry, called *Segre varieties*, and a substructure, called an *exterior set*, associated to these Segre varieties [448]. This again illustrates the many links between coding theory and finite geometry.

Definition 14.5.35 The **Segre mapping** $\sigma : \mathrm{PG}(n-1, q) \times \mathrm{PG}(n-1, q) \to \mathrm{PG}(n^2-1, q)$ maps two points $x = (x_1, x_2, \ldots, x_n)$, $y = (y_1, y_2, \ldots, y_n)$ of $\mathrm{PG}(n-1, q)$ to the point $(x_1 y_1, x_1 y_2, x_1 y_3, \ldots, x_n y_n)$ of $\mathrm{PG}(n^2-1, q)$. The image of this Segre mapping σ is called the **Segre variety** $S_{n-1, n-1}$.

When we order the n^2 coordinates of a point of $\mathrm{PG}(n^2-1, q)$ in an $n \times n$ matrix, then the points of $\mathrm{PG}(n^2-1, q)$ correspond to matrices from $\mathbb{F}_q^{n \times n}$. In particular, for the Segre variety $S_{n-1, n-1}$, we have

$$(x_1 y_1, x_1 y_2, x_1 y_3, \ldots, x_n y_n) \mapsto \begin{bmatrix} x_1 y_1 & x_1 y_2 & \cdots & x_1 y_n \\ x_2 y_1 & x_2 y_2 & \cdots & x_2 y_n \\ \vdots & \vdots & & \vdots \\ x_n y_1 & x_n y_2 & \cdots & x_n y_n \end{bmatrix}.$$

Theorem 14.5.36 ([963, Theorem 25.1.2]) *The Segre variety $S_{n-1, n-1}$ of $\mathrm{PG}(n^2-1, q)$ is the set of points of $\mathrm{PG}(n^2-1, q)$ for which the corresponding matrix has rank 1.*

Definition 14.5.37 An **exterior set** of points of $\mathrm{PG}(n^2-1, q)$ with respect to the Segre variety $S_{n-1, n-1}$ in $\mathrm{PG}(n^2-1, q)$ is a set of points $\mathcal{E}$ of $\mathrm{PG}(n^2-1, q) \setminus S_{n-1, n-1}$ of size θ_{n^2-n-1} such that every line through two different points of $\mathcal{E}$ is skew to the Segre variety $S_{n-1, n-1}$.

Theorem 14.5.38 ([448]) *An exterior set with respect to $S_{n-1, n-1}$ defines an MRD code in $\mathbb{F}_q^{n \times n}$ with minimum distance 2, which is closed under scalar multiplication in $\mathbb{F}_q$. Conversely, every MRD code in $\mathbb{F}_q^{n \times n}$ with minimum distance 2, which is closed under scalar multiplication in $\mathbb{F}_q$, defines an exterior set with respect to $S_{n-1, n-1}$.*

The Gabidulin code $\mathcal{G}_{n, n, n-1}$ is the best known MRD code with minimum distance 2 and so defines an exterior set with respect to $S_{n-1, n-1}$.

Theorem 14.5.39 ([448]) *A Gabidulin code $\mathcal{G}_{n, n, n-1}$ over $\mathbb{F}_q$ is an $(n^2 - n - 1)$-space X of $\mathrm{PG}(n^2-1, q)$ which defines an exterior set with respect to $S_{n-1, n-1}$.*

Cossidente, Marino, and Pavese [448] constructed an exterior set with respect to the Segre variety $S_{2, 2}$. This construction was generalized to arbitrary $n \geq 3$ by Durante and Siciliano [653].

14.6 A Geometric Result Arising from a Coding Theoretic Result

In this chapter, many links between finite geometry and coding theory were presented, and many geometric results were presented which are of relevance for coding theory. To end this chapter, we present a result in coding theory that immediately implies a result in finite geometry.

Definition 14.6.1

(a) Let S be a set of points in $\mathrm{PG}(n,q)$ and let $P \in S$. A **tangent hyperplane** $T_P(S)$ **to the set** S **in the point** P is a hyperplane only sharing P with S.

(b) A **strong representative system** S in $\mathrm{PG}(n,q)$ is a set of points such that every point $P \in S$ belongs to at least one tangent hyperplane $T_P(S)$ to S.

Definition 14.6.2 A **unital** $\mathcal{U}$ of $\mathrm{PG}(2,q)$, q square, is a set of $q\sqrt{q}+1$ points of $\mathrm{PG}(2,q)$ such that every line of $\mathrm{PG}(2,q)$ intersects $\mathcal{U}$ in either 1 or $\sqrt{q}+1$ points.

Example 14.6.3 The classical example of a unital in $\mathrm{PG}(2,q)$, q square, is the set of points of a **nonsingular Hermitian curve** $\mathcal{H}$. This is an absolutely irreducible algebraic curve of degree $\sqrt{q}+1$ over $\mathbb{F}_q$ having standard equation $X_0^{\sqrt{q}+1} + X_1^{\sqrt{q}+1} + X_2^{\sqrt{q}+1} = 0$. Also other examples of unitals in $\mathrm{PG}(2,q)$, q square, called **Buekenhout–Metz** unitals, are known [318, 1383].

Definition 14.6.4 An **ovoid** $\mathcal{O}$ of $\mathrm{PG}(3,q)$ is a set of q^2+1 points of $\mathrm{PG}(3,q)$, such that every plane of $\mathrm{PG}(3,q)$ intersects $\mathcal{O}$ in either 1 or $q+1$ points.

Example 14.6.5 The classical example of an ovoid in $\mathrm{PG}(3,q)$ is the set of points of a **nonsingular elliptic quadric** $\mathcal{Q}$. This is a nonsingular quadric in $\mathrm{PG}(3,q)$ having standard equation $f(X_0, X_1) + X_2 X_3 = 0$, where $f(X_0, X_1)$ is a homogeneous quadratic polynomial, irreducible over $\mathbb{F}_q$. Next to the elliptic quadric, one other type of ovoid is known. This is the **Tits ovoid** in $\mathrm{PG}(3,q)$, $q = 2^{2h+1}$, $h \geq 1$ [1809].

Theorem 14.6.6 ([958, 959])

(a) *The $(q+1)$-arcs of $\mathrm{PG}(2,q)$ are strong representative systems of $\mathrm{PG}(2,q)$.*

(b) *The unitals of $\mathrm{PG}(2,q)$, q square, are strong representative systems of $\mathrm{PG}(2,q)$.*

(c) *The ovoids of $\mathrm{PG}(3,q)$ are strong representative systems of $\mathrm{PG}(3,q)$.*

The next theorem presents upper bounds on the size of strong representative systems of $\mathrm{PG}(n,q)$.

Theorem 14.6.7 (Bruen–Thas [316])

(a) *A strong representative system in $\mathrm{PG}(2,q)$ has size at most $q\sqrt{q}+1$, and if q is square, then the strong representative systems of $\mathrm{PG}(2,q)$ of size $q\sqrt{q}+1$ are unitals.*

(b) *A strong representative system in $\mathrm{PG}(3,q)$ has size at most q^2+1, and in $\mathrm{PG}(3,q)$ every strong representative system of size q^2+1 is equal to an ovoid.*

(c) *For $n \geq 4$, every strong representative system S in $\mathrm{PG}(n,q)$ with $q > 2$ satisfies the bound $|S| < q^{(n+1)/2} + 1$.*

The preceding upper bounds were obtained via geometric arguments. However, there is an alternative way to obtain upper bounds on the size of strong representative systems in $\mathrm{PG}(n,q)$ by using coding theoretic results. The next theorem follows from the ideas developed in [225, Section 4].

Theorem 14.6.8 *Let A be the point-hyperplane incidence matrix of $\mathrm{PG}(n,q)$, $q = p^h$, p prime, $h \geq 1$. If S is a strong representative system of $\mathrm{PG}(n,q)$, then $|S| \leq \mathrm{rank}_p(A)$.*

By applying Theorem 14.4.4(c) to the result of the preceding theorem, the following result on strong representative systems is obtained.

Corollary 14.6.9 ([225, Theorem 1.6]) *Let* S *be a strong representative system of* $\mathrm{PG}(n, q)$, $q = p^h$, *p prime*, $h \geq 1$. *Then*

$$|S| \leq \binom{n + p - 1}{n}^h + 1.$$

The preceding theorem and corollary motivate the title of this subsection. The result of this theorem and corollary is, for particular primes p and exponents h, better than the geometric bound of Theorem 14.6.7. So, here, results from coding theory, applied to a geometric problem, lead to better results than the geometric techniques.

The preceding theorem and corollary are a particular example of the relevance of coding theoretic research for Galois geometries, again showing the close relationship between the research areas of coding theory and Galois geometries.

Chapter 15

Algebraic Geometry Codes and Some Applications

Alain Couvreur

Inria and LIX, École Polytechnique

Hugues Randriambololona

ANSSI Laboratoire de cryptographie and Télécom Paris

Introduction

Algebraic geometry codes is a fascinating topic at the confluence of number theory and algebraic geometry on one side and computer science involving coding theory, combinatorics, information theory, and algorithms, on the other side.

History

The beginning of the story dates back to the early 1970s where the Russian mathematician V. D. Goppa proposed code constructions from algebraic curves, using rational functions or differential forms on algebraic curves over finite fields [836, 837, 838]. In particular, this geometric point of view permits one to regard Goppa's original construction of codes from rational functions [834, 835] (see [168] for a description of these codes in English) as codes from differential forms on the projective line.

Shortly after, there appeared one of the most striking results in the history of coding theory, which is probably at the origin of the remarkable success of algebraic geometry codes. In 1982, Tsfasman, Vlăduţ, and Zink [1822] related the existence of a sequence of curves whose numbers of rational points go to infinity and grow linearly with respect to the curves' genera to the existence of sequences of asymptotically good codes. Next, using sequences of modular curves and Shimura curves, they proved the existence of sequences of codes over a field $\mathbb{F}_q$ where $q = p^2$ or p^4, for p a prime number, whose asymptotic rate R and asymptotic relative distance δ satisfy

$$R \geq 1 - \delta - \frac{1}{\sqrt{q} - 1}.$$

In an independent work and using comparable arguments, Ihara [1021] proved a similar result over any field $\mathbb{F}_q$ where q is a square. For this reason, the ratio

$$\limsup_{g \to +\infty} \frac{\max |X(\mathbb{F}_q)|}{g},$$

where the maximum is taken over the set of curves X of genus g over $\mathbb{F}_q$, is usually referred to as the *Ihara constant* and denoted by $A(q)$. Further, Vlăduţ and Drinfeld [1857] proved an upper bound for the Ihara constant showing that the families of curves exhibited in [1021, 1822] are optimal.

The groundbreaking aspect of such a result appears when q is a square and $q \geq 49$, since, for such a parameter, the Tsfasman–Vlăduţ–Zink Bound is better than the asymptotic Gilbert–Varshamov Bound. Roughly speaking, this result asserts that some algebraic geometry codes are better than random codes, while the opposite statement was commonly believed in the coding theory community.

For this breakthrough, Tsfasman, Vlăduţ, and Zink received the prestigious *Information Theory Society Paper Award*, and their result motivated an intense development of the theory of algebraic geometry codes. The community explored various sides of this theory in the following decades. First, the question of producing sequences of curves with maximal Ihara constant, or the estimate of the Ihara constant $A(q)$ when q is not a square, became a challenging new problem in number theory. In particular, some constructions based on class field theory gave lower bounds for $A(q)$ when q is no longer a square. Second, in 1995, García and Stichtenoth [790] obtained new optimal sequences of curves (i.e., reaching the Drinfeld–Vlăduţ Bound) using a much more elementary construction called *recursive towers*. Apart from the asymptotic questions, many authors improved, in some specific cases, Goppa's estimate for the minimum distance of algebraic geometry codes. Such results permitted the construction of new codes of given length whose parameters beat the tables of best known codes; see, for instance, [846]. Third, another fruitful direction is on the algorithmic side with the development of polynomial time decoding algorithms correcting up to half the designed distance and even further using list decoding. This chapter presents known results on improved bounds on the minimum distance and discusses unique and list decoding. The asymptotic performances of algebraic geometry codes are also quickly surveyed without providing an in-depth study of the Ihara constant and the construction of optimal towers.

The latter topic, being much too rich, would require a separate treatment which we decided not to develop in the present survey.

It should also be noted that codes may be constructed from higher dimensional varieties. This subject, also of deep interest, will not be discussed in the present chapter. We refer the interested reader to [1276] for a survey on this question, to [1474] for a decoding algorithm, and to [460] for a first attempt toward good asymptotic constructions of codes from surfaces.

Applications of Algebraic Geometry Codes

Algebraic geometry (AG) codes admit many interesting properties that make them suitable for a very wide range of applications. Most of these properties are inherited from Reed–Solomon (RS) codes and their variants,[1] of which AG codes are a natural extension:

- AG codes can be explicitly constructed.

- AG codes can be efficiently decoded.

- AG codes admit good bounds on their parameters: although they might not be MDS, they remain close to the Singleton Bound [1323, Th. 1.11].

- AG codes behave well under duality: the dual of an AG code is an AG code.

- AG codes behave well under multiplication: the ⋆-product of two AG codes is included in, and in many situations, is equal to an AG code.

- AG codes may have automorphisms, reflecting the geometry of the underlying objects.

However, as already discussed above, AG codes enjoy an additional property over their Reed–Solomon genus 0 counterparts, which was perhaps the main motivation for their introduction:

- For a given q, the length of an (extended) RS code over $\mathbb{F}_q$ cannot exceed $q + 1$ while one can construct arbitrarily long AG codes over a given fixed field $\mathbb{F}_q$.

It would be an endless task to list all applications of AG codes. Below we focus on a selection of those we find most meaningful. Basically, in every situation where Reed–Solomon codes are used, replacing them by algebraic geometry codes is natural and frequently leads to improvements. These application topics include symmetric cryptography, public-key cryptography, algebraic complexity theory, multiparty computation and secret sharing, distributed storage, and so on. The present survey aims at presenting various aspects of the theory of algebraic geometry codes together with several applications.

Organization of the Chapter

We start by fixing some general notation in Section 15.1. The background on algebraic geometry and number theory is recalled in Section 15.2, and the construction and first properties of algebraic geometry codes are presented in Section 15.3. In particular, the Goppa Bound for the minimum distance is given. In Section 15.4, we discuss the asymptotic performance of algebraic geometry codes and relate this question to the Ihara constant. As said earlier, the construction and study of good families of curves is too rich to be

---

[1]Beware that the terminology on Reed–Solomon codes varies in the literature with several names for variants: *generalized Reed–Solomon codes, extended Reed–Solomon codes, doubly extended Reed–Solomon codes*, etc. In this chapter, we refer to *Reed–Solomon* or *generalized Reed–Solomon codes* as the algebraic geometry codes from a curve of genus 0. See Section 15.3.2 for further details.

developed in the present survey and would require a separate treatment. Next, Section 15.5 is devoted to various improvements of the Goppa designed distance. Section 15.6 surveys the various decoding algorithms. Starting with algorithms correcting up to half the designed distance and then moving to the more recent developments of list decoding permits one to exceed this threshold. Finally, the last three sections present various applications of algebraic geometry codes. Namely, we study their possible use for post-quantum public key cryptography in Section 15.7. Thanks to their nice behavior with respect to the so-called $\star$-*product*, algebraic geometry codes have applications to algebraic complexity theory, secret sharing and multiparty computation, which are presented in Section 15.8. Finally, applications to distributed storage with the algebraic geometric constructions of *locally recoverable codes* are presented in Section 15.9.

15.1 Notation

For any prime power q, the finite field with q elements is denoted by $\mathbb{F}_q$ and its algebraic closure by $\overline{\mathbb{F}}_q$. Given any ring R, the group of invertible elements of R is denoted by $R^\times$. In particular, for a field $\mathbb{F}$, the group $\mathbb{F}^\times$ is nothing but $\mathbb{F} \setminus \{0\}$.

Unless otherwise specified, any code in this chapter is linear. The vector space $\mathbb{F}_q^n$ is equipped with the **Hamming weight** denoted by $\mathrm{wt}_H(\cdot)$, and the **Hamming distance** between two vectors $\mathbf{x}, \mathbf{y}$ is denoted by $d_H(\mathbf{x}, \mathbf{y})$. Given a linear code $\mathcal{C} \subseteq \mathbb{F}_q^n$, as is usual in the literature, the fundamental parameters of $\mathcal{C}$ are listed as a triple of the form $[n, k, d]$, where n denotes its block length, k its dimension as an $\mathbb{F}_q$-space, and d its minimum Hamming distance, which is sometimes also referred to as $d_H(\mathcal{C})$. When the minimum distance is unknown, we sometimes denote the known parameters by $[n, k]$.

Another important notion in the sequel is that of the $\star$-*product*, also called the *Hadamard product*. The $\star$-**product** is nothing but the componentwise multiplication in $\mathbb{F}_q^n$: for $\mathbf{x} = (x_1, \ldots, x_n)$ and $\mathbf{y} = (y_1, \ldots, y_n)$, we have

$$\mathbf{x} \star \mathbf{y} \overset{\text{def}}{=} (x_1 y_1, \ldots, x_n y_n).$$

This notion extends to codes: given two linear codes $\mathcal{C}, \mathcal{C}' \subseteq \mathbb{F}_q^n$ we let

$$\mathcal{C} \star \mathcal{C}' \overset{\text{def}}{=} \mathrm{span}_{\mathbb{F}_q}\{\mathbf{c} \star \mathbf{c}' \mid \mathbf{c} \in \mathcal{C}, \, \mathbf{c}' \in \mathcal{C}'\}$$

be the *linear span* of the pairwise products of codewords from $\mathcal{C}$ and $\mathcal{C}'$. Observe that $\mathcal{C} \star \mathcal{C}' \subseteq \mathbb{F}_q^n$ is again a linear code since we take the linear span. Then the **square** of $\mathcal{C}$ is defined as

$$\mathcal{C}^{\star 2} \overset{\text{def}}{=} \mathcal{C} \star \mathcal{C}.$$

Recall that $\mathbb{F}_q^n$ is equipped with the Euclidean inner product (or bilinear form) defined as

$$
\begin{array}{rccc}
\langle \cdot, \cdot \rangle_{\text{Eucl}} : & \mathbb{F}_q^n \times \mathbb{F}_q^n & \longrightarrow & \mathbb{F}_q \\
& (\mathbf{x}, \mathbf{y}) & \longmapsto & \sum_{i=1}^n x_i y_i.
\end{array}
$$

The $\star$-product and the Euclidean inner product are related by the following adjunction property.

Lemma 15.1.1 *Let* $\mathbf{x}, \mathbf{y}, \mathbf{z} \in \mathbb{F}_q^n$, *then* $\langle \mathbf{x} \star \mathbf{y}, \mathbf{z} \rangle_{\text{Eucl}} = \langle \mathbf{x}, \mathbf{y} \star \mathbf{z} \rangle_{\text{Eucl}}$.

15.2   Curves and Function Fields

Algebraic geometry, the study of geometric objects defined by polynomial equations, has a long history. In the last century it received rigorous foundations, in several waves, each bringing its own language. For *most* applications to coding theory, we will only need to work with some of the simplest geometric objects, namely curves. These can be described equally well in the following languages:

- the language of *algebraic function fields*, for which a recommended reference is [1755];

- the language of *varieties over an algebraically closed field*, as in [758] or [1657];

- the language of *schemes*, for which we refer to [915].

As long as one works with one fixed curve, these languages have the same power of expression.

We briefly recall some of the basic notions and results that we need, and explain how they correspond in these different languages. For more details the reader should look in the references given above.

15.2.1   Curves, Points, Function Fields, and Places

Definition 15.2.1 An **algebraic function field** F with constant field $\mathbb{F}_q$ is a finite extension of a purely transcendental extension of $\mathbb{F}_q$ of transcendence degree 1, in which $\mathbb{F}_q$ is *algebraically closed*, i.e., any element $\alpha \in F$ which is algebraic over $\mathbb{F}_q$ is actually in $\mathbb{F}_q$.

Any such F is of the form $F = \mathrm{Frac}(\mathbb{F}_q[x,y]/(P(x,y)))$ where $P \in \mathbb{F}_q[x,y]$ is *absolutely irreducible*, i.e., irreducible even when regarded as an element of $\overline{\mathbb{F}}_q[x,y]$.

Definition 15.2.2 A **curve** over $\mathbb{F}_q$ is a geometrically irreducible smooth projective variety of dimension 1 defined over $\mathbb{F}_q$.

Any such curve can be obtained as the projective completion and desingularization of an affine plane curve of the form $\{P(x,y) = 0\}$ where P is an absolutely irreducible polynomial in the two indeterminates x and y over $\mathbb{F}_q$.

From this observation we see that Definitions 15.2.1 and 15.2.2 are essentially equivalent. This can be made more precise.

Theorem 15.2.3 *There is an equivalence of categories between:*

- *algebraic function fields, with field morphisms, over $\mathbb{F}_q$, and*

- *curves, with dominant (surjective) morphisms, over $\mathbb{F}_q$.*

The proof can be found for example in [915, § I.6]. In one direction, to each curve X one associates its field of rational functions $F = \mathbb{F}_q(X)$, which is an algebraic function field. We then have a natural correspondence between:

- places, or discrete valuations of F;

- Galois orbits in the set $X(\overline{\mathbb{F}}_q)$ of points of X with coordinates in the algebraic closure of $\mathbb{F}_q$;

- closed points of the topological space of X seen as a scheme.

If P is a closed point of X, or a place of F, we denote by $v_P : F \to \mathbb{Z} \cup \{\infty\}$ the corresponding discrete valuation, by $\mathcal{O}_P = \{f \in F \mid v_P(f) \geq 0\}$ its valuation ring (the local ring of X at P), and by $\mathfrak{m}_P = \{f \in F \mid v_P(f) > 0\}$ its maximal ideal. An element $t_P \in \mathfrak{m}_P$ is called a **local parameter**, or a **uniformizer**, at P if it satisfies $v_P(t_P) = 1$. The residue field $k_P = \mathcal{O}_P/\mathfrak{m}_P$ is a finite extension of $\mathbb{F}_q$. We define the **degree** of P as the degree of that field extension

$$\deg(P) = [k_P : \mathbb{F}_q].$$

This degree is equal to the cardinality of the Galois orbit corresponding to P in $X(\overline{\mathbb{F}}_q)$. Conversely, for any finite extension $\mathbb{F}_{q^d}$ of $\mathbb{F}_q$, the set $X(\mathbb{F}_{q^d})$ of $\mathbb{F}_{q^d}$-rational points of X (i.e., points of X with coordinates in $\mathbb{F}_{q^d}$) identifies with the set of degree 1 places in the base field extension $\mathbb{F}_{q^d}F$ seen as a function field over $\mathbb{F}_{q^d}$.

15.2.2 Divisors

The **divisor group** $\mathrm{Div}(X)$ is the free abelian group generated by the set of closed points of X or equivalently by the set of places of its function field $\mathbb{F}_q(X)$. Thus, a divisor is a formal sum

$$D = \sum_P n_P P$$

where P ranges over closed points and $v_P(D) \overset{\mathrm{def}}{=} n_P \in \mathbb{Z}$ are almost all zero. The **support** of D is the finite set $\mathrm{supp}(D)$ of such P with $n_P \neq 0$. The **degree** of D is

$$\deg(D) \overset{\mathrm{def}}{=} \sum_P n_P \deg(P).$$

We say that D is **effective** if $n_P \geq 0$ for all P. We write $D_1 \geqslant D_2$ if $D_1 - D_2$ is effective.

A divisor is **principal** if it is of the form

$$\mathrm{div}(f) \overset{\mathrm{def}}{=} \sum_P v_P(f)P$$

for $f \in F^\times = F \setminus \{0\}$. So we can write $\mathrm{div}(f) = (f)_0 - (f)_\infty$ where $(f)_0 = \sum_{v_P(f)>0} v_P(f)P$ is the divisor of zeros of f, and $(f)_\infty = \sum_{v_P(f)<0} -v_P(f)P$ is its divisor of poles. Principal divisors have degree zero ("a rational function on a curve has as many poles as zeros"), and they form a subgroup of $\mathrm{Div}(X)$. We say that two divisors D_1 and D_2 are **linearly equivalent**, and we write

$$D_1 \sim D_2, \tag{15.1}$$

when $D_1 - D_2$ is principal. Passing to the quotient we get the **divisor class group** $\mathrm{Cl}(X) = \mathrm{Div}(X)/\sim$ of X, together with a **degree map** $\mathrm{Cl}(X) \to \mathbb{Z}$ that is a surjective group morphism [1401, Th. 3.2(i)].

The **Riemann–Roch space** of a divisor D is the vector space

$$L(D) = \{f \in F^\times \mid \mathrm{div}(f) \geqslant -D\} \cup \{0\}.$$

It has finite dimension $\ell(D) = \dim_{\mathbb{F}_q} L(D)$. Actually $\ell(D)$ only depends on the linear equivalence class of D in $\mathrm{Cl}(X)$.

To a divisor $D = \sum_P n_P P$, one can associate an invertible sheaf (or line bundle) $\mathcal{O}(D)$ on X, generated locally at each P by $t_P^{-n_P}$, for t_P a uniformizer at P. There is then a natural identification

$$L(D) = \Gamma(X, \mathcal{O}(D))$$

between the Riemann–Roch space of D and the space of global sections of $\mathcal{O}(D)$. Conversely, given an invertible sheaf $\mathcal{L}$ on X, any choice of a nonzero rational section s of $\mathcal{L}$ defines a divisor $D = \operatorname{div}(s)$ with $\mathcal{L} \simeq \mathcal{O}(D)$. Another choice of s gives a linearly equivalent D. From this we get an isomorphism

$$\operatorname{Cl}(X) \simeq \operatorname{Pic}(X)$$

where $\operatorname{Pic}(X)$, the **Picard group** of X, is the group of isomorphism classes of invertible sheaves on X equipped with the tensor product.

15.2.3 Morphisms of Curves and Pullbacks

A **morphism of curves** is a map $\phi : X \to Y$ that is componentwise described by polynomials or rational functions. To such a map is associated a function field extension $\phi^* : \mathbb{F}_q(Y) \hookrightarrow \mathbb{F}_q(X)$: given a rational function f on Y, one defines the **pullback** of f by ϕ, denoted $\phi^* f$, to be the function $f \circ \phi$ on X. The **degree** of ϕ is the extension degree $[\mathbb{F}_q(X) : \mathbb{F}_q(Y)]$ induced by the ϕ^* field extension.

Definition 15.2.4 Given a divisor $D = \sum_{i=1}^r n_i P_i$ on Y, one defines the **pullback of the divisor D by** ϕ and denotes it by $\phi^* D$:

$$\phi^* D \stackrel{\text{def}}{=} \sum_{i=1}^r \sum_{Q \stackrel{\phi}{\to} P_i} n_i \cdot e_{Q|P_i} \cdot Q,$$

where $e_{Q|P_i}$ denotes the **ramification index** at Q; see [1755, Def. 3.1.5].

The divisor $\phi^* D$ is sometimes also called the **conorm** of D; see [1755, Def. 3.1.8]. In addition, it is well-known that

$$\deg(\phi^* D) = \deg(\phi) \cdot \deg(D). \tag{15.2}$$

15.2.4 Differential Forms

The space of **rational differential forms** on X is the one-dimensional F-vector space Ω_F whose elements are of the form

$$\omega = u \mathrm{d} v$$

for $u, v \in F$, subject to the usual Leibniz rule $\mathrm{d}(u_1 u_2) = u_1 \mathrm{d} u_2 + u_2 \mathrm{d} u_1$. Given $\omega \in \Omega_F$ and t_P a uniformizer at P, we can write $\omega = f_P \mathrm{d} t_P$ for some $f_P \in F$ and we define the valuation of ω at P as

$$v_P(\omega) = v_P(f_P).$$

One can prove that the definition does not depend on the choice of t_P. In the same spirit as functions, to any nonzero rational differential form ω, one associates its divisor

$$\operatorname{div}(\omega) \stackrel{\text{def}}{=} \sum_P v_P(\omega) P.$$

Equivalently, Ω_F is the space of rational sections of the invertible sheaf Ω_X^1, called the **canonical sheaf,** or the sheaf of differentials of X, generated locally at each P by the differential $\mathrm{d} t_P$, for t_P a uniformizer at P.

For any divisor D we set

$$\Omega(D) \stackrel{\text{def}}{=} \Gamma(X, \Omega_X^1 \otimes \mathcal{O}(-D)) = \{\omega \in \Omega_F \setminus \{0\} \mid \operatorname{div}(\omega) \geqslant D\} \cup \{0\}.$$

Remark 15.2.5 Beware of the sign change compared to the definition of the $L(D)$ space. This choice of notation could seem unnatural, but it is somehow standard in the literature on algebraic geometry codes (see, e.g., [1755]), and is related to Serre duality.

15.2.4.1 Canonical Divisors

A **canonical divisor** is a divisor K_X on X such that $\Omega_X^1 \simeq \mathcal{O}(K_X)$. Thus a canonical divisor is of the form

$$K_X = \operatorname{div}(\omega)$$

for any choice of $\omega = u dv \in \Omega_F \setminus \{0\}$. More explicitly, we have $K_X = \sum_P v_P(f_P)P$ where locally at each P we write $\omega = f_P dt_P$. Any two canonical divisors are linearly equivalent; see, for instance, [1755, Prop. 1.5.13(b)].

15.2.4.2 Residues

Given a rational differential ω_P at P and a uniformizer t_P we have a local Laurent series expansion

$$\omega_P = a_{-N} t_P^{-N} dt_P + \cdots + a_{-1} t_P^{-1} dt_P + \eta_P$$

where N is the order of the pole of ω_P at P and η_P is regular at P, i.e., $v_P(\eta_P) \geq 0$. Then

$$\operatorname{res}_P(\omega_P) = a_{-1}$$

is independent of the choice of t_P and is called the **residue** of ω_P at P. In particular, if ω_P is regular at P, then we have $\operatorname{res}_P(\omega_P) = 0$. We refer to [1755, Chap. IV] for further details.

15.2.5 Genus and the Riemann–Roch Theorem

An important numerical invariant of a curve X is its **genus**

$$g = \ell(K_X).$$

We then also have $\deg(K_X) = 2g - 2$ [1755, Cor. 1.5.16].

The following, which is proved, for instance, in [1755, Th. 1.5.15], is a central result in the theory of curves.

Theorem 15.2.6 (Riemann–Roch) *For a divisor D on X, we have*

$$\ell(D) - \ell(K_X - D) = \deg(D) + 1 - g.$$

In particular, we always have

$$\ell(D) \geq \deg(D) + 1 - g.$$

In addition, Riemann–Roch spaces satisfy the following properties; see, e.g., Cor. 1.4.12(b) and Th. 1.5.17 of [1755].

Proposition 15.2.7 *Let D be a divisor on a curve X such that $\deg(D) < 0$. Then $L(D) = \{0\}$.*

Corollary 15.2.8 *When $\deg(D) > 2g - 2$, we have $\ell(K_X - D) = 0$, and then*

$$\ell(D) = \deg(D) + 1 - g.$$

15.3 Basics on Algebraic Geometry Codes

15.3.1 Algebraic Geometry Codes, Definitions, and Elementary Results

Let X be a curve over $\mathbb{F}_q$ and fix a divisor G and an ordered sequence of n distinct rational points $\mathcal{P} = (P_1, \ldots, P_n)$ disjoint from $\operatorname{supp}(G)$. The latter sequence $\mathcal{P}$ will be referred to as the **evaluation points sequence** and a divisor $D_\mathcal{P}$ is associated to it, namely:

$$D_\mathcal{P} \stackrel{\text{def}}{=} P_1 + \cdots + P_n. \tag{15.3}$$

With this data, we can define two codes.

Definition 15.3.1 The **evaluation code**, or **function code**, $\mathcal{C}_L(X, \mathcal{P}, G)$ is the image of the map

$$
\begin{array}{ccc}
L(G) & \longrightarrow & \mathbb{F}_q^n \\
f & \longmapsto & (f(P_1), \ldots, f(P_n)).
\end{array}
$$

Definition 15.3.2 The **residue code**, or **differential code**, $\mathcal{C}_\Omega(X, \mathcal{P}, G)$ is the image of the map

$$
\begin{array}{ccc}
\Omega(G - D_\mathcal{P}) & \longrightarrow & \mathbb{F}_q^n \\
\omega & \longmapsto & (\operatorname{res}_{P_1}(\omega), \ldots, \operatorname{res}_{P_n}(\omega))
\end{array}
$$

where again $D_\mathcal{P} = P_1 + \cdots + P_n$.

The two constructions are dual to each other; see, for instance, [1755, Th. 2.2.8]).

Theorem 15.3.3 *The evaluation code and the residue code are dual of each other:*

$$\mathcal{C}_\Omega(X, \mathcal{P}, G) = \mathcal{C}_L(X, \mathcal{P}, G)^\perp.$$

Recall that the $\star$-product denotes the componentwise multiplication in $\mathbb{F}_q^n$; so if $\mathbf{x} = (x_1, \ldots, x_n)$ and $\mathbf{y} = (y_1, \ldots, y_n)$, then $\mathbf{x} \star \mathbf{y} = (x_1 y_1, \ldots, x_n y_n)$.

Definition 15.3.4 Two codes $\mathcal{C}_1, \mathcal{C}_2 \subseteq \mathbb{F}_q^n$ are **diagonally equivalent** (under $\mathbf{a} \in (\mathbb{F}_q^\times)^n$) if

$$\mathcal{C}_2 = \mathcal{C}_1 \star \mathbf{a}$$

or equivalently if

$$\mathbf{G}_2 = \mathbf{G}_1 \mathbf{D}_\mathbf{a}$$

where $\mathbf{G}_1, \mathbf{G}_2$ are generator matrices of $\mathcal{C}_1, \mathcal{C}_2$ respectively, and $\mathbf{D}_\mathbf{a}$ is the $n \times n$ diagonal matrix whose diagonal entries are the entries of $\mathbf{a}$.

Remark 15.3.5 Diagonally equivalent codes are isometric with respect to the Hamming distance.

Lemma 15.3.6 *Let $G_1 \sim G_2$ be two linearly equivalent divisors on X, both with support disjoint from $\mathcal{P} = \{P_1, \ldots, P_n\}$. Then $\mathcal{C}_L(X, \mathcal{P}, G_1)$ and $\mathcal{C}_L(X, \mathcal{P}, G_2)$ are diagonally equivalent under $\mathbf{a} = (h(P_1), \ldots, h(P_n))$ where h is any choice of function with $\operatorname{div}(h) = G_1 - G_2$. Likewise $\mathcal{C}_\Omega(X, \mathcal{P}, G_1)$ and $\mathcal{C}_\Omega(X, \mathcal{P}, G_2)$ are diagonally equivalent under $\mathbf{a}^{-1}$.*

See [1755, Prop. 2.2.14] for a proof of the latter statement.

Remark 15.3.7 In Lemma 15.3.6, the choice of h is not unique. More precisely, h may be replaced by any nonzero scalar multiple λh of h. This would replace $\mathbf{a}$ by $\lambda \mathbf{a}$, which has no consequence, since the codes are linear and hence globally invariant by a scalar multiplication.

Remark 15.3.8 If we accept codes defined only up to diagonal equivalence, then we can relax the condition that $\operatorname{supp}(G)$ is disjoint from $\mathcal{P}$ in Definitions 15.3.1 and 15.3.2. Indeed, if $\operatorname{supp}(G)$ is not disjoint from $\mathcal{P}$, then by the weak approximation theorem [1755, Th. 1.3.1] we can find $G' \sim G$ with support disjoint from $\mathcal{P}$, and then we can use $\mathcal{C}_L(X, \mathcal{P}, G')$ in place of $\mathcal{C}_L(X, \mathcal{P}, G)$, and $\mathcal{C}_\Omega(X, \mathcal{P}, G')$ in place of $\mathcal{C}_\Omega(X, \mathcal{P}, G)$. Lemma 15.3.6 then shows that, up to diagonal equivalence, these codes do not depend on the choice of G'.

In summary, the usual restriction "the support of G should avoid the P_i's" in the definition of $\mathcal{C}_L(X, \mathcal{P}, G)$ can always be ruled out at the cost of some technical clarifications.

A slightly more general construction is the following.

Construction 15.3.9 Begin with a curve X over $\mathbb{F}_q$, an invertible sheaf $\mathcal{L}$ on X, and an ordered sequence of n distinct rational points $\mathcal{P} = (P_1, \ldots, P_n)$. After some choice of a trivialization $\mathcal{L}|_{P_i} \simeq \mathbb{F}_q$ for the fibres of $\mathcal{L}$ at the $P_i \in \mathcal{P}$, the code $\mathcal{C}(X, \mathcal{P}, \mathcal{L})$ is the image of the map

$$
\begin{array}{ccc}
\Gamma(X, \mathcal{L}) & \longrightarrow & \bigoplus_{1 \le i \le n} \mathcal{L}|_{P_i} \simeq \mathbb{F}_q^n \\
s & \longmapsto & (s|_{P_1}, \ldots, s|_{P_n}).
\end{array}
$$

Choosing another trivialization of the fibres, and also replacing the invertible sheaf $\mathcal{L}$ with an isomorphic one, leaves $\mathcal{C}(X, \mathcal{P}, \mathcal{L})$ unchanged up to diagonal equivalence.

Definition 15.3.1 is a special case of this construction with $\mathcal{L} = \mathcal{O}(G)$ together with the natural trivialization $\mathcal{O}(G)|_{P_i} = \mathbb{F}_q$ when $P_i \notin \operatorname{supp}(G)$. Relaxing this last condition, we can use a trivialization $\mathcal{O}(G)|_{P_i} = \mathbb{F}_q \cdot h_i|_{P_i}$ depending on the choice of a local function h_i at P_i with minimal valuation $v_{P_i}(h_i) = -v_{P_i}(G)$. This defines $\mathcal{C}_L(X, \mathcal{P}, G)$ as the image of the map

$$
\begin{array}{ccc}
L(G) & \longrightarrow & \mathbb{F}_q^n \\
f & \longmapsto & ((f/h_1)(P_1), \ldots, (f/h_n)(P_n)).
\end{array}
$$

A possible choice for h_i is $h_i = t_{P_i}^{-v_{P_i}(G)}$ where t_{P_i} is a uniformizer. Alternatively, given $G' \sim G$ with $\operatorname{supp}(G') \cap \mathcal{P} = \emptyset$, one can find h with $\operatorname{div}(h) = G' - G$ and set $h_i = h$ for all i. Doing so, we obtain Remark 15.3.8.

Likewise Definition 15.3.2 is a special case of Definition 15.3.9 with $\mathcal{L} = \Omega_X^1 \otimes \mathcal{O}(D_\mathcal{P} - G)$ and trivialization given by the residue map when $\operatorname{supp}(G) \cap \mathcal{P} = \emptyset$, and can be relaxed in a similar way when this condition is relaxed. This also gives the following result.

Lemma 15.3.10 *The codes $\mathcal{C}_\Omega(X, \mathcal{P}, G)$ and $\mathcal{C}_L(X, \mathcal{P}, K_X + D_\mathcal{P} - G)$ are diagonally equivalent for any canonical divisor K_X on X.*

Actually, if $\operatorname{supp}(G)$ is disjoint from $\mathcal{P}$, then there is a choice of a canonical divisor K_X, with support disjoint from $\operatorname{supp}(G)$ and $\mathcal{P}$, that turns this diagonal equivalence into an equality: $\mathcal{C}_\Omega(X, \mathcal{P}, G) = \mathcal{C}_L(X, \mathcal{P}, K_X - D_\mathcal{P} + G)$.

Remark 15.3.11 The canonical divisors providing the equality between the $\mathcal{C}_\Omega$ and the $\mathcal{C}_L$ is the divisor of a differential form having simple poles with residue equal to 1 at all the P_i's. The existence of such a differential is a consequence of the weak approximation theorem; see [1755, Lem. 2.2.9 & Prop. 2.2.10].

The parameters of AG codes satisfy the following basic estimates.

Theorem 15.3.12 ([1755, Th. 2.2.2 & 2.2.7]) *The evaluation code* $\mathcal{C}_L(X, \mathcal{P}, G)$ *is a linear code of length* $n = |\mathcal{P}| = \deg(D_\mathcal{P})$ *and dimension*

$$k = \ell(G) - \ell(G - D_\mathcal{P}).$$

In particular, if $\deg(G) < n$, *then*

$$k = \ell(G) \geq \deg(G) + 1 - g,$$

and if moreover $2g - 2 < \deg(G) < n$, *then*

$$k = \ell(G) = \deg(G) + 1 - g,$$

where g *is the genus of* X. *Its minimum distance* $d = d_H(\mathcal{C}_L(X, \mathcal{P}, G))$ *satisfies*

$$d \geq d^*_{\mathrm{Gop}} \overset{\mathrm{def}}{=} n - \deg(G)$$

where d^*_{Gop} *is the so-called* **Goppa designed distance** *of* $\mathcal{C}_L(X, \mathcal{P}, G)$.

Joint with Lemma 15.3.10 we likewise have the following.

Corollary 15.3.13 *If* $2g - 2 < \deg(G) < n$, *the residue code* $\mathcal{C}_\Omega(X, \mathcal{P}, G)$ *has dimension*

$$k = n + g - 1 - \deg(G)$$

and minimum distance

$$d \geq d_{\mathrm{Gop}} \overset{\mathrm{def}}{=} \deg(G) + 2 - 2g$$

where d_{Gop} *is the* **Goppa designed distance** *of* $\mathcal{C}_\Omega(X, \mathcal{P}, G)$.

Another important consequence of these bounds is the following statement, providing a comparison of these bounds with the well-known Singleton Bound of Theorem 1.9.10; the following is sometimes known as the Singleton Bound for algebraic geometry codes.

Corollary 15.3.14 *Let* $\mathcal{C} = \mathcal{C}_L(X, \mathcal{P}, G)$ *with* $\deg(G) < n$, *or* $\mathcal{C} = \mathcal{C}_\Omega(X, \mathcal{P}, G)$ *with* $\deg(G) > 2g - 2$. *Then*

$$k + d \geq n + 1 - g,$$

i.e., the **Singleton defect** *of* $\mathcal{C}$ *is at most* g.

Remark 15.3.15 In the sequel, both quantities d^*_{Gop} and d_{Gop} are referred to as the **Goppa Bound** or the **Goppa designed distance**. They do not provide *a priori* the actual minimum distance but yield a lower bound. In addition, as we will see in Section 15.6, correcting errors up to half these lower bounds will be considered as good "targets" for decoding.

Finally, automorphisms of X give rise to automorphisms of evaluation codes on it.

Proposition 15.3.16 *Assume* $\mathrm{supp}(G)$ *is disjoint from* $\mathcal{P}$, *and let* σ *be an automorphism of* X *such that* $\sigma(\mathcal{P}) = \mathcal{P}$ *and* $\sigma^*G \sim G$ *(see Definition 15.2.4). Let* $\mathbf{P}_\sigma$ *be the permutation matrix given by* $(\mathbf{P}_\sigma)_{i,j} = 1$ *if* $P_i = \sigma(P_j)$ *and* $(\mathbf{P}_\sigma)_{i,j} = 0$ *otherwise. Also set* $\mathbf{v} = (h(P_1), \ldots, h(P_n))$, *where* $\mathrm{div}(h) = \sigma^*G - G$. *Then the map*

$$\mathbf{c} \mapsto \mathbf{c}\mathbf{P}_\sigma \star \mathbf{v}$$

defines a linear automorphism of $\mathcal{C}_L(X, \mathcal{P}, G)$.

The proof of Proposition 15.3.16 uses the following lemma.

Lemma 15.3.17 *In the context of Proposition 15.3.16, the map*

$$\varphi_\sigma : \quad \mathbb{F}_q(X) \quad \longrightarrow \quad \mathbb{F}_q(X)$$
$$f \quad \longmapsto \quad f \circ \sigma$$

induces an isomorphism $L(G) \xrightarrow{\sim} L(\sigma^*G)$.

Proof: The map φ_σ is clearly an isomorphism with inverse $h \mapsto h \circ \sigma^{-1}$. Hence, we only need to prove that $\varphi_\sigma(L(G)) \subseteq L(\sigma^*G)$. From Definition 15.2.4, we have

$$\sigma^*G = \sum_P v_P(G)\sigma^{-1}(P). \tag{15.4}$$

Next, for any place P of $\mathbb{F}_q(X)$ and any $f \in \mathbb{F}_q(X)$,

$$v_P(f) = v_{\sigma^{-1}(P)}(f \circ \sigma). \tag{15.5}$$

Combining (15.4) and (15.5), for any place P of $\mathbb{F}_q(X)$, we have

$$v_P(f) \geq -v_P(G) \implies v_{\sigma^{-1}(P)}(f \circ \sigma) \geq v_{\sigma^{-1}(P)}(\sigma^*G).$$

This yields the result. $\qquad\square$

Proof of Proposition 15.3.16: The map φ_σ of Lemma 15.3.17 induces an isomorphism

$$\phi_\sigma : \quad \mathcal{C}_L(X, \mathcal{P}, G) \quad \longrightarrow \quad \mathcal{C}_L(X, \mathcal{P}, \sigma^*G)$$
$$(f(P_1), \ldots, f(P_n)) \quad \longmapsto \quad (f(\sigma(P_1)), \ldots, f(\sigma(P_n))),$$

which, by definition of $\mathbf{P}_\sigma$, is nothing but the map $\mathbf{c} \mapsto \mathbf{c}\mathbf{P}_\sigma$. Next, Lemma 15.3.6 yields an isomorphism

$$\psi : \quad \mathcal{C}_L(X, \mathcal{P}, \sigma^*G) \quad \longrightarrow \quad \mathcal{C}_L(X, \mathcal{P}, G)$$
$$\mathbf{c} \quad \longmapsto \quad \mathbf{c} \star \mathbf{v}.$$

The composition map $\psi \circ \phi_\sigma$ provides an automorphism of $\mathcal{C}_L(X, \mathcal{P}, G)$ which is the map $\mathbf{c} \mapsto \mathbf{c}\mathbf{P}_\sigma \star \mathbf{v}$. $\qquad\square$

15.3.2 Genus 0, Generalized Reed–Solomon and Classical Goppa Codes

In this section, we focus on algebraic geometry codes from the projective line $\mathbb{P}^1$. We equip this line with homogeneous coordinates $(X : Y)$. We denote by x the rational function $x \overset{\text{def}}{=} \frac{X}{Y}$, and for any $x_i \in \mathbb{F}_q$ we associate the point $P_i = (x_i : 1)$. Finally, we denote by $P_\infty \overset{\text{def}}{=} (1 : 0)$.

Remark 15.3.18 In this chapter, according to the usual notation in algebraic geometry, the projective space of dimension m is denoted as $\mathbb{P}^m$. In particular, its set of rational points $\mathbb{P}^m(\mathbb{F}_q)$ is the finite set sometimes denoted as $PG(m, q)$ in the literature of finite geometries and combinatorics.

15.3.2.1   The $\mathcal{C}_L$ Description

One of the most famous families of codes is probably that of Reed–Solomon codes.

Definition 15.3.19 Let $\mathbf{x} = (x_1, \ldots, x_n)$ be an n-tuple of distinct elements of $\mathbb{F}_q$ and $\mathbf{y} = (y_1, \ldots, y_n)$ be an n-tuple of nonzero elements of $\mathbb{F}_q$. Let $k < n$, the **generalized Reed–Solomon (GRS) code** of dimension k associated to the pair $(\mathbf{x}, \mathbf{y})$ is defined as

$$\mathcal{GRS}_k(\mathbf{x}, \mathbf{y}) \stackrel{\text{def}}{=} \{(y_1 f(x_1), \ldots, y_n f(x_n)) \mid f \in \mathbb{F}_q[X],\ \deg(f) < k\},$$

where, by convention, the zero polynomial has degree $-\infty$. A **Reed–Solomon (RS) code** is a GRS code with $\mathbf{y} = (1, \ldots, 1)$ and is denoted as $\mathcal{RS}_k(\mathbf{x})$.

Remark 15.3.20 In terms of diagonal equivalence, any *generalized* Reed–Solomon code is diagonally equivalent to a Reed–Solomon code thanks to the obvious relation

$$\mathcal{GRS}_k(\mathbf{x}, \mathbf{y}) = \mathcal{RS}_k(\mathbf{x}) \star \mathbf{y}.$$

Remark 15.3.21 Beware that our definition of *Reed–Solomon codes* slightly differs from the most usual one in the literature and in particular may differ from that of other chapters of this encyclopedia. Indeed, most of the references define Reed–Solomon codes as cyclic codes of length $q - 1$, i.e., as a particular case of BCH codes. For instance, see Definition 1.14.8, [1008, § 5.2], [1323, § 10.2], [1602, § 5.2], or [1836, Def. 6.8.1]. Note that this commonly accepted definition is not exactly the historical one made by Reed and Solomon themselves in [1582], where they introduced a code of length $n = 2^m$ over $\mathbb{F}_{2^m}$ which is not cyclic.

Further, Reed–Solomon codes "of length q" are sometimes referred to as *extended Reed–Solomon codes*. Next, using Remark 15.3.8, one can actually define generalized Reed–Solomon codes of length $q+1$, corresponding to the codes called (*generalized*) *doubly extended Reed–Solomon codes* in the literature; see also Definition 14.2.6.

Generalized Reed–Solomon codes are known to have length n, dimension k, and minimum distance $d = n - k + 1$; see Lemma 24.1.2. That is to say, such codes are maximum distance separable (MDS), i.e., they meet the Singleton Bound of Theorem 1.9.10 asserting that for any code of length n and dimension k and minimum distance d, we always have $k + d \leq n + 1$. In addition, many algebraic constructions of codes such as BCH codes, Goppa codes, or Srivastava codes derive from some particular GRS codes by applying the *subfield subcode operation*.

Definition 15.3.22 Consider a finite field $\mathbb{F}_q$ and its degree m extension $\mathbb{F}_{q^m}$ for some positive integer m. Let $\mathcal{C} \subseteq \mathbb{F}_{q^m}^n$ be a linear code; the **subfield subcode** of $\mathcal{C}$ is the code $\mathcal{C} \cap \mathbb{F}_q^n$.

The operation defined above is of particular interest for public-key cryptography applications. All these codes fit in a broader class called *alternant codes*. See for instance [1323, Chap. 12, Fig. 12.1].

In some sense, algebraic geometry codes are natural generalizations of generalized Reed–Solomon codes, the latter being algebraic geometry codes from the projective line $\mathbb{P}^1$. Let us start with the case of Reed–Solomon codes.

Proposition 15.3.23 *Let* $\mathbf{x} = (x_1, \ldots, x_n) \in \mathbb{F}_q^n$ *be an n-tuple of distinct elements. Set* $\mathcal{P} = ((x_1 : 1), \ldots, (x_n : 1)) \in \mathbb{P}^1$ *and* $P_\infty = (1 : 0)$. *Then the code* $\mathcal{C}_L(\mathbb{P}^1, \mathcal{P}, (k-1)P_\infty)$ *is the Reed–Solomon code* $\mathcal{RS}_k(\mathbf{x})$.

Proof: By definition, the Riemann–Roch space $L((k-1)P_\infty)$ is the space of rational functions with a pole of order less than k at infinity, which is nothing but the space of polynomials of degree less than k. $\square$

More generally, the equivalence between GRS codes and algebraic geometry codes from $\mathbb{P}^1$ is summarized in the next theorem.

Theorem 15.3.24 *Any generalized Reed–Solomon code is an AG code from $\mathbb{P}^1$ whose evaluation points avoid P_∞. Conversely, any such AG code is a GRS code. More precisely:*

(a) *For any pair $(\mathbf{x}, \mathbf{y})$ as in Definition 15.3.19, we have $\mathcal{GRS}_k(\mathbf{x}, \mathbf{y}) = \mathcal{C}_L(\mathbb{P}^1, \mathcal{P}, G)$, with*

$$\mathcal{P} = ((x_1 : 1), \ldots, (x_n : 1)) \quad and \quad G = (k-1)P_\infty - \mathrm{div}(h),$$

where h is the Lagrange interpolation polynomial satisfying $\deg(h) < n$ and $h(x_i) = y_i$ for any $1 \le i \le n$.

(b) *Conversely, for any ordered n-tuple $\mathcal{P}$ of distinct rational points of $\mathbb{P}^1 \setminus \{P_\infty\}$ with coordinates $(x_1 : 1), \ldots, (x_n : 1)$ and any divisor G of $\mathbb{P}^1$ of degree $k-1$ whose support avoids $\mathcal{P}$, we have $\mathcal{C}_L(\mathbb{P}^1, \mathcal{P}, G) = \mathcal{GRS}_k(\mathbf{x}, \mathbf{y})$ where*

$$\mathbf{x} = (x_1, \ldots, x_n) \quad and \quad \mathbf{y} = (f(x_1), \ldots, f(x_n))$$

for some function $f \in L(G - (k-1)P_\infty) \setminus \{0\}$.

Proof: Using Remark 15.3.20, it suffices to prove that $\mathcal{C}_L(\mathbb{P}^1, \mathcal{P}, G) = \mathcal{RS}_k(\mathbf{x}) \star \mathbf{y}$. This last equality is deduced from Proposition 15.3.23 together with Lemma 15.3.6 since $(k-1)P_\infty - G = \mathrm{div}(h)$, proving (a).

Conversely, since $\deg(G) = k-1$ and the genus of $\mathbb{P}^1$ is 0, then from Corollary 15.2.8, we deduce that the space $L(G - (k-1)P_\infty)$ has dimension 1. Let us take any nonzero function f in this space. Then by definition,

$$\mathrm{div}(f) \geqslant -G + (k-1)P_\infty$$

and, for degree reasons, the above inequality is an equality. Set $\mathbf{y} = (f(x_1), \ldots, f(x_n))$. Again from Proposition 15.3.23 and Lemma 15.3.6, we deduce that

$$\mathcal{C}_L(\mathbb{P}^1, \mathcal{P}, G) = \mathcal{C}_L(\mathbb{P}^1, \mathcal{P}, (k-1)P_\infty) \star \mathbf{y} = \mathcal{RS}_k(\mathbf{x}) \star \mathbf{y} = \mathcal{GRS}_k(\mathbf{x}, \mathbf{y}).$$

This proves (b). $\square$

Remark 15.3.25 Theorem 15.3.24 asserts that any AG code from the projective line is diagonally equivalent to a *one-point code*, i.e., an AG code whose divisor G is supported by one point. This fact can be directly observed without introducing the notation of Reed–Solomon codes, by using a classical result in algebraic geometry asserting that on $\mathbb{P}^1$, two divisors are linearly equivalent if and only if they have the same degree.

15.3.2.2 The $\mathcal{C}_\Omega$ Description

Another way to describe codes from the projective line is to use the differential description. Note that the equivalence between the $\mathcal{C}_L$ and the $\mathcal{C}_\Omega$ description can be made easily explicit in the $\mathbb{P}^1$ case as follows.

Proposition 15.3.26 *Let* $\mathcal{P} = (P_1, \ldots, P_n)$ *be an ordered n-tuple of distinct points of* $\mathbb{P}^1 \setminus \{P_\infty\}$ *with respective homogeneous coordinates* $(x_1 : 1), \ldots, (x_n : 1)$, *and let G be a divisor on* $\mathbb{P}^1$ *whose support avoids the* P_i's. *Then*

$$\mathcal{C}_\Omega(\mathbb{P}^1, \mathcal{P}, G) = \mathcal{C}_L(\mathbb{P}^1, \mathcal{P}, \mathrm{div}(\omega) - D_\mathcal{P} + G),$$

with $\omega \overset{def}{=} \frac{dh}{h}$, *where* $h(x) \overset{def}{=} \prod_{i=1}^n (x - x_i)$.

Proof: A classical result on logarithmic differential forms asserts that $\frac{dh}{h} = \sum_i \frac{dx}{x - x_i}$ has simple poles at the P_i's with residue 1 at them. The proof follows using Lemma 15.3.10 and Remark 15.3.11. $\qquad\qquad\qquad\qquad\qquad\qquad\qquad\qquad\qquad\qquad\qquad\qquad\qquad\qquad\qquad\square$

Even though the $\mathcal{C}_L$ and $\mathcal{C}_\Omega$ definitions are equivalent thanks to Lemma 15.3.10, the differential description is of interest since it allows one to redefine the so-called *classical Goppa codes* [834, 835]. Given an n-tuple $\mathbf{x} = (x_1, \ldots, x_n) \in \mathbb{F}_q$ with distinct entries, a polynomial $f \in \mathbb{F}_q[X]$ which does not vanish at any of the entries of $\mathbf{x}$, and a subfield $\mathbb{K}$ of $\mathbb{F}_q$, then the **classical Goppa code** associated to $(\mathbf{x}, f, \mathbb{K})$ is defined as

$$\Gamma(\mathbf{x}, f, \mathbb{K}) \overset{def}{=} \left\{ \mathbf{c} = (c_1, \ldots, c_n) \in \mathbb{K}^n \; \middle| \; \sum_{i=1}^n \frac{c_i}{X - x_i} \equiv 0 \mod (f) \right\}. \quad (15.6)$$

The major interest of this definition lies in the case of a proper subfield $\mathbb{K} \subsetneq \mathbb{F}_q$, for which the corresponding code benefits from a better estimate of its parameters compared to general alternant codes [1323, 1763]. This estimate comes with an efficient decoding algorithm called the *Patterson Algorithm* [1493] correcting up to half the designed minimum distance. However, in what follows, we are mainly interested in the relationship between this construction and that of $\mathcal{C}_\Omega$ codes and, for this reason, we will focus on the case $\mathbb{F}_q = \mathbb{K}$.

Remark 15.3.27 Note that the terminology might be misleading here. Algebraic geometry codes are sometimes referred to as *Goppa codes*, while the *classical Goppa codes* are **not** algebraic geometry codes from $\mathbb{P}^1$ in general since $\mathbb{K}$ may be different from $\mathbb{F}_q$. These codes are subfield subcodes (see Definition 15.3.22) of some algebraic geometry codes from $\mathbb{P}^1$.

For this reason and to avoid any confusion, in the present chapter, we refer to *algebraic geometry codes* when speaking about $\mathcal{C}_L$ and $\mathcal{C}_\Omega$ codes and to *Goppa codes* or *classical Goppa codes* when dealing with codes as defined in (15.6) with $\mathbb{K} \subsetneq \mathbb{F}_q$.

Theorem 15.3.28 *Denote by* $\mathcal{P} = (P_1, \ldots, P_n) = ((x_1 : 1), \ldots, (x_n : 1))$. *The code* $\Gamma(\mathbf{x}, f, \mathbb{F}_q)$ *equals* $\mathcal{C}_\Omega(\mathbb{P}^1, \mathcal{P}, (f)_0 - P_\infty)$ *where* $(f)_0$ *is the effective divisor given by the zeroes of the polynomial f counted with multiplicity.*

Proof: Let $\mathbf{c} \in \Gamma(\mathbf{x}, f, \mathbb{F}_q)$ and set

$$\omega_\mathbf{c} \overset{def}{=} \sum_{i=1}^n \frac{c_i dx}{x - x_i}.$$

By definition of $\Gamma(\mathbf{x}, f, \mathbb{F}_q)$, the form $\omega_\mathbf{c}$ vanishes on $(f)_0$. In addition, it has simple poles at all the P_i's and is regular at any other point of $\mathbb{P}^1 \setminus \{P_\infty\}$. It remains to check its valuation at infinity. This can be done by replacing x by $1/u$ obtaining

$$\omega_\mathbf{c} = \sum_{i=1}^n \frac{-c_i \frac{du}{u^2}}{\frac{1}{u} - x_i} = -\sum_{i=1}^n \frac{c_i du}{u(1 - x_i u)}.$$

We deduce that $\omega_{\mathbf{c}}$ has valuation ≥ -1 at P_∞ and hence $\omega_{\mathbf{c}} \in \Omega((f)_0 - P_\infty - D_{\mathcal{P}})$, which yields $\Gamma(\mathbf{x}, f, \mathbb{F}_q) \subseteq \mathcal{C}_\Omega(\mathbb{P}^1, \mathcal{P}, (f)_0 - P_\infty)$.

Conversely, given $\omega \in \Omega((f)_0 - P_\infty - D_{\mathcal{P}})$, consider the differential form

$$\eta \overset{\text{def}}{=} \sum_{i=1}^n \frac{\operatorname{res}_{P_i}(\omega)dx}{x - x_i}.$$

The previous discussion shows that the poles of η are simple and are contained in $\{P_1, \ldots, P_n, P_\infty\}$. In addition, the two forms have simple poles and the same residue at any of the points $P_1, \ldots, P_n$, and by the residue formula [1755, Cor. 4.3.3] they also should have the same residue at P_∞. Therefore, the differential form $\eta - \omega$ has no pole on $\mathbb{P}^1$. Moreover, since the degree of a canonical divisor is $2g - 2 = -2$, a nonzero rational differential form on $\mathbb{P}^1$ should have poles. Therefore $\eta = \omega$, and we deduce that the rational function $\sum_i \frac{\operatorname{res}_{P_i}(\omega)}{x - x_i}$ vanishes on $(f)_0$ or, equivalently, that

$$\sum_{i=1}^n \frac{\operatorname{res}_{P_i}(\omega)}{x - x_i} \equiv 0 \mod (f).$$

This yields the inclusion $\mathcal{C}_\Omega(\mathbb{P}^1, \mathcal{P}, (f)_0 - P) \subseteq \Gamma(\mathbf{x}, f, \mathbb{F}_q)$ and concludes the proof. $\qquad\square$

15.4 Asymptotic Parameters of Algebraic Geometry Codes

15.4.1 Preamble

A nonconstructive argument shows that **asymptotically good codes**, i.e., $[n, k, d]$ codes with

- $n \to \infty$,

- $\liminf \frac{k}{n} \geq R > 0$ (positive asymptotic rate), and

- $\liminf \frac{d}{n} \geq \delta > 0$ (positive asymptotic relative minimum distance)

exist over any given finite field $\mathbb{F}_q$. More precisely, the asymptotic version of the Gilbert–Varshamov Bound (see Theorem 1.9.28(e)) shows that this is possible for $0 < \delta < 1 - 1/q$ and $R = 1 - H_q(\delta)$ where

$$H_q(x) \overset{\text{def}}{=} x \log_q(q - 1) - x \log_q(x) - (1 - x) \log_q(1 - x)$$

is the q-ary entropy function.

For a long time it remained an open question to give an explicit description, or even better an effectively computable construction, of asymptotically good codes. The first such construction was proposed by Justesen in [1063]. Let us briefly recall how it works (in a slightly modified version).

Let m be an integer, and assume that we have an explicit $\mathbb{F}_q$-linear identification of $\mathbb{F}_{q^m}$ with $\mathbb{F}_q^m$. Set $n = q^m - 1$ and let $\mathbf{x} = (x_1, \ldots, x_n)$ be an n-tuple consisting of all the nonzero elements in $\mathbb{F}_{q^m}$. For an integer $k < n$, consider the evaluation map

$$\begin{array}{ccc} \mathbb{F}_{q^m}[X]_{<k} & \longrightarrow & (\mathbb{F}_{q^m})^{2n} \\ f(X) & \longmapsto & (f(X), Xf(X))(\mathbf{x}) \end{array},$$

where $\mathbb{F}_{q^m}[X]_{<k}$ denotes the subspace of polynomials of degree less than k, and

$$(f(X), Xf(X))(\mathbf{x}) \overset{\text{def}}{=} (f(x_1), x_1f(x_1), f(x_2), x_2f(x_2), \ldots, f(x_n), x_nf(x_n)).$$

The image of this evaluation map is a $[2n, k]$ code over $\mathbb{F}_{q^m}$, and identifying $\mathbb{F}_{q^m}$ with $\mathbb{F}_q^m$ this becomes a $[2nm, km]$ code $\mathcal{C}$ over $\mathbb{F}_q$.

Theorem 15.4.1 *Let $0 < R < \frac{1}{2}$. Then as $m \to \infty$ and $\frac{k}{n} \to 2R$, the codes $\mathcal{C}$ are asymptotically good, with asymptotic rate R and asymptotic relative minimum distance at least $(1 - 2R)H_q^{-1}(\frac{1}{2})$.*

The idea of the proof is that, for $\epsilon > 0$ and $m \to \infty$, the number of vectors in $\mathbb{F}_q^{2m}$ of relative weight less than $H_q^{-1}(\frac{1}{2} - \epsilon)$ is roughly $q^{2m(\frac{1}{2} - \epsilon)} = q^{m(1-2\epsilon)}$, which is exponentially negligible compared to $n = q^m - 1$. Moreover, if such a vector, seen in $\mathbb{F}_{q^m} \times \mathbb{F}_{q^m}$, is of the form $(\alpha, x_i\alpha)$, then it uniquely determines x_i. Now if f has degree $k \leq 2Rn$, there are at least $(1 - 2R)n$ values of x_i such that $f(x_i) \neq 0$, and then, except for a negligible fraction of them, $(f(x_i), x_if(x_i))$ has weight at least $2mH_q^{-1}(\frac{1}{2} - \epsilon)$ in $\mathbb{F}_q^{2m}$. The conclusion follows.

15.4.2   The Tsfasman–Vlăduţ–Zink Bound

Algebraic geometry also provides asymptotically good codes. Dividing by n in the basic estimate of Corollary 15.3.14 and setting $R = \frac{k}{n}$ and $\delta = \frac{d}{n}$ gives codes whose rate R and relative minimum distance δ satisfy $R + \delta \geq 1 + \frac{1}{n} - \frac{g}{n}$. Letting $n \to \infty$ motivates the following definition.

Definition 15.4.2 The **Ihara constant** of $\mathbb{F}_q$ is

$$A(q) = \limsup_{g(X) \to \infty} \frac{n(X)}{g(X)}$$

where X ranges over all curves over $\mathbb{F}_q$, and $n(X) = |X(\mathbb{F}_q)|$ is the number of rational points of X.

We then readily get:

Theorem 15.4.3 *Assume $A(q) > 1$. Then, for any $R, \delta > 0$ satisfying*

$$R + \delta = 1 - \frac{1}{A(q)},$$

there exist asymptotically good algebraic geometry codes with asymptotic rate at least R and asymptotic relative minimum distance at least δ.

For this result to be meaningful, we need estimates on $A(q)$. Let us start with upper bounds. First, the well-known Hasse–Weil Bound [1755, Th. 5.2.3] implies $A(q) \leq 2\sqrt{q}$. Several improvements were proposed, culminating with the following, known as the **Drinfeld–Vlăduţ Bound**.

Theorem 15.4.4 ([1857]) *For any q we have $A(q) \leq \sqrt{q} - 1$.*

On the other hand, lower bounds on $A(q)$ combine with Theorem 15.4.3 to give lower bounds on codes.

- When $q = p^{2m}$ $(m \geq 1)$ is a square, we have $A(q) \geq \sqrt{q} - 1$. This was first proved by Ihara in [1021], then independently in [1822], using modular curves if $m = 1$, and Shimura curves for general m. Observe that Ihara's lower bound matches the Drinfeld–Vlăduţ Bound, so we actually get equality: $A(q) = \sqrt{q} - 1$. Other more effective constructions matching the Drinfeld–Vlăduţ Bound were later proposed, for instance in [790]. These constructions use recursive towers of curves, although it was observed by Elkies [676, 677] that they in fact yield modular curves.

  Combined with Theorem 15.4.3, Ihara's lower bound gives asymptotically good codes with
  $$R + \delta \geq 1 - \frac{1}{\sqrt{q} - 1}.$$
  This is known as the **Tsfasman–Vlăduţ–Zink (TVZ) Bound** [1822]. For $q \geq 49$, it is shown that this TVZ Bound beats the Gilbert–Varshamov Bound on a certain interval as illustrated in Figure 15.1

- When $q = p^{2m+1}$ $(m \geq 1)$ is a non-prime odd power of a prime, Bassa, Beelen, García, and Stichtenoth [135] show
  $$A(p^{2m+1}) \geq \frac{2(p^{m+1} - 1)}{p + 1 + \frac{p-1}{p^m - 1}} = \left(\frac{1}{2}((p^m - 1)^{-1} + (p^{m+1} - 1)^{-1}) \right)^{-1}.$$

  The proof is constructive and uses a recursive tower of curves, although these curves can also be interpreted in terms of Drinfeld modular varieties.

- For general q, Serre [1653] shows $A(q) \geq c \log(q) > 0$ for a certain constant c which can be taken as $c = \frac{1}{96}$ (see [1437, Th. 5.2.9]). When $q = p$ is prime, this is often the best one knows.

FIGURE 15.1: Tsfasman–Vlăduţ–Zink Bound for $q = 64$

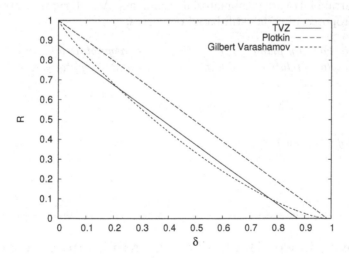

There are some explicit lower bounds for $A(p)$ where $p = 2, 3, 5, \dots$ is a small prime. However these bounds cannot be used in Theorem 15.4.3, which requires $A(q) > 1$. Indeed, the Drinfeld–Vlăduţ Bound actually shows $A(p) < 1$ for $p = 2$ or 3.

15.4.3 Subfield Subcodes and the Katsman–Tsfasman–Wirtz Bound

The Tsfasman–Vlăduţ–Zink Bound provides remarkable results when $q \geq 49$ but, on the other hand, it turns out to be inefficient for smaller values of q and is in particular irrelevant for $q \leq 4$ where it does not even prove the existence of asymptotically good families of algebraic geometry codes.

Very often, lower bounds on codes over a small base field can be obtained by considering good codes over an extension field, and then using either a concatenation argument or a subfield subcode argument; see Definition 15.3.22. In [1896], and independently in [1092], the parameters of subfield subcodes of algebraic geometric codes and asymptotic parameters are studied. More recently, a slightly different construction called *Cartier codes* [458] using the Cartier operator has been proposed providing the same asymptotic parameters leading to the following result.

Theorem 15.4.5 *For any even positive integer ℓ, there exists a sequence of codes over $\mathbb{F}_q$ defined as subfield subcodes of codes over $\mathbb{F}_{q^\ell}$ whose asymptotic parameters (R, δ) satisfy*

$$R \geq 1 - \frac{2(q-1)\ell}{q(q^{\ell/2}-1)} - \frac{(q-1)\ell}{q}\delta \quad \text{for} \quad \frac{q-2}{q^{\ell/2}-1} \leq \delta \leq \frac{q}{m(q-1)} - \frac{2}{q^{\ell/2}-1}.$$

In particular, when $\ell \to \infty$ the asymptotic parameters of such codes reach the Gilbert–Varshamov Bound for $\delta \sim 0$.

15.4.4 Nonlinear Codes

Works of Xing [1916], Elkies [678], and Niederreiter and Özbudak [1436] show that it is possible to construct codes with asymptotic parameters better than those in Theorem 15.4.3, and in particular better than the TVZ Bound, if one turns to *nonlinear* codes. Observe then that, strictly speaking, these codes are not AG codes of the form $\mathcal{C}_L(X, \mathcal{P}, G)$. However they are still obtained by evaluation of functions on an algebraic curve. More precisely, the original construction of these codes uses derivative evaluation of functions on curves, combined with certain intricate combinatorial arguments. An alternative, arguably simpler, construction is proposed in [1756] still based on curves.

Theorem 15.4.6 *For any prime power q, there exist asymptotically good nonlinear codes over $\mathbb{F}_q$ with asymptotic relative minimum distance δ and asymptotic rate*

$$R \geq 1 - \delta - \frac{1}{A(q)} + \log_q\left(1 + \frac{1}{q^3}\right).$$

This holds for any $\delta > 0$ such that this quantity is positive.

15.5 Improved Lower Bounds for the Minimum Distance

Given a curve X over $\mathbb{F}_q$ of genus g, a divisor G, and an ordered set of n rational points $\mathcal{P}$ such that $n > \deg(G)$ (respectively, $\deg(G) > 2g - 2$), from Corollary 15.3.14

$$k + d \geq n + 1 - g. \tag{15.7}$$

Therefore, the parameters of an algebraic geometry code are "at distance at most g" from the Singleton Bound. However, the Goppa Bound is not always reached and improvements may exist under some hypotheses. The present section is devoted to such possible improvements. It should be emphasized that this concerns only "finite length" codes; the statements to follow do not provide any improvement on the asymptotic performances of (linear) algebraic geometry codes presented in Section 15.4.

First, let us try to understand why the Goppa Bound may be not reached by proving the following lemma.

Lemma 15.5.1 *The code $C_L(X, \mathcal{P}, G)$ has minimum distance $d = d^*_{\mathrm{Gop}} = n - \deg(G)$ if and only if there exists $f \in L(G)$ such that the positive divisor $\mathrm{div}(f) + G$ is of the form $P_{i_1} + \cdots + P_{i_s}$ where the P_{i_j}'s are distinct points among $P_1, \ldots, P_n$.*

Proof: Suppose there exists such an $f \in L(G)$ satisfying $\mathrm{div}(f) + G = P_{i_1} + \cdots + P_{i_s}$. Since principal divisors have degree 0, then $s = \deg(G)$. Consequently, the corresponding codeword vanishes at positions with index $i_1, \ldots, i_s$ and hence has weight $n - \deg(G)$. Thus, such a code reaches the Goppa Bound.

Conversely, if the Goppa Bound is reached, then there exists $f \in L(G)$ vanishing at $s = \deg(G)$ distinct points $P_{i_1}, \ldots, P_{i_s}$ among the P_i's. Hence

$$\mathrm{div}(f) \geqslant -G + P_{i_1} + \cdots + P_{i_s}.$$

For degree reasons, the above inequality is an equality. □

Remark 15.5.2 In algebraic geometry, the **complete linear system** or **complete linear series** associated to a divisor G, denoted by $|G|$, is the set of positive divisors linearly equivalent to G. Such a set is parameterized by the projective space $\mathbb{P}(L(G))$. Using this language, one can claim that *the code $C_L(X, \mathcal{P}, G)$ reaches the Goppa Bound if and only if $|G|$ contains a reduced divisor supported by the P_i's.*

For instance, one can provide examples of codes from **elliptic curves** (i.e., curves with $g = 1$) that are MDS, i.e., reach the Singleton Bound.

Example 15.5.3 Recall that given an elliptic curve E over a finite field $\mathbb{F}_q$ with a fixed rational point O_E, the set of rational points has a natural structure of an abelian group with zero element O_E given by the chord-tangent group law; see [1713, § III.2]. The addition law will be denoted by $\oplus$. This group law can also be deduced from the linear equivalence of divisors (see (15.1)) as follows:

$$P \oplus Q = R \quad \Longleftrightarrow \quad (P - O_E) + (Q - O_E) \sim (R - O_E) \tag{15.8}$$

(see [1713, Prop. 3.4]).

From [1821, Th. 3.3.15.5a], there exists an elliptic curve E over a field $\mathbb{F}_q$ whose group of rational points $E(\mathbb{F}_q)$ is cyclic of cardinality $q+1$. Let $P \in E(\mathbb{F}_q)$ be a generator of the group of points and set $\mathcal{P} = (P_1, P_2, \ldots, P_n)$ for some positive integer n such that $\frac{n(n+1)}{2} < q$ and where for any $i \in \{1, \ldots, n\}$, $P_i = iP = P \oplus \cdots \oplus P$ (i times). Now, let $q + 1 > r > \frac{n(n+1)}{2}$ and set $Q = rP$. Choose an integer $n > k > 0$ and consider $Q_1, \ldots, Q_k \in E(\mathbb{F}_q)$ (possibly non-distinct) such that $Q_1 \oplus \cdots \oplus Q_k = Q$. Let $G \sim Q_1 + \cdots + Q_k$ be a divisor. From (15.8), we have

$$(G - kO_E) \sim Q - O_E. \tag{15.9}$$

Consider the code $C \overset{\mathrm{def}}{=} C_L(E, \mathcal{P}, G)$. Its length equals n. Moreover, since $\deg(G) > 0 = 2g - 2$, from Theorem 15.3.12, we have $\dim C = k + 1 - g = k$.

We claim that $\mathcal{C}$ is MDS. Indeed, from the Singleton Bound and (15.7), the minimum distance of the code is either $n - k$ or $n - k + 1$. Suppose the minimum distance is $n - k$; then, from Lemma 15.5.1, there exists $f \in L(G)$ and k distinct points $P_{i_1}, \ldots, P_{i_k}$ among $P_1, \ldots, P_n$ such that

$$\mathrm{div}(f) = P_{i_1} + \cdots + P_{i_k} - G.$$

Therefore

$$G \sim P_{i_1} + \cdots + P_{i_k} \quad \Longleftrightarrow \quad G - kO_E \sim (P_{i_1} - O_E) + \cdots + (P_{i_k} - O_E)$$
$$\Longleftrightarrow \quad Q = rP = P_{i_1} \oplus \cdots \oplus P_{i_k},$$

where the last equivalence is a consequence of (15.8) and (15.9), but contradicts the assertion $r > \frac{n(n+1)}{2}$. Thus, $\mathcal{C}$ is MDS.

Remark 15.5.4 The existence of MDS codes from elliptic curves is further discussed in [1821, § 4.4.2].

A significant part of the literature on algebraic geometry codes proposes improvements of the Goppa Bound under some assumptions on the pair $(\mathcal{P}, G)$ and applies them to examples in order to beat records. These improved bounds can roughly be split into two main categories:

1. *Floor bounds* rest on the use of *base points* of divisors related to G.

2. *Order bounds* rest on *filtrations* of the code

$$\mathcal{C}_L(X, \mathcal{P}, G) \supset \mathcal{C}_L(X, \mathcal{P}, G_1) \supset \mathcal{C}_L(X, \mathcal{P}, G_2) \supset \cdots$$

   for some strictly decreasing sequence of divisors $(G_i)_i$ and the iterated estimates of the minimum distance of the sets $\mathcal{C}_L(X, \mathcal{P}, G_i) \setminus \mathcal{C}_L(X, \mathcal{P}, G_{i+1})$.

A nice overview of many improved bounds in the literature is given in [656].

Thanks to Lemma 15.3.10, we know that any $\mathcal{C}_L$ code is a $\mathcal{C}_\Omega$ one. Therefore, always choosing the most convenient point of view, we alternate between improved lower bounds for the minimum distance of $\mathcal{C}_L$ and $\mathcal{C}_\Omega$ codes. Our point is to provide lower bounds for the minimum distance that improve the Goppa designed distance

$$d_{\mathrm{Gop}} \stackrel{\mathrm{def}}{=} \deg(G - K_X) = \deg(G) + 2 - 2g,$$

where K_X denotes a canonical divisor on X.

15.5.1 Floor Bounds

Given a divisor A on a curve X, a point P of X is said to be a **base point** of A if $L(A) = L(A - P)$.

Remark 15.5.5 Actually, the notion of base point depends only on the divisor class. Thus, if $A' \sim A$, then P is also a base point of A'.

Remark 15.5.6 From the Riemann–Roch Theorem, any divisor A such that $\deg(A) > 2g - 1$ has no base points.

If a divisor G has a base point P outside the set $\mathcal{P}$, then $\mathcal{C}_L(X, \mathcal{P}, G) = \mathcal{C}_L(X, \mathcal{P}, G - P)$ which implies that the minimum distance of this code satisfies $d \geq n - \deg(G) + 1$ instead of $n - \deg(G)$. Similarly, for $\mathcal{C}_\Omega$ codes, if $G - K_X$ has a base point, we have the following.

Lemma 15.5.7 ([656, Lem. 1.3]) *Let P be a base point of $G - K_X$ where P is disjoint from the elements of $\mathcal{P}$. Then*

$$d_H(\mathcal{C}_\Omega(X, \mathcal{P}, G)) \geq d_{\mathrm{Gop}} + 1.$$

Remark 15.5.8 According to Remark 15.3.8, one can get rid of the hypothesis that P is disjoint from the elements of $\mathcal{P}$. In such a situation, the codes $\mathcal{C}_L(X, \mathcal{P}, G)$ and $\mathcal{C}_L(X, \mathcal{P}, G - P)$ are only diagonally equivalent, but the result on the minimum distance still holds.

More generally, the **floor** $\lfloor A \rfloor$ of a divisor A is the divisor A' of smallest degree such that $L(A') = L(A)$. Such a divisor satisfies $\lfloor A \rfloor \leqslant A$ [1325, Prop. 2.1] and we have these bounds:

Theorem 15.5.9 (Maharaj, Matthews, Pirsic [1325, Th. 2.9]) *The code $\mathcal{C}_L(X, \mathcal{P}, G)$ has dimension k and minimum distance d which satisfy*

$$k \geq \deg(G) + 1 - g \quad and \quad d \geq n - \deg(\lfloor G \rfloor).$$

This principle has been used and improved in several references such as [658, 872, 1304, 1325]. The so-called **ABZ Bound** due to Duursma and Park [658, Th. 2.4], inspired from the *AB Bound* of van Lint and Wilson [1838, Th. 5], permits one to deduce many other bounds.

Theorem 15.5.10 (Duursma, Park [658, Th. 2.4]) *Let $G = A + B + Z$ for a divisor $Z \geqslant 0$ whose support is disjoint from $\mathcal{P}$. Then*

$$d_H(\mathcal{C}_\Omega(X, \mathcal{P}, G)) \geq \ell(A) - \ell(A - G + K_X) + \ell(B) - \ell(B - G + K_X).$$

Remark 15.5.11 The Goppa designed distance can be deduced from the ABZ Bound by choosing $A = G$ and $B = Z = 0$. Indeed, we get

$$d_H(\mathcal{C}_\Omega(X, \mathcal{P}, G)) \geq \ell(G) - \ell(K_X) + \ell(0) - \ell(K_X - G).$$

From the Riemann–Roch Theorem, we have $\ell(G) - \ell(K_X - G) = \deg(G) + 1 - g$ and $\ell(0) - \ell(K_X) = 1 - g$. Therefore,

$$d_H(\mathcal{C}_\Omega(X, \mathcal{P}, G)) \geq \deg(G) + 2 - 2g = d_{\mathrm{Gop}}.$$

Several floor bounds can be deduced from Theorem 15.5.10, such as the **Lundell–McCullough (LM) Bound** and the **Güneri–Stichtenoth–Taşkin (GST) Bound**.

Theorem 15.5.12 (Lundell, McCullough [1304, Th. 3]) *Let $G = A + B + Z$ where $Z \geqslant 0$, the support of Z is disjoint from $\mathcal{P}$, $L(A + Z) = L(A)$, and $L(B + Z) = L(B)$. Then*

$$d_H(\mathcal{C}_\Omega(X, \mathcal{P}, G)) \geq d_{\mathrm{LM}} \overset{def}{=} \deg(G) + 2 - 2g + \deg(Z) = d_{\mathrm{Gop}} + \deg(Z).$$

Proof: This is a consequence of Theorem 15.5.10. See [656, Cor. 2.5]. $\square$

Remark 15.5.13 The case $A = B$ was previously proved by Maharaj, Matthews, and Pirsic in [1325, Th. 2.10].

Theorem 15.5.14 (Güneri, Stichtenoth, Taşkin [872, Th. 2.4]) *Let A, B, C, and Z be divisors on X satisfying:*

- *The support of $A + B + C + Z$ is disjoint from $\mathcal{P}$.*

- $L(A) = L(A - Z)$.

- $L(B) = L(B + Z) = L(C)$.

If $G = A + B$, then

$$d_H(\mathcal{C}_\Omega(X, \mathcal{P}, G)) \geq d_{\text{GST}} \overset{\text{def}}{=} d_{\text{Gop}} + \deg(Z) + \ell(K_X - A) - \ell(K_X - G + C).$$

Proof: This is proved in [656, Cor. 2.6] as another consequence of Theorem 15.5.10. $\square$

Example 15.5.15 This example is borrowed from [656]. Calculations have been verified using **Magma** [250]. Consider the *Suzuki curve* X over $\mathbb{F}_8$ defined by the affine equation

$$y^8 + y = x^2(x^8 + x).$$

This curve is known to have genus 14 and 65 rational points. We set P and Q to be the places above the points with respective homogeneous coordinates $(0 : 1 : 0)$ and $(0 : 0 : 1)$. Let $\mathcal{P}$ contain all the rational points of X except P and Q. Set $G = 22P + 6Q$. The code $\mathcal{C}_\Omega(X, \mathcal{P}, G)$ has length 63 and dimension 48. According to the Goppa Bound, its minimum distance satisfies

$$d \geq d_{\text{Gop}} = 2.$$

If we set $A = 16P$, $B = 5P + 4Q$, and $Z = P + 2Q$, a calculation gives $\ell(A) = \ell(A + Z) = 6$ and $\ell(B) = \ell(B + Z) = 1$. Then the Lundell–McCullough Bound may be applied and yields

$$d \geq d_{\text{LM}} = 5.$$

Next, taking $A = 14P + 2Q$, $B = 8P + 4Q$, $C = 8P$, and $Z = 2Q$, one can check that the conditions of Theorem 15.5.14 are satisfied and that

$$d \geq d_{\text{GST}} = 6.$$

Actually, Theorem 15.5.10 gives this lower bound $d \geq 6$ using $A = 14P$, $B = 8P$, and $Z = 6Q$.

15.5.2 Order Bounds

In [713], Feng and Rao propose a new decoding algorithm for algebraic geometry codes associated to a divisor G supported by a single rational point. This algorithm, further discussed in Section 15.6, permits the correction of errors up to half the designed distance. For this sake, they introduced a new method to bound the minimum distance from below. The lower bound they obtained turned out to always be at least as good as the Goppa designed distance. Their approach is at the origin of the so-called *order bounds*.

The Feng–Rao Bound and the corresponding algorithm applied only to codes of the form $\mathcal{C}_\Omega(X, \mathcal{P}, G)$ where the divisor G is supported by one point, that is, $G = rP$ for some rational point P and some positive integer r. Next, some generalizations to arbitrary divisors appeared. One of the most general order bounds is due to Beelen [147]. Before stating it, we need to introduce some notation and definitions.

Definition 15.5.16 Given a divisor F on a curve X and a point P, the **non-gaps semigroup** of F at P denoted by $\nu(F, P)$ is defined as

$$\nu(F, P) \overset{\text{def}}{=} \{j \in \mathbb{Z} \mid L(F + (j - 1)P) \neq L(F + jP)\}.$$

Remark 15.5.17 When $F = 0$, we find the classical notion of *Weierstrass gaps*.

Lemma 15.5.18 (See, e.g., [147, Rem. 2] or [1124, Rem. 3.2]) *The semi-group* $\nu(F, P)$ *satisfies the following conditions.*

(a) $\nu(F, P) \subseteq \{n \in \mathbb{Z} \mid n \geq \deg(F)\}$.

(b) *The set* $\{n \in \mathbb{Z} \mid n \geq \deg(F)\} \setminus \nu(F, P)$, *usually referred to as the set of* F-***gaps*** *at* P, *is finite of cardinality g and contained in* $\{-\deg(F), \ldots, -\deg(F) + 2g - 1\}$.

Now let F_1, F_2, G be divisors on a curve X such that $F_1 + F_2 = G$. We aim to give a lower bound for the minimum distance of $\mathcal{C}_\Omega(X, \mathcal{P}, G)$. Fix a rational point P outside $\mathcal{P}$, and let $(\mu_i)_{i \in \mathbb{N}}$, $(\nu_i)_{i \in \mathbb{N}}$, and $(\rho_i)_{i \in \mathbb{Z}}$ denote, respectively, the F_1-, F_2-, and G-non-gap sequences at P. In addition, for any $r \geq 0$, set

$$n_r \stackrel{\text{def}}{=} \left| \left\{ (i, j) \in \mathbb{N}^2 \mid \mu_i + \nu_j = \rho_{r+1} \right\} \right|. \tag{15.10}$$

Proposition 15.5.19 ([147, Prop. 4]) *Let r be a positive integer. The minimum weight of a codeword in* $\mathcal{C}_\Omega(X, \mathcal{P}, G + rP) \setminus \mathcal{C}_\Omega(X, \mathcal{P}, G + (r+1)P)$ *is bounded from below by* n_r.

A direct consequence of this result is the following.

Theorem 15.5.20 *Let* F_1, F_2, G *be divisors on* X *with* $G = F_1 + F_2$, $\mathcal{P}$ *be an ordered set of rational points of* X, *and* P *be a fixed rational point outside* $\mathcal{P}$. *Then the minimum distance* d *of the code* $\mathcal{C}_\Omega(X, \mathcal{P}, G)$ *satisfies*

$$d \geq d_{\text{Ord}} \stackrel{\text{def}}{=} \min_{r \geq 0} \{n_r\},$$

where n_r *is defined in* (15.10).

Proposition 15.5.21 ([147, Prop. 10]) *The lower bound given in Theorem 15.5.20 is at least as good as the Goppa Bound:*

$$d_{\text{Ord}} \geq d_{\text{Gop}}.$$

Remark 15.5.22 Of course, the previous result may be refined in order to get the minimum distance of a code of the form $\mathcal{C}_\Omega(X, \mathcal{P}, G + sP)$ for some positive s. In particular, Feng and Rao's original approach [713] can be interpreted by choosing $F_1 = F_2 = G = 0$ in order to estimate the minimum distance (and decode) a code of the form $\mathcal{C}_\Omega(X, \mathcal{P}, mP)$ for some positive integer m.

Remark 15.5.23 In [147], Beelen proposes a slightly more general statement in which, instead of considering a filtration of the form

$$\mathcal{C}_\Omega(X, \mathcal{P}, G) \supseteq \cdots \supseteq \mathcal{C}_\Omega(X, \mathcal{P}, G + iP) \supseteq \mathcal{C}_\Omega(X, \mathcal{P}, G + (i+1)P) \supseteq \cdots$$

associated to a single point P, he considers a filtration associated to various points $Q_1, \ldots, Q_s$ where two consecutive terms of the filtration are of the form

$$\mathcal{C}_\Omega(X, \mathcal{P}, G_i) \supseteq \mathcal{C}_\Omega(X, \mathcal{P}, G_{i+1})$$

where $G_{i+1} - G_i \in \{Q_1, \ldots, Q_s\}$. The choice of the optimal path is obtained by a kind of tree exploration using some classical backtracking tricks.

Example 15.5.24 Here we refer to [147, Ex. 8] for an application of this bound to the Klein quartic, the genus 3 curve with affine equation

$$x^3 y + y^3 + x = 0.$$

15.5.3 Further Bounds

The literature provides further more technical and involved bounds that are not discussed in the present chapter. The best reference surveying them and explaining in a clear manner how these bounds are related is due to Duursma, Kirov, and Park [656].

15.5.4 Geometric Bounds for Codes from Embedded Curves

To conclude this section, let us notice that all the previous bounds arise from the intrinsic geometry of the curve and independently from any particular embedding. In another direction, bounds deduced from the geometry of the ambient space for a given embedding have been studied. In [457], the following statement is proved.

Theorem 15.5.25 *Let $X \subseteq \mathbb{P}^N$ be a smooth curve which is a complete intersection in $\mathbb{P}^N$. Let $m \geq 2$ and G_m be a divisor obtained from the intersection of X and a hypersurface of degree m (the points being counted with the intersection multiplicities). Finally, let $\mathcal{P} = (P_1, \ldots, P_n)$ be an ordered n-tuple of rational points of X. Then the minimum distance d of the code $\mathcal{C}_\Omega(X, \mathcal{P}, G_m)$ satisfies:*

(a) $d = m + 2$ *if and only if $m + 2$ of the P_i's are collinear;*

(b) $d = 2m + 2$ *if and only if* (a) *is not satisfied and $2m + 2$ of the P_i's lie on a plane conic;*

(c) $d = 3m$ *if and only if both* (a) *and* (b) *are not satisfied and $3m$ of the P_i's lie at the intersection of a plane cubic and a plane curve of degree m with no common components;*

(d) $d > 3m$ *if and only if none of the previous configurations exist.*

Remark 15.5.26 Actually, Theorem 15.5.25 applies not only to codes from embedded curves but to duals of $\mathcal{C}_L$ codes from arbitrary dimensional complete intersections in a projective space.

In [457, Th. 4.1], it is proved that, for plane curves, Theorem 15.5.25 provides a nontrivial lower bound even in cases where the Goppa Bound is negative and hence irrelevant.

Example 15.5.27 This example is borrowed from [457, Ex. 4.3]. Consider the finite field $\mathbb{F}_{64}$ and the curve X with homogeneous equation

$$w^{24}x^{11} + w^{44}x^6y^2z^3 + w^{24}x^5yz^5 + w^{20}x^4y^6z + w^{33}x^2z^9 +$$
$$w^{46}xy^5z^5 + w^{46}xz^{10} + w^{39}y^{11} + w^{30}y^2z^9 = 0,$$

where w is a primitive element of $\mathbb{F}_{64}$ over $\mathbb{F}_2$ with minimal polynomial $x^6 + x^4 + x^3 + x + 1$. This curve has genus 45 and 80 rational points in the affine chart $\{z \neq 0\}$ and 1 rational point at infinity. We set $\mathcal{P}$ to be the whole set of affine points with some arbitrary ordering. The divisor G_m has degree $11m$; hence the Goppa designed distance applied to $\mathcal{C}_\Omega(X, \mathcal{P}, G_m)$ gives

$$d_{\mathrm{Gop}} = 11m - 88$$

which is negative for $m \leq 8$. Using Theorem 15.5.25 together with a computer aided calculation, we prove that the codes $\mathcal{C}_\Omega(X, D_\mathcal{P}, G_m)$ for $m \in \{1, \ldots, 8\}$ have parameters, respectively, of the form: $[80, 77, \geq 3]$, $[80, 74, \geq 4]$, $[80, 70, \geq 5]$, $[80, 65, \geq 6]$, $[80, 59, \geq 7]$, $[80, 52, \geq 8]$, $[80, 46, \geq 9]$, and $[80, 35, \geq 10]$.

15.6 Decoding Algorithms

While AG codes appeared in the very early 1980s, the first decoding algorithm was not proposed until 1989 by Justesen et al. in [1065] for codes from smooth plane curves. Then, Skorobogatov and Vlăduţ gave a generalization to arbitrary AG codes in [1720]. Further, Pellikaan and independently Kötter gave an abstract version of the algorithm expurgated from algebraic geometry [1155, 1498, 1500]. All these algorithms correct errors up to half the designed distance minus some defect proportional to the curve's genus. In the 1990s many authors tried to fill this gap [654, 665, 1499, 1720, 1856].

In the late 1990s, after Sudan's breakthrough [1762] showing that, at the cost of possibly returning a list of codewords instead of a single codeword, it was possible to correct errors in Reed–Solomon codes beyond half the designed distance, a generalization of Sudan's algorithm was proposed by Shokrollahi and Wasserman in [1693]. Further, Guruswami and Sudan gave an improved algorithm correcting errors up to the so-called Johnson Bound [880].

For a detailed survey on decoding algorithms up to half the designed distance, see [971]. For a more recent survey including list decoding, see [148] and also Chapter 24.

Remark 15.6.1 In the sequel, we suppose that for any divisor A on the given curve X, bases of the spaces $L(A)$ and $\Omega(A)$ can be efficiently computed. It is worth noting that the effective computation of Riemann–Roch spaces is a difficult algorithmic problem of deep interest but which requires an independent treatment. For references on this topic, we refer the reader to, for instance, [950, 1210].

15.6.1 Decoding Below Half the Designed Distance

15.6.1.1 The Basic Algorithm

We first present what is sometimes referred to as the *basic algorithm* in the literature. However, compared to the usual approach, we present this algorithm for a $\mathcal{C}_L$ code instead of a $\mathcal{C}_\Omega$ one. For this reason and as it does not represent a high degree of difficulty, we detail the proofs in the sequel instead of referring to the literature. This algorithm corrects any error pattern of weight t satisfying

$$t \leq \frac{d^*_{\text{Gop}} - 1}{2} - \frac{g}{2}.$$

Let X be a curve of genus g, let $\mathcal{P} = (P_1, \ldots, P_n)$ an ordered n-tuple of distinct rational points of X, and let G be a divisor whose support avoids the P_i's. Let $\mathcal{C}$ be the algebraic geometry code $\mathcal{C}_L(X, \mathcal{P}, G)$. For $\mathbf{c} \in \mathcal{C}$ a codeword and $\mathbf{e} \in \mathbb{F}_q^n$ of Hamming weight $\text{wt}_{\text{H}}(\mathbf{e}) = w \leq t$ for some positive integer t, consider the **received word**

$$\mathbf{y} \overset{\text{def}}{=} \mathbf{c} + \mathbf{e}.$$

Recall from Theorem 15.3.12 that $d^*_{\text{Gop}} = n - \deg(G)$ denotes the designed distance of $\mathcal{C}$.

By definition, there exists a function $f \in L(G)$ such that $\mathbf{c} = (f(P_1), \ldots, f(P_n))$. Denote by $\{i_1, \ldots, i_w\} \in \{1, \ldots, n\}$ the **support** of $\mathbf{e}$, that is to say, the set of indexes corresponding to the nonzero entries of $\mathbf{e}$, i.e., the positions at which errors occurred.

A decoding algorithm correcting t errors takes as inputs $(\mathcal{C}, \mathbf{y})$ and returns either $\mathbf{c}$ (or equivalently $\mathbf{e}$) if $\text{wt}_{\text{H}}(\mathbf{e}) \leq t$ or "?" if $\text{wt}_{\text{H}}(\mathbf{e})$ is too large. The first algorithms in the literature [1065, 1720] rest on the calculation of an *error-locating function*. For this sake, one introduces an extra divisor F whose support avoids $\mathcal{P}$ and whose additional properties are to be decided later. The point is to compute a nonzero function $\lambda \in L(F)$ "locating the error positions", i.e., such that $\lambda(P_{i_1}) = \cdots = \lambda(P_{i_w}) = 0$. Once such a function is computed, its zero locus provides a subset of indexes $J \subseteq \{1, \ldots, n\}$ such that $\{i_1, \ldots, i_w\} \subseteq J$. If this set J is small enough, then its knowledge allows the decoding to proceed by solving a linear system as suggested by the following statement. This is a classical result of coding theory: once the errors are located, decoding reduces to correcting erasures.

Proposition 15.6.2 *Let* $\mathbf{H}$ *be a parity check matrix for* C *and* $J \subseteq \{1, \ldots, n\}$ *such that* $|J| < d_H(C)$ *and which contains the support of* $\mathbf{e}$. *Then* $\mathbf{e}$ *is the unique solution of the system*

$$\begin{cases} \mathbf{H}\mathbf{e}^T = \mathbf{H}\mathbf{y}^T \\ e_i = 0 \qquad \text{for all } i \in \{1, \ldots, n\} \setminus J. \end{cases} \tag{15.11}$$

Proof: Clearly $\mathbf{e}$ is a solution. If $\mathbf{e}'$ is another solution, then $\mathbf{e} - \mathbf{e}' \in \ker \mathbf{H} = C$ and has support included in J. This word has weight less than the code's minimum distance and hence is $\mathbf{0}$. $\qquad \square$

For a function $\lambda \in L(F)$ vanishing at the error positions, the fundamental observation is that

$$\lambda(P_i)y_i = \lambda(P_i)f(P_i) \quad \text{for all } i \in \{1, \ldots, n\}. \tag{15.12}$$

Indeed, either there is no error at position i, i.e., $e_i = 0$ and hence $y_i = f(P_i)$, or there is an error but in this situation $\lambda(P_i) = 0$, making the above equality obviously true. Next, since $\lambda f \in L(G + F)$, we deduce that $(\lambda(P_1)y_1, \ldots, \lambda(P_n)y_n) \in \mathcal{C}_L(X, \mathcal{P}, G + F)$. This motivates the introduction of the following space:

$$K_{\mathbf{y}} \overset{\text{def}}{=} \{\lambda \in L(F) \mid (\lambda(P_1)y_1, \ldots, \lambda(P_n)y_n) \in \mathcal{C}_L(X, \mathcal{P}, G + F)\}. \tag{15.13}$$

Lemma 15.6.3 *Let* $D_{\mathbf{e}} \overset{\text{def}}{=} P_{i_1} + \cdots + P_{i_w}$ *be the sum of points at which an error occurs. Then*

$$\mathcal{C}_L(X, \mathcal{P}, F - D_{\mathbf{e}}) \subseteq K_{\mathbf{y}}.$$

Proof: If $\lambda \in L(F - D_{\mathbf{e}})$, then λ vanishes at the error points and the result is a consequence of (15.12). $\qquad \square$

Proposition 15.6.4 *If* $t \leq d_{\text{Gop}}^* - \deg(F) - 1 = n - \deg(G - F) - 1$, *then*

$$K_{\mathbf{y}} = L(F - D_{\mathbf{e}}).$$

Proof: Inclusion $\supseteq$ is given by Lemma 15.6.3. Conversely, if $(\lambda(P_1)y_1, \ldots, \lambda(P_n)y_n) \in \mathcal{C}_L(X, \mathcal{P}, G + F)$, then, since $(\lambda(P_1)f(P_1), \ldots, \lambda(P_n)f(P_n)) \in \mathcal{C}_L(X, \mathcal{P}, G + F)$, this implies that $\mathbf{u} \overset{\text{def}}{=} (\lambda(P_1)e_1, \ldots, \lambda(P_n)e_n)$ also lies in this code. In addition, $\text{wt}_H(\mathbf{u}) \leq \text{wt}_H(\mathbf{e}) \leq t$. On the other hand, from Theorem 15.3.12, $d_H(\mathcal{C}_L(X, \mathcal{P}, G + F)) \geq n - \deg(G + F)$. Therefore, $\text{wt}_H(\mathbf{u})$ is less than the code's minimum distance, and hence $\mathbf{u} = (\lambda(P_1)e_1, \ldots, \lambda(P_n)e_n) = \mathbf{0}$. Thus, λ vanishes at any position where $\mathbf{e}$ does not. Hence $\lambda \in L(F - D_{\mathbf{e}})$. $\qquad \square$

The previous statements provide the necessary material to describe a decoding algorithm for the code $\mathcal{C}_L(X, \mathcal{P}, G)$. Mainly, the algorithm consists of:

1. computing $K_{\mathbf{y}}$;

2. taking a nonzero function λ in $K_{\mathbf{y}}$;

3. computing the zeroes of λ among the P_i's, which, hopefully, should provide a set localizing the errors; and

4. finding **e** by solving a linear system using Proposition 15.6.2.

More precisely the pseudo-code of the complete procedure is given in Algorithm 15.6.5.

Algorithm 15.6.5 (Basic Decoding)

Input: A code $\mathcal{C} = \mathcal{C}_L(X, \mathcal{P}, G)$, a vector $\mathbf{y} \in \mathbb{F}_q^n$, and an integer $t > 0$

Output: A vector **e** such that $\mathrm{wt_H}(\mathbf{e}) \leq t$ and $\mathbf{y} - \mathbf{e} \in \mathcal{C}$ if it exists; otherwise "?"

1: Compute the space $K_\mathbf{y}$ defined in (15.13);
2: **if** $K_\mathbf{y} = \{0\}$ **then**
3:    **return** "?"
4: **else**
5:    Take $\lambda \in K_\mathbf{y} \setminus \{0\}$;
6:    Compute $i_1, \ldots, i_s \in \{1, \ldots, n\}$ such that $\lambda(P_{i_j}) = 0$;
7:    Let S be a affine space of solutions of the system (15.11);
8:    **if** $S = \emptyset$ or $|S| > 2$ **then**
9:       **return** "?"
10:    **else**
11:       **return** **e** (the unique solution of (15.11))
12:    **end if**
13: **end if.**

Theorem 15.6.6 *If* $t \leq \frac{d^*_{\mathrm{Gop}}-1}{2} - \frac{g}{2}$ *and* $\deg(F) = t + g$, *then Algorithm* 15.6.5 *is correct and returns the solution* **e** *in* $O(n^\omega)$ *operations in* $\mathbb{F}_q$, *where* ω *is the complexity exponent of linear algebraic operations (in particular* $\omega \leq 3$).

Proof: If $\deg(F) \geq t + g$, then $\deg(F - D_\mathbf{e}) \geq g$ and hence, by the Riemann–Roch Theorem, $\ell(F - D_\mathbf{e}) > 0$. In particular, from Lemma 15.6.3, $K_\mathbf{y} \neq \{0\}$. Next, by assumption on t and $\deg(F)$, we get

$$2t + g \leq d^*_{\mathrm{Gop}} - 1 \implies t \leq d^*_{\mathrm{Gop}} - \deg(F) - 1 \tag{15.14}$$

which, from Proposition 15.6.4, yields the equality $K_\mathbf{y} = L(F - D_\mathbf{e})$. Therefore, one can compute $L(F - D_\mathbf{e})$. Take any nonzero function λ in this space. It remains to prove that the conditions of Proposition 15.6.2 are satisfied, i.e., that the zero locus of λ in $\{P_1, \ldots, P_n\}$ is not too large. But, from (15.14), $\lambda \in L(F)$, implying its zero divisor $(\lambda)_0$ has degree at most $\deg(F)$. Since $0 \leq t \leq d^*_{\mathrm{Gop}} - \deg(F) - 1$, we immediately see that $\deg((\lambda)_0) \leq \deg(F) < d^*_{\mathrm{Gop}}$ which proves that the solution of the system (15.11) will provide **e** as the unique possible solution.

About the complexity, the computation of $K_\mathbf{y}$ and the solution of the system (15.11) are nothing but the solution of linear systems with $O(n)$ equations and $O(n)$ unknowns. The other operations in the algorithm are negligible. $\qquad\square$

15.6.1.2   Getting Rid of Algebraic Geometry: Error-Correcting Pairs

As observed by Pellikaan [1498, 1500] and independently by Kötter [1155], the Basic Decoding Algorithm can be defined on the level of codes without involving any algebraic geometric notion. To do that, observe that

$$\mathbf{a} \overset{\text{def}}{=} (\lambda(P_1), \ldots, \lambda(P_n)) \in \mathcal{C}_L(X, \mathcal{P}, F).$$

Next, on the level of codes, we search for a vector $\mathbf{a}$ such that $\mathbf{a} \star \mathbf{y} \in \mathcal{C}_L(X, \mathcal{P}, G + F)$. Thus, the space $K_{\mathbf{y}}$ can be redefined on the level of codes as the space

$$\widehat{K}_{\mathbf{y}} \overset{\text{def}}{=} \left\{ \mathbf{a} \in \mathcal{C}_L(X, \mathcal{P}, F) \mid \text{for all } \mathbf{b} \in \mathcal{C}_L(X, \mathcal{P}, G + F)^\perp, \ \langle \mathbf{a} \star \mathbf{y}, \mathbf{b} \rangle_{\text{Eucl}} = 0 \right\}.$$

Next, one shows easily that the adjunction property noticed in Lemma 15.1.1 yields an equivalent reformulation of the above definition as

$$\widehat{K}_{\mathbf{y}} = \left\{ \mathbf{a} \in \mathcal{C}_L(X, \mathcal{P}, F) \mid \text{for all } \mathbf{b} \in \mathcal{C}_L(X, \mathcal{P}, G + F)^\perp, \ \langle \mathbf{a} \star \mathbf{b}, \mathbf{y} \rangle_{\text{Eucl}} = 0 \right\}.$$

If we set

$$\mathcal{A} = \mathcal{C}_L(X, \mathcal{P}, F) \ \text{ and } \ \mathcal{B} \overset{\text{def}}{=} \mathcal{C}_L(X, \mathcal{P}, G + F)^\perp = \mathcal{C}_\Omega(X, \mathcal{P}, G + F),$$

then one can prove in particular that

$$\mathcal{A} \star \mathcal{B} \subseteq \mathcal{C}_\Omega(X, \mathcal{P}, G) = \mathcal{C}^\perp,$$

and this material turns out to be sufficient to provide a decoding algorithm.

Definition 15.6.7 Let $\mathcal{C} \subseteq \mathbb{F}_q^n$ be a code and t a positive integer. A pair of codes $(\mathcal{A}, \mathcal{B})$ is said to be a t-**Error-Correcting Pair (ECP)** for $\mathcal{C}$ if it satisfies the following conditions:

(ECP1) $\mathcal{A} \star \mathcal{B} \subseteq \mathcal{C}^\perp$,

(ECP2) $\dim \mathcal{A} > t$,

(ECP3) $\dim \mathcal{B}^\perp > t$, and

(ECP4) $d_H(\mathcal{A}) + d_H(\mathcal{C}) > n$.

Theorem 15.6.8 (See, e.g., [1500, Th. 2.14]) *Let $\mathcal{C} \subseteq \mathbb{F}_q^n$ be a code equipped with a t-error-correcting pair $(\mathcal{A}, \mathcal{B})$ with $t \leq \frac{d_H(\mathcal{C}) - 1}{2}$. Then there is a decoding algorithm for $\mathcal{C}$ correcting any error pattern of weight less than or equal to t in $O(n^\omega)$ operations in $\mathbb{F}_q$.*

Remark 15.6.9 Using Theorem 15.3.12 and Corollary 15.3.13, one can easily observe that the codes $\mathcal{A} = \mathcal{C}_L(X, \mathcal{P}, F)$ and $\mathcal{B} = \mathcal{C}_\Omega(X, \mathcal{P}, G + F)$ satisfy the conditions of Definition 15.6.7. Next, the algorithm mentioned in Theorem 15.6.8 is nothing but Algorithm 15.6.5 described at the level of codes, in particular, replacing elements $\lambda \in L(F)$ by the corresponding evaluation vectors $(\lambda(P_1), \ldots, \lambda(P_n)) \in \mathcal{C}_L(X, \mathcal{P}, F)$.

Remark 15.6.10 As noticed earlier, the tradition in the literature is to present the basic algorithm using error-correcting pairs to decode $\mathcal{C}_\Omega(X, \mathcal{P}, G)$. This code has a very similar decoding algorithm correcting up to $\frac{d_{\text{Gop}} - 1}{2} - \frac{g}{2}$ errors using the error-correcting pair

$$\mathcal{A} = \mathcal{C}_L(X, \mathcal{P}, F) \ \text{ and } \ \mathcal{B} = \mathcal{C}_L(X, \mathcal{P}, G - F)$$

where F is an extra divisor with $\deg(F) \geq t + g$.

Remark 15.6.11 Note that the defect $\frac{g}{2}$ corresponds actually to a *worst case*; it is observed for instance in [971, Rem. 4.6] that this algorithm can actually correct a uniformly random error pattern of weight $t = \lfloor \frac{d_{\text{Gop}}^* - 1}{2} \rfloor$ with a failure probability of $O(1/q)$.

15.6.1.3 Reducing the Gap to Half the Designed Distance

After the basic algorithm, several attempts appeared in the literature to reduce the gap between the decoding radius of this algorithm and half the designed distance.

- The *modified algorithm* [1720] and the *extended modified algorithm* [654] combine the basic algorithm with an iterative search for a relevant choice for the extra divisor F. These algorithms permit a reduction in the gap $\frac{g}{2}$ to about $\frac{g}{4}$; see [971, Rem. 4.11 & 4.15].

- In [1499, 1856] the question of the existence of an extra divisor F for which the basic algorithm corrects up to half the designed distance is discussed. It is, in particular, proved that such an F exists and can be chosen in a set of $O(n)$ divisors as soon as $q \geq 37$.

- Finally, an iterative approach to finding an F achieving half the designed distance was proposed by Ehrhard [665].

All the above contributions are summarized with further details in [971, § 4 to 7].

15.6.1.4 Decoding Up to Half the Designed Distance: The Feng–Rao Algorithm and Error-Correcting Arrays

The first success in getting a decoding algorithm correcting exactly up to half the designed distance is due to Feng and Rao [713]. Their algorithm consists of using a filtration of codes and to iterate *coset decoding*. Note that Feng–Rao's original algorithm applied to codes $\mathcal{C}_\Omega(X, \mathcal{P}, rP)$ for some rational point P of X. Later, their approach was extended to arbitrary divisors [1124, 1501].

Similar to the basic algorithm which led to the abstract and algebraic geometry free formulation of error-correcting pairs, Feng and Rao's approach led to a purely coding theoretic formulation called *error-correcting arrays*. We present this approach in the sequel using the notation of [1501] together with refinements appearing in [462].

Definition 15.6.12 An **array of codes** for a code $\mathcal{C}$ is a triple of sequences of codes $\left((\mathcal{A}_i)_{0 \leq i \leq u}, (\mathcal{B}_j)_{0 \leq j \leq v}, (\mathcal{C}_r)_{w \leq r \leq n}\right)$ such that

(A1) $\mathcal{C} = \mathcal{C}_w$;

(A2) for all i, j, r, $\dim \mathcal{A}_i = i$, $\dim \mathcal{B}_j = j$, and $\dim \mathcal{C}_r = n - r$;

(A3) the sequences $(\mathcal{A}_i)_i$ and $(\mathcal{B}_j)_j$ are increasing, while $(\mathcal{C}_r)_r$ is decreasing; and

(A4) for all i, j, we define $\widehat{r}(i, j)$ to be the least integer r such that $\mathcal{A}_i \star \mathcal{B}_j \subseteq \mathcal{C}_r^\perp$. For any $i, j \geq 1$, if $\mathbf{a} \in \mathcal{A}_i \setminus \mathcal{A}_{i-1}$ and $\mathbf{b} \in \mathcal{B}_j \setminus \mathcal{B}_{j-1}$ and $r = \widehat{r}(i, j) > w$, then $\mathbf{a} \star \mathbf{b} \in \mathcal{C}_r^\perp \setminus \mathcal{C}_{r-1}^\perp$.

Note that the function $\widehat{r}(i, j)$ is increasing in the variables i and j but not necessarily strictly increasing. This motivates the following definition of a *well-behaving pair*, a terminology borrowed from that of *well-behaving sequences* [799].

Definition 15.6.13 A pair $(i, j) \in \{1, \ldots, u\} \times \{1, \ldots, v\}$ is said to be **well-behaving (WB)** if for any pair (i', j') such that $i' \leq i$, $j' \leq j$ and $(i', j') \neq (i, j)$, we have $\widehat{r}(i', j') < \widehat{r}(i, j)$. For any $r \in \{w, \ldots, n-1\}$, define

$$\widehat{n}_r \overset{\text{def}}{=} \left|\{(i, j) \in \{1, \ldots, u\} \times \{1, \ldots, v\} \mid (i, j) \text{ is WB and } \widehat{r}(i, j) = r + 1\}\right|.$$

Theorem 15.6.14 (See [1501, Th. 4.2]) *For any array of codes* $\left((\mathcal{A}_i)_i, (\mathcal{B}_j)_j, (\mathcal{C}_r)_r\right)$, *we have for all* $w \leq r \leq n - 1$

$$d_H(\mathcal{C}_r) \geq \min\{\widehat{n}_{r'} \mid r \leq r' \leq n - 1\}.$$

Remark 15.6.15 Of course, Theorem 15.6.14 is almost the same as Theorem 15.5.20. The slight difference is in the quantity $\widehat{n}_r$, which is very close to n_r introduced in (15.10) from Section 15.5.2. The difference lies in the fact that n_r is defined on the level of function algebras while $\widehat{n}_r$ is defined on the level of codes. On function algebras over curves, valuations assert a strict growth of the function $r(i, j)$ that could be defined in this context. When dealing on the level of codes, we should restrict to a subset of pairs (the well-behaving ones) to keep this strict increasing property.

This requirement of strict increasing is necessary to prove Theorem 15.6.14. This bound on the minimum distance is obtained by relating the Hamming weight of some codeword to the rank of a given matrix. The well-behaving pairs provide the position of some pivots in this matrix. Thus, their number give a lower bound for the rank. Of course two pivots cannot lie on the same row or column, hence the requirement of strict decreasing. Note that in [1501, Th. 4.2], the strict increasing requirement is lacking, which makes the proof not completely correct. Replacing general pairs by well-behaving ones, as is done in [462], fixes the proof.

Definition 15.6.16 An array of codes $\left((\mathcal{A}_i)_{1 \leq i \leq u}, (\mathcal{B}_j)_{1 \leq j \leq v}, (\mathcal{C}_r)_{w \leq r \leq n}\right)$ for a code $\mathcal{C}$ is said to be a **t-error-correcting array for** $\mathcal{C}$ if

(i) $\mathcal{C}_w = \mathcal{C}$ and

(ii) $t \leq \frac{d_H(\mathcal{C}_w) - 1}{2}$.

Theorem 15.6.17 (See [1501, Th. 4.4]) *If a code $\mathcal{C}$ has a t-error-correcting array, then it has a decoding algorithm correcting up to t errors in $O(n^\omega)$ operations in $\mathbb{F}_q$.*

The spirit of the algorithm is to start from a received vector $\mathbf{y} = \mathbf{c} + \mathbf{e}$ with $\mathbf{c} \in \mathcal{C} = \mathcal{C}_w$ and $\mathrm{wt}_H(\mathbf{e}) \leq t$, then use the array to "guess an additional syndrome" using a process called *majority voting*. This transforms the problem into decoding

$$\mathbf{y}_{w+1} = \mathbf{c}_{w+1} + \mathbf{e}$$

where $\mathbf{c}_{w+1} \in \mathcal{C}_{w+1}$ noting that *the error is unchanged*. Applying this process iteratively until $\mathbf{c}_n = \mathbf{0}$ yields the error.

Remark 15.6.18 Actually, as noticed in [1501], the process may be stopped before reaching $\mathbf{c}_n$. Indeed, after $r \geq g$ iterations, one obtains a new decoding problem $\mathbf{y}_r = \mathbf{c}_r + \mathbf{e}$ where $\mathcal{C}_r$ benefits from a t-error-correcting pair; one can at this step switch to the error-correcting pair algorithm, which corrects up to $\frac{d_H(\mathcal{C}_r) - 1}{2} - \frac{g}{2} \geq \frac{d_H(\mathcal{C}_w) - 1}{2}$ errors, and get $\mathbf{e}$.

The impact of the previous constructions on algebraic geometry codes is summarized in the following statement.

Theorem 15.6.19 ([1501, Th. 4.5]) *Let $\mathcal{C} = \mathcal{C}_\Omega(X, \mathcal{P}, G)$ be a code on a curve X of genus g such that $2g - 2 < \deg(G) < n - g$. Then $\mathcal{C}$ has a t-error-correcting array with*

$$t = \frac{d_{\mathrm{Gop}} - 1}{2}.$$

Remark 15.6.20 Of course, a similar result holds for evaluation codes by replacing G by $K - G + D_{\mathcal{P}}$.

Remark 15.6.21 A simple choice of an error-correcting array for $\mathcal{C}_\Omega(X, \mathcal{P}, G)$ can be obtained by choosing a rational point P of X and considering the following sequences to construct the $\mathcal{A}_i$'s, the $\mathcal{B}_j$'s, and the $\mathcal{C}_r$'s:

$$\cdots \subseteq \mathcal{C}_L(X, \mathcal{P}, \mu_i P) \subseteq \mathcal{C}_L(X, \mathcal{P}, \mu_{i+1} P) \subseteq \cdots$$
$$\cdots \subseteq \mathcal{C}_L(X, \mathcal{P}, G + \nu_j P) \subseteq \mathcal{C}_L(X, \mathcal{P}, G + \nu_{j+1} P) \subseteq \cdots$$
$$\cdots \supseteq \mathcal{C}_\Omega(X, \mathcal{P}, G + \nu_j P) \supseteq \mathcal{C}_\Omega(X, \mathcal{P}, G + \nu_{j+1} P) \supseteq \cdots$$

where $(\mu_i)_i$ and $(\nu_j)_j$, respectively, denote the 0- and G-non-gap sequences at P; see Definition 15.5.16.

15.6.2 List Decoding and the Guruswami–Sudan Algorithm

The previous algorithms permit error-correction up to half the designed distance. Note that Shannon theory essentially asserts that, for a random code, decoding up to the minimum distance is almost always possible. A manner to fill the gap between Hamming and Shannon's point of view is to use *list decoding*. That is to say, given a code $C \subseteq \mathbb{F}_q^n$, a received word $\mathbf{y} \in \mathbb{F}_q^n$, and a positive integer t, list decoding consists in returning the whole list of codewords $\mathbf{c} \in C$ at distance less than or equal to t from $\mathbf{y}$.

Let $C = \mathcal{C}_L(X, \mathcal{P}, G)$, $\mathbf{y} \in \mathbb{F}_q^n$, and t be a positive integer. The Guruswami–Sudan Algorithm may be regarded as a generalization of Algorithm 15.6.5 when reformulated as follows. In Section 15.6.1.1, we were looking for a function $Q \in \mathbb{F}_q(X)[Y]$ of the form $Q = Q_0 + Q_1 Y$ with $Q_0 \in L(G + F)$ and $Q_1 \in L(F)$ such that the function $Q(f)$ is identically 0. In this reformulation, the function Q_1 is nothing but our former error-locating function λ. Note that, regarded as a polynomial in the variable Y, the polynomial Q has degree 1 and hence has a unique root. Sudan's key idea is to consider a polynomial of larger degree in Y possibly providing a list of solutions instead of a single one.

Fix a positive integer ℓ which will be our maximal list size, a positive integer s called the *multiplicity* and an extra divisor F whose support avoids the points of $\mathcal{P}$ and whose additional properties are to be decided later. The algorithm is divided in two main steps: *interpolation* and *root finding* which are described in Algorithm 15.6.22. (Compare this algorithm to the Guruswami–Sudan Algorithm 24.4.6 for generalized Reed–Solomon codes.)

Algorithm 15.6.22 (Guruswami–Sudan Decoding for AG Codes)

Interpolation: Compute a nonzero polynomial $Q \in \mathbb{F}_q(X)[Y]$ of the form

$$Q = Q_0 + Q_1 Y + \cdots + Q_\ell Y^\ell$$

satisfying:

(i) for any $j \in \{0, \ldots, \ell\}$, $Q_j \in L(F + (\ell - j)G)$, and

(ii) for any $i \in \{1, \ldots, n\}$, the function Q vanishes at (P_i, y_i) with multiplicity at least s.

Root finding: Compute the roots $f_1, \ldots, f_m$ $(m \leq \ell)$ of $Q(Y)$ lying in $\mathbb{F}_q(X)$ and output the list of codewords of the form $(f_i(P_1), \ldots, f_i(P_n))$ that are at distance at most t from $\mathbf{y}$.

Remark 15.6.23 Geometrically speaking, the function Q in Algorithm 15.6.22 can be interpreted as a rational function on the surface $X \times \mathbb{P}^1$. In this context, saying that Q *vanishes at (P_i, y_i) with multiplicity at least s* means that $Q \in \mathfrak{m}^s_{(P_i, y_i)}$, where $\mathfrak{m}_{(P_i, y_i)}$ denotes the maximal ideal of the local ring of the surface at the point (P_i, y_i). From a more computational point of view, given a local parameter $t \in \mathbb{F}_q(X)$ at P_i, the function Q has a Taylor series expansion $\sum_{u,v} q_{uv} t^u (Y - y_i)^v$ and the vanishing requirement means that $q_{uv} = 0$ for any pair (u, v) such that $u + v < s$.

Remark 15.6.24 In [876, § 6.3.4], a polynomial time algorithm to perform the root finding step is presented. This algorithm consists in a reduction of Q modulo some place of $\mathbb{F}_q(X)$ of large enough degree, then factoring the reduced polynomial using the Berlekamp or the Cantor–Zassenhaus Algorithm, followed by a "lifting" of the roots in $\mathbb{F}_q(Y)$, under the assumption that they lie in a given Riemann–Roch space.

Theorem 15.6.25 (See, e.g., [148, Lem. 2.3.2 & 2.3.3]) *Let*

$$t \le n - \frac{n(s+1)}{2(\ell+1)} - \frac{\ell \deg(G)}{2s} - \frac{g}{s}.$$

Then, for any extra divisor F satisfying

$$\frac{ns(s+1)}{2(\ell+1)} + \frac{\ell \deg(G)}{2} + g - 1 < \deg(F + \ell G) < s(n - t),$$

the Guruswami–Sudan Algorithm 15.6.22 succeeds in returning in polynomial time the full list of codewords in $\mathcal{C}_L(X, \mathcal{P}, G)$ at distance less than or equal to t from $\mathbf{y}$.

Remark 15.6.26 The case with no multiplicity, i.e., the case $s = 1$ was considered before Guruswami and Sudan by Sudan for Reed–Solomon codes [1762] (see Algorithm 24.3.1) and Shokrollahi and Wasserman [1693] for algebraic geometry codes.

Remark 15.6.27 When $\ell = s = 1$, the algorithm should yield the basic algorithm. However, Theorem 15.6.25 yields a decoding radius of $\left\lfloor \frac{d^*_{\text{Gop}} - 1}{2} - g \right\rfloor$: a defect of g instead of $\frac{g}{2}$. Further analysis shows that, whenever $g > 0$, the linear system whose solution is the polynomial Q never has full rank. On the other hand, the analysis of the decoding radius is done without taking this rank defect into account, leading to a slightly pessimistic estimate. A further analysis of the rank of the system in the general case might provide a slight improvement of the Guruswami–Sudan decoding radius.

15.7 Application to Public-Key Cryptography: A McEliece-Type Cryptosystem

15.7.1 History

In [171] Berlekamp, McEliece, and van Tilborg proved that the following problem is NP-complete.

Problem 15.7.1 Let $\mathcal{C} \subseteq \mathbb{F}_q^n$ be a code, $t \le n$ be a positive integer, and $\mathbf{y} \in \mathbb{F}_q^n$. Decide whether or not there exists $\mathbf{c} \in \mathcal{C}$ whose Hamming distance to $\mathbf{y}$ is less than or equal to t.

Note that NP-completeness may only assert hardness in the worst case. However, this problem is commonly believed by the community to be "difficult in average". By this, we mean that most of the instances of the problem seem difficult.

This result motivated McEliece to propose a code-based encryption scheme[2] whose security was relying on the hardness of this problem [1361]. Roughly speaking, the McEliece scheme can be described follows.

- The public key is a code $\mathcal{C}$.

- The secret key is a secret element related to $\mathcal{C}$ that permits decoding.

- Encryption consists in encoding the plain text and then including errors in it.

- Decryption is decoding.

More formally, let $\mathcal{F}_{n,k}$ be a family of codes of fixed length n and dimension k. Consider a set $\mathcal{S}$ of *secrets* together with a surjective map $\mathcal{C} : \mathcal{S} \to \mathcal{F}_{n,k}$ such that for any $s \in \mathcal{S}$, the code $\mathcal{C}(s)$ has a decoding algorithm $\mathbf{Dec}(s)$ *depending on* s that corrects up to t errors. See Figure 15.2.

FIGURE 15.2: McEliece encryption scheme

Key generation: Draw a uniformly random element $s \in \mathcal{S}$.

    Secret key: The secret s

    Public key: A pair $(\mathbf{G}, t)$ where $\mathbf{G}$ is a $k \times n$ generator matrix of the code $\mathcal{C}(s) \in \mathcal{F}_{n,k}$ and t is the number of errors that the decoder $\mathbf{Dec}(s)$ can correct for the code $\mathcal{C}(s)$

Encryption: The plain text is a vector $\mathbf{m} \in \mathbb{F}_q^k$. Pick $\mathbf{e}$ a uniformly random element of $\mathbb{F}_q^n$ of weight t and define the cipher text as

$$\mathbf{y}_{\text{cipher}} \overset{\text{def}}{=} \mathbf{mG} + \mathbf{e}.$$

Decryption: Apply $\mathbf{Dec}(s)(\mathbf{y}_{\text{cipher}})$ to recover the plain text $\mathbf{m}$ from $\mathbf{y}_{\text{cipher}}$.

Example 15.7.2 (Algebraic geometry codes) Let X be a curve of genus g and $\mathcal{S}$ be the set of pairs $(\mathcal{P}, G)$ where $\mathcal{P}$ is an ordered n-tuple of rational points of X and G a divisor of degree $k - 1 + g$ with $2g - 2 < \deg(G) < n$.

Then the corresponding family $\mathcal{F}_{n,k}$ is nothing but the family of any $[n, k]$ algebraic geometry code $\mathcal{C}_L(X, \mathcal{P}, G)$. For such codes, according to the results of Section 15.6.1.4, using error-correcting arrays, these codes have an efficient decoding algorithm correcting up to $t = \frac{d^*_{\text{Gop}} - 1}{2}$ errors.

Example 15.7.3 (Generalized Reed–Solomon codes) A subcase of the previous example consists in considering generalized Reed–Solomon codes, i.e., algebraic geometry codes from $\mathbb{P}^1$. According to Definition 15.3.19, the set $\mathcal{S}$ can be constructed as the set

---

[2]Also see Chapter 34 which examines the broader topic of code-based cryptography including the McEliece scheme.

of pairs $(\mathbf{x}, \mathbf{y})$ where $\mathbf{x}$ is an ordered n-tuple of distinct elements of $\mathbb{F}_q$ and $\mathbf{y}$ and ordered n-tuple of nonzero elements of $\mathbb{F}_q$. Then the map $\mathcal{C}$ is $\mathcal{C} : (\mathbf{x}, \mathbf{y}) \mapsto \mathcal{GRS}_k(\mathbf{x}, \mathbf{y})$.

15.7.2 McEliece's Original Proposal Using Binary Classical Goppa Codes

As explained further in Section 15.7.5.2, neither Example 15.7.2 nor Example 15.7.3 provides secure instantiations for the McEliece scheme. An efficient, and up to now secure, way to instantiate the scheme is to use subfield subcodes (see Definition 15.3.22) of GRS codes instead of genuine GRS ones. Such codes are usually called *alternant codes*; see [1323, Chap. 12]. McEliece's historic proposal was based on classical Goppa codes defined in (15.6) of Section 15.3.2.2. These are codes of the form

$$\Gamma(\mathbf{x}, f, \mathbb{F}_{q_0}) = \mathcal{C}_\Omega(\mathbb{P}^1, \mathcal{P}, G) \cap \mathbb{F}_{q_0}^n$$

where $\mathcal{P}$ is the sequence of points $((x_1 : 1), \dots, (x_n : 1))$ and $G = (f)_0 + P_\infty$. Therefore, classical Goppa codes are subfield subcodes of AG codes from $\mathbb{P}^1$.

15.7.3 Advantages and Drawbacks of the McEliece Scheme

Since the work of Shor [1694], it is known that if a quantum computer exists in the future, then the currently used cryptographic primitives based on number theoretic problems such as integer factoring or the discrete logarithm problem would become insecure. Code-based cryptography (see also Chapter 34) is currently of central interest since it belongs to the few cryptographic paradigms that are believed to be resistant to a quantum computer attack.

On the other hand, a major drawback of McEliece's original proposal [1361] is the size of the public key: about 32.7 kBytes to claim 65 bits of security[3]. Note that, at the current (2020) time, the limits in terms of computation lie around 2^{80} operations and the new standards require at least 128 bits security. To reach this latter level, the recent National Institute of Standards and Technology (NIST) submission *Classic McEliece* [177] suggests public keys of at least 261 kBytes. As a comparison, according to NIST recommendations for key managements [130, § 5.6.1.1, Tab. 2], to reach a similar security level with RSA, a public key of 384 Bytes (3072 bits) would be sufficient!

15.7.4 Janwa and Moreno's Proposals Using AG Codes

Because of this drawback, many works subsequent to McEliece consisted of proposing other code families in order to reduce the size of the keys. In particular, Niederreiter suggests using Reed–Solomon codes in [1432], i.e., AG codes from $\mathbb{P}^1$. Later, the use of AG codes from curves of arbitrary genus was suggested in Janwa and Moreno's article [1038]. More precisely, Janwa and Moreno made three proposals involving algebraic geometry codes:

(JM1) a proposal based on concatenated AG codes,

(JM2) a proposal based on AG codes, and

(JM3) a proposal based on subfield subcodes of AG codes.

---

[3] A cryptosystem is claimed to have λ *bits of security* if the best known attack would cost more than 2^λ operations to the attacker.

15.7.5 Security

For any instantiation of the McEliece scheme, one should distinguish two kinds of attacks:

1. **Message recovery attacks** consist in trying to recover the plain text from the data of the cipher text. Such an attack rests on generic decoding algorithms, such as Prange's *Information Set Decoding* [1540] and its improvements [146, 336, 646, 1212, 1356, 1357, 1752].

2. **Key recovery attacks** consist in recovering the secret key from the data of the public key and rest on ad hoc methods depending on the family of codes composing the set of public keys.

We conclude this section with a discussion of the security of Janwa and Moreno's three proposals with respect to key-recovery attacks. It is explained in particular that proposals (JM1) and (JM2) are not secure.

15.7.5.1 Concatenated Codes Are Not Secure

In [1646], Sendrier showed that concatenated codes have an inherent weakness which makes them insecure for public key encryption. In particular, proposal (JM1) should not be used.

15.7.5.2 Algebraic Geometry Codes Are Not Secure

The raw use of algebraic geometry codes (JM2) has been subject to two kinds of attacks.

- The case of curves of genus 0 was proved to be insecure by Sidelńikov and Shestakov [1705] in 1992. Note that actually a procedure to recover the structure of a generalized Reed–Solomon code from the data of a generator matrix was already known by Roth and Seroussi [1603].

- An extension of Sidelńikov and Shestakov's attack due to Minder permits one to break AG codes defined on elliptic curves [1388]. This attack has been extended to genus 2 curves by Faure and Minder in [708]. Actually, Faure and Minder's attack can be extended to any AG code constructed from a hyperelliptic curve but its cost is exponential in the curve's genus.

- Finally, an attack due to Couvreur, Márquez-Corbella, and Pellikaan [462] permits one to recover an efficient decoding algorithm in polynomial time from the public key. This attack can be extended to subcodes of small codimension.

  The attack is based on a distinguisher which rests on the $\star$-product operation (see Section 15.1) and on the computation of a filtration of the public code composed of algebraic geometry subcodes. This filtration leads to the construction of an error-correcting array for the public code (see Section 15.6.1.4). Similar approaches have been used to attack the McEliece scheme based on variants of GRS codes [459] and on classical Goppa codes when the subfield has index 2 [464].

15.7.5.3 Conclusion: Only Subfield Subcodes of AG Codes Remain Unbroken

The raw use of algebraic geometry codes is not secure for public key encryption. On the other hand, subfield subcodes of AG codes (JM3) are still resistant to any known attack. Some recent proposals make a step in this direction such as [120, Chap. 5]. Recall that these codes can be regarded as an extension to arbitrary genus of classical Goppa codes which remain unbroken forty years after McEliece's original proposal.

15.8 Applications Related to the $\star$-Product: Frameproof Codes, Multiplication Algorithms, and Secret Sharing

Recall that $\star$ denotes componentwise multiplication in $\mathbb{F}_q^n$, and that the $\star$-product of two linear codes $\mathcal{C}, \mathcal{C}' \subseteq \mathbb{F}_q^n$ is the linear span of the pairwise products of codewords from $\mathcal{C}$ and $\mathcal{C}'$:

$$\mathcal{C} \star \mathcal{C}' \overset{\text{def}}{=} \operatorname{span}_{\mathbb{F}_q}\{\mathbf{c} \star \mathbf{c}' \mid \mathbf{c} \in \mathcal{C},\ \mathbf{c}' \in \mathcal{C}'\}.$$

Recall also that the square of $\mathcal{C}$ is $\mathcal{C}^{\star 2} = \mathcal{C} \star \mathcal{C}$, and likewise its higher powers are defined by induction: $\mathcal{C}^{\star(t+1)} = \mathcal{C}^{\star t} \star \mathcal{C}$.

The link between AG codes and the theory of $\star$-products essentially comes from the obvious inclusion

$$\mathcal{C}_L(X, \mathcal{P}, A) \star \mathcal{C}_L(X, \mathcal{P}, B) \subseteq \mathcal{C}_L(X, \mathcal{P}, A + B). \tag{15.15}$$

It is also observed that in many cases this inclusion is an equality. For instance a sufficient condition for equality [462, Cor. 9] (as a consequence of [1410, Th. 6]) is that A, B satisfy $\deg(A) \geq 2g$ and $\deg(B) \geq 2g + 1$.

Somehow surprisingly, although the $\star$-product is a very simple operation, it turns out to have many interesting applications. We already saw in Sections 15.6 and 15.7 how it can be used for decoding and cryptanalysis. In this section we will focus on a selection of further applications in which AG codes play a prominent role: frameproof codes, multiplication algorithms, and arithmetic secret sharing. A common feature of these constructions was pointed out in [361]: they involve codes $\mathcal{C}_L(X, \mathcal{P}, G)$ where G is a solution to so-called *Riemann–Roch equations* arising from (15.15).

Eventually, the numerous applications of the $\star$-product of codes made it desirable to have a better understanding of this operation for its own sake from a purely coding theoretical perspective. Departing from chronology, we will start with these aspects and then come back to applications later.

15.8.1 The $\star$-Product from the Perspective of AG Codes

Works studying the $\star$-product of codes for itself include the research articles [357, 359, 362, 1390, 1561, 1562, 1563] as well as the expository paper [1564] which organizes the theory in a more systematic way.

We will survey these works with emphasis on the results in which AG codes are involved.

15.8.1.1 Basic Properties

Proposition 15.8.1 *Let $\mathcal{C} \subseteq \mathbb{F}_q^n$ be a nonzero linear code.*

(a) *([1561, Prop. 11]) For any $t \geq 1$ we have*

$$\dim(\mathcal{C}^{\star(t+1)}) \geq \dim(\mathcal{C}^{\star t})$$

and

$$d_H(\mathcal{C}^{\star(t+1)}) \leq d_H(\mathcal{C}^{\star t}).$$

(b) *([1564, Cor. 2.33]) If for some r we have $\dim(\mathcal{C}^{\star(r+1)}) = \dim(\mathcal{C}^{\star r})$, then also $\dim(\mathcal{C}^{\star(r+i)}) = \dim(\mathcal{C}^{\star r})$ for all $i \geq 0$.*

The smallest $r = r(\mathcal{C})$ such that $\dim(\mathcal{C}^{\star(r+1)}) = \dim(\mathcal{C}^{\star r})$ is called the **regularity** of $\mathcal{C}$ [1564, Def. 1.5]. Thus, $\dim(\mathcal{C}^{\star t})$ strictly increases for $t < r$ and then it stabilizes.

If $\mathcal{C}$ is an $[n, k]$ code and $\mathcal{C}'$ is an $[n, k']$ code, we have a short exact sequence [1564, § 1.10]

$$0 \longrightarrow I(\mathcal{C}, \mathcal{C}') \longrightarrow \mathcal{C} \otimes \mathcal{C}' \xrightarrow{\pi_\Delta} \mathcal{C} \star \mathcal{C}' \longrightarrow 0 \tag{15.16}$$

where $\mathcal{C} \otimes \mathcal{C}'$ is the usual $[n^2, kk']$ (tensor) product code that identifies with the space of $n \times n$ matrices all of whose columns are in $\mathcal{C}$ and all of whose rows are in $\mathcal{C}'$, $I(\mathcal{C}, \mathcal{C}')$ is the subspace of such matrices that are zero on the diagonal, and π_Δ is the projection onto the diagonal.

Equivalently, let $\mathbf{p}_1, \ldots, \mathbf{p}_n$ (respectively, $\mathbf{p}'_1, \ldots, \mathbf{p}'_n$) be the columns of a full rank generator matrix of $\mathcal{C}$ (respectively, of $\mathcal{C}'$), and consider the evaluation map

$$\begin{array}{ccc} \mathrm{Bilin}(\mathbb{F}_q^k \times \mathbb{F}_q^{k'} \to \mathbb{F}_q) & \longrightarrow & \mathbb{F}_q^n \\ b & \longmapsto & (b(\mathbf{p}_1, \mathbf{p}'_1), \ldots, b(\mathbf{p}_n, \mathbf{p}'_n)), \end{array}$$

where $\mathrm{Bilin}(\mathbb{F}_q^k \times \mathbb{F}_q^{k'} \to \mathbb{F}_q)$ denotes the space of $\mathbb{F}_q$-bilinear forms on $\mathbb{F}_q^k \times \mathbb{F}_q^{k'}$. Then $\mathcal{C} \star \mathcal{C}'$ is the image of this evaluation map, and $I(\mathcal{C}, \mathcal{C}')$ is its kernel.

Similar to (15.16), for any $t \geq 2$, we have a natural short exact sequence [1332, Sec. 3]

$$0 \longrightarrow I_t(\mathcal{C}) \longrightarrow S^t \mathcal{C} \longrightarrow \mathcal{C}^{\star t} \longrightarrow 0 \tag{15.17}$$

where $S^t \mathcal{C}$ is the tth symmetric power of $\mathcal{C}$, which actually defines $I_t(\mathcal{C})$ as the kernel of the natural map $S^t \mathcal{C} \to \mathcal{C}^{\star t}$.

Equivalently and more concretely, let $\mathbf{p}_1, \ldots, \mathbf{p}_n$ be the columns of a generator matrix of $\mathcal{C}$. Denote by $\mathbb{F}_q[x_1, \ldots, x_k]_t$ the space of degree t homogeneous polynomials in k variables and consider the evaluation map

$$\begin{array}{ccc} \mathbb{F}_q[x_1, \ldots, x_k]_t & \longrightarrow & \mathbb{F}_q^n \\ f & \longmapsto & (f(\mathbf{p}_1), \ldots, f(\mathbf{p}_n)). \end{array}$$

Then $\mathcal{C}^{\star t}$ is the image of this evaluation map, and $I_t(\mathcal{C})$ is its kernel, the space of degree t homogeneous forms vanishing at $\mathbf{p}_1, \ldots, \mathbf{p}_n$.

When $\mathcal{C}$ is an AG code, Márquez-Corbella, Martínez-Moro, and Pellikaan showed how $I_2(\mathcal{C})$ allows one to retrieve the underlying curve of an AG code:

Proposition 15.8.2 ([1332, Prop. 8 & Th. 12]) *Let $\mathcal{C} = \mathcal{C}_L(X, \mathcal{P}, G)$ be an evaluation code, and assume that $g = g(X)$, $n = |\mathcal{P}|$, and $m = \deg(G)$ satisfy*

$$2g + 2 \leq m < \frac{n}{2}.$$

Then $I_2(\mathcal{C})$ is a set of quadratic equations defining X embedded in $\mathbb{P}^{k-1}$.

15.8.1.2 Dimension of $\star$-Products

Let $\mathcal{C}$ be an $[n, k]$ code and $\mathcal{C}'$ an $[n, k']$ code. From (15.16) it follows that

$$\dim(\mathcal{C} \star \mathcal{C}') \leq \min\{n, kk'\},$$

and one expects that if $\mathcal{C}$ and $\mathcal{C}'$ are independent random codes, then, with high probability, this inequality becomes an equality: $\dim(\mathcal{C} \star \mathcal{C}') = \min\{n, kk'\}$. We refer to [1563, Th. 16–18] for precise statements and proofs. By contrast, we observe that for AG codes $\mathcal{C} = \mathcal{C}_L(X, \mathcal{P}, G)$ and $\mathcal{C}' = \mathcal{C}_L(X, \mathcal{P}, G')$, with the same evaluation points sequence $\mathcal{P}$, we can get much stronger upper bounds. For instance, if $2g - 2 < \deg(G + G') < n$, then Theorem 15.3.12 together with (15.15) gives $\dim(\mathcal{C} \star \mathcal{C}') \leq k + k' - 1 + g$.

Likewise from (15.17) it follows that

$$\dim(\mathcal{C}^{\star t}) \leq \min\left\{n, \binom{k + t - 1}{t}\right\},$$

and, at least for $t = 2$, one expects that if $\mathcal{C}$ is a random code, then with high probability this inequality becomes an equality: $\dim(\mathcal{C}^{\star 2}) = \min\left\{n, \frac{k(k+1)}{2}\right\}$. We refer to [359, Th. 2.2–2.3] for precise statements and proofs.

Remark 15.8.3 ([1563, Prop. 19]) When $t > q$, we always have the strict inequality $\dim(\mathcal{C}^{\star t}) < \binom{k+t-1}{t}$ because of the extra relations $\mathbf{c}^{\star q} \star \mathbf{c}' = \mathbf{c} \star \mathbf{c}'^{\star q}$.

Concerning lower bounds, from now on we will make the simplifying assumption that *all our codes have full support*, i.e., any generator matrix has no zero column.

Proposition 15.8.4 ([1564, § 3.5]) *Assume $\mathcal{C}$ or $\mathcal{C}'$ is MDS. Then*

$$\dim(\mathcal{C} \star \mathcal{C}') \geq \min\{n, k + k' - 1\}.$$

The **stabilizing algebra** of a code is introduced in [1564, Def. 2.6] as

$$\text{Stab}(\mathcal{C}) \overset{\text{def}}{=} \{\mathbf{x} \in \mathbb{F}_q^n \mid \mathbf{x} \star \mathcal{C} \subseteq \mathcal{C}\}.$$

It is a subalgebra of $\mathbb{F}_q^n$ and admits a basis made of the characteristic vectors of the supports of the indecomposable components of $\mathcal{C}$ ([1129, Th. 1.2] and [1564, Prop. 2.11]). In particular, $\dim(\text{Stab}(\mathcal{C}))$ is equal to the number of indecomposable components of $\mathcal{C}$.

Mirandola and Zémor [1390] then established a coding analog of Kneser's Theorem from additive combinatorics [1790, Th. 5.5] in which $\text{Stab}(\mathcal{C})$ plays the same role as the stabilizer of a subgroup.

Theorem 15.8.5 ([1390, Th. 18]) *We have*

$$\dim(\mathcal{C} \star \mathcal{C}') \geq k + k' - \dim(\text{Stab}(\mathcal{C} \star \mathcal{C}')).$$

Pursuing the analogy with additive combinatorics, they also obtained the following characterization of cases of equality in Proposition 15.8.4, which one might see as an analog of Vosper's Theorem [1790, Th. 5.9].

Theorem 15.8.6 ([1390, Th. 23]) *Assume **both** $\mathcal{C}$ and $\mathcal{C}'$ are MDS, with $k, k' \geq 2$, $k + k' \leq n - 1$, and*

$$\dim(\mathcal{C} \star \mathcal{C}') = k + k' - 1.$$

Then $\mathcal{C}$ and $\mathcal{C}'$ are (generalized) Reed–Solomon codes with a common evaluation point sequence.

Remark 15.8.7 The symmetric version ($\mathcal{C} = \mathcal{C}'$) of Theorem 15.8.6 can actually be regarded as a direct consequence of Castelnuovo's Lemma [61, § III.2, p. 120] asserting that for $n \geq 2k + 1$, any n points in general position in $\mathbb{P}^{k-1}$ imposing $2r - 1$ independent conditions on quadrics lie on a Veronese embedding of $\mathbb{P}^1$.

15.8.1.3 Joint Bounds on Dimension and Distance

A fundamental problem in coding theory is to find lower bounds (existence results) or upper bounds (nonexistence results) relating the possible dimension and minimum distance of a code. The analog for $\star$-products is to find similar bounds relating the dimensions of a given number of codes and the minimum distance of their product (or the dimension of a code and the minimum distance of some power).

Concerning lower bounds, a raw use of AG codes easily shows that for a given t, if q is large enough (depending on t), there are asymptotically good codes over $\mathbb{F}_q$ whose t^{th} powers are also asymptotically good. For $t = 2$ the following theorem shows that this result actually holds for *all* q. The proof still uses AG codes but in combination with a specially devised concatenation argument.

Theorem 15.8.8 ([1561]) *Over any finite field $\mathbb{F}_q$, there exist asymptotically good codes $\mathcal{C}$ whose squares $\mathcal{C}^{\star 2}$ are also asymptotically good.*

In particular for $q = 2$ and for any $0 < \delta < 0.003536$ and $R \leq 0.001872 - 0.5294\delta$, there exist codes $\mathcal{C}$ of asymptotic rate at least R whose squares $\mathcal{C}^{\star 2}$ have asymptotic relative minimum distance at least δ.

In finite length, other constructions have been studied by Cascudo and co-authors that give codes such that both $\mathcal{C}$ and $\mathcal{C}^{\star 2}$ have good parameters. This includes cyclic codes [357] and matrix-product codes [362].

Concerning upper bounds, the following result is sometimes called the **Product Singleton Bound**.

Theorem 15.8.9 ([1562]) *Let $t \geq 2$ be an integer and $\mathcal{C}_1, \ldots, \mathcal{C}_t \subseteq \mathbb{F}_q^n$ linear codes with full support. Then there exist codewords $\mathbf{c}_1 \in \mathcal{C}_1, \ldots, \mathbf{c}_t \in \mathcal{C}_t$ whose product has Hamming weight*

$$0 < \mathrm{wt}_{\mathrm{H}}(\mathbf{c}_1 \star \cdots \star \mathbf{c}_t) \leq \min\left\{ t - 1, n - (k_1 + \cdots + k_t) + t \right\}$$

where $k_i = \dim(\mathcal{C}_i)$. In particular we have

$$d_H(\mathcal{C}_1 \star \cdots \star \mathcal{C}_t) \leq \min\left\{ t - 1, n - (k_1 + \cdots + k_t) + t \right\}.$$

Mirandola and Zémor describe cases of equality for $t = 2$. Essentially these are either pairs of (generalized) Reed–Solomon codes with a common evaluation point sequence, or pairs made of a code and its dual (up to diagonal equivalence). See [1390, Sec. V] for further details.

15.8.1.4 Automorphisms

Let $\mathcal{C}$ be a linear code of regularity $r(\mathcal{C})$, and let $t \leq t'$ be two integers. Assume one of the following two conditions holds:

(i) $t \mid t'$ or

(ii) $t' \geq r(\mathcal{C})$.

Then, from [1564, § 2.50–2.55], it follows that $|\mathrm{Aut}(\mathcal{C}^{\star t})|$ divides $|\mathrm{Aut}(\mathcal{C}^{\star t'})|$.

This makes one wonder whether one could compare $\mathrm{Aut}(\mathcal{C}^{\star t})$ and $\mathrm{Aut}(\mathcal{C}^{\star t'})$ for arbitrary $t \leq t'$. For instance, do we always have $|\mathrm{Aut}(\mathcal{C}^{\star 2})| \leq |\mathrm{Aut}(\mathcal{C}^{\star 3})|$?

Motivated by Proposition 15.3.16, Couvreur and Ritzenthaler tested this question against AG codes and eventually showed that the answer is negative.

Example 15.8.10 ([466]) Let E be the elliptic curve defined by the Weierstrass equation $y^2 = x^3 + 1$ over $\mathbb{F}_7$ with point at infinity O_E. Set $\mathcal{P} = E(\mathbb{F}_7)$, $G = 3O_E$, and consider the evaluation code $\mathcal{C} = \mathcal{C}_L(E, E(\mathbb{F}_7), 3O_E)$. So $\mathcal{C}$ has generator matrix

$$
\begin{bmatrix}
0 & 0 & 1 & 1 & 2 & 2 & 3 & 4 & 4 & 5 & 6 & 0 \\
1 & 6 & 3 & 4 & 3 & 4 & 0 & 3 & 4 & 0 & 0 & 1 \\
1 & 1 & 1 & 1 & 1 & 1 & 1 & 1 & 1 & 1 & 1 & 0
\end{bmatrix}.
$$

Computer aided calculations show that $\mathcal{C}^{\star 2} = \mathcal{C}_L(E, E(\mathbb{F}_7)), 6O_E)$ has

$$|\mathrm{Aut}(\mathcal{C}^{\star 2})| = 432,$$

and $\mathcal{C}^{\star 3} = \mathcal{C}_L(E, E(\mathbb{F}_7), 9O_E)$ has

$$|\mathrm{Aut}(\mathcal{C}^{\star 3})| = 108.$$

So in particular

$$|\mathrm{Aut}(\mathcal{C}^{\star 2})| > |\mathrm{Aut}(\mathcal{C}^{\star 3})|.$$

15.8.2   Frameproof Codes and Separating Systems

Frameproof codes were introduced in the context of traitor tracing schemes [233, 406, 1757]. Slightly differing definitions of this notion can be found. Here we work only with linear codes, and we say such a code $\mathcal{C}$ is s-**frameproof**, or s-**wise intersecting** if the supports of any s nonzero codewords have a nonempty common intersection. In terms of $\star$-products, this means that for any $\mathbf{c}_1, \ldots, \mathbf{c}_s \in \mathcal{C}$,

$$\mathbf{c}_1, \ldots, \mathbf{c}_s \neq \mathbf{0} \quad \Longrightarrow \quad \mathbf{c}_1 \star \cdots \star \mathbf{c}_s \neq \mathbf{0}.$$

Some elementary combinatorial constructions of frameproof codes can be found in [206, 419].

In [1915] Xing considers asymptotic bounds, and in particular, constructions from AG codes. The starting observation of his main result is that a sufficient condition for an evaluation code $\mathcal{C}_L(X, \mathcal{P}, G)$ to be s-frameproof is

$$\ell(sG - D_\mathcal{P}) = 0, \tag{15.18}$$

where $D_\mathcal{P}$ is defined in (15.3). Condition (15.18) is perhaps the simplest instance of what was later called a *Riemann–Roch equation* [361].

Xing uses a counting argument in the divisor class group of the curve to prove the existence of solutions to (15.18) with $\deg(sG - D_\mathcal{P}) \approx (1 - 2\log_q s)g$. This leads to the following result.

Theorem 15.8.11 ([1915]) *For $2 \leq s \leq A(q)$, there exist s-frameproof codes over $\mathbb{F}_q$ of length going to infinity and asymptotic rate at least*

$$\frac{1}{s} - \frac{1}{A(q)} + \frac{1 - 2\log_q s}{sA(q)}.$$

In this bound, the $2\log_q s$ term reflects the possible s-torsion in the class group that hinders the counting argument.

For $s = 2$, an alternative method is proposed in [1560] that allows one to construct solutions to (15.18) up to $\deg(2G - D_\mathcal{P}) \approx g - 1$, which is best possible. This gives:

Theorem 15.8.12 ([1560]) *If $A(q) \geq 4$, one can construct 2-frameproof codes over $\mathbb{F}_q$ of length going to infinity and asymptotic rate at least*

$$\frac{1}{2} - \frac{1}{2A(q)}.$$

This result has a somehow unexpected application. In [683], Erdös and Füredi studied a certain problem in combinatorial geometry. It led them to introduce certain configurations, which admit the following equivalent descriptions:

- sets of M vertices of the unit cube $\{0,1\}^n$ in the Euclidean space $\mathbb{R}^n$, any three $\mathbf{x}, \mathbf{y}, \mathbf{z}$ of which form an acute angle: $\langle \mathbf{x} - \mathbf{z}, \mathbf{y} - \mathbf{z} \rangle_{\mathrm{Eucl}} > 0$;

- sets of M points in the binary Hamming space $\mathbb{F}_2^n$, any three $\mathbf{x}, \mathbf{y}, \mathbf{z}$ of which satisfy the strict triangle inequality: $d_H(\mathbf{x}, \mathbf{y}) < d_H(\mathbf{x}, \mathbf{z}) + d_H(\mathbf{y}, \mathbf{z})$;

- sets of M binary vectors of length n, any three $\mathbf{x}, \mathbf{y}, \mathbf{z}$ of which have a position i where $x_i = y_i \neq z_i$;

- sets of M subsets in an n-set, no three A, B, C of which satisfy $A \cap B \subseteq C \subseteq A \cup B$.

In the context of coding theory, such a configuration is also called a **binary $(2,1)$-separating system** of length n and size M. A random coding argument shows that there exist such separating systems of length $n \to \infty$ and size

$$M \approx (2/\sqrt{3})^n \approx 1.1547005^n.$$

However, combining Theorem 15.8.12 with a convenient concatenation argument provides a dramatic(?) improvement.

Theorem 15.8.13 ([1560]) *One can construct binary $(2,1)$-separating systems of length $n \to \infty$ and size*

$$M \approx (11^{\frac{3}{50}})^n \approx 1.1547382^n.$$

This is an instance of a construction based on AG codes beating a random coding argument *over the binary field*!

15.8.3 Multiplication Algorithms

The theory of bilinear complexity started with the celebrated algorithms of Karatsuba [1082] that multiplies two 2-digit numbers with 3 elementary multiplications instead of 4, and of Strassen [1759] that multiplies two 2×2 matrices with 7 field multiplications instead of 8. Used recursively, these algorithms then allow one to multiply numbers with a large number of digits, or matrices of large size, with a dramatic improvement in complexity over naive methods.

Authors subsequently studied the complexity of various multiplication maps such as the multiplication of two polynomials modulo a given polynomial [725, 1895]. This includes, in particular, the multiplication map $m_{\mathbb{F}_{q^k}/\mathbb{F}_q}$ in an extension of finite fields $\mathbb{F}_{q^k}$ over $\mathbb{F}_q$.

Definition 15.8.14 A **bilinear multiplication algorithm** of length n for $\mathbb{F}_{q^k}$ over $\mathbb{F}_q$ is the data of linear maps $\alpha, \beta : \mathbb{F}_{q^k} \to \mathbb{F}_q^n$ and $\omega : \mathbb{F}_q^n \to \mathbb{F}_{q^k}$ such that the following diagram commutes:

$$
\begin{array}{ccc}
\mathbb{F}_{q^k} \times \mathbb{F}_{q^k} & \xrightarrow{m_{\mathbb{F}_{q^k}/\mathbb{F}_q}} & \mathbb{F}_{q^k} \\
{\scriptstyle \alpha \times \beta} \downarrow & & \uparrow {\scriptstyle \omega} \\
\mathbb{F}_q^n \times \mathbb{F}_q^n & \xrightarrow{\quad \star \quad} & \mathbb{F}_q^n.
\end{array}
$$

Equivalently, it is the data of linear forms $\alpha_1, \ldots, \alpha_n, \beta_1, \ldots, \beta_n : \mathbb{F}_{q^k} \to \mathbb{F}_q$ and elements $\omega_1, \ldots, \omega_n \in \mathbb{F}_{q^k}$ such that for all $x, y \in \mathbb{F}_{q^k}$ we have

$$xy = \sum_{i=1}^{n} \alpha_i(x)\beta_i(y)\omega_i.$$

Definition 15.8.15 A multiplication algorithm as above is **symmetric** if $\alpha = \beta$, or equivalently, if $\alpha_i = \beta_i$ for all i.

Definition 15.8.16 The **bilinear complexity** $\mu_q(k)$ (respectively, the **symmetric bilinear complexity** $\mu_q^{\mathrm{sym}}(k)$) of $\mathbb{F}_{q^k}$ over $\mathbb{F}_q$ is the smallest possible length of a bilinear multiplication algorithm (respectively, a symmetric bilinear multiplication algorithm) for $\mathbb{F}_{q^k}$ over $\mathbb{F}_q$.

Obviously we always have $\mu_q(k) \leq \mu_q^{\mathrm{sym}}(k) \leq k^2$. For $k \leq \frac{q}{2} + 1$ this is easily improved to $\mu_q(k) \leq \mu_q^{\mathrm{sym}}(k) \leq 2k - 1$, using the Fourier transform, or equivalently, Reed–Solomon codes.

In [408], Chudnovsky and Chudnovsky highlighted further links between multiplication algorithms and codes. Then, using a construction similar to AG codes, they were able to prove the first linear asymptotic upper bound on $\mu_q(k)$.

Theorem 15.8.17 ([408, Th. 7.7]) *If $q \geq 25$ is a square, then*

$$\limsup_{k \to \infty} \frac{1}{k}\mu_q(k) \leq 2\left(1 + \frac{1}{\sqrt{q} - 3}\right).$$

This result was originally stated for bilinear complexity, but the proof also works for symmetric bilinear complexity, since it provides symmetric algorithms. So we actually get

$$\limsup_{k \to \infty} \frac{1}{k}\mu_q^{\mathrm{sym}}(k) \leq 2\left(1 + \frac{1}{\sqrt{q} - 3}\right)$$

for $q \geq 25$ a square.

Chudnovsky and Chudnovsky proceeded by *evaluation-interpolation* on curves. Suppose we are given a curve X over $\mathbb{F}_q$, together with a collection $\mathcal{P} = \{P_1, \ldots, P_n\}$ of n distinct rational points, a point Q of degree k, and a suitably chosen auxiliary divisor G. Also assume:

(Eval1) the evaluation-at-Q map $L(G) \longrightarrow \mathbb{F}_{q^k}$ is surjective; and

(Eval2) the evaluation-at-$\mathcal{P}$ map $L(2G) \longrightarrow \mathbb{F}_q^n$ is injective.

Then, in order to multiply x, x' in $\mathbb{F}_{q^k}$ with only n multiplications in $\mathbb{F}_q$, one can proceed as follows.

Step 1: Thanks to (Eval1), lift x, x' to functions $f_x, f_{x'}$ in $L(G)$ such that $f_x(Q) = x$, $f_{x'}(Q) = x'$, and then evaluate these functions at $\mathcal{P}$ to get codewords $\mathbf{c}_x = (f_x(P_1), \ldots, f_x(P_n))$, $\mathbf{c}_{x'} = (f_{x'}(P_1), \ldots, f_{x'}(P_n))$ in $\mathcal{C}_L(X, \mathcal{P}, G)$.

Step 2: Compute $\mathbf{c}_x \star \mathbf{c}_{x'} = (y_1, \ldots, y_n)$ in $\mathcal{C}_L(X, \mathcal{P}, 2G)$, i.e., $y_i = f_x(P_i)f_{x'}(P_i)$ for $1 \leq i \leq n$.

Step 3: By Lagrange interpolation, find a function h in $L(2G)$ that takes these values $h(P_1) = y_1, \ldots, h(P_n) = y_n$, and then evaluate h at Q.

Then (Eval2) ensures that we necessarily have $h = f_x f_{x'}$; so the last evaluation in Step 3 gives $h(Q) = xx'$ as desired.

In [1697], Shparlinski, Tsfasman, and Vlăduţ propose several improvements. First, they correct certain imprecise statements in [408] or give additional details to the proofs. For instance, on the choice of the curves to which the method is applied in order to get Theorem 15.8.17, they provide an explicit description of a family of Shimura curves X_i over $\mathbb{F}_q$, for q a square, that satisfy

- (optimality) $|X_i(\mathbb{F}_q)|/g(X_i) \to A(q) = \sqrt{q} - 1$ and
- (density) $g(X_{i+1})/g(X_i) \to 1$.

They further show how to deduce linearity of the complexity over an arbitrary finite field.

Lemma 15.8.18 ([1697, Cor. 1.3]) *Set $M_q \overset{def}{=} \limsup_{k \to \infty} \frac{1}{k} \mu_q(k)$. Then, for any prime power q and integer m, we have*

$$M_q \leq \mu_q(m) M_{q^m}.$$

From this and Theorem 15.8.17 it readily follows that

$$M_q < +\infty$$

for all q. As above, this result was originally stated only for bilinear complexity, but the proof also works for symmetric bilinear complexity; so we likewise get

$$M_q^{\mathrm{sym}} \overset{def}{=} \limsup_{k \to \infty} \frac{1}{k} \mu_q^{\mathrm{sym}}(k) < +\infty.$$

Finally, they devise a possible improvement on Theorem 15.8.17. For this they observe that for given $\mathcal{P}$ and Q, the choice of the auxiliary divisor G satisfying (Eval1) and (Eval2) above reduces to the solution of the following system of two Riemann–Roch equations:

$$\begin{cases} \ell(K_X - G + Q) = 0, \\ \ell(2G - D_{\mathcal{P}}) = 0. \end{cases} \tag{15.19}$$

In order to get the best possible parameters, one would like to set $\mathcal{P} = X(\mathbb{F}_q)$, and find a solution with $\deg(K_X - G + Q)$ and $\deg(2G - D_{\mathcal{P}})$ close to $g - 1$. The authors propose a method to achieve this, but it was later observed that their argument is incomplete. A corrected method was then proposed in [1559], relying on the tools introduced in the proof of Theorem 15.8.12. Combining all these ideas then gives:

Theorem 15.8.19 ([1559, Th. 6.4]) *If $q \geq 49$ is a square, then*

$$M_q^{\mathrm{sym}} \leq 2\left(1 + \frac{1}{\sqrt{q} - 2}\right).$$

In this same work, it is also observed that if one does not insist on having symmetric algorithms, then (15.19) can be replaced with an asymmetric variant: for given $\mathcal{P}$ and Q, find two divisors G and G' satisfying

$$\begin{cases} \ell(K_X - G + Q) = 0, \\ \ell(K_X - G' + Q) = 0, \\ \ell(G + G' - D_{\mathcal{P}}) = 0. \end{cases}$$

This asymmetric system is easier to solve, leading to the following result.

Theorem 15.8.20 ([1559, Th. 6.4]) *If $q \geq 9$ is a square, then*

$$M_q \leq 2 \left(1 + \frac{1}{\sqrt{q} - 2} \right).$$

The study of bilinear multiplication algorithms over finite fields is a very active area of research and we covered only one specific aspect. Other research directions include:

- Give asymptotic bounds for non-square q, or for very small $q = 2, 3, 4, \ldots$.

- Instead of asymptotic bounds valid for $k \to \infty$, give uniform bounds valid for all k.

- Study multiplication algorithms in more general finite dimensional algebras, not only extension fields.

- Give effective constructions of multiplication algorithms.

For a more exhaustive survey of recent results, see [108].

15.8.4   Arithmetic Secret Sharing

Our starting point here will be Shamir's secret sharing scheme.[4] Suppose Alice has a secret $s \in \mathbb{F}_q$, and she wants to distribute it among n players. We will assume $n < q$, and the n players are labelled by n distinct nonzero elements $x_1, \ldots, x_n$ in $\mathbb{F}_q$. Given a certain threshold $t \leq n$, Alice picks $t - 1$ random elements $c_1, \ldots, c_{t-1}$ in $\mathbb{F}_q$ and considers the polynomial $P(x) = s + c_1 x + c_2 x^2 + \cdots + c_{t-1} x^{t-1}$. Then each player x_i receives their share $y_i = P(x_i)$. The main properties of this scheme are the following:

(Sha1) Any coalition of at least t players can use Lagrange interpolation to reconstruct the polynomial P, hence also the secret $s = P(0)$, from their shares.

(Sha2) Any coalition of up to $t - 1$ players has no information at all about the secret, i.e., all possible values for s appear equiprobable to them.

Another property of Shamir's scheme is its *linearity*: suppose Alice has two secrets s and $\widetilde{s}$, and distributes them to the same players. Let P be the polynomial used to distribute s, with corresponding shares $y_1, \ldots, y_n$, and let $\widetilde{P}$ be the polynomial used to distribute $\widetilde{s}$, with corresponding shares $\widetilde{y}_1, \ldots, \widetilde{y}_n$. Then $y_1 + \widetilde{y}_1, \ldots, y_n + \widetilde{y}_n$ are shares corresponding to the sum $s + \widetilde{s}$ of the two secrets: indeed, these are the shares obtained when distributing $s + \widetilde{s}$ with the polynomial $P + \widetilde{P}$.

Shamir's scheme also enjoys a *multiplicative* property, but it is more subtle. We have $(P\widetilde{P})(0) = s\widetilde{s}$ and $(P\widetilde{P})(x_i) = y_i \widetilde{y}_i$; so in some sense, $y_1 \widetilde{y}_1, \ldots, y_n \widetilde{y}_n$ are the shares obtained when distributing $s\widetilde{s}$ with the polynomial $P\widetilde{P}$. A drawback is that $P\widetilde{P}$ may have degree up to $2t - 2$, instead of $t - 1$ in the original scheme. Still we can say that any coalition of $2t - 1$ players can reconstruct the secret product $s\widetilde{s}$ from their product shares $y_i \widetilde{y}_i$. On the other hand, $P\widetilde{P}$ is not uniformly distributed in the set of polynomials R of degree up to $2t - 2$ satisfying $R(0) = s\widetilde{s}$; so it is unclear what information a smaller coalition can get.

There are close links between linear secret sharing schemes and linear codes, under which Shamir's scheme corresponds to Reed–Solomon codes. Indeed, observe that share vectors $(y_1, \ldots, y_n)$ in Shamir's scheme are precisely the codewords of an RS code. Properties (Sha1) and (Sha2) then reflect the MDS property of Reed–Solomon codes. And a common limitation to Shamir's scheme and to RS codes is that, for a given q, the number n of players, or the length of the code, remains bounded essentially by q.

---

[4]Secret sharing schemes are the topic of Chapter 33.

The importance of multiplicative linear secret sharing schemes perhaps comes from a result in [468] that shows that these schemes can serve as a basis for secure multiparty computation protocols. In [1026] it is also shown that certain two-party protocols, for instance a zero-knowledge proof, admit communication-efficient implementations in which one player, for instance the verifier, has to simulate "in her head" a multiparty computation with a large number of players. This last result makes it very desirable to overcome the limitation on the number of players in Shamir's scheme.

In the same way that AG codes provide a generalization of RS codes of arbitrary length, one can construct linear secret sharing schemes with an arbitrary number of players by using evaluation of functions on an algebraic curve. Moreover, under certain conditions, these schemes also admit good multiplicative properties. This was first studied by Chen and Cramer in [394] and then refined and generalized in several works such as [358] and [360]. We follow the presentation of the latter.

Given a finite field $\mathbb{F}_q$ and integers $k, n \geq 1$, we equip the vector space $\mathbb{F}_q^k \times \mathbb{F}_q^n$ with the linear projection maps

$$\pi_0 : \quad \begin{array}{ccc} \mathbb{F}_q^k \times \mathbb{F}_q^n & \longrightarrow & \mathbb{F}_q^k \\ \mathbf{v} = (s_1, \ldots, s_k, c_1, \ldots, c_n) & \longmapsto & \mathbf{v}_0 \overset{\text{def}}{=} (s_1, \ldots, s_k) \end{array}$$

and, for any subset $B \subseteq \{1, \ldots, n\}$,

$$\pi_B : \quad \begin{array}{ccc} \mathbb{F}_q^k \times \mathbb{F}_q^n & \longrightarrow & \mathbb{F}_q^{|B|} \\ \mathbf{v} = (s_1, \ldots, s_k, c_1, \ldots, c_n) & \longmapsto & \mathbf{v}_B \overset{\text{def}}{=} (c_i)_{i \in B} \end{array}.$$

Definition 15.8.21 A (n, t, d, r)-**arithmetic secret sharing scheme** for $\mathbb{F}_q^k$ over $\mathbb{F}_q$ is a linear subspace $\mathcal{C} \subseteq \mathbb{F}_q^k \times \mathbb{F}_q^n$ with the following properties:

(i) (t-**disconnectedness**) for any subset $B \subseteq \{1, \ldots, n\}$ of cardinality $|B| = t$, the projection map

$$\pi_{0,B} : \quad \begin{array}{ccc} \mathcal{C} & \longrightarrow & \mathbb{F}_q^k \times \pi_B(\mathcal{C}) \\ \mathbf{v} & \longmapsto & (\mathbf{v}_0, \mathbf{v}_B) \end{array}$$

is surjective

(ii) (d^{th} **power** r-**reconstruction**) for any subset $B \subseteq \{1, \ldots, n\}$ of cardinality $|B| = r$ we have

$$(\ker \pi_B) \cap \mathcal{C}^{\star d} \subseteq (\ker \pi_0) \cap \mathcal{C}^{\star d}.$$

Moreover we say that $\mathcal{C}$ has **uniformity** if in (i) we have $\pi_B(\mathcal{C}) = \mathbb{F}_q^{|B|}$ for all B with $|B| = t$.

Such a scheme permits distribution of a secret $\mathbf{s} = (s_1, \ldots, s_k) \in \mathbb{F}_q^k$ among n players. To do this, one chooses a random $\mathbf{v} \in \mathcal{C}$ such that $\mathbf{v}_0 = \mathbf{s}$, and for each $i \in \{1, \ldots, n\}$, the ith player receives their share $c_i = \mathbf{v}_{\{i\}}$. Now condition (i) means that for each coalition B of t adversary players, the secret vector $\mathbf{v}_0$ is independently distributed from their share vector $\mathbf{v}_B$. On the other hand, condition (ii) means that any coalition of r honest players can reconstruct the $\star$-product of d secret vectors from the $\star$-product of their corresponding d share vectors.

It turns out that the two conditions of Definition 15.8.21 can be captured by Riemann–Roch equations.

Lemma 15.8.22 *Let X be an algebraic curve over $\mathbb{F}_q$, together with a collection of $k+n$ distinct rational points $\mathcal{S} = \{Q_1, \ldots, Q_k, P_1, \ldots, P_n\}$. Define the divisors $Q = Q_1 + \cdots + Q_k$, and for each subset $B \subseteq \{1, \ldots, n\}$, $P_B = \sum_{i \in B} P_i$. Let G be a divisor on X that satisfies the following system:*

$$\begin{cases} \ell(K_X - G + P_B + Q) = 0 & \text{for all } B \subseteq \{1, \ldots, n\} \text{ with } |B| = t, \\ \ell(dG - P_B) = 0 & \text{for all } B \subseteq \{1, \ldots, n\} \text{ with } |B| = r. \end{cases} \tag{15.20}$$

Then $\mathcal{C} = \mathcal{C}_L(X, \mathcal{S}, G)$ is an (n, t, d, r)-arithmetic secret sharing scheme for $\mathbb{F}_q^k$ with uniformity.

In [360] a method is developed to solve (15.20) using some control on the d-torsion of the class group of the curve. It is not known if this method is optimal, but it gives arithmetic secret sharing schemes with the best parameters up to now.[5] It leads to:

Theorem 15.8.23 ([360]) *For all prime powers q except perhaps for $q = 2, 3, 4, 5, 7, 11, 13$, there is an infinite family of $(n, t, 2, n-t)$-arithmetic secret sharing schemes for $\mathbb{F}_q^k$ over $\mathbb{F}_q$ with uniformity, where n is unbounded, $k = \Omega(n)$, and $t = \Omega(n)$.*

It should be noted that, if one is ready to drop the uniformity condition, then the corresponding result holds for *all* q. This can be proved using a concatenation argument [358].

The literature on arithmetic secret sharing is rapidly evolving, and we cannot cover all recent developments. For further references on the topic, together with details on the connection with multiparty computation, we recommend the book [469].

15.9 Application to Distributed Storage Systems: Locally Recoverable Codes

15.9.1 Motivation

The impressive development of cloud computing and distributed storage in the last decade motivated new paradigms and new questions in coding theory yielding an impressive number of works studying the construction, the features, and the limitations of codes having "good local properties". For more on this topic, we refer the reader to Chapter 31, which focuses entirely on codes for distributed storage.

While the literature provides many definitions, such as *locally correctable codes*, *locally decodable codes*, etc., in this section we only focus on so-called *locally recoverable codes* (LRC). To define them, let us first define the notion of the *restriction* of a code.

Definition 15.9.1 Let $\mathcal{C} \subseteq \mathbb{F}_q^n$ be a code and $I \subseteq \{1, \ldots, n\}$. The **restriction of $\mathcal{C}$ to I**, denoted by $\mathcal{C}_{|I}$, is the image of $\mathcal{C}$ under the projection

$$\begin{array}{ccc} \mathbb{F}_q^n & \longrightarrow & \mathbb{F}_q^{|I|} \\ (c_1, \ldots, c_n) & \longmapsto & (c_i)_{i \in I}. \end{array}$$

Remark 15.9.2 Classically in the literature, this operation is referred to as *puncturing* $\mathcal{C}$ at $\{1, \ldots, n\} \setminus I$. In the sequel, the term *restriction* seems more relevant since we will deal with evaluation codes and the restriction of the code will be obtained by evaluating restrictions of some given functions.

---

[5] Observe that the method of [1560], that gives optimal solutions to (15.18) and (15.19), does not operate with (15.20), even in the case $d = 2$, because the number of equations in the system is too high.

Definition 15.9.3 A code $\mathcal{C} \subseteq \mathbb{F}_q^n$ is **locally recoverable** with **locality** ℓ if, for any $i \in \{1, \ldots, n\}$, there exists at least one subset $A(i) \subseteq \{1, \ldots, n\}$ containing i such that $|A(i)| \leq \ell + 1$ and $\mathcal{C}_{|A(i)}$ has minimum distance ≥ 2.

Definition 15.9.4 In the context of Definition 15.9.3, a subset $A(i)$ is called a **recovery set of $\mathcal{C}$ for** i.

Remark 15.9.5 Note that the sets $A(1), \ldots, A(n)$ need not be distinct. In addition, we emphasize that, for a given position $i \in \{1, \ldots, n\}$, there might exist more than one recovery set for i; this is actually the point of the notion of *availability* discussed further in Section 15.9.6.

On the other hand, in most of the examples presented in Sections 15.9.3, 15.9.4, and 15.9.5, we consider a partition $A_1 \sqcup \cdots \sqcup A_s$ of $\{1, \ldots, n\}$ such that for any $i \in \{1, \ldots, n\}$, the unique recovery set for i is the unique subset A_j containing i.

Remark 15.9.6 One can prove that Definition 15.9.3 is equivalent to the following one. For any $i \in \{1, \ldots, n\}$, there exists at least one codeword $\mathbf{c}^{(i)} \in \mathcal{C}^{\perp}$ of weight less than or equal to $\ell + 1$ and whose i^{th} entry is nonzero.

Let us give some practical motivation for this definition. Suppose we distribute data on n servers. Our file (or a part of it) is an element $\mathbf{m} \in \mathbb{F}_q^k$ that has been encoded as a codeword $\mathbf{c} = (c_1, \ldots, c_n) \in \mathcal{C}$, where $\mathcal{C}$ is a code with locality ℓ. For any $i \in \{1, \ldots, n\}$ the element $c_i \in \mathbb{F}_q$ is stored in the ith server. In distributed storage systems, data should be recoverable at any moment, even if failures or maintenance operations are performed. For this reason, when a machine fails, the data it contains should be recovered and saved on another machine. To perform such operations efficiently, we wish to limit the number of machines from which data is downloaded. Here comes the interest of codes having a small locality! Suppose the ith server failed. Then we need to reconstruct c_i from the knowledge of the c_j's for $j \neq i$. The objective is to recover c_i from the knowledge of an as small as possible number of c_j's. From Remark 15.9.6, there exists $\mathbf{d} = (d_1, \ldots, d_n) \in \mathcal{C}^{\perp}$ of weight less than or equal to $\ell + 1$ with $d_i \neq 0$. Then the support of $\mathbf{d}$ is $\{i, i_1, \ldots, i_s\}$ with $s \leq \ell$ and

$$c_i = -\frac{1}{d_i} \sum_{j=1}^{s} c_{i_j} d_{i_j}.$$

Consequently, c_i can be recovered after downloading data from at most ℓ distinct servers. The smaller the ℓ, the more efficient the recovery process.

Remark 15.9.7 Note that the literature on distributed storage actually involves two distinct families of codes:

- *locally recoverable codes* which are the purpose of the present section (see also Sections 20.7, 31.3, and 31.4);

- *regenerating codes* which are not discussed in the present chapter but are examined in Section 31.2.

We also refer the interested reader to [542] for an introduction to regenerating codes and to [1430] for a more geometric presentation of them.

15.9.2 A Bound on the Parameters Involving Locality

The most classical bound, which can be regarded as a Singleton Bound for locally recoverable codes, is due to Gopalan, Haung, Simitci, and Yekhanin.

Theorem 15.9.8 ([830, 1489]) *Let $C \subseteq \mathbb{F}_q^n$ be a locally recoverable code of dimension k, minimum distance d, and locality ℓ. Then*

$$d \leq n - k - \left\lceil \frac{k}{\ell} \right\rceil + 2. \tag{15.21}$$

15.9.3 Tamo–Barg Codes

The original proposal of *optimal locally recoverable codes*, i.e., codes reaching the bound (15.21) is due to Barg and Tamo [1772] and are codes derived from Reed–Solomon codes; see also Sections 20.7 and 31.3.2. The construction is as follows.

- Consider a subset $A \subseteq \mathbb{F}_q$ of cardinality n such that $\ell + 1$ divides n and a partition of A:

$$A = A_1 \sqcup \cdots \sqcup A_{\frac{n}{\ell+1}}$$

  into disjoint subsets of size $\ell + 1$. Denote by $x_1, \ldots, x_n$ the elements of A.

- Consider a polynomial[6] g of degree $\ell + 1$ which is constant on any element of the partition, i.e.,

$$g(x) = g(y) \text{ for all } 1 \leq i \leq \frac{n}{\ell+1} \text{ and all } x, y \in A_i.$$

Then, for k divisible by ℓ, define the $[n, k]$ **Tamo–Barg code** of locality ℓ as the code

$$C \stackrel{\text{def}}{=} \left\{ (f(x_1), \ldots, f(x_n)) \;\middle|\; f(X) = \sum_{i=0}^{\ell-1} \sum_{j=0}^{\frac{k}{\ell}-1} a_{ij} X^i g(X)^j \right\}. \tag{15.22}$$

This code has length n and dimension k. In addition, the polynomials that are evaluated to generate codewords have degree at most

$$\deg(f) \leq \ell - 1 + (\ell + 1)\left(\frac{k}{\ell} - 1\right) = k + \frac{k}{\ell} - 2.$$

To get a lower bound for its minimum distance, it suffices to embed this code into a larger code whose minimum distance is known, namely a Reed–Solomon code

$$C \subseteq \left\{ (f(x_1), \ldots, f(x_n)) \;\middle|\; \deg(f) \leq k + \frac{k}{\ell} - 2 \right\} = C_L\left(\mathbb{P}^1, A, \left(k + \frac{k}{\ell} - 2\right)P_\infty\right),$$

whose minimum distance equals $n - k - \frac{k}{\ell} + 2$. Finally, it remains to be shown that the code has locality ℓ which will explain the rationale behind the construction. Suppose we wish to recover a symbol c_r from a given codeword $\mathbf{c} \in C$. The index $r \in A_s$ for some integer s; since g is constant on A_s and identically equal to some constant γ_s, the restriction to A_s of any polynomial f as in (15.22) coincides with the polynomial $\sum_{i,j} a_{ij} \gamma_s^j x^i$, which is a

---

[6]Such polynomials are called ℓ-*good* or (ℓ, n)-*good polynomials*; they are discussed extensively in Section 20.7 where numerous examples are given.

polynomial of degree $< \ell$. By Lagrange interpolation, this polynomial is entirely determined by its evaluations at the ℓ locations of $A_s \setminus \{r\}$, and hence its evaluation at location r can be deduced from these evaluations.

In summary, this code has parameters $\left[n, k, n - k - \frac{k}{\ell} + 2\right]$ and locality ℓ; thus it is optimal with respect to the bound of (15.21).

Example 15.9.9 An explicit example of a polynomial g which is constant on each element of a given partition can be obtained from a polynomial which is invariant under some group action on the affine line and using the partition given by the cosets with respect to this action. For instance, suppose that $(\ell+1) \mid (q-1)$; then $\mathbb{F}_q$ contains a primitive $(\ell+1)^{\text{th}}$ root of unity ζ. The cyclic subgroup of $\mathbb{F}_q^{\times}$ generated by ζ acts multiplicatively on the affine line via the map $z \mapsto \zeta z$ which splits $\mathbb{F}_q^{\times}$ into $\frac{q-1}{\ell+1}$ cosets. Next, the polynomial $g(X) = X^{\ell+1}$ is obviously constant on these cosets. This provides optimal $[n, k]$ codes of locality ℓ for any $n \leq q - 1$ and any $k < n$ such that $(\ell+1) \mid n$ and $\ell \mid k$. Also see Section 20.7.1.1 for a similar construction.

15.9.4 Locally Recoverable Codes from Coverings of Algebraic Curves: Barg–Tamo–Vlăduţ Codes

The discussion to follow requires the introduction of the following definition.

Definition 15.9.10 A morphism $\phi : Y \to X$ is a **Galois cover** if the induced field extension $\phi^* : \mathbb{F}_q(X) \hookrightarrow \mathbb{F}_q(Y)$ (see Section 15.2.3) is Galois. The **Galois group of the cover** is nothing but the Galois group of the extension.

Remark 15.9.11 It is well known that, given a Galois cover $\phi : Y \to X$, the Galois group Γ acts on Y and the orbits are the pre-images of points of X.

Tamo and Barg's construction can be generalized in terms of curve morphisms. Indeed, the situation of Example 15.9.9 can be interpreted as follows. The elements $x_1, \ldots, x_n$ at which polynomials are evaluated are now regarded as a sequence of rational points $P_1, \ldots, P_n$ of $\mathbb{P}^1$ that are from a disjoint union of orbits under the action of the automorphism $\sigma : (x : y) \longmapsto (\zeta x : y)$. Next the polynomial g induces a cyclic cover $\phi : \mathbb{P}^1 \xrightarrow{g} \mathbb{P}^1$ with Galois group spanned by σ. The orbits can be regarded as fibres of rational points of $\mathbb{P}^1$ that split totally.

Similarly to Reed–Solomon codes, for a given ground field, the Tamo–Barg approach permits one to generate optimal codes with respect to (15.21), but the code length will be bounded from above by the number of rational points of $\mathbb{P}^1$, i.e., by $q + 1$. If one wishes to create longer good locally recoverable codes, the construction can be generalized as proposed in [128]. Consider

- two curves X, Y and a Galois cover $\phi : Y \to X$ of degree $\ell + 1$ with Galois group Γ;

- rational points $Q_1, \ldots, Q_{\frac{n}{(\ell+1)}}$ of X which split completely in the cover ϕ;

- their pre-images by ϕ, the points $P_{1,1}, \ldots, P_{1,\ell+1}, \ldots, P_{\frac{n}{\ell+1},1}, \ldots, P_{\frac{n}{\ell+1},\ell+1} \in Y(\mathbb{F}_q)$ which are grouped by orbits of size $\ell + 1$ under the action of Γ. These orbits are the recovery sets.

Equivalently, the recovery sets are *fibres* of ϕ; that is to say, such a set is the pre-image set of a given totally split rational point of X. Let $x \in \mathbb{F}_q(Y)$ be a primitive element of the extension $\mathbb{F}_q(Y)/\mathbb{F}_q(X)$ whose pole locus avoids[7] the points $P_{i,j}$, and let G be a divisor on X. Then one can construct the code

$$\mathcal{C} \stackrel{\text{def}}{=} \left\{ \left(f\left(P_{1,1}\right), \ldots, f(P_{\frac{n}{\ell+1},\ell+1}) \right) \;\middle|\; f = \sum_{i=0}^{\ell-1} (\phi^* f_i) \cdot x^i, \; f_i \in L(G) \right\}, \qquad (15.23)$$

where $L(G) \subseteq \mathbb{F}_q(X)$.

Note that ϕ is constant on any recovery set and hence so are the functions $\phi^* f_i$. Therefore, we have the following statement.

Lemma 15.9.12 *The restriction of* $f = \sum(\phi^* f_i)x^i$ *to a recovery set coincides with a polynomial in* x *with constant coefficients. Consequently, the restriction of* $\mathcal{C}$ *to a recovery set is a Reed–Solomon code of length* $\ell + 1$ *and dimension* ℓ.

Remark 15.9.13 Since x is a primitive element of the extension $\mathbb{F}_q(Y)/\mathbb{F}_q(X)$, then its restriction to a fibre is injective. Indeed, suppose that for two distinct points R_1, R_2 in a given fibre, we have $x(R_1) = x(R_2)$. For any $f \in \mathbb{F}_q(X)$, $\phi^* f(R_1) = \phi^* f(R_2)$; also $\mathbb{F}_q(Y)$ is generated as an algebra by $\phi^* \mathbb{F}_q(X)$ and x. Together these two facts imply that no function in $\mathbb{F}_q(Y)$ takes distinct values at R_1 and R_2, a contradiction.

Similarly to the Reed–Solomon-like construction, these codes have locality ℓ. To estimate the other parameters, denote by $\deg(\phi)$ the degree of the morphism, which is nothing but the extension degree $[\mathbb{F}_q(Y) : \mathbb{F}_q(X)]$. This degree is also the degree of the pole locus divisor $(x)_\infty$ of the function x [1401, Lem. 2.2].

Theorem 15.9.14 ([128, Th. 3.1]) *The Barg–Tamo–Vlăduţ code defined in* (15.23) *has locality* ℓ *and parameters* $[n, k, d]$ *with*

$$k \geq \ell(\deg(G) + 1 - g) \quad \text{and} \quad d \geq n - (\ell - 1)\deg(\phi) - (\ell + 1)\deg(G).$$

Proof: The dimension is a consequence of the definition and of the Riemann–Roch Theorem. The proof of the locality is the very same as that of Tamo–Barg codes given in Section 15.9.3. For the minimum distance, observe that the code $\mathcal{C}$ is contained in the code $\mathcal{C}_L(Y, \mathcal{P}, (\ell - 1)(x)_\infty + \phi^* G)$. Therefore, it suffices to bound from below the minimum distance of this code to get a lower bound for the minimum distance of $\mathcal{C}$. From (15.2), $\deg(\phi^* G) = (\ell + 1)\deg(G)$ and, from Theorem 15.3.12, the code $\mathcal{C}_L(Y, \mathcal{P}, (\ell - 1)(x)_\infty + \phi^* G)$ has minimum distance at least $n - (\ell - 1)\deg((x)_\infty) - (\ell + 1)\deg(G)$. Finally, from [1401, Lem. 2.2], we get $\deg((x)_\infty) = \deg(\phi)$, which concludes the proof. $\qquad\square$

Remark 15.9.15 The above proof is more or less that of [128, Th. 3.1]. We chose to reproduce it here in order describe the general strategy of evaluation of the parameters of such locally recoverable codes constructed from curves. Namely:

- the dimension is obtained by applying the Riemann–Roch Theorem on X together with an elementary count of monomials;

- the minimum distance is obtained by observing that the constructed LRC is contained in an actual algebraic geometry code to which the Goppa Bound (Theorem 15.3.12) can be applied.

Remark 15.9.16 See [128, § IV.A] for examples of LRC from the Hermitian curve.

---

[7]Here again, this avoiding condition can be relaxed thanks to Remark 15.3.8.

15.9.5 Improvement: Locally Recoverable Codes with Higher Local Distance

Until now, we introduced codes whose restriction to any recovery set is nothing but a parity code, that is to say a code of minimum distance 2 which permits only a recovery of one symbol from the other ones. One can expect more, such as being able to correct errors for such a *local code*. Thus one could look for codes whose restrictions to recovery sets have a minimum distance larger than 2. Also see Section 31.3.4.5.

Definition 15.9.17 The **local distance** of a locally recoverable code C is defined as

$$\min_i \min_{A(i)} \left\{ d_H(C_{|A(i)}) \right\},$$

where $i \in \{1, \dots, n\}$ and $A(i)$ ranges over all the recovery sets of C for i (which may be non-unique according to Remark 15.9.5).

The codes introduced in (15.23) have local distance 2. Actually, improving the local distance permits a reduction in the amount of requested symbols for the recovery of a given symbol as suggested by the following statement.

Lemma 15.9.18 *Let $C \subseteq \mathbb{F}_q^n$ be a locally recoverable code with local distance ρ and recovery sets of cardinality $\ell + 1$. For a codeword $\mathbf{c} \in C$, any symbol c_i of $\mathbf{c}$ can be recovered from any $(\ell - \rho + 2)$-tuple of other symbols in the same recovery set.*

Proof: Let $A(i) \subseteq \{1, \dots, n\}$ be a recovery set for the position i. The restricted code $C_{|A(i)}$ has length $\ell + 1$ and minimum distance ρ. Therefore, by definition of the minimum distance for any $I \subseteq A_i \setminus \{i\}$ with $|I| = (\ell + 1) - (\rho - 1) = \ell - \rho + 2$, the restriction map $C_{|A(i)} \longrightarrow C_{|I}$ is injective. $\qquad \square$

This can be done by reducing the degree in x of the evaluated functions, that is to say, by considering a code

$$C' = \left\{ \left(f(P_{1,1}, \dots, f(P_{\frac{n}{\ell+1}, \ell+1}) \right) \ \middle| \ f = \sum_{i=0}^{s-1} (\phi^* f_i) x^i, \ f_i \in (L(G)) \right\} \tag{15.24}$$

for some integer $0 \leq s \leq \ell$. Here again, the code restricted to a recovery set is nothing but an $[\ell + 1, s, \ell - s + 2]$ Reed–Solomon code. In such a code, a codeword is entirely determined by any s-tuple of its entries.

Theorem 15.9.19 *The Tamo–Barg–Vlăduţ code C' defined in (15.24) has length n and locality ℓ, with local distance ρ, dimension k, and minimum distance d satisfying*

$$\rho = \ell - s + 2, \quad k \geq s(\deg(G) + 1 - g), \quad \text{and} \quad d \geq n - (s-1)\deg(x) - (\ell+1)\deg(G).$$

15.9.6 Fibre Products of Curves and the Availability Problem

Still motivated by distributed storage applications, another parameter called *availability* is of interest. This topic is also covered briefly in Section 31.3.4.3.

Definition 15.9.20 The **availability** of a locally recoverable code is the minimum over all positions $i \in \{1, \dots, n\}$ of the number of recovery sets for i.

Practically, a distributed storage system with a large availability benefits from more flexibility in choosing the servers contacted for recovering the contents of a given one.

The availability of an LRC is a positive integer and all the previous constructions had availability 1. Constructing LRC with multiple recovery sets is a natural problem.

- For Reed–Solomon-like LRC, this problem is discussed in [1772, § IV] by considering two distinct group actions on $\mathbb{F}_q$. The cosets with respect to these group actions provide two partitions of the support yielding LRC with availability 2.

- In the curve case, constructions of LRC with availability 2 from a curve X together with two distinct morphisms from this curve to other curves $Y^{(1)}, Y^{(2)}$ is considered in [122, § 6] and [128, § V]; the case of availability $t \geq 2$ is treated in [926].

The construction can be realized from the bottom using the notion of a *fibre product*. Given three curves $Y^{(1)}, Y^{(2)}$, and X with morphisms

$$Y^{(1)} \xrightarrow{\phi_1} X \quad \text{and} \quad Y^{(2)} \xrightarrow{\phi_2} X,$$

then the **fibre product** $Y^{(1)} \times_X Y^{(2)}$ is defined as

$$(Y^{(1)} \times_X Y^{(2)})(\overline{\mathbb{F}}_q) \overset{\text{def}}{=} \left\{ (P_1, P_2) \in (Y^{(1)} \times Y^{(2)})(\overline{\mathbb{F}}_q) \mid \phi_1(P_1) = \phi_2(P_2) \right\}.$$

It comes with two canonical projections ψ_1 and ψ_2 onto $Y^{(1)}$ and $Y^{(2)}$ respectively:

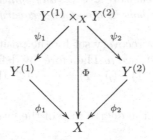

where we denote by Φ the morphism $\Phi = \phi_1 \circ \psi_1 = \phi_2 \circ \psi_2 : Y^{(1)} \times_X Y^{(2)} \longrightarrow X$.

The construction of LRC with availability 2 can be done as follows. Let ℓ_1, ℓ_2 be integers such that $\deg(\phi_1) = \ell_1 + 1$, $\deg(\phi_2) = \ell_2 + 1$ (hence $\deg(\psi_1) = \ell_2 + 1$ and $\deg(\psi_2) = \ell_1 + 1$), and let

- x_1 and x_2 be respective primitive elements of the extensions $\mathbb{F}_q(Y^{(1)})/\mathbb{F}_q(X)$ and $\mathbb{F}_q(Y^{(2)})/\mathbb{F}_q(X)$;

- G be a divisor on X;

- $Q_1, \ldots, Q_s$ be rational points of X that are totally split in Φ and denote by $P_1, \ldots, P_n$ their pre-images.

Similar to the previous cases, either we suppose that the supports of G and the points $Q_1, \ldots, Q_s$ avoid the image under ϕ_1 of the pole locus of x_1 and the image under ϕ_2 of the pole locus of x_2, or we use Remark 15.3.8.

With the above data, we define a locally recoverable code as follows:

$$\mathcal{C} \overset{\text{def}}{=} \left\{ (f(P_1), \ldots, f(P_n)) \;\middle|\; f = \sum_{i=0}^{\ell_1-1} \sum_{j=0}^{\ell_2-1} (\Phi^* h_{i,j}) \psi_1^*(x_1)^i \psi_2^*(x_2)^j, \; h_{i,j} \in L(G) \right\}. \quad (15.25)$$

Let $i \in \{1, \ldots, n\}$. The point P_i is associated to two recovery sets, namely, $\psi_1^{-1}(\{\psi_1(P_1)\})$ and $\psi_2^{-1}(\{\psi_2(P_2)\})$, which have respective cardinalities $\ell_2 + 1$ and $\ell_1 + 1$.

Theorem 15.9.21 *The code C of* (15.25) *has availability 2 with localities (ℓ_1, ℓ_2). Its parameters $[n, k, d]$ satisfy*

$$n = s(\ell_1 + 1)(\ell_2 + 1),$$
$$k \geq (\deg(G) + 1 - g_X)\ell_1\ell_2, \quad and$$
$$d \geq n - \deg(G)(\ell_1 + 1)(\ell_2 + 1)$$
$$- (\ell_1 - 1)(\ell_2 + 1)\deg((x_2)_\infty) - (\ell_2 - 1)(\ell_1 + 1)\deg((x_2)_\infty),$$

where g_X denotes the genus of X.

Proof: The minimum distance comes from the fact that the code is a subcode of $C_L(Y^{(1)} \times_X Y^{(2)}, \mathcal{P}, \Phi^*(G) + (\ell_1 - 1)(\psi_1^*((x_1)_\infty)) + (\ell_2 - 1)(\psi_2^*(x_2)_\infty))$. The dimension is a direct consequence of the definition of the code. For further details, see, for instance, [926, Th. 3.1]. $\qquad\square$

Remark 15.9.22 See [926, § 5, 6, 7] for examples of LRC with availability ≥ 2 from Giulietti–Korchmaros curves, Suzuki curves, and Artin–Schreier curves.

Acknowledgement

To write Section 15.5, the authors followed some advice of Iwan Duursma. Remark 15.6.27 was transmitted by Peter Beelen. The authors warmly thank these colleagues for their help.

Chapter 16

Codes in Group Algebras

Wolfgang Willems

Otto-von-Guericke Universität and *Universidad del Norte*

16.1 Introduction

Group codes, i.e., ideals in a group algebra $\mathbb{F}G$, where $\mathbb{F}$ is a finite field and G a finite group, serve as a fairly good illustration of the fact that the more algebraic structure the ambient space carries the better one understands the code. In the sixties of the last century group codes came into the game. Already in 1967 Berman [174] discovered that binary Reed–Muller codes are ideals in a group algebra where the underlying group is an elementary abelian 2-group. At the same time MacWilliams investigated ideals in a group algebra over a dihedral group of order $2n$ [1320]. In [1321] she proposed that one should "look for a class of groups, not cyclic, which produce codes with some desirable practical properties."

Since that time many papers have dealt with group codes, and there are mainly two reasons for that. First of all, there are numerous good codes in the class of group codes. Second, methods from representation theory seem to be extremely powerful when dealing with coding theoretic properties.

Throughout this chapter we always denote by G a finite group and by $\mathbb{F} = \mathbb{F}_q$ a finite field of size q and characteristic p if nothing else is said. The notation we shall use in both, coding theory and representation theory, is standard. For coding theory we refer to Chapter 1 in this encyclopedia and for representation theory to the textbooks [43, 1200] and [1013, Chapter VII].

In the following we denote by S_n, A_n the symmetric, respectively alternating, group on n letters; by C_n, D_n a cyclic, respectively dihedral, group of order n. The notion $n \mid m$ for

$n, m \in \mathbb{Z}$ means that n divides m. Furthermore, for $n \in \mathbb{N}$ we write $n = n_p n_{p'}$ where n_p is a power of p and $p \nmid n_{p'}$.

In the next two sections we summarize basic facts from finite dimensional algebras, in particular group algebras. The remaining sections then deal with particular classes of group codes.

16.2   Finite Dimensional Algebras

Throughout this section $\mathbb{F}$ is an arbitrary field.

Definition 16.2.1 A ring A with a unit element 1 is called an $\mathbb{F}$-**algebra**, or simply an **algebra**, if A is a vector space over an arbitrary field $\mathbb{F}$ and $\lambda(ab) = (\lambda a)b = a(\lambda b)$ for all $a, b \in A$ and $\lambda \in \mathbb{F}$.

In the following, A always denotes a *finite dimensional algebra* over the field $\mathbb{F}$. By an A-**module** M we mean a *right A-module of finite dimension* over $\mathbb{F}$; by an **ideal** I of A a *right ideal* in A. To be brief we write $I \leq A$. Recall that an A-module M satisfies

- $m1 = m$ for all $m \in M$,
- $(m + m')a = ma + m'a$ for all $m, m' \in M$ and $a \in A$,
- $m(a + a') = ma + ma'$ for all $m \in M$ and $a, a' \in A$,
- $(ma)a' = m(aa')$ for all $m \in M$ and $a, a' \in A$.

Note that the algebra structure of A makes A into an A-module which we call the **regular module** of A. By a **submodule** of an A-module M we mean an $\mathbb{F}$-linear subspace of M which is invariant under the action of A.

Definition 16.2.2 Let $0 \neq M$ be an A-module.

(a) M is called **irreducible** or **simple** if there does not exist a submodule N of M such that $0 < N < M$; i.e., 0 and M are the only submodules of M.

(b) M is called **completely reducible** or **semisimple** if M is a direct sum of irreducible A-modules.

(c) M is called **indecomposable** if there does not exist a direct decomposition $M = M_1 \oplus M_2$ with A-modules $M_1 \neq 0 \neq M_2$.

Definition 16.2.3 Let M and N be A-modules. We call an $\mathbb{F}$-linear map $\alpha : M \longrightarrow N$ which satisfies $(ma)\alpha = (m\alpha)a$ for all $m \in M$ and $a \in A$ an A-**homomorphism** or simply a **homomorphism**. If in addition α is an $\mathbb{F}$-isomorphism, we say that M and N are **isomorphic** and write $M \cong N$. (Note that we always write homomorphisms on the right.)

In the following we do not distinguish between isomorphic A-modules. This means that we usually suppress 'up to isomorphism'.

Theorem 16.2.4 ([1200]) *A has only finitely many irreducible modules; i.e., the number of pairwise non-isomorphic irreducible A-modules is finite.*

Definition 16.2.5 The **Jacobson radical** $J(A)$ of A is defined by

$$J(A) = \{a \in A \mid Ma = 0 \text{ for all irreducible } A\text{-modules } M\}.$$

$J(A)$ is a 2-sided nilpotent ideal in A where **nilpotent** means that there exists an $n \in \mathbb{N}$ such that $J(A)^n = 0$. It can also be characterized as the intersection of all maximal right or left ideals in A.

Definition 16.2.6 A is called **semisimple** if the regular A-module is completely reducible.

Theorem 16.2.7 ([1200]) *The following statements are equivalent.*

(a) *Every A-module is completely reducible.*

(b) *A is semisimple.*

(c) *$J(A) = 0$.*

Example 16.2.8 We consider the 3-dimensional algebra

$$A = \left\{ \left[\begin{array}{cc} a & 0 \\ b & c \end{array} \right] \;\middle|\; a, b, c \in \mathbb{F} \right\}.$$

The Jacobson radical of A is

$$J(A) = \left\{ \left[\begin{array}{cc} 0 & 0 \\ b & 0 \end{array} \right] \;\middle|\; b \in \mathbb{F} \right\}.$$

Furthermore, A has two irreducible modules, both of dimension 1, namely $V_1 = \{(x,0) \mid x \in \mathbb{F}\} \le \mathbb{F}^2$ and $V_2 = \mathbb{F}^2 / V_1$. All minimal ideals of A are isomorphic to V_1. Thus V_2 is not isomorphic to a minimal ideal of A.

Theorem 16.2.9 ([1200]) *Let $M \neq 0$ be an A-module.*

(a) (Jordan–Hölder) *For all composition series*

$$M_0 = 0 < M_1 < \cdots < M_t = M$$

of M with M_i/M_{i-1} irreducible, the number t is the same and the composition factors M_i/M_{i-1} are the same up to isomorphism including multiplicities.

(b) (Krull–Schmidt) *For all decompositions*

$$M = M_1 \oplus \cdots \oplus M_s$$

of M into a direct sum of indecomposable modules M_i, the number s is the same and the M_i are unique up to isomorphism and labeling.

Definition 16.2.10 For any subset $C \subseteq M$, where M is an A-module, the **right annihilator** $\mathrm{Ann}_r(C)$ is defined by

$$\mathrm{Ann}_r(C) = \{a \in A \mid ca = 0 \text{ for all } c \in C\}.$$

Analogously, if M is a left A-module, the **left annihilator** of C is given by

$$\mathrm{Ann}_l(C) = \{a \in A \mid ac = 0 \text{ for all } c \in C\}.$$

Note that the right (left) annihilators are right (left) ideals in A.

Since the Jacobson radical is the intersection of all maximal right ideals of A, we have $J(A/J(A)) = 0$. Thus, by Theorem 16.2.7, $A/J(A)$ is a completely reducible $A/J(A)$-module, hence a completely reducible A-module since $J(A)$ annihilates the irreducible A-modules. Moreover, each irreducible A-module occurs as a factor module of $A/J(A)$. This can be seen as follows: If M is an irreducible A-module and $0 \neq m \in M$, then the A-linear map $\alpha : A \longrightarrow M$ defined by $a \mapsto ma \in M$ for $a \in A$ is an epimorphism where the kernel $\mathrm{Ann}_r(m)$ is a maximal right ideal of A. Thus $A/\mathrm{Ann}_r(m) \cong M$ and M is an epimorphic image of $A/J(A)$ since $J(A) \leq \mathrm{Ann}_r(m)$.

In general, an irreducible A-module does not need to be isomorphic to a minimal ideal of A as Example 16.2.8 shows.

Definition 16.2.11

(a) An element $0 \neq e \in A$ is called an **idempotent** if $e^2 = e$.

(b) Two idempotents $e, f \in A$ are **orthogonal** if $ef = 0 = fe$.

(c) If an idempotent e cannot be written as $e = e_1 + e_2$ with orthogonal idempotents e_1 and e_2, then we say that e is **primitive**.

Note that by definition an idempotent is always different from 0, but $e = 0$ satisfies $e^2 = e$. The next two results are due to Fitting [1200, Chapter I, Theorem 1.4].

Proposition 16.2.12 *The following are equivalent.*

(a) $A = P_1 \oplus \cdots \oplus P_n$ *with A-modules P_i.*

(b) $1 = e_1 + \cdots + e_n$ *with pairwise orthogonal idempotents e_i and $e_i A = P_i$.*

Proposition 16.2.13 *Let $0 \neq e = e^2 \in A$. Then eA is indecomposable if and only if e is primitive.*

Example 16.2.14 Let A be the algebra of Example 16.2.8. Then $e_1 = \begin{bmatrix} 1 & 0 \\ 0 & 0 \end{bmatrix}$ and $e_2 = \begin{bmatrix} 0 & 0 \\ 0 & 1 \end{bmatrix}$ are primitive orthogonal idempotents with $1 = e_1 + e_2$. Thus $A = e_1 A \oplus e_2 A$ with indecomposable A-modules $e_i A$. Note that $e_1 A$ is irreducible and isomorphic to V_1. The A-module $e_2 A$ contains the submodule $J(A)$ which is isomorphic to V_1. Furthermore $e_2 A/J(A) \cong V_2$.

Definition 16.2.15

(a) An A-module P is called **projective** if P is a direct summand of A^n for some n, i.e., if there exists an A-module P' such that

$$P \oplus P' \cong A \oplus \cdots \oplus A$$

as A-modules.

(b) If $e = e^2 \in A$ is a primitive idempotent, then we call $P = eA$ a **PIM** (**projective indecomposable module**). Note that the module eA is indeed projective as $A = eA \oplus (1 - e)A$.

Proposition 16.2.16 ([1200]) *Let $P = eA$ be a PIM of A. Then we have the following.*

(a) *P contains a unique maximal submodule, namely $PJ(A)$.*

(b) $P/PJ(A) \cong M$ *where M is an irreducible A-module.*

(c) *If P' is also a PIM, then $P \cong P'$ if and only if $P/PJ(A) \cong P'/P'J(A)$.*

We say that P is <u>the</u> **projective cover** of M since P and M determine each other up to isomorphism. In particular, there is a one-to-one correspondence between the isomorphism classes of PIM's of A and the isomorphism classes of irreducible A-modules.

Definition 16.2.17

(a) Since A is of finite dimension over $\mathbb{F}$, we may decompose A into a direct sum

$$A = B_0 \oplus B_1 \oplus \cdots \oplus B_s$$

with 2-sided indecomposable ideals $B_i \neq 0$ of A (i.e., indecomposable as 2-sided ideals). Up to the labeling the B_i are uniquely determined by A and called the **blocks**, or more precisely the p-**blocks** of A if the characteristic of $\mathbb{F}$ is p.

(b) If we write $1 = f_0 + \cdots + f_s$ with $f_i \in B_i$, then the f_i are orthogonal primitive idempotents in the **center**

$$Z(A) = \{b \in A \mid ab = ba \text{ for all } a \in A\}$$

of A. The idempotent f_i is uniquely determined by B_i and called the **block idempotent** of B_i.

(c) If M is an indecomposable A-module, then the decomposition

$$M = Mf_0 \oplus \cdots \oplus Mf_s$$

with A-modules Mf_i implies that there is exactly one index i such that $M = Mf_i$ and $Mf_j = 0$ for $j \neq i$. We say that M **belongs to the block** B_i.

Note that a block of A is an algebra over $\mathbb{F}$ where the block idempotent serves as the identity.

16.3   Group Algebras

Definition 16.3.1 Let G be a finite group and let $\mathbb{F}$ be an arbitrary field. The $\mathbb{F}$-vector space

$$\mathbb{F}G = \left\{ a = \sum_{g \in G} a_g g \mid a_g \in \mathbb{F} \right\}$$

with basis $\{g \in G\}$ and multiplication

$$ab = \left(\sum_{g \in G} a_g g \right) \left(\sum_{g \in G} b_g g \right) = \sum_{g \in G} \left(\sum_{h \in G} a_h b_{h^{-1}g} \right) g$$

is an $\mathbb{F}$-algebra which is called a **group algebra** or more precisely the **group algebra of G over $\mathbb{F}$**.

Observe that $\mathbb{F}G$ always has an irreducible module of dimension 1, namely the **trivial module** $\mathbb{F}\sum_{g\in G} g$, which is a minimal ideal in $\mathbb{F}G$. The block to which the trivial module belongs is called the **principal block** of $\mathbb{F}G$.

The number of irreducible $\mathbb{F}G$-modules depends on G and $\mathbb{F}$. If $\mathbb{F}$ has characteristic p and is large enough, for instance if $\mathbb{F}$ is a splitting field for G which means that $\mathbb{F}G/J(\mathbb{F}G)$ is a direct sum of full matrix algebras over $\mathbb{F}$, then this number is equal to the number of p'-conjugacy classes of G; see [1013, Chapter VII, Section 3]. A p'**-conjugacy class** is a conjugacy class $g^G = \{g^x = x^{-1}gx \mid x \in G\}$ of G where g is of order prime to p. If $\mathbb{F}$ is not a splitting field, then there are less irreducible modules. Finally, note that algebraically closed fields are splitting fields, but there are always finite splitting fields for any G and any characteristic $p > 0$ of the underlying field $\mathbb{F}$; see [1013, Chapter VII, Section 2].

Example 16.3.2 Suppose that the characteristic of $\mathbb{F}$ is p and G is a p-group; i.e., the **order** $|G|$ of G is a power of p. Then the trivial module is the only irreducible $\mathbb{F}G$-module. Furthermore, the regular $\mathbb{F}G$-module is indecomposable and in the case $|G| > 1$ not irreducible.

Example 16.3.3 Let $G = A_5$ be the alternating group on 5 letters. Representatives of the conjugacy classes are $1, (1,2)(3,4), (1,2,3), (1,2,3,4,5)$, and $(1,3,5,2,4)$. Apart from $(1,2)(3,4)$ all classes are $2'$-conjugacy classes. If the characteristic of $\mathbb{F}$ is 2, then $\mathbb{F}G$ has two blocks, the principal block B_0 generated by f_0 and another block B_1 generated by $f_1 = 1 - f_0$ which is the sum of all elements of order 3 and 5. For any $\mathbb{F} \geq \mathbb{F}_2$ the block B_1 contains only one irreducible module of dimension 4 and $J(B_1) = 0$. The Jacobson radical $J(B_0)$ of the principal block has dimension 35. If $\mathbb{F} = \mathbb{F}_2$, then B_0 contains two irreducible modules, the trivial one and one of dimension 4. If $\mathbb{F} \geq \mathbb{F}_4$, then $\mathbb{F}$ is a splitting field for G and B_0 contains three irreducible modules, the trivial one and two modules of dimension 2.

Suppose that char $\mathbb{F} = p$ and $p \mid |G|$. If $a = \sum_{g\in G} g$, then $a^2 = |G|a = 0$, and $J = a\mathbb{F}G$ is a nilpotent ideal since $a \in Z(\mathbb{F}G)$. If M is an irreducible $\mathbb{F}G$-module, then obviously $MJ = 0$ or $MJ = M$ since MJ is a submodule of M. As J is nilpotent the second case also leads to $MJ = 0$. Thus $0 \neq J \leq J(\mathbb{F}G)$ by Definition 16.2.5, which means that $\mathbb{F}G$ is not semisimple.

Theorem 16.3.4 (Maschke [43]) *The following are equivalent.*

(a) $\mathbb{F}G$ *is semisimple.*

(b) char $\mathbb{F} \nmid |G|$.

Thus, if the characteristic p of the field $\mathbb{F}$ divides $|G|$, then there are irreducible $\mathbb{F}G$-modules whose PIM's are indecomposable but not irreducible.

Again let char $\mathbb{F} = p$, and let B be a p-block of $\mathbb{F}G$ with block idempotent $f = \sum_{g\in G} a_g g \in Z(\mathbb{F}G)$. By a result of R. Brauer [1200], there exists a $g \in G$ with $a_g \neq 0$ and a Sylow p-subgroup $\delta(B)$ of $C_G(g) = \{x \in G \mid xg = gx\}$ such that for all $h \in G$ with $a_h \neq 0$ the Sylow p-subgroup of $C_G(h)$ is contained in a G-conjugate of $\delta(B)$, i.e., in $\delta(B)^x = x^{-1}\delta(B)x$ for some $x \in G$. The p-subgroup $\delta(B)$ of G is unique up to conjugation.

Definition 16.3.5 Up to conjugation the p-subgroup $\delta(B)$ of G is called the **defect group** of B. If $|\delta(B)| = p^{d(B)}$, then we call $d(B)$ the **defect** of B.

Recall that for any prime p a finite group G always has subgroups of order $|G|_p$, which are called **Sylow p-subgroups**, and that all of them are conjugate [43, Section 7]. By definition $|G|_p$ is the highest power of p dividing $|G|$.

Remark 16.3.6 Let char $\mathbb{F} = p$.

(a) The defect group of the principal p-block $B_0(G)$ of $\mathbb{F}G$ is a Sylow p-subgroup of G. In particular $|G|_p = p^{d(B_0(G))}$.

(b) The defect group of a p-block B of $\mathbb{F}G$ is the trivial group if and only if B is a full matrix algebra over some finite extension field of $\mathbb{F}$. In this case B contains only one irreducible $\mathbb{F}G$-module M and $B \cong M \oplus \cdots \oplus M$ as right modules. In particular, M is a projective $\mathbb{F}G$-module since B is a direct summand of $\mathbb{F}G$.

Definition 16.3.7 Let M be an $\mathbb{F}G$-module. The vector space $\mathrm{Hom}_{\mathbb{F}}(M, \mathbb{F})$ of all $\mathbb{F}$-linear maps from M to $\mathbb{F}$ becomes an $\mathbb{F}G$-module by

$$m(\alpha g) = (mg^{-1})\alpha$$

for $m \in M, g \in G$ and $\alpha \in \mathrm{Hom}_{\mathbb{F}}(M, \mathbb{F})$. This module is denoted by M^* and called the **dual module** of M. If $M \cong M^*$, we say that M is a **self-dual $\mathbb{F}G$-module**.

Observe that the trivial $\mathbb{F}G$-module is always self-dual. Furthermore, for any $\mathbb{F}G$-module M we have $\dim M = \dim M^*$.

Example 16.3.8 Let $\mathbb{F} = \mathbb{F}_4$ and let G be a cyclic group of order 3 generated by g. Clearly, $\mathbb{F}$ contains a primitive third root of unity, say ϵ. The 1-dimensional vector-space $M = \langle m \rangle$ becomes an $\mathbb{F}G$-module via the action $mg = \epsilon m$. On $M^* = \langle m^* \rangle$ the generator g acts via $m^*g = \epsilon^{-1}m^*$. Thus $M \not\cong M^*$.

16.4 Group Codes

From now on $\mathbb{F}$ always denotes a finite field of characteristic p.

Definition 16.4.1 For $a = \sum_{g \in G} a_g g \in \mathbb{F}G$, where $a_g \in \mathbb{F}$, we define the **support** of a by

$$\mathrm{supp}(a) = \{g \in G \mid a_g \neq 0\}$$

and the **Hamming weight** of a by $\mathrm{wt}_{\mathrm{H}}(a) = |\mathrm{supp}(a)|$.

Definition 16.4.2 On the group algebra $\mathbb{F}G$ there is a **distance** d_{H}, also called the **Hamming distance**, which is given by

$$\mathrm{d}_{\mathrm{H}}(a, b) = \mathrm{wt}_{\mathrm{H}}(a - b)$$

for $a, b \in \mathbb{F}G$. Note that d_{H} satisfies the axioms of a metric; see Remark 1.6.3.

Definition 16.4.3 A right ideal $\mathcal{C}$ in $\mathbb{F}G$ endowed with the distance d_{H} is called a **group code**, or in case we want to refer to the underlying group, a **G-code**. If G is abelian, we call $\mathcal{C}$ an **abelian group code**.

To focus on right ideals as group codes is only for convention. All results easily transfer to left ideals.

In general it is hard to determine all ideals, i.e., group codes, in a given group algebra $\mathbb{F}G$.

Example 16.4.4 Let $G = S_3$ be the symmetric group on three letters.

(a) If char $\mathbb{F} \nmid |G|$, then $\mathbb{F}G$ is semisimple (by Maschke's Theorem 16.3.4) and has a direct decomposition $\mathbb{F}G \cong M \oplus M^- \oplus V \oplus V$, where M is the trivial module, M^- is a one-dimensional module on which an involution acts by -1 and V is an irreducible module of dimension 2. One easily computes that $\mathbb{F}_q G$ contains exactly $4 \cdot q + 12$ group codes.

(b) If char $\mathbb{F} = 2$, then the group algebra $\mathbb{F}G$ has a direct sum decomposition $\mathbb{F}G \cong \genfrac{}{}{0pt}{}{M}{M} \oplus V \oplus V$, where V is again an irreducible module of dimension 2 and the first component is the projective cover $P(M)$ of the trivial module M. The notation $\genfrac{}{}{0pt}{}{M}{M}$ means that $P(M)$ is a non-split extension of M by M. Furthermore, $P(M)$ is the principal 2-block B_0 and $V \oplus V$ is a 2-block B_1 of defect 0. Since each ideal I in $\mathbb{F}G$ can be written as $I = (B_0 \cap I) \oplus (B_1 \cap I)$, we see that $\mathbb{F}_{2^n} G$ contains exactly $3 \cdot 2^n + 9$ group codes.

(c) If char $\mathbb{F} = 3$, then $\mathbb{F}G$ can be written as

$$\mathbb{F}G = \begin{matrix} M \\ M^- \\ M \end{matrix} \oplus \begin{matrix} M^- \\ M \\ M^- \end{matrix} = P(M) \oplus P(M^-)$$

where M is again the trivial module and M^- is a 1-dimensional module on which an involution acts by -1. Determining all group codes in this case is more difficult.

Proposition 16.4.5 ([1050]) *Let $\mathbb{F} = \mathbb{F}_q$ be of characteristic p and let $G = A \times C_{p^k}$ where A is an abelian p'-group and C_{p^k} is cyclic of order p^k. Then $\mathbb{F}G$ contains exactly $(p^k + 1)^t$ group codes where t is the number of q-cyclotomic cosets modulo $|A|$ or equivalently the number of irreducible $\mathbb{F}A$-modules.*

Remark 16.4.6 In this remark a G-code or group code means a 2-sided ideal in $\mathbb{F}G$.

(a) [175] If $G = AB$ with abelian subgroups $A, B \leq G$, then any G-code is an abelian group code (up to equivalence). In particular, a given G-code may also be realized as an H-code where G and H are non-isomorphic groups.

(b) [791, 792, 793] For $|G| < 128$ and $|G| \notin \{24, 48, 50, 60, 64, 72, 96, 108, 120\}$ all G-codes are abelian group codes. Over $\mathbb{F}_5$ there are S_4-codes which are not abelian group codes. Over the binary field there are $SL_2(3)$-codes which are optimal and non-abelian group codes.

Next we consider the natural $\mathbb{F}$-linear action of the symmetric group S_n on $\mathbb{F}^n$ defined on the standard basis $e_i = (0, \ldots, 0, 1, 0, \ldots, 0)$ (where 1 is at position i) by

$$e_i \sigma = e_{i\sigma}$$

for $\sigma \in S_n$ and $i = 1, \ldots, n$. For an $\mathbb{F}$-linear subspace $\mathcal{C}$ of $\mathbb{F}^n$, we denote by

$$\mathrm{PAut}(\mathcal{C}) = \{\sigma \in S_n \mid \sigma \in \mathrm{Aut}(\mathcal{C})\}$$

the **group of permutation automorphisms** of $\mathcal{C}$. Clearly, if $\mathcal{C} \leq \mathbb{F}G$ is a group code, then $G \leq \mathrm{PAut}(\mathcal{C})$ and G is a regular subgroup of $S_{|G|}$. Recall that a subgroup H of S_n is called **regular** if H acts transitively on $\{1, \ldots, n\}$ and $|H| = n$.

Theorem 16.4.7 ([175]) *Let C be a linear code over $\mathbb{F}$ of length n, and let G be a group of order n. Then, up to permutation equivalence, C is a G-code if and only if G is isomorphic to a regular subgroup of* $\mathrm{PAut}(C)$.

In order to see the 'if' part we may assume that $G \leq \mathrm{PAut}(C)$. Since G is transitive of order n there exists to each $j \in \{1, \ldots, n\}$ exactly one $g \in G$ such that $j = 1g$. The $\mathbb{F}$-linear map $\phi : \mathbb{F}^n \longrightarrow \mathbb{F}G$ defined by $\phi(e_j) = \phi(e_{1g}) = g$ shows that $\phi(C)$ is a (right) ideal of $\mathbb{F}G$.

There is a slight extension of Theorem 16.4.7 to monomial equivalence in [1424].

Example 16.4.8 The linear code $C = \{(c_1, \ldots, c_n) \mid c_i \in \mathbb{F}, \sum_{i=1}^n c_i = 0\}$ is a G-code for any group G of order n. This may be seen as follows. First note that $\alpha : \mathbb{F}G \to \mathbb{F}$ defined by $\alpha(\sum_{g \in G} a_g g) = \sum_{g \in G} a_g$ is G-linear. Since C is permutation equivalent to the kernel of α, which is an ideal in $\mathbb{F}G$, we are done.

Example 16.4.9 A cyclic code of length n is an ideal in $\mathbb{F}[x]/\langle x^n - 1 \rangle \cong \mathbb{F}C_n$, where C_n is a cyclic group of order n. Thus cyclic codes are abelian group codes. In particular, BCH codes, Reed–Solomon codes, and quadratic residue codes are group codes.

Example 16.4.10 Let $\mathcal{H}_{m,q}$ denote the Hamming code of length $n = \frac{q^m - 1}{q - 1}$ over the field $\mathbb{F}_q$; see Section 1.10.

(a) [1008] If $\gcd(m, q - 1) = 1$, then $\mathcal{H}_{m,q}$ is a BCH code, hence a group code.

(b) The ternary Hamming code $C = \mathcal{H}_{2,3}$ which is of dimension 2 and has length 4 is not a group code: Suppose that $C \leq \mathbb{F}_3 G$ where G is cyclic of order 4 or a Klein four group. Note that these groups are the only groups of order 4. Since $\mathbb{F}G$ is semisimple by Maschke's Theorem 16.3.4, one easily computes $d_{\mathrm{H}}(C) = 2$, a contradiction as $d_{\mathrm{H}}(\mathcal{H}_{2,3}) = 3$.

(c) For all m and q, the Hamming code $\mathcal{H}_{m,q}$ is a right ideal in a twisted group algebra of a cyclic group.

Example 16.4.11 Let p be a prime.

(a) [174, 385] Let $G = C_p \times \cdots \times C_p$ be an elementary abelian p-group of order p^m. Furthermore let $J = J(\mathbb{F}_p G)$ denote the Jacobson radical of $\mathbb{F}_p G$; i.e., $J = \sum_{g \in G}(1 - g)\mathbb{F}_p$. Then, for $r = -1, 0, 1, \ldots, N = m(p - 1)$ the ideal $J^{N-r} \leq \mathbb{F}_p G$ defines the Reed–Muller code $\mathcal{RM}_p(r, m)$ of order r and length p^m over $\mathbb{F}_p$. Thus Reed–Muller codes over prime fields are group codes.

(b) In [1201] it is shown that generalized Reed–Muller codes of length p^m over $\mathbb{F}_{p^s}$, with $s \mid m$, are ideals in $\mathbb{F}_{p^s} G$ where G is again elementary abelian of order p^m. If $s > 1$, then, in general, the codes are not powers of the Jacobson radical.

Instead of looking at elementary abelian p-groups, we may look more generally at arbitrary p-groups. Using Jennings' description [1041] of the radical powers, Faldum proved that this larger class of group codes does not lead to better parameters. More precisely, he proved the following.

Theorem 16.4.12 ([701]) *Suppose that $\mathbb{F}$ is of characteristic p. Let G be an elementary abelian p-group of order p^m and let H be any p-group with $|G| = |H|$. If $\dim J(\mathbb{F}H)^{l'} \begin{Bmatrix} \geq \\ > \end{Bmatrix} \dim J(\mathbb{F}G)^l \neq 0$, then $d_{\mathrm{H}}(J(\mathbb{F}H)^{l'}) \begin{Bmatrix} \leq \\ < \end{Bmatrix} d_{\mathrm{H}}(J(\mathbb{F}G)^l)$. Moreover, if equality holds for the dimension and the minimum distance, then $l = l'$ and $G \cong H$ for $l \notin \{0, 1, m(p - 1)\}$.*

Remark 16.4.13 Many codes which are optimal in the class of linear codes are group codes. Here we mention a few examples over the binary field.

(a) [792] There is a $[24, 6, 10]$ group code for $G = \mathrm{SL}_2(3)$.

(b) [1858] There is a $[32, 17, 8]$ group code for $G = \mathrm{C}_4 \wr \mathrm{C}_2$, $G = \mathrm{D}_8 \mathrm{Y} \mathrm{D}_8$ with amalgamated centers, and $G = \mathrm{C}_2 \times \mathrm{C}_2 \times \mathrm{C}_2 : \mathrm{C}_2 \times \mathrm{C}_2$.

(c) [435] There is a $[45, 13, 16]$ abelian group code for $G = \mathrm{C}_3 \times \mathrm{C}_{15}$.

Remark 16.4.14 Geometric Goppa codes arising from curves with many rational points often turn out to be group codes.

(a) Using the Klein quartic $x^3 y + y^3 z + z^3 x = 0$, Hansen proved in [900] that a $[21, 10, 9]_8$ code exists as a group code in $\mathbb{F}_8 G$, where G is a Frobenius group of order 21. Note that 9 is the largest minimum distance of a known $[21, 10]_8$ code as stated in [845].

(b) In [901] there are geometric Goppa codes constructed from Hermitian curves. Some of them are group codes in $\mathbb{F}_{q^2} G$, where G is a non-abelian group of order q^3 and q is a power of p.

(c) According to [902] a similar result holds true for the Suzuki curve in which the codes are group codes over $\mathbb{F}_q$ for the Sylow 2-subgroup of the simple Suzuki group $Sz(q)$ of order $q^2(q^2 + 1)(q - 1)$ where $q = 2^{2n+1}$ and $n \in \mathbb{N}$.

Definition 16.4.15 The group algebra $\mathbb{F}G$ carries a non-degenerate symmetric **bilinear form** $\langle \cdot, \cdot \rangle$ defined by

$$\langle a, b \rangle = \sum_{g \in G} a_g b_g \in \mathbb{F}$$

for $a = \sum_{g \in G} a_g g$ and $b = \sum_{g \in G} b_g g$ in $\mathbb{F}G$. Note that $\langle \cdot, \cdot \rangle$ is G-**invariant**, i.e., $\langle ag, bg \rangle = \langle a, b \rangle$ for all $g \in G$ and all $a, b \in \mathbb{F}G$.

Definition 16.4.16 For a group code $\mathcal{C} \le \mathbb{F}G$ we call

$$\mathcal{C}^\perp = \{x \in \mathbb{F}^n \mid \langle x, c \rangle = 0 \text{ for all } c \in \mathcal{C}\}$$

the **dual code** of $\mathcal{C}$. If $\mathcal{C} \subseteq \mathcal{C}^\perp$, we say that $\mathcal{C}$ is **self-orthogonal**. In the case $\mathcal{C} = \mathcal{C}^\perp$, we call $\mathcal{C}$ a **self-dual code**.

Observe that with $\mathcal{C}$ the dual code $\mathcal{C}^\perp$ is a group code as well since

$$\langle c, c^\perp g \rangle = \langle cg^{-1}, c^\perp \rangle = 0$$

for all $c \in \mathcal{C}$, $c^\perp \in \mathcal{C}^\perp$, and $g \in G$.

Definition 16.4.17 On $\mathbb{F}G$ we define an $\mathbb{F}$-linear map $\widehat{}$ by $\widehat{g} = g^{-1}$ for $g \in G$.

Note that this map is an anti-algebra automorphism of $\mathbb{F}G$. Thus, if $\mathcal{C}$ is a right ideal in $\mathbb{F}G$, then $\widehat{\mathcal{C}} = \{\widehat{c} \mid c \in \mathcal{C}\}$ is a left ideal.

Proposition 16.4.18 ([1320]) *If $\mathcal{C} \le \mathbb{F}G$ is a group code, then $\mathcal{C}^\perp = \widehat{\mathrm{Ann}_l(\mathcal{C})}$.*

As an immediate consequence we get the following.

Lemma 16.4.19 *If $\mathcal{C} = a\mathbb{F}G$ with $\widehat{a}a = 0$, then $\mathcal{C}$ is self-orthogonal.*

For a group code $\mathcal{C}$ there exists a powerful connection between the dual code $\mathcal{C}^\perp$ and the dual module $\mathcal{C}^*$ given in Definition 16.3.7.

Proposition 16.4.20 ([482, 1893]) *If $\mathcal{C} \leq \mathbb{F}G$ is a group code, then $\mathbb{F}G/\mathcal{C}^\perp \cong \mathcal{C}^*$ as $\mathbb{F}G$-modules.*

Corollary 16.4.21 *If $\mathcal{C}^\perp = \mathcal{C} \leq \mathbb{F}G$ is a self-dual group code, then the multiplicity of every irreducible self-dual $\mathbb{F}G$-module as a composition factor in a Jordan–Hölder series of $\mathbb{F}G$ is even.*

Proof: If X is a composition factor of $\mathcal{C} = \mathcal{C}^\perp$, then X^* is a composition factor of $\mathcal{C}^*$. Since $\mathbb{F}G/\mathcal{C} \cong \mathcal{C}^*$ by Proposition 16.4.20, we see that X has even multiplicity in $\mathbb{F}G$ if $X \cong X^*$. Note, that if X is a composition factor of $\mathbb{F}G/\mathcal{C} \cong \mathcal{C}^*$, then X^* is a composition factor of $\mathcal{C}^{**} \cong \mathcal{C}$. $\qquad\square$

The statement of Corollary 16.4.21 is one of the main tools when studying self-dual group codes.

Proposition 16.4.22 ([1464]) *If $\mathcal{C} = a\mathbb{F}G$, then $\mathcal{C}^* \cong \widehat{a}\mathbb{F}G$. In particular, $\dim a\mathbb{F}G = \dim \mathbb{F}Ga$.*

16.5 Self-Dual Group Codes

In coding theory much attention is paid to self-dual codes, and they occur in all characteristics. In contrast to codes which are only linear vector spaces, self-dual group codes exist only in characteristic 2 (Theorem 16.5.4). The next example shows that also for self-dual group codes the isomorphism type of the underlying group is in general not unique.

Example 16.5.1

(a) The binary self-dual $[24, 12, 8]$ extended Golay code $\mathcal{G}_{24}$ is a group code for $G = S_4$ in [176], for $G = D_{24}$ in [1372], and for $G = A_4 \times C_2, D_8 \times C_3$, and $(C_3 \times C_2) \rtimes C_2$ in [623].

(b) The binary self-dual $[48, 24, 12]$ extended quadratic residue code is a group code for $G = D_{48}$; see [1371].

Remark 16.5.2 The binary self-dual $[24, 12, 8]$ extended Golay code $\mathcal{G}_{24}$ does not occur as an ideal in $\mathbb{F}_2 C_{24}$ according to Theorem 16.4.7, since the cyclic group C_{24} is not a subgroup of $\mathrm{PAut}(\mathcal{G}_{24}) = M_{24}$. As shown in [470], it is the image of a derivation on $\mathbb{F}_2 C_{24}$.

There is the long standing question [1726] whether a binary self-dual $[72, 36, 16]$ code $\mathcal{C}$ exists. If there is such a code $\mathcal{C}$, then $\mathcal{C}$ cannot be a group code since $|\mathrm{PAut}(\mathcal{C})| \leq 5$; see for instance [239] and Remark 4.3.11.

Example 16.5.3 Let $\mathbb{F}$ be of characteristic p and let $G = C_{p^a}$ be a cyclic group of order p^a. Suppose that $\mathcal{C} = \mathcal{C}^\perp \leq \mathbb{F}G$ is a self-dual group code. Since the multiplicity of the trivial module in $\mathbb{F}G$ is p^a we get $p = 2$ according to Corollary 16.4.21. If J is the Jacobson radical of $\mathbb{F}G$, then J^i $(i = 0, \dots, 2^a)$ are the only ideals in $\mathbb{F}G$ and

$$0 = J^{2^a} < J^{2^a - 1} < \cdots < J < J^0 = \mathbb{F}G.$$

Now $2 \dim \mathcal{C} = \dim \mathcal{C} + \dim \mathcal{C}^\perp = |G| = 2^a$ forces $\mathcal{C} = J^{2^{a-1}}$.

Theorem 16.5.4 ([1893]) *The group algebra $\mathbb{F}G$ contains a self-dual group code if and only if the characteristic of $\mathbb{F}$ is 2 and $2 \mid |G|$.*

Remark 16.5.5 ([176]) The ternary extended self-dual Golay code $\mathcal{G}_{12}$ is not a group code, but an ideal in a twisted group algebra of the alternating group $G = A_4$ over $\mathbb{F}_3$.

In [1727] Sloane and Thompson proved that a binary self-dual group code is never doubly-even provided the Sylow 2-subgroup of the underlying group is cyclic. More generally, we have the following.

Theorem 16.5.6 ([1342]) *Let $\mathbb{F} = \mathbb{F}_2$ and let T be a Sylow 2-subgroup of G. Then $\mathbb{F}G$ contains a self-dual doubly-even group code if and only if T is neither cyclic nor a Klein four group.*

If $\mathcal{C}$ is a G-code, then G is a subgroup of the symmetric group $S_{|G|}$. In the case that $\mathbb{F}$ is binary and $\mathcal{C} = \mathcal{C}^\perp$ is doubly-even, we can say more.

Theorem 16.5.7 ([873]) *If $\mathcal{C} = \mathcal{C}^\perp \leq \mathbb{F}_2^n$ is a doubly-even linear code, then $\mathrm{PAut}(\mathcal{C})$ is contained in the alternating group A_n. In particular, a binary self-dual doubly-even G-code implies $G \leq A_{|G|}$ (up to isomorphism).*

Remark 16.5.8 In [1043] an enumeration of self-dual cyclic codes over $\mathbb{F}_{2^m}$ is given. This has been extended in [1050] to the case where the underlying group is abelian with a cyclic Sylow 2-subgroup.

16.6 Idempotent Group Codes

Definition 16.6.1 A group code $\mathcal{C} \leq \mathbb{F}G$ is called an **idempotent code** if there exists an idempotent $e = e^2 \in \mathbb{F}G$ such that $\mathcal{C} = e\mathbb{F}G$.

Remark 16.6.2 If $\mathrm{char}\,\mathbb{F} \nmid |G|$, i.e., $\mathbb{F}G$ is a semisimple group algebra by Maschke's Theorem 16.3.4, then all G-codes over $\mathbb{F}$ are idempotent codes.

Note that an idempotent code $\mathcal{C} = e\mathbb{F}G$ is a projective $\mathbb{F}G$-module since $\mathbb{F}G = e\mathbb{F}G \oplus (1-e)\mathbb{F}G$.

Theorem 16.6.3 ([1013, Dickson]) *If $\mathrm{char}\,\mathbb{F} = p$ and $\mathcal{C}$ is an idempotent code in $\mathbb{F}G$, then $|G|_p \mid \dim \mathcal{C}$.*

Let $\mathrm{char}\,\mathbb{F} = p$. For a normal p'-subgroup H of G we put

$$\widetilde{H} = \frac{1}{|H|} \sum_{h \in H} h \in \mathbb{F}G.$$

Note that $\frac{1}{|H|} \in \mathbb{F}$ exists since H is a p'-group. Furthermore, $\widetilde{H}$ is an idempotent lying in the center $Z(\mathbb{F}G)$ of $\mathbb{F}G$. If $e = e^2 \in \mathbb{F}G$ is a primitive idempotent, then $\widetilde{H}e = e$ or $\widetilde{H}e = 0$. The second case means that $e\mathbb{F}G \cap \widetilde{H}\mathbb{F}G = 0$. In order to understand the first case let T be a right transversal of H in G; i.e., $G = \bigcup_{t \in T} Ht$ (disjoint union). Thus

$$e\mathbb{F}G = \widetilde{H}e\mathbb{F}G \subseteq \widetilde{H}\mathbb{F}G = \widetilde{H}\left(\bigoplus_{t \in T} \mathbb{F}t \right).$$

In particular, a coordinate in T of a codeword in $e\mathbb{F}G$ is repeated $|H|$ times.

Definition 16.6.4 Let $C \leq \mathbb{F}G$ be a group code. If there exists $T \subseteq G$ with $|T| \mid |G|$ such that, for every $c \in C$, each coordinate c_t of c with $t \in T$ is repeated $|G : T|$ times, then we call C a $|G : T|$-**repetition code**.

In the next result we use the notion $O_{p'}(G)$ which denotes the largest normal subgroup of G whose order is not divisible by p.

Theorem 16.6.5 ([482]) *Let* $\operatorname{char} \mathbb{F} = p$ *and let* $e \in \mathbb{F}G$ *be a primitive idempotent. If* $\widetilde{O_{p'}(G)}e = e$, *then* $e\mathbb{F}G$ *is an* $|O_{p'}(G)|$-*repetition code.*

Remark 16.6.6 If $G = HT$ where H is a p'-subgroup and T is a Sylow p-subgroup of G, then $\widetilde{H}$ is a primitive idempotent in $\mathbb{F}G$ where $p = \operatorname{char} \mathbb{F}$. This is a consequence of Dickson's Theorem 16.6.3 and Proposition 16.2.13. Note that $\widetilde{H}$ may not be central.

Theorem 16.6.7 ([378]) *Let* A *be an abelian group with* $\operatorname{char} \mathbb{F} \nmid |A|$ *and let* e *be a primitive idempotent in* $\mathbb{F}A$. *If* $\widetilde{H}e = 0$ *for all subgroups* $H \neq 1$ *of* A, *then* A *is cyclic.*

Corollary 16.6.8 *Let* A *be an abelian group with* $\operatorname{char} \mathbb{F} \nmid |A|$, *but not cyclic. Then any minimal group code in* $\mathbb{F}A$ *is a* $|U|$-*repetition code for some subgroup* $1 \neq U \leq A$.

Theorem 16.6.9 ([378]) *Every minimal abelian group code in a semisimple group algebra is permutation equivalent to a cyclic code.*

16.7   LCP and LCD Group Codes

Definition 16.7.1

(a) A pair (C, D) of linear codes $C, D \leq \mathbb{F}^n$ is called a <u>l</u>inear <u>c</u>omplementary <u>p</u>air of codes, or simply an **LCP** of codes, if $C \oplus D = \mathbb{F}^n$.

(b) If $D = C^{\perp}$, then C is called a <u>l</u>inear <u>c</u>omplementary <u>d</u>ual code, or simply an **LCD** code.

LCD codes were first considered by Massey in [1347]. The present-day interest in LCP and LCD codes arises from the fact that they can be used in protection against side channel and fault injection attacks; see [304, 353, 1431]. In this context the security of a linear complementary pair (C, D) can be measured by its **security parameter** $\min \{d_{\mathrm{H}}(C), d_{\mathrm{H}}(D^{\perp})\}$. Clearly, if $D = C^{\perp}$, i.e., C is an LCD code, then the security parameter is $d_{\mathrm{H}}(C)$.

LCD codes are asymptotically good [1347], and they achieve the Gilbert–Varshamov Bound [1648]. Applying Proposition 16.2.12 we immediately get the following.

Proposition 16.7.2 *If* (C, D) *is an LCP of group codes, then* $C = e\mathbb{F}G$ *and* $D = (1 - e)\mathbb{F}G$ *for some* $e = e^2 \in \mathbb{F}G$.

Proposition 16.7.3 *If* $\mathbb{F}G = C \oplus D$ *with group codes* C *and* D, *i.e.,* (C, D) *is an LCP of group codes, then the following hold.*

(a) C *and* D *are projective* $\mathbb{F}G$-*modules.*

(b) $|G|_p$ *divides both* $\dim C$ *and* $\dim D$.

(c) *If C is an LCD code, i.e., $D = C^\perp$, then $C \cong C^*$ as $\mathbb{F}G$-modules.*

Proof: Part (a) is just Definition 16.2.15; part (b) is Dickson's Theorem 16.6.3 using Proposition 16.7.2; and part (c) follows from Proposition 16.4.20.   □

Theorem 16.7.4 ([242]) *Let (C, D) be an LCP of codes in $\mathbb{F}G$, where C and D are 2-sided ideals of $\mathbb{F}G$. Then C and $D^\perp$ are permutation equivalent. In particular (C, D) has security parameter $d_H(C)$.*

If C and D are nD cyclic codes, then the codes can be expressed as ideals in $\mathbb{F}[x_1, \dots, x_n]/\langle x_1^{m_1} - 1, \dots, x_n^{m_n} - 1 \rangle$. In this case a proof of Theorem 16.7.4, which uses only polynomials, can be found in [354, 870].

In the case that C and D are only right ideals, $D^\perp$ is in general not permutation equivalent to C. It may even happen that $d_H(C) \neq d_H(D^\perp)$ as the next example shows.

Example 16.7.5 Let $\mathbb{F} = \mathbb{F}_2$ be the binary field and let

$$G = D_{14} = \langle a, b \mid a^7 = 1 = b^2, a^b = a^{-1} \rangle$$

be a dihedral group of order 14. If we put

$$e = 1 + a + a^2 + a^4 + b + a^2 b + a^5 b + a^6 b,$$

then $e = e^2$. Furthermore $d_H((1 - e)\mathbb{F}G) = 2$ and $d_H(\mathbb{F}G(1 - e)) = 3$.

Now let $C = (1 - e)\mathbb{F}G$ and $D = e\mathbb{F}G$. Since $D^\perp = (1 - \hat{e})\mathbb{F}G$ we get

$$d_H(D^\perp) = d_H((1 - \hat{e})\mathbb{F}G) = d_H(\mathbb{F}G(1 - e)) = 3,$$

but $d_H(C) = d_H((1 - e)\mathbb{F}G) = 2$. We would like to mention here that C and D are quasi-cyclic codes.

Theorem 16.7.6 ([502]) *A group code $C \leq \mathbb{F}G$ is an LCD code if and only if $C = e\mathbb{F}G$ with $e^2 = e = \hat{e} \in \mathbb{F}G$. In this case $C^\perp = (1 - e)\mathbb{F}G$.*

Example 16.7.7 Suppose that $\mathrm{char}\,\mathbb{F} = p$ and H is a subgroup of G with $p \nmid |H|$. Then $e = \frac{1}{|H|} \sum_{h \in H} h$ satisfies $e^2 = e = \hat{e}$. Thus $C = e\mathbb{F}G$ is an LCD group code. In the case that H is a p-complement in G, i.e., $G = HT$ where T is a Sylow p-subgroup of G, we have $\dim C = |G|_p$ and $d_H(C) = |H| = |G|_{p'}$. This follows directly from $C = e\mathbb{F}G = e\mathbb{F}T$.

Theorem 16.7.8 ([502]) *Let $\mathrm{char}\,\mathbb{F} = 2$ and let $C = e\mathbb{F}G$ with $e^2 = e = \hat{e}$. Then C is symplectic, i.e., $\langle c, c \rangle = 0$ for all $c \in C$, if and only if $1 \notin \mathrm{supp}(e)$.*

Thus, if C is a binary symplectic LCD group code, then C is singly-even. Observe that C is never doubly-even. This can be seen as follows. If C were doubly-even, then $C \subseteq C^\perp$, a contradiction to the assumption that $C \neq 0$ is an LCD group code.

Definition 16.7.9 Let $C \leq \mathbb{F}G$ be a group code. We call $\hat{C} = \{\hat{c} \mid c \in C\}$ the **adjoint code** of C. In the case $C = \hat{C}$ we say that C is **self-adjoint**.

For a group code C the vector space $\hat{C}$ is a left ideal. Thus, if C is self-adjoint, then C is a 2-sided ideal.

Theorem 16.7.10 ([502]) *A group code $C \leq \mathbb{F}G$ is a self-adjoint LCD code if and only if $C = e\mathbb{F}G$ where $e^2 = e = \hat{e}$ and $e \in Z(\mathbb{F}G)$.*

Note that a self-adjoint group code $\mathcal{C} \neq 0$ is a direct sum of blocks of the group algebra; see Definition 16.2.17.

Furthermore, as a consequence of Theorem 16.7.10 we immediately obtain an early result of Yang and Massey.

Corollary 16.7.11 ([1920]) *If $g(x)$ is the generator polynomial of an $[n, k]$ cyclic code $\mathcal{C}$ (the characteristic of $\mathbb{F}$ and n not necessarily coprime), then $\mathcal{C}$ is an LCD code if and only if $g(x)$ is a self-reciprocal polynomial and $\gcd\left(g(x), \frac{x^n - 1}{g(x)}\right) = 1$.*

Example 16.7.12 Let $G = A_5$ and let $\mathbb{F} = \mathbb{F}_2$. Furthermore, let e denote the sum of all elements of G of order 3 and 5. Obviously $\widehat{e} = e^2 = e \in Z(\mathbb{F}_2 G)$. Thus $\mathcal{C} = e\mathbb{F}_2 G$ is a self-adjoint LCD group code. It is the unique 2-block of defect 0 in $\mathbb{F}G$. Its parameters are $[60, 16, 18]$. According to Grassl's list [845] a best known binary $[60, 16]$ code has minimum distance 20. The principal 2-block $B_0(G) = (1 - e)\mathbb{F}_2 G$ has parameters $[60, 44, 5]$ whereas the best known binary $[60, 44]$ code has minimum distance 6.

Remark 16.7.13 Let $q = p^m$ where p is a prime and let $n = q - 1$. Then there are Reed–Solomon LCD codes over $\mathbb{F}_q$ of length n for all dimensions k with $0 < k < n$. Using Corollary 16.7.11, a construction is given in [353] in the case $p = 2$. It also for works for p odd.

Recall that a linear code $\mathcal{C}$ of dimension $k \geq 1$ in $\mathbb{F}^n$ is called an MDS code if the minimum distance of $\mathcal{C}$ reaches the Singleton Bound; i.e., $d_H(\mathcal{C}) = n - k + 1$.

Theorem 16.7.14 *If $\mathcal{C} = e\mathbb{F}G$ with $e^2 = e = \widehat{e} \neq 1$ is an MDS code, then we have $\mathrm{wt}_H(e) = |\mathrm{supp}(e)| > \frac{|G|}{2}$. In particular $\langle \mathrm{supp}(e) \rangle = G$.*

Proof: First note that the conditions $\mathcal{C} < \mathbb{F}G$ and $\mathcal{C}$ MDS imply that $\mathcal{C}^\perp$ is MDS. Thus $d_H(\mathcal{C}) = |G| - \dim \mathcal{C} + 1$ and $d_H(\mathcal{C}^\perp) = \dim \mathcal{C} + 1$. Furthermore, $\mathcal{C}^\perp = (1 - e)\mathbb{F}G$. Hence

$$|G| + 2 = d_H(\mathcal{C}) + d_H(\mathcal{C}^\perp) \leq \mathrm{wt}_H(e) + \mathrm{wt}_H(1 - e) \leq 2\mathrm{wt}_H(e) + 1$$

which proves the first part of the theorem. The second statement follows by elementary group theory. $\square$

Question 16.7.15 Does the existence of an LCD MDS group code $\mathcal{C} \leq \mathbb{F}G$ imply that $\mathrm{char}\,\mathbb{F} \nmid |G|$? The answer is yes if G is abelian; see [502].

16.8 Divisible Group Codes

Definition 16.8.1 For a group code $\mathcal{C} \leq \mathbb{F}G$ the **monomial kernel** $K_M(\mathcal{C})$ of $\mathcal{C}$ is defined by

$$K_M(\mathcal{C}) = \{g \in G \mid gc = a(g)c \text{ with } a(g) \in \mathbb{F} \text{ for all } c \in \mathcal{C}\}.$$

Clearly, $K_M(\mathcal{C})$ is a subgroup of G but in general not a normal subgroup.

Remember that a linear code $\mathcal{C}$ is r-**divisible** if $r \mid \mathrm{wt}_H(c)$ for all $c \in \mathcal{C}$. In the following we denote by $\Delta(\mathcal{C})$ the greatest common divisor of all weights of codewords in $\mathcal{C}$. Sometimes, $\Delta(\mathcal{C})$ is called the **divisor** of $\mathcal{C}$.

If $|\mathbb{F}| = q = p^f$ with p a prime and $\mathcal{C}$ is a G-code over $\mathbb{F}$ of dimension $k \geq 1$, then the average weight equation says that

$$\sum \text{wt}_H(c) = |G|q^{k-1}$$

where the sum runs over representatives of the one dimensional subspaces of $\mathcal{C}$. Thus $\Delta(\mathcal{C})_{p'} \,|\, |G|$.

Example 16.8.2 Let $\mathcal{C}$ be the $\left[\frac{p^k-1}{p-1}, k, p^{k-1}\right]$ simplex code over the prime field $\mathbb{F}_p$ where $k \geq 2$ and $\gcd(k, p-1) = 1$. Then $\mathcal{C}$ is a group code in $\mathbb{F}_p G$ where G is cyclic of order $\frac{p^k-1}{p-1}$ (see Example 16.4.10(a)). Clearly, $\Delta(\mathcal{C}) = \Delta(\mathcal{C})_p = p^{k-1}$, but $\Delta(\mathcal{C}) \nmid |G|$.

Theorem 16.8.3 ([482, 1873]) *Let* $\operatorname{char} \mathbb{F} = p$ *and let* $0 \neq \mathcal{C} \leq \mathbb{F}G$ *be a group code. Then the following hold true.*

(a) $|K_M(\mathcal{C})|$ *divides the weights of all codewords* $c \in \mathcal{C}$.

(b) $\Delta(\mathcal{C})_{p'} = |K_M(\mathcal{C})|_{p'}$.

In contrast to the p'-part, the p-part of $\Delta(\mathcal{C})$ does not seem to be easy to calculate. As a first result we like to mention a theorem of McEliece.

Theorem 16.8.4 ([1360]) *Let* $\mathcal{C}$ *be a cyclic code over* $\mathbb{F}_p$ *of length* n *where the prime* p *does not divide* n. *Let* x *denote the cyclic shift of order* n, *and let* $\mathcal{E}$ *be the set of eigenvalues of the action of* x *on* $\mathcal{C}$. *Then* $\Delta(\mathcal{C})_p = p^e$, *where* $m = (p-1)(e+1)$ *is the smallest multiple of* $p-1$ *for which a product of* m *members of* $\mathcal{E}$ *(repetitions allowed) is equal to* 1.

To state a more general result which holds for arbitrary groups we need the following definition. For $f \in \mathbb{F}G^* = \operatorname{Hom}_\mathbb{F}(\mathbb{F}G, \mathbb{F})$ we put $a_f = \sum_{g \in G} f(g)g^{-1}$. Thus the map $\eta : \mathbb{F}G \longrightarrow \mathbb{F}G$ defined by $z\eta = a_f z$ for $z \in \mathbb{F}G$ is an element of $\operatorname{End}_{\mathbb{F}G}(\mathbb{F}G)$. Furthermore

$$f(z) = \langle a_f z, 1 \rangle = \langle z\eta, 1 \rangle,$$

for all $z \in \mathbb{F}G$. We say that a group code $\mathcal{C} \leq \mathbb{F}G$ satisfies Condition (E) if the following holds true.

Condition (E): Whenever $f \in \mathcal{C}^*$ there exists $\eta \in \operatorname{End}_{\mathbb{F}G}(\mathcal{C})$ such that $f(c) = \langle c\eta, 1 \rangle$ for all $c \in \mathcal{C}$.

In general it is hard to see whether a given right ideal $\mathcal{C}$ satisfies (E) or not. A 2-sided ideal in $\mathbb{F}G$ always satisfies the condition according to [1874].

Theorem 16.8.5 ([1874]) *Let* $\mathcal{C} = e\mathbb{F}_p G$ *where* p *is again a prime and* $e = e^2 \neq 0$. *Suppose that* $\mathcal{C}$ *satisfies Condition* (E). *Then* $\Delta(\mathcal{C})_p = p^{r-1}$ *where* r *is the least positive integer for which* $\mathcal{C}$ *has a nontrivial* G-*invariant multilinear form of degree* $r(p-1)$.

For an extension of Theorem 16.8.5 to fields $\mathbb{F}_{p^s}$, which requires more definitions, we refer the reader directly to [1874].

Example 16.8.6

(a) Let $\mathcal{C} = P_0$ be the projective cover of the trivial module in $\mathbb{F}G$ where $\operatorname{char} \mathbb{F} = 2$. Since the restriction $\langle \cdot, \cdot \rangle|_{P_0}$ of the bilinear form to P_0 is non-degenerate, we see that P_0 is an LCD code. Thus $P_0 = e\mathbb{F}G$ where $e^2 = e = \widehat{e}$, according to Theorem 16.7.6. It follows that

$$\langle e, e \rangle = \langle e\widehat{e}, 1 \rangle = \langle e, 1 \rangle.$$

By [502, Proposition 3.6], we know that $\langle e, 1 \rangle = 1_{\mathbb{F}}$. Suppose for a moment that $\mathbb{F} = \mathbb{F}_2$. Thus $\text{wt}_H(e)$ is odd since $\text{wt}_H(e)1_{\mathbb{F}} = \langle e, e \rangle$. This implies $\Delta(P_0)_2 = 1$ if we are working over the prime field $\mathbb{F}_2$. Since the codewords of the projective cover of the trivial module over $\mathbb{F}_2$ are also codewords in the projective cover of the trivial module over any extension field $\mathbb{F} = \mathbb{F}_{2^s}$, we get $\Delta(P_0)_2 = 1$ for every field $\mathbb{F}$ of characteristic 2.

(b) Suppose in (a) that $\text{char}\,\mathbb{F} = p$ and that G has a p-complement H. Let T be a Sylow p-subgroup of G. If $e = \frac{1}{|H|} \sum_{h \in H} h$, then $e = e^2 = \hat{e}$ and $P_0 = e\mathbb{F}G = e\mathbb{F}T$ is the projective cover of the trivial module. Thus $\Delta(P_0) = |H| = |G|_{p'}$.

(c) [482] Let $\mathcal{C} = B_0$ be the principal p-block of $\mathbb{F}G$ where again $\text{char}\,\mathbb{F} = p$. Let $K(B_0) = \{g \in G \mid gb = b \text{ for all } b \in B_0\}$. By [1013, Chapter VII, Section 14], we have $K(B_0) = O_{p'}(G)$, where $O_{p'}(G)$ is the largest normal subgroup of G of order prime to p. Since B_0 is a left $\mathbb{F}G$-module, $K_M(B_0)$ is a normal subgroup of G. Furthermore, $K_M(B_0)/K(B_0)$ is a p'-group. Thus we obtain $K_M(B_0) = O_{p'}(G)$. Suppose again for a moment that $\mathbb{F} = \mathbb{F}_p$ and note that B_0 satisfies Condition (E). Thus, by Theorem 16.8.5, we have $r = 1$. Therefore $\Delta(B_0)_p = 1$ in this case. Now recall that the principal p-block over $\mathbb{F}_p$ is contained in the principal p-block over any extension field. Thus we get $\Delta(B_0) = |O_{p'}(G)|$ for every field of characteristic p.

Question 16.8.7 Does $\Delta(P_0)_p = 1$ for the projective cover P_0 of the trivial module in odd characteristic p?

Based on a theorem of Ax [78] the following explicit example has been carried out.

Example 16.8.8 ([526, 1875]) Let $\mathcal{RM}_q(r, m)$ denote the rth order generalized Reed–Muller code over $\mathbb{F}_q$ where q is a power of the prime p, as described in Example 16.4.11. Then $\Delta(\mathcal{RM}_q(r, m))_p = q^{\lceil \frac{m}{r} \rceil - 1}$.

For irreducible group codes $\mathcal{C}$ in semisimple abelian group algebras over fields of characteristic p, the p'-part of the divisor determines the equivalence class of $\mathcal{C}$.

Theorem 16.8.9 ([1873]) *Let $\mathcal{C}_1$ and $\mathcal{C}_2$ be irreducible group codes in a semisimple abelian group algebra $\mathbb{F}G$ where $\text{char}\,\mathbb{F} = p$. Then $\mathcal{C}_1$ and $\mathcal{C}_2$ are equivalent if and only if $\Delta(\mathcal{C}_1)_{p'} = \Delta(\mathcal{C}_2)_{p'}$.*

Remark 16.8.10 The assertion in Theorem 16.8.9 is not true in general for non-abelian groups. For instance, we may take a field $\mathbb{F}$ of characteristic $p \neq 2, 3$ and $G = S_3$. There are two irreducible $\mathbb{F}G$-modules of dimension 1, namely $T = (\sum_{g \in G} g)\mathbb{F}$ and $S = (\sum_{g \in G} \text{sgn}(g)g)\mathbb{F}$. Clearly, $T \not\cong S$, but $\Delta(T) = \Delta(S) = 6$.

16.9   Checkable Group Codes

Definition 16.9.1 ([1049]) Let A be a finite dimensional algebra over the field $\mathbb{F}$.

(a) A right ideal $I \leq A$ is called **checkable** if there exists an $a \in A$ such that $I = \text{Ann}_r(a)$. Analogously, a left ideal I is checkable if $I = \text{Ann}_l(a)$.

(b) We say that A is **code-checkable** if all right ideals of A are checkable.

Example 16.9.2

(a) Let $e = e^2 \in A$. Then the ideal eA is checkable which can be seen as follows. Clearly, $eA \leq \mathrm{Ann}_r(A(1-e))$. Since any

$$0 \neq (1-e)b \in (1-e)A$$

is not in $\mathrm{Ann}_r(A(1-e))$ we have $eA = \mathrm{Ann}_r(A(1-e)) = \mathrm{Ann}_r(1-e)$.

(b) LCD group codes $\mathcal{C}$ are checkable since $\mathcal{C} = e\mathbb{F}G$ with $e = e^2 = \widehat{e}$.

(c) All cyclic codes are checkable via the control polynomial.

(d) Maximal group codes in $\mathbb{F}G$ are checkable since they are right annihilators of minimal left ideals which obviously are principal.

(e) A semisimple algebra is code-checkable, in particular group algebras $\mathbb{F}G$ if char $\mathbb{F} \nmid |G|$.

(f) A p-block B of defect 0 in $\mathbb{F}G$ is code-checkable, since B is a simple algebra.

Theorem 16.9.3 ([243]) *A group code $\mathcal{C} \leq \mathbb{F}G$ is checkable if and only if $\mathcal{C}^\perp$ is a principal ideal in $\mathbb{F}G$.*

Example 16.9.4 The extended binary Golay code is checkable in $\mathbb{F}_2 G$ for all groups G mentioned in Example 16.5.1, since it is always constructed as a principal ideal. The same holds true for the binary self-dual $[48, 24, 12]$ code.

Example 16.9.5 Let char $\mathbb{F} = p$ and let G be a finite non-cyclic p-group. Let $\mathcal{C} = (\sum_{g \in G} g)\mathbb{F}$ be the trivial $\mathbb{F}G$-module in $\mathbb{F}G$. One easily computes that $\mathcal{C}^\perp = \sum_{g \in G}(1-g)\mathbb{F}$, which is the Jacobson radical of $\mathbb{F}G$. Since G is not cyclic, it has a factor group of type (p, p) which implies that $\mathcal{C}^\perp$ is not a principal ideal. Thus $\mathcal{C}$ is not checkable. Note that in this example $\mathbb{F}G$ consists only of the principal p-block $B_0(G)$ which is the projective cover of the trivial module and not uniserial.

Definition 16.9.6 For an $\mathbb{F}$-algebra A a module M is called **uniserial** if M has only one Jordan–Hölder series; i.e., the submodules M_i in Theorem 16.2.9(a) are unique.

Theorem 16.9.7 ([243]) *Let char $\mathbb{F} = p$ and let B be a p-block of $\mathbb{F}G$. Then the following are equivalent.*

(a) *All right ideals in B are checkable; i.e., B is code-checkable.*

(b) *All left ideals in B are principal.*

(c) *B contains only one irreducible left module, say L, and its projective cover $P(L)$ is uniserial.*

The condition that B contains only one irreducible left module whose projective cover is uniserial is obviously equivalent to the fact that B contains only one irreducible right module whose projective cover is uniserial. Thus we may interchange left and right in (a) and (b) of Theorem 16.9.7.

Definition 16.9.8 Let p be a prime. A finite group G is called p-**nilpotent** if G has a normal p'-subgroup N such that G/N is a p-group. In particular, G/N is isomorphic to a Sylow p-subgroup of G and $N = O_{p'}(G)$.

Theorem 16.9.9 ([243]) *Let* char $\mathbb{F} = p$ *and let* $B_0(G)$ *be the principal p-block of* $\mathbb{F}G$. *Then the following are equivalent.*

(a) $B_0(G)$ *is code-checkable.*

(b) G *is p-nilpotent with cyclic Sylow p-subgroups.*

(c) $\mathbb{F}G$ *is code-checkable.*

The equivalence of (b) and (c) is already contained in [1492], and in [734] for nilpotent groups, i.e., for groups which are p-nilpotent for all primes p dividing the order of the group.

Definition 16.9.10 Let p be a prime. A finite group G is called p-**solvable** if all non-abelian composition factors of G are p'-groups.

Corollary 16.9.11 ([243]) *Let* G *be a p-solvable group and let* B *be a p-block of* G. *Then the following are equivalent.*

(a) B *is code-checkable.*

(b) *The defect group of* B *is cyclic and* B *contains only one irreducible left (right) module.*

It turned out that in numerous cases the parameters of checkable group codes are as good as the best known linear codes [845]. Even more, some earlier bounds have been improved via group codes.

Remark 16.9.12 ([1049]) There is a checkable $[36, 28, 6]$ group code in $\mathbb{F}_5(C_6 \times C_6)$ and a checkable $[72, 62, 6]$ group code in $\mathbb{F}_5(C_6 \times C_{12})$. In both cases the minimum distance is improved by 1 from an earlier lower bound given in [845].

16.10 Decoding Group Codes

Clearly, if $d_H(\mathcal{C}) = 1$ for a group code $\mathcal{C}$ in $\mathbb{F}G$, then $\mathcal{C} = \mathbb{F}G$. Furthermore, for a 2-sided ideal $\mathcal{C}$ of FG we have $\text{Ann}_l(\mathcal{C}) = \text{Ann}_r(\mathcal{C})$, which we denote by $\text{Ann}(\mathcal{C})$; see [482, Section 2].

Theorem 16.10.1 ([482]) *Let* $\mathcal{C}$ *be a 2-sided ideal in* $\mathbb{F}G$ *with* $0 \neq \mathcal{C} \neq \mathbb{F}G$. *Then* $d_H(\mathcal{C}) \geq 3$ *if and only if* $K_M(\text{Ann}(\mathcal{C})) = 1$ *(see Definition 16.8.1).*

Corollary 16.10.2 *Let* G *be a simple group. If* $\mathbb{F}G$ *has more than one block, then any sum of blocks different from* $\mathbb{F}G$ *has minimum distance at least 3.*

Note that all simple non-abelian groups have more than one p-block unless $p = 2$ and $G = M_{22}$ or M_{24} (Mathieu groups).

For cyclic codes, i.e., group codes for cyclic groups G, and Reed–Muller codes, i.e., group codes for elementary abelian p-groups, there are several decoding algorithms which work quite effectively. To our knowledge, almost nothing seems to be known for general groups G.

Syndrome Decoding 16.10.3 Let $\mathcal{C} = \bigcap_{i=1}^{m} \operatorname{Ann}_r(a_i)$ with $a_i \in \mathbb{F}G$. Furthermore suppose that $d_H(\mathcal{C}) \geq 2t + 1$. Suppose that $c \in \mathcal{C}$ is transmitted and $v = c + e$ is received with an error vector e of weight $\operatorname{wt}_H(e) \leq t$. If we write $v = c' + e'$ with $c' \in \mathcal{C}$ and $\operatorname{wt}_H(e') \leq t$, then we get $a_i v = a_i e = a_i e'$ for all i. Thus

$$e - e' \in \bigcap_{i=1}^{m} \operatorname{Ann}_r(a_i) = \mathcal{C},$$

implying $e - e' = 0$. This shows that the **syndromes** $a_i v$ uniquely determine the error vector, hence the transmitted codeword c.

Remark 16.10.4 The above decoding algorithm has been proposed in [176] to decode the binary extended Golay code, and in [900] to decode a $[21, 16, 3]_8$ group code in $\mathbb{F}_8 G$ where G is a Frobenius group of order 21.

A **quasi-group code** $\mathcal{C}$ of index ℓ over the field $\mathbb{F}$ is a submodule of the $\mathbb{F}G$-module $\mathbb{F}G^\ell$ where G is a any finite group. Obviously, for $\ell = 1$, $\mathcal{C}$ is a group code. If H is a subgroup of G, then we may write $G = \bigcup_{i=1}^{s} g_i H$ where $s = |G : H|$. Thus

$$\mathbb{F}G = \bigoplus_{i=1}^{s} g_i \mathbb{F}H \cong \mathbb{F}H^s.$$

In particular, a G-code $\mathcal{C}$ is a quasi-group code for any subgroup H of index $|G : H|$. In the case that H is a cyclic group, $\mathcal{C}$ is a quasi-cyclic code in the sense of classical coding theory. For decoding quasi-cyclic codes we refer to [116] and [1940].

In [251] Bossert proposed a decoding algorithm for cyclic codes using codewords of the dual code. A modification to general group codes is possible and straightforward. How efficiently such an algorithm works is admittedly open.

Research Problem 16.10.5 Find a decoding algorithm which works effectively for group codes or at least some classes of group codes, which are not cyclic.

16.11 Asymptotic Results

There is still the long standing question whether the class of cyclic codes is asymptotically good [74].

Theorem 16.11.1 ([173]) *Let $\mathbb{F}$ be of characteristic p. If in the lengths of a sequence of cyclic codes over $\mathbb{F}$ only finitely many primes p_i all different from p are involved, then the sequence is asymptotically bad.*

A partial answer in the positive to the above question has recently been given by Shi, Wu, and Solé for additive cyclic codes. By this we mean a code in $\mathbb{F}_{q^l}^n$ ($l \geq 2$) which is linear over the field $\mathbb{F}_q$ and invariant by cyclic shifts.

Theorem 16.11.2 ([1678]) *The class of additive cyclic codes over $\mathbb{F}_{q^l}$ ($l \geq 2$) of rate $\geq \frac{1}{l}$ is asymptotically good.*

Stichtenoth constructed in [1754] an asymptotically good sequence of linear codes over $\mathbb{F}_{q^2}$ in which all automorphism groups are transitive. But it is not clear whether the occurring groups contain regular subgroups in order to be group codes.

Using dihedral groups of order $2m$ with m odd, Bazzi and Mitter proved the following theorem in the special case that $\mathbb{F} = \mathbb{F}_2$.

Theorem 16.11.3 ([145, 244]) *The class of group codes in code-checkable group algebras is asymptotically good over every finite field.*

16.12 Group Codes over Rings

Similar to group algebras over fields we may define group rings RG over commutative rings R with identity and consider ideals in RG as group codes. As rings, $\mathbb{Z}_m = \mathbb{Z}/m\mathbb{Z}$, Galois rings $\mathrm{GR}(p^r, k)$ which include $\mathbb{Z}_{p^r}$ for $k = 1$ and $\mathbb{F}_{p^k}$ for $r = 1$, finite chain rings, and finite Frobenius rings are of particular interest; see Chapter 6.

Definition 16.12.1 Let R be a finite commutative ring.

(a) A **group code over** R is a right ideal in RG.

(b) A **cyclic code of length** n **over** R is an ideal in the ring $R[x]/\langle x^n - 1 \rangle \cong RC_n$ where C_n is a cyclic group of order n. Thus cyclic codes over R are group codes.

In contrast to the field case a cyclic code in $R[x]/\langle x^n - 1 \rangle$ may not be generated by $g \in R[x]$ with $g \mid (x^n - 1)$.

Theorem 16.12.2 ([851]) *Let $\mathcal{C} \leq R[x]/\langle x^n - 1 \rangle$ be a cyclic code. Then the following are equivalent.*

(a) *There exists a divisor g of $x^n - 1$ in RG such that $\mathcal{C} = gR[x]/\langle x^n - 1 \rangle$.*

(b) *$\mathcal{C}$ has a complement in $R[x]/\langle x^n - 1 \rangle$ as an R-module.*

Remark 16.12.3 Already in 1972 Blake constructed in [211] group codes in $\mathbb{Z}_m C_n$ where m is a product of different primes and $\gcd(m, n) = 1$. More on cyclic codes over rings can be found in [213, 332, 441, 1742, 1743] and Chapter 6.

Lemma 16.12.4 ([1197]) *RG is a semisimple ring if and only if R is a semisimple ring and $|G|$ is a unit in R.*

Example 16.12.5 $\mathbb{Z}_m C_n$ is semisimple if and only if m is a product of different primes and $\gcd(m, n) = 1$.

Theorem 16.12.6 ([1893]) *Let $m = 2^s p_1^{n_1} \cdots p_r^{n_r}$ with pairwise different odd primes p_i. Then $\mathbb{Z}_m G$ contains a self-dual group code if and only if all the n_i are even and s or $|G|$ is even.*

Remark 16.12.7 In [1012] structure theorems for group ring codes are presented via cohomology. In [1556] a characterization of abelian group codes over $\mathbb{Z}_{p^r}$ is given by using a generalized Discrete Fourier Transform. As a consequence in both articles a weaker version of Theorem 16.12.6 is obtained.

If π is a set of primes, then $O_{\pi'}(G)$ denotes the largest normal subgroup of G whose order has no prime factor in π.

Theorem 16.12.8 ([7, 617]) *Let R be a finite commutative semisimple ring and let G be abelian. If π denotes the set of non-invertible primes in R, then the following are equivalent.*

(a) *Every ideal in RG is principal.*

(b) *RG is code-checkable.*

(c) *$G/O_{\pi'}(G)$ is a cyclic π-group.*

Chapter 17

Constacyclic Codes over Finite Commutative Chain Rings

Hai Q. Dinh

Kent State University

Sergio R. López-Permouth

Ohio University

17.1 Introduction

Noise is unavoidable. You cannot get rid of it; all you can do is manage it. Codes are used for data compression, cryptography, error-correction, and more recently for network coding. The study of codes has grown into an important subject that lies at the intersection of various disciplines, including information theory, electrical engineering, mathematics, and computer science. Its purpose is designing efficient, reliable, and secure data transmission methods.

The expression 'algebraic coding theory' refers to the approach to coding theory where alphabets and ambient spaces are enhanced with algebraic structures to facilitate the design, analysis, and decoding of the codes produced. As with coding theory in general, algebraic coding theory is divided in two major subcategories of study: block codes and convolutional codes. The study of block codes focuses on three important parameters: code length, total number of codewords, and minimum distance between codewords. Originally, work centered on the Hamming distance. Recently, fueled by an incremental use of finite rings as alphabets, the use of other distances such as the Lee distance, the uniform distance, and Euclidean distance has increased.

The algebraic theory of error-correcting codes has traditionally taken place in the setting of vector spaces over finite fields. Codes over finite rings were first considered in the early

1970s. But they might have initially been considered mostly as a mathematical curiosity and their study was limited to only a handful of publications.

Some of the highlights of that period include the work of Blake [211], who, in 1972, showed how to construct codes over $\mathbb{Z}_m$ from cyclic codes over $\mathbb{F}_p$ where p is a prime factor of m. He then focused on studying the structure of codes over $\mathbb{Z}_{p^r}$; see [213]. In 1977, Spiegel [1742, 1743] generalized those results to codes over $\mathbb{Z}_m$, where m is an arbitrary positive integer.

A boost for codes over finite rings came when connections with ring-linearity were found for well-known families of nonlinear codes that at the time of the discovery had more codewords than all comparable linear codes. The nonlinear codes by Kerdock, Preparata, Nordstrom–Robinson, Goethals, and Delsarte–Goethals [312, 524, 816, 817, 1107, 1323, 1446, 1541] exhibited great error-correcting capabilities and remarkable structures. For instance, the fact that the weight distributions of Kerdock and Preparata codes relate to each other as if they were linear duals (via the MacWilliams Identities) became a tantalizing enigma that was only settled when ring-alphabet linear codes were reintroduced for the purpose. Several researchers have investigated these and other codes with the same weight distributions [94, 342, 1076, 1077, 1078, 1080, 1834]. The families of Kerdock and Preparata codes exist for all length $n = 4^m \geq 16$, and at length 16, they coincide, providing the Nordstrom–Robinson code [817, 1446, 1732]; this code is the unique binary code of length 16 consisting of 256 codewords with Hamming distance 6. In [329, 898] (see also [441, 442]), it has also been shown that the Nordstrom–Robinson code is equivalent to a quaternary code which is self-dual. From that point on, codes over finite rings in general and over $\mathbb{Z}_4$ in particular, have gained considerable prominence in the literature. There are now numerous research papers on this subject and there is at least one book [1861] devoted to the study of codes over $\mathbb{Z}_4$.

The discovery in the early 1990s that these nonlinear binary codes are actually equivalent to linear codes over the ring of integers modulo four, the so-called **quaternary codes**[1], propelled the importance of codes over rings and prompted many more publications such as [329, 441, 898, 1425, 1426, 1522, 1524]. A code of length n over a ring R is **linear** provided it is a submodule of R^n. Nechaev pointed out that the Kerdock codes are, in fact, linear cyclic codes over $\mathbb{Z}_4$ in [1426]. Furthermore, the intriguing relationship between the weight distributions of Kerdock and Preparata codes, a relation that is akin to that between the weight distributions of a linear code and its dual, was explained by Calderbank, Hammons, Kumar, Sloane, and Solé [329, 898] when they showed in 1993 that these well-known codes are in fact equivalent to a pair of mutually dual linear codes over the ring $\mathbb{Z}_4$.

More precisely, the relation between the Kerdock and Preparata codes and the corresponding ring-linear counterparts over $\mathbb{Z}_4$ is due to an isometry between them, induced by the Gray map $\mu : \mathbb{Z}_4 \to \mathbb{Z}_2^2$ sending 0 to 00, 1 to 01, 2 to 11, and 3 to 10. The isometry relates codes over $\mathbb{Z}_4$ equipped with the so-called Lee metric to the Kerdock and Preparata codes with the standard Hamming metric. Consequently, from its inception, the theory of codes over rings was not only about the introduction of an alternate algebraic structure for the alphabet of the codes under consideration but also of a different metric for them. In addition to the Lee metric, other alternative metrics have been considered; see Chapter 22.

Cyclic codes have been one of the most important classes of codes for at least two reasons. First, cyclic codes can be efficiently encoded using shift registers; this explains their preferred role in engineering. Secondly, from a more theoretical perspective, cyclic codes are easily characterized as the ideals of a specific ring: $\frac{\mathbb{F}[x]}{\langle x^n - 1 \rangle}$. This ring is a quotient of the (infinite)

---

[1] In the coding theory literature, the term 'quaternary codes' sometimes is used for codes over the finite field $\mathbb{F}_4$. Throughout this chapter, including references, unless otherwise stated, by 'quaternary codes' we mean codes over $\mathbb{Z}_4$.

ring $\mathbb{F}[x]$ of polynomials with coefficients in the alphabet field $\mathbb{F}$, a principal ideal domain. It is this characterization that makes the notion of cyclic codes suitable to be generalized in various ways. The concepts of negacyclic and constacyclic codes, for example, are the result of focusing, respectively, on codes that correspond to ideals of the quotient rings $\frac{\mathbb{F}[x]}{\langle x^n+1 \rangle}$ and $\frac{\mathbb{F}[x]}{\langle x^n - \lambda \rangle}$ (where $\lambda \in \mathbb{F} \setminus \{0\}$) of $\mathbb{F}[x]$. In fact, the most general such generalization is the notion of a polycyclic code, namely, those codes that correspond to ideals of some quotient ring $\frac{\mathbb{F}[x]}{\langle f(x) \rangle}$ of $\mathbb{F}[x]$; see [1287].

All of the notions above can easily be extended to the finite ring alphabet case by replacing the finite field $\mathbb{F}$ by the finite ring R in each definition. Those concepts, when R is a chain ring, are the main subject of our survey. Section 17.2 discusses chain rings and their archetypical example, the Galois rings, in some detail; that section also serves as an opportunity to briefly visit various alternative metrics which often serve the purposes of ring-linear coding theory in a better way. Section 17.3 surveys what can be said about constacyclic codes in full generality over finite commutative rings without further assumptions on the alphabet. Staying true to the purpose of this chapter, we focus on chain rings alphabets in the next three sections 17.4, 17.5, and 17.6. Section 17.4 deals with the simple-root case for arbitrary chain rings. Considering the repeated-root case requires specific techniques for the various chain rings considered. With one remarkable exception ($\mathbb{F}_2 + u\mathbb{F}_2$ where $u^2 = 0$), Section 17.5 focuses mostly on Galois rings. Section 17.6 deals with the case of the chain ring $\mathcal{R} = \mathbb{F}_{p^m} + u\mathbb{F}_{p^m}$ where $u^2 = 0$. $\mathcal{R}$ consists of all p^m-ary polynomials of degree 0 and 1 in indeterminate u; it is closed under p^m-ary polynomial addition and multiplication modulo u^2. Thus, $\mathcal{R} = \frac{\mathbb{F}_{p^m}[u]}{\langle u^2 \rangle} = \{a + ub \mid a, b \in \mathbb{F}_{p^m}\}$ is a chain ring with maximal ideal $u\mathbb{F}_{p^m}$. The ring $\mathcal{R}$ has precisely $p^m(p^m - 1)$ units, which are of the forms $\alpha + u\beta$ and γ, where α, β, γ are nonzero elements of the field $\mathbb{F}_{p^m}$. The analysis in that section requires independent considerations of various subcases which are thus presented in corresponding subsections. In Section 17.7, we close by considering various directions in which the study of constacyclic codes has been extended. Notions considered include polycyclic and sequential codes; Theorem 17.7.3 highlights a theoretical reason for the centrality of the notion of constacyclicity.

This chapter updates and extends [585]. Note that general codes over rings are examined in Chapter 6.

17.2   Chain Rings, Galois Rings, and Alternative Distances

Let R be a finite commutative ring. An ideal I of R is called **principal** if it is generated by a single element. A ring R is a **principal ideal ring** if all of its ideals are principal. R is called a **local ring** if R has a unique maximal ideal. Furthermore, a ring R is called a **chain ring** if the set of all ideals of R is a chain under set-theoretic inclusion. It can be shown easily that chain rings are principal ideal rings. Examples of finite commutative chain rings include the ring $\mathbb{Z}_{p^k}$ of integers modulo p^k, for a prime p and positive integer k. A similar construction yields the so-called Galois rings: $\mathrm{GR}(p^k, m)$, the Galois extension of degree m of $\mathbb{Z}_{p^k}$; this is a common generalization of Galois fields and the rings in the previous sentence; see [1008, 1359].[2] These classes of rings have been used widely as an

---

[2]Although we only consider finite commutative chain rings in this chapter, it is worth noting that a finite chain ring need not be commutative. The smallest noncommutative chain ring has order 16 [1127] and can

alphabet for constacyclic codes. Various decoding schemes for codes over Galois rings have been considered in [322, 323, 324, 325].

The following equivalent conditions are well-known for the class of finite commutative chain rings; see [584, Proposition 2.1].

Proposition 17.2.1 *For a finite commutative ring R the following conditions are equivalent.*

(a) *R is a local ring and the maximal ideal M of R is principal,*

(b) *R is a local principal ideal ring,*

(c) *R is a chain ring.*

Let ζ be a fixed generator of the maximal ideal M of a finite commutative chain ring R. Then ζ is nilpotent, and we denote its nilpotency index by t. The ideals of R form a chain:

$$R = \langle \zeta^0 \rangle \supsetneq \langle \zeta^1 \rangle \supsetneq \cdots \supsetneq \langle \zeta^{t-1} \rangle \supsetneq \langle \zeta^t \rangle = \langle 0 \rangle.$$

Let $\overline{R}$ denote the field $\frac{R}{M}$, called the **residue field of R**. The natural epimorphism of R onto $\overline{R}$ extends naturally to $^{-} : R[x] \to \overline{R}[x]$, such that, for all $r \in R$, $r \mapsto r + M$ and the variable $x \in R[x]$ maps to $x \in \overline{R}[x]$.

The following theorem summarizes various known general facts about finite commutative chain rings; see [1359].

Proposition 17.2.2 *Let R be a finite commutative chain ring with maximal ideal $M = \langle \zeta \rangle$, and let t be the nilpotency of ζ. Then*

(a) *there exists a prime p such that $|R|$, $|\overline{R}|$, and the characteristics of R and $\overline{R}$ are powers of p,*

(b) *for $i = 1, 2, \ldots, t$, the quotients $\frac{\langle \zeta^{i-1} \rangle}{\langle \zeta^i \rangle}$ are isomorphic as R-modules and as $\overline{R}$-vector spaces, and*

(c) *for $i = 0, 1, \ldots, t$, $|\langle \zeta^i \rangle| = |\overline{R}|^{t-i}$. In particular, $|R| = |\overline{R}|^t$.*

We continue with a family of rings that constitute a general yet rather concrete family of finite commutative chain rings; they are the so-called Galois rings. While they can be characterized in pure theoretical terms, they also have a very specific construction which resembles that of finite (Galois) fields; undoubtedly, that is the reason for their name.

The theory of Galois rings was first developed by W. Krull in 1924. A finite ring R with identity 1 is said to be a **Galois ring** if the set of its zero divisors forms a principal ideal $\langle p \cdot 1 \rangle$, where p is a prime integer and $p \cdot 1 = 1 + \cdots + 1$ denotes the p-fold addition of copies of 1. Chapter 14 of [1864] is devoted to a characterization of Galois rings which amounts to showing that they must be one of the rings in the following definition.

Definition 17.2.3 *The **Galois ring of characteristic p^a and dimension m**, denoted by $\mathrm{GR}(p^a, m)$, is the ring*

$$\mathrm{GR}(p^a, m) = \frac{\mathbb{Z}_{p^a}[z]}{\langle h(z) \rangle}$$

---

be represented as $R = \mathbb{F}_4 \oplus \mathbb{F}_4$, where the operations $+, \cdot$ are given by

$$(a_1, b_1) + (a_2, b_2) = (a_1 + a_2, b_1 + b_2),$$
$$(a_1, b_1) \cdot (a_2, b_2) = (a_1 a_2, a_1 b_2 + b_1 a_2^2).$$

where $h(z)$ is a monic basic irreducible polynomial of degree m in $\mathbb{Z}_{p^a}[z]$. A polynomial $h(z)$ is said to be **basic irreducible** if $\overline{h}(z)$ is irreducible over the residue field of $\mathrm{GR}(p^a, m)$.

Note that if $a = 1$, then $\mathrm{GR}(p, m) = \mathrm{GF}(p^m) = \mathbb{F}_{p^m}$, and if $m = 1$, then $\mathrm{GR}(p^a, 1) = \mathbb{Z}_{p^a}$. We gather here some well-known facts about Galois rings; see [898, 1008, 1359].

Proposition 17.2.4 *Let* $\mathrm{GR}(p^a, m) = \frac{\mathbb{Z}_{p^a}[z]}{\langle h(z) \rangle}$ *be a Galois ring. Then the following hold.*

(a) *Each ideal of* $\mathrm{GR}(p^a, m)$ *is of the form* $\langle p^k \rangle = p^k \mathrm{GR}(p^a, m)$, *for* $0 \le k \le a$. *In particular,* $\mathrm{GR}(p^a, m)$ *is a chain ring with maximal ideal* $\langle p \rangle = p\mathrm{GR}(p^a, m)$, *and residue field* $\mathbb{F}_{p^m}$.

(b) *For* $0 \le i \le a$, $|p^i \mathrm{GR}(p^a, m)| = p^{m(a-i)}$.

(c) *Each element of* $\mathrm{GR}(p^a, m)$ *can be represented as* up^k, *where* u *is a unit and* $0 \le k \le a$; *in this representation* k *is unique and* u *is unique modulo* $\langle p^{n-k} \rangle$.

(d) $h(z)$ *has a root* ξ, *which is also a primitive* $(p^m - 1)^{th}$ *root of unity. The set*

$$\mathcal{T}_m = \{0, 1, \xi, \xi^2, \dots, \xi^{p^m - 2}\}$$

is a complete set of representatives of the cosets $\frac{\mathrm{GR}(p^a, m)}{p\mathrm{GR}(p^a, m)} = \mathbb{F}_{p^m}$ *in* $\mathrm{GR}(p^a, m)$. *Each element* $r \in \mathrm{GR}(p^a, m)$ *can be written uniquely as*

$$r = \xi_0 + \xi_1 p + \cdots + \xi_{a-1} p^{a-1},$$

with $\xi_i \in \mathcal{T}_m$, $0 \le i \le a - 1$.

(e) *There is a natural injective ring homomorphism* $\mathrm{GR}(p^a, m) \to \mathrm{GR}(p^a, md)$ *for any positive integer* d.

(f) *There is a natural surjective ring homomorphism* $\mathrm{GR}(p^a, m) \to \mathrm{GR}(p^{a-1}, m)$ *with kernel* $\langle p^{a-1} \rangle$.

(g) *Each subring of* $\mathrm{GR}(p^a, m)$ *is a Galois ring of the form* $\mathrm{GR}(p^a, l)$, *where* l *divides* m. *Conversely, if* l *divides* m *then* $\mathrm{GR}(p^a, m)$ *contains a unique copy of* $\mathrm{GR}(p^a, l)$. *That means, the number of subrings of* $\mathrm{GR}(p^a, m)$ *is the number of positive divisors of* m.

The set $\mathcal{T}_m$ in Theorem 17.2.4(d) is called a **Teichmüller set** of $\mathrm{GR}(p^a, m)$.

We conclude this section by mentioning three important distances used for codes over specific classes of finite chain rings, namely, the homogeneous, Lee, and Euclidean distances. We will provide the exact values of these distances for several classes of codes in Section 17.5. *In this chapter, we make the convention that the zero code has distance 0, for any distance.*

The homogeneous weight was first introduced in [430] (see also [431, 432]) over integer residue rings, and later over finite Frobenius rings. This weight has numerous applications for codes over finite rings, such as constructing extensions of the Gray isometry to finite chain rings [853, 932, 978], or providing a combinatorial approach to MacWilliams equivalence theorems [1317, 1318, 1906] for codes over finite Frobenius rings [854]. The homogeneous distance of codes over the Galois rings $\mathrm{GR}(2^a, m)$ is defined as follows.

Let $a \ge 2$; the **homogeneous weight** on $\mathrm{GR}(2^a, m)$ is a weight function $\mathrm{wt}_{\mathrm{hom}} : \mathrm{GR}(2^a, m) \to \mathbb{N}$ given by

$$\mathrm{wt}_{\mathrm{hom}}(r) = \begin{cases} 0 & \text{if } r = 0, \\ (2^m - 1)\, 2^{m(a-2)} & \text{if } r \in \mathrm{GR}(2^a, m) \setminus 2^{a-1}\mathrm{GR}(2^a, m), \\ 2^{m(a-1)} & \text{if } r \in 2^{a-1}\mathrm{GR}(2^a, m) \setminus \{0\}. \end{cases}$$

The homogeneous weight of a vector $(c_0, c_1, \ldots, c_{n-1})$ of length n over $\mathrm{GR}(2^a, m)$ is the rational sum of the homogeneous weights of its components:

$$\mathrm{wt}_{\mathrm{hom}}(c_0, c_1, \ldots, c_{n-1}) = \mathrm{wt}_{\mathrm{hom}}(c_0) + \mathrm{wt}_{\mathrm{hom}}(c_1) + \cdots + \mathrm{wt}_{\mathrm{hom}}(c_{n-1}).$$

The **homogeneous distance** (or **minimum homogeneous weight**) d_{hom} of a linear code $\mathcal{C}$ is the minimum homogeneous weight of nonzero codewords of $\mathcal{C}$:

$$d_{hom}(\mathcal{C}) = \min\left\{\mathrm{wt}_{\mathrm{hom}}(\mathbf{x} - \mathbf{y}) \mid \mathbf{x}, \mathbf{y} \in \mathcal{C}, \ \mathbf{x} \neq \mathbf{y}\right\} = \min\left\{\mathrm{wt}_{\mathrm{hom}}(\mathbf{c}) \mid \mathbf{c} \in \mathcal{C}, \ \mathbf{c} \neq \mathbf{0}\right\}.$$

The Lee distance, named after its originator [1211], is a good alternative to the Hamming distance in algebraic coding theory, especially for codes over $\mathbb{Z}_4$. For instance, the Lee distance plays an important role in constructing an isometry between binary and quaternary codes via the Gray map in landmark papers [329, 898] of the theory of codes over rings. Classically, for codes over finite fields, Berlekamp's negacyclic codes [167, 170], the class of cyclic codes investigated in [404], and the class of alternant codes discussed in [1605], are examples of codes designed with the Lee metric in mind.

Let $z \in \mathbb{Z}_{2^a}$; the **Lee value** of z, denoted by $|z|_L$, is given as

$$|z|_L = \begin{cases} z & \text{if } 0 \leq z \leq 2^{a-1}, \\ 2^a - z & \text{if } 2^{a-1} < z \leq 2^a - 1. \end{cases}$$

The **Lee weight** of a vector $(c_0, c_1, \ldots, c_{n-1})$ of length n over $\mathbb{Z}_{2^a}$ is the rational sum of the Lee values of its components:

$$\mathrm{wt}_{\mathrm{L}}(c_0, c_1, \ldots, c_{n-1}) = |c_0|_L + |c_1|_L + \cdots + |c_{n-1}|_L.$$

The **Lee distance** (or **minimum Lee weight**) d_L of a linear code $\mathcal{C}$ is the minimum Lee weight of nonzero codewords of $\mathcal{C}$:

$$d_L(\mathcal{C}) = \min\left\{\mathrm{wt}_{\mathrm{L}}(\mathbf{x} - \mathbf{y}) \mid \mathbf{x}, \mathbf{y} \in \mathcal{C}, \ \mathbf{x} \neq \mathbf{y}\right\} = \min\left\{\mathrm{wt}_{\mathrm{L}}(\mathbf{c}) \mid \mathbf{c} \in \mathcal{C}, \ \mathbf{c} \neq \mathbf{0}\right\}.$$

As codes over $\mathbb{Z}_4$ have gained more prominence, interesting connections with binary codes and unimodular lattices were found with relations to codes over $\mathbb{Z}_{2k}$; see [115]. The connection between codes over $\mathbb{Z}_4$ and unimodular lattices prompted the definition of the Euclidean weight of codewords of length n over $\mathbb{Z}_4$ [235, 236], and more generally, over $\mathbb{Z}_{2k}$ [115, 624, 625].

Let $z \in \mathbb{Z}_{2^a}$, the **Euclidean weight** of z, denoted by $|z|_E$, is given as

$$|z|_E = \begin{cases} z^2 & \text{if } 0 \leq z \leq 2^{a-1}, \\ (2^a - z)^2 & \text{if } 2^{a-1} < z \leq 2^a - 1. \end{cases}$$

The **Euclidean weight** of a vector $(c_0, c_1, \ldots, c_{n-1})$ of length n over $\mathbb{Z}_{2^a}$ is the rational sum of the Euclidean weights of its components:

$$\mathrm{wt}_{\mathrm{E}}(c_0, c_1, \ldots, c_{n-1}) = |c_0|_E + |c_1|_E + \cdots + |c_{n-1}|_E.$$

The **Euclidean distance** (or **minimum Euclidean weight**) d_E of a linear code $\mathcal{C}$ is the minimum Euclidean weight of nonzero codewords of $\mathcal{C}$:

$$d_E(\mathcal{C}) = \min\left\{\mathrm{wt}_{\mathrm{E}}(\mathbf{x} - \mathbf{y}) \mid \mathbf{x}, \mathbf{y} \in \mathcal{C}, \ \mathbf{x} \neq \mathbf{y}\right\} = \min\left\{\mathrm{wt}_{\mathrm{E}}(\mathbf{c}) \mid \mathbf{c} \in \mathcal{C}, \ \mathbf{c} \neq \mathbf{0}\right\}.$$

17.3   Constacyclic Codes over Arbitrary Commutative Finite Rings

Given an n-tuple $(x_0, x_1, \ldots, x_{n-1}) \in R^n$, the **cyclic shift** τ and **negashift** ν on R^n are defined as usual, i.e.,

$$\tau(x_0, x_1, \ldots, x_{n-1}) = (x_{n-1}, x_0, x_1, \ldots, x_{n-2}),$$

and

$$\nu(x_0, x_1, \ldots, x_{n-1}) = (-x_{n-1}, x_0, x_1, \ldots, x_{n-2}).$$

A code $\mathcal{C}$ is called **cyclic** if $\tau(\mathcal{C}) = \mathcal{C}$, and $\mathcal{C}$ is called **negacyclic** if $\nu(\mathcal{C}) = \mathcal{C}$. More generally, if λ is a unit of the ring R, then the λ-**constacyclic** (λ-**twisted**) **shift** τ_λ on R^n is the shift

$$\tau_\lambda(x_0, x_1, \ldots, x_{n-1}) = (\lambda x_{n-1}, x_0, x_1, \ldots, x_{n-2}),$$

and a code $\mathcal{C}$ is said to be λ-**constacyclic** (or λ-**twisted**) if $\tau_\lambda(\mathcal{C}) = \mathcal{C}$, i.e., if $\mathcal{C}$ is closed under the λ-constacyclic shift τ_λ. Equivalently, $\mathcal{C}$ is a λ-constacyclic code if and only if

$$\mathcal{C}S_\lambda \subseteq \mathcal{C},$$

where S_λ is the λ-constacyclic shift matrix given by

$$S_\lambda = \begin{bmatrix} 0 & 1 & \cdots & 0 \\ \vdots & \vdots & \ddots & \vdots \\ 0 & 0 & \cdots & 1 \\ \lambda & 0 & \cdots & 0 \end{bmatrix} = \begin{bmatrix} 0 & & & \\ \vdots & & I_{n-1} & \\ 0 & & & \\ \lambda & 0 & \cdots & 0 \end{bmatrix} \in R^{n \times n}.$$

In light of this definition, when $\lambda = 1$, λ-constacyclic codes are cyclic codes, and when $\lambda = -1$, λ-constacyclic codes are just negacyclic codes.

Each codeword $\mathbf{c} = (c_0, c_1, \ldots, c_{n-1})$ is customarily identified with its polynomial representation $c(x) = c_0 + c_1 x + \cdots + c_{n-1} x^{n-1}$, and the code $\mathcal{C}$ is in turn identified with the set of all polynomial representations of its codewords. Then in the ring $\frac{R[x]}{\langle x^n - \lambda \rangle}$, $xc(x)$ corresponds to a λ-constacyclic shift of $c(x)$. From that, the following fact is well-known and straightforward.

Proposition 17.3.1 *A linear code $\mathcal{C}$ of length n is λ-constacyclic over R if and only if $\mathcal{C}$ is an ideal of $\frac{R[x]}{\langle x^n - \lambda \rangle}$.*

Under the ordinary inner product, the dual of a cyclic code is a cyclic code, and the dual of a negacyclic code is a negacyclic code. In general, we have the following implication of the dual of a λ-constacyclic code.

Proposition 17.3.2 ([578]) *Under the ordinary inner product, the dual of a λ-constacyclic code is a λ^{-1}-constacyclic code.*

For a nonempty subset S of the ring R, the **annihilator** of S, denoted by $\text{ann}(S)$, is the set

$$\text{ann}(S) = \{f \mid fg = 0 \text{ for all } g \in R\}.$$

Note that $\text{ann}(S)$ is always an ideal of R.

Customarily, for a polynomial f of degree k, its **reciprocal polynomial** $x^k f(x^{-1})$ will be denoted by f^*. Thus, for example, if $f(x) = a_0 + a_1 x + \cdots + a_{k-1} x^{k-1} + a_k x^k$, then

$$f^*(x) = x^k(a_0 + a_1 x^{-1} + \cdots + a_{k-1} x^{-(k-1)} + a_k x^{-k})$$
$$= a_k + a_{k-1} x + \cdots + a_1 x^{k-1} + a_0 x^k.$$

Note that $(f^*)^* = f$ if and only if the constant term of f is nonzero if and only if $\deg(f) = \deg(f^*)$. For $A \subseteq R[x]$, we let $A^* = \{f^*(x) \mid f(x) \in A\}$. It is easy to see that if A is an ideal of $R[x]$, then A^* is also an ideal of $R[x]$.

Proposition 17.3.3 ([591, Propositions 3.3, 3.4]) *Let R be a finite commutative ring, and λ a unit of R.*

(a) *Let $a(x) = a_0 + a_1 x + \cdots + a_{n-1} x^{n-1}$, $b(x) = b_0 + b_1 x + \cdots + b_{n-1} x^{n-1}$ be in $R[x]$. Then $a(x)b(x) = 0$ in $\frac{R[x]}{\langle x^n - \lambda \rangle}$ if and only if $(a_0, a_1, \ldots, a_{n-1})$ is orthogonal to $(b_{n-1}, b_{n-2}, \ldots, b_0)$ and all its λ^{-1}-constacyclic shifts.*

(b) *If $\mathcal{C}$ is a λ-constacyclic code of length n over R, then the dual $\mathcal{C}^\perp$ of $\mathcal{C}$ is $(\operatorname{ann}(\mathcal{C}))^*$.*

When studying λ-constacyclic codes over finite fields, most researchers assume that the code length n is not divisible by the characteristic p of the field. This ensures that $x^n - \lambda$, and hence the generator polynomial of any λ-constacyclic code, will have no multiple factors, and hence no repeated roots in an extension field. The case when the code length n is divisible by the characteristic p of the field yields the so-called **repeated-root codes**, which were first studied in 1967 by Berman [173], and then in the 1970s and 1980s by several authors such as Massey et al. [1350], Falkner et al. [702], and Roth and Seroussi [1604]. However, repeated-root codes over finite fields were investigated in the most generality in the 1990s by Castagnoli [369] and van Lint [1835], where they showed that repeated-root cyclic codes have a concatenated construction, and are asymptotically bad. Nevertheless, such codes are optimal in a few cases and that motivates further study of the class.

Repeated-root constacyclic codes over a class of finite chain rings have been extensively studied over the last few years by many researchers, such as Abualrub and Oehmke [9, 10, 11], Blackford [207, 208], Dinh [573, 574, 575, 576, 578, 580], Ling et al. [631, 1112, 1268], Sălăgean et al. [1449, 1611], etc. To distinguish the two cases, codes where the code length is not divisible by the characteristic p of the residue field $\overline{R}$ are called **simple-root codes**. We will consider this class of codes in Section 17.4 and the class of repeated-root codes in Sections 17.5 and 17.6.

The dual notions of polycyclic and sequential codes were introduced in [1287]. In addition to being generalizations of constacyclicity, they serve to characterize precisely that concept in terms of a symmetry criterion. We mention this result as Theorem 17.7.3 at the end of this chapter.

17.4 Simple-Root Cyclic and Negacyclic Codes over Finite Chain Rings

All codes considered in this section are simple-root codes over a finite chain ring R; i.e., the code length n is not divisible by the characteristic p of the residue field $\overline{R}$. The structure of cyclic codes over $\mathbb{Z}_{p^a}$ was obtained by Calderbank and Sloane [332] in 1995, and later on with a different proof by Kanwar and López-Permouth [1081] in 1997. In 1999, with a different technique, Norton and Sălăgean [1448] extended the structure theorems given in [332] and [1081] to cyclic codes over finite chain rings; they used an elementary approach which did not appeal to commutative algebra as [332] and [1081] did.

Let R be a finite chain ring with maximal ideal $\langle \zeta \rangle$, and let t be the nilpotency of ζ. For a linear code $\mathcal{C}$ of length n over R, the **submodule quotient** of $\mathcal{C}$ by $r \in R$ is the code

$$(\mathcal{C} : r) = \{ e \in R^n \mid er \in \mathcal{C} \} .$$

Thus we have a tower of linear codes over R

$$\mathcal{C} = (\mathcal{C} : \zeta^0) \subseteq \cdots \subseteq (\mathcal{C} : \zeta^i) \subseteq \cdots \subseteq (\mathcal{C} : \zeta^{t-1}).$$

Its projection to $\overline{R}$ forms a tower of linear codes over $\overline{R}$

$$\overline{\mathcal{C}} = \overline{(\mathcal{C} : \zeta^0)} \subseteq \cdots \subseteq \overline{(\mathcal{C} : \zeta^i)} \subseteq \cdots \subseteq \overline{(\mathcal{C} : \zeta^{t-1})}.$$

If $\mathcal{C}$ is a cyclic code over R, then for $0 \le i \le t - 1$, $(\mathcal{C} : \zeta^i)$ is cyclic over R, and $\overline{(\mathcal{C} : \zeta^i)}$ is cyclic over $\overline{R}$. For codes over $\mathbb{Z}_4$, $\overline{\mathcal{C}} = \overline{(\mathcal{C} : \zeta^0)} \subseteq \overline{(\mathcal{C} : \zeta)}$ where $\zeta = 2$, were first introduced by Conway and Sloane in [441], and later were generalized to codes over any chain ring by Norton and Sălăgean [1448].

For a code $\mathcal{C}$ of length n over R, a matrix G is called a **generator matrix** of $\mathcal{C}$ if the rows of G span $\mathcal{C}$, and none of them can be written as a linear combination of other rows of G. A generator matrix G is said to be in **standard form** if after a suitable permutation of the coordinates,

$$
G = \begin{bmatrix}
I_{k_0} & A_{0,1} & A_{0,2} & A_{0,3} & \cdots & A_{0,t-1} & A_{0,t} \\
0 & \zeta I_{k_1} & \zeta A_{1,2} & \zeta A_{1,3} & \cdots & \zeta A_{1,t-1} & \zeta A_{1,t} \\
0 & 0 & \zeta^2 I_{k_2} & \zeta^2 A_{2,3} & \cdots & \zeta^2 A_{2,t-1} & \zeta^2 A_{2,t} \\
\vdots & \vdots & \vdots & \vdots & \ddots & \vdots & \vdots \\
0 & 0 & 0 & 0 & \cdots & \zeta^{t-1} I_{k_{t-1}} & \zeta^{t-1} A_{t-1,t}
\end{bmatrix}
$$

$$
= \begin{bmatrix}
A_0 \\
\zeta A_1 \\
\zeta^2 A_2 \\
\vdots \\
\zeta^{t-1} A_{t-1}
\end{bmatrix} , \tag{17.1}
$$

where the columns are grouped into blocks of sizes $k_0, k_1, \ldots, k_{t-1}, n - \sum_{i=0}^{t-1} k_i$. The generator matrix in standard form G is associated to the matrix

$$
A = \begin{bmatrix}
A_0 \\
A_1 \\
A_2 \\
\vdots \\
A_{t-1}
\end{bmatrix} .
$$

We denote by $\gamma(\mathcal{C})$ the number of rows of a generator matrix in standard form of $\mathcal{C}$, and $\gamma_i(\mathcal{C})$ the number of rows divisible by ζ^i but not by ζ^{i+1}. Equivalently, $\gamma_0(\mathcal{C}) = \dim(\overline{\mathcal{C}})$, and $\gamma_i(\mathcal{C}) = \dim \overline{(\mathcal{C} : \zeta^i)} - \dim \overline{(\mathcal{C} : \zeta^{i-1})}$, for $1 \le i \le t - 1$. Obviously, $\gamma(\mathcal{C}) = \sum_{i=0}^{t-1} \gamma_i(\mathcal{C})$.

For a linear code $\mathcal{C}$ of length n over a finite chain ring R, the information on generator matrices, parity check matrices, and sizes of $\mathcal{C}$, its dual $\mathcal{C}^\perp$, its projection $\overline{\mathcal{C}}$ to the residue field $\overline{R}$, is given as follows.

Theorem 17.4.1 ([1448, Lemma 3.4, Theorems 3.5, 3.10]) *Let $\mathcal{C}$ be a linear code of length n over a finite chain ring R with generator matrix G in standard form (17.1) associated to the matrix*

$$A = \begin{bmatrix} A_0 \\ A_1 \\ A_2 \\ \vdots \\ A_{t-1} \end{bmatrix}.$$

Then

(a) *For $0 \le i \le t-1$, $\overline{(\mathcal{C} : \zeta^i)}$ has generator matrix*

$$\begin{bmatrix} \overline{A_0} \\ \overline{A_1} \\ \vdots \\ \overline{A_i} \end{bmatrix},$$

and $\dim \overline{(\mathcal{C} : \zeta^i)} = k_0 + k_1 + \cdots + k_i$.

(b) *If $E_0 \subseteq E_1 \subseteq \cdots \subseteq E_{t-1}$ are linear codes of length n over $\overline{R}$, then there is a code D of length n over R such that $\overline{(D : \zeta^i)} = E_i$, for $0 \le i \le t-1$.*

(c) *The parameters $k_0, k_1, \ldots, k_{t-1}$ are the same for any generator matrix G in standard form for $\mathcal{C}$.*

(d) *Any codeword $\mathbf{c} \in \mathcal{C}$ can be written uniquely as*

$$\mathbf{c} = (\mathbf{v}_0, \mathbf{v}_1, \ldots, \mathbf{v}_{t-1}) G,$$

where $\mathbf{v}_i \in (R/\zeta^{t-i}R)^{k_i} \cong (\zeta^i R)^{k_i}$.

(e) *The number of codewords in $\mathcal{C}$ is*

$$|\mathcal{C}| = |\overline{R}|^{\sum_{i=0}^{t-1}(t-i)k_i}.$$

(f) *If, for $0 \le i < j \le t$, $B_{i,j} = -\sum_{l=i+1}^{j-1} B_{i,l} A_{t-j,t-l}^{\mathsf{T}} - A_{t-j,t-i}^{\mathsf{T}}$, then*

$$H = \begin{bmatrix} B_{0,t} & B_{0,t-1} & \cdots & B_{0,1} & I_{n-\gamma(\mathcal{C})} \\ \zeta B_{1,t} & \zeta B_{1,t-1} & \cdots & \zeta I_{\gamma_{t-1}(\mathcal{C})} & 0 \\ \vdots & \vdots & \ddots & \vdots & \vdots \\ \zeta^{t-1} B_{t-1,t} & \zeta^{t-1} I_{\gamma_1(\mathcal{C})} & \cdots & 0 & 0 \end{bmatrix} = \begin{bmatrix} B_0 \\ \zeta B_1 \\ \vdots \\ \zeta^{t-1} B_{t-1} \end{bmatrix}$$

is a generator matrix for $\mathcal{C}^\perp$ and a parity check matrix for $\mathcal{C}$.

(g) *For $0 \le i \le t-1$, $\overline{(\mathcal{C}^\perp : \zeta^i)} = \overline{(\mathcal{C} : \zeta^i)}^\perp$, $\gamma_0(\mathcal{C}^\perp) = n - \gamma(\mathcal{C})$, and $\gamma_i(\mathcal{C}^\perp) = \gamma_{t-i}(\mathcal{C})$.*

(h) *$|\mathcal{C}^\perp| = |R^n|/|\mathcal{C}|$, and $(\mathcal{C}^\perp)^\perp = \mathcal{C}$.*

(i) *Associate the generator matrix H of $\mathcal{C}^\perp$ with the matrix*

$$B = \begin{bmatrix} B_0 \\ B_1 \\ \vdots \\ B_{t-1} \end{bmatrix}.$$

Then $\overline{C}$ has generator matrix $\overline{A_0}$, and parity check matrix

$$\overline{B} = \begin{bmatrix} \overline{B_0} \\ \overline{B_1} \\ \vdots \\ \overline{B_{t-1}} \end{bmatrix}.$$

The set $\{\zeta^{a_0} g_{a_0}, \zeta^{a_1} g_{a_1}, \ldots, \zeta^{a_k} g_{a_k}\}$ is said to be a **generating set in standard form** of the cyclic code C if the following conditions hold:

- $C = \langle \zeta^{a_0} g_{a_0}, \zeta^{a_1} g_{a_1}, \ldots, \zeta^{a_k} g_{a_k} \rangle$,

- $0 \leq k < t$,

- $0 \leq a_0 < a_1 < \cdots < a_k < t$,

- $g_{a_i} \in R[x]$ is monic for $0 \leq i \leq k$,

- $\deg(g_{a_i}) > \deg(g_{a_{i+1}})$ for $0 \leq i \leq k - 1$, and

- $g_{a_k} \mid g_{a_{k-1}} \mid \cdots \mid g_{a_0} \mid (x^n - 1)$.

The existence and uniqueness of a generating set in standard form of a cyclic code were proven by Calderbank and Sloane [332] in 1995 for the alphabet $\mathbb{Z}_{p^a}$; in 2000, that was extended to the general case of any chain ring R by Norton and Sălăgean [1448].

Proposition 17.4.2 ([332, Theorem 6], [1448, Theorem 4.4]) *Any nonzero cyclic code C over a finite chain ring R has a unique generating set in standard form.*

If the constant term f_0 of f is a unit, we set $f^\# = f_0^{-1} f^*$. In particular, the constant term of any factor of $x^n - 1$ is a unit. Moreover, if $f(x)$ is a factor of $x^n - 1$, let $\widehat{f}(x) = \frac{x^n - 1}{f(x)}$.

The generating set in standard form of a cyclic code is related to its generator matrix and the generating set in standard form of its dual as follows.

Theorem 17.4.3 ([1448, Theorems 4.5, 4.9]) *Let C be a cyclic code with generating set $\{\zeta^{a_0} g_{a_0}, \zeta^{a_1} g_{a_1}, \ldots, \zeta^{a_k} g_{a_k}\}$ in standard form. The following hold.*

(a) *If, for $0 \leq i \leq k$, $d_i = \deg(g_{a_i})$, and by convention, $d_{-1} = n$, $d_{k+1} = 0$, and*

$$T = \bigcup_{i=0}^{k} \left\{ \zeta^{a_i} g_{a_i} x^{d_{i-1} - d_i - 1}, \ldots, \zeta^{a_i} g_{a_i} x, \zeta^{a_i} g_{a_i} \right\},$$

then T defines a generator matrix for C.

(b) *Any $c \in C$ can be uniquely represented as $c = \sum_{i=0}^{k} h_i g_{a_i} \zeta^{a_i}$, where*

$$h_i \in \left(R / R \zeta^{t-a_i} \right) [x] \cong \left(R \zeta^{a_i} \right) [x],$$

and $\deg(h_i) < d_{i-1} - d_i$.

(c)

$$\gamma_j(C) = \begin{cases} d_{i-1} - d_i & \text{if } j = a_i \text{ for some } i, \\ 0 & \text{otherwise,} \end{cases}$$

and

$$|C| = |\overline{R}|^{\sum\limits_{i=0}^{k} (t-a_i)(d_{i-1}-d_i)}.$$

(d) Let $a_{k+1} = t$, and $g_{a-1} = x^n - 1$. For $0 \leq i \leq k+1$, set $b_i = t - a_{k+1-i}$, and $g'_{b_i} = \widehat{g}^{\#}_{a_{k-i}}$. Then $\{\zeta^{b_0} g'_{b_0}, \zeta^{b_1} g'_{b_1}, \ldots, \zeta^{b_k} g'_{b_k}\}$ is the generating set in standard form for $\mathcal{C}^{\perp}$.

In 2004, Dinh and López-Permouth [584] generalized the methods of [332, 1081] for simple-root cyclic codes over $\mathbb{Z}_{p^a}$ to obtain the structures of simple-root cyclic and self-dual cyclic codes over finite chain rings R. The strategy was independent from the approach in [1448] and the results were more detailed.

Since the code length n and the characteristic p of the residue field $\overline{R}$ are coprime, $x^n - 1$ factors uniquely to a product of monic basic irreducible pairwise coprime polynomials in $R[x]$. The ambient ring $\frac{R[x]}{\langle x^n - 1 \rangle}$ can be decomposed as a direct sum of chain rings. So any cyclic code of length n over R, viewed as an ideal of this ambient ring $\frac{R[x]}{\langle x^n - 1 \rangle}$, is represented as a direct sum of ideals from those chain rings. A polynomial $f \in R[x]$ is said to be **regular** if it is not a zero divisor.

Theorem 17.4.4 ([584, Lemma 3.1, Theorem 3.2, Corollary 3.3]) *Let R be a finite chain ring with maximal ideal $\langle \zeta \rangle$ and t the nilpotency of ζ. The following hold.*

(a) *If f is a regular basic irreducible polynomial of the ring $R[x]$, then $\frac{R[x]}{\langle f \rangle}$ is also a chain ring whose ideals are $\langle \zeta^i \rangle$, $0 \leq i \leq t$.*

(b) *Let $x^n - 1 = f_1 f_2 \cdots f_r$ be a representation of $x^n - 1$ as a product of monic basic irreducible pairwise coprime polynomials in $R[x]$. Then $\frac{R[x]}{\langle x^n - 1 \rangle}$ can be represented as a direct sum of chain rings $\frac{R[x]}{\langle f_i \rangle}$:*

$$\frac{R[x]}{\langle x^n - 1 \rangle} \cong \bigoplus_{i=1}^{r} \frac{R[x]}{\langle f_i \rangle}.$$

(c) *Each cyclic code of length n over R, i.e., each ideal of $\frac{R[x]}{\langle x^n - 1 \rangle}$, is a direct sum of ideals of the form $\langle \zeta^j \widehat{f}_i \rangle$, where $0 \leq j \leq t$, $1 \leq i \leq r$.*

(d) *The number of cyclic codes over R of length n is $(t + 1)^r$.*

For each cyclic code $\mathcal{C}$, using the decomposition above, a unique set of pairwise coprime monic polynomials that generates $\mathcal{C}$ is constructed, which in turn provides the sizes of $\mathcal{C}$ and its dual $\mathcal{C}^{\perp}$, and a set of generators for $\mathcal{C}^{\perp}$. The set of pairwise coprime monic polynomial generators of $\mathcal{C}$ also gives a single generator of $\mathcal{C}$; that implies $\frac{R[x]}{\langle x^n - 1 \rangle}$ is a principle ideal ring. Two elements a, b of a ring R are said to be **associate** if there is an invertible element u of R such that $a = bu$, i.e., $\langle a \rangle = \langle b \rangle$.

Theorem 17.4.5 ([584, Theorems 3.4, 3.5, 3.6, 3.8, 3.10, 4.1]) *Let R be a finite chain ring with maximal ideal $\langle \zeta \rangle$ and t the nilpotency of ζ. Let $\mathcal{C}$ be a cyclic code of length n over R. The following hold.*

(a) *There exists a unique family of pairwise coprime monic polynomials $F_0, F_1, \ldots, F_t$ in $R[x]$ such that $F_0 F_1 \cdots F_t = x^n - 1$ and $\mathcal{C} = \langle \widehat{F}_1, \zeta \widehat{F}_2, \ldots, \zeta^{t-1} \widehat{F}_t \rangle$.*

(b) *The number of codewords in $\mathcal{C}$ is*

$$|\mathcal{C}| = |\overline{R}|^{\sum_{i=0}^{t-1}(t-i)\deg F_{i+1}}.$$

(c) *There exist polynomials* $g_0, g_1, \ldots, g_{t-1}$ *in* $R[x]$ *such that* $\mathcal{C} = \langle g_0, \zeta g_1, \ldots, \zeta^{t-1} g_{t-1} \rangle$ *and*

$$g_{t-1} \,|\, g_{t-2} \,|\, \cdots \,|\, g_1 \,|\, g_0 \,|\, (x^n - 1).$$

(d) *Let* $F = \widehat{F}_1 + \zeta \widehat{F}_2 + \cdots + \zeta^{t-1} \widehat{F}_t$. *Then* F *is a generating polynomial of* $\mathcal{C}$, *i.e.,* $\mathcal{C} = \langle F \rangle$. *In particular,* $\frac{R[x]}{\langle x^n - 1 \rangle}$ *is a principal ideal ring.*

(e) *The dual* $\mathcal{C}^\perp$ *of* $\mathcal{C}$ *is the cyclic code*

$$\mathcal{C}^\perp = \langle \widehat{F}_0^*, \zeta \widehat{F}_t^*, \ldots, \zeta^{t-1} \widehat{F}_2^* \rangle,$$

and

$$|\mathcal{C}^\perp| = |\overline{R}|^{\sum\limits_{i=1}^{t} i \deg F_{i+1}}$$

where $F_{t+1} = F_0$.

(f) *Let* $G = \widehat{F}_0^* + \zeta \widehat{F}_t^* + \cdots + \zeta^{t-1} \widehat{F}_2^*$. *Then* G *is a generating polynomial of* $\mathcal{C}^\perp$, *i.e.,* $\mathcal{C}^\perp = \langle G \rangle$.

(g) $\mathcal{C}$ *is self-dual if and only if* F_i *is an associate of* F_j^* *for all* $i, j \in \{0, \ldots, t\}$ *such that* $i + j \equiv 1 \pmod{t+1}$.

If the nilpotency t of ζ is even, then $\langle \zeta^{t/2} \rangle$ is a cyclic self-dual code, which is the so-called trivial self-dual code. Using the structure of cyclic codes above, a necessary and sufficient condition for the existence of nontrivial self-dual cyclic codes was obtained.

Theorem 17.4.6 ([584, Theorems 4.3, 4.4]) *Assuming that* t *is an even integer, the following conditions are equivalent.*

(a) *Nontrivial self-dual cyclic codes exist.*

(b) *There exists a basic irreducible factor* $f \in R[x]$ *of* $x^n - 1$ *such that* f *and* f^* *are not associates.*

(c) $p^i \not\equiv -1 \pmod{n}$ *for all positive integers* i.

When p is an odd prime, a characterization of integers n where $p^i \not\equiv -1 \pmod{n}$ for all positive integers i, is still unknown. However when $p = 2$, the integers n where $2^i \not\equiv -1 \pmod{n}$ for all positive integers i, were completely characterized by Moree in [1524, Appendix B] with more details in [1400].

Theorem 17.4.7 ([1524, Theorem 4], [584, Theorem 4.5]) *Let* R *be a finite chain ring with maximal ideal* $\langle \zeta \rangle$ *where* $|R| = 2^{lt}$, $|\overline{R}| = 2^l$, *and* t *is the nilpotency of* ζ. *If* t *is even and* n *is odd, then nontrivial self-dual cyclic codes of length* n *over* R *exist if and only if* n *is divisible by any of the following:*

* *a prime* $\tau \equiv 7 \pmod{8}$, *or*

* *a prime* $\tau \equiv 1 \pmod{8}$, *where the order of* $2 \pmod{\tau}$ *is odd, or*

* *different odd primes* ϱ *and* σ *such that the order of* $2 \pmod{\varrho}$ *is* $2^\varsigma i$ *and the order of* $2 \pmod{\sigma}$ *is* $2^\varsigma j$, *where* i *is odd,* j *is even, and* $\varsigma \geq 1$.

There are cases where $p^i \equiv -1 \pmod{n}$ for some integer i, which leads to the nonexistence of nontrivial self-dual cyclic codes for certain values of n and p. Recall that for relatively prime integers a and m, a is called a **quadratic residue** or **quadratic non-residue** of m according to whether the congruence $x^2 \equiv a \pmod{m}$ has a solution or not. We refer to [584, Appendix] for important properties of quadratic residues and related concepts.

Theorem 17.4.8 ([584, Corollaries 4.6, 4.7, 4.8]) *Let R be a finite chain ring with maximal ideal $\langle \zeta \rangle$, $|R| = p^{lt}$, where $|\bar{R}| = p^l$, and t is the nilpotency of ζ such that t is even. The following hold.*

(a) *If n is a prime, then nontrivial self-dual cyclic codes of length n over R do not exist in the following cases:*

- $p = 2$, $n \equiv 3, 5 \pmod{8}$,
- $p = 3$, $n \equiv 5, 7 \pmod{12}$,
- $p = 5$, $n \equiv 3, 7, 13, 17 \pmod{20}$,
- $p = 7$, $n \equiv 5, 11, 13, 15, 17, 23 \pmod{28}$,
- $p = 11$, $n \equiv 3, 13, 15, 17, 21, 23, 27, 29, 31, 41 \pmod{44}$.

(b) *If n is an odd prime different from p, and p is a quadratic non-residue of n^k, where $k \geq 1$, then nontrivial self-dual cyclic codes of length n over R do not exist.*

(c) *If n is an odd prime, then nontrivial self-dual cyclic codes of length n over R do not exist in the following cases:*

- $p \equiv 1 \pmod{4}$, *and there exists a positive integer k such that $\gcd(p, 4n^k) = 1$ and p is a quadratic non-residue of $4n^k$,*
- $p \equiv 1 \pmod{8}$, *and there exist positive integers i, j with $i > 2$ such that $\gcd(p, 2^i n^j) = 1$ and p is a quadratic non-residue of $2^i n^j$.*

Furthermore, let $m = 2^{k_0} p_1^{k_1} \cdots p_r^{k_r}$ be the prime factorization of some integer $m > 1$ with $k_0 \geq 2$. Assume that $\gcd(p, m) = 1$, p is a quadratic non-residue of m, and

$$p \equiv \begin{cases} 1 \pmod{4} & \text{if } k_0 = 2, \\ 1 \pmod{8} & \text{if } k_0 \geq 3. \end{cases}$$

Then there exists an integer $i \in \{1, 2, \ldots, r\}$ such that nontrivial self-dual cyclic codes of length p_i over R do not exist.

Remark 17.4.9

(a) All results in this section for simple-root cyclic codes also hold for simple-root negacyclic codes, reformulated accordingly. We obtain valid results if we replace "cyclic" by "negacyclic" and "$x^n - 1$" by "$x^n + 1$".

(b) Most of the techniques that Dinh and López-Permouth [584] used for simple-root cyclic codes over finite chain rings (Theorems 17.4.4 − 17.4.8) are the most general form of the techniques that were first introduced by Pless et al. [1522, 1524] in 1996 for simple-root cyclic codes over $\mathbb{Z}_4$. Those were previously extended to the setting of simple-root cyclic codes over $\mathbb{Z}_{p^m}$ by Kanwar and López-Permouth [1081] in 1997, and simple-root cyclic codes over Galois rings $\mathrm{GR}(p^a, m)$ by Wan [1862] in 1999.

(c) As shown by Hammons et al. [898], well-known nonlinear binary codes can be constructed from quaternary linear codes using the Gray map. The **Gray map** $\mathcal{G} : \mathbb{Z}_4^n \to \mathbb{Z}_2^{2n}$ is defined as follows: since each $\mathbf{c} \in \mathbb{Z}_4^n$ is uniquely represented as $\mathbf{c} = \mathbf{a} + 2\mathbf{b}$ with $\mathbf{a}, \mathbf{b} \in \mathbb{Z}_2^n$, define $\mathcal{G}(\mathbf{c}) = (\mathbf{b}, \mathbf{a} \oplus \mathbf{b})$ where $\oplus$ is the componentwise addition of vectors modulo 2. The Gray map is significant because it is an isometry in the sense that the Lee weight of $\mathbf{c}$ is equal to the Hamming weight of $\mathcal{G}(\mathbf{c})$. The Gray map also preserves duality, since for any linear code $\mathcal{C}$ over $\mathbb{Z}_4$, $\mathcal{G}(\mathcal{C})$ and $\mathcal{G}(\mathcal{C}^\perp)$ are formally dual, i.e., their Hamming weight enumerators are MacWilliams transforms of each other.

However, the Gray map does not preserve linearity; in fact the Gray image of a linear code is usually not linear. It was shown in [898] that for a $\mathbb{Z}_4$-linear cyclic code of odd length $\mathcal{C}$, its Gray image $\mathcal{G}(\mathcal{C})$ is linear if and only if for any codewords $\mathbf{c}_1, \mathbf{c}_2 \in \mathcal{C}$, $2(\mathbf{c}_1 \star \mathbf{c}_2) \in \mathcal{C}$, where $\star$ is the componentwise multiplication of vectors called the **Hadamard product**, that is, $\mathbf{a} \star \mathbf{b} = (a_0 b_0, \ldots, a_{n-1} b_{n-1})$. Indeed, binary nonlinear codes having better parameters than their linear counterparts have been constructed via the Gray map.

Wolfmann [1901, 1902] showed that the Gray image of a simple-root linear negacyclic code over $\mathbb{Z}_4$ is a (not necessarily linear) cyclic binary code. He classified all $\mathbb{Z}_4$-linear negacyclic codes of odd length and provided a method to determine all linear binary cyclic codes of length $2n$ (n is odd) that are images of negacyclic codes under the Gray map. Therefore, the Gray image of a simple-root negacyclic code over $\mathbb{Z}_4$ is permutation equivalent to a binary cyclic code under the Nechaev permutation.

17.5 Repeated-Root Constacyclic Codes over Galois Rings

Except otherwise stated, all codes in this section are repeated-root codes over a finite chain ring R, i.e., the code length n is divisible by the characteristic p of the residue field $\overline{R}$.

When the code length n is odd, there is a one-to-one correspondence between cyclic and negacyclic codes (single-root or repeated-root) over any finite commutative ring.

Proposition 17.5.1 ([584, Proposition 5.1]) *Let R be a finite commutative ring and n be an odd integer. The map $\xi : \frac{R[x]}{\langle x^n - 1 \rangle} \to \frac{R[x]}{\langle x^n + 1 \rangle}$ defined by $\xi(f(x)) = f(-x)$ is a ring isomorphism. In particular, $\mathcal{A}$ is an ideal of $\frac{R[x]}{\langle x^n - 1 \rangle}$ if and only if $\xi(\mathcal{A})$ is an ideal of $\frac{R[x]}{\langle x^n + 1 \rangle}$. Equivalently, $\mathcal{A}$ is a cyclic code of length n over R if and only if $\xi(\mathcal{A})$ is a negacyclic code of length n over R.*

It was shown by Sălăgean in [1611] that repeated-root cyclic and negacyclic codes over finite chain rings are not in general principally generated.

Proposition 17.5.2 ([1611, Theorem 3.4]) *Let R be a finite chain ring whose residue field has characteristic p. For $p \mid n$ the following hold.*

(a) $\frac{R[x]}{\langle x^n - 1 \rangle}$ *is not a principal ideal ring.*

(b) *If p is odd or $p = 2$ and R is not a Galois ring, then $\frac{R[x]}{\langle x^n + 1 \rangle}$ is not a principal ideal ring.*

(c) *If $p = 2$ and R is a Galois ring, then $\frac{R[x]}{\langle x^n + 1 \rangle}$ is a principal ideal ring.*

The description of generators of ideals of $R[x]$ by Gröbner bases was developed in [1427, 1428, 1449] for a chain ring R. Sălăgean et al. [1449, 1611] used Gröbner bases to obtain the structure of repeated-root cyclic codes over finite chain rings, and furthermore provide generating matrices, sizes, and Hamming distances of such codes.

Theorem 17.5.3 ([1449, Theorem 4.2], [1611, Theorems 4.1, 5.1, 6.1]) *Let R be a finite chain ring with maximal ideal $\langle \zeta \rangle$ and t be the nilpotency of ζ. If C is a nonzero cyclic code of length n over R, then*

(a) *C admits a set of generators $C = \langle \zeta^{a_0} g_{a_0}, \zeta^{a_1} g_{a_1}, \ldots, \zeta^{a_k} g_{a_k} \rangle$ such that*

    (i) *$0 \leq k < t$,*

    (ii) *$0 \leq a_0 < a_1 < \cdots < a_k < t$,*

    (iii) *$g_{a_i} \in R[x]$ is monic for $0 \leq i \leq k$,*

    (iv) *$\deg(g_{a_i}) > \deg(g_{a_{i+1}})$ for $0 \leq i \leq k - 1$,*

    (v) *For $0 \leq i \leq k$, $\zeta^{a_{i+1}} g_{a_i} \in \langle \zeta^{a_{i+1}} g_{a_{i+1}}, \ldots, \zeta^{a_k} g_{a_k} \rangle$ in $R[x]$, and*

    (vi) *$\zeta^{a_0}(x^n - 1) \in \langle \zeta^{a_0} g_{a_0}, \zeta^{a_1} g_{a_1}, \ldots, \zeta^{a_k} g_{a_k} \rangle$ in $R[x]$.*

(b) *This set $\{\zeta^{a_0} g_{a_0}, \zeta^{a_1} g_{a_1}, \ldots, \zeta^{a_k} g_{a_k}\}$ of generators is a strong Gröbner basis. It is not necessarily unique. However, the cardinality $k + 1$ of the basis, the degrees of its polynomials and the exponents $a_0, a_1, \ldots, a_k$ are unique.*

(c) *Let $d_i = \deg(g_{a_i})$ for $0 \leq i \leq k$, and $d_{-1} = n$. Then the matrix consisting of the rows corresponding to the codewords $\zeta^{a_i} x^j g_{a_i}$, with $0 \leq i \leq k$ and $0 \leq j \leq d_{i-1} - d_i - 1$, is a generator matrix for C.*

(d) *The number of codewords in C is*

$$|C| = |\overline{R}|^{\sum_{i=0}^{k}(t - a_i)(d_{i-1} - d_i)}.$$

(e) *The Hamming distance of C equals the Hamming distance of $\langle \overline{g_{a_k}} \rangle$.*

(f) *The results in parts (a), (b), (c), (d), (e) hold for negacyclic codes, reformulated by replacing $x^n - 1$ with $x^n + 1$.*

In fact, Theorem 17.5.3(a) provides a structure theorem for both simple-root and repeated-root cyclic codes. Conditions (v) and (vi) imply that $g_{a_k} \mid g_{a_{k-1}} \mid \cdots \mid g_{a_0} \mid (x^n - 1)$. In the simple-root case, conditions (v) and (vi) can be replaced by the stronger condition $g_{a_k} \mid g_{a_{k-1}} \mid \cdots \mid g_{a_0} \mid (x^n - 1)$, as in Proposition 17.4.2, giving a structure theorem for simple-root cyclic codes. For repeated-root cyclic codes, conditions (v) and (vi) cannot be improved in general; [1449, Example 3.3] gave cyclic codes for which no set of generators of the form given in Theorem 17.5.3(a) has the property $g_{a_k} \mid g_{a_{k-1}} \mid \cdots \mid g_{a_0} \mid (x^n - 1)$.

Most of the research on repeated-root codes concentrated on the situation where the chain ring is a Galois ring, i.e., $R = \mathrm{GR}(p^a, m)$. In this case, using polynomial representation, it is easy to show that the ideals $\langle x - 1, p \rangle$ and $\langle x + 1, p \rangle$ are the sets of non-invertible elements of $\frac{\mathrm{GR}(p^a, m)[x]}{\langle x^{p^s} - 1 \rangle}$ and $\frac{\mathrm{GR}(p^a, m)[x]}{\langle x^{p^s} + 1 \rangle}$, respectively. Therefore, $\frac{\mathrm{GR}(p^a, m)[x]}{\langle x^{p^s} - 1 \rangle}$ and $\frac{\mathrm{GR}(p^a, m)[x]}{\langle x^{p^s} + 1 \rangle}$ are local rings whose maximal ideals are $\langle x - 1, p \rangle$ and $\langle x + 1, p \rangle$, respectively. When $a \geq 2$, $\mathrm{GR}(p^a, m)$ is not a field; Proposition 17.5.2 gives us information on the ambient rings of cyclic and negacyclic codes of length p^s over $\mathrm{GR}(p^a, m)$.

Proposition 17.5.4 ([583, 592]) *If $a \geq 2$, then the following hold.*

(a) $\frac{\mathrm{GR}(p^a, m)[x]}{\langle x^{p^s} - 1 \rangle}$ *is a local ring with maximal ideal $\langle x - 1, p \rangle$, but it is not a chain ring.*

(b) *If p is odd,* $\frac{\mathrm{GR}(p^a, m)[x]}{\langle x^{p^s} + 1 \rangle}$ *is a local ring with maximal ideal $\langle x + 1, p \rangle$, but it is not a chain ring.*

(c) *If $p = 2$,* $\frac{\mathrm{GR}(p^a, m)[x]}{\langle x^{p^s} + 1 \rangle}$ *is a chain ring with maximal ideal $\langle x + 1 \rangle$.*

When $a = 1$, the Galois ring $\mathrm{GR}(p^a, m)$ is the Galois field $\mathbb{F}_{p^m}$. Dinh [577] showed that the ambient rings $\frac{\mathbb{F}_{p^m}[x]}{\langle x^{p^s} - 1 \rangle}$ and $\frac{\mathbb{F}_{p^m}[x]}{\langle x^{p^s} + 1 \rangle}$ are chain rings, and used this to establish the structure of cyclic and negacyclic codes of length p^s over $\mathbb{F}_{p^m}$, as well as the Hamming distances of all such codes. These results were then generalized by Dinh [579] to all constacyclic codes.

Theorem 17.5.5 ([577, 579]) *For any a nonzero element λ of $\mathbb{F}_{p^m}$, there exists $\lambda_0 \in \mathbb{F}_{p^m}$ such that $\lambda_0^{p^s} = \lambda$. The ambient ring $\frac{\mathbb{F}_{p^m}[x]}{\langle x^{p^s} - \lambda \rangle}$ is a chain ring with maximal ideal $\langle x - \lambda_0 \rangle$. The λ-constacyclic codes of length p^s over $\mathbb{F}_{p^m}$ are precisely the ideals $C_i = \langle (x - \lambda_0)^i \rangle$ of $\frac{\mathbb{F}_{p^m}[x]}{\langle x^{p^s} - \lambda_0 \rangle}$, for $i \in \{0, 1, \ldots, p^s\}$. The λ-constacyclic code C_i has $p^{m(p^s - i)}$ codewords. Its dual code is the λ^{-1}-constacyclic code $\langle (x - \lambda_0^{-1})^{p^s - i} \rangle \subseteq \frac{\mathbb{F}_{p^m}[x]}{\langle x^{p^s} - \lambda^{-1} \rangle}$ with p^{mi} codewords. The λ-constacyclic code C_i has Hamming distance*

$$
d_H(C_i) = \begin{cases}
1 & \text{if } i = 0, \\
\beta + 2 & \text{if } \beta p^{s-1} + 1 \leq i \leq (\beta + 1) p^{s-1} \text{ where } 0 \leq \beta \leq p - 2, \\
(t + 1) p^k & \text{if } p^s - p^{s-k} + (t - 1) p^{s-k-1} + 1 \leq i \leq p^s - p^{s-k} + t p^{s-k-1} \\
& \text{where } 1 \leq t \leq p - 1 \text{ and } 1 \leq k \leq s - 1, \\
0 & \text{if } i = p^s.
\end{cases}
$$

When $p = 2$, there is no one-to-one correspondence between cyclic and negacyclic codes of length 2^s over $\mathrm{GR}(2^a, m)$ (Proposition 17.5.1 does not hold when the code length is even). In 2005, Dinh gave the structure of such negacyclic codes, and the Hamming distances of most of them in [573], and later on, in [580], obtained the Hamming and homogeneous distances of all of them, using their structure in [573] and the Hamming distances of 2^m-ary cyclic codes in Theorem 17.5.5.

Theorem 17.5.6 ([573, 580]) *The ring $\frac{\mathrm{GR}(2^a, m)[x]}{\langle x^{2^s} + 1 \rangle}$ is a chain ring with maximal ideal $\langle x + 1 \rangle$ and residue field $\mathbb{F}_{2^m}$. Negacyclic codes of length 2^s over the Galois ring $\mathrm{GR}(2^a, m)$ are precisely the ideals $C_i = \langle (x + 1)^i \rangle$, $0 \leq i \leq 2^s a$, of $\frac{\mathrm{GR}(2^a, m)[x]}{\langle x^{2^s} + 1 \rangle}$. Each negacyclic code C_i has $2^{m(2^s a - i)}$ codewords; its dual is the negacyclic code $C_{2^s a - i}$ with 2^{mi} codewords. The Hamming distance $d_H(C_i)$ and homogeneous distances $d_{hom}(C_i)$ are completely determined as follows:*

$$
d_H(C_i) = \begin{cases}
0 & \text{if } i = 2^s a, \\
1 & \text{if } 0 \leq i \leq 2^s(a - 1), \\
2 & \text{if } 2^s(a - 1) + 1 \leq i \leq 2^s(a - 1) + 2^{s-1}, \\
2^{k+1} & \text{if } 2^s(a - 1) + 2^s - 2^{s-k} + 1 \leq i \leq \\
& \quad 2^s(a - 1) + 2^s - 2^{s-k} + 2^{s-k-1}, \\
& \text{i.e., } 2^s(a - 1) + 1 + \sum_{l=1}^{k} 2^{s-l} \leq i \leq \\
& \quad 2^s(a - 1) + \sum_{l=1}^{k+1} 2^{s-l} \text{ for } 1 \leq k \leq s - 1.
\end{cases}
$$

$$d_{hom}(\mathcal{C}_i) = \begin{cases} 0 & \text{if } i = 2^s a, \\ (2^m - 1)2^{m(a-2)} & \text{if } 0 \le i \le 2^s(a-2), \\ 2^{m(a-1)} & \text{if } 2^s(a-2)+1 \le i \le 2^s(a-1), \\ 2^{m(a-1)+1} & \text{if } 2^s(a-1)+1 \le i \le 2^s(a-1)+2^{s-1}, \\ 2^{m(a-1)+k+1} & \text{if } 2^s(a-1)+2^s-2^{s-k}+1 \le i \le \\ & \qquad 2^s(a-1)+2^s-2^{s-k}+2^{s-k-1}, \\ & \text{i.e., } 2^s(a-1)+1+\sum_{l=1}^{k} 2^{s-l} \le i \le \\ & \qquad 2^s(a-1)+\sum_{l=1}^{k+1} 2^{s-l} \text{ for } 1 \le k \le s-1. \end{cases}$$

If the dimension $m = 1$, the Galois ring $\mathrm{GR}(2^a, m)$ is the ring $\mathbb{Z}_{2^a}$. The Hamming, homogeneous, Lee, and Euclidean distances of all negacyclic codes of length 2^s over $\mathbb{Z}_{2^a}$ were established in [576].

Theorem 17.5.7 ([576]) *Let $\mathcal{C}$ be a negacyclic code of length 2^s over $\mathbb{Z}_{2^a}$. Then $\mathcal{C} = \mathcal{C}_i = \langle (x+1)^i \rangle \subseteq \frac{\mathbb{Z}_{2^a}[x]}{\langle x^{2^s}+1 \rangle}$, for $i \in \{0, 1, \ldots, 2^s a\}$, and the Hamming distance $d_H(\mathcal{C}_i)$, homogeneous distance $d_{hom}(\mathcal{C}_i)$, Lee distance $d_L(\mathcal{C}_i)$, and Euclidean distance $d_E(\mathcal{C}_i)$ of $\mathcal{C}$ are determined as follows:*

$$d_H(\mathcal{C}_i) = \begin{cases} 0 & \text{if } i = 2^s a, \\ 1 & \text{if } 0 \le i \le 2^s(a-1), \\ 2 & \text{if } 2^s(a-1)+1 \le i \le 2^s(a-1)+2^{s-1}, \\ 2^{k+1} & \text{if } 2^s(a-1)+1+\sum_{j=1}^{k} 2^{s-j} \le i \le 2^s(a-1)+\sum_{j=1}^{k+1} 2^{s-j} \\ & \quad \text{for } 1 \le k \le s-1. \end{cases}$$

$$d_{hom}(\mathcal{C}_i) = \begin{cases} 0 & \text{if } i = 2^s a, \\ 2^{a-2} & \text{if } 0 \le i \le 2^s(a-2), \\ 2^{a-1} & \text{if } 2^s(a-2)+1 \le i \le 2^s(a-1), \\ 2^a & \text{if } 2^s(a-1)+1 \le i \le 2^s(a-1)+2^{s-1}, \\ 2^{a+k} & \text{if } 2^s(a-1)+1+\sum_{j=1}^{k} 2^{s-j} \le i \le 2^s(a-1)+\sum_{j=1}^{k+1} 2^{s-j} \\ & \quad \text{for } 1 \le k \le s-1. \end{cases}$$

$$d_L(\mathcal{C}_i) = \begin{cases} 0 & \text{if } i = 2^s a, \\ 1 & \text{if } i = 0, \\ 2 & \text{if } 1 \le i \le 2^s, \\ 2^{l+1} & \text{if } 2^s l + 1 \le i \le 2^s(l+1) \text{ for } 1 \le l \le a-2, \\ 2^a & \text{if } 2^s(a-1)+1 \le i \le 2^s(a-1)+2^{s-1}, \\ 2^{a+k} & \text{if } 2^s(a-1)+1+\sum_{j=1}^{k} 2^{s-j} \le i \le 2^s(a-1)+\sum_{j=1}^{k+1} 2^{s-j} \\ & \quad \text{for } 1 \le k \le s-1. \end{cases}$$

$$d_E(\mathcal{C}_i) = \begin{cases} 0 & \textit{if } i = 2^s a, \\ 1 & \textit{if } i = 0, \\ 2^{2l+1} & \textit{if } 2^s l + 1 \leq i \leq 2^s l + 2^{s-1} \textit{ for } 0 \leq l \leq a-2, \\ 2^{2l+2} & \textit{if } 2^s l + 2^{s-1} + 1 \leq i \leq 2^s(l+1) \textit{ for } 0 \leq l \leq a-2, \\ 2^{2a-1} & \textit{if } 2^s(a-1) + 1 \leq i \leq 2^s(a-1) + 2^{s-1}, \\ 2^{2a+k-1} & \textit{if } 2^s(a-1) + 1 + \sum_{j=1}^{k} 2^{s-j} \leq i \leq 2^s(a-1) + \sum_{j=1}^{k+1} 2^{s-j} \\ & \textit{for } 1 \leq k \leq s-1. \end{cases}$$

In the special case when the alphabet is $\mathbb{Z}_4$, or its Galois extension $\mathrm{GR}(4,m)$, repeated-root cyclic and negacyclic codes have been studied in more detail. Among other partial results, the structures of negacyclic and cyclic codes over $\mathbb{Z}_4$ of any length were provided, respectively, by Blackford in 2003 [208] and by Dougherty and Ling in 2006 [631].

The Discrete Fourier Transform is a useful tool to study structures of codes; for instance, it was used by Blackford [207, 208], and Dougherty and Ling [631] to recover a tuple $\mathbf{c}$ from its Mattson–Solomon polynomial. In 2003, Blackford used the Discrete Fourier Transform to give a decomposition of the ambient ring $\frac{\mathbb{Z}_4[x]}{\langle x^{2^a n}+1 \rangle}$ of cyclic codes of length $2^a n$ over $\mathbb{Z}_4$ as a direct sum of $\frac{\mathrm{GR}(4,m_i)[u]}{\langle u^{2^a}+1 \rangle}$. The rings $\frac{\mathrm{GR}(4,m_i)[u]}{\langle u^{2^a}+1 \rangle}$ are the ambient rings of negacyclic codes of length 2^a over $\mathrm{GR}(4,m_i)$, which were shown to be chain rings by Blackford, and later by Dinh [573] for the more general case over $\mathrm{GR}(2^z, m_i)$.

Theorem 17.5.8 ([208, Lemma 2, Theorem 1]) *Let n be an odd positive integer and a be any nonnegative integer. Let I denote a complete set of representatives of the 2-cyclotomic cosets modulo n, and for each $i \in I$, let m_i be the size of the 2-cyclotomic coset containing i. The following hold.*

(a) *For any $m \geq 1$, the ring $\frac{\mathrm{GR}(4,m)[u]}{\langle u^{2^a}+1 \rangle}$ is a chain ring with maximal ideal $\langle u+1 \rangle$ and residue field $\mathbb{F}_{2^m}$. Its ideals, i.e., negacyclic codes of length 2^a over $\mathrm{GR}(4,m)$, are $\langle 0 \rangle$, $\langle 1 \rangle$, $\langle (u+1)^i \rangle$, and $\langle 2(u+1)^i \rangle$, where $1 \leq i \leq 2^a - 1$.*

(b) *The map*

$$\phi : \frac{\mathbb{Z}_4[x]}{\langle x^{2^a n}+1 \rangle} \longrightarrow \bigoplus_{i \in I} \frac{\mathrm{GR}(4,m_i)[u]}{\langle u^{2^a}+1 \rangle},$$

given by

$$\phi(c(x)) = [\widehat{c}_i]_{i \in I},$$

where $(\widehat{c}_0, \widehat{c}_1, \ldots, \widehat{c}_{n-1})$ is the Discrete Fourier Transform of $c(x)$, is a ring isomorphism.

(c) *Each negacyclic code of length $2^a n$ over $\mathbb{Z}_4$, i.e., an ideal of the ring $\frac{\mathbb{Z}_4[x]}{\langle x^{2^a n}+1 \rangle}$, is isomorphic to $\bigoplus_{i \in I} \mathcal{C}_i$, where $\mathcal{C}_i$ is an ideal of $\frac{\mathrm{GR}(4,m_i)[u]}{\langle u^{2^a}+1 \rangle}$ (such ideals are given in part (a)).*

Using this, Blackford went on to show that $\frac{\mathbb{Z}_4[x]}{\langle x^{2^a n}+1 \rangle}$ is a principal ideal ring and established a concatenated structure of negacyclic codes over $\mathbb{Z}_4$.

Theorem 17.5.9 ([208, Theorems 2, 3]) *Let $\mathcal{C}$ be a negacyclic code of length $2^a n$ over $\mathbb{Z}_4$ with n odd, i.e., an ideal of the ring $\frac{\mathbb{Z}_4[x]}{\langle x^{2^a n}+1 \rangle}$. The following hold.*

(a) $\mathcal{C} = \langle g(x) \rangle$, where $g(x) = \prod_{i=0}^{2^{a+1}} [g_i(x)]^i$ and $\{g_i(x)\}$ are monic pairwise coprime divisors of $x^n - 1$ in $\mathbb{Z}_4[x]$.

(b) Any codeword of $\mathcal{C}$ is equivalent to a $(2^a n)$-tuple of the form $(\mathbf{b}_0 \,|\, \mathbf{b}_1 \,|\, \cdots \,|\, \mathbf{b}_{2^a-1})$, where

$$\mathbf{b}_i = \sum_{j=0}^{2^a-1} \overline{\binom{j}{i}} \, \mathbf{a}_j \quad \text{with} \quad \overline{\binom{j}{i}} = \binom{j}{i} \bmod 2$$

and

$$\mathbf{a}_j \in \langle g_{j+1} \cdots g_{2^{a+1}} + 2g_{j+2^a+1} \cdots g_{2^{a+1}} \rangle \subseteq \frac{\mathbb{Z}_4[x]}{\langle x^n - 1 \rangle}.$$

We now turn our attention to repeated-root cyclic codes over $\mathbb{Z}_4$. In 2003, Abualrub and Oehmke [10, 11] classified cyclic codes of length 2^k over $\mathbb{Z}_4$ by their generators, and in 2004 they derived a mass formula for the number of such codes [9]. In 2006, Dougherty and Ling [631] generalized that to give a classification of cyclic codes of length 2^k over the Galois ring $\mathrm{GR}(4, m)$.

Theorem 17.5.10 ([631, Lemma 2.3, Theorem 2.6]) *Let η be a primitive $(2^m - 1)^{th}$ root of unity in $\mathrm{GR}(4, m)$ with Teichmüller set $\mathcal{T}_m = \{0, 1, \eta, \eta^2, \ldots, \eta^{2^m-2}\}$. Then the ambient ring $\frac{\mathrm{GR}(4,m)[u]}{\langle u^{2^k}-1 \rangle}$ is a local ring with maximal ideal $\langle 2, u - 1 \rangle$ and residue field $\mathbb{F}_{2^m}$. Cyclic codes of length 2^k over $\mathrm{GR}(4, m)$, i.e., ideals of $\frac{\mathrm{GR}(4,m)[u]}{\langle u^{2^k}-1 \rangle}$, are as follows:*

- $\langle 0 \rangle$, $\langle 1 \rangle$,

- $\langle 2(u-1)^i \rangle$ *where* $0 \le i \le 2^k - 1$,

- $\left\langle (u-1)^i + 2 \sum_{j=0}^{i-1} s_j (x-1)^j \right\rangle$ *where* $1 \le i \le 2^k - 1$ *and* $s_j \in \mathcal{T}_m$, *or equivalently,* $\langle (u-1)^i + 2(u-1)^t h(u) \rangle$ *where* $0 \le t \le i - 1$ *and* $h(u)$ *is 0 or a unit,*

- $\left\langle 2(u-1)^l, \ (u-1)^i + 2 \sum_{j=0}^{i-1} s_j (u-1)^j \right\rangle$ *where* $1 \le i \le 2^k - 1$, $l < i$, *and* $s_j \in \mathcal{T}_m$, *or equivalently,* $\langle 2(u-1)^l, \ (u-1)^i + 2(u-1)^t h(u) \rangle$, *where* $0 \le t \le i - 1$ *and* $h(u)$ *is 0 or a unit.*

Furthermore, the number of such cyclic codes is

$$\mathcal{N}(m) = 5 + 2^{2^{k-1}m} + 2^m (5 \cdot 2^m - 1) \frac{2^{m(2^{k-1}-1)} - 1}{(2^m - 1)^2} - 4 \cdot \frac{2^{k-1} - 1}{2^m - 1}.$$

In 2003, using the Discrete Fourier Transform, Blackford [207] gave the structure of cyclic codes of length $2n$ (n is odd) over $\mathbb{Z}_4$. Later, in 2006, Dougherty and Ling [631] generalized that to obtain a description of cyclic codes of any length over $\mathbb{Z}_4$ as a direct sum of cyclic codes of length 2^k over $\mathrm{GR}(4, m_i)$.

Theorem 17.5.11 ([207, Theorem 2], [631, Theorem 3.2, Corollaries 3.3, 3.4]) *Let n be an odd positive integer and k any nonnegative integer. Let I denote a complete set of representatives of the 2-cyclotomic cosets modulo n, and for each $i \in I$, let m_i be the size of the 2-cyclotomic coset containing i. The following hold.*

(a) *The map*

$$\gamma : \frac{\mathbb{Z}_4[x]}{\langle x^{2^k n} - 1 \rangle} \to \bigoplus_{i \in I} \frac{\mathrm{GR}(4, m_i)[u]}{\langle u^{2^k} - 1 \rangle}$$

given by

$$\gamma(c(x)) = [\widehat{c}_i]_{i \in I},$$

where $(\widehat{c}_0, \widehat{c}_1, \ldots, \widehat{c}_{n-1})$ *is the Discrete Fourier Transform of* $c(x)$, *is a ring isomorphism.*

(b) *Each cyclic code of length* $2^k n$ *over* $\mathbb{Z}_4$, *i.e., an ideal of the ring* $\frac{\mathbb{Z}_4[x]}{\langle x^{2^k n} - 1 \rangle}$, *is isomorphic to* $\bigoplus_{i \in I} C_i$, *where* C_i *is an ideal of* $\frac{\mathrm{GR}(4, m_i)[u]}{\langle u^{2^k} - 1 \rangle}$ (*such ideals are classified in Theorem 17.5.10*).

(c) *The number of distinct cyclic code of length* $2^k n$ *over* $\mathbb{Z}_4$ *is* $\prod_{i \in I} \mathcal{N}(m_i)$, *where* $\mathcal{N}(m_i)$ *is the number of cyclic codes of length* 2^k *over* $\mathrm{GR}(4, m_i)$, *which is given in Theorem 17.5.10.*

This decomposition of cyclic codes was then used to completely determine the generators of all cyclic codes and their sizes; see [631] for specific details about the polynomials involved in the next result.

Theorem 17.5.12 ([631, Theorems 4.2, 4.3]) *Let* n *be an odd positive integer and* k *any nonnegative integer, and let* $\mathcal{C}$ *be a cyclic code of length* $2^k n$ *over* $\mathbb{Z}_4$, *i.e.,* $\mathcal{C}$ *is an ideal of the ring* $\frac{\mathbb{Z}_4[x]}{\langle x^{2^k n} - 1 \rangle}$. *The following hold.*

(a) $\mathcal{C}$ *is of the form*

$$\left\langle p(x^{2^k}) \prod_{i=0}^{2^k - 1} q_i(x^{2^k}) \prod_{i=1}^{2^k - 1} \left(\prod_T \widetilde{r_{i,T}(x)} \right)^i \prod_{i=1}^{2^k - 1} \left(\prod_{l=0}^{i-1} \widetilde{s_{i,l}(x)} \right)^i, \right.$$

$$\left. 2p(x^{2^k}) \prod_{i=0}^{2^k - 1} q_i(x)^i \prod_{i=1}^{2^k - 1} \left(\prod_T r_{i,T}(x)^T \right) \prod_{i=1}^{2^k - 1} \left(\prod_{l=0}^{i-1} s_{i,l}(x)^l \right) \right\rangle,$$

where

$$x^n - 1 = p(x) \left(\prod_{i=0}^{2^k - 1} q_i(x) \right) \left(\prod_{i=1}^{2^k - 1} \left(\prod_T r_{i,T}(x) \right) \right) \left(\prod_{i=1}^{2^k - 1} \left(\prod_{l=0}^{i-1} s_{i,l}(x) \right) \right) y(x),$$

and $\widetilde{r_{i,T}(x)} = r_{i,T}(x) \bmod 2$, $\widetilde{s_{i,l}(x)} = s_{i,l}(x) \bmod 2$, *and for each* i, *the product* $\prod_T$ *is taken over all possible values of* T *as follows:*

- *if* $1 \leq i \leq 2^{k-1}$, *then* $T = i$,
- *if* $2^{k-1} < i < 2^{k-1} + t$ $(t > 0)$, *then* $T = 2^{k-1}$,
- *if* $i = 2^{k-1} + t$ $(t > 0)$, *then* $2^{k-1} \leq T \leq i$,
- *if* $i > 2^{k-1} + t$ $(t > 0)$, *then* $T = 2^{k-1}$ *or* $2^k - i + t$.

(b) *The number of codewords in C is*

$$|\mathcal{C}| = 4^{2^k \deg(p)} \prod_{i=0}^{2^k-1} 2^{(2^k-i)\deg(q_i)} \prod_{i=1}^{2^k-1} \left(\prod_T 2^{(2^{k+1}-i-T)\deg(r_{i,T})} \right) \times$$

$$\prod_{i=1}^{2^k-1} \left(\prod_{l=0}^{i-1} 2^{(2^{k+1}-i-l)\deg(s_{i,l})} \right).$$

There are four finite commutative rings of four elements, namely, the Galois field $\mathbb{F}_4$, the ring $\mathbb{Z}_4$ of integers modulo 4, the ring $\mathbb{F}_2 + u\mathbb{F}_2$ where $u^2 = 0$, and the ring $\mathbb{F}_2 + v\mathbb{F}_2$ where $v^2 = v$. The first three are chain rings, while the last one, $\mathbb{F}_2 + v\mathbb{F}_2$, is not. Indeed, $\mathbb{F}_2 + v\mathbb{F}_2 \cong \mathbb{F}_2 \times \mathbb{F}_2$, which is not even a local ring. The ring $\mathbb{F}_2 + u\mathbb{F}_2$ consists of all binary polynomials of degree 0 and 1 in indeterminate u; it is closed under binary polynomial addition and multiplication modulo u^2. Thus, $\mathbb{F}_2 + u\mathbb{F}_2 = \frac{\mathbb{F}_2[u]}{\langle u^2 \rangle} = \{0, 1, u, \overline{u} = u + 1\}$ is a chain ring with maximal ideal $\{0, u\}$. The addition in $\mathbb{F}_2 + u\mathbb{F}_2$ is similar to that of the Galois field $\mathbb{F}_4 = \{0, 1, \xi, \xi^2 = \xi + 1\}$, where u is replaced by ξ. The multiplication in $\mathbb{F}_2 + u\mathbb{F}_2$ is similar to the multiplication in the ring $\mathbb{Z}_4$, where u is replaced by 2. In fact, $(\mathbb{F}_2 + u\mathbb{F}_2, +) \cong (\mathbb{F}_4, +)$, and $(\mathbb{F}_2 + u\mathbb{F}_2, *) \cong (\mathbb{Z}_4, *)$. Thus, $\mathbb{F}_2 + u\mathbb{F}_2$ lies between $\mathbb{F}_4$ and $\mathbb{Z}_4$, in the sense that it is additively analogous to $\mathbb{F}_4$, and multiplicatively analogous to $\mathbb{Z}_4$. In 2009, Dinh [578] established the structure of all constacyclic codes of length 2^s over $\mathbb{F}_{2^m} + u\mathbb{F}_{2^m}$, for any positive integer m. Of course, over $\mathbb{F}_{2^m} + u\mathbb{F}_{2^m}$, cyclic and negacyclic codes coincide; their structure and sizes are as follows.

Theorem 17.5.13 ([578])

(a) *The ring $\frac{(\mathbb{F}_{2^m}+u\mathbb{F}_{2^m})[x]}{\langle x^{2^s}+1 \rangle}$ is a local ring with maximal ideal $\langle u, x+1 \rangle$, but it is not a chain ring.*

(b) *Cyclic codes of length 2^s over $\mathbb{F}_{2^m} + u\mathbb{F}_{2^m}$ are precisely the ideals of the ring $\frac{(\mathbb{F}_{2^m}+u\mathbb{F}_{2^m})[x]}{\langle x^{2^s}+1 \rangle}$, which are*

Type 1: (trivial ideals) $\langle 0 \rangle$, $\langle 1 \rangle$,

Type 2: (principal ideals with non-monic polynomial generators)

$$\langle u(x+1)^i \rangle$$

where $0 \le i \le 2^s - 1$,

Type 3: (principal ideals with monic polynomial generators)

$$\langle (x+1)^i + u(x+1)^t h(x) \rangle$$

where $1 \le i \le 2^s - 1$, $0 \le t < i$, and either $h(x)$ is 0 or $h(x)$ is a unit that can be represented as $h(x) = \sum_j h_j(x+1)^j$ with $h_j \in \mathbb{F}_{2^m}$ and $h_0 \ne 0$,

Type 4: (non-principal ideals)

$$\left\langle (x+1)^i + u \sum_{j=0}^{\kappa-1} c_j(x+1)^j, \ u(x+1)^\kappa \right\rangle$$

where $1 \le i \le 2^s - 1$, $c_j \in \mathbb{F}_{2^m}$, and $\kappa < T$ with T the smallest integer such that $u(x+1)^T \in \left\langle (x+1)^i + u \sum_{j=0}^{i-1} c_j(x+1)^j \right\rangle$, or equivalently,

$$\langle (x+1)^i + u(x+1)^t h(x), \ u(x+1)^\kappa \rangle$$

with $h(x)$ as in Type 3 and $\deg(h) \le \kappa - t - 1$.

(c) *The number of distinct cyclic codes of length 2^s over $\mathbb{F}_{2^m} + u\mathbb{F}_{2^m}$ is*

$$\frac{2^{m(2^{s-1}-1)}(2^{2m} + 2^m + 2) - 2^{2m+1} - 2}{(2^m - 1)^2} + \frac{6 \cdot 2^{m(2^s-1)} - 2^{s+1} - 1}{2^m - 1} +$$

$$2^{m2^{s-1}} + 4 \cdot 2^{m(2^{s-1}-1)} + 3 \cdot 2^{s-1} - 1.$$

(d) *Let $\mathcal{C}$ be a cyclic code of length 2^s over $\mathbb{F}_{2^m} + u\mathbb{F}_{2^m}$, as classified in (b). Then the number of codewords $n_{\mathcal{C}}$ of $\mathcal{C}$ is given as follows.*

- *If $\mathcal{C} = \langle 0 \rangle$, then $n_{\mathcal{C}} = 1$.*
- *If $\mathcal{C} = \langle 1 \rangle$, then $n_{\mathcal{C}} = 2^{m2^{s+1}}$.*
- *If $\mathcal{C} = \langle u(x+1)^i \rangle$ where $0 \leq i \leq 2^s - 1$, then $n_{\mathcal{C}} = 2^{m(2^s-i)}$.*
- *If $\mathcal{C} = \langle (x+1)^i \rangle$ where $1 \leq i \leq 2^s - 1$, then $n_{\mathcal{C}} = 2^{2m(2^s-i)}$.*
- *If $\mathcal{C} = \langle (x+1)^i + u(x+1)^t h(x) \rangle$ where $1 \leq i \leq 2^s - 1$, $0 \leq t < i$, and $h(x)$ is a unit, then*

$$n_{\mathcal{C}} = \begin{cases} 2^{2m(2^s-i)} & \text{if } 1 \leq i \leq 2^{s-1} + \frac{t}{2}, \\ 2^{m(2^s-t)} & \text{if } 2^{s-1} + \frac{t}{2} < i \leq 2^s - 1. \end{cases}$$

- *If $\mathcal{C} = \langle (x+1)^i + u(x+1)^t h(x), \, u(x+1)^\kappa \rangle$ where $1 \leq i \leq 2^s - 1$, $0 \leq t < i$, either $h(x)$ is 0 or $h(x)$ is a unit, and*

$$\kappa < T = \begin{cases} i & \text{if } h(x) = 0, \\ \min\{i, 2^s - i + t\} & \text{if } h(x) \neq 0, \end{cases}$$

then $n_{\mathcal{C}} = 2^{m(2^{s+1}-i-\kappa)}$.

17.6 Repeated-Root Constacyclic Codes over $\mathcal{R} = \mathbb{F}_{p^m} + u\mathbb{F}_{p^m}$, $u^2 = 0$

Throughout this section p is an odd prime, except in Subsection 17.6.4. The ring $\mathcal{R} = \mathbb{F}_{p^m} + u\mathbb{F}_{p^m}$ with $u^2 = 0$ consists of all p^m-ary polynomials of degree 0 and 1 in indeterminate u; it is closed under p^m-ary polynomial addition and multiplication modulo u^2. Thus, $\mathcal{R} = \frac{\mathbb{F}_{p^m}[u]}{\langle u^2 \rangle} = \{a + ub \mid a, b \in \mathbb{F}_{p^m}\}$ is a chain ring with maximal ideal $u\mathbb{F}_{p^m}$. The ring $\mathcal{R}$ has precisely $p^m(p^m - 1)$ units, which are of the forms $\alpha + u\beta$ and γ, where α, β, γ are nonzero elements of the field $\mathbb{F}_{p^m}$.

17.6.1 All Constacyclic Codes of Length p^s over $\mathcal{R}$

In 2010, Dinh [579] gave the structure of all constacyclic codes of length p^s over $\mathcal{R}$ as follows.

Theorem 17.6.1 ([579]) *Let λ be a unit of the ring $\mathcal{R}$, i.e., λ is of the form $\alpha + u\beta$ or γ, where α, β, γ are nonzero elements of the field $\mathbb{F}_{p^m}$.*

(a) *If* $\lambda = \alpha + u\beta$, *then the ambient ring* $\Re(p^s, \alpha + u\beta) = \frac{\mathcal{R}[x]}{\langle x^{p^s} - (\alpha + u\beta)\rangle}$ *is a chain ring with maximal ideal* $\langle \alpha_0 x - 1\rangle$, *and* $\langle (\alpha_0 x - 1)^{p^s}\rangle = \langle u\rangle$.[3] *The* $(\alpha + u\beta)$-*constacyclic codes of length* p^s *over* $\mathcal{R}$ *are the ideals* $\langle (\alpha_0 x - 1)^i\rangle$, $0 \le i \le 2p^s$, *of the chain ring* $\Re(p^s, \alpha + u\beta)$.

(b) *If* $\lambda = \gamma \in \mathbb{F}_{p^m}^*$, *then the ambient ring* $\Re(p^s, \gamma) = \frac{\mathcal{R}[x]}{\langle x^{p^s} - \gamma\rangle}$ *is a local ring with maximal ideal* $\langle u, x - \gamma_0\rangle$, *but it is not a chain ring.*[4] *The* γ-*constacyclic codes of length* p^s *over* $\mathcal{R}$, *i.e., ideals of the ring* $\Re(p^s, \gamma)$, *are as follows:*

Type 1: (trivial ideals) $\langle 0\rangle$, $\langle 1\rangle$,

Type 2: (principal ideals with non-monic polynomial generators)

$$\langle u(x - \gamma_0)^i\rangle$$

where $0 \le i \le p^s - 1$,

Type 3: (principal ideals with monic polynomial generators)

$$\langle (x - \gamma_0)^i + u(x - \gamma_0)^t h(x)\rangle$$

where $1 \le i \le p^s - 1$, $0 \le t < i$, *and either* $h(x)$ *is* 0 *or* $h(x)$ *is a unit in* $\frac{\mathbb{F}_{p^m}[x]}{\langle x^{p^s} - \gamma\rangle}$,

Type 4: (non-principal ideals)

$$\langle (x - \gamma_0)^i + u(x - \gamma_0)^t h(x), \ u(x - \gamma_0)^\kappa\rangle$$

with $h(x)$ *as in Type 3,* $\deg(h) \le \kappa - t - 1$, *and* $\kappa < T$, *where* T *is the smallest integer such that* $u(x - \gamma_0)^T \in \langle (x - \gamma_0)^i + u(x - \gamma_0)^t h(x)\rangle$; *i.e., such* T *can be determined as*

$$T = \begin{cases} i & \text{if } h(x) = 0, \\ \min\{i, p^s - i + t\} & \text{if } h(x) \ne 0. \end{cases}$$

Using the structure in Theorem 17.6.1, the Hamming distances of all constacyclic codes of length p^s over $\mathcal{R}$ were established in [579, 586]. Recall that $\mathbb{F}_{p^m}$ is a subring of $\mathcal{R}$. In this subsection, we denote $\langle (x - \gamma_0)^i\rangle_F$ as the code of length p^s generated by $(x - \gamma_0)^i$ over $\mathbb{F}_{p^m}$; i.e., $\langle (x - \gamma_0)^i\rangle_F$ is the ideal $\langle (x - \gamma_0)^i\rangle$ of $\frac{\mathbb{F}_{p^m}[x]}{\langle x^{p^s} - \gamma\rangle}$.

Theorem 17.6.2 ([579, 586]) *Let the notations and* λ-*constacyclic codes* $\mathcal{C}$ *of length* p^s *over* $\mathcal{R}$ *be as given in Theorem 17.6.1.*

(a) *If* $\lambda = \alpha + u\beta$, *then*

$$d_H(\mathcal{C}) = \begin{cases} 1 & \text{if } 0 \le i \le p^s, \\ (e+1)p^k & \text{if } 2p^s - p^{s-k} + (e-1)p^{s-k-1} + 1 \le i \le 2p^s - p^{s-k} + ep^{s-k-1} \\ & \quad \text{for } 1 \le e \le p - 1 \text{ and } 0 \le k \le s - 1, \\ 0 & \text{if } i = 2p^s. \end{cases}$$

---

[3]Here $\alpha_0 \in \mathbb{F}_{p^m}$ such that $\alpha_0^{p^s} = \alpha^{-1}$, which is determined as follows. Since α is a nonzero element of the field $\mathbb{F}_{p^m}$, $\alpha^{-p^m} = \alpha^{-1}$, implying $\alpha^{-p^{km}} = \alpha^{-1}$, for any nonnegative integer k. Using the positive integers s and m as dividend and divisor, by the division algorithm, there exist nonnegative integers q_m and r_m such that $s = q_m m + r_m$, and $0 \le r_m \le m - 1$. Let $\alpha_0 = \alpha^{-p^{(q_m + 1)m - s}} = \alpha^{-p^{m - r_m}}$. Then $\alpha_0^{p^s} = \alpha^{-p^{(q_m + 1)m}} = \alpha^{-1}$.

[4]Here $\gamma_0 \in \mathbb{F}_{p^m}$ such that $\gamma_0^{p^s} = \gamma$, which is determined similarly to α_0 in part (a) as follows. Let $\gamma_0 = \gamma^{p^{(q_m + 1)m - s}} = \gamma^{p^{m - r_m}}$. Then $\gamma_0^{p^s} = \gamma^{p^{(q_m + 1)m}} = \gamma$.

(b) *If $\lambda = \gamma \in \mathbb{F}_{p^m}^*$, then $d_H(\mathcal{C})$ is as follows:*

Type 1: (trivial ideals) $d_H(\langle 0 \rangle) = 0$, $d_H(\langle 1 \rangle) = 1$,

Type 2: (principal ideals with non-monic polynomial generators)

$$d_H(\mathcal{C}) = d_H(\langle (x - \gamma_0)^i \rangle_F)$$
$$= \begin{cases} 1 & \text{if } i = 0, \\ (e+1)p^k & \text{if } p^s - p^{s-k} + (e-1)p^{s-k-1} + 1 \leq i \leq \\ & p^s - p^{s-k} + ep^{s-k-1} \\ & \text{for } 1 \leq e \leq p-1 \text{ and } 0 \leq k \leq s-1, \end{cases}$$

Type 3: (principal ideals with monic polynomial generators)

$$d_H(\mathcal{C}) = d_H(\langle (x - \gamma_0)^i \rangle_F) = (e+1)p^k$$

for $p^s - p^{s-k} + (e-1)p^{s-k-1} + 1 \leq i \leq p^s - p^{s-k} + ep^{s-k-1}$, $1 \leq e \leq p-1$, and $0 \leq k \leq s-1$,

Type 4: (non-principal ideals)

$$d_H(\mathcal{C}) = d_H(\langle (x - \gamma_0)^\kappa \rangle_F) = (e+1)p^k$$

for $p^s - p^{s-k} + (e-1)p^{s-k-1} + 1 \leq \kappa \leq p^s - p^{s-k} + ep^{s-k-1}$, $1 \leq e \leq p-1$, and $0 \leq k \leq s-1$.

With the development of high-density data storage technologies, symbol-pair codes are proposed to protect efficiently against a certain number of pair-errors. Let Ξ be the code alphabet consisting of q elements. Then each element $c \in \Xi$ is called a **symbol**. In symbol-pair read channels, a codeword $(c_0, c_1, \ldots, c_{n-1})$ is read as $((c_0, c_1), (c_1, c_2), \ldots, (c_{n-1}, c_0))$. Suppose that $\mathbf{x} = (x_0, x_1, \ldots, x_{n-1})$ is a vector in Ξ^n. The **symbol-pair vector** of $\mathbf{x}$ is defined as $\pi(\mathbf{x}) = ((x_0, x_1), (x_1, x_2), \ldots, (x_{n-1}, x_0))$. Hence, each vector has a unique symbol-pair vector $\pi(\mathbf{x}) \in (\Xi, \Xi)^n$. An important parameter of symbol-pair codes is symbol-pair distance. In 2010, Cassuto and Blaum [365] gave the definition of the symbol-pair distance as the Hamming distance over the alphabet (Ξ, Ξ). Given $\mathbf{x} = (x_0, x_1, \ldots, x_{n-1}), \mathbf{y} = (y_0, y_1, \ldots, y_{n-1})$, the **symbol-pair distance** between $\mathbf{x}$ and $\mathbf{y}$ is defined as

$$d_{sp}(\mathbf{x}, \mathbf{y}) = d_H(\pi(\mathbf{x}), \pi(\mathbf{y})) = \left| \{ i \mid (x_i, x_{i+1}) \neq (y_i, y_{i+1}) \} \right|.$$

A q-ary code of length n is a nonempty subset $\mathcal{C} \subseteq \Xi^n$. The **symbol-pair distance** of a **symbol-pair code** $\mathcal{C}$ is defined as

$$d_{sp}(\mathcal{C}) = \min_{\mathbf{x}, \mathbf{y} \in \mathcal{C}, \mathbf{x} \neq \mathbf{y}} \{ d_{sp}(\mathbf{x}, \mathbf{y}) \}.$$

The **symbol-pair weight** of a vector $\mathbf{x}$ is defined as the Hamming weight of its symbol-pair vector $\pi(\mathbf{x})$:

$$\text{wt}_{sp}(\mathbf{x}) = \text{wt}_H(\pi(\mathbf{x})) = \left| \{ i \mid (x_i, x_{i+1}) \neq (0, 0), \ 0 \leq i \leq n-1, \ x_n = x_0 \} \right|.$$

If the code $\mathcal{C}$ is linear, its symbol-pair distance equals the minimum symbol-pair weight of nonzero codewords of $\mathcal{C}$, namely

$$d_{sp}(\mathcal{C}) = \min \{ \text{wt}_{sp}(\mathbf{x}) \mid \mathbf{x} \neq \mathbf{0}, \ \mathbf{x} \in \mathcal{C} \}.$$

In [365], Cassuto and Blaum studied the model of symbol-pair read channels. They provided constructions and decoding methods of symbol-pair codes. Then in [366] they considered lower and upper bounds on such codes; moreover, they gave bounds and asymptotics on the size of optimal symbol-pair codes. They also established the relationship between the symbol-pair distance and the Hamming distance in [366, Theorem 2]. In 2011, by using algebraic methods, Cassuto and Litsyn [367] constructed cyclic symbol-pair codes. Applying the Discrete Fourier Transform to coefficients of codeword polynomials $c(x) \in \frac{\mathbb{F}_q[x]}{\langle x^n - 1 \rangle}$ and the BCH Bound, Cassuto and Litsyn proved that the symbol-pair distance of a cyclic code $\mathcal{C}$ with Hamming distance $d_H(\mathcal{C})$ is at least $d_H(\mathcal{C}) + 2$ [367, Theorem 10]. This implies that the lower bounds on symbol-pair distances of cyclic codes can be computed by determining the Hamming distances of the cyclic codes. In addition, if $g(x)$ is a generator polynomial of a cyclic code $\mathcal{C}$ with prime length n and $g(x)$ has at least m roots in $\mathbb{F}_{q^t}$ and $d_H(\mathcal{C}) \le \min\{2m - n + 2, m - 1\}$, then the lower bounds on symbol-pair distance of the cyclic code $\mathcal{C}$ is at least $d_H(\mathcal{C}) + 3$ [367, Theorem 11]. Cassuto and Litsyn also constructed symbol-pair codes from cyclic codes and showed that there exist symbol-pair codes with rates strictly higher than such codes in the Hamming metric with the same relative distance.

More recently, Yaakobi et al. [1917] considered and gave a lower bound on the symbol-pair distances for binary cyclic codes as follows: for a given linear cyclic code $\mathcal{C}$ with Hamming distance $d_H(\mathcal{C})$, the symbol-pair distance is at least $d_H(\mathcal{C}) + \left\lceil \frac{d_H(\mathcal{C})}{2} \right\rceil$ [1917, Theorem 4]. This shows that the lower bound on the symbol-pair distances proved by Yaakobi et al. [1917, Theorem 4] is better than the result of Cassuto and Litsyn [367, Theorem 10] for binary cyclic codes. However, the algorithms introduced by Cassuto et al. and Yaakobi et al. for decoding symbol-pair codes could not construct all pair-errors within the pair-error-correcting capability. By giving the definition of a parity check matrix for symbol-pair codes, Hirotomo et al. [956] proposed a new syndrome decoding algorithm of symbol-pair codes that improved the algorithms of Cassuto et al. [367] and Yaakobi et al. [1917] for decoding symbol-pair codes. In particular, in 2015, Kai et al. [1071] extended the result of Cassuto and Litsyn [367, Theorem 10] for the case of simple-root constacyclic codes.

Recently, the symbol-pair distance distribution of all constacyclic code of length p^s over $\mathbb{F}_{p^m}$ was established by Dinh et al. in [587].

Theorem 17.6.3 ([587]) *Given a λ-constacyclic code of length p^s over $\mathbb{F}_{p^m}$, then it has the form $\mathcal{C}_i = \langle (x - \lambda_0)^i \rangle \subseteq \frac{\mathbb{F}_{p^m}[x]}{\langle x^{p^s} - \lambda \rangle}$ for $i \in \{0, 1, \ldots, p^s\}$, where $\lambda_0 \in \mathbb{F}_{p^m}$ such that $\lambda_0^{p^s} = \lambda$ (such λ_0 exists and can be determined as in Theorem 17.6.1). The symbol-pair distance $d_{sp}(\mathcal{C}_i)$ of $\mathcal{C}_i$ is completely determined as follows:*

$$
d_{sp}(\mathcal{C}_i) = \begin{cases}
2 & \text{if } i = 0, \\
3p^k & \text{if } i = p^s - p^{s-k} + 1 \text{ for } 0 \le k \le s - 2, \\
4p^k & \text{if } p^s - p^{s-k} + 2 \le i \le p^s - p^{s-k} + p^{s-k+1} \\
& \quad \text{for } 0 \le k \le s - 2, \\
2(\delta + 2)p^k & \text{if } p^s - p^{s-k} + \delta\, p^{s-k-1} + 1 \le i \le \\
& \quad p^s - p^{s-k} + (\delta + 1)\, p^{s-k-1} \\
& \quad \text{for } 0 \le k \le s - 2 \text{ and } 1 \le \delta \le p - 2, \\
(\delta + 2)p^{s-1} & \text{if } i = p^s - p + \delta \text{ for } 0 \le \delta \le p - 2, \\
p^s & \text{if } i = p^s - 1, \\
0 & \text{if } i = p^s.
\end{cases}
$$

Using the Hamming and symbol-pair distances of constacyclic codes of length p^s over

$\mathbb{F}_{p^m}$ (see Theorems 17.5.5 and 17.6.3), Dinh et al. [586] obtained the symbol-pair distances of all such codes over $\mathcal{R}$.

Theorem 17.6.4 ([586]) *Let the notations and λ-constacyclic codes $\mathcal{C}$ of length p^s over $\mathcal{R}$ be as listed in Theorem 17.6.1.*

(a) *If $\lambda = \alpha + u\beta$, then*

$$
d_{sp}(\mathcal{C}) = \begin{cases}
2 & \text{if } 0 \le i \le p^s, \\
3p^k & \text{if } i = 2p^s - p^{s-k} + 1 \text{ for } 0 \le k \le s - 2, \\
4p^k & \text{if } 2p^s - p^{s-k} + 2 \le i \le 2p^s - p^{s-k} + p^{s-k+1} \\
& \quad \text{for } 0 \le k \le s - 2, \\
2(\delta + 2)p^k & \text{if } 2p^s - p^{s-k} + \delta p^{s-k-1} + 1 \le i \le \\
& \quad 2p^s - p^{s-k} + (\delta + 1)p^{s-k-1} \\
& \quad \text{for } 0 \le k \le s - 2 \text{ and } 1 \le \delta \le p - 2, \\
(\delta + 2)p^{s-1} & \text{if } i = 2p^s - p + \delta \text{ for } 0 \le \delta \le p - 2, \\
p^s & \text{if } i = 2p^s - 1, \\
0 & \text{if } i = 2p^s.
\end{cases}
$$

(b) *If $\lambda = \gamma \in \mathbb{F}_{p^m}^*$, then $d_{sp}(\mathcal{C})$ is as follows:*

Type 1: (trivial ideals) $d_{sp}(\langle 0 \rangle) = 0$, $d_{sp}(\langle 1 \rangle) = 2$.

Type 2: (principal ideals with non-monic polynomial generators)

$$
\begin{aligned}
d_{sp}(\mathcal{C}) = d_{sp}\left(\langle (x - \gamma_0)^i \rangle_F\right) \\
= \begin{cases}
2 & \text{if } i = 0, \\
3p^k & \text{if } i = p^s - p^{s-k} + 1 \text{ for } 0 \le k \le s - 2, \\
4p^k & \text{if } p^s - p^{s-k} + 2 \le i \le p^s - p^{s-k} + p^{s-k+1} \\
& \quad \text{for } 0 \le k \le s - 2, \\
2(\delta + 2)p^k & \text{if } p^s - p^{s-k} + \delta p^{s-k-1} + 1 \le i \le \\
& \quad p^s - p^{s-k} + (\delta + 1)p^{s-k-1} \\
& \quad \text{for } 0 \le k \le s - 2 \text{ and } 1 \le \delta \le p - 2, \\
(\delta + 2)p^{s-1} & \text{if } i = p^s - p + \delta \text{ for } 0 \le \delta \le p - 2, \\
p^s & \text{if } i = p^s - 1.
\end{cases}
\end{aligned}
$$

Type 3: (principal ideals with monic polynomial generators)

$$
\begin{aligned}
d_{sp}(\mathcal{C}) = d_{sp}\left(\langle (x - \gamma_0)^i \rangle_F\right) \\
= \begin{cases}
3p^k & \text{if } i = p^s - p^{s-k} + 1 \text{ for } 0 \le k \le s - 2, \\
4p^k & \text{if } p^s - p^{s-k} + 2 \le i \le p^s - p^{s-k} + p^{s-k+1} \\
& \quad \text{for } 0 \le k \le s - 2, \\
2(\delta + 2)p^k & \text{if } p^s - p^{s-k} + \delta p^{s-k-1} + 1 \le i \le \\
& \quad p^s - p^{s-k} + (\delta + 1)p^{s-k-1} \\
& \quad \text{for } 0 \le k \le s - 2 \text{ and } 1 \le \delta \le p - 2 \\
(\delta + 2)p^{s-1} & \text{if } i = p^s - p + \delta \text{ for } 0 \le \delta \le p - 2 \\
p^s & \text{if } i = p^s - 1.
\end{cases}
\end{aligned}
$$

Type 4: (non-principal ideals)

$$d_{sp}(\mathcal{C}) = d_{sp}\left(\langle (x - \gamma_0)^\kappa \rangle_F\right)$$

$$= \begin{cases} 3p^k & \text{if } \kappa = p^s - p^{s-k} + 1 \text{ for } 0 \le k \le s - 2, \\ 4p^k & \text{if } p^s - p^{s-k} + 2 \le \kappa \le p^s - p^{s-k} + p^{s-k+1} \\ & \quad \text{for } 0 \le k \le s - 2, \\ 2(\delta + 2)p^k & \text{if } p^s - p^{s-k} + \delta\, p^{s-k-1} + 1 \le \kappa \le \\ & \quad p^s - p^{s-k} + (\delta + 1)\, p^{s-k-1} \\ & \quad \text{for } 0 \le k \le s - 2 \text{ and } 1 \le \delta \le p - 2, \\ (\delta + 2)p^{s-1} & \text{if } \kappa = p^s - p + \delta \text{ for } 0 \le \delta \le p - 2, \\ p^s & \text{if } \kappa = p^s - 1. \end{cases}$$

17.6.2 All Constacyclic Codes of Length $2p^s$ over $\mathcal{R}$

For codes of length $2p^s$ over $\mathcal{R}$, in 2015, Dinh et al. [598] studied negacyclic codes and then generalized the results to all constacyclic codes in 2016 [390]. The algebraic structures of all λ-constacyclic codes of length $2p^s$ over $\mathcal{R}$ were determined as follows.

Theorem 17.6.5 ([390, Section 3])

(a) *If λ is a square in $\mathcal{R}$ and $\lambda = \delta^2$, then it follows from the Chinese Reminder Theorem that*

$$\Re(2p^s, \lambda) = \frac{\mathcal{R}[x]}{\langle x^{2p^s} - \lambda \rangle} \cong \frac{\mathcal{R}[x]}{\langle x^{p^s} + \delta \rangle} \oplus \frac{\mathcal{R}[x]}{\langle x^{p^s} - \delta \rangle} = \Re(p^s, -\delta) \oplus \Re(p^s, \delta).$$

So a λ-constacyclic code of length $2p^s$ over $\mathcal{R}$, i.e., an ideal $\mathcal{C}$ of $\Re(2p^s, \lambda)$, is represented as a direct sum $\mathcal{C} = \mathcal{C}_+ \oplus \mathcal{C}_-$, where $\mathcal{C}_+$ and $\mathcal{C}_-$ are ideals of $\Re(p^s, -\delta)$ and $\Re(p^s, \delta)$, respectively. Thus, the classification, detailed structure, and number of codewords of constacyclic codes $\mathcal{C}$ of length $2p^s$ over $\mathcal{R}$ can be obtained from that of the direct summands $\mathcal{C}_+$ and $\mathcal{C}_-$ in Theorem 17.6.1. Also $\mathcal{C}^\perp = \mathcal{C}_+^\perp \oplus \mathcal{C}_-^\perp$.

(b) *If λ is not a square in $\mathcal{R}$, and $\lambda = \alpha + u\beta$, $\alpha, \beta \in \mathbb{F}_{p^m}^*$, then the ring $\Re(2p^s, \alpha + u\beta) = \frac{\mathcal{R}[x]}{\langle x^{2p^s} - (\alpha + u\beta) \rangle}$ is a chain ring whose ideals are*

$$\Re(2p^s, \alpha + u\beta) = \langle 1 \rangle \supsetneq \langle x^2 - \alpha_0 \rangle \supsetneq \cdots \supsetneq \langle (x^2 - \alpha_0)^{2p^s - 1} \rangle \supsetneq \langle (x^2 - \alpha_0)^{2p^s} \rangle = \langle 0 \rangle$$

where $\alpha_0^{p^s} = \alpha$. In other words, $(\alpha + u\beta)$-constacyclic codes of length $2p^s$ over $\mathcal{R}$ are precisely the ideals $\langle (x^2 - \alpha_0)^i \rangle \subseteq \Re(2p^s, \alpha + u\beta)$, where $0 \le i \le 2p^s$. Each $(\alpha + u\beta)$-constacyclic code $\mathcal{C} = \langle (x^2 - \alpha_0)^i \rangle$ has $p^{2m(2p^s - i)}$ codewords; its dual $\mathcal{C}^\perp$ is the $(\alpha^{-1} - u\alpha^{-2}\beta)$-constacyclic code

$$\mathcal{C}^\perp = \left\langle (x^2 - \alpha_0^{-1})^{2p^s - i} \right\rangle \subseteq \Re(2p^s, \alpha^{-1} - u\alpha^{-2}\beta),$$

which contains p^{2mi} codewords. Moreover, the ideal $\langle u \rangle_{\Re(2p^s, \alpha + u\beta)}$ is the unique self-dual $(\alpha + u\beta)$-constacyclic code of length $2p^s$ over $\mathcal{R}$.

(c) *If λ is not a square in $\mathcal{R}$ and $\lambda = \gamma \in \mathbb{F}_{p^m}^*$, then γ-constacyclic codes are classified by categorizing the ideals of the local ring $\Re(2p^s, \gamma) = \frac{\mathcal{R}[x]}{\langle x^{2p^s} - \gamma \rangle}$ into 4 distinct types, where $\gamma_0^{p^s} = \gamma$, as follows:*

Type 1: (trivial ideals) $\langle 0 \rangle$, $\langle 1 \rangle$,

Type 2: (principal ideals with non-monic polynomial generators)

$$\langle u(x^2 - \gamma_0)^i \rangle$$

where $0 \le i \le p^s - 1$,

Type 3: (principal ideals with monic polynomial generators)

$$\langle (x^2 - \gamma_0)^i + u(x^2 - \gamma_0)^t h(x) \rangle$$

where $1 \le i \le p^s - 1, 0 \le t < i$, *and either* $h(x)$ *is* 0 *or* $h(x)$ *is a unit in* $\Re(2p^s, \gamma)$ *which can be represented as* $h(x) = \sum_j (h_{0j}x + h_{1j})(x^2 - \gamma_0)^j$ *with* $h_{0j}, h_{1j} \in \mathbb{F}_{p^m}$ *and* $h_{00}x + h_{10} \ne 0$,

Type 4: (non-principal ideals)

$$\langle (x^2 - \gamma_0)^i + u(x^2 - \gamma_0)^t h(x), \ u(x^2 - \gamma_0)^\omega \rangle$$

with $h(x)$ *as in Type 3*, $\deg h(x) \le \omega - t - 1$, *and* $\omega < T$, *where* T *is the smallest integer such that* $u(x^2 - \gamma_0)^T \in \langle (x^2 - \gamma_0)^i + u(x^2 - \gamma_0)^t h(x) \rangle$; *i.e., such* T *can be determined as*

$$T = \begin{cases} i & \text{if } h(x) = 0, \\ \min\{i, p^s - i + t\} & \text{if } h(x) \ne 0. \end{cases}$$

Furthermore, the number of distinct γ-*constacyclic codes of length* $2p^s$ *over* $\mathcal{R}$, *i.e., distinct ideals of the ring* $\Re(2p^s, \gamma)$, *is*

$$\frac{2(p^{2m} + 1)p^{m(p^s - 1)} - 2p^{4m} - 2}{(p^{2m} - 1)^2} + \frac{(2p^{2m} + 3)p^{m(p^s - 1)} - 2p^s - 1}{p^{2m} - 1} + p^{m(p^s - 1)} + 2.$$

When $\gamma \in \mathbb{F}_{p^m}^*$, it is easy to see that for any γ-constacyclic code $\mathcal{C}$ of length $2p^s$ over $\mathcal{R}$, its residue code $\text{Res}(\mathcal{C})$ and torsion code $\text{Tor}(\mathcal{C})$ are γ-constacyclic codes of length $2p^s$ over $\mathbb{F}_{p^m}$. By [581], each γ-constacyclic code of length $2p^s$ over $\mathbb{F}_{p^m}$ is an ideal of the form $\langle (x^2 - \gamma_0)^i \rangle$ of the finite chain ring $\frac{\mathbb{F}_{p^m}[x]}{\langle x^{2p^s} - \gamma \rangle}$, where $0 \le i \le p^s$, and each code $\langle (x^2 - \gamma_0)^i \rangle$ contains $p^{2m(p^s - i)}$ codewords. Therefore, we can determine the size of all γ-constacyclic codes of length $2p^s$ over $\mathcal{R}$ in Theorem 17.6.5(c) by multiplying the sizes of $\text{Res}(\mathcal{C})$ and $\text{Tor}(\mathcal{C})$ in each case.

Theorem 17.6.6 ([390, Section 3]) *Let* $\gamma \in \mathbb{F}_{p^m}^*$ *and* $\mathcal{C}$ *be a* γ-*constacyclic code of length* $2p^s$ *over* $\mathcal{R}$ *as in Theorem 17.6.5(c). Then the number of codewords of* $\mathcal{C}$, *denoted by* $n_\mathcal{C}$, *is determined as follows.*

- *If* $\mathcal{C} = \langle 0 \rangle$, *then* $n_\mathcal{C} = 1$.

- *If* $\mathcal{C} = \langle 1 \rangle$, *then* $n_\mathcal{C} = p^{4mp^s}$.

- *If* $\mathcal{C} = \langle u(x^2 - \gamma_0)^i \rangle$ *where* $0 \le i \le p^s - 1$, *then* $n_\mathcal{C} = p^{2m(p^s - i)}$.

- *If* $\mathcal{C} = \langle (x^2 - \gamma_0)^i \rangle$ *where* $1 \le i \le p^s - 1$, *then* $n_\mathcal{C} = p^{4m(p^s - i)}$.

- *If* $\mathcal{C} = \langle (x^2 - \gamma_0)^i + u(x^2 - \gamma_0)^t h(x) \rangle$ *where* $1 \le i \le p^s - 1, 0 \le t < i$, *and* $h(x)$ *is a unit, then*

$$n_\mathcal{C} = \begin{cases} p^{4m(p^s - i)} & \text{if } 1 \le i \le p^{s-1} + \frac{t}{2}, \\ p^{2m(p^s - i)} & \text{if } p^{s-1} + \frac{t}{2} < i \le p^s - 1. \end{cases}$$

- *If $\mathcal{C} = \langle (x^2 - \gamma_0)^i + u(x^2 - \gamma_0)^t h(x), u(x^2 - \gamma_0)^\omega \rangle$ where $1 \le i \le p^s - 1$, $0 \le t < i$, and either $h(x)$ is 0 or $h(x)$ is a unit, then $n_{\mathcal{C}} = p^{2m(2p^s - i - \omega)}$.*

Theorem 17.6.7 ([390, Section 3]) *Let $\mathcal{C}$ be a γ-constacyclic code as in Theorem 17.6.5(c). Then the dual $\mathcal{C}^\perp$ of $\mathcal{C}$ is determined as follows.*

Type 1: (trivial ideals) $\langle 0 \rangle^\perp = \langle 1 \rangle$, and $\langle 1 \rangle^\perp = \langle 0 \rangle$.

Type 2: (principal ideals with non-monic polynomial generators)

    If $\mathcal{C} = \langle u(x^2 - \gamma_0)^i \rangle$, then $\mathcal{C}^\perp = \langle (x^2 - \gamma_0^{-1})^{p^s - i}, u \rangle$.

Type 3: (principal ideals with monic polynomial generators)

    If $\mathcal{C} = \langle (x^2 - \gamma_0)^i + u(x^2 - \gamma_0)^t h(x) \rangle$ where $h(x)$ is 0 or $h(x)$ is a unit, then the dual code $\mathcal{C}^\perp$ is associated to the ideal $\mathrm{ann}(\mathcal{C})^*$, determined as follows.

- *If $h(x)$ is 0, then $\mathrm{ann}(\mathcal{C})^* = \langle (x^2 - \gamma_0^{-1})^{p^s - i} \rangle$.*

- *If $h(x)$ is a unit and $1 \le i \le \frac{p^s + t}{2}$, then $\mathrm{ann}(\mathcal{C})^* = \langle a(x) \rangle$ where*

$$a(x) = (-\gamma_0)^{i-t}(x^2 - \gamma_0^{-1})^{p^s - i} - u(x^2 - \gamma_0^{-1})^{p^s - 2i + t} \times$$
$$\sum_{j=0}^{i-t-1} (b_j x + a_j)(-\gamma_0)^j (x^2 - \gamma_0^{-1})^j x^{2i - 2t - 2j - 1}.$$

- *If $h(x)$ is a unit and $\frac{p^s + t}{2} < i \le p^s - 1$, then $\mathrm{ann}(\mathcal{C})^* = \langle b(x), u(x^2 - \gamma_0^{-1})^{p^s - i} \rangle$ where*

$$b(x) = (-\gamma_0)^{i-t}(x^2 - \gamma_0^{-1})^{i-t} - u \sum_{j=0}^{p^s - i - 1} (b_j x + a_j)(-\gamma_0)^j (x^2 - \gamma_0^{-1})^j x^{2i - 2t - 2j - 1}.$$

Type 4: (non-principal ideals)

    If $\mathcal{C} = \langle (x^2 - \gamma_0)^i + u(x^2 - \gamma_0)^t h(x), u(x^2 - \gamma_0)^\omega \rangle$ where $h(x)$ is 0 or $h(x)$ is a unit, then the dual code $\mathcal{C}^\perp$ is associated to the ideal $\mathrm{ann}(\mathcal{C})^*$, determined as follows.

- *If $h(x) = 0$, then $\mathrm{ann}(\mathcal{C})^* = \langle (x^2 - \gamma_0^{-1})^{p^s - \omega}, u(x^2 - \gamma_0^{-1})^{p^s - i} \rangle$.*

- *If $h(x)$ is a unit, then $\mathrm{ann}(\mathcal{C})^* = \langle c(x), u(x^2 - \gamma_0^{-1})^{p^s - i} \rangle$ where*

$$c(x) = (-\gamma_0)^{i-t}(x^2 - \gamma_0^{-1})^{p^s - \omega} - u(x^2 - \gamma_0^{-1})^{p^s - i - \omega + t} \times$$
$$\sum_{j=0}^{\omega - t - 1} (b_j x + a_j)(-\gamma_0)^j (x^2 - \gamma_0^{-1})^j x^{2i - 2t - 2j - 1}.$$

17.6.3 All Constacyclic Codes of Length $4p^s$ over $\mathcal{R}$

For any odd prime p such that $p^m \equiv 1 \pmod 4$, Dinh et al. [582] provided the structure of all λ-constacyclic codes of length $4p^s$ over $\mathcal{R}$ in terms of their generator polynomials. If the unit λ is a square, each λ-constacyclic code of length $4p^s$ is expressed as a direct sum of a $-\alpha$-constacyclic code and an α-constacyclic code of length $2p^s$. In the main case that the unit λ is not a square, if $\lambda = \alpha + u\beta$ for nonzero elements α, β of $\mathbb{F}_{p^m}$, the ambient ring $\frac{\mathcal{R}[x]}{\langle x^{4p^s} - (\alpha + u\beta) \rangle}$ is a chain ring with maximal ideal $\langle x^4 - \alpha_0 \rangle$, and so the $(\alpha + u\beta)$-constacyclic codes are $\langle (x^4 - \alpha_0)^i \rangle$ for $0 \le i \le 2p^s$. For the remaining case that the unit $\lambda = \gamma$ is not a square for a nonzero element γ of $\mathbb{F}_{p^m}$, the ambient ring $\frac{(\mathbb{F}_{p^m} + u\mathbb{F}_{p^m})[x]}{\langle x^{4p^s} - \gamma \rangle}$ is a local ring with unique maximal ideal $\langle x^4 - \gamma_0, u \rangle$. Such λ-constacyclic codes are then classified into 4 distinct types of ideals, and the detailed structures of ideals in each type are provided.

The key result in [582] was that, when the unit λ is not a square in $\mathcal{R}$, any nonzero polynomial of degree < 4 over $\mathbb{F}_{p^m}$ is invertible in the ambient ring $\frac{\mathcal{R}[x]}{\langle x^{4p^s} - \lambda \rangle}$ of constacyclic codes of length $4p^s$; see [582, Propositions 4.1 and 5.1]. This fact was then used to obtain the algebraic structure of all constacyclic codes of length $4p^s$ over $\mathbb{F}_{p^m} + u\mathbb{F}_{p^m}$ and their duals. However, [582, Section 6] explained that this fact is no longer true for the case $p^m \equiv 3 \pmod 4$.

In 2018, Dinh et al. [589, 590] completed the case $p^m \equiv 3 \pmod 4$ by showing that, if the unit λ is not a square, then $x^4 - \lambda_0$ can be decomposed into a product of two irreducible coprime quadratic polynomials which are $x^2 + \gamma x + \frac{\gamma^2}{2}$ and $x^2 - \gamma x + \frac{\gamma^2}{2}$, where $\lambda_0^{p^s} = \lambda$ and $\gamma^4 = -4\lambda_0$. The quotient rings $\frac{\mathcal{R}[x]}{\left\langle (x^2 + \gamma x + \frac{\gamma^2}{2})^{p^s} \right\rangle}$ and $\frac{\mathcal{R}[x]}{\left\langle (x^2 - \gamma x + \frac{\gamma^2}{2})^{p^s} \right\rangle}$ are local non-chain rings, and their ideals, which are completely classified, provide the structure of all λ-constacyclic codes and their duals.

Theorem 17.6.8 ([582, 589, 590, 593])

(a) *λ is a square in $\mathcal{R}$, i.e., $\lambda = \delta^2$. Then it follows from the Chinese Remainder Theorem that*

$$\Re(4p^s, \lambda) = \frac{\mathcal{R}[x]}{\langle x^{4p^s} - \lambda \rangle} \cong \frac{\mathcal{R}[x]}{\langle x^{2p^s} + \delta \rangle} \oplus \frac{\mathcal{R}[x]}{\langle x^{2p^s} - \delta \rangle} = \Re(2p^s, -\delta) \oplus \Re(2p^s, \delta).$$

So a λ-constacyclic code of length $4p^s$ over $\mathcal{R}$, i.e., an ideal $\mathcal{C}$ of $\Re(4p^s, \lambda)$, is represented as a direct sum $\mathcal{C} = \mathcal{C}_+ \oplus \mathcal{C}_-$ where $\mathcal{C}_+$ and $\mathcal{C}_-$ are ideals of $\Re(2p^s, -\delta)$ and $\Re(2p^s, \delta)$, respectively. Thus, the classification, detailed structure, and number of codewords of constacyclic codes $\mathcal{C}$ of length $4p^s$ over $\mathcal{R}$ can be obtained from that of the direct summands $\mathcal{C}_+$ and $\mathcal{C}_-$ in Theorem 17.6.5. Also $\mathcal{C}^\perp = \mathcal{C}_+^\perp \oplus \mathcal{C}_-^\perp$.

(b) *$\lambda = \alpha + u\beta$, with $\alpha, \beta \in \mathbb{F}_{p^m}^*$, is not a square in $\mathcal{R}$ where $p^m \equiv 1 \pmod 4$. Let $\alpha_0^{p^s} = \alpha$. Then the ambient ring*

$$\Re(4p^s, \alpha + u\beta) = \frac{\mathcal{R}[x]}{\langle x^{4p^s} - (\alpha + u\beta) \rangle}$$

is a chain ring with maximal ideal $\langle x^4 - \alpha_0 \rangle$ whose ideals are

$$\Re(4p^s, \alpha + u\beta) = \langle 1 \rangle \supsetneq \langle x^4 - \alpha_0 \rangle \supsetneq \cdots \supsetneq \langle (x^4 - \alpha_0)^{2p^s - 1} \rangle \supsetneq \langle (x^4 - \alpha_0)^{2p^s} \rangle = \langle 0 \rangle,$$

and $\langle (x^4 - \alpha_0)^{p^s} \rangle = \langle u \rangle$. In other words, the $(\alpha + u\beta)$-constacyclic codes of length $4p^s$ over $\mathcal{R}$ are the ideals $\langle (x^4 - \alpha_0)^i \rangle$, $0 \le i \le 2p^s$, of the chain ring $\Re(4p^s, \alpha + u\beta)$. Each $(\alpha + u\beta)$-constacyclic code $\mathcal{C} = \langle (x^4 - \alpha_0)^i \rangle$ has $p^{4m(2p^s - i)}$ codewords. Its dual $\mathcal{C}^\perp$ is the $(\alpha^{-1} - u\alpha^{-2}\beta)$-constacyclic code

$$\mathcal{C}^\perp = \left\langle (x^4 - \alpha_0^{-1})^{2p^s - i} \right\rangle \subseteq \Re(4p^s, \alpha^{-1} - u\alpha^{-2}\beta),$$

which contains p^{4mi} codewords.

(c) $\lambda = \gamma \in \mathbb{F}_{p^m}^*$ *is not a square in* $\mathcal{R}$ *where* $p^m \equiv 1 \pmod 4$. *Let* $\gamma_0^{p^s} = \gamma$. *Then the ambient ring*

$$\Re(4p^s, \gamma) = \frac{\mathcal{R}[x]}{\langle x^{4p^s} - \gamma \rangle}$$

is a local ring with maximal ideal $\langle x^4 - \gamma_0, u \rangle$, *but it is not a chain ring. The* γ-*constacyclic codes of length* $4p^s$ *over* $\mathcal{R}$, *i.e., ideals of the ring* $\Re(4p^s, \gamma)$, *are*

Type 1: (trivial ideals) $\langle 0 \rangle$, $\langle 1 \rangle$,

Type 2: (principal ideals with non-monic polynomial generators)

$$\langle u(x^4 - \gamma_0)^i \rangle$$

where $0 \le i \le p^s - 1$,

Type 3: (principal ideals with monic polynomial generators)

$$\langle (x^4 - \gamma_0)^i + u(x^4 - \gamma_0)^t h(x) \rangle$$

where $1 \le i \le p^s - 1, 0 \le t < i$, *and either* $h(x)$ *is* 0 *or* $h(x)$ *is a unit in* $\Re(4p^s, \gamma)$ *which can be represented as* $h(x) = \sum_j (h_{3j} x^3 + h_{2j} x^2 + h_{1j} x + h_{00})(x^4 - \gamma_0)^j$ *with* $h_{0j}, h_{1j}, h_{2j}, h_{3j} \in \mathbb{F}_{p^m}$ *and* $h_{3,0} x^3 + h_{2,0} x^2 + h_{1,0} x + h_{0,0} \ne 0$,

Type 4: (non-principal ideals)

$$\langle (x^4 - \gamma_0)^i + u(x^4 - \gamma_0)^t h(x), u(x^4 - \gamma_0)^\omega \rangle$$

with $h(x)$ *as in Type 3,* $\deg h(x) \le \omega - t - 1$, *and* $\omega < T$, *where* T *is the smallest integer such that* $u(x^4 - \gamma_0)^T \in \langle (x^4 - \gamma_0)^i + u(x^4 - \gamma_0)^t h(x) \rangle$; *i.e., such* T *can be determined as*

$$T = \begin{cases} i & \text{if } h(x) = 0, \\ \min\{i, p^s - i + t\} & \text{if } h(x) \ne 0. \end{cases}$$

(d) $\lambda = \alpha + u\beta$, *with* $\alpha, \beta \in \mathbb{F}_{p^m}^*$, *is not a square in* $\mathcal{R}$ *where* $p^m \equiv 3 \pmod 4$. *Let* $\alpha_0^{p^s} = \alpha$ *and* $\alpha_0 = -4\eta^4$. *Then* $x^4 - \alpha_0 = (x^2 + 2\eta x + 2\eta^2)(x^2 - 2\eta x + 2\eta^2)$ *is a factorization of* $x^4 - \alpha_0$ *into monic coprime irreducible factors. The ambient ring*

$$\Re(4p^s, \alpha + u\beta) = \frac{\mathcal{R}[x]}{\langle x^{4p^s} - (\alpha + u\beta) \rangle}$$

is a principal ideal ring with ideals $\langle (x^2 + 2\eta x + 2\eta^2)^i (x^2 - 2\eta x + 2\eta^2)^j \rangle$ *where* $0 \le i, j \le 2p^s$. *Equivalently, each* $(\alpha + u\beta)$-*constacyclic code of length* $4p^s$ *over* $\mathcal{R}$ *has the form*

$$\mathcal{C} = \langle (x^2 + 2\eta x + 2\eta^2)^i (x^2 - 2\eta x + 2\eta^2)^j \rangle$$

for $0 \le i, j \le 2p^s$, *which contains* $p^{m(8p^s - 2i - 2j)}$ *codewords. Its dual* $\mathcal{C}^\perp$ *is the* $(\alpha^{-1} - u\alpha^{-2}\beta)$-*constacyclic code*

$$\mathcal{C}^\perp = \left\langle \left(x^2 + \eta^{-1} x + \frac{\eta^{-2}}{2} \right)^{2p^s - i} \left(x^2 - \eta^{-1} x + \frac{\eta^{-2}}{2} \right)^{2p^s - j} \right\rangle$$

contained in $\Re(4p^s, \alpha^{-1} - u\alpha^{-2}\beta)$, *which has* $p^{m(2i + 2j)}$ *codewords.*

(e) $\lambda \in \mathbb{F}_{p^m}^*$ *is not a square in* $\mathcal{R}$ *where* $p^m \equiv 3 \pmod 4$. *Let* $\lambda_0^{p^s} = \lambda$ *and* $\gamma^4 = -4\lambda_0$. *Then* $x^4 - \lambda_0 = (x^2 + \gamma x + \frac{\gamma^2}{2})(x^2 - \gamma x + \frac{\gamma^2}{2})$ *is a factorization of* $x^4 - \lambda_0$ *into irreducible coprime quadratic polynomials. The ambient ring*

$$\Re(4p^s, \lambda) = \frac{\mathcal{R}[x]}{\langle x^{4p^s} - \lambda \rangle} \cong \bigoplus_{\delta \in \{-1,1\}} \frac{\mathcal{R}[x]}{\left\langle \left(x^2 + \delta\gamma x + \frac{\gamma^2}{2} \right)^{p^s} \right\rangle}.$$

The λ-*constacyclic codes of length* $4p^s$ *over* $\mathcal{R}$ *can be expressed as* $\mathcal{C} = \mathcal{C}_1 \oplus \mathcal{C}_{-1}$ *where* $\mathcal{C}_\delta$ *are ideals of the quotient ring* $\frac{\mathcal{R}[x]}{\left\langle \left(x^2 + \delta\gamma x + \frac{\gamma^2}{2} \right)^{p^s} \right\rangle}$. *For* $\delta \in \{-1, 1\}$, *the ring* $\frac{\mathcal{R}[x]}{\left\langle \left(x^2 + \delta\gamma x + \frac{\gamma^2}{2} \right)^{p^s} \right\rangle}$ *is a local, non-chain ring, whose ideals are as follows:*

Type 1: (trivial ideals) $\langle 0 \rangle$, $\langle 1 \rangle$,

Type 2: (principal ideals with non-monic polynomial generators)

$$\left\langle u \left(x^2 + \delta\gamma x + \frac{\gamma^2}{2} \right)^i \right\rangle$$

where $0 \le i \le p^s - 1$,

Type 3: (principal ideals with monic polynomial generators)

$$\left\langle \left(x^2 + \delta\gamma x + \frac{\gamma^2}{2} \right)^i + u \left(x^2 + \delta\gamma x + \frac{\gamma^2}{2} \right)^t h(x) \right\rangle$$

where $1 \le i \le p^s - 1, 0 \le t < i$, *and either* $h(x)$ *is* 0 *or* $h(x)$ *is a unit in* $\frac{\mathcal{R}[x]}{\left\langle \left(x^2 + \delta\gamma x + \frac{\gamma^2}{2} \right)^{p^s} \right\rangle}$ *which can be represented as* $h(x) = \sum_j (h_{0j}x + h_{1j})(x^2 + \delta\gamma x + \frac{\gamma^2}{2})^j$ *with* $h_{0j}, h_{1j} \in \mathbb{F}_{p^m}$ *and* $h_{00}x + h_{10} \ne 0$,

Type 4: (non-principal ideals)

$$\left\langle \left(x^2 + \delta\gamma x + \frac{\gamma^2}{2} \right)^i + u \left(x^2 + \delta\gamma x + \frac{\gamma^2}{2} \right)^t h(x), u \left(x^2 + \delta\gamma x + \frac{\gamma^2}{2} \right)^\omega \right\rangle$$

with $h(x)$ *as in Type 3,* $\deg h(x) \le \omega - t - 1$, *and* $\omega < T$, *where* T *is the smallest integer such that*

$$u(x^2 + \delta\gamma x + \frac{\gamma^2}{2})^T \in \left\langle (x^2 + \delta\gamma x + \frac{\gamma^2}{2})^i + u(x^2 + \delta\gamma x + \frac{\gamma^2}{2})^t h(x) \right\rangle;$$

such T *can be determined as*

$$T = \begin{cases} i & \text{if } h(x) = 0, \\ \min\{i, p^s - i + t\} & \text{if } h(x) \ne 0. \end{cases}$$

Theorem 17.6.9 ([589]) *Let* $\gamma \in \mathbb{F}_{p^m}^*$, *and* $\mathcal{C}$ *be a* γ-*constacyclic code of length* $4p^s$ *over* $\mathcal{R}$ *in Theorem 17.6.8(c). Then the number of codewords of* $\mathcal{C}$, *denoted by* $n_\mathcal{C}$, *is determined as follows.*

- *If* $\mathcal{C} = \langle 0 \rangle$, *then* $n_\mathcal{C} = 1$.

- If $\mathcal{C} = \langle 1 \rangle$, then $n_{\mathcal{C}} = p^{8mp^s}$.

- If $\mathcal{C} = \langle u(x^4 - \gamma_0)^i \rangle$ where $0 \leq i \leq p^s - 1$, then $n_{\mathcal{C}} = p^{4m(p^s - i)}$.

- If $\mathcal{C} = \langle (x^4 - \gamma_0)^i \rangle$, where $1 \leq i \leq p^s - 1$, then $n_{\mathcal{C}} = p^{8m(p^s - i)}$.

- If $\mathcal{C} = \langle (x^4 - \gamma_0)^i + u(x^4 - \gamma_0)^t h(x) \rangle$ where $1 \leq i \leq p^s - 1$, $0 \leq t < i$, and $h(x)$ is a unit, then

$$
n_{\mathcal{C}} = \begin{cases} p^{8m(p^s - i)} & \text{if } 1 \leq i \leq p^{s-1} + \frac{t}{2}, \\ p^{4m(p^s - i)} & \text{if } p^{s-1} + \frac{t}{2} < i \leq p^s - 1. \end{cases}
$$

- If $\mathcal{C} = \langle (x^4 - \gamma_0)^i + u(x^4 - \gamma_0)^t h(x), u(x^4 - \gamma_0)^\omega \rangle$ where $1 \leq i \leq p^s - 1$, $0 \leq t < i$, and either $h(x)$ is 0 or $h(x)$ is a unit, then $n_{\mathcal{C}} = p^{4m(2p^s - i - \omega)}$.

17.6.4 λ-Constacyclic Codes of Length np^s over $\mathcal{R}$, $\lambda \in \mathbb{F}_{p^m}^*$

In this subsection, p is a prime including $p = 2$. For any unit $\lambda \in \mathbb{F}_{p^m}^*$, [341] provides an explicit representation for all distinct λ-constacyclic codes of length np^s over $\mathcal{R}$ by a canonical form decomposition for each code, where s and n are arbitrary positive integers with $\gcd(p, n) = 1$.

Since $\lambda \in \mathbb{F}_{p^m}^*$ and $\mathbb{F}_{p^m}^*$ is a multiplicative cyclic group of order $p^m - 1$, there is an element $\lambda_0 \in \mathbb{F}_{p^m}^*$ such that $\lambda_0^{p^s} = \lambda$. This implies that $x^{np^s} - \lambda = (x^n - \lambda_0)^{p^s}$ in $\mathbb{F}_{p^m}[x]$. As $\gcd(p, n) = 1$, there are pairwise coprime monic irreducible polynomials $f_1(x), \dots, f_r(x)$ in $\mathbb{F}_{p^m}[x]$ such that $x^n - \lambda_0 = f_1(x) \cdots f_r(x)$. Hence

$$
x^{np^s} - \lambda = (x^n - \lambda_0)^{p^s} = f_1(x)^{p^s} \cdots f_r(x)^{p^s}.
$$

For any integer j, $1 \leq j \leq r$, we assume $\deg(f_j(x)) = d_j$ and denote $F_j(x) = \frac{x^n - \lambda_0}{f_j(x)}$. Then $F_j(x)^{p^s} = \frac{x^{np^s} - \lambda}{f_j(x)^{p^s}}$ and $\gcd(F_j(x), f_j(x)) = 1$. Hence there exist $v_j(x), w_j(x) \in \mathbb{F}_{p^m}[x]$ such that $\deg(v_j(x)) < \deg(f_j(x)) = d_j$ and $v_j(x) F_j(x) + w_j(x) f_j(x) = 1$. This implies

$$
v_j(x)^{p^s} F_j(x)^{p^s} + w_j(x)^{p^s} f_j(x)^{p^s} = (v_j(x) F_j(x) + w_j(x) f_j(x))^{p^s} = 1.
$$

Denote

- $\Re(np^s, \lambda) = \mathcal{R}[x] / \langle x^{np^s} - \lambda \rangle$;

- $\mathcal{A} = \mathbb{F}_{p^m}[x] / \langle x^{np^s} - \lambda \rangle$ and $\mathcal{A}[u] / \langle u^2 \rangle = \mathcal{A} + u\mathcal{A}$ $(u^2 = 0)$;

- $\mathcal{K}_j = \mathbb{F}_{p^m}[x] / \langle f_j(x)^{p^s} \rangle$ and $\mathcal{K}_j[u] / \langle u^2 \rangle = \mathcal{K}_j + u\mathcal{K}_j$ $(u^2 = 0)$;

- $\varepsilon_j(x) \in \mathcal{A}$ defined by

$$
\varepsilon_j(x) = v_j(x)^{p^s} F_j(x)^{p^s} = 1 - w_j(x)^{p^s} f_j(x)^{p^s} \pmod{x^{np^s} - \lambda}.
$$

A direct sum decomposition for a λ-constacyclic code over $\mathcal{R}$ is derived as follows.

Theorem 17.6.10 ([341, Theorem 2.5]) *Let $\mathcal{C} \subseteq \Re(np^s, \lambda)$. Then the following statements are equivalent.*

(a) *$\mathcal{C}$ is a λ-constacyclic code over $\mathcal{R}$ of length np^s, i.e. an ideal of $\Re(np^s, \lambda)$.*

(b) *$\mathcal{C}$ is an ideal of $\mathcal{A} + u\mathcal{A}$.*

(c) *For each integer j, $1 \le j \le r$, there is a unique ideal $\mathcal{C}_j$ of $\mathcal{K}_j + u\mathcal{K}_j$ such that*
$$\mathcal{C} = \bigoplus_{j=1}^{r} \varepsilon_j(x)\mathcal{C}_j \pmod{x^{np^s} - \lambda}.$$

Hence, in order to determine all distinct λ-constacyclic codes of length np^s over $\mathcal{R}$, by Theorem 17.6.10, it is sufficient to list all distinct ideals of the ring $\mathcal{K}_j + u\mathcal{K}_j$ ($u^2 = 0$) for all $j = 1, \ldots, r$.

Theorem 17.6.11 ([341, Theorem 3.8]) *All distinct ideals $\mathcal{C}_j$ of the ring $\mathcal{K}_j + u\mathcal{K}_j$ ($u^2 = 0$) are given by one of the following five cases:*

Case I: $p^{\left(p^s - \lceil \frac{p^s}{2} \rceil\right)md_j}$ *ideals:*
$$\mathcal{C}_j = \langle f_j(x)b(x) + u \rangle \text{ with } |\mathcal{C}_j| = p^{md_j p^s},$$

where $b(x) \in f_j(x)^{\lceil \frac{p^s}{2} \rceil - 1}(\mathcal{K}_j / \langle f_j(x)^{p^s - 1} \rangle)$.

Case II: $\sum_{k=1}^{p^s - 1} p^{\left(p^s - k - \lceil \frac{1}{2}(p^s - k) \rceil\right)md_j}$ *ideals:*
$$\mathcal{C}_j = \langle f_j(x)^{k+1}b(x) + uf_j(x)^k \rangle \text{ with } |\mathcal{C}_j| = p^{md_j(p^s - k)},$$

where $b(x) \in f_j(x)^{\lceil \frac{p^s - k}{2} \rceil - 1}(\mathcal{K}_j / \langle f_j(x)^{p^s - k - 1} \rangle)$ and $1 \le k \le p^s - 1$.

Case III: $p^s + 1$ *ideals: $\mathcal{C}_j = \langle f_j(x)^k \rangle$ with $|\mathcal{C}_j| = p^{2md_j(p^s - k)}$, $0 \le k \le p^s$.*

Case IV: $\sum_{t=1}^{p^s - 1} p^{\left(t - \lceil \frac{t}{2} \rceil\right)md_j}$ *ideals:*
$$\mathcal{C}_j = \langle f_j(x)b(x) + u, f_j(x)^t \rangle \text{ with } |\mathcal{C}_j| = p^{md_j(2p^s - t)},$$

where $b(x) \in f_j(x)^{\lceil \frac{t}{2} \rceil - 1}(\mathcal{K}_j / \langle f_j(x)^{t-1} \rangle)$, $1 \le t \le p^s - 1$.

Case V: $\sum_{k=1}^{p^s - 2} \sum_{t=1}^{p^s - k - 1} p^{\left(t - \lceil \frac{t}{2} \rceil\right)md_j}$ *ideals:*
$$\mathcal{C}_j = \langle f_j(x)^{k+1}b(x) + uf_j(x)^k, f_j(x)^{k+t} \rangle \text{ with } |\mathcal{C}_j| = p^{md_j(2p^s - 2k - t)},$$

where $b(x) \in f_j(x)^{\lceil \frac{t}{2} \rceil - 1}(\mathcal{K}_j / \langle f_j(x)^{t-1} \rangle)$, $1 \le t \le p^s - k - 1$ and $1 \le k \le p^s - 2$.

Moreover, let $N_{(p^m, d_j, p^s)}$ be the number of ideals in $\mathcal{K}_j + u\mathcal{K}_j$. Then
$$N_{(p^m, d_j, p^s)} = \begin{cases} \sum_{i=0}^{2^{s-1}} (1 + 4i)2^{(2^{s-1} - i)md_j} & \text{if } p = 2, \\ \sum_{i=0}^{\frac{p^s - 1}{2}} (3 + 4i)p^{(\frac{p^s - 1}{2} - i)md_j} & \text{if } p \text{ is odd.} \end{cases}$$

For the rest of this subsection, we adopt the following notations and definitions. Let $\lambda, \lambda_0 \in \mathbb{F}_{p^m}^*$ satisfying $\lambda_0^{p^s} = \lambda$. Recall that $f^*(x) = x^{\deg(f)}f(x^{-1})$ is the reciprocal polynomial of $f(x) \in \mathbb{F}_{p^m}[x]$; see Section 17.3. Note that $(fg)^* = f^*g^*$. Denote

- $\widehat{\mathcal{A}} = \mathbb{F}_{p^m}[x]/\langle x^{np^s} - \lambda^{-1} \rangle$ and $\widehat{\mathcal{A}}[u]/\langle u^2 \rangle = \widehat{\mathcal{A}} + u\widehat{\mathcal{A}}$ ($u^2 = 0$);

- $\widehat{\mathcal{K}}_j = \mathbb{F}_{p^m}[x]/\langle f_j^*(x)^{p^s} \rangle$ and $\widehat{\mathcal{K}}_j[u]/\langle u^2 \rangle = \widehat{\mathcal{K}}_j + u\widehat{\mathcal{K}}_j$ ($u^2 = 0$);

- $\widehat{\Psi} : \widehat{\mathcal{A}} + u\widehat{\mathcal{A}} \to \mathcal{R}[x]/\langle x^{np^s} - \lambda^{-1} \rangle$ via

$$\widehat{\Psi} : g_0(x) + ug_1(x) \mapsto \sum_{i=0}^{np^s - 1} (g_{i,0} + ug_{i,1})x^i$$

for all $g_k(x) = \sum_{i=0}^{np^s - 1} g_{i,k}x^i \in \widehat{\mathcal{A}}$ with $g_{i,k} \in \mathbb{F}_{p^m}$, $0 \le i \le np^s - 1$ and $k = 0, 1$.

It can be easily verified that $\widehat{\Psi}$ is a ring isomorphism from $\widehat{\mathcal{A}} + u\widehat{\mathcal{A}}$ onto $\Re(np^s, \lambda^{-1}) = \mathcal{R}[x]/\langle x^{np^s} - \lambda^{-1}\rangle$. Then we will identify $\widehat{\mathcal{A}} + u\widehat{\mathcal{A}}$ with $\Re(np^s, \lambda^{-1})$ under $\widehat{\Psi}$. As $x^{np^s} = \lambda^{-1}$ in the ring $\widehat{\mathcal{A}}$, we have

$$x^{-1} = \lambda x^{np^s-1} \text{ in } \widehat{\mathcal{A}} \subset \Re(np^s, \lambda^{-1}).$$

Now we define a map $\tau : \mathcal{A} \to \widehat{\mathcal{A}}$ via

$$\tau : a(x) \mapsto a(x^{-1}) = a(\lambda x^{np^s-1}) \ (\text{mod } x^{np^s} - \lambda^{-1})$$

for all $a(x) \in \mathcal{A}$. Then one can easily verify that τ is a ring isomorphism from $\mathcal{A}$ onto $\widehat{\mathcal{A}}$ and can be extended to a ring isomorphism from $\mathcal{A} + u\mathcal{A}$ onto $\widehat{\mathcal{A}} + u\widehat{\mathcal{A}}$ in the natural way, namely

$$\tau : \rho(x) \mapsto \rho(x^{-1}) = a(x^{-1}) + ub(x^{-1})$$

for all $\rho(x) = a(x) + ub(x)$ with $a(x), b(x) \in \mathcal{A}$.

Clearly, for $1 \leq j \leq r$, $f_j^*(x)^{p^s}$ is a divisor of $x^{np^s} - \lambda^{-1}$ in $\mathbb{F}_{p^m}[x]$. This implies $x^{np^s} \equiv \lambda^{-1} \ (\text{mod } f_j^*(x)^{p^s})$. Hence $x^{-1} = \lambda x^{np^s-1}$ in the rings $\widehat{\mathcal{K}}_j$ and $\widehat{\mathcal{K}}_j + u\widehat{\mathcal{K}}_j$ as well. Moreover, we define

- $\widehat{\varepsilon}_j(x) = \tau(\varepsilon_j(x)) = \varepsilon_j(x^{-1}) = \varepsilon_j(\lambda x^{np^s-1}) \ (\text{mod } x^{np^s} - \lambda^{-1})$;

- $\widehat{\Phi} : (\widehat{\mathcal{K}}_1 + u\widehat{\mathcal{K}}_1) \times \cdots \times (\widehat{\mathcal{K}}_r + u\widehat{\mathcal{K}}_r) \to \widehat{\mathcal{A}} + u\widehat{\mathcal{A}}$ via

$$\widehat{\Phi} : (\xi_1 + u\eta_1, \ldots, \xi_r + u\eta_r) \mapsto \sum_{j=1}^{r} \widehat{\varepsilon}_j(x)(\xi_j + u\eta_j) \ (\text{mod } x^{np^s} - \lambda^{-1})$$

  for all $\xi_j, \eta_j \in \widehat{\mathcal{K}}_j$, $j = 1, \ldots, r$;

- $\tau_j : \mathcal{K}_j + u\mathcal{K}_j \to \widehat{\mathcal{K}}_j + u\widehat{\mathcal{K}}_j$ via

$$\tau_j : \xi \mapsto a(x^{-1}) + ub(x^{-1}) = a(\lambda x^{np^s-1}) + ub(\lambda x^{np^s-1}) \ (\text{mod } f_j^*(x)^{p^s}),$$

  for all $\xi = a(x) + ub(x) \in \mathcal{K}_j + u\mathcal{K}_j$ with $a(x), b(x) \in \mathcal{K}_j$.

It follows that $\widehat{\Phi}$ is a ring isomorphism from $(\widehat{\mathcal{K}}_1 + u\widehat{\mathcal{K}}_1) \times \cdots \times (\widehat{\mathcal{K}}_r + u\widehat{\mathcal{K}}_r)$ onto $\widehat{\mathcal{A}} + u\widehat{\mathcal{A}}$. Moreover, we have the following.

Lemma 17.6.12 ([341, Lemmas 4.2, 4.3])

(a) τ_j *is a ring isomorphism from* $\mathcal{K}_j + u\mathcal{K}_j$ *onto* $\widehat{\mathcal{K}}_j + u\widehat{\mathcal{K}}_j$ *satisfying* $\tau(\varepsilon_j(x)\xi) = \widehat{\varepsilon}_j(x)\tau_j(\xi)$ *for all* $\xi \in \mathcal{K}_j + u\mathcal{K}_j$.

(b) *Let* $\mathbf{a} = (a_0, a_1, \ldots, a_{p^s n-1})$ *and* $\mathbf{b} = (b_0, b_1, \ldots, b_{p^s n-1})$ *be in* $\mathcal{R}^{np^s}$. *Set*

$$a(x) = \sum_{i=0}^{np^s-1} a_i x^i \in \mathcal{A} + u\mathcal{A} \quad \text{and} \quad b(x) = \sum_{i=0}^{np^s-1} b_i x^i \in \widehat{\mathcal{A}} + u\widehat{\mathcal{A}}.$$

If $\tau(a(x))b(x) = 0$ *in* $\widehat{\mathcal{A}} + u\widehat{\mathcal{A}}$, *then* $\mathbf{a} \cdot \mathbf{b} = \sum_{i=0}^{np^s-1} a_i b_i = 0$.

The dual of each λ-constacyclic code of length np^s over $\mathcal{R}$ can be obtained from its canonical form decomposition.

Theorem 17.6.13 ([341, Theorem 4.4]) *Let $\mathcal{C}$ be a λ-constacyclic code of length np^s over $\mathcal{R}$ with canonical form decomposition $\mathcal{C} = \bigoplus_{j=1}^{r} \varepsilon_j(x)\mathcal{C}_j$ (mod $x^{p^s n} - \lambda$), where $\mathcal{C}_j$ is an ideal of $\mathcal{K}_j + u\mathcal{K}_j$ given in Theorem 17.6.11. Then the dual code $\mathcal{C}^\perp$ of $\mathcal{C}$ is a λ^{-1}-constacyclic code of length np^s over $\mathcal{R}$ with canonical form decomposition*

$$\mathcal{C}^\perp = \bigoplus_{j=1}^{r} \widehat{\varepsilon}_j(x)\widehat{\mathcal{D}}_j \pmod{x^{np^s} - \lambda^{-1}},$$

where $\widehat{\mathcal{D}}_j$ is an ideal of $\widehat{\mathcal{K}}_j + u\widehat{\mathcal{K}}_j$ given by one of the following five cases:

Case I: $\widehat{\mathcal{D}}_j = \langle -\lambda x^{p^s n - d_j} f_j^*(x) b(x^{-1}) + u \rangle$, *if* $\mathcal{C}_j = \langle f_j(x)b(x) + u \rangle$ *where*

$$b(x) \in f_j(x)^{\lceil \frac{p^s}{2} \rceil - 1}(\mathcal{K}_j / \langle f_j(x)^{p^s - 1} \rangle).$$

Case II: $\widehat{\mathcal{D}}_j = \langle -\lambda x^{p^s n - d_j} f_j^*(x) b(x^{-1}) + u, f_j^*(x)^{p^s - k} \rangle$, *if* $\mathcal{C}_j = \langle f_j(x)^{k+1} b(x) + u f_j(x)^k \rangle$ *where* $b(x) \in f_j(x)^{\lceil \frac{p^s - k}{2} \rceil - 1}(\mathcal{K}_j / \langle f_j(x)^{p^s - k - 1} \rangle)$ *and* $1 \leq k \leq p^s - 1$.

Case III: $\widehat{\mathcal{D}}_j = \langle f_j^*(x)^{p^s - k} \rangle$, *if* $\mathcal{C}_j = \langle f_j(x)^k \rangle$ *where* $0 \leq k \leq p^s$.

Case IV: $\widehat{\mathcal{D}}_j = \langle -\lambda x^{p^s n - d_j} f_j^*(x)^{p^s - t + 1} b(x^{-1}) + u f_j^*(x)^{p^s - t} \rangle$, *if* $\mathcal{C}_j = \langle f_j(x)b(x) + u, f_j(x)^t \rangle$ *where* $1 \leq t \leq p^s - 1$ *and* $b(x) \in f_j(x)^{\lceil \frac{t}{2} \rceil - 1}(\mathcal{K}_j / \langle f_j(x)^{t-1} \rangle)$.

Case V: $\widehat{\mathcal{D}}_j = \langle -\lambda x^{p^s n - d_j} f_j^*(x)^{p^s - k - t + 1} b(x^{-1}) + u f_j^*(x)^{p^s - k - t}, f_j^*(x)^{p^s - k} \rangle$, *if*

$$\mathcal{C}_j = \langle f_j(x)^{k+1} b(x) + u f_j(x)^k, f_j(x)^{k+t} \rangle$$

where $1 \leq t \leq p^s - k - 1$, $1 \leq k \leq p^s - 2$, *and* $b(x) \in f_j(x)^{\lceil \frac{t}{2} \rceil - 1}(\mathcal{K}_j / \langle f_j(x)^{t-1} \rangle)$.

In the rest of this subsection we determine all self-dual cyclic codes and negacyclic codes of length np^s over $\mathcal{R}$. Let $\nu \in \{1, -1\}$ and $\Re(np^s, \nu) = \mathcal{R}[x]/\langle x^{np^s} - \nu \rangle$. Then $\mathcal{C}$ is a cyclic, respectively negacyclic, code if and only if $\mathcal{C}$ is an ideal of the ring $\Re(np^s, \nu)$, i.e., a ν-constacyclic code over $\mathcal{R}$ of length np^s, where $\nu = 1$, respectively $\nu = -1$.

Using the notations above, as $\nu^{-1} = \nu$, we have $\Re(np^s, \nu) = \Re(np^s, \nu^{-1})$ and $\mathcal{A} = \widehat{\mathcal{A}} = \mathbb{F}_{p^m}[x]/\langle x^{p^s n} - \nu \rangle$. Hence the map $\tau : \mathcal{A} \to \mathcal{A}$ defined by $\tau(a(x)) = a(x^{-1}) = a(\nu x^{np^s - 1})$ (mod $x^{np^s} - \nu$) for all $a(x) \in \mathcal{A}$ is a ring automorphism on $\mathcal{A}$ satisfying $\tau^{-1} = \tau$.

We then have

$$x^{np^s} - \nu = f_1(x)^{p^s} \cdots f_r(x)^{p^s} \quad \text{and}$$
$$x^{np^s} - \nu = -\nu(x^{p^s n} - \nu)^* = -\nu f_1^*(x)^{p^s} \cdots f_r^*(x)^{p^s}.$$

Since $f_1(x), \ldots, f_r(x)$ are pairwise coprime monic irreducible polynomials in $\mathbb{F}_{p^m}[x]$, for each $1 \leq j \leq r$ there is a unique integer j', $1 \leq j' \leq r$, such that $f_j^*(x) = \delta_j f_{j'}(x)$ where $\delta_j = f_j(0)^{-1} \in \mathbb{F}_{p^m}^*$. In the following, we still denote the bijection $j \mapsto j'$ on the set $\{1, \ldots, r\}$ by τ, i.e.,

$$f_j^*(x) = \delta_j f_{\tau(j)}(x).$$

Whether τ denotes the automorphism of $\mathcal{A}$ or this map on $\{1, \ldots, r\}$ is determined by context. The next lemma shows the compatibility of the two uses of τ.

Lemma 17.6.14 ([341, Lemma 5.1]) *The following hold.*

(a) *τ is a permutation on the set $\{1, \ldots, r\}$ satisfying $\tau^{-1} = \tau$.*

(b) *After a suitable rearrangement of $f_1(x), \ldots, f_r(x)$, there are nonnegative integers ρ, ϵ such that $\rho + 2\epsilon = r$, $\tau(j) = j$ for all $j = 1, \ldots, \rho$, $\tau(\rho + i) = \rho + \epsilon + i$ and $\tau(\rho + \epsilon + i) = \rho + i$ for all $i = 1, \ldots, \epsilon$.*

(c) *For any $1 \leq j \leq r$, $\tau(\varepsilon_j(x)) = \widehat{\varepsilon}_j(x) = \varepsilon_j(x^{-1}) = \varepsilon_{\tau(j)}(x)$ in the ring $\mathcal{A}$.*

(d) *For any $1 \leq j \leq r$, the map $\tau_j : \mathcal{K}_j + u\mathcal{K}_j \to \mathcal{K}_{\tau(j)} + u\mathcal{K}_{\tau(j)}$ defined by*

$$\tau_j(a(x) + ub(x)) = a(x^{-1}) + ub(x^{-1}),$$

for all $a(x), b(x) \in \mathcal{K}_j$, is a ring isomorphism from $\mathcal{K}_j + u\mathcal{K}_j$ onto $\mathcal{K}_{\tau(j)} + u\mathcal{K}_{\tau(j)}$ with inverse $\tau_j^{-1} = \tau_{\tau(j)}$. Moreover, for any $\xi \in \mathcal{K}_j + u\mathcal{K}_j$ we have

$$\tau(\varepsilon_j(x)\xi) = \varepsilon_{\tau(j)}(x)\tau_j(\xi).$$

By Theorem 17.6.13 and Lemma 17.6.14, we have the following decomposition for dual codes of cyclic and negacyclic codes over $\mathcal{R}$ of length np^s.

Corollary 17.6.15 ([341, Corollary]) *Let $\nu \in \{1, -1\}$ and $\mathcal{C}$ be a ν-constacyclic code over $\mathcal{R}$ of length np^s with canonical form decomposition $\mathcal{C} = \bigoplus_{j=1}^{r} \varepsilon_j(x)\mathcal{C}_j \pmod{x^{p^s n} - \nu}$, where $\mathcal{C}_j$ is an ideal of $\mathcal{K}_j + u\mathcal{K}_j$ given in Theorem 17.6.11. Then the dual code $\mathcal{C}^{\perp}$ of $\mathcal{C}$ is a ν-constacyclic code over $\mathcal{R}$ of length np^s with canonical form decomposition*

$$\mathcal{C}^{\perp} = \bigoplus_{j=1}^{r} \varepsilon_{\tau(j)}(x)\mathcal{D}_{\tau(j)} \pmod{x^{np^s} - \nu},$$

where $\mathcal{D}_{\tau(j)}$ is an ideal of $\mathcal{K}_{\tau(j)} + u\mathcal{K}_{\tau(j)}$ given by one of the following five cases:

Case I: *If $\mathcal{C}_j = \langle f_j(x)b(x) + u \rangle$ where $b(x) \in f_j(x)^{\lceil \frac{p^s}{2} \rceil - 1}(\mathcal{K}_j / \langle f_j(x)^{p^s - 1} \rangle)$, then*

$$\mathcal{D}_{\tau(j)} = \langle -\nu\delta_j x^{p^s n - d_j} f_{\tau(j)}(x)b(x^{-1}) + u \rangle.$$

Case II: *If $\mathcal{C}_j = \langle f_j(x)^{k+1}b(x) + uf_j(x)^k \rangle$ where $b(x) \in f_j(x)^{\lceil \frac{p^s - k}{2} \rceil - 1}(\mathcal{K}_j / \langle f_j(x)^{p^s - k - 1} \rangle)$ and $1 \leq k \leq p^s - 1$, then*

$$\mathcal{D}_{\tau(j)} = \langle -\nu\delta_j x^{p^s n - d_j} f_{\tau(j)}(x)b(x^{-1}) + u, \ f_{\tau(j)}(x)^{p^s - k} \rangle.$$

Case III: *If $\mathcal{C}_j = \langle f_j(x)^k \rangle$ where $0 \leq k \leq p^s$, then $\mathcal{D}_{\tau(j)} = \langle f_{\tau(j)}(x)^{p^s - k} \rangle$.*

Case IV: *If $\mathcal{C}_j = \langle f_j(x)b(x) + u, \ f_j(x)^t \rangle$ where $b(x) \in f_j(x)^{\lceil \frac{t}{2} \rceil - 1}(\mathcal{K}_j / \langle f_j(x)^{t-1} \rangle)$ and $1 \leq t \leq p^s - 1$, then*

$$\mathcal{D}_{\tau(j)} = \langle -\nu\delta_j x^{p^s n - d_j} f_{\tau(j)}(x)^{p^s - t + 1} b(x^{-1}) + uf_{\tau(j)}(x)^{p^s - t} \rangle.$$

Case V: *If $\mathcal{C}_j = \langle f_j(x)^{k+1}b(x) + uf_j(x)^k, \ f_j(x)^{k+t} \rangle$ where $1 \leq t \leq p^s - k - 1$, $1 \leq k \leq p^s - 2$, and $b(x) \in f_j(x)^{\lceil \frac{t}{2} \rceil - 1}(\mathcal{K}_j / \langle f_j(x)^{t-1} \rangle)$, then*

$$\mathcal{D}_{\mu(j)} = \langle -\nu\delta_j x^{p^s n - d_j} f_{\tau(j)}(x)^{p^s - k - t + 1} b(x^{-1}) + uf_{\tau(j)}(x)^{p^s - k - t}, \ f_{\tau(j)}(x)^{p^s - k} \rangle.$$

Using these structures, all self-dual cyclic and negacyclic codes of length $p^s n$ over $\mathcal{R}$ can be determined as follows.

Theorem 17.6.16 ([341, Theorem 5.3]) *Let $\nu \in \{-1, 1\}$ and $x^{-1} = \nu x^{np^s - 1}$. All distinct self-dual ν-constacyclic codes of length $p^s n$ over $\mathcal{R}$ have decomposition*

$$\mathcal{C} = \bigoplus_{j=1}^{r} \varepsilon_j(x)\mathcal{C}_j \pmod{x^{np^s} - \nu},$$

where $\mathcal{C}_j$ is an ideal of $\mathcal{K}_j + u\mathcal{K}_j$ given by one of the following two cases.

Case I: *If* $1 \le j \le \rho$, $\mathcal{C}_j$ *is given by one of the following three subcases.*

I-1: $\mathcal{C}_j = \langle f_j(x)b(x) + u \rangle$ *where* $b(x) \in f_j(x)^{\lceil \frac{p^s}{2} \rceil - 1}(\mathcal{K}_j/\langle f_j(x)^{p^s-1} \rangle)$ *satisfying* $b(x) + \nu\delta_j x^{p^s n - d_j} b(x^{-1}) \equiv 0 \pmod{f_j(x)^{p^s - 1}}$.

I-2: $\mathcal{C}_j = \langle f_j(x)^k \rangle$ *where* k *is an integer satisfying* $2k = p^s$.

I-3: $\mathcal{C}_j = \langle f_j(x)^{k+1}b(x) + uf_j(x)^k, f_j(x)^{k+t} \rangle$ *where* $1 \le t \le p^s - k - 1$, $1 \le k \le p^s - 2$, *and* $b(x) \in f_j(x)^{\lceil \frac{t}{2} \rceil - 1}(\mathcal{K}_j/\langle f_j(x)^{t-1} \rangle)$ *satisfying* $p^s = 2k + t$ *and* $b(x) + \nu\delta_j x^{p^s n - d_j} b(x^{-1}) \equiv 0 \pmod{f_j(x)^{t-1}}$.

Case II: *If* $j = \rho + i$ *where* $1 \le i \le \epsilon$, *the pair* $(\mathcal{C}_j, \mathcal{C}_{j+\epsilon})$ *of ideals is given by one of the following five subcases.*

II-1: $\mathcal{C}_j = \langle f_j(x)b(x) + u \rangle$ *and* $\mathcal{C}_{j+\epsilon} = \langle -\nu\delta_j x^{p^s n - d_j} f_{j+\epsilon}(x)b(x^{-1}) + u \rangle$ *where* $b(x) \in f_j(x)^{\lceil \frac{p^s}{2} \rceil - 1}(\mathcal{K}_j/\langle f_j(x)^{p^s-1} \rangle)$.

II-2: $\mathcal{C}_j = \langle f_j(x)^{k+1}b(x) + uf_j(x)^k \rangle$ *and*

$$\mathcal{C}_{j+\epsilon} = \langle -\nu\delta_j x^{p^s n - d_j} f_{j+\epsilon}(x)b(x^{-1}) + u, \ f_{j+\epsilon}(x)^{p^s - k} \rangle$$

where $b(x) \in f_j(x)^{\lceil \frac{p^s - k}{2} \rceil - 1}(\mathcal{K}_j/\langle f_j(x)^{p^s - k - 1} \rangle)$ *and* $1 \le k \le p^s - 1$.

II-3: $\mathcal{C}_j = \langle f_j(x)^k \rangle$ *and* $\mathcal{C}_{j+\epsilon} = \langle f_{j+\epsilon}(x)^{p^s - k} \rangle$ *where* $0 \le k \le p^s$.

II-4: $\mathcal{C}_j = \langle f_j(x)b(x) + u, \ f_j(x)^t \rangle$ *and*

$$\mathcal{C}_{j+\epsilon} = \langle -\nu\delta_j x^{p^s n - d_j} f_{j+\epsilon}(x)^{p^s - t + 1} b(x^{-1}) + uf_{j+\epsilon}(x)^{p^s - t} \rangle$$

where $b(x) \in f_j(x)^{\lceil \frac{t}{2} \rceil - 1}(\mathcal{K}_j/\langle f_j(x)^{t-1} \rangle)$ *and* $1 \le t \le p^s - 1$.

II-5: $\mathcal{C}_j = \langle f_j(x)^{k+1}b(x) + uf_j(x)^k, \ f_j(x)^{k+t} \rangle$ *and*

$$\mathcal{C}_{j+\epsilon} = \langle -\nu\delta_j x^{p^s n - d_j} f_{j+\epsilon}(x)^{p^s - k - t + 1} b(x^{-1}) + uf_{j+\epsilon}(x)^{p^s - k - t}, \ f_{j+\epsilon}(x)^{p^s - k} \rangle$$

where $b(x) \in f_j(x)^{\lceil \frac{t}{2} \rceil - 1}(\mathcal{K}_j/\langle f_j(x)^{t-1} \rangle)$, $1 \le t \le p^s - k - 1$, *and* $1 \le k \le p^s - 2$.

17.7 Extensions

In this section we briefly mention a few alternative directions in which the theories studied here have been extended.

In information theory, the traditional way to analyze noisy channels is to divide the message into information units called symbols. The research on the process of writing and reading is often presumed to be performed on individual symbols. However, in some of today's emerging storage technologies, one finds that symbols can only be written and read in possibly overlapping groups. Then, symbol-pair read channels were first studied by Cassuto and Blaum [365, 366], where the outputs of the read process are pairs of consecutive symbols. After that, their results were generalized by Yaakobi et al. [1918] to b-symbol read channels, where the read operation is performed as a consecutive sequence of b symbols.

Let Ξ be an alphabet of size q, whose elements are called **symbols**. Suppose that $\mathbf{x} = (x_0, x_1, \ldots, x_{n-1})$ is a vector in Ξ^n; in [543], Ding et al. defined the b-**symbol read vector** of $\mathbf{x}$ as

$$\pi_b(\mathbf{x}) = \big((x_0, \ldots, x_{b-1}), (x_1, \ldots, x_b), \ldots, (x_{n-1}, x_0, \ldots, x_{b-2})\big) \in (\Xi^b)^n$$

where $b \geq 1$. For any two vectors $\mathbf{x}$ and $\mathbf{y}$, the b-**distance** between $\mathbf{x}$ and $\mathbf{y}$ is defined as

$$d_b(\mathbf{x}, \mathbf{y}) = \big|\{0 \leq i \leq n-1 \mid (x_i, \ldots, x_{i+b-1}) \neq (y_i, \ldots, y_{i+b-1})\}\big|$$

where the subscripts are reduced modulo n. Accordingly, the b-**weight** of a vector $\mathbf{x}$ is defined as

$$\mathrm{wt}_b(\mathbf{x}) = \big|\{0 \leq i \leq n-1 \mid (x_i, \ldots, x_{i+b-1}) \neq \mathbf{0}\}\big|$$

where the subscripts are reduced modulo n.

In light of this definition, if $b = 1$, the b-distance is the Hamming distance, and if $b = 2$, such b-distance is the symbol-pair distance. The b-**distance** of a code $\mathcal{C}$ is defined to be

$$d_b(\mathcal{C}) = \min\{d_b(\mathbf{c_1}, \mathbf{c_2}) \mid \mathbf{c_1}, \mathbf{c_2} \in \mathcal{C}, \mathbf{c_1} \neq \mathbf{c_2}\}.$$

Clearly, if $\mathcal{C}$ is a linear b-symbol code, the minimum b-weight and b-distance are the same, i.e., $d_b(\mathcal{C}) = \min\{\mathrm{wt}_b(\mathbf{c}) \mid \mathbf{c} \in \mathcal{C}, \mathbf{c} \neq \mathbf{0}\}$.

Recently, Dinh et al. [599] provided the b-distance for all constacyclic codes of length p^s over $\mathbb{F}_{p^m}$ when $1 \leq b \leq \lfloor \frac{p}{2} \rfloor$.

Theorem 17.7.1 ([599]) *Let $1 \leq b \leq \lfloor \frac{p}{2} \rfloor$ and $1 \leq \beta \leq p - 1$. A λ-constacyclic code of length p^s over $\mathbb{F}_{p^m}$ has the form $\mathcal{C}_i = \langle (x - \lambda_0)^i \rangle \subseteq \frac{\mathbb{F}_{p^m}[x]}{\langle x^{p^s} - \lambda \rangle}$ for $i \in \{0, 1, \ldots, p^s\}$. For $0 \leq \theta \leq s - 2$, the symbol-pair distance $d_b(\mathcal{C}_i)$ of $\mathcal{C}_i$ is*

$$d_b(\mathcal{C}_i) = \begin{cases} b & \text{if } i = 0, \\ (\beta + b)(\varphi + 1)p^\theta & \text{if } i = p^s - p^{s-\theta} + \varphi p^{s-\theta-1} + \beta \\ & \quad \text{for } 0 \leq \varphi \leq p - 2 \text{ and } b > \beta(\varphi + 1), \\ b(\varphi + 2)p^\theta & \text{if } p^s - p^{s-\theta} + \varphi p^{s-\theta-1} + \beta \leq i \leq p^s - p^{s-\theta} + (\varphi + 1)p^{s-\theta-1} \\ & \quad \text{for } 0 \leq \varphi \leq p - 2 \text{ and } b \leq \beta(\varphi + 1), \\ (\varphi + b)p^{s-1} & \text{if } i = p^s - p + \varphi \text{ for } 0 \leq \varphi \leq p - b, \\ p^s & \text{if } p^s - b + 1 \leq i \leq p^s - 1, \\ 0 & \text{if } i = p^s. \end{cases}$$

Note that when $b = 1$ and $b = 2$, the b-symbol distance is the Hamming distance and symbol-pair distance of codes. Namely, Theorem 17.7.1 is exactly the Hamming distance and symbol-pair distance of all λ-constacyclic codes of length p^s over $\mathbb{F}_{p^m}$, presented in Theorems 17.5.5 and 17.6.3, respectively.

Certain specific cyclic codes can be constructed as cyclic DNA codes to be used in DNA computing. The idea of DNA computing, introduced by Head in [927], is to link genetic data analysis with scientific computations in order to handle computationally difficult problems. However, Adleman [13] initiated the studies on DNA computing by solving an instance of an NP-complete problem over DNA molecules. Four different constraints on DNA codes are considered.

- The **Hamming constraint**: For all $\mathbf{x}, \mathbf{y} \in \mathcal{C}$, $d_H(\mathbf{x}, \mathbf{y}) \geq d$ with $\mathbf{x} \neq \mathbf{y}$, for some prescribed minimum distance d.

- The **reverse constraint**: For all $\mathbf{x}, \mathbf{y} \in \mathcal{C}$, $d_H(\mathbf{x}^r, \mathbf{y}) \geq d$, where $\mathbf{x}^r$ is the reverse of a codeword $\mathbf{x}$.

- The **reverse-complement constraint**: For all $\mathbf{x}, \mathbf{y} \in \mathcal{C}$, $d_H(\mathbf{x}^{rc}, \mathbf{y}) \geq d$, where $\mathbf{x}^{rc}$ is the reverse-complement of $\mathbf{x}$ obtained by taking the reverse $\mathbf{x}^r$ of $\mathbf{x}$, and performing the symbol interchanges $A \leftrightarrow T$, $C \leftrightarrow G$; this is called taking Watson–Crick complements.

- The **fixed GC-content constraint**: Any codeword $\mathbf{x} \in \mathcal{C}$ has the same number of G and C elements.

The objective of the first three constraints is to avoid undesirable hybridization between different strands. The fixed GC-content constraint ensures that all codewords have similar thermodynamic characteristics. Furthermore, cyclic DNA computing has generated great interest because it provides more storage capacity than silicon-based computing systems. In [12], DNA codes over the finite field of four elements were studied by Abualrub and Şiap. Later, Şiap et al. [1700] considered DNA codes over the finite ring $\mathbb{F}_2[u]/\langle u^2 - 1 \rangle$ with four elements. Algebraic structure of DNA cyclic codes over finite rings with 16 elements was given by Dinh et al. in [597, 594]. Later, Singh and Kumar [1175] considered cyclic codes over certain Frobenius rings with 16 elements for DNA computing. DNA cyclic codes over non-chain rings with 64 elements have also been recently discussed by Dinh et al. in [595, 596].

As mentioned in Section 17.4, for a unit λ in the ring R, the λ-constacyclic (or λ-twisted) shift τ_λ on R^n is the shift

$$\tau_\lambda(x_0, x_1, \ldots, x_{n-1}) = (\lambda x_{n-1}, x_0, x_1, \ldots, x_{n-2}).$$

A code $\mathcal{C}$ is said to be a **quasi-cyclic code of index** l if $\mathcal{C}$ is closed under the cyclic shift of l symbols τ^l, i.e., if $\tau^l(\mathcal{C}) = \mathcal{C}$, and $\mathcal{C}$ is called a λ-**quasi-twisted code of index** l if it is closed under the λ-twisted shift of l symbols, i.e., $\tau_\lambda^l(\mathcal{C}) = \mathcal{C}$. Of course, when $\lambda = 1$, a λ-quasi-twisted code of index l is just a quasi-cyclic code of index l, and it becomes a λ-constacyclic code if $l = 1$. It is easy to see that a code of length n is λ-quasi-twisted (quasi-cyclic) of index l if and only if it is λ-quasi-twisted (quasi-cyclic) of index $\gcd(l, n)$. Therefore, without loss of generality, one only need to consider λ-quasi-twisted (quasi-cyclic) codes of index l where l is a divisor of the length n.

Quasi-cyclic codes over finite fields, the topic of Chapter 7, have a rich history in and of themselves. Connections between quasi-cyclic block codes and convolutional codes, the subject of Chapter 10, are found in [686, 1736] and Section 7.6. Beginning in the 1990s many new quasi-cyclic linear codes over finite fields and finite rings have been discovered; see, for example, [79, 398, 485, 486, 862, 863, 1269, 1271, 1701].

Another variation that yields interesting results both for codes over fields and codes over rings is when one starts with a noncommutative ambient space for codes rather than the usual commutative setting of quotient rings of the polynomial ring $\mathbb{F}[x]$. Specifically, consider the codes that are ideals of quotient rings of the skew polynomial ring $R[x; \sigma]$ (where σ is an automorphism of the ring R). These are the **skew-cyclic codes**. They have the property that if $(a_0, a_1, \ldots, a_{n-1})$ is a codeword in a skew-cyclic code $\mathcal{C}$, then $(\sigma(a_{n-1}), \sigma(a_0), \ldots, \sigma(a_{n-2}))$ is also a codeword in $\mathcal{C}$. Of course when σ is the identity, this produces the normal cyclic shift. This approach, introduced in [256] for skew-cyclic codes over finite fields, was later extended to skew-cyclic and skew-constacyclic codes over finite rings; see for example [257, 588]. Chapter 8 is devoted to the study of codes over skew polynomial rings.

If quotients of a multivariable polynomial ring $R[x_1, \ldots, x_n]$ are used as ambient spaces for codes, one gets the so-called **multivariable codes**. The study of multivariable codes goes back to the work of Poli in [1531, 1532] where multivariable codes over finite fields were first introduced and studied. There, ideals of $\frac{R[x,y,z]}{\langle t_1(x), t_2(y), t_3(z)\rangle}$, where R is a finite field, were considered. This notion then was extended by Martínez-Moro and Rúa in [1337, 1338] where R is assumed to be a finite chain ring.

Finally, there are the notions of polycyclic codes and sequential codes, which were introduced in [1287] and [987], respectively. A linear code $\mathcal{C}$ of length n is **right polycyclic** if there exists an n-tuple $\mathbf{c} = (c_0, c_1, \ldots, c_{n-1}) \in \mathbb{F}^n$, where $\mathbb{F}$ is a finite field, such that for every codeword $(a_0, a_1, \ldots, a_{n-1}) \in \mathcal{C}$, $(0, a_0, a_1, \ldots, a_{n-2}) + a_{n-1}(c_0, c_1, \ldots, c_{n-1}) \in \mathcal{C}$. Left polycyclic is defined similarly. $\mathcal{C}$ is **bi-polycyclic** if it is both left and right polycyclic. Polycyclicity of codes is clearly a generalization of cyclicity, as a λ-constacyclic code is right polycyclic induced by $\mathbf{c} = (\lambda, 0, \ldots, 0)$, and left polycyclic using $\mathbf{d} = (0, \ldots, 0, \lambda^{-1})$. So indeed a λ-constacyclic code is bi-polycyclic.

As with cyclic and constacyclic codes, polycyclic codes may be understood in terms of ideals in a quotient ring of a polynomial ring. Given $\mathbf{c} = (c_0, c_1, \ldots, c_{n-1}) \in \mathbb{F}^n$, let $f(x) = x^n - c(x)$ where $c(x) = c_0 + c_1 x + \cdots + c_{n-1}x^{n-1}$. Then the $\mathbb{F}$-linear isomorphism $\rho : \mathbb{F}^n \to \frac{\mathbb{F}[x]}{\langle f(x)\rangle} = R_n$ sending $\mathbf{a} = (a_0, a_1, \ldots, a_{n-1}) \in \mathbb{F}^n$ to the polynomial $a_0 + a_1 x + \cdots + a_{n-1}x^{n-1}$ associates the right polycyclic codes induced by $\mathbf{c}$ with the ideals of R_n.

Similarly, when $\mathcal{C}$ is a left polycyclic code, a slightly different isomorphism gives the identification of the left polycyclic codes induced by $\mathbf{c}$ as ideals of the corresponding ambient ring. As before, let $\mathbf{c} = (c_0, c_1, \ldots, c_{n-1}) \in \mathbb{F}^n$ but this time let $c'(x) = c_0 x^{n-1} + c_1 x^{n-2} + \cdots + c_{n-1}$. Then let $f'(x) = x^n - c'(x)$ and consider $\gamma : \mathbb{F}^n \to \frac{\mathbb{F}[x]}{\langle f'(x)\rangle} = L_n$ defined via $\gamma : (a_0, a_1, \ldots, a_{n-1}) \mapsto a_0 x^{n-1} + \cdots + a_{n-2}x + a_{n-1}$. In this setting, very much like before, one can see that $\gamma(\mathcal{C})$ is an ideal of L_n.

Since all ideals of $\mathbb{F}[x]$ are principal, the same is true in $\frac{\mathbb{F}[x]}{\langle f(x)\rangle}$; thus the ambient space $\frac{\mathbb{F}[x]}{\langle f(x)\rangle}$ is a principal ideal ring. Furthermore, following the usual arguments used in the theory of cyclic codes, one easily sees that every polycyclic code $\mathcal{C}$ of dimension k has a monic polynomial $g(x)$ of minimum degree $n - k$ belonging to the code. This polynomial, called a **generator polynomial** of $\mathcal{C}$, is a factor of $f(x)$. Also, a generator polynomial of a code is unique up to associates in the sense that if $g_1(x) \in \mathbb{F}[x]$ has degree $n - k$, it is easy to show that $g_1(x)$ is in the code generated by $g(x)$ if and only if $g_1(x) = ag(x)$ for some $0 \neq a \in \mathbb{F}$.

As with cyclic codes, using the generator polynomial of a polycyclic code $\mathcal{C}$, one can readily construct a generator matrix for it. It turns out that this property in fact characterizes polycyclic codes, as pointed out in [1287, Theorem 2.3].

Theorem 17.7.2 *A code $\mathcal{C} \subseteq \mathbb{F}^n$ is right polycyclic if and only if it has a $k \times n$ generator matrix of the form*

$$G = \begin{bmatrix} g_0 & g_1 & \cdots & g_{n-k} & 0 & \cdots & 0 \\ 0 & g_0 & \cdots & g_{n-k-1} & g_{n-k} & \cdots & 0 \\ \vdots & \vdots & \ddots & \vdots & \vdots & \ddots & \vdots \\ 0 & 0 & \cdots & g_0 & g_1 & \cdots & g_{n-k} \end{bmatrix}$$

with $g_{n-k} \neq 0$. In this case $\rho(\mathcal{C}) = \langle g_0 + g_1 x + \cdots + g_{n-k}x^{n-k}\rangle$ is an ideal of $R_n = \frac{\mathbb{F}[x]}{\langle f(x)\rangle}$. The same criterion, but requiring $g_0 \neq 0$ instead of $g_{n-k} \neq 0$, serves to characterize left polycyclic codes. In the latter case, $\gamma(\mathcal{C}) = \langle g_{n-k} + g_{n-k-1}x + \cdots + g_0 x^{n-k}\rangle$ is an ideal of $L_n = \frac{\mathbb{F}[x]}{\langle f(x)\rangle}$.

A code $\mathcal{C}$ is **right sequential** if there is a function $\phi : \mathbb{F}^n \to \mathbb{F}$ such that for every $(a_0, a_1, \ldots, a_{n-1}) \in \mathcal{C}$, $(a_1, \ldots, a_{n-1}, b) \in \mathcal{C}$ where $b = \phi(a_0, a_1, \ldots, a_{n-1})$. Left sequential is defined similarly. $\mathcal{C}$ is **bi-sequential** if it is both right and left sequential. Examples to illustrate the promise of sequential codes as a source for good (even optimal) codes are found in [987, Examples 6.3, 6.4].

It has been shown in [1287] that a code $\mathcal{C}$ over a field $\mathbb{F}$ is right sequential if and only if its dual $\mathcal{C}^{\perp}$ is right polycyclic. Also, $\mathcal{C}$ is sequential and polycyclic if and only if $\mathcal{C}$ and $\mathcal{C}^{\perp}$ are both sequential if and only if $\mathcal{C}$ and $\mathcal{C}^{\perp}$ are both polycyclic. Furthermore, any one of these equivalent statements characterizes the family of constacyclic codes. In fact, the following result of [1287, Theorems 3.2, 3.5] gives several equivalences.

Theorem 17.7.3 *Let $\mathcal{C}$ be a code of length n over the finite field $\mathbb{F}$. Then*

(a) *The following conditions are equivalent.*

  (i) $\mathcal{C}$ *is right (respectively left, bi-) sequential.*

  (ii) $\mathcal{C}^{\perp}$ *is right (respectively left, bi-) polycyclic.*

(b) *The following conditions are equivalent.*

 (1-R) $\mathcal{C}$ *and $\mathcal{C}^{\perp}$ are right sequential.*

 (2-R) $\mathcal{C}$ *and $\mathcal{C}^{\perp}$ are right polycyclic.*

 (3-R) $\mathcal{C}$ *is right sequential and right polycyclic.*

 (4-R) $\mathcal{C}$ *is right sequential and bi-polycyclic.*

 (5-R) $\mathcal{C}$ *is right sequential and left polycyclic with generator polynomial not a monomial of the form x^t $(t \geq 1)$.*

 (1-L) $\mathcal{C}$ *and $\mathcal{C}^{\perp}$ are left sequential.*

 (2-L) $\mathcal{C}$ *and $\mathcal{C}^{\perp}$ are left polycyclic.*

 (3-L) $\mathcal{C}$ *is left sequential and left polycyclic.*

 (4-L) $\mathcal{C}$ *is left sequential and bi-polycyclic.*

 (5-L) $\mathcal{C}$ *is left sequential and right polycyclic with generator polynomial not a monomial of the form x^t $(t \geq 1)$.*

 (A) $\mathcal{C}$ *is right polycyclic and bi-sequential.*

 (B) $\mathcal{C}$ *is left polycyclic and bi-sequential.*

 (C) $\mathcal{C}$ *is constacyclic.*

In particular, this theorem highlights in theoretical terms the significance of constacyclic codes as a central notion in coding theory.

Chapter 18

Weight Distribution of Trace Codes over Finite Rings

Minjia Shi

Anhui University

18.1   Introduction

Few-weight codes, codes with a small number of codeword weights, were studied first for their intrinsic mathematical appeal. The study of various two-weight codes is described in Chapter 19; few-weight codes appear prominently in Chapter 20. Few-weight codes are also widely used in secret sharing schemes as introduced in Chapter 33. Ding et al. in [560, 568, 570, 572, 1935, 1946] constructed several few-weight trace codes over finite fields and studied their applications. Inspired by these articles, many scholars have further studied the trace codes over some finite rings. Using the trace function to construct few-weight codes is an effective method. In different alphabets, under different defining sets (the definition of defining sets will be introduced later), many scholars have constructed many classes of interesting few-weight codes. Selecting the suitable Gray map, some image codes are optimal and their parameters are completely new.

18.2   Preliminaries

The classes of trace codes play an important role in the theory of error-correcting codes. Many scholars considered trace codes over some finite rings [1307, 1308, 1669, 1670, 1671, 1672, 1673, 1674, 1675, 1676, 1677, 1679]. These finite rings are

$$R_k = \mathbb{F}_2[u_1, u_2, \ldots, u_k]/\langle u_i^2 = 0, u_i u_j = u_j u_i \rangle, \text{ (Type I)}$$
$$R_k(p) = \mathbb{F}_p[u_1, u_2, \ldots, u_k]/\langle u_i^2 = 0, u_i u_j = u_j u_i \rangle, \text{ and}$$
$$R(k, p, u^k = a) = \mathbb{F}_p + u\mathbb{F}_p + \cdots + u^{k-1}\mathbb{F}_p, \ u^k = a, \ a \in \mathbb{F}_p \text{ (Type II)}.$$

The ring R_k is also considered in Chapter 6.

Let R denote one of the three finite rings given above. Let $\mathcal{R}$ be the ring extension of R obtained by replacing the field $\mathbb{F}_p$ in R by $\mathbb{F}_{p^m}$. These extensions will be denoted by $\mathcal{R}_k$, $\mathcal{R}_k(p)$ and $\mathcal{R}(k, p, u^k = a)$, respectively. For q a prime power, the **trace function** $\mathrm{Tr}_{q^s/q}(\cdot)$ from $\mathbb{F}_{q^s}$ down to $\mathbb{F}_q$ is defined by $\mathrm{Tr}_{q^s/q}(x) = \sum_{i=0}^{s-1} x^{q^i}$ for all $x \in \mathbb{F}_{q^s}$. For each of the three extensions we can define a **generalized trace function** $\mathrm{Tr}(\cdot)$ from $\mathcal{R}$ down to R, as given in the above references. For example, when we consider the ring $R(k, p, u^k = a)$, the generalized trace function from $\mathcal{R}(k, p, u^k = a)$ down to $R(k, p, u^k = a)$ is defined as $\mathrm{Tr}(a_0 + a_1 u + \cdots + a_{k-1}u^{k-1}) = \mathrm{Tr}_{p^m/p}(a_0) + \mathrm{Tr}_{p^m/p}(a_1)u + \cdots + \mathrm{Tr}_{p^m/p}(a_{k-1})u^{k-1}$ for $a_i \in \mathbb{F}_{p^m}$, $i = 0, 1, \ldots, k-1$. In the other two cases, the definition of the generalized trace function is similar.

Let $L \subseteq \mathcal{R}^*$ where $\mathcal{R}^*$ denotes the group of units in $\mathcal{R}$. For $r \in \mathcal{R}$, the vector $\mathrm{Ev}(r)$ is obtained by applying the **evaluation map** to r: $\mathrm{Ev}(r) = (\mathrm{Tr}(rx))_{x \in L}$. Define the code $\mathcal{C}(m, p, L)$ of length $|L|$ by the formula

$$\mathcal{C}(m, p, L) = \{\mathrm{Ev}(r) \mid r \in \mathcal{R}\} = \{(\mathrm{Tr}(rx))_{x \in L} \mid r \in \mathcal{R}\},$$

where the subset L is called the **defining set** of $\mathcal{C}(m, p, L)$.

18.3   A Class of Special Finite Rings R_k (Type I)

Take $R_k = \mathbb{F}_2[u_1, u_2, \ldots, u_k]/\langle u_i^2 = 0, u_i u_j = u_j u_i \rangle$ with $k \geq 1$. R_k can also be described recursively as $R_k = R_{k-1}[u_k]/\langle u_k^2 = 0, u_k u_j = u_j u_k, j = 1, 2, \ldots, k-1 \rangle$ [638]. This ring is the ring of Boolean functions in k variables [346]. For convenience, we define $u_A := \prod_{i \in A} u_i$ for any subset $A \subseteq \{1, 2, \ldots, k\}$. Furthermore, $u_A = 1$ when $A = \emptyset$ by convention. Then $r \in R_k$ is written as $r = \sum_A c_A' u_A$, $c_A' \in \mathbb{F}_2$. Assume that $\sum_A c_A' u_A$ and $\sum_B d_B' u_B$ with $A, B \subseteq \{1, 2, \ldots, k\}$ are two elements of the ring R_k; then their product is defined as

$$\left(\sum_A c_A' u_A \right) \left(\sum_B d_B' u_B \right) = \sum_{\substack{A, B \subseteq \{1, 2, \ldots, k\}, \\ A \cap B = \emptyset}} c_A' d_B' u_{A \cup B}.$$

For a given positive integer $m \geq 2$, the ring R_k can be extended to

$$\mathcal{R}_k = \mathbb{F}_{2^m}[u_1, u_2, \ldots, u_k]/\langle u_i^2 = 0, u_i u_j = u_j u_i, i, j \in \{1, 2, \ldots, k\} \rangle.$$

The elements of $\mathcal{R}_k$ are in the form of $\sum_{A \subseteq \{1, 2, \ldots, k\}} c_A u_A$, $c_A \in \mathbb{F}_{2^m}$.

18.3.1 Case (i) $k = 1$

Shi et al. [1672] utilized exponential sums to present the Lee weight enumerator of trace codes over R_k when $k = 1$ and the defining set of $C(m, 2, L)$ is $\mathcal{R}_1^*$. To simplify notation, let $u = u_1$.

Define the **Lee weight** $\mathrm{wt_L}(\cdot)$ as the Hamming weight of the image of the codeword under the Gray map. The **Lee distance** $\mathrm{d_L}(\cdot, \cdot)$ from $\mathbf{x}$ to $\mathbf{y}$ is $\mathrm{d_L}(\mathbf{x}, \mathbf{y}) = \mathrm{wt_L}(\mathbf{x} - \mathbf{y})$. The **Gray map** ϕ^1 from R_1 to $\mathbb{F}_2^2$ is given by $\phi^1(r + uq) = (q, r + q)$. Thus ϕ^1 is a bijection. This map can be extended to R_1^n in the natural way. For any $\mathbf{x} = (x_1, x_2, \ldots, x_n) \in R_1^n$, where $x_j = r_j + uq_j$, $1 \le j \le n$, $\phi^1(\mathbf{x}) = (q(\mathbf{x}), r(\mathbf{x}) + q(\mathbf{x}))$, where $r(\mathbf{x}) = (r_1, r_2, \ldots, r_n)$, $q(\mathbf{x}) = (q_1, q_2, \ldots, q_n)$, and they are unique. Then ϕ^1 is a distance preserving map from $(R_1^n, \mathrm{d_L})$ to $(\mathbb{F}_2^{2n}, \mathrm{d_H})$; that is, $\mathrm{d_L}(\mathbf{x}, \mathbf{y}) = \mathrm{d_H}(\phi^1(\mathbf{x}), \phi^1(\mathbf{y}))$, where $\mathrm{d_H}(\phi^1(\mathbf{x}), \phi^1(\mathbf{y}))$ denotes the number of places $\phi^1(\mathbf{x})$ and $\phi^1(\mathbf{y})$ differ.

The ring $\mathcal{R}_1$ has a unique maximal ideal $M = \{ru \mid r \in \mathbb{F}_{2^m}\}$. $\mathcal{R}_1$ has size 2^{2m} and M has size 2^m. Furthermore, $\mathcal{R}_1^* = \mathcal{R}_1 \setminus M$.

Theorem 18.3.1 ([1672]) *For $a \in \mathcal{R}_1$, the Lee weight distribution of the code $C(m, 2, \mathcal{R}_1^*)$ is as follows.*

(a) *If $a = 0$, then $\mathrm{wt_L}(\mathrm{Ev}(a)) = 0$.*

(b) *If $a \in M \setminus \{0\}$, then $\mathrm{wt_L}(\mathrm{Ev}(a)) = 2^{2m}$.*

(c) *If $a \in \mathcal{R}_1^*$, then $\mathrm{wt_L}(\mathrm{Ev}(a)) = 2^{2m} - 2^m$.*

Theorem 18.3.1 shows that $C(m, 2, \mathcal{R}_1^*)$ is a two-weight code of length $2^{2m} - 2^m$ over R_1. By Theorem 18.3.1 together with the sizes of M and $\mathcal{R}_1^*$, $\phi^1(C(m, 2, \mathcal{R}_1^*))$ is a binary two-weight linear code of length $n = 2^{2m+1} - 2^{m+1}$ and dimension $2m$ with Hamming weight enumerator

$$\mathrm{Hwe}_{\phi^1(C(m,2,\mathcal{R}_1^*))}(x, y) = y^n + (2^{2m} - 2^m)x^{n/2}y^{n/2} + (2^m - 1)x^{2^{2m}}y^{n-2^{2m}}.$$

Remark 18.3.2 These parameters are reminiscent of those of the MacDonald code [1312, Table 6, page 54] but they are not the same. However, if we make the concatenation of a MacDonald code of type 13 in the sense of [1312, Table 6, page 54] ($k = 2m$, $u = m$) with a length 2 repetition code (i.e., replace 0 by 00 and 1 by 11), then we obtain a code with the same parameters as $\phi^1(C(m, 2, \mathcal{R}_1^*))$. It is in fact an equivalent code, obtained by the Gray map from a simplex code over $\mathbb{F}_2 + u\mathbb{F}_2 = R_1$. See the construction in [981, page 18, equation (31)] with $K = 2m + 1$, $U = m + 1$.

By the Griesmer Bound ([855] and Theorem 1.9.18) the parameters of an $[n, k, d]_q$ linear code satisfy $\sum_{i=0}^{k-1} \lceil \frac{d}{q^i} \rceil \le n$. Theorem 18.3.1 and the Griesmer Bound yield the following result.

Theorem 18.3.3 ([1672]) *For any $m \ge 2$, the code $\phi^1(C(m, 2, \mathcal{R}_1^*))$ is optimal for a given length and dimension.*

We find that $\phi^1(C(2, 2, \mathcal{R}_1^*))$ is a $[24, 4, 12]$ binary code. This code is optimal, for a given length and dimension, by [845], but different from the `Magma` $BKLC(\mathbb{F}_2, 24, 4)$ [250], which has a different Hamming weight distribution, namely

$$\mathrm{Hwe}_{BKLC(\mathbb{F}_2, 24, 4)}(x, y) = y^{24} + 14x^{12}y^{12} + x^{16}y^8,$$

and a very complicated construction [845].

We also see that $\phi^1(\mathcal{C}(3,2,\mathcal{R}_1^*))$ is a $[112,6,56]_2$ binary code with Hamming weight enumerator

$$\mathrm{Hwe}_{\phi^1(\mathcal{C}(3,2,\mathcal{R}_1^*))}(x,y) = y^{112} + 56x^{56}y^{56} + 7x^{64}y^{48}.$$

This code is optimal for a given length and dimension, but it is different from the `Magma` $BKLC(\mathbb{F}_2, 112, 6)$ code as $\mathrm{Hwe}_{BKLC(\mathbb{F}_2,112,6)}(x,y) = y^{112} + 60x^{56}y^{56} + 3x^{64}y^{48}$. The `Magma` code has a fourteen point construction in [845].

Theorem 18.3.4 ([1672]) *The minimum Lee distance of* $\mathcal{C}(m,2,\mathcal{R}_1^*)^\perp$ *is 2, for all* $m \geq 2$.

We say that a vector $\mathbf{x}$ **covers** a vector $\mathbf{y}$ if $\mathrm{supp}(\mathbf{x})$ contains $\mathrm{supp}(\mathbf{y})$. A **minimal codeword** of a linear code $\mathcal{C}$ is a nonzero codeword $\mathbf{x}$ that does not cover any other nonzero codeword except multiples of $\mathbf{x}$. A code in which every nonzero codeword is minimal is called a **minimal code**. In Chapter 33, minimal codes are discussed in relation to secret sharing schemes.

In general determining the minimal codewords of a given linear code is a difficult task; determining minimal codewords is called the **covering problem**. However, there is a numerical condition, derived in [67], bearing on the weights of the code, that is easy to check.

Theorem 18.3.5 ([67]) *Denote by w_0 and w_∞ the minimum and maximum nonzero Hamming weights of a linear code $\mathcal{C}$ over $\mathbb{F}_q$. If $\frac{w_0}{w_\infty} > \frac{q-1}{q}$, then every nonzero codeword of $\mathcal{C}$ is minimal; hence $\mathcal{C}$ is minimal.*

Theorem 18.3.6 ([1672]) $\phi^1(\mathcal{C}(m,2,\mathcal{R}_1^*))$ *is a minimal code.*

18.3.2　Case (ii) $k = 2$

The code $\mathcal{C}(m,2,L)$ with $L = \mathcal{R}_2^*$ is examined in [1673]. When $k = 2$, simplifying notation, $R_2 = \mathbb{F}_2 + u\mathbb{F}_2 + v\mathbb{F}_2 + uv\mathbb{F}_2$, with $u^2 = v^2 = 0$, $uv = vu$. Let $M = \{bu + cv + duv \mid b,c,d \in \mathbb{F}_{2^m}\}$. Obviously, for any $m \in M$, m is a non-unit in $\mathcal{R}_2$. The group of units in $\mathcal{R}_2$, denoted by $\mathcal{R}_2^*$, is $\{a + bu + cv + duv \mid a \in \mathbb{F}_{2^m}^*, b,c,d \in \mathbb{F}_{2^m}\}$. It is obvious that $\mathcal{R}_2^*$ is not a cyclic group under multiplication and $\mathcal{R}_2 = \mathcal{R}_2^* \cup M$.

The Gray map $\phi^2 : R_2^n \rightarrow \mathbb{F}_2^{4n}$ is given by

$$\phi^2(\mathbf{a} + \mathbf{b}u + \mathbf{c}v + \mathbf{d}uv) = (\mathbf{d}, \mathbf{c} + \mathbf{d}, \mathbf{b} + \mathbf{d}, \mathbf{a} + \mathbf{b} + \mathbf{c} + \mathbf{d}),$$

where $\mathbf{a}, \mathbf{b}, \mathbf{c}, \mathbf{d} \in \mathbb{F}_2^n$. As was observed in [1072], ϕ^2 is a distance preserving isometry from $(R_2^n, \mathrm{d_L})$ to $(\mathbb{F}_2^{4n}, \mathrm{d_H})$.

Theorem 18.3.7 ([1673]) *For $a \in \mathcal{R}_2$, the Lee weight of codewords in the code $\mathcal{C}(m,2,\mathcal{R}_2^*)$ is given below.*

(a) *If $a = 0$, then* $\mathrm{wt_L}(\mathrm{Ev}(a)) = 0$.

(b) *If $a = \alpha uv$, where $\alpha \in \mathbb{F}_{2^m}^*$, then* $\mathrm{wt_L}(\mathrm{Ev}(a)) = 2^{4m+1}$.

(c) *If $a \in M \setminus \{\alpha uv \mid \alpha \in \mathbb{F}_{2^m}\}$, then* $\mathrm{wt_L}(\mathrm{Ev}(a)) = 2^{4m+1} - 2^{3m+1}$.

(d) *If $a \in \mathcal{R}_2^*$, then* $\mathrm{wt_L}(\mathrm{Ev}(a)) = 2^{4m+1} - 2^{3m+1}$.

In Theorem 18.3.7, we have constructed a binary code $\phi^2(\mathcal{C}(m,2,\mathcal{R}_2^*))$ of length $n = 2^{3m+2}(2^m - 1)$, dimension $4m$, with two nonzero weights $\omega_1 < \omega_2$ of values

$$\omega_1 = 2^{3m+1}(2^m - 1), \quad \omega_2 = 2^{4m+1},$$

and respective frequencies f_1, f_2 given by

$$f_1 = 2^{4m} - 2^m, \quad f_2 = 2^m - 1.$$

Remark 18.3.8 These parameters are reminiscent of those of the family $SU1$ in [330]. However, identifying the weights we found with that of $SU1$ forces $\ell = 4m + 2$, $t = 3m + 2$. These values lead to a different f_1. Besides, the code dimension is only $4m < \ell$. Thus, our family is not of the form $SU1$.

Theorem 18.3.9 ([1673]) *For any $m \geq 2$, the code $\phi^2(\mathcal{C}(m, 2, \mathcal{R}_2^*))$ is optimal for a given length and dimension.*

Theorem 18.3.10 ([1673]) *The minimum Lee distance of $\mathcal{C}(m, 2, \mathcal{R}_2^*)^\perp$ is 2, for all $m \geq 2$.*

Theorem 18.3.11 ([1673]) *For $m \geq 2$, $\phi^2(\mathcal{C}(m, 2, \mathcal{R}_2^*))$ is a minimal code.*

18.3.3 Case (iii) $k > 2$

For the general case $k > 2$, Shi et al. [1670] studied two different defining sets of $\mathcal{C}(m, 2, L)$. This construction is very powerful: many interesting codes [944, 945] can be obtained by choosing different defining sets L.

Let α be a fixed primitive element of $\mathbb{F}_{2^m}$ and N_0 be a positive integer such that $N_0 \mid (2^m - 1)$. Define $D = \langle \alpha^{N_0} \rangle \subseteq \mathbb{F}_{2^m}$, where $\langle \alpha^{N_0} \rangle$ denotes the subgroup of $\mathbb{F}_{2^m}^* = \langle \alpha \rangle$ of order $\frac{2^m - 1}{N_0}$.

We present the following two different defining sets:

- $L_1 = \mathcal{R}_k^* \cong \mathbb{F}_{2^m}^* \times \underbrace{\mathbb{F}_{2^m} \times \cdots \times \mathbb{F}_{2^m}}_{2^k - 1}$;

- $L_2 \subset \mathcal{R}_k^*$ with $L_2 \cong D \times \underbrace{\mathbb{F}_{2^m} \times \cdots \times \mathbb{F}_{2^m}}_{2^k - 1}$.

For $\mathbf{c} \in R_k^n$ define the Lee weight of $\mathbf{c}$ as $\mathrm{wt}_H(\phi_k^3(\mathbf{c}))$ where $\phi_k^3 : R_k^n \to \mathbb{F}_2^{2^k n}$ is the distance preserving Gray map defined recursively in [638]. When $k = 1$, ϕ_1^3 is the map defined previously; i.e., $\phi_1^3 = \phi^1$. For $k \geq 2$, write $\mathbf{c} = \mathbf{c}_1 + u_k \mathbf{c}_2$ with $\mathbf{c}_1, \mathbf{c}_2 \in R_{k-1}^n$. Define $\phi_k^3(\mathbf{c}) = (\phi_{k-1}^3(\mathbf{c}_2), \phi_{k-1}^3(\mathbf{c}_1) + \phi_{k-1}^3(\mathbf{c}_2))$. Notice that $\phi_2^3 = \phi^2$ as defined earlier.

Let M be the maximal ideal of $\mathcal{R}_k$ given by

$$M = \left\{ \sum_A r_A u_A \mid A \subseteq \{1, \ldots, k\}, r_\emptyset = 0 \right\}.$$

As before, $\mathcal{R}_k^* = \mathcal{R}_k \setminus M$.

Theorem 18.3.12 ([1670]) *For any integer $k \geq 3$, denote the integer range $\{1, 2, \ldots, k\}$ by $[k]$. For $a \in \mathcal{R}_k$, the weight distribution of the trace code $\mathcal{C}(m, 2, L_1)$ is as follows.*

(a) *If $a = 0$, then $\mathrm{wt}_L(\mathrm{Ev}(a)) = 0$.*

(b) *If $a = \alpha u_{[k]}$, where $\alpha \in \mathbb{F}_{2^m}^*$, then $\mathrm{wt}_L(\mathrm{Ev}(a)) = 2^{2^k m + k - 1}$.*

(c) *If $a \in M \setminus \{\alpha u_{[k]} \mid \alpha \in \mathbb{F}_{2^m}\}$, then $\mathrm{wt}_L(\mathrm{Ev}(a)) = (2^m - 1)2^{m(2^k - 1) + k - 1}$.*

(d) *If $a \in \mathcal{R}_k^*$, then $\mathrm{wt}_L(\mathrm{Ev}(a)) = (2^m - 1)2^{m(2^k - 1) + k - 1}$.*

Therefore, we obtained a binary two-weight code $\phi_k^3(\mathcal{C}(m, 2, L_1))$ of length $(2^m - 1)2^{m(2^k - 1) + k}$ and dimension $2^k m$. More details are shown in Table 18.1.

TABLE 18.1: Hamming weight distribution of $\phi_k^3(\mathcal{C}(m,2,L_1))$; see Theorem 18.3.12

Hamming weight	Frequency
0	1
$(2^m-1)2^{m(2^k-1)+k-1}$	$2^{m2^k}-2^m$
$2^{2^k m+k-1}$	2^m-1

Remark 18.3.13 We compare the two-weight code $\phi_k^3(\mathcal{C}(m,2,L_1))$ with MacDonald codes in [1312]. It can be seen as a concatenation of the MacDonald code of type 13 ($k=2^km$, $u=2^km-m$) in [1312, Table 6, page 54] with the $[2^k,1,2^k]_2$ repetition code. In fact, it is an equivalent code obtained by the Gray map ϕ_k^3 from a simplex code over R_k. Note that the parameters in Table 18.1 includes references [1672, 1673] as special cases.

Theorem 18.3.14 ([1670]) *Let m be even. If $1 \leq N_0 < \sqrt{2^m}+1$, then $\mathcal{C}(m,2,L_2)$ is a $(|L_2|, 2^{m2^k}, d_L)$ linear code over R_k which has at most N_0+1 different nonzero Lee weights, where $\frac{1}{N_0}2^{m(2^k-1)+k-1}\big(2^m-(N_0-1)2^{\frac{m}{2}}\big) \leq d_L \leq \frac{1}{N_0}2^{m(2^k-1)+k-1}(2^m-1)$.*

Theorem 18.3.15 ([1670]) *Let m be even and $N_0 > 2$. Assume that there exists a positive integer l such that $2^l \equiv -1 \pmod{N_0}$ and $2l$ divides m. Set $t = \frac{m}{2l}$. The linear code $\phi_k^3(\mathcal{C}(m,2,L_2))$ is a three-weight code provided that $2^{\frac{m}{2}}+(-1)^t(N_0-1) > 0$, and its weight distribution is given in Table 18.2.*

TABLE 18.2: Hamming weight distribution of $\phi_k^3(\mathcal{C}(m,2,L_2))$ in Theorem 18.3.15

Hamming weight	Frequency
0	1
$\dfrac{2^{m(2^k-1)+k-1}\big(2^m+(-1)^t(N_0-1)2^{\frac{m}{2}}\big)}{N_0}$	$\dfrac{2^m-1}{N_0}$
$\dfrac{2^{m(2^k-1)+k-1}(2^m-1)}{N_0}$	$2^{2^k m}-2^m$
$\dfrac{2^{m(2^k-1)+k-1}\big(2^m-(-1)^t 2^{\frac{m}{2}}\big)}{N_0}$	$\dfrac{(N_0-1)(2^m-1)}{N_0}$

The choice of the defining set L_2 is inspired from [945], but $\phi_k^3(\mathcal{C}(m,2,L_2))$ is different from codes constructed in [945]. As the length, minimum distance and weight distribution of the three-weight code $\phi_k^3(\mathcal{C}(m,2,L_2))$ are determined by the values of N_0, k and m, it is difficult to compare $\phi_k^3(\mathcal{C}(m,2,L_2))$ with other three-weight codes completely.

Note that when $N_0 = 1$, $L_1 = L_2$ and so $\mathcal{C}(m,2,L_2) = \mathcal{C}(m,2,L_1)$.

Theorem 18.3.16 ([1670]) *For $k \geq m \geq 2$, the minimum Lee distance of $\mathcal{C}(m,2,L_i)^{\perp}$ is 2 when $i = 1,2$.*

Theorem 18.3.17 ([1670]) *The following hold.*

(a) *For $m \geq 2$, $\phi_k^3(\mathcal{C}(m,2,L_1))$ is a minimal code.*

(b) *With the conditions of Theorem 18.3.15, when t is odd, $\phi_k^3(\mathcal{C}(m,2,L_2))$ is a minimal code for $2 < N_0 < 2^{\frac{m}{2}-1}+\frac{1}{2}$; when t is even, $\phi_k^3(\mathcal{C}(m,2,L_2))$ is a minimal code for $2 < N_0 < 2^{\frac{m}{2}}-1$.*

18.3.4   Case (iv) $R_k(p)$, p an Odd Prime

Codes over $R_k(p) = \mathbb{F}_p[u_1, u_2, \ldots, u_k]/\langle u_i^2 = 0, u_i u_j = u_j u_i, i, j \in \{1, 2, \ldots, k\}\rangle$ for p an odd prime were considered in [1308, 1674, 1675, 1676] when $k = 1, 2$.

Trace codes over the ring $R_1(p) = \mathbb{F}_p + u\mathbb{F}_p$, with $u^2 = 0$, were considered in [1308, 1676]. Given an integer m we can construct the ring extension $\mathcal{R}_1(p) = \mathbb{F}_{p^m} + u\mathbb{F}_{p^m}$. Denote by $\mathcal{Q}$ and $\mathcal{N}$, respectively, the nonzero squares and the non-squares of $\mathbb{F}_{p^m}$. Many interesting defining sets have been considered in [1307, 1308, 1676]. Similar to case (i), the Gray map ϕ^1 from $R_1(p)$ to $\mathbb{F}_p^2$ is also given by $\phi^1(a + ub) = (b, a + b)$.

Theorem 18.3.18 ([1676]) *Let m be even and $\epsilon(p) = (-1)^{\frac{p+1}{2}}$. For $a \in \mathcal{R}_1(p)$, the Lee weight of codewords of $\mathcal{C}(m, p, \mathcal{Q} + u\mathbb{F}_{p^m})$ is as follows.*

(a) *If $a = 0$, then $\mathrm{wt_L}(\mathrm{Ev}(a)) = 0$.*

(b) *If $a = \alpha u$, where $\alpha \in \mathcal{Q}$, then $\mathrm{wt_L}(\mathrm{Ev}(a)) = (p-1)\big((p^{2m-1} - p^{m-1}) + p^{m-1}(\epsilon(p)p^{m/2} - 1)\big)$.*

(c) *If $a = \alpha u$, where $\alpha \in \mathcal{N}$, then $\mathrm{wt_L}(\mathrm{Ev}(a)) = (p-1)\big((p^{2m-1} - p^{m-1}) - p^{m-1}(\epsilon(p)p^{m/2} + 1)\big)$.*

(d) *If $a \in \mathcal{R}_1(p)^*$, then $\mathrm{wt_L}(\mathrm{Ev}(a)) = (p-1)(p^{2m-1} - p^{m-1})$.*

Thus we have constructed a p-ary code $\phi^1(\mathcal{C}(m, p, \mathcal{Q} + u\mathbb{F}_{p^m}))$ of length $p^{2m} - p^m$, dimension $2m$, with three nonzero weights $w_1 < w_2 < w_3$ of values

$$w_1 = (p-1)\big((p^{2m-1} - p^{m-1}) - p^{m-1}(p^{m/2} + 1)\big),$$
$$w_2 = (p-1)(p^{2m-1} - p^{m-1}),$$
$$w_3 = (p-1)\big((p^{2m-1} - p^{m-1}) + p^{m-1}(p^{m/2} - 1)\big),$$

and respective frequencies f_1, f_2, f_3 given by

$$f_1 = \frac{p^m - 1}{2}, \quad f_2 = p^{2m} - p^m, \quad f_3 = \frac{p^m - 1}{2}.$$

(Note that taking $\epsilon(p) = 1$ or -1 leads to the same values of w_1 and w_3.)

Theorem 18.3.19 ([1676]) *Assume m is odd and $p \equiv 3 \pmod 4$. For $a \in \mathcal{R}_1(p)$, the Lee weight of codewords of $\mathcal{C}(m, p, \mathcal{Q} + u\mathbb{F}_{p^m})$ is the following.*

(a) *If $a = 0$, then $\mathrm{wt_L}(\mathrm{Ev}(a)) = 0$.*

(b) *If $a = \alpha u$, where $\alpha \in \mathbb{F}_{p^m}^*$, then $\mathrm{wt_L}(\mathrm{Ev}(a)) = (p-1)p^{2m-1}$.*

(c) *If $a \in \mathcal{R}_1(p)^*$, then $\mathrm{wt_L}(\mathrm{Ev}(a)) = (p-1)(p^{2m-1} - p^{m-1})$.*

Thus we obtain a family of p-ary two-weight codes $\phi^1(\mathcal{C}(m, p, \mathcal{Q} + u\mathbb{F}_{p^m}))$ of length $p^{2m} - p^m$, dimension $2m$, with nonzero weights $w_1 < w_2$ given by

$$w_1 = (p-1)(p^{2m-1} - p^{m-1}), \quad w_2 = (p-1)p^{2m-1}$$

and respective frequencies f_1, f_2 given by

$$f_1 = p^{2m} - p^m, \quad f_2 = p^m - 1.$$

Theorem 18.3.20 ([1676]) *If m is odd and $p \equiv 3 \pmod{4}$, then $\phi^1(\mathcal{C}(m, p, \mathcal{Q} + u\mathbb{F}_{p^m}))$ meets the Griesmer Bound with equality.*

Theorem 18.3.21 ([1676]) *When $m \geq 4$ is even, $\phi^1(\mathcal{C}(m, p, \mathcal{Q} + u\mathbb{F}_{p^m}))$ is a minimal code.*

Theorem 18.3.22 ([1676]) *For $m \geq 1$ odd and $p \equiv 3 \pmod{4}$, $\phi^1(\mathcal{C}(m, p, \mathcal{Q} + u\mathbb{F}_{p^m}))$ is a minimal code.*

Theorem 18.3.23 ([1308]) *Let D_l be an l-subset of $\mathbb{F}_q^*$, where $0 < l < q$. Assume that $K_1 = D_l + u\mathbb{F}_r$ with $u^2 = 0$ and $r = q^m$. Then $\mathcal{C}(m, q, K_1)$ is a linear code of length lr over $\mathbb{F}_q + u\mathbb{F}_q$, and the Lee weight distribution is listed in Table 18.3.*

TABLE 18.3: Lee weight distribution of $\mathcal{C}(m, q, K_1)$ in Theorem 18.3.23

Lee weight	Frequency
0	1
$2rl$	$q - 1$
$\frac{2(q-1)rl}{q}$	$q(r - 1)$

Theorem 18.3.24 ([1308]) *With the notation of Theorem 18.3.23, then $\phi^1(\mathcal{C}(m, q, K_1))$ is a $[2q^m l, m + 1, 2q^{m-1} l(q - 1)]$ linear code over $\mathbb{F}_q$ with the weight distribution in Table 18.4. Furthermore, $\phi^1(\mathcal{C}(m, q, K_1))$ meets the Griesmer Bound with equality if $2l < q$, and $\phi^1(\mathcal{C}(m, q, K_1))$ is optimal for a given length and dimension if $2l \geq q$.*

TABLE 18.4: Hamming weight distribution of $\phi^1(\mathcal{C}(m, q, K_1))$ in Theorem 18.3.24

Hamming weight	Frequency
0	1
$2q^m l$	$q - 1$
$2q^{m-1} l(q - 1)$	$q^{m+1} - q$

Theorem 18.3.25 ([1279]) *Let $K_2 = \langle \xi^e \rangle + u\mathbb{F}_r \subseteq \mathcal{R}_1(q)^*$, where ξ is a primitive element of $\mathbb{F}_r$, $u^2 = 0$, $r = q^m$ and e is a divisor of $r - 1$.*

(a) *If $\gcd(e, m) = 1$, the Hamming weight distribution of $\phi^1(\mathcal{C}(m, q, K_2))$ is listed in Table 18.5.*

(b) *If $\gcd(e, m) = 2$, the Hamming weight distribution of $\phi^1(\mathcal{C}(m, q, K_2))$ is listed in Table 18.6.*

Let $r = q^m$ and $s = q^k$, where $m > 1$ and $k > 1$ are integers with $k \,|\, m$. In [1307], Luo et al. considered the following defining sets:

(i) $K_3 = D_c + u\mathbb{F}_r$, $D_c = \{x \in \mathbb{F}_s \mid \mathrm{Tr}_{q^s/q}(x) = c\}$ for all $c \in \mathbb{F}_q$;

(ii) $K_4 = \mathbb{F}_s \setminus \mathbb{F}_q + u\mathbb{F}_r$;

(iii) $K_5 = SQ_{\mathbb{F}_s \setminus \mathbb{F}_q} + u\mathbb{F}_r$, where $SQ_{\mathbb{F}_s \setminus \mathbb{F}_q}$ denotes the set of all squares in $\mathbb{F}_s \setminus \mathbb{F}_q$;

TABLE 18.5: Hamming weight distribution of $\phi^1(\mathcal{C}(m, q, K_2))$ in Theorem 18.3.25 with $\gcd(e, m) = 1$

Hamming weight	Frequency
0	1
$2\frac{q-1}{eq}(r^2 - r)$	$r^2 - r$
$2\frac{q-1}{eq}r^2$	$r - 1$

TABLE 18.6: Hamming weight distribution of $\phi^1(\mathcal{C}(m, q, K_2))$ in Theorem 18.3.25 with $\gcd(e, m) = 2$

Hamming weight	Frequency
0	1
$2\frac{q-1}{eq}(r^2 - r^{\frac{3}{2}})$	$\frac{r-1}{2}$
$2\frac{q-1}{eq}(r^2 - r)$	$r^2 - r$
$2\frac{q-1}{eq}(r^2 + r^{\frac{3}{2}})$	$\frac{r-1}{2}$

(iv) $K_6 = \langle \xi^e \rangle + u\mathbb{F}_r$ where ξ is a primitive element of $\mathbb{F}_r$.

Using the above four defining sets, five classes of optimal two-weight trace codes are obtained. For more details, please refer to [1307].

We now turn to codes over $R_2(p)$. Shi et al. considered the finite ring $R_2(p) = \mathbb{F}_p + u\mathbb{F}_p + v\mathbb{F}_p + uv\mathbb{F}_p$, where $u^2 = 0$, $v^2 = 0$, $uv = vu$ in [1674, 1675]. They chose different defining sets of $\mathcal{C}(m, p, L)$ as follows:

- $L_1 = \{a + bu + cv + duv \mid a \in \mathcal{Q}, b, c, d \in \mathbb{F}_{p^m}\}$;

- $L_2 = \{a + bu + cv + duv \mid a \in D, b, c, d \in \mathbb{F}_{p^m}\} \subseteq R_2(p)^*$ (where the definition of D can be found in (18.1));

- $L_3 = \{a + bu + cv + duv \mid a \in \mathbb{F}_{p^m}^*, b, c, d \in \mathbb{F}_{p^m}\} = R_2(p)^*$.

Let ξ be a primitive element of $\mathbb{F}_{p^m}$. For N a positive integer such that $N \mid (p^m - 1)$, let $N_1 = \mathrm{lcm}\left(N, \frac{p^m - 1}{p - 1}\right)$ and $N_2 = \gcd\left(N, \frac{p^m - 1}{p - 1}\right)$. Define $n = \frac{N_1}{N}$ and

$$D = \{\xi^{N(j-1)} \mid j = 1, 2, \ldots, n\}. \tag{18.1}$$

Similar to case (ii), the Gray map ϕ^2 from $R_2(p)$ to $\mathbb{F}_p^4$ is also given by

$$\phi^2(a + ub + vc + uvd) = (d, c + d, b + d, a + b + c + d),$$

where $a, b, c, d \in \mathbb{F}_p$.

Theorem 18.3.26 ([1674]) *Assume $m \equiv 2 \pmod 4$. Let $\epsilon(p) = (-1)^{\frac{p+1}{2}}$. Let $\mathcal{Q}$ be the nonzero squares in $\mathbb{F}_{p^m}$ and $\mathcal{N}$ the non-squares in $\mathbb{F}_{p^m}$. For $a \in R_2(p)$, the Lee weight of codewords of $\mathcal{C}(m, p, L_1)$ is given below.*

(a) *If $a = 0$, then $\mathrm{wt}_L(\mathrm{Ev}(a)) = 0$.*

(b) *If $a = \alpha uv$, where $\alpha \in \mathcal{Q}$, then $\mathrm{wt}_L(\mathrm{Ev}(a)) = 2(p-1)\left(p^{4m-1} - \epsilon(p)p^{\frac{7m-2}{2}}\right)$.*

(c) *If* $a = \alpha uv$, *where* $\alpha \in \mathcal{N}$, *then* $\mathrm{wt_L}(\mathrm{Ev}(a)) = 2(p-1)\left(p^{4m-1} + \epsilon(p)p^{\frac{7m-2}{2}}\right)$.

(d) *If* $a \in \mathcal{R}_2(p) \setminus \{\alpha uv \mid \alpha \in \mathbb{F}_{p^m}\}$, *then* $\mathrm{wt_L}(\mathrm{Ev}(a)) = 2(p-1)\left(p^{4m-1} - p^{3m-1}\right)$.

We have constructed a class of p-ary linear codes of length $2p^{4m} - 2p^{3m}$, dimension $4m$, with three nonzero weights $w_1 < w_2 < w_3$ of values

$$w_1 = 2(p-1)(p^{4m-1} - p^{\frac{7m-2}{2}}),$$
$$w_2 = 2(p-1)(p^{4m-1} - p^{3m-1}),$$
$$w_3 = 2(p-1)(p^{4m-1} + p^{\frac{7m-2}{2}}),$$

and respective frequencies f_1, f_2, f_3 given by

$$f_1 = \frac{p^m - 1}{2}, \quad f_2 = p^{4m} - p^m, \quad f_3 = \frac{p^m - 1}{2}.$$

(Note that taking $\epsilon(p) = 1$ or -1 leads to the same values of w_1 and w_3.)

Theorem 18.3.27 ([1674]) *Assume m is odd and $p \equiv 3 \pmod 4$. For $a \in \mathcal{R}_2(p)$, the Lee weight of codewords of $\mathcal{C}(m, p, L_1)$ is given below.*

(a) *If* $a = 0$, *then* $\mathrm{wt_L}(\mathrm{Ev}(a)) = 0$.

(b) *If* $a = \alpha uv$, *where* $\alpha \in \mathbb{F}_{p^m}^*$, *then* $\mathrm{wt_L}(\mathrm{Ev}(a)) = 2(p^{4m} - p^{4m-1})$.

(c) *If* $a \in \mathcal{R}_2(p) \setminus \{\alpha uv \mid \alpha \in \mathbb{F}_{p^m}\}$, *then* $\mathrm{wt_L}(\mathrm{Ev}(a)) = 2(p-1)(p^{4m-1} - p^{3m-1})$.

We obtain a class of p-ary two-weight linear $[2p^{4m} - 2p^{3m}, 4m]$ codes with two nonzero weights $w_1 < w_2$ given by

$$w_1 = 2(p-1)(p^{4m-1} - p^{3m-1}), \quad w_2 = 2(p^{4m} - p^{4m-1}),$$

and respective frequencies f_1, f_2 given by

$$f_1 = p^m - 1, \quad f_2 = p^{4m} - p^m.$$

Theorem 18.3.28 ([1674]) *If m is odd and $p \equiv 3 \pmod 4$, then $\phi^2(\mathcal{C}(m, p, L_1))$ is optimal, meeting the Griesmer Bound.*

Theorem 18.3.29 ([1675]) *If $N_2 = 1$, suppose either m is even or m is odd with $p \equiv 3 \pmod 4$. Let $M = \{\alpha u + \beta v + \gamma uv \mid \alpha, \beta, \gamma \in \mathbb{F}_{p^m}\}$.*

(a) *For $a \in \mathcal{R}_2(p)$, the Lee weight of codewords in $\mathcal{C}(m, p, L_2)$ is given below.*

   (i) *If* $a = 0$, *then* $\mathrm{wt_L}(\mathrm{Ev}(a)) = 0$.

   (ii) *If* $a = \gamma uv$ *where* $\gamma \in \mathbb{F}_{p^m}^*$, *then* $\mathrm{wt_L}(\mathrm{Ev}(a)) = 4p^{4m-1}$.

   (iii) *If* $a \in M \setminus \{\gamma uv \mid \gamma \in \mathbb{F}_{p^m}\}$, *then* $\mathrm{wt_L}(\mathrm{Ev}(a)) = 4p^{4m-1} - 4p^{3m-1}$.

   (iv) *If* $a \in \mathcal{R}_2(p)^*$, *then* $\mathrm{wt_L}(\mathrm{Ev}(a)) = 4p^{4m-1} - 4p^{3m-1}$.

(b) *For $a \in \mathcal{R}_2(p)$, the Lee weight of codewords in $\mathcal{C}(m, p, L_3)$ is given below.*

   (i) *If* $a = 0$, *then* $\mathrm{wt_L}(\mathrm{Ev}(a)) = 0$.

   (ii) *If* $a = \gamma uv$ *where* $\gamma \in \mathbb{F}_{p^m}^*$, *then* $\mathrm{wt_L}(\mathrm{Ev}(a)) = 4(p-1)p^{4m-1}$.

   (iii) *If* $a \in M \setminus \{\gamma uv \mid \gamma \in \mathbb{F}_{p^m}\}$, *then* $\mathrm{wt_L}(\mathrm{Ev}(a)) = 4(p-1)(p^{4m-1} - 4p^{3m-1})$.

(iv) *If $a \in \mathcal{R}_2(p)^*$, then* $\mathrm{wt}_L(\mathrm{Ev}(a)) = 4(p-1)(p^{4m-1} - 4p^{3m-1})$.

In fact, by Theorem 18.3.29, we can obtain the MacDonald codes [981] with parameters $\left[\frac{p^{K'} - p^{U'}}{p-1}, K', p^{K'-1} - p^{U'-1}\right]$. As $N_2 = 1$, $|L_2| = np^{3m} = \frac{N_1}{N}p^{3m} = \frac{p^m - 1}{p-1}p^{3m}$. Thus $\phi^2(\mathcal{C}(m, p, L_2))$ is a p-ary code with parameters $\left[\frac{4p^{4m} - 4p^{3m}}{p-1}, 4m, 4p^{4m-1} - 4p^{3m-1}\right]$; these parameters are 4-fold of the MacDonald codes with $K' = 4m$ and $U' = 3m$. Also $\phi^2(\mathcal{C}(m, p, L_3))$ is a p-ary code with parameters $[4p^{4m} - 4p^{3m}, 4m, 4(p-1)(p^{4m-1} - p^{3m-1})]$; these parameters are $4(p-1)$-fold of the MacDonald codes with $K' = 4m$ and $U' = 3m$. And these trace codes can be described geometrically as analogs of the simplex codes over $\mathbb{F}_p + u\mathbb{F}_p + v\mathbb{F}_p + uv\mathbb{F}_p$ [981].

By Theorem 18.3.29, $\phi^2(\mathcal{C}(m, p, L_2))$ is a two-weight linear code over $\mathbb{F}_p$ of length $\frac{4(p^m - 1)p^{3m}}{p-1}$ and dimension $4m$. The Hamming weight distribution of $\phi^2(\mathcal{C}(m, p, L_2))$ is given in Table 18.7.

TABLE 18.7: Hamming weight distribution of $\phi^2(\mathcal{C}(m, p, L_2))$; see Theorem 18.3.29

Hamming weight	Frequency
0	1
$4p^{4m-1}$	$p^m - 1$
$4p^{4m-1} - 4p^{3m-1}$	$p^{4m} - p^m$

By Theorem 18.3.29, $\phi^2(\mathcal{C}(m, p, L_3))$ is a two-weight linear code over $\mathbb{F}_p$ of length $4(p^m - 1)p^{3m}$ and dimension $4m$. The Hamming weight distribution of $\phi^2(\mathcal{C}(m, p, L_3))$ is given in Table 18.8.

TABLE 18.8: Hamming weight distribution of $\phi^2(\mathcal{C}(m, p, L_3))$; see Theorem 18.3.29

Hamming weight	Frequency
0	1
$4(p-1)p^{4m-1}$	$p^m - 1$
$4(p-1)(p^{4m-1} - p^{3m-1})$	$p^{4m} - p^m$

Comparing parameters in [1674], it is not hard to see that the corresponding dimension and frequencies are the same. However, the corresponding length and weights are different; for example, the weights are double those of parameters in [1674, Theorem 5].

Next, we consider the case $N_2 > 1$.

Theorem 18.3.30 ([1675]) *Let m be even or m be odd and $p \equiv 3 \pmod 4$. If $1 < N_2 < \sqrt{p^m} + 1$, then $\mathcal{C}(m, p, L_2)$ is a $(|L_2|, p^{4m'}, d_L)$ linear code over $\mathcal{R}_2(p)$, which has at most $N_2 + 1$ nonzero Lee weights, where $m' \le m$ and*

$$\frac{4p^{3m-1}\left(p^m - (N_2 - 1)p^{\frac{m}{2}}\right)}{N_2} \le d_L(\mathcal{C}(m, p, L_2)) \le \frac{4p^{3m-1}(p^m - 1)}{N_2}.$$

If there exists a positive integer l such that $p^l \equiv -1 \pmod{N_2}$, then the weight distribution of $\mathcal{C}(m, p, L_2)$ is given in the following result.

Theorem 18.3.31 ([1675]) *Let m be even and $N_2 = \gcd\left(N, \frac{p^m - 1}{p-1}\right) > 2$ with $N \mid (p^m - 1)$. Assume that there exists a positive integer l such that $p^l \equiv -1 \pmod{N_2}$ and $2l$ divides m. Set $t = \frac{m}{2l}$.*

(a) *Assuming that p, t, and $\frac{p^l+1}{N_2}$ are odd, then the linear code $C(m, p, L_2)$ is a three-weight linear code, where N_2 is even and $N_2 < p^{\frac{m}{2}} + 1$. The weights of $C(m, p, L_2)$ are presented in Table 18.9.*

(b) *In all other cases, the linear code $C(m, p, L_2)$ is a three-weight linear code, where $p^{\frac{m}{2}} + (-1)^t(N_2 - 1) > 0$. The weights of $C(m, p, L_2)$ are presented in Table 18.10.*

TABLE 18.9: Lee weight distribution of $\phi^2(C(m, p, L_2))$ in Theorem 18.3.31(a)

Lee weight	Frequency
0	1
$\dfrac{4p^{3m-1}\left(p^m-(N_2-1)p^{\frac{m}{2}}\right)}{N_2}$	$\dfrac{p^m-1}{N_2}$
$\dfrac{4p^{3m-1}(p^m-1)}{N_2}$	$p^{4m}-p^m$
$\dfrac{4p^{3m-1}\left(p^m+p^{\frac{m}{2}}\right)}{N_2}$	$\dfrac{(N_2-1)(p^m-1)}{N_2}$

TABLE 18.10: Lee weight distribution of $\phi^2(C(m, p, L_2))$ in Theorem 18.3.31(b)

Lee weight	Frequency
0	1
$\dfrac{4p^{3m-1}\left(p^m+(-1)^t(N_2-1)p^{\frac{m}{2}}\right)}{N_2}$	$\dfrac{p^m-1}{N_2}$
$\dfrac{4p^{3m-1}(p^m-1)}{N_2}$	$p^{4m}-p^m$
$\dfrac{4p^{3m-1}\left(p^m-(-1)^tp^{\frac{m}{2}}\right)}{N_2}$	$\dfrac{(N_2-1)(p^m-1)}{N_2}$

Theorem 18.3.32 ([1675]) *The minimum Lee distance of $C(m, p, L_1)^\perp$ is 2 for all $m > 1$.*

Theorem 18.3.33 ([1675]) *The Gray image $\phi^2(C(m, p, L_1))$ satisfies the following properties.*

(a) *For $m > 2$ with m even, $\phi^2(C(m, p, L_1))$ is a minimal code.*

(b) *For $m > 1$ with m odd and $p \equiv 3 \pmod 4$, $\phi^2(C(m, p, L_1))$ is a minimal code.*

Theorem 18.3.34 ([1675]) *If $N_2 = 1$, m is even or m is odd and $p \equiv 3 \pmod 4$, the minimum Lee distance of $C(m, p, L_i)^\perp$ is 2 for $i = 2, 3$.*

Theorem 18.3.35 ([1675]) *Let $m > 1$ and p be an odd prime. We have the following.*

(a) *If m is even or m is odd and $p \equiv 3 \pmod 4$, then $\phi^2(C(m, p, L_2))$ is a minimal code when $N_2 = 1$.*

(b) *$\phi^2(C(m, p, L_3))$ is a minimal code.*

18.4   A Class of Special Finite Rings $R(k, p, u^k = a)$ (Type II)

Let $R(k, p, u^k = a) = \mathbb{F}_p + u\mathbb{F}_p + \cdots + u^{k-1}\mathbb{F}_p$ with $u^k = a$. The trace code over $R(k, p, u^k = 0)$ has been studied in [1677]. Let $\mathcal{R}(k, p, u^k = 0) = \mathbb{F}_{p^m} + u\mathbb{F}_{p^m} + \cdots + u^{k-1}\mathbb{F}_{p^m}$, which is a ring extension of $R(k, p, u^k = 0)$ of degree m where m is a positive integer.

Any integer z can be written uniquely in base p as $z = p_0(z) + p p_1(z) + p^2 p_2(z) + \cdots$, where $0 \le p_i(z) \le p - 1$, $i = 0, 1, 2, \ldots$. The Gray map $\phi^4 : R(k, p, u^k = 0) \to \mathbb{F}_p^{p^{k-1}}$ is defined as follows for $a = a_0 + a_1 u + \cdots + a_{k-1} u^{k-1}$:

$$\phi^4(a) = (b_0, b_1, b_2, \ldots, b_{p^{k-1}-1}),$$

where for all $0 \le i \le p^{k-2} - 1$, $0 \le \epsilon \le p - 1$, we have

$$b_{ip+\epsilon} = \begin{cases} a_{k-1} + \sum_{l=1}^{k-2} p_{l-1}(i) a_l + \epsilon a_0 & \text{if } k \ge 3, \\ a_1 + \epsilon a_0 & \text{if } k = 2. \end{cases}$$

For instance, when $p = k = 2$, it is easy to check that the Gray map ϕ^4 is identified with [1672]. As an additional example, when $p = k = 3$, we write $\phi^4(a_0 + a_1 u + a_2 u^2) = (b_0, b_1, b_2, \ldots, b_8)$. According to the above definition, we have $0 \le i \le 2$, $0 \le \epsilon \le 2$, and $\sum_{l=1}^{k-2} p_{l-1}(i) a_l = p_0(i) a_1 = i a_1$. Then we get

$$b_0 = a_2, \quad b_1 = a_2 + a_0, \quad b_2 = a_2 + 2a_0, \quad b_3 = a_2 + a_1, \quad b_4 = a_2 + a_1 + a_0$$
$$b_5 = a_2 + a_1 + 2a_0, \quad b_6 = a_2 + 2a_1, \quad b_7 = a_2 + 2a_1 + a_0, \quad \text{and} \quad b_8 = a_2 + 2a_1 + 2a_0.$$

It is easy to extend the Gray map ϕ^4 from $R(k, p, u^k = 0)^n$ to $\mathbb{F}_p^{p^{k-1}n}$, and we also know from [1680] that ϕ^4 is injective and linear.

The **homogeneous weight** of an element $x \in R(k, p, u^k = 0)$ is defined as follows:

$$\text{wt}_{\text{hom}}(x) = \begin{cases} 0 & \text{if } x = 0, \\ p^{k-1} & \text{if } x = \alpha u^{k-1} \text{ with } \alpha \in \mathbb{F}_p^*, \\ (p-1)p^{k-2} & \text{otherwise.} \end{cases}$$

The homogeneous weight of a codeword $\mathbf{c} = (c_1, c_2, \ldots, c_n)$ of $R(k, p, u^k = 0)^n$ is defined as $\text{wt}_{\text{hom}}(\mathbf{c}) = \sum_{i=1}^n \text{wt}_{\text{hom}}(c_i)$. For any $\mathbf{x}, \mathbf{y} \in R(k, p, u^k = 0)^n$, the **homogeneous distance** d_{hom} is given by $d_{\text{hom}}(\mathbf{x}, \mathbf{y}) = \text{wt}_{\text{hom}}(\mathbf{x} - \mathbf{y})$. As was observed in [1680], ϕ^4 is a distance preserving isometry from $(R(k, p, u^k = 0)^n, d_{\text{hom}})$ to $(\mathbb{F}_p^{p^{k-1}n}, d_H)$, where d_{hom} and d_H denote the homogeneous and Hamming distance in $R(k, p, u^k = 0)^n$ and $\mathbb{F}_p^{p^{k-1}n}$, respectively. This means if $\mathcal{C}$ is a linear code over $R(k, p, u^k = 0)$ with parameters (n, p^t, d), then $\phi^4(\mathcal{C})$ is a linear code of parameters $[p^{k-1}n, t, d]$ over $\mathbb{F}_p$. Note that when $p = k = 2$, the homogeneous weight is none other than the Lee weight, which was considered in [1672].

Shi et al. [1677] studied three different defining sets of $\mathcal{C}(m, p, L)$ over $\mathcal{R}(k, p, u^k = 0)$. For the first defining set, $\mathcal{Q}$ denotes the nonzero squares in $\mathbb{F}_{p^m}$ and $\mathcal{N}$ the non-squares in $\mathbb{F}_{p^m}$. For the third defining set, we need additional notation (analogous to that found in Section 18.3.4). Let ξ be a primitive element of $\mathbb{F}_{p^m}$. For N' a positive integer such that $N' \mid (p^m - 1)$, let $N_1' = \text{lcm}(N', \frac{p^m-1}{p-1})$ and $N_2' = \gcd(N', \frac{p^m-1}{p-1})$. Define $n' = \frac{N_1'}{N'}$ and $D' = \{\xi^{N'(j-1)} \mid j = 1, 2, \ldots, n'\}$. The three defining sets are as follows:

- $L_1 = \mathcal{Q} + u\mathbb{F}_{p^m} + \cdots + u^{k-1}\mathbb{F}_{p^m}$;

- $L_2 = \mathcal{R}(k, p, u^k = 0)^*$;

- $L_3 = D' + u\mathbb{F}_{p^m} + \cdots + u^{k-1}\mathbb{F}_{p^m}$. (See also [945].)

Theorem 18.4.1 ([1677]) *Let m be even, $p \equiv 1 \,(\mathrm{mod}\ 4)$, and $N_1 = (p^m - 1)p^{(k-1)(m+1)}/2$. For $a \in \mathcal{R}(k, p, u^k = 0)$, the homogeneous weight distribution of codewords in $\mathcal{C}(m, p, L_1)$ is given below.*

(a) *If $a = 0$, then $\mathrm{wt}_{\mathrm{hom}}(\mathrm{Ev}(a)) = 0$.*

(b) *If $a = \alpha u^{k-1}$, where $\alpha \in \mathcal{Q}$, then $\mathrm{wt}_{\mathrm{hom}}(\mathrm{Ev}(a)) = \frac{p-1}{p}\left(N_1 + p^{(k-1)(m+1)}(p^{\frac{m}{2}}+1)/2\right)$.*

(c) *If $a = \alpha u^{k-1}$, where $\alpha \in \mathcal{N}$, then $\mathrm{wt}_{\mathrm{hom}}(\mathrm{Ev}(a)) = \frac{p-1}{p}\left(N_1 - p^{(k-1)(m+1)}(p^{\frac{m}{2}}+1)/2\right)$.*

(d) *If $a \in \mathcal{R}(k, p, u^k = 0) \setminus \{\alpha u^{k-1} \mid \alpha \in \mathbb{F}_{p^m}\}$, then $\mathrm{wt}_{\mathrm{hom}}(\mathrm{Ev}(a)) = \frac{p-1}{p}N_1$.*

Theorem 18.4.2 ([1677]) *Let m be odd, $p \equiv 3 \,(\mathrm{mod}\ 4)$, and $N_1 = (p^m - 1)p^{(k-1)(m+1)}/2$. For $a \in \mathcal{R}(k, p, u^k = 0)$, the homogeneous weight distribution of codewords in $\mathcal{C}(m, p, L_1)$ is given below.*

(a) *If $a = 0$, then $\mathrm{wt}_{\mathrm{hom}}(\mathrm{Ev}(a)) = 0$.*

(b) *If $a = \alpha u^{k-1}$, where $\alpha \in \mathbb{F}_{p^m}^*$, then $\mathrm{wt}_{\mathrm{hom}}(\mathrm{Ev}(a)) = \frac{p-1}{p}\left(N_1 + p^{(k-1)(m+1)}/2\right)$.*

(c) *If $a \in \mathcal{R}(k, p, u^k = 0) \setminus \{\alpha u^{k-1} \mid \alpha \in \mathbb{F}_{p^m}\}$, then $\mathrm{wt}_{\mathrm{hom}}(\mathrm{Ev}(a)) = \frac{p-1}{p}N_1$.*

It is necessary to distinguish the difference between the case when $k = 2$ in Theorems 18.4.1 and 18.4.2 and the case in [1676]. Although the ring and the defining set are the same as [1676], the codes are not the same, because the Gray maps are different. We list their homogeneous weight distributions in Tables 18.11 and 18.12 to show the difference.

TABLE 18.11: Homogeneous weight distribution of the three-weight case ($k = 2$)

Weight in [1676, Theorem 1]	Weight in Theorem 18.4.1	Frequency
0	0	1
$(p^m - p^{m-1})(p^m - p^{\frac{m}{2}})$	$\frac{(p^{m+1}-p^m)(p^m - p^{\frac{m}{2}})}{2}$	$\frac{p^m - 1}{2}$
$(p^m - p^{m-1})(p^m - 1)$	$\frac{(p^{m+1}-p^m)(p^m - 1)}{2}$	$p^{2m} - p^m$
$(p^m - p^{m-1})(p^m + p^{\frac{m}{2}})$	$\frac{(p^{m+1}-p^m)(p^m + p^{\frac{m}{2}})}{2}$	$\frac{p^m - 1}{2}$

TABLE 18.12: Homogeneous weight distribution of the two-weight case ($k = 2$)

Weight in [1676, Theorem 2]	Weight in Theorem 18.4.2	Frequency
0	0	1
$(p^m - p^{m-1})(p^m - 1)$	$\frac{(p^{m+1}-p^m)(p^m - 1)}{2}$	$p^{2m} - p^m$
$p^{m-1}(p^{m+1} - p^m)$	$\frac{p^m(p^{m+1}-p^m)}{2}$	$p^m - 1$

According to Tables 18.11 and 18.12, it is easy to see that the corresponding dimension and frequency are the same. However, the nonzero weights of the codes are different. Furthermore, we can check that the corresponding nonzero weights and lengths have constant ratio in both tables. For example, in Table 18.12, we have

$$\frac{\frac{(p^{m+1}-p^m)(p^m-1)}{2}}{(p^m-p^{m-1})(p^m-1)} = \frac{\frac{p^m(p^{m+1}-p^m)}{2}}{p^{m-1}(p^{m+1}-p^m)} = \frac{p}{2},$$

and we know the length of the codes has the same proportional relationship.

Theorem 18.4.3 ([1677]) *Let* $N_2 = (p^m - 1)p^{(k-1)(m+1)}$. *For* $a \in \mathcal{R}(k, p, u^k = 0)$, *the homogeneous weight distribution of codewords in* $\mathcal{C}(m, p, L_2)$ *is given below.*

(a) *If* $a = 0$, *then* $\mathrm{wt}_{\mathrm{hom}}(\mathrm{Ev}(a)) = 0$.

(b) *If* $a = \alpha u^{k-1}$, *where* $\alpha \in \mathbb{F}_{p^m}^*$, *then* $\mathrm{wt}_{\mathrm{hom}}(\mathrm{Ev}(a)) = \frac{p-1}{p}\left(N_2 + p^{(k-1)(m+1)}\right)$.

(c) *If* $a \in \mathcal{R}(k, p, u^k = 0) \setminus \{\alpha u^{k-1} \mid \alpha \in \mathbb{F}_{p^m}\}$, *then* $\mathrm{wt}_{\mathrm{hom}}(\mathrm{Ev}(a)) = \frac{p-1}{p}N_2$.

Theorem 18.4.4 ([1677]) *Suppose* $N_2' = 1$ *and either* m *is even or* m *is odd with* $p \equiv 3$ (mod 4). *For* $a \in \mathcal{R}(k, p, u^k = 0)$, *the homogeneous weight distribution of codewords in* $\mathcal{C}(m, p, L_3)$ *is given below.*

(a) *If* $a = 0$, *then* $\mathrm{wt}_{\mathrm{hom}}(\mathrm{Ev}(a)) = 0$.

(b) *If* $a = \alpha u^{k-1}$, *where* $\alpha \in \mathbb{F}_{p^m}^*$, *then* $\mathrm{wt}_{\mathrm{hom}}(\mathrm{Ev}(a)) = p^{k(m+1)-2}$.

(c) *If* $a \in \mathcal{R}(k, p, u^k = 0) \setminus \{\alpha u^{k-1} \mid \alpha \in \mathbb{F}_{p^m}\}$, *then*

$$\mathrm{wt}_{\mathrm{hom}}(\mathrm{Ev}(a)) = (p^m - 1)p^{(k-1)(m+1)-1}.$$

Theorem 18.4.5 ([1677]) *Suppose* $1 < N_2' < p^{\frac{m}{2}} + 1$. *Assume* m *is even or* m *is odd and* $p \equiv 3$ (mod 4). *Let* $N_3 = n'p^{(k-1)(m+1)}$. *Then* $\mathcal{C}(m, p, L_3)$ *is a* $(N_3, p^{km}, d_{\mathrm{hom}})$ *linear code over* $R(k, p, u^k = 0)$ *which has at most* $N_2' + 1$ *nonzero homogeneous weights, and*

$$p^{(k-1)(m+1)-1} \cdot \frac{p^m - (N_2' - 1)p^{\frac{m}{2}}}{N_2'} \leq d_{\mathrm{hom}} \leq p^{(k-1)(m+1)-1} \cdot \frac{p^m - 1}{N_2'}.$$

Theorem 18.4.6 ([1677]) *Let* m *be even and* $N_2' > 2$. *Assume there exists a positive integer* k' *such that* $p^{k'} \equiv -1$ (mod N_2') *and* $2k'$ *divides* m. *Let* $t = \frac{m}{2k'}$.

(a) *If* N_2' *is even,* p, t *and* $\frac{p^{k'}+1}{N_2'}$ *are odd, then the code* $\mathcal{C}(m, p, L_3)$ *is a three-weight linear code provided that* $N_2' < p^{\frac{m}{2}} + 1$, *and its homogeneous weight distribution is given in Table 18.13.*

(b) *In all other cases, the code* $\mathcal{C}(m, p, L_3)$ *is a three-weight linear code provided that* $p^{\frac{m}{2}} + (-1)^t(N_2' - 1) > 0$ *and its homogeneous weight distribution is given in Table 18.14.*

Theorem 18.4.7 ([1677]) *If* m *is odd and* $p \equiv 3$ (mod 4), *then the code* $\phi^4(\mathcal{C}(m, p, L_1))$ *is optimal for a given length and dimension whenever*

$$m \geq \max\left\{k, \left\lfloor \frac{p^{k-1} - 2k + 1}{2(k-1)} \right\rfloor + 1\right\}.$$

TABLE 18.13: Homogeneous weight distribution of $\mathcal{C}(m,p,L_3)$ in Theorem 18.4.6(a)

Homogeneous weight	Frequency
0	1
$p^{(k-1)(m+1)-1} \cdot \dfrac{p^m - (N_2'-1)p^{\frac{m}{2}}}{N_2'}$	$\dfrac{p^m - 1}{N_2'}$
$p^{(k-1)(m+1)-1} \cdot \dfrac{p^{\frac{m}{2}} - 1}{N_2'}$	$p^{km} - p^m$
$p^{(k-1)(m+1)-1} \cdot \dfrac{p^m + p^{\frac{m}{2}}}{N_2'}$	$\dfrac{(N_2'-1)(p^m-1)}{N_2'}$

TABLE 18.14: Homogeneous weight distribution of $\mathcal{C}(m,p,L_3)$ in Theorem 18.4.6(b)

Homogeneous weight	Frequency
0	1
$p^{(k-1)(m+1)-1} \cdot \dfrac{p^m + (-1)^t (N_2'-1)p^{\frac{m}{2}}}{N_2'}$	$\dfrac{p^m - 1}{N_2'}$
$p^{(k-1)(m+1)-1} \cdot \dfrac{p^{\frac{m}{2}} - 1}{N_2'}$	$p^{km} - p^m$
$p^{(k-1)(m+1)-1} \cdot \dfrac{p^m - (-1)^t p^{\frac{m}{2}}}{N_2'}$	$\dfrac{(N_2'-1)(p^m-1)}{N_2'}$

Theorem 18.4.8 ([1677]) *The code $\phi^4(\mathcal{C}(m,p,L_2))$ is optimal for a given length and dimension if*

$$m \geq \max\left\{ k, \left\lfloor \frac{p^{k-1} - k}{k-1} \right\rfloor + 1 \right\}.$$

Theorem 18.4.9 ([1677]) *Assume $N_2' = 1$, m is even or m is odd and $p \equiv 3 \pmod 4$. Then the code $\phi^4(\mathcal{C}(m,p,L_3))$ is optimal for a given length and dimension if*

$$m \geq \max\left\{ k, \left\lfloor \frac{p^{k-1} - p(k-1) + k - 2}{(p-1)(k-1)} \right\rfloor + 1 \right\}.$$

Theorem 18.4.10 ([1677]) *For $m \geq 2$, then the minimum homogeneous distance of $\mathcal{C}(m,p,L_i)^\perp$ with $i = 1,2,3$ is $2(p-1)p^{k-2}$.*

Theorem 18.4.11 ([1677]) *If either $m \geq 3$ is odd with $p \equiv 3 \pmod 4$, or $m \geq 4$ is even, then $\phi^4(\mathcal{C}(m,p,L_1))$ is a minimal code.*

Theorem 18.4.12 ([1677]) *For $m \geq 2$, $\phi^4(\mathcal{C}(m,p,L_2))$ and $\phi^4(\mathcal{C}(m,p,L_3))$ are minimal codes.*

18.5 Three Special Rings

18.5.1 $R(2,p,u^2 = u)$

Shi et al. [1669] studied the trace from $\mathcal{R}(2,p,u^2 = u) = \mathbb{F}_{p^m} + u\mathbb{F}_{p^m}$ to $R(2,p,u^2 = u) = \mathbb{F}_p + u\mathbb{F}_p$ where $u^2 = u$. Under the Gray map ϕ^5, they obtained two infinite families of linear p-ary codes. The Gray map ϕ^5 from $R(2,p,u^2 = u)$ to $\mathbb{F}_p^2$ is defined by $\phi^5(a + ub) = (-b, 2a + b)$ for $a,b \in \mathbb{F}_p$. It is a one-to-one map which extends naturally into a map from

$R(2, p, u^2 = u)^n$ to $\mathbb{F}_p^{2n}$. As earlier, $\mathcal{Q}$ is the nonzero squares in $\mathbb{F}_{p^m}$, and $\mathcal{N}$ is the non-squares in $\mathbb{F}_{p^m}$.

Theorem 18.5.1 ([1669]) *Let $m \equiv 2 \pmod 4$ and $\epsilon(p) = (-1)^{\frac{p+1}{2}}$. For $a \in \mathcal{R}(2, p, u^2 = u)$, the Lee weight of codewords of $\mathcal{C}(m, p, \mathcal{Q} + u\mathbb{F}_{p^m}^*)$ is given below.*

(a) *If $a = 0$, then $\mathrm{wt_L}(\mathrm{Ev}(a)) = 0$.*

(b) *If $a = \alpha u$, where $\alpha \in \mathcal{Q}$, then*

$$\mathrm{wt_L}(\mathrm{Ev}(a)) = (p-1)\left(p^{2m-1} - p^{m-1} - \epsilon(p)p^{3m/2-1} + \epsilon(p)p^{m/2-1}\right).$$

(c) *If $a = \alpha u$, where $\alpha \in \mathcal{N}$, then*

$$\mathrm{wt_L}(\mathrm{Ev}(a)) = (p-1)\left(p^{2m-1} - p^{m-1} + \epsilon(p)p^{3m/2-1} - \epsilon(p)p^{m/2-1}\right).$$

(d) *If $a = \beta(1-u)$, where $\beta \in \mathbb{F}_{p^m}^*$, then $\mathrm{wt_L}(\mathrm{Ev}(a)) = (p-1)(p^{2m-1} - p^{m-1})$.*

(e) *If $a = \alpha u + \beta(1-u) \in \mathcal{R}(2, p, u^2 = u)^*$, where $\alpha \in \mathcal{Q}$, then*

$$\mathrm{wt_L}(\mathrm{Ev}(a)) = (p-1)\left(p^{2m-1} - 2p^{m-1} + \epsilon(p)p^{m/2-1}\right).$$

(f) *If $a = \alpha u + \beta(1-u) \in \mathcal{R}(2, p, u^2 = u)^*$, where $\alpha \in \mathcal{N}$, then*

$$\mathrm{wt_L}(\mathrm{Ev}(a)) = (p-1)\left(p^{2m-1} - 2p^{m-1} - \epsilon(p)p^{m/2-1}\right).$$

Theorem 18.5.2 ([1669]) *Assume m is odd and $p \equiv 3 \pmod 4$. For $a \in \mathcal{R}(2, p, u^2 = u)$, the Lee weight of codewords of $\mathcal{C}(m, p, \mathcal{Q} + u\mathbb{F}_{p^m}^*)$ is given below.*

(a) *If $a = 0$, then $\mathrm{wt_L}(\mathrm{Ev}(a)) = 0$.*

(b) *If $a = \beta u$ with $\beta \in \mathbb{F}_{p^m}^*$, then $\mathrm{wt_L}(\mathrm{Ev}(a)) = (p-1)(p^{2m-1} - p^{m-1})$.*

(c) *If $a = \beta(1-u)$ with $\beta \in \mathbb{F}_{p^m}^*$, then $\mathrm{wt_L}(\mathrm{Ev}(a)) = (p-1)(p^{2m-1} - p^{m-1})$.*

(d) *If $a \in \mathcal{R}(2, p, u^2 = u)^*$, then $\mathrm{wt_L}(\mathrm{Ev}(a)) = (p-1)(p^{2m-1} - 2p^{m-1})$.*

Thus we obtain a family of p-ary two-weight codes of parameters $[p^{2m} - 2p^m + 1, 2m]$, with weight distribution as given in Table 18.15. The parameters are different from those in [330, 1672, 1673, 1676].

TABLE 18.15: Lee weight distribution of $\mathcal{C}(m, p, \mathcal{Q} + u\mathbb{F}_{p^m}^*)$ in Theorem 18.5.2

Lee weight	Frequency
0	1
$(p-1)(p^{2m-1} - 2p^{m-1})$	$(p^m - 1)^2$
$(p-1)(p^{2m-1} - p^{m-1})$	$2(p^m - 1)$

Theorem 18.5.3 ([1669]) *For $a \in \mathcal{R}(2, p, u^2 = u)$, the Lee weight distribution of codewords of $\mathcal{C}(m, p, \mathbb{F}_{p^m} + u\mathbb{F}_{p^m}^*)$ is given below.*

(a) *If $a = 0$, then $\mathrm{wt_L}(\mathrm{Ev}(a)) = 0$.*

(b) *If $a = \alpha u$ with $\alpha \in \mathbb{F}_{p^m}^*$, then $\mathrm{wt_L}(\mathrm{Ev}(a)) = 2(p-1)(p^{2m-1} - p^{m-1})$.*

(c) *If $a = \beta(1 - u)$ with $\beta \in \mathbb{F}_{p^m}^*$, then $\mathrm{wt_L}(\mathrm{Ev}(a)) = 2(p-1)(p^{2m-1} - p^{m-1})$.*

(d) *If $a \in \mathcal{R}(2, p, u^2 = u)^*$, then $\mathrm{wt_L}(\mathrm{Ev}(a)) = 2(p-1)(p^{2m-1} - 2p^{m-1})$.*

The code $\phi^5(\mathcal{C}(m, p, \mathbb{F}_{p^m}^* + u\mathbb{F}_{p^m}^*))$ is a two-weight linear code over $\mathbb{F}_p$ of length $2(p^m - 1)^2$ and dimension $2m$. The Hamming weight distribution is given in Table 18.16. Note that the parameters are different from those in [330, 1307, 1308, 1312].

TABLE **18.16**: Hamming weight distribution of $\phi^5(\mathcal{C}(m, p, \mathbb{F}_{p^m}^* + u\mathbb{F}_{p^m}^*))$; see Theorem 18.5.3

Hamming weight	Frequency
0	1
$2(p-1)(p^{2m-1} - 2p^{m-1})$	$p^{2m} - 2p^m + 1$
$2(p-1)(p^{2m-1} - p^{m-1})$	$2p^m - 2$

Theorem 18.5.4 ([1669]) *Assume $m \geq 3$ is odd and $p \equiv 3$ (mod 4). Then $\phi^5(\mathcal{C}(m, p, \mathcal{Q} + u\mathbb{F}_{p^m}^*))$ is optimal for a given length and dimension.*

Theorem 18.5.5 ([1669]) *Assume $m \geq 3$ with $p = 3$ or $m \geq 4$ with $p \geq 5$. Then $\phi^5(\mathcal{C}(m, p, \mathbb{F}_{p^m}^* + u\mathbb{F}_{p^m}^*))$ is optimal for a given length and dimension.*

18.5.2 $R(3, 2, u^3 = 1)$

Shi et al. [1679] considered the ring $R(3, 2, u^3 = 1) = \mathbb{F}_2 + u\mathbb{F}_2 + u^2\mathbb{F}_2$ with $u^3 = 1$. The Gray map ϕ^6 from $R(3, 2, u^3 = 1)$ to $\mathbb{F}_2^3$ is defined by

$$\phi^6(a_1 + a_2u + a_3u^2) = (a_1, a_2, a_3),$$

where $a_1, a_2, a_3 \in \mathbb{F}_2$. It is a one-to-one map from $R(3, 2, u^3 = 1)$ to $\mathbb{F}_2^3$, which can be extended naturally to a map from $R(3, 2, u^3 = 1)^n$ to $\mathbb{F}_2^{3n}$. As usual, the extension ring $\mathcal{R}(3, 2, u^3 = 1)$ is $\mathbb{F}_{2^m} + u\mathbb{F}_{2^m} + u^2\mathbb{F}_{2^m}$ with $u^3 = 1$.

Theorem 18.5.6 ([1679]) *Suppose m is odd. For $a \in \mathcal{R}(3, 2, u^3 = 1)$, the Lee weight distribution of codewords in $\mathcal{C}(m, 2, \mathcal{R}(3, 2, u^3 = 1)^*)$ is given below.*

(a) *If $a = 0$, then $\mathrm{wt_L}(\mathrm{Ev}(a)) = 0$.*

(b) *If $a \in \mathcal{R}(3, 2, u^3 = 1)^*$, then $\mathrm{wt_L}(\mathrm{Ev}(a)) = 3(2^{3m-1} - 2^{2m-1} - 2^{m-1})$.*

(c) *If $a \in \mathcal{R}(3, 2, u^3 = 1) \setminus \{\mathcal{R}(3, 2, u^3 = 1)^* \cup \{0\}\}$, $a = a_1 + a_2u + a_3u^2$, and*

    (i) *if $(a_1+a_2, a_1+a_3) = (0, 0)$, $a_1+a_2+a_3 \neq 0$, then $\mathrm{wt_L}(\mathrm{Ev}(a)) = 3(2^{3m-1} - 2^{m-1})$;*

    (ii) *if $(a_1+a_2, a_1+a_3) \neq (0, 0)$, $a_1+a_2+a_3 = 0$, then $\mathrm{wt_L}(\mathrm{Ev}(a)) = 3(2^{3m-1} - 2^{2m-1})$.*

Theorem 18.5.7 ([1679]) *Suppose m is even. Let $\mathbb{F}_4 = \{0, 1, \omega, \omega^2\} \subseteq \mathbb{F}_{2^m}$ where $\omega^2 = 1 + \omega$ and $\omega^3 = 1$. For $a \in \mathcal{R}(3, 2, u^3 = 1)$, the Lee weight distribution of codewords in $\mathcal{C}(m, 2, \mathcal{R}(3, 2, u^3 = 1)^*)$ is given below.*

(a) *If $a = 0$, then $\mathrm{wt_L}(\mathrm{Ev}(a)) = 0$.*

(b) *If $a \in \mathcal{R}(3, 2, u^3 = 1)^*$, then $\mathrm{wt_L}(\mathrm{Ev}(a)) = 3(2^{3m-1} - 3 \cdot 2^{2m-1} + 3 \cdot 2^{m-1})$.*

(c) *If* $a = a_1 + a_2 u + a_3 u^2 \in R(3, 2, u^3 = 1) \setminus \{R(3, 2, u^3 = 1)^* \cup \{0\}\}$, *then*

  (i) *if exactly one of* $a_1 + a_2 + a_3$, $a_1 + a_2 \omega + a_3 \omega^2$, *or* $a_1 + a_2 \omega^2 + a_3 \omega$ *equals 0, then*
  $$\text{wt}_L(\text{Ev}(a)) = 3(2^{3m-1} - 3 \cdot 2^{2m-1} + 2^m);$$

  (ii) *if exactly two of* $a_1 + a_2 + a_3$, $a_1 + a_2 \omega + a_3 \omega^2$, *or* $a_1 + a_2 \omega^2 + a_3 \omega$ *equals 0, then*
  $$\text{wt}_L(\text{Ev}(a)) = 3(2^{3m-1} - 2^{2m} + 2^{m-1}).$$

Next, the optimality of their binary images is given in the following theorem.

Theorem 18.5.8 ([1679]) *For any odd* $m > 1$, *the code* $\phi^6(\mathcal{C}(m, 2, \mathcal{R}(3, 2, u^3 = 1)^*))$ *is optimal for a given length and dimension.*

18.5.3 $R(3, 3, u^3 = 1)$

Using the same Gray map as ϕ^6, Shi et al. [1671] considered the ring $R(3, 3, u^3 = 1) = \mathbb{F}_3 + u\mathbb{F}_3 + u^2\mathbb{F}_3$ and extension ring $\mathcal{R}(3, 3, u^3 = 1) = \mathbb{F}_{3^m} + u\mathbb{F}_{3^m} + u^2\mathbb{F}_{3^m}$ with $u^3 = 1$. As usual, $\mathcal{Q}$ is the nonzero squares in $\mathbb{F}_{3^m}$, and $\mathcal{N}$ is the non-squares in $\mathbb{F}_{3^m}$.

Theorem 18.5.9 ([1671]) *Suppose* $m \equiv 2 \pmod 4$. *For* $a \in \mathcal{R}(3, 3, u^3 = 1)$, *the Lee weight distribution of codewords in* $\mathcal{C}(m, 3, \mathcal{Q} + u\mathbb{F}_{3^m} + u^2\mathbb{F}_{3^m})$ *is given below.*

  (a) *If* $a = 0$, *then* $\text{wt}_L(\text{Ev}(a)) = 0$.

  (b) *If* $a = \alpha(u - 1)^2$, *where* $\alpha \in \mathcal{Q}$, *then* $\text{wt}_L(\text{Ev}(a)) = 3^{3m} - 3^{5m/2}$.

  (c) *If* $a = \alpha(u - 1)^2$, *where* $\alpha \in \mathcal{N}$, *then* $\text{wt}_L(\text{Ev}(a)) = 3^{3m} + 3^{5m/2}$.

  (d) *If* $a \in \mathcal{R}(3, 3, u^3 = 1) \setminus \langle (u - 1)^2 \rangle$, *then* $\text{wt}_L(\text{Ev}(a)) = 3^{3m} - 3^{2m}$.

Theorem 18.5.10 ([1671]) *Suppose* m *is odd. For* $a \in \mathcal{R}(3, 3, u^3 = 1)$, *the Lee weight distribution of codewords in* $\mathcal{C}(m, 3, \mathcal{Q} + u\mathbb{F}_{3^m} + u^2\mathbb{F}_{3^m})$ *is given below.*

  (a) *If* $a = 0$, *then* $\text{wt}_L(\text{Ev}(a)) = 0$.

  (b) *If* $a = \alpha(u - 1)^2$ *with* $\alpha \in \mathbb{F}_{3^m}^*$, *then* $\text{wt}_L(\text{Ev}(a)) = 3^{3m}$.

  (c) *If* $a \in \mathcal{R}(3, 3, u^3 = 1) \setminus \langle (u - 1)^2 \rangle$, *then* $\text{wt}_L(\text{Ev}(a)) = 3^{3m} - 3^{2m}$.

Theorem 18.5.11 ([1671]) *For* $a \in \mathcal{R}(3, 3, u^3 = 1)$, *the Lee weight distribution of codewords in* $\mathcal{C}(m, 3, \mathcal{R}(3, 3, u^3 = 1)^*)$ *is given below.*

  (a) *If* $a = 0$, *then* $\text{wt}_L(\text{Ev}(a)) = 0$.

  (b) *If* $a = \alpha(u - 1)^2$ *with* $\alpha \in \mathbb{F}_{3^m}^*$, *then* $\text{wt}_L(\text{Ev}(a)) = 2 \cdot 3^{3m}$.

  (c) *If* $a \in \mathcal{R}(3, 3, u^3 = 1) \setminus \langle (u - 1)^2 \rangle$, *then* $\text{wt}_L(\text{Ev}(a)) = 2(3^{3m} - 3^{2m})$.

Theorem 18.5.12 ([1671]) *The codes* $\phi^6(\mathcal{C}(m, 3, L))$ *of length* $3|L|$ *are optimal given length and dimension for the following* L:

  (a) m *is odd and* $L = \mathcal{Q} + u\mathbb{F}_{3^m} + u^2\mathbb{F}_{3^m}$ *in Theorem 18.5.10;*

  (b) m *is a positive integer and* $L = \mathcal{R}(3, 3, u^3 = 1)^*$ *in Theorem 18.5.11.*

18.6   Conclusion

Inspired by trace codes over finite fields, many scholars have studied several classes of few-weight codes over finite rings. This chapter has explored some of these codes. With the proper definition of the Gray map, images of these trace codes under the Gray map have led to few-weight linear codes over finite fields, many of which turn out to be optimal.

Chapter 19

Two-Weight Codes

Andries E. Brouwer

Eindhoven University of Technology

19.1   Generalities

A **linear code** C with **length** n, **dimension** m, and **minimum distance** d over the field $\mathbb{F}_q$ (in short, an $[n, m, d]_q$ code) is an m-dimensional subspace of the vector space $\mathbb{F}_q^n$ such that any two distinct codewords (elements of C) differ in at least d coordinates. A **generator matrix** for C is an $m \times n$ matrix M such that its rows span C. The **dual code**

$C^\perp$ of C is the $(n-m)$-dimensional code consisting of the vectors orthogonal to all of C for the inner product $(u, v) = \sum u_i v_i$.

The **weight** $\mathrm{wt}_H(c)$ of the codeword c is its number of nonzero coordinates. A weight of C is the weight of some codeword in C. A **two-weight code** is a linear code with exactly two nonzero weights. (Codes with few weights appear in Chapters 18 and 20.) The **weight enumerator** of C is the polynomial $\sum f_i x^i$ where the coefficient f_i of x^i is the number of words of weight i in C. (This equals $\mathrm{Hwe}_C(x, 1)$ from Definition 1.15.1.)

19.2   Codes as Projective Multisets

Let C be a linear code of length n and dimension m over the field $\mathbb{F}_q$. Let the $m \times n$ matrix M be a generator matrix of C. The columns of M are elements of $V = \mathbb{F}_q^m$, the m-dimensional vector space over $\mathbb{F}_q$, and up to coordinate permutation the code C is uniquely determined by the n-multiset of column vectors consisting of the columns of M. Let us call a coordinate position where C is identically zero a **zero position**. The code C will have zero positions if and only if the dual code $C^\perp$ has codewords of weight 1. Usually, zero positions are uninteresting and can be discarded.

Let PV be the **projective space** of which the points are the 1-spaces in V. If C does not have zero positions, then each column c of M determines a projective point $\langle c \rangle$ in PV, and we find a projective multiset X of size n in PV. Note that PV is spanned by X.

In this way we get a 1-1 correspondence between linear codes C without zero positions (up to equivalence) and projective multisets X (up to nonsingular linear transformations): If C' is a linear code equivalent to C, and M' a generator matrix for C', then $M' = AMB$, where A is a nonsingular matrix of order m (so that $AC = C$), and B is a monomial matrix of order n (a matrix with a single nonzero entry in each row and column), so that $XB = X$.

The code C is called **projective** when no two coordinate positions are dependent, i.e., when the dual code $C^\perp$ has minimum distance at least 3. This condition says that the multiset does not contain repeated points, i.e., is a set.

19.2.1   Weights

Let Z be the (multi)set of columns of M, so that $V = \langle Z \rangle$. We can extend any codeword $u = (u(z))_{z \in Z} \in C$ to a linear functional on V. Let X be the projective (multi)set in PV determined by the columns of M. For $u \neq 0$, let H_u be the hyperplane in PV defined by $u(z) = 0$. The weight of the codeword u is $\mathrm{wt}_H(u) = n - |X \cap H_u|$.

Therefore searching for a code with large minimum distance is the same as searching for a projective (multi)set such that all hyperplane intersections are small. Searching for a code with few different weights is the same as searching for a projective (multi)set such that its hyperplane sections only have a few different sizes.

Example 19.2.1 Consider codes with dimension $m = 3$ and minimum distance $d = n - 2$. According to the above, these correspond to subsets X of the projective plane $\mathrm{PG}(2, q)$ such that each line meets X in at most 2 points. It follows that X is an arc (or a double point). For odd q the best one can do is to pick a conic (of size $q + 1$) and one finds $[q + 1, 3, q - 1]_q$ codes. For even q one can pick a hyperoval (of size $q + 2$) and one finds $[q + 2, 3, q]_q$ codes. The $[6, 3, 4]_4$ code is the famous **hexacode** [442].

19.3   Graphs

Let Γ be a graph with vertex set S of size s, undirected, without loops or multiple edges. For $x, y \in S$ we write $x = y$ or $x \sim y$ or $x \not\sim y$ when the vertices x and y are equal, adjacent, or nonadjacent, respectively. The **adjacency matrix** of Γ is the matrix A of order s with rows and columns indexed by S, where $A_{xy} = 1$ if $x \sim y$ and $A_{xy} = 0$ otherwise. The **spectrum** of Γ is the spectrum of A, that is, its (multi)set of eigenvalues.

19.3.1   Difference Sets

Given an abelian group G and a subset D of G such that $D = -D$ and $0 \notin D$, we can define a graph Γ with vertex set G by letting $x \sim y$ whenever $y - x \in D$. This graph is known as the **Cayley graph** on G with **difference set** D.

If A is the adjacency matrix of Γ, and χ is a character of G, then $(A\chi)(x) = \sum_{y \sim x} \chi(y) = \sum_{d \in D} \chi(x + d) = (\sum_{d \in D} \chi(d))\chi(x)$. It follows that the eigenvalues of Γ are the numbers $\sum_{d \in D} \chi(d)$, where χ runs through the characters of G. In particular, the trivial character yields the eigenvalue $|D|$, the valency of Γ.

19.3.2   Using a Projective Set as a Difference Set

Let V be a vector space of dimension m over the finite field $\mathbb{F}_q$. Let X be a subset of size n of the point set of the projective space PV. Let D be the set of $(q-1)n$ vectors spanning the points of X. Define a graph Γ with vertex set V by letting $x \sim y$ whenever $y - x \in D$. This graph has $v = q^m$ vertices, and is regular of valency $k = (q-1)n$.

Let q be a power of the prime p, let $\zeta = e^{2\pi i/p}$ be a primitive p^{th} root of unity, and let $\text{tr} : \mathbb{F}_q \to \mathbb{F}_p$ be the trace function (if $q = p^r$, then $\text{tr}(x) = \sum_{i=0}^{r-1} x^{p^i}$). Let V^* be the dual vector space to V, that is, the space of $\mathbb{F}_q$-linear maps from V to $\mathbb{F}_q$. Then the characters χ of V are of the form $\chi_a(x) = \zeta^{\text{tr}(a(x))}$, with $a \in V^*$. Now

$$\sum_{\lambda \in \mathbb{F}_q} \chi_a(\lambda x) = \begin{cases} q & \text{if } a(x) = 0, \\ 0 & \text{otherwise.} \end{cases}$$

It follows that $\sum_{d \in D} \chi_a(d) = q\,|H_a \cap X| - |X|$ where H_a is the hyperplane $\{\langle x \rangle \mid a(x) = 0\}$ in PV.

This can be formulated in terms of coding theory. To the set X corresponds a (projective) linear code $\mathcal{C}$ of length n and dimension m. Each $a \in V^*$ gives rise to the vector $(a(x))_{x \in X}$ indexed by X, and the collection of all these vectors is the code $\mathcal{C}$. A codeword a of weight w corresponds to a hyperplane H_a that meets X in $n - w$ points, and hence to an eigenvalue $q(n - w) - n = k - qw$. The number of codewords of weight w in $\mathcal{C}$ equals the multiplicity of the eigenvalue $k - qw$ of Γ.

19.3.3   Strongly Regular Graphs

A **strongly regular graph** with parameters (v, k, λ, μ) is a graph on v vertices, regular of valency k, where $0 < k < v - 1$ (there are both edges and non-edges), such that the number of common neighbors of any two distinct vertices equals λ if they are adjacent, and μ if they are nonadjacent. For the adjacency matrix A of the graph this means that $A^2 = kI + \lambda A + \mu(J - I - A)$, where J is the all-1 matrix. A regular graph is strongly regular precisely when apart from the valency it has precisely two distinct eigenvalues. The eigenvalues of Γ, that is, the eigenvalues of A, are the valency k and the two solutions of $x^2 + (\mu - \lambda)x + \mu - k = 0$.

In the above setting, with a graph Γ on a vector space, with adjacency defined by a projective set X as difference set, the graph Γ will be strongly regular precisely when $|H \cap X|$ takes only two different values for hyperplanes H, that is, when the code corresponding to X is a two-weight code.

This 1-1-1 correspondence between projective two-weight codes, projective sets that meet the hyperplanes in two cardinalities (these are known as **2-character sets**), and strongly regular graphs defined on a vector space by a projective difference set, is due to Delsarte [518].

The more general case of a code $\mathcal{C}$ with dual $\mathcal{C}^\perp$ of minimum distance at least 2 corresponds to a multiset X. Brouwer and van Eupen [313] gives a 1-1 correspondence between arbitrary projective codes and arbitrary two-weight codes. See Section 19.9 below.

A survey of two-weight codes was given by Calderbank and Kantor [330]. Additional families and examples are given in [446, 447]. In [270] numerical data (such as the number of non-isomorphic codes and the order of the automorphism group) is given for small cases.

19.3.4   Parameters

Let V be a vector space of dimension m over $\mathbb{F}_q$. Let X be a subset of size n of the point set of PV that meets hyperplanes in either m_1 or m_2 points, where $m_1 > m_2$. Let f_1 and f_2 be the numbers of such hyperplanes. Then f_1 and f_2 satisfy

$$f_1 + f_2 = \frac{q^m - 1}{q - 1},$$

$$f_1 m_1 + f_2 m_2 = n \frac{q^{m-1} - 1}{q - 1},$$

$$f_1 m_1 (m_1 - 1) + f_2 m_2 (m_2 - 1) = n(n - 1) \frac{q^{m-2} - 1}{q - 1}$$

and it follows that

$$(q^m - 1)m_1 m_2 - n(q^{m-1} - 1)(m_1 + m_2 - 1) + n(n - 1)(q^{m-2} - 1) = 0,$$

so that in particular $n \mid (q^m - 1)m_1 m_2$.

The corresponding two-weight code is a q-ary linear code with dimension m, length n, weights $w_1 = n - m_1$ and $w_2 = n - m_2$, minimum distance w_1, and weight enumerator $1 + (q - 1)f_1 x^{w_1} + (q - 1)f_2 x^{w_2}$. Here $(q - 1)f_1 = \frac{1}{w_2 - w_1}(w_2(q^m - 1) - nq^{m-1}(q - 1))$.

The corresponding strongly regular graph Γ has parameters determined by

$$v = q^m, \qquad k = (q - 1)n, \qquad r = qm_1 - n, \qquad s = qm_2 - n,$$
$$\lambda = \mu + r + s, \qquad \mu = rs + k = \frac{w_1 w_2}{q^{m-2}}, \qquad f = (q - 1)f_1, \qquad g = (q - 1)f_2,$$

where r, s are the eigenvalues of Γ other than k (with $r \geq 0 > s$) and f, g their multiplicities.

Example 19.3.1 The hyperoval in $\mathrm{PG}(2,4)$ ($m = 3, q = 4, n = 6$) gives the linear $[6, 3, 4]_4$ code (with weight enumerator $1 + 45x^4 + 18x^6$), but also corresponds to a strongly regular graph with parameters $(v, k, \lambda, \mu) = (64, 18, 2, 6)$ and spectrum $18^1\ 2^{45}\ (-6)^{18}$, where exponents denote multiplicities.

It is not often useful, but one can also check the definition of strong regularity directly. The graph Γ defined by the difference set X will be strongly regular with constants λ, μ if and only if each point outside X is collinear with μ ordered pairs of points of X, while each point p inside X is collinear with $\lambda - (q - 2)$ ordered pairs of points of $X \setminus \{p\}$.

19.3.5 Complement

Passing from a 2-character set X to its complement corresponds to passing from the strongly regular graph to its complement. The two-weight codes involved have a more complicated relation and will look very different, with different lengths and minimum distances.

Example 19.3.2 The dual of the ternary Golay code is a $[11, 5, 6]_3$-code with weights 6 and 9. It corresponds to an 11-set in $PG(4, 3)$ such that hyperplanes meet it in 5 or 2 points. Its complement is a 110-set in $PG(4, 3)$ such that hyperplanes meet it in 35 or 38 points. It corresponds to a $[110, 5, 72]_3$-code with weights 72 and 75.

19.3.6 Duality

Suppose X is a subset of the point set of PV that meets hyperplanes in either m_1 or m_2 points. We find a subset Y of the point set of the dual space PV^* consisting of the hyperplanes that meet X in m_1 points. Also Y is a 2-character set. If each point of PV is on n_1 or n_2 hyperplanes in Y, with $n_1 > n_2$, then $n_2 = \frac{n(q^{m-2}-1)-m_2(q^{m-1}-1)}{(q-1)(m_1-m_2)}$ and $(m_1 - m_2)(n_1 - n_2) = q^{m-2}$. It follows that the difference of the weights in a projective two-weight code is a power of the characteristic. (This is a special case of the duality for translation association schemes. See [519, Section 2.6] and [309, Section 2.10B].)

To a pair of complementary sets or graphs belongs a dual pair of complementary sets or graphs. The valencies k, $v - k - 1$ of the dual graph are the multiplicities f_1, f_2 of the graph.

Let $\mathcal{C}$ be the two-weight code belonging to X. Then the graph belonging to Y has vertex set $\mathcal{C}$, where codewords are joined when their difference has weight w_1.

Example 19.3.3 For the $[11, 5, 6]_3$-code of Example 19.3.2 (with weight enumerator $1 + 132x^6 + 110x^9$) the corresponding strongly regular graph has parameters $(v, k, \lambda, \mu) = (243, 22, 1, 2)$ and spectrum $22^1\ 4^{132}\ (-5)^{110}$. One of the two dual graphs has parameters $(v, k, \lambda, \mu) = (243, 110, 37, 60)$ and spectrum $110^1\ 2^{220}\ (-25)^{22}$. The corresponding two-weight code is a $[55, 5, 36]_3$-code with weights 36 and 45.

19.3.7 Field Change

Our graphs are defined by a difference set in an abelian group, and are independent of a multiplicative field structure we put on that additive group.

Suppose V is a vector space of dimension m over F, where F has a subfield F_0 with $[F : F_0] = e$, say $F = \mathbb{F}_q$, $F_0 = \mathbb{F}_r$, with $q = r^e$. Let V_0 be V, but regarded as a vector space (of dimension me) over F_0. Each projective point in PV corresponds to $\frac{q-1}{r-1}$ projective points in PV_0. If our graph belonged to a projective subset X of size n of PV, it also belongs to a set X_0 of size $n\frac{q-1}{r-1}$ of PV_0. If the intersection numbers were m_i before, they will be $\frac{r^e-1}{r-1}m_i + \frac{r^{e-1}-1}{r-1}(n - m_i)$ now. We see that a q-ary code of dimension m, length n, and weights w_i becomes an r-ary code of dimension me, length $n\frac{q-1}{r-1}$ and weights $w_i\frac{q}{r}$.

19.4 Irreducible Cyclic Two-Weight Codes

In the case of a vector space that is a field F, one conjectures that all examples are known of difference sets that are subgroups of the multiplicative group F^* containing the multiplicative group of the base field.

Conjecture 19.4.1 (Schmidt–White [1627, Conj. 4.4]; see also [798, Conj. 1.2])
Let F be a finite field of order $q = p^f$, p prime. Suppose $1 < e \mid (q-1)/(p-1)$ and let D be the subgroup of F^* of index e. If the Cayley graph on F with difference set D is strongly regular, then one of the following holds:

(a) **Subfield case:** D is the multiplicative group of a subfield of F.

(b) **Semiprimitive case:** There exists a positive integer l such that $p^l \equiv -1 \pmod{e}$.

(c) **Exceptional cases:** $|F| = p^f$, and (e, p, f) takes one of the following eleven values: $(11, 3, 5)$, $(19, 5, 9)$, $(35, 3, 12)$, $(37, 7, 9)$, $(43, 11, 7)$, $(67, 17, 33)$, $(107, 3, 53)$, $(133, 5, 18)$, $(163, 41, 81)$, $(323, 3, 144)$, $(499, 5, 249)$.

In each of the exceptional cases the graph is strongly regular. These graphs correspond to two-weight codes over $\mathbb{F}_p$.

Since F^* has a partition into cosets of D, the point set of the projective space PF is partitioned into isomorphic copies of the two-intersection set $X = \{\langle d \rangle \mid d \in D\}$. See also [568, 1565, 1848].

19.5 Cyclotomy

More generally, the difference set D can be a union of cosets of a subgroup of F^*, for some finite field F. Let $F = \mathbb{F}_q$ where $q = p^f$, p is prime, and let $e \mid (q-1)$, say $q = em + 1$. Let $K \subseteq \mathbb{F}_q^*$ be the subgroup of the eth powers (so that $|K| = m$). Let α be a primitive element of $\mathbb{F}_q$. For $J \subseteq \{0, 1, \ldots, e-1\}$ put $u := |J|$ and $D := D_J := \bigcup\{\alpha^j K \mid j \in J\} = \{\alpha^{ie+j} \mid j \in J, 0 \le i < m\}$. Define a graph $\Gamma = \Gamma_J$ with vertex set $\mathbb{F}_q$ and edges (x, y) whenever $y - x \in D$. Note that Γ will be undirected if q is even or $e \mid (q-1)/2$.

As before, the eigenvalues of Γ are the sums $\theta(\chi) = \sum_{d \in D} \chi(d)$ for the characters χ of F. Their explicit determination requires some theory of Gauss sums. Let us write $A\chi = \theta(\chi)\chi$. Clearly, $\theta(1) = mu$, the valency of Γ. Now assume $\chi \ne 1$. Then $\chi = \chi_g$ for some g, where

$$\chi_g(\alpha^j) = \exp\left(\frac{2\pi i}{p} \operatorname{tr}(\alpha^{j+g})\right)$$

and $\operatorname{tr} : \mathbb{F}_q \to \mathbb{F}_p$ is the trace function. If μ is any multiplicative character of order e (say, $\mu(\alpha^j) = \zeta^j$, where $\zeta = \exp(\frac{2\pi i}{e})$), then

$$\sum_{i=0}^{e-1} \mu^i(x) = \begin{cases} e & \text{if } \mu(x) = 1, \\ 0 & \text{otherwise.} \end{cases}$$

Hence,

$$\theta(\chi_g) = \sum_{d \in D} \chi_g(d) = \sum_{j \in J} \sum_{y \in K} \chi_{j+g}(y) = \frac{1}{e} \sum_{j \in J} \sum_{x \in \mathbb{F}_q^*} \chi_{j+g}(x) \sum_{i=0}^{e-1} \mu^i(x)$$

$$= \frac{1}{e} \sum_{j \in J} \left(-1 + \sum_{i=1}^{e-1} \sum_{x \neq 0} \chi_{j+g}(x) \mu^i(x) \right) = \frac{1}{e} \sum_{j \in J} \left(-1 + \sum_{i=1}^{e-1} \mu^{-i}(\alpha^{j+g}) G_i \right)$$

where G_i is the Gauss sum $\sum_{x \neq 0} \chi_0(x) \mu^i(x)$. In a few cases these sums can be evaluated.

Proposition 19.5.1 (Stickelberger–Davenport–Hasse; see [1366]) *Suppose $e > 2$ and p is semiprimitive mod e; i.e., there exists an l such that $p^l \equiv -1 \pmod{e}$. Choose l minimal and write $f = 2lt$. Then*

$$G_i = (-1)^{t+1} \varepsilon^{it} \sqrt{q},$$

where

$$\varepsilon = \begin{cases} -1 & \text{if } e \text{ is even and } (p^l + 1)/e \text{ is odd,} \\ +1 & \text{otherwise.} \end{cases}$$

Under the hypotheses of this proposition, we have

$$\sum_{i=1}^{e-1} \mu^{-i}(\alpha^{j+g}) G_i = \sum_{i=1}^{e-1} \zeta^{-i(j+g)} (-1)^{t+1} \varepsilon^{it} \sqrt{q} = \begin{cases} (-1)^t \sqrt{q} & \text{if } r \neq 1, \\ (-1)^{t+1} \sqrt{q}(e-1) & \text{if } r = 1, \end{cases}$$

where $r = r_{g,j} = \zeta^{-j-g} \varepsilon^t$ (so that $r^e = \varepsilon^{et} = 1$), and hence

$$\theta(\chi_g) = \frac{u}{e}(-1 + (-1)^t \sqrt{q}) + (-1)^{t+1} \sqrt{q} \cdot \#\{j \in J \mid r_{g,j} = 1\}.$$

Noting that if $\varepsilon^t = -1$, then e is even and p is odd, we have

$$\#\{j \in J \mid r_{g,j} = 1\} = \begin{cases} 1 & \text{if } \varepsilon^t = 1 \text{ and } g \in -J \pmod{e}, \\ 1 & \text{if } \varepsilon^t = -1 \text{ and } g \in \frac{1}{2}e - J \pmod{e}, \\ 0 & \text{otherwise.} \end{cases}$$

We have proved the following.

Theorem 19.5.2 ([141, 314]) *Let $q = p^f$, p prime, $f = 2lt$ and $e \mid (p^l + 1) \mid (q - 1)$. Let $u = |J|$, $1 \leq u \leq e - 1$. Then the graphs Γ_J are strongly regular with eigenvalues*

$$k = \frac{q-1}{e} u \quad \text{with multiplicity 1,}$$
$$\frac{u}{e}(-1 + (-1)^t \sqrt{q}) \quad \text{with multiplicity } q - 1 - k,$$
$$\frac{u}{e}(-1 + (-1)^t \sqrt{q}) + (-1)^{t+1} \sqrt{q} \quad \text{with multiplicity } k.$$

This will yield two-weight codes over $\mathbb{F}_r$ in case K is invariant under multiplication by nonzero elements in $\mathbb{F}_r$, i.e., when $e \mid \frac{q-1}{r-1}$. This is always true for $r = p^l$, but also happens, for example, when $q = p^{2lt}$, $r = p^{lt}$, $e \mid (p^l + 1)$, and t is odd.

19.5.1 The Van Lint–Schrijver Construction

Van Lint and Schrijver [1837] used the above setup in the case e is an odd prime and p is primitive mod e (so that $l = (e-1)/2$ and $f = (e-1)t$); notice that the group G consisting of the maps $x \mapsto ax^{p^i} + b$, where $a \in K$ and $b \in F$ and $i \geq 0$ acts as a rank 3 group on F.

19.5.2 The De Lange Graphs

De Lange [507] found that one gets strongly regular graphs in the following three cases (that are not semiprimitive).

p	f	e	J
3	8	20	$\{0, 1, 4, 8, 11, 12, 16\}$
3	8	16	$\{0, 1, 2, 8, 10, 11, 13\}$
2	12	45	$\{0, 5, 10\}$

One finds two-weight codes over $\mathbb{F}_r$ for $r = 9, 3, 8$, respectively.

This last graph can be viewed as a graph with vertex set $\mathbb{F}_q^3$ for $q = 16$ such that each vertex has a unique neighbor in each of the $q^2 + q + 1 = 273$ directions.

19.5.3 Generalizations

The examples given by de Lange and by Ikuta and Munemasa [1022, 1023] ($p = 2$, $f = 20$, $e = 75$, $J = \{0, 3, 6, 9, 12\}$ and $p = 2$, $f = 21$, $e = 49$, $J = \{0, 1, 2, 3, 4, 5, 6\}$) and the sporadic cases of the Schmidt–White Conjecture 19.4.1 were generalized by Feng and Xiang [717], Ge, Xiang, and Yuan [798], Momihara [1394], and Wu [1910], who found several further infinite families of strongly regular graphs. See also [1395].

19.6 Rank 3 Groups

Let Γ be a graph and G a group of automorphisms of Γ. The group G is called **rank 3** when it is transitive on vertices, edges, and non-edges. In this case, the graph Γ is strongly regular (or complete or empty).

All rank 3 groups have been classified in a series of papers by Foulser, Kallaher, Kantor, Liebler, Liebeck, Saxl and others. The affine case that interests us here was finally settled by Liebeck [1256].

19.6.1 One-Dimensional Affine Rank 3 Groups

Let $q = p^r$ be a prime power, where p is prime. Consider the group $A\Gamma L(1, q)$ consisting of the affine semilinear maps $x \mapsto ax^\sigma + b$ on $\mathbb{F}_q$. Let T be the subgroup of size q consisting of the translations $x \mapsto x + b$. We classify the rank 3 subgroups R of $A\Gamma L(1, q)$ that contain T. They are the groups generated by T and H, where H fixes 0 and has two orbits on the nonzero elements.

After deleting the fixed point 0, the group H is a subgroup of the semilinear group $G = \Gamma L(1, q)$ acting on the nonzero elements of $\mathbb{F}_q$. It consists of the maps $t_{a,i} : x \mapsto ax^\sigma$, where $a \neq 0$ and $\sigma = p^i$. Foulser and Kallaher [749, Section 3] determined which subgroups H of G have precisely two orbits.

Lemma 19.6.1 *Let H be a subgroup of $\Gamma L(1, q)$. Then $H = \langle t_{b,0} \rangle$ for suitable b, or $H = \langle t_{b,0}, t_{c,s} \rangle$ for suitable b, c, s, where $s \mid r$ and $c^{(q-1)/(p^s - 1)} \in \langle b \rangle$.*

Proof: The subgroup of all elements $t_{a,0}$ in H is cyclic and has a generator $t_{b,0}$. If this was not all of H, then $H/\langle t_{b,0} \rangle$ is cyclic again, and has a generator $t_{c,s}$ with $s \mid r$. Since

$t_{c,s}{}^i = t_{c^j,is}$ where $j = 1+p^s+p^{2s}+\cdots+p^{(i-1)s}$, it follows for $i = r/s$ that $c^{(q-1)/(p^s-1)} \in \langle b \rangle$. $\square$

Theorem 19.6.2 $H = \langle t_{b,0} \rangle$ *has two orbits if and only if q is odd and H consists precisely of the elements $t_{a,0}$ with a a square in $\mathbb{F}_q^*$.*

Proof: Let b have multiplicative order m. Then $m \mid (q-1)$, and $\langle t_{b,0} \rangle$ has d orbits, where $d = (q-1)/m$. $\square$

Let b have order m and put $d = (q-1)/m$. Choose a primitive element $\omega \in \mathbb{F}_q^*$ with $b = \omega^d$. Let $c = \omega^e$.

Theorem 19.6.3 $H = \langle t_{b,0}, t_{c,s} \rangle$, *where $s \mid r$ and $d \mid e(q-1)/(p^s-1)$, has exactly two orbits of different lengths n_1, n_2, where $n_1 < n_2$, $n_1 + n_2 = q - 1$, if and only if $n_1 = m_1 m$, with*

(a) *the prime divisors of m_1 divide $p^s - 1$,*

(b) *$v := (q-1)/n_1$ is an odd prime, and $p^{m_1 s}$ is a primitive root mod v,*

(c) *$\gcd(e, m_1) = 1$, and*

(d) *$m_1 s(v-1) \mid r$.*

That settled the case of two orbits of different lengths. Next consider that of two orbits of equal length. As before, let b have order m and put $d = (q-1)/m$. Choose a primitive element $\omega \in \mathbb{F}_q^*$ with $b = \omega^d$. Let $c = \omega^e$.

Theorem 19.6.4 $H = \langle t_{b,0}, t_{c,s} \rangle$, *where $s \mid r$ and $d \mid e(q-1)/(p^s-1)$, has exactly two orbits of the same length $(q-1)/2$ where $(q-1)/2 = m_1 m$ if and only if*

(a) *the prime divisors of $2m_1$ divide $p^s - 1$,*

(b) *no odd prime divisor of m_1 divides e,*

(c) *$m_1 s \mid r$, and*

(d) *one of the following cases applies:*

    (i) *m_1 is even, $p^s \equiv 3 \pmod 8$, and e is odd,*

    (ii) *$m_1 \equiv 2 \pmod 4$, $p^s \equiv 7 \pmod 8$, and e is odd,*

    (iii) *m_1 is even, $p^s \equiv 1 \pmod 4$, and $e \equiv 2 \pmod 4$, or*

    (iv) *m_1 is odd and e is even.*

The graphs from Theorem 19.6.2 are the Paley graphs. The Van Lint–Schrijver Construction from Section 19.5.1 is the special case of Theorem 19.6.3 where $s = 1$, $e = 0$, $m_1 = 1$.

19.7 Two-Character Sets in Projective Space

Since projective two-weight codes correspond to 2-character sets in projective space, we want to classify the latter. The surrounding space will always be the projective space PV, where V is an m-dimensional vector space over $\mathbb{F}_q$.

19.7.1   Subspaces

(i) Easy examples are subspaces of PV. A subspace with vector space dimension i (projective dimension $i-1$), where $1 \leq i \leq m-1$, has size $n = \frac{q^i-1}{q-1}$ and meets hyperplanes in either $m_1 = \frac{q^i-1}{q-1}$ or $m_2 = \frac{q^{i-1}-1}{q-1}$ points.

Here $m_1 - m_2 = q^{i-1}$ can take many values.

(ii) If $m = 2l$ is even, we can take the union of any family of pairwise disjoint l-subspaces. A hyperplane will contain either 0 or 1 of these, so that $n = \frac{q^l-1}{q-1}u$, $m_1 = \frac{q^{l-1}-1}{q-1}u+q^{l-1}$, $m_2 = \frac{q^{l-1}-1}{q-1}u$ where u is the size of the family, $1 \leq u \leq q^l$.

Clearly, one has a lot of freedom choosing this family of pairwise disjoint l-subspaces, and one obtains exponentially many non-isomorphic graphs with the same parameters (cf. [1079]). There are many further constructions with these parameters; see, e.g., Section 19.7.2(ii) below, the alternating forms graphs on $\mathbb{F}_q^5$ (with $u = q^2+1$, see [309, Theorem 9.5.6]), and [223, 224, 453, 497, 513].

19.7.2   Quadrics

(i) Let $X = Q$ be the point set of a non-degenerate quadric in PV, and let H denote a hyperplane. Intersections $Q \cap H$ are quadrics in H, and in the cases where there is only one type of non-degenerate quadric in H, there are two intersection sizes, dependent on whether H is tangent or not.

In more detail, if m is even, then $n = |Q| = \frac{q^{m-1}-1}{q-1} + \varepsilon q^{m/2-1}$ with $\varepsilon = 1$ for a hyperbolic quadric, and $\varepsilon = -1$ for an elliptic quadric. A non-degenerate hyperplane meets Q in $m_1 = \frac{q^{m-2}-1}{q-1}$ points, and a tangent hyperplane meets Q in $m_2 = \frac{q^{m-2}-1}{q-1} + \varepsilon q^{m/2-1}$ points. (Here we dropped the convention that $m_1 > m_2$.) The corresponding weights are $w_1 = q^{m-2} + \varepsilon q^{m/2-1}$ and $w_2 = q^{m-2}$. The corresponding graphs are known as the affine polar graphs $VO^\varepsilon(m,q)$.

In the special case $m = 4$, $\varepsilon = -1$ one has $n = q^2 + 1$, $m_1 = q + 1$, $m_2 = 1$, and not only the elliptic quadrics but also the Tits ovoids have these parameters.

(ii) The above construction with $\varepsilon = 1$ has the same parameters as the subspaces construction in Section 19.7.1(ii) with $u = q^{m/2-1}+1$. Brouwer et al. [310] gave a common generalization of both by taking (for $m = 2l$) the disjoint union of pairwise disjoint l-spaces and non-degenerate hyperbolic quadrics, where possibly a number of pairwise disjoint l-spaces contained in some of the hyperbolic quadrics is removed.

(iii) For odd q and even m, consider a non-degenerate quadric Q of type $\varepsilon = \pm 1$ in V, the m-dimensional vector space over $\mathbb{F}_q$. The non-isotropic points fall into two classes of equal size, depending on whether $Q(x)$ is a square or not. Both sets are (isomorphic) 2-character sets.

Let X be the set of non-isotropic projective points x where $Q(x)$ is a nonzero square (this is well-defined). Then $|X| = \frac{1}{2}(q^{m-1}-\varepsilon q^{m/2-1})$ and $m_1, m_2 = \frac{1}{2}q^{m/2-1}(q^{m/2-1}\pm 1)$ (independent of ε). The corresponding graphs are known as $VNO^\varepsilon(m,q)$.

(iv) In Brouwer [307] a construction for two-weight codes is given by taking a quadric defined over a small field and cutting out a quadric defined over a larger field. Let $F_1 = \mathbb{F}_r$, and $F = \mathbb{F}_q$, where $r = q^e$ for some $e > 1$. Let V_1 be a vector space of dimension d over F_1, where d is even, and write V for V_1 regarded as a vector space of dimension de over F. Let tr : $F_1 \to F$ be the trace map. Let $Q_1 : V_1 \to F_1$ be a non-degenerate quadratic form on V_1. Then $Q = \text{tr} \circ Q_1$ is a non-degenerate quadratic form on V. Let $X = \{x \in PV \mid Q(x) = 0 \text{ and } Q_1(x) \neq 0\}$. Write $\varepsilon = 1$ ($\varepsilon = -1$) if Q is hyperbolic (elliptic).

Proposition 19.7.1 *In the situation described in* (iv), *the corresponding two-weight code has dimension* de, *length* $n = |X| = (q^{e-1} - 1)(q^{de-e} - \varepsilon q^{de/2-e})/(q - 1)$, *and weights* $w_1 = (q^{e-1} - 1)q^{de-e-1}$ *and* $w_2 = (q^{e-1} - 1)q^{de-e-1} - \varepsilon q^{de/2-1}$.

For example, when $q = e = 2$, $d = 4$, $\varepsilon = -1$, this yields a projective binary $[68, 8]$ code with weights 32, 40. This construction was generalized in Hamilton [894].

19.7.3 Maximal Arcs and Hyperovals

A maximal arc in a projective plane $\mathrm{PG}(2, q)$ is a 2-character set with intersection numbers $m_1 = a$, $m_2 = 0$, for some constant a ($1 < a < q$). Clearly, maximal arcs have size $n = qa - q + a$, and necessarily $a \mid q$. For $a = 2$ these objects are called hyperovals, and exist for all even q. Denniston [531] constructed maximal arcs for all even q and all divisors a of q. Ball et al. [105] showed that there are no maximal arcs in $\mathrm{PG}(2, q)$ when q is odd.

These arcs show that the difference between the intersection numbers need not be a power of q. Also for a unital, one has intersection sizes 1 and $\sqrt{q} + 1$.

19.7.4 Baer Subspaces

Let $q = r^2$ and let m be odd. Then $\mathrm{PG}(m - 1, q)$ has a partition into pairwise disjoint Baer subspaces $\mathrm{PG}(m - 1, r)$. Each hyperplane hits all of these in a $\mathrm{PG}(m - 3, r)$, except for one which is hit in a $\mathrm{PG}(m - 2, r)$. Let X be the union of u such Baer subspaces, $1 \leq u < (r^m + 1)/(r + 1)$. Then $n = |X| = u(r^m - 1)/(r - 1)$, $m_2 = u(r^{m-2} - 1)/(r - 1)$, $m_1 = m_2 + r^{m-2}$.

19.7.5 Hermitian Quadrics

Let $q = r^2$ and let V be provided with a non-degenerate Hermitian form. Let X be the set of isotropic projective points. Then

$$n = |X| = (r^m - \varepsilon)(r^{m-1} + \varepsilon)/(q - 1),$$
$$w_2 = r^{2m-3},$$
$$w_1 - w_2 = \varepsilon r^{m-2},$$

where $\varepsilon = (-1)^m$. If we view V as a vector space of dimension $2m$ over $\mathbb{F}_r$, the same set X now has $n = (r^m - \varepsilon)(r^{m-1} + \varepsilon)/(r - 1)$, $w_2 = r^{2m-2}$, $w_1 - w_2 = \varepsilon r^{m-1}$, as expected, since the form is a non-degenerate quadratic form in $2m$ dimensions over $\mathbb{F}_r$. Thus, the graphs that one gets here are also graphs one gets from quadratic forms, but the codes here are defined over a larger field.

19.7.6 Sporadic Examples

In Table 19.1 we give some small sporadic examples (or series of parameters for which examples are known, some of which are sporadic). Many of these also have a cyclotomic description.

TABLE 19.1: Sporadic examples

q	m	n	w_1	w_2-w_1	Comments
2	9	73	32	8	Fiedler & Klin [726]; [1136]
2	9	219	96	16	dual
2	10	198	96	16	Kohnert [1136]
2	11	276	128	16	Conway-Smith $2^{11}.M_{24}$ rank 3 graph
2	11	759	352	32	dual; [819]
2	12	$65i$	$32i$	32	Kohnert [1136] ($12 \le i \le 31, i \ne 19$)
2	24	98280	47104	2048	Rodrigues [1593]
4	5	$11i$	$8i$	8	Dissett [600] ($7 \le i \le 14, i \ne 8$)
4	6	78	56	8	Hill [952]
4	6	429	320	32	dual
4	6	147	96	16	[307]; Cossidente et al. [445]
4	6	210	144	16	Cossidente et al. [445]
4	6	273	192	16	Section 19.7.1; De Wispelaere and Van Maldeghem [512]
4	6	315	224	16	[307]; Cossidente et al. [445]
8	4	39	32	4	De Lange [507]
8	4	273	224	16	dual
16	3	78	72	4	De Resmini and Migliori [510]
3	5	11	6	3	dual of the ternary Golay code
3	5	55	36	9	dual
3	6	56	36	9	Games graph, Hill cap [951]
3	6	84	54	9	Gulliver [861]; [1343]
3	6	98	63	9	Gulliver [861]; [1343]
3	6	154	99	9	Van Eupen [1831]; [860]
3	8	$82i$	$54i$	27	Kohnert [1136] ($8 \le i \le 12$)
3	8	$41i$	$27i$	27	Kohnert [1136] ($26 \le i \le 39$)
3	8	1435	945	27	de Lange [507]
3	12	32760	21627	243	Liebeck [1256] $3^{12}.2.\mathrm{Suz}$ rank 3 graph
9	3	35	30	3	De Resmini [509]
9	3	42	36	3	Penttila and Royle [1502]
9	4	287	252	9	de Lange [507]
5	4	39	30	5	Dissett [600]; [270]
5	6	1890	1500	25	Liebeck [1256] $5^6.4.J_2$ rank 3 graph
125	3	829	820	5	Batten and Dover [138]
125	3	7461	7400	25	dual
343	3	3189	3178	7	Batten and Dover [138]
343	3	28701	28616	49	dual

19.8 Nonprojective Codes

When the code C is not projective (which is necessarily the case when $n > \frac{q^m - 1}{q-1}$) the set X is a multiset. Still, it allows a geometric description of the code, which is very helpful. For example, see Cheon et al. [400].

Two-weight $[n, m, d]_q$ codes with the two weights d and n were classified in Jungnickel and Tonchev [1060]—the corresponding multiset X is either a multiple of a plane maximal arc, or a multiple of the complement of a hyperplane.

Part of the literature is formulated in terms of the complement Z of X in PV (or the multiset containing some fixed number t of copies of each point of PV). The code C will have minimum distance at least d when $|X \cap H| \leq n - d$ for all hyperplanes H. For Z that says $|Z \cap H| \geq t\frac{q^{m-1}-1}{q-1} - n + d$ for all hyperplanes H. Such sets Z are studied under the name minihypers, especially when they correspond to codes meeting the Griesmer Bound $n \geq \sum_{i=0}^{m-1} \lceil \frac{d}{q^i} \rceil$. See for example Hamada and Deza [890], Storme [1758], and Hill and Ward [955].

For projective two-weight codes we saw that $w_2 - w_1$ is a power of the characteristic. So whenever this does not hold, the code must be non-projective. (This settles a question in [1306].)

19.9 Brouwer–Van Eupen Duality

Brouwer and van Eupen [313] give a correspondence between arbitrary projective codes and arbitrary two-weight codes. The correspondence can be said to be 1-1, even though there are choices to be made in both directions.

19.9.1 From Projective Code to Two-Weight Code

Given a linear code C with length n, let n_C be its effective length, that is, the number of coordinate positions where C is not identically zero.

Let C be a projective $[n, m, d]_q$ code with nonzero weights $w_1, \ldots, w_t$. In a subcode D of codimension 1 in C these weights occur with frequencies $f_1, \ldots, f_t$, where $\sum f_i = q^{m-1} - 1$ and $\sum (n_D - w_i) f_i = n_D (q^{m-2} - 1)$. It follows that for arbitrary choice of α, β the sum $\sum (\alpha w_i + \beta) f_i$ does not depend on D but only on n_D. Since C is projective, we have $n_D = n - 1$ for n subcodes D, and $n_D = n$ for the remaining $\frac{q^m-1}{q-1} - n$ subcodes of codimension 1. Therefore, the above sum takes only two values.

Fix α, β in such a way that all numbers $\alpha w_i + \beta$ are nonnegative integers, and consider the multiset Y in PC consisting of the 1-spaces $\langle c \rangle$ with $c \in C$ taken $\alpha w + \beta$ times, where w is the weight of c. Since an arbitrary hyperplane D meets Y in $\alpha q^{m-2} n_D + \beta \frac{q^{m-1}-1}{q-1}$ points, the set Y defines a two-weight code of length $|Y| = \beta \frac{q^{m-1}-1}{q-1} + q^{m-1} \alpha n$, dimension m, and weights $w = |Y| - \frac{|Y| - \beta}{q}$ and $w' = w + \alpha q^{m-2}$.

Example 19.9.1 If we start with the unique $[16, 5, 9]_3$-code, with weight enumerator $1 + 116x^9 + 114x^{12} + 12x^{15}$, and take $\alpha = 1/3$, $\beta = -3$, we find a $[69, 5, 45]_3$-code with weight enumerator $1 + 210x^{45} + 32x^{54}$. With $\alpha = -1/3$, $\beta = 5$, we find a $[173, 5, 108]_3$-code with weight enumerator $1 + 32x^{108} + 210x^{117}$.

19.9.2   From Two-Weight Code to Projective Code

Let $\mathcal{C}$ be a two-weight $[n, m, d]_q$-code with nonzero weights w_1 and w_2. Let X be the corresponding projective multiset. Let Y be the set of hyperplanes meeting X in $|X| - w_2$ points. Then Y defines a projective code of length $|H| = \frac{1}{w_2 - w_1}(nq^{m-1} - w_1\frac{q^m - 1}{q - 1})$ and dimension m, and with a number of distinct weights equal to the number of distinct multiplicities in X.

Remark 19.9.2

- In both directions there is a choice: pick α, β or pick $w_2 \in \{w_1, w_2\}$. The correspondence is 1-1 in the sense that if $\mathcal{C}^*$ is a Brouwer and van Eupen dual (BvE-dual) of $\mathcal{C}$, then $\mathcal{C}$ is a BvE-dual of $\mathcal{C}^*$. If the projective code $\mathcal{C}$ one starts with has only two different weights, then one can choose α, β so that Y becomes a set and the BvE-dual coincides with the Delsarte dual.

- For another introduction and further examples, see Hill and Kolev [954].

- In Section 19.9.1, the degree 1 polynomial $p(w) = \alpha w + \beta$ was used. One can use higher degree polynomials when more information about subcodes is available. See the last section of [313] and Dodunekov and Simonis [611].

- See also [267, 1032] and [1768, Lemma 5.1].

Chapter 20

Linear Codes from Functions

Sihem Mesnager

University of Paris VIII and Télécom Paris

20.1   Introduction

Vectorial multi-output Boolean functions are functions from the finite field $\mathbb{F}_{2^n}$ of order 2^n to the finite field $\mathbb{F}_{2^m}$, for given positive integers n and m. These functions are called (n, m)-*functions* and include the single-output Boolean functions, which correspond to the case $m = 1$. The (n, m)-functions are a mathematical object but naturally appear in different areas of computer science and engineering since any such mapping can be understood as a transformation substituting a sequence of n bits (zeros and ones) with a sequence of m bits according to a given prescription. Moreover, (n, m)-functions play an important role in symmetric cryptography, usually referred to as an "S-box" or "substitution box" in this context; they are fundamental parts of block ciphers by providing confusion, a requirement mentioned by C. E. Shannon [1662], which is necessary to withstand known (and hopefully future) attacks. When they are used as S-boxes in block ciphers, the number m of their output bits is less than or equal to the number n of input bits, most often. Such functions can also be used in stream ciphers, with m significantly smaller than n, in the place of Boolean functions to speed up the ciphers. Researchers have defined various properties which measure the resistance of an (n, m)-function to different kinds of cryptanalysis, including nonlinearity, differential uniformity, boomerang uniformity, algebraic degree, and so forth. Vectorial Boolean functions have been extended and studied in any characteristic. An excellent survey on Boolean and vectorial Boolean functions can be found in [346, 347]. We also deeply recommend the outstanding book of C. Carlet [349] dealing with Boolean and vectorial functions for cryptography and coding theory.

Linear codes are important in communication systems, consumer electronics, data storage systems, etc. Those with good parameters can be employed in data storage systems and communication systems. Determining the (Hamming) weights of linear codes is useful for secret sharing, authentication codes, association schemes, strongly regular graphs, etc. Cryptographic functions and codes have important applications in data communication and data storage. These two areas are closely related and have had fascinating interplay. Cryptographic functions (e.g., highly nonlinear functions, PN, APN, bent, AB, plateaued) have important applications in coding theory. For instance, bent functions and almost bent (AB) functions have been employed to construct optimal linear codes.

In the past decades, a lot of progress has been made in the construction of linear codes from cryptographic functions and polynomials defined over finite fields. The main aim of this chapter is to give an overview of linear codes from functions and polynomials using different approaches. It is quite fascinating that vectorial functions can be nicely involved to construct codes with good parameters and play a major role in these constructions. A recent survey in this direction is [1244a].

The chapter is structured as follows. In Section 20.2, we present some background on vectorial functions and related notions (some parameters of vectorial functions and tools for handling vectorial functions). In Section 20.3, generic constructions of linear codes and their extensions are presented. The core of the chapter starts in Section 20.4. We shall present constructions of binary linear codes and highlight the role of some Boolean and vectorial Boolean functions in the construction of codes with few weights. Section 20.5 is devoted to the construction of cyclic codes using a sequence approach. We shall see for example that APN functions play an important role in deriving such codes. Section 20.6 deals with p-ary codes from functions over finite fields in odd characteristic p. We shall present constructions of p-ary codes from specific functions based on the algebraic approaches presented in Section 20.3. Finally, in Section 20.7, we present linear locally recoverable codes (LRC) and highlight the role of special functions in the design of optimal LRC.

20.2 Preliminaries

For a finite set A, $\#A$ denotes the cardinality of A and A^* denotes the set $A \setminus \{0\}$. For a complex number z, $|z|$ denotes its absolute value.

20.2.1 The Trace Function

Let q be a power of a prime p and r be a positive integer. The **(relative) trace function** $\mathrm{Tr}_{q^r/q} : \mathbb{F}_{q^r} \to \mathbb{F}_q$ is defined as

$$\mathrm{Tr}_{q^r/q}(x) := \sum_{i=0}^{r-1} x^{q^i} = x + x^q + x^{q^2} + \cdots + x^{q^{r-1}}.$$

The trace function from $\mathbb{F}_{q^r} = \mathbb{F}_{p^n}$ to its prime subfield $\mathbb{F}_p$ is called the **absolute trace function**.

Recall that the (relative) trace function $\mathrm{Tr}_{q^r/q}$ is $\mathbb{F}_q$-linear. It satisfies the transitivity property in a chain of extension fields. More precisely, let $\mathbb{F}_{p^h}$ be a finite field, let $\mathbb{F}_{p^m}$ be a finite extension of $\mathbb{F}_{p^h}$ and $\mathbb{F}_{p^r}$ be a finite extension of $\mathbb{F}_{p^m}$. Then $\mathrm{Tr}_{p^r/p^h}(x) = \mathrm{Tr}_{p^m/p^h}(\mathrm{Tr}_{p^r/p^m}(x))$ for all $x \in \mathbb{F}_{p^r}$.

20.2.2 Vectorial Functions

In this chapter, we shall identify the Galois field $\mathbb{F}_{p^n}$ of order p^n with the vector space $\mathbb{F}_p^n$ by using $\mathbb{F}_{p^n}$ as an n-dimensional vector space over $\mathbb{F}_p$ and writing $\mathbb{F}_p^n$ instead of $\mathbb{F}_{p^n}$ when the field structure is not really used. In the following, "$\cdot$" denotes the standard inner (dot) product of two vectors, that is, $\lambda \cdot x := \lambda_1 x_1 + \cdots + \lambda_n x_n$, where $\lambda, x \in \mathbb{F}_p^n$. If we identify the vector space $\mathbb{F}_p^n$ with the finite field $\mathbb{F}_{p^n}$, we use the trace bilinear form $\mathrm{Tr}_{p^n/p}(\lambda x)$ instead of the dot product, that is, $\lambda \cdot x = \mathrm{Tr}_{p^n/p}(\lambda x)$, where $\lambda, x \in \mathbb{F}_{p^n}$.

Definition 20.2.1 Let $q = p^r$ where p is a prime. A vectorial function $\mathbb{F}_q^n \to \mathbb{F}_q^m$ (or $\mathbb{F}_{q^n} \to \mathbb{F}_{q^m}$) is called an (n, m)-q-**ary function**. When $q = 2$, an (n, m)-2-ary function will be simply denoted an (n, m)-**function**. A **Boolean function** is an $(n, 1)$-function, i.e., a function $\mathbb{F}_2^n \to \mathbb{F}_2$ (or $\mathbb{F}_{2^n} \to \mathbb{F}_2$).

20.2.2.1 Representations of p-Ary Functions

There exist several kinds of possible *univariate representations* (also called trace, or polynomial, representations) of p-ary functions which are not all unique and use the identification between the vector space $\mathbb{F}_p^n$ and the field $\mathbb{F}_{p^n}$.

A p-**ary function** is a function from $\mathbb{F}_p^n$ to $\mathbb{F}_p$. If we identify $\mathbb{F}_p^n$ with $\mathbb{F}_{p^n}$, all p-ary functions can be described in the so-called *univariate form*, which can be put in the (non-unique) **trace form** $\mathrm{Tr}_{p^n/p}(F(x))$ for some polynomial function F from $\mathbb{F}_{p^n}$ to $\mathbb{F}_{p^n}$ of degree at most $p^n - 1$. A unique univariate form of a p-ary function, called the **trace representation**, is given by

$$f(x) = \sum_{j \in \Gamma_n} \mathrm{Tr}_{p^{o(j)}/p}(A_j x^j) + A_{p^n-1} x^{p^n - 1}$$

where

- Γ_n is the set of integers obtained by choosing the smallest element, called the **coset leader**, in each p-cyclotomic coset modulo $p^n - 1$;

- $o(j)$ is the size of the p-cyclotomic coset containing j;

- $A_j \in \mathbb{F}_{p^{o(j)}}$;

- $A_{p^n-1} \in \mathbb{F}_p$.

Note that $o(j)$ divides n. The **algebraic degree** of f, coming from its unique trace representation, is $\max\{w_p(j) \mid A_j \neq 0\}$ where $w_p(j)$ is the number of nonzero entries in the p-ary expansion of j.

Example 20.2.2 Let $p = 3$. Denote by $C(j)$ the cyclotomic coset of j modulo $3^n - 1$. We have
$$C(j) = \{j, 3j, 3^2 j, \dots, 3^{o(j)-1} j\},$$
where $o(j)$ is the smallest positive integer such that $j \equiv j3^{o(j)} \pmod{3^n - 1}$. For example, if $n = 3$, the cyclotomic cosets modulo 26 are: $C(0) = \{0\}$, $C(1) = \{1, 3, 9\}$, $C(2) = \{2, 6, 18\}$, $C(4) = \{4, 10, 12\}$, $C(5) = \{5, 15, 19\}$, $C(7) = \{7, 11, 21\}$, $C(8) = \{8, 20, 24\}$, $C(13) = \{13\}$, $C(14) = \{14, 16, 22\}$ and $C(17) = \{17, 23, 25\}$. So $\Gamma_3 = \{0, 1, 2, 4, 5, 7, 8, 13, 14, 17\}$; also $w_3(0) = 0$, $w_3(1) = w_3(2) = 1$, $w_3(4) = w_3(5) = w_3(7) = w_3(8) = 2$, and $w_3(13) = w_3(14) = w_3(17) = 3$. Any nonzero function f from $\mathbb{F}_{3^n}$ into $\mathbb{F}_3$ can be represented as $f(x) = \sum_{j \in \Gamma_3} \mathrm{Tr}_{3^{o(j)}/3}(A_j x^j) + A_{3^n-1} x^{3^n-1}$ where $A_j \in \mathbb{F}_{3^{o(j)}}$, $A_{3^n-1} \in \mathbb{F}_3$.

If we do not identify $\mathbb{F}_p^n$ with $\mathbb{F}_{p^n}$, a p-ary function f has a representation as a unique multinomial in $x_1, \dots, x_n$, where the variables x_i occur with exponent at most $p - 1$. This is called the **multivariate representation** or **algebraic normal form (ANF)** of f. The **algebraic degree** of f is the global (total) degree of its multivariate representation.

Example 20.2.3 Let $p = 2$. An n-variable Boolean function $f(x_1, \dots, x_n)$ can be written as
$$f(x_1, \dots, x_n) = \sum_{u \in \mathbb{F}_2^n} \lambda_u \left(\prod_{i=1}^{n} x_i^{u_i} \right)$$
for $\lambda_u \in \mathbb{F}_2$ and $u = (u_1, \dots, u_n)$.

The **bivariate representation** of a p-ary function f of even dimension $n = 2m$ is defined as follows: we identify $\mathbb{F}_p^n$ with $\mathbb{F}_{p^m} \times \mathbb{F}_{p^m}$ and consider the argument of f as an ordered pair (x, y) of elements $x, y \in \mathbb{F}_{p^m}$. Since f takes on its values in $\mathbb{F}_p$, the bivariate representation can be written in the form $f(x, y) = \mathrm{Tr}_{p^m/p}(P(x, y))$ where
$$P(x, y) = \sum_{0 \leq i, j \leq p^m - 1} A_{i,j} x^i y^j$$
is a unique polynomial in two variables, representing f, with $A_{i,j} \in \mathbb{F}_{p^m}$. The **algebraic degree** of f is $\max\{w_p(i) + w_p(j) \mid A_{i,j} \neq 0\}$.

A p-ary **linear function** $f(x)$ over $\mathbb{F}_{p^n}$ has the form $\mathrm{Tr}_{p^n/p}(ax)$ for some unique $a \in \mathbb{F}_{p^n}$; linear functions include the identically zero function. A p-ary function whose algebraic degree equals 1 (respectively 2) is called **affine** (respectively **quadratic**).

Example 20.2.4 Let $p = 2$. Any Boolean affine function $f(x)$ over $\mathbb{F}_{2^n}$ can be written uniquely as $f(x) = \mathrm{Tr}_{2^n/2}(ax + b)$ where $a, b \in \mathbb{F}_{2^n}$ with $a \neq 0$; these include the nonzero

Boolean linear functions. Any Boolean quadratic function $f(x)$ over $\mathbb{F}_{2^n}$ can be written uniquely in the form

$$f(x) = \mathrm{Tr}_{2^n/2}\left(\sum_{k=1}^{\lfloor \frac{n}{2} \rfloor} a_k x^{2^k+1}\right) + \mathrm{Tr}_{2^n/2}(ax+b)$$

where $a, b, a_k \in \mathbb{F}_{2^n}$ with $a_k \neq 0$ for at least one k.

20.2.2.2 The Walsh Transform of a Vectorial Function

Let f be a function from $\mathbb{F}_{p^n}$ to $\mathbb{F}_p$, and let $\xi_p = e^{\frac{2\pi i}{p}}$ which is a complex primitive p^{th} root of unity. The elements of $\mathbb{F}_p$ are considered as integers modulo p.

Definition 20.2.5 The function $\xi_p^{f(x)}$ from $\mathbb{F}_{p^n}$ to $\mathbb{C}$ is usually called the **sign** function of f. The Fourier transform of the function $\xi_p^{f(x)}$ is called the Walsh transform. The **Walsh transform of f** at $b \in \mathbb{F}_{p^n}$, which we denote by $W_f(b)$, is defined by

$$W_f(b) = \sum_{x \in \mathbb{F}_{p^n}} \xi_p^{f(x)-\mathrm{Tr}_{p^n/p}(bx)}.$$

Note that the notion of a Walsh transform refers to a scalar product; it is convenient to choose the isomorphism such that the canonical scalar product "$\cdot$" in $\mathbb{F}_{p^n}$ coincides with the trace of the product bx, i.e., $b \cdot x := \mathrm{Tr}_{p^n/p}(bx)$.

Example 20.2.6 Let $p = 2$. Consider a quadratic Boolean function from $\mathbb{F}_{2^m}$ to $\mathbb{F}_2$ of the form

$$f(x) = \mathrm{Tr}_{2^m/2}\left(\sum_{i=0}^{\lfloor m/2 \rfloor} f_i x^{2^i+1}\right)$$

where $f_i \in \mathbb{F}_{2^m}$. The **rank of f**, denoted by r_f, is defined to be the codimension of the $\mathbb{F}_2$-vector space

$$V_f = \{x \in \mathbb{F}_{2^m} \mid f(x+z) - f(x) - f(z) = 0 \text{ for all } z \in \mathbb{F}_{2^m}\}.$$

It is an even integer. The Walsh spectrum of f is known [337] and given in Table 20.1.

TABLE 20.1: Walsh spectrum of quadratic Boolean functions

$W_f(w)$	Number of w's
0	$2^m - 2^{r_f}$
$2^{m-r_f/2}$	$2^{r_f-1} + 2^{(r_f-2)/2}$
$-2^{m-r_f/2}$	$2^{r_f-1} - 2^{(r_f-2)/2}$

From its Walsh transform, the function f can be recovered by the inverse transform

$$\xi_p^{f(x)} = \frac{1}{p^n} \sum_{b \in \mathbb{F}_{p^n}} W_f(b)\xi_p^{\mathrm{Tr}_{p^n/p}(bx)}$$

For every function f from $\mathbb{F}_{p^n}$ to $\mathbb{F}_p$, we have

$$\sum_{w \in \mathbb{F}_{p^n}} W_f(w) = p^n \xi_p^{f(0)}.$$

eval's Identity states that

$$\sum_{w \in \mathbb{F}_{p^n}} |W_f(w)|^2 = p^{2n}. \tag{20.1}$$

nition 20.2.7 Let F be a vectorial function from $\mathbb{F}_{p^n}$ into $\mathbb{F}_{p^m}$. The linear combina-
of the coordinates of F are the p-ary functions $F_u : x \mapsto u \cdot F(x) := \mathrm{Tr}_{p^m/p}(uF(x))$,
$\mathbb{F}_{p^m}$, where F_0 is the null function. The functions F_u $(u \neq 0)$ are called the **components**

nition 20.2.8 For a vectorial function $F : \mathbb{F}_{p^n} \to \mathbb{F}_{p^m}$, the **Walsh–Hadamard**
sform of F, denoted by W_F, is the function which maps any ordered pair $(u, v) \in$
$\times \mathbb{F}_{p^m}$ to the value at u of the Walsh transform of the p-ary function $v \cdot F(x) :=$
$_p(vF(x))$; i.e.,

$$W_F(u, v) = \sum_{x \in \mathbb{F}_{p^n}} \xi_p^{\mathrm{Tr}_{p^m/p}(vF(x)) - \mathrm{Tr}_{p^n/p}(ux)}.$$

.3 Nonlinearity of Vectorial Boolean Functions and Bent Boolean Functions

he **nonlinearity of the Boolean function** f defined over $\mathbb{F}_2^n$ is the minimum Ham-
distance to the set A_n of all affine functions (see Example 20.2.4), i.e.,

$$\mathrm{nl}(f) = \min_{g \in A_n} \mathrm{d_H}(f, g),$$

$_e$ $\mathrm{d_H}(f, g)$ is the Hamming distance between f and g; that is, $\mathrm{d_H}(f, g) := \{x \in \mathbb{F}_2^n \mid$
$\neq g(x)\}$. The relationship between nonlinearity and the Walsh spectrum of f is

$$\mathrm{nl}(f) = 2^{n-1} - \frac{1}{2} \max_{\omega \in \mathbb{F}_2^n} |W_f(\omega)|.$$

arseval's Identity (20.1) with $p = 2$, it can be shown that $\max\{|W_f(\omega)| \mid \omega \in \mathbb{F}_2^n\} \geq 2^{\frac{n}{2}}$,
$_1$ implies that $\mathrm{nl}(f) \leq 2^{n-1} - 2^{\frac{n}{2}-1}$.

nition 20.2.9 Let n be an even integer. An n-variable Boolean function is said to be
if the upper bound $2^{n-1} - 2^{n/2-1}$ on its nonlinearity $\mathrm{nl}(f)$ is achieved with equality.

ent Boolean functions f defined over $\mathbb{F}_{2^n}$ are then those of maximum Hamming distance
$_e$ set of affine Boolean functions. They exist only when n is even.
he notion of bent function was introduced by Rothaus [1606] and attracted a lot of
rch for more than four decades. Such functions are extremal combinatorial objects
several areas of application, such as coding theory, maximum length sequences, and
ography. A recent survey on bent functions can be found in [355] as well as the book
].
ent functions can be characterized by means of the Walsh transform as follows; this
cterization is independent of the choice of the inner product on $\mathbb{F}_{2^n}$.

osition 20.2.10 (see, e.g., [538]) *Let n be an even integer and function f be a*
an function defined over $\mathbb{F}_{2^n}$. Then f is bent if and only if $W_f(\omega) \in \{2^{\frac{n}{2}}, -2^{\frac{n}{2}}\}$,
$_l$ $\omega \in \mathbb{F}_{2^n}$.

Bent Boolean functions can also be characterized in terms of *difference sets*.

Definition 20.2.11 Let G be a finite abelian group of order μ. A subset D of G of cardinality k is called a (μ, k, λ)-**difference set** in G if every element $g \in G$, different from the identity, can be written as $d_1 - d_2$, $d_1, d_2 \in D$, in exactly λ different ways.

Let $p = 2$. It is observed in Rothaus' paper [1606] and developed in Dillon's thesis [538] that a Boolean function $f : \mathbb{F}_{2^m} \to \mathbb{F}_2$ is bent if and only if its support is a difference set. More precisely, a function f from $\mathbb{F}_{2^m}$ to $\mathbb{F}_2$ is bent if and only if its support $D_f := \{x \in \mathbb{F}_{2^m} \mid f(x) = 1\}$ is a difference set in $(\mathbb{F}_{2^m}, +)$ with parameters

$$\left(2^m, 2^{m-1} \pm 2^{\frac{m-2}{2}}, 2^{m-2} \pm 2^{\frac{m-2}{2}}\right).$$

A standard notion of **nonlinearity of Boolean vectorial functions** $F : \mathbb{F}_2^n \to \mathbb{F}_2^m$ is defined as

$$\mathcal{N}(F) = \min_{v \in \mathbb{F}_2^m \setminus \{0\}} \mathrm{nl}(v \cdot F),$$

where $v \cdot F$ denotes the usual inner product on $\mathbb{F}_2^m$ and $\mathrm{nl}(\cdot)$ denotes the nonlinearity of Boolean functions. From the Covering Radius Bound, we have $\mathcal{N}(F) \leq 2^{n-1} - 2^{n/2-1}$. The functions achieving this bound are called (n, m)-**bent** functions and are characterized by the fact that all nonzero component functions of F are bent Boolean functions. In [1451], it is shown that (n, m)-bent functions exist only if n is even and $m \leq n/2$.

The generalization to vectorial functions of the notion of nonlinearity, as introduced by Nyberg [1450, 1452] and further studied by Chabaud and Vaudenay [377], can be expressed by means of the Walsh transform as follows:

$$\mathcal{N}(F) = 2^{n-1} - \frac{1}{2} \max_{\substack{v \in \mathbb{F}_2^m \setminus \{0\} \\ u \in \mathbb{F}_2^n}} |W_F(u,v)| \quad \text{where} \quad W_F(u,v) = \sum_{x \in \mathbb{F}_2^n} (-1)^{v \cdot F(x) + u \cdot x}.$$

Bent (n, m)-functions do not exist if $m > \frac{n}{2}$. For $m \geq n - 1$, the following bound has been found by Sidelńikov and re-discovered by Chabaud and Vaudenay in [377] and is now called the *Sidelńikov–Chabaud–Vaudenay Bound*.

Theorem 20.2.12 (Sidelńikov–Chabaud–Vaudenay (SCV) Bound) *Let* n *and* m *be any positive integers such that* $m \geq n - 1$. *Let* F *be any* (n, m)-function. *Then*

$$\mathcal{N}(F) \leq 2^{n-1} - \frac{1}{2} \sqrt{3 \times 2^n - 2 - 2 \frac{(2^n - 1)(2^{n-1} - 1)}{2^m - 1}}.$$

The Sidelńikov–Chabaud–Vaudenay Bound can be tight only if $n = m$ with n odd, in which case it states that

$$\mathcal{N}(F) \leq 2^{n-1} - 2^{\frac{n-1}{2}}. \tag{20.2}$$

Definition 20.2.13 ([377]) The (n, n)-functions F which achieve (20.2) with equality are called **almost bent (AB)**.

AB functions can be characterized as follows.

Proposition 20.2.14 *Let* F *be a function from* $\mathbb{F}_{2^n}$ *to* $\mathbb{F}_{2^n}$. *Then* F *is AB if and only if* $\mathcal{W}_F(a, b) := \sum_{x \in \mathbb{F}_{2^n}} (-1)^{\mathrm{Tr}_{2^n/2}(aF(x) + bx)}$ *takes the values* 0 *or* $\pm 2^{(n+1)/2}$ *for every pair* $(a, b) \in \mathbb{F}_{2^n}^* \times \mathbb{F}_{2^n}$.

.4 Plateaued Functions and More about Bent Functions

lateaued p-ary functions can be defined as follows.

nition 20.2.15 Let s and n be two nonnegative integers such that $0 \le s \le n$. A p-ary
ion f defined over $\mathbb{F}_{p^n}$ is said to be s-**plateaued** of **amplitude** μ if $|W_f(\omega)|^2 \in \{0, \mu^2\}$
ery $\omega \in \mathbb{F}_{p^n}$ where $\mu^2 = p^{n+s}$.

rial plateaued functions are defined as follows.

nition 20.2.16 A vectorial function F from $\mathbb{F}_{p^n}$ to $\mathbb{F}_{p^m}$ is called **vectorial plateaued**
its components F_u from $\mathbb{F}_{p^n}$ to $\mathbb{F}_p$ (defined by $F_u(x) = \mathrm{Tr}_{p^m/p}(uF(x))$ for every $x \in$
are s_u-plateaued for every $u \in \mathbb{F}_{p^m}^*$ with possibly different amplitudes. In particular,
called **vectorial s-plateaued** if F_u are s-plateaued with the same amplitude μ for
$u \in \mathbb{F}_{p^m}^*$.

lateaued Boolean functions include four important classes of Boolean functions. Indeed
aracteristic two, 0-plateaued functions are the bent functions, 1-plateaued functions
alled **near-bent** functions, 2-plateaued functions are the **semi-bent** functions, and
rth sub-class is the so-called **partially bent** functions. Partially bent functions were
duced by Carlet in [343, 344]. They are defined as Boolean functions whose derivatives
) := $f(x) + f(x + a)$, for $a \in \mathbb{F}_{2^n}$, are all either balanced (see Definition 20.2.23) or
ant. Notice that, in characteristic two, 0-plateaued and 2-plateaued functions exist
n is even, while 1-plateaued functions exist when n is odd. These functions have been
y studied by several researchers; a complete survey can be found in [1377].
fact, in characteristic two, the Walsh distribution of plateaued Boolean functions can
rived easily. More specifically, let $f : \mathbb{F}_{2^n} \to \mathbb{F}_2$ be an s-plateaued Boolean function
$f(0) = 0$ and $n + s$ an even integer. Then the Walsh distribution of f is given by Table

LE 20.2: Walsh spectrum of s-plateaued Boolean functions f for $f(0) = 0$ and $n + s$

$W_f(w)$	Number of w's
$2^{\frac{n+s}{2}}$	$2^{n-s-1} + 2^{\frac{n-s-2}{2}}$
0	$2^n - 2^{n-s}$
$-2^{\frac{n+s}{2}}$	$2^{n-s-1} - 2^{\frac{n-s-2}{2}}$

odd characteristic, bent functions are classified into *regular bent* and *weakly regular*
functions.

nition 20.2.17 A bent function f is called **regular bent** if for every $b \in \mathbb{F}_{p^m}$,
$W_f(b)$ equals $\xi_p^{f^\star(b)}$ for some p-ary function $f^\star : \mathbb{F}_{p^m} \to \mathbb{F}_p$. The bent function f
led **weakly regular bent** if there exists a complex number u with $|u| = 1$ and a p-ary
ion $f^\star$ such that $up^{-\frac{m}{2}} W_f(b) = \xi_p^{f^\star(b)}$ for all $b \in \mathbb{F}_{p^m}$. The function $f^\star(x)$ is called the
of $f(x)$.

om [937, 938], a weakly regular bent function $f(x)$ satisfies

$$W_f(b) = \epsilon_f \sqrt{p^*}^m \xi_p^{f^\star(b)}, \tag{20.3}$$

$\epsilon_f = \pm 1$ is called the **sign of the Walsh transform** of $f(x)$ and $p^* = (-1)^{(p-1)/2}p$.

Moreover, the Walsh transform coefficients of a p-ary bent function f with odd p satisfy

$$p^{-\frac{m}{2}}W_f(b) = \begin{cases} \pm\xi_p^{f^\star(b)} & \text{if } m \text{ is even or } m \text{ is odd and } p \equiv 1 \pmod 4, \\ \pm i\xi_p^{f^\star(b)} & \text{if } m \text{ is odd and } p \equiv 3 \pmod 4, \end{cases}$$

where i is a complex primitive 4th root of unity; see [1176]. Therefore, regular bent functions can only be found for even m and for odd m with $p \equiv 1 \pmod 4$. Moreover, for a weakly regular bent function, the constant u in Definition 20.2.17 can only be equal to ± 1 or $\pm i$. For more information about those functions, we invite the reader to consult the book [1378].

We summarize in Table 20.3 all known weakly regular bent functions over $\mathbb{F}_{p^m}$ with odd characteristic p.

TABLE 20.3: Known weakly regular bent functions over $\mathbb{F}_{p^m}$ for p odd

Bent functions	m	p
$\sum_{i=0}^{\lfloor m/2 \rfloor} \mathrm{Tr}_{p^m/p}(a_i x^{p^i+1})$	arbitrary	arbitrary
$\sum_{i=0}^{p^k-1} \mathrm{Tr}_{p^m/p}(a_i x^{i(p^k-1)}) + \mathrm{Tr}_{p^l/p}(\delta x^{\frac{p^m-1}{e}})$ with $e \mid (p^k+1)$, $l = \min\{i \mid e \mid (p^i-1) \text{ and } i \mid m\}$	$m = 2k$	arbitrary
$\mathrm{Tr}_{p^m/p}(a x^{\frac{3^m-1}{4}+3^k+1})$	$m = 2k$	$p = 3$
$\mathrm{Tr}_{p^m/p}(x^{p^{3k}+p^{2k}-p^k+1}+x^2)$	$m = 4k$	arbitrary
$\mathrm{Tr}_{p^m/p}(a x^{\frac{3^i+1}{2}})$, with i odd, $\gcd(i,m)=1$	arbitrary	$p = 3$

Similarly to bent functions, plateaued functions are classified in odd characteristic into *regular plateaued* and *weakly regular plateaued* functions. The absolute Walsh distribution of plateaued functions follows from Parseval's Identity (20.1); see, e.g., [1376]. More precisely, we have the following result.

Proposition 20.2.18 *Let $f : \mathbb{F}_{p^n} \to \mathbb{F}_p$ be an s-plateaued function. Then for $\omega \in \mathbb{F}_{p^n}$, $W_f(\omega)|^2$ takes p^{n-s} times the value p^{n+s} and $p^n - p^{n-s}$ times the value 0.*

Definition 20.2.19 *The **support of a vectorial function** $f : \mathbb{F}_{q^n} \to \mathbb{F}_{q^m}$ is the set* $\mathrm{supp}(f) := \{x \in \mathbb{F}_{q^n} \mid f(x) \neq 0\}$.

In 2016, Hyun, Lee, and Lee [1018] showed that the Walsh transform coefficients of p-ary s-plateaued f satisfy

$$W_f(\omega) = \begin{cases} \pm p^{\frac{n+s}{2}}\xi_p^{g(\omega)}, 0 & \text{if } n+s \text{ is even, or } n+s \text{ is odd and } p \equiv 1 \pmod 4, \\ \pm i p^{\frac{n+s}{2}}\xi_p^{g(\omega)}, 0 & \text{if } n+s \text{ is odd and } p \equiv 3 \pmod 4, \end{cases}$$

where i is a complex primitive 4th root of unity and g is a p-ary function over $\mathbb{F}_{p^n}$ with $g(\omega) = 0$ for all $\omega \in \mathbb{F}_{p^n} \setminus \mathrm{supp}(W_f)$. It is worth noting that by the definition of $g : \mathbb{F}_{p^n} \to \mathbb{F}_p$, it can be regarded as a mapping from $\mathrm{supp}(W_f)$ to $\mathbb{F}_p$ such that $g(\omega) = 0$ for all $\omega \in \mathbb{F}_{p^n} \setminus \mathrm{supp}(W_f)$. In 2017, Mesnager, Özbudak, and Sınak [1380, 1381] introduced the notion of *weakly regular plateaued functions* in odd characteristic, which covers a nontrivial subclass of the class of plateaued functions.

Definition 20.2.20 ([1380, 1381]) *Let p be an odd prime and $f : \mathbb{F}_{p^n} \to \mathbb{F}_p$ be a p-ary s-plateaued function, where s is an integer with $0 \leq s \leq n$. Then f is called **weakly***

lar p-ary s-plateaued if there exists a complex number u, not dependent on ω, such $|u| = 1$ and $W_f(\omega) \in \left\{ 0, u p^{\frac{n+s}{2}} \xi_p^{g(\omega)} \right\}$ for all $\omega \in \mathbb{F}_{p^n}$, where g is a p-ary function $\mathbb{F}_{p^n}$ with $g(\omega) = 0$ for all $\omega \in \mathbb{F}_{p^n} \setminus \operatorname{supp}(W_f)$; otherwise, f is called **(non)-weakly lar p-ary s-plateaued**. In particular, a weakly regular p-ary s-plateaued function is l regular p-ary s-plateaued when $u = 1$.

he following lemma has, in particular, a significant role in finding the Hamming weights e codewords of a linear code. It can be proven in a similar way to the case of $s = 0$; ·.g., [1379].

ma 20.2.21 *Let p be an odd prime and $f : \mathbb{F}_{p^n} \to \mathbb{F}_p$ be a weakly regular s-plateaued ion. For all $\omega \in \operatorname{supp}(W_f)$,*

$$W_f(\omega) = \epsilon_f \sqrt{p^*}^{n+s} \xi_p^{g(\omega)},$$

$\epsilon_f = \pm 1$ is the sign of W_f, $p^* = (-1)^{(p-1)/2} p$, and g is a p-ary function over (W_f).

ark 20.2.22 Notice that the notion of (non)-weakly regular 0-plateaued functions ides with the one of (non)-weakly regular bent functions. Indeed, if we have $|W_f(\omega)|^2 \in$ $^{\prime}\}$ for all $\omega \in \mathbb{F}_{p^n}$, then by Parseval's Identity, we have $p^{2n} = p^n \# \operatorname{supp}(W_f)$, and by, $\# \operatorname{supp}(W_f) = p^n$. Hence, a (non)-weakly regular 0-plateaued function is a (non)-ly regular bent function.

nition 20.2.23 A function $f : \mathbb{F}_{p^m} \to \mathbb{F}_p$ is said to be **balanced** over $\mathbb{F}_p$ if f takes value of $\mathbb{F}_p$ the same number p^{n-1} of times; otherwise, f is called **unbalanced**.

nple 20.2.24 Let $p = n = 3$. Let $\mathbb{F}_{3^3}^* = \langle \zeta \rangle$ with $\zeta^3 + 2\zeta + 1 = 0$.

The function $f(x) = \operatorname{Tr}_{3^3/3}(\zeta x^{13} + \zeta^7 x^4 + \zeta^7 x^3 + \zeta x^2)$ is weakly regular 3-ary 1-plateaued with $W_f(\omega) \in \{0, -9\xi_3^{g(\omega)}\}$ for all $\omega \in \mathbb{F}_{3^3}$, where g is an unbalanced 3-ary function.

The function $f(x) = \operatorname{Tr}_{3^3/3}(\zeta^5 x^{11} + \zeta^{20} x^5 + \zeta^{11} x^4 + \zeta^2 x^3 + \zeta x^2)$ is regular 3-ary 1-plateaued over $\mathbb{F}_{3^3}$.

The function $\operatorname{Tr}_{3^3/3}(\zeta^{16} x^{13} + \zeta^2 x^4 + \zeta^2 x^3 + \zeta x^2)$ is (non)-weakly regular 3-ary 2-plateaued over $\mathbb{F}_{3^3}$.

.5 Differential Uniformity of Vectorial Boolean Functions, PN, and APN Functions

·ifferentially uniform functions are defined as follows.

nition 20.2.25 **([1451, 1452])** Let n and m be any positive integers such that $m \leq n$ et δ be any positive integer. An (n, m)-function F is called **differentially δ-uniform** r every nonzero $a \in \mathbb{F}_2^n$ and every $b \in \mathbb{F}_2^m$, the equation $F(x) + F(x + a) = b$ has at δ solutions. The minimum of such values δ for a given function F is denoted by δ_F :alled the **differential uniformity** of F.

The expression $F(x) + F(x + a)$ denoted by $D_a F(x)$ is called the **derivative of F in the direction of** a. The differential uniformity is necessarily even since the solutions of equation $D_a F(x) = b$ arise in pairs; if x is a solution of $F(x) + F(x + a) = b$, then $x + a$ is also a solution.

When F is used as an S-box inside a cryptosystem, the differential uniformity measures its resistance to the differential attack: the smaller the value δ_F, the better the contribution of F used as S-box in a cryptosystem to the resistance against this attack.

The differential uniformity δ_F of any (n, m)-function F is bounded below by 2^{n-m}. The differential uniformity δ_F equals 2^{n-m} if and only if F is **perfect nonlinear (PN)**. PN functions are also called *bent functions* [1451]. In fact, we have that an (n, m)-function is bent if and only if all the component functions of F are bent. Recall also that an (n, m)-function is bent if and only if all its derivatives $D_a F(x) = F(x) + F(x + a)$, $a \in \mathbb{F}_{2^n}^*$ are balanced. The equivalence between these two characteristic properties, called respectively bentness and perfect nonlinearity (that is, *bent* and *perfect nonlinear* are equivalent), is a direct consequence of the derivatives being balanced. As we already recalled, perfect nonlinear (n, n)-functions do not exist; but they do exist in other characteristics than 2 (see, e.g., [351]); they are then often called **planar** functions instead of *perfect nonlinear*.

Definition 20.2.26 ([186, 1453, 1454]) An (n, n)-function F is called **almost perfect nonlinear (APN)** if it is differentially 2-uniform; that is, if for every $a \in \mathbb{F}_2^{n*}$ and every $b \in \mathbb{F}_2^n$, the equation $F(x) + F(x + a) = b$ has 0 or 2 solutions.

Since (n, m)-functions have differential uniformity at least 2^{n-m} when $m \leq n/2$ (n even) and strictly larger when n is odd or $m > n/2$, we shall use the term 'APN function' only when $m = n$. An interesting article dealing with Boolean, vectorial plateaued, and APN functions is [348].

20.2.6 APN and Planar Functions over $\mathbb{F}_{q^m}$

Let q be a positive power of a prime p, and m be a positive integer. A function $F : \mathbb{F}_{q^m} \to \mathbb{F}_{q^m}$ is called **almost perfect nonlinear (APN)** if

$$\max_{a \in \mathbb{F}_{q^m}^*} \max_{b \in \mathbb{F}_{q^m}} \#\{x \in \mathbb{F}_{q^m} \mid F(x + a) - F(x) = b\} = 2.$$

And F is said to be **perfect nonlinear** or **planar** if

$$\max_{a \in \mathbb{F}_{q^m}^*} \max_{b \in \mathbb{F}_{q^m}} \#\{x \in \mathbb{F}_{q^m} \mid F(x + a) - F(x) = b\} = 1.$$

Both planar and APN functions over $\mathbb{F}_{q^m}$ for odd q exist. The following is a summary of known APN monomials $F(x) = x^d$ over $\mathbb{F}_{2^m}$:

- [822, 1453] $d = 2^i + 1$ with $\gcd(i, m) = 1$ (Gold function – degree 2);

- [1039, 1087] $d = 2^{2i} - 2^i + 1$ with $\gcd(i, m) = 1$ (Kasami function – degree $i + 1$);

- [607] $d = 2^t + 3$ with $m = 2t + 1$ (Welch function – degree 3);

- [606] $d = 2^t + 2^{t/2} - 1$ if t is even and $d = 2^t + 2^{(3t+1)/2} - 1$ if t is odd where $m = 2t + 1$ (Niho function – degree equals $(t + 2)/2$ if t is even and $t + 1$ if t is odd);

- [186, 1453] $d = 2^{2t} - 1$ with $m = 2t + 1$ (Inverse function – degree $m - 1$);

- [609] $d = 2^{4t} + 2^{3t} + 2^{2t} + 2^t - 1$ with $m = 5t$ (Dobbertin function – degree $t + 3$).

The following is a summary of known APN monomials $F(x) = x^d$ over $\mathbb{F}_{p^m}$ where p is odd:

- $d = 3$, $p > 3$;

- $d = p^m - 2$, $p > 2$, and $p \equiv 2 \pmod 3$;

- $d = \frac{p^m - 3}{2}$, $p \equiv 3, 7 \pmod{20}$, $p^m > 7$, $p^m \neq 27$, and m is odd;

- $d = \frac{p^m + 1}{4} + \frac{p^m - 1}{2}$, $p^m \equiv 3 \pmod 8$;

- $d = \frac{p^m + 1}{4}$, $p^m \equiv 7 \pmod 8$;

- $d = \frac{2p^m - 1}{3}$, $p^m \equiv 2 \pmod 3$;

- $d = 3^m - 3$, $p = 3$, and m is odd;

- $d = p^h + 2$, $p^h \equiv 1 \pmod 3$, and $m \equiv 2h \pmod{17}$;

- $d = \frac{5^h + 1}{2}$, $p = 5$, and $\gcd(2m, h) = 1$;

- $d = \left(3^{(m+1)/4} - 1\right)\left(3^{(m+1)/2} + 1\right)$, $m \equiv 3 \pmod 4$, and $p \equiv 3 \pmod{33}$;

- $p = 3$ and $d = \frac{3^{(m+1)/2} - 1}{2}$ if $m \equiv 3 \pmod 4$ or $d = \frac{3^{(m+1)/2} - 1}{2} + \frac{3^m - 1}{2}$ if $m \equiv 1 \pmod 4$;

- $p = 3$ and $d = \frac{3^{m+1} - 1}{8}$ if $m \equiv 3 \pmod 4$ or $d = \frac{3^{m+1} - 1}{8} + \frac{3^m - 1}{2}$ if $m \equiv 1 \pmod 4$;

- $d = \frac{5^m - 1}{4} + \frac{5^{(m+1)/2} - 1}{2}$, $p = 5$, and m is odd.

The following is a list of some known planar functions over $\mathbb{F}_{p^m}$ where p is odd:

- $f(x) = x^2$;

- $f(x) = x^{p^h + 1}$ where $m / \gcd(m, h)$ is odd;

- $f(x) = x^{(3^h + 1)/2}$ where $p = 3$ and $\gcd(m, h) = 1$;

- $f(x) = x^{10} - ux^6 - u^2 x^2$, where $p = 3$, $u \in \mathbb{F}_{p^m}$, and m is odd.

20.2.7 Dickson Polynomials

Let p be prime, q be a positive power of p, and m be a positive integer. In 1896 Dickson [537] introduced the following family of polynomials over $\mathbb{F}_{q^m}$:

$$D_{h,a}(x) = \sum_{i=0}^{\lfloor \frac{h}{2} \rfloor} \frac{h}{h - i} \binom{h - i}{i} (-a)^i x^{h-2i},$$

where $a \in \mathbb{F}_{q^m}$, and $h \geq 0$ is called the **order** of the polynomial. This family is referred to as the **Dickson polynomials of the first kind**.

Dickson polynomials of the second kind over $\mathbb{F}_{q^m}$ are defined by

$$E_{h,a}(x) = \sum_{i=0}^{\lfloor \frac{h}{2} \rfloor} \binom{h}{h - i} (-a)^i x^{h-2i},$$

where $a \in \mathbb{F}_{q^m}$, and $h \geq 0$ is called the **order** of the polynomial. We invite the reader to consult the excellent book [1253] for detailed information about Dickson polynomials.

20.3 Generic Constructions of Linear Codes

In coding theory, Boolean functions, or more generally p-ary functions have been used to construct linear codes. Historically, the Reed–Muller codes and Kerdock codes have been, for a long time, the two famous classes of binary codes derived from Boolean functions. A lot of progress has been made in this direction, and further codes have been derived from more general and complex functions. Nevertheless, as highlighted by Ding in his excellent survey [552], despite the advances in the past two decades, one can isolate essentially two generic constructions of linear codes from functions; see, e.g., [1378, Chapter 18].

20.3.1 The First Generic Construction

The first generic construction is obtained by considering a code $\mathcal{C}(f)$ over $\mathbb{F}_p$ involving a polynomial $f : \mathbb{F}_q \to \mathbb{F}_q$, where $q = p^m$. Such a code is defined by

$$\mathcal{C}(f) = \big\{ \mathbf{c} = \big(\mathrm{Tr}_{q/p}(af(x) + bx)\big)_{x \in \mathbb{F}_q} \mid a \in \mathbb{F}_q, b \in \mathbb{F}_q \big\}.$$

The resulting code $\mathcal{C}(f)$ is a linear code over $\mathbb{F}_p$ of length q, and its dimension is bounded above by $2m$; this bound is reached when the nonlinearity of the vectorial function f is larger than 0, which happens in many cases.

One can also define a code $\mathcal{C}^*(f)$ over $\mathbb{F}_p$ involving a polynomial f from $\mathbb{F}_q$ to $\mathbb{F}_q$, where $q = p^m$, which vanishes at 0 defined by

$$\mathcal{C}^*(f) = \big\{ \mathbf{c} = \big(\mathrm{Tr}_{q/p}(af(x) + bx)\big)_{x \in \mathbb{F}_q^*} \mid a \in \mathbb{F}_q, b \in \mathbb{F}_q \big\}.$$

The resulting code $\mathcal{C}^*(f)$ is a linear code of length $q - 1$, and its dimension is also bounded above by $2m$, which is reached in many cases.

As mentioned in the literature (see, e.g., [552]), the first generic construction has a long history and its importance is supported by Delsarte's Theorem [522]. In the binary case, that is, when $p = 2$, the first generic construction allows a kind of coding-theoretic characterization of special cryptographic functions such as APN functions, almost bent functions, and semi-bent functions; see [350].

More generally, but in a more complex way, for any $\alpha, \beta \in \mathbb{F}_{q^r}$, define

$$
\begin{aligned}
f_{\alpha,\beta} \;:\; \mathbb{F}_{q^r} &\longrightarrow \mathbb{F}_q \\
x &\longmapsto f_{\alpha,\beta}(x) := \mathrm{Tr}_{q^r/q}(\alpha\Psi(x) - \beta x)
\end{aligned}
$$

where Ψ is a mapping from $\mathbb{F}_{q^r}$ to $\mathbb{F}_{q^r}$ such that $\Psi(0) = 0$. One can define a linear code $\mathcal{C}_\Psi$ over $\mathbb{F}_q$ as

$$\mathcal{C}_\Psi = \big\{ \widetilde{c}_{\alpha,\beta} = \big(f_{\alpha,\beta}(\zeta_1), f_{\alpha,\beta}(\zeta_2), \dots, f_{\alpha,\beta}(\zeta_{q^r-1})\big) \mid \alpha, \beta \in \mathbb{F}_{q^r} \big\} \qquad (20.4)$$

where $\zeta_1, \dots, \zeta_{q^r-1}$ denote the nonzero elements of $\mathbb{F}_{q^r}$.

Proposition 20.3.1 ([1379]) *The linear code $\mathcal{C}_\Psi$ has length $q^r - 1$. If the mapping Ψ has no linear component $\Psi_u : x \mapsto \mathrm{Tr}_{q^r/q}(u\Psi(x))$ over $\mathbb{F}_p$, then $\mathcal{C}_\Psi$ has dimension $2r$.*

Proof: It is clear that $\mathcal{C}_\Psi$ is of length $q^r - 1$. Now, let us compute the cardinality of $\mathcal{C}_\Psi$. Let $\widetilde{c}_{\alpha,\beta}$ be a codeword of $\mathcal{C}_\Psi$. We have

$$\widetilde{c}_{\alpha,\beta} = 0 \iff \mathrm{Tr}_{q^r/q}(\alpha\Psi(\zeta_i) - \beta\zeta_i) = 0, \text{ for all } i \in \{1, \dots, q^r - 1\}$$
$$\iff \mathrm{Tr}_{q^r/q}(\alpha\Psi(x) - \beta x) = 0, \text{ for all } x \in \mathbb{F}_{q^r}^*$$
$$\implies \mathrm{Tr}_{q^r/p}(\alpha\Psi(x) - \beta x) = 0, \text{ for all } x \in \mathbb{F}_{q^r}^*$$
$$\implies \mathrm{Tr}_{q^r/p}(\alpha\Psi(x) - \beta x) = 0, \text{ for all } x \in \mathbb{F}_{q^r}$$
$$\implies \mathrm{Tr}_{q^r/p}(\alpha\Psi(x)) = \mathrm{Tr}_{q^r/p}(\beta x), \text{ for all } x \in \mathbb{F}_{q^r}.$$

Hence, $\widetilde{c}_{\alpha,\beta} = 0$ implies that the component of Ψ associated with $\alpha \neq 0$ is linear or null and coincides with $x \mapsto \mathrm{Tr}_{q^r/p}(\beta x)$. Therefore, to ensure that the null codeword appears only once (for $\alpha = \beta = 0$), it suffices that no component function of Ψ is identically equal to 0 or linear. Furthermore, this implies that all the codewords $\widetilde{c}_{\alpha,\beta}$ are pairwise distinct. In this case, the size of the code is q^{2r} and its dimension thus equals $2r$. $\qquad\square$

The following statement shows that the weight distribution of the code $\mathcal{C}_\Psi$ of length $q^r - 1$ can be expressed by means of the Walsh transform of some absolute trace functions over $\mathbb{F}_{p^m}$ involving the map Ψ.

Proposition 20.3.2 ([1379]) *We keep the notation above. Set $q = p^m$. Let $a \in \mathbb{F}_{p^m}$ and let Ψ be a mapping from $\mathbb{F}_{q^r}$ to $\mathbb{F}_{q^r}$ such that $\Psi(0) = 0$. Let us denote by ψ_a a mapping from $\mathbb{F}_{p^m}$ to $\mathbb{F}_p$ defined as*

$$\psi_a(x) = \mathrm{Tr}_{p^m/p}(a\Psi(x)).$$

For $\widetilde{c}_{\alpha,\beta} \in \mathcal{C}_\Psi$, we have

$$\mathrm{wt}_{\mathrm{H}}(\widetilde{c}_{\alpha,\beta}) = p^m - \frac{1}{q} \sum_{\omega \in \mathbb{F}_q} W_{\psi_{\omega\alpha}}(\omega\beta).$$

Proof: Note that $\Psi(0) = 0$ implies that $f_{\alpha,\beta}(0) = 0$. Now let $\widetilde{c}_{\alpha,\beta}$ be a codeword of $\mathcal{C}_\Psi$. Then

$$\mathrm{wt}_{\mathrm{H}}(\widetilde{c}_{\alpha,\beta}) = \#\{x \in \mathbb{F}_{q^r}^* \mid f_{\alpha,\beta}(x) \neq 0\} = \#\{x \in \mathbb{F}_{q^r} \mid f_{\alpha,\beta}(x) \neq 0\}$$
$$= p^m - \#\{x \in \mathbb{F}_{q^r} \mid f_{\alpha,\beta}(x) = 0\} = p^m - \sum_{x \in \mathbb{F}_{q^r}} \frac{1}{q} \sum_{\omega \in \mathbb{F}_q} \xi_p^{\mathrm{Tr}_{q/p}(\omega f_{\alpha,\beta}(x))}.$$

The latter equality comes from the fact that the sum of characters equals q if $f_{\alpha,\beta}(x) = 0$ and 0 otherwise. Moreover, using the transitivity property of the trace function $\mathrm{Tr}_{q^r/p}$ and the fact that $\mathrm{Tr}_{q^r/q}$ is $\mathbb{F}_q$-linear, we have

$$\mathrm{wt}_{\mathrm{H}}(\widetilde{c}_{\alpha,\beta}) = p^m - \frac{1}{q} \sum_{\omega \in \mathbb{F}_q} \sum_{x \in \mathbb{F}_{q^r}} \xi_p^{\mathrm{Tr}_{q/p}(\omega f_{\alpha,\beta}(x))}$$
$$= p^m - \frac{1}{q} \sum_{\omega \in \mathbb{F}_q} \sum_{x \in \mathbb{F}_{q^r}} \xi_p^{\mathrm{Tr}_{q/p}(\omega \mathrm{Tr}_{q^r/q}(\alpha\Psi(x) - \beta x))}$$
$$= p^m - \frac{1}{q} \sum_{\omega \in \mathbb{F}_q} \sum_{x \in \mathbb{F}_{q^r}} \xi_p^{\mathrm{Tr}_{q^r/p}(\omega\alpha\Psi(x) - \omega\beta x)}$$
$$= p^m - \frac{1}{q} \sum_{\omega \in \mathbb{F}_q} \sum_{x \in \mathbb{F}_{q^r}} \xi_p^{\psi_{\omega\alpha}(x) - \mathrm{Tr}_{q^r/p}(\omega\beta x)} = p^m - \frac{1}{q} \sum_{\omega \in \mathbb{F}_q} W_{\psi_{\omega\alpha}}(\omega\beta).$$

$\square$

20.3.2 The Second Generic Construction

In this subsection, we introduce the second generic construction of linear codes from cryptographic functions and its generalizations, and we also present recent results and some open problems regarding this construction. We shall see that many new linear codes with few nonzero weights from cryptographic functions, such as PN functions, APN functions, bent functions, AB functions, etc., have been obtained from this construction, including some optimal linear codes.

20.3.2.1 The Defining Set Construction of Linear Codes

The second generic construction of linear codes from functions is obtained by fixing a set $D = \{d_1, d_2, \ldots, d_n\}$ in $\mathbb{F}_q$, where $q = p^k$, and by defining a linear code involving D as follows:
$$\mathcal{C}_D = \left\{ \left(\mathrm{Tr}_{q/p}(xd_1), \mathrm{Tr}_{q/p}(xd_2), \ldots, \mathrm{Tr}_{q/p}(xd_n) \right) \mid x \in \mathbb{F}_q \right\}.$$
The set D is usually called the **defining set** of the code $\mathcal{C}_D$. The resulting code $\mathcal{C}_D$ is linear over $\mathbb{F}_p$ of length n with dimension at most k. This construction is generic in the sense that any linear code of dimension k over $\mathbb{F}_p$ and whose generator matrix G is made of distinct columns (i.e., does not allow repetitions)[1] can be obtained this way. Indeed, the function $d \in \mathbb{F}_q \mapsto \mathrm{Tr}_{q/p}(xd)$ matches once with each linear form over $\mathbb{F}_q$ when x ranges over $\mathbb{F}_q$, and the codewords of the code are obtained as $x \times G$, where x ranges over $\mathbb{F}_q$. The defining set construction idea was introduced by J. Wolfmann [1898] in 1975. The idea also goes back to the earlier description of the trace construction of irreducible cyclic codes given by L. D. Baumert and R. J. McEliece [140]. However, it is hard to trace who used the idea first, as the idea is based on the simple fact (known 150 years ago, much earlier than the beginning of coding theory in 1949) that every linear function from $\mathbb{F}_{q^m}$ to $\mathbb{F}_q$ must be of the form $\mathrm{Tr}_{q^m/q}(ax)$ for some $a \in \mathbb{F}_{q^m}$.

The defining set construction has been revisited many times since then; see, e.g., [563]. The code quality in terms of parameters is closely related to the choice of the set D. The Hamming weights of the codewords can be expressed by means of exponential sums. For each $x \in \mathbb{F}_q$, define
$$\mathbf{c}_x = \left(\mathrm{Tr}_{q/p}(xd_1), \mathrm{Tr}_{q/p}(xd_1), \ldots, \mathrm{Tr}_{q/p}(xd_n) \right).$$
The Hamming weight $\mathrm{wt}_\mathrm{H}(\mathbf{c}_x)$ of $\mathbf{c}_x$ is $n - N_x(0)$, where
$$N_x(0) = \#\{1 \leq i \leq n \mid \mathrm{Tr}_{q/p}(xd_i) = 0\}, \quad x \in \mathbb{F}_q.$$
Note that
$$p N_x(0) = \sum_{i=1}^{n} \sum_{y \in \mathbb{F}_p} \xi_p^{y \mathrm{Tr}_{q/p}(xd_i)} = \sum_{i=1}^{n} \sum_{y \in \mathbb{F}_p} \chi_1(yxd_i) = n + \sum_{y \in \mathbb{F}_p^*} \chi_1(yxD),$$
where χ_1 is the canonical additive character[2] of $\mathbb{F}_q$, aD denotes the set $\{ad \mid d \in D\}$, and $\chi_1(S) := \sum_{x \in S} \chi_1(x)$ for any subset S of $\mathbb{F}_q$. Therefore,
$$\mathrm{wt}_\mathrm{H}(\mathbf{c}_x) = \frac{(p-1)}{p} n - \frac{1}{p} \sum_{y \in \mathbb{F}_p^*} \chi_1(yxD).$$

---

[1]Such codes are sometimes called **projective codes**.

[2]The characters of a finite field $\mathbb{F}_q$ (where q is a power of a prime p) are parameterized by an element $a \in \mathbb{F}_q$ and are given by the formula $\chi_a(x) = \xi^{\mathrm{Tr}_{q/p}(ax)}$ where ξ is a primitive complex p^th root of unity. The choice $a = 0$ gives the trivial character, equal to the constant function 1. The character corresponding to $a = 1$ is often called the *canonical character*, but this term is not in universal use. It is well known that any additive character of $\mathbb{F}_q$ can be written as $\chi_a(x) = \chi_1(ax)$.

In practice, the defining set could be the support or the co-support (i.e., the complement of the support) of a function defined over a finite field.

20.3.2.2 Generalizations of the Defining Set Construction of Linear Codes

The defining set construction of linear codes is an efficient method to produce linear codes with few weights, including optimal linear codes; so it has been extensively studied in recent years. To the best of our knowledge, this method has been extended in the following three directions:

I. Generalization I: Let $f : \mathbb{F}_{2^k} \to \mathbb{F}_2$ with $f(ax) = f(x)$, where $a \in \mathbb{F}_{2^t}^*$, $t \mid k$ and $x \in \mathbb{F}_{2^k}$. Let $T = \{t_1, t_2, \ldots, t_n\}$ satisfying $S = \{x \in \mathbb{F}_{2^k}^* \mid f(x) = 0\} = \bigcup_{i=1}^n t_i \mathbb{F}_{2^t}^*$. Define $D = T$ or $D = S$, and a linear code $\mathcal{C}_D$ over $\mathbb{F}_{2^t}$ as follows:

$$\mathcal{C}_D = \left\{ \left(\mathrm{Tr}_{2^k/2^t}(xd) \right)_{d \in D} \mid x \in \mathbb{F}_{2^k} \right\}.$$

II. Generalization II: Let $F : \mathbb{F}_{2^k} \to \mathbb{F}_{2^s}$ and D be the support of $\mathrm{Tr}_{2^s/2}(\lambda F(x))$ where $\lambda \in \mathbb{F}_{2^s}^*$. A linear code $\mathcal{C}_D$ over $\mathbb{F}_2$ is defined by

$$\mathcal{C}_D = \left\{ \left(\mathrm{Tr}_{2^k/2}(xd) + \mathrm{Tr}_{2^s/2}(yF(d)) \right)_{d \in D} \mid x \in \mathbb{F}_{2^k}, y \in \mathbb{F}_{2^s} \right\}.$$

III. Generalization III: Let $F_i : \mathbb{F}_{p^k} \to \mathbb{F}_{p^k}$ for $i = 1, 2, \ldots, t$, where t is a positive integer. Define a linear code $\mathcal{C}_D$ as follows:

$$\mathcal{C}_D = \left\{ \left(\mathrm{Tr}_{p^k/p}(a_1 x_1 + \cdots + a_t x_t) \right)_{(a_1, \ldots, a_t) \in D} \mid x_1, \ldots, x_t \in \mathbb{F}_{p^k} \right\},$$

where the defining set D is given by

$$D = \{(x_1, x_2, \ldots, x_t) \mid \mathrm{Tr}_{p^k/p}(F_1(x_1) + F_2(x_2) + \cdots + F_t(x_t)) = 0\}.$$

20.3.2.3 A Modified Defining Set Construction of Linear Codes

Let $F(x)$ be a mapping from $\mathbb{F}_{p^k}$ to $\mathbb{F}_{p^k}$ and $D = \{d_1, d_2, \ldots, d_n\} \subseteq \mathbb{F}_{p^k}$. A linear code $\mathcal{C}_{F(D)}$ then can be obtained as follows:

$$\mathcal{C}_{F(D)} = \left\{ \left(\mathrm{Tr}_{p^k/p}\left(xF(d_1) \right), \mathrm{Tr}_{p^k/p}\left(xF(d_2) \right), \ldots, \mathrm{Tr}_{p^k/p}\left(xF(d_n) \right) \right) \mid x \in \mathbb{F}_{p^k} \right\}. \quad (20.5)$$

Note that the linear code defined as (20.5) coincides with the one given in the original defining set construction if $F(x) = x$.

20.4 Binary Codes with Few Weights from Boolean Functions and Vectorial Boolean Functions

20.4.1 A First Example of Codes from Boolean Functions: Reed–Muller Codes

Reed–Muller codes, introduced by D. E. Muller and I. S. Reed in 1954, are one of the best-understood families of codes; see, e.g., [1323]. Except for first-order Reed–Muller codes and for codes of small lengths, their minimum distance is lower than that of BCH codes. But they have very efficient decoding algorithms, they contain nonlinear subcodes with optimal parameters together with efficient decoding algorithms, and they give a useful framework for the study of Boolean functions in cryptography. Reed–Muller codes can be defined in terms of Boolean functions defined over $\mathbb{F}_2^m$.

For every $0 \leq r \leq m$, the Reed–Muller code of order r is the set of all Boolean functions of algebraic degrees at most r. More precisely, it is the linear code of all binary words of length 2^m corresponding to the last columns of the truth-tables of these functions; see [1323]. The Reed–Muller codes are nested: $\mathcal{RM}(0,m) \subset \mathcal{RM}(1,m) \subset \cdots \subset \mathcal{RM}(m,m)$. For every $0 \leq r \leq m-1$, the dual code of $\mathcal{RM}(r,m)$ is the $(m-r-1)^{\text{th}}$-order Reed–Muller code $\mathcal{RM}(m-r-1,m)$; see Theorem 1.11.6.

20.4.2 A General Construction of Binary Linear Codes from Boolean Functions

We start with a simple but interesting construction of binary linear codes from Boolean functions; see, e.g., [556]. Let f be a Boolean function from $\mathbb{F}_{2^m}$ to $\mathbb{F}_2$ such that $f(0) = 0$ but $f(b) = 1$ for at least one $b \in \mathbb{F}_{2^m}$. We define a binary linear code based on a modified version of the first generic construction by

$$\mathcal{C}_f = \left\{ \left(uf(x) + \mathrm{Tr}_{2^m/2}(vx) \right)_{x \in \mathbb{F}_{2^m}^*} \mid u \in \mathbb{F}_2, v \in \mathbb{F}_{2^m} \right\}.$$

The binary code $\mathcal{C}_f$ has length $2^m - 1$ and dimension $m + 1$ if $f(x) \neq \mathrm{Tr}_{2^m/2}(wx)$ for all $w \in \mathbb{F}_{2^m}$. In addition, the weight distribution of $\mathcal{C}_f$ is given by the multiset union $\left\{ \frac{1}{2}(2^m - W_f(w)) \mid w \in \mathbb{F}_{2^m} \right\} \cup \left\{ 2^{m-1} \mid w \in \mathbb{F}_{2^m}^* \right\} \cup \{0\}$.

20.4.3 Binary Codes from the Preimage $f^{-1}(b)$ of Boolean Functions f

Let f be a function from $\mathbb{F}_{2^m}$ to $\mathbb{F}_2$, and let D be any subset of the preimage $f^{-1}(b)$ for any $b \in \mathbb{F}_2$. In general, it is very hard to determine the parameters of the code $\mathcal{C}_D$ whose defining set is D. Let $n_f = \#D_f$ be the cardinality of the support D_f of f. Ding [550] has been interested in the binary code $\mathcal{C}_{D_f}$, obtained from the second generic construction and whose defining set is D_f, with length n_f and dimension at most m. We will present his results on the weight distribution of the code $\mathcal{C}_{D_f}$ for several classes of Boolean functions f.

The following result of Ding [550] establishes a connection between the set of Boolean functions f such that $2n_f + W_f(w) \neq 0$ for all $w \in \mathbb{F}_{2^m}^*$ and a class of binary linear codes.

Theorem 20.4.1 ([550]) *Let f be a function from $\mathbb{F}_{2^m}$ to $\mathbb{F}_2$, and let D_f be the support of f. If $2n_f + W_f(w) \neq 0$ for all $w \in \mathbb{F}_{2^m}^*$, then $\mathcal{C}_{D_f}$ is a binary linear code with length n_f and dimension m, and its weight distribution is given by the multiset*

$$\left\{ \frac{2n_f + W_f(w)}{4} \mid w \in \mathbb{F}_{2^m}^* \right\} \cup \{0\}. \tag{20.6}$$

The determination of the weight distribution of the binary linear code $\mathcal{C}_{D_f}$ is equivalent to that of the Walsh spectrum of the Boolean function f satisfying $2n_f + W_f(w) \neq 0$ for all $w \in \mathbb{F}_{2^m}^*$. As highlighted by Ding, when the Boolean function f is selected properly, the code $\mathcal{C}_{D_f}$ could have only a few weights and may have good parameters.

Theorem 20.4.1 can be generalized into the following statement whose proof is the same as that of Theorem 20.4.1.

Theorem 20.4.2 ([550]) *Let f be a function from $\mathbb{F}_{2^m}$ to $\mathbb{F}_2$, and let D_f be the support of f. Let e_w denote the multiplicity of the element $\frac{2n_f + W_f(w)}{4}$ and e the multiplicity of 0 in the multiset $\{2n_f + W_f(w) \mid w \in \mathbb{F}_{2^m}^*\}$. Then $\mathcal{C}_{D_f}$ is a binary linear code with length n_f and dimension $m - \log_2 e$, and the weight distribution of $\mathcal{C}_{D_f}$ is given by*

$$\frac{2n_f + W_f(w)}{4} \quad \text{with frequency} \quad \frac{e_w}{e}$$

for all $\frac{2n_f + W_f(w)}{4}$ in the multiset of (20.6).

20.4.4 Codes with Few Weights from Bent Boolean Functions

This subsection is devoted to codes from bent functions. For every even m, the Kerdock code $\mathcal{K}_{m/2}$ [1107] of size 2^{2m} contains the Reed–Muller code $\mathcal{RM}(1,m)$ and is a subcode of $\mathcal{RM}(2,m)$. It is constructed from bent functions; see, e.g., [349, 346]. The codewords of the Kerdock code are the lists of values of the functions

$$(x, x_m) \mapsto \operatorname{Tr}_{2^{m-1}/2}\left(\sum_{j=1}^{m/2-1} (ux)^{2^j+1} \right) + x_m \operatorname{Tr}_{2^{m-1}/2}(ux) + \operatorname{Tr}_{2^{m-1}/2}(ax) + \eta x_m + \epsilon,$$

where $u, a, x \in \mathbb{F}_{2^{m-1}}$ and $\epsilon, \eta, x_m \in \mathbb{F}_2$. $\mathcal{K}_{m/2}$ is a union of cosets $f_u + \mathcal{RM}(1,m)$ of $\mathcal{RM}(1,m)$, where the functions

$$f_u(x, x_m) = \operatorname{Tr}_{2^{m-1}/2}\left(\sum_{j=1}^{m/2-1} (ux)^{2^j+1} \right) + x_m \operatorname{Tr}_{2^{m-1}/2}(ux)$$

are bent quadratic, except for $u = 0$. Moreover, the sum of any two distinct functions f_u is bent. Since we know that the sum of a bent function and an affine function is bent, the code $\mathcal{K}_{m/2}$ has for minimum distance $2^{m-1} - 2^{\frac{m}{2}-1}$, which is the best possible minimum distance for a code equal to a union of cosets of $\mathcal{RM}(1,m)$, according to the Covering Radius Bound. In fact, as shown by Delsarte [520], $2^{m-1} - 2^{\frac{m}{2}-1}$ is the best possible minimum distance for any code of length 2^m and size 2^{2m}. Interesting developments about these nonlinear codes with their relationship to bent Boolean functions can be found in [345, 346]. Later, more codes, especially with few weights, were constructed from bent Boolean functions. We highlight that Wolfmann [1900] was the first to construct two-weight codes from bent functions.

Recall that a function f from $\mathbb{F}_{2^m}$ to $\mathbb{F}_2$ is bent if and only if its support $D_f = \{x \in \mathbb{F}_{2^m} \mid f(x) = 1\}$ is a difference set in $(\mathbb{F}_{2^m}, +)$ with parameters

$$\left(2^m, 2^{m-1} \pm 2^{\frac{m-2}{2}}, 2^{m-2} \pm 2^{\frac{m-2}{2}} \right).$$

Hence, when f is bent, we have

$$n_f := \#D_f = 2^{m-1} \pm 2^{\frac{m-2}{2}}.$$

Wolfmann established for the first time a link between bent Boolean functions and two-weight binary codes.

Theorem 20.4.3 ([1900]) *Let f be a Boolean function from $\mathbb{F}_{2^m}$ to $\mathbb{F}_2$ with $f(0) = 0$ where m is even and $m \geq 4$. Then the code $\mathcal{C}_{D_f}$, obtained from the second generic construction and whose defining set is D_f, is an $\left[n_f, m, (n_f - 2^{(m-2)/2})/2 \right]_2$ two-weight binary code with weight distribution given by Table 20.4 if and only if f is bent.*

TABLE 20.4: Weight distribution of $\mathcal{C}_{D_f}$ in Theorem 20.4.3

Hamming weight w	Multiplicity A_w
0	1
$\frac{n_f}{2} - 2^{(m-4)/2}$	$\frac{2^m - 1 - n_f 2^{-(m-2)/2}}{2}$
$\frac{n_f}{2} + 2^{(m-4)/2}$	$\frac{2^m - 1 + n_f 2^{-(m-2)/2}}{2}$

Consequently, any bent function can be plugged into Theorem 20.4.3 to obtain a two-weight binary linear code.

Now we are interested in constructing binary linear codes based on the first generic construction. With $\Psi : \mathbb{F}_{2^m} \to \mathbb{F}_{2^m}$ where $\Psi(0) = 0$, let $\mathcal{C}_\Psi$ be the linear code defined by (20.4). Set $\psi_1(x) = \text{Tr}_{2^m/2}(\Psi(x))$. Let us define a subcode $\mathcal{C}_{\psi_1}$ of $\mathcal{C}_\Psi$ as follows:

$$\mathcal{C}_{\psi_1} = \left\{ \widetilde{c}_{\alpha,\beta} = \left(f_{\alpha,\beta}(\zeta_1), f_{\alpha,\beta}(\zeta_2), \dots, f_{\alpha,\beta}(\zeta_{2^m-1}) \right) \mid \alpha \in \mathbb{F}_2, \ \beta \in \mathbb{F}_{2^m} \right\}, \qquad (20.7)$$

where $\zeta_1, \dots, \zeta_{2^m-1}$ denote the nonzero elements of $\mathbb{F}_{2^m}$.

Theorem 20.4.4 ([1379]) *Let $\mathcal{C}_{\psi_1}$ be the linear code defined by (20.7) whose codewords are denoted by $\widetilde{c}_{\alpha,\beta}$. Let m be an even integer. Assume that the function $\psi_1 = \text{Tr}_{2^m/2}(\Psi)$ is bent. We denote by $\psi_1^\star$ its dual function. Then $\mathcal{C}_{\psi_1}$ is a three-weight code with weights as follows. We have $\text{wt}_\text{H}(\widetilde{c}_{0,0}) = 0$, and for $\beta \neq 0$ $\text{wt}_\text{H}(\widetilde{c}_{0,\beta}) = 2^m - 2^{m-1}$. Moreover, the Hamming weight of $\widetilde{c}_{1,\beta}$ for $\beta \in \mathbb{F}_{2^m}$ is $\text{wt}_\text{H}(\widetilde{c}_{1,\beta}) = 2^{m-1} - (-1)^{\psi_1^\star(\beta)} 2^{\frac{m}{2}-1}$. The weight distribution of $\mathcal{C}_{\psi_1}$, which is of dimension $m + 1$, is in Table 20.5.*

TABLE 20.5: Weight distribution of $\mathcal{C}_{\psi_1}$ in Theorem 20.4.4

Hamming weight w	Multiplicity A_w
0	1
$2^m - 2^{m-1}$	$2^m - 1$
$2^{m-1} - 2^{\frac{m}{2}-1}$	$2^{m-1} + 2^{\frac{m}{2}-1}$
$2^{m-1} + 2^{\frac{m}{2}-1}$	$2^{m-1} - 2^{\frac{m}{2}-1}$

20.4.5 Codes with Few Weights from Semi-Bent Boolean Functions

Let f be a semi-bent Boolean function on $\mathbb{F}_{2^m}$ with m odd. Denote by D_f its support and by n_f the cardinality of D_f. Recall that in this case the Walsh spectrum W_f of f is equal to $\{0, \pm 2^{\frac{m+1}{2}}\}$. Let $\mathcal{C}_{D_f}$ be the linear code obtained via the second generic construction. Ding [550] proved the following result on $\mathcal{C}_{D_f}$.

Theorem 20.4.5 ([550]) *Let f be a Boolean function on $\mathbb{F}_{2^m}$ with $f(0) = 0$ where m is odd. Then $\mathcal{C}_{D_f}$ is an $\left[n_f, m, (n_f - 2^{(m-1)/2})/2 \right]_2$ three-weight binary code with the weight distribution as given in Table 20.6 where n_f is defined by*

$$n_f = \#D_f = \begin{cases} 2^{m-1} - 2^{\frac{m-1}{2}} & \text{if } W_f(0) = 2^{(m+1)/2}, \\ 2^{m-1} + 2^{\frac{m-1}{2}} & \text{if } W_f(0) = -2^{(m+1)/2}, \\ 2^{m-1} & \text{if } W_f(0) = 0 \end{cases}$$

if and only if the function f is semi-bent.

All semi-bent functions can be plugged into Theorem 20.4.5 to obtain three-weight binary linear codes. The known semi-bent functions are given in [1378, Chapter 17].

TABLE 20.6: Weight distribution of $\mathcal{C}_{D_f}$ in Theorem 20.4.5

Hamming weight w	Multiplicity A_w
0	1
$\frac{n_f - 2^{(m-1)/2}}{2}$	$n_f(2^m - n_f)2^{-m} - n_f 2^{-(m+1)/2}$
$\frac{n_f}{2}$	$2^m - 1 - n_f(2^m - n_f)2^{-(m-1)}$
$\frac{n_f + 2^{(m-1)/2}}{2}$	$n_f(2^m - n_f)2^{-m} + n_f 2^{-(m+1)/2}$

20.4.6 Linear Codes from Quadratic Boolean Functions

Recalling Example 20.2.6, let

$$f(x) = \mathrm{Tr}_{2^m/2}\left(\sum_{i=0}^{\lfloor m/2 \rfloor} f_i x^{2^i+1}\right) \tag{20.8}$$

be a quadratic Boolean function from $\mathbb{F}_{2^m}$ to $\mathbb{F}_2$, where $f_i \in \mathbb{F}_{2^m}$, and the rank of f is r_f. Let D_f be the support of f. By definition, we have

$$n_f = \#D_f = 2^{m-1} - \frac{W_f(0)}{2} = \begin{cases} 2^{m-1} & \text{if } W_f(0) = 0, \\ 2^{m-1} - 2^{m-1-r_f/2} & \text{if } W_f(0) = 2^{m-r_f/2}, \\ 2^{m-1} + 2^{m-1-r_f/2} & \text{if } W_f(0) = -2^{m-r_f/2}. \end{cases} \tag{20.9}$$

Let $\mathcal{C}_{D_f}$ be the code based on the second generic construction with defining set D_f. From Theorem 20.4.1 and Table 20.1, Ding has deduced the following result.

Theorem 20.4.6 ([550]) *Let f be a quadratic Boolean function of the form in (20.8) such that $r_f > 2$. Then $\mathcal{C}_{D_f}$ is a three-weight binary code with length n_f given in (20.9), dimension m, and the weight distribution in Table 20.7, where*

$$(\epsilon_1, \epsilon_2, \epsilon_3) = \begin{cases} (1,0,0) & \text{if } W_f(0) = 0, \\ (0,1,0) & \text{if } W_f(0) = 2^{m-1-r_f/2}, \\ (0,0,1) & \text{if } W_f(0) = -2^{m-1-r_f/2}. \end{cases}$$

TABLE 20.7: Weight distribution of $\mathcal{C}_{D_f}$ in Theorem 20.4.6

Hamming weight w	Multiplicity A_w
0	1
$\frac{n_f}{2}$	$2^m - 2^{r_f} - \epsilon_1$
$\frac{n_f + 2^{m-1-r_f/2}}{2}$	$2^{r_f-1} + 2^{(r_f-2)/2} - \epsilon_2$
$\frac{n_f - 2^{m-1-r_f/2}}{2}$	$2^{r_f-1} - 2^{(r_f-2)/2} - \epsilon_3$

As mentioned by Ding in [552], the code $\mathcal{C}_{D_f}$ in Theorem 20.4.6 defined by any quadratic Boolean function f is different from any subcode of the second-order Reed–Muller code, due to the difference in their lengths as well as their weight distributions.

20.4.7 Binary Codes $\mathcal{C}_{D_f}$ with Three Weights

Let $f(x) = Tr_{2^m/2}(x^d)$ for the following d:

- [822] $d = 2^h + 1$ where $\gcd(m, h)$ is odd and $1 \leq h \leq m/2$;

- [1087] $d = 2^{2h} - 2^h + 1$ where $\gcd(m, h)$ is odd and $1 \leq h \leq m/2$;

- [477] $d = 2^{m/2} + 2^{(m+2)/4} + 1$ where $m \equiv 2 \pmod 4$;

- [477] $d = 2^{(m+2)/2} + 3$ where $m \equiv 2 \pmod 4$.

For all the values of d listed above, one has $\gcd(d, 2^m - 1) = 1$. Therefore $n_f = \#D_f = 2^{m-1}$. Ding [552] computed the Walsh spectrum of the functions f, given in Table 20.8, and derived three-weight codes based on the second generic construction.

TABLE 20.8: Boolean functions with three-valued Walsh spectrum

$W_f(w)$	Number of w's
0	$2^m - 2^{m-d}$
$2^{(m+d)/2}$	$2^{m-d-1} + 2^{(m-d-2)/2}$
$-2^{(m+d)/2}$	$2^{m-d-1} - 2^{(m-d-2)/2}$

Theorem 20.4.7 ([552]) *Let $m \geq 4$ be even. Let $f(x) = \mathrm{Tr}_{2^m/2}(x^d)$ where d are given above. Then the code $\mathcal{C}_{D_f}$ with defining set D_f is a three-weight code and has parameters $\left[2^{m-1}, m, 2^{m-2} - 2^{(m-2)/2}\right]_2$ and weight distribution given in Table 20.9.*

TABLE 20.9: Weight distribution of $\mathcal{C}_{D_f}$ in Theorem 20.4.7

Hamming weight w	Multiplicity A_w
0	1
2^{m-2}	$2^m - 2^{m-2} - 1$
$2^{m-2} + 2^{(m-2)/2}$	$2^{m-3} + 2^{(m-4)/2}$
$2^{m-2} - 2^{(m-2)/2}$	$2^{m-3} - 2^{(m-4)/2}$

20.4.8 Binary Codes $\mathcal{C}_{D_f}$ with Four Weights

Let f be the Boolean function given by $f(x) = \mathrm{Tr}_{2^m/2}(x^d)$ where d is given in the following list:

- [1438] When $d = 2^{(m+2)/2} - 1$ and $m \equiv 0 \pmod 4$, the code $\mathcal{C}_{D_f}$ has length 2^{m-1} and dimension m, and the weight distribution of $\mathcal{C}_{D_f}$ is deduced from Theorem 20.4.1 and Table 20.10.

- [1438] When $d = 2^{(m+2)/2} - 1$ and $m \equiv 2 \pmod 4$, the code $\mathcal{C}_{D_f}$ has length $2^{m-1} - 2^{m/2}$ and dimension m, and the weight distribution of $\mathcal{C}_{D_f}$ is deduced from Theorem 20.4.1 and Table 20.12. Note that in this case, $\gcd(d, 2^m - 1) = 3$.

- [1438] When $d = (2^{m/2} + 1)(2^{m/4} - 1) + 2$ and $m \equiv 0 \pmod 4$, the code $\mathcal{C}_{D_f}$ has length 2^{m-1} and dimension m, and the weight distribution of $\mathcal{C}_{D_f}$ is deduced from Theorem 20.4.1 and Table 20.11.

- [605] When $d = \frac{2^{(m+2)h/2}-1}{2^h-1}$ and $m \equiv 0 \pmod 4$, where $1 \le h < m$ and $\gcd(h, m) = 1$, the code $\mathcal{C}_{D_f}$ has length 2^{m-1} and dimension m, and the weight distribution of $\mathcal{C}_{D_f}$ is deduced from Theorem 20.4.1 and Table 20.10.

- [1438] When $d = \frac{2^{(m+2)h/2}-1}{2^h-1}$ and $m \equiv 2 \pmod 4$, where $1 \le h < m$ and $\gcd(h, m) = 1$, the code $\mathcal{C}_{D_f}$ has length $2^{m-1} - 2^{m/2}$ and dimension m, and the weight distribution of $\mathcal{C}_{D_f}$ is deduced from Theorem 20.4.1 and Table 20.12. Note that in this case, $\gcd(d, 2^m - 1) = 3$.

- [941] When $d = \frac{2^m + 2^{h+1} - 2^{m/2+1} - 1}{2^h - 1}$ and $m \equiv 0 \pmod 4$, where $2h$ divides $m/2$, the code $\mathcal{C}_{D_f}$ has length 2^{m-1} and dimension m, and the weight distribution of $\mathcal{C}_{D_f}$ is deduced from Theorem 20.4.1 and Table 20.10.

- When $d = (2^{m/2} - 1)s + 1$ with $s = 2^h(2^h \pm 1)^{-1} \pmod{2^{m/2} + 1}$, where $e_2(h) < e_2(m/2)$ and $e_2(h)$ denotes the highest power of 2 dividing h, the parameters and the weight distribution of the code $\mathcal{C}_{D_f}$ can be deduced from Theorem 20.4.1 and the results in [610].

- Let d be any integer such that $1 \le d \le 2^m - 2$ and $d(2^\ell + 1) \equiv 2^h \pmod{2^m - 1}$ for some positive integers ℓ and h. Then the parameters and the weight distribution of the code $\mathcal{C}_{D_f}$ can be deduced from Theorem 20.4.1 and the results in [936].

All these cases of d above are derived from the cross-correlation of a binary maximum-length sequence with its d-decimation version.

TABLE 20.10: Boolean functions with four-valued Walsh spectrum: Case 1

$W_f(w)$	Number of w's
$-2^{m/2}$	$(2^m - 2^{m/2})/3$
0	$2^{m-1} - 2^{(m-2)/2}$
$2^{m/2}$	$2^{m/2}$
$2^{(m+2)/2}$	$(2^{m-1} - 2^{(m-2)/2})/3$

TABLE 20.11: Boolean functions with four-valued Walsh spectrum: Case 2

$W_f(w)$	Number of w's
$-2^{m/2}$	$2^{m-1} - 2^{(3m-4)/4}$
0	$2^{3m/4} - 2^{m/4}$
$2^{m/2}$	$2^{m-1} - 2^{(3m-4)/4}$
$2^{3m/4}$	$2^{m/4}$

20.4.9 Binary Codes $\mathcal{C}_{D_f}$ with at Most Five Weights

The code $\mathcal{C}_{D_f}$ has at most five weights for the following f:

TABLE 20.12: Boolean functions with four-valued Walsh spectrum: Case 3

$W_f(w)$	Number of w's
$-2^{m/2}$	$(2^m - 2^{m/2} - 2)/3$
0	$2^{m-1} - 2^{(m-2)/2} + 2$
$2^{m/2}$	$2^{m/2} - 2$
$2^{(m+2)/2}$	$(2^{m-1} - 2^{(m-2)/2} + 2)/3$

- When $f(x) = \mathrm{Tr}_{2^m/2}(x^{2^{m/2}+3})$ where $m \geq 6$ and is even, $\mathcal{C}_{D_f}$ is a five-weight code with length 2^{m-1} and dimension m, and its weight distribution can be derived from [934].

- When $f(x) = \mathrm{Tr}_{2^m/2}(ax^{(2^m-1)/3})$ with $\mathrm{Tr}_{2^m/4}(a) \neq 0$ where m is even, $\mathcal{C}_{D_f}$ is a two-weight code with length $(2^{m+2} - 4)/6$ and dimension m, and its weight distribution can be derived from [1237].

- [340] When $f(x) = \mathrm{Tr}_{2^m/2}(\lambda x^{2^{m/2}+1}) + \mathrm{Tr}_{2^m/2}(x)\mathrm{Tr}_{2^m/2}(\mu x^{2^{m/2}-1})$ where m is even, $\mu \in \mathbb{F}_{2^{m/2}}^*$, and $\lambda \in \mathbb{F}_{2^m}$ with $\lambda + \lambda^{2^m} = 1$, $\mathcal{C}_{D_f}$ is a five-weight code.

- [340] When $f(x) = (1 + \mathrm{Tr}_{2^m/2}(x))\mathrm{Tr}_{2^m/2}(\lambda x^{2^{m/2}+1}) + \mathrm{Tr}_{2^m/2}(x)\mathrm{Tr}_{2^m/2}(\mu x^{2^{m/2}-1})$ where m is even, $\mu \in \mathbb{F}_{2^{m/2}}^*$, and $\lambda \in \mathbb{F}_{2^m}$ with $\lambda + \lambda^{2^m} = 1$, $\mathcal{C}_{D_f}$ is a five-weight code.

Some Boolean functions f documented in [1549] also give binary linear codes $\mathcal{C}_{D_f}$ with five weights.

20.4.10   A Class of Two-Weight Binary Codes from the Preimage of a Type of Boolean Function

Let m be a positive integer and let r be a prime such that 2 is a primitive root modulo r^m. Let $q = 2^{r^{m-1}(r-1)}$. Define

$$D = \left\{ x \in \mathbb{F}_q^* \mid \mathrm{Tr}_{2^m/2}\big(x^{\frac{q-1}{r^m}}\big) = 0 \right\}. \tag{20.10}$$

Theorem 20.4.8 ([1872]) *Let $r^m \geq 9$ and let D be defined in (20.10). Then $\mathcal{C}_D$ is a two-weight binary code of length $(q-1)(r^m - r + 1)/r^m$, dimension $(r-1)r^{m-1}$, and weight distribution given in Table 20.13.*

TABLE 20.13: Weight distribution of $\mathcal{C}_D$ in Theorem 20.4.8

Hamming weight w	Multiplicity A_w
0	1
$\frac{q-\sqrt{q}}{4} + \frac{q+\sqrt{q}}{4r^m}(r^m - 2r + 2)$	$\frac{(q-1)(r^m-r+1)}{r^m}$
$\frac{q+\sqrt{q}}{4} + \frac{q+\sqrt{q}}{4r^m}(r^m - 2r + 2)$	$\frac{(q-1)(r-1)}{r^m}$

20.4.11 Binary Codes from Boolean Functions Whose Supports are Relative Difference Sets

Definition 20.4.9 Let $(A, +)$ be an abelian group of order $t\ell$ and $(N, +)$ a subgroup of A of order ℓ. An n-subset D of A is called a (t, ℓ, n, λ) **relative difference set**, if the multiset $\{d_1 - d_2 \mid d_1, d_2 \in D, \ d_1 \neq d_2\}$ does not contain any element in N, but every element in $A \setminus N$ exactly λ times.

It is known (see, e.g., [552, 1900]) that any relative difference set D of size n in $(\mathbb{F}_2^m, +)$ defines a binary code $\mathcal{C}_D$, where D is the support of a Boolean function on $\mathbb{F}_{2^m}$, with at most the following four weights:

$$\frac{n \pm \sqrt{n}}{2}, \ \frac{n \pm \sqrt{n - \lambda \ell}}{2}.$$

20.4.12 Binary Codes with Few Weights from Plateaued Boolean Functions

Let $\Psi : \mathbb{F}_{2^m} \to \mathbb{F}_{2^m}$ be a mapping with $\Psi(0) = 0$. Set $\psi_1(x) = \mathrm{Tr}_{2^m/2}(\Psi(x))$. Let $f_{\alpha,\beta}(x)$ be the Boolean function defined by

$$f_{\alpha,\beta}(x) = \mathrm{Tr}_{2^m/2}(\alpha \Psi(x) - \beta x).$$

Recalling (20.4) and (20.7), the subcode $\mathcal{C}_{\psi_1}$ of $\mathcal{C}_\Psi$ is defined by

$$\mathcal{C}_{\psi_1} = \left\{ \widetilde{c}_{\alpha,\beta} = \left(f_{\alpha,\beta}(\zeta_1), f_{\alpha,\beta}(\zeta_2), \ldots, f_{\alpha,\beta}(\zeta_{2^m-1}) \right) \mid \alpha \in \mathbb{F}_2, \ \beta \in \mathbb{F}_{2^m} \right\} \tag{20.11}$$

where $\zeta_1, \ldots, \zeta_{2^m-1}$ are the elements of $\mathbb{F}_{2^m}^*$. The linear code $\mathcal{C}_{\psi_1}$ of length $2^m - 1$ over $\mathbb{F}_2$ defined by (20.11) is a k-dimensional subspace of $\mathbb{F}_2^m$ where $k = 2m$ if we assume that Ψ has no linear component.

Assume that ψ_1 is an s-plateaued Boolean function where $m + s$ is an even integer with $0 \leq s \leq m - 2$. The following theorem (compare to Theorem 20.4.4) gives the Hamming weights of the codewords and the weight distribution of the binary three-weight linear code $\mathcal{C}_{\psi_1}$ defined by (20.11)

Theorem 20.4.10 ([1381]) *With $\mathcal{C}_{\psi_1}$ as in (20.11), under the above assumptions, the Hamming weights of the codewords and the weight distribution of $\mathcal{C}_{\psi_1}$ are as in Table 20.14.*

TABLE 20.14: Weight distribution of $\mathcal{C}_{\psi_1}$ in Theorem 20.4.10 for $m + s$ even

Hamming weight w	Multiplicity A_w
0	1
2^{m-1}	$2^{m+1} - 2^{m-s} - 1$
$2^{m-1} - 2^{(m+s-2)/2}$	$2^{m-s-1} + 2^{(m-s-2)/2}$
$2^{m-1} + 2^{(m+s-2)/2}$	$2^{m-s-1} - 2^{(m-s-2)/2}$

Example 20.4.11 Let $\Psi(x) = \zeta^{18} x^5 + \zeta^2 x^3$ be the mapping over $\mathbb{F}_{2^5}$, where $\mathbb{F}_{2^5}^* = \langle \zeta \rangle$ with $\zeta^5 + \zeta^2 + 1 = 0$. Then $\psi_1(x) = \mathrm{Tr}_{2^5/2}(\Psi(x))$ is a 3-plateaued Boolean function, and the set $\mathcal{C}_{\psi_1}$ in (20.11) is the binary three-weight linear code with parameters $[31, 6, 8]_2$ and weight enumerator $\mathrm{Hwe}_{\mathcal{C}_{\psi_1}}(x, y) = y^{31} + 3x^8 y^{23} + 59 x^{16} y^{15} + x^{24} y^7$; see Definition 1.15.1.

20.4.13   Binary Codes with Few Weights from Almost Bent Functions

Recall by Proposition 20.2.14 that a function F from $\mathbb{F}_{2^m}$ to $\mathbb{F}_{2^m}$ is *almost bent* (AB) if and only if $\mathcal{W}_F(a,b) = \sum_{x \in \mathbb{F}_{2^m}} (-1)^{\mathrm{Tr}_{2^m/2}(aF(x)+bx)}$ takes the values 0 or $\pm 2^{(m+1)/2}$ for every pair $(a,b) \in \mathbb{F}_{2^m}^* \times \mathbb{F}_{2^m}$.

A characterization of AB functions by the weight distribution of related codes has been obtained in [350].

Theorem 20.4.12 ([350]) *Let F be any vectorial (Boolean) function from $\mathbb{F}_{2^m}$ to $\mathbb{F}_{2^m}$ such that $F(0) = 0$. Let H be the matrix*

$$\begin{bmatrix} 1 & \alpha & \alpha^2 & \cdots & \alpha^{2^m-2} \\ F(1) & F(\alpha) & F(\alpha^2) & \cdots & F(\alpha^{2^m-2}) \end{bmatrix}$$

where α is a primitive element of the field $\mathbb{F}_{2^m}$, and where each symbol stands for the column of its coordinates with respect to a basis of the $\mathbb{F}_2$-vector space $\mathbb{F}_{2^m}$. Let $\mathcal{C}_F$ be the linear code admitting H as a parity check matrix. Then F is AB if and only if $\mathcal{C}_F^{\perp}$ (i.e., the code admitting H as a generator matrix) has Hamming weights

$$0, \ 2^{m-1} - 2^{(m-1)/2}, \ 2^{m-1}, \ 2^{m-1} + 2^{(m-1)/2}.$$

By definition, almost bent functions over $\mathbb{F}_{2^m}$ exist only for m odd. Moreover, by definition, for every almost bent function F, the sum $\mathcal{W}_F(1,0) \in \{0, \pm 2^{(m+1)/2}\}$. Ding has deduced the following result.

Theorem 20.4.13 ([550]) *Let F be an almost bent function from $\mathbb{F}_{2^m}$ to $\mathbb{F}_{2^m}$ (where m is odd) such that $F(0) = 0$. Define $f := \mathrm{Tr}_{2^m/2}(F)$. Then $\mathcal{C}_{D_f}$ is an $[n_f, m, (n_f - 2^{(m-1)/2})/2]_2$ three-weight binary code with the weight distribution as given by Table 20.15 where*

$$n_f = \#D_f = \begin{cases} 2^{m-1} + 2^{(m-1)/2} & \text{if } \mathcal{W}_F(1,0) = -2^{(m+1)/2} \\ 2^{m-1} - 2^{(m-1)/2} & \text{if } \mathcal{W}_F(1,0) = 2^{(m+1)/2} \\ 2^{m-1} & \text{if } \mathcal{W}_F(1,0) = 0. \end{cases}$$

TABLE 20.15: Weight distribution of $\mathcal{C}_{D_f}$ in Theorem 20.4.13

Hamming weight w	Multiplicity A_w
0	1
$\dfrac{n_f - 2^{(m-1)/2}}{2}$	$n_f(2^m - n_f)2^{-m} - n_f 2^{-(m+1)/2}$
$\dfrac{n_f}{2}$	$2^m - 1 - n_f(2^m - n_f)2^{-(m-1)}$
$\dfrac{n_f + 2^{(m-1)/2}}{2}$	$n_f(2^m - n_f)2^{-m} + n_f 2^{-(m+1)/2}$

The list of the known AB power functions $F(x) = x^d$ on $\mathbb{F}_{2^m}$ for odd m is given below (see, e.g., [1378, Chapter 12]):

- [822] $d = 2^h + 1$ where $\gcd(m,h) = 1$;
- [1087] $d = 2^{2h} - 2^h + 1$ where $h \geq 2$ and $\gcd(m,h) = 1$;
- [1087] $d = 2^{(m-1)/2} + 3$;
- [974, 986] $d = 2^{(m-1)/2} + 2^{(m-1)/4} - 1$ where $m \equiv 1 \pmod 4$;
- [974, 986] $d = 2^{(m-1)/2} + 2^{(3m-1)/4} - 1$ where $m \equiv 3 \pmod 4$.

The length of each code $\mathcal{C}_{D_f}$ obtained from this list is equal to $n_f = 2^{m-1}$, and the weight distribution of the code is given in Table 20.15.

20.4.14 Binary Codes $\mathcal{C}_{D(G)}$ from Functions on $\mathbb{F}_{2^m}$ of the Form $G(x) = F(x) + F(x+1) + 1$

Definition 20.4.14 Let A and B be two finite sets, and let f be a mapping from A to B. Then f is called a **2-to-1 mapping** if one of the following two cases hold:

(i) $\#A$ is even, and for any $b \in B$, b has either 2 or 0 preimages under f, or

(ii) $\#A$ is odd, and for all but one $b \in B$, b has either 2 or 0 preimages under f, and the exception element has exactly one preimage.

Let F be any function on $\mathbb{F}_{2^m}$. Define

$$G(x) = F(x) + F(x+1) + 1.$$

By the definition of APNness, G is 2-to-1 for APN functions $F(x)$ over $\mathbb{F}_{2^m}$. For example, $G(x) = x^{2^{2h}-2^h+1} + (x+1)^{2^{2h}-2^h+1} + 1$ is 2^s-to-1, where $s = \gcd(h,m)$; see [948]. We define

$$D(G) = \{G(x) \mid x \in \mathbb{F}_{2^m}\}.$$

In this subsection, we consider the code $\mathcal{C}_{D(G)}$ based on the second generic construction whose defining set is $D(G)$.

Theorem 20.4.15 ([552]) *Let $F(x) = x^{2^h+1}$, and let $\gcd(h,m) = 1$. Then $\mathcal{C}_{D(G)}$ is a one-weight code with parameters $[2^{m-1}, m-1, 2^{m-2}]_2$.*

From Ding [552], we have the following comments on other APN monomials.

- Let $F(x) = x^{2^m-2}$. Then $\mathcal{C}_{D(G)}$ is a binary code of length 2^{m-1}, dimension m, and has at most m weights. The weights are determined by the Kloosterman sums.[3]

- For the Niho function $F(x) = x^{2^{(m-1)/2}+2^{(m-1)/4}-1}$ where $m \equiv 1 \pmod 4$, the code $\mathcal{C}_{D(G)}$ has length 2^{m-1} and dimension m, but many weights.

- For the Niho function $F(x) = x^{2^{(m-1)/2}+2^{(3m-1)/4}-1}$ where $m \equiv 3 \pmod 4$, the code $\mathcal{C}_{D(G)}$ has length 2^{m-1} and dimension m, but many weights.

According to Ding, it would be extremely difficult to determine the weight distribution of the code $\mathcal{C}_{D(G)}$ for these three classes of APN monomials.

20.4.15 Binary Codes from the Images of Certain Functions on $\mathbb{F}_{2^m}$

Let $F(x)$ be a function from $\mathbb{F}_{2^m}$ to $\mathbb{F}_{2^m}$. We define

$$D(F) = \{F(x) \mid x \in \mathbb{F}_{2^m}\} \quad \text{and} \quad D(F)^* = \{F(x) \mid x \in \mathbb{F}_{2^m}\} \setminus \{0\}.$$

We consider again the codes $\mathcal{C}_{D(F)}$ and $\mathcal{C}_{D(F)^*}$ based on the second generic construction. As highlighted by Ding [552], it is difficult to determine in general the parameters of these codes except in certain special cases.

If $0 \notin D(F)$, then the two codes $\mathcal{C}_{D(F)}$ and $\mathcal{C}_{D(F)^*}$ are the same. Otherwise, the length of the code $\mathcal{C}_{D(F)^*}$ is one less than that of the code $\mathcal{C}_{D(F)}$, but the two codes have the same weight distribution. Ding has provided more information about the codes $\mathcal{C}_{D(F)^*}$ in [552].

---

[3]The binary Kloosterman sum $K_m : \mathbb{F}_{2^m} \to \{-1, 1\}$ is defined as follows: For $a \in \mathbb{F}_{2^m}$, $K_m(a) = \sum_{x \in \mathbb{F}_{2^m}} (-1)^{\mathrm{Tr}_{2^m/2}(ax + \frac{1}{x})}$.

Moreover he provided a remarkable presentation in [552, Section 6] on binary codes $\mathcal{C}_{D(F)}$ in the case where F is an oval polynomial on $\mathbb{F}_{2^m}$. Recall that a bijection F on $\mathbb{F}_{2^m}$ is called a **permutation polynomial**. A permutation polynomial F is an **oval** or **o-polynomial** if $F(0) = 0$, and for each $y \in \mathbb{F}_{2^m}$,

$$F_y(x) = (F(x + y) + F(y))x^{2^m - 2}$$

is also a permutation polynomial. In finite geometry, o-polynomials lead to the construction of hyperovals, which in turn lead to the construction of MDS codes as described in Section 14.2.3. The simplest example of an o-polynomial is the Frobenius automorphism $F(z) = z^{2^i}$ where $\gcd(i, m) = 1$. Other known examples are the following; the last three are not polynomials but they are still bijections on $\mathbb{F}_{2^m}$:

- [1641] $F(z) = z^6$ where m is odd;

- [810] $F(z) = z^{3 \cdot 2^k + 4}$ where $m = 2k - 1$;

- [810] $F(z) = z^{2^k + 2^{2k}}$ where $m = 4k - 1$;

- [810] $F(z) = z^{2^{2k+1} + 2^{3k+1}}$ where $m = 4k + 1$;

- [401] $F(z) = z^{2^k} + z^{2^k + 2} + z^{3 \cdot 2^k + 4}$ where $m = 2k - 1$;

- [1496] $F(z) = z^{\frac{1}{6}} + z^{\frac{1}{2}} + z^{\frac{5}{6}}$ where m is odd (note that $F(z) = D_5(z^{\frac{1}{6}})$ where D_5 is the Dickson polynomial of index 5 [1253]);

- [403] $F(z) = \frac{\delta^2(z^4 + z) + \delta^2(1 + \delta + \delta^2)(z^3 + z^2)}{z^4 + \delta^2 z^2 + 1} + z^{1/2}$ where $\mathrm{Tr}_{2^m/2}(1/\delta) = 1$ and, if $m \equiv 2$ (mod 4), then $\delta \notin \mathbb{F}_4$.

- [402] For m even, $r = \pm \frac{2^m - 1}{3}$, $v \in \mathbb{F}_{2^{2m}}$, $v^{2^m + 1} = 1$, and $v \neq 1$

$$F(z) = z^{1/2} + \frac{1}{\mathrm{Tr}_{2^n/2^m}(v)} \times$$

$$\left[\mathrm{Tr}_{2^n/2^m}(v^r)(z + 1) + \mathrm{Tr}_{2^n/2^m}\left((vz + v^{2^m})^r\right) \left(z + \mathrm{Tr}_{2^n/2^m}(v)z^{1/2} + 1\right)^{1-r} \right].$$

20.5 Constructions of Cyclic Codes from Functions: The Sequence Approach

In the following we focus on cyclic codes, and we present constructions of such codes using the sequence approach. The obtained codes are often optimal or almost optimal for given length and dimension.

20.5.1 A Generic Construction of Cyclic Codes with Polynomials

Cyclic codes, as an important class of linear codes, have significant applications in data storage systems, communication systems, and consumer electronics products. This is due to their cycle structure, which is easy to implement in hardware, and their efficient encoding and decoding algorithms. Cyclic codes can also be used to construct other interesting structures such as quantum codes [1799], frequency hopping sequences [569], etc. Thus, it is of great interest to construct cyclic codes with good parameters.

Let q be a positive power of a prime p. One way of constructing cyclic codes over $\mathbb{F}_q$ of length n is to use the generator polynomial

$$\frac{x^n - 1}{\gcd(S^n(x), x^n - 1)} \tag{20.12}$$

where $S^n(x) = \sum_{i=0}^{n-1} s_i x^i \in \mathbb{F}_q[x]$ and $s^\infty = (s_i)_{i=0}^\infty$ is a sequence of period n over $\mathbb{F}_q$. We call the cyclic code $\mathcal{C}_s$ with generator polynomial (20.12) the **code defined by the sequence** s^∞, and the sequence s^∞ the **defining sequence** of $\mathcal{C}_s$.

It can be seen that every cyclic code of length n over $\mathbb{F}_q$ can be expressed as $\mathcal{C}_s$ for some sequence s^∞ of period n over $\mathbb{F}_q$. For this reason, this construction of cyclic codes is said to be *fundamental*. In the past decade, impressive progress, in particular thanks to C. Ding, in the construction of cyclic codes with this approach has been made; see, e.g., [545, 547, 548, 549, 571, 1779, 1886].

Let $s^\infty = (s_i)_{i=0}^\infty$ be a sequence of period n over $\mathbb{F}_q$. The polynomial $c(x) = \sum_{i=0}^{\ell} c_i x^i$ over $\mathbb{F}_q$, where $c_0 = 1$, is called the **characteristic polynomial** of s^∞ if

$$-c_0 s_i = c_1 s_{i-1} + c_2 s_{i-2} + \cdots + c_l s_{i-\ell} \text{ for all } i \geq \ell.$$

The characteristic polynomial with the smallest degree is called the **minimal polynomial** of s^∞. The degree of the minimal polynomial is referred to as the **linear span** or **linear complexity** of s^∞. Since we require that the constant term of any characteristic polynomial be 1, the minimal polynomial of any periodic sequence s^∞ must be unique. In addition, any characteristic polynomial must be a multiple of the minimal polynomial.

For periodic sequences, there are a few ways to determine their linear span and minimal polynomials. One of them is given in the following lemma [567, page 87, Theorem 5.3].

Lemma 20.5.1 ([567]) *Let s^∞ be a sequence of period n over $\mathbb{F}_q$. Define $S^n(x) = \sum_{i=0}^{n-1} s_i x^i \in \mathbb{F}_q[x]$. Then the minimal polynomial $\mathbb{M}_s(x)$ of s^∞ is given by*

$$\frac{x^n - 1}{\gcd(x^n - 1, S^n(x))}$$

and the linear span $\mathbb{L}_s$ of s^∞ is given by $n - \deg(\gcd(x^n - 1, S^n(x)))$.

The other one is given in the following lemma ([52, Theorem 3], [940]).

Lemma 20.5.2 ([571]) *Any sequence s^∞ over $\mathbb{F}_q$ of period $q^m - 1$ has a unique expansion of the form*

$$s_t = \sum_{i=0}^{q^m - 2} c_i \alpha^{it} \text{ for all } t \geq 0$$

where α is a generator of $\mathbb{F}_{q^m}^*$ and $c_i \in \mathbb{F}_{q^m}$. If $I = \{i \mid c_i \neq 0\}$, then the minimal polynomial $\mathbb{M}_s(x)$ of s^∞ is $\mathbb{M}_s(x) = \prod_{i \in I}(1 - \alpha^i x)$, and the linear span $\mathbb{L}_s$ of s^∞ is $\#I$.

It should be noted that in some references the reciprocal of $\mathbb{M}_s(x)$ is called the minimal polynomial of the sequence s^∞. Lemma 20.5.2 is a modified version of the original one [940].

Given a polynomial $F(x)$ over $\mathbb{F}_{q^m}$, one can define its associate sequence s^∞ by

$$s_i = \text{Tr}_{q^m/q}(F(\alpha^i + 1)) \text{ for all } i \geq 0 \tag{20.13}$$

where α is a generator of $\mathbb{F}_q^*$. Any method for constructing an $[n, k]$ cyclic code over $\mathbb{F}_q$ corresponds to the selection of a divisor F over $\mathbb{F}_q$ of $x^n - 1$ of degree $n - k$, which is employed as the generator polynomial. The minimum weight $d_H(\mathcal{C}_s)$ and other parameters of this cyclic code $\mathcal{C}_s$ are determined by the generator polynomial $F(x)$. It is unnecessary to require that F is highly nonlinear in order to obtain cyclic codes $\mathcal{C}_s$ defined by F with good parameters. Both linear and highly nonlinear polynomials F could give optimal cyclic codes $\mathcal{C}_s$ when they are plugged into the above generic construction. As demonstrated by Ding (see, e.g., [544, 548]), it is possible to construct optimal cyclic codes meeting some bound on parameters of linear cyclic codes or cyclic codes with good parameters. But this generic construction may produce also codes with bad parameters; see for instance the discussion of Ding in [548].

20.5.2   Binary Cyclic Codes from APN Functions

Let m be a positive integer and $n = 2^m - 1$. Recall from Section 20.2.6 that a polynomial $F(x)$ over $\mathbb{F}_{2^m}$ is almost perfect nonlinear (APN) if

$$\max_{a \in \mathbb{F}_{2^m}^*} \max_{b \in \mathbb{F}_{2^m}} \#\{x \in \mathbb{F}_{2^m} \mid F(x + a) - F(x) = b\} = 2.$$

A characterization of APN functions by the minimum distance of related codes has been obtained in [350].

Theorem 20.5.3 ([350]) *Let F be any vectorial (Boolean) function from $\mathbb{F}_{2^m}$ to $\mathbb{F}_{2^m}$ such that $F(0) = 0$. Let H be the matrix*

$$\begin{bmatrix} 1 & \alpha & \alpha^2 & \cdots & \alpha^{2^m-2} \\ F(1) & F(\alpha) & F(\alpha^2) & \cdots & F(\alpha^{2^m-2}) \end{bmatrix}$$

where α is a primitive element of the field $\mathbb{F}_{2^m}$, and where each symbol stands for the column of its coordinates with respect to a basis of the $\mathbb{F}_2$-vector space $\mathbb{F}_{2^m}$. Let $\mathcal{C}_F$ be the linear code with parity check matrix H. Then F is APN if and only if $\mathcal{C}_F$ has minimum distance 5.

Note that APN and perfect nonlinear (PN) functions over finite fields have been employed to construct a number of classes of cyclic codes. APN monomials over $\mathbb{F}_{2^m}$ were employed to construct binary $[2^m - 1, 2^m - 1 - 2m, 5]_2$ cyclic codes by Carlet, Charpin, and Zinoviev [350], and PN polynomials over $\mathbb{F}_p$, with p an odd prime, were used to define $[p^m - 1, p^m - 1 - 2m, d]_p$ non-binary codes by Carlet, Ding, and Yuan [352]. The dimension of the codes obtained from these two constructions in [350] and [352] is always equal to $p^m - 1 - 2m$. In [548] Ding employed both PN polynomials and APN monomials to construct cyclic codes. His constructions using these polynomials are different from those in [352] as the dimension of the codes obtained in [548] is usually different from $p^m - 1 - 2m$. In fact, the parameters of the cyclic codes from APN or PN polynomials obtained in [548] do not depend on the APN or PN property of these polynomials. Further, Ding and Zhou have constructed in [571] a number of families of binary cyclic codes with monomials and trinomials of special types.

Ding [544, 548, 554] and Ding and Zhou [571] provided several constructions of cyclic codes from polynomials based on the generic construction of cyclic codes with polynomials; see Section 20.5.1. Below we give a brief description of state-of-the-art cyclic codes constructed from some APN functions. To get more information about those codes, we invite the reader to consult the excellent survey [554].

Recalling Section 20.5.1, we adopt the following notation unless otherwise stated:

- p is a prime, q is a positive power of p, and m is a positive integer.

- $\mathbb{N}_p(x)$ is a function defined by $\mathbb{N}_p(x) = 0$ if $x \equiv 0 \pmod{p}$ and $\mathbb{N}_p(x) = 1$ otherwise.

- α is a generator of $\mathbb{F}_{q^m}^*$, the multiplicative group of $\mathbb{F}_{q^m}$.

- $m_a(x)$ is the minimal polynomial of $a \in \mathbb{F}_{q^m}$ over $\mathbb{F}_q$.

- s^∞ is the sequence given by (20.13).

- $\mathbb{L}_s$ is the linear span of s^∞.

- $\mathbb{M}_s$ is the minimal polynomial of s^∞.

- $\mathcal{C}_s$ is the cyclic code over $\mathbb{F}_q$ of length $n = q^m - 1$ with defining sequence s^∞.

Theorem 20.5.4 ([544, 571]) *Let* $m = 2t + 1 \geq 7$ *and let* $F(x)$ *be the APN function* $F(x) = x^{2^t+3}$ *defined over* $\mathbb{F}_{2^{2t+1}}$. *The sequence* s^∞ *has* $\mathbb{L}_s = 5m + 1$ *and*

$$\mathbb{M}_s(x) = (x-1)m_{\alpha^{-1}}(x)m_{\alpha^{-3}}(x)m_{\alpha^{-(2^t+1)}}(x)m_{\alpha^{-(2^t+2)}}(x)m_{\alpha^{-(2^t+3)}}(x).$$

$\mathcal{C}_s$ *is a binary* $[2^m - 1, 2^m - 2 - 5m, d]_2$ *cyclic code with* $d \geq 8$ *and generator polynomial* $\mathbb{M}_s(x)$.

In the next three theorems, let

$$\kappa_{2j+1}^{(h)} = \begin{cases} 1 & \text{if } j = 2^{h-1} - 1, \\ \left\lceil \log_2 \frac{2^h - 1}{2j+1} \right\rceil \bmod 2 & \text{if } 0 \leq j < 2^{h-1} - 1. \end{cases}$$

Theorem 20.5.5 ([544, 571]) *Let* $F(x)$ *be the APN function* $F(x) = x^{2^h-1}$ *with* $2 \leq h \leq \left\lceil \frac{m}{2} \right\rceil$. *The sequence* s^∞ *has linear span*

$$\mathbb{L}_s = \begin{cases} \dfrac{m\left(2^h + (-1)^{h-1}\right)}{3} & \text{if } m \text{ is even}, \\ \dfrac{m\left(2^h + (-1)^{h-1}\right)+3}{3} & \text{if } m \text{ is odd} \end{cases}$$

and minimal polynomial

$$\mathbb{M}_s(x) = (x-1)^{\mathbb{N}_2(m)} \prod_{\substack{1 \leq 2j+1 \leq 2^h - 1 \\ \kappa_{2j+1}^{(h)} = 1}} m_{\alpha^{-(2j+1)}}(x).$$

Also $\mathcal{C}_s$ *is a binary* $[2^m - 1, 2^m - 1 - \mathbb{L}_s, d]_2$ *cyclic code with*

$$d \geq \begin{cases} 2^{h-2} + 2 & \text{if } m \text{ is odd and } h > 2, \\ 2^{h-2} + 1 & \text{otherwise} \end{cases}$$

and generator polynomial $\mathbb{M}_s(x)$.

Theorem 20.5.6 ([544, 571]) *Let* $m \equiv 1 \pmod 4$ *with* $m \geq 9$. *Let* $F(x)$ *be the APN function* $F(x) = x^e$ *where* $e = 2^{(m-1)/2} + 2^{(m-1)/4} - 1$. *The sequence* s^∞ *has linear span*

$$\mathbb{L}_s = \begin{cases} \dfrac{m\left(2^{(m+7)/4} + (-1)^{(m-5)/4}\right)+3}{3} & \text{if } m \equiv 1 \pmod 8, \\ \dfrac{m\left(2^{(m+7)/4} + (-1)^{(m-5)/4} - 6\right)+3}{3} & \text{if } m \equiv 5 \pmod 8 \end{cases}$$

and minimal polynomial

$$\mathbb{M}_s(x) = (x-1) \prod_{i=0}^{2^{\frac{m-1}{4}}-1} m_{\alpha^{-i-2^{\frac{m-1}{2}}}}(x) \prod_{\substack{1 \leq 2j+1 \leq 2^{\frac{m-1}{4}}-1 \\ \kappa_{2j+1}^{((m-1)/4)}=1}} m_{\alpha^{-2j-1}}(x)$$

if $m \equiv 1 \pmod 8$ *and*

$$\mathbb{M}_s(x) \doteq (x-1) \prod_{i=1}^{2^{\frac{m-1}{4}}-1} m_{\alpha^{-i-2^{\frac{m-1}{2}}}}(x) \prod_{\substack{3 \leq 2j+1 \leq 2^{\frac{m-1}{4}}-1 \\ \kappa_{2j+1}^{((m-1)/4)}=1}} m_{\alpha^{-2j-1}}(x)$$

if $m \equiv 5 \pmod 8$. *Also* $\mathcal{C}_s$ *is a binary* $[2^m-1, 2^m-1-\mathbb{L}_s, d]_2$ *cyclic code with*

$$d \geq \begin{cases} 2^{(m-1)/4}+2 & \text{if } m \equiv 1 \pmod 8, \\ 2^{(m-1)/4} & \text{if } m \equiv 5 \pmod 8 \end{cases}$$

and generator polynomial $\mathbb{M}_s(x)$.

Theorem 20.5.7 ([544, 571]) *Let* $F(x)$ *be the APN function* $F(x) = x^{2^{2h}-2^h+1}$ *where* h *satisfies* $\gcd(m,h) = 1$ *and*

$$1 \leq h \leq \begin{cases} \frac{m-1}{4} & \text{if } m \equiv 1 \pmod 4, \\ \frac{m-3}{4} & \text{if } m \equiv 3 \pmod 4, \\ \frac{m-4}{4} & \text{if } m \equiv 0 \pmod 4, \\ \frac{m-2}{4} & \text{if } m \equiv 2 \pmod 4. \end{cases}$$

The sequence s^∞ *has linear span*

$$\mathbb{L}_s = \begin{cases} \dfrac{m(2^{(h+2)}+(-1)^{h-1})+3}{3} & \text{if } h \text{ is even}, \\ \dfrac{m(2^{h+2}+(-1)^{h-1}-6)+3}{3} & \text{if } h \text{ is odd} \end{cases}$$

and minimal polynomial

$$\mathbb{M}_s(x) = (x-1) \prod_{i=0}^{2^h-1} m_{\alpha^{-i-2^{m-h}}}(x) \prod_{\substack{1 \leq 2j+1 \leq 2^h-1 \\ \kappa_{2j+1}^{(h)}=1}} m_{\alpha^{-2j-1}}(x)$$

if h *is even and*

$$\mathbb{M}_s(x) = (x-1) \prod_{i=1}^{2^h-1} m_{\alpha^{-i-2^{m-h}}}(x) \prod_{\substack{3 \leq 2j+1 \leq 2^h-1 \\ \kappa_{2j+1}^{(h)}=1}} m_{\alpha^{-2j-1}}(x)$$

if h *is odd. Also* $\mathcal{C}_s$ *is a binary* $[2^m-1, 2^m-1-\mathbb{L}_s, d]_2$ *cyclic code with*

$$d \geq \begin{cases} 2^h+2 & \text{if } h \text{ is even}, \\ 2^h & \text{if } h \text{ is odd} \end{cases}$$

and generator polynomial $\mathbb{M}_s(x)$.

The code C_s in Theorems 20.5.5, 20.5.6, and 20.5.7 could be optimal in some cases [571].

We examine one final APN monomial.

Theorem 20.5.8 ([544, 571]) *Let F be the APN function $F(x) = x^{2^m-2}$ over $\mathbb{F}_{2^m}$. Let C_i be the 2-cyclotomic coset modulo $2^m - 1$ containing i, ρ_i the total number of even integers in C_i, and $\ell_i = \#C_i$. Let Γ be the set of coset leaders modulo $n = 2^m - 1$. Define*

$$\nu_i = \frac{m\rho_i}{\ell_i} \bmod 2.$$

The sequence s^∞ has linear span $\mathbb{L}_s = (n+1)/2$ and minimal polynomial

$$\mathbb{M}_s(x) = \prod_{j \in \Gamma,\ \nu_j = 1} m_{\alpha^{-j}}(x).$$

C_s *is a binary $[2^m - 1, 2^{m-1} - 1, d]_2$ cyclic code with generator polynomial $\mathbb{M}_s(x)$. If m is odd, then $d \geq d_1$ where d_1 is the smallest even integer satisfying $d_1^2 - d_1 + 1 \geq 2^m - 1$; moreover, the dual code $C_s^\perp$ is a binary $[2^m - 1, 2^{m-1}, d^\perp]_2$ cyclic code where $(d^\perp)^2 - d^\perp + 1 \geq 2^m - 1$.*

When $F(x) = x^{q^m-2}$ and $q > 2$, the dimension of the code C_s over $\mathbb{F}_q$ was settled in [1779]. But no lower bound on the minimum distance of C_s is developed.

20.5.3 Non-Binary Cyclic Codes from Monomials and Trinomials

All the results presented in this subsection come from [548]. C. Ding has obtained very significant results on non-binary cyclic codes from monomials and trinomials over finite fields; some of these are optimal or almost optimal among all linear codes.

Throughout this subsection, m is a positive integer and q is a power of an <u>odd</u> prime p. Also $\mathbb{N}_p(x)$, α, $m_a(x)$, s^∞, $\mathbb{L}_s$, $\mathbb{M}_s$, and C_s remain as in the previous subsection.

Ding [548] studied the cyclic codes obtained from the monomial $F(x) = x^{q^\kappa+1}$ in the case that m is odd.

Theorem 20.5.9 ([548]) *Let m and q be odd, and let $\kappa \geq 0$. Let C_s be the code defined by the sequence s^∞ given by (20.13) where $F(x) = x^{q^\kappa+1}$. Then the linear span $\mathbb{L}_s$ of s^∞ is equal to $2m + \mathbb{N}_p(m)$ and the minimal polynomial $\mathbb{M}_s(x)$ of s^∞ is*

$$\mathbb{M}_s(x) = (x-1)^{\mathbb{N}_p(m)} m_{\alpha^{-1}}(x) m_{\alpha^{-(p^\kappa+1)}}(x).$$

Moreover the code C_s has parameters $[q^m - 1, q^m - 1 - 2m - \mathbb{N}_p(m), d]_q$ where

$$\begin{cases} d = 4 & \text{if } q = 3 \text{ and } m \equiv 0 \pmod{p}, \\ 4 \leq d \leq 5 & \text{if } q = 3 \text{ and } m \not\equiv 0 \pmod{p}, \\ d = 3 & \text{if } q > 3 \text{ and } m \equiv 0 \pmod{p}, \\ 3 \leq d \leq 4 & \text{if } q > 3 \text{ and } m \not\equiv 0 \pmod{p} \end{cases}$$

and generator polynomial $\mathbb{M}_s(x)$.

Example 20.5.10 Let $(m, \kappa, q) = (3, 1, 3)$ in Theorem 20.5.9. If α be a generator of $\mathbb{F}_{3^3}^*$ with $\alpha^3 + 2\alpha + 1 = 0$, then C_s is a $[26, 20, 4]_3$ ternary code with generator polynomial $\mathbb{M}_s(x) = x^6 + 2x^5 + 2x^4 + x^3 + x^2 + 2x + 1$. This cyclic code is an optimal linear code according to [845].

Before proceeding, we need a notation that will be used in the next result and later in the chapter:

- For $x \in \mathbb{F}_{q^m}$, $\delta(x) = \begin{cases} 0 & \text{if } \mathrm{Tr}_{q^m/q}(x) = 0, \\ 1 & \text{otherwise.} \end{cases}$

Ding [548] studied the code $\mathcal{C}_s$ obtained from the trinomial $F(x) = x^{10} - ux^6 - u^2x^2$ when $q = 3$, m is odd, and $u \in \mathbb{F}_{3^m}$. We present the results in the next theorem and example. The lower bounds on d in the theorem come from the BCH Bound while the upper bounds on d follow from the Sphere Packing Bound.

Theorem 20.5.11 ([548]) *Let m be odd and $q = 3$. Let s^∞ be the sequence in (20.13) where $F(x) = x^{10} - ux^6 - u^2x^2$ with $u \in \mathbb{F}_{3^m}$. Then the linear span $\mathbb{L}_s$ of s^∞ is*

$$\mathbb{L}_s = \begin{cases} 2m + \delta(u^2 + u - 1) & \text{if } u^6 + u = 0, \\ 3m + \delta(u^2 + u - 1) & \text{otherwise,} \end{cases}$$

and the minimal polynomial $\mathbb{M}_s(x)$ of s^∞ is

$$\mathbb{M}_s(x) = \begin{cases} (x-1)^{\delta(u^2+u-1)} m_{\alpha^{-1}}(x) m_{\alpha^{-10}}(x) & \text{if } u^6 + u = 0, \\ (x-1)^{\delta(u^2+u-1)} m_{\alpha^{-1}}(x) m_{\alpha^{-2}}(x) m_{\alpha^{-10}}(x) & \text{otherwise.} \end{cases}$$

The code $\mathcal{C}_s$ defined by the sequence s^∞ has generator polynomial $\mathbb{M}_s(x)$ and parameters $[3^m - 1, 3^m - 1 - \mathbb{L}_s, d]_3$ where

$$\begin{cases} 5 \le d \le 8 & \text{if } u^6 + u \ne 0 \text{ and } \delta(u^2 + u - 1) = 1, \\ 4 \le d \le 6 & \text{if } u^6 + u \ne 0 \text{ and } \delta(u^2 + u - 1) = 0, \\ 3 \le d \le 6 & \text{if } u^6 + u = 0 \text{ and } \delta(u^2 + u - 1) = 1, \\ 3 \le d \le 4 & \text{if } u^6 + u = 0 \text{ and } \delta(u^2 + u - 1) = 0. \end{cases}$$

Example 20.5.12 Let $(m, q, u) = (3, 3, 1)$ in Theorem 20.5.11. If α be a generator of $\mathbb{F}_{3^3}^*$ with $\alpha^3 + 2\alpha + 1 = 0$, then $\mathcal{C}_s$ is a $[26, 17, 5]_3$ ternary code with generator polynomial $\mathbb{M}_s(x) = x^9 + x^8 + 2x^7 + 2x^6 + 2x^5 + x^4 + x^3 + x^2 + 2x + 1$. This cyclic code is an optimal linear code according to [845].

Let h be a positive integer satisfying $1 \le h \le \lceil \frac{m}{2} \rceil$. Ding [548] studied the code $\mathcal{C}_s$ defined by the sequence s^∞ of (20.13) with $F(x) = x^{(q^h-1)/(q-1)}$. When $h = 1$, the code $\mathcal{C}_s$ has parameters $[q^m - 1, q^m - 1 - m - \mathbb{N}_p(m), 2 + \mathbb{N}_p(m)]_q$. When $h = 2$, the code $\mathcal{C}_s$ becomes a special case of the code presented in Theorem 20.5.9. When $h \ge 3$, Ding provided the following information on the code $\mathcal{C}_s$. In Theorems 20.5.13 and 20.5.15, we need

- $\mathbb{N}(u, t) = \begin{cases} 1 & \text{if } t = 1,\ u \ge 1, \\ \#\{(i_1, \ldots, i_{t-1}) \in \mathbb{Z}^{t-1} \mid 1 \le i_1 < i_2 < \cdots < i_{t-1} < u\} & \text{if } t \ge 2. \end{cases}$

Theorem 20.5.13 ([548]) *Let s^∞ be the sequence in (20.13) where $F(x) = x^{(q^h-1)/(q-1)}$ with $3 \le h \le \lceil \frac{m}{2} \rceil$. Then the linear span $\mathbb{L}_s$ and minimal polynomial $\mathbb{M}_s(x)$ of s^∞ are*

$$\mathbb{L}_s = \left(\mathbb{N}_p(h) + \sum_{t=1}^{h-1} \sum_{u=1}^{h-1} \mathbb{N}_p(h-u)\mathbb{N}(u,t) \right) m + \mathbb{N}_p(m)$$

and

$$\mathbb{M}_s(x) = (x-1)^{\mathbb{N}_p(m)} m_{\alpha^{-1}}(x)^{\mathbb{N}_p(h)} \prod_{\substack{1 \le u \le h-1 \\ \mathbb{N}_p(h-u)=1}} m_{\alpha^{-(q^0+q^u)}}(x) \times$$

$$\prod_{t=2}^{h-1} \prod_{\substack{t \le u \le h-1 \\ \mathbb{N}_p(h-u)=1}} \prod_{1 \le i_1 < \cdots < i_{t-1} < u} m_{\alpha^{-(q^0 + \sum_{j=1}^{t-1} q^{i_j} + q^u)}}(x).$$

The code $\mathcal{C}_s$ defined by the sequence s^∞ has parameters $[q^m - 1, q^m - 1 - \mathbb{L}_s, d]_q$ and generator polynomial $\mathbb{M}_s(x)$.

As a corollary of Theorem 20.5.13, Ding has proved the following result when $h = 3$.

Corollary 20.5.14 ([548]) *Let $h = 3$ and $m \geq 6$. The code $\mathcal{C}_s$ of Theorem 20.5.13 has parameters $[q^m - 1, q^m - 1 - \mathbb{L}_s, d]_q$ where*

$$\mathbb{L}_s = \begin{cases} 4m + \mathbb{N}_p(m) & \text{if } p \neq 3, \\ 3m + \mathbb{N}_p(m) & \text{if } p = 3 \end{cases}$$

and generator polynomial $\mathbb{M}_s(x)$ given by

$$\mathbb{M}_s(x) = \begin{cases} (x - 1)^{\mathbb{N}_p(m)} m_{\alpha^{-1}}(x) m_{\alpha^{-1-q}}(x) m_{\alpha^{-1-q^2}}(x) m_{\alpha^{-1-q-q^2}}(x) & \text{if } p \neq 3, \\ (x - 1)^{\mathbb{N}_p(m)} m_{\alpha^{-1-q}}(x) m_{\alpha^{-1-q^2}}(x) m_{\alpha^{-1-q-q^2}}(x) & \text{if } p = 3. \end{cases}$$

In addition,

$$\begin{cases} 3 \leq d \leq 8 & \text{if } p = 3 \text{ and } \mathbb{N}_p(m) = 1, \\ 3 \leq d \leq 6 & \text{if } p = 3 \text{ and } \mathbb{N}_p(m) = 0, \\ 3 \leq d \leq 8 & \text{if } p > 3. \end{cases}$$

In the case $q = 3$, Ding [548] also studied the code $\mathcal{C}_s$ defined by the sequence s^∞ of (20.13) with $F(x) = x^{(3^h+1)/2}$ where h is odd with $1 \leq h \leq \lceil \frac{m}{2} \rceil$ and $\gcd(m, h) = 1$. When $h = 1$, the code $\mathcal{C}_s$ becomes a special case of a code in Theorem 20.5.9. The following theorem provides information on the code $\mathcal{C}_s$ when $h \geq 3$.

Theorem 20.5.15 *With $q = 3$, let h be odd where $3 \leq h \leq \lceil \frac{m}{2} \rceil$ and $\gcd(m, h) = 1$. Let s^∞ be the sequence in (20.13) where $F(x) = x^{(3^h+1)/2}$. The code $\mathcal{C}_s$ defined by s^∞ has parameters $[3^m - 1, 3^m - 1 - \mathbb{L}_s, d]_3$ and generator polynomial $\mathbb{M}_s(x)$ where $\mathbb{L}_s$ and $\mathbb{M}_s(x)$ are*

$$\mathbb{L}_s = \mathbb{N}_3(m) + \left(\sum_{i=0}^{h} \mathbb{N}_3(h - i + 1) \right) m + \left(\sum_{t=2}^{h} \mathbb{N}(h, t) + \sum_{t=2}^{h-1} \sum_{i_t=t}^{h-1} \mathbb{N}_3(h - i_t + 1) \mathbb{N}(i_t, t) \right) m$$

and

$$\mathbb{M}_s(x) = (x - 1)^{\mathbb{N}_3(m)} m_{\alpha^{-1}}(x)^{\mathbb{N}_3(h+1)} m_{\alpha^{-2}}(x) \prod_{t=1}^{h-1} \prod_{1 \leq i_1 < \cdots < i_t \leq h-1} m_{\alpha^{-(2+\sum_{j=1}^{t} 3^{i_j})}}(x) \times$$

$$\prod_{\substack{1 \leq u \leq h-1 \\ \mathbb{N}_3(h-u+1)=1}} m_{\alpha^{-(1+3^u)}}(x) \prod_{t=2}^{h-1} \prod_{\substack{t \leq i_t \leq h-1 \\ \mathbb{N}_3(h-i_t+1)=1}} \prod_{1 \leq i_1 < \cdots < i_{t-1} < i_t} m_{\alpha^{-(1+\sum_{j=1}^{t} 3^{i_j})}}(x).$$

As a corollary of Theorem 20.5.15, Ding proved the following result on the code $\mathcal{C}_s$ when $h = 3$; the lower bound on d follows from the BCH Bound and the upper bound follows from the Sphere Packing Bound.

Corollary 20.5.16 ([548]) *The code $\mathcal{C}_s$ of Theorem 20.5.15 with $h = 3$ has parameters $[3^m - 1, 3^m - 1 - \mathbb{L}_s, d]_3$ where $\mathbb{L}_s = 7m + \mathbb{N}_3(m)$ and $4 + \mathbb{N}_3(m) \leq d \leq 16$. $\mathcal{C}_s$ has generator polynomial $\mathbb{M}_s(x)$ given by*

$$\mathbb{M}_s(x) = (x - 1)^{\mathbb{N}_3(m)} m_{\alpha^{-1}}(x) m_{\alpha^{-2}}(x) m_{\alpha^{-5}}(x) m_{\alpha^{-10}}(x) m_{\alpha^{-11}}(x) m_{\alpha^{-13}}(x) m_{\alpha^{-14}}(x).$$

In [548], Ding also provided a list of nice open problems related to cyclic codes from monomials and trinomials over finite fields.

20.5.4   Cyclic Codes from Dickson Polynomials

Cyclic codes from Dickson polynomials of the first kind have been studied by Ding [553, 554]. We shall present them in this subsection in a compact form. As indicated by Ding, cyclic codes from Dickson polynomials of the second kind can be developed in a similar way. All the results presented in this subsection come from [553] and the excellent recent survey of Ding [554]. We continue using the notation presented in Sections 20.5.2 and 20.5.3.

Ding studied the cyclic codes $\mathcal{C}_s$ over $\mathbb{F}_q$ from the Dickson polynomial $D_{p^u,a}(x) = x^{p^u}$ and provided the following complete result.

Theorem 20.5.17 ([553, 554]) *The cylic code $\mathcal{C}_s$ over $\mathbb{F}_q$ defined by the sequence in* (20.13) *arising from the Dickson polynomial $F(x) = D_{p^u,a}(x) = x^{p^u}$ has generator polynomial $\mathbb{M}_s(x) = (x-1)^{\delta(1)} m_{\alpha^{-p^u}}(x)$ and parameters $[q^m - 1, q^m - 1 - m - \delta(1), d]_q$ where*

$$d = \begin{cases} 4 & \text{if } q = 2 \text{ and } \delta(1) = 1, \\ 3 & \text{if } q = 2 \text{ and } \delta(1) = 0, \\ 3 & \text{if } q > 2 \text{ and } \delta(1) = 1, \\ 2 & \text{if } q > 2 \text{ and } \delta(1) = 0. \end{cases}$$

As highlighted by Ding, when $q = 2$, the code in Theorem 20.5.17 is equivalent to the binary Hamming code or its even-weight subcode and is thus optimal. For $q > 2$, the code is either optimal or almost optimal with respect to the Sphere Packing Bound.

Ding studied the cyclic codes $\mathcal{C}_s$ coming from the Dickson polynomial $D_{2,a}(x) = x^2 - 2a$ over $\mathbb{F}_{q^m}$. When $p = 2$, this code was treated in Theorem 20.5.17. When $p > 2$, the following theorem is a variant of Theorem 5.2 in [545] but has a much stronger conclusion on the minimum distance of the code.

Theorem 20.5.18 ([553, 554]) *Let $p > 2$ and $m \geq 3$. The code $\mathcal{C}_s$ defined by the sequence in* (20.13) *arising from the Dickson polynomial $F(x) = D_{2,a}(x) = x^2 - 2a$ with $a \in \mathbb{F}_{q^m}$ has parameters $[q^m - 1, q^m - 1 - 2m - \delta(1 - 2a), d]_q$ and generator polynomial*

$$\mathbb{M}_s(x) = (x-1)^{\delta(1-2a)} m_{\alpha^{-1}}(x) m_{\alpha^{-2}}(x),$$

where

$$d = \begin{cases} 4 & \text{if } q = 3 \text{ and } \delta(1 - 2a) = 0, \\ 5 & \text{if } q = 3 \text{ and } \delta(1 - 2a) = 1, \\ 3 & \text{if } q > 3 \text{ and } \delta(1 - 2a) = 0, \\ 4 & \text{if } q > 3 \text{ and } \delta(1 - 2a) = 1. \end{cases}$$

Again using the Sphere Packing Bound, the code in Theorem 20.5.18 is either optimal or almost optimal for all $m \geq 2$.

The cyclic codes $\mathcal{C}_s$ arising from the Dickson polynomial $D_{3,a}(x) = x^3 - 3ax$ were studied by Ding. In his study, Ding has distinguished among the three cases: $p = 2$, $p = 3$, and $p \geq 5$. The case $p = 3$ was addressed previously in Theorem 20.5.17. So we give information for only the two remaining cases. The following theorem provides information on the code $\mathcal{C}_s$ when $q = p = 2$.

Theorem 20.5.19 ([553, 554]) *Let $q = p = 2$ and let $m \geq 4$. Let s^∞ be the sequence in* (20.13) *where $F(x) = D_{3,a}(x) = x^3 - 3ax = x^3 + ax$ with $a \in \mathbb{F}_{2^m}$. Then the minimal polynomial $\mathbb{M}_s(x)$ of s^∞ is*

$$\mathbb{M}_s(x) = \begin{cases} (x-1)^{\delta(1)} m_{\alpha^{-3}}(x) & \text{if } a = 0, \\ (x-1)^{\delta(1+a)} m_{\alpha^{-1}}(x) m_{\alpha^{-3}}(x) & \text{if } a \neq 0, \end{cases}$$

and the linear span $\mathbb{L}_s$ *of* s^∞ *is*

$$\mathbb{L}_s = \begin{cases} \delta(1) + m & \text{if } a = 0, \\ \delta(1+a) + 2m & \text{if } a \neq 0. \end{cases}$$

Moreover, the binary code $\mathcal{C}_s$ *defined by the sequence* s^∞ *has generator polynomial* $\mathbb{M}_s(x)$ *and parameters* $[2^m - 1, 2^m - 1 - \mathbb{L}_s, d]_2$ *where*

$$d = \begin{cases} 2 & \text{if } a = 0 \text{ and } \delta(1) = 0, \\ 4 & \text{if } a = 0 \text{ and } \delta(1) = 1, \\ 5 & \text{if } a \neq 0 \text{ and } \delta(1+a) = 0, \\ 6 & \text{if } a \neq 0 \text{ and } \delta(1+a) = 1. \end{cases}$$

As highlighted by Ding, when $a = 0$ and $\delta(1) = 1$, the code is equivalent to the even-weight subcode of the Hamming code. When $a = 1$, the code $\mathcal{C}_s$ is a double-error correcting binary BCH code or its even-weight subcode. Theorem 20.5.19 shows that well-known classes of cyclic codes can be constructed with Dickson polynomials of order 3. The code is either optimal or almost optimal.

In the case where $q = p^t$ with $p \geq 5$ or $p = 2$ and $t \geq 2$, Ding has given the following result.

Theorem 20.5.20 ([553, 554]) *Let* $q = p^t$ *where* $p \geq 5$ *or* $p = 2$ *and* $t \geq 2$. *Let* s^∞ *be the sequence in* (20.13) *where* $F(x) = D_{3,a}(x) = x^3 - 3ax$ *with* $a \in \mathbb{F}_{q^m}$. *Then the minimal polynomial* $\mathbb{M}_s(x)$ *of* s^∞ *is*

$$\mathbb{M}_s(x) = \begin{cases} (x-1)^{\delta(-2)} m_{\alpha^{-3}}(x) m_{\alpha^{-2}}(x) & \text{if } a = 1, \\ (x-1)^{\delta(1-3a)} m_{\alpha^{-3}}(x) m_{\alpha^{-2}}(x) m_{\alpha^{-1}}(x) & \text{if } a \neq 1, \end{cases}$$

and the linear span $\mathbb{L}_s$ *of* s^∞ *is*

$$\mathbb{L}_s = \begin{cases} \delta(-2) + 2m & \text{if } a = 1, \\ \delta(1+a) + 3m & \text{if } a \neq 1. \end{cases}$$

Moreover, the code $\mathcal{C}_s$ *defined by the sequence* s^∞ *has generator polynomial* $\mathbb{M}_s(x)$ *and parameters* $[q^m - 1, q^m - 1 - \mathbb{L}_s, d]_q$ *where*

$$d \geq \begin{cases} 3 & \text{if } a = 1, \\ 4 & \text{if } a \neq 1 \text{ and } \delta(1 - 3a) = 0, \\ 5 & \text{if } a \neq 1 \text{ and } \delta(1 - 3a) = 1, \\ 5 & \text{if } a \neq 1 \text{ and } \delta(1 - 3a) = 0 \text{ and } q = 4, \\ 6 & \text{if } a \neq 1 \text{ and } \delta(1 - 3a) = 1 \text{ and } q = 4. \end{cases}$$

The code $\mathcal{C}_s$ of Theorem 20.5.20 is either a BCH code or the even-like subcode of a BCH code. Ding has showed that the code is either optimal or almost optimal.

Similarly Ding studied cyclic codes coming from the Dickson polynomial $D_{4,a}(x) = x^4 - 4ax^2 + 2a^2$ where he distinguished among the three cases: $p = 2$, $p = 3$, and $p \geq 5$. The case $p = 2$ was covered by a code presented in Theorem 20.5.17.

The following result gives information about the cyclic code $\mathcal{C}_s$ when $q = p = 3$

Theorem 20.5.21 ([553, 554]) *Let* $q = p = 3$ *and* $m \geq 3$. *Let* s^∞ *be the sequence in* (20.13) *where* $F(x) = D_{4,a}(x) = x^4 - 4ax^2 + 2a^2$ *with* $a \in \mathbb{F}_{3^m}$. *Then the minimal polynomial* $\mathbb{M}_s(x)$ *of* s^∞ *is*

$$\mathbb{M}_s(x) = \begin{cases} (x-1)^{\delta(1)} m_{\alpha^{-4}}(x) m_{\alpha^{-1}}(x) & \text{if } a = 0, \\ (x-1)^{\delta(1)} m_{\alpha^{-4}}(x) m_{\alpha^{-2}}(x) & \text{if } a = 1, \\ (x-1)^{\delta(1-a-a^2)} m_{\alpha^{-4}}(x) m_{\alpha^{-2}}(x) m_{\alpha^{-1}}(x) & \text{otherwise,} \end{cases}$$

and the linear span $\mathbb{L}_s$ of s^∞ is

$$\mathbb{L}_s = \begin{cases} \delta(1) + 2m & \text{if } a = 0, \\ \delta(1) + 2m & \text{if } a = 1, \\ \delta(1 - a - a^2) + 3m & \text{otherwise.} \end{cases}$$

Moreover, the code $\mathcal{C}_s$ defined by the sequence s^∞ has generator polynomial $\mathbb{M}_s(x)$ and parameters $[3^m - 1, 3^m - 1 - \mathbb{L}_s, d]_3$ where

$$\begin{cases} d = 2 & \text{if } a = 1, \\ d = 3 & \text{if } a = 0 \text{ and } m \equiv 0 \pmod 6, \\ d \geq 4 & \text{if } a = 0 \text{ and } m \not\equiv 0 \pmod 6, \\ d \geq 5 & \text{if } a^2 \neq a \text{ and } \delta(1 - a - a^2) = 0, \\ d = 6 & \text{if } a^2 \neq a \text{ and } \delta(1 - a - a^2) = 1. \end{cases}$$

Ding showed that when $a = 1$, the code in Theorem 20.5.21 is neither optimal nor almost optimal. The code is either optimal or almost optimal in all other cases.

In the case where $q = p^t$, with $p \geq 5$ or $p = 3$ and $t \geq 2$, Ding proved the following result about the code $\mathcal{C}_s$.

Theorem 20.5.22 ([553, 554]) *Let $m \geq 2$ and $q = p^t$ where $p \geq 5$ or $p = 3$ and $t \geq 2$. Let s^∞ be the sequence in (20.13) where $F(x) = D_{4,a}(x) = x^4 - 4ax^2 + 2a^2$ with $a \in \mathbb{F}_{q^m}$. Then the minimal polynomial $\mathbb{M}_s(x)$ of s^∞ is*

$$\mathbb{M}_s(x) = \begin{cases} (x-1)^{\delta(1)} m_{\alpha^{-4}}(x) m_{\alpha^{-3}}(x) m_{\alpha^{-1}}(x) & \text{if } a = \frac{3}{2}, \\ (x-1)^{\delta(1)} m_{\alpha^{-4}}(x) m_{\alpha^{-3}}(x) m_{\alpha^{-2}}(x) & \text{if } a = \frac{1}{2}, \\ (x-1)^{\delta(1-4a+2a^2)} \prod_{i=1}^4 m_{\alpha^{-i}}(x) & \text{if } a \notin \{\frac{3}{2}, \frac{1}{2}\}, \end{cases}$$

and the linear span $\mathbb{L}_s$ of s^∞ is

$$\mathbb{L}_s = \begin{cases} \delta(1) + 3m & \text{if } a \in \{\frac{3}{2}, \frac{1}{2}\}, \\ \delta(1 - 4a + 2a^2) + 4m & \text{otherwise.} \end{cases}$$

Moreover, the code $\mathcal{C}_s$ defined by the sequence s^∞ has generator polynomial $\mathbb{M}_s(x)$ and parameters $[q^m - 1, q^m - 1 - \mathbb{L}_s, d]_q$ where

$$\begin{cases} d \geq 3 & \text{if } a = \frac{3}{2}, \\ d \geq 4 & \text{if } a = \frac{1}{2}, \\ d \geq 5 & \text{if } a \notin \{\frac{3}{2}, \frac{1}{2}\} \text{ and } \delta(1 - 4a + a^2) = 0, \\ d = 6 & \text{if } a \notin \{\frac{3}{2}, \frac{1}{2}\} \text{ and } \delta(1 - 4a + a^2) = 1. \end{cases}$$

As indicated by Ding, the code $\mathcal{C}_s$ of Theorem 20.5.22 is either optimal or almost optimal, except in the cases that $a \in \{\frac{3}{2}, \frac{1}{2}\}$.

Ding studied cyclic codes $\mathcal{C}_s$ determined by the Dickson polynomial $D_{5,a}(x) = x^5 - 5ax^3 + 5a^2 x$ distinguishing among the three cases: $p = 2$ (with three subcases), $p = 3$ (with two subcases), and $p \geq 7$. The case $p = 5$ was included in Theorem 20.5.17.

When $q = p = 2$, the following result describes parameters of the code $\mathcal{C}_s$.

Theorem 20.5.23 ([553, 554]) *Let $q = p = 2$ and $m \geq 5$. Let s^∞ be the sequence in (20.13) where $F(x) = D_{5,a}(x) = x^5 - 5ax^3 + 5a^2 x$ with $a \in \mathbb{F}_{2^m}$. Then the minimal polynomial $\mathbb{M}_s(x)$ of s^∞ is*

$$\mathbb{M}_s(x) = \begin{cases} (x-1)^{\delta(1)} m_{\alpha^{-5}}(x) & \text{if } a = 0, \\ (x-1)^{\delta(1)} m_{\alpha^{-5}}(x) m_{\alpha^{-3}}(x) & \text{if } 1 + a + a^3 = 0, \\ (x-1)^{\delta(1)} \prod_{i=0}^2 m_{\alpha^{-(2i+1)}}(x) & \text{if } a + a^2 + a^4 \neq 0, \end{cases}$$

and the linear span $\mathbb{L}_s$ *of* s^∞ *is*

$$\mathbb{L}_s = \begin{cases} \delta(1) + m & if\ a = 0, \\ \delta(1) + 2m & if\ 1 + a + a^3 = 0, \\ \delta(1) + 3m & if\ a + a^2 + a^4 \neq 0. \end{cases}$$

Moreover, the code $\mathcal{C}_s$ *defined by the sequence* s^∞ *has generator polynomial* $\mathbb{M}_s(x)$ *and parameters* $[2^m - 1, 2^m - 1 - \mathbb{L}_s, d]_2$ *where*

$$\begin{cases} d = 2 & if\ a = 0\ and\ \delta(1) = 0\ and\ 5\,|\,(2^m - 1), \\ d = 3 & if\ a = 0\ and\ \delta(1) = 0\ and\ 5 \nmid (2^m - 1), \\ d = 4 & if\ a = 0\ and\ \delta(1) = 1, \\ d \geq 3 & if\ 1 + a + a^3 = 0\ and\ \delta(1) = 0, \\ d \geq 4 & if\ 1 + a + a^3 = 0\ and\ \delta(1) = 1, \\ d \geq 7 & if\ a + a^2 + a^4 \neq 0\ and\ \delta(1) = 0, \\ d = 8 & if\ a + a^2 + a^4 \neq 0\ and\ \delta(1) = 1. \end{cases}$$

As showed by Ding, the code in Theorem 20.5.23 is either optimal or almost optimal. The code is not a BCH code when $1 + a + a^3 = 0$ and is a BCH code in the remaining cases.

For the case where $(p, q) = (2, 4)$, Ding has proved the following result on the code $\mathcal{C}_s$.

Theorem 20.5.24 ([553, 554]) *Let* $p = 2$, $q = 4$, *and* $m \geq 3$. *Let* s^∞ *be the sequence in* (20.13) *where* $F(x) = D_{5,a}(x) = x^5 - 5ax^3 + 5a^2 x$ *with* $a \in \mathbb{F}_{4^m}$. *Then the minimal polynomial* $\mathbb{M}_s(x)$ *of* s^∞ *is*

$$\mathbb{M}_s(x) = \begin{cases} (x - 1)^{\delta(1)} m_{\alpha^{-5}}(x) & if\ a = 0, \\ (x - 1)^{\delta(1)} m_{\alpha^{-5}}(x) m_{\alpha^{-3}}(x) m_{\alpha^{-2}}(x) & if\ a = 1, \\ (x - 1)^{\delta(1 + a + a^2)} m_{\alpha^{-5}}(x) m_{\alpha^{-3}}(x) m_{\alpha^{-2}}(x) m_{\alpha^{-1}}(x) & if\ a + a^2 \neq 0, \end{cases}$$

and the linear span $\mathbb{L}_s$ *of* s^∞ *is*

$$\mathbb{L}_s = \begin{cases} \delta(1) + m & if\ a = 0, \\ \delta(1) + 3m & if\ a = 1, \\ \delta(1) + 4m & if\ a + a^2 \neq 0. \end{cases}$$

Moreover, the code $\mathcal{C}_s$ *defined by the sequence* s^∞ *has generator polynomial* $\mathbb{M}_s(x)$ *and parameters* $[4^m - 1, 4^m - 1 - \mathbb{L}_s, d]_4$ *where*

$$\begin{cases} d = 2 & if\ a = 0\ and\ \delta(1) = 0\ and\ 5\,|\,(4^m - 1), \\ d = 3 & if\ a = 0\ and\ 5 \nmid (4^m - 1), \\ d \geq 3 & if\ a = 1, \\ d \geq 6 & if\ a + a^2 \neq 0\ and\ \delta(1) = 0, \\ d \geq 7 & if\ a + a^2 \neq 0\ and\ \delta(1) = 1. \end{cases}$$

Examples of the code in Theorem 20.5.24 are documented in [553], and many of them are optimal.

We turn to the case where $(p, q) = (2, 2^t)$ with $t \geq 3$.

Theorem 20.5.25 ([553, 554]) *Let* $p = 2$, $q = 2^t$ *with* $t \geq 3$, *and* $m \geq 3$. *Let* s^∞ *be the sequence in* (20.13) *where* $F(x) = D_{5,a}(x) = x^5 - 5ax^3 + 5a^2 x$ *with* $a \in \mathbb{F}_{q^t}$. *Then the minimal polynomial* $\mathbb{M}_s(x)$ *of* s^∞ *is*

$$\mathbb{M}_s(x) = \begin{cases} (x - 1)^{\delta(1)} m_{\alpha^{-5}}(x) m_{\alpha^{-4}}(x) m_{\alpha^{-1}}(x) & if\ a = 0, \\ \prod_{i=2}^5 m_{\alpha^{-i}}(x) & if\ 1 + a + a^2 = 0, \\ (x - 1)^{\delta(1 + a + a^2)} \prod_{i=1}^5 m_{\alpha^{-i}}(x) & if\ a + a^2 + a^3 \neq 0, \end{cases}$$

and the linear span $\mathbb{L}_s$ *of* s^∞ *is*

$$\mathbb{L}_s = \begin{cases} \delta(1) + 3m & \text{if } a = 0, \\ \delta(1) + 4m & \text{if } 1 + a + a^2 = 0, \\ \delta(1) + 5m & \text{if } a + a^2 + a^3 \neq 0. \end{cases}$$

Moreover, the code $\mathcal{C}_s$ *defined by the sequence* s^∞ *has generator polynomial* $\mathbb{M}_s(x)$ *and parameters* $[q^m - 1, q^m - 1 - \mathbb{L}_s, d]_q$ *where*

$$d \geq \begin{cases} 3 & \text{if } a = 0 \text{ and } \delta(1) = 0, \\ 4 & \text{if } a = 0 \text{ and } \delta(1) = 1, \\ 5 & \text{if } 1 + a + a^2 = 0, \\ 6 & \text{if } a + a^2 + a^3 \neq 0 \text{ and } \delta(1) = 0, \\ 7 & \text{if } a + a^2 + a^3 \neq 0 \text{ and } \delta(1) = 1. \end{cases}$$

Examples of the code in Theorem 20.5.25 can be found in [553], and many of them are optimal. The code in Theorem 20.5.25 is not a BCH code when $a = 0$ and is a BCH code otherwise.

When $q = p = 3$, Ding has stated the following result.

Theorem 20.5.26 ([553, 554]) *Let* $q = p = 3$ *and* $m \geq 3$. *Let* s^∞ *be the sequence in* (20.13) *where* $F(x) = D_{5,a}(x) = x^5 - 5ax^3 + 5a^2x$ *with* $a \in \mathbb{F}_{3^m}$. *Then the minimal polynomial* $\mathbb{M}_s(x)$ *of* s^∞ *is*

$$\mathbb{M}_s(x) = \begin{cases} (x-1)^{\delta(1+a+2a^2)} m_{\alpha^{-5}}(x) m_{\alpha^{-4}}(x) m_{\alpha^{-2}}(x) & \text{if } a - a^6 = 0, \\ (x-1)^{\delta(1+a+2a^2)} \prod_{i=2}^5 m_{\alpha^{-i}}(x) & \text{if } a - a^6 \neq 0, \end{cases}$$

and the linear span $\mathbb{L}_s$ *of* s^∞ *is*

$$\mathbb{L}_s = \begin{cases} \delta(1 + a + 2a^2) + 3m & \text{if } a - a^6 = 0, \\ \delta(1 + a + 2a^2) + 4m & \text{if } a - a^6 \neq 0. \end{cases}$$

Moreover, the code $\mathcal{C}_s$ *defined by the sequence* s^∞ *has generator polynomial* $\mathbb{M}_s(x)$ *and parameters* $[3^m - 1, 3^m - 1 - \mathbb{L}_s, d]_3$ *where*

$$d \geq \begin{cases} 4 & \text{if } a - a^6 = 0, \\ 7 & \text{if } a - a^6 \neq 0 \text{ and } \delta(1 + a + 2a^2) = 0, \\ 8 & \text{if } a - a^6 \neq 0 \text{ and } \delta(1 + a + 2a^2) = 1. \end{cases}$$

Examples of the code in Theorem 20.5.26 are described in [553], and some of them are optimal.

The case $(p, q) = (3, 3^t)$ where $t \geq 2$ is stated in the following theorem.

Theorem 20.5.27 ([553, 554]) *Let* $p = 3$, $q = 3^t$ *with* $t \geq 2$, *and* $m \geq 2$. *Let* s^∞ *be the sequence in* (20.13) *where* $F(x) = D_{5,a}(x) = x^5 - 5ax^3 + 5a^2x$ *with* $a \in \mathbb{F}_{q^m}$. *Then the minimal polynomial* $\mathbb{M}_s(x)$ *of* s^∞ *is*

$$\mathbb{M}_s(x) = \begin{cases} (x-1)^{\delta(1)} m_{\alpha^{-5}}(x) m_{\alpha^{-4}}(x) m_{\alpha^{-2}}(x) m_{\alpha^{-1}}(x) & \text{if } 1 + a = 0, \\ (x-1)^{\delta(a-1)} m_{\alpha^{-5}}(x) m_{\alpha^{-4}}(x) m_{\alpha^{-3}}(x) m_{\alpha^{-2}}(x) & \text{if } 1 + a^2 = 0, \\ (x-1)^{\delta(1+a+2a^2)} \prod_{i=1}^5 m_{\alpha^{-i}}(x) & \text{if } (a+1)(a^2+1) \neq 0, \end{cases}$$

and the linear span $\mathbb{L}_s$ *of* s^∞ *is*

$$\mathbb{L}_s = \begin{cases} \delta(1) + 4m & \text{if } a + 1 = 0, \\ \delta(a-1) + 4m & \text{if } a^2 + 1 = 0, \\ \delta(1 + a + 2a^2) + 5m & \text{if } (a+1)(a^2+1) \neq 0. \end{cases}$$

Moreover, the code $\mathcal{C}_s$ defined by the sequence s^∞ has generator polynomial $\mathbb{M}_s(x)$ and parameters $[q^m - 1, q^m - 1 - \mathbb{L}_s, d]_q$ where

$$d \geq \begin{cases} 3 & \text{if } a = -1 \text{ and } \delta(1) = 0, \\ 4 & \text{if } a = -1 \text{ and } \delta(1) = 1, \\ 5 & \text{if } a^2 = -1 \text{ and } \delta(a-1) = 0, \\ 6 & \text{if } a^2 = -1 \text{ and } \delta(a-1) = 1, \\ 6 & \text{if } (a+1)(a^2+1) \neq 0 \text{ and } \delta(1 + a + 2a^2) = 0, \\ 7 & \text{if } (a+1)(a^2+1) \neq 0 \text{ and } \delta(1 + a + 2a^2) = 1. \end{cases}$$

Examples of the code in Theorem 20.5.27 are available in [553], and some of them are optimal. The code $\mathcal{C}_s$ is a BCH code, except in the case that $a = -1$.

Finally, in the case $p \geq 7$, Ding has shown the following result.

Theorem 20.5.28 ([553, 554]) *Let $p \geq 7$ and $m \geq 2$. Let s^∞ be the sequence in (20.13) where $F(x) = D_{5,a}(x) = x^5 - 5ax^3 + 5a^2x$ with $a \in \mathbb{F}_{q^m}$. Then the minimal polynomial $\mathbb{M}_s(x)$ of s^∞ is*

$$\mathbb{M}_s(x) = \begin{cases} (x-1)^{\delta(1-5a+5a^2)} m_{\alpha^{-5}}(x) m_{\alpha^{-4}}(x) m_{\alpha^{-2}}(x) m_{\alpha^{-1}}(x) & \text{if } a = 2, \\ (x-1)^{\delta(1-5a+5a^2)} m_{\alpha^{-5}}(x) m_{\alpha^{-4}}(x) m_{\alpha^{-3}}(x) m_{\alpha^{-1}}(x) & \text{if } a = \frac{2}{3}, \\ (x-1)^{\delta(1-5a+5a^2)} m_{\alpha^{-5}}(x) m_{\alpha^{-4}}(x) m_{\alpha^{-3}}(x) m_{\alpha^{-2}}(x) & \text{if } a^2 - 3a + 1 = 0, \\ (x-1)^{\delta(1-5a+5a^2)} \prod_{i=1}^{5} m_{\alpha^{-i}}(x) & \text{otherwise,} \end{cases}$$

and the linear span $\mathbb{L}_s$ of s^∞ is

$$\mathbb{L}_s = \begin{cases} \delta(1 - 5a + 5a^2) + 4m & \text{if } (a^2 - 3a + 1)(a - 2)(3a - 2) = 0, \\ \delta(1 - 5a + 5a^2) + 5m & \text{otherwise.} \end{cases}$$

Moreover, the code $\mathcal{C}_s$ defined by the sequence s^∞ has generator polynomial $\mathbb{M}_s(x)$ and parameters $[q^m - 1, q^m - 1 - \mathbb{L}_s, d]_q$ where

$$d \geq \begin{cases} 3 & \text{if } a = 2 \text{ and } \delta(1 - 5a + 5a^2) = 0, \\ 4 & \text{if } a = 2 \text{ and } \delta(1 - 5a + 5a^2) = 1, \\ 4 & \text{if } a = \frac{2}{3} \text{ and } \delta(1 - 5a + 5a^2) = 0, \\ 5 & \text{if } a = \frac{2}{3} \text{ and } \delta(1 - 5a + 5a^2) = 1, \\ 5 & \text{if } 1 - 3a + a^2 = 0 \text{ and } \delta(1 - 5a + 5a^2) = 0, \\ 6 & \text{if } 1 - 3a + a^2 = 0 \text{ and } \delta(1 - 5a + 5a^2) = 1, \\ 6 & \text{if } (a^2 - 3a + 1)(a - 2)(3a - 2) \neq 0 \text{ and } \delta(1 - 5a + 5a^2) = 0, \\ 7 & \text{if } (a^2 - 3a + 1)(a - 2)(3a - 2) \neq 0 \text{ and } \delta(1 - 5a + 5a^2) = 1. \end{cases}$$

Examples of the code in Theorem 20.5.28 can be found in [553], and some of them are optimal. The code $\mathcal{C}_s$ is a BCH code, except in the cases $a \in \{2, 2/3\}$.

20.6 Codes with Few Weights from p-Ary Functions with p Odd

20.6.1 Codes with Few Weights from p-Ary Weakly Regular Bent Functions Based on the First Generic Construction

Let p be an <u>odd</u> prime, $\Psi : \mathbb{F}_{p^m} \to \mathbb{F}_{p^m}$ with $\Psi(0) = 0$, and $\psi_1(x) := \mathrm{Tr}_{p^m/p}(\Psi(x))$. Let $\mathcal{C}_\Psi$ be the code defined by (20.4). Analogous to (20.11), define the subcode $\mathcal{C}_{\psi_1}$ of $\mathcal{C}_\Psi$ by

$$\mathcal{C}_{\psi_1} := \{\widetilde{c}_{\alpha,\beta} = (f_{\alpha,\beta}(\zeta_1), f_{\alpha,\beta}(\zeta_2), \ldots, f_{\alpha,\beta}(\zeta_{p^m-1})) \mid \alpha \in \mathbb{F}_p, \beta \in \mathbb{F}_{p^m}\} \quad (20.14)$$

TABLE 20.16: Weight distribution of $\mathcal{C}_{\psi_1}$ in Theorem 20.6.2 for m even and p odd

Hamming weight w	Multiplicity A_w
0	1
$p^m - p^{m-1}$	$p^m - 1$
$p^m - p^{m-1} - p^{\frac{m}{2}-1}(p-1)$	$p^m - p^{m-1} + p^{\frac{m}{2}-1}(p-1)^2$
$p^m - p^{m-1} + p^{\frac{m}{2}-1}$	$(p^m - p^{m-1})(p-1) - p^{\frac{m}{2}-1}(p-1)^2$

TABLE 20.17: Weight distribution of $\mathcal{C}_{\psi_1}$ in Theorem 20.6.2 for m and p odd

Hamming weight w	Multiplicity A_w
0	1
$p^m - p^{m-1}$	$2p^m - p^{m-1} - 1$
$p^m - p^{m-1} - \left(\frac{-1}{p}\right)^{\frac{m+1}{2}} p^{\frac{m-1}{2}}$	$\left(p^{m-1} + p^{\frac{m-1}{2}}\right)\frac{(p-1)^2}{2}$
$p^m - p^{m-1} + \left(\frac{-1}{p}\right)^{\frac{m+1}{2}} p^{\frac{m-1}{2}}$	$\left(p^{m-1} - p^{\frac{m-1}{2}}\right)\frac{(p-1)^2}{2}$

where $\zeta_1, \ldots, \zeta_{p^m-1}$ are the nonzero elements of $\mathbb{F}_{p^m}$.

In the next theorem and in Table 20.17 $\left(\frac{a}{p}\right)$ denotes the Legendre symbol when $a \in \mathbb{F}_p^*$; specifically $\left(\frac{-1}{p}\right) = (-1)^{(p-1)/2}$.

Theorem 20.6.1 ([1379]) *Let p an odd prime. Let $\mathcal{C}_{\psi_1}$ be the linear code defined by (20.14) whose codewords are denoted by $\widetilde{c}_{\alpha,\beta}$. Assume that the function $\psi_1 := \mathrm{Tr}_{p^m/p}(\Psi)$ is weakly regular. Denote by ψ_1^* its dual function. Let ϵ be the sign ± 1 of the Walsh transform of ψ_1. For $\alpha \in \mathbb{F}_p^*$, $\overline{\alpha}$ satisfies $\overline{\alpha}\alpha = 1$. The weight distribution of $\mathcal{C}_{\psi_1}$ is given as follows.*

(a) $\mathrm{wt}_\mathrm{H}(\widetilde{c}_{0,0}) = 0$, and for $\beta \neq 0$, $\mathrm{wt}_\mathrm{H}(\widetilde{c}_{0,\beta}) = p^m - p^{m-1}$.

(b) *If $\alpha \neq 0$ and m is even, then*

$$\mathrm{wt}_\mathrm{H}(\widetilde{c}_{\alpha,\beta}) = \begin{cases} p^m - p^{m-1} - \epsilon p^{\frac{m}{2}-1}(p-1) & \text{if } \psi_1^*(\overline{\alpha}\beta) = 0, \\ p^m - p^{m-1} + \epsilon p^{\frac{m}{2}-1} & \text{if } \psi_1^*(\overline{\alpha}\beta) \neq 0. \end{cases}$$

(c) *If $\alpha \neq 0$ and m is odd, then*

$$\mathrm{wt}_\mathrm{H}(\widetilde{c}_{\alpha,\beta}) = \begin{cases} p^m - p^{m-1} & \text{if } \psi_1^*(\overline{\alpha}\beta) = 0, \\ p^m - p^{m-1} - \epsilon \left(\frac{-1}{p}\right)^{\frac{m+1}{2}} p^{\frac{m-1}{2}} \left(\frac{\psi_1^*(\overline{\alpha}\beta)}{p}\right) & \text{if } \psi_1^*(\overline{\alpha}\beta) \neq 0. \end{cases}$$

The following theorem presents the weight distribution of the code $\mathcal{C}_{\psi_1}$.

Theorem 20.6.2 ([1379]) *Let $\mathcal{C}_{\psi_1}$ be the three-weight code in Theorem 20.6.1. Then $\mathcal{C}_{\psi_1}$ has dimension $m + 1$; its Hamming weight distribution is given in Table 20.16 when m is even and in Table 20.17 when m is odd.*

20.6.2 Linear Codes with Few Weights from Cyclotomic Classes and Weakly Regular Bent Functions

Let p be an <u>odd</u> prime and $q = p^m$. Recently (2020), Y. Wu, N. Li and X. Zeng [1911] generalized the construction of linear codes of length n defined by

$$\mathcal{C} = \left\{ \left(\mathrm{Tr}_{q/p}(ax)\right)_{x \in D} \mid a \in \mathbb{F}_q \right\}.$$

where D is the defining set of $\mathcal{C}$. More specifically, they investigated codes $\mathcal{C}_D$ of the form

$$\mathcal{C}_D = \left\{ c(a,b) = \left(\mathrm{Tr}_{q/p}(ax + by) \right)_{(x,y) \in D} \mid a, b \in \mathbb{F}_q \right\} \tag{20.15}$$

where $D \subseteq \mathbb{F}_q^2$ by employing weakly regular bent functions (Definition 20.2.17) and cyclotomic classes over finite fields.

Let α be a generator of $\mathbb{F}_q^*$.

Definition 20.6.3 Let e be a positive integer with $e \mid (q-1)$. Define $\mathfrak{C}_i = \alpha^i \langle \alpha^e \rangle$ for $i = 0, 1, \ldots, e-1$, where $\langle \alpha^e \rangle$ denotes the subgroup of $\mathbb{F}_q^*$ generated by α^e. The cosets $\mathfrak{C}_i$ are called the **cyclotomic classes of order** e with respect to $\mathbb{F}_q$. The **Gauss periods** are defined by

$$\eta_i = \sum_{x \in \mathfrak{C}_i} \xi_p^{\mathrm{Tr}_{q/p}(x)}.$$

where ξ_p is the complex primitive p^{th} root of unity $e^{\frac{2\pi i}{p}}$.

Concretely, Wu et al. first considered the linear codes of the form (20.15) by selecting the defining set as

$$D = \{ (x,y) \in \mathbb{F}_q^2 \mid x \in \mathfrak{C}_i, \, y \in \mathfrak{C}_j \}, \tag{20.16}$$

where $\mathfrak{C}_i$, $\mathfrak{C}_j$ are any two cyclotomic classes of order e. It has been seen that the weights of $\mathcal{C}_D$ depend on the values of Gauss periods of order e which are normally difficult to calculate. By using the semi-primitive case of cyclotomic classes of order e, they showed that $\mathcal{C}_D$ is a five-weight linear code and determined its weight distribution according to Gauss sums. As a special case, when $e = 2$ and m is odd, $\mathcal{C}_D$ is proved to be a two-weight linear code. Moreover, Wu et al. derived new two-weight linear codes and three-weight linear codes of the form (20.15) from weakly regular bent functions. The defining set D for this case is of the form

$$D = \left\{ (x,y) \in \mathbb{F}_q^2 \setminus \{(0,0)\} \mid f(x) + g(y) = 0 \right\}, \tag{20.17}$$

and the following two cases are considered to construct linear codes:

(1) $f(x) = \mathrm{Tr}_{q/p}(x)$ and $g(y)$ is a weakly regular bent function, or

(2) both $f(x)$ and $g(y)$ are weakly regular bent functions.

By calculating Weil sums and using the properties of weakly regular bent functions, it was shown that a class of three-weight linear codes can be obtained from the first case, and the second case leads to both two-weight and three-weight linear codes. We present below the main results in [1911].

We first consider linear codes obtained from cyclotomic classes. When D is as defined in (20.16), let

$$N_{a,b}(0) = \#\{ (x,y) \in D \mid \mathrm{Tr}_{q/p}(ax + by) = 0 \}.$$

Then the Hamming weight of the codeword $c(a,b) \in \mathcal{C}_D$ can be expressed as

$$\mathrm{wt_H}(c(a,b)) = n - N_{a,b}(0).$$

Theorem 20.6.4 ([1911]) *Let e be an integer and $q = p^m$ with $m = 2d\gamma$, where d is the least positive integer satisfying $p^d \equiv -1 \pmod{e}$. Then the linear code $\mathcal{C}_D$ defined by (20.15) and (20.16) is a five-weight $\left[\frac{(p^m-1)^2}{e^2}, 2m \right]_p$ code over $\mathbb{F}_p$ with the weight distribution in Table 20.18.*

For the special case $e = 2$, Wu et al. have derived the following result.

TABLE 20.18: Weight distribution of $\mathcal{C}_D$ in Theorem 20.6.4 for p odd

Hamming weight w	Multiplicity A_w
0	1
$\frac{(p-1)(p^m-1)}{e^2}\left(p^{m-1}+(-1)^\gamma(e-1)p^{d\gamma-1}\right)$	$\frac{2(p^m-1)}{e}$
$\frac{(p-1)(p^m-1)}{e^2}\left(p^{m-1}-(-1)^\gamma p^{d\gamma-1}\right)$	$(e-1)\frac{2(p^m-1)}{e}$
$\frac{(p-1)}{e^2}\left(p^{2m-1}-(e^2-2e+3)p^{m-1}+2(-1)^{\gamma+1}(e-1)p^{d\gamma-1}\right)$	$\frac{(p^m-1)^2}{e^2}$
$\frac{(p-1)}{e^2}\left(p^{2m-1}-3p^{m-1}+2(-1)^\gamma p^{d\gamma-1}\right)$	$(e-1)^2\frac{(p^m-1)^2}{e^2}$
$\frac{(p-1)}{e^2}\left(p^{2m-1}+(e-3)p^{m-1}+(-1)^{\gamma+1}(e-2)p^{d\gamma-1}\right)$	$2(e-1)\frac{(p^m-1)^2}{e^2}$

TABLE 20.19: Weight distribution of $\mathcal{C}_D$ in Theorem 20.6.5 for m even, p odd, and $e=2$

Hamming weight w	Multiplicity A_w
0	1
$\frac{(p-1)(p^m-1)p^{m-1}}{4}$	$\frac{(p^m-1)^2}{2}$
$\frac{(p-1)(p^{2m-1}-3p^{m-1}-2p^{\frac{m}{2}-1})}{4}$	$\frac{(p^m-1)^2}{4}$
$\frac{(p-1)(p^{2m-1}-3p^{m-1}+2p^{\frac{m}{2}-1})}{4}$	$\frac{(p^m-1)^2}{4}$
$\frac{(p-1)(p^m-1)(p^{m-1}-p^{\frac{m}{2}-1})}{4}$	p^m-1
$\frac{(p-1)(p^m-1)(p^{m-1}+p^{\frac{m}{2}-1})}{4}$	p^m-1

Theorem 20.6.5 ([1911]) *Let the linear code $\mathcal{C}_D$ be defined as (20.15) and (20.16) with $e=2$. Then $\mathcal{C}_D$ is a five-weight $\left[\frac{(p^m-1)^2}{4},2m\right]_p$ code over $\mathbb{F}_p$ with the weight distribution in Table 20.19 if m is even, and otherwise $\mathcal{C}_D$ is a two-weight $\left[\frac{(p^m-1)^2}{4},2m\right]_p$ code over $\mathbb{F}_p$ with the weight distribution in Table 20.20 if m is odd.*

Now we consider linear codes derived from weakly regular bent functions. Let $\mathcal{RF}$ be the set of p-ary weakly regular bent functions $g:\mathbb{F}_q\to\mathbb{F}_p$ with $g(0)=0$ and $g(ax)=a^{h_g}g(x)$ for any $a\in\mathbb{F}_p^*$ and $x\in\mathbb{F}_q$, where h_g is a positive even integer with $\gcd(h_g-1,p-1)=1$. It is known that if $g(x)\in\mathcal{RF}$ and $g^\star(x)$ is its dual function, then $g^\star(0)=0$ and there exists a positive even integer l_g with $\gcd(l_g-1,p-1)=1$ such that $g^\star(ax)=a^{l_f}g^\star(x)$ for any $a\in\mathbb{F}_p^*$ and $x\in\mathbb{F}_q$.

Using this notation we present two results on linear codes $\mathcal{C}_D$ defined in (20.15) and (20.17), where the first theorem is valid for $f(x)=\mathrm{Tr}_{q/p}(x)$ and $g(y)\in\mathcal{RF}$, and the second

TABLE 20.20: Weight distribution of $\mathcal{C}_D$ in Theorem 20.6.5 for m odd, p odd, and $e=2$

Hamming weight w	Multiplicity A_w
0	1
$\frac{(p-1)(p^{2m-1}-3p^{m-1})}{4}$	$\frac{(p^m-1)^2}{2}$
$\frac{(p-1)(p^{2m-1}-p^{m-1})}{4}$	$\frac{(p^m-1)(p^m+3)}{2}$

TABLE 20.21: Weight distribution of $\mathcal{C}_D$ in Theorem 20.6.6 for m even and p odd

Hamming weight w	Multiplicity A_w
0	1
$(p-1)(p^{2m-2} - \epsilon_g p^{m-2}\sqrt{p^{*m}})$	$(p-1)(p^{m-1} + \frac{p-1}{p}\epsilon_g\sqrt{p^{*m}})$
$(p-1)p^{2m-2}$	$(p^m - p + 1)p^m - 1$
$(p-1)p^{2m-2} + \epsilon_g p^{m-2}\sqrt{p^{*m}}$	$(p-1)^2(p^{m-1} - \frac{1}{p}\epsilon_g\sqrt{p^{*m}})$

TABLE 20.22: Weight distribution of $\mathcal{C}_D$ in Theorem 20.6.6 for m and p odd

Hamming weight w	Multiplicity A_w
0	1
$(p-1)p^{2m-2} - p^{m-2}p^{\frac{m+1}{2}}$	$\frac{(p-1)^2}{2}(p^{m-1} + p^{\frac{m-1}{2}})$
$(p-1)p^{2m-2}$	$p^{2m} - 1 - (p-1)^2 p^{m-1}$
$(p-1)p^{2m-2} + p^{m-2}p^{\frac{m+1}{2}}$	$\frac{(p-1)^2}{2}(p^{m-1} - p^{\frac{m-1}{2}})$

is valid when both $f(x)$ and $g(y)$ are in $\mathcal{RF}$. Let ϵ_g denote the sign of the Walsh transform of $g \in \mathcal{RF}$ (see (20.3)) and set $p^* = (-1)^{(p-1)/2}p$.

Theorem 20.6.6 ([1911]) *Let $f(x) = \mathrm{Tr}_{q/p}(x)$ and assume $g(y)$ is a weakly regular bent function in $\mathcal{RF}$. Then the linear code $\mathcal{C}_D$ defined by (20.15) and (20.17) is a three-weight $\left[p^{2m-1} - 1, 2m\right]_p$ code over $\mathbb{F}_p$ with the weight distribution in Table 20.21 when m is even and in Table 20.22 when m is odd.*

Theorem 20.6.7 ([1911]) *Let $f(x)$ and $g(y)$ be any two weakly bent functions in $\mathcal{RF}$ where l_f, l_g, ϵ_f, and ϵ_g are defined as before. If $l_f \neq l_g$ and m is even, then the linear code $\mathcal{C}_D$ defined by (20.15) and (20.17) is a three-weight $\left[p^{2m-1} + \frac{p-1}{p}\epsilon_f\epsilon_g(p^{*m} - 1, 2m\right]_p$ code over $\mathbb{F}_p$ with the weight distribution given in Table 20.23. If $l_f = l_g$, then $\mathcal{C}_D$ is a two-weight $\left[p^{2m-1} + \frac{p-1}{p}\epsilon_f\epsilon_g p^{*m} - 1, 2m\right]_p$ code over $\mathbb{F}_p$ with the weight distribution in Table 20.24.*

20.6.3 Codes with Few Weights from p-Ary Weakly Regular Bent Functions Based on the Second Generic Construction

Zhou, Li, Fan and Helleseth [1947] derived several classes of p-ary linear codes with two or three weights constructed from quadratic bent functions over $\mathbb{F}_p$ where p is an <u>odd</u> prime.

TABLE 20.23: Weight distribution of $\mathcal{C}_D$ in Theorem 20.6.7 for $l_f \neq l_g$ and p odd

i	Hamming weight ω_i	Multiplicity A_{ω_i}
0	0	1
1	$(p-1)p^{2m-2}$	$(p^{m-1} + \frac{p-1}{p}\epsilon_f\sqrt{p^{*m}})(p^{m-1} + \frac{p-1}{p}\epsilon_g\sqrt{p^{*m}}) - 1$
2	$(p-1)p^{2m-2} + \frac{p-3}{p}\epsilon_f\epsilon_g p^{*m}$	$\frac{(p-1)^2}{2}(p^{m-1} - \frac{1}{p}\epsilon_f\sqrt{p^{*m}})(p^{m-1} - \frac{1}{p}\epsilon_g\sqrt{p^{*m}})$
3	$(p-1)p^{2m-2} + \frac{p-1}{p}\epsilon_f\epsilon_g p^{*m}$	$p^{2m} - A_{\omega_1} - A_{\omega_2} - 1$

TABLE 20.24: Weight distribution of $\mathcal{C}_D$ in Theorem 20.6.7 for $l_f = l_g$ and p odd

Hamming weight w	Multiplicity A_w
0	1
$(p-1)p^{2m-2}$	$p^{2m-1} + \frac{p-1}{p}\epsilon_f\epsilon_g p^{*m} - 1$
$(p-1)p^{2m-2} + \frac{p-1}{p}\epsilon_f\epsilon_g p^{*m}$	$(p-1)(p^{2m-1} - \frac{1}{p}\epsilon_f\epsilon_g p^{*m})$

Let Q be a quadratic bent function from $\mathbb{F}_{p^m}$ to $\mathbb{F}_p$. Define $D_Q = \{x \in \mathbb{F}_{p^m}^* \mid Q(x) = 0\}$. Then if m is odd, we have $\#D_Q = p^{m-1} - 1$ and if m is even, we have $\#D_Q = p^{m-1} + \epsilon(p-1)p^{\frac{m-2}{2}}$ where $\epsilon \in \{-1, 1\}$. Denote by $\mathcal{C}_{D_Q}$ the linear code based on the second generic construction whose defining set is D_Q.

Theorem 20.6.8 ([1947]) *With the above notation the following hold. If m is odd, then $\mathcal{C}_{D_Q}$ is a three-weight linear code with parameters $[p^{m-1} - 1, m]_p$. If m is even, then $\mathcal{C}_{D_Q}$ is a two-weight linear code with parameters $\left[p^{m-1} + \epsilon(p-1)p^{\frac{m-2}{2}} - 1, m\right]_p$. The weight distributions and the description of ϵ are given in [1947].*

Tang, Li, Qi, Zhou and Helleseth [1778] generalized their approach to weakly regular bent functions. More precisely, they derived linear codes with two or three weights from the subclass $\mathcal{RF}$ of p-ary weakly regular bent functions defined in Section 20.6.2.

Given a p-ary function $f : \mathbb{F}_{p^m} \to \mathbb{F}_p$ in $\mathcal{RF}$ define $D_f := \{x \in \mathbb{F}_{p^m} \mid f(x) = 0\}$ and denote by $\mathcal{C}_{D_f}$ the related linear code based on the second generic construction whose defining set is D_f.

Theorem 20.6.9 ([1778]) *Let f be a function in $\mathcal{RF}$. If m is odd, then $\mathcal{C}_{D_f}$ is a three-weight linear code with parameters $[p^{m-1} - 1, m]_p$. If m is even, then $\mathcal{C}_{D_f}$ is a two-weight linear code with parameters $\left[p^{m-1} - 1 + \epsilon(p-1)p^{(m-2)/2}, m\right]_p$ where ϵ is the sign of the Walsh transform of f. The weight distributions are given in [1778].*

20.6.4 Codes with Few Weights from p-Ary Weakly Regular Plateaued Functions Based on the First Generic Construction

Let p be an <u>odd</u> prime. Let $\Psi : \mathbb{F}_{p^m} \to \mathbb{F}_{p^m}$ be a mapping with $\Psi(0) = 0$. We examine the code $\mathcal{C}_{\psi_1}$ given by (20.14) where $\psi_1(x) = \mathrm{Tr}_{p^m/p}(\Psi(x))$ is a weakly regular s-plateaued p-ary function with $0 \leq s \leq m - 2$.

Throughout this subsection $\eta_0 = (-1)^{(p-1)/2}$. The following theorem describes the Hamming weights of the codewords of $\mathcal{C}_{\psi_1}$.

Theorem 20.6.10 ([1381]) *Let $\mathcal{C}_{\psi_1}$ be the linear p-ary code defined by (20.14). Assume that $\psi_1 = \mathrm{Tr}_{p^m/p}(\Psi)$ is weakly regular p-ary s-plateaued with $0 \leq s \leq m - 2$. Let g be a p-ary function over $\mathrm{supp}(W_{\psi_1})$ as described in [1381] and [1018, Theorem 2]. Let $\epsilon = \pm 1$ be the sign of W_{ψ_1}. For $\alpha \in \mathbb{F}_p^*$, $\overline{\alpha}$ satisfies $\overline{\alpha}\alpha = 1$. The weight distribution of $\mathcal{C}_{\psi_1}$ is as follows.*

(a) $\mathrm{wt_H}(\widetilde{c}_{0,0}) = 0$ *and* $\mathrm{wt_H}(\widetilde{c}_{0,\beta}) = p^m - p^{m-1}$ *for* $\beta \in \mathbb{F}_{p^m}^*$.

(b) *For* $\alpha \in \mathbb{F}_p^*$ *and* $\beta \in \mathbb{F}_{p^m}$, *if* $\overline{\alpha}\beta \notin \mathrm{supp}(W_{\psi_1})$, *then* $\mathrm{wt_H}(\widetilde{c}_{\alpha,\beta}) = p^m - p^{m-1}$.

(c) *For* $\alpha \in \mathbb{F}_p^*$ *and* $\beta \in \mathbb{F}_{p^m}$, *if* $\overline{\alpha}\beta \in \mathrm{supp}(W_{\psi_1})$ *and* $m + s$ *is odd, then* $\mathrm{wt_H}(\widetilde{c}_{\alpha,\beta}) = p^m - p^{m-1} - \epsilon\eta_0^{\frac{m+s+1}{2}} p^{\frac{m+s-1}{2}}\left(\frac{g(\overline{\alpha}\beta)}{p}\right)$.

(d) *For $\alpha \in \mathbb{F}_p^*$ and $\beta \in \mathbb{F}_{p^m}$, if $\overline{\alpha}\beta \in \mathrm{supp}(W_{\psi_1})$ and $m + s$ is even, then*

$$\mathrm{wt_H}(\widetilde{c}_{\alpha,\beta}) = \begin{cases} p^m - p^{m-1} - \epsilon\eta_0^{\frac{m+s}{2}} p^{\frac{m+s-2}{2}} (p-1) & \text{if } g(\overline{\alpha}\beta) = 0, \\ p^m - p^{m-1} + \epsilon\eta_0^{\frac{m+s}{2}} p^{\frac{m+s-2}{2}} & \text{if } g(\overline{\alpha}\beta) \neq 0. \end{cases}$$

One can determine the weight distribution of the code $\mathcal{C}_{\psi_1}$ addressed by Theorem 20.6.10. We present separately the even case of $m + s$ and its odd case.

Theorem 20.6.11 ([1381]) *Let $\mathcal{C}_{\psi_1}$ be the three-weight code of Theorem 20.6.10. If $m + s$ is even, then $\mathcal{C}_{\psi_1}$ is a $[p^m - 1, m + 1, d]_p$ code over $\mathbb{F}_p$ where*

$$d = \begin{cases} p^m - p^{m-1} - (p-1)p^{\frac{m+s-2}{2}} & \text{if } \epsilon\eta_0^{(m+s)/2} = 1, \\ p^m - p^{m-1} - p^{\frac{m+s-2}{2}} & \text{if } \epsilon\eta_0^{(m+s)/2} = -1 \end{cases}$$

and weight distribution given in Table 20.25, noting again that $\epsilon = \pm 1$ is the sign of W_{ψ_1}.

TABLE 20.25: Weight distribution of $\mathcal{C}_{\psi_1}$ in Theorem 20.6.11 for $m + s$ even and p odd

Hamming weight w	Multiplicity A_w
0	1
$p^m - p^{m-1}$	$p^{m+1} - p^{m-s}(p-1) - 1$
$p^m - p^{m-1} - \epsilon\eta_0^{(m+s)/2} p^{\frac{m+s-2}{2}}(p-1)$	$p^{m-s-1}(p-1) + \epsilon p^{\frac{m-s-2}{2}}(p-1)^2$
$p^m - p^{m-1} + \epsilon\eta_0^{(m+s)/2} p^{\frac{m+s-2}{2}}$	$(p^{m-s} - p^{m-s-1})(p-1) - \epsilon p^{\frac{m-s-2}{2}}(p-1)^2$

As an example of Theorems 20.6.10 and 20.6.11, we give a weakly regular ternary 1-plateaued function and the corresponding ternary linear code. Recall that a function $g : \mathbb{F}_{p^m} \to \mathbb{F}_p$ is *balanced* over $\mathbb{F}_p$ if g takes every value of $\mathbb{F}_p$ the same number p^{m-1} times; otherwise, g is *unbalanced*.

Example 20.6.12 Let $\Psi : \mathbb{F}_{3^3} \to \mathbb{F}_{3^3}$ be the map defined by $\Psi(x) = \zeta^{22}x^{13} + \zeta^7 x^4 + \zeta x^2$ where $\mathbb{F}_{3^3}^* = \langle \zeta \rangle$ with $\zeta^3 + 2\zeta + 1 = 0$. The function $\psi_1(x) = \mathrm{Tr}_{3^3/3}(\Psi(x))$ is weakly regular ternary 1-plateaued with

$$W_{\psi_1}(\omega) \in \left\{ 0, \epsilon 3^2 \xi_3^{g(\omega)} \right\} = \{0, -9, -9\xi_3, -9\xi_3^2\}$$

for all $\omega \in \mathbb{F}_{3^3}$, where $\epsilon = -1$ and g is an unbalanced function over $\mathbb{F}_{3^3}$. The code $\mathcal{C}_{\psi_1}$ of Theorem 20.6.11 is a three-weight ternary linear $[26, 4, 15]_3$ code with Hamming weight enumerator $\mathrm{Hwe}_{\mathcal{C}_{\psi_1}}(x, y) = y^{26} + 16x^{15}x^{11} + 62x^{18}y^8 + 2x^{24}y^2$. This was verified by Magma [250].

The following theorem determines the weight distribution of $\mathcal{C}_{\psi_1}$ in Theorem 20.6.10 when $m + s$ is an odd integer.

Theorem 20.6.13 ([1381]) *Let $\mathcal{C}_{\psi_1}$ be the code of Theorem 20.6.10. If $m + s$ is odd, then $\mathcal{C}_{\psi_1}$ is a three-weight $[p^m - 1, m + 1, p^m - p^{m-1} - p^{(m+s-1)/2}]_p$ code over $\mathbb{F}_p$. $\mathcal{C}_{\psi_1}$ has weight distribution given in Table 20.26, noting again that $\epsilon = \pm 1$ is the sign of W_{ψ_1}. The number of minimum weight codewords depends on the sign of $\epsilon\eta_0^{(m+s+1)/2} = \pm 1$.*

TABLE 20.26: Weight distribution of $\mathcal{C}_{\psi_1}$ in Theorem 20.6.13 for $m + s$ and p odd

Hamming weight w	Multiplicity A_w
0	1
$p^m - p^{m-1}$	$p^{m+1} - p^{m-s-1}(p-1)^2 - 1$
$p^m - p^{m-1} - \epsilon\eta_0^{(m+s+1)/2}p^{\frac{m+s-1}{2}}$	$\frac{1}{2}\left(p^{m-s-1} + \epsilon p^{\frac{m-s-1}{2}}\right)(p-1)^2$
$p^m - p^{m-1} + \epsilon\eta_0^{(m+s+1)/2}p^{\frac{m+s-1}{2}}$	$\frac{1}{2}\left(p^{m-s-1} - \epsilon p^{\frac{m-s-1}{2}}\right)(p-1)^2$

20.6.5   Codes with Few Weights from p-Ary Weakly Regular Plateaued Functions Based on the Second Generic Construction

Let f be a p-ary function from $\mathbb{F}_q$ to $\mathbb{F}_p$ where $q = p^n$ and p is an <u>odd</u> prime. Define the set

$$D_f = \{x \in \mathbb{F}_q^* \mid f(x) = 0\}. \tag{20.18}$$

Let us denote $m = \#D_f$ and $D_f = \{d_1, d_2, \ldots, d_m\}$. Recall that the second generic construction of a linear code $\mathcal{C}_{D_f}$ with defining set D_f is

$$\mathcal{C}_{D_f} = \{\mathbf{c}_\beta = \left(\mathrm{Tr}_{p^n/p}(\beta d_1), \mathrm{Tr}_{p^n/p}(\beta d_2), \ldots, \mathrm{Tr}_{p^n/p}(\beta d_m)\right) \mid \beta \in \mathbb{F}_q\}.$$

Let us define the subset of the set of weakly regular unbalanced plateaued functions that we shall use to construct linear codes. Similar to the class $\mathcal{RF}$, denote by $\mathcal{WRP}$ the set of those weakly regular p-ary s-plateaued unbalanced functions $f : \mathbb{F}_q \to \mathbb{F}_p$, where $0 \leq s \leq n$, with $f(0) = 0$ such that there exists an even positive integer h_f with $\gcd(h_f - 1, p - 1) = 1$ where $f(ax) = a^{h_f}f(x)$ for any $a \in \mathbb{F}_p^*$ and $x \in \mathbb{F}_q$.

Throughout this subsection, as earlier, $\eta_0 = (-1)^{(p-1)/2}$ and $p^* = (-1)^{(p-1)/2}p$.

Theorem 20.6.14 ([1382]) *Let $n + s$ be an even integer and $f \in \mathcal{WRP}$. Then $\mathcal{C}_{D_f}$ is a three-weight linear code over $\mathbb{F}_p$ with parameters $\left[p^{n-1} - 1 + \epsilon\eta_0(p-1)\sqrt{p^*}^{n+s-2}, n\right]_p$. The Hamming weights of the codewords and the weight distribution of $\mathcal{C}_{D_f}$ are as in Table 20.27, where $\epsilon = \pm 1$ is the sign of W_f.*

TABLE 20.27: Weight distribution of $\mathcal{C}_{D_f}$ in Theorem 20.6.14 for $n + s$ even and p odd

Hamming weight w	Multiplicity A_w
0	1
$(p-1)\left(p^{n-2} + \epsilon(p-1)\sqrt{p^*}^{n+s-4}\right)$	$p^n - p^{n-s}$
$(p-1)p^{n-2}$	$p^{n-s-1} + \epsilon\eta_0^{n+1}(p-1)\sqrt{p^*}^{n-s-2} - 1$
$(p-1)\left(p^{n-2} + \epsilon\eta_0\sqrt{p^*}^{n+s-2}\right)$	$(p-1)\left(p^{n-s-1} - \epsilon\eta_0^{n+1}\sqrt{p^*}^{n-s-2}\right)$

Example 20.6.15 The function $f : \mathbb{F}_{3^8} \to \mathbb{F}_3$ defined by $f(x) = \mathrm{Tr}_{3^8/3}(\zeta x^4 + \zeta^{816}x^2)$, where $\mathbb{F}_{3^8}^* = \langle\zeta\rangle$ with $\zeta^8 + 2\zeta^5 + \zeta^4 + 2\zeta^2 + 2\zeta + 2 = 0$, is a quadratic 2-plateaued unbalanced function in the set $\mathcal{WRP}$ such that

$$W_f(\beta) \in \{0, \epsilon\eta_0^5 3^5 \xi_3^{g(\beta)}\} = \{0, 243, 243\xi_3, 243\xi_3^2\} \quad \text{for all } \beta \in \mathbb{F}_{3^8}$$

where g is an unbalanced 3-ary function with $g(0) = 0$. Also $\epsilon = \eta_0 = -1$. Then $\mathcal{C}_{D_f}$ is a three-weight ternary linear $[2348, 8, 1458]_3$ code with Hamming weight enumerator $\mathrm{Hwe}_{\mathcal{C}_{D_f}}(x, y) = y^{2348} + 260x^{1458}y^{890} + 5832x^{1566}y^{782} + 468x^{1620}y^{728}$, verified by Magma [250].

The following is the odd case of Theorem 20.6.14.

Theorem 20.6.16 ([1382]) *Let $n+s$ be an odd integer with $0 \leq s \leq n-3$ and $f \in \mathcal{WRP}$. Then $\mathcal{C}_{D_f}$ is a three-weight linear $\left[p^{n-1} - 1, n, (p-1)\left(p^{n-2} - p^{(n+s-3)/2}\right)\right]_p$ code over $\mathbb{F}_p$. The Hamming weights of the codewords and the weight distribution of $\mathcal{C}_{D_f}$ are as in Table 20.28, where $\epsilon = \pm 1$ is the sign of W_f.*

TABLE 20.28: Weight distribution of $\mathcal{C}_{D_f}$ in Theorem 20.6.16 for $n + s$ and p odd

Hamming weight w	Multiplicity A_w
0	1
$(p-1)p^{n-2}$	$p^n + p^{n-s-1} - p^{n-s} - 1$
$(p-1)\left(p^{n-2} - \epsilon\sqrt{p^*}^{n+s-3}\right)$	$\frac{p-1}{2}\left(p^{n-s-1} + \epsilon\eta_0^n\sqrt{p^*}^{n-s-1}\right)$
$(p-1)\left(p^{n-2} + \epsilon\sqrt{p^*}^{n+s-3}\right)$	$\frac{p-1}{2}\left(p^{n-s-1} - \epsilon\eta_0^n\sqrt{p^*}^{n-s-1}\right)$

Example 20.6.17 The function $f : \mathbb{F}_{3^3} \to \mathbb{F}_3$ defined by $f(x) = \mathrm{Tr}_{3^3/3}(x^4 + x^2)$, where $\mathbb{F}_{3^3}^* = \langle \zeta \rangle$ with $\zeta^3 + 2\zeta + 1 = 0$, is a quadratic bent function ($s = 0$) in the set $\mathcal{WRP}$ with

$$W_f(\beta) \in \{i3\sqrt{3}, i3\sqrt{3}\xi_3, i3\sqrt{3}\xi_3^2\} = \{6\xi_3 + 3, -3\xi_3 - 6, -3\xi_3 + 3\} \quad \text{for all} \quad \beta \in \mathbb{F}_{3^3},$$

where $\epsilon = \eta_0 = -1$. Then $\mathcal{C}_{D_f}$ is a three-weight ternary linear $[8, 3, 4]_3$ code with Hamming weight enumerator $\mathrm{Hwe}_{\mathcal{C}_{D_f}}(x, y) = y^8 + 12x^4y^4 + 8x^6y^2 + 6x^8$, verified by Magma [250].

Let $f \in \mathcal{WRP}$. For any $x \in \mathbb{F}_q$, $f(x) = 0$ if and only if $f(ax) = 0$ for every $a \in \mathbb{F}_p^*$. Then one can choose a subset $\overline{D}_f$ of the defining set D_f defined by (20.18) such that $\bigcup_{a\in\mathbb{F}_p^*} a\overline{D}_f$ is a partition of D_f; namely, $D_f = \mathbb{F}_p^*\overline{D}_f = \{a\overline{d} \mid a \in \mathbb{F}_p^* \text{ and } \overline{d} \in \overline{D}_f\}$, where for each pair of distinct elements $\overline{d}_1, \overline{d}_2 \in \overline{D}_f$ we have $\frac{\overline{d}_1}{\overline{d}_2} \notin \mathbb{F}_p^*$. This implies that the linear code $\mathcal{C}_{D_f}$ can be punctured into a shorter linear code $\mathcal{C}_{\overline{D}_f}$, which is based on the second generic construction with the defining set $\overline{D}_f$. Notice that for $\beta \in \mathbb{F}_q^*$, $\#\{x \in D_f \mid f(x) = 0 \text{ and } \mathrm{Tr}_{q/p}(\beta x) = 0\} = (p-1)\#\{x \in \overline{D}_f \mid f(x) = 0 \text{ and } \mathrm{Tr}_{q/p}(\beta x) = 0\}$.

Corollary 20.6.18 ([1382]) *The punctured version $\mathcal{C}_{\overline{D}_f}$ of the code $\mathcal{C}_{D_f}$ in Theorem 20.6.14 is a three-weight linear code with parameters $\left[(p^{n-1}-1)/(p-1) + \epsilon\eta_0\sqrt{p^*}^{n+s-2}, n\right]_p$ whose weight distribution is listed in Table 20.29, where $\epsilon = \pm 1$ is the sign of W_f.*

TABLE 20.29: Weight distribution of $\mathcal{C}_{\overline{D}_f}$ in Corollary 20.6.18 for $n + s$ even and p odd

Hamming weight w	Multiplicity A_w
0	1
$p^{n-2} + \epsilon(p-1)\sqrt{p^*}^{n+s-4}$	$p^n - p^{n-s}$
p^{n-2}	$p^{n-s-1} + \epsilon\eta_0^{n+1}(p-1)\sqrt{p^*}^{n-s-2} - 1$
$p^{n-2} + \epsilon\eta_0\sqrt{p^*}^{n+s-2}$	$(p-1)\left(p^{n-s-1} - \epsilon\eta_0^{n+1}\sqrt{p^*}^{n-s-2}\right)$

Example 20.6.19 The punctured version $\mathcal{C}_{\overline{D}_f}$ of $\mathcal{C}_{D_f}$ in Example 20.6.15 is a three-weight ternary $[1174, 8, 729]_3$ linear code with Hamming weight enumerator $\mathrm{Hwe}_{\mathcal{C}_{\overline{D}_f}}(x, y) = y^{1174} + 260x^{729}y^{445} + 5832x^{783}y^{391} + 468x^{810}y^{364}$.

Corollary 20.6.20 ([1382]) *The punctured version $\mathcal{C}_{\overline{D}_f}$ of the code $\mathcal{C}_{D_f}$ in Theorem 20.6.16 is a three-weight linear code with parameters $\left[(p^{n-1} - 1)/(p - 1), n, p^{n-2} - p^{(n+s-3)/2}\right]_p$ whose weight distribution is listed in Table 20.30, where $\epsilon = \pm 1$ is the sign of W_f.*

TABLE 20.30: Weight distribution of $\mathcal{C}_{\overline{D}_f}$ in Corollary 20.6.20 for $n + s$ and p odd

Hamming weight w	Multiplicity A_w
0	1
p^{n-2}	$p^n + p^{n-s-1} - p^{n-s} - 1$
$p^{n-2} - \epsilon\sqrt{p^*}^{n+s-3}$	$\frac{p-1}{2}\left(p^{n-s-1} + \epsilon\eta_0^n\sqrt{p^*}^{n-s-1}\right)$
$p^{n-2} + \epsilon\sqrt{p^*}^{n+s-3}$	$\frac{p-1}{2}\left(p^{n-s-1} - \epsilon\eta_0^n\sqrt{p^*}^{n-s-1}\right)$

Example 20.6.21 The punctured version $\mathcal{C}_{\overline{D}_f}$ of $\mathcal{C}_{D_f}$ in Example 20.6.17 is a three-weight ternary $[4, 3, 2]_3$ linear code with Hamming weight enumerator $\mathrm{Hwe}_{\mathcal{C}_{\overline{D}_f}}(x, y) = 1 + 12x^2y^2 + 8x^3y + x^4$, as verified by `Magma` [250]. This code is optimal by the Singleton Bound.

We can work on the Walsh support of a weakly regular plateaued function f to define a subcode of each code constructed above. Consider therefore a linear code $\overline{\mathcal{C}}_{D_f}$ involving D_f defined by

$$\overline{\mathcal{C}}_{D_f} = \left\{\mathbf{c}_\beta = \left(Tr_{p^n/p}(\beta d_1), Tr_{p^n/p}(\beta d_2), \ldots, Tr_{p^n/p}(\beta d_m)\right) \mid \beta \in \mathrm{supp}(W_f)\right\} \quad (20.19)$$

which is a subcode of $\mathcal{C}_{D_f}$. Hence, the following codes in Corollaries 20.6.22, 20.6.23, 20.6.24, and 20.6.25 are subcodes of the codes in Theorems 20.6.14, 20.6.16 and Corollaries 20.6.18, and 20.6.20, respectively. Notice that their parameters are directly derived from those of the corresponding codes.

Corollary 20.6.22 ([1382]) *The subcode $\overline{\mathcal{C}}_{D_f}$ of the code $\mathcal{C}_{D_f}$ in Theorem 20.6.14 is a two-weight linear code with parameters $\left[p^{n-1} - 1 + \epsilon\eta_0(p - 1)\sqrt{p^*}^{n+s-2}, n - s\right]_p$ whose weight distribution is listed in Table 20.31.*

TABLE 20.31: Weight distribution of $\overline{\mathcal{C}}_{D_f}$ in Corollary 20.6.22 for $n + s$ even and p odd

Hamming weight w	Multiplicity A_w
0	1
$(p - 1)p^{n-2}$	$p^{n-s-1} + \epsilon\eta_0^{n+1}(p - 1)\sqrt{p^*}^{n-s-2} - 1$
$(p - 1)\left(p^{n-2} + \epsilon\eta_0\sqrt{p^*}^{n+s-2}\right)$	$(p - 1)\left(p^{n-s-1} - \epsilon\eta_0^{n+1}\sqrt{p^*}^{n-s-2}\right)$

TABLE 20.32: Weight distribution of $\overline{\mathcal{C}}_{D_f}$ in Corollary 20.6.23 for $n + s$ and p odd

Hamming weight w	Multiplicity A_w
0	1
$(p-1)p^{n-2}$	$p^{n-s-1} - 1$
$(p-1)\left(p^{n-2} - \epsilon\sqrt{p^*}^{n+s-3}\right)$	$\frac{p-1}{2}\left(p^{n-s-1} + \epsilon\eta_0^n\sqrt{p^*}^{n-s-1}\right)$
$(p-1)\left(p^{n-2} + \epsilon\sqrt{p^*}^{n+s-3}\right)$	$\frac{p-1}{2}\left(p^{n-s-1} - \epsilon\eta_0^n\sqrt{p^*}^{n-s-1}\right)$

Corollary 20.6.23 ([1382]) *The subcode $\overline{\mathcal{C}}_{D_f}$ of the code $\mathcal{C}_{D_f}$ in Theorem 20.6.16 is the three-weight linear code with parameters $\left[p^{n-1} - 1, n - s, (p-1)\left(p^{n-2} - p^{(n+s-3)/2}\right)\right]_p$ whose weight distribution is listed in Table 20.32.*

Corollary 20.6.24 ([1382]) *The subcode $\overline{\mathcal{C}}_{\overline{D}_f}$ of the code $\mathcal{C}_{\overline{D}_f}$ in Corollary 20.6.18 is a two-weight linear code with parameters $\left[(p^{n-1} - 1)/(p-1) + \epsilon\eta_0\sqrt{p^*}^{n+s-2}, n - s\right]_p$ whose weight distribution is listed in Table 20.33.*

TABLE 20.33: Weight distribution of $\overline{\mathcal{C}}_{\overline{D}_f}$ in Corollary 20.6.24 for $n + s$ even and p odd

Hamming weight w	Multiplicity A_w
0	1
p^{n-2}	$p^{n-s-1} + \epsilon\eta_0^{n+1}(p-1)\sqrt{p^*}^{n-s-2} - 1$
$p^{n-2} + \epsilon\eta_0\sqrt{p^*}^{n+s-2}$	$(p-1)\left(p^{n-s-1} - \epsilon\eta_0^{n+1}\sqrt{p^*}^{n-s-2}\right)$

Corollary 20.6.25 ([1382]) *The subcode $\overline{\mathcal{C}}_{\overline{D}_f}$ of the code $\mathcal{C}_{\overline{D}_f}$ in Corollary 20.6.20 is a three-weight linear code with parameters $\left[(p^{n-1} - 1)/(p-1), n - s, p^{n-2} - p^{(n+s-3)/2}\right]_p$ whose weight distribution is listed in Table 20.34.*

TABLE 20.34: Weight distribution of $\overline{\mathcal{C}}_{\overline{D}_f}$ in Corollary 20.6.25 for $n + s$ and p odd

Hamming weight w	Multiplicity A_w
0	1
p^{n-2}	$p^{n-s-1} - 1$
$p^{n-2} - \epsilon\sqrt{p^*}^{n+s-3}$	$\frac{p-1}{2}\left(p^{n-s-1} + \epsilon\eta_0^n\sqrt{p^*}^{n-s-1}\right)$
$p^{n-2} + \epsilon\sqrt{p^*}^{n+s-3}$	$\frac{p-1}{2}\left(p^{n-s-1} - \epsilon\eta_0^n\sqrt{p^*}^{n-s-1}\right)$

Below, we show that one can construct other linear codes. We shall push further the use of weakly regular plateaued functions in the recent construction methods proposed by Tang et al. [1778]. Let $f : \mathbb{F}_q \to \mathbb{F}_p$ be a p-ary function. Define the sets $D_{f,sq} = \{x \in \mathbb{F}_q \mid f(x) \in SQ\}$ and $D_{f,nsq} = \{x \in \mathbb{F}_q \mid f(x) \in NSQ\}$ where SQ and NSQ denote, respectively, the sets of squares and non-squares in $\mathbb{F}_p^*$. With a similar definition to the linear code $\mathcal{C}_{D_f}$, define a linear code $\mathcal{C}_{D_{f,sq}}$ involving $D_{f,sq} = \{d_1', d_2', \dots, d_m'\}$ where

$$\mathcal{C}_{D_{f,sq}} = \left\{ \mathbf{c}_\beta = \left(Tr_{p^n/p}(\beta d_1'), Tr_{p^n/p}(\beta d_2'), \dots, Tr_{p^n/p}(\beta d_m') \right) \mid \beta \in \mathbb{F}_q \right\} \tag{20.20}$$

and a linear code $\mathcal{C}_{D_{f,nsq}}$ involving $D_{f,nsq} = \{d_1'', d_2'', \dots, d_t''\}$ where

$$\mathcal{C}_{D_{f,nsq}} = \left\{ \mathbf{c}_\beta = \left(Tr_{p^n/p}(\beta d_1''), Tr_{p^n/p}(\beta d_2''), \dots, Tr_{p^n/p}(\beta d_t'') \right) \mid \beta \in \mathbb{F}_q \right\}. \tag{20.21}$$

Clearly, the codes $\mathcal{C}_{D_{f,sq}}$ and $\mathcal{C}_{D_{f,nsq}}$ have length m and t, respectively, and dimension at most n.

Theorem 20.6.26 ([1382]) *Let $n + s$ be an even integer and $f \in \mathcal{WRP}$. Then $\mathcal{C}_{D_{f,sq}}$ is a three-weight linear code with parameters $\left[\frac{p-1}{2} \left(p^{n-1} - \epsilon \eta_0 \sqrt{p^*}^{n+s-2} \right), n \right]_p$ where $\epsilon = \pm 1$ is the sign of W_f. The Hamming weights of the codewords and the weight distribution of $\mathcal{C}_{D_{f,sq}}$ are in Table 20.35.*

TABLE 20.35: Weight distribution of $\mathcal{C}_{D_{f,sq}}$ in Theorem 20.6.26 for $n + s$ even and p odd

Hamming weight w	Multiplicity A_w
0	1
$\frac{(p-1)^2}{2} \left(p^{n-2} - \epsilon \sqrt{p^*}^{n+s-4} \right)$	$p^n - p^{n-s}$
$\frac{(p-1)^2}{2} p^{n-2}$	$p^{n-s-1} + \frac{p-1}{2} \left(p^{n-s-1} + \epsilon \eta_0^{n+1} \sqrt{p^*}^{n-s-2} \right) - 1$
$(p-1) \left(\frac{p-1}{2} p^{n-2} - \epsilon \eta_0 \sqrt{p^*}^{n+s-2} \right)$	$\frac{p-1}{2} \left(p^{n-s-1} - \epsilon \eta_0^{n+1} \sqrt{p^*}^{n-s-2} \right)$

Remark 20.6.27 In Theorem 20.6.26, if $\epsilon \eta_0^{(n+s)/2} = 1$ and $p = 3$, then we have the condition $0 \leq s \leq n - 4$; otherwise, $0 \leq s \leq n - 2$ and $n \geq 3$. This guarantees that we have $\text{wt}_H(\mathbf{c}_\beta) > 0$ for each $\beta \in \mathbb{F}_q^*$, which confirms that the code $\mathcal{C}_{D_{f,sq}}$ has dimension n.

Example 20.6.28 The function $f : \mathbb{F}_{3^5} \to \mathbb{F}_3$ defined as $f(x) = Tr_{3^5/3}(\zeta^{19} x^4 + \zeta^{238} x^2)$, where $\mathbb{F}_{3^5}^* = \langle \zeta \rangle$ with $\zeta^5 + 2\zeta + 1 = 0$, is a quadratic 1-plateaued unbalanced function in the set $\mathcal{WRP}$ with

$$W_f(\beta) \in \left\{ 0, \epsilon \eta_0^3 (-1) 3^3 \xi_3^{g(\beta)} \right\} = \left\{ 0, -27, -27\xi_3, -27\xi_3^2 \right\} \text{ for all } \beta \in \mathbb{F}_{3^5}$$

where $\epsilon = 1$, $\eta_0 = -1$, and g is an unbalanced ternary function with $g(0) = 0$. Then $\mathcal{C}_{D_{f,sq}}$ is a three-weight ternary $[90, 5, 54]_3$ linear code with Hamming weight enumerator $\text{Hwe}_{\mathcal{C}_{D_{f,sq}}}(x, y) = y^{90} + 50x^{54}y^{36} + 162x^{60}y^{30} + 30x^{72}y^{18}$ as verified by Magma [250].

Theorem 20.6.29 ([1382]) *Let $n + s$ be an odd integer and $f \in \mathcal{WRP}$. Then $\mathcal{C}_{D_{f,sq}}$ is a three-weight linear code with parameters $\left[\frac{p-1}{2} \left(p^{n-1} + \epsilon \sqrt{p^*}^{n+s-1} \right), n \right]_p$ where $\epsilon = \pm 1$ is the sign of W_f. The Hamming weights of the codewords and the weight distribution of $\mathcal{C}_{D_{f,sq}}$ are in Table 20.36 and Table 20.37 when $p \equiv 1 \pmod 4$ and $p \equiv 3 \pmod 4$, respectively.*

Remark 20.6.30 In Theorem 20.6.29, if $p \equiv 3 \pmod 4$ and $\epsilon \eta_0^{(n+s-1)/2} = -1$ or $p \equiv 1 \pmod 4$ and $\epsilon = -1$, then we have the condition $0 \leq s \leq n - 3$; otherwise, $0 \leq s \leq n - 1$ and $n \geq 2$. This guarantees that we have $\text{wt}_H(\mathbf{c}_\beta) > 0$ for each $\beta \in \mathbb{F}_q^*$, which confirms that the code $\mathcal{C}_{D_{f,sq}}$ has dimension n.

TABLE 20.36: Weight distribution of $\mathcal{C}_{D_{f,sq}}$ in Theorem 20.6.29 for $p \equiv 1 \pmod 4$ and $n + s$ odd

Hamming weight w	Multiplicity A_w
0	1
$\frac{(p-1)^2}{2}p^{n-2}$	$p^{n-s-1} - 1$
$\frac{p-1}{2}\left((p-1)p^{n-2} + \epsilon(p+1)\sqrt{p}^{n+s-3}\right)$	$\frac{p-1}{2}\left(p^{n-s-1} + \epsilon\sqrt{p}^{n-s-1}\right)$
$\frac{(p-1)^2}{2}\left(p^{n-2} + \epsilon\sqrt{p}^{n+s-3}\right)$	$p^n - p^{n-s} + \frac{p-1}{2}\left(p^{n-s-1} - \epsilon\sqrt{p}^{n-s-1}\right)$

TABLE 20.37: Weight distribution of $\mathcal{C}_{D_{f,sq}}$ in Theorem 20.6.29 for $p \equiv 3 \pmod 4$ and $n + s$ odd

Hamming weight w	Multiplicity A_w
0	1
$\frac{(p-1)^2}{2}p^{n-2}$	$p^{n-s-1} - 1$
$\frac{(p-1)^2}{2}\left(p^{n-2} - \epsilon\sqrt{p^*}^{n+s-3}\right)$	$p^n - p^{n-s} + \frac{p-1}{2}\left(p^{n-s-1} + \epsilon(-1)^n\sqrt{p^*}^{n-s-1}\right)$
$\frac{p-1}{2}\left((p-1)p^{n-2} - \epsilon(p+1)\sqrt{p^*}^{n+s-3}\right)$	$\frac{p-1}{2}\left(p^{n-s-1} - \epsilon(-1)^n\sqrt{p^*}^{n-s-1}\right)$

Theorem 20.6.31 ([1382]) *Let $n + s$ be an odd integer and $f \in \mathcal{WRP}$. Then $\mathcal{C}_{D_{f,nsq}}$ is a three-weight linear code with parameters $\left[\frac{p-1}{2}\left(p^{n-1} - \epsilon\sqrt{p^*}^{n+s-1}\right), n\right]_p$ where $\epsilon = \pm 1$ is the sign of W_f. The Hamming weights of the codewords and the weight distribution of $\mathcal{C}_{D_{f,nsq}}$ are in Table 20.38 and Table 20.39 when $p \equiv 1 \pmod 4$ and $p \equiv 3 \pmod 4$, respectively.*

TABLE 20.38: Weight distribution of $\mathcal{C}_{D_{f,nsq}}$ in Theorem 20.6.31 for $p \equiv 1 \pmod 4$ and $n + s$ odd

Hamming weight w	Multiplicity A_w
0	1
$\frac{(p-1)^2}{2}p^{n-2}$	$p^{n-s-1} - 1$
$\frac{(p-1)^2}{2}\left(p^{n-2} - \epsilon\sqrt{p}^{n+s-3}\right)$	$p^n - p^{n-s} + \frac{p-1}{2}\left(p^{n-s-1} + \epsilon\sqrt{p}^{n-s-1}\right)$
$\frac{p-1}{2}\left((p-1)p^{n-2} - \epsilon(p+1)\sqrt{p}^{n+s-3}\right)$	$\frac{p-1}{2}\left(p^{n-s-1} - \epsilon\sqrt{p}^{n-s-1}\right)$

TABLE 20.39: Weight distribution of $\mathcal{C}_{D_{f,nsq}}$ in Theorem 20.6.31 for $p \equiv 3 \pmod 4$ and $n + s$ odd

Hamming weight w	Multiplicity A_w
0	1
$\frac{(p-1)^2}{2}p^{n-2}$	$p^{n-s-1} - 1$
$\frac{p-1}{2}\left((p-1)p^{n-2} + \epsilon(p+1)\sqrt{p^*}^{n+s-3}\right)$	$\frac{p-1}{2}\left(p^{n-s-1} + \epsilon(-1)^n\sqrt{p^*}^{n-s-1}\right)$
$\frac{(p-1)^2}{2}\left(p^{n-2} + \epsilon\sqrt{p^*}^{n+s-3}\right)$	$p^n - p^{n-s} + \frac{p-1}{2}\left(p^{n-s-1} - \epsilon(-1)^n\sqrt{p^*}^{n-s-1}\right)$

Remark 20.6.32 In Theorem 20.6.31, if $p \equiv 3 \pmod 4$ and $\epsilon\eta_0^{(n+s-1)/2} = 1$ or $p \equiv 1$ (mod 4) and $\epsilon = 1$, then we have the condition $0 \leq s \leq n - 3$; otherwise, $0 \leq s \leq n - 1$ and $n \geq 2$. This guarantees that we have $\mathrm{wt}_H(\mathbf{c}_\beta) > 0$ for each $\beta \in \mathbb{F}_q^*$, which confirms that the code $\mathcal{C}_{D_{f,nsq}}$ has dimension n.

Example 20.6.33 The function $f : \mathbb{F}_{3^6} \to \mathbb{F}_3$ defined as $f(x) = \mathrm{Tr}_{3^6/3}(\zeta x^4 + \zeta^{27}x^2)$, where $\mathbb{F}_{3^6}^* = \langle \zeta \rangle$ with $\zeta^6 + 2\zeta^4 + \zeta^2 + 2\zeta + 2 = 0$, is a quadratic 1-plateaued unbalanced function in the set $\mathcal{WRP}$ with

$$W_f(\beta) \in \{0, i27\sqrt{3}, i27\sqrt{3}\xi_3, i27\sqrt{3}\xi_3^2\} = \{0, 54\xi_3 + 27, -27\xi_3 - 54, -27\xi_3 + 27\}$$

for all $\beta \in \mathbb{F}_{3^6}$ where $\epsilon = \eta_0 = -1$ and g is an unbalanced ternary function with $g(0) = 0$. Then $\mathcal{C}_{D_{f,nsq}}$ is a three-weight ternary linear $[216, 6, 126]_3$ code with Hamming weight enumerator $\mathrm{Hwe}_{\mathcal{C}_{D_{f,nsq}}}(x, y) = y^{216} + 72x^{126}y^{90} + 576x^{144}y^{72} + 80x^{162}y^{54}$ again verified by Magma [250].

Remark 20.6.34 Let $n + s$ be an even integer and $f \in \mathcal{WRP}$. Then $\mathcal{C}_{D_{f,nsq}}$ is a three-weight linear code with the same parameters and weight distribution as $\mathcal{C}_{D_{f,sq}}$ of Theorem 20.6.26.

From each of the codes constructed above, we can obtain a shorter linear code as done earlier in this section. Indeed, the code $\mathcal{C}_{D_{f,sq}}$ can be punctured into a shorter one whose weight distribution is derived from that of $\mathcal{C}_{D_{f,sq}}$. Let $f \in \mathcal{WRP}$. For any $x \in \mathbb{F}_q$, $f(x)$ is a quadratic residue (respectively quadratic non-residue) in $\mathbb{F}_p^*$ if and only if $f(ax)$ is a quadratic residue (respectively quadratic non-residue) in $\mathbb{F}_p^*$ for every $a \in \mathbb{F}_p^*$. Thus one can choose a subset $\overline{D}_{f,sq}$ of $D_{f,sq}$ such that $\bigcup_{a \in \mathbb{F}_p^*} a\overline{D}_{f,sq}$ is a partition of $D_{f,sq}$. A similar subset $\overline{D}_{f,nsq}$ of $D_{f,nsq}$ can be chosen. So one can easily obtain the punctured versions $\mathcal{C}_{\overline{D}_{f,sq}}$ and $\mathcal{C}_{\overline{D}_{f,nsq}}$ of the codes $\mathcal{C}_{D_{f,sq}}$ and $\mathcal{C}_{D_{f,nsq}}$, respectively, whose parameters are derived directly from those of the original codes. Corollaries 20.6.35, 20.6.37, and 20.6.38 below follow directly from Theorems 20.6.26, 20.6.29, and 20.6.31, respectively.

Corollary 20.6.35 ([1382]) *The punctured version* $\mathcal{C}_{\overline{D}_{f,sq}}$ *of the code* $\mathcal{C}_{D_{f,sq}}$ *in Theorem 20.6.26 is a three-weight linear code with parameters* $\left[\frac{1}{2}(p^{n-1} - \epsilon\eta_0\sqrt{p^*}^{n+s-2}), n\right]_p$ *and weight distribution in Table 20.40.*

TABLE 20.40: Weight distribution of $\mathcal{C}_{\overline{D}_{f,sq}}$ in Corollary 20.6.35 for $n + s$ even and p odd

Hamming weight w	Multiplicity A_w
0	1
$\frac{p-1}{2}(p^{n-2} - \epsilon\sqrt{p^*}^{n+s-4})$	$p^n - p^{n-s}$
$\frac{(p-1)}{2}p^{n-2}$	$p^{n-s-1} + \frac{p-1}{2}(p^{n-s-1} + \epsilon\eta_0^{n+1}\sqrt{p^*}^{n-s-2}) - 1$
$\frac{p-1}{2}p^{n-2} - \epsilon\eta_0\sqrt{p^*}^{n+s-2}$	$\frac{p-1}{2}(p^{n-s-1} - \epsilon\eta_0^{n+1}\sqrt{p^*}^{n-s-2})$

Example 20.6.36 The punctured version $\mathcal{C}_{\overline{D}_{f,sq}}$ of $\mathcal{C}_{D_{f,sq}}$ in Example 20.6.28 is a three-weight ternary linear $[45, 5, 27]_3$ code with Hamming weight enumerator $\mathrm{Hwe}_{\mathcal{C}_{\overline{D}_{f,sq}}}(x, y) = y^{45} + 50x^{27}y^{18} + 162x^{30}y^{15} + 30x^{36}y^9$. This code is minimal by Theorem 20.6.47 and is almost optimal by the Griesmer Bound.

Corollary 20.6.37 ([1382]) *The punctured version $\mathcal{C}_{\overline{D}_{f,sq}}$ of the code $\mathcal{C}_{D_{f,sq}}$ in Theorem 20.6.29 is a three-weight linear code with parameters $\left[\frac{1}{2}\left(p^{n-1}+\epsilon\sqrt{p^{*}}^{n+s-1}\right),n\right]_{p}$ and weight distribution in Table 20.41 and Table 20.42 when $p\equiv 1\pmod 4$ and $p\equiv 3\pmod 4$, respectively.*

TABLE 20.41: Weight distribution of $\mathcal{C}_{\overline{D}_{f,sq}}$ in Corollary 20.6.37 for $p\equiv 1\pmod 4$ and $n+s$ odd

Hamming weight w	Multiplicity A_w
0	1
$\frac{p-1}{2}p^{n-2}$	$p^{n-s-1}-1$
$\frac{1}{2}\left((p-1)p^{n-2}+\epsilon(p+1)\sqrt{p}^{n+s-3}\right)$	$\frac{p-1}{2}\left(p^{n-s-1}+\epsilon\sqrt{p}^{n-s-1}\right)$
$\frac{(p-1)}{2}\left(p^{n-2}+\epsilon\sqrt{p}^{n+s-3}\right)$	$p^{n}-p^{n-s}+\frac{p-1}{2}\left(p^{n-s-1}-\epsilon\sqrt{p}^{n-s-1}\right)$

TABLE 20.42: Weight distribution of $\mathcal{C}_{\overline{D}_{f,sq}}$ in Corollary 20.6.37 for $p\equiv 3\pmod 4$ and $n+s$ odd

Hamming weight w	Multiplicity A_w
0	1
$\frac{p-1}{2}p^{n-2}$	$p^{n-s-1}-1$
$\frac{p-1}{2}\left(p^{n-2}-\epsilon\sqrt{p^{*}}^{n+s-3}\right)$	$p^{n}-p^{n-s}+\frac{p-1}{2}\left(p^{n-s-1}+\epsilon(-1)^{n}\sqrt{p^{*}}^{n-s-1}\right)$
$\frac{1}{2}\left((p-1)p^{n-2}-\epsilon(p+1)\sqrt{p^{*}}^{n+s-3}\right)$	$\frac{p-1}{2}\left(p^{n-s-1}-\epsilon(-1)^{n}\sqrt{p^{*}}^{n-s-1}\right)$

Corollary 20.6.38 ([1382]) *The punctured version $\mathcal{C}_{\overline{D}_{f,nsq}}$ of the code $\mathcal{C}_{D_{f,nsq}}$ in Theorem 20.6.31 is a three-weight linear code with parameters $\left[\frac{1}{2}\left(p^{n-1}-\epsilon\sqrt{p^{*}}^{n+s-1}\right),n\right]_{p}$ and weight distribution in Table 20.43 and Table 20.44 when $p\equiv 1\pmod 4$ and $p\equiv 3\pmod 4$, respectively.*

TABLE 20.43: Weight distribution of $\mathcal{C}_{\overline{D}_{f,nsq}}$ in Corollary 20.6.38 for $p\equiv 1\pmod 4$ and $n+s$ odd

Hamming weight w	Multiplicity A_w
0	1
$\frac{(p-1)}{2}p^{n-2}$	$p^{n-s-1}-1$
$\frac{p-1}{2}\left(p^{n-2}-\epsilon\sqrt{p}^{n+s-3}\right)$	$p^{n}-p^{n-s}+\frac{p-1}{2}\left(p^{n-s-1}+\epsilon\sqrt{p}^{n-s-1}\right)$
$\frac{1}{2}\left((p-1)p^{n-2}-\epsilon(p+1)\sqrt{p}^{n+s-3}\right)$	$\frac{p-1}{2}\left(p^{n-s-1}-\epsilon\sqrt{p}^{n-s-1}\right)$

Example 20.6.39 The punctured version $\mathcal{C}_{\overline{D}_{f,nsq}}$ of $\mathcal{C}_{D_{f,nsq}}$ in Example 20.6.33 is a three-weight ternary linear $[108,6,63]_{3}$ code with Hamming weight enumerator $\mathrm{Hwe}_{\mathcal{C}_{\overline{D}_{f,nsq}}}(x,y)=y^{108}+72x^{63}y^{45}+576x^{72}y^{36}+80x^{81}y^{27}$. This code is minimal by Theorem 20.6.47 and is almost optimal by the Griesmer Bound.

TABLE 20.44: Weight distribution of $\mathcal{C}_{\overline{D}_{f,nsq}}$ in Corollary 20.6.38 for $p \equiv 3 \pmod 4$ and $n+s$ odd

Hamming weight w	Multiplicity A_w
0	1
$\frac{p-1}{2}p^{n-2}$	$p^{n-s-1}-1$
$\frac{1}{2}\big((p-1)p^{n-2}+\epsilon(p+1)\sqrt{p^*}^{n+s-3}\big)$	$\frac{p-1}{2}\big(p^{n-s-1}+\epsilon(-1)^n\sqrt{p^*}^{n-s-1}\big)$
$\frac{p-1}{2}\big(p^{n-2}+\epsilon\sqrt{p^*}^{n+s-3}\big)$	$p^n-p^{n-s}+\frac{p-1}{2}\big(p^{n-s-1}-\epsilon(-1)^n\sqrt{p^*}^{n-s-1}\big)$

With a definition similar to that of the subcode $\overline{\mathcal{C}}_{D_f}$, we have a linear code involving $D_{f,sq}$ defined by

$$\overline{\mathcal{C}}_{D_{f,sq}} = \big\{ \mathbf{c}_\beta = \big(\mathrm{Tr}_{p^n/p}(\beta d'_1), \mathrm{Tr}_{p^n/p}(\beta d'_2), \ldots, \mathrm{Tr}_{p^n/p}(\beta d'_m)\big) \mid \beta \in \mathrm{supp}(W_f) \big\}$$

which is the subcode of $\mathcal{C}_{D_{f,sq}}$ defined by (20.20). The following codes in Corollaries 20.6.40, 20.6.41, 20.6.42, and 20.6.43 are subcodes of the codes in Theorems 20.6.26, 20.6.29 and Corollaries 20.6.35, and 20.6.37, respectively.

Corollary 20.6.40 ([1382]) *The subcode $\overline{\mathcal{C}}_{D_{f,sq}}$ of the code $\mathcal{C}_{D_{f,sq}}$ in Theorem 20.6.26 is a two-weight linear code with parameters $\big[\frac{p-1}{2}\big(p^{n-1}-\epsilon\eta_0\sqrt{p^*}^{n+s-2}\big), n-s\big]_p$ and weight distribution in Table 20.45.*

TABLE 20.45: Weight distribution of $\overline{\mathcal{C}}_{D_{f,sq}}$ in Corollary 20.6.40 for $n+s$ even and p odd

Hamming weight w	Multiplicity A_w
0	1
$\frac{(p-1)^2}{2}p^{n-2}$	$p^{n-s-1}+\frac{p-1}{2}\big(p^{n-s-1}+\epsilon\eta_0^{n+1}\sqrt{p^*}^{n-s-2}\big)-1$
$(p-1)\big(\frac{p-1}{2}p^{n-2}-\epsilon\eta_0\sqrt{p^*}^{n+s-2}\big)$	$\frac{p-1}{2}\big(p^{n-s-1}-\epsilon\eta_0^{n+1}\sqrt{p^*}^{n-s-2}\big)$

Corollary 20.6.41 ([1382]) *The subcode $\overline{\mathcal{C}}_{D_{f,sq}}$ of the code $\mathcal{C}_{D_{f,sq}}$ in Theorem 20.6.29 is a three-weight linear code with parameters $\big[\frac{p-1}{2}\big(p^{n-1}+\epsilon\sqrt{p^*}^{n+s-1}\big), n-s\big]_p$ and weight distribution in Table 20.46.*

TABLE 20.46: Weight distribution of $\overline{\mathcal{C}}_{D_{f,sq}}$ in Corollary 20.6.41 for $n+s$ and p odd

Hamming weight w	Multiplicity A_w
0	1
$\frac{(p-1)^2}{2}p^{n-2}$	$p^{n-s-1}-1$
$\frac{p-1}{2}\big((p-1)p^{n-2}+\epsilon(p^*+1)\sqrt{p^*}^{n+s-3}\big)$	$\frac{p-1}{2}\big(p^{n-s-1}+\epsilon\eta_0^n\sqrt{p^*}^{n-s-1}\big)$
$\frac{p-1}{2}\big((p-1)p^{n-2}+\epsilon(p^*-1)\sqrt{p^*}^{n+s-3}\big)$	$\frac{p-1}{2}\big(p^{n-s-1}-\epsilon\eta_0^n\sqrt{p^*}^{n-s-1}\big)$

TABLE 20.47: Weight distribution of $\overline{\mathcal{C}}_{\overline{D}_{f,sq}}$ in Corollary 20.6.42 for $n+s$ even and p odd

Hamming weight w	Multiplicity A_w
0	1
$\frac{(p-1)}{2}p^{n-2}$	$p^{n-s-1} + \frac{p-1}{2}\left(p^{n-s-1} + \epsilon\eta_0^{n+1}\sqrt{p^*}^{n-s-2}\right) - 1$
$\frac{p-1}{2}p^{n-2} - \epsilon\eta_0\sqrt{p^*}^{n+s-2}$	$\frac{p-1}{2}\left(p^{n-s-1} - \epsilon\eta_0^{n+1}\sqrt{p^*}^{n-s-2}\right)$

Corollary 20.6.42 ([1382]) *The subcode $\overline{\mathcal{C}}_{\overline{D}_{f,sq}}$ of the code $\mathcal{C}_{\overline{D}_{f,sq}}$ in Corollary 20.6.35 is a two-weight linear code with parameters $\left[\frac{1}{2}\left(p^{n-1} - \epsilon\eta_0\sqrt{p^*}^{n+s-2}\right), n-s\right]_p$ and weight distribution in Table 20.47.*

Corollary 20.6.43 ([1382]) *The subcode $\overline{\mathcal{C}}_{\overline{D}_{f,sq}}$ of the code $\mathcal{C}_{\overline{D}_{f,sq}}$ in Corollary 20.6.37 is a three-weight linear code with parameters $\left[\frac{1}{2}\left(p^{n-1} + \epsilon\sqrt{p^*}^{n+s-1}\right), n-s\right]_p$ and weight distribution in Table 20.48.*

TABLE 20.48: Weight distribution of $\overline{\mathcal{C}}_{\overline{D}_{f,sq}}$ in Corollary 20.6.43 for $n+s$ and p odd

Hamming weight w	Multiplicity A_w
0	1
$\frac{(p-1)}{2}p^{n-2}$	$p^{n-s-1} - 1$
$\frac{1}{2}\left((p-1)p^{n-2} + \epsilon(p^*+1)\sqrt{p^*}^{n+s-3}\right)$	$\frac{p-1}{2}\left(p^{n-s-1} + \epsilon\eta_0^n\sqrt{p^*}^{n-s-1}\right)$
$\frac{1}{2}\left((p-1)p^{n-2} + \epsilon(p^*-1)\sqrt{p^*}^{n+s-3}\right)$	$\frac{p-1}{2}\left(p^{n-s-1} - \epsilon\eta_0^n\sqrt{p^*}^{n-s-1}\right)$

With a similar definition for the subcode $\overline{\mathcal{C}}_{D_f}$ defined by (20.19), we have a linear code $\overline{\mathcal{C}}_{D_{f,nsq}}$ involving $D_{f,nsq}$ defined by

$$\overline{\mathcal{C}}_{D_{f,nsq}} = \left\{ \mathbf{c}_\beta = \left(\mathrm{Tr}_{p^n/p}(\beta d_1''), \mathrm{Tr}_{p^n/p}(\beta d_2''), \ldots, \mathrm{Tr}_{p^n/p}(\beta d_t'')\right) \mid \beta \in \mathrm{supp}(W_f)\right\},$$

which is the subcode of $\mathcal{C}_{D_{f,nsq}}$ defined by (20.21). Hence, the following codes in Corollaries 20.6.44 and 20.6.45 are subcodes of the codes constructed in Theorem 20.6.31 and Corollary 20.6.38, respectively.

Corollary 20.6.44 ([1382]) *The subcode $\overline{\mathcal{C}}_{D_{f,nsq}}$ of the code $\mathcal{C}_{D_{f,nsq}}$ in Theorem 20.6.31 is a three-weight linear code with parameters $\left[\frac{p-1}{2}\left(p^{n-1} - \epsilon\sqrt{p^*}^{n+s-1}\right), n-s\right]_p$ and weight distribution in Table 20.49.*

Corollary 20.6.45 ([1382]) *The subcode $\overline{\mathcal{C}}_{\overline{D}_{f,nsq}}$ of the code $\mathcal{C}_{\overline{D}_{f,nsq}}$ in Corollary 20.6.38 is a three-weight linear code with parameters $\left[\frac{1}{2}\left(p^{n-1} - \epsilon\sqrt{p^*}^{n+s-1}\right), n-s\right]_p$ and weight distribution in Table 20.50.*

When we assume only the weakly regular bent-ness in this subsection, we can obviously recover the linear codes obtained by Tang et al. [1778].

Codes with few weights from functions have some applications to the design of minimal codes for secret sharing schemes and for two-party computation.

TABLE 20.49: Weight distribution of $\overline{C}_{D_{f,nsq}}$ in Corollary 20.6.44 for $n+s$ and p odd

Hamming weight w	Multiplicity A_w
0	1
$\frac{p-1}{2}(p-1)p^{n-2}$	$p^{n-s-1}-1$
$\frac{p-1}{2}\left((p-1)p^{n-2}-\epsilon(p^*-1)\sqrt{p^*}^{\,n+s-3}\right)$	$\frac{p-1}{2}\left(p^{n-s-1}+\epsilon\eta_0^n\sqrt{p^*}^{\,n-s-1}\right)$
$\frac{p-1}{2}\left((p-1)p^{n-2}-\epsilon(p^*+1)\sqrt{p^*}^{\,n+s-3}\right)$	$\frac{p-1}{2}\left(p^{n-s-1}-\epsilon\eta_0^n\sqrt{p^*}^{\,n-s-1}\right)$

TABLE 20.50: Weight distribution of $\overline{C}_{\overline{D}_{f,nsq}}$ in Corollary 20.6.45 for $n+s$ and p odd

Hamming weight w	Multiplicity A_w
0	1
$\frac{1}{2}(p-1)p^{n-2}$	$p^{n-s-1}-1$
$\frac{1}{2}\left((p-1)p^{n-2}-\epsilon(p^*-1)\sqrt{p^*}^{\,n+s-3}\right)$	$\frac{p-1}{2}\left(p^{n-s-1}+\epsilon\eta_0^n\sqrt{p^*}^{\,n-s-1}\right)$
$\frac{1}{2}\left((p-1)p^{n-2}-\epsilon(p^*+1)\sqrt{p^*}^{\,n+s-3}\right)$	$\frac{p-1}{2}\left(p^{n-s-1}-\epsilon\eta_0^n\sqrt{p^*}^{\,n-s-1}\right)$

Definition 20.6.46 A vector **a** **covers** another vector **b** if the support of **a** contains the support of **b**. A nonzero codeword **a** of the code $\mathcal{C}$ defined over $\mathbb{F}_q$ is said to be **minimal** if **a** covers only the codewords $\beta\mathbf{a}$ for every $\beta \in \mathbb{F}_q$, but no other nonzero codeword of $\mathcal{C}$. The code $\mathcal{C}$ is said to be **minimal** if every nonzero codeword of $\mathcal{C}$ is minimal.

It is worth noting that determining the minimality of a linear code over finite fields has been an attractive research topic in coding theory. Almost all the codes presented in the previous sections are minimal, since they satisfy the sufficient condition of minimality given by Ashikhmin and Barg in the theorem below.

Theorem 20.6.47 ([67]) *All nonzero codewords of a linear code $\mathcal{C}$ over $\mathbb{F}_q$ are minimal if*

$$\frac{q-1}{q} < \frac{\mathrm{wt}_{\min}}{\mathrm{wt}_{\max}},$$

where $\mathrm{wt}_{\min}$ and $\mathrm{wt}_{\max}$ denote the minimum and maximum weights of nonzero codewords in $\mathcal{C}$, respectively.

In [943], Ding, Heng, and Zhou presented a necessary and sufficient condition for linear codes over finite fields to be minimal; see also [556] for minimal binary codes.

Theorem 20.6.48 ([943]) *Let $\mathcal{C}$ be a linear code in $\mathbb{F}_q^n$. Then $\mathcal{C}$ is minimal if and only if* $\sum_{\beta\in\mathbb{F}_q^*}\mathrm{wt}_{\mathrm{H}}(\mathbf{a}+\beta\mathbf{b}) \neq (q-1)\mathrm{wt}_{\mathrm{H}}(\mathbf{a}) - \mathrm{wt}_{\mathrm{H}}(\mathbf{b})$ *for any $\mathbb{F}_q$-linearly independent codewords* **a**, **b** $\in \mathcal{C}$.

Minimal codes are also explored in Chapters 18 and 19. In [556, 943] the authors presented an infinite family of minimal linear codes not satisfying the sufficient condition given by Ashikhmin and Barg [67] showing then that this condition is not necessary. See more on the connection between linear codes and secret sharing in Section 33.3.1.

Since minimal codes for secret sharing schemes have been nicely developed in [942], we only deal here with minimal codes for two-party computation.

Secure two-party computation (2PC) is a sub-problem of secure multi-party computation

(MPC) that has received special attention by researchers because of its close relation to many cryptographic tasks. The goal of 2PC is to create a generic protocol that allows two parties to jointly compute an arbitrary function on their inputs without sharing the value of their inputs with the opposing party. One of the most well-known examples of 2PC is Yao's millionaire problem, in which two parties, Alice and Bob, are millionaires who wish to determine who is wealthier without revealing their wealth. Formally, Alice has wealth a, Bob has wealth b, and they wish to compute $a \geq b$ without revealing the values a or b.

In [374], Chabanne, Cohen and Patey have introduced a protocol for secure two-party computation of linear functions in the semi-honest model, based on coding techniques. Their protocol uses q-ary minimal codes. We refer the reader to [374] for the actual details.

20.7 Optimal Linear Locally Recoverable Codes from p-Ary Functions

A code $\mathcal{C}$ is called a *locally recoverable code* (LRC) if every coordinate of the codeword $\mathbf{c}$ can be recovered from some subset of r other coordinates of $\mathbf{c}$. Such an LRC is said to have *locality* r. Mathematically, we have the following definition.

Definition 20.7.1 A code $\mathcal{C} \subseteq \mathbb{F}_{q^n}$ is a **locally recoverable code (LRC)** with **locality** r if for every $i \in \{1, \ldots, n\}$, there exists a subset $R_i \subseteq \{1, \ldots, n\} \setminus \{i\}$ with $|R_i| \leq r$ and a function ϕ_i such that, for every codeword $\mathbf{c} = (c_1, c_2, \ldots, c_n) \in \mathcal{C}$,

$$c_i = \phi_i(\{c_j \mid j \in R_i\}).$$

An $(n, k, r)_q$ LRC $\mathcal{C}$ over $\mathbb{F}_q$ is of code length n, cardinality q^k, and locality r.

LRCs have recently been a very attractive subject in research in coding theory, due to their theoretical appeal, and to applications in large-scale distributed storage systems, where a single storage node erasure is considered as a frequent error event. Distributed storage systems are the subject of Chapter 31 with applications of LRCs in such systems detailed in Section 31.3. Distributed storage systems and LRCs involving algebraic geometry codes are presented in Section 15.9.

The parameters of an $(n, k, r)_q$ LRC have been studied; the upper bound on d is sometimes called the Singleton Bound for LRC.

Theorem 20.7.2 ([830, 1489]) *Let $\mathcal{C}$ be an $(n, k, r)_q$ LRC of cardinality q^k over an alphabet of size q; then the minimum distance d and rate $\frac{k}{n}$ of $\mathcal{C}$ satisfy*

$$d \leq n - k - \lceil k/r \rceil + 2 \quad and \quad \frac{k}{n} \leq \frac{r}{r+1}.$$

Note that if $r = k$, then the upper bound given in the above theorem coincides with the well-known Singleton Bound, $d \leq n - k + 1$.

In view of the above upper bound on the minimal distance, *optimal/almost optimal* LRC are defined as follows.

Definition 20.7.3 LRCs for which $d = n - k - \lceil k/r \rceil + 2$ are called **optimal** codes. The code is **almost optimal** when the minimum distance differs by at most 1 from the optimal value.

We now recall the concept of *r-good polynomials*, which is the key ingredient for constructing optimal linear LRCs.

Definition 20.7.4 Let q be a power of a prime and n be a positive integer such that $q \geq n$. A polynomial F over $\mathbb{F}_q$ is said to be an **r-good polynomial**[4] if

1. the degree of F is $r + 1$, and

2. there exists a partition $A = \{A_1, \ldots, A_{\frac{n}{r+1}}\}$ of a set $A \subseteq \mathbb{F}_q$ of size n into sets of size $r + 1$ such that F is constant on each set A_i in the partition.

Good polynomials produce optimal LRCs according to the construction due to Tamo and Barg [1772], yielding so-called **Tamo–Barg codes**, given in the next theorem.

Theorem 20.7.5 ([1772]) *For $r \geq 1$, let $g(x)$ be an r-good polynomial over $\mathbb{F}_{p^s}$. Set $n = (r + 1)l$ and $k = rt$, where $t \leq l$. For $a = (a_{ij})_{i=0,\ldots,r-1;\ j=0,\ldots,t-1} \in \mathbb{F}_{p^s}^{rt} = \mathbb{F}_{p^s}^k$, let*

$$f_a(x) = \sum_{i=0}^{r-1}\sum_{j=0}^{t-1} a_{ij} g(x)^j x^i.$$

If $\{A_1, \ldots, A_l\}$ is the partition associated to $g(x)$, set $A = \bigcup_{i=1}^{l} A_i$ and define

$$\mathcal{C} = \left\{ \left(f_a(\beta)\right)_{\beta \in A} \mid a \in \mathbb{F}_{p^s}^k \right\}.$$

Then $\mathcal{C}$ is an optimal linear $(n, k, r)_{p^s}$ LRC over $\mathbb{F}_{p^s}$.

In the above theorem, the elements of the set A are called **localizations**, $a \in \mathbb{F}_{p^s}^k$ is the **information vector** for $\left(f_a(\beta)\right)_{\beta \in A}$, and the elements $\left(f_a(\beta)\right)_{\beta \in A}$ are called **symbols** of the codeword. The local recovery is accomplished as follows. Let $\mathbf{c} = (c_\beta)_{\beta \in A} \in \mathcal{C}$ where $c_\beta = f_a(\beta)$ for some $a \in A$. Suppose that the erased symbol in $\mathbf{c}$ corresponds to the localization $\alpha \in A_j$ where A_j is one of the sets in the partition A. $\{c_\beta \mid \beta \in A_j \setminus \{\alpha\}\}$ denote the remaining r symbols in the localizations of the set A_j. To find the value $c_\alpha = f_a(\alpha)$, find the unique polynomial $\delta(x)$ of degree less than r such that $\delta(\beta) = c_\beta$ for all $\beta \in A_j \setminus \{\alpha\}$, that is,

$$\delta(x) = \sum_{\beta \in A_j \setminus \{\alpha\}} c_\beta \prod_{\beta' \in A_j \setminus \{\alpha, \beta\}} \frac{x - \beta'}{\beta - \beta'} \tag{20.22}$$

and set $c_\alpha = \delta(\alpha)$. Hence, to find one erased symbol, we need to perform polynomial interpolation from r known symbols in its recovery set.

Example 20.7.6 ([1772]) We construct a $(9, 4, 2)_{13}$ LRC over $\mathbb{F}_{13}$ with $g(x) = x^3$. Since $g(1) = g(3) = g(9) = 1$, $g(2) = g(6) = g(5) = 8$, $g(4) = g(12) = g(10) = 12$, let $A = A_1 \cup A_2 \cup A_3$ where $A_1 = \{1, 3, 9\}$, $A_2 = \{2, 6, 5\}$, $A_3 = \{4, 12, 10\}$. If the information vector is $a = (a_{00}, a_{01}, a_{10}, a_{11})$, then $f_a(x) = \left(a_{00} + a_{01}g(x)\right) + x\left(a_{10} + a_{11}g(x)\right) = a_{00} + a_{10}x + a_{01}x^3 + a_{11}x^4$. Let $\mathbf{c} = \left(f_a(\beta)\right)_{\beta \in A}$ with $a = (1, 1, 1, 1)$. So $\mathbf{c} = \left(f_a(1), f_a(3), f_a(9) \mid f_a(2), f_a(6), f_a(5) \mid f_a(4), f_a(12), f_a(10)\right) = (4, 8, 7 \mid 1, 11, 2 \mid 0, 0, 0)$. Suppose $f_a(1)$ is erased. This erased symbol can be recovered by accessing the 2 other codeword symbols at the localization 3 and 9. Using (20.22), we find that $\delta(x) = 2x + 2$ and compute $\delta(1) = 4$, which is the required value.

---

[4]For simplicity, we write r-good polynomial instead of (r, n)-good polynomial.

20.7.1 Constructions of r-Good Polynomials for Optimal LRC Codes

In this subsection, we present r-good polynomials over $\mathbb{F}_{p^s}$ with $r = mp^t - 1$ where $\gcd(m, p) = 1$. We first remark that if F is an r-good polynomial over $\mathbb{F}_{p^s}$, then γF and $F - \alpha$ are again r-good polynomials over $\mathbb{F}_{p^s}$ for every $\gamma \in \mathbb{F}_{p^s}^*$ and every $\alpha \in \mathbb{F}_{p^s}$.

20.7.1.1 Good Polynomials from Power Functions

If $t = 0$ and $p^s \equiv 1 \pmod{m}$, then the p^s-ary power function[5]

$$G_\gamma(x) = \gamma x^m$$

is an r-good polynomial, where $\gamma \in \mathbb{F}_{p^s}^*$. Note that $\mathbb{F}_{p^s}^*$ can be split into pairwise disjoint multiplicative cosets of the form bU_m, where $b \in \mathbb{F}_{p^s}^*$ and $U_m = \{x \in \mathbb{F}_{p^s} \mid x^m = 1\}$. Observe next that, for every $x \in bU_m$, $G_\gamma(x) = G_\gamma(b)$; see [1772, Proposition 3.2].

20.7.1.2 Good Polynomials from Linearized Functions

If $t > 0$ and $m = 1$, then the p^s-ary linearized function

$$F_a(x) = \sum_{i=0}^{t} a_i x^{p^i} \tag{20.23}$$

is an r-good polynomial where $a = (a_0, \ldots, a_t) \in (\mathbb{F}_{p^s})^{t+1}$, $a_0 \neq 0$, and $a_t \neq 0$. Note that $\mathbb{F}_{p^s}$ can be split into pairwise disjoint additive cosets of the form $b + E_a$ where $b \in \mathbb{F}_{p^s}$ and $E_a = \{x \in \mathbb{F}_{p^s} \mid F_a(x) = 0\}$. Observe next that, for every $x \in b + E_a$, $F_a(x) = F_a(b)$; again see [1772, Proposition 3.2].

20.7.1.3 Good Polynomials from Function Composition

We present three r-good polynomials F, H, and I using function composition of linearized polynomials where again $r = mp^t - 1$. For each polynomial we assume $p^s \equiv 1 \pmod{m}$ where $m > 1$ and $U_m = \{x \in \mathbb{F}_{p^s} \mid x^m = 1\} = \{\alpha_1, \ldots, \alpha_m\}$.

If $t > 0$ and $p^t \equiv 1 \pmod{m}$, then the p^s-ary function

$$F(x) = \left(\sum_{i=0}^{t/e} a_i x^{p^{ei}} \right)^m$$

is an r-good polynomial where e is a divisor of t such that $p^e \equiv 1 \pmod{m}$, and $a_i \in \mathbb{F}_{p^s}$ satisfying $\sum_{i=0}^{t/e} a_i = 0$ with $a_0 \neq 0$ and $a_{t/e} \neq 0$. In fact, this $(mp^t - 1)$-good polynomial can be written as

$$F(x) = \prod_{i=1}^{m} \prod_{k \in K} (x + k + \alpha_i) = \prod_{k \in K} (x + k)^m = \left(\sum_{i=0}^{t/e} a_i x^{p^{ei}} \right)^m$$

where K is the set of roots of the linearized polynomial $\sum_{i=0}^{t/e} a_i x^{p^{ei}}$. Note that K is an additive subgroup of $\mathbb{F}_{p^s}$ that is closed under multiplication by $\mathbb{F}_{p^e}$, and $U_m \subseteq \mathbb{F}_{p^e} \subseteq K$; $\mathbb{F}_{p^e} \subseteq K$ follows because $1 \in K$ as $\sum_{i=0}^{t/e} a_i = 0$. See [1772, Theorem 3.3] for more details.

For the functions H and I, let $G_\gamma(x) = \gamma x^m$ and F_a be defined as in (20.23), and assume that $\mathbb{F}_{p^s}$ contains all the roots of F_a.

---

[5]See Example 15.9.9 for a similar use of a power function to construct an algebraic geometry LRC.

Set $H(x) = F_a(G_\gamma(x)) = \sum_{i=0}^{t} a_i \gamma^{p^i} x^{mp^i}$ where $t > 0$. Then H is an r-good polynomial over $\mathbb{F}_{p^s}$ if and only if $\{b \in \mathbb{F}_{p^s} \setminus E_a \mid b + E_a \subseteq \mathrm{Im}(G_\gamma)\}$ is nonempty where $E_a = \{x \in \mathbb{F}_{p^s} \mid F_a(x) = 0\}$ and $\mathrm{Im}(G_\gamma) = \{G_\gamma(x) \mid x \in \mathbb{F}_{p^s}\}$. Let cU_m be a multiplicative coset of U_m where $c \in \mathbb{F}_{p^s}^*$. Then the r-good polynomial $H(x)$ is constant on $\bigcup_{i=1}^{p^t} x_i U_m$, where each $x_i \in \mathbb{F}_{p^s}^*$, and $\{\gamma x_1^m, \dots, \gamma x_{p^t}^m\}$ is an additive coset of the form $b + E_a$. See [1280, Theorem 8] for more details.

Finally set $I(x) = G_1(F_a(x)) = \left(\sum_{i=0}^{t} a_i x^{p^i}\right)^m$. Then I is an r-good polynomial over $\mathbb{F}_{p^s}$ if and only if $\{b \in \mathbb{F}_{p^s}^* \mid bU_m \subseteq \mathrm{Im}(F_a)\}$ is nonempty where $\mathrm{Im}(F_a) = \{F_a(x) \mid x \in \mathbb{F}_{p^s}\}$. The r-good polynomial I is constant on $\bigcup_{i=1}^{m}(x_i + E_a)$ where $E_a = \{x \in \mathbb{F}_{p^s} \mid F_a(x) = 0\}$, and $\{F_a(x_1), \dots, F_a(x_m)\}$ is a coset of U_m. Again see [1280] for details.

20.7.1.4 Good Polynomials from Dickson Polynomials of the First Kind

For $b \in \mathbb{F}_q$ and integer $m \geq 1$, recall that the Dickson polynomial of the first kind of degree m over $\mathbb{F}_q$ is defined as follows:

$$D_{m,b}(x) = \sum_{j=0}^{\lfloor \frac{m}{2} \rfloor} \frac{m}{m-j} \binom{m-j}{j} (-b)^j x^{m-2j}. \tag{20.24}$$

In what follows let $U_m = \{x \in \mathbb{F}_q \mid x^m = 1\}$. We consider separately the cases q odd and q even; see [1281].

For q odd, let $b \in \mathbb{F}_q^*$ and let $m \geq 3$ be an integer satisfying $m \mid (q-1)$. Then the Dickson polynomial $D_{m,b}(x)$ is an $(m-1)$-good polynomial as we now describe. Suppose $b \in \xi^l U_m$ for some $l \in S = \{0, 1, \dots, \frac{q-1}{m} - 1\}$ where ξ is a primitive element of $\mathbb{F}_q$ and $\xi^i U_m$ is the multiplicative coset of U_m in $\mathbb{F}_q^*$. Then pairwise disjoint subsets of $\mathbb{F}_q$ with cardinality m such that $D_{m,b}$ is constant on each subset include

$$D_i = \{u + b \cdot u^{-1} \mid u \in \xi^i U_m\} \quad \text{for } i \in I \subseteq S$$

where

$$I = \begin{cases} \{0, 1, \dots, \frac{l}{2} - 1, l+1, l+2, \dots, \frac{l}{2} + \frac{q-1}{2m} - 1\} & \text{if } l \text{ is even}, \frac{q-1}{m} \text{ is even}, \\ \{0, 1, \dots, \frac{l}{2} - 1, l+1, l+2, \dots, \frac{l}{2} + \frac{q-1}{2m} - \frac{1}{2}\} & \text{if } l \text{ is even}, \frac{q-1}{m} \text{ is odd}, \\ \{0, 1, \dots, \frac{l-1}{2}, l+1, l+2, \dots, \frac{l}{2} + \frac{q-1}{2m} - \frac{1}{2}\} & \text{if } l \text{ is odd}, \frac{q-1}{m} \text{ is even}, \\ \{0, 1, \dots, \frac{l-1}{2}, l+1, l+2, \dots, \frac{l}{2} + \frac{q-1}{2m} - 1\} & \text{if } l \text{ is odd}, \frac{q-1}{m} \text{ is odd}. \end{cases} \tag{20.25}$$

The Dickson polynomial $D_{m,b}(x)$ is constant on exactly $l_{D_{m,b}}$ pairwise disjoint subsets with cardinality m where

$$l_{D_{m,b}} = \begin{cases} \frac{q-1}{2m} - 1 & \text{if } l \text{ is even}, \frac{q-1}{m} \text{ is even}, \\ \frac{q-1-m}{2m} & \text{if } l \text{ is even}, \frac{q-1}{m} \text{ is odd}, \\ \frac{q-1}{2m} & \text{if } l \text{ is odd}, \frac{q-1}{m} \text{ is even}, \\ \frac{q-1-m}{2m} & \text{if } l \text{ is odd}, \frac{q-1}{m} \text{ is odd}. \end{cases}$$

For q even, let $b \in \mathbb{F}_q^*$ and let $m \geq 2$ be an integer satisfying $m \mid (q-1)$. Then the Dickson polynomial $D_{m,b}(x)$ is an $(m-1)$-good polynomial as follows. Suppose $b \in \xi^l U_m$ for some $l \in S = \{0, 1, \dots, \frac{q-1}{m} - 1\}$ where ξ is a primitive element of $\mathbb{F}_q$ and $\xi^i U_m$ is the multiplicative coset of U_m in $\mathbb{F}_q^*$. Then pairwise disjoint subsets of $\mathbb{F}_q$ with cardinality m such that $D_{m,b}$ is constant on each subset include

$$D_i = \{u + b \cdot u^{-1} \mid u \in \xi^i U_m\} \quad \text{for } i \in I \subseteq S$$

where

$$I = \begin{cases} \{0,1,\ldots,\frac{l}{2}-1,l+1,l+2,\ldots,\frac{l}{2}+\frac{q-1}{2m}-\frac{1}{2}\} & \text{if } l \text{ is even}, \\ \{0,1,\ldots,\frac{l-1}{2},l+1,l+2,\ldots,\frac{l}{2}+\frac{q-1}{2m}-1\} & \text{if } l \text{ is odd}. \end{cases} \qquad (20.26)$$

20.7.1.5 Good Polynomials from the Composition of Functions Involving Dickson Polynomials

Let $q = p^s$. If $F : \mathbb{F}_q \to \mathbb{F}_q$, let $\text{Im}(F) = \{F(x) \mid x \in \mathbb{F}_q\}$ denote the image set of F. Let $D_{m,b}$ and F_a be defined as in (20.24) and (20.23), respectively. For $i \in \{0,1,\ldots,\frac{q-1}{m}-1\}$, let $D_i = \{u + b \cdot u^{-1} \mid u \in \xi^i U_m\}$ where ξ is a primitive element of $\mathbb{F}_q$. Suppose that $m \mid (p^s-1)$, $m \geq 3$, and $\mathbb{F}_q$ contains all the roots of F_a. We have the following [1281]:

- $H(x) = D_{m,b} \circ F_a(x)$ is an (mp^t-1)-good polynomial over $\mathbb{F}_q$ if and only if the set $\{i \in \{0,1,\ldots,q-2\} \mid i \bmod \frac{q-1}{m} \in I, \ D_i \subseteq \text{Im}(F_a)\}$ is nonempty, where I is defined in (20.25) and (20.26) for q odd and q even, respectively.

- $H'(x) = F_a \circ D_{m,b}(x)$ is an (mp^t-1)-good polynomial over $\mathbb{F}_q$ if and only if the set $\{c \in \mathbb{F}_q \mid c + E_a \subseteq D_{m,b}(\bigcup_{i \in I} D_i)\}$ is nonempty where $E_a = \{x \in \mathbb{F}_{p^s} \mid F_a(x) = 0\}$ and I is defined in (20.25) and (20.26) for q odd and q even, respectively. Here $D_{m,b}(\bigcup_{i \in I} D_i) = \{D_{m,b}(x) \mid x \in \bigcup_{i \in I} D_i\}$.

We now present several examples of new r-good polynomials H over $\mathbb{F}_{p^s}$ with $r = mp^t-1$ where $m > 1$, $\gcd(m,p) = 1$, $p^t \not\equiv 1 \pmod{m}$, and $p^s \equiv 1 \pmod{m}$. Let l_H be the number of pairwise disjoint subsets with cardinality $r+1$ on which H is constant. The examples were found using `Magma` [250].

Example 20.7.7 Let $F_a(x) = x^{p^t} - a^{p^t-1}x$ where a is a primitive element of $\mathbb{F}_{p^s}$. Let $p = 3$, $t = 1$, and $m = 4$. Define

$$H(x) = D_{m,-a} \circ F_a(x) = (x^3 - a^2 x)^4 + a(x^3 - a^2 x)^2 + 2a^2.$$

- If $s = 4$, then $H(x)$ is an 11-good polynomial on $\mathbb{F}_{3^4}$ and $l_H = 2$.

- If $s = 6$, then $H(x)$ is an 11-good polynomial on $\mathbb{F}_{3^6}$ and $l_H = 11$.

- If $s = 8$, then $H(x)$ is an 11-good polynomial on $\mathbb{F}_{3^8}$ and $l_H = 95$.

- If $s = 10$, then $H(x)$ is an 11-good polynomial on $\mathbb{F}_{3^{10}}$ and $l_H = 803$.

Example 20.7.8 Let $F_a(x) = x^{p^t} - a^{p^t-1}x$ where a is a primitive element of $\mathbb{F}_{p^s}$. Let $p = 3$, $t = 1$, and $m = 4$. Define

$$H(x) = F_a \circ D_{m,-a}(x) = (x^4 + ax^2 + 2a^2)^3 - a^2(x^4 + ax^2 + 2a^2).$$

- If $s = 8$, then $H(x)$ is an 11-good polynomial on $\mathbb{F}_{3^8}$ and $l_H = 6$.

- If $s = 10$, then $H(x)$ is an 11-good polynomial on $\mathbb{F}_{3^{10}}$ and $l_H = 34$.

Very recently, Micheli [1385] provided a Galois theoretical framework which allows production of good polynomials and showed that the construction of good polynomials can be reduced to a Galois theoretical problem over global function fields.

Acknowledgement

The author is deeply grateful to Claude Carlet for his very careful reading, valuable comments and precious information (in particular on Kerdock codes) and to Cunsheng Ding for his great help in bringing material in this chapter and suggesting improvements. She is also indebted to Cary Huffman for his precious comments and great help in improving this chapter.

Chapter 21

Codes over Graphs

Christine A. Kelley

University of Nebraska-Lincoln

21.1   Introduction

Due to advances in graph-based decoding algorithms, codes over graphs have been shown to be capacity-achieving on many communication channels and are now ubiquitous in industry applications such as hard disk drives, flash drives, wireless standards (IEEE 802.11n, IEEE 802.16e, WiMax) [292, 409, 1589, 1590, 1681, 1712], and deep space communication. While any linear code gives rise to a natural graph representation, the phrase "codes over graphs" is typically reserved for codes whose corresponding graph representation aids in the decoding or understanding of those codes. Perhaps the most common example is the family of low-density parity check (LDPC) codes [787] that have become standard error-correcting codes in modern wireless and storage devices. LDPC codes are characterized by having sparse graph representations that allow for efficient and practical implementation of their iterative decoders. Although the concepts of iterative decoding on sparse graph representations of codes was introduced much earlier [787, 1784], they remained largely unexplored until [1316] when improvements in computational power made them more attractive for longer practical block lengths. Other prominent families of graph-based codes include turbo codes, repeat accumulate codes, fountain codes, Raptor codes, low-density generator-matrix codes, and others [183, 604, 1044, 1292, 1296, 1687]. Fountain and Raptor codes are described in Chapter 30.

The following definition presents terminology that we will use in this chapter.

Definition 21.1.1 A **graph** $\mathcal{G}$ is a pair (V, E) where V is a set of points called **vertices**, or **nodes**, and E is a collection of sets of cardinality two of vertices from V called **edges**. Unless otherwise specified, the graphs in this chapter are assumed to be **simple**, meaning that there are no loops or multiedges. Namely, if $\{v_i, v_j\} \in E$, then $v_i \neq v_j$, and the pairs in E are distinct. A graph is **bipartite** if the vertices can be partitioned in two sets V_1 and V_2 such that any edge in E contains one element from V_1 and one from V_2. A vertex v has **degree** r if v belongs to r pairs in E. A graph is r-**regular**, or **regular**, if every vertex has the same degree r. A **subgraph** of $\mathcal{G}$ is a graph (V', E') where $V' \subseteq V$ and E' is a set of unordered pairs of V' such that $E' \subseteq E$. If S is a subset of V, the subgraph **induced** by S in $\mathcal{G}$ is the pair (S, E_S) where $\{v_i, v_j\} \in E_S$ if and only if $\{v_i, v_j\} \in E$ and $v_i, v_j \in S$.

21.2 Low-Density Parity Check Codes

The idea of using sparse matrices for codes dates back to Gallager in 1963 [787] when he analyzed random ensembles of regular LDPC codes and provided the first iterative decoding algorithm for these codes. This iterative principle has since been extended to many other engineering problems; see for example [886].

Definition 21.2.1 A **low-density parity check (LDPC) code** $\mathcal{C}$ is a linear block code that is specified by a parity check matrix H that is sparse in the number of its nonzero entries (thereby giving its name). An LDPC code is said to be (j, r)-**regular**, or simply **regular**) if the parity check matrix H has a fixed number j of nonzero entries per column and a fixed number r of nonzero entries per row, and is said to be **irregular** otherwise.

Example 21.2.2 (Gallager ensemble) A regular (n, j, r) ensemble of LDPC codes is defined by a family of $m \times n$ parity check matrices H (where $m = nj/r$)

$$H = \begin{bmatrix} 1 & 1 & \cdots & 1 & 0 & 0 & \cdots & 0 & 0 & 0 & \cdots & 0 & 0 & \cdots & 0 \\ 0 & 0 & \cdots & 0 & 1 & 1 & \cdots & 1 & 0 & 0 & \cdots & 0 & 0 & \cdots & 0 \\ 0 & 0 & \cdots & 0 & 0 & 0 & \cdots & 0 & 1 & 1 & \cdots & 1 & 0 & \cdots & 0 \\ \vdots & \vdots & \vdots & \vdots & \vdots & \vdots & & \vdots & \vdots & \vdots & & \vdots & \vdots & & \vdots \\ & & & & & & P_2 & & & & & & & & \\ & & & & & & P_3 & & & & & & & & \\ & & & & & & \vdots & & & & & & & & \\ & & & & & & P_j & & & & & & & & \end{bmatrix},$$

where the first row has r ones in the first r columns, the second row has r ones in the next r columns, and so on until the $(n/r)^{\text{th}}$ row that has r ones in the last r columns. The second set of n/r rows of a matrix from this ensemble are a random column permutation of the first n/r rows of the matrix and denoted by P_2. The third set of n/r rows are defined by another random column permutation of the first n/r rows, and denoted by P_3, and so on, until P_j.

Gallager showed that a code chosen at random from this ensemble is asymptotically good with high probability – meaning, the minimum distance of such a code is increasing linearly with the block length n and the design rate of the code is $1 - \frac{j}{r} > 0$, as $n \to \infty$. Specifically, he provided the rate/distance tradeoff in Figure 21.1.

In another landmark paper, Tanner [1784] initiated the area of designing codes from graphs and generalized the constraints that could be used in linear codes. Consequently, the resulting graph representation for linear codes is commonly referred to as a *Tanner graph*, described next.

FIGURE 21.1: Relative minimum distance and code rate tradeoff

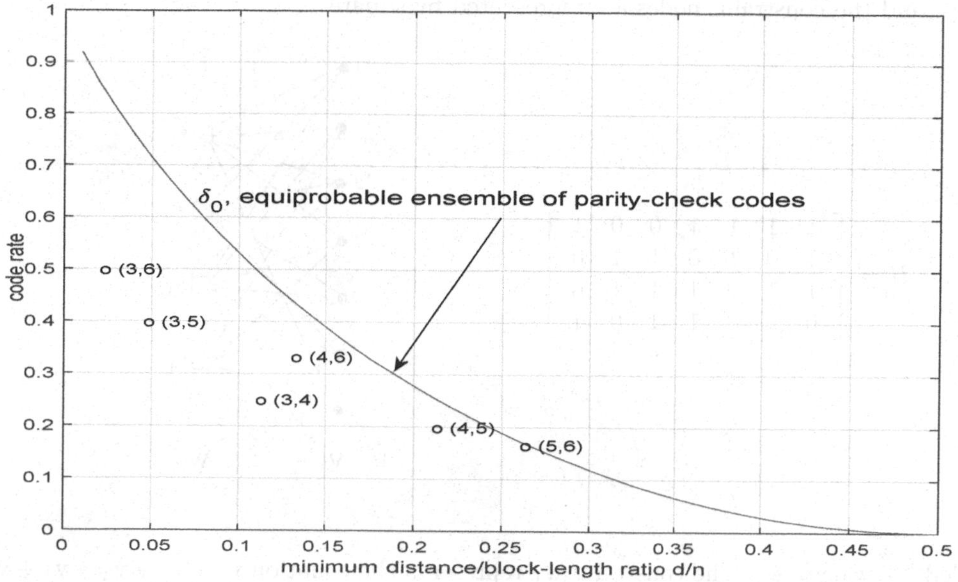

Definition 21.2.3 An LDPC code $\mathcal{C}$ with a parity check matrix H may be represented by a bipartite graph $\mathcal{G}$, called a **Tanner graph**, with vertex set $V \cup W$ where $V \cap W = \emptyset$. The vertices in V are the **variable nodes** and correspond to the columns of H; the vertices in W are the **check nodes** or **constraint nodes** and correspond to the rows of H. Two nodes $v_t \in V$ and $w_s \in W$ are connected by an edge if and only if the $(s,t)^{\text{th}}$ entry of H is nonzero. The value of the $(s,t)^{\text{th}}$ entry acts as a weight for the corresponding edge. We will restrict our attention to binary LDPC codes, where each edge implicitly has 1 as a weight. If the LDPC code is (j,r)-regular, then the degree of each variable node is j and the degree of each check node is r. The check nodes of an LDPC code represent simple parity check constraints; i.e., the values of the adjacent variable nodes sum to 0 mod 2.

While early constructions of LDPC codes focused on the regular case, irregular codes have been shown to be better in decoding performance. The *degree distribution* of a Tanner graph is used to analyze the code.

Definition 21.2.4 The **degree distribution** of the variable nodes and the constraint nodes of a Tanner graph $\mathcal{G}$ are denoted by polynomials $\lambda(x)$ and $\rho(x)$, respectively. The variable node degree distribution $\lambda(x)$ is defined by $\lambda(x) = \sum_{i=2}^{d_v} \lambda_i x^{i-1}$, and the constraint node degree distribution $\rho(x)$ is defined by $\rho(x) = \sum_{i=2}^{d_c} \rho_i x^{i-1}$, where λ_i (respectively ρ_i) is the fraction of edges in the Tanner graph that are connected to degree i variable (respectively check) nodes, and d_v and d_c are the maximum variable node and check node degrees, respectively. For example, a $(3,6)$-regular LDPC code has degree distribution $\lambda(x) = x^2$ and $\rho(x) = x^5$. Since degree 1 nodes have been shown to cause poor performance, we assume there are no degree 1 nodes in the Tanner graph and omit them in the polynomials described above.

Example 21.2.5 For an LDPC code with a Tanner graph shown in Figure 21.2, the variable nodes on the left are labeled $v_1, v_2, \ldots, v_7$ and the constraint nodes on the right are

FIGURE 21.2: A parity check matrix H for a code $\mathcal{C}$ on the left with its corresponding Tanner graph representation on the right. The variable nodes are represented by shaded circles, and the constraint nodes are represented by squares.

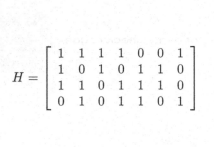

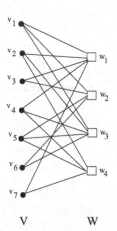

labeled w_1, w_2, w_3, w_4. The constraint w_1 represents the equation $v_1 + v_2 + v_3 + v_4 + v_7 = 0 \bmod 2$, the constraint w_2 represents the equation $v_1 + v_3 + v_5 + v_6 = 0 \bmod 2$, the constraint w_3 represents the constraint $v_1 + v_2 + v_4 + v_5 + v_6 = 0 \bmod 2$, and the constraint w_4 represents $v_2 + v_4 + v_5 + v_7 = 0 \bmod 2$. The degree distribution of the Tanner graph is $\lambda(x) = \frac{1}{3}x + \frac{2}{3}x^2$ and $\rho(x) = \frac{4}{9}x^3 + \frac{5}{9}x^4$.

Remark 21.2.6 For non-binary LDPC codes over $\mathbb{F}_{p^s}$, the edge weights act as coefficients in the corresponding parity check equations, and the sum is set to 0 in the field. For example, if the first row of a parity check matrix for an LDPC code over $\mathbb{F}_3$ was $\begin{bmatrix} 0 & 1 & 2 & 0 & 0 & 1 \end{bmatrix}$, then the corresponding equation at the first constraint w_1 would be $v_2 + 2v_3 + v_6 = 0 \bmod 3$. For more detail on non-binary LDPC codes, we refer the reader to [490].

Remark 21.2.7 The Tanner graph may be seen as representing the corresponding code $\mathcal{C}$ in the following way. Any binary vector $\mathbf{x} = x_1 x_2 \cdots x_n$ that is input to the variable node set (i.e., v_i is assigned value x_i for $i = 1, \ldots, n$) is a codeword of $\mathcal{C}$ if and only if all of the constraints $w_1, \ldots, w_m$ are simultaneously satisfied. This coincides with the definition of parity check matrix since $\mathbf{x}H^{\mathsf{T}} = \mathbf{0}$ if and only if $\mathbf{x} \in \mathcal{C}$. Moreover, while a parity check matrix (or equivalently, a Tanner graph) represents a unique code, a code may have many parity check matrix (and Tanner graph) representations.

Definition 21.2.8 A **generalized LDPC** (or **GLDPC**) code is described by a sparse bipartite graph similar to a Tanner graph, but where each constraint node of degree r represents a specified linear code of block length r. The variable nodes adjacent to a particular constraint node (with some fixed ordering on the edges) must form a codeword in this linear code in order to satisfy the constraint. Note that LDPC codes, as in Definition 21.2.3, correspond to the choice of using simple parity codes at the constraint nodes and are sometimes referred to as **simple** LDPC codes.

Example 21.2.9 Suppose the Tanner graph in Figure 21.2 represents a GLDPC code so that the parity check constraints represent more general constraints. The constraint w_1 has degree 5 and variable node neighbors v_1, v_2, v_3, v_4 and v_7. Suppose w_1 represents a linear code $\mathcal{C}_{w_1}$ of length 5 with check equations $v_1+v_2+v_7 = 0$ mod 2 and $v_1+v_3+v_4 = 0$ mod 2. Then, assuming some ordering on the edges to v_1, v_2, v_3, v_4, v_7, the constraint node w_1 is satisfied if and only if the values assigned to v_1, v_2, v_3, v_4, v_7 form a codeword in $\mathcal{C}_{w_1}$ (i.e., satisfy the two equations). Thus, for a generalized LDPC code, there are more overall constraints imposed by the same set of constraint nodes in the Tanner graph. Consequently, while the matrix in Figure 21.2 is still an incidence matrix of the Tanner graph for the GLDPC code, it is no longer a parity check matrix of the GLDPC code. Note further that the code rate of a GLDPC code is typically smaller than the code rate of a simple LDPC code for the same Tanner graph representation.

With the Tanner graph representation of a code's parity check matrix, the iterative decoder from [787] may be seen as a *message-passing* graph-based decoder that exchanges information between variable and check nodes in the graph [1784]. The structure of the graph affects the success of the decoder. Recall that the **girth** of a graph is the smallest number of edges in any cycle in the graph, and in a bipartite graph the girth is always even. Moreover, the **diameter** of a graph is the largest distance between any two vertices, where distance is measured by the number of edges in the shortest path between them. For decoding, the girth corresponds to the number of iterations for which the messages propagated are independent; thus much research has focused on designing codes with large girth so as to prevent bad information from reinforcing variable node beliefs. Indeed, iterative decoding is optimal on cycle-free graphs [1889]. In contrast, a small graph diameter may be desired to allow information from any node in the graph to reach any other node in the fewest iterations. Despite this intuition, there exist special cases of codes with small girth and/or large diameter that still perform well, suggesting that other factors such as overall cycle structure or graph topology may also play a role.

Remark 21.2.10 Since a girth of 4 is the smallest possible in a bipartite (simple) graph, it is common practice to remove all 4-cycles in the construction of LDPC codes. Linear codes that have Tanner graph representations without 4-cycles are characterized in [1181] in terms of the parameters and minimum distance of the dual code. Moreover, while cycle-free representations ensure independent messages during iterative decoding, codes having such representations were shown to have poor minimum distance [693]. In particular, Etzion, Trachtenberg, and Vardy showed the following.

Theorem 21.2.11 ([693]) *If an* $[n, k, d]$ *linear code* $\mathcal{C}$ *with code rate* R *has a cycle-free Tanner graph representation, then for* $R \geq 0.5$, $d \leq 2$, *and for* $R < 0.5$,

$$d \leq \left\lfloor \frac{n}{k+1} \right\rfloor + \left\lfloor \frac{n+1}{k+1} \right\rfloor < \frac{2}{R}.$$

Using the girth, the minimum distance of LDPC codes with column weight j (equivalently, having variable node degree j) may be bounded as follows.

Theorem 21.2.12 (Tree Bound [1784]) *Suppose a Tanner graph* $\mathcal{G}$ *of an LDPC code* $\mathcal{C}$ *has girth* g *and variable node degree* j. *Then the minimum distance* d *satisfies*

$$d \geq \begin{cases} 1 + j + j(j-1) + j(j-1)^2 + \cdots + j(j-1)^{\frac{g-6}{4}} & \text{if } g/2 \text{ is odd,} \\ 1 + j + j(j-1) + j(j-1)^2 + \cdots + j(j-1)^{\frac{g-8}{4}} + (j-1)^{\frac{g-4}{4}} & \text{if } g/2 \text{ is even.} \end{cases}$$

A similar result for GLDPC codes was also given in [1784]. These results are shown by enumerating the Tanner graph of the code from a variable node as a tree until the layer in which nodes may not be distinct (due to girth), and examining how many variable nodes must have a value 1 in a minimum weight codeword.

Remark 21.2.13 A special case of simple LDPC codes with $j = 2$ are referred to as **cycle codes**. Their minimum distance d can be shown to be $g/2$, where g is the girth of the corresponding Tanner graph. Since the girth is known to increase only logarithmically in the block length [516], cycle codes are known to be bad (i.e., $d/n \to 0$ as $n \to \infty$). Thus, if the variable nodes have degree 2, it is typical to use generalized LDPC codes with constraints from a strong linear code at the check nodes to improve the minimum distance and performance.

LDPC codes are often constructed by randomly choosing the positions and values of the nonzero entries of H. In a random construction, many desirable features for practical implementation are lost, such as efficient representation, simple analysis, and easy encoding. However, with some algebraic structure in the code, some of these features may be retained. The theory of designing and decoding LDPC codes has attracted widespread interest in the coding theory community; see for example [409, 1121, 1209, 1589, 1590, 1591]. For example, designing graphs for codes that have properties such as large girth, good expansion (see Section 21.5), small diameter, or optimizing other combinatorial objects in a graph to obtain better codes is of fundamental interest.

21.3 Decoding

A codeword transmitted over a communication channel is typically corrupted by noise. A decoder takes the received (noisy) word $\mathbf{r}$ as input and tries to estimate the codeword that was transmitted. A maximum likelihood decoder (see Section 1.17) finds the most likely transmitted codeword as follows:

$$\widehat{\mathbf{c}}^*_{ML} = \arg \max_{\mathbf{c} \in \mathcal{C}} \text{Prob}(\mathbf{r} \text{ received} \mid \mathbf{c} \text{ transmitted}).$$

Typically, performing ML decoding is hard as the number of codewords to search over is exponential in the dimension of the code and the codes that are of interest in practical applications have block lengths in 1000s or 10000s, making ML decoding infeasible. In contrast, the iterative decoding algorithms that were initially developed in [787] and later generalized by others work remarkably well at low decoding complexity on LDPC codes of large block lengths. These iterative decoders can be seen as message-passing algorithms that exchange messages between the variable and check nodes of a code's Tanner graph $\mathcal{G}$. The sparsity yields linear complexity decoders for the corresponding LDPC codes.

Iterative message-passing decoders may either be hard decision decoders that use only the symbol values of the codeword alphabet during decoding or soft decision decoders that use the symbol values and probability values associated with those symbol values during decoding; see Definition 1.17.1. While hard decision decoders are much easier to implement and take less time to run, their performance is quite inferior compared to soft decision decoders. Here, we provide some well known hard and soft decision decoders for LDPC codes. Before we give the algorithms, we start with some notation.

Let H be an $m \times n$ parity check matrix. Iterative decoding starts by taking the channel information (i.e., the received message) and assigning those values to the variable nodes in the graph. A single iteration in a **message-passing algorithm** consists of passing messages from variable nodes to their neighboring check nodes, determined by a **variable node update rule**, and passing messages back from check nodes to their neighboring variable nodes, determined by a **check node update rule**. At the end of each iteration, variable node symbols are estimated according to an **estimation rule**. If the estimates of all of the variable nodes corresponds to a codeword, the iterative decoder stops and outputs that codeword; otherwise another iteration is performed. This continues until the decoder either produces an output or some specified number of iterations is reached.

Let $\mu_{ch}(v)$ be the channel/received message for the variable node v. Let $\mu_{c \to v}^{(t)}$ be the message from check node c to variable node v at a particular decoding iteration t. Let $\mu_{v \to c}^{(t)}$ be the message from variable node v to check node c at a particular decoding iteration t. Let $N(v)$ denote the set of check node neighbors of a variable node v in the Tanner graph $\mathcal{G}$ representing the parity check matrix H, and let $N(c)$ denote the set of variable node neighbors of a check node c in $\mathcal{G}$. Let $N(v) \setminus \{c\}$ denote the set of check node neighbors of v excluding check node c. We define $N(c) \setminus \{v\}$ similarly.

Algorithm 21.3.1 (Gallager A/B [787])

Gallager A and B are hard decision algorithms and the messages $\mu_{ch}(v)$, $\mu_{v \to c}^{(t)}(v)$, and $\mu_{c \to v}^{(t)}$ take on values from the code's alphabet. For binary codes, they take values 0 or 1.

Initialization: For the first decoding iteration, $t = 0$, set $\mu_{v \to c}^{(0)} = \mu_{ch}(v)$ for all $c \in N(v)$.

Check node update rule: The message sent from a check node c to a variable node v in iteration t is given by the modulo 2 sum (for binary codes)

$$\mu_{c \to v}^{(t)} = \bigoplus_{v' \in N(c) \setminus \{v\}} \mu_{v' \to c}^{(t)}.$$

Variable node update rule (Gallager A): The message sent from a variable node v to a check node c in iteration $t + 1$ is given by

$$\mu_{v \to c}^{(t+1)} = \begin{cases} \mu_{c' \to v}^{(t)} & \text{if } \mu_{c' \to v}^{(t)} \text{ are identical for all } c' \in N(v) \setminus \{c\}, \\ \mu_{ch}(v) & \text{otherwise.} \end{cases}$$

Variable node update rule (Gallager B): The variable node update sets $\mu_{v \to c}^{(t+1)} = x$ if, for some chosen $b \in \{1, 2, \ldots, |N(v)| - 1\}$, at least b of the check nodes in $N(v) \setminus \{c\}$ are equal to value x.

Estimation: The variable nodes are estimated similarly to the variable node update rule at the end of each iteration t. If the estimated values of the variable nodes correspond to a valid codeword in the LDPC code, decoding stops. Otherwise, another round of decoding is performed until a maximum specified limit of decoding iterations is reached.

Example 21.3.2 In Figure 21.3, the update rule at a check node (on the left) and at a variable node (on the right) are shown. At the check node, the outgoing message is simply the XOR of the incoming messages. At the variable node, since all the incoming check node to variable node messages agree and equal to 1 in this example, the outgoing message is also 1.

We now turn to the soft decision Sum-Product Algorithm (SPA) [787, 1170, 1590]. A soft decision decoder such as this uses likelihood values for the messages. For binary codes, these likelihood values may be of the form of probability pairs (p_0, p_1), where p_0 is the probability of a message symbol representing value 0 and p_1 is the probability of a message symbol

FIGURE 21.3: Check node update on the left and variable node update on the right for the Gallager A Algorithm

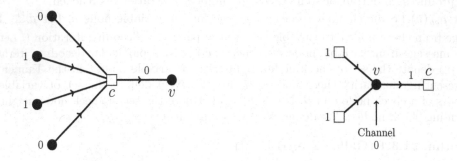

representing value 1, or the likelihood values may be of the form of **log-likelihood ratios (LLRs)** (for example $\log(p_0/p_1)$). While either representation may be used, for practical implementation, the log representation, referred to as **log-domain decoding**, is preferred over the probability representation, referred to as **probability-domain decoding**, due to numerical precision issues that occur in practice. A log-domain description of the SPA is presented below. We refer the reader to [787] for a probability-domain description of the SPA. Before we proceed, we modify our notation to represent soft information.

Let $\mu_{ch}(v) = \log\left(\frac{\text{Prob}(v=0\,|\,\mathbf{r})}{\text{Prob}(v=1\,|\,\mathbf{r})}\right)$ be the LLR associated with the variable node v based on the received word $\mathbf{r}$ from the channel. Let $\mu_{v\to c}^{(t)}$ be the LLR message sent from a variable node v to a check node c in decoding iteration t and let $\mu_{c\to v}^{(t)}$ be the LLR message sent from a check node c to a variable node v in decoding iteration t. Note that $\mu_{v\to c}^{(t)}$ represents the LLR for the value associated with variable node v based on information available at node v from all neighboring nodes excluding node c and $\mu_{c\to v}^{(t)}$ represents the LLR for the value associated with node v based on information available at node c from all its neighboring nodes excluding node v.

Algorithm 21.3.3 (Sum-Product (SPA))

<u>Initialization</u>: For the first decoding iteration, $t = 0$, set $\mu_{v\to c}^{(0)} = \mu_{ch}(v)$ for all $c \in N(v)$.

<u>Check node update rule</u>: The absolute value of the message sent from a check node c to a variable node v in iteration t is given by

$$\tanh\left(\frac{|\mu_{c\to v}^{(t)}|}{2}\right) = \prod_{v'\in N(c)\setminus\{v\}} \tanh\left(\frac{|\mu_{v'\to c}^{(t)}|}{2}\right).$$

The sign value of the message sent from a check node c to a variable node v in iteration t is given by

$$\text{sign}\left(\mu_{c\to v}^{(t)}\right) = \prod_{v'\in N(c)\setminus\{v\}} \text{sign}\left(\mu_{v'\to c}^{(t)}\right).$$

<u>Variable node update rule</u>: The message sent from a variable node v to a check node c in iteration $t + 1$ is given by

$$\mu_{v\to c}^{(t+1)} = \mu_{ch}(v) + \sum_{c'\in N(v)\setminus\{c\}} \mu_{c'\to v}^{(t)}.$$

FIGURE 21.4: Check node update on the left and variable node update on the right for the Sum-Product Algorithm in the probability-domain

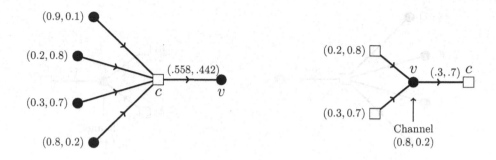

Estimation: The variable nodes are estimated at the end of each iteration t as follows:

$$\hat{v} = \begin{cases} 0 & \text{if } \mu_{ch}(v) + \sum_{c' \in N(v)} \mu_{c' \to v}^{(t)} > 0, \\ 1 & \text{otherwise.} \end{cases}$$

If the estimated values of the variable nodes correspond to a valid codeword in the LDPC code, decoding stops. Otherwise, another round of decoding is performed until a maximum specified limit of decoding iterations is reached.

Example 21.3.4 In Figure 21.4, the update rule at a check node (on the left) and at a variable node (on the right) are shown. Instead of using LLRs as messages, the messages are shown in the probability-domain. The first coordinate of the message represents that probability of the adjacent variable node being a 0 and the second coordinate represents the probability of the adjacent variable node being a 1. At the check node, the outgoing message is a probability pair that depends on the probabilities on the incoming messages. In the example shown, the probability that the variable node v is a 0 given the incoming messages at c (as shown in the figure) is 0.558 and the probability that it is a 1 is 0.442. Specifically, the probability of the outgoing message being a 0 (respectively a 1) is the sum of the product of the probabilities of an even (respectively odd) number of incoming messages being 1. The variable node update is also shown using messages in the probability-domain. The outgoing message is obtained by a component-wise product of the incoming messages and the channel message and scaled by a normalization factor.

The Min-Sum Algorithm is a simplification of the SPA.

Algorithm 21.3.5 (Min-Sum (MSA))

Initialization, the variable node update rule, and estimation are as in the SPA.

Check node update rule: $|\mu_{c \to v}^{(t)}| = \min_{v' \in N(c) \setminus \{v\}} |\mu_{v' \to c}^{(t)}|$. The sign value of the check node to variable node message remains the same as for the SPA.

Most practical implementations of the MSA [396, 1591] use this approximation in the check node updates along with scaling factors to improve decoder convergence.

FIGURE 21.5: Check node update on the left and variable node update on the right for the Min-Sum Algorithm using LLRs

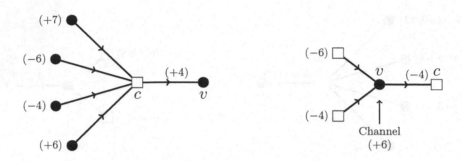

Example 21.3.6 In Figure 21.5, the update rule at a check node (on the left) and at a variable node (on the right) are shown with the messages as LLRs. At the check node, the outgoing message has magnitude that is the minimum of the magnitudes of the incoming message and has a sign that is the product of the signs of the incoming messages. At a variable node, the outgoing message is simply a sum of the incoming messages and the channel message.

Linear programming (LP) decoding for LDPC codes was proposed in [710, 711] and can be applied to any linear code, with complexity depending on the sparsity of the representation. By taking the convex hull of all codewords, a codeword polytope may be obtained whose vertices are precisely the codewords of the code. ML decoding can be expressed as an integer linear programming problem, and its output is a vertex of this polytope. For the additive white Gaussian noise channel, the ML decoding essentially finds the codeword that minimizes the cost function

$$\widehat{\mathbf{c}} = \arg\min_{\mathbf{c}\in\mathcal{C}} \mathbf{cr}^{\mathsf{T}}.$$

When the code constraints are relaxed, the convex hull results in a new polytope, called the **pseudocodeword polytope** $\mathcal{P}$. In this case, the linear programming decoder complexity is significantly reduced, and the LP decoding problem finds

$$\widehat{\mathbf{x}} = \arg\min_{\mathbf{x}\in\mathcal{P}} \mathbf{xr}^{\mathsf{T}}.$$

The vertices of this polytope with all integer coordinates are codewords of $\mathcal{C}$. The solution of LP decoding is a vertex of this polytope $\mathcal{P}$, and when the LP decoder finds an all integer coordinate vertex as the solution, its estimate coincides with that of the ML decoder. Thus, although LP decoding is more complex in general over other iterative decoding methods such as the SPA, it has the advantage that when it decodes to a codeword, the codeword is guaranteed to be the maximum-likelihood codeword [710, 711].

Remark 21.3.7 When decoding GLDPC codes, iterative decoding is performed between the variable and constraint nodes, but at the constraint nodes, typically ML decoding of the code representing the constraint node is performed. This is feasible since the degree of the constraint node, and hence the block length of the code represented by the constraint, is typically small. For soft decision decoding, the decoding at the constraint node may be performed efficiently using trellis-based decoding [1897].

FIGURE 21.6: Decoder performance plot of an LDPC code

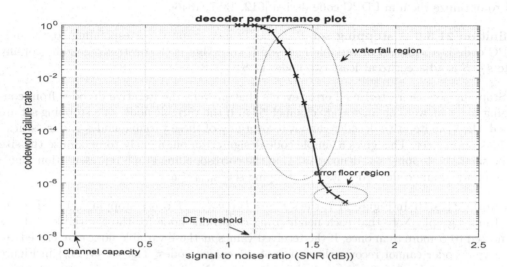

FIGURE 21.6: Decoder performance plot of an LDPC code

21.3.1 Decoder Analysis

The success of graph-based codes lies in their efficient decoding algorithms and capacity approaching performance. However, the iterative decoder is sub-optimal since it uses only the local Tanner graph structure for decoding. In many cases, the iterative decoding performance is inferior to the optimal maximum likelihood (ML) performance which is, however, too complex to implement for long block length codes.

There are two main approaches for analyzing iterative decoders for codes over graphs; one is an asymptotic analysis applicable to long block length codes, and the other examines the combinatorial structure of the graph and its effect on the decoding performance for finite length codes. The former is known as *density evolution* and tracks the probability density function of the messages exchanged during iterations of the decoder assuming a cycle-free graph. The special case for the binary erasure channel will be summarized in Section 21.7. In the rest of this section, we will summarize *finite length decoder* analysis.

Remark 21.3.8 A typical performance plot of an iterative decoder for an LDPC code is shown in Figure 21.6. The plot shows the failure rate of the decoder as a function of the channel quality. In the case of the AWGN channel, the channel quality is measured in terms of **signal to noise ratio (SNR)**. As the SNR increases, the decoder failure rate decreases. There are two prominent regions: the **waterfall region** where the decoder failure rate drops steeply with increasing SNR and the **error floor region**, where the decoder failure rate drops slowly with increasing SNR. While the waterfall region typically starts at an SNR value that is dependent on the degree distribution of the LDPC code [1590] and the chosen iterative decoder, the error floor region tends to start at an SNR value that is dependent on the distance properties of the code, the combinatorial structures in the Tanner graph, and the chosen iterative decoder.

Finite length LDPC codes when decoded via iterative message-passing decoders may sometimes fail to converge to a codeword, or sometimes converge to an incorrect codeword. Depending on the channel and choice of iterative decoder, these types of failure events have been attributed to combinatorial structures in the underlying LDPC Tanner graph called *stopping sets* [534], *trapping sets* [1387, 1846], *pseudocodewords* [710, 1099, 1100, 1102, 1158], and *absorbing sets* [612]. Substantial research has been conducted to analyze these structures and to optimize them in LDPC code design [612, 1387, 1846].

Definition 21.3.9 A **stopping set** S in the Tanner graph $\mathcal{G}$ corresponding to a simple LDPC code is a subset S of variable nodes in $\mathcal{G}$ such that each constraint neighbor of any node $s \in S$ is adjacent to at least two nodes in S.

Stopping sets characterize all error events that prevent the iterative decoder from converging to a codeword on an erasure channel [534]. If the variable nodes of a stopping set are erased, then the iterative decoder cannot recover any of these variable nodes and is "stuck" at the stopping set. This gives a simple code design criterion, namely to maximize the size of the smallest stopping set, denoted $s_{\min}$, in the corresponding graph representation. Since the support of any codeword is itself a stopping set, $s_{\min} \leq d$.

Example 21.3.10 In Figure 21.7, the set $S = \{v_1, v_3, v_6\}$ forms a stopping set of size 3 in the Tanner graph. Note that each neighboring constraint node for these variable nodes is connected to S more than once. If the received values at these variable nodes are erased, an iterative decoder cannot recover the values of the erased nodes. The set $\{v_2, v_4\}$ in Figure 21.7 is an example of a stopping set of smallest size. Note that $\{v_1, v_6\}$ is not a stopping set since w_1 is only adjacent to v_1.

FIGURE 21.7: A stopping set $\{v_1, v_3, v_6\}$ of size 3 in an LDPC Tanner graph

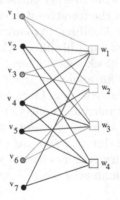

Example 21.3.11 Consider the Tanner graph in Figure 21.7. In Figure 21.8, the received word $\mathbf{r} = 011???0$ is input to the variable nodes, shown in the left-most graph. Recall that any bit that is received is guaranteed to be correct for the binary erasure channel (BEC). In the second graph, the information at w_1 is used to recover the value at v_4. This allows w_4 to have enough information to recover the value at v_5, shown in the third graph, which in turn enables w_3 to recover the value at v_6. It is worth noting that if the set of erased variable nodes does not contain a stopping set, the decoder will successfully recover the transmitted word regardless of the number of initially erased nodes; in particular, this number can exceed $d - 1$. Indeed, the minimum distance of the corresponding code in this example is 2.

For other channels such as the binary symmetric channel (BSC) or the additive white Gaussian noise (AWGN) channel, if certain subsets of variable nodes receive incorrect messages from the channel, the iterative decoder may eventually get stuck at a so-called trapping

FIGURE 21.8: Binary erasure channel (BEC) decoding example

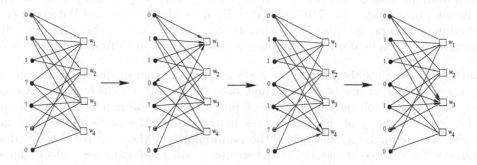

set and not be able to correct the errors on the variable nodes of this trapping set. Combinatorially, an (a, b)-**trapping set** is a subset S of variable nodes in the Tanner graph $\mathcal{G}$ where $|S| = a$ and whose induced subgraph in $\mathcal{G}$ contains b odd degree check nodes. However, whether this set causes an error or not depends on the corresponding choice of decoder [1846]. To remove this dependence on the decoder choice, absorbing sets were introduced in [612] to characterize failure on these channels in a purely combinatorial way.

Definition 21.3.12 An (a, b)-**absorbing set** is an (a, b)-trapping set in a Tanner graph $\mathcal{G}$ with the additional property that every variable node in S has more even degree than odd degree neighbors in the subgraph induced by S in $\mathcal{G}$ [612]. More refined notions of absorbing sets, such as *fully absorbing* and *elementary absorbing* sets, have also been defined to describe varying degrees of harmfulness [612].

Intuitively, if the information at the variable nodes in an absorbing set is incorrect, their estimates are unlikely to change since each neighboring check node of the absorbing set connects an even number of times to these erroneous nodes.

Remark 21.3.13 In general, small absorbing sets are regarded as the most problematic. Removing these is one method to obtain low error-floor performance [151, 612, 1182]. The smallest (a, b)-absorbing set of the Tanner graph of a code is the size of the smallest a, and the corresponding smallest b for that given a, for which an absorbing set exists [612].

During message-passing iterative decoding, if the shortest path length from variable node v_i to another variable node v_j is $2t$, then it takes t iterations for information to propagate from v_i to v_j. This message propagation can be better visualized using the decoder's computation tree for each node.

Definition 21.3.14 A **computation tree** with root node v is a representation of the Tanner graph $\mathcal{G}$ with v as the root node, where instead of having cycles in the Tanner graph, the tree may have multiple copies of each variable node. For t iterations of iterative decoding, the tree is enumerated up to a depth of $2t$ layers.

Remark 21.3.15 Recall that iterative decoding is optimal on a cycle-free graph [1889]. Using this observation, Wiberg and others showed that the iterative decoder attempts to converge to the closest codeword solution for codewords on the tree. The codewords on the computation tree contain all the codewords of the original LDPC code as well as other valid configurations that are not codewords of the LDPC code. Valid configurations on the computation tree are referred to as **pseudocodewords**. For the case of LP decoding, the set of pseudocodewords is shown to be all codewords existing in Tanner graphs that are obtained from all possible lifts (for any lift degree and choice of permutations in the lift) of the original Tanner graph; see [710, 711, 1158]. Pseudocodewords for an LP decoder can be described more accurately as belonging to polyhedra and bounds on the parameters of the pseudocodewords can be derived using the structure of the LDPC Tanner graph.

Remark 21.3.16 Since the parity check matrix for a code is not unique, the choice of which parity check matrix to use is an important one. Depending on which matrix is chosen, the corresponding graph may have different properties. Several papers have investigated the effect of redundancy of the parity check matrix on decoder performance. In general, the higher the redundancy, the better the performance of the iterative decoder due to the reduction of bad combinatorial structures (e.g., small stopping sets, absorbing sets, pseudocodewords) that typically results [1636, 1950]. For example, in [1636] it is shown that with enough redundancy, the smallest stopping set size can match the minimum distance of the code. However, higher redundancy comes at a cost of increased decoder complexity.

21.4   Codes from Finite Geometries

In contrast to random constructions, one design methodology for LDPC codes is to take incidence matrices of combinatorial or algebraic structures as the parity check matrices for the LDPC codes; the properties of the structure may be used to derive properties of the corresponding code. In [1159], Kou, Lin, and Fossorier introduced LDPC codes with parity check matrices determined by the incidence structure of points and lines in finite Euclidean and projective geometries (often referred to as finite geometry LDPC, or FG-LDPC, codes). These matrices alone can be used as parity check matrices of LDPC codes and have been shown to perform remarkably well in comparison to random LDPC codes. Alternatively, they can also be modified by a column splitting process that results in codes of longer block lengths [1159, 1780].

Definition 21.4.1 The m-**dimensional finite projective geometry** $\mathrm{PG}(m, q)$ over $\mathbb{F}_q$ has the following parameters: There are $\frac{q^{m+1}-1}{q-1}$ points and $\frac{(q^m+\cdots+q+1)(q^{m-1}+\cdots+q+1)}{(q+1)}$ lines. Each line contains $q + 1$ points, and each point is on $\frac{q^m-1}{q-1}$ lines. Moreover, any two points have exactly one line in common and each pair of lines has exactly one point in common. More on projective geometry can be found in Chapter 14.

An LDPC code, denoted a **PG-LDPC** code, is obtained by taking the points to lines incidence matrix of $\mathrm{PG}(m, q)$ and using that for its parity check matrix. For $m = 2$ and $q = 2^s$, the corresponding PG-LDPC code has block length $n = 2^{2s} + 2^s + 1$, dimension $k = 2^{2s} + 2^s - 3^s$, and minimum distance $d_{\min} = 2^s + 2$.

Example 21.4.2 An incidence matrix for the points and lines of $\mathrm{PG}(2, 2)$ over $\mathbb{F}_2$ (i.e., the projective plane) is shown in Figure 21.9, along with the corresponding Tanner graph. The geometry is commonly known as the Fano plane. The corresponding PG-LDPC code is the $[7, 3, 4]_2$ simplex code, the dual of the Hamming code $\mathcal{H}_{3,2}$.

FIGURE 21.9: A Fano plane PG(2,2) on the left, the corresponding incidence matrix in the middle, and the LDPC Tanner graph on the right

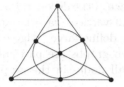
$$H = \begin{bmatrix} 1 & 1 & 0 & 1 & 0 & 0 & 0 \\ 0 & 1 & 1 & 0 & 1 & 0 & 0 \\ 0 & 0 & 1 & 1 & 0 & 1 & 0 \\ 0 & 0 & 0 & 1 & 1 & 0 & 1 \\ 1 & 0 & 0 & 0 & 1 & 1 & 0 \\ 0 & 1 & 0 & 0 & 0 & 1 & 1 \\ 1 & 0 & 1 & 0 & 0 & 0 & 1 \end{bmatrix}$$
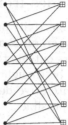

Definition 21.4.3 The m-**dimensional finite Euclidean geometry** over $\mathbb{F}_q$ without the origin point and the lines containing it, denoted $\text{EG}_0(m,q)$, has the following parameters [1159]: There are $q^m - 1$ points and $\frac{q^{m-1}(q^m-1)}{q-1}$ lines. Each line contains q points and each point is on $\frac{q^m-1}{q-1}$ lines. Moreover, any two points have exactly one line in common and any two lines either have one point in common or are parallel (i.e., they have no points in common).

An LDPC code, denoted an **EG-LDPC** code, is obtained by taking the points to lines incidence matrix of $\text{EG}_0(m,q)$ and using that for its parity check matrix. For $m = 2$ and $q = 2^s$, the corresponding EG-LDPC code has block length $n = 2^{2s} - 1$, dimension $k = 2^{2s} - 3^s$, and minimum distance $d_{\min} = 2^s + 1$.

There are a wide variety of finite geometry codes. For example, a μ-dimensional subspace of a finite geometry is called a μ-**flat**. By taking the incidence matrix of μ_1-flats and μ_2-flats from an m-dimensional finite geometry, where $0 \le \mu_1 < \mu_2 \le m$, LDPC codes have also been obtained [1780].

Remark 21.4.4 FG-LDPC codes have the advantage of being cyclic or quasi-cyclic; see Section 1.12 and Chapters 2 and 7. In a cyclic code, each row of the parity check matrix is a cyclic shift of the preceding row by one position. For quasi-cyclic codes, the parity check matrix can be seen as an array of submatrices, each of which is cyclic. The cyclic or quasi-cyclic structure of these codes is particularly desirable for practical implementation where identical processing units can be run in parallel to improve the efficiency of encoding and decoding operations. Many codes used in industry have this property.

Remark 21.4.5 The parity check matrices for FG-LDPC codes have a large number of redundant rows, which may positively impact their decoding performance; see Remark 21.3.16. In particular, the redundancy allows decoding to proceed even when some nodes become faulty. Moreover, the geometric structure of these codes has led to concrete results on stopping sets and pseudocodewords [1101, 1912], and trapping sets of FG-LDPC codes [536]. More recently, absorbing sets of FG-LDPC codes were analyzed in [150, 1261, 1278].

Remark 21.4.6 Incidence matrices from a wide variety of discrete structures may be used to obtain LDPC codes that have practical advantages. This may include a more compact representation, easier implementation, and/or theoretical guarantees on the corresponding code parameters using the discrete structure's properties. For example, Steiner systems, combinatorial block designs (examined in Chapter 5), partial geometries, generalized quadrangles, generalized polygons, and Latin squares have all been explored for LDPC code design [1052, 1053, 1054, 1101, 1284, 1859].

21.5   Codes from Expander Graphs

The connectivity properties of a graph also affect the performance of the decoder. Loosely speaking, a graph is a good expander if small (nonempty) sets of vertices have large sets of neighbors outside the set; alternatively, if small sets of vertices define large "edge cuts". They may be thought of as highly connected sparse graphs. Such graphs are particularly suited for iterative decoders in that messages are dispersed to all nodes in the graph as quickly as possible.

Definition 21.5.1 Let $\mathcal{G}$ be a graph on n vertices. The **adjacency matrix** $A(\mathcal{G})$ of $\mathcal{G}$ is an $n \times n$ matrix where the $(s,t)^{\text{th}}$ entry $A_{s,t}$ is equal to the number of edges between vertices v_s and v_t. Thus, for simple graphs, each diagonal element of $A(\mathcal{G})$ is 0, and each nonzero entry is 1. As $A(\mathcal{G})$ is symmetric, it has real eigenvalues $\nu_1, \nu_2, \ldots, \nu_n$, which we order so that $|\nu_1| \geq |\nu_2| \geq \cdots \geq |\nu_n|$. We will use $\mu_i = |\nu_i|$ to represent the absolute value of the ith eigenvalue for $i = 1, 2, \ldots, n$.

Graphs whose second largest eigenvalue μ_2 (in absolute value) of the corresponding adjacency matrix is small compared to the largest eigenvalue μ_1 are known to be good expanders. Recall that the adjacency matrix of a r-regular connected graph has r as its largest eigenvalue, and for a (j,r)-regular bipartite graph, the largest eigenvalue is $\sqrt{jr}$.

Definition 21.5.2 A connected simple graph $\mathcal{G}$ is said to be an (n, r, μ_2)-**expander** if $\mathcal{G}$ has n vertices, is r-regular, and the second largest eigenvalue of $A(\mathcal{G})$ (in absolute value) is μ_2.

We start by mentioning two bounds on the minimum distance, called the *Bit-Oriented* and *Parity-Oriented* Bounds, that rely on the connectivity properties of the Tanner graph.

Theorem 21.5.3 (Bit-Oriented Bound [1787]) *Let $\mathcal{G}$ be a regular connected graph with n variable nodes of uniform degree j and m constraint nodes of uniform degree r, and let each constraint node represent constraints from a $[r, kr, \epsilon r]$ code[1]. Then the minimum distance d of the GLDPC code $\mathcal{C}$ with parity check matrix H represented by the Tanner graph $\mathcal{G}$ satisfies*

$$d \geq \frac{n(jr\epsilon - \mu_2)}{jr - \mu_2},$$

where μ_2 is the second largest eigenvalue of $H^T H$.

Theorem 21.5.4 (Parity-Oriented Bound [1787]) *Let $\mathcal{G}$ be a regular connected graph with n variable nodes of uniform degree j and m constraint nodes of uniform degree r, and let each constraint node represent constraints from a $[r, kr, \epsilon r]$ code. Then the minimum distance d of the GLDPC code $\mathcal{C}$ with parity check matrix H represented by the Tanner graph $\mathcal{G}$ satisfies*

$$d \geq \frac{n\epsilon(\epsilon jr + r - \epsilon r - \mu_2)}{jr - \mu_2},$$

where μ_2 is the second largest eigenvalue of HH^T.

---

[1]Recall that an $[a, b, s]$ linear code is a code of block length a, dimension b, and minimum distance s. The code rate of the code is b/a. For simple LDPC codes, the check nodes represent simple $[r, r-1, 2]$ parity check codes.

In both cases, the smaller the second largest eigenvalue μ_2, the better the bound on the minimum distance.

Constructing graphs with good expansion is a hard problem [982]. Since most known constructions yield regular graphs, we start by showing a common technique to convert a regular graph into a bipartite graph that may be used to represent a code.

Definition 21.5.5 Let $\mathcal{G}$ be a graph with vertex set V and edge set E. Then the **edge-vertex incidence graph** of $\mathcal{G}$ is the bipartite graph with vertex set $V \cup E$ and edge set $\{\{v, e\} \mid v \in V \text{ and } e \in E \text{ are incident in } \mathcal{G}\}$.

Remark 21.5.6 If $\mathcal{G}$ is r-regular on n vertices, then the edge-vertex incidence graph of $\mathcal{G}$ has $(nr)/2$ vertices in the part corresponding to the edges of $\mathcal{G}$, n vertices in the other part, and is $(2, r)$-regular since each edge in $\mathcal{G}$ is incident to two vertices in $\mathcal{G}$ and each vertex in $\mathcal{G}$ has degree r.

Example 21.5.7 A 3-regular graph on 8 vertices and its corresponding $(2, 3)$-regular edge-vertex incidence graph are shown in Figure 21.10. Since edge-vertex incidence graphs are 2-left regular by construction, it is standard to use edge-vertex incidence graphs as Tanner graphs for GLDPC codes; see Remark 21.2.13.

FIGURE 21.10: A $(2, 3)$-regular edge-vertex incidence graph on the right corresponding to the 3-regular graph on the left

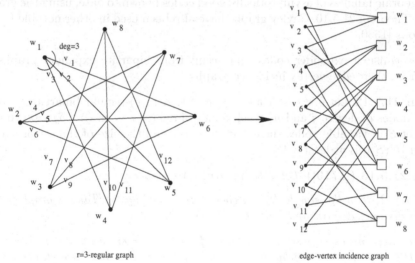

r=3-regular graph   edge-vertex incidence graph

Definition 21.5.8 Expander codes are families of graph-based codes whose corresponding Tanner graphs are obtained from expander graphs. If the expander graph is not bipartite, then the code is represented by the edge-vertex incidence graph of the expander graph (and inherits some level of expansion). If the expander graph is bipartite, then the code may be a simple LDPC code, or a GLDPC code with code constraints used at the check nodes, or a GLDPC code represented by the edge-vertex incidence graph of the bipartite expander graph with possibly multiple types of code constraints at the vertices.

We now state a particularly useful result describing the expansion of an r-regular graph due to Alon and Chung.

Theorem 21.5.9 ([40]) *Let $\mathcal{G}$ be an r-regular graph on n vertices and let μ_2 be the second largest eigenvalue (in absolute value) of its adjacency matrix. Then every subset S of γn vertices contains at most $\frac{nr}{2}\left(\gamma^2 + \frac{\mu_2}{r}(\gamma - \gamma^2)\right)$ edges in the subgraph induced by S in $\mathcal{G}$.*

Using Theorem 21.5.9, Sipser and Spielman show that the guaranteed error correction capabilities of a code may be improved when the underlying graph is a good expander.

Theorem 21.5.10 ([1719]) *Let C be a GLDPC code formed by taking the edge-vertex incidence graph of an (n, r, μ_2)-expander. Let the check nodes represent constraints of an $[r, kr, \epsilon r]$ linear block code. Then the resulting GLDPC code has block length $N = nr/2$, rate $R \geq 2k - 1$, and minimum distance*

$$d \geq N \frac{(\epsilon - \mu_2/r)^2}{(1 - \mu_2/r)^2}.$$

Remark 21.5.11 A family of highly expanding graphs known as Ramanujan graphs [1291] was constructed with excellent graph properties that surpassed the parameters predicted for random graphs. One notable family is obtained by taking the Cayley graph of projective special linear groups over $\mathbb{F}_q$ with a particular choice of generators [1291]. The description of these graphs and their analysis relies on deep results from mathematics using tools from graph theory, number theory, and representation theory of groups [491]. Sipser and Spielman show that graphs with good expansion lead to LDPC codes with minimum distance growing linearly with the block length [1719]. Indeed, [1719] gives a construction of one of the few explicit algebraic families of asymptotically good codes known to date, using the graphs from [1291] and Theorem 21.5.10. Cayley graphs have also been used in other notable LDPC code constructions [1330].

We now consider expander codes that result from bipartite expander graphs directly (and not from their edge-vertex incidence graphs).

Definition 21.5.12 Let $0 < \alpha < 1$ and $0 < \delta < 1$. A (j, r)-regular bipartite graph $\mathcal{G}$ with n variable nodes of degree j and m check nodes of degree r is an $(\alpha n, \delta j)$-**expander** if for every subset S of variable nodes such that $|S| < \alpha n$ and the size of the set of neighbors of S, denoted $|N(S)|$, is at least $\delta j |S|$.

Theorem 21.5.13 ([1719]) *Let $\mathcal{G}$ be an $(\alpha n, \delta j)$-expander.*

(a) *If $\delta > 1/2$, then the simple LDPC code represented by the Tanner graph $\mathcal{G}$ has minimum distance $d \geq \alpha n$ and code rate $R \geq 1 - \frac{j}{r}$.*

(b) *If the check nodes represent constraints from a $[r, kr, \epsilon r]$ code and if $\delta > 1/(\epsilon r)$, the GLDPC code represented by the Tanner graph $\mathcal{G}$ has minimum distance $d \geq \alpha n$ and code rate $R \geq 1 - j(1 - k)$.*

Finally, we consider the case of taking the edge-vertex incidence graph of a bipartite expander graph for a code.

Definition 21.5.14 A (j, r)-regular bipartite graph $\mathcal{G}$ on n left vertices and m right vertices is a (j, r, n, m, μ_2)-**expander** if the second largest eigenvalue of $A(\mathcal{G})$ (in absolute value) is μ_2.

Remark 21.5.15 Let $\mathcal{G}$ be an (j, r, n, m, μ_2)-expander and consider the edge-vertex incidence graph of $\mathcal{G}$. The corresponding GLDPC code is formed by using constraints of a $[j, k_1 j, \epsilon_1 j]$ linear block code at each degree j vertex, and using the constraints of an $[r, k_2 r, \epsilon_2 r]$ linear block code at each degree r vertex. The resulting LDPC code has block length $N = nj = mr$ and rate $R \geq k_1 + k_2 - 1$.

The edge-expansion of a regular bipartite graph may be described as follows.

Theorem 21.5.16 ([1037]) *Let $\mathcal{G}$ be a (j, r)-regular bipartite graph with n left nodes of degree j and m right nodes of degree r, and let μ_2 be the second largest eigenvalue (in absolute value) of the adjacency matrix of $\mathcal{G}$. Then for subsets S and T of the left and right vertices, respectively, the number of edges in the subgraph induced by $S \cup T$, denoted $|E(S, T)|$ is at most*

$$|E(S, T)| \leq \frac{r}{n}|S||T| + \frac{\mu_2}{2}(|S| + |T|).$$

Using this, Janwa and Lal obtain the following result.

Theorem 21.5.17 ([1037]) *If $\epsilon_2 r \geq \epsilon_1 j > \mu_2/2$, then the GLDPC code obtained from the (j, r, n, m, μ_2)-expander graph $\mathcal{G}$ as in Remark 21.5.15 has minimum distance*

$$d \geq nj\left(\epsilon_1\epsilon_2 - \frac{\mu_2}{2\sqrt{jr}}\left(\epsilon_1\sqrt{j/r} + \epsilon_2\sqrt{r/j}\right)\right).$$

Remark 21.5.18 Expander graphs are of fundamental interest in mathematics and engineering and have several applications in computer science, complexity theory, derandomization, designing communication networks, and coding theory [982]. Since Sipser and Spielman's notable result, there have been several papers addressing the design and analysis of codes using expander graphs [1037, 1941]. Other authors have investigated non-algebraic approaches to designing expander graphs; one such construction takes an appropriately defined product of small component expander graphs to construct a larger expander graph [982, 1586]. The distance bounds in this section indicate that LDPC codes constructed from expander graphs provide a certain guaranteed level of performance and error correction capability with graph-based iterative decoding. Similar lower bounds have been obtained on other decoding-related parameters [1096, 1098].

21.6 Protograph Codes

Since codes with long block lengths are desirable in practice, one avenue of code design is to construct families of LDPC codes by taking random lifts of a specially chosen bipartite base graph, called a *protograph*. The graph lift is used as a Tanner graph for the resulting so-called **protograph code** [601, 602, 603, 1802].

Definition 21.6.1 Let $\mathcal{G} = (V, E)$ be a graph with $V = \{v_1, \ldots, v_n\}$ and $E = \{e_1, \ldots, e_m\}$. A degree ℓ **lift** of $\mathcal{G}$ is a graph $\widehat{\mathcal{G}} = (\widehat{V}, \widehat{E})$ with ℓn vertices and ℓm edges, where each $v_i \in V$ is replaced by ℓ copies $\{v_{i,1}, v_{i,2}, \ldots, v_{i,\ell}\} \in \widehat{V}$ (called a **cloud** of v_i) and each edge $\{v_i, v_j\} \in E$ is replaced by ℓ edges in $\widehat{E}$ that connect the vertices in the clouds v_i and v_j in a 1-1 manner. That is, each $v_{i,k}$ connects to a unique $v_{j,k'}$, for $k, k' \in \{1, 2, \ldots, \ell\}$. $\mathcal{G}$ is called the **protograph** of $\widehat{\mathcal{G}}$

Recall that given a permutation σ on ℓ elements, the **permutation matrix** P corresponding to σ is the $\ell \times \ell$ binary matrix having a 1 in position (s, t) if $\sigma(s) = t$ and 0 otherwise. Parity check matrices for protograph codes may be obtained from the incidence matrix of the protograph as follows: 0s are replaced by $\ell \times \ell$ all-zero matrices, and a nonzero entry of p in the incidence matrix is replaced by a sum of p (possibly different) $\ell \times \ell$ permutation matrices. In general, permutations may be chosen randomly or algebraically.

FIGURE 21.11: A protograph and a degree-three graph lift

$$
\begin{bmatrix} 2 & 0 & 1 & 1 & 1 \\ 1 & 1 & 1 & 1 & 1 \\ 1 & 1 & 1 & 1 & 0 \end{bmatrix} \longrightarrow
\left[\begin{array}{ccc|ccc|ccc|ccc|ccc}
1 & 1 & 0 & 0 & 0 & 0 & 0 & 1 & 0 & 1 & 0 & 0 & 0 & 0 & 1 \\
0 & 1 & 1 & 0 & 0 & 0 & 0 & 0 & 1 & 0 & 1 & 0 & 1 & 0 & 0 \\
1 & 0 & 1 & 0 & 0 & 0 & 1 & 0 & 0 & 0 & 0 & 1 & 0 & 1 & 0 \\
\hline
1 & 0 & 0 & 0 & 0 & 1 & 1 & 0 & 0 & 0 & 0 & 1 & 0 & 0 & 1 \\
0 & 1 & 0 & 0 & 1 & 0 & 0 & 1 & 0 & 1 & 0 & 0 & 1 & 0 & 0 \\
0 & 0 & 1 & 1 & 0 & 0 & 0 & 0 & 1 & 0 & 1 & 0 & 0 & 1 & 0 \\
\hline
1 & 0 & 0 & 0 & 1 & 0 & 1 & 0 & 0 & 1 & 0 & 0 & 0 & 0 & 0 \\
0 & 1 & 0 & 0 & 0 & 1 & 0 & 1 & 0 & 0 & 1 & 0 & 0 & 0 & 0 \\
0 & 0 & 1 & 1 & 0 & 0 & 0 & 0 & 1 & 0 & 0 & 1 & 0 & 0 & 0 \\
\end{array}\right]
$$

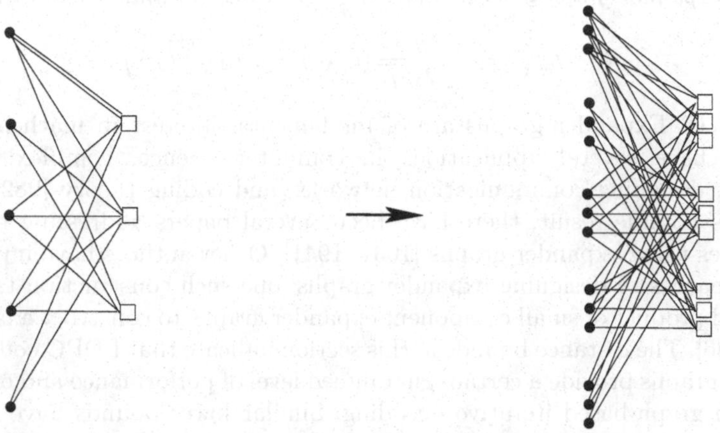

Remark 21.6.2 Note that the protograph used in a protograph code construction may have multiple edges between the same pair of vertices. Hence, the incidence matrix may have entries of value greater than one. Typically, the lifted graph that is used to represent the code no longer retains the multiple edges by the choice of permutations used.

Example 21.6.3 In Figure 21.11, a protograph is shown on the left with its corresponding incidence matrix. Note that the top right node (represented by the first row) has 2 edges to the top left node, no edges to the second left node, and a single edge to each of the remaining left nodes. An example of a degree 3 lift is shown on the right with its corresponding parity check matrix. The permutation matrices correspond to the assignments made to the edges of the protograph while constructing the lift, and indicate how the vertices in the clouds connect to each other. If the permutation σ is assigned to the edge $\{v_i, c_j\}$ in the protograph (with v_i a left node and c_j a right node), then $v_{i,k}$ is connected to $c_{j,\sigma(k)}$ in the lift for $k = 1, \ldots, 3$. Note that the local structure at each node belonging to the cloud of v (or the cloud of c) in the lift is the same as that for v (or for c) in the protograph.

An assignment of algebraically chosen permutations to the edges of a protograph is the same as a **permutation voltage assignment** which was introduced in the 1970s [856] to construct covering spaces for graphs. The corresponding labeled protograph is called a **permutation voltage graph** where it is assumed there is an orientation on the edges for the purpose of construction. In the case of protograph codes, we can assume, without loss of generality, that edges are oriented from variable nodes to check nodes to obtain the edge connections in the lift. Using theory from voltage graphs, LDPC codes were constructed in [227, 1095, 1097, 1103].

Definition 21.6.4 Array-based LDPC codes are defined as LDPC codes with parity check matrices composed of blocks of circulant matrices and were introduced in [703]. That is, the parity check matrix H is composed of $j \times r$ blocks of circulant matrices[2], each of size $\ell \times \ell$. The overall block length is $r\ell$ and the code rate of the corresponding LDPC code is $R \geq 1 - j/r$. Typically, each circulant matrix in an array is a cyclically shifted identity matrix, where the shift is chosen either randomly or algebraically.

Example 21.6.5 The **Tanner–Sridhara–Fuja (TSF) codes** introduced in [1788] are one family of array-based codes with algebraically chosen circulant matrices. Let a and b be two elements from the multiplicative group of $\mathbb{F}_\ell$, with orders j and r, respectively, and ℓ a prime. Then the parity check matrix H is defined as

$$H = \begin{bmatrix} I_{1,\ell} & I_{b,\ell} & I_{b^2,\ell} & \cdots & I_{b^{r-1},\ell} \\ I_{a,\ell} & I_{ab,\ell} & I_{ab^2,\ell} & \cdots & I_{ab^{r-1},\ell} \\ \vdots & \vdots & \vdots & \cdots & \vdots \\ I_{a^{j-1},\ell} & I_{a^{j-1}b,\ell} & I_{a^{j-1}b^2,\ell} & \cdots & I_{a^{j-1}b^{r-1},\ell} \end{bmatrix},$$

where $I_{a^i b^k, \ell}$ is the $\ell \times \ell$ identity matrix cyclically shifted to the left by $a^i b^k$ positions, where $a^i b^k$ is interpreted as an element in the integers $\mathbb{Z}_\ell = \{0, 1, \ldots, \ell - 1\}$.

One popular code is the $[155, 64, 20]$ code that is obtained by choosing $\ell = 31, a = 5, b = 2$ and $j = 3, r = 5$ in the above construction. The matrix is given by

$$H = \begin{bmatrix} I_{1,31} & I_{2,31} & I_{4,31} & I_{8,31} & I_{16,31} \\ I_{5,31} & I_{10,31} & I_{20,31} & I_{9,31} & I_{18,31} \\ I_{25,31} & I_{19,31} & I_{7,31} & I_{14,31} & I_{28,31} \end{bmatrix},$$

where $I_{x,31}$ is the 31×31 identity matrix cyclically shifted to the left by x positions. This code has block length 155, dimension 64, minimum distance 20, and the corresponding Tanner graph has a girth of 8.

Furthermore, several graph automorphism properties were shown to exist within the Tanner graph for these codes [1785] that provide an algebraic framework for analyzing and understanding the structure and properties of the codes.

Remark 21.6.6 It was shown in [1103] that all array-based LDPC codes and several other classes of protograph LDPC codes may be obtained and analyzed in the voltage graph framework. Moreover, tools from voltage graph theory can be used to obtain (often simpler) proofs of code graph properties.

The array-based LDPC codes as defined above also belong to the class of quasi-cyclic (QC) codes. One interesting feature of quasi-cylic array-based codes is that they may be used to obtain convolutional codes. A binary $\ell \times \ell$ circulant matrix may be represented by a polynomial in the ring $\mathbb{F}_2[X]$ mod $(X^\ell - 1)$. Viewing these polynomials as elements in $\mathbb{F}_2[X]$, a convolutional code is obtained. In terms of the Tanner graph, the convolutional code can be seen as an "unwrapping" of the Tanner graph of the QC code within each circulant [1789]. Also, by Theorem 7.6.3, the free distance of the convolutional code is bounded below by the minimum distance of the original QC code. LDPC convolutional codes were later shown to belong to the family of spatially-coupled LDPC codes [1172]. The latter are currently the most interesting class of LDPC codes capable of achieving near capacity performance on many communication channels and will be discussed in Section 21.8.

Array-based codes built from circulant matrices have a limited minimum distance. Specifically, the following result has been shown.

---

[2]A circulant matrix is a square matrix with each row of the matrix being a cyclic shift of its previous row by one position.

Theorem 21.6.7 ([1315]) *If an LDPC code is represented by a parity check matrix H composed of a $j \times r$ array of $\ell \times \ell$ permutation matrices and there is a $j\ell \times (j+1)\ell$ submatrix containing a grid of $j(j+1)$ nonzero permutation matrices that commute with each other, then the corresponding code has minimum distance less than or equal to $(j+1)!$.*

It is well known that circulant matrices commute under multiplication. Thus, the minimum distance bound holds for all array-based LDPC codes with the property defined in Theorem 21.6.7 that the parity check matrix contains a $j\ell \times (j+1)\ell$ submatrix having a $j(j+1)$ grid of nonzero circulants. A generalization of this theorem is given next [1730]. Protograph codes whose permutation assignments generate a commutative permutation group also have limitations on properties such as Tanner graph girth [1103].

Theorem 21.6.8 ([1730]) *Let C be an LDPC code with a parity check matrix H composed of a $j \times r$ grid of $\ell \times \ell$ permutation matrices that commute and is derived from a degree ℓ lift of a protograph represented by a $j \times r$ parity check matrix B. Then the minimum distance of C is bounded above by*

$$d \leq \min_{S \subseteq \{1,2,\ldots,r\}, \, |S|=j+1}^{*} \sum_{i \in S} \mathrm{perm}(B_{S \setminus \{i\}}),$$

where $\mathrm{perm}(B_{S \setminus \{i\}})$ denotes the permanent of the matrix consisting of the j columns of B in the set $S \setminus \{i\}$ and the $\overset{*}{\min}$ operator returns the smallest nonzero value from a set.

Despite the upper bounds on the minimum distance in Theorems 21.6.7 and 21.6.8, array-based LDPC codes are prominent in industry applications as their structure lends itself to efficient practical implementation.

21.7 Density Evolution

In this section, we explain the process of density evolution, which is an analytical technique to test how well a graph-based code may perform. Details will be given for the case of the binary erasure channel. For other channels, we refer to [1590]. An important result, known as the "Concentration Theorem", was presented in [1589, 1590] and essentially says the following.

Theorem 21.7.1 (Concentration Theorem [1590]) *Consider a random ensemble of LDPC codes in which each has the same variable node degree distribution and check node degree distribution in the corresponding Tanner graph. The following hold.*

(a) *A randomly chosen graph from this ensemble has iterative decoder performance "very close" (in terms of the difference in probabilities of sending an incorrect message at a specified iteration t) to the average iterative decoder performance of the random ensemble, and the difference between the two converges to 0 exponentially fast in the block length n.*

(b) *The average iterative decoder performance at iteration t is very close to the iterative decoder performance on a cycle free graph up to t decoding iterations (with the same degree distributions on the variable and check nodes) and the difference between the two converges to 0 exponentially fast in the block length n.*

(c) *There is a deterministic way of computing the probability density function of the messages exchanged in a cycle-free Tanner graph for t decoding iterations.*

Remark 21.7.2 Regarding (c), if the probability density function (pdf) of wrong/incorrect messages[3] in the graph tends to a delta function at $\pm\infty$ after $t \to \infty$ iterations, then it means that the iterative decoder is likely to converge for the specified channel quality. Hence, a *density evolution (DE) threshold* is defined as the worst channel quality that allows the pdfs of the incorrect messages to tend to a delta function at $\pm\infty$ as $t, n \to \infty$. This quantity is sometimes referred to as the *capacity* of the LDPC graph with the specified degree distribution on the variable and check nodes of the graph under the chosen iterative message-passing decoder. Thus, the performance of LDPC codes via iterative sum-product decoding is typically limited by their density evolution thresholds [1589, 1590].

Definition 21.7.3 The **density evolution (DE) threshold** of an LDPC code is the worst case channel condition at which the code may still converge under sum-product decoding, and relates primarily to the degree distribution $(\lambda(x), \rho(x))$ of the vertices in the code's Tanner graph.

Remark 21.7.4 In Figure 21.6, the DE threshold for a long block length LDPC code may be seen as the channel SNR where the waterfall region begins. The channel capacity is an SNR value that is at most the DE threshold. Closing the gap between the DE threshold and the channel capacity requires optimizing the degree distribution of the Tanner graph for the code [1591].

The binary erasure channel (BEC) is the simplest case to analyze. For the case of the BEC, let ϵ_{ch} be the probability of a symbol being erased by the channel; see Figure 30.2 where $\epsilon_{ch} = p$. Let $\epsilon(t)$ be the overall probability of sending an erasure message from a variable node to a check node in iteration t, and let $\gamma(t)$ be the overall probability of sending an erasure message from a check node to a variable node in iteration t. Further, let $\gamma_i(t)$ be the probability of sending an erasure message from a degree i check node to a neighboring variable node in iteration t. Similarly, let $\epsilon_i(t)$ be the probability of sending an erasure message from a degree i variable node to a neighboring check node in iteration t. Then we can derive the following:

$$\gamma_i(t) = 1 - \left(1 - \epsilon(t)\right)^{i-1},$$

since if any other incoming message to a check node is an erasure message, the outgoing message from that check node will also be an erasure message. Averaging over all degrees of a check node, we get

$$\gamma(t) = \sum_i \rho_i \gamma_i(t) = \sum_i \rho_i \left(1 - \left(1 - \epsilon(t)\right)^{i-1}\right) = 1 - \rho\left(1 - \epsilon(t)\right).$$

At the variable node of degree i, we have

$$\epsilon_i(t+1) = \epsilon_{ch}\left(\gamma(t)\right)^{i-1},$$

since only when all other of the incoming check node messages and the channel message at a variable node are erased will the outgoing message at that variable node also be an erasure message. Averaging over all degrees of a variable node, we get

$$\epsilon(t+1) = \sum_i \lambda_i \epsilon_i(t+1) = \sum_i \lambda_i \epsilon_{ch}\left(\gamma(t)\right)^{i-1} = \epsilon_{ch} \lambda(\gamma(t)).$$

---

[3] For example, a wrong message would be one with the incorrect sign compared to the true value of the corresponding variable node in the Sum-Product Algorithm implemented in the log-domain.

TABLE 21.1: DE thresholds for (j, r)-regular LDPC codes on the BEC

(j, r)	$(3, 4)$	$(3, 5)$	$(3, 6)$	$(4, 8)$	$(5, 10)$
DE threshold ϵ^*	0.64	0.52	0.423	0.38	0.34
Capacity ϵ_{cap}	0.75	0.6	0.5	0.5	0.5

Combining the above equations yields the general recursion for the probability of sending an erasure message during iteration $t + 1$ as

$$\epsilon(t + 1) = \epsilon_{ch}\lambda(1 - \rho(1 - \epsilon(t))).$$

Thus, the DE threshold is derived as the largest value for ϵ_{ch} for which $\epsilon(t + 1) \to 0$ as $t \to \infty$. Sometimes the DE threshold is written as

$$\epsilon^* = \arg\max_{\epsilon_{ch}}\{\epsilon_{ch}\lambda(1 - \rho(1 - \epsilon(t))) \to 0 \text{ as } t \to \infty\}.$$

A fixed point in the DE analysis is defined by the solutions to the following equation

$$x = \epsilon_{ch}\lambda(1 - \rho(1 - x)).$$

These are the points where the pdfs do not change with increasing iterations. A table of the DE thresholds for different regular LDPC codes over the binary erasure channel (BEC) are given in Table 21.1. The Shannon capacity ϵ_{cap} of the channel for the specified LDPC code rate is also given to compare against their DE threshold.

Remark 21.7.5 As noted in Table 21.1, regular LDPC codes typically have DE thresholds that are not close to the channel capacity. To reverse the problem, one can ask what is the distribution $\lambda(\cdot)$ and $\rho(\cdot)$ for which the DE threshold approaches the channel capacity. By optimizing irregular degree distributions and designing irregular codes, [1589] showed that we can come very close to the channel capacity on many communication channels.

21.8   Other Families of Codes over Graphs

In this section, we briefly describe a few other prominent families of codes over graphs. The family of fountain codes, including LT codes and Raptor codes, is treated in Chapter 30.

21.8.1   Turbo Codes

Turbo codes were introduced in [183] and perform very well under iterative decoding despite having relatively low minimum distance. A **turbo code** is a parallel concatenation of two or more convolutional codes in which information bits are permuted prior to being encoded by the individual convolutional codes. Graphically, turbo codes are represented by m constraint nodes, each representing a convolutional code component, and n variable nodes. Iterative decoding is performed by doing trellis computation on the component convolutional codes and exchanging that information among them. For more detail, we refer the reader to [183]. The main design problem for optimizing the performance of turbo codes is in the choice of component codes and permutations (called **interleavers**) used. For example, the free distances of the component convolutional codes limit the minimum distance of the overall turbo codes; consequently, these codes tend to have a poor distance distribution. A decoding threshold similar to a DE threshold may be obtained using a process called EXtrinsic Information Transfer (EXIT) analysis [1797, 1798] that measures how the quality of each component code's output improves based on the information gained from the other component codes during iterative decoding. Typically for block lengths less than 1000 bits, turbo codes tend to outperform LDPC codes, whereas for longer block lengths, LDPC codes tend to be superior.

21.8.2 Repeat Accumulate Codes

In contrast to turbo codes, a **repeat accumulate (RA) code** is a *serial* concatenation of a repetition code (where each symbol is repeated n times) and a simple rate 1 linear code such as an accumulator code, separated by a random permuter/interleaver. Since the encoding for repetition and accumulator codes is a fixed deterministic process, the design problem for RA codes is in the choice of permutation used. Similar to LDPC codes, RA codes have a simple graph representation that may be used for iterative decoding.

Moreover, certain subgraphs in the graph representation were shown in [397, 1786] to restrict the minimum distance of RA codes. While ensembles of RA codes are provably not asymptotically good [144], ensembles of **repeat accumulate accumulate (RAA) codes** were shown to be asymptotically good when the interleavers are chosen randomly [144]. RAA codes consist of a serial concatenation of a repetition code and *two* accumulators each following an interleaver. The design aspect of RAA codes again lies in the choice of the two interleavers. In [877], a construction of asymptotically good RAA codes is presented that uses one explicit interleaver with one random. Constructing asymptotically good RAA codes with two explicit interleavers remains an open problem.

21.8.3 Spatially-Coupled LDPC Codes

Among LDPC codes, regular LDPC codes typically have a poor DE threshold (with a large gap to channel capacity) but a larger minimum distance, whereas irregular LDPC codes with a certain fraction of degree two variable nodes typically have a DE threshold closer to the channel capacity but a smaller minimum distance [535]. In a breakthrough in further understanding LDPC codes and their decoding behavior, it was discovered that introducing a certain convolutional-like structure within the code's graph can improve the DE threshold without significantly affecting its minimum distance [1172, 1173, 1216, 1745]. The structure essentially has a regular subgraph repeated in a special way, wherein the repeated copies are interconnected by switching edges while preserving the original degree structure within each copy. Finally a few additional check node vertices of small degree, called terminating vertices, are added to the ends of the graph as necessary. This yields the SC-protograph, which is then lifted to obtain the graph for the code; see Figure 21.12. This design method is called **spatial-coupling**, and during graph-based decoding there is a *wave-effect* of good information moving from the boundaries to the center of the spatially-coupled graph. Certain fixed points of the Sum-Product Algorithm in the uncoupled case are eliminated when decoding over the spatially-coupled graph due to the structure and lower degree check nodes at the boundary [1172].

It was shown in [1173] that by having this spatial-coupling structure and also allowing for variable node degrees to increase, the DE threshold of the underlying spatially-coupled code reaches the MAP threshold of the code – a phenomenon termed as **threshold saturation**,

FIGURE 21.12: On the top, a base Tanner graph to be coupled to form an SC-protograph. Variable nodes are denoted by •, and check nodes are denoted by ◇. On the bottom, an SC-protograph resulting from randomly edge-spreading L copies of the Tanner graph on top, with coupling width $w = 1$, and applying the same map at each position.

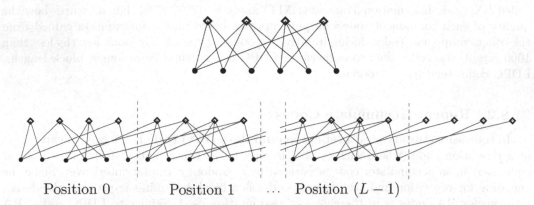

Position 0   Position 1   ⋯   Position $(L - 1)$

meaning the DE threshold is the best possible. Further, increasing the variable node degrees allows the MAP threshold to approach capacity of the channel. While the results were rigorously proven for the binary erasure channel (BEC), spatial-coupling has been shown to be effective on a wide range of other channels [682, 1172, 1173, 1216, 1745]. While spatially-coupled LDPC codes may be designed using several methods including edge-spreading an initial Tanner graph to obtain an SC-protograph (that is then lifted to obtain an SC-LDPC code) as in Figure 21.12, or using a so-called edge-cutting vector [685], all of these methods can be seen as special cases of a single protograph code with algebraically chosen permutations [149].

Part III

Applications

Chapter 22

Alternative Metrics

Marcelo Firer

State University of Campinas (Unicamp)

22.1   Introduction

The main scope of this chapter is metrics defined for coding and decoding purposes, mainly for block codes. Despite the fact that metrics are nearly ubiquitous in coding theory, it is very common to find them as a general background, with the role of metric invariants not always clearly stated. As a simple but eloquent example, the role of the minimum distance and the packing radius of a code are commonly interchanged. While the minimum distance $d(\mathcal{C})$ of a code $\mathcal{C}$ ensures that any error of size at most $d(\mathcal{C}) - 1$ can be detected, the packing radius $R(\mathcal{C})$ ensures that any error of size at most $R(\mathcal{C})$ can be corrected. These statements are true for any metric (considering additive errors) and the interchange between these two figures of merits follows from the fact that, in the most relevant metric in coding theory, the Hamming metric, the packing radius is determined by the minimum distance in the famous formula $R(\mathcal{C}) = \left\lfloor \frac{d_H(\mathcal{C})-1}{2} \right\rfloor$. This is not the case for general metrics, where not only one of the invariants may not be determined by the other, but determining the packing radius may be an intractable work, even in the case of a code with only two codewords; see, for example, [613]. As a second example of such a situation is the description of equivalence of codes, found in nearly every textbook on the subject, that seldom mentions the fact that differing by a combination of a permutation of the coordinates and the product by an invertible diagonal matrix means that the two considered codes are images one of the other by a linear isometry of the Hamming space. What is natural to the Hamming metric may become a "sloppy" approach when considering different metrics.

For this reason, besides properly defining a (non-exhaustive) list of metrics that are alternative to the Hamming metric, we will focus our attention on the *structural* aspects of the metric space $(\mathcal{X}, d)$. By structural, in this context, we mean aspects that are relevant for coding theory: the relation between the minimum distance and the packing radius, the weight (or distance) enumerator of the ambient space (or codes), the existence of a MacWilliams-type identity, a useful description of the group of (linear) isometries of $(\mathcal{X}, d)$ and, of course, bounds for the existence of codes with given parameters. What will be mainly missing is the search for codes attaining or approaching the bounds and particular interesting decoding algorithms, among other things, mainly due to space restrictions.

The most important metric in coding theory is the Hamming metric d_H. Being an *alternative metric* means (in the context of coding theory) being an alternative to the Hamming metric (or to the Lee metric). The prominence of the Hamming metric is due mainly to the fact that it is **matched** to the binary symmetric channel, in the sense that (considering codewords to be equiprobable)

$$\arg \min_{\mathbf{c} \in \mathcal{C}} \{d_H(\mathbf{x}, \mathbf{c})\} = \arg \max_{\mathbf{c} \in \mathcal{C}} \{P_{BSC}(\mathbf{c} \mid \mathbf{x})\},$$

for every code $\mathcal{C} \subseteq \mathbb{F}_2^n$ and every $\mathbf{x} \in \mathbb{F}_2^n$, where $P_{BSC}(\mathbf{c} \mid \mathbf{x}) := \varrho^{d_H(\mathbf{x},\mathbf{c})}(1 - \varrho)^{n-d_H(\mathbf{x},\mathbf{c})}$ is the probabilistic model of the channel; see Section 1.17.

Since errors are always described in a probabilistic model, *being matched* to a channel is a crucial property of a metric. Despite the fact that any metric can be matched to a channel (the reverse affirmation is not true - see [732]), only part of the alternative metrics arose matched to a specific channel. Others are considered to be "suitable" to correct some types of errors that are characteristic of the channel, despite the fact that the metric and the channel are not matched. Finally, some of the metrics were studied for their intrinsic interest, shedding some light on the role of metric invariants and waiting on a theoretical shelf to acquire more practical application. This recently happened to the rank metric, introduced by È. M. Gabidulin in 1985 [760] which became outstandingly relevant after the burst in research of network coding; see Chapter 29.

All the distance functions surveyed in this chapter actually determine a **metric**, in the sense that they satisfy the symmetry and positivity axioms (what entitles them to be called a **distance**, according to the terminology used in [532, Chapter 1]) and also the triangular inequality. At this point we should remark that the triangle inequality, many times the unique property that is not obvious, is a somehow superfluous property in the context of coding theory. Given two distances (or metrics) d_1 and d_2 defined on $\mathbb{F}_q^n$, they should be considered equivalent if they determine the same minimum distance decoding for every possible code $\mathcal{C}$ and every possible received message $\mathbf{x}$, in the sense that $\arg \min_{\mathbf{c} \in \mathcal{C}} \{d_1(\mathbf{x}, \mathbf{c})\} = \arg \min_{\mathbf{c} \in \mathcal{C}} \{d_2(\mathbf{x}, \mathbf{c})\}$. Since $\mathbb{F}_q^n$ is a finite space, given a distance d it is always possible to find a metric d' that is equivalent to d. For more details on this and related subjects see [614].

Nearly every metric to be presented in this chapter is a generalization of the Hamming or the Lee metric. Each generalization preserves some relevant properties, while it may fail to preserve others. In order to understand the way they depart from the Hamming metric, we should list and name some of these properties.

WD Weight defined: It is **defined by a weight**: $d(\mathbf{x}, \mathbf{y}) = \text{wt}(\mathbf{x} - \mathbf{y})$. This means that the metric is **invariant by translation**, in the sense that $d(\mathbf{x} + \mathbf{z}, \mathbf{y} + \mathbf{z}) = d(\mathbf{x}, \mathbf{y})$ for all $\mathbf{x}, \mathbf{y}, \mathbf{z} \in \mathbb{F}_q^n$. The invariance by translation property makes the metric suitable for decoding linear codes, allowing, for instance, syndrome decoding.

AP Additive Property: It is **additive**, in the sense that the weight wt may be considered to be defined on the alphabet $\mathbb{F}_q$ and $\text{wt}(\mathbf{x}) = \sum \text{wt}(x_i)$, for $\mathbf{x} = x_1 x_2 \cdots x_n \in \mathbb{F}_q^n$. The additive property is a condition for a metric to be matched to a memoryless channel.

RS Respect to support: This means that whenever $\text{supp}(\mathbf{x}) \subseteq \text{supp}(\mathbf{y})$, we have that $\text{wt}(\mathbf{x}) \leq \text{wt}(\mathbf{y})$. Respecting the support is an essential condition for a metric being suitable for decoding over a channel with additive noise.

Nearly all the alternative metrics presented in this chapter are considered in the context of block coding over an alphabet, and they may be essentially distinguished in the way they differ from the Hamming metric in the alphabet, in the spreading over the coordinates or in both ways. They can be classified into some different types. There are metrics that look at the way errors spread over the bits, ignoring any additional structure on the alphabet. This is the case of the *metrics generated by subspaces* and the *poset metrics*, introduced in Sections 22.2 and 22.3, respectively. In Section 22.4 we move into a different direction, considering metrics that "dig" into the structure of the alphabet and spread additively over the bits. These are natural generalizations of the Lee metric. In Section 22.5 we consider two families of metrics that move in these two directions simultaneously; they consider different weights on the alphabet but are non-additive. Up to this point, all the metrics are defined by a weight, hence invariant by translations. In the next three sections, we approach metrics that are not weight-defined: the *metrics for asymmetric channels* (Section 22.6), *editing metrics*, defined over strings of different length (Section 22.7) and *metrics defined on a permutation group* (Section 22.8).

Before we start presenting the alternative metrics, we shall establish some notations and conventions:

- Given $\mathbf{x} = x_1 x_2 \cdots x_n \in \mathbb{F}_q^n$, its **support** is denoted by $\text{supp}(\mathbf{x}) := \{i \in [n] \mid x_i \neq 0\}$, where $[n] := \{1, 2, \ldots, n\}$.

- $\mathbf{e}_i$ is the vector with $\text{supp}(\mathbf{e}_i) = \{i\}$ and the ith coordinate equals 1. We shall refer to $\beta = \{\mathbf{e}_1, \mathbf{e}_2, \ldots, \mathbf{e}_n\}$ as the **standard basis** of $\mathbb{F}_q^n$.

- Since we are considering many different metrics, we will denote each **distance function** by a subscript: the Hamming metric is denoted by d_H, the Lee metric by d_L, and so on. The subscript is used also in the notation for the **minimum distance** of a code $\mathcal{C}$: $d_H(\mathcal{C}), d_L(\mathcal{C})$, etc.

- Given a metric d on $\mathbb{F}_q^n$, we denote by $B_d^n(r, \mathbf{x}) = \{\mathbf{y} \in \mathbb{F}_q^n \mid d(\mathbf{x}, \mathbf{y}) \leq r\}$ and $S_d^n(r, \mathbf{x}) = \{\mathbf{y} \in \mathbb{F}_q^n \mid d(\mathbf{x}, \mathbf{y}) = r\}$ the metric **ball** and **sphere**, with cardinality $b_d^n(r)$ and $s_d^n(r)$ respectively, the center $\mathbf{x}$ being immaterial if the metric is weight-defined.

- The **packing radius** $R_d(\mathcal{C})$ of a code $\mathcal{C}$ contained in a space (for instance $\mathbb{F}_q^n$) endowed with a metric d is defined by $R_d(\mathcal{C}) = \max \{r \mid B_d^n(r, \mathbf{x}) \cap B_d^n(r, \mathbf{y}) = \emptyset$ for all $\mathbf{x}, \mathbf{y} \in \mathcal{C}, \mathbf{x} \neq \mathbf{y}\}$.

- When considering a metric d, we shall say that a code is **capable of correcting** t **errors** if the packing radius of the code, according to the metric d, equals t.

- Since we will consider many different metrics, we may explicitly indicate it by the suffix d$_*$, writing, for example, d$_*$-weight enumerator, d$_*$-perfect, d$_*$-isometry, $\mathrm{GL}(\mathcal{X}^n, \mathrm{d}_*)$, and so on.

22.2   Metrics Generated by Subspaces

This is a large family (increasing exponentially with the length n) of non-additive metrics. Metrics generated by subspaces, or simply, **subspace metrics**, were introduced by Gabidulin and Simonis in 1998 [767], generalizing some pre-existing families of metrics.

Given a set $X \subseteq \mathbb{F}_q^n$, we denote by $\mathrm{span}(X)$ the linear subspace of $\mathbb{F}_q^n$ spanned by X. Let $\mathcal{F} = \{F_1, F_2, \ldots, F_m\}$ be a family of subsets of $\mathbb{F}_q^n$ which **generates** the ambient space, in the sense that $\mathrm{span}(\bigcup_{i=1}^m F_i) = \mathbb{F}_q^n$. The elements of $\mathcal{F}$ are called **basic sets**.

Definition 22.2.1 The $\mathcal{F}$-weight $\mathrm{wt}_{\mathcal{F}}(\mathbf{x})$ of $\mathbf{0} \neq \mathbf{x} \in \mathbb{F}_q^n$ is the minimum size of a set $I \subseteq [n]$ such that $\mathbf{x} \in \mathrm{span}(\bigcup_{i \in I} F_i)$ and $\mathrm{wt}_{\mathcal{F}}(\mathbf{0}) = 0$. The $\mathcal{F}$-distance is $\mathrm{d}_{\mathcal{F}}(\mathbf{x}, \mathbf{y}) = \mathrm{wt}_{\mathcal{F}}(\mathbf{x} - \mathbf{y})$.

Without loss of generality, since $\mathrm{span}(\bigcup_{i \in [n]} F_i) = \mathrm{span}(\bigcup_{i \in [n]} \mathrm{span}(F_i))$, we may assume that each F_i is a vector subspace of $\mathbb{F}_q^n$; hence

$$\mathrm{wt}_{\mathcal{F}}(\mathbf{x}) = \min\left\{ |I| \ \middle| \ I \subseteq [m], \ \mathbf{x} = \sum_{i \in I} \mathbf{x}^i, \ \mathbf{x}^i \in F_i \right\}.$$

The Hamming metric is a particular case of subspace metric, which happens when $m = n$ and $F_i = \mathrm{span}(\mathbf{e}_i)$ for each $i \in [n]$. The well known **rank metric** on the space of all $k \times n$ matrices over $\mathbb{F}_q$, which was introduced in [760], can also be viewed as a particular instance of a projective metric (a special type of subspace metric, to be defined in Section 22.2.1), where $\mathcal{F}$ is the set of all $k \times n$ matrices of rank 1. This metric became nearly ubiquitous in network coding after Silva, Kschischang, and Koetter's work [1711]. Despite its importance, the rank metric will not be explored in this chapter. Many of its important features can be found in Chapter 29, devoted to network coding, and in Chapter 11, where rank-metric codes are examined.

As far as this author was able to track, the subspace metrics, in their full generality, are unexplored in the professional literature. However, something is known about some particular cases, namely projective and combinatorial metrics, most of it due to the work of Gabidulin and co-authors. We present some of these results on particular subclasses in the sequence.

22.2.1   Projective Metrics

Projective metrics were introduced in the same paper [767] where Gabidulin and Simonis first defined the subspace metrics. When each F_i is a vector or a one-dimensional vector subspace of $\mathbb{F}_q^n$, it may be seen as a point in projective space; so in this case the $\mathbb{F}_q^n$-metric is called a **projective metric**. In this situation, we can identify each F_i with a nonzero vector $\mathbf{f}_i \in F_i$. We denote by

$$d_{\mathcal{F}}(\mathcal{C}) = \min\{d_{\mathcal{F}}(\mathbf{x}, \mathbf{y}) \mid \mathbf{x}, \mathbf{y} \in \mathcal{C}, \mathbf{x} \neq \mathbf{y}\}$$

the $d_{\mathcal{F}}$-**minimum distance** of a code $\mathcal{C}$. For an $[n, k]_q$ linear code $\mathcal{C}$, the usual Singleton Bound $d_{\mathcal{F}}(\mathcal{C}) \leq n - k + 1$ was proved to hold (see [765]).

We remark that, for a projective metric, since $\{\mathbf{f}_1, \mathbf{f}_2, \ldots, \mathbf{f}_m\}$ generates $\mathbb{F}_q^n$, we have that $m \geq n$. By considering the linear map $\varphi : \mathbb{F}_q^m \to \mathbb{F}_q^n$ determined by $\varphi(\mathbf{e}_i) = \mathbf{f}_i$, its kernel $\mathcal{P} := \ker(\varphi) \subseteq \mathbb{F}_q^m$ is an $[m, m - n]_q$ linear code called the **parent code** of $\mathcal{F}$. Given an $[m, m - n]_q$ linear code $\mathcal{C}$, the columns of a parity check matrix determine a family $\mathcal{F}_{\mathcal{C}}$ of basic sets such that $\mathcal{C}$ is the parent code of $\mathcal{F}_{\mathcal{C}}$. The parent code helps to compute the $\mathcal{F}$-weight of a vector. Given $\mathbf{x} \in \mathbb{F}_q^n$, the inverse image $\varphi^{-1}(\mathbf{x})$ is a coset of the parent code $\mathcal{P}$. If $\mathbf{y} \in \varphi^{-1}(\mathbf{x})$ is a coset leader (a vector of minimal Hamming weight), we have that $\mathrm{wt}_{\mathcal{F}}(\mathbf{x}) = \mathrm{wt}_{\mathrm{H}}(\mathbf{y})$ [767, Proposition 2].

This simple relation between the $\mathcal{F}$-weight of a vector and the Hamming weight of the parental coset is used to produce a class of Gilbert–Varshamov Bounds for codes with an $\mathcal{F}$-metric. We let $L_i(\mathcal{P})$ be the number of cosets of $\mathcal{P}$ with Hamming weight i. Then we have that $L_i(\mathcal{P}) = s_{d_{\mathcal{F}}}^m(i)$ for every i less than or equal to the packing radius of the parent code $\mathcal{P}$. This simple remark is the key to prove the following Gilbert–Varshamov Bound.

Theorem 22.2.2 ([767, Theorem 3.1.]) *Let $\mathcal{P}_m$ be a sequence of $[m, n]_2$ linear codes with $L_i(\mathcal{P}_m) \simeq \binom{m}{i}$, where $\simeq$ means asymptotic behavior. Suppose that $\mathcal{P}_m$ attains the Gilbert–Varshamov Bound, and let $\mathcal{F}_m$ be the family of basic sets determined by $\mathcal{P}_m$. Then there exists a code $\mathcal{C} \subseteq \mathbb{F}_q^n$ with $\mathcal{F}_m$-distance $d = d_{\mathcal{F}_m}(\mathcal{C})$ and M elements where*

$$M \simeq \frac{2^n}{\sum_{i=0}^{d} \binom{m}{i}}.$$

The parent code of $\mathcal{F}$ is used to produce a family of codes attaining the Singleton Bound, a generalization of Reed–Solomon codes, using concatenation of Vandermonde matrices to generate the basic sets (see [765, Section 3]) and also to produce a family of $d_{\mathcal{F}}$-perfect codes (see [768] for the construction).

The **phase-rotation** metric is a special instance of a projective metric when $m = n + 1$, $F_i = \mathrm{span}(\mathbf{e}_i)$ for $i \leq n$ and $F_{n+1} = \mathrm{span}(\mathbf{1})$ where $\mathbf{1} := 11 \cdots 1$. It is suitable for decoding in a binary channel in which errors are a cascade of a phase-inversion channel (where every bit of the input message is flipped) and a binary symmetric channel. The actual model of the channel depends on the probability of the two types of errors, and hence the term "suitable" is not to be confused with the matching property.

It is easy to see that, in this case, $\mathrm{wt}_{\mathcal{F}}(\mathbf{x}) = \min\{\mathrm{wt}_{\mathrm{H}}(\mathbf{x}), n + 1 - \mathrm{wt}_{\mathrm{H}}(\mathbf{x})\}$. In [763], the authors provide constructions of perfect codes relative to the phase-rotation metric, based on the construction of perfect codes with the Hamming metric and also show how to reduce the decision decoding for the phase-rotation metric to decoding for the Hamming metric.

22.2.2 Combinatorial Metrics

The family of **combinatorial metrics** was introduced in 1973 by Gabidulin [759]. It is a particular instance of subspace metrics that is obtained as follows: we let $\mathcal{A} = \{A_1, A_2, \ldots, A_m \mid A_i \subseteq [n]\}$ be a **covering** of $[n]$, in the sense that $[n] = \bigcup_{i=1}^{m} A_i$ and define $F_i = \{\mathbf{x} \in \mathbb{F}_q^n \mid \mathrm{supp}(\mathbf{x}) \subseteq A_i\}$, for $i \in [m]$, and $\mathcal{F}_{\mathcal{A}} = \{F_i \mid 1 \leq i \leq m\}$.

We remark that the combinatorial metric can be expressed in a simpler way, as introduced in [759]:

$$\mathrm{wt}_{\mathcal{F}}(\mathbf{x}) = \min\left\{ |I| \;\Big|\; I \subseteq [m],\ \mathrm{supp}(\mathbf{x}) \subseteq \bigcup_{i \in I} A_i \right\}.$$

When considering a combinatorial metric, we shall identify a basic set with its support; hence call each A_i also a basic set and denote $\mathcal{F} = \{A_1, A_2, \ldots, A_m\}$. We note that, if $\mathcal{F} = \big\{ \{i\} \mid i \in [n] \big\}$, then $\mathrm{d}_{\mathcal{F}}$ is the Hamming metric.

There are many particular instances of combinatorial metrics that have been used for particular coding necessities, and there is a substantial literature devoted to codes with "good" properties according to such particular instances, despite the fact that the actual metric is hardly mentioned. However, since its introduction in 1973 by Gabidulin, the structural aspects of the metric were left nearly untouched.

A Singleton-type bound was determined in 1996 [252] for codes which are not necessarily linear. We present here its linear version: given an $[n,k]_q$ linear code $\mathcal{C}$ and a covering $\mathcal{F}$, its minimum distance $\mathrm{d}_{\mathcal{F}}(\mathcal{C})$ is bounded by

$$n\frac{\mathrm{d}_{\mathcal{F}}(\mathcal{C}) - 1}{D} \leq \left\lceil n\frac{\mathrm{d}_{\mathcal{F}}(\mathcal{C}) - 1}{D} \right\rceil \leq n - k,$$

where $\lceil \cdot \rceil$ is the ceiling function and D is the minimum number of basic sets needed to cover $[n]$. We remark that $D \leq n$ and equality in the bounds holds if and only if $\mathrm{d}_{\mathcal{F}}$ is the Hamming metric, and, in this case, we get the usual Singleton Bound.

Recently, combinatorial metrics started to be explored in a more systematic way. In [1510] the authors give a necessary and sufficient condition for the existence of a MacWilliams Identity, that is, conditions on $\mathcal{F}$ to ensure that the $\mathrm{d}_{\mathcal{F}}$-weight distribution of a code $\mathcal{C}$ determines the $\mathrm{d}_{\mathcal{F}}$-weight distribution of the dual code $\mathcal{C}^{\perp}$.

Theorem 22.2.3 *Let $\mathcal{F} = \{A_1, A_2, \ldots, A_m\}$ be a covering of $[n]$. Then the combinatorial metric $\mathrm{d}_{\mathcal{F}}$ admits a MacWilliams Identity if and only if $|A_i|$ is constant and $A_i \cap A_j = \emptyset$, for all $i \neq j$.*

In the same work, there is a description of the group $\mathrm{GL}(\mathbb{F}_q^n, \mathrm{d}_{\mathcal{F}})$ of linear $\mathrm{d}_{\mathcal{F}}$-isometries. We briefly explain how it is obtained.

First of all, we say that a permutation $\sigma \in \mathrm{S}_n$ **preserves the covering** if $\sigma(A) \in \mathcal{F}$, for all $A \in \mathcal{F}$. Each such permutation induces a linear isometry $T_\sigma(\mathbf{x}) := x_{\sigma(1)} x_{\sigma(2)} \cdots x_{\sigma(n)}$. The set of all such isometries is a group denoted by $G_{\mathcal{F}}$.

Next, let $\mathcal{F}^i := \{A \in \mathcal{F} \mid i \in A\}$. This defines an equivalence relation on $[n]$: $i \sim_{\mathcal{F}} j$ if and only if $\mathcal{F}^i = \mathcal{F}^j$. Let s be the number of equivalence classes and denote by $[\![i]\!]$ the equivalence class of i. We construct an $s \times m$ matrix $M = [m_{[\![i]\!]j}]$, the **incidence matrix** of this equivalence relation:

$$m_{[\![i]\!]j} = \begin{cases} 1 & \text{if } [\![i]\!] \subseteq A_j, \\ 0 & \text{otherwise.} \end{cases}$$

Given an $n \times n$ matrix $B = [b_{xy}]$ with entries in $\mathbb{F}_q$, let us consider the (i,j) block (sub-matrix) $B_{ij} = [b_{xy}]_{x \in [\![i]\!], y \in [\![j]\!]}$. We say that B **respects** M if, for $[\![i]\!] = [\![j]\!]$, the block $B_{ij} = B_{ii} = [b_{xy}]_{x,y \in [\![i]\!]}$ is an invertible matrix and, for $[\![i]\!] \neq [\![j]\!]$, then $B_{ij} \neq \mathbf{0}$ (B_{ij} is a nonzero matrix) implies that $\mathrm{supp}(\mathbf{v}^j) \subseteq \mathrm{supp}(\mathbf{v}^i)$, where $\mathbf{v}^k$ is the kth row of M. Each matrix B respecting M determines a linear $\mathrm{d}_{\mathcal{F}}$-isometry, and the set K_M of all such matrices is also a group.

Theorem 22.2.4 *For $q > 2$, the group $\mathrm{GL}(\mathbb{F}_q^n, \mathrm{d}_\mathcal{F})$ of linear $\mathrm{d}_\mathcal{F}$-isometries is the semi-direct product $\mathrm{GL}(\mathbb{F}_q^n, \mathrm{d}_\mathcal{F}) = G_\mathcal{F} \ltimes K_M$. For $q = 2$, we have the inclusion $G_\mathcal{F} \ltimes K_M \subseteq \mathrm{GL}(\mathbb{F}_2^n, \mathrm{d}_\mathcal{F})$ and equality may or not hold, depending of $\mathcal{F}$.*

There are some special instances of combinatorial metrics that deserve to be mentioned, namely the block metrics, burst metrics, and 2-dimensional burst metrics.

22.2.2.1 Block Metrics

This is the case when $\mathcal{F}$ determines a partition of $[n]$, that is, if $A_i \cap A_j = \emptyset$ for any $A_i, A_j \in \mathcal{F}, i \neq j$. In this context, each basic set is usually called a **block**. In [716] we can find many classical questions of coding theory approached with the block metrics, starting with the Hamming and Singleton Bounds. If $n_i = |A_i|$ ordered so that $n_1 \geq n_2 \geq \cdots \geq n_m$, the Singleton Bound is just $n - k \geq n_1 + n_2 + \cdots + n_{d-1}$, where d is the $\mathrm{d}_\mathcal{F}$-minimum distance. The Hamming Bound is provided by simply counting the number of elements in a metric ball. Both the Singleton and the Hamming Bounds are proved in [716, Theorem 2.1]. We note that, as a particular instance of a combinatorial metric, the MacWilliams Identity holds for a block metric only if the basic sets (blocks) all have equal size, and in this case, Feng et al. provide an explicit expression for the identity.

Theorem 22.2.5 ([716, Theorem 4.3]) *Let $\mathcal{C} \subseteq \mathbb{F}_q^n$ be a linear code and $\mathcal{F}$ be a partition of $[n]$ with m basic sets each having cardinality r. Let $\mathcal{F}\mathrm{we}_\mathcal{C}(x, y) = \sum_{c \in \mathcal{C}} x^{\mathrm{wt}_\mathcal{F}(\mathcal{C})} y^{m - \mathrm{wt}_\mathcal{F}(\mathcal{C})}$ be the $\mathcal{F}$-weight enumerator of the code $\mathcal{C}$. Then,*

$$\mathcal{F}\mathrm{we}_{\mathcal{C}^\perp}(x, y) = \frac{1}{|\mathcal{C}|} \mathcal{F}\mathrm{we}_\mathcal{C}(y - x, y + (q^r - 1)x).$$

The authors also establish necessary and sufficient conditions on a parity check matrix for a code to be perfect or MDS. Also, algebraic geometry codes are explored with this metric.

22.2.2.2 b-Burst Metrics

The b-**burst metric** is the particular case of a combinatorial metric where the basic sets are all the sequences of b consecutive elements in $[n]$, that is, $\mathcal{F}_{[b]} = \{[b], [b] + 1, [b] + 2, \ldots, [b] + (n - b)\}$, with $[b] + i := \{1 + i, 2 + i, \ldots, b + i\}$. A variation will consider all the n cyclic (modulus n) sequences of b consecutive elements. It was first introduced in 1970 [303]. It is suitable for predicting error-correcting capabilities of errors which occur in bursts of length at most b.

Theorem 22.2.6 ([303, Theorems 1,2 and 4]) *Let $\mathcal{C}$ be a linear code with minimum burst distance $d_\mathcal{F}(\mathcal{C})$ where $\mathcal{F} = \mathcal{F}_{[b]}$. Then $\mathcal{C}$ corrects*

(a) *any pattern of $\left\lfloor \frac{d_\mathcal{F}(\mathcal{C}) - 1}{2} \right\rfloor$ error bursts of length b;*

(b) *any pattern of $d_\mathcal{F}(\mathcal{C}) - 1$ erasure bursts of length b;*

(c) *any pattern of m_1 error bursts and m_2 erasure bursts of length $e_1, e_2, \ldots, e_{m_2}$, with every erasure preceded and followed by a burst error of length $b_{1,i}$ and $b_{2,i}$, respectively, provided that $2m_1 + m_2 < d_\mathcal{F}(\mathcal{C})$ and $b_{1,i} + e_i + b_{2,i} \leq b$.*

As an example, it is possible to prove that a code $\mathcal{C}$ can correct any burst error of length b if and only if $3 \leq \min \{\mathrm{d}_{\mathcal{F}_{[b]}}(\mathbf{x}, \mathbf{y}) \mid \mathbf{x}, \mathbf{y} \in \mathcal{C}, \mathbf{x} \neq \mathbf{y}\}$; that is, the error correction capability (the packing radius) is determined by the minimum distance and this happens in a similar way to the Hamming metric. The Reiger Bound [1585], for example, can be expressed in terms of the packing radius.

A different instance where the b-burst metric arises is in high density storage media. In these storage devices, the density does not permit reading the bits one-by-one, but only in sequences of b-bits at a time. This is what is called the **b-symbol read channel**. Recently, Cassuto and Blaum ([366], for the case $b = 2$) and Yaakobi et al. ([1918], for $b > 2$), proposed coding and decoding schemes for the b-symbol read channel. The decoding scheme is a nearest-neighbor decoding, which considers a metric that is actually the b-burst metric. Many bounds on such coding schemes and construction of MDS codes can be found in the literature; see for example [543]. We should remark that in all these works, the authors consider as undistinguished a particular encoding approach and the metric decoding.

22.2.2.3 $b_1 \times b_2$-**Burst Metrics**

When considering codewords to be represented as a matrix, a way to couple bursts of size b_1 and b_2 in the rows and columns, respectively, is to consider as the basic sets the $(b_2 + 1) \times (b_1 + 1)$ matrices

$$
T_{r,s} = \begin{bmatrix}
(r,s) & (r,s+1) & \cdots & (r,s+b_1) \\
(r+1,s) & (r+1,s+1) & \cdots & (r+1,s+b_1) \\
\vdots & \vdots & \ddots & \vdots \\
(r+b_2,s) & (r+b_2,s+1) & \cdots & (r+b_2,s+b_1)
\end{bmatrix},
$$

where $r \in [n_1], s \in [n_2]$ and the entries are considered $(\bmod\ n_1, \bmod\ n_2)$. A large account on the subject can be found in the survey [217] on array codes.

22.3   Poset Metrics

Poset metrics, in their full generality, were introduced by Brualdi, Graves and Lawrence in 1995 [315], as a generalization of Niederreiter's metric [1434]. Few of these metrics are matched to actual relevant channels, but they are relatively well understood in their intrinsic aspects and so become an interesting contribution to the understanding of the role of metrics in coding theory. As with the subspace metrics presented in Section 22.2, poset metrics are a family of non-additive metrics, respecting support and defined by a weight (properties RS and WD in Section 22.1).

In what follows we shall introduce the poset metrics as presented in [315], then consider some particular cases of posets and later move on to the many generalizations that arose in recent years. A recent account on poset metrics and its generalizations can be found in [731].

Let $P = ([n], \preceq_P)$ be a partial order (**poset**, for short) over $[n]$. A subset $I \subseteq [n]$ is called an **ideal** if $i \in I$ and $j \preceq_P i$ implies $j \in I$. For $X \subseteq [n]$, we denote by $\langle X \rangle_P$ the **ideal generated by** X (the smallest ideal containing it). Given $\mathbf{x} \in \mathbb{F}_q^n$, the P-**weight of** $\mathbf{x}$ is $\mathrm{wt}_P(\mathbf{x}) = |\langle \mathrm{supp}(\mathbf{x}) \rangle_P|$, where $|A|$ is the cardinality of A and $\mathrm{supp}(\mathbf{x})$ is the support of the vector $\mathbf{x}$. The function wt_P is a weight and the **poset metric** d_P is the weight defined metric $\mathrm{d}_P(\mathbf{x}, \mathbf{y}) = \mathrm{wt}_P(\mathbf{x} - \mathbf{y})$ We should remark that for the particular case of an **anti-chain** ($i \preceq_P j \iff i = j$) we have that the poset metric coincides with the Hamming metric. It is worth remarking that this family is complementary to the family of metrics generated by subspaces (Section 22.2), in the sense that the Hamming metric is the only intersection of these two families.

Codes with a poset metric, or simply poset codes, were investigated along many different paths, including conditions (on the poset) to turn a given family of codes into perfect or MDS codes. The fact that there are relatively many codes with such properties is what first attracted attention to poset metrics. For perfect or MDS codes with poset metrics see [315, 637, 1016, 1019, 1035, 1118, 1550], among others.

Considering more structural results, one can find in [1487] a description of the group of linear isometries of the metric space $(\mathbb{F}_q^n, d_P)$ as a semi-direct product of groups. Each element σ of the group $\mathrm{Aut}(P)$ of poset automorphisms (permutations of $[n]$ that preserve the order) determines a linear isometry $T_\sigma(\mathbf{x}) = x_{\sigma(1)} x_{\sigma(2)} \cdots x_{\sigma(n)}$. The set of all such isometries is a group, also denoted by $\mathrm{Aut}(P)$. We let M_P be the order matrix of P, that is, $M_P = [m_{ij}]$ is an $n \times n$ matrix and $m_{ij} = 1$ if $i \preceq j$ and $m_{ij} = 0$ otherwise. An $n \times n$ matrix $A = [a_{ij}]$ with entries in $\mathbb{F}_q$ such that $a_{ii} \neq 0$ and $\mathrm{supp}(A) \subseteq \mathrm{supp}(M_P)$ determines a linear isometry, and the family of all such maps is a group denoted by G_P.

Theorem 22.3.1 ([1487, Corollary 1.3]) *The group* $\mathrm{GL}(\mathbb{F}_q^n, d_P)$ *of linear isometries of* $(\mathbb{F}_q^n, d_P)$ *is the semi-direct product* $\mathrm{GL}(\mathbb{F}_q^n, d_P) = \mathrm{Aut}(P) \ltimes G_P$.

MacWilliams-type identities are explored in [405], revealing a deep understanding about the most important duality result in coding theory. First of all, it was noticed since [1035] that, when looking at the dual code $\mathcal{C}^\perp$, we are actually not interested in the d_P-weight distribution but in its $d_{P^\perp}$-weight distribution, determined by the **opposite poset** $P^\perp$ defined by $i \preceq_{P^\perp} j \iff j \preceq_P i$. The innovation proposed in [405] is to note that the metric invariant is not necessarily the weight distribution, but rather a distribution of elements in classes determined by a suitable equivalence relation on ideals of P. The authors identify three very basic and different equivalence relations of ideals. Given two poset ideals $I, J \subseteq [n]$ we establish the following equivalence relations.

E_C: $(I, J) \in E_C$ if they have the same cardinality.

E_S: $(I, J) \in E_S$ if they are isomorphic as sub-posets.

E_H: Given a subgroup $H \subseteq \mathrm{Aut}(P)$, $(I, J) \in E_H$ if there is $\sigma \in H$ such that $\sigma(I) = J$.

It is easy to see that $E_H \subseteq E_S \subseteq E_C$.

Given an equivalence relation E over the set $\mathcal{I}(P)$ of ideals of P, for each coset $\overline{I} \in \mathcal{I}(P)/E$, the $\overline{I}$-sphere $S_{\overline{I}, E}^P$ centered at $\mathbf{0}$ with respect to E is

$$S_{\overline{I}, E}^P := \{ \mathbf{x} \in \mathbb{F}_q^n \mid (\langle \mathrm{supp}(\mathbf{x}) \rangle_P, I) \in E \}.$$

If E is an equivalence relation on $\mathcal{I}(P)$, we denote by $E^\perp$ the equivalence relation on $\mathcal{I}(P^\perp)$, where E may be either E_C, E_S, or E_H. Given an ideal $I \in \mathcal{I}(P)$, we denote its complement by $I^c = [n] \setminus I$. We say that $E^\perp$ is the **dual relation** of E.

We define $A_{\overline{I}, E}(\mathcal{C}) := |S_{\overline{I}, E}^P \cap \mathcal{C}|$ and call $W_{P, E}(\mathcal{C}) := [A_{\overline{I}, E}(\mathcal{C})]_{\overline{I} \in \mathcal{I}(P)/E}$ the **spectrum** of $\mathcal{C}$ with respect to E. We remark that, for $E = E_C$, the spectrum of $\mathcal{C}$ consists of the coefficients of the weight enumerator polynomial of $\mathcal{C}$.

Definition 22.3.2 Let P be a poset, E an equivalence relation on $\mathcal{I}(P)$ and suppose that $E^\perp$ is the dual relation on $\mathcal{I}(P^\perp)$. The relation E is a **MacWilliams equivalence relation** if, given linear codes $\mathcal{C}_1$ and $\mathcal{C}_2$, $W_{P,E}(\mathcal{C}_1) = W_{P,E}(\mathcal{C}_2)$ implies $W_{P^\perp, E^\perp}(\mathcal{C}_1^\perp) = W_{P^\perp, E^\perp}(\mathcal{C}_2^\perp)$.

For the case $E = E_C$, to be a MacWilliams equivalence relation implies the existence of a MacWilliams Identity in the usual sense. In general, for E being a MacWilliams equivalence relation depends on the choice of P.

Theorem 22.3.3 ([405, Theorem 3])

(a) E_H is a MacWilliams equivalence relation, for every poset P and every subgroup $H \subseteq \mathrm{Aut}(P)$.

(b) E_S is a MacWilliams equivalence relation if and only if P is such that $I \sim J$ implies $I^c \sim J^c$, for every $I, J \in \mathcal{I}(P)$.

(c) E_C is a MacWilliams equivalence relation if and only if P is a hierarchical poset.

As for the last item, a poset P is called **hierarchical** if $[n]$ can be partitioned into a disjoint union $[n] = H_1 \cup H_2 \cup \cdots \cup H_r$ in such a way that $i \preceq_P j$ if and only if $i \in H_{h_i}, j \in H_{h_j}$ and $h_i < h_j$. This family of posets includes the minimal poset (an anti-chain) and maximal (a chain, defined shortly). Hierarchical poset metrics are a true generalization of the Hamming metric, in the sense that many well known results concerning the Hamming metric are actually a characteristic of hierarchical poset metrics.

Theorem 22.3.4 ([1313]) *Let $P = ([n], \preceq_P)$ be a poset. Then P is hierarchical if and only if any of the following (equivalent) properties holds.*

(a) *P admits a MacWilliams Identity (E_C is a MacWilliams equivalence relation).*

(b) *The MacWilliams Extension Theorem holds: given two $[n, k]_q$ linear codes C_1, C_2, any linear isometry $f : C_1 \to C_2$ can be extended to a linear isometry $F : \mathbb{F}_q^n \to \mathbb{F}_q^n$.*

(c) *The packing radius of a code is a function of the minimum distance.*

(d) *Linear isometries act transitively on metric spheres: for every $\mathbf{x}, \mathbf{y} \in \mathbb{F}_q^n$, if $\mathrm{wt}_P(\mathbf{x}) = \mathrm{wt}_P(\mathbf{y})$, there is $\sigma \in \mathrm{GL}(\mathbb{F}_q^n, \mathrm{d}_P)$ such that $\sigma(\mathbf{x}) = \mathbf{y}$.*

All these properties (and some more in [1313]) are a direct consequence of the canonical decomposition of codes with hierarchical posets [712], which allows the transformation of problems with hierarchical poset metrics into problems with the well-studied Hamming metric.

Another relevant class of poset metrics is the family of NRT metrics, defined by a poset which is a disjoint union of chains. A poset P is a **chain**, or total order, when (up to relabeling) $1 \preceq_P 2 \preceq_P \cdots \preceq_P n$. Let $(P, [n])$ be a disjoint union of m chains, each of length l; that is, $n = ml$ and the relations in P are $jl + 1 \preceq_P jl + 2 \preceq_P \cdots \preceq_P (j+1)l$, for $j = 0, 1, \ldots, m - 1$. This is the original instance explored by Niederreiter in a sequence of three papers [1433, 1434, 1435] and later, in 1997, by Rosenbloom and Tsfasman [1595], with an information-theoretic approach. After these three main authors, it is known as the **NRT metric**. Many coding-theoretic questions have been investigated with respect to the NRT metric, including MacWilliams Duality [636], MDS codes [635, 1721], structure and decoding of linear codes [1480, 1486], and coverings [370]. In [127] the authors show the connection of NRT metric codes with simple models of information transmission channels and introduce a simple invariant that characterizes the orbits of vectors under the group of linear isometries (the **shape** of a vector). For the particular case of one single chain ($n = l, m = 1$), d_P is an *ultra-metric* and, given a code C with minimum distance $d_{\mathrm{d}_P}(C)$, it is possible to detect and correct every error of P-weight at most $d_{\mathrm{d}_P}(C) - 1$. Despite some strangeness that may be caused by ultra-metrics, this instance is simple and well understood; see [1486].

An extensive and updated survey of poset metrics in the context of coding theory can be found in [731]. In the following two subsections we will present some recent generalizations of the poset metrics.

22.3.1   Poset-Block Metrics

In 2006, Feng et al. [716] started to explore block metrics, a special instance of combinatorial metrics, presented in Section 22.2.2.1. In 2008, Alves et al. [45] combined the poset and the block structure, giving rise to the so-called *poset-block metrics*.

We let

$$\mathbb{F}_q^N := V_1 \oplus V_2 \oplus \cdots \oplus V_n$$

be a decomposition of $\mathbb{F}_q^N$ as a direct sum of subspaces. We denote $k_i = \dim(V_i) > 0$, $\pi = (k_1, k_2, \ldots, k_n)$, and we call this decomposition a **block structure**. Being a direct sum, we have that each $\mathbf{x} \in \mathbb{F}_q^N$ has a unique decomposition as $\mathbf{x} = \mathbf{x}_1 + \mathbf{x}_2 + \cdots + \mathbf{x}_n$ with $\mathbf{x}_i \in V_i$. Considering such a decomposition, the π-**support** of $\mathbf{x}$ is defined as $\mathrm{supp}_\pi(\mathbf{x}) := \{i \in [n] \mid \mathbf{x}_i \neq 0\}$. Counting the π-support gives rise to the π-**weight** $\mathrm{wt}_\pi(\mathbf{x}) := |\mathrm{supp}_\pi(\mathbf{x})|$.

If $P = ([n], \preceq_P)$ is a poset over $[n]$, we may combine the block structure π and the poset structure P into a single one, by counting the P-ideal generated by the π-support of a vector, obtaining the **poset-block weight** and the **poset-block metric**:

$$\mathrm{wt}_{(P,\pi)}(\mathbf{x}) := |\langle \mathrm{supp}_\pi(\mathbf{x})\rangle_P| \quad \text{and} \quad \mathrm{d}_{(P,\pi)}(\mathbf{x}, \mathbf{y}) := \mathrm{wt}_{(P,\pi)}(\mathbf{x} - \mathbf{y}).$$

In [45], the authors give a characterization of the group of linear isometries of $(\mathbb{F}_q^n, \mathrm{d}_{(P,\pi)})$. Necessary and sufficient conditions are provided for such a metric to admit a MacWilliams Identity. Perfect and MDS codes with poset-block metrics are investigated in [45] and [487]. It is worth noting that an interesting feature of this family of metrics is combining together a poset metric, which increases the weight of vectors (hence "shrinking" the metric balls), with a block metric, which decreases the weights (hence "blowing-up" the balls).

22.3.2   Graph Metrics

The Hasse diagram of a poset P can be seen as special kind of directed graph, thinking of an edge of the diagram connecting $i, j \in [n]$ to be a directed edge with initial point at j if $i \preceq_P j$. Inasmuch, it is natural to generalize the poset metrics to digraphs. This family of metrics was introduced in 2016 in [689] and explored in [690].

Consider a finite **directed graph** (or simply **digraph**) $G = G(V, E)$ consisting of a finite set of **vertices** $V = \{v_1, v_2, \ldots, v_n\}$ and a set of **directed edges** $E \subseteq V \times V$ (parallel edges are not allowed). A **directed path** is a sequence $v_{i_1}, v_{i_2}, \ldots, v_{i_r}$ of vertices where every two consecutive vertices determine a directed edge. A **cycle** is a directed path in which $v_{i_1} = v_{i_r}$. We say that a vertex u **dominates** a vertex v if there is a directed path starting at u and passing through v. A set $X \subseteq V$ is called a **closed set** if $u \in X$ and u dominates $v \in V$ implies that $v \in X$. The **closure** $\langle X\rangle_G$ of a set $X \subseteq V$ is the smallest closed subset containing X.

We identify $V = \{v_1, v_2, \ldots, v_n\}$ with $[n] = \{1, 2, \ldots, n\}$ and define the G-weight $\mathrm{wt}_G(\mathbf{x})$ of $\mathbf{x} \in \mathbb{F}_q^n$ as the number of vertices in G dominated by the vertices in the support of $\mathbf{x}$:

$$\mathrm{wt}_G(\mathbf{x}) := |\langle \mathrm{supp}(\mathbf{x})\rangle_G|.$$

The G-**distance** between $\mathbf{x}, \mathbf{y} \in \mathbb{F}_q^n$ is defined by $\mathrm{d}_G(\mathbf{x}, \mathbf{y}) := \mathrm{wt}_G(\mathbf{y} - \mathbf{x})$.

In the case that G is **acyclic** (that is, contains no cycles), the metric d_G is actually a poset metric: the existence of a nontrivial cycle would contradict the anti-symmetry axiom of a poset ($i \preceq_P j$ and $j \preceq_P i$ implies $i = j$). Among the few things that are known about such metrics, there are two canonical forms of a digraph which are able to establish whether two different graphs determine the same metric. Also, similarly to the poset-block case, in [690] there is a reasonable description of the group of linear isometries and some sufficient (not necessary) conditions for a digraph metric to admit the MacWilliams Extension Theorem and the MacWilliams Identity. It is worth noting that the conditions for the validity of the Extension Theorem and the MacWilliams Identity do not coincide. As far as the author was able to look, this is the first instance where the hypotheses for the MacWilliams Extension Theorem and the MacWilliams Identity are different.

A different formulation of this digraph metric can be obtained by considering weights on a poset, as defined in [1017], where perfect codes start to be explored in this context.

22.4 Additive Generalizations of the Lee Metric

In this section we present some families of metrics defined over an alphabet $\mathcal{X}$, generally a ring $\mathbb{Z}_q$ or a field $\mathbb{F}_q$, which generalize the Lee metric. They are all additive metrics, depending on the definition of a weight on the base ring or field.

To study different weights on $\mathbb{Z}_q$ (or other algebraic structures) means to describe (and possibly classify) all possible weights, up to an equivalence relation that arises naturally in the context of coding theory: two weights (or metrics) are considered to be equivalent if they determine the same collection of metric balls (see [615] for a precise definition). The problem of describing all inequivalent weights on $\mathbb{Z}_q$ was raised by Gabidulin in [762], and discussed for $q = 4$. A general approach to this problem was made in [1852, Theorem 2.3.8] where all possible weights are classified, assuming the alphabet to be just an abelian group.

22.4.1 Metrics over Rings of Integers

Metrics on the quotient rings of Gaussian integers were first introduced by Huber in [998], and they admit many variations and generalizations. They are considered suitable for signal constellations as QAM. We introduce them here in their simplest and most usual setting.

Given $x + iy \in \mathbb{Z}[i]$, we consider the norm $\mathcal{N}(x + iy) = x^2 + y^2$. Let $\alpha = a + bi \in \mathbb{Z}[i]$ be a Gaussian prime, that is, either $a^2 + b^2 = p$ is a prime and $p \equiv 1 \pmod 4$ or $\alpha = p + 0i$ with $p \equiv 3 \pmod 4$. Given $\beta \in \mathbb{Z}[i]$, there are $q, r \in \mathbb{Z}[i]$ such that $\beta = q\alpha + r$, with $\mathcal{N}(r) < \mathcal{N}(\alpha)$. We denote by $\mathbb{Z}[i]/\langle\alpha\rangle$ the ring of the equivalence classes of $\mathbb{Z}[i]$ modulo the ideal generated by α. For $\beta \in \mathbb{Z}[i]/\langle\alpha\rangle$, let $x + iy$ be a representative of the class β with $|x| + |y|$ minimum. The α-**weight** wt_α and the α-**distance** d_α are defined by $\mathrm{wt}_\alpha(\beta) := |x| + |y|$ and $\mathrm{d}_\alpha(\beta, \gamma) = \mathrm{wt}_\alpha(\beta - \gamma)$, respectively.

The α-weight and α-distance are extended additively to the product $(\mathbb{Z}[i]/\langle\alpha\rangle)^n$ by setting

$$\mathrm{wt}_\alpha(\beta) = \sum_{i=1}^{n} \mathrm{wt}_\alpha(\beta_i) \ \text{ and } \ \mathrm{d}_\alpha(\beta, \gamma) = \mathrm{wt}_\alpha(\beta - \gamma),$$

for $\beta = \beta_1\beta_2 \cdots \beta_n \in (\mathbb{Z}[i]/\langle\alpha\rangle)^n$. The distance d_α is known as the **Mannheim metric**.

There are many variations of the Mannheim metric, obtained by considering other rings of integers in a cyclotomic field. The same kind of construction can be done, for example in the ring of **Eisenstein-Jacobi integers**; see [997]. We let $\zeta = (-1 + i\sqrt{3})/2$ and consider $\mathbb{Z}[\zeta] := \{x + y\zeta \mid x, y \in \mathbb{Z}\}$. We consider a prime $p \equiv 1 \pmod 6$, which can be expressed as $p = a^2 + 3b^2 = \alpha\alpha^*$, where $\alpha = a + b + 2b\zeta$ and $\alpha^* = a + b + 2b\zeta^2$, and let $\mathbb{Z}[\zeta]/\langle\alpha\rangle$ be the quotient ring. Given $\beta \in \mathbb{Z}[\zeta]/\langle\alpha\rangle$, we define $\mathrm{wt}_{\zeta,\alpha}(\beta)$ to be the minimum of $|x_1| + |x_2|$, where $\beta = x_1\epsilon_1 + x_2\epsilon_2$ with $\epsilon_1, \epsilon_2 \in \{\pm1, \pm\zeta, \pm(1 + \zeta)\}$. The definition of a metric on $(\mathbb{Z}[\zeta]/\langle\alpha\rangle)^n$ is made in a similar way, considering the distance on $\mathbb{Z}[\zeta]/\langle\alpha\rangle$ determined by the weight and extending it additively to the n-fold product of $\mathbb{Z}[\zeta]/\langle\alpha\rangle$.

We remark that codes over rings of integers can be considered as codes on a flat torus, a generalization of spherical codes; see, for example, [454, 1336].

Most of the results using metrics over rings of integers are constructions of codes that are able to correct a certain number of errors, along with decoding algorithms; see [997, 998]. It is also worth noting that these metrics can be obtained as path-metrics on appropriate circulant graphs, as presented in [1336].

22.4.2 *l*-Dimensional Lee Weights

This is a family of metrics defined by S. Nishimura and T. Hiramatsu in 2008 [1440]. It is an interesting construction that has not yet been explored in the literature. We let $\beta = \{\mathbf{e}_1, \mathbf{e}_2, \ldots, \mathbf{e}_l\}$ be the standard basis of $\mathbb{R}^l$, and we consider the finite field $\mathbb{F}_q$, $q = p^m$. Let $\phi : \beta \to \mathbb{F}_q$ be a map and denote $a_i := \phi(\mathbf{e}_i)$. Given $\mathbf{u} = \sum_{i=1}^l u_i \mathbf{e}_i \in \mathbb{Z}^l$, we define $\phi(\mathbf{u}) = \sum_{i=1}^l a_i u_i \in \mathbb{F}_q$. We shall assume that the a_i's are chosen in such a way that ϕ is surjective, so that given $a \in \mathbb{F}_q$ there is $\mathbf{u}$ such that $\phi(\mathbf{u}) = a$.

The *l*-dimensional ϕ-weight $\mathrm{wt}_{l,\phi}(a)$ of $a \in \mathbb{F}_q$ is defined by

$$\mathrm{wt}_{l,\phi}(a) := \min \left\{ \sum_{i=1}^l |u_i| \;\middle|\; \mathbf{u} \in \phi^{-1}(a) \right\}.$$

The *l*-**dimensional Lee weight** of $\mathbf{x} \in \mathbb{F}_q^n$ is defined additively by $\mathrm{wt}_{l,\phi}(\mathbf{x}) := \sum_{i=1}^n \mathrm{wt}_{l,\phi}(x_i)$. The weight of the difference $d_{l,\phi}(\mathbf{x}, \mathbf{y}) = \mathrm{wt}_{l,\phi}(\mathbf{x} - \mathbf{y})$ is the *l*-**dimensional Lee distance** determined by ϕ.

The *l*-dimensional Lee distance is a metric which generalizes some known metrics. This is the case for the Lee metric, which is obtained by setting $l = 1$ and $\phi(\mathbf{e}_1) = 1$. Also the Mannheim metric can be obtained as a 2-dimensional Lee metric: for $p \equiv 1 \pmod 4$, the equation $x^2 \equiv -1 \pmod p$ has a solution $x = a \in \mathbb{F}_p$. For $l = 2$, we define $\phi(\mathbf{e}_1) = 1$ and $\phi(\mathbf{e}_2) = a$ and one gets the Mannheim metric (Section 22.4.1) as a particular instance of a 2-dimensional Lee metric.

In [1440] there are constructions of codes correcting any error $\mathbf{e}$ of weight $\mathrm{wt}_{l,\phi}(\mathbf{e}) = 1$ (for q odd) and $\mathrm{wt}_{l,\phi}(\mathbf{e}) = 2$ (for the case $q = p^m = 4n + 1$ with $p > 5$).

22.4.3 Kaushik–Sharma Metrics

A **Kaushik–Sharma metric** is an additive metric over $\mathbb{Z}_q$, for an integer $q > 1$, which generalizes the Lee metric. It is named after M. L. Kaushik, who introduced it in [1094] and B. D. Sharma, who studied it in subsequent works. We say that a partition $\mathcal{B} = \{B_0, B_1, \ldots, B_{m-1}\}$ of $\mathbb{Z}_q$ is a **KS-partition** if it satisfies the following conditions:

1. $B_0 = \{0\}$ and, for $i \in \mathbb{Z}_q \setminus \{0\}$, $i \in B_s \iff q - i \in B_s$;

2. if $i \in B_s, j \in B_t$ and $s < t$, then $\min\{i, q - i\} < \min\{j, q - j\}$;

3. $|B_0| \leq |B_1| \leq \cdots \leq |B_{m-2}|$ and $\frac{1}{2}|B_{m-2}| \leq |B_{m-1}|$.

Given $j \in \mathbb{Z}_q$, there is a unique $s \leq m - 1$ such that $j \in B_s$ and we write $\mathrm{wt}_{\mathcal{B}}(j) = s$. The **KS-weight** and **KS-distance** determined by the KS-partition $\mathcal{B}$ are defined, respectively, by

$$\mathrm{wt}_{\mathcal{B}}(\mathbf{x}) = \sum_{i=1}^n \mathrm{wt}_{\mathcal{B}}(x_i) \quad \text{and} \quad d_{\mathcal{B}}(\mathbf{x}, \mathbf{y}) = \mathrm{wt}_{\mathcal{B}}(\mathbf{x} - \mathbf{y}),$$

for $\mathbf{x}, \mathbf{y} \in \mathbb{Z}_q^n$. The KS-distance determines a metric which generalizes the Hamming metric (for the partition $\mathcal{B}_H$ defined by $B_1 = \{1, 2, \ldots, q-1\}$) and the Lee metric (for the partition $\mathcal{B}_L$ defined by $B_i = \{i, q-i\}$, for every $1 \leq i \leq \lfloor 1 + q/2 \rfloor$).

Considering $\mathbb{Z}_q^n$ as a module over $\mathbb{Z}_q$, we say that $\mathcal{C} \subseteq \mathbb{Z}_q^n$ is **linear over** $\mathbb{Z}_q$ if it is a $\mathbb{Z}_q$-submodule of $\mathbb{Z}_q^n$; see Chapter 6. Since $\mathbb{Z}_q$ is commutative, it has invariant basis number, and we denote by k the rank of $\mathcal{C}$. Considering a KS-metric, the main existing bound is a generalization of the Hamming Bound, and it applies for linear codes over $\mathbb{Z}_q$.

Theorem 22.4.1 ([1094, Theorem 1]) *Let $\mathcal{C} \subset \mathbb{Z}_q^n$ be a linear code of length n over $\mathbb{Z}_q$ that corrects all errors $\mathbf{e}$ with $\mathrm{wt}_{\mathcal{B}}(\mathbf{e}) \leq r_1$ and all errors $\mathbf{f}$ consisting of bursts of length at most $b < n/2$ and $\mathrm{wt}_{\mathcal{B}}(\mathbf{f}) \leq r_2$, with $1 < r_1 < r_2 < (m-1)b$. Then*

$$n - k \geq \log_q b_{\mathrm{d}_{\mathcal{B}}}^n(r_1) + \sum_{i=1}^{b}(n - i + 1)(b_{\mathrm{d}_{\mathcal{B}}}^i(r_2) - b_{\mathrm{d}_{\mathcal{B}}}^i(r_1)),$$

where $b_{\mathrm{d}_{\mathcal{B}}}^i(r)$ is the cardinality of a metric ball of radius r in $\mathbb{Z}_q^i$.

In more recent works [796, 797], there is a refinement of the bounds, by considering errors of limited pattern, that is, by possibly limiting the range of errors in each coordinate: $\mathrm{wt}_{\mathcal{B}}(x_i) \leq d$. It is worth noting that a generalization of the Kaushik–Sharma metric can be done for any finite group (not necessarily cyclic), as presented in [137].

22.5 Non-Additive Metrics Digging into the Alphabet

In the previous sections, we considered metrics that either ignore different weights on the alphabet and look for the way it spreads over different positions (as in Sections 22.2 and 22.3) or metrics that look into the alphabet and spread it additively over a finite product of the alphabet (as in Section 22.4).

In this section we consider two families of metrics that dig into the alphabet, i.e., consider different weights on the alphabet, and are not necessarily additive, in the sense that different positions of errors may count differently.

22.5.1 Pomset Metrics

Considering partially ordered multisets (pomsets), one can generalize simultaneously both the poset metrics (which essentially do not care about the alphabet structure) and the Lee metric (which is solely concerned with the cyclic structure of the alphabet). This was achieved by Sudha and Selvaraj [812] in 2017; all the content of this section refers to this paper.

Multiset is a generalization of the concept of a set that allows multiple instances of the elements, and it had been used earlier in the context of coding theory, mainly to explore duality issues, as an alternative to the use of matroids.

A **multiset** (or simply **mset**) is a pair $M = (\mathcal{X}, \mathfrak{c})$, where $\mathfrak{c} : \mathcal{X} \to \mathbb{N}$ is the **counting function** (also called the **weight function**): $\mathfrak{c}(i)$ counts the number of occurrences of i in M. We define the cardinality of M as $|M| = \sum_{x \in \mathcal{X}} \mathfrak{c}(x)$, and we denote by $M = \{k_1/a_1, \ldots, k_n/a_n\}$ the multiset with $\mathfrak{c}(a_i) = k_i$. Let $\mathcal{M}^m(\mathcal{X})$ be the family of all multisets underlying $\mathcal{X}$ such that any element occurs at most m times.

To define an order relation on a multiset one should consider the **Cartesian product of multisets** $M_1(\mathcal{X}, \mathfrak{c}_1)$ and $M_2(\mathcal{X}, \mathfrak{c}_2)$: it is the mset with underlying set $\mathcal{X} \times \mathcal{X}$ and counting function $\mathfrak{c}$ defined as the product, $\mathfrak{c}((a, b)) = \mathfrak{c}_1(a)\mathfrak{c}_2(b)$, that is,

$$M_1 \times M_2 := \{mn/(m/a, n/b) \mid m/a \in M_1, \, n/b \in M_2\}.$$

A submultiset (**submset**) of $M = (\mathcal{X}, \mathfrak{c})$ is a multiset $S = (\mathcal{X}, \mathfrak{c}')$ such that $\mathfrak{c}'(x) \leq \mathfrak{c}(x)$, for all $x \in \mathcal{X}$. We denote it by $S \ll M$. A submset $R = (M \times M, \mathfrak{c}_\times) \ll M \times M$ is called a **mset relation** on M if $\mathfrak{c}_\times(m/a, n/b) = mn$ for all $(m/a, n/b) \in R$. If we consider, for example, the multiset $M = \{4/a, 2/b\}$, we have that $S = \{5/(4/a, 2/a), 8/(4/a, 2/b)\}$ is a submset of $M \times M$, but it is not a mset relation, since $5 \neq 4 \cdot 2$, whereas $R = \{2/(2/a, 1/b), 6/(2/b, 3/a)\}$ is a mset relation.

Definition 22.5.1 Let M be a multiset. A **partially ordered mset relation** (or **pomset relation**) R on M is a mset relation satisfying

(a) (**reflexivity**) $(m/a)R(m/a)$, for all $m/a \in M$;

(b) (**anti-symmetry**) $(m/a)R(n/b)$ and $(n/b)R(m/a)$ implies $m = n$ and $a = b$;

(c) (**transitivity**) $(m/a)R(n/b)R(k/c)$ implies $(m/a)R(k/c)$.

The pair $\mathbb{P} = (M, R)$, where M is an mset and R is a pomset relation, is a **partially ordered mset** (or **pomset**). Given $\mathbb{P} = (M, R)$, a subset $I \ll M$ is called a **pomset ideal** if $m/a \in I$ and $(n/b)R(k/a)$, with $k > 0$ and $b \neq a$, implies $n/b \in I$. Given a submset $S \ll M$, we denote by $\langle S \rangle_\mathbb{P}$ the ideal generated by S, that is, the smallest ideal of $\mathbb{P}$ containing S.

Consider the mset $M = \{r/1, r/2, \ldots, r/n\} \in \mathcal{M}^r([n])$ where $r := \lfloor q/2 \rfloor$ is the integer part of $q/2$. The **Lee support** of a vector $\mathbf{x} \in \mathbb{Z}_q^n$ is $\mathrm{supp}_L(\mathbf{x}) := \{k/i \mid k = \mathrm{wt}_L(x_i), \, k \neq 0\}$, where $\mathrm{wt}_L(x_i) = \min\{x_i, q - x_i\}$ is the usual Lee weight.

The $\mathbb{P}$-**weight** and $\mathbb{P}$-**distance** on $\mathbb{Z}_q^n$ are defined as

$$\mathrm{wt}_\mathbb{P}(\mathbf{x}) := |\langle \mathrm{supp}_L(\mathbf{x}) \rangle_\mathbb{P}| \quad \text{and} \quad \mathrm{d}_\mathbb{P}(\mathbf{x}, \mathbf{y}) := \mathrm{wt}_\mathbb{P}(\mathbf{x} - \mathbf{y}),$$

respectively.

The $\mathbb{P}$-distance satisfies the metric axioms. In the case that $\mathbb{P}$ is an *anti-chain*, that is, any pair of distinct points $m/a, n/b \in M$ with $a \neq b$ are not comparable (neither $(m/a)R(n/b)$ nor $(n/b)R(m/a)$), we have that $\langle \mathrm{supp}_L(\mathbf{x}) \rangle_\mathbb{P} = \mathrm{supp}_L(\mathbf{x})$ and so $\mathrm{wt}_\mathbb{P}(\mathbf{x}) = \sum_i \mathrm{wt}_L(x_i) = \mathrm{wt}_L(\mathbf{x})$; therefore the pomset metric is a generalization of the Lee metric. In the case $M = \{1/1, 1/2, \ldots, 1/n\}$ we have a poset, hence generalizing also the poset metrics.

Not much is known about pomset metrics, only what is presented in [812]: the authors generalize for pomsets the basic operations known for posets (direct and ordinal sum, direct and ordinal products), and study the behavior of the minimum distance under some of these operations, producing either closed expressions or bounds.

22.5.2 m-Spotty Metrics

The m-spotty weights were introduced by Suzuki et al. in [1765], considering only the binary case. It was extended later for any finite field [1481] and rings [1665] as an extension of both the Hamming metric and the Lee metric in [1699].

Let $\mathbf{x}^i = x_1^i x_2^i \cdots x_b^i \in \mathbb{F}_q^b$ and $\mathbf{x} = \mathbf{x}^1 \mathbf{x}^2 \cdots \mathbf{x}^n \in \mathbb{F}_q^{bn}$. We call $\mathbf{x}^i$ the ith b-byte of $\mathbf{x}$. A **spotty byte error** is defined as t if t or fewer bit errors occur within a b-byte, for $1 \leq t \leq b$.

Given $\mathbf{x} = \mathbf{x}^1 \mathbf{x}^2 \cdots \mathbf{x}^n$, $\mathbf{y} = \mathbf{y}^1 \mathbf{y}^2 \cdots \mathbf{y}^n$, $\mathbf{x}, \mathbf{y} \in \mathbb{F}_q^{bn}$, the $(m, *)$-**spotty weight** and $(m, *)$-**spotty distance** are defined as

$$\mathrm{wt}_{m,*}(\mathbf{x}) = \sum_{i=1}^n \left\lceil \frac{\mathrm{wt}_*(\mathbf{x}^i)}{t} \right\rceil \quad \text{and} \quad \mathrm{d}_{m,*}(\mathbf{x}, \mathbf{y}) = \mathrm{wt}_{m,*}(\mathbf{x} - \mathbf{y})$$

where $\lceil x \rceil$ is the ceiling function and $*$ stands for any of the following cases:

- the Hamming structure, with $\mathrm{wt}_* = \mathrm{wt}_\mathrm{H}$;

- the Lee structure, with $\mathrm{wt}_* = \mathrm{wt}_\mathrm{L}$;

- the Niederreiter–Rosenbloom chain structure, with $\mathrm{wt}_*(\mathbf{x}^i) = \mathrm{wt}_\mathrm{NR}(\mathbf{x}^i) = \max\{j \mid x_j^i \neq 0\}$.

Considering the m-spotty metrics, most of the attention was devoted to the weight distribution of a code: MacWilliams Identities were obtained for the Hamming m-spotty weight $\mathrm{wt}_{m,\mathrm{H}}$ ([1765] for the binary case, [1481] for finite fields in general, and [1665] over rings), the Lee spotty weight $\mathrm{wt}_{m,\mathrm{L}}$ in [1699], and the RT spotty weight $\mathrm{wt}_{m,\mathrm{RT}}$ in [1482].

22.6 Metrics for Asymmetric Channels

The **binary asymmetric channel** is a memoryless channel with transition probabilities $\mathrm{Prob}(0 \mid 1) = \varrho_{01}$ and $\mathrm{Prob}(1 \mid 0) = \varrho_{10}$, where it is assumed that $0 \leq \varrho_{01} \leq \varrho_{10} \leq 1/2$. The binary asymmetric channel is a model for non-volatile memories, where errors occur due to leakage of charge. The extreme cases, $\varrho_{01} = \varrho_{10}$ and $\varrho_{01} = 0$ correspond to the binary symmetric channel (see Section 1.17) and the Z-channel, respectively.

It is possible to generalize it for a q-ary alphabet $\mathcal{X} = \{x_0, x_1, \dots, x_{q-1}\}$, by assuming that $\varrho_{ij} := \mathrm{Prob}(x_i \mid x_j)$ is not constant for $x_i, x_j \in \mathcal{X}$, $x_i \neq x_j$. Sometimes, as in [1420], the word asymmetric is reserved for a generalization of the Z-channel, where $\varrho_{ij} = 0$ for $i > j$.

22.6.1 The Asymmetric Metric

In a binary asymmetric channel we distinguish the two types of errors: when a transmitted 0 is received as a 1, it is called a **0-error**, and when a transmitted 1 is received as a 0, it is referred to as a **1-error**. The asymmetric metric is reported in [433] to have been introduced in 1975 by Rao and Chawla ([1567], a reference difficult to find). Given $\mathbf{x}, \mathbf{y} \in \mathbb{F}_2^n$, we define $N(\mathbf{x}, \mathbf{y}) = |\{i \in [n] \mid x_i = 0 \text{ and } y_i = 1\}|$. We remark that the Hamming distance d_H can be expressed as $\mathrm{d}_\mathrm{H}(\mathbf{x}, \mathbf{y}) = N(\mathbf{x}, \mathbf{y}) + N(\mathbf{y}, \mathbf{x})$. The **asymmetric distance** d_a between $\mathbf{x}$ and $\mathbf{y}$ is $\mathrm{d}_a(\mathbf{x}, \mathbf{y}) = \max\{N(\mathbf{x}, \mathbf{y}), N(\mathbf{y}, \mathbf{x})\}$. It is worth noting that the asymmetric metric is not matched to the binary asymmetric channel, in the sense presented in Section 22.1. We say that a code $\mathcal{C} \subseteq \mathbb{F}_2^n$ corrects r or fewer 0-errors if

$$\{\mathbf{x} \mid N(\mathbf{c}_1, \mathbf{x}) \leq r\} \cap \{\mathbf{x} \mid N(\mathbf{c}_2, \mathbf{x}) \leq r\} = \emptyset$$

whenever $\mathbf{c}_1, \mathbf{c}_2 \in \mathcal{C}$ with $\mathbf{c}_1 \neq \mathbf{c}_2$, and similarly for 1-errors, changing $N(\mathbf{c}_i, \mathbf{x}) \leq r$ to $N(\mathbf{x}, \mathbf{c}_i) \leq r$ for $i = 1, 2$.

Theorem 22.6.1 ([433, Theorem 3]) *Let $\mathcal{C} \subset \mathbb{F}_2^n$ be a code with minimum distance $d_a(\mathcal{C})$. Then $\mathcal{C}$ is capable of correcting r_0 or fewer 0-errors and r_1 or fewer 1-errors, where r_0 and r_1 are fixed and $r_0 + r_1 < d_a(\mathcal{C})$. In particular, it can correct $(d_a(\mathcal{C}) - 1)$ or fewer 0-errors or $(d_a(\mathcal{C}) - 1)$ or fewer 1-errors.*

If a message $\mathbf{x}$ is sent and $\mathbf{y}$ is received, we say that the error is t-**symmetric** if $\mathrm{d}_\mathrm{H}(\mathbf{x}, \mathbf{y}) = t$. The error is said to be t-**unidirectional** if $\mathrm{d}_\mathrm{H}(\mathbf{x}, \mathbf{y}) = t$ and either $N(\mathbf{x}, \mathbf{y}) = t$

or $N(\mathbf{y}, \mathbf{x}) = t$. A code is called a t-**EC-AUED** code if it can correct t or fewer symmetric errors, detect $t+1$ symmetric errors and detect all ($t+2$ or more) unidirectional errors. It is possible to prove that a code $\mathcal{C}$ is a t-EC-AUED code if $N(\mathbf{x}, \mathbf{y}) \geq t+1$ and $N(\mathbf{y}, \mathbf{x}) \geq t+1$ for all distinct $\mathbf{x}, \mathbf{y} \in \mathcal{C}$ [246, Theorem 8].

There are many bounds for the size of a t-EC-AUED code, the lower bounds generally presented as parameters of specific codes. Many of these bounds can be found in a comprehensive survey, with bibliography updated in 1995, due to Kløve [1128]. We present just two of the most classical upper bounds, whose proofs can also be found in [1128].

Theorem 22.6.2 *Let $A(n,t)$ denote the maximal size of a t-EC-AUED code of length n. The following bounds hold.*

$$\text{Lin–Bose [1258, Theorem 2.5]}: \quad A(n,1) \leq \frac{2}{n}\binom{n}{\lfloor n/2 \rfloor},$$

$$\text{Varshamov [1845]}: \quad A(n,t) \leq \frac{2^{n+1}}{\sum_{j=1}^{t}\left\{\binom{\lfloor n/2 \rfloor}{j} + \binom{\lceil n/2 \rceil}{j}\right\}}.$$

22.6.2 The Generalized Asymmetric Metric

The asymmetric metric, originally defined over a binary alphabet, can be generalized over a q-ary alphabet $[\![q]\!] := \{0, 1, \ldots, q-1\}$, or the alphabet $\mathbb{Z}_{\geq}$, consisting of nonnegative integers.

Given $\mathbf{x}, \mathbf{y} \in (\mathbb{Z}_{\geq})^n$, we define the **generalized asymmetric weight** and the **generalized asymmetric metric**, respectively, as

$$N_g(\mathbf{x}, \mathbf{y}) := \sum_{i=1}^{n} \max\{x_i - y_i, 0\} \quad \text{and} \quad d_g(\mathbf{x}, \mathbf{y}) = \max\{N_g(\mathbf{x}, \mathbf{y}), N_g(\mathbf{y}, \mathbf{x})\}.$$

This distance was recently introduced in [1113], where it is used to define a metric on the so-called *l-gramm profile* of $\mathbf{x} \in [\![q]\!]^n$. It turns out it is an interesting distance for the types of errors that may occur in the DNA storage channel, which includes substitution errors (in the synthesis and sequencing processes) and coverage errors (which may happen during the DNA fragmentation process). For details see [1113].

22.7 Editing Metrics

An **editing distance** is a measure of similarity between strings (not necessarily of the same length) based on the minimum number of operations required to transform one into the other. Different types of operations lead to different metrics, but the most common ones are insertions, deletions, substitution and transpositions.

Consider an alphabet $\mathcal{X}$ and let $\mathcal{X}^*$ be the space of all finite sequences (strings) in $\mathcal{X}$. Given two strings $\mathbf{x} = x_1 x_2 \cdots x_m$ and $\mathbf{y} = y_1 y_2 \cdots y_n$, there are four basic errors, described as operations from which $\mathbf{y}$ can be obtained from $\mathbf{x}$.

(*I*) <u>Insertion</u>: In case **y** is obtained by inserting a single letter into **x** ($n = m + 1$);

(*D*) <u>Deletion</u>: In case **y** is obtained by deleting a single letter from **x** ($n = m - 1$);

(*S*) <u>Substitution</u>: In case **y** is obtained by substituting a single letter of **x** ($n = m$);

(*T*) <u>Transposition</u>: In case **y** is obtained by transposing two adjacent letters of **x**; that is, for some i we have $y_i = x_{i+1}$, $y_{i+1} = x_i$ ($n = m$).

These operations are considered to be an appropriate measure to describe human typeset misspellings. In [481], Damerau claims that 80 percent of the misspelling errors have distance one in the metric admitting all the four types of basic errors. It also has applications in genomics, since DNA duplication is commonly disturbed by the considered operations, each operation occurring with similar probabilities.

Definition 22.7.1 Given a set $\mathcal{E}$ of operations, $\{I, D\} \subseteq \mathcal{E} \subseteq \{I, D, S, T\}$, the distance $d_{\mathcal{E}}$ between strings **x** and **y** is the minimal number of operations in $\mathcal{E}$ needed to get **y** from **x**. For simplicity, we will write d_{ID} instead of $d_{\{I,D\}}$, and similarly for other subsets $\mathcal{E}$.

For $\mathcal{E} = \{I, D\}$, we have **Levenshtein's insertion-deletion metric**, introduced in [1225] for the case of binary alphabets. The case $\mathcal{E} = \{I, D, S\}$ is known as **Levenshtein's editing metric**, also introduced in [1225] for the binary case. In this work, Levenshtein proved [1225, Lemma 1] that a code capable of correcting t deletions and separately correcting t insertions can correct a mixture of t insertions and deletions.

Denoting by $|\mathbf{x}|$ the length of the string, in [973] it was proved that $d_{ID}(\mathbf{x}, \mathbf{y}) = |\mathbf{x}| + |\mathbf{y}| - 2\rho(\mathbf{x}, \mathbf{y})$, where $\rho(\mathbf{x}, \mathbf{y})$ is the maximum length of a common subsequence of **x** and **y**. As observed in [973], it is important to distinguish between the minimum number of insertions/deletions and minimum number of insertions and deletions. If one defines $d_{ID}^*(\mathbf{x}, \mathbf{y}) = 2e$ meaning that e is the smallest number in which **x** can be transformed into **y** by at most e insertions and at most e deletions, we get a distinct metric, with $d_{ID}(\mathbf{x}, \mathbf{y}) \leq d_{ID}^*(\mathbf{x}, \mathbf{y})$.

Theorem 22.7.2 ([973, Theorems 4.1 and 4.2]) *A code $\mathcal{C} \subseteq \mathcal{X}^*$ is capable of correcting e insertions/deletions if and only if $d_{ID}(\mathcal{C}) > 2e$, and it is capable of correcting i insertions and d deletions if and only if $d_{ID}^*(\mathcal{C}) > 2(d + i)$.*

The structure of the edit space, that is, the structure of $\mathcal{X}^*$ equipped with the metric d_{IDS}, is studied in [990]. The group $\mathrm{GL}(\mathcal{X}^*, d_{IDS})$ is described as the product $\mathrm{GL}(\mathcal{X}^*, d_{IDS}) = \langle \gamma \rangle \times S_q$ where $\gamma : \mathcal{X}^* \to \mathcal{X}^*$ is the reversion map $\gamma(x_1 x_2 \cdots x_m) = x_m x_{m-1} \cdots x_1$, $\langle \gamma \rangle$ is the group generated by γ and S_q is the group of permutations of the alphabet $\mathcal{X}$ ($|\mathcal{X}| = q$), acting as usual on every position of a string, $\sigma(x_1 x_2 \cdots x_m) = x_{\sigma(1)} x_{\sigma(2)} \cdots x_{\sigma(m)}$.

22.7.1 Bounds for Editing Codes

Denote by $A_{\mathcal{E}}(n, t)_q$ the maximal size of a q-ary code of length n capable of correcting t errors of the types belonging to $\mathcal{E}$. The asymptotic behavior of $A_{\mathcal{E}}(n, t)_q$ was first studied by Levenshtein in [1225], showing that $A_{ID}(n, 1)_2$ behaves asymptotically as $2^n/n$. Lower and upper bounds for the q-ary case were given in [1233]:

$$\frac{q^{n+t}}{\left(\sum_{i=0}^{t} \binom{n}{i}(q-1)^i \right)^2} \leq A_{ID}(n, t)_q \leq \frac{q^{n-t}}{\sum_{i=0}^{t} \binom{n-t}{i}} + \sum_{i=0}^{n-2} \binom{n-1}{i}(q-1)^i.$$

It was proved in [476, Theorem 2] that a code capable of correcting a deletions and b insertions has size asymptotically bounded above by

$$\frac{q^{n+b}}{(q-1)^{a+b}\binom{n}{a+b}\binom{a+b}{b}}.$$

The case $\mathcal{E} = \{I, D, S\}$ is also approached in [1225] where Levenshtein gives the bounds for codes capable of correcting one insertion-deletion-substitution error:

$$\frac{2^{n-1}}{n} \leq A_{IDS}(n,1)_2 \leq \frac{2^n}{n+1}.$$

As a general relation, useful for studying bounds, we can find in [990]:

$$A_{IDS}(n,t)_q \leq q \cdot A_{IDS}(n-1,t)_q.$$

Tables of known values of $A_{IDS}(n,t)_q$ and $A_{ID}(n,t)_q$ (for small values of n,t) can be found in [988] for $q = 2$ and [989] for $q = 3$.

22.8 Permutation Metrics

The symmetric group S_n acts on $\mathcal{X}^n$ permuting the coordinates:

$$\sigma(\mathbf{x}) = x_{\sigma(1)}x_{\sigma(2)}\cdots x_{\sigma(n)},$$

for $\sigma \in S_n$, $\mathbf{x} \in \mathcal{X}^n$. Considering the group structure of S_n or a subgroup $G \subset S_n$ is a common procedure to handle algorithm difficulties, a topic explored in Chapter 23. It is used, for example, when considering a code to be an orbit (best if with no fixed-points) of such a group. Considering permutations as codewords is what is known as **code in permutations**, and it is relevant in coding for flash memories and to handle problems concerning rankings comparisons. It was in the context of ranking comparisons that, in 1938, Kendall introduced [1104] the metric which became known as the *Kendall-tau metric*.

We denote a permutation $\sigma \in S_n$ as $[\sigma(1), \sigma(2), \ldots, \sigma(n)]$. The (group) product of permutations $\sigma, \pi \in S_n$ is defined by $\pi \circ \sigma(i) = \sigma(\pi(i))$ for $i \in [n]$. A transposition $\sigma := (i, k)$ is a permutation such that

$$\sigma(i) = k, \ \sigma(k) = i \ \text{and} \ \sigma(j) = j \ \text{for} \ j \neq i, \ k.$$

We say the transposition is adjacent if $k = i + 1$. The **Kendall-tau metric** $d_\tau(\sigma, \pi)$ is the minimal number of adjacent transpositions needed to transform σ into π. The formula

$$d_\tau(\sigma, \pi) = \left|\{(i,j) \mid \sigma^{-1}(i) < \sigma^{-1}(j) \text{ and } \pi^{-1}(i) > \pi^{-1}(j)\}\right|$$

is well-known and is actually the original formulation of Kendall.

We denote by $A_{d_\tau}(n,t)$ the maximum size of a code $\mathcal{C} \subseteq S_n$ with minimum distance $d_\tau(\mathcal{C}) = t$. In [124] we find various bounds for $A_{d_\tau}(n,t)$, among others the Singleton and Sphere Packing Bounds.

Theorem 22.8.1 ([124]) *Let $B_{d_\tau}(R)$ be the d_τ-ball of radius R. The following hold.*

(a) *For $n - 1 < t < n(n-1)/2$, $A_{d_\tau}(n, t) \leq \lfloor 3/2 + \sqrt{n(n-1) - 2t + 1/4} \rfloor!$.*

(b) $\dfrac{n!}{|B_{d_\tau}(2r)|} \leq A_{d_\tau}(n, 2r+1) \leq \dfrac{n!}{|B_{d_\tau}(r)|}$.

Obstruction conditions for the existence of perfect codes with the Kendall-tau metric can be found in [320].

It is worth remarking that the set $S = \{(1,2), (2,3), \ldots, (n-1,n)\}$ of adjacent transpositions is a set of generators of S_n, minimal if considering only transpositions as generators. Moreover, a transposition $\sigma = (i, j)$ is its own inverse ($\sigma = \sigma^{-1}$). It follows that the Kendall-tau metric is just the graph metric in the Cayley graph of S_n determined by the generator set S; the vertices are the elements of the group and two vertices are connected by an edge if and only if they differ by a generator. This is actually the Coxeter group structure. The group S_n acts with no fixed points on the subspace $V = \{\mathbf{x} \in \mathbb{R}^n \mid x_1 + x_2 + \cdots + x_n = 0\}$ and there is a simplicial structure on the unit sphere of V, known as the Coxeter complex, on which S_n acts simply transitively. This action makes Coxeter groups suitable for spherical coding; see Definition 12.4.2 for the definition of a spherical code. The study of such group codes, a generalization of Slepian's permutation modulation codes, is studied in [1393]. See also Chapter 16.

There are other relevant metrics on S_n. Let $G_n := \mathbb{Z}_2 \times \mathbb{Z}_3 \times \cdots \times \mathbb{Z}_n$ and consider the embedding : $S_n \to G_n$ which associates to $\sigma \in S_n$ the element $\mathbf{x}_\sigma = x_\sigma(1) x_\sigma(2) \cdots x_\sigma(n-1) \in G_n$, where x_σ is defined by

$$x_\sigma(i) = \left| \{ j \mid j < i + 1, \, \sigma(j) > \sigma(i+1) \} \right|,$$

for $i = 1, 2, \ldots, n-1$. The ℓ_1 metric on G_n, $d_{\ell_1}(\mathbf{x}, \mathbf{y}) = \sum_{i=1}^{n-1} |x(i) - y(i)|$, induces a metric on S_n: $d_{\ell_1}(\sigma, \pi) := d_{\ell_1}(\mathbf{x}_\sigma, \mathbf{x}_\pi)$. This is a metric related to the Kendall-tau metric by the relation $d_\tau(\sigma, \pi) \geq d_{\ell_1}(\sigma, \pi)$. More on permutation metrics can be found in [124].

Chapter 23

Algorithmic Methods

Alfred Wassermann
Universität Bayreuth

23.1   Introduction

In this chapter, we study algorithms for computer construction of "good" linear $[n,k]_q$ codes and methods to determine the minimum distance and the weight distribution of a linear code. While we will focus on construction methods in this chapter, for methods for decoding received words, we refer to Section 1.17 of Chapter 1 and Chapters 15, 21, 24, and 32.

In the search for good codes coding theorists always want to satisfy two conflicting goals:

1. The redundancy of the code should be as small as possible, i.e., the ratio k/n should be as large as possible, and

2. the code should allow one to correct as many errors as possible; i.e., the minimum distance should be as big as possible.

As a consequence, a typical approach to find good codes over a finite field $\mathbb{F}_q$ is to fix two parameters from $[n, k, d]$ and optimize the third one.

First, we note that in this chapter we claim to have found a certain code as soon as we have a generator matrix and know the parameters n, k, d. Fortunately, the construction method described in this chapter also reveals the weight distribution as soon as a code has been found. In Section 23.8 we study a state-of-the-art algorithm for computing the minimum distance of linear codes in general. In special cases, this algorithm can be modified easily to determine the weight distribution of the code.

When it comes to code construction by searching for a suitable generator matrix, we immediately realize the following: a code in general will have many generator matrices, and permuting columns and multiplying columns by nonzero scalars do not change fundamental properties of the code. One possible approach to find a suitable generator matrix for given parameters n and d is to build up the generator matrix row by row, thereby avoiding isomorphic copies. This method is described in detail in [266, 268, 1180]. In a sense, by doing this we fix n and d and try to maximize k.

In this chapter we will focus on an alternative approach and fix k and d and try to minimize the length of n, which means find a minimal set of columns of a generator matrix such that the resulting code has at least minimum distance d. Another formulation of the problem would be to fix k and n and try to maximize d.

Both approaches—row-by-row or column-by-column—have been used successfully. The latter approach, which is the approach that will be described in this chapter, is to select the columns of a generator matrix appropriately. In order to do this an equivalence between the existence of a linear code with a prescribed minimum distance and the existence of a solution of a certain system of Diophantine linear equations is exploited. Since the number of candidates for the possible columns of the generator matrix is equal to $(q^k - 1)/(q - 1)$, for interesting cases of linear codes the search has to be restricted, for example to codes having a prescribed symmetry. Thus, by applying group theory the number of Diophantine linear equations often reduces to a size which can be handled by today's computer hardware. Additionally, the search can be tailored to special types of codes like two-weight, self-orthogonal or LCD codes; see Section 23.6.

As before, a linear $[n, k]_q$ code is a k-dimensional subspace of the n-dimensional vector space $\mathbb{F}_q^n$ over the finite field $\mathbb{F}_q$ with q elements. The q^k codewords of length n are the elements of the subspace; they are written as row vectors. The (Hamming) weight $\mathrm{wt_H}(\mathbf{c})$ of a codeword $\mathbf{c}$ is defined to be the number of nonzero entries of $\mathbf{c}$, and the minimum distance $d_H(\mathcal{C})$ of a code $\mathcal{C}$ is the minimum of all weights of the nonzero codewords in $\mathcal{C}$.

For the purpose of error correcting codes we are interested in codes with high minimum distance d as these allow the correction of $\lfloor (d-1)/2 \rfloor$ errors. On the other hand we are interested in codes of small length n. As already stated, high minimum distance and small length are contrary goals for the optimization of codes. A linear code $\mathcal{C}$ is called **optimal** if there is no linear $[n, k, d_H(\mathcal{C}) + 1]$ code.

The definitive reference on lower and upper limits for the maximum minimum distance of a linear code of fixed length n and dimension k are the online tables [845]. In most cases the best upper bound is given by the Griesmer Bound [855]; see Theorem 1.9.18 and Remark 1.9.19. In many cases there is a gap between the minimum distance of the best known linear $[n, k]_q$ code and the limit given by these upper bounds.

23.2 Linear Codes with Prescribed Minimum Distance

To start with we give a characterization of the codewords of a code with the standard inner product. If G denotes a generator matrix of an $[n, k]_q$ code $\mathcal{C}$, then the codewords of the code consist of the set

$$\mathcal{C} = \{\mathbf{v}G \mid \mathbf{v} \in \mathbb{F}_q^k\}.$$

As we desire optimal codes, we assume G has no zero columns. If $\mathbf{G}_j$ indicates the jth column of the generator matrix G, then each codeword $\mathbf{v}G$ can be written as the vector

$$\mathbf{v}G = \mathbf{v}\mathbf{G}_1 \, \mathbf{v}\mathbf{G}_2 \, \cdots \, \mathbf{v}\mathbf{G}_n \, .$$

Let $\mathbf{v} \cdot \mathbf{w} = \sum_i v_i w_i$ denote the standard **inner product**, and let $\Sigma_{\mathbf{v}}$ be the numbers of columns $\mathbf{u}$ of the generator matrix G for which $\mathbf{v} \cdot \mathbf{u} = 0$; then the codeword $\mathbf{v}G$ has weight d' if and only if $\Sigma_{\mathbf{v}} = n - d'$.

Hence, G is a generator matrix of an $[n, k, d]_q$ code over $\mathbb{F}_q$ if and only if

$$\max\{\Sigma_{\mathbf{v}} \mid \mathbf{v} \in \mathbb{F}_q^k, \ \mathbf{v} \neq \mathbf{0}\} = n - d \, .$$

It is clear that the Hamming distance of a codeword $\mathbf{v}G$ is not changed if we multiply a column of G by a nonzero scalar $\lambda \in \mathbb{F}_q^*$ or permute the columns of G. Thus, it is appropriate to view a linear code $\mathcal{C}$ with generator matrix G as the *multiset*

$$\{\mathbf{G}_1, \mathbf{G}_2, \ldots, \mathbf{G}_n\}$$

of column vectors $\mathbf{G}_i \in \mathbb{F}_q^k$. Since we do not exclude that columns can be scalar multiples of other columns, we view the generator matrix as a multiset. This view of a code as a multiset of points makes it worthwhile to have a short excursion into finite projective geometry.

23.3 Linear Codes as Sets of Points in Projective Geometry

Let $V = \mathbb{F}_q^k$ be the k-dimensional vector space over $\mathbb{F}_q$. We take the view of projective geometry and call the 1-dimensional subspaces of V **points**, the 2-dimensional subspaces **lines**, and the $(k-1)$-dimensional subspaces **hyperplanes**.

$\mathbb{F}_q^k$ can be considered as a $(k-1)$-dimensional projective geometry $\mathrm{PG}(k-1, q)$. But in order to make things easier, we strictly adhere to the vector space language, that is "dimension" of a subspace means dimension as a vector space. Nevertheless, it is convenient to use the term *point* for subspaces of dimension 1. Additionally, instead of saying a 1-dimensional subspace $P = \langle \mathbf{P} \rangle$, $\mathbf{P} \in V$, is a subspace of a t-dimensional subspace H, we will say P is an element of H. Consequently, instead of $P \leq H$, we write $P \in H$. More information about the connection between coding theory and projective geometry can be found in Chapter 14.[1]

The number of t-dimensional subspaces of V is given by the **Gaussian binomial coefficient**

$$\begin{bmatrix} k \\ t \end{bmatrix}_q = \prod_{i=0}^{t-1} \frac{q^{k-i} - 1}{q^{i+1} - 1} \, .$$

Similarly, the set of all t-dimensional subspaces of V is denoted by

$$\begin{bmatrix} V \\ t \end{bmatrix}_q = \{U \leq V \mid \dim U = t\} \, .$$

---

[1] In Chapter 14, "dimension" of a projective subspace is the projective dimension, which is one less than the vector space dimension.

Let $\mathbf{h} = h_1 h_2 \cdots h_k \in V$, $\mathbf{h} \neq \mathbf{0}$. The set

$$\mathbf{h}^{\perp} = \left\{ \mathbf{x} \in V \mid \mathbf{h} \cdot \mathbf{x} = \sum_{i=1}^{k} h_i x_i = 0 \right\}$$

is a $(k-1)$-dimensional subspace of V; i.e., the vector $\mathbf{h}$ is the **normal vector** of the hyperplane $\mathbf{h}^{\perp}$.

A geometry can be described by the **incidence matrix** $M_{k,q}$ between points and hyperplanes. Let $H = \mathbf{h}^{\perp}$ be a hyperplane of V and $\mathbf{P} \in V$ where $P = \langle \mathbf{P} \rangle$ is a point of V; the entry $m_{H,P}$ in the row labeled by H and in the column labeled by P of $M_{k,q}$ is defined by

$$m_{P,H} := \begin{cases} 1 & P \in H, \\ 0 & \text{otherwise,} \end{cases}$$

where a point $P = \langle \mathbf{P} \rangle$ is **incident** with the hyperplane $H = \mathbf{h}^{\perp}$, i.e., $P \in H$, if and only if $\mathbf{P} \cdot \mathbf{h} = 0$ in $\mathbb{F}_q$.

Example 23.3.1 $\mathbb{F}_2^3$ has $\begin{bmatrix} 3 \\ 1 \end{bmatrix}_2 = 7$ points and 7 hyperplanes, i.e., 1-dimensional subspaces and 2-dimensional subspaces. An incidence matrix between 2-dimensional subspaces (rows) and 1-dimensional subspaces (columns) is:

		P_1	P_2	P_3	P_4	P_5	P_6	P_7	
		$\langle 100 \rangle$	$\langle 010 \rangle$	$\langle 001 \rangle$	$\langle 110 \rangle$	$\langle 011 \rangle$	$\langle 101 \rangle$	$\langle 111 \rangle$	
	$100^{\perp}$	0	1	1	0	1	0	0	
	$010^{\perp}$	1	0	1	0	0	1	0	
$M_{3,2} =$	$001^{\perp}$	1	1	0	1	0	0	0	(23.1)
	$110^{\perp}$	0	0	1	1	0	0	1	
	$011^{\perp}$	1	0	0	0	1	0	1	
	$101^{\perp}$	0	1	0	0	0	1	1	
	$111^{\perp}$	0	0	0	1	1	1	0	

Now, we are ready to construct linear $[n,k]_q$ codes by searching for suitable (multi)sets of points in V.

The codewords of an $[n,k]_q$ code $\mathcal{C}$ are the linear combinations of rows of a generator matrix G of $\mathcal{C}$; i.e., for every codeword $\mathbf{c} = c_1 c_2 \cdots c_n \in \mathcal{C}$, there exists a vector $\mathbf{h} = h_1 h_2 \cdots h_k \in V$ such that

$$\mathbf{c} = c_1 c_2 \cdots c_n = \mathbf{h} G$$

If $\mathbf{G_i}$ denotes the ith column of the matrix G, then for $i = 1, \ldots, n$ we have the inner products

$$\mathbf{h} \cdot \mathbf{G_i^{\mathsf{T}}} = c_i.$$

If we take $\mathbf{G_i^{\mathsf{T}}}$ as base vector of the point $\langle \mathbf{G_i^{\mathsf{T}}} \rangle$ in V, it follows that

$$c_i = 0 \text{ if and only if } \langle \mathbf{G_i^{\mathsf{T}}} \rangle \in \mathbf{h}^{\perp}.$$

The Hamming weight of the codeword $\mathbf{c}$ is the number of its nonzero entries. Now the following lemma is obvious.

Lemma 23.3.2 *If a generator matrix of a $[n,k]_q$ code $\mathcal{C}$ is viewed as a multiset $\mathcal{P}$ of points in V, then the following hold.*

(a) *Every codeword $\mathbf{c}$ of $\mathcal{C}$ corresponds to a hyperplane $\mathbf{h}^{\perp}$ of V.*

(b) *The Hamming weight of* **c** *is equal to the number of points in* $\mathcal{P}$ *which are* <u>*not*</u> *contained in* $\mathbf{h}^{\perp}$; *i.e., it is equal to the number of zeros in the row labeled by* $\mathbf{h}^{\perp}$ *of the columns of the matrix* $M_{k,q}$ *labeled by* $\mathcal{P}$.

(c) *The* $[n, k]_q$ *code defined by a multiset* $\mathcal{P}$ *has minimum weight at least* d *if every hyperplane contains at most* $n - d$ *points of* $\mathcal{P}$.

Remark 23.3.3 We speak of $\mathcal{P}$ as a **projective point set**; compare [611].

Example 23.3.4 Continuing Example 23.3.1, we can take the projective point set $\mathcal{P} = \{P_1, P_2, \ldots, P_7\}$ of *all* points of $\mathbb{F}_2^3$ as columns for our generator matrix. Since the row sum of each row in the incidence matrix (23.1) is equal to 3, for every hyperplane there are $7 - 3 = 4$ points of $\mathcal{P}$ *not* contained in that hyperplane. It follows that every nonzero codeword has weight 4, and we have constructed a $[7, 3, 4]_2$ code. Taking the base vectors of the points as columns of the generator matrix G gives

$$G = \begin{bmatrix} 1 & 0 & 0 & 1 & 0 & 1 & 1 \\ 0 & 1 & 0 & 1 & 1 & 0 & 1 \\ 0 & 0 & 1 & 0 & 1 & 1 & 1 \end{bmatrix}.$$

This is the well known *simplex code*; see also Definition 1.10.2.

Choosing points P_1, P_2, P_3, and P_7, we get a code with the generator matrix

$$G = \begin{bmatrix} 1 & 0 & 0 & 1 \\ 0 & 1 & 0 & 1 \\ 0 & 0 & 1 & 1 \end{bmatrix}.$$

Summing up the corresponding columns of the incidence matrix, we immediately see that this gives a code with 6 codewords of weight 2 and one codeword of weight 4; i.e., it is a $[4, 3, 2]_2$ code.

We note that every linear code $\mathcal{C}$ can be described by projective point sets in many different ways. For example, each step of the Gaussian elimination process applied on a generator matrix will give a different projective point set without changing the code. On the other hand, if a projective point set and its order as columns of the generator matrix is given, the **row reduced echelon form** of the generator matrix defines a canonical projective point set describing the code.

The special case when a code $\mathcal{C}$ is described by a set of projective points instead of a multiset can be read from the weight distribution of the dual code $\mathcal{C}^{\perp}$. There is a codeword of weight 2 in the dual code if and only if there is a linear combination of two columns of the generator matrix of $\mathcal{C}$ which equals the zero vector; i.e., two columns of the generator matrix are nonzero scalar multiples of each other. If a code $\mathcal{C}$ is described by a *set* rather than a *multiset* of projective points, $\mathcal{C}$ is called **projective**.

Proposition 23.3.5 *For a linear* $[n, k]_q$-*code* $\mathcal{C}$ *the following statements are equivalent.*

(a) $\mathcal{C}$ *is a projective code.*

(b) $\mathcal{C}^{\perp}$ *does not contain words of weight 2.*

(c) *The columns of a generator matrix of* $\mathcal{C}$ *correspond to different projective points.*

Remark 23.3.6 The relation between projective point sets and linear codes is well studied; see [611]. MacWilliams and Sloane mention the connection in the special case of MDS codes [1323, Chapter 11, Section 6].

The complement of a projective point set defining an $[n, k, d]_q$ code is called a minihyper. A (b, t)-**minihyper** is a multiset of b points in $\begin{bmatrix} V \\ 1 \end{bmatrix}_q$ such that every hyperplane in $\begin{bmatrix} V \\ k-1 \end{bmatrix}_q$ contains at least t of these points and there is at least one hyperplane which contains exactly t points; see further [293, 720, 891, 1344].

The above considerations lead us to the following theorem which shows the equivalence between the construction of codes with a prescribed minimum distance and solving a system of Diophantine linear equations. In the next theorem and subsequently I_t is the $t \times t$ identity matrix.

Theorem 23.3.7 *There is a $[n, k, d']_q$ code with minimum distance $d' \geq d$ if and only if there are column vectors $\mathbf{x} \in \{0, \ldots, n\}^{\begin{bmatrix} k \\ 1 \end{bmatrix}_q}$ and $\mathbf{y} \in \{0, \ldots, n-d\}^{\begin{bmatrix} k \\ 1 \end{bmatrix}_q}$ satisfying simultaneously*

- $$\begin{bmatrix} M_{k,q} & \Big| & I_{\begin{bmatrix} k \\ 1 \end{bmatrix}_q} \end{bmatrix} \begin{bmatrix} \mathbf{x} \\ \mathbf{y} \end{bmatrix} = (n-d)\mathbf{1}^\mathsf{T}, \tag{23.2}$$

- $$\sum_{i=1}^{\begin{bmatrix} k \\ 1 \end{bmatrix}_q} x_i = n, \tag{23.3}$$

where $\mathbf{1}$ is the all-one vector.

Remark 23.3.8 The constructed codes in Theorem 23.3.7 are projective if the solution vector $\mathbf{x}$ is restricted to $\{0, 1\}^{\begin{bmatrix} k \\ 1 \end{bmatrix}_q}$.

Example 23.3.9 Continuing with Example 23.3.4, the $[7, 3, 4]_2$ code is obtained by choosing $x_1 = x_2 = \cdots = x_7 = 1$ and $y_1 = y_2 = \cdots = y_7 = 0$ in Theorem 23.3.7. The $[4, 3, 2]_2$ code is obtained with the choice $x_1 = x_2 = x_3 = x_7 = 1$, $x_4 = x_5 = x_6 = 0$, $y_1 = y_2 = \cdots = y_6 = 2$, and $y_7 = 0$.

For this brute force approach to the code search, clearly the number of rows, respectively columns, of $M_{k,q}$, namely $\begin{bmatrix} k \\ 1 \end{bmatrix}_q = (q^k - 1)/(q - 1)$, is the limiting factor of the construction. Solving the system of Diophantine linear equations is only possible for small values of k and q. Therefore, we need to apply a well-known method, also described in [293], to shrink the matrix $M_{k,q}$ to the much smaller $M_{k,q}^\Gamma$ by prescribing a subgroup Γ of the general linear group $\mathrm{GL}_k(q)$ as a group of automorphisms. This approach will be described in the following sections.

23.3.1   Automorphisms of Projective Point Sets

The geometric view describes codes up to monomial equivalence: permutations of columns and multiplications of column vectors by nonzero scalars do not change the corresponding projective point set.

A mapping $\pi : \mathbb{F}_q^k \to \mathbb{F}_q^k$ is an **automorphism** of a set of t-dimensional subspaces $\mathcal{U} = \{U_1, U_2, \ldots, U_n\} \subseteq \begin{bmatrix} \mathbb{F}_q^k \\ t \end{bmatrix}_q$, $1 \leq t \leq k$, if $\pi(\mathcal{U}) = \mathcal{U}$. We will not further discuss this, but one requirement on π is that the dimension of a subspace does not change under this mapping. The famous fundamental theorem of projective geometry states that all such automorphisms are elements of $\mathrm{P\Gamma L}_k(q)$; see for example the book by Artin [66].

Theorem 23.3.10 (Fundamental Theorem of Projective Geometry) *If $\mathbb{F}$ is a field and $k \geq 3$, then the full group of automorphisms that map each t-dimensional subspace of $\mathbb{F}^k$ to a t-dimensional subspace is $\mathrm{P\Gamma L}_k(\mathbb{F})$.*

Leaving the field automorphisms aside, which are trivial if q is a prime, the fundamental theorem of projective geometry says that the automorphism $\pi : \mathbb{F}_q^k \to \mathbb{F}_q^k$ is an element of $\mathrm{GL}_k(q)$; i.e., it can be expressed as an invertible $k \times k$ matrix M_π over $\mathbb{F}_q$. For more details on the action of $\mathrm{GL}_k(q)$ on the subspace lattice of $\mathrm{PG}(k, q)$, see [296, 297] and [190, Chapter 8].

The set of all automorphisms of a set of subspaces $\mathcal{U}$ forms a group Γ. Let $\Gamma \leq \mathrm{GL}_k(q)$ be a group represented by $k \times k$ matrices over $\mathbb{F}_q$. An **action** of the group Γ (from the left) on a nonempty set $\mathcal{U}$ is defined by a mapping

$$\Gamma \times \mathcal{U} \to \mathcal{U} : (\pi, U) \mapsto \pi(U)$$

with the properties

$$(\pi \pi')(U) = (\pi(\pi'(U)) \quad \text{and} \quad 1(U) = U$$

for $U \in \mathcal{U}$, $\pi, \pi' \in \Gamma$ and the identity element 1 of Γ. The crucial point is that the action of Γ induces an equivalence relation on $\mathcal{U}$ defined by

$$U \sim_\Gamma U'$$

if and only if there exists $\pi \in \Gamma$ such that $\pi(U) = U'$. The proof relies on the fact $\pi(U) = U'$ if and only if $U = \pi^{-1}(U')$. The equivalence class

$$\Gamma(U) = \{\pi(U) \mid \pi \in \Gamma\}$$

of $U \in \mathcal{U}$ is called the Γ-**orbit** or **orbit** of U. Since the action of Γ on $\mathcal{U}$ induces an equivalence relation on $\mathcal{U}$, the set of orbits $\{\Gamma(U) \mid U \in \mathcal{U}\}$ is a partition of $\mathcal{U}$. A minimal but complete set of orbit representatives is called a **transversal** of the orbits.

Let π be an automorphism of the projective point multiset $\mathcal{P}$, and let the points of $\mathcal{P}$ for the moment be represented by vectors $\mathbf{P}_1, \mathbf{P}_2, \ldots, \mathbf{P}_n \in V$. Let the $k \times n$ matrix P consist of the columns $\mathbf{P}_1^\mathsf{T}, \mathbf{P}_2^\mathsf{T}, \ldots, \mathbf{P}_n^\mathsf{T}$. Then the induced action $\pi(\mathcal{P})$ of π on the projective point set $\mathcal{P}$ can be realized as the matrix multiplication

$$M_\pi P = \begin{bmatrix} M_\pi \mathbf{P}_1^\mathsf{T} & M_\pi \mathbf{P}_2^\mathsf{T} & \cdots & M_\pi \mathbf{P}_n^\mathsf{T} \end{bmatrix}.$$

Since $\pi(\mathcal{P}) = \mathcal{P}$, the base vectors of points in $\mathcal{P}$ are mapped to base vectors of $\mathcal{P}$. That means there are scalars $\alpha_1, \alpha_2, \ldots, \alpha_n \in \mathbb{F}_q$ and a permutation $\sigma \in \mathrm{Sym}(n)$ of the indices $1, 2, \ldots, n$, such that

$$M_\pi P = \begin{bmatrix} \alpha_1 \mathbf{P}_{\sigma(1)}^\mathsf{T} & \alpha_2 \mathbf{P}_{\sigma(2)}^\mathsf{T} & \cdots & \alpha_n \mathbf{P}_{\sigma(n)}^\mathsf{T} \end{bmatrix}.$$

That is, $M_\pi P$ can be expressed as

$$M_\pi P = P P_\sigma \mathrm{diag}(\alpha),$$

where P_σ is the permutation matrix induced by the permutation σ and $\mathrm{diag}(\alpha)$ is the diagonal matrix with diagonal entries $\alpha_1, \alpha_2, \ldots, \alpha_n$. But that means $P_\sigma \mathrm{diag}(\alpha)$ is an automorphism of the code $\mathcal{C}$ (see Definition 1.8.7), and we have shown the following theorem.

Theorem 23.3.11 *Let $\Gamma \leq \mathrm{GL}_k(q)$ be a group of automorphisms of a projective point (multi)set $\mathcal{P}$ where $\mathcal{P}$ defines a linear $[n, k]_q$ code $\mathcal{C}$. Then Γ induces an automorphism group Γ' of $\mathcal{C}$.*

Lemma 23.3.12 *Let Γ be a subgroup of $\mathrm{GL}_k(q)$.*

(a) *The group Γ is a group of automorphisms of a projective point set $\mathcal{P} \subseteq \begin{bmatrix} V \\ 1 \end{bmatrix}_q$ if and only if $\mathcal{P}$ is a union of Γ-orbits on $\begin{bmatrix} V \\ 1 \end{bmatrix}_q$.*

(b) *The incidence between a point P and a hyperplane H is invariant under the action of Γ. That means if $P \in H$, then $\pi(P) \in \pi(H)$ for all $\pi \in \Gamma$.*

(c) *The number of Γ-orbits and their sizes on the set of points is equal to the number of Γ-orbits on the set of hyperplanes.*

(d) *If $\Gamma(H)$ is a Γ-orbit on hyperplanes and $\Gamma(P)$ is a Γ-orbit on points, then the cardinality of $\{Q \in \Gamma(P) \mid Q \in H\}$ is independent of the choice of the representative H of the orbit $\Gamma(H)$.*

(e) *If the action of the group Γ on a projective point set consists of a single orbit, i.e., the action is **transitive**, then the corresponding code is* cyclic.

There are several methods available to compute the full automorphism group of a linear code, to test two codes for isomorphism, or to partition a set of linear codes into isomorphism classes.

One possibility to tackle these problems is to use the general methods described in Section 3.2. For linear codes, specific algorithms have been developed; see [265, 266, 1217, 1219, 1650] and the references therein. A very fast algorithm, which is available in various computer algebra systems as well as a stand-alone program, is codecan by Feulner [721, 722].

23.4   Projective Point Sets with Prescribed Automorphism Groups

In this section we attempt to overcome the limits of the construction algorithm of Section 23.3. By prescribing a group of automorphisms $\Gamma \leq \mathrm{GL}_k(q)$ on the projective point sets we can reduce the search space considerably. Instead of selecting n individual points as columns of a generator matrix, we search for suitable orbits of Γ on the projective points whose union consists of n projective points.

Using Lemma 23.3.12 the construction theorem for linear codes with a prescribed group of automorphisms is as follows. Note that by a group of automorphisms of a generator matrix we mean a group of automorphisms of the corresponding projective point set.

Theorem 23.4.1 *Let Γ be a subgroup of $\mathrm{GL}_k(q)$, let $\omega_1, \ldots, \omega_m$ be the orbits of Γ on the projective points of V, and let $\Omega_1, \ldots, \Omega_m$ be the orbits of Γ on the set of hyperplanes of V together with a transversal $H_1 \in \Omega_1$, $H_2 \in \Omega_2$, $\ldots$, $H_m \in \Omega_m$. Let $M_{k,q}^{\Gamma} = [m_{i,j}^{\Gamma}]$ be the $m \times m$ matrix with integer entries*

$$m_{i,j}^{\Gamma} := \big|\{P \in \omega_j \mid P \in H_i\}\big|,$$

for $1 \leq i, j \leq m$.

There is an $[n, k, d']_q$ code with minimum distance $d' \geq d$ such that a generator matrix of this code has Γ as a group of automorphisms if and only if there are vectors $\mathbf{x}$ with $x_i \in \{0, \ldots, \lfloor n/|\omega_i| \rfloor\}$, $1 \leq i \leq m$, and $\mathbf{y} \in \{0, \ldots, n-d\}^m$ satisfying simultaneously

$$\bullet \quad \left[M_{k,q}^{\Gamma} \ \Big| \ I_m \right] \begin{bmatrix} \mathbf{x} \\ \mathbf{y} \end{bmatrix} = (n-d)\mathbf{1}^{\mathsf{T}}, \tag{23.4}$$

$$\bullet \quad \sum_{j=1}^{m} |\omega_j| \cdot x_j = n. \tag{23.5}$$

Remark 23.4.2 As in Section 23.3, the codes constructed in Theorem 23.4.1 are projective if the solution vector $\mathbf{x}$ is restricted to $\{0, 1\}^m$.

For algorithms to compute the matrix $M_{k,q}^\Gamma$, see [1090, 1162]; a more advanced approach is described in [190, Chapter 8] and [1106, Chapter 9].

Note that for the computation of the hyperplane orbits the group action for the group element $M \in \Gamma$ is realized by a multiplication from the right of the matrix M to a row vector $\mathbf{h}$, i.e., $\mathbf{h}M$.

Example 23.4.3 Continuing Example 23.3.4, we again search for a $[4, 3, 2]_2$ code. But now we prescribe a cyclic group Γ of order 3, generated by the matrix

$$\begin{bmatrix} 0 & 1 & 0 \\ 1 & 0 & 0 \\ 0 & 0 & 1 \end{bmatrix}.$$

The action of Γ on the points $\begin{bmatrix} \mathbb{F}_2^3 \\ 1 \end{bmatrix}_2$ yields the following three orbits:

$$\{\langle 100\rangle, \langle 010\rangle, \langle 001\rangle\}, \{\langle 110\rangle, \langle 101\rangle, \langle 011\rangle\}, \{\langle 111\rangle\}$$

as well as the following three orbits on the set of hyperplanes:

$$\{100^\perp, 010^\perp, 001^\perp\}, \{110^\perp, 101^\perp, 011^\perp\}, \{111^\perp\}.$$

The reduced 3×3 matrix $M_{3,2}^\Gamma$ is equal to

$$M_{3,2}^\Gamma = \begin{bmatrix} 2 & 1 & 0 \\ 1 & 1 & 1 \\ 0 & 3 & 0 \end{bmatrix}$$

For example, entry $m_{1,2}^\Gamma = 1$ is computed as follows. $\omega_2 = \{110, 101, 110\}$ and one representative of Ω_1 is the hyperplane $H = 100^\perp$. For the members of ω_2 we have that $110, 101 \notin H$ and $010 \in H$. Therefore it follows that $m_{1,2}^\Gamma = 1$.

The resulting system of Diophantine equations (23.4) and (23.5) is

$$\left[\begin{array}{ccc|ccc} 2 & 1 & 0 & 1 & 0 & 0 \\ 1 & 1 & 1 & 0 & 1 & 0 \\ 0 & 3 & 0 & 0 & 0 & 1 \\ \hline 3 & 3 & 1 & 0 & 0 & 0 \end{array}\right] \begin{bmatrix} x_1 \\ x_2 \\ x_3 \\ y_1 \\ y_2 \\ y_3 \end{bmatrix} = \begin{bmatrix} 2 \\ 2 \\ 2 \\ 4 \end{bmatrix}$$

where $x_1, x_2 \in \{0, 1\}$, $x_2 \in \{0, 1\}$, $x_3 \in \{0, 1, \ldots, 4\}$ and $y_i \in \{0, 1, 2\}$, $1 \le i \le 3$. The unique solution vector of this system of equation is

$$[1, 0, 1; 0, 0, 2]^\mathsf{T},$$

which means that the columns of the generator matrix will be composed of the union of the orbits ω_1 and ω_3. As in Example 23.3.4, we obtain the following generator matrix

$$\begin{bmatrix} 1 & 0 & 0 & 1 \\ 0 & 1 & 0 & 1 \\ 0 & 0 & 1 & 1 \end{bmatrix}$$

of an optimal linear $[4, 3, 2]_2$-code. From the solution vector $\mathbf{y}$, we immediately can read off the weight distribution of the code. There are two entries of $\mathbf{y}$ equal to zero. The corresponding orbits on hyperplanes both have size 3. Therefore, the code has $(q-1) \cdot (3+3) = 6$ codewords of weight $d+0 = 2$. Further, $y_3 = 2$. The hyperplane orbit Ω_3 has size 1, which means that the code has exactly $(q-1) \cdot 1 = 1$ codeword of weight $d+2 = 4$.

Since the solution vector $x_1 x_2 x_3$ has only entries from $\{0, 1\}$ we know that the resulting code is projective.

Example 23.4.4 Here, we are searching for a linear $[14, 3, 9]_3$ code over $\mathbb{F}_3$. Such a code is optimal. First note that a code with these parameters cannot be projective, since there are exactly $\begin{bmatrix} 3 \\ 1 \end{bmatrix}_3 = (3^3 - 1)/(3 - 1) = 13$ 1-dimensional subspaces of $\mathbb{F}_3^3$. Therefore, at least one column has to occur at least two times in a generator matrix of such a code. The incidence matrix $M_{3,3}$ has size 13×13. The 1-dimensional subspaces are $\langle 001 \rangle$, $\langle 010 \rangle$, $\langle 011 \rangle$, $\langle 012 \rangle$, $\langle 100 \rangle$, $\langle 101 \rangle$, $\langle 102 \rangle$, $\langle 110 \rangle$, $\langle 111 \rangle$, $\langle 112 \rangle$, $\langle 120 \rangle$, $\langle 121 \rangle$, $\langle 122 \rangle$.

Now we prescribe the complete monomial group $\Gamma = \mathrm{Mon}_3(3)$; see [294]. This group is generated by the matrices

$$\begin{bmatrix} 0 & 1 & 0 \\ 1 & 0 & 0 \\ 0 & 0 & 1 \end{bmatrix}, \begin{bmatrix} 0 & 1 & 0 \\ 0 & 0 & 1 \\ 1 & 0 & 0 \end{bmatrix}, \begin{bmatrix} 2 & 0 & 0 \\ 0 & 1 & 0 \\ 0 & 0 & 1 \end{bmatrix}, \begin{bmatrix} 1 & 0 & 0 \\ 0 & 2 & 0 \\ 0 & 0 & 1 \end{bmatrix}, \begin{bmatrix} 1 & 0 & 0 \\ 0 & 1 & 0 \\ 0 & 0 & 2 \end{bmatrix},$$

and has order 48. The action of Γ on $\begin{bmatrix} \mathbb{F}_3^3 \\ 1 \end{bmatrix}_3$ yields the following three orbits:

$$\{\langle 001 \rangle, \langle 010 \rangle, \langle 100 \rangle\}, \ \{\langle 011 \rangle, \langle 101 \rangle, \langle 110 \rangle, \langle 012 \rangle, \langle 102 \rangle, \langle 120 \rangle\},$$

$$\{\langle 111 \rangle, \langle 122 \rangle, \langle 121 \rangle, \langle 112 \rangle\}.$$

In addition, we obtain the following three orbits on the set of hyperplanes:

$$\{001^\perp, 010^\perp, 100^\perp\}, \ \{011^\perp, 101^\perp, 110^\perp, 012^\perp, 102^\perp, 120^\perp\},$$

$$\{111^\perp, 122^\perp, 121^\perp, 112^\perp\}.$$

The reduced 3×3 matrix $M_{3,3}^\Gamma$ is equal to

$$M_{3,3}^\Gamma = \begin{bmatrix} 2 & 2 & 0 \\ 1 & 1 & 2 \\ 0 & 3 & 1 \end{bmatrix}$$

As an example, we show how to compute the entry $m_{1,1}^\Gamma = 2$: $\omega_1 = \{\langle 001 \rangle, \langle 010 \rangle, \langle 100 \rangle\}$ and one representative of Ω_1 is the hyperplane $H = 001^\perp$. For the members of ω_1 we have that $\langle 001 \rangle \notin H$ and $\langle 010 \rangle, \langle 100 \rangle \in H$. It follows that $m_{1,1}^\Gamma = 2$.

Having computed $M_{3,3}^\Gamma$, the corresponding system of Diophantine equations (23.4) and (23.5) is

$$\begin{bmatrix} 2 & 2 & 0 & 1 & 0 & 0 \\ 1 & 1 & 2 & 0 & 1 & 0 \\ 0 & 3 & 1 & 0 & 0 & 1 \\ \hline 3 & 6 & 4 & 0 & 0 & 0 \end{bmatrix} \begin{bmatrix} x_1 \\ x_2 \\ x_3 \\ y_1 \\ y_2 \\ y_3 \end{bmatrix} = \begin{bmatrix} 5 \\ 5 \\ 5 \\ 14 \end{bmatrix}$$

where $x_1 \in \{0, 1, 2, 3, 4\}$, $x_2 \in \{0, 1, 2\}$, $x_3 \in \{0, 1, 2, 3\}$ and $y_i \in \{0, 1, 2, 3, 4, 5\}$, $1 \leq i \leq 3$. The unique solution vector of this system of equations is equal to

$$[0, 1, 2; 3, 0, 0]^\mathsf{T},$$

which means that the orbit ω_2 occurs once in the corresponding generator matrix and the orbit ω_3 occurs two times in the generator matrix. Thus, we obtain the generator matrix

$$\begin{bmatrix} 0 & 1 & 1 & 0 & 1 & 1 & 1 & 1 & 1 & 1 & 1 & 1 & 1 & 1 \\ 1 & 0 & 1 & 1 & 0 & 2 & 1 & 1 & 2 & 2 & 2 & 2 & 1 & 1 \\ 1 & 1 & 0 & 2 & 2 & 0 & 1 & 1 & 2 & 2 & 1 & 1 & 2 & 2 \end{bmatrix}$$

of an optimal linear $[14, 3, 9]_3$ code. From the solution vector $\mathbf{y}$, we immediately can read off the weight distribution of the code: there are $(q - 1) \cdot 3 = 6$ codewords of weight $d + 3 = 12$ and $(q - 1) \cdot (6 + 4) = 20$ codewords of weight $d + 0 = 9$.

Since $x_3 > 1$, we know that the code is not projective.

23.4.1 Strategies for Choosing Groups

Subgroups of $\mathrm{GL}_k(q)$ which are promising candidates to allow construction of good codes should be large enough to produce a small number of orbits, so that we can solve the corresponding system of Diophantine linear equations, but also "nice" enough to be the group of automorphisms of a good linear code. The least requirement is that the group action on the points of V has orbits whose sizes add up to the length n of the code.

So far, several classes of subgroups of $\mathrm{GL}_k(q)$ proved to be successful. In [1483] the authors went through the online *Atlas of Finite Group Representations* [1894] and searched for group representations generating at least one orbit of a given size n on the points of V. Classes of subgroups used in [293, 299, 1141] are the following:

- Permutation groups:

  As the permutation matrices are elements of $\mathrm{GL}_k(q)$ we have that the symmetric group S_k is a subgroup of $\mathrm{GL}_k(q)$.

- Prime fields:

  In the case when q is a proper prime power, and particularly in the cases of fields with 4, 8, and 9 elements, the general linear group over the prime field is a subgroup of $\mathrm{GL}_k(q)$. This class of subgroups contains the permutation groups.

- Cyclic groups:

  Conjugated subgroups of $\mathrm{GL}_k(q)$ produce equivalent codes. If all conjugacy classes of cyclic codes of $\mathrm{GL}_k(q)$ have been tested, we know that all possible cyclic subgroups of all possible groups of automorphisms of a code are covered. This may enable a systematic, exhaustive search. Another successful approach is to use random cyclic generators.

- $\mathrm{GL}_l(q)$ for $l < k$:

  Subgroups which are direct products of $\mathrm{GL}_l(q)$ for $l < k$ are also candidates.

23.4.2 Observations for Permutation Groups

This class of groups is a good source for codes. The symmetric group was used in [293] to produce series of new codes. The alternating group for example was used in the case of $q = 8$ and $k = 4$ to reduce the size of the system of equations to 57 rows and to construct five new $[n, 4, d]_8$ codes (the bold face minimum distance means that the code is optimal):

n	d
96	81
97	82
102	87
103	**88**
108	92

In [299] in more than 30 cases a code with a minimum distance as good as the current lower bound could be constructed. In situations where the number of orbits generated by a permutation group were too big, the group was enlarged by adding a diagonal matrix as a generator to get fewer orbits. For example in the case of $q = 9$ and $k = 4$ a double transposition and a diagonal matrix were used to reduce the system of equations to 59 rows which gave the following 3 new $[n, 4, d]_9$ codes:

n	d
100	86
128	110
130	112

23.4.3   Observations for Cyclic Groups

For a random generator of a cyclic subgroup the typical situation is as follows: there is only a small number of different orbit sizes. Thus one can only hope for a good code in the cases where some orbit sizes add up to n. In [299] a random generator was found in the case of $q = 4$ and $k = 6$ which produced 43 orbits, one of size 1 and the others of sizes 17 and 51. The group produced a new best code with parameters $[136, 6, 96]_4$ and in eight further lengths a code as good as the best known. An even more extreme situation was in the case of another random element, again in $q = 4$ and $k = 6$. It produced 65 orbits of length 21; i.e., there could only be codes for n being multiples of 21. But that generator was good enough to produce a new best $[189, 6, 136]_4$ code and for several lengths a code as good as the best known codes.

With the method presented in this chapter the gap between upper and lower bounds could be shortened in hundreds of cases by constructing codes with a higher minimum distance than previously known; see [293, 299, 1141].

Remark 23.4.5 The approach to prescribe an automorphism group and solve a resulting Diophantine linear system to construct a linear code is very general and can be used in a wide variety of other contexts.

For example, in [1116, 1117] it was used to find very good linear codes over rings. In [300, 1139] subspace codes were found with the same approach; see Chapter 14 for more information about subspace codes.

Even more general, a code can be regarded as an incidence structure, represented by an incidence matrix whose row sums are within a certain range. Now the theory of incidence preserving group actions [1106] allows one to shrink the incidence matrix to a smaller one whose rows, respectively columns, correspond to orbits of a group and whose row sums stay in the same range. This construction is a general approach that works for many discrete structures. Kramer and Mesner were among the first to apply this method in design theory [1161]. Subsequently, it has become known as **Kramer–Mesner Method**. It allowed construction of record breaking structures such as, for example, combinatorial designs [192], q-analogs of designs [295, 296, 297], and (n, r)-arcs in $PG(2, q)$ [298].

23.5 Solving Strategies

So far, we have reduced the problem of finding good linear codes to the solution of Diophantine linear systems arising in Theorems 23.3.7 and 23.4.1. This problem is also known as an **integer linear programming** problem. A natural candidate for a solving algorithm for this problem class is integer linear programming software like CPLEX [1020] and Gurobi [875] among others. Since the integer linear programming problem is known to be NP-complete, see [1633], the problem can be reduced to any other NP-complete problem and can be attempted to be solved by problem specific algorithms.

The problem formulation in Theorems 23.3.7 and 23.4.1 is without objective function and for fixed values of n and k. This can be changed to replace, for example, the equation (23.5) by the objective function

$$\min_{\mathbf{x}} \sum_{j=1}^{m} |\omega_j| \cdot x_j$$

in order to find the shortest codes for fixed values of k and d.

Further promising strategies are generic backtrack algorithms [1091] and SAT solvers [1134]. However, the most successful strategies besides integer linear programming seem to be heuristic algorithms like [1141, 1951] and methods based on lattice basis reduction [1879, 1880]. In the latter case, the Diophantine linear system is transformed into the problem of finding certain small vectors in a lattice. The search for these vectors is done with lattice basis reduction followed by exhaustive enumeration.

23.6 Construction of Codes with Additional Restrictions

In this section, we construct certain classes of codes, like projective codes, two-weight codes, self-orthogonal codes, and linear codes with a complementary dual (LCD). These restrictions can be accomplished (and combined) by adding certain additional constraints to Theorems 23.3.7 and 23.4.1.

It would be desirable to construct $[n, k]_q$ codes with prescribed minimum distance as well as prescribed dual minimum distance. First attempts to accomplish this were made in [1137].

23.6.1 Projective Codes

As already pointed out in Remark 23.3.8, the search for codes can be restricted to projective codes by restricting the solution vectors $\mathbf{x}$ in Theorems 23.3.7 and 23.4.1 to $\{0, 1\}$-vectors.

23.6.2 Codes with Few Weights

If the nonzero codewords of a linear code attain only two different values, the code is called a linear **two-weight code**; see [330] and Chapters 18, 19, and 20.

The search for two-weight codes is much more restrictive than the search for general linear codes, since in Theorem 23.3.7 the solutions can be restricted to $\{0, 1\}$ vectors. In the language of projective point sets, a projective point set corresponding to a two-weight code has exactly two intersection numbers with the hyperplanes. Let w_1 and w_2 be the two possible weights of the linear code; without restriction let's assume that $w_1 < w_2$. Then, if q is a power of the prime p, it is well known that there are integers j and t such that $w_1 = p^j t$ and $w_2 = p^j(t+1)$; see [330, Corollary 5.5].

It is easy to generalize Theorem 23.3.7 to linear two-weight codes; see [1136].

Theorem 23.6.1 *Let $w_1 < w_2$. There is a two-weight $[n, k, w_1]_q$ code with nonzero weights w_1 and w_2 if and only if there is a column vector $\mathbf{x} \in \{0, \ldots, n\}^{\begin{bmatrix} k \\ 1 \end{bmatrix}_q}$ and a column vector $\mathbf{y} \in \{0, 1\}^{\begin{bmatrix} k \\ 1 \end{bmatrix}_q}$, where $00 \cdots 0 \neq \mathbf{y} \neq 11 \cdots 1$, satisfying simultaneously*

$$\bullet \quad \left[M_{k,q} \mid \mathrm{diag}(w_2 - w_1) \right] \begin{bmatrix} \mathbf{x} \\ \mathbf{y} \end{bmatrix} = (n - w_1)\mathbf{1}^\mathsf{T},$$

$$\bullet \quad \sum_{i=1}^{\begin{bmatrix} k \\ 1 \end{bmatrix}_q} x_i = n,$$

where $\mathrm{diag}(w_2 - w_1)$ is the $\begin{bmatrix} k \\ 1 \end{bmatrix}_q \times \begin{bmatrix} k \\ 1 \end{bmatrix}_q$ diagonal matrix with diagonal entries $w_2 - w_1$.

Remark 23.6.2 Prescribing a group of automorphisms in the search for two-weight codes is also straightforward: In Theorem 23.4.1 the identity matrix I_m in equation (23.4) has to be replaced by the $m \times m$ diagonal matrix $\mathrm{diag}(w_2 - w_1)$ and the entries of the column vector $\mathbf{y}$ from Theorem 23.4.1 have to be restricted to $\{0, 1\}$.

Another straightforward modification of the equations allows one to restrict the search to linear **three-weight codes**, i.e., linear codes attaining three different nonzero weights $w_1 < w_2 < w_3$. In the special case that the three weights are symmetric around w_2, i.e., there is a $\delta > 0$ such that $w_1 + \delta = w_2 = w_3 - \delta$, condition (23.2) reduces to

$$\bullet \quad \left[M_{k,q} \mid \mathrm{diag}(\delta) \right] \begin{bmatrix} \mathbf{x} \\ \mathbf{y} \end{bmatrix} = (n - w_1)\mathbf{1}^\mathsf{T},$$

and $\mathbf{y}$ is required to be a vector of length $\begin{bmatrix} k \\ 1 \end{bmatrix}_q$ with entries from $\{0, 1, 2\}$, whereas $\mathbf{x}$ has the same restrictions as in Theorem 23.6.1; for more see [1114].

Prescribing a group of automorphisms Γ to the search for two-weight or three-weight codes is straightforward; in the above condition $M_{k,q}$ has to be replaced by the matrix $M_{k,q}^\Gamma$ and the bounds on $\mathbf{x}$ and $\mathbf{y}$ have to be adjusted accordingly.

23.6.3   Divisible Codes

A linear code is called a Δ-**divisible** code if all weights of codewords are divisible by some given integer Δ. Divisible codes were introduced by Ward [1873]; for a survey and further related literature, see [1877]. The divisibility constant Δ is severely constrained [1873, Theorem 1]: If $\gcd(\Delta, q) = 1$, then a Δ-divisible code over $\mathbb{F}_q$ (with no zero coordinates) is a Δ-fold replicated code. Thus, the interesting cases of divisible codes are those where Δ is a multiple of p with q being a power of the prime p. The corresponding point set is also called a projective Δ-**divisible point set**.

It is an easy task to restrict the search for projective point sets to Δ-divisible point sets. We modify condition (23.2) to

$$\bullet \quad \left[M_{k,q} \mid \mathrm{diag}(\Delta) \right] \begin{bmatrix} \mathbf{x} \\ \mathbf{y} \end{bmatrix} = (n - d)\mathbf{1}^\mathsf{T},$$

where $\mathbf{x}$ has the same restrictions as in Theorem 23.6.1 and $\mathbf{y}$ is required to be a vector of length $\begin{bmatrix} k \\ 1 \end{bmatrix}_q$ with entries from $\{0, 1, \ldots, \lfloor (n-d)/\Delta \rfloor\}$.

Prescribing a group of automorphisms Γ to the search for divisible codes is straightforward: in the above condition $M_{k,q}$ has to be replaced by the matrix $M_{k,q}^\Gamma$.

23.6.4 Codes with Prescribed Gram Matrix

In the important cases $q = 2$ and $q = 3$ it is also possible to use the construction approach of this section to search for even more restricted codes.

Let $\mathcal{C}$ be a linear $[n, k]_q$ code. Continuing our view of codes as projective point sets, the columns of a generator matrix G of C are transposed base vectors $\mathbf{G}_1, \mathbf{G}_2, \ldots, \mathbf{G}_n$ of the points of a projective point set $\{\langle \mathbf{G}_1 \rangle, \langle \mathbf{G}_2 \rangle, \ldots, \langle \mathbf{G}_n \rangle\}$. The $k \times k$ matrix GG^T over $\mathbb{F}_q$ is called the **Gram matrix** of the generator matrix G. Note that the Gram matrix is *symmetric*.

The goal of this section is to restrict the search to codes with a prescribed Gram matrix. Two important applications will be discussed in the next sections: if the prescribed Gram matrix is equal to zero, then the codes will be self-orthogonal, and if the prescribed Gram matrix is nonsingular, then LCD codes will be constructed.

Obviously, a permutation of the columns of the generator matrix affects the Gram matrix. Moreover, a multiplication of the generator matrix G from the right with a diagonal matrix $\mathrm{diag}(\alpha)$ consisting of nonzero diagonal entries from $\mathbb{F}_q$, i.e., $\alpha = \alpha_1 \alpha_2 \cdots \alpha_n \in (\mathbb{F}_q^*)^n$, may change the rank of the Gram matrix. The Gram matrix of the generator matrix $G\,\mathrm{diag}(\alpha)$ is equal to

$$(G\,\mathrm{diag}(\alpha))(G\,\mathrm{diag}(\alpha))^\mathsf{T} = G\,\mathrm{diag}(\alpha)^2 G^\mathsf{T},$$

where $\mathrm{diag}(\alpha)^2$ is the diagonal matrix with entries $\alpha_1^2 \alpha_2^2 \cdots \alpha_n^2 \in (\mathbb{F}_q^*)^n$. Therefore, the Gram matrix is not invariant under multiplication by a diagonal matrix with nonzero entries. Even worse, depending on the vector α, the determinant of the Gram matrix may be zero or nonzero over $\mathbb{F}_q$ if $q > 3$.

Example 23.6.3 Let $\begin{bmatrix} 1 & 1 \end{bmatrix}$ be a generator matrix of a $[2,1]_5$ code. The Gram matrix is the 1×1 matrix $\begin{bmatrix} 2 \end{bmatrix}$ which is obviously nonzero. However, the monomial equivalent code with generator matrix $\begin{bmatrix} 1 & 2 \end{bmatrix}$ has Gram matrix $\begin{bmatrix} 0 \end{bmatrix}$ over $\mathbb{F}_5$.

The situation is different in the binary and ternary case. For these two fields, $\alpha_i^2 = 1$ if and only if $\alpha_i \neq 0$. Therefore, for every choice of α, it follows that

$$\mathrm{diag}(\alpha)^2 = I$$

and multiplication of the generator matrix from the right by $\mathrm{diag}(\alpha)$ does not change the Gram matrix.

Now, as before let G be a $k \times n$ generator matrix, and let Q be a symmetric $k \times k$ matrix over $\mathbb{F}_q$. Let $\mathbf{v}^{(j)}$, $1 \le j \le k$, denote the jth entry of the vector $\mathbf{v}$. The equation $GG^\mathsf{T} = Q$ holds over $\mathbb{F}_q$ if and only if

$$\sum_{s=1}^n \mathbf{G}_s^{(i)} \mathbf{G}_s^{(j)} = Q_{i,j} \quad \text{over } \mathbb{F}_q \text{ for all } 1 \le i \le j \le k.$$

Note that, since the Gram matrix is symmetric, it suffices to consider the $\binom{k+1}{2}$ entries $Q_{i,j}$, $1 \le i \le j \le k$. With this notation the additional equations over $\mathbb{F}_q$ can therefore be written as

$$\sum_{s=1}^{\begin{bmatrix} k \\ 1 \end{bmatrix}_q} x_s \mathbf{P}_s^{(i)} \mathbf{P}_s^{(j)} = Q_{i,j} \quad \text{for } 1 \le i \le j \le k,$$

where $\langle \mathbf{P}_s \rangle$, $1 \leq s \leq \begin{bmatrix} k \\ 1 \end{bmatrix}_q$, runs through all points. These equations guarantee that the pairwise inner products of rows of a generator matrix built from the columns of $\begin{bmatrix} V \\ 1 \end{bmatrix}_q$ have the required values. Combining this with the result from Section 23.3 we get the following theorem.

Theorem 23.6.4 *Let $q \in \{2,3\}$ and Q be a symmetric $k \times k$ matrix over $\mathbb{F}_q$. There exists a $[n,k,d']_q$ code with minimum distance $d' \geq d$ if and only if there are column vectors $\mathbf{x} \in \{0,\dots,n\}^{\begin{bmatrix} k \\ 1 \end{bmatrix}_q}$ and $\mathbf{y} \in \{0,\dots,n-d\}^{\begin{bmatrix} k \\ 1 \end{bmatrix}_q}$ satisfying the system of equations*

- $\left[M_{k,q} \ \middle| \ I_{\begin{bmatrix} k \\ 1 \end{bmatrix}_q} \right] \begin{bmatrix} \mathbf{x} \\ \mathbf{y} \end{bmatrix} = (n-d)\mathbf{1}^\mathsf{T}$,

- $\displaystyle\sum_{i=1}^{\begin{bmatrix} k \\ 1 \end{bmatrix}_q} x_i = n$,

- *the inner product $[P_{k,q}]_{i,j} \cdot \mathbf{x} = Q_{i,j}$ for $1 \leq i \leq j \leq k$,*

where the $\binom{k+1}{2} \times \begin{bmatrix} k \\ 1 \end{bmatrix}_q$ matrix $P_{k,q} = [p_{(i,j),\mathbf{P}}]$ is defined by

$$p_{(i,j),\mathbf{P}} = \mathbf{P}^{(i)}\mathbf{P}^{(j)}$$

for $1 \leq i \leq j \leq k$ and $\langle \mathbf{P} \rangle \in \begin{bmatrix} V \\ 1 \end{bmatrix}_q$.

Additionally prescribing a group of automorphisms in the search for linear codes with a prescribed Gram matrix is straightforward.

Theorem 23.6.5 *Let Γ be a subgroup of $\mathrm{GL}_k(q)$ and Q be a symmetric $k \times k$ matrix over $\mathbb{F}_q$. Let $\omega_1,\dots,\omega_m$ be the orbits of Γ on the projective points of V, and let $\Omega_1,\dots,\Omega_m$ be the orbits of Γ on the set of hyperplanes of V together with a transversal $H_1, H_2, \dots, H_m$. Let $M_{k,q}^\Gamma = [m_{i,j}^\Gamma]$ be the $m \times m$ matrix with integer entries*

$$m_{i,j}^\Gamma := \left| \{ P \in \omega_j \mid P \in H_i \} \right|,$$

for $1 \leq i,j \leq m$.

Further, let $P_{k,q}^\Gamma = [p_{(i,j),s}^\Gamma]$ be the $\binom{k+1}{2} \times m$ matrix with entries

$$p_{(i,j),s}^\Gamma = \sum_{\mathbf{P} \in \omega_s} \mathbf{P}^{(i)}\mathbf{P}^{(j)} \quad \text{for } 1 \leq i \leq j \leq k \text{ and } 1 \leq s \leq m.$$

There is an $[n,k,d']_q$ code with minimum distance $d' \geq d$ such that a generator matrix of this code has Γ as a group of automorphisms and a generator matrix has Gram matrix Q if and only if there are vectors $\mathbf{x}$ with $x_i \in \{0,\dots,\lfloor n/|\omega_i| \rfloor\}$ for $1 \leq i \leq m$, $\mathbf{y} \in \{0,\dots,n-d\}^m$, and $\mathbf{z} \in \mathbb{Z}^{\binom{k+1}{2}}$ satisfying

$$\begin{bmatrix} M_{k,q}^\Gamma & I_m & \mathbf{0}_{m \times \binom{k+1}{2}} \\ P_{k,q}^\Gamma & \mathbf{0}_{\binom{k+1}{2} \times m} & -qI_{\binom{k+1}{2}} \\ |\omega_1| \quad \cdots \quad |\omega_m| & \mathbf{0}_{1 \times m} & \mathbf{0}_{1 \times \binom{k+1}{2}} \end{bmatrix} \begin{bmatrix} \mathbf{x} \\ \mathbf{y} \\ \mathbf{z} \end{bmatrix} = \begin{bmatrix} (n-d)\mathbf{1}^T \\ \mathbf{Q} \\ n \end{bmatrix},$$

where $\mathbf{0}_{a \times b}$ is the $a \times b$ zero matrix and $\mathbf{Q}$ is the column vector of length $\binom{k+1}{2}$ consisting of the entries $Q_{i,j}$, $1 \leq i \leq j \leq k$.

Note that in the above system of equations the integer variables z_i, $0 \leq i < \binom{k+1}{2}$ are implicitly bounded by the restrictions on the vector $\mathbf{x}$ and $\mathbf{y}$.

As for Theorem 23.6.4, in the case of $q \notin \{2,3\}$ the above conditions involving the entries of the Gram matrix are sufficient but not necessary for a code together with a generator matrix with given Gram matrix to exist.

23.6.5 Self-Orthogonal Codes

Recall that for an $[n, k]_q$ code $\mathcal{C}$ the dual code

$$\mathcal{C}^{\perp} = \{\mathbf{w} \in \mathbb{F}_q^n \mid \mathbf{w} \cdot \mathbf{c} = 0 \text{ for all } \mathbf{c} \in \mathcal{C}\}.$$

is an $[n, n-k]_q$ code. A **self-orthogonal** linear $[n, k]_q$ code is a k-dimensional subspace of the n-dimensional vector space $\mathbb{F}_q^n$ over the finite field $\mathbb{F}_q$ with the additional requirement that $\mathcal{C} \subseteq \mathcal{C}^{\perp}$.

This means that a code $\mathcal{C}$ is self-orthogonal if and only if it holds over $\mathbb{F}_q$ that

$$\mathbf{v} \cdot \mathbf{w} = 0 \quad \text{for all } \mathbf{v}, \mathbf{w} \in \mathcal{C}.$$

It is known that if G is a generator matrix of $\mathcal{C}$ and $\mathbf{G}^{(0)}, \mathbf{G}^{(1)}, \ldots, \mathbf{G}^{(k-1)}$ are the rows of G then $\mathcal{C}$ is self-orthogonal if and only if

$$\mathbf{G}^{(i)} \cdot \mathbf{G}^{(j)} = \sum_{0 \leq s < n} \mathbf{G}_s^{(i)} \mathbf{G}_s^{(j)} = 0 \quad \text{for all } 1 \leq i \leq j \leq k.$$

In other words, a linear $[n, k]_q$ code $\mathcal{C}$ with generator matrix G is self-orthogonal if and only if the Gram matrix $GG^{\mathsf{T}} = \mathbf{0}_{k \times k}$ over $\mathbb{F}_q$. This enables us to apply the method from Section 23.6.4 to the search for self-orthogonal codes for $q = 2, 3$.

Example 23.6.6 In [1140] the described method allowed the authors to construct self-orthogonal codes for the following parameters, which were improvements of the bounds (for general linear codes) in [845].

- Parameters of optimal codes: $[177, 10, 84]_2$, $[38, 7, 21]_3$, $[191, 6, 126]_3$, $[202, 6, 132]_3$, $[219, 6, 144]_3$, $[60, 7, 36]_3$.

- Parameters of codes which are improvements to the previous bounds in [845] but which are not optimal codes: $[175, 10, 82]_2$, $[140, 11, 64]_2$, $[61, 7, 36]_3$, $[188, 7, 120]_3$, $[243, 7, 156]_3$.

The time needed to compute the matrices $M_{k,q}^{\Gamma}$ of Section 23.4 is small compared to the time needed to solve the corresponding system of Diophantine equations. The computation times depend heavily on the number of orbits and the number $n - d$ which is the upper bound of parts of the variables. Also, the number $\binom{k}{2}$ of equations to ensure self-orthogonal solutions comes into play. All this is shown in Table 23.1 which gives detailed information for the six optimal codes.

TABLE 23.1: Code parameters, number of orbits, and solving time for optimal self-orthogonal codes

Code	# Orbits	Time	$n - d$	$\binom{k}{2}$
$[177, 10, 84]_2$	51	$< 3h$	93	45
$[38, 7, 21]_3$	101	$< 100s$	17	21
$[60, 7, 36]_3$	67	$< 100s$	24	21
$[202, 6, 132]_3$	44	$< 10s$	70	15
$[219, 6, 144]_3$	38	$< 10s$	75	15
$[191, 6, 126]_3$	44	$< 10s$	65	15

One further challenge is to choose a group such that the reduction is large enough to

TABLE 23.2: Parameters of all self-orthogonal $[n, 9, d]_2$ codes which could be constructed in [1140] with the method described in this chapter and whose minimum distance is as least as high as the best known codes in [845]. An entry $30 - 33$ means that codes of length 30, 31, 32, and 33 could be constructed.

n	d	n	d	n	d	n	d	n	d
$21 - 25$	8	$70 - 74$	32	$118 - 121$	56	$175 - 176$	84	$222 - 223$	108
$30 - 33$	12	$80 - 81$	36	$127 - 128$	60	$182 - 184$	88	226	110
$38 - 42$	16	84	38	$135 - 138$	64	189, 191	92	$228 - 233$	112
45	18	$86 - 90$	40	142	66	194	94	$238 - 240$	116
$47 - 50$	20	93	42	$144 - 145$	68	$196 - 201$	96	243	118
53	22	$95 - 97$	44	148	70	205	98	$245 - 248$	120
$55 - 58$	24	100	46	$150 - 154$	72	$207 - 209$	100	250	122
$63 - 65$	28	$102 - 106$	48	$159 - 160$	76	212	102	252	124
68	30	$111 - 113$	52	$166 - 170$	80	$214 - 216$	104	256	128

get a system which can be handled by the solving algorithm but which, on the other hand, is also a group of automorphisms of a self-orthogonal code with high minimum distance. The least requirement on the group is that there exist point orbits under the action of the group with length at most n.

For more details on promising groups for construction of general linear codes, see [299, 1483]. However, in the case of self-orthogonal codes, cyclic groups (i.e., generated by one element) seem to be especially good. For example in the case of $q = 2$ and $k = 9$ all the distance-optimal self-orthogonal codes in [1140] were found by using cyclic groups.

In many cases it is possible to find self-orthogonal codes which meet the minimum weight of the best known linear codes. Of course this is possible only in the case of even weight (in the binary case) or weight d with $d \equiv 0 \bmod 3$ (in the ternary case). This situation is reflected in Table 23.2 for the case $q = 2$ and $k = 9$ from [1140].

An alternative approach is to classify self-orthogonal codes by reversing the operation of going from a code to its residual code; see [271]. This means that instead of searching for suitable columns of a generator matrix, one builds up generator matrices row by row. In Chapter 4 construction methods for self-dual codes are studied.

23.6.6   LCD Codes

In general, a code $\mathcal{C}$ and its dual code $\mathcal{C}^\perp$ over the finite field $\mathbb{F}_q$ can have intersection larger than $\{\mathbf{0}\}$. The extreme case $\mathcal{C} \subseteq \mathcal{C}^\perp$ is well studied, as we saw in Section 23.6.5.

However, in 1993 Massey [1347] studied the other extreme situation, namely the codes $\mathcal{C}$ with $\mathcal{C} \cap \mathcal{C}^\perp = \{\mathbf{0}\}$. An $[n, k, d]_q$ code $\mathcal{C}$ with the property that $\mathcal{C} \cap \mathcal{C}^\perp = \{\mathbf{0}\}$ is called a **linear code with a complementary dual (LCD code)**. For an LCD code $\mathcal{C}$ it follows that $\mathbb{F}_q^n = \mathcal{C} \oplus \mathcal{C}^\perp$.

Theorem 23.6.7 (Massey [1347]) *If G is a generator matrix for the linear $[n, k]_q$ code $\mathcal{C}$, then $\mathcal{C}$ is an LCD code if and only if the $k \times k$ Gram matrix GG^T is nonsingular. Moreover, if $\mathcal{C}$ is an LCD code, then $G^T(GG^T)^{-1}G$ is the orthogonal projection from $\mathbb{F}_q^n$ to $\mathcal{C}$.*

Remark 23.6.8 $G^T(GG^T)^{-1}$ is known as the **Penrose inverse** of the matrix G. It does exist if and only if the Gram matrix GG^T is invertible.

Carlet and Guilley [353] used LCD codes as a counter-measure to *side-channel attacks* in cryptography. LCD codes are also used in the construction of self-dual codes and lattices [1421]. In [687] decoding properties of LCD codes were studied.

There are many constructions of LCD codes. For example, Yang and Massey [1920] proved that $\mathcal{C}$ is a cyclic LCD code if and only if the generator polynomial is self-reciprocal plus some condition on the irreducible factors of the polynomial; see also Section 2.10. Sendrier [1648] showed that LCD codes meet the Gilbert–Varshamov Bound. Dougherty et al. [629] describe constructions of LCD codes by codes over rings, orthogonal matrices and block designs. For other constructions, see Chapter 16 and Galindo et al. [785].

At first sight it appears that in an exhaustive search for all optimal LCD codes we have to prescribe all symmetric matrices from $\mathrm{GL}_k(q)$. Fortunately, the set of Gram matrices which have to be examined can be reduced to a much smaller set. To see this we make a short excursion to quadratic forms.

A symmetric matrix $A \in \mathbb{F}_q^{k \times k}$ gives rise to a **quadratic form** Q_A, which is a mapping

$$Q_A : \mathbb{F}_q^k \times \mathbb{F}_q^k \to \mathbb{F}_q \quad \text{where} \quad (\mathbf{x}, \mathbf{x}) \mapsto \mathbf{x} A \mathbf{x}^\mathsf{T}.$$

The matrix A is called the **coefficient matrix** of the quadratic form Q_A. Two quadratic forms with coefficient matrices A and B are called **equivalent** if there is an invertible matrix $C \in \mathbb{F}_q^{k \times k}$ such that

$$C A C^\mathsf{T} = B.$$

In this case we also call the coefficient matrices A and B equivalent. It is easy to check that this indeed is an equivalence relation.

Suppose $\mathcal{P}$ and $\mathcal{P}'$ are isomorphic projective point sets of size n which give the $k \times n$ matrices P and P', and there is a matrix $M \in \mathrm{GL}_k(q)$ such that

$$M\mathcal{P} = \mathcal{P}'.$$

Thus $MP = P'$. Then it follows from

$$P'P'^\mathsf{T} = (MP)(MP)^\mathsf{T} = M(PP^\mathsf{T})M^\mathsf{T}$$

that the Gram matrices of P and P' are equivalent.

In [34], Albert classified quadratic forms up to equivalence. The classification can be summarized in the following theorem for our context.

Theorem 23.6.9 (Albert [34]) *Let A be a symmetric matrix in $\mathbb{F}_q^{k \times k}$. If q is even and A has rank $2t + 1$ or if q is odd, then A is equivalent to a diagonal matrix.*

If q is even and A has rank $2t$, then A is either equivalent to a diagonal matrix or to the matrix

$$\begin{bmatrix} E_1 & & & 0 \\ & \ddots & & \vdots \\ & & E_t & \\ 0 & \cdots & & 0 \end{bmatrix}, \tag{23.6}$$

where $E_i = \left[\begin{smallmatrix} 0 & 1 \\ 1 & 0 \end{smallmatrix}\right]$ for $i = 1, \ldots, t$.

Corollary 23.6.10 *Let $q = 2$ and the symmetric $k \times k$ matrix A be invertible over $\mathbb{F}_q$. If k is odd, then A is equivalent to I_k. If k is even, then A is either equivalent to I_k or to the matrix (23.6) with $2t = k$.*

This classification enables an exhaustive search for the $[n, k]_2$ LCD codes with the largest possible minimum distance for small values of k, i.e., $k < 6$.

In the search for LCD codes for larger values of k one strategy could be to restrict the set of prescribed invertible Gram matrices to, for example, permutation matrices.

TABLE 23.3: Table of parameters of binary LCD codes with $n < 50$. The entries in the table are the largest values of d for which an $[n, k, d]_2$ LCD code has been found with the methods of this chapter. The search was exhaustive for $k < 6$. Bold face entries indicate that the code is an optimal $[n, k, d]_2$ code for general linear codes.

```
n/k| 1  2  3  4  5 | 6  7  8  9 10 11 12 13 14 15 16 17 18 19 20 21 22 23 24 25 26 27 28 29 30 31 32 33 34 35 36 37 38 39 40 41 42 43 44 45 46 47 48 49
 1 | 1
 2 | 1  1
 3 | 3  2  1
 4 | 3  2  1  1
 5 | 5  2  2  2  1
 6 | 3  3  2  2  1 | 1
 7 | 7  4  3  2  2 | 2  1
 8 | 7  5  3  3  2 | 2  1  1
 9 | 9  6  4  4  3 | 2  2  1
10 | 9  6  5  4  3 | 3  2  1  1
11 |11  6  5  4  4 | 4  2  2  2  1
12 |11  7  6  5  4 | 4  3  2  2  1  1
13 |13  8  6  6  5 | 4  4  3  2  2  2
14 |13  9  7  6  5 | 5  4  4  3  2  2
15 |15 10  7  6  6 | 6  4  4  4  3  2  2  2
16 |15 10  8  7  6 | 6  5  4  4  3  3  2  2
17 |17 10  9  8  7 | 6  6  4  4  4  3  3  2  2  2
18 |17 11  9  8  7 | 7  6  4  5  3  2  3  3  2  2
19 |19 12 10  9  8 | 8  6  6  6  4  4  3  3  2  2  2  2
20 |19 13 10 10  9 | 8  6  6  6  5  2  4  3  3  3  2  2
21 |21 14 11 10  9 | 8  8  7  6  6  5  2  4  4  3  3  2  2  2
22 |21 14 11 10 10 | 9  8  7  6  6  5  4  2  4  3  3  2  2  2
23 |23 14 12 11 10 |10  9  8  6  6  6  6  3  2  4  4  3  2  2  2  2
24 |23 15 13 12 11 |10  9  8  6  6  6  6  2  2  4  3  3  3  2  2  2
25 |25 16 13 12 11 |10 10  8  7  6  6  2  2  2  3  4  4  4  2  2  2  2  2
26 |25 17 14 12 12 |11 10 10  8  6  6  6  2  2  4  4  2  3  2  2  2  2
27 |27 18 14 13 12 |12 11 10  8  8  6  4           3  2  4  4  3  3  3  2  2  2  2
28 |27 18 15 14 13 |12 11 10  9  8  6  8           4  2  4  4  2  3  3  2  2  2  2
29 |29 18 15 14 13 |12 12 10 10  8  8  4              2  2  3  4  3  3  2  2  2  2  2
30 |29 19 16 14 14 |13 12 12 10  9  8  6              4  3  4  4  4  2  3  2  2  2  2
31 |31 20 17 15 14 |14 12 12 10 10  9  8           3  4  3  4  3  2  3  2  2  2  2
32 |31 21 17 16 15 |14 13 12 10 10 10  8              2  4  4  3  2  2  2  2  2  2
33 |33 22 18 16 15 |14 13 12 11 10 10  8           3  4  4  3  3  2  2  2  2  2  2
34 |33 22 18 17 16 |14 14 12 12 12 11  8              2  5  4  3  2  2  2  2  2  2
35 |35 22 19 18 16 |16 14 14 12 12 12  8           3  2  4  2  3  3  2  2  2  2  2  2
36 |35 23 19 18 17 |16 15 14 13 12 12  8              2  2  4  4  4  3  2  2  2  2  2
37 |37 24 20 18 17 |16 15 15 13 12 12  8           2  2  3  4  3  2  3  2  2  2  2  2
38 |37 25 21 19 18 |17 16 15 14 12 12  8              2  2  4  2  2  3  3  2  2  2  2
39 |39 26 21 20 18 |18 17 16 14 14 12  8           2  2  3  3  3  3  8  2  2  2  2  2  2
40 |39 26 22 20 19 |18 17 16 14 13 12  8              2  2  3  3  3  2  2  2  2  2
41 |41 26 22 20 19 |19 18 16 14 15 12  8                    2  2  3  2  3  3  8  2  2  2  2
42 |41 27 23 21 20 |20 18 18 16 15 12  8              2  2  4  3  3  3  8  2  2  2  2
43 |43 28 23 22 20 |20 19 18 16 14 14  8           2  4  2  3  3  3  2  2  2  2  2  2
44 |43 29 24 22 21 |20 19 18 16 14 14  8              2  4  2  2  2  2  2  2  2  2
45 |45 30 25 22 21 |20 20 18 18 16 14  8           2  2  4  4  2  2  2  2  2  2  2
46 |45 30 25 23 22 |21 20 18 18 17 16  8              2  2  4  3  2  2  3  2  2  2  2
47 |47 30 26 24 22 |22 21 18 18 16 16  8           2  2  3  3  3  2  2  2  2  2  2  2
48 |47 31 26 24 23 |22 21 20 19 18 16  8              2  2  2  3  3  3  2  2  2  2  2
49 |49 32 27 25 23 |22 22 20 20 16 16  8           2  2  2  4  3  3  2  2  2  2  2  2
```

Example 23.6.11 Table 23.3 contains the results of a computer search for $q = 2$, $n < 50$, $k < 13$ carried out for this chapter. The search was exhaustive for $k < 6$ using Corollary 23.6.10. The entries in the table are the largest values of d for which an $[n, k, d]_2$ LCD code has been found with the methods of this chapter. Bold face entries indicate that the code meets the general upper bound for linear codes; i.e., it is an optimal $[n, k, d]_2$ code. The entries for values of $k \geq 13$ were determined by additionally computing the minimum distance of the corresponding dual codes.

23.7 Extensions of Codes

If the minimum distance d of a binary linear $[n, k, d]_2$ code is odd, the extended code has parameters $[n + 1, k, d + 1]_2$.

Hill [953] studied sufficient conditions for $[n, k, d]_q$ codes which can be extended. We speak of an ℓ-extension if for an $[n, k, d]_q$ code there is an **extended** $[n + \ell, k, d + 1]_q$ code.

The construction approach described in this chapter can easily be adapted to find ℓ-extensions of a linear code. If the $[n, k, d]_q$ code $\mathcal{C}$ is extended to a code of length $n + \ell$ and minimum distance $d + 1$ this means we search for an additional ℓ projective points such that all minimum weight codewords of $\mathcal{C}$ are increased to weight at least $d+1$. This is the case if the system of equations (23.2) and (23.3) is adapted to

- $\left[M'_{k,q} \mid I_{[\frac{k}{1}]_q} \right] \begin{bmatrix} \mathbf{x} \\ \mathbf{y} \end{bmatrix} = (\ell-1)\mathbf{1}^\mathsf{T},$

- $\sum x_i = \ell,$

where $M'_{k,q}$ is the submatrix of $M_{k,q}$ whose rows are labeled by the hyperplanes corresponding to weight d codewords of $\mathcal{C}$. The columns are labeled by the points which are allowed as additional columns of a generator matrix. In other words, we search for an additional ℓ columns of a matrix such that for every minimum weight codeword at most $\ell-1$ zeros are added. See [1135, 1138] for further information.

Above is the formulation of the problem as a *packing problem*. If the ones and zeros of the matrix $M'_{k,q}$ are flipped, then one arrives at a *covering problem*, i.e., for every hyperplane corresponding to a minimum weight codeword, the weight has to increase by at least 1. In this formulation, the very efficient approach of the *Dancing Links* Algorithm [1133] can be applied.

23.8   Determining the Minimum Distance and Weight Distribution

The problem of determining the minimum distance of a general linear code is known to be NP-complete [1843]. Nevertheless, there are algorithms which are able to compute the minimum distance exactly for reasonable code sizes. In the sequel we study the algorithm of Zimmermann [191] which is the minimum distance algorithm found for example in Magma [250]. Moreover, with a small variation of this method, the number of codewords of minimum weight can be counted. The algorithm was first published in German [191]; see [190] for a detailed description in English.

Let $\mathcal{C}$ be a linear $[n,k,d]_q$ code with generator matrix G. In the first step, we compute from the generator matrix G a set of generator matrices $G^{(1)}, G^{(2)}, \ldots, G^{(m)}$ with mutually disjoint information sets. This is achieved with Gaussian elimination and possibly column permutations, but only permutations of columns from the right of the pivot column in the Gaussian elimination. That is $G^{(i)}$, with $1 \le i \le m$, has the form

$$G^{(i)} = \left[\begin{array}{c|c|c} & I_{r_i} & B^{(i)} \\ A^{(i)} & & \\ & 0 & 0 \end{array} \right].$$

This means that the ranks of the information sets of the matrices $G^{(1)}, G^{(2)}, \ldots, G^{(m)}$ are $r_1, r_2, \ldots, r_m$, respectively, and the matrices $A^{(i)}$ consist of $\sum_{j=1}^{i-1} r_j$ columns. So from now on we will work with a code equivalent to $\mathcal{C}$ which has m consecutive disjoint information sets. Nevertheless, we will call this code $\mathcal{C}$.

Let $\mathbf{G}_j^{(i)}$ denote the jth row of $G^{(i)}$. In the second step, for each $t=1,2,\ldots,k$, we do the following. For each of the matrices $G^{(i)}$, $1 \le i \le m$, we compute all possible linear combinations of t rows of $G^{(i)}$ and determine their weights. That means that we search for the codewords with minimum weight in the set of all codewords of $\mathcal{C}$ which are linear combinations of t rows $j_1, j_2, \ldots, j_t$ of $G^{(i)}$:

$$\mathbf{c} = a_1 \mathbf{G}_{j_1}^{(i)} + a_2 \mathbf{G}_{j_2}^{(i)} + \cdots + a_t \mathbf{G}_{j_t}^{(i)}$$

and $a_j \in \mathbb{F}_q \setminus \{0\}$ for $1 \leq j \leq t$. In the case of binary codes, $a_j = 1$ for $1 \leq j \leq t$. Thus, the codewords of minimum weight which are linear combinations of t rows of $G^{(i)}$ give an upper bound $\overline{d}_t$ on the overall minimum distance of the code $\mathcal{C}$. It is clear that

$$\overline{d}_1 \geq \overline{d}_2 \geq \cdots \geq \overline{d}_t \geq \overline{d}_k.$$

However, we want to avoid enumerating all linear combinations of up to k rows of the generator matrices because this would mean enumerating all codewords of the code for each of the m generator matrices.

Why can we stop this enumeration scheme before reaching $t = k$? To see this recall that after completing stage t we have enumerated for all matrices $G^{(i)}$, $1 \leq i \leq m$, all codewords of $\mathcal{C}$ which are linear combinations of at most t rows of $G^{(i)}$, and we have determined an upper bound $\overline{d}_t$ on the minimum distance of $\mathcal{C}$. That is, if there is a codeword $\mathbf{c} \in \mathcal{C}$ with weight $< \overline{d}_t$, then $\mathbf{c}$ has to be a linear combination with nonzero coefficients of at least $t + 1$ rows of any of the generator matrices $G^{(i)}$. But then, for every i with $1 \leq i \leq m$ the codeword $\mathbf{c}$ contains at least $t + 1 - (k - r_i)$ nonzero entries in the positions which correspond to the information set I_{r_i}. Therefore, after completing stage t,

$$\underline{d}_t = \sum_{i=1}^{m} t + 1 - k + r_i$$

is a lower bound for the weight of codewords which are the linear combination (with nonzero coefficients) of at least $t + 1$ rows of the generator matrices $G^{(i)}$. The enumeration can be stopped as soon as $\underline{d}_t \geq \overline{d}_t$.

This algorithm is a generalization of an idea of Brouwer for cyclic codes. Also, for self-dual codes this algorithm seems to be common knowledge. In that case there are exactly two disjoint information sets, each of rank k.

Example 23.8.1 We apply the minimum distance algorithm to the $[7,3]_2$ code with generator matrix

$$G^{(1)} = \begin{bmatrix} 1 & 0 & 0 & 1 & 0 & 1 & 1 \\ 0 & 1 & 0 & 1 & 1 & 0 & 1 \\ 0 & 0 & 1 & 0 & 1 & 1 & 1 \end{bmatrix}.$$

Computing consecutive disjoint information sets gives the following two generator matrices $G^{(2)}$ and $G^{(3)}$:

$$G^{(2)} = \begin{bmatrix} 0 & 1 & 1 & 1 & 0 & 0 & 1 \\ 1 & 1 & 0 & 0 & 1 & 0 & 1 \\ 1 & 1 & 1 & 0 & 0 & 1 & 0 \end{bmatrix} \quad \text{and}$$

$$G^{(3)} = \begin{bmatrix} 0 & 1 & 1 & 1 & 0 & 0 & 1 \\ 1 & 0 & 1 & 1 & 1 & 0 & 0 \\ 1 & 1 & 1 & 0 & 0 & 1 & 0 \end{bmatrix}.$$

In the first enumeration step $t = 1$ of the algorithm we simply have to determine the weights of the rows of the three generator matrices. All the rows have weight four, whence $\overline{d}_1 = 4$. For the lower bound after completing $t = 1$ we have

$$\underline{d}_1 = (2 - 0) + (2 - 0) + (2 - (3 - 1)) = 4.$$

Hence $d = \overline{d}_1 = 4$ is the minimum distance of $\mathcal{C}$.

If $n = 2k$ and if the two disjoint information sets have full rank k, a slight variation of this algorithm enables one to count the number of codewords of a fixed weight w of the code $\mathcal{C}$ as follows.

Algorithm 23.8.2 (Counting Codewords)

Use the previous notation and assumptions.

Step 1: For all values of t with $1 \leq t \leq w/2$ and all possible choices of $a_j \in \mathbb{F}_q \setminus \{0\}$, $1 \leq j \leq t$, count the number of codewords $\mathbf{c}$ of weight w which are the linear combination of t rows $j_1, j_2, \ldots, j_t$ of $G^{(1)}$:

$$\mathbf{c} = a_1 \mathbf{G}_{j_1}^{(1)} + a_2 \mathbf{G}_{j_2}^{(1)} + \cdots + a_t \mathbf{G}_{j_t}^{(1)}.$$

Step 2: For all values of t with $1 \leq t < w/2$ and all possible choices of $a_j \in \mathbb{F}_q \setminus \{0\}$, $1 \leq j \leq t$, count the number of codewords $\mathbf{c}$ of weight w which are the linear combination of t rows of $G^{(2)}$:

$$\mathbf{c} = a_1 \mathbf{G}_{j_1}^{(2)} + a_2 \mathbf{G}_{j_2}^{(2)} + \cdots + a_t \mathbf{G}_{j_t}^{(2)}.$$

The sum of these two numbers is the number of codewords in $\mathcal{C}$ of weight w.

When it comes to implementation of the minimum distance algorithm some details are important:

- For a fixed selection $j_1, j_2, \ldots, j_t$ of rows of a generator matrix, all linear combinations $a_1 \mathbf{G}_{j_1}^{(1)} + a_2 \mathbf{G}_{j_2}^{(1)} + \cdots + a_t \mathbf{G}_{j_t}^{(1)}$ and $a_1 \mathbf{G}_{j_1}^{(2)} + a_2 \mathbf{G}_{j_2}^{(2)} + \cdots + a_t \mathbf{G}_{j_t}^{(2)}$ with $a_j \in \mathbb{F}_q \setminus \{0\}$, $1 \leq j \leq t$, have to be tested. An obvious optimization is to enumerate the linear combinations of the rows of the generator matrices up to scalar multiples; i.e., we can fix $a_1 = 1$ and for each selection of t rows $(q-1)^{t-1}$ linear combinations have to be tested. The resulting number of codewords then has to be multiplied by $(q-1)$.

- For a fast implementation of this algorithm, especially for the cases $q = 2$ or 3, working with bit vectors is crucial. Counting the number of 1-bits in a vector is usually called **popcount**. In some processors it is available in hardware; for software implementations see for example [1132, 1878].

- The generating of all combinations of t rows $j_1, j_2, \ldots, j_t$ out of k rows of the generator matrices $G^{(1)}$ and $G^{(2)}$, respectively, can be realized with a revolving door algorithm like the algorithm of Nijenhuis and Wilf [1439]. It has the property that in each step only one row in $j_1, j_2, \ldots, j_t$ is exchanged. The resulting sequence of t-tuples has the interesting property that the combination $j_1, j_2, \ldots, j_t$ of the rows of the generator matrix is visited after exactly

$$N = \binom{j_t + 1}{t} - \binom{j_{t-1} + 1}{t - 1} + \cdots$$
$$+ (-1)^t \binom{j_2 + 1}{2} - (-1)^t \binom{j_1 + 1}{1} - [t \text{ odd}]$$

other combinations have been visited; see Lüneburg [1305] and Knuth [1132]. This enables an easy parallelization of the problem: For $1 \leq t \leq w/2$ the work of $\binom{k}{t}$ enumeration steps can be divided into $\lfloor \binom{k}{t}/T \rfloor$ chunks of size T. The rth job for $1 \leq r \leq \lfloor \binom{k}{t}/T \rfloor$ then starts at enumeration step $(r-1) \cdot T$. With the above formula the corresponding combination of rows $j_1, j_2, \ldots, j_t$ of the generator matrix can be determined.

The minimum distance algorithm of this section has been generalized to nonlinear codes [1853] and to linear codes over rings [1115].

An alternative approach to compute the minimum distance of a linear code for $\mathbb{F}_q$, $q \in \{2, 3\}$, is based on lattice basis reduction [1881]. While this method may not be able to compute the exact minimum distance for large codes, it usually determines good upper bounds on the minimum distance of the code. This can be of importance in cryptographic applications. The complexity of approximating the minimum distance of linear codes is discussed in [650].

Chapter 24

Interpolation Decoding

Swastik Kopparty

Rutgers University

24.1   Introduction

In this chapter we will see some beautiful algorithmic ideas based on *polynomial interpolation* for decoding algebraic codes. Here we will focus only on (Generalized) Reed–Solomon codes, but these ideas extend very naturally to the important class of algebraic geometry codes studied in Chapter 15.

Let us quickly recall the *polynomial-evaluation* based definition of (narrow-sense) Reed–Solomon codes from Theorem 1.14.11. Following the notation from that chapter, let q be a prime power, let $\alpha \in \mathbb{F}_q$ be a generator of the multiplicative group $\mathbb{F}_q^* = \mathbb{F}_q \setminus \{0\}$ of $\mathbb{F}_q$, and define[1]

$$\mathcal{P}_{k,q} = \{p(x) \in \mathbb{F}_q[x] \mid \deg(p) < k\} \quad \text{and}$$

$$\mathcal{RS}_k(\boldsymbol{\alpha}) = \{(p(\alpha^0), p(\alpha^1), p(\alpha^2), \ldots, p(\alpha^{q-2})) \mid p(x) \in \mathcal{P}_{k,q}\} \subseteq \mathbb{F}_q^{q-1}.$$

Informally, the codewords of $\mathcal{RS}_k(\boldsymbol{\alpha})$ are the vectors of all evaluations of a polynomial of degree $< k$ at all the nonzero points of $\mathbb{F}_q$. See specifically Theorem 1.14.11.

More generally, we can take an arbitrary $\mathbf{s} = (s_1, \ldots, s_n) \in \mathbb{F}_q^n$ with $n \leq q$ and the s_i distinct. Consider evaluations of polynomials of degree $< k$ at all the points of $\mathbf{s}$. This

---

[1] We will follow the convention that the degree of the 0 polynomial is $-\infty$.

length n code is known as the **Generalized Reed–Solomon code**[2]:

$$\mathcal{GRS}_k(\mathbf{s}) = \{(p(s_1), p(s_2), \ldots, p(s_n)) \mid p(x) \in \mathcal{P}_{k,q}\}.$$

Abusing notation, we will sometimes refer to the polynomials $p(x)$ of degree $< k$ as *codewords*. When $\mathbf{s} = \boldsymbol{\alpha} = (\alpha^0, \alpha^1, \ldots, \alpha^{q-2})$, $\mathcal{GRS}_k(\mathbf{s}) = \mathcal{RS}_k(\boldsymbol{\alpha})$.

Working with this more general family of codes greatly clarifies and even motivates the important ideas in this chapter. We begin with the most basic and useful fact about polynomials.

Lemma 24.1.1 *Let $p(x) \in \mathbb{F}_q[x]$ be a nonzero polynomial of degree d. Then the number of $u \in \mathbb{F}_q$ such that $p(u) = 0$ is at most d.*

This lemma lets us compute the dimension and minimum distance of $\mathcal{GRS}_k(\mathbf{s})$.

Lemma 24.1.2 *Let $\mathbf{s} = (s_1, \ldots, s_n) \in \mathbb{F}_q^n$ with $n \leq q$ and the s_i distinct. Suppose $1 \leq k \leq n$. Then $\mathcal{GRS}_k(\mathbf{s})$ is an $[n, k, n - k + 1]_q$ code.*

Indeed, by Lemma 24.1.1 every nonzero codeword of $\mathcal{GRS}_k(\mathbf{s})$ has at most $k-1$ coordinates equal to 0, and thus at least $n - k + 1$ nonzero coordinates. This implies that the minimum distance is at least $n-k+1$. Finally, we can check that this bound is achieved by the codeword coming from the polynomial $p(x) = \prod_{i=1}^{k-1}(x - u_i)$, where $u_1, \ldots, u_{k-1}$ are arbitrary distinct elements of $S = \{s_1, \ldots, s_n\}$. A similar argument gives that $\mathcal{GRS}_k(\mathbf{s})$ has dimension k.

For generalized Reed–Solomon codes $\mathcal{GRS}_k(\mathbf{s})$, the question of decoding up to half the minimum distance takes on the following pleasant form. Let $e = \lfloor (n - k)/2 \rfloor$ and $S = \{s_1, \ldots, s_n\}$. We are given a function $r : S \to \mathbb{F}_q$ representing the received vector. We would like to find the unique polynomial $p(x) \in \mathbb{F}_q[x]$ of degree $< k$ such that $p(u) = r(u)$ for all but at most e values of $u \in S$.

In this formulation, the problem looks like one of *error-tolerant* polynomial interpolation. Classical polynomial interpolation is the problem of finding a low degree polynomial taking desired values at certain points. Here we need to do interpolation despite some of the values being wrong.

There is a naive brute force search algorithm for this problem, namely to try all $\binom{n}{\leq e}$ possible subsets of S as candidates for the set of error locations E, and to do standard polynomial interpolation through the remaining points. This takes time exponential in e. Remarkably, there are polynomial time algorithms for this problem! In time $(n \log q)^{O(1)}$ one can find the polynomial p. The first such algorithm was found by Peterson [1504]. In Section 24.2 we will see the ingenious algorithm of Berlekamp and Welch [172] for this problem. Later in Sections 24.3 and 24.4 we will see powerful generalizations and extensions of this algorithm by Sudan [1762] and Guruswami–Sudan [880], respectively. Finally, in Section 24.5 we introduce interleaved generalized Reed–Solomon codes and present an algorithm, due to Bleichenbacher, Kiayias and Yung [220] modeled on the Berlekamp–Welch Algorithm, to decode them.

24.2 The Berlekamp–Welch Algorithm

Let $p(x) \in \mathbb{F}_q[x]$ be the polynomial that we are trying to find, namely the polynomial with degree $< k$ such that $p(u) = r(u)$ for all but at most $e = \lfloor (n - k)/2 \rfloor$ values of u. Let

---

[2]Sometimes the word 'generalized' is omitted. Also note that $\mathcal{GRS}_k(\mathbf{s})$ in this chapter is the same as $\mathcal{RS}_k(\mathbf{s})$ in Definition 15.3.19.

$E = \{u \in S \mid p(u) \neq r(u)\}$ be the set of error locations. Let $Z(x) \in \mathbb{F}_q[x]$ be the polynomial given by

$$Z(x) = \prod_{u \in E} (x - u).$$

$Z(x)$ is called the **error-locating polynomial**. We do not know $Z(x)$; indeed, finding $Z(x)$ is as hard as finding $p(x)$. Nevertheless thinking about $Z(x)$ will motivate our algorithm.

The crux of the Berlekamp–Welch Algorithm is the following identity. For each $u \in S$, we have

$$Z(u) \cdot r(u) = Z(u) \cdot p(u). \tag{24.1}$$

Indeed, if $u \notin E$, then $r(u) = p(u)$ and so $Z(u) \cdot r(u) = Z(u) \cdot p(u)$. Otherwise if $u \in E$, then $Z(u) = 0$, and so (24.1) holds.

Let $W(x) \in \mathbb{F}_q[x]$ be the polynomial $Z(x) \cdot p(x)$. Observe that $\deg(Z) \leq e$ and $\deg(W) < k + e$. This motivates the following idea: let us try to search for the polynomials Z and W by solving the linear equations

$$Z(u) \cdot r(u) = W(u)$$

for each $u \in S$. Hopefully we will then recover $p(x)$ as $W(x)/Z(x)$. This algorithm actually works, but its analysis is more subtle than the naive arguments given above suggest. We now formally present the algorithm.

Algorithm 24.2.1 (Berlekamp–Welch RSDecode [172])

Input: $r : S \to \mathbb{F}_q$, e, and k.

Step 1: **Interpolation:** Let $a_0, a_1, \ldots, a_e$ and $b_0, b_1, \ldots, b_{k+e-1}$ be indeterminates. Consider the following system of homogeneous linear equations in these indeterminates, one equation for each $u \in S$:

$$\left(\sum_{i=0}^{e} a_i u^i \right) \cdot r(u) = \sum_{j=0}^{k+e-1} b_j u^j. \tag{24.2}$$

Solve this system to get a *nonzero* solution $(a_0, \ldots, a_e, b_0, \ldots, b_{k+e-1}) \in \mathbb{F}_q^{k+2e+1}$. If there is no nonzero solution, then return FAIL.

Step 2: **Polynomial Algebra:** For the solution found in the previous step, define $A(x), B(x) \in \mathbb{F}_q[x]$ by

$$A(x) = \sum_{i=0}^{e} a_i x^i \quad \text{and}$$

$$B(x) = \sum_{j=0}^{k+e-1} b_j x^j.$$

If $A(x)$ divides $B(x)$, then return $p(x) = B(x)/A(x)$; otherwise return FAIL.

The Berlekamp–Welch Algorithm will correctly decode a Generalized Reed–Solomon code under the conditions of the following theorem.

Theorem 24.2.2 *Suppose $r(x)$ is within Hamming distance[3] $e = \lfloor (n - k)/2 \rfloor$ of some codeword $p(x)$ of $\mathcal{GRS}_k(\mathbf{s})$ where $\mathbf{s} = (s_1, \ldots, s_n) \in \mathbb{F}_q^n$ with the s_i distinct. Then with $S = \{s_1, \ldots, s_n\}$ the Berlekamp–Welch Algorithm RSDecode will output that codeword $p(x)$.*

Before analyzing the correctness of the algorithm, we make some observations about its running time. The main operations involved are solving systems of linear equations and polynomial division. By well-known algorithms, both of these can be solved in polynomial time (in fact, time $O(n^3 \log^2 q)$). With more advanced algebraic algorithms, the running time can even be made $O(n(\log q \cdot \log n)^3)$. This is nearly-linear time, and almost as fast as noiseless polynomial interpolation!

24.2.1   Correctness of the Algorithm RSDecode

We prove Theorem 24.2.2. Suppose $p(x)$ is the desired polynomial whose distance from $r(x)$ is at most $e = \lfloor (n - k)/2 \rfloor$. We need to show that the algorithm outputs $p(x)$. We do this through two claims.

Claim 1: *In Step 1 there does indeed exist a nonzero solution $(a_0, \ldots, a_e, b_0, \ldots, b_{k+e-1}) \in \mathbb{F}_q^{k+2e+1}$.*

Let $Z(x)$ be the error locating polynomial, and let $W(x) = Z(x) \cdot p(x)$. Note that $Z(x)$ is a nonzero polynomial. Taking $a_0, \ldots, a_e$ to be the coefficients of $Z(x)$, and taking $b_0, \ldots, b_{k+e-1}$ to be the coefficients of $W(x)$, the identity (24.1) immediately implies that this $(a_0, \ldots, a_e, b_0, \ldots, b_{k+e-1})$ is a nonzero solution to the system of equations (24.2).

Claim 2: *In Step 2 the polynomial $A(x)$ does divide $B(x)$ and $B(x)/A(x) = p(x)$.*

Take $A(x), B(x)$ and consider the polynomial $H(x) = B(x) - p(x)A(x) \in \mathbb{F}_q[x]$. We know that for every $u \notin E$, we have

$$H(u) = B(u) - p(u)A(u) = B(u) - r(u)A(u) = 0,$$

where the last equality follows from the fact that $(a_0, \ldots, a_e, b_0, \ldots, b_{k+e-1})$ satisfies (24.2). Observe that $H(x)$ is a polynomial of degree at most $k + e - 1$. By the above, we have that $H(u)$ vanishes for at least $n - e > k + e - 1$ values of $u \in S$. By Lemma 24.1.1, this implies that $H(x)$ is the identically zero polynomial. Thus $B(x) = p(x)A(x)$ and the claim follows.

It is worth noting that there may be multiple nonzero solutions to the system of equations (24.2). Claim 2 is about every such nonzero solution!

Another way of viewing the Berlekamp–Welch Algorithm is through the lens of *rational function interpolation*. Roughly, we start off trying to find a rational function $B(x)/A(x)$ such that $\deg(A), \deg(B)$ are both small, and $B(u)/A(u) = r(u)$ for *all* $u \in S$. We have to be careful about what we mean by division when the denominator is 0.

Definition 24.2.3 *Let $S \subseteq \mathbb{F}_q$ and let $r : S \to \mathbb{F}_q$ be a function. We say that the rational function $B(x)/A(x) \in \mathbb{F}_q(x)$ **interpolates** $r(x)$ if for every $u \in S$, either*

(i) $A(u) \neq 0$ *and* $B(u)/A(u) = r(u)$, *or*

(ii) $A(u) = 0$ *and* $B(u) = 0$.

---

[3]The Hamming distance between $p_1(x)$ and $p_2(x)$ is the number of $\beta \in S$ where $p_1(\beta) \neq p_2(\beta)$ as in Definition 1.6.1. This is different from the Hamming distance between polynomials in Chapter 10.

The Berlekamp–Welch Algorithm is essentially based on the fact that rational interpolation can be efficiently solved using linear algebra, and the analysis basically shows any low degree rational interpolation of $r(x)$ is (after division) the nearby polynomial $p(x)$.

24.3 List-decoding of Reed–Solomon Codes

We now come to the most spectacular application of interpolation ideas: to a more general decoding problem called *list-decoding*. Just as in the classical decoding problem for a code $C \subseteq \Sigma^n$ over an alphabet Σ, we are given a received string $r \in \Sigma^n$. Now we are also given a radius parameter e, and we would like to find the list $\mathcal{L}$ of all codewords $c \in C$ such that the distance from r to c is at most e. In classical decoding, the radius parameter e is always at most half the minimum distance of C where we are guaranteed that $|\mathcal{L}| \leq 1$. In list-decoding, we can allow a larger radius e. Now the list size $|\mathcal{L}|$ may be larger than 1, but as long as it is not too big, it is reasonable to ask for a fast algorithm that finds $\mathcal{L}$.

In this section we will see how interpolation based ideas, vastly generalizing the ideas in the Berlekamp–Welch Algorithm, lead to efficient list-decoding algorithms for Reed–Solomon codes to surprisingly large radii. This is known as the *Sudan Algorithm*.

At the high level, this algorithm has two parts. The first step is interpolation. The goal of the interpolation step is to find a *bivariate polynomial* $Q(x, y)$ such that $Q(u, r(u)) = 0$ for *all* $u \in S$. Pictorially, if $\Sigma = \mathbb{F}_q$, this finds an algebraic curve (which looks like a union of irreducible algebraic curves) in the $\mathbb{F}_q \times \mathbb{F}_q$ plane that passes through all the points $(u, r(u))$.

For the second step, the key insight is to consider a polynomial $p(x) \in \mathbb{F}_q[x]$ that is close to $r(x)$ and see how the graph of the relation $y = p(x)$ looks in relation to the above picture. We see that the two graphs $y = p(x)$ and $Q(x, y) = 0$ have many points of intersection. The classical Bezout Theorem[4] implies that low degree curves $P(x, y) = 0$ and $Q(x, y) = 0$ cannot have too many points of intersection unless P and Q have a common factor. Thus in our case, the polynomials $y - p(x)$ and $Q(x, y)$ must have a common factor, and the irreducibility of $y - p(x)$ implies that $y - p(x)$ must be a factor of $Q(x, y)$. The second step of the algorithm is to factor the polynomial $Q(x, y)$ and to thus find $p(x)$.

To gain some insight into this algorithm, we remark that it *is* a generalization of the Berlekamp–Welch Algorithm. Indeed, the first step of the Berlekamp–Welch Algorithm is to find a polynomial $Q(x, y)$ of the form $A(x)y - B(x)$ such that $Q(u, r(u)) = 0$ for all $u \in S$.

24.3.1 The Sudan Algorithm

We now give a formal description of (one version of) the Sudan Algorithm.

Algorithm 24.3.1 (Sudan RSListDecodeV1 [1762])

Input: $r : S \to \mathbb{F}_q$, $n = |S|$, e, and k.

Step 1: **Interpolation:** Let $I = \lceil \sqrt{nk} \rceil$ and $J = \lceil \sqrt{\frac{n}{k}} \rceil$. For each i, j with $0 \leq i \leq I$ and $0 \leq j \leq J$, we let a_{ij} be an indeterminate. Consider the system of homogenous linear equations in these indeterminates, where for each $u \in S$, we have the equation

$$\sum_{i \leq I, j \leq J} a_{ij} u^i (r(u))^j = 0.$$

---

[4]When we formally analyze the algorithm, everything will be elementary and we will not use the Bezout Theorem. We mention the Bezout Theorem only to provide motivation.

Solve this system to find a *nonzero* solution $(a_{ij})_{i \leq I, j \leq J} \in \mathbb{F}_q^{(I+1)(J+1)}$. If there is no nonzero solution, then return FAIL.

Step 2: **Polynomial Algebra:** Let $Q(x, y) \in \mathbb{F}_q[x, y]$ be the polynomial given by

$$Q(x, y) = \sum_{i \leq I, j \leq J} a_{ij} x^i y^j.$$

Factor $Q(x, y)$ into its irreducible factors over $\mathbb{F}_q$. Let $\mathcal{L}$ be the set of all $p(x)$ for which $y - p(x)$ is an irreducible factor of $Q(x, y)$. These are our candidate codewords. Now output those codewords of $\mathcal{L}$ which are within distance e of the received word r.

We will prove the following theorem.

Theorem 24.3.2 *Let* $\mathbf{s} = (s_1, \ldots, s_n) \in \mathbb{F}_q^n$ *with the* s_i *distinct and* $S = \{s_1, \ldots, s_n\}$. *If* $e < n - 2\sqrt{nk} - k$, *then the Sudan Algorithm RSListDecodeV1 outputs the list of all codewords in* $\mathcal{GRS}_k(\mathbf{s})$ *that are within Hamming distance* e *of* r. *Furthermore, this list has size at most* $\lceil \sqrt{\frac{n}{k}} \rceil$.

To understand how remarkable the above theorem is, consider the setting $k = (0.01)n$ and $e = (0.75)n$. Then the theorem says that the Sudan Algorithm can find all codewords within distance $(0.75)n$ of any given received word r. Note that most entries of r may be wrong; yet we can find the corrected codeword!

Before analyzing the correctness of the algorithm, we discuss the running time. Factoring of bivariate polynomials of degree d over $\mathbb{F}_q$ can be done in time $(d \log q)^{O(1)}$ by a randomized algorithm (or $(dq)^{O(1)}$ by a deterministic algorithm). Thus the entire Sudan Algorithm can be made to run in polynomial time.

24.3.2   Correctness of the Algorithm RSListDecodeV1

We prove Theorem 24.3.2. Suppose $p(x)$ is a codeword of $\mathcal{GRS}_k(\mathbf{s})$ whose Hamming distance from $r(x)$ is at most e. We want to show that $p(x)$ is one of the codewords output by the algorithm. Let $E \subseteq S$ be the set of $u \in S$ such that $p(u) \neq r(u)$. This is the error set for $p(x)$. By hypothesis, $|E| \leq e$.

First we show that the interpolation step of the algorithm does not FAIL. In this step, we have to solve a system of homogeneous linear equations to find a nonzero solution. There are n equations in $(I+1)(J+1)$ unknowns. It is well known, despite ample intuition to the contrary, that for general systems of linear equations we cannot deduce anything about the solvability based on the number of equations and the number of unknowns. But for *homogeneous* linear systems we can! If the number of unknowns is greater than the number of equations, there is a *nonzero* solution. In our case, the number of unknowns is $(I+1)(J+1)$, which is greater than the number n of equations because

$$(I+1)(J+1) > \sqrt{nk} \cdot \sqrt{\frac{n}{k}} = n.$$

Thus the first step of the algorithm succeeds in finding a nonzero solution.

Now we examine the second step. By construction, the polynomial $Q(x, y)$ has the property that $Q(u, r(u)) = 0$ for all $u \in S$. Consider the polynomial $H(x) = Q(x, p(x))$. For any point $u \in S \setminus E$, we have $H(u) = Q(u, p(u)) = Q(u, r(u)) = 0$. Furthermore, the degree of H is at most $I + (k - 1)J$. Thus $H(x)$ is a polynomial of degree at most

$$I + (k - 1)J < (\sqrt{nk} + 1) + (k - 1) \left(\sqrt{\frac{n}{k}} + 1 \right) < 2\sqrt{nk} + k,$$

whose number of roots is at least

$$|S| - |E| \geq n - e > n - \left(n - 2\sqrt{nk} - k \right) = 2\sqrt{nk} + k.$$

By Lemma 24.1.1, we conclude that $H(x) = Q(x, p(x))$ must be the zero polynomial.

Finally, we use the fact that if $Q(x, y)$ is such that $Q(x, p(x)) = 0$, it means that the bivariate polynomial $y - p(x)$ divides $Q(x, y)$. This is a form of the so-call *Factor Theorem*.[5] So $y - p(x)$ will appear in the list of factors of Q, and thus $p(x)$ will appear in the output of the Sudan Algorithm RSListDecodeV1, as desired. The number of such factors is at most the y-degree of Q, which is at most $J = \lceil \sqrt{\frac{n}{k}} \rceil$. This completes the proof of Theorem 24.3.2.

We make some remarks about the argument that we just saw.

Remark 24.3.3

- The precise algebraic problem that we need to solve in the second step of the algorithm is *root finding* rather than *factoring*. There are faster and simpler algorithms for root finding than for general factoring.

- A slightly cleverer choice of monomials[6] used in the polynomial $Q(x, y)$ leads to an improvement of the decoding radius to $n - \sqrt{2nk}$. This is the error-correction performance of the original Sudan Algorithm. This is larger than half the minimum distance for all $k < n/3$.

- The significance of the Sudan Algorithm is that it showed for the first time that it is possible to efficiently list-decode positive rate codes beyond what is possible for unique decoding. For a code of rate R, the classical Singleton Bound Theorem 1.9.10 implies that it is not possible to uniquely decode from more than $((1 - R)/2)$-fraction errors. We just saw that Reed–Solomon codes of rate R can be efficiently decoded from $(1 - 2\sqrt{R} - R)$-fraction errors, which for small R is larger than $(1 - R)/2$.

24.4 List-decoding of Reed–Solomon Codes Using Multiplicities

Now we come to a powerful new tool that greatly strengthens the reach of the interpolation method: *multiplicities*. We will see the algorithm of Guruswami and Sudan for list-decoding Reed–Solomon codes from $n - \sqrt{nk}$ errors. This is larger than half the minimum distance for *all* k. It correspondingly shows that it is possible to have codes of rate R for which efficient list-decoding from $(1 - \sqrt{R})$-fraction errors is possible – this is larger than the unique decoding limit of $(1 - R)/2$ for *all* R.

We begin with a definition of *multiplicity of vanishing*.

---

[5]In more detail, write $Q(x, z + p(x))$ as a $h_0(x) + zh_1(x) + z^2h_2(x) + \cdots$. Then $Q(x, p(x)) = 0$ means that $h_0(x) = 0$. Thus z divides $Q(x, z + p(x))$. Setting $y = z + p(x)$, we get the claim.

[6]We will see this cleverer choice when we discuss the Guruswami–Sudan Algorithm in the next section.

Definition 24.4.1 Over a field $\mathbb{F}$, let $Q(x_1, \ldots, x_m) \in \mathbb{F}[x_1, \ldots, x_m]$ be a polynomial. Let $\mathbf{u} = (u_1, \ldots, u_m) \in \mathbb{F}^m$ be a point. We define the **multiplicity of vanishing** of Q at $\mathbf{u}$, denoted $\mathrm{mult}(Q, \mathbf{u})$, to be the smallest integer M such that the polynomial $Q(\mathbf{u} + \mathbf{x}) \in \mathbb{F}[x_1, \ldots, x_m]$ has no monomials $\prod_{j=1}^m x_j^{b_j}$ of degree $< M$. If Q is the zero polynomial, we define $\mathrm{mult}(Q, \mathbf{u}) = -\infty$ by convention.

Classically, over fields of characteristic 0 multiplicity is defined using derivatives. Over finite fields one has to be careful. Everything works well if we use the notion of the Hasse derivative. See [960] for more on this.

Example 24.4.2 The following multiplicity calculations are easy to check.

- $\mathrm{mult}((x-1)^2(3+2x), 1) = 2$.

- $\mathrm{mult}(y^2 - x^3 - x^4, (0,0)) = 2$.

The Guruswami–Sudan Algorithm for list-decoding of Reed–Solomon codes is based on interpolating a bivariate polynomial $Q(x, y)$ that vanishes at each point $(u, r(u))$ for $u \in S$ *with high multiplicity*. Asking that a polynomial vanishes somewhere with high multiplicity imposes more linear constraints on the coefficients of Q than simply asking that Q vanishes there. To accommodate this, we need to enlarge the space where we search for Q, and thus we end up with such a Q of higher degree than before.

We then benefit from the increased vanishing multiplicity to deduce that for any $p(x)$ that is near the received word $r(x)$, the univariate polynomial $H(x) = Q(x, p(x))$ vanishes *with high multiplicity* at many points, and thus must be identically 0. This lets us recover $p(x)$ by factoring Q, as before. *A priori* it is not clear that there will be an improvement in the decoding radius: the larger degree of Q is traded off against the increased vanishing multiplicity. Nevertheless there is an improvement. We discuss the philosophical reason to expect an improvement later in Section 24.4.4.

24.4.1 Preparations

We state below some simple properties of multiplicity. We omit the (easy) proofs.

Lemma 24.4.3 *Suppose $H(x) \in \mathbb{F}[x]$ and $u \in \mathbb{F}$ are such that $\mathrm{mult}(H, u) \geq M$. Then*

$$(x - u)^M \mid H(x).$$

As an immediate consequence of this lemma, we get the multiplicity analog of Lemma 24.1.1.

Lemma 24.4.4 *Let $H(x) \in \mathbb{F}[x]$ be a nonzero polynomial of degree at most d. Then*

$$\sum_{u \in \mathbb{F}} \mathrm{mult}(H, u) \leq d.$$

Lemma 24.4.5 *Suppose $Q(x, y) \in \mathbb{F}[x, y]$ and $p(x) \in \mathbb{F}[x]$. Suppose $u \in \mathbb{F}$ is such that $\mathrm{mult}(Q, (u, p(u))) \geq M$. Then, letting $H(x) = Q(x, p(x))$, we have $\mathrm{mult}(H, u) \geq M$.*

We will be dealing with certain weighted degrees of bivariate polynomials. For a monomial $x^i y^j$, its (α, β)-**weighted degree** is defined to be $\alpha i + \beta j$.

24.4.2 The Guruswami–Sudan Algorithm

With this preparation in hand we can now formally describe the Guruswami–Sudan Algorithm. (Compare this algorithm to the Guruswami–Sudan Algorithm 15.6.22 for algebraic geometry codes.)

Algorithm 24.4.6 (Guruswami–Sudan RSListDecodeV2 [880])

Input: $r : S \to \mathbb{F}_q$, $n = |S|$, e, and k.

Step 1: **Interpolation:** Let $M = k$ and $D = M\sqrt{nk}$. Let $\mathcal{M}_D$ be the set of all monomials $x^i y^j$ for which the $(1, k-1)$-weighted degree is at most D. Thus

$$\mathcal{M}_D = \{x^i y^j \mid i + (k-1)j \leq D\}.$$

We also consider the corresponding set of exponents

$$T_D = \{(i,j) \mid i, j \geq 0,\ i + (k-1)j \leq D\}.$$

For each $(i,j) \in T_D$, we let a_{ij} be an indeterminate. Let $Q(x,y) \in \mathbb{F}_q[x,y]$ be the polynomial all of whose monomials, with coefficients a_{ij}, are in $\mathcal{M}_D$:

$$Q(x,y) = \sum_{(i,j) \in T_D} a_{ij} x^i y^j.$$

Consider the system of homogeneous linear equations on the a_{ij} by imposing the condition, for each $u \in S$,

$$\mathrm{mult}(Q, (u, r(u))) \geq M.$$

Find a *nonzero* solution $(a_{ij})_{(i,j) \in T_D} \in \mathbb{F}_q^{|T_D|}$. If no nonzero solution exists, then return FAIL.

Step 2: **Polynomial Algebra:** Factor $Q(x,y)$ into its irreducible factors over $\mathbb{F}_q$. Let $\mathcal{L}$ be the set of all $p(x)$ for which $y - p(x)$ is an irreducible factor of $Q(x,y)$. These are our candidate codewords. Now output those codewords of $\mathcal{L}$ which are within distance e of the received word r.

We will prove the following theorem.

Theorem 24.4.7 *Let* $\mathbf{s} = (s_1, \ldots, s_n) \in \mathbb{F}_q^n$ *with the* s_i *distinct and* $S = \{s_1, \ldots, s_n\}$. *If* $e < n - \sqrt{nk}$, *then the Guruswami–Sudan Algorithm RSListDecodeV2 outputs the list of all codewords in* $\mathcal{GRS}_k(\mathbf{s})$ *that are within Hamming distance* e *of* r. *Furthermore, this list has size at most* $2\sqrt{nk}$.

24.4.3   Correctness of the Algorithm RSListDecodeV2

We prove Theorem 24.4.7. Suppose $p(x)$ is a codeword of $\mathcal{GRS}_k(\mathbf{s})$ whose distance from $r(x)$ is at most e. We want to show that $p(x)$ is one of the codewords output by the algorithm. Let $E \subseteq S$ be the set of $u \in S$ such that $p(u) \neq r(u)$. This is the error set for $p(x)$. By hypothesis, $|E| \leq e$.

First we look at the interpolation step. Since it is a system of *homogeneous* linear equations, we can show existence of a nonzero solution by counting equations and unknowns. The total number of unknowns is $|T_D|$, which can be bounded below by

$$|T_D| = \sum_{j \leq D/(k-1)} \left((D+1) - (k-1)j \right) \geq \frac{D^2}{2(k-1)}.$$

How many equations are there? Asking that Q vanishes at a point $(u, r(u))$ with multiplicity at least M is the same as asking that $\binom{M+1}{2}$ coefficients of the polynomial $Q(x+u, y+r(u))$ vanish. Thus the total number of equations in the system is $n \cdot \binom{M+1}{2}$. We now check that the number of unknowns is larger than the number of constraints as follows:

$$\frac{D^2}{2(k-1)} = \frac{nkM^2}{2(k-1)} = n \cdot \frac{M^2}{2} \cdot \frac{k}{k-1} > n \cdot \frac{M(M+1)}{2},$$

where the last inequality uses the fact that $\frac{M+1}{M} < \frac{k}{k-1}$ since $M = k$. Thus the system of equations has a *nonzero* solution, and the first step of the algorithm succeeds.

Now we consider the second step. By construction, the polynomial $Q(x, y)$ has the property that $\text{mult}(Q, (u, r(u))) \geq M$ for all $u \in S$. Consider the polynomial $H(x) = Q(x, p(x))$. For any point $u \in S \setminus E$, we have that $p(u) = r(u)$, and thus by Lemma 24.4.5, $\text{mult}(H, u) \geq M$. Furthermore, the degree of H is at most D. Thus $H(x)$ is a polynomial of degree at most D whose total multiplicity of vanishing at all points in S is at least

$$(|S| - |E|) \cdot M \geq (n - e) \cdot M > (n - (n - \sqrt{nk})) \cdot M = \sqrt{nk} \cdot M = D.$$

By Lemma 24.4.4, we conclude that $H(x)$ must be the zero polynomial.

Again, since $Q(x, p(x)) = H(x) = 0$, we get that $y - p(x)$ divides $Q(x, y)$. So $y - p(x)$ will appear in the list of factors of Q, and thus $p(x)$ will appear in the output of the Guruswami–Sudan Algorithm RSListDecodeV2, as desired. The number of such factors is at most the y-degree of Q, which is at most $\frac{D}{k-1} \leq \frac{k\sqrt{nk}}{k-1} \leq 2\sqrt{nk}$, completing the proof of Theorem 24.4.7.

24.4.4   Why Do Multiplicities Help?

The role of multiplicities in the previous argument seems mysterious. Why did multiplicities improve the decoding radius of the Sudan Algorithm?

In the univariate case, if we have a set $T \subseteq \mathbb{F}_q$, then we have the simple fact that any polynomial $Q(x)$ that vanishes at all points of T must be a multiple of $Z_T(x) = \prod_{u \in T}(x - u)$. Analogously, any polynomial $Q(x)$ that vanishes with multiplicity M at each point of T must be a multiple of $Z_T(x)^M$. Informally, this means that in the univariate case, we don't get more information by asking Q to vanish with high multiplicity at the points of a set. In the multivariate case, there is no similar tight connection between simply vanishing at all points of a set $T \subseteq \mathbb{F}_q^m$ and vanishing with high multiplicity at all points of T.

The following facts shed further light on this phenomenon. If we take a *typical* polynomial $Q(x, y) \in \mathbb{F}_q[x, y]$ of low degree, it will vanish at approximately q points in $\mathbb{F}_q \times \mathbb{F}_q$. On the other hand, a typical Q will vanish with multiplicity ≥ 2 at only $O(1)$ points of $\mathbb{F}_q \times \mathbb{F}_q$.

This means that a polynomial that vanishes with high multiplicity at all the points of a set of interest is a very special polynomial, and presumably its fate is more strongly tied to that of the set.

There have been many successful uses of the idea of interpolation in several areas of mathematics, and taking multiplicities into account is often useful [660, 1628, 1629].

24.5 Decoding of Interleaved Reed–Solomon Codes under Random Error

We give one final example of the power of interpolation ideas for decoding algebraic codes. This example concerns *interleaved Reed–Solomon codes*, which we now define.

Let q be a prime power, $S \subseteq \mathbb{F}_q$, and let $k \geq 0$ and $t \geq 1$ be integers. We define the **interleaved Reed–Solomon code** $\mathcal{IRS}_q(k, S, t)$ as follows. The alphabet Σ of this code equals $\mathbb{F}_q^t$, and the coordinates of each codeword are indexed by S; thus the block length equals $|S|$. For each t-tuple of polynomials $(p_1(x), \ldots, p_t(x)) \in (\mathbb{F}_q[x])^t$ with $\deg(p_i) < k$ for each i, there is a codeword

$$\left((p_1(\alpha), p_2(\alpha), \ldots, p_t(\alpha)) \right)_{\alpha \in S} \in \Sigma^{|S|}.$$

Another way to view this is as follows. To get codewords of the interleaved Reed–Solomon code $\mathcal{IRS}_q(k, S, t)$, we take a $t \times |S|$ matrix whose rows are codewords of $\mathcal{GRS}_k(\mathbf{s})$, with $\mathbf{s} = \{s_1, \ldots, s_n\}$ where $S = \{s_1, \ldots, s_n\}$, and view each column of the matrix as a single symbol in $\Sigma = \mathbb{F}_q^t$. The strings in $\Sigma^{|S|}$ so obtained are the codewords of $\mathcal{IRS}_q(k, S, t)$.

Clearly, interleaved Reed–Solomon codes are very closely related to Reed–Solomon codes. It is easy to see that the rate of $\mathcal{IRS}_q(k, S, t)$ equals $k/|S|$, and the minimum Hamming distance of $\mathcal{IRS}_q(k, S, t)$ equals $|S| - k + 1$, exactly as in the case of Reed–Solomon codes.

We now turn our attention to decoding algorithms. The problem of decoding interleaved Reed–Solomon codes from e errors is the following natural-looking question. Let $r : S \to \mathbb{F}_q^t$ be the received word. We will sometimes think of r as a tuple of t functions $r^{(1)}, r^{(2)}, \ldots, r^{(t)}$, where $r^{(i)} : S \to \mathbb{F}_q$ is the ith output coordinate of r. Our goal is to find one/many/all t-tuples of polynomials $p = (p_1, \ldots, p_t) \in (\mathbb{F}_q[x])^t$ with $\deg(p_i) < k$ such that $r(u) = (p_1(u), \ldots, p_t(u))$ for all but at most e values of $u \in S$; in other words $d_H(p, r) \leq e$ where $d_H(\cdot, \cdot)$ denotes Hamming distance on $\Sigma^{|S|}$.

Clearly, if $d_H(p, r) \leq e$, then we have that for all i, $d_H(p_i, r^{(i)}) \leq e$ where this latter $d_H(\cdot, \cdot)$ denotes Hamming distance on $\mathbb{F}_q^{|S|}$. Note that the converse does not hold: the sets of coordinates $u \in S$ where p_i and $r^{(i)}$ agree can look very different for different i. When e is at most $(|S| - k)/2$ (half the minimum distance of the code), then the above observation gives an easy algorithm to decode interleaved Reed–Solomon codes from adversarial errors: for each i, decode $r^{(i)}$, using a standard Reed–Solomon decoder such as the Berlekamp–Welch Algorithm, to find the unique polynomial p_i of degree $< k$ that is within Hamming distance e of it. Taking these p_i together to form a t-tuple p, we get a candidate codeword p of $\mathcal{IRS}_q(k, S, t)$, and then we check whether or not $d_H(p, r) \leq e$.[7]

We will now describe an algorithm that can decode from a significantly larger number of *random* errors. Specifically, the model for generating r is as follows. There is some unknown

---

[7] A more sophisticated variation using the Guruswami–Sudan Algorithm RSListDecodeV2 can extend this algorithm to decode interleaved Reed–Solomon codes from any $e \leq n - \sqrt{nk}$ errors. Further improvements seem difficult: decoding from an even larger number of errors in the worst case would need a breakthrough on decoding of standard Reed–Solomon codes.

t-tuple of polynomials $p = (p_1, \ldots, p_t) \in (\mathbb{F}_q[x])^t$ where $\deg(p_i) < k$. There is an unknown set $J \subseteq S$ (the set of error locations) of size at most e. Based on these unknowns, the received word $r : S \to \mathbb{F}_q^t$ is generated by setting

$$r(u) = \begin{cases} (p_1(u), p_2(u), \ldots, p_t(u)) & \text{if } u \notin J, \\ \text{a uniformly random element of } \mathbb{F}_q^t & \text{if } u \in J. \end{cases} \tag{24.3}$$

Note that the number of errors (i.e., the number of $u \in S$ for which $p(u) \neq r(u)$) is always at most $|J|$, and is with high probability equal to $|J|$, but it could be smaller.

In this setting, we would like to design a decoding algorithm that takes r as input and recovers p, the underlying codeword. Below we give a natural (given the previous sections) algorithm for this due to Bleichenbacher, Kiayias, and Yung [220]. A clever and nontrivial analysis shows that this algorithm succeeds with high probability even with the number of errors e being as large as $\frac{t}{t+1} \cdot (n - k)$, which for large t is close to the minimum distance of the code (*twice the worst-case unique decoding radius!*). This result is very surprising.

The key idea is to run a Berlekamp–Welch type decoding algorithm for each of the $r^{(i)}$, while taking into account the fact that *the error locations, and hence the error-locating polynomials, are the same*.

Algorithm 24.5.1 (Bleichenbacher–Kiayias–Yung IRSDecode [220])

Input: $r : S \to \mathbb{F}_q^t$, e, and k.

Step 1: Interpolation: Let $a_0, a_1, \ldots, a_e$ be indeterminates, and for each $\ell \in \{1, \ldots, t\}$, let $b_{\ell,0}, b_{\ell,1}, \ldots, b_{\ell,k+e-1}$ be indeterminates. Consider the following system of homogeneous linear equations in these indeterminates, one equation for each $(u, \ell) \in S \times \{1, \ldots, t\}$:

$$\left(\sum_{i=0}^{e} a_i u^i \right) \cdot r^{(\ell)}(u) = \sum_{j=0}^{k+e-1} b_{\ell,j} u^j.$$

Solve the system to find any *nonzero* solution $(a_0, \ldots, a_e, b_{1,0}, \ldots, b_{t,k+e-1}) \in \mathbb{F}_q^{e+t(k+e)}$. If there is no nonzero solution, then return FAIL.

Step 2: Polynomial Algebra: For the solution found in the previous step, define $A(x), B_1(x), B_2(x), \ldots, B_t(x) \in \mathbb{F}_q[x]$ by

$$A(x) = \sum_{i=0}^{e} a_i x^i \quad \text{and}$$

$$B_\ell(x) = \sum_{j=0}^{k+e-1} b_{\ell,j} x^j.$$

If $A(x)$ divides $B_\ell(x)$ for each ℓ, then return the t-tuple

$$(B_1(x)/A(x), B_2(x)/A(x), \ldots, B_s(x)/A(x));$$

otherwise return FAIL.

The correctness guarantee of this algorithm is given by the following theorem.

Theorem 24.5.2 *Suppose $e < \frac{t}{t+1} \cdot (n - k)$, and let $J \subseteq S$ be of size at most e. Suppose $p = (p_1(x), \ldots, p_t(x)) \in (\mathbb{F}_q[x])^t$ is such that $\deg(p_i) < k$. Finally, let r be generated based on p and J by the random process described in (24.3). Then with probability at least $1 - \frac{|S|}{q}$, the Bleichenbacher–Kiayias–Yung Algorithm IRSDecode on input r will return p.*

A proof of this theorem is a bit too involved to give here. We instead just give a high-level overview of what we expect will happen.

Our hope is that $A(x)$ ends up equalling the error-locating polynomial for this setting, namely

$$Z(x) = \prod_{\alpha \in J} (x - \alpha).$$

This error-locating polynomial has degree at most e. Consider also the polynomial $W_\ell(x) = Z(x) \cdot p_\ell(x)$, which has degree $< k + e$. We clearly have

$$Z(x) r^{(i)}(x) = W_i(x)$$

for each $x \in S$. Thus $(Z(x), W_1(x), W_2(x), \ldots, W_t(x))$ is a valid solution to the system of linear equations in Step 1, and then the correct solution $p_i(x) = W_i(x)/Z(x)$ does get returned.

There may be other valid solutions to the system of linear equations. For example, if the error set J has size $< e$, then $A(x) = Z(x) \cdot (x + 1)$ and $B_i(x) = W_i(x) \cdot (x + 1)$ is also a valid solution, but the final output of the algorithm is still the same.

The actual proof of Bleichenbacher, Kiayias, and Yung is very interesting, and shows that these are the only possibilities with high probability. It is based on treating the random variables $(r^{(i)}(u))_{u \in J}$ as formal variables, and careful studying the multivariate polynomials that arise as sub-determinants of the matrix underlying the system of linear equations. Most crucially, one needs to identify appropriate Vandermonde matrices inside this matrix, and then apply the Schwartz–Zippel Lemma, which says that nonzero multivariate polynomials evaluate to nonzero at a random point with high probability.

24.6   Further Reading

Algorithms for decoding interleaved Reed–Solomon codes from random errors were given by Bleichenbacher, Kiayas, and Yung [220] (Algorithm 24.5.1 we saw here) and Coppersmith and Sudan [443] (a different but related algorithm).

There have been many advances on the list-decoding of error-correcting codes, and list-decoding of algebraic codes in particular, in recent years. The most important result is the construction and decoding of *capacity achieving* list-decodable codes by Guruswami and Rudra [878], based on a breakthrough by Parvaresh and Vardy [1491]. These codes, called *folded Reed–Solomon codes*, are a variation on Reed–Solomon codes, and achieve the optimal tradeoff between rate and number of errors correctable by list-decoding with polynomially bounded list size.

Another family of algebraic codes called *multiplicity codes*, based on evaluating polynomials and their derivatives, was also shown to achieve list-decoding capacity by Guruswami and Wang [881] and Kopparty [1145].

All these list-decoding algorithms use extensions of the interpolation decoding technique using high-variate interpolation. Very recently, Kopparty, Ron-Zewi, Saraf and Wootters [1147] showed that both folded Reed–Solomon codes and multiplicity codes achieve list-decoding capacity KRSW

The list-decodability of Reed–Solomon codes is still open. For all we know, Reed–Solomon codes themselves may be list-decodable up to list decoding capacity; this corresponds to list-decoding from $O(k)$ agreements, instead of the $\sqrt{nk}$ agreements that the Guruswami–Sudan Algorithm needs. The best negative result in this direction is by Ben-Sasson, Kopparty, and Radhakrishnan [159], who show that the list size may become superpolynomial for Reed–Solomon codes of vanishing rate.

Interpolation methods have a long history in mathematics. Some striking classical applications include the Thue–Siegel–Roth theorems on diophantine approximation, the Gelfond–Schneider–Baker theorems on transcendental numbers [93], and the Stepanov–Bombieri–Schmidt proofs of the Weil Bounds [1629]. More recent applications include the results of Dvir, Lev, Kopparty, Saraf, and Sudan [659, 660, 1146, 1612] on the Kakeya problems over finite fields, the work of Guth and Katz [884, 885] in combinatorial geometry, and the results of Croot, Lev, Pach, Ellenberg and Giswijit on the cap set problem [472, 679].

Chapter 25

Pseudo-Noise Sequences

Tor Helleseth
University of Bergen

Chunlei Li
University of Bergen

25.1 Introduction

Pseudo-noise sequences (PN sequences), also referred to as pseudo-random sequences, are sequences that are deterministically generated but possess some favorable properties that one would expect to find in randomly generated sequences. Applications of PN sequences include signal synchronization, navigation, radar ranging, random number generation, spread-spectrum communications, multi-path resolution, cryptography, and signal identification in multiple-access communication systems [824, 825]. Interested readers are referred to comprehensive investigations and surveys of the design and applications of PN sequences in [794, 824, 825, 940, 1616, 1617, 1850]. This chapter aims to provide a brief overview of the topics of sequences with low correlation and/or maximal periods.

Correlation of sequences is a measure of their similarity or relatedness. In communication and engineering systems, there are a large number of problems that require sets of sequences with one or both of the following properties: (i) each sequence in the set is easy to distinguish from a time-shifted version of itself; (ii) each sequence is easy to distinguish from (a possibly time-shifted version of) every other sequence in the set. The first property is important for applications such as ranging systems, radar systems, and spread spectrum communications systems; and the second is desirable for simultaneous ranging to several targets, multiple-terminal system identification, and code-division multiple-access (CDMA) communications systems [1616, 1617]. There are strong ties between low correlation sequence design and the theory of error-correcting codes. Interestingly, the dual of an efficient error-correcting code will often yield a family of sequences with desirable correlation properties and vice versa [940]. The first part of this chapter will be dedicated to individual sequences and sets of sequences with low periodic correlation.

Feedback shift registers are one of the most popular methods to generate PN sequences in an easy, efficient, and low-cost manner. Due to the intrinsic combinatorial structures, shift register sequences with maximal possible length are of particular interest. They are widely used in the areas of error-correcting codes, combinatorics, graph theory, and cryptography [824]. For linear feedback shift registers, the maximum-length sequences are used in a variety of engineering applications as well as in the construction of optimal error-correcting codes [940]. For nonlinear cases, de Bruijn sequences, which are deemed as the nonlinear counterpart of m-sequences, are considered as a model example of the interaction of discrete mathematics and computer science [1558]. They have been used in building error-correcting codes for storage systems [381], stream ciphers [933], and in DNA sequences correction and assemblies [429]. Techniques from combinatorics, graph theory, and abstract algebra have yielded a number of constructions and generations of de Bruijn sequences. The second part of this chapter will introduce some approaches in these areas to efficiently generate de Bruijn sequences.

25.2   Sequences with Low Correlation

This section will give a brief introduction of sequences with low correlation, with the focus on established correlation measures of sequences and some constructions of individual sequences and sequence families with low correlation. It can be considered as a simplified version of the comprehensive chapter [940] by Helleseth and Kumar and the recent survey [794] by Gargantuas, Helleseth, and Kumar. Interested readers can refer to them and references therein for more in-depth discussions on low-correlation sequences.

25.2.1   Correlation Measures of Sequences

In this subsection we will introduce some correlation measures of sequences for their applications in communication systems.

A sequence $\{s(t)\}$ over an alphabet $\mathcal{A}$ is said to be **periodic** if for a certain positive integer N, $s(t) = s(t + N)$ for all integers $t \geq 0$. The smallest integer N such that $s(t) = s(t + N)$ for all integers $t \geq 0$ is called the **least period** of $\{s(t)\}$, and $\{s(t)\}$ is said to be a sequence of (least) **period** N. For instance, the sequence $011011011\cdots$ is a periodic sequence with least period 3.

PN sequences used in digital communications are commonly sequences over the complex unit circle. Namely, they are often of the form $\{\omega_q^{s(t)}\}$, where $\omega_q = e^{2\pi\sqrt{-1}/q}$ is a complex primitive q^{th} root of unity and $\{s(t)\}$ takes on values in $\mathbb{Z}_q$, which is the set of integers modulo q. Sequences over $\mathbb{Z}_q$ are commonly referred to as **q-ary sequences**. Throughout this section, we restrict ourselves to q-ary sequences due to their one-to-one correspondence with complex-valued sequences. In particular, a q-ary sequence is termed a **binary sequence** when $q = 2$, and is termed a **quaternary sequence** when $q = 4$. We denote periodic PN sequences with lower-case letters such as $\{s(t)\}, \{u(t)\}, \{v(t)\}$.

The **correlation** of two sequences is the inner product of the first sequence in the complex-valued version with a shifted version of the second sequence in the complex-valued version. According to different shift operations, the correlation of two sequences is called a *periodic correlation* if the shift is a cyclic shift; an *odd-periodic correlation* if the shift is a cyclic shift with an additional sign; an *aperiodic correlation* if the shift is not cyclic, and a *partial-period correlation* if the inner product involves only a partial segment of the two sequences. Sequences based on the periodic correlation constraint have stronger ties to coding theory. Partly for this reason and partly on account of our own research interests, the emphasis in this chapter is on the design of sequences with low periodic correlation. We will first introduce some definitions of correlation measures of sequences in this subsection, and restrict our discussion to sequences with periodic correlation in later subsections.

Let $\{u(t)\}$ and $\{v(t)\}$ be two q-ary sequences of period N, not necessarily distinct. The **periodic correlation**, sometimes also called *even-periodic correlation*, of the sequences $\{u(t)\}$ and $\{v(t)\}$ is the collection $\{\theta_{u,v}(\tau) \mid 0 \leq \tau \leq N - 1\}$ with

$$\theta_{u,v}(\tau) = \sum_{t=0}^{N-1} \omega_q^{u(t+\tau)-v(t)},$$

where the sum $t+\tau$ is computed modulo N. The correlation is called **in-phase correlation** when $\tau = 0$ and **out-of-phase correlation** when $\tau \neq 0$. When the q-ary sequences $\{u(t)\}$, $\{v(t)\}$ are the same, we speak of the **autocorrelation function** of the sequence $\{u(t)\}$, simply denoted as $\theta_u(\tau)$, and speak of the **crosscorrelation function** of them if they are distinct.

For a family $\mathcal{F}$ of M periodic sequences over $\mathbb{Z}_q$, we define the **peak crosscorrelation magnitude** by

$$\theta_c = \max\left\{|\theta_{u,v}(\tau)| \mid 0 \leq \tau < N,\ u, v \in \mathcal{F} \text{ and } u \neq v\right\},$$

the **peak autocorrelation magnitude** by

$$\theta_a = \max\left\{|\theta_u(\tau)| \mid 1 \leq \tau < N,\ u \in \mathcal{F}\right\},$$

and the **correlation magnitude** by

$$\theta_{\max} = \max\left\{\theta_c, \theta_a\right\}.$$

A sequence family is said to have **low correlation** if it has a relatively small magnitude $\theta_{\max}$ when compared with other families of the same size and period [940].

From the definition of the periodic correlation function, it can be verified [1616, 1617] that

$$\sum_{\tau=0}^{N-1} |\theta_{u,v}(\tau)|^2 = \sum_{\tau=0}^{N-1} \theta_u(\tau)\theta_v^*(\tau).$$

Then, given a sequence family $\mathcal{F}$, we have

$$\sum_{u,v\in\mathcal{F}} \sum_{\tau=0}^{N-1} |\theta_{u,v}(\tau)|^2 = \sum_{u,v\in\mathcal{F}} \sum_{\tau=0}^{N-1} \theta_u(\tau)\theta_v^*(\tau) = \sum_{\tau=0}^{N-1} \left|\sum_{u\in\mathcal{F}} \theta_u(\tau)\right|^2 \geq \left|\sum_{u\in\mathcal{F}} \theta_u(0)\right|^2 = M^2 N^2.$$

On the other hand,

$$\sum_{u,v \in \mathcal{F}} \sum_{\tau=0}^{N-1} |\theta_{u,v}(\tau)|^2 = \sum_{u \in \mathcal{F}} \sum_{\tau=0}^{N-1} |\theta_u(\tau)|^2 + \sum_{u \neq v} \sum_{\tau=0}^{N-1} |\theta_{u,v}(\tau)|^2$$

$$= \sum_{u \in \mathcal{F}} |\theta_u(0)|^2 + \sum_{u \in \mathcal{F}} \sum_{\tau=1}^{N-1} |\theta_u(\tau)|^2 + \sum_{u \neq v} \sum_{\tau=0}^{N-1} |\theta_{u,v}(\tau)|^2$$

$$\leq MN^2 + M(N-1)\theta_a^2 + M(M-1)N\theta_c^2.$$

Hence we obtain the inequality

$$M^2 N^2 \leq MN^2 + M(N-1)\theta_a^2 + M(M-1)N\theta_c^2.$$

This gives the Sarwate Bound [1616, 1617]:

$$\left(\frac{\theta_c^2}{N}\right) + \frac{N-1}{N(M-1)}\left(\frac{\theta_a^2}{N}\right) \geq 1,$$

which leads to the following lower bound on the peak correlation magnitude:

$$\theta_{\max}^2 \geq \frac{N^2(M-1)}{MN-1}. \tag{25.1}$$

This bound is known as the Welch Bound, and it is shown to be tight in [1884]. A sequence family is commonly considered to have **optimal correlation** if it achieves the Welch Bound.

The Sidelńikov Bound [940] is asymptotic for q-ary sequence families with periodic correlation. The Sidelńikov Bound is derived from a successive estimation of the ratio of even moments and can be approximated by

$$\theta_{\max}^2 \geq \begin{cases} (2s+1)(N-s) + \frac{s(s+1)}{2} - \frac{2^s N^{2s+1}}{M(2s)!\binom{N}{s}} & \text{for } q = 2 \text{ and } 0 \leq s < \frac{2N}{5}, \\ \frac{1}{2}(s+1)(2N-s) - \frac{2^s N^{2s+1}}{M(s!)^2\binom{2N}{s}} & \text{for } q > 2 \text{ and } s \geq 0. \end{cases} \tag{25.2}$$

This bound is significantly tighter than the Welch Bound when $q > 2$.

In addition, by applying the Cauchy inequality to the sum on the right-hand side of the equality

$$\sum_{\tau=0}^{N-1} |\theta_{u,v}(\tau)|^2 = \sum_{\tau=0}^{N-1} \theta_u(\tau)\theta_v^*(\tau) = N^2 + \sum_{\tau=1}^{N-1} \theta_u(\tau)\theta_v^*(\tau),$$

we have

$$\left|\sum_{\tau=1}^{N-1} \theta_u(\tau)\theta_v^*(\tau)\right|^2 \leq \left(\sum_{\tau=1}^{N-1} |\theta_u(\tau)|^2\right)\left(\sum_{\tau=1}^{N-1} |\theta_v(\tau)|^2\right).$$

Therefore, the upper and lower bounds on the **mean square values of periodic correlation functions** can be given by

$$\left|\sum_{\tau=0}^{N-1} |\theta_{u,v}(\tau)|^2 - N^2\right| \leq \left(\sum_{\tau=1}^{N-1} |\theta_u(\tau)|^2\right)^{\frac{1}{2}}\left(\sum_{\tau=1}^{N-1} |\theta_v(\tau)|^2\right)^{\frac{1}{2}} \leq (N-1)\theta_a^2.$$

Frequently in practice, there is greater interest in the mean square correlation distribution of a sequence family than in the parameter $\theta_{\max}$. Quite often in sequence design, the sequence family is derived from a binary cyclic code of length N by picking a set of cyclically distinct sequences of period N. The families of Gold and Kasami sequences are constructed in this way [821, 1086]. In this case, the mean square correlation of the family can be shown to be either optimum or close to optimum, under certain easily satisfied conditions, imposed on the minimum distance of the dual code. A similar situation holds even when the sequence family does not come from a linear cyclic code. In this sense, mean square correlation is not a very discriminating measure of the correlation properties of a family of sequences.

For the case when there is approximate, but not perfect, synchronism present in a CDMA communication system, there is interest in the design of families of sequences whose correlations, both autocorrelation functions and crosscorrelation functions, are zero or small for small values of relative time shift [1286]. In this context, a **low correlation zone (LCZ)** sequence family with parameters (N, M, L, δ) is a family $\mathcal{F}$ of M sequences of period N such that for each time shift τ in the low correlation zone $(-L, L)$, the magnitude of the periodic crosscorrelation function of any two distinct sequences at the shift τ and that of the out-of-phase autocorrelation function of any sequence at the shift τ are bounded by δ, i.e., for $u, v \in \mathcal{F}$

$$|\theta_{u,v}(\tau)| \leq \delta \text{ for } |\tau| < L, \ u \neq v, \text{ and for } 0 < |\tau| < L \text{ when } u = v.$$

It is shown in [1782] that the parameters of an (N, M, L, δ) LCZ sequence family are mutually constrained by the following inequality

$$\delta^2 \geq \frac{N(ML - N)}{ML - 1}.$$

It is worth noting that if L is taken as N, the above lower bound on δ matches the Welch Bound (25.1).

The **odd-periodic correlation** of two q-ary sequences $\{u(t)\}$ and $\{v(t)\}$ is the collection $\{\widehat{\theta}_{u,v}(\tau) \mid -(N-1) \leq \tau \leq N-1\}$ given by

$$\widehat{\theta}_{u,v}(\tau) = \sum_{t=0}^{N-1} (-1)^{\lfloor (t+\tau)/N \rfloor} \omega_q^{u(t+\tau)-v(t)}.$$

The odd-periodic correlation of the sequences $\{u(t)\}$ and $\{v(t)\}$ is termed **odd-periodic crosscorrelation** when they are different, and termed **odd-periodic autocorrelation** when they are identical, where $\widehat{\theta}_u(\tau)$ is used for simplicity. Similar to the (even-)periodic correlation, the odd-periodic correlation properties of sequences can play an equally important role in applications for extracting desired information from the signals at the receiver. Hence, sequences with low odd-periodic correlation magnitude are favorable. Nevertheless, the odd-periodic correlation properties of sequences are little explored in general. Only a few constructions of sequences with low autocorrelations have been proposed in the literature [1302, 1921]. Lüke and Schotten in [1302] showed that there exists no binary sequence of period greater than 2 that are perfect with respect to odd-periodic autocorrelation (namely, all out-of-phase odd-periodic autocorrelations equal zero). They proposed a class of odd-periodic almost perfect binary sequences with period $q + 1$ and with one position taking a special element, where q is an odd prime power. As we shall see, these sequences can be used to construct quaternary sequences with low periodic correlation. Yang and Tang [1921] recently proposed several families of quaternary sequences with low odd-periodic autocorrelation.

The **aperiodic correlation** of two q-ary sequences $\{u(t)\}$ and $\{v(t)\}$ is the collection $\{\rho_{u,v}(\tau) \mid -(N-1) \leq \tau \leq N-1\}$ given by

$$\rho_{u,v}(\tau) = \sum_{t=\max\{0,-\tau\}}^{\min\{N-1, N-1-\tau\}} \omega_q^{u(t+\tau)-v(t)},$$

where the range of t in the summation is the overlapped area of a shifted-version of $\{u(t)\}$ with $\{v(t)\}$. The aperiodic correlation of the sequences $\{u(t)\}$ and $\{v(t)\}$ is termed **aperiodic crosscorrelation** when they are different and termed **aperiodic autocorrelation** when they are identical, where $\rho_u(\tau)$ is used for simplicity. For a family of M sequences of common period N, let $\rho_{\max}$ be the maximum magnitude of the aperiodic crosscorrelation of any two different sequences and the out-of-phase aperiodic autocorrelation of any sequence. In practice sequences with small $\rho_{\max}$ are desirable. A lower bound on $\rho_{\max}$ due to Welch [1884] is

$$\rho_{\max}^2 \geq \frac{N^2(M-1)}{M(2N-1)-1}.$$

For large families of sequences, the best bound on $\rho_{\max}$ is due to Levenshtein [1232]. The Levenshtein Bound is based on linear programming theory as well as the theory of association schemes. When $M \geq 4$ and $N \geq 2$, the Levenshtein Bound is

$$\rho_{\max}^2 \geq \frac{(2N^2+1)M - 3N^2}{3(MN-1)}.$$

Aperiodic correlations are used more often than periodic correlations in a CDMA communication system. However, the problem of designing sequence sets, particularly the crosscorrelation functions, with low aperiodic correlation is difficult. The conventional approach has been to design sequences based on periodic correlation properties and to subsequently analyze the aperiodic correlation properties of the resulting design.

For a sequence $\{u(t)\}$ of period N, the **merit factor** is defined as

$$F = \frac{\rho_u(0)^2}{\sum_{0<|\tau|\leq N-1}\rho_u^2(\tau)}.$$

The merit factor is the ratio of the square of the in-phase autocorrelation to the sum of the squares of the out-of-phase aperiodic autocorrelation values. Hence it can be deemed as a measure of the aperiodic autocorrelation properties of a sequence. It is also closely connected with the signal to self-generated noise ratio of a communication system in which coded pulses are transmitted and received. It is desirable that the merit factor of a sequence be high, corresponding to the requirement that the sequence should have low aperiodic autocorrelation magnitude. Due to the intractability of aperiodic correlation, progress on the merit factor was only made for binary sequences in recent years [1040, 1093]. Let F_N denote the largest merit factor for any binary sequence of length N. Exhaustive computer searches for $N \leq 40$ have revealed that $F_{11} = 12.1$, $F_{13} = 14.08$ and $3.3 \leq F_N \leq 9.85$ for other positive integers N up to 40. The values F_{11} and F_{13} are achieved by Barker sequences of corresponding lengths. From partial searches, for lengths up to 117, the highest known merit factor is between 8 and 9.56; for lengths from 118 to 200, the best-known factor is close to 6. For lengths > 200, statistical search methods have failed to yield a sequence having merit factor exceeding 5. Researchers have also looked into the asymptotic behavior of the merit factor of binary sequences of infinitely large length N. It is proved that skew-symmetric sequences exhibit the best known asymptotic merit factor as large as $6.342\cdots$ [1040, 1093].

The **partial-period correlation** of two sequences $\{u(t)\}$ and $\{v(t)\}$ is the collection $\{\Delta_{u,v}(l,\tau,t_0) \mid 1 \leq l \leq N,\ 0 \leq \tau \leq N-1,\ 0 \leq t_0 \leq N-1\}$ given by

$$\Delta_{u,v}(l,\tau,t_0) = \sum_{t=t_0}^{t=t_0+l-1} \omega_q^{u(t+\tau)-v(t)},$$

where l is the length of the partial-period and the sum $t+\tau$ is again computed modulo N. In direct-sequence CDMA systems, the pseudo-random signature sequences used by the various users are often very long for reasons of data security. In such situations, to minimize receiver hardware complexity, correlation over a partial period of the signature sequence is often used to demodulate data as well as to achieve synchronization [825, 940]. For this reason, the partial-period correlation properties of a sequence are of interest. Researchers have attempted to determine the moments of the partial-period correlation. Here, the main tool is the application of the Pless Power Moments of coding theory; see [1513] and Section 1.15. These identities often allow the first and second partial-period correlation moments to be completely determined. For example, this is true in the case of m-sequences.

We have recalled basic correlation measures of sequences used in digital communication systems. Other correlation metrics related to transmitted signal energy, like peak-to-sidelobes level (PSL) and peak-to-average power ratio (PAPR) of sequences can be found in a recent survey [1850].

25.2.2   Sequences with Low Periodic Autocorrelation

For a q-ary sequence $\{s(t)\}$ of period N, it is clear that the in-phase autocorrelation is equal to N. In many applications it is desirable to have sequences with low out-of-phase autocorrelation magnitude, which is given by $\Theta = \max_{\tau \neq 0} |\theta_s(\tau)|$. A sequence $\{s(t)\}$ of length N is said to have **perfect autocorrelation** if $\Theta = 0$. However, such sequences are rare in general, particularly for the binary and quaternary sequences, which are of more practical interest for simplicity of implementation. Binary and quaternary sequences with perfect autocorrelation are believed to exist only for a few lengths, namely length 4 for binary sequences and lengths 2, 4, 8, 16 for quaternary sequences [1302]. It is thus natural and interesting to consider sequences of other lengths and low autocorrelation. Sequences with lowest possible out-of-phase autocorrelation magnitude are commonly said to have **optimal autocorrelation** [59, 328]. There exist many constructions of q-ary sequences with optimal autocorrelation properties. It is beyond our intention to cover all the known constructions here. Instead, we choose to introduce only some constructions for binary sequences and quaternary sequences, which are of particularly practical interest. Other q-ary sequences with good periodic autocorrelation property can be found in Arasu's recent survey [59].

Binary Sequences with Optimal Autocorrelation

For a binary sequence $\{s(t)\}$ of period N, its **support** is defined as $C_s = \{0 \leq t < N \mid s(t) = 1\}$. A binary sequence $\{s(t)\}$ is commonly termed the **characteristic sequence** of a given set that is taken as support of $\{s(t)\}$. It is well known (e.g., in [1057]) that the autocorrelation function of $\{s(t)\}$ at any time shift τ satisfies $\theta_s(\tau) \equiv N \pmod{4}$. The optimal values of Θ can be classified into the following four types:

(i) If $N \equiv 0 \pmod 4$, then the smallest possible value of Θ is equal to 0. However, the only known such example is 0001 when $N = 4$, and it is conjectured that perfect sequences with $\Theta = 0$ do not exist for periods greater than 4. Hence the optimal value of Θ in this case is commonly considered to be 4.

(ii) If $N \equiv 1 \pmod 4$, then the smallest possible value of Θ is equal to 1. There exist examples with $\Theta = 1$ for $N = 5$ and $N = 13$. However, it is proved that there are no such sequences of lengths from 14 up to 20201. Some evidence [1057] shows that there exist no binary sequences of period $N > 13$ with $\Theta = 1$. In this case the optimal value of Θ is believed to be 3.

(iii) If $N \equiv 2$ (mod 4), then the smallest possible value of Θ is equal to 2. In this case, the out-of-phase autocorrelation function $\theta_s(\tau)$ may take values from $\{-2, 2\}$. Jungnickel and Pott in [1057] showed that there exist no binary sequences with the out-of-phase autocorrelation function $\theta_s(\tau)$ always taking the value 2 or always taking the value -2 with lengths up to 10^9, except for 4 length values; in addition, there exist binary sequences with out-of-phase autocorrelation values $\{-2, 2\}$, which are said to have optimal autocorrelation.

(iv) If $N \equiv 3$ (mod 4), then the smallest possible value of Θ is again equal to 1 since the out-of-phase autocorrelation function may equal -1. As a matter of fact, there exist binary sequences with out-of-phase autocorrelation function always taking the value -1. These sequences are said to have ideal two-level autocorrelation.

Due to their intrinsic connection, many known families of binary sequences with optimal autocorrelation are constructed from combinatorial objects [60]. Here we recall basic definitions of *difference sets* and *almost difference sets* for self-completeness. Let $(A, +)$ be an abelian group of order v and C be a k-subset of A. Let $d_C(w) = |C \cap (C + w)|$ be the difference function at any nonzero element w in A. The subset C is called a (v, k, λ) **difference set (DS)** in A if $d_C(w) = \lambda$ for any nonzero element w in A, and is called a (v, k, λ, t) **almost difference set (ADS)** if $d_C(w) = \lambda$ for t nonzero elements w and $d_C(w) = \lambda + 1$ for $v - 1 - t$ nonzero elements w. Difference sets and almost difference sets in an abelian group A are said to be *cyclic* if the group itself is cyclic.

Let $\mathbb{Z}_N$ be the set of integers modulo N. A binary sequence $\{s(t)\}$ with a support C_s is said to have optimal autocorrelation if one of the following is satisfied:

(i) If $N \equiv 0$ (mod 4), C_s is an $(N, k, k - (N+4)/4, Nk - k^2 - (N-1)N/4)$ ADS in $\mathbb{Z}_N$.

(ii) If $N \equiv 1$ (mod 4), C_s is an $(N, k, k - (N+3)/4, Nk - k^2 - (N-1)^2/4)$ ADS in $\mathbb{Z}_N$.

(iii) If $N \equiv 2$ (mod 4), C_s is an $(N, k, k - (N+2)/4, Nk - k^2 - (N-1)(N-2)/4)$ ADS in $\mathbb{Z}_N$.

(iv) If $N \equiv 3$ (mod 4), C_s is an $(N, (N+1)/2, (N+1)/4)$ or $(N, (N-1)/2, (N-3)/4)$ DS in $\mathbb{Z}_N$.

Among the above four cases, the last one when $N \equiv 3$ (mod 4) is the family of Paley–Hadamard difference sets with flexible parameters and ideal two-value autocorrelation. The conditions for the first three cases are more restrictive. Here we shall introduce one example for each of the first three cases and then discuss the last case in more depth.

The examples for cases (i)-(iii) we introduce here all pertain to square and non-square elements in finite fields. A nonzero element x in the finite field $\mathbb{F}_q$ is called a **square** if there exists an element $y \in \mathbb{F}_q$ satisfying $x = y^2$. Equivalently, a nonzero element $x \in \mathbb{F}_q$ with $x = \alpha^k$ is a square if and only if the exponent k is even, where α is a primitive element of $\mathbb{F}_q$. For case (i), let q be a prime power with $q \equiv 1$ (mod 4); then the set $\{t \mid \alpha^t + 1 \text{ is a non-square in } \mathbb{F}_q, 0 \le t < q-1\}$ is a $(q-1, (q-1)/2, (q-5)/4, (q-1)/4)$ ADS in $\mathbb{Z}_{q-1}$, of which the binary sequence has optimal autocorrelation values $\{0, -4\}$. For case (iii), let q be a prime power with $q \equiv 3$ (mod 4); then the set $\{t \mid \alpha^t + 1 \text{ is a non-square in } \mathbb{F}_q, 0 \le t < q-1\}$ is a $(q-1, (q-1)/2, (q-3)/4, (3q-5)/4)$ ADS in $\mathbb{Z}_{q-1}$, of which the binary sequence has optimal autocorrelation values $\{0, -2, 2\}$ [1702]. For case (ii), let $q = p$ be a prime with $p \equiv 1$ (mod 4); then the set of squares modulo p forms an ADS in $\mathbb{Z}_p$. The binary sequence having this ADS as its support is the well-known Legendre sequence [704], and it has optimal autocorrelation values $\{0, -3, 1\}$. Other families of binary sequences with optimal autocorrelation for N (mod 4) $\in \{0, 1, 2\}$ are mostly constructed from cyclotomic classes. More instances of such sequences can be found in [328, 1163, 1783, 1934].

For the last case of Paley–Hadamard difference sets, we cover one important case $N = 2^n - 1$, for which n can be freely chosen and the corresponding binary sequences can be converted to good binary cyclic codes of the same length. Known infinite families of Paley–Hadamard difference sets fall into two classes:

(1) difference sets with Singer parameters, and

(2) difference sets from cyclotomic classes.

These correspond to sequences of period N of the form $N = 2^n - 1$, $n \geq 1$ for the first class, and N is a prime, or of the form $N = p(p + 2)$ with both p and $p + 2$ being prime, for the second class.

For the first class with $N = 2^n - 1$, the parameters $(2^n - 1, 2^{n-1}, 2^{n-2})$ or their complements $(2^n - 1, 2^{n-1} - 1, 2^{n-2} - 1)$ are known as Singer parameters after Singer's work in [1716]. The Singer construction corresponds to the well known binary maximum-length sequences, called m-sequences, which will be discussed in more depth in the context of shift register sequences. The construction was later generalized by Gordon, Mills, and Welch [839]. Some 40 years after the appearance of the GMW construction [839], new and interesting cyclic Hadamard difference sets with Singer parameters were discovered in the late 1990s by several research groups. In 1998, Maschietti [1345] constructed a new family of cyclic Hadamard difference sets that were based on the theory of hyperovals in finite projective geometry. Around the same time, several difference sets were postulated by No et al. in [1444] based on the results of a numeric search and these led to a subsequent postulate by Dobbertin [608]. The difference sets postulated in [608, 1444] were later shown by Dillon and Dobbertin [540] to be examples of difference sets constructed by considering the image set of two mappings in the finite field $\mathbb{F}_{2^n}$. From the aforementioned difference sets in the multiplicative group $\mathbb{F}_{2^n}^*$, we obtain binary sequences $\{s(t)\}$ of period $2^n - 1$ with optimal autocorrelation given by

$$s(t) = 1 \text{ if and only if } \alpha^t \in D,$$

where α is a generator of $\mathbb{F}_{2^n}^*$ and D is one of the difference sets with parameters $(2^n - 1, 2^{n-1}, 2^{n-1})$. We summarize these difference sets below.

Construction 25.2.1 (Trace Function) Define the trace function $\mathrm{Tr}_{q^r/q} : \mathbb{F}_{q^r} \to \mathbb{F}_q$ by $\mathrm{Tr}_{q^r/q}(x) = \sum_{i=0}^{r-1} x^{q^i}$. Let n_0 be a divisor of n and $R = \{x \in \mathbb{F}_{2^{n_0}} \mid \mathrm{Tr}_{2^{n_0}/2}(x) = 1\}$, which is a difference set [1716]. Gordon, Mills, and Welch showed the set $D = \{x \in \mathbb{F}_{2^n} \mid \mathrm{Tr}_{2^n/2^{n_0}}(x) \in R\}$ is a difference set in $\mathbb{F}_{2^n}^*$ with Singer parameters [839]. It is a generalization of the original difference set $\{x \in \mathbb{F}_{2^n} \mid \mathrm{Tr}_{2^n/2}(x) = 1\}$ by Singer [1716].

Construction 25.2.2 (Hyperoval) Maschietti [1345] showed that $M_k = \mathbb{F}_{2^n} \setminus \{x^k + x \mid x \in \mathbb{F}_{2^n}\}$ is a difference set in $\mathbb{F}_{2^n}^*$ with Singer parameters for an odd integer $n > 0$ and the following integers k:

- $k = 2$ (the Singer case),

- $k = 6$ (the Segre case),

- $k = 2^\sigma + 2^\pi$ with $\sigma = (n + 1)/2$ and $4\pi \equiv 1 \pmod{n}$ (the Glynn I case), and

- $k = 3 \cdot 2^\sigma + 4$ with $\sigma = (n + 1)/2$ (the Glynn II case).

Construction 25.2.3 (Kasami–Welch Exponent) Let k be an integer coprime to n and $d = 2^{2k} - 2^k + 1$ (known as the Kasami–Welch exponent). Dillon and Dobbertin [540] showed that under the mappings $\phi_k^{(0)}(x) = (x+1)^d + x^d$ and $\phi_k^{(1)}(x) = (x+1)^d + x^d + 1$, the set

$$D_k = \begin{cases} \phi_k^{(0)}(\mathbb{F}_{2^n}) & \text{if } n \text{ is odd,} \\ \mathbb{F}_{2^n} \setminus \phi_k^{(0)}(\mathbb{F}_{2^n}) & \text{if } n \text{ is even,} \end{cases}$$

for $3k \equiv 1 \pmod{n}$ and the set $\Delta_k = \mathbb{F}_{2^n} \setminus \phi_k^{(1)}(\mathbb{F}_{2^n})$ for $\gcd(n, k) = 1$ are difference sets in $\mathbb{F}_{2^n}^*$ with Singer parameters. This unified the results in [1444] and [539].

For the second class of Paley–Hadamard difference sets, cyclotomy plays a critical role in the known constructions. Let $q = df + 1$ be a power of prime and α a primitive element of $\mathbb{F}_q$. The cosets $D_i^{(d,q)} = \alpha^i \langle \alpha^d \rangle$, where $\langle \alpha^d \rangle$ is the cyclic group generated by α^d in $\mathbb{F}_q^*$, are called **cyclotomic classes** of order d with respect to $\mathbb{F}_q$. Clearly $\mathbb{F}_q^* = \bigcup_{i=0}^{d-1} D_i^{(d,q)}$. Known constructions of Paley–Hadamard difference sets are summarized as follows.

Construction 25.2.4 (Hall [887]) Let p be a prime of the form $p = 4s^2 + 27$ for some s. The Hall difference set is given by $D = D_0^{(6,p)} \cup D_1^{(6,p)} \cup D_3^{(6,p)}$.

Construction 25.2.5 (Paley [1484]) Let p be a prime with $p \equiv 3 \pmod 4$. Then the cyclotomic class $D_1^{(\frac{p-1}{2}, p)}$ is a $(p, \frac{p-1}{2}, \frac{p-3}{4})$ difference set in $\mathbb{Z}_p$, which is the well-known Legendre sequence [704].

Construction 25.2.6 (Two-Prime [142, Chapter 2]) Let $N = p(p+2)$ with p and $p+2$ being two primes. The two-prime difference set is given by

$$(\mathbb{Z}_p \times \{0\}) \bigcup \left(D_0^{(\frac{p-1}{2}, p)} \times D_0^{(\frac{p+1}{2}, p+2)} \right) \bigcup \left(D_1^{(\frac{p-1}{2}, p)} \times D_1^{(\frac{p+1}{2}, p+2)} \right),$$

and it corresponds to the two-prime sequence of period $p(p+2)$.

Quaternary Sequences with Optimal Autocorrelation

Let $\{s(t)\}$ be a quaternary sequence of period N, namely, each $s(t)$ takes a value from $\{0, 1, 2, 3\}$. With $\omega_4 = i = \sqrt{-1}$, the periodic autocorrelation of $\{s(t)\}$ at a shift τ is

$$\theta_s(\tau) = \sum_{t=0}^{N-1} \omega_4^{s(t+\tau)-s(t)}.$$

Let $n_j(s, \tau)$ denote the number of occurrences of $j \in \{0, 1, 2, 3\}$ in the difference $s(t+\tau) - s(t)$ (mod 4) as t ranges over $\mathbb{Z}_N$. The autocorrelation of $\{s(t)\}$ can be rewritten as

$$\theta_s(\tau) = (n_0(s, \tau) - n_2(s, \tau)) + \omega_4(n_1(s, \tau) - n_3(s, \tau)),$$

and the maximum magnitude of out-of-phase autocorrelation is given by

$$\Theta = \max \left\{ \sqrt{(n_0(s, \tau) - n_2(s, \tau))^2 + (n_1(s, \tau) - n_3(s, \tau))^2} \mid 1 \leq \tau < N \right\}.$$

The magnitude Θ could be as small as zero, which is achieved by perfect quaternary sequences. Nevertheless, perfect quaternary sequences have only been found for lengths 2, 4, 8, and 16, and it is believed that there is no perfect quaternary sequence with lengths greater than 16 [704]. Quaternary sequences of period N that have $\Theta = 1$ for odd N or have $\Theta = 2$ for even N were found, which are considered to be the best possible autocorrelation for quaternary sequences. Hence quaternary sequences with $\Theta = 1$ for odd N or $\Theta = 2$ for even N are said to have **optimal autocorrelation**. Lüke and Schotten in [1302] gave a comprehensive survey on both binary and quaternary sequences with optimal autocorrelation properties. In addition to binary and quaternary sequences, they also introduced *almost binary* and *almost quaternary* sequences, which have a single position take a certain special value and are used to construct optimal sequences. There are several constructions of optimal (almost) quaternary sequences. Below we recall the generalized Sidelńikov sequence proposed in [1301] and related constructions of sequences with optimal autocorrelation.

Lüke, Schotten, and Mahram in [1301] generalized the binary Sidelńikov sequence [1702] as follows. Let α be a primitive element in $\mathbb{F}_q$. The quaternary Sidelńikov sequence of length $q - 1$ can be given by $s(0) = c$ for certain $c \in \mathbb{Z}_4$ and $s(t) = \log_\alpha(\alpha^t - 1) \pmod 4$ for $t = 1, \ldots, q - 2$, where $\log_\alpha(\cdot)$ denotes the discrete logarithm function in $\mathbb{F}_q$. The authors in [1301] showed that the quaternary Sidelńikov sequences have $\Theta = 2$ when $N = q - 1 \equiv 0 \pmod 4$ and $c = 0, 2$, or when $N = q - 1 \equiv 2 \pmod 4$ and $c = 1, 3$.

When N is odd, the optimal autocorrelation $\Theta = 1$ is attained by a so-called complementary-based method due to Schotten [1631]. The complementary-based method utilizes odd-perfect binary sequences proposed in [1300] as building blocks. The construction can be described as below, using the trace function defined in Construction 25.2.1.

Construction 25.2.7 (Complementary-Based) Let β be a primitive element of $\mathbb{F}_{q^2}$ with odd q and $f(t) = \text{Tr}_{q^2/q}(\beta^{tq+k})$ with $k = -\log_\beta(\beta^q - \beta)$. Let $\{a(t)\}$ be a binary sequence of length $N_a = q + 1$ defined as follows: for $t = 0, 1, \ldots, q$, $a(t) = 0$ if $f(t)$ is 0 or a square in $\mathbb{F}_q$, and $a(t) = 1$ otherwise. Then the sequence $\{a(t)\}$ has odd perfect autocorrelation, and also satisfies the property $a(t) = (t + \frac{N_a}{2}) + a(N_a - t) \pmod 2$ for $t = 1, \ldots, \frac{N_a}{2} - 1$. In the case of $N_a \equiv 2 \pmod 4$, define two sequences $\{x(t)\}$ and $\{y(t)\}$ of length $N = \frac{N_a}{2}$ by

$$x(t) = (-1)^{t + \widehat{a}(2t+1+N)} \quad \text{and} \quad y(t) = (-1)^{t + \widehat{a}(2t+1)} \quad \text{for } t = 0, 1, \ldots, N - 1,$$

where $\widehat{a} = a \| (a \oplus 1)$ is a sequence of length $2N_a$ obtained by concatenating the sequence a and its complement $a \oplus 1$. A quaternary sequence $\{s(t)\}$ is then defined by

$$s(t) = \frac{2x(t) + x(t)y(t) - 1}{2} \quad \text{for } t = 0, 1, \ldots, N - 1.$$

It can be verified that the periodic correlation function of $\{s(t)\}$ has out-of-phase autocorrelation magnitude $\Theta = 1$.

We give an example of the Complementary-Based Construction.

Example 25.2.8 Let $q = p = 17$ with primitive element β of $\mathbb{F}_{q^2}$ satisfying $(\beta^q - \beta)^{-1} = \beta^7$. From the definition, one gets the binary sequence $\{a(t)\}$: 000010000110100010 of length 18. Then the quaternary sequence $\{s(t)\}$ of length 9 is given by 201030103, which has the periodic autocorrelation spectrum

$$\theta_s(\tau) : (9, -1, -1, -1, 1, 1, -1, -1, -1).$$

In addition, when $N \equiv 1 \pmod 4$, the binary Legendre sequences of odd prime length can be modified to yield optimal quaternary sequences [1302]. Take $N = p$ as an odd prime with $N \equiv 1 \pmod 4$. The quaternary sequence $\{s(t)\}$ is given as follows: $s(0) = 1$, $s(t) = 0$ if t is a square modulo p, and $s(t) = 2$ if t is a non-square modulo p. For instance, when $p = 5$, the modified Legendre sequence is defined as 10220, which has periodic autocorrelation spectrum $\theta_s(\tau) : (5, -1, -1, -1, -1)$.

Using optimal binary and quaternary sequences with odd length N and $\Theta = 1$, one can further obtain more optimal quaternary sequences of length $2N$ and $\Theta = 2$ by multiplying them with the perfect quaternary sequence 01. The aforementioned optimal quaternary sequences are constructed by slightly modifying binary sequences with optimal autocorrelation. Hence they cannot have good balance over $\mathbb{Z}_4$. (Here by *balance* we mean the values of $\mathbb{Z}_4$ are (almost) uniformly distributed in a quaternary sequence.) Tang and Ding [1781] employed the interleaving technique and the inverse Gray mapping in constructing balanced quaternary sequences for even length $N \equiv 2 \pmod 4$.

25.2.3 Sequence Families with Low Periodic Correlation

The crosscorrelation function is important in code-division multiple-access (CDMA) communication systems where each user is assigned a distinct signature sequence. In order to minimize interference among users, it is desirable that the signature sequences have, pairwise, low values of their crosscorrelation function. In addition, to provide the system with a self-synchronizing capability, it is desirable that the signature sequences have low values of their autocorrelation function as well. Hence, it is of great interest to construct a family of sequences with correlation magnitude $\theta_{\max}$ as low as possible. In addition, the optimal value of $\theta_{\max}$ is affected by the total number of sequences in the family and the period of the sequences as well as the size of the alphabet of the sequences. Bounds on the minimum possible value of $\theta_{\max}$ for given period N, family size M, and alphabet size q are available, which are usually used to judge the merits of a particular sequence design. The most efficient bounds are those due to Welch (25.1) and Sidelńikov (25.2). In CDMA systems, there is greatest interest in designing sequence families with $\theta_{\max}$ in the range $\sqrt{N} \leq \theta_{\max} \leq 2\sqrt{N}$, where N is the common period of all sequences in the family.

Consistent with our treatment of autocorrelation, this subsection will introduce some sequence families having low correlation magnitude $\theta_{\max}$ for $q = 2$ and $q = 4$.

Binary Sequence Families with Low Correlation

The binary maximum-length sequences, called *m*-sequences, are an important family of sequences with optimal autocorrelation. A binary *m*-**sequence** with period $2^n - 1$ can be defined by $\{s(t)\}$ with $s(t) = \mathrm{Tr}_{2^n/2}(\alpha^t)$, where $\mathrm{Tr}_{2^n/2}$ is the trace function defined in Construction 25.2.1 and α is a primitive element in $\mathbb{F}_{2^n}$. A left cyclic shift of $\{s(t)\}$ by l positions is the sequence $s(l)s(l+1) \cdots s(l + 2^n - 2)$, where the indices are taken modulo $2^n - 1$. In other words, all cyclic shifts of a binary *m*-sequence can be simply given by $\{s'(t)\}$ with $s'(t) = \mathrm{Tr}_{2^n/2}(\alpha^l \alpha^t)$, where l is an integer in the interval $[0, 2^n - 2]$. Given the optimal autocorrelation of an *m*-sequence and its simplicity, it is natural to attempt to construct families of low correlation sequences from *m*-sequences. Many families of low correlation sequences have been constructed from pairs of an *m*-sequence $\{s(t)\}$ and its *d*-decimated sequence $\{s(dt)\}$, which is again an *m*-sequence when d is coprime to the period $N = 2^n - 1$. Interestingly, the study of the crosscorrelation between *m*-sequences $\{s(t)\}$ and $\{s(dt)\}$ also determines the weight distribution of binary cyclic codes with zeros α, α^d, i.e., with the generator polynomial $g(x) = \prod_{i=0}^{n-1}(x - \alpha^{2^i})(x - \alpha^{d2^i})$. For certain decimations d, all possible values, and even the occurrences of values, of the crosscorrelation function can be determined. From those decimations, many constructions of sequence families with low correlation are proposed in the literature; see, e.g., [794, 821, 934, 940, 1086]. Among the families constructed from *m*-sequences, two of the better known constructions are the families of Gold and Kasami sequences.

Construction 25.2.9 (Gold sequences [821]) Let $\{s(t)\}$ be a cyclic shift of the *m*-sequence with period $N = 2^n - 1$. Let $d = 2^k + 1$ with $\gcd(k, n) = 1$ and $\{s(dt)\} \neq \{0\}$.

The Gold sequence family is given by

$$\mathcal{G} = \{s(t)\} \cup \{s(dt)\} \cup \big\{\{s(t) + s(d(t + \tau))\}\big\} \mid 0 \le \tau < N\}.$$

The maximum correlation magnitude $\theta_{\max}$ of $\mathcal{G}$ satisfies $\theta_{\max} \le 1 + \sqrt{2^{n+1}}$. An application of the Sidelńikov Bound (25.2) coupled with the information that $\theta_{\max}$ must be an odd integer yields that $\theta_{\max}$ is as small as possible for $\mathcal{G}$. In this sense, the sequence family $\mathcal{G}$ has optimal correlation. With the same technique and argument, the above sequence family also has optimal correlation when $d = 2^k + 1$ replaced by $d = 2^{2k} - 2^k + 1$ with the same condition $\gcd(k, n) = 1$. The Gold family is one of the best-known families of m-sequences having low crosscorrelation, and is used to design the signals in the global positioning system (GPS) [940].

Construction 25.2.10 (Kasami sequences [1086]) The family of Kasami sequences has a similar description. Let n be an even integer, $d = 2^{\frac{n}{2}} + 1$. Let $\{s(t)\}$ be a cyclic shift of an m-sequence with period $2^n - 1$ and $\{s(dt)\} \ne \{0\}$. The Kasami sequence family is defined by

$$\mathcal{K} = \{s(t)\} \cup \big\{\{s(t) + s(d(t + \tau))\}\big\} \mid 0 \le \tau < 2^{\frac{n}{2}} - 1\}.$$

It can be shown that the Kasami family has $2^{\frac{n}{2}}$ sequences of period $2^n - 1$ and maximum correlation magnitude $\theta_{\max} = 1 + 2^{\frac{n}{2}}$. In this case, the Kasami family has optimal correlation with respect to the Welch Bound (25.1).

Quaternary Sequence Families with Low Correlation

According to the Welch Bound (25.1) and Sidelńikov Bound (25.2) on $\theta_{\max}$ for a family of sequences, there may exist q-ary sequence families with $q > 2$ that allow for better performance. Below we introduce two families of quaternary sequences that outperform the Gold and Kasami sequences.

We first introduce some notation for the construction. Let $\mathrm{GR}(4, n)$ be the Galois ring with 4^n elements, an extension of $\mathbb{Z}_4$ of degree n. There exists an element $\beta \in \mathrm{GR}(4, n)^*$ of order $2^n - 1$. Let

$$\mathcal{T}_n = \big\{0, 1, \beta, \dots, \beta^{2^n - 2}\big\}$$

be the set of so-called **Teichmüller representatives** of $\mathrm{GR}(4, n)$. Every element $z \in \mathrm{GR}(4, n)$ has the unique 2-adic expansion $z = z_0 + 2z_1$ where $z_i \in \mathcal{T}_n$. This induces an automorphism σ on $\mathrm{GR}(4, n)$ with $\sigma(z) = z_0^2 + 2z_1^2$ and a trace function $T_n : \mathrm{GR}(4, n) \to \mathbb{Z}_4$ given by

$$T_n(z) = \sum_{i=0}^{n-1} \sigma^i(z).$$

Let η_i, $1 \le i \le 2^n$, be an ordering of the elements of $\mathcal{T}_n$. Let $\mathcal{A}$ be a family of $\mathbb{Z}_4$ sequences with period $N = 2^n - 1$ given by

$$\mathcal{A} = \{s_i(t) \mid 1 \le i \le 2^n + 1\},$$

where

$$s_i(t) = T_n\big((1 + 2\eta_i)\beta^t\big) \text{ for } 1 \le i \le 2^n \quad \text{and} \quad s_{2^n + 1}(t) = T_n(2\beta^t).$$

Then the correlation values

$$\theta_{i,j}(\tau) = \sum_{t=0}^{n-1} \omega_4^{s_i(t+\tau) - s_j(t)}$$

of Family $\mathcal{A}$ are all of the form

$$\theta_{i,j}(\tau) = -1 + \sum_{x \in \mathcal{T}_n} \omega_4^{T_n(ax)}$$

for some $a \in \mathrm{GR}(4, n)$ depending upon i, j, and τ. The correlation distribution of Family $\mathcal{A}$ was determined in [940], which indicates that

$$\theta_{\max} \le 1 + \sqrt{2^n}\,.$$

Compared with the binary Gold sequences, Family $\mathcal{A}$ has the same family size and better correlation.

With the same notation as in Family $\mathcal{A}$, we define another sequence family $\mathcal{L}$ as follows:

$$\mathcal{L} = \left\{ T_n\big((1+2\eta)\beta^t + 2\mu\beta^{3t}\big) \mid \eta,\, \mu \in \mathcal{T}_n \right\} \cup \left\{ T_n\big(2\beta^t + 2\mu\beta^{3t}\big) \mid \mu \in \mathcal{T}_n \right\} \cup \left\{ T_n\big(2\beta^{3t}\big) \right\}$$

when n is odd and by

$$\mathcal{L} = \left\{ T_n\big((1+2\eta)\beta^t + 2\mu\beta^{3t}\big) \mid \eta,\, \mu \in \mathcal{T}_n \right\} \cup \left\{ T_n\big(2\beta^t + 2\mu\beta^{3t}\big) \mid \mu \in \mathcal{T}_n \right\}$$

when n is even. Each sequence Family $\mathcal{L}$ can be verified to have period $N = 2^n - 1$. Clearly the family has size

$$|\mathcal{L}| = \begin{cases} 2^{2n} + 2^n + 1 & \text{when } n \text{ is odd,} \\ 2^{2n} + 2^n & \text{when } n \text{ is even.} \end{cases}$$

Furthermore, it can be shown [940] that $\theta_{\max}$ for Family $\mathcal{L}$ satisfies

$$\theta_{\max} \le 1 + 2\sqrt{2^n}\,.$$

By the Sidelńikov Bound, Family $\mathcal{L}$ is superior to any possible family of binary sequences having the same size. Moreover, when n is odd, $\mathcal{L}$ is simultaneously a superset of Family $\mathcal{A}$ as well as of the family of binary of Gold sequences.

25.3   Shift Register Sequences

As introduced in Golomb's landmark book [824], the theory of shift register sequences has found major applications in a wide variety of technological situations, including secure, reliable and efficient communications, digital ranging and tracking systems, deterministic simulation of random processes, and computer sequencing and timing schemes. While the theory is very well worked out and fully ready for use from an engineering point of view, there are many unresolved problems worthy of further study from a mathematical standpoint. In the study of shift register sequences, two families of sequences are of particular interest. One is the family of maximum-length sequences for linear shift registers, and the other is the family of de Bruijn sequences for nonlinear shift registers. These sequences have maximal possible periods and interesting combinatorial properties. The generation and properties of linear maximum-length sequences are well-understood. In his book [1903], Wolfram commented that these sequences are the most-used mathematical construct in history: "An octillion. A billion billion billion. That's a fairly conservative estimate of the number of times a cellphone or other device somewhere in the world has generated a bit using a maximum-length linear-feedback shift register sequence." Nevertheless, our understanding of nonlinear shift registers, even of the extreme de Bruijn sequences, is rather limited [824, 935]. In this section, we will briefly introduce some basics of shift register sequences and summarize common approaches to generating de Bruijn sequences. Different from the previous section, we will denote a shift register sequence by $\{s_t\}$ (instead of $\{s(t)\}$) in this section for simpler presentation. In addition, binary n-tuples will be denoted by upper-case letters in bold, e.g., $\mathbf{X}, \mathbf{Y}, \mathbf{A}, \mathbf{B}$.

25.3.1 Feedback Shift Registers

An n-stage **feedback shift register (FSR)** consists of n consecutive storage units $x_0, x_1, \ldots, x_{n-1}$ each of which holds a value and a feedback function $f(x_0, x_1, \ldots, x_{n-1})$. The content in the n storage units is called the **state** of the shift register. Figure 25.1 displays a diagram of an n-stage FSR.

FIGURE 25.1: Diagram of a feedback shift register (FSR)

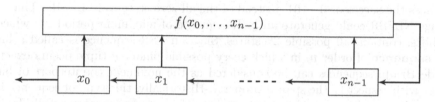

At every clock pulse, the content of x_0 is output, the content of x_{i+1} is transferred into x_i for $0 \le i \le n-2$ and the last unit x_{n-1} is filled with $f(x_0, x_1, \ldots, x_{n-1})$. Namely, given an initial state $\mathbf{S}_0 = s_0 s_1 \cdots s_{n-1}$, a feedback shift register outputs a sequence $\mathbf{s} = \{s_t\} = s_0 s_1 \cdots s_{n-1} s_n s_{n+1} \cdots$ according to the following recurrence

$$s_{n+t} = f(s_t, s_{t+1}, \ldots, s_{t+n-1}) \ \text{ for } t = 0, 1, 2, \ldots.$$

The sequence $\mathbf{s} = \{s_t\}$ is called an **FSR sequence** of order n. It exhibits a sequence of state transitions $\mathbf{S}_0 \to \mathbf{S}_1 \to \mathbf{S}_2 \to \cdots$ in the shift register, where $\mathbf{S}_t = s_t s_{t+1} \cdots s_{t+n-1}$ is called the tth state of $\mathbf{s}$ for $t = 0, 1, 2, \ldots$. The states $\mathbf{S}_{t-1}$ and $\mathbf{S}_{t+1}$ are called the **predecessor** and the **successor** of $\mathbf{S}_t$, respectively.

A shift register is called a **linear feedback shift register (LFSR)** when the feedback function is of linear form

$$f(x_0, x_1, \ldots, x_{n-1}) = c_0 x_0 + c_1 x_1 + \cdots + c_{n-1} x_{n-1},$$

and is called a **nonlinear feedback shift register (NLFSR)** when the feedback function is nonlinear.

In LFSRs, the all-zero state $\mathbf{0}_n = \overbrace{0 \cdots 0}^{n}$ repeatedly produces the all-zero successor state. Hence, a nonzero binary LFSR sequence of order n has length at most $2^n - 1$ since $\mathbf{0}_n$ as an initial state will yield the all-zero sequence. LFSR sequences of order n with maximum possible length $2^n - 1$ are called **maximum-length sequences**, or **m-sequences** for short. It is well-known that m-sequences of order n have a one-to-one correspondence with primitive polynomials of degree n. This indicates that there are a total of $\phi(2^n - 1)/n$ cyclically distinct binary m-sequences of order n, where $\phi(2^n - 1)$ is the Euler totient of $2^n - 1$, namely, the number of positive integers that are less than and coprime to $2^n - 1$. Binary m-sequences possess many good random properties [824]:

- **two-level ideal autocorrelation**: the autocorrelation function of an m-sequence takes the value -1 at all out-of-phase shifts;

- **cycle-and-add property**: the addition of an m-sequence and any of its cyclic shifts is again an m-sequence;

- **span-n property**: any nonzero n-tuple binary string occurs exactly once in an m-sequence of order n;

- **run property**: the runs of length i take up $1/2^i$ portion of all runs in an m-sequence, where a run of length i is a block of i consecutive identical 1's or 0's.

The above properties of m-sequences are favorable in radar systems, synchronization of data, Global Positioning Systems (GPS), coding theory, and code-division multiple-access (CDMA) communication systems [824]. On the other hand, the regularities of m-sequences are not desirable in cryptography. Nonlinearity and irregularity need to be introduced when m-sequences are used in cryptographic applications.

For n-stage binary NLFSRs, there are a total of 2^{2^n} possible Boolean feedback functions. When the feedback function of an NLFSR has a nonzero constant term, the all-zero state $\mathbf{0}_n$ produces the successor $0\cdots01$, instead of the all-zero state as in LFSRs. That is to say, an n-stage NLFSR could generate an FSR sequence of maximum period 2^n, which, under cyclic shifts, contains all possible 2^n states. Such an FSR sequence is called a binary **de Bruijn sequence** of order n, in which every possible binary n-tuple occurs exactly once. Binary de Bruijn sequences can be considered as the nonlinear counterpart of binary m-sequences with respect to the span-n property. Historically, this type of sequence has been 'rediscovered' several times in different contexts. They became more generally known after the work [496] of Nicolaas S. de Bruijn, whose name was later associated with this type of sequence.

De Bruijn sequences showcase nice combinatorial and random properties. They have found applications in a variety of areas, particularly in control mechanisms, coding and communication systems, pseudo-random number generators, and cryptography [824, 1558]. It was shown that binary de Bruijn sequences exist for all positive integers n, and there are a total of $2^{2^{n-1}-n}$ cyclically distinct binary de Bruijn sequences of order n [496]. The number of de Bruijn sequences is huge compared to that of m-sequences, but it takes up only a small portion $2^{(2^{n-1}-n)}/2^{2^n}$ in the whole space of all possible binary FSR sequences of order n. Therefore, efficient generation of de Bruijn sequences has been an interesting and challenging topic. As we shall see, this problem can be approached from perspectives of different mathematics branches, such as combinatorics, graph theory, and abstract algebra [755, 824, 1558].

25.3.2   Cycle Structure

For an n-stage FSR with feedback function f, let $\Omega(f)$ denote the set of sequences generated from all possible 2^n initial states. Under a cyclic shift operator, these sequences can be partitioned into equivalence classes, in which one FSR sequence can be transformed to all other sequences by cyclic shifts. All periodic sequences in an equivalence class form a cycle of distinct states in the FSR. We denote by [$\mathbf{s}$] the **cycle of states** formed by an equivalence class of sequences, where $\mathbf{s}$ is a representative in the equivalence class. If $\Omega(f)$ has exactly r cycles $[\mathbf{s}_1], [\mathbf{s}_2], \ldots, [\mathbf{s}_r]$, the **cycle structure** of $\Omega(f)$ is given by

$$\Omega(f) = [\mathbf{s}_1] \cup [\mathbf{s}_2] \cup \cdots \cup [\mathbf{s}_r].$$

In some contexts, it is sufficient to express the cycle structure of $\Omega(f)$ by the distribution of lengths of cycles for simplicity, where we will denote the cycle structure as $[d_i(\lambda_i)]_{i \geq 1}$, where $d_i(\lambda_i)$ means that there are d_i cycles of length λ_i in the cycle structure.

FIGURE 25.2: The state graph of the pure cycling register (PCR) for $n = 3$

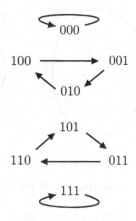

FIGURE 25.3: The de Bruijn graphs G_n for $n = 1, 2, 3$

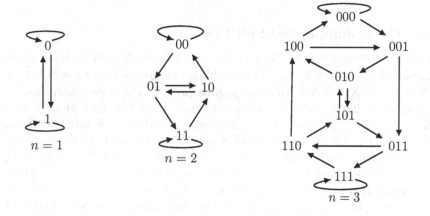

Graphically, the cycle structure of $\Omega(f)$ exhibits a graph of state transitions in the FSR. As an instance, Figure 25.2 displays the state graph for the simple **pure cycling register** **(PCR)** with feedback function $f(x_0, x_1, \ldots, x_{n-1}) = x_0$ for $n = 3$. This state graph has four disjoint cycles and the cycle structure can be given by $\Omega(x_0) = [\mathbf{0}] \cup [\mathbf{1}] \cup [\mathbf{001}] \cup [\mathbf{011}]$.

There are 2^{2^n} possible feedback functions for n-stage binary FSRs. The superposition of the state graphs of all these FSRs yields the **de Bruijn graph** G_n of order n. It is a directed graph composed of 2^n vertices and 2^{n+1} edges, where each vertex denotes an n-tuple state and one vertex $x_0 x_1 \cdots x_{n-1}$ is directed to another vertex $y_0 y_1 \cdots y_{n-1}$ if and only if $x_1 \cdots x_{n-1} = y_0 \cdots y_{n-2}$. Figure 25.3 displays the binary de Bruijn graphs for $n = 1, 2, 3$.

The de Bruijn graph provides an overview of state transitions in all possible FSRs. Choosing a specific feedback function simply means that some edges in the graph will be removed. Clearly, an FSR with any feedback function transits a state to a unique successor state, in which the last bit is uniquely determined by the feedback function. While a state does not necessarily have a unique predecessor state, Golomb [824] gave a necessary and sufficient condition on feedback functions that restricts each state to have a unique

FIGURE 25.4: The process of cycle joining and cycle splitting

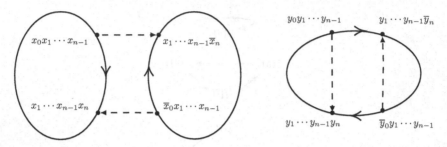

predecessor. An FSR is said to be **nonsingular** if its feedback function has the form

$$f(x_0, x_1, \ldots, x_{n-1}) = x_0 + g(x_1, \ldots, x_{n-1}).$$

The state graph of a nonsingular FSR consists of pure cycles that are disjoint from each other (as in Figure 25.2). Such cycle structures are branchless and of more practical interest [824]. We will restrict our discussion to nonsingular FSRs in the remainder of this section.

25.3.3   Cycle Joining and Splitting

Given an n-stage FSR, let $\mathbf{X} = x_0 x_1 \cdots x_{n-1}$ be a state in its state graph. The state $\widehat{\mathbf{X}} = \overline{x}_0 x_1 \cdots x_{n-1}$, where $\overline{x}_i = x_i \oplus 1$ is the binary complement of x_i, is called the **conjugate** of $\mathbf{X}$. The states $\mathbf{X}$ and $\widehat{\mathbf{X}}$ form a **conjugate pair**. In the state graph, two cycles are said to be **adjacent** if there exists a conjugate pair $\mathbf{X}$ and $\widehat{\mathbf{X}}$ such that $\mathbf{X}$ is on one cycle and $\widehat{\mathbf{X}}$ is on the other cycle. Two adjacent cycles with a conjugate pair $\mathbf{X}$ and $\widehat{\mathbf{X}}$ can be joined into one cycle if the successors of $\mathbf{X}$ and $\widehat{\mathbf{X}}$ are swapped. As shown by the left part of Figure 25.4, there are originally two cycles with a conjugate pair $\mathbf{X} = x_0 x_1 \cdots x_{n-1}$ and $\widehat{\mathbf{X}} = \overline{x}_0 x_1 \cdots x_{n-1}$, which have successors $x_1 \cdots x_{n-1} x_n$ and $x_1 \cdots x_{n-1} \overline{x}_n$, respectively, indicated by the arrows. If we replace the successors of $\mathbf{X}$, $\widehat{\mathbf{X}}$ as $x_0 x_1 \cdots x_{n-1} \to x_1 \cdots x_{n-1} \overline{x}_n$ and $\overline{x}_0 x_1 \cdots x_{n-1} \to x_1 \cdots x_{n-1} x_n$, as indicated by the dashed arrow, we join these two adjacent cycles into one larger cycle. The process of joining adjacent cycles in this way is known as **cycle joining** [824]. Conversely, **cycle splitting** refers to the process in which a single cycle is split into two disjoint cycles by swapping the successors of a conjugate pair that is on the same cycle, as displayed by the right part of Figure 25.4.

Given an FSR with multiple cycles, one can merge these cycles into a full-length cycle if the conjugate pairs between all disjoint cycles can be determined. In addition, one can also transform a full-length cycle to another full-length cycle by first splitting into two cycles with one conjugate pair and then joining the two cycles again in accordance with another conjugate pair. Such a process is known as the **cross joining method** [824]. In the above operations on cycles, it is critical to know adjacent cycles in the state graph and the conjugate pairs between adjacent cycles. The notion of the adjacency graph of an FSR was proposed to facilitate the cycle joining method in generating de Bruijn sequences. For an FSR with feedback function f, its **adjacency graph** $G(f)$ is an undirected multi-graph, in which the vertices correspond to the cycles in $\Omega(f)$, and there exists an edge between two vertices if and only if they share a conjugate pair [925]. Moreover, when two vertices are connected by exactly m edges, it is convenient to denote them by an edge labeled with m. For instance, Figure 25.5 shows the adjacency graph of the PCR for $n = 3$, from which one can see the cycle [**001**] shares one conjugate pair $(100, 000)$ with the cycle [**0**] and two

FIGURE 25.5: The adjacency graph of the pure cycling register (PCR) for $n = 3$

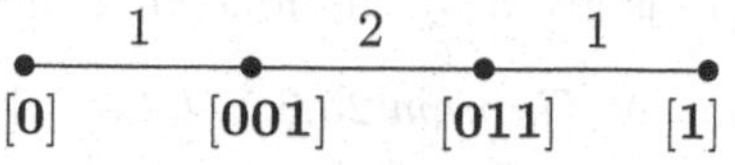

$$
\begin{array}{ccccccc}
 & 1 & & 2 & & 1 & \\
\bullet & \!\!\!\!\!\!\!\!\rule[0.5ex]{2cm}{0.4pt}\!\!\!\!\!\!\!\! & \bullet & \!\!\!\!\!\!\!\!\rule[0.5ex]{2cm}{0.4pt}\!\!\!\!\!\!\!\! & \bullet & \!\!\!\!\!\!\!\!\rule[0.5ex]{2cm}{0.4pt}\!\!\!\!\!\!\!\! & \bullet \\
[0] & & [001] & & [011] & & [1]
\end{array}
$$

conjugate pairs $(001, 101)$, $(010, 110)$ with the cycle $[011]$, and the cycle $[011]$ shares one conjugate pair $(011, 111)$ with the cycle $[1]$.

For an FSR with feedback function f, the cycle joining and splitting methods change the cycle structure of $\Omega(f)$, which results in changes to the values of f at the conjugate pairs accordingly. When the successors of a conjugate pair $a_0 a_1 \cdots a_{n-1}$ and $\bar{a}_0 a_1 \cdots a_{n-1}$ are swapped in its state graph, the resulting feedback function is given by

$$
f'(x_0, x_1, \ldots, x_{n-1}) = x_0 + g(x_1, \ldots, x_{n-1}) + \prod_{i=1}^{n-1} (x_i + a_i + 1)
$$

$$
= x_0 + g'(x_1, \ldots, x_{n-1}),
$$

where the function g' takes the same value as g at all $(n-1)$-tuples except $(a_1, \ldots, a_{n-1})$. In other words, each cycle joining or cycle splitting has a one-to-one correspondence to the change of $g(x_1, \ldots, x_{n-1})$ at one position, causing the Hamming weight of the truth table of g to change by one. It can be shown that the parity of the number of cycles is the same as the parity of the weight of the truth table of $g(x_1, \ldots, x_{n-1})$ [755].

25.3.4 Cycle Structure of LFSRs

For the feedback function $f(x_0, x_1, \ldots, x_{n-1}) = \sum_{i=0}^{n-1} c_i x_i$, where $c_0 = 1$, of an LFSR, its associated characteristic polynomial is given by

$$
f(x) = x^n + c_{n-1} x^{n-1} + \cdots + c_1 x + c_0.
$$

Here we use the same notation f for the associated polynomial and the feedback function for simplicity. Let $f^*(x) = x^n f(\frac{1}{x})$ denote the reciprocal polynomial of $f(x)$. Below we present some important results of LFSRs from [1642] together with auxiliary examples.

Theorem 25.3.1 *Let* $\mathbf{s} = \{s_t\}$ *be a sequence in* $\Omega(f)$ *and* $G(x) = \sum_{i=0}^{\infty} s_i x^i$ *be its generating function. Then*

$$
G(x) f^*(x) = \varphi^*(x),
$$

where

$$
\varphi(x) = s_0 x^{n-1} + (s_1 + c_{n-1} s_0) x^{n-2} + \cdots + (s_{n-1} + c_{n-1} s_{n-2} + \cdots + c_1 s_0).
$$

From the above theorem, all sequences $\mathbf{s}$ in $\Omega(f)$ can be given by $\varphi^*(x)/f^*(x)$ for some polynomial $\varphi^*(x)$ with $\deg(\varphi^*(x)) < n$.

Example 25.3.2 The sequence 0010111 of period 7 is generated by $f(x) = x^3 + x + 1$. The generating function of the sequence is

$$
G(x) = (x^2 + x^4 + x^5 + x^6) + (x^9 + x^{11} + x^{12} + x^{13}) + \cdots
$$

In this case we find $\varphi(x) = 1$ and its reciprocal to be $\varphi^*(x) = x^2$. The reciprocal of $f(x) = x^3 + x + 1$ is $f^*(x) = x^3 + x^2 + 1 = (x^7 + 1)/(x^4 + x^3 + x^2 + 1)$. Hence,

$$
\frac{\varphi^*(x)}{f^*(x)} = \frac{x^2(1 + x^2 + x^3 + x^4)}{1 + x^7} = (x^2 + x^4 + x^5 + x^6) + (x^9 + x^{11} + x^{12} + x^{13}) + \cdots.
$$

This matches the equality $G(x)f^*(x) = \varphi^*(x)$ in Theorem 25.3.1.

The period of a sequence in $\Omega(f)$ is given by the following theorem.

Theorem 25.3.3 *With the notation of Theorem 25.3.1, let* $\mathbf{s} = \{s_t\}$ *have period ϵ. Then*

$$(x^\epsilon - 1)\varphi(x) = \sigma(x)f(x),$$

where $\sigma(x) = s_0 x^{\epsilon-1} + s_1 x^{\epsilon-2} + \cdots + s_{\epsilon-1}$.

The identity in Theorem 25.3.3 is the key to studying the periodic properties of LFSRs. To study the periods of the sequences in $\Omega(f)$, it is very useful to define the period of the characteristic polynomial $f(x)$. Let $f(x)$ be a polynomial with nonzero constant term. The **period** of a polynomial $f(x)$, denoted by per(f), is defined as the smallest positive integer e such that $f(x) \mid (x^e - 1)$. From the key identity in Theorem 25.3.3, the period of a sequence $\mathbf{s}$ in $\Omega(f)$ and the period of $f(x)$ are closely connected.

Theorem 25.3.4 *Let* $\mathbf{s} = \{s_t\}$ *be a nonzero sequence in* $\Omega(f)$. *Then the period of* $\mathbf{s}$ *divides the period of the characteristic polynomial* $f(x)$. *Moreover, if* $f(x)$ *is irreducible, then the period of* $\mathbf{s}$ *equals the period of* $f(x)$.

Theorem 25.3.4 explains the essential reason that m-sequences of order n are generated from primitive polynomials of degree n, which by definition have period $2^n - 1$.

Example 25.3.5 Let $n = 4$. Below we choose three different polynomials of degree 4 that illustrate the result of Theorem 25.3.4.

- We first take a polynomial $f(x) = x^4 + x^3 + x^2 + 1 = (x+1)(x^3 + x + 1)$. It is easily seen (from Example 25.3.2) that $f(x)$ divides $x^7 - 1$ and $f(x)$ actually has per$(f) = 7$. Moreover, the cycle structure of $\Omega(f)$ is given by

$$\Omega(f) = [0] \cup [1] \cup [0010111] \cup [1101000].$$

  The length of each cycle in $\Omega(f)$ is a divisor of per(f).

- We then take a polynomial $f(x) = x^4 + x^3 + x^2 + x + 1$. It can be verified that $f(x)$ is irreducible and has period 5. In this case, the cycle structure of $\Omega(f)$ is given by

$$\Omega(f) = [0] \cup [00011] \cup [00101] \cup [01111].$$

  All the above nonzero cycles have the same length 5.

- Lastly, we take $f(x) = x^4 + x + 1$. It can be verified that $f(x)$ is primitive and has period 15. In this case, the cycle structure of $\Omega(f)$ is given by

$$\Omega(f) = [0] \cup [000100110101111].$$

  The nonzero cycle has length 15, which is the same as per(f).

By Theorems 25.3.3 and 25.3.4, we can rewrite $\Omega(f)$ with per$(f) = e$ as follows:

$$\Omega(f) = \left\{ \frac{x^e - 1}{f(x)} \varphi(x) \;\middle|\; \deg(\varphi(x)) < n \right\}$$

since all sequences in $\Omega(f)$ repeat with period e. Therefore, for two polynomials $g(x)$ and $h(x)$, we have

$$\Omega(g) \cap \Omega(h) = \Omega(\gcd(g,h)) \quad \text{and} \quad \Omega(g) + \Omega(h) = \Omega(\mathrm{lcm}(g,h)).$$

In particular, when $\gcd(g(x), h(x)) = 1$,

$$\Omega(gh) = \Omega(g) \oplus \Omega(h) = \{\mathbf{u} + \mathbf{v} \mid \mathbf{u} \in \Omega(g), \mathbf{v} \in \Omega(h)\},$$

and

$$\text{per}(\mathbf{u} + \mathbf{v}) = \text{lcm}(\text{per}(\mathbf{u}), \text{per}(\mathbf{v})).$$

Thus, the full cycle structure of $\Omega(gh)$ for co-prime polynomials g and h can be determined in a relatively easy way.

Example 25.3.6 Take $g(x) = x^2 + x + 1$ and $h(x) = x^3 + x + 1$. We have $\Omega(g) = [\mathbf{0}] \cup [\mathbf{011}]$ and $\Omega(h) = [\mathbf{0}] \cup [\mathbf{0010111}]$. The set $\Omega(gh)$ can be given as follows:

$$\Omega(gh) = \Omega(g) \oplus \Omega(h) = \{\mathbf{u} + \mathbf{v} \mid \mathbf{u} \in \Omega(g), \mathbf{v} \in \Omega(h)\}$$
$$= [\mathbf{0}] \cup [\mathbf{011}] \cup [\mathbf{0010111}] \cup [\mathbf{010000111110101001100}],$$

where the sum of any cycle $[\mathbf{u}]$ and $[\mathbf{0}]$ is the cycle $[\mathbf{u}]$ itself, and the last cycle $[\mathbf{010000111110101001100}]$ is the sum of $[\mathbf{011}]$ and $[\mathbf{0010111}]$ formed by adding 7 replications of $[\mathbf{011}]$ and 3 replications of $[\mathbf{0010111}]$.

Theorem 25.3.7 *Let* $\gcd(g, h) = 1$. *Let* $\Omega(g)$ *contain* d_1 *cycles of length* λ_1 *denoted as* $[d_1(\lambda_1)]$, *and let* $\Omega(h)$ *contain* d_2 *cycles of length* λ_2 *denoted as* $[d_2(\lambda_2)]$. *Combine by adding in all possible ways the corresponding sequences. This leads to* $d_1 \lambda_1 d_2 \lambda_2$ *sequences, all of period* $\text{lcm}(\lambda_1, \lambda_2)$. *The number of cycles obtained in this way is* $d_1 d_2 \gcd(\lambda_1, \lambda_2)$.

Note that both $\Omega(g)$ and $\Omega(h)$ may contain many cycles. By Theorem 25.3.7 we can repeatedly combine all sequences from $\Omega(g)$ and $\Omega(h)$.

Example 25.3.8 Let $f(x) = (x^2 + x + 1)(x + 1)^2$ and define $g(x) = x^2 + x + 1$ and $h(x) = (x + 1)^2$. Then the cycle structure of $\Omega(g)$ is given by $[\mathbf{0}] \cup [\mathbf{011}]$, denoted as $[1(1), 1(3)]$, and the cycle structure of $\Omega(h)$ is $[\mathbf{0}] \cup [\mathbf{1}] \cup [\mathbf{01}]$, denoted as $[2(1), 1(2)]$. The cycle structure of $\Omega(f)$ is therefore obtained by combining the two cycle structures. By Theorem 25.3.7, the cycle structure of $\Omega(f)$ has the form

$$[2(1), 1(2), 2(3), 1(6)].$$

More specifically, the cycles structure of $\Omega(f)$ is given by

$$\Omega(f) = [\mathbf{0}] \cup [\mathbf{1}] \cup [\mathbf{01}] \cup [\mathbf{011}] \cup [\mathbf{001}] \cup [\mathbf{000111}].$$

The following result determines the cycle structure of polynomials having the form $f(x)^k$, where $f(x)$ is an irreducible polynomial.

Theorem 25.3.9 *Let* $f(x)$ *be an irreducible polynomial of degree* n *and period* e. *For a positive integer* k, *determine* κ *such that* $2^\kappa < k \leq 2^{\kappa+1}$. *Then* $\Omega(f^k)$ *contains the following number of sequences with the following periods:*

Number	1	$2^n - 1$	$2^{2n} - 2^n$	$\cdots$	$2^{2^\kappa n} - 2^{2^{\kappa-1}n}$	$2^{kn} - 2^{2^\kappa n}$
Period	1	e	$2e$	$\cdots$	$2^\kappa e$	$2^{\kappa+1} e$

Example 25.3.10 Let $f(x) = x^2 + x + 1$, i.e., $n = 2$, $e = 3$. Let $[d(\lambda)]$ denote that there are d cycles of period λ in $\Omega(f)$. Then Theorem 25.3.9 gives the cycle structure of $\Omega(f^2)$ as $[1(1), 1(3), 2(6)]$. In fact, the cycle structure of $\Omega(f^2)$ is

$$\Omega(f^2) = [\mathbf{0}] \cup [\mathbf{011}] \cup [\mathbf{000101}] \cup [\mathbf{001111}].$$

For a general polynomial $f(x) = \prod_{i=1}^{r} f_i(x)^{k_i}$ with $f_i(x)$ irreducible, we can obtain the cycle structure of $\Omega(f)$ by repeatedly applying Theorem 25.3.7 and Theorem 25.3.9. We provide a simple example.

Example 25.3.11 Let $f(x) = (x^2 + x + 1)^2(x + 1)^2$. The cycle structure of $(x^2 + x + 1)^2$ is $[1(1), 1(3), 2(6)]$ and the cycle structure of $(x + 1)^2$ is $[2(1), 1(2)]$. Hence, by Theorem 25.3.7, the cycle structure of $\Omega(f)$ is given by $[2(1), 1(2), 2(3), 7(6)]$. More specifically, the cycle structure of $\Omega(f)$ is

$$\Omega(f) = [0] \cup [1] \cup [01] \cup [001] \cup [011] \cup [000101] \cup [001111] \cup [010111]$$
$$\cup [000011] \cup [000111] \cup [000001] \cup [001101].$$

25.4 Generation of De Bruijn Sequences

De Bruijn sequences have interesting combinatorial and random properties. People are particularly interested in algorithms and methods to generate de Bruijn sequences for a given order. In his survey paper [755], Fredricksen alluded to an interesting phenomenon: "When a mathematician on the street is presented with the problem of generating a de Bruijn sequence, one of three things happens. He gives up, or produces a sequence based on a primitive polynomial, or produces the sequence by the 'prefer one' algorithm. Only rarely is a new algorithm proposed." Fredricksen's description roughly summarizes the known simple approaches to generating de Bruijn sequences, and also indicates the intractability of generating de Bruijn sequences. A variety of approaches from combinatorics, graph theory, and abstract algebra were proposed for the problem in the past decades. They have advantages and disadvantages in terms of performance, time and/or memory cost, and complexity of the approaches themselves. This subsection will introduce some of those approaches proposed in the literature.

25.4.1 Graphical Approach

Recall that the de Bruijn graph G_n is a directed graph composed of 2^n vertices and 2^{n+1} edges that leave every vertex $x_0 x_1 \cdots x_{n-1}$ and enter the two vertices $x_1 \cdots x_{n-1}0$ and $x_1 \cdots x_{n-1}1$. It is the superposition of all state graphs of n-stage FSRs. The existence of binary de Bruijn sequences can be easily confirmed by simple graphical methods.

In graph theory, an **Eulerian path** in a finite graph G is a path that goes through all of the edges of G exactly once (allowing revisits to vertices); a **Hamiltonian path** in a graph G is a path that traverses all of the vertices of G exactly once. It is easily seen that a Hamiltonian path of the de Bruijn graph G_n corresponds to a de Bruijn sequence of order n. Let's now take a closer look at the de Bruijn graph G_n. It consists of 2^n vertices, of which each has in-degree and out-degree 2. If we label the 2^{n+1} edges of G_n with the $(n + 1)$-tuple corresponding to the pair of n-tuples on the ends of edges, then G_{n+1} can be made by interpreting the edges of G_n as vertices of G_{n+1} and connecting two vertices of G_{n+1} if their corresponding edges enter and leave the same node in G_n. Hence a Eulerian path in G_n defines a Hamiltonian path in G_{n+1}. It is known that a Eulerian path exists for a directed graph G if and only if G is connected and the number of edges leaving a vertex v equals the number of edges entering v. Such a condition is clearly met by the de Bruijn graph G_n for any positive integer n. Since the cycle (01) is a Hamiltonian path through G_1, there exist Hamiltonian paths through G_n for all positive integers n. Moreover, it can be shown [496] that if the de Bruijn graph G_n has N complete Hamiltonian cycles, then there exist $2^{2^{n-1}-1} \cdot N$ complete Hamiltonian cycles in G_{n+1}. Starting from the unique Hamiltonian path through the de Bruijn graph G_1, by induction one can easily prove the fact that there are a total of $2^{2^{n-1}-n}$ de Bruijn sequences of order n.

The above result was proved by a method known as the *BEST Theorem* due to de Bruijn, Ehrenfest, Smith, and Tutte [496]. For a given graph G without loops, the BEST Theorem determines the number of subgraphs of G that are rooted trees. In graph theory, a **rooted tree** is a connected graph without cycles, and there exists a vertex that is not the initial vertex of any edge. The number of rooted trees of a graph without loops is determined by the following BEST Theorem.

Theorem 25.4.1 (BEST [496]) *Given a graph G of t nodes without loops, let A be its associated matrix in which $A_{i,i}$ equals the degree of vertex i, and $A_{i,j}$ denotes (-1) times the number of edges between vertex i and vertex j. Then the number of rooted trees of G is equal to the minor of any element of A.*

Denote by G_n^* the modified de Bruijn graph, which is obtained by dropping the cycles at all-zero and all-one vertices. We shall apply the BEST Theorem to a directed graph G_n^* with certain adjustments: we use $A_{i,i}$ to denote the number of edges exiting the vertex numbered i and $A_{i,j}$ to denote (-1) times the number of edges from vertex i to vertex j. For the directed graph G_n^*, denote any integer $\mathbf{x} \in \{0, 1, \ldots, 2^n - 1\}$ by a binary n-tuple $x_0 x_1 \cdots x_{n-1}$. It is relatively easy to verify that the associated matrix of G_n^* has the form:

- $A_{0,0} = A_{2^n-1,2^n-1} = 1$;
- $A_{0,1} = A_{2^n-1,2^n-2} = -1$;
- $A_{i,i} = 2$ for $i = 1, 2, \ldots, 2^n - 2$;
- $A_{i,j} = -1$ for $i = 1, 2, \ldots, 2^n - 2$, $j = 2i, 2i+1$ modulo 2^n;
- $A_{i,j} = 0$ otherwise.

From the structure of the associated matrix of G_n^*, it can be shown by induction that there are $2^{2^n-(n+1)}$ trees with root 0 by evaluating the determinant of the minor of the matrix A at $(0, 0)$. Furthermore, each directed rooted tree of G_n^* at vertex 0 can be used to find a Hamiltonian path of G_{n+1} by the following Graphical Algorithm.

Algorithm 25.4.2 (Graphical)

Step 1: Add the loop at vertex 0 to any given tree rooted at 0 in G_n.

Step 2: Place a label (i, j) on every edge from vertex i to vertex j of the augmented tree in G_n.

Step 3: Choose vertex 0 as the starting vertex.

Step 4: When at vertex i, leave the vertex by the unlabeled edge if it has not been used; otherwise, leave the vertex by the labeled edge.

Step 5: Denote the edges traversed in the process as the state of G_{n+1} and connect two edges if they enter and leave the same node.

Step 6: When we enter a vertex, if both edges out of that vertex have been used, terminate the algorithm. The sequence of states visited is a Hamiltonian path through G_{n+1}.

In is shown in [757] that the Graphical Algorithm can generate all de Bruijn sequences of order n. As an illustration, we recall an example from [757].

Example 25.4.3 Let T be a rooted tree in G_3^* given by vertices $0, 1, \ldots, 7$ and edges

$$(1,2), (2,4), (3,6), (4,0), (5,2), (6,4), (7,6),$$

where the label (i, j) denotes an edge from vertex i to vertex j. These edges together with $(0, 0)$ will be labeled in Step 2. Thus, the unlabeled edges in G_3 are

$$(0,1), (1,3), (2,5), (3,7), (4,1), (5,3), (6,5), (7,7).$$

Starting from vertex 0, with $(0, 0)$ labeled, $(0, 1)$ satisfies Step 4. Following the steps in the Graphical Algorithm, we get a sequence of edges

$$0 \to 1 \to 3 \to 7 \to 7 \to 6 \to 5 \to 3 \to 6 \to 4 \to 1 \to 2 \to 5 \to 2 \to 4 \to 0 \to 0.$$

This corresponds to a Hamiltonian path of G_4:

$$0001, 0011, 0111, 1111, 1110, 1101, 1011, 0110,$$
$$1100, 1001, 0010, 0101, 1010, 0100, 1000, 0000.$$

In a similar way, all Hamiltonian paths of G_4 can be generated by all trees rooted at vertex 0 in G_3^*.

The preceding process using a graphical approach can generate all de Bruijn sequences of order n. However, we have to obtain all trees in G_{n-1}. Another drawback of this approach is that we need to store the tree and generated path in some way. These drawbacks can be overcome by some approaches from a combinatorial or algebraic perspective.

25.4.2   Combinatorial Approach

Recall that in a de Bruijn sequence of order n, every n-tuple binary string occurs exactly once. The main idea of a combinatorial approach is to generate a de Bruijn sequence one symbol at a time based on certain successor rules, which can be as simple as preferring one and checking that the newly formed n-tuple has not previously appeared. These approaches mostly generate one or a few de Bruijn sequences for a given order. Below we introduce known algorithms with simple successor rules, followed by an example in the case $n = 5$ to demonstrate those algorithms. Recall that $\overline{x}_i = x_i \oplus 1$ is the binary complement of x_i.

There are three well-known greedy approaches: the prefer-one [1333], the prefer-same [670], and the prefer-opposite [37]. Each of them has a straightforward successor rule, but they all have the drawback of requiring an exponential amount of memory.

Algorithm 25.4.4 (Prefer-One [1333])

Step 1: Start with the all-zero n-tuple $x_0 \cdots x_{n-1} = 0 \cdots 0$.

Step 2: For an integer $t \geq 0$, if the newly formed n-tuple $x_{t+1} \cdots x_{t+n}$ with $x_{t+n} = 1$ has not previously appeared in the sequence, then take $x_{t+n} = 1$, increment t by one and repeat Step 2; otherwise, take $x_{t+n} = 0$ and repeat Step 2.

Step 3: When Step 2 can no longer be repeated to produce new n-tuples, remove the last $n - 1$ symbols of the sequence generated by Step 2.

Steps 1 - 2 of the Prefer-One Algorithm ignore the wraparound sequences. When Step 2 terminates, it outputs a de Bruijn sequence appended with $(n-1)$ extra bits. However, Step 2 is sufficient to produce a de Bruijn sequence. This elegant algorithm, as noted in Fredricksen's survey [755], was proposed from different areas several times.

Algorithm 25.4.5 (Prefer-Same [670])

Step 1: Write n zeros followed by n ones to form $x_0 \cdots x_{n-1} x_n \cdots x_{2n-1} = 0 \cdots 0 1 \cdots 1$.

Step 2: For an integer $t \geq n$, if $x_{t+n-1} = v$ has been placed and the following two conditions

  (i) the n-tuple $x_{t+1} \cdots x_{t+n}$ with $x_{t+n} = x_{t+n-1} = v$ has not previously appeared; and

  (ii) if placing $x_{t+n} = v$ makes a run $\overbrace{v \cdots v}^{i}$, which means a string of i consecutive v's with both the preceding bit (if any) and succeeding bit (if any) of the string different from v, then we have not generated 2^{n-2-i} runs of i consecutive v's in the sequence for $1 \leq i \leq n-2$,

  are not violated, then take $x_{t+n} = x_{t+n-1} = v$, increment t by one and repeat Step 2.

Step 3: If either Condition (i) or (ii) is violated, take $x_{t+n} = \overline{x}_{t+n-1}$ as long as it does not violate either condition (i) or (ii), increase t by one and continue with Step 2. If we cannot place $x_t = \overline{x}_{t+n-1}$ either, the algorithm terminates.

In the Prefer-Same Algorithm, the condition Step 2(ii) on the run length is a necessary requirement for producing de Bruijn sequences.

Algorithm 25.4.6 (Prefer-Opposite [37])

Step 1: Start with the all-zero n-tuple $x_0 \cdots x_{n-1} = 0 \cdots 0$.

Step 2: For an integer $t \geq 0$, if the newly formed n-tuple $x_{t+1} \cdots x_{t+n}$ with $x_{t+n} = \overline{x}_{t+n-1}$ has not previously appeared in the sequence, then take $\overline{x}_{t+n} = x_{t+n-1}$, increment t by one and repeat Step 2. Otherwise, take $x_{t+n} = x_{t+n-1}$, increment t by one. If $x_{t+1} \cdots x_{t+n-1} \neq \overbrace{1 \cdots 1}^{n-1}$, repeat Step 2; otherwise, go to Step 3;

Step 3: Take $x_{t+n} = 1$ and exit.

In the Prefer-Opposite Algorithm, repeating Step 2 produces a periodic sequence of length $2^n - 1$ and the all-one $(n-1)$-tuple will occur at the end of the sequence. Step 3 is therefore needed to detect this pattern and terminate the process.

The reader may notice that the successor rules in the above algorithms are surprisingly simple. However, in each of these algorithms, the previously generated sequence needs to be stored for use in checking whether the preferred value can be taken or not, leading to exponential memory cost. With the property of necklaces or co-necklaces, some combinatorial approaches with less memory cost than the greedy algorithms were proposed [756, 757, 1557].

A **Lyndon word** is the lexicographically smallest string among all its circular shifts. A necklace of a string $\mathbf{x}$, denoted by $\text{Neck}(\mathbf{x})$, is the set consisting of $\mathbf{x}$ along with all its circular shifts. The necklace representative of $\text{Neck}(\mathbf{x})$ is the unique Lyndon word in $\text{Neck}(\mathbf{x})$. We say $\mathbf{x}$ is a **co-Lyndon word** if $\mathbf{x}\overline{\mathbf{x}}$ is a Lyndon word. Let $\text{coNeck}(\mathbf{x})$ denote the set containing all length-n substrings of the circular string $\mathbf{x}\overline{\mathbf{x}}$. The representative of $\text{coNeck}(\mathbf{x})$ is the lexicographically smallest string in $\text{coNeck}(\mathbf{x})$. For instance, let $\mathbf{x} = 01010$; we have $\text{Neck}(01010) = \{01010, 10100, 01001, 10010, 00101\}$ and its representative is 00101; $\text{coNeck}(00111) = \{00111, 01111, 11111, 11110, 11100, 11000, 10000, 00000, 00001, 00011\}$. The co-necklace representative is 00000, and it is a co-Lyndon word.

The Lyndon word and co-Lyndon word are closely related to the pure cycling register (PCR) and its complement. Recall that the feedback function of the PCR is given by $f(x_0, x_1, \ldots, x_{n-1}) = x_0$. Its complement, called the **complement cycling register** (**CCR**), has feedback function $f_c(x_0, x_1, \ldots, x_{n-1}) = \overline{x}_0$. It can be easily verified that the state graph of the PCR actually consists of necklaces and the state graph of the CCR consists of co-necklaces [781]. For instance, for the PCR with $n = 3$, the cycles $[0], [1], [001], [011]$ as in Figure 25.2 are necklaces; for the CCR with $n = 3$, the cycle $[000111]$ forms a co-necklace $\{000, 001, 011, 111, 110, 100\}$ and the cycle $[010]$ is a co-necklace $\{010, 101\}$.

We introduce some combinatorial algorithms for generating de Bruijn sequences by investigating Lyndon or co-Lyndon words. We start with an efficient necklace concatenation algorithm, and then present several follow-up algorithms by setting rules on Lyndon words.

Algorithm 25.4.7 (Granddaddy [754])

Step 1: Start with the all-zero n-tuple $x_0 \cdots x_{n-1} = 0 \cdots 0$.

Step 2: For the tth n-tuple $\mathbf{x} = x_t \cdots x_{t+n-1}$, $t \geq 0$, denote by j the smallest index such that $x_j = 0$ and $j > t + 1$, or let $j = t + n$ if no such index exists.

Step 3: Let $\mathbf{x}^* = x_j \cdots x_{t+n-1} 0 x_{t+1} \cdots x_{j-1} = x_j \cdots x_{t+n-1} 01 \cdots 1$. If $\mathbf{x}^*$ is a Lyndon word, take $x_{t+n} = \overline{x}_t$; otherwise, take $x_{t+n} = x_t$.

The name *Granddaddy* for the above algorithm originated in [1132], where Knuth referred to the lexicographically least de Bruijn sequence for each positive integer n as the *granddaddy*. Different from the preceding greedy algorithms, the Granddaddy Algorithm does not require storing the previously generated sequence. Instead, it only requires checking whether the n-tuple $\mathbf{x}^*$ is a Lyndon word or not, hence reducing memory cost to $O(n)$. The reverse version was recently proposed in [642], where the authors termed it the Grandma Algorithm corresponding to the playful name Granddaddy Algorithm.

Algorithm 25.4.8 (Grandma [642])

Step 1: Start with the all-zero n-tuple $x_0 \cdots x_{n-1} = 0 \cdots 0$.

Step 2: For the tth n-tuple $\mathbf{x} = x_t \cdots x_{t+n-1}$, $t \geq 0$, let j be the largest index such that $x_j = 1$, or let $j = t - 1$ if no such index exists.

Step 3: Let $\mathbf{x}^* = x_{j+1} \cdots x_{t+n-1} 1 x_{t+1} \cdots x_j = 0 \cdots 01 x_{t+1} \cdots x_j$. If $\mathbf{x}^*$ is a Lyndon word, take $x_{t+n} = \overline{x}_t$; otherwise, take $x_{t+n} = x_t$.

From the above algorithms, one may already notice that the process essentially provides ways of concatenating the necklaces in the state graph of PCRs. In fact, we can have even simpler algorithms for creating possible Lyndon words [1036, 1625].

Algorithm 25.4.9 (PCR 1 [1036, 1625])

Step 1: Start with the all-zero n-tuple $x_0 \cdots x_{n-1} = 0 \cdots 0$.

Step 2: For the tth n-tuple $\mathbf{x} = x_t \cdots x_{t+n-1}$, $t \geq 0$, if $\mathbf{x}^* = x_{t+1} \cdots x_{t+n-1}1$ is a Lyndon word, take $x_{t+n} = \overline{x}_t$; otherwise, take $x_{t+n} = x_t$.

Algorithm 25.4.10 (PCR 2 [1625])

Step 1: Start with the all-zero n-tuple $x_0 \cdots x_{n-1} = 0 \cdots 0$.

Step 2: For the tth n-tuple $\mathbf{x} = x_t \cdots x_{t+n-1}$, $t \geq 0$, if $\mathbf{x}^* = 0x_{t+1} \cdots x_{t+n-1}$ is a Lyndon word, take $x_{t+n} = \overline{x}_t$; otherwise, take $x_{t+n} = x_t$.

Since the necklace Neck($\mathbf{x}$) has a unique Lyndon word, the above two algorithms essentially generate PCRs with the successor rule $x_{t+n} = x_t$ and join the cycles by swapping the successors of that Lyndon word with $x_{t+n} = \overline{x}_t$. The preceding approaches that work with PCRs can be applied to CCRs with moderate modifications [781].

Algorithm 25.4.11 (CCR 1 [781])

Step 1: Start with the all-zero n-tuple $x_0 \cdots x_{n-1} = 0 \cdots 0$.

Step 2: For the tth n-tuple $\mathbf{x} = x_t \cdots x_{t+n-1}$, $t \geq 0$, denote by j the smallest index such that $x_j = 0$ and $j > t$, or $j = t + n$ if no such index exists.

Step 3: Let $\mathbf{x}^* = x_j \cdots x_{t+n-1}1\overline{x}_{t+1} \cdots \overline{x}_{j-1} = x_j \cdots x_{t+n-1}10 \cdots 0$. If $\mathbf{x}^*$ is a co-Lyndon word, take $x_{t+n} = x_t$; otherwise, take $x_{t+n} = \overline{x}_t$.

Algorithm 25.4.12 (CCR 2 [781])

Step 1: Start with the all-zero n-tuple $x_0 \cdots x_{n-1} = 0 \cdots 0$.

Step 2: For the tth n-tuple $\mathbf{x} = x_t \cdots x_{t+n-1}$, $t \geq 0$, let j be the largest index such that $x_j = 1$, or $j = t - 1$ if no such index exists.

Step 3: Let $\mathbf{x}^* = x_{j+1} \cdots x_{t+n-1}1\overline{x}_{t+1} \cdots \overline{x}_j = 0 \cdots 01\overline{x}_{t+1} \cdots \overline{x}_j$. If $\mathbf{x}^*$ is a co-Lyndon word, take $x_{t+n} = x_t$; otherwise, take $x_{t+n} = \overline{x}_t$.

Algorithm 25.4.13 (CCR 3 [781, 1036])

Step 1: Start with the all-zero n-tuple $x_0 \cdots x_{n-1} = 0 \cdots 0$.

Step 2: For the tth n-tuple $\mathbf{x} = x_t \cdots x_{t+n-1}$, $t \geq 0$, if $\mathbf{x}^* = x_{t+1} \cdots x_{t+n-1}0$ is a co-Lyndon word and $\mathbf{x}^* \neq 0 \cdots 0$, take $x_{t+n} = x_t$; otherwise, take $x_{t+n} = \overline{x}_t$.

The proofs of the preceding algorithms are given in the corresponding references [37, 641, 754, 755, 1036, 1333], and PCR1 and CCR3, despite different presentations, are essentially the same as the ones in [1036]. Observe that these algorithms heavily depend on the simple cycle structures of PCRs and CCRs. They have the advantages of simplicity and low memory cost. Nevertheless, they only generate a few de Bruijn sequences. From an algebraic point of view, they are among the simplest LFSRs with manageable cycle structures. After the following example we will introduce some other LFSRs that have been used to generate de Bruijn sequences.

Example 25.4.14 Table 25.1 illustrates the process of these algorithms for $n = 5$ starting with all-zero string 00000. Note that in this example, the algorithms CCR1 and CCR2 produce the same de Bruijn sequence. This is not true for $n > 5$. For instance, if $n = 6$, the algorithms CCR1 and CCR2 generate the following de Bruijn sequences, respectively:

0000001111110001101110011001010110101001000100111011000010111101;
0000001111110001001110110011000010111101001010110101000110111001.

TABLE 25.1: De Bruijn sequences of order 5 from the combinatorial algorithms

Algorithm	De Bruijn sequence
Prefer-One [1333]	00000111110111001101011000101001
Prefer-Same [670]	00000111110001001100101110110101
Prefer-Opposite [37]	00000101011010010001100111011111
Granddaddy [754]	00000100011001010011101011011111
Grandma [642]	00000100101000110101100111011111
PCR 1 [1036]	00000111110111001100010110101001
PCR 2 [781]	00000111110110101001011100110001
CCR 1 [781]	00000111110010011011000101110101
CCR 2 [781]	00000111110010011011000101110101
CCR 3 [641]	00000100110110010101110100011111

25.4.3 Algebraic Approach

Compared with NLFSRs, the mathematical properties of LFSRs are fairly well understood. Thanks to their combinatorial structure, LFSRs are also important components for constructing de Bruijn sequences. Recall that m-sequences are the sequences of maximum length from primitive LFSRs. The state graph of an n-stage primitive LFSR consists of two cycles: the all-zero cycle and the cycle of m-sequences. One can simply join these two cycles by swapping the successors of the conjugate pair $0 \cdots 0$ and $10 \cdots 0$. Hence we can obtain de Bruijn sequences from all m-sequences of order n, which accounts for $\phi(2^n - 1)/n$ de Bruijn sequences. Primitive LFSRs only take up a small portion of all possible LFSRs. The reader may raise a natural question here. Can de Bruijn sequences be obtained from other non-primitive LFSRs? The short answer to the question is "yes" but more complicated. Here we introduce some work on constructing de Bruijn sequences from LFSRs by the cycle joining approach.

FIGURE 25.6: The adjacency graph of LFSRs from a product of primitive polynomials $p_1(x), p_2(x)$ that have relatively co-prime degrees

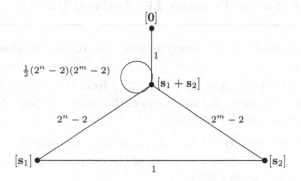

De Bruijn Sequences from LFSRs

As mentioned in the preceding subsection, the simplest way to construct de Bruijn sequences from LFSRs is to use m-sequences. More specifically, given a primitive LFSR with feedback function $f(x_0, \ldots, x_{n-1})$, one can derive a de Bruijn sequence of order n by simply appending 0 to the $(n-1)$-tuple of all zeros. In this way, the new feedback function is given by

$$f'(x) = f(x_0, \ldots, x_{n-1}) + \prod_{i=1}^{n-1} (x_i + 1).$$

Beyond this simplest case, a natural step would be to consider FSRs with characteristic polynomial the product of two primitive polynomials.

Proposition 25.4.15 ([1240, 1244]) *Let* $p_1(x) = x^n + \sum_{i=0}^{n-1} c_i x^i$ *and* $p_2(x) = x^m + \sum_{j=0}^{m-1} d_j x^j$ *be two primitive polynomials and* $\gcd(n, m) = 1$. *The cycle structure of an* $(n+m)$-*stage LFSR with characteristic polynomial* $f(x) = p_1(x)p_2(x)$ *is given by*

$$\Omega(f) = [\mathbf{0}] \cup [\mathbf{s_1}] \cup [\mathbf{s_2}] \cup [\mathbf{s_1} + \mathbf{s_2}],$$

where $\mathbf{s_1}$ *is the* m-*sequence of length* $2^n - 1$ *corresponding to* $p_1(x)$, $\mathbf{s_2}$ *is the* m-*sequence of length* $2^m - 1$ *corresponding to* $p_2(x)$, *and the sequence* $\mathbf{s_1} + \mathbf{s_2}$ *of length* $(2^n-1)(2^m-1)$ *is the sum of the sequences* $\mathbf{s_1}$ *and* $\mathbf{s_2}$. *Moreover, the adjacency graph* $G(f)$ *can be characterized as follows:*

(a) *The cycle* $[\mathbf{s_1} + \mathbf{s_2}]$ *has one conjugate pair with* $[\mathbf{0}]$, $2^n - 2$ *conjugate pairs with* $[\mathbf{s_1}]$ *and* $2^m - 2$ *conjugate pairs with* $[\mathbf{s_2}]$, *and*

(b) *the cycle* $[\mathbf{s_1}]$ *has one conjugate pair with* $[\mathbf{s_2}]$.

Figure 25.6 displays the adjacency graph of $\Omega(f)$ [1240].

From Proposition 25.4.15, one can derive a full cycle from $\Omega(f)$ using the following algorithm.

Algorithm 25.4.16 (Simple Primitive LFSRs-Based [1244])

Step 1: Swap the successors of the conjugate pair $(0\cdots 0, 10\cdots 0)$ shared by the cycles $[\mathbf{0}]$ and $[\mathbf{s}_1 + \mathbf{s}_2]$.

Step 2: Pick an $(n + m)$-tuple $\mathbf{A} = a_0 a_1 \cdots a_{n+m-1}$ from the cycle $[\mathbf{s}_1]$, with its conjugate $\widehat{\mathbf{A}}$ on $[\mathbf{s}_1 + \mathbf{s}_2]$. Swapping the successors of the conjugate pair $(\mathbf{A}, \widehat{\mathbf{A}})$ joins the cycles $[\mathbf{s}_1]$ and $[\mathbf{s}_1 + \mathbf{s}_2]$ (as in Figure 25.4).

Step 3: Pick an $(n + m)$-tuple $\mathbf{B} = b_0 b_1 \cdots b_{n+m-1}$ from the cycle $[\mathbf{s}_2]$, with its conjugate $\widehat{\mathbf{B}}$ on $[\mathbf{s}_1 + \mathbf{s}_2]$. Swapping the successors of the conjugate pair $(\mathbf{B}, \widehat{\mathbf{B}})$ joins the cycles $[\mathbf{s}_2]$ and $[\mathbf{s}_1 + \mathbf{s}_2]$ (as in Figure 25.4).

Step 4: A new feedback function that generates de Bruijn sequences of order $(n + m)$ is given by

$$f'(x_0, \ldots, x_{n+m-1}) = \sum_{i=0}^{n-1} \sum_{j=0}^{m-1} c_i d_j x_{i+j} + \sum_{\delta \in \Delta} \prod_{i=1}^{n+m-1} (x_i + \delta_i + 1),$$

where c_i, d_j are the coefficients in $p_1(x)$ and $p_2(x)$, respectively, as given in Proposition 25.4.15, and Δ is a set consisting of $\overbrace{0 \cdots 0}^{n+m-1}, a_1 \cdots a_{n+m-1}$, and $b_1 \cdots b_{n+m-1}$.

Example 25.4.17 Let $p_1(x) = x^2 + x + 1$ and $p_2(x) = x^3 + x + 1$. The m-sequences from $p_1(x)$ and $p_2(x)$ are $\mathbf{s}_1 = 011$ and $\mathbf{s}_2 = 0010111$, respectively. For the polynomial $f(x) = p_1(x)p_2(x)$, the cycle structure of $\Omega(f)$ is given by

$$\Omega(f) = [\mathbf{0}] \cup [\mathbf{s}_1] \cup [\mathbf{s}_2] \cup [\mathbf{s}_1 + \mathbf{s}_2],$$

where $\mathbf{s}_1 + \mathbf{s}_2 = 010000111110101001100$. It is easily seen that the cycle $[\mathbf{s}_1 + \mathbf{s}_2]$ shares the conjugate pair $(10000, 00000)$ with $[\mathbf{0}]$. Furthermore, picking one 5-tuple 01101 from the cycle $[\mathbf{s}_1]$, we get the corresponding conjugate state 11101 from $[\mathbf{s}_1 + \mathbf{s}_2]$; picking one 5-tuple 00101 from $[\mathbf{s}_2]$, we get the conjugate state 10101 from $[\mathbf{s}_1 + \mathbf{s}_2]$. From these three pairs, we derive a new feedback function $f'(x_0, \ldots, x_4)$ from the original feedback function $f(x_0, \ldots, x_4) = x_0 + x_4$ as follows

$$f'(x_0, \ldots, x_4) = (x_0 + x_4) + \prod_{i=1}^{4} (x_i + 1) + x_1 x_2 (x_3 + 1) x_4 + (x_1 + 1) x_2 (x_3 + 1) x_4$$

$$= \prod_{i=1}^{4} (x_i + 1) + x_2 (x_3 + 1) x_4 + x_0 + x_4.$$

Given the initial state 00000, an FSR with feedback function f' generates the de Bruijn sequence $00000111110110101110010100110001$.

The previous example shows the process of joining smaller cycles into a full cycle for a relatively simple characteristic polynomial, and shows how the corresponding feedback function is obtained. It is proven in [1239] that when different conjugate pairs are chosen, the resulting de Bruijn sequences will be different. This indicates that the number of de Bruijn sequences obtained in this way can be determined by explicitly counting the number of conjugate pairs among adjacent cycles [1239].

Following the idea from Proposition 25.4.15, we can further consider LFSRs that have the characteristic polynomial as a product of multiple primitive polynomials.

Proposition 25.4.18 ([1240]) *For an n-stage LFSR with characteristic polynomial of the form $f(x) = \prod_{i=0}^{k} p_i(x)$, where $p_i(x)$ is a primitive polynomial of degree d_i such that $d_i < d_j$ and $\gcd(d_i, d_j) = 1$ for $0 \le i < j \le k$ and $i \ne j$, let $\mathbf{s}_i$ be the m-sequence of length $2^{d_i} - 1$ generated by $p_i(x)$. Then the cycle structure of $\Omega(f)$ is given by*

$$\Omega(f) = [\mathbf{0}] \cup \bigcup_{l=1}^{k+1} \bigcup_{\mathcal{I}_l \subseteq \mathbb{Z}_{k+1}} [\mathbf{s}_{i_1} + \mathbf{s}_{i_2} + \cdots + \mathbf{s}_{i_l}],$$

where $\mathcal{I}_l = \{i_1, i_2, \ldots, i_l\}$ and $\mathbb{Z}_{k+1} = \{0, 1, \ldots, k\}$. Furthermore, the adjacency graph $G(f)$ is given as follows.

(a) *The cycle $[\mathbf{0}]$ is adjacent to the cycle $[\mathbf{s}_0 + \mathbf{s}_1 + \cdots + \mathbf{s}_k]$.*

(b) *For any set $\mathcal{I}_l \subseteq \{0, 1, \ldots, k\}$, the cycle $\left[\sum_{j \in \mathcal{I}_l} \mathbf{s}_j\right]$ is adjacent to all cycles $\left[\sum_{j \in M} \mathbf{s}_j\right]$ where M is any set satisfying $\mathcal{I}_l \cup M = \mathbb{Z}_{k+1}$.*

(c) *For any set M given in (b), the cycles $\left[\sum_{j \in \mathcal{I}_l} \mathbf{s}_j\right]$ and $\left[\sum_{j \in M} \mathbf{s}_j\right]$ share only one conjugate pair if $M \cap \mathcal{I}_l = \emptyset$, and exactly $\prod_{i_j \in \mathcal{I}_l \cap M} (2^{d_{i_j}} - 2)$ conjugate pairs otherwise.*

In particular, the cycle $[\mathbf{s}_0 + \mathbf{s}_1 + \cdots + \mathbf{s}_k]$ contains $\frac{1}{2} \prod_{i=0}^{k} (2^{d_i} - 2)$ conjugate pairs.

By the above proposition, an algorithm can be developed to join smaller cycles in $\Omega(f)$ into a full-length cycle [1240]. In particular, when $p_0(x) = 1 + x$, the adjacency graph $G(f)$ for $f(x) = \prod_{i=0}^{k} p_i(x)$ has no loop. In this case, the number of de Bruijn sequences from $f(x)$ can be determined by the BEST Theorem and the theory of cyclotomy [1240].

A further extension for constructing de Bruijn sequences from LFSRs is to consider non-primitive irreducible polynomials or even arbitrary characteristic polynomials. However, the study of LFSRs with more flexible choices of characteristic polynomials also involves much more complex investigation of adjacency graphs and heavier notation. (Readers may already notice the heavy notation and complexity from Proposition 25.4.18 and Theorem 25.3.9.) For a deep understanding of this topic, please refer to [382, 383, 1238, 1239, 1244] for recent work in this direction. In the following, we will introduce a simple case for non-primitive irreducible polynomials followed by an auxiliary example.

Note that a non-primitive irreducible polynomial f of degree n has its period e as a divisor of $2^n - 1$. According to Theorem 25.3.4, all nonzero sequences in $\Omega(f)$ have period e. That is to say, the state diagram has the cycle structure

$$\Omega(f) = [\mathbf{0}] \cup [\mathbf{s}_0] \cup [\mathbf{s}_1] \cup \cdots \cup [\mathbf{s}_{d-1}],$$

where $d = (2^n - 1)/e$ and each cycle $[\mathbf{s}_i]$ has length e.

In order to investigate the conjugate pairs between adjacent cycles in $\Omega(f)$, we need to first determine the states on each cycle. A handy approach is to associate $f(x)$ to a primitive polynomial $q(x)$ as follows. Let α be a primitive element in the finite field $\mathbb{F}_{2^n} = \mathbb{F}_2[x]/(f(x))$. Then $\beta = \alpha^d$ is a root of $f(x)$ in $\mathbb{F}_{2^n}$. The minimal polynomial $q(x)$ of α over

$\mathbb{F}_2$ is defined as the **associated primitive polynomial** of $f(x)$. Since $1, \beta, \ldots, \beta^{n-1}$ is a basis of $\mathbb{F}_{2^n}$ over $\mathbb{F}_2$, any element α^j can be written uniquely as $\alpha^j = \sum_{t=0}^{n-1} a_{j,t}\beta^t$ where $a_{j,t} \in \mathbb{F}_2$. Hauge and Helleseth in [924] introduced a mapping $\phi : \mathbb{F}_{2^n} \to \mathbb{F}_2^n$ as

$$\phi(0) = 0 \cdots 0 \quad \text{and} \quad \phi(\alpha^l) = a_{l,0} a_{l+d,0} \cdots a_{l+(n-1)d,0}.$$

Observe that the mapping ϕ satisfies $\phi(x) + \phi(y) = \phi(x+y)$ for any $x, y \in \mathbb{F}_{2^n}$. This statement is clearly valid when at least one of x, y is the zero element. Indeed, for nonzero x, y in $\mathbb{F}_{2^n}$, suppose $x = \alpha^k$, $y = \alpha^l$, and $\alpha^k + \alpha^l = \alpha^m$. Then the ith coordinate of $\phi(x) + \phi(y)$ is $a_{k+id,0} + a_{l+id,0}$. It is the starting coordinate of $\alpha^{k+id} + \alpha^{l+id} = \alpha^{m+id}$, which is actually the ith coordinate of α^m. It thus follows that $\phi(\alpha^k) + \phi(\alpha^l) = \phi(\alpha^m) = \phi(\alpha^k + \alpha^l)$. In addition, it is readily seen that $\phi(x) = 0$ if and only if $x = 0$. Hence, ϕ is a bijective mapping from $\mathbb{F}_{2^n}$ to $\mathbb{F}_2^n$, and $\phi(x)$, $\phi(x+1)$ for any $x \in \mathbb{F}_{2^n}$ always form a conjugate pair [924].

The cyclotomic class with respect to d is given by $C_i = \{\alpha^i \beta^j \mid 0 \le j < e\}$, where $0 \le i < d$. It was shown in [924, Theorem 3] that the mapping ϕ induces a one-to-one correspondence between cycles in $\Omega(f)$ and cyclotomic classes. As a result, the number of conjugate pairs between two cycles $[\mathbf{s}_i]$ and $[\mathbf{s}_j]$ is equal to the cyclotomic number $(i,j)_d$ given by $(i,j)_d = |C_i \cap (C_j + 1)|$. In cyclotomy theory, the cyclotomic numbers $(i,j)_d$ for small integers d are determined. Hence, the adjacency graph $G(f)$ can be explicitly given for those integers d with known cyclotomic numbers.

Example 25.4.19 Let $n = 4$ and $f(x) = x^4 + x^3 + x^2 + x + 1$. For the finite field $\mathbb{F}_{2^4} = \mathbb{F}_2[x]/(f(x))$, we take a primitive element α that is a root of the primitive polynomial $q(x) = x^4 + x + 1$. Then $\beta = \alpha^3$. The FSR with characteristic polynomial $f(x)$ is the so-called pure summing register and has cycle structure as follows

$$\Omega(f) = [\mathbf{0}] \cup [\mathbf{00011}] \cup [\mathbf{00101}] \cup [\mathbf{01111}]$$
$$= [\mathbf{0}] \cup C_0 \cup C_1 \cup C_2,$$

where $C_i = \{\alpha^i, \alpha^i\beta, \alpha^i\beta^2, \alpha^i\beta^3, \alpha^i\beta^4\}$ for $i = 0, 1, 2$. The number of conjugate pairs between these cycles can be explicitly given by the cyclotomic numbers $(i,j)_3$, which in this case are

$$(0,0)_3 = \left(2^n + (-2)^{n/2+1} - 8\right)/9 = 0;$$
$$(0,1)_3 = (1,0)_3 = (2,2)_3 = \left(2^{n+1} - (-2)^{n/2+1} - 4\right)/18 = 2;$$
$$(0,2)_3 = (2,0)_3 = (1,1)_3 = \left(2^{n+1} - (-2)^{n/2+1} - 4\right)/18 = 2;$$
$$(1,2)_3 = (2,1)_3 = \left(2^n + (-2)^{n/2+1} + 1\right)/9 = 1.$$

In the above we introduced the generation of de Bruijn sequences from LFSRs with simple characteristic polynomials. For LFSRs with more general characteristic polynomials, such as a product of two different irreducible polynomials or even an arbitrary polynomial, Chang et al. in [382, 384] intensively investigated the property of their cycle structure and adjacency graph, and also developed a `python` program to generate de Bruijn sequences from products of irreducible polynomials, which is available at `https://github.com/adamasstokhorst/deBruijn`.

Chapter 26

Lattice Coding

Frédérique Oggier

Nanyang Technological University

26.1   Introduction

A **lattice** Λ in $\mathbb{R}^n$ is a set of points in $\mathbb{R}^n$ composed of all integer linear combinations of independent vectors $\mathbf{b}_1, \mathbf{b}_2, \ldots, \mathbf{b}_m$:

$$\Lambda = \{a_1\mathbf{b}_1 + \cdots + a_m\mathbf{b}_m \mid a_1, \ldots, a_m \in \mathbb{Z}\}.$$

The basis vectors may be collected into a **generator matrix** B, i.e.,

$$B = \begin{bmatrix} \mathbf{b}_1 \\ \mathbf{b}_2 \\ \vdots \\ \mathbf{b}_m \end{bmatrix}$$

such that a vector $\mathbf{x} = (x_1, \ldots, x_n) \in \mathbb{R}^n$ is in Λ if and only if it can be written as

$$(x_1, \ldots, x_n) = (a_1, \ldots, a_m) \begin{bmatrix} \mathbf{b}_1 \\ \mathbf{b}_2 \\ \vdots \\ \mathbf{b}_m \end{bmatrix} \quad \text{for } a_1, \ldots, a_m \in \mathbb{Z}.$$

We thus define a lattice Λ in terms of a generator matrix as $\Lambda = \{\mathbf{a}B \mid \mathbf{a} \in \mathbb{Z}^m\}$. Note that B is a matrix of rank m which is not unique. We call $G = BB^\mathsf{T}$ a **Gram matrix** for Λ. The **volume** of a lattice Λ, denoted by $V(\Lambda)$, is the (positive) square root of the determinant of G. The integer $m \leq n$ is called the **rank** of Λ. We will focus on the case $m = n$, where the lattice is **full-rank**.

FIGURE 26.1: An AWGN channel model.

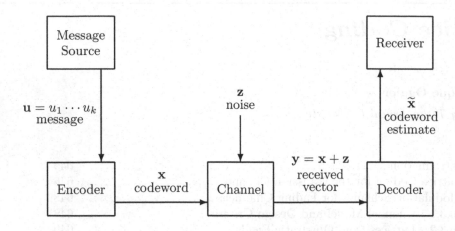

Lattices have been extensively studied; see, e.g., [442] for an encyclopedic reference, [661, 1335] for mathematical theories, [1938] for an information theoretic approach, [1384, 1497] for applications of lattices to cryptography, and [455] for a recent tutorial on lattices and some of their applications.

This chapter introduces some aspects of lattice coding, namely coding for the Gaussian channel (see Section 26.2), modulation schemes for fading channels (in Section 26.3) and constructions of lattices from linear codes (in Section 26.4). Some other aspects of lattice coding will be reviewed in Section 26.5.

26.2   Lattice Coding for the Gaussian Channel

We consider the transmission of a vector $\mathbf{x}$ from a discrete set of points $S \subset \mathbb{R}^n$ (S does not have to form a lattice for now) over an **additive white Gaussian noise (AWGN) channel**, so that the received signal $\mathbf{y}$ is of the form

$$\mathbf{y} = \mathbf{x} + \mathbf{z} \in \mathbb{R}^n \tag{26.1}$$

where $\mathbf{z}$ is a random vector whose components are independent Gaussian random variables with mean 0 and variance σ^2, as shown in Figure 26.1.

The Gaussian channel coding problem consists of constructing the set S such that $\mathbf{x}$ can be decoded from $\mathbf{y}$ despite the presence of the noise $\mathbf{z}$. If the transmitter had infinite power, given σ^2, it could just scale the set S, so that the distance between any two vectors in S is much larger than $2\sigma^2$. Then points in S will be far enough apart that the decoder can just choose the point $\mathbf{x}$ which is closest to the received point $\mathbf{y}$. However we do have a power constraint: all the points of S must lie within a sphere of radius $\sqrt{nP}$ around the origin, where P defines an average power constraint.

Suppose now that S is a subset from a lattice Λ. The receiver will make a correct decision to choose the closest lattice point $\mathbf{x} \in S$ from $\mathbf{y}$ as the decoded point exactly if the noise vector $\mathbf{z}$ falls in the **Voronoi region** $\mathcal{V}_\Lambda = \mathcal{V}_\Lambda(\mathbf{x})$ of $\mathbf{x} = \mathbf{0}$ given by

$$\mathcal{V}_\Lambda(\mathbf{x}) = \{\mathbf{y} \in \mathbb{R}^n \mid \|\mathbf{x} - \mathbf{y}\| \leq \|\mathbf{v} - \mathbf{y}\| \text{ for all } \mathbf{v} \in \Lambda\},$$

an event of probability

$$\frac{1}{(\sigma\sqrt{2\pi})^n} \int_{\mathcal{V}_\Lambda} e^{-\|\mathbf{x}\|^2/(2\sigma^2)} d\mathbf{x}.$$

Assuming that all points $\mathbf{x}$ are equally likely to be sent, the **error probability** P_e of decoding a lattice point $\widetilde{\mathbf{x}} \neq \mathbf{x}$ when $\mathbf{x}$ is sent is 1 minus the above probability. Given σ, we are thus looking for the n-dimensional lattice of volume normalized to 1 for which P_e is minimized. Using the so-called **Union Bound**, the probability of an error event is bounded above by

$$\frac{1}{2} \sum_{\mathbf{x} \in \Lambda \setminus \{\mathbf{0}\}} e^{-\|\mathbf{x}\|^2/(8\sigma^2)}.$$

The dominant terms in the sum correspond to vectors with **minimum norm**[1]

$$\lambda = \min_{\mathbf{0} \neq \mathbf{x} \in \Lambda} \|\mathbf{x}\|.$$

Therefore, dropping all terms except the ones of minimum norm, the upper bound can be approximated by

$$\frac{\kappa e^{-\rho^2/(2\sigma^2)}}{2},$$

where $\rho = \lambda/2$ is the **packing radius** of Λ, and κ is the **kissing number** of Λ, that is, the number of packing balls that touch a fixed one, which corresponds to the number of lattice points having the minimum nonzero norm.

This expression is minimized if ρ is maximized (once this is done, one may then minimize κ). Intuitively, we expect the number of points in S to be close to the ratio between the volume of the sphere $\mathcal{B}^n(\sqrt{nP})$ of radius $\sqrt{nP}$ and the volume $V(\Lambda)$ of Λ, i.e.,

$$|S| \approx \frac{\text{vol } \mathcal{B}^n(1)(nP)^{n/2}}{V(\Lambda)} = \frac{\Delta(nP)^{n/2}}{\rho^n}$$

where Δ is the **packing density** of Λ defined by

$$\Delta(\Lambda) = \frac{\text{vol } \mathcal{B}^n(\rho)}{V(\Lambda)}.$$

Then $\rho^2 \approx (nP)\left(\frac{\Delta}{|S|}\right)^{2/n}$, and for a fixed number of points, minimizing the probability of error can be achieved by maximizing the packing density of Λ.

We give some examples of famous lattices, with some of their parameters introduced above; more are found in [442, Chapter 4].

- **The cubic lattice** $\mathbb{Z}^n$: It has minimal norm $\lambda(\mathbb{Z}^n) = \min_{\mathbf{x} \in \mathbb{Z}^n \setminus \{0\}} \|\mathbf{x}\| = 1$, packing radius $\rho(\mathbb{Z}^n) = 1/2$, and kissing number $\kappa(\mathbb{Z}^n) = 2n$. Its packing density is $\Delta(\mathbb{Z}^n) = \frac{\text{vol } \mathcal{B}^n(1)}{2^n}$.

- **The checkerboard lattice** D_n: It is a sublattice of $\mathbb{Z}^n$ where the sum of coordinates is even. It has basis $\{\mathbf{b}_1, \mathbf{b}_2, \mathbf{b}_3, \ldots, \mathbf{b}_n\}$ with $\mathbf{b}_1 = (-1, -1, 0, \ldots, 0)$, $\mathbf{b}_2 = (1, -1, 0, \ldots, 0)$, $\mathbf{b}_3 = (0, 1, -1, 0, \ldots, 0)$, $\ldots$, $\mathbf{b}_n = (0, 0, \ldots, 0, 1, -1)$. Its minimal norm is $\lambda(D_n) = \sqrt{2}$, its volume is $V(D_n) = 2$, and its kissing number is $\kappa(D_n) = 2n(n-1)$.

---

[1] In several textbooks and papers the minimum norm is defined as the square of this number.

- **The lattice** A_n: It is a sublattice of $\mathbb{Z}^{n+1}$ lying in the hyperplane H where the sum of the coordinates is zero ($A_n \subset D_{n+1} \subset \mathbb{Z}^{n+1}$). It has basis $\{\mathbf{b}_1, \ldots, \mathbf{b}_n\}$ given by $\mathbf{b}_1 = (-1, 1, 0, \ldots, 0)$, $\mathbf{b}_2 = (0, -1, 1, 0, \ldots, 0)$, $\mathbf{b}_3 = (0, 0, -1, 1, 0, \ldots, 0), \ldots, \mathbf{b}_n = (0, 0, \ldots, 0, -1, 1)$. Its volume is $V(A_n) = n + 1$. It has minimal norm $\lambda(A_n) = \sqrt{2}$ and kissing number $\kappa(A_n) = n(n + 1)$.

- **The lattices** E_6, E_7 **and** E_8: The Gosset lattice E_8 is defined as

$$E_8 = \left\{ \mathbf{x} = (x_1, \ldots, x_8) \in \mathbb{Z}^8 \ \middle| \ \mathbf{x} \in D_8 \text{ or } \mathbf{x} + \left(\frac{1}{2}, \ldots, \frac{1}{2} \right) \in D_8 \right\}.$$

  It has minimal norm $\lambda(E_8) = \sqrt{2}$. The lattices E_7 and E_6 are lattices of rank 7 and 6 naturally defined in $\mathbb{R}^8$ as:

$$E_7 = \{ \mathbf{x} = (x_1, \ldots, x_8) \in E_8 \mid x_1 = x_2 \},$$
$$E_6 = \{ \mathbf{x} = (x_1, \ldots, x_8) \in E_8 \mid x_1 = x_2 = x_3 \}.$$

- **The Barnes–Wall lattice** Λ_{16}: The so-called Barnes–Wall lattices BW_n are defined in dimension $n = 2^k$, $k \geq 2$. In dimension 16, $\Lambda_{16} = BW_{16}$ has the best known packing density in dimension 16.

26.3   Modulation Schemes for Fading Channels

26.3.1   Channel Model and Design Criteria

We now consider transmission over fast fading channels, where communication is modelled by

$$\mathbf{y} = \mathbf{x}\mathbf{H} + \mathbf{z} \in \mathbb{R}^n \quad \text{with} \quad \mathbf{H} = \begin{bmatrix} h_1 & 0 & 0 \\ 0 & \ddots & 0 \\ 0 & 0 & h_n \end{bmatrix}$$

where $\mathbf{z}$ is a random vector whose components are independent Gaussian random variables with mean 0 and variance σ^2, and $h_1, \ldots, h_n$ are independently Rayleigh distributed. This model is similar to (26.1), except for the matrix H which takes into account fading in a wireless environment.

Assuming the receiver knows $\mathbf{H}$ (this is called a **coherent** channel), the receiver is facing a Gaussian channel, with as transmitted lattice constellation a version of the original one twisted by the fading $\mathbf{H}$: if $\mathbf{x}$ is a lattice point of the form $\mathbf{a}B$, then it is as if the lattice used for transmission had in fact $B\mathbf{H}$ for a generator matrix. As explained in the above section, lattice constellations for a Gaussian channel will make sure that lattice points are separated enough to resist the channel noise. However even if two lattice points $\mathbf{x}, \mathbf{x}'$ are designed to be apart, $h_j x_j$ and $h_j x_j'$ could be arbitrarily close, depending on h_j, $j = 1, 2, \ldots, n$.

We thus want to first make sure that the lattices used are such that for any two points $\mathbf{x} \neq \mathbf{x}'$, they differ in all their components, the intuition being that if the fading affects some components, the lattice points will still be distinguishable on their other nonzero components [264]. By linearity, this means that any of their vectors $\mathbf{x} = (x_1, x_2, \ldots, x_n)$ have $x_i \neq 0$, for any i. We say that such lattices in $\mathbb{R}^n$ have **full diversity** n. The relevant distance is then the **product distance**, which is the minimum of the absolute value of the product of the coordinates of nonzero vectors in the lattice.

FIGURE 26.2: A fast fading channel model.

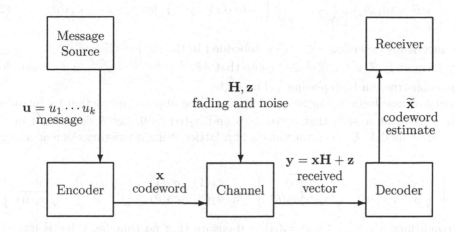

The intuition follows the formal derivation of the error probability of decoding $\widetilde{\mathbf{x}} \neq \mathbf{x}$ when $\mathbf{x}$ is sent, given that $\mathbf{H}$ is known, and assuming that the decoder looks for the closest lattice point, in which case this error probability is bounded above by

$$\frac{1}{2} \prod_{x_i \neq \widetilde{x}_i} \frac{1}{\frac{(x_i - \widetilde{x}_i)^2}{8\sigma^2}}.$$

The actual lattice decoder used to find the closest lattice point is called the *sphere decoder* [479, 922, 1855].

Algebraic constructions (see, e.g., [143, 1055]) provide a systematic method to obtain full diversity, and also to compute the minimum distance. Such an algebraic construction is explained below.

26.3.2 Lattices from Quadratic Fields

We will briefly explain how to construct lattices using quadratic fields, and refer to [1462] for more general constructions.

For $d > 1$ a square-free integer, a **quadratic number field** is of the form

$$\mathbb{Q}(\sqrt{d}) = \{a + b\sqrt{d} \mid a, b \in \mathbb{Q}\},$$

where $\{1, \sqrt{d}\}$ forms a basis, and we have two embeddings of $\mathbb{Q}(\sqrt{d})$ into $\mathbb{R}$:

$$\sigma_1 : \quad a + b\sqrt{d} \mapsto a + b\sqrt{d}$$
$$\sigma_2 : \quad a + b\sqrt{d} \mapsto a - b\sqrt{d}.$$

Therefore $\sigma = (\sigma_1, \sigma_2)$ gives an embedding of $\mathbb{Q}(\sqrt{d})$ into $\mathbb{R}^2$.

Now

$$\mathbb{Z}[\sqrt{d}] = \{a + b\sqrt{d} \mid a, b, \in \mathbb{Z}\} \subset \mathbb{Q}(\sqrt{d})$$

has a $\mathbb{Z}$-basis, given for example by $\{1, \sqrt{d}\}$. Embedding this basis using σ gives

$$B = \begin{bmatrix} 1 & 1 \\ \sigma_1(\sqrt{d}) & \sigma_2(\sqrt{d}) \end{bmatrix} = \begin{bmatrix} 1 & 1 \\ \sqrt{d} & -\sqrt{d} \end{bmatrix}$$

which is a lattice generator matrix. Note that

$$\mathbf{a}B = (a_1, a_2) \begin{bmatrix} 1 & 1 \\ \sqrt{d} & -\sqrt{d} \end{bmatrix} = (\sigma_1(x), \sigma_2(x)) \quad \text{for } x = a_1 + a_2\sqrt{d}, \qquad (26.2)$$

which geometrically describes $x \in \mathbb{Z}[\sqrt{d}]$ embedded in the lattice $\sigma(\mathbb{Z}[\sqrt{d}])$.

If $d \equiv 1 \pmod 4$, $\mathbb{Z}[\sqrt{d}] \subset \mathbb{Z}[\frac{1+\sqrt{d}}{2}]$ (note that $\frac{1+\sqrt{d}}{2} - \frac{1-\sqrt{d}}{2} = \sqrt{d}$), and we can do the exact same construction by replacing $\sqrt{d}$ by $\frac{\sqrt{d}+1}{2}$.

A variety of new lattices can be obtained from the above construction by introducing a "twisting" element α such that $\sigma_1(\alpha) > 0$ and $\sigma_2(\alpha) > 0$. Let θ denote $\sqrt{d}$ or $\frac{1+\sqrt{d}}{2}$ depending on $d \bmod 4$. A generator matrix of a lattice using a twisting element α is given by

$$B = \begin{bmatrix} \sqrt{\sigma_1(\alpha)} & \sqrt{\sigma_2(\alpha)} \\ \sqrt{\sigma_1(\alpha)}\sigma_1(\theta) & \sqrt{\sigma_2(\alpha)}\sigma_2(\theta) \end{bmatrix} = \begin{bmatrix} \sigma_1(1) & \sigma_2(1) \\ \sigma_1(\theta) & \sigma_2(\theta) \end{bmatrix} \begin{bmatrix} \sqrt{\sigma_1(\alpha)} & 0 \\ 0 & \sqrt{\sigma_2(\alpha)} \end{bmatrix}$$

and the conditions $\sigma_1(\alpha) > 0$ and $\sigma_2(\alpha) > 0$ ensure that no complex value is introduced when taking the square root.

The volume of this lattice is given by the square root of

$$\det \begin{bmatrix} \sigma_1(1) & \sigma_2(1) \\ \sigma_1(\theta) & \sigma_2(\theta) \end{bmatrix}^2 \det \begin{bmatrix} \sqrt{\sigma_1(\alpha)} & 0 \\ 0 & \sqrt{\sigma_2(\alpha)} \end{bmatrix}^2 = \sigma_1(\alpha)\sigma_2(\alpha)(\sigma_2(\theta) - \sigma_1(\theta))^2.$$

We are left to exhibit full diversity. Recall from (26.2) that points in lattices obtained from quadratic fields are of the general form

$$(\sqrt{\sigma_1(\alpha)}\sigma_1(x), \sqrt{\sigma_2(\alpha)}\sigma_2(x)) \quad \text{and} \quad (\sqrt{\sigma_1(\alpha)}\sigma_1(y), \sqrt{\sigma_2(\alpha)}\sigma_2(y))$$

and their difference belongs to the lattice. Then for $x \neq y$

$$\sqrt{\sigma_1(\alpha)}\sigma_1(x) - \sqrt{\sigma_1(\alpha)}\sigma_1(y) = \sqrt{\sigma_1(\alpha)}\sigma_1(x - y) \neq 0,$$
$$\sqrt{\sigma_2(\alpha)}\sigma_2(x) - \sqrt{\sigma_2(\alpha)}\sigma_2(y) = \sqrt{\sigma_2(\alpha)}\sigma_2(x - y) \neq 0,$$

since $x' = x - y = a_1 + a_2\sqrt{d} = 0$ if and only if $a_1 = a_2 = 0$.

Example 26.3.1 We embed $\mathbb{Z}[\frac{1+\sqrt{5}}{2}]$ into $\mathbb{R}^2$ using $\sigma = (\sigma_1, \sigma_2)$ where

$$\sigma_1 : \sqrt{5} \mapsto \sqrt{5} \quad \text{and} \quad \sigma_2 : \sqrt{5} \mapsto -\sqrt{5}$$

to get a generator matrix

$$B = \begin{bmatrix} 1 & 1 \\ \sigma_1(\frac{1+\sqrt{5}}{2}) & \sigma_2(\frac{1+\sqrt{5}}{2}) \end{bmatrix}.$$

This lattice has for Gram matrix

$$G = BB^\mathsf{T} = \begin{bmatrix} 1 & 1 \\ \sigma_1(\frac{1+\sqrt{5}}{2}) & \sigma_2(\frac{1+\sqrt{5}}{2}) \end{bmatrix} \begin{bmatrix} 1 & \sigma_1(\frac{1+\sqrt{5}}{2}) \\ 1 & \sigma_2(\frac{1+\sqrt{5}}{2}) \end{bmatrix} = \begin{bmatrix} 2 & 1 \\ 1 & 3 \end{bmatrix}.$$

We next introduce a twisting element α (set $\theta = \frac{1+\sqrt{5}}{2}$):

$$\alpha = 3 - \tfrac{1+\sqrt{5}}{2}, \quad \alpha\theta = -1 + 2\tfrac{1+\sqrt{5}}{2}, \quad \alpha\theta^2 = 2 + \tfrac{1+\sqrt{5}}{2}.$$

Then

$$B = \begin{bmatrix} \sqrt{\alpha} & \sqrt{\sigma_2(\alpha)} \\ \sqrt{\alpha}\frac{1+\sqrt{5}}{2} & \sqrt{\sigma_2(\alpha)}\frac{1-\sqrt{5}}{2} \end{bmatrix}$$

with corresponding Gram matrix G equalling

$$\begin{bmatrix} \sigma_1\left(3 - \frac{1+\sqrt{5}}{2}\right) + \sigma_2\left(3 - \frac{1+\sqrt{5}}{2}\right) & \sigma_1\left(-1 + 2\frac{1+\sqrt{5}}{2}\right) + \sigma_2\left(-1 + 2\frac{1+\sqrt{5}}{2}\right) \\ \sigma_1\left(-1 + 2\frac{1+\sqrt{5}}{2}\right) + \sigma_2\left(-1 + 2\frac{1+\sqrt{5}}{2}\right) & \sigma_1\left(2 + \frac{1+\sqrt{5}}{2}\right) + \sigma_2\left(2 + \frac{1+\sqrt{5}}{2}\right) \end{bmatrix}$$

yielding

$$G = \begin{bmatrix} 5 & 0 \\ 0 & 5 \end{bmatrix}$$

and volume

$$\sqrt{|\sigma_1(\alpha)\sigma_2(\alpha)|}\,|\sigma_2(\theta) - \sigma_1(\theta)| = \sqrt{5}\,|\sqrt{5}| = 5.$$

Up to a change of basis, this lattice is in fact a scaled version of $\mathbb{Z}^2$.

26.4 Lattices from Linear Codes

Any choice of n linearly independent vectors in $\mathbb{R}^n$ generates a lattice, but some lattices are more interesting than others. We listed some of those above. In this section, we show how families of lattices can be obtained using linear codes. There are several such constructions, called Constructions A, B, C, D, E [442]. We will present two of the (arguably) most famous ones, Constructions A and D.

26.4.1 Construction A

For $q \geq 2$ a positive integer, consider the set $\mathbb{Z}_q = \{0, 1, \dots, q-1\}$ of integers modulo q. We will use the notation $\mathbb{F}_p$ to denote $\mathbb{Z}_p$ to emphasize the case $q = p$ is prime. A linear code $\mathcal{C}$ in $\mathbb{Z}_q^n$ is by definition an additive subgroup of $\mathbb{Z}_q^n$, whose vectors are called codewords. If $q = p$ is prime, we have more structure, and a linear code is a subspace of dimension k of the vector space $\mathbb{F}_p^n$ (called an $[n, k]$ code).

Next we establish a connection between linear codes in $\mathbb{Z}_q^n$ and lattices. Consider the map ρ defined over n copies of $\mathbb{Z}_q$ by taking the reduction modulo q component-wise:

$$\rho : \mathbb{Z}^n \to \mathbb{Z}_q^n \quad \text{where} \quad \mathbf{x} \mapsto \rho(\mathbf{x}).$$

One may take any arbitrary subset S of $\mathbb{Z}_q^n$ and compute $\rho^{-1}(S)$, but we will start with $S \subset \mathbb{Z}_q^n$ a linear code, to understand how this code structure is carried over to $\rho^{-1}(S)$.

The key result is as follows (see, e.g., [455] for a proof):

> Given a subset $S \subset \mathbb{Z}_q^n$, $\rho^{-1}(S)$ is a lattice in $\mathbb{R}^n \iff S$ is a linear code.

For $\mathcal{C} \subset \mathbb{Z}_q^n$ a linear code, the lattice $\Lambda_{\mathcal{C}} = \rho^{-1}(\mathcal{C})$ is said to have been obtained via **Construction A**. Construction A generates good (capacity-achieving) codes for the Gaussian channel, for some channels with side information [1937]. It is further used for wiretap coding, as will be explained below. It also appears in lattice-based cryptographic schemes, linked to the computational difficulty of the shortest and closest vector problems [1384].

Some properties of Construction A of lattices include (see [455] for a proof):

- We have the lattice inclusions $q\mathbb{Z}^n \subset \Lambda_\mathcal{C} \subset \mathbb{Z}_q^n$.

- We have $\left| \dfrac{\Lambda_\mathcal{C}}{q\mathbb{Z}^n} \right| = \dfrac{q^n}{V(\Lambda_\mathcal{C})} = |\mathcal{C}|$, where $|\mathcal{C}|$ is the number of codewords of $\mathcal{C}$.

- If q is prime, a linear code $\mathcal{C}$ of dimension $k \leq n$ has q^k codewords, and $V(\Lambda_\mathcal{C}) = q^{n-k}$.

A generator matrix for the lattice $\Lambda_\mathcal{C}$ can be computed from that of $\mathcal{C}$. If $q = p$ is prime, the linear code $\mathcal{C}$ of dimension k over $\mathbb{F}_p$ is a subspace and has a basis, formed by k vectors. These k vectors are stacked as row vectors to form a $k \times n$ matrix with elements in $\mathbb{F}_p$. In this case, up to coordinate permutation, a generator matrix may always be reduced to its systematic form $\begin{bmatrix} I_k & A \end{bmatrix}$, where I_k is the $k \times k$ identity matrix, and A is a $k \times (n-k)$ matrix.

When q is a composite number, even though we still have a generator matrix, which contains vectors $\mathbf{v}_i = (v_{i1}, \ldots, v_{in})$ for $i = 1, \ldots, l$ ($l = k$ when $q = p$) that generate $\mathcal{C}$ as its rows, these vectors do not always form a basis and the generator matrix may not always be reduced to its systematic form. But each codeword $\mathbf{a} \in \mathcal{C}$ can be written as $\mathbf{a} = \sum_{i=1}^{l} a_i \mathbf{v}_i$, so that

$$\mathbf{a} = \sum_{i=1}^{l} a_i \mathbf{v}_i \in \mathcal{C} \iff \rho^{-1}(\mathbf{a}) = \sum_{i=1}^{l} a_i \mathbf{v}_i + \sum_{i=1}^{n} q h_i \mathbf{e}_i \in \mathbb{R}^n$$

where $\mathbf{e}_i$, $i = 1, \ldots, n$, form the canonical basis of $\mathbb{R}^n$ and $h_1, \ldots, h_n \in \mathbb{Z}$.

Stacking the vectors as row vectors of an $(n+l) \times n$ matrix yields an expanded generator matrix. Since we are working over $\mathbb{Z}$, a generator matrix for the lattice $\Lambda_\mathcal{C}$ is given by the $n \times n$ full rank matrix H obtained by computing the **Hermite normal form (HNF)** of

$$\begin{bmatrix} \mathbf{v}_1 \\ \vdots \\ \mathbf{v}_l \\ q\mathbf{e}_1 \\ \vdots \\ q\mathbf{e}_n \end{bmatrix}.$$

Recall that being in a Hermite normal form means that $H = [h_{ij}]$ is a square matrix satisfying:

- $h_{ij} = 0$ for $i < j$, which means the matrix H will be upper triangular, and

- $0 \leq h_{ij} < h_{ii}$ for $i > j$, that is, entries are nonnegative and each row has a maximum entry on the diagonal.

If the generator matrix of $\mathcal{C}$ can be put in systematic form, then a generator matrix of $\Lambda_\mathcal{C}$ is

$$\begin{bmatrix} I_l & A \\ \mathbf{0}_{(n-l) \times l} & q I_{n-l} \end{bmatrix}.$$

Example 26.4.1 We consider the two linear codes

$$\mathcal{C}_1 = \{(2a, 2b, a+b) \mid a, b \in \mathbb{F}_3\} \quad \text{and} \quad \mathcal{C}_2 = \{(2a, 2b, a+b) \mid a, b \in \mathbb{Z}_4\}.$$

The code over $\mathbb{F}_3$ has dimension 2, length $n = 3$, and has for generator matrices

$$M = \begin{bmatrix} 2 & 0 & 1 \\ 0 & 2 & 1 \end{bmatrix} \quad \text{and} \quad R = \begin{bmatrix} 1 & 0 & 2 \\ 0 & 1 & 2 \end{bmatrix}$$

where R is the systematic form of M.

The matrix M is also a generator matrix for C_2, but it cannot be put in the form R, and the vectors $(2,0,1)$ and $(0,2,1)$ do not form a basis because they are not linearly independent. Indeed $\lambda_1(2,0,1) + \lambda_2(0,2,1) = 0$ does not imply $\lambda_1 = \lambda_2 = 0$; we could also have $\lambda_1 = \lambda_2 = 2$.

A generator matrix for the lattices Λ_{C_2} is obtained by computing the Hermite normal form of

$$B_2 = \begin{bmatrix} 2 & 0 & 1 \\ 0 & 2 & 1 \\ 4 & 0 & 0 \\ 0 & 4 & 0 \\ 0 & 0 & 4 \end{bmatrix}$$

which yields

$$H_2 = \begin{bmatrix} 2 & 0 & 1 \\ 0 & 2 & 1 \\ 0 & 0 & 2 \end{bmatrix}.$$

Famous lattices have been shown to be obtained using Construction A. For example, the linear code $C = \left\{ (a_1, \ldots, a_{n-1}, \sum_{i=1}^{n-1} a_i) \mid a_1, \ldots, a_{n-1} \in \mathbb{F}_2 \right\}$ of length n and dimension $n-1$ gives the lattice Λ_C with generator matrix

$$\begin{bmatrix} I_{n-1} & \begin{matrix} 1 \\ \vdots \\ 1 \end{matrix} \\ \mathbf{0}_{1\times(n-1)} & 2 \end{bmatrix}.$$

Thus every vector in Λ_C is such that the sum of its entries is even and we have just constructed the lattices

$$D_n = \left\{ (x_1, \ldots, x_n) \in \mathbb{Z}^n \mid \sum_{i=1}^{n} x_i \text{ is even} \right\}.$$

We note that the lattice construction from number fields from the previous section can in turn be combined with Construction A to obtain further lattices; see, e.g., [1154] and references therein.

Furthermore, a popular use of Construction A is the construction of lattices which are particularly structured with respect to their dual; they are called modular lattices (see, e.g., [115, 236]). We will not give more details on these since they have little application to coding theory. We will however mention in Section 26.5.2 how the weight enumerator of C is interpreted in the context of Construction A.

26.4.2 Constructions D and $\overline{\text{D}}$

Let ψ be the natural embedding of $\mathbb{F}_2^n$ into $\mathbb{Z}^n$. Construction D is another construction of lattices from codes given as follows [442].

Definition 26.4.2 (Construction D (scaled)) Let $C_0 \subseteq C_1 \subseteq \cdots \subseteq C_{a-1} \subseteq C_a = \mathbb{F}_2^n$ be a family of nested binary linear codes where the minimum distance of C_i is at least $4^{a-i}/\gamma$ with $\gamma = 1$ or 2. Let $k_i = \dim(C_i)$ and let $\mathbf{b}_1, \mathbf{b}_2, \ldots, \mathbf{b}_n$ be a basis of $\mathbb{F}_2^n$ such that

$\mathbf{b}_1, \ldots, \mathbf{b}_{k_i}$ span $\mathcal{C}_i$. The lattice Λ_D consists of all vectors of the form

$$\sum_{i=0}^{a-1} 2^i \sum_{j=1}^{k_i} \alpha_j^{(i)} \psi(\mathbf{b}_j) + 2^a \mathbf{1}$$

where $\alpha_j^{(i)} \in \{0, 1\}$ and $\mathbf{1} \in \mathbb{Z}^n$.

The minimum distance condition does not play a role in the following discussion. A related construction is used in [913] to construct Barnes–Wall lattices from Reed–Muller codes. It was called Construction $\overline{\mathrm{D}}$ in [1153].

Definition 26.4.3 (Construction $\overline{\mathbf{D}}$) Let $\mathcal{C}_0 \subseteq \mathcal{C}_1 \subseteq \cdots \subseteq \mathcal{C}_{a-1} \subseteq \mathcal{C}_a = \mathbb{F}_2^n$ be a family of nested binary linear codes. Let

$$\Gamma_{\overline{D}} = \psi(\mathcal{C}_0) \oplus 2\psi(\mathcal{C}_1) \oplus \cdots \oplus 2^{a-1}\psi(\mathcal{C}_{a-1}) \oplus 2^a \mathbb{Z}^n.$$

It was shown in [1153] that the set $\Gamma_{\overline{D}}$ itself may not necessarily be a lattice. Let us denote by $\Lambda_{\overline{D}}$ the smallest lattice that contains $\Gamma_{\overline{D}}$, and by $\star$ the componentwise multiplication (known also as the **Schur product** or Hadamard product): for $\mathbf{x} = (x_1, \ldots, x_n)$, $\mathbf{y} = (y_1, \ldots, y_n) \in \mathbb{F}_2^n$, we have

$$\mathbf{x} \star \mathbf{y} := (x_1 y_1, \ldots, x_n y_n) \in \mathbb{F}_2^n.$$

A family of nested binary linear codes $\mathcal{C}_0 \subseteq \mathcal{C}_1 \subseteq \cdots \subseteq \mathcal{C}_{a-1} \subseteq \mathcal{C}_a = \mathbb{F}_2^n$ is said to be **closed under the Schur product** if and only if the Schur product of any two codewords of $\mathcal{C}_i$ is contained in $\mathcal{C}_{i+1}$ for all i. In other words, if $\mathbf{c}_1, \mathbf{c}_2 \in \mathcal{C}_i$, then $\mathbf{c}_1 \star \mathbf{c}_2 \in \mathcal{C}_{i+1}$ for all $i = 0, \ldots, a-1$.

The following equivalences that describe in particular the connection between Constructions D and $\overline{\mathrm{D}}$ was proved in [1153].

Theorem 26.4.4 *Given a family of nested binary linear codes $\mathcal{C}_0 \subseteq \mathcal{C}_1 \subseteq \cdots \subseteq \mathcal{C}_{a-1} \subseteq \mathcal{C}_a = \mathbb{F}_2^n$, the following are equivalent.*

(a) $\Gamma_{\overline{D}}$ *is a lattice.*

(b) $\Gamma_{\overline{D}} = \Lambda_{\overline{D}}$.

(c) $\mathcal{C}_0 \subseteq \mathcal{C}_1 \subseteq \cdots \subseteq \mathcal{C}_{a-1} \subseteq \mathcal{C}_a = \mathbb{F}_2^n$ *is closed under the Schur product.*

(d) $\Gamma_{\overline{D}} = \Lambda_D$.

26.5 Variations of Lattice Coding Problems

As mentioned in the introduction, there are many more applications of lattice codes, e.g., for quantization and cryptography. Underlying lattice structures also appear for designing space-time codes (see, e.g., [154, 155]) and spherical codes (see, e.g., [455]). We will briefly review two other applications of lattice codes to design index codes and wiretap codes.

26.5.1 Index Codes

This scenario considers a wireless broadcast channel, where a transmitted signal is received not only by the intended recipient but also by other terminals within the transmission range which may even receive the transmitted signal earlier or better than the intended recipient. We thus have a transmitter having to deliver a finite set of messages to finitely many receivers, each interested in a subset of the transmitted messages while having prior knowledge, called **side information**, of the values of a different subset of messages, e.g., obtained by overhearing previous transmissions.

Index coding refers to coding for the broadcast channel where messages are jointly encoded to meet the demands of all the receivers with the highest rate by exploiting side information.

We saw in the first section of this chapter that lattices can be used for transmission over the Gaussian channel. Consider a lattice $\Lambda \in \mathbb{R}^n$ and let Λ' be a sublattice of Λ, that is Λ' is a lattice included in Λ. If B is a generator matrix of Λ, MB for M an integer matrix with $\det(M) \neq 0$ is a generator matrix of Λ'. The set Λ/Λ' is called a **Voronoi constellation**.

For a Gaussian broadcast channel having a single transmitter with K messages $(w_1, \ldots, w_K) \in \mathcal{W}_1 \times \cdots \times \mathcal{W}_K$ to be sent to every receiver, where every finite set $\mathcal{W}_i$ is an alphabet for w_i, $i = 1, \ldots, K$, a code $\mathcal{C} \subset \mathbb{R}^n$ is used and the vector $(w_1, \ldots, w_K)$ is mapped to an n-dimensional vector $\mathbf{x} \in \mathcal{C}$. This is a labelling of every $\mathbf{x} \in \mathcal{C}$ by a distinct message tuple $(w_1, \ldots, w_K)$. Every receiver is getting $\mathbf{y} = \mathbf{x} + \mathbf{z}$ with $\mathbf{z}$ the Gaussian noise it experiences, and $\mathbf{w}_S$ its side information; that is $\mathbf{w}_S$ is the set of exact realizations of the message symbols w_k for $k \in S$. This means that each receiver decodes from $\mathbf{y}$ by looking for the closest $(w_1, \ldots, w_K)$ knowing $\mathbf{w}_S$. Effectively, the receiver is thus decoding a subcode of $\mathcal{C}$, and therefore the decoding performance will depend on the minimum distance d_S which is the smallest value the minimum distance can take over all choices of side information $\mathbf{w}_S$. The parameters at play are thus d_S, d_0 which is the minimum distance of $\mathcal{C}$ (with no side information), the rate $R_k = \frac{1}{n} \log_2 |\mathcal{W}_k|$ of the kth message for $k = 1, \ldots, K$, and the side information gain $\Gamma(\mathcal{C}) = \min_{S \subset \{1, \ldots, K\}} \frac{10 \log_{10}(d_S^2/d_0^2)}{\sum_{k=1}^K R_k}$.

Consider K lattices $\Lambda_1, \ldots, \Lambda_K$ with a common sublattice $\Lambda' \subset \Lambda_i$, $i = 1, \ldots, K$, and define the alphabet $\mathcal{W}_i$ to be the Voronoi constellation Λ_i/Λ', $i = 1, \ldots K$. A lattice index code for K messages is defined [1416] as a bijective map that sends $(\mathbf{x}_1, \ldots, \mathbf{x}_K) \in \Lambda_1/\Lambda' \times \cdots \times \Lambda_K/\Lambda' = \mathcal{W}_1 \times \cdots \times \mathcal{W}_K$ to $\mathbf{x}_1 + \cdots + \mathbf{x}_K \mod \Lambda'$. We would like the resulting code to have large minimum distance d_0 and large information gain Γ. Further analysis and constructions are found in [1416].

26.5.2 Wiretap Codes

Given a Gaussian channel used by honest transmitter (Alice) and receiver (Bob) which communicate in the presence of an eavesdropper (Eve), one may ask the question of designing lattice codes which achieve both reliability and confidentiality. They are referred to as **wiretap codes**. Recalling from (26.1) the transmission model of a vector $\mathbf{x}$ over a Gaussian channel, the wiretap model is [1224]

$$\mathbf{y}_B = \mathbf{x} + \mathbf{z}_B \quad \text{and} \quad \mathbf{y}_E = \mathbf{x} + \mathbf{z}_E$$

for the respective received vectors at Bob and Eve. To obtain confidentiality, randomness is introduced at the transmitter, and coset coding gives a practical way to handle this randomness.

Let us look again at the lattice $\Lambda_{\mathcal{C}} = \rho^{-1}(\mathcal{C})$ obtained from a linear code $\mathcal{C} \subset \mathbb{Z}_q^n$ via Construction A geometrically. It is obtained by considering the lattice $q\mathbb{Z}^n$ and its translations by the codewords of $\mathcal{C}$. In other words, $\rho^{-1}(\mathcal{C})$ is the union of cosets of $q\mathbb{Z}^n$, and codewords of $\mathcal{C}$ form coset representatives. This makes Construction A particularly suitable for a coding strategy called **coset coding**, which is used in particular in the context of wiretap coding.

The secret information is encoded into cosets, while $\mathbf{x}$ is then chosen randomly within this coset. If we consider for example the code $\{(0,0),(1,1)\} \subset \mathbb{Z}_2^2$, one bit of secret can be transmitted using coset coding: to send 0, choose the coset $2\mathbb{Z}^2$, and to send 1, choose the coset $2\mathbb{Z}^2 + (1,1)$. The geometric intuition is that when Bob receives a noisy codeword, his channel is such that in the radius around his received point, only the codeword that was sent is present, while Eve, who is assumed to have a stronger noise than Bob, will find in her radius points from different cosets, such that each coset is equally likely to have been sent. See [1290] for a practical implementation of coset coding over a USRP testbed.

Coset encoding uses two nested lattices $\Lambda_E \subset \Lambda_B$, where Λ_B is the lattice from which a signal constellation is carved for transmission to Bob, while Λ_E is the sublattice used to partition Λ_B. For Construction A, as explained above, Λ_B is partitioned using $\Lambda_E = q\mathbb{Z}^n$. More general versions of Construction A can be used for the same purpose (see, e.g., [661, 1154] and references therein), and one may ask for a design criterion to choose nested pairs of lattices $\Lambda_E \subset \Lambda_B$ that bring most confidentiality. The **(strong) secrecy gain** $\chi_{\Lambda, strong}$ of an n-dimensional lattice Λ defined by

$$\chi_{\Lambda, strong} = \sup_{y>0} \frac{\Theta_{\nu\mathbb{Z}^n}(y)}{\Theta_\Lambda(y)}$$

for $y > 0$ was proposed as one design criterion in [1461], where Θ_Λ is the **theta series** of Λ [442] defined by

$$\Theta_\Lambda(z) = \sum_{\mathbf{x} \in \Lambda} q^{\|\mathbf{x}\|^2} \quad \text{for } q = e^{i\pi z}, \ \text{Im}(z) > 0$$

(we set $y = -iz$). Note the normalization factor ν which ensures that both $\mathbb{Z}^n$ and Λ have the same volume, for fair comparison. The theta series of an integral lattice keeps track of the different norms of lattice points. The coefficient $N(m)$ of q^m in this series tells how many points in the lattice are at squared distance m from the origin. This series always starts with 1, corresponding to the zero vector. The second term corresponds to the squared minimum norm λ^2, and thus the coefficient $N(\lambda^2)$ of q^{λ^2} is the kissing number κ of the lattice.

The theta series of a lattice is not that easy to compute in full generality. If the lattice is obtained using Construction A over $\mathbb{F}_2$, then $\Theta_{\Lambda_\mathcal{C}}(z) = \text{Hwe}_\mathcal{C}(\theta_3(2z), \theta_2(2z))$ where $\text{Hwe}_\mathcal{C}(x, y)$ is the Hamming weight enumerator of $\mathcal{C}$ and θ_2, θ_3 are Jacobi theta functions [442].

The role of the theta series Θ_{Λ_E} at the point $y = \frac{1}{2\pi\sigma_E^2}$ has been independently confirmed in [1266], where it was shown for the mod-Λ Gaussian channel that the mutual information $I(\mathbf{S}; \mathbf{Z})$, an information theoretic measure of the amount of information that Eve gets about the secret message $\mathbf{S}$ by receiving $\mathbf{Z}$, is bounded by a function that depends of the channel parameters and of $\Theta_{\Lambda_E}\left(\frac{1}{2\pi\sigma_E^2}\right)$.

Chapter 27

Quantum Error-Control Codes

Martianus Frederic Ezerman

Nanyang Technological University

27.1   Introduction

Information is physical. It is sensible to use quantum mechanics as a basis of computation and information processing [724]. Here at the intersection of information theory, computing, and physics, mathematicians and computer scientists must think in terms of the quantum physical realizations of messages. The often philosophical debates among physicists over the nature and interpretations of quantum mechanics shift to harnessing its power for information processing and testing the theory for completeness.

One cannot directly access information stored and processed in massively entangled quantum systems without destroying the content. Turning large-scale quantum computing into practical reality is massively challenging. To start with, it requires techniques for error control that are much more complex than those implemented effectively in classical systems. As a quantum system grows in size and circuit depth, error control becomes ever more important.

Quantum error-control is a set of methods to protect quantum information from unwanted environmental interactions, known as *decoherence*. Classically, one encodes information-carrying vectors into a larger space to allow for sufficient redundancy for error detection and correction. In the quantum setup, information is stored in a subspace embedded in a larger Hilbert space, which is a finite dimensional, normed, vector space over the field of complex numbers $\mathbb{C}$. Codewords are quantum states and errors are operators.

The good news is that noise, if it can be kept below a certain level, is not an obstacle to resilient quantum computation. This crucial insight is arrived at based on seminal results that form the so-called threshold theorems. Theoretical references include the exposition of Knill et al. in [1131], the work of Preskill in [1542], the results shown by Steane in [1751], and the paper of Ahoronov and Ben-Or [24]. A comprehensive review on related experiments is available in [334].

The possibility of correcting errors in quantum systems was shown, e.g., in the early works of Shor [1695], Steane [1749], and Laflamme et al. [1187]. While the quantum codes that these pioneers proposed may nowadays seem to be rather trivial in performance, their construction highlighted both the main obstacles and their respective workarounds. Measurement collapses the information contained in the state into something useless. One should measure the error, not the data. Since repetition is ruled out due to the *No-cloning Theorem* [1907], we use redundancy from spreading the states to avoid repetition. There are multiple types of errors, as we will soon see. The key is to start by correcting the phase errors, and then use the Hadamard transform to exchange the bit flips and the phase errors. Quantum errors are continuous. Controlling them seemed to be too daunting a task. It turned out that handling a set of discrete *error operators*, represented by the tensor product of Pauli matrices, allows for the control of every $\mathbb{C}$-linear combination of these operators.

Advances continue to be made as effort intensifies to scale quantum computing up. Research in **quantum error-correcting codes (QECs)** has attracted the sustained attention of established researchers and students alike. Several excellent online lecture notes, surveys, and books are available. Developments up to 2011 have been well-documented in [1252]. It is impossible to describe the technical details of every important research direction in QECs. We focus on **quantum stabilizer codes** and their variants. The decidedly biased take here is for the audience with more applied mathematics background, including coding theorists, information theorists, and researchers in discrete mathematics and finite geometries. No knowledge of quantum mechanics is required beyond the very basic. This chapter is meant to serve as an entry point for those who want to understand and get involved in building upon this foundational aspect of quantum information processing and computation, which have been proven to be indispensable in future technologies.

A quantum stabilizer code is designed so that errors with high probability of occuring transform information-carrying states to an error space which is orthogonal to the original space. The beauty lies in how natural the determination of the location and type of each error in the system is. Correction becomes a simple application of the type of error at the very location.

27.2 Preliminaries

Consider the field extensions $\mathbb{F}_p$ to $\mathbb{F}_{q=p^r}$ to $\mathbb{F}_{q^m=p^{rm}}$, for prime p and positive integers r and m. For $\alpha \in \mathbb{F}_{q^m}$, the **trace mapping** from $\mathbb{F}_{q^m}$ to $\mathbb{F}_q$ is given by

$$\mathrm{Tr}_{\mathbb{F}_{q^m}/\mathbb{F}_q}(\alpha) = \alpha + \alpha^q + \ldots + \alpha^{q^{m-1}} \in \mathbb{F}_q.$$

The trace of α is the sum of its **conjugates**. If the extension $\mathbb{F}_q$ of $\mathbb{F}_p$ is contextually clear, the notation Tr is sufficient. Properties of the trace map can be found in standard textbooks, e.g., [1254, Chapter 2]. The key idea is that $\mathrm{Tr}_{\mathbb{F}_{q^m}/\mathbb{F}_q}$ serves as a *description* for all linear transformations from $\mathbb{F}_{q^m}$ to $\mathbb{F}_q$.

Let G be a finite *abelian* group, for now written multiplicatively, with identity 1_G. Let U be the multiplicative group of complex numbers of modulus 1, i.e., the unit circle of radius 1 on the complex plane $\mathbb{C}$. A **character** $\chi : G \to U$ is a homomorphism. For any $g \in G$, the images of χ are $|G|^{\mathrm{th}}$ roots of unity since $(\chi(g))^{|G|} = \chi\left(g^{|G|}\right) = \chi(1_G) = 1$. Let $\bar{c}$ denote the complex conjugate of c. Then $\chi(g^{-1}) = (\chi(g))^{-1} = \overline{\chi(g)}$. The only trivial character is $\chi_0 : g \mapsto 1$ for all $g \in G$. One can associate χ and $\overline{\chi}$ by using $\overline{\chi}(g) = \overline{\chi(g)}$. The set of all characters of G forms, under function multiplication, an abelian group $\widehat{G}$. For $g, h \in G$ and

$\chi, \Psi \in \widehat{G}$ we have two **orthogonality relations**

$$\sum_{g \in G} \chi(g) \, \overline{\Psi(g)} = \begin{cases} 0 & \text{if } \chi \neq \Psi, \\ |G| & \text{if } \chi = \Psi \end{cases} \quad \text{and} \quad \sum_{\chi \in \widehat{G}} \chi(g) \, \chi(h^{-1}) = \begin{cases} 0 & \text{if } g \neq h, \\ |G| & \text{if } g = h. \end{cases} \quad (27.1)$$

Consider the additive abelian group $(\mathbb{F}_q, +)$. The additive character $\chi_1 : (\mathbb{F}_q, +) \to U$ given by $\chi_1 : c \mapsto e^{\frac{2\pi}{p} \text{Tr}(c)}$, for all $c \in \mathbb{F}_q$, with $\text{Tr} = \text{Tr}_{\mathbb{F}_q/\mathbb{F}_p}$, is called the **canonical character** of $(\mathbb{F}_q, +)$. For a chosen $b \in \mathbb{F}_q$ and for all $c \in \mathbb{F}_q$,

$$\chi_b := \mathbb{F}_q \to U \quad \text{where} \quad c \mapsto \chi_1(b \cdot c) = e^{\frac{2\pi}{p} \text{Tr}(b \cdot c)}$$

is a character of $(\mathbb{F}_q, +)$. Every character of $(\mathbb{F}_q, +)$ can, in fact, be expressed in this manner. The extension to $(\mathbb{F}_q^n, +)$ is straightforward.

Theorem 27.2.1 *Let* $\zeta := e^{\frac{2\pi}{p}}$ *and* $\text{Tr} = \text{Tr}_{\mathbb{F}_q/\mathbb{F}_p}$ *be the trace map with* $q = p^m$. *Let* $\mathbf{a} = (a_1, a_2, \ldots, a_n)$ *and* $\mathbf{b} = (b_1, b_2, \ldots, b_n)$ *be vectors in* $\mathbb{F}_q^n$. *For each* $\mathbf{a}$,

$$\lambda_{\mathbf{a}} : \mathbb{F}_q^n \mapsto \{1, \zeta, \zeta^2, \ldots, \zeta^{p-1}\} \quad \text{where} \quad \mathbf{b} \mapsto \zeta^{\text{Tr}(\mathbf{a} \cdot \mathbf{b})} = \zeta^{\text{Tr}(a_1 b_1 + \ldots + a_n b_n)} \text{ for all } \mathbf{b} \in \mathbb{F}_q$$

is a character of $(\mathbb{F}_q^n, +)$. *Hence* $\widehat{\mathbb{F}_q^n} = \{\lambda_{\mathbf{a}} \mid \mathbf{a} \in \mathbb{F}_q^n\}$.

A *qubit*, a term coined by Schumacher in [1635], is the *canonical quantum system* consisting of two distinct levels. The states of a qubit live in $\mathbb{C}^2$ and are defined by their continuous amplitudes. A *qudit* refers to a system of $q \geq 3$ distinct levels, with a *qutrit* commonly used when $q = 3$. Physicists prefer the "**bra**" $\langle \cdot |$ and "**ket**" $| \cdot \rangle$ notation to describe quantum systems. A $| \varphi \rangle$ is a (column) vector while $\langle \psi |$ is the vector dual of $| \psi \rangle$.

Definition 27.2.2 (Quantum systems) A **qubit** is a nonzero vector in $\mathbb{C}^2$, usually with basis $\{|0\rangle, |1\rangle\}$. It is written in vector form as $|\varphi\rangle := \alpha |0\rangle + \beta |1\rangle$, or in matrix form as $\begin{bmatrix} \alpha \\ \beta \end{bmatrix}$, with $\|\alpha\|^2 + \|\beta\|^2 = 1$.

An n-**qubit system** or vector is a nonzero element in $\left(\mathbb{C}^2\right)^{\otimes n} \cong \mathbb{C}^{2^n}$. Let $\mathbf{a} = (a_1, \ldots, a_n) \in \mathbb{F}_2^n$. The standard $\mathbb{C}$-basis is

$$\{|a_1 a_2 \cdots a_n\rangle := |a_1\rangle \otimes |a_2\rangle \otimes \cdots \otimes |a_n\rangle \mid \mathbf{a} \in \mathbb{F}_2^n\}.$$

An arbitrary nonzero vector in $\mathbb{C}^{2^n}$ is written

$$|\psi\rangle = \sum_{\mathbf{a} \in \mathbb{F}_2^n} c_{\mathbf{a}} |\mathbf{a}\rangle \quad \text{with } c_{\mathbf{a}} \in \mathbb{C} \text{ and } \frac{1}{2^n} \sum_{\mathbf{a} \in \mathbb{F}_2^n} \|c_{\mathbf{a}}\|^2 = 1.$$

The normalization is optional since $|\psi\rangle$ and $\alpha |\psi\rangle$ are considered the same state for nonzero $\alpha \in \mathbb{C}$.

The **inner product** of $|\psi\rangle := \sum_{\mathbf{a} \in \mathbb{F}_2^n} c_{\mathbf{a}} |\mathbf{a}\rangle$ and $|\varphi\rangle := \sum_{\mathbf{a} \in \mathbb{F}_2^n} b_{\mathbf{a}} |\mathbf{a}\rangle$ is

$$\langle \psi | \varphi \rangle = \sum_{\mathbf{a} \in \mathbb{F}_2^n} \overline{c_{\mathbf{a}}} \, b_{\mathbf{a}}.$$

Their **(Kronecker) tensor product** is written as $|\varphi\rangle \otimes |\psi\rangle$ and is often abbreviated to $|\varphi \psi\rangle$. The states $|\psi\rangle$ and $|\varphi\rangle$ are **orthogonal** or **distinguishable** if $\langle \psi | \varphi \rangle = 0$. Let A be a $2^n \times 2^n$ complex unitary matrix with conjugate transpose $A^\dagger$. The **(Hermitian) inner product** of $|\varphi\rangle$ and $A|\psi\rangle$ is equal to that of $A^\dagger |\varphi\rangle$ and $|\psi\rangle$. Henceforth, $i := \sqrt{-1}$.

Definition 27.2.3 A **qubit error operator** is a unitary $\mathbb{C}$-linear operator acting on $\mathbb{C}^{2^n}$ qubit by qubit. It can be expressed by a unitary matrix with respect to the basis $\{|0\rangle, |1\rangle\}$. The three *nontrivial* errors acting on a qubit are known as the **Pauli matrices**:

$$\sigma_x = \begin{bmatrix} 0 & 1 \\ 1 & 0 \end{bmatrix}, \quad \sigma_z = \begin{bmatrix} 1 & 0 \\ 0 & -1 \end{bmatrix}, \quad \sigma_y = i\,\sigma_x\sigma_z = \begin{bmatrix} 0 & -i \\ i & 0 \end{bmatrix}.$$

The actions of the error operators on a qubit $|\varphi\rangle = \alpha|0\rangle + \beta|1\rangle \in \mathbb{C}^2$ can be considered based on their types. The **trivial operator** I_2 leaves the qubit unchanged. The **bit-flip error** σ_x flips the probabilities

$$\sigma_x\,|\varphi\rangle = \beta|0\rangle + \alpha|1\rangle \quad \text{or} \quad \sigma_x \begin{bmatrix} \alpha \\ \beta \end{bmatrix} = \begin{bmatrix} \beta \\ \alpha \end{bmatrix}.$$

The **phase-flip error** σ_z modifies the angular measures

$$\sigma_z\,|\varphi\rangle = \alpha|0\rangle - \beta|1\rangle \quad \text{or} \quad \sigma_z \begin{bmatrix} \alpha \\ \beta \end{bmatrix} = \begin{bmatrix} \alpha \\ -\beta \end{bmatrix}.$$

The **combination error** σ_y contains both a bit-flip and a phase-flip, implying

$$\sigma_y\,|\varphi\rangle = -i\beta|0\rangle + i\alpha|1\rangle \quad \text{or} \quad \sigma_y \begin{bmatrix} \alpha \\ \beta \end{bmatrix} = \begin{bmatrix} -i\beta \\ i\alpha \end{bmatrix}.$$

It is immediate to confirm that $\sigma_x^2 = \sigma_y^2 = \sigma_z^2 = I_2$ and $\sigma_x\sigma_z = -\sigma_z\sigma_x$. The Pauli matrices generate a group of order 16. Each of its elements can be uniquely represented as $i^\lambda w$, with $\lambda \in \{0,1,2,3\}$ and $w \in \{I_2, \sigma_x, \sigma_z, \sigma_y\}$.

27.3 The Stabilizer Formalism

The most common route from classical coding theory to QEC is via the *stabilizer formalism*, from which numerous specific constructions emerge. Classical codes *cannot* be used as quantum codes but can *model* the error operators in some quantum channels. The capabilities of a QEC can then be inferred from the properties of the corresponding classical codes. The main tools come from character theory and symplectic geometry over finite fields. Our focus is on the qubit setup since it is the most deployment-feasible and because the general qudit case naturally follows from it.

Let $\mathbf{a} = (a_1, a_2, \ldots, a_n) \in \mathbb{F}_2^n$, $\lambda \in \{0,1,2,3\}$, and $w_j \in \{I_2, \sigma_x, \sigma_z, \sigma_y\}$. A **quantum error operator** on $\mathbb{C}^{2^n}$ is of the form $\mathrm{E} := i^\lambda w_1 \otimes w_2 \otimes \cdots \otimes w_n$. It is a $\mathbb{C}$-linear unitary operator acting on a $\mathbb{C}^{2^n}$-basis $\{|\mathbf{a}\rangle = |a_1\rangle \otimes |a_2\rangle \otimes \cdots \otimes |a_n\rangle\}$ by $\mathrm{E}|\mathbf{a}\rangle := i^\lambda (w_1|a_1\rangle \otimes w_2|a_2\rangle \otimes \cdots \otimes w_n|a_n\rangle)$. The set of error operators

$$\mathcal{E}_n := \{i^\lambda w_1 \otimes w_2 \otimes \cdots \otimes w_n\}$$

is a non-abelian group of cardinality 4^{n+1}. Given $\mathrm{E} := i^\lambda w_1 \otimes w_2 \otimes \cdots \otimes w_n$ and $\mathrm{E}' := i^{\lambda'} w_1' \otimes w_2' \otimes \cdots \otimes w_n'$ in $\mathcal{E}_n$, we have

$$\mathrm{E}\,\mathrm{E}' = i^{\lambda + \lambda'} (w_1 w_1') \otimes (w_2 w_2') \otimes \cdots \otimes (w_n w_n')$$

$$= i^{\lambda + \lambda' + \lambda''} w_1'' \otimes w_2'' \otimes \cdots \otimes w_n'' \quad \text{where } w_j w_j' = i^{\lambda_j''} w_j'' \text{ and } \lambda'' = \sum_{j=1}^{n} \lambda_j''.$$

Expanding $\mathrm{E}'\mathrm{E}$ makes it clear that $\mathrm{E}\,\mathrm{E}' = \pm 1\,\mathrm{E}'\mathrm{E}$.

Example 27.3.1 Given $n = 2$, $E = I_2 \otimes \sigma_x$ and $E' = \sigma_z \otimes \sigma_y$, we have $E\,E' = \sigma_z \otimes \sigma_x \sigma_y = \sigma_z \otimes i\sigma_z = i\sigma_z \otimes \sigma_z$ and $E'\,E = \sigma_z \otimes \sigma_y \sigma_x = \sigma_z \otimes i^3 \sigma_z = i^3 \sigma_z \otimes \sigma_z$.

The **center** of $\mathcal{E}_n$ is $\mathcal{Z}(\mathcal{E}_n) := \{i^\lambda I_2 \otimes I_2 \otimes \ldots \otimes I_2\}$. Let $\overline{\mathcal{E}_n}$ denote the quotient group $\mathcal{E}_n / \mathcal{Z}(\mathcal{E}_n)$ of cardinality $|\overline{\mathcal{E}_n}| = 4^n$. This group is an elementary abelian 2-group $\cong (\mathbb{F}_2^{2n}, +)$, since $\overline{E}^2 = I_2 \otimes \cdots \otimes I_2 = I_{2^n}$ for any $\overline{E} \in \overline{\mathcal{E}_n}$.

We switch notation to define the product of error operators in terms of an inner product of their vector representatives. We write $E = i^\lambda w_1 \otimes w_2 \otimes \cdots \otimes w_n$ as $E = i^{\lambda + \epsilon} X(\mathbf{a}) Z(\mathbf{b})$, where $\mathbf{a} = (a_1, \ldots, a_n)$, $\mathbf{b} = (b_1, \ldots, b_n) \in \mathbb{F}_2^n$ and $\epsilon := \left| \{i \mid w_i = \sigma_y, 1 \leq i \leq n\} \right|$, by setting (a_i, b_i) equal to $(0,0)$ if $w_i = I_2$, equal to $(1,0)$ if $w_i = \sigma_x$, equal to $(0,1)$ if $w_i = \sigma_z$, and equal to $(1,1)$ if $w_i = \sigma_y$.

The respective actions of $X(\mathbf{a})$ and $Z(\mathbf{b})$ on any vector $|\mathbf{v}\rangle \in \mathbb{C}^{2^n}$, for $\mathbf{v} \in \mathbb{F}_2^n$, are $X(\mathbf{a})|\mathbf{v}\rangle = |\mathbf{a} + \mathbf{v}\rangle$ and $Z(\mathbf{b})|\mathbf{v}\rangle = (-1)^{\mathbf{b} \cdot \mathbf{v}} |\mathbf{v}\rangle$. The matrix for $X(\mathbf{a})$ is a symmetric $\{0,1\}$ matrix. It represents a permutation consisting of 2^{n-1} transpositions. The matrix for $Z(\mathbf{b})$ is diagonal with diagonal entries ± 1. Hence, writing the operators in $\mathcal{E}_n$ as $E := i^\lambda X(\mathbf{a}) Z(\mathbf{b})$ and $E' := i^{\lambda'} X(\mathbf{a}') Z(\mathbf{b}')$, one gets $E\,E' = (-1)^{\mathbf{a} \cdot \mathbf{b}' + \mathbf{a}' \cdot \mathbf{b}} E'\,E$. The **symplectic inner product** of $(\mathbf{a}|\mathbf{b})$ and $(\mathbf{a}'|\mathbf{b}')$ in $\mathbb{F}_2^{2n}$ is

$$\langle (\mathbf{a}|\mathbf{b}), (\mathbf{a}'|\mathbf{b}') \rangle_s = \mathbf{a} \cdot \mathbf{b}' + \mathbf{a}' \cdot \mathbf{b} \tag{27.2}$$

or, in matrix form,

$$\langle (\mathbf{a}|\mathbf{b}), (\mathbf{a}'|\mathbf{b}') \rangle_s = \begin{bmatrix} \mathbf{a} & \mathbf{b} \end{bmatrix} \begin{bmatrix} 0 & I_n \\ I_n & 0 \end{bmatrix} \begin{bmatrix} \mathbf{a}'^T \\ \mathbf{b}'^T \end{bmatrix}.$$

The **symplectic dual** of $\mathcal{C} \subseteq \mathbb{F}_2^{2n}$ is $\mathcal{C}^{\perp_s} = \{\mathbf{u} \in \mathbb{F}_2^{2n} \mid \langle \mathbf{u}, \mathbf{c} \rangle_s = 0 \text{ for all } \mathbf{c} \in \mathcal{C}\}$. Thus, a subgroup G of $\mathcal{E}_n$ is abelian if and only if $\overline{G}$ is a symplectic self-orthogonal subspace of $\overline{\mathcal{E}_n} \cong \mathbb{F}_2^{2n}$.

Example 27.3.2 Continuing from Example 27.3.1, we write $E = X((0,1)) Z((0,0))$ and $E' = iX((0,1)) Z((1,1))$. We choose the ordering $(0,0), (0,1), (1,0), (1,1)$ of $\mathbb{F}_2^2$ and the corresponding ordering for the basis of $\mathbb{C}^4$. The matrix for $X((0,1))$ agrees with $I_2 \otimes \sigma_x$, the matrix for $Z((0,0))$ is I_4, and the matrix for $Z((1,1))$ is diagonal with diagonal entries $1, -1, -1, 1$. Multiplying matrices confirms that $\sigma_z \otimes \sigma_y$ is indeed $iX((0,1)) Z((1,1))$.

The **quantum weight** of an error operator $E = i^\lambda X(\mathbf{a}) Z(\mathbf{b}) \in \mathcal{E}_n$ is

$$\mathrm{wt}_Q(E) := \mathrm{wt}_Q(\overline{E}) = \mathrm{wt}_Q(\mathbf{a}|\mathbf{b}) = \left| \{i \mid a_i = 1 \text{ or } b_i = 1, \ 1 \leq i \leq n\} \right|$$
$$= \left| \{i \mid w_i \neq I_2, \ 1 \leq i \leq n\} \right|.$$

By definition, $\mathrm{wt}_Q(E\,E') \leq \mathrm{wt}_Q(E) + \mathrm{wt}_Q(E')$, for any $E, E' \in \mathcal{E}_n$. We can define the set of all error operators of weight at most δ in $\mathcal{E}_n$ and determine its cardinality. Let

$$\mathcal{E}_n(\delta) := \{E \in \mathcal{E}_n \mid \mathrm{wt}_Q(E) \leq \delta\} \quad \text{and} \quad \overline{\mathcal{E}_n}(\delta) = \{\overline{E} \in \overline{\mathcal{E}_n} \mid \mathrm{wt}_Q(\overline{E}) \leq \delta\}.$$

Then $|\mathcal{E}_n(\delta)| = 4 \sum_{j=0}^\delta 3^j \binom{n}{j}$ and $|\overline{\mathcal{E}_n}(\delta)| = \sum_{j=0}^\delta 3^j \binom{n}{j}$.

In the classical setup, both errors and codewords are vectors over the same field. In the quantum setup, errors are linear combinations of the tensor products of Pauli matrices. A qubit code $Q \subseteq \mathbb{C}^{2^n}$ has three parameters: its **length** n, **dimension** K over $\mathbb{C}$, and **minimum distance** $d = d(Q)$. We use

$$((n, K, d)) \quad \text{or} \quad [[n, k, d]] \text{ with } k = \log_2 K$$

to signify that Q describes the encoding of k logical qubits as n physical qubits, with d being the smallest number of simultaneous errors that can transform a valid codeword into another.

Definition 27.3.3 (Knill–Laflamme condition [1130]) A quantum code Q can correct up to ℓ quantum errors if the following hold. If $|\varphi\rangle, |\psi\rangle \in Q$ are distinguishable, i.e., $\langle\varphi|\psi\rangle = 0$, then $\langle\varphi|E_1 E_2|\psi\rangle = 0$, i.e., $E_1|\varphi\rangle$ and $E_2|\psi\rangle$ must remain distinguishable, for all $E_1, E_2 \in \mathcal{E}_n(\ell)$. The minimum distance of Q is $d := d(Q)$ if $\langle\varphi|E|\psi\rangle = 0$ for all $E \in \mathcal{E}_n(d-1)$ and for all distinguishable $|\varphi\rangle, |\psi\rangle \in Q$.

Given an $((n, K, d))$-qubit code Q and an $E \in \mathcal{E}_n$, EQ is a subspace of $\mathbb{C}^{2^n}$. The fact that Q corrects errors of weight up to $\ell = \lfloor\frac{d-1}{2}\rfloor$ does *not* imply that the subspaces in $\{\overline{E}Q \mid \overline{E} \in \overline{\mathcal{E}_n}(\ell)\}$ are orthogonal to each other. It is possible that a codeword $|v\rangle$ is fixed by some $E \neq I_{2^n}$, say, when $|v\rangle$ is an **eigenvector** of E satisfying $E|v\rangle = \alpha|v\rangle$ for some nonzero $\alpha \in \mathbb{C}$. If the subspaces $\{\overline{E}Q \mid \overline{E} \in \overline{\mathcal{E}_n}(\ell)\}$ are orthogonal to each other, then Q is said to be **pure**. Otherwise, the code is **degenerate** or **impure**.

To formally define a **qubit stabilizer code**, we choose an abelian group G, which is a subgroup of $\mathcal{E}_n$, and associate G with a classical code $\mathcal{C} \subset \mathbb{F}_2^{2n}$, which is *self-orthogonal* under the symplectic inner product. The action of G partitions $\mathbb{C}^{2^n}$ into a direct sum of χ-eigenspaces $Q(\chi)$ with $\chi \in \widehat{G}$. The properties of $Q := Q(\chi)$ follow from the properties of $\mathcal{C}$ and $\mathcal{C}^{\perp_s}$. The *stabilizer formalism*, first introduced by Gottesman in his thesis [843] and described in the language of group algebra by Calderbank et al. in [331], remains the most widely-studied approach to control quantum errors. Ketkar et al. generalized the formalism to qudit codes derived from classical codes over $\mathbb{F}_{q^2}$ in [1109].

Let G be a finite abelian group acting on a finite dimensional $\mathbb{C}$-vector space V. Each $g \in G$ is a Hermitian operator of V and, for any $g, g' \in G$ and for all $|v\rangle \in V$, $(gg')|v\rangle = g(g'(|v\rangle))$ and $gg^{-1}(|v\rangle) = |v\rangle$. Let $\widehat{G}$ be the character group of G. For any $\chi \in \widehat{G}$, the map $L_\chi := \frac{1}{|G|}\sum_{g \in G} \overline{\chi}(g)\, g$ is a linear operator over V. The set $\{L_\chi \mid \chi \in \widehat{G}\}$ is the **system of orthogonal primitive idempotent operators**.

Proposition 27.3.4 L_χ *is idempotent, i.e.,* $L_\chi^2 = L_\chi$ *and* $L_\chi L_{\chi'} = 0$ *if* $\chi \neq \chi'$. *The operators in the system sum to the identity* $\sum_{\chi \in \widehat{G}} L_\chi = 1$. *For all* $g \in G$, *we have* $g\,L_\chi = \chi(g)\,L_\chi$.

Proof: In G, let $gh = a$, i.e., $h = ag^{-1}$. Using $\overline{\chi} = \chi^{-1}$ and the orthogonality of characters, we write

$$L_\chi L_{\chi'} = \frac{1}{|G|^2}\sum_{g \in G}\overline{\chi}(g)\, g\,\sum_{h \in G}\overline{\chi'}(h)\, h = \frac{1}{|G|^2}\sum_{a,g \in G}\overline{\chi}(g)\,\overline{\chi'}(ag^{-1})\, a$$

$$= \frac{1}{|G|^2}\sum_{a \in G}\overline{\chi'}(a)\, a\,\sum_{g \in G}(\overline{\chi}\chi')(g).$$

The third equality comes from collecting terms that contain only a and only g. By the first orthogonality relation in (27.1), one arrives at

$$L_\chi L_{\chi'} = \frac{1}{|G|}\sum_{a \in G}\overline{\chi'}(a)a = \begin{cases} 0 & \text{if } \chi \neq \chi', \\ L_\chi & \text{if } \chi = \chi'. \end{cases}$$

We verify the second assertion by using the second orthogonality relation in (27.1). Since $\overline{\chi}(1) = 1$, we obtain

$$\sum_{\chi \in \widehat{G}} L_\chi = \frac{1}{|G|}\sum_{\chi \in \widehat{G}}\sum_{g \in G}\overline{\chi}(g)\, g = \frac{1}{|G|}\sum_{g \in G} g\,\sum_{\chi \in \widehat{G}}\overline{\chi}(g)\,\overline{\chi}(1) = 1.$$

Using $g\,h = a$, the definition of L_χ, and the equality $\overline{\chi}(g^{-1}) = \chi(g)$, one gets

$$g\,L_\chi = \frac{1}{|G|} \sum_{h \in G} \overline{\chi}(h)\,g\,h = \frac{1}{|G|} \sum_{a \in G} \overline{\chi}(ag^{-1})\,a = \frac{1}{|G|}\overline{\chi}(g^{-1}) \sum_{a \in G} \overline{\chi}(a)\,a = \chi(g)\,L_\chi. \qquad \square$$

Proposition 27.3.5 *For each* $\chi \in \widehat{G}$, *let* $V(\chi) := L_\chi V = \{L_\chi(|\mathbf{v}\rangle) \mid |\mathbf{v}\rangle \in V\}$. *For* $|\mathbf{v}\rangle \in V(\chi)$ *and* $g \in G$, *we have* $g|\mathbf{v}\rangle = \chi(g)\,|\mathbf{v}\rangle$. *Thus* $V(\chi)$ *is a common eigenspace of all operators in* G. *A direct decomposition* $V = \bigoplus_{\chi \in \widehat{G}} V(\chi)$ *ensures that each* $|\mathbf{v}\rangle \in V$ *has a unique expression*

$$|\mathbf{v}\rangle = \sum_{\chi \in \widehat{G}} |\mathbf{v}\rangle_\chi \quad \text{where } |\mathbf{v}\rangle_\chi \in V(\chi).$$

Proof: For $|\mathbf{v}\rangle \in V(\chi)$ and $g \in G$, there exists $|\mathbf{w}\rangle \in V$ such that

$$g|\mathbf{v}\rangle = g\,L_\chi(|\mathbf{w}\rangle) = \chi(g)\,L_\chi(|\mathbf{w}\rangle) = \chi(g)|\mathbf{v}\rangle,$$

confirming the first assertion, where the second equality follows from Proposition 27.3.4.

Using Proposition 27.3.4 again, if $|\mathbf{v}\rangle \in V$, we write $|\mathbf{v}\rangle = 1|\mathbf{v}\rangle = \left(\sum_{\chi \in \widehat{G}} L_\chi\right)|\mathbf{v}\rangle = \sum_{\chi \in \widehat{G}} |\mathbf{v}\rangle_\chi$, where $|\mathbf{v}\rangle_\chi := L_\chi(|\mathbf{v}\rangle) \in V(\chi)$. On the other hand, if $|\mathbf{v}\rangle = \sum_{\chi \in \widehat{G}} |\mathbf{u}\rangle_\chi$ for $|\mathbf{u}\rangle_\chi \in V(\chi)$, then $|\mathbf{u}\rangle_\chi = L_\chi(|\mathbf{w}\rangle_\chi)$ for some $|\mathbf{w}\rangle_\chi \in V$. Since $\{L_\chi \mid \chi \in \widehat{G}\}$ has the orthogonality property, for every $\chi \in \widehat{G}$, we have

$$|\mathbf{v}\rangle_\chi = L_\chi(|\mathbf{v}\rangle) = L_\chi\left(\sum_{\chi' \in \widehat{G}} |\mathbf{u}\rangle_{\chi'}\right) = L_\chi\left(\sum_{\chi' \in \widehat{G}} L_{\chi'}(|\mathbf{w}\rangle_{\chi'})\right)$$
$$= \sum_{\chi' \in \widehat{G}} L_\chi L_{\chi'}(|\mathbf{w}\rangle_{\chi'}) = L_\chi(|\mathbf{w}\rangle_\chi) = |\mathbf{u}\rangle_\chi.$$

Thus, $V = \bigoplus_{\chi \in \widehat{G}} V(\chi)$. $\qquad \square$

It is also a well-known fact that $V(\chi)$ and $V(\chi')$ are Hermitian orthogonal for all $\chi \neq \chi' \in \widehat{G}$ when all $g \in G$ are unitary linear operators on V.

All the tools to connect qubit stabilizer codes to classical codes are now in place. We choose $G := \langle g_1, g_2, \dots, g_k \rangle$ to be an abelian subgroup of $\mathcal{E}_n$ with $g_j := i^{\lambda_j} X(\mathbf{a}_j)\,Z(\mathbf{b}_j)$ for $1 \le j \le k$, where $\mathbf{a}_j, \mathbf{b}_j \in \mathbb{F}_2^n$ and $\lambda_j \equiv \mathbf{a}_j \cdot \mathbf{b}_j \pmod 2$. Since σ_x and σ_z are Hermitian unitary matrices, $X(\mathbf{a}_j)$ and $Z(\mathbf{b}_j)$ are also Hermitian matrices. The basis element g_j is Hermitian since

$$g_j^\dagger := \overline{g}_j^{\mathsf{T}} = (-i)^{\lambda_j}\,Z(\mathbf{b}_j)^{\mathsf{T}}\,X(\mathbf{a}_j)^{\mathsf{T}} = (-i)^{\lambda_j}\,Z(\mathbf{b}_j)\,X(\mathbf{a}_j)$$
$$= i^{\lambda_j}\,(-1)^{\mathbf{a}_j \cdot \mathbf{b}_j}\,(-1)^{\mathbf{a}_j \cdot \mathbf{b}_j}\,X(\mathbf{a}_j)\,Z(\mathbf{b}_j) = i^{\lambda_j}\,X(\mathbf{a}_j)\,Z(\mathbf{b}_j).$$

Theorem 27.3.6 *Let* $\mathcal{C}$ *be an* $(n - k)$-*dimensional self-orthogonal subspace of* $\mathbb{F}_2^{2n}$ *under the symplectic inner product. Let* $d := w_Q\left(\mathcal{C}^{\perp_s} \setminus \mathcal{C}\right) = \min\left\{\operatorname{wt}_Q(\mathbf{v}) \mid \mathbf{v} \in \mathcal{C}^{\perp_s} \setminus \mathcal{C}\right\}$. *Then there is an* $[\![n, k, d]\!]$-*qubit stabilizer code* Q.

Proof: We lift $\mathcal{C} := \overline{G} \subset \overline{\mathcal{E}_n}$ to an abelian subgroup G of $\mathcal{E}_n$, with $G \cong \mathbb{F}_2^{n-k}$. Then $\mathbb{C}^{2^n} = \bigoplus_{\chi \in \widehat{G}} Q(\chi)$, where $Q(\chi) = L_\chi(\mathbb{C}^{2^n})$, is the subspace

$$Q(\chi) = \left\{|\mathbf{v}\rangle \in \mathbb{C}^{2^n} \mid g|\mathbf{v}\rangle = \chi(g)\,|\mathbf{v}\rangle \text{ for all } g \in G\right\}.$$

Showing that each $Q(\chi)$ is an $[\![n, k, d]\!]$-qubit code means proving

$$\dim_{\mathbb{C}} Q(\chi) = 2^k \quad \text{and} \quad d(Q(\chi)) \ge w_Q\left(\mathcal{C}^{\perp_s} \setminus \mathcal{C}\right).$$

Consider the action of $\mathcal{E}_n$ on $\{Q(\chi) \mid \chi \in \widehat{G}\}$. For any $|\mathbf{v}\rangle \in Q(\chi)$ and $g \in G$, we have $g|\mathbf{v}\rangle = \chi(g)|\mathbf{v}\rangle$. Thus, for any $\mathrm{E} \in \mathcal{E}_n$ and any $g \in G$,

$$g(\mathrm{E}|\mathbf{v}\rangle) = (-1)^{\langle \overline{g}, \overline{\mathrm{E}}\rangle_s} \mathrm{E}(g|\mathbf{v}\rangle) = (-1)^{\langle \overline{g}, \overline{\mathrm{E}}\rangle_s} \chi(g)\, \mathrm{E}|\mathbf{v}\rangle.$$

Since $\chi_{\overline{\mathrm{E}}} : G \to \{\pm 1\}$ and $\chi_{\overline{\mathrm{E}}}(g) = (-1)^{\langle \overline{g}, \overline{\mathrm{E}}\rangle_s}$ is a character of G, we have

$$g(\mathrm{E}|\mathbf{v}\rangle) = \chi_{\overline{\mathrm{E}}}(g)\, \chi(g)\, \mathrm{E}|\mathbf{v}\rangle = \chi'(g)\, \mathrm{E}|\mathbf{v}\rangle \quad \text{for all } g \in G.$$

This implies $\mathrm{E}|\mathbf{v}\rangle \in Q(\chi')$ and $\mathrm{E} : Q(\chi) \to Q(\chi')$ where $\chi' := \chi_{\overline{\mathrm{E}}}\chi$. Since $\mathcal{E}_n$ is a group, E is a bijection, making $\dim_{\mathbb{C}} Q(\chi) = \dim_{\mathbb{C}} Q(\chi')$. As E runs through $\mathcal{E}_n$, $\chi_{\overline{\mathrm{E}}}$ takes all characters of G, ensuring that $\dim_{\mathbb{C}} Q(\chi)$ is the same for all $\chi \in \widehat{G}$. Thus, $\dim_{\mathbb{C}} Q(\chi) = 2^{n-(n-k)} = 2^k$ for any $\chi \in \widehat{G}$ using Proposition 27.3.5.

We now show that, if $\mathrm{E} \in \mathcal{E}_{d-1}$ and $|\mathbf{v}_1\rangle, |\mathbf{v}_2\rangle \in Q(\chi)$ with $\langle \mathbf{v}_1|\mathbf{v}_2\rangle = 0$, then $\langle \mathbf{v}_1|\mathrm{E}_1\, \mathrm{E}_2|\mathbf{v}_2\rangle = 0$, where $\mathrm{E} := \mathrm{E}_1\, \mathrm{E}_2$. If $\overline{\mathrm{E}} \in \overline{G} = \mathcal{C}$, then $\langle \mathbf{v}_1|\mathrm{E}|\mathbf{v}_2\rangle = \chi(\overline{\mathrm{E}})(\langle \mathbf{v}_1|\mathbf{v}_2\rangle) = 0$. Otherwise, $\overline{\mathrm{E}} \notin \mathcal{C}$. From $w_Q(\overline{\mathrm{E}}) = w_Q(\mathrm{E}) \leq d-1$ and the assumption $(\mathcal{C}^{\perp_s} \setminus \mathcal{C}) \cap \overline{\mathcal{E}}_{d-1} = \emptyset$, we know $\overline{\mathrm{E}} \notin \mathcal{C}^{\perp_s}$. Hence, there exists $\overline{\mathrm{E}}' \in \overline{G}$ such that $\mathrm{E}\,\mathrm{E}' = -\mathrm{E}'\,\mathrm{E}$. Then, for $|\mathbf{v}_2\rangle \in Q(\chi)$, we have

$$\mathrm{E}'\,\mathrm{E}|\mathbf{v}_2\rangle = -\mathrm{E}\,\mathrm{E}'|\mathbf{v}_2\rangle = -\chi(\mathrm{E}')\mathrm{E}|\mathbf{v}_2\rangle \quad \text{with } -\chi(\mathrm{E}') \neq \chi(\mathrm{E}').$$

Therefore, $\mathrm{E}|\mathbf{v}_2\rangle \in Q(\chi')$ for some $\chi' \neq \chi$. Since $|\mathbf{v}_1\rangle \in Q(\chi)$ and $Q(\chi)$ is orthogonal to $Q(\chi')$, we confirm $\langle \mathbf{v}_1|\mathrm{E}|\mathbf{v}_2\rangle = 0$. $\qquad\square$

Example 27.3.7 We exhibit a $[\![5,1,3]\!]$-qubit stabilizer code Q. Consider a subspace $\mathcal{C} \subset \mathbb{F}_2^{10}$ with generator matrix

$$\begin{bmatrix} 1 & 1 & 0 & 0 & 0 & 0 & 0 & 1 & 0 & 1 \\ 0 & 1 & 1 & 0 & 0 & 1 & 0 & 0 & 1 & 0 \\ 0 & 0 & 1 & 1 & 0 & 0 & 1 & 0 & 0 & 1 \\ 0 & 0 & 0 & 1 & 1 & 1 & 0 & 1 & 0 & 0 \end{bmatrix} = \begin{bmatrix} \mathbf{v}_1 \\ \mathbf{v}_2 \\ \mathbf{v}_3 \\ \mathbf{v}_4 \end{bmatrix}.$$

One reads $\mathbf{v}_1 = (\mathbf{a}|\mathbf{b})$ as having $\mathbf{a} = (1,1,0,0,0)$ and $\mathbf{b} = (0,0,1,0,1)$. The code $\mathcal{C}$ is symplectic self-orthogonal, with $\dim_{\mathbb{F}_2} \mathcal{C} = 4$ and $\dim_{\mathbb{F}_2} \mathcal{C}^\perp = 6$; i.e., the codimension is 2. To extend the basis for $\mathcal{C}$ to a basis for $\mathcal{C}^{\perp_s}$ we use $(0,0,0,0,0,\,1,1,1,1,1)$ and $(1,1,1,1,1,\,0,0,0,0,0)$. Since $w_Q(\mathcal{C}) = 4$ and $w_Q(\mathcal{C}^{\perp_s}) = 3$, one obtains $w_Q(\mathcal{C}^{\perp_s} \setminus \mathcal{C}) = 3$. We can write the $[\![5,1,3]\!]$ code $Q = Q(\chi_0)$ explicitly by using $G = \langle g_1, g_2, g_3, g_4 \rangle$, with $g_1 = \sigma_x \otimes \sigma_x \otimes \sigma_z \otimes I_2 \otimes \sigma_z$, $g_2 = \sigma_z \otimes \sigma_x \otimes \sigma_x \otimes \sigma_z \otimes I_2$, $g_3 = I_2 \otimes \sigma_z \otimes \sigma_x \otimes \sigma_x \otimes \sigma_z$, and $g_4 = \sigma_z \otimes I_2 \otimes \sigma_z \otimes \sigma_x \otimes \sigma_x$.

Since $k = 1$ and $\dim_{\mathbb{C}} Q = 2^k = 2$, two independent vectors in $\mathbb{C}^{32}$ form a basis of $Q = \{|\mathbf{v}\rangle \in \mathbb{C}^{32} \mid g|\mathbf{v}\rangle = \chi_0(g)|\mathbf{v}\rangle \text{ for all } g \in G\}$. Q consists of vectors which are fixed by all $g \in G$. After some computation, we conclude that Q can be generated by $|\mathbf{v}_0\rangle = \sum_{g \in G} g|00000\rangle$ and $|\mathbf{v}_1\rangle = \sum_{g \in G} g|11111\rangle$.

With minor modifications, the qubit stabilizer formalism extends to the general qudit case. A complete treatment is available in [1109]. We outline the main steps here. An n-**qudit system** is a nonzero element in $(\mathbb{C}^q)^{\otimes n} \cong \mathbb{C}^{q^n}$. Let $\mathbf{a} = (a_1, \ldots, a_n) \in \mathbb{F}_q^n$. The standard $\mathbb{C}$-basis is

$$\{|a_1 a_2 \ldots a_n\rangle := |a_1\rangle \otimes |a_2\rangle \otimes \ldots \otimes |a_n\rangle \mid \mathbf{a} \in \mathbb{F}_q^n\},$$

and an arbitrary vector in $\mathbb{C}^{q^n}$ is written $|\psi\rangle = \sum_{\mathbf{a} \in \mathbb{F}_q^n} c_{\mathbf{a}}|\mathbf{a}\rangle$ with $c_{\mathbf{a}} \in \mathbb{C}$ such that $q^{-n} \sum_{\mathbf{a} \in \mathbb{F}_q^n} \|c_{\mathbf{a}}\|^2 = 1$.

Let $\mathbf{a} = (a_1, \ldots, a_n)$, $\mathbf{b} = (b_1, \ldots, b_n) \in \mathbb{F}_q^n$, and $\omega := e^{\frac{2\pi i}{p}}$, where $q = p^m$ with p a prime. The error operators form $\mathcal{E}_n := \{\omega^\beta X(\mathbf{a})Z(\mathbf{b}) \mid \mathbf{a}, \mathbf{b} \in \mathbb{F}_q^n, \beta \in \mathbb{F}_p\}$ of cardinality pq^{2n}. The respective actions of $X(\mathbf{a})$ and $Z(\mathbf{b})$ on $|\mathbf{v}\rangle \in \mathbb{C}^{q^n}$ are $X(\mathbf{a})|\mathbf{v}\rangle = |\mathbf{a} + \mathbf{v}\rangle$ and $Z(\mathbf{b})|\mathbf{v}\rangle = (\omega)^{\mathrm{Tr}(\mathbf{b} \cdot \mathbf{v})}|\mathbf{v}\rangle$ where $\mathrm{Tr} = \mathrm{Tr}_{\mathbb{F}_q/\mathbb{F}_p}$. Hence, for $\mathrm{E} := \omega^\beta X(\mathbf{a})Z(\mathbf{b})$ and $\mathrm{E}' := \omega^{\beta'} X(\mathbf{a}')Z(\mathbf{b}')$ in $\mathcal{E}_n$, one gets $\mathrm{E}\,\mathrm{E}' = \omega^{\mathrm{Tr}(\mathbf{b} \cdot \mathbf{a}' - \mathbf{b}' \cdot \mathbf{a})}\,\mathrm{E}'\,\mathrm{E}$. The **symplectic weight** of $(\mathbf{a}|\mathbf{b})$ is the quantum weight of E.

The **(trace) symplectic inner product** of $(\mathbf{a}|\mathbf{b})$ and $(\mathbf{a}'|\mathbf{b}')$ in $\mathbb{F}_q^{2n}$, generalizing (27.2), is

$$\langle (\mathbf{a}|\mathbf{b}), (\mathbf{a}'|\mathbf{b}') \rangle_s = \mathrm{Tr}(\mathbf{b} \cdot \mathbf{a}' - \mathbf{b}' \cdot \mathbf{a}).$$

The **symplectic dual** of $\mathcal{C} \subseteq \mathbb{F}_q^{2n}$ is $\mathcal{C}^{\perp_s} = \{\mathbf{u} \in \mathbb{F}_q^{2n} \mid \langle \mathbf{u}, \mathbf{c} \rangle_s = 0 \text{ for all } \mathbf{c} \in \mathcal{C}\}$. As in the qubit case, in the general qudit setup, a subgroup G of $\mathcal{E}_n$ is abelian if and only if $\overline{G}$ is a symplectic self-orthogonal subspace of $\overline{\mathcal{E}_n} \cong \mathbb{F}_q^{2n}$. The analog of Theorem 27.3.6 follows.

Theorem 27.3.8 *Let $\mathcal{C}$ be an $n - k$-dimensional self-orthogonal subspace of $\mathbb{F}_q^{2n}$ under the (trace) symplectic inner product. Let $d := w_Q\left(\mathcal{C}^{\perp_s} \setminus \mathcal{C}\right) = \min\{\mathrm{wt}_Q(\mathbf{v}) \mid \mathbf{v} \in \mathcal{C}^{\perp_s} \setminus \mathcal{C}\}$. Then there is an $[\![n, k, d]\!]$-qudit stabilizer code Q.*

> "With group and eigenstate, we've learned to fix
> Your quantum errors with our quantum tricks."

> Daniel Gottesman
> in *Quantum Error Correction Sonnet*

27.4 Constructions via Classical Codes

Any stabilizer code Q is fully characterized by its stabilizer group that specifies the set of errors that Q can correct. Any linear combination of the operators in the error set is correctable, allowing Q to correct a continuous set of operators. For this reason, the best-known qubit codes in the online table [847] maintained by M. Grassl are given in terms of their stabilizer generators. The stabilizer approach has massive advantages over other frameworks, some of which will be mentioned below. It describes a large set of QECs, complete with their encoding and decoding mechanism, in a very compact form.

A valid codeword of Q is a $+1$ eigenvector of *all* the stabilizer generators. An error E, expressed as a tensor product of Pauli operators, anticommutes with some of the stabilizer generators and commutes with others. It sends a codeword to an **eigenstate** of the stabilizer generators. The eigenvalue remains $+1$ for all operators that commute with E but becomes -1 for those generators that anticommute with E. From the resulting error syndrome, one knows which Pauli operators act on which qubits. Applying the respective Pauli operators on the corresponding locations corrects the error. Suppose that the location of the error is known but the type is not, then this is a **quantum erasure**. By the Knill–Laflamme condition, correcting ℓ general errors means correcting 2ℓ erasures.

The encoding and syndrome reading circuits can be written using only three quantum gates, namely the **Hadamard gate**, the **phase S gate**, and the **CNOT gate**, whose respective matrices are

$$\mathrm{H} = \frac{1}{\sqrt{2}} \begin{bmatrix} 1 & 1 \\ 1 & -1 \end{bmatrix}, \quad \mathrm{S} = \begin{bmatrix} 1 & 0 \\ 0 & i \end{bmatrix}, \quad \mathrm{CNOT} = \begin{bmatrix} 1 & 0 & 0 & 0 \\ 0 & 1 & 0 & 0 \\ 0 & 0 & 0 & 1 \\ 0 & 0 & 1 & 0 \end{bmatrix}.$$

A treatment on the circuit implementations is available, for example, in [848].

We now look into suitable classical codes that fully describe the set of correctable errors. All constructions are applications of the stabilizer formalism. Since all of the inner products used are non-degenerate, i.e., $(\mathcal{C}^\perp)^\perp = \mathcal{C}$, one can interchange self-orthogonality and dual-containment, provided that the derived parameters are adjusted accordingly.

First, we consider a generic construction of q-ary quantum codes via additive (i.e., $\mathbb{F}_q$-linear) codes over $\mathbb{F}_{q^2}$. Let $\{1, \gamma\}$ be a basis of $\mathbb{F}_{q^2}$ over $\mathbb{F}_q$. The map $\Phi : \overline{\mathcal{E}_n} \cong \mathbb{F}_q^{2n} \to \mathbb{F}_{q^2}^n$ that sends $\overline{\mathrm{E}} := \mathbf{v} = (\mathbf{a}|\mathbf{b}) = (a_1, \ldots, a_n | b_1, \ldots, b_n)$ to $(a_1 + \gamma b_1, \ldots, a_n + \gamma b_n)$ is an isomorphism of $\mathbb{F}_q$-vector spaces. It is also an isometry, since $\mathrm{wt}_Q(\mathrm{E}) = \mathrm{wt}_H(\Phi((\mathbf{a}|\mathbf{b})))$. For any $\mathbf{u} := (u_1, \ldots, u_n)$ and $\mathbf{v} := (v_1, \ldots, v_n) \in \mathbb{F}_{q^2}^n$, we use $\mathbf{u}^q$ to denote $(u_1^q, \ldots, u_n^q)$ and define the

trace alternating inner product of $\mathbf{u}$ and $\mathbf{v}$ as $\langle \mathbf{u}, \mathbf{v} \rangle_{\mathrm{alt}} := \mathrm{Tr}_{\mathbb{F}_{q^2}/\mathbb{F}_q} \left(\dfrac{\mathbf{u} \cdot \mathbf{v}^q - \mathbf{u}^q \cdot \mathbf{v}}{\gamma - \gamma^q} \right)$.

When $q = 2$, $\langle \mathbf{u}, \mathbf{v} \rangle_{\mathrm{alt}}$ coincides with the **trace Hermitian inner product** $\langle \mathbf{u}, \mathbf{v} \rangle_{\mathrm{Tr_H}} := \sum_{j=1}^n \left(u_j v_j^2 + u_j^2 v_j \right)$, since $\gamma - \gamma^2 = 1$. For any $(\mathbf{a}, \mathbf{b})$ and $(\mathbf{a}', \mathbf{b}')$ in $\overline{\mathcal{E}_n}$, we immediately verify that $\langle (\mathbf{a}, \mathbf{b}), (\mathbf{a}', \mathbf{b}') \rangle_s = \langle \Phi((\mathbf{a}, \mathbf{b})), \Phi((\mathbf{a}', \mathbf{b}')) \rangle_{\mathrm{alt}}$. Hence, a linear code $\mathcal{C} \subseteq \mathbb{F}_q^{2n}$ is symplectic self-orthogonal if and only if the additive code $\Phi(\mathcal{C})$ is trace alternating self-orthogonal. The **Hermitian inner product** of any $\mathbf{u}, \mathbf{v} \in \mathbb{F}_{q^2}^n$ is $\langle \mathbf{u}, \mathbf{v} \rangle_{\mathrm{H}} := \sum_{j=1}^n u_j v_j^q$. If $\Phi(\mathcal{C})$ is $\mathbb{F}_{q^2}$-linear, instead of being strictly additive, then $\Phi(\mathcal{C}) \subseteq (\Phi(\mathcal{C}))^{\perp_{\mathrm{alt}}}$ if and only if $\Phi(\mathcal{C}) \subseteq (\Phi(\mathcal{C}))^{\perp_{\mathrm{H}}}$. Thus, Theorem 27.3.8 has the following equivalent statement.

Theorem 27.4.1 *Let $\mathcal{C} \subseteq \mathbb{F}_{q^2}^n$ be an $\mathbb{F}_q$-additive code such that $\mathcal{C} \subseteq \mathcal{C}^{\perp_{\mathrm{alt}}}$, with $|\mathcal{C}| = q^{n-k}$. Then there exists an $[\![n, k, d]\!]_q$ quantum code Q with*

$$d(Q) = wt_H(\mathcal{C}^{\perp_{\mathrm{alt}}} \setminus \mathcal{C}) = \min \{ wt_H(\mathbf{v}) \mid \mathbf{v} \in \mathcal{C}^{\perp_{\mathrm{alt}}} \setminus \mathcal{C} \}.$$

If $\mathcal{C}$ is $\mathbb{F}_{q^2}$-linear, we can conveniently replace the trace alternating inner product by the Hermitian inner product, which is easier to compute.

If $\mathcal{C} \subseteq \mathbb{F}_4^n$ is additive (i.e., $\mathbb{F}_2$-linear) and is even (i.e., $\mathrm{wt}_H(\mathbf{c})$ is even for all $\mathbf{c} \in \mathcal{C}$), then $\mathcal{C}$ is trace Hermitian self-orthogonal. If $\mathcal{C}$ is trace Hermitian self-orthogonal and $\mathcal{C}$ is $\mathbb{F}_4$-linear, then $\mathcal{C}$ is an even code.

The quantum codes in Theorem 27.4.1 are modeled after classical codes with an additive structure, but the error operators are in fact multiplicative. An error E may have the same effect as ES where $\mathrm{S} \neq I$ is an element of the stabilizer group. A QEC is degenerate or **impure** if the set of correctable errors contains degenerate errors. Studies on impure codes have been rather scarce. The existence of two inequivalent $[\![6, 1, 3]\!]$ impure qubit codes was shown in [331, Section IV]. Remarkably, there is no $[\![6, 1, 3]\!]$ pure qubit code. A systematic construction based on duadic codes and further discussion on the advantages of degenerate quantum codes are supplied in [46].

A very popular construction is based on nested classical codes. We denote the Euclidean dual of $\mathcal{C}$ by $\mathcal{C}^{\perp_{\mathrm{E}}}$. As in the above theorem, for the remainder of the chapter, if $\mathcal{S} \subseteq \mathbb{F}_q^n$, the **Hamming weight** of $\mathcal{S}$ is $wt_H(\mathcal{S}) = \min \{ \mathrm{wt}_H(\mathbf{v}) \mid \mathbf{v} \in \mathcal{S} \}$.

Theorem 27.4.2 (Calderbank–Shor–Steane (CSS) Construction) *For $j \in \{1, 2\}$ let $\mathcal{C}_j$ be an $[n, k_j, d_j]_q$ code with $\mathcal{C}_1^{\perp_{\mathrm{E}}} \subseteq \mathcal{C}_2$. Then there is an*

$$[\![n, k_1 + k_2 - n, \min \{ wt_H(\mathcal{C}_2 \setminus \mathcal{C}_1^{\perp_{\mathrm{E}}}), wt_H(\mathcal{C}_1 \setminus \mathcal{C}_2^{\perp_{\mathrm{E}}}) \}]\!]_q$$

code Q. The code is pure whenever $\min\{wt_H(\mathcal{C}_2 \setminus \mathcal{C}_1^{\perp E}), wt_H(\mathcal{C}_1 \setminus \mathcal{C}_2^{\perp E})\} = \min\{d_1, d_2\}$.

Proof: Let G_j and H_j be the generator and parity check matrices of $\mathcal{C}_j$. Consider the linear code $\mathcal{C} \subseteq \mathbb{F}_q^{2n}$ with generator matrix $\begin{bmatrix} H_1 & 0 \\ 0 & H_2 \end{bmatrix}$. Since $\mathcal{C}_1^{\perp E} \subseteq \mathcal{C}_2$, we have $H_1 H_2^\mathsf{T} = 0$. Similarly, from $\mathcal{C}_2^{\perp E} \subseteq \mathcal{C}_1$, we know $H_2 H_1^\mathsf{T} = 0$. Define $\mathcal{C}^{\perp s}$ to be the code with parity check and generator matrices, respectively, $\begin{bmatrix} H_2 & 0 \\ 0 & H_1 \end{bmatrix}$ and $\begin{bmatrix} G_2 & 0 \\ 0 & G_1 \end{bmatrix}$. We verify that $\mathcal{C} \subseteq \mathcal{C}^{\perp s}$ with $\dim_{\mathbb{F}_q} \mathcal{C} = 2n - (k_1 + k_2)$. By Theorem 27.3.8, $\dim_{\mathbb{C}} Q = n - \dim_{\mathbb{F}_q} \mathcal{C} = k_1 + k_2 - n$. The distance computation is clear. $\qquad\square$

A special case of the CSS Construction comes via a Euclidean dual-containing code $\mathcal{C}^{\perp E} \subseteq \mathcal{C}$. From such an $[n, k, d]_q$ code $\mathcal{C}$, one obtains an $[\![n, 2k - n, \geq d]\!]_q$ code Q. The next method allows for most qubit CSS codes to be enlarged while avoiding a significant drop in the distance. The choice of the extra vectors in the generator matrix of $\mathcal{C}'$ is detailed in [1750, Section III].

Theorem 27.4.3 (Steane Enlargement of CSS Codes) *Let $\mathcal{C}$ be an $[n, k, d]_2$ code that contains its Euclidean dual $\mathcal{C}^{\perp E} \subseteq \mathcal{C}$. Suppose that $\mathcal{C}$ can be enlarged to an $[n, k' > k+1, d']_2$ code $\mathcal{C}'$. Then there is a pure qubit code with parameters $[\![n, k + k' - n, \min\{d, \lceil 3d'/2 \rceil\}]\!]$.*

A generalization to the qudit case was subsequently given in [1267], where the distance is $\min\{d, \lceil \frac{q+1}{q} d' \rceil\}$. Comparing the minimum distances in the resulting codes, the enlargement offers a better chance of relative gain in the qubit case as compared with the $q > 2$ cases.

Lisoněk and Singh, inspired by the classical *Construction X*, proposed a modification to qubit stabilizer codes in [1273]. The construction generalizes naturally to qudit codes. Here $\mathcal{C}^{\perp H}$ is the Hermitian dual of $\mathcal{C}$.

Theorem 27.4.4 (Quantum Construction X) *For an $[n, k]_{q^2}$ linear code $\mathcal{C}$, let $e := k - \dim(\mathcal{C} \cap \mathcal{C}^{\perp H})$. Then there exists an $[\![n + e, n - 2k + e, d]\!]_q$ code Q, with $d := d(Q) \geq \min\{d_H(\mathcal{C}^{\perp H}), d_H(\mathcal{C} + \mathcal{C}^{\perp H}) + 1\}$, where $\mathcal{C} + \mathcal{C}^{\perp H} := \{\mathbf{u} + \mathbf{v} \mid \mathbf{u} \in \mathcal{C}, \mathbf{v} \in \mathcal{C}^{\perp H}\}$.*

The case $e = 0$ is the usual stabilizer construction. To prevent a sharp drop in d, we want small e, i.e., large **Hermitian hull** $\mathcal{C} \cap \mathcal{C}^{\perp H}$.

We shift our attention now to *propagation rules* and *bounds*. Most of them are direct consequences of the propagation rules and bounds on the classical codes used as ingredients in the above constructions.

Proposition 27.4.5 (see [331, Theorem 6] for the binary case) *From an $[\![n, k, d]\!]_q$ code, the following codes can be derived: an $[\![n, k-1, \geq d]\!]_q$ code by **subcode construction** if $k > 1$ or if $k = 1$ and the initial code is pure, an $[\![n+1, k, \geq d]\!]_q$ code by **lengthening** if $k > 0$, and an $[\![n-1, k, \geq d-1]\!]_q$ code by **puncturing** if $n \geq 2$.*

The analog of **shortening** is less straightforward. It requires the construction of an auxiliary code and, then, a check on whether this code has codewords of a given length. The details on how to shorten quantum codes are available in [849, Section 4], building upon the initial idea of Rains in [1554].

How can we measure the goodness of a QEC? For stabilizer codes, given their classical ingredients and constructions, there are numerous bounds.

Theorem 27.4.6 (Quantum Hamming Bound; see [331] for the binary case) *Let Q be a pure $[\![n, k, d]\!]_q$ code with $d \geq 2\ell + 1$ and $k > 0$. Then*

$$q^{n-k} \geq \sum_{j=0}^{\ell} (q^2 - 1)^j \binom{n}{j}.$$

Q is **perfect** if it meets the bound. The proof comes from the observation that

$$q^n \geq \sum_{\overline{\mathrm{E}} \in \overline{\mathcal{E}_n}(\ell)} \dim_{\mathbb{C}}(\overline{\mathrm{E}}\, Q) = \dim_{\mathbb{C}} Q \cdot |\overline{\mathcal{E}_n}(\ell)| = q^k \sum_{j=0}^{\ell} (q^2 - 1)^j \binom{n}{j}.$$

The code in Example 27.3.7 is perfect. It has $2^{n-k} = 16 = \sum_{j=0}^{1} 3^j \binom{5}{j} = 1 + 15$ codewords.

Here is another bound which had been established as a necessary condition for pure stabilizer codes.

Theorem 27.4.7 (Quantum Gilbert–Varshamov Bound [715]) *Let $n > k \geq 2$, $d \geq 2$, and $n \equiv k$ (mod 2). A pure $[\![n, k, d]\!]_q$ code exists if*

$$\frac{q^{n-k+2} - 1}{q^2 - 1} > \sum_{j=1}^{d-1} (q^2 - 1)^{j-1} \binom{n}{j}.$$

An upper bound, which is well-suited for computer search, is the *Quantum Linear Programming (LP) Bound*; see Theorem 12.5.3. In the qubit case, the bound is explained in detail in [331, Section VII]. The same routine adjusts immediately to the general qudit case, as was shown in [1109, Section VI]. The main tool is the MacWilliams Identities of Theorem 1.15.3 (and [1319]); these are linear relations between the weight distribution of a classical code and its dual. They hold for all of the inner products we are concerned with here and have been very useful in ruling out the existence of quantum codes of certain ranges of parameters.

Rains supplied a nice proof for the next bound, which is a corollary to the Quantum LP Bound, using the *quantum weight enumerator* in [1554]. A quantum code that reaches the equality in the bound is said to be **quantum MDS (QMDS)**.

Theorem 27.4.8 (Quantum Singleton Bound) *An $[\![n, k, d]\!]_q$ code with $k > 0$ satisfies $k \leq n - 2d + 2$.*

Nearly all known families of classical codes over finite fields, especially those with well-studied algebraic and combinatorial structures, have been used in each of the constructions above. A partial list, compiled in mid 2005 as [1109, Table II], already showed a remarkable breadth. The large family of cyclic-type codes, whose corresponding structures in the rings of polynomials are ideals or modules, has been a rich source of ingredients for QECs with excellent parameters. This includes the BCH, cyclic, constacyclic, quasi-cyclic, and quasi-twisted codes; see, e.g., Chapter 9. In the family, the nestedness property, very useful in the CSS construction, comes for free. A great deal is known about their dual codes under numerous applicable inner products. For small q, the structures allow for extensive computer algebra searches for reasonable lengths, aided by their minimum distance bounds.

The most comprehensive record for best-known qubit codes is Grassl's online table [847]. Numerous entries have been certified optimal, while still more entries indicate gaps between the best-possible and the best-known. It is a two-fold challenge to contribute meaningfully to the table. First, for $n \leq 100$, many researchers have attempted exhaustive searches. Better codes are unlikely to be found without additional clever strategies. Second, for $n > 100$, computing the actual distance $d(Q)$ tends to be prohibitive. As the length and dimension grow, computing the minimum distances of the relevant classical codes to derive the quantum distance is hard [1843]. Improvements remain possible, with targeted searches. Recent examples include the works of Galindo et al. on quasi-cyclic constructions of quantum codes [786] where Steane enlargement is deployed, the search reported in [1273] on cyclic codes over $\mathbb{F}_4$ where Construction X is used with $e \in \{1, 2, 3\}$, and similar random searches on quasi-cyclic codes done in [699] for qubit and qutrit codes.

Less attention has been given to record-holding qutrit codes, for which there is no publicly available database of comparative extent. A table listing numerous qutrit codes is kept by Y. Edel in [662] based on their explicit construction as quantum twisted codes in [201]. Better codes than many of those in the table have since been found.

Attempts to derive new quantum codes by shortening good stabilizer codes motivate closer studies on the weight distribution of the classical auxiliary codes, in particular when the stabilizer codes are QMDS. Shortening is very effective in constructing qudit MDS codes of lengths up to $q^2 + 2$ and minimum distances up to $q + 1$.

There has been a large literature on QMDS codes. All of the above constructions via classical codes as well as the propagation rules have been applied to families of classical MDS codes, particularly Generalized Reed–Solomon (see Chapter 24) and constacyclic MDS codes. Since the dual of an MDS code is MDS, the dual distance is evident, leaving only the orthogonality property to investigate. While the theoretical advantages are clearly abundant, there are practical limitations. The length of such codes is bounded above by $q^2 + 2$, when q is even, and by $q^2 + 1$, when q is odd, assuming the MDS Conjecture holds; see Conjecture 3.3.21 of Chapter 3.

For qubit codes, the only nontrivial QMDS codes are those with parameters $[\![5, 1, 3]\!]$, $[\![6, 0, 4]\!]$, and $[\![2m, 2m - 2, 2]\!]$. As q grows larger, the practical value of QMDS codes quickly diminishes, since controlling qudit systems with $q > 3$ is currently prohibitive. A list for q-ary QMDS codes, with $2 \leq q \leq 17$, is available in [850]. Another list that covers families of QMDS codes and their references can be found in [391, Table V]. More works on QMDS codes continue to appear, with detailed analysis on the self-orthogonality conditions supplied from number theoretic and combinatorial tools.

Taking algebraic geometry (AG) codes, examined in Chapter 15, as the classical ingredients is another route. A wealth of favourable results had already been available prior to the emergence of QECs. Curves with many rational points often lead, via the Goppa construction, to codes with good parameters. Their duals are well-understood, via the residue formula. Their designed distances can be computed from the Riemann–Roch Theorem. Chen et al. showed how to combine Steane enlargement and concatenated AG codes to derive excellent qubit codes in [395]. A quantum asymptotic bound was established in [714]. Construction of QECs from AG codes was initially a very active line of inquiry. It has somewhat lessened in the last decade, mostly due to lack of practical values to add to the quest as q grows.

Using codes over rings to construct QECs has also been tried. This route, however, does not usually lead to parameter improvements over QECs constructed from codes over fields. The absence of a direct connection from codes over rings to QECs necessitates the use of a Gray map, which often causes a significant drop in the minimum distance.

27.5   Going Asymmetric

So far we have been working on the assumption that the bit-flips and the phase-flips are equiprobable. Quantum systems, however, have noise properties that are dynamic and asymmetric. The fault-tolerant threshold is improved when asymmetry is considered [38]. It was Steane who first hinted at the idea of adjusting error-correction to the particular characteristics of the channel in [1749]. Designing error control methods to suit the noise profile, which can be hardware-specific, is crucial. The study of *asymmetric quantum codes* (*AQCs*) gained traction when the ratios of how often σ_z occurs over the occurrence of σ_x were discussed in [1025], with follow-up constructions offered soon after in [1615]. Wang et al. established a mathematical model of AQCs in the general qudit system in [1871].

As in the symmetric case, $\mathcal{E}_n := \{\omega^\beta X(\mathbf{a})Z(\mathbf{b}) \mid \mathbf{a}, \mathbf{b} \in \mathbb{F}_q^n, \beta \in \mathbb{F}_p\}$. An error $\mathrm{E} := \omega^\beta X(\mathbf{a})Z(\mathbf{b}) \in \mathcal{E}_n$ has $\mathrm{wt_X(E)} := \mathrm{wt}_H(\mathbf{a})$ and $\mathrm{wt_Z(E)} := \mathrm{wt}_H(\mathbf{b})$.

Definition 27.5.1 Let d_x and d_z be positive integers. A qudit code Q with dimension $K \geq q$ is called an **asymmetric quantum code (AQC)** with parameters $((n, K, d_z, d_x))_q$ or $[\![n, k, d_z, d_x]\!]_q$, where $k = \log_q K$, if Q detects $d_x - 1$ qudits of X-errors and, at the same time, $d_z - 1$ qudits of Z-errors. The code Q is **pure** if $|\varphi\rangle$ and $\mathrm{E}|\psi\rangle$ are orthogonal for any $|\varphi\rangle, |\psi\rangle \in Q$ and any $\mathrm{E} \in \mathcal{E}_n$ such that $1 \leq \mathrm{wt_X(E)} \leq d_x - 1$ or $1 \leq \mathrm{wt_Z(E)} \leq d_z - 1$. Any code Q with $K = 1$ is assumed to be pure.

An $[\![n, k, d, d]\!]_q$ AQC is symmetric, with parameters $[\![n, k, d]\!]_q$, but the converse is not true since, for $\mathrm{E} \in \mathcal{E}_n$ with $\mathrm{wt_X(E)} \leq d-1$ and $\mathrm{wt_Z(E)} \leq d-1$, $\mathrm{wt_Q(E)}$ may be bigger than $d - 1$.

To date, most known families of AQCs come from the asymmetric version of the CSS Construction and its generalization in [697].

Theorem 27.5.2 (CSS-like Constructions for AQCs) *Let $\mathcal{C}_j$ be an $[n, k_j, d_j]_q$ code for $j \in \{1, 2\}$. Let $\mathcal{C}_j^{\perp*}$ be the dual of $\mathcal{C}_j$ under one of the Euclidean, the trace Euclidean, the Hermitian, and the trace Hermitian inner products. Let $\mathcal{C}_1^{\perp*} \subseteq \mathcal{C}_2$, with*

$$d_z := \max\left\{wt_H(\mathcal{C}_2 \setminus \mathcal{C}_1^{\perp*}), \ wt_H(\mathcal{C}_1 \setminus \mathcal{C}_2^{\perp*})\right\} \ and$$
$$d_x := \min\left\{wt_H(\mathcal{C}_2 \setminus \mathcal{C}_1^{\perp*}), \ wt_H(\mathcal{C}_1 \setminus \mathcal{C}_2^{\perp*})\right\}.$$

Then there exists an $[\![n, k_1 + k_2 - n, d_z, d_x]\!]_q$ code Q, which is pure whenever $\{d_z, d_x\} = \{d_1, d_2\}$. If we have $\mathcal{C} \subseteq \mathcal{C}^{\perp*}$ where $\mathcal{C}$ is an $[n, k, d]_q$ code, then Q is an $[\![n, n - 2k, d', d']\!]_q$ code, where $d' = wt_H(\mathcal{C}^{\perp*} \setminus \mathcal{C})$. The code Q is pure whenever $d' = d^{\perp*} := d(\mathcal{C}^{\perp*})$.

The propagation rules and bounds for AQCs follow from the relevant rules and bounds on the nested codes and their respective duals. Details on how to derive new AQCs from already known ones were discussed by La Guardia in [1183, 1184]. The *asymmetric version of the quantum Singleton Bound* reads $k \leq n - (d_z + d_x) + 2$. To benchmark codes of large lengths, one can use the *asymmetric versions of the quantum Gilbert–Varshamov Bound* established by Matsumoto in [1354].

Many best-performing AQCs with small d_x and moderate n and k were derived in [697]. The optimal ones reach the upper bounds certified by an improved LP Bound, called the *Triangle Bound* in [697, Section V]. Recent results, covering also AQCs of large lengths, came from an interesting family of nested codes defined from multivariate polynomials and Cartesian product point sets due to Gallindo et al. in [784].

Most known **asymmetric quantum MDS (AQMDS)** codes, i.e., codes that meet the asymmetric quantum Singleton Bound, are pure CSS. Assuming the validity of the MDS Conjecture, all possible parameters that pure CSS AQMDS codes can have were established in [696].

Theorem 27.5.3 *Assuming the MDS Conjecture holds, there is a pure CSS AQMDS code with parameters $[\![n, k, d_z, d_x]\!]_q$, where $\{d_z, d_x\} = \{n - j - k + 1, j + 1\}$ if and only if one of the followings holds:*

(a) *q is arbitrary, $n \geq 2$, $j \in \{1, n - 1\}$, and $k \in \{0, n - j\}$;*

(b) $q = 2$, n *is even,* $j = 1$, *and* $k = n - 2$;

(c) $q \geq 3$, $n \geq 2$, $j = 1$, *and* $k = n - 2$;

(d) $q \geq 3$, $2 \leq n \leq q$, $j \leq n - 1$, *and* $0 \leq k \leq n - j$;

(e) $q \geq 3$, $n = q + 1$, $j \leq n - 1$, *and* $k \in \{0, 2, \ldots, n - j\}$;

(f) $q = 2^m$, $n = q + 1$, $k = 1$, *and* $j \in \{2, 2^m - 2\}$;

(g) $q = 2^m$ *where* $m \geq 2$, $n = q + 2$,

$$\begin{cases} j = 1, \ and \ k \in \{2, 2^m - 2\}, \ or \\ j = 3, \ and \ k \in \{0, 2^m - 4, 2^m - 1\}, \ or \\ j = 2^m - 1, \ and \ k \in \{0, 3\}. \end{cases}$$

Going forward, three general challenges can be identified. First, find better AQCs, particularly in qubit systems, than the currently best-known. More tools to determine or to lower bound d_z and d_x remain to be explored if we are to improve on the parameters. Second, construct codes with very high d_z to d_x ratio, since experimental results suggest that this is typical in qubit channels. Third, find conditions on the nested classical codes that yield impure codes.

27.6 Other Approaches and a Conclusion

We briefly mention other approaches to quantum error control before concluding.

Successful small-scale hardware implementations often rely on *topological codes*, first put forward by Kitaev [1125]. This family of codes includes *surface codes* [301] and *color codes* [231]. Topological codes encode information in, mostly 2-dimensional, lattices. They are CSS codes with a clever design. The lattice, on which the stabilizer generators act locally, has a bounded weight. The extra restrictions make the error syndrome easier to infer.

Instead of block quantum codes, studies have been done on *convolutional qubit codes*; see, e.g., [746] and subsequent works that cited it. The logical qubits are encoded and transmitted as soon as they arrive in a steady stream. The rate k over n is fixed, but the length is not. This type of codes, like their classical counterparts, may be useful in quantum communication.

An approach, that does not require self-orthogonality, constructs *entanglement-assisted quantum codes (EA-QECs)* [317]. The price to pay is the need for a number of maximally entangled states, called **ebits** for *entangled qubits*, to increase either the rate or the ability to handle errors. Producing and maintaining ebits, however, tend to be costly, which offset their efficacy. Pairs of classical codes, whose intersections have some prescribed dimensions, were shown to result in EA-QECs in [859, Section 4]. A formula on the optimal number of ebits that an EA-QEC requires is given in [1891].

Theorem 27.6.1 *Given a linear* $[n, k, d]_{q^2}$ *code* C *with parity check matrix* H, *the code* $C^{\perp_H}$ *stabilizes an EA-QEC with parameters* $[\![n, 2k - n + c, d; c]\!]_q$, *where* $c := \mathrm{rank}(HH^T)$ *is the number of ebits required.*

A larger class of QECs that includes all stabilizer codes is the *codeword stabilized (CWS) codes*. The framework was proposed by Cross et al. in [473] to unify stabilizer (additive) codes and known examples of good non-additive codes. General constructions for large CWS codes are yet to be devised. Also currently unavailable are efficient encoding and decoding algorithms.

The bridge between classical coding theory and quantum error-control was firmly put in place via the stabilizer formalism. Various generalizations and modifications have been studied since, benefitting both the classical and quantum sides of the error-control theory. Well-researched tools and the wealth of results in classical coding theory translate almost effortlessly to the design of good quantum codes, moving far beyond what is currently practical to implement in actual quantum devices. Research problems triggered by error-control issues in the quantum setup revive and expand studies on specific aspects of classical codes, which were previously overlooked or deemed not so interesting. This fruitful cross-pollination between the classical and the quantum, in terms of error-control, is set to continue.

Acknowledgement

This work is supported by Nanyang Technological University Research Grant no. M4080456.

Chapter 28

Space-Time Coding

Frédérique Oggier

Nanyang Technological University

28.1 Introduction

Space-time coding refers to an area of coding theory that appears in the context of transmission of information over a wireless channel where both the transmitter and the receiver are equipped with multiple antennas. The goal is to design space-time codebooks, that is families of codewords to be transmitted over this wireless channel, that achieve high reliability, and high rate (to be formally defined below). When the wireless channel is such that it may be assumed constant over a given period of time, the codewords are transmitted via multiple antennas ("space") over this time period ("time"), which explains the terminology.

This chapter presents typical channel models for which space-time codes are designed (see Section 28.2), out of which design criteria are extracted. Then some families of space-time codes with examples are provided in Section 28.3. Finally different variations of space-time coding problems are reviewed in Section 28.4.

FIGURE 28.1: A multiple antenna communication channel

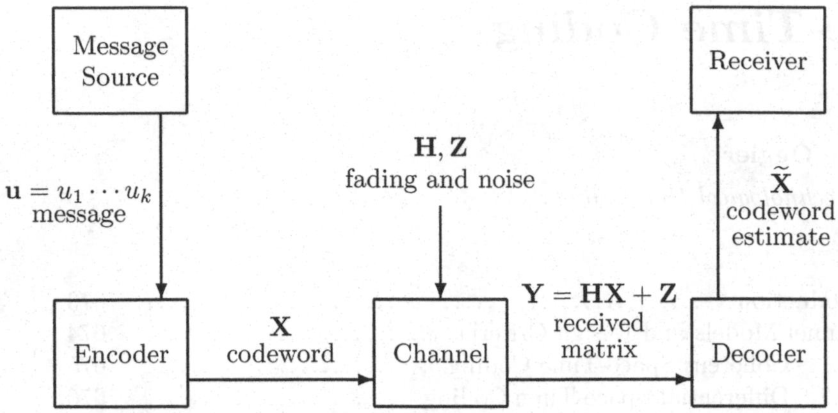

28.2 Channel Models and Design Criteria

Consider a communication system with M transmit antennas and N receive antennas over a fading block of T symbol duration given by

$$\mathbf{Y}_{N \times T} = \mathbf{H}_{N \times M} \mathbf{X}_{M \times T} + \mathbf{Z}_{N \times T} \in \mathbb{C}^{N \times T}. \tag{28.1}$$

The $N \times M$ **fading matrix H** models the wireless environment: its coefficient h_{ij} corresponds to the channel coefficient between the jth transmit and the ith receive antenna and is a complex Gaussian random variable with zero mean and unit variance. The $N \times T$ matrix **Z** corresponds to the channel noise, whose independent entries are complex Gaussian random variables with zero mean and variance N_0 (see, e.g., [204] for more details). The x_{il} entry of the **space-time codeword X** corresponds to the signal transmitted from the ith antenna during the lth symbol interval for $1 \le l \le T$. It is a function of the message symbols, and the mapping between the message symbols and the codeword is done by an encoder, as usual (see Figure 28.1). A **space-time block code** is a finite set $\mathcal{C}$ of $M \times T$ complex matrices **X**, whose cardinality is denoted by $|\mathcal{C}|$.

The channel model (28.1) is sometimes referred to as a **multiple-input multiple-output (MIMO) channel**. Multiple antenna systems have been extensively studied since they are known to support high data rates [748, 1796].

Design criteria for space-time codes depend on the type of receiver that is considered. Two major classes of receivers have been considered in the literature:

- **coherent**: it is assumed that the channel matrix **H** is known to the receiver.

- **non-coherent**: the channel matrix **H** is not known to the receiver, in which case different solutions are available (see, e.g., [152, 967]), including a technique called *differential space-time coding*, which we will explain below.

28.2.1 Coherent Space-Time Coding

Assuming that $\mathbf{H}$ is known at the receiver, **maximum likelihood** (ML) decoding corresponds to choosing the codeword $\mathbf{X}$ that minimizes the squared Frobenius norm:

$$\min_{\mathbf{X} \in \mathcal{C}} \|\mathbf{Y} - \mathbf{HX}\|^2 \tag{28.2}$$

where $\|\mathbf{A}\| = \sqrt{\sum_{i=1}^{m} \sum_{j=1}^{n} |a_{ij}|^2}$ for an $m \times n$ matrix $\mathbf{A} = [a_{ij}]$. Then it can be shown that the **pairwise error probability** $\mathrm{Prob}(\mathbf{X} \to \widehat{\mathbf{X}})$, i.e., the probability that, when a codeword $\mathbf{X}$ is transmitted, the ML receiver decides erroneously in favor of another codeword $\widehat{\mathbf{X}}$, assuming only $\mathbf{X}$ and $\widehat{\mathbf{X}}$ are in the codebook, is bounded above by

$$\mathrm{Prob}(\mathbf{X} \to \widehat{\mathbf{X}}) \leq \det \left[\mathbf{I}_M + \frac{(\mathbf{X} - \widehat{\mathbf{X}})(\mathbf{X} - \widehat{\mathbf{X}})^*}{4N_0} \right]^{-N},$$

where $\mathbf{I}_M$ is the $M \times M$ identity matrix. Let r denote the rank of the codeword difference matrix $\mathbf{X} - \widehat{\mathbf{X}}$, with nonzero eigenvalues λ_j, $j = 1, \ldots, r$. For small N_0, we have

$$\mathrm{Prob}(\mathbf{X} \to \widehat{\mathbf{X}}) \leq \left(\frac{(\prod_{j=1}^{r} \lambda_j)^{1/r}}{4N_0} \right)^{-rN}.$$

When $\mathbf{X}$ and $\widehat{\mathbf{X}}$ vary through distinct codewords in the codebook, we can look at the minimum value of r, and we call $\min\{rN\}$ the **diversity gain** of the code. If $r = M$, we say that the code has **full diversity**. This means that we can exploit all the MN independent channels available in the MIMO system. Maximizing r is sometimes called the **rank** or **diversity criterion**. Furthermore, we want to maximize the **coding gain** [1793], given by $\left(\min_{\mathbf{X} \neq \widehat{\mathbf{X}}} \det((\mathbf{X} - \widehat{\mathbf{X}})(\mathbf{X} - \widehat{\mathbf{X}})^*) \right)^{1/M}$.

When a space-time block code $\mathcal{C}$ is linear, meaning that

$$\mathbf{X} \pm \mathbf{X}' \in \mathcal{C} \quad \text{for all } \mathbf{X}, \mathbf{X}' \in \mathcal{C},$$

$\min_{\mathbf{X} \neq \widehat{\mathbf{X}}} \det((\mathbf{X} - \widehat{\mathbf{X}})(\mathbf{X} - \widehat{\mathbf{X}})^*)$ reduces to $\min_{\mathbf{X} \neq \mathbf{0}_{M \times T}} \det(\mathbf{XX}^*)$; maximizing this quantity is referred to as the **determinant criterion**.

Considering linear space-time codes not only simplifies the design problem, it also endows the code of a lattice structure and enables the application of the sphere decoder [479, 922, 1855] at the receiver.

The space-time coding problem.

We would like to design a linear family $\mathcal{C}$ of $M \times T$ complex matrices with full diversity such that $\min_{\mathbf{X} \in \mathcal{C}, \mathbf{X} \neq \mathbf{0}_{M \times T}} \det(\mathbf{XX}^*)$ is maximized.

The message symbols $u_1, \ldots, u_k$ to be encoded in the space-time codeword $\mathbf{X}$ are assumed to be QAM or HEX symbols [745] (more precisely, information bits are labeling QAM or HEX symbols), which are represented by finite subsets from respectively $\mathbb{Z}[i]$ and $\mathbb{Z}[\zeta_3]$ where $\zeta_3 = e^{2\pi i/3}$ (see [154] for more details on scaling and translation of these subsets to get the actual signal constellations).

Let k denote the number of information symbols (QAM or HEX) that are encoded in the space-time codewords. We say that a code has **full rate** when $k = NT$.

There are further space-time design considerations.

- One may prefer full rate space-time codes, in which case it will be requested to have $k = NT$.

- One may take into consideration the notion of **shaping gain** [747], which suggests considering constellations with a cubic shaping, in order to have both a practical way of labeling QAM/HEX symbols, and low average energy. We note that the best shaping gain is obtained by a spherical shaping, whose labeling becomes complex for large constellations. See also [1174] for a nonlinear mapping encoder with a good shaping gain.

- Cubic shaping is related to the concept of *information lossless*. A space-time code is **information lossless** if the mutual information of the equivalent channel obtained by including the encoder in the channel is equal to the mutual information of the MIMO channel (see, e.g., [1666]).

- Another space-time coding property which may be desirable is that of having a **non-vanishing determinant** (NVD) [155]. Adaptive transmission schemes may require the use of different size constellations, in which case it is important to ensure that coding gain does not depend on the constellation size, even if the code is infinite. The NVD property is a necessary and sufficient condition for a space-time code to achieve the **diversity-multiplexing tradeoff** [671, 1945] (see also [1923]).

28.2.2   Differential Space-Time Coding

We again consider a MIMO channel with M transmit antennas and N receive antennas, but this time with no known channel information. The channel is used in blocks of M channel uses, so that the transmitted signal can be represented as an $M \times M$ matrix $\mathbf{S}_t$, where $t = 0, 1, 2, \ldots$ represents the block channel use. Assuming that the channel is constant over M channel uses, we may write it as

$$\mathbf{Y}_t = \mathbf{S}_t \mathbf{H}_t + \mathbf{Z}_t \ \text{ for } t = 0, 1, 2, \ldots. \tag{28.3}$$

The main difference with the coherent case is that we immediately assume that the transmitted matrix $\mathbf{S}_t$ is square, and the index t emphasizes a time dependency, which will be needed in the **differential unitary space-time modulation** scheme [967, 1011] explained next.

Assuming $\mathbf{S}_0 = \mathbf{I}$, the transmitted signal $\mathbf{S}_t$ is encoded as follows:

$$\mathbf{S}_t = \mathbf{V}_{u_t} \mathbf{S}_{t-1} \ \text{ for } t = 1, 2, \ldots, \tag{28.4}$$

where $u_t \in \{0, \ldots, L-1\}$ is the data to be transmitted, and $\mathcal{C} = \{\mathbf{V}_0, \ldots, \mathbf{V}_{L-1}\}$ the constellation to be designed. By definition, the matrices in $\mathcal{C}$ have to be unitary (so that their product does not go to zero or infinity). Since the channel is used M times, the transmission rate is $R = \frac{1}{M} \log_2 L$. The size $|\mathcal{C}|$ of the constellation is thus $L = 2^{MR}$. We remark that in the coherent case, since the number of codewords could be infinite, we use as rate the ratio of information symbols per transmitted coefficients in one codeword.

By further assuming the channel constant for $2M$ consecutive uses ($\mathbf{H}_{t-1} = \mathbf{H}_t = \mathbf{H}$), we get from (28.3) and (28.4) that

$$\begin{aligned} \mathbf{Y}_t &= \mathbf{V}_{u_t} \mathbf{S}_{t-1} \mathbf{H} + \mathbf{Z}_t \\ &= \mathbf{V}_{u_t} (\mathbf{Y}_{t-1} - \mathbf{Z}_{t-1}) + \mathbf{Z}_t \\ &= \mathbf{V}_{u_t} \mathbf{Y}_{t-1} + \mathbf{Z}'_t. \end{aligned}$$

Since the matrix $\mathbf{H}$ does not appear in the last equation, this means differential modulation allows decoding without knowledge of the channel.

The maximum likelihood decoder is thus given by

$$\widehat{u}_t = \arg \min_{l=0,\dots,|\mathcal{C}|-1} \|\mathbf{Y}_t - \mathbf{V}_l \mathbf{Y}_{t-1}\|.$$

The pairwise block probability of error can then be bounded above [967, 1011], similarly to what was done in the coherent case, to obtain the so-called **diversity product** given by

$$\frac{1}{2} \min_{\mathbf{V}_i \neq \mathbf{V}_j} |\det(\mathbf{V}_i - \mathbf{V}_j)|^{1/M} \quad \text{for all } \mathbf{V}_i \neq \mathbf{V}_j \in \mathcal{C} \tag{28.5}$$

as a design criterion. We say that **diversity** is achieved when

$$\det(\mathbf{V}_i - \mathbf{V}_j) \neq 0 \quad \text{for all } \mathbf{V}_i \neq \mathbf{V}_j \in \mathcal{C}$$

and the goal is to maximize the diversity product.

The differential space-time coding problem.
We would like to design a family $\mathcal{C}$ of $M \times M$ unitary complex matrices with full diversity such that $\min_{\mathbf{V}_i \neq \mathbf{V}_j \in \mathcal{C}} |\det(\mathbf{V}_i - \mathbf{V}_j)|$ is maximized.

28.3   Some Examples of Space-Time Codes

There is a wealth of literature on designing space-time block codes. We propose a few examples below: the Alamouti code is arguably one of the most well known of the space-time block codes, linear dispersion codes give a generic framework for building space-time codes, and the golden code illustrates the type of codes that can be obtained using noncommutative algebra. Some other classical constructions may be found in [478, 480, 666, 897, 985, 1249, 1251, 1289], or in [1030, 1249, 1792] for orthogonal space-time block codes. Space-time trellis codes have also been extensively studied in the literature developing from [1793]. Techniques involving noncommutative algebras include [972, 1547].

Differential space-time coding has also been well-studied, and early constructions include, e.g., [1045, 1046, 1250, 1689]. We will choose to give as an example Cayley codes, because they provide a systematic way to obtain unitary codes, and do not involve algebraic techniques, such as representation theory; see [1455] for a survey of algebraic space-time codes for the differential case.

28.3.1   The Alamouti Code

The first example of a space-time code we present is the Alamouti code [31], to be used with QAM symbols for $N = 1$ receive and $M = 2$ transmit antennas.

An Alamouti codeword is of the form

$$\mathbf{X} = \begin{bmatrix} u_1 & -\overline{u_2} \\ u_2 & \overline{u_1} \end{bmatrix},$$

where u_1, u_2 are QAM symbols and $\bar{}$ denotes complex conjugation. We thus transmit 2 symbols for $NM = 2$; this code is thus full rate. It is also clearly linear.

The Alamouti code is fully diverse since

$$\det(\mathbf{X}) = |u_1|^2 + |u_2|^2 > 0,$$

for any $u_1, u_2 \in \mathbb{C}$ nonzero. For arbitrary complex numbers, it does not have a non-vanishing determinant; however it does have it for QAM symbols ($|u| \geq 1$ if $u \in \mathbb{Z}[i]$, $u \neq 0$). The Alamouti code is one of the most well known and popular space-time codes, thanks to its excellent performance and easy decoding.

28.3.2   Linear Dispersion Codes

A linear dispersion code [921] contains codewords $\mathbf{X}$ of the form

$$\mathbf{X} = \sum_{q=1}^{Q} (u_q C_q + u_q^* D_q)$$

where $u_1, \ldots, u_Q$ are typically chosen from a QAM constellation, and where C_q, D_q are fixed complex matrices to be designed. Alternatively, one may consider the real and imaginary parts of u_q, $\mathrm{Re}(u_q)$ and $\mathrm{Im}(u_q)$, respectively, and write

$$\mathbf{X} = \sum_{q=1}^{Q} (\mathrm{Re}(u_q) A_q + \mathrm{Im}(u_q) B_q)$$

where $A_q = C_q + D_q$ and $B_q = C_q - D_q$. For example, the Alamouti code is a linear dispersion code with

$$A_1 = \begin{bmatrix} 1 & 0 \\ 0 & 1 \end{bmatrix}, \quad A_2 = \begin{bmatrix} 0 & 1 \\ -1 & 0 \end{bmatrix}, \quad B_1 = \begin{bmatrix} 1 & 0 \\ 0 & -1 \end{bmatrix}, \quad B_2 = \begin{bmatrix} 0 & 1 \\ 1 & 0 \end{bmatrix}.$$

In [921], the matrices $\{A_q, B_q\}$ are optimized to yield space-time codes that approach capacity. They were proposed before the introduction of the properties of non-vanishing determinant, or diversity-multiplexing trade-off.

28.3.3   The Golden Code

The golden code is a 2×2 space-time code [156, 493] to be used in a MIMO channel with $M = N = 2$ antennas. Its name comes from its algebraic construction [156], which involves the Golden number $\theta = \frac{1+\sqrt{5}}{2}$.

We start with space-time codewords of the form

$$\begin{bmatrix} a + b\theta & c + d\theta \\ i(c + d\sigma(\theta)) & a + b\sigma(\theta) \end{bmatrix},$$

with $a, b, c, d \in \mathbb{Z}[i]$ and $\sigma(a + b\theta) = a + b\frac{1-\sqrt{5}}{2}$.

By definition, the codebook obtained is both linear and full rate (since it contains the 4 information symbols a, b, c, d, and $MN = 4$).

Since the code is linear, to check that it has full diversity, it is enough to show that the determinant of any nonzero codeword is nonzero. We have

$$\det \begin{bmatrix} a + b\theta & c + d\theta \\ i(c + d\sigma(\theta)) & a + b\sigma(\theta) \end{bmatrix} = (a + b\theta)\sigma(a + b\theta) - i(c + d\theta)\sigma(c + d\theta)$$

which is zero if and only if $i = \alpha\sigma(\alpha)$ for some $\alpha = \frac{a+b\theta}{c+d\theta}$. It can be shown [154, 156] that such an α does not exist (the problem is equivalent to showing that i is not an algebraic norm in the field extension $\mathbb{Q}(i, \sqrt{5})$).

We now further restrict the above codewords to obtain golden space-time codewords as follows:

$$\mathbf{X} = \frac{1}{\sqrt{5}} \begin{bmatrix} \alpha(a + b\theta) & \alpha(c + d\theta) \\ i\sigma(\alpha)(c + d\sigma(\theta)) & \sigma(\alpha)(a + b\sigma(\theta)) \end{bmatrix}$$

where a, b, c, d are QAM symbols, and $\alpha = 1 + i - i\theta$.

Consider now the matrix

$$\mathbf{M} = \begin{bmatrix} \alpha & \alpha\theta \\ \sigma(\alpha) & \sigma(\alpha\theta) \end{bmatrix}.$$

A direct computation shows that $\mathbf{MM}^* = 5\mathbf{I}_2$. Thus $\frac{1}{\sqrt{5}}\mathbf{M}$ is a unitary matrix, and it turns out that the diagonal of a golden codeword is obtained by multiplying the vector containing the symbols a, b with $\mathbf{M}$. The same holds with the antidiagonal and the symbols c, d, and an extra multiplication by i which is such that $|i|^2 = 1$.

This illustrates the property of cubic shaping discussed above: an encoding process for the golden code consists of having a column vector containing the information symbols a, b, c, d, and the encoding is done by multiplying the vector by a block matrix which is unitary. Geometrically, this means that the constellation is extracted with a cubic shaping (and in fact, without change in the energy allocated to the constellation).

Let us now compute the minimum determinant of the (infinite) golden code. Since $\alpha\bar{\alpha} = 2 + i$, we have

$$\det(\mathbf{X}) = \frac{2+i}{5}\left((a + b\theta)(a + b\bar{\theta}) - i(c + d\theta)(c + d\bar{\theta})\right)$$

$$= \frac{1}{2-i}\left((a^2 + ab - b^2) - i(c^2 + cd - d^2)\right).$$

By definition of a, b, c, d, we have that the nontrivial minimum of $|a^2 + ab - b^2 - i(c^2 + cd - d^2)|^2$ is 1; thus

$$\min_{0 \neq \mathbf{X} \in \mathcal{C}} |\det(\mathbf{X})|^2 = \frac{1}{5}.$$

This illustrates the property of non-vanishing determinant.

The golden code is built using cyclic division algebras. The interested reader may refer for example to [154, 166, 671, 673, 972, 1122, 1458, 1547, 1656, 1666] for learning more about space-time codes from division algebras, and to [1655] for an interpretation of the Alamouti code in terms of division algebras.

28.3.4 Cayley Codes

Cayley codes for differential space-time coding have been introduced in [920]. They are based on the Cayley transform, which maps the nonlinear Stiefel manifold to the linear space of Hermitian or skew-Hermitian matrices (and vice versa). Let $\mathbf{A}$ be a Hermitian matrix, and thus $i\mathbf{A}$ be skew-Hermitian. The **Cayley transform** of $i\mathbf{A}$ is given by

$$\mathbf{V} = (\mathbf{I} + i\mathbf{A})^{-1}(\mathbf{I} - i\mathbf{A}).$$

It is easy to check that $\mathbf{V}$ is unitary. The Cayley transform of $i\mathbf{A}$ is preferred since all its eigenvalues are strictly imaginary, thus different from 1, so that the inverse of $\mathbf{I} + i\mathbf{A}$ always exists.

In order to encode information, the Hermitian matrix $\mathbf{A}$ is defined by

$$\mathbf{A} = \sum_{q=1}^{Q} u_q \mathbf{A}_q,$$

where $u_1, \ldots, u_Q \in \mathbb{R}$ are the information symbols, chosen from a finite set, and where $\mathbf{A}_q$ are fixed $M \times M$ complex Hermitian matrices. We remark that the encoding technique is similar to that of linear dispersion codes.

The performance of a Cayley code depends on Q, the Hermitian basis matrices $\{\mathbf{A}_q\}$, and the set from which each u_q is chosen. To choose the $\{\mathbf{A}_q\}$, the approach of [920] consists of optimizing a mutual information criterion, instead of the traditional diversity criterion (28.5), since it is argued that at high rate, checking the diversity may become intractable. However, one could combine the Cayley code approach explained above with that of division algebras to ensure full diversity, as was tried in [1456].

28.4 Variations of Space-Time Coding Problems

We provide a non-exhaustive list of space-time coding problems, whose design criteria (and code design) are variations of those discussed above.

28.4.1 Distributed Space-Time Coding

Given a wireless network where nodes are equipped with one or several antennas, cooperative coding strategies see the nodes as forming a virtual antenna array (see, e.g., [1202]) to obtain the diversity known to be achieved by MIMO systems. They are broadly classified between amplify-and-forward and decode-and-forward protocols. Both protocols consist of a two-step transmission: a broadcast phase during which the transmitter broadcasts its message to the neighbor relay nodes, and a forward phase, where in the amplify-and-forward protocol, relay nodes forward the amplified version of their received signals to the receiver, while in the decode-and-forward protocol, relay nodes first try to decode the received signal, and those which manage then forward the decoded signal to the receiver.

In the decode-and-forward case, relays which performed the decoding can cooperate in re-encoding a space-time code [494, 1202]. In the amplify-and-forward case, a way of getting cooperation is to use *distributed space-time coding* [1047], as we explain next; we ignore on purpose normalization factors for the sake of simplicity, and all random variables for noise and fading are assumed to be complex Gaussian with zero mean and unit variance.

The transmitter broadcasts its signal $\mathbf{s}$ to each relay, and the ith relay gets

$$\mathbf{r}_i = f_i \mathbf{s} + \mathbf{z}_i,$$

where $\mathbf{z}_i$ is the noise vector and f_i is the fading at the ith relay. Now each relay transmits

$$\mathbf{t}_i = A_i \mathbf{r}_i,$$

where A_i is a unitary matrix, so that the receiver gets

$$\mathbf{x} = \sum_{i=1}^{R} g_i \mathbf{t}_i + \mathbf{w} = \mathbf{S}H + W,$$

with

$$\mathbf{S} = \begin{bmatrix} A_1\mathbf{s} & \cdots & A_R\mathbf{s} \end{bmatrix}, \quad H = \begin{bmatrix} f_1g_1 \\ \vdots \\ f_Rg_R \end{bmatrix}, \quad \text{and} \quad W = \sum_{i=1}^{R} g_iA_i\mathbf{z}_i + \mathbf{w}.$$

The matrix $\mathbf{S}$ is called a **distributed space-time codeword**. Analyzing the behavior of the pairwise error probability shows [1047] that the diversity criterion still holds. Adaptations of space-time codes have thus been used for wireless networks for example in [672, 673, 1123, 1457].

The diversity-multiplexing gain tradeoff (DMT) has naturally been studied as well in the context of wireless networks; see [80] for the DMT of the amplify-and-forward protocol, assuming a direct link from the transmitter to the receiver, unlike in the distributed space-time code model. It was shown that in order to reach the tradeoff, the transmitter node always transmits, which yields to so-called non-orthogonal amplify-and-forward protocol. In [1919], the protocol has further been extended to the case of relays equipped with multiple antennas.

28.4.2 Space-Time Coded Modulation

If the channel (28.1) is constant over a long time period LT (that is, we have a slow fading MIMO channel), a sequence of space-time codes may be used for transmission; that is, a codeword $\mathbf{X}$ is of the form $\mathbf{X} = (\mathbf{X}_1, \ldots, \mathbf{X}_L)$ where every $\mathbf{X}_i$ is an $M \times T$ space-time codeword. Slow MIMO fading induces a loss in diversity, which is known to be compensated for by using concatenated coding schemes, enabling us to distinguish the two main design criteria, namely the diversity and determinant criteria: an inner code guarantees full diversity (this is the choice of $\mathbf{X}_i$), an outer code brings coding gain, which is characterized by the minimum determinant:

$$\min_{\mathbf{X} \neq 0} \det(\mathbf{X}\mathbf{X}^*) = \min_{\mathbf{X} \neq 0} \det(\mathbf{X}_1\mathbf{X}_1^* + \cdots + \mathbf{X}_L\mathbf{X}_L^*) \geq \min_{\mathbf{X} \neq 0} \left(\sum_{i=1}^{L} |\det(\mathbf{X}_i)| \right)^2$$

to be maximized.

In [976], a concatenated scheme is considered, where the inner code is the golden code (see Subsection 28.3.3) and the outer code is a trellis code. This can be viewed as a multidimensional trellis coded modulation, where the golden code serves as a signal set to be partitioned, which is done based on lattice set partitioning to increase the minimum determinant between codewords. A first attempt to design such a scheme was made in [379], whose ad-hoc scheme suffered from a high trellis complexity. In [1407], the golden code is serially concatenated with a convolutional code. In [1309], coset coding is used to design an outer code for the golden code, and the coding problem becomes that of designing a suitable outer code over the given ring of matrices ($\mathbb{F}_2^{2\times 2}$ and $\mathbb{F}_2[i]^{2\times 2}$): the repetition code of length 2 and a construction using Reed–Solomon codes are proposed. In [1460], codes over matrix rings were more generally considered for some space-time codes built over division algebras.

28.4.3 Fast Decodable Space-Time Codes

Suppose we have twice the number of transmit antennas than the number of receive antennas in (28.1), namely $M = 2N$; the explanation made below can be adapted to M a multiple of N without any difficulty. A full rate $M \times M$ space-time codeword $\mathbf{X}$ transmits up to $2N^2$ complex (say QAM) information symbols, or equivalently $4N^2$ real (say PAM) information symbols.

We rewrite a space-time codeword $\mathbf{X}$ in terms of basis matrices $\mathbf{B}_i$, $i = 1, \ldots, 4N^2$, and a PAM vector $\mathbf{u} = (u_1, \ldots, u_{4N^2})^{\mathsf{T}}$ as

$$\mathbf{X} = \sum_{i=1}^{4N^2} u_i \mathbf{B}_i,$$

as was done for linear dispersion codes. Then (28.2) in Euclidean norm becomes

$$d(X) = \|\mathbf{y} - \mathbf{B}\mathbf{u}\|_E^2, \tag{28.6}$$

where $\mathbf{y} \in \mathbb{R}^{4N^2}$ is the vectorized channel output, with real and imaginary parts separated, and a $4N^2 \times 4N^2$ real matrix

$$\mathbf{B} = (\mathbf{b}_1, \mathbf{b}_2, \ldots, \mathbf{b}_{4N^2}) \in \mathbb{R}^{4N^2 \times 4N^2},$$

also obtained by vectorizing and separating the real and imaginary parts of $\mathbf{H}\mathbf{B}_i$ to obtain $\mathbf{b}_i$ for $i = 1, \ldots, 4N^2$.

From (28.6), a QR decomposition of $\mathbf{B}$, $\mathbf{B} = \mathbf{Q}\mathbf{R}$, with $\mathbf{Q}$ unitary, reduces to computing

$$d(X) = \|\mathbf{y} - \mathbf{Q}\mathbf{R}\mathbf{u}\|_E^2 = \|\mathbf{Q}^*\mathbf{y} - \mathbf{R}\mathbf{u}\|_E^2$$

where $\mathbf{R}$ is an upper right triangular matrix. The number and position of nonzero elements in the upper right part of $\mathbf{R}$ will determine the complexity of the real sphere decoding process [922, 1855].

The worst case is given when the matrix $\mathbf{R}$ is a full upper right triangular matrix, yielding the complexity of the exhaustive-search ML decoder, that is here $O(|S|^{4N^2})$, where $|S|$ is the number of PAM symbols in use. If the structure of the code is such that the decoding complexity has an exponent of $|S|$ smaller than $4N^2$, we say that the code is **fast-decodable** [203].

We list three shapes for the matrix $\mathbf{R}$ which have been shown [1048, 1417] to result in reduced decoding complexity. Suppose a space-time codeword has matrix $\mathbf{R}$ of the form:

$$\mathbf{R} = \begin{bmatrix} \Delta & * \\ \mathbf{0} & \mathbf{R}' \end{bmatrix},$$

where $\Delta, \mathbf{R}', \mathbf{0}, *$ are respectively a $d \times d$ diagonal, upper-triangular, zero, and an arbitrary matrix. Then the exponent of $|S|$ in the decoding complexity of the code is reduced by $d - 1$. The zero structure of $\mathbf{R}$ is actually related [1048] to the zero structure of the matrix M defined by $\mathbf{M}_{k,l} = \|\mathbf{B}_k\mathbf{B}_l^* + \mathbf{B}_l\mathbf{B}_k^*\|_F$ which captures the orthogonality relations of the basis elements B_i. Furthermore, M has the advantage that its zero structure is stable under pre-multiplication of $\mathbf{B}_i$ by a channel matrix $\mathbf{H}$ (the same does not generally hold for $\mathbf{R}$). It was shown in [1048, Lemma 2] that

$$\mathbf{M} = \begin{bmatrix} \Delta & * \\ * & * \end{bmatrix} \Rightarrow \mathbf{R} = \begin{bmatrix} \Delta & * \\ \mathbf{0} & \mathbf{R}' \end{bmatrix},$$

where Δ is as above a d-dimensional diagonal matrix. It is thus sufficient to compute $\mathbf{M}$ to deduce fast-decodability.

A space-time code is called g-**group decodable** if there exists a partition of nonempty subsets $\Gamma_1, \ldots, \Gamma_g$, so that $\mathbf{M}_{k,l} = 0$ when k, l are in disjoint subsets Γ_i, Γ_j.

In this case, the matrix $\mathbf{R}$ has the form $\mathbf{R} = \mathrm{diag}(\mathbf{R}_1, \ldots, \mathbf{R}_g)$, where each $\mathbf{R}_i$ is a square upper triangular matrix [1048] and the symbols u_k and u_l can be decoded independently when their corresponding basis matrices $\mathbf{B}_k$ and $\mathbf{B}_l$ belong to disjoint subsets of the induced partition of the basis, yielding a decoding complexity of $O(|S|^{\max_{1 \le i \le g} |\Gamma_i|})$.

Finally, a code which is not g-group decodable but could become one assuming that a set of symbols are already decoded gives another characterization of fast-decodability [1417]. A code is called **conditionally g-group decodable** if there exists a partition into $g + 1$ disjoint subsets $\Gamma_1, \ldots, \Gamma_g, \Gamma^C$ such that

$$\|\mathbf{B}_l\mathbf{B}_m^* + \mathbf{B}_m\mathbf{B}_l^*\|_F = 0 \quad \text{for all } l \in \Gamma_i \text{ and for all } m \in \Gamma_j \text{ with } i \ne j.$$

The sphere decoding complexity order then reduces to $|S|^{|\Gamma^C| + \max_{1 \le i \le g} |\Gamma_i|}$.

By the above discussion, in order to demonstrate fast-decodability, respectively (conditional) g-group decodability, it suffices to find an ordering on the basis elements $\mathbf{B}_i$, which results in the desired zero structure of $\mathbf{M}$.

As an example, we may consider the Alamouti code:

$$\mathbf{X} = \begin{bmatrix} u_1 & -u_2^* \\ u_2 & u_1^* \end{bmatrix} = \begin{bmatrix} u_{11} + iu_{12} & -u_{21} + iu_{22} \\ u_{21} + iu_{22} & u_{11} - iu_{12} \end{bmatrix},$$

where u_1, u_2 are QAM symbols and $\mathbf{u} = (u_{11}, u_{12}, u_{21}, u_{22})$ is the PAM symbol vector. A decomposition into basis matrices $\mathbf{B}_1, \mathbf{B}_2, \mathbf{B}_3, \mathbf{B}_4$ is given by

$$\mathbf{X} = u_{11}\mathbf{B}_1 + u_{12}\mathbf{B}_2 + u_{21}\mathbf{B}_3 + u_{22}\mathbf{B}_4,$$

where

$$\mathbf{B}_1 = \begin{bmatrix} 1 & 0 \\ 0 & 1 \end{bmatrix}, \ \mathbf{B}_2 = \begin{bmatrix} i & 0 \\ 0 & -i \end{bmatrix}, \ \mathbf{B}_3 = \begin{bmatrix} 0 & -1 \\ 1 & 0 \end{bmatrix}, \ \mathbf{B}_4 = \begin{bmatrix} 0 & i \\ i & 0 \end{bmatrix}.$$

We assume transmission through a channel with $M = 2, N = 1$ antennas, described by the vector $\mathbf{H} = (h_1, h_2)$ so that $\mathbf{B} = [\mathbf{b}_1, \mathbf{b}_2, \mathbf{b}_3, \mathbf{b}_4]$ with

$$\mathbf{b}_1 = (\text{Re}(h_1), \text{Im}(h_1), \text{Re}(h_2), \text{Im}(h_2))^\mathsf{T},$$
$$\mathbf{b}_2 = (-\text{Im}(h_1), \text{Re}(h_1), \text{Im}(h_2), -\text{Re}(h_2))^\mathsf{T},$$
$$\mathbf{b}_3 = (\text{Re}(h_2), \text{Im}(h_2), -\text{Re}(h_1), -\text{Im}(h_1))^\mathsf{T},$$
$$\mathbf{b}_4 = (-\text{Im}(h_2), \text{Re}(h_2), -\text{Im}(h_1), \text{Re}(h_1))^\mathsf{T}.$$

Since $\langle \mathbf{b}_i, \mathbf{b}_j \rangle = 0$ for $i \ne j$, the QR decomposition of $\mathbf{B}$ yields a matrix $\mathbf{R}$ which is in fact diagonal. Thus the worst case decoding complexity of such a code is the size of the QAM alphabet, that is, of linear order. In fact, the design of fast-decodable space-time codes is mimicking that of the Alamouti code, to reduce their decoding complexity. For design of fast-decodable space-time codes, see, e.g., [1331, 1418, 1419, 1718, 1747, 1849].

28.4.4 Secure Space-Time Coding

Given an honest transmitter and an honest receiver which communicate in the presence of an eavesdropper, one may ask the question of designing secure space-time codes. In [947], the so-called constraints of low probability of detection, and low probability of intercept were introduced, considering the scenario where the transmitter and the receiver are both informed about their channel while the eavesdropper is only informed about its own channel. Such assumptions are not standard: the eavesdropper is usually knowledgeable about everything, and in this case, codes that achieve both reliability and confidentiality are referred to as **wiretap codes**.

We are next interested in deriving a code design criterion for MIMO wiretap codes (there is an important literature on designing wiretap codes, in particular capacity achieving codes from an information theoretic view point, which we voluntarily skip here). We know from (28.1) what is the channel to the honest user (which we will index by M for main); the channel to the eavesdropper is similar:

$$\mathbf{Y}_M = \mathbf{H}_M \mathbf{X} + \mathbf{Z}_M \in \mathbb{C}^{n_M \times T}$$
$$\mathbf{Y}_E = \mathbf{H}_E \mathbf{X} + \mathbf{Z}_E \in \mathbb{C}^{n_E \times T},$$

where the transmitter has M transmit antennas, while the receiver and the eavesdropper have respectively n_M and n_E receive antennas. Ignoring the eavesdropper, we know how to design space-time codes, as reviewed above. We are left with deriving a code design criterion for confidentiality. The noise coefficients for both receivers have different variances and the covariance matrices of the fading matrices are different (we do not normalize the covariance matrices for deriving code design criteria).

To achieve confidentiality, we use an encoding scheme called **coset coding**, which is a well accepted encoding to provide confusion in the presence of an eavesdropper, whose original version is already present in Wyner's work, where the notion of wiretap codes was first proposed. The idea of coset encoding is to partition the set of codewords into subsets (called cosets), label each coset by information symbols, and then pick a codeword uniformly at random within a coset for the actual transmission. Concretely, the transmitter partitions the group of linear space-time codewords into a union of disjoint cosets $\Lambda_E + C$, where Λ_E is a subgroup and C encodes the data. The transmitter then chooses a random $\mathbf{R} \in \Lambda_E$ so that $\mathbf{X} = \mathbf{R} + \mathbf{C} \in \Lambda_E + \mathbf{C}$.

We mimic the usual pairwise error probability computations to obtain Eve's probability of correctly obtaining the secret codeword, in an attempt to derive a confidentiality code design criterion. It remains an open question whether the obtained criterion can be linked to the information theoretical characterization of secrecy; see [1266] for a discussion in the context of Gaussian channels.

We have that Eve's probability $P_{c,e}$ of correctly decoding a space-time code $\mathbf{X}$ sent is [153]

$$P_{c,e} \leq C_{\text{MIMO}} \gamma_E^{TM} \sum_{\mathbf{X} \in \Lambda_E} \det\left(\mathbf{I}_M + \gamma_E \mathbf{X}\mathbf{X}^*\right)^{-n_E - T} \tag{28.7}$$

where γ_E is Eve's signal to noise ratio and C_{MIMO} is a constant that depends on different channel parameters.

To derive a code design criterion from (28.7), similarly to the way diversity and coding gain were obtained for reliability, we rewrite:

$$P_{c,e} \leq C_{\text{MIMO}} \gamma_E^{TM} \left[1 + \sum_{\mathbf{X} \in \Lambda_E \setminus \{0\}} \det\left(\mathbf{I}_M + \gamma_E \mathbf{X}\mathbf{X}^*\right)^{-n_E - T} \right].$$

The space-time code used to transmit data to Bob is assumed to be fully diverse, since a wiretap code needs to ensure reliability for Bob, namely, if $\mathbf{X} \neq 0$ and $T \geq M$ then the rank of $\mathbf{X}$ is M. To minimize Eve's average probability of correct decoding, a first simplified design criterion is then

$$\boxed{\min_{\Lambda_E} \sum_{\mathbf{X} \in \Lambda_E \setminus \{0\}} \frac{1}{\det(\mathbf{X}\mathbf{X}^*)^{n_E + T}}.}$$

For works on MIMO wiretap codes achieving the secrecy capacity and satisfying Wyner's original criterion for secrecy given in terms of mutual information, see [1111].

Chapter 29

Network Codes

Frank R. Kschischang
University of Toronto

29.1 Packet Networks

Communication networks generalize the simple communication channel—having a single transmitter, a single receiver, and a single channel as described in Chapter 1—to multiple nodes interconnected by multiple communication channels. In **packet networks**, information to be transmitted from one node to another is segmented into relatively short pieces called **packets**, which travel over various communication channels, often passing through various intermediate nodes which serve as relays. The channels between nodes are typically modelled as being error-free, since errors caused by noise are assumed to be corrected using an appropriate type of error-correcting code, as described elsewhere in this encyclopedia. In addition to their information **payload**, packets are typically augmented with a **header**, which contains information such as the network address of the information source, the network address of the destination, the protocol type, a sequence number, and other similar useful metadata. An often-abused analogy is that packets travelling over networks are similar to cars travelling along highways, proceeding from source to destination. The design and analysis of communication networks is covered in many textbooks; see, e.g., [1221, 1746] .

In this chapter, communication networks will be modelled as directed multigraphs, with certain nodes designated as information sources and certain nodes designated as information sinks. For example, the directed graph shown in Figure 29.1(a)—the so-called **butterfly network**—models a packet network with a single source s and two sinks t and u. Each directed edge represents an error-free packet channel capable of delivering, in each time slot, a single packet of some fixed number of bits. The communication objective is to transfer a common set of packets from the source node to the two sink nodes at as high a rate as possible. Note that the butterfly network contains a bottleneck link: the edge (x, y) in Figure 29.1(a).

FIGURE 29.1: (a) the butterfly network with bottleneck channel (x, y), in which source node s wishes to send fixed-length packets as quickly as possible to two sink nodes t and u; (b) a routing solution, achieving the transfer of three packets a, b, c over two time slots (here $-$ indicates an idle time slot); (c) a more efficient network coding solution, achieving the transfer of two packets a, b in a single channel time slot, at the expense of a coding operation at node x and decoding operations at the sink nodes

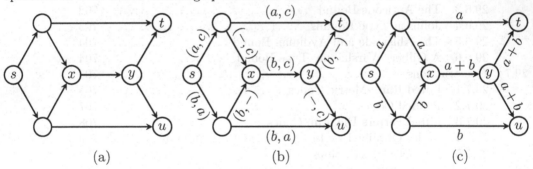

In the conventional store-and-forward **routing solution**, information-flow is treated as commodity-flow; i.e., the contents of packets are not permitted to be altered at the intermediate nodes, but are simply forwarded (routed) from incoming to outgoing channels. For this reason, the intermediate nodes are referred to as **routers**. Although the source can transmit two distinct packets in each time slot, only one of them may pass through the bottleneck link, and thus only one of the two sinks has the benefit of receiving two distinct packets in a single time slot. By alternating this benefit between the sink nodes, it is possible to achieve a transmission rate of 3 packets per 2 time-slots, i.e., 1.5 packets per time slot. One of several possible means of achieving this throughput by routing is shown in Figure 29.1(b), where each edge is labelled with a packet pair having components corresponding to time slots.

In the **network coding** solution shown in Figure 29.1(c), the contents of packets are indeed permitted to be altered at the intermediate nodes, i.e., the intermediate nodes may transmit packets that are a function of the packets that they receive, so that, in effect, routers become **coders**. The benefit of this is that node x, which receives two packets a and b, can transmit their sum $a + b$ (here we view packets of n bits as binary vectors, and add them as elements of $\mathbb{F}_2^n$). In a single time slot, node t now receives a and $a + b$, and can recover b by forming $a + (a + b)$ in $\mathbb{F}_2^n$. Likewise, node u recovers packet a by forming $(a + b) + b$. Thus at the expense of an encoding operation at intermediate node x and decoding operations at the sinks, a transmission rate of 2 packets per time slot can be achieved. Since the source can emit at most 2 distinct packets per time slot and the sinks can receive at most 2 distinct packets per time slot, this rate is clearly the best possible.

The butterfly network, introduced in the seminal paper [26] of Ahlswede, Cai, Li, and Yeung (in which the field of network coding originated), illustrates the important conclusion that routing alone is, in general, insufficient to achieve the full information-carrying capability of packet networks. Stated differently, information-flow is not the same as commodity-flow: communication networks are not the same as highways, since, unlike cars on the highway, packets can in fact be combined to squeeze through network bottlenecks, provided that sufficient side information is available at the sinks to be able to reverse such combinations.

29.2 Multicasting from a Single Source

Our interest in this chapter will be to define and study the **multicast problem** in combinational packet networks. Although this problem is not the most general network information flow problem, it is perhaps the easiest, as it admits simple and elegant linear network coding solutions. We will first consider an error-free approach, and later consider coding strategies to deal with an adversary who can inject erroneous packets into the network.

29.2.1 Combinational Packet Networks

We assume that the reader is somewhat familiar with the terminology of graph theory at the level of, say, the first chapter of [1888], as we will use such terminology often. A **combinational packet network** $\mathcal{N} = (V, E, S, T, A)$ comprises a finite directed acyclic multigraph $G = (V, E)$ where V is the set of vertices and E is the multiset of directed edges, a distinguished nonempty set $S \subset V$ of **sources**, a distinguished nonempty set $T \subset V$ of **sinks**, and a finite **packet alphabet** A, with $|A| \geq 2$.

Vertices in V model communication nodes within the packet network, while directed edges in E model error-free communication channels between the nodes. Each edge $(u, v) \in E$ has **unit capacity**, in the sense that, in each channel use, it can deliver a single packet (i.e., a symbol drawn from A) from u to v without any errors. It is also possible that an edge remains idle (no packet is transmitted), and, in this case, it is assumed that v is aware that the edge is idle. Equivalently, idle edges can be removed from E. Idle edges were denoted with '$-$' in the routing solution of Figure 29.1(b). To allow for greater (but still integral) capacity from u to v, parallel edges between u and v are permitted; thus G is in general a **multigraph** (as opposed to a simple graph, where such parallel edges are not permitted).

The network of Figure 29.1, for example, has 7 vertices, 9 directed edges (but no parallel edges), a single source vertex s, 2 sink vertices t and u, and a packet alphabet $A = \mathbb{F}_2^n$.

At each vertex $v \in V$, we denote by $I(v)$ the set of (incoming) edges incident **to** v and by $O(v)$ the set of (outgoing) edges incident **from** v. Possibly by introducing "virtual nodes" in a given network we may, without loss of generality, assume that $S \cap T = \emptyset$, that $I(s) = \emptyset$ for every source node $s \in S$, and that $O(t) = \emptyset$ for every sink node $t \in T$.

As usual in graph theory, a (directed) **path** in G from a vertex $u \in V$ to a vertex $v \in V$ is a finite sequence of directed edges $(u, v_1), (v_1, v_2), \ldots, (v_{\ell-1}, v)$, where each element of the sequence is an element of E. A cycle in G is a directed path from some vertex to itself. The term **acyclic** in the definition of a packet network means that the multigraph G contains no cycles. A vertex $v \in V$ is said to be **reachable** from $u \in V$ if $v = u$ or if there is a directed path in G from u to v.

As in combinational logic (hence the terminology **combinational packet network**), it is assumed that packets transmitted on the non-idle edges of $O(v)$ are **functions** of the packets received on the non-idle edges of $I(v)$, or, if v is a source, of packets internally generated at v. Acyclicity of G ensures that each packet sent on any edge is indeed a well-defined function of the packets sent by the sources. In an actual network implementation, packets must be received (and possibly buffered) at v before outgoing packets can be computed and transmitted. Delays caused by transmission, buffering, and processing of packets are, however, not explicitly modelled. We refer to the function applied by a particular node v as the **local encoding function** at v. For example, if v were acting as a router, then each packet sent on an edge of $O(v)$ would be a copy of one of the packets received on an edge of $I(v)$. In Figure 29.1(c), node x receives two packets a and b on the two edges of $I(x)$, and transmits the packet $a + b$ on the single edge in $O(x)$.

By a single **channel use** of a combinational packet network $\mathcal{N}$ we mean a particular assignment of packets drawn from A to each of the (non-idle) edges of E, consistent with the particular local encoding functions implemented by the nodes of the network. In other words, in a single channel use, the local encoding functions are fixed, and each edge of the network is used at most once. Of course it is possible, as in the routing example of Figure 29.1(b), that the local encoding functions may vary from channel use to channel use.

Transmission efficiency is measured in units of **packets per channel use**. For example, the butterfly network routing solution of Fig. 29.1(b) achieves a multicast rate of 1.5 packets per channel use, while the network coding solution of Fig. 29.1(c) achieves a multicast rate of 2 packets per channel use. It should be noted that the measure of information ("packets") used here depends on the alphabet size $|A|$. It is straightforward to convert a rate from packets per channel use to bits per channel use, simply by multiplying by $\log_2 |A|$ bits per packet.

This combinational model is not the most general possible. For example, if the underlying transmission channels were wireless, then the network model would need to be modified to take into account the broadcast nature of the channel (i.e., the possibility that a single transmission might be received simultaneously by multiple receivers) and the multiple-access nature of the channel (i.e., that simultaneous transmissions from multiple transmitters may interfere with each other at a given receiver). The model also does not explicitly incorporate outage (the possibility that links may not be perfectly reliable), delay (or any notion of timing as in sequential logic), or the possibility of feedback (cycles in the graph). Nevertheless, the combinational model given is sufficiently rich that it will well illustrate the main ideas of network coding.

29.2.2 Network Information Flow Problems

Let the set of source vertices be given as $S = \{s_1, s_2, \ldots, s_{|S|}\}$. We assume that each source vertex $s_i \in S$ has an infinite supply of packets to send. We also assume that each sink vertex $t \in T$ is interested in reconstructing the packets sent by some subset $D_t \subseteq S$ of sources. We refer to the elements of D_t as the sources **demanded** by sink t. A network information flow problem is said to be a **single-source** problem if $|S| = 1$, i.e., if there is just a single source in the network.

Suppose that a source s_i has packets $p_1, p_2, p_3, \ldots$ to send. If, over n channel uses, a sink t can reconstruct (as a function of the packets it receives on $I(t)$ over those channel uses) k source packets $p_1, p_2, \ldots, p_k$, then we say that sink t can decode source s_i at **rate** $r_i = k/n$ packets per channel use. We will require that each sink demanding a particular source $s_i \in S$ be served at the same transmission rate r_i. A particular collection of demands $\{D_t \mid t \in T\}$ is said to be **achievable** with rates $(r_1, r_2, \ldots, r_{|S|})$ over a given network if, for some choice of (possibly time-varying) local encoding operations, and some n, each sink can reconstruct the output of the sources it demands at the specified rate.

An important special case of this general network information flow problem is the **multicast problem**, in which $D_t = S$ for all $t \in T$, i.e., the case where all sources are demanded by all sinks. Other important special cases include the **unicast problem** (with a single source s, a single sink t, and $D_t = \{s\}$) or the problem of **multiple unicasts** (for some integer $L > 1$, $S = \{s_1, \ldots, s_L\}$, $T = \{t_1, \ldots, t_L\}$, and $D_{t_i} = \{s_i\}$). Of course many further variations are possible.

29.2.3 The Unicast Problem

As described in the previous subsection, in the single-source multicast problem, the network has a single source node s and, in general, multiple sink nodes, all of which demand $\{s\}$. We start by considering the special case where there is also just a single sink t.

Suppose that vertex $t \in V$ is reachable in G from vertex $s \in V$. An $\{s, t\}$-**separating cut** is a collection C of edges such that every path from s to t contains an edge from C. If the edges of C were removed from E, then there would be no path from s to t. A **minimum cut** separating s and t is a cut of smallest possible cardinality (among all $\{s, t\}$-separating cuts). We denote the cardinality of an $\{s, t\}$-separating minimum cut as $\mathrm{mincut}(s, t)$. If t is not reachable from s, then $\mathrm{mincut}(s, t) = 0$.

Let $R(s, t)$ denote the information rate (in packets per channel use) that flows between s and t in any fixed transmission scheme. If C is any $\{s, t\}$-separating cut, then since at most $|C|$ distinct packets can flow over the edges of C in each channel use, we have the trivial upper bound $R(s, t) \leq |C|$. Minimizing $|C|$ over all cuts yields the upper bound

$$R(s, t) \leq \mathrm{mincut}(s, t).$$

To achieve $R(s, t) = \mathrm{mincut}(s, t)$, it is necessary that in each channel use, through each minimum cut exactly $\mathrm{mincut}(s, t)$ distinct packets must flow, from which the $\mathrm{mincut}(s, t)$ information packets emitted by the source must be recoverable.

It is well known from the edge-connectivity version of Menger's Theorem [1374] or from the theory of commodity flows [675, 738] that $\mathrm{mincut}(s, t)$ is equal to the maximum number of mutually edge-disjoint paths from s to t. Such a collection of mutually edge-disjoint paths can be found using, e.g., the Ford–Fulkerson method [444, Section 26.2] [738]. Thus the solution of the single-source unicast problem, i.e., transmission of $R(s, t) = \mathrm{mincut}(s, t)$ packets per channel use from the source s to a **single** sink t can be achieved by **routing**, sending a distinct packet along each of the edge-disjoint paths between s and t in each channel use.

For example, in the butterfly network of Figure 29.1, $\text{mincut}(s,t) = 2$, and indeed it is easy to find a pair of edge-disjoint paths from s to t over which 2 packets may be routed in each channel use. Note that $\text{mincut}(s,u) = 2$ also. However, as we have already seen, in this network the paths over which packets are routed to t conflict with the paths over which packets are routed to u, making it impossible to achieve, using routing alone, a transmission rate of 2 packets per channel use to *both* t and u simultaneously.

29.2.4   Linear Network Coding Achieves Multicast Capacity

Suppose now, that the set of sinks, T, contains more than one node. It is clear that the multicast rate $R(s,T)$ from s to T cannot exceed the transmission rate that can be achieved from s to any particular element of T; i.e., the multicast rate $R(s,T)$ must satisfy

$$R(s,T) \leq \min_{t \in T} \text{mincut}(s,t). \tag{29.1}$$

The main theorem of network multicasting is that the upper bound (29.1) is achievable (with equality) via network coding [26]. Moreover, the upper bound is achievable with **linear** network coding, provided that the packet alphabet A is a vector space over a sufficiently large finite field [1248]. The quantity $\min_{t \in T} \text{mincut}(s,t)$ is referred to as the **multicast capacity** of the given combinational packet network.

In linear network coding, we assume that the packet alphabet A is an n-dimensional vector space $\mathbb{F}_q^n$ over the finite field $\mathbb{F}_q$. We assume that the source will communicate at some (integer) rate r to all of the sink nodes $t \in T$. This means that the source vertex s has r information packets $p_1, p_2, \ldots, p_r$ to send in any given channel use, which we view as "incoming" to s. We will gather these information packets to form the rows of an $r \times n$ matrix denoted as P. Each sink node $t \in T$ must generate r output packets, which we view as "outgoing" from t. These will be gathered into the rows of an $r \times n$ matrix denoted, with slight abuse of notation, as $X_{O(t)}$. We require, for all $t \in T$, that $X_{O(t)} = P$. In other words, the communication problem is to reproduce the source matrix P at the output $X_{O(t)}$ of each sink node $t \in T$.

We will assume that no edges of the network are idle. This means that a vertex $v \in V \setminus \{s\}$ will observe exactly $|I(v)|$ incoming packets, forming the rows of an $|I(v)| \times n$ matrix denoted as $X_{I(v)}$. Vertex v (including, now, each sink node $t \in T$) will then apply its local encoding function to produce $|O(v)|$ outgoing packets, forming the rows of an $|O(v)| \times n$ matrix denoted as $X_{O(v)}$.

In **linear network coding**, these local encoding functions are **linear** over $\mathbb{F}_q$, i.e.,

$$X_{O(v)} = L_v X_{I(v)},$$

for some $|O(v)| \times |I(v)|$ matrix L_v with entries from $\mathbb{F}_q$, called the **local transfer matrix** at v. In other words, each outgoing packet (a row of $X_{O(v)}$) is an $\mathbb{F}_q$-linear combination of the incoming packets at node v (the rows of $X_{I(v)}$). The row of L_v corresponding to a particular edge $e \in O(v)$ is called the **local encoding vector** associated with e.

Note that, since only linear operations are performed in the network, every packet transmitted along an edge is a linear combination of the source packets $p_1, p_2, \ldots, p_r$. In particular, for every vertex $v \in V$, we have

$$X_{I(v)} = G_v P$$

where G_v is an $|I(v)| \times r$ matrix of coefficients from $\mathbb{F}_q$, called the **global transfer matrix** at v. The row of G_v corresponding to a particular edge $e \in I(v)$ is called the **global encoding vector** associated with e.

Now, in order for a sink node t to be able to recover P, it is necessary for the $|I(t)| \times r$ global transfer matrix G_t at t to have rank r, since only then does G_t have a *left* inverse G_t^{-1} satisfying $G_t^{-1} G_t = I_r$, where I_r is the $r \times r$ identity matrix. (Note that G_t does not need to be *square* to have a left inverse, merely full rank.) Forming $G_t^{-1} X_{I(t)}$ at node t produces P, as desired. The following theorem claims that if $r \leq \min_{t \in T} \text{mincut}(s, t)$ and q is sufficiently large, then it is possible to make G_t an $r \times r$ invertible matrix **simultaneously** at each sink node t, thereby achieving a multicast rate of r with linear network coding.

Theorem 29.2.1 (Linear Network Multicasting Theorem) *If $\mathcal{N} = (V, E, \{s\}, T, \mathbb{F}_q)$, then a multicast rate of*

$$R(s, T) = \min_{t \in T} \text{mincut}(s, t) \tag{29.2}$$

is achievable, for sufficiently large q, with linear network coding. In fact, $q \geq |T|$ suffices for (29.2) to hold.

This theorem was established in [1248], and an elegant algebraic proof was given in [1157], in which the bound $q > |T|$ was proved to be sufficient for the theorem to hold. That $q = |T|$ also suffices follows from the **linear information flow algorithm** of [1034], which provides a polynomial-time construction of capacity-achieving network codes. See [750, 1169] for tutorial introductions.

29.2.5 Multicasting from Multiple Sources

The Linear Network Multicasting Theorem can be extended to the case of multicasting from multiple sources ($|S| > 1$) by augmenting the network with a virtual "supersource" connected to each actual source via an appropriate number of parallel edges.

Specifically, let $S = \{s_1, \ldots, s_{|S|}\}$. Let $(r_1, r_2, \ldots, r_{|S|})$ be any $|S|$-tuple of integers with the property that, for each sink node $t \in T$,

1. it is possible to establish r_i paths from s_i to t for all $i \in \{1, 2, \ldots, |S|\}$, and

2. these $r = r_1 + r_2 + \cdots + r_{|S|}$ paths are mutually edge-disjoint.

Under these conditions it would be possible for the source nodes to communicate—using routing alone—with any *single* sink node t at a rate of r packets per channel use, with r_i packets originating at source node s_i.

With linear network coding, it is possible to achieve the rate-tuple $(r_1, \ldots, r_{|S|})$ *simultaneously* at all sink nodes $t \in T$. To see this, simply augment the network with a "virtual source" s^*, connected to s_i with r_i parallel edges, and note that each sink $t \in T$ is reachable from s^* by r edge-disjoint paths. Theorem 29.2.1 implies that s^* can, using linear network coding, multicast to all sinks at a rate of r packets per channel use. Let us fix one such linear network coding solution. We then have, for some $r \times r$ matrix L_{s^*},

$$X_{O(s^*)} = L_{s^*} P,$$

where, as above, the rows of P consist of the r packets to be sent. In other words, the packets transmitted on the edges incident from s^* are linear combinations of the r super-source packets. However, since $O(s^*)$ is a cut separating s^* from the rest of the network, the supersource packets must be decodable from $X_{O(s^*)}$, which means that L_{s^*} must be invertible. Premultiplying at s^* by $L_{s^*}^{-1}$ does not affect the ability of any of the sinks to decode, but translates a general network coding solution into a solution where the global encoding vectors on the edges of $O(s^*)$ are unit vectors. Thus source s_i receives r_i "uncoded" super-source packets. By supposing that these r_i supersource packets are actually generated at

s_i, we see that the given linear network coding solution has the effect of delivering packets from s_i to each sink at a rate r_i packets per channel use, thereby achieving the rate-tuple $(r_1, \ldots, r_{|S|})$.

29.3 Random Linear Network Coding

Theorem 29.2.1 shows that the single-source multicast capacity of a given network can be achieved via linear network coding over a finite field $\mathbb{F}_q$, with $q \geq |T|$. Indeed, a feasible linear network coding design can be produced in a computation time that is a polynomial function of $|E|$, $|T|$, and $\log_q$ using the linear information flow algorithm [1034]. An alternative to the linear information flow algorithm is a matrix-completion approach due to Harvey, et al. [916].

In fact, a careful deterministic linear network coding design is usually not needed in practice, since a completely **random** choice of the elements of the local encoding vector will, with high probability, produce a feasible solution provided that the network coding field size q is sufficiently large. This approach, which has a number of practical benefits, was first investigated by Ho, et al. [964].

Suppose that the given network has multicast capacity r. Using, say, the Ford–Fulkerson method [444, Section 26.2] [738], it is then possible to find r mutually edge-disjoint paths from the source s to any sink node $t \in T$. Edges not involved in any of these paths to any of the sinks can safely be discarded without reducing the multicast capacity. In some cases (for example, if $|I(v)| = 1$) it may be possible for a node v to determine the local encoding vector for an outgoing edge deterministically with no reduction in the achievable multicast rate. We call an edge a **random coding edge** if its local encoding vector is not fixed deterministically, but instead is chosen randomly. Let c_t denote the number of random coding edges encountered on the r edge-disjoint paths from the source s to a sink node t as found above, and let $\eta = \max_{t \in T} c_t$. A key result, [964, Theorem 3], is the following.

Theorem 29.3.1 *Fix $\mathbb{F}_q$, with $q > |T|$. The probability that local encoding coefficients chosen independently and uniformly at random from $\mathbb{F}_q$ result in a feasible network code, assuming there exists a feasible solution with the same values for the fixed local encoding vectors, is at least*
$$\mathrm{Prob}[success] \geq (1 - |T|/q)^{\eta}.$$

Stated differently, and making the practical assumption that $q = 2^m$ (so that each field element comprises m bits), the probability $\mathrm{Prob}[failure]$ that such a random choice of local encoding coefficients fails to produce a feasible network code satisfies

$$\mathrm{Prob}[failure] \leq |T| \eta 2^{-m},$$

which, for any fixed network, decreases exponentially with the number of bits per field element.

The random linear network coding approach has many practical benefits, as it can be used not only to design a fixed network code, but also as a **communication protocol** [407, 964]. In this approach, communication between source and sinks occurs in a series of rounds or **generations**; during each generation, the source injects a number of fixed-length packets into the network, each of which is a fixed length row vector over $\mathbb{F}_q$. These packets propagate throughout the network, possibly passing through a number of intermediate nodes between transmitter and receiver. Whenever an intermediate node is granted a transmission opportunity, it creates a random $\mathbb{F}_q$-linear combination of the packets it has available and transmits this random combination. Finally, the receiver collects such randomly generated packets and tries to infer the set of packets injected into the network. This model is robust in the sense that we make no assumption that the network operates synchronously or without delay or even that the network is acyclic or has a fixed time-invariant topology. Even the number of intermediate nodes participating in the protocol from generation to generation can be time-varying.

The set of successful packet transmissions in a generation induces a directed multigraph with the same vertex set as the network, in which edges denote successful packet transmissions. As we have seen, the rate of information transmission (packets per generation) between the source and the sinks is upper-bounded by the smallest mincut between these nodes (the multicast capacity in each generation), and, furthermore, Theorem 29.3.1 implies that random linear network coding in $\mathbb{F}_q$ is able to achieve a transmission rate equal to this upper bound with probability approaching one as $q \to \infty$ [964].

Recovery of the transmitted packets from the random linear combinations received at a given terminal node remains an issue. As above, let the transmitted packets $p_1, \ldots, p_r \in \mathbb{F}_q^n$ form the rows of a matrix

$$P = \begin{bmatrix} p_1 \\ \vdots \\ p_r \end{bmatrix} \in \mathbb{F}_q^{r \times n}. \tag{29.3}$$

Suppose that the sink node t collects ℓ packets $y_1, \ldots, y_\ell \in \mathbb{F}_q^n$ in a generation, forming the rows of a matrix $Y_t \in \mathbb{F}_q^{\ell \times n}$. Since the network performs linear operations, we have

$$Y_t = G_t P, \tag{29.4}$$

where the $G_t \in \mathbb{F}_q^{\ell \times r}$ is the global transfer matrix induced at sink node t by the random choice of local encoding coefficients in the network. If we assume that G_t is not known to either the source or the sink t, then we refer to the network operation as **non-coherent network coding**.

How can the sink recover P? One obvious method is to endow the transmitted packets with so-called **headers** that can be used to record the particular linear combination of the components of the message present in each received packet. In wireless communication, the transmission of known symbols for the purpose of determining the transfer characteristics of an unknown channel is known as **channel sounding**. Here, we let packet $p_i = (e_i, m_i)$, where e_i is the ith unit vector and m_i is the packet **payload**. We then have $P = \begin{bmatrix} I_r & M \end{bmatrix}$, so that

$$Y_t = G_t P = G_t \begin{bmatrix} I_r & M \end{bmatrix} = \begin{bmatrix} G_t & G_t M \end{bmatrix}.$$

The first r columns of Y_t thus yield G_t, and, assuming G_t has a left inverse G_t^{-1}—which, as we have seen, occurs with high probability if the field size is sufficiently large—sink node t can recover $M = G_t^{-1}(G_t M)$.

Of course, the use of packet headers incurs a transmission **overhead** relative to a fixed network (with the relevant global encoding vectors known *a priori* by all sinks). For many practical transmission scenarios, where packet lengths can potentially be large, this overhead can be held to a small percentage of the total information transferred in a generation. For example, take $q = 2^8 = 256$ (field elements are bytes), a generation size $r = 50$, and packets of length $2^{11} = 2048$ bytes. The packet header then must be 50 bytes, representing an overhead of just $50/2048 = 2.4\%$.

29.4   Operator Channels

The channel sounding approach, though very convenient, is not fundamentally necessary in order to communicate over a non-coherent random linear network coding channel. Although the network will certainly impose structure on G_t, one robust approach is to simply ignore such structure, treating G_t as random (but with full rank r with high probability, where r is the number of packets in the current generation). When G_t has full rank, then $Y_t = G_t P$ has a **row space** equal to the row space of P. This suggests an interesting approach for transmission of information over the non-coherent random linear network coding channel [1156]: transmit information by the choice of the **vector space** spanned by the rows of P. Even when G_t is not full rank, what is received is a subspace of the row space of P; such a dimensional **erasure** is potentially even correctable, if we can design a codebook (of vector spaces) in which such a lower-dimensional subspace is contained within just one codeword. It is further possible to design codebooks to correct **errors**, i.e., injection into the network (by an adversary or by channel noise) of packets *not* sent by the source.

The remainder of this chapter will describe this approach and various code constructions. In this section, we will begin by establishing some notation and stating some useful facts about subspaces of $\mathbb{F}_q^n$. We will then define the operator channel, which will serve as a convenient abstraction of a random linear network coding channel.

29.4.1   Vector Space, Matrix, and Combinatorial Preliminaries

We start with subspaces. Let $\mathcal{P}_q(n)$ denote the collection of all subspaces of $\mathbb{F}_q^n$. The dimension of a subspace $V \in \mathcal{P}_q(n)$ is denoted as $\dim(V)$; clearly $0 \leq \dim(V) \leq n$. The set of subspaces $\mathcal{P}_q(n)$ can be partitioned (by dimension) into so-called **Grassmannians**; i.e.,

$$\mathcal{P}_q(n) = \bigcup_{k=0}^{n} \mathcal{G}_q(n, k),$$

where the Grassmannian $\mathcal{G}_q(n, k)$, simply the set of all k-dimensional subspaces of $\mathbb{F}_q^n$, is given as

$$\mathcal{G}_q(n, k) = \{V \in \mathcal{P}_q(n) \mid \dim(V) = k\}.$$

Let U and V be two subspaces of $\mathbb{F}_q^n$. Their intersection $U \cap V$ is then also a subspace of $\mathbb{F}_q^n$, as is their sum

$$U + V = \{u + v \mid u \in U, v \in V\},$$

which is the smallest subspace of $\mathbb{F}_q^n$ containing both U and V. It is well known that

$$\dim(U + V) + \dim(U \cap V) = \dim(U) + \dim(V). \tag{29.5}$$

When $U \cap V = \{0\}$, i.e., when U and V intersect trivially, then their sum $U + V$ is a direct sum, denoted as $U \oplus V$, and in this case $\dim(U \oplus V) = \dim(U) + \dim(V)$.

Turning now to matrices, we let $\mathbb{F}_q^{m \times n}$ denote the set of all $m \times n$ matrices over $\mathbb{F}_q$. The zero matrix is denoted as $\mathbf{0}_{m \times n}$; however, the subscript is often omitted when the dimensions are clear from context. As above, the $r \times r$ identity matrix is denoted as I_r.

The row space of a matrix X is denoted as $\langle X \rangle$, and the rank of X is denoted as $\text{rank}(X)$; of course, $\text{rank}(X) = \dim(\langle X \rangle)$. If $X \in \mathbb{F}_q^{m_1 \times n}$ and $Y \in \mathbb{F}_q^{m_2 \times n}$, then

$$\left\langle \begin{bmatrix} X \\ Y \end{bmatrix} \right\rangle = \langle X \rangle + \langle Y \rangle;$$

therefore

$$\text{rank} \begin{bmatrix} X \\ Y \end{bmatrix} = \dim(\langle X \rangle + \langle Y \rangle).$$

A vector in $\mathbb{F}_q^n$ will be considered as an element of $\mathbb{F}_q^{1 \times n}$, i.e., as a row vector. For $v = (v_1, \ldots, v_n) \in \mathbb{F}_q^n$, let $\text{supp}(v) = \{i \in \{1, \ldots, n\} \mid v_i \neq 0\}$ denote its support and let $\text{wt}_H(v) = |\text{supp}(v)|$ denote its Hamming weight.

Given two vectors $u, v \in \mathbb{F}_q^n$, let $u \star v$ denote their **Hadamard product**, i.e., the vector obtained by multiplying u and v componentwise so that $(u \star v)_i = u_i v_i$. The Hamming distance between *binary* vectors $u, v \in \mathbb{F}_2^n$ can be expressed as

$$d_H(u, v) = \text{wt}_H(u) + \text{wt}_H(v) - 2\text{wt}_H(u \star v).$$

It will also be useful to define the **asymmetric distance** [1567] between two binary vectors $u, v \in \mathbb{F}_2^n$ as

$$d_A(u, v) = \max\{\text{wt}_H(u), \text{wt}_H(v)\} - \text{wt}_H(u \star v).$$

In the special case when $\text{wt}_H(u) = \text{wt}_H(v)$, we have $d_H(u, v) = 2d_A(u, v)$.

If $X \in \mathbb{F}_q^{m \times n}$ is nonzero, then its reduced row echelon form is denoted as $\text{rref}(X)$. Associated with a nonzero X is a vector $\text{id}(X) \in \mathbb{F}_2^n$, called the **identifying vector** of X, in which $\text{supp}(\text{id}(X))$ is the set of column positions of the leading ones in the rows of $\text{rref}(X)$. For example in $\mathbb{F}_2^{3 \times 5}$,

$$\text{if } X = \begin{bmatrix} 1 & 1 & 0 & 1 & 0 \\ 0 & 1 & 1 & 1 & 1 \\ 1 & 0 & 1 & 1 & 0 \end{bmatrix}, \text{ then } \text{rref}(X) = \begin{bmatrix} \underline{1} & 0 & 1 & 0 & 1 \\ 0 & \underline{1} & 1 & 0 & 0 \\ 0 & 0 & 0 & \underline{1} & 1 \end{bmatrix}$$

and $\text{id}(X) = (1, 1, 0, 1, 0)$, where the leading ones in $\text{rref}(X)$ have been indicated by underlining. If $X = \mathbf{0}$, then we set $\text{id}(X)$ to the zero vector. Note that $\text{rank}(X) = \text{wt}_H(\text{id}(X))$.

Associated with a vector space $V \in \mathcal{G}_q(n, k)$, $k > 0$, is a unique $k \times n$ matrix G_V in reduced row echelon form (i.e., with $G_V = \text{rref}(G_V)$) having the property that $\langle G_V \rangle = V$. We call G_V the **basis matrix** for V. With a slight abuse of notation we extend the id function to vector spaces by defining

$$\text{id}(V) = \text{id}(G_V),$$

where $\text{id}(V)$ is the zero vector if $\dim(V) = 0$. Clearly we have $\dim(V) = \text{wt}_H(\text{id}(V))$.

The following bounds on the dimension of the sum and product of spaces in terms of their identifying vectors will be useful.

Theorem 29.4.1 *For any pair of spaces $U, V \in \mathcal{P}_q(n)$, let $u = \text{id}(U)$ and let $v = \text{id}(V)$ be their respective identifying vectors. Then*

$$\dim(U + V) \geq \text{wt}_H(u) + \text{wt}_H(v) - \text{wt}_H(u \star v),$$
$$\dim(U \cap V) \leq \text{wt}_H(u \star v).$$

Proof: To show the first inequality, let $P_U = \text{supp}(u)$ be the set of pivot positions in G_U and similarly let $P_V = \text{supp}(v)$. Since every pivot position of G_U is a pivot position of

$$X = \begin{bmatrix} G_U \\ G_V \end{bmatrix},$$

and similarly for G_V, it follows that $P_U \cup P_V \subseteq \text{supp}(\text{id}(X))$. Therefore, since $\dim(U + V) = |\text{supp}(\text{id}(X))|$, we get

$$\dim(U + V) \geq |P_U \cup P_V|$$
$$= |P_U| + |P_V| - |P_U \cap P_V| = \text{wt}_H(u) + \text{wt}_H(v) - \text{wt}_H(u \star v).$$

The second inequality follows from (29.5).   □

Given any binary identifying vector $a \in \{0,1\}^n$, the so-called **Schubert cell** [237] in $\mathcal{P}_q(n)$ corresponding to a is the set

$$\mathcal{S}_q(a) = \mathrm{id}^{-1}(a) = \{V \in \mathcal{P}_q(n) \mid \mathrm{id}(V) = a\}.$$

If $\mathrm{wt}_H(a) = k$, then $\mathcal{S}_q(a) \subseteq \mathcal{G}_q(n,k)$. Thus binary identifying vectors (in general) induce a *partition* of $\mathcal{P}_q(n)$ into 2^n distinct Schubert cells, while binary identifying vectors of weight k (in particular) induce a partition of $\mathcal{G}_q(n,k)$ into $\binom{n}{k}$ Schubert cells. These partitions will become useful in Section 29.7.

The basis matrices corresponding to the elements of any fixed Schubert cell $\mathcal{S}_q(a)$ all have a similar form: their pivot elements are in the same positions and take value 1; non-pivot entries in the same column as a pivot take value zero; non-pivot entries in the same row as, but to the left of, a pivot also take value zero. We refer to the remaining matrix entries as **free**, as they can take on any arbitrary value in $\mathbb{F}_q$. Free positions will be denoted with the symbol '•'. If the fixed zero and ones of a basis matrix are removed, and the free positions are right-justified, a so-called **Ferrers diagram** is obtained. A row without any free positions is denoted as '·' in the Ferrers diagram.

For example, the matrices and corresponding Ferrers diagram forms for a few sample identifying vectors are given as follows:

$$a = 111000 \leftrightarrow \begin{bmatrix} 1 & 0 & 0 & \bullet & \bullet & \bullet \\ 0 & 1 & 0 & \bullet & \bullet & \bullet \\ 0 & 0 & 1 & \bullet & \bullet & \bullet \end{bmatrix} \leftrightarrow \begin{matrix} \bullet & \bullet & \bullet \\ \bullet & \bullet & \bullet \\ \bullet & \bullet & \bullet \end{matrix}$$

$$a = 101010 \leftrightarrow \begin{bmatrix} 1 & \bullet & 0 & \bullet & 0 & \bullet \\ 0 & 0 & 1 & \bullet & 0 & \bullet \\ 0 & 0 & 0 & 0 & 1 & \bullet \end{bmatrix} \leftrightarrow \begin{matrix} \bullet & \bullet & \bullet \\ \bullet & \bullet \\ \bullet \end{matrix}$$

$$a = 010011 \leftrightarrow \begin{bmatrix} 0 & 1 & \bullet & \bullet & 0 & 0 \\ 0 & 0 & 0 & 0 & 1 & 0 \\ 0 & 0 & 0 & 0 & 0 & 1 \end{bmatrix} \leftrightarrow \begin{matrix} \bullet & \bullet \\ \cdot \\ \cdot \end{matrix}$$

Note that the number of free positions is a non-increasing function of the row index; i.e., the number of free positions in row i cannot exceed the number of free positions in row $i-1$, and of course the number of such positions must be a nonnegative integer. These are the defining properties of a Ferrers diagram (also known as a Young diagram), which arise in the study of integer partitions. It is easy to go from an identifying vector to the corresponding Ferrers diagram. In fact, when n is given it is also possible to do the reverse, and obtain an identifying vector from a Ferrers diagram. We leave this as an exercise for the reader (*hint*: start from the bottom row).

Associated with $\mathcal{G}_q(n,k)$ is a distance-regular graph (called a **Grassmann graph**) whose vertices correspond to the elements of $\mathcal{G}_q(n,k)$ and where two vertices are adjacent if the corresponding subspaces intersect in a space of dimension $k-1$ [309]. We mention in passing that the Grassmannian $\mathcal{G}_q(n,k)$ also forms an association scheme, the so-called q-Johnson scheme [1839, Chapter 30], in which two spaces are i^{th} associates if they intersect in a space of dimension $k-i$, or, equivalently, if they are separated by graph distance i in the Grassmann graph.

The number of k-dimensional subspaces of $\mathbb{F}_q^n$, i.e., the cardinality of $\mathcal{G}_q(n,k)$, or the number of vertices of the corresponding Grassmann graph, is given by a Gaussian coefficient (a so-called "q-analog" of the binomial coefficient), namely,

$$|\mathcal{G}_q(n,k)| = \begin{bmatrix} n \\ k \end{bmatrix}_q = \frac{(q^n - 1)(q^{n-1} - 1) \cdots (q^{n-k+1} - 1)}{(q^k - 1)(q^{k-1} - 1) \cdots (q - 1)} = \prod_{i=0}^{k-1} \frac{q^{n-i} - 1}{q^{k-i} - 1};$$

see, e.g., [1839, Chapter 24]. The subscript q will be omitted when there is no possibility of confusion. Note that $\begin{bmatrix} n \\ k \end{bmatrix} = \begin{bmatrix} n \\ n-k \end{bmatrix}$ and $\begin{bmatrix} n \\ 0 \end{bmatrix} = \begin{bmatrix} n \\ n \end{bmatrix} = 1$.

Let $V \in \mathcal{G}_q(n, k)$ be a fixed k-dimensional subspace of $\mathbb{F}_q^n$. Among all j-dimensional subspaces W of $\mathbb{F}_q^n$, how many intersect with V in some subspace U of dimension ℓ? It turns out that this number, denoted as $N_q(n, k, j, \ell)$, depends on V only through its dimension k, and is given as

$$N_q(n, k, j, \ell) = q^{(k-\ell)(j-\ell)} \begin{bmatrix} k \\ \ell \end{bmatrix} \begin{bmatrix} n-k \\ j-\ell \end{bmatrix}. \tag{29.6}$$

To see this, observe that the space U of intersection can be chosen in $\begin{bmatrix} k \\ \ell \end{bmatrix}$ ways. This subspace can be extended to a j-dimensional subspace in

$$\frac{(q^n - q^k)(q^n - q^{k+1})(q^n - q^{k+2}) \cdots (q^n - q^{k+j-\ell-1})}{(q^j - q^\ell)(q^j - q^{\ell+1})(q^j - q^{\ell+2}) \cdots (q^j - q^{j-1})} = q^{(j-\ell)(k-\ell)} \begin{bmatrix} n-k \\ j-l \end{bmatrix}$$

ways, since we can extend U by adjoining any of the $q^n - q^k$ vectors not in V, then adjoining any of the $q^n - q^{k+1}$ vectors not in the resulting $(k+1)$-space, etc., but any specific choice is in an equivalence class of size $(q^j - q^\ell)(q^j - q^{\ell+1}) \cdots (q^j - q^{j-1})$.

The quantity $N_q(n, k, j, \ell)$ is very useful. For example, $N_q(n, n, k, k) = \begin{bmatrix} n \\ k \end{bmatrix}$ (the number of k-subspaces of an n-space, i.e., $|\mathcal{G}_q(n, k)|$), $N_q(n, k, j, k) = \begin{bmatrix} n-k \\ j-k \end{bmatrix}$ (the number of j-dimensional spaces containing the k-space V), $N_q(n, k, k, k - i) = q^{i^2} \begin{bmatrix} k \\ i \end{bmatrix} \begin{bmatrix} n-k \\ i \end{bmatrix}$ (the number of k-spaces at graph distance i from the k-space V in the Grassmann graph), etc.

Let us also mention here two additional properties of the Gaussian coefficient. We have [760]

$$\begin{bmatrix} m \\ n \end{bmatrix} \begin{bmatrix} n \\ t \end{bmatrix} = \begin{bmatrix} m \\ t \end{bmatrix} \begin{bmatrix} m-t \\ n-t \end{bmatrix} \quad \text{for } t \leq n \leq m, \tag{29.7}$$

and [1156, Lemma 5]

$$q^{i(n-i)} \leq \begin{bmatrix} n \\ i \end{bmatrix} \leq h(q) q^{i(n-i)} \quad \text{where } h(q) = \prod_{j=1}^{\infty} \frac{1}{1 - q^{-j}}. \tag{29.8}$$

It is shown in [1156] that $h(q)$ decreases monotonically with q, approaching $q/(q - 1)$ for large q. The infinite product for $h(q)$ converges rapidly; the following table lists $h(q)$ for various values of q.

q	2	3	4	5	7	8	16	32	64	128	256
$h(q)$	3.46	1.79	1.45	1.32	1.20	1.16	1.07	1.03	1.02	1.01	1.004

29.4.2 The Operator Channel

Let us return now to the random linear network coding channel defined by (29.3) and (29.4). The transmitter sends the r rows of a matrix $P \in \mathbb{F}_q^{r \times n}$ and any particular sink node t gathers the ℓ rows of a matrix $Y_t = G_t P$, where G_t is a random $\ell \times r$ global transfer matrix induced at t by the random choice of local encoding coefficients in the network. Even though G_t is random, it is clear that

$$\langle Y_t \rangle = \langle G_t P \rangle \subseteq \langle P \rangle,$$

i.e., even though the network linearly mixes the transmitted packets, the received packets span a space contained in the space spanned by the transmitted packets. This motivates a convenient abstraction—termed the *operator channel*—in which the input and output of the channel are considered simply as subspaces of $\mathbb{F}_q^n$, and any more particular relationship between the input and output matrices (induced, say, by a particular network topology) is ignored.

For integer $k \geq 0$, define a stochastic operator $\mathcal{H}_k$, called an **erasure operator**, that operates on $\mathcal{P}_q(n)$. If $\dim(U) > k$, then $\mathcal{H}_k(U)$ returns a randomly chosen k-dimensional subspace of U; otherwise, $\mathcal{H}_k(U)$ returns U. For the purposes of code constructions, the actual probability distribution of $\mathcal{H}_k$ is unimportant; for example, it could be chosen to be uniform. Given a channel input U and a channel output V, it is always possible to write V as

$$V = \mathcal{H}_k(U) \oplus E$$

for some subspace $E \in \mathcal{P}_q(n)$, assuming that $k = \dim(U \cap V)$, and that $\mathcal{H}_k(U) = U \cap V$.

Definition 29.4.2 An **operator channel** associated with $\mathbb{F}_q^n$ is a channel with input and output alphabet $\mathcal{P}_q(n)$. The channel input U and channel output V are related as $V = \mathcal{H}_k(U) \oplus E$, where $k = \dim(U \cap V)$ and E is an error space satisfying $E \cap U = \{0\}$. In transforming U to V, we say that the channel commits $\rho = \dim(U) - k$ erasures and $t = \dim(E)$ errors.

It is important to note that the concept of an "error" differs from that in classical coding theory: here it refers to the insertion into the network (by an adversary, say) of vectors not in the span of the transmitted vectors, so that the received space may contain vectors outside of that span. Errors (in the absence of erasures) will make the dimension of the received space larger than that of the transmitted one. The concept of an "erasure" also differs from that in classical coding theory: here it refers to the possibility that the intersection of the received subspace with the span of the transmitted vectors may not include all of the vectors in that span. Erasures (in the absence of errors) will make the dimension of the received space smaller than the transmitted one.

Note that, according to this definition, the error space E is assumed to intersect trivially with U; thus the choice of E is not independent of the choice of U. However, if one were to model the received space as $V = \mathcal{H}_k(U) + E$ for an arbitrary error space E, then, since E always decomposes for some space E' as $E = (E \cap U) \oplus E'$, one would get

$$V = \mathcal{H}_k(U) + ((E \cap U) \oplus E') = \mathcal{H}_{k'}(U) \oplus E'$$

for some $k' \geq k$. In other words, components of an error space that intersect with the transmitted space U would only be helpful, possibly decreasing the number of erasures seen by the receiver.

In summary, for any sink node t, we consider the operator channel as a point-to-point channel between s and t, with input and output alphabet equal to $\mathcal{P}_q(n)$. The channel takes in a vector space U and puts out another vector space V, possibly with erasures (deletion of vectors from U) or errors (addition of vectors to U). Of course, in actuality, the source node s transmits a basis for U into the network, and the sink node t gathers a basis for V.

29.5 Codes and Metrics in $\mathcal{P}_q(n)$

29.5.1 Subspace Codes

The operator channel given in Definition 29.4.2 concisely captures the effects of random linear network coding over networks with erasures, varying mincuts, and/or erroneous packets. A code for the operator channel is known as a subspace code. Although one can consider multi-shot codes [1413, 1445], we will mainly be interested in so-called **one-shot** codes, that convey information via a single channel use, sending just a single subspace from an appropriate codebook of subspaces.

Definition 29.5.1 A (one-shot) **subspace code** Ω of packet-length n over $\mathbb{F}_q$ is a nonempty collection of subspaces of $\mathbb{F}_q^n$, i.e., a nonempty subset of $\mathcal{P}_q(n)$.

Unlike in classical coding theory, where a codeword is a vector, here a codeword of Ω is itself an entire *space* of vectors. A code in which each codeword has the same dimension, i.e., a code contained within a single Grassmannian, is called a **constant-dimension** code. For example,

$$\Psi = \left\{ \left\langle \begin{bmatrix} 1\,0\,0\,0 \\ 0\,1\,1\,0 \end{bmatrix} \right\rangle, \left\langle \begin{bmatrix} 0\,1\,0\,0 \\ 0\,0\,1\,1 \end{bmatrix} \right\rangle, \left\langle \begin{bmatrix} 1\,1\,0\,1 \\ 0\,0\,1\,0 \end{bmatrix} \right\rangle, \left\langle \begin{bmatrix} 1\,0\,1\,0 \\ 0\,0\,0\,1 \end{bmatrix} \right\rangle, \left\langle \begin{bmatrix} 1\,0\,0\,1 \\ 0\,1\,0\,1 \end{bmatrix} \right\rangle \right\} \qquad (29.9)$$

is a binary constant-dimension code of packet-length 4.

29.5.2 Coding Metrics on $\mathcal{P}_q(n)$

As in classical coding theory, it is important to define a distance measure between codewords. A natural distance measure, which treats errors and erasures symmetrically, is the so-called **subspace distance**[1] [1156]

$$\begin{aligned} d_S(U, V) &= \dim(U + V) - \dim(U \cap V) \\ &= \dim(U) + \dim(V) - 2\dim(U \cap V) \\ &= 2\dim(U + V) - \dim(U) - \dim(V), \end{aligned}$$

where the latter equalities follow from (29.5). It can be shown [1156, Lemma 1] that this distance measure does indeed satisfy all of the axioms (including the triangle inequality) that make it a metric. That $d_S(U, V)$ is a metric also follows from the fact that this quantity represents a graph distance (the length of a geodesic) in the undirected Hasse graph representing the lattice of subspaces of $\mathbb{F}_q^n$ partially ordered by inclusion.

It is easy to show that

$$d_S(U, V) = d_S(U, U \cap V) + d_S(U \cap V, V) = d_S(U, U + V) + d_S(U + V, V).$$

The first equality can be interpreted as saying that the subspace distance between U and V is equal to the number of dimensions that must be removed from U to produce $U \cap V$, plus the number of additional dimensions that must be adjoined to $U \cap V$ to produce V. The second equality says that $d_S(U, V)$ is equal to the number of dimensions that must be adjoined to U to produce $U + V$, plus the number of dimensions that must be removed from $U + V$ to produce V. Evidently, at least one geodesic between U and V in the Hasse graph passes through $U \cap V$ while another passes through $U + V$.

The **minimum subspace distance** between distinct codewords in a code Ω is denoted as $d_S(\Omega)$, i.e.,

$$d_S(\Omega) = \min \left\{ d_S(U, V) \mid U, V \in \Omega, \ U \neq V \right\}.$$

---

[1]When speaking of subspace distance in this chapter, dimension is measured in the non-projective sense. In Section 14.5, when speaking of subspace codes and subspace distance, dimension is measured in the projective sense; fortunately the numerical values of subspace distance agree in both senses. See Footnote 1 of Section 14.5.

For example, the binary constant-dimension subspace code Ψ defined in (29.9) has $d_S(\Psi) = 4$, since every pair of codewords intersects trivially.

Consider now information transmission with a subspace code Ω of packet-length n over $\mathbb{F}_q$ using an operator channel that takes a codeword $U \in \Omega$ to the received space

$$V = \mathcal{H}_{\dim(U)-\rho}(U) \oplus E$$

where $E \cap U = \{0\}$ and $\rho \geq 0$. As above, in transforming U to V, the channel commits ρ erasures and $t = \dim(E)$ errors. Given V, a nearest-subspace-distance decoder for Ω returns a codeword $U \in \Omega$ having smallest subspace distance to V.

Theorem 29.5.2 *When $U \in \Omega$ is sent, and V is received with ρ erasures and t errors, then a nearest-subspace-distance decoder for Ω is guaranteed to return U if $2(\rho + t) < d_S(\Omega)$.*

Proof: Let $U' = \mathcal{H}_{\dim(U)-\rho}(U)$. From the triangle inequality we have $d_S(U, V) \leq d_S(U, U') + d_S(U', V) = \rho + t$. If $T \neq U$ is any other codeword in Ω, then $d_S(\Omega) \leq d_S(U, T) \leq d_S(U, V) + d_S(V, T)$, from which it follows that $d_S(V, T) \geq d_S(\Omega) - d_S(U, V) \geq d_S(\Omega) - (\rho + t) > \rho + t \geq d_S(U, V)$. Since $d_S(V, T) > d_S(U, V)$, a nearest-subspace-distance decoder must produce U. $\square$

Conversely, consider two codewords $U, T \in \Omega$ with $d_S(U, T) = d_S(\Omega)$, where (without loss of generality) $\dim(U) \leq \dim(T)$. Let $\rho = d_S(U, U \cap T)$, and let t be the smallest integer such that $2(t + \rho) \geq d_S(\Omega)$. Note that $\dim(U) \leq \dim(T)$ implies that t is nonnegative. Writing $T = (U \cap T) \oplus T'$, let $V = (U \cap T) \oplus E$, where E is any subspace of T' of dimension t. By construction, we then have $d_S(U, V) = \rho + t$, while $d_S(T, V) \leq d_S(\Omega) - (\rho + t) \leq \rho + t$. Thus, if U is sent and V is received, since both U and T are within subspace distance $\rho + t$ of V, there is no guarantee that the receiver will produce U.

Another distance measure, introduced in the later paper [1710], is the so-called **injection distance**

$$d_I(U, V) = \max\{\dim(U), \dim(V)\} - \dim(U \cap V)$$
$$= \dim(U + V) - \min\{\dim(U), \dim(V)\}.$$

The **minimum injection distance** between distinct codewords in a code Ω is denoted as $d_I(\Omega)$, i.e.,

$$d_I(\Omega) = \min\{d_I(U, V) \mid U, V \in \Omega, \ U \neq V\}.$$

Again it can be shown [1710] that this distance measure is indeed a metric. That $d_I(U, V)$ is a metric also follows from the fact that this quantity represents a graph distance in the undirected Hasse graph representing the lattice of subspaces of $\mathbb{F}_q^n$ partially ordered by inclusion, adjoined with the edges of the Grassmann graphs in each dimension.

As explained in [1710], the injection distance is designed to measure adversarial effort in networks, counting the minimum number of packets that an adversary would need to inject into the network to transform a given input space U to a given output space V. Whereas the subspace distance treats errors and erasures symmetrically, the injection distance reflects the fact that the injection of a single error packet may cause the replacement of a dimension (i.e., a simultaneous error and erasure); thus errors and erasures should not be treated symmetrically in network coding applications of subspace codes.

The injection distance and the subspace distance are, however, closely related, as

$$2d_I(U, V) = d_S(U, V) + |\dim(V) - \dim(U)|. \tag{29.10}$$

In fact, the two metrics are equivalent when U and V have the same dimension, i.e., if $\dim(U) = \dim(V)$, then $d_S(U, V) = 2d_I(U, V)$. It follows from (29.10) that

$$2d_I(\Omega) \geq d_S(\Omega),$$

with equality if (but not only if) Ω is a constant-dimension code.

Given a received space V, a nearest-injection-distance decoder for a code Ω returns a codeword $U \in \Omega$ having smallest injection distance to V. In general, as illustrated in [1710], a nearest-injection-distance decoder does not return the same codeword as a nearest-subspace-distance decoder. However, if Ω is a constant-dimension code then, given V, both decoders will return a codeword U maximizing the intersection dimension $\dim(U \cap V)$, i.e., the two decoding rules are equivalent for constant-dimension codes.

Before moving to a discussion of bounds on the parameters of subspace codes and various means for their construction, it is interesting to note that every subspace code Ω is associated with a natural **complementary code** $\Omega^{\perp}$, defined as

$$\Omega^{\perp} = \{U^{\perp} \mid U \in \Omega\},$$

where

$$U^{\perp} = \left\{ w \in \mathbb{F}_q^n \ \middle| \ \sum_{i=1}^{n} w_i u_i = 0 \text{ for all } u \in U \right\}$$

is simply the (orthogonal) dual of the subspace U, as in linear coding theory. Note that $|\Omega^{\perp}| = |\Omega|$. It is easily seen that $d_S(U^{\perp}, V^{\perp}) = d_S(U, V)$ and $d_I(U^{\perp}, V^{\perp}) = d_I(U, V)$; thus the subspace distance and the injection distance between subspaces U and V are perfectly mirrored by the subspace and injection distances between the orthogonal subspaces $U^{\perp}$ and $V^{\perp}$. It follows immediately that $d_S(\Omega^{\perp}) = d_S(\Omega)$ and $d_I(\Omega^{\perp}) = d_I(\Omega)$.

In the remainder of this chapter, we will mainly be interested in working with constant-dimension codes, where the distinction between subspace distance and injection distance is irrelevant. To avoid extraneous factors of two, we adopt the injection distance in all that follows.

29.6 Bounds on Constant-Dimension Codes

A constant-dimension code $\Omega \subseteq \mathcal{G}_q(n, k)$, in which each codeword has dimension k, is denoted as an $(n, d, k)_q$ code if $d_I(\Omega) = d$. We denote by $A_q(n, d, k)$ the size of a largest $(n, d, k)_q$ code. We need to consider only $k \leq n/2$, since the complement $\Omega^{\perp}$ of an $(n, d, k)_q$ code Ω is an $(n, d, n - k)_q$ code of the same cardinality, which implies that $A_q(n, d, k) = A_q(n, d, n - k)$.

29.6.1 The Sphere Packing Bound

For $V \in \mathcal{G}_q(n, k)$, let

$$\mathcal{B}_V(n, k, q, t) = \{U \in \mathcal{G}_q(n, k) \mid d_I(V, U) \leq t\}$$

be the set of all k-dimensional subspaces of $\mathbb{F}_q^n$ at injection distance at most t from V, i.e., a ball of injection radius t centered at V in $\mathcal{G}_q(n, k)$. For any $V \in \mathcal{G}_q(n, k)$, the size of $\mathcal{B}_V(n, k, q, t)$ is [1156]

$$|\mathcal{B}_V(n, k, q, t)| = \sum_{i=0}^{t} q^{i^2} \begin{bmatrix} k \\ i \end{bmatrix} \begin{bmatrix} n - k \\ i \end{bmatrix},$$

which follows easily from (29.6). Note that the size of a ball in $\mathcal{G}_q(n, k)$ is independent of its center, but this would not be true if subspaces from $\mathcal{P}_q(n)$ of dimension other than k were included. For convenience, let us denote $|\mathcal{B}_V(n, k, q, t)|$ as $B(n, k, q, t)$.

Since balls of radius $\lfloor (d-1)/2 \rfloor$ centered at the codewords of an $(n, d, k)_q$ code must be disjoint, the total number of points included in the union of such balls cannot exceed the number of points $|\mathcal{G}_q(n, k)| = \begin{bmatrix} n \\ k \end{bmatrix}$ in the space as a whole. Thus we get the Sphere Packing Bound [1156].

Theorem 29.6.1 (Sphere Packing Bound)

$$A_q(n, d, k) \leq \frac{\begin{bmatrix} n \\ k \end{bmatrix}}{B(n, k, q, \lfloor (d-1)/2 \rfloor)}.$$

29.6.2   The Singleton Bound

In [1156] a natural puncturing operation in $\mathcal{G}_q(n, k)$ is defined that reduces by one the dimension of the ambient space and the dimension of each subspace in $\mathcal{G}_q(n, k)$. According to this puncturing operation, a punctured code obtained by puncturing an $(n, d, k)_q$ code is itself an $(n-1, d', k-1)_q$ code, where $d' \geq d-1$. If an $(n, d, k)_q$ code is punctured $d-1$ times repeatedly, an $(n-d+1, d'', k-d+1)_q$ code (with $d'' \geq 1$) is obtained, which may have size no greater than $|\mathcal{G}_q(n-d+1, k-d+1)|$. Thus the following Singleton-type bound is established [1156].

Theorem 29.6.2 (Singleton Bound)

$$A_q(n, d, k) \leq |\mathcal{G}_q(n-d+1, k-d+1)| = \begin{bmatrix} n-d+1 \\ k-d+1 \end{bmatrix} = \begin{bmatrix} n-d+1 \\ n-k \end{bmatrix}.$$

From (29.8) it follows[2] that

$$A_q(n, d, k) \leq h(q)q^{(n-k)(k-d+1)}. \tag{29.11}$$

29.6.3   The Anticode Bound

Since $\mathcal{G}_q(n, k)$ is an association scheme, the Anticode Bound of Delsarte [519] can be applied. Let Ω be an $(n, d, k)_q$ code. Then the Anticode Bound implies that

$$|\Omega| \leq \frac{|\mathcal{G}_q(n, k)|}{|\mathcal{A}|},$$

where $\mathcal{A} \subseteq \mathcal{G}_q(n, k)$ is any set with maximum distance $d-1$ (called an **anticode**).

Note that, for all $U, V \in \mathcal{G}_q(n, k)$, $d_I(U, V) \leq d-1$ if and only if $\dim(U \cap V) \geq k-d+1$. Thus, we can take $\mathcal{A}$ as a set in which any two elements intersect in a space of dimension at least $k-d+1$. From the results of Frankl and Wilson [753], it follows that, for $k \leq n/2$, the maximum value of $|\mathcal{A}|$ is equal to $\begin{bmatrix} n-k+d-1 \\ d-1 \end{bmatrix}$. Hence, we have the following bound.

Theorem 29.6.3 (Anticode Bound)

$$A_q(n, d, k) \leq \frac{\begin{bmatrix} n \\ k \end{bmatrix}}{\begin{bmatrix} n-k+d-1 \\ d-1 \end{bmatrix}} = \frac{\begin{bmatrix} n \\ k-d+1 \end{bmatrix}}{\begin{bmatrix} k \\ k-d+1 \end{bmatrix}}.$$

---

[2]Compare the upper bound of (29.11) to the lower bound of Theorem 14.5.5 taking Footnote 1 of Chapter 14 into account.

The equality in this theorem follows by observing from (29.7) that

$$
\begin{bmatrix} n \\ k \end{bmatrix} \begin{bmatrix} k \\ k-d+1 \end{bmatrix} = \begin{bmatrix} n \\ k-d+1 \end{bmatrix} \begin{bmatrix} n-k+d-1 \\ d-1 \end{bmatrix}.
$$

The Anticode Bound is also useful for bounding the parameters of rank-metric codes; see Chapter 11.

It is easy to observe that the Anticode Bound also implies the Sphere Packing Bound as a special case, since a ball $\mathcal{B}_V(n,k,q,\lfloor (d-1)/2 \rfloor)$ is (by the triangle inequality) an anticode of maximum distance $d-1$. However, a ball is not an optimal anticode in $\mathcal{G}_q(n,k)$, and therefore the bound of Theorem 29.6.3 is always tighter for nontrivial codes.

The bound in Theorem 29.6.3 was first obtained by Wang, Xing and Safavi-Naini in [1870] using a different argument. The proof that Theorem 29.6.3 follows from Delsarte's Anticode Bound is due to Etzion and Vardy [695]. As observed in [1913], the Anticode Bound is always stronger than the Singleton Bound for nontrivial codes in $\mathcal{G}_q(n,k)$.

29.6.4 Johnson-Type Bounds

Let $U \in \mathcal{G}_q(n, n-1)$ be any $(n-1)$-dimensional subspace $\mathbb{F}_q^n$, and let $\phi_U : U \to \mathbb{F}_q^{n-1}$ be an isomorphism identifying U with $\mathbb{F}_q^{n-1}$. For any subspace W of U, let $\phi_U(W)$ denote the image of W under transformation by ϕ_U. Since ϕ_U is an isomorphism, $\dim(\phi_U(W)) = \dim(W)$.

Now let Ω be an $(n,d,k)_q$ code with $A_q(n,d,k)$ codewords and consider the subcode $\Omega_U = \{V \in \Omega \mid V \subseteq U\}$ consisting of those codewords of Ω contained entirely within U. Let $\phi_U(\Omega_U) = \{\phi_U(W) \mid W \in \Omega_U\}$ and observe that the elements of $\phi_U(\Omega_U)$ are k-dimensional subspaces of $\mathbb{F}_q^{n-1}$. Thus $\phi_U(\Omega_U)$ is an $(n-1,d',k)_q$ code with $|\Omega_U|$ codewords. It is easy to see, since ϕ_U is an isomorphism, that $d' \geq d$; thus $\phi_U(\Omega_U)$ cannot have cardinality greater than $A_q(n-1,d,k)$. If we now form the summation of such cardinalities, ranging over all possible U, we obtain

$$
\sum_{U \in \mathcal{G}_q(n,n-1)} |\phi_U(\Omega_U)| = \left(\frac{q^{n-k}-1}{q-1} \right) A_q(n,d,k) \leq \left(\frac{q^n-1}{q-1} \right) A_q(n-1,d,k),
$$

where the first equality follows from the fact that each codeword of Ω will appear as a codeword in exactly $N_q(n,k,n-1,k) = (q^{n-k}-1)/(q-1)$ of the Ω_U's. This argument yields the following theorem [695].

Theorem 29.6.4 ([695])

$$
A_q(n,d,k) \leq \frac{q^n-1}{q^{n-k}-1} A_q(n-1,d,k).
$$

Since $A_q(n,d,k) = A_q(n,d,n-k)$, we also get the following.

Theorem 29.6.5 ([695, 1913])

$$
A_q(n,d,k) \leq \frac{q^n-1}{q^k-1} A_q(n-1,d,k-1).
$$

Theorems 29.6.4 and 29.6.5 may be iterated to give an upper bound for $A_q(n,d,k)$. However, as in the classical case of the Johnson space, the order in which the two bounds should be iterated in order to get the tightest bound is unclear. By iterating Theorem 29.6.5 with itself, the following bound is established in [695, 1913].

Theorem 29.6.6 ([695, 1913])

$$A_q(n,d,k) \leq \left\lfloor \frac{q^n - 1}{q^k - 1} \left\lfloor \frac{q^{n-1} - 1}{q^{k-1} - 1} \cdots \left\lfloor \frac{q^{n-k+d} - 1}{q^d - 1} \right\rfloor \cdots \right\rfloor \right\rfloor.$$

It is shown in [1913] that Theorem 29.6.4 improves on the Anticode Bound.

29.6.5   The Ahlswede and Aydinian Bound

Let D be a nonempty subset of $\{1, \ldots, n\}$ and let $\Omega \subseteq \mathcal{G}_q(n,k)$ be a code. If, for all $U, V \in \Omega$, with $U \neq V$, we have $d_{\mathrm{I}}(U,V) \in D$, then we say that Ω is a code with distances in D. The following lemma is given in [25].

Lemma 29.6.7 ([25]) *Let $\Omega_D \subseteq \mathcal{G}_q(n,k)$ be a code with distances in a set D. Then, for a nonempty subset $S \subseteq \mathcal{G}_q(n,k)$, there exists a code $\Omega_D^*(S) \subseteq S$ with distances in D such that*

$$\frac{|\Omega_D^*(S)|}{|S|} \geq \frac{|\Omega_D|}{\left[\begin{smallmatrix} n \\ k \end{smallmatrix}\right]},$$

where, if $|\Omega_D^*(S)| = 1$, then $\Omega_D^*(S)$ is a code with distances in D by convention.

In particular when Ω_D is an $(n,d,k)_q$ code and S is an anticode of maximum distance $d-1$, then $|\Omega_D^*(S)| = 1$ and the Anticode Bound on $\mathcal{G}_q(n,k)$ is obtained.

Using Lemma 29.6.7 Ahlswede and Aydinian obtain the following bound.

Theorem 29.6.8 (Ahlswede–Aydinian Bound [25]) *For integers $0 \leq t \leq d \leq k$, $k - t \leq m \leq n$,*

$$A_q(n,d,k) \leq \frac{\left[\begin{smallmatrix} n \\ k \end{smallmatrix}\right] A_q(m, d-t, k-t)}{\displaystyle\sum_{i=0}^{t} q^{i(m-i)} \begin{bmatrix} m \\ k-i \end{bmatrix} \begin{bmatrix} n-m \\ i \end{bmatrix}}$$

It is shown in [25] that for $t = 0$ and $m = n - 1$, Theorem 29.6.8 gives Theorem 29.6.4.

29.6.6   A Gilbert–Varshamov-Type Bound

The counterpart of the classical Gilbert–Varshamov lower bound for $A_q(n,d,k)$ is easily obtained, via the standard argument. To obtain a code of distance d one can start by taking any space in $\mathcal{G}_q(n,k)$ as a codeword, and then eliminate all subspaces within injection distance $d-1$ of the chosen codeword, as these subspaces are too close to the chosen codeword. If there are any subspaces remaining (i.e., not yet eliminated, or already a codeword), one can select one of these as a codeword, and eliminate all subspaces within injection distance $d-1$ of that subspace, continuing in this fashion until all subspaces have been eliminated. In the worst case, the subspaces eliminated in each step are disjoint, but even in this case the number of codewords obtained is guaranteed to be at least as large as the ratio of the size of the space to the number of words eliminated in each step. Since the number of words eliminated is the size of a ball in $\mathcal{G}_q(n,k)$, the argument sketched above gives the following lower bound on $A_q(n,d,k)$.

Theorem 29.6.9 ([1156])

$$A_q(n,d,k) \geq \frac{\left[\begin{smallmatrix} n \\ k \end{smallmatrix}\right]}{B(n,k,q,d-1)}.$$

Unlike the case of classical coding theory in the Hamming metric, the best lower bounds on $A_q(n, d, k)$ result from code constructions, the subject of the next section. Indeed, given in Figure 29.2 are plots of the Singleton, Anticode, and Johnson upper bounds, together with the Gilbert–Varshamov (GV) lower bound for $(2048, d, k)_{256}$ constant-dimension codes, with $k = 50$ and $k = 512$, expressed in terms of rate versus relative distance. The rate R of an $(n, d, k)_q$ code Ω is defined as

$$R = \frac{1}{nk} \log_q |\Omega|;$$

this is the number of q-ary information symbols that are sent in the transmission of k packets of length n symbols from $\mathbb{F}_q$ using code Ω. The relative distance is the ratio d/k; since $1 \leq d \leq k$, the relative distance is bounded between $1/k$ and 1. As can be seen in Figure 29.2, the upper bounds all give essentially the same result (when expressed on a logarithmic scale). Also shown in the figure are points achieved by the lifted maximum rank distance (MRD) codes described in the next section. Although there is a gap between the Gilbert–Varshamov Bound and the various upper bounds on $A_q(n, d, k)$, the lifted MRD codes come very close to the upper bounds, and hence are essentially optimal (at least for the reasonably realistic case of packet lengths of 2K bytes, with network coding performed on bytes, i.e., in $\mathbb{F}_{256}$).

FIGURE 29.2: The GV (lower) Bound and various upper bounds on $A_q(n, d, k)$ expressed in terms of rate versus relative distance for $n = 248$, $q = 256$ and $k \in \{50, 512\}$. Also shown are achievable rates for $(n, d, k)_q$ lifted maximum rank distance codes, with $d \in \{1, 2, 3, \ldots, 50\}$ for $k = 50$ and $d \in \{1, 8, 15, \ldots, 512\}$ for $k = 512$.

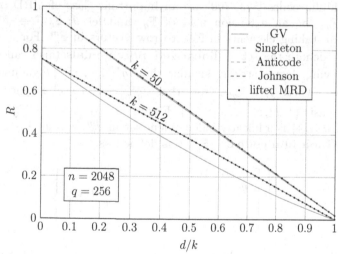

29.7 Constructions

29.7.1 Lifted Rank-Metric Codes

In this section, we describe the simplest construction of asymptotically good subspace codes, which uses rank-metric codes as building blocks. This construction was first proposed in [1870], and then rediscovered in [1156] for the special case where the rank-metric code is a Delsarte–Gabidulin code. The construction was later explained in [1708, 1711] in the context of the subspace/injection distance. The latter description is reviewed below. Rank-metric codes are described in more detail in Chapter 11; here we review only some basic properties needed for the purposes of this chapter.

For matrices $X, Y \in \mathbb{F}_q^{n \times m}$, the **rank distance** is defined as

$$d_{\mathrm{R}}(X, Y) = \mathrm{rank}(Y - X).$$

As observed in [760], the rank distance is indeed a metric.

A **rank-metric code** $\mathcal{C} \subseteq \mathbb{F}_q^{n \times m}$ is a matrix code (i.e., a nonempty set of matrices) used in the context of the rank metric. We use $d_R(\mathcal{C})$ to denote the **minimum rank distance** of $\mathcal{C}$, i.e., the minimum rank distance between distinct codewords. Taking the transpose of every codeword in a rank-metric code over $\mathbb{F}_q^{n \times m}$ yields a rank-metric code of the same cardinality and minimum rank distance in $\mathbb{F}_q^{m \times n}$. Thus there is no loss of generality in considering, as we will do, only the case where $n \leq m$ (and thus the codewords are in general "wide," not "tall").

The Singleton Bound for the rank metric [523, 760] (see also [783, 1285, 1711] and Theorem 11.3.5) states that

$$|\mathcal{C}| \leq q^{m(n-d+1)}$$

for every code $\mathcal{C} \subseteq \mathbb{F}_q^{n \times m}$ with $d_R(\mathcal{C}) = d$ and $n \leq m$. Codes that achieve this bound are called **maximum rank distance** (MRD) codes and linear MRD codes are known to exist for all choices of parameters q and $d \leq n \leq m$ [760] (and hence, by transposition, for *all* choices of q, n, m, and $d \leq \min\{m, n\}$).

Delsarte–Gabidulin codes [523, 760] are an important class of MRD codes, described as follows. Let $\mathbb{F}_{q^m}$ be an extension field of $\mathbb{F}_q$, and let $\theta \colon \mathbb{F}_{q^m} \to \mathbb{F}_q^m$ be a vector space isomorphism, taking elements in $\mathbb{F}_{q^m}$ to row vectors in $\mathbb{F}_q^m$. For any positive integer δ, let $\mathbb{F}_{q,m}^{<\delta}[x]$ denote the set of **linearized polynomials** (also called q-**linearized polynomials**) having degree strictly smaller than q^δ, i.e., all polynomials of the form $f(x) = \sum_{i=0}^{\delta-1} f_i x^{q^i}$, where $f_i \in \mathbb{F}_{q^m}$. Note that $\mathbb{F}_{q,m}^{<\delta}[x]$ is a vector space of dimension δ over $\mathbb{F}_{q^m}$, having basis $\{1, x^q, x^{q^2}, \dots, x^{q^{\delta-1}}\}$. Finally, let $\alpha_1, \dots, \alpha_n \in \mathbb{F}_{q^m}$ be elements for which $\{\theta(\alpha_1), \dots, \theta(\alpha_n)\}$ is a linearly independent set in $\mathbb{F}_q^m$, and let $1 \leq d \leq n$.

A **Delsarte–Gabidulin code** $\mathcal{C} \subseteq \mathbb{F}_q^{n \times m}$ is defined as

$$\mathcal{C} = \left\{ \begin{bmatrix} \theta(f(\alpha_1)) \\ \theta(f(\alpha_2)) \\ \vdots \\ \theta(f(\alpha_n)) \end{bmatrix} \;\middle|\; f(x) \in \mathbb{F}_{q,m}^{<n-d+1}[x] \right\}.$$

It is shown in [760] that such a code has $d_R(\mathcal{C}) = d$ and $q^{m(n-d+1)}$ codewords, and so it is indeed an MRD code.

Given a rank-metric code $\mathcal{C} \subseteq \mathbb{F}_q^{n \times m}$, a minimum-rank-distance decoder for $\mathcal{C}$ takes a matrix $r \in \mathbb{F}_q^{n \times m}$ and returns a codeword $c \in \mathcal{C}$ that minimizes the rank distance $d_{\mathrm{R}}(c, r)$. It is easy to see that, if $d_{\mathrm{R}}(c, r) < d_R(\mathcal{C})/2$ for some $c \in \mathcal{C}$, then c is the unique solution to the above problem. A bounded-distance decoder for $\mathcal{C}$ returns $c \in \mathcal{C}$ if $d_{\mathrm{R}}(c, r) < d_R(\mathcal{C})/2$, or declares a failure if no such codeword can be found. For Delsarte–Gabidulin codes, very efficient bounded-distance decoders exist; see, e.g., [760, 1711].

Let us return, now, to the construction of constant-dimension subspace codes. For a matrix $X \in \mathbb{F}_q^{k \times m}$, let the subspace

$$\Lambda(X) = \left\langle \begin{bmatrix} I_k & X \end{bmatrix} \right\rangle \in \mathcal{G}_q(k + m, k)$$

be called the **lifting** of X. Similarly, for a matrix code $\mathcal{C} \subseteq \mathbb{F}_q^{k \times m}$, let the subspace code

$$\Lambda(\mathcal{C}) = \{\Lambda(X) \mid X \in \mathcal{C}\}$$

be called the **lifting** of $\mathcal{C}$. Since every subspace corresponds to a unique matrix in reduced row echelon form, we have that the mapping $X \to \Lambda(X)$ is injective, and therefore $|\Lambda(\mathcal{C})| = |\mathcal{C}|$. Note that $\Lambda(\mathcal{C})$ is a constant-dimension code, i.e., $\Lambda(\mathcal{C}) \subseteq \mathcal{G}_q(k + m, k)$.

Lemma 29.7.1 (Lifting Lemma [1711]) *For all $X, X' \in \mathbb{F}_q^{k \times m}$ and all $\mathcal{C} \subseteq \mathbb{F}_q^{k \times m}$,*

$$d_I(\Lambda(X), \Lambda(X')) = d_R(X, X'),$$
$$d_I(\Lambda(\mathcal{C})) = d_R(\mathcal{C}).$$

Proof: We have

$$d_I(\Lambda(X), \Lambda(X')) = \dim(\Lambda(X) + \Lambda(X')) - \min\{\dim(\Lambda(X)), \dim(\Lambda(X'))\}$$
$$= \mathrm{rank}\begin{bmatrix} I_k & X \\ I_k & X' \end{bmatrix} - k$$
$$= \mathrm{rank}\begin{bmatrix} I_k & X \\ 0 & X' - X \end{bmatrix} - k$$
$$= \mathrm{rank}(X' - X).$$

The second statement immediately follows from the first. □

Lemma 29.7.1 shows that a subspace code constructed by lifting inherits the distance properties of its underlying rank-metric code.

In particular, let $\mathcal{C} \subseteq \mathbb{F}_q^{k \times (n-k)}$ be an MRD code with $d_R(\mathcal{C}) = d$ and, without loss of generality, let $k \leq n - k$. Then $\Lambda(\mathcal{C})$ is an $(n, d, k)_q$ code with cardinality

$$|\Lambda(\mathcal{C})| = q^{(n-k)(k-d+1)}. \tag{29.12}$$

Note that (29.12) gives a lower bound on $A_q(n, d, k)$; see also Theorem 14.5.5. Comparing with the upper bound of (29.11), we see that the ratio of the upper and lower bounds is a constant depending only on q, thus demonstrating that this construction yields asymptotically optimal codes.

We now mention a particular way of constructing lifted rank-metric codes. When $m \geq 2k$, it is convenient to construct an MRD code $\mathcal{C} \subseteq \mathbb{F}_q^{k \times m}$ as a Cartesian product of simpler MRD codes. Let $m_1, \ldots, m_r \geq k$ be such that $\sum_{i=1}^r m_i = m$, and let $\mathcal{C}_i \subseteq \mathbb{F}_q^{k \times m_i}$, $i = 1, \ldots, r$, be MRD codes with minimum rank distance d. Then it is easy to see that the Cartesian product $\mathcal{C} = \mathcal{C}_1 \times \cdots \times \mathcal{C}_r$ is also an MRD code with $d_R(\mathcal{C}) = d$, where a specific element $(X_1, \ldots, X_r)$ in the Cartesian product is interpreted as the $k \times m$ matrix $[X_1 \ X_2 \ \cdots \ X_r]$. Clearly, we have $|\mathcal{C}| = \prod_{i=1}^r q^{m_i(k-d+1)} = q^{m(k-d+1)}$. Note the importance of choosing $m_i \geq k$ for the resulting code to be MRD. Now, since $d_R(\mathcal{C}) = d$, it follows that $\Lambda(\mathcal{C})$ is a $(k + m, k, d)_q$ code.

29.7.2 Padded Codes

Padded codes are a set of constant-dimension codes obtained as a union of lifted product rank-metric codes. Let $n = (r + 1)k + s$, where $r, s \in \mathbb{N}$ and $s < k$. Let $\mathcal{C} \subseteq \mathbb{F}_q^{k \times k}$, and $\mathcal{C}' \subseteq \mathbb{F}_q^{k \times (k+s)}$ be rank-metric codes of minimum rank distance d. Define a padded code as $\Omega = \Omega_0 \cup \Omega_1 \cup \cdots \cup \Omega_{r-1}$ where

$$\Omega_i = \{\langle [\overbrace{\mathbf{0}_{k \times k} \cdots \mathbf{0}_{k \times k}}^{i} \ I_k \ c_{i+1} \cdots c_r] \rangle \mid c_{i+1}, \ldots, c_{r-1} \in \mathcal{C}, c_r \in \mathcal{C}'\}.$$

Each codeword U of Ω_i has an identifying vector $\mathrm{id}(U) = (0, \ldots, 0, 1, \ldots, 1, 0, \ldots, 0)$ of weight k, in which the ones occur in k consecutive positions starting at position $ik + 1$. Consider now the injection distance between two distinct codewords $U \in \Omega_i$ and $V \in \Omega_j$. If $i = j$, then we have $\mathrm{d_I}(U, V) \geq d_I(\Omega_i) = d$. If $i \neq j$, observe that $\mathrm{wt_H}(\mathrm{id}(U) \star \mathrm{id}(V)) = 0$ which, by Theorem 29.4.1, implies that $\dim(U \cap V) = 0$, and hence $\mathrm{d_I}(U, V) = k - \dim(U \cap V) = k \geq d$. Thus we obtain $d_I(\Omega) = d$.

When $\mathcal{C} = \mathcal{C}'$ are Delsarte–Gabidulin codes with $d_R(\mathcal{C}) = k$, then the construction above results in the "spread codes" of [764] and [1328].

29.7.3   Lifted Ferrers Diagram Codes

In [692], Etzion and Silberstein provide a multi-level construction for codes in $\mathcal{P}_q(n)$. The basic idea of this construction is to generalize the lifting construction to Schubert cells (as defined in Section 29.4.1) so that a lifted rank-metric code is contained completely within any given cell. A code can then be constructed by taking a union of such lifted rank-metric codes in suitably well-separated Schubert cells. We now give a detailed description of this construction.

The key idea is the following theorem, which bounds the injection distance between two spaces in terms of the asymmetric distance between their identifying vectors.

Theorem 29.7.2 ([1110]) *For any pair of spaces* $U, V \in \mathcal{P}_q(n)$,

$$\mathrm{d_I}(U, V) \geq \mathrm{d_A}(\mathrm{id}(U), \mathrm{id}(V)).$$

Proof: This follows from Theorem 29.4.1, from which we know that $\dim(U \cap V) \leq \mathrm{wt_H}(\mathrm{id}(U) \star \mathrm{id}(V))$. If $\dim(U) \geq \dim(V)$, then $\mathrm{wt_H}(\mathrm{id}(U)) \geq \mathrm{wt_H}(\mathrm{id}(V))$ and

$$\begin{aligned}
\mathrm{d_I}(U, V) &= \dim(U) - \dim(U \cap V) \\
&\geq \mathrm{wt_H}(\mathrm{id}(U)) - \mathrm{wt_H}(\mathrm{id}(U) \star \mathrm{id}(V)) \\
&= \mathrm{d_A}(\mathrm{id}(U), \mathrm{id}(V)).
\end{aligned}$$

If $\dim(V) \geq \dim(U)$, simply interchange the roles of U and V.   □

Now in case U and V have the same dimension, then asymmetric distance and Hamming distance are the same up to scale, in which case we get

$$2\mathrm{d_I}(U, V) \geq \mathrm{d_H}(\mathrm{id}(U), \mathrm{id}(V)).$$

This also follows from the work of Etzion and Silberstein [692], who give the subspace distance counterpart of Theorem 29.7.2.

Theorem 29.7.3 ([692]) *For any pair of spaces* $U, V \in \mathcal{P}_q(n)$,

$$\mathrm{d_S}(U, V) \geq \mathrm{d_H}(\mathrm{id}(U), \mathrm{id}(V)).$$

Proof: Similar to the proof of Theorem 29.7.2.   □

Note that both lower bounds are achieved with equality when the non-pivot elements of G_U and G_V are all zero.

Consider now a Schubert cell with identifying vector a and corresponding Ferrers diagram $F(a)$ having n rows and m columns. A **Ferrers diagram (FD) code** over $\mathbb{F}_q$ for a is a matrix code $\mathcal{C}_a$ with codewords in $\mathbb{F}_q^{n \times m}$ such that all matrix entries not in free positions are fixed to zero. The minimum rank distance of an FD code is defined in the usual way. The lifting $\Lambda_a(\mathcal{C}_a)$ is defined in a manner analogous to the usual lifting of rectangular matrix codes, however with a reduced row-echelon form conforming to matrices having identifying vector a.

For example, the identifying vector $a = (101010)$ corresponds to the 3×3 Ferrers diagram

$$F(a) = \begin{matrix} \bullet & \bullet & \bullet \\ \bullet & \bullet & \\ \bullet & & \end{matrix} \ .$$

A codeword $v \in \mathcal{C}_a$ must then have the general form

$$v = \begin{bmatrix} v_{11} & v_{12} & v_{13} \\ 0 & v_{22} & v_{23} \\ 0 & 0 & v_{33} \end{bmatrix},$$

for some elements $v_{11}, v_{12}, v_{13}, v_{22}, v_{23}, v_{33} \in \mathbb{F}_q$. The codeword v is lifted to the subspace

$$\Lambda_a(v) = \left\langle \begin{bmatrix} 1 & v_{11} & 0 & v_{12} & 0 & v_{13} \\ 0 & 0 & 1 & v_{22} & 0 & v_{23} \\ 0 & 0 & 0 & 0 & 1 & v_{23} \end{bmatrix} \right\rangle \in \Lambda_a(\mathcal{C}_a).$$

Theorem 29.7.4 *Let $\mathcal{A}$ be a binary code of length n with constant codeword weight k and minimum Hamming distance $2d$, or equivalently, minimum asymmetric distance d. For each codeword $a \in \mathcal{A}$, let $\mathcal{C}_a$ be an FD code over $\mathbb{F}_q$ for a with minimum rank distance d. Finally, let*

$$\Omega = \bigcup_{a \in \mathcal{A}} \Lambda_a(\mathcal{C}_a).$$

Then Ω is an $(n, d, k)_q$ constant-dimension code of cardinality

$$|\Omega| = \sum_{a \in \mathcal{A}} |\mathcal{C}_a|.$$

Proof: Distinct codewords of Ω in the same Schubert cell are separated by an injection distance of at least d, as they are codewords of the same lifted FD code. Codewords of Ω in different Schubert cells are separated by an injection distance of at least d due to Theorems 29.7.2 and 29.7.3. The packet-length, codeword dimension, and cardinality of Ω are obvious. □

We refer to a code obtained by this multi-level construction as a **union of lifted FD codes**.

The size of such constant-dimension codes depends on the size of the corresponding FD codes. Bounds on the size of linear FD codes with prescribed minimum distance are given in [692]; see also [1110]. The construction of maximal Ferrers diagram rank-metric codes is considered in [51, 691].

It is interesting to observe the padded codes of Section 29.7.2 are in fact a union of lifted FD codes, corresponding to the constant-weight-k code $\mathcal{A}$ in which the codewords have k consecutive nonzero positions starting in position $1, k + 1, 2k + 1, \ldots$.

In this construction a naive choice for $\mathcal{A}$ would be one with a high information rate. However, a high information rate would only result in a large number of selected Schubert cells, and does not necessarily guarantee a high overall rate for the resulting $(n, d, k)_q$ code. This is due to the fact that the size of a lifted FD code depends on the size of its underlying $\mathcal{C}_a$ code, which in turn depends on a. In particular, the largest Schubert cell, with the largest corresponding lifted FD code, is selected when $a = 11 \cdots 100 \cdots 0$. This vector is actually the first to occur in reverse lexicographic order. In [692] constant-weight lexicodes, always including $11 \cdots 100 \cdots 0$, are used to select well-separated Schubert cells in the Grassmannian.

This construction idea can be extended to obtain mixed-dimension subspace codes. If $\mathcal{A}$ has minimum asymmetric distance d and each $\mathcal{C}_a$ code is designed to have minimum rank distance d, then the resulting subspace code is guaranteed to have minimum injection distance d. Similarly, if $\mathcal{A}$ has minimum Hamming distance $2d$ and each $\mathcal{C}_a$ code is designed to have minimum rank distance d, then the resulting subspace code is guaranteed to have minimum subspace distance $2d$.

Further constructions of subspace codes based on Ferrers diagrams and related concepts are given in [1707].

29.7.4   Codes Obtained by Integer Linear Programming

In [1139], Kohnert and Kurtz view the construction of constant-dimension subspace codes as an optimization problem involving integer variables. Their aim is to construct an $(n, d, k)_q$ code Ω with as many codewords as possible while satisfying that for all $U, V \in \Omega$ we have $d_{\mathrm{I}}(U, V) = k - \dim(U \cap V) \geq d$. This means that no pair of codewords in Ω may have a $(k - d + 1)$-dimensional subspace in common.

Let $M \in \{0, 1\}^{\left[\begin{smallmatrix} n \\ k-d+1 \end{smallmatrix}\right] \times \left[\begin{smallmatrix} n \\ k \end{smallmatrix}\right]}$ be an incidence matrix whose rows correspond to elements $W \in \mathcal{G}_q(n, k - d + 1)$ and whose columns correspond to elements $V \in \mathcal{G}_q(n, k)$. In the row corresponding to W and column corresponding to V, let M have entry

$$M_{W,V} = \begin{cases} 1 & \text{if } W \subseteq V, \\ 0 & \text{otherwise.} \end{cases}$$

Every $(n, \delta, k)_q$ code Ω can be associated with a $\{0, 1\}$-valued column vector x with $\left[\begin{smallmatrix} n \\ k \end{smallmatrix}\right]$ rows corresponding to elements $V \in \mathcal{G}_q(n, k)$. The entry x_V in the row corresponding to subspace V takes value 1 if and only if $V \in \Omega$, so that $|\Omega| = \sum_{V \in \mathcal{G}_q(n,k)} x_V$. Furthermore, Mx is then a column vector with $\left[\begin{smallmatrix} n \\ k-d+1 \end{smallmatrix}\right]$ rows corresponding to elements $W \in \mathcal{G}_q(n, k - d + 1)$, where the entry in the row corresponding to subspace W takes value i if subspace W occurs exactly i times as a subspace of some codeword of Ω. In order to ensure that $\delta \geq d$, no W can occur more than once (otherwise two codewords of Ω would have a common $(k-d+1)$-dimensional subspace); thus no entry of Mx may exceed unity.

The code construction problem then becomes the following integer optimization problem:

$$\text{maximize:} \quad \sum_{V \in \mathcal{G}_q(n,k)} x_V$$

$$\text{subject to:} \quad x_V \in \{0, 1\} \text{ for all } V \in \mathcal{G}_q(n, k)$$

$$Mx \leq [1\ 1\ \cdots\ 1]^\mathsf{T}.$$

It is possible to significantly reduce the size of the problem by prescribing a group of automorphisms for the code, and then using the induced symmetry to reduce the number of equations. See [1139] for details.

29.7.5   Further Constructions

A variety of further constructions, both for constant-dimension subspace codes and for mixed-dimension subspace codes, have been proposed; see, e.g., [984] and references therein. Similar to the construction of group codes for the Gaussian channel [1725], one prominent approach is to consider the orbit of a particular subspace under a group acting on the

Grassmannian in which the subspace is contained; see, e.g., [158, 808, 1476, 1599, 1818]. A generalization of subspace codes to so-called "flag codes" is given in [1257]. Subspace code constructions based on concepts of discrete geometry are given in [449, 450, 451, 452]. Constructions and bounds for mixed-dimension subspace codes are considered in [977].

29.8 Encoding and Decoding

Let $\Omega \in \mathcal{P}_q(n)$ be a subspace code with $d_I(\Omega) = d$. Throughout this section, let $t = \lfloor (d-1)/2 \rfloor$. In this section, we consider two problems related to the use of Ω for error control in non-coherent linear network coding. The *encoding problem* is how to efficiently map an integer in $\{0, \dots, |\Omega| - 1\}$ into a codeword of Ω (and back). The *decoding problem* is how to efficiently find a codeword of Ω that is closest (in injection distance) to a given subspace $U \in \mathcal{P}_q(n)$. More specifically, we focus on a **bounded-distance decoder**, which returns a codeword $V \in \Omega$ if V is the unique codeword that satisfies $d_I(V, U) \le t$ and returns a failure otherwise.

29.8.1 Encoding a Union of Lifted FD Codes

Let

$$\Omega = \bigcup_{a \in \mathcal{A}} \Lambda_a(\mathcal{C}_a)$$

be an $(n, d, k)_q$ union of lifted FD-codes, constructed as described in Section 29.7.3, and suppose that $\mathcal{A}$ is given as $\{a_1, a_2, \dots, a_{|\mathcal{A}|}\}$. Let $c_1 = 0$, and, for $2 \le i \le |\mathcal{A}|$, let $c_i = \sum_{j=1}^{i-1} |\mathcal{C}_{a_j}|$.

Codewords are numbered starting at zero. To map an integer m in the range $\{0, \dots, |\Omega| - 1\}$ to a codeword: (a) find the largest index i such that $c_i \le m$, (b) map the integer $j = m - c_i$ to a codeword of $\mathcal{C}_{a_i}$ (using an encoder for the corresponding rank-metric code), which can then be lifted to the corresponding subspace. Note that $0 \le j < |\mathcal{C}_{a_i}|$. Conversely, the jth codeword of $\mathcal{C}_{a_i}$ maps back to the message $m = c_i + j$. Assuming efficient encoding of the underlying rank-metric codes, the main complexity of the encoding algorithm, given m, is to determine the corresponding c_i, which can be done using a binary search in time at worst proportional to $\log |\mathcal{A}|$.

29.8.2 Decoding Lifted Delsarte–Gabidulin Codes

Let $\mathcal{C} \subseteq \mathbb{F}_q^{k \times m}$ be a Delsarte–Gabidulin code with $d_R(\mathcal{C}) = d$. Recall that $\Lambda(\mathcal{C})$ is a $(k + m, d, k)_q$ code.

A bounded-distance decoder for $\Lambda(\mathcal{C})$ is a function dec: $\mathcal{P}_q(n) \to \mathcal{C} \cup \{\varepsilon\}$ such that $\mathrm{dec}(U) = c$ for all $U \in \mathcal{B}_{\Lambda(c)}(k + m, k, q, t)$ and all $c \in \mathcal{C}$, and such that $\mathrm{dec}(U) = \varepsilon$ for all other U.

We note first that decoding of $\Lambda(\mathcal{C})$ is not a straightforward application of rank-distance decoding. To see this, let $A \in \mathbb{F}_q^{\ell \times k}$, $y \in \mathbb{F}_q^{\ell \times (n-k)}$ and $Y = \begin{bmatrix} A & y \end{bmatrix}$ be such that $\langle Y \rangle = U$ is the received subspace. If $\ell = k$ and A is nonsingular, then

$$d_I(\Lambda(c), U) = d_R(c, A^{-1}y)$$

and therefore decoding of Ω reduces to rank-distance decoding of $\mathcal{C}$. In general, however, A may not be invertible, in which case the argument above does not hold.

Several algorithms have been proposed for implementing the function $\text{dec}(\cdot)$. The first such algorithm was proposed by Kötter and Kschischang in [1156] and is a version of Sudan's "list-of-1" decoding for Delsarte–Gabidulin codes. The time complexity of the algorithm is $O((k+m)^2 m^2)$ operations in $\mathbb{F}_q$. A faster algorithm was proposed in [1711] which is a generalization of the standard ("time-domain") decoding algorithm for Delsarte–Gabidulin codes. The complexity of this algorithm is $O(dm^3)$ operations in $\mathbb{F}_q$. As shown in [1709], the algorithm in [1711] can significantly benefit from the use of optimal (or low-complexity) normal bases, further reducing the decoding complexity to $(11t^2 + 13t + m)m^2/2$ multiplications in $\mathbb{F}_q$ (and a similar number of additions). Finally, a transform-domain decoding algorithm was proposed in [1708, 1709], which is slightly faster than that of [1711] for low-rate codes.

As we will see, a bounded-distance decoder for a lifted Delsarte–Gabidulin code can be used as a black box for decoding many other subspace codes. For instance, consider $\mathcal{C}^r = \mathcal{C} \times \cdots \times \mathcal{C}$, the rth Cartesian power of a Delsarte–Gabidulin code $\mathcal{C} \subseteq \mathbb{F}_q^{k \times m}$. Recall that $\Lambda(\mathcal{C}^r)$ is a $(k + rm, d, k)_q$ code, where $d = d_R(\mathcal{C})$. Let $\text{dec}(\cdot)$ be a bounded-distance decoder for $\Lambda(\mathcal{C})$. Then a bounded-distance decoder for $\Lambda(\mathcal{C}^r)$ can be obtained as the map $\mathcal{P}_q(k + rm) \to \mathcal{C}^r \cup \{\varepsilon\}$ given by

$$U \mapsto \begin{cases} [\widehat{c}_1 \quad \cdots \quad \widehat{c}_r] = \widehat{c} & \text{if } \widehat{c}_i \neq \varepsilon,\ i = 1, \ldots, r,\ \text{and } d_I(\Lambda(\widehat{c}), U) \leq t, \\ \varepsilon & \text{otherwise} \end{cases}$$

where $\widehat{c}_i = \text{dec}(\langle [A \quad y_i] \rangle)$, $i = 1, \ldots, r$, and $A \in \mathbb{F}_q^{\ell \times k}$ and $y_1, \ldots, y_r \in \mathbb{F}_q^{\ell \times m}$ are such that $\langle [A \quad y_1 \quad \cdots \quad y_r] \rangle = U$. In other words, we can decode $\Lambda(\mathcal{C}^r)$ by decoding each Cartesian component individually (using the same matrix A on the left) and then checking whether the resulting subspace codeword is within the bounded distance from U.

29.8.3   Decoding a Union of Lifted FD Codes

Let $\mathcal{A}$ be a binary code with $d_A(\mathcal{A}) \geq d$. For all $a \in \mathcal{A}$, let $\mathcal{C}_a$ be an FD code for a with $d_R(\mathcal{C}_a) \geq d$. Let

$$\Omega = \bigcup_{a \in \mathcal{A}} \Lambda_a(\mathcal{C}_a).$$

Note that Ω need not be a constant-dimension code. Let $t = \lfloor (d-1)/2 \rfloor$.

A bounded-distance decoder for Ω is a function $\text{dec} \colon \mathcal{P}_q(n) \to \left(\mathcal{A} \times \bigcup_{a \in \mathcal{A}} \mathcal{C}_a \right) \cup \{\varepsilon\}$ such that $\text{dec}(V) = (a, c)$ whenever $d_I(V, \Lambda_a(c)) \leq t$, for all $c \in \mathcal{C}_a$, and all $a \in \mathcal{A}$, and such that $\text{dec}(V) = \varepsilon$ for all other V. In other words, the decoding function returns both the identifying vector $a \in \mathcal{A}$ and the corresponding matrix codeword $c \in \mathcal{C}_a$ whenever the injection distance between V and $\Lambda_a(c)$ is not greater than t, and it returns a decoding failure otherwise.

We will show that we can efficiently decode Ω, provided that we have efficient decoders for $\mathcal{A}$ and for each $\mathcal{C}_a$, $a \in \mathcal{A}$. The basic procedure was proposed in [692] for decoding in the subspace metric; here we adapt it for the injection metric.

Let us first consider the decoding of $\Lambda_a(\mathcal{C}_a)$, for some $a \in \mathcal{A}$. Let $c \in \mathcal{C}_a$ and $V \in \mathcal{P}_q(n)$. For every identifying vector a define a linear projection map $P(a)$ so that

$$\Lambda_a(c)P(a) = \Lambda(c)$$

is satisfied. Note that $P(a)$ simply rearranges the columns of a matrix so that an FD-restricted lifting is transformed to a usual lifting. Since $P(a)$ is a nonsingular linear transformation, and therefore preserves dimensions, we have

$$d_I(\Lambda_a(c), V) = d_I(\Lambda_a(c)P(a), VP(a)) = d_I(\Lambda(c), VP(a)).$$

It follows that bounded-distance decoding of $\Lambda_a(\mathcal{C}_a)$ can be performed by first computing $\widehat{c} = \text{dec}(VP(a))$, and then returning $(a, \widehat{c})$ unless $\widehat{c} = \varepsilon$.

Now, consider the decoding of Ω. Let $V \in \mathcal{P}_q(n)$ be such that $d_I(U, V) \leq t$, for some (unique) $U \in \Omega$. The first step is to compute the identifying vector a corresponding to the Schubert cell containing U. Let a' denote the identifying vector corresponding to the Schubert cell containing V. Since by Theorem 29.7.2

$$d_A(a, a') \leq d_I(U, V) \leq t,$$

it follows that a can be found by inputting a' to a bounded-asymmetric-distance decoder for $\mathcal{A}$. Then the actual $c \in \mathcal{C}_a$ such that $V = \Lambda_a(c)$ can be found by using the decoder for $\mathcal{C}_a$ described above.

29.9 Conclusions

As we have seen in this chapter, the ideas that underlie network coding are simple, yet potentially far-reaching. Allowing the nodes in a network to perform coding (not just routing), permits the multicast capacity of a network to be achieved with linear network coding. A careful network coding design is not required, as a completely random choice of network coding coefficients will, with high probability of success when the field size is sufficiently large, achieve capacity. In practice, the random linear network coding approach results in multicast protocols that are robust to changes in the underlying network topology.

The random linear network coding approach also opens an intriguing set of problems in coding theory, centered on the design of subspace codes, but strongly connected with the design of codes in the rank metric, and problems in geometry. As we have seen in this chapter, many bounds and constructions of classical coding theory have subspace-code analogs. The book [852] collects together many contributions from experts in this area, going well beyond the results presented in this chapter. A table of bounds and constructions for subspace codes and constant-dimension codes is maintained at

> http://subspacecodes.uni-bayreuth.de/.

From a practical standpoint, at least for applications in network coding, it seems that the main problems are largely solved, as constant-dimension lifted rank-metric codes contain close to the maximum possible number of codewords (at least on a logarithmic scale), and efficient encoding and decoding algorithms have been developed. It is unlikely that codes with marginally larger codebooks (even though they exist) will justify the additional complexity needed to process them.

From a mathematical standpoint, however, much remains open and—as in coding theory in general—finding new approaches for constructing subspace codes and decoding them efficiently as well as finding new applications for such codes remains an active area of research.

Chapter 30

Coding for Erasures and Fountain Codes

Ian F. Blake

University of British Columbia

30.1 Introduction

The chapter addresses the problem of the efficient correction of erasures for such applications as the dissemination of large downloads on the internet. Two aspects of the problem are considered. In the next section a class of block codes designed for erasure-correction are considered, the Tornado codes. The construction of these introduced several concepts, in particular the notion of probabilistic construction of bipartite graphs, that play a prominent role in the other type of code considered, the fountain codes examined in Sections 30.3 and 30.4. In both classes of codes, linear encoding and decoding complexity is achieved.

Remark 30.1.1 The term *packet* will be used interchangeably with *symbol* as a sequence of bits, which in the sequel will also be identified with nodes of a code bipartite graph. For uniformity, the left nodes of such a graph will be referred to as *information symbols* or packets and the right nodes as coded or *check symbols* or packets. Since the only operation on such symbols is an XOR with other symbols, sometimes referred to as addition (as in the finite field $\mathbb{F}_2$), the length of the symbols is not an issue.

Example 30.1.2 For $0110, 1101 \in \mathbb{F}_2^4$ the XOR is $0110 \oplus 1101 = 1011$.

Remark 30.1.3 There are two commonly used bipartite graphs to represent codes. In one, the k left nodes are **information nodes** and the $(n - k)$ right nodes are **check nodes**. The alternative representation is for the left nodes of the graph to represent the n codeword positions or parity check matrix columns and the right nodes the parity check matrix rows. Thus left node j is connected to right node i if the element h_{ij} of the parity check matrix H is 1. This second representation is usually referred to as the *Tanner graph* of the code and when the left nodes represent a codeword, the right nodes are all zero. This representation is commonly used in the analysis of LDPC codes where the left nodes are usually referred to as *variable nodes* and the right nodes as *check nodes*; see Chapter 21 for the study of

LDPC codes. The first representation will be used in this work. The representations are shown in Figure 30.1 for the $[7,4,3]_2$ Hamming code.

FIGURE 30.1: (a) Code parity check matrix of the $[7,4,3]_2$ Hamming code; (b) first representation of the code; (c) Tanner graph of the code

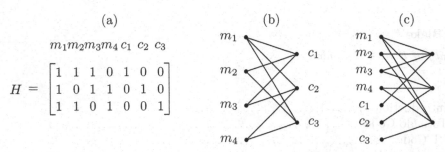

(a)   (b)   (c)

$$m_1 m_2 m_3 m_4\, c_1\ c_2\ c_3$$

$$H = \begin{bmatrix} 1 & 1 & 1 & 0 & 1 & 0 & 0 \\ 1 & 0 & 1 & 1 & 0 & 1 & 0 \\ 1 & 1 & 0 & 1 & 0 & 0 & 1 \end{bmatrix}$$

Remark 30.1.4 In fountain codes, coded packets are generated as random XOR's of selected information packets and transmitted on the channel. A client or receiver accepts coded packets as they arrive until a sufficient number are received that allows for decoding (solving a matrix equation) to retrieve the information packets. In this context an *erasure* is simply a missing packet. All randomly generated coded packets are equivalent in this regard and such a system is ideal for a multicast environment where the possibility of feedback from the client to the transmitter is not allowed to prevent the server from being overwhelmed by retransmission requests for particular missing packets.

FIGURE 30.2: The binary erasure channel with erasure probability p

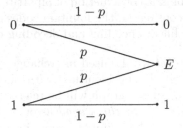

Remark 30.1.5 In preparation for the discussion of Tornado codes in the next section, some elements of erasure-correction for linear block codes are recalled. The **binary erasure channel (BEC)**, shown in Figure 30.2, has a binary input $\{0,1\}$ and ternary output $\{0,1,E\}$ where E represents an erasure, which is an error whose location is known. If the erasure probability is p, the **capacity** of such a channel is $C_{BEC} = 1 - p$. Consider the use of a binary linear $[n,k,d]_2$ code. Suppose the codeword $\mathbf{c} \in \mathbb{F}_2^n$ is transmitted and the word $\mathbf{r}$ received. This word may have up to $d-1$ erasures in it which must be solved for. All vectors will be row vectors and displayed in bold face. Suppose $e \le d-1$ erasures are made in transmission, and let H be the $(n-k) \times n$ parity check matrix of the code. Suppose the columns of the parity check matrix are labelled $\mathbf{h}_i^\mathsf{T}$, $i = 1, 2, \ldots, n$. Assume the e erasures occurred in positions $i_1, i_2, \ldots, i_e$ and that the received word is $\mathbf{r} = r_1 r_2 \cdots r_n$.

It is sought to determine the values of the erasure such that

$$H\mathbf{r}^\mathsf{T} = \mathbf{0}^\mathsf{T}$$

or equivalently to solve the matrix equation

$$H'\mathbf{e}^\mathsf{T} = \mathbf{r}'^\mathsf{T}$$

where $\mathbf{e}$ is the erasure vector of length e whose values are sought, H' is the $(n-k) \times e$ matrix with columns $\mathbf{h}_{i_1}^\mathsf{T}, \mathbf{h}_{i_2}^\mathsf{T}, \ldots, \mathbf{h}_{i_e}^\mathsf{T}$, and $\mathbf{r}'^\mathsf{T}$ is the sum of the columns of H corresponding to 1's in $\mathbf{r}$. The complexity of this process is at most $O(d^3)$.

The aim of this work is to show how, with the careful choice of the matrix H, the encoding and decoding complexity can be reduced to linear in the code length n.

30.2 Tornado Codes

The idea of Tornado codes arose in the works [321, 1295]. The reference [321] also introduced the notion of a fountain code studied in the next two sections. Tornado codes, which are linear block codes, were further studied in the more complete version in [1298] (while not actually being referred to as Tornado codes there). Only binary codes are considered. As noted, since the only coding and decoding operations involve XOR, the information and coded symbols can be viewed as binary packets of a fixed length. For the discussion of Tornado codes it is assumed the code is described by a bipartite graph with left and right nodes. The term *Tornado* [321] arose from the behavior of the following algorithm that typically proceeds regularly until a certain stage of decoding when the arrival of a single symbol generates a rapid succession of substitutions and decoding is completed quickly.

The codes that will be designed will use the following very simple decoding algorithm:

Algorithm 30.2.1

Given the value of a check symbol and all but one of the information symbols on which it depends, set the missing information symbol to be the XOR of the check symbol and its known attached information symbols.

Remark 30.2.2 The success of this algorithm depends on there being a check node with the desired property at each stage of the decoding process. The algorithm requires that all check nodes be received without erasure. A cascade of codes is constructed as follows. At the first stage consider a bipartite graph with k information nodes on the left and βk check nodes on the right for some $\beta < 1$. Label this code $\mathcal{C}(B_0)$. For the next stage, create another bipartite graph using the check nodes of $\mathcal{C}(B_0)$ as left nodes and add $\beta^2 k$ check nodes on the right—a code graph labelled $\mathcal{C}(B_1)$. Continue in this manner, at each stage the check (right) nodes of the code graph $\mathcal{C}(B_i)$ serving as the input (left) nodes of $\mathcal{C}(B_{i+1})$. The algorithm applied to each of these stages requires all check nodes to be received without erasures. To complete the construction a small erasure-correction code C is added with $k\beta^{m+2}/(1-\beta)$ check bits. The overall code is referred to as $\mathcal{C}(B_0, B_1, \ldots, B_m, C)$. It has k information bits and

$$\sum_{i=1}^{m+1} \beta^i k + \frac{\beta^{m+2}k}{1-\beta} = \frac{k\beta}{1-\beta}$$

check bits and is of rate $(1 - \beta)$.

Remark 30.2.3 If the last code C is able to correct any fraction of $\beta(1 - \epsilon)$ of erasures (in either check or information bits - left or right nodes of C) with high probability and if each of the intermediate codes $\mathcal{C}(B_i)$ is able to correct a fraction $\beta(1 - \epsilon)$ of left nodes with high probability, given all check or right nodes are correct, then overall the code should be able to correct a fraction of $\beta(1 - \epsilon')$ of erasures with high probability. The relationship between ϵ and ϵ' is complex and the reader is referred to the proof of Theorem 2 in [1299] for details.

The cascade construction of Remark 30.2.2 gives an overall structure of a code and its decoding but has so far not specified how the bipartite graphs are to be constructed. The construction of the graphs will each be randomized according to certain distributions. For a given bipartite graph, an edge in the graph is said to be of **left edge degree** i if it is connected to a node on the left of degree i, and similarly for **right edge degree**. Consider two edge degree distributions, a left distribution $(\lambda_1, \lambda_2, \ldots)$ and a right distribution $(\rho_1, \rho_2, \ldots)$ where λ_i (respectively ρ_i) is the fraction of edges in the graph of left (respectively right) degree i. Note that each stage of the cascade construction requires a bipartite graph and each bipartite graph will be chosen randomly according to the (λ, ρ) distributions. Define the polynomials $\lambda(x) = \sum_i \lambda_i x^{i-1}$ and $\rho(x) = \sum_i \rho_i x^{i-1}$. These will be referred to as a (λ, ρ) **edge degree distribution**.

Several properties of a bipartite graph constructed according to these distributions follow immediately. Let E be the total number of edges in the graph. Then the number of left (respectively right) nodes of degree i is $E\lambda_i/i$ (respectively $E\rho_i/i$). If a_L (respectively a_R) is the average degree of the left (respectively right) nodes, then

$$a_L = \frac{1}{\sum_i \lambda_i/i} = \frac{1}{\int_0^1 \lambda(x)\,dx} \quad \text{and} \quad a_R = \frac{1}{\sum_i \rho_i/i} = \frac{1}{\int_0^1 \rho(x)\,dx}.$$

The polynomials ([1684, Lemma 1])

$$\Lambda(x) = \frac{\int_0^x \lambda(t)\,dt}{\int_0^1 \lambda(t)\,dt} \quad \text{and} \quad R(x) = \frac{\int_0^x \rho(t)\,dt}{\int_0^1 \rho(t)\,dt}$$

give the left and right node degree distributions, respectively.

It is also clear that $\beta \int_0^1 \rho(x)\,dx = \int_0^1 \lambda(x)\,dx$. It is assumed that where necessary the constants needed are integers. Furthermore ([1684, Theorem 1]), the rate of the code can be bounded by

$$R \geq 1 - \frac{a_L}{a_R} = 1 - \frac{\int_0^1 \rho(x)\,dx}{\int_0^1 \lambda(x)\,dx}.$$

Remark 30.2.4 Given a left and a right edge degree distribution the following technique is one possibility for constructing a bipartite graph with the required properties. It is assumed all the quantities needed are integral; suitable approximations could be obtained for the cases when this is not true. Consider four columns of nodes. Choose an appropriate number of edges E for the graph. The number of left nodes in the first column of degree i will be $E\lambda_i/i$. These nodes have i edges emanating from each of them and these, for all i, terminate in the second column of nodes. Similarly for the right nodes: create $E\rho_j/j$ right nodes, with j edges and these form the fourth column with the nodes at the other end of their edges forming the third column. So the second and third columns have E nodes each. Select a random permutation on the integers $1, 2, \ldots, E$ and connect the nodes of the second and third columns, according to this permutation, to create the bipartite graph. The nodes in the first column are the left nodes of the bipartite graph, and the nodes of the fourth column are the right nodes; the nodes in columns two and three are absorbed into the edges in the obvious way. The resulting bipartite graph has the required left and right edge degrees. There may be multiple edges between two nodes which can be corrected by deleting edges. Such a process will have little effect on the probabilities involved.

Algorithm 30.2.1 is equivalent to the next algorithm, which is in a slightly more convenient form for the sequel. It is applied to each stage of the cascade of graphs.

Algorithm 30.2.5

Associate with the nodes of the code graph a register initially loaded with received values. For each left information node that is not erased, add the associated value to all neighbor registers on the right and erase all edges emanating from that node. If after this operation there is a check node of degree one, replace the erased left neighbor with the value of the check node register, XOR that value to all other attached nodes of the left neighbor and erase all edges from the two nodes. The left information neighbor node is said to be **resolved**. Continue the process until all left nodes are resolved or there are no check nodes of degree one. If there is no check node of degree one and there are remaining unresolved information nodes, the decoding fails.

FIGURE 30.3: A small example of Algorithm 30.2.5 using the Hamming $[7, 4, 3]_2$ code of Figure 30.1

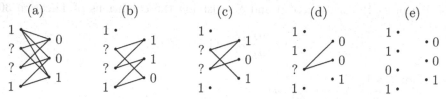

| (a) | (b) | (c) | (d) | (e) |

Example 30.2.6 The various steps of Algorithm 30.2.5 are shown in Figure 30.3. Notice in Step (d) the two uppermost check digits are the same and give the corresponding information bit the value 0. In this circumstance the two check bits could not differ if the set of information and check bits corresponds to a valid codeword and the number of erasures does not exceed 2, the erasure-correcting capability of the code.

With the probabilistic construction of the graphs according to the above edge distributions, one is able to set up differential equations for the number of left and right degree edges as time evolves and solve them at each stage of the algorithm to give an analysis of the decoding algorithm; see [1295, 1298, 1299]. It is desired to choose the left and right edge degree distributions to maximize the probability that the process ends successfully, i.e., with all information nodes resolved. The analysis is straightforward but involved. The results are reported in the following two impressive theorems.

Theorem 30.2.7 ([1298, Theorem 2]) *Let k be an integer and suppose that*

$$\mathcal{C} = \mathcal{C}(B_0, B_1, \ldots, B_m, C)$$

is a cascade of bipartite graphs from Remark 30.2.2 where B_0 has k left (information) nodes. Suppose that each B_i is chosen at random with edge degrees specified by $\lambda(x)$ and $\rho(x)$ such that $\lambda_1 = \lambda_2 = 0$, and suppose δ is such that

$$\rho(1 - \delta\lambda(x)) > 1 - x \quad \text{for } 0 < x \leq 1. \tag{30.1}$$

Then if at most a δ fraction of the encoded word in C is erased independently at random, the erasure decoding Algorithm 30.2.5 terminates successfully with probability $1 - O(k^{-3/4})$ and does so in $O(k)$ steps.

Theorem 30.2.8 ([1298, Theorem 3]) *For any rate R with $0 < R < 1$ and any ϵ with $0 < \epsilon < 1$ and sufficiently large block length n, there is a linear code and a decoding algorithm that, with probability $1 - O(n^{-3/4})$, is able to correct a random $(1 - R)(1 - \epsilon)$-fraction of erasures in time proportional to $n \ln(1/\epsilon)$.*

Remark 30.2.9 It is noted [1298] that there is a dual version of the condition on the distributions in the Theorem 30.2.7:

$$\delta\lambda(1 - \rho(y)) < 1 - y \quad \text{for } 0 \leq y < 1,$$

which is sometimes useful. It is obtained from (30.1) by substituting $y = \rho^{-1}(1 - x)$ which is valid since the function ρ is monotonically increasing on $(0, 1]$ noting that a polynomial with positive coefficients has a positive derivative on $(0, 1]$.

Remark 30.2.10 Theorem 30.2.7 establishes a probabilistic encoding and decoding algorithm that is capacity achieving. Three properties of the construction are:

(i) [1684, Theorem 1] For $\lambda(x), \rho(x)$ and δ satisfying the conditions of Theorem 30.2.7 above,

$$\delta \leq \frac{a_L}{a_R}\left(1 - (1 - \delta)^{a_R}\right).$$

(ii) [1684, Corollary 1]

$$\delta \leq \frac{a_L}{a_R}\left(1 - (1 - a_L/a_R)^{a_R}\right),$$

a result that follows from (i) since $\delta \leq a_L/a_R$.

(iii) [1684, Lemma 2] If $(\lambda(x), \rho(x))$ satisfy the condition that $\delta\lambda(1 - \rho(1 - x)) < x$ for all $x \in (0, \delta]$, then $\delta \leq \rho'(1)/\lambda'(0)$.

Remark 30.2.11 Given the theory developed for achieving capacity on the BEC with linear decoding complexity, it is of interest to derive distributions that satisfy the conditions of Theorem 30.2.7 and, equivalently, Remark 30.2.9.

Example 30.2.12 ([1684, 1685]) Two distributions with a parameter a are given by:

$$\rho_a(x) = x^{a-1} \quad \text{and} \quad \lambda_{\alpha,N}(x) = \alpha\frac{\sum_{k=1}^{N-1}\binom{\alpha}{k}(-1)^{k+1}x^k}{\alpha - N\binom{\alpha}{N}(-1)^{N+1}},$$

where the generalized binomial coefficient for real $\alpha = 1/(a - 1)$ is

$$\binom{\alpha}{N} = \frac{\alpha(\alpha - 1)\cdots(\alpha - N + 1)}{N!}.$$

The numerator of the right-hand side of the equation for $\lambda_{\alpha,N}(x)$ is a truncated version of

$1 - (1-x)^\alpha$ for real α, and is a version of the general binomial theorem. Notice that for $0 < \alpha < 1$ the terms in the numerator are positive—the binomials alternate in sign as does $(-1)^{k+1}$. With the distributions $(\lambda, \rho) = (\lambda_{a,N}, \rho_a)$, the condition required for asymptotic optimality reduces to ([1684, Theorem 2])

$$\delta\lambda\big(1 - \rho(1-x)\big) \leq \frac{\delta\alpha}{\alpha - N\binom{\alpha}{N}(-1)^{N+1}}x.$$

So to satisfy the condition $\delta\lambda\big(1 - \rho(1-x)\big) < x$ (from Remark 30.2.9) it is required that

$$\delta \leq \frac{\alpha - \binom{\alpha}{N}(-1)^{N+1}}{\alpha},$$

and from this equation it can be shown [1684] that the distribution pair (λ, ρ) is asymptotically optimal. Notice the nodes on the right are all of degree $(a-1)$ and the graph is referred to as right regular.

Example 30.2.13 ([1685, Section 3.4]) Let D be a positive integer (that will be an indicator for how close δ can be made to $1 - R$ for the sequences obtained). Let $H(D) = \sum_{i=1}^{D} \frac{1}{i}$ denote the harmonic sum truncated at D and note that $H(D) \approx \ln(D)$. Two distributions parameterized by the positive integer D are

$$\lambda_D(x) = \frac{1}{H(D)} \sum_{i=1}^{D} \frac{x^i}{i} \quad \text{and} \quad \rho_D(x) = e^{\mu(x-1)} \text{ (truncated)}$$

where $\lambda_D(x)$ is a truncated version of $\ln(x)$ and $\rho_D(x)$ is a truncated version of the Poisson distribution by which is meant that the sequence of probabilities

$$\rho_{D,i} = e^{-\mu}\frac{\mu^{i-1}}{(i-1)!}$$

is truncated at some high power so that the following development remains valid. The parameter μ is the solution to the equation

$$\frac{1}{\mu}(1 - e^{-\mu}) = \frac{1-R}{H(D)}\left(1 - \frac{1}{D+1}\right). \tag{30.2}$$

Such a distribution gives a code of rate at least R. Note the average left degree is $a_L = H(D)(D+1)/D$ and $\int_0^1 \lambda_D(x)\,dx = (1/H(D))\big(1 - 1/(D+1)\big)$. The right degree edge distribution is approximated by the Poisson distribution. It can also be established that

$$\delta\lambda_D\big(1 - \rho_D(1-x)\big) = \delta\lambda_D(1 - e^{-\mu x}) \leq \frac{-\delta}{H(D)}\ln(e^{-\mu x}) = \frac{\delta\mu x}{H(D)}.$$

For the right hand side of the last equation to be at most x it is required that

$$\delta \leq H(D)/\mu.$$

By (30.2)

$$H(D)/\mu = (1-R)\big(1 - 1/(D+1)\big)/(1 - e^{-\mu}).$$

Thus this pair of distributions (λ_D, ρ_D) satisfy the condition

$$(1-R)(1-1/D)\lambda_D\big(1 - \rho_D(1-x)\big) < x \quad \text{for } 0 < x < (1-R)(1-1/D),$$

as shown in [1685], and hence is asymptotically capacity achieving.

Example 30.2.14 A technique is established in [1295, 1298] using linear programming to find distribution sequences satisfying the capacity achieving equation (30.1). An example in that paper is the following:

$$\lambda(x) = 0.430034x^2 + 0.237331x^{12} + 0.007979x^{13} + 0.119493x^{47} +$$
$$0.052153x^{48} + 0.079630x^{161} + 0.073380x^{162}$$
$$\rho(x) = 0.713788x^9 + 0.122494x^{10} + 0.163718x^{199}$$

The technique involves, for a given $\lambda(x)$ and δ, using the condition of Theorem 30.2.7 to generate linear constraints for the ρ_i and uses linear programming to determine if the suitable ρ_i exist. The reader is referred to the numerous references on capacity achieving sequences, including [1297, 1298, 1299, 1475, 1589, 1684, 1690].

30.3 LT Codes

The random construction idea of bipartite graphs for erasure coding of the previous section was contained in the original contribution of [321], which also mentioned the idea of fountain codes. The first incarnation of fountain codes was the work of Luby [1292] called **LT codes** (for Luby Transform) there. It seems a natural consequence of the work in [321], and they are the precursors of the Raptor codes which feature in so many commercial standards for big downloads. The original paper [1292] and the articles [1688, 1691] are excellent references for the definition and analysis of these codes used for this presentation. Indeed much of what is known about the construction and performance of fountain codes derives from the work of Luby and Shokrollahi and their colleagues.

Consider k packets of fixed size (since, as noted previously, the only operation used is that of XOR and the extension from bit operations to packet operations is trivial). The notation used will be that of information packets and coded or check packets. The idea of **fountain codes** is to create check packets by choosing at random in a carefully chosen manner a subset of information packets and XOR'ing them together to form a check packet. How this is done is discussed below. The receiver will collect from the channel a sufficient number of check packets before starting the decoding algorithm. It will be assumed here the receiver collects $k(1 + \epsilon)$ packets and ϵ is referred to as the **overhead** of the code, the number (fraction) of coded packets beyond k, the minimum required to retrieve the original k information packets with high probability.

With the $k(1 + \epsilon)$ collected coded packets, the receiver forms a bipartite graph with the left nodes the unknown information packets (which are to be solved for) and the right nodes the coded packets (received) with edges between them designating which information packets are included in the coded packet, information contained in a coded packet header. Decoding will be unique only if the associated $k \times k(1 + \epsilon)$ matrix is of full rank over $\mathbb{F}_2$.

Algorithm 30.3.1

Initially the code graph has k unknown information (left) nodes corresponding to the information packets and $k(1 + \epsilon)$ right nodes corresponding to received coded packets. At each stage of decoding, a coded node of degree one is sought. Suppose coded node u has degree 1 and its (unique) information node neighbor is v. The packet associated with u is then associated with attached information node v and XOR'ed to each of v's other coded neighbors. All edges involved with this process emanating from nodes u and v are removed and the process is repeated. It is said the decoded information node v is **recovered** or **resolved**. If at any stage before all information nodes are resolved there is no coded packet of degree 1, decoding fails.

FIGURE 30.4: Simple example of LT decoding

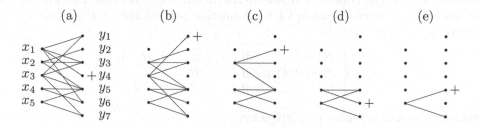

Example 30.3.2 The LT code of interest is in Figure 30.4 (a). At each stage of decoding the chosen coded symbol of degree one is designated by $+$. At the first stage (a) the coded symbols y_2 and y_4 have degree 1 - choose symbol y_4 and delete node x_1 and all its other neighbors to give graph (b). Coded symbols y_1, y_2 and y_6 have degree 1 - choose y_1 and repeat the process to give graph (c). Coded symbols y_2, y_3, y_6 and y_7 have degree 1 - choose y_2 to give graph (d). Finally choose node y_6 and then y_5 to resolve all input symbols.

An **LT code** is formed in the following manner. A **degree distribution** $\Omega(x) = \sum_{i \geq 1} \Omega_i x^i$ is sought which with high probability ensures a degree one coded packet will be available at each stage to the completion of the process. To form a coded packet the distribution is sampled for a value. If i is the result, (with probability Ω_i) i information packets are chosen uniformly at random from the k information packets and XOR'ed to form a coded packet. An LT code with k information symbols and distribution $\Omega(x)$ will be denoted a $(k, \Omega(x))$ LT code.

Given a $(k, \Omega(x))$ LT code, the **output ripple** at step i of the decoding process is defined as the set of coded symbols at that step of reduced degree 1, and we say that a coded symbol **is released** at step $i + 1$ if its degree is larger than 1 at step i and 1 at step $i + 1$. At least one coded symbol is released at each step of the algorithm. One is able to set up differential discrete time equations to model the decoding process, and the analysis [1688, 1691] shows that if the expected number of coded symbols in the output ripple is to be 1 at step k of the algorithm, the final step, the distribution should satisfy the equation

$$(1 - x)\Omega''(x) = 1 \quad \text{for } 0 < x < 1,$$

which, with the constraint $\Omega(1) = 1$ (the sum of all probabilities is to add to 1), has the solution

$$\Omega(x) = \sum_{i \geq 2} \frac{x^i}{i(i-1)}.$$

This distribution is unsuitable as it has no nodes of degree 1. The modified result suggested in [1292] is the **soliton distribution** given by

$$\Omega(x) = \frac{x}{k} + \frac{x^2}{1 \cdot 2} + \cdots + \frac{x^k}{(k-1) \cdot k}. \tag{30.3}$$

That this distribution sums to unity is easily established by induction on k. If $r(L)$ denotes the probability a coded symbol is released when L information symbols are unresolved (as yet unknown), this soliton distribution has the interesting property ([1292, Proposition 10]) that $r(L) = 1/k$ for all $L = 1, 2, \ldots, k$, i.e., a uniform release probability.

Remark 30.3.3 The soliton distribution does not work well in practice as the variance of the ripple size for finite k will be too large resulting in too high a probability the output ripple will vanish. To remedy this, the following distribution is proposed.

Definition 30.3.4 The **robust soliton distribution** [1292] is defined as follows: Let δ be the probability of decoder failure for k information packets and let $R = c \ln(k/\delta)\sqrt{k}$ for some suitable constant $c > 0$. Let

$$
\tau_i = \begin{cases}
R/(ik) & \text{if } i = 1, 2, \ldots, k/R - 1, \\
R \ln(R/\delta)/k & \text{if } i = k/R, \\
0 & \text{if } i = k/R + 1, \ldots, k.
\end{cases}
$$

The robust soliton distribution μ_i is given by

$$
\mu_i = (\Omega_i + \tau_i)/\beta \quad \text{for } i = 1, 2, \ldots, k \text{ where } \beta = \sum_{i=1}^{k} (\Omega_i + \tau_i)
$$

and where Ω_i is the soliton distribution of (30.3).

Remark 30.3.5 The intuition of this distribution is that the random walk represented by the output ripple size deviating from its mean by more than $\ln(k/\delta)\sqrt{k}$ is at most δ resulting in a higher probability of complete decoding.

Theorem 30.3.6 ([1292, Theorems 12 and 13]) *With the above notation the average degree of a coded symbol is $D = O(\ln(k/\delta))$ and the number of coded symbols required to achieve a decoding failure of at most δ is $K = k + O(\sqrt{k} \ln^2(k/\delta))$.*

Remark 30.3.7 For obvious reasons LT codes and their relatives are often referred to as **fountain codes** and sometimes **rateless codes** since there is no notion of code rate involved. While the parameter of packet loss rate on the internet is an important parameter in the overall performance of the coded system, it does not figure in the code design problem.

Definition 30.3.8 ([1687]) A **reliable decoding algorithm** for a fountain code is one which can recover the k input packets from n coded packets and errs (fails to complete decoding) with a probability that is at most inversely polynomial in k (of the form $1/k^c$).

Proposition 30.3.9 ([1687, Proposition 1]) *If an LT code with k information symbols possesses a reliable decoding algorithm, then there is a constant c such that the graph associated with the decoder has at least $ck \ln(k)$ edges.*

Proof: An informal overview of the proof is given. Consider the LT code $(k, \Omega(x))$ and consider the probability an information node u is not attached to any coded node, in which case it is impossible for decoding to be successful. Consider a coded node v of degree d. The probability that u is not a neighbor to v is

$$
\binom{k-1}{d} \Big/ \binom{k}{d} = 1 - d/k.
$$

If $a = \Omega'(1)$ is the average degree of a coded node, then averaging this expression over d

gives the probability u is not a neighbor of v as $1 - a/k$. Similarly the probability u is not a neighbor of any of the $n = k(1 + \epsilon)$ output nodes is $(1 - a/k)^n$. For large k this expression is approximated as

$$(1 - a/k)^n = \left[(1 - a/k)^k\right]^{n/k} \approx (\exp(-a))^{n/k} = \exp(-\alpha)$$

where $\alpha = an/k$ is the average degree of an information node. If the probability of failure of the algorithm is assumed to be $1/k^c$ and the fact that an information symbol not being attached to any coded node is only one way the decoding could fail, then

$$\exp(-\alpha) \leq 1/k^c \ \text{ or } \ \alpha \geq c\ln(k)$$

as required. $\qquad\qquad\qquad\qquad\qquad\qquad\qquad\qquad\qquad\qquad\qquad\qquad\qquad\qquad\square$

Remark 30.3.10 For very large downloads it will be important to have a linear (in the number of information packets, k) decoding complexity and the above discussion shows that this is not achievable for LT codes since the average node degree of the soliton distribution is $O(\ln(k))$. The modification of LT codes to achieve this linear complexity resulted in the Raptor codes introduced in the next section. The important modification was to obtain a distribution with a constant average degree that allowed recovery of almost all information packets (not all), with an acceptable overhead, and to introduce a precoding stage to recover all the information packets. The process is described in the next section.

An excellent analysis of the LT decoding process is contained in [1688, Theorem 1] that gathers and extends analyses available in the literature; see, e.g., [1084, 1311, 1327]. Define the **state** of an LT decoder by the triple (c, r, u) where u is the number of information symbols not yet resolved at some stage, c is the size of the **cloud** (the coded symbols at that stage of reduced degree ≥ 2), and r is the size of the output ripple (the coded symbols of degree 1). Let $P_{c,r,u}$ be the probability the decoder is in that state. Then a recursion is developed for the generating function of these probabilities in terms of p_u, the probability a randomly chosen information symbol is attached to a coded symbol of degree 1 after the current state, given that it was of degree ≥ 2 before the transition and there are currently u undecoded information symbols. The analysis is quite involved and only a brief outline of the results is given.

Let

$$P_u(x, y) = \sum_{r \geq 1, c} P_{c,r,u} x^c y^{r-1}$$

be the generating function. The argument to derive these functions is combinatorial. The reader is referred to the proof in [1688, Theorem 1], due to Karp et al. [1084], for details. A recursion is given to compute $P_{u-1}(x, y)$ knowing $P_u(x, y)$ and p_u. For $u = k + 1, k, \ldots, 1$, the recursion is given by

$$P_{u-1}(x, y) = \frac{P_u\big(x(1 - p_u) + yp_u, (1/u) + y(1 - 1/u)\big) - P_u\big(x(1 - p_u), 1/u\big)}{y},$$

where a complicated expression for the quantity p_u is given; as noted, it gives the probability a random cloud element joins the ripple after the current transition. The initial conditions include

$$P_{c,r,k} = \begin{cases} \binom{n}{r}\Omega_1^r(1 - \Omega_1)^c & \text{if } c + r = n, \\ 0 & \text{otherwise,} \end{cases}$$

$$P_k(x, y) = \sum_{r \geq 1, c} P_{c,r,k} x^{r-1} y^c = \frac{\big(x(1 - \Omega_1) + y\Omega_1\big)^n - x(1 - \Omega_1)}{y},$$

$$P_{k+1}(x, y) \overset{\text{def}}{=} x^n \quad \text{and} \quad p_{k+1} \overset{\text{def}}{=} \Omega_1.$$

These definitions allow $P_k(x, y)$ to be computed from $P_{k+1}(x, y)$ and, subsequently, all $P_u(x, y)$, $u = k - 1, \ldots, 1$.

These expressions are important, beyond their intrinsic interest, in that they allow the probability of error (decoding failure for LT codes) to be computed (see, e.g., [1311]) in that

$$P_{err}(u) = \sum_{c \geq 0} P_{c,0,u} = 1 - \sum_{c \geq 0, r \geq 1} P_{c,r,u} = 1 - P_u(1, 1)$$

and the overall probability of error is

$$P_{err} = \sum_{u=1}^{k} P_{err}(u).$$

Let $R_u(x, y) = \partial P_u / \partial y$. Then $R_u(1, 1) = R(u)$ is the expected ripple size, given the decoding process has continued up to u undecoded information symbols. It can be shown [1311, 1688] that

$$R(u) = u\big(\Omega'(1 - u/k) + u \ln(u/k)\big) + O(1).$$

An expression for the expected size of the cloud is also given [1311, Theorem 2] where it is also shown that the standard deviation of the ripple size is $O(\sqrt{k})$.

Remark 30.3.11 An important consequence of the above analysis is the behavior of the probability of decoding failure as the decoding process evolves. It is shown [1688, Figure 9] that decoding failure typically occurs towards the end of the process when there are relatively few information symbols unresolved, an observation that matches behavior observed in simulations. This behavior has important implications for the formulation of the Raptor codes of the next section.

Remark 30.3.12 It is shown in [1292, Theorem 12], and commented on in [1687], that in order to have a reliable decoder for LT codes, it is necessary for the decoder to gather at least $k + O\big(\sqrt{k} \cdot \ln^2(k/\delta)\big)$ coded symbols for a probability of error of less than δ, and hence an overhead of $O\big(\ln^2(k/\delta)/\sqrt{k}\big)$. Many other aspects of the analysis of the performance of LT codes and their decoding algorithm are given in [1294, 1295, 1298, 1299, 1327, 1475, 1684, 1686, 1688, 1691].

Remark 30.3.13 It should be noted that the decoding algorithm for the LT codes is an instance of a **belief propagation (BP)** or **message-passing** decoding algorithm. It is clear that it is not **maximum likelihood** (ML). An algorithm that is ML is **Gaussian Elimination (GE)** where the decoder, on gathering N coded symbols from the fountain, forms the $N \times k$ matrix G of coded symbols and information symbols and solves the matrix equation

$$G\mathbf{x}^{\mathsf{T}} = \mathbf{r}^{\mathsf{T}}$$

using GE on the matrix G, where $\mathbf{x}$ is the k-tuple of information symbols and $\mathbf{r}$ the N-tuple of received coded symbols. The complexity of such an algorithm is $O(k^3)$ and is far too complex for the application to large downloads of interest here.

That BP is not ML is clear since one can imagine a situation where the matrix G is of full rank (and hence the above equation has a unique solution), yet the arrangement of checks is such that the decoding algorithm fails to complete.

30.4 Raptor Codes

The genesis of Raptor (RAPid TORnado) codes stems from Remark 30.3.11 that the LT decoding process, when it fails, tends to fail towards the end of the decoding process where a large fraction of the input symbols have been decoded. This suggests relaxing the constraint that the LT code decode all information symbols to requiring to decode all but a fraction of them and making up for the incompleteness by precoding the information symbols with a weak erasure-correcting code. This simple artifice will allow both decoding to completion with high probability and a linear decoding complexity. Raptor codes were invented in late 2000 and a patent was filed in 2001.

Definition 30.4.1 For the construction of a **Raptor code** $(k, \mathcal{C}_n, \Omega(x))$ imagine three vertical sequences of nodes. The first is the set of k information nodes. The second, referred to as **intermediate nodes** is the set of k information nodes with $n - k$ parity nodes added above, generated according to the simple erasure code $\mathcal{C}_n$. The third is a set of $n(1 + \epsilon)$ coded nodes generated according to the $(n, \Omega(x))$ LT code acting on the set of n intermediate nodes. The process is illustrated in Figure 30.5.

FIGURE 30.5: Structure of a Raptor code

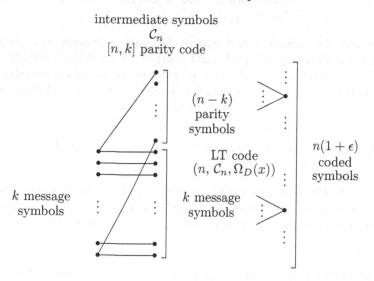

Remark 30.4.2 There is an interesting trivial case for Raptor codes, that of the **precode only (PCO)** codes [1687] where the distribution $\Omega(x) = x$ is used; i.e., for each coded symbol, a random intermediate symbol is chosen and no XOR operations are used. Such a code is equivalent to converting a block code ($\mathcal{C}_n$) into a fountain code. The performance of such codes is actually quite good. The reader is referred to [1687] for more information.

The distribution that will be used for the LT code is a modified and truncated version of the soliton distribution (see [1687]):

$$\Omega_D(x) = \frac{1}{1 + \mu} \left(\mu x + \sum_{i=2}^{D} \frac{x^i}{(i-1)i} + \frac{x^{D+1}}{D} \right)$$

where $D = \lceil 4(1+\epsilon)/\epsilon \rceil$ for ϵ a positive number and $\mu = (\epsilon/2) + (\epsilon/2)^2$. That this distribution sums to unity is easy to establish using the soliton distribution. Notice that the average degree of a coded node is $O(1)$ for D a constant. Recall the average degree of a coded node for the soliton distribution was $O(\ln(k/\delta))$.

The following result is crucial to the development of Raptor codes.

Lemma 30.4.3 ([1687, Lemma 4]) *There exists a positive real number c (depending on ϵ) such that, with an error probability of at most e^{-cn}, any set of $(1+\epsilon/2)n+1$ output symbols of the LT code $(n, \Omega_D(x))$ is sufficient to recover at least $(1-\delta)n$ of the intermediate symbols via BP decoding where $\delta = (\epsilon/4)/(1+\epsilon)$.*

Remark 30.4.4 Suppose the LT code $(n, \Omega(x))$ is used. The edge distribution of the check nodes of the LT codes then is, as shown in the previous section, $\omega(x) = \Omega'(x)/\Omega'(1)$. Let $a = \Omega'(1)$ be the average degree of a coded node. The probability an intermediate node is the neighbor of exactly ℓ coded nodes is given by the binomial distribution $\binom{N}{n}(a/\ell)^\ell(1-a/n)^{N-\ell}$ where $N = n(1+\epsilon/2)+1$ coded nodes. The intermediate node edge degree distribution then is

$$\iota(x) = \left(1 - \frac{a(1-x)}{n}\right)^{(1+\epsilon/2)n} \approx e^{a(x-1)} = \sum_{j=0}^{\infty} e^{-a}(a^j/j!)x^j$$

where the last approximation is the standard approximation to the binomial by the Poisson distribution. Notice that asymptotically this is also the degree distribution of the intermediate nodes since $\iota(x) = \iota'(x)/\iota'(1)$.

Remark 30.4.5 As in Section 30.2, part of the proof of Lemma 30.4.3 includes the observation that if, for any constant δ, $\iota(1 - \omega(1-x)) < x$ for $x \in [\delta, 1]$, then the probability the decoder cannot recover at least $(1-\delta)n$ or more of the input (intermediate) nodes is less than e^{-cn}. It is straightforward to show for the distributions noted [1687, proof of Lemma 4] that

$$\iota(1 - \omega(1-x)) < e^{-(1+\epsilon/2)\Omega'(1-x)}$$

and it is shown that this is indeed less than x on the interval $[\delta, 1]$, a crucial part of the proof of Lemma 30.4.3.

Remark 30.4.6 The requirements of the code C_n of the Raptor code are quite minimal and not specified strictly. It is assumed of such codes [1687] that, for a fixed real number ϵ and for every length n, there is a linear code C_n with the properties:

(i) The rate of C_n is $(1 + \epsilon/2)/(1 + \epsilon)$.

(ii) The BP decoder can decode C_n on a BEC with erasure probability

$$\delta = (\epsilon/4)(1 + \epsilon) = (1 - R)/2$$

with $O(n \log(1/\epsilon))$ arithmetic operations. Recall that the capacity of the BEC is $1 - R$ and the conditions on this code are not severe.

Examples of such codes are Tornado codes of Section 30.2.

Theorem 30.4.7 ([1687, Theorem 5]) *Let ϵ be a positive real number, k an integer, $D = \lceil 4(1+\epsilon)/\epsilon \rceil$, $R = (1+\epsilon/2)/(1+\epsilon)$, $n = \lceil k/(1-R) \rceil$, and let C_n be a code with the above properties. Then the Raptor code with parameters $(k, C_n, \Omega_D(x))$ has space consumption $1/R$, overhead ϵ, and a cost of $O(\log(1/\epsilon))$ with respect to BP decoding of both the precode and the LT code.*

Remark 30.4.8 It is noted that a similar construction to Raptor codes was discovered independently by Maymounkov [1358].

Inherent in Theorem 30.4.7 is the notion of a reliable decoder and hence the probability of error of the Raptor code is inversely polynomial in k. However, if the C_n used has an exponentially small error probability in k, the Raptor code that uses it will also.

It is also noted that the codes of Theorem 30.4.7 have a constant overhead while the LT codes have a vanishingly small overhead with k. This is a consequence of our insistence on a linear decoding complexity. As noted earlier, the Raptor codes achieve linear decoding complexity by requiring only a fraction of information symbols to be decoded, relying on the linear code C_n to decode the remainder.

Definition 30.4.9 The **input ripple** at a given stage of decoding is defined as the set of all input symbols that are released at that stage.

The expected size of the input ripple is of interest as the decoding algorithm develops. A slight change of notation from Section 30.2 is adopted in that the BP messages sent on the graph edges (from/to intermediate and coded nodes) are 0 and 1, a 0 from a coded symbol to an intermediate neighbor if and only if the intermediate symbol cannot be recovered at that point and from an intermediate symbol to a neighbor coded symbol if its value is not yet determined.

An overview of the analysis is given [1687, Section VIIA]. It makes use of some arguments of Tornado codes. Let

$$p_{i+1} = \omega\big(1 - \iota(1 - p_i)\big)$$

denote the probability an edge of the graph carries a 1 value from a coded node to an attached intermediate node, where $p_0 = \delta$, the design erasure probability of the code C_n. As in Section 30.2, this assumes independent messages (a tree to the depth of interest) where $\omega(x)$ and $\iota(x)$ are the edge distributions of the coded and intermediate edges, respectively. The reference [1294] approaches this problem from a general point of view, from which the above result follows easily.

Let u_i be the probability that an intermediate symbol is recovered at round i (which is the probability the corresponding edge from a coded symbol to that intermediate symbol carried a 1). If the intermediate symbol node had a degree of d, the probability it is recovered at round i, given the degree d, is 1 minus the probability all zeros were transmitted to it, or $1 - (1 - p_i)^d$; averaging over the distribution $\iota(x)$ this probability is $u_i = 1 - \iota(1 - p_i) = 1 - e^{-\alpha p_i}$. It follows that $p_i = -\ln(1 - u_i)/\alpha$ where α is the average degree of an intermediate symbol. Thus the fraction of symbols recovered by round $(i + 1)$ is

$$u_{i+1} = 1 - e^{-\alpha\omega(u_i)}.$$

So if x is the fraction of intermediate symbols recovered by some round, then $1 - e^{-\alpha\omega(x)}$ is the fraction recovered by the next stage. Therefore the fraction recovered during this last stage is

$$1 - x - e^{-\alpha\omega(x)} = 1 - x - e^{-\Omega'(x)(1+\epsilon)}.$$

The latter equality follows from two observations: $\Omega'(1)$ is the average degree of a check node which also equals $\alpha/(1 + \epsilon)$, and $\omega(x) = \Omega'(x)/\Omega'(1)$ by Remark 30.4.4.

The size of the input ripple at any stage of the decoding algorithm is a discrete random variable, and it is desired to choose the coded distribution $\Omega(x)$ so that this size never goes below 1 at any stage of decoding. It is suggested [1687] that the expected size of the input ripple when a fraction x of the inputs has been recovered be kept at least as large as

$c\sqrt{(1-x)k}$ for some positive constant c, which, from the above argument leads to designing distributions $\Omega(x)$ such that

$$1 - x - \exp\left(-\Omega'(x)(1+\epsilon)\right) \geq c\sqrt{(1-x)/k}$$

or equivalently

$$\Omega'(x) \geq -\ln\left(1 - x - c\sqrt{(1-x)/k}\right)\bigg/(1+\epsilon) \qquad (30.4)$$

for $x \in [0, 1-\delta]$.

The reader is referred to the works of [1687, Section VIIA] and [1084, 1311] for further details of the analysis.

Example 30.4.10 As with the Tornado codes of Section 30.2 a linear programming technique is suggested to design distributions that satisfy (30.4). Impressive results on Raptor codes are given in [1687, Section VII A, Table 1]. One of these examples is given here for $k = 100,000$:

$$\Omega(x) = 0.006495x + 0.495044x^2 + 0.168010x^3 + 0.067900x^4 +$$
$$0.089209x^5 + \cdots + 0.010777x^{67}$$

The code has an overhead of 0.028 and an average coded node degree of 5.85. The δ value is 0.01 and it is noted that for small values of d the distribution is approximately soliton, i.e., $\Omega_d \approx 1/(d(d-1))$.

Inactivation Decoding of LT Codes

It has been observed that the BP decoding algorithm, while very efficient computationally, is not ML and may fail due to a lack of coded nodes of degree 1 at some stage of decoding. GE decoding is ML but expensive computationally, typically $O(k^3)$. Inactivation decoding is a hybrid of the two techniques, using BP decoding as far as it can go and then using GE on a hopefully small set of variables. This typically occurs towards the end of the process for a well designed code. The algorithm will be discussed for LT codes, assuming k information nodes and n coded nodes. The description of the algorithm below is slightly unconventional compared to that in the literature.

Algorithm 30.4.11

It as assumed that each node is associated with a register which can be updated with information passed between nodes as the algorithm runs. The BP algorithm runs on the code bipartite graph until there is no coded node of degree 1. If there is a coded node u of degree 2 at this stage, attached to information nodes v and w, and the register associated with u, r_u contains the symbol b_u, then add a variable x_1 to the register r_v of v and $x_1 \oplus b_u$ to r_w. These operations are recorded in the register in some suitable manner. The variable x_1 is referred to as an **inactivation variable**. Continue the algorithm by removing all edges incident with the visited nodes u, v, and w. Each time in the decoding process there is no degree 1 coded node available, a sufficient number of inactivation variables are introduced and the information node registers and attached coded node registers are updated and the process continues. At the end of the process, some of the left nodes will contain inactivation variables in their 'solutions'. If ℓ inactivation variables have been introduced, a matrix equation can be established relating the left nodes containing inactivation variables and the originally received coded symbols. Solving this set of equations is achieved by using GE on an $\ell \times \ell$ matrix. The information symbols are then resolved. The complexity of this stage is $O(\ell^3)$ and typically ℓ is quite small. The result is an ML algorithm with a complexity that is considerably reduced compared to ML on the originally received coded symbols.

The above algorithm may be viewed as an equivalent matrix reduction on the $n \times k$ matrix G that relates the coded and information symbols. This matrix can first be subjected to row and column permutations (no other arithmetic operations) to place as large an identity matrix as possible in the upper left corner of the matrix. Suppose the process results in a $\tau \times \tau$ identity matrix. The $(n - \tau) \times \tau$ matrix under the identity I_τ matrix, say A, can be zeroed using row and XOR operations. The $(n-\tau) \times (k-\tau)$ matrix to the right of A, say B, which, due to the previous operations has higher density than the original relatively sparse matrix, can then be solved by GE. This part relates to the inactivation variables in the previous description. The matrix above B can then be used to give all information symbols. It is clear that if the original matrix is of full rank, this algorithm yields the solution and is ML.

There are numerous studies on inactivation decoding including [215, 216, 1488, 1690].

Systematic Encoding of Raptor Codes

There are situations where it is advantageous to use systematic Raptor codes, for example where a client does not have the correct version of the decoding software. The obvious solution is to simply broadcast the information packets in some order among the coded packets. An argument is given in [1688, Section 6], that this approach is not a good solution as it results in a significant performance loss and increased overhead. The relationship between the LT distribution used to achieve low overhead is delicate and the systematic coding procedure quite involved. The argument below is based on [1687, Section VIII] and [1688, Section 6].

It is assumed the Raptor code is a $(k, \mathcal{C}_n, \Omega(x))$ code and that the code $\mathcal{C}_n$ has the $k \times n$ binary generator matrix G. All vectors are row vectors—some are binary and some are vectors of symbols, each symbol viewed as a vector of bits that allows the XOR of symbols to be formed. The systematic coding procedure for Raptor codes involves forming, from the k information symbols denoted by $\mathbf{x}_1, \mathbf{x}_2, \ldots, \mathbf{x}_k$, k **processed symbols** $\mathbf{y}_1, \mathbf{y}_2, \ldots, \mathbf{y}_k$, n **intermediate symbols** $\mathbf{u}_1, \mathbf{u}_2, \ldots, \mathbf{u}_n$, and from these N **coded symbols** $\mathbf{z}_1, \mathbf{z}_2, \ldots, \mathbf{z}_N$. The corresponding vectors of these sets of symbols are denoted $\mathbf{x}, \mathbf{y}, \mathbf{u}$, and $\mathbf{z}$ respectively. The processing of these layers of symbols is given in Algorithm 30.4.12 from which it will be clear that the k information symbols will appear among the N coded symbols.

As a preliminary, the $N \times n$ matrix S is constructed as follows. Let each row $\mathbf{v}$ of S be a binary row of weight corresponding to a sampling of the distribution $\Omega(x)$, a row being of weight d if the sample of $\Omega(x)$ was d, and the d nonzero elements of $\mathbf{v}$ chosen uniformly at random among the $\binom{n}{d}$ possibilities. The procedure is repeated N times corresponding to the N rows of S and the N coded symbols produced. The following algorithm is from Algorithms 7 and 11 of [1687, Section VIIIC]. It consists of a slight variation of the procedure described there.

Algorithm 30.4.12

Step 1: Sample the distribution $\Omega(x)$, $N = k(1 + \epsilon)$ times and produce an $N \times n$ binary matrix S with rows $\mathbf{v}_1, \mathbf{v}_2, \ldots, \mathbf{v}_N, \mathbf{v}_i \in \mathbb{F}_2^n$.

Step 2: Assuming the $N \times k$ binary matrix SG^{T} is of full rank, over $\mathbb{F}_2$, find indices $i_1, i_2, \ldots, i_k$ such that the corresponding rows of SG^{T} form the nonsingular matrix R with inverse R^{-1}. Thus $R = AG^{\mathsf{T}}$ where A, a submatrix of S, is formed by rows $\mathbf{v}_{i_1}, \ldots, \mathbf{v}_{i_k}$.

Step 3: Compute the k processed symbols $\mathbf{y} = \mathbf{y}_1 \cdots \mathbf{y}_k$ by $\mathbf{y}^{\mathsf{T}} = R^{-1} \mathbf{x}^{\mathsf{T}}$.

Step 4: Compute the intermediate symbols $\mathbf{u} = \mathbf{u}_1 \cdots \mathbf{u}_n$ as $\mathbf{u}^{\mathsf{T}} = G^{\mathsf{T}} \mathbf{y}^{\mathsf{T}}$.

Step 5: Compute the N coded symbols $\mathbf{z}_i = \mathbf{v}_i \cdot \mathbf{u}^{\mathsf{T}}$, $i = 1, 2, \ldots, N$, with the LT code $(n, \Omega(x))$.

It is clear that in Algorithm 30.4.12 the coded symbols $\mathbf{z}_{i_1} = \mathbf{x}_1, \mathbf{z}_{i_2} = \mathbf{x}_2, \ldots, \mathbf{z}_{i_k} = \mathbf{x_k}$ since $R = AG^{\mathsf{T}}$ and $R\mathbf{y}^{\mathsf{T}} = RR^{-1}\mathbf{x}^{\mathsf{T}} = \mathbf{x}^{\mathsf{T}}$. Thus among the LT encoded symbols are the information symbols.

Standardized Raptor Codes

Raptor codes have been incorporated into a large number of commercial standards that include mobile applications and broadcast/multicast file delivery to the more high end streaming applications. Two forms of Raptor code design have been designed to meet the requirements of these systems, the code Raptor R10 (Raptor version 10) and RaptorQ; an excellent detailed discussion of these codes is given in [1691, Chapter 3]. Both of these codes use inactivation decoding and are systematic and attempt to achieve the performance of a random binary fountain code with greater efficiency.

The R10 code supports the range of source symbols of between 4 and 8192 with up to 65,536 coded blocks with a small overhead and error probability of less than 10^{-6}. It is noted the design of the LT distribution is a challenging task for such an application. The design discussion of [1691] includes a detailed look at the overhead achieved and the number of inactivations expected.

The RaptorQ or RQ code supports up to 56,403 information symbols and up to 16,777,216 coded symbols, figures derived from the applications intended. It achieves a performance of a random fountain code over the finite field $\mathbb{F}_{256}$, on the order of an error probability of 10^{-6}. The notion of using non-binary alphabets for fountain codes is also considered [1691, Section 3.3.1]. Further information is contained in the RFC's [1389, 1692].

Other Comments

The term **online codes** has been used [1358] synonymously with the term fountain codes. More recently [368] the term has been introduced to mean a fountain code that adapts its code distribution in response to learning of the 'state' of the decoder. Thus as the decoding proceeds, such a decoder can offer superior performance. Of course this decoder would require the characterization of the state and the transmission of this information back to the encoder, in contrast to the multicast situation. Using the dynamics of random graph processes, it is shown in [368] how such a strategy can result in a considerable lowering of the overhead required for a given performance compared to the usual fountain codes.

The performance of fountain and in particular Raptor codes over the past decade has been spectacular, essentially achieving capacity on a BEC with linear encoding and decoding complexity. An essential feature of such codes is that it is unnecessary for the encoder to learn the characteristics of the channel to achieve such performance. It is natural to consider the performance of such codes on other channels such as the BSC and AWGN using an appropriate form of BP decoding. The issue is addressed in [688] where in particular the fraction of nodes of degree 2 is shown to be of importance for the use of Raptor codes in error-correction on the BSC. Other studies are available in the literature but are not commented on here.

Numerous other works on interesting questions related to these erasure-correcting codes include [33, 42, 215, 216, 1293, 1389, 1488, 1589, 1684, 1685, 1686, 1690, 1692, 1744].

Chapter 31

Codes for Distributed Storage

Vinayak Ramkumar

Indian Institute of Science

Myna Vajha

Indian Institute of Science

S. B. Balaji

Qualcomm

M. Nikhil Krishnan

University of Toronto

Birenjith Sasidharan

Govt. Engineering College, Barton Hill

P. Vijay Kumar

Indian Institute of Science

The traditional means of ensuring reliability in data storage is to store multiple copies of the same file in different storage units. Such a replication strategy is clearly inefficient in terms of storage overhead. By **storage overhead** we mean the ratio of the amount of data stored pertaining to a file to the size of the file itself. Modern-day data centers can store several Exabytes of data, and there are enormous costs associated with such vast amounts of storage not only in terms of hardware, software and maintenance, but also in terms of the power and water consumed. For this reason, reduction in storage overhead is of paramount importance. An efficient approach to reduce storage overhead, without compromising on reliability, is to employ erasure codes instead of replication. When an $[n, k]$ erasure code is employed to store a file, the file is first broken into k fragments. To this, $n - k$ redundant fragments are then added, and the resultant n fragments are stored across n storage units. Thus the storage overhead incurred is given by $\frac{n}{k}$. In terms of minimizing storage overhead, a class of codes known as maximum distance separable (MDS) codes, of which Reed–Solomon (RS) codes are the principal example, are the most efficient. MDS codes have the property that the entire file can be obtained by connecting to any k storage units, which means that an MDS codes can handle the failure of $n - k$ storage units without suffering data loss. RS codes are widely used in current-day **distributed storage systems**. Examples include, the $[9, 6]$ RS code in the Hadoop Distributed File System with Erasure Coding (HDFS-EC), the $[14, 10]$ RS code employed in Facebook's f4 BLOB storage, and the $[12, 8]$ RS code employed in Baidu's Atlas cloud storage [489].

A second important consideration governing choice of the erasure code is the ability of the code to efficiently handle the failure of an individual storage unit, as such failures are a common occurrence [1568, 1624]. From here on, we will interchangeably refer to a storage unit as a **node**, as we will, at times, employ a graphical representation of the code. We use the term **node failure** to encompass not only the actual physical failure of a storage unit, but also instances when a storage unit is down for maintenance or else is unavailable on account of competing serving requests. Thus, the challenge is to design codes which are efficient not only with respect to storage overhead, but which also have the ability to efficiently handle the repair of a failed node. The nodes from which data is downloaded for the repair of a failed node are termed **helper nodes**, and the number of helper nodes contacted is termed the **repair degree**. The amount of data downloaded to a replacement node from the helper nodes during node repair is called the **repair bandwidth**.

If the employed erasure code is an $[n, k]$ MDS code, then the conventional approach towards repair of a failed node is as follows. A set of k helper nodes would be contacted and the entire data from these k nodes would then be downloaded. This would then permit reconstruction of the entire data file and hence, in particular, enable repair of the failed node. Thus under this approach, the repair bandwidth is k times the amount of data stored in the replacement node, which is clearly wasteful of network bandwidth resources and can end up clogging the storage network.

This is illustrated in Figure 31.1 in the case of Facebook's $[14, 10]$ RS code. If each node stores 100MB of data, then the total amount of data download needed for the repair of a node storing 100MB from the 10 helper nodes equals 1GB. Thus in general, in the case of

FIGURE 31.1: Conventional repair of a failed node in $[14, 10]$ RS code

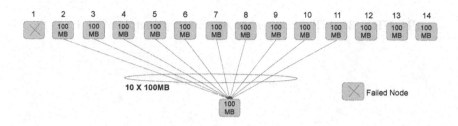

an $[n, k]$ MDS code, the repair degree equals k and repair bandwidth is the entire file size which is k times the amount of data stored in a node. As will be seen, under conventional node repair, MDS codes are inefficient both in terms of repair degree and repair bandwidth.

Coding theorists responded to the challenge of node repair by coming up with two classes of codes, known as *ReGenerating Codes* (RGCs) [541] and *Locally Recoverable Codes* (LRCs) [830]. The goal of an RGC is to reduce the repair bandwidth, whereas with an LRC, one aims to minimize the repair degree. Recently much progress has also been made in coming up with more efficient approaches for the repair of RS codes [882, 1660]. This chapter will primarily focus on RGCs and LRCs. We will also briefly discuss novel methods of RS repair, as well as a class of codes known as *Locally ReGenerating Codes* (LRGCs) that combine the desirable properties of RGCs and LRCs.

A brief overview of the topic of codes for distributed storage, and from the perspective of the authors, is presented here. For further details, the readers are referred to surveys such as can be found in [96, 488, 542, 1241, 1283]. Distributed systems related to algebraic geometry codes are described in Section 15.9. We would like to thank the editors of *Advanced Computing and Communications* for permitting the reuse of some material from [1165].

We begin with a brief primer on RS codes.

31.1 Reed–Solomon Codes

We provide here a brief description of an $[n, k]$ Reed–Solomom (RS) code.[1] Let $a_i \in \mathbb{F}_q$, $i = 0, 1, \ldots, k - 1$, represent the k message symbols. Let $x_0, x_1, \ldots, x_{n-1}$ be an arbitrary collection of n distinct elements from $\mathbb{F}_q$ and the polynomial f be defined by

$$f(x) = \sum_{i=0}^{k-1} a_i \prod_{\substack{j=0 \\ j \neq i}}^{k-1} \frac{x - x_j}{x_i - x_j} := \sum_{i=0}^{k-1} b_i x^i.$$

It follows that f is the interpolation polynomial of degree $k - 1$ that satisfies

$$f(x_i) = a_i \text{ for } 0 \leq i \leq k - 1.$$

The n code symbols in the RS codeword corresponding to message symbols $\{a_i\}_{i=0}^{k-1}$ are precisely the n ordered symbols $(f(x_0), f(x_1), \ldots, f(x_{n-1}))$; see Figure 31.2 for an illustra-

---

[1]Throughout this chapter the term 'Reed–Solomon codes' will include what are often called 'generalized Reed–Solomon codes', defined specifically in Section 31.5.

tion. Of these, the first k symbols are message symbols while the remaining are redundant symbols.

FIGURE 31.2: Illustrating the operating principle of an RS code

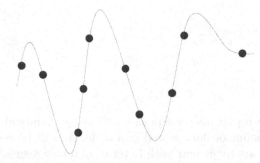

The RS code derives its MDS property from the fact that the polynomial f can be determined from the knowledge of any k evaluations simply by solving the following nonsingular set of k equations in the k unknown coefficients $\{b_i\}_{i=0}^{k-1}$:

$$\begin{bmatrix} f(x_{i_1}) \\ f(x_{i_2}) \\ \vdots \\ f(x_{i_k}) \end{bmatrix} = \underbrace{\begin{bmatrix} 1 & x_{i_1} & \cdots & x_{i_1}^{k-1} \\ 1 & x_{i_2} & \cdots & x_{i_2}^{k-1} \\ \vdots & \vdots & \vdots & \vdots \\ 1 & x_{i_k} & \cdots & x_{i_k}^{k-1} \end{bmatrix}}_{\substack{\text{a Vandermonde matrix} \\ \text{and therefore invertible}}} \begin{bmatrix} b_0 \\ b_1 \\ \vdots \\ b_{k-1} \end{bmatrix},$$

where $i_1, \ldots, i_k$ are any set of k distinct indices drawn from $\{0, \ldots, n-1\}$. From this it follows that an RS code can recover from the erasure of any $n-k$ code symbols. The conventional approach of repairing a failed node corresponding to code symbol $f(x_j)$ would be to use the contents of any k nodes (i.e., any k code symbols) to recover the polynomial f and then evaluate the polynomial f at x_j to recover the value of $f(x_j)$.

31.2   Regenerating Codes

Traditional erasure codes are scalar codes, i.e., each code symbol corresponds to a single symbol from a finite field. It turns out, however, that the design of codes with improved repair bandwidth calls for codes that have an underlying vector alphabet. Thus, each code symbol is now a vector. The process of moving from a scalar symbol to a vector symbol is referred to as **sub-packetization**. The reason for this choice of terminology is that we view a scalar symbol over the finite field $\mathbb{F}_{q^\alpha}$ of size q^α as being replaced by a vector of α symbols drawn from $\mathbb{F}_q^\alpha$. An example is presented in Figure 31.3.

FIGURE 31.3: Showing how breaking up a single scalar symbol into two smaller symbols helps improve the repair bandwidth. This breaking up of a symbol is referred to as sub-packetization. The sub-packetization level equals 2 here.

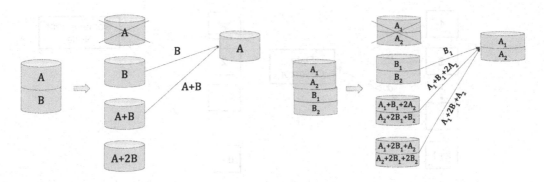

31.2.1 An Example of a Regenerating Code and Sub-Packetization

In Figure 31.3, the setup on the left represents a $[4, 2]$ MDS code. The symbols stored in the 4 nodes are, respectively, $A, B, A + B, A + 2B$, all drawn from a finite field of suitable size, for example $\mathbb{F}_{3^2}$. To repair the failed node (disk) 1 that previously stored A, we have to download 2 symbols. Consider next the setup to the right. Here the sub-packetization level is 2; each symbol drawn from $\mathbb{F}_{3^2}$ is replaced by 2 'half-symbols', each drawn from $\mathbb{F}_3$. Thus A is replaced by A_1, A_2 and B by B_1, B_2. Note that if the data stored in the remaining two parity nodes is as shown in the figure, then node 1 can be repaired by downloading 3 half-symbols in place of two full symbols, thereby achieving a reduction in repair bandwidth. Note however that we have contacted all the remaining nodes, 3 in this case, as opposed to $k = 2$ in the case of the MDS code. Thus while regenerating codes reduce the repair bandwidth, they do in general, result in increased repair degree.

31.2.2 General Definition of a Regenerating Code

Definition 31.2.1 ([541]) Given a file of size B, as measured in number of symbols over $\mathbb{F}_q$, an $\{(n, k, d), (\alpha, \beta), B, \mathbb{F}_q\}$ **regenerating code** $\mathcal{C}$ stores data pertaining to this file across n nodes, where each node stores α symbols from the field $\mathbb{F}_q$. The code $\mathcal{C}$ is required to have the following properties (see Figure 31.4):

(i) **Data Collection:** The entire file can be obtained by downloading contents of any k nodes.

(ii) **Node Repair:** If a node fails, then the replacement node can connect to any subset of d helper nodes, where $k \le d \le n - 1$, download β symbols from each of these helper nodes, and repair the failed node.

With respect to the definition above, there could be two interpretations as to what it means to repair a failed node. One interpretation of repair is that the failed node is replaced by a replacement node which, upon repair, stores precisely the same data as did the failed node. This is called **Exact Repair** (E-R). This is not however, a requirement and there is an alternative, more general definition of node repair. Under this more general interpretation, the failed node is replaced by a replacement node in such a way that the resultant collection of n nodes once again satisfies the requirements of a regenerating code. Such a replacement

FIGURE 31.4: Data collection and node repair properties of an $\{(n, k, d),\ (\alpha, \beta),\ B,\ \mathbb{F}_q\}$ regenerating code

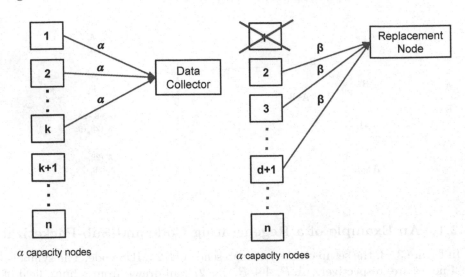

of a failed node is termed as **Functional Repair** (F-R). Clearly, E-R is a special case of F-R. E-R is preferred in practice as it simplifies management of the storage network.

It is easily verified that the storage overhead of a regenerating code equals $\frac{n\alpha}{B}$, the repair degree is d, and the repair bandwidth is $d\beta$. The parameter β is typically much smaller than α, resulting in savings in repair bandwidth. The parameter α is called the **sub-packetization level** of the regenerating code, and having low sub-packetization level is preferred from a practical perspective; see [1825]. A regenerating code is said to possess the **optimal-access** property if no computation is required at the helper nodes during node repair. If repair of a failed node can be done without any computation at either the helper nodes or the replacement node, then the regenerating code is said to have the **Repair-By-Transfer** (RBT) property.

31.2.3 Bound on File Size

It turns out by using the cut-set bound of network coding one can show that the size B of the file encoded by a regenerating code must satisfy the following inequality [541]:

$$B \leq \sum_{i=0}^{k-1} \min\{\alpha, (d-i)\beta\}, \tag{31.1}$$

which we will refer to as the **Cut-Set Bound**. This bound holds for F-R (and hence also for E-R).

A regenerating code is said to be **optimal** if the Cut-Set Bound is satisfied with equality and if, in addition, decreasing either α or β would cause the bound to be violated. There are many flavors of optimality in the sense that for a fixed (B, k, d), inequality (31.1) can be met with equality by several different pairs (α, β). The parameter α determines the storage overhead $\frac{n\alpha}{B}$, whereas β is an indicator of normalized repair bandwidth $\frac{d\beta}{B}$. The various pairs of (α, β) which satisfy the Cut-Set Bound with equality present a tradeoff between storage overhead and normalized repair bandwidth. The existence of codes achieving the Cut-Set Bound for all possible parameters is known again from network coding in the F-R case. Thus in the F-R case, the storage-repair bandwidth (S-RB) tradeoff is fully characterized. An example of this tradeoff is presented for the case $(B = 5400, k = 6, d = 10)$. The discussion above suggests that the tradeoff is a discrete collection of optimal pairs (α, β). However, in the plot, these discrete pairs are connected by straight lines, giving rise to the piecewise linear graph in Figure 31.5. The straight line connections have a physical interpretation and correspond to a space sharing solution to the problem of node repair; the reader is referred to [1658, 1804] for details.

FIGURE 31.5: S-RB tradeoff for F-R with $(B = 5400, k = 6, d = 10)$

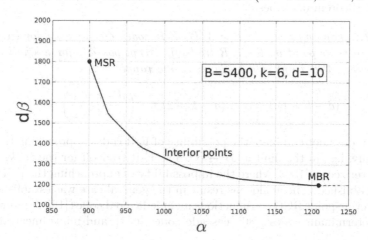

31.2.4   MSR and MBR Codes

There are two extreme choices of (α, β) in this tradeoff. In the first, α is the least possible, as a result of which $\alpha \leq (d - k + 1)\beta$, thereby forcing $B = k\alpha$. Under this condition β is minimized by setting $\alpha = (d - k + 1)\beta$. Regenerating codes with this flavor of optimality are called **Minimum Storage Regenerating** (MSR) codes. In the case of an MSR code, the size of the code is given by $q^B = q^{k\alpha} = (q^\alpha)^k = (q^\alpha)^{(n - d_{\min} + 1)}$, where $d_{\min}$ is the minimum distance. It follows that MSR codes achieve the Singleton Bound on code size over the vector alphabet $\mathbb{F}_q^\alpha$, and hence MSR codes belong to the class of MDS codes. At the other extreme, we have the case $\alpha \geq d\beta$, and file size $B = \sum_{i=0}^{k-1}(d - i)\beta = kd\beta - \binom{k}{2}\beta$. Here α is minimized by setting $\alpha = d\beta$. Regenerating codes having these parameters are termed as **Minimum Bandwidth Regenerating** (MBR) codes. MBR codes have the minimum possible repair bandwidth but are not MDS. The storage overhead of an MBR code can be shown to be bounded below by 2, whereas MSR codes can have storage overhead arbitrarily close to 1.

Several constructions of E-R MSR [832, 1243, 1569, 1578, 1619, 1621, 1764, 1774, 1824, 1925, 1926] and MBR [1164, 1263, 1569, 1570] codes can be found in the literature. A selected subset of these constructions is discussed here.

31.2.5   Storage Bandwidth Tradeoff for Exact-Repair

The Cut-Set Bound (31.1) may be not achievable under E-R, and hence the S-RB tradeoff for E-R may vary from that for F-R. For brevity, we will refer to the S-RB tradeoff in the case of E-R as the *E-R tradeoff* and similarly *F-R tradeoff* in the case of functional repair. Since E-R codes are a special case of F-R codes, the file size B in the case of E-R codes cannot exceed (31.1). Constructions of exact-repair MSR and MBR codes are known and hence the E-R tradeoff coincides with the F-R tradeoff at the MSR and MBR points. Points on the S-RB tradeoff, other than the MSR and MBR points, are referred to as interior points. In [1658], it was shown that it is impossible to achieve, apart from one exceptional subset, any interior point of the F-R tradeoff using E-R. The exceptional set of interior points corresponds to a small region of the F-R tradeoff curve, adjacent to the MSR point, as the following theorem describes.

Theorem 31.2.2 *For any given* $(n, k \geq 3, d)$, *E-R codes do not exist for* (α, β, B) *corresponding to an interior point on the F-R tradeoff, except possibly for a small region in the neighborhood of the MSR point with* α *values in the range*

$$(d - k + 1)\beta \leq \alpha \leq (d - k + 2)\beta - \left(\frac{d - k + 1}{d - k + 2} \right) \beta.$$

This theorem does not eliminate the possibility of E-R codes approaching the F-R tradeoff asymptotically, i.e., in the limit as $B \to \infty$. The E-R tradeoff for the $(n, k, d) = (4, 3, 3)$ case was characterized in [1803] where the impossibility of approaching the F-R tradeoff under E-R was established. The analogous result in the general case was established in [1620]. Examples of interior-point RGC constructions include layered codes [1804], improved layered codes [1652], determinant codes [680], cascade codes [681], and multi-linear-algebra-based codes [657].

31.2.6   Polygon MBR Codes

We present a simple construction of an MBR code [1570] possessing the RBT property through an example known as the Pentagon MBR code. The parameters of the example construction are

$$\{(n = 5, k = 3, d = 4), (\alpha = 4, \beta = 1), B = 9, \mathbb{F}_2\}.$$

Note that as required of an MBR code,

$$B = kd\beta - \binom{k}{2}\beta = 9 \quad \text{and} \quad \alpha = d\beta = 4.$$

The file to be stored consists of the 9 symbols $\{a_1, a_2, \ldots, a_9\}$. We first generate a parity symbol a_P given by $a_P = a_1 + a_2 + \cdots + a_9 \pmod{2}$. Next, set up a complete graph with $n = 5$ nodes, i.e., form a fully-connected pentagon. The pentagon has $\binom{5}{2} = 10$ edges, and we assign each of the 10 code symbols $\{a_i \mid 1 \leq i \leq 9\} \cup \{a_P\}$ to a distinct edge; see Figure 31.6. Each node stores all the symbols appearing on edges incident on that node, giving $\alpha = 4$.

Data Collection: The data collection property requires that the entire data file be recoverable by connecting to any $k = 3$ nodes. It can be easily seen that any collection of 3 nodes contains 9 distinct symbols from $\{a_i \mid 1 \leq i \leq 9\} \cup \{a_P\}$, which is sufficient to recover the entire file.

FIGURE 31.6: An illustration of the pentagon MBR code

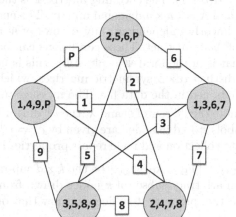

Node Repair: Now suppose one of the nodes failed. The repair is accomplished by downloading, from each of the remaining $d = 4$ nodes, the code symbol it has in common with the failed node. Since each helper node passes on a single symbol to aid in node repair, we have $\beta = 1$.

This construction can be generalized by replacing the pentagon by a polygon with n vertices. The parameters of the general construction are

$$\left\{ (n, k, d = n - 1),\ (\alpha = n - 1, \beta = 1),\ B = k(n - 1) - \binom{k}{2},\ \mathbb{F}_q \right\},$$

where $q = O(n^2)$. In the initial step of this construction, an $\left[\binom{n}{2}, B \right]$ MDS code is used to generate the code symbols, and these code symbols are then assigned to distinct edges of a complete graph on n vertices. Each node stores the symbols appearing on incident edges, resulting in $\alpha = n - 1$. The data collection and node-repair properties in the case of this general construction are easily verified.

31.2.7 The Product-Matrix MSR and MBR Constructions

The Product-Matrix (PM) framework introduced in [1569] allows for the construction of both MBR and MSR codes. For codes constructed under this framework, the parameter β is the smallest value possible, i.e., is always equal to 1. The framework allows for the construction of MBR code with $\alpha = d$ for all parameter sets (n, k, d). In the MSR case, the framework yields constructions for all (n, k, d) with $\beta = 1$ and $d \geq 2k - 2$.

The mathematical setting under this framework is as follows. There is an $n \times \alpha$ code matrix C whose i^{th} row, denoted by c_i^{T}, contains the α symbols stored in node i. The code matrix C is the product of an $n \times d$ encoding matrix Ψ and a $d \times \alpha$ message matrix M, i.e., $C = \Psi M$. The entries of the encoding matrix Ψ are independent of the file to be stored. The message matrix M contains the B symbols in the file, with some symbols repeated. Let ψ_i^{T} denote the ith row of Ψ; then the content of node i is given by $c_i^{\mathsf{T}} = \psi_i^{\mathsf{T}} M$.

31.2.7.1 PM-MSR Codes

We will begin by identifying the encoding and message matrices that will yield an MSR code with $\beta = 1$ [1569] for the case $d = 2k - 2$. This construction can be extended to

Concise Encyclopedia of Coding Theory

$d > 2k - 2$ through the mechanism of shortening a code [1569]. When $d = 2k - 2$, we have $\alpha = k-1$, $d = 2\alpha$, and $B = \alpha(\alpha+1)$. The encoding matrix Ψ is then given by $\Psi = \begin{bmatrix} \Phi & \Lambda\Phi \end{bmatrix}$, where Φ is a $n \times \alpha$ matrix and Λ is a $n \times n$ diagonal matrix. The entries of Ψ are chosen such that any d rows of Ψ are linearly independent, any α rows of Φ are linearly independent, and the diagonal entries of Λ are distinct. These conditions can be met by picking Ψ to be a Vandermonde matrix that is of the form $\Psi = \begin{bmatrix} \Phi & \Lambda\Phi \end{bmatrix}$ (this is not difficult).

Let S_1 and S_2 be two distinct $\alpha \times \alpha$ symmetric matrices, which together contain all the $B = \alpha(\alpha + 1)$ elements contained in the data file. The message matrix M is then given by $M = \begin{bmatrix} S_1 & S_2 \end{bmatrix}^\mathsf{T}$. Let ϕ_i^T denote the ith row of Φ and λ_i the ith diagonal element of the diagonal matrix Λ. Then the α symbols stored in node i are given by $c_i^\mathsf{T} = \psi_i^\mathsf{T} M = \phi_i^\mathsf{T} S_1 + \lambda_i \phi_i^\mathsf{T} S_2$. We will now show that the data collection and node repair properties hold for this construction.

Data Collection: Let $\Psi_{\text{DC}} = \begin{bmatrix} \Phi_{\text{DC}} & \Lambda_{\text{DC}}\Phi_{\text{DC}} \end{bmatrix}$ be the $k \times d$ sub-matrix of the $n \times d$ matrix Ψ and corresponding to an arbitrary subset of k nodes drawn from the totality of n nodes. To establish the data collection property it suffices to show that one can recover S_1 and S_2 from the matrix

$$\Psi_{\text{DC}} M = \Phi_{\text{DC}} S_1 + \Lambda_{\text{DC}} \Phi_{\text{DC}} S_2,$$

given Φ_{DC} and Λ_{DC}. The first step is to multiply both sides of the equation on the right by the matrix $\Phi_{\text{DC}}^\mathsf{T}$ to obtain

$$\Psi_{\text{DC}} M \Phi_{\text{DC}}^\mathsf{T} = \Phi_{\text{DC}} S_1 \Phi_{\text{DC}}^\mathsf{T} + \Lambda_{\text{DC}} \Phi_{\text{DC}} S_2 \Phi_{\text{DC}}^\mathsf{T}.$$

Set $P = \Phi_{\text{DC}} S_1 \Phi_{\text{DC}}^\mathsf{T}$ and $Q = \Phi_{\text{DC}} S_2 \Phi_{\text{DC}}^\mathsf{T}$, so that $\Psi_{\text{DC}} M \Phi_{\text{DC}}^\mathsf{T} = P + \Lambda_{DC} Q$. It can be seen that P and Q are symmetric matrices. The $(i, j)^{\text{th}}$ element of $P + \Lambda_{DC} Q$ is $P_{ij} + \lambda_i Q_{ij}$, whereas the $(j, i)^{\text{th}}$ element is $P_{ji} + \lambda_j Q_{ji} = P_{ij} + \lambda_j Q_{ij}$. Since all the $\{\lambda_i\}$ are distinct, one can solve for P_{ij} and Q_{ij} for all $i \neq j$, thus obtaining all the non-diagonal entries of both matrices P and Q. The ith row of P excluding the diagonal element is given by $\phi_i^\mathsf{T} S_1 [\phi_1 \cdots \phi_{i-1} \phi_{i+1} \cdots \phi_{\alpha+1}]$, from which the vector $\phi_i^\mathsf{T} S_1$ can be obtained, since the matrix on the right is invertible. Next, one can form $[\phi_1 \cdots \phi_\alpha]^\mathsf{T} S_1$. Since the matrix on the left of S_1 is invertible, we can then recover S_1. In a similar manner, the entries of the matrix S_2 can be recovered from the non-diagonal entries of Q.

Node Repair: Let f be the index of the failed node and $\{h_j \mid j = 1, \ldots, d\}$ denote the arbitrary set of d helper nodes chosen. The helper node h_i computes $\psi_{h_i}^\mathsf{T} M \phi_f$ and passes it on to the replacement node. Set $\Psi_{\text{rep}} = [\psi_{h_1} \psi_{h_2} \cdots \psi_{h_d}]^\mathsf{T}$. Then the d symbols obtained by the replacement node from the repair node can be aggregated into the form $\Psi_{\text{rep}} M \phi_f$. From the properties of Ψ, it can be seen that Ψ_{rep} is invertible. Thus the replacement node can recover $M \phi_f = \begin{bmatrix} S_1 \phi_f & S_2 \phi_f \end{bmatrix}^\mathsf{T}$. Since S_1 and S_2 are symmetric matrices, $\phi_f^\mathsf{T} S_1$ and $\phi_f^\mathsf{T} S_2$ can be obtained by simply taking the transpose. Now $\phi_f^\mathsf{T} S_1 + \lambda_f \phi_f^\mathsf{T} S_2 = \phi_f^\mathsf{T} M = c_f^\mathsf{T}$, completing the repair process.

31.2.7.2 PM-MBR Codes

For the sake of brevity, we describe here only the structure of the encoding and message matrices under the product-matrix framework that will result in an MBR code with $\beta = 1$. A proof of the data collection and node-repair properties can be found in [1569].

From the properties described earlier in Section 31.2.4 of an MBR code, it follows that $\alpha = d\beta = d$ and $B = kd - \binom{k}{2}$. The $n \times d$ encoding matrix Ψ takes on the form $\Psi = \begin{bmatrix} \Phi & \Delta \end{bmatrix}$, where Φ is an $n \times k$ matrix and Δ is an $n \times (d - k)$ matrix. The matrices are chosen such that any d rows of Ψ are linearly independent and any k rows of Φ are linearly independent. We remark that these requirements can be satisfied by choosing Ψ to be a Vandermonde matrix. This places an $O(n)$ requirement on the field size. The number of message symbols

$B = kd - \binom{k}{2}$ can be expressed in the form $B = \binom{k+1}{2} + k(d-k)$. Accordingly, let the B message symbols be divided into two subsets A_1, A_2 of respective sizes $\binom{k+1}{2}$ and $k(d-k)$. Let S be a $k \times k$ symmetric matrix whose distinct entries correspond precisely to the set A_1, placed in any order. The symbols in A_2 are used to fill up, again in any order, a $k \times (d-k)$ matrix T.

Given the matrices S and T, the $d \times d$ symmetric message matrix M is then formed by setting

$$M = \begin{bmatrix} S & T \\ T^\mathsf{T} & 0 \end{bmatrix}.$$

For the repair of failed node i, the jth helper node sends $\psi_j^\mathsf{T} M \psi_i$.

31.2.8 The Clay Code

In this subsection, we present the construction of an optimal-access MSR code with minimum possible level of sub-packetization, having a coupled-layer structure and which is therefore sometimes referred to as the *Clay code*. The Clay code was first introduced by Ye and Barg [1926] and then independently discovered in [1621]. Work that is very closely related to the Clay code can be found in [1243]. A systems implementation and evaluation of the Clay code appears in [1825]. Clay codes are MSR codes that possess the optimal-access property, have optimal sub-packetization level, and can be constructed over a finite field $\mathbb{F}_q$ of any size $q \geq n$. The parameters of a Clay code construction are of the form

$$\{(n = rt, k = r(t-1), d = n-1), \ (\alpha = r^t, \beta = r^{t-1}), \ B = k\alpha, \ \mathbb{F}_q\},$$

where $q \geq n$, for some $t > 1$ and $r \geq 1$. Each codeword in the Clay code is composed of a total of $n\alpha = r \times t \times r^t$ symbols over the finite field $\mathbb{F}_q$. We will refer to these finite field symbols as code symbols of the codeword. Using the notation $[a : b] = \{a, a+1, \ldots, b\}$, these $n\alpha$ code symbols will be indexed by the three tuple

$$(x, y; \mathbf{z}) \quad \text{where} \quad x \in [0 : r-1], \ y \in [0 : t-1], \ \text{and} \ \mathbf{z} = z_0 z_1 \cdots z_{t-1} \in \mathbb{Z}_r^t.$$

Such an indexing allows us to identify each code symbol with an interior or exterior point of an $r \times t \times r^t$ three-dimensional (3D) cube, and an example is shown in Figure 31.7. For a code symbol $A(x, y; \mathbf{z})$ indexed by $(x, y; \mathbf{z})$, the pair (x, y) indicates the node to which the code symbol belongs, while $\mathbf{z}$ is used to uniquely identify the specific code symbol within the set of $\alpha = r^t$ code symbols stored in that node.

Uncoupled code: As an intermediate step in describing the construction of the Clay code $\mathcal{A}$, we introduce a second code $\mathcal{B}$ that has the same length $n = rt$, the same level $\alpha = r^t$ of sub-packetization, and which possesses a simpler description. For reasons that will shortly become clear, we shall refer to the code $\mathcal{B}$ as the *uncoupled code*. The uncoupled code $\mathcal{B}$ is simply described in terms of a set of $r\alpha$ parity check (p-c) equations. Let $\{B(x, y; \mathbf{z}) \mid x \in [0 : r-1], y \in [0 : t-1], \mathbf{z} \in \mathbb{Z}_r^t\}$ be the $n\alpha$ code symbols corresponding to the code $\mathcal{B}$.

The $r\alpha$ p-c equations satisfied by the code symbols that make up each codeword in code $\mathcal{B}$ are given by

$$\sum_{x=0}^{r-1} \sum_{y=0}^{t-1} \theta_{x,y}^\ell B(x, y; \mathbf{z}) = 0 \quad \text{for all } \ell \in [0 : r-1] \text{ and } \mathbf{z} \in \mathbb{Z}_r^t, \tag{31.2}$$

where the $\{\theta_{x,y}\}$ are all distinct. Such a $\{\theta_{x,y}\}$ assignment can always be carried out with a field of size $q \geq n = rt$. The uncoupled code is also an MDS code as it is formed by simply stacking α codewords, each belonging to the same $[n, k]_q$ MDS code.

FIGURE 31.7: Illustrating the data cube associated with a Clay code having parameters $(r = 2, t = 2)$. Hence $n = rt = 4$ and $\alpha = r^t = 4$. We associate, with each codeword in this example Clay code, a $2 \times 2 \times 4$ data cube. Each of the 16 points within the data cube is thus associated with a distinct code symbol of the codeword. The data cube is made up of $r^t = 4$ planes and each plane is represented by a vector $\mathbf{z}$. The vector $\mathbf{z}$ associated with a plane in the cube is identified by the location of dots within the plane, which are placed at the coordinates $(x, y; \mathbf{z})$ satisfying $x = z_y$.

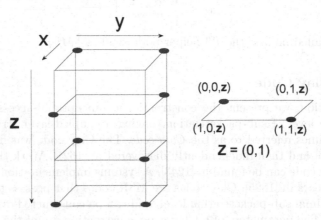

Pairwise Coupling: Next, we introduce a pairing among the $n\alpha$ code symbols (see Figure 31.8) associated with uncoded codeword $B(x, y; \mathbf{z})$. The pair of a code symbol $B(x, y; \mathbf{z})$, for the case $x \neq z_y$ is given by $B(z_y, y; \mathbf{z}(x \to z_y))$ where the notation $\mathbf{z}(x \to z_y)$ denotes the vector $\mathbf{z}$ with the yth component z_y, replaced by x, i.e.,

$$\mathbf{z}(x \to z_y) = z_0 z_1 \cdots z_{y-1} x z_{y+1} \cdots z_{t-1} \in \mathbb{Z}_r^t.$$

The code symbols $B(x, y; \mathbf{z})$, for the case $x = z_y$ will remain unpaired. One can alternately view this subset of code symbols as being paired with themselves, i.e., the pair of $B(x, y; \mathbf{z})$, for the case $x = z_y$, is $B(x, y; \mathbf{z})$ itself.

We now introduce a pairwise transformation, which we will refer to as the *coupling transformation*, which will lead from a codeword

$$\left(B(x, y; \mathbf{z}) \mid x \in [0 : r - 1], y \in [0 : t - 1], \mathbf{z} \in \mathbb{Z}_r^t\right),$$

in the uncoupled code to a codeword

$$\left(A(x, y; \mathbf{z}) \mid x \in [0 : r - 1], y \in [0 : t - 1], \mathbf{z} \in \mathbb{Z}_r^t\right),$$

in the coupled code. For $x \neq z_y$ the pairwise transformation takes on the form

$$\begin{bmatrix} A(x, y; \mathbf{z}) \\ A(z_y, y, \mathbf{z}(x \to z_y)) \end{bmatrix} = \begin{bmatrix} 1 & \gamma \\ \gamma & 1 \end{bmatrix}^{-1} \begin{bmatrix} B(x, y; \mathbf{z}) \\ B(z_y, y; \mathbf{z}(x \to z_y)) \end{bmatrix}, \qquad (31.3)$$

where γ is selected such that $\gamma^2 \neq 1$. This causes the 2×2 linear transformation above to be invertible. For the case $x = z_y$, we simply set

$$A(x, y; \mathbf{z}) = B(x, y; \mathbf{z}).$$

FIGURE 31.8: Figure illustrating the data cube with Clay code symbols on the left and the uncoupled code symbols on the right. A, A^* and B, B^* indicate the paired symbols in the respective cubes.

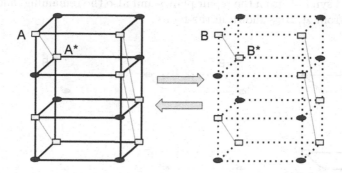

Parity Check Equations: Combining equations (31.3) and (31.2) gives us the p-c equations satisfied by the code symbols of the Clay code $\mathcal{A}$:

$$\sum_{x=0}^{r-1}\sum_{y=0}^{t-1} \theta^\ell_{x,y}\left(A(x,y;\mathbf{z}) + 1_{\{x \neq z_y\}}A(x,y;\mathbf{z}(x \to z_y))\right) = 0, \tag{31.4}$$

for all $\ell \in [0:r-1]$ and all $\mathbf{z} \in \mathbb{Z}_r^t$, where $1_{\{x \neq z_y\}}$ is equal to 1 if $x \neq z_y$ or else takes the value 0.

Optimal-Access Node Repair: We will show how repair of a single node (x_0, y_0) in the Clay code is accomplished by downloading β symbols from each of the remaining $n-1$ nodes. Since no computation is required at a helper node, this will also establish that the Clay code possesses the optimal-access property. The $\beta = r^{t-1}$ symbols passed on by a helper node $(x,y) \neq (x_0, y_0)$ are precisely the subset $\{A(x,y;\mathbf{z}) \mid \mathbf{z} \in P\}$ of the $\alpha = r^t$ symbols contained in that node, in which

$$P = \{\mathbf{z} \in \mathbb{Z}_r^t \mid z_{y_0} = x_0\}$$

identifies an r^{t-1}-sized subset of the totality of r^t planes in the cube. We will refer to P as the set of *repair planes*. Pictorially, these are precisely the planes that have a dot in the location of the failed node; see Figure 31.9.

Consider the r p-c equations given by (31.4) for a fixed repair plane $\mathbf{z}$, i.e., a plane $\mathbf{z}$ belonging to P. The code symbols appearing in these equations either belong to the plane $\mathbf{z}$ itself or else are paired to a code symbol belonging to $\mathbf{z}$. For $y \neq y_0$, the pair of a code symbol $A(x,y,\mathbf{z})$ lying in $\mathbf{z}$ is a second code symbol $A(z_y, y, \mathbf{z}(x \to z_y))$ that does not belong to $\mathbf{z}$ but lies, however, in a different plane $\hat{\mathbf{z}}$ that is also a member of P. It follows that, for $y \neq y_0$, the replacement for failed node (x_0, y_0) has access to all the code symbols with $y \neq y_0$ that appear in (31.4). For the case $y = y_0$ with $x \neq x_0$, the replacement node has access to code symbols that belong to the plane $\mathbf{z}$ but not to their pairs. Based on the above, the p-c equations corresponding to the plane $\mathbf{z} \in P$ can be expressed in the form

$$\theta^\ell_{x_0,y_0}A(x_0,y_0;\mathbf{z}) + \sum_{x \neq z_{y_0}} \gamma\theta^\ell_{x,y_0}A(x_0,y_0;\mathbf{z}(x \to z_{y_0})) = \kappa^*,$$

where κ^* indicates a quantity that can be computed at a replacement node based on the

FIGURE 31.9: Figure illustrating node repair of $(x_0, y_0) = (1, 0)$. The helper data sent corresponds to repair planes $\mathbf{z}$ with a dot at failed node $(1, 0)$. The Uncoupled code symbols corresponding to repair planes are recovered as described in the text. We can therefore recover the failed symbols from the repair planes and also the remaining failed node symbols of the form A^* from A, B as shown in the figure.

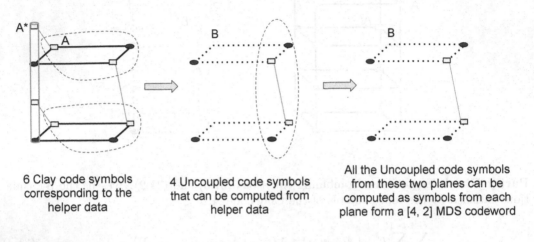

| 6 Clay code symbols corresponding to the helper data | 4 Uncoupled code symbols that can be computed from helper data | All the Uncoupled code symbols from these two planes can be computed as symbols from each plane form a [4, 2] MDS codeword |

code symbols supplied by the $n - 1$ helper nodes. Thus we have a set of r equations in r unknowns. From the theory of generalized RS codes, it is known that these equations suffice to recover the symbols

$$\{A(x_0, y_0; \mathbf{z}(x \to z_{y_0})) \mid \mathbf{z} \in P, x \in [0 : r - 1]\} = \{A(x_0, y_0, \mathbf{z}) \mid \mathbf{z} \in \mathbb{Z}_r^t\}.$$

This completes the proof of node repair. We refer the readers to [1621] for a proof of the data collection property.

31.2.9   Variants of Regenerating Codes

Cooperative RGC Regenerating Codes were introduced initially with the aim of efficiently dealing with single-node failures; subsequent work extended their applicability to the simultaneous failure of $t > 1$ nodes. Two approaches were adopted. Under the first approach, there is a *single repair center* that collects all the helper information pertinent to the repair of all the failed nodes and then subsequently distributes the requisite information to the replacement nodes. In [326], the notion of interference alignment is applied to establish that the total amount of data downloaded to the repair center, for the case when the code is an MDS code, is at least $\frac{\alpha dt}{d+t-k}$.

Under the second approach, there are t repair centers, each corresponding to a replacement node. In the first phase of the repair process, helper data is transferred to the t repair centers from the $d \leq n - t$ helper nodes. In the second phase, data is transferred between repair centers. This latter approach introduced in [992] is known as *cooperative regeneration*. A storage-repair-bandwidth tradeoff for cooperative regenerating codes was derived in [1108] and [1698] using Cut-Set-Bound-based techniques. The extreme points of the tradeoff are known as *minimum-bandwidth cooperative regenerating* (MBCR) points and *minimum-storage cooperative regenerating* (MSCR) points. Constructions of optimal codes that operate at these extreme points can be found in [1865] and [1927].

Near Optimal Bandwidth MDS Codes are yet another variant of RGCs explored in the literature. They are vector MDS codes that trade between sub-packetization level and savings in repair bandwidth. As demonstrated in [1825], a large sub-packetization level is not a desirable feature in a distributed storage system. Though MSR codes have optimal repair bandwidth, they necessarily incur a high sub-packetization level as established in [44, 98, 833, 1775]. The piggybacking framework [1571], ϵ-MSR framework [1580], and the transformation in [1242] are examples of construction methods for vector MDS codes that have a small sub-packetization level while ensuring substantial reduction in repair bandwidth.

Fractional-Repetition Codes, introduced in [1607], may be regarded as a generalization of the RBT polygon MBR code [1570] presented in Section 31.2.6. In a fractional-repetition code, the symbols corresponding to a data file are first encoded using an MDS code. Each code symbol is then replicated ρ times. These replicated code symbols are stored across n nodes, with each node storing α symbols and each code symbol appearing in exactly ρ distinct nodes. The definition of MBR codes requires that any set of d surviving nodes can be called upon to serve as helper nodes for node repair. In contrast, in the case of fractional-repetition codes, it is only guaranteed that there is at least one choice of d helper nodes corresponding to which RBT is possible. A fractional-repetition code with ρ-repetition allows repair without any computation for up to $\rho-1$ node failures. Constructions of fractional-repetition codes can be found in [1465, 1494, 1607, 1706].

Secure RGCs, introduced in [1495], are a variant of RGCs where a passive but curious eavesdropper is present, who has access to the data stored in a subset A of size $|A| = \ell < k$ of the storage nodes. The aim here is to prevent the eavesdropper from gaining any information pertinent to the stored data. Here again, there is a notion of exact and functional repair and there are corresponding storage-repair bandwidth tradeoffs. Codes that operate at extreme ends of the tradeoff are called *secure MBR* and *secure MSR* codes, respectively. In [1495], an upper bound on file size for the case of F-R secure RGC is provided along with a matching construction corresponding to the MBR point for the case $d = n - 1$. In [1664], the authors study the E-R storage-bandwidth tradeoff for this model.

This eavesdropper model was subsequently extended in [1572] to the case where the passive eavesdropper also has access to data received during the repair of a subset of nodes $A_1 \subseteq A$ of size ℓ_1. In [1572], the authors provide a secure MBR construction that holds for any value of the parameter d. The file size under this construction matches the upper bound shown in [1495]. In [1576], for the extended model, the authors provide an upper bound on the file size corresponding to the secure MSR point and a matching construction. In [1924], the authors study the E-R storage-bandwidth tradeoff for the extended model.

31.3 Locally Recoverable Codes

In the case of the class of regenerating codes discussed in the previous section, the aim was to reduce the repair bandwidth. In contrast, the focus in the case of **Locally Recoverable Codes** (LRCs) introduced in [899, 994] and discussed in the present section, is on reducing the repair degree, i.e., on reducing the number of helper nodes contacted for the purpose of node repair.[2] Two comments are in order here. First, reducing the repair degree does tend to lower the repair bandwidth. Second, the storage overhead in the case of

---

[2]LRCs arising from algebraic geometry are discussed in Section 15.9. Constructions of LRCs from functions can be found in Section 20.7.

a nontrivial LRC is necessarily larger than that of an MDS code. There are two broad classes of LRCs. LRCs with **Information Symbol Locality** (ISL) are systematic linear codes in which the repair degree is reduced only for the repair of nodes corresponding to message symbols. In an LRC with **All-Symbol Locality** (ASL), the repair degree is reduced for the repair of all n nodes, regardless of whether the node corresponds to a message or parity symbol. Clearly, the class of LRCs with ASL is a sub-class of the set of all LRCs with ISL.

31.3.1 Information Symbol Locality

Unless otherwise specified, when we speak of an LRC in this section, the reference will be to an LRC with ISL. Recall that a linear code is *systematic* if the k message symbols are explicitly present among the n code symbols. An (n, k, r) LRC $\mathcal{C}$ over a field $\mathbb{F}_q$ is a systematic $[n, k]$ linear block code having the property that every message symbol c_t can be recovered by computing a linear combination of the form

$$c_t = \sum_{j \in S_t} a_j c_j \quad \text{where } a_j \in \mathbb{F}_q$$

involving at most r other code symbols c_j with $j \in S_t$. Thus the set S_t in the equation above has size at most r. The minimum distance of an (n, k, r) LRC [830] must necessarily satisfy the Singleton Bound for LRC:

$$d_{\min} \leq (n - k + 1) - \left(\left\lceil \frac{k}{r} \right\rceil - 1 \right). \tag{31.5}$$

Thus for the same values of $[n, k]$, an LRC has $d_{\min}$ which is smaller by an amount equal to $\left(\left\lceil \frac{k}{r} \right\rceil - 1 \right)$ in comparison to an MDS code. The quantity $\left(\left\lceil \frac{k}{r} \right\rceil - 1 \right)$ may thus be regarded as the *penalty* associated with imposing the locality requirement. An LRC whose minimum distance satisfies the above bound with equality is said to be **optimal**. The class of pyramid codes [994] are an example of a class of optimal LRCs and are described below. Analysis of nonlinear LRCs can be found in [737, 1489].

31.3.1.1 Pyramid Codes

We introduce the pyramid code [994] construction of an LRC with ISL through an illustrative example corresponding to parameter set $(n = 9, k = 6, r = 3)$. The starting point is the generator matrix of an RS code. Let G_{RS} be the generator matrix of an $[n_{\text{RS}} = 8, k_{\text{RS}} = 6]$ RS code $\mathcal{C}_{\text{RS}}$ in systematic form, i.e.,

$$G_{\text{RS}} = \begin{bmatrix} 1 & 0 & 0 & 0 & 0 & 0 & g_{11} & g_{12} \\ 0 & 1 & 0 & 0 & 0 & 0 & g_{21} & g_{22} \\ 0 & 0 & 1 & 0 & 0 & 0 & g_{31} & g_{32} \\ 0 & 0 & 0 & 1 & 0 & 0 & g_{41} & g_{42} \\ 0 & 0 & 0 & 0 & 1 & 0 & g_{51} & g_{52} \\ 0 & 0 & 0 & 0 & 0 & 1 & g_{61} & g_{62} \end{bmatrix}.$$

The generator matrix of the associated pyramid code is obtained by splitting a single parity column in G_{RS} and then rearranging columns as shown below:

$$G_{\text{pyr}} = \begin{bmatrix} 1 & 0 & 0 & g_{11} & 0 & 0 & 0 & & g_{12} \\ 0 & 1 & 0 & g_{21} & 0 & 0 & 0 & & g_{22} \\ 0 & 0 & 1 & g_{31} & 0 & 0 & 0 & & g_{32} \\ 0 & 0 & 0 & 0 & 1 & 0 & 0 & g_{41} & g_{42} \\ 0 & 0 & 0 & 0 & 0 & 1 & 0 & g_{51} & g_{52} \\ 0 & 0 & 0 & 0 & 0 & 0 & 1 & g_{61} & g_{62} \end{bmatrix}.$$

This yields the generator matrix G_{pyr} of an $(n = 9, k = 6, r = 3)$ optimal LRC $\mathcal{C}_{\text{pyr}}$. The proof that the above code is an optimal LRC with ISL is as follows. It is clear that the code $\mathcal{C}_{\text{pyr}}$ is an LRC and that the minimum distance $d_{\min}$ of the code $\mathcal{C}_{\text{pyr}}$ is at least the minimum distance of the RS code $\mathcal{C}_{\text{RS}}$. This follows from the fact that the minimum Hamming distance of a linear code equals its minimum Hamming weight. The minimum distance of the $\mathcal{C}_{\text{RS}}$ equals $n_{\text{RS}} - k_{\text{RS}} + 1 = 8 - 6 + 1 = 3$ from the Singleton Bound. It follows that the minimum distance of the pyramid code is at least 3. On the other hand, from (31.5), we have that

$$d_{\min} \leq (n - k + 1) - \left(\left\lceil \frac{k}{r} \right\rceil - 1 \right) = 9 - 6 + 1 - \left(\left\lceil \frac{6}{3} \right\rceil - 1 \right) = 3.$$

It follows that the code is an optimal LRC. In the general case, if we start with an $[n, k]$ RS code and split a single parity column, we will obtain an optimal

$$(n_{\text{pyr}} = n + \lceil k/r \rceil - 1, \; k_{\text{pyr}} = k, \; r)$$

pyramid LRC.

31.3.1.2 Windows Azure LRC

Figure 31.10 shows the $(n = 18, k = 14, r = 7)$ LRC employed in the Windows Azure cloud storage system [995] and which is related in structure to the pyramid code. The dotted boxes indicate a collection of symbols that satisfy an overall parity check. This code has minimum distance 4 which is the same as that of the $[n = 9, k = 6]$ RS code.

FIGURE 31.10: The LRC employed in Windows Azure cloud storage

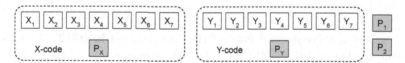

In terms of reliability, the $(n = 18, k = 14, r = 7)$ Windows Azure code and the $[9, 6]$ RS code are comparable as they both have the same minimum distance $d_{\min} = 4$. In terms of repair degree, the two codes are again comparable, having respective repair degrees of 7 (Windows Azure LRC) and 6 (RS). The major difference is in the storage overhead, which stands at $\frac{18}{14} = 1.29$ in the case of the Azure LRC and $\frac{9}{6} = 1.5$ in the case of the $[9, 6]$ RS code. This reduction in storage overhead has reportedly saved Microsoft millions of dollars [1386].

31.3.2 All Symbol Locality

An LRC in which every code symbol can be recovered from a linear combination of at most r other code symbols is called an LRC with ASL. An LRC with ASL will be said to be **optimal** if it has minimum distance that satisfies (31.5) with equality. A construction for optimal ASL LRCs can be found in [1772]; see also Section 15.9.3 and discussion surrounding Theorem 20.7.5. The codes in the construction may be regarded as subcodes of RS codes. The idea behind the construction is illustrated in Figure 31.11 for the case $r = 2$. As noted in Section 31.1, the code symbols within a codeword of an $[n.k]$ RS code over $\mathbb{F}_q$ may be regarded as evaluations of a polynomial associated with the message symbols. More specifically, the codeword $(f(P_1), \ldots, f(P_n)) \in \mathbb{F}_q^n$, where $f(x) = \sum_{i=0}^{k-1} m_i x^i$ and where $P_1, \ldots, P_n$ are distinct elements from $\mathbb{F}_q$, is associated to the set $\{m_i\}_{i=0}^{k-1}$ of message symbols. The construction, depicted in Figure 31.11, is one in which code symbols are obtained by evaluating a subclass of this set of polynomials. This subclass of polynomials has the property that given any code symbol corresponding to the evaluation $f(P_a)$, there exist two other code symbols $f(P_b), f(P_c)$ such that the three values lie on a straight line and hence satisfy an equation of the form

$$u_a f(P_a) + u_b f(P_b) + u_c f(P_c) = 0,$$

where $u_a, u_b, u_c \in \mathbb{F}_q$. Thus the value of an example code symbol $f(P_a)$ can be recovered from the values of two other code symbols, $f(P_b)$ and $f(P_c)$ in the present case. Thus this construction represents an LRC with $r = 2$ which structurally is a subcode of an RS code.

FIGURE 31.11: Illustrating the construction of an optimal ASL LRC

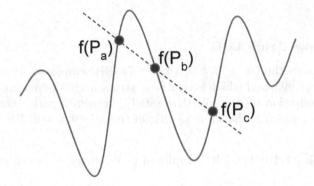

We now present a more general form of the construction in [1772] of an (n, k, r) ASL LRC, sometimes called a Tamo–Barg code, for the case $n = q - 1$ and $(r + 1) \mid (q - 1)$. It will be convenient to express k in the form

$$k = \ell r + a \quad \text{where} \quad \ell = \left\lceil \frac{k}{r} \right\rceil - 1 \text{ and } 1 \leq a \leq r.$$

In an RS code, we evaluate all polynomials of degree $\leq k - 1$; here we restrict attention to the subset Q of polynomials that can be expressed in the form

$$f(x) = \sum_{i=0}^{\ell} x^{(r+1)i} f_i(x),$$

where the polynomials

$$f_i(x) = \sum_{j=0}^{r-1} b_{i,j} x^j \quad \text{for } 0 \leq i \leq \ell - 1$$

have degree at most $r - 1$ and where

$$f_\ell(x) = \sum_{j=0}^{a-1} b_{\ell,j} x^j$$

has degree at most $a-1$. Clearly by counting the number of coefficients $b_{i,j} \in \mathbb{F}_q$, we see that the number of polynomials in the set Q equals $q^{\ell r + a} = q^k$ and hence this code has dimension k. Let $\mathbb{F}_q^*$ denote the set of $q-1$ nonzero elements in $\mathbb{F}_q$. Code symbols are obtained by evaluating each polynomial in Q at all the elements in $\mathbb{F}_q^*$. Let $\mathcal{C}$ be the resultant code, i.e.,

$$\mathcal{C} = \left\{ (f(u) \mid u \in \mathbb{F}_q^*) \mid f \in Q \right\}.$$

We will next establish that $\mathcal{C}$ is an LRC. To see this, let H denote the set of $(r+1)^{\text{th}}$ roots of unity contained in $\mathbb{F}_q$. Then the $\frac{q-1}{r+1}$ multiplicative cosets of H partition $\mathbb{F}_q^*$. We first note that for any $\beta \in H$ and any $b \in \mathbb{F}_q^*$, the product $b\beta$ is a zero of the polynomial $x^{r+1} - b^{r+1}$. It follows then that for $f \in Q$,

$$f(b\beta) = f(x)|_{x=b\beta} = \left(f(x) \pmod{x^{r+1} - b^{r+1}} \right) \Big|_{x=b\beta}$$

$$= \left(\sum_{i=0}^{\ell} x^{(r+1)i} f_i(x) \pmod{x^{r+1} - b^{r+1}} \right) \Big|_{x=b\beta}$$

$$= \left(\sum_{i=0}^{\ell} b^{(r+1)i} f_i(x) \right) \Big|_{x=b\beta}. \tag{31.6}$$

Since each polynomial $f_i(x)$ is of degree $\leq r-1$, the polynomial appearing in (31.6) is also of degree $\leq r-1$. As a result, we can recover the value $f(b\beta)$ from the r evaluations $\{f(b\theta) \mid \theta \in H \setminus \{\beta\}\}$. Thus this construction results in an LRC with locality parameter r.

We will now show that $\mathcal{C}$ is an optimal LRC with respect to the minimum distance bound in (31.5). We bound the minimum distance of the code by computing the maximum degree of a polynomial in Q. We see that

$$\deg(f) \leq \ell(r+1) + a - 1 = k + \ell - 1 = k + \left(\left\lceil \frac{k}{r} \right\rceil - 1 \right) - 1.$$

Since a polynomial of degree d can have at most d zeros and the minimum Hamming weight of a linear code equals its minimum distance, it follows that

$$d_{\min} \geq n - \max \deg(f) \geq n - k + 1 - \left(\left\lceil \frac{k}{r} \right\rceil - 1 \right). \tag{31.7}$$

Comparing (31.5) and (31.7), we see that the code $\mathcal{C}$ is an optimal ASL LRC. Note that the field size needed for this construction is $O(n)$. A different construction of optimal LRC with $O(n)$ field size that is based on cyclic codes can be found in [1773].

It turns out that for parameter sets where $(r+1) \nmid n$, the bound (31.5) cannot be achieved with equality by any ASL LRC. Improved versions of the bound (31.5) can be found in [1373, 1867, 1942]. A construction achieving the improved bound in [1866] with equality and with exponential field size can be found in the same paper for the case $n_1 > n_2$ where $n_1 = \left\lceil \frac{n}{r+1} \right\rceil$ and $n_2 = n_1(r+1) - n$.

31.3.3 LRCs over Small Field Size

For field size $q < n$, it is challenging to construct optimal LRCs. Upper bounds on the minimum distance of an (n, k, r) LRC over $\mathbb{F}_q$ that take into account the field size q can be found in [14, 97, 327, 831, 996, 1773, 1868]. Example constructions that are optimal with respect to these improved bounds can be found in [831, 996, 1868]. Asymptotic upper bounds on an LRC, i.e., upper bounds on the rate $\frac{k}{n}$ as a function of relative minimum distance $\frac{d_{\min}}{n}$ in the limit as $n \to \infty$, which take into account the field size q can be found in [14]. Asymptotic lower bounds can be found in [128, 327].

31.3.4   Recovery from Multiple Erasures

There are several approaches towards designing an LRC that can recover from more than one erasure. A classification of these approaches is presented in Figure 31.12.

FIGURE 31.12: Classification of various approaches for LRCs for multiple erasures

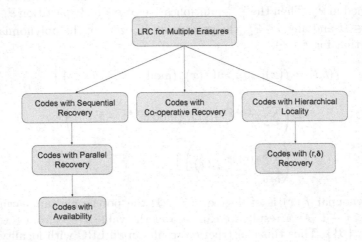

31.3.4.1   Codes with Sequential Recovery

The most general approach, by which we mean the approach that imposes the least constraint in terms of how recovery is to be accomplished, is sequential recovery [95, 1535]. An example of a code with sequential recovery is shown in Figure 31.13. In the figure, the numbers shown correspond to the indices of the 8 code symbols. The 4 vertices correspond to the 4 parity checks. Each parity check represents the equation that the sum of code symbols, corresponding to the numbers attached to it, is equal to 0. It can be seen that if the code symbols 1 and 5 are erased, and one chooses to decode using locality, then one must first decode code symbol 5 before decoding symbol 1. Hence, this code can recover sequentially from two erasures where each erasure is recovered by contacting $r = 2$ code symbols with block length $n = 8$ and dimension $k = 4$.

FIGURE 31.13: An example code with sequential recovery which can recover from 2 erasures with $n = 8$, $k = 4$, $r = 2$

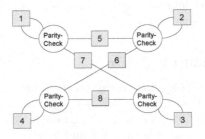

More formally a code with **sequential recovery** from t erasures is an $[n, k]$ linear code

over $\mathbb{F}_q$ such that an arbitrary set of t erased symbols $c_{j_1}, \ldots, c_{j_t}$ can be recovered as follows:

$$c_{j_i} = \sum_{m \in S_i} a_m c_m \text{ with } a_m \in \mathbb{F}_q$$

where $S_i \subseteq \{1, 2, \ldots, n\}$, $|S_i| \le r$, and $\{j_i, j_{i+1}, \ldots, j_t\} \cap S_i = \emptyset$. The example given in the Figure 31.13 corresponds to the parameter set $(n = 8, k = 4, r = 2, t = 2)$. Sequential recovery was introduced in [1535], where a detailed analysis for the $t = 2$ case can be found. Characterization of the maximum possible rate for given n, r, and $t = 3$ can be found in [1739]. The maximum possible rate of codes with sequential recovery for a given r, t is characterized in [95]. The construction of codes having high rate, but rate that is less in comparison to the rate of the construction in [95], can be found in [1579, 1737]. The construction in [1737] however has lower block length $O(r^{O(\log(t))})$ in comparison to the construction in [95], which has $O(r^{O(t)})$ block length.

31.3.4.2 Codes with Parallel Recovery

If in the definition of sequential recovery, we impose the stronger requirement that $\{j_1, \ldots, j_t\} \cap S_i = \emptyset$ for $1 \le i \le t$, in place of $\{j_i, j_{i+1}, \ldots, j_t\} \cap S_i = \emptyset$, we will obtain the definition of a code with **parallel recovery**. Thus under parallel recovery each of the t erased code symbols can be recovered in any desired order. Please see [1485] for additional details on parallel recovery.

31.3.4.3 Codes with Availability

Codes with availability [1866, 1869], also discussed briefly in Section 15.9.6, cater to the situation when a node, containing a code symbol that is desired to be accessed, is unavailable as the particular node is busy serving other requests. To handle such situations, an availability code is designed so that the same code symbol can be recovered in multiple ways, as a linear combination of a small and disjoint subset of the remaining code symbols. The binary product code is one example of an availability code. Consider a simple example of a product code in which code symbols are arranged in the form of an $(r + 1) \times (r + 1)$ array and the code symbols are such that each row and column satisfies even parity; see Figure 31.14 for an example. Thus the code symbols within any row or column of the array sum to zero. It follows that each code symbol can be recovered in 3 distinct ways: directly from the node storing the code symbol or else by computing the sum of the remaining entries in either the row or the column containing the desired symbol.

Formally, a code with t-**availability**, is an $[n, k]$ linear over $\mathbb{F}_q$ such that each code symbol c_i can be recovered in t disjoint ways as follows:

$$c_i = \sum_{m \in S_j^i} a_m c_m \text{ with } a_m \in \mathbb{F}_q$$

where for $1 \le j \ne j' \le t$, we have that

$$i \notin S_j^i, \quad |S_j^i| \le r, \quad \text{and} \quad S_j^i \cap S_{j'}^i = \emptyset.$$

The sets $\{S_j^i\}$ will be referred to as *recovery sets*. The example product code described above corresponds to the parameter set $(n = (r+1)^2, k = r^2, t = 2)$ and the sets S_1^i, S_2^i to symbols lying within the same row and column respectively.

Codes with t-availability can recover from t erasures. This can be seen as follows. If there are t erased symbols including c_i, then apart from c_i there are $t - 1$ other erased symbols. These however, can be present in at most $t - 1$ out of the t disjoint recovery sets $S_1^i, \ldots, S_t^i$.

FIGURE 31.14: The binary product code as an example of a code with availability. In this example, the code symbol 5 can be recovered either directly from the node storing it or else by computing either the row sum or the column sum.

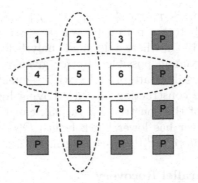

Hence, there must exist at least one recovery set S_j^i in which none of the erased symbols is present, and this recovery set can be used to recover c_i. It can be verified that a code with t-availability is also a code that can recover for t erasures in parallel.

Upper bounds on the rate of codes with availability for a given (r, t) can be found in [97, 1070, 1771]. A construction of high-rate codes with availability is presented in [1869]. An upper bound on the minimum distance of a code with availability that is independent of field size q can be found in [97, 1771, 1866]. Field size dependent upper bounds on minimum distance can be found in [97, 996]. Asymptotic lower bounds, i.e., lower bounds on the rate $\frac{k}{n}$ as a function of the relative minimum distance $\frac{d_{\min}}{n}$ as $n \to \infty$, for fixed (r, t, q) can be found in [128, 1771].

31.3.4.4   Codes with Cooperative Recovery

In all the different types of t-erasure LRCs that we have encountered thus far, the constraint placed has always been on the number r of unerased symbols contacted for the repair of a single erased symbol. In cooperative recovery, a constraint is placed, instead, on the total number of unerased code symbols contacted for the recovery of all t erased symbols. Formally, a code with **cooperative recovery** is an $[n, k]$ linear code over a field $\mathbb{F}_q$ such that an arbitrary set $\{c_{j_1}, \ldots, c_{j_t}\}$ of t erased symbols can be recovered from a set of t equations

$$c_{j_i} = \sum_{m \in S} a_{m,i} c_m \quad \text{where } a_{m,i} \in \mathbb{F}_q \text{ for } 1 \le i \le t$$

that involve a common set $\{c_m \mid m \in S\}$ of r unerased code symbols, where $S \subseteq \{1, 2, \ldots, n\}$ and $|S| \le r$. Further details including constructions and performance bounds can be found in [1579].

31.3.4.5   Codes with (r, δ) Locality

The definition of an LRC required that each code symbol be part of a single parity check code of length $\le r + 1$. If it was required, instead, that each code symbol be part of an $[r + \delta - 1, r, \delta]$ MDS code, then the resultant code would be an example of a code with (r, δ) locality. Thus each local code is stronger in terms of minimum distance, allowing local recovery from a larger number of erasures. Also see Section 15.9.5.

More formally, a code with (r, δ) **locality** $\mathcal{C}$ over $\mathbb{F}_q$ is an $[n, k]$ linear code over $\mathbb{F}_q$ such that for each code symbol c_i there is an index set $S_i \subseteq \{1, 2, \ldots, n\}$ such that $d_{\min}(\mathcal{C}_{|S_i}) \geq \delta$ and $|S_i| \leq r + \delta - 1$ where $\mathcal{C}_{|S_i}$ is the *restriction* of the code to the coordinates corresponding to the set S_i. Alternately, we may regard the code $\mathcal{C}_{|S_i}$ as being obtained from $\mathcal{C}$ by *puncturing* $\mathcal{C}$ in the locations corresponding to index set $\{1, 2, \ldots, n\} \setminus S_i$. Note that an LRC is an instance of an (r, δ) code with $\delta = 2$.

The classification of this class of codes into information symbol and all symbol (r, δ) locality codes follows in the same way as was carried out in the case of an LRC. There is an analogous minimum distance bound [1534] given by

$$d_{\min}(\mathcal{C}) \leq (n - k + 1) - \left(\left\lceil \frac{k}{r} \right\rceil - 1 \right)(\delta - 1). \tag{31.8}$$

A code with (r, δ) locality satisfying the above bound with equality is said to be **optimal**. Optimal codes with (r, δ) information symbol locality can be obtained from pyramid codes by extending the approach described in Section 31.3.1.1 and splitting a larger number $\delta - 1$ of the parity columns in the generator matrix of a systematic RS code. Optimal codes with (r, δ) ASL can be obtained by employing the construction in [1772] as described in Section 31.3.2 for the case when $(r + \delta - 1) \mid n$ and $q = O(n)$. Optimal (r, δ) cyclic codes with $q = O(n)$ can be found in [392] for the case when $(r + \delta - 1) \mid n$. A detailed analysis as to when the upper bound on minimum distance appearing in (31.8) is achievable can be found in [1738]. Characterization of binary codes achieving the bound in (31.8) with equality can be found in [903]. A field size dependent upper bound on dimension k for fixed $(r, \delta, n, d_{\min})$ appears in [14]. Asymptotic lower bounds, i.e., lower bounds on the rate $\frac{k}{n}$ as a function of the relative minimum distance $\frac{d_{\min}}{n}$ as $n \to \infty$, for a fixed (r, δ, q) can be found in [128].

31.3.4.6 Hierarchical Codes

From a certain perspective, the idea of an LRC is not scalable. Consider for instance, a $[24, 14]$ linear code which is made up of the union of 6 disjoint $[4, 3]$ local codes (see Figure 31.15 (left)). These local codes are single parity check codes and ensure that the code has locality 3. However if there are 2 or more erasures within a single local code, then local recovery is no longer possible and one has to resort to decoding the entire code as a whole to recover the two erasures. Clearly, this problem becomes more acute as the block length n increases. One option to deal with this situation would be to build codes with (r, δ) locality, but even in this case, if there are more than $\delta - 1$ erasures within a local code, local decoding is no longer possible. Codes with hierarchical locality [107, 1618] (see Figure 31.15 (right)) seek to overcome this by building a hierarchy of local codes having increasing block length to ensure that in the event that a local code at the lowest level is overcome by a larger number of erasures than it can handle, then the local code at the next level in the hierarchy can take over. As one goes up the hierarchy, both block length and minimum distance increase. An example hierarchical code is presented in Figure 31.15.

31.3.5 Maximally Recoverable Codes

Let $\mathcal{C}$ be an $[n, k]$ linear code with (r, δ) locality such that every local code has disjoint support and where further, each local code is an $[r + \delta - 1, r]$ MDS code. Let $E \subseteq \{1, 2, \ldots, n\}$ be formed by picking $\delta - 1$ coordinates from each of the local codes within $\mathcal{C}$. Then $\mathcal{C}$ is said to be **maximally recoverable** (MR) [829] if the code obtained by puncturing $\mathcal{C}$ on coordinates defined by E is an MDS code. An MR code can correct all possible erasure patterns that are information-theoretically correctable given the locality constraints. MR codes were originally introduced as partial MDS codes in [218]. The notion of maximal

FIGURE 31.15: The code on the left is an LRC in which each code symbol is protected by a $[n = 4, k = 3, d_{\min} = 2]$ local code and each local code is contained in an overall $[24, 14, 7]$ global code. In the hierarchical locality code appearing to the right, each local code is a part of a $[12, 8, 3]$ so-called middle code, and the middle codes, in turn, are contained in an overall $[24, 14, 6]$ global code.

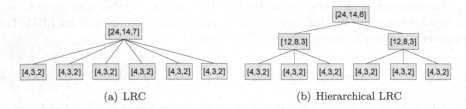

(a) LRC   (b) Hierarchical LRC

recoverability finds particular application in the design of sector-disk codes [1512] that are used in RAID storage systems to combat simultaneous sector and disk erasures.

31.4   Locally Regenerating Codes

We have seen earlier that while RGCs minimize repair bandwidth, LRCs minimize the repair degree. **Locally Regenerating Codes** (LRGCs) [1075, 1577] are codes which simultaneously possess low repair bandwidth as well as low repair degree. LRGCs are perhaps best viewed as vector codes with locality, in which the local codes are themselves RGCs. In Figure 31.16, we illustrate an LRGC where each local code is a repair-by-transfer, pentagon MBR code.

FIGURE 31.16: An LRGC in which each of the three local codes is a pentagon MBR code. The set of 30 scalar symbols that make up the LRGC form a scalar ASL LRC in which there are three disjoint local codes, each of block length $r + 1 = 10$. The contents of each of the three pentagons are obtained from the 10 scalar symbols making up the respective local code by following the same procedure employed to construct a pentagon MBR code from a set of 10 scalar symbols that satisfy an overall parity check.

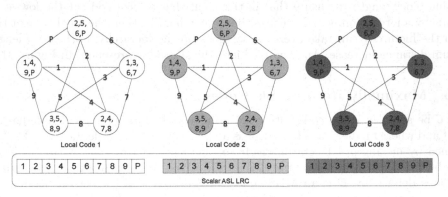

31.5   Efficient Repair of Reed–Solomon Codes

In an $[n, k]$ MDS code, each code symbol is traditionally considered as an indivisible unit over $\mathbb{F}_q$. As any k code symbols of an MDS code form an information set, the conventional approach to the recovery of an erased code symbol is to access an arbitrary set of k other code symbols, use these to decode the code and in this way, recover the erased symbol. In the context of a distributed storage system, where each code symbol of a codeword is stored in a distinct node, this implies a repair bandwidth which is k times the amount of data stored in a failed node. The first step in developing a more efficient repair strategy, in the case of an MDS code, is to employ a finite field $\mathbb{F}_q$ that is a degree-t extension of a base field $\mathbb{B}$. Thus, if $|\mathbb{B}| = p$, where p is some prime power, we will have $|\mathbb{F}_q| = q = p^t$. In this setting, the authors of [1660] took the important next step of regarding each code symbol as a vector over the base field $\mathbb{B}$ and showing that the repair bandwidth can be reduced by carrying out repair operations over the base field. The paper [1660] dealt with the specific case $n - k = 2$, where only the repair of systematic nodes was considered. This approach was subsequently generalized in [882] to present an efficient all-node-repair scheme for Generalized Reed–Solomon (GRS) (see [1323, Chapter 10]) codes as described below.

GRS Codes: Let $\Theta := \{\theta_i\}_{i=1}^{n} \subseteq \mathbb{F}_q$ denote a subset of $\mathbb{F}_q$ of size n. Let $\mathcal{F}, \mathcal{H}$ denote the set of all polynomials, including the zero polynomial, in $\mathbb{F}_q[x]$ having degree bounded above by $k - 1$ and $n - k - 1$, respectively. Each codeword in an $[n, k]$ GRS code $\mathcal{GRS}_k(\Theta, \mathbf{u})$ is obtained by evaluating a polynomial in $\mathcal{F}$, along with scaling coefficients $\{u_i\}_{i=1}^{n} \subseteq \mathbb{F}_q^*$ where $\mathbf{u} = (u_1, \ldots, u_n)$, at the elements of $\Theta = (\theta_1, \ldots, \theta_n)$, i.e.,

$$\mathcal{GRS}_k(\Theta, \mathbf{u}) = \left\{ \big(u_1 f(\theta_1), u_2 f(\theta_2), \ldots, u_n f(\theta_n) \big) \mid f \in \mathcal{F} \right\}.$$

The dual code of $\mathcal{GRS}_k(\Theta, \mathbf{u})$ is then of the form

$$\mathcal{GRS}_k(\Theta, \mathbf{u})^{\perp} = \left\{ \big(v_1 h(\theta_1), v_2 h(\theta_2), \ldots, v_n h(\theta_n) \big) \mid h \in \mathcal{H} \right\} = \mathcal{GRS}_{n-k}(\Theta, \mathbf{v}).$$

Like the $\{u_i\}$, the $\{v_i\}_{i=1}^{n} \subseteq \mathbb{F}_q^*$ are also a set of scaling coefficients where $\mathbf{v} = (v_1, \ldots, v_n)$. The scaling coefficients $\{u_i\}, \{v_j\}$ do not, however, play any role in determining the repair bandwidth, and for this reason, in the text below, we assume all the scaling coefficients u_i, v_j are equal to 1.

Trace Function and Trace-Dual Basis: The trace function $\mathrm{Tr}_{q/p} : \mathbb{F}_q \to \mathbb{B}$ is given by

$$\mathrm{Tr}_{q/p}(x) = \sum_{i=0}^{t-1} x^{p^i},$$

where $x \in \mathbb{F}_q$. For every basis $\Gamma = \{\gamma_1, \gamma_2, \ldots, \gamma_t\}$ of $\mathbb{F}_q$ over $\mathbb{B}$, there exists a second basis $\Delta = \{\delta_1, \delta_2, \ldots, \delta_t\}$, termed the **trace-dual basis**, satisfying

$$\mathrm{Tr}_{q/p}(\gamma_i \delta_j) = \left\{ \begin{array}{ll} 1 & \text{if } i = j, \\ 0 & \text{otherwise.} \end{array} \right.$$

It can be verified that each element $x \in \mathbb{F}_q$ has the basis expansion

$$x = \sum_{i=1}^{t} \mathrm{Tr}_{q/p}(x \gamma_i) \delta_i.$$

Thus given $\{\mathrm{Tr}_{q/p}(x \gamma_i)\}_{i=1}^{t}$, the element x can be uniquely recovered.

Node Repair via the Dual Code: Recall that $\mathcal{GRS}_k(\Theta, \mathbf{u})$ and its dual $\mathcal{GRS}_{n-k}(\Theta, \mathbf{v})$ are scaled evaluations of polynomials of degree at most $k - 1$ and at most $n - k - 1$, respectively. Hence for $f, h \in \mathcal{F}, \mathcal{H}$, respectively, we have $\sum_{i=1}^{n} f(\theta_i)h(\theta_i) = 0$ (assuming that each u_i and each v_j equals 1 for reasons explained earlier). Suppose the code symbol $f(\theta_i)$ has been erased. We have

$$f(\theta_i)h(\theta_i) = -\sum_{j=1, j\neq i}^{n} f(\theta_j)h(\theta_j).$$

Thus,

$$\mathrm{Tr}_{q/p}(f(\theta_i)h(\theta_i)) = -\sum_{j=1, j\neq i}^{n} \mathrm{Tr}_{q/p}(f(\theta_j)h(\theta_j)). \tag{31.9}$$

Next, let us assume that it is possible to select a subset $\mathcal{H}_i$ of $\mathcal{H}$ in such a way that $\{h(\theta_i)\}_{h\in\mathcal{H}_i}$ forms a basis for $\mathbb{F}_q$ over $\mathbb{B}$. It follows from (31.9) and the existence of a trace-dual basis that $f(\theta_i)$ can be recovered from the set $\left\{\sum_{j=1, j\neq i}^{n} \mathrm{Tr}_{q/p}(f(\theta_j)h(\theta_j))\right\}_{h\in\mathcal{H}_i}$. In [882], the authors carefully choose the subsets $\{\mathcal{H}_i\}_{i=1}^{n}$ so as to not only satisfy the above basis requirement, but also reduce the repair bandwidth associated with the recovery of $f(\theta_i)$ via (31.9).

The Repair Scheme in [882]: Let $n - k \geq p^{t-1}$ for a GRS code $\mathcal{C} = \mathcal{GRS}_k(\Theta, 1)$. Let $\Gamma = \{\gamma_1, \gamma_2, \ldots, \gamma_t\}$ be a basis for $\mathbb{F}_q$ over $\mathbb{B}$. Each codeword in $\mathcal{C}$ corresponds to the scaled evaluation at the elements in Θ of a polynomial $f \in \mathcal{F}$. With respect to the scheme for failed-node recovery described above, consider the set

$$\mathcal{H}_i = \left\{\frac{\mathrm{Tr}_{q/p}(\gamma_j(x - \theta_i))}{(x - \theta_i)}\right\}_{j=1}^{t}.$$

It is straightforward to verify that $\{h(\theta_i)\}_{h\in\mathcal{H}_i} \equiv \Gamma$ and for $j \neq i$, $\{h(\theta_j)\}_{h\in\mathcal{H}_i}$ is a set consisting of scalar multiples (over $\mathbb{B}$) of $\frac{1}{\theta_j-\theta_i}$. Hence, from the $\mathbb{B}$-linearity of the trace function $\mathrm{Tr}_{q/p}$, it is possible to compute all elements in the set $\{\mathrm{Tr}_{q/p}(f(\theta_j)h(\theta_j))\}_{h\in\mathcal{H}_i}$ from $\mathrm{Tr}_{q/p}\left(\frac{f(\theta_j)}{\theta_j-\theta_i}\right)$. Clearly, in order for the replacement node to be able to compute $\left\{\sum_{j=1, j\neq i}^{n} \mathrm{Tr}_{q/p}(f(\theta_j)h(\theta_j))\right\}_{h\in\mathcal{H}_i}$, each node j ($j \neq i$) needs only provide the single symbol $\mathrm{Tr}_{q/p}\left(\frac{f(\theta_j)}{\theta_j-\theta_i}\right) \in \mathbb{B}$. This results in a repair bandwidth of $n - 1$ symbols over $\mathbb{B}$ to recover each $f(\theta_i)$. In contrast, as noted earlier, the traditional approach for recovering a code symbol incurs a repair bandwidth of k symbols over $\mathbb{F}_q$ or equivalently kt symbols over $\mathbb{B}$.

There has been much subsequent work on the repair of codes dealing with issues such as repairing RS codes in the presence of multiple erasures, achieving the Cut-Set Bound on node repair, enabling optimal access, etc. The reader is referred to [399, 489, 1776] and the references therein for details. RS repair schemes specific to the $[n = 14, k = 10]_{q=256}$ RS code employed by HDFS have been provided in [655].

31.6   Codes for Distributed Storage in Practice

Given the clear-cut storage-overhead advantage that erasure codes provide over replication, popular distributed systems such as Hadoop, Google File System (GFS), Windows

Azure, Ceph, and Openstack have enabled support for erasure codes within their systems. These erasure coding options were initially limited to RS codes. It was subsequently realized that the frequent node-repair operations taking place in the background and the consequent network traffic, and helper-node distraction, were hampering front-end operations. This motivated the development of the RGCs, LRCs, and the improved repair of RS codes. As noted in Section 31.3.1.2, LRCs are very much a part of the Windows Azure cloud-storage system. Hadoop EC has made Piggybacked RS codes available as an option. Both LRC and MSR (Clay) codes are available as erasure coding options in Ceph.

Chapter 32

Polar Codes

Noam Presman

Tel Aviv University

Simon Litsyn

Tel Aviv University

32.1 Introduction

Introduced by Arıkan [63], polar codes are error-correcting codes (ECCs) that achieve the symmetric capacity of discrete input memoryless channels (DMCs) with polynomial encoding and decoding complexities. Arguably the most useful member of this family is Arıkan's $(u+v, v)$ polar code. It was also the first one that was discovered; however generalizations were soon to follow [1149, 1150, 1406, 1403].

In this chapter we attempt to provide a general presentation of polar codes and their associated algorithms. At the same time, many of the examples in the chapter use Arıkan's original construction due to its simplicity and wide applicability. Our journey begins in Section 32.2 by introducing a code construction based on a recursive concatenation of short codes (a.k.a. kernels) and by predetermining some of their information input symbols (a.k.a. symbol freezing). The theory of polar codes is based on such constructions and on associating with each information symbol of the ECC a synthetic channel. Such a synthetic channel has as input the information symbol and as outputs the original communication channel outputs and some of the other information symbols. In Section 32.3 we describe the notion of those channels using the channel-combining and channel-splitting operations. A kernel is polarizing if almost all of its resultant synthetic channels are either perfect (i.e., having capacity $\to 1$) or useless (i.e., having capacity $\to 0$) as the code length grows. This polarization property is studied in Section 32.4 and used in Section 32.5 to specify polar codes. Based on their recursive structure we describe efficient encoding and decoding algorithms in Sections 32.6 and 32.7, respectively.

Notations

Throughout we use the following notations. For a natural number ℓ, we denote $[\ell] = \{1, 2, 3, \ldots, \ell - 1\}$ and $[\ell]_- = \{0, 1, 2, \ldots, \ell - 1\}$. Vectors are represented by bold lowercase letters and matrices by bold uppercase letters. For $i \geq j$, let $\mathbf{u}_j^i = [u_j \; u_{j+1} \cdots \; u_i]$ be the sub-vector of $\mathbf{u}$ of length $i - j + 1$ (if $i < j$ we say that $\mathbf{u}_j^i = [\;]$, the empty vector, and its length is 0). For a set of indices $\mathcal{I} \subseteq [n]_-$, the vector $\mathbf{u}_\mathcal{I}$ is a sub-vector of $\mathbf{u}$ with indices corresponding to $\mathcal{I}$ (the order of elements in $\mathbf{u}_\mathcal{I}$ is the same as in $\mathbf{u}$), where n is the length of $\mathbf{u}$.

32.2 Kernel Based ECCs

We begin our polar codes study by introducing a general construction framework of which polar codes are instances. ECCs generated by this construction are defined by providing a kernel-based transformation and by specifying constraints on the inputs to the transformation, thereby expurgating some of the transformation outputs. In Section 32.2.1 we see that, in fact, such codes are recursive generalized concatenated codes (GCCs). Later in this chapter, we show that properly selecting the kernel and constraints yields good and practical ECCs that we call polar codes.

The basic component of any polar code is a bijective mapping $g(\cdot)$ over $\mathbb{F}^\ell$, $g(\cdot) : \mathbb{F}^\ell \to \mathbb{F}^\ell$. This means that $g(\mathbf{u}) = \mathbf{x}$ for $\mathbf{u}, \mathbf{x} \in \mathbb{F}^\ell$. Such a mapping is called a **kernel** of ℓ dimensions. Both the inputs and outputs of $g(\cdot)$ are from some field $\mathbb{F}$, and therefore we say that $g(\cdot)$ is homogenous. Symbols from $\mathbb{F}$ are called $\mathbb{F}$-symbols in this chapter. The homogenous kernel $g(\cdot)$ may generate a larger code of length ℓ^m $\mathbb{F}$-symbols by inducing a larger mapping from it, as Definition 32.2.1 specifies [1405].

Definition 32.2.1 Given an ℓ dimensions kernel $g(\cdot)$, we construct a **kernel based transformation** $g^{(n)}(\cdot)$ of $N = \ell^n$ dimensions (i.e., $g^{(n)}(\cdot) : \mathbb{F}^{\ell^n} \to \mathbb{F}^{\ell^n}$) in the following recursive fashion:

$$g^{(1)}(\mathbf{u}_0^{\ell-1}) = g(\mathbf{u}_0^{\ell-1});$$
$$g^{(n)}(\mathbf{u}_0^{N-1}) = \Big[g\left(\gamma_{0,0}, \gamma_{1,0}, \gamma_{2,0}, \ldots, \gamma_{\ell-1,0}\right), g\left(\gamma_{0,1}, \gamma_{1,1}, \gamma_{2,1}, \ldots, \gamma_{\ell-1,1}\right), \ldots,$$
$$g\left(\gamma_{0,N/\ell-1}, \gamma_{1,N/\ell-1}, \gamma_{2,N/\ell-1}, \ldots, \gamma_{\ell-1,N/\ell-1}\right)\Big] \quad \text{for} \quad n > 1$$

where

$$[\gamma_{i,j}]_{j=0}^{j=N/\ell-1} = g^{(n-1)}\left(\mathbf{u}_{i \cdot (N/\ell)}^{(i+1) \cdot (N/\ell)-1}\right) \quad \text{for} \quad i \in [\ell]_- .$$

If $g(\cdot)$ is a linear mapping over $\mathbb{F}$, it can be specified by a generating matrix $\mathbf{G}$ of ℓ dimensions such that $g(\mathbf{u}) = \mathbf{u} \cdot \mathbf{G}$. Moreover $g^{(n)}(\cdot)$ is also linear for every $n > 1$.

Example 32.2.2 (The $(u+v,v)$ kernel) In his seminal paper [63] Arıkan proposed the following linear binary kernel $g(u,v) = [u+v,v] = [u,v] \cdot \mathbf{G}$ and $\mathbf{G} = \begin{bmatrix} 1 & 0 \\ 1 & 1 \end{bmatrix}$, where $u, v, G_{i,j} \in \mathbb{F}_2$. Such a mapping is referred to as the $(u+v,v)$ kernel. It can be seen that

$$g^{(2)}(u_0, u_1, u_2, u_3) = [u_0 + u_1 + u_2 + u_3, u_2 + u_3, u_1 + u_3, u_3]$$
$$= [u_0, u_1, u_2, u_3] \cdot \mathbf{G}^{(2)}$$

where $\mathbf{G}^{(2)} = \begin{bmatrix} 1 & 0 & 0 & 0 \\ 1 & 0 & 1 & 0 \\ 1 & 1 & 0 & 0 \\ 1 & 1 & 1 & 1 \end{bmatrix}$. Moreover, we have $g^{(3)}(\mathbf{u}_0^7) = \mathbf{u}_0^7 \cdot \mathbf{G}^{(3)}$ where

$$\mathbf{G}^{(3)} = \begin{bmatrix} 1 & 0 & 0 & 0 & 0 & 0 & 0 & 0 \\ 1 & 0 & 0 & 0 & 1 & 0 & 0 & 0 \\ 1 & 0 & 1 & 0 & 0 & 0 & 0 & 0 \\ 1 & 0 & 1 & 0 & 1 & 0 & 1 & 0 \\ 1 & 1 & 0 & 0 & 0 & 0 & 0 & 0 \\ 1 & 1 & 0 & 0 & 1 & 1 & 0 & 0 \\ 1 & 1 & 1 & 1 & 0 & 0 & 0 & 0 \\ 1 & 1 & 1 & 1 & 1 & 1 & 1 & 1 \end{bmatrix}.$$

Generally speaking, $g^{(n)}(\mathbf{u}_0^{N-1}) = \mathbf{u} \cdot \mathbf{G}^{(n)}$ where $\mathbf{G}^{(n)} = \mathbf{R}_N \bigotimes \mathbf{G}^{\otimes n}$, $\mathbf{R}_N$ is the bit-reversal permutation matrix and $\mathbf{G}^{\otimes n}$ is the Kronecker n^{th} power of $\mathbf{G}$ [63, Section VII].

The input to the transformation $g^{(n)}(\cdot)$ is called the **input information vector** $\mathbf{u} \in \mathbb{F}^{\ell^n}$. Application of $g^{(n)}(\cdot)$ on the set of information vectors uniquely transforms all these vectors to other codeword vectors $\mathbf{x} \in \mathbb{F}^{\ell^n}$. By constraining $\mathbf{u}$ to a subset of $\mathbb{F}^{\ell^n}$ the set of possible codewords $\mathbf{x}$ is expurgated. The common way to do so is to partition the information vector indices into two subsets $\mathcal{A}, \mathcal{A}^c \subseteq [N]_-$ and have the information vector values at indices in $\mathcal{A}^c$ predetermined and known both to the encoder and the decoder. As a consequence, a code $\mathcal{C}$ is defined as $\{g^{(n)}(\mathbf{u}) \mid \mathbf{u}_\mathcal{A} \in \mathbb{F}^k\}$ where the code dimension is $k = |\mathcal{A}|$, resulting in a code rate of $R = k/N$. We call the u_i's for which $i \in \mathcal{A}^c$, **frozen-symbols** of the code. In the case of a linear kernel, $\mathcal{C}$ is a coset code, where the frozen symbol values determine the vector shift of the associated linear code. For symmetric channels the values of the frozen symbols do not affect the quality of the ECC and as such we typically set $\mathbf{u}_{\mathcal{A}^c} = \mathbf{0}$ which results in a linear code.

Example 32.2.3 Consider the code of length 8 bits and rate $1/2$ induced by the transformation $g^{(3)}(\cdot)$ from Example 32.2.2. Let the frozen indices set be $\mathcal{A}^c = \{0,1,2,4\}$. Assume that we set all the frozen symbols to zero. Therefore the induced code is a linear code with generating matrix

$$B = \begin{bmatrix} 1 & 0 & 1 & 0 & 1 & 0 & 1 & 0 \\ 1 & 1 & 0 & 0 & 1 & 1 & 0 & 0 \\ 1 & 1 & 1 & 1 & 0 & 0 & 0 & 0 \\ 1 & 1 & 1 & 1 & 1 & 1 & 1 & 1 \end{bmatrix}.$$

Example 32.2.4 (Reed–Muller codes) Consider $g^{(n)}(\cdot)$ from Example 32.2.2. Selecting $\mathcal{A}$ to be the indices of the rows of $\mathbf{G}^{(n)}$ with Hamming weight at least 2^{n-r} results in the rth order Reed–Muller (RM) code $\mathcal{RM}(r,n)$ of length $N = 2^n$. Note that row i of $\mathbf{G}^{(n)}$ has weight $2^{\text{wt}_H(i)}$ where $\text{wt}_H(i)$ is the Hamming weight of the binary representation of i. For a general indices selection $\mathcal{A}$, the minimum distance of the code is $d_{min} = \min_{i \in \mathcal{A}} 2^{\text{wt}_H(i)}$ [1148, Chapter 6].

Codes constructed as kernel-based transformations are members of a broader family of general concatenated codes. Studying this affiliation is instructive because it provides both terminology and insights into this code construction and its related algorithms.

32.2.1   Kernel Based ECCs are Recursive GCCs

GCCs are error-correcting codes constructed by a technique introduced by Blokh and Zyabolov [222] and Zinoviev [1948]. In this construction, we have ℓ **outer-codes** $\{C_i\}_{i=0}^{\ell-1}$, where C_i is a code of length N_{out} and size M_i over alphabet F_i. We also have an **inner-code** of length N_{in} and size $\prod_{i=0}^{\ell-1} |F_i|$ over alphabet F, with a nested encoding function $\phi : F_0 \times F_1 \times \cdots \times F_{\ell-1} \to F^{N_{in}}$. The GCC that is generated by these components is a code of length $N_{out} \cdot N_{in}$ symbols and of size $\prod_{i=0}^{\ell-1} M_i$. It is created by taking an $\ell \times N_{out}$ matrix, in which the ith row is a codeword from C_i, and applying the inner mapping ϕ on each of the N_{out} columns of the matrix.

Some of the GCCs have good code parameters for short length codes when using appropriate combinations of outer-codes and a nested inner-code [648]. In fact, there exist GCCs with the best parameters known per their length and dimension. Moreover, decoding algorithms may utilize their structure by performing local decoding steps on the outer-codes and utilizing the inner-code layer for exchanging decisions between the outer-codes.

Arıkan identified polar codes as instances of recursive GCCs [63, Section I.D]. This observation is useful as it allows us to formalize the construction of any long polar code as a concatenation of several shorter polar codes (outer-codes) by using a kernel mapping (an inner-code). Therefore, applying this notion to Definition 32.2.1, we observe that a polar code of length $N = \ell^n$ symbols may be regarded as a collection of ℓ outer polar codes of length ℓ^{n-1}; the ith outer-code is $[\gamma_{i,j}]_{j=0}^{j=N/\ell-1} = g^{(n-1)}\left(\mathbf{u}_{i \cdot N/\ell}^{(i+1) \cdot N/\ell-1}\right)$ for $i \in [\ell]_-$. These codes are then joined together by employing an inner-code, defined by the kernel function $g(\cdot)$, on the outputs of these mappings. There are N/ℓ instances of the inner mapping such that instance number $j \in [N/\ell]_-$ is applied on the jth symbol from each outer-code.

FIGURE 32.1: A GCC representation of a polar code of length $N = \ell^n$ symbols constructed by a homogenous kernel according to Definition 32.2.1

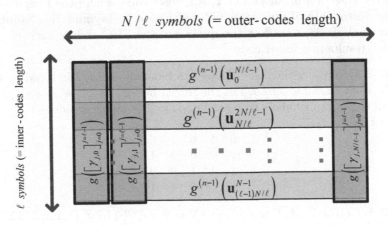

The above GCC formalization is illustrated in Figure 32.1. In this figure, we see the ℓ outer-code codewords of length ℓ^{n-1} depicted as horizontal rectangles (similar to rows of a matrix). The instances of the inner-codeword mapping are depicted as vertical rectangles that are located on top of the outer-codes rows (resembling columns of a matrix). This is appropriate, as this mapping operates on columns of the matrix whose rows are the outer-code codewords. Note that for brevity we only drew three instances of the inner mapping, but there should be ℓ^{n-1} instances of it, one for each column of this matrix. In the homogenous case, the outer-codes themselves are constructed in the same manner. Note, however, that even though these outer-codes have the same structure, they form different codes in the general case. The reason is that they may have different sets of frozen symbols.

Example 32.2.5 (GCC structure for Arıkan's Construction) Figure 32.2 depicts the GCC block diagram for the Arıkan $(u+v, v)$ Construction specified in Example 32.2.2. Note that $[\gamma_{0,j}]_{j=0}^{N/2-1} = g^{(n-1)}\left(\mathbf{u}_0^{N/2-1}\right)$ and $[\gamma_{1,j}]_{j=0}^{N/2-1} = g^{(n-1)}\left(\mathbf{u}_{N/2}^{N-1}\right)$ are the two outer-codes, each of length $N/2$ bits, and the kernel $g(\cdot)$ is the inner mapping.

FIGURE 32.2: Example 32.2.5's GCC representation (Arıkan's Construction)

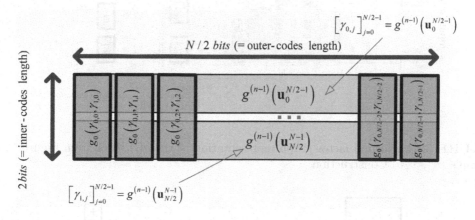

The GCC structure of polar codes can be also represented by a layered[1] Forney's normal factor graph (NFG) [744]. Layer #0 of this graph contains the inner mappings (represented as sets of vertices), and therefore we refer to it as the **inner layer**. Layer #1 contains the vertices of the inner layers of all the outer-codes that are concatenated by layer #0. We may continue and generate layer #i by considering the outer-codes that are concatenated by layer #$(i-1)$ and include in this layer all the vertices describing their inner mappings. This recursive construction process may continue until we obtain outer-codes that cannot be decomposed into nontrivial inner-codes and outer-codes. Edges (representing variables) connect between outputs of the outer-codes to the input of the inner mappings. This representation can be understood as viewing the GCC structure in Figure 32.1 from its side.

Example 32.2.6 (Layered NFG for Arıkan's Construction) Figure 32.3 depicts a layered factor graph representation for a polar code of length $N = 2^n$ symbols with kernel of $\ell = 2$ dimensions. We have two outer-codes (#0 and #1) each one of length $N/2$ bits that are connected by the inner layer (note the similarities to the GCC block diagram in Figure 32.2). Half edges represent the inputs $\mathbf{u}_0^{N-1}$ and the outputs $\mathbf{x}_0^{N-1}$ of the transformation. The edges, denoted by $\gamma_{i,j}$ for $j \in [N/2]_-$ and $i \in [2]_-$, connect the outputs of the two outer-codes to the inputs of the inner mapping blocks, $g(\cdot)$. To demonstrate the recursive

---

[1]In a layered graph, the vertices set can be partitioned into a sequence of sub-sets called layers and denoted by $L_0, L_1, \cdots, L_{k-1}$. The edges of the graph connect only vertices within the layer or in successive layers.

FIGURE 32.3: Representation of a polar code with kernel of $\ell = 2$ dimensions as a layered NFG

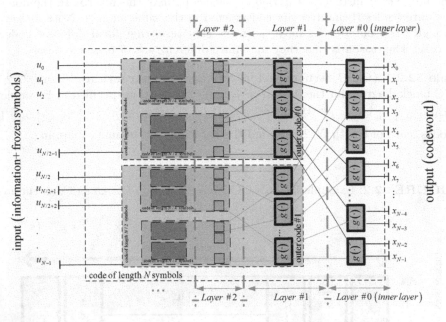

FIGURE 32.4: Normal factor graph representation of the $g(\cdot)$ block from Figure 32.3 for Arıkan's $(u + v, v)$ Construction

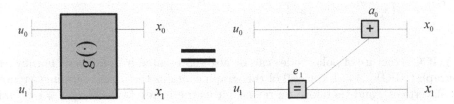

structure of the code we also unfolded the interior organization of outer-codes #0 and #1, thereby uncovering their inner-codes ($g(\cdot)$) and outer-codes (each one of length $N/4$ symbols).

Strictly speaking, the vertical blocks that represent the $g(\cdot)$ inner mapping are themselves factor graphs (i.e., collections of vertices and edges). An example of a normal factor graph specifying such a block is given in Figure 32.4 for Arıkan's $(u + v, v)$ Construction (see Example 32.2.5). Vertex a_0 represents a parity constraint and vertex e_1 represents an equivalence constraint. The half edges u_0, u_1 represent the inputs of the mapping, and the half edges x_0, x_1 represent its outputs. This graphical structure is probably the most popular visual representation of polar codes (see, e.g., [63, Figure 12] and [1148, Figure 5.2]) and is also known as the "butterflies" graph because of the edges arrangement.

Note that it is also possible to employ several types of kernels in the same construction and even kernels that have different types of input alphabets (non-homogenous kernels). The authors introduced such mixed-kernels construction that is advantageous over homogenous kernel constructions [1545].

32.3 Channel Splitting and Combining and the SC Algorithm

Let us consider a length $N = \ell^n$ symbols code that was constructed using a recursive transformation $g^{(n)}(\cdot)$ and information symbol freezing. Assume that any codeword $\mathbf{x} = g^{(n)}(\mathbf{u})$ is transmitted over a memoryless channel $W : \mathbb{F} \to \mathcal{Y}$ with channel transition function $W(y \mid x) = \mathrm{Prob}(Y = y \mid X = x)$ indicating the probability that $y \in \mathcal{Y}$ was received given that a symbol $x \in \mathbb{F}$ was transmitted. Transmitting a codeword of length N symbols on the channel means that we use the channel N times, each time transmitting a symbol x_i and receiving y_i. The vector $\mathbf{x}$ is transmitted over N independent copies of the channel W. However, $\mathbf{x}$ is a result of a transformation of $\mathbf{u}$ and as such its components are statistically dependant. Arıkan defined a sequence of N mathematical channels, each one having as input a different symbol of the information vector u_i for each $i \in [N]_-$. Thus, the transform induces new N (dependent) channels between the individual components of $\mathbf{u}$ and the outputs of the channels.

The operation of transforming N independent channels to dependent channels is called **channel combining**. Let $W^N : \mathbb{F}^N \to \mathcal{Y}^N$ denote the vector channel between the vector $\mathbf{x}$ and the output vector $\mathbf{y}$, and let $\mathcal{W}_N : \mathbb{F}^N \to \mathcal{Y}^N$ denote the synthesized vector channel between $\mathbf{u}$ and $\mathbf{y}$. Furthermore, let us define the channel transition functions of W^N and $\mathcal{W}_N$ as $W^N(\mathbf{y} \mid \mathbf{x}) = \mathrm{Prob}\,(\mathbf{Y} = \mathbf{y} \mid \mathbf{X} = \mathbf{x})$ and $\mathcal{W}_N(\mathbf{y} \mid \mathbf{u}) = \mathrm{Prob}\,(\mathbf{Y} = \mathbf{y} \mid \mathbf{U} = \mathbf{u})$, respectively. Therefore, it is easy to see that

$$\mathcal{W}_N(\mathbf{y} \mid \mathbf{u}) = W^N(\mathbf{y} \mid \mathbf{x} = g^{(n)}(\mathbf{u})) = \prod_{i=0}^{N-1} W(y_i \mid x_i).$$

The next step Arıkan takes is called **channel splitting**. As its name implies, the vector channel $\mathcal{W}_N$ is split into a set of N $\mathbb{F}$-symbol input coordinate channels denoted by $\mathcal{W}_N^{(i)} : \mathbb{F} \to \mathbb{F}^i \times \mathcal{Y}^N$, having the corresponding transition probabilities

$$\mathcal{W}_N^{(i)}(\mathbf{y}, \mathbf{u}_0^{i-1} \mid u_i) = \frac{1}{q^{N-1}} \sum_{\mathbf{u}_{i+1}^{N-1} \in \mathbb{F}^{N-i}} \mathcal{W}_N(\mathbf{y} \mid \mathbf{u}),$$

where $|\mathbb{F}| = q$. The outputs of a channel $\mathcal{W}_N^{(i)}$ $i \in [N]_-$ consist of the output vector of the physical channel, $\mathbf{y}$, and the ith prefix of the input vector $\mathbf{u}$, i.e., $\mathbf{u}_0^{i-1}$. It is further asserted that symbols $\mathbf{u}_{i+1}^{N-1}$ are unknown and uniformly distributed. The set of coordinate channels $\left\{\mathcal{W}_N^{(i)}\right\}_{i \in [N]_-}$ is referred to as the **synthetic channels** generated by the transformation.

The channel splitting induces the decoding method specified in Algorithm 32.3.1, referred to as Successive-Cancellation (SC) decoding. According to this method the decoder decides on symbols of input information sequentially according to (32.1). If i is a frozen index, then it uses the fixed and predetermined frozen value u_i. Otherwise it applies maximum-likelihood on the ith synthetic channel, in which it assumes the correctness of its previous decisions $\widehat{\mathbf{u}}_0^{i-1}$. The outputs of the algorithm are the estimated information vector $\widehat{\mathbf{u}}$ and its corresponding codeword $\widehat{\mathbf{x}}$.

The reader may raise two concerns with regards to the specification of Algorithm 32.3.1: (i) The complexity of the method seems to be exponential in the code length. It turns out that a recursive and efficient method exists for polar codes with time complexity of $O(q^\ell \cdot N \cdot \log N)$ and space complexity of $O(q \cdot N)$. (ii) Failure in step i of the algorithm causes the entire algorithm to fail. While error propagation is indeed a problem for any sequential method, it turns out that there exists a method to select $\mathcal{A}$ such that the code rate is approaching the symmetric capacity of the channel $\mathcal{W}$ with vanishing error probability (in the code length) in SC decoding.

Algorithm 32.3.1 (SC Decoding Method)

[▶▶▶] **Input:** $\mathbf{y}$.

//Initialization:
   ▷ Allocate decision vector $\widehat{\mathbf{u}} = \mathbf{0}$.

//Sequential decision:
   ▷ For each $i = 0, 1, 2, \ldots, N - 1$ do

$$\widehat{u}_i = \begin{cases} u_i & \text{if } i \in \mathcal{A}^c, \\ \operatorname{argmax}_{f \in \mathbb{F}} W_N^{(i)}(\mathbf{y}, \widehat{\mathbf{u}}_0^{i-1} \mid u_i = f) & \text{if } i \in \mathcal{A}, \end{cases} \tag{32.1}$$

[◀◀◀] **Output:** $\widehat{\mathbf{u}}$ and $\widehat{\mathbf{x}} = g(\widehat{\mathbf{u}})$.

We further consider a variant of the SC scheme called **genie-aided** (GA) SC decoding. According to this method, for deciding on u_i the decoder is informed by the genie on the true values of $\mathbf{u}_0^{i-1}$ (while in the standard SC it uses only the estimated values $\widehat{\mathbf{u}}_0^{i-1}$ that may be erroneous). While the GA SC description seems to be highly impractical, it turns out that it forms the theoretical foundation of polar codes. By proving that GA SC achieves capacity with vanishing probability of error it is possible to deduce that standard SC also achieves capacity with vanishing error probability. The reasoning behind this assertion is that until the first step i in which an error occurs, standard SC and GA SC behave the same. Specifically, both GA and the standard algorithm have the same first error i and therefore have the same frame error rate.

Let us return once again to the synthetic channels. The recursive mathematical construction took N independent channel usages of $\mathcal{W}$ and obtained N dependent synthetic channel $\left\{ \mathcal{W}_N^{(i)} \right\}_{i \in [N]_-}$ usages corresponding to the GA SC decoding algorithm. For a channel $\mathcal{W}$ we may define its symmetric capacity as $I(\mathcal{W}) = -1/q \sum_{x \in \mathcal{X}, y \in \mathcal{Y}} \operatorname{Prob}(Y = y \mid X = x) \cdot \log_q (W(Y = y \mid X = x))$. Similarly, for each of the synthetic channels we may attach a capacity as well:

$$I\left(\mathcal{W}_N^{(i)} \right) = - \sum_{\mathbf{u}_0^i \in \mathbb{F}^i, \mathbf{y} \in \mathcal{Y}^N} \frac{\operatorname{Prob}(\mathbf{Y} = \mathbf{y} \mid \mathbf{U}_0^i = \mathbf{u}_0^i)}{q^i} \cdot \log_q \left(W_N^{(i)}(\mathbf{y}, \mathbf{u}_0^{i-1} \mid u_i) \right).$$

Note that $I(\mathcal{W}), I\left(\mathcal{W}_N^{(i)} \right) \in [0, 1]$ and by the information chain-rule we have that $I(\mathcal{W}) = 1/N \cdot \sum_{i=0}^{N-1} I\left(\mathcal{W}_N^{(i)} \right)$. Finally, let us denote the GA SC error probability on step i as $P_e(\mathcal{W}_N^{(i)}) = \operatorname{Prob}(\widehat{U}_i \neq U_i)$ where $i \in [N]_-$.

Recall that Shannon measures the channel quality by the channel capacity, indicating the maximum code rate that can be used to transmit data on the channel with vanishing error probability. Similarly, $I\left(\mathcal{W}_N^{(i)} \right)$ indicates the quality of the ith synthetic channel. Specifically, as $I\left(\mathcal{W}_N^{(i)} \right) \to 0$ the channel worsens and the probability of error in (32.1) GA

SC grows to 1 (assuming that $i \notin \mathcal{A}$), i.e., $P_e(\mathcal{W}_N^{(i)}) \to 1$. Conversely, as $I\left(\mathcal{W}_N^{(i)}\right) \to 1$ the channel improves and the probability of error in (32.1) tends to zero ($P_e(\mathcal{W}_N^{(i)}) \to 0$). We say that a kernel $g(\cdot)$ is *polarizing* if almost all the synthetic channels have their symmetric capacities go to zero or to one as N grows. Let us formalize this.

For a length N transformation let us induce a uniform probability distribution on the synthetic channels and define I_n and $P_{e,n}$ to be random variables, respectively, standing for the capacity and error-probability of the nth synthetic channel selected uniformly at random. In other words, we pick each one of the N synthetic channels with probability $1/N$ and I_n and $P_{e,n}$ are the capacity and error-probability of this channel, respectively. For a subset of channels $\Gamma \subseteq \left\{\mathcal{W}_N^{(i)}\right\}_{i \in [N]_-}$, we understand the probabilistic notation $\text{Prob}(I_n \in \Gamma)$ as the chance to select any channel in Γ uniformly at random; therefore $\text{Prob}(I_n \in \Gamma) = |\Gamma|/N$.

Definition 32.3.2 A kernel $g(\cdot)$ **polarizes** a channel $\mathcal{W}$ if and only if for every $\delta \in (0, 0.5)$ we have

$$\lim_{n \to \infty} \text{Prob}(\iota_0 \le I_n \le \iota_1) = \begin{cases} I(\mathcal{W}) & \text{if } \iota_0 = 1 - \delta \text{ and } \iota_1 = 1, \\ 1 - I(\mathcal{W}) & \text{if } \iota_0 = 0 \text{ and } \iota_1 = \delta. \end{cases}$$

In other words, polarization indicates that as the code length grows almost all the channels either tend to be perfect ($I\left(\mathcal{W}_N^{(i)}\right) \approx 1$) or useless ($I\left(\mathcal{W}_N^{(i)}\right) \approx 0$), hence its name. Moreover, the proportion of perfect channels equals the original channel capacity. In the next section we turn to study conditions for kernel polarization.

Example 32.3.3 Figure 32.5 depicts the polarization phenomenon of the bi-additive white Gaussian noise (bi-AWGN) channel at $E_s/N_0 = -1.54\,\text{dB}$ ($I(\mathcal{W}) = 0.6$ bits) by the $(u+v, v)$ kernel. On the left, we took $g^{(11)}(\cdot)$ of length 2048 bits and estimated using density evolution the capacities of the synthetic channels $I\left(\mathcal{W}_{2048}^{(i)}\right)$. It is easy to see that most of channels are either with capacity close to 1 or to 0, even for this intermediate length of 2048 bits. The right figure indicates the trend of polarization: as the length of transformation grows the proportion of channels with capacity close to 1 tends to $I(\mathcal{W})$ and those with capacity close to 0 tends to $1 - I(\mathcal{W})$.

32.4 Polarization Conditions

In this section we describe conditions on a kernel $g(\cdot)$ to be polarizing according to Definition 32.3.2. Before doing that we define a useful kernel property called partial distances sequence.

Definition 32.4.1 For an ℓ dimensions kernel $g(\cdot)$ over $\mathbb{F}$ we define the following **sequences of partial distances**, where $\text{d}_{\text{H}}(\cdot, \cdot)$ is Hamming distance:

$$D_{min}^{(i)}\left(\mathbf{u}_0^{i-1}\right) = \min \left\{ \text{d}_{\text{H}}\left(g(\mathbf{u}_0^{i-1}, \omega, \mathbf{v}_{i+1}^{\ell-1}), g(\mathbf{u}_0^{i-1}, \theta, \mathbf{w}_{i+1}^{\ell-1})\right) \mid \right.$$
$$\left. \mathbf{v}_{i+1}^{\ell-1}, \mathbf{w}_{i+1}^{\ell-1} \in \mathbb{F}^{N-i-1}, \omega, \theta \in \mathbb{F}, \omega \neq \theta \right\} \text{ with } \mathbf{u}_0^{i-1} \in \mathbb{F}^i \quad \text{and} \tag{32.2}$$

$$D_{min}^{(i)} = \min \left\{ D_{min}^{(i)}\left(\mathbf{u}_0^{i-1}\right) \mid \mathbf{u}_0^{i-1} \in \mathbb{F}^i \right\}. \tag{32.3}$$

FIGURE 32.5: Polarization property for the $(u+v,v)$ kernel based transformation of length 2048 bits, operated on the bi-AWGN channel $\mathcal{W}$ with $E_s/N_0 = -1.54$ dB, $I(\mathcal{W}) = 0.6$ bits: (a) the synthetic channel capacity $I\left(\mathcal{W}_{2048}^{(i)}\right)$ as a function of the channel index i; (b) proportion of polarizing channels as function of the block length ($\delta = 0.2$)

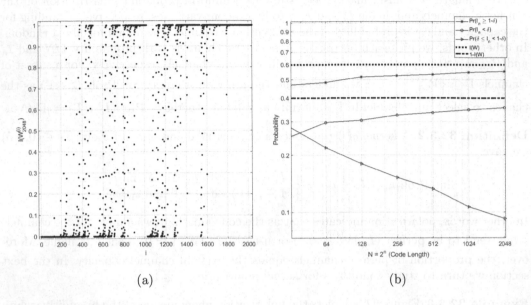

(a)   (b)

A partial distance element $D_{min}^{(i)}\left(\mathbf{u}_0^{i-1}\right)$ corresponds to the ith step of SC, in which we have to decide on u_i after we have already determined the prefix $\mathbf{u}_0^{i-1}$ of the input $\mathbf{u}$ to the kernel $g(\cdot)$. In this case the event $u_i = \omega$ corresponds to the subset $\left\{g(\mathbf{u}_0^{i-1}, \omega, \mathbf{v}_{i+1}^{\ell-1}) \mid \mathbf{v}_{i+1}^{\ell-1} \in \mathbb{F}^{N-i-1}, \omega \in \mathbb{F}\right\}$. Generally speaking, the larger the gap between any two codewords when each one of them corresponds to a different u_i value, the smaller the chance that a decoder will perform erroneous decisions. The minimum gap corresponding to a certain prefix is defined in (32.2) and the minimum across all prefixes is defined in (32.3).

Example 32.4.2 Consider the following binary kernel

$$g(\mathbf{u}_0^3) = [u_0, u_1, u_2, u_3] \cdot \begin{bmatrix} 1 & 0 & 0 & 0 \\ 1 & 1 & 0 & 0 \\ 1 & 0 & 1 & 0 \\ 1 & 1 & 1 & 1 \end{bmatrix}.$$

We have that $D_{min}^{(0)} = 1$, $D_{min}^{(1)} = 2$, $D_{min}^{(2)} = 2$, $D_{min}^{(3)} = 4$. We also have that $D_{min}^{(i)} = D_{min}^{(i)}\left(\mathbf{u}_0^{i-1}\right)$ for all $i \in [4]_-$ and $\mathbf{u}_0^{i-1} \in \mathbb{F}_2^i$.

SC on this kernel may be interpreted as sequentially deciding on the prefix of the input vector $\mathbf{u}_0^3$, thereby eliminating some of the possible candidates. As we proceed in the decision sequence the minimum partial distance grows (except between step 1 and step 2). This property suggests that $P_e\left(\mathcal{W}_4^{(0)}\right) \geq P_e\left(\mathcal{W}_4^{(1)}\right) \geq P_e\left(\mathcal{W}_4^{(3)}\right)$, for good enough (e.g., having sufficient signal to noise ratio (SNR)) input channel $\mathcal{W}$.

Example 32.4.3 (RS kernels [1406]) Let the Reed–Solomon kernel over $\mathbb{F}_q$ be defined as $g(\mathbf{u}_0^{q-1}) = \mathbf{u} \cdot \mathbf{G}_{RS}(q)$ where $\mathbf{G}_{RS}(q)$ is a $q \times q$ matrix given by

$$
\mathbf{G}_{RS}(q) = \begin{bmatrix}
1 & 1 & 1 & \cdots & 1 & 0 \\
1 & \alpha & \alpha^2 & \cdots & \alpha^{q-2} & 0 \\
1 & \alpha^2 & \alpha^4 & \cdots & \alpha^{2(q-2)} & 0 \\
1 & \vdots & \vdots & \cdots & \vdots & \vdots \\
1 & \alpha^{q-2} & \alpha^{2(q-2)} & \cdots & \alpha^{(q-2)(q-2)} & 0 \\
1 & 1 & 1 & \cdots & 1 & 1
\end{bmatrix}
$$

with α a primitive element of $\mathbb{F}_q$. We have that for each $i \in [q]$, the i lowest rows of $\mathbf{G}_{RS}(q)$ form a generating matrix of the $[q, i, q - i + 1]$ extended RS code. As a consequence, we can easily observe that $D_{min}^{(i)} = i + 1$ for all $i \in [q]_-$.

Arıkan defined the polarization term, and provided the first example of a polarizing kernel $(u + v, v)$ [63]. Korada et al. found conditions for polarizing binary and linear kernels [1149, 1150]. Mori and Tanaka considered the general case of a mapping $g(\cdot)$ that is not necessarily linear and binary [1406, 1403]. For brevity, we provide in this chapter two useful necessary and sufficient conditions for polarization based on the latter works. The first condition deals with binary kernels that are not necessarily linear.

Theorem 32.4.4 (Polarization of Binary Kernels [1403]) *Let $g(\cdot)$ be a binary kernel of ℓ dimensions. If there exists $i \in [\ell]_-$ and $\mathbf{u}_0^{i-1} \in \{0, 1\}^i$ where $D_{min}^{(i)}(\mathbf{u}_0^{i-1}) \geq 2$, then $g(\cdot)$ is polarizing according to Definition 32.3.2.*

The next theorem deals with linear kernels over $\mathbb{F}_q$. Such kernels are specified by an $\ell \times \ell$ kernel generating matrix $\mathbf{G}$. We say that two kernel matrices are **polarization equivalent** if the synthetic channels induced by them have the same capacity sequences for any symmetric channel $\mathcal{W}$. Lemma 32.4.5 specifies an equivalence condition.

Lemma 32.4.5 (Polarization Equivalence Conditions [1406]) *Two matrices $\mathbf{G}$ and $\widetilde{\mathbf{G}}$ are polarization equivalent if there exists an invertible upper-triangular matrix $\mathbf{V}$ such that $\mathbf{G}$ can be derived from $\mathbf{V} \cdot \widetilde{\mathbf{G}}$ by permuting its columns and multiplying each column by elements from $\mathbb{F}_q \backslash \{0\}$.*

Definition 32.4.6 A lower triangular matrix with a unit diagonal that is equivalent to a matrix $\mathbf{G}$ is called a **standard form** of $\mathbf{G}$.

In general, a standard form of a matrix $\mathbf{G}$ is not necessarily unique. If there exists a standard form of $\mathbf{G}$ that is the identity matrix, then it is unique.

Example 32.4.7 Let α be a primitive element of $\mathbb{F}_4$. In the notation of Example 32.4.3, $\mathbf{G}_{RS}(4) = \mathbf{V} \cdot \widetilde{\mathbf{G}}$ where

$$
\widetilde{\mathbf{G}} = \begin{bmatrix}
1 & 0 & 0 & 0 \\
1 & 1 & 0 & 0 \\
\alpha^2 & \alpha & 1 & 0 \\
1 & 1 & 1 & 1
\end{bmatrix} \text{ and } \mathbf{V} = \begin{bmatrix}
1 & \alpha^2 & 1 & 0 \\
0 & \alpha^2 & \alpha^2 & 0 \\
0 & 0 & \alpha & 0 \\
0 & 0 & 0 & 1
\end{bmatrix}.
$$

So $\widetilde{\mathbf{G}}$ is a standard form of $\mathbf{G}_{RS}(4)$.

Theorem 32.4.8 (Polarization of Linear Kernels [1406]) *For a linear kernel $g(\cdot)$ over $\mathbb{F}_q$ specified by an invertible matrix $\mathbf{G}$ with non-identity standard form, the following conditions are equivalent.*

(a) *Any q-ary input symmetric channel $\mathcal{W}$ is polarized by $g(\cdot)$.*

(b) *There exists a standard form $\widetilde{\mathbf{G}}$ of $\mathbf{G}$ that contains at least one element $\widetilde{G}_{i,j} \in \mathbb{F}_q$ that is not a member of any proper subfield of $\mathbb{F}_q$.*

(c) *All the standard forms $\widetilde{\mathbf{G}}$ of $\mathbf{G}$ contain at least one element $\widetilde{G}_{i,j} \in \mathbb{F}_q$ that is not a member of any proper subfield of $\mathbb{F}_q$.*

Example 32.4.9 The kernel from Example 32.4.2 is polarizing based on Theorem 32.4.4. Following Example 32.4.7 and Theorem 32.4.8 we have that the kernel defined by $\mathbf{G}_{RS}(4)$ is polarizing. In fact, for any q that is a power of a prime, we have that the kernel defined by $\mathbf{G}_{RS}(q)$ is polarizing [1406].

32.4.1 Polarization Rate

The partial distances sequence $\left\{ D_{min}^{(i)} \right\}_{i=0}^{\ell-1}$ of a polarizing kernel $g(\cdot)$ influences a performance property of SC, called the **polarization rate**. Roughly speaking, we may expect that as the distances in the partial distances sequence grow, the kernel should have better correction performance.

In this subsection we elaborate on the polarization rate. We begin by defining a **kernel exponent**, which is a numerical measure of how fast the error probability of the good synthetic channels decays.

Definition 32.4.10 Let $g(\cdot)$ be an ℓ dimensions polarizing kernel over $\mathbb{F}$ with partial distances sequence $\left\{ D_{min}^{(i)} \right\}_{i=0}^{\ell-1}$. The **polarizing exponent** of $g(\cdot)$ is defined as $E(g) = \frac{1}{\ell} \cdot \sum_{i=0}^{\ell-1} \log_\ell D_{min}^{(i)}$.

Example 32.4.11 The Arıkan kernel as well as the kernel in Example 32.4.2 have a polarization exponent of 0.5. Linear kernels with $\mathbf{G}_{RS}(q)$ have a polarization exponent $\log_q(q!)/q$. Specifically, the kernel based on $\mathbf{G}_{RS}(4)$ has the exponent 0.573120.

The following theorem associates the error-correction performance of the GA SC and exponent.

Theorem 32.4.12 (Polarization Rate [1406, 1403]) *Let $g(\cdot)$ be an ℓ dimensions kernel over $\mathbb{F}$ that polarizes a channel $\mathcal{W}$. Then for any $\epsilon > 0$*

$$\lim_{n \to \infty} \text{Prob}\left(P_{e,n} < 2^{-\ell^{(E(g)-\epsilon)n}} \right) = I(\mathcal{W}), \tag{32.4}$$

and

$$\lim_{n \to \infty} \text{Prob}\left(P_{e,n} < 2^{-\ell^{(E(g)+\epsilon)n}} \right) = 0. \tag{32.5}$$

Roughly speaking, (32.4) and (32.5) suggest that as N grows the SC GA decoding error rate of almost all the good channels (the $N \cdot I(\mathcal{W})$ channels with capacity approaching 1) is in the range of $\left[2^{-N^{E(g)+\epsilon}}, 2^{-N^{E(g)-\epsilon}} \right]$.

Note that the exponent is always smaller than 1. As a consequence the error probability decays sub-exponentially in the length of the code N. Korada et al. showed that as ℓ grows, the maximum achievable polarization exponent of binary linear kernels of ℓ dimensions approaches 1 [1150, Theorem 22]. The exponent of binary kernels is ≤ 0.5 for $\ell \leq 13$ [1259, 1546]. For $\ell = 16$ dimensions the maximum exponents for binary linear and nonlinear kernels are 0.51828 [1150, Example 28] and 0.52742 [1546], respectively. The RS kernels have the maximum exponent per their number of dimensions.

32.5 Polar Codes

In this section we incorporate the kernel-based ECCs construction from Section 32.2 with the polarization theory of Section 32.4 to specify polar codes. Polar codes are constructed using polarizing kernel-based transforms where the frozen symbols are selected in a way that the frame error rate under SC is vanishing with the block length. Moreover, such codes can achieve the symmetric capacity of the channel using SC. The existence of such codes is guaranteed by Theorem 32.5.1 that was first conceived by Arıkan for the $(u+v, v)$ kernel [63] and later was generalized by Korada et al. [1149, 1150] and Mori and Tanaka [1406, 1403] for other kernels.

Theorem 32.5.1 (Polar Codes Achieve Capacity with SC Decoding) *Let $g(\cdot)$ be an ℓ dimensions kernel over $\mathbb{F}_q$ polarizing a channel $\mathcal{W}$ with symmetric capacity $I(\mathcal{W})$. For each $\epsilon > 0$ there exists a sequence of codes $\{\mathcal{C}^{(i)}\}_{i=1}^{\infty}$ where $\mathcal{C}^{(n)}$ is of length ℓ^n, rate $\geq I(\mathcal{W}) - \epsilon$, constructed from $g^{(n)}(\cdot)$ and non-frozen symbols set $\mathcal{A}^{(n)}$, such that the frame error of the code under SC decoding is $P_{e,n} = o\left(2^{-\ell^{(E(g)-\epsilon)n}}\right)$.*

Proof: Given $\epsilon > 0$, for each n let us specify $\mathcal{A}^{(n)}$ to be the set of channels with GA SC error probability below $2^{-\ell^{(E(g)-2\epsilon)n}}$. According to Theorem 32.4.12 there exists n_0 such that for each $n > n_0$ we have $|\mathcal{A}^{(n)}| \geq I(\mathcal{W}) - \epsilon$. In that case the probability of frame error under SC is bounded from above by

$$P_{e,n} < \sum_{i \in \mathcal{A}^{(n)}} P_e(\mathcal{W}_N^{(i)}) \leq (I(\mathcal{W}) - \epsilon) \cdot \ell^n \cdot 2^{-\ell^{(E(g)-2\epsilon)n}} = o\left(2^{-\ell^{(E(g)-\epsilon)n}}\right).$$

$\square$

Remark 32.5.2 A more refined notion of polarization that links the gap to capacity and the length of a polar code achieving it (with certain error probability) was studied by Guruswami and Xia [883]. The scaling behavior of polar codes is a related phenomenon to that [823, 918, 919, 1151, 1397, 1398]. Sasoglu et al. showed how to achieve the Shannon capacity of a non-symmetric channel by a suitable extension of the input alphabet that depends on the capacity achieving input probability distribution [1622].

The proof of Theorem 32.5.1 suggests that in order to specify a good polar code, the quality of the synthetic channels has to be assessed and graded. Bad channels are assigned to be frozen while good channels are used to carry the information. This channel organization operation is referred to as **polar code design**.

32.5.1 Polar Code Design

Given a memoryless channel $\mathcal{W}$ and a polarizing kernel $g(\cdot)$, the **polar code design** problem of a code of length $N = \ell^n$ symbols and rate R (such that $R \cdot N$ is an integer) is to select the optimal subset of synthetic channel indices $\mathcal{A} \subseteq [N]_-$ ($|\mathcal{A}| = R \cdot N$) assigned to carry the information. Information symbols with indices in $\mathcal{A}^c$ are frozen. The optimality here is typically in the sense of bringing to minimum the code's frame error rate (FER) or symbol error rate under SC decoding over $\mathcal{W}$.

Since the selection criterion for a good channel is based on the GA SC algorithm, the performance of this algorithm has to be evaluated. In some cases running the SC GA algorithm based on the channel $\mathcal{W}$ using, e.g., Monte-Carlo simulation and estimating properties of the channels (e.g., symbol-error-rate, Bhattacharyya, capacity, etc.) may allow us to assess their ordering [63].

Alternatively, it may be possible to assess the qualities of the synthetic channels by performing estimation of their log likelihood ratio (LLR) distribution or parameter thereof. Mori and Tanaka followed that approach by employing the density evolution (DE) algorithm for Arıkan's Construction [1404, 1403]. This algorithm recursively computes the probability density function of the LLR messages assuming that the zero codeword was transmitted. However, as the block length grows the computation precision has to increase and thus the complexity. An efficient approximation to DE is done by assuming that all computed distributions are Gaussian, thereby only the expectation and the standard deviation of them have to be tracked [1819].

Tal and Vardy developed a quantization method that allows performing DE analysis while still maintaining manageable complexity [1769]. They devised two approximation methods that bound the bit-channel capacity from above and from below by applying the so-called **channel-upgrading** and **channel-degrading** methods, respectively. A fidelity parameter μ controls the tightness of these bounds. As μ increases the bounds are tighter while complexity also increases. Generalizations of this approach for multiple access channels (MACs) and non-binary input alphabets were studied by Pereg and Tal [1503].

So far we have seen that polar codes achieve the symmetric capacity of channels. Because of their kernel based recursive structure, they also have efficient implementations of their encoding and decoding algorithms. This is explored in the next section.

32.6   Polar Codes Encoding Algorithms

Let $g(\cdot)$ be an ℓ dimensions kernel over $\mathbb{F}_q$. Let us assume that the computation of the mapping $g(\mathbf{u})$ for any $\mathbf{u} \in \mathbb{F}_q^\ell$ has time and space complexities of τ and s, respectively. It is easy to see that for any $\mathbf{u} \in \mathbb{F}_q^N$ ($N = \ell^n$) it is possible to compute $g^{(n)}(\mathbf{u})$ in time complexity of $O(\tau \cdot N/\ell \cdot \log_\ell(N))$ and space complexity of $O(N \cdot \log(q) + s)$. Let us illustrate this by the recursive GCC construction depicted in Figure 32.1.

Let $\mathbf{u}$ be vector of length N over $\mathbb{F}_q$, serving as the encoder input. Each polar code of length N is defined by its frozen-indicator vector $\mathbf{z}$ (of length N), such that $z_i = 1$ if and only if the ith input of the encoder is frozen (i.e., fixed and known to both the encoder and the decoder) and $z_i = 0$ otherwise. Let $\mathcal{C}$ be a polar code of length $N = \ell^n$ symbols and dimension k. We have that $\sum_{i=0}^{N-1} z_i = N - k$. Given a user message $\breve{\mathbf{u}} \in \mathbb{F}_q^k$, it is the role of the encoder to output a corresponding codeword $\mathbf{x} \in \mathbb{F}_q^N$. Given a $\breve{\mathbf{u}}$ and $\mathbf{z}$, it is easy to generate $\mathbf{u} \in \mathbb{F}^n$, the encoder input, such that values of $\breve{\mathbf{u}}$ are sequentially assigned to the non-frozen components of $\mathbf{u}$ and elements corresponding to frozen indices are set to predetermined values (here, we arbitrarily decided to set the frozen values of $\mathbf{u}$ to zero), i.e.,

$$u_i = \begin{cases} 0 & \text{if } z_i = 1, \\ \breve{u}_{i - \sum_{m=0}^{i} z_m} & \text{if } z_i = 0. \end{cases}$$

If the code length is $N = \ell$, then the encoder's output is $g(\mathbf{u})$. Otherwise we compute the ℓ outer-codes codewords by employing the transformation $g^{(n-1)}(\cdot)$ on the sub-vectors $\left\{ \mathbf{u}_{i \cdot N/\ell}^{(i+1) \cdot N/\ell - 1} \right\}_{i \in [\ell]_-}$. We then employ the inner mapping on the outer-code codewords by using $g(\cdot)$ on the columns of the outer-code codewords matrix.

The method described here results in non-systematic encoding. Systematic encoding of polar codes has the same (asymptotic) complexity but may require more effort in order to determine the assignment to the vector $\mathbf{u}$ [65, 1544, 1614]. Note that being systematic or non-systematic doesn't affect the frame error rate performance of any decoder of polar codes; however it turns out that the symbol-error-rate of systematically encoded codewords is better than that of non-systematic systems under SC decoding [65].

32.7 Polar Codes Decoding Algorithms

The recursive GCC structure of polar codes enables recursive formalizations of the decoding algorithms associated with them. These algorithms benefit from simple and clear descriptions, which support elegant analysis. There are two types of equivalent recursion notions that are useful to describe the decoding algorithm: (i) The recursion presented by Arıkan [63] is a **left-to-right** or **outer-to-inner** recursion, meaning that in order to calculate an LLR of the leftmost layer of the normal factor graph of Figure 32.3 the algorithm asks for calculations of certain LLRs of edges of its adjacent right layer. This request triggers a calculation of LLRs of edges at the next layer to the right. The recursion continues until we reach the input to the algorithm (the channel LLRs). (ii) A **right-to-left** or **inner-to-outer** recursion in which we first calculate LLRs resulting from the decoder input and then move to the left by recursively decoding the outer-codes of the construction. The way we describe such recursions is by the scheme given in Algorithm 32.7.1, meaning that for each code we first prepare the LLR input to its outer-code decoder (by using the inner-code structures) and then recursively call the outer-code decoder [1544].

Consider the GCC structure of Figure 32.1. In this construction we have a code of length N symbols that is composed of ℓ outer-codes, denoted by $\{\mathcal{C}_i\}_{i=0}^{\ell-1}$, each one of length N/ℓ symbols. The decoding algorithms that are considered here are composed of ℓ pairs of steps. The ith pair is dedicated to decoding $\mathcal{C}_i$ as described in Algorithm 32.7.1. Typically, the codes $\{\mathcal{C}_i\}_{i=0}^{\ell-1}$ are polar codes of length N/ℓ symbols, thereby creating the recursive structure of the decoding algorithm.

Algorithm 32.7.1 (Decoding Outer-code $\mathcal{C}_i$, $i \in [\ell]_-$)

//**STEP** $2 \cdot i$:
> ▷ Using the previous steps, prepare the inputs to the decoder of outer-code $\mathcal{C}_i$.

//**STEP** $2 \cdot i + 1$:
> ▷ Run the decoder of code $\mathcal{C}_i$ on the inputs you prepared.
> ▷ Process the output of this decoder, together with the outputs of the previous steps.

The structure of Algorithm 32.7.1 is typical for decoding algorithms of GCCs. For example, see the decoding algorithms in Dumer's survey on GCCs [648] and his RM recursive decoding algorithms [649, 651]. In fact, Dumer's simplified decoding algorithm for RM codes [649, Section IV] is the SC decoding for Arıkan's structure we describe in Subsection 32.7.1.

32.7.1   The SC Decoding Algorithm

We already introduced SC in Algorithm 32.3.1. In this section we study efficient inner-to-outer recursive implementations of it. We begin by considering the SC decoder for Arıkan's $(u+v, v)$ Construction. A brief description of the algorithm for generalized arbitrary kernels then follows.

32.7.1.1   SC for $(u+v, v)$

The inputs of the SC algorithm for Arıkan's Construction are listed below.

- A vector of N input LLRs, $\boldsymbol{\lambda}$, such that $\lambda_j = \ln \frac{\text{Prob}(Y_j = y_j \mid X_j = 0)}{\text{Prob}(Y_j = y_j \mid X_j = 1)}$ for $j \in [N]_-$, where Y_j is the observation of the jth realization of the channel $\mathcal{W}$, $X_j \to Y_j$.

- Frozen indicator vector $\mathbf{z} \in \{0, 1\}^N$, defined in Section 32.6.

The algorithm outputs two binary vectors of length N bits, $\widehat{\mathbf{u}}$ and $\widehat{\mathbf{x}}$, containing the estimated information word (including frozen bits placed at appropriate locations) and its corresponding codeword, respectively. The SC function signature is defined as $[\widehat{\mathbf{u}}, \widehat{\mathbf{x}}] = \text{SCDecoder}(\boldsymbol{\lambda}, \mathbf{z})$.

First, let us describe the decoding algorithm for a code of length $N = 2$ bits, i.e., for the basic kernel $g^{(1)}(u, v) = [u + v, v]$. Note that this is the base case of the recursion. We get as input $\boldsymbol{\lambda} = [\lambda_0, \lambda_1]$ which are the LLRs of the output of the channel (λ_0 corresponds to the first output of the channel and λ_1 corresponds to the second output). The procedure has four steps as described in Algorithm 32.7.2.

Algorithm 32.7.2 (SC of the $(u+v, v)$ Kernel)

[▶▶▶] **Input:** $[\lambda_0, \lambda_1]$; $[z_0, z_1]$.

//STEP 0:
  ▷ Compute the LLR of u: $\widehat{\lambda} = 2 \tanh^{-1} \left(\tanh(\lambda_0 / 2) \tanh(\lambda_1 / 2) \right)$.

//STEP 1:
  ▷ Decide on u; denote the decision by $\widehat{u}$.

//STEP 2:
  ▷ Compute the LLR of v, given the estimate of $\widehat{u}$: $\widehat{\lambda} = (-1)^{\widehat{u}} \cdot \lambda_0 + \lambda_1$.

//STEP 3:
  ▷ Decide on v; denote the decision by $\widehat{v}$.

[◀◀◀] **Output:** $\widehat{\mathbf{u}} = [\widehat{u}, \widehat{v}]$; $\widehat{\mathbf{x}} = [\widehat{u} + \widehat{v}, \widehat{v}]$.

Note that STEPs 1 and 3 may be done based on the LLRs computed in STEPs 0 and 2, respectively (i.e., by their sign), or using additional side information (for example, if u is frozen, then the decision is based on its known value). A decoder of a polar code of length N bits is described in Algorithm 32.7.3.

The reader may observe the similarities of the computations in STEPs 0 and 2 in Algorithms 32.7.2 and 32.7.3. Those steps correspond to the LLR calculation of u and v respectively for the base case or the LLR preparations for outer-code #0 and #1 decoders in the longer code. They both employ the constraint manifested by $g(\cdot)$. Specifically, for Algorithm 32.7.3 those computations are separately employed on columns of Figure 32.2 or the inner layer of Figure 32.3. The input to the computations are the pair of LLRs corresponding to the output of $g(\cdot)$ (i.e., λ_{2j} and λ_{2j+1} corresponding to x_{2j} and x_{2j+1}, respectively) and the results of it are LLRs corresponding to inputs of $g(\cdot)$ (i.e., $\widehat{\lambda}_j$ corresponding to $\gamma_{0,j}$ and $\gamma_{1,j}$ in STEPs 0 and 2, respectively).

Algorithm 32.7.3 (SC Recursive Description for a $(u+v, v)$ Polar Code of Length $N = 2^n$ bits)

[▶▶▶] **Input:** λ; z.

//STEP 0:

▷ Compute the LLR input vector, $\widehat{\lambda}_0^{N/2-1}$, for the first outer-code such that

$$\widehat{\lambda}_i = 2\tanh^{-1}\left(\tanh(\lambda_{2i}/2)\tanh(\lambda_{2i+1}/2)\right) \quad \text{for all } i \in [N/2]_- .$$

//STEP 1:

▷ Give the vector $\widehat{\lambda}$ as an input to the polar code decoder of length $N/2$. Also provide to the decoding algorithm, the indices of the frozen bits from the first half of the codeword (corresponding to the first outer-code); i.e., run

$$\left[\widehat{\mathbf{u}}^{(0)}, \ \widehat{\mathbf{x}}^{(0)}\right] = \text{SCDecoder}\left(\widehat{\lambda}, \ \mathbf{z}_0^{N/2-1}\right).$$

where $\widehat{\mathbf{u}}^{(0)}$ is the information word estimation for the first outer-code, and $\widehat{\mathbf{x}}^{(0)}$ is its corresponding codeword.

//STEP 2:

▷ Using λ and $\widehat{\mathbf{x}}^{(0)}$, prepare the LLR input vector, $\widehat{\lambda}_0^{N/2-1}$, for the second outer-code, such that

$$\widehat{\lambda}_i = (-1)^{\widehat{x}_i^{(0)}} \cdot \lambda_{2i} + \lambda_{2i+i} \quad \text{for all } i \in [N/2]_- .$$

//STEP 3:

▷ Give the vector $\widehat{\lambda}$ as an input to the polar code decoder of length $N/2$. In addition, provide the indices of the frozen bits from the second half of the codeword (corresponding to the second outer-code); i.e., run

$$\left[\widehat{\mathbf{u}}^{(1)}, \ \widehat{\mathbf{x}}^{(1)}\right] = \text{SCDecoder}\left(\widehat{\lambda}, \ \mathbf{z}_{N/2}^{N-1}\right),$$

where $\widehat{\mathbf{u}}^{(1)}$ and $\widehat{\mathbf{x}}^{(1)}$ are the estimations of the information word and its corresponding codeword of the second outer-code.

[◀◀◀] **Output:**

- $\widehat{\mathbf{u}} = \left[\widehat{\mathbf{u}}^{(0)}, \widehat{\mathbf{u}}^{(1)}\right]$;

- $\widehat{\mathbf{x}} = \left[\widehat{x}_i^{(0)} + \widehat{x}_i^{(1)}, \widehat{x}_i^{(1)}\right]_{i=0}^{N/2-1} .$

32.7.1.2   SC for General Kernels

Let us now generalize this decoding algorithm for a polar code with a general kernel. In this case for a code of length N symbols, we have an ℓ dimension kernel $g(\cdot)$ over $\mathbb{F}_q$. The inputs and the outputs of the decoding algorithm are the same as in the $(u + v, v)$ case, except that here the LLRs may correspond to a non-binary alphabet. As a consequence, we need to have $q - 1$ LLR input vectors $\left\{ \boldsymbol{\lambda}^{(t)} \right\}_{t \in \mathbb{F}_q \backslash \{0\}}$ each one of length N and defined such that

$$\lambda_j^{(t)} = \ln\left(\frac{\operatorname{Prob}\left(Y_j = y_j \mid X_j = 0\right)}{\operatorname{Prob}\left(Y_j = y_j \mid X_j = t\right)} \right)$$

for $j \in [N]_-$, where Y_j is the measurement of the jth channel realization $\mathcal{W}$, $X_j \to Y_j$. Furthermore $\mathbf{u}$ and $\mathbf{x}$ are in $\mathbb{F}_q^N$. By convention we have $\lambda_j^{(0)} = 0$ and therefore it doesn't have to be calculated. The general SC decoder signature is

$$[\widehat{\mathbf{u}}, \ \widehat{\mathbf{x}}] = \operatorname{SCDecoder}\left(\left\{ \boldsymbol{\lambda}^{(t)} \right\}_{t \in \mathbb{F}_q \backslash \{0\}}, \ \mathbf{z} \right). \tag{32.6}$$

In the GCC structure of this polar code there are ℓ outer-codes $\{\mathcal{C}_i\}_{i=0}^{\ell-1}$, each one of length N/ℓ symbols. We assume that each outer-code has a decoding algorithm associated with it. This decoding algorithm is assumed to receive as input the channel observations on the outer-code symbols (usually formatted as LLR vectors). If the outer-code is a polar code, then this algorithm should also receive the indices of the frozen symbols of the outer-code. We require that the algorithm outputs its estimation on the information vector and its corresponding outer-code codeword.

Let us first consider the recursion base-case of a code of length ℓ generated by a single application of the kernel, i.e., $\mathbf{x} = g(\mathbf{u})$. Assuming that we already decided on symbols $\mathbf{u}_0^{i-1}$ (denote this decision by $\widehat{\mathbf{u}}_0^{i-1}$), computing the LLRs $\left\{ \widehat{\lambda}^{(t)} \right\}_{t \in \mathbb{F}_q}$ corresponding to the ith input of the transformation (i.e., u_i) is done according to the following rule

$$\widehat{\lambda}^{(t)} = \ln\left(\frac{\sum_{\mathbf{u}_{i+1}^{\ell-1} \in \mathbb{F}_q^{\ell-i-1}} R_g\left(\widehat{\mathbf{u}}_0^{i-1}, 0, \mathbf{u}_{i+1}^{\ell-1} \right)}{\sum_{\mathbf{u}_{i+1}^{\ell-1} \in \mathbb{F}_q^{\ell-i-1}} R_g\left(\widehat{\mathbf{u}}_0^{i-1}, t, \mathbf{u}_{i+1}^{\ell-1} \right)} \right), \tag{32.7}$$

where $R_g(\mathbf{u}_0^{\ell-1}) = \exp\left(-\sum_{r=0}^{\ell-1} \lambda_r^{(x_r)} \right)$, such that $\mathbf{x} = g(\mathbf{u})$.

Consequently, SC decoding for a polar code of length ℓ includes sequential calculations of the likelihood values $\left\{ \widehat{\lambda}^{(t)} \right\}_{t \in \mathbb{F}_q \backslash \{0\}}$ corresponding to non-frozen u_i according to (32.7) followed by a decision on u_i (denoted by $\widehat{u}_i$) for $i \in [\ell]_-$. If u_i is frozen, we set $\widehat{u}_i$ to be equal to its predetermined value. Finally, in (32.6) we output $\widehat{\mathbf{u}} = [\widehat{u}_0 \ \widehat{u}_1 \ \cdots \ \widehat{u}_{\ell-1}]$, and $\widehat{\mathbf{x}} = g(\widehat{\mathbf{u}})$.

For longer codes ($N > \ell$) we follow the recipe of Algorithm 32.7.1, using the LLR preparation rule of (32.7) applied on the columns of the matrix in Figure 32.1. Before applying step $2 \cdot i$ we already decoded outer-codes $\{\mathcal{C}_m\}_{m=0}^{i-1}$ and stored the outer-codes information words and codeword in $\{\widehat{\mathbf{u}}^{(m)}\}_{m=0}^{i-1}$ and $\{\widehat{\mathbf{x}}^{(m)}\}_{m=0}^{i-1}$, respectively. We use these outcomes to generate an LLR input vector to the decoder of $\mathcal{C}_i$ denoted as $\left\{ \left[\widehat{\lambda}_j^{(t)} \right]_{j=0}^{N/\ell-1} \right\}_{t \in \mathbb{F}_q \backslash \{0\}}$. In order to compute $\widehat{\lambda}_j^{(t)}$ we use (32.7) with input LLRs corresponding to the jth column of the GCC matrix and known values $\widehat{\mathbf{u}}_0^{i-1} = \left[\widehat{x}_j^{(m)} \right]_{m=0}^{i-1}$. Then we call the SC decoder for outer-code $\mathcal{C}_i$, using the computed LLRs and $\mathbf{z}_{i \cdot N/\ell}^{(i+1) \cdot N/\ell - 1}$ as inputs to (32.6), and receive

$\widehat{\mathbf{u}}^{(i)}$ and $\widehat{\mathbf{x}}^{(i)}$ as outputs. Finally, we use the ℓ decided outer-codes codewords $\{\widehat{\mathbf{x}}^{(m)}\}_{m=0}^{\ell-1}$ to generate $\mathcal{C}$'s estimated codeword (employing the inner mapping) and the outer-codes information words $\{\widehat{\mathbf{u}}^{(m)}\}_{m=0}^{\ell-1}$ to generate the information word (by serially concatenating them).

32.7.1.3 SC Complexity

The SC decoding steps can be classified into three categories: **(SC.a)** LLR calculations; **(SC.b)** decision making based on these LLRs; **(SC.c)** partial encoding of the decided symbols using the kernel. The time and space complexities of **(SC.a)** category operations dominate the time complexity of the entire SC algorithm. This is our justification for regarding the number of operations of **(SC.a)** as a good measure of the SC decoder time complexity.

For a kernel of ℓ dimensions over the field $\mathbb{F}_q$, the straight-forward calculation of the likelihoods performed on the ith decoding step (32.7) ($i \in [\ell]_-$) requires $O(\ell \cdot q^{\ell-i})$ operations. For linear kernels it is possible to apply trellis decoding based on the zero-coset's parity check matrix which yield an upper bound of $O(\ell \cdot q^i)$ operations. Let τ and s be the total time and space complexities for LLR computations of a kernel $g(\cdot)$ (note that $\tau = O(\ell \cdot q^\ell)$ and $\tau = O(\ell \cdot q^{\ell/2})$ for general and linear codes respectively and $s = O(\ell \cdot q)$). Inspection of the recursive SC decoding algorithm yields total time complexity of $O(\tau \cdot N/\ell \cdot \log_\ell(N))$ and total space complexity of $O(N \cdot q + s)$.

Although being advantageous due to its light complexity, SC suffers from two inherent deficiencies resulting in error-correction performance sub-optimality: (i) Decisions on information symbols are final and cannot be changed after being made. (ii) When the SC algorithm decides on u_i, it assumes that all the assignments to $\mathbf{u}_i^{N-1}$ are possible, although some of the components may be frozen. The SCL decoder is a generalization of SC that aims to mitigate these shortcomings by allowing multiple decision options to exist at any given point in time.

32.7.2 The SCL Decoding Algorithm

Introduced by Tal and Vardy [1770], SCL provides significant improvement to the error-correction performance of SC by supporting a list of decision candidates at any given time. On step i of SC, the algorithm assumes a single assignment to its $i - 1$ prefix, i.e., its previous decision $\widehat{\mathbf{u}}_0^{i-1}$. Such an assignment is referred to as a **decoding path**. On the other hand, in SCL with list size L, the decoder considers at most L different assignments to its $i - 1$ prefix $\widehat{\mathbf{u}}_0^{<0>\,i-1}$, $\widehat{\mathbf{u}}_0^{<1>\,i-1}, \ldots, \widehat{\mathbf{u}}_0^{<L-1>\,i-1}$. SCL computes likelihoods associated with extending those options in one symbol, resulting in at most $q \cdot L$ possibilities. From these candidates we keep L decoding paths which are the most likely and move to step $i+1$.

For each decoding path, SCL performs the same operations that would have been employed in an SC decoding instance for that prefix. In addition to that, SCL has to determine the best L paths from the $q \cdot L$ extended paths, which can be done in $O(q \cdot L)$ time and space. As a consequence, for fixed list size L the complexity of SCL is bounded from above by L times the complexity of SC. Two additional challenges of SCL are efficiently keeping track of the L different prefixes and the fact that comparison between different SCL decoding paths requires retrieval of the log-likelihood of the paths (and not just their LLRs) [99, 1543].

As we increase the list size L the error-correction performance of SCL improves. However, the relative gain due to L growth is decreasing since we approach the maximum-likelihood (ML) performance. Tal and Vardy [1770] found that concatenating a cyclic redundancy check (CRC) to the information word, such that the SCL final decoded codeword must meet the CRC constraints, significantly improves the performance, making it comparable to the state of the art coding systems [1770, Section V].

In order to attach a CRC of b symbols we have to unfreeze b information symbols in the inner polar code such that the total information rate of the concatenated system remains the same. This technique is characterized by the following tradeoff. Increasing the CRC size b causes the performance of the inner-code SCL to deteriorate. On the other hand, it may improve the performance of the concatenated system as a larger proportion of the polar code codewords is eliminated. Therefore, the CRC should be carefully selected based on the SNR and the list size. As a general rule of thumb, as the list size increases, larger CRC should be considered.

Example 32.7.4 Figure 32.6 provides FER curves of several polar codes of length $N = 2048$ bits and rate $R = 0.5$ simulated on a bi-AWGN channel. Two kernels are used here: Arıkan's $(u + v, v)$ and the RS(4) kernel. The codes were designed using GA simulation at $E_b/N_0 = 2$ dB. We considered SC and SCL decoders with list size (L) in $\{2, 4, 8, 16, 32\}$. For SCL with $L = 16, 32$ we also examined CRC of 16 bits concatenated to the information part. The reader may notice the following phenomena: (i) SCL outperforms SC. Moreover, as the list size grows the decoding performance improves; however it eventually saturates, in the limit of ML decoding. (ii) Attaching CRC significantly lowers the FER. (iii) RS(4) based polar codes outperform $(u + v, v)$ polar codes due to their better partial distances distribution.

FIGURE 32.6: FER curves of polar codes of length $N = 2048$ bits and rate $R = 0.5$ simulated on the bi-AWGN channel

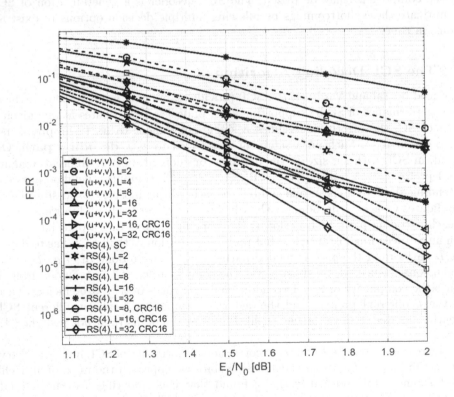

Remark 32.7.5 Additional polar code decoding algorithms are the iterative belief-propagation (BP) [62, 63, 1148, 1543] and the sequential SC Stack (SCS) [1442, 1443]. Many useful modifications to SCL were proposed over the years; see, e.g., [234, 917, 1235, 1613]. For more detailed discussion on SC and SCL complexity the reader may refer to [1543, Section 5.5].

32.8   Summary and Concluding Remarks

Polar codes are capacity achieving GCCs constructed by short kernels recursively forming the constituent codes. The kernel of the code is the fundamental key for the behavior of the system. A polarizing kernel transforms a physical channel W into an ensemble of (almost only) extremal synthetic channels that are either perfect or useless. A polar code is constructed by carrying information on the inputs of the perfect channels and freezing the inputs to the useless ones. Sequential decoding algorithms such as SC and SCL can be efficiently implemented by taking advantage of the recursive kernel structure.

In this chapter we studied the essentials of polar codes as efficient solutions to the channel-coding problem. This was indeed their first usage. However, it was soon discovered that they are applicable to many other problems in information theory as well. For all of these problems the polar coding solution transforms the probabilistic structures of the problems to a set of a few extremal structures which are trivial to handle. As a consequence, the solutions inherit the same polynomial time complexity of the encoding and decoding algorithms that we saw for the ECC case. We mention here just a few examples: lossless source coding [3, 4, 64, 471], rate-distortion and lossy source coding [1085, 1148, 1152], wiretap channels [49, 969, 1326], broadcast channels [815, 1148, 1396] and the multi-access channel (MAC) problem [5, 6, 1623].

Chapter 33

Secret Sharing with Linear Codes

Cunsheng Ding
The Hong Kong University of Science and Technology

33.1   Introduction to Secret Sharing Schemes

Secret sharing is an interesting topic of cryptography[1] and has a number of applications in real-world security systems. The objective of this chapter is to give a brief introduction to secret sharing schemes and their connections to codes.

Definition 33.1.1 A **secret sharing scheme** consists of

- a dealer;

- a group $\mathcal{P} = \{P_1, P_2, \ldots, P_\ell\}$ of ℓ participants;

- a secret space $\mathcal{S}$ with the uniform probability distribution;

- ℓ share spaces $\mathcal{S}_1, \mathcal{S}_2, \ldots, \mathcal{S}_\ell$;

- a share computing procedure F; and

- a secret recovering procedure G.

The dealer randomly chooses an element $s \in \mathcal{S}$ as the secret to be shared among the ℓ participants in $\mathcal{P}$, then computes $(s_1, s_2, \ldots, s_\ell) \in \mathcal{S}_1 \times \mathcal{S}_2 \times \cdots \times \mathcal{S}_\ell$ with F, s and possibly some other random parameters, and finally distributes s_i to Participant P_i as his/her share of the secret s for each i with $1 \le i \le \ell$.

If a subset $\{P_{i_1}, P_{i_2}, \ldots, P_{i_t}\}$ of t participants meets together with their shares $\{s_{i_1}, s_{i_2}, \ldots, s_{i_t}\}$ and is able to determine s from G, $(i_1, i_2, \ldots, i_t)$, and $(s_{i_1}, s_{i_2}, \ldots, s_{i_t})$,

---

[1]Code-based cryptography is the topic of Chapter 34.

where $1 \leq t \leq \ell$, then the subset $\{P_{i_1}, P_{i_2}, \ldots, P_{i_t}\}$ is called an **access set**. The set Γ of all access sets is called the **access structure**, which is an element of the power set $2^{\mathcal{P}}$ of $\mathcal{P}$. An access set is **minimal** if any proper subset of it is not an access set. The set of all minimal access sets is called the **minimal access structure**.

A secret sharing scheme is said to be t-**democratic** if every group of t participants is in the same number of minimal access sets, where $t \geq 1$. A participant is called a **dictator** if she/he is a member of every minimal access set.

A secret sharing scheme is said to be **perfect** if any subset of participants and their shares either gives all information about the secret s or does not contain any information about s. A secret sharing scheme is said to be **ideal** if each share has the same size as the corresponding secret.

The following example illustrates the building blocks of secret sharing schemes and related concepts.

Example 33.1.2 Let $\ell \geq 2$ be an integer, and let $\mathcal{P} := \{P_1, P_2, \ldots, P_\ell\}$ be the set of participants. Let $\mathcal{S} := \mathbb{F}_2^{256}$ be the secret set. The share sets $\mathcal{S}_i$ are the same as the secret set $\mathcal{S}$. The secret sharing procedure is the following:

- The dealer chooses a random vector $s \in \mathcal{S}$ (which can be obtained by flipping a balanced coin 256 times) as the secret.

- The dealer then chooses $\ell - 1$ random vectors $s_1, \ldots, s_{\ell-1}$ in $\mathbb{F}_2^{256}$, and then computes $s_\ell = s + s_1 + \cdots + s_{\ell-1}$; this computation is the share computing procedure F.

- The dealer finally gives s_i to Participant P_i for all i.

The secret sharing recovering procedure goes as follows. When all ℓ participants meet together with their shares, they recover the secret by computing the sum of all their shares, i.e.,

$$s := s_1 + s_2 + \cdots + s_{\ell-1} + s_\ell.$$

This is the secret recovering procedure G. It is easily seen that the access structure $\Gamma = \{\{P_1, P_2, \ldots, P_\ell\}\}$. Hence, this secret sharing scheme has only one access set, which is the only minimal access set. This scheme is both perfect and ideal. It is also t-democratic for all $1 \leq t \leq l$, and every participant is a dictator.

Definition 33.1.3 A secret sharing scheme is called a (t, ℓ)-**threshold scheme**, where $1 \leq t \leq \ell$, if the following conditions are satisfied:

1. the number of participants is ℓ;

2. any subset of t or more participants can obtain all information about the secret with their shares; and

3. any subset of $t - 1$ or fewer participants cannot get any information about the secret with their shares.

By definition, any (t, ℓ)-threshold scheme is perfect and t-democratic. The scheme described in Example 33.1.2 is an (ℓ, ℓ)-threshold scheme.

The first secret sharing schemes appeared in the public literature in 1979, and were developed by Shamir [1659] and Blakley [214] independently. The Blakley scheme has a geometric description, while the Shamir scheme has an algebraic expression. The two schemes are in fact equivalent. The following example describes the Shamir secret sharing scheme.[2]

---

[2]The Shamir scheme is also presented in the context of algebraic geometry codes in Section 15.8.4.

Example 33.1.4 Let q be a prime power. Let ℓ and t be positive integers with $2 \le t \le \ell < q$. In the **Shamir scheme**, the secret space is $\mathcal{S} := \mathbb{F}_q$ with the uniform probability distribution, the set of participants is $\mathcal{P} := \{P_1, P_2, \ldots, P_\ell\}$. The share spaces are the same as the secret space. The sharing computing procedure is the following:

- The dealer randomly chooses an element $s \in \mathcal{S}$ as the secret to be shared among the ℓ participants.

- The dealer then randomly and independently chooses $t - 1$ elements $a_1, \ldots, a_{t-1}$ from $\mathbb{F}_q$, and constructs the polynomial

$$a(x) = s + \sum_{i=1}^{t-1} a_i x^i \in \mathbb{F}_q[x].$$

- The dealer finally randomly and independently chooses ℓ pairwise distinct elements $u_1, \ldots, u_\ell$ from $\mathbb{F}_q^* := \mathbb{F}_q \setminus \{0\}$, computes $s_i = a(u_i)$ and distributes (u_i, s_i) to P_i as the share for all i.

When t participants $P_{i_1}, \ldots, P_{i_t}$ meet together with their shares, they can recover the secret s by solving the following system of t linear equations to obtain $a(x)$ and hence s:

$$s_{i_j} = a(u_{i_j}) \quad \text{for } j = 1, 2, \ldots, t.$$

It can be easily proved that any set of $t - 1$ or fewer participants get no information about s with their shares. Hence, the Shamir scheme is a (t, ℓ)-threshold scheme.

33.2 The First Construction of Secret Sharing Schemes

Two constructions of secret sharing schemes from linear codes have been investigated in the literature. The objective of this section is to introduce the first construction, which can be viewed as a variant of the second construction considered later in this chapter.

Construction 33.2.1 Let $\mathcal{C}$ be an $[n, k, d]_q$ code with generator matrix G, where $k \ge 2$. In the **secret sharing scheme based on G with respect to the first construction**, the secret space and share spaces are $\mathbb{F}_q$, and the set of participants is $\mathcal{P} = \{P_1, P_2, \ldots, P_n\}$. The share computing procedure is the following:

- The dealer chooses randomly an element s_1 from $\mathbb{F}_q$ as the secret to be shared among the n participants.

- The dealer then randomly and independently chooses $s_2, \ldots, s_k$ from $\mathbb{F}_q$, and computes

$$\mathbf{t} = (t_1, t_2, \ldots, t_n) := \mathbf{s}G,$$

where $\mathbf{s} = (s_1, s_2, \ldots, s_k)$.

- The dealer gives t_i to Participant P_i as the share.

It is easily seen that a set of shares $\{t_{i_1}, t_{i_2}, \ldots, t_{i_m}\}$ determines the secret s_1 if and only if the vector $\mathbf{e}_1 = (1, 0, \ldots, 0)^\mathsf{T}$ is a linear combination of $\mathbf{g}_{i_1}, \mathbf{g}_{i_2}, \ldots, \mathbf{g}_{i_m}$, where $\mathbf{g}_i$ is the ith column of G. Furthermore, the secret sharing scheme is perfect and ideal.

The secret recovering procedure goes as follows. Given a set of shares $\{t_{i_1}, \ldots, t_{i_m}\}$ which can determine the secret, the subgroup of m participants solves the equation

$$\mathbf{e}_1 = \sum_{j=1}^{m} x_j \mathbf{g}_{i_j}$$

to obtain all x_j. They then recover the secret with

$$s_1 = \mathbf{s} \cdot \mathbf{e}_1 = \sum_{j=1}^{m} x_j \mathbf{s} \cdot \mathbf{g}_{i_j} = \sum_{j=1}^{m} x_j t_{i_j},$$

where $\mathbf{s} \cdot \mathbf{e}_1$ is the standard inner product of the two vectors.

Definition 33.2.2 The **support** of $\mathbf{c} = (c_0, c_1, \ldots, c_{n-1}) \in \mathbb{F}_q^n$ is defined by

$$\text{supp}(\mathbf{c}) = \{i \mid 0 \le i \le n-1, \, c_i \ne 0\}.$$

Theorem 33.2.3 *Let $\mathcal{C}$ be an $[n, k, d]_q$ code with generator matrix $G = \begin{bmatrix} \mathbf{g}_1 & \cdots & \mathbf{g}_n \end{bmatrix}$. The access structure of the secret sharing scheme based on G with respect to the first construction is given by*

$$\Gamma = \left\{ \{P_i \mid i \in \text{supp}((x_1, x_2, \ldots, x_n))\} \,\middle|\, \mathbf{e}_1 = \sum_{j=1}^{n} x_j \mathbf{g}_j \right\}.$$

Remark 33.2.4 Theorem 33.2.3 indicates that the access structure of the secret sharing scheme based on a linear code with respect to the first construction depends on the choice of the underlying generator matrix G.

McEliece and Sarwate were the first to observe a connection between secret sharing and error-correcting codes [1367]. The secret sharing schemes based on linear codes with respect to the first construction were considered by McEliece and Sarwate [1367] where the Shamir scheme formulated in terms of polynomial interpolation was generalized in terms of Reed–Solomon codes. In terms of vector spaces, Brickell also studied this kind of secret sharing schemes [302]. A generalization of this approach was given by Bertilsson according to [1830] and also by van Dijk [1830].

Secret sharing schemes based on MDS codes over $\mathbb{F}_q$ with respect to the first construction were investigated by Renvall and Ding in [1587]. Their access structures are known and documented in the following theorems.

Theorem 33.2.5 *Let G be a generator matrix of an $[n, k, n - k + 1]_q$ MDS code $\mathcal{C}$. In the secret sharing scheme based on G with respect to the first construction, any k shares determine the secret.*

Theorem 33.2.6 *Let G be a generator matrix of an $[n, k, n-k+1]_q$ MDS code $\mathcal{C}$ such that its ith column is a scalar multiple of $\mathbf{e}_1$. In the secret sharing scheme based on G with respect to the first construction, a set of shares determines the secret if and only if it contains the ith share or its cardinality is no less than k.*

The access structure of the secret sharing schemes described by Theorem 33.2.6 could be interesting in applications where the president of some organization alone should know the secret, and a set of other participants can determine the secret if and only if the number of participants is no less than k.

A practical question concerning such a secret sharing scheme is whether each $[n, k, n - k + 1]$ MDS code has such a generator matrix and how to find them if there are some. The answer is "yes" and such a generator matrix is found as follows. First, take any generator matrix G'. Then find a coordinate of the first row vector of G' and add multiples of this row to other rows to make the coordinate of other rows equal zero. The obtained matrix is the desired one.

Theorem 33.2.7 *Let G be a generator matrix of an $[n, k, n - k + 1]_q$ MDS code C such that $\mathbf{e}_1$ is a linear combination of $\mathbf{g}_1$ and $\mathbf{g}_2$, but not a multiple of either of them, where $\mathbf{g}_i$ denotes the ith column vector of G. The access structure of the secret sharing scheme based on G with respect to the first construction is then described as follows:*

(a) *any k shares determine the secret;*

(b) *a set of shares with cardinality $k - 2$ or less determines the secret if and only if the set contains both t_1 and t_2;*

(c) *a set of $k - 1$ shares $\{t_{i_1}, t_{i_2}, \ldots, t_{i_{k-1}}\}$*

    (i) *determines the secret when it contains both t_1 and t_2;*

    (ii) *cannot determine the secret when it contains one and only one of t_1 and t_2;*

    (iii) *determines the secret if and only if $\mathrm{rank}(\mathbf{e}_1, \mathbf{g}_{i_1}, \ldots, \mathbf{g}_{i_{k-1}}) = k - 1$ when it contains neither of t_1 and t_2.*

Theorem 33.2.8 *Let G be a generator matrix of an $[n, k, n - k + 1]_q$ MDS code C. The secret sharing scheme based on G with respect to the first construction is a (k, n)-threshold scheme if and only if $\mathbf{e}_1$ is not a linear combination of any $k - 1$ columns of G.*

Example 33.2.9 Let $u_1, u_2, \ldots, u_n$ be n pairwise distinct elements of $\mathbb{F}_q^*$. Let t be an integer with $1 \leq t \leq n$. Denote by C the linear code over $\mathbb{F}_q$ with generator matrix

$$
G = \begin{bmatrix}
1 & 1 & \cdots & 1 \\
u_1 & u_2 & \cdots & u_n \\
\vdots & \vdots & \ddots & \vdots \\
u_1^{t-1} & u_2^{t-1} & \cdots & u_n^{t-1}
\end{bmatrix}.
$$

Then C is a generalized Reed–Solomon code and is an $[n, t, n - t + 1]_q$ MDS code. It is easily seen that $\mathbf{e}_1$ is not a linear combination of any $t - 1$ columns of G. It then follows from Theorem 33.2.8 that the secret sharing scheme based on G with respect to the first construction is a (t, n)-threshold scheme. It is clearly the Shamir scheme.

The minimum distance of a linear code is one of the important parameters of the code since it determines the error-correcting capacity of the code. The minimum distance of the dual code has an implication on the access structure of the secret sharing scheme based on the code. The following is such a result developed in [1587].

Theorem 33.2.10 *Let G be a generator matrix of an $[n, k, d]_q$ code C. In the secret sharing scheme based on G with respect to the first construction, if each of two sets of shares*

$\{t_{i_1}, \ldots, t_{i_l}\}$ and $\{t_{j_1}, \ldots, t_{j_m}\}$ *determines the secret, but no subset of any of them can do so, then*

$$|\{i_1, \ldots, i_l\} \cup \{j_1, \ldots, j_m\}| - |\{i_1, \ldots, i_l\} \cap \{j_1, \ldots, j_m\}| \geq d^{\perp},$$

where $d^{\perp}$ is the minimum distance of the dual code $\mathcal{C}^{\perp}$.

The access structures of the secret sharing scheme based on linear codes with respect to the first construction are very complex in general. Not much work in this direction has been done.

Remark 33.2.11 Any secret s can be encoded into a binary string $s_1 s_2 \cdots s_u$ with an encoding scheme. Then s can be shared bit by bit by a group of participants. Hence, binary linear codes or linear codes over a small field $\mathbb{F}_q$ can be employed for sharing a secret with large size. This remark applies also to the second construction to be treated later.

33.3 The Second Construction of Secret Sharing Schemes

The first construction of secret sharing schemes was dealt with in Section 33.2. This section treats the second construction. As will be seen later, the second construction looks more flexible and interesting.

33.3.1 Minimal Linear Codes and the Covering Problem

Minimal linear codes are a class of special linear codes that give secret sharing schemes with interesting access structures. The objective of this subsection is to introduce minimal linear codes and the covering problem of linear codes.

Definition 33.3.1 Two vectors $\mathbf{a}$ and $\mathbf{b}$ in $\mathbb{F}_q^n$ are said to be **scalar multiples** of each other if $\mathbf{a} = h\mathbf{b}$ for some $h \in \mathbb{F}_q^*$.

Two vectors in $\mathbb{F}_2^n$ have the same support if and only if they are identical. However, two distinct vectors in $\mathbb{F}_q^n$ may have the same support for $q > 2$. If $\mathbf{a} = h\mathbf{b}$ for some $h \in \mathbb{F}_q^*$, then $\mathbf{a}$ and $\mathbf{b}$ have the same support.

Definition 33.3.2 A vector $\mathbf{u} \in \mathbb{F}_q^n$ **covers** a vector $\mathbf{v} \in \mathbb{F}_q^n$ if $\operatorname{supp}(\mathbf{u})$ contains $\operatorname{supp}(\mathbf{v})$. We write $\mathbf{v} \preceq \mathbf{u}$ if $\mathbf{v}$ is **covered** by $\mathbf{u}$, and $\mathbf{v} \prec \mathbf{u}$ if $\operatorname{supp}(\mathbf{v})$ is a proper subset of $\operatorname{supp}(\mathbf{u})$.

It is now time to define minimal codewords and minimal linear codes.

Definition 33.3.3 A codeword $\mathbf{u}$ in a linear code $\mathcal{C}$ is said to be **minimal** if $\mathbf{u}$ covers only the codeword $a\mathbf{u}$ for all $a \in \mathbb{F}_q$, but no other codewords in $\mathcal{C}$. A linear code $\mathcal{C}$ is said to be **minimal** if every codeword in $\mathcal{C}$ is minimal.

Example 33.3.4 The dual of the $[7, 4, 3]$ binary Hamming code is minimal. Its codewords and supports are the following:

Codeword	Support	Codeword	Support
0000000	$\emptyset$	1000111	$\{0, 4, 5, 6\}$
0101011	$\{1, 3, 5, 6\}$	0011101	$\{2, 3, 4, 6\}$
1101100	$\{0, 1, 3, 4\}$	1011010	$\{0, 2, 3, 5\}$
0110110	$\{1, 2, 4, 5\}$	1110001	$\{0, 1, 2, 6\}$

Definition 33.3.5 The **covering problem** of a linear code is to determine the set of all minimal codewords of the code.

The covering problem of linear codes is very hard in general. Minimal linear codes are special in the sense that all their codewords are minimal. It has been a hot topic in the past decades to search for minimal linear codes. The following is a useful tool developed by Ashikmin and Barg [67].

Lemma 33.3.6 *A linear code C over $\mathbb{F}_q$ is minimal if $w_{\min}/w_{\max} > (q-1)/q$. Herein and hereafter, $w_{\min}$ and $w_{\max}$ denote the minimum and maximum nonzero Hamming weights in C, respectively.*

Many families of minimal linear codes have been identified with Lemma 33.3.6. Most of these codes are one-weight codes, two-weight codes [330], and three-weight codes [557]. These codes usually have few weights and have relatively small dimensions. Minimal codes were also investigated in [28] and [69].

Lemma 33.3.6 presents a sufficient condition, which is not necessary. This was demonstrated with two examples by Cohen, Mesnager, and Patey [420]. Recently, the first infinite family of minimal binary linear codes with $w_{\min}/w_{\max} \leq 1/2$ was discovered in [380]. The reader is informed that necessary and sufficient conditions for a linear code to be minimal have been just derived and more infinite families of minimal linear codes with $w_{\min}/w_{\max} \leq (q-1)/q$ have been constructed [556, 943]. Families of minimal linear codes are also presented in Chapter 18.

33.3.2 The Second Construction of Secret Sharing Schemes

We are now ready to introduce the second construction of secret sharing schemes from linear codes.

Construction 33.3.7 Let C be an $[n, k, d]_q$ code with generator matrix G where $G = [\mathbf{g}_0^\mathsf{T} \ \mathbf{g}_1^\mathsf{T} \ \cdots \ \mathbf{g}_{n-1}^\mathsf{T}]$ with $\mathbf{g}_i \in \mathbb{F}_q^k$. The **secret space** S **of the secret sharing scheme based on C with respect to the second construction** is $\mathbb{F}_q$ with the uniform probability distribution. The participant set $\mathcal{P} = \{P_1, P_2, \ldots, P_{n-1}\}$. The secret sharing procedure goes as follows:

- The dealer chooses randomly $\mathbf{u} = (u_0, \ldots, u_{k-1})$ such that $s = \mathbf{u} \cdot \mathbf{g}_0$, where $\mathbf{u} \cdot \mathbf{g}_0$ is the standard inner product of the vectors $\mathbf{u}$ and $\mathbf{g}_0$.

- The dealer then treats $\mathbf{u}$ as an information vector and computes the corresponding codeword

$$\mathbf{t} = (t_0, t_1, \ldots, t_{n-1}) := \mathbf{u}G.$$

- The dealer finally gives t_i to Participant P_i as the share for each $i \geq 1$.

Note that $t_0 = \mathbf{u} \cdot \mathbf{g}_0 = s$. A set of shares $\{t_{i_1}, t_{i_2}, \ldots, t_{i_m}\}$ determines the secret if and only if $\mathbf{g}_0$ is a linear combination of $\mathbf{g}_{i_1}, \ldots, \mathbf{g}_{i_m}$. Hence, the access structure

$$\Gamma = \left\{ \{P_{i_1}, \ldots, P_{i_m}\} \ \middle| \ \begin{array}{c} 1 \leq i_1 < \cdots < i_m \leq n-1 \\ \mathbf{g}_0 \text{ is a linear combination of } \mathbf{g}_{i_1}, \ldots, \mathbf{g}_{i_m} \end{array} \right\}.$$

It is easily seen that a set of shares $\{t_{i_1}, t_{i_2}, \ldots, t_{i_m}\}$ gives no information about s if $\mathbf{g}_0$ is not a linear combination of $\mathbf{g}_{i_1}, \ldots, \mathbf{g}_{i_m}$. Hence, the secret sharing scheme is perfect.

Suppose that

$$\mathbf{g}_0 = \sum_{j=1}^{m} x_j \mathbf{g}_{i_j}.$$

Then the secret s is recovered by the group $\{P_{i_1}, \ldots, P_{i_m}\}$ of participants after computing

$$s = \sum_{j=1}^{m} x_j t_{i_j}.$$

The second construction above was considered by Massey [1348, 1349]. Van Dijk pointed out in [1830] that Massey's approach is a special case of the construction introduced by Bertilsson and Ingemarson in [184]. The following theorem was observed by Massey. It documents the access structure of the secret sharing scheme based on $\mathcal{C}$; see [1348, 1349].

Theorem 33.3.8 *Let $\mathcal{C}$ be an $[n, k, d]_q$ code. Then the access structure of the secret sharing scheme based on $\mathcal{C}$ with respect to the second construction is given by*

$$\Gamma = \left\{ \{P_i \mid i \in \mathrm{supp}((c_1, \ldots, c_{n-1}))\} \mid (1, c_1, \ldots, c_{n-1}) \in \mathcal{C}^{\perp} \right\}.$$

Remark 33.3.9 Theorem 33.3.8 shows that the access structure of the secret sharing scheme based on a linear code with respect to the second construction is independent of the choice of the underlying generator matrix G. This is a major difference between the first and second constructions.

The minimum distance $d^{\perp}$ of the dual code $\mathcal{C}^{\perp}$ gives a lower bound on the size of any minimal access set in the secret sharing scheme based on $\mathcal{C}$ with respect to the second construction. Specifically, we have the following which was proved by Renvall and Ding in [1587].

Theorem 33.3.10 *Let $\mathcal{C}$ be an $[n, k, d]_q$ code and let $d^{\perp}$ denote the minimum distance of the dual code. In the secret sharing scheme based on $\mathcal{C}$ with respect to the second construction, any set of $d^{\perp} - 2$ or fewer shares does not give any information on the secret, and there is at least one set of $d^{\perp}$ shares that determines the secret.*

33.3.3 Secret Sharing Schemes from the Duals of Minimal Codes

The following theorem describes the access structure of the secret sharing scheme based on the dual of a minimal linear code and was developed by Ding and Yuan in [570].

Theorem 33.3.11 *Let $\mathcal{C}$ be an $[n, k, d]_q$ code, and let $H = \begin{bmatrix} \mathbf{h}_0 & \mathbf{h}_1 & \cdots & \mathbf{h}_{n-1} \end{bmatrix}$ be its parity check matrix. If $\mathcal{C}^{\perp}$ is minimal, then in the secret sharing scheme based on $\mathcal{C}$ with respect to the second construction, the set of participants is $\mathcal{P} = \{P_1, P_2, \ldots, P_{n-1}\}$, and there are altogether q^{n-k-1} minimal access sets.*

(a) *When $d = 2$, the access structure is as follows.*

If $\mathbf{h}_i$ is a scalar multiple of $\mathbf{h}_0$, $1 \leq i \leq n - 1$, then Participant P_i must be in every minimal access set.

If $\mathbf{h}_i$ is not a scalar multiple of $\mathbf{h}_0$, $1 \leq i \leq n - 1$, then Participant P_i must be in $(q - 1)q^{n-k-2}$ out of q^{n-k-1} minimal access sets.

(b) *When $d \geq 3$, for any fixed $1 \leq t \leq \min\{n - k - 1, d - 2\}$ every set of t participants is involved in $(q - 1)^t q^{n-k-(t+1)}$ out of q^{n-k-1} minimal access sets.*

Example 33.3.12 Let C be the linear code with parameters $[13, 10, 2]_3$ and generator matrix

$$\begin{bmatrix}
1 & 0 & 0 & 0 & 0 & 0 & 0 & 0 & 0 & 1 & 0 & 2 & 1 \\
0 & 1 & 0 & 0 & 0 & 0 & 0 & 0 & 0 & 0 & 0 & 0 & 1 \\
0 & 0 & 1 & 0 & 0 & 0 & 0 & 0 & 0 & 0 & 0 & 1 & 1 \\
0 & 0 & 0 & 1 & 0 & 0 & 0 & 0 & 0 & 2 & 0 & 1 & 0 \\
0 & 0 & 0 & 0 & 1 & 0 & 0 & 0 & 0 & 2 & 0 & 1 & 1 \\
0 & 0 & 0 & 0 & 0 & 1 & 0 & 0 & 0 & 0 & 0 & 2 & 1 \\
0 & 0 & 0 & 0 & 0 & 0 & 1 & 0 & 0 & 1 & 0 & 1 & 2 \\
0 & 0 & 0 & 0 & 0 & 0 & 0 & 1 & 0 & 2 & 0 & 1 & 2 \\
0 & 0 & 0 & 0 & 0 & 0 & 0 & 0 & 1 & 1 & 0 & 0 & 1 \\
0 & 0 & 0 & 0 & 0 & 0 & 0 & 0 & 0 & 0 & 1 & 1 & 2
\end{bmatrix}.$$

Then $C^\perp$ is a $[13, 3, 7]_3$ code. In the secret sharing scheme based on C with respect to the second construction, the secret space is $\mathcal{S} = \mathbb{F}_3$, the participant set is $\mathcal{P} = \{P_1, \ldots, P_{12}\}$, and there are $q^{n-k-1} = 3^2 = 9$ minimal access sets which are the following:

$$\{1,2,3,5,7,8,9,10,12\}, \ \{1,2,3,6,7,9,11,12\}, \quad \{3,4,6,7,8,9\},$$
$$\{1,4,5,7,9,10,11,12\}, \ \{2,3,4,5,6,7,10,11\}, \quad \{1,2,4,6,7,8,9,11,12\},$$
$$\{1,2,4,5,6,7,8,10,12\}, \ \{1,3,5,6,7,8,10,11,12\}, \ \{2,3,4,5,7,8,9,10,11\}.$$

Participant P_7 is a dictator. Each other participant is in 6 ($= (q-1)q^{n-k-2}$) minimal access sets.

Example 33.3.13 Let $C^\perp$ be the $[12, 4, 6]_2$ code with generator matrix

$$H = \begin{bmatrix}
0 & 0 & 0 & 0 & 1 & 1 & 1 & 1 & 1 & 1 & 1 & 1 \\
1 & 1 & 1 & 1 & 0 & 0 & 0 & 0 & 1 & 1 & 1 & 1 \\
0 & 0 & 1 & 1 & 0 & 0 & 1 & 1 & 0 & 0 & 1 & 1 \\
0 & 1 & 0 & 1 & 0 & 1 & 0 & 1 & 0 & 1 & 0 & 1
\end{bmatrix}$$

The secret sharing scheme based on C with respect to the second construction has the secret space $\mathbb{F}_2$, the participant set $\mathcal{P} = \{P_1, \ldots, P_{11}\}$, and the following minimal access sets:

$$\{1,6,7,8,9\}, \quad \{2,5,7,8,10\}, \quad \{1,4,5,10,11\}, \quad \{2,4,6,9,11\},$$
$$\{3,5,6,8,11\}, \quad \{3,4,7,9,10\}, \quad \{1,2,3,4,5,6,7\}, \quad \{1,2,3,8,9,10,11\}.$$

Every participant is in 4 minimal access sets. Therefore, the secret sharing scheme is 1-democratic.

33.3.4 Other Works on the Second Construction

The objective of this subsection is to give a summary of secret sharing schemes based on linear codes with respect to the second construction. The reader is referred to specific references for details.

Let $C^\perp$ be an $[n, k, d]_q$ code with only nonzero weights d and n. The secret sharing schemes based on C with respect to the second construction were investigated by Anderson, Ding, Helleseth, and Kløve in [48]. The access structures of these schemes are related to t-designs. Specifically, simplex codes, Hamming codes and the first-order Reed–Muller codes were employed for secret sharing in [48].

Planar functions were employed to construct linear codes over $\mathbb{F}_q$ with three and five weights by Carlet, Ding, and Yuan in [352], where the duals of these codes were used to construct secret sharing schemes with special access structures.

The duals of some irreducible codes and a class of linear codes from quadratic forms were used to construct secret sharing schemes by Yuan and Ding in [352]. Several families of trace codes were utilized for secret sharing by Yuan and Ding in [1935]. A family of group character codes was employed for secret sharing by Ding, Kohel, and Ling in [557]. Some linear codes were also considered for secret sharing in [564, 566].

Secret sharing schemes based on additive codes over $\mathbb{F}_4$ and their connections with t-designs were considered by Kim and Lee in [1120]. Self-dual codes were employed to construct secret sharing schemes by Dougherty, Mesnager, and Solé in [633].

33.4   Multisecret Sharing with Linear Codes

To describe some multisecret sharing schemes with linear codes, we introduce basic notation and notions of information theory. The **entropy** of a random variable X with a probability distribution $p(x)$ is defined by

$$H(X) = -\sum_x p(x) \log p(x).$$

The entropy is a measure of the average uncertainty in the random variable. Entropy is the uncertainty of a single random variable. We can define conditional entropy $H(X \mid Y)$, which is the entropy of a random variable X conditional on the knowledge of another random variable Y. The reduction in uncertainty due to another random variable is called the mutual information. For two random variables X and Y this reduction is the **mutual information**

$$I(X;Y) = H(X) - H(X \mid Y) = \sum_{x,y} p(x,y) \log \frac{p(x,y)}{p(x)p(y)},$$

where $p(x,y)$ is the joint probability distribution of X and Y. The mutual information $I(X;Y)$ is a measure of the dependence between the two random variables. It is symmetric in X and Y and always nonnegative and is equal to zero if and only if X and Y are independent.

In [1083] Karnin, Greene, and Hellman considered the situation where there are k secrets $s_1, s_2, \ldots, s_k$ from secret spaces $\mathcal{S}_i$ with the uniform probability distribution to be shared. The following are required for any $1 \le j \le k$:

- Any set of m shares determines the secret s_j; i.e., for any set of m indices $1 \le i_1 < \cdots < i_m \le n$, $H(s_j \mid (t_{i_1}, \ldots, t_{i_m})) = 0$. Here and hereafter $t_i \in \mathcal{T}_i$ denotes the share of Participant P_i and $H(s_j \mid (t_{i_1}, \ldots, t_{i_m}))$ denotes the uncertainty of s_j given $(t_{i_1}, \ldots, t_{i_m})$.

- Any set of $m - 1$ shares gives no information about the secret s_j; i.e., for any set of $m - 1$ indices $1 \le i_1 < \cdots < i_{m-1} \le n$, $H(s_j \mid (t_{i_1}, \ldots, t_{i_{m-1}})) = H(s_j)$, or in terms of mutual information $I(s_j; (t_{i_1}, \ldots, t_{i_{m-1}})) = 0$, where $H(s_j)$ denotes the uncertainty of s_j, $H(a \mid b)$ the uncertainty of a when event b happened, and $I(a; b)$ the amount of mutual information between a and b.

Such schemes are necessary in applications where a number of secrets should be shared at the same time by a number of participants. We will refer to such systems as $[k, m, n]$ **(multisecret sharing) threshold schemes**. Another important fact is that each $[k, m, n]$ threshold scheme for multisecret sharing gives naturally k (m, n)-threshold schemes for single secret sharing. Thus, the importance of multisecret sharing follows also from that of single secret sharing.

Multisecret sharing schemes were also studied by Jackson, Martin, and O'Keefe [1027], where they considered the case in which each subset of k participants is associated with a secret which is protected by a (t, k)-threshold access structure and lower bounds on the size of a participant's share. Some information aspects of multisecret sharing schemes were also studied by Blundo, De Santis, Di Crescenzo Gaggia, and Vaccaro [226], where they tried to work out a general theory of multisecret sharing schemes and to establish some lower bounds on the size of information held by each participant for various access structures. Ding, Laihonen, and Renvall [558] have established some connections between linear multisecret sharing schemes and linear codes, which will be introduced in the next subsections.

33.4.1   The Relation Between Multisecret Sharing and Codes

We assume that each element of the secret space $\mathcal{S}_i$ is equally likely to be the ith secret for each i. Let the secret spaces $\mathcal{S}_i$ for each $1 \leq i \leq k$ and the share spaces $\mathcal{T}_i$ for each $1 \leq i \leq n$ be vector spaces over a field $\mathbb{F}$. Then the product spaces $\mathcal{S} = \mathcal{S}_1 \times \cdots \times \mathcal{S}_k$ and $\mathcal{T} = \mathcal{T}_1 \times \cdots \times \mathcal{T}_n$ are also vector spaces over $\mathbb{F}$. In a multisecret sharing scheme a dealer uses a **share function** $f : \mathcal{S} \to \mathcal{T}$ to compute the shares for the n participants; i.e., let $\mathbf{s} = (s_1, \ldots, s_k)$ be the vector consisting of k secrets s_i and $\mathbf{t} = (t_1, \ldots, t_n) = f(\mathbf{s})$, where the share given only to the ith participant is t_i. Clearly, the share function must be one-to-one. A multisecret sharing scheme is said to be **linear** if for all $a, a' \in \mathbb{F}$ and all $s, s' \in \mathcal{S}$

$$f(as + a's') = af(s) + a'f(s').$$

If there is a constant $t \in \mathcal{T}$ such that $f(x) - t$ is linear, then the secret sharing scheme is said to be **affine**.

In what follows we consider the case that $\mathcal{S}_i = \mathbb{F}_q$ and $\mathcal{T}_i = \mathbb{F}_q$. Thus, $\mathcal{T} = \mathbb{F}_q^n$ and $\mathcal{S} = \mathbb{F}_q^k$ both are vector spaces over $\mathbb{F}_q$.

Theorem 33.4.1 *A multisecret sharing scheme defined over the above secret and share spaces is linear if and only if its share function is of the form*

$$f(\mathbf{s}) = \mathbf{s}G, \tag{33.1}$$

where $\mathbf{s} = (s_1, \ldots, s_k) \in \mathcal{S}$, *and* G *is a* $k \times n$ *matrix over* $\mathbb{F}_q$ *with rank* k.

Theorem 33.4.1 clearly shows that a linear multisecret sharing scheme gives an $[n, k, d]_q$ code $\mathcal{C}$ with generator matrix G, and each generator matrix G of an $[n, k, d]_q$ linear code $\mathcal{C}$ gives a linear multisecret sharing scheme. The share function is an encoding mapping of a linear code.

For a linear multisecret sharing scheme with the share function f of (33.1), recovering the original multisecret $\mathbf{s}$ is carried out as follows. Let $G(i_1, \ldots, i_u)$ denote the submatrix consisting of columns $i_1, i_2, \ldots, i_u$ of the matrix G, where $1 \leq u \leq n$, and $1 \leq i_1 < \cdots < i_u \leq n$. Suppose that the shares $t_{i_1}, t_{i_2}, \ldots, t_{i_u}$ are known; then recovering the multisecret becomes solving the linear equation

$$\mathbf{s}G(i_1, \ldots, i_u) = (t_{i_1}, \ldots, t_{i_u}).$$

The complexity of solving such a linear equation is $O(u^3)$ with a method like the Gaussian elimination method. Thus, recovering the multisecret is much simpler than decoding linear codes.

33.4.2   Linear Threshold Schemes and MDS Codes

Since we have assumed that the secrets from each secret space S_i are equally likely, without the knowledge of any share the uncertainty (denoted as $H(s_i)$) or self-information (denoted as $I(s_i)$) of each s_i is $H(s_i) = I(s_i) = \log_2 q$ bits, and the uncertainty or self-information of each $\mathbf{s} = (s_1, \ldots, s_k)$ is $I(\mathbf{s}) = H(\mathbf{s}) = k \log_2 q$ bits; here we assume that all secrets are independent. To recover the multisecret, a set of shares must provide $I(\mathbf{s})$ bits of information about the secret.

Theorem 33.4.2 *Let a multisecret sharing scheme have the share function of* (33.1). *Then*

$$I(\mathbf{s}; (t_{i_1}, \ldots, t_{i_u})) = r \log_2 q = \begin{cases} < I(\mathbf{s}) & \text{if } r < k, \\ = I(\mathbf{s}) & \text{if } r = k, \end{cases}$$

$$H(\mathbf{s} \mid (t_{i_1}, \ldots, t_{i_u})) = (k - r) \log_2 q = \begin{cases} > 0 & \text{if } r < k, \\ = 0 & \text{if } r = k, \end{cases}$$

where $r = \text{rank } G(i_1, \ldots, i_u)$.

By Theorem 33.4.2 the amount of information about the multisecret $\mathbf{s}$ given by a set of shares $\{t_{i_1}, \ldots, t_{i_u}\}$ is completely determined by the rank of the submatrix $G(i_1, \ldots, i_u)$ of G. Thus, each share gives information about the multisecret $\mathbf{s}$; however we shall prove this could not be true for each individual secret s_j.

Theorem 33.4.3 *Let* $\mathbf{e}_j$ *denote the vector where the jth coordinate is 1 and the remaining coordinates are 0. Let a multisecret sharing scheme have the share function of* (33.1). *If* $r = \text{rank } G(i_1, \ldots, i_u) = k$, *then*

$$I(s_j; (t_{i_1}, \ldots, t_{i_u})) = \log_2 q = I(s_j),$$
$$H(s_j \mid (t_{i_1}, \ldots, t_{i_u})) = 0.$$

If $r < k$, *then* $I(s_j; (t_{i_1}, \ldots, t_{i_u})) = \log_2 q = I(s_j)$ *if and only if the vector* $\mathbf{e}_j$ *is a linear combination of the column vectors of the submatrix* $G(i_1, \ldots, i_u)$; *otherwise* $I(s_j; (t_{i_1}, \ldots, t_{i_u})) = 0$.

Below we study only $[k, k, n]$ linear multisecret sharing schemes.

Theorem 33.4.4 *A multisecret sharing scheme with the share function of* (33.1) *is a $[k, k, n]$ threshold scheme if and only if*

(a) *the linear code* C *with generator matrix* G *is MDS, and*

(b) *any set of $k - 1$ column vectors of G generates a $[k, k - 1, 2]$ MDS code.*

By Theorem 33.4.4, a $[k, k, n]$ threshold scheme for multisecret sharing gives one $[n, k, n - k + 1]$ MDS code and k MDS codes with parameters $[k, k - 1, 2]$.

Theorem 33.4.5 *For any linear $[k, k, n]$ threshold scheme with the share function of* (33.1)

$$I(\mathbf{s}; (t_{i_1}, \ldots, t_{i_u})) = \min \{k, u\} \log_2 q.$$

This theorem clearly shows the information hierarchy about the multisecret of $[k, k, n]$ threshold schemes, i.e., $I(\mathbf{s}; B) = |B| \log_2 q$, where B is a set of shares. Thus, linear $[k, k, n]$ threshold schemes are the most democratic schemes in that each share contains the same amount of information about the multisecret, and two sets of shares give the same amount of information about the multisecret if and only if the numbers of shares in the two sets are equal. However, the information hierarchy about each individual secret s_j is quite different, i.e., $I(s_j; B) = 0$ if $|B| < k$ and $I(s_j; B) = I(s_j)$ otherwise.

The relation between linear $[k, k, m]$ threshold schemes and linear $[n, k, n - k + 1]$ MDS codes is now clear. To construct such multisecret sharing schemes, we need to find linear MDS codes. It is obvious that not every generator matrix of an MDS code satisfies Theorem 33.4.4(b). So our task is first to find MDS codes, and then to find generator matrices of those codes satisfying Theorem 33.4.4(b). Below we give an example.

Let $\alpha = (\alpha_1, \ldots, \alpha_n)$ where the α_i are distinct elements of $\mathbb{F}_q$. The $[n, k, n - k + 1]$ generalized Reed–Solomon code has generator matrix

$$
G = \begin{bmatrix}
1 & 1 & \cdots & 1 \\
\alpha_1 & \alpha_2 & \cdots & \alpha_n \\
\vdots & \vdots & \vdots & \vdots \\
\alpha_1^{k-1} & \alpha_2^{k-1} & \cdots & \alpha_n^{k-1}
\end{bmatrix}. \tag{33.2}
$$

To construct linear multisecret sharing $[k, k, n]$ threshold schemes based on the Reed–Solomon code, we need the following lemma.

Lemma 33.4.6 *Any $(k - 1) \times (k - 1)$ submatrix of the G of (33.2) has rank $k - 1$ if and only if for any set of indices $1 \le i_1 < \cdots < i_{k-1} \le n$*

$$
\sum_{1 \le u_1 < \cdots < u_j \le k-1} \alpha_{i_{u_1}} \alpha_{i_{u_2}} \cdots \alpha_{i_{u_j}} \ne 0 \quad \text{for all } j = 1, 2, \ldots, k - 2. \tag{33.3}
$$

The linear multisecret sharing scheme based on the Reed–Solomon codes is constructed as follows. Choose the α_i such that (33.3) holds. Then we use the matrix of (33.2) as the one of (33.1) to construct the share function for the multisecret sharing scheme.

Theorem 33.4.7 *The multisecret sharing scheme based on the Reed–Solomon code with generator matrix G of (33.2) satisfying (33.3) is a linear $[k, k, n]$ threshold scheme.*

To illustrate the above linear multisecret sharing $[k, k, n]$ threshold scheme based on the Reed–Solomon codes, we take the following example.

Example 33.4.8 Consider the field $\mathbb{F}_{11}$ and $k = 3$. Let $\alpha_i = i$ for $i = 1, 2, 3, 4, 5$. Then the matrix in (33.2) becomes

$$
G' = \begin{bmatrix}
1 & 1 & 1 & 1 & 1 \\
1 & 2 & 3 & 4 & 5 \\
1 & 4 & 9 & 5 & 3
\end{bmatrix},
$$

which generates a $[5, 3, 3]$ MDS code over $\mathbb{F}_{11}$. It is easily checked that each 2×2 submatrix of G' is invertible; so G' gives a $[3, 3, 5]$ threshold scheme for multisecret sharing.

Connections between single-secret sharing and multisecret sharing can be found in [558]. Multisecret sharing schemes based on codes were investigated in [333, 1740].

Chapter 34

Code-Based Cryptography

Philippe Gaborit

University of Limoges

Jean-Christophe Deneuville

University of Toulouse

Introduction

Code-based cryptography refers to asymmetric cryptography based on coding theory assumptions where the underlying primitives can be reduced to solving decoding problems. This chapter focuses on the essentials of code-based cryptography and covers some reminders on coding theory and the associated hard problems useful for cryptography, historical constructions such as the McEliece [1361] and Niederreiter [1432] constructions, as well as more recent proposals for public-key cryptography, key-exchange protocols, and digital signatures.

With the announced downfall of number theoretic based approaches to provide secure public-key primitives against quantum adversaries [858, 1696], the field of post-quantum cryptography has received growing interest with a lot of research effort. Among the possible alternative tools such as lattices, hash functions, multivariate polynomials, or isogenies of elliptic curves, coding theory stands as a credible mature candidate for several reasons that will be discussed in the next sections: it allows one to design most of the usual (most used) primitives, the complexity of the best known attacks is well-studied and rather stable, and it offers time/memory/security trade-offs.

In particular, in recent years a lot of progress has been made. In addition to the classical McEliece framework published in 1978, with its advantages and drawbacks, some new approaches have introduced the possibility of having code-based cryptosystems with small key sizes and with security reductions to generic problems as a way to avoid the classical structural attacks associated to the McEliece framework.

The recent standardization process launched by the National Institute of Standards and Technology (NIST) in 2017 [1441] has thrown code-based cryptography into a new era by forcing it to consider real applications, which had not really been done before because of the practical efficiency of number theory-based cryptosystems like RSA.

34.1   Preliminaries

34.1.1   Notation

Throughout this chapter, $\mathbb{Z}$ denotes the ring of integers and $\mathbb{F}_q$ denotes the finite field of q elements for q a power of a prime. Let $\mathcal{R} = \mathbb{F}_q[X]/(X^n - 1)$ denote the quotient ring of polynomials modulo $X^n - 1$ whose coefficients lie in $\mathbb{F}_q$. Elements of $\mathcal{R}$ will be interchangeably considered as row vectors or polynomials. Vectors and polynomials (respectively, matrices) will be represented by lower-case (respectively, upper-case) bold letters.[1]

The typical case is $q = 2$ and n a prime number chosen such that 2 is primitive modulo n; indeed in the typical case $X^n - 1$ factors as the product of two irreducible polynomials $X^n - 1 = (X + 1)(X^{n-1} + X^{n-2} + \cdots + 1)$ [779]. Hence $X^n - 1$ can be considered as *almost* irreducible, which is considered to be better for security and makes it easy to find invertible elements of $\mathcal{R}$; modulo $X^n - 1$, the invertible elements are exactly polynomials with an odd number of terms.

---

[1] Vectors are usually considered in row representation in coding theory, against column representation in lattice-based cryptography.

For any two elements $\mathbf{x}, \mathbf{y} \in \mathcal{R}$, their product, defined as the usual product of two polynomials modulo $X^n - 1$, is as follows: $\mathbf{x} \cdot \mathbf{y} = \mathbf{z} \in \mathcal{R}$ with

$$z_k = \sum_{i+j \equiv k \pmod{n}} x_i y_j \quad \text{for } i, j, k \in \{0, \dots, n-1\}.$$

Notice that as $\mathcal{R}$ is a *commutative* ring, we have $\mathbf{x} \cdot \mathbf{y} = \mathbf{y} \cdot \mathbf{x}$.

For any finite set $\mathcal{S}$, $x \xleftarrow{\$} \mathcal{S}$ denotes a uniformly random element x sampled from $\mathcal{S}$. For any $x \in \mathbb{R}$, let $\lfloor x \rfloor$ denote the biggest integer smaller than or equal to x. Finally, all logarithms $\log(\cdot)$ will be base 2 unless explicitly mentioned.

Definition 34.1.1 Let $\mathbf{x} = (x_0, \dots, x_{n-1}) \in \mathbb{F}_q^n$. The **circulant matrix** induced by $\mathbf{x}$ is defined and denoted as follows:

$$\mathbf{rot}(\mathbf{x}) = \begin{bmatrix} x_0 & x_1 & \cdots & x_{n-1} \\ x_{n-1} & x_0 & \cdots & x_{n-2} \\ \vdots & \vdots & \ddots & \vdots \\ x_1 & x_2 & \cdots & x_0 \end{bmatrix} \in \mathbb{F}_q^{n \times n}.$$

It is easy to see that the product of any two elements $\mathbf{x}, \mathbf{y} \in \mathcal{R}$ can be expressed as a usual vector-matrix (or matrix-vector) product using the $\mathbf{rot}(\cdot)$ operator as

$$\mathbf{x} \cdot \mathbf{y} = \mathbf{x} \times \mathbf{rot}(\mathbf{y}) = \left(\mathbf{rot}(\mathbf{x})^\mathsf{T} \times \mathbf{y}^\mathsf{T}\right)^\mathsf{T} = \mathbf{y} \times \mathbf{rot}(\mathbf{x}) = \mathbf{y} \cdot \mathbf{x}. \tag{34.1}$$

34.1.2 Background on Coding Theory

We now recall some basic definitions and properties about coding theory that are regularly used in code-based cryptography. More details about coding theory can be found in Chapter 1 and dedicated books such as [879, 1008, 1323].

Definition 34.1.2 A **linear code** $\mathcal{C}$ of length n and dimension k (denoted an $[n, k]$ code) is a subspace of the vector space $\mathbb{F}_q^n$ of dimension k. Elements of $\mathcal{C}$ are referred to as **codewords**.

Definition 34.1.3 If $\mathbf{x} = (x_1, \dots, x_n)$ and $\mathbf{y} = (y_1, \dots, y_n)$ are two elements of $\mathbb{F}_q^n$, then the **Euclidean inner product** of $\mathbf{x}$ and $\mathbf{y}$ is defined as

$$\langle \mathbf{x}, \mathbf{y} \rangle = \sum_{i=1}^{n} x_i y_i.$$

Definition 34.1.4 The **dual code** $\mathcal{C}^\perp$ of a linear $[n, k]$ code $\mathcal{C}$ is the linear $[n, n-k]$ code defined by

$$\mathcal{C}^\perp = \{\mathbf{x} \in \mathbb{F}_q^n \mid \langle \mathbf{x}, \mathbf{c} \rangle = 0 \text{ for all } \mathbf{c} \in \mathcal{C}\}.$$

Definition 34.1.5 We say that $\mathbf{G} \in \mathbb{F}_q^{k \times n}$ is a **generator matrix** for the $[n, k]$ code $\mathcal{C}$ if

$$\mathcal{C} = \{\mathbf{mG} \mid \mathbf{m} \in \mathbb{F}_q^k\}.$$

Definition 34.1.6 Given an $[n, k]$ code $\mathcal{C}$, we say that $\mathbf{H} \in \mathbb{F}_q^{(n-k) \times n}$ is a **parity check matrix** for $\mathcal{C}$ if $\mathbf{H}$ is a generator matrix of the dual code $\mathcal{C}^\perp$, or more formally, if

$$\mathcal{C} = \{\mathbf{x} \in \mathbb{F}_q^n \mid \mathbf{Hx}^\mathsf{T} = \mathbf{0}\}, \quad \text{or equivalently } \mathcal{C}^\perp = \{\mathbf{yH} \mid \mathbf{y} \in \mathbb{F}_q^{n-k}\},$$

where $\mathbf{Hx}^\mathsf{T}$ is called the **syndrome** of $\mathbf{x}$.

Remark 34.1.7 It follows from the previous definitions that $\mathbf{GH}^\mathsf{T} = \mathbf{0}_{k \times (n-k)}$.

Code-based cryptography was originally proposed using the standard Hamming metric. Other metrics, such as the rank metric, allow for the design of cryptographic primitives with different properties. We focus on the Hamming metric for most of this chapter and refer the reader to Section 34.10 for a brief overview of code-based cryptography with the rank metric.

Definition 34.1.8 Let $\mathbf{x}, \mathbf{y} \in \mathbb{F}_q^n$. The **Hamming weight** of $\mathbf{x} = (x_1, \ldots, x_n)$ is the number of its nonzero coordinates. Formally $\mathrm{wt}_H(\mathbf{x}) = |\{i \mid x_i \neq 0\}|$. The **Hamming distance** between $\mathbf{x}$ and $\mathbf{y}$ is $d_H(\mathbf{x}, \mathbf{y}) = \mathrm{wt}_H(\mathbf{x} - \mathbf{y})$.

Definition 34.1.9 Let $\mathcal{C}$ be an $[n, k]$ linear code over $\mathbb{F}_q$. The **minimum distance** of $\mathcal{C}$ is

$$d_H(\mathcal{C}) = \min_{\mathbf{x}, \mathbf{y} \in \mathcal{C}, \, \mathbf{x} \neq \mathbf{y}} d_H(\mathbf{x}, \mathbf{y}) = \min_{\mathbf{x}, \mathbf{y} \in \mathcal{C}, \, \mathbf{x} \neq \mathbf{y}} \mathrm{wt}_H(\mathbf{x} - \mathbf{y}).$$

Remark 34.1.10 Notice that by definition, the minimum distance of a linear code is exactly the minimum weight of a nonzero codeword over all possible codewords.

When it is possible to decode a code $\mathcal{C}$, there is a unique decoding up to $\delta = \left\lfloor \frac{d_H(\mathcal{C})-1}{2} \right\rfloor$ errors. When $d = d_H(\mathcal{C})$ is known, the code parameters of $\mathcal{C}$ are denoted $[n, k, d]$.

Code-based cryptography usually suffers from huge keys. In order to keep the key sizes reasonable, several proposals use the strategy introduced by Gaborit [769]. This results in quasi-cyclic codes, as defined below. Quasi-cyclic codes are the topic of Chapter 7.

Definition 34.1.11 ([1323, Chap. 16, §7]) View a vector $\mathbf{c} = (\mathbf{c}_0, \ldots, \mathbf{c}_{s-1})$ of $\mathbb{F}_2^{sn}$ as s successive blocks (n-tuples). An $[sn, k, d]$ linear code $\mathcal{C}$ is **quasi-cyclic (QC)** of index s if for any $\mathbf{c} = (\mathbf{c}_0, \ldots, \mathbf{c}_{s-1}) \in \mathcal{C}$, the vector obtained after applying a simultaneous circular shift to every block $\mathbf{c}_0, \ldots, \mathbf{c}_{s-1}$ is also a codeword.

More formally, by considering each block $\mathbf{c}_i$ as a polynomial in $\mathcal{R} = \mathbb{F}_q[X]/(X^n - 1)$, the code $\mathcal{C}$ is QC of index s if for any $\mathbf{c} = (\mathbf{c}_0, \ldots, \mathbf{c}_{s-1}) \in \mathcal{C}$, then $(X \cdot \mathbf{c}_0, \ldots, X \cdot \mathbf{c}_{s-1}) \in \mathcal{C}$.

Definition 34.1.12 A **systematic** quasi-cyclic $[sn, n]$ code of index s and rate $1/s$ is a quasi-cyclic code with an $(s-1)n \times sn$ parity check matrix of the form

$$\mathbf{H} = \begin{bmatrix} \mathbf{I}_n & \mathbf{0}_{n \times n} & \cdots & \mathbf{0}_{n \times n} & \mathbf{A}_0 \\ \mathbf{0}_{n \times n} & \mathbf{I}_n & & & \mathbf{A}_1 \\ \vdots & & \ddots & & \vdots \\ \mathbf{0}_{n \times n} & & \cdots & \mathbf{I}_n & \mathbf{A}_{s-2} \end{bmatrix}$$

where $\mathbf{A}_0, \ldots, \mathbf{A}_{s-2}$ are circulant $n \times n$ matrices, $\mathbf{I}_n$ is the $n \times n$ identity matrix, and $\mathbf{0}_{n \times n}$ is the $n \times n$ zero matrix.

Remark 34.1.13 The definition of systematic quasi-cyclic codes of index s can of course be generalized to all rates ℓ/s, $\ell = 1, \ldots, s-1$, but we shall only use systematic QC codes of rates $1/2$ and $1/3$ and wish to lighten notation with the above definition. In the sequel, referring to a systematic QC code will imply by default that it is of rate $1/s$. Note that arbitrary QC codes are not necessarily equivalent to a systematic QC code.

Definition 34.1.14 For an $[n, k]$ code over $\mathbb{F}_q$, the **Gilbert–Varshamov distance** is the smallest value of d which satisfies the inequality

$$q^{n-k} \leq \sum_{i=0}^{d} \binom{n}{i} (q-1)^i.$$

This distance corresponds to the average minimum distance of a random $[n, k]$ code. Notice that for quasi-cyclic codes, this value can be a little higher [779].

Definition 34.1.15 We denote by $\mathcal{S}_w^n(\mathbb{F}_q)$ the set of all **words of weight** w in $\mathbb{F}_q^n$.

34.2 Difficult Problems for Code-Based Cryptography: The Syndrome Decoding Problem and Its Variations

The main problem considered in code-based cryptography is the *Syndrome Decoding (SD) problem* with different variations.

Problem 34.2.1 (Computational Syndrome Decoding (CSD)) Given a random $\mathbf{H} \in \mathbb{F}_q^{(n-k) \times n}$, a syndrome $\mathbf{y} \in \mathbb{F}_q^{n-k}$, and an integer w, is it possible to find $\mathbf{x} \in \mathbb{F}_q^n$ such that $\mathbf{H}\mathbf{x}^\mathsf{T} = \mathbf{y}$ and $\mathrm{wt}_\mathrm{H}(\mathbf{x}) \leq w$?

Problem 34.2.2 (Ideal-Decision Syndrome Decoding (IDSD)) Given a random $\mathbf{H} \in \mathbb{F}_q^{(n-k) \times n}$, a syndrome $\mathbf{y} \in \mathbb{F}_q^{n-k}$, and an integer w, does there exist a vector $\mathbf{x} \in \mathbb{F}_q^n$ such that $\mathbf{H}\mathbf{x}^\mathsf{T} = \mathbf{y}$ and $\mathrm{wt}_\mathrm{H}(\mathbf{x}) \leq w$?

Recall that an adversary's advantage is a measure of how successfully an adversary can distinguish a cryptographic value from a idealized random value. A **negligible advantage** means an advantage within $O(2^{-\lambda})$ for λ the value of the security parameter, in bits.

Problem 34.2.3 (Decision Syndrome Decoding (DSD)) Let $\mathbf{H}$ be a random $(n - k) \times n$ matrix over $\mathbb{F}_q$ and let w be a positive integer. Given $\mathbf{x} \in \mathbb{F}_q^n$ of weight w, is it possible to distinguish between $\mathbf{H}\mathbf{x}^\mathsf{T}$ and $\mathbf{r}$ random in $\mathbb{F}_q^{n-k}$ with a non-negligible advantage?

The Ideal-Decision Syndrome Decoding problem was proven NP-complete in [171] under the name of the Coset Weight problem; it is also sometimes called the Maximum Likelihood Decoding problem. The computational version CSD is obviously a harder problem. In practice the Ideal-Decision Syndrome Decoding problem assumes that the answer to the question can only be "yes" or "no"; it corresponds to an ideal case, which does not fit very well the usual attacker models. In practice, for security proofs in cryptography, it is easier to consider a more precise notion of advantage and indistinguishability for an attacker, which motivates the introduction of the Decision Syndrome Decoding problem. The DSD problem is easier than the IDSD problem, but it was proven in [54, 733] that over $\mathbb{F}_2$ the DSD problem is harder than the CSD problem, so that overall for random binary codes these three problems are equivalent. But in practice the DSD problem appears more naturally.

One can draw a distinction with the Diffie–Hellman problems, for which there is a computational version CDH (Computational Diffie–Hellman) and a decisional version DDH (Decisional Diffie–Hellman); in practice proofs are often based on the DDH assumption. In the case of the Diffie–Hellman problem, the CDH problem is harder than the DDH problem, and it is not known whether or not the two problems are equivalent.

Another related problem is the *Minimum Weight* problem.

Problem 34.2.4 (Minimum Weight) Let $\mathcal{C}$ be a random $[n, k]$ code over $\mathbb{F}_q$ and let w be a positive integer. Does there exist a nonzero codeword $\mathbf{x}$ in $\mathcal{C}$ with $\mathrm{wt}_\mathrm{H}(\mathbf{x}) \leq w$?

This problem was proven NP-complete in [1843].

As will be discussed in Section 34.4, code-based cryptography usually suffers from large public key sizes. In order to mitigate this issue, several constructions were proposed using structured codes, for instance, quasi-cyclic codes. We now explicitly define the syndrome decoding problem for quasi-cyclic codes with computational and decision variations and discuss their relative hardness.

Problem 34.2.5 (Computational s-Quasi-Cyclic Syndrome Decoding (QCSD)) Let $\mathcal{C}$ be a random systematic quasi-cyclic $[sn, n]$ code of index s over $\mathbb{F}_q$ with parity check matrix $\mathbf{H}$, and let $\mathbf{y}$ be a syndrome in $\mathbb{F}_q^{(s-1)n}$. Is it possible to find $\mathbf{x} \in \mathbb{F}_q^{sn}$ with weight w such that $\mathbf{y} = \mathbf{H}\mathbf{x}^{\mathsf{T}}$?

Problem 34.2.6 (Decision s-Quasi-Cyclic Syndrome Decoding (DQCSD)) Let $\mathbf{H}$ be a parity check matrix of a $[sn, n]$ quasi-cyclic code of index s over $\mathbb{F}_q$, and let w be a positive integer. Given $\mathbf{x} \in \mathbb{F}_q^{sn}$ of weight w, is it possible to distinguish between $\mathbf{H}\mathbf{x}^{\mathsf{T}}$ and $\mathbf{r}$ random in $\mathbb{F}_q^{(s-1)n}$ with a non-negligeable advantage?

There is no known reduction for the QCSD problem, but the problem is considered hard by the cryptographic community. In the typical case for n as described in Section 34.1.1, there is no known attack using the quasi-cyclic structure which drastically reduces the cost of the attack for the QCSD problem. The best improvement in that case, the DOOM approach [1649], only permits a gain in $\sqrt{n}$. Concerning the reduction between the decision and the computational QCSD problems, different from of the previously mentioned case of random binary codes, it is not known whether or not the two problems are polynomially equivalent.

Problem 34.2.7 (Learning Parity with Noise (LPN)) Fix a secret $\mathbf{s} \in \mathbb{F}_2^n$ and an error probability p. A sample t is defined by the binary value $t = \langle \mathbf{r}, \mathbf{s} \rangle + e$ for $\mathbf{r}$ a random element in $\mathbb{F}_2^n$ and e an error value which is 1 with probability p and 0 otherwise. Given N samples, is it possible to recover $\mathbf{s}$?

The complexity of the LPN problem depends on the number of samples considered [1310]. For a fixed number of samples the LPN problem corresponds exactly to the SD problem for a $[N, n]$ binary code with an error weight depending on the probability p. To the best of our knowledge, except for the case of the Hopper–Blum (HB) authentication scheme [983] where the notion of unlimited samples appears naturally, for encryption schemes based on LPN, one has to fix the number of samples. The fact that the weight of the error may vary because of the probability p makes it hard to have efficient parameters, so that sometimes one considers LPN problems with a fixed weight distribution. Hence, in practice, efficient cryptosystems based on LPN or its variations are equivalent to cryptosystems based on the classical SD (or QCSD) problems.

34.3 Best-Known Attacks for the Syndrome Decoding Problem

The best known solvers for the syndrome decoding problem are Information Set Decoding (ISD) algorithms: a combination of linear algebra and collision search algorithms. For cryptographic applications, an adversary has to decode a noisy version $\mathbf{u} = \mathbf{c} + \mathbf{e}$ of a codeword $\mathbf{c}$ where $\mathrm{wt}_{\mathrm{H}}(\mathbf{e}) = w$. ISD algorithms try to find an **information set**: a subset

$\mathcal{I}$ of indexes for which $e_i = 0$ for $i \in \mathcal{I}$, i.e., an error-free subset of positions. The average running time is hence a function of n, k, and w. Let $\mathcal{C}$ be a code with generator matrix $\mathbf{G}$ and parity check matrix $\mathbf{H}$. Remember that, by multiplying by a parity check matrix $\mathbf{H}$, decoding a noisy codeword $\mathbf{y} = \mathbf{mG} + \mathbf{e}$ for a small weight error $\mathbf{e}$ and a codeword $\mathbf{c} = \mathbf{mG}$ is equivalent to solving $\mathbf{eH}^\mathsf{T} = \mathbf{yH}^\mathsf{T}$.

The first ISD dates back to Prange [1540] and is presented in Algorithm 34.3.1. The general idea consists in guessing k coordinates positions of a noisy vector and hoping that these positions do not contain errors. If the guess is correct, one can reconstruct the codeword and check that the error has the desired small weight. Otherwise the attacker considers a new set of coordinates. The algorithm essentially relies on the fact that, for an invertible matrix $\mathbf{U} \in \mathbb{F}_2^{(n-k) \times (n-k)}$ and a permutation matrix $\mathbf{P} \in \mathbb{F}_2^{n \times n}$, if $\mathbf{H}' = \mathbf{UHP}$, solving $\mathbf{eH}^\mathsf{T} = \mathbf{s}$ is equivalent to solving $\mathbf{e}'\mathbf{H}'^\mathsf{T} = \mathbf{s}'$, with $\mathbf{e}' = \mathbf{eP}$ and $\mathbf{s}' = \mathbf{sU}^\mathsf{T}$. Using Gaussian elimination, an information set can be identified in polynomial time by writing the resulting parity check matrix in systematic form. The algorithm succeeds when $\mathbf{sU}^\mathsf{T}$ has a low enough weight.

Algorithm 34.3.1 (Prange-ISD)

Input: $\mathbf{H} \in \mathbb{F}_2^{(n-k) \times n}$, $\mathbf{s} \in \mathbb{F}_2^{n-k}$, and target weight w

Output: $\mathbf{e} \in \mathbb{F}_2^n$ such that $\mathbf{eH}^\mathsf{T} = \mathbf{s}$ and $\mathrm{wt_H}(\mathbf{e}) \leq w$

1: **repeat**
2:     Sample a uniformly random permutation $\mathbf{P} \in \mathbb{F}_2^{n \times n}$;
3:     Compute $\mathbf{HP}$;
4:     When it exists, find $\mathbf{U} \in \mathbb{F}_2^{(n-k) \times (n-k)}$ such that $\mathbf{UHP} = \begin{bmatrix} \mathbf{I}_{n-k} & | & \widetilde{\mathbf{H}} \end{bmatrix}$;
5:     Compute $\mathbf{sU}^\mathsf{T}$
6: **until** $\mathrm{wt_H}(\mathbf{sU}^\mathsf{T}) \leq w$;
7: **return** $(\mathbf{sU}^\mathsf{T}, \mathbf{0})\mathbf{P}^{-1}$.

The complexity of the Prange-ISD Algorithm to decode an error of weight w associated to a codeword in a random $[n, k]$ code is the cost of matrix inversion times the average work factor (the inverse of the probability of finding an adequate set of columns which does not intersect with the error). More precisely, in the algorithm, we want the image under $\mathbf{P}^{-1}$ of the set of the last k positions of a vector of length n to not contain any error position. Since $\mathbf{P}$ is random, this probability is hence the number of possible cases for which the algorithm succeeds divided by the number of possible choices. The number of cases for which the algorithm succeeds corresponds to the number of choices of k positions among $n - w$ positions (the total number of positions minus the w forbidden positions of the error); the total number of choices is choosing k positions among n. The probability that the attack succeeds is therefore $\dfrac{\binom{n-w}{k}}{\binom{n}{k}} = \dfrac{\binom{n-k}{w}}{\binom{n}{w}}$, and the whole complexity of the attack is hence $O\left((n-k)^3 \dfrac{\binom{n}{w}}{\binom{n-k}{w}} \right)$.

Starting from Prange [1540], a long line of research papers improved the complexity of these solvers [146, 178, 179, 182, 254, 335, 336, 338, 339, 375, 376, 413, 417, 418, 647, 730, 893, 1069, 1167, 1212, 1218, 1356, 1357, 1752, 1840, 1841]. As we saw with the Prange-ISD Algorithm formula, the general complexity formulae depend on three parameters: n, k, and the weight w. In order to compare the exponent part of the complexity of different approaches, one usually compares the asymptotic exponent of the attack for searching for a codeword of weight the Gilbert–Varshamov distance of an $[n, n/2]$ code, which in that case is approximately $\frac{n}{9}$. For this special set of parameters, the complexity depends linearly on a unique parameter n. In Table 34.1 we give the values of the exponent depending on the attack. In the table, the last three attacks also use nearest neighbor techniques together with the ISD approach; these approaches require extensive memory and are not necessarily the most efficient for cryptographic parameters. All these attacks are probabilistic. There also exist deterministic methods, but they are too expensive and in practice probabilistic methods are used. The Prange algorithm can be adapted in a straightforward manner to the case of finding minimum weight codewords of a code. In that case the syndrome is null, but it is possible to use the Prange algorithm by guessing an error position of the small weight vector and use the associated column of the parity check matrix as the new syndrome.

TABLE 34.1: Evolution of the value of the coefficient α in the exponential part $2^{\alpha n}$ of the complexity of attacks for decoding a random $[n, n/2, d]$ code for d the Gilbert–Varshamov distance of the code (approximately $\frac{n}{9}$).

Name	Date	α
Exhaustive search		0.386
Prange [1540]	1962	0.1207
Stern [1752]	1988	0.1164
Dumer [647]	1991	0.1162
May, Meurer, Thomas [1356]	2011	0.1114
Becker, Joux, May, Meurer [146]	2012	0.1019
May, Ozerov [1357]	2015	0.1019
Both, May [253]	2017	0.0953
Both, May [254]	2018	0.0885

In the case where w is very small compared to n and k (for instance, typically, $k = n/2$, $w = O(\sqrt{n})$), all these attacks have the same complexity, approximately $2^{-w \log\left(1 - \frac{k}{n}\right)(1 + o(1))}$, which is the complexity of the Prange-ISD Algorithm [339].

Remark 34.3.2 The best known quantum attack consists in using the Grover Algorithm through the Prange-ISD Algorithm and gives a complexity in $O\left(\sqrt{\binom{n}{w} / \binom{n-k}{w}}\right)$ [858]; see also [1069] for small exponential improvements with other improved ISD attacks. This complexity roughly means to divide by a factor of 2 in the exponential term of the attack, and that a 256 bits classical security level (the complexity needed for an adversary to attack a problem, that is, 2^{256} binary operations for a 256 bits security) would give a 128 quantum bits security. In contrast to number theory-based problems like the factorization problem or the discrete logarithm problem, there is no known quantum attack with logarithmic gain for the Syndrome Decoding problem; the *a priori* reason for which the community does not believe in it, is that the SD problem is NP-complete, and the general rationale is that this class of complexity remains hard even in front of a quantum adversary. Meanwhile the fact that there may exist a better quantum attack for the QCSD problem is still an open question.

To gauge the gap between theory and practice, a challenge website was created in 2019 [58]. It gathers instances of generic problems such as the SD problem or the small codeword finding problem, as well as instances more specific to the NIST post-quantum cryptography standardization process.

34.4 Public-Key Encryption from Coding Theory with Hidden Structure

34.4.1 The McEliece and Niederreiter Frameworks

The first code-based scheme was proposed by McEliece in 1978 [1361].[2] This scheme is usually seen as a cryptosystem in itself, when it is actually a more general encryption framework, in the sense that the security reduction of the scheme depends on the family of codes considered in the instantiation. A generic description of McEliece's cryptosystem is depicted in Figure 34.1, where sk is the **secret key** and pk the **public key**.

FIGURE 34.1: Generic presentation of the McEliece framework [1361]

KeyGen
Generate a (private) random $[n, k]$ linear code $\mathcal{C}$ and its associated generator matrix $\mathbf{G} \in \mathbb{F}_2^{k \times n}$, along with an efficient decoding algorithm $\mathcal{D}$. Sample uniformly at random an invertible matrix $\mathbf{S} \in \mathbb{F}_2^{k \times k}$ and a permutation matrix $\mathbf{P} \in \mathbb{F}_2^{n \times n}$. Return $(\mathbf{sk}, \mathbf{pk}) = \left((\mathbf{S}, \mathbf{G}, \mathbf{P}), \widetilde{\mathbf{G}} = \mathbf{SGP} \right)$.

Encrypt
To encrypt $\mathbf{m} \in \mathbb{F}_2^k$, sample uniformly $\mathbf{e} \in \mathbb{F}_2^n$ with $\mathrm{wt}_H(\mathbf{e}) = w$ and return $\mathbf{c} = \mathbf{m}\widetilde{\mathbf{G}} + \mathbf{e}$.

Decrypt
To decrypt, using $\mathcal{D}$, first decode $\mathbf{d} = \mathbf{cP}^{-1}$ to retrieve $\mathbf{mS}$, then $\mathbf{m}$ from $\mathbf{mS}$.

Why decryption works
$\mathbf{d} = (\mathbf{mS})\mathbf{G} + \mathbf{eP}^{-1}$; decoding $\mathbf{d}$ permits recovery of $\mathbf{mS}$, then $\mathbf{m} = (\mathbf{mS})\mathbf{S}^{-1}$.

McEliece parameters: The size of the public key is $(n - k)k$; the size of the ciphertext is n.

Security of the McEliece framework: The OW-CPA (One-Wayness against Chosen Plaintext Attack) security of the scheme relies on two problems. First, the hidden code structure is indistinguishable from a random code, which assumes that the **SGP** matrix cannot be distinguished from a random one. Second, based on this assumption, the security is then the security of decoding random codes (hence the SD problem). From the OW-CPA security one can obtain IND-CCA2 (Indistinguishable under Chosen Plaintext Attacks) security with a small ciphertext overhead using a general conversion as, for instance, in [970].

Structural attacks of McEliece: The main security assumption is the indistinguishability of the public matrix from random codes. The particular type of attack, called a *structural attack*, consists in trying to recover directly the private key **G** from the public key **SGP**. There exist two type of hidden families. In the first case, the attacker knows that the public key **SGP** is obtained from a code belonging to a large family of codes but does not know the specific code in the large family (for instance Goppa codes). Another possibility consists

---

[2]Section 15.7 examines the McEliece scheme in the context of algebraic geometry codes.

in considering a unique code but one which has to be resistant to recovering the primary structure from a permuted structure (for instance Reed–Muller or Reed–Solomon codes). The best generic attacks to recover a permutation from a permuted code and the code are the Leon Algorithm [1217] (with complexity: enumerating sufficiently many small weight codewords) and the Support Splitting Algorithm [1647] (with complexity $\tilde{O}\big(2^{\dim(\mathcal{C}\cap\mathcal{C}^{\perp})}\big)$). In the case of specific families of codes, for instance Reed–Solomon codes with monomial transformations (permutation matrices where each column is moreover multiplied by a nonzero scalar), it is possible to break the system in polynomial time [1705].

Niederreiter approach: Niederreiter's approach which can be seen as a dual approach of McEliece, is presented in Figure 34.2. The security is equivalent to the security of the McEliece framework. The main interest is that the ciphertext has size $n-k$ (the dimension of the dual) rather than n. The drawback is that the message has to be expressed in terms of a small weight vector. In practice, with the KEM/DEM (Key Encapsulation Mechanism/Data Encapsulation Mechanism) approach, the latter drawback becomes less important.

FIGURE 34.2: Neiderreiter's cryptosystem [1432]

KeyGen

Generate a (private) random $[n,k]$ linear code $\mathcal{C}$ and its associated parity check matrix $\mathbf{H}_{\mathrm{sec}} \in \mathbb{F}_2^{(n-k)\times n}$, along with an efficient decoding algorithm $\mathcal{D}$. Sample uniformly at random an invertible matrix $\mathbf{S} \in \mathbb{F}_2^{(n-k)\times(n-k)}$ and a permutation $\mathbf{P} \in \mathbb{F}_2^{n\times n}$. Return $(\mathrm{sk},\mathrm{pk}) = ((\mathbf{S},\mathbf{H}_{\mathrm{sec}},\mathbf{P}),\mathbf{H}_{\mathrm{pub}} = \mathbf{S}\mathbf{H}_{\mathrm{sec}}\mathbf{P})$.

Encrypt

To encrypt $\mathbf{m} \in \mathbb{F}_2^k$, first encode it into a vector $\widetilde{\mathbf{m}}$ of length n and weight w, then return $\mathbf{c} = \mathbf{H}_{\mathrm{pub}}\widetilde{\mathbf{m}}^\mathsf{T}$.

Decrypt

To decrypt, first decode $\mathbf{d} = \mathcal{D}\left(\mathbf{S}^{-1}\mathbf{c}\right)$, then retrieve $\mathbf{m}$ from $\widetilde{\mathbf{m}} = \mathbf{d}\mathbf{P}^{-1}$.

Instantiations: In 1978, McEliece originally proposed to instantiate this framework using binary Goppa codes ($n = 1024, k = 524, w = 50$). The encryption and decryption procedures are fast (of complexity $O(n^2)$) compared to RSA. However, the public key sizes are also large in $O(n^2)$, making it impractical compared to RSA. Other proposals were later made to either improve efficiency or reduce the key sizes [102, 103, 163, 164, 165, 180, 766, 769, 1038, 1391, 1392, 1432, 1704]. Since the framework works for any decodable code, many families of code were proposed and many were broken by structural attacks; in particular Reed–Solomon codes are an optimal family of codes with a lot of structure difficult to mask. The *square attack* proposed in [459] is an efficient tool for distinguishing codes related to Reed–Solomon codes (Reed–Solomon codes to which random columns have been added, or subcodes of Reed–Solomon codes among many possible variations) from random codes. Even if the original McEliece cryptosystem, based on the family of Goppa codes, is still considered secure today by the cryptography community, the large size of the public key of order 260kByte (for 128 bits of security) remains a big drawback. Many variants based on alternative families of codes (Reed–Solomon codes, Reed–Muller codes, or some alternant codes [163, 1391]) were broken by recovering in polynomial time the hidden structure [706]. Moreover, high rate Goppa codes have been proved not to behave like random codes [705]. The fact that the hidden code structure may be uncovered (even possibly for Goppa codes, see [463, 464, 707]) lies like a sword of Damocles over the system, and finding a practical alternative cryptosystem based on the difficulty of decoding unstructured or random codes has always been a major issue in code-based cryptography.

Generalized McEliece framework: The original McEliece framework described in Figure 34.1 has a straightforward generalization in a scheme which can be roughly described as:

- **Secret Key:** a decodable code $\mathcal{C}$
- **Public Key:** a matrix $\mathbf{G}' = \text{TrapDoor}(\mathcal{C})$, for $\text{TrapDoor}(\mathcal{C})$ a transformation hiding $\mathcal{C}$ in $\mathbf{G}'$
- **Encryption:** $\mathbf{c} = \mathbf{m}\mathbf{G}' + \mathbf{e}$, for $\mathbf{e}$ a small weight error
- **Decryption:** $\mathcal{C}.\text{Decode}(\text{TrapDoor}^{-1}(\mathbf{c}))$, for $\mathcal{C}.\text{Decode}$ a decoding algorithm of $\mathcal{C}$

This more general point of view permits one to not only consider the original McEliece permutation hiding, but also more general hiding as, for instance, the group-structured McEliece or the MDPC cryptosystem variations of the next sections.

This generalized framework also contains other natural variations on the scheme, such as the Wieschbrink variation in which one adds random columns to a code to hide it [1890], or using a subcode of a code [165], or a mix of the two approaches. There also exist variations [101] where the permutation is replaced by a matrix with a very small numbers of 1s in each row and column or by considering a monomial transformation rather than a basic permutation (typically when considering non-binary codes). But this type of system was also attacked [465]. All these variations do not fit in the original McEliece framework, but have in common the notion of TrapDoor and hiding a decodable code in a public matrix, with a security relying on the indistinguishability of the public matrix from a random code. Most of these variations were also broken [119, 459, 461, 1222].

Research Problem 34.4.1 An interesting question would be to find a hiding TrapDoor with a proven indistinguishability. Such constructions exist for lattices (see [800]) but are still not known in code-based cryptography. Among many broken families in the McEliece framework, it is interesting to notice that up to now there is no specific attack when considering the smaller BCH code family rather than Goppa codes.

34.4.2   Group-Structured McEliece Framework

In order to reduce the size of the public key, it is possible to consider a group-structured McEliece scheme. A quasi-cyclic McEliece framework was proposed in 2005 [769]. The main idea of this approach is to consider a compact representation of a code through the action of a group (for instance, a quasi-cyclic code) and then consider a hiding procedure compatible with the compact representation of the code (for instance, block permutations rather than general permutations). Quasi-cyclic codes allow one to drastically reduce the size of the public key from millions of bits to a few thousand. A first approach based on quasi-cyclic BCH codes was broken in [1478], and an improved approach based on alternant codes [163] had their proposed parameters broken in [706]. Eventually this approach led to the BigQuake submission (based on quasi-cyclic Goppa codes) [117] to the NIST competition, whose parameters have never been attacked while gaining a factor of roughly 10 on the classical McEliece Goppa instantiation. Another approach based on a similar idea but with the action of another group (the dyadic group) [1391] had some of their parameters broken in [706] and led to the NIST submission [110] whose parameters were broken in [121]. Overall while this approach should lead to a decrease of the size of the public key by some factor, the underlying quasi-cyclic structure may now be seen as a weakness.

Security and parameters: Everything is similar to the McEliece framework except that the size of the public key is smaller by a factor of roughly 10 and more structure is added on the hidden code. The reduction is done with the QCSD problem and not with the SD problem.

34.4.3   Moderate-Density Parity Check (MDPC) Codes

The MDPC cryptosystem [1392] introduced in 2013 uses a hidden code approach with the McEliece framework, but in that case no hiding permutation is used. The masking comes from the knowledge of small weight codewords of the code which permit one to decode the MDPC code.

Since LDPC codes, discussed in Chapter 21, are a very efficient class of decodable codes, it is very natural to consider them within the McEliece framework. However the very reason why these codes are efficient is the fact that the dual code is built from very small weight vectors, and this is precisely why they cannot be used as such in the McEliece framework, since recovering very small weight codewords in the dual code is easy. A first reasonable approach using LDPC codes for cryptography was done in [102] through a masking of the LDPC structure. This type of approach eventually led to the LEDA submission to the NIST standardization process [100], eventually attacked in [53].

The main idea from [1392] is to consider a generalization of LDPC codes: LDPC codes have very small weight vectors of weight $O(1)$ in their dual but decode in $O(n)$. Moderate-density parity check (MDPC) codes are generated from higher weight codewords in $O(\sqrt{n})$ but will also be less efficient and decode errors of weight $O(\sqrt{n})$. Now, while LDPC codes have been studied for a long time, MDPC codes still raise questions regarding the analysis of their decoding rate, especially in a cryptographic context for which a **decryption failure rate (DFR)** of 2^{-128} is expected.

In order to get smaller key sizes, it is also possible to consider a quasi-cyclic version of MDPC codes (QC-MDPC codes). The main advantage of QC-MDPC codes is that the knowledge of one small weight vector is enough to decode them. In practice the QC-MDPC cryptosystem (usually simply denoted by MDPC although the codes are quasi-cyclic) can be seen as being close to the NTRU (N^{th} degree Truncated polynomial Ring Units) cryptosystem but in Hamming weight version. The basic decoding BitFlipping Algorithm 34.4.2, based on the classical LDPC decoder, is very simple to describe (the value of the threshold T plays a central role), but usually optimized variations are considered [644, 1651]. The original scheme was described in a McEliece-like version; we describe, in Figure 34.3, a polynomial key-exchange version similar to what is presented in [15]; see also Section 34.1.1 for the polynomial representation. Notice that turning the key-exchange protocol into an encryption scheme is straightforward.

Algorithm 34.4.2 (BitFlipping(h_0, h_1, s, T, t))

Input: h_0, h_1, and $s = h_1 e_1 + h_0 e_0$, threshold value T required to flip a bit, weight t of e_0 and e_1.

Output: (e_0, e_1) if the algorithm succeeds, "?" otherwise.

1: $(\mathbf{u}, \mathbf{v}) \leftarrow (\mathbf{0}, \mathbf{0}) \in (\mathbb{F}_2^n)^2$, $\mathbf{H} \leftarrow (\mathbf{rot}(h_0)^{\mathsf{T}}, \mathbf{rot}(h_1)^{\mathsf{T}}) \in \mathbb{F}_2^{n \times 2n}$, syndrome $\leftarrow \mathbf{s}$;
2: **while** $\big[\mathrm{wt}_{\mathrm{H}}(\mathbf{u}) \neq t$ **or** $\mathrm{wt}_{\mathrm{H}}(\mathbf{v}) \neq t\big]$ **and** syndrome $\neq \mathbf{0}$ **do**
3:     sum $\leftarrow$ syndrome $\times \mathbf{H}$;   /* No modular reduction, values in $\mathbb{Z}$ /*

4:     flipped_positions $\leftarrow \mathbf{0} \in \mathbb{F}_2^{2n}$;

5:     for $0 \leq i \leq 2n - 1$ do

6:         if sum$[i] \geq T$ then flipped_positions$[i]$ = flipped_positions$[i] \oplus 1$;

7:     $(\mathbf{u}, \mathbf{v}) = (\mathbf{u}, \mathbf{v}) \oplus$ flipped_positions;

8:     syndrome = syndrome $- \mathbf{H} \times$ flipped_positions^{T}

9: **end while**;

10: **if** $\mathbf{s} - \mathbf{H} \times (\mathbf{u}, \mathbf{v})^{\mathsf{T}} \neq \mathbf{0}$

11:     **then return** "?"

12: **else return** $(\mathbf{u}, \mathbf{v})$.

FIGURE 34.3: Description of QC-MDPC Key-Exchange protocol in a Niederreiter form. n usually follows the typical case described in Section 34.1.1, w is odd, $\mathbf{h}_0$ and $\mathbf{h}_1$ are small weight vectors considered as elements of $\mathcal{R}$. $\mathbf{h}_0$ is taken invertible in $\mathcal{R}$. The products follow the definition of (34.1), with simplified notation $\mathbf{xy}$ rather than $\mathbf{x} \cdot \mathbf{y}$. An encryption version can be obtained by considering $\mathbf{h}$ as a public key and adding a second part in Bob's response: $\mathbf{c}_0 = \mathbf{m}$ XOR Hash$(\mathbf{e}_0, \mathbf{e}_1)$. w and t are in $O(\sqrt{n})$.

Alice		Bob
$\mathbf{h}_0, \mathbf{h}_1 \xleftarrow{\$} \mathcal{S}_w^n(\mathbb{F}_2),\ \mathbf{h} \leftarrow \mathbf{h}_1 \mathbf{h}_0^{-1}$	$\xrightarrow{\ \mathbf{h}\ }$	
$\mathbf{s} \leftarrow \mathbf{h}_0 \mathbf{c}_1$	$\xleftarrow{\ \mathbf{c}_1\ }$	$\mathbf{e}_0, \mathbf{e}_1 \xleftarrow{\$} \mathbb{F}_2^n$ s.t. $\mathrm{wt}_H(\mathbf{e}_0) + \mathrm{wt}_H(\mathbf{e}_1) = 2t$
$(\mathbf{e}_0, \mathbf{e}_1) \leftarrow$ BitFlipping$(\mathbf{h}_0, \mathbf{h}_1, \mathbf{s}, t, w)$		$\mathbf{c}_1 \leftarrow \mathbf{e}_0 + \mathbf{h}\mathbf{e}_1$
$\boxed{\text{Hash}(\mathbf{e}_0, \mathbf{e}_1)}$	SHARED SECRET	$\boxed{\text{Hash}(\mathbf{e}_0, \mathbf{e}_1)}$

Why it works: $\mathbf{h}_0 \mathbf{c}_1 = \mathbf{e}_0 \mathbf{h}_0 + \mathbf{e}_1 \mathbf{h}_1$, a syndrome for the small weight vector $(\mathbf{e}_0, \mathbf{e}_1)$ associated to the MDPC matrix derived from $\mathbf{h}_0$ and $\mathbf{h}_1$, which can be decoded by the BitFlipping Algorithm 34.4.2.

Parameters: For security level λ, the length of the code is in $O(\lambda^2)$ and the weight of the generating vector is in $O(\lambda)$.

Security: The original presentation of the MDPC scheme had an OW-CPA security, based not only on the QCSD problem but also on the indistinguishability between the public generator matrix of a QC-MDPC code (generated by a small weight vector) and a random quasi-cyclic code. The Niederreiter version presented in Figure 34.3 follows [15] and can be proven IND-CCA2 (under the two previous hypotheses) with a generic transformation of type [970] at the condition that the DFR is proven sufficiently low in $2^{-\lambda}$, for λ the security parameter in bits.

Attacks and DFR: The main attacks are related to the decryption failure rate [874]. The MDPC scheme can be used for key exchange in a one-time process with a DFR as sufficiently small as 2^{-30}; using it for encryption necessitates a stronger analysis of the decoder to reach a DFR of at least 2^{-128} [1651]. A strong drawback of the cryptosystem for one-time key exchange is the cost of inverting a polynomial, which is drastically improved in [643].

Variations on the scheme: The original MDPC scheme [1392] is in a McEliece-like form. It is possible to consider it in a Niederreiter-like form as in the BIKE-2 version of the NIST standardization process, and also a version which does not necessitate a costly inversion [132] (the BIKE-1 version of the NIST standardization process); see the Bike submission [15] for details.

34.5 PKE Schemes with Reduction to Decoding Random Codes without Hidden Structure

34.5.1 Alekhnovich's Approach

In 2003, Alekhnovich proposed an innovative approach based on the difficulty of decoding purely random codes [36]. In this system the trapdoor (or secret key) is a random error vector that has been added to a random codeword of a random code. Recovering the secret key is therefore equivalent to solving the problem of decoding a random code – with no hidden structure. Alekhnovich also proved that breaking the system in any way, not necessarily by recovering the secret key, involves decoding a random linear code. The single bit encryption scheme is presented in Figure 34.4. The scheme can be turned into a multiple bits encryption scheme and necessitates the use of error-correcting codes to handle the decryption failure rate induced by the system. Typically one encrypts a codeword (seen as a sequence of single bits) obtained as an encoding of the plaintext by a given decodable code. The decryption results in a noisy codeword which can be decoded to recover the codeword and then from it, the plaintext.

FIGURE 34.4: Alekhnovich encryption scheme for a single bit [36]

KeyGen
   Let $\mathbf{A} \overset{\$}{\leftarrow} \mathbb{F}_2^{k \times n}$, $\mathbf{x} \overset{\$}{\leftarrow} \mathbb{F}_2^k$, $w = O(\sqrt{n})$, $\mathbf{e} \overset{\$}{\leftarrow} \mathcal{S}_w^n(\mathbb{F}_2)$, $\mathbf{y} \leftarrow \mathbf{xA} + \mathbf{e}$, and $\mathbf{H} \leftarrow [\mathbf{A}^\mathsf{T} \mid \mathbf{y}^\mathsf{T}]^\mathsf{T}$. Let $\mathcal{C}$ be the code with parity check matrix $\mathbf{H}$ and generator matrix $\mathbf{G}$. Return $(\mathrm{sk}, \mathrm{pk}) = (\mathbf{e}, \mathbf{G})$.

Encrypt
   To encrypt $m \in \{0,1\}$, if $m = 0$, $\mathbf{c} \leftarrow \mathbf{c}' + \mathbf{e}'$, otherwise if $m = 1$, $\mathbf{c} \leftarrow \mathbf{u}$, for $\mathbf{c}'$ a random codeword of $\mathcal{C}$, $\mathbf{e}' \overset{\$}{\leftarrow} \mathcal{S}_w^n$, and $\mathbf{u} \overset{\$}{\leftarrow} \mathbb{F}_2^n$. Return $\mathbf{c}$.

Decrypt
   To decrypt, return $b \leftarrow \langle \mathbf{c}, \mathbf{e} \rangle$.

Why decryption works: If $m = 0$, $\langle \mathbf{c}, \mathbf{e} \rangle = \langle \mathbf{c}' + \mathbf{e}', \mathbf{e} \rangle = \langle \mathbf{e}', \mathbf{e} \rangle = 0$ with overwhelming probability (since $w = O(\sqrt{n})$). If $m = 1$, decryption succeeds with probability $1/2$. The scheme is probabilistic.

Even if the system was not totally practical, the approach in itself was a breakthrough for code-based cryptography. Its inspiration was provided in part by the Ajtai–Dwork cryptosystem [27] which is based on solving hard lattice problems. The Ajtai–Dwork cryptosystem also inspired the Learning With Errors (LWE) lattice-based cryptosystem by Regev [1583] which generated a huge amount of research in lattice-based cryptography.

34.5.2 HQC: Efficient Encryption from Random Quasi-Cyclic Codes

The previous Alekhnovich cryptosystem, although not quite efficient, was a first step towards efficient cryptosystems based on random codes. In 2018, Aguilar et al. proposed an efficient cryptosystem based on the difficulty of decoding random quasi-cyclic codes and on the notion of noisy Diffie–Hellman; see also [770] for the first version of the scheme in 2010. The main novelty of the system, different from the McEliece framework, is that there is not only one masked code which does both encryption **and** decryption (through decoding), but two codes: a first random double circulant code which is used for encryption and a second code which is used for decryption. The novelty in the scheme is that this second code is public with no masking; it just needs to have an efficient decoding algorithm. In particular, considering other decoding codes does not change the security of the scheme. The scheme is very simple and expressed in polynomial form (see Section 34.1.1 for the relation between polynomial representation and matrices) and described in Figure 34.5. The Hamming Quasi-Cyclic (HQC) scheme was submitted to the NIST standardization process [17].

FIGURE 34.5: Description of the HQC cryptosystem. The multiplication $\mathbf{xy}$ of two elements $\mathbf{x}$ and $\mathbf{y}$ of $\mathcal{R}$ is a simplified notation for $\mathbf{x} \cdot \mathbf{y}$ of (34.1); the choice of n follows the typical case of Section 34.1.1.

- **KeyGen(param)**: Samples $\mathbf{h} \xleftarrow{\$} \mathcal{R}$, a generator matrix $\mathbf{G} \in \mathbb{F}_2^{k \times n}$ of $\mathcal{C}$ which decodes $O(n)$ errors, $\mathrm{sk} = (\mathbf{x}, \mathbf{y}) \xleftarrow{\$} \mathcal{R}^2$ such that $\mathrm{wt}_\mathrm{H}(\mathbf{x}) = \mathrm{wt}_\mathrm{H}(\mathbf{y}) = O(\sqrt{n})$. Sets $\mathrm{pk} = (\mathbf{h}, \mathbf{s} = \mathbf{x} + \mathbf{hy})$, and returns $(\mathrm{pk}, \mathrm{sk})$.

- **Encrypt(pk, m)**: Generates $\mathbf{e} \xleftarrow{\$} \mathcal{R}$, $\mathbf{r} = (\mathbf{r}_1, \mathbf{r}_2) \xleftarrow{\$} \mathcal{R}^2$ such that $\mathrm{wt}_\mathrm{H}(\mathbf{e}) = w_\mathbf{e}$, and $\mathrm{wt}_\mathrm{H}(\mathbf{r}_1) = \mathrm{wt}_\mathrm{H}(\mathbf{r}_2) = w_\mathbf{r}$, for $w_\mathbf{r} = w_\mathbf{e} = O(\sqrt{n})$. Sets $\mathbf{u} = \mathbf{r}_1 + \mathbf{hr}_2$ and $\mathbf{v} = \mathbf{mG} + \mathbf{sr}_2 + \mathbf{e}$, and returns $\mathbf{c} = (\mathbf{u}, \mathbf{v})$.

- **Decrypt(sk, c)**: Returns $\mathcal{C}.\mathrm{Decode}(\mathbf{v} - \mathbf{uy})$.

Why decryption works: $\mathbf{v} - \mathbf{uy} = \mathbf{mG} - \mathbf{yr}_1 + \mathbf{xr}_2 + \mathbf{e}$, a word decodable by $\mathcal{C}$, with a decodable error of weight $O(n)$ for chosen parameters.

Parameters: Similarly to MDPC codes, for λ the security level, the length of the code is in $O(\lambda^2)$ and the weight of the small weight generating vectors is in $O(\lambda)$. In practice the parameters are slightly larger than for MDPC and two blocks are needed for the ciphertext.

Security: The IND-CPA security of HQC relies on the QCSD problem and the decision DQCSD problem; no indistinguishability hypothesis for a hidden code is needed. The IND-CCA2 security can be reached through a generic and efficient transformation such as [970] depending on the DFR of the system.

Instantiations: In the original HQC submission, the use of a tensor code obtained from BCH codes and a repetition code was proposed for decoding. Recently a more efficient concatenated code based on Reed–Muller and Reed–Solomon codes was proposed in [57]. Typically one needs to decode codes with a very low rate of order 1% and an error rate of order 30%.

Decryption Failure Rate analysis: The main point to keep in mind for this system is that since the decoding is probabilistic, one needs to have a precise DFR analysis for the security proof which requires a DFR in $2^{-\lambda}$ for λ the security level, which is the case with the proposed decoding codes. One could consider more efficient iterative decoding algorithms which exist in the literature, but in that case having a precise DFR analysis which is able to go as a low as 2^{-128} is trickier to obtain.

A weakness of the system is its relatively low encryption rate, but this is not a major issue for classical applications of public-key encryption schemes such as authentication or key exchange. In the NIST standardization process, it was possible to consider a message size of only 256 bits.

34.5.3 Ouroboros Key-Exchange Protocol

While HQC has the advantage of relying only on the hardness of the (decisional) syndrome decoding problem for random quasi-cyclic codes, it still features big keys for low encryption rates. For most concrete applications, such as key exchange over the internet through TLS for instance, having a small plaintext space big enough to encrypt a session key is sufficient.

Based upon this observation, lighter parameters (and some other tricks) were proposed for a key-exchange protocol, resulting in the Ouroboros key-exchange protocol [530] and presented in Figure 34.6. The protocol works similarly as HQC, except that no message needs to be encrypted (hence there is no public code $\mathcal{C}$), and the decoding algorithm, which is very close to the bit flipping algorithm [787] used for LDPC and MDPC codes, is tweaked to handle noisy syndromes; see Algorithm 34.5.1. The Ouroboros protocol was submitted to the NIST standardization process as BIKE-3.

Algorithm 34.5.1 (xBitFlipping($\mathbf{h}_0, \mathbf{h}_1, \mathbf{s}, T, t$))

Input: $\mathbf{h}_0, \mathbf{h}_1$, and $\mathbf{s} = \mathbf{h}_1 \mathbf{e}_1 - \mathbf{h}_0 \mathbf{e}_0 + \mathbf{e}$, a threshold value T required to flip a bit, weight t of $\mathbf{e}_0$, $\mathbf{e}_1$, and $\mathbf{e}$.

Output: $(\mathbf{e}_0, \mathbf{e}_1)$ if the algorithm succeeds, "?" otherwise.

1: $(\mathbf{u}, \mathbf{v}) \leftarrow (\mathbf{0}, \mathbf{0}) \in (\mathbb{F}_2^n)^2$, $\mathbf{H} \leftarrow (\mathbf{rot}(-\mathbf{h}_0)^\mathsf{T}, \mathbf{rot}(\mathbf{h}_1)^\mathsf{T}) \in \mathbb{F}_2^{n \times 2n}$, syndrome $\leftarrow \mathbf{s}$;

2: **while** $[\mathrm{wt}_H(\mathbf{u}) \neq t$ **or** $\mathrm{wt}_H(\mathbf{v}) \neq t]$ **and** $\mathrm{wt}_H(\text{syndrome}) > t$ **do**

3:      sum $\leftarrow$ syndrome $\times \mathbf{H}$;   /* No modular reduction, values in $\mathbb{Z}$ /*

4:      flipped_positions $\leftarrow \mathbf{0} \in \mathbb{F}_2^{2n}$;

5:      **for** $0 \leq i \leq 2n - 1$ **do**

6:          **if** sum$[i] \geq T$ **then** flipped_positions$[i]$ = flipped_positions$[i] \oplus 1$;

7:      $(\mathbf{u}, \mathbf{v}) = (\mathbf{u}, \mathbf{v}) \oplus$ flipped_positions;

8:      syndrome = syndrome $- \mathbf{H} \times$ flipped_positions$^\mathsf{T}$;

9: **end while**;

10: **if** $\mathrm{wt}_H\left(\mathbf{s} - \mathbf{H} \times (\mathbf{u}, \mathbf{v})^\mathsf{T}\right) > t$

11:     **then return** "?"

12: **else return** $(\mathbf{u}, \mathbf{v})$.

FIGURE 34.6: Description of the Ouroboros Key-Exchange protocol. f_0 and f_1 constitute the public key. Alternatively f_1 can be recovered by sending/publishing only the λ bits of the seed seed_{f_1} (instead of the n coordinates of f_1). The choice of n follows the typical case of Section 34.1.1.

Alice		Bob

$$\text{seed}_{f_1} \xleftarrow{\$} \{0,1\}^\lambda, \ f_1 \xleftarrow{\text{seed}_{f_1}} \mathbb{F}_2^n$$

$$h_0, h_1 \xleftarrow{\$} \mathcal{S}_w^n(\mathbb{F}_2), \ f_0 \leftarrow h_1 + f_1 h_0 \qquad \xrightarrow{\quad f_0, f_1 \quad} \qquad e \xleftarrow{\$} \mathcal{S}_t^n(\mathbb{F}_2)$$

$$e_0, e_1 \xleftarrow{\$} \mathbb{F}_2 \text{ s.t. } \text{wt}_H(e_0) + \text{wt}_H(e_1) = 2t$$

$$s \leftarrow c_0 - h_0 c_1 \qquad \xleftarrow{\quad c_0, c_1 \quad} \qquad c_0 \leftarrow f_0 e_1 + e, \quad c_1 \leftarrow e_0 + f_1 e_1$$

$$(e_0, e_1) \leftarrow \text{xBitFlipping}(h_0, h_1, s, t, w)$$

$\boxed{\text{Hash}(e_0, e_1)}$ SHARED SECRET $\boxed{\text{Hash}(e_0, e_1)}$

Why it works: $s = -e_0 h_0 + e_1 h_1 + e$, a syndrome for the small weight vector (e_0, e_1) associated to the MDPC matrix derived from h_0 and h_1 to which one adds a small weight vector e. w and t are in $O(\sqrt{n})$. The modified syndrome s can be decoded by the xBitFlipping Algorithm 34.5.1.

Security and parameters: The Ouroboros protocol can be seen as intermediate to HQC and MDPC schemes: it benefits from the security reduction of HQC with reduction to attacking random quasi-cyclic instances on one side (the QCSD and DQCSD problems), and on the other side, it uses an extended bitflip-like decoding algorithm, xBitFlipping of Algorithm 34.5.1. As with the HQC protocol, the Ouroboros protocol necessitates sending two blocks for the encryption. Similarly to MDPC and HQC, for λ the security level, the length of the code is in $O(\lambda^2)$, and the weight of the small weight generating vector is in $O(\lambda)$. In practice the parameters lie between those of HQC and MDPC. Notice that since the value f_1 is random, a seed to generate it is sufficient. As with MDPC, the Ouroboros key-exchange protocol can be turned into an encryption scheme which can be proven IND-CCA2 through a generic transformation like [970] depending on a low DFR, but without an additional public matrix indistinguishability hypothesis.

Decryption Failure Rate analysis: The decoding of Ouroboros is very close to the decoding of MDPC codes as the properties of the decoder are very close in practice. Basically for a small DFR such as 2^{-30}, like MDPC, the Ouroboros protocol can be used as a KEM for key exchange. It can be used for encryption but with a lower DFR.

Key exchange and encryption: Similarly to MDPC the protocol is presented as a key-exchange protocol but it can be easily turned into an encryption scheme for a message m by adding a ciphertext $c = m$ XOR $\text{Hash}(e_0, e_1)$.

34.6 Examples of Parameters for Code-Based Encryption and Key Exchange

In this section we provide a table with practical examples for 128 bits security.

TABLE 34.2: Comparison between several 1st and 2nd round code-based submissions to the NIST Post-Quantum Cryptography (PQC) standardization process. All sizes are expressed in bytes. The targeted security level is NIST Level-1, approximately equivalent to 128 bits of classical security. The notation ct stands for ciphertext.

Cryptosystem name	pk size	ct size	DFR
Classic McEliece	261120	128	0
Big Quake	25482	201	0
HQC-original	3024	6017	2^{-128}
HQC-RMRS	2607	5191	2^{-128}
BIKE-II (MDPC)	1473	1505	2^{-128}
BIKE-III (Ouroboros)	1566	3100	2^{-128}
BIKE-I (Cake)	2945	2945	2^{-128}

34.7   Authentication: The Stern Zero-Knowledge Protocol

Authentication is an important cryptographic primitive. When there is an obvious generic transformation to turn a public key encryption scheme or a signature scheme into a public key authentication scheme, there also exist specific authentication schemes and, in particular, zero-knowledge authentication schemes. After the seminal work of Fiat and Shamir, there were a series of papers related to code-based authentication schemes which culminated in efficiency with the Stern authentication algorithm.

We now recall *Stern's zero-knowledge*[3] *authentication protocol* [1753]; see Figure 34.7. Given a random public parity check matrix $\mathbf{H} \in \mathbb{F}_2^{(n-k) \times n}$, the secret key is a small weight vector $\mathbf{x}$ of length n and of weight w. The public key is constructed as $\mathbf{s} = \mathbf{H}\mathbf{x}^\mathsf{T}$. In Stern's identification protocol, a prover P wants to convince a verifier V that she/he is indeed the person corresponding to the public identifier $\mathbf{s}$.

Stern's protocol has a cheating probability of 2/3 for each round. This implies that the protocol has to be repeated $\lceil -\lambda / \log_2 (2/3) \rceil$ times to achieve a negligible (in the security parameter λ) cheating probability. For instance, for $\lambda = 128$ bits of security, this results in 219 rounds.

Security and parameters: The security of the protocol is the generic SD problem. There are two main drawbacks to the scheme: the public key, a random matrix, is very large and the cheating probability is rather high, which implies a large number of rounds to get small overall cheating probabilities. In particular, the signature is very large on the order of the square of the security level. Indeed increasing the security level increases both the length of the code and the number of necessary rounds. Typically, for parameters one considers, a secret word is of weight just below the Gilbert–Varshamov distance for a rate $\frac{1}{2}$ code; this leads to code lengths on the order of 1000 for 128 bits security.

Variations and improvements: There are two main improvements on the scheme when quasi-cyclic codes are used. Their use decreases the size of the public matrix [771], and the cheating probability can be decreased to $\frac{1}{2}$ [23]. There exists a dual version of the protocol that is slightly more efficient [1851]. It is also possible to obtain a $\frac{1}{2}$ cheating probability

---

[3]Often due to the original Stern's protocol description, the protocol is presented without the random seeds $\mathbf{r}_1$, $\mathbf{r}_2$, and $\mathbf{r}_3$. In that case the protocol is not zero-knowledge but only testable weak zero-knowledge [30]; random seeds $\mathbf{r}_i$ need to be added as in Figure 34.7.

FIGURE 34.7: Stern's protocol

Secret key: $\mathbf{x} \in \mathbb{F}_2^n$ of weight w

Public key: A random $(n - k) \times n$ binary matrix $\mathbf{H}$, $\mathbf{s} = \mathbf{H}\mathbf{x}^\mathsf{T}$

1. (Commitment Step) P samples uniformly at random a commitment $\mathbf{y} \in \mathbb{F}_2^n$, a permutation σ of $\{1, 2, \ldots, n\}$, and three random seeds $\mathbf{r}_i$ for $i \in \{1, 2, 3\}$ of $\mathbb{F}_2^\lambda$, for λ the security level in bits. Then P sends to V the commitments $\mathbf{c}_1$, $\mathbf{c}_2$, and $\mathbf{c}_3$ such that:

$$\mathbf{c}_1 = h(\mathbf{r}_1|\sigma|\mathbf{H}\mathbf{y}^\mathsf{T}); \quad \mathbf{c}_2 = h(\mathbf{r}_2|\sigma(\mathbf{y})); \quad \mathbf{c}_3 = h(\mathbf{r}_3|\sigma(\mathbf{y} + \mathbf{x})),$$

where $a|b$ denotes the concatenation of a and b, and h is a hash function.

2. (Challenge Step) V sends $b \in \{0, 1, 2\}$ to P.

3. (Answer Step) Three possibilities:
   - If $b = 0$, P reveals $\mathbf{r}_1$, $\mathbf{r}_2$, $\mathbf{y}$, and σ.
   - If $b = 1$, P reveals $\mathbf{r}_1$, $\mathbf{r}_3$, $(\mathbf{y} + \mathbf{x})$ and σ.
   - If $b = 2$, P reveals $\mathbf{r}_2$, $\mathbf{r}_3$, $\sigma(\mathbf{y})$ and $\sigma(\mathbf{x})$.

4. (Verification Step) Three possibilities:
   - If $b = 0$, V verifies that $\mathbf{c}_1, \mathbf{c}_2$ have been honestly calculated.
   - if $b = 1$, V verifies that $\mathbf{c}_1, \mathbf{c}_3$ have been honestly calculated.
   - if $b = 2$, V verifies that $\mathbf{c}_2, \mathbf{c}_3$ have been honestly calculated and that the weight of $\sigma(\mathbf{x})$ is w.

5. Iterate Steps 1, 2, 3, 4 until the expected security level is reached.

Why it works: For $b = 1$, $\mathbf{H}\mathbf{y}^\mathsf{T}$ can be obtained as $\mathbf{H}\mathbf{y}^\mathsf{T} = \mathbf{H}(\mathbf{x} + \mathbf{y})^\mathsf{T} + \mathbf{s}$

---

by considering q-ary codes [372]. An interesting question would be to try to decrease the cheating probability for one round, below the $\frac{1}{2}$ best known cheating probability.

34.8 Digital Signatures from Coding Theory

Designing a secure and efficient signature scheme based on coding theory is a long-standing open problem. A first approach consists in designing a zero-knowledge identification scheme and then turning it into a digital signature scheme by applying the Fiat–Shamir transform to the identification transcript, using a collision-resistant hash function h (not necessarily based on coding assumptions). This kind of approach usually yields large signatures that are impractical for real-life use.

The other approach is known as the hash-and-sign paradigm and will be described later in this section. While this latter approach can yield much shorter and efficient signature schemes, it is rather hard to obtain securely.

34.8.1 Signature from a Zero-Knowledge Authentication Scheme with the Fiat–Shamir Heuristic

The Fiat–Shamir heuristic permits one to turn a zero-knowledge protocol into a signature scheme. The main idea is that the new protocol is no longer interactive anymore, and the prover proves itself through the use of a hash function. We recall the Fiat–Shamir heuristic in Figure 34.8.

FIGURE 34.8: Signature through the Fiat–Shamir heuristic

- **KeyGen**(param): KeyGen parameters of the zero-knowledge scheme, returns $(\mathsf{pk}, \mathsf{sk})$.

- **Signature of a message** M: Fix a security level and a number N of associated rounds. Generate the N commitments C_i at once, and compute the general commitment C as a concatenation of all the C_i. Compute a sequence of N challenges r_i from a hash of the concatenation of C and the message M. Compute the N answers A_i associated to the r_i and concatenate the A_i as A. The signature is the two supersets (C, A).

- **Verification:** Compute the sequence of challenges r_i from the received C and M. Check that the answers A_i correspond to the challenges r_i and commitments C_i. Accept the signature if all the answers fit. Reject otherwise.

Security: The Fiat–Shamir heuristic is proven secure in the random oracle model [32, 1528]. The security is based on the SD problem for the original Stern algorithm and under QCSD for its quasi-cyclic improvement. The question of the extension of the proof in a quantum oracle is actively considered by researchers.

Parameters: The main advantage of such signatures is that the public key is rather small (in the quasi-cyclic version), a few hundred bits, and the security is really strong. Now the signature length in itself is rather large (a few hundred thousand bits).

Variations and applications: Because of the simplicity of the scheme and the really strong security of the scheme, the Fiat–Shamir heuristic-based signature has been used to introduce many of the classical signatures used in classical number theory-based cryptography: blind signatures [219], group signatures [698], undeniable signatures [18], ring signatures [20, 1944], identity-based signatures [371], etc.

34.8.2 The CFS Signature Scheme

A classical way to build a hash-and-sign signature, like the RSA signature for instance, is to consider a full domain hash in which it is possible to associate to any hash (an element of a domain) a pre-image through a particular hard to invert function. For coding theory, this approach is basically somehow difficult; indeed if one considers a given code $\mathcal{C}$, the set of decodable vectors is exactly the whole space $\mathbb{F}_2^n$ if and only if the code is perfect. There are not many such codes and typically their structure is hard to hide. The idea of the CFS scheme, proposed by Courtois, Finiasz, and Sendrier in [456], is to consider a classical McEliece scheme built with Goppa codes, but then to consider very dense Goppa codes in order to obtain a high density of decodable vectors. The algorithm then consists in considering hash values obtained from the message to sign together with a counter, and check whether the hash value considered as a syndrome can be decoded, and repeat it until one obtains a decodable hash value.

For a t-correcting binary Goppa code $\mathcal{C}$ of length $n = 2^m$ over $\mathbb{F}_2$, only $\binom{n}{t}$ among the 2^{mt} syndromes are decodable, yielding an asymptotic density of $\frac{\binom{n}{t}}{2^{mt}} \approx \frac{1}{t!}$. The scheme is described in Figure 34.9; all details can be found in the original paper [456].

FIGURE 34.9: CFS signature scheme

KeyGen

> Use the Niederreiter KeyGen Algorithm (see Figure 34.2) to generate a public parity check matrix $\mathbf{H}_{\mathrm{pub}}$ and a secret parity check matrix $\mathbf{H}_{\mathrm{sec}}$.

Sign

> Compute $\mathbf{s}_i \leftarrow h\left(h\left(\mathbf{m}\right)|i\right)$ until $\mathbf{s}_{i_0}$ can be decoded in $\mathbf{z}$ using $\mathbf{H}_{\mathrm{sec}}$ where h is a hash function. The signature is $(\mathbf{z}, i_0)$.

Verify

> Accept the signature if $\mathbf{H}_{\mathrm{pub}}\mathbf{z}^{\mathsf{T}} = h\left(h\left(\mathbf{m}\right)|i_0\right)$. Reject otherwise.

Security: The security proof of the scheme relies on the indistinguishability of a dense permuted Goppa code. It was proven in [705] that this assumption did not hold and that it was possible to distinguish between dense Goppa codes and random codes; this property holds for very dense Goppa codes but not for Goppa codes used in the classical encryption scheme for which the rate of the code is often of order $\frac{1}{2}$. It is interesting to notice that even if it is possible to distinguish between dense Goppa codes and random codes, in practice there is no known attack on the scheme from this distinguisher.

Parameters: The density of $\frac{1}{t!}$ obliges one to consider on average $t!$ trials; hence t cannot be too large. Moreover decoding t errors for a random code has to be difficult. Overall these two constraints lead to considering very long Goppa codes so that the size of the public key is super-polynomial in the security parameter. In practice, the scheme can be used for levels of security that are not too high, such as 80 bits of security, in which case $n = 2^{16}$, $n - k = 144$, and $t = 9$; higher security levels like 128 bits seem out of reach in practice. Because the CFS scheme was the only existing hash-and-sign signature scheme for a long time, it was used in many code-based constructions.

34.8.3 The WAVE Signature

Recently a new hash-and-sign signature was proposed in [514]. This scheme considers a ternary generalized $(\mathbf{U}, \mathbf{U} + \mathbf{V})$ code construction, an extension of the $(\mathbf{u} \mid \mathbf{u} + \mathbf{v})$ construction of Definition 1.11.1. This construction is in the spirit of the lattice construction [800] in the sense that a proof is given that the signature is indistinguishable from a random distribution by using rejection sampling. Moreover, in contrast to the CFS scheme, for which in the range of parameters under consideration there is a unique pre-image for syndromes, in the case of WAVE, the trapdoor is used to construct a pre-image of a syndrome but in a range where the pre-images may not be unique. In the end, the paper introduces a very interesting new notion of decoding for high weights. The security of the scheme relies both on the hardness of generic decoding for high weights and the indistinguishability of generalized $(\mathbf{U}, \mathbf{U} + \mathbf{V})$ codes. Even if the parameters are rather large, they do not increase as dramatically as the CFS signature scheme, and hence the WAVE signature scheme can be considered as the first hash-and-sign code-based signature scheme. For 128 bits of classical security, signature sizes are on the order of 15 thousand bits, the public key size on the order of 4 megabytes, and the rejection rate is limited to one rejection every 10 to 12 signatures.

34.8.4 Few-Times Signature Schemes and Variations

One of the main problems in designing a signature scheme is the fact that there has to be no leak of information when a signature is given. Some schemes exist which permit one to obtain only a few signatures. The main idea of these schemes is to give, as the public key, a set of syndromes associated to small weight secret vectors. The signature is then a pre-image (built with the secret vectors) of a linear combination of the public syndromes. The first proposed scheme was the Kabatianskii–Krouk–Smeets (KKS) scheme wrongly proposed at first as being a general signature scheme; see [1066] and attacks [1477]. More recently proposed full signature schemes adapting ideas from lattices were also broken or found to be much too inefficient [529, 1213], eventually becoming few-times signatures schemes. A few-times signature scheme based on the action of an automorphism was also presented in [777].

34.9 Other Primitives

There exist many other cryptographic primitives; not all of them possess a code-based equivalent, but some do, including the following two.

Code-based pseudo-random number generators: It is possible to construct pseudo-random number generators based on the SD problem. The first scheme was proposed in [733]; a more efficient variation based on quasi-cyclic codes was proposed in [773].

Code-based hash functions: Hash functions are a very important tool in cryptography. Following what was done for lattices, an early code-based hash was proposed in [77]. It was followed by quasi-cyclic optimized versions [76, 729], submitted to the SHA3 (Secure Hash Algorithm 3) competition in 2008, as well as other variations described in [181].

Research Problem 34.9.1 Among primitives which are still not known but are of real interest, one can cite homomorphic encryption. It is trivially possible to add $O(n)$ ciphertexts by a code-based cryptosystem and still be able to decrypt the sum of them, but it is not known if it is possible with $O(n^2)$ ciphertexts or even more. Also it is not known how to obtain fully homomorphic encryption. Lastly, identity-based encryption is also an open question.

34.10 Rank-Based Cryptography

Rank-based cryptography is based on codes using the rank metric rather than the Hamming metric; see Chapters 11 and 22. When considering codes over an extension field $\mathbb{F}_{q^m}$ of $\mathbb{F}_q$, a vector $\mathbf{v} \in \mathbb{F}_{q^m}^n$ can be unfolded, using an $\mathbb{F}_q$-basis of $\mathbb{F}_{q^m}$, as an $m \times n$ matrix V. The rank weight of $\mathbf{v}$ is then the rank of the matrix V. Alternatively the rank weight of a vector $\mathbf{v}$ is the dimension of the $\mathbb{F}_q$-space generated by its coordinates. There is a natural dictionary of many notions in Hamming distance when they are considered with the rank metric. For instance the notion of permutation is turned into a notion of invertible matrix over the base field $\mathbb{F}_q$, and the notion of support in the Hamming metric is turned into the space generated by coordinates in the rank metric. Also the SD problem has an analog in the rank metric, namely, the Rank Syndrome Decoding (RSD) problem, where the Hamming distance is replaced by the rank distance. Changing the metric implies several changes to the geometric properties of the codes. Overall some cryptographic schemes can be easily adapted; for instance, the GPT (Gabidulin–Paramonov–Tretjakov) cryptosystem [766] is an adaptation of the McEliece cryptosystem with Gabidulin codes replacing Goppa codes. However, Gabidulin codes have a structure that is hard to mask, which leads to many attacks on the scheme or its variations [1479]. There also exist equivalent systems to MDPC, namely the LRPC (Low Rank Parity Check) cryptosystem [774]; to HQC, namely the RQC (Rank Quasi-Cyclic) cryptosystem [19]; and to Ouroboros, namely the Ouroboros-R scheme [16]. These schemes were presented in the ROLLO 2nd round submission of the NIST standardization process [55].

After a first attempt in 2014 at a signature scheme using the rank metric, namely the RankSign signature [776] that was attacked in [515], the efficient signature scheme Durandal was proposed in 2019 [56]. It is possible to adapt the Stern authentication scheme to the rank metric [778]. Other primitives with rank metric formulations, e.g., [772], are also known. But some others, e.g., hash functions, seem to resist because of properties of the metric.

In general the rank metric leads to cryptosystems with smaller key sizes. The main difficulty of the problem is not yet completely fixed even though, in recent years, there has developed better knowledge of the security of the general Rank Syndrome Decoding problem, to which the Syndrome Decoding problem can be probabilistically reduced [118, 775, 780].

Acknowledgement

The first author thanks Olivier Blazy, Alain Couvreur, Thomas Debris-Alazard, and Gilles Zémor for helpful comments and discussions.

Bibliography

[1] M. Aaltonen. Linear programming bounds for tree codes. *IEEE Trans. Inform. Theory*, 25:85–90, 1979.

[2] M. Aaltonen. A new upper bound on nonbinary block codes. *Discrete Math.*, 83:139–160, 1990.

[3] E. Abbe. Randomness and dependencies extraction via polarization. In *Proc. Information Theory and Applications Workshop*, La Jolla, California, 7 pages, February 2011.

[4] E. Abbe. Randomness and dependencies extraction via polarization, with applications to Slepian-Wolf coding and secrecy. *IEEE Trans. Inform. Theory*, 61:2388–2398, 2015.

[5] E. Abbe and E. Telatar. Polar codes for the m-user MAC. arXiv:1002.0777 [cs.IT], February 2010.

[6] E. Abbe and E. Telatar. Polar codes for the m-user multiple access channel. *IEEE Trans. Inform. Theory*, 58:5437–5448, 2012.

[7] N. Abdelghany and N. Megahed. Code-checkable group rings. *J. Algebra Comb. Discrete Struct. Appl.*, 4:115–122, 2017.

[8] M. Abramowitz and I. A. Stegun. *Handbook of Mathematical Functions with Formulas, Graphs, and Mathematical Tables*, volume 55 of *National Bureau of Standards Applied Mathematics Series*. For sale by the Superintendent of Documents, U.S. Government Printing Office, Washington, D.C.; 9th Dover printing, 10th GPO printing, 1964.

[9] T. Abualrub, A. Ghrayeb, and R. H. Oehmke. A mass formula and rank of $\mathbb{Z}_4$ cyclic codes of length 2^e. *IEEE Trans. Inform. Theory*, 50:3306–3312, 2004.

[10] T. Abualrub and R. H. Oehmke. On the generators of $\mathbb{Z}_4$ cyclic codes of length 2^e. *IEEE Trans. Inform. Theory*, 49:2126–2133, 2003.

[11] T. Abualrub and R. H. Oehmke. Correction to: "On the generators of $\mathbb{Z}_4$ cyclic codes of length 2^e" [*IEEE Trans. Inform. Theory* 49:2126–2133, 2003]. *IEEE Trans. Inform. Theory*, 51:3009, 2005.

[12] T. Abualrub and I. Şiap. Cyclic codes over the rings $\mathbb{Z}_2 + u\mathbb{Z}_2$ and $\mathbb{Z}_2 + u\mathbb{Z}_2 + u^2\mathbb{Z}_2$. *Des. Codes Cryptogr.*, 42:273–287, 2007.

[13] L. M. Adleman. Molecular computation of solutions to combinatorial problems. *Science*, 266:1021–1024, 1994.

[14] A. Agarwal, A. Barg, S. Hu, A. Mazumdar, and I. Tamo. Combinatorial alphabet-dependent bounds for locally recoverable codes. *IEEE Trans. Inform. Theory*, 64:3481–3492, 2018.

[15] C. Aguilar-Melchor, N. Aragon, P. Barreto, S. Bettaieb, L. Bidoux, O. Blazy, J.-C. Deneuville, P. Gaborit, S. Gueron, T. Güneysu, R. Misoczki, E. Persichetti, N. Sendrier, J.-P. Tillich, and G. Zémor. BIKE. First round submission to the NIST post-quantum cryptography call: https://bikesuite.org, November 2017.

[16] C. Aguilar-Melchor, N. Aragon, S. Bettaieb, L. Bidoux, O. Blazy, J.-C. Deneuville, P. Gaborit, A. Hauteville, and G. Zémor. Ouroboros-R. First round submission to the NIST post-quantum cryptography call: https://pqc-ouroborosr.org, November 2017.

[17] C. Aguilar-Melchor, N. Aragon, S. Bettaieb, L. Bidoux, O. Blazy, J.-C. Deneuville, P. Gaborit, E. Persichetti, and G. Zémor. HQC. First round submission to the NIST post-quantum cryptography call: https://pqc-hqc.org, November 2017.

[18] C. Aguilar-Melchor, S. Bettaieb, P. Gaborit, and J. Schrek. A code-based undeniable signature scheme. In M. Stam, editor, *Cryptography and Coding*, volume 8308 of *Lecture Notes in Comput. Sci.*, pages 99–119. Springer, Berlin, Heidelberg, 2013.

[19] C. Aguilar-Melchor, O. Blazy, J.-C. Deneuville, P. Gaborit, and G. Zémor. Efficient encryption from random quasi-cyclic codes. *IEEE Trans. Inform. Theory*, 64:3927–3943, 2018.

[20] C. Aguilar-Melchor, P.-L. Cayrel, P. Gaborit, and F. Laguillaumie. A new efficient threshold ring signature scheme based on coding theory. *IEEE Trans. Inform. Theory*, 57:4833–4842, 2011.

[21] C. Aguilar-Melchor and P. Gaborit. On the classification of extremal [36, 18, 8] binary self-dual codes. *IEEE Trans. Inform. Theory*, 54:4743–4750, 2008.

[22] C. Aguilar-Melchor, P. Gaborit, J.-L. Kim, L. Sok, and P. Solé. Classification of extremal and *s*-extremal binary self-dual codes of length 38. *IEEE Trans. Inform. Theory*, 58:2253–2262, 2012.

[23] C. Aguilar-Melchor, P. Gaborit, and J. Schrek. A new zero-knowledge code based identification scheme with reduced communication. In *Proc. IEEE Information Theory Workshop*, pages 648–652, Paraty, Brazil, October 2011.

[24] D. Aharonov and M. Ben-Or. Fault-tolerant quantum computation with constant error rate. *SIAM J. Comput.*, 38:1207–1282, 2008.

[25] R. Ahlswede and H. Aydinian. On error control codes for random network coding. In *Workshop on Network Coding, Theory, and Applications*, pages 68–73, Lausanne, Switzerland, June 2009.

[26] R. Ahlswede, N. Cai, S.-Y. R. Li, and R. W. Yeung. Network information flow. *IEEE Trans. Inform. Theory*, 46:1204–1216, 2000.

[27] M. Ajtai and C. Dwork. A public-key cryptosystem with worst-case/average-case equivalence. In *Proc. 29th Annual ACM Symposium on the Theory of Computing*, pages 284–293, El Paso, Texas, May 1997.

[28] A. Alahmadi, R. E. L. Aldred, R. de la Cruz, S. Ok, P. Solé, and C. Thomassen. The minimum number of minimal codewords in an $[n, k]$-code and in graphic codes. *Discrete Appl. Math.*, 184:32–39, 2015.

[29] A. Alahmadi, C. Güneri, B. Özkaya, H. Shohaib, and P. Solé. On self-dual double negacirculant codes. *Discrete Appl. Math.*, 222:205–212, 2017.

[30] Q. Alamélou, O. Blazy, S. Cauchie, and P. Gaborit. A code-based group signature scheme. *Des. Codes Cryptogr.*, 82:469–493, 2017.

[31] S. M. Alamouti. A simple transmit diversity technique for wireless communications. *IEEE J. Selected Areas Commun.*, 16:1451–1458, 1998.

[32] S. M. E. Y. Alaoui, Ö Dagdelen, P. Véron, D. Galindo, and P.-L. Cayrel. Extended security arguments for signature schemes. In A. Mitrokotsa and S. Vaudenay, editors, *Progress in Cryptology – AFRICACRYPT 2012*, volume 7374 of *Lecture Notes in Comput. Sci.*, pages 19–34. Springer, Berlin, Heidelberg, 2012.

[33] A. Albanese, J. Blömer, J. Edmonds, M. G. Luby, and M. Sudan. Priority encoding transmission. *IEEE Trans. Inform. Theory*, 42:1737–1744, 1996.

[34] A. A. Albert. Symmetric and alternate matrices in an arbitrary field. I. *Trans. Amer. Math. Soc.*, 43:386–436, 1938.

[35] T. L. Alderson. (6,3)-MDS codes over an alphabet of size 4. *Des. Codes Cryptogr.*, 38:31–40, 2006.

[36] M. Alekhnovich. More on average case vs. approximation complexity. In *Proc. 44th IEEE Symposium on Foundations of Computer Science*, pages 298–307, Cambridge, Massachusetts, October 2003.

[37] A. M. Alhakim. A simple combinatorial algorithm for de Bruijn sequences. *Amer. Math. Monthly*, 117:728–732, 2010.

[38] P. Aliferis and J. Preskill. Fault-tolerant quantum computation against biased noise. *Physical Review A*, 78(5):052331, November 2008.

[39] P. Almeida, D. Napp, and R. Pinto. A new class of superregular matrices and MDP convolutional codes. *Linear Algebra Appl.*, 439:2145–2157, 2013.

[40] N. Alon and F. R. K. Chung. Explicit construction of linear sized tolerant networks. *Discrete Math.*, 72:15–19, 1988.

[41] N. Alon, O. Goldreich, J. Håstad, and R. Peralta. Simple constructions of almost k-wise independent random variables. *Random Structures Algorithms*, 3:289–304, 1992.

[42] N. Alon and M. G. Luby. A linear time erasure-resilient code with nearly optimal recovery. *IEEE Trans. Inform. Theory*, 42:1732–1736, 1996.

[43] J. L. Alperin and R. B. Bell. *Groups and Representations*, volume 162 of *Graduate Texts in Mathematics*. Springer-Verlag, New York, 1995.

[44] O. Alrabiah and V. Guruswami. An exponential lower bound on the sub-packetization of MSR codes. In *Proc. 51st Annual ACM Symposium on the Theory of Computing*, pages 979–985, Phoenix, Arizona, June 2019.

[45] M. M. S. Alves, L. Panek, and M. Firer. Error-block codes and poset metrics. *Adv. Math. Commun.*, pages 95–111, 2008.

[46] S. A. Aly, A. Klappenecker, and P. K. Sarvepalli. Remarkable degenerate quantum stabilizer codes derived from duadic codes. In *Proc. IEEE International Symposium on Information Theory*, pages 1105–1108, Seattle, Washington, July 2006.

[47] S. A. Aly, A. Klappenecker, and P. K. Sarvepalli. On quantum and classical BCH codes. *IEEE Trans. Inform. Theory*, 53:1183–1188, 2007.

[48] R. J. Anderson, C. Ding, T. Helleseth, and T. Kløve. How to build robust shared control systems. *Des. Codes Cryptogr.*, 15:111–124, 1998.

[49] M. Andersson, V. Rathi, R. Thobaben, J. Kliewer, and M. Skoglund. Nested polar codes for wiretap and relay channels. *IEEE Commun. Letters*, 14:752–754, 2010.

[50] M. F. Anjos and J. B. Lasserre, editors. *Handbook on Semidefinite, Conic and Polynomial Optimization.* Springer-Verlag, New York, 2012.

[51] J. Antrobus and H. Gluesing-Luerssen. Maximal Ferrers diagram codes: constructions and genericity considerations. *IEEE Trans. Inform. Theory*, 65:6204–6223, 2019.

[52] M. Antweiler and L. Bömer. Complex sequences over $GF(p^M)$ with a two-level autocorrelation function and a large linear span. *IEEE Trans. Inform. Theory*, 38:120–130, 1992.

[53] D. Apon, R. Perlner, A. Robinson, and P. Santini. Cryptanalysis of LEDAcrypt. Cryptology ePrint Archive, Report 2020/455, https://eprint.iacr.org/2020/455, 2020.

[54] B. Applebaum, Y. Ishai, and E. Kushilevitz. Cryptography with constant input locality. In A. Menezes, editor, *Advances in Cryptology – CRYPTO 2007*, volume 4622 of *Lecture Notes in Comput. Sci.*, pages 92–110. Springer, Berlin, Heidelberg, 2007.

[55] N. Aragon, O. Blazy, J.-C. Deneuville, P. Gaborit, A. Hauteville, O. Ruatta, J.-P. Tillich, G. Zémor, C. Aguilar-Melchor, S. Bettaieb, L. Bidoux, Bardet M., and A. Otmani. ROLLO (merger of Rank-Ouroboros, LAKE and LOCKER). Second round submission to the NIST post-quantum cryptography call: https://pqc-rollo.org, March 2019.

[56] N. Aragon, O. Blazy, P. Gaborit, A. Hauteville, and G. Zémor. Durandal: a rank metric based signature scheme. In Y. Ishai and V. Rijmen, editors, *Advances in Cryptology – EUROCRYPT 2019, Part III*, volume 11478 of *Lecture Notes in Comput. Sci.*, pages 728–758. Springer, Berlin, Heidelberg, 2019.

[57] N. Aragon, P. Gaborit, and G. Zémor. HQC-RMRS, an instantiation of the HQC encryption framework with a more efficient auxiliary error-correcting code. arXiv:2005.10741 [cs.CR], May 2020.

[58] N. Aragon, J. Lavauzelle, and M. Lequesne. Decoding challenge. http://decodingchallenge.org, 2019.

[59] K. T. Arasu. Sequences and arrays with desirable correlation properties. In D. Crnković and V. Tonchev, editors, *Information Security, Coding Theory and Related Combinatorics*, volume 29 of *NATO Sci. Peace Secur. Ser. D Inf. Commun. Secur.*, pages 136–171. IOS, Amsterdam, 2011.

[60] K. T. Arasu, C. Ding, T. Helleseth, P. V. Kumar, and H. M. Martinsen. Almost difference sets and their sequences with optimal autocorrelation. *IEEE Trans. Inform. Theory*, 47:2934–2943, 2001.

[61] E. Arbarello, M. Cornalba, P. A. Griffiths, and J. Harris. *Geometry of Algebraic Curves. Volume I*, volume 267 of *Grundlehren der Mathematischen Wissenschaften [Fundamental Principles of Mathematical Sciences]*. Springer-Verlag, New York, 1985.

[62] E. Arıkan. A performance comparison of polar codes and Reed-Muller codes. *IEEE Commun. Letters*, 12:447–449, 2008.

[63] E. Arıkan. Channel polarization: a method for constructing capacity-achieving codes for symmetric binary-input memoryless channels. *IEEE Trans. Inform. Theory*, 55:3051–3073, 2009.

[64] E. Arıkan. Source polarization. In *Proc. IEEE International Symposium on Information Theory*, pages 899–903, Austin, Texas, June 2010.

[65] E. Arıkan. Systematic polar coding. *IEEE Commun. Letters*, 15:860–862, 2011.

[66] E. Artin. *Geometric Algebra*. Wiley Classics Library. John Wiley & Sons, Inc., New York, 1988. Reprint of the 1957 original, a Wiley-Interscience Publication.

[67] A. Ashikhmin and A. Barg. Minimal vectors in linear codes. *IEEE Trans. Inform. Theory*, 44:2010–2017, 1998.

[68] A. Ashikhmin and A. Barg. Binomial moments of the distance distribution: bounds and applications. *IEEE Trans. Inform. Theory*, 45:438–452, 1999.

[69] A. Ashikhmin, A. Barg, G. D. Cohen, and L. Huguet. Variations on minimal codewords in linear codes. In G. D. Cohen, M. Giusti, and T. Mora, editors, *Applied Algebra, Algebraic Algorithms and Error-Correcting Codes*, volume 948 of *Lecture Notes in Comput. Sci.*, pages 96–105. Springer, Berlin, Heidelberg, 1995.

[70] A. Ashikhmin and S. Litsyn. Upper bounds on the size of quantum codes. *IEEE Trans. Inform. Theory*, 45:1206–1215, 1999.

[71] E. F. Assmus, Jr. and J. D. Key. *Designs and Their Codes*, volume 103 of *Cambridge Tracts in Mathematics*. Cambridge University Press, Cambridge, UK, 1992.

[72] E. F. Assmus, Jr. and H. F. Mattson, Jr. New 5-designs. *J. Combin. Theory*, 6:122–151, 1969.

[73] E. F. Assmus, Jr. and H. F. Mattson, Jr. Coding and combinatorics. *SIAM Rev.*, 16:349–388, 1974.

[74] E. F. Assmus, Jr., H. F. Mattson, Jr., and R. Turyn. *Cyclic Codes*. AFCRL-66-348, Air Force Cambridge Research Labs, Bedford, Massachusetts, 1966.

[75] J. Astola. The Lee-scheme and bounds for Lee-code. *Cybernetics and Systems*, 13:331–343, 1982.

[76] D. Augot, M. Finiasz, P. Gaborit, S. Manuel, and N. Sendrier. SHA-3 proposal: FSB. Submission to the SHA3 NIST competition; see `https://www.rocq.inria.fr/secret/CBCrypto/fsbdoc.pdf`, 2008.

[77] D. Augot, M. Finiasz, and N. Sendrier. A family of fast syndrome based cryptographic hash functions. In E. Dawson and S. Vaudenay, editors, *Progress in Cryptology – Mycrypt 2005*, volume 3715 of *Lecture Notes in Comput. Sci.*, pages 64–83. Springer, Berlin, Heidelberg, 2005.

[78] J. Ax. Zeroes of polynomials over finite fields. *Amer. J. Math.*, 86:255–261, 1964.

[79] N. Aydin and D. K. Ray-Chaudhuri. Quasi-cyclic codes over $\mathbb{Z}_4$ and some new binary codes. *IEEE Trans. Inform. Theory*, 48:2065–2069, 2002.

[80] K. Azarian, H. El Gamal, and P. Schniter. On the achievable diversity-multiplexing tradeoff in half-duplex cooperative channels. *IEEE Trans. Inform. Theory*, 51:4152–4172, 2005.

[81] A. H. Baartmans, I. Bluskov, and V. D. Tonchev. The Preparata codes and a class of 4-designs. *J. Combin. Des.*, 2:167–170, 1994.

[82] C. Bachoc. On harmonic weight enumerators of binary codes. *Des. Codes Cryptogr.*, 18:11—28, 1999.

[83] C. Bachoc. Linear programming bounds for codes in Grassmannian spaces. *IEEE Trans. Inform. Theory*, 52:2111–2125, 2006.

[84] C. Bachoc. Semidefinite programming, harmonic analysis and coding theory. arXiv:0909.4767 [cs.IT], September 2010.

[85] C. Bachoc, V. Chandar, G. D. Cohen, P. Solé, and A. Tchamkerten. On bounded weight codes. *IEEE Trans. Inform. Theory*, 57:6780–6787, 2011.

[86] C. Bachoc and P. Gaborit. Designs and self-dual codes with long shadows. *J. Combin. Theory Ser. A*, 105:15–34, 2004.

[87] C. Bachoc, D. C. Gijswijt, A. Schrijver, and F. Vallentin. Invariant semidefinite programs. In M. F. Anjos and J. B. Lasserre, editors, *Handbook on Semidefinite, Conic and Polynomial Optimization*, pages 219–269. Springer-Verlag, New York, 2012.

[88] C. Bachoc, A. Passuello, and F. Vallentin. Bounds for projective codes from semidefinite programming. *Adv. Math. Comm.*, 7:127–145, 2013.

[89] C. Bachoc and F. Vallentin. New upper bounds for kissing numbers from semidefinite programming. *J. Amer. Math. Soc.*, 21:909–924, 2008.

[90] C. Bachoc and G. Zémor. Bounds for binary codes relative to pseudo-distances of k points. *Adv. Math. Commun.*, 4:547–565, 2010.

[91] B. Bagchi. On characterizing designs by their codes. In N. S. N. Sastry, editor, *Buildings, Finite Geometries and Groups*, volume 10 of *Springer Proc. Math.*, pages 1–14. Springer, New York, 2012.

[92] B. Bagchi and S. P. Inamdar. Projective geometric codes. *J. Combin. Theory Ser. A*, 99:128–142, 2002.

[93] A. Baker. *Transcendental Number Theory*. Cambridge Mathematical Library. Cambridge University Press, Cambridge, UK, 2nd edition, 1990.

[94] R. D. Baker, J. H. van Lint, and R. M. Wilson. On the Preparata and Goethals codes. *IEEE Trans. Inform. Theory*, 29:342–345, 1983.

[95] S. B. Balaji, G. R. Kini, and P. V. Kumar. A tight rate bound and matching construction for locally recoverable codes with sequential recovery from any number of multiple erasures. *IEEE Trans. Inform. Theory*, 66:1023–1052, 2020.

[96] S. B. Balaji, M. N. Krishnan, M. Vajha, V. Ramkumar, B. Sasidharan, and P. V. Kumar. Erasure coding for distributed storage: an overview. *SCIENCE CHINA Information Sciences*, 61(10):100301:1–100301:45, 2018.

[97] S. B. Balaji and P. V. Kumar. Bounds on the rate and minimum distance of codes with availability. In *Proc. IEEE International Symposium on Information Theory*, pages 3155–3159, Aachen, Germany, June 2017.

[98] S. B. Balaji and P. V. Kumar. A tight lower bound on the sub-packetization level of optimal-access MSR and MDS codes. In *Proc. IEEE International Symposium on Information Theory*, pages 2381–2385, Vail, Colorado, June 2018.

[99] A. Balatsoukas-Stimming, M. B. Parizi, and A. Burg. LLR-based successive cancellation list decoding of polar codes. In *Proc. IEEE International Conference on Acoustics, Speech and Signal Processing (ICASSP)*, pages 3903–3907, Florence, Italy, May 2014.

[100] M. Baldi, A. Barenghi, F. Chiaraluce, G. Pelosi, and P. Santini. LEDAkem. First round submission to the NIST post-quantum cryptography call: https://csrc.nist.gov/CSRC/media/Projects/Post-Quantum-Cryptography/documents/round-1/submissions/LEDAkem.zip, November 2017.

[101] M. Baldi, M. Bianchi, F. Chiaraluce, J. Rosenthal, and D. Schipani. Enhanced public key security for the McEliece cryptosystem. *J. Cryptology*, 29:1–27, 2016.

[102] M. Baldi, M. Bodrato, and F. Chiaraluce. A new analysis of the McEliece cryptosystem based on QC-LDPC codes. In R. Ostrovsky, R. De Prisco, and I. Visconti, editors, *Security and Cryptography for Networks*, volume 5229 of *Lecture Notes in Comput. Sci.*, pages 246–262. Springer, Berlin, Heidelberg, 2008.

[103] M. Baldi and F. Chiaraluce. Cryptanalysis of a new instance of McEliece cryptosystem based on QC-LDPC codes. In *Proc. IEEE International Symposium on Information Theory*, pages 2591–2595, Nice, France, June 2007.

[104] S. Ball. On sets of vectors of a finite vector space in which every subset of basis size is a basis. *J. Eur. Math. Soc. (JEMS)*, 14:733–748, 2012.

[105] S. Ball, A. Blokhuis, and F. Mazzocca. Maximal arcs in Desarguesian planes of odd order do not exist. *Combinatorica*, 17:31–41, 1997.

[106] S. Ball and J. De Beule. On sets of vectors of a finite vector space in which every subset of basis size is a basis II. *Des. Codes Cryptogr.*, 65:5–14, 2012.

[107] S. Ballentine, A. Barg, and S. G. Vlăduţ. Codes with hierarchical locality from covering maps of curves. *IEEE Trans. Inform. Theory*, 65:6056–6071, 2019.

[108] S. Ballet, J. Chaumine, J. Pieltant, M. Rambaud, H. Randriambololona, and R. Rolland. On the tensor rank of multiplication in finite extensions of finite fields and related issues in algebraic geometry. arXiv:1906.07456 [math.AG], June 2019, to appear *Russian Math. Surveys*.

[109] B. Ballinger, G. Blekherman, H. Cohn, N. Giansiracusa, E. Kelly, and A. Schürmann. Experimental study of energy-minimizing point configurations on spheres. *Experiment. Math.*, 18:257–283, 2009.

[110] G. Banegas, P. S. L. M. Barreto, B. O. Boidje, P.-L. Cayrel, G. N. Dione, K. Gaj, C. T. Gueye, R. Haeussler, J. B. Klamti, O. N'diaye, D. T. Nguyen, E. Persichetti, and J. E. Ricardini. DAGS: key encapsulation for dyadic GS codes. First round submission to the NIST post-quantum cryptography call: https://csrc.nist.gov/CSRC/media/Projects/Post-Quantum-Cryptography/documents/round-1/submissions/DAGS.zip, November 2017.

[111] Ei. Bannai and Et. Bannai. A survey on spherical designs and algebraic combinatorics on spheres. *European J. Combin.*, 30:1392–1425, 2009.

[112] Ei. Bannai, Et. Bannai, H. Tanaka, and Y. Zhu. Design theory from the viewpoint of algebraic combinatorics. *Graphs Combin.*, 33:1–41, 2017.

[113] Ei. Bannai and R. M. Damerell. Tight spherical designs, I. *J. Math. Soc. Japan*, 31:199–207, 1979.

[114] Ei. Bannai and R. M. Damerell. Tight spherical disigns, II. *J. London Math. Soc. (2)*, 21:13–30, 1980.

[115] Ei. Bannai, S. T. Dougherty, M. Harada, and M. Oura. Type II codes, even unimodular lattices, and invariant rings. *IEEE Trans. Inform. Theory*, 45:1194–1205, 1999.

[116] M. Barbier, C. Chabot, and G. Quintin. On quasi-cyclic codes as a generalization of cyclic codes. *Finite Fields Appl.*, 18:904–919, 2012.

[117] M. Bardet, É. Barelli, O. Blazy, R. Canto Torres, A. Couvreur, P. Gaborit, A. Otmani, N. Sendrier, and J.-P. Tillich. BIG QUAKE. First round submission to the NIST post-quantum cryptography call: https://bigquake.inria.fr, November 2017.

[118] M. Bardet, M. Bros, D. Cabarcas, P. Gaborit, R. Perlner, D. Smith-Tone, J.-P. Tillich, and J. Verbel. Improvements of algebraic attacks for solving the rank decoding and MinRank problems. arXiv:2002.08322 [cs.CR], February 2020.

[119] M. Bardet, J. Chaulet, V. Dragoi, A. Otmani, and J.-P. Tillich. Cryptanalysis of the McEliece public key cryptosystem based on polar codes. In T. Takagi, editor, *Post-Quantum Cryptography*, volume 9606 of *Lecture Notes in Comput. Sci.*, pages 118–143. Springer, Berlin, Heidelberg, 2016.

[120] É. Barelli. *Étude de la Sécurité de Certaines Clés Compactes pour le Schéma de McEliece Utilisant des Codes Géométriques*. PhD thesis, Université Paris-Saclay, 2018.

[121] É Barelli and A. Couvreur. An efficient structural attack on NIST submission DAGS. In T. Peyrin and S. D. Galbraith, editors, *Advances in Cryptology – ASIACRYPT 2018, Part 1*, volume 11272 of *Lecture Notes in Comput. Sci.*, pages 93–118. Springer, Berlin, Heidelberg, 2018.

[122] A. Barg, K. Haymaker, E. W. Howe, G. L. Matthews, and A. Várilly-Alvarado. Locally recoverable codes from algebraic curves and surfaces. In E. W. Howe, K. E. Lauter, and J. L. Walker, editors, *Algebraic Geometry for Coding Theory and Cryptography*, volume 9 of *Assoc. Women Math. Ser.*, pages 95–127. Springer, Cham, 2017.

[123] A. Barg and D. B. Jaffe. Numerical results on the asymptotic rate of binary codes. In A. Barg and S. Litsyn, editors, *Codes and Association Schemes: DIMACS Workshop Codes and Association Schemes, November 9-12, 1999, DIMACS Center*, volume 56 of *Discrete Mathematics and Theoretical Computer Science*, pages 25–32. Amer. Math. Soc., Providence, Rhode Island, 2001.

[124] A. Barg and A. Mazumdar. Codes in permutations and error correction for rank modulation. *IEEE Trans. Inform. Theory*, 56:3158–3165, 2010.

[125] A. Barg and D. Y. Nogin. Bounds on packings of spheres in the Grassmann manifold. *IEEE Trans. Inform. Theory*, 48:2450–2454, 2002.

[126] A. Barg and D. Y. Nogin. Spectral approach to linear programming bounds on codes (in Russian). *Problemy Peredači Informacii*, 42(2):12–25, 2006. (English translation in *Probl. Inform. Transm.*, 42(2):77–89, 2006).

[127] A. Barg and W. Park. On linear ordered codes. *Mosc. Math. J.*, 15:679–702, 2015.

[128] A. Barg, I. Tamo, and S. G. Vlăduţ. Locally recoverable codes on algebraic curves. *IEEE Trans. Inform. Theory*, 63:4928–4939, 2017.

[129] A. Barg and W.-H. Yu. New bounds for spherical two-distance sets. *Experiment. Math.*, 22:187–194, 2013.

[130] E. Barker. Recommendation for key management, 2019. Draft NIST special publication 800-57 Part 1. https://doi.org/10.6028/NIST.SP.800-57pt1r5-draft.

[131] A. Barra and H. Gluesing-Luerssen. MacWilliams extension theorems and the local-global property for codes over Frobenius rings. *J. Pure Appl. Algebra*, 219:703–728, 2015.

[132] P. S. L. M. Barreto, S. Gueron, T. Güneysu, R. Misoczki, E. Persichetti, N. Sendrier, and J.-P. Tillich. CAKE: Code-based Algorithm for Key Encapsulation. In M. O'Neill, editor, *Cryptography and Coding*, volume 10655 of *Lecture Notes in Comput. Sci.*, pages 207–226. Springer, Berlin, Heidelberg, 2017.

[133] D. Bartoli, A. Sboui, and L. Storme. Bounds on the number of rational points of algebraic hypersurfaces over finite fields, with applications to projective Reed-Muller codes. *Adv. Math. Commun.*, 10:355–365, 2016.

[134] A. Barvinok. *A Course in Convexity*. Amer. Math. Soc., Providence, Rhode Island, 2002.

[135] A. Bassa, P. Beelen, A. Garcia, and H. Stichtenoth. Towers of function fields over non-prime finite fields. *Mosc. Math. J.*, 15:1–29, 181, 2015.

[136] L. A. Bassalygo, V. A. Zinov'ev, V. K. Leont'ev, and N. I. Fel'dman. Nonexistence of perfect codes for certain composite alphabets (in Russian). *Problemy Peredači Informacii*, 11(3):3–13, 1975 (English translation in *Probl. Inform. Transm.*, 11(3):181–189, 1975).

[137] V. Batagelj. Norms and distances over finite groups. Department of Mathematics, University of Ljubljana, Yugoslavia, 1990.

[138] L. M. Batten and J. M. Dover. Some sets of type (m, n) in cubic order planes. *Des. Codes Cryptogr.*, 16:211–213, 1999.

[139] H. Bauer, B. Ganter, and E. Hergert. Algebraic techniques for nonlinear codes. *Combinatorica*, 3:21–33, 1983.

[140] L. D. Baumert and R. J. McEliece. Weights of irreducible cyclic codes. *Inform. and Control*, 20:158–175, 1972.

[141] L. D. Baumert, W. H. Mills, and R. L. Ward. Uniform cyclotomy. *J. Number Theory*, 14:67–82, 1982.

[142] Leonard D. Baumert. *Cyclic Difference Sets*, volume 182 of *Lecture Notes in Math.* Springer-Verlag, Berlin-New York, 1971.

[143] E. Bayer-Fluckiger, F. Oggier, and E. Viterbo. New algebraic constructions of rotated $\mathbb{Z}^n$-lattice constellations for the Rayleigh fading channel. *IEEE Trans. Inform. Theory*, 50:702–714, 2004.

[144] L. M. J. Bazzi, M. Mahdian, and D. Spielman. The minimum distance of turbo-like codes. *IEEE Trans. Inform. Theory*, 55:6–15, 2009.

[145] L. M. J. Bazzi and S. K. Mitter. Some randomized code constructions from group actions. *IEEE Trans. Inform. Theory*, 52:3210–3219, 2006.

[146] A. Becker, A. Joux, A. May, and A. Meurer. Decoding random binary linear codes in $2^{n/20}$: how $1 + 1 = 0$ improves information set decoding. In D. Pointcheval and T. Johansson, editors, *Advances in Cryptology – EUROCRYPT 2012*, volume 7237 of *Lecture Notes in Comput. Sci.*, pages 520–536. Springer, Berlin, Heidelberg, 2012.

[147] P. Beelen. The order bound for general algebraic geometric codes. *Finite Fields Appl.*, 13:665–680, 2007.

[148] P. Beelen and T. Høholdt. The decoding of algebraic geometry codes. In *Advances in Algebraic Geometry Codes*, volume 5 of *Ser. Coding Theory Cryptol.*, pages 49–98. World Sci. Publ., Hackensack, New Jersey, 2008.

[149] A. Beemer, S. Habib, C. A. Kelley, and J. Kliewer. A general approach to optimizing SC-LDPC codes. In *Proc. 55th Allerton Conference on Communication, Control, and Computing*, pages 672–679, Monticello, Illinois, October 2017.

[150] A. Beemer, K. Haymaker, and C. A. Kelley. Absorbing sets of codes from finite geometries. *Cryptogr. Commun.*, 11:1115–1131, 2019.

[151] A. Beemer and C. A. Kelley. Avoiding trapping sets in SC-LDPC codes under windowed decoding. In *Proc. IEEE International Symposium on Information Theory and its Applications*, pages 206–210, Monterey, California, October-November 2016.

[152] J.-C. Belfiore and A. M. Cipriano. Space-Time Coding for Non-Coherent Channels. In H. Boelcskei, D. Gesbert, C. Papadias, and A. J. van der Veen, editors, *Space-Time Wireless Systems: From Array Processing to MIMO Communications*, chapter 10. Cambridge University Press, Cambridge, UK, 2006.

[153] J.-C. Belfiore and F. Oggier. An error probability approach to MIMO wiretap channels. *IEEE Trans. Commun.*, 61:3396–3403, 2013.

[154] J.-C. Belfiore, F. Oggier, and E. Viterbo. Cyclic division algebras: a tool for space-time coding. *Found. Trends Commun. and Inform. Theory*, 4:1–95, 2007.

[155] J.-C. Belfiore and G. Rekaya. Quaternionic lattices for space-time coding. In *Proc. IEEE Information Theory Workshop*, pages 267–270, Paris, France, April 2003.

[156] J.-C. Belfiore, G. Rekaya, and E. Viterbo. The golden code: a 2×2 full-rate space-time code with non-vanishing determinants. *IEEE Trans. Inform. Theory*, 51:1432–1436, 2005.

[157] B. I. Belov, V. N. Logačev, and V. P. Sandimirov. Construction of a class of linear binary codes that attain the Varšamov-Griesmer bound (in Russian). *Problemy Peredači Informacii*, 10(3):36–44, 1974. (English translation in *Probl. Inform. Transm.*, 10(3):211–217, 1974).

[158] E. Ben-Sasson, T. Etzion, A. Gabizon, and N. Raviv. Subspace polynomials and cyclic subspace codes. *IEEE Trans. Inform. Theory*, 62:1157–1165, 2016.

[159] E. Ben-Sasson, S. Kopparty, and J. Radhakrishnan. Subspace polynomials and limits to list decoding of Reed-Solomon codes. *IEEE Trans. Inform. Theory*, 56:113–120, 2010.

[160] A. Ben-Tal and A. Nemirovski. *Lectures on Modern Convex Optimization*. Society for Industrial and Applied Mathematics (SIAM); Mathematical Programming Society (MPS), Philadelphia, Pennsylvania, 2001.

[161] T. P. Berger. The automorphism group of double-error-correcting BCH codes. *IEEE Trans. Inform. Theory*, 40:538–542, 1994.

[162] T. P. Berger. Isometries for rank distance and permutation group of Gabidulin codes. *IEEE Trans. Inform. Theory*, 49:3016–3019, 2002.

[163] T. P. Berger, P.-L. Cayrel, P. Gaborit, and A. Otmani. Reducing key length of the McEliece cryptosystem. In B. Preneel, editor, *Progress in Cryptology – AFRICACRYPT 2009*, volume 5580 of *Lecture Notes in Comput. Sci.*, pages 77–97. Springer, Berlin, Heidelberg, 2009.

[164] T. P. Berger and P. Loidreau. Designing an efficient and secure public-key cryptosystem based on reducible rank codes. In A. Canteaut and K. Viswanathan, editors, *Progress in Cryptology – INDOCRYPT 2004*, volume 3348 of *Lecture Notes in Comput. Sci.*, pages 218–229. Springer, Berlin, Heidelberg, 2005.

[165] T. P. Berger and P. Loidreau. How to mask the structure of codes for a cryptographic use. *Des. Codes Cryptogr.*, 35:63–79, 2005.

[166] G. Berhuy and F. Oggier. On the existence of perfect space–time codes. *IEEE Trans. Inform. Theory*, 55:2078–2082, 2009.

[167] E. R. Berlekamp. Negacyclic codes for the Lee metric. In R. C. Bose and T. A. Dowling, editors, *Combinatorial Mathematics and its Applications (Proc. Conf., Univ. North Carolina, Chapel Hill, NC, 1967)*, pages 298–316. Univ. North Carolina Press, Chapel Hill, NC, 1969.

[168] E. R. Berlekamp. Goppa codes. *IEEE Trans. Inform. Theory*, 19:590–592, 1973.

[169] E. R. Berlekamp, editor. *Key Papers in the Development of Coding Theory.* IEEE Press, New York, 1974.

[170] E. R. Berlekamp. *Algebraic Coding Theory.* World Scientific Publishing Co. Pte. Ltd., Hackensack, New Jersey, revised edition, 2015.

[171] E. R. Berlekamp, R. J. McEliece, and H. C. A. van Tilborg. On the inherent intractability of certain coding problems. *IEEE Trans. Inform. Theory*, 24:384–386, 1978.

[172] E. R. Berlekamp and L. Welch. Error correction of algebraic block codes. US Patent Number 4,633,470.

[173] S. D. Berman. Semisimple cyclic and abelian codes. II (in Russian). *Kibernetika (Kiev)*, 3:21–30, 1967. (English translation in *Cybernetics*, 3:17–23, 1967).

[174] S. D. Berman. On the theory of group codes. *Cybernetics*, 3:25–31, 1969.

[175] J. J. Bernal, Á. del Río, and J. J. Simón. An intrinsical description of group codes. *Des. Codes Cryptogr.*, 51:289–300, 2009.

[176] F. Bernhardt, P. Landrock, and O. Manz. The extended Golay codes considered as ideals. *J. Combin. Theory Ser. A*, 55:235–246, 1990.

[177] D. J. Bernstein, T. Chou, T. Lange, I. von Maurich, R. Mizoczki, R. Niederhagen, E. Persichetti, C. Peters, P. Schwabe, N. Sendrier, J. Szefer, and W. Wen. Classic McEliece: conservative code-based cryptography. Second round submission to the NIST post-quantum cryptography call. https://classic.mceliece.org, March 2019.

[178] D. J. Bernstein, T. Lange, and C. Peters. Attacking and defending the McEliece cryptosystem. In J. Buchmann and J. Ding, editors, *Post-Quantum Cryptography*, volume 5299 of *Lecture Notes in Comput. Sci.*, pages 31–46. Springer, Berlin, Heidelberg, 2008.

[179] D. J. Bernstein, T. Lange, and C. Peters. Smaller decoding exponents: ball-collision decoding. In P. Rogaway, editor, *Advances in Cryptology – CRYPTO 2011*, volume 6841 of *Lecture Notes in Comput. Sci.*, pages 743–760. Springer, Berlin, Heidelberg, 2011.

[180] D. J. Bernstein, T. Lange, and C. Peters. Wild McEliece. In A. Biryukov, G. Gong, and D. R. Stinson, editors, *Selected Areas in Cryptography – SAC 2010*, volume 6544 of *Lecture Notes in Comput. Sci.*, pages 143–158. Springer, Berlin, Heidelberg, 2011.

[181] D. J. Bernstein, T. Lange, C. Peters, and P. Schwabe. Really fast syndrome-based hashing. In A. Nitaj and D. Pointcheval, editors, *Progress in Cryptology – AFRICACRYPT 2011*, volume 6737 of *Lecture Notes in Comput. Sci.*, pages 134–152. Springer, Berlin, Heidelberg, 2011.

[182] D. J. Bernstein, T. Lange, C. Peters, and H.C.A. van Tilborg. Explicit bounds for generic decoding algorithms for code-based cryptography. In *Pre-Proc. International Workshop on Coding and Cryptography*, pages 168–180, Ullensvang, Norway, May 2009.

[183] C. Berrou, A. Glavieux, and P. Thitimajshima. Near Shannon limit error correcting coding and decoding. In *Proc. International Communications Conference*, pages 1064–1070, Geneva, Switzerland, May 1983.

[184] M. Bertilsson and I. Ingemarsson. A construction of practical secret sharing schemes using linear block codes. In J. Seberry and Y. Zheng, editors, *Advances in Cryptology – AUSCRYPT '92*, volume 718 of *Lecture Notes in Comput. Sci.*, pages 67–79. Springer, Berlin, Heidelberg, 1993.

[185] M. R. Best. Perfect codes hardly exist. *IEEE Trans. Inform. Theory*, 29:349–351, 1983.

[186] T. Beth and C. Ding. On almost perfect nonlinear permutations. In T. Helleseth, editor, *Advances in Cryptology – EUROCRYPT '93*, volume 765 of *Lecture Notes in Comput. Sci.*, pages 65–76. Springer, Berlin, Heidelberg, 1994.

[187] T. Beth, D. Jungnickel, and H. Lenz. *Design Theory*. Cambridge University Press, Cambridge, UK, 2nd edition, 1999.

[188] K. Betsumiya, T. A. Gulliver, and M. Harada. On binary extremal formally self-dual even codes. *Kyushu J. Math.*, 53:421—-430, 1999.

[189] K. Betsumiya, M. Harada, and A. Munemasa. A complete classification of doubly even self-dual codes of length 40. *Electronic J. Combin.*, 19:#P18 (12 pages), 2012.

[190] A. Betten, M. Braun, H. Fripertinger, A. Kerber, A. Kohnert, and A. Wassermann. *Error-Correcting Linear Codes*, volume 18 of *Algorithms and Computation in Mathematics*. Springer-Verlag, Berlin, 2006. Classification by isometry and applications, with 1 CD-ROM (Windows and Linux).

[191] A. Betten, H. Fripertinger, A. Kerber, A. Wassermann, and K-H. Zimmermann. *Codierungstheorie – Konstruktion und Anwendung linearer Codes*. Springer-Verlag, Heidelberg, 1998.

[192] A. Betten, R. Laue, and A. Wassermann. Simple 7-designs with small parameters. *J. Combin. Des.*, 7:79–94, 1999.

[193] A. Beutelspacher. Partial spreads in finite projective spaces and partial designs. *Math. Z.*, 145:211–229, 1975.

[194] A. Beutelspacher. Correction to: "Partial spreads in finite projective spaces and partial designs" [*Math. Z.*, 145:211–229, 1975]. *Math. Z.*, 147:303, 1976.

[195] A. Beutelspacher. On t-covers in finite projective spaces. *J. Geom.*, 12:10–16, 1979.

[196] J. Bierbrauer. The theory of cyclic codes and a generalization to additive codes. *Des. Codes Cryptogr.*, 25:189–206, 2002.

[197] J. Bierbrauer. Cyclic additive and quantum stabilizer codes. In C. Carlet and B. Sunar, editors, *Arithmetic of Finite Fields*, volume 4547 of *Lecture Notes in Comput. Sci.*, pages 276–283. Springer, Berlin, Heidelberg, 2007.

[198] J. Bierbrauer. Cyclic additive codes. *J. Algebra*, 372:661–672, 2012.

[199] J. Bierbrauer. *Introduction to Coding Theory*. Discrete Mathematics and its Applications. Chapman & Hall/CRC, Boca Raton, Florida, 2nd edition, 2017.

[200] J. Bierbrauer, D. Bartoli, S. Marcugini, and F. Pambianco. Geometric constructions of quantum codes. In A. A. Bruen and D. L. Wehlau, editors, *Error-Correcting Codes, Finite Geometries and Cryptography*, volume 523 of *Contemp. Math.*, pages 149–154. Amer. Math. Soc., Providence, Rhode Island, 2010.

[201] J. Bierbrauer and Y. Edel. Quantum twisted codes. *J. Combin. Des.*, 8:174–188, 2000.

[202] J. Bierbrauer, Y. Edel, G. Faina, S. Marcugini, and F. Pambianco. Short additive quaternary codes. *IEEE Trans. Inform. Theory*, 55:952–954, 2009.

[203] E. Biglieri, Y. Hong, and E. Viterbo. On fast-decodable space-time block codes. *IEEE Trans. Inform. Theory*, 55:524–530, 2009.

[204] E. Biglieri, J. Proakis, and S. Shamai. Fading channels: information-theoretic and communications aspects. *IEEE Trans. Inform. Theory*, 44:2619–2692, 1998.

[205] R. T. Bilous. Enumeration of the binary self-dual codes of length 34. *J. Combin. Math. Combin. Comput.*, 59:173–211, 2006.

[206] S. R. Blackburn. Frameproof codes. *SIAM J. Discrete Math.*, 16:499–510, 2003.

[207] T. Blackford. Cyclic codes over $\mathbb{Z}_4$ of oddly even length. *Discrete Appl. Math.*, 128:27–46, 2003.

[208] T. Blackford. Negacyclic codes over $\mathbb{Z}_4$ of even length. *IEEE Trans. Inform. Theory*, 49:1417–1424, 2003.

[209] R. E. Blahut. *Theory and Practice of Error Control Codes*. Addison-Wesley, Reading, Massachusetts, 1983.

[210] R. E. Blahut. The Gleason–Prange theorem. *IEEE Trans. Inform. Theory*, 37:1269–1273, 1991.

[211] I. F. Blake. Codes over certain rings. *Inform. and Control*, 20:396–404, 1972.

[212] I. F. Blake, editor. *Algebraic Coding Theory History and Development*. Dowden, Hutchinson, & Ross, Inc. Stroudsburg, Pennsylvania, 1973.

[213] I. F. Blake. Codes over integer residue rings. *Inform. and Control*, 29:295–300, 1975.

[214] G. R. Blakley. Safeguarding cryptographic keys. In *Proc. National Computer Conference*, pages 313–317, New York, June 1979. AFIPS Press.

[215] F. L. Blasco. *Fountain codes under maximum likelihood decoding*. PhD thesis, Technischen Universität Hamburg-Harburg, 2017.

[216] F. L. Blasco, G. Liva, and G. Bauch. LT code design for inactivation decoding. In *Proc. IEEE Information Theory Workshop*, pages 441–445, Hobart, Tasmania, November 2014.

[217] M. Blaum, P. G. Farrell, and H. C. A. van Tilborg. Array Codes. In V. S. Pless and W. C. Huffman, editors, *Handbook of Coding Theory, Vol. I, II*, chapter 22, pages 1855–1909. North-Holland, Amsterdam, 1998.

[218] M. Blaum, J. L. Hafner, and S. Hetzler. Partial-MDS codes and their application to RAID type of architectures. *IEEE Trans. Inform. Theory*, 59:4510–4519, 2013.

[219] O. Blazy, P. Gaborit, J. Schrek, and N. Sendrier. A code-based blind signature. In *Proc. IEEE International Symposium on Information Theory*, pages 2718–2722, Aachen, Germany, June 2017.

[220] D. Bleichenbacher, A. Kiayias, and M. Yung. Decoding interleaved Reed-Solomon codes over noisy channels. *Theoret. Comput. Sci.*, 379:348–360, 2007.

[221] G. Blekherman, P. A. Parrilo, and R. R. Thomas, editors. *Semidefinite Optimization and Convex Algebraic Geometry*. Society for Industrial and Applied Mathematics (SIAM); Mathematical Optimization Society (MPS), Philadelphia, Pennsylvania, 2013.

[222] È. L. Blokh and V. V. Zyablov. Coding of generalized concatenated codes (in Russian). *Problemy Peredači Informacii*, 10(3):45–50, 1974 (English translation in *Probl. Inform. Transm.*, 10(3):218–222, 1974).

[223] A. Blokhuis and M. Lavrauw. Scattered spaces with respect to a spread in PG(n, q). *Geom. Dedicata*, 81:231–243, 2000.

[224] A. Blokhuis and M. Lavrauw. On two-intersection sets with respect to hyperplanes in projective spaces. *J. Combin. Theory Ser. A*, 99:377–382, 2002.

[225] A. Blokhuis and G. E. Moorhouse. Some p-ranks related to orthogonal spaces. *J. Algebraic Combin.*, 4:295–316, 1995.

[226] C. Blundo, A. De Santis, G. Di Cresceno, A. Gaggia, and U. Vaccaro. Multi-secret sharing schemes. In D. R. Stinson, editor, *Advances in Cryptology – CRYPTO '93*, volume 773 of *Lecture Notes in Comput. Sci.*, pages 126–135. Springer, Berlin, Heidelberg, 1994.

[227] I. Bocharova, F. Hug, R. Johannesson, B. Kudryashov, and R. Satyukov. Some voltage graph-based LDPC tailbiting codes with large girth. In A. Kuleshov, V. M. Blinovsky, and A. Ephremides, editors, *Proc. IEEE International Symposium on Information Theory*, pages 732–736, St. Petersburg, Russia, July-August 2011.

[228] M. Bogaerts and P. Dukes. Semidefinite programming for permutation codes. *Discrete Math.*, 326:34–43, 2014.

[229] K. Bogart, D. Goldberg, and J. Gordon. An elementary proof of the MacWilliams theorem on equivalence of codes. *Inform. and Control*, 37:19–22, 1978.

[230] G. T. Bogdanova, A. E. Brouwer, S. N. Kapralov, and P. R. J. Östergård. Error-correcting codes over an alphabet of four elements. *Des. Codes Cryptogr.*, 23:333–342, 2001.

[231] H. Bombín. Gauge color codes: optimal transversal gates and gauge fixing in topological stabilizer codes. *New J. Phys.*, 17:083002, August 2015.

[232] A. Bondarenko, D. Radchenko, and M. Viazovska. Optimal asymptotic bounds for spherical designs. *Ann. of Math. (2)*, 178:443–452, 2013.

[233] D. Boneh and J. Shaw. Collusion-secure fingerprinting for digital data. *IEEE Trans. Inform. Theory*, 44:1897–1905, 1998.

[234] G. Bonik, S. Goreinov, and N. Zamarashkin. A variant of list plus CRC concatenated polar code. arXiv:1207.4661 [cs.IT], July 2012.

[235] A. Bonnecaze, P. Solé, C. Bachoc, and B. Mourrain. Type II codes over $\mathbb{Z}_4$. *IEEE Trans. Inform. Theory*, 43:969–976, 1997.

[236] A. Bonnecaze, P. Solé, and A. R. Calderbank. Quaternary quadratic residue codes and unimodular lattices. *IEEE Trans. Inform. Theory*, 41:366–377, 1995.

[237] A. Borel. *Linear Algebraic Groups*, volume 126 of *Graduate Texts in Mathematics*. Springer-Verlag, New York, 2nd edition, 1991.

[238] M. Borello. The automorphism group of a self-dual [72, 36, 16] binary code does not contain elements of order 6. *IEEE Trans. Inform. Theory*, 58:7240–7245, 2012.

[239] M. Borello. The automorphism group of a self-dual [72, 36, 16] code is not an elementary abelian group of order 8. *Finite Fields Appl.*, 25:1–7, 2014.

[240] M. Borello. On automorphism groups of binary linear codes. In G. Kyureghyan, G. L. Mullen, and A. Pott, editors, *Topics in Finite Fields*, volume 632 of *Contemp. Math.*, pages 29–41. Amer. Math. Soc., Providence, Rhode Island, 2015.

[241] M. Borello, F. Dalla Volta, and G. Nebe. The automorphism group of a self-dual [72, 36, 16] code does not contain S_3, A_4 or D_8. *Adv. Math. Commun.*, 7:503–510, 2013.

[242] M. Borello, J. de la Cruz, and W. Willems. A note on linear complementary pairs of group codes. arXiv:1907.07506 [cs.IT], July 2019.

[243] M. Borello, J. de la Cruz, and W. Willems. On checkable codes in group algebras. arXiv:1901.10979 [cs.IT], January 2019.

[244] M. Borello and W. Willems. Group codes over fields are asymptotically good. arXiv:1904.10885 [cs.IT], April 2019.

[245] S. V. Borodachov, D. P. Hardin, and E. B. Saff. *Discrete Energy on Rectifiable Sets*. Springer-Verlag, Berlin, Heidelberg, 2019.

[246] B. Bose and T. R. N. Rao. Theory of unidirectional error correcting/detecting codes. *IEEE Trans. Comput.*, 31:521–530, 1982.

[247] R. C. Bose. Mathematical theory of the symmetrical factorial design. *Sankhya*, 8:107–166, 1947.

[248] R. C. Bose and D. K. Ray-Chaudhuri. On a class of error correcting binary group codes. *Inform. and Control*, 3:68–79, 1960. (Also reprinted in [169] pages 75–78 and [212] pages 165–176).

[249] R. C. Bose and D. K. Ray-Chaudhuri. Further results on error correcting binary group codes. *Inform. and Control*, 3:279–290, 1960. (Also reprinted in [169] pages 78–81 and [212] pages 177–188).

[250] W. Bosma, J. Cannon, and C. Playoust. The Magma algebra system. I. The user language. Computational algebra and number theory (London, 1993). *J. Symbolic Comput.*, 24:235–265, 1997.

[251] M. Bossert. On decoding using codewords of the dual code. arXiv:2001.02956 [cs.IT], January 2020.

[252] M. Bossert and V. Sidorenko. Singleton-type bounds for blot-correcting codes. *IEEE Trans. Inform. Theory*, 42:1021–1023, 1996.

[253] L. Both and A. May. Optimizing BJMM with nearest neighbors: full decoding in $2^{2/21n}$ and McEliece security. In *Proc. 10th International Workshop on Coding and Cryptography*, Saint Petersburg, Russia, September 2017. Available at http://cits.rub.de/imperia/md/content/may/paper/bjmm+.pdf.

[254] L. Both and A. May. Decoding linear codes with high error rate and its impact for LPN security. In T. Lange and R. Steinwandt, editors, *Post-Quantum Cryptography*, volume 10786 of *Lecture Notes in Comput. Sci.*, pages 25–46. Springer, Berlin, Heidelberg, 2018.

[255] D. Boucher. Construction and number of self-dual skew codes over $\mathbb{F}_{p^2}$. *Adv. Math. Commun.*, 10:765–795, 2016.

[256] D. Boucher, W. Geiselmann, and F. Ulmer. Skew-cyclic codes. *Appl. Algebra Engrg. Comm. Comput.*, 18:379–389, 2007.

[257] D. Boucher, P. Solé, and F. Ulmer. Skew constacyclic codes over Galois rings. *Adv. Math. Commun.*, 2:273–292, 2008.

[258] D. Boucher and F. Ulmer. Codes as modules over skew polynomial rings. In M. G. Parker, editor, *Cryptography and Coding*, volume 5921 of *Lecture Notes in Comput. Sci.*, pages 38–55. Springer, Berlin, Heidelberg, 2009.

[259] D. Boucher and F. Ulmer. Coding with skew polynomial rings. *J. Symbolic Comput.*, 44:1644–1656, 2009.

[260] D. Boucher and F. Ulmer. Linear codes using skew polynomials with automorphisms and derivations. *Des. Codes Cryptogr.*, 70:405–431, 2014.

[261] D. Boucher and F. Ulmer. Self-dual skew codes and factorization of skew polynomials. *J. Symbolic Comput.*, 60:47–61, 2014.

[262] M. Boulagouaz and A. Leroy. (σ, δ)-codes. *Adv. Math. Commun.*, 7:463–474, 2013.

[263] S. Boumova, P. G. Boyvalenkov, and D. P. Danev. Necessary conditions for existence of some designs in polynomial metric spaces. *European J. Combin.*, 20:213–225, 1999.

[264] J. Boutros, E. Viterbo, C. Rastello, and J.-C. Belfiore. Good lattice constellations for both Rayleigh fading and Gaussian channels. *IEEE Trans. Inform. Theory*, 42:502–518, 1996.

[265] I. Bouyukliev. About the code equivalence. In T. Shaska, W. C. Huffman, D. Joyner, and V. Ustimenko, editors, *Advances in Coding Theory and Cryptography*, volume 3 of *Ser. Coding Theory Cryptol.*, pages 126–151. World Sci. Publ., Hackensack, New Jersey, 2007.

[266] I. Bouyukliev. What is Q-extension? *Serdica J. Comput.*, 1:115–130, 2007.

[267] I. Bouyukliev. Classification of Griesmer codes and dual transform. *Discrete Math.*, 309:4049–4068, 2009.

[268] I. Bouyukliev, S. Bouyuklieva, and S. Kurz. Computer classification of linear codes. arXiv:2002.07826 [cs.IT], February 2020.

[269] I. Bouyukliev, M. Dzhumalieva-Stoeva, and V. Monev. Classification of binary self-dual codes of length 40. *IEEE Trans. Inform. Theory*, 61:4253–4258, 2015.

[270] I. Bouyukliev, V. Fack, W. Willems, and J. Winne. Projective two-weight codes with small parameters and their corresponding graphs. *Des. Codes Cryptogr.*, 41:59–78, 2006.

[271] I. Bouyukliev and P. R. J. Östergård. Classification of self-orthogonal codes over $\mathbb{F}_3$ and $\mathbb{F}_4$. *SIAM J. Discrete Math.*, 19:363–370, 2005.

[272] S. Bouyuklieva. A method for constructing self-dual codes with an automorphism of order 2. *IEEE Trans. Inform. Theory*, 46:496–504, 2000.

[273] S. Bouyuklieva. On the automorphisms of order 2 with fixed points for the extremal self-dual codes of length $24m$. *Des. Codes Cryptogr.*, 25:5–13, 2002.

[274] S. Bouyuklieva. On the automorphism group of a doubly-even $(72, 36, 16)$ code. *IEEE Trans. Inform. Theory*, 50:544–547, 2004.

[275] S. Bouyuklieva and I. Bouyukliev. Classification of the extremal formally self-dual even codes of length 30. *Adv. Math. Commun.*, 4:433–439, 2010.

[276] S. Bouyuklieva and I. Bouyukliev. An algorithm for classification of binary self-dual codes. *IEEE Trans. Inform. Theory*, 58:3933–3940, 2012.

[277] S. Bouyuklieva, I. Bouyukliev, and M. Harada. Some extremal self-dual codes and unimodular lattices in dimension 40. *Finite Fields Appl.*, 21:67–83, 2013.

[278] S. Bouyuklieva and M. Harada. Extremal $[50, 25, 10]$ codes with automorphisms of order 3 and quasi-symmetric 2-$(49, 9, 6)$ designs. *Des. Codes Cryptogr.*, 28:163–169, 2003.

[279] S. Bouyuklieva, M. Harada, and A. Munemasa. Determination of weight enumerators of binary extremal self-dual $[42, 21, 8]$ codes. *Finite Fields Appl.*, 14:177–187, 2008.

[280] S. Bouyuklieva, E. A. O'Brien, and W. Willems. The automorphism group of a binary self-dual doubly even $[72, 36, 16]$ code is solvable. *IEEE Trans. Inform. Theory*, 52:4244–4248, 2006.

[281] S. Bouyuklieva, R. Russeva, and N. Yankov. On the structure of binary self-dual codes having an automorphism a square of an odd prime. *IEEE Trans. Inform. Theory*, 51:3678–3686, 2005.

[282] S. Bouyuklieva and Z. Varbanov. Some connections between self-dual codes, combinatorial designs and secret sharing schemes. *Adv. Math. Commun.*, 5:191–198, 2011.

[283] S. Bouyuklieva and W. Willems. Singly-even self-dual codes with minimal shadow. *IEEE Trans. Inform. Theory*, 58:3856–3860, 2012.

[284] S. Bouyuklieva, N. Yankov, and R. Russeva. Classification of the binary self-dual $[42, 21, 8]$ codes having an automorphism of order 3. *Finite Fields Appl.*, 13:605–615, 2007.

[285] S. Bouyuklieva, N. Yankov, and R. Russeva. On the classification of binary self-dual $[44, 22, 8]$ codes with an automorphism of order 3 or 7. *Internat. J. Inform. Coding Theory*, 2, 2011.

[286] P. G. Boyvalenkov and D. P. Danev. On maximal codes in polynomial metric spaces. In T. Mora and H. F. Mattson, Jr., editors, *Applied Algebra, Algebraic Algorithms and Error-Correcting Codes*, volume 1255 of *Lecture Notes in Comput. Sci.*, pages 29–38. Springer, Berlin, Heidelberg, 1997.

[287] P. G. Boyvalenkov and D. P. Danev. On linear programming bounds for codes in polynomial metric spaces (in Russian). *Problemy Peredači Informacii*, 34(2):16–31, 1998. (English translation in *Probl. Inform. Transm.*, 34(2):108–120, 1998).

[288] P. G. Boyvalenkov, D. P. Danev, and S. P. Bumova. Upper bounds on the minimum distance of spherical codes. *IEEE Trans. Inform. Theory*, 42:1576–1581, 1996.

[289] P. G. Boyvalenkov, D. P. Danev, and M. Stoyanova. Refinements of Levenshtein bounds in q-ary Hamming spaces (in Russian). *Problemy Peredači Informacii*, 54(4):35–50, 2018. (English translation in *Probl. Inform. Transm.*, 54(4):329–342, 2018).

[290] P. G. Boyvalenkov, P. D. Dragnev, D. P. Hardin, E. B. Saff, and M. M. Stoyanova. Universal lower bounds for potential energy of spherical codes. *Constr. Approx.*, 44:385–415, 2016.

[291] P. G. Boyvalenkov, P. D. Dragnev, D. P. Hardin, E. B. Saff, and M. M. Stoyanova. Energy bounds for codes and designs in Hamming spaces. *Des. Codes Cryptogr.*, 82:411–433, 2017.

[292] T. Brack, M. Alles, F. Kienle, and N. Wehn. A synthesizable IP core for WIMAX 802.16E LDPC Code Decoding. In *Proc. IEEE 17th International Symposium on Personal, Indoor and Mobile Radio Communications*, Helsinki, Finland, 5 pages, September 2006.

[293] M. Braun. Construction of linear codes with large minimum distance. *IEEE Trans. Inform. Theory*, 50:1687–1691, 2004.

[294] M. Braun. Designs over the binary field from the complete monomial group. *Australas. J. Combin.*, 67:470–475, 2017.

[295] M. Braun, T. Etzion, P. R. J. Östergård, A. Vardy, and A. Wassermann. Existence of q-analogs of Steiner systems. *Forum Math. Pi*, 4:e7, 14, 2016.

[296] M. Braun, M. Kiermaier, and A. Wassermann. Computational methods in subspace designs. In *Network Coding and Subspace Designs*, Signals Commun. Technol., pages 213–244. Springer, Cham, 2018.

[297] M. Braun, M. Kiermaier, and A. Wassermann. q-analogs of designs: subspace designs. In *Network Coding and Subspace Designs*, Signals Commun. Technol., pages 171–211. Springer, Cham, 2018.

[298] M. Braun, A. Kohnert, and A. Wassermann. Construction of (n, r)-arcs in PG(2, q). *Innov. Incidence Geom.*, 1:133–141, 2005.

[299] M. Braun, A. Kohnert, and A. Wassermann. Optimal linear codes from matrix groups. *IEEE Trans. Inform. Theory*, 51:4247–4251, 2005.

[300] M. Braun, P. R. J. Östergård, and A. Wassermann. New lower bounds for binary constant-dimension subspace codes. *Experiment. Math.*, 27:179–183, 2018.

[301] S. Bravyi, M. Suchara, and A. Vargo. Efficient algorithms for maximum likelihood decoding in the surface code. *Physical Review A*, 90(3):032326, September 2014.

[302] E. F. Brickell. Some ideal secret sharing schemes. In J.-J. Quisquater and J. Vandewalle, editors, *Proc. Advances in Cryptology – EUROCRYPT '89*, volume 434 of *Lecture Notes in Comput. Sci.*, pages 468–475. Springer, Berlin, Heidelberg, 1990.

[303] J. D. Bridwell and J. K. Wolf. Burst distance and multiple-burst correction. *Bell System Tech. J.*, 49:889–909, 1970.

[304] J. Bringer, C. Carlet, H. Cabanne, S. Guilley, and H. Maghrebi. Orthogonal direct sum masking: a smartcard friendly computation paradigm in a code, with builtin protection against side-channel and fault attacks. In D. Naccache and D. Sauveron, editors, *Information Security Theory and Practice. Securing the Internet of Things*, volume 8501 of *Lecture Notes in Comput. Sci.*, pages 40–56. Springer, Berlin, Heidelberg, 2014.

[305] A. E. Brouwer. Personal homepage. `http://www.win.tue.nl//~aeb/`. Accessed: 2018-08-31.

[306] A. E. Brouwer. Some unitals on 28 points and their embeddings in projective planes of order 9. In M. Aigner and D. Jungnickel, editors, *Geometries and Groups*, volume 893 of *Lecture Notes in Math.*, pages 183–188. Springer-Verlag, Berlin, Heidelberg, 1981.

[307] A. E. Brouwer. Some new two-weight codes and strongly regular graphs. *Discrete Appl. Math.*, 10:111–114, 1985.

[308] A. E. Brouwer. Strongly regular graphs. In C. J. Colbourn and J. H. Dinitz, editors, *The CRC Handbook of Combinatorial Designs*, chapter VII.11, pages 852–868. Chapman & Hall/CRC Press, 2nd edition, 2007.

[309] A. E. Brouwer, A. M. Cohen, and A. Neumaier. *Distance-Regular Graphs*, volume 18 of *Ergebnisse der Mathematik und ihrer Grenzgebiete (3) [Results in Mathematics and Related Areas (3)]*. Springer-Verlag, Berlin, 1989.

[310] A. E. Brouwer, A. V. Ivanov, and M. H. Klin. Some new strongly regular graphs. *Combinatorica*, 9:339–344, 1989.

[311] A. E. Brouwer and S. C. Polak. Uniqueness of codes using semidefinite programming. *Des. Codes Cryptogr.*, 87:1881–1895, 2019.

[312] A. E. Brouwer and L. M. G. M. Tolhuizen. A sharpening of the Johnson bound for binary linear codes and the nonexistence of linear codes with Preparata parameters. *Des. Codes Cryptogr.*, 3:95–98, 1993.

[313] A. E. Brouwer and M. van Eupen. The correspondence between projective codes and 2-weight codes. *Des. Codes Cryptogr.*, 11:261–266, 1997.

[314] A. E. Brouwer, R. M. Wilson, and Q. Xiang. Cyclotomy and strongly regular graphs. *J. Algebraic Combin.*, 10:25–28, 1999.

[315] R. A. Brualdi, J. S. Graves, and K. M. Lawrence. Codes with a poset metric. *Discrete Math.*, 147:57–72, 1995.

[316] A. A. Bruen and J. A. Thas. Hyperplane coverings and blocking sets. *Math. Z.*, 181:407–409, 1982.

[317] T. A. Brun, I. Devetak, and M.-H. Hsieh. Correcting quantum errors with entanglement. *Science*, 314(5798):436–439, 2006.

[318] F. Buekenhout. Existence of unitals in finite translation planes of order q^2 with a kernel of order q. *Geom. Dedicata*, 5:189–194, 1976.

[319] K. A. Bush. Orthogonal arrays of index unity. *Ann. Math. Stat.*, 23:279–290, 1952.

[320] S. Buzaglo and T. Etzion. Bounds on the size of permutation codes with the Kendall τ-metric. *IEEE Trans. Inform. Theory*, 61:3241–3250, 2015.

[321] J. W. Byers, M. G. Luby, M. Mitzenmacher, and A. Rege. A digital fountain approach to reliable distribution of bulk data. In *Proc. ACM SIGCOMM Conference on Applications, Technologies, Architectures, and Protocols for Computer Communication*, pages 56–67, Vancouver, Canada, August-September 1998. ACM Press, New York.

[322] E. Byrne. Lifting decoding schemes over a Galois ring. In S. Boztaş and I. E. Shparlinski, editors, *Applied Algebra, Algebraic Algorithms and Error-Correcting Codes*, volume 2227 of *Lecture Notes in Comput. Sci.*, pages 323–332. Springer, Berlin, Heidelberg, 2001.

[323] E. Byrne. Decoding a class of Lee metric codes over a Galois ring. *IEEE Trans. Inform. Theory*, 48:966–975, 2002.

[324] E. Byrne and P. Fitzpatrick. Gröbner bases over Galois rings with an application to decoding alternant codes. *J. Symbolic Comput.*, 31:565–584, 2001.

[325] E. Byrne and P. Fitzpatrick. Hamming metric decoding of alternant codes over Galois rings. *IEEE Trans. Inform. Theory*, 48:683–694, 2002.

[326] V. R. Cadambe, S. A. Jafar, H. Maleki, K. Ramchandran, and C. Suh. Asymptotic interference alignment for optimal repair of MDS codes in distributed storage. *IEEE Trans. Inform. Theory*, 59:2974–2987, 2013.

[327] V. R. Cadambe and A. Mazumdar. Bounds on the size of locally recoverable codes. *IEEE Trans. Inform. Theory*, 61:5787–5794, 2015.

[328] Y. Cai and C. Ding. Binary sequences with optimal autocorrelation. *Theoret. Comput. Sci.*, 410:2316–2322, 2009.

[329] A. R. Calderbank, A. R. Hammons, Jr., P. V. Kumar, N. J. A. Sloane, and P. Solé. A linear construction for certain Kerdock and Preparata codes. *Bull. Amer. Math. Soc. (N.S.)*, 29(2):218–222, 1993.

[330] A. R. Calderbank and W. M. Kantor. The geometry of two-weight codes. *Bull. London Math. Soc.*, 18:97–122, 1986.

[331] A. R. Calderbank, E. M. Rains, P. W. Shor, and N. J. A. Sloane. Quantum error correction via codes over GF(4). *IEEE Trans. Inform. Theory*, 44:1369–1387, 1998.

[332] A. R. Calderbank and N. J. A. Sloane. Modular and p-adic cyclic codes. *Des. Codes Cryptogr.*, 6:21–35, 1995.

[333] S. Calkavur and P. Solé. Multisecret-sharing schemes and bounded distance decoding of linear codes. *Internat. J. Compt. Math.*, 94:107–114, 2017.

[334] E. T. Campbell, B. M. Terhal, and C. Vuillot. Roads towards fault-tolerant universal quantum computation. *Nature*, 549:172–179, 2017.

[335] A. Canteaut and H. Chabanne. A further improvement of the work factor in an attempt at breaking McEliece's cryptosystem. In P. Charpin, editor, *EUROCODE 94 – Livre des résumés*, pages 169–173. INRIA, 1994.

[336] A. Canteaut and F. Chabaud. A new algorithm for finding minimum-weight words in a linear code: application to McEliece's cryptosystem and to narrow-sense BCH codes of length 511. *IEEE Trans. Inform. Theory*, 44:367–378, 1998.

[337] A. Canteaut, P. Charpin, and G. M. Kyureghyan. A new class of monomial bent functions. *Finite Fields Appl.*, 14:221–241, 2008.

[338] A. Canteaut and N. Sendrier. Cryptanalysis of the original McEliece cryptosystem. In K. Ohta and D. Pei, editors, *Advances in Cryptology – ASIACRYPT '98*, volume 1514 of *Lecture Notes in Comput. Sci.*, pages 187–199. Springer, Berlin, Heidelberg, 1998.

[339] R. Canto-Torres and N. Sendrier. Analysis of information set decoding for a sub-linear error weight. In T. Takagi, editor, *Post-Quantum Cryptography*, volume 9606 of *Lecture Notes in Comput. Sci.*, pages 144–161. Springer, Berlin, Heidelberg, 2016.

[340] X. Cao and L. Hu. Two Boolean functions with five-valued Walsh spectra and high nonlinearity. *Internat. J. Found. Comput. Sci.*, 26:537–556, 2015.

[341] Y. Cao, Y. Cao, H. Q. Dinh, F.-W. Fu, J. Gao, and S. Sriboonchitta. Constacyclic codes of length np^s over $\mathbb{F}_{p^m} + u\mathbb{F}_{p^m}$. *Adv. Math. Commun.*, 12:231–262, 2018.

[342] C. Carlet. A simple description of Kerdock codes. In G. D. Cohen and J. Wolfmann, editors, *Coding Theory and Applications*, volume 388 of *Lecture Notes in Comput. Sci.*, pages 202–208. Springer, Berlin, Heidelberg, 1989.

[343] C. Carlet. Partially-bent functions. In E. F. Brickell, editor, *Advances in Cryptology – CRYPTO '92*, volume 740 of *Lecture Notes in Comput. Sci.*, pages 280–291. Springer, Berlin, Heidelberg, 1993.

[344] C. Carlet. Partially-bent functions. *Des. Codes Cryptogr.*, 3:135–145, 1993.

[345] C. Carlet. On Kerdock codes. In *Finite Fields: Theory, Applications, and Algorithms (Waterloo, ON, 1997)*, volume 225 of *Contemp. Math.*, pages 155–163. Amer. Math. Soc., Providence, Rhode Island, 1999.

[346] C. Carlet. Boolean Functions for Cryptography and Error-Correcting Codes. In Y. Crama and P. L. Hammer, editors, *Boolean Models and Methods in Mathematics, Computer Science, and Engineering*, volume 134 of *Encyclopedia of Mathematics and its Applications*, chapter 8, pages 257–397. Cambridge University Press, Cambridge, UK, 2010.

[347] C. Carlet. Vectorial Boolean Functions for Cryptography. In Y. Crama and P. L. Hammer, editors, *Boolean Models and Methods in Mathematics, Computer Science, and Engineering*, volume 134 of *Encyclopedia of Mathematics and its Applications*, chapter 9, pages 398–469. Cambridge University Press, Cambridge, UK, 2010.

[348] C. Carlet. Boolean and vectorial plateaued functions and APN functions. *IEEE Trans. Inform. Theory*, 61:6272–6289, 2015.

[349] C. Carlet. *Boolean Functions for Cryptography and Coding Theory.* Cambridge University Press, Cambridge, UK, 2020.

[350] C. Carlet, P. Charpin, and V. A. Zinoviev. Codes, bent functions and permutations suitable for DES-like cryptosystems. *Des. Codes Cryptogr.*, 15:125–156, 1998.

[351] C. Carlet and C. Ding. Highly nonlinear mappings. *J. Complexity*, 20:205–244, 2004.

[352] C. Carlet, C. Ding, and J. Yuan. Linear codes from perfect nonlinear mappings and their secret sharing schemes. *IEEE Trans. Inform. Theory*, 51:2089–2102, 2005.

[353] C. Carlet and S. Guilley. Complementary dual codes for counter-measures to side-channel attacks. In R. Pinto, P. R. Malonek, and P. Vettori, editors, *Coding Theory and Applications*, volume 3 of *CIM Ser. Math. Sci.*, pages 97–105. Springer, Cham, 2015.

[354] C. Carlet, C. Güneri, F. Özbudak, B. Özkaya, and P. Solé. On linear complementary pairs of codes. *IEEE Trans. Inform. Theory*, 64:6583–6589, 2018.

[355] C. Carlet and S. Mesnager. Four decades of research on bent functions. *Des. Codes Cryptogr.*, 78:5–50, 2016.

[356] X. Caruso and J. Le Borgne. A new faster algorithm for factoring skew polynomials over finite fields. *J. Symbolic Comput.*, 79:411–443, 2017.

[357] I. Cascudo. On squares of cyclic codes. *IEEE Trans. Inform. Theory*, 65:1034–1047, 2019.

[358] I. Cascudo, H. Chen, R. Cramer, and C. Xing. Asymptotically good ideal linear secret sharing with strong multiplication over *any* fixed finite field. In S. Halevi, editor, *Advances in Cryptology—CRYPTO 2009*, volume 5677 of *Lecture Notes in Comput. Sci.*, pages 466–486. Springer, Berlin, Heidelberg, 2009.

[359] I. Cascudo, R. Cramer, D. Mirandola, and G. Zémor. Squares of random linear codes. *IEEE Trans. Inform. Theory*, 61:1159–1173, 2015.

[360] I. Cascudo, R. Cramer, and C. Xing. The torsion-limit for algebraic function fields and its application to arithmetic secret sharing. In P. Rogaway, editor, *Advances in Cryptology—CRYPTO 2011*, volume 6841 of *Lecture Notes in Comput. Sci.*, pages 685–705. Springer, Heidelberg, Heidelberg, 2011.

[361] I. Cascudo, R. Cramer, and C. Xing. Torsion limits and Riemann-Roch systems for function fields and applications. *IEEE Trans. Inform. Theory*, 60:3871–3888, 2014.

[362] I. Cascudo, J. S. Gundersen, and D. Ruano. Squares of matrix-product codes. *Finite Fields Appl.*, 62:101606, 21, 2020.

[363] L. R. A. Casse and D. G. Glynn. The solution to Beniamino Segre's problem $I_{r,q}$, $r = 3$, $q = 2^h$. *Geom. Dedicata*, 13:157–163, 1982.

[364] L. R. A. Casse and D. G. Glynn. On the uniqueness of $(q + 1)_4$-arcs of PG(4, q), $q = 2^h$, $h \geq 3$. *Discrete Math.*, 48:173–186, 1984.

[365] Y. Cassuto and M. Blaum. Codes for symbol-pair read channels. In *Proc. IEEE International Symposium on Information Theory*, pages 988–992, Austin, Texas, June 2010.

[366] Y. Cassuto and M. Blaum. Codes for symbol-pair read channels. *IEEE Trans. Inform. Theory*, 57:8011–8020, 2011.

[367] Y. Cassuto and S. Litsyn. Symbol-pair codes: algebraic constructions and asymptotic bounds. In A. Kuleshov, V. M. Blinovsky, and A. Ephremides, editors, *Proc. IEEE International Symposium on Information Theory*, pages 2348–2352, St. Petersburg, Russia, July-August 2011.

[368] Y. Cassuto and M. A. Shokrollahi. Online fountain codes with low overhead. *IEEE Trans. Inform. Theory*, 61:3137–3149, June 2015.

[369] G. Castagnoli, J. L. Massey, P. A. Schoeller, and N. von Seeman. On repeated-root cyclic codes. *IEEE Trans. Inform. Theory*, 37:337–342, 1991.

[370] A. G. Castoldi and E. L. M. Carmelo. The covering problem in Rosenbloom-Tsfasman spaces. *Electron. J. Combin.*, 22(3):Paper 3.30, 18 pages, 2015.

[371] P.-L. Cayrel, P. Gaborit, and M. Girault. Identity-based identification and signature schemes using correcting codes. In D. Augot, N. Sendrier, and J.-P. Tillich, editors, *Proc. 5th International Workshop on Coding and Cryptography*, pages 69–78, Versailles, France, April 2007.

[372] P.-L. Cayrel, P. Véron, and S. M. E. Y. Alaoui. A zero-knowledge identification scheme based on the q-ary syndrome decoding problem. In A. Biryukov, G. Gong, and D. R. Stinson, editors, *Selected Areas in Cryptography – SAC 2010*, volume 6544 of *Lecture Notes in Comput. Sci.*, pages 171–186. Springer, Berlin, Heidelberg, 2011.

[373] Y. Cengellenmis, A. Dertli, and S. T. Dougherty. Codes over an infinite family of rings with a Gray map. *Des. Codes Cryptogr.*, 72:559–580, 2014.

[374] H. Chabanne, G. D. Cohen, and A. Patey. Towards secure two-party computation from the wire-tap channel. In H.-S. Lee and D.-G. Han, editors, *Information Security and Cryptology – ICISC 2013*, volume 8565 of *Lecture Notes in Comput. Sci.*, pages 34–46. Springer, Berlin, Heidelberg, 2014.

[375] H. Chabanne and B. Courteau. Application de la méthode de décodage itérative d'omura a la cryptanalyse du système de McEliece. Technical Report 122, University of Sherbrooke 1993, 1993.

[376] F. Chabaud. Bounds for self-complementary codes and their applications. In P. Camion, P. Charpin, and S. Harari, editors, *Proc. International Symposium on Information Theory – EUROCODE 1992*, volume 339 of *CISM Courses and Lect.*, pages 175–183. Springer, Vienna, 1993.

[377] F. Chabaud and S. Vaudenay. Links between differential and linear cryptanalysis. In A. De Santis, editor, *Advances in Cryptology – EUROCRYPT '94*, volume 950 of *Lecture Notes in Comput. Sci.*, pages 356–365. Springer, Berlin, Heidelberg, 1995.

[378] G. Chalom, R. A. Ferraz, and C. P. Milies. Essential idempotents and simplex codes. *J. Algebra Comb. Discrete Struct. Appl.*, 4:181–188, 2017.

[379] D. Champion, J.-C. Belfiore, G. Rekaya, and E. Viterbo. Partitionning the golden code: a framework to the design of space-time coded modulation. In *Proc. 9th Canadian Workshop on Information Theory*, Montreal, Canada, 2 pages, June 2005.

[380] S. Chang and J. Y. Hyun. Linear codes from simplicial complexes. *Des. Codes Cryptogr.*, 86:2167–2181, 2018.

[381] Z. Chang, J. Chrisnata, M. F. Ezerman, and H. M. Kiah. Rates of DNA sequence profiles for practical values of read lengths. *IEEE Trans. Inform. Theory*, 63:7166–7177, 2017.

[382] Z. Chang, M. F. Ezerman, S. Ling, and H. Wang. Construction of de Bruijn sequences from product of two irreducible polynomials. *Cryptogr. Commun.*, 10:251–275, 2018.

[383] Z. Chang, M. F. Ezerman, S. Ling, and H. Wang. The cycle structure of LFSR with arbitrary characteristic polynomial over finite fields. *Cryptogr. Commun.*, 10:1183–1202, 2018.

[384] Z. Chang, M. F. Ezerman, S. Ling, and H. Wang. On binary de Bruijn sequences from LFSRs with arbitrary characteristic polynomials. *Des. Codes Cryptogr.*, 87:1137–1160, 2019.

[385] P. Charpin. *Codes Idéaux de Certaines Algèbres Modulaires*, Thèse de 3ième cycle. Université de Paris VII, 1982.

[386] P. Charpin. Weight distributions of cosets of two-error-correcting binary BCH codes, extended or not. *IEEE Trans. Inform. Theory*, 40:1425–1442, 1994.

[387] P. Charpin, T. Helleseth, and V. A. Zinov'ev. The coset distribution of triple-error-correcting binary primitive BCH codes. *IEEE Trans. Inform. Theory*, 52:1727–1732, 2006.

[388] P. Charpin and V. A. Zinov'ev. On coset weight distributions of the 3-error-correcting BCH-codes. *SIAM J. Discrete Math.*, 10:128–145, 1997.

[389] L. Chaussade, P. Loidreau, and F. Ulmer. Skew codes of prescribed distance or rank. *Des. Codes Cryptogr.*, 50:267–284, 2009.

[390] B. Chen, H. Q. Dinh, H. Liu, and L. Wang. Constacyclic codes of length $2p^s$ over $\mathbb{F}_{p^m} + u\mathbb{F}_{p^m}$. *Finite Fields Appl.*, 37:108–130, 2016.

[391] B. Chen, S. Ling, and G. Zhang. Application of constacyclic codes to quantum MDS codes. *IEEE Trans. Inform. Theory*, 61:1474–1484, 2015.

[392] B. Chen, S.-T. Xia, J. Hao, and F.-W. Fu. Constructions of optimal cyclic (r, δ) locally repairable codes. *IEEE Trans. Inform. Theory*, 64:2499–2511, 2018.

[393] C. L. Chen, W. W. Peterson, and E. J. Weldon, Jr. Some results on quasi-cyclic codes. *Inform. and Control*, 15:407–423, 1969.

[394] H. Chen and R. Cramer. Algebraic geometric secret sharing schemes and secure multi-party computations over small fields. In C. Dwork, editor, *Advances in Cryptology—CRYPTO 2006*, volume 4117 of *Lecture Notes in Comput. Sci.*, pages 521–536. Springer, Berlin, Heidelberg, 2006.

[395] H. Chen, S. Ling, and C. Xing. Quantum codes from concatenated algebraic-geometric codes. *IEEE Trans. Inform. Theory*, 51:2915–2920, 2005.

[396] J. Chen, A. Dholakia, E. Eleftheriou, M. P. C. Fossorier, and X. Hu. Reduced-complexity decoding of LDPC codes. *IEEE Trans. Commun.*, 53:1288–1299, 2005.

[397] J. Chen, R. M. Tanner, J. Zhang, and M. P. C. Fossorier. Construction of irregular LDPC codes by quasi-cyclic extension. *IEEE Trans. Inform. Theory*, 53:1479–1483, 2007.

[398] Z. Chen. Six new binary quasi-cyclic codes. *IEEE Trans. Inform. Theory*, 40:1666–1667, 1994.

[399] Z. Chen, M. Ye, and A. Barg. Enabling optimal access and error correction for the repair of Reed-Solomon codes. arXiv:2001.07189 [cs.IT], January 2020.

[400] E. J. Cheon, Y. Kageyama, S. J. Kim, N. Lee, and T. Maruta. A construction of two-weight codes and its applications. *Bull. Korean Math. Soc.*, 54:731–736, 2017.

[401] W. E. Cherowitzo. α-flocks and hyperovals. *Geom. Dedicata*, 72:221–246, 1998.

[402] W. E. Cherowitzo, C. M. O'Keefe, and T. Penttila. A unified construction of finite geometries associated with q-clans in characteristic 2. *Adv. Geom.*, 3:1–21, 2003.

[403] W. E. Cherowitzo, T. Penttila, I. Pinneri, and G. F. Royle. Flocks and ovals. *Geom. Dedicata*, 60:17–37, 1996.

[404] J. C. Y. Chiang and J. K. Wolf. On channels and codes for the Lee metric. *Inform. and Control*, 19:159–173, 1971.

[405] S. Choi, J. Y. Hyun, H. K. Kim, and D. Y. Oh. MacWilliams-type equivalence relations. arXiv:1205.1090 [math.CO], May 2012.

[406] B. Chor, A. Fiat, and M. Naor. Tracing traitors. In Y. G. Desmedt, editor, *Advances in Cryptology – CRYPTO '94*, volume 839 of *Lecture Notes in Comput. Sci.*, pages 257–270. Springer, Berlin, Heidelberg, 1994.

[407] P. A. Chou, Y. Wu, and K. Jain. Practical network coding. In *Proc. 41st Allerton Conference on Communication, Control, and Computing*, pages 40–49, Monticello, Illinois, October 2003.

[408] D. V. Chudnovsky and G. V. Chudnovsky. Algebraic complexities and algebraic curves over finite fields. *J. Complexity*, 4:285–316, 1988.

[409] S.-Y. Chung, G. D. Forney, Jr., T. J. Richardson, and R. L. Urbanke. On the design of low-density parity-check codes within 0.0045 dB of the Shannon limit. *IEEE Commun. Letters*, 5:58–60, 2001.

[410] D. Clark, D. Jungnickel, and V. D. Tonchev. Affine geometry designs, polarities, and Hamada's conjecture. *J. Combin. Theory Ser. A*, 118:231–239, 2011.

[411] D. Clark, D. Jungnickel, and V. D. Tonchev. Exponential bounds on the number of designs with affine parameters. *J. Combin. Des.*, 19:131–140, 2011.

[412] D. Clark and V. D. Tonchev. A new class of majority-logic decodable codes derived from finite geometry. *Adv. Math. Commun.*, 7:175–186, 2013.

[413] G. C. Clark, Jr. and J. B. Cain. *Error-Correction Coding for Digital Communications*. Applications of Communications Theory. Plenum Press, New York-London, 1981.

[414] J.-J. Climent, D. Napp, C. Perea, and R. Pinto. A construction of MDS 2D convolutional codes of rate $1/n$ based on superregular matrices. *Linear Algebra Appl.*, 437:766–780, 2012.

[415] J.-J. Climent, D. Napp, C. Perea, and R. Pinto. Maximum distance separable 2D convolutional codes. *IEEE Trans. Inform. Theory*, 62:669–680, 2016.

[416] J.-J. Climent, D. Napp, R. Pinto, and R. Simões. Series concatenation of 2D convolutional codes. In *Proc. IEEE 9th International Workshop on Multidimensional (nD) Systems (nDS)*, Vila Real, Portugal, 6 pages, September 2015.

[417] J. T. Coffey and R. M. Goodman. The complexity of information set decoding. *IEEE Trans. Inform. Theory*, 36:1031–1037, 1990.

[418] J. T. Coffey, R. M. Goodman, and P. G. Farrell. New approaches to reduced-complexity decoding. *Discrete Appl. Math.*, 33:43–60, 1991.

[419] G. D. Cohen and S. Encheva. Efficient constructions of frameproof codes. *Electron. Lett.*, 36:1840–1842, 2000.

[420] G. D. Cohen, S. Mesnager, and A. Patey. On minimal and quasi-minimal linear codes. In M. Stam, editor, *Cryptography and Coding*, volume 8308 of *Lecture Notes in Comput. Sci.*, pages 85–98. Springer, Berlin, Heidelberg, 2013.

[421] H. Cohn, J. H. Conway, N. D. Elkies, and A. Kumar. The D_4 root system is not universally optimal. *Experiment. Math.*, 16:313–320, 2007.

[422] H. Cohn and M. de Courcy-Ireland. The Gaussian core model in high dimensions. *Duke Math. J.*, 167:2417–2455, 2018.

[423] H. Cohn and N. D. Elkies. New upper bounds on sphere packings I. *Ann. of Math.*, 157:689–714, 2003.

[424] H. Cohn and A. Kumar. Universally optimal distribution of points on spheres. *J. Amer. Math. Soc.*, 20:99–148, 2007.

[425] H. Cohn and J. Woo. Three-point bounds for energy minimization. *J. Amer. Math. Soc.*, 25:929–958, 2012.

[426] H. Cohn and Y. Zhao. Energy-minimizing error-correcting codes. *IEEE Trans. Inform. Theory*, 60:7442–7450, 2014.

[427] P. M. Cohn. *Free Rings and Their Relations*, volume 19 of *London Math. Soc. Monographs*. Academic Press, Inc. [Harcourt Brace Jovanovich, Publishers], London, 2nd edition, 1985.

[428] C. J. Colbourn and J. F. Dinitz, editors. *The CRC Handbook of Combinatorial Designs*. Chapman & Hall/CRC Press, 2nd edition, 2007.

[429] P. Compeau, P. Pevzner, and G. Tesler. How to apply de Bruijn graphs to genome assembly. *Nature Biotechnology*, 29:987–991, 2011.

[430] I. Constaninescu. *Lineare Codes über Restklassenringen ganzer Zahlen und ihre Automorphismen bezüglich einer verallgemeinerten Hamming-Metrik*. PhD thesis, Technische Universität, München, Germany, 1995.

[431] I. Constaninescu and W. Heise. A metric for codes over residue class rings of integers (in Russian). *Problemy Peredači Informacii*, 33(3):22–28, 1997. (English translation in *Probl. Inform. Transm.*, 33(3):208–213, 1997).

[432] I. Constaninescu, W. Heise, and T. Honold. Monomial extensions of isometries between codes over $\mathbb{Z}_m$. In *Proc. 5th International Workshop on Algebraic and Combinatorial Coding Theory*, pages 98–104, Sozopol, Bulgaria, June 1996.

[433] S. D. Constantin and T. R. N. Rao. On the theory of binary asymmetric error correcting codes. *Inform. and Control*, 40:20–36, 1979.

[434] J. H. Conway. Three lectures on exceptional groups. In M. B. Powell and G. Higman, editors, *Finite Simple Groups*, pages 215–247. Academic Press, New York, 1971.

[435] J. H. Conway, S. J. Lomonaco, Jr., and N. J. A. Sloane. A [45, 13] code with minimal distance 16. *Discrete Math.*, 83:213–217, 1990.

[436] J. H. Conway and V. S. Pless. On the enumeration of self-dual codes. *J. Combin. Theory Ser. A*, 28:26–53, 1980.

[437] J. H. Conway and V. S. Pless. On primes dividing the group order of a doubly-even (72, 36, 16) code and the group order of a quaternary (24, 12, 10) code. *Discrete Math.*, 38:143–156, 1982.

[438] J. H. Conway, V. S. Pless, and N. J. A. Sloane. Self-dual codes over GF(3) and GF(4) of length not exceeding 16. *IEEE Trans. Inform. Theory*, 25:312–322, 1979.

[439] J. H. Conway, V. S. Pless, and N. J. A. Sloane. The binary self-dual codes of length up to 32: a revised enumeration. *J. Combin. Theory Ser. A*, 60:183–195, 1992.

[440] J. H. Conway and N. J. A. Sloane. A new upper bound on the minimal distance of self-dual codes. *IEEE Trans. Inform. Theory*, 36:1319–1333, 1990.

[441] J. H. Conway and N. J. A. Sloane. Self-dual codes over the integers modulo 4. *J. Combin. Theory Ser. A*, 62:30–45, 1993.

[442] J. H. Conway and N. J. A. Sloane. *Sphere Packings, Lattices and Groups*, volume 290 of *Grundlehren der Mathematischen Wissenschaften [Fundamental Principles of Mathematical Sciences]*. Springer-Verlag, New York, 3rd edition, 1999. With additional contributions by Ei. Bannai, R. E. Borcherds, J. Leech, S. P. Norton, A. M. Odlyzko, R. A. Parker, L. Queen and B. B. Venkov.

[443] D. Coppersmith and M. Sudan. Reconstructing curves in three (and higher) dimensional spaces from noisy data. In *Proc. 35th Annual ACM Symposium on the Theory of Computing*, pages 136–142, San Diego, California, June 2003. ACM Press, New York.

[444] T. H. Cormen, C. E. Leiserson, R. L. Rivest, and C. Stein. *Introduction to Algorithms*. MIT Press, Cambridge, Massachusetts; McGraw-Hill Book Co., Boston, 2nd edition, 2001.

[445] A. Cossidente, N. Durante, G. Marino, T. Penttila, and A. Siciliano. The geometry of some two-character sets. *Des. Codes Cryptogr.*, 46:231–241, 2008.

[446] A. Cossidente and O. H. King. Some two-character sets. *Des. Codes Cryptogr.*, 56:105–113, 2010.

[447] A. Cossidente and G. Marino. Veronese embedding and two-character sets. *Des. Codes Cryptogr.*, 42:103–107, 2007.

[448] A. Cossidente, G. Marino, and F. Pavese. Non-linear maximum rank distance codes. *Des. Codes Cryptogr.*, 79:597–609, 2016.

[449] A. Cossidente and F. Pavese. On subspace codes. *Des. Codes Cryptogr.*, 78:527–531, 2016.

[450] A. Cossidente and F. Pavese. Veronese subspace codes. *Des. Codes Cryptogr.*, 81:445–457, 2016.

[451] A. Cossidente and F. Pavese. Subspace codes in PG(2N − 1, Q). *Combinatorica*, 37:1073–1095, 2017.

[452] A. Cossidente, F. Pavese, and L. Storme. Geometrical aspects of subspace codes. In M. Greferath, M. Pavčević, N. Silberstein, and M. Á. Vázquez-Castro, editors, *Network Coding and Subspace Designs. Signals and Communication Technology*, Signals Commun. Technol., pages 107–129. Springer, Cham, 2018.

[453] A. Cossidente and H. Van Maldeghem. The simple exceptional group $G_2(q)$, q even, and two-character sets. *J. Combin. Theory Ser. A*, 114:964–969, 2007.

[454] S. I. R. Costa, M. M. S. Alves, E. Agustini, and R. Palazzo. Graphs, tessellations, and perfect codes on flat tori. *IEEE Trans. Inform. Theory*, 50:2363–2377, 2004.

[455] S. I. R. Costa, F. Oggier, A. Campello, J.-C. Belfiore, and E. Viterbo. *Lattices Applied to Coding for Reliable and Secure Communications*. SpringerBriefs in Mathematics. Springer, Cham, 2017.

[456] N. T. Courtois, M. Finiasz, and N. Sendrier. How to achieve a McEliece-based digital signature scheme. In C. Boyd, editor, *Advances in Cryptology – ASIACRYPT 2001*, volume 2248 of *Lecture Notes in Comput. Sci.*, pages 157–174. Springer, Berlin, Heidelberg, 2001.

[457] A. Couvreur. The dual minimum distance of arbitrary-dimensional algebraic-geometric codes. *J. Algebra*, 350:84–107, 2012.

[458] A. Couvreur. Codes and the Cartier operator. *Proc. Amer. Math. Soc.*, 142(6):1983–1996, 2014.

[459] A. Couvreur, P. Gaborit, V. Gauthier-Umaña, A. Otmani, and J.-P. Tillich. Distinguisher-based attacks on public-key cryptosystems using Reed-Solomon codes. *Des. Codes Cryptogr.*, 73:641–666, 2014.

[460] A. Couvreur, P. Lebacque, and M. Perret. Toward good families of codes from towers of surfaces. arXiv:2002.02220 [math.AG], February 2020, to appear *AMS Contemp. Math.*

[461] A. Couvreur, M. Lequesne, and J.-P. Tillich. Recovering short secret keys of RLCE in polynomial time. In J. Ding and R. Steinwandt, editors, *Post-Quantum Cryptography*, volume 11505 of *Lecture Notes in Comput. Sci.*, pages 133–152. Springer, Berlin, Heidelberg, 2019.

[462] A. Couvreur, I. Márquez-Corbella, and R. Pellikaan. Cryptanalysis of McEliece cryptosystem based on algebraic geometry codes and their subcodes. *IEEE Trans. Inform. Theory*, 63:5404–5418, 2017.

[463] A. Couvreur, A. Otmani, and J.-P. Tillich. Polynomial time attack on wild McEliece over quadratic extensions. In P. Q. Nguyen and E. Oswald, editors, *Advances in Cryptology – EUROCRYPT 2014*, volume 8441 of *Lecture Notes in Comput. Sci.*, pages 17–39. Springer, Berlin, Heidelberg, 2014.

[464] A. Couvreur, A. Otmani, and J.-P. Tillich. Polynomial time attack on wild McEliece over quadratic extensions. *IEEE Trans. Inform. Theory*, 63:404–427, 2017.

[465] A. Couvreur, A. Otmani, J.-P. Tillich, and V. Gauthier-Umaña. A polynomial-time attack on the BBCRS scheme. In J. Katz, editor, *Public-Key Cryptography – PKC 2015*, volume 9020 of *Lecture Notes in Comput. Sci.*, pages 175–193. Springer, Berlin, Heidelberg, 2015.

[466] A. Couvreur and C. Ritzenthaler. Oral presentation, Day 2 of ANR Manta First Retreat (Lacapelle-Biron, France), April 12, 2016.

[467] T. M. Cover and J. A. Thomas. *Elements of Information Theory*. J. Wiley & Sons, New York, 1991.

[468] R. Cramer, I. B. Damgård, and U. Maurer. General secure multi-party computation from any linear secret-sharing scheme. In B. Preneel, editor, *Advances in Cryptology—EUROCRYPT 2000 (Bruges)*, volume 1807 of *Lecture Notes in Comput. Sci.*, pages 316–334. Springer, Berlin, Heidelberg, 2000.

[469] R. Cramer, I. B. Damgård, and J. Nielsen. *Secure Multiparty Computation and Secret Sharing*. Cambridge University Press, Cambridge, UK, 2015.

[470] L. Creedon and K. Hughes. Derivations on group algebras with coding theory applications. *Finite Fields Appl.*, 56:247–265, 2019.

[471] H. S. Cronie and S. B. Korada. Lossless source coding with polar codes. In *Proc. IEEE International Symposium on Information Theory*, pages 904–908, Austin, Texas, June 2010.

[472] E. Croot, V. F. Lev, and P. P. Pach. Progression-free sets in $\mathbb{Z}_4^n$ are exponentially small. *Ann. of Math. (2)*, 185:331–337, 2017.

[473] A. Cross, G. Smith, J. A. Smolin, and B. Zeng. Codeword stabilized quantum codes. *IEEE Trans. Inform. Theory*, 55:433–438, 2009.

[474] B. Csajbók, G. Marino, O. Polverino, and F. Zullo. Maximum scattered linear sets and MRD-codes. *J. Algebraic Combin.*, 46:517–531, 2017.

[475] B. Csajbók and A. Siciliano. Puncturing maximum rank distance codes. *J. Algebraic Combin.*, 49:507–534, 2019.

[476] D. Cullina and N. Kiyavash. An improvement to Levenshtein's upper bound on the cardinality of deletion correcting codes. *IEEE Trans. Inform. Theory*, 60:3862–3870, 2014.

[477] T. W. Cusick and H. S. Dobbertin. Some new three-valued crosscorrelation functions for binary m-sequences. *IEEE Trans. Inform. Theory*, 42:1238–1240, 1996.

[478] M. O. Damen, K. Abed-Meraim, and J.-C. Belfiore. Diagonal algebraic space-time block codes. *IEEE Trans. Inform. Theory*, 48:628–636, 2002.

[479] M. O. Damen, A. Chkeif, and J.-C. Belfiore. Lattice code decoder for space-time codes. *IEEE Commun. Letters*, 4:161–163, 2000.

[480] M. O. Damen, A. Tewfik, and J.-C. Belfiore. A construction of a space-time code based on number theory. *IEEE Trans. Inform. Theory*, 48:753–760, 2002.

[481] F. J. Damerau. A technique for computer detection and correction of spelling errors. *Commun. ACM*, 7:171–176, 1964.

[482] I. B. Damgård and P. Landrock. *Ideals and Codes in Group Algebras*. Aarhus Preprint Series, 1986.

[483] L. E. Danielsen and M. G. Parker. Edge local complementation and equivalence of binary linear codes. *Des. Codes Cryptogr.*, 49:161–170, 2008.

[484] L. E. Danielsen and M. G. Parker. Directed graph representation of half-rate additive codes over GF(4). *Des. Codes Cryptogr.*, 59:119–130, 2011.

[485] R. N. Daskalov and T. A. Gulliver. New good quasi-cyclic ternary and quaternary linear codes. *IEEE Trans. Inform. Theory*, 43:1647–1650, 1997.

[486] R. N. Daskalov, T. A. Gulliver, and E. Metodieva. New ternary linear codes. *IEEE Trans. Inform. Theory*, 45:1687–1688, 1999.

[487] B. K. Dass, N. Sharma, and R. Verma. Perfect codes in poset spaces and poset block spaces. *Finite Fields Appl.*, 46:90–106, 2017.

[488] A. Datta and F. Oggier. An overview of codes tailor-made for better repairability in networked distributed storage systems. *ACM SIGACT News*, 44:89–105, 2013.

[489] H. Dau, I. M. Duursma, H. M. Kiah, and O. Milenkovic. Repairing Reed-Solomon codes with multiple erasures. *IEEE Trans. Inform. Theory*, 64:6567–6582, 2018.

[490] M. C. Davey and D. J. C. MacKay. Low density parity check codes over GF(q). *IEEE Commun. Letters*, 2:165–167, 1998.

[491] G. Davidoff, P. Sarnak, and A. Valette. *Elementary Number Theory, Group Theory, and Ramanujan Graphs*, volume 55 of *London Math. Soc. Student Texts*. Cambridge University Press, Cambridge, UK, 2003.

[492] P. J. Davis. *Circulant Matrices*. John Wiley & Sons, New York-Chichester-Brisbane, 1979.

[493] P. Dayal and M. K. Varanasi. An optimal two transmit antenna space-time code and its stacked extensions. In *Proc. 37th Asilomar Conference on Signals, Systems and Computers*, volume 1, pages 987–991, Pacific Grove, California, November 2003. IEEE, Piscataway, New Jersey.

[494] P. Dayal and M. K. Varanasi. Distributed QAM-based space-time block codes for efficient cooperative multiple-access communication. *IEEE Trans. Inform. Theory*, 54:4342–4354, 2008.

[495] J. De Beule, K. Metsch, and L. Storme. Characterization results on weighted mini-hypers and on linear codes meeting the Griesmer bound. *Adv. Math. Commun.*, 2:261–272, 2008.

[496] N. G. de Bruijn. A combinatorial problem. *Nederl. Akad. Wetensch., Proc.*, 49:758–764, 1946 (See also *Indagationes Math.*, 8:461–467, 1946).

[497] F. De Clerck and M. Delanote. Two-weight codes, partial geometries and Steiner systems. *Des. Codes Cryptogr.*, 21:87–98, 2000.

[498] E. de Klerk. *Aspects of Semidefinite Programming*. Kluwer Acad. Publ., Dordrecht, Netherlands, 2002.

[499] E. de Klerk and F. Vallentin. On the Turing model complexity of interior point methods for semidefinite programming. *SIAM J. Optim.*, 26:1944–1961, 2016.

[500] J. de la Cruz, E. Gorla, H. Lopez, and A. Ravagnani. Weight distribution of rank-metric codes. *Des. Codes Cryptogr.*, 86:1–16, 2018.

[501] J. de la Cruz, M. Kiermaier, A. Wassermann, and W. Willems. Algebraic structures of MRD codes. *Adv. Math. Commun.*, 10:499–510, 2016.

[502] J. de la Cruz and W. Willems. On group codes with complementary duals. *Des. Codes Cryptogr.*, 86:2065–2073, 2018.

[503] D. de Laat. Moment methods in energy minimization: new bounds for Riesz minimal energy problems. arXiv:1610.04905 [math.OC], October 2016.

[504] D. de Laat, F. M. de Oliveira Filho, and F. Vallentin. Upper bounds for packings of spheres of several radii. *Forum Math. Sigma*, 2, 42 pages, 2014.

[505] D. de Laat, F. C. Machado, F. M. de Oliveira Filho, and F. Vallentin. k-point semidefinite programming bounds for equiangular lines. arXiv:1812.06045 [math.OC], December 2018.

[506] D. de Laat and F. Vallentin. A semidefinite programming hierarchy for packing problems in discrete geometry. *Math. Program. Ser. B*, 151:529–553, 2015.

[507] C. L. M. de Lange. Some new cyclotomic strongly regular graphs. *J. Algebraic Combin.*, 4:329–330, 1995.

[508] F. M. de Oliveira Filho and F. Vallentin. Computing upper bounds for packing densities of congruent copies of a convex body. In G. Ambrus, I. Bárány, K. J. Böröczky, G. F. Tóth, and J. Pach, editors, *New Trends in Intuitive Geometry*, volume 27 of *Bolyai Society Mathematical Studies*, pages 155–188. Springer-Verlag, Berlin, 2018.

[509] M. J. De Resmini. A 35-set of type $(2,5)$ in PG$(2,9)$. *J. Combin. Theory Ser. A*, 45:303–305, 1987.

[510] M. J. De Resmini and G. Migliori. A 78-set of type $(2,6)$ in PG$(2,16)$. *Ars Combin.*, 22:73–75, 1986.

[511] C. de Seguins Pazzis. The classification of large spaces of matrices with bounded rank. *Israel J. Math.*, 208:219–259, 2015.

[512] A. De Wispelaere and H. Van Maldeghem. Codes from generalized hexagons. *Des. Codes Cryptogr.*, 37:435–448, 2005.

[513] A. De Wispelaere and H. Van Maldeghem. Some new two-character sets in PG$(5, q^2)$ and a distance-2 ovoid in the generalized hexagon H(4). *Discrete Math.*, 308:2976–2983, 2008.

[514] T. Debris-Alazard, N. Sendrier, and J.-P. Tillich. WAVE: a new family of trapdoor one-way preimage sampleable functions based on codes. In S. D. Galbraith and S. Moriai, editors, *Advances in Cryptology – ASIACRYPT 2019, Part 1*, volume 11921 of *Lecture Notes in Comput. Sci.*, pages 21–51. Springer, Berlin, Heidelberg, 2019.

[515] T. Debris-Alazard and J.-P. Tillich. Two attacks on rank metric code-based schemes: RankSign and an IBE scheme. In T. Peyrin and S. D. Galbraith, editors, *Advances in Cryptology – ASIACRYPT 2018, Part 1*, volume 11272 of *Lecture Notes in Comput. Sci.*, pages 62–92. Springer, Berlin, Heidelberg, 2018.

[516] L. Decreusefond and G. Zémor. On the error-correcting capabilities of cycle codes of graphs. In *Proc. IEEE International Symposium on Information Theory*, page 307, Trondheim, Norway, June-July 1994.

[517] P. Delsarte. Bounds for unrestricted codes, by linear programming. *Philips Res. Rep.*, 27:272–289, 1972.

[518] P. Delsarte. Weights of linear codes and strongly regular normed spaces. *Discrete Math.*, 3:47–64, 1972.

[519] P. Delsarte. An algebraic approach to the association schemes of coding theory. *Philips Res. Rep. Suppl.*, 10:vi+97, 1973.

[520] P. Delsarte. *An Algebraic Approach to the Association Schemes of Coding Theory*. PhD thesis, Université Catholique de Louvain, 1973.

[521] P. Delsarte. Four fundamental parameters of a code and their combinatorial significance. *Inform. and Control*, 23:407–438, 1973.

[522] P. Delsarte. On subfield subcodes of modified Reed-Solomon codes. *IEEE Trans. Inform. Theory*, 21:575–576, 1975.

[523] P. Delsarte. Bilinear forms over a finite field, with applications to coding theory. *J. Combin. Theory Ser. A*, 25:226–241, 1978.

[524] P. Delsarte and J.-M. Goethals. Alternating bilinear forms over $GF(q)$. *J. Combin. Theory Ser. A*, 19:26–50, 1975.

[525] P. Delsarte and J.-M. Goethals. Unrestricted codes with the Golay parameters are unique. *Discrete Math.*, 12:211–224, 1975.

[526] P. Delsarte, J.-M. Goethals, and F. J. MacWilliams. On generalized Reed-Muller codes and their relatives. *Inform. and Control*, 16:403–442, 1970.

[527] P. Delsarte, J.-M. Goethals, and J. J. Seidel. Spherical codes and designs. *Geom. Dedicata*, 6:363–388, 1977.

[528] P. Delsarte and V. I. Levenshtein. Association schemes and coding theory. *IEEE Trans. Inform. Theory*, 44:2477–2504, 1998.

[529] J.-C. Deneuville and P. Gaborit. Cryptanalysis of a code-based one-time signature. *Des. Codes Cryptogr.*, 88:to appear, 2020.

[530] J.-C. Deneuville, P. Gaborit, and G. Zémor. Ouroboros: a simple, secure and efficient key exchange protocol based on coding theory. In T. Lange and T. Takagi, editors, *Post-Quantum Cryptography*, volume 10346 of *Lecture Notes in Comput. Sci.*, pages 18–34. Springer, Berlin, Heidelberg, 2017.

[531] R. H. F. Denniston. Some maximal arcs in finite projective planes. *J. Combin. Theory*, 6:317–319, 1969.

[532] M. M. Deza and E. Deza. *Encyclopedia of Distances*. Springer-Verlag, Berlin, 2009.

[533] A. Dholakia. *Introduction to Convolutional Codes with Applications*. International Series in Engineering and Computer Science. Springer Science & Business Media, New York, 1994.

[534] C. Di, D. Proietti, T. J. Richardson, and R. L. Urbanke. Finite length analysis of low-density parity-check codes. *IEEE Trans. Inform. Theory*, 48:1570–1579, 2002.

[535] C. Di, T. J. Richardson, and R. L. Urbanke. Weight distribution of low-density parity-check codes. *IEEE Trans. Inform. Theory*, 52:4839–4855, 2006.

[536] Q. Diao, Y. Y. Tai, S. Lin, and K. Abdel-Ghaffar. Trapping set structure of finite geometry LDPC codes. In *Proc. IEEE International Symposium on Information Theory*, pages 3088–3092, Cambridge, Massachusetts, July 2012.

[537] L. E. Dickson. The analytic representation of substitutions on a power of a prime number of letters with a discussion of the linear group. *Ann. of Math.*, 11:65–120, 161–183, 1896/97.

[538] J. F. Dillon. *Elementary Hadamard Difference Sets*. PhD thesis, University of Maryland, 1974.

[539] J. F. Dillon. Multiplicative difference sets via additive characters. *Des. Codes Cryptogr.*, 17:225–235, 1999.

[540] J. F. Dillon and H. S. Dobbertin. New cyclic difference sets with Singer parameters. *Finite Fields Appl.*, 10:342–389, 2004.

[541] A. G. Dimakis, P. B. Godfrey, Y. Wu, M. J. Wainwright, and K. Ramchandran. Network coding for distributed storage systems. *IEEE Trans. Inform. Theory*, 56:4539–4551, 2010.

[542] A. G. Dimakis, K. Ramchandran, Y. Wu, and C. Suh. A survey on network codes for distributed storage. *Proc. IEEE*, 99:476–489, 2011.

[543] B. K. Ding, T. Zhang, and G. N. Ge. Maximum distance separable codes for b-symbol read channels. *Finite Fields Appl.*, 49:180–197, 2018.

[544] C. Ding. Cyclic codes from APN and planar functions. arXiv:1206.4687 [cs.IT], June 2012.

[545] C. Ding. Cyclic codes from the two-prime sequences. *IEEE Trans. Inform. Theory*, 58:3881–3891, 2012.

[546] C. Ding. Cyclotomic constructions of cyclic codes with length being the product of two primes. *IEEE Trans. Inform. Theory*, 58:2231–2236, 2012.

[547] C. Ding. Cyclic codes from cyclotomic sequences of order four. *Finite Fields Appl.*, 23:8–34, 2013.

[548] C. Ding. Cyclic codes from some monomials and trinomials. *SIAM J. Discrete Math.*, 27:1977–1994, 2013.

[549] C. Ding. *Codes from Difference Sets*. World Scientific Publishing Co. Pte. Ltd., Hackensack, New Jersey, 2015.

[550] C. Ding. Linear codes from some 2-designs. *IEEE Trans. Inform. Theory*, 61:3265–3275, 2015.

[551] C. Ding. Parameters of several classes of BCH codes. *IEEE Trans. Inform. Theory*, 61:5322–5330, 2015.

[552] C. Ding. A construction of binary linear codes from Boolean functions. *Discrete Math.*, 339:2288–2303, 2016.

[553] C. Ding. Cyclic codes from Dickson polynomials. arXiv:1206.4370 [cs.IT], November 2016.

[554] C. Ding. A sequence construction of cyclic codes over finite fields. *Cryptogr. Commun.*, 10:319–341, 2018.

[555] C. Ding, C. Fan, and Z. Zhou. The dimension and minimum distance of two classes of primitive BCH codes. *Finite Fields Appl.*, 45:237–263, 2017.

[556] C. Ding, Z. Heng, and Z. Zhou. Minimal binary linear codes. *IEEE Trans. Inform. Theory*, 64:6536–6545, 2018.

[557] C. Ding, D. Kohel, and S. Ling. Secret sharing with a class of ternary codes. *Theoret. Comput. Sci.*, 246:285–298, 2000.

[558] C. Ding, T. Laihonen, and A. Renvall. Linear multisecret-sharing schemes and error-correcting codes. *J. Universal Comput. Sci.*, 3:1023–1036, 1997.

[559] C. Ding, K. Y. Lam, and C. Xing. Construction and enumeration of all binary duadic codes of length p^m. *Fund. Inform.*, 38:149–161, 1999.

[560] C. Ding, C. Li, N. Li, and Z. Zhou. Three-weight cyclic codes and their weight distributions. *Discrete Math.*, 339:415–427, 2016.

[561] C. Ding, C. Li, and Y. Xia. Another generalisation of the binary Reed-Muller codes and its applications. *Finite Fields Appl.*, 53:144–174, 2018.

[562] C. Ding and S. Ling. A q-polynomial approach to cyclic codes. *Finite Fields Appl.*, 20:1–14, 2013.

[563] C. Ding and H. Niederreiter. Cyclotomic linear codes of order 3. *IEEE Trans. Inform. Theory*, 53:2274–2277, 2007.

[564] C. Ding, D. Pei, and A. Salomaa. *Chinese Remainder Theorem: Applications in Computing, Coding, Cryptography*. World Scientific Publishing Co., Inc., River Edge, New Jersey, 1996.

[565] C. Ding and V. S. Pless. Cyclotomy and duadic codes of prime lengths. *IEEE Trans. Inform. Theory*, 45:453–466, 1999.

[566] C. Ding and A. Salomaa. Secret sharing schemes with nice access structures. *Fund. Inform.*, 73:51–62, 2006.

[567] C. Ding, G. Xiao, and W. Shan. *The Stability Theory of Stream Ciphers*, volume 561 of *Lecture Notes in Comput. Sci.* Springer, Berlin, Heidelberg, 1991.

[568] C. Ding and J. Yang. Hamming weights in irreducible cyclic codes. *Discrete Math.*, 313:434–446, 2013.

[569] C. Ding, Y. Yang, and X. Tang. Optimal sets of frequency hopping sequences from linear cyclic codes. *IEEE Trans. Inform. Theory*, 56:3605–3612, 2010.

[570] C. Ding and J. Yuan. Covering and secret sharing with linear codes. In C. S. Calude, M. J. Dinneen, and V. Vajnovszki, editors, *Discrete Mathematics and Theoretical Computer Science*, volume 2731 of *Lecture Notes in Comput. Sci.*, pages 11–25. Springer, Berlin, Heidelberg, 2003.

[571] C. Ding and Z. Zhou. Binary cyclic codes from explicit polynomials over GF(2^m). *Discrete Math.*, 321:76–89, 2014.

[572] K. Ding and C. Ding. A class of two-weight and three-weight codes and their applications in secret sharing. *IEEE Trans. Inform. Theory*, 61:5835–5842, 2015.

[573] H. Q. Dinh. Negacyclic codes of length 2^s over Galois rings. *IEEE Trans. Inform. Theory*, 51:4252–4262, 2005.

[574] H. Q. Dinh. Repeated-root constacyclic codes of length 2^s over $\mathbb{Z}_{2^a}$. In *Algebra and its Applications*, volume 419 of *Contemp. Math.*, pages 95–110. Amer. Math. Soc., Providence, Rhode Island, 2006.

[575] H. Q. Dinh. Structure of some classes of repeated-root constacyclic codes over integers modulo 2^m. In A. Giambruno, C. P. Milies, and S. K. Sehgal, editors, *Groups, Rings and Group Rings*, volume 248 of *Lecture Notes Pure Appl. Math.*, pages 105–117. Chapman & Hall/CRC, Boca Raton, Florida, 2006.

[576] H. Q. Dinh. Complete distances of all negacyclic codes of length 2^s over $\mathbb{Z}_{2^a}$. *IEEE Trans. Inform. Theory*, 53:147–161, 2007.

[577] H. Q. Dinh. On the linear ordering of some classes of negacyclic and cyclic codes and their distance distributions. *Finite Fields Appl.*, 14:22–40, 2008.

[578] H. Q. Dinh. Constacyclic codes of length 2^s over Galois extension rings of $\mathbb{F}_2 + u\mathbb{F}_2$. *IEEE Trans. Inform. Theory*, 55:1730–1740, 2009.

[579] H. Q. Dinh. Constacyclic codes of length p^s over $\mathbb{F}_{p^m} + u\mathbb{F}_{p^m}$. *J. Algebra*, 324:940–950, 2010.

[580] H. Q. Dinh. On some classes of repeated-root constacyclic codes of length a power of 2 over Galois rings. In *Advances in Ring Theory*, Trends Math., pages 131–147. Birkhäuser/Springer Basel AG, Basel, 2010.

[581] H. Q. Dinh. Repeated-root constacyclic codes of length $2p^s$. *Finite Fields Appl.*, 18:133–143, 2012.

[582] H. Q. Dinh, S. Dhompongsa, and S. Sriboonchitta. On constacyclic codes of length $4p^s$ over $\mathbb{F}_{p^m} + u\mathbb{F}_{p^m}$. *Discrete Math.*, 340:832–849, 2017.

[583] H. Q. Dinh, H. Liu, X.-S. Liu, and S. Sriboonchitta. On structure and distances of some classes of repeated-root constacyclic codes over Galois rings. *Finite Fields Appl.*, 43:86–105, 2017.

[584] H. Q. Dinh and S. R. López-Permouth. Cyclic and negacyclic codes over finite chain rings. *IEEE Trans. Inform. Theory*, 50:1728–1744, 2004.

[585] H. Q. Dinh, S. R. López-Permouth, and S. Szabo. On the structure of cyclic and negacyclic codes over finite chain rings. In P. Solé, editor, *Codes Over Rings*, volume 6 of *Series on Coding Theory and Cryptology*, pages 22–59. World Sci. Publ., Hackensack, New Jersey, 2009.

[586] H. Q. Dinh, B. T. Nguyen, A. K. Singh, and S. Sriboonchitta. Hamming and symbol-pair distances of repeated-root constacyclic codes of prime power lengths over $\mathbb{F}_{p^m} + u\mathbb{F}_{p^m}$. *IEEE Commun. Letters*, 22:2400–2403, 2018.

[587] H. Q. Dinh, B. T. Nguyen, A. K. Singh, and S. Sriboonchitta. On the symbol-pair distance of repeated-root constacyclic codes of prime power lengths. *IEEE Trans. Inform. Theory*, 64:2417–2430, 2018.

[588] H. Q. Dinh, B. T. Nguyen, and S. Sriboonchitta. Skew constacyclic codes over finite fields and finite chain rings. *Math. Probl. Eng.*, Art. ID 3965789, 17 pages, 2016.

[589] H. Q. Dinh, B. T. Nguyen, S. Sriboonchitta, and T. M. Vo. On a class of constacyclic codes of length $4p^s$ over $\mathbb{F}_{p^m} + u\mathbb{F}_{p^m}$. *J. Algebra Appl.*, 18(2):1950022, 2019.

[590] H. Q. Dinh, B. T. Nguyen, S. Sriboonchitta, and T. M. Vo. On $(\alpha + u\beta)$-constacyclic codes of length $4p^s$ over $\mathbb{F}_{p^m} + u\mathbb{F}_{p^m}$. *J. Algebra Appl.*, 18(2):1950023, 2019.

[591] H. Q. Dinh and H. D. T. Nguyen. On some classes of constacyclic codes over polynomial residue rings. *Adv. Math. Commun.*, 6:175–191, 2012.

[592] H. Q. Dinh, H. D. T. Nguyen, S. Sriboonchitta, and T. M. Vo. Repeated-root constacyclic codes of prime power lengths over finite chain rings. *Finite Fields Appl.*, 43:22–41, 2017.

[593] H. Q. Dinh, A. Sharma, S. Rani, and S. Sriboonchitta. Cyclic and negacyclic codes of length $4p^s$ over $\mathbb{F}_{p^m} + u\mathbb{F}_{p^m}$. *J. Algebra Appl.*, 17(9):1850173, 22, 2018.

[594] H. Q. Dinh, A. K. Singh, S. Pattanayak, and S. Sriboonchitta. On cyclic DNA codes over the ring $\mathbb{Z}_4 + u\mathbb{Z}_4$. Unpublished.

[595] H. Q. Dinh, A. K. Singh, S. Pattanayak, and S. Sriboonchitta. Cyclic DNA codes over the ring $\mathbb{F}_2 + u\mathbb{F}_2 + v\mathbb{F}_2 + uv\mathbb{F}_2 + v^2\mathbb{F}_2 + uv^2\mathbb{F}_2$. *Des. Codes Cryptogr.*, 86:1451–1467, 2018.

[596] H. Q. Dinh, A. K. Singh, S. Pattanayak, and S. Sriboonchitta. DNA cyclic codes over the ring $\mathbb{F}_2[u, v]/\langle u^2 - 1, v^3 - v, uv - vu \rangle$. *Internat. J. Biomath.*, 11(3):1850042, 19, 2018.

[597] H. Q. Dinh, A. K. Singh, S. Pattanayak, and S. Sriboonchitta. Construction of cyclic DNA codes over the ring $\mathbb{Z}_4[u]/\langle u^2 - 1 \rangle$ based on the deletion distance. *Theoret. Comput. Sci.*, 773:27–42, 2019.

[598] H. Q. Dinh, L. Wang, and S. Zhu. Negacyclic codes of length $2p^s$ over $\mathbb{F}_{p^m} + u\mathbb{F}_{p^m}$. *Finite Fields Appl.*, 31:178–201, 2015.

[599] H. Q. Dinh, X. Wang, H. Liu, and S. Sriboonchitta. On the b-distance of repeated-root constacyclic codes of prime power lengths. *Discrete Math.*, 343(4):111780, 2020.

[600] L. A. Dissett. *Combinatorial and Computational Aspects of Finite Geometries*. PhD thesis, University of Toronto, 2000.

[601] D. Divsalar, S. Dolinar, and C. Jones. Protograph LDPC codes over burst erasure channels. In *Proc. MILCOM '06 – IEEE Military Communications Conference*, Washington, DC, 7 pages, October 2006.

[602] D. Divsalar, S. Dolinar, C. Jones, and J. Thorpe. Protograph LDPC codes with minimum distance linearly growing with block size. In *Proc. GLOBECOM '05 – IEEE Global Telecommunications Conference*, volume 3, pages 1152–1156, St. Louis, Missouri, November-December 2005.

[603] D. Divsalar and L. Dolocek. On the typical minimum distance of protograph-based non-binary LDPC codes. In *Proc. Information Theory and Applications Workshop*, pages 192–198, San Diego, California, February 2012.

[604] D. Divsalar, H. Jin, and R. J. McEliece. Coding theorems for turbo-like codes. In *Proc. 36th Allerton Conference on Communication, Control, and Computing*, pages 201–210, Monticello, Illinois, September 1998.

[605] H. S. Dobbertin. One-to-one highly nonlinear power functions on GF(2^n). *Appl. Algebra Engrg. Comm. Comput.*, 9:139–152, 1998.

[606] H. S. Dobbertin. Almost perfect nonlinear power functions on GF(2^n): the Niho case. *Inform. and Comput.*, 151:57–72, 1999.

[607] H. S. Dobbertin. Almost perfect nonlinear power functions on GF(2^n): the Welch case. *IEEE Trans. Inform. Theory*, 45:1271–1275, 1999.

[608] H. S. Dobbertin. Kasami power functions, permutation polynomials and cyclic difference sets. In *Difference Sets, Sequences and Their Correlation Properties (Bad Windsheim, 1998)*, volume 542 of *NATO Adv. Sci. Inst. Ser. C Math. Phys. Sci.*, pages 133–158. Kluwer Acad. Publ., Dordrecht, Netherlands, 1999.

[609] H. S. Dobbertin. Almost perfect nonlinear power functions on GF(2^n): a new case for n divisible by 5. In D Jungnickel and H. Niederreiter, editors, *Proc. 5th International Conference on Finite Fields and Applications (Augsburg 1999)*, pages 113–121. Springer-Verlag, Berlin, Heidelberg, 2001.

[610] H. S. Dobbertin, P. Felke, T. Helleseth, and P. Rosendahl. Niho type cross-correlation functions via Dickson polynomials and Kloosterman sums. *IEEE Trans. Inform. Theory*, 52:613–627, 2006.

[611] S. Dodunekov and J. Simonis. Codes and projective multisets. *Electron. J. Combin.*, 5:Research Paper 37, 23 pages, 1998.

[612] L. Dolecek, Z. Zhang, V. Vnantharam, M. J. Wainwright, and B. Nikolic. Analysis of absorbing sets and fully absorbing sets of array-based LDPC codes. *IEEE Trans. Inform. Theory*, 56:181–201, 2010.

[613] R. G. L. D'Oliveira and M. Firer. The packing radius of a code and partitioning problems: the case for poset metrics on finite vector spaces. *Discrete Math.*, 338:2143–2167, 2015.

[614] R. G. L. D'Oliveira and M. Firer. Channel metrization. *European J. Combin.*, 80:107–119, 2019.

[615] R. G. L. D'Oliveira and M. Firer. A distance between channels: the average error of mismatched channels. *Des. Codes Cryptogr.*, 87:481–493, 2019.

[616] R. A. Dontcheva, A. J. van Zanten, and S. M. Dodunekov. Binary self-dual codes with automorphisms of composite order. *IEEE Trans. Inform. Theory*, 50:311–318, 2004.

[617] T. J. Dorsey. Morphic and principal-ideal group rings. *J. Algebra*, 318:393–411, 2007.

[618] M. Dostert, C. Guzman, F. M. de Oliveira Filho, and F. Vallentin. New upper bounds for the density of translative packings of three-dimensional convex bodies with tetrahedral symmetry. *Discrete Comput. Geom.*, 58:449–481, 2017.

[619] S. T. Dougherty. *Algebraic Coding Theory over Finite Commutative Rings*. Springer-Briefs in Mathematics. Springer International Publishing, New York, 2017.

[620] S. T. Dougherty and C. Fernández-Córdoba. Codes over $\mathbb{Z}_{2^k}$, Gray map and self-dual codes. *Adv. Math. Commun.*, 5:571–588, 2011.

[621] S. T. Dougherty, P. Gaborit, M. Harada, A. Munemasa, and P. Solé. Type IV self-dual codes over rings. *IEEE Trans. Inform. Theory*, 45:2345–2360, 1999.

[622] S. T. Dougherty, P. Gaborit, M. Harada, and P. Solé. Type II codes over $\mathbb{F}_2 + u\mathbb{F}_2$. *IEEE Trans. Inform. Theory*, 45:32–45, 1999.

[623] S. T. Dougherty, J. Gildea, R. Taylor, and A. Tylyshchak. Group rings, G-codes and constructions of self-dual and formally self-dual codes. *Des. Codes Cryptogr.*, 86:2115–2138, 2018.

[624] S. T. Dougherty, T. A. Gulliver, and M. Harada. Type II self-dual codes over finite rings and even unimodular lattices. *J. Algebraic Combin.*, 9:233–250, 1999.

[625] S. T. Dougherty, M. Harada, and P. Solé. Self-dual codes over rings and the Chinese remainder theorem. *Hokkaido Math. J.*, 28:253–283, 1999.

[626] S. T. Dougherty, S. Karadeniz, and B. Yildiz. Cyclic codes over R_k. *Des. Codes Cryptogr.*, 63:113–126, 2012.

[627] S. T. Dougherty, J.-L. Kim, and H. Kulosman. MDS codes over finite principal ideal rings. *Des. Codes Cryptogr.*, 50:77–92, 2009.

[628] S. T. Dougherty, J.-L. Kim, H. Kulosman, and H. Liu. Self-dual codes over commutative Frobenius rings. *Finite Fields Appl.*, 16:14–26, 2010.

[629] S. T. Dougherty, J.-L. Kim, B. Özkaya, L. Sok, and P. Solé. The combinatorics of LCD codes: linear programming bound and orthogonal matrices. *Internat. J. Inform. Coding Theory*, 4:116–128, 2017.

[630] S. T. Dougherty, J.-L. Kim, and P. Solé. Open problems in coding theory. In S. T. Dougherty, A. Facchini, A. Leroy, E. Puczyłowski, and P. Solé, editors, *Noncommutative Rings and Their Applications*, volume 634 of *Contemp. Math.*, pages 79–99. Amer. Math. Soc., Providence, Rhode Island, 2015.

[631] S. T. Dougherty and S. Ling. Cyclic codes over $\mathbb{Z}_4$ of even length. *Des. Codes Cryptogr.*, 39:127–153, 2006.

[632] S. T. Dougherty and H. Liu. Independence of vectors in codes over rings. *Des. Codes Cryptogr.*, 51:55–68, 2009.

[633] S. T. Dougherty, S. Mesnager, and P. Solé. Secret-sharing schemes based on self-dual codes. In F. R. Kschischang and E.-H. Yang, editors, *Proc. IEEE International Symposium on Information Theory*, pages 338–342, Toronto, Canada, July 2008.

[634] S. T. Dougherty and K. Shiromoto. Maximum distance codes over rings of order 4. *IEEE Trans. Inform. Theory*, 47:400–404, 2001.

[635] S. T. Dougherty and K. Shiromoto. Maximum distance codes in $\text{Mat}_{n,s}(\mathbb{Z}_k)$ with a non-Hamming metric and uniform distributions. *Des. Codes Cryptogr.*, 33:45–61, 2004.

[636] S. T. Dougherty and M. M. Skriganov. MacWilliams duality and the Rosenbloom-Tsfasman metric. *Mosc. Math. J.*, 2:81–97, 2002.

[637] S. T. Dougherty and M. M. Skriganov. Maximum distance separable codes in the ρ metric over arbitrary alphabets. *J. Algebraic Combin.*, 16:71–81, 2002.

[638] S. T. Dougherty, B. Yildiz, and S. Karadeniz. Codes over R_k, Gray maps and their binary images. *Finite Fields Appl.*, 17:205–219, 2011.

[639] S. T. Dougherty, B. Yildiz, and S. Karadeniz. Self-dual codes over R_k and binary self-dual codes. *Eur. J. Pure Appl. Math.*, 6:89–106, 2013.

[640] J. Doyen, X. Hubaut, and M. Vandensavel. Ranks of incidence matrices of steiner triple systems. *Math. Z.*, 163:251–259, 1978.

[641] P. B. Dragon, O. I. Hernandez, J. Sawada, A. Williams, and D. Wong. Constructing de Bruijn sequences with co-lexicographic order: the k-ary Grandmama sequence. *European J. Combin.*, 72:1–11, 2018.

[642] P. B. Dragon, O. I. Hernandez, and A. Williams. The Grandmama de Bruijn sequence for binary strings. In E. Kranakis, G. Navarro, and E. Chávez, editors, *LATIN 2016: Theoretical Informatics*, volume 9644 of *Lecture Notes in Comput. Sci.*, pages 347–361. Springer, Berlin, Heidelberg, 2016.

[643] N. Drucker, S. Gueron, and D. Kostic. Fast polynomial inversion for post quantum QC-MDPC cryptography. In S. Dolev, V. Kolesnikov, S. Lodha, and G. Weiss, editors, *Cyber Security Cryptography and Machine Learning*, volume 12161 of *Lecture Notes in Comput. Sci.*, pages 110–127. Springer, Berlin, Heidelberg, 2020.

[644] N. Drucker, S. Gueron, and D. Kostic. QC-MDPC decoders with several shades of gray. In J. Ding and J.-P. Tillich, editors, *Post-Quantum Cryptography*, volume 12100 of *Lecture Notes in Comput. Sci.*, pages 35–50. Springer, Berlin, Heidelberg, 2020.

[645] J. Ducoat. Generalized rank weights: a duality statement. In G. Kyureghyan, G. Mullen, and A. Pott, editors, *Topics in Finite Fields*, volume 632 of *Contemp. Math.*, pages 101–109. Amer. Math. Soc., Providence, Rhode Island, 2015.

[646] I. Dumer. Two decoding algorithms for linear codes (in Russian). *Problemy Peredači Informacii*, 25(1):24–32, 1989 (English translation in *Probl. Inform. Transm.*, 25(1):17–23, 1989).

[647] I. Dumer. On minimum distance decoding of linear codes. In *Proc. 5th Joint Soviet-Swedish Workshop on Information Theory*, pages 50–52, Moscow, USSR, 1991.

[648] I. Dumer. Concatenated Codes and Their Multilevel Generalizations. In V. S. Pless and W. C. Huffman, editors, *Handbook of Coding Theory, Vol. I, II*, chapter 23, pages 1911–1988. North-Holland, Amsterdam, 1998.

[649] I. Dumer. Soft-decision decoding of Reed-Muller codes: a simplified algorithm. *IEEE Trans. Inform. Theory*, 52:954–963, 2006.

[650] I. Dumer, D. Micciancio, and M. Sudan. Hardness of approximating the minimum distance of a linear code. *IEEE Trans. Inform. Theory*, 49:22–37, 2003.

[651] I. Dumer and K. Shabunov. Soft-decision decoding of Reed-Muller codes: recursive lists. *IEEE Trans. Inform. Theory*, 52:1260–1266, 2006.

[652] C. F. Dunkl. Discrete quadrature and bounds on t-designs. *Michigan Math. J.*, 26:81–102, 1979.

[653] N. Durante and A. Siciliano. Non-linear maximum rank distance codes in the cyclic model for the field reduction of finite geometries. *Electron. J. Combin.*, 24:Paper 2.33, 18, 2017.

[654] I. M. Duursma. Algebraic decoding using special divisors. *IEEE Trans. Inform. Theory*, 39:694–698, 1993.

[655] I. M. Duursma and H. Dau. Low bandwidth repair of the RS(10, 4) Reed-Solomon code. In *Proc. Information Theory and Applications Workshop*, San Diego, California, 10 pages, February 2017.

[656] I. M. Duursma, R. Kirov, and S. Park. Distance bounds for algebraic geometric codes. *J. Pure Appl. Algebra*, 215:1863–1878, 2011.

[657] I. M. Duursma, X. Li, and H. Wang. Multilinear algebra for distributed storage. arXiv:2006.08911 [cs.IT], June 2020.

[658] I. M. Duursma and S. Park. Coset bounds for algebraic geometric codes. *Finite Fields Appl.*, 16:36–55, 2010.

[659] Z. Dvir. On the size of Kakeya sets in finite fields. *J. Amer. Math. Soc.*, 22:1093–1097, 2009.

[660] Z. Dvir, S. Kopparty, S. Saraf, and M. Sudan. Extensions to the method of multiplicities, with applications to Kakeya sets and mergers. In *Proc. 50th IEEE Symposium on Foundations of Computer Science*, pages 181–190, Atlanta, Georgia, October 2009.

[661] W. Ebeling. *Lattices and Codes*. Advanced Lectures in Mathematics. Springer Spektrum, Wiesbaden, 3rd edition, 2013. A course partially based on lectures by Friedrich Hirzebruch.

[662] Y. Edel. Parameters of some GF(3)-linear quantum twisted codes. `https://www.mathi.uni-heidelberg.de/~yves/Matritzen/QTBCH/QTBCHTab3.html`. Accessed on 2019-01-17.

[663] Y. Edel and J. Bierbrauer. Twisted BCH-codes. *J. Combin. Des.*, 5:377–389, 1997.

[664] J. Egan and I. M. Wanless. Enumeration of MOLS of small order. *Math. Comp.*, 85:799–824, 2016.

[665] D. Ehrhard. Achieving the designed error capacity in decoding algebraic-geometric codes. *IEEE Trans. Inform. Theory*, 39:743–751, 1993.

[666] H. El Gamal and M. Damen. Universal space-time coding. *IEEE Trans. Inform. Theory*, 49:1097–1119, 2003.

[667] S. I. El-Zanati, O. Heden, G. F. Seelinger, P. A. Sissokho, L. E. Spence, and C. Vanden Eynden. Partitions of the 8-dimensional vector space over GF(2). *J. Combin. Des.*, 18:462–474, 2010.

[668] S. I. El-Zanati, H. Jordon, G. F. Seelinger, P. A. Sissokho, and L. E. Spence. The maximum size of a partial 3-spread in a finite vector space over GF(2). *Des. Codes Cryptogr.*, 54:101–107, 2010.

[669] S. I. El-Zanati, G. F. Seelinger, P. A. Sissokho, L. E. Spence, and C. Vanden Eynden. On partitions of finite vector spaces of low dimension over GF(2). *Discrete Math.*, 309:4727–4735, 2009.

[670] C. Eldert, H. M. Gurk, H. J. Gray, and M. Rubinoff. Shifting counters. *Trans. Amer. Inst. of Electrical Engineers, Part I: Communication and Electronics*, 77:70–74, 1958.

[671] P. Elia, K. R. Kumar, S. A. Pawar, P. V. Kumar, and H.-F. Lu. Explicit space-time codes achieving the diversity-multiplexing gain tradeoff. *IEEE Trans. Inform. Theory*, 52:3869–3884, 2006.

[672] P. Elia, F. Oggier, and P. V. Kumar. Asymptotically optimal cooperative wireless networks with reduced signaling complexity. *IEEE J. Selected Areas Commun.*, 25:258–267, 2007.

[673] P. Elia, B. A. Sethuraman, and P. V Kumar. Perfect space-time codes with minimum and non-minimum delay for any number of antennas. In *International Conference on Wireless Networks, Communications and Mobile Computing*, volume 1, pages 722–727. IEEE, Piscataway, New Jersey, 2005.

[674] P. Elias. Coding for noisy channels. *IRE Conv. Rec.*, 4:37–46, 1955.

[675] P. Elias, A. Feinstein, and C. E. Shannon. A note on the maximum flow through a network. *IRE Trans. Inform. Theory*, 2:117–119, 1956.

[676] N. D. Elkies. Explicit modular towers. In T. Basar and A. Vardy, editors, *Proc. 35th Allerton Conference on Communication, Control, and Computing*, pages 23–32, Monticello, Illinois, September 1997.

[677] N. D. Elkies. Explicit towers of Drinfeld modular curves. In C. Casacuberta, R. M. Miró-Roig, J. Verdera, and S. Xambó-Descamps, editors, *European Congress of Mathematics, Vol. II (Barcelona, 2000)*, volume 202 of *Progress in Mathematics*, pages 189–198. Birkhäuser, Basel, 2001.

[678] N. D. Elkies. Still better nonlinear codes from modular curves. arXiv:math/0308046 [math.NT], August 2003.

[679] J. S. Ellenberg and D. C. Gijswijt. On large subsets of $\mathbb{F}_q^n$ with no three-term arithmetic progression. *Ann. of Math. (2)*, 185:339–343, 2017.

[680] M. Elyasi and S. Mohajer. Determinant coding: a novel framework for exact-repair regenerating codes. *IEEE Trans. Inform. Theory*, 62:6683–6697, 2016.

[681] M. Elyasi and S. Mohajer. Cascade codes for distributed storage systems. arXiv:1901.00911 [cs.IT], January 2019.

[682] K. Engdahl and K. Sh. Zigangirov. To the theory of low-density convolutional codes I (in Russian). *Problemy Peredači Informacii*, 35(4):12–28, 1999. (English translation in *Probl. Inform. Transm.*, 35(4):295–310, 1999).

[683] P. Erdős and Z. Füredi. The greatest angle among n points in the d-dimensional Euclidean space. In *Combinatorial Mathematics (Marseille-Luminy, 1981)*, volume 75 of *North-Holland Math. Stud.*, pages 275–283. North-Holland, Amsterdam, 1983.

[684] T. Ericson and V. A. Zinov'ev. *Codes on Euclidean Spheres*. North-Holland Mathematical Library. Elsevier Science, Amsterdam, 2001.

[685] H. Esfahanizadeh, A. Hareedy, and L. Dolecek. Finite-length construction of high performance spatially-coupled codes via optimized partitioning and lifting. *IEEE Trans. Commun.*, 67:3–16, 2019.

[686] M. Esmaeili, T. A. Gulliver, N. P. Secord, and S. A. Mahmoud. A link between quasi-cyclic codes and convolutional codes. *IEEE Trans. Inform. Theory*, 44:431–435, 1998.

[687] J. Etesami, F. Hu, and W. Henkel. LCD codes and iterative decoding by projections, a first step towards an intuitive description of iterative decoding. In *Proc. GLOBECOM '11 – IEEE Global Telecommunications Conference*, Houston, Texas, December 2011, 4 pages.

[688] O. Etesami and M. A. Shokrollahi. Raptor codes on binary memoryless symmetric channels. *IEEE Trans. Inform. Theory*, 52:2033–2051, 2006.

[689] T. Etzion and M. Firer. Metrics based on finite directed graphs. In *Proc. IEEE International Symposium on Information Theory*, pages 1336–1340, Barcelona, Spain, July 2016.

[690] T. Etzion, M. Firer, and R. A. Machado. Metrics based on finite directed graphs and coding invariants. *IEEE Trans. Inform. Theory*, 64:2398–2409, 2018.

[691] T. Etzion, E. Gorla, A. Ravagnani, and A. Wachter-Zeh. Optimal Ferrers diagram rank-metric codes. *IEEE Trans. Inform. Theory*, 62:1616–1630, 2016.

[692] T. Etzion and N. Silberstein. Error-correcting codes in projective spaces via rank-metric codes and Ferrers diagrams. *IEEE Trans. Inform. Theory*, 55:2909–2919, 2009.

[693] T. Etzion, A. Trachtenberg, and A. Vardy. Which codes have cycle-free Tanner graphs? *IEEE Trans. Inform. Theory*, 45:2173–2181, 1999.

[694] T. Etzion and A. Vardy. Perfect binary codes: constructions, properties and enumeration. *IEEE Trans. Inform. Theory*, 40:754–763, 1994.

[695] T. Etzion and A. Vardy. Error-correcting codes in projective space. *IEEE Trans. Inform. Theory*, 57:1165–1173, 2011.

[696] M. F. Ezerman, S. Jitman, H. M. Kiah, and S. Ling. Pure asymmetric quantum MDS codes from CSS construction: a complete characterization. *Internat. J. Quantum Inf.*, 11(3):1350027, 2013.

[697] M. F. Ezerman, S. Jitman, S. Ling, and D. V. Pasechnik. CSS-like constructions of asymmetric quantum codes. *IEEE Trans. Inform. Theory*, 59:6732–6754, 2013.

[698] M. F. Ezerman, H. T. Lee, S. Ling, K. Nguyen, and H. Wang. A provably secure group signature scheme from code-based assumptions. In T. Iwata and J. H. Cheon, editors, *Advances in Cryptology – ASIACRYPT 2015, Part 1*, volume 9452 of *Lecture Notes in Comput. Sci.*, pages 260–285. Springer, Berlin, Heidelberg, 2015.

[699] M. F. Ezerman, S. Ling, B. Özkaya, and P. Solé. Good stabilizer codes from quasi-cyclic codes over $\mathbb{F}_4$ and $\mathbb{F}_9$. In *Proc. IEEE International Symposium on Information Theory*, pages 2898–2902, Paris, France, July 2019.

[700] M. F. Ezerman, S. Ling, B. Özkaya, and J. Tharnnukhroh. Spectral bounds for quasi-twisted codes. In *Proc. IEEE International Symposium on Information Theory*, pages 1922–1926, Paris, France, July 2019.

[701] A. Faldum. *Radikalcodes*. Diploma Thesis, Mainz, 1993.

[702] G. Falkner, B. Kowol, W. Heise, and E. Zehendner. On the existence of cyclic optimal codes. *Atti Sem. Mat. Fis. Univ. Modena*, 28(2):326–341 (1980), 1979.

[703] J. L. Fan. Array codes as low-density parity-check codes. In *Proc. 2ⁿᵈ International Symposium on Turbo Codes and Related Topics*, pages 543–546, Brest, France, September 2000.

[704] P. Fan and M. Darnell. *Sequence Design for Communications Applications*. Communications Systems, Techniques, and Applications Series. J. Wiley, New York, 1996.

[705] J.-C. Faugère, V. Gauthier-Umaña, A. Otmani, L. Perret, and J.-P. Tillich. A distinguisher for high-rate McEliece cryptosystems. *IEEE Trans. Inform. Theory*, 59:6830–6844, 2013.

[706] J.-C. Faugère, A. Otmani, L. Perret, and J.-P. Tillich. Algebraic cryptanalysis of McEliece variants with compact keys. In H. Gilbert, editor, *Advances in Cryptology – EUROCRYPT 2010*, volume 6110 of *Lecture Notes in Comput. Sci.*, pages 279–298. Springer, Berlin, Heidelberg, 2010.

[707] J.-C. Faugère, L. Perret, and F. de Portzamparc. Algebraic attack against variants of McEliece with Goppa polynomial of a special form. In P. Sarkar and T. Iwata, editors, *Advances in Cryptology – ASIACRYPT 2014, Part 1*, volume 8873 of *Lecture Notes in Comput. Sci.*, pages 21–41. Springer, Berlin, Heidelberg, 2014.

[708] C. Faure and L. Minder. Cryptanalysis of the McEliece cryptosystem over hyperelliptic codes. In *Proc. 11ᵗʰ International Workshop on Algebraic and Combinatorial Coding Theory*, pages 99–107, Pamporovo, Bulgaria, June 2008.

[709] G. Fazekas and V. I. Levenshtein. On upper bounds for code distance and covering radius of designs in polynomial metric spaces. *J. Combin. Theory Ser. A*, 70:267–288, 1995.

[710] J. Feldman. *Decoding Error-Correcting Codes via Linear Programming*. PhD thesis, Massachusetts Institute of Technology, 2003.

[711] J. Feldman, M. J. Wainwright, and D. R. Karger. Using linear programming to decode binary linear codes. *IEEE Trans. Inform. Theory*, 51:954–972, 2005.

[712] L. V. Felix and M. Firer. Canonical-systematic form for codes in hierarchical poset metrics. *Adv. Math. Commun.*, 6:315–328, 2012.

[713] G.-L. Feng and T. R. N. Rao. Decoding algebraic-geometric codes up to the designed minimum distance. *IEEE Trans. Inform. Theory*, 39:37–45, 1993.

[714] K. Feng, S. Ling, and C. Xing. Asymptotic bounds on quantum codes from algebraic geometry codes. *IEEE Trans. Inform. Theory*, 52:986–991, 2006.

[715] K. Feng and Z. Ma. A finite Gilbert-Varshamov bound for pure stabilizer quantum codes. *IEEE Trans. Inform. Theory*, 50:3323–3325, 2004.

[716] K. Feng, L. Xu, and F. J. Hickernell. Linear error-block codes. *Finite Fields Appl.*, 12:638–652, 2006.

[717] T. Feng and Q. Xiang. Strongly regular graphs from unions of cyclotomic classes. *J. Combin. Theory Ser. B*, 102:982–995, 2012.

[718] C. Fernández-Córdoba, J. Pujol, and M. Villanueva. On rank and kernel of $\mathbb{Z}_4$-linear codes. In Á. Barbero, editor, *Coding Theory and Applications*, volume 5228 of *Lecture Notes in Comput. Sci.*, pages 46–55. Springer, Berlin, Heidelberg, 2008.

[719] C. Fernández-Córdoba, J. Pujol, and M. Villanueva. $\mathbb{Z}_2\mathbb{Z}_4$-linear codes: rank and kernel. *Des. Codes Cryptogr.*, 56:43–59, 2010.

[720] S. Ferret and L. Storme. Minihypers and linear codes meeting the Griesmer bound: improvements to results of Hamada, Helleseth and Maekawa. *Des. Codes Cryptogr.*, 25:143–162, 2002.

[721] T. Feulner. The automorphism groups of linear codes and canonical representatives of their semilinear isometry classes. *Adv. Math. Commun.*, 3:363–383, 2009.

[722] T. Feulner. Classification and nonexistence results for linear codes with prescribed minimum distances. *Des. Codes Cryptogr.*, 70:127–138, 2014.

[723] T. Feulner and G. Nebe. The automorphism group of an extremal [72, 36, 16] code does not contain Z_7, $Z_3 \times Z_3$, or D_{10}. *IEEE Trans. Inform. Theory*, 58:6916–6924, 2012.

[724] R. P. Feynman. Simulating physics with computers. *Int. J. Theor. Phys.*, 21:467–488, 1982.

[725] C. M. Fiduccia and Y. Zalcstein. Algebras having linear multiplicative complexities. *J. ACM*, 24:311–331, 1977.

[726] F. Fiedler and M. Klin. A strongly regular graph with the parameters (512, 73, 438, 12, 10) and its dual graph. Preprint MATH-AL-7-1998, Technische Universität Dresden, 23 pages, July 1998.

[727] J. E. Fields, P. Gaborit, W. C. Huffman, and V. S. Pless. On the classification of extremal even formally self-dual codes. *Des. Codes Cryptogr.*, 18:125—148, 1999.

[728] J. E. Fields, P. Gaborit, W. C. Huffman, and V. S. Pless. On the classification of extremal even formally self-dual codes of lengths 20 and 22. *Discrete Appl. Math.*, 111:75—-86, 2001.

[729] M. Finiasz, P. Gaborit, and N. Sendrier. Improved fast syndrome based crypto-graphic hash functions. In V. Rijmen, editor, *ECRYPT Hash Workshop*, 2007.

[730] M. Finiasz and N. Sendrier. Security bounds for the design of code-based cryptosys-tems. In M. Matsui, editor, *Advances in Cryptology – ASIACRYPT 2009*, volume 5912 of *Lecture Notes in Comput. Sci.*, pages 88–105. Springer, Berlin, Heidelberg, 2009.

[731] M. Firer, M. M. S. Alves, J. A. Pinheiro, and L. Panek. *Poset Codes: Partial Orders, Metrics and Coding Theory*. SpringerBriefs in Mathematics. Springer, Cham, 2018.

[732] M. Firer and J. L. Walker. Matched metrics and channels. *IEEE Trans. Inform. Theory*, 62:1150–1156, 2016.

[733] J.-B. Fischer and J. Stern. An efficient pseudo-random generator provably as secure as syndrome decoding. In U. Maurer, editor, *Advances in Cryptology – EUROCRYPT '96*, volume 1070 of *Lecture Notes in Comput. Sci.*, pages 245–255. Springer, Berlin, Heidelberg, 1996.

[734] J. L. Fisher and S. K. Sehgal. Principal ideal group rings. *Comm. Algebra*, 4:319–325, 1976.

[735] N. L. Fogarty. *On Skew-Constacyclic Codes*. PhD thesis, University of Kentucky, 2016.

[736] N. L. Fogarty and H. Gluesing-Luerssen. A circulant approach to skew-constacyclic codes. *Finite Fields Appl.*, 35:92–114, 2015.

[737] M. A. Forbes and S. Yekhanin. On the locality of codeword symbols in non-linear codes. *Discrete Math.*, 324:78–84, 2014.

[738] L. R. Ford, Jr. and D. R. Fulkerson. Maximal flow through a network. *Canad. J. Math.*, 8:399–404, 1956.

[739] E. Fornasini and G. Marchesini. Structure and properties of two dimensional systems. In S. G. Tzafestas, editor, *Multidimensional Systems, Techniques and Applications*, pages 37–88. CRC Press, Boca Raton, Florida, 1986.

[740] G. D. Forney, Jr. Convolutional codes I: algebraic structure. *IEEE Trans. Inform. Theory*, 16:720–738, 1970.

[741] G. D. Forney, Jr. Correction to: "Convolutional codes I: algebraic structure" [*IEEE Trans. Inform. Theory* 16:720–738, 1970]. *IEEE Trans. Inform. Theory*, 17:360, 1971.

[742] G. D. Forney, Jr. Structural analysis of convolutional codes via dual codes. *IEEE Trans. Inform. Theory*, 19:512–518, 1973.

[743] G. D. Forney, Jr. Convolutional codes II: maximum-likelihood decoding. *Inform. and Control*, 25:222–266, 1974.

[744] G. D. Forney, Jr. Codes on graphs: normal realizations. *IEEE Trans. Inform. Theory*, 47:520–548, 2001.

[745] G. D. Forney, Jr., R. G. Gallager, G. R. Lang, F. M. Longstaff, and S. U. Qureshi. Efficient modulation for band-limited channels. *IEEE J. Selected Areas Commun.*, 2:632–647, 1984.

[746] G. D. Forney, Jr., M. Grassl, and S. Guha. Convolutional and tail-biting quantum error-correcting codes. *IEEE Trans. Inform. Theory*, 53:865–880, 2007.

[747] G. D. Forney, Jr. and L.-F. Wei. Multidimensional constellations I: Introduction, figures of merit, and generalized cross constellations. *IEEE J. Selected Areas Commun.*, 7:877–892, 1989.

[748] G. J. Foschini and M. J. Gans. On limits of wireless communication in a fading environment when using multiple antennas. *Wireless Personal Communications*, 6:311–335, 1998.

[749] D. A. Foulser and M. J. Kallaher. Solvable, flag-transitive, rank 3 collineation groups. *Geom. Dedicata*, 7:111–130, 1978.

[750] C. Fragouli and E. Soljanin. Network coding fundamentals. *Found. Trends Networking*, 2:1–133, 2007.

[751] J. B. Fraleigh. *A First Course in Abstract Algebra*. Addison-Wesley, New York, 7th edition, 2003.

[752] P. Frankl and N. Tokushige. Invitation to intersection problems for finite sets. *J. Combin. Theory Ser. A*, 144:157–211, 2016.

[753] P. Frankl and R. M. Wilson. The Erdős-Ko-Rado theorem for vector spaces. *J. Combin. Theory Ser. A*, 43:228–236, 1986.

[754] H. Fredricksen. Generation of the Ford sequence of length 2^n, n large. *J. Combinatorial Theory Ser. A*, 12:153–154, 1972.

[755] H. Fredricksen. A survey of full length nonlinear shift register cycle algorithms. *SIAM Rev.*, 24:195–221, 1982.

[756] H. Fredricksen and I. J. Kessler. An algorithm for generating necklaces of beads in two colors. *Discrete Math.*, 61:181–188, 1986.

[757] H. Fredricksen and J. Maiorana. Necklaces of beads in k colors and k-ary de Bruijn sequences. *Discrete Math.*, 23:207–210, 1978.

[758] W. Fulton. *Algebraic Curves*. Advanced Book Classics. Addison-Wesley Publishing Company, Advanced Book Program, Redwood City, California, 1989. An introduction to algebraic geometry. Notes written with the collaboration of Richard Weiss; reprint of 1969 original.

[759] È. M. Gabidulin. Combinatorial metrics in coding theory. In B. N. Petrov and F. Csádki, editors, *Proc. 2nd IEEE International Symposium on Information Theory, Tsahkadsor, Armenia, USSR, September 2-8, 1971*, pages 169–176. Akadémiai Kiadó, Budapest, 1973.

[760] È. M. Gabidulin. Theory of codes with maximum rank distance (in Russian). *Problemy Peredači Informacii*, 21(1):3–16, 1985 (English translation in *Probl. Inform. Transm.*, 21(1):1–12, 1985).

[761] È. M. Gabidulin. Rank q-cyclic and pseudo-q-cyclic codes. In *Proc. IEEE International Symposium on Information Theory*, pages 2799–2802, Seoul, Korea, June-July 2009.

[762] È. M. Gabidulin. A brief survey of metrics in coding theory. *Math. Distances and Appl.*, 66:66–84, 2012.

[763] È. M. Gabidulin and M. Bossert. Hard and soft decision decoding of phase rotation invariant block codes. In *Proc. International Zurich Seminar on Broadband Communications: Accessing, Transmission, Networking. (Cat. No.98TH8277)*, pages 249–251, February 1998.

[764] È. M. Gabidulin and M. Bossert. Codes for network coding. In F. R. Kschischang and E.-H. Yang, editors, *Proc. IEEE International Symposium on Information Theory*, pages 867–870, Toronto, Canada, July 2008.

[765] È. M. Gabidulin and V. A. Obernikhin. Codes in the Vandermonde F-metric and their application (in Russian). *Problemy Peredači Informacii*, 39(2):3–14, 2003. (English translation in *Probl. Inform. Transm.*, 39(2):159–169, 2003).

[766] È. M. Gabidulin, A. V. Paramonov, and O. V. Tretjakov. Ideals over a noncommutative ring and their application in cryptology. In D. W. Davies, editor, *Advances in Cryptology – EUROCRYPT '91*, volume 547 of *Lecture Notes in Comput. Sci.*, pages 482–489. Springer, Berlin, Heidelberg, 1991.

[767] È. M. Gabidulin and J. Simonis. Metrics generated by families of subspaces. *IEEE Trans. Inform. Theory*, 44:1336–1341, 1998.

[768] È. M. Gabidulin and J. Simonis. Perfect codes for metrics generated by primitive 2-error-correcting binary BCH codes. In *Proc. IEEE International Symposium on Information Theory*, page 68, Cambridge, Massachussets, August 1998.

[769] P. Gaborit. Shorter keys for code based cryptography. In *Proc. 4th International Workshop on Coding and Cryptography, Book of Extended Abstracts*, pages 81–91, Bergen, Norway, March 2005. `https://www.unilim.fr/pages\_perso/philippe.gaborit/shortIC.ps`.

[770] P. Gaborit and C. Aguilar-Melchor. Cryptographic method for communicating confidential information. `https://patents.google.com/patent/EP2537284B1/en`, 2010.

[771] P. Gaborit and M. Girault. Lightweight code-based identification and signature. In *Proc. IEEE International Symposium on Information Theory*, pages 191–195, Nice, France, June 2007.

[772] P. Gaborit, A. Hauteville, and J.-P. Tillich. RankSynd: a PRNG based on rank metric. In T. Takagi, editor, *Post-Quantum Cryptography*, volume 9606 of *Lecture Notes in Comput. Sci.*, pages 18–28. Springer, Berlin, Heidelberg, 2016.

[773] P. Gaborit, C. Lauradoux, and N. Sendrier. SYND: a fast code-based stream cipher with a security reduction. In *Proc. IEEE International Symposium on Information Theory*, pages 186–190, Nice, France, June 2007.

[774] P. Gaborit, G. Murat, O. Ruatta, and G. Zémor. Low rank parity check codes and their application to cryptography. In *Proc. 8th International Workshop on Coding and Cryptography*, Bergen, Norway, April 2013.

[775] P. Gaborit, O. Ruatta, and J. Schrek. On the complexity of the rank syndrome decoding problem. *IEEE Trans. Inform. Theory*, 62:1006–1019, 2016.

[776] P. Gaborit, O. Ruatta, J. Schrek, and G. Zémor. RankSign: an efficient signature algorithm based on the rank metric. In M. Mosca, editor, *Post-Quantum Cryptography*, volume 8772 of *Lecture Notes in Comput. Sci.*, pages 88–107. Springer, Berlin, Heidelberg, 2014; extended version at arXiv:1606.00629 [cs.CR].

[777] P. Gaborit and J. Schrek. Efficient code-based one-time signature from automorphism groups with syndrome compatibility. In *Proc. IEEE International Symposium on Information Theory*, pages 1982–1986, Cambridge, Massachussets, July 2012.

[778] P. Gaborit, J. Schrek, and G. Zémor. Full cryptanalysis of the Chen identification protocol. In B.-Y. Yang, editor, *Post-Quantum Cryptography*, volume 7071 of *Lecture Notes in Comput. Sci.*, pages 35–50. Springer, Berlin, Heidelberg, 2011.

[779] P. Gaborit and G. Zémor. Asymptotic improvement of the Gilbert-Varshamov bound for linear codes. *IEEE Trans. Inform. Theory*, 54:3865–3872, 2008.

[780] P. Gaborit and G. Zémor. On the hardness of the decoding and the minimum distance problems for rank codes. *IEEE Trans. Inform. Theory*, 62:7245–7252, 2016.

[781] D. Gabric, J. Sawada, A. Williams, and D. Wong. A framework for constructing de Bruijn sequences via simple successor rules. *Discrete Math.*, 341:2977–2987, 2018.

[782] M. Gadouleau and Z. Yan. MacWilliams identities for the rank metric. In *Proc. IEEE International Symposium on Information Theory*, pages 36–40, Nice, France, June 2007.

[783] M. Gadouleau and Z. Yan. Packing and covering properties of rank metric codes. *IEEE Trans. Inform. Theory*, 54:3873–3883, 2008.

[784] C. Galindo, O. Geil, F. Hernando, and D. Ruano. Improved constructions of nested code pairs. *IEEE Trans. Inform. Theory*, 64:2444–2459, 2018.

[785] C. Galindo, O. Geil, F. Hernando, and D. Ruano. New binary and ternary LCD codes. *IEEE Trans. Inform. Theory*, 65:1008–1016, 2019.

[786] C. Galindo, F. Hernando, and R. Matsumoto. Quasi-cyclic constructions of quantum codes. *Finite Fields Appl.*, 52:261–280, 2018.

[787] R. G. Gallager. *Low Density Parity Check Codes*. M.I.T. Press, Cambridge, Massachusetts, 1963.

[788] F. R. Gantmacher. *The Theory of Matrices*, volume I. AMS Chelsea Publishing, Providence, Rhode Island, 1977.

[789] J. Gao, L. Shen, and F.-W. Fu. A Chinese remainder theorem approach to skew generalized quasi-cyclic codes over finite fields. *Cryptogr. Commun.*, 8:51–66, 2016.

[790] A. García and H. Stichtenoth. A tower of Artin-Schreier extensions of function fields attaining the Drinfeld-Vlăduţ bound. *Invent. Math.*, 121:211–222, 1995.

[791] C. García Pillado, S. González, V. T. Markov, C. Martínez, and A. A. Nechaev. When are all group codes of a noncommutative group abelian (a computational approach)? *J. Math. Sci. (N. Y.)*, 186:578–585, 2012.

[792] C. García Pillado, S. González, V. T. Markov, C. Martínez, and A. A. Nechaev. New examples of non-abelian group codes. *Adv. Math. Commun.*, 10:1–10, 2016.

[793] C. García Pillado, S. González, C. Martínez, V. T. Markov, and A. A. Nechaev. Group codes over non-abelian groups. *J. Algebra Appl.*, 12(7):1350037, 20, 2013.

[794] G. Garg, T. Helleseth, and P. V. Kumar. Recent advances in low-correlation sequences. In V. Tarokh, editor, *New Directions in Wireless Communications Research*, pages 63–92. Springer, Boston, Massachusetts, 2009.

[795] B. Gärtner and J. Matoušek. *Approximation Algorithms and Semidefinite Programming*. Springer, Heidelberg, 2012.

[796] A. Gaur and B. D. Sharma. Codes correcting limited patterns of random error using SK-metric. *Cybern. Inf. Technol.*, 13:34–45, 2013.

[797] A. Gaur and B. D. Sharma. Upper bound on correcting partial random errors. *Cybern. Inf. Technol.*, 13:41–49, 2013.

[798] G. Ge, Q. Xiang, and T. Yuan. Constructions of strongly regular Cayley graphs using index four Gauss sums. *J. Algebraic Combin.*, 37:313–329, 2013.

[799] O. Geil, R. Matsumoto, and D. Ruano. Feng-Rao decoding of primary codes. *Finite Fields Appl.*, 23:35–52, 2013.

[800] C. Gentry, C. Peikert, and V. Vaikuntanathan. Trapdoors for hard lattices and new cryptographic constructions. In *Proc. 40th Annual ACM Symposium on the Theory of Computing*, pages 197–206, Victoria, British Columbia, Canada, May 2008.

[801] D. N. Gevorkyan, A. M. Avetisyan, and G. A. Tigranyan. On the structure of two-error-correcting in Hamming metric over Galois fields (in Russian). In *Computational Techniques*, volume 3, pages 19–21. Kuibyshev, 1975.

[802] M. Giesbrecht. Factoring in skew-polynomial rings over finite fields. *J. Symbolic Comput.*, 26:463–486, 1998.

[803] D. C. Gijswijt. Block diagonalization for algebras associated with block codes. arXiv:0910.4515 [math.OC], October 2009.

[804] D. C. Gijswijt, H. D. Mittelmann, and A. Schrijver. Semidefinite code bounds based on quadruple distances. *IEEE Trans. Inform. Theory*, 58:2697–2705, 2012.

[805] D. C. Gijswijt, A. Schrijver, and H. Tanaka. New upper bounds for nonbinary codes based on the Terwilliger algebra and semidefinite programming. *J. Combin. Theory Ser. A*, 113:1719–1731, 2006.

[806] E. N. Gilbert. A comparison of signaling alphabets. *Bell System Tech. J.*, 31:504–522, 1952. (Also reprinted in [169] pages 14–19 and [212] pages 24–42).

[807] H. Gluesing-Luerssen and B. Langfeld. A class of one-dimensional MDS convolutional codes. *J. Algebra Appl.*, 5:505–520, 2006.

[808] H. Gluesing-Luerssen, K. Morrison, and C. Troha. Cyclic orbit codes and stabilizer subfields. *Adv. Math. Commun.*, 9:177–197, 2015.

[809] H. Gluesing-Luerssen, J. Rosenthal, and R. Smarandache. Strongly-MDS convolutional codes. *IEEE Trans. Inform. Theory*, 52:584–598, 2006.

[810] D. G. Glynn. Two new sequences of ovals in finite Desarguesian planes of even order. In L. R. A. Casse, editor, *Combinatorial Mathematics X*, volume 1036 of *Lecture Notes in Math.*, pages 217–229. Springer, Berlin, Heidelberg, 1983.

[811] D. G. Glynn. The nonclassical 10-arc of PG(4, 9). *Discrete Math.*, 59:43–51, 1986.

[812] I. Gnana Sudha and R. S. Selvaraj. Codes with a pomset metric and constructions. *Des. Codes Cryptogr.*, 86:875–892, 2018.

[813] C. D. Godsil. Polynomial spaces. *Discrete Math.*, 73:71–88, 1988.

[814] C. D. Godsil. *Algebraic Combinatorics*. Chapman and Hall Mathematics Series. Chapman & Hall, New York, 1993.

[815] N. Goela, E. Abbe, and M. Gastpar. Polar codes for broadcast channels. *IEEE Trans. Inform. Theory*, 61:758–782, 2015.

[816] J.-M. Goethals. Two dual families of nonlinear binary codes. *Electron. Lett.*, 10:471–472, 1974.

[817] J.-M. Goethals. The extended Nadler code is unique. *IEEE Trans. Information Theory*, 23:132–135, 1977.

[818] J.-M. Goethals and P. Delsarte. On a class of majority-logic decodable cyclic codes. *IEEE Trans. Inform. Theory*, 14:182–188, 1968.

[819] J.-M. Goethals and J. J. Seidel. Strongly regular graphs derived from combinatorial designs. *Canad. J. Math.*, 22:597–614, 1970.

[820] M. J. E. Golay. Notes on digital coding. *Proc. IRE*, 37:657, 1949. (Also reprinted in [169] page 13 and [212] page 9).

[821] R. Gold. Optimal binary sequences for spread spectrum multiplexing. *IEEE Trans. Inform. Theory*, 13:619–621, 1967.

[822] R. Gold. Maximal recursive sequences with 3-valued recursive cross-correlation functions. *IEEE Trans. Inform. Theory*, 14:154–156, 1968.

[823] D. Goldin and D. Burshtein. Improved bounds on the finite length scaling of polar codes. *IEEE Trans. Inform. Theory*, 60:6966–6978, 2014.

[824] S. W. Golomb. *Shift Register Sequences*. World Scientific, Hackensack, New Jersey, 3rd edition, 2017.

[825] S. W. Golomb and G. Gong. *Signal Design for Good Correlation: For Wireless Communication, Cryptography, and Radar*. Cambridge University Press, Cambridge, UK, 2005.

[826] S. W. Golomb and E. C. Posner. Rook domains, Latin squares, affine planes, and error-distributing codes. *IEEE Trans. Inform. Theory*, 10:196–208, 1964.

[827] J. Gómez-Torrecillas, F. J. Lobillo, and G. Navarro. Peterson-Gorenstein-Zierler algorithm for skew RS codes. *Linear Multilinear Algebra*, 66:469–487, 2018.

[828] J. Gómez-Torrecillas, F. J. Lobillo, G. Navarro, and A. Neri. Hartmann-Tzeng bound and skew cyclic codes of designed Hamming distance. *Finite Fields Appl.*, 50:84–112, 2018.

[829] P. Gopalan, C. Huang, B. Jenkins, and S. Yekhanin. Explicit maximally recoverable codes with locality. *IEEE Trans. Inform. Theory*, 60:5245–5256, 2014.

[830] P. Gopalan, C. Huang, H. Simitci, and S. Yekhanin. On the locality of codeword symbols. *IEEE Trans. Inform. Theory*, 58:6925–6934, 2012.

[831] S. Goparaju and A. R. Calderbank. Binary cyclic codes that are locally repairable. In *Proc. IEEE International Symposium on Information Theory*, pages 676–680, Honolulu, Hawaii, June-July 2014.

[832] S. Goparaju, A. Fazeli, and A. Vardy. Minimum storage regenerating codes for all parameters. *IEEE Trans. Inform. Theory*, 63:6318–6328, 2017.

[833] S. Goparaju, I. Tamo, and A. R. Calderbank. An improved sub-packetization bound for minimum storage regenerating codes. *IEEE Trans. Inform. Theory*, 60:2770–2779, 2014.

[834] V. D. Goppa. A new class of linear correcting codes (in Russian). *Problemy Peredači Informacii*, 6(3):24–30, 1970 (English translation in *Probl. Inform. Transm.*, 6(3):207–212, 1970).

[835] V. D. Goppa. Rational representation of codes and (L, g)-codes (in Russian). *Problemy Peredači Informacii*, 7(3):41–49, 1971 (English translation in *Probl. Inform. Transm.*, 7(3):223–229, 1971).

[836] V. D. Goppa. Codes associated with divisors (in Russian). *Problemy Peredači Informacii*, 13(1):33–39, 1977 (English translation in *Probl. Inform. Transm.*, 13(1):22–27, 1977).

[837] V. D. Goppa. Codes on algebraic curves (in Russian). *Dokl. Akad. Nauk SSSR*, 259:1289–1290, 1981.

[838] V. D. Goppa. Algebraic-geometric codes (in Russian). *Izv. Akad. Nauk SSSR Ser. Mat.*, 46(4):762–781, 896, 1982 (English translation in *Mathematics of the USSR-Izvestiya*, 21(1):75–91, 1983).

[839] B. Gordon, W. H. Mills, and L. R. Welch. Some new difference sets. *Canadian J. Math.*, 14:614–625, 1962.

[840] D. C. Gorenstein and N. Zierler. A class of error-correcting codes in p^m symbols. *J. Soc. Indust. Appl. Math.*, 9:207–214, 1961. (Also reprinted in [169] pages 87–89 and [212] pages 194–201).

[841] E. Gorla, R. Jurrius, H. H. López, and A. Ravagnani. Rank-metric codes and q-polymatroids. arXiv:1803.10844 [cs.IT], April 2019.

[842] E. Gorla and A. Ravagnani. Codes Endowed with the Rank Metric. In M. Greferath, M. Pavčević, N. Silberstein, and M. Vázquez-Castro, editors, *Network Coding and Subspace Designs. Signals and Communication Technology*, chapter 1, pages 3–23. Springer, 2018.

[843] D. E. Gottesman. *Stabilizer Codes and Quantum Error Correction*. PhD thesis, California Institute of Technology, 1997.

[844] R. L. Graham and F. J. MacWilliams. On the number of information symbols in difference-set cyclic codes. *Bell System Tech. J.*, 45:1057–1070, 1966.

[845] M. Grassl. Code tables: Bounds on the parameters of various types of codes. `http://www.codetables.de`.

[846] M. Grassl. Code tables: Bounds on the parameters of various types of codes. `http://www.codetables.de`. Accessed on 2020-07-05.

[847] M. Grassl. Code tables: Bounds on the parameters of various types of codes: Quantum error-correcting codes. `http://www.codetables.de`. Accessed on 2020-02-20.

[848] M. Grassl. Variations on encoding circuits for stabilizer quantum codes. In Y. M. Chee, Z. Guo, S. Ling, F. Shao, Y. Tang, H. Wang, and C. Xing, editors, *Coding and Cryptology*, volume 6639 of *Lecture Notes in Comput. Sci.*, pages 142–158. Springer, Berlin, Heidelberg, 2011.

[849] M. Grassl, T. Beth, and M. Rotteler. On optimal quantum codes. *Int. J. Quantum Inform.*, 2:55–64, 2004.

[850] M. Grassl and M. Rötteler. Quantum MDS codes over small fields. In *Proc. IEEE International Symposium on Information Theory*, pages 1104–1108, Hong Kong, China, June 2015.

[851] M. Greferath. Cyclic codes over finite rings. *Discrete Math.*, 177:273–277, 1997.

[852] M. Greferath, M. Pavčević, N. Silberstein, and M. Á. Vázquez-Castro, editors. *Network Coding and Subspace Designs. Signals and Communication Technology*. Springer, Cham, 2018.

[853] M. Greferath and S. E. Schmidt. Gray isometries for finite chain rings and a nonlinear ternary (36, 3^{12}, 15) code. *IEEE Trans. Inform. Theory*, 45:2522–2524, 1999.

[854] M. Greferath and S. E. Schmidt. Finite-ring combinatorics and MacWilliams' equivalence theorem. *J. Combin. Theory Ser. A*, 92:17–28, 2000.

[855] J. H. Griesmer. A bound for error-correcting codes. *IBM J. Research Develop.*, 4:532–542, 1960.

[856] J. L. Gross and T. W. Tucker. *Topological Graph Theory*. Dover Publications, Inc., Mineola, NY, 2001.

[857] M. Grötschel, L. Lovász, and A. Schrijver. *Geometric Algorithms and Combinatorial Optimization*. Springer-Verlag, Berlin, 1988.

[858] L. K. Grover. A fast quantum mechanical algorithm for database search. In *Proc. 28th Annual ACM Symposium on the Theory of Computing*, pages 212–219, Phildelphia, Pennsylvania, May 1996.

[859] K. Guenda, T. A. Gulliver, S. Jitman, and S. Thipworawimon. Linear ℓ-intersection pairs of codes and their applications. *Des. Codes Cryptogr.*, 88:133–152, 2020.

[860] T. A. Gulliver. A new two-weight code and strongly regular graph. *Appl. Math. Lett.*, 9(2):17–20, 1996.

[861] T. A. Gulliver. Two new optimal ternary two-weight codes and strongly regular graphs. *Discrete Math.*, 149:83–92, 1996.

[862] T. A. Gulliver and V. K. Bhargava. Nine good rate $(m-1)/pm$ quasi-cyclic codes. *IEEE Trans. Inform. Theory*, 38:1366–1369, 1992.

[863] T. A. Gulliver and V. K. Bhargava. New good rate $(m-1)/pm$ ternary and quaternary quasi-cyclic codes. *Des. Codes Cryptogr.*, 7:223–233, 1996.

[864] T. A. Gulliver and M. Harada. Classification of extremal double circulant self-dual codes of lengths 64 to 72. *Des. Codes and Cryptogr.*, 13:257–269, 1998.

[865] T. A. Gulliver and M. Harada. Classification of extremal double circulant self-dual codes of lengths 74–88. *Discrete Math.*, 306:2064–2072, 2006.

[866] T. A. Gulliver, M. Harada, and H. Miyabayashi. Double circulant and quasi-twisted self-dual codes over $\mathbb{F}_5$ and $\mathbb{F}_7$. *Adv. Math. Commun.*, 1:223–238, 2007.

[867] T. A. Gulliver and P. R. J. Östergård. Binary optimal linear rate 1/2 codes. *Discrete Math.*, 283:255–261, 2004.

[868] C. Güneri and F. Özbudak. A bound on the minimum distance of quasi-cyclic codes. *SIAM J. Discrete Math.*, 26:1781–1796, 2012.

[869] C. Güneri and F. Özbudak. The concatenated structure of quasi-cyclic codes and an improvement of Jensen's bound. *IEEE Trans. Inform. Theory*, 59:979–985, 2013.

[870] C. Güneri, B. Özkaya, and S. Sayici. On linear complementary pair of nD cyclic codes. *IEEE Commun. Lett.*, 22:2404–2406, 2018.

[871] C. Güneri, B. Özkaya, and P. Solé. Quasi-cyclic complementary dual codes. *Finite Fields Appl.*, 42:67–80, 2016.

[872] C. Güneri, H. Stichtenoth, and İ. Taşkın. Further improvements on the designed minimum distance of algebraic geometry codes. *J. Pure Appl. Algebra*, 213:87–97, 2009.

[873] A. Günther and G. Nebe. Automorphisms of doubly even self-dual binary codes. *Bull. Lond. Math. Soc.*, 41:769–778, 2009.

[874] Q. Guo, T. Johansson, and P. Stankovski. A key recovery attack on MDPC with CCA security using decoding errors. In J. H. Cheon and T. Takagi, editors, *Advances in Cryptology – ASIACRYPT 2016, Part 1*, volume 10031 of *Lecture Notes in Comput. Sci.*, pages 789–815. Springer, Berlin, Heidelberg, 2009.

[875] Gurobi Optimizer Reference Manual. gurobi.com.

[876] V. Guruswami. *List Decoding of Error-Correcting Codes: Winning Thesis of the 2002 ACM Doctoral Dissertation Competition*, volume 3282 of *Lecture Notes in Comput. Sci.* Springer, Berlin, Heidelberg, 2005.

[877] V. Guruswami and W. Machmouchi. Explicit interleavers for a repeat accumulate accumulate (RAA) code construction. In F. R. Kschischang and E.-H. Yang, editors, *Proc. IEEE International Symposium on Information Theory*, pages 1968–1972, Toronto, Canada, July 2008.

[878] V. Guruswami and A. Rudra. Explicit codes achieving list decoding capacity: error-correction with optimal redundancy. *IEEE Trans. Inform. Theory*, 54:135–150, 2008.

[879] V. Guruswami, A. Rudra, and M. Sudan. *Essential Coding Theory*. Draft available at https://cse.buffalo.edu/faculty/atri/courses/coding-theory/book/web-coding-book.pdf, 2012.

[880] V. Guruswami and M. Sudan. Improved decoding of Reed-Solomon and algebraic-geometry codes. *IEEE Trans. Inform. Theory*, 45:1757–1767, 1999.

[881] V. Guruswami and C. Wang. Linear-algebraic list decoding for variants of Reed-Solomon codes. *IEEE Trans. Inform. Theory*, 59:3257–3268, 2013.

[882] V. Guruswami and M. Wootters. Repairing Reed-Solomon codes. *IEEE Trans. Inform. Theory*, 63:5684–5698, 2017.

[883] V. Guruswami and P. Xia. Polar codes: speed of polarization and polynomial gap to capacity. *IEEE Trans. Inform. Theory*, 61:3–16, 2015.

[884] L. Guth and N. H. Katz. Algebraic methods in discrete analogs of the Kakeya problem. *Adv. Math.*, 225:2828–2839, 2010.

[885] L. Guth and N. H. Katz. On the Erdős distinct distances problem in the plane. *Ann. of Math. (2)*, 181:155–190, 2015.

[886] J. Hagenauer, E. Offer, and L. Papke. Iterative decoding of binary block and convolutional codes. *IEEE Trans. Inform. Theory*, 42:429–445, 1996.

[887] M. Hall, Jr. A survey of difference sets. *Proc. Amer. Math. Soc.*, 7:975–986, 1956.

[888] M. Hall, Jr. *Combinatorial Theory*. Blaisdell Publishing Company, 1967.

[889] N. Hamada. On the p-rank of the incidence matrix of a balanced or partially balanced incomplete block design and its application to error correcting codes. *Hiroshima Math. J.*, 3:154–226, 1973.

[890] N. Hamada and M. Deza. A survey of recent works with respect to a characterization of an $(n, k, d; q)$-code meeting the Griesmer bound using a min·hyper in a finite projective geometry. *Discrete Math.*, 77:75–87, 1989.

[891] N. Hamada and T. Helleseth. Arcs, blocking sets, and minihypers. *Comput. Math. Appl.*, 39:159–168, 2000.

[892] N. Hamada and H. Ohmori. On the BIB-design having the minimum p-rank. *J. Combin. Theory Ser. A*, 18:131–140, 1975.

[893] Y. Hamdaoui and N. Sendrier. A non asymptotic analysis of information set decoding. IACR Cryptology ePrint Archive, Report 2013/162, 2013. `https://eprint.iacr.org/2013/162`, 2013.

[894] N. Hamilton. Strongly regular graphs from differences of quadrics. *Discrete Math.*, 256:465–469, 2002.

[895] R. W. Hamming. Error detecting and error correcting codes. *Bell System Tech. J.*, 29:10–23, 1950. (Also reprinted in [169] pages 9–12 and [212] pages 10–23).

[896] R. W. Hamming. *Coding and Information Theory*. Prentice Hall, Englewood Cliffs, New Jersey, 2nd edition, 1986.

[897] A. R. Hammons and H. El Gamal. On the theory of space-time codes for PSK modulation. *IEEE Trans. Inform. Theory*, 46:524–542, 2000.

[898] A. R. Hammons, Jr., P. V. Kumar, A. R. Calderbank, N. J. A. Sloane, and P. Solé. The $\mathbb{Z}_4$-linearity of Kerdock, Preparata, Goethals, and related codes. *IEEE Trans. Inform. Theory*, 40:301–319, 1994.

[899] J. Han and L. A. Lastras-Montano. Reliable memories with subline accesses. In *Proc. IEEE International Symposium on Information Theory*, pages 2531–2535, Nice, France, June 2007.

[900] J. P. Hansen. Codes on the Klein quartic, ideals, and decoding. *IEEE Trans. Inform. Theory*, 33:923–925, 1987.

[901] J. P. Hansen. *Group Codes on Algebraic Curves*. Mathematica Gottingensis, heft 9, 1987.

[902] J. P. Hansen and H. Stichtenoth. Group codes on certain algebraic curves with many rational points. *Appl. Algebra Engrg. Comm. Comput.*, 1:67–77, 1990.

[903] J. Hao, S. T. Xia, and B. Chen. Some results on optimal locally repairable codes. In *Proc. IEEE International Symposium on Information Theory*, pages 440–444, Barcelona, Spain, July 2016.

[904] M. Harada. The existence of a self-dual [70, 35, 12] code and formally self-dual codes. *Finite Fields Appl.*, 3:131–139, 1997.

[905] M. Harada, T. A. Gulliver, and H. Kaneta. Classification of extremal double circulant self-dual codes of length up to 62. *Discrete Math.*, 188:127–136, 1998.

[906] M. Harada, W. H. Holzmann, H. Kharaghani, and M. Khorvash. Extremal ternary self-dual codes constructed from negacirculant matrices. *Graphs Combin.*, 23:401–417, 2007.

[907] M. Harada, C. Lam, and V. D. Tonchev. Symmetric (4, 4)-nets and generalized Hadamard matrices over groups of order 4. *Des. Codes Cryptogr.*, 34:71–87, 2005.

[908] M. Harada and A. Munemasa. Database of self-dual codes. http://www.math.is.tohoku.ac.jp/~munemasa/selfdualcodes.htm.

[909] M. Harada and A. Munemasa. A quasi-symmetric 2-(49, 9, 6) design. *J. Combin. Des.*, 10:173–179, 2002.

[910] M. Harada and A. Munemasa. Classification of self-dual codes of length 36. *Adv. Math. Commun.*, 6:229–235, 2012.

[911] M. Harada, E. Novak, and V. D. Tonchev. The weight distribution of the self-dual [128, 64] polarity design code. *Adv. Math. Commun.*, 10:643–648, 2016.

[912] M. Harada and P. R. J. Östergård. Classification of extremal formally self-dual even codes of length 22. *Graphs Combin.*, 18:507–516, 2002.

[913] J. Harshan, E. Viterbo, and J.-C. Belfiore. Practical encoders and decoders for Euclidean codes from Barnes-Wall lattices. *IEEE Trans. Commun.*, 61:4417–4427, 2013.

[914] C. R. P. Hartmann and K. K. Tzeng. Generalizations of the BCH bound. *Inform. and Control*, 20:489–498, 1972.

[915] R. Hartshorne. *Algebraic Geometry*. Springer-Verlag, New York-Heidelberg, 1977. Graduate Texts in Mathematics, No. 52.

[916] N. J. A. Harvey, D. R. Karger, and K. Murota. Deterministic network coding by matrix completion. In *Proc. 16th Annual ACM-SIAM Symposium on Discrete Algorithms*, pages 489–498, Vancouver, Canada, January 2005.

[917] S. A. Hashemi, C. Condo, and W. J. Gross. Fast and flexible successive-cancellation list decoders for polar codes. *IEEE Trans. Signal Process.*, 65:5756–5769, 2017.

[918] S. H. Hassani. *Polarization and Spatial Coupling: Two Techniques to Boost Performance.* PhD thesis, École Polytechnique Fédérale de Lausanne, 2013.

[919] S. H. Hassani, K. Alishahi, and R. L. Urbanke. Finite-length scaling for polar codes. *IEEE Trans. Inform. Theory*, 60:5875–5898, 2014.

[920] B. Hassibi and B. M. Hochwald. Cayley differential unitary space-time codes. *IEEE Trans. Inform. Theory*, 48:1485–1503, 2002.

[921] B. Hassibi and B. M. Hochwald. High-rate codes that are linear in space and time. *IEEE Trans. Inform. Theory*, 48:1804–1824, 2002.

[922] B. Hassibi and H. Vikalo. On the sphere decoding algorithm I. Expected complexity. *IEEE Trans. Signal Process.*, 53:2806–2818, 2005.

[923] J. Håstad. Clique is hard to approximate within $n^{1-\epsilon}$. *Acta Math.*, 182:105–142, 1999.

[924] E. R. Hauge and T. Helleseth. De Bruijn sequences, irreducible codes and cyclotomy. *Discrete Math.*, 159:143–154, 1996.

[925] E. R. Hauge and J. Mykkeltveit. On the classification of de Bruijn sequences. *Discrete Math.*, 148:65–83, 1996.

[926] K. Haymaker, B. Malmskog, and G. L. Matthews. Locally recoverable codes with availability $t \geq 2$ from fiber products of curves. *Adv. Math. Commun.*, 12:317–336, 2018.

[927] T. Head. Formal language theory and DNA: an analysis of the generative capacity of specific recombinant behaviors. *Bull. Math. Biol.*, 49:737–759, 1987.

[928] A. S. Hedayat, N. J .A. Sloane, and J. Stufken. *Orthogonal Arrays.* Springer Nature, New York, 1999.

[929] O. Heden. A survey of perfect codes. *Adv. Math. Commun.*, 2:223–247, 2008.

[930] O. Heden. On perfect codes over non prime power alphabets. In A. A. Bruen and D. L. Wehlau, editors, *Error-Correcting Codes, Finite Geometries and Cryptography*, volume 523 of *Contemp. Math.*, pages 173–184. Amer. Math. Soc., Providence, Rhode Island, 2010.

[931] O. Heden and C. Roos. The non-existence of some perfect codes over non-prime power alphabets. *Discrete Math.*, 311:1344–1348, 2011.

[932] W. Heise, T. Honold, and A. A. Nechaev. Weighted modules and representations of codes. In *Proc. 6th International Workshop on Algebraic and Combinatorial Coding Theory*, pages 123–129, Pskov, Russia, September 1998.

[933] M. Hell, T. Johansson, A. Maximov, and W. Meier. The Grain family of stream ciphers. In M. Robshaw and O. Billet, editors, *New Stream Cipher Designs*, volume 4986 of *Lecture Notes in Comput. Sci.*, pages 179–190. Springer, Berlin, Heidelberg, 2008.

[934] T. Helleseth. Some results about the cross-correlation function between two maximal linear sequences. *Discrete Math.*, 16:209–232, 1976.

[935] T. Helleseth. Nonlinear shift registers—a survey and challenges. In *Algebraic Curves and Finite Fields*, volume 16 of *Radon Ser. Comput. Appl. Math.*, pages 121–144. De Gruyter, Berlin, 2014.

[936] T. Helleseth, L. Hu, A. Kholosha, X. Zeng, N. Li, and W. Jiang. Period-different m-sequences with at most four-valued cross correlation. *IEEE Trans. Inform. Theory*, 55:3305–3311, 2009.

[937] T. Helleseth and A. Kholosha. Monomial and quadratic bent functions over the finite fields of odd characteristic. *IEEE Trans. Inform. Theory*, 52:2018–2032, 2006.

[938] T. Helleseth and A. Kholosha. New binomial bent functions over the finite fields of odd characteristic. *IEEE Trans. Inform. Theory*, 56:4646–4652, 2010.

[939] T. Helleseth, T. Kløve, and J. Mykkeltveit. The weight distribution of irreducible cyclic codes with block lengths $n_1((q^\ell - 1)/n)$. *Discrete Math.*, 18:179–211, 1977.

[940] T. Helleseth and P. V. Kumar. Sequences with Low Correlation. In V. S. Pless and W. C. Huffman, editors, *Handbook of Coding Theory, Vol. I, II*, chapter 21, pages 1765–1853. North-Holland, Amsterdam, 1998.

[941] T. Helleseth and P. Rosendahl. New pairs of m-sequences with 4-level cross-correlation. *Finite Fields Appl.*, 11:674–683, 2005.

[942] Z. L. Heng and C. Ding. A construction of q-ary linear codes with irreducible cyclic codes. *Des. Codes Cryptogr.*, 87:1087–1108, 2019.

[943] Z. L. Heng, C. Ding, and Z. Zhou. Minimal linear codes over finite fields. *Finite Fields Appl.*, 54:176–196, 2018.

[944] Z. L. Heng and Q. Yue. A class of binary linear codes with at most three weights. *IEEE Commun. Letters*, 19:1488–1491, 2015.

[945] Z. L. Heng and Q. Yue. A class of q-ary linear codes derived from irreducible cyclic codes. arXiv:1511.09174 [cs.IT], April 2016.

[946] F. Hergert. The equivalence classes of the Vasil'ev codes of length 15. In D. Jungnickel and K. Vedder, editors, *Combinatorial Theory*, volume 969 of *Lecture Notes in Math.*, pages 176–186. Springer-Verlag, Berlin, Heidelberg, 1982.

[947] A. O. Hero. Secure space-time communication. *IEEE Trans. Inform. Theory*, 49:3235–3249, 2003.

[948] D. Hertel. A note on the Kasami power function. Cryptology ePrint Archive, Report 2005/436, 2005. https://eprint.iacr.org/2005/436.

[949] M. Herzog and J. Schönheim. Group partition, factorization and the vector covering problem. *Canad. Math. Bull.*, 15:207–214, 1972.

[950] F. Hess. Computing Riemann-Roch spaces in algebraic function fields and related topics. *J. Symbolic Comput.*, 33:425–445, 2002.

[951] R. Hill. On the largest size of cap in $S_{5,3}$. *Atti Accad. Naz. Lincei Rend. Cl. Sci. Fis. Mat. Nat. (8)*, 54:378–384 (1974), 1973.

[952] R. Hill. Caps and groups. In *Colloquio Internazionale sulle Teorie Combinatorie (Rome, 1973), Tomo II*, pages 389–394. Atti dei Convegni Lincei, No. 17. Accad. Naz. Lincei, Rome, 1976.

[953] R. Hill. An extension theorem for linear codes. *Des. Codes Cryptogr.*, 17:151–157, 1999.

[954] R. Hill and E. Kolev. A survey of recent results on optimal linear codes. In *Combinatorial Designs and their Applications (Milton Keynes, 1997)*, volume 403 of *Chapman & Hall/CRC Res. Notes Math.*, pages 127–152. Chapman & Hall/CRC, Boca Raton, Florida, 1999.

[955] R. Hill and H. N. Ward. A geometric approach to classifying Griesmer codes. *Des. Codes Cryptogr.*, 44:169–196, 2007.

[956] M. Hirotomo, M. Takita, and M. Morii. Syndrome decoding of symbol-pair codes. In *Proc. IEEE Information Theory Workshop*, pages 162–166, Hobart, Tasmania, November 2014.

[957] J. W. P. Hirschfeld. Rational curves on quadrics over finite fields of characteristic two. *Rend. Mat. (6)*, 4:773–795, 1971.

[958] J. W. P. Hirschfeld. *Finite Projective Spaces of Three Dimensions*. Oxford Mathematical Monographs. The Clarendon Press, Oxford University Press, New York, 1985. Oxford Science Publications.

[959] J. W. P. Hirschfeld. *Projective Geometries over Finite Fields*. Oxford Mathematical Monographs. The Clarendon Press, Oxford University Press, New York, 2nd edition, 1998. Oxford Science Publications.

[960] J. W. P. Hirschfeld, G. Korchmáros, and F. Torres. *Algebraic Curves Over a Finite Field*. Princeton Series in Applied Mathematics. Princeton University Press, Princeton, New Jersey, 2008.

[961] J. W. P. Hirschfeld and L. Storme. The packing problem in statistics, coding theory and finite projective spaces: Update 2001. In *Finite Geometries*, volume 3 of *Dev. Math.*, pages 201–246. Kluwer Acad. Publ., Dordrecht, Netherlands, 2001.

[962] J. W. P. Hirschfeld and J. A. Thas. Open problems in finite projective spaces. *Finite Fields Appl.*, 32:44–81, 2015.

[963] J. W. P. Hirschfeld and J. A. Thas. *General Galois Geometries*. Springer Monographs in Mathematics. Springer, London, 2016.

[964] T. Ho, M. Médard, R. Kötter, D. R. Karger, M. Effros, J. Shi, and B. Leong. A random linear network coding approach to multicast. *IEEE Trans. Inform. Theory*, 52:4413–4430, 2006.

[965] S. A. Hobart. Bounds on subsets of coherent configurations. *Michigan Math. J.*, 58:231–239, 2009.

[966] S. A. Hobart and J. Williford. The absolute bound for coherent configurations. *Linear Algebra Appl.*, 440:50–60, 2014.

[967] B. M. Hochwald and T. L. Marzetta. Unitary space-time modulation for multiple-antenna communications in Rayleigh flat fading. *IEEE Trans. Inform. Theory*, 46:543–564, 2000.

[968] A. Hocquenghem. Codes correcteurs d'erreurs. *Chiffres (Paris)*, 2:147–156, 1959 (Also reprinted in [169] pages 72–74 and [212] pages 155–164).

[969] E. Hof and S. Shamai. Secrecy-achieving polar-coding for binary-input memoryless symmetric wire-tap channels. arXiv:1005.2759 [cs.IT], August 2010.

[970] D. Hofheinz, K. Hövelmanns, and E. Kiltz. A modular analysis of the Fujisaki-Okamoto transformation. In Y. Kalai and L. Reyzin, editors, *Theory of Cryptography, Part 1*, volume 10677 of *Lecture Notes in Comput. Sci.*, pages 341–371. Springer, Berlin, Heidelberg, 2017.

[971] T. Høholdt and R. Pellikaan. On the decoding of algebraic-geometric codes. *IEEE Trans. Inform. Theory*, 41:1589–1614, 1995.

[972] C. Hollanti, J. Lahtonen, and H.-F. Lu. Maximal orders in the design of dense space-time lattice codes. *IEEE Trans. Inform. Theory*, 54:4493–4510, 2008.

[973] H. D. L. Hollmann. A relation between Levenshtein-type distances and insertion-and-deletion correcting capabilities of codes. *IEEE Trans. Inform. Theory*, 39:1424–1427, 1993.

[974] H. D. L. Hollmann and Q. Xiang. A proof of the Welch and Niho conjectures on cross-correlations of binary m-sequences. *Finite Fields Appl.*, 7:253–286, 2001.

[975] Y. Hong. On the nonexistence of unknown perfect 6- and 8-codes in Hamming schemes $H(n,q)$ with q arbitrary. *Osaka J. Math.*, 21:687–700, 1984.

[976] Y. Hong, E. Viterbo, and J.-C. Belfiore. Golden space-time trellis coded modulation. *IEEE Trans. Inform. Theory*, 53:1689–1705, 2007.

[977] T. Honold, M. Kiermaier, and S. Kurz. Constructions and bounds for mixed-dimension subspace codes. *Adv. Math. Commun.*, 10:649–682, 2016.

[978] T. Honold and I. Landjev. Linear representable codes over chain rings. In *Proc. 6th International Workshop on Algebraic and Combinatorial Coding Theory*, pages 135–141, Pskov, Russia, September 1998.

[979] T. Honold and I. Landjev. Linearly representable codes over chain rings. *Abh. Math. Sem. Univ. Hamburg*, 69:187–203, 1999.

[980] T. Honold and I. Landjev. Linear codes over finite chain rings. *Electron. J. Combin.*, 7:Research Paper 11, 22 pages, 2000.

[981] T. Honold and I. Landjev. Linear codes over finite chain rings and projective Hjelm-slev geometries. In P. Solé, editor, *Codes Over Rings*, volume 6 of *Series on Coding Theory and Cryptology*, pages 60–123. World Sci. Publ., Hackensack, New Jersey, 2009.

[982] S. Hoory, N. Linial, and A. Wigderson. Expander graphs and their applications. *Bull. Amer. Math. Soc. (N.S.)*, 43:439–561, 2006.

[983] N. J. Hopper and M. Blum. Secure human identification protocols. In C. Boyd, editor, *Advances in Cryptology – ASIACRYPT 2001*, volume 2248 of *Lecture Notes in Comput. Sci.*, pages 52–66. Springer, Berlin, Heidelberg, 2001.

[984] A.-L. Horlemann-Trautmann and J. Rosenthal. Constructions of constant dimension codes. In M. Greferath, M. Pavčević, N. Silberstein, and M. Á. Vázquez-Castro, editors, *Network Coding and Subspace Designs. Signals and Communication Technology*, pages 25–42. Springer, Cham, 2018.

[985] A. Hottinen, O. Tirkkonen, and R. Wichman. *Multi-Antenna Transceiver Techniques for 3G and Beyond*. John Wiley & Sons, Ltd., Chichester, UK, 2003.

[986] X.-D. Hou. A note on the proof of Niho's conjecture. *SIAM J. Discrete Math.*, 18:313–319, 2004.

[987] X.-D. Hou, S. R. López-Permouth, and B. R. Parra-Avila. Rational power series, sequential codes and periodicity of sequences. *J. Pure Appl. Algebra*, 213:1157–1169, 2009.

[988] S. K. Houghten. A table of bounds on optimal variable-length binary edit-metric codes. https://www.cosc.brocku.ca/~houghten/binaryeditvar.html.

[989] S. K. Houghten. A table of bounds on optimal variable-length ternary edit-metric codes. https://www.cosc.brocku.ca/~houghten/ternaryeditvar.html.

[990] S. K. Houghten, D. Ashlock, and J. Lenarz. Construction of optimal edit metric codes. In *Proc. IEEE Information Theory Workshop*, pages 259–263, Chengdu, China, October 2006.

[991] S. K. Houghten, C. W. H. Lam, L. H. Thiel, and J. A. Parker. The extended quadratic residue code is the only (48, 24, 12) self-dual doubly-even code. *IEEE Trans. Inform. Theory*, 49:53–59, 2003.

[992] Y. Hu, Y. Xu, X. Wang, C. Zhan, and P. Li. Cooperative recovery of distributed storage systems from multiple losses with network coding. *IEEE J. Selected Areas Commun.*, 28:268–276, 2010.

[993] L. K. Hua. A theorem on matrices over a field and its applications. *Acta Math. Sinica*, 1:109–163, 1951.

[994] C. Huang, M. Chen, and J. Li. Pyramid codes: flexible schemes to trade space for access efficiency in reliable data storage systems. In *Proc. 6th IEEE International Symposium on Network Computing and Applications*, pages 79–86, Cambridge, Massachusetts, July 2007.

[995] C. Huang, H. Simitci, Y. Xu, A. Ogus, B. Calder, P. Gopalan, J. Li, and S. Yekhanin. Erasure coding in Windows Azure Storage. In *Proc. USENIX Annual Technical Conference (ATC)*, pages 15–26, Boston, Massachusetts, June 2012.

[996] P. Huang, E. Yaakobi, H. Uchikawa, and P. H. Siegel. Cyclic linear binary locally repairable codes. In *Proc. IEEE Information Theory Workshop*, pages 259–263, Jerusalem, Israel, April-May 2015.

[997] K. Huber. Codes over Eisenstein-Jacobi integers. *Contemp. Math.*, 168:165–165, 1994.

[998] K. Huber. Codes over Gaussian integers. *IEEE Trans. Inform. Theory*, 40:207–216, 1994.

[999] W. C. Huffman. Automorphisms of codes with applications to extremal doubly-even codes of length 48. *IEEE Trans. Inform. Theory*, 28:511–521, 1982.

[1000] W. C. Huffman. Decomposing and shortening codes using automorphisms. *IEEE Trans. Inform. Theory*, 32:833–836, 1986.

[1001] W. C. Huffman. On extremal self-dual quaternary codes of lengths 18 to 28, I. *IEEE Trans. Inform. Theory*, 36:651–660, 1990.

[1002] W. C. Huffman. The automorphism groups of the generalized quadratic residue codes. *IEEE Trans. Inform. Theory*, 41:378–386, 1995.

[1003] W. C. Huffman. Characterization of quaternary extremal codes of lengths 18 and 20. *IEEE Trans. Inform. Theory*, 43:1613–1616, 1997.

[1004] W. C. Huffman. Decompositions and extremal Type II codes over $\mathbb{Z}_4$. *IEEE Trans. Inform. Theory*, 44:800–809, 1998.

[1005] W. C. Huffman. On the classification and enumeration of self-dual codes. *Finite Fields Appl.*, 11:451–490, 2005.

[1006] W. C. Huffman. Additive self-dual codes over $\mathbb{F}_4$ with an automorphism of odd prime order. *Adv. Math. Commun.*, 1:357–398, 2007.

[1007] W. C. Huffman. Self-dual codes over $\mathbb{F}_2 + u\mathbb{F}_2$ with an automorphism of odd order. *Finite Fields Appl.*, 15:277–293, 2009.

[1008] W. C. Huffman and V. S. Pless. *Fundamentals of Error-Correcting Codes*. Cambridge University Press, Cambridge, UK, 2003.

[1009] W. C. Huffman and V. D. Tonchev. The existence of extremal self-dual $[50, 25, 10]$ codes and quasi-symmetric 2-$(49, 9, 6)$ designs. *Des. Codes Cryptogr.*, 6:97–106, 1995.

[1010] W. C. Huffman and V. Y. Yorgov. A $[72, 36, 16]$ doubly even code does not have an automorphism of order 11. *IEEE Trans. Inform. Theory*, 33:749–752, 1987.

[1011] B. L. Hughes. Differential space-time modulation. *IEEE Trans. Inform. Theory*, 46:2567–2578, 2000.

[1012] G. Hughes. Structure theorems for group ring codes with an application to self-dual codes. *Des. Codes Cryptogr.*, 24:5–14, 2001.

[1013] B. Huppert and N. Blackburn. *Finite Groups II*, volume 242 of *Grundlehren der Mathematischen Wissenschaften [Fundamental Principles of Mathematical Sciences]*. Springer-Verlag, Berlin-New York, 1982.

[1014] R. Hutchinson. The existence of strongly MDS convolutional codes. *SIAM J. Control Optim.*, 47:2812–2826, 2008.

[1015] R. Hutchinson, J. Rosenthal, and R. Smarandache. Convolutional codes with maximum distance profile. *Systems Control Lett.*, 54:53–63, 2005.

[1016] J. Y. Hyun and H. K. Kim. The poset structures admitting the extended binary Hamming code to be a perfect code. *Discrete Math.*, 288:37–47, 2004.

[1017] J. Y. Hyun, H. K. Kim, and J. Rye Park. The weighted poset metrics and directed graph metrics. arXiv:1703.00139 [math.CO], March 2017.

[1018] J. Y. Hyun, J. Lee, and Y. Lee. Explicit criteria for construction of plateaued functions. *IEEE Trans. Inform. Theory*, 62:7555–7565, 2016.

[1019] J. Y. Hyun and Y. Lee. MDS poset-codes satisfying the asymptotic Gilbert-Varshamov bound in Hamming weights. *IEEE Trans. Inform. Theory*, 57:8021–8026, 2011.

[1020] IBM ILOG CPLEX Optimization Studio. ibm.com.

[1021] Y. Ihara. Some remarks on the number of rational points of algebraic curves over finite fields. *J. Fac. Sci. Univ. Tokyo Sect. IA Math.*, 28:721–724 (1982), 1981.

[1022] T. Ikuta and A. Munemasa. A new example of non-amorphous association schemes. *Contrib. Discrete Math.*, 3(2):31–36, 2008.

[1023] T. Ikuta and A. Munemasa. Pseudocyclic association schemes and strongly regular graphs. *European J. Combin.*, 31:1513–1519, 2010.

[1024] S. P. Inamdar and N. S. Narasimha Sastry. Codes from Veronese and Segre embeddings and Hamada's formula. *J. Combin. Theory Ser. A*, 96:20–30, 2001.

[1025] L. Ioffe and M. Mézard. Asymmetric quantum error-correcting codes. *Physical Review A*, 75(3):032345, March 2007.

[1026] Y. Ishai, E. Kushilevitz, R. Ostrovsky, and A. Sahai. Zero-knowledge proofs from secure multiparty computation. *SIAM J. Comput.*, 39:1121–1152, 2009.

[1027] W. A. Jackson, K. M. Martin, and C. M. O'Keefe. Multi-secret sharing schemes. In Y. G. Desmedt, editor, *Advances in Cryptology – CRYPTO '94*, volume 839 of *Lecture Notes in Comput. Sci.*, pages 150–163. Springer, Berlin, Heidelberg, 1994.

[1028] N. Jacobson. *The Theory of Rings*, volume II of *Amer. Math. Soc. Mathematical Surveys*. Amer. Math. Soc., New York, 1943.

[1029] N. Jacobson. *Finite-Dimensional Division Algebras over Fields*. Springer-Verlag, Berlin, 1996.

[1030] H. Jafarkhani. A quasi-orthogonal space-time block code. *IEEE Trans. Commun.*, 49:1–4, 2001.

[1031] D. B. Jaffe. Optimal binary linear codes of length ≤ 30. *Discrete Math.*, 223:135–155, 2000.

[1032] D. B. Jaffe and J. Simonis. New binary linear codes which are dual transforms of good codes. *IEEE Trans. Inform. Theory*, 45:2136–2137, 1999.

[1033] D. B. Jaffe and V. D. Tonchev. Computing linear codes and unitals. *Des. Codes Cryptogr.*, 14:39–52, 1998.

[1034] S. Jaggi, P. Sanders, P. A. Chou, M. Effros, S. Egner, K. Jain, and L. M. G. M. Tolhuizen. Polynomial time algorithms for multicast network code construction. *IEEE Trans. Inform. Theory*, 51:1973–1982, 2005.

[1035] Y. Jang and J. Park. On a MacWilliams type identity and a perfectness for a binary linear $(n, n − 1, j)$-poset code. *Discrete Math.*, 265:85–104, 2003.

[1036] C. J. A. Jansen, W. G. Franx, and D. E. Boekee. An efficient algorithm for the generation of de Bruijn cycles. *IEEE Trans. Inform. Theory*, 37:1475–1478, 1991.

[1037] H. L. Janwa and A. K. Lal. On expander graphs: parameters and applications. arXiv:cs/0406048 [cs.IT], June 2004.

[1038] H. L. Janwa and O. Moreno. McEliece public key cryptosystems using algebraic-geometric codes. *Des. Codes Cryptogr.*, 8:293–307, 1996.

[1039] H. L. Janwa and R. M. Wilson. Hyperplane sections of Fermat varieties in P^3 in char. 2 and some applications to cyclic codes. In G. D. Cohen, T. Mora, and O. Moreno, editors, *Applied Algebra, Algebraic Algorithms and Error-Correcting Codes*, volume 673 of *Lecture Notes in Comput. Sci.*, pages 180–194. Springer, Berlin, Heidelberg, 1993.

[1040] J. Jedwab, D. J. Katz, and K.-U. Schmidt. Advances in the merit factor problem for binary sequences. *J. Combin. Theory Ser. A*, 120:882–906, 2013.

[1041] S. A. Jennings. The structure of the group ring of a p-group over a modular field. *Trans. Amer. Math. Soc.*, 50:175–185, 1941.

[1042] J. M. Jensen. The concatenated structure of cyclic and abelian codes. *IEEE Trans. Inform. Theory*, 31:788–793, 1985.

[1043] Y. Jia, S. Ling, and C. Xing. On self-dual cyclic codes over finite fields. *IEEE Trans. Inform. Theory*, 57:2243–2251, 2011.

[1044] H. Jin, R. Khandekar, and R. J. McEliece. Irregular repeat-accumulate codes. In *Proc. 2nd International Symposium on Turbo Codes and Related Topics*, pages 1–8, Brest, France, September 2000.

[1045] Y. Jing and B. Hassibi. Design of full-diverse multiple-antenna codes based on Sp(2). *IEEE Trans. Inform. Theory*, 50:2639–2656, 2004.

[1046] Y. Jing and B. Hassibi. Three-transmit-antenna space-time codes based on SU(3). *IEEE Trans. Signal Process.*, 53:3688–3702, 2005.

[1047] Y. Jing and B. Hassibi. Distributed space-time coding in wireless relay networks. *IEEE Trans. Wireless Commun.*, 5:3524–3536, 2006.

[1048] G. R. Jithamitra and B. Sundar Rajan. Minimizing the complexity of fast sphere decoding of STBCs. In A. Kuleshov, V. M. Blinovsky, and A. Ephremides, editors, *Proc. IEEE International Symposium on Information Theory*, pages 1846–1850, St. Petersburg, Russia, July-August 2011.

[1049] S. Jitman, S. Ling, H. Liu, and X. Xie. Checkable codes from group rings. arXiv:1012.5498 [cs.IT], December 2010.

[1050] S. Jitman, S. Ling, H. Liu, and X. Xie. Abelian codes in principal ideal group algebras. *IEEE Trans. Inform. Theory*, 59:3046–3058, 2013.

[1051] R. Johannesson and K. S. Zigangirov. *Fundamentals of Convolutional Coding*. IEEE Series on Digital & Mobile Communication. IEEE Press, New York, 1999.

[1052] S. J. Johnson and S. R. Weller. Regular low-density parity-check codes from combinatorial designs. In *Proc. IEEE Information Theory Workshop*, pages 90–92, Cairns, Australia, September 2001.

[1053] S. J. Johnson and S. R. Weller. Codes for low-density parity check codes from Kirkman triple systems. In *Proc. GLOBECOM '02 – IEEE Global Telecommunications Conference*, volume 2, pages 970–974, Taipei, Taiwan, November 2002.

[1054] S. J. Johnson and S. R. Weller. Codes for iterative decoding from partial geometries. *IEEE Trans. Commun.*, 52:236–243, 2004.

[1055] G. C. Jorge, A. A. de Andrade, S. I. R. Costa, and J. E. Strapasson. Algebraic constructions of densest lattices. *J. Algebra*, 429:218–235, 2015.

[1056] D. Jungnickel. Characterizing geometric designs, II. *J. Combin. Theory Ser. A*, 118:623–633, 2011.

[1057] D. Jungnickel and A. Pott. Perfect and almost perfect sequences. *Discrete Appl. Math.*, 95:331–359, 1999.

[1058] D. Jungnickel and V. D. Tonchev. Polarities, quasi-symmetric designs, and Hamada's conjecture. *Des. Codes Cryptogr.*, 51:131–140, 2009.

[1059] D. Jungnickel and V. D. Tonchev. The number of designs with geometric parameters grows exponentially. *Des. Codes Cryptogr.*, 55:131–140, 2010.

[1060] D. Jungnickel and V. D. Tonchev. The classification of antipodal two-weight linear codes. *Finite Fields Appl.*, 50:372–381, 2018.

[1061] R. Jurrius and R. Pellikaan. On defining generalized rank weights. *Adv. Math. Commun.*, 11:225–235, 2017.

[1062] R. Jurrius and R. Pellikaan. Defining the q-analogue of a matroid. *Electron. J. Combin.*, 25(3):Paper 3.2, 32, 2018.

[1063] J. Justesen. A class of constructive asymptotically good algebraic codes. *IEEE Trans. Inform. Theory*, 18:652–656, 1972.

[1064] J. Justesen. An algebraic construction of rate $1/v$ convolutional codes. *IEEE Trans. Inform. Theory*, 21:577–580, 1975.

[1065] J. Justesen, K. J. Larsen, H. E. Jensen, A. Havemose, and T. Høholdt. Construction and decoding of a class of algebraic geometry codes. *IEEE Trans. Inform. Theory*, 35:811–821, 1989.

[1066] G. A. Kabatianskii, E. A. Krouk, and B. Smeets. A digital signature scheme based on random error-correcting codes. In M. Darnell, editor, *Crytography and Coding*, volume 1355 of *Lecture Notes in Comput. Sci.*, pages 161–167. Springer, Berlin, Heidelberg, 1997.

[1067] G. A. Kabatianskii and V. I. Levenshtein. On bounds for packings on a sphere and in space (in Russian). *Problemy Peredači Informacii*, 14(1):3–25, 1978. (English translation in *Probl. Inform. Transm.*, 14(1):1–17, 1978).

[1068] G. A. Kabatianskii and V. I. Panchenko. Packings and coverings of the Hamming space by balls of unit radius (in Russian). *Problemy Peredači Informacii*, 24(4):3–16, 1988 (English translation in *Probl. Inform. Transm.*, 24(4):261–272, 1988).

[1069] G. Kachigar and J.-P. Tillich. Quantum information set decoding algorithms. In T. Lange and T. Takagi, editors, *Post-Quantum Cryptography*, volume 10346 of *Lecture Notes in Comput. Sci.*, pages 69–89. Springer, Berlin, Heidelberg, 2017.

[1070] S. Kadhe and A. R. Calderbank. Rate optimal binary linear locally repairable codes with small availability. In *Proc. IEEE International Symposium on Information Theory*, pages 166–170, Aachen, Germany, June 2017.

[1071] X. Kai, S. Zhu, and P. Li. A construction of new MDS symbol-pair codes. *IEEE Trans. Inform. Theory*, 61:5828–5834, 2015.

[1072] X. Kai, S. Zhu, and L. Wang. A family of constacyclic codes over $\mathbb{F}_2 + u\mathbb{F}_2 + v\mathbb{F}_2 + uv\mathbb{F}_2$. *J. Syst. Sci. Complex.*, 25:1032–1040, 2012.

[1073] T. Kailath. *Linear Systems*. Prentice-Hall, Inc., Englewood Cliffs, New Jersey, 1980.

[1074] J. G. Kalbfleisch and R. G. Stanton. A combinatorial problem in matching. *J. London Math. Soc.*, 44:60–64, 1969. Corrigendum appears in [*J. London Math Soc. (2) 1* (1969), 398].

[1075] G. M. Kamath, N. Prakash, V. Lalitha, and P. V. Kumar. Codes with local regeneration and erasure correction. *IEEE Trans. Inform. Theory*, 60:4637–4660, 2014.

[1076] W. M. Kantor. An exponential number of generalized Kerdock codes. *Inform. and Control*, 53:74–80, 1982.

[1077] W. M. Kantor. Spreads, translation planes and Kerdock sets. I. *SIAM J. Algebraic Discrete Methods*, 3:151–165, 1982.

[1078] W. M. Kantor. Spreads, translation planes and Kerdock sets. II. *SIAM J. Algebraic Discrete Methods*, 3:308–318, 1982.

[1079] W. M. Kantor. Exponential numbers of two-weight codes, difference sets and symmetric designs. *Discrete Math.*, 46:95–98, 1983.

[1080] W. M. Kantor. On the inequivalence of generalized Preparata codes. *IEEE Trans. Inform. Theory*, 29:345–348, 1983.

[1081] P. Kanwar and S. R. López-Permouth. Cyclic codes over the integers modulo p^m. *Finite Fields Appl.*, 3:334–352, 1997.

[1082] A. Karatsuba and Y. Ofman. Multiplication of multi-digit numbers on automata. *Soviet Physics Doklady*, 7:595–596, 1963.

[1083] E. D. Karnin, J. W. Greene, and M. E. Hellman. On secret sharing systems. *IEEE Trans. Inform. Theory*, 29:35–41, 1983.

[1084] R. Karp, M. G. Luby, and M. A. Shokrollahi. Finite length analysis of LT codes. In *Proc. IEEE International Symposium on Information Theory*, page 39, Chicago, Illinois, June-July 2004.

[1085] M. Karzand and E. E. Telatar. Polar codes for q-ary source coding. In *Proc. IEEE International Symposium on Information Theory*, pages 909–912, Austin, Texas, June 2010.

[1086] T. Kasami. Weight distribution formula for some class of cyclic codes. Technical Report No. R-285, Coordinated Science Laboratory, University of Illinois at Urbana-Champaign, 1966.

[1087] T. Kasami. The weight enumerators for several classes of subcodes of the 2nd order binary Reed-Muller codes. *Inform. and Control*, 18:369–394, 1971.

[1088] T. Kasami. A Gilbert-Varshamov bound for quasi-cyclic codes of rate 1/2. *IEEE Trans. Inform. Theory*, 20:679, 1974.

[1089] T. Kasami, S. Lin, and W. Peterson. New generalizations of the Reed–Muller codes. Part I: primitive codes. *IEEE Trans. Inform. Theory*, 14:189–199, 1968. (Also reprinted in [212] pages 323–333).

[1090] P. Kaski and P. R. J. Östergård. *Classification Algorithms for Codes and Designs*, volume 15 of *Algorithms and Computation in Mathematics*. Springer-Verlag, Berlin, 2006. With 1 DVD-ROM (Windows, Macintosh and UNIX).

[1091] P. Kaski and O. Pottonen. Libexact User's Guide: Version 1.0. Number 2008-1 in HIIT technical reports. Helsinki Institute for Information Technology HIIT, Finland, 2008.

[1092] G. L. Katsman and M. A. Tsfasman. A remark on algebraic geometric codes. In *Representation Theory, Group Rings, and Coding Theory*, volume 93 of *Contemp. Math.*, pages 197–199. Amer. Math. Soc., Providence, Rhode Island, 1989.

[1093] D. J. Katz. Sequences with low correlation. In L. Budaghyan and F. Rodríguez-Henríquez, editors, *Arithmetic of Finite Fields*, volume 11321 of *Lecture Notes in Comput. Sci.*, pages 149–172. Springer, Berlin, Heidelberg, 2018.

[1094] M. L. Kaushik. Necessary and sufficient number of parity checks in codes correcting random errors and bursts with weight constraints under a new metric. *Information Sciences*, 19:81–90, 1979.

[1095] C. A. Kelley. On codes designed via algebraic lifts of graphs. In *Proc. 46th Allerton Conference on Communication, Control, and Computing*, pages 1254–1261, Monticello, Illinois, September 2008.

[1096] C. A. Kelley. Minimum distance and pseudodistance lower bound for generalized LDPC codes. *Internat. J. Inform. Coding Theory*, 1:313–333, 2010.

[1097] C. A. Kelley. Algebraic design and implementation of protograph codes using non-commuting permutation matrices. *IEEE Trans. Commun.*, 61:910–918, 2013.

[1098] C. A. Kelley and D. Sridhara. Eigenvalue bounds on the pseudocodeword weight of expander codes. *Adv. Math. Commun.*, 1:287–307, 2007.

[1099] C. A. Kelley and D. Sridhara. Pseudocodewords of Tanner graphs. *IEEE Trans. Inform. Theory*, 53:4013–4038, 2007.

[1100] C. A. Kelley, D. Sridhara, and J. Rosenthal. Pseudocodeword weights for non-binary LDPC codes. In *Proc. IEEE International Symposium on Information Theory*, pages 1379–1383, Seattle, Washington, July 2006.

[1101] C. A. Kelley, D. Sridhara, and J. Rosenthal. Tree-based construction of LDPC codes having good pseudocodeword weights. *IEEE Trans. Inform. Theory*, 53:287–307, 2007.

[1102] C. A. Kelley, D. Sridhara, J. Xu, and J. Rosenthal. Pseudocodeword weights and stopping sets. In *Proc. IEEE International Symposium on Information Theory*, page 67, Chicago, Illinois, June-July 2004.

[1103] C. A. Kelley and J. Walker. LDPC codes from voltage graphs. In F. R. Kschischang and E.-H. Yang, editors, *Proc. IEEE International Symposium on Information Theory*, pages 792–796, Toronto, Canada, July 2008.

[1104] M. G. Kendall. A new measure of rank correlation. *Biometrika*, 30:81–93, 1938.

[1105] G. Kennedy and V. S. Pless. On designs and formally self-dual codes. *Des. Codes Cryptogr.*, 4:43—-55, 1994.

[1106] A. Kerber. *Applied Finite Group Actions*, volume 19 of *Algorithms and Combinatorics*. Springer-Verlag, Berlin, 2nd edition, 1999.

[1107] A. M. Kerdock. A class of low-rate nonlinear binary codes. *Inform. and Control*, 20:182–187, 1972; ibid, 21:395, 1972.

[1108] A. M. Kermarrec, N. L. Scouarnec, and G. Straub. Repairing multiple failures with coordinated and adaptive regenerating codes. In *Proc. International Symposium on Networking Coding*, Beijing, China, 6 pages, July 2011.

[1109] A. Ketkar, A. Klappenecker, S. Kumar, and P. K. Sarvepalli. Nonbinary stabilizer codes over finite fields. *IEEE Trans. Inform. Theory*, 52:4892–4914, 2006.

[1110] A. Khaleghi, D. Silva, and F. R. Kschischang. Subspace codes. In M. G. Parker, editor, *Cryptography and Coding*, volume 5921 of *Lecture Notes in Comput. Sci.*, pages 1–21. Springer, Berlin, Heidelberg, 2009.

[1111] A. Khina, Y. Kochman, and A. Khisti. Decomposing the MIMO wiretap channel. In *Proc. IEEE International Symposium on Information Theory*, pages 206–210, Honolulu, Hawaii, June-July 2014.

[1112] H. M. Kiah, K. H. Leung, and S. Ling. Cyclic codes over $GR(p^2, m)$ of length p^k. *Finite Fields Appl.*, 14:834–846, 2008.

[1113] H. M. Kiah, G. J. Puleo, and O. Milenkovic. Codes for DNA sequence profiles. *IEEE Trans. Inform. Theory*, 62:3125–3146, 2016.

[1114] M. Kiermaier, S. Kurz, P. Solé, and A. Wassermann. On strongly walk regular graphs, triple sum sets and their codes. In preparation, 2020.

[1115] M. Kiermaier and A. Wassermann. Minimum weights and weight enumerators of $\mathbb{Z}_4$-linear quadratic residue codes. *IEEE Trans. Inform. Theory*, 58:4870–4883, 2012.

[1116] M. Kiermaier and J. Zwanzger. A $\mathbb{Z}_4$-linear code of high minimum Lee distance derived from a hyperoval. *Adv. Math. Commun.*, 5:275–286, 2011.

[1117] M. Kiermaier and J. Zwanzger. New ring-linear codes from dualization in projective Hjelmslev geometries. *Des. Codes Cryptogr.*, 66:39–55, 2013.

[1118] H. K. Kim and D. S. Krotov. The poset metrics that allow binary codes of codimension m to be m, $(m - 1)$, or $(m - 2)$-perfect. *IEEE Trans. Inform. Theory*, 54:5241–5246, 2008.

[1119] J.-L. Kim. New extremal self-dual codes of lengths 36, 38 and 58. *IEEE Trans. Inform. Theory*, 47:1575–1580, 2001.

[1120] J.-L. Kim and N. Lee. Secret sharing schemes based on additive codes over GF(4). *Appl. Algebra Engrg. Comm. Comput.*, 28:79–97, 2017.

[1121] J.-L. Kim, U. Peled, I. Perepelitsa, V. S. Pless, and S. Friedland. Explicit construction of families of LDPC codes with no 4-cycles. *IEEE Trans. Inform. Theory*, 50:2378–2388, 2004.

[1122] T. Kiran and B. S. Rajan. STBC-schemes with non-vanishing determinant for certain number of transmit antennas. *IEEE Trans. Inform. Theory*, 51:2984–2992, 2005.

[1123] T. Kiran and B. S. Rajan. Distributed space-time codes with reduced decoding complexity. In *Proc. IEEE International Symposium on Information Theory*, pages 542–546, Seattle, Washington, July 2006.

[1124] C. Kirfel and R. Pellikaan. The minimum distance of codes in an array coming from telescopic semigroups. *IEEE Trans. Inform. Theory*, 41:1720–1732, 1995.

[1125] A. Y. Kitaev. Fault-tolerant quantum computation by anyons. *Ann. Physics*, 303:2–30, 2003.

[1126] B. Kitchens. Symbolic dynamics and convolutional codes. In *Codes, Systems, and Graphical Models (Minneapolis, MN, 1999)*, volume 123 of *IMA Vol. Math. Appl.*, pages 347–360. Springer, New York, 2001.

[1127] E. Kleinfeld. Finite Hjelmslev planes. *Illinois J. Math.*, 3:403–407, 1959.

[1128] T. Kløve. Error Correction Codes for the Asymmetric Channel. Department of Informatics, University of Bergen, Norway, 1983, bibliography updated 1995.

[1129] W. Knapp and P. Schmid. Codes with prescribed permutation group. *J. Algebra*, 67:415–435, 1980.

[1130] E. Knill and R. Laflamme. Theory of quantum error-correcting codes. *Physical Review A*, 55:900–911, February 1997.

[1131] E. Knill, R. Laflamme, and W. H. Zurek. Resilient quantum computation. *Science*, 279(5349):342–345, 1998.

[1132] D. E. Knuth. *The Art of Computer Programming. Volume 4A: Combinatorial Algorithms, Part 1*. Addison-Wesley Professional, 2011.

[1133] D. E. Knuth. *The Art of Computer Programming. Volume 4. Fascicle 5: Mathematical Preliminaries Redux; Introduction to Backtracking; Dancing Links*. Addison-Wesley, 2015.

[1134] D. E. Knuth. *The Art of Computer Programming. Volume 4: Satisfiability*. Addison-Wesley Professional, 2015.

[1135] A. Kohnert. Update on the extension of good linear codes. In *Combinatorics 2006*, volume 26 of *Electron. Notes Discrete Math.*, pages 81–85. Elsevier Sci. B. V., Amsterdam, 2006.

[1136] A. Kohnert. Constructing two-weight codes with prescribed groups of automorphisms. *Discrete Appl. Math.*, 155:1451–1457, 2007.

[1137] A. Kohnert. Construction of linear codes having prescribed primal-dual minimum distance with applications in cryptography. *Albanian J. Math.*, 2:221–227, 2008.

[1138] A. Kohnert. (l, s)-extension of linear codes. *Discrete Math.*, 309:412–417, 2009.

[1139] A. Kohnert and S. Kurz. Construction of large constant dimension codes with a prescribed minimum distance. In J. Calmet, W. Geiselmann, and J. Müller-Quade, editors, *Mathematical Methods in Computer Science: Essays in Memory of Thomas Beth*, volume 5393 of *Lecture Notes in Comput. Sci.*, pages 31–42. Springer, Berlin, Heidelberg, 2008.

[1140] A. Kohnert and A. Wassermann. Construction of binary and ternary self-orthogonal linear codes. *Discrete Appl. Math.*, 157:2118–2123, 2009.

[1141] A. Kohnert and J. Zwanzger. New linear codes with prescribed group of automorphisms found by heuristic search. *Adv. Math. Commun.*, 3:157–166, 2009.

[1142] J. I. Kokkala, D. S. Krotov, and P. R. J. Östergård. On the classification of MDS codes. *IEEE Trans. Inform. Theory*, 61:6485–6492, 2015.

[1143] J. I. Kokkala and P. R. J. Östergård. Further results on the classification of MDS codes. *Adv. Math. Commun.*, 10:489–498, 2016.

[1144] D. König. Graphok és matrixok. *Mat. Fiz. Lapok*, 38:115–119, 1931.

[1145] S. Kopparty. List-decoding multiplicity codes. In *Electronic Colloquium on Computational Complexity (ECCC)*, volume 19, Report No. 44, 26 pages, 2012.

[1146] S. Kopparty, V. F. Lev, S. Saraf, and M. Sudan. Kakeya-type sets in finite vector spaces. *J. Algebraic Combin.*, 34:337–355, 2011.

[1147] S. Kopparty, N. Ron-Zewi, S. Saraf, and M. Wootters. Improved decoding of folded Reed-Solomon and multiplicity codes. In M. Thorup, editor, *Proc. 59th IEEE Symposium on Foundations of Computer Science*, pages 212–223, Paris, France, October 2018.

[1148] S. B. Korada. *Polar Codes for Channel and Source Coding*. PhD thesis, École Polytechnique Fédérale de Lausanne, 2009.

[1149] S. B. Korada, E. Şaşoğlu, and R. L. Urbanke. Polar codes: characterization of exponent, bounds, and constructions. arXiv:0901.0536 [cs.IT], January 2009.

[1150] S. B. Korada, E. Şaşoğlu, and R. L. Urbanke. Polar codes: characterization of exponent, bounds, and constructions. *IEEE Trans. Inform. Theory*, 56:6253–6264, 2010.

[1151] S. B. Korada, A. Montanari, E. Telatar, and R. L. Urbanke. An empirical scaling law for polar codes. In *Proc. IEEE International Symposium on Information Theory*, pages 884–888, Austin, Texas, June 2010.

[1152] S. B. Korada and R. L. Urbanke. Polar codes are optimal for lossy source coding. *IEEE Trans. Inform. Theory*, 56:1751–1768, 2010.

[1153] W. Kositwattanarerk and F. Oggier. Connections between Construction D and related constructions of lattices. *Des. Codes Cryptogr.*, 73:441–455, 2014.

[1154] W. Kositwattanarerk, S. S. Ong, and F. Oggier. Construction A of lattices over number fields and block fading (wiretap) coding. *IEEE Trans. Inform. Theory*, 61:2273–2282, 2015.

[1155] R. Kötter. A unified description of an error locating procedure for linear codes. In D. Yorgov, editor, *Proc. 3rd International Workshop on Algebraic and Combinatorial Coding Theory*, pages 113–117, Voneshta Voda, Bulgaria, June 1992. Hermes.

[1156] R. Kötter and F. R. Kschischang. Coding for errors and erasures in random network coding. *IEEE Trans. Inform. Theory*, 54:3579–3591, 2008.

[1157] R. Kötter and M. Médard. An algebraic approach to network coding. *IEEE/ACM Trans. Networking*, 11:782–795, 2003.

[1158] R. Kötter and P. O. Vontobel. Graph covers and iterative decoding of finite-length codes. In *Proc. 3rd International Symposium on Turbo Codes and Related Topics*, pages 75–82, Brest, France, September 2003.

[1159] Y. Kou, Lin. S., and M. P. C. Fossorier. Low-density parity-check codes based on finite geometries: a rediscovery and more. *IEEE Trans. Inform. Theory*, 47:2711–2736, 2001.

[1160] I. Kra and S. R. Simanca. On circulant matrices. *Notices Amer. Math. Soc.*, 59(3):368–377, 2012.

[1161] E. S. Kramer and D. M. Mesner. *t*-designs on hypergraphs. *Discrete Math.*, 15:263–296, 1976.

[1162] D. L. Kreher and D. R. Stinson. *Combinatorial Algorithms: Generation, Enumeration, and Search*. CRC Press, Boca Raton, Florida, 1999.

[1163] E. I. Krengel and P. V. Ivanov. Two constructions of binary sequences with optimal autocorrelation magnitude. *Electron. Lett.*, 52:1457–1459, 2016.

[1164] M. N. Krishnan and P. V. Kumar. On MBR codes with replication. In *Proc. IEEE International Symposium on Information Theory*, pages 71–75, Barcelona, Spain, July 2016.

[1165] M. N. Krishnan, M. Vajha, V. Ramkumar, B. Sasidharan, S. B. Balaji, and P. V. Kumar. Erasure coding for big data. *Advanced Computing and Communications (ACCS)*, 3(1):6 pages, 2019.

[1166] D. S. Krotov, P. R. J. Östergård, and O. Pottonen. On optimal binary one-error-correcting codes of lengths $2^m - 4$ and $2^m - 3$. *IEEE Trans. Inform. Theory*, 57:6771–6779, 2011.

[1167] E. A. Krouk. Decoding complexity bound for linear block codes (in Russian). *Problemy Peredači Informacii*, 25(3):103–107, 1989 (English translation in *Probl. Inform. Transm.*, 25(3):251–254, 1989).

[1168] V. Krčadinac. Steiner 2-designs $S(2, 4, 28)$ with nontrivial automorphisms. *Glasnik Matematički*, 37:259–268, 2002.

[1169] F. R. Kschischang. An introduction to network coding. In M. Médard and A. Sprintson, editors, *Network Coding: Fundamentals and Applications*, chapter 1. Academic Press, Waltham, Massachusetts, 2012.

[1170] F. R. Kschischang, B. Frey, and H.-A. Loeliger. Factor graphs and the sum-product algorithm. *IEEE Trans. Inform. Theory*, 47:498–519, 2001.

[1171] A. Kshevetskiy and È. M. Gabidulin. The new construction of rank codes. In *Proc. IEEE International Symposium on Information Theory*, pages 2105–2108, Adelaide, Australia, September 2005.

[1172] S. Kudekar, T. J. Richardson, and R. L. Urbanke. Threshold saturation via spatial coupling: why convolutional LDPC ensembles perform so well over the BEC. *IEEE Trans. Inform. Theory*, 57:803–834, 2011.

[1173] S. Kudekar, T. J. Richardson, and R. L. Urbanke. Spatially coupled ensembles universally achieve capacity under belief propagation. *IEEE Trans. Inform. Theory*, 59:7761–7783, 2013.

[1174] K. R. Kumar and G. Caire. Construction of structured LaST codes. In *Proc. IEEE International Symposium on Information Theory*, pages 2834–2838, Seattle, Washington, July 2006.

[1175] N. Kumar and A. K. Singh. DNA computing over the ring $\mathbb{Z}_4[v]/\langle v^2 - v\rangle$. *Internat. J. Biomath.*, 11(7):1850090, 18, 2018.

[1176] P. V. Kumar, R. A. Scholtz, and L. R. Welch. Generalized bent functions and their properties. *J. Combin. Theory Ser. A*, 40:90–107, 1985.

[1177] S.-Y. Kung, B. C. Lévy, M. Morf, and T. Kailath. New results in 2-D systems theory, part II: 2-D state-space models—realization and the notions of controllability, observability, and minimality. *Proc. of IEEE*, 65:945–961, 1977.

[1178] J. Kurihara, R. Matsumoto, and T. Uyematsu. Relative generalized rank weight of linear codes and its applications to network coding. *IEEE Trans. Inform. Theory*, 61:3912–3936, 2015.

[1179] S. Kurz. Improved upper bounds for partial spreads. *Des. Codes Cryptogr.*, 85:97–106, 2017.

[1180] S. Kurz. LinCode – computer classification of linear codes. arXiv:1912.09357 [math.CO], December 2019.

[1181] D. K. Kythe and P. K. Kythe. *Algebraic and Stochastic Coding Theory*. CRC Press, Boca Raton, Florida, 2012.

[1182] G. B. Kyung and C. C. Wang. Finding the exhaustive list of small fully absorbing sets and designing the corresponding low error-floor decoder. *IEEE Trans. Commun.*, 60:1487–1498, 2012.

[1183] G. G. La Guardia. Asymmetric quantum codes: new codes from old. *Quantum Inform. Process.*, 12:2771–2790, 2013.

[1184] G. G. La Guardia. Erratum to: Asymmetric quantum codes: new codes from old. *Quantum Inform. Process.*, 12:2791, 2013.

[1185] H. Laakso. Nonexistence of nontrivial perfect codes in the case $q = p_1^a p_2^b p_3^c$, $e \geq 3$. *Ann. Univ. Turku. Ser. A I*, 177, 43 pages, 1979. Dissertation, University of Turku, Turku, 1979.

[1186] A. Laaksonen and P. R. J. Östergård. Constructing error-correcting binary codes using transitive permutation groups. *Discrete Appl. Math.*, 233:65–70, 2017.

[1187] R. Laflamme, C. Miquel, J. P. Paz, and W. H. Zurek. Perfect quantum error correcting code. *Physical Review Lett.*, 77:198–201, July 1996.

[1188] C. Lai and A. Ashikhmin. Linear programming bounds for entanglement-assisted quantum error-correcting codes by split weight enumerators. *IEEE Trans. Inform. Theory*, 64:622–639, 2018.

[1189] K. Lally. Quasicyclic codes of index l over $\mathbb{F}_q$ viewed as $\mathbb{F}_q[x]$-submodules of $\mathbb{F}_{q^l}[x]/\langle x^m - 1\rangle$. In M. Fossorier, T. Høholdt, and A. Poli, editors, *Applied Algebra, Algebraic Algorithms and Error-Correcting Codes*, volume 2643 of *Lecture Notes in Comput. Sci.*, pages 244–253. Springer, Berlin, Heidleberg, 2003.

[1190] K. Lally. Algebraic lower bounds on the free distance of convolutional codes. *IEEE Trans. Inform. Theory*, 52:2101–2110, 2006.

[1191] K. Lally and P. Fitzpatrick. Algebraic structure of quasi-cyclic codes. *Discrete Appl. Math.*, 111:157–175, 2001.

[1192] C. W. H. Lam and L. Thiel. Backtrack search with isomorph rejection and consistency check. *J. Symbolic Comput.*, 7:473–485, 1989.

[1193] T. Y. Lam. A general theory of Vandermonde matrices. *Exposition. Math.*, 4:193–215, 1986.

[1194] T. Y. Lam and A. Leroy. Vandermonde and Wronskian matrices over division rings. *J. Algebra*, 119:308–336, 1988.

[1195] T. Y. Lam and A. Leroy. Wedderburn polynomials over division rings. I. *J. Pure Appl. Algebra*, 186:43–76, 2004.

[1196] T. Y. Lam, A. Leroy, and A. Ozturk. Wedderburn polynomials over division rings. II. In S. K. Jain and S. Parvathi, editors, *Noncommutative Rings, Group Rings, Diagram Algebras and Their Applications*, volume 456 of *Contemp. Math.*, pages 73–98. Amer. Math. Soc., Providence, Rhode Island, 2008.

[1197] J. Lambek. *Lectures on Rings and Modules*. Blaisdell Publishing Co. Ginn and Co., Waltham, Mass.-Toronto, Ont.-London, 1966.

[1198] I. N. Landjev. The geometric approach to linear codes. In *Finite Geometries*, volume 3 of *Dev. Math.*, pages 247–256. Kluwer Acad. Publ., Dordrecht, Netherlands, 2001.

[1199] I. N. Landjev and P. Vandendriessche. A study of (xv_t, xv_{t-1})-minihypers in $\mathrm{PG}(t, q)$. *J. Combin. Theory Ser. A*, 119:1123–1131, 2012.

[1200] P. Landrock. *Finite Group Algebras and Their Modules*, volume 84 of *London Math. Soc. Lecture Note Ser.* Cambridge University Press, Cambridge, UK, 1983.

[1201] P. Landrock and O. Manz. Classical codes as ideals in group algebras. *Des. Codes Cryptogr.*, 2:273–285, 1992.

[1202] J. N. Laneman and G. W. Wornell. Distributed space-time-coded protocols for exploiting cooperative diversity in wireless network. *IEEE Trans. Inform. Theory*, 49:2415–2425, 2003.

[1203] S. Lang. *Algebra*, volume 211 of *Graduate Texts in Mathematics*. Springer-Verlag, New York, revised 3rd edition, 2002.

[1204] J. B. Lasserre. An explicit equivalent positive semidefinite program for nonlinear 0-1 programs. *SIAM J. Optim.*, 12:756–769, 2002.

[1205] M. Laurent. A comparison of the Sherali-Adams, Lovász-Schrijver, and Lasserre relaxations for 0-1 programming. *Math. Oper. Res.*, 28:470–496, 2003.

[1206] M. Laurent. Strengthened semidefinite programming bounds for codes. *Math. Program. Ser. B*, 109:239–261, 2007.

[1207] M. Laurent and F. Vallentin. *A Course on Semidefinite Optimization*. Cambridge University Press, Cambridge, UK, in preparation.

[1208] M. Lavrauw, L. Storme, and G. Van de Voorde. Linear codes from projective spaces. In A. A. Bruen and D. L. Wehlau, editors, *Error-Correcting Codes, Finite Geometries and Cryptography*, volume 523 of *Contemp. Math.*, pages 185–202. Amer. Math. Soc., Providence, Rhode Island, 2010.

[1209] F. Lazebnik and V. A. Ustimenko. Explicit construction of graphs with arbitrary large girth and of large size. *Discrete Appl. Math.*, 60:275–284, 1997.

[1210] D. Le Brigand and J.-J. Riesler. Algorithme de Brill–Noether et codes de Goppa. *Bull. Soc. Math. France*, 116:231–253, 1988.

[1211] C. Y. Lee. Some properties of nonbinary error-correcting codes. *IRE Trans. Inform. Theory*, 4:77–82, 1958.

[1212] P. J. Lee and E. F. Brickell. An observation on the security of McEliece's public-key cryptosystem. In J. Stoer, N. Wirth, and C. G. Günther, editors, *Advances in Cryptology – EUROCRYPT '88*, volume 330 of *Lecture Notes in Comput. Sci.*, pages 275–280. Springer, Berlin, Heidelberg, 1988.

[1213] W. Lee, Y.-S. Kim, Y.-W. Lee, and J.-S. No. Post quantum signature scheme based on modified Reed-Muller code pqsigRM. First round submission to the NIST post-quantum cryptography call: `https://csrc.nist.gov/CSRC/media/Projects/Post-Quantum-Cryptography/documents/round-1/submissions/pqsigRM.zip`, November 2017.

[1214] J. Lehmann and O. Heden. Some necessary conditions for vector space partitions. *Discrete Math.*, 312:351–361, 2012.

[1215] H. W. Lenstra, Jr. Two theorems on perfect codes. *Discrete Math.*, 3:125–132, 1972.

[1216] M. Lentmaier, D. Truhachev, and K. Sh. Zigangirov. To the Theory of Low-Density Convolutional Codes II. *IEEE Trans. Inform. Theory*, 37:288–306, 2001.

[1217] J. S. Leon. Computing automorphism groups of error-correcting codes. *IEEE Trans. Inform. Theory*, 28:496–511, 1982.

[1218] J. S. Leon. A probabilistic algorithm for computing minimum weights of large error-correcting codes. *IEEE Trans. Inform. Theory*, 34:1354–1359, 1988.

[1219] J. S. Leon. Partitions, refinements, and permutation group computation. In L. Finkelstein and W. M. Kantor, editors, *Groups and Computation, II (New Brunswick, NJ, 1995)*, volume 28 of *DIMACS Ser. Discrete Math. Theoret. Comput. Sci.*, pages 123–158. Amer. Math. Soc., Providence, Rhode Island, 1997.

[1220] J. S. Leon, J. M. Masley, and V. S. Pless. Duadic codes. *IEEE Trans. Inform. Theory*, 30:709–714, 1984.

[1221] A. Leon-Garcia and I. Widjaja. *Communication Networks: Fundamental Concepts and Key Architectures*. McGraw-Hill, New York, 2004.

[1222] M. Lequesne and J.-P. Tillich. Attack on the EDON-$\mathcal{K}$ key encapsulation mechanism. In *Proc. IEEE International Symposium on Information Theory*, pages 981–985, Vail, Colorado, June 2018.

[1223] A. Leroy. Noncommutative polynomial maps. *J. Algebra Appl.*, 11(4):1250076, 16, 2012.

[1224] S. Leung-Yan-Cheong and M. Hellman. Concerning a bound on undetected error probability. *IEEE Trans. Inform. Theory*, 22:235–237, 1976.

[1225] V. I. Levenshtein. Binary codes capable of correcting deletions, insertions, and reversals. *Soviet Physics Dokl.*, 10:707–710, 1965.

[1226] V. I. Levenshtein. On bounds for packings in n-dimensional Euclidean space (in Russian). *Soviet Math. Doklady*, 20:417–421, 1979.

[1227] V. I. Levenshtein. Bounds for packings in metric spaces and certain applications (in Russian). *Probl. Kibernet.*, 40:44–110, 1983.

[1228] V. I. Levenshtein. Designs as maximum codes in polynomial metric spaces. *Acta Appl. Math.*, 29:1–82, 1992.

[1229] V. I. Levenshtein. Krawtchouk polynomials and universal bounds for codes and designs in Hamming spaces. *IEEE Trans. Inform. Theory*, 41:1303–1321, 1995.

[1230] V. I. Levenshtein. Universal Bounds for Codes and Designs. In V. S. Pless and W. C. Huffman, editors, *Handbook of Coding Theory, Vol. I, II*, chapter 6, pages 499–648. North-Holland, Amsterdam, 1998.

[1231] V. I. Levenshtein. Equivalence of Delsarte's bounds for codes and designs in symmetric association schemes, and some applications. *Discrete Math.*, 197/198:515–536, 1999.

[1232] V. I. Levenshtein. New lower bounds on aperiodic crosscorrelation of binary codes. *IEEE Trans. Inform. Theory*, 45:284–288, 1999.

[1233] V. I. Levenshtein. Bounds for deletion/insertion correcting codes. In *Proc. IEEE International Symposium on Information Theory*, page 370, Lausanne, Switzerland, June-July 2002.

[1234] B. C. Lévy. *2-D Polynomial and Rational Matrices, and their Applications for the Modeling of 2-D Dynamical Systems*. PhD thesis, Standford University, 1981.

[1235] B. Li, H. Shen, and D. Tse. An adaptive successive cancellation list decoder for polar codes with cyclic redundancy check. *IEEE Commun. Letters*, 16:2044–2047, 2012.

[1236] C. Li, C. Ding, and S. Li. LCD cyclic codes over finite fields. *IEEE Trans. Inform. Theory*, 63:4344–4356, 2017.

[1237] C. Li and Q. Yue. The Walsh transform of a class of monomial functions and cyclic codes. *Cryptogr. Commun.*, 7:217–228, 2015.

[1238] C. Li, X. Zeng, T. Helleseth, C. Li, and L. Hu. The properties of a class of linear FSRs and their applications to the construction of nonlinear FSRs. *IEEE Trans. Inform. Theory*, 60:3052–3061, 2014.

[1239] C. Li, X. Zeng, C. Li, and T. Helleseth. A class of de Bruijn sequences. *IEEE Trans. Inform. Theory*, 60:7955–7969, 2014.

[1240] C. Li, X. Zeng, C. Li, T. Helleseth, and M. Li. Construction of de Bruijn sequences from LFSRs with reducible characteristic polynomials. *IEEE Trans. Inform. Theory*, 62:610–624, 2016.

[1241] J. Li and B. Li. Erasure coding for cloud storage systems: a survey. *Tsinghua Science and Technology*, 18:259–272, 2013.

[1242] J. Li and X. Tang. A systematic construction of MDS codes with small sub-packetization level and near optimal repair bandwidth. arXiv:1901.08254 [cs.IT], January 2019.

[1243] J. Li, X. Tang, and C. Tian. A generic transformation to enable optimal repair in MDS codes for distributed storage systems. *IEEE Trans. Inform. Theory*, 64:6257–6267, 2018.

[1244] M. Li and D. Lin. De Bruijn sequences, adjacency graphs, and cyclotomy. *IEEE Trans. Inform. Theory*, 64:2941–2952, 2018.

[1244a] N. Li and S. Mesnager. Recent results and problems on constructions of linear codes from cryptographic functions. *Cryptogr. Commun.*, 12:965–986, 2020.

[1245] S. Li. The minimum distance of some narrow-sense primitive BCH codes. *SIAM J. Discrete Math.*, 31:2530–2569, 2017.

[1246] S. Li, C. Ding, M. Xiong, and G. Ge. Narrow-sense BCH codes over GF(q) with length $n = \frac{q^m-1}{q-1}$. *IEEE Trans. Inform. Theory*, 63:7219–7236, 2017.

[1247] S. Li, C. Li, C. Ding, and H. Liu. Two families of LCD BCH codes. *IEEE Trans. Inform. Theory*, 63:5699–5717, 2017.

[1248] S.-Y. R. Li, R. W. Yeung, and N. Cai. Linear network coding. *IEEE Trans. Inform. Theory*, 49:371–381, 2003.

[1249] X.-B. Liang. Orthogonal designs with maximal rates. *IEEE Trans. Inform. Theory*, 49:2468–2503, 2003.

[1250] X.-B. Liang and X.-G. Xia. Unitary signal constellations for differential space-time modulation with two transmit antennas: parametric codes, optimal designs, and bounds. *IEEE Trans. Inform. Theory*, 48:2291–2322, 2002.

[1251] H. Liao and X.-G. Xia. Some designs of full rate space-time codes with nonvanishing determinant. *IEEE Trans. Inform. Theory*, 53:2898–2908, 2007.

[1252] D. A. Lidar and T. A. Brun, editors. *Quantum Error Correction*. Cambridge University Press, Cambridge, UK, 2013.

[1253] R. Lidl, G. L. Mullen, and G. Turnwald. *Dickson Polynomials*, volume 65 of *Pitman Monographs and Surveys in Pure and Applied Mathematics*. Longman Scientific & Technical, Harlow; copublished in the United States with John Wiley & Sons, Inc., New York, 1993.

[1254] R. Lidl and H. Niederreiter. *Finite Fields*, volume 20 of *Encyclopedia of Mathematics and its Applications*. Cambridge University Press, Cambridge, UK, 2nd edition, 1997.

[1255] J. Lieb. Complete MDP convolutional codes. *J. Algebra Appl.*, 18(6):1950105, 13, 2019.

[1256] M. W. Liebeck. The affine permutation groups of rank three. *Proc. London Math. Soc. (3)*, 54:477–516, 1987.

[1257] D. Liebhold, G. Nebe, and M. Á. Vázquez-Castro. Generalizing subspace codes to flag codes using group actions. In M. Greferath, M. Pavčević, N. Silberstein, and M. Á. Vázquez-Castro, editors, *Network Coding and Subspace Designs. Signals and Communication Technology*, Signals Commun. Technol., pages 67–89. Springer, Cham, 2018.

[1258] D. J. Lin and B. Bose. On the maximality of the group theoretic single error correcting and all unidirectional error detecting (SEC-AUED) codes. In R. M. Capocelli, editor, *Sequences: Combinatorics, Compression, Security, and Transmission*, pages 506–529. Springer-Verlag, New York, 1990.

[1259] H.-P. Lin, S. Lin, and K. A. S. Abdel-Ghaffar. Linear and nonlinear binary kernels of polar codes of small dimensions with maximum exponents. *IEEE Trans. Inform. Theory*, 61:5253–5270, 2015.

[1260] S. Lin and D. J. Costello, Jr. *Error Control Coding: Fundamentals and Applications*. Prentice-Hall, Upper Saddle River, New Jersey, 2nd edition, 2004.

[1261] S. Lin, Q. Diao, and I. F. Blake. Error floors and finite geometries. In *Proc. 8th International Symposium on Turbo Codes and Iterative Information Processing*, pages 42–46, Bremen, Germany, August 2014.

[1262] S. Lin and E. J. Weldon, Jr. Long BCH codes are bad. *Inform. and Control*, 11:445–451, 1967.

[1263] S.-J. Lin and W. Chung. Novel repair-by-transfer codes and systematic exact-MBR codes with lower complexities and smaller field sizes. *IEEE Trans. Parallel Distrib. Syst.*, 25:3232–3241, 2014.

[1264] D. Lind and B. Marcus. *An Introduction to Symbolic Dynamics and Coding*. Cambridge University Press, Cambridge, UK, 1995.

[1265] B. Lindström. On group and nongroup perfect codes in q symbols. *Math. Scand.*, 25:149–158, 1969.

[1266] C. Ling, L. Luzzi, J.-C. Belfiore, and D. Stehlé. Semantically secure lattice codes for the Gaussian wiretap channel. *IEEE Trans. Inform. Theory*, 60:6399–6416, 2014.

[1267] S. Ling, J. Luo, and C. Xing. Generalization of Steane's enlargement construction of quantum codes and applications. *IEEE Trans. Inform. Theory*, 56:4080–4084, 2010.

[1268] S. Ling, H. Niederreiter, and P. Solé. On the algebraic structure of quasi-cyclic codes IV: repeated roots. *Des. Codes Cryptogr.*, 38:337–361, 2006.

[1269] S. Ling and P. Solé. On the algebraic structure of quasi-cyclic codes I: finite fields. *IEEE Trans. Inform. Theory*, 47:2751–2760, 2001.

[1270] S. Ling and P. Solé. Good self-dual quasi-cyclic codes exist. *IEEE Trans. Inform. Theory*, 49:1052–1053, 2003.

[1271] S. Ling and P. Solé. On the algebraic structure of quasi-cyclic codes II: chain rings. *Des. Codes Cryptogr.*, 30:113–130, 2003.

[1272] S. Ling and P. Solé. On the algebraic structure of quasi-cyclic codes III: generator theory. *IEEE Trans. Inform. Theory*, 51:2692–2700, 2005.

[1273] P. Lisoněk and V. Singh. Quantum codes from nearly self-orthogonal quaternary linear codes. *Des. Codes Cryptogr.*, 73:417–424, 2014.

[1274] B. Litjens. Semidefinite bounds for mixed binary/ternary codes. *Discrete Math.*, 341:1740–1748, 2018.

[1275] B. Litjens, S. Polak, and A. Schrijver. Semidefinite bounds for nonbinary codes based on quadruples. *Des. Codes Cryptogr.*, 84:87–100, 2017.

[1276] J. B. Little. Algebraic geometry codes from higher dimensional varieties. In E. Martínez-Moro, C. Munuera, and D. Ruano, editors, *Advances in Algebraic Geometry Codes*, volume 5 of *Ser. Coding Theory Cryptol.*, pages 257–293. World Sci. Publ., Hackensack, New Jersey, 2008.

[1277] H. Liu, C. Ding, and C. Li. Dimensions of three types of BCH codes over GF(q). *Discrete Math.*, 340:1910–1927, 2017.

[1278] H. Liu, Y. Li, L. Ma, and J. Chen. On the smallest absorbing sets of LDPC codes from finite planes. *IEEE Trans. Inform. Theory*, 58:4014–4020, 2012.

[1279] H. Liu and Y. Maouche. Two or few-weight trace codes over $\mathbb{F}_q + u\mathbb{F}_q$. *IEEE Trans. Inform. Theory*, 65:2696–2703, 2019.

[1280] J. Liu, S. Mesnager, and L. Chen. New constructions of optimal locally recoverable codes via good polynomials. *IEEE Trans. Inform. Theory*, 64:889–899, 2018.

[1281] J. Liu, S. Mesnager, and D. Tang. Constructions of optimal locally recoverable codes via Dickson polynomials. *Des. Codes Cryptogr.*, 88:1759–1780, 2020.

[1282] S. Liu, F. Manganiello, and F. R. Kschischang. Construction and decoding of generalized skew-evaluation codes. In *Proc. 14th Canadian Workshop on Information Theory*, pages 9–13, St. John's, Canada, July 2015.

[1283] S. Liu and F. Oggier. An overview of coding for distributed storage systems. In *Network Coding and Subspace Designs*, Signals Commun. Technol., pages 363–383. Springer, Cham, 2018.

[1284] Z. Liu and D. A. Pados. LDPC codes from generalized polygons. *IEEE Trans. Inform. Theory*, 51:3890–3898, 2005.

[1285] P. Loidreau. *Étude et optimisation de cryptosystèmes à clé publique fondés sur la théorie des codes correcteurs.* PhD thesis, École Polytechnique, Paris, 2001.

[1286] B. Long, P. Zhang, and J. Hu. A generalized QS-CDMA system and the design of new spreading codes. *IEEE Trans. Vehicular Tech.*, 47:1268–1275, 1998.

[1287] S. R. López-Permouth, B. R. Parra-Avila, and S. Szabo. Dual generalizations of the concept of cyclicity of codes. *Adv. Math. Commun.*, 3:227–234, 2009.

[1288] L. Lovász. On the Shannon capacity of a graph. *IEEE Trans. Inform. Theory*, 25:1–7, 1979.

[1289] H.-F. Lu and P. V. Kumar. A unified construction of space-time codes with optimal rate-diversity tradeoff. *IEEE Trans. Inform. Theory*, 51:1709–1730, 2005.

[1290] J. Lu, J. Harshan, and F. Oggier. Performance of lattice coset codes on Universal Software Radio Peripherals. *Physical Commun.*, 24:94–102, 2017.

[1291] A. Lubotzky, R. Phillips, and P. Sarnak. Ramanujan graphs. *Combinatorica*, 8:261–277, 1988.

[1292] M. G. Luby. LT codes. In *Proc. 43rd Annual IEEE Symposium on Foundations of Computer Science*, pages 271–280, Vancouver, Canada, November 2002.

[1293] M. G. Luby and M. Mitzenmacher. Verification-based decoding for packet-based low-density parity-check codes. *IEEE Trans. Inform. Theory*, 51:120–127, 2005.

[1294] M. G. Luby, M. Mitzenmacher, and M. A. Shokrollahi. Analysis of random processes via And-Or tree evaluation. In *Proc. 9th Annual ACM-SIAM Symposium on Discrete Algorithms*, pages 364–373, San Francisco, California, January 1998. ACM Press, New York.

[1295] M. G. Luby, M. Mitzenmacher, M. A. Shokrollahi, D. A. Speileman, and V. Stenman. Practical loss-resilient codes. In *Proc. 29th Annual ACM Symposium on the Theory of Computing*, pages 150–159, El Paso, Texas, May 1997. ACM Press, New York.

[1296] M. G. Luby, M. Mitzenmacher, M. A. Shokrollahi, and D. A. Speilman. Improved low-density parity-check codes using irregular graphs. *IEEE Trans. Inform. Theory*, 47:585–598, 2001.

[1297] M. G. Luby, M. Mitzenmacher, M. A. Shokrollahi, and D. A. Spielman. Analysis of low density codes and improved designs using irregular graphs. In *Proc. 30th Annual ACM Symposium on the Theory of Computing*, pages 249–258, Dallas, Texas, May 1998. ACM Press, New York.

[1298] M. G. Luby, M. Mitzenmacher, M. A. Shokrollahi, and D. A. Spielman. Efficient erasure correcting codes. *IEEE Trans. Inform. Theory*, 47:569–584, 2001.

[1299] M. G. Luby, M. Mitzenmacher, M. A. Shokrollahi, and D. A. Spielman. Improved low-density parity-check codes using irregular graphs. *IEEE Trans. Inform. Theory*, 47:585–598, 2001.

[1300] H. D. Lüke and H. D. Schotten. Odd-perfect, almost binary correlation sequences. *IEEE Trans. Aerospace and Electronic Systems*, 31:495–498, 1995.

[1301] H. D. Lüke, H. D. Schotten, and H. Hadinejad-Mahram. Generalised Sidelńikov sequences with optimal autocorrelation properties. *Electron. Lett.*, 36:525–527, 2000.

[1302] H. D. Lüke, H. D. Schotten, and H. Hadinejad-Mahram. Binary and quadriphase sequences with optimal autocorrelation properties: a survey. *IEEE Trans. Inform. Theory*, 49:3271–3282, 2003.

[1303] G. Lunardon, R. Trombetti, and Y. Zhou. Generalized twisted Gabidulin codes. *J. Combin. Theory Ser. A*, 159:79–106, 2018.

[1304] B. Lundell and J. McCullough. A generalized floor bound for the minimum distance of geometric Goppa codes. *J. Pure Appl. Algebra*, 207:155–164, 2006.

[1305] H. Lüneburg. Gray-codes. *Abh. Math. Sem. Univ. Hamburg*, 52:208–227, 1982.

[1306] G. Luo and X. Cao. A construction of linear codes and strongly regular graphs from q-polynomials. *Discrete Math.*, 340:2262–2274, 2017.

[1307] G. Luo and X. Cao. Five classes of optimal two-weight linear codes. *Cryptogr. Commun.*, 10:1119–1135, 2018.

[1308] G. Luo, X. Cao, G. Xu, and S. Xu. A new class of optimal linear codes with flexible parameters. *Discrete Appl. Math.*, 237:126–131, 2018.

[1309] L. Luzzi, G. R.-B. Othman, J.-C. Belfiore, and E. Viterbo. Golden space-time block coded modulation. *IEEE Trans. Inform. Theory*, 55:584–597, 2009.

[1310] V. Lyubashevsky. Lattice signatures without trapdoors. In D. Pointcheval and T. Johansson, editors, *Advances in Cryptology – EUROCRYPT 2012*, volume 7237 of *Lecture Notes in Comput. Sci.*, pages 738–755. Springer, Berlin, Heidelberg, 2012.

[1311] G. Maatouk and M. A. Shokrollahi. Analysis of the second moment of the LT decoder. In *Proc. IEEE International Symposium on Information Theory*, pages 2326–2330, Seoul, Korea, June-July 2009.

[1312] J. E. MacDonald. Design methods for maximum minimum-distance error-correcting codes. *IBM J. Research Develop.*, 4:43–57, 1960.

[1313] R. A. Machado, J. A. Pinheiro, and M. Firer. Characterization of metrics induced by hierarchical posets. *IEEE Trans. Inform. Theory*, 63:3630–3640, 2017.

[1314] D. J. C. MacKay. *Information Theory, Inference, and Learning Algorithms.* Cambridge University Press, Cambridge, UK, 2003.

[1315] D. J. C. MacKay and M. C. Davey. Evaluation of Gallager codes for short block length and high rate applications. In B. Marcus and J. Rosenthal, editors, *Codes, Systems, and Graphical Models (Minneapolis, MN, 1999)*, volume 123 of *IMA Vol. Math. Appl.*, pages 113–130. Springer, New York, 2001.

[1316] D. J. C. MacKay and R. M. Neal. Near Shannon limit performance of low density parity check codes. *IEEE Electronic Letters*, 32:1645–1646, 1996.

[1317] F. J. MacWilliams. Error-correcting codes for multiple-level transmission. *Bell System Tech. J.*, 40:281–308, 1961.

[1318] F. J. MacWilliams. *Combinatorial Problems of Elementary Abelian Groups.* PhD thesis, Harvard University, 1962.

[1319] F. J. MacWilliams. A theorem on the distribution of weights in a systematic code. *Bell System Tech. J.*, 42:79–94, 1963. (Also reprinted in [169] pages 261–265 and [212] pages 241–257).

[1320] F. J. MacWilliams. Codes and ideals in group algebras. In R. C. Bose and T. A. Dowling, editors, *Combinatorial Mathematics and its Applications (Proc. Conf., Univ. North Carolina, Chapel Hill, NC, 1967)*, pages 317–328. Univ. North Carolina Press, Chapel Hill, NC, 1969.

[1321] F. J. MacWilliams. Binary codes which are ideals in the group algebra of an abelian group. *Bell System Tech. J.*, 49:987–1011, 1970.

[1322] F. J. MacWilliams and H. B. Mann. On the p-rank of the design matrix of a difference set. *Inform. and Control*, 12:474–488, 1968.

[1323] F. J. MacWilliams and N. J. A. Sloane. *The Theory of Error-Correcting Codes.* North-Holland Publishing Co., Amsterdam-New York-Oxford, 1977.

[1324] F. J. MacWilliams, N. J. A. Sloane, and J. G. Thompson. Good self-dual codes exist. *Discrete Math.*, 3:153–162, 1972.

[1325] H. Maharaj, G. L. Matthews, and G. Pirsic. Riemann-Roch spaces of the Hermitian function field with applications to algebraic geometry codes and low-discrepancy sequences. *J. Pure Appl. Algebra*, 195:261–280, 2005.

[1326] H. Mahdavifar and A. Vardy. Achieving the secrecy capacity of wiretap channels using polar codes. *IEEE Trans. Inform. Theory*, 57:6428–6443, 2011.

[1327] E. Maneva and M. A. Shokrollahi. New model for rigorous analysis of LT-codes. In *Proc. IEEE International Symposium on Information Theory*, pages 2677–2679, Seattle, Washington, July 2006.

[1328] F. Manganiello, E. Gorla, and J. Rosenthal. Spread codes and spread decoding in network coding. In F. R. Kschischang and E.-H. Yang, editors, *Proc. IEEE International Symposium on Information Theory*, pages 881–885, Toronto, Canada, July 2008.

[1329] B. Marcus. Symbolic dynamics and connections to coding theory, automata theory and system theory. In A. R. Calderbank, editor, *Different Aspects of Coding Theory (San Francisco, CA, 1995)*, volume 50 of *Proc. Sympos. Appl. Math.*, pages 95–108. Amer. Math. Soc., Providence, Rhode Island, 1995.

[1330] G. A. Margulis. Explicit construction of graphs without short cycles and low density codes. *Combinatorica*, 2:71–78, 1982.

[1331] N. Markin and F. Oggier. Iterated space-time code constructions from cyclic algebras. *IEEE Trans. Inform. Theory*, 59:5966–5979, 2013.

[1332] I. Márquez-Corbella, E. Martínez-Moro, and R. Pellikaan. Evaluation of public-key cryptosystems based on algebraic geometry codes. In J. Borges and M. Villanueva, editors, *Proc. 3rd International Castle Meeting on Coding Theory and Applications*, pages 199–204, Cardona Castle in Catalonia, Spain, September 2011.

[1333] M. H. Martin. A problem in arrangements. *Bull. Amer. Math. Soc.*, 40:859–864, 1934.

[1334] W. J. Martin and H. Tanaka. Commutative association schemes. *European J. Combin.*, 30:1497–1525, 2009.

[1335] J. Martinet. *Perfect Lattices in Euclidean Spaces*, volume 327 of *Grundlehren der Mathematischen Wissenschaften [Fundamental Principles of Mathematical Sciences]*. Springer-Verlag, Berlin, 2003.

[1336] C. Martínez, M. Moreto, R. Beivide, and È. M. Gabidulin. A generalization of perfect Lee codes over Gaussian integers. In *Proc. IEEE International Symposium on Information Theory*, pages 1070–1074, Seattle, Washington, July 2006.

[1337] E. Martínez-Moro and I. F. Rúa. Multivariable codes over finite chain rings: serial codes. *SIAM J. Discrete Math.*, 20:947–959, 2006.

[1338] E. Martínez-Moro and I. F. Rúa. On repeated-root multivariable codes over a finite chain ring. *Des. Codes Cryptogr.*, 45:219–227, 2007.

[1339] U. Martínez-Peñas. On the roots and minimum rank distance of skew cyclic codes. *Des. Codes Cryptogr.*, 83:639–660, 2017.

[1340] U. Martínez-Peñas. Skew and linearized Reed-Solomon codes and maximum sum rank distance codes over any division ring. *J. Algebra*, 504:587–612, 2018.

[1341] U. Martínez-Peñas and R. Matsumoto. Relative generalized matrix weights of matrix codes for universal security on wire-tap networks. *IEEE Trans. Inform. Theory*, 64:2529–2549, 2018.

[1342] C. Martínez-Pérez and W. Willems. Self-dual codes and modules for finite groups in characteristic two. *IEEE Trans. Inform. Theory*, 50:1798–1803, 2004.

[1343] M. Martis, J. Bamberg, and S. Morris. An enumeration of certain projective ternary two-weight codes. *J. Combin. Des.*, 24:21–35, 2016.

[1344] T. Maruta. A characterization of some minihypers and its application to linear codes. *Geom. Dedicata*, 74:305–311, 1999.

[1345] A. Maschietti. Difference sets and hyperovals. *Des. Codes Cryptogr.*, 14:89–98, 1998.

[1346] J. L. Massey. Reversible codes. *Inform. and Control*, 7:369–380, 1964.

[1347] J. L. Massey. Linear codes with complementary duals. *Discrete Math.*, 106/107:337–342, 1992.

[1348] J. L. Massey. Minimal codewords and secret sharing. In *Proc. 6th Joint Swedish-Russian Workshop on Information Theory*, pages 276–279, Molle, Sweden, August 1993.

[1349] J. L. Massey. Some applications of coding theory. In P. G. Ferell, editor, *Cryptography, Codes and Ciphers: Cryptography and Coding IV*, pages 33–47. Formara Ltd, Essex, 1995.

[1350] J. L. Massey, D. J. Costello, Jr., and J. Justesen. Polynomial weights and code constructions. *IEEE Trans. Inform. Theory*, 19:101–110, 1973.

[1351] J. L. Massey and M. K. Sain. Codes, automata, and continuous systems: explicit interconnections. *IEEE Trans. Automat. Control*, 12:644–650, 1967.

[1352] É. Mathieu. Mémoire sur l'étude des fonctions de plusieurs quantités, sur la manière de les former et sur les substitutions qui les laissent invariables. *J. de Mathématiques Pures et Appliquées*, 6:241–323, 1861.

[1353] É. Mathieu. Sur la fonction cinq fois transitive de 24 quantités. *J. de Mathématiques Pures et Appliquées*, 18:25–46, 1873.

[1354] R. Matsumoto. Two Gilbert-Varshamov-type existential bounds for asymmetric quantum error-correcting codes. *Quantum Inform. Process.*, 16(12):Paper No. 285, 7 pages, 2017.

[1355] V. C. Mavron, T. P. McDonough, and V. D. Tonchev. On affine designs and Hadamard designs with line spreads. *Discrete Math.*, 308:2742–2750, 2008.

[1356] A. May, A. Meurer, and E. Thomae. Decoding random linear codes in $\widetilde{O}(2^{0.054n})$. In D. H. Lee and X. Wang, editors, *Advances in Cryptology – ASIACRYPT 2011*, volume 7073 of *Lecture Notes in Comput. Sci.*, pages 107–124. Springer, Berlin, Heidelberg, 2011.

[1357] A. May and I. Ozerov. On computing nearest neighbors with applications to decoding of binary linear codes. In E. Oswald and M. Fischlin, editors, *Advances in Cryptology – EUROCRYPT 2015*, volume 9056 of *Lecture Notes in Comput. Sci.*, pages 203–228. Springer, Berlin, Heidelberg, 2015.

[1358] P. Maymounkov. Online codes. *New York University Technical Report TR2002-833*, 2002.

[1359] B. R. McDonald. *Finite Rings with Identity*. Marcel Dekker, Inc., New York, 1974. Pure and Applied Mathematics, Vol. 28.

[1360] R. J. McEliece. Weight congruences of p-ary cyclic codes. *Discrete Math.*, 3:177–192, 1972.

[1361] R. J. McEliece. A public-key system based on algebraic coding theory. DSN Progress Report 44, pages 114–116, Jet Propulsion Laboratory, Pasadena, California, 1978.

[1362] R. J. McEliece. *Finite Fields for Computer Scientists and Engineers*. Kluwer Acad. Pub., Boston, 1987.

[1363] R. J. McEliece. The Algebraic Theory of Convolutional Codes. In V. S. Pless and W. C. Huffman, editors, *Handbook of Coding Theory, Vol. I, II*, chapter 12, pages 1065–1138. North-Holland, Amsterdam, 1998.

[1364] R. J. McEliece, E. R. Rodemich, and H. Rumsey, Jr. The Lovász bound and some generalizations. *J. Combin. Inform. System Sci.*, 3:134–152, 1978.

[1365] R. J. McEliece, E. R. Rodemich, H. Rumsey, Jr., and L. Welch. New upper bounds on the rate of a code via the Delsarte–MacWilliams inequalities. *IEEE Trans. Inform. Theory*, 23:157–166, 1977.

[1366] R. J. McEliece and H. Rumsey, Jr. Euler products, cyclotomy, and coding. *J. Number Theory*, 4:302–311, 1972.

[1367] R. J. McEliece and D. V. Sarwate. On sharing secrets and Reed-Solomon codes. *Commun. ACM*, 24:583–584, 1981.

[1368] G. McGuire, V. D. Tonchev, and H. N. Ward. Characterizing the Hermitian and Ree unitals on 28 points. *Des. Codes Cryptogr.*, 13:57–61, 1998.

[1369] B. D. McKay. Isomorph-free exhaustive generation. *J. Algorithms*, 26:306–324, 1998.

[1370] B. D. McKay and A. Piperno. Practical graph isomorphism, II. *J. Symbolic Comput.*, 60:94–112, 2014.

[1371] I. McLoughlin. A group ring construction of the [48,24,12] type II linear block code. *Des. Codes Cryptogr.*, 63:29–41, 2012.

[1372] I. McLoughlin and T. Hurley. A group ring construction of the extended binary Golay code. *IEEE Trans. Inform. Theory*, 54:4381–4383, 2008.

[1373] M. Mehrabi and M. Ardakani. On minimum distance of locally repairable codes. In *Proc. 15th Canadian Workshop on Information Theory*, Quebec, Canada, 5 pages, June 2017.

[1374] K. Menger. Zur allgemeinen Kurventheorie. *Fund. Math.*, 10:96–115, 1927.

[1375] R. Meshulam. On the maximal rank in a subspace of matrices. *Q. J. Math.*, 36:225–229, 1985.

[1376] S. Mesnager. Characterizations of plateaued and bent functions in characteristic p. In K.-U. Schmidt and A. Winterhof, editors, *Sequences and Their Applications – SETA 2014*, volume 8865 of *Lecture Notes in Comput. Sci.*, pages 72–82. Springer, Berlin, Heidelberg, 2014.

[1377] S. Mesnager. On semi-bent functions and related plateaued functions over the Galois field $\mathbb{F}_{2^n}$. In Ç. K. Koç, editor, *Open Problems in Mathematics and Computational Science*, pages 243–273. Springer, Cham, 2014.

[1378] S. Mesnager. *Bent Functions: Fundamentals and Results*. Springer, Cham, 2016.

[1379] S. Mesnager. Linear codes with few weights from weakly regular bent functions based on a generic construction. *Cryptogr. Commun.*, 9:71–84, 2017.

[1380] S. Mesnager, F. Özbudak, and A. Sınak. A new class of three-weight linear codes from weakly regular plateaued functions. In *Proc. 10th International Workshop on Coding and Cryptography*, Saint Petersburg, Russia, September 2017.

[1381] S. Mesnager, F. Özbudak, and A. Sınak. Linear codes from weakly regular plateaued functions and their secret sharing schemes. *Des. Codes Cryptogr.*, 87:463–480, 2019.

[1382] S. Mesnager and A. Sınak. Several classes of minimal linear codes with few weights from weakly regular plateaued functions. arXiv:1808.03877 [cs.IT], August 2018.

[1383] R. Metz. On a class of unitals. *Geom. Dedicata*, 8:125–126, 1979.

[1384] D. Micciancio and O. Regev. Lattice-based cryptography. In *Post-Quantum Cryptography*, pages 147–191. Springer, Berlin, 2009.

[1385] G. Micheli. Constructions of locally recoverable codes which are optimal. *IEEE Trans. Inform. Theory*, 66:167–175, 2020.

[1386] Microsoft Research Blog: A Better Way to Store Data. `https://www.microsoft.com/en-us/research/blog/better-way-store-data/`. Accessed: 2018-Mar-29.

[1387] O. Milenkovic, E. Soljanin, and P. Whiting. Asymptotic trapping set spectra in regular and irregular LDPC code ensembles. *IEEE Trans. Inform. Theory*, 53:39–55, 2007.

[1388] L. Minder. *Cryptography Based on Error Correcting Codes*. PhD thesis, École Polytechnique Fédérale de Lausanne, 2007.

[1389] L. Minder, M. A. Shokrollahi, M. Watson, M. G. Luby, and T. Stockhammer. RaptorQ Forward Error Correction Scheme for Object Delivery. RFC 6330, August 2011.

[1390] D. Mirandola and G. Zémor. Critical pairs for the product Singleton bound. *IEEE Trans. Inform. Theory*, 61:4928–4937, 2015.

[1391] R. Misoczki and P. S. L. M. Barreto. Compact McEliece keys from Goppa codes. In M. J. Jacobson, Jr., V. Rijmen, and R. Safavi-Naini, editors, *Selected Areas in Cryptography*, volume 5867 of *Lecture Notes in Comput. Sci.*, pages 376–392. Springer, Berlin, Heidelberg, 2009.

[1392] R. Misoczki, J.-P. Tillich, N. Sendrier, and P. S. L. M. Barreto. MDPC-McEliece: new McEliece variants from moderate density parity-check codes. In *Proc. IEEE International Symposium on Information Theory*, pages 2069–2073, Istanbul, Turkey, July 2013.

[1393] T. Mittelholzer and J. Lahtonen. Group codes generated by finite reflection groups. *IEEE Trans. Inform. Theory*, 42:519–528, 1996.

[1394] K. Momihara. Strongly regular Cayley graphs, skew Hadamard difference sets, and rationality of relative Gauss sums. *European J. Combin.*, 34:706–723, 2013.

[1395] Koji Momihara. Certain strongly regular Cayley graphs on $\mathbb{F}_{2^{2(2s+1)}}$ from cyclotomy. *Finite Fields Appl.*, 25:280–292, 2014.

[1396] M. Mondelli, S. H. Hassani, I. Sason, and R. L. Urbanke. Achieving Marton's region for broadcast channels using polar codes. *IEEE Trans. Inform. Theory*, 61:783–800, 2015.

[1397] M. Mondelli, S. H. Hassani, and R. L. Urbanke. Scaling exponent of list decoders with applications to polar codes. In *Proc. IEEE Information Theory Workshop*, Seville, Spain, 5 pages, September 2013.

[1398] M. Mondelli, S. H. Hassani, and R. L. Urbanke. Scaling exponent of list decoders with applications to polar codes. arXiv:1304.5220 [cs.IT], September 2014.

[1399] J. V. S. Morales and H. Tanaka. An Assmus-Mattson theorem for codes over commutative association schemes. *Des. Codes Cryptogr.*, 86:1039–1062, 2018.

[1400] P. Moree. On the divisors of $a^k + b^k$. *Acta Arith.*, 80(3):197–212, 1997.

[1401] C. Moreno. *Algebraic Curves over Finite Fields*, volume 97 of *Cambridge Tracts in Mathematics*. Cambridge University Press, Cambridge, UK, 1991.

[1402] M. Morf, B. C. Lévy, and S.-Y. Kung. New results in 2-D systems theory, part I: 2-D polynomial matrices, factorization, and coprimeness. *Proc. of IEEE*, 65:861–872, 1977.

[1403] R. Mori and T. Tanaka. Performance and construction of polar codes on symmetric binary-input memoryless channels. In *Proc. IEEE International Symposium on Information Theory*, pages 1496–1500, Seoul, Korea, June-July 2009.

[1404] R. Mori and T. Tanaka. Performance of polar codes with the construction using density evolution. *IEEE Commun. Letters*, 13:519–521, 2009.

[1405] R. Mori and T. Tanaka. Channel polarization on q-ary discrete memoryless channels by arbitrary kernels. In *Proc. IEEE Information Theory Workshop*, pages 894–898, Austin, Texas, June 2010.

[1406] R. Mori and T. Tanaka. Source and channel polarization over finite fields and Reed-Solomon matrices. *IEEE Trans. Inform. Theory*, 60:2720–2736, 2014.

[1407] L. Mroueh, S. Rouquette, and J.-C. Belfiore. Application of perfect space time codes: PEP bounds and some practical insights. *IEEE Trans. Commun.*, 60:747–755, 2012.

[1408] G. L. Mullen and D. Panario, editors. *Handbook of Finite Fields*. Discrete Mathematics and its Applications. CRC Press, Boca Raton, Florida, 2013.

[1409] D. E. Muller. Application of Boolean algebra to switching circuit design and to error detection. *IEEE Trans. Comput.*, 3:6–12, 1954. (Also reprinted in [169] pages 20–26 and [212] pages 43–49).

[1410] D. Mumford. Varieties defined by quadratic equations. In E. Marchionna, editor, *Questions on Algebraic Varieties (C.I.M.E., III Ciclo, Varenna, 1969)*, pages 29–100. Edizioni Cremonese, Rome, 1970.

[1411] O. R. Musin. The kissing number in four dimensions. *Ann. of Math. (2)*, 168:1–32, 2008.

[1412] D. Napp, C. Perea, and R. Pinto. Input-state-output representations and constructions of finite support 2D convolutional codes. *Adv. Math. Commun.*, 4:533–545, 2010.

[1413] D. Napp and F. Santana. Multi-shot network coding. In M. Greferath, M. Pavčević, N. Silberstein, and M. Á. Vázquez-Castro, editors, *Network Coding and Subspace Designs. Signals and Communication Technology*, Signals Commun. Technol., pages 91–104. Springer, Cham, 2018.

[1414] D. Napp and R. Smarandache. Constructing strongly-MDS convolutional codes with maximum distance profile. *Adv. Math. Commun.*, 10:275–290, 2016.

[1415] E. L. Năstase and P. A. Sissokho. The maximum size of a partial spread in a finite projective space. *J. Combin. Theory Ser. A*, 152:353–362, 2017.

[1416] L. P. Natarajan, Y. Hong, and E. Viterbo. Lattice index coding. *IEEE Trans. Inform. Theory*, 61:6505–6525, 2015.

[1417] L. P. Natarajan and B. S. Rajan. Fast group-decodable STBCs via codes over GF(4). In *Proc. IEEE International Symposium on Information Theory*, pages 1056–1060, Austin, Texas, June 2010.

[1418] L. P. Natarajan and B. S. Rajan. Asymptotically-optimal, fast-decodable, full-diversity STBCs. In *Proc. IEEE International Conference on Communications*, Kyoto, Japan, 6 pages, June 2011.

[1419] L. P. Natarajan and B. S. Rajan. Fast-group-decodable STBCs via codes over GF(4): further results. In *Proc. IEEE International Conference on Communications*, Kyoto, Japan, 6 pages, June 2011.

[1420] I. P. Naydenova. *Error Detection and Correction for Symmetric and Asymmetric Channels*. PhD thesis, University of Bergen, 2007.

[1421] G. Nebe. A generalisation of Turyn's construction of self-dual codes. In *RIMS Workshop Report: Research into Vector Operator Algebras, Finite Groups and Combinatorics*, volume 1756, pages 51–59, Kyoto, Japan, December 2010.

[1422] G. Nebe. An extremal [72, 36, 16] binary code has no automorphism group containing $Z_2 \times Z_4$, Q_8, or Z_{10}. *Finite Fields Appl.*, 18:563–566, 2012.

[1423] G. Nebe, E. Rains, and N. J. A. Sloane. *Self-Dual Codes and Invariant Theory*. Springer-Verlag, Berlin, Heidelberg, 2006.

[1424] G. Nebe and A. Schäfer. A nilpotent non abelian group code. *Algebra Discrete Math.*, 18:268–273, 2014.

[1425] A. A. Nechaev. Trace functions in Galois rings and noise-stable codes. In *Proc. 5th All-Union Symposium on the Theory of Rings, Algebras and Modules*, page 97, 1982.

[1426] A. A. Nechaev. Kerdock's code in cyclic form (in Russian). *Diskret. Mat.*, 1(4):123–139, 1989. (English translation in *Discrete Math. Appl.* 1(4):365–384, 1991).

[1427] A. A. Nechaev. Linear recurrence sequences over commutative rings (in Russian). *Diskret. Mat.*, 3(4):105–127, 1991. (English translation in *Discrete Math. Appl.* 2(6):659–683, 1992).

[1428] A. A. Nechaev and D. A. Mikhaĭlov. A canonical system of generators of a unitary polynomial ideal over a commutative Artinian chain ring (in Russian). *Diskret. Mat.*, 13(4):3–42, 2001. (English translation in *Discrete Math. Appl.* 11(6):545–586, 2001).

[1429] A. Nemirovski. Advances in convex optimization: conic programming. In M. Sanz-Solé, J. Soria, J. L. Varona, and J. Verdera, editors, *Proc. International Congress of Mathematicians (Madrid, 2006)*, pages 413–444. Eur. Math. Soc., Zürich, 2007.

[1430] S.-L. Ng and M. B. Paterson. Functional repair codes: a view from projective geometry. *Des. Codes Cryptogr.*, 87:2701–2722, 2019.

[1431] X. T. Ngo, S. Bahsin, J.-L. Danger, S. Guilley, and Z. Najim. Linear complementary dual code improvement to strengthen encoded circuit against hardware Trojan horses. In *Proc. IEEE International Symposium on Hardware Oriented Security and Trust (HOST)*, pages 82–87, McLean, Virginia, May 2015.

[1432] H. Niederreiter. Knapsack-type cryptosystems and algebraic coding theory. *Problems Control Inform. Theory/Problemy Upravlen. Teor. Inform.*, 15(2):159–166, 1986.

[1433] H. Niederreiter. Point sets and sequences with small discrepancy. *Monatsh. Math.*, 104:273–337, 1987.

[1434] H. Niederreiter. A combinatorial problem for vector spaces over finite fields. *Discrete Math.*, 96:221–228, 1991.

[1435] H. Niederreiter. Orthogonal arrays and other combinatorial aspects in the theory of uniform point distributions in unit cubes. *Discrete Math.*, 106/107:361–367, 1992.

[1436] H. Niederreiter and F. Özbudak. Constructive asymptotic codes with an improvement on the Tsfasman-Vlăduţ-Zink and Xing bounds. In K. Feng, H. Niederreiter, and C. Xing, editors, *Coding, Cryptography and Combinatorics*, volume 23 of *Progr. Comput. Sci. Appl. Logic*, pages 259–275. Birkhäuser Verlag Basel, 2004.

[1437] H. Niederreiter and C. Xing. *Rational Points on Curves over Finite Fields: Theory and Applications*, volume 285 of *London Math. Soc. Lecture Note Ser.* Cambridge University Press, Cambridge, UK, 2001.

[1438] Y. Niho. *Multi-Valued Cross-Correlation Functions Between Two Maximal Linear Recursive Sequences*. PhD thesis, University of Southern California, 1972.

[1439] A. Nijenhuis and H. S. Wilf. *Combinatorial Algorithms*. Academic Press, Inc. [Harcourt Brace Jovanovich, Publishers], New York-London, 2nd edition, 1978. For computers and calculators, Computer Science and Applied Mathematics.

[1440] S. Nishimura and T. Hiramatsu. A generalization of the Lee distance and error correcting codes. *Discrete Appl. Math.*, 156:588–595, 2008.

[1441] NIST. Post-quantum cryptography standardization. https://csrc.nist.gov/Projects/post-quantum-cryptography/Post-Quantum-Cryptography-Standardization, 2017.

[1442] K. Niu and K. Chen. CRC-aided decoding of polar codes. *IEEE Commun. Letters*, 16:1668–1671, 2012.

[1443] K. Niu and K. Chen. Stack decoding of polar codes. *Electron. Lett.*, 48:695–697, 2012.

[1444] J.-S. No, S. W. Golomb, G. Gong, H.-K. Lee, and P. Gaal. Binary pseudorandom sequences of period $2^n - 1$ with ideal autocorrelation. *IEEE Trans. Inform. Theory*, 44:814–817, 1998.

[1445] R. W. Nóbrega and B. F. Uchôa-Filho. Multishot codes for network coding: bounds and a multilevel construction. In *Proc. IEEE International Symposium on Information Theory*, pages 428–432, Seoul, Korea, June-July 2009.

[1446] A. W. Nordstrom and J. P. Robinson. An optimum nonlinear code. *Inform. and Control*, 11:613–616, 1967.

[1447] G. H. Norton and A. Sălăgean. On the Hamming distance of linear codes over a finite chain ring. *IEEE Trans. Inform. Theory*, 46:1060–1067, 2000.

[1448] G. H. Norton and A. Sălăgean. On the structure of linear and cyclic codes over a finite chain ring. *Appl. Algebra Engrg. Comm. Comput.*, 10:489–506, 2000.

[1449] G. H. Norton and A. Sălăgean. Cyclic codes and minimal strong Gröbner bases over a principal ideal ring. *Finite Fields Appl.*, 9:237–249, 2003.

[1450] K. Nyberg. Constructions of bent functions and difference sets. In I. B. Damgård, editor, *Advances in Cryptology – EUROCRYPT '90*, volume 473 of *Lecture Notes in Comput. Sci.*, pages 151–160. Springer, Berlin, Heidelberg, 1991.

[1451] K. Nyberg. Perfect nonlinear S-boxes. In D. W. Davies, editor, *Advances in Cryptology – EUROCRYPT '91*, volume 547 of *Lecture Notes in Comput. Sci.*, pages 378–386. Springer, Berlin, Heidelberg, 1991.

[1452] K. Nyberg. On the construction of highly nonlinear permutations. In R. A. Rueppel, editor, *Advances in Cryptology – EUROCRYPT '92*, volume 658 of *Lecture Notes in Comput. Sci.*, pages 92–98. Springer, Berlin, Heidelberg, 1993.

[1453] K. Nyberg. Differentially uniform mappings for cryptography. In T. Helleseth, editor, *Advances in Cryptology – EUROCRYPT '93*, volume 765 of *Lecture Notes in Comput. Sci.*, pages 55–64. Springer, Berlin, Heidelberg, 1994.

[1454] K. Nyberg and L. R. Knudsen. Provable security against differential cryptanalysis. In E. F. Brickell, editor, *Advances in Cryptology – CRYPTO '92*, volume 740 of *Lecture Notes in Comput. Sci.*, pages 566–574. Springer, Berlin, Heidelberg, 1993.

[1455] F. Oggier. A survey of algebraic unitary codes. In Y. M. Chee, C. Li, S. Ling, H. Wang, and C. Xing, editors, *Coding and Cryptology*, volume 5557 of *Lecture Notes in Comput. Sci.*, pages 171–187. Springer, Berlin, Heidelberg, 2009.

[1456] F. Oggier and B. Hassibi. Algebraic Cayley differential space-time codes. *IEEE Trans. Inform. Theory*, 53:1911–1919, 2007.

[1457] F. Oggier and B. Hassibi. An algebraic coding scheme for wireless relay networks with multiple-antenna nodes. *IEEE Trans. Signal Process.*, 56:2957–2966, 2008.

[1458] F. Oggier, G. Rekaya, J.-C. Belfiore, and E. Viterbo. Perfect space-time block codes. *IEEE Trans. Inform. Theory*, 52:3885–3902, 2006.

[1459] F. Oggier and A. Sboui. On the existence of generalized rank weights. In *Proc. IEEE International Symposium on Information Theory and its Applications*, pages 406–410, Honolulu, Hawaii, October 2012.

[1460] F. Oggier, P. Solé, and J.-C. Belfiore. Codes over matrix rings for space-time coded modulations. *IEEE Trans. Inform. Theory*, 58:734–746, 2012.

[1461] F. Oggier, P. Solé, and J.-C. Belfiore. Lattice codes for the wiretap Gaussian channel: construction and analysis. *IEEE Trans. Inform. Theory*, 62:5690–5708, 2016.

[1462] F. Oggier and E. Viterbo. Algebraic number theory and code design for Rayleigh fading channels. *Found. Trends Commun. and Inform. Theory*, 1:333–415, 2004.

[1463] C. M. O'Keefe and T. Penttila. A new hyperoval in $PG(2, 32)$. *J. Geom.*, 44:117–139, 1992.

[1464] T. Okuyama and Y. Tsushima. On a conjecture of P. Landrock. *J. Algebra*, 104:203–208, 1986.

[1465] O. Olmez and A. Ramamoorthy. Fractional repetition codes with flexible repair from combinatorial designs. *IEEE Trans. Inform. Theory*, 62:1565–1591, 2016.

[1466] O. Ore. Theory of non-commutative polynomials. *Ann. of Math. (2)*, 34:480–508, 1933.

[1467] P. R. J. Östergård. Classifying subspaces of Hamming spaces. *Des. Codes Cryptogr.*, 27:297–305, 2002.

[1468] P. R. J. Östergård. On the size of optimal three-error-correcting binary codes of length 16. *IEEE Trans. Inform. Theory*, 57:6824–6826, 2011.

[1469] P. R. J. Östergård. The sextuply shortened binary Golay code is optimal. *Des. Codes Cryptogr.*, 87:341–347, 2019.

[1470] P. R. J. Östergård and M. K. Kaikkonen. New single-error-correcting codes. *IEEE Trans. Inform. Theory*, 42:1261–1262, 1996.

[1471] P. R. J. Östergård and O. Pottonen. There exists no Steiner system $S(4, 5, 17)$. *J. Combin. Theory Ser. A*, 115:1570–1573, 2008.

[1472] P. R. J. Östergård and O. Pottonen. The perfect binary one-error-correcting codes of length 15: Part I—Classification. *IEEE Trans. Inform. Theory*, 55:4657–4660, 2009.

[1473] P. R. J. Östergård, O. Pottonen, and K. T. Phelps. The perfect binary one-error-correcting codes of length 15: Part II—Properties. *IEEE Trans. Inform. Theory*, 56:2571–2582, 2010.

[1474] M. E. O'Sullivan. Decoding of codes on surfaces. In *Proc. IEEE Information Theory Workshop*, pages 33–34, Killarney, Ireland, June 1998.

[1475] P. Oswald and M. A. Shokrollahi. Capacity-achieving sequences for the erasure channel. *IEEE Trans. Inform. Theory*, 48:3017–3028, 2002.

[1476] K. Otal and F. Özbudak. Cyclic subspace codes via subspace polynomials. *Des. Codes Cryptogr.*, 85:191–204, 2017.

[1477] A. Otmani and J.-P. Tillich. An efficient attack on all concrete KKS proposals. In B.-Y. Yang, editor, *Post-Quantum Cryptography*, volume 7071 of *Lecture Notes in Comput. Sci.*, pages 98–116. Springer, Berlin, Heidelberg, 2011.

[1478] A. Otmani, J.-P. Tillich, and L. Dallot. Cryptanalysis of two McEliece cryptosystems based on quasi-cyclic codes. *Math. Comput. Sci.*, 3:129–140, 2010.

[1479] R. Overbeck. A new structural attack for GPT and variants. In E. Dawson and S. Vaudenay, editors, *Progress in Cryptology – Mycrypt 2005*, volume 3715 of *Lecture Notes in Comput. Sci.*, pages 50–63. Springer, Berlin, Heidelberg, 2005.

[1480] M. Özen and I. Şiap. Linear codes over $\mathbb{F}_q[u]/(u^s)$ with respect to the Rosenbloom-Tsfasman metric. *Des. Codes Cryptogr.*, 38:17–29, 2006.

[1481] M. Özen and V. Şiap. The MacWilliams identity for m-spotty weight enumerators of linear codes over finite fields. *Comput. Math. Appl.*, 61:1000–1004, 2011.

[1482] M. Özen and V. Şiap. The MacWilliams identity for m-spotty Rosenbloom-Tsfasman weight enumerator. *J. Franklin Inst.*, 351:743–750, 2014.

[1483] N. Pace and A. Sonnino. On linear codes admitting large automorphism groups. *Des. Codes Cryptogr.*, 83:115–143, 2017.

[1484] R. E. A. C. Paley. On orthogonal matrices. *J. Mathematics and Physics*, 12:311–320, 1933.

[1485] L. Pamies-Juarez, H. D. L. Hollmann, and F. Oggier. Locally repairable codes with multiple repair alternatives. In *Proc. IEEE International Symposium on Information Theory*, pages 892–896, Istanbul, Turkey, July 2013.

[1486] L. Panek, M. Firer, and M. M. S. Alves. Classification of Niederreiter-Rosenbloom-Tsfasman block codes. *IEEE Trans. Inform. Theory*, 56:5207–5216, 2010.

[1487] L. Panek, M. Firer, H. K. Kim, and J. Y. Hyun. Groups of linear isometries on poset structures. *Discrete Math.*, 308:4116–4123, 2008.

[1488] E. Paolini, G. Liva, B. Matuz, and M. Chiani. Maximum likelihood erasure decoding of LDPC codes: pivoting algorithms and code design. *IEEE Trans. Commun.*, 60:3209–3220, 2012.

[1489] D. S. Papailiopoulos and A. G. Dimakis. Locally repairable codes. *IEEE Trans. Inform. Theory*, 60:5843–5855, 2014.

[1490] Y. H. Park. Modular independence and generator matrices for codes over $\mathbb{Z}_m$. *Des. Codes Cryptogr.*, 50:147–162, 2009.

[1491] F. Parvaresh and A. Vardy. Correcting errors beyond the Guruswami-Sudan radius in polynomial time. In *Proc. 46th IEEE Symposium on Foundations of Computer Science*, pages 285–294, Pittsburgh, Pennsylvania, October 2005.

[1492] D. S. Passman. Observations on group rings. *Comm. Algebra*, 5:1119–1162, 1977.

[1493] N. J. Patterson. The algebraic decoding of Goppa codes. *IEEE Trans. Inform. Theory*, 21:203–207, 1975.

[1494] S. Pawar, N. Noorshams, S. Y. E. Rouayheb, and K. Ramchandran. DRESS codes for the storage cloud: simple randomized constructions. In A. Kuleshov, V. M. Blinovsky, and A. Ephremides, editors, *Proc. IEEE International Symposium on Information Theory*, pages 2338–2342, St. Petersburg, Russia, July-August 2011.

[1495] S. Pawar, S. Y. E. Rouayheb, and K. Ramchandran. Securing dynamic distributed storage systems against eavesdropping and adversarial attacks. *IEEE Trans. Inform. Theory*, 57:6734–6753, 2011.

[1496] S. E. Payne. A new infinite family of generalized quadrangles. *Congr. Numer.*, 49:115–128, 1985.

[1497] C. Peikert. A decade of lattice cryptography. *Found. Trends Commun. and Inform. Theory*, 10:283–424, 2016.

[1498] R. Pellikaan. On decoding linear codes by error correcting pairs. Preprint, Technical University Eindhoven, 1988.

[1499] R. Pellikaan. On a decoding algorithm for codes on maximal curves. *IEEE Trans. Inform. Theory*, 35:1228–1232, 1989.

[1500] R. Pellikaan. On decoding by error location and dependent sets of error positions. *Discrete Math.*, 106/107:369–381, 1992.

[1501] R. Pellikaan. On the efficient decoding of algebraic-geometric codes. In P. Camion, P. Charpin, and S. Harari, editors, *Proc. International Symposium on Information Theory – EUROCODE 1992*, volume 339 of *CISM Courses and Lect.*, pages 231–253. Springer, Vienna, 1993.

[1502] T. Penttila and G. F. Royle. Sets of type (m, n) in the affine and projective planes of order nine. *Des. Codes Cryptogr.*, 6:229–245, 1995.

[1503] U. Pereg and I. Tal. Channel upgradation for non-binary input alphabets and MACs. In *Proc. IEEE International Symposium on Information Theory*, pages 411–415, Honolulu, Hawaii, June-July 2014.

[1504] W. W. Peterson. Encoding and error-correction procedures for the Bose-Chaudhuri codes. *IRE Trans. Inform. Theory*, 6:459–470, 1960.

[1505] W. W. Peterson. *Error-Correcting Codes*. MIT Press, Cambridge, Massachusetts, 1961.

[1506] W. W. Peterson and E. J. Weldon, Jr. *Error-Correcting Codes*. MIT Press, Cambridge, Massachusetts, 2nd edition, 1972.

[1507] K. T. Phelps. A combinatorial construction of perfect codes. *SIAM J. Algebraic Discrete Methods*, 4:398–403, 1983.

[1508] K. T. Phelps. A general product construction for error correcting codes. *SIAM J. Algebraic Discrete Methods*, 5:224–228, 1984.

[1509] K. T. Phelps. A product construction for perfect codes over arbitrary alphabets. *IEEE Trans. Inform. Theory*, 30:769–771, 1984.

[1510] J. A. Pinheiro, R. A. Machado, and M. Firer. Combinatorial metrics: MacWilliams-type identities, isometries and extension property. *Des. Codes Cryptogr.*, 87:327–340, 2019.

[1511] P. Piret. Structure and constructions of cyclic convolutional codes. *IEEE Trans. Information Theory*, 22:147–155, 1976.

[1512] J. S. Plank and M. Blaum. Sector-disk (SD) erasure codes for mixed failure modes in RAID systems. *ACM Trans. on Storage*, 10(1):4:1–4:17, 2014.

[1513] V. S. Pless. Power moment identities on weight distributions in error correcting codes. *Inform. and Control*, 6:147–152, 1963. (Also reprinted in [169] pages 266–267 and [212] pages 257–262).

[1514] V. S. Pless. On the uniqueness of the Golay codes. *J. Combin. Theory*, 5:215–228, 1968.

[1515] V. S. Pless. A classification of self-orthogonal codes over GF(2). *Discrete Math.*, 3:209–246, 1972.

[1516] V. S. Pless. 23 does not divide the order of the group of a $(72, 36, 16)$ doubly even code. *IEEE Trans. Inform. Theory*, 28:113–117, 1982.

[1517] V. S. Pless. Q-codes. *J. Combin. Theory Ser. A*, 43:258–276, 1986.

[1518] V. S. Pless. More on the uniqueness of the Golay code. *Discrete Math.*, 106/107:391–398, 1992.

[1519] V. S. Pless. Duadic codes and generalizations. In P. Camion, P. Charpin, and S. Harari, editors, *Proc. International Symposium on Information Theory – EUROCODE 1992*, volume 339 of *CISM Courses and Lect.*, pages 3–15. Springer, Vienna, 1993.

[1520] V. S. Pless. *Introduction to the Theory of Error-Correcting Codes*. J. Wiley & Sons, New York, 3rd edition, 1998.

[1521] V. S. Pless and W. C. Huffman, editors. *Handbook of Coding Theory, Vol. I, II.* North-Holland, Amsterdam, 1998.

[1522] V. S. Pless and Z. Qian. Cyclic codes and quadratic residue codes over $\mathbb{Z}_4$. *IEEE Trans. Inform. Theory*, 42:1594–1600, 1996.

[1523] V. S. Pless and N. J. A. Sloane. On the classification and enumeration of self-dual codes. *J. Combin. Theory Ser. A*, 18:313–335, 1975.

[1524] V. S. Pless, P. Solé, and Z. Qian. Cyclic self-dual $\mathbb{Z}_4$-codes. *Finite Fields Appl.*, 3:48–69, 1997.

[1525] V. S. Pless and J. G. Thompson. 17 does not divide the order of the group of a $(72, 36, 16)$ doubly even code. *IEEE Trans. Inform. Theory*, 28:537–541, 1982.

[1526] M. Plotkin. *Binary Codes with Specified Minimum Distance*. Master's thesis, Moore School of Electrical Engineering, University of Pennsylvania, 1952.

[1527] M. Plotkin. Binary codes with specified minimum distances. *IRE Trans. Inform. Theory*, 6:445–450, 1960. (Also reprinted in [169] pages 238–243).

[1528] D. Pointcheval and J. Stern. Security proofs for signature schemes. In U. Maurer, editor, *Advances in Cryptology – EUROCRYPT '96*, volume 1070 of *Lecture Notes in Comput. Sci.*, pages 387–398. Springer, Berlin, Heidelberg, 1996.

[1529] S. C. Polak. Semidefinite programming bounds for Lee codes. arXiv:1810.05066 [math.CO], October 2018.

[1530] S. C. Polak. Semidefinite programming bounds for constant weight codes. *IEEE Trans. Inform. Theory*, 65:28–39, 2019.

[1531] A. Poli. Important algebraic calculations for n-variables polynomial codes. *Discrete Math.*, 56:255–263, 1985.

[1532] A. Poli and L. Huguet. *Error Correcting Codes*. Prentice Hall International, Hemel Hempstead; Masson, Paris, 1992; Translated from the 1989 French original by Iain Craig.

[1533] O. Polverino and F. Zullo. Codes arising from incidence matrices of points and hyperplanes in PG(n,q). *J. Combin. Theory Ser. A*, 158:1–11, 2018.

[1534] N. Prakash, G. M. Kamath, V. Lalitha, and P. V. Kumar. Optimal linear codes with a local-error-correction property. In *Proc. IEEE International Symposium on Information Theory*, pages 2776–2780, Cambridge, Massachusetts, July 2012.

[1535] N. Prakash, V. Lalitha, S. B. Balaji, and P. V. Kumar. Codes with locality for two erasures. *IEEE Trans. Inform. Theory*, 65:7771–7789, 2019.

[1536] E. Prange. *Cyclic Error-Correcting Codes in Two Symbols*. TN-57-103, Air Force Cambridge Research Labs, Bedford, Massachusetts, September 1957.

[1537] E. Prange. *Some Cyclic Error-Correcting Codes with Simple Decoding Algorithms*. TN-58-156, Air Force Cambridge Research Labs, Bedford, Massachusetts, April 1958.

[1538] E. Prange. *An Algorithm for Factoring $x^n - 1$ over a Finite Field*. TN-59-175, Air Force Cambridge Research Labs, Bedford, Massachusetts, October 1959.

[1539] E. Prange. *The Use of Coset Equivalence in the Analysis and Decoding of Groups Codes*. TN-59-164, Air Force Cambridge Research Labs, Bedford, Massachusetts, 1959.

[1540] E. Prange. The use of information sets in decoding cyclic codes. *IRE Trans. Inform. Theory*, 8:S 5–S 9, 1962.

[1541] F. P. Preparata. A class of optimum nonlinear double-error-correcting codes. *Inform. and Control*, 13:378–400, 1968 (Also reprinted in [212] pages 366–388).

[1542] J. Preskill. Fault-tolerant quantum computation. In H.-K. Lo, S. Popescu, and T. Spiller, editors, *Introduction to Quantum Computation and Information*, pages 213–269. World Sci. Publ., River Edge, New Jersey, 1998.

[1543] N. Presman. *Methods in Polar Coding*. PhD thesis, Tel Aviv University, 2015.

[1544] N. Presman and S. Litsyn. Recursive descriptions of polar codes. *Adv. Math. Commun.*, 11:1–65, 2017.

[1545] N. Presman, O. Shapira, and S. Litsyn. Mixed-kernels constructions of polar codes. *IEEE J. Selected Areas Commun.*, 34:239–253, 2016.

[1546] N. Presman, O. Shapira, S. Litsyn, T. Etzion, and A. Vardy. Binary polarization kernels from code decompositions. *IEEE Trans. Inform. Theory*, 61:2227–2239, 2015.

[1547] S. Pumplün and T. Unger. Space-time block codes from nonassociative division algebras. *Adv. Math. Commun.*, 5:449–471, 2011.

[1548] M. Putinar. Positive polynomials on compact semi-algebraic sets. *Indiana Univ. Math. J.*, 42:969–984, 1993.

[1549] L.-J. Qu, Y. Tan, and C. Li. On the Walsh spectrum of a family of quadratic APN functions with five terms. *Sci. China Inf. Sci.*, 57(2):028104, 7, 2014.

[1550] J. Quistorff. On Rosenbloom and Tsfasman's generalization of the Hamming space. *Discrete Math.*, 307:2514–2524, 2007.

[1551] B. Qvist. Some remarks concerning curves of the second degree in a finite plane. *Ann. Acad. Sci Fenn.*, 134:1–27, 1952.

[1552] M. Rahman and I. F. Blake. Majority logic decoding using combinatorial designs. *IEEE Trans. Inform. Theory*, 21:585–587, 1975.

[1553] E. M. Rains. Shadow bounds for self-dual codes. *IEEE Trans. Inform. Theory*, 44:134–139, 1998.

[1554] E. M. Rains. Nonbinary quantum codes. *IEEE Trans. Inform. Theory*, 45:1827–1832, 1999.

[1555] E. M. Rains and N. J. A. Sloane. Self-Dual Codes. In V. S. Pless and W. C. Huffman, editors, *Handbook of Coding Theory, Vol. I, II*, chapter 3, pages 177–294. North-Holland, Amsterdam, 1998.

[1556] B. S. Rajan and M. U. Siddiqi. A generalized DFT for abelian codes over Z_m. *IEEE Trans. Inform. Theory*, 40:2082–2090, 1994.

[1557] A. Ralston. A new memoryless algorithm for de Bruijn sequences. *J. Algorithms*, 2:50–62, 1981.

[1558] A. Ralston. De Bruijn sequences—a model example of the interaction of discrete mathematics and computer science. *Math. Mag.*, 55:131–143, 1982.

[1559] H. Randriambololona. Bilinear complexity of algebras and the Chudnovsky-Chudnovsky interpolation method. *J. Complexity*, 28:489–517, 2012.

[1560] H. Randriambololona. (2,1)-separating systems beyond the probabilistic bound. *Israel J. Math.*, 195:171–186, 2013.

[1561] H. Randriambololona. Asymptotically good binary linear codes with asymptotically good self-intersection spans. *IEEE Trans. Inform. Theory*, 59:3038–3045, 2013.

[1562] H. Randriambololona. An upper bound of Singleton type for componentwise products of linear codes. *IEEE Trans. Inform. Theory*, 59:7936–7939, 2013.

[1563] H. Randriambololona. Linear independence of rank 1 matrices and the dimension of ∗-products of codes. In *Proc. IEEE International Symposium on Information Theory*, pages 196–200, Hong Kong, China, June 2015.

[1564] H. Randriambololona. On products and powers of linear codes under componentwise multiplication. In S. Ballet, M. Perret, and A. Zaytsev, editors, *Algorithmic Arithmetic, Geometry, and Coding Theory*, volume 637 of *Contemp. Math.*, pages 3–77. Amer. Math. Soc., Providence, Rhode Island, 2015.

[1565] A. Rao and N. Pinnawala. A family of two-weight irreducible cyclic codes. *IEEE Trans. Inform. Theory*, 56:2568–2570, 2010.

[1566] C. R. Rao. Factorial experiments derivable from combinatorial arrangements of arrays. *Suppl. to J. Roy. Statist. Soc.*, 9:128–139, 1947.

[1567] T. R. N. Rao and A. S. Chawla. Asymmetric error codes for some LSI semiconductor memories. In *Proc. 7th Southeastern Symposium on System Theory*, pages 170–171, Auburn-Tuskegee, Alabama, March 1975.

[1568] K. V. Rashmi, N. B. Shah, D. Gu, H. Kuang, D. Borthakur, and K. Ramchandran. A solution to the network challenges of data recovery in erasure-coded distributed storage systems: a study on the Facebook Warehouse cluster. In *Proc. 5th USENIX Workshop on Hot Topics in Storage and File Systems*, San Jose, California, 5 pages, June 2013.

[1569] K. V. Rashmi, N. B. Shah, and P. V. Kumar. Optimal exact-regenerating codes for distributed storage at the MSR and MBR points via a product-matrix construction. *IEEE Trans. Inform. Theory*, 57:5227–5239, 2011.

[1570] K. V. Rashmi, N. B. Shah, P. V. Kumar, and K. Ramchandran. Explicit construction of optimal exact regenerating codes for distributed storage. In *Proc. 47th Allerton Conference on Communication, Control, and Computing*, pages 1243–1249, Monticello, Illinois, September-October 2009.

[1571] K. V. Rashmi, N. B. Shah, and K. Ramchandran. A piggybacking design framework for read-and download-efficient distributed storage codes. *IEEE Trans. Inform. Theory*, 63:5802–5820, 2017.

[1572] K. V. Rashmi, N. B. Shah, K. Ramchandran, and P. V. Kumar. Information-theoretically secure erasure codes for distributed storage. *IEEE Trans. Inform. Theory*, 64:1621–1646, 2018.

[1573] A. Ravagnani. Generalized weights: an anticode approach. *J. Pure Appl. Algebra*, 220:1946–1962, 2016.

[1574] A. Ravagnani. Rank-metric codes and their duality theory. *Des. Codes Cryptogr.*, 80:197–216, 2016.

[1575] A. Ravagnani. Duality of codes supported on regular lattices, with an application to enumerative combinatorics. *Des. Codes Cryptogr.*, 86:2035–2063, 2018.

[1576] A. S. Rawat. Secrecy capacity of minimum storage regenerating codes. In *Proc. IEEE International Symposium on Information Theory*, pages 1406–1410, Aachen, Germany, June 2017.

[1577] A. S. Rawat, O. O. Koyluoglu, N. Silberstein, and S. Vishwanath. Optimal locally repairable and secure codes for distributed storage systems. *IEEE Trans. Inform. Theory*, 60:212–236, 2014.

[1578] A. S. Rawat, O. O. Koyluoglu, and S. Vishwanath. Progress on high-rate MSR codes: enabling arbitrary number of helper nodes. In *Proc. Information Theory and Applications Workshop*, La Jolla, California, 6 pages, January-February 2016.

[1579] A. S. Rawat, A. Mazumdar, and S. Vishwanath. Cooperative local repair in distributed storage. *EURASIP J. Adv. Signal Process.*, 2015:107, 2015.

[1580] A. S. Rawat, I. Tamo, V. Guruswami, and K. Efremenko. MDS code constructions with small sub-packetization and near-optimal repair bandwidth. *IEEE Trans. Inform. Theory*, 64:6506–6525, 2018.

[1581] I. S. Reed. A class of multiple-error-correcting codes and the decoding scheme. *IRE Trans. Inform. Theory*, 4:38–49, 1954. (Also reprinted in [169] pages 27–38 and [212] pages 50–61).

[1582] I. S. Reed and G. Solomon. Polynomial codes over certain finite fields. *J. Soc. Indust. Appl. Math.*, 8:300–304, 1960. (Also reprinted in [169] pages 70–71 and [212] pages 189–193).

[1583] O. Regev. New lattice based cryptographic constructions. In *Proc. 35th Annual ACM Symposium on the Theory of Computing*, pages 407–416, San Diego, California, June 2003. ACM Press, New York.

[1584] G. Regts. *Upper Bounds for Ternary Constant Weight Codes from Semidefinite Programming and Representation Theory*. Master's thesis, University of Amsterdam, 2009.

[1585] S. Reiger. Codes for the correction of "clustered" errors. *IRE Trans. Inform. Theory*, 6:16–21, 1960.

[1586] O. Reingold, S. Vadhan, and A. Wigderson. Entropy waves, the zig-zag graph product, and new constant degree expanders and extractors. In *Proc. 41st IEEE Symposium on Foundations of Computer Science*, pages 3–13, Redondo Beach, California, November 2000.

[1587] A. Renvall and C. Ding. The access structure of some secret-sharing schemes. In J. Pieprzyk and J. Seberry, editors, *Information Security and Privacy*, volume 1172 of *Lecture Notes in Comput. Sci.*, pages 67–78. Springer, Berlin, Heidelberg, 1996.

[1588] H. F. H. Reuvers. *Some Non-existence Theorems for Perfect Codes over Arbitrary Alphabets*. Technische Hogeschool Eindhoven, Eindhoven, 1977. Written for conferment of a Doctorate in Technical Sciences at the Technische Hogeschool, Eindhoven, 1977.

[1589] T. J. Richardson, M. A. Shokrollahi, and R. L. Urbanke. Design of capacity-approaching irregular low-density parity-check codes. *IEEE Trans. Inform. Theory*, 47:619–637, 2001.

[1590] T. J. Richardson and R. L. Urbanke. The capacity of low-density parity-check codes under message-passing decoding. *IEEE Trans. Inform. Theory*, 47:599–618, 2001.

[1591] T. J. Richardson and R. L. Urbanke. *Modern Coding Theory*. Cambridge University Press, Cambridge, UK, 2008.

[1592] P. Rocha. *Structure and Representation of 2-D Systems*. PhD thesis, University of Groningen, 1990.

[1593] B. G. Rodrigues. A projective two-weight code related to the simple group Co$_1$ of Conway. *Graphs Combin.*, 34:509–521, 2018.

[1594] C. Roos. A new lower bound for the minimum distance of a cyclic code. *IEEE Trans. Inform. Theory*, 29:330–332, 1983.

[1595] M. Y. Rosenbloom and M. A. Tsfasman. Codes for the m-metric (in Russian). *Problemy Peredači Informacii*, 33(1):55–63, 1997. (English translation in *Probl. Inform. Transm.*, 33(1):45–52, 1997).

[1596] J. Rosenthal. Connections between linear systems and convolutional codes. In B. Marcus and J. Rosenthal, editors, *Codes, Systems, and Graphical Models (Minneapolis, MN, 1999)*, volume 123 of *IMA Vol. Math. Appl.*, pages 39–66. Springer, New York, 2001.

[1597] J. Rosenthal, J. M. Schumacher, and E. V. York. On behaviors and convolutional codes. *IEEE Trans. Inform. Theory*, 42:1881–1891, 1996.

[1598] J. Rosenthal and R. Smarandache. Maximum distance separable convolutional codes. *Appl. Algebra Engrg. Comm. Comput.*, 10:15–32, 1999.

[1599] J. Rosenthal and A.-L. Trautmann. A complete characterization of irreducible cyclic orbit codes and their Plücker embedding. *Des. Codes Cryptogr.*, 66:275–289, 2013.

[1600] J. Rosenthal and E. V. York. BCH convolutional codes. *IEEE Trans. Inform. Theory*, 45:1833–1844, 1999.

[1601] R. M. Roth. Maximum-rank array codes and their application to criss-cross error correction. *IEEE Trans. Inform. Theory*, 37:328–336, 1991.

[1602] R. M. Roth. *Introduction to Coding Theory*. Cambridge University Press, Cambridge, UK, 2006.

[1603] R. M. Roth and G. Seroussi. On generator matrices of MDS codes. *IEEE Trans. Inform. Theory*, 31:826–830, 1985.

[1604] R. M. Roth and G. Seroussi. On cyclic MDS codes of length q over GF(q). *IEEE Trans. Inform. Theory*, 32:284–285, 1986.

[1605] R. M. Roth and P. H. Siegel. Lee-metric BCH codes and their application to constrained and partial-response channels. *IEEE Trans. Inform. Theory*, 40:1083–1096, 1994.

[1606] O. S. Rothaus. On "bent" functions. *J. Combinatorial Theory Ser. A*, 20:300–305, 1976.

[1607] S. Y. E. Rouayheb and K. Ramchandran. Fractional repetition codes for repair in distributed storage systems. arXiv:1010.2551 [cs.IT], October 2010.

[1608] G. F. Royle. An orderly algorithm and some applications in finite geometry. *Discrete Math.*, 185:105–115, 1998.

[1609] L. D. Rudolph. A class of majority-logic decodable codes. *IEEE Trans. Inform. Theory*, 13:305–307, 1967.

[1610] SAGE. Delsarte, a.k.a. linear programming (LP), upper bounds. http://doc.sagemath.org/html/en/reference/coding/sage/coding/delsarte_bounds.html. Accessed: 2018-08-31.

[1611] A. Sălăgean. Repeated-root cyclic and negacyclic codes over a finite chain ring. *Discrete Appl. Math.*, 154:413–419, 2006.

[1612] S. Saraf and M. Sudan. An improved lower bound on the size of Kakeya sets over finite fields. *Anal. PDE*, 1:375–379, 2008.

[1613] G. Sarkis, P. Giard, A. Vardy, C. Thibeault, and W. J. Gross. Fast polar decoders: algorithm and implementation. *IEEE J. Selected Areas Commun.*, 32:946–957, 2014.

[1614] G. Sarkis, I. Tal, P. Giard, A. Vardy, C. Thibeault, and W. J. Gross. Flexible and low-complexity encoding and decoding of systematic polar codes. *IEEE Trans. Commun.*, 64:2732–2745, 2016.

[1615] P. K. Sarvepalli, A. Klappenecker, and M Rötteler. Asymmetric quantum codes: constructions, bounds and performance. *Proc. Roy. Soc. London Ser. A*, 465(2105):1645–1672, 2009.

[1616] D. V. Sarwate and M. B. Pursley. Correction to "Crosscorrelation properties of pseudorandom and related sequences" [*Proceedings of the IEEE*, 68:593–619, 1980]. *Proceedings of the IEEE*, 68:1554, 1980.

[1617] D. V. Sarwate and M. B. Pursley. Crosscorrelation properties of pseudorandom and related sequences. *Proceedings of the IEEE*, 68:593–619, 1980.

[1618] B. Sasidharan, G. K. Agarwal, and P. V. Kumar. Codes with hierarchical locality. In *Proc. IEEE International Symposium on Information Theory*, pages 1257–1261, Hong Kong, China, June 2015.

[1619] B. Sasidharan, G. K. Agarwal, and P. V. Kumar. A high-rate MSR code with polynomial sub-packetization level. In *Proc. IEEE International Symposium on Information Theory*, pages 2051–2055, Hong Kong, China, June 2015.

[1620] B. Sasidharan, K. Senthoor, and P. V. Kumar. An improved outer bound on the storage repair-bandwidth tradeoff of exact-repair regenerating codes. In *Proc. IEEE International Symposium on Information Theory*, pages 2430–2434, Honolulu, Hawaii, June-July 2014.

[1621] B. Sasidharan, M. Vajha, and P. V. Kumar. An explicit, coupled-layer construction of a high-rate MSR code with low sub-packetization level, small field size and all-node repair. arXiv:1607.07335 [cs.IT], 2016.

[1622] E. Saşoğlu, E. Telatar, and E. Arıkan. Polarization for arbitrary discrete memoryless channels. In *Proc. IEEE Information Theory Workshop*, pages 144–148, Taormina, Italy, October 2009.

[1623] E. Saşoğlu, E. Telatar, and E. Yeh. Polar codes for the two-user binary-input multiple-access channel. In *Proc. IEEE Information Theory Workshop*, Cairo, Egypt, 5 pages, January 2010.

[1624] M. Sathiamoorthy, M. Asteris, D. S. Papailiopoulos, A. G. Dimakis, R. Vadali, S. Chen, and D. Borthakur. XORing elephants: novel erasure codes for big data. *Proc. VLDB Endowment (PVLDB)*, 6:325–336, 2013.

[1625] J. Sawada, A. Williams, and D. Wong. A surprisingly simple de Bruijn sequence construction. *Discrete Math.*, 339:127–131, 2016.

[1626] A. Sboui. Special numbers of rational points on hypersurfaces in the n-dimensional projective space over a finite field. *Discrete Math.*, 309:5048–5059, 2009.

[1627] B. Schmidt and C. White. All two-weight irreducible cyclic codes? *Finite Fields Appl.*, 8:1–17, 2002.

[1628] W. M. Schmidt. *Diophantine Approximation*, volume 785 of *Lecture Notes in Math.* Springer, Berlin, 1980.

[1629] W. M. Schmidt. *Equations Over Finite Fields: An Elementary Approach.* Kendrick Press, Heber City, Utah, 2nd edition, 2004.

[1630] J. Schönheim. On linear and nonlinear single-error-correcting q-nary perfect codes. *Inform. and Control*, 12:23–26, 1968.

[1631] H. D. Schotten. Optimum complementary sets and quadriphase sequences derived from q-ary m-sequences. In *Proc. IEEE International Symposium on Information Theory*, page 485, Ulm, Germany, June-July 1997.

[1632] A. Schrijver. A comparison of the Delsarte and Lovász bounds. *IEEE Trans. Inform. Theory*, 25:425–429, 1979.

[1633] A. Schrijver. *Theory of Linear and Integer Programming.* Wiley-Interscience Series in Discrete Mathematics. John Wiley & Sons, Ltd., Chichester, UK, 1986.

[1634] A. Schrijver. New code upper bounds from the Terwilliger algebra and semidefinite programming. *IEEE Trans. Inform. Theory*, 51:2859–2866, 2005.

[1635] B. Schumacher. Quantum coding. *Physical Review A*, 51:2738–2747, April 1995.

[1636] M. Schwartz and A. Vardy. On the stopping distance and stopping redundancy of codes. In *Proc. IEEE International Symposium on Information Theory*, pages 975–979, Adelaide, Australia, September 2005.

[1637] G. F. Seelinger, P. A. Sissokho, L. E. Spence, and C. Vanden Eynden. Partitions of $V(n, q)$ into 2- and s-dimensional subspaces. *J. Combin. Des.*, 20:467–482, 2012.

[1638] B. Segre. Curve razionali normali e k-archi negli spazi finiti. *Ann. Mat. Pura Appl. (4)*, 39:357–379, 1955.

[1639] B. Segre. Ovals in a finite projective plane. *Canad. J. Math.*, 7:414–416, 1955.

[1640] B. Segre. Sui k-archi nei piani finiti di caratteristica due. *Rev. Math. Pures Appl.*, 2:289–300, 1957.

[1641] B. Segre. Ovali e curve σ nei piani di Galois di caratteristica due (in Italian). *Atti Accad. Naz. Lincei Rend. Cl. Sci. Fis. Mat. Nat. (8)*, 32:785–790, 1962.

[1642] E. S. Selmer. *Linear Recurrence Relations Over Finite Fields.* Department of Informatics, University of Bergen, 1966.

[1643] N. V. Semakov and V. A. Zinov'ev. Equidistant q-ary codes and resolved balanced incomplete block designs (in Russian). *Problemy Peredači Informacii*, 4(2):3–10, 1968. (English translation in *Probl. Inform. Transm.*, 4(2):1–7, 1968).

[1644] N. V. Semakov, V. A. Zinov'ev, and G. V. Zaitsev. Class of maximal equidistant codes (in Russian). *Problemy Peredači Informacii*, 5(2):84–87, 1969. (English translation in *Probl. Inform. Transm.*, 5(2):65–68, 1969).

[1645] P. Semenov and P. Trifonov. Spectral method for quasi-cyclic code analysis. *IEEE Commun. Letters*, 16:1840–1843, 2012.

[1646] N. Sendrier. On the structure of a randomly permuted concatenated code. In P. Charpin, editor, *EUROCODE 94 – Livre des résumés*, pages 169–173. INRIA, 1994.

[1647] N. Sendrier. Finding the permutation between equivalent linear codes: the support splitting algorithm. *IEEE Trans. Inform. Theory*, 46:1193–1203, 2000.

[1648] N. Sendrier. Linear codes with complementary duals meet the Gilbert-Varshamov bound. *Discrete Math.*, 285:345–347, 2004.

[1649] N. Sendrier. Decoding one out of many. In B.-Y. Yang, editor, *Post-Quantum Cryptography*, volume 7071 of *Lecture Notes in Comput. Sci.*, pages 51–67. Springer, Berlin, Heidelberg, 2011.

[1650] N. Sendrier and D. E. Simos. The hardness of code equivalence over $\mathbb{F}_q$ and its application to code-based cryptography. In P. Gaborit, editor, *Post-Quantum Cryptography*, volume 7932 of *Lecture Notes in Comput. Sci.*, pages 203–216. Springer, Berlin, Heidelberg, 2013.

[1651] N. Sendrier and V. Vasseur. On the decoding failure rate of QC-MDPC bit-flipping decoders. In J. Ding and R. Steinwandt, editors, *Post-Quantum Cryptography*, volume 11505 of *Lecture Notes in Comput. Sci.*, pages 404–416. Springer, Berlin, Heidelberg, 2019.

[1652] K. Senthoor, B. Sasidharan, and P. V. Kumar. Improved layered regenerating codes characterizing the exact-repair storage-repair bandwidth tradeoff for certain parameter sets. In *Proc. IEEE Information Theory Workshop*, pages 224–228, Jerusalem, Israel, April-May 2015.

[1653] J.-P. Serre. Nombres de points des courbes algébriques sur $\mathbb{F}_q$. In *Seminar on Number Theory, 1982–1983 (Talence, 1982/1983)*, pages Exp. No. 22, 8. Univ. Bordeaux I, Talence, 1983.

[1654] J.-P. Serre. Lettre à M. Tsfasman. *Astérisque*, 198-200(11):351–353, 1991/1992.

[1655] B. A. Sethuraman and B. S. Rajan. An algebraic description of orthogonal designs and the uniqueness of the Alamouti code. In *Proc. GLOBECOM '02 – IEEE Global Telecommunications Conference*, volume 2, pages 1088–1093, Taipei, Taiwan, November 2002.

[1656] B. A. Sethuraman, B. S. Rajan, and V. Shashidhar. Full-diversity, high-rate space-time block codes from division algebras. *IEEE Trans. Inform. Theory*, 49:2596–2616, 2003.

[1657] I. R. Shafarevich. *Basic Algebraic Geometry 1: Varieties in Projective Space.* Springer-Verlag, Berlin, 2nd edition, 1994. Translated from the 1988 Russian edition and with notes by Miles Reid.

[1658] N. B. Shah, K. V. Rashmi, P. V. Kumar, and K. Ramchandran. Distributed storage codes with repair-by-transfer and nonachievability of interior points on the storage-bandwidth tradeoff. *IEEE Trans. Inform. Theory*, 58:1837–1852, 2012.

[1659] A. Shamir. How to share a secret. *Commun. ACM*, 22:612–613, 1979.

[1660] K. Shanmugam, D. S. Papailiopoulos, A. G. Dimakis, and G. Caire. A repair framework for scalar MDS codes. *IEEE J. Selected Areas Commun.*, 32:998–1007, 2014.

[1661] C. E. Shannon. A mathematical theory of communication. *Bell System Tech. J.,* 27:379–423 and 623–656, 1948.

[1662] C. E. Shannon. Communication theory of secrecy systems. *Bell System Tech. J.,* 28:656–715, 1949.

[1663] C. E. Shannon. The zero error capacity of a noisy channel. *IRE Trans. Inform. Theory,* 2:8–19, 1956.

[1664] S. Shao, T. Liu, C. Tian, and C. Shen. On the tradeoff region of secure exact-repair regenerating codes. *IEEE Trans. Inform. Theory,* 63:7253–7266, 2017.

[1665] A. Sharma and A. K. Sharma. MacWilliams identities for m-spotty weight enumerators of codes over rings. *Australas. J. Combin.,* 58:67–105, 2014.

[1666] V. Shashidhar, B. S. Rajan, and B. A. Sethuraman. Information-lossless space-time block codes from crossed-product algebras. *IEEE Trans. Inform. Theory,* 52:3913–3935, 2006.

[1667] J. Sheekey. A new family of linear maximum rank distance codes. *Adv. Math. Commun.,* 10:475–488, 2016.

[1668] M. Shi, A. Alahmadi, and P. Solé. *Codes and Rings: Theory and Practice.* Academic Press, 2017.

[1669] M. Shi, Y. Guan, and P. Solé. Two new families of two-weight codes. *IEEE Trans. Inform. Theory,* 63:6240–6246, 2017.

[1670] M. Shi, Y. Guan, C. Wang, and P. Solé. Few-weight codes from trace codes over R_k. *Bull. Aust. Math. Soc.,* 98:167–174, 2018.

[1671] M. Shi, D. Huang, and P. Solé. Optimal ternary cubic two-weight codes. *Chinese J. Electronics,* 27:734–738, 2018.

[1672] M. Shi, Y. Liu, and P. Solé. Optimal two-weight codes from trace codes over $\mathbb{F}_2 + u\mathbb{F}_2$. *IEEE Commun. Letters,* 20:2346–2349, 2016.

[1673] M. Shi, Y. Liu, and P. Solé. Optimal binary codes from trace codes over a non-chain ring. *Discrete Appl. Math.,* 219:176–181, 2017.

[1674] M. Shi, Y. Liu, and P. Solé. Two-weight and three-weight codes from trace codes over $\mathbb{F}_p + u\mathbb{F}_p + v\mathbb{F}_p + uv\mathbb{F}_p$. *Discrete Math.,* 341:350–357, 2018.

[1675] M. Shi, L. Q. Qian, and P. Solé. Few-weight codes from trace codes over a local ring. *Appl. Algebra Engrg. Comm. Comput.,* 29:335–350, 2018.

[1676] M. Shi, R. S. Wu, Y. Liu, and P. Solé. Two and three weight codes over $\mathbb{F}_p + u\mathbb{F}_p$. *Cryptogr. Commun.,* 9:637–646, 2017.

[1677] M. Shi, R. S. Wu, L. Q. Qian, L. Sok, and P. Solé. New classes of p-ary few weight codes. *Bull. Malays. Math. Sci. Soc.,* 42:1393–1412, 2019.

[1678] M. Shi, R. S. Wu, and P. Solé. Asymptotically good additive cyclic codes exist. *IEEE Commun. Letters,* 22:1980–1983, 2018.

[1679] M. Shi, H. Zhu, and P. Solé. Optimal three-weight cubic codes. *Appl. Comput. Math.,* 17:175–184, 2018.

[1680] M. Shi, S. Zhu, and S. Yang. A class of optimal p-ary codes from one-weight codes over $\mathbb{F}_p[u]/(u^m)$. *J. Franklin Inst.*, 350:929–937, 2013.

[1681] K. Shimizu. A parallel LSI architecture for LDPC decoder improving message-passing schedule. In *Proc. IEEE International Symposium on Circuits and Systems*, pages 5099–5102, Island of Kos, Greece, May 2006.

[1682] K. Shiromoto. Singleton bounds for codes over finite rings. *J. Algebraic Combin.*, 12:95–99, 2000.

[1683] K. Shiromoto. Matroids and codes with the rank metric. arXiv:1803.06041 [math.CO], March 2018.

[1684] M. A. Shokrollahi. New sequences of linear time erasure codes approaching the channel capacity. In M. Fossorier, H. Imai, S. Shu Lin, and A. Poli, editors, *Applied Algebra, Algebraic Algorithms and Error-Correcting Codes*, volume 1719 of *Lecture Notes in Comput. Sci.*, pages 65–76. Springer, Berlin, Heidelberg, 1999.

[1685] M. A. Shokrollahi. Codes and graphs. In H. Reichel and S. Tison, editors, *STACS 2000 – Symposium on Theoretical Aspects of Computer Science*, volume 1770 of *Lecture Notes in Comput. Sci.*, pages 1–12. Springer, Berlin, Heidelberg, 2000.

[1686] M. A. Shokrollahi. Capacity-achieving sequences. In M. Brian and J. Rosenthal, editors, *Codes, Systems, and Graphical Models (Minneapolis, MN 1999)*, volume 123 of *The IMA Volumes in Mathematics and its Applications*, pages 153–166. Springer, New York, 2001.

[1687] M. A. Shokrollahi. Raptor codes. *IEEE Trans. Inform. Theory*, 52:2551–2567, 2006.

[1688] M. A. Shokrollahi. Theory and applications of Raptor codes. In A. Quarteroni, editor, *MATHKNOW—Mathematics, Applied Sciences and Real Life*, volume 3 of *MS&A. Model. Simul. Appl.*, pages 59–89. Springer-Verlag Italia, Milan, 2009.

[1689] M. A. Shokrollahi, B. Hassibi, B. M. Hochwald, and W. Sweldens. Representation theory for high-rate multiple-antenna code design. *IEEE Trans. Inform. Theory*, 47:2335–2367, 2001.

[1690] M. A. Shokrollahi, S. Lassen, and R. Karp. Systems and processes for decoding chain reaction codes through inactivation. http://www.freepatentsonline.com/6856263.html, February 2005.

[1691] M. A. Shokrollahi and M. G. Luby. Raptor codes. *Found. Trends Commun. and Inform. Theory*, 6:213–322, 2009.

[1692] M. A. Shokrollahi, T. Stockhammer, M. G. Luby, and M. Watson. Raptor Forward Error Correction Scheme for Object Delivery. RFC 5053, October 2007.

[1693] M. A. Shokrollahi and H. Wasserman. List decoding of algebraic-geometric codes. *IEEE Trans. Inform. Theory*, 45:432–437, 1999.

[1694] P. W. Shor. Algorithms for quantum computation: discrete logarithms and factoring. In S. Goldwasser, editor, *Proc. 35th IEEE Symposium on Foundations of Computer Science*, pages 124–134, Santa Fe, New Mexico, 1994. IEEE Comput. Soc. Press, Los Alamitos, California.

[1695] P. W. Shor. Scheme for reducing decoherence in quantum computer memory. *Physical Review A*, 52:R2493–R2496, October 1995.

[1696] P. W. Shor. Polynomial-time algorithms for prime factorization and discrete logarithms on a quantum computer. *SIAM J. Comput.*, 26:1484–1509, 1997.

[1697] I. E. Shparlinski, M. A. Tsfasman, and S. G. Vlăduţ. Curves with many points and multiplication in finite fields. In H. Stichtenoth and M. A. Tsfasman, editors, *Coding Theory and Algebraic Geometry*, volume 1518 of *Lecture Notes in Math.*, pages 145–169. Springer, Berlin, Heidelberg, 1992.

[1698] K. W. Shum and Y. Hu. Cooperative regenerating codes. *IEEE Trans. Inform. Theory*, 59:7229–7258, 2013.

[1699] I. Şiap. MacWilliams identity for m-spotty Lee weight enumerators. *Appl. Math. Lett.*, 23:13–16, 2010.

[1700] I. Şiap, T. Abualrub, and A. Ghrayeb. Cyclic DNA codes over the ring $\mathbb{F}_2[u]/(u^2-1)$ based on the deletion distance. *J. Franklin Inst.*, 346:731–740, 2009.

[1701] I. Şiap, N. Aydin, and D. K. Ray-Chaudhuri. New ternary quasi-cyclic codes with better minimum distances. *IEEE Trans. Inform. Theory*, 46:1554–1558, 2000.

[1702] V. M. Sidelńikov. Some k-valued pseudo-random sequences and nearly equidistant codes (in Russian). *Problemy Peredači Informacii*, 5(1):16–22, 1969 (English translation in *Probl. Inform. Transm.*, 5(1):12–16, 1969).

[1703] V. M. Sidelńikov. On extremal polynomials used in bounds of code volume (in Russian). *Problemy Peredači Informacii*, 16(3):17–30, 1980. (English translation in *Probl. Inform. Transm.*, 16(3):174–186, 1980).

[1704] V. M. Sidelńikov. A public-key cryptosytem based on Reed-Muller codes (in Russian). *Diskret. Mat.*, 6(2):3–20, 1994 (English translation in *Discrete Math. Appl.*, 4(3):191–207, 1994).

[1705] V. M. Sidelńikov and S. O. Shestakov. On the insecurity of cryptosystems based on generalized Reed-Solomon codes (in Russian). *Diskret. Mat.*, 4(3):57–63, 1992 (English translation in *Discrete Math. Appl.*, 2(4):439–444, 1992).

[1706] N. Silberstein and T. Etzion. Optimal fractional repetition codes based on graphs and designs. *IEEE Trans. Inform. Theory*, 61:4164–4180, 2015.

[1707] N. Silberstein and A.-L. Trautmann. Subspace codes based on graph matchings, Ferrers diagrams, and pending blocks. *IEEE Trans. Inform. Theory*, 61:3937–3953, 2015.

[1708] D. Silva. *Error Control for Network Coding*. PhD thesis, University of Toronto, 2009.

[1709] D. Silva and F. R. Kschischang. Fast encoding and decoding of Gabidulin codes. In *Proc. IEEE International Symposium on Information Theory*, pages 2858–2862, Seoul, Korea, June-July 2009.

[1710] D. Silva and F. R. Kschischang. On metrics for error correction in network coding. *IEEE Trans. Inform. Theory*, 55:5479–5490, 2009.

[1711] D. Silva, F. R. Kschischang, and R. Kötter. A rank-metric approach to error control in random network coding. *IEEE Trans. Inform. Theory*, 54:3951–3967, 2008.

[1712] F. G. Silva, V. Sousa, and Marinho L. High coded data rate and multicodeword WiMAX LDPC decoding on the Cell/BE. *IET Electronics Letters*, 44:1415–1417, 2008.

[1713] J. H. Silverman. *The Arithmetic of Elliptic Curves*, volume 106 of *Graduate Texts in Mathematics*. Springer, Dordrecht, 2nd edition, 2009.

[1714] B. Simon. *A Course in Convexity—An Analytic Viewpoint*. Cambridge University Press, Cambridge, UK, 2011.

[1715] J. Simonis. The [18, 9, 6] code is unique. *Discrete Math.*, 106/107:439—-448, 1992.

[1716] J. Singer. A theorem in finite projective geometry and some applications to number theory. *Trans. Amer. Math. Soc.*, 43:377–385, 1938.

[1717] R. C. Singleton. Maximum distance q-ary codes. *IEEE Trans. Inform. Theory*, 10:116–118, 1964.

[1718] M. Sinnokrot, J. R. Barry, , and V. Madisetti. Embedded Alamouti space-time codes for high rate and low decoding complexity. In *Proc. 42nd Asilomar Conference on Signals, Systems and Computers*, pages 1749–1753, Pacific Grove, California, October 2008.

[1719] M. Sipser and D. Spielman. Expander codes. *IEEE Trans. Inform. Theory*, 42:1710–1722, 1996.

[1720] A. N. Skorobogatov and S. G. Vlăduţ. On the decoding of algebraic-geometric codes. *IEEE Trans. Inform. Theory*, 36:1051–1060, 1990.

[1721] M. M. Skriganov. Coding theory and uniform distributions. *St. Petersburg Math. J.*, 13:301–337, 2002.

[1722] D. Slepian. A class of binary signaling alphabets. *Bell System Tech. J.*, 35:203–234, 1956. (Also reprinted in [169] pages 56–65 and [212] pages 83–114).

[1723] D. Slepian. A note on two binary signaling alphabets. *IRE Trans. Inform. Theory*, 2:84–86, 1956. (Also reprinted in [212] pages 115–117).

[1724] D. Slepian. Some further theory of group codes. *Bell System Tech. J.*, 39:203–234, 1960. (Also reprinted in [212] pages 118–151).

[1725] D. Slepian. Group codes for the Gaussian channel. *Bell System Tech. J.*, 47:575–602, 1968.

[1726] N. J. A. Sloane. Is there a (72, 36) $d = 16$ self-dual code? *IEEE Trans. Inform. Theory*, 19:251, 1973.

[1727] N. J. A. Sloane and J. G. Thompson. Cyclic self-dual codes. *IEEE Trans. Inform. Theory*, 29:364–366, 1983.

[1728] R. Smarandache, H. Gluesing-Luerssen, and J. Rosenthal. Constructions of MDS-convolutional codes. *IEEE Trans. Inform. Theory*, 47:2045–2049, 2001.

[1729] R. Smarandache and J. Rosenthal. A state space approach for constructing MDS rate $1/n$ convolutional codes. In *Proc. IEEE Information Theory Workshop*, pages 116–117, Killarney, Ireland, June 1998.

[1730] R. Smarandache and P. O. Vontobel. Quasi-cyclic LDPC codes: influence of proto- and Tanner-graph structure on minimum Hamming distance upper bounds. *IEEE Trans. Inform. Theory*, 58:585–607, 2012.

[1731] K. J. C. Smith. On the p-rank of the incidence matrix of points and hyperplanes in a finite projective geometry. *J. Combin. Theory*, 7:122–129, 1969.

[1732] S. L. Snover. *The Uniqueness of the Nordstrom–Robinson and the Golay Binary Codes*. PhD thesis, Michigan State University, 1973.

[1733] P. Solé. Asymptotic bounds on the covering radius of binary codes. *IEEE Trans. Inform. Theory*, 36:1470–1472, 1990.

[1734] P. Solé. The covering radius of spherical designs. *European J. Combin.*, 12:423–431, 1991.

[1735] P. Solé, editor. *Codes Over Rings*, volume 6 of *Series on Coding Theory and Cryptology*. World Sci. Publ., Hackensack, New Jersey, 2009.

[1736] G. Solomon and H. C. A. van Tilborg. A connection between block and convolutional codes. *SIAM J. Appl. Math.*, 37:358–369, 1979.

[1737] W. Song, K. Cai, C. Yuen, K. Cai, and G. Han. On sequential locally repairable codes. *IEEE Trans. Inform. Theory*, 64:3513–3527, 2017.

[1738] W. Song, S. H. Dau, C. Yuen, and T. J. Li. Optimal locally repairable linear codes. *IEEE J. Selected Areas Commun.*, 32:1019–1036, 2014.

[1739] W. Song and C. Yuen. Locally repairable codes with functional repair and multiple erasure tolerance. arXiv:1507.02796 [cs.IT], 2015.

[1740] Y. Song, Y. Li, Z. Li, and J. Li. A new multi-use multi-secret sharing scheme based on the duals of minimal linear codes. *Security and Commun. Networks*, 8:202–211, 2015.

[1741] A. B. Sørensen. Projective Reed-Muller codes. *IEEE Trans. Inform. Theory*, 37:1567–1576, 1991.

[1742] E. Spiegel. Codes over Z_m. *Inform. and Control*, 35:48–51, 1977.

[1743] E. Spiegel. Codes over Z_m, revisited. *Inform. and Control*, 37:100–104, 1978.

[1744] D. A. Spielman. Linear-time encodable and decodable error-correcting codes. *IEEE Trans. Inform. Theory*, 42:1723–1731, 1996.

[1745] A. Sridharan, M. Lentmaier, D. J. Costello, Jr., and K. Sh. Zigangirov. Terminated LDPC convolutional codes with thresholds close to capacity. In *Proc. IEEE International Symposium on Information Theory*, pages 1372–1376, Adelaide, Australia, September 2005.

[1746] R. Srikant and L. Ying. *Communication Networks: An Optimization, Control, and Stochastic Networks Perspective*. Cambridge University Press, Cambridge, UK, 2014.

[1747] K. P. Srinath and B. S. Rajan. Low ML-decoding complexity, large coding gain, full-diversity STBCs for 2×2 and 4×2 MIMO systems. *IEEE J. Selected Topics Signal Process.*, 3:916–927, 2010.

[1748] M. K. Srinivasan. Symmetric chains, Gelfand-Tsetlin chains, and the Terwilliger algebra of the binary Hamming scheme. *J. Algebraic Combin.*, 34:301–322, 2011.

[1749] A. M. Steane. Multiple-particle interference and quantum error correction. *Proc. Roy. Soc. London Ser. A*, 452(1954):2551–2577, 1996.

[1750] A. M. Steane. Enlargement of Calderbank-Shor-Steane quantum codes. *IEEE Trans. Inform. Theory*, 45:2492–2495, 1999.

[1751] A. M. Steane. Overhead and noise threshold of fault-tolerant quantum error correction. *Physical Review A*, 68(4):042322, October 2003.

[1752] J. Stern. A method for finding codewords of small weight. In G. D. Cohen and J. Wolfmann, editors, *Coding Theory and Applications*, volume 388 of *Lecture Notes in Comput. Sci.*, pages 106–113. Springer, Berlin, Heidelberg, 1989.

[1753] J. Stern. A new identification scheme based on syndrome decoding. In D. R. Stinson, editor, *Advances in Cryptology – CRYPTO '93*, volume 773 of *Lecture Notes in Comput. Sci.*, pages 13–21. Springer, Berlin, Heidelberg, 1994.

[1754] H. Stichtenoth. Transitive and self-dual codes attaining the Tsfasman-Vlăduţ-Zink bound. *IEEE Trans. Inform. Theory*, 52:2218–2224, 2006.

[1755] H. Stichtenoth. *Algebraic Function Fields and Codes*, volume 254 of *Graduate Texts in Mathematics*. Springer-Verlag, Berlin, 2nd edition, 2009.

[1756] H. Stichtenoth and C. Xing. Excellent nonlinear codes from algebraic function fields. *IEEE Trans. Inform. Theory*, 51:4044–4046, 2005.

[1757] D. R. Stinson and R. Wei. Combinatorial properties and constructions of traceability schemes and frameproof codes. *SIAM J. Discrete Math.*, 11:41–53, 1998.

[1758] L. Storme. Linear codes meeting the Griesmer bound, minihypers and geometric applications. *Le Matematiche*, 59:367–392, 2004.

[1759] V. Strassen. Gaussian elimination is not optimal. *Numer. Math.*, 13:354–356, 1969.

[1760] S. Suda and H. Tanaka. A cross-intersection theorem for vector spaces based on semidefinite programming. *Bull. London Math. Soc.*, 46:342–348, 2014.

[1761] S. Suda, H. Tanaka, and N. Tokushige. A semidefinite programming approach to a cross-intersection problem with measures. *Math. Program. Ser. A*, 166:113–130, 2017.

[1762] M. Sudan. Decoding of Reed-Solomon codes beyond the error-correction bound. *J. Complexity*, 13:180–193, 1997.

[1763] Y. Sugiyama, M. Kasahara, S. Hirasawa, and T. Namekawa. Further results on Goppa codes and their applications to constructing efficient binary codes. *IEEE Trans. Inform. Theory*, 22:518–526, 1976.

[1764] C. Suh and K. Ramchandran. Exact-repair MDS code construction using interference alignment. *IEEE Trans. Inform. Theory*, 57:1425–1442, 2011.

[1765] K. Suzuki, H. Kaneko, and E. Fujiwara. MacWilliams identity for m-spotty weight enumerator. In *Proc. IEEE International Symposium on Information Theory*, pages 31–35, Nice, France, June 2007.

[1766] G. Szegö. *Orthogonal Polynomials*. Colloquium publications. Amer. Math. Soc., New York, 1939.

[1767] T. Szőnyi and Z. Weiner. Stability of k mod p multisets and small weight codewords of the code generated by the lines of PG(2, q). *J. Combin. Theory Ser. A*, 157:321–333, 2018.

[1768] M. Takenaka, K. Okamoto, and T. Maruta. On optimal non-projective ternary linear codes. *Discrete Math.*, 308:842–854, 2008.

[1769] I. Tal and A. Vardy. How to construct polar codes. *IEEE Trans. Inform. Theory*, 59:6562–6582, 2013.

[1770] I. Tal and A. Vardy. List decoding of polar codes. *IEEE Trans. Inform. Theory*, 61:2213–2226, 2015.

[1771] I. Tamo and A. Barg. Bounds on locally recoverable codes with multiple recovering sets. In *Proc. IEEE International Symposium on Information Theory*, pages 691–695, Honolulu, Hawaii, June-July 2014.

[1772] I. Tamo and A. Barg. A family of optimal locally recoverable codes. *IEEE Trans. Inform. Theory*, 60:4661–4676, 2014.

[1773] I. Tamo, A. Barg, S. Goparaju, and A. R. Calderbank. Cyclic LRC codes, binary LRC codes, and upper bounds on the distance of cyclic codes. *Internat. J. Inform. Coding Theory*, 3:345–364, 2016.

[1774] I. Tamo, Z. Wang, and J. Bruck. Zigzag codes: MDS array codes with optimal rebuilding. *IEEE Trans. Inform. Theory*, 59:1597–1616, 2013.

[1775] I. Tamo, Z. Wang, and J. Bruck. Access versus bandwidth in codes for storage. *IEEE Trans. Inform. Theory*, 60:2028–2037, 2014.

[1776] I. Tamo, M. Ye, and A. Barg. The repair problem for Reed-Solomon codes: optimal repair of single and multiple erasures with almost optimal node size. *IEEE Trans. Inform. Theory*, 65:2673–2695, 2019.

[1777] H. Tanaka. New proofs of the Assmus-Mattson theorem based on the Terwilliger algebra. *European J. Combin.*, 30:736–746, 2009.

[1778] C. Tang, N. Li, Y. Qi, Z. Zhou, and T. Helleseth. Linear codes with two or three weights from weakly regular bent functions. *IEEE Trans. Inform. Theory*, 62:1166–1176, 2016.

[1779] C. Tang, Y. Qi, and M. Xu. A note on cyclic codes from APN functions. *Appl. Algebra Engrg. Comm. Comput.*, 25:21–37, 2014.

[1780] H. Tang, J. Xu, Lin. S., and K. Abdel-Ghaffar. Codes on finite geometries. *IEEE Trans. Inform. Theory*, 52:572–596, 2005.

[1781] X. Tang and C. Ding. New classes of balanced quaternary and almost balanced binary sequences with optimal autocorrelation value. *IEEE Trans. Inform. Theory*, 56:6398–6405, 2010.

[1782] X. Tang, P. Fan, and S. Matsufuji. Lower bounds on correlation of spreading sequence set with low or zero correlation zone. *Electron. Lett.*, 36:551–552, 2000.

[1783] X. Tang and G. Gong. New constructions of binary sequences with optimal autocorrelation value/magnitude. *IEEE Trans. Inform. Theory*, 56:1278–1286, 2010.

[1784] R. M. Tanner. A recursive approach to low complexity codes. *IEEE Trans. Inform. Theory*, 27:533–547, 1981.

[1785] R. M. Tanner. A transform theory for a class of group-invariant codes. *IEEE Trans. Inform. Theory*, 34:752–775, 1988.

[1786] R. M. Tanner. On quasi-cyclic repeat accumulate codes. In *Proc. 37th Allerton Conference on Communication, Control, and Computing*, pages 249–259, Monticello, Illinois, September 1999.

[1787] R. M. Tanner. Minimum distance bounds by graph analysis. *IEEE Trans. Inform. Theory*, 47:808–821, 2001.

[1788] R. M. Tanner, D. Sridhara, and T. E. Fuja. A class of group structured LDPC codes. In *Proc. 6th International Symposium on Communication Theory and Applications*, pages 365–370, Ambleside, UK, July 2001.

[1789] R. M. Tanner, D. Sridhara, A. Sridharan, D. J. Costello, Jr., and T. E. Fuja. LDPC block and convolutional codes from circulant matrices. *IEEE Trans. Inform. Theory*, 50:2966–2984, 2004.

[1790] T. Tao and V. H. Vu. *Additive Combinatorics*, volume 105 of *Cambridge Studies in Advanced Mathematics*. Cambridge University Press, Cambridge, UK, 2006.

[1791] L. F. Tapia Cuitiño and A. L. Tironi. Some properties of skew codes over finite fields. *Des. Codes Cryptogr.*, 85:359–380, 2017.

[1792] V. Tarokh, H. Jafarkhani, and A. R. Calderbank. Space-time block codes from orthogonal design. *IEEE Trans. Inform. Theory*, 45:1456–1466, 1999.

[1793] V. Tarokh, N. Seshadri, and A. R. Calderbank. Space-time codes for high data rate wireless communication: performance criterion and code construction. *IEEE Trans. Inform. Theory*, 44:744–765, 1998.

[1794] O. Taussky and J. Todd. Covering theorems for groups. *Ann. Soc. Polon. Math.*, 21:303–305, 1948.

[1795] L. Teirlinck. On projective and affine hyperplanes. *J. Combin. Theory Ser. A*, 28:290–306, 1980.

[1796] E. Telatar. Capacity of multi-antenna Gaussian channels. *European Trans. on Telecommun.*, 10:585–596, 1999.

[1797] S. Ten Brink. Convergence of iterative decoding. *IEEE Electronic Letters*, 35:806–808, 1999.

[1798] S. Ten Brink. Convergence behavior of iteratively decoded parallel concatenated codes. *IEEE Trans. Inform. Theory*, 49:1727–1737, 2001.

[1799] A. Thangaraj and S. W. McLaughlin. Quantum codes from cyclic codes over GF(4^m). *IEEE Trans. Inform. Theory*, 47:1176–1178, 2001.

[1800] J. G. Thompson. Weighted averages associated to some codes. *Scripta Math.*, 29:449–452, 1973.

[1801] T. M. Thompson. *From Error-Correcting Codes Through Sphere Packings to Simple Groups*, volume 21, Carus Mathematical Monographs. Mathematical Association of America, Washington DC, 1983.

[1802] J. Thorpe. Low-density parity-check codes constructed from protographs. *IPN Progress Report*, pages 42–154, 2003.

[1803] C. Tian. Characterizing the rate region of the $(4,3,3)$ exact-repair regenerating codes. *IEEE J. Selected Areas Commun.*, 32:967–975, 2014.

[1804] C. Tian, B. Sasidharan, V. Aggarwal, V. A. Vaishampayan, and P. V. Kumar. Layered exact-repair regenerating codes via embedded error correction and block designs. *IEEE Trans. Inform. Theory*, 61:1933–1947, 2015.

[1805] A. Tietäväinen. On the nonexistence of perfect codes over finite fields. *SIAM J. Appl. Math.*, 24:88–96, 1973.

[1806] A. Tietäväinen. Nonexistence of nontrivial perfect codes in case $q = p_1^r p_2^s$, $e \geq 3$. *Discrete Math.*, 17:199–205, 1977.

[1807] A. Tietäväinen. An upper bound on the covering radius as a function of the dual distance. *IEEE Trans. Inform. Theory*, 36:1472–1474, 1990.

[1808] A. Tietäväinen. Covering radius and dual distance. *Des. Codes Cryptogr.*, 1:31–46, 1991.

[1809] J. Tits. Ovoïdes à translations. *Rend. Mat. e Appl. (5)*, 21:37–59, 1962.

[1810] V. Tomás, J. Rosenthal, and R. Smarandache. Decoding of convolutional codes over the erasure channel. *IEEE Trans. Inform. Theory*, 58:90–108, 2012.

[1811] V. D. Tonchev. Quasi-symmetric 2-(31,7,7) designs and a revision of Hamada's conjecture. *J. Combin. Theory Ser. A*, 42:104–110, 1986.

[1812] V. D. Tonchev. *Combinatorial Configurations, Designs, Codes, Graphs*. John Wiley & Sons, Inc. New York, 1988.

[1813] V. D. Tonchev. A class of Steiner 4-wise balanced designs derived from Preparata codes. *J. Combin. Des.*, 4:203–204, 1996.

[1814] V. D. Tonchev. Codes and Designs. In V. S. Pless and W. C. Huffman, editors, *Handbook of Coding Theory, Vol. I, II*, chapter 15, pages 1229–1267. North-Holland, Amsterdam, 1998.

[1815] V. D. Tonchev. Linear perfect codes and a characterization of the classical designs. *Des. Codes Cryptogr.*, 17:121–128, 1999.

[1816] V. D. Tonchev. Codes. In C. J. Colbourn and J. H. Dinitz, editors, *The CRC Handbook of Combinatorial Designs*, chapter VII.1, pages 677–702. Chapman & Hall/CRC Press, 2nd edition, 2007.

[1817] V. D. Tonchev. On resolvable Steiner 2-designs and maximal arcs in projective planes. *Des. Codes Cryptogr.*, 84:165–172, 2017.

[1818] A.-L. Trautmann, F. Manganiello, M. Braun, and J. Rosenthal. Cyclic orbit codes. *IEEE Trans. Inform. Theory*, 59:7386–7404, 2013.

[1819] P. Trifonov. Efficient design and decoding of polar codes. *IEEE Trans. Commun.*, 60:3221–3227, 2012.

[1820] H. Trinker. The triple distribution of codes and ordered codes. *Discrete Math.*, 311:2283–2294, 2011.

[1821] M. A. Tsfasman, S. G. Vlăduţ, and D. Nogin. *Algebraic Geometric Codes: Basic Notions*, volume 139 of *Mathematical Surveys and Monographs*. Amer. Math. Soc., Providence, Rhode Island, 2007.

[1822] M. A. Tsfasman, S. G. Vlăduţ, and T. Zink. Modular curves, Shimura curves, and Goppa codes, better than Varshamov-Gilbert bound. *Math. Nachr.*, 109:21–28, 1982.

[1823] L. Tunçel. *Polyhedral and Semidefinite Programming Methods in Combinatorial Optimization*. Amer. Math. Soc., Providence, Rhode Island; Fields Institute for Research in Mathematical Sciences, Toronto, Canada, 2010.

[1824] M. Vajha, S. B. Balaji, and P. V. Kumar. Explicit MSR codes with optimal access, optimal sub-packetization and small field size for $d = k+1, k+2, k+3$. In *Proc. IEEE International Symposium on Information Theory*, pages 2376–2380, Vail, Colorado, June 2018.

[1825] M. Vajha, V. Ramkumar, B. Puranik, G. R. Kini, E. Lobo, B. Sasidharan, P. V. Kumar, A. Barg, M. Ye, S. Narayanamurthy, S. Hussain, and S. Nandi. Clay codes: moulding MDS codes to yield an MSR code. In *Proc. 16th USENIX Conference on File and Storage Technologies*, pages 139–154, Oakland, California, February 2018.

[1826] M. E. Valcher and E. Fornasini. On 2D finite support convolutional codes: an algebraic approach. *Multidimens. Systems Signal Process.*, 5:231–243, 1994.

[1827] A. E. A. Valdebenito and A. L. Tironi. On the duals codes of skew constacyclic codes. *Adv. Math. Commun.*, 12:659–679, 2018.

[1828] F. Vallentin. Lecture notes: Semidefinite programming and harmonic analysis. arXiv:0809.2017 [math.OC], September 2008.

[1829] F. Vallentin. Symmetry in semidefinite programs. *Linear Algebra Appl.*, 430:360–369, 2009.

[1830] M. van Dijk. A linear construction of perfect secret sharing schemes. In A. DeSantis, editor, *Advances in Cryptology – EUROCRYPT '94*, volume 950 of *Lecture Notes in Comput. Sci.*, pages 23–34. Springer, Berlin, Heidelberg, 1995.

[1831] M. van Eupen. Some new results for ternary linear codes of dimension 5 and 6. *IEEE Trans. Inform. Theory*, 41:2048–2051, 1995.

[1832] J. H. van Lint. Recent results on perfect codes and related topics. In M. Hall, Jr. and J. H. van Lint, editors, *Combinatorics. Part 1: Theory of Designs, Finite Geometry and Coding Theory*, volume 55 of *Math. Centre Tracts*, pages 158–178. Math. Centrum, Amsterdam, 1974.

[1833] J. H. van Lint. A survey on perfect codes. *Rocky Mountain J. Math.*, 5:199–224, 1975.

[1834] J. H. van Lint. Kerdock codes and Preparata codes. *Congr. Numer.*, 39:25–41, 1983.

[1835] J. H. van Lint. Repeated-root cyclic codes. *IEEE Trans. Inform. Theory*, 37:343–345, 1991.

[1836] J. H. van Lint. *Introduction to Coding Theory*. Springer-Verlag, Berlin, Heidelberg, 3rd edition, 1999.

[1837] J. H. van Lint and A. Schrijver. Construction of strongly regular graphs, two-weight codes and partial geometries by finite fields. *Combinatorica*, 1:63–73, 1981.

[1838] J. H. van Lint and R. M. Wilson. On the minimum distance of cyclic codes. *IEEE Trans. Inform. Theory*, 32:23–40, 1986.

[1839] J. H. van Lint and R. M. Wilson. *A Course in Combinatorics*. Cambridge University Press, Cambridge, UK, 2nd edition, 2001.

[1840] J. van Tilburg. On the McEliece public-key cryptosystem. In S. Goldwasser, editor, *Advances in Cryptology – CRYPTO '88*, volume 403 of *Lecture Notes in Comput. Sci.*, pages 119–131. Springer, Berlin, Heidelberg, 1990.

[1841] J. van Tilburg. *Security-Analysis of a Class of Cryptosystems Based on Linear Error-Correcting Codes*. PhD thesis, Technical University Eindhoven, 1994.

[1842] G. J. M. van Wee. On the nonexistence of certain perfect mixed codes. *Discrete Math.*, 87:323–326, 1991.

[1843] A. Vardy. The intractability of computing the minimum distance of a code. *IEEE Trans. Inform. Theory*, 43:1757–1766, 1997.

[1844] R. R. Varshamov. The evaluation of signals in codes with correction of errors (in Russian). *Dokl. Akad. Nauk SSSR*, 117:739–741, 1957. (English translation in [212] pages 68–71).

[1845] R. R. Varshamov. Some features of linear codes that correct asymmetric errors. *Soviet Physics Doklady*, 9:538, 1965.

[1846] B. Vasic, S. K. Chilappagari, D. V. Nguyen, and S. K. Planjery. Trapping set ontology. In *Proc. 47th Allerton Conference on Communication, Control, and Computing*, pages 1–7, Monticello, Illinois, September-October 2009.

[1847] J. L. Vasil'ev. On nongroup close-packed codes (in Russian). *Probl. Kibernet.*, 8:337–339, 1962. (English translation in [212] pages 351–357).

[1848] G. Vega. A critical review and some remarks about one- and two-weight irreducible cyclic codes. *Finite Fields Appl.*, 33:1–13, 2015.

[1849] R. Vehkalahti, C. Hollanti, and F. Oggier. Fast-decodable asymmetric space-time codes from division algebras. *IEEE Trans. Inform. Theory*, 58:2362–2385, 2012.

[1850] J. M. Velazquez-Gutierrez and C. Vargas-Rosales. Sequence sets in wireless communication systems: a survey. *IEEE Commun. Surveys & Tutorials*, 19:1225–1248, 2017.

[1851] P. Véron. A fast identification scheme. In *Proc. IEEE International Symposium on Information Theory*, page 359, Whistle, British Columbia, Canada, September 1995.

[1852] M. G. Vides. *Métricas Sobre Grupos y Anilloscon Aplicaciones a la Teoría de Códigos*. PhD thesis, Universidad Nacional de Córdoba, 2018.

[1853] M. Villanueva, F. Zeng, and J. Pujol. Efficient representation of binary nonlinear codes: constructions and minimum distance computation. *Des. Codes Cryptogr.*, 76:3–21, 2015.

[1854] A. Viterbi. Error bounds for convolutional codes and an asymptotically optimum decoding algorithm. *IEEE Trans. Inform. Theory*, 13:260–269, 1967.

[1855] E. Viterbo and J. Boutros. A universal lattice code decoder for fading channels. *IEEE Trans. Inform. Theory*, 45:1639–1642, 1999.

[1856] S. G. Vlăduţ. On the decoding of algebraic-geometric codes over $\mathbb{F}_q$ for $q \geq 16$. *IEEE Trans. Inform. Theory*, 36:1461–1463, 1990.

[1857] S. G. Vlăduţ and V. G. Drinfeld. The number of points of an algebraic curve (in Russian). *Funktsional. Anal. i Prilozhen.*, 17:68–69, 1983 (English translation in *Funct. Anal. Appl.*, 17:53–54, 1983).

[1858] A. vom Felde. A new presentation of Cheng-Sloane's $[32, 17, 8]$-code. *Arch. Math. (Basel)*, 60:508–511, 1993.

[1859] P. Vontobel and R. M. Tanner. Construction of codes based on finite generalized quadrangles for iterative decoding. In *Proc. IEEE International Symposium on Information Theory*, page 223, Washington, DC, June 2001.

[1860] Z.-X. Wan. A proof of the automorphisms of linear groups over a field of characteristic 2. *Scientia Sinica*, 11:1183–1194, 1962.

[1861] Z.-X. Wan. *Quaternary Codes*, volume 8 of *Series on Applied Mathematics*. World Scientific Publishing Co., Inc., River Edge, New Jersey, 1997.

[1862] Z.-X. Wan. Cyclic codes over Galois rings. *Algebra Colloq.*, 6:291–304, 1999.

[1863] Z.-X. Wan. The MacWilliams identity for linear codes over Galois rings. In I. Althöfer, N. Cai, G. Dueck, L. Khachatrian, M. S. Pinsker, A. Sárközy, I. Wegener, and Z. Zhang, editors, *Numbers, Information and Complexity*, pages 333–338. Kluwer Acad. Publ., Boston, 2000.

[1864] Z.-X. Wan. *Lectures on Finite Fields and Galois Rings*. World Scientific Publishing Co., Inc., River Edge, New Jersey, 2003.

[1865] A. Wang and Z. Zhang. Exact cooperative regenerating codes with minimum-repair-bandwidth for distributed storage. In *Proc. 32nd IEEE International Conference on Computer Communications*, pages 400–404, Turin, Italy, April 2013.

[1866] A. Wang and Z. Zhang. Repair locality with multiple erasure tolerance. *IEEE Trans. Inform. Theory*, 60:6979–6987, 2014.

[1867] A. Wang and Z. Zhang. An integer programming-based bound for locally repairable codes. *IEEE Trans. Inform. Theory*, 61:5280–5294, 2015.

[1868] A. Wang, Z. Zhang, and D. Lin. Bounds and constructions for linear locally repairable codes over binary fields. In *Proc. IEEE International Symposium on Information Theory*, pages 2033–2037, Aachen, Germany, June 2017.

[1869] A. Wang, Z. Zhang, and M. Liu. Achieving arbitrary locality and availability in binary codes. In *Proc. IEEE International Symposium on Information Theory*, pages 1866–1870, Hong Kong, China, June 2015.

[1870] H. Wang, C. Xing, and R. Safavi-Naini. Linear authentication codes: bounds and constructions. *IEEE Trans. Inform. Theory*, 49:866–872, 2003.

[1871] L. Wang, K. Feng, S. Ling, and C. Xing. Asymmetric quantum codes: characterization and constructions. *IEEE Trans. Inform. Theory*, 56:2938–2945, 2010.

[1872] Q. Wang, K. Ding, and R. Xue. Binary linear codes with two weights. *IEEE Commun. Letters*, 19:1097–1100, 2015.

[1873] H. N. Ward. Divisible codes. *Arch. Math. (Basel)*, 36:485–494, 1981.

[1874] H. N. Ward. Multilinear forms and divisors of codeword weights. *Quart. J. Math. Oxford Ser. (2)*, 34(133):115–128, 1983.

[1875] H. N. Ward. Divisors of codes of Reed-Muller type. *Discrete Math.*, 131:311–323, 1994.

[1876] H. N. Ward. Quadratic Residue Codes and Divisibility. In V. S. Pless and W. C. Huffman, editors, *Handbook of Coding Theory, Vol. I, II*, chapter 9, pages 827–870. North-Holland, Amsterdam, 1998.

[1877] H. N. Ward. Divisible codes—a survey. *Serdica Math. J.*, 27:263–278, 2001.

[1878] H. S. Warren. *Hacker's Delight*. Addison-Wesley Professional, 2nd edition, 2012.

[1879] A. Wassermann. Finding simple t-designs with enumeration techniques. *J. Combin. Des.*, 6:79–90, 1998.

[1880] A. Wassermann. Attacking the market split problem with lattice point enumeration. *J. Comb. Optim.*, 6:5–16, 2002.

[1881] A. Wassermann. Computing the minimum distance of linear codes. In *Proc. 8th International Workshop on Algebraic and Combinatorial Coding Theory*, pages 254–257, Tsarskoe Selo, Russia, September 2002.

[1882] V. K. Wei. Generalized Hamming weights for linear codes. *IEEE Trans. Inform. Theory*, 37:1412–1418, 1991.

[1883] P. Weiner. *Multidimensional Convolutional Codes*. PhD thesis, University of Notre Dame, 1998.

[1884] L. R. Welch. Lower bounds on the maximum cross correlation of signals. *IEEE Trans. Inform. Theory*, 20:397–399, 1974.

[1885] L. R. Welch, R. J. McEliece, and H. Rumsey, Jr. A low-rate improvement on the Elias bound. *IEEE Trans. Inform. Theory*, 20:676–678, 1974.

[1886] E. J. Weldon, Jr. Difference-set cyclic codes. *Bell System Tech. J.*, 45:1045–1055, 1966.

[1887] E. J. Weldon, Jr. New generalizations of the Reed–Muller codes. Part II: nonprimitive codes. *IEEE Trans. Inform. Theory*, 14:199–205, 1968. (Also reprinted in [212] pages 334–340).

[1888] D. B. West. *Introduction to Graph Theory*. Prentice Hall, Inc., Upper Saddle River, New Jersey, 2nd edition, 2001.

[1889] N. Wiberg. *Codes and Decoding on General Graphs*. PhD thesis, Electrical Engineering, Linköping University, Sweden, 1996.

[1890] C. Wieschebrink. Two NP-complete problems in coding theory with an application in code based cryptography. In *Proc. IEEE International Symposium on Information Theory*, pages 1733–1737, Seattle, Washington, July 2006.

[1891] M. M. Wilde and T. A. Brun. Optimal entanglement formulas for entanglement-assisted quantum coding. *Physical Review A*, 77(6):064302, June 2008.

[1892] J. C. Willems. From time series to linear system I: finite-dimensional linear time invariant systems. *Automatica J. IFAC*, 22:561–580, 1986.

[1893] W. Willems. A note on self-dual group codes. *IEEE Trans. Inform. Theory*, 48:3107–3109, 2002.

[1894] R. A. Wilson. An Atlas of sporadic group representations. In R. T. Curtis and R. A. Wilson, editors, *The Atlas of Finite Groups: Ten Years On (Birmingham, 1995)*, volume 249 of *London Math. Soc. Lecture Note Ser.*, pages 261–273. Cambridge University Press, Cambridge, UK, 1998.

[1895] S. Winograd. Some bilinear forms whose multiplicative complexity depends on the field of constants. *Mathematical Systems Theory*, 10:169–180, 1977.

[1896] M. Wirtz. On the parameters of Goppa codes. *IEEE Trans. Inform. Theory*, 34:1341–1343, 1988.

[1897] J. K. Wolf. Efficient maximum-likelihood decoding of linear block codes using a trellis. *IEEE Trans. Inform. Theory*, 24:76–80, 1978.

[1898] J. Wolfmann. Codes projectifs à deux ou trois poids associés aux hyperquadriques d'une géométrie finie. *Discrete Math.*, 13:185–211, 1975.

[1899] J. Wolfmann. New bounds on cyclic codes from algebraic curves. In G. D. Cohen and J. Wolfmann, editors, *Coding Theory and Applications*, volume 388 of *Lecture Notes in Comput. Sci.*, pages 47–62. Springer, Berlin, Heidelberg, 1989.

[1900] J. Wolfmann. Bent functions and coding theory. In A. Pott, V. Kumaran, T. Helleseth, and D. Jungnickel, editors, *Difference Sets, Sequences and Their Correlation Properties*, volume 542 of *NATO Adv. Sci. Inst. Ser. C Math. Phys. Sci.*, pages 393–418. Kluwer Acad. Publ., Dordrecht, Netherlands, 1999.

[1901] J. Wolfmann. Negacyclic and cyclic codes over $\mathbb{Z}_4$. *IEEE Trans. Inform. Theory*, 45:2527–2532, 1999.

[1902] J. Wolfmann. Binary images of cyclic codes over $\mathbb{Z}_4$. *IEEE Trans. Inform. Theory*, 47:1773–1779, 2001.

[1903] S. Wolfram. *Idea Makers: Personal Perspectives on the Lives & Ideas of Some Notable People*. Wolfram Media, Inc., Champaign, Illinois, 2016.

[1904] H. Wolkowicz, R. Saigal, and L. Vandenberghe, editors. *Handbook of Semidefinite Programming*. Kluwer Acad. Pub., Boston, 2000.

[1905] J. A. Wood. Extension theorems for linear codes over finite rings. In T. Mora and H. F. Mattson, Jr., editors, *Applied Algebra, Algebraic Algorithms and Error-Correcting Codes*, volume 1255 of *Lecture Notes in Comput. Sci.*, pages 329–340. Springer, Berlin, Heidelberg, 1997.

[1906] J. A. Wood. Duality for modules over finite rings and applications to coding theory. *Amer. J. Math.*, 121:555–575, 1999.

[1907] W. K. Wootters and W. H. Zurek. A single quantum cannot be cloned. *Nature*, 299:802–803, 1982.

[1908] S. J. Wright. *Primal-dual Interior-point Methods.* Society for Industrial and Applied Mathematics (SIAM), Philadelphia, 1997.

[1909] B. Wu and Z. Liu. Linearized polynomials over finite fields revisited. *Finite Fields Appl.*, 22:79–100, 2013.

[1910] F. Wu. Constructions of strongly regular Cayley graphs using even index Gauss sums. *J. Combin. Des.*, 21:432–446, 2013.

[1911] Y. Wu, N. Li, and X. Zeng. Linear codes with few weights from cyclotomic classes and weakly regular bent functions. *Des. Codes Cryptogr.*, 88:1255–1272, 2020.

[1912] S.-T. Xia and F.-W. Fu. On the stopping distance of finite geometry LDPC codes. *IEEE Commun. Letters*, 10:381–383, 2006.

[1913] S.-T. Xia and F.-W. Fu. Johnson type bounds on constant dimension codes. *Des. Codes Cryptogr.*, 50:163–172, 2009.

[1914] C. Xiang. It is indeed a fundamental construction of all linear codes. arXiv:1610.06355 [cs.IT], October 2016.

[1915] C. Xing. Asymptotic bounds on frameproof codes. *IEEE Trans. Inform. Theory*, 48:2991–2995, 2002.

[1916] C. Xing. Nonlinear codes from algebraic curves improving the Tsfasman-Vlăduţ-Zink bound. *IEEE Trans. Inform. Theory*, 49:1653–1657, 2003.

[1917] E. Yaakobi, J. Bruck, and P. H. Siegel. Decoding of cyclic codes over symbol-pair read channels. In *Proc. IEEE International Symposium on Information Theory*, pages 2891–2895, Cambridge, Massachussets, July 2012.

[1918] E. Yaakobi, J. Bruck, and P. H. Siegel. Constructions and decoding of cyclic codes over b-symbol read channels. *IEEE Trans. Inform. Theory*, 62:1541–1551, 2016.

[1919] S. Yang and J.-C. Belfiore. Optimal space-time codes for the MIMO amplify-and-forward cooperative channel. *IEEE Trans. Inform. Theory*, 53:647–663, 2007.

[1920] X. Yang and J. L. Massey. The condition for a cyclic code to have a complementary dual. *Discrete Math.*, 126:391–393, 1994.

[1921] Y. Yang and X. Tang. Generic construction of binary sequences of period $2N$ with optimal odd correlation magnitude based on quaternary sequences of odd period N. *IEEE Trans. Inform. Theory*, 64:384–392, 2018.

[1922] N. Yankov. A putative doubly even [72, 36, 16] code does not have an automorphism of order 9. *IEEE Trans. Inform. Theory*, 58:159–163, 2012.

[1923] H. Yao and G. W. Wornell. Achieving the full MIMO diversity-multiplexing frontier with rotation-based space-time codes. In *Proc. 41st Allerton Conference on Communication, Control, and Computing*, pages 400–409, Monticello, Illinois, October 2003.

[1924] F. Ye, K. W. Shum, and R. W. Yeung. The rate region for secure distributed storage systems. *IEEE Trans. Inform. Theory*, 63:7038–7051, 2017.

[1925] M. Ye and A. Barg. Explicit constructions of high-rate MDS array codes with optimal repair bandwidth. *IEEE Trans. Inform. Theory*, 63:2001–2014, 2017.

[1926] M. Ye and A. Barg. Explicit constructions of optimal-access MDS codes with nearly optimal sub-packetization. *IEEE Trans. Inform. Theory*, 63:6307–6317, 2017.

[1927] M. Ye and A. Barg. Cooperative repair: constructions of optimal MDS codes for all admissible parameters. *IEEE Trans. Inform. Theory*, 65:1639–1656, 2019.

[1928] V. Y. Yorgov. Binary self-dual codes with an automorphism of odd order (in Russian). *Problemy Peredači Informacii*, 19(4):11–24, 1983. (English translation in *Probl. Inform. Transm.*, 19(4):260–270, 1983).

[1929] V. Y. Yorgov. A method for constructing inequivalent self-dual codes with applications to length 56. *IEEE Trans. Inform. Theory*, 33:77–82, 1987.

[1930] V. Y. Yorgov. On the automorphism group of a putative code. *IEEE Trans. Inform. Theory*, 52:1724–1726, 2006.

[1931] V. Y. Yorgov and D. Yorgov. The automorphism group of a self-dual [72, 36, 16] code does not contain $\mathbb{Z}_4$. *IEEE Trans. Inform. Theory*, 60:3302–3307, 2014.

[1932] E. V. York. *Algebraic Description and Construction of Error Correcting Codes: A Linear Systems Point of View*. PhD thesis, University of Notre Dame, 1997.

[1933] D. C. Youla and P. F. Pickel. The Quillen-Suslin theorem and the structure of n-dimensional elementary polynomial matrices. *IEEE Trans. Circuits and Systems*, 31:513–518, 1984.

[1934] N. Y. Yu and G. Gong. New binary sequences with optimal autocorrelation magnitude. *IEEE Trans. Inform. Theory*, 54:4771–4779, 2008.

[1935] J. Yuan and C. Ding. Secret sharing schemes from three classes of linear codes. *IEEE Trans. Inform. Theory*, 52:206–212, 2006.

[1936] V. A. Yudin. Minimum potential energy of a point system of charges (in Russian). *Diskret. Mat.*, 4(2):115–121, 1992. (English translation in *Discrete Math. Appl.*, 3(1):75–81, 1993).

[1937] R. Zamir. Lattices are everywhere. In *Proc. Information Theory and Applications Workshop*, pages 392–421, San Diego, California, February 2009.

[1938] R. Zamir. *Lattice Coding for Signals and Networks: A Structured Coding Approach to Quantization, Modulation, and Multiuser Information Theory*. Cambridge University Press, Cambridge, UK, 2014.

[1939] S. K. Zaremba. Covering problems concerning abelian groups. *J. London Math. Soc.*, 27:242–246, 1952.

[1940] A. Zeh and S. Ling. Decoding of quasi-cyclic codes up to a new lower bound for the minimum distance. In *Proc. IEEE International Symposium on Information Theory*, pages 2584–2588, Honolulu, Hawaii, June-July 2014.

[1941] G. Zémor. Minimum distance bounds by graph analysis. *IEEE Trans. Inform. Theory*, 47:835–837, 2001.

[1942] J. Zhang, X. Wang, and G. Ge. Some improvements on locally repairable codes. arXiv:1506.04822 [cs.IT], June 2015.

[1943] S. Zhang. On the nonexistence of extremal self-dual codes. *Discrete Appl. Math.*, 91:277–286, 1999.

[1944] D. Zheng, X. Li, and K. Chen. Code-based ring signature scheme. *Internat. J. Network Security*, 5:154–157, 2007.

[1945] L. Zheng and D. N. C. Tse. Diversity and multiplexing: a fundamental tradeoff in multiple-antenna channels. *IEEE Trans. Inform. Theory*, 49:1073–1096, 2003.

[1946] Z. Zhou and C. Ding. A class of three-weight cyclic codes. *Finite Fields Appl.*, 25:79–93, 2014.

[1947] Z. Zhou, N. Li, C. Fan, and T. Helleseth. Linear codes with two or three weights from quadratic bent functions. *Des. Codes Cryptogr.*, 81:283–295, 2016.

[1948] V. A. Zinov'ev. Generalized cascade codes (in Russian). *Problemy Peredači Informacii*, 12(1):5–15, 1976. (English translation in *Probl. Inform. Transm.*, 12(1):2–9, 1976).

[1949] V. A. Zinov'ev and V. K. Leont'ev. On non-existence of perfect codes over Galois fields (in Russian). *Probl. Control and Inform. Theory*, 2:123–132, 1973.

[1950] J. Zumbragel, M. F. Flanagan, and V. Skachek. On the pseudocodeword redundancy. In *Proc. IEEE International Symposium on Information Theory*, pages 759–763, Austin, Texas, June 2010.

[1951] J. Zwanzger. *Computergestützte Suche Nach Optimalen Linearen Codes Über Endlichen Kettenringen Unter Verwendung Heuristischer Methoden*. PhD thesis, University of Bayreuth, 2011.

Index

Printed in the United States
By Bookmasters